TV & Video Engineer's Reference Book

TV & Video Engineer's Reference Book

Edited by
K G Jackson
G B Townsend
With specialist contributors

Butterworth-Heinemann Ltd
Linacre House, Jordan Hill, Oxford OX2 8DP

A member of the Reed Elsevier group

OXFORD LONDON BOSTON
MUNICH NEW DELHI SINGAPORE SYDNEY
TOKYO TORONTO WELLINGTON

First published 1991
Paperback edition 1994

British Library Cataloguing in Publication Data
Television and video engineer's reference book
1. Television equipment 2. Video equipment
I. Jackson, K. G. (Kenneth George) *1930-*
II. Townsend, Boris

Library of Congress Cataloguing in Publication Data
Television and video engineer's reference book/edited by
Kenneth G. Jackson, Boris Townsend
p. cm.
Includes bibliographical references and index
ISBN 0 7506 1953 8
1. Television - handbooks, manuals, etc.
I. Jackson, Kenneth George II. Townsend, Boris
TK6642.T436 1990
621.388-dc20 90-2107

ISBN 0 7506 1953 8

Printed and bound in Great Britain by Hartnolls Ltd, Bodmin, Cornwall

Preface

'The compilers of this book would be wanting in courtesy if they did not expressly say what might otherwise be safely left to the reader's discernment.' So wrote the brothers Fowler in the preface to their famous work on the King's English. It is a precept which any editor could observe with advantage, though perhaps disciplining himself to avoid too slavish a repetition of the Contents List.

The most superficial examination of this volume will indicate that it is a comprehensive survey of television technology and that it is authoritative. Eminent engineers of international stature and from many countries have spared time from their development work to share their knowledge and experience with the rest of us.

It is, indeed, in the best traditions of science that research and teaching should go hand-in-hand; and many a student has discovered that the originator of new ideas was more comprehensible in his writings than the popularizing scribes who attempted to explain what the innovator was talking about.

Television is currently in turmoil. It has always been so and, like it or not, since it is engineering-led it will always be in turmoil. As electronics and materials science develop at an ever-increasing rate, so will television. In the United Kingdom, de-regulation is in vogue and the economics of programme making will press ever more heavily on equipment design, though the basic engineering principles will persist.

To say that television is international is a platitude – and is incorrect. It is nowhere near as international as film. Countries insist on using their own technical standards and their own language. Direct broadcasting from geostationary satellites across national frontiers is an everyday occurrence yet governments cannot agree on a world standard. Even in the European Economic Community the directives defining which scanning and coding standards should be used for broadcasting from space are flouted by entrepreneurs. But the inexorable march of engineering developments, based on the principles and factors detailed in this book, have their own logic. Conversion from one picture standard to another in real time has been solved by the engineers, and machine translation between spoken languages is a process already working in the laboratory. When this is coupled with higher definition, seeing and hearing at a distance will become an even greater pleasure and even more instructive.

The world is growing smaller, even for those of us who do not travel.

GBT
Wrea Green, England
1991

Contents

Part 5 · Distribution of Broad-band Signals

Part 6 · Direct Broadcasting by Satellite

Part 7 · TV Studios and Studio Equipment

Part 15 · Industrial, Commercial, Medical and Defence Applications of Television

Part 16 · Performance Measurements and Electromagnetic Compatibility

List of Contributors

I G Aizlewood
Managing Director, Continental Microwave Ltd

C P Arbuthnot BEng
British Telecom Research Laboratories

P Audemars MA
Senior Film & Video Editor, London Weekend Television

J L E Baldwin BSc, MInstP, FRTS, FSMPTE
Consultant

J Barron BA, MA (Cantab)
University of Cambridge

P G J Barten
Barten Consultancy

D J Bradshaw BSc(Eng), AMIEE
Design and Equipment Department, BBC

D Bryan
Formerly Technical Director, Michael Cox Electronics Ltd

J H Causebrook BSc, PhD, MIEE, CEng, AMIOP
Service Area Planning Section, IBA

C K P Clarke
Senior Engineer, BBC Research Department

P A Crozier-Cole
Head of Telemetry and Automation Section, IBA

K Davison
Manager Communications, Thames Television

C Dawkins BA (Cantab)
British Telecom Research Laboratories

C Debnam
Quality Assurance Manager, PAG Ltd

R Elen
Creative Technology Associates

S R Ely PhD, CEng, MIEE
Head of Carrier Systems Section, BBC Research Department

B Flowers MRTS
Head of Eurovision Control Centre, European Broadcasting Union

D Hardy
Design Manager, PAG Ltd

S Hirata
Senior Specialist, Toshiba Corp

R G Hunt DSc, FRPS, FRSA, MRTS
Professor of Physiological Optics, City University

Y Imahori
Chief Engineer, NHK

G A Johnson BSc, EEng, MIEE
Deputy Head Engineering Services, ITV Association

E A Jones BEng, AMIEE
Assistant Technical Support Manager, Philips Communications and Security

J Kelleher MIEE
Formerly Chief Engineer, Dynamic Technology Ltd

P Kemble CEng, MIEE, BSc
Principal Engineer, IBA

W H Klemmer
Broadcast Television Systems GmbH

K Komada
General Manager, Toshiba Corp

S J Lent CEng, MIEE
Engineering Research Department, BBC

S Lowe
Manager International Technical Training, Ampex Ltd

R G Manton BSc(Eng), PhD, CEng, MIEE
Transmission Engineering Department, BBC

C A Marshman BTech, CEng, MIEF
York Electronics Centre, University of York

J D Millward BSc, CEng, MIEE
Head of Research, Rank Cintel Ltd

P L Mothersole FEng, CEng, FIEE
VG Electronics Ltd

K C Quinton MBE, BSc, FEng, FIEE, FRTS
Formerly Director of Research, British Cable Services Ltd

A F Reekie AMIEE
Formerly Senior Engineer, European Broadcasting Union

F M Remley Jr FSMPTE, MBKSTS
Technical Director Broadcasting Services, University of Michigan

R S Roberts CEng, FIEE, SenMIEEE
Consultant Electronics Engineer

J T P Robinson
MVC Crow Ltd

J G Sawdy BSc, CEng, MIEE
Independent Broadcasting Authority

B L Smith
Chief Technical Writer, Thomson-CSF

P Sproxton
Alpha Image Ltd

R Stevens BSc, CEng, MIEE
Engineering Project Supervisor, Thames Television

L Strashun MSc, CEng, MIEE, MBKS, MRTS
Senior Manager, Sony Broadcast Ltd

J Summers
Formerly Lighting Director, BBC

M Talbot-Smith BSc, CPhys, MInstP
Formerly BBC Engineering Training Department

D G Thompson BSc
Philips Components

E C Thomson
Mullard Application Laboratory

A Todorović
Director, Televigija Beograd

G P Tozer BSc
Principal Lecturer, Sony Broadcast and Communications

E Trundle MSERT, MRTS, MISTC
Chief Engineer, RNF Services Ltd

L W Turner FIEE
Consultant Engineer

N Wassiczek
European Broadcasting Union

G T Waters
European Broadcasting Union

I M Waters CEng, MIEE
Product Manager Transmitters, Varian TVT Ltd

R Watson CEng, MIEE, MRTS, MBKS
Consultant

P W Wayne
Formerly Marketing Director, Vinten Broadcast Ltd

L E Weaver BSc, CEng, MIEE
Formerly Head of Measurements Laboratory, BBC Designs Department

J P Whiting MSc, CEng, FIEE
Head of Power Systems, IBA

R Wilson BSc
Continental Microwave

G W Wiskin BSc, CEng, MICE, MIStructE
Architectural and Civil Engineering Department, BBC

D Wood
European Broadcasting Union

J M Woodgate BSc(Eng), CEng, MIEE, MAES, MInstSCE
Electronics Design Consultant

Part 1
Basic Reference Material

R S Roberts C Eng, FIEE, Sen MIEEE
Consultant Electronics Engineer

1 Television Standards and Broadcasting Spectrum

Every colour television channel consists of three modulated carriers:

- The *vision* information, derived from a camera or other signal source, is used to amplitude modulate a carrier with the electrical equivalents of the basic 'black and white' variations that are encountered during transmission of the scene.
- A subcarrier, situated within the bandwidth of the vision modulated carrier, is itself modulated with information related to the *colour* information in the scene.
- A separate adjacent carrier is modulated with the *sound* information contained in the scene.

The eye, as a visual communication system, 'sees' a large amount of detail simultaneously, by virtue of the fact that it has several million communication channels operating in parallel at any instant. The electrical signals that are generated by the millions of sensors in the eye are partly processed in the retina at the back of the eye, and further processed in the brain to provide the familiar human experience of normal vision. The mass of detail forming the visual scene consists of variations in light and shade, colour and, because we have two eyes, perspective.

Picture transmission, using electronic means to convey information of a scene, cannot be carried out as a simultaneous process embracing the total field of view. Any telecommunication system can process only a single item of information at a time, and hence the data relating to any visual scene must be analysed in such a way that the complete scene can be transmitted as separate items of electrical information. At the receiver, the individual bits of information are recovered and processed for display.

1.1 Scanning and aspect ratio

The visual scene is explored by examining the small areas of detail that are contained in it, a process known as *scanning*. When we read the page of a book, our eyes scan it line by line to extract the total visual information. Electronic scanning carries out a similar line by line scan process, the detail encountered being translated into voltage variations that can be used to modulate a radio transmitter. At the receiver, the received signals are demodulated and used to vary the beam current(s) of a display tube, the beam of which is sweeping in synchronism with the transmitter scanning beam.

A constraint of the electronic scanning system is the need to put a *frame* round the field of view to be transmitted. In the human seeing process the eye is quite unrestricted in its movements, and it roams freely over a very wide angular range which, with head and body movements, provides an unlimited field of view. In the electronic process, a finite limitation must be imposed by means of a frame, within which the picture can be analysed line by line.

The cinema industry has been in the business of presenting pictures as visual information for very many years, and has established a large number of basic principles. The newcomers to the art are the television engineers who did not 're-invent the wheel' but, very wisely, adopted many of the principles and standards that have evolved in film presentation. One of these concerns the shape of the frame. In the film industry, a standard rectangular shape with an *aspect ratio* (ar) of 4 (horizontal):3 (vertical) is the norm. This standard is still in general use for the main products of the film industry, despite the use of various 'wide screen' and other ratios. If a system has a standard aspect ratio at both the transmitter and the receiver, picture size is irrelevant. The relative dimensions of objects in the field will be correct.

The first television engineers concerned with the need to establish standards had no reason to depart from the 4:3 ar, particularly as it was realized that film would constitute a large proportion of programme material. These engineers were the team that created the standards for the world's first regular broadcast service of television programmes in 1936. This ratio has been adopted by all the systems that followed, and only recently has a change been considered in the 4:3 ratio.

Earlier experimental systems by Baird scanned the frame vertically with an aspect ratio of 1:2.

1.2 Still and moving pictures

A variety of picture transmission methods have been in use for many years. Still pictures have been communicated over telecommunication links by means of a *facsimile* (*fax*) system. The picture to be transmitted is wrapped round a drum, and scanned line by line as the drum is rotated and advanced with each turn. At the receiver, a photo-sensitive paper, wrapped round a similar drum, is synchronously rotated past a light beam which is modulated by the received signals. A high-quality 250 × 200 mm picture can be transmitted over a voice communication circuit in about 12 minutes. (Some modern fax scanners use linear flat scanning, similar to an office duplicator.)

The difference between scanning and transmission of a still or moving picture is one of time. Transmission of a still picture can take as long as we wish, but a moving field of view must be totally scanned in a time that is very short compared with the time being taken by any movements in the field of view. In other words, complete scanning of the moving picture must be so fast that we are concerned with what is, virtually, a still picture.

One of the properties exhibited by the human eye is *persistence of vision*. When the image of a still picture is impressed on the eye, removal of the visual stimulus does not result in an immediate cessation of the signals passed to the brain. An exponential lag takes place with a relatively long time for a total decay of the image. The cinema exploits this effect by presenting to the eye a succession of still pictures (or *frames*) one after the other, each frame differing from the previous one only by the change in position of any moving objects in the field of view. The presentation of one frame after another must not allow time for the image decay to become obvious and, provided that the presentation is sufficiently rapid and not too bright, an illusion of continuous movement is maintained.

The still pictures are projected in succession onto a screen. A frame is drawn into position with light cut off by means of a rotating shutter. As the shutter opens, the frame is stationary and the projected image illuminates the screen. The shutter cuts off the light, the next frame is drawn into position, the light is re-exposed through the frame, and so on in a continuous sequence. A great deal of early work with film showed that, for most people, a projection rate as low as 10–12 frames/second is adequate to present a complete illusion of movement.

However, at this projection rate another property of the eye becomes significant. The eye is extremely sensitive to the interruption of light at this rate, and the viewer would be very aware of *flicker*. As a consequence, the standard rate adopted for projection was 16 frames/second, well above that necessary for the presentation of continuous movement and, with the low-level illuminants of those days, adequate to minimize any awareness of flicker.

The eye sensitivity to flicker is a function of picture brightness and interruption rate, the rate needing to be higher if brightness is increased. With improvements in projector lamps over the years, flicker became a problem. Raising the projection rate would reduce flicker, but would result in a larger quantity of expensive film being required. An ingenious solution of this problem became a standard feature of all film projection systems. The film is drawn into position with the light cut off by the shutter, as previously explained. The shutter exposes the light through the film, cuts it off, re-exposes the light through the same frame, cuts it off and only then is the next frame drawn into position for the next double exposure. The light is projected and cut off twice during each frame. For a picture projection rate of 16 frames/second, the interruption frequency is raised to 32 per second, and the visibility of flicker is very much reduced without doubling the length of film.

As time passed, better illuminants came into use, and flicker reappeared. This problem was solved along with another concerned with the sound track that films now required. Film was not passing through the projector system fast enough for good sound quality. The standard was changed to that in use today. The frame rate was raised from 16 to 24 frames/second. This raised the interruption frequency to 48, and increased the sound track length by 50 per cent.

1.3 Television picture frequency

The engineers who had the task of establishing standards for the first broadcast system (system A, see *Table 1.1*) adopted the aspect ratio of 4:3, but were concerned by the film projection rate of 24 frames/second. It was feared that, with a 50 Hz power supply frequency, any residual 50 Hz or 100 Hz power supply ripple in the receiver might modulate the beam current of the display tube with a sub-harmonic at 25 Hz which would produce a visible 'bar' across the picture. If film was being transmitted at the film standard rate of 24 frames/second, the difference frequency of 1 Hz would result in the bar sweeping down the picture once per second.

It was decided that the picture rate would be 25 pictures/second instead of the film rate of 24. It was considered that the effect on sound would not be serious. It was further considered that, in the event of interference from the power supply, a stationary bar across the picture would be less offensive than a bar sweeping down the picture at the 'beat' frequency of 1 per second.

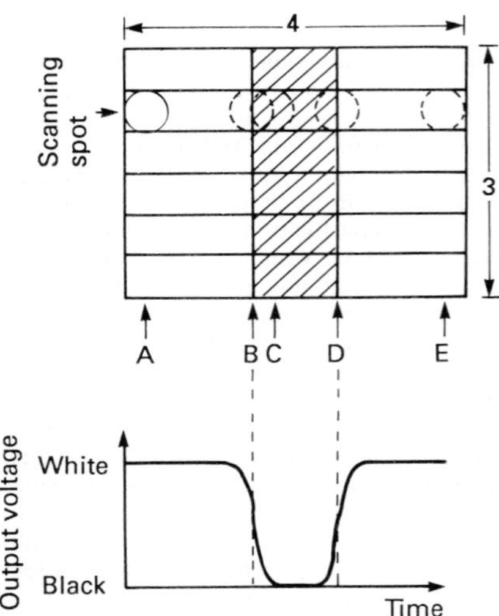

Figure 1.1 A six-line scan of a 4:3 field, showing aperture distortion

Figure 1.1 shows a simple six-line picture consisting of a black bar on a white background and the voltage output from the scanning system during a one-line scan. The scanning spot has a diameter equal to the width of one line, and sweeps across the picture from right to left along line 1, returns to its starting point displaced vertically by one line, sweeps line 2, and so on down the field of view to line 6. It is then returned to the top of the picture for the second picture scan, and so on. The voltage output may be as shown with maximum voltage indicating peak white, and minimum voltage corresponding to black, termed *positive modulation*. In a system, the polarity may be inverted, so that minimum voltage equals white and maximum is black. This would be *negative modulation*, as exemplified in the current UK system I (see *Table 1.1*). This method of scanning is termed *sequential scanning*, and results in a very obvious flicker because the field rate and picture frequency are the same, i.e. 25 per second.

A scanning system was standardized which provided a similar effect to the double-shuttering used in film projection. Instead of scanning the lines in sequence, the picture field is scanned by one field using lines 1, 3 and 5, and a second scan fills the gaps by re-scanning the field using lines 2, 4 and 6. This is *interlaced scanning*, and constitutes two sweeps of each field for a complete picture field. This has the same effect as double-shuttering of film, and raises the flicker frequency to 50 per second.

The interruption rate imposes a limitation on the brightness level at which the display tube can be operated before flicker becomes visible. The relationship between flicker and brightness is expressed by the Ferry-Porter law:

$$f_c = F + 12.6 \log_{10} B$$

where f_c is the critical frequency below which flicker is observed, F is a constant related to viewing conditions, and B is the luminance of picture highlights.

Tests on the viewing conditions of television pictures have suggested a value of about 37 for F and, with $f_c = 50$ (as in all European and some other systems), a picture highlight value of about 10 foot-lamberts is obtained.

Television standards in the USA adopted the same general principles. The picture rate was related to a power supply frequency of 60 Hz, resulting in a picture rate of 30 per second and a light-interruption rate of 60 per second. This increase of interruption frequency, compared with the UK rate, results in a permissible increase in highlight value by 6.8 times.

The six-line system illustrated in *Figure 1.1* would have very poor picture quality. The picture has a sharp transition at the edges of the black bar, but the voltage output does not change instantly from the white value to the black value. At position A (*Figure 1.1*) the scanning spot 'sees' peak white. As the spot reaches the bar at B it 'sees' half white and half black, the output being a half of the peak value, as shown. The output only reaches the value due to black at position C. At D, the half value is derived as shown, and the remainder of the scan gives a white output. The resulting effect is termed *aperture distortion*; it prevents any small detail in the picture being reproduced accurately.

In any practical television system, the picture quality will depend on the ability of the system to reproduce at the display tube all the sharp edges and fine detail. This requires the scanning spot size to be reduced, with a consequent increase in the number of lines necessary for a complete scan of the field of view. It is shown in section *1.5* that the channel bandwidth is determined by the scanning spot size. The smaller the spot, the more lines that are required for a complete scan, and channel space is at a premium. This means that a 'standard' spot size has to be a compromise between the ability of the system to provide a picture quality that is acceptable, and the minimum demand for channel space. Such standards are quoted in terms of the number of lines that are required in the vertical dimension for complete scanning of the entire field of view.

1.4 The video signal

The video signal derived by the camera or other scanning device for a practical black and white system will consist of random

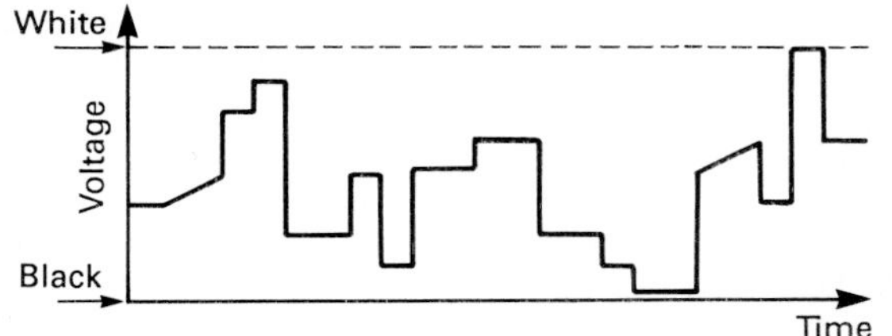

Figure 1.2 A possible video output from a camera during a one-line scan

voltages generated by the scanning of black, white and grey images during the line scan. A possible scan output is shown in *Figure 1.2*.

Two important conclusions arise from consideration of this type of voltage waveform:

- Voltage variations will generally consist of 'step' changes from one value to another. Smooth transitions from white to black, or from black to white, will be rare.
- AC voltage variations are extremely unlikely, and their rare appearance might arise from a scan across regular bars, such as the black and white bars on a test card.

Figure 1.3 illustrates another important feature that arises from this type of waveform. The two line scans each show the same signal voltage variation. In (a) a white bar is shown on a grey background, and in (b) the same output voltage variation shows a grey bar on a black background. The difference between the two identical signal variations is due to there being, in each, an average dc voltage component. The dc level determines picture brightness.

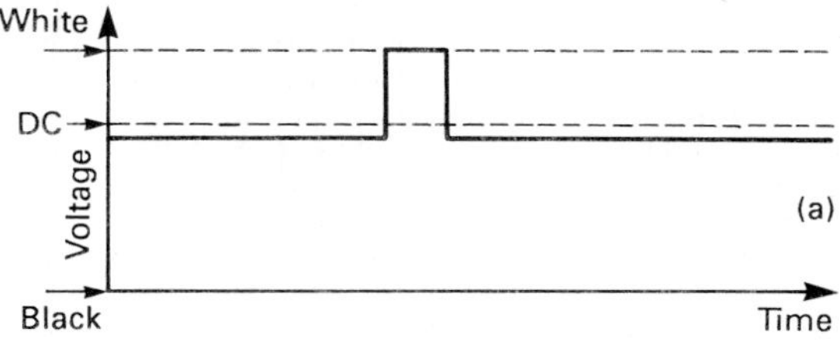

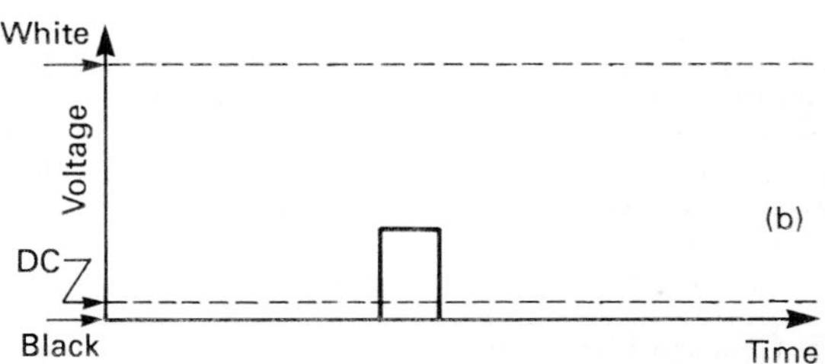

Figure 1.3 Similar video output signals are shown in (a) and (b), but with different dc levels

The video signals resulting from the scanning operation are processed and used to amplitude modulate the transmitter output. All standard broadcast transmitters use am, but fm is employed for certain links, and for satellite systems.

The principles of amplitude modulation are well known, but there are some important differences between sound and video as modulating signals. *Figure 1.4*(a) shows an alternating current variation that might be measured in the antenna system of a sound transmitter. Initially, the carrier is not modulated, then one cycle of an audio modulating tone is applied. The familiar features of this process are:

- The unmodulated carrier is radiated, and its mean level is constant with or without modulation being present.
- The carrier peak level is varied at the modulating frequency. The audio variation in carrier peak values during modulation is termed the *envelope*.
- An obvious limit exists in the modulating process whereby the carrier peak must not exceed twice the unmodulated level if distortion due to *clipping* is to be avoided.

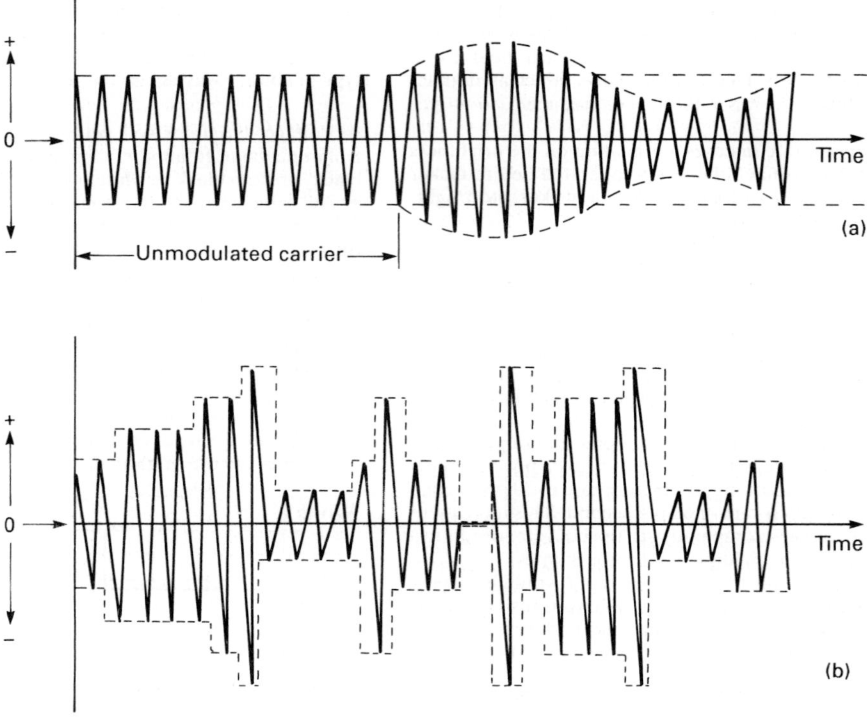

Figure 1.4 A carrier is shown amplitude modulated (a) by an audio tone and (b) by a video signal

Figure 1.4(b) shows a similar situation; it is amplitude modulation by a video signal similar to *Figure 1.2*. The envelope is now of a random character, and there is no mean carrier of constant level during modulation. When no modulating signal is present, no transmitter output is radiated.

1.5 Channel bandwidth

If a carrier is amplitude modulated with, for example, a 1 kHz tone, three frequencies are produced: the carrier, a frequency lower than the carrier by 1 kHz (a *lower side frequency*), and a frequency higher than the carrier by 1 kHz. If the modulating signal is a band of frequencies such as voice, music or video, a band of frequencies is generated each side of the carrier, termed *sidebands*, extending on each side of the carrier frequency to a limit determined by the highest frequency in the range of modulating frequencies.

There are several ways in which the highest frequency component in a modulating video signal may be determined, and thereby the channel bandwidth. One is indicated in *Figure 1.5* in which (a) is the top left-hand corner of a picture consisting of a regular pattern of alternate black and white squares. The sides of the squares are equal in length to the scanning spot diameter, and consequently the output signal generated will be a sine-wave, the frequency being the highest that will be generated at full amplitude. Any detail of smaller dimensions will not generate maximum output.

Consider now the six-line picture discussed in section *1.3*, where this picture is of the pattern shown in *Figure 1.5*, i.e. alternate squares of black and white, each with a width and height equal to the scanning spot diameter. The total number of squares in a 4:3 ar frame will be (6 × 6)4/3=48. If we transmit 25 complete pictures per second, the squares will be scanned in 1/25 s so that 48 × 25 = 1200 squares will be scanned in one second.

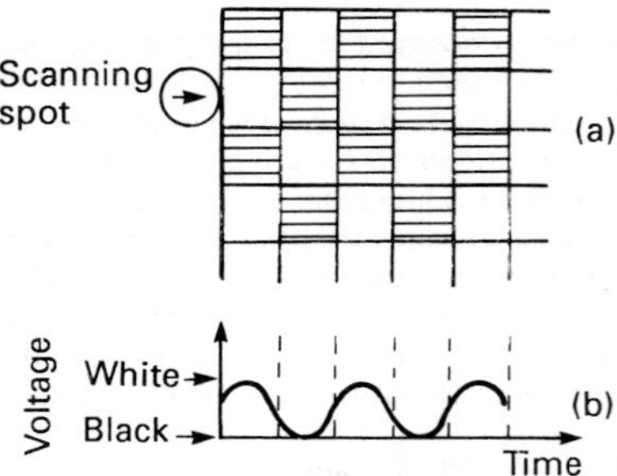

Figure 1.5 The smallest detail than can be resolved at full amplitude

The resulting ac cycle corresponds to the scan of one black and one white square. Thus the highest modulating frequency is 1200/2 = 600 Hz, and the two sidebands would require an overall rf channel bandwidth of 2 × 600 = 1200 Hz.

We have seen that the picture quality would be poor, and severely lacking in the resolution of edges and fine detail. We know that, to improve resolution, we must reduce the scanning spot diameter and increase the number of scanning lines appropriately. For example, if we halved the spot diameter of our *Figure 1.1* model, we would have to double the number of lines for a complete picture scan, and then the total number of squares would be 12 × 12 × $^4/_3$ = 192 and the number of squares scanned in 1 s would be 192 × 25 = 4800. This would result in a channel bandwidth of 4.8 kHz, four times that of the original six-line system. The bandwidth increases as the square of the number of lines.

A standard adopted for a good quality television system has to be a compromise between the need for acceptable definition of edges and fine detail in a picture and the overall bandwidth of the channel.

There are two historical examples of 'line standards' that are worth consideration. The first UK television broadcast system developed by Baird used an aspect ratio of 1:2 and 12½ pictures per second. Scanning was sequential and vertical. Thus, the number of squares would be (30 × 30) × 2 = 1800, and the number scanned in 1 s was 1800 × 12.5 = 22 500. The highest modulating frequency was thus 11 250 Hz.

The second example was the first ever 'high definition' broadcasts that commenced in 1936 in the UK with a picture frequency of 25 per second, interlaced scanning and an aspect ratio of 4:3. The compromise on definition and bandwidth was decided on the basis of the scanning spot being of such a size that 405 lines would be required to cover the picture area, and give acceptable picture quality. The highest video frequency is 405 × 405 × 4/3 × 25/2 = 2.7 MHz. The output of detail smaller than 1/405 of picture height would be less than maximum. Detail generating 3.0 MHz, for example, would be about −3 dB. The total video channel rf bandwidth becomes 3.4 MHz and the lowest modulating frequency is zero or dc. The total radio spectrum space occupied by this first *system A* (as it became known) is shown in *Figure 1.6*.

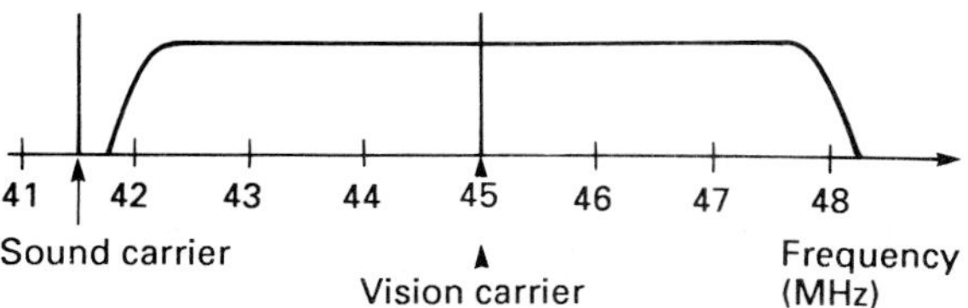

Figure 1.6 The full bandwidth of a dsb system A channel

1.6 Synchronism between scanning systems

The waveform in *Figure 1.4*(b) could represent the transmitter modulation during a one-line scan. At the receiver, the modulating signal is recovered and, after suitable processing, is used to modulate the beam current of the display tube, thus re-creating the detail seen by the scanner during the one-line scan.

Two further important items of information are necessary to ensure that the scanning spot at the transmitter, and the beam position on the face of the display tube, occupy identical positions in their 4:3 frame. One determines the position of the scanning spot in the vertical plane, and the other ensures the correct position in the horizontal plane.

It would appear from *Figure 1.4*(b) that the video waveform is very complete, and there is no way in which any additional information can be provided, but a development of system A showed how it could be done, in a manner that forms part of any television standard today. *Figure 1.7*(a) shows how the video signal modulation can be established between the two limits: the maximum transmitter output, and the carrier level corresponding to black. This leaves a region between zero carrier output and the black level into which we can put extra information.

However, the entire transmitter/receiver system can deal with only one bit of information at any instant, and we must remove the video information while we provide any extra information. A *blanking pulse* blanks out all video information down to the black level, at the start of the line scan. A narrow *line synchronizing pulse* is then inserted from the foot of the blanking pulse down to zero. The leading edge of this pulse is used to start the line scan on its traverse across the field from left to right. At the receiver, the leading edge of the demodulated narrow pulse is used to start the sweep of the display tube beam across the face of the tube. At the end of the line scan, the next blanking/synchronizing pulse triggers the return of the spot to the left-hand side, and the scanning cycle starts again, this time slightly displaced vertically, to trace a new line path.

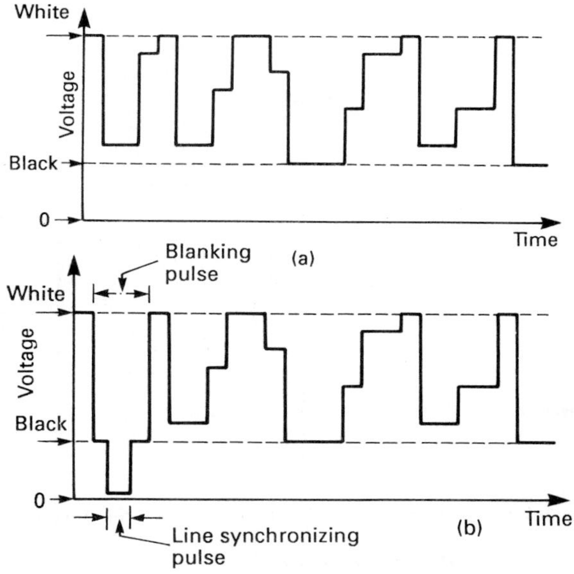

Figure 1.7 A typical camera output (a) before and (b) after the insertion of blanking and line sync pulses

The waveform of a line of the UK system I is shown in *Figure 1.8*. The heavy black line is the waveform of a black and white system, and the shaded areas are concerned with the colour information. The timing of all the pulses and other features is with reference to the leading edge of the narrow sync pulse. All television standards use the same type of waveform, but their timing and pulse widths may differ.

To effect vertical displacement of the scanning spot in synchronism, it is necessary to use a different type of pulse. The line sync pulses are narrow and relatively infrequent, so we can suppress video information for several lines, and transmit either a single, long pulse or a train of relatively broad pulses to effect vertical synchronism. *Figure 1.9* shows the four-field sequence of field sync pulses used in the UK system I. The 'burst' sequence is necessary for the PAL colour system.

At the receiver, the two types of pulse can be separated from the composite waveform by relatively simple forms of amplitude discrimination, and they can then be separated from each other by the passive circuitry shown in *Figures 1.10* and *1.11*.

Figure 1.10 shows one form of a *differentiation circuit* in which the time constant *CR* is short compared with *t*. It will be recalled that synchronizing information is required from the leading edge of the line sync pulse if all the timing is to be correct, and this type of circuit provides this discrimination between leading and lagging edges.

Figure 1.11 shows an *integrating circuit*. The time constant *CR* is long compared with the duration of the train of pulses provided at the input. Successive pulses build up the voltage on C to the value V_p and, at the end of the pulse train, C

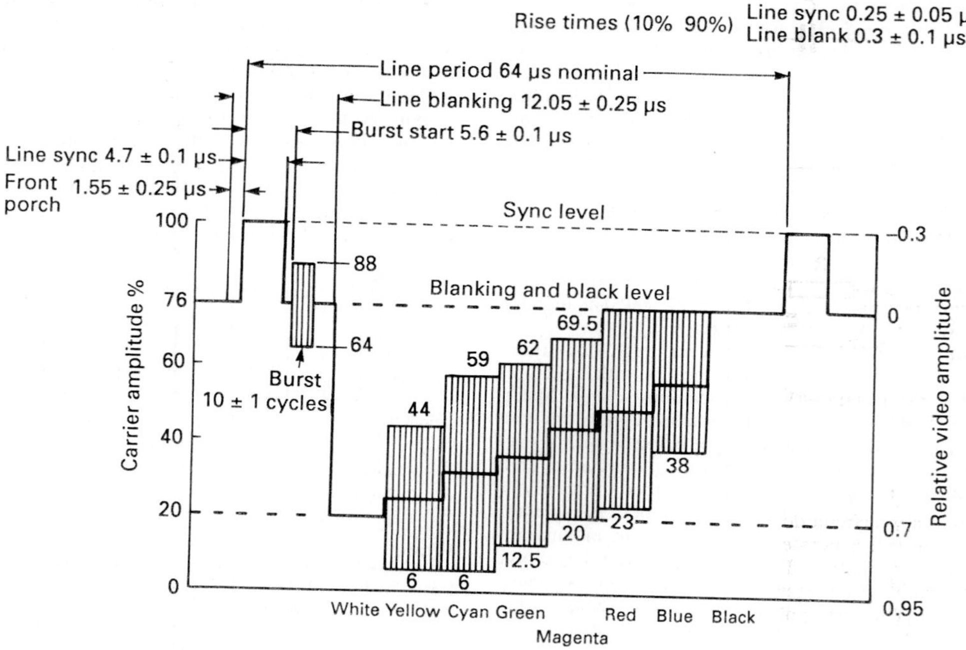

Figure 1.8 The thick line show a possible signal output resulting from a single line scan of a black and white system, the image being a grey 'staircase' from white to black. The shaded areas are concerned with the addition of colour (BBC)

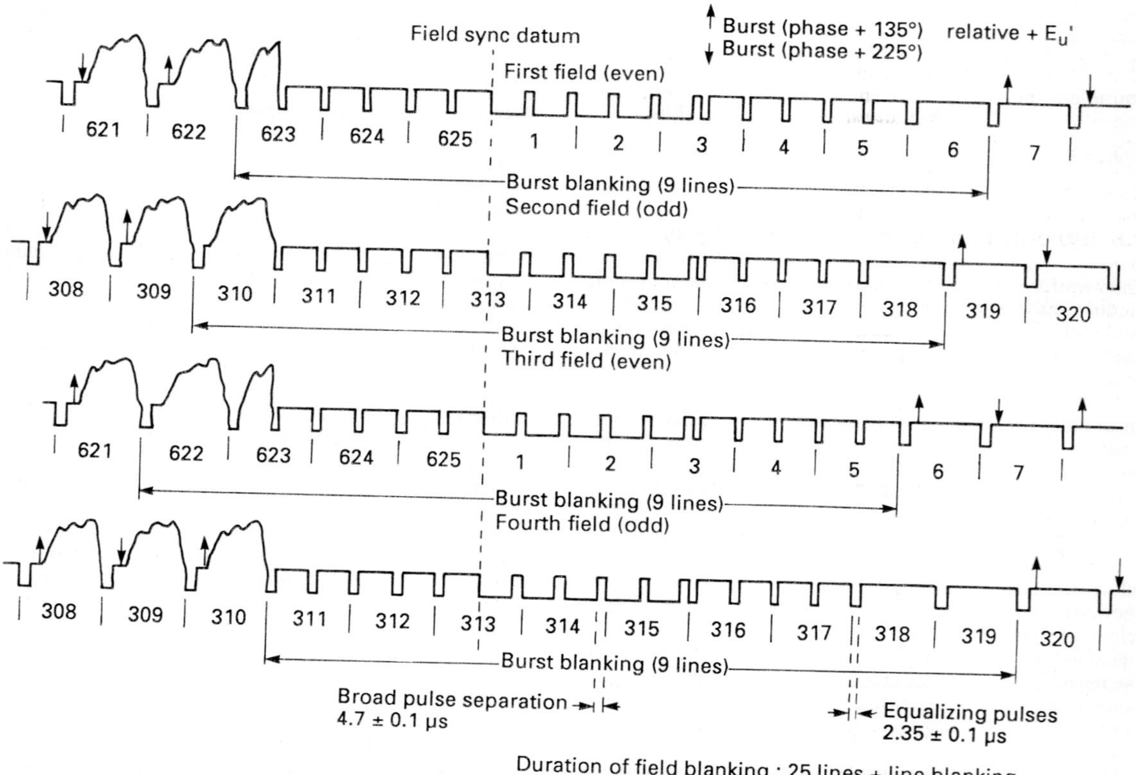

Field sync datum

Burst (phase + 135°) relative + E_u'

Burst (phase + 225°)

First field (even)

621 622 623 624 625 1 2 3 4 5 6 7

Burst blanking (9 lines)

Second field (odd)

308 309 310 311 312 313 314 315 316 317 318 319 320

Burst blanking (9 lines)

Third field (even)

621 622 623 624 625 1 2 3 4 5 6 7

Burst blanking (9 lines)

Fourth field (odd)

308 309 310 311 312 313 314 315 316 317 318 319 320

Burst blanking (9 lines)

Broad pulse separation 4.7 ± 0.1 μs

Equalizing pulses 2.35 ± 0.1 μs

Duration of field blanking : 25 lines + line blanking

Figure 1.9 The broad pulses used for synchronism of the vertical scan during blanking of 25 lines

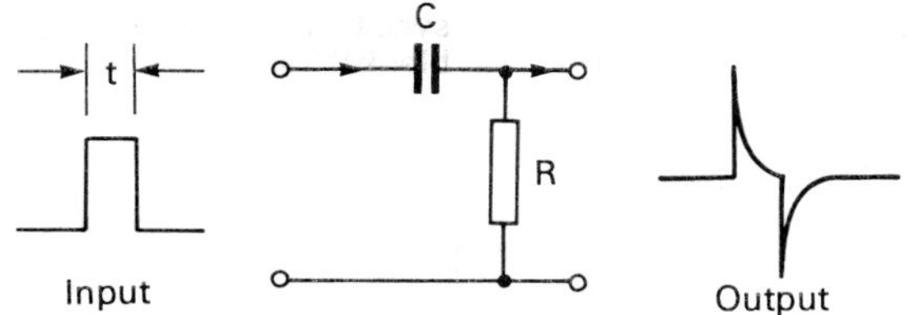

Figure 1.10 A differentiation circuit that provides an output when the input voltage changes in value

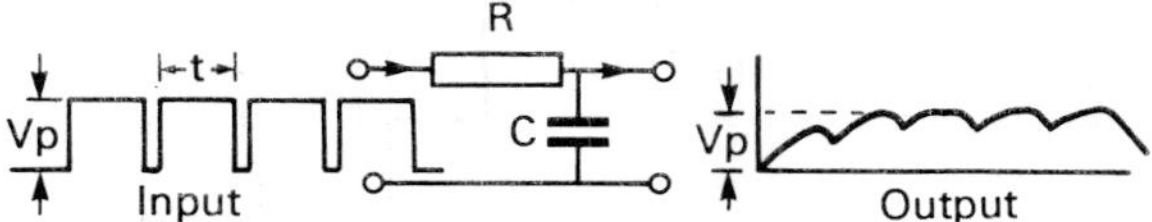

Figure 1.11 An integration system that enables a train of broad pulses to build up to a peak value

discharges. The output waveform thus constitutes a single broad pulse, which operates at a relatively slow speed to trigger the vertical sweep system.

The differentiation circuit does not distinguish between line or field pulses; it will generate 'spike' pulse output from the edges of any pulse fed into it. The integration circuit is the one that *does* discriminate between the two types of pulse. The narrow, infrequent line sync pulses will provide no output from the integrator.

1.7 Porches

Figure 1.8 shows two other features that are concerned with the line sync pulse region. Note that the pulse is not in the centre of the blanking interval.

In any system that contains inductance, capacitance and resistance, voltage or current changes cannot take place instantaneously. A video signal change from black to peak white, or white to black, takes time for completion, requiring possibly a capacitor to charge or discharge. A black/white edge results in a voltage change as shown in *Figure 1.12* (not to be confused with the aperture distortion shown in *Figure 1.1*). Ahead of the pulse in *Figure 1.8* is a narrow plateau, the *front porch*. Its purpose is to allow the video signal of the previous line (which might have been at peak white, for example) to reach black level before the sync pulse starts.

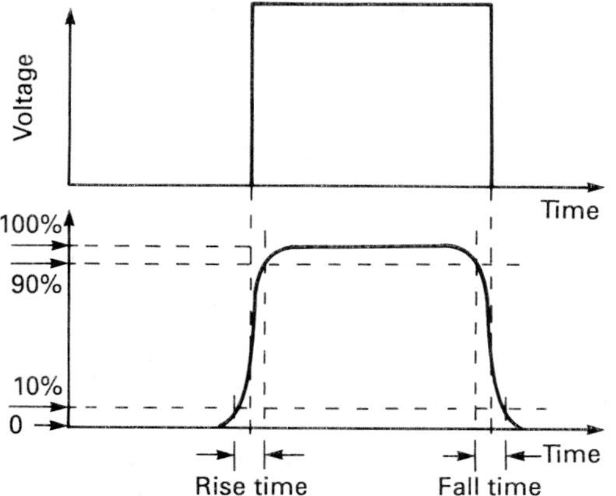

Figure 1.12 Voltage and current values cannot change instantly. Rise or fall times are determined by the values at 10 per cent and 90 per cent of peak values

Behind the sync pulse is another plateau, the *back porch*. The original purpose of the back porch, when first used in system A, was to make sure that there was plenty of time for the receiver line scan circuitry to effect complete retrace of the line scan, and for the beam to be in its correct position for commencement of the next line scan. By comparison with current technology, receiver scan circuits in 1936 were crude, ponderous and extravagant of power, and they required a lot of time for retrace. Today, the back porch provides plenty of time for retrace and it also provides the space necessary for colour information to be transmitted and extracted at the receiver.

1.8 DSB, ssb, asb and vsb

The UK system A operated from 1936 to 1939, and then had to close down due to the outbreak of World War II. The USA standardized its television system and immediately found a serious problem. The double sideband (dsb) system was very extravagant of channel space and, as many channels were required, a new system of modulation was devised. It was known as a *vestigial sideband* (vsb) system, sometimes termed an *asymmetric sideband* (asb) system, and is now used throughout the world because it saves a considerable amount of channel space.

To appreciate the operation of vsb, it is necessary to digress into some important differences between sound and video amplitude modulation. Single sideband (ssb) am systems have been used for radio transmission of speech and music for at least 60 years. Why cannot ssb be used for video modulation and thus save half the dsb bandwidth?

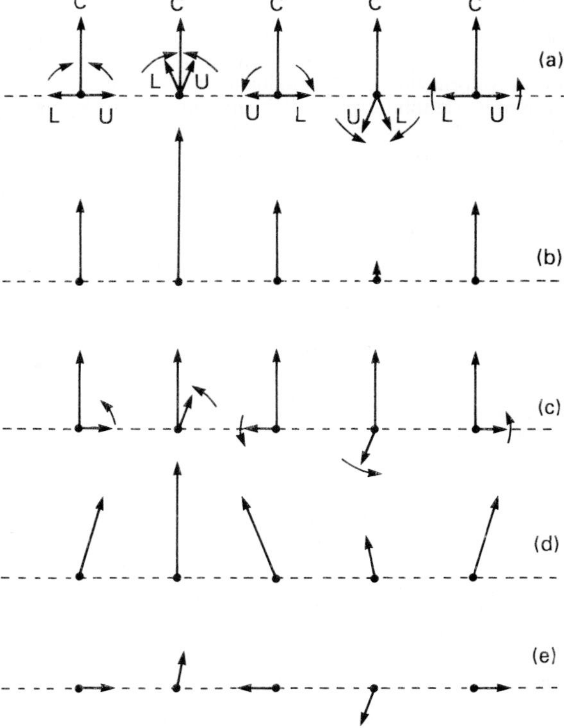

Figure 1.13 The vector relationships of dsb (a) and (b) ssb (c) and (d), and a single side frequency (e)

There are two sources of distortion in a dsb speech or music system. An amplitude constraint exists whereby any attempt at over-modulation results in severe waveform distortion, with unacceptable audible quality. The second effect occurs in single-channel sound reproduction where the quality of the reproduced sound is unaffected by changes of phase response. Many do not accept that our normal binaural experience distorts. Nevertheless, simply turning one's head, for example, results in a severe phase change of, say, a 5 kHz tone, where about 4 cm movement represents about 180°. We live with this effect and do not notice it, even when we use it for directional information.

Figure 1.13(a) shows the vectors relating to the carrier side frequencies during one cycle of a modulating tone of dsb am at maximum modulation; (b) shows the resultant vector addition of the vectors in (a), causing the rise and fall of carrier amplitude, and the phase of the resultant vector, which remains that of the carrier at all times during the modulating cycle.

Figure 1.13(c) is as (a), but with the lower side frequency suppressed, and (d) is the new resultant. Two features can be seen:

- The modulation, considered in terms of carrier amplitude, is halved.
- The angle of the resultant vector is now swinging with respect to the carrier vector.

We have produced phase modulation.

To determine the requirements of a video signal used for amplitude modulation, consider the simple video modulating signal shown in *Figure 1.14*(a). Such a pulse can be analysed into a number of discrete harmonic components, which extend out to infinity with a descending order of amplitude. The square waveform of (a) could be synthesized by addition of the frequency components of (b) at the correct amplitudes and phases. In a television system, we require the pulse signal to progress through the system in such a manner that the shape of the waveform that finally modulates the display tube beam current is a faithful copy of the signal derived by the scanning process at the transmitter.

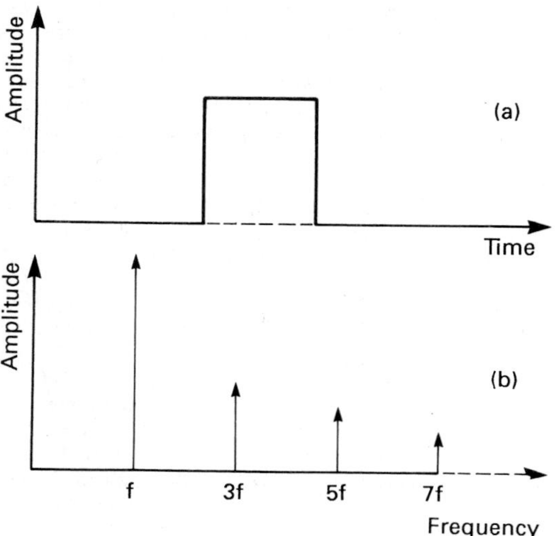

Figure 1.14 A theoretical rectangular pulse contains a fundamental frequency *f*, and all the odd harmonic frequencies. The practical pulse loses some of the higher harmonics due to the finite bandwith of the system, and has sloping sides with rounded corners

A pulse can experience degradation of its waveform in a number of ways as it goes through the system:

1 The bandwidth of the system is finite, and therefore some of the higher frequency harmonics will not be present.
2 An inadequate hf response at some point in the signal path can change the pulse shape by, for example, rounding the corners and sloping the sides of the pulse.
3 Inadequate lf response can produce a 'tilt' at the top of a pulse. The dc voltage may not hold up for the duration of the pulse.
4 The various discrete frequency components that constitute the signal waveform may experience differing transit times as they travel through the system. Some may go through faster or slower than others, distorting the wave shapes in varying degrees at the point where they are intended to arrive together, i.e. at the point where the display tube beam current is modulated.

Of the above, 1 is determined by a definition compromise as discussed in section *1.5*, and 2 and 3 are concerned with circuit behaviour, and any deficiencies can be resolved. The timing effect of 4 is much more important. The phase response of the system, clearly, determines the preservation of pulse shape. Each of the signal frequency components must travel through the system in the same time, although the actual time of transit is, within reason, of no importance.

If a fundamental frequency f passes through the system in time t, this time can be interpreted as a phase change ϕ. The frequency $2f$ will go through twice this phase angle in the same time. $3f$ will shift three times the fundamental phase angle, and so on for each harmonic component. If one of the harmonic components goes through too great a phase shift in the time t, it means that it is travelling too fast and will arrive early at the tube beam current control. It is seen that the phase response of the system should be such that the angle ϕ must be proportional to frequency. Any one of the lines shown in *Figure 1.15* would indicate a satisfactory phase response. The linear relationship is essential, and the only significance of the differing slopes is that they relate to different transit times. (The horizontal line would indicate zero time which is, of course, not possible.)

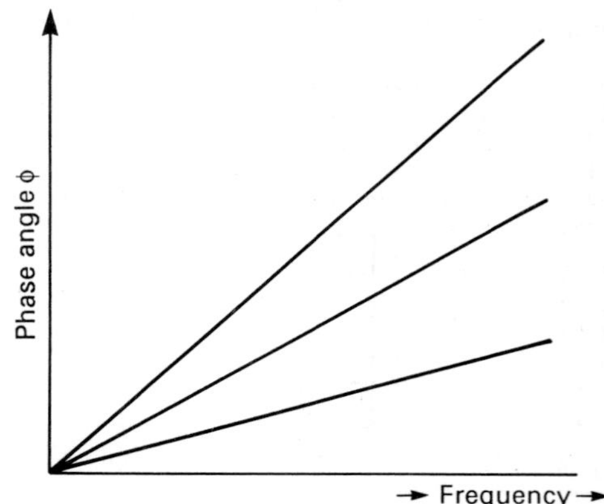

Figure 1.15 A video system phase response must be proportional to frequency

The main differences between an audio channel and video signal processing are now more clearly seen. Audio signal channels can tolerate a high degree of phase distortion, but amplitude distortion must be kept to very low limits. Video signal processing must ensure that phase distortion is minimized, but overload effects are far less serious. The peak value of *Figure 1.14* may represent peak white, for example, and any increase above this value would not be visually of much significance, although it may drive a transmitter into an overload condition. Alternatively, the peak value may correspond to a black signal, in which case 'blacker than black' is of no visual consequence.

The phase distortion produced by ssb (*Figure 1.13*(c) and (d)) may be acceptable for audio, but is quite unacceptable for video. However, it *is* possible to use partial suppression of one sideband (vsb) for a television system and save considerable channel space. Whereas phase distortion can be quite unacceptable where large picture areas are involved, phase errors become difficult to see on small detail in the picture. Thus, a practical system must have a very low phase distortion at low video frequencies, while the high frequencies generated by the fine detail in the picture can be severely distorted, but still acceptable.

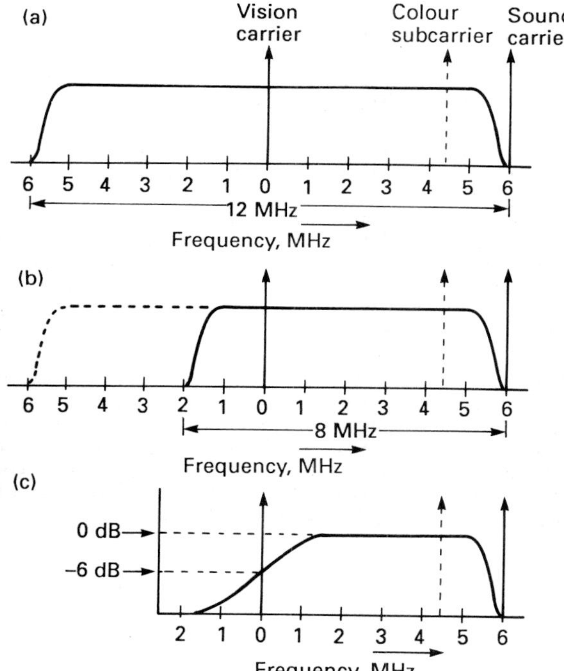

Figure 1.16 A full channel bandwith is shown with, at (b), the radiated signal of vsb modulation, and at (c), the required receiver if response

Figure 1.16 shows, at (a), the full dsb channel bandwidth that would result from video amplitude modulation, together with the sound channel, for the UK system I. The transmitted signals are shown in (b) where the lower sideband has had about 4 MHz filtered off, leaving a 'vestigial' 1.25 MHz. The 12 MHz channel has been reduced to 8 MHz.

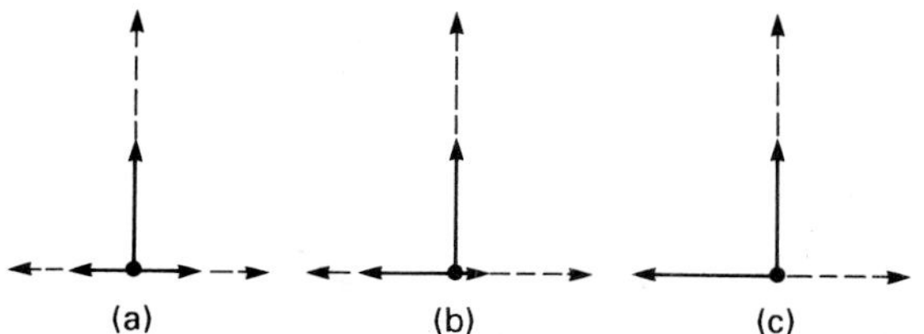

Figure 1.17 Demodulated output from the vsb input to a receiver. See description in text

For vsb to function correctly, it is necessary for the received signals to be processed before being presented to the demodulator. The pass-band of the receiver if channel must be shaped as shown in *Figure 1.16*(c). Consider, for example, a 50 Hz modulating frequency, and its position in *Figure 1.16*. Such a frequency would result from the scan of a large area, indeed a complete field, and the resulting side frequencies would be very close to the carrier. The signals presented to the demodulator by the receiver if amplifier are shown as vectors in *Figure 1.17*(a). The carrier and side frequencies are cut by 6 dB, but no phase distortion would result because the signal is a full double sideband am signal at 100 per cent modulation.

Consider next a modulating frequency resulting from the scan of a smaller detail, producing a frequency around, say, 500 kHz. *Figure 1.17*(b) shows the vector relationship between side frequencies and carrier. The overall length of the upper plus lower side vectors add to the same value as in (a), showing that the modulation factor has not changed. However, one vector is longer than the other, and some phase distortion is beginning to appear in the resultant addition of the three vectors. At frequencies of about 1.5 MHz and above, the vector situation is as shown in (c). The single side frequency resulting from the scan of a small detail has an amplitude equal to the attenuated carrier and, again, the level of modulation is unchanged. Phase distortion is present, but the detail is small and the errors become more difficult to see as the modulating frequency increases.

The US problem of minimizing channel space was eased considerably by the use of vsb, and all standard systems now use vsb. All UK system A stations that followed Channel I used vsb and, when the first station moved from its original site to Crystal Palace, the opportunity was taken to bring it into line and change it from dsb to vsb. The vestigial band was about 1 MHz, and the overall channel width was 5 MHz.

1.9 National standards

Many of the world's standards use 625 lines, the exceptions being the 525-line system M, used by North and South America, Japan and a few other countries, and the French system E on 619 lines. System E was developed before World War II in an attempt to provide a better picture quality than the 405-line standard then prevailing. This virtual doubling of the line structure does, indeed, provide an excellent picture quality, although at the expense of channel space. Consideration is currently being given to the adoption of new line standards that will approximately double the number of lines.

The almost universal adoption of a 625-line standard resulted from the first international conference on television standards after World War II. All countries planned to start a television service, and it was hoped that the 625-line standard would permit international links between European and other countries that used 625 lines. Unfortunately, widespread adoption of 625 lines did not result in a universal standard. There are many 625-line systems, but few are compatible. Some have positive modulation, others negative. Some have am sound, and others fm sound. There are further differences in the colour.

The US system M was the first practical broadcast colour system. A later European conference attempted to standardize a European colour system. It was considered that the NTSC system M could be improved, and several proposals were made for systems based on the NTSC principle that used a subcarrier for colour information. Germany proposed a modified NTSC system known as PAL (*Phase Alternation Line*). The French proposal was for the subcarrier to be frequency-modulated, the system being termed SECAM (*Sequentiel á Mémoire*). Other variants of the NTSC system were considered for a European standard, but the final outcome was that PAL was preferred

Table 1.1 Standard television systems

	A	*B*	*C*	*D(K)*	*E*	*G*	*H*	*I*	*L*	*M*	*N*
Lines per picture	405	625	625	625	819	625	625	625	625	525	625
Field frequency, Hz	50	50	50	50	50	50	50	50	50	60	50
Line frequency, kHz	10.125	15.625	15.625	15.625	20.475	15.625	15.625	15.625	15.625	15.734	15.625
Video bandwidth, MHz	3	5	5	6	10	5	5	5.5	6	4.2	4.2
Channel bandwidth, MHz	5	7	7	8	14	8	8	8	8	6	6
Sound/vision carrier spacing, MHz	3.5	5.5	5.5	6.5	11.15	5.5	5.5	6	6.5	4.5	4.5
Vestigial sideband width, MHz	0.75	0.75	0.75	0.75	2	0.75	1.25	1.25	1.25	0.75	0.75
Vision modulation polarity	+ ve	− ve	+ ve	− ve	+ ve	− ve	− ve	− ve	+ ve	− ve	− ve
Sound modulation	am	fm	am	fm	am	fm	fm	fm	am	fm	fm
Deviation, k Hz		50		50		50	50	50		25	25
Pre-emphasis, μs		50		50		50	50	50		75	75

and adopted by most countries, while France and Russia decided to use SECAM. (They actually have different versions of SECAM.)

A common feature of all colour systems is that they use a subcarrier for colour information, and transmit it in the form of a 'burst' of a few cycles for use by the receiver as a reference. The position for the burst is on the back porch, as shown in *Figure 1.8*.

Table 1.1 gives details of most of the standards in use throughout the world. (System A has been included, although it is no longer in use in the UK.) These are not the only differences between systems, but the radio frequencies on which they operate also differ between countries. The frequency allocations on which radio services operate are decided by the International Telecommunications Union (ITU). The world is divided by pole-to-pole boundaries into three regions:

Region 1 includes Europe, Africa and Russia. The eastern boundary includes the whole of Russia and Mongolia.
Region 2 includes North and South America, Greenland and Alaska.
Region 3 includes Australia, New Zealand, India and Pakistan, China and Japan.

This partition is intended to ensure that the various radio services should operate with minimum interference to each other and other services. However, this does not ensure that all television services will operate on the same frequency bands. Considerable differences exist, not only between regions, but within regions.

For example, *Table 1.2* shows the UK and European allocations for television services, but Africa and Scandinavia, also in Region 1, differ. South Africa has two bands, one on vhf 175–255 MHz and the other a uhf band on 471–632 MHz. The Danish bands are 55–68 MHz, 175–216 MHz and 615–856 MHz. Both countries use PAL. South Africa uses system I, as in the UK, and Denmark uses PAL on systems B and G which differ only in channel bandwidth.

Many similar examples can be quoted. A notable one is that Japan uses NTSC system M, but a receiver manufactured by the Japanese for home use could not be used in the USA because the two countries are in different regions and their radio frequencies are different.

Further examination of *Table 1.1* shows several very similar standards. Many have derived from basic systems, and differences are small. Systems D and K, for example, are the same for the listed parameters, but the D version uses PAL and the K version uses SECAM. The K system, in turn, has spawned the K′ system, which is similar to the K system but with a wider vestigial sideband.

1.10 Bands and channels

The radio frequency spectrum is divided up into bands as follows:

vlf	<30 kHz
lf	30–300 kHz
mf	300–3000 kHz
hf	3–30 MHz
vhf	30–300 MHz
uhf	300–3000 MHz
shf	3–30 GHz
ehf	30–300 GHz

The band classifications are international, and broadcasting takes place in some of them. Medium-wave broadcasting, for example, occurs in the mf band. The '3–30' range of each band is not so arbitrary as it may appear (see section *12.2.1*). Each band is characterized by its own propagation and antenna features.

Table 1.2 European bands and designations

Band	European band	Frequency range
vhf	I	41-68 MHz
vhf	II	88-108 MHz (fm sound)
vhf	III	174-216 MHz
uhf	IV	470-582 MHz
uhf	V	614-854 MHz
shf		11.7-12.5 GHz

Notes:
1. The UK used bands I and III for a television service, on standard A. This has now been discontinued, but other parts of Europe still use these bands for television.
2. Band II is not used for television, but for fm sound only.
3. The shf band has been allocated for satellite television services.

The sections of the spectrum that are allocated to television are in the vhf and uhf bands. There are three world zones with different frequency allocations for broadcasting, and *Table 1.2* shows the European bands and their designations.

The UK television service functions in Bands IV and V only. Channels 21–34 are in Band IV, and channels 39–68 are in Band V. The various standards in use have different channel bandwidths so, for example, a UK channel 40 does not necessarily occupy the same frequency space as a European channel 40.

Table 1.3 UK shf frequencies

SHF channel	Frequency GHz	
4	11.78502	all left-hand circular polarization
8	11.86174	
12	11.93846	
16	12.01518	
20	12.09190	

The actual frequencies in the shf band allocated to television broadcasting in the UK are shown in *Table 1.3*. There are 40 channels, each 20 MHz wide. The precise form of the transmission has yet to be decided.

1.11 Adding colour to a monochrome system

Colour television was demonstrated in 1928 by Baird, using red, green and blue filters before each field in a sequential manner, with appropriate synchronized filters before a modulated white light source at the receiver. Various unsatisfactory features appeared, such as *colour break-up* (a fast-moving object in the field was trailed by a colour), and flicker, requiring an increase in field rate.

A more sophisticated version of a mechanical sequential colour field system was proposed by CBS when the US sought a standard for a colour system, and this proposal was adopted in 1951, only to be subsequently rejected when the US National Television Standards Committee (NTSC) was formed to derive a wholly electronic standard. The NTSC system was adopted by the US in 1953, and had its first broadcast in 1955.

Modern colour television has resulted from the addition of suitable techniques to existing monochrome systems. The transmission, reception and processing of black and white signals require three separate items of information: the video information and the synchronizing information for two-dimensional scanning.

It is shown in Part *3* that colour requires three further types of information. It is necessary to determine *luminance*, *hue* and *saturation* of any colour at any instant. Fortunately, existing monochrome systems are luminance-only systems, and they can supply the luminance information. This leaves two extra items of information to be transmitted and received which, when suitably processed, can determine the hue and saturation encountered during scanning.

Apart from the obvious difficulties, some severe constraints exist on the addition of colour information:

- It must not require extra channel space.
- There must be no interference between the existing luminance information and the added information.
- It must be totally compatible, i.e. a monochrome receiver must display a colour transmission in black and white, and a colour receiver must be capable of displaying a monochrome transmission in black and white.

1.11.1 Colour bandwidth

There are some aspects of human colour perception that can be used to engineering advantage. One of these concerns the bandwidth required for colour display. Our ability to perceive fine detail in colour is considerably inferior to our awareness of fine detail as a brightness or luminance variation, thus colour information does not require the same wide bandwidth that is needed by a luminance system.

The eye can resolve small variations of luminance detail over the very small angle of 0.5–1.0′ (the precise figure varies with individuals). Visual acuity for colour is far less sensitive than for luminance, and depends to some degree on the range and saturation of the colours involved. Over an orange/cyan range of colours the resolution angle becomes 1.5–3.0′, and over a green/magenta range, detail resolution requires about 4–8′. The bandwidth required for colour information can thus be reduced to between 1/3 and 1/8 of the width required for luminance. Whatever methods we are going to use to process colour, it is only necessary to use about 1 MHz for the relatively low-detail colour information.

1.11.2 Spectrum utilization

Let us examine more closely the distribution of energy over the band of frequencies occupied by a carrier which is modulated by a video waveform. Without any video modulation, the carrier is already modulated by a number of frequencies. The line scanning rate is at a constant frequency (15 625 lines/second in system I), and it establishes side frequencies on each side of the carrier, spaced at line frequency and from each other, as shown in *Figure 1.18*. Interlaced scanning modulates the carrier at 50 Hz, and the side frequency of 50 Hz and the harmonics are spaced on both sides of the carrier and the line scanning harmonics. The picture frequency is another modulating signal at 25 Hz, and its energy joins the clusters around each line frequency harmonic.

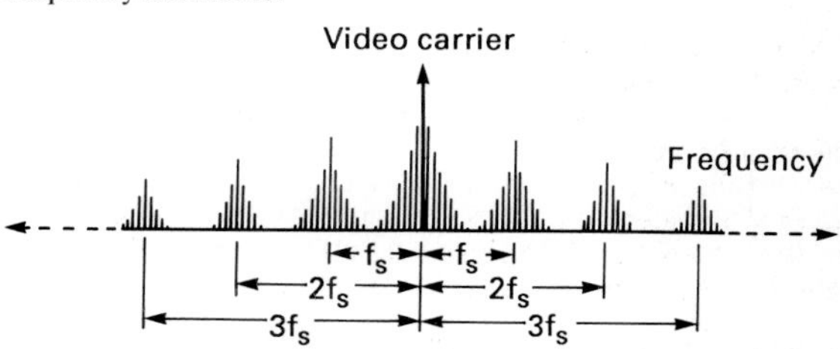

Figure 1.18 The signal components resulting from amplitude modulation. f_s is the line scanning frequency

The line frequency harmonics with their side frequencies extend out to the bandwidth limits of the video channel, and they are of a descending order of energy level as they become more remote from the carrier. Over the channel bandwidth, the energy generated across the band, without video modulation, is contained in clusters, spaced at line-scanning frequency from the carrier and from each other. Large regions of the channel space contain little or no energy.

In 1934, long before any broadcast television system existed, two mathematicians, Merz and Gray, studied the situation when video signals were included in the modulation process and concluded that the video energy simply joined the existing clusters round each line-scan harmonic, leaving the overall pattern substantially unchanged. The only differences between one picture and another, or whether one is moving or stationary, are relatively small variations in the magnitude of the energy in the clusters. The low or zero energy gaps between the line frequency harmonics remain with all types of video modulation.

The original work of Merz and Gray was directed at reducing interference between two adjacent transmitters, operating on nominally the same frequency. If their carrier frequencies were 'offset' by half the line frequency, the line harmonics of each would drop into the low-energy gaps of the other, thus minimizing mutual interference. However, in the years 1951–1953, the gaps were seen by the NTSC as the place for colour information.

A subcarrier can be inserted in the video frequency band, at a frequency about 1 MHz lower than the top of the video band, where the line-scan harmonics are low in amplitude. This subcarrier can then be modulated with colour information in such a manner that, when extracted and processed in the receiver, the colour display tube is driven with appropriate display information.

All existing standard colour systems are derived from the concept of the use of a subcarrier in the video band to supply colour information.

Acknowledgement

This section derives from a series of lectures given by the author during courses on Television Engineering, organized for the Royal Television Society. The RTS course lectures were arranged into a book *Television Engineering* and published by Pentech Press. We acknowledge permission given by Pentech Press for the use of some diagrams and text from the RTS book.

L.W.Turner FIEE,
Consultant Engineer

2 Quantities and Units

2.1 International unit system

The International System of Units (SI) is the modern form of the metric system agreed at an international conference in 1960. It has been adopted by the International Standards Organisation (ISO) and the International Electrotechnical Commission (IEC) and its use is recommended wherever the metric system is applied. It is now being adopted throughout most of the world and is likely to remain the primary world system of units of measurement for a very long time. The indications are that SI units will supersede the units of existing metric systems and all systems based on Imperial units.

SI units and the rules for their application are contained in *ISO Resolution* R1000 (1969, updated 1973) and an informatory document *SI-Le Système International d'Unités,* published by the Bureau International des Poids et Mesures (BIPM). An abridged version of the former is given in British Standards Institution (BSI) publication PD 5686 *The use of SI Units* (1969, updated 1973) and BS 3763 *International System (SI) Units;* BSI (1964) incorporates information from the BIPM document.

The adoption of SI presents less of a problem to the electronics engineer and the electrical engineer than to those concerned with other engineering disciplines as all the practical electrical units were long ago incorporated in the metre-kilogram-second (MKS) unit system and these remain unaffected in SI.

The SI was developed from the metric system as a fully coherent set of units for science, technology and engineering. A coherent system has the property that corresponding equations between quantities and between numerical values have exactly the same form, because the relations between units do not involve numerical conversion factors. In constructing a coherent unit system, the starting point is the selection and definition of a minimum set of independent 'base' units. From these, 'derived' units are obtained by forming products or quotients in various combinations, again without numerical factors. Thus the base units of length (metre), time (second) and mass (kilogram) yield the SI units of velocity (metre/second), force (kilogram-metre/second-squared) and so on. As a result there is, for any given physical quantity, only one SI unit with no alternatives and with no numerical conversion factors. A single SI unit (joule = kilogram metre-squared/second-squared) serves for energy of any kind, whether it be kinetic, potential, thermal, electrical, chemical..., thus unifying the usage in all branches of science and technology.

The SI has seven base units, and two supplementary units of angle. Certain important derived units have special names and can themselves be employed in combination to form alternative names for further derivations.

Each physical quantity has a quantity-symbol (e.g., m for mass) that represents it in equations, and a unit-symbol (e.g., kg for kilogram) to indicate its SI unit of measure.

2.1.1 Base units

Definitions of the seven base units have been laid down in the following terms. The quantity-symbol is given in italics, the unit-symbol (and its abbreviation) in roman type.

Length: l; metre (m). The length equal to 1 650 763.73 wavelengths in vacuum of the radiation corresponding to the transition between the levels $2p_{10}$ and $5d_5$ of the krypton-86 atom.

Mass: m; kilogram (kg). The mass of the international prototype kilogram (a block of platinum preserved at the International Bureau of Weights and Measures at Sèvres).

Time: t; second (s). The duration of 9 192 631 770 periods of the radiation corresponding to the transition between the two hyperfine levels of the ground state of the caesium-133 atom.

Electric current: i; ampere (A). The current which, maintained in two straight parallel conductors of infinite length, of negligible circular cross-section and 1 m apart in vacuum, produces a force equal to 2×10^{-7} newton per metre of length.

Thermodynamic temperature: T; kelvin (K). The fraction 1/273.16 of the thermodynamic (absolute) temperature of the triple point of water.

Luminous intensity: I; candela (cd). The luminous intensity in the perpendicular direction of a surface of 1/600 000 m^2 of a black body at the temperature of freezing platinum under a pressure of 101 325 newtons per square metre.

Amount of substance: Q; mole (mol). The amount of substance of a system which contains as many elementary entities as there are atoms in 0.012 kg of carbon-12. The elementary entity must be specified and may be an atom, a molecule, an ion, an electron, etc., or a specified group of such entities.

2.1.2 Supplementary units

Plane angle: α, β...; radian (rad). The plane angle between two radii of a circle which cut off on the circumference an arc of length equal to the radius.
Solid angle: **Ω**; steradian (sr). The solid angle which, having its vertex at the centre of a sphere, cuts off an area of the surface of the sphere equal to a square having sides equal to the radius.
Force: The base SI unit of electric current is in terms of force in newtons (N). A force of 1 N is that which endows unit mass (1 kg) with unit acceleration (1 m/s^2). The newton is thus not only a coherent unit; it is also devoid of any association with gravitational effects.

2.1.3 Temperature

The base SI unit of thermodynamic temperature is referred to a point of 'absolute zero' at which bodies possess zero thermal energy. For practical convenience two points on the Kelvin temperature scale, namely 273.15 K and 373.15 K, are used to define the Celsius (or Centigrade) scale (0°C and 100°C). Thus in terms of temperature *intervals*, 1 K = 1°C; but in terms of temperature *levels*, a Celsius temperature *θ* corresponds to a Kelvin temperature (θ+273.15)K.

2.1.4 Derived units

Nine of the more important SI derived units with their definitions are given below.

Quantity	*Unit name*	*Unit symbol*
Force	newton	N
Energy	joule	J
Power	watt	W
Electric charge	coulomb	C
Electrical potential difference and EMF	volt	V
Electric resistance	ohm	Ω
Electric capacitance	farad	F
Electric inductance	henry	H
Magnetic flux	weber	Wb

Newton That force which gives to a mass of 1 kilogram an acceleration of 1 metre per second squared.
Joule The work done when the point of application of 1 newton is displaced a distance of 1 metre in the direction of the force.
Watt The power which gives rise to the production of energy at the rate of 1 joule per second.
Coulomb The quantity of electricity transported in 1 second by a current of 1 ampere.
Volt The difference of electric potential between two points of a conducting wire carrying a constant current of 1 ampere, when the power dissipated between these points is equal to 1 watt.
Ohm The electric resistance between two points of a conductor when a constant difference of potential of 1 volt, applied between these two points, produces in this conductor a current of 1 ampere, this conductor not being the source of any electromotive force.
Farad The capacitance of a capacitor between the plates of which there appears a difference of potential of 1 volt when it is charged by a quantity of electricity equal to 1 coulomb.
Henry The inductance of a closed circuit in which an electromotive force of 1 volt is produced when the electric current in the circuit varies uniformly at a rate of 1 ampere per second.
Weber The magnet flux which, linking a circuit of one turn, produces in it an electromotive force of 1 volt as it is reduced to zero at a uniform rate in 1 second.

Some of the simpler derived units are expressed in terms of the seven basic and two supplementary units directly. Examples are listed in Table 2.1.

Table 2.1 Directly derived units

Quantity	*Unit name*	*Unit symbol*
Area	square metre	m^2
Volume	cubic metre	m^3
Mass density	kilogram per cubic metre	kg/m^3
Linear velocity	metre per second	m/s
Linear acceleration	metre per second squared	m/s^2
Angular velocity	radian per second	rad/s
Angular acceleration	radian per second squared	rad/s^2
Force	kilogram metre per second squared	kg m/s^2
Magnetic field strength	ampere per metre	A/m
Concentration	mole per cubic metre	mol/m^3
Luminance	candela per square metre	cd/m^2

Units in common use, particularly those for which a statement in base units would be lengthy or complicated, have been given special shortened names (see Table 2.2). Those that are named from scientists and engineers are abbreviated to an initial capital letter: all others are in lower-case letters.

Table 2.2 Named derived units

Quantity	*Unit name*	*Unit symbol*	*Derivation*
Force	newton	N	kg m/s^2
Pressure	pascal	Pa	N/m^2
Power	watt	W	J/s
Energy	joule	J	N m, W s
Electric charge	coulomb	C	A s
Electric flux	coulomb	C	A s
Magnetic flux	weber	Wb	V s
Magnetic flux density	tesla	T	Wb/m^2
Electric potential	volt	V	J/C, W/A
Resistance	ohm	Ω	V/A
Conductance	siemens	S	A/V
Capacitance	farad	F	A s/V, C/V
Inductance	henry	H	V s/A, Wb/A
Luminous flux	lumen	lm	cd sr
Illuminance	lux	lx	lm/m^2
Frequency	hertz	Hz	l/s

The named derived units are used to form further derivations. Examples are given in Table 2.3.

Names of SI units and the corresponding EMU and ESU CGS units are given in Table 2.4.

2.1.5 Gravitational and absolute systems

There may be some difficulty in understanding the difference between SI and the Metric Technical System of units which has been used principally in Europe. The main difference is that while mass is expressed in kg in both systems, weight (representing a force) is expressed as kgf, a gravitational unit, in the MKSA system and as N in SI. An absolute unit of force differs from a gravitational unit of force because it induces unit acceleration in a unit mass whereas a gravitational unit imparts gravitational acceleration to a unit mass.

A comparison of the more commonly known systems and SI is shown in Table 2.5.

2.1.6 Expressing magnitudes of SI units

To express magnitudes of a unit, decimal multiples and submultiples are formed using the prefixes shown in Table 2.6. This method of expressing magnitudes ensures complete adherence to a decimal system.

Table 2.3 Further derived units

Quantity	*Unit name*	*Unit symbol*
Torque	newton metre	N m
Dynamic viscosity	pascal second	Pa s
Surface tension	newton per metre	N/m
Power density	watt per square metre	W/m^2
Energy density	joule per cubic metre	J/m^3
Heat capacity	joule per kelvin	J/K
Specific heat capacity	joule per kilogram kelvin	J/(kg K)
Thermal conductivity	watt per metre kelvin	W/(m K)
Electric field strength	volt per metre	V/m
magnetic field strength	ampere per metre	A/m
Electric flux density	coulomb per square metre	C/m^2
Current density	ampere per square metre	A/m^2
Resistivity	ohm metre	Ω m
Permittivity	farad per metre	F/m
Permeability	henry per metre	H/m

Table 2.4 Unit names

Quantity	*Symbol*	*SI*	*EMU & ESU*
Length	*l*	metre (m)	centimetre (cm)
Time	*t*	second (s)	second
Mass	*m*	kilogram (kg)	gram (g)
Force	*F*	newton (N)	dyne (dyn)
Frequency	*f, v*	hertz (Hz)	hertz
Energy	*E, W*	joule (J)	erg (erg)
Power	*P*	watt (W)	erg/second (erg/s)
Pressure	*p*	newton/metre² (N/m^2)	dyne/centimetre² ($dyne/cm^2$)
Electric charge	*Q*	coulomb (C)	coulomb
Electric potential	*V*	volt (V)	volt
Electric current	*I*	ampere (A)	ampere
Magnetic flux	φ	weber (Wb)	maxwell (Mx)
Magnetic induction	*B*	tesla (T)	gauss (G)
Magnetic field strength	*H*	ampere turn/metre (At/m)	oersted (Oe)
Magnetomotive force	Fm	ampere turn (At)	gilbert (Gb)
Resistance	*R*	ohm (Ω)	ohm
Inductance	*L*	henry (H)	henry
Conductance	*G*	mho (Ω^{-1}) (siemens)	mho
Capacitance	*C*	farad (F)	farad

Table 2.5 Commonly used units of measurement

		SI (absolute)	*FPS (gravitational)*	*FPS (absolute)*	*cgs (absolute)*	*Metric technical units (gravitational)*
Length		metre (m)	ft	ft	cm	metre
Force		newton (N)	lbf	poundal (pdl)	dyne	kgf
Mass		kg	lb or slug	lb	gram	kg
Time		s	s	s	s	s
Temperature		°C K	°F	°F °R	°C K	°C K
Energy	mech.	joule*	ft lbf	ft pdl	dyne cm = erg	kgf m
	heat		Btu	Btu	calorie	kcal
Power	mech.	watt	hp	hp	erg/s	metric hp
	elec.		watt	watt		watt
Electric current		amp	amp	amp	amp	amp
Pressure		N/m^2	lbf/ft^2	pdl/ft^2	$dyne/cm^2$	kgf/cm^2

* 1 joule = 1 newton metre or 1 watt second.

Table 2.6 The internationally agreed multiples and submultiples

Factor by which the unit is multiplied		*Prefix*	*Symbol*	*Common everyday examples*
One million million (billion)	10^{12}	tera	T	
One thousand million	10^9	giga	G	gigahertz (GHz)
One million	10^6	mega	M	megawatt (MW)
One thousand	10^3	kilo	k	kilometre (km)
One hundred	10^2	hecto*	h	
Ten	10^1	deca*	da	decagram (dag)
UNITY	1			
One tenth	10^{-1}	deci*	d	decimetre (dm)
One hundredth	10^{-2}	centi*	c	centimetre (cm)
One thousandth	10^{-3}	milli	m	milligram (mg)
One millionth	10^{-6}	micro	μ	microsecond (μs)
One thousand millionth	10^{-9}	nano	n	nanosecond (ns)
One million millionth	10^{-12}	pico	p	picofarad (pF)
One thousand million millionth	10^{-15}	femto	f	
One million million millionth	10^{-18}	atto	a	

* To be avoided wherever possible.

2.1.7 Auxiliary units

Certain auxiliary units may be adopted where they have application in special fields. Some are acceptable on a temporary basis, pending a more widespread adoption of the SI system. Table 2.7 lists some of these.

Table 2.7 Auxiliary units

Quantity	*Unit symbol*	*SI equivalent*
Day	d	86 400 s
Hour	h	3600 s
Minute (time)	min	60 s
Degree (angle)	°	π/180 rad
Minute (angle)	′	π/10 800 rad
Second (angle)	″	π/648 000 rad
Acre	a	$1\ dam^2 = 10^2\ m^2$
Hectare	ha	$1\ hm^2 = 10^4\ m^2$
Barn	b	$100\ fm^2 = 10^{-28}\ m^2$
Standard atmosphere	atm	101 325 Pa
Bar	bar	$0.1\ MPa = 10^5\ Pa$
Litre	l	$1\ dm^3 = 10^{-3} m^3$
Tonne	t	$10^3\ kg = 1\ Mg$
Atomic mass unit	u	$1.660\ 53 \times 10^{-27}\ kg$
Angström	A	$0.1\ nm = 10^{-10}\ m$
Electron-volt	eV	$1.602\ 19 \times 10^{-19}\ J$
Curie	Ci	$3.7 \times 10^{10}\ s^{-1}$
Röntgen	R	$2.58 \times 10^{-4}\ C/kg$

2.2 Universal constants in SI units

Table 2.8 Universal constants

The digits in parentheses following each quoted value represent the standard deviation error in the final digits of the quoted value as computed on the criterion of internal consistency. The unified scale of atomic weights is used throughout ($^{12}C=12$). C=coulomb; G=gauss; Hz=hertz; J=joule; N=newton; T=tesla; u=unified nuclidic mass unit; W=watt; Wb=weber. For result multiply the numerical value by the SI unit.

Constant	*Symbol*	*Numerical value*	*SI unit*
Speed of light in vacuum	c	2.997 925(1)	10^{8} m/s
Gravitational constant	G	6.670(5)*	10^{-11} N m^{2} kg^{2}
Elementary charge	e	1.602 10(2)	10^{-19} C
Avogadro constant	N_A	6.022 52(9)	10^{26} kmol^{-1}
Mass unit	u	1.660 43(2)	10^{-27} kg
Electron rest mass	m_e	9.109 08(13)	10^{-31} kg
		5.485 97(3)	10^{-4} u
Proton rest mass	m_p	1.672 52(3)	10^{-27} kg
		1.007 276 63(8)	u
Neutron rest mass	m_n	1.674 82(3)	10^{-27} kg
		1.008 665 4(4)	u
Faraday constant	F	9.684 70(5)	10^{4} C/mol
Planck constant	h	6.625 59(16)	10^{-34} J s
	$h/2\pi$	1.054 494(25)	10^{-34} J s
Fine-structure constant	$\propto$	7.297 20(3)	10^{-3}
	$1/\alpha$	137.038 8(6)	
Charge-to-mass ratio for electron	e/m_e	1.758 796(6)	10^{11} C/kg
Quantum of magnetic flux	hc/e	4.135 56(4)	$10\$^{11}$ Wb
Rydberg constant	$R_\propto$	1.097 373 1(1)	10^{7} m^{-1}
Bohr radius	a_0	5.291 67(2)	10^{-11} m
Compton wavelength of electron	$h/m_e c$	2.426 21(2)	10^{-12} m
	$\lambda C/2\pi$	3.861 44(3)	10^{-13} m
Electron radius	$e^2/m_e c^2 = r_e$	2.817 77(4)	10^{-15} m
Thomson cross-section	$8\pi r_e^2/3$	6.651 6(2)	10^{-29} m^{2}
Compton wavelength of proton	$\lambda c,p$	1.321 398(13)	10^{-15} m
	$\lambda c,p/2\pi$	2.103 07(2)	10^{-16} m
Gyromagnetic ratio of proton	γ	2.675 192(7)	10^{8} rad/(s T)
	$\gamma/2\pi$	4.257 70(1)	10^{7} Hz/T
(uncorrected for diamagnetism of H_2O)	γ'	2.675 123(7)	10^{8} rad/(s T)
	$\gamma'/2\pi$	4.257 59(1)	10^{7} Hz/T
Bohr magneton	μB	9.273 2(2)	10^{-24} J/T
Nuclear magneton	μN	5.050 50(13)	10^{-27} J/T
Proton magnetic moment	μp	1.410 49(4)	10^{-26} J/T
	$\mu p/\mu N$	2.792 76(2)	
(uncorrected for diamagnetism in H_2O sample)	$\mu' p/\mu N$	2.792 68(2)	
Gas constant	R_0	8.314 34(35)	J/K mol
Boltzmann constant	k	1.380 54(6)	10^{-23} J/K
First radiation constant ($2\pi hc^2$)	c_1	3.741 50(9)	10^{-16} W/m^{2}
Second radiation constant (hc/k)	c_2	1.438 79(6)	10^{-2} m K
Stefan-Boltzmann constant	σ	5.669 7(10)	10^{-8} W/m^{2} K^{4}

* The universal gravitational constant is not, and cannot in our present state of knowledge, be expressed in terms of other fundamental constants. The value given here is a direct determination by P.R. Heyland and P. Chrzanowski, *J. Res. Natl. Bur. Std.* (U.S.) 29, 1 (1942).

The above values are extracts from *Review of Modern Physics* Vol. 37 No. 4 October 1965 published by the American Institute of Physics.

2.3 Metric to Imperial conversion factors

Table 2.9 Conversion factors

SI units	*British units*
SPACE AND TIME	
Length:	
1 μm (micron)	= 39.37 × 10^{-6} in
1 mm	= 0.039 370 1 in
1 cm	= 0.393 701 in
1 m	= 3.280 84 ft
1 m	= 1.093 61 yd
1 km	= 0.621 371 mile
Area:	
1 mm^{2}	= 1.550 × 10^{-3} in^{2}

Table 2.9 *continued*

SI units	*British units*
1 cm^{2}	= 0.155 0 in^{2}
1 m^{2}	= 10.763 9 ft^{2}
1 m^{2}	= 1.195 99 yd^{2}
1 ha	= 2.471 05 acre
Volume:	
1 mm^{3}	= 61.023 7 × 10^{-6} in^{3}
1 cm^{3}	= 61.023 7 × 10^{-3} in^{3}
1 m^{3}	= 35.314 7 ft^{3}
1 m^{3}	= 1.307 95 yd^{3}
Capacity:	
10^{6} m^{3}	= 219.969 × 10^{6} gal
1 m^{3}	= 219.969 gal
1 litre (l)	= 0.219 969 gal
	= 1.759 80 pint

Table 2.9 *continued*

SI units	*British units*
Capacity flow:	
10^3 m³/s	= 791.9 × 10^6 gal/h
1 m³/s	= 13.20 × 10^3 gal/min
1 litre/s	= 13.20 gal/min
1 m³/kW h	= 219.969 gal/kW h
1 m³/s	= 35.314 7 ft³/s (cusecs)
1 litre/s	= 0.588 58 × 10^{-3} ft³/min (cfm)
Velocity:	
1 m/s	= 3.280 84 ft/s = 2.236 94 mile/h
1 km/h	= 0.621 371 mile/h
Acceleration:	
1 m/s²	= 3.280 84 ft/s²
MECHANICS	
Mass:	
1 g	= 0.035 274 oz
1 kg	= 2.204 62 lb
1 t	= 0.984 207 ton = 19.684 1 cwt
Mass flow:	
1 kg/s	= 2.204 62 lb/s = 7.936 64 klb/h
Mass density:	
1 kg/m³	= 0.062 428 lb/ft³
1 kg/litre	= 10.022 119 lb/gal
Mass per unit length:	
1 kg/m	= 0.671 969 lb/ft = 2.015 91 lb/yd
Mass per unit area:	
1 kg/m²	= 0.204 816 lb/ft²
Specific volume:	
1 m³/kg	= 16.018 5 ft³/lb
1 litre/tonne	= 0.223 495 gal/ton
Momentum:	
1 kg m/s	= 7.233 01 lb ft/s
Angular momentum:	
1 kg m²/s	= 23.730 4 lb ft²/s
Moment of inertia:	
1 kg m²	= 23.730 4 lb ft²
MECHANICS	
Force:	
1 N	= 0.224 809 lbf
Weight (force) per unit length:	
1 N/m	= 0.068 521 lbf/ft
	= 0.205 566 lbf/yd
Moment of force (or torque):	
1 N m	= 0.737 562 lbf ft
Weight (force) per unit area:	
1 N/m²	= 0.020 885 lbf/ft²
Pressure:	
1 N/m²	= 1.450 38 × 10^{-4} lbf/in²
1 bar	= 14.503 8 lbf/in²
1 bar	= 0.986 923 atmosphere
1 mbar	= 0.401 463 in H_2O
	= 0.029 53 in Hg
Stress:	
1 N/mm²	= 6.474 90 × 10^{-2} tonf/in²
1 MN/m²	= 6.474 90 × 10^{-2} tonf/in²
1 hbar	= 0.647 490 tonf/in²
Second moment of area:	
1 cm⁴	= 0.024 025 in⁴
Section modulus:	
1 m³	= 61 023.7 in³
1 cm³	= 0.061 023 7 in³
Kinematic viscosity:	
1 m²/s	= 10.762 75 ft²/s = 10^6 cSt
1 cSt	= 0.038 75 ft²/h
Energy, work:	
1 J	= 0.737 562 ft lbf
1 MJ	= 0.372 5 hph
1 MJ	= 0.277 78 kW h
Power:	
1 W	= 0.737 562 ft lbf/s
1 kW	= 1.341 hp = 737.562 ft lbf/s

Table 2.9 *continued*

SI units	*British units*
Fluid mass:	
(Ordinary) 1 kg/s	= 2.204 62 lb/s = 793 6.64 lb/h
(Velocity) 1 kg/m²s	= 0.204 815 lb/ft²s
HEAT	
Temperature:	
(Interval) 1 K	= 9/5 deg R (Rankine)
1°C	= 9/5 deg F
(Coefficient)1°R^{-1}	= 1 deg F^{-1} = 5/9 deg C
1°C^{-1}	= 5/9 deg F^{-1}
Quantity of heat:	
1 J	= 9.478 17 × 10^{-4} Btu
1 J	= 0.238 846 cal
1 kJ	= 947.817 Btu
1 GJ	= 947.817 × 10^3 Btu
1 kJ	= 526.565 CHU
1 GJ	= 526.565 × 10^3 CHU
1 GJ	= 9.478 17 therm
Heat flow rate:	
1 W(J/s)	= 3.412 14 Btu/h
1 W/m²	= 0.316 998 Btu/ft² h
Thermal conductivity:	
1 W/m °C	= 6.933 47 Btu in/ft² h °F
Coefficient and heat transfer:	
1 W/m² °C	= 0.176 110 Btu/ft² h °F
Heat capacity:	
1 J/°C	= 0.526 57 × 10^{-3} Btu/°R
Specific heat capacity:	
1 J/g °C	= 0.238 846 Btu/lb °F
1 kJ/kg °C	= 0.238 846 Btu/lb °F
Entropy:	
1 J/K	= 0.526 57 × 10^{-3} Btu/°R
Specific entropy:	
1 J/kg °C	= 0.238 846 × 10^{-3} Btu/lb °F
1 J/kg K	= 0.238 846 × 10^{-3} Btu/lb °R
Specific energy/specific latent heat:	
1 J/g	= 0.429 923 Btu/lb
1 J/kg	= 0.429 923 × 10^{-3} Btu/lb
Calorific value:	
1 kJ/kg	= 0.429 923 Btu/lb
1 kJ/kg	= 0.773 861 4 CHU/lb
1 J/m³	= 0.026 839 2 × 10^{-3} Btu/ft³
1 kJ/m³	= 0.026 839 2 Btu/ft³
1 kJ/litre	= 4.308 86 Btu/gal
1 kJ/kg	= 0.009 630 2 therm/ton
ELECTRICITY	
Permeability:	
1 H/m	= $10^7/4\pi\ \mu_0$
Magnetic flux density:	
1 tesla	= 10^4 gauss = 1 Wb/m²
Conductivity:	
1 mho	= 1 reciprocal ohm
1 siemens	= 1 reciprocal ohm
Electric stress:	
1 kV/mm	= 25.4 kV/in
1 kV/m	= 0.025 4 kV/in

2.4 Symbols and abbreviations

Table 2.10 Quantities and units of periodic and related phenomena (based on ISO Recommendation R31)

Symbol	*Quantity*
T	periodic time
τ, (*T*)	time constant of an exponentially varying quantity
f, *v*	frequency

Table 2.10 *continued*

Symbol	*Quantity*
η	rotational frequency
ω	angular frequency
λ	wavelength
σ ($\tilde{\nu}$)	wavenumber
K	circular wavenumber
$\log_e (A_1/A_2)$	natural logarithm of the ratio of two amplitudes
$10 \log_{10} (P_1/P_2)$	ten times the common logarithm of the ratio of two powers
δ	damping coefficient
Λ	logarithmic decrement
α	attenuation coefficient
β	phase coefficient
γ	propagation coefficient

Table 2.11 Symbols for quantities and units of electricity and magnetism (based on ISO Recommendation R31)

Symbol	*Quantity*
I	electric current
Q	electric charge, quantity of electricity
ρ	volume density of charge, charge density (Q/V)
σ	surface density of charge (Q/A)
E, (K)	electric field strength
V, (φ)	electric potential
U, (V)	potential difference, tension
E	electromotive force
D	displacement (rationalised displacement)
D'	non-rationalised displacement
ψ	electric flux, flux of displacement (flux of rationalised displacement)
ψ'	flux of non-rationalised displacement
C	capacitance
ε	permittivity
ε_o	permittivity of vacuum
ε'	non-rationalised permittivity
ε'_o	non-rationalised permittivity of vacuum
ε_r	relative permittivity
χ_e	electric susceptibility
χ'_e	non-rationalised electric susceptibility
P	electric polarisation
p, (P_e)	electric dipole moment
J, (S)	current density
A, (α)	linear current density
H	magnetic field strength
H'	non-rationalised magnetic field strength
U_m	magnetic potential difference
F, (F_m)	magnetomotive force
B	magnetic flux density, magnetic induction
Φ	magnetic flux
A	magnetic vector potential
L	self-inductance
M, (L)	mutual inductance
k, (x)	coupling coefficient
σ	leakage coefficient
μ	permeability
μ_o	permeability of vacuum
μ'	non-rationalised permeability
μ'_o	non-rationalised permeability of vacuum
μ_r	relative permeability
k, (χ_m)	magnetic susceptibility
k', (χ'_m)	non-rationalised magnetic susceptibility
m	electromagnetic moment (magnetic moment)
H_v (M)	magnetisation
J, (B_i)	magnetic polarisation
J'	non-rationalised magnetic polarisation
w	electromagnetic energy density
S	Poynting vector

Table 2.11 *continued*

Symbol	*Quantity*
c	velocity of propagation of electromagnetic waves *in vacuo*
R	resistance (to direct current)
G	conductance (to direct current)
ρ	resistivity
y, σ	conductivity
R, R_m	reluctance
Λ, (P)	permeance
N	number of turns in winding
m	number of phases
p	number of pairs of poles
φ	phase displacement
Z	impedance (complex impedance)
[Z]	modulus of impedance (impedance)
X	reactance
R	resistance
Q	quality factor
Y	admittance (complex admittance)
[Y]	modulus of admittance (admittance)
B	susceptance
G	conductance
P	active power
S, (P_s)	apparent power
Q, (P_q)	reactive power

Table 2.12 Symbols for quantities and units of acoustics (based on ISO Recommendation R31)

Symbol	*Quantity*
T	period, periodic time
f, ν	frequency, frequency interval
ω	angular frequency, circular frequency
λ	wavelength
k	circular wavenumber
ρ	density (mass density)
P_s	static pressure
p	(instantaneous) sound pressure
ε, (x)	(instantaneous) sound particle displacement
u, v	(instantaneous) sound particle velocity
a	(instantaneous) sound particle acceleration
q, U	(instantaneous) volume velocity
c	velocity of sound
E	sound energy density
P, (N, W)	sound energy flux, sound power
I, J	sound intensity
Z_s, (W)	specific acoustic impedance
Z_a, (Z)	acoustic impedance
Z_m, (w)	mechanical impedance
L_p, (L_N, L_W)	sound power level
L_p, (L)	sound pressure level
δ	damping coefficient
Λ	logarithmic decrement
α	attenuation coefficient
β	phase coefficient
γ	propagation coefficient
δ	dissipation coefficient
r, τ	reflection coefficient
γ	transmission coefficient
α, (α_a)	acoustic absorption coefficient
R	sound reduction index sound transmission loss
A	equivalent absorption area of a surface or object
T	reverberation time
L_N, (Λ)	loudness level
N	loudness

Table 2.13 Some technical abbreviations and symbols

Quantity	*Abbreviation*	*Symbol*
Alternating current	ac	
Ampere	A or amp	
Amplification factor		μ
Amplitude modulation	am	
Angular velocity		ω
Audio frequency	af	
Automatic frequency control	afc	
Automatic gain control	agc	
Bandwidth		Δf
Beat frequency oscillator	bfo	
British thermal unit	Btu	
Cathode-ray oscilloscope	cro	
Cathode-ray tube	crt	
Celsius	C	
Centi-	c	
Centimetre	cm	
Square centimetre	cm^2 or sq cm	
Cubic centimetre	cm^3 or cu cm or cc	
Centimetre-gram-second	cgs	
Continuous wave	cw	
Coulomb	C	
Deci-	d	
Decibel	dB	
Direct current	dc	
Direction finding	df	
Double sideband	dsb	
Efficiency		η
Equivalent isotropic radiated power	eirp	
Electromagnetic unit	emu	
Electromotive force instantaneous value	emf	E or V, e or v
Electron-volt	eV	
Electrostatic unit	esu	
Fahrenheit	F	
Farad	F	
Frequency	freq.	f
Frequency modulation	fm	
Gauss	G	
Giga-	G	
Gram	g	
Henry	H	
Hertz	Hz	
High frequency	hf	
Independent sideband	isb	
Inductance-capacitance		L-C
Intermediate frequency	if	
Kelvin	K	
Kilo-	k	
Knot	kn	
Length		l
Local oscillator	lo	
Logarithm, common		log or $\log_{10}$
Logarithm, natural		ln or $\log_e$
Low frequency	lf	
Low tension	lt	
Magnetomotive force	mmf	F or M
Mass		m
Medium frequency	mf	
Mega-	M	
Metre	m	
Metre-kilogram-second	mks	
Micro-	μ	
Micromicro-	p	
Micron		μ
Milli-	m	
Modulated continuous wave	mcw	
Nano-	n	
Neper	N	
Noise factor		N
Ohm		Ω
Peak to peak	p-p	
Phase modulation	pm	
Pico-	p	
Plan-position indication	PPI	
Potential difference	pd	V
Power factor	pf	
Pulse repetition frequency	prf	
Radian	rad	
Radio frequency	rf	
Radio telephony	R/T	
Root mean square	rms	
Short-wave	sw	
Single sideband	ssb	
Signal frequency	sf	
Standing wave ratio	swr	
Super-high frequency	shf	
Susceptance		B
Travelling-wave tube	twt	
Ultra-high frequency	uhf	
Very high frequency	vhf	
Very low frequency	vlf	
Volt	V	
Voltage standing wave ratio	vswr	
Watt	W	
Weber	Wb	
Wireless telegraphy	W/T	

Table 2.14 Greek alphabet and symbols

Name	*Symbol*		*Quantities used for*
alpha	A	α	angles, coefficients, area
beta	B	β	angles, coefficients
gamma	Γ	γ	specific gravity
delta	Δ	δ	density, increment, finite difference operator
epsilon	E	ε	Napierian logarithm, linear strain, permittivity, error, small quantity
zeta	Z	ζ	coordinates, coefficients, impedance (capital)
eta	H	η	magnetic field strength, efficiency
theta	Θ	θ	angular displacement, time
iota	I	ι	inertia
kappa	K	κ	bulk modulus, magnetic susceptibility
lambda	Λ	λ	permeance, conductivity, wavelength
mu	M	μ	bending moment, coefficient of friction, permeability
nu	N	ν	kinematic viscosity, frequency, reluctivity
xi	Ξ	ζ	output coefficient
omicron	O	ο	
pi	Π	π	circumference ÷ diameter
rho	P	ρ	specific resistance

Table 2.14 *continued*

Name	*Symbol*		*Quantities used for*
sigma	Σ	σ	summation (capital), radar cross-section, standard deviation
tau	T	τ	time constant, pulse length
upsilon	Y	*u*	
phi	Φ	φ	flux, phase
chi	X	χ	reactance (capital)
psi	Ψ	ψ	angles
omega	Ω	ω	angular velocity, ohms

References

1 COHEN, E.R. and TAYLOR, B.N., *Journal of Physical and Chemical Reference Data*, vol. 2, 663 (1973)
2 'Recommended values of physical constants'. CODATA (1973)
3 McGLASHAN, M. L., *Physicochemical quantities and units*, London: The Royal Institute of Chemistry (1971)

P Sproxton,
Alpha Image Ltd

3 Analogue and Digital Circuit Theory

3.1 Analogue circuit theory

3.1.1 Resistors

Resistors are the simplest linear components encountered when applying circuit theory techniques. The symbol for a resistor is shown in *Figure 3.1* (a). If an emf V (measured in volts) is applied across the terminals of the resistor (resistance R, measured in ohms), it can be shown, as in (b), that the current flowing through the resistor is linearly proportional to the current I (measured in amperes).

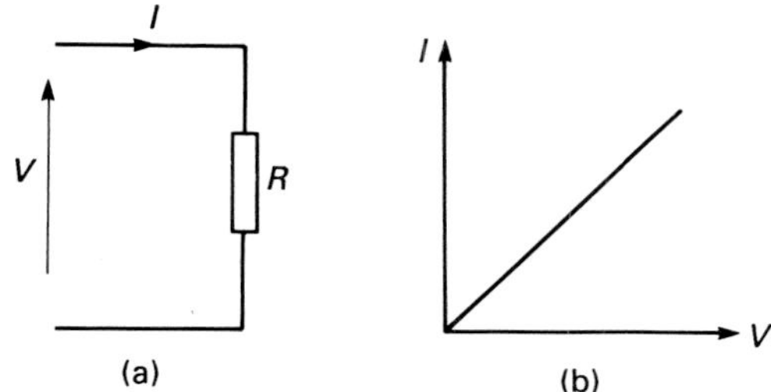

Figure 3.1 Resistor and voltage/current characteristic

This relationship is better known as Ohm's law, i.e.:

$$I = \frac{V}{R} \tag{3.1}$$

Another important relationship here is the power P (in watts) which is dissipated in the resistor. This is given by:

$$\begin{aligned} P &= VI \\ &= I^2R \end{aligned} \tag{3.2}$$

3.1.2 Series resistance circuits

A series configuration for resistors is illustrated in *Figure 3.2*. In this circuit, the current flowing through all three resistors is the same. Therefore according to equation (3.1) the potential difference across each resistance is given by:

$$V_1 = R_1,I \; ; V_2 = R_2I \; ; V_3 = R_3I$$

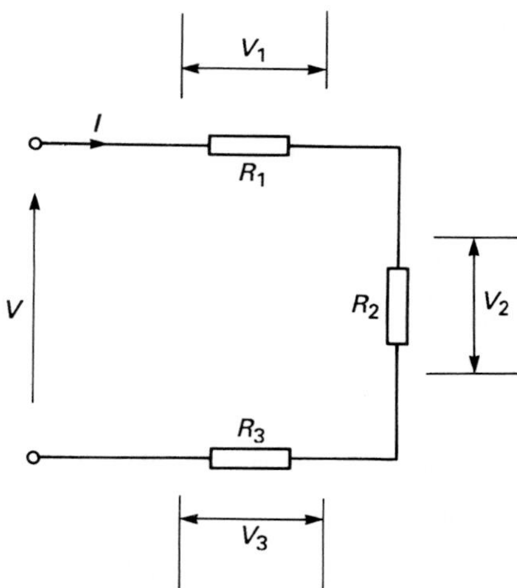

Figure 3.2 Series resistance circuit

The sum of these three potential differences is equal to the applied voltage V, i.e.:

$$V = V_1 + V_2 + V_3 \tag{3.3}$$

Equation (3.3) indicates that the algebraic sum of the potential differences around any complete circuit is equal to zero.

Using equation (3.1) we can re-arrange equation (3.3) to give:

$$\begin{aligned} V &= IR_1 + IR_2 + IR_3 \\ &= I(R_1 + R_2 + R_3) \end{aligned}$$

Therefore

$$I = \frac{V}{R_1 + R_2 + R_3} = \frac{V}{R_e} \tag{3.4}$$

The equivalent resistance of the circuit in *Figure 3.2* is therefore R_e, which is given by

$$R_e = R_1 + R_2 + R_3 \tag{3.5}$$

Equation (3.5) can be stated more generally: If any number of resistors are connected in series, then the equivalent resistance is the sum of the individual values.

3.1.3 Parallel resistance circuits

A second way of configuring resistors is shown in *Figure 3.3*. In this circuit, the voltage across each resistor is the same, but the current in each is given by:

$$I_1 = \frac{V}{R_1} \; ; \; I_2 = \frac{V}{R_2} \; ; \; I_3 = \frac{V}{R_3} \quad (3.6)$$

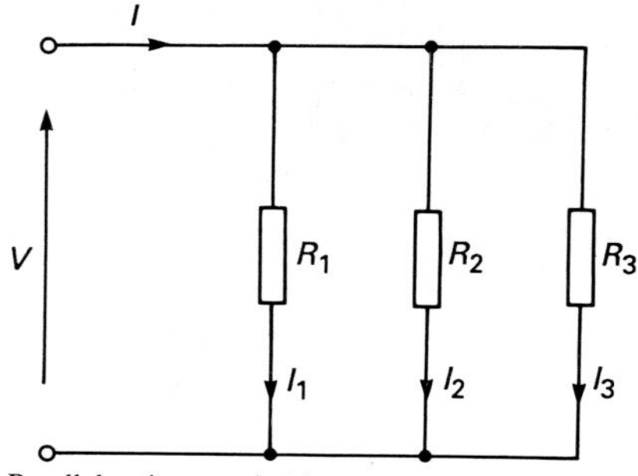

Figure 3.3 Parallel resistance circuit

Here, the total supplied current equals the sum of the currents through each resistor. i.e.:

$$I = I_1 + I_2 + I_3 \quad (3.7)$$

Substituting for currents in equation (3.6):

$$I = \frac{V}{R_1} + \frac{V}{R_2} + \frac{V}{R_3} = V\left(\frac{1}{R_1} + \frac{1}{R_2} + \frac{1}{R_3}\right) \quad (3.8)$$

From Ohm's law (equation (3.1)) we have

$$I = \frac{V}{R_e} = V\left(\frac{1}{R_1} + \frac{1}{R_2} + \frac{1}{R_3}\right) \quad (3.9)$$

Therefore

$$\frac{1}{R_e} = \frac{1}{R_1} + \frac{1}{R_2} + \frac{1}{R_3} \quad (3.10)$$

This can be more generally stated by:

$$\frac{1}{R_e} = \frac{1}{R_1} + \frac{1}{R_2} + \frac{1}{R_3} + \ldots + \frac{I}{R_n} \quad (3.11)$$

i.e. the reciprocal of the equivalent resistance is equal to the sum of the reciprocals of the component resistances.

3.1.4 Voltage dividers

One very common implementation of resistor networks is the voltage divider (see *Figure 3.4*). For this circuit, the output is always smaller than the input, in a ratio which is determined by the value of R_1 and R_2. The current through R_1 and R_2 is given by:

$$I = \frac{V_i}{R_1 + R_2} \quad (3.12)$$

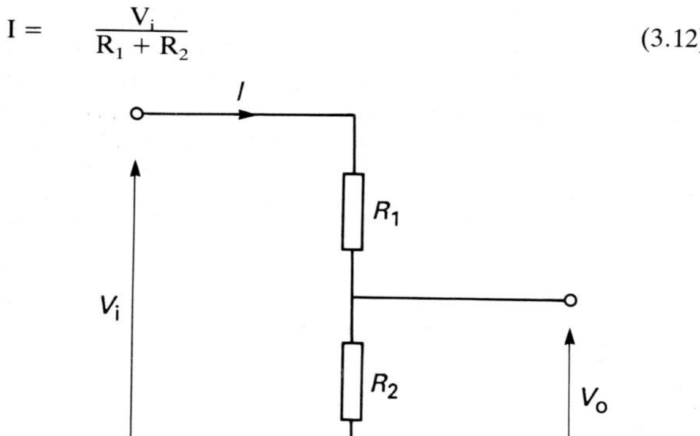

Figure 3.4 Voltage divider

and, as $V_o = IR_2$:

$$V_o = \frac{R_2}{R_1 + R_2} V_i \quad (3.13)$$

3.1.5 Kirchoff's laws

The circuit in *Figure 3.5* (a) contains a parallel set of resistances with two *nodes* indicated. A node is a point where three or more conductors are joined. Kirchoff's first law states: The total current flowing towards a node is equal to the total current flowing away from that node, i.e. the algebraic sum of currents flowing towards a node is zero. Thus in *Figure 3.5* (b) at node B:

$$I_1 + I_2 + I_3 = I$$

or

$$I_1 + I_2 + I_3 - I = O$$

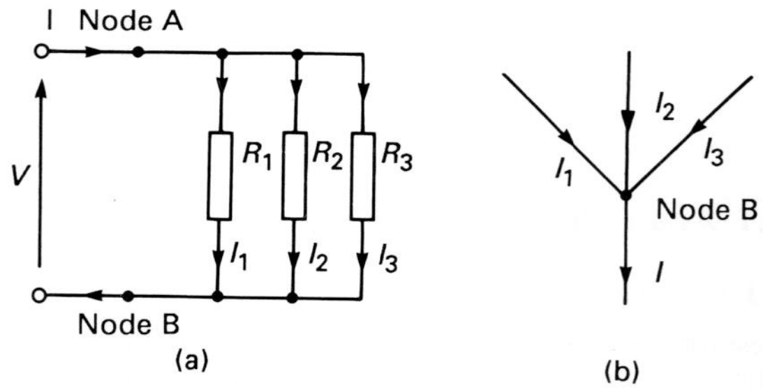

Figure 3.5 Kirchoff's current law

In general, for any node,

$$\Sigma I = O \quad (3.14)$$

The circuit in *Figure 3.6* illustrates Kirchoff's second law, which states: The algebraic sum of the potential differences is zero, i.e.:

$$V = V_1 + V_2 + V_3$$

or

$$V_1 + V_2 + V_3 - V = O$$

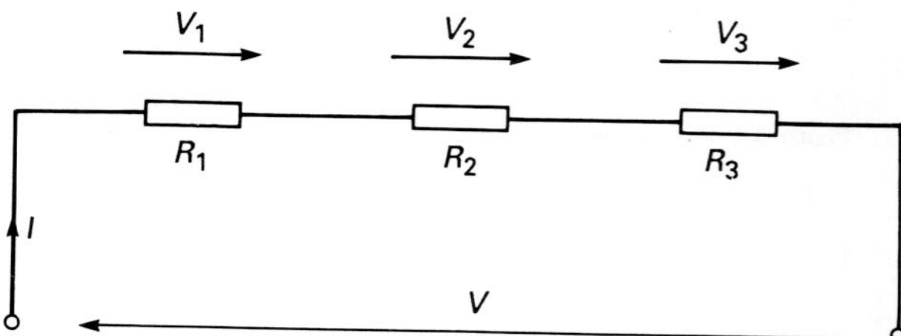

Figure 3.6 Kirchoff's voltage law

Generally, for any node,

$$\Sigma V = O \quad (3.15)$$

3.1.6 Equivalent circuits

To simplify circuit analysis, it is often necessary to reduce portions of a circuit to a simpler equivalent form. This is in order to clarify areas of the circuit that are of particular interest. The following two sections describe two theorems that enable some networks to be simplified.

3.1.6.1 Thevenin's theorem

Consider the network and load resistance in *Figure 3.7* (a). Thevenin's theorem can be stated as follows: An active network, having two terminals, A and B, can be replaced by a constant voltage source having an emf V_o and an internal resistance r. The value of V_o is equal to the open circuit potential between A and B, and r is the resistance of the network measured between A and B, with the load disconnected and the sources of emf replaced by their internal resistances. On this basis, the network in *Figure 7*(a) can be redrawn as in (b).

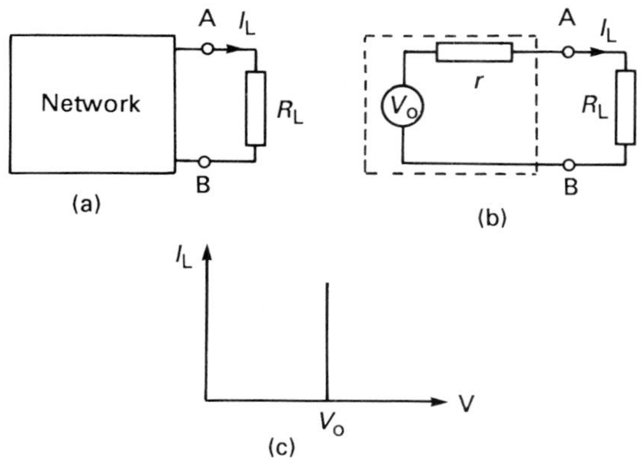

Figure 3.7 Thevenin's equivalent circuit

3.1.6.2 Norton's theorem

This is another theorem for an equivalent circuit. Norton's theorem states that: Any two-terminal network consisting of dc voltage sources and resistors can be replaced by a parallel combination of a current source I_e and a resistance R_e (see *Figure 3.8*). The current source is the short-circuit current at the output terminals and R_e is the same as for Thevenin's theorem. The Norton equivalent is illustrated in *Figure 3.8*, with the characteristic of the current generator in (c).

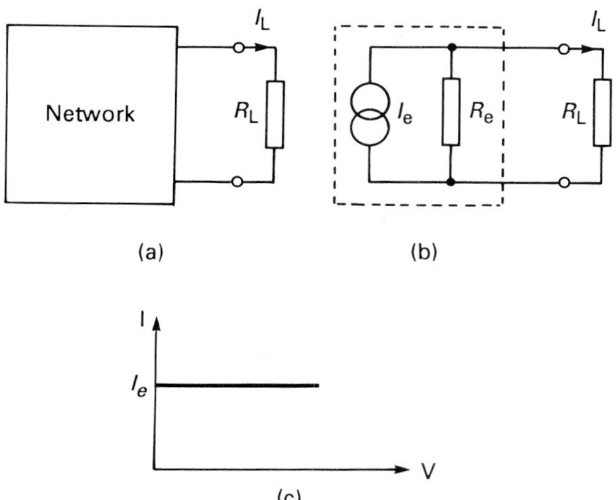

Figure 3.8 Norton's equivalent circuit

3.2 Alternating current circuits

In practice, the currents and voltages most often found in electronic circuits vary with time. The simplest time-varying waveform is one where the current or voltage changes direction periodically; this is termed an *alternating current* (ac).

The simplest ac waveform is the sine-wave, where current or voltage vary sinusoidally with time. A sinusoidal waveform (*Figure 3.9*) is generated by the variation of the vertical component of a vector rotating counter-clockwise with uniform angular velocity. A single revolution is termed a *cycle*, where the elapsed time interval for one revolution is the period T. The number of cycles per second is the *frequency*, f, of the sine-wave and the *period* is the reciprocal of the frequency.

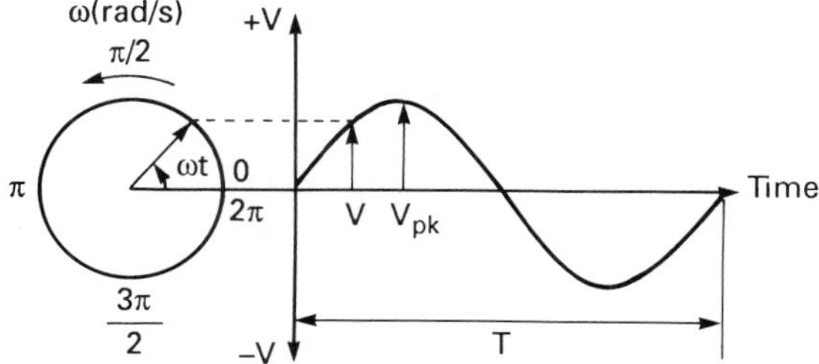

Figure 3.9 Rotating vector representation of a sinusoidal quantity

For a complete cycle, the vector will turn through 2π radians, therefore the angular frequency (in radians per second) is given by:

$$\omega = 2\pi f \quad (3.16)$$

If the magnitude of the rotating vector is V_{pk}, the instantaneous value of v, at any time t, is given by:

$$v = V_{pk}\sin\omega t \quad (3.17)$$

To produce a more generalized expression for the instantaneous voltage, the *phase* of the sine-wave must be considered. *Figure 3.10* illustrates two sinusoidal voltage waveforms with different phases.

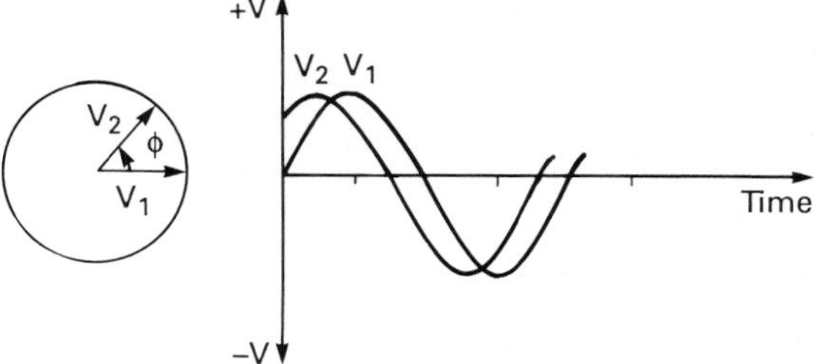

Figure 3.10 Vector representation of phase sinusoidal quantity

This shows that voltage V_2 is leading V_1, because V_2 has passed through zero in advance of V_1. In fact the voltage V_2 is *leading* voltage V_1 by the phase angle ø. The phase angle can only be specified between sine-waves of the same frequency. It is not possible to completely describe a sine-wave in terms of its amplitude and frequency unless it is being compared to a reference waveform of the same frequency. Hence, a more general expression for instantaneous voltage can be stated thus:

$$v = V_{pk}\sin(\omega t + \phi) \quad (3.18)$$

In working with ac circuits, there are two circuit elements that become particularly significant. These are capacitors and inductors.

3.2.1 Capacitors

The circuit in *Figure 3.11*(a) contains a capacitor C consisting of two parallel plates separated by an insulator (e.g. air) with a dc voltage applied across the plates.

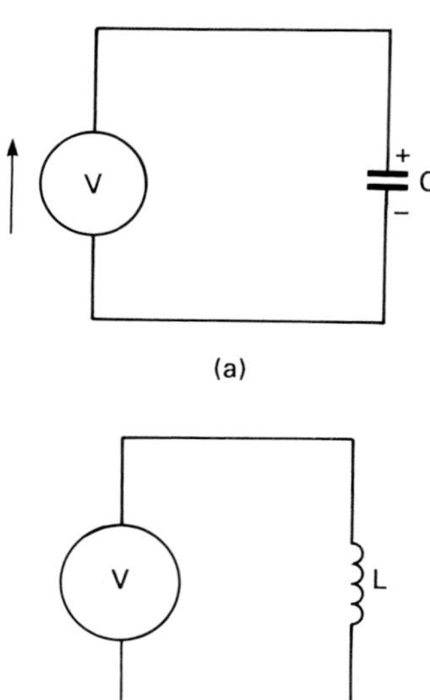

Figure 3.11 Capacitor and inductor circuits

A potential difference exists between the plates of the capacitor, and hence a positive charge will develop on the plate connected to the positive battery terminal and a negative charge on the plate connected to the negative terminal. The charge, q, on the plates is proportional to the voltage across them, i.e.:

$$q = CV \tag{3.19}$$

where C is a constant, called the capacitance, which depends upon the size, shape and separation of the plates, and the type of insulator between the plates.

In an ac circuit, the effect of the voltage changing with time will give rise to a time-varying charge on the capacitor plates. This is equivalent to a current through the circuit:

$$i = \frac{dq}{dt} \tag{3.20}$$

Substituting for q from equation (3.19),

$$i = C\frac{dv}{dt} \tag{3.21}$$

For a sinusoidal exciting voltage v, equations (3.19) and (3.20) give the current i in the circuit as:

$$\begin{aligned} i &= C\frac{d}{dt}(V_{pk}\sin\omega t) \\ &= \omega CV_{pk}\cos\omega t \\ &= \omega CV_{pk}\sin(\omega t + \pi/2) \end{aligned} \tag{3.22}$$

Equation (3.22) shows the current to be sinusoidal and leading the voltage by a phase $\pi/2$.

3.2.2 Inductors

An inductor consists of a coil of wire around a magnetic circuit, a circuit which may, for example, consist of iron or air. The symbol for the inductor is shown connected to a voltage source V in *Figure 3.11*(b). The current in an inductor produces a magnetic flux around that inductor. The magnetic field will vary as the current changes, which will, in turn, induce an emf in the circuit. This emf is given by:

$$v = L\frac{di}{dt} \tag{3.23}$$

Equation (3.23) indicates that the voltage v across the inductance is proportional to the rate of change of current through it. The proportional constant, L, is determined by the size, shape and magnetic properties of the inductor. Substituting a sinusoidal exciting current for the current in equation (3.23).

$$\begin{aligned} V &= L\frac{d}{dt}(I_{pk}\sin\omega t) \\ &= \omega LI_{pk}\cos\omega t \\ &= \omega LI_{pk}\sin(\omega t + \pi/2) \end{aligned} \tag{3.24}$$

Equation (3.24) shows that the current through the inductor *lags* the voltage by a phase angle of $\pi/2$.

3.2.2.1 *Complex representation of sinusoidal quantities*

We have seen in section *3.2* that sinusoidally varying voltages or currents can be represented by a rotating vector (*Figure 3.9*). By representing this vector on an Argand diagram, it can be described by a complex number.

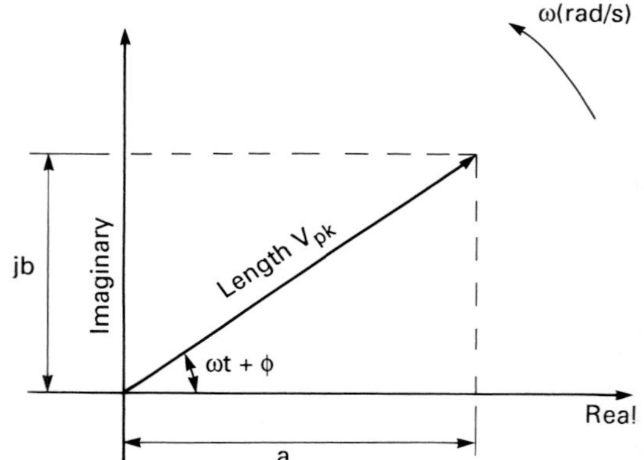

Figure 3.12 Complex representation of a sinusoidal quantity

For a complex number of the form $a + jb$, *Figure 3.12* shows that

$$a = V_{pk}\cos(\omega t + \phi)$$
$$b = V_{pk}\sin(\omega t + \phi)$$

Therefore

$$v = V_{pk}\cos(\omega t + \phi) + jV_{pk}\sin(\omega t + \phi) \tag{3.25}$$

This result can also be represented exponentially:

$$v = V_{pk}e^{j(wt + \phi)} \tag{3.26}$$

By this means, the inductor and capacitor can be represented in terms of their complex impedance. Consider first the inductor;

$$\begin{aligned} V_{pk}e^{j(wt+\phi)} &= L\frac{d}{dt}(I_{pk}e^{jwt}) \\ &= j\omega LI_{pk}e^{jwt} \\ \therefore V &= j\omega LI \end{aligned} \tag{3.27}$$

By comparing equation (3.27) with Ohm's law (equation (3.1)), it can be reduced to

$$V = ZI \tag{3.28}$$

This is the ac form of Ohm's law, where Z is the complex impedance. Similarly, for the capacitor;

$$I_{pk}e^{j(w+\phi)} = C\frac{d}{dt}(V_{pk}e^{jwt})$$

$$= j\omega C V_{pk} e^{jwt}$$
$$\therefore I = j\omega C V \qquad (3.29)$$

Here the complex impedance for the capacitor is $Z = 1/j\omega C$. From equation (3.29) it can be seen that the impedance for the capacitor is $1/j\omega C$, and from equation (3.27) the impedance for the inductor is $j\omega L$. This indicates that they are both imaginary quantities. This means that the voltage and current are always 90° out of phase, i.e. these circuits are purely *reactive*. In practice circuits always have some resistance, which results in a real part to the impedance.

Consider the circuit in *Figure 3.13*.

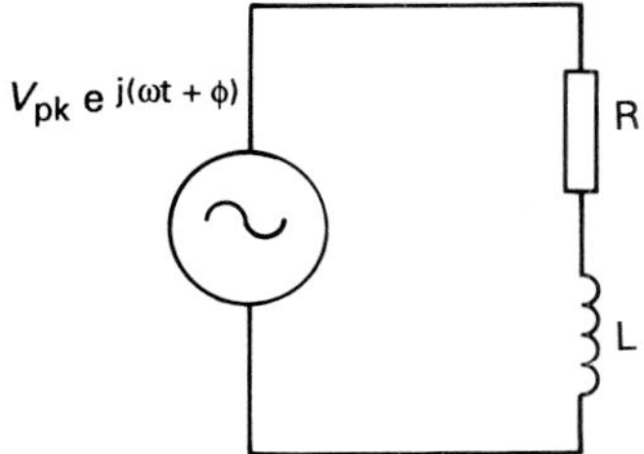

Figure 3.13 RL series circuit

The differential equation for this circuit is:

$$v = R_i + \frac{di}{dt} \qquad (3.30)$$

Allowing for a phase angle between the current and voltage, equation (3.30) can be expressed in its complex form, i.e.:

$$V_{pk} e^{j(wt+\phi)} = RI_{pk} e^{jwt} + j\omega L I_{pk} e^{jwt}$$
$$= (R + j\omega L) I_{pk} e^{jwt}$$
$$\therefore V = (R + j\omega L) I \qquad (3.31)$$

Equation (3.31) shows that the impedance for the RL series circuit is given by

$$Z = R + j\omega L \qquad (3.32)$$

which has a real part *R* and an imaginery part $j\omega L$. The impedance phase angle can be found for equation (3.32), as generally, for a complex number $a + jb$, the angle is arctan a/b, or for equation (3.32):

$$\phi = \arctan \frac{\omega L}{R}$$

or more generally:

$$\phi = \arctan \frac{X}{R} \qquad (3.33)$$

where *X* is the pure reactance, and *R* is the pure resistance.

To summarize, sinusoidally excited circuits, consisting of inductance, capacitance and resistance can be analysed using the component complex impedance:

pure resistive impedance $Z_r = R$
pure capacitive impedance $Z_c = 1/j\omega c$
pure inductive impedance $Z_l = j\omega L$

For complex impedance, Ohm's law can be rewritten:

$$V = IZ$$

where *Z* is the equivalent impedance of the circuit. The complex impedance obeys the same rules as for parallel and series resistance circuits, i.e.:

series: $Z = Z_1 + Z_2 + Z_3$
parallel: $Z = 1(1/Z_1 + 1/Z_2 + 1/Z_3)$

3.2.2.2 *RLC series circuits*

Consider the circuit in *Figure 3.14*. The complex impedance method of determining the total equivalent impedance in the circuit gives:

$$Z = R + j\omega L + \frac{1}{j\omega C} \qquad (3.34)$$

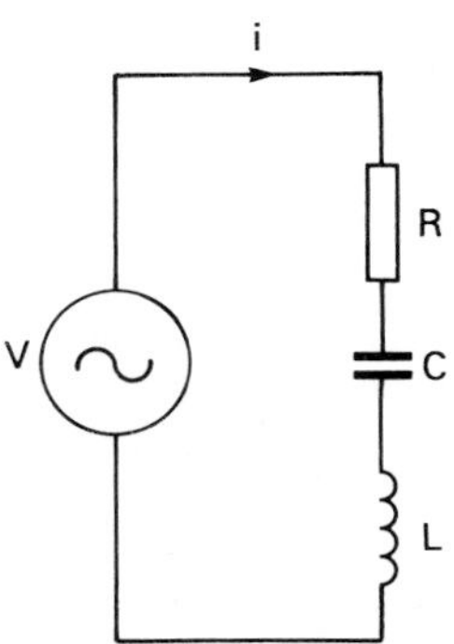

Figure 3.14 RLC series circuit

Separating into real and imaginary parts:

$$Z = R + j(\omega L - \frac{1}{\omega C}) \qquad (3.35)$$

Therefore the current in the circuit is:

$$I = \frac{V}{Z} = \frac{V}{\{R + j[\omega L - (1/\omega c)]\}} \qquad (3.36)$$

and the impedance angle is given by:

$$\phi = \arctan \frac{[\omega L - (1/\omega c)]}{R} \qquad (3.37)$$

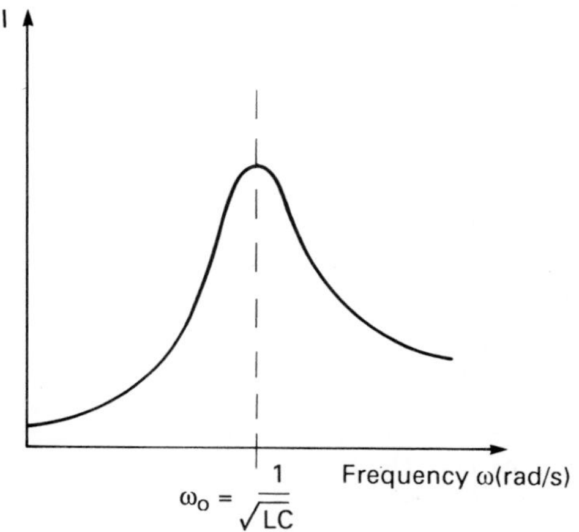

Figure 3.15 Resonance condition for the series circuit

Observation of equation (3.36) indicates that, as $\omega \rightarrow 0$, the current will be very small due to a high capacitive reactance. Also, as $\omega \rightarrow$ infinity, the inductive reactance becomes very high and hence the current very small. However, at a value of ω where the capacitive reactance is equal to the inductive reactance, the current is only dependent on the pure resistance

R, i.e.:

$$\text{for } \omega L = \frac{1}{\omega C}\ , \quad I = \frac{V}{R}$$

The value of ω for this condition is the *resonant frequency*, ω_o; i.e.:

$$\omega_o = \frac{1}{\sqrt{LC}} \qquad (3.38)$$

In *Figure 3.15* the RLC series current is plotted against frequency, showing the current peaking at resonant frequency. Two points to notice about the series resonance condition are that the circuit appears as a pure resistance, and its current is in phase with the applied voltage. Also, in calculating the voltage drops across C and L, Kirchoff's laws would appear to be breached. This is explained, however, by indicating that the voltages are in antiphase, and hence cancel each other.

3.2.2.3 Parallel RLC circuits

In the circuit shown in *Figure 3.16* (a), the impedance for the parallel LC combination is given by:

$$Z = j\,\frac{\omega L}{1 - \omega^2 LC} \qquad (3.39)$$

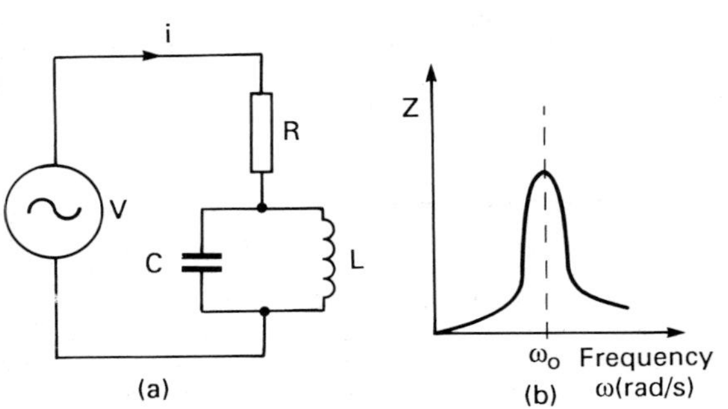

Figure 3.16 RLC parallel circuit

Equation (3.39) shows that the impedance is infinite when

$$\omega_o^2 LC = 1 \qquad (3.40)$$

Here, ω_o is the resonant frequency. Furthermore, equation (3.40) can be rewritten:

$$\omega_o = \frac{1}{\sqrt{LC}} \qquad (3.41)$$

Equation (3.41) is identical to equation (3.38) for the series LC combination. The difference between series and parallel resonance, however, is that the impedance is a *minimum* for series circuits at resonance, whereas it is a *maximum* for parallel circuits at resonance. The impedance characteristic for the parallel circuit is shown in *Figure 3.16* (b).

3.2.2.4 Q-factor

About the resonant frequency for a series circuit, there is a voltage magnification across the inductor (or capacitor), which is given by (voltage across inductor)/(supply voltage).

Thus

$$Q = \frac{\omega_o LI}{RI}$$

$$= \frac{\omega_o L}{R} \qquad (3.42)$$

Equation (3.42) indicates that the voltage magnification factor (the Q-factor) is equal to the ratio of the inductive reactance to the resistance. This means that near resonance, the resistance inherent in all inductors (due to multiple turns of wire) becomes significant in determining the sharpness or selectivity of the resonant circuit (*Figure 3.17*).

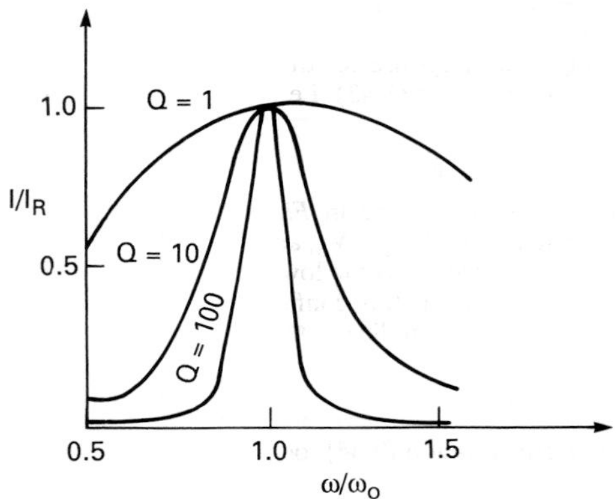

Figure 3.17 Q-factor

For circuits that require very high frequency selectivity, crystals are usually used instead of inductors and capacitors. This is because the Q-factor for LC circuits is typically in the range 10–100, whereas the Q for crystals can be as high as several thousand.

3.2.3 RC circuits as filters

RC circuits can be arranged to discriminate selected frequency bands by using low-pass and high-pass configurations.

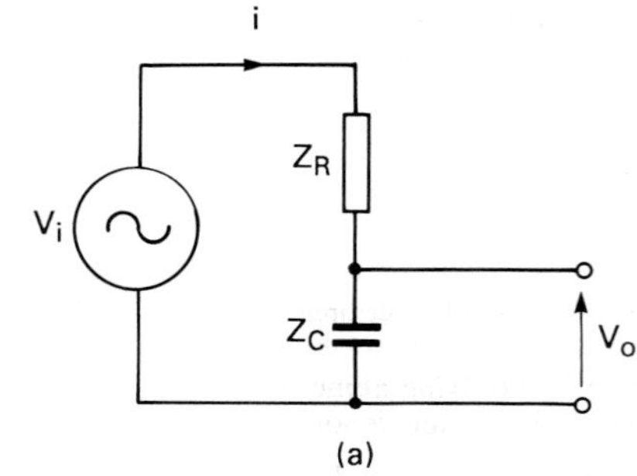

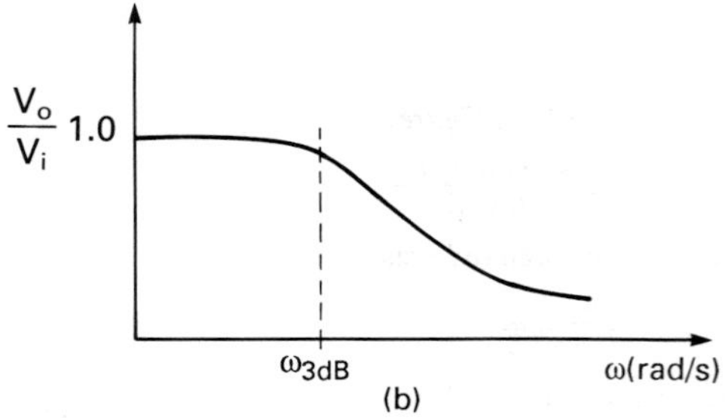

Figure 3.18 Low-pass RC filter circuit and characteristic

3.2.3.1 Low-pass RC filters

Consider the circuit in *Figure 3.18* (a). The output, V_o, of this circuit can be defined as a proportion of the input V_i:

$$V_o = \frac{Z_C}{Z_R + Z_C} V_i = \frac{1/j\omega C}{R + 1/j\omega C} V_i = \frac{1}{1 + j\omega CR} V_i \quad (3.43)$$

The amplitude response for this circuit is found by taking the modulus of equation (3.43), i.e.:

$$\frac{V_o}{V_i} = \frac{1}{\sqrt{1 + \omega^2 C^2 R^2}} \quad (3.44)$$

This response is plotted in *Figure 3.18*(b). Equation (3.44) shows that as $\omega \rightarrow 0$, $V_i = V_o$, and that as $\omega \rightarrow \infty, V_o = 0$. Thus, the larger the frequency the lower the magnitude of the output V_o. For this reason, this configuration is called the *low-pass filter*. *Figure 3.18*(b) indicates the *half power* frequency, where:

$$V_o = \frac{1}{\sqrt{2}} V_i \quad (3.45)$$

which, from equation (3.44), occurs when $\omega CR = 1$, or

$$\omega_{3dB} = \frac{1}{RC} \quad (3.46)$$

3.2.3.2 High-pass RC filters

The configuration for the high-pass RC filter is shown in *Figure 3.19*(a). The current in this circuit is given by:

$$I = \frac{V_i}{Z} = \frac{V_i}{R + \frac{1}{j\omega C}} \quad (3.47)$$

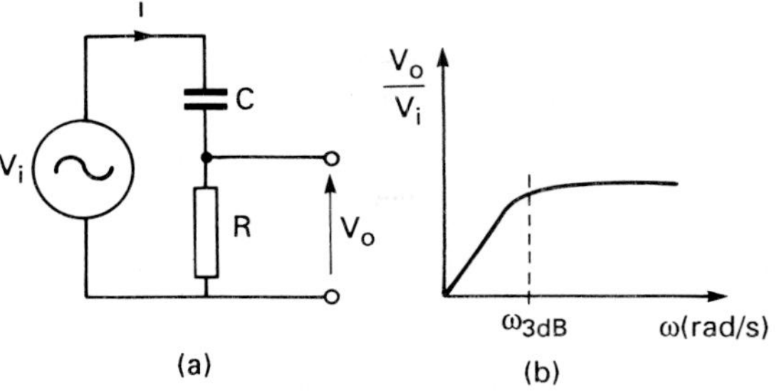

Figure 3.19 High-pass RC filter circuit and characteristic

Therefore, multiplying numerator and denominator by the complex conjugate of the denominator

$$I = V_i \left\{ \frac{j\omega C}{1 + j\omega CR} \cdot \frac{1 - j\omega CR}{1 - j\omega CR} \right\}$$

$$= V_i \frac{[R + (j/\omega C)]}{R^2 + (1/\omega^2 C^2)} \quad (3.48)$$

The voltage across R in *Figure 3.19*(a) is given by:

$$V_o = IR = V_i \frac{[R + (j/\omega C)]R}{R^2 + (1/\omega^2 C^2)} \quad (3.49)$$

Therefore, the magnitude response is given by

$$\frac{V_o}{V_i} = \frac{R}{\sqrt{R^2 + (1/\omega^2 C^2)}} \quad (3.50)$$

Equation (3.50) results in the response in *Figure 3.19*(b), which shows a zero amplitude response at dc ($\omega = 0$). For high frequencies, there is no attenuation at the output, V_o. This figure also indicates the half power frequency, which is again given by:

$$\omega_{3db} = \frac{I}{RC}$$

3.3 Digital circuit theory

Electronic circuits used for digital systems are designed to generate only two recognized output voltage levels, and probably the most common definition for these voltage levels is 5V for the high level and 0V for the lower level. In practice, a certain range is allowed for each of the two levels, again the most common being 0–0.8V for 0V, and 2.7–5V for 5V. Any voltage levels that exist outside these two ranges are invalid, and if present in a digital system will give rise to error conditions.

These voltage ranges are as defined for the ttl (transistor-transistor logic) family of digital circuits; other digital circuit families can use different voltage ranges, e.g. ecl (emitter coupled logic) uses levels -2.1 – -1.7V and -1.3 – -0.9V.

Ensuring digital circuits operate on two distinct voltage ranges, these two ranges can be equated to the two binary conditions 1 and 0 which are used in logic circuits. For ttl type circuits, binary 1 can be equated to the range 2.7V – 5V and the binary 0 condition can be equated to the range 0 – 0.8V. This is known as the *positive logic convention*, where the negative logic definition would equate 1 to the range 0 – 0.8V and 0 to the range 2.7 – 5V. Most digital systems use positive or mixed logic conventions.

3.3.1 Logic gates

The construction of the logic gate from transistor or fet type devices is beyond the scope of this section, as the aim is to describe the use of logic gates, not their construction. Any logic gate manufacturer provides this information in his data sheets.

All digital systems consist, at the most fundamental level, of individual logic gates. Although there are a number of different gate characteristics, there are three types from which all other logic functions can be synthesized: OR, AND and NOT gates.

Logic gates are circuits with one output and one or more inputs. The gate in *Figure 3.20* illustrates a two input logic gate. The truth table in (b) lists the gate output for all possible combinations of binary inputs.

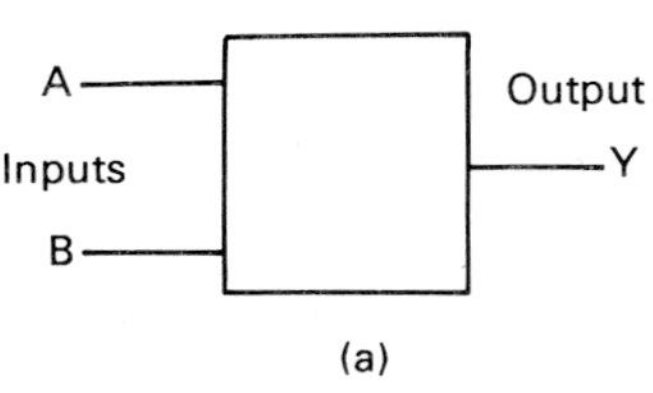

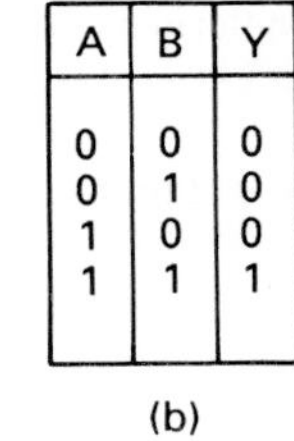

A	B	Y
0	0	0
0	1	0
1	0	0
1	1	1

(b)

Figure 3.20 Logic gate and truth table

From this truth table we can see that the output of the gate will be binary 0 unless both input A and input B are binary 1. Manufacturers will normally quote their truth tables in terms of positive logic.

The truth table is used in the following descriptions of the basic logic gates.

3.3.1.1 OR gates

The symbol for the OR gate along with its truth table is shown in *Figure 3.21*. Considering the truth table for the two input OR gate, it can be seen that the output Y will be a binary 1 if input A is 1, *or* input B is 1, *or* both are 1, hence its name.

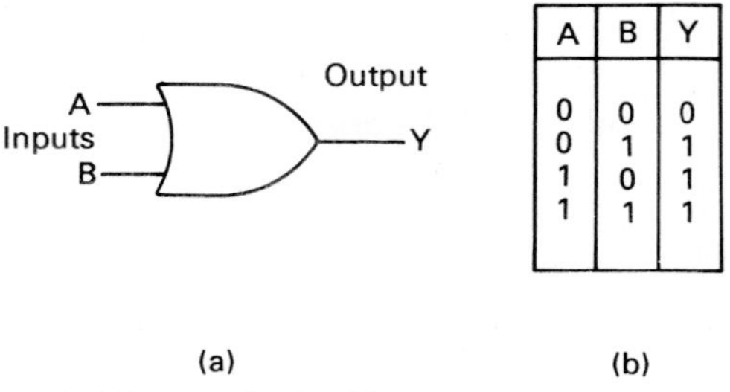

A	B	Y
0	0	0
0	1	1
1	0	1
1	1	1

(a) (b)

Figure 3.21 OR gate and truth table

3.3.1.2 AND gates

The symbol for the AND gate along with its truth table is shown in *Figure 3.22*. Observing the truth table for this device, we see that the output Y will only be binary 1 when input A is 1 *and* input B is 1.

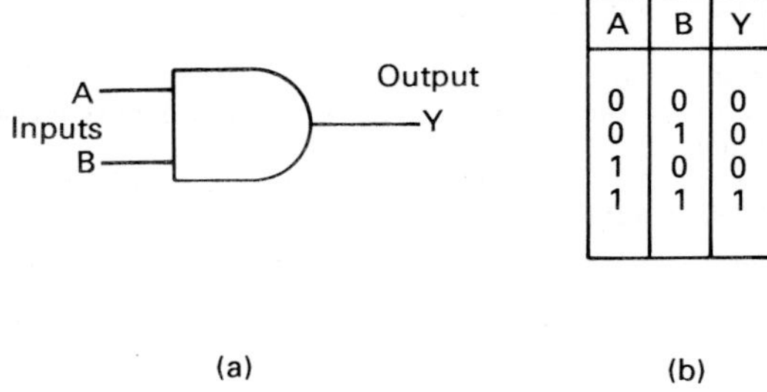

A	B	Y
0	0	0
0	1	0
1	0	0
1	1	1

(a) (b)

Figure 3.22 AND gate and truth table

3.3.1.3 The inverter

The symbol for the NOT gate along with its truth table is shown in *Figure 3.23*. Observation of the truth table indicates the output Y to be the logical inverse of whichever state is present on the input A.

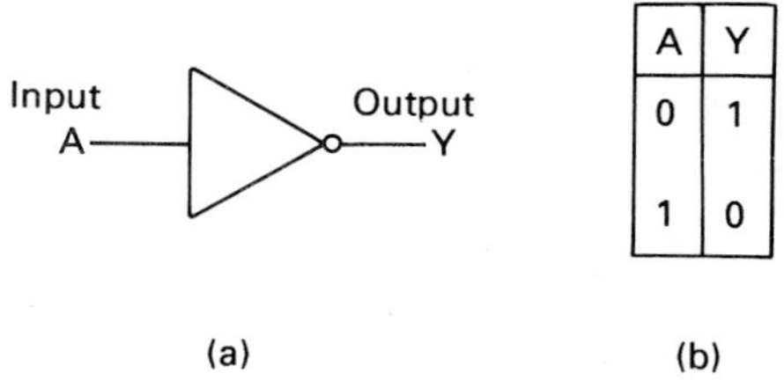

A	Y
0	1
1	0

(a) (b)

Figure 3.23 INVERTER and truth table

3.3.1.4 NOR gates

The symbol for the NOR gate along with its truth table is shown in *Figure 3.24*. The NOR gate is simply an OR gate followed by an inverter; this can be seen from the truth table in *Figure 3.24*(b) where the output Y is the inverse of the output Y for the OR gate in *Figure 3.21*(b).

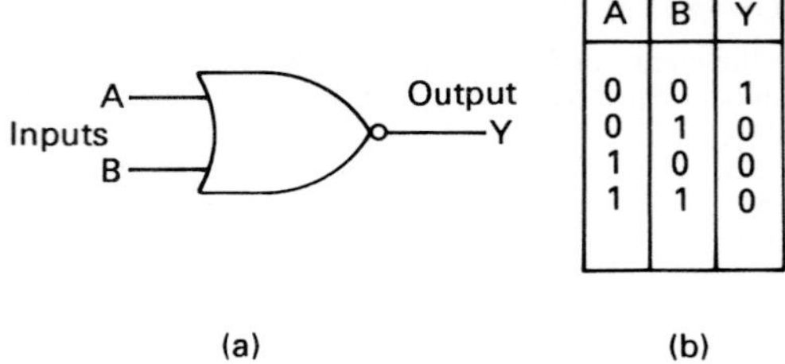

A	B	Y
0	0	1
0	1	0
1	0	0
1	1	0

(a) (b)

Figure 3.24 NOR gate and truth table

3.3.1.5 NAND gates

The symbol for the NAND gate, and its truth table, is shown in *Figure 3.25*. As with the NOR function, the NAND gate is an AND gate followed by an inverter as can be seen from the two truth tables which show the output Y of the NAND gate to be the inverse of the output of the AND gate.

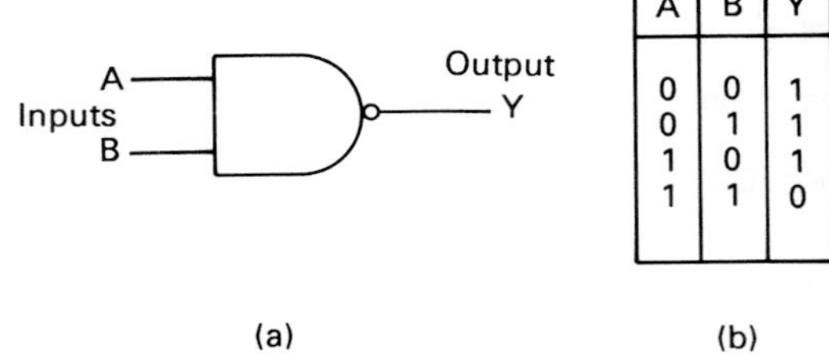

A	B	Y
0	0	1
0	1	1
1	0	1
1	1	0

(a) (b)

Figure 3.25 NAND gate and truth table

3.3.2 Implementing AND / OR functions from NAND / NOR gates

Both the NAND and the NOR gates can be used to implement inverters (see *Figure 3.26*). In *Figure 3.27* a NAND gate has been followed by a second NAND configured as an inverter. The truth table demonstrates that the function implemented is that of an AND gate. This means that NAND elements can be used to implement AND functions, and that NOR functions can be used to implement OR functions, hence the NAND and NOR gates are functionally complete. Thise becomes particularly important when considering that most manufacturers provide multiple gates within a single integrated circuit.

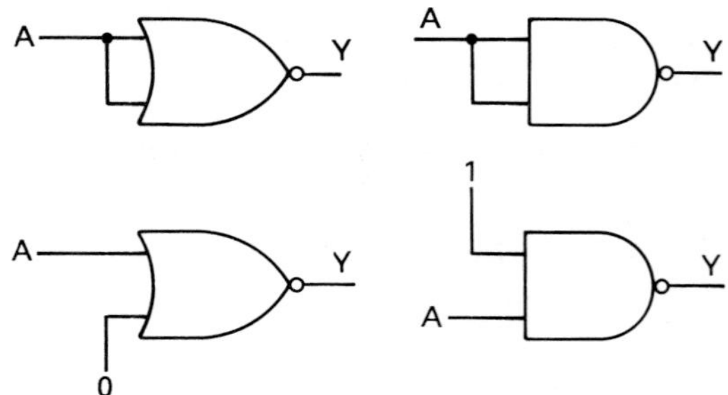

Figure 3.26 Implementing inverters with NAND and NOR gates

3.4 Boolean algebra

3.4.1 Combinational and sequential circuits

All logic circuits can be subdivided into two types: combinational logic and sequential logic. A *combinational* circuit can be described by stating that its output will be true for only certain combinations of input variables; all other input combinations will cause the output to be false. The output(s) for a *sequential* circuit depend upon current input variables, time and past input variables. Sequential circuits use combinational circuits as building blocks, so it is essential to understand these elements. Examples of the implementation of sequential logic circuits are covered in some later chapters.

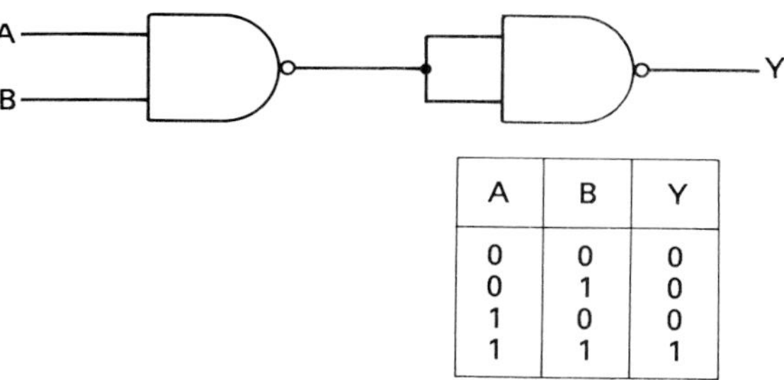

A	B	Y
0	0	0
0	1	0
1	0	0
1	1	1

Figure 3.27 Synthesizing an AND gate with NAND gates

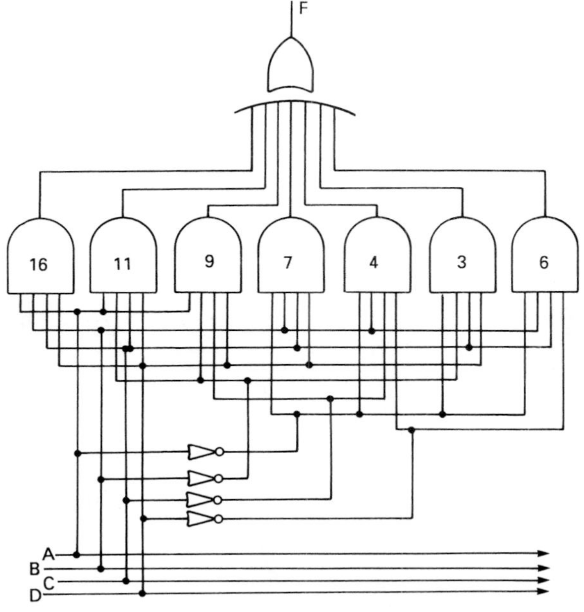

Figure 3.28 Logic system example

For effective design of digital systems, the designer must be able to specify clearly the function of the system, ensure the design will be reliable, and ensure the minimum number of logic elements are used. This section covers a method of defining a logic system in terms of an algebraic equation, and this method will be extended in the discussion of Karnaugh maps for minimizing logic circuit resources.

Consider the logic circuit in *Figure 3.28*. In this example, we have a four bit parallel data transmission link, where there is a requirement to recognize the presence of particular binary codes. The detection circuit is designed such that its output will be logic 1 for the presence of any of the decimal codes 16,11,9,7,4,3 or 6 and a logic 0 for any other combination of codes on the link.

Each of the logic system input variables are designated an alphabetic character (in this case A,B,C,D). Each of these variables implies its complement, i.e. $\overline{A}$, $\overline{B}$, $\overline{C}$, $\overline{D}$. For a positive logic convention, A = 1 (true) and $\overline{A}$ = 0 (false). Thus, for the binary code 0100 on the transmission link in *Figure 3.28*, it is required to recognize $\overline{A}B\overline{C}\,\overline{D}$.

To recognize the code 0100, in this example, we can use a four input AND gate to detect the presence of $\overline{A}B\overline{C}\,\overline{D}$. The other required codes are similarly detected.

Instead of using unwieldy grammar to describe this example, Boolean algebra can be used. The basic rules for Boolean algebra are covered in the following sections.

3.4.2 Boolean OR/AND identities

The truth table for the logical OR relation is shown in *Figure 3.21*(b). It can be represented in terms of a Boolean expression:

$$A + B = X \tag{3.51}$$

where the + symbol indicates the Boolean OR operation, A and B are the input variables, and X is the output. There are a number of important Boolean identities associated with the OR function, which can all be verified using the OR truth table:

$$A + 0 = A \tag{3.52}$$

$$A + 1 = 1 \tag{3.53}$$

$$A + A = A \tag{3.54}$$

$$A + B + C = (A+B)+C = A + (B+C) \tag{3.55}$$

$$A + B = B + A \tag{3.56}$$

All these relations can be directly realized using OR, as illustrated in *Figure 3.29*.

The truth table for the AND gate is shown in *Figure 3.22*(b). It can be written in the Boolean expression

$$A.B = X \tag{3.57}$$

As with the OR gate, there are a number of important AND identities, these are:

$$AO = O \tag{3.58}$$
$$A1 = A \tag{3.59}$$
$$AA = A \tag{3.60}$$
$$ABC = (AB)C = A(BC) \tag{3.61}$$
$$AB = BA \tag{3.62}$$

All these relations can be directly realized using AND, as illustrated in *Figure 3.30*.

There are a number of other very important Boolean identities:

$$\overline{\overline{A}} = A \tag{3.63}$$
$$A + \overline{A} = 1 \text{ (OR complement)} \tag{3.64}$$
$$A\overline{A} = O \text{ (AND complement)} \tag{3.65}$$
$$A(B+C) = AB+AC \tag{3.66}$$

3.4.3 De Morgan's theorem

De Morgan's theorem indicates a useful relationship between AND and OR functions. It can be stated in the form of two laws, i.e.:

$$\overline{A} + \overline{B} + \overline{C} + \ldots + \overline{N} = \overline{ABC\ldots N} \tag{3.67}$$
$$\overline{A}\,\overline{B}\,\overline{C}\ldots\overline{N} = \overline{A + B + C + \ldots + N} \tag{3.68}$$

Figure 3.31 illustrates the physical realization of De Morgan's laws. In terms of gates, it can be seen that a NAND gate is equivalent to an OR gate with inverted inputs, and that a NOR gate is equivalent to an AND gate with inverted inputs.

Application of these laws can enable a designer to implement OR/NOR functions when there are only AND/NAND functions available, and vice versa.

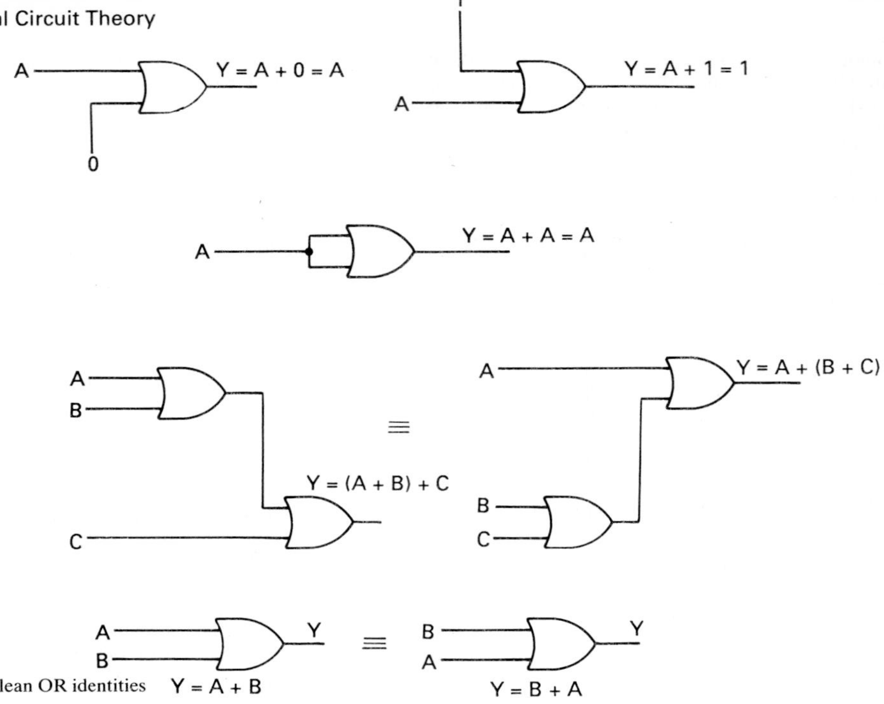

Figure 3.29 Realization of Boolean OR identities

3.5 Karnaugh maps

There are a number of different methods of minimizing Boolean expressions, but for functions of up to six variables, the Karnaugh map provides a method most suitable for manipulating expressions by hand. Functions above six variables are best processed using computer algorithms. Such algorithms are often supplied by the manufacturers of programmable logic devices.

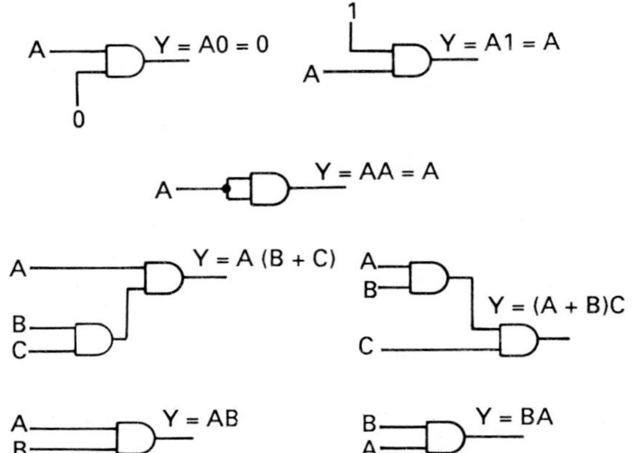

Figure 3.30 Realization of Boolean AND identities

3.5.1 Preparing a Boolean expression for plotting on a Karnaugh map

Boolean expressions will often exist in two distinct forms: the standard sum of products (*SOP*) form and the product of sums (*POS*). Expression (3.69) shows an example of the SOP form, whereas expression (3.70) gives an example of the POS form.

Before a Boolean expression can be plotted on the Karnaugh map, it must be converted into the standard sum of products form (*SSOP*).

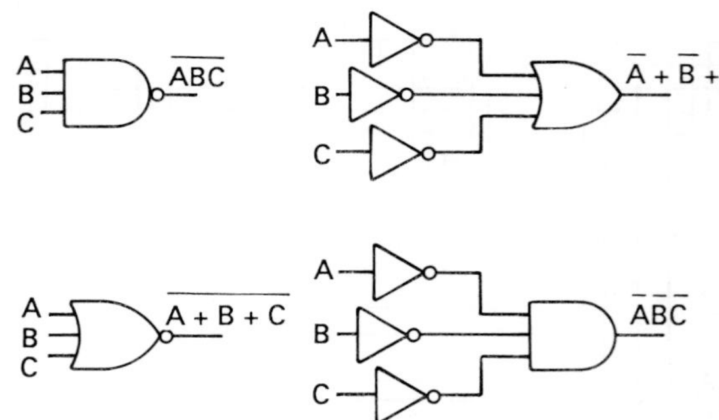

Figure 3.31 Realization of De Morgan's laws

$$AB + BC + \overline{B}D \tag{3.69}$$

$$(A+B+C)(\overline{B} + \overline{C}) \tag{3.70}$$

Using the distributive law (equation (3.66)), the POS expression (3.70) can be converted into the SOP form:

$$A\overline{B} + B\overline{B} + C\overline{B} + A\overline{C} + B\overline{C} + C\overline{C}$$

$$= A\overline{B} + C\overline{B} + A\overline{C} + B\overline{C} \tag{3.71}$$

This form, now in SOP form, needs one further conversion into SSOP form. The original expression contains three variables A, B and C; for SSOP form, each product term must include all of these variables or their complements. This can be done by taking each product term in equation (3.71) with a missing variable, and ANDing that term with the sum of the missing variable and its complement, i.e.:

$$A\overline{B}(C+\overline{C}) + (A+\overline{A})\overline{B}C + A(B+\overline{B})\overline{C} + (A+\overline{A})B\overline{C}$$

$$= A\overline{B}C + A\overline{B}\,\overline{C} + A\overline{B}C + \overline{A}\,\overline{B}C + AB\overline{C} + A\overline{B}\,\overline{C} + AB\overline{C} + \overline{A}B\overline{C} \tag{3.72}$$

This operation does not affect the expression, as the sum of a variable and its own complement is 1 (see equation (3.64)). Duplicate terms can now be removed from equation (3.72), giving:

$$A\overline{B}C + A\overline{B}\,\overline{C} + \overline{A}\,\overline{B}C + AB\overline{C} + \overline{A}B\overline{C} \qquad (3.73)$$

The SSOP terms in an expression are called *minterms*. Minterms are numbered according to the decimal code they represent, e.g.:

$$F = \overline{A}\,\overline{B}C + \overline{A}BC + A\overline{B}\,\overline{C} \qquad (3.74)$$

contains the minterms m1, m3 and m4. This shows that F = 1 when minterms m1, m2 or m4 are present and that F = 0 for all other possible minterms. (For equation (3.74), F will be 1 when, for example, the third term, minterm 4 is 1 i.e. A=1, B=0 and C=0.)

3.5.2 Entering an expression on the Karnaugh map

A Karnaugh map is plotted by entering each term of the SSOP expression in one of the map locations. The map for a two variable function is shown in *Figure 3.32*. It contains four locations, as there are four possible combinations of the two input variables, A and B. The minterm numbers are also shown in *Figure 3.32*, although they are not normally drawn in.

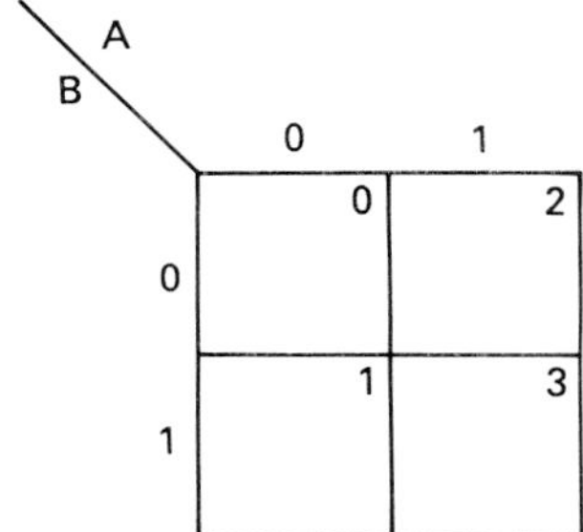

Figure 3.32 Two variable Karnaugh map

Consider the Boolean function

$$F = AB + \overline{A}B \qquad (3.75)$$

which indicates that F will be 1 when A is 1 and B is 1 and when A is 1 and B is 0. F will be 0 for the other two combinations of variables A and B. This can be plotted on the Karnaugh map as in *Figure 3.33*. Note the minterms present and their number and position on the Karnaugh map.

3.5.3 Reducing an expression using a Karnaugh map

Once an expression has been plotted, it can be reduced by forming the entries into logically adjacent groups. Consider *Figure 3.33*, where the function $F = AB + A\overline{B}$ has been plotted. In this example, the bottom left and right entries are logically adjacent, forming a *couple*. This couple indicates that the function will be a 1 regardless of the state of A, because A is present in its complemented and uncomplemented form. Therefore

$$F_R = B \qquad (3.76)$$

Diagonal coupling is not valid, as it would mean more than one variable changing state at one time, e.g., minterm 2 is not

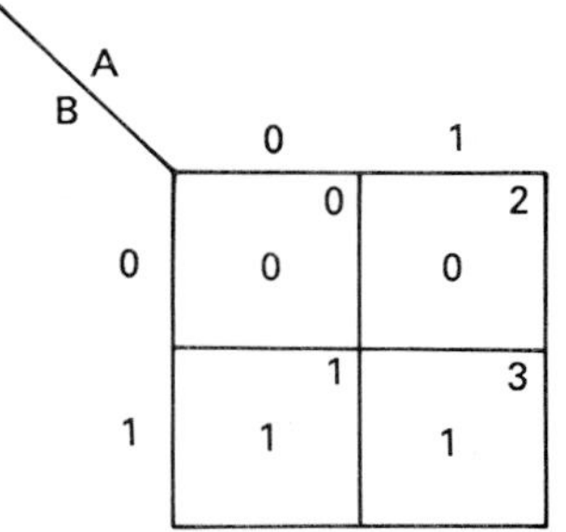

Figure 3.33 Karnaugh map of the function $F = AB + A\overline{B}$

logically adjacent to minterm 1. The result in equation (3.76) can be verified using the Boolean identities, i.e.:

$$F = BA + B\overline{A} = B(A+\overline{A}) = B \qquad (3.77)$$

where $(A+\overline{A})$ is the OR complement.

A three variable Karnaugh map is shown in *Figure 3.34*. It contains eight locations to represent each of the three variable combinations. The locations on the map are arranged so that all entries are logically adjacent. This means that only one variable will change state between any two adjacent map locations. Logical adjacency also exists between the extreme opposite squares for a row or column; for the example in *Figure 3.34*, location $A\overline{BC}$ (minterm 4) is adjacent to the location $\overline{ABC}$ (minterm 0). Note that due to the logical adjacency, the minterm numbers do not follow in order.

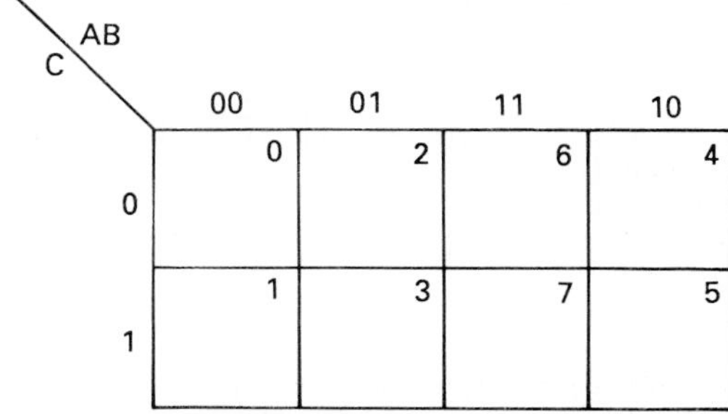

Figure 3.34 Three variable Karnaugh map

An example of a three variable expression reduction is shown in *Figure 3.35*, which maps the following expression:

$$F = \overline{A}\,\overline{B}C + A\overline{B}\,\overline{C} + A\overline{B}C + AB\overline{C} + ABC \qquad (3.78)$$

Grouping logically adjacent locations to form the largest possible groups, the map reveals a *quad* and a *couple* of adjacent terms. The quad indicates the constituent terms to be independent of the variables B and C because the complemented and non-complemented form exists for this quad. The couple indicates its constituent terms to be independent of A, because its complemented and non-complemented forms appear. Thus the function F will depend only upon A for the quad, and $\overline{B}C$ for the couple. Thus the expression in equation (3.78) can be reduced to:

$$F_R = A + \overline{B}C \qquad (3.79)$$

Using Boolean algebraic reduction, the result for the quad can be verified, i.e.:

$$\begin{aligned} &A\overline{B}\,\overline{C} + A\overline{B}C + AB\overline{C} + ABC \\ &\quad = A(\overline{B}\,\overline{C}+\overline{B}C+B\overline{C}+BC) \\ &\quad = A[\overline{B}(\overline{C}+C) + B(\overline{C}+C)] \\ &\quad = A(\overline{B}+B) \\ &\quad = A \end{aligned} \qquad (3.80)$$

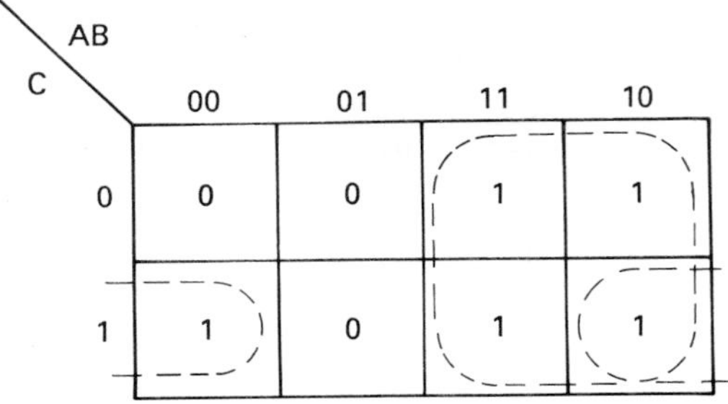

Figure 3.35 Karnaugh map of the function $F = \bar{A}\bar{B}C + A\bar{B}\bar{C} + A\bar{B}C + AB\bar{C} + ABC$

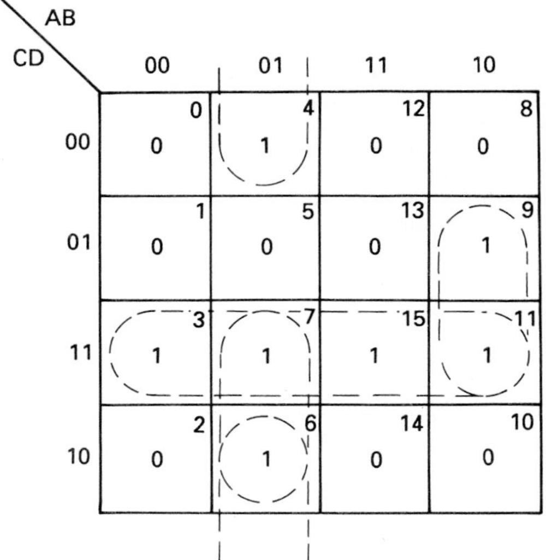

Figure 3.36 Karnaugh map of the function $F = ABCD + A\bar{B}CD + A\bar{B}\bar{C}D + \bar{A}BCD + \bar{A}B\bar{C}\bar{D} + \bar{A}\bar{B}CD + \bar{A}BC\bar{D}$

The example in *Figure 3.28* describes a logic system whose function can be described by the Boolean expression:

$$F = ABCD + A\bar{B}CD + A\bar{B}\bar{C}D + \bar{A}BCD + \bar{A}B\bar{C}\bar{D} + \bar{A}\bar{B}CD + \bar{A}BC\bar{D} \qquad (3.81)$$

This can be plotted on the four variable Karnaugh map as in *Figure 3.36*. In this example there are 16 possible input combinations for the four variables, so the map contains 16 locations. From this map it can be seen that the largest group of terms that can be formed is the quad; the other entries on the map can be grouped into two quads, as shown. The quad indicates that both A and B exist in their complemented and non-complemented forms; therefore it reduces to CD. Reducing the two couples, the resultant reduced expression becomes:

$$F_R = CD + \bar{A}B\bar{C} + A\bar{B}D \qquad (3.82)$$

The power of this technique can be illustrated by considering the implementation of the unreduced expression in equation (3.81) and comparing it with the reduced expression in equation (3.82). The unreduced implementation is shown in *Figure 3.28* and the reduced implementation in *Figure 3.37*; the reduction in gates can be clearly seen.

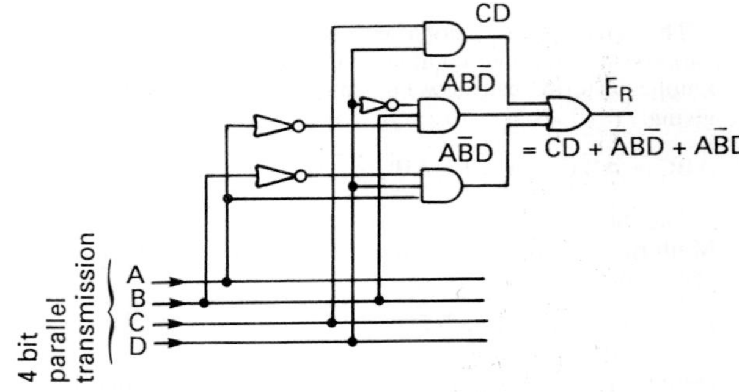

Figure 3.37 Implementation of function $F_R = CD + \bar{A}B\bar{C} + A\bar{3}D$

The operation of using a Karnaugh map for reducing an expression can be summarized as follows:

- Form standard sum of products of the unreduced expression.
- Plot the minterms on the map.
- Form the largest and least number of groups of logically adjacent entries (these groups will always contain a number which is a power of 2, e.g. 2,4,8).

Generally, a couple allows two original terms to be reduced to one smaller term, a quad allows four original terms to be reduced to one smaller term, etc. A map entry surrounded by 0s cannot be reduced, and will appear in the final expression without any reduction.

3.5.4 Prime implicants

Given the reduction method described in section *3.53*, it is still possible to produce a non-minimal result. This can be illustrated by considering the *prime implicants*.

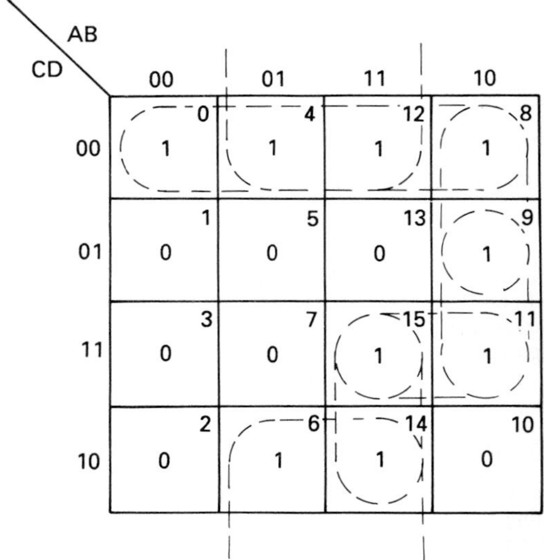

Figure 3.38 Function with non-essential prime implicants

Once the Karnaugh map has been plotted, logically adjacent entries are grouped together; these groups are the prime implicants. The four variable example in *Figure 3.38* shows the following prime implicants are formed:

$\bar{C}\bar{D}$	$A\bar{B}D$
$B\bar{D}$	ACD
$A\bar{B}\bar{C}$	ABC

The figure shows that soon entries on the map are covered by more than one prime implicant, e.g. minterm 15. If any of the overlapping prime implicants results in all entries of another prime implicant being covered, then that prime implicant is non-essential. For example, the couple covering minterms 15 and 11 is covered by couple $^{9}/_{11}$ and couple $^{14}/_{15}$, hence it is not an essential implicant. For a minimal expression all of these non-essential implicants must be removed. *Figure 3.38* indicates three possible results, depending upon which essential implicants are chosen. These are:

$$F_R = \overline{C}\,\overline{D} + B\overline{D} + A\overline{B}D + ABC$$
$$F_R = \overline{C}\,\overline{D} + B\overline{D} + A\overline{B}\,\overline{C} + ACD$$
$$F_R = \overline{C}\,\overline{D} + B\overline{D} + A\overline{B}D + ACD$$

The results of these three equations indicate that there can be more than one result for a minimized expression; each one is valid.

J Barron BA, MA (Cantab)
University of Cambridge

4 Engineering Mathematics, Formulas and Calculations

4.1 Mathematical signs and symbols

Sign, symbol	*Quantity*
$=$	equal to
$\neq$ $\neq$	not equal to
$\equiv$	identically equal to
Δ	corresponds to
$\approx$	approximately equal to
$\rightarrow$	approaches
$\simeq$	asymptotically equal to
$\sim$	proportional to
∞	infinity
$<$	smaller than
$>$	larger than
$\leqslant$ $\leq$ $\leqq$	smaller than or equal to
$\geqslant$ $\geq$ $\geqq$	larger than or equal to
	much smaller than
	much larger than
$+$	plus
$-$	minus
. $\times$	multiplied by
$\frac{a}{b}$ a/b	a divided by b
$\lvert a\rvert$	magnitude of a
a^n	a raised to the power of n
$a^{1/2}$ $\sqrt{a}$	square root of a
$a^{1/n}$ $\sqrt[n]{a}$	nth root of a
a $\langle a\rangle$	mean value of a
$p!$	factorial p, $1 \times 2 \times 3 \times \ldots \times p$
$\binom{n}{p}$	binomial coefficient, $\frac{n(n-1)\ldots(n-p+1)}{1 \times 2 \times 3 \times \ldots \times p}$
Σ	sum
Π	product
$f(x)$	function f of the variable x
$[f(x)]_a^b$	$f(b) - f(a)$
$\lim_{x\rightarrow a} f(x)$; $\lim_{x\rightarrow a}f(x)$	the limit to which $f(x)$ tends as x approaches a
Δx	delta x= finite increment of x
δx	delta x=variation of x
$\frac{df}{dx}$; df/dx; $f'(x)$	differential coefficient of $f(x)$ with respect to x

Sign, symbol	*Quantity*
$\frac{d^nf}{dx^n}$; $f^{(n)}(x)$	differential coefficient of order n of $f(x)$
$\frac{\delta f(x, y, \ldots)}{\delta x}$; $\left(\frac{\delta f}{\delta x}\right)_{y,\ldots}$	partial differential coefficient of $f(x, y, \ldots)$ with respect to x, when $y, \ldots$ are held constant
df	the total differential of f
$\int f(x)dx$	indefinite integral of $f(x)$ with respect to x
$\int_a^b f(x)dx$	definite integral of $f(x)$ from $x = a$ to $x = b$
e	base of natural logarithms
e^x, $\exp x$	e raised to the power x
$\log_a x$	logarithm to the base a of x
$\lg x$; $\log x$; $\log_{10} x$	common (Briggsian) logarithm of x
$\text{lb}\, x$; $\log_2 x$	binary logarithm of x
$\sin x$	sine of x
$\cos x$	cosine of x
$\tan x$; $\text{tg}\, x$	tangent of x
$\cot x$; $\text{ctg}\, x$	cotangent of x
$\sec x$	secant of x
$\text{cosec}\, x$	cosecant of x
$\arcsin x$, etc.	arc sine of x, etc.
$\sinh x$, etc.	hyperbolic sine of x, etc.
$\text{arsinh}\, x$, etc.	inverse hyperbolic sine of x, etc.
i, j	imaginary unity, $i^2 = -1$
$\text{Re}\, z$	real part of z
$\text{Im}\, z$	imaginary part of z
$\lvert z\rvert$	modulus of z
$\arg z$	argument of z
z^*	conjugate of z, complex conjugate of z
$\tilde{A}$, A', A^t	transpose of matrix A
A, a	vector
$\lvert \mathbf{A}\rvert$,A	magnitude of vector
$\mathbf{A \cdot B}$	scalar product
$\mathbf{A \times B}$, $\mathbf{A \wedge B}$	vector product
∇	differential vector operator
$\nabla \varphi$, grad φ	gradient of φ
$\nabla^2 \varphi$, $\triangle \varphi$	Laplacian of φ

4.2 Trigonometric formulas

$\sin^2 A + \cos^2 A = \sin A \operatorname{cosec} A = 1$

$\sin A = \dfrac{\cos A}{\cot A} = \dfrac{1}{\operatorname{cosec} A} = (1 - \cos^2 A)^{1/2}$

$\cos A = \dfrac{\sin A}{\tan A} = \dfrac{1}{\sec A} = (1 - \sin^2 A)^{1/2}$

$\tan A = \dfrac{\sin A}{\cos A} = \dfrac{1}{\cot A}$

$1 + \tan^2 A = \sec^2 A$

$1 + \cot^2 A = \operatorname{cosec}^2 A$

$1 - \sin A = \operatorname{coversin} A$

$1 - \cos A = \operatorname{versin} A$

$\tan \tfrac{1}{2}\theta = t$; $\sin\theta = 2t/(1 + t^2)$; $\cos\theta = (1 - t^2)/(1 + t^2)$

$\cot A = 1/\tan A$

$\sec A = 1/\cos A$

$\operatorname{cosec} A = 1/\sin A$

$\cos(A \pm B) = \cos A \cos B \mp \sin A \sin B$

$\sin(A \pm B) = \sin A \cos B \pm \cos A \sin B$

$\tan(A \pm B) = \dfrac{\tan A \pm \tan B}{1 \mp \tan A \tan B}$

$\cot(A \pm B) = \dfrac{\cot A \cot B \mp 1}{\cot B \pm \cot A}$

$\sin A \pm \sin B = 2 \sin \tfrac{1}{2}(A \pm B) \cos \tfrac{1}{2}(A \mp B)$

$\cos A + \cos B = 2 \cos \tfrac{1}{2}(A + B) \cos \tfrac{1}{2}(A - B)$

$\cos A - \cos B = 2 \sin \tfrac{1}{2}(A + B) \sin \tfrac{1}{2}(B - A)$

$\tan A \pm B = \dfrac{\sin(A \pm B)}{\cos A \cos B}$

$\cot A \pm \cot B = \dfrac{\sin(B \pm A)}{\sin A \sin B}$

$\sin 2A = 2 \sin A \cos A$

$\cos 2A = \cos^2 A - \sin^2 A = 2\cos^2 A - 1 = 1 - 2\sin^2 A$

$\cos^2 A - \sin^2 B = \cos(A + B)\cos(A - B)$

$\tan 2A = 2\tan A/(1 - \tan^2 A)$

$\sin \tfrac{1}{2}A = \left(\dfrac{1 - \cos A}{2}\right)^{1/2}$

$\cos \tfrac{1}{2}A = \pm\left(\dfrac{1 + \cos A}{2}\right)^{1/2}$

$\tan \tfrac{1}{2}A = \dfrac{\sin A}{1 + \cos A}$

$\sin^2 A = \tfrac{1}{2}(1 - \cos 2A)$

$\cos^2 A = \tfrac{1}{2}(1 + \cos 2A)$

$\tan^2 A = \dfrac{1 - \cos 2A}{1 + \cos 2A}$

$\tan \tfrac{1}{2}(A \pm B) = \dfrac{\sin A \pm \sin B}{\cos A + \cos B}$

$\cot \tfrac{1}{2}(A \pm B) = \dfrac{\sin A \pm \sin B}{\cos B - \cos A}$

4.3 Trigonometric values

Angle	*0°*	*30°*	*45°*	*60°*	*90°*	*180°*	*270°*	*360°*
Radians	0	$\pi/6$	$\pi/4$	$\pi/3$	$\pi/2$	π	$3\pi/2$	2π
Sine	0	$\tfrac{1}{2}$	$\tfrac{1}{2}\sqrt{2}$	$\tfrac{1}{2}\sqrt{3}$	1	0	-1	0
Cosine	1	$\tfrac{1}{3}\sqrt{3}$	$\tfrac{1}{2}\sqrt{2}$	$\tfrac{1}{2}$	0	-1	0	1
Tangent	0	$\tfrac{1}{3}\sqrt{3}$	1	$\sqrt{3}$	$\propto$	0	$\propto$	0

4.4 Approximations for small angles

$\sin\theta = \theta^3/6$; $\cos\theta = 1 - \theta^2/2$; $\tan\theta = \theta + \theta^3/3$;
(θ in radians)

4.5 Solution of triangles

$\dfrac{\sin A}{a} = \dfrac{\sin B}{b} = \dfrac{\sin C}{c}$ $\quad\cos A = \dfrac{b^2 + c^2 - a^2}{2bc}$

$\cos B = \dfrac{c^2 + a^2 - b^2}{2ca}$ $\quad\cos C = \dfrac{a^2 + b^2 - c^2}{2ab}$

where *A*, *B*, *C* and *a*, *b*, *c* are shown in *Figure 4.1*. If $s = \tfrac{1}{2}(a + b + c)$,

$\sin \dfrac{A}{2} = \sqrt{\dfrac{(s-b)(s-c)}{bc}}$ $\quad\sin \dfrac{B}{2} = \sqrt{\dfrac{(s-c)(s-a)}{ca}}$

$\sin \dfrac{C}{2} = \sqrt{\dfrac{(s-a)(s-b)}{ab}}$

$\cos \dfrac{A}{2} = \sqrt{\dfrac{s(s-a)}{bc}}$ $\quad\cos \dfrac{B}{2} = \sqrt{\dfrac{s(s-b)}{ca}}$

$\cos \dfrac{C}{2} = \sqrt{\dfrac{s(s-c)}{ab}}$

$\tan \dfrac{A}{2} = \sqrt{\dfrac{(s-b)(s-c)}{s(s-a)}}$ $\quad\tan \dfrac{B}{2} = \sqrt{\dfrac{(s-c)(s-a)}{s(s-b)}}$

$\tan \dfrac{C}{2} = \sqrt{\dfrac{(s-a)(s-b)}{s(s-c)}}$

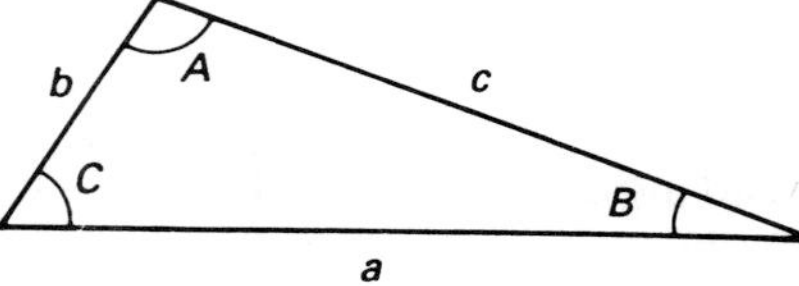

Figure 4.1 Triangle

4.6. Spherical triangle

$\dfrac{\sin A}{\sin a} = \dfrac{\sin B}{\sin b} = \dfrac{\sin C}{\sin c}$

$\cos a = \cos b \cos c + \sin b \sin c \cos A$

$\cos b = \cos c \cos a + \sin c \sin a \cos B$

$\cos c = \cos a \cos b + \sin a \sin b \cos C$

where *A*, *B*, *C* and *a*, *b*, *c* are now as in *Figure 4.2*.

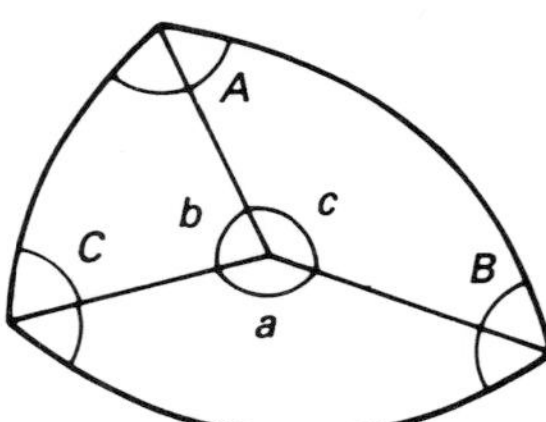

Figure 4.2 Spherical triangle

4.7 Exponential form

$\sin\theta = \dfrac{e^{i\theta} - e^{-i\theta}}{2i}$ $\quad\cos\theta = \dfrac{e^{i\theta} + e^{-\theta}}{2}$

$e^{i\theta} = \cos\theta + i\sin\theta$ $\quad e^{-i\theta} = \cos\theta - i\sin\theta$

4.8 De Moivre's theorem

$(\cos A + i\sin A)(\cos B + i\sin B) = \cos(A + B) + i\sin(A + B)$

4.9 Euler's relation

$(\cos\theta + i\sin\theta)^n = \cos n\theta + i\sin n\theta = e^{in\theta}$

4.10 Hyperbolic functions

$\sinh x=(e^x-e^{-x})/2$ $\qquad$ $\cosh x=(e^x+e^{-x})/2$

$\tanh x=\sinh x/\cosh x$

Relations between hyperbolic functions can be obtained from the corresponding relations between trigonometric functions by reversing the sign of any term containing the product or implied product of two sines, e.g.:

$\cosh^2 A-\sinh^2 A=1$

$$\cosh 2A=2\cosh^2 A-1=1+2\sinh^2 A$$
$$=\cosh^2 A+\sinh^2 A$$

$\cosh(A\pm B)=\cosh A\cosh B\pm\sinh A\sinh B$

$\sinh(A\pm B)=\sinh A\cosh B\pm\cosh A\sinh B$

$e^x=\cosh x+\sinh x$ $\qquad$ $e^{-x}=\cosh x-\sinh x$

4.11 Complex variable

If $z=x+iy$, where x and y are real variables, z is a complex variable and is a function of x and y. z may be represented graphically in an Argand diagram (*Figure 4.3*).

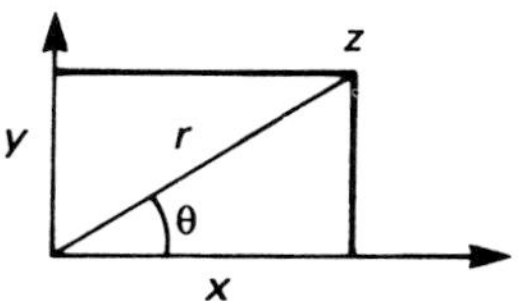

Figure 4.3 Argand diagram

Polar form:

$z=x+iy=|z|e^{i\theta}=|z|(\cos\theta+i\sin\theta)$

$x=r\cos\theta$ $\qquad$ $y=r\sin\theta)$

where $r=|z|$.

Complex arithmetic:

$z_1=x_1+iy_1$ $\qquad$ $z_2=x_2+iy_2$

$z_1\pm z_2=(x_1\pm x_2)+i(y_1\pm y_2)$

$z_1.z_2=(x_1x_2-y_1y_2)+i(x_1y_2+x_2y_1)$

Conjugate:

$z^*=x-iy$ $\qquad$ $z.z^*=x^2+y^2=|z|^2$

Function: another complex variable $w=u+iv$ may be related functionally to z by

$w=u+iv=f(x+iy)=f(z)$

which implies

$u=u(x,y)$ $\qquad$ $v=v(x,y)$

e.g.,

$$\cosh z=\cosh(x+iy)=\cosh x\cosh iy+\sinh x\sinh iy$$
$$=\cosh x\cos y+i\sinh x\sin y$$

$u=\cosh x\cos y$ $\qquad$ $v=\sinh x\sin y$

4.12 Cauchy–Riemann equations

If $u(x,y)$ and $v(x,y)$ are continuously differentiable with respect to x and y,

$$\frac{\delta u}{\delta x}=\frac{\delta v}{\delta y} \qquad \frac{\delta u}{\delta y}=-\frac{\delta v}{\delta x}$$

$w=f(z)$ is continuously differentiable with respect to z and its derivative is

$$f'(z)=\frac{\delta u}{\delta x}+i\frac{\delta v}{\delta x}=\frac{\delta v}{\delta y}-i\frac{\delta u}{\delta y}=\frac{1}{i}\left(\frac{\delta u}{\delta y}+i\frac{\delta v}{\delta y}\right)$$

It is also easy to show that $\nabla^2u=\nabla^2v=0$. Since the transformation from z to w is conformal, the curves $u=$ constant and $v=$ constant intersect each other at right angles, so that one set may be used as equipotentials and the other as field lines in a vector field.

4.13 Cauchy's theorem

If $f(z)$ is analytic everywhere inside a region bounded by C and a is a point within C

$$f(a)=\frac{1}{2\pi i}\int_c\frac{f(z)}{z-a}\,dz$$

This formula gives the value of a function at a point in the interior of a closed curve in terms of the values on that curve.

4.14 Zeros, poles and residues

If $f(z)$ vanishes at the point z_0 the Taylor series for z in the region of z_0 has its first two terms zero, and perhaps others also: $f(z)$ may then be written

$f(z)=(z-z_0)^n g(z)$

where $g(z_0)\neq 0$. Then $f(z)$ has a *zero* of order n at z_0. The reciprocal

$q(z)=1/f(z)=h(z)/(z-z_0)^n$

where $h(z)=1/g(z)\neq 0$ at z_o. $q(z)$ becomes infinite at $z=z_o$ and is said to have a *pole* of order n at z_o. $q(z)$ may be expanded in the form.

$q(z)=c_{-n}(z-z_0)^n+\ldots+c_{-1}(z-z_0)^{-1}+c_0+\ldots$

where c_{-1} is the *residue* of $q(z)$ at $z=z_o$. From Cauchy's theorem, it may be shown that if a function $f(z)$ is analytic throughout a region enclosed by a curve C except at a finite number of poles, the integral of the function around C has a value of $2\pi i$ times the sum of the residues of the function at its poles within C. This fact can be used to evaluate many definite integrals whose indefinite form cannot be found.

4.15 Some standard forms

$$\int_0^{2\pi} e^{\cos\theta}\cos(n\theta-\sin\theta)d\theta=2\pi/n!$$

$$\int_0^{\alpha}\frac{x^{a-1}}{1+x}dx=\pi\operatorname{cosec}a\pi$$

$$\int_0^{\alpha}\frac{\sin\theta}{\theta}d\theta=\frac{\pi}{2}$$

$$\int_0^{\alpha}x\exp(-h^2x^2)dx=\frac{1}{2h^2}$$

$$\int_0^{\alpha}\frac{x^{a-1}}{1-x}dx=\pi\cot a\pi$$

$$\int_0^{\alpha}\exp(-h^2x^2)dx=\frac{\sqrt{\pi}}{2h}$$

$$\int_0^\alpha x^2 \exp(-h^2x^2)\mathrm{d}x = \frac{\sqrt{\pi}}{4h^3}$$

4.16 Coordinate systems

The basic system is the rectangular Cartesian system (x, y, z) to which all other systems are referred. Two other commonly used systems are as follows.

4.16.1 Cylindrical coordinates

Coordinates of point P are (x, y, z) or (r, θ, z) (see *Figure 4.4*), where

$x=r\cos\theta$ $\quad y=r\sin\theta$ $\quad z=z$

In these coordinates the volume element is $r\,\mathrm{d}r\,\mathrm{d}\theta\,\mathrm{d}z$.

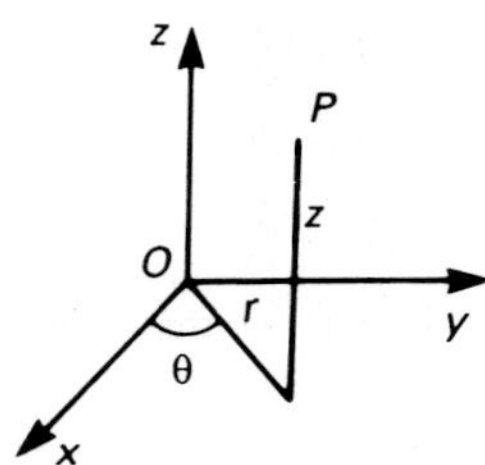

Figure 4.4 Cylindrical coordinates

4.16.2 Spherical polar coordinates

Coordinates of point P are (x, y, z) or (r, θ, φ) *(see Figure 4.5), where*

$x=r\sin\theta\cos\phi$ $\quad y=r\sin\theta\sin\phi$ $\quad z=r\cos\theta$

In these coordinates the volume element is $r^2\sin\theta\,\mathrm{d}r\,\mathrm{d}\theta\,\mathrm{d}\phi$.

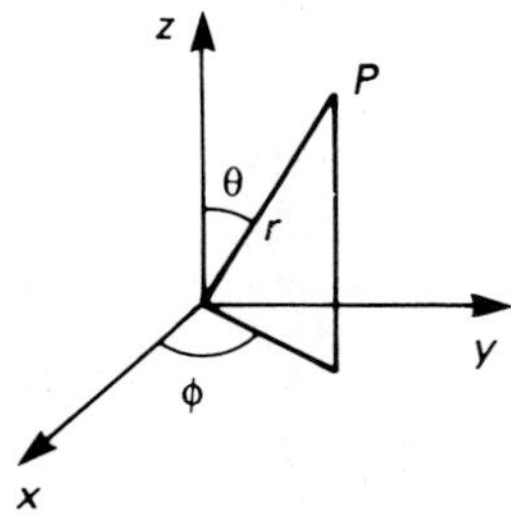

Figure 4.5 Spherical polar coordinates

4.17 Transformation of integrals

$$\iiint f(x, y, z)\mathrm{d}x\,\mathrm{d}y\,\mathrm{d}z = \iiint \varphi(u, v, w)|J|\mathrm{d}u\,\mathrm{d}v\,\mathrm{d}w$$

where

$$J = \begin{vmatrix} \frac{\delta x}{\delta u} & \frac{\delta y}{\delta u} & \frac{\delta z}{\delta u} \\ \frac{\delta x}{\delta v} & \frac{\delta y}{\delta v} & \frac{\delta z}{\delta v} \\ \frac{\delta x}{\delta w} & \frac{\delta y}{\delta w} & \frac{\delta z}{\delta w} \end{vmatrix} = \frac{\delta(x,y,z)}{\delta(u,v,w)}$$

is the Jacobian of the transformation of coordinates. For Cartesian to cylindrical coordinates, $J=r$, and for Cartesian to spherical polars, it is $r^2\sin\theta$.

4.18 Laplace's equation

The equation satisfied by the scalar potential from which a vector field may be derived by taking the gradient is Laplace's equation, written as:

$$\Delta^2\phi = \frac{\delta^2\phi}{\delta x^2} + \frac{\delta^2\phi}{\delta y^2} + \frac{\delta^2\phi}{\delta z^2} = 0$$

In cylindrical coordinates:

$$\Delta^2\phi = \frac{1}{r}\frac{\delta}{\delta r}\left(r\frac{\delta\phi}{\delta r}\right) + \frac{1}{r^2}\frac{\delta^2\phi}{\delta\theta^2} + \frac{\delta^2\phi}{\delta z^2}$$

In spherical polars:

$$\nabla^2\phi = \frac{1}{r^2}\frac{\delta}{\delta r}\left(r^2\frac{\delta\phi}{\delta r}\right) + \frac{1}{r^2\sin\theta}\frac{\delta\phi}{\delta\theta} + \frac{1}{r^2\sin^2\theta}\frac{\delta^2\phi}{\delta\phi^2}$$

The equation is solved by setting

$\phi=U(u)V(v)W(w)$

in the appropriate form of the equation, separating the variables and solving separately for the three functions, where (u,v,w) is the coordinate system in use.

In Cartesian coordinates, typically the functions are trigonometric, hyperbolic and exponential; in cylindrical coordinates the function of z is exponential, that of θ trigonometric and that of r is a Bessel function. In spherical polars, typically the function of r is a power of r, that of φ is trigonometric, and that of θ is a Legendre function of $\cos\theta$.

4.19 Solution of equations

4.19.1 Quadratic equation

$ax^2+bx+c=0$

$$x = -\frac{b}{2a} \pm \frac{\sqrt{b^2-4ac}}{2a}$$

In practical calculations if $b^2>4ac$, so that the roots are real and unequal, calculate the root of larger modulus first, using the same sign for both terms in the formula, then use the fact that $x_1x_2=c/a$ where x_1 and x_2 are the roots. This avoids the severe cancellation of significant digits which may otherwise occur in calculating the smaller root.

For polynomials other than quadratics, and for other functions, several methods of successive approximation are available.

4.19.2 Bisection method

By trial find x_0 and x_1 such that $f(x0)$ and $f(x_1)$ have opposite signs (see *Figure 4.6*). Set $x_2=(x_0+x_1)/2$ and calculate $f(x_2)$. If $f(x_0)f(x_2)$ is positive, the root lies in the interval (x_1,x_2); if

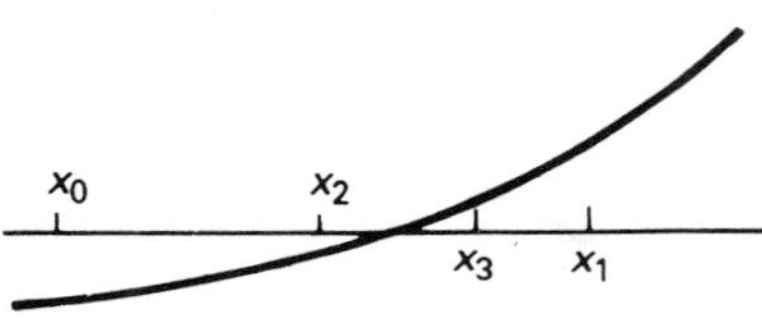

Figure 4.6

negative in the interval (x_0,x_2); and if zero, x_2 is the root. Continue if necessary using the new interval.

4.19.3 Regula falsi

By trial, find x_0 and x_1 as for the bisection method; these two values define two points $(x_0,f(x_0))$ and $(x_1,f(x_1))$. The straight line joining these two points cuts the x-axis at the point (see *Figure 4.7*):

$$x_2=\frac{x_0f(x_1)-x_1f(x_0)}{f(x_1)-f(x_0)}$$

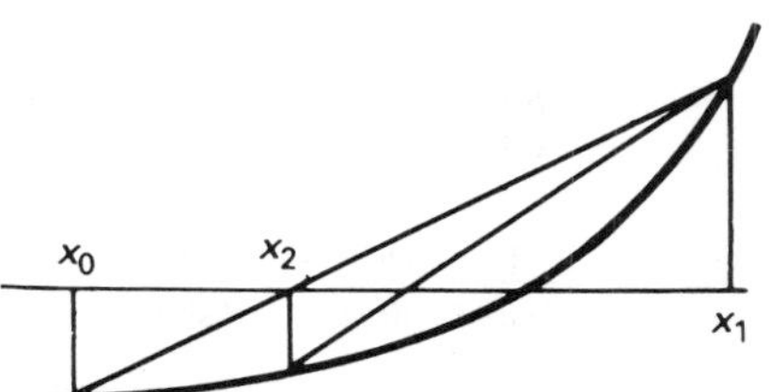

Figure 4.7 Regula falsi

Evaluate $f(x_2)$ and repeat the process for whichever of the intervals (x_0,x_2) or (x_1,x_2) contains the root. This method can be accelerated by halving at each step the function value at the retained end of the interval, as shown in *Figure 4.8*.

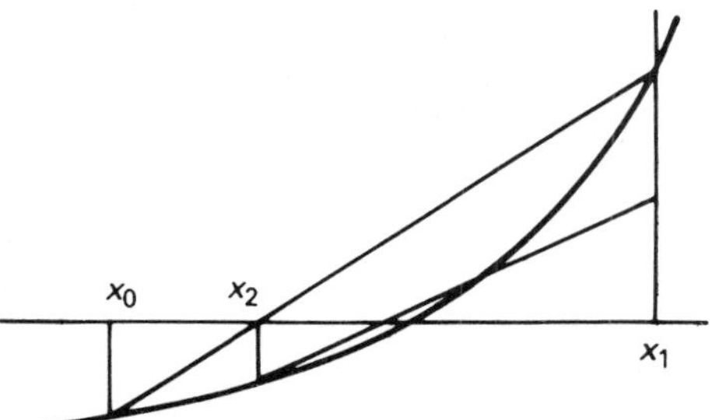

Figure 4.8 Accelerated method

4.19.4 Fixed-point iteration

Arrange the equation in the form

$x=f(x)$

Choose an initial value of x by trial, and calculate repetitively

$x_{k+1}=f(x_k)$

This process will not always converge.

4.19.5 Newton's method

Calculate repetitively (*Figure 4.9*)

$x_{k+1}=x_k-f(x_k)/f'(x_k)$

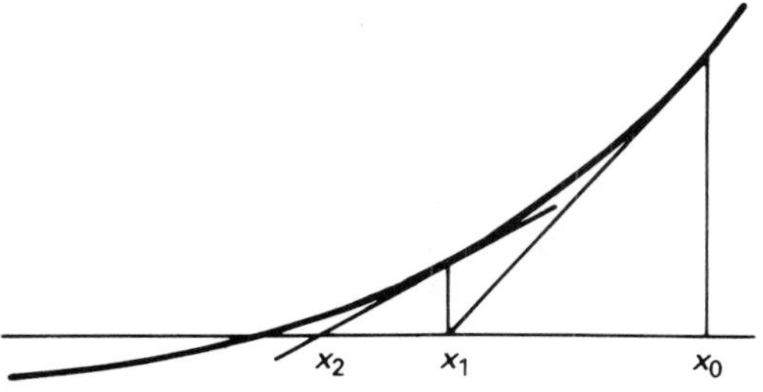

Figure 4.9 Newton's method

This method will converge unless: (a) x_k is near a point of inflexion of the function; or (b) x_k is near a local minimum; or (c) the root is multiple. If one of these cases arises, most of the trouble can be overcome by checking at each stage that

$f(x_{k+1})>f(x_k)$

and, if not, halving the preceding value of $|x_{k+1}-x_k|$.

4.20 Method of least squares

To obtain the best fit between a straight line $ax+by=1$ and several points (x_1,y_1), (x_2,y_2),..., (x_n,y_n) found by observation, the coefficients a and b are to be chosen so that the sum of the squares of the errors

$e_i=ax_i+by_i-1$

is a minimum. To do this, first write the set of inconsistent equations

$$ax_1+by_1-1=0$$
$$ax_2+by_2-1=0$$
$$\vdots$$
$$ax_n+by_n-1=0$$

Multiply each equation by the value of x it contains, and add, obtaining

$$a\sum_{i=1}^{n} x_i^2+b\sum_{i=1}^{n} x_iy_i-\sum_{i=1}^{n} x_i=0$$

Similarly multiply by y and add, obtaining

$$a\sum_{i=1}^{n} x_iy_i+b\sum_{i=1}^{n} y_i^2-\sum_{i=1}^{n} y_i=0$$

Lastly, solve these two equations for a and b, which will be the required values giving the least squares fit.

4.21 Relation between decibels, current and voltage ratio, and power ratio

$$\mathrm{dB}=10\log\frac{P_1}{P_2}=20\log\frac{V_1}{V_2}=20\log\frac{I_1}{I_2}$$

dB	I_1/I_2 or V_1/V_2	I_2/I_1 or V_2/V_1	P_1/P_2	P_2/P_1
0.1	1.012	0.989	1.023	0.977
0.2	1.023	0.977	1.047	0.955
0.3	1.035	0.966	1.072	0.933
0.4	1.047	0.955	1.096	0.912
0.5	1.059	0.944	1.122	0.891
0.6	1.072	0.933	1.148	0.871
0.7	1.084	0.923	1.175	0.851
0.8	1.096	0.912	1.202	0.832
0.9	1.109	0.902	1.230	0.813
1.0	1.122	0.891	1.259	0.794
1.1	1.135	0.881	1.288	0.776
1.2	1.148	0.871	1.318	0.759
1.3	1.162	0.861	1.349	0.741
1.4	1.175	0.851	1.380	0.724
1.5	1.188	0.841	1.413	0.708
1.6	1.202	0.832	1.445	0.692
1.7	1.216	0.822	1.479	0.676

dB	I_1/I_2 or V_1/V_2	I_2/I_1 or V_2/V_1	P_1/P_2	P_2/P_1
1.8	1.230	0.813	1.514	0.661
1.9	1.245	0.804	1.549	0.645
2.0	1.259	0.794	1.585	0.631
2.5	1.334	0.750	1.778	0.562
3.0	1.413	0.708	1.995	0.501
3.5	1.496	0.668	2.24	0.447
4.0	1.585	0.631	2.51	0.398
4.5	1.679	0.596	2.82	0.355
5.0	1.778	0.562	3.16	0.316
5.5	1.884	0.531	3.55	0.282
6.0	1.995	0.501	3.98	0.251
6.5	2.11	0.473	4.47	0.224
7.0	2.24	0.447	5.01	0.200
7.5	2.37	0.422	5.62	0.178
8.0	2.51	0.398	6.31	0.158
8.5	2.66	0.376	7.08	0.141
9.0	2.82	0.355	7.94	0.126
9.5	2.98	0.335	8.91	0.112
10.0	3.16	0.316	10.00	0.100
10.5	3.35	0.298	11.2	0.0891
11.0	3.55	0.282	12.6	0.0794
15.0	5.62	0.178	31.6	0.0316
15.5	5.96	0.168	35.5	0.0282
16.0	6.31	0.158	39.8	0.0251
16.5	6.68	0.150	44.7	0.0224
17.0	7.08	0.141	50.1	0.0200
17.5	7.50	0.133	56.2	0.0178
18.0	7.94	0.126	63.1	0.0158
18.5	8.41	0.119	70.8	0.0141
19.0	8.91	0.112	79.4	0.0126
19.5	9.44	0.106	89.1	0.0112
20.0	10.00	0.1000	100	0.0100
20.5	10.59	0.0944	112	0.00891
21.0	11.22	0.0891	126	0.00794
21.5	11.88	0.0841	141	0.00708
22.0	12.59	0.0794	158	0.00631
22.5	13.34	0.0750	178	0.00562
23.0	14.13	0.0708	200	0.00501
23.5	14.96	0.0668	224	0.00447
24.0	15.85	0.0631	251	0.00398
24.5	16.79	0.0596	282	0.00355
25.0	17.78	0.0562	316	0.00316
25.5	18.84	0.0531	355	0.00282
26.0	19.95	0.0501	398	0.00251
26.5	21.1	0.0473	447	0.00224
27.0	22.4	0.0447	501	0.00200
27.5	23.7	0.0422	562	0.00178
28.0	25.1	0.0398	631	0.00158
28.5	26.6	0.0376	708	0.00141
29.0	28.2	0.0355	794	0.00126
29.5	29.8	0.0335	891	0.00112
30.0	31.6	0.031.6	1000	0.00100
31.0	35.5	0.0282	1260	7.94×10^{-4}
32.0	39.8	0.0251	1580	6.31×10^{-4}
33.0	44.7	0.0224	2000	5.01×10^{-4}
34.0	50.1	0.0200	2510	3.98×10^{-4}
35.0	56.2	0.0178		3.16×10^{-4}
36.0	63.1	0.0158	3980	2.51×10^{-4}
37.0	70.8	0.0141	5010	2.00×10^{-4}

4.22 Calculus

4.22.1 Derivative

$$f'(x)=\lim_{\delta\to 0}\frac{f(x+\delta x)-f(x)}{\delta x}$$

If u and v are functions of x,

$$(uv)'=u'v+uv'$$

$$\left(\frac{u}{v}\right)'=\frac{u'v-uv'}{v^2}$$

$$(uv)^{(n)}=u^{(n)}v+nu^{(n-1)}v^{(1)}+\ldots+{}^nC_pu^{(n-p)}v^{(p)}+\ldots+uv^{(n)}$$

where

$${}^nC_p=\frac{n!}{p!(n-p)!}$$

If $z=f(x)$ and $y=g(z)$, then

$$\frac{dy}{dx}=\frac{dy}{dz}\frac{dz}{dx}$$

4.22.2 Maxima and minima

$f(x)$ has a stationary point wherever $f'(x)=0$: the point is a maximum, minimum or point of inflexion according as $f''(x)<$, $>$ or $=0$.

$f(x,y)$ has a stationary point wherever

$$\frac{\delta f}{\delta x}=\frac{\delta f}{\delta y}=0$$

Let (a,b) be such a point, and let

$$\frac{\delta^2 f}{\delta x^2}=A,\qquad \frac{\delta^2 f}{\delta x\delta y}=H,\qquad \frac{\delta^2 f}{\delta y^2}=B$$

all at that point, then:

If $H^2-AB>0$, $f(x,y)$ has a saddle point at (a,b).
If $H^2-AB<0$ and if $A<0$, $f(x,y)$ has a maximum at (a,b), but if $A>0$, $f(x,y)$ has a minimum at (a,b).
If $H^2=AB$, higher derivatives need to be considered.

4.22.3 Integral

$$\int_a^b f(x)\mathrm{d}x=\lim_{N\to\infty}\sum_{n=0}^{n-1} f\left(a+\frac{n(b-a)}{N}\right)\left(\frac{b-a}{N}\right)$$

$$=\lim_{N\to\infty if}\sum_{n=1}^{N} f(a+(n-1)\delta x)\delta x$$

where $\delta x=(b-a)/N$.

If u and v are functions of x, then

$$\int uv'=uv-\int u'v\mathrm{d}x \text{ (integration by parts)}$$

4.22.4 Derivatives and integrals

y	$\frac{dy}{dx}$	$\int y dx$
x^n	nx^{n-1}	$x^{n+1}/(n+1)$
$1/x$	$-1/x^2$	$\ln(x)$
e^{ax}	ae^{ax}	e^{ax}/a
$\ln(x)$	$1/x$	$x[\ln(x)-1]$
$\log_a x$	$\frac{1}{x}\log_a e$	$x \log_a\left(\frac{x}{e}\right)$
$\sin ax$	$a \cos ax$	$-\frac{1}{a}\cos ax$
$\cos ax$	$-a \sin ax$	$\frac{1}{a}\sin ax$
$\tan ax$	$a \sec^2 ax$	$-\frac{1}{a}\ln(\cos ax)$
$\cot ax$	$-a\,\mathrm{cosec}^2 ax$	$\frac{1}{a}\ln(\sin ax)$
$\sec ax$	$a \tan ax \sec ax$	$\frac{1}{a}\ln(\sec ax+\tan ax)$
$\mathrm{cosec}\, ax$	$-a \cot ax\, \mathrm{cosec}\, ax$	$\frac{1}{a}\ln(\mathrm{cosec}\, ax-\cot ax)$
$\arcsin(x/a)$	$1/(a^2-x^2)^{1/2}$	$x \arcsin(x/a)+(a^2-x^2)^{1/2}$
$\arccos(x/a)$	$-1/(a^2-x^2)^{1/2}$	$x \arccos(x/a)-(a^2-x^2)^{1/2}$
$\arctan(x/a)$	$a/(a^2+x^2)$	$x \arctan(x/a)-\frac{1}{2}a\ln(a^2+x^2)$
$\mathrm{arccot}(x/a)$	$-a/(a^2+x^2)$	$x\,\mathrm{arccot}(x/a)+\frac{1}{2}a\ln(a^2+x^2)$
$\mathrm{arcsec}(x/a)$	$a(x^2-a^2)^{-1/2}/x$	$x\,\mathrm{arcsec}(x/a)-a\ln[x+(x^2-a^2)^{1/2}]$
$\mathrm{arccosec}(x/a)$	$-a(x^2-a^2)^{-1/2}/x$	$x\,\mathrm{arccosec}(x/a)+a\ln[x+(x^2-a^2)^{1/2}]$
$\sinh ax$	$a \cosh ax$	$\frac{1}{a}\cosh ax$
$\coth ax$	$a \sinh ax$	$\frac{1}{a}\sinh ax$
$\tanh ax$	$a\,\mathrm{sech}^2 ax$	$\frac{1}{a}\ln(\cosh ax)$
$\coth ax$	$-a\,\mathrm{cosech}^2 ax$	$\frac{1}{a}\ln(\sinh ax)$
$\mathrm{sech}\, ax$	$-a \tanh ax\, \mathrm{sech}\, ax$	$\frac{2}{a}\arctan(e^{ax})$
$\mathrm{cosech}\, ax$	$-a \coth ax\, \mathrm{cosech}\, ax$	$\frac{1}{a}\ln\left(\tanh\frac{ax}{2}\right)$
$\mathrm{arsinh}(x/a)$	$(x^2+a^2)^{-1/2}$	$x\,\mathrm{arsinh}(x/a)-(x^2+a^2)^{1/2}$
$\mathrm{arcosh}(x/a)$	$(x^2-a^2)^{-1/2}$	$x\,\mathrm{arcosh}(x/a)-(x^2-a^2)^{1/2}$
$\mathrm{artanh}(x/a)$	$a(a^2-x^2)^{-1}$	$x\,\mathrm{artanh}(x/a)+\frac{1}{2}a\ln(a^2-x^2)$
$\mathrm{arcoth}(x/a)$	$-a(x^2-a^2)^{-1}$	$x\,\mathrm{arcoth}(x/a)+\frac{1}{2}a\ln(x^2-a^2)$
$\mathrm{arsech}(x/a)$	$-a(a^2-x^2)^{-1/2}/x$	$x\,\mathrm{arsech}(x/a)+a \arcsin(x/a)$
$\mathrm{arcosech}(x/a)$	$-a(x^2+a^2)^{-1/2}/x$	$x\,\mathrm{arcosech}(x/a)+a\,\mathrm{arsinh}(x/a)$
$(x^2\pm a^2)^{1/2}$		$\frac{1}{2}x(x^2\pm a^2)^{1/2}\pm\frac{1}{2}a^2\,\mathrm{arsinh}(x/a)$
$(a^2-x^2)^{1/2}$		$\frac{1}{2}x(a^2-x^2)^{1/2}+\frac{1}{2}a^2\arcsin(x/a)$
$(x^2\pm a^2)^p x$		$\frac{1}{2}(x^2\pm a^2)^{p+1}/(p+1)$ $(p\neq-1)$ $\frac{1}{2}\ln(x^2\pm a^2)$ $(p=-1)$
$(a^2-x^2)^p x$		$-\frac{1}{2}(a^2-x^2)^{p+1}/(p+1)$ $(p\neq-1)$ $-\frac{1}{2}\ln(a^2-x^2)$ $(p=-1)$
$x(ax^2+b)^p$		$(ax^2+b)^{p+1}/2a(p+1)$ $(p\neq-1)$ $[\ln(ax^2+b)]/2a$ $(p=-1)$
$(2ax-x^2)^{-1/2}$		$\arccos\left(\frac{a-x}{a}\right)$
$(a^2\sin^2 x+b^2\cos^2 x)^{-1}$		$\frac{1}{ab}\arctan\left(\frac{a}{b}\tan x\right)$
$(a^2\sin^2 x-b^2\cos^2 x)^{-1}$		$-\frac{1}{ab}\mathrm{artanh}\left(\frac{a}{b}\tan x\right)$
$e^{ax}\sin bx$		$e^{ax}\frac{a \sin bx-b \cos bx}{a^2+b^2}$
$e^{ax}\cos bx$		$e^{ax}\frac{(a \cos bx+b \sin bx)}{a^2+b^2}$

y	$\int y\,dx$
$\sin mx \sin nx$	$\frac{1}{2}\frac{\sin(m-n)x}{m-n} - \frac{1}{2}\frac{\sin(m+n)x}{m+n} \quad (m \neq n)$
	$\frac{1}{2}\left(x - \frac{\sin 2mx}{2m}\right) \quad (m=n)$
$\sin mx \cos nx$	$-\frac{1}{2}\frac{\cos(m+n)x}{m+n} - \frac{1}{2}\frac{\cos(m-n)x}{m-n} \quad (m \neq n)$
	$-\frac{1}{2}\frac{\cos 2mx}{2m} \quad (m=n)$
$\cos mx \cos nx$	$\frac{1}{2}\frac{\sin(m+n)x}{m+n} + \frac{1}{2}\frac{\sin(m-n)x}{m-n} \quad (m \neq n)$
	$\frac{1}{2}\left(x + \frac{\sin 2mx}{2m}\right) \quad (m=n)$

4.22.5 Standard substitutions

Integral a function of	*Substitute*
a^2-x^2	$x=a\sin\theta$ or $x=a\cos\theta$
a^2+x^2	$x=a\tan\theta$ or $x=a\sinh\theta$
x^2-a^2	$x=a\sec\theta$ or $x=a\cosh\theta$

4.22.6 Reduction formulas

$$\int \sin^m x\,dx = -\frac{1}{m}\sin^{m-1}x\cos x + \frac{m-1}{m}\int \sin^{m-2}x\,dx$$

$$\int \cos^m x\,dx = \frac{1}{m}\cos^{m-1}x\sin x + \frac{m-1}{m}\int \cos^{m-2}x\,dx$$

$$\int \sin^m x\cos^n x\,dx = \frac{\sin^{m+1}x\cos^{n-1}x}{m+n} + \frac{n-1}{m+n}\int \sin^m x\cos^{n-2}x\,dx$$

If the integrand is a rational function of $\sin x$ and/or $\cos x$, substitute $t=\tan \frac{1}{2}x$, then

$$\sin x = \frac{1}{1+t^2}, \quad \cos x = \frac{1-t^2}{1+t^2}, \quad dx = \frac{2dt}{1+t^2}$$

4.22.7 Numerical integration

4.22.7.1 Trapezoidal rule (Figure 4.10)

$$\int_{x1}^{x2} y\,dx = \tfrac{1}{2}h(y_1+y_2)+O(h^3)$$

4.22.7.2 Simpson's rule (Figure 4.10)

$$\int_{x1}^{x2} y\,dx = 2h(y_1+4y_2+y_3)/6 + O(h^5)$$

4.22.7.3 Change of variable in double integral

$$\iint f(x,y)\,dx\,dy = \iint F(u,v)|J|\,du\,dv$$

where

$$J = \frac{\delta(x,y)}{\delta(u,v)} = \begin{vmatrix} \frac{dx}{\delta u} & \frac{dx}{\delta v} \\ \frac{\delta y}{\delta u} & \frac{\delta y}{\delta v} \end{vmatrix} = \begin{vmatrix} \frac{dx}{\delta u} & \frac{dy}{\delta u} \\ \frac{\delta x}{\delta v} & \frac{\delta y}{\delta v} \end{vmatrix}$$

is the Jacobian of the transformation.

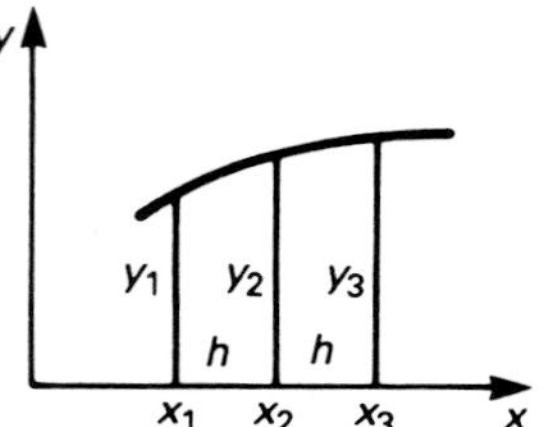

Figure 4.10 Numerical integration

4.22.7.4 Differential mean value theorem

$$\frac{f(x+h)-f(x)}{h} = f'(x+\theta h) \qquad 0<\theta<1$$

4.22.7.5 Integral mean value theorem

$$\int_a^b f(x)g(x)dx = g(a+\theta h)\int_a^b f(x)dx$$

$h=b-a,\ 0<\theta<1$

4.22.8 Vector calculus

Let $s(x,y,z)$ be a scalar function of position and let

$\mathbf{v}(x,y,z)=\mathbf{i}v_x(x,y,z)+\mathbf{j}v_y(x,y,z)+\mathbf{k}v_z(x,y,z)$

be a vector function of position. Define

$$\nabla = \mathbf{i}\frac{\delta}{\delta x} + \mathbf{j}\frac{\delta}{\delta y} + \mathbf{k}\frac{\delta}{\delta z}$$

so that

$$\nabla.\nabla = \nabla^2 = \frac{\delta^2}{\delta x^2} + \frac{\delta^2}{\delta y^2} + \frac{\delta^2}{\delta z^2}$$

then

$$\text{grad } s = \nabla s = \mathbf{i}\frac{\delta s}{\delta x} + \mathbf{j}\frac{\delta s}{\delta y} + \mathbf{k}\frac{\delta s}{\delta z}$$

$$\text{div } \mathbf{v} = \nabla.\mathbf{v} = \frac{\delta v_x}{\delta x} + \frac{\delta v_y}{\delta y} + \frac{\delta v_z}{\delta z}$$

$$\text{curl } \mathbf{v} = \nabla\times\mathbf{v} = \mathbf{i}\left(\frac{\delta v_z}{\delta y} - \frac{\delta v_y}{\delta z}\right) + \mathbf{j}\left(\frac{\delta v_x}{\delta z} - \frac{\delta v_z}{\delta x}\right) + \mathbf{k}\left(\frac{\delta v_y}{\delta x} - \frac{\delta v_x}{\delta y}\right)$$

The following identities are then true:

div(s**v**)=s *div* **v**+(grad s).**v**

curl(s**v**)=$s$curl **v**+(grad s)×**v**

div(**u**×**v**) =**v**.curl **u**−**u**.curl **v**

curl(**u**×**v**)=**u** div **v**−**v** div **u**+(**v**.∇)**u**−(**u**.∇)**v**

div grad $s=\nabla^2 s$

div curl **v**=0

curl grad s=0

curl curl **v**=grad(div **v**)−∇²**v**

where ∇² operates on each component of **v**.

vx curl **v**+(**v**.∇)**v**=grad½**v**²

Potentials:

If curl **v**=0,**v**=grad φ where φ is a scalar potential.
If div **v**=0,**v**=curl **A** where **A** is a vector potential.

Part 2
Materials, Components and Construction

J M Woodgate B Sc(Eng), C Eng, MIEE, MAES, M Inst SCE
Electronics design consultant

5

Conductors and Insulators, Passive Components, Printed Circuit Boards

5.1 Conductors, semiconductors and insulators

5.1.1 Atomic and molecular structure

A reasonably accurate qualitative picture of electrical conduction and insulation can be obtained by assuming that an *atom* (the smallest possible particle of a chemical element) consists of a tiny massive core, the positively charged *nucleus*, around which circulate a number of rather diffuse 'blobs', called *electrons*. These are negatively charged and much less massive, but considerably larger, than the particles which make up the nucleus (*protons* and *neutrons*). The diffuseness of the electron is an expression of its quantum-mechanical nature.

In a *molecule* (the smallest possible particle of a chemical compound), atoms are bound together by forces which arise from transfer or sharing of electrons between adjacent atoms. Substantially complete transfer of electrons from one atom to another creates *ionic bonds* between them, while the sharing of electrons creates *covalent bonds*. Chemical compounds can thus be divided into two structural types, ionic and covalent, with very different electrical properties: there are also compounds of mixed nature. A third structure, in which electrons can move about fairly freely in the material, not being attached to any particular atom, is known as the *metallic structure*. This is exhibited by metallic elements, carbon in the graphite form and some non-metallic compounds. Due to the presence of mobile electrons, these materials conduct electricity. A fourth structure, in which the free electrons experience no interference whatever with their movement, gives rise to the property of *superconductivity*, where there is no resistance at all to the flow of electric current. There is evidence that more than one type of structure may in fact result in this property.

The model of the atom described above is not consistent with classical mechanics. The circulating electrons are electrically charged, and their curved paths indicate that they are accelerating (Newton's first law), so they should continuously radiate energy. To explain the fact that this radiation does not occur, Bohr (1913) postulated that electrons can gain or lose energy only in discrete amounts, called *quanta*, and that each energy state corresponds to a different orbital radius in the atom. In one of these 'permitted' energy states, an orbiting electron is quasi-stable, and does not radiate continuously. If it is in a higher energy state than the lowest possible, or ground state, it will release its excess energy, in the form of one or more quanta of radiation, as soon as conditions are favourable. Electrons can enter 'forbidden' energy states, with energies intermediate between permitted states, only under special circumstances, and normally give up their excess energy very quickly to return to a permitted state of lower energy.

This hypothesis explains a number of physical phenomena, such as the line structure of the spectra of elements. The electron orbits are not all spherical, but for most purposes in electronics the more complex shapes do not require consideration. Quantum mechanics requires that the 'orbital radius' is not a definite value, but rather a 'most probable' value, for each energy level. It also stipulates that each energy level can contain a maximum of only two electrons. However, the difference in energy between two different levels may be very small.

5.1.2 Energy levels

The study of energy levels is essential in order to understand how solid-state electronic devices work. For a single, isolated atom, the permitted levels are (nearly) single-valued (i.e. the differences in energy between adjacent levels are not very small), and consist of three groups. The lowest levels are occupied by *core electrons* which do not take part in chemical reactions or electronic phenomena except under extreme conditions of temperature, pressure and/or electric stress. The outermost electrons of the atom, the *valence electrons*, form a group of higher energy levels, which are responsible for chemical reactions. At yet higher energies are a group of levels known as the *conduction band*, where the electron cloud responsible for electrical conductivity exists in conductors.

In a bulk material, whether gaseous, liquid or solid, the presence of vast numbers of atoms, each having an effect on others near it, results in the splitting of the discrete energy levels of the isolated atom into *bands* of levels so close together that energy transitions no longer occur in clearly discrete steps, but appear to occur almost smoothly. This effect is naturally most prevalent in solids, where the interatomic distances are smallest. However, between the bands of permitted levels there may, depending on the material, still remain bands of forbidden levels, or *gaps*, where there are no electrons with corresponding energies. If the temperature of the material is raised, electrons will absorb energy and move into higher levels. The energy levels of the bands and gaps themselves are characteristic of the material and do not change with temperature unless the structure changes.

5.1.3 Conductors and superconductors

In a material which conducts electricity, there is no band gap between the valence band and the conduction band. Electronics can pass more or less freely through the material. In many non-metallic conductors, electron mobility is very different in different directions relative to the molecular structure; often the conducting behaviour is practically confined to one or two axes.

The variation of metallic conductivity with temperature is a complex matter. The increased numbers of electrons forced into the conduction band by the absorption of thermal energy tend to increase the conductivity, but the interference with electron flow due to the fixed atoms also increases, due to their more energetic vibration. In most cases, conductivity decreases above room temperature. Special alloys, such as Eureka, Manganin and Nichrome, have been developed to have conductivity which is rather low for a metal, and varies very little with temperature. Such alloys are thus used for making wirewound resistors and heating elements.

At very low temperatures, some materials exhibit superconductivity, where the resistance to electron flow disappears entirely. For metals, metal alloys and compounds, the temperature has to be very low indeed, mostly below 10 K. The original explanation of this effect was that pairs of electrons could be formed by the interaction of electrons with vibrations of the material structure or 'lattice'. Such paired electrons are almost confined to a single energy state, above which is a gap representing the dissociation energy of the pairs. Because there is only one permitted state, the paired electrons cannot absorb from, or give up energy to, electrons of different energy. Since this absorption is the primary cause of resistive losses in conductors, there is no resistance at all in this case. Recent discoveries in connection with so-called *high temperature superconductors*, which are superconducting at temperatures of the order of 100 K or above, have suggested that, at least for these materials, lattice vibration does not occur, and a different explanation of lossless propagation is required.

5.1.4 Semiconductors

A semiconductor may be an element, silicon (Si) or germanium (Ge), or a chemical compound. Silicon and germanium are members, together with carbon (C), tin (Sn) and lead (Pb), of group IV of the periodic table of elements. Compound semiconductors may consist of elements from groups III and V, or groups II and VI, and the compound silicon carbide (SiC, a IV/IV semiconductor) is used for some devices (*Table 5.1*).

In a semiconductor, the band gap between the valence and conduction bands is fairly narrow, and the valence band is almost full of electrons. A full band does not permit conduction, because for this to occur, there must be vacant energy levels for electrons, accelerated by the applied electric field, to move into. At room temperature, only a few electrons are (thermally) energetic enough to enter the conduction band, so the resistivity of a typical pure, or *intrinsic*, semiconductor is high compared with that of a typical metallic conductor. It should be understood that materials can now be made with intermediate properties, i.e. high-resistivity metallic conductors and low-resistivity intrinsic semiconductors. Since the resistivity of an intrinsic semiconductor depends on the thermal energy of the electrons, it is strongly temperature dependent. In fact the conductivity varies exponentially with temperature.

Group	*Element*	*Symbol*	*Uses*
	Zinc	Zn	ZnS, ZnSe: Camera tube targets, phosphor screens
II	Cadmium	Cd	CdS, CdSe, CdTe: optoelectronic and microwave devices
	Boron	B	Dopant
	Aluminium	Al	Dopant, conductor Al_2O_3: insulator, substrate
III	Gallium	Ga	Dopant GaP: optoelectric devices GaAs: semiconductor substrate, especially for microwave GaAsP: light-emitting diodes GaAlAsP: semiconductor laser diodes
	Indium	In	Dopant InP, InAs: microwave devices InSb: Hall-effect devices
	Thallium	Tl	Dopant
	Carbon	C	Possible future uses
	Silicon	Si	Major semiconductor substrate SiC: blue LEDs, special high-temperature devices
IV	Germanium	Ge	Semiconductor substrate: now mainly for diodes
	Tin	Sn	SnO_2: transparent conductor for optoelectronic devices
	Lead	Pb	PbS, PbSe, PbTe: camera tube targets
	Nitrogen	N	Si_3N_4: insulator in semiconductors
	Phosphorus	P	Dopant, compounds
V	Arsenic	As	Dopant, compounds
	Antimony	Sb	Dopant, compounds
	Bismuth	Bi	Bi_4Te_3: cooling devices (reverse Peltier effect)
	Oxygen	O	SiO_2: insulator in semiconductors
VI	Sulphur	S	Compounds (see above)
	Selenium	Se	Optoelectronic devices: diodes (formerly)
	Tellurium	Te	Compounds (see above)

Table 5.1 Examples of applications of elements in semiconductor technology

The conductivity and other properties of a semiconductor can be changed profoundly by adding a minute amount of impurity or *dopant*. The dopant may have one more electron in the valence band than the semiconductor, and a superfluous electron is attracted only weakly by the fixed atoms and is easily promoted to the conduction band. A vacancy for an electron, produced by a dopant with one fewer electron per atom than the semiconductor, is called a *hole*, and behaves like a positively charged electron. In semiconductors with useful levels of doping, the room temperature conduction is largely due to these impurity electrons and holes, and varies little with temperature. At higher temperatures the greatly increased intrinsic conduction becomes dominant.

Semiconductor junctions are explained in section *6.1*.

5.1.5 Insulators

Most practical insulators are covalent compounds, where all the electrons are tightly bound to the atoms and none are available for conduction. The energy band theory which explains the properties of metals and semiconductors is unsatisfactory for many insulators, if applied in any simple form, due to structural complexities. The conductivity of most insulators increases with temperature, in some cases very rapidly.

Strongly ionic compounds are also insulators in the solid state, where the electrons are strongly bound to the ionized atoms which in turn are fixed in the crystal lattice, but are good conductors in the liquid state. This has considerable implications when failure modes are concerned.

5.1.5.1 Failure of insulators

Failure of covalent insulators occurs under high voltage stress if the internal electric field is high enough to disrupt the chemical structure of the material, a process known as *ionization* (stripping electrons from one or more atoms). This in itself may cause failure, but there may be secondary effects, due to the production of conducting and/or corrosive degradation products. Voids, usually air bubbles, in moulded thermoplastic insulators can lead to rapid failure, due to the production of oxygen ions in the air, which rapidly attack the insulating material. Many of these materials are also rapidly attacked by ozone, the triatomic form of oxygen which is produced by corona discharge and electric sparks.

A special case of some importance is that of glasses and ceramics, which are ionic compounds. Glasses are not crystalline, but are actually supercooled liquids, so viscous that they will not crystallize even at temperatures well below the melting point. Both of these materials can be very good insulators (although conducting types can also be made), and are particularly useful at high temperatures, e.g. as structural elements of resistors. However, if the material melts, e.g. under fault conditions, catastrophic failure may occur, because the conductivity decreases rapidly to a very low value. This may result in a large fault current and the generation of extremely high temperatures.

Long-term effects occur in insulators under combined electric and thermal stress. Known as *dielectric degradation*, these effects lead to an indefinite rise in conductivity after a stress duration which may vary from a few hours to (predicted values of) thousands of years. Mechanisms include electrolysis, migration and segregation of impurities and the development of defects in crystal structure, apart from possible chemical changes not primarily due to the electric stress.

A further failure mechanism, which is not due to the insulator itself, is known as *ionic drift* or *electromigration*. This is caused by a conductive film, derived from conductors in contact with the insulator, growing across the surface, or possibly within the bulk, of the insulator. This effect is aggravated by the presence of moisture, and some metal/insulator combinations seem more prone to exhibit it than others.

A similar effect can occur where, although a stable conducting film is not formed, surface conduction takes place through a film of moisture, even when no surface moisture is apparent. This effect can occur on the surface of printed wiring boards, if not protected by insulating varnish, especially if moisture is trapped in a capillary slit between a component and the board surface. For this reason, such capillary slits should always be avoided.

5.1.5.2 Dielectric properties of insulators

Most of the dielectric properties of insulators can be described by two characteristics, the relative permittivity and the loss factor, both of which may vary with temperature and frequency. The *relative permittivity* of an insulator indicates the ratio of the value of a capacitor using the insulator as dielectric to the value of a capacitor of the same dimensions having a vacuum dielectric. The *loss factor* is the ratio of the resistive component of the impedance of a capacitor using the insulator as dielectric to the capacitive component, when other resistive losses (due to the electrodes, for example) have been eliminated.

However, some insulators exhibit an effect known as *dielectric hysteresis*. This is due to the retention of internal electric stress within the material after the applied field has been reduced to zero, and the residual stress may not relax for several seconds after the external field has vanished. This can result, for example, in the retention of charge on an apparently discharged capacitor, which may cause pulse waveform distortion or give rise to a shock hazard.

5.2 Resistors, capacitors and inductors

5.2.1 Passive and active components

A *passive component* is one that requires no energy for its operation, other than the applied signal. Thus resistors, capacitors and inductors are passive. Diodes, including varactors but not diodes showing negative resistance characteristics, are also passive components.

An *active component* is one that requires an external energy source. The external energy source may be a second signal, such as in a balanced modulator using transistors as switches. It is usual, but not essential, for an active component to cause energy to be transferred from the external source to the signal. For example, while an active device is often used as an amplifier, it may be a biassing device or even act as an attenuator or switch. An *active network*, however, is defined as a network that transfers energy from an external source to an output signal, which may be generated within the network (as in an oscillator) or be controlled by an input signal.

Figure 5.1 Metal film precision resistors 'bandoliered' for use with automatic technology (Maplin Electronics Ltd)

5.2.2 Resistors

In principle, resistors are the simplest components to understand, and their selection and specification is easier now than it used to be, because high quality components are freely available at low cost. On the other hand, the greater precision in design that is made practicable by the improved component quality requires factors to be taken into account that once were negligible.

Resistors may be usefully classified according to the materials used in their construction and the intended application of the finished component. It is possible to deal with both fixed and variable resistors together; important differences are noted.

5.2.2.1 Wirewound resistors

As the name implies, the resistive element in these components is in the form of a wire or tape, usually wound on an insulating former. These may be sub-classified by intended application, as precision, power, precision power or special.

Precision wirewound resistors are used in measuring instruments, the emphasis being on precision and stability of resistance value. It is often required in addition that the parasitic series inductance and/or parallel capacitance should be negligible, i.e. that the impedance of the component is a pure resistance. The permitted power dissipation and temperature rise are limited.

Power wirewound resistors are usually required to have good stability of resistance value and a low temperature coefficient of resistance, while being capable of dissipating a significant amount of power. With the continuous trend towards smaller components, high operating temperatures, and consequently high surface temperatures, at maximum dissipation are inevitable.

Precision power wirewound resistors combine the attributes of the previous two types, and are usually of metal-cased construction, requiring an external heat-sink to achieve rated maximum power dissipation without exceeding the permitted temperature rating. The metal case results in a parasitic capacitance (either in parallel with the resistor or to the heat-sink potential, normally common) which may not be negligible.

Special types of resistor include those with specified significant temperature coefficients (e.g. copper wire resistors for the temperature compensation of transistor bias circuits) and types with specified thermoelectric potentials. Special variable wirewound resistors include types with a precise 'law' relating resistance value to slider position or spindle rotation (e.g. sine/cosine potentiometers), types with more than one element, with characteristics approaching constant-resistance attenuators (e.g. for loudspeaker volume controls) and types with 'balanced' parallel capacitance, where the attenuation/frequency characteristic is improved with respect to those of standard types.

5.2.2.2 Carbon resistors

These have been the most common types until recently, but many types are rapidly becoming obsolete. They may be sub-classified according to the form in which the carbon element is made, as carbon composition, carbon film, sprayed, conductive plastic or thick film.

Carbon composition resistors have the resistive element in the form of a rod of carbon mixed with inert filler and binder. The resistivity is adjusted by controlling the proportions of carbon and filler. Largely obsolete, they have unpredictable stability and a wide spread of resistance value in manufacture. They also generate noise, which, for higher resistance values, may be considerably in excess of the inevitable Johnson noise.

Carbon film resistors have as their resistance element a thin film of carbon deposited on a glass or ceramic rod. For high value resistors, the effective length of the element is increased by cutting a spiral track in the film. These components show fair stability and reproducability in manufacture, but are subject to large, unpredictable changes in resistance (usually increases) under heat and voltage stress. For most applications, except in values above about 1 megohm, they are obsolescent.

Sprayed resistors are low-cost variable elements in which the 'tracks' or resistive elements are sprayed. Their precision and stability are not good but billions of these components have been found satisfactory in service, and they are still being used in large quantities.

Conductive plastic resistors use either a plastic substrate into which carbon is diffused, or a printing (thick film) technique with a carbon-containing ink. The former technique is employed in high quality potentiometers, such as audio faders.

Thick film resistors are produced by a screen-printing technique, using a carbon-loaded ink, followed by a firing process. Resistor networks, comprising several carbon thick film resistors in one package, are available for noncritical applications in digital circuits, such as pull-up resistors and current-limiters for display devices.

5.2.2.3 Metal oxide and metal film resistors

Most of the low- and medium-power resistor requirements are now met by these types, which offer better characteristics than earlier types at comparable, or lower, cost. They are also available in physically smaller types. They may be metal oxide, metal glaze, metal film or cermet.

Metal oxide resistors have a resistance element composed of a mixture of metallic oxides (mostly stannic oxide, SnO_2). They can be made in both low-power and high-power types, and feature good stability and other properties. Low-power types have largely been superseded by metal film types (see below).

Metal glaze resistors have their resistive element formed from a metal alloy film, fused and chemically bonded to the surface of a ceramic rod or plate. The resistivity of such a film can be made high, to allow the achievement of high resistance values in physically small components without excessively fine spiralling of the element. Metal glaze technology is also used for surface-mounting 'chip' resistors which are very resistant to external heating, and may actually be immersed in molten solder during circuit board manufacture.

Metal film resistors are the general-purpose resistors of the present time. They are available in a wide range of values and power ratings, and in two groups of tolerances. Standard components are available in tolerances down to 1 per cent, while in precision types tolerances as close as 0.01 per cent can be obtained. Chip resistors for surface-mounting are also available. Fusible resistors, which fail on overload without fire risk, are obtainable. (It should be noted that other types of resistor using ceramic or glass for support may fail hazardously on overload due to conduction of the molten ceramic former. See section *5.1.5.1*). Noise generation is very low in all metal film resistors of reliable manufacture: indeed, noise generation is a sign of incipient failure.

Cermet resistors have as a resistive material a mixture of metal and metal oxide particles with glass powder. A wide range of resistivity can be achieved, with good thermal endurance. This allows the technique to be used for small power resistors and for high resistance values. High quality preset and variable resistors can also be made, together with thick film resistor networks.

5.2.2.4 Selecting and specifying resistors

In selecting resistors, or any components for that matter, the intention should be to obtain the required characteristics for the lowest cost, and without buying too many attributes that are of no importance. It is not good engineering to select components with specifications far in excess of those required, unless by chance there is no cost penalty in doing so. The selection process normally involves the following characteristics in order, information on which should be provided by the manufacturer.

5.2.2.4.1 Value

Most resistors are available in a fixed logarithmic series of internationally standardized values. These logarithmic series are designated E12, E24, E96, etc. (the number signifying the number of different values in each decade), and were originally related to selection tolerance; there is no point in specifying a value to a greater precision than the selection tolerance allows (e.g. 109 045 ohms ±20 %, as a facetious example). However, for many purposes it is unnecessary to have so many values available in 1 per cent tolerance components, and only values in the E24 series are stock items. This also avoids the use of the rather unpopular five-band colour code necessary to indicate E96 values. While values outside the standard series are made to order, it is preferable to use standard values wherever possible.

Most ranges of resistor cover values between 12 ohms and 100 kilohms; outside this range it is necessary to check availability and/or the possibility of a price premium.

5.2.2.4.2 Tolerance

Tolerance engineering is a subject in itself, and cannot be treated here. Luckily, 1 per cent tolerance components are now available at low cost, and it is usually very clearly evident during circuit design if a closer tolerance than this is necessary. For extremely low and high resistance values, however, close tolerance is still expensive and careful consideration is required of means to avoid its necessity.

5.2.2.4.3 Power dissipation

It is necessary for the component to dissipate the required power without exceeding its own permitted temperature and without overheating adjacent components. Careful consideration is necessary regarding ventilation or other means of cooling, the position and orientation of the hot component (e.g. long thin components mounted vertically may overheat at one end only, which may be either end) and whether it is better to have 'high grade' heat from a small component with a high surface temperature, or 'low grade' heat from a larger, cooler component. Frequently the latter is preferable, as ageing processes tend to double in rate for every 10 K rise in temperature. While there is usually a size penalty if a component of higher permitted dissipation is chosen, the cost penalty may be small or nil.

5.2.2.4.4 Maximum voltage

This is a relatively new factor for the selection of general-purpose resistors. Some types now available have maximum voltage ratings below 250 V,which may be exceeded in what would not be considered as 'high voltage' circuits. Note that the peak value of 240 V mains is 339 V! Special resistor types are available for applied voltages in excess of 1 kV. These are preferable to series chains of ordinary resistors, where an increase in resistance may result in cumulative unequal voltage sharing, and rapid failure to open-circuit.

5.2.2.4.5 Stability

The above characteristics are all that are usually necessary to consider for most applications. Stability and the following characteristics are nevertheless vitally important in particular cases.

Stability of resistance value with time is affected by ambient temperatures and internally generated heat. Carbon resistors are not as good in this respect as metal film and wirewound types, with metal oxide types showing intermediate characteristics.

5.2.2.4.6 Temperature coefficient

This is more significant in carbon types than in others. It is possible to obtain resistors of almost any desired temperature coefficient where essential.

5.2.2.4.7 Noise

Excess noise in resistors is related to the applied voltage, and is least in metal film types. The Johnson noise voltage is proportional to the square root of the absolute temperature of the resistor, so sensitive circuits should be kept cool.

5.2.2.4.8 Voltage coefficient

The dependence of resistance value on applied voltage is normally significant only for carbon composition resistors, but it is a non-linearity which can give rise to unwanted signal distortion, albeit at very low levels (<0.1 per cent). The effect is naturally observed to a significant extent only in high value components, as for lower resistance values the applied voltage is limited by the power rating.

5.2.2.4.9 Operating frequency

For most resistor types the limit to operating frequency is set by the parallel capacitance, but for power wirewound types of medium resistance the series inductance may become important at low radio frequencies. At higher frequencies wirewound resistors are not usable, and unspiralled metal film or metal glaze resistors are preferred. Special types are made for attenuators and for uhf and higher frequency applications.

5.2.2.4.10 Thermoelectric voltage

An emf is generated at the junction of almost any pair of dissimilar conductors. In a loop at constant temperature throughout, these emfs add to zero, but a practical circuit may well not be all at the same temperature. Since the junction emf may be some tens of microvolts per kelvin, it may be significant in high-gain dc amplifiers. Since resistors use conductors other than copper, they may give rise to these thermal emfs. Carbon film types are particularly prone, but other types may also cause problems, and it is necessary to check individual specifications or consult manufacturers on this subject.

Figure 5.2 A selection of modern fixed capacitors, ranging from low-value cermic types at right foreground, through polystyrene and polyester types to high-value electrolytics at the rear (Maplin Electronics Ltd)

5.2.3 Capacitors

All capacitors consist of a pair of conductors, or electrodes, separated by an insulating layer, the *dielectric*. They are produced in a variety of physical forms, and may be classified by the nature of the dielectric material. Sub-classification depends on the dielectric, and may be based on physical form or the precise composition of the dielectric material.

5.2.3.1 Ceramic capacitors

These used to be available in a vast number of different types, which, however, have now been rationalized. Metal electrodes are evaporated onto a ceramic insulator, which may take one of several physical forms. The precise type of ceramic material may be chosen to give various desired properties. At present, three types are common, one (NPO or COG) with a low permittivity and near-zero temperature coefficient of permittivity, one (X7R) with an intermediate permittivity and a higher, rather non-linear temperature coefficient, and one (Z5U) with high permittivity and a very non-linear temperature characteristic. Ceramic capacitors can be obtained in values from about 1 pF to 10 μF, and even larger values are available.

Types include tube, disc and plate, multilayer, surface-mounting, lead-through and stand-off.

Tube capacitors have, as the name implies, a tubular dielectric, with electrodes on the interior and exterior surfaces. This type has been largely replaced by the plate type.

Disc and plate capacitors have a classical style of construction, with electrodes on each side of a disc or plate. While they have low losses and low series inductance, the larger disc types tend to act as short antennas.

Multilayer capacitors are constructed from a stack of plate capacitors connected in parallel and pressed into a solid block. They offer relatively large capacitance in a small volume, with low losses and inductance.

Surface-mounting capacitors are 'chip' capacitors of the plate or multilayer types, with terminations and thermal endurance suitable for surface-mounting applications.

Lead-through and *stand-off* capacitors are made for vhf and uhf applications, and consist of a centre conductor surrounded by a ceramic tube, which carries a metal collar forming the other electrode termination. This construction ensures very low inductance, and allows, for example, power supply conductors to pass through a screening box without compromising the screening.

5.2.3.2 Plastic film capacitors

These components may consist of metal foil electrodes separated by dielectric film and rolled into a cylinder. Alternatively, the dielectric film may be metallized in a vacuum, and the capacitor may consist of a stack of rectangular sheets connected in parallel, instead of a continuous spiral. Sub-classification by dielectric material is the most useful. These are polystyrene, polypropylene, polyester, polycarbonate, mixed and special types.

Polystyrene capacitors have a low, stable permittivity and very low losses even at high radio frequencies. Capacitors can be made with tolerances closer than 1 per cent, and in values from about 10 pF to an economic limit of 100 nF. However, the dielectric melts at about 85°C, so operating temperature is restricted, and care is necessary in soldering during manufacture. Because of the chemical properties and low melting point of the film, metallization is not normally possible, and foil electrodes are used.

Polypropylene capacitors are suitable for applications where a high voltage is applied at up to low radio frequencies, and for dc operation above 500 V. Foil electrodes are normal, although metallized film is increasingly used, and capacitance values run from 1 nF to above 1 μF.

Polyester capacitors include a dielectric that can be produced in very thin films, with a high permittivity. Consequently it is used for general-purpose components in the capacitance range from 1 nF to 10 μF. However, the dielectric loss is only moderately low, and its temperature coefficient is variable. Metallized film is usually used, although foil construction is also obtainable. Cylindrical, flattened cylinder and stacked constructions are used.

Polycarbonate capacitors have better loss and leakage resistance characteristics than polyester, but are a little higher in price. Polycarbonate tends to be used only in 'premium quality' components, but lower cost types are also available. Metallized foil construction is usual, in cylindrical or flattened cylinder form. Unlike metallized polyester capacitors, however, metallized polycarbonate capacitors do not reliably 'self-heal' (by local evaporation of the electrode film) if a pinhole short-circuit occurs. Consequently, they should not be used in positions where failure could be hazardous.

Mixed dielectric capacitors are obsolescent except for special applications involving high-voltage pulses of fast rise-time, and those detailed below. In most cases, the dielectric is composed of a polyester film, bonded to paper. Polypropylene capacitors are now used for this type of service.

Special types of plastic film capacitor are available. For applications where a short-circuit might give rise to a safety hazard, or an undetected fault condition, special metallized polyester capacitors can be used. In these, local failure of the dielectric, due to an imperfection or weak spot failing under voltage stress which may be outside the normal voltage range (e.g. caused by lightning striking power lines), causes local evaporation of the metallizing and consequently a 'healing' of the short-circuit. Two grades of component are made. *X type* will self-heal but may do so rather often in service, and must not be used where failure could create a shock hazard, e.g. between the mains supply and common rail of a Class II (double-insulated) equipment (see BS 2754:1976 or IEC Publication 356; Class II is defined differently in North America). For this more critical service, *Y type* mixed dielectric capacitors are made, and the use of X type and Y type components, in appropriate circuit positions, in such applications as interference filters is required by safety standards in many countries. The components themselves also have to carry recognized safety approvals.

5.2.3.3 Electrolytic capacitors

In these components the dielectric is formed by an extremely thin layer of oxide on the surface of the positive electrode. Both aluminium and tantalum have oxides suitable for this purpose. Unlike ceramic and film types, electrolytic capacitors have specific positive and negative terminals, and, except for special 'reversible' types (see below), require a dc polarizing voltage to be applied. They also pass appreciable leakage current when correctly polarized, and are likely to be destroyed if incorrectly polarized. For these reasons, they are restricted to high capacitance values (generally exceeding 1 μF), which are difficult or expensive to manufacture with other techniques.

Reversible electrolytic capacitors are made by forming an oxide layer on both electrodes. Because of the two dielectrics in series, the volume for a given capacitance and rated voltage is twice that of a polarized capacitor. These components are often used for passive dividing networks in multi-way loudspeakers, and are available in two types, a low-loss type capable of passing high currents, and a more compact type for less critical positions.

Sub-classification of electrolytic capacitors is by dielectric (and therefore electrode) material, and construction. They include aluminium foil, solid aluminium, solid tantalum and very high capacitance types.

Aluminium foil capacitors have two aluminium foil electrodes, separated by an absorbent paper spacer impregnated with an electrolyte solution. The assembly is wound in cylindrical form and encased. Oxide dielectric is formed on the surface of the positive electrode by applying a dc voltage to cause electrolysis of the electrolyte. In order to increase the capacitance achievable in a given volume, the electrode surfaces are acid-etched. The negative electrode has an extremely thin film of oxide on its surface, formed on exposure to air. This allows the capacitor to withstand a small reverse polarizing voltage (i.e. the leakage current remains low).

Solid aluminium capacitors have not proved very popular, but they show many characteristics improved over those of aluminium foil and tantalum types. There is (ideally) no liquid in the capacitor at all; the 'electrolyte' is semiconducting manganese dioxide, usually dispersed in an inert matrix such as glass fibre. These capacitors will withstand considerable reverse polarization and high ripple current, are highly reliable and have a long life.

Solid tantalum capacitors consist of a porous 'slug' of sintered (compressed and partially fused) tantalum, the surface of which is oxidized electrolytically, surrounded by an electrolyte gel and a metal case which forms the negative electrode. These capacitors had size advantages over aluminium types when first introduced, but this is now less evident. They are subject to failure due to current surges (e.g. surges must be limited to 3 A or less under all conditions), and end-of-life failure is normally due to the penetration of the oxide dielectric by a crystalline form of oxide which is a conductor. Permissible reverse polarization is often limited to a very low voltage (e.g. 0.5 V), contrary to popular opinion, which also holds, incorrectly, that the leakage current of tantalum capacitors is much lower than that of aluminium types of similar value, working voltage and size. In fact, the specifications of aluminium and tantalum components are not dissimilar in this respect, and in practice, the aluminium components often show lower leakage. However, tantalum components show lower losses and are available in temperature ratings up to 125°C.

Very high capacitance types are made for energy storage at low voltage, mostly for the maintenance of semiconductor memories. The electrolyte is in liquid form, and the dielectric is an electrical double layer, formed on the surface of an oxide semiconductor. This layer is extremely thin, and has a very high permittivity, so that capacitances up to several farads can be made in very small volumes.

5.2.3.4 Mica, glass, vacuum and air dielectrics

Capacitors made with these dielectrics are now used for special purposes, such as in transmitters where high rf voltages are present. Glass dielectrics used to be popular for precision capacitors in filters, and mica capacitors are used for this service as well.

Mica is a natural silicate mineral with a pronounced laminar structure. When pure, it can be split into very thin sheets and has a high dielectric strength. Compared with organic insulators its permittivity (around 7) is quite high, and its losses are low. Consequently it was extensively used in capacitors for the first 50 years of electronics. Subsequently its use has become more specialized, taking advantage of the low losses and stable performance that can be achieved, together with the high voltage capability. Capacitors are normally made from stacked metallized sheets for low-current applications, and with foil electrodes for high-current types.

Glass capacitors have similar characteristics to mica capacitors, and may be hermetically sealed. They appear to have decreased in volume of use, for cost reasons. A very common glass capacitor is formed by the internal and external coatings of cathode ray display tubes, together with the glass of the envelope. This capacitor often forms the reservoir capacitor for the final anode voltage supply.

Vacuum capacitors are used in high-voltage technology, where the penalty of large size, and possible mechanical fragility, have to be accepted in order to achieve low losses and very high breakdown voltage.

Air dielectric fixed capacitors are now used mostly in transmitters. Losses can be low, although only quite low capacitance values are achievable in reasonable volumes.

5.2.3.5 Variable capacitors

Variable capacitors may be divided into operational control types and preset or trimmer types, and sub-classified by dielectric: air, plastic or ceramic.

Air dielectric variable capacitors, as operational controls, are used mainly for tuning radio receivers: in other former applications, electronic tuning has all but eliminated the variable capacitor. Air-spaced low-value preset capacitors, however, are still in wide use.

Plastic film dielectric capacitors are available with vane-type electrodes and with foil electrodes. Practically all the comments on air dielectric types apply also to plastic film types, but larger capacitance can naturally be achieved in a given volume. Dielectric materials include polypropylene, polycarbonate, polytetrafluoroethylene (ptfe) and polyimide, the last two being especially suitable for higher temperatures. A further advantage of plastic film variable capacitors over air dielectric types is that they do not suffer from microphony (modulation of the capacitance by mechanical vibration).

Vane-type variable capacitors are bulky if capacitance values exceeding a few hundred picofards are required. For the relatively few applications (such as some types of filter) where larger value preset variable capacitors are essential, foil and film variables are available, in which a rolled cylinder of foils and dielectric sheets is mechanically deformed to produce a fairly small percentage change in capacitance value.

Ceramic variable capacitors are extensively used as preset components in all types of circuit. They feature small size and low losses.

5.2.3.6 Selecting and specifying capacitors

Practical capacitor values range from 1 pF to several farads, and working voltages from a few volts to tens of kilovolts. Operating frequency is a very important influence on the choice of capacitor type. It is important to use a logical and exhaustive selection procedure in order to arrive at an appropriate choice from the many types available. Six selection criteria are detailed below.

Type	*Normal range*	*Extended range*
Ceramic	1pF — 100nF	<1pF — 10μF+
Plastic films:		
Polystyrene	10pF — 10nF	<p10F — 10μF
Other	1nF — 10μF	to order
Electrolytics:		
Aluminium foil	1μF — 100mF	100nF — >1F
Solid tantalum	100nF — 100μF	

Table 5.2 Value ranges of fixed capacitors

5.2.3.6.1 Value

Capacitors of various construction are available in wide ranges of values, with a great deal of overlap, so consideration of value will usually lead to more than one possible solution. *Table 5.2* may be used as a guide, but some of the limit values given refer to rather expensive products and, conversely, special 'out of range' products are made.

5.2.3.6.2 Tolerance

It is important to consider tolerance because for some types the selection tolerance is very wide, such as +80 per cent to -20 per cent for high value ceramics, and even larger for some electrolytics. It should be noted that the relative cost of close tolerance types varies with the dielectric; polystyrene is the first type to consider for close tolerance applications.

5.2.3.6.3 Applied voltages

Where the applied voltage is almost entirely dc, it is usually only necessary to consider the rated voltage (specified by the manufacturer), taking into account any derating for temperature or other reasons. However, tantalum capacitors (see section *5.2.3.3*) require surge-current limitation as well.

Where the applied alternating voltage and frequency or frequencies are such that appreciable current flows through the capacitor, it is also necessary to ensure that this can occur without damage or deterioration. The current, in conjunction with resistive losses which may be represented as a resistor in series with the capacitor, will generate heat inside the body of the component, and may cause chemical changes as well. This effect is particularly important in power supply reservoir capacitors, and extensive *ripple current ratings* are given for aluminium electrolytic capacitors intended for this service. Similar considerations apply when plastic film types are required to pass significant current, but manufacturer's data may be sparse. Sometimes alternating voltage ratings are given, particularly for mixed-dielectric types (see section *5.2.3.2*), and these are often frequency-dependent. If in doubt, ask.

5.2.3.6.4 Operating frequency

Two factors have to be taken into account when considering selection on this basis; these are losses, expressed as series or parallel loss resistance, and series inductance. For low frequencies, and in non-resonant circuits, the *equivalent series resistance* (esr) sets a lower limit to the impedance of the component, and often controls the maximum current limit, dictated by internal heating. For higher frequencies, the *parallel loss resistance*, or the Q of the capacitor, may be more convenient for calculation and measurement.

The *series inductance* of a capacitor sets an upper limit to the useful operating frequency range, because it forms a series resonant circuit with the capacitance. Above the resonant frequency the impedance of the capacitor rises, i.e. it behaves as a low-Q inductor.

The intrinsic inductance of a capacitor depends on the construction, and is lowest for disc and plate ceramic types; it is approximately proportional to the distance between the points of attachment of the terminations. Extrinsic inductance is due to the lead wires (if any), and is lowest for chip and lead-through ceramic types. Stacked-foil film capacitors also have low inductance.

For very high value capacitors, resonance with lead inductance may occur at low frequencies; for example, a 1 F capacitor resonates with 1 cm leads at about 10 kHz.

Special low-inductance types of electrolytic capacitor are necessary for reservoir and smoothing applications in switched-mode power supplies operating at low radio frequencies. Where the applied voltage has a pulse waveform it may be necessary to consider the minimum permitted rise-time of the applied voltage, both in terms of the consequent surge current and also the mechanical stresses that may occur in the body of the component due to electrostatic forces and piezoelectric effects. Manufacturers usually supply data on such applications, but conditions are so varied that consultation is often necessary. If there is no mention of pulse operation in the data sheet, this may mean either that there are no special limitations or that *the component is not suitable for such applications*.

5.2.3.6.5 Operating temperature

Maximum temperature ratings are always given by manufacturers. It is necessary to take into account both external and internal heating. All time-dependent deteriorations accelerate exponentially as the temperature increases, usually doubling in rate for each 10 K rise. Thus for long term reliability, it is essential to keep components as cool as possible.

5.2.3.6.6 Temperature coefficient of capacitance

In timing and oscillator circuits this is of major importance. Values of temperature coefficient are given in manufacturers' data, and it should be noted that it is possible to balance the positive temperature coefficient of resistors in RC timing circuits by choosing film (not polyester, which has a variable temperature characteristic) or ceramic capacitors with appropriate negative coefficients. Ceramic capacitors are available in values up to about 1 nF with a range of specified temperature coefficients. Larger values have non-linear temperature characteristics, and it should be particularly noted that the common Y5V and Z5U dielectrics suffer a steep loss of permittivity above the relatively low temperature of 10°C.

5.2.3.6.7 Other considerations

Because of the wide range of types of capacitor, and of their applications, a considerable number of other factors may have to be taken into account in special circumstances. Examples are the tendency for high value ceramic capacitors to be microphonic (because the dielectric is piezoelectric), the possibility of film capacitors burning or spreading fire under fault conditions, and the possibility of a shock hazard to service personnel due to charge recovery in high voltage capacitors through dielectric hysteresis.

5.2.4 Inductors and transformers

Inductors are now available, at least in values up to a few henrys, as stock items like resistors and capacitors. In the past this was, except in a few cases, not so, and inductors were always designed and manufactured to order.

Inductors are classified by core material and the physical form.

5.2.4.1 Air-cored inductors

As stock items, air-cored inductors are available in low values, as axial-lead components and as 'can types' for board mounting with a screening can. Can types may have provision for adjustment of the inductance by means of a brass or aluminium core. Also available are rather larger values, wound in sections to reduce self-capacitance, which are intended for use as 'rf chokes'. Much larger values may be specially designed and manufactured.

5.2.4.2 Iron dust cores (rod and pot cores)

These are now used only where suitable ferrite cores (see section *5.2.4.3*) are not available. Rod types are available as axial-lead components, but most inductors of this kind are can types, which are usually adjustable.

5.2.4.3 Ferrite cores (rod and pot cores)

These are available as stock axial-lead components in a range of values up to about 5 mH, and in miniature can types up to about 2 H. Adjustable types are available in the lower inductance values. Components for specialized applications, such as transformers for if amplifiers in receivers, fm discriminators, narrow-band filters and balanced modulators are also available. Larger values, and larger cores for higher current applications, are normally wound to order. Some types are adjustable.

5.2.4.4 Ferrite cores (EI, F and U cores)

These cores are used for high voltage and/or high current transformers in switching power supplies and cathode ray tube deflection circuits. Inductors are designed and made to order.

5.2.4.5 Toroidal ferrite and iron dust cores

These are used for applications where the low external magnetic field and/or freedom from the need to screen against external fields or the influence of adjacent metal parts is advantageous. Very small toroidal inductors are available for surface-mounting applications.

Special lossy grades of iron dust toroid are available for interference filters. These filters are resonant circuits and should have controlled Q, otherwise they may generate interference voltages far higher than that of the source, and consequently increase, rather than reduce, interference radiation.

5.2.4.6 Laminated iron and nickel-iron cores (EI, TU and F cores)

These are used for inductors and transformers for frequencies up to low radio frequencies. Inductors and transformers are available as stock items, but many are made to order.

5.2.4.7 Toroidal iron cores

Toroidal iron cores are used for power transformers and inductors, which are available in stock ranges, but many types are made to order.

5.2.4.8 Selecting and specifying inductors

The selection and/or specification of an inductor or transformer is likely to be a much more complex matter than for other types of passive component. The criteria listed below are those which are relevant in most cases, but there may be other essential requirements in particular cases. Component prices, both for stock items and for 'specials', vary widely, so it is essential to specify closely and to obtain a number of quotations.

5.2.4.8.1 Value

Even this characteristic is not as simple as for other components, as non-linearity is significant in most types of inductor under some conditions. Except for air-cored inductors it is necessary to specify the value at a stated applied voltage and frequency. It should also be made clear whether the stated value is the *series* inductance (where the simplified equivalent circuit of the component is considered as a pure inductance in series with a pure resistance), or the *parallel* inductance (where the equivalent circuit is an inductance and resistance in parallel). It may even be necessary to specify values under different operating conditions, e.g. with and without a specified direct current flowing.

5.2.4.8.2 Tolerance

This aspect is fairly straightforward. However, care is necessary not to impose impracticable tolerances, especially on iron-cored inductors. Often, only a minimum inductance need be specified, leaving an unspecified positive tolerance.

5.2.4.8.3 Applied voltages

Direct voltage will clearly result in direct current flowing through the winding. This will cause heating due to resistance losses, and will affect (generally reduce) the permeability of any solid core material. It is usual to specify the dc conditions in terms of the current rather than the applied voltage.

The applied peak alternating voltage controls the peak flux density in the core, as well as the current in the winding. Care is necessary to avoid saturation of the core, possibly under permitted over-voltage conditions, as this will greatly increase losses and give rise to waveform distortion and possibly mechanical noise. In resonant circuits, particularly in the series configuration, the voltage across an inductor may greatly exceed the supply rail voltages in the equipment, and large currents may flow in quite small inductors.

5.2.4.8.4 Resistive losses

For high frequency operation, resistive losses are usually expressed as the Q of the inductor. It is essential to indicate the frequency, or the tuning capacitance value, at which the Q is specified. For lower frequencies, where copper loss usually predominates, losses can be conveniently expressed as the dc resistance of the winding(s), preferably as a target value with tolerances. This will automatically also control the *regulation* of a transformer (the difference between no-load and full-load voltage, expressed as a percentage of no-load voltage).

5.2.4.8.5 Leakage inductance

This is a measure of the amount of primary flux in a transformer which does not link with the secondary windings(s). It is important that it should be minimized in power and audio transformers, and this can be achieved by interleaving primary and secondary windings. In some other applications it may be required to have a definite value, with tolerances.

5.2.4.8.6 Self-capacitance

This affects the impedance/frequency characteristic of the inductor, and inevitably results in parallel resonance at some frequency, above which the impedance is capacitive, not inductive. It can be reduced in air and ferrite rod cored inductors by splitting the winding into sections. In iron cored transformers, interwinding capacitance may be controlled by choosing insulation materials and thicknesses, and also the disposition of the windings.

5.2.4.8.7 Magnetizing current

This is a very useful quality-monitoring characteristic of iron cored components, as it is very dependent on peak flux and on the quality and condition of the core material. It is thus subject to rather large variations between samples, but a maximum permitted value can be deduced from measurements on a representative batch. The applied voltage and frequency must be specified.

5.2.4.8.8 Turns ratio

This applies to power and audio transformers and to ferrite cored components. It may conveniently be specified and measured as no-load voltage ratio.

5.2.4.8.9 External magnetic field

This is sometimes a problem with power transformers, but it should also be noted in rod cored inductors, which make good magnetic transmitting antennas. For power transformers, the external field is usually most economically reduced by means of a thick copper or aluminium screen, acting as a short-circuited turn linked with the leakage flux, and thus generating a nearly equal and opposite flux. This is often more effective, and much cheaper, than nickel-iron magnetic screening.

The converse problem, interference from an external field, can require the use of nickel-iron screening. An alternative technique is the use of *astatic* winding, where windings of opposite sense are arranged to link equally with the interfering flux and are connected in series.

5.2.4.9 Designing inductors and transformers

The subject of inductor design is very large, and only an outline can be given. The best way to design one is to have designed many others previously; in other words, experience is very valuable. It is well worth taking a systematic approach from the beginning, by keeping records of all design procedures and data in one notebook, which will become a valuable reference in future.

For high-frequency inductors, the first step is usually to decide the optimum core type, which can be deduced from the catalogues of stock items. It is then necessary to choose the number of turns and wire size to obtain the specified inductance and Q. Manufacturers usually give appropriate design data. In their absence a trial winding on the chosen core must be measured and the results used to deduce a closer solution if necessary.

For low frequency and power transformers, the first step is the choice of a suitable core size. For power transformers, manufacturers now indicate the maximum VA ratings available from each core size, which is a valuable guide. It is then necessary to design the primary winding, taking into account the maximum permissible applied voltage and the peak flux density allowed in the core. The primary winding normally occupies half the winding space, and each secondary winding takes space proportional to its VA rating. Care is necessary to ensure compliance with relevant safety requirements. Once designed, the primary winding will be suitable for most transformers using that core size, so, provided that its details are properly recorded, the design need only be done once.

5.3 Printed circuits

Early attempts to find alternatives to wired circuits include a household radio receiver made by GEC in the 1930s, in which copper strips were riveted to an srbp (synthetic resin bonded paper) board. The strips were large enough to form the contacts for the valve (tube) pins directly. Further attempts included the work of Sargrove, far in advance of its time, on the automatic production of radio receivers, using printed conductors and resistors. Some military devices from World War II used printed conductors made of silver-loaded ink. Practical printed circuits, using etched copper-clad sheets of srbp, were developed in the 1950s. Development would have been less troublesome if the transistor had come into general use a little earlier, because heat dispersal from valve (tube) circuits presented a significant problem when there was no metal chassis to conduct and radiate it away.

5.3.1 Substrate materials

There are only two choices of rigid substrate material for general application. These are srbp which is used almost exclusively in household products, and glass-fibre reinforced epoxy, which is used for all other general applications. Special substrates are available for such devices as printed switches and commutators, and for shf circuits.

Flexible substrates include polyimide, polysulphone and polyphenylene oxide. Substrates which are less resistant to high temperatures are used for printed flexible connectors which do not require to be soldered.

5.3.2 Conductor materials

Copper is the normal conductor material, and is available in a standard thickness of about 35 μm, and a double thickness for vhf/uhf and high-current applications. Resistive materials are also available for special purposes. The substrate may be clad on one or both sides.

5.3.3 Dielectric and thermal properties

The application of srbp is limited. It has a lower permittivity and a lower maximum operating temperature (85°C under normal conditions and 110°C under fault conditions) than glass-epoxy (120°C under normal conditions and 150°C under fault conditions). The relative permittivity of glass-epoxy varies with grade and thickness, but is usually about 3.3. Flame-retardant grades are available, and are demanded by some safety regulations.

5.3.4 Through-hole and surface-mounting techniques

When printed circuits were first introduced, all components were designed for conventional wiring techniques, and the components industry was slow to change designs, partly because the existing products were not actually unusable. These components require at least two holes to be punched or drilled in the substrate, so that the lead wires can pass through to the conductor pattern. The holes must not be very much larger than the wires, or soldering will be unreliable, and they must be accurately positioned with respect to the conductor pattern. There may be 5000 holes or more on a single board. Thus a complicated and fragile press tool is necessary, or a time-consuming drilling procedure, requiring a skilled operator to do rather unpleasant work. Nevertheless, many millions of boards have been made by these techniques.

Surface-mounting techniques eliminate the costly and inconvenient holes, but require components that will withstand the higher temperatures involved in the necessary soldering processes. An additional operation, to apply solder paste and possibly adhesive, is also necessary. The component packing density can be increased, thus saving board area. Stray capacitances and inductances are reduced.

5.3.5 Layout design

It is quite surprising that printed circuit layouts can be, and often are, prepared by draughtsmen or others without formal training in electronic technology. However, these layouts are often less than optimal. The introduction of computer aided techniques may allow the circuit designer to become more involved in the layout at the design stage, rather than in overcoming problems at a later stage.

The layout should be based on the flow of signals through the circuit, with the signal paths being made as short as possible. Particular attention should be paid to sensitive input circuits of linear amplifiers, where interference may be introduced or stability compromised. The conductors directly connected to the inverting input of such an amplifier must be made very short indeed, and the area of the input circuit minimized. Attention is also required in order to avoid earth loops, particularly when double-clad board is used. These are not only a problem in audio circuits, but may affect other types as well. In particular, resonant loops may occur, which greatly increase the sensitivity of the circuit to rf interference. Power supply leads can conveniently be formed on the ground-plane side of the board, with decoupling components mounted close to the active devices.

A systematic approach to *earthing* (common rail routeing) is essential for reliable design, but is very often lacking. The best way of avoiding unwanted interactions via the common rail is to consider each active device and its power supply as forming a *current loop*, and to arrange the connections to the common rail (and the supply rails as well, although this seems to be done instinctively in most cases) so as to avoid the intersection, i.e. sharing of lengths of conductor, of loops carrying large currents with loops associated with sensitive circuits. It is also necessary to ensure that multiple current paths, i.e. 'earth loops', are avoided so that current paths are defined and do not, perhaps, vary between samples of pcb due to variations of track width or resistivity.

5.3.6 Heat dissipation

Individual hot components, such as power resistors, should be supported on pillars or otherwise reliably spaced away from the board. Where the board is horizontal, the provision of ventilation holes in the board below the hot component can be very helpful in reducing its surface temperature. The use of heat-sink structures for power semiconductors is well established. Care is necessary to ensure that the heat-sink(s) do not impress unacceptable mechanical stresses onto the board, especially if the equipment could be dropped, either in use or in transit. Failure to do this may well result in unacceptable repair costs. Some integrated circuits are designed to use areas of copper on the pcb to conduct and radiate heat away. It is essential to check the performance of such arrangements by temperature measurements under worst case conditions, because the cooling effect is very difficult to predict.

The orientation of the board (horizontal or vertical) profoundly affects the temperature distribution; this is likely to be more uneven when the board is vertical. For equipment which may be used in different positions, the design must be proved by temperature measurements with the board(s) in all operating positions.

5.3.7 Production methods

The majority of pcbs are produced by a photomechanical process. The cleaned copper surface of the board is coated with photoresist material, which is then exposed to ultraviolet light through a positive film of the conductor pattern. After exposure, the photoresist is chemically developed, and the unwanted copper is etched away in an acid solution of ferric chloride (or ammonium persulphate in some cases). The board is then washed, and may be coated with solder resist by a similar process. Finally, component designations (circuit references) may be printed on the board by a silk-screening process. Some pcbs are produced throughout by silk-screening, but it is difficult to cope with the small dimensions of modern components with this technique.

5.3.8 Designing to ease production and servicing

Production and servicing can be considerably eased by following these guidelines:

- Print component designations (circuit references) legibly on the board.
- Ensure as far as possible that all component type numbers, values and tolerance codes are visible when components have been mounted on the board.
- Orient all polarized, but physically reversible, components in the same direction.
- Ensure that functions are split between boards in such a way that each board can be adequately evaluated by testing in isolation, i.e. without having to be connected to other boards in the equipment.
- Ensure that each board can be assessed for servicing without presenting the service engineer with a difficult mechanical problem (e.g. complex disassembly or having to balance an assembly in an unstable position).

J M Woodgate B Sc(Eng), C Eng, MIEE, MAES, M Inst SCE
Electronics design consultant

6 Semiconductors and Microelectronics

6.1 Semiconductors

6.1.1 Junctions

A very brief introduction to semiconductor materials is given in section *5.1.4*. In order to discuss the operation, selection and specification of devices, we have to consider first the effects which occur at the junction between two different materials, conductors or semiconductors. To do this, we need two terms connected with electron (or hole) energy levels (see section *5.1.2*).

The *Fermi level* is the average energy level of electrons inside the material. There may be no electrons actually having this energy because it is forbidden, but it is equally probable that an electron will have a given amount more energy than this level as that it will have the same amount less. The *work function* is the difference in energy between an electron at the Fermi level and one at the surface of the material. The energy level at the surface is usually normalized to zero. Fermi levels also apply to holes; these levels are not necessarily the same as those of electrons.

6.1.2 Metallic junctions

If two metals with different values of work function are placed in contact, electrons will diffuse from the one having the lower work function into the other, until the space charge set up by the ionized atoms fixed in the lattice near the junction prevents any further net flow. This produces a potential difference between the surfaces of the two metals, equal to the difference between the work functions and known as the *contact potential*. The potential is independent of any externally imposed current flowing through the junction. Inside the materials, the Fermi level becomes the same on each side of the junction.

6.1.3 Metal/semiconductor junctions

In order to connect semiconductor devices to other components, junctions of this type are essential, and they also occur within some types of device. If a metal contacts an n-type semiconductor with a higher work function, or a p-type semiconductor with a lower work function, the situation is similar to that at the junction of two metals, and an *ohmic contact* is established. This is just what is required for attaching lead out wires to a discrete device or conductors between devices in an integrated circuit.

However, if the work function of the metal is greater than that of an n-type semiconductor (or less than that of a p-type semiconductor) with which it is in contact, carriers (electrons or holes) diffuse from the semiconductor to the metal, leaving a very thin layer of semiconductor with no free carriers in it; this is known as a *depletion layer*. Carriers with sufficient thermal energy can diffuse across this layer, but at room temperature they pass through it very quickly.

The potential difference between the material surfaces is equal to the difference in work functions, and is called the *diffusion potential*. This represents an energy barrier which carriers have to overcome if current is to flow through the junction.

If an external voltage is applied, almost all of it appears across the depletion layer, and if it is opposite in sign to the diffusion potential, the barrier is lowered and the junction conducts freely. However, if the applied voltage is in the same sense as the diffusion potential, the barrier height is increased and only a small thermal leakage current can flow. Thus a *rectifying contact* has been established. The reverse biased depletion layer forms a capacitor, and if the layer is very narrow, the capacitance may be hundreds of picofarads per square millimetre.

In the special case where an n-type semiconductor is so heavily doped that its band-gap (see section *5.1.2*) disappears (so-called *degenerate n^+ type*), the situation is exactly as for the contact of two metals, i.e. ohmic contact is made. This is of great practical importance in the formation of internal connections in integrated circuits.

6.1.4 Semiconductor junctions

By far the most important type of semiconductor junction is the p-n junction, which is usually formed within a single crystal of the substrate material, by alloying or diffusing donor and acceptor impurities towards each other through the crystal. Most junctions are now formed by diffusion.

Across such a junction, electrons and holes diffuse in opposite directions until a depletion layer is formed which encloses the actual junction region. The corresponding diffusion potential depends on the material, being in the order of 0.7 V for silicon, 0.4 V for germanium, and 1.2 V for gallium arsenide.

If the p region is made positive with respect to the n region by an external voltage, the potential barrier is lowered and current will flow. Above the diffusion potential the current rises exponentially with the applied voltage, being limited by the resistance of the external circuit and that of the material of the device between the external connections and the junction, the *extrinsic* resistance. If the external voltage is reversed, however, the barrier will be increased in height, and only a thermal leakage current can result. This current can be small at room temperature, but rises exponentially as the temperature increases, approximately doubling for every 10 K increase.

Increasing the external voltage will eventually result in a carrier velocity sufficient to ionize substrate atoms by collision. This is a cumulative effect, known as *avalanche breakdown*, and results in an increase in current limited only by the extrinsic and external circuit resistance. Other types of breakdown are discussed below in connection with Zener and tunnel diodes.

6.2 Diodes

6.2.1 Principles

A diode consists of a single rectifying junction, which may be a p-n junction or a metal/semiconductor junction, enclosed in some form of substantially gas-tight enclosure or *encapsulation*, to prevent surface contamination or corrosion. For external connections, one or two ohmic contacts must be made to the semiconductor element. Some types of high voltage 'diode stack' consist of a series chain of junctions in a single encapsulation. Power rectifier diodes are available in many different types of multiple-junction assemblies, containing from two to many tens of interconnected junctions.

6.2.2 Types of diode

Diodes may be broadly classified into small signal, power, regulator, reference, variable capacitance, photosensitive, photoemissive and special purpose types; further sub-classification may depend on construction or intended application.

6.2.2.1 Small signal diodes

These are used for rectification (including detection), signal routeing, pulse shaping, modulation and very many other applications. In this context, 'small signal' may include peak inverse voltages up to 100 V and/or forward currents of 100 mA. Germanium point-contact diodes are used where their low diffusion potential and/or low shunt capacitance is vital (such as in low level, high frequency detectors), although they are being replaced by silicon Schottky (metal/semiconductor junction) types (see section *6.2.2.7*). However, gold-bonded germanium diodes still offer the lowest forward voltage at low currents (about 300 μA).

Silicon small signal diodes are available as point-contact types, but are mostly of 'whiskerless' construction, where contact with the device crystal or *die* is made by axial mechanical compression.

'Implosion diodes' are made by a process wherein the compression is achieved by imploding the (softened) glass encapsulation. Special types are made with very low forward resistance and parallel capacitance for electronic bandswitching in LC tuned receivers.

6.2.2.2 Power diodes

Power diodes are almost exclusively made of silicon. They may have peak inverse voltage ratings ranging from 50 V to well over 1000 V, and forward current ratings from 100 mA to hundreds of amps. For the highest currents, for low voltage supplies and for those derived from high frequency ac, Schottky diodes offer lower forward voltage, lower resistance and lower shunt capacitance, all but the latter attribute formerly being available only in germanium junction types, which have a much lower maximum permitted junction temperature.

Low power types are normally air cooled, with some recourse to heat dissipation via connecting leads to the printed circuit board, but higher power types are designed for mounting on a heat-sink, or may be supplied in a heat-dissipating assembly.

6.2.2.3 Regulator and reference diodes

If a diode is operated under avalanche breakdown conditions, with a suitable current-limiting impedance in series, any rise in applied voltage will result in a sharp rise in avalanche current, and the voltage across the diode will remain practically constant. This indicates that the device can be used as a voltage regulator.

If the doping level (see section *5.1.4*) on both sides of the junction is made large, the depletion layer is very narrow and the electric field across it may be hundreds of millions of volts per metre. Under these conditions the quantum mechanical nature of the carriers (see section *5.1.1*) becomes significant and there is a finite probability of carriers, with insufficient thermal energy to cross the layer, nevertheless appearing on the other side of it. This process is known as *quantum mechanical tunnelling*, and the resulting current in this type of device is known as *Zener current*. In common usage, both avalanche and Zener diodes are called Zener diodes.

The breakdown mechanisms occur together in silicon diodes with breakdown voltages of about 4–7 V: below 4 V Zener breakdown predominates, and above 7 V avalanche current is much greater. Since Zener breakdown voltage falls with temperature, while avalanche voltage increases, diodes with a breakdown voltage of about 5.6 V have nearly zero temperature coefficient and are therefore particularly useful for establishing stable reference voltages.

Avalanche and Zener currents have considerable wide-band noise components, which vary rapidly and in a complex manner with the average current. Parallel capacitance and/or series inductance, together with physical spacing and even screening, may be necessary to keep this noise out of nearby circuits.

It is also possible to use the forward voltage (diffusion potential) of a diode as a reference voltage, but it is necessary to use diodes specially designed for this application, as the spread of forward voltage between samples of standard types is rather large. Even the special types have much larger variations between samples than do the reverse current types. Some types of forward reference device have two or three junctions in series. The temperature coefficient of reference voltage is quite large at about 2.2 mV/K per junction, but the noise current is very much lower than for reverse-breakdown diodes.

6.2.2.4 Variable capacitance diodes

The capacitance of a reverse-biased depletion layer depends inversely on the applied voltage, because this affects the width of the layer. Unfortunately the dependence usually follows an inverse fractional power law. It is, however, possible to use these devices for voltage controlled tuning in receivers and filters, for frequency and phase modulation and for automatic frequency control. At high radio frequencies, varactor diodes can also be used in 'parametric amplifiers' and other related applications.

For tuning purposes, varactor diodes are often supplied in matched sets, because the capacitance and the capacitance/voltage law vary between samples.

6.2.2.5 *Photosensitive diodes*

If electromagnetic radiation falls on a semiconductor junction of a diode, there is no effect (except heating) unless the quantum energy of the radiation exceeds the band-gap energy of the semiconductor. If the wavelength of the radiation is short enough (i.e. the frequency is high enough), the energy of each quantum is sufficient to create an electron/hole pair, and the charged particles (if they do not recombine shortly after creation) diffuse away from the depletion layer. If the diode is in a circuit, this diffusion forms a component of reverse current additional to the thermal leakage current. For both germanium and silicon the quantum energy equal to the band-gap energy corresponds to radiation in the near infrared part of the spectrum.

For radiation at wavelengths near the minimum energy, the efficiency of conversion of radiant to electrical energy is low, because there is only a limited number of permitted states with the necessary energy difference.

For radiation of shorter wavelengths, the efficiency increases to a maximum, and then falls as the quantum energy becomes sufficient, first to excite electrons into the surface layers of the material, where there are imperfections in the structure allowing many opportunities for recombination, and then to excite the electrons so much that they are actually emitted from the material.

In such a *p-n photodiode*, the reverse current in excess of the leakage current is proportional to the incident radiation power. If the reverse bias voltage is high enough to bring the diode into the reverse breakdown region, both the thermal leakage current and the photo current are increased considerably by avalanche multiplication.

Although the avalanche process is itself noisy, a net gain in signal/noise ratio can be obtained compared with operation outside the avalanche region.

In the above modes of operation, the diode is reverse biased. In addition, the diode generates a useful emf (up to about 0.5 V) if operated in the forward conducting mode. A diode specifically intended for this application (in which the thermal leakage current may not be specified or controlled) is called a *solar cell*. The cell does not, however, behave as a voltage generator like a primary cell, but is better regarded as equivalent to a current generator in parallel with a diode.

For high speed operation, the capacitance of the reverse-biased p-n junction is a disadvantage, and a useful improvement can be obtained by using a p-i-n diode, in which a thin layer of intrinsic (or, in practice, very lightly doped) material is included. The capacitance change with change of bias voltage of such diodes is low, but the useful modulation bandwidth is large.

Photosensitivity is an inherent property of semiconductor diodes, so that means have to be provided to prevent light from affecting diodes which are not intended to respond to it.

6.2.2.6 *Photoemissive diodes*

Photoemissive diodes are otherwise knows as *leds* (light emitting diodes).

When an electron and a hole recombine in a semiconductor material, they may give up their excess energy as visible light. In germanium and silicon this does not occur to a significant extent, because the most probable kinetic energy of a conduction band electron is too high to allow a direct transition to the valence band (which is what happens in recombination). In order to recombine directly, the electron has to transfer a critical amount of vibrational energy to the crystal lattice, and simultaneously emit a photon, which is a very improbable combination of events. In these semiconductors, recombination normally occurs through 'traps', which are intermediate energy levels in the forbidden band, made accessible by defects in the crystal.

In compound semiconductors such as gallium arsenide, however, it is possible, indeed very likely, for an electron to fall directly from the conduction band to the valence band, while emitting a photon whose quantum energy is (approximately) equal to the band-gap energy. For pure gallium arsenide this energy corresponds to a wavelength of 0.87 μm, in the near infrared. Doping with phosphorus, zinc oxide or nitrogen increases the band-gap, and thus shortens the wavelength, producing green and yellow light emitters. The production of blue light requires the use of more exotic semiconductors, such as silicon carbide, and the relative cost of such devices is high. (However, filament lamp simulations of blue leds are available.)

The light is produced at the junction, and in order to minimize absorption the anode layer is made very thin. Because of the high refractive index of the semiconductor material (of the order of 3.5), total internal reflection occurs at the top surface for rays at an angle exceeding about 15° to the normal, so that the emission is confined to a narrow cone. The viewing angle may be increased by coating the surface with a material of intermediate refractive index (ideally 1.9), and by the use of lenses.

The high band-gap energy of leds is reflected in their forward voltages, which are in the order of 2 V, being greater for yellow emitters. Blue emitters have forward voltages in the order of 5 V. LEDs are not usually characterized for reverse breakdown voltage or current, and reverse voltage must be limited to less than 1 V by a parallel, reversed polarity silicon diode.

6.2.2.7 *Special types*

Diodes for more specialized applications include Gunn diodes, IMPATT and TRAPATT diodes, laser leds, p-i-n diodes, Schottky diodes and tunnel diodes.

6.2.2.7.1 *Gunn diodes*

These devices are not really diodes at all, since the action takes place in the bulk of a semiconductor material and does not depend on the presence of a junction. However, they are usually characterized only for a preferred direction of current flow.

Gunn devices have the valuable property of converting direct current to current at microwave frequencies. In use, an electric field exceeding about 300 kV/m is established across a piece of gallium arsenide, which may be up to 3 mm long. It is found that, superimposed on the resulting direct current, there are pulses of current at a frequency proportional to the length of the specimen and corresponding to a transit speed in the region of 100 km/s.

The Gunn effect occurs only in semiconductors which have two conduction bands, separated by a gap, in which the mobility of electrons in the higher energy band is less than in the lower. In such a case, if a group of a few electrons enter the higher conduction band, due for example to the action of a crystal defect, they will drift to the anode more slowly than the electrons in the lower conduction band. This causes the electric field associated with the group of electrons to increase, thus allowing more electrons to enter the group in the higher conduction band. Thus the group becomes a *domain* of increased electric field, and results in a pulse of current in the external circuit. Once the domain has left the semiconductor, another can form near the cathode, and the process repeats continuously.

Gunn devices can be made to operate in the *transit-time* mode, where the frequency of oscillation is controlled by the device dimensions, or in the *limited space-charge accumulation* (lsa) mode, where the frequency is determined by external components.

6.2.2.7.2 IMPATT and TRAPATT diodes
These, also, are not really diodes, being three-layer (p^+-n-n^+) or four-layer (p^+-p-n-n^+) two-terminal devices for direct conversion of dc power to microwave frequencies. Unlike Gunn diodes, however, they operate in a reverse-current mode. Three-layer IMPATT (IMPact Avalanche Transit Time) diodes are biased so that the p^+-n junction operates in the avalanche mode. The avalanche ionization process is relatively slow; the avalanche current therefore lags the field potential and the avalanche region behaves as an inductor. This, with the shunt capacitance of the junction forms a tuned circuit, which can be tuned by varying the applied bias voltage.

For frequencies above the natural resonant frequency of the avalanche region, it can be shown that the n-type drift region behaves as a negative resistance. The device will therefore sustain oscillations at some frequency where the net circuit resistance is zero. Four-layer devices have two drift regions, and can therefore operate more efficiently.

If the reverse bias of a suitable device is increased until the avalanche region fills the former drift region, operation in the TRAPATT (TRApped Plasma Avalanche Transit Time) mode occurs. This operates somewhat as a relaxation oscillator; the terminal voltage rises as the plasma extends and the shunt capacitance charges. When the plasma reaches the end of the drift region, a large current pulse occurs, discharging the capacitance and destroying the plasma. The terminal voltage falls suddenly, and then begins to build up again.

6.2.2.7.3 Laser leds
If photons with sufficient energy pass through a material, some will be absorbed by atoms, which are thus raised to a higher energy level. Normally, these excited atoms quickly return to the ground state by re-emitting a photon at random. In some materials, however, such a higher energy level may be metastable, so that it is possible to collect large numbers of excited atoms in it. The incidence of a few photons of the right energy can then trigger a chain reaction, whereby the photon released from one atom causes another to relax, and so on. If there are more atoms at the higher energy than in the ground state (i.e. a population inversion), the number of photons produced exceeds the number incident, and laser (light amplification by stimulated emission of radiation) action is produced.

A laser led, or injection laser, consists of a p-n junction in a crystal (usually of gallium arsenide), whose faces perpendicular to the junction plane are polished. These act as semi-transparent mirrors, feeding back into the junction region some of the photons emitted by normal led action. Above a critical bias current, a population inversion is established, and coherent light, or infrared radiation, is emitted along the junction plane.

Lower critical currents, and therefore fewer problems of heat dissipation, can be obtained by reducing the loss of photons in directions approaching the normal to the junction plane. This may be done by the introduction of appropriately doped layers of gallium aluminium arsenide, which has a lower refractive index than gallium arsenide and therefore produces internal reflection at the interface. Such a device is called a *heterostructure* or *heterojunction laser*.

Still further improvement can be obtained by confining the bias current to a narrow strip of the crystal, producing a stripe laser. Such devices are capable of continuous-wave operation.

6.2.2.7.4 p-i-n diodes
As the name implies, these are three-layer devices with a central instrinsic (or, in practice, very lightly doped) layer. Such a device has a low reverse-bias capacitance which is nearly independent of bias voltage and a forward resistance inversely proportional to current. It can therefore be used as a switch or a component of a voltage controlled attenuator.

6.2.2.7.5 Schottky diodes
These devices have a metal/semiconductor junction, the metal having a higher work function than the semiconductor. They are increasingly widely used, as discrete diodes, even for general purposes, in integrated circuits and as microwave devices. They have lower forward voltage than p-n diodes of similar rating, and no carrier-storage effects because they are majority-carrier devices.

6.2.2.7.6 Tunnel diodes
If a p-n diode is constructed with very heavy doping, the junction may be in the breakdown region at room temperature even with a small forward bias. As the bias is increased, the forward current falls to normal values. This produces a negative resistance characteristic.

As in the Zener diode, the breakdown current is due to quantum-mechanical tunnelling of electrons through the very narrow potential barrier. This is a very fast process, so that operation at extreme frequencies is possible, in switching, amplifier and oscillator circuits.

6.2.3 Selecting and specifying diodes

For all diodes except some special types, the basic characteristics are the forward voltage and average forward current, the reverse leakage current and the reverse breakdown voltage. These characteristics should not be forgotten, even when other characteristics are of more immediate interest (e.g. in the case of a light emitting diode). For rectifier diodes, other important characteristics include the peak forward current, the reverse recovery time and, in the case of power rectifiers, the thermal resistance. See also section *6.14*.

6.3 Bipolar junction transistors

6.3.1 Principles

The bipolar junction transistor is a three-layer (n-p-n or p-n-p) device, consisting of a narrow central region, the base, between a heavily doped emitter region and a less heavily doped collector region. In normal use, the emitter/base junction is forward biased, and the collector/base is reverse biased. The following explanation applies to n-p-n transistors; for p-n-p devices the same mode of operation applies with 'electrons' and 'holes' interchanged.

The base width is made much less than the diffusion length for electrons, so that most electrons entering from the emitter do not combine with holes in the base. In a practical transistor (*graded base* device), the doping level is not constant in the base region, becoming less towards the collector. There is thus an accelerating field for electrons across the base width. When they approach the collector, they are quickly swept across the reverse biased junction by the strong field, and form a collector current. This current can be controlled by varying the base-to-emitter voltage, which in turn controls the emitter current. The base current is composed of the reverse leakage current of the collector/base junction and the fraction of emitter current which does not reach the collector. Both of these currents are normally much smaller than the collector current. A small power in the base emitter circuit can therefore control a larger power in the collector circuit.

6.3.2 Types of bipolar junction transistor

These devices are made for a very wide range of applications, but the categories most used in short-form catalogues, device selectors and the like are as follows:

Small signal general-purpose devices can be used for most low frequency (say up to 3 MHz) applications, and most of them are equally suitable in linear and switching circuits.

Low frequency, low noise devices are designed to have low noise at low audio frequencies, when used in correctly designed circuits. It is wise to use these types in any application where low frequency noise could be a problem as there is a wide variation in the noise performance of general-purpose types.

Small signal, high frequency devices are designed for use in rf and if amplifiers at frequencies up to several hundred megahertz. These transistors are characterized and tested specifically for the application. Some types are characterized for large-signal linearity, where low intermodulation distortion is required, e.g. for antenna amplifiers.

Low power switching devices are intended for fast switching and are characterized accordingly. Generally, it is unwise to use these types for linear amplification, as linearity and noise may not be satisfactory.

High voltage transistors are further classified by circuit application, such as switching, video amplification and crt deflection.

Audio power devices are specifically designed for high power audio amplifiers, with controlled high frequency characteristics for use in circuits with large amounts of negative feedback and a large 'safe operating area' (i.e. region of collector current/collector voltage characteristics free from second breakdown effects) to cope with reactive load impedances.

R.F. power and *microwave* devices are specifically designed for the application, and characterized accordingly. They are often designed for use in a specific circuit configuration.

Darlington devices. If two transistors are connected so that the input device is an emitter follower and the output device operates in the common-emitter mode, the combination behaves more or less as a single transistor (except at high frequencies, where the behaviour is often quite different) known as a Darlington device. While the configuration is capable of use in many applications, Darlingtons are usually made for power linear and switching applications, where the high power gain is an advantage.

There are, of course, many other categories; generally the manufacturer provides circuit application data for these also.

6.3.3 Selecting and specifying bipolar junction transistors

Generally, lowest cost and greatest availability will be obtained if a general-purpose device can be used. In some cases, general-purpose devices actually offer the best characteristics available. However, care should be taken not to depend on unspecified characteristics of a device. Where unconventional use is contemplated, the advice of the supplier should be sought. See also section *6.14*.

6.4 Field effect transistors

6.4.1 Principles

Field effect transistors (fets) depend on conduction through a strip or *channel* of semiconductor provided with (usually) ohmic contacts at both ends, the *source* and *drain*. The conduction can be controlled by applying an electric field to the middle region of the bar, via a *gate* electrode. This gate may be formed by a junction on the surface of the region, or by a very thin insulating layer on the surface, covered with a conducting layer of metal or semiconductor.

Figure 6.1 Field effect transistors, ranging from small-signal junction typesto high power vmos types. Note the Japanese practice of abbreviating the type numbers: the TO-3 cased devices are 2SJ49 and 2SK134 (Maplin Electronics Ltd)

In the first case, the device is known as a *junction gate field effect transistor* (jfet), while the latter are insulated gate or metal oxide field effect transistors (*mosfets*) and silicon gate (normally the semiconductor is silicon) field effect transistors respectively.

In normal operation of a jfet, the bias voltages are arranged so that the junction is sufficiently reverse biased at the drain end for the depletion layer, which extends into the channel, to control the current flow from source to drain. Under these conditions, it is found that the drain current is largely independent of the drain-to-source voltage, and can be controlled by the gate-to-source voltage. If the gate-to-source voltage reverse biases the junction even at the source end, the gate voltage and current are very small compared to the drain-to-source voltage and drain current, so the device gives useful power gain.

In metal oxide semiconductor transistors (MOSTs) and silicon gate devices, the current through the channel is controlled by the electric field set up between the gate electrode and the channel. In *enhancement mode* devices, the source and drain regions are of opposite doping to the body of the device, and the channel is created by the field, which, when sufficiently strong, attracts enough minority carriers into the channel region to change it from n-type to p-type, or vice versa. The channel is then effectively of the same doping type as the source and drain, so that conduction occurs. In a *depletion mode* device, a conducting channel is formed by diffusion and the gate bias voltage polarity is chosen to reduce the drain current from its value at zero gate-to-source voltage. Because the gate is insulated from the rest of the device, the input impedance is extremely high, and consequently so is the power gain.

MOSTs with two gates are, naturally, called 'dual gate MOSTs', and are particularly intended for vhf linear applications. They can have low feedback capacitance (C_{rs}) and high transconductance (y_{fs}), and are therefore useful in high gain amplifiers. The transconductance can be varied by varying the bias voltage on the gate which is not used for signal input, and this allows automatic gain control to be applied simply.

A device has been developed which combines the characteristics of a MOST at the input and a bipolar transistor at the output. This is known as an insulated gate bipolar transistor (igbt), and offers much greater power gain than a bipolar transistor with higher voltage operation and lower losses than a MOST. It may be regarded as a MOST integrated with a bipolar transistor in a form of Darlington configuration (see section *6.3.2*).

The dimensions of field effect devices have to be small to give useful characteristics. In particular, the insulating layer in

MOSTs is very thin, and can easily be damaged by static electric discharge. The majority of devices include protective diodes to minimize this effect, but it is still necessary to take precautions to minimize the generation and retention of static charge when these devices are being handled.

The characteristics of field effect devices depend in a complex manner on details of the device geometry, and various special forms of construction, such as 'vmos' (vertical metal oxide silicon), have been developed in order to give greater breakdown voltage, greater power and better linearity. Care is needed to understand the specific advantages and disadvantages of each variety, as the latter are rarely mentioned by the manufacturer.

6.4.2 Selecting and specifying field effect devices

It should be noted that there is usually a large spread of values for each characteristic of field effect devices. It is advisable to check the full data sheets of the devices and to carry out worst-case analyses of circuit designs using them. See also section *6.14*.

6.5 Other discrete solid-state semiconductor devices

6.5.1 Thyristors and triacs

These are four-layer (p-n-p-n) devices which are used for switching applications, especially in power control and conversion. There are three electrodes, those of the thyristor being termed *anode*, *cathode* and *gate*. Some devices have two gates, one connected to the layer next to the cathode layer, and one to the layer next to the anode layer. If there is only one gate, it is usually a cathode gate.

The thyristor is a unidirectional device, which blocks current from anode to cathode until a trigger current is applied to the gate electrode. Once main circuit current is triggered, the gate loses control, and main circuit current only ceases when it is forced below some critical value, the *holding current*. This may be done, for example, by reducing the main circuit supply voltage.

A slightly different construction allows the *gate turn-off* (gto) thyristor to be produced. In this device, main circuit current can be turned both on and off by controlling the gate voltage. However, these devices are often not characterized for reverse conduction or reverse blocking, and supplementary diodes are therefore needed. Some devices have an anti-parallel diode integrated with the main device.

The triac is a bi-directional device, which may be considered as two thyristors connected in inverse-parallel. Its electrodes are usually called *main terminal 1*, *main terminal 2* and *gate*, the gate being closer in potential to main terminal 1. The triac can be used for controlling power directly in ac mains circuits.

The triggering conditions for these devices have to be carefully controlled to minimize gate dissipation while ensuring full triggering of the main circuit, since incorrect circuit design will lead to rapid failure of the device. There are many special-purpose devices, such as diacs, and integrated circuits designed for this application.

6.5.2 Programmable unijunction transistors

Programmable unijunction transistors are similar to thyristors with only an anode gate. The anode-to-cathode breakdown voltage can be controlled by the gate voltage, and the anode current at the breakdown voltage is controlled by the gate current (i.e. by the resistance between the gate and its bias voltage source). With a suitable value of load resistance in the anode circuit, when breakdown occurs the anode-to-cathode voltage falls to a low value and the current rises to a high value, which can also be controlled by the external gate resistance. The device is useful in simple timing and relaxation oscillator circuits, and in trigger circuits for thyristors and triacs.

6.5.3 Hall effect devices

The current flow through a broad, thin conductor is deflected by a magnetic field perpendicular to the plane of the conductor, in accordance with the 'motor rule'. This effect is known as the Hall effect, after its discoverer, and gives rise to a potential difference between the opposite edges of the conductor which are parallel to the current flow. It has, for this reason, also been described as the transverse magnetoelectric effect. The voltage generated is very small in metallic conductors but can be quite large in some semiconductors.

Hall effect devices may be used for detecting and measuring magnetic fields and for electronic commutation in brushless motors.

6.5.4 Light-dependent resistors

Unlike Hall effect devices, which are constructed of single crystals of semiconductor material, light-dependent resistors (ldrs) are polycrystalline devices. The usual material is cadmium sulphide, which gives a spectral response similar to that of the human eye. Devices made of lead sulphide are sensitive to infrared radiation. The current through the cell is due to the electron-hole pairs created by the incident photons, and is therefore proportional to the illumination. The response of these devices is of the order of tens or even hundreds of milliseconds, especially at low levels of illumination.

6.5.5 Thermistors

Thermistors are resistive elements which are ohmically linear at constant temperature, but which have strongly negative or strongly positive temperature coefficients of resistance. They are made of semiconducting combinations of metal oxides; those with negative temperature coefficients behave approximately as intrinsic semiconductors. The relation between resistance and temperature is often based on an exponential law.

These devices are used for temperature measurement and current control.

6.5.6 Surge suppressors

Surge suppressors are made of metallic oxide semiconductors which exhibit very little conduction below a well defined threshold voltage, and a greatly increased conduction above this voltage. Such devices are able to absorb very large amounts of power in short duration pulses, and are useful in limiting surges and interference pulses on power supply lines.

6.5.7 Peltier effect devices

When electrons flow across the junction of two dissimilar materials, there is also a flow of thermal energy. This is known as the Peltier, or longitudinal electrothermal, effect and is due to the preferential transfer of high-energy electrons across the junction, reducing the average energy and hence the temperature on one side. If the materials are a metal and a non-degenerate semiconductor, this effect may produce significant cooling of one side of the junction, considerably greater than the heating due to resistance. As the packing density of electronic equipment increases, so the importance of such cooling devices is likely to rise.

The best semiconductor material for this application known at present is bismuth telluride, and advanced methods of preparation of this material have made the cost of devices using it economically acceptable for some applications.

Figure 6.2 Integrated circuit. All these devices are audio amplifiers, in different encapsulations. Low power devices can be built in 8-pin DIL (dual in line) form, with a copper lead frame with heat-sink tabs. Other encapsulations are the 5-lead TO-220 style and the 11-lead package for twin devices, which is of SIL (single in line) form but with staggered leads (Maplin Electronics Ltd)

6.6 Microelectronics

6.6.1 Principles

The epitaxial method of manufacturing transistors, in which layers of semiconductor materials may be deposited from the gas phase onto a substrate while preserving an ordered crystal structure, allows also the production on a single *die* or *chip* of much more complex structures, equivalent to hundreds, thousands or even millions of transistors and other devices. This is done by combining the growth of epitaxial layers with processes of masking the surface of the die and diffusing dopants into selected areas, and by coating areas with insulators such as silicon oxides and nitride. By these means, bipolar and field effect transistors, diodes, Zener diodes and thyristors may be made, together with resistors and capacitors.

There is a very large number of standard and proprietary processes which offer various advantages in performance, space, cost and reliability. Developments are being made at such a rate that it would be pointless to describe current processes in any detail in this work, still less those used in the past.

6.6.2 Design constraints

The versatility of microelectronics manufacturing is continually increasing, but there are three major factors which, relatively unaffected by new developments, affect the feasibility and cost of realizing a given device specification. These are the quantity of devices, the number of diffusions and the area of semiconductor substrate required.

The first factor is very significant because of the high costs of initially designing the circuit, verifying the design and preparing the diffusion masks and other specialized manufacturing equipment. Because of this, and the great demand for small and medium volumes of diverse products, user-configurable and semi-custom devices have been made available. This applies to logic and microprocessor devices, but for linear applications it is usually necessary, and possible, to use standard volume-produced ('commodity') devices.

The second factor depends on the detailed design of the device and the manufacturing processes used. It can be controlled only by the device designer and manufacturer.

The third factor clearly depends on the device complexity, but the type of functional element also has a considerable effect. Resistors and capacitors require large areas of substrate, while field effect transistors can be made very small. Bipolar transistors are intermediate in size, and diodes and Zener diodes are made as variants of bipolar transistors. Special means have to be used to make p-n-p transistors, and formerly only slow devices with very low current gain were available, but devices with conventional performance are now available in integrated form.

Because of the above considerations, circuits optimally designed for integration are considerably different from those for discrete execution. The use of resistors and capacitors is avoided as far as possible. Many circuit functions, such as current mirrors, are not readily available in discrete form but are easily realized in integrated circuits, because of the close matching of the characteristics of adjacent devices.

Much use can also be made of devices with multiple electrodes, such as multi-emitter or multi-collector bipolar transistors, and matched devices, which are made to the same dimensions and placed close together on the die.

6.7 Linear integrated devices

By far the greatest volume of linear device production is devoted to *operational amplifiers*. The name derives from the approximation of the performance to that of an ideal amplifier, with infinite gain, infinite input impedance and zero output impedance. A variant is the *operational transconductance amplifier*, which has an output whose impedance approaches infinity, i.e. a current source. These devices do not necessarily have a high input impedance; this also applies to some voltage-source output devices (*Norton amplifiers*) designed particularly for current-mode input and for single supply rail operation.

Other linear devices include voltage and current regulators, audio preamplifers and power amplifiers, video and rf amplifiers, voltage controlled gain/attenuation blocks, tone generators and detectors, analogue multipliers, companders, and colour television encoders and decoders.

6.7.1 Selecting and specifying linear devices

There is a very large number of devices catalogued for many of the common linear applications, and it is often by no means easy to make an 'optimum' choice. Attempting to do so may well take more time than is justifiable, especially when the optimum choice is found to be a device which is not used in large volume and is consequently expensive and not freely available. Except for applications where extreme performance is required in respect of one or more characteristics, it is usually best to choose the lowest priced 'industry standard' device which meets all the requirements. See also section *6.14*.

6.8 Digital integrated devices

There is a vast range of digital devices, the principal types being general-purpose logic, microprocessors, primary memory devices, peripherals and arrays.

General-purpose, or *glue*, logic devices comprise both sequential and combinational logic functions on a relatively small scale of complexity, in contrast to that of processors and similar devices. There are numerous 'families' of logic devices, and more are continually being developed with new combinations of advantages in terms of speed, low power consumption, high noise immunity, etc.

Microprocessors include not only general-purpose computing devices but also devices designed for such diverse special purposes as calculators, television receiver tuning and remote control and automatic photographic cameras.

Primary memory devices are intended for use with microprocessors and are of two kinds, *random access memory* (RAM) and *read only memory* (ROM). Within these major divisions, there are many varieties, such as dynamic and static RAM and various types of user-programmable and user-alterable ROM.

Peripherals are a large category including input/output devices for use with microprocessors, such as analogue/digital (ADC) and digital/analogue (DAC) converters, many types of data bus driver and receiver and video display drivers. Other devices of this type include disc drive and graphics controllers (which are also processors) and direct memory access (DMA) and memory management devices.

Arrays are intended for user-specific adaptation for applications where the development of special-purpose devices (application-specific integrated circuits, asics) cannot be justified. There are a number of types, some of whose usual identifiers are actually trade names, and new types are continually being introduced. Some types can be adapted by the user, while others (e.g. 'mask-programmable' devices) require the services of the manufacturer for this purpose.

6.8.1 Selecting and specifying digital devices

For general-purpose logic, the usual considerations are speed, power consumption and noise immunity. If used in conjunction with a processor or its peripherals, logic-level compatibility is also important. With the continual introduction of new ranges, it is important to avoid devices that may not be available during the whole planned service life of the equipment.

For processors and peripherals, the (very complex) choice is often based largely on previous experience with software; few designers are equally happy writing and de-bugging code for several types of processor. See also section *6.14*.

6.9 Miscellaneous integrated devices

Some of the more widely used devices in this category are described below.

6.9.1 Analogue switches

These devices allow analogue signal routeing to be controlled by digital signals. Important characteristics are the *on* and *off* resistances, the analogue and digital voltage ranges and, since most devices are multi-channel, the isolation between channels.

6.9.2 Optocouplers

Electrical isolation between circuits is, in many cases, much more cheaply and conveniently achieved by the use of these devices than with the traditional transformer or relay. They are available in several forms, differing in the type of output device, which may be a bipolar transistor or Darlington, a thyristor or triac or a logic gate. Additional logic or other functions may be included in the device.

It is important to ensure that the isolation provided is sufficient to meet the relevant safety requirements, particularly in Europe where the mains voltages are high. Some specifications also require devices to be free from ionic migration across the isolation.

6.9.3 Switched-capacitor devices

These can be used for a growing number of applications, such as the construction of filters for analogue signals, precision dc amplifiers and voltage/frequency and frequency/voltage converters. They depend for their action on the transfer of charge from input to output through one or more capacitors. In precision amplifier applications, there may be no need for external high precision components, as gains can be arranged to depend only on precise ratios of integrated capacitor values.

6.9.4 Clocks, counters, digital voltmeters and similar devices

These devices include oscillator, divider and, often, display driver functions (see section *6.10.6*). They can be used for many applications which are not precisely what they were originally intended for, and can provide low cost and elegant solutions in many cases. For example, they can be used in automatic sensing equipment to detect time intervals or signal levels but without any display function being required.

6.10 Display devices

Cathode ray tubes for picture display are dealt with separately in Part *3*. With the exception of liquid-crystal picture displays, the following types of device are mainly intended for the display of text and graphics rather than television pictures. Devices for the latter purpose are dealt with in detail in Part *10*.

6.10.1 Liquid-crystal devices

Liquid crystals are solutions of, or pure liquid, substances which have long, thin molecules carrying unlike electric charges on the ends. The liquid has a structure that is less ordered than that of a crystalline solid but more ordered than that of a normal liquid. In *nematic* liquids, the molecules are arranged with the long axes parallel, while in *smectic* liquids the molecules are additionally arranged in layers perpendicular to the long axes of the molecules. *Cholesteric* crystals are arranged in sheets, parellel to the long axes of the molecules.

In practical devices, a very thin layer (typically 10 μm) of liquid is trapped between two glass plates provided with transparent conducting electrodes. These allow an electric field to be applied to the liquid. In the *twisted nematic* device, the molecules tend to align with the field, but also align with a preferred axis in the surface of the electrodes (determined by surface treatment of the electrode). The electrodes are arranged with their preferred directions at right-angles, and this causes the device to rotate the plane of polarization of polarized light.

In use, the device is sandwiched between crossed polarizing filters and backed by a mirror. With no applied field, incident light passes with little attenuation through to the mirror and back again; thus the device appears transparent. When a field is applied, the plane of polarization is rotated in the device, which thus appears opaque.

The field is applied by means of a low alternating voltage to avoid electrolysis, together with a small dc bias. The power requirements are very low, so that battery operation is possible, and high contrast displays are available. However, the viewing angle is rather limited and at some angles a spurious display may appear.

Higher contrast displays, with wide viewing angles, are achieved by mixing a suitable dye with the liquid crystal. This type of device is known as a *guest-host* cell, and is opaque when not energized but transparent when energized. Colour filters can be used to give colour displays; a coloured backing for the cell may be provided as well.

Some substances that form smectic liquid crystals are ferroelectric, i.e. the molecules have a strong electrical polarization. These materials can be used to make display devices which have bistable elements, remaining in the transparent or opaque state until switched. Such displays have considerably reduced complication in the element addressing compared with conventional displays, and consequently may be less costly, more reliable and larger.

6.10.2 Light emitting diode displays

The operation of the led itself is dealt with in section *6.2.2.6*. In order to obtain a readable display size, each die may be equipped with a prismatic lens, which distributes the light over a wide angle in one direction, thus producing a linear light source. Displays are available with various numbers of line segments, allowing numeric (7-segment), alphanumeric (16-segment) and special symbols to be displayed. Another type of display uses an array of small circular leds, which is driven, under the control of a microprocessor, to produce fixed or moving displays of symbols. Such displays can be multi-coloured and of dynamically variable brightness.

6.10.3 Vacuum incandescent devices

Introduced to provide high-brightness numerical displays, these devices are obsolescent but are still used for applications where there is no suitable alternative. They share with individual and matrix lamp displays the drive problems caused by high inrush currents when switching current into cold filaments. The continuous power requirements are quite large compared with those of newer types of display, and the devices generally respond too slowly to allow multiplex operation (see section *6.10.6*).

6.10.4 Vacuum fluorescent devices

These are used for numeric and symbolic displays. Electrons from a heated cathode electrode are collimated by grid structures and accelerated towards a fluorescent screen by an anode structure. The fluorescence is controlled by the grid-to-cathode voltages. These devices are being superseded to some extent by liquid-crystal and led displays.

6.10.5 Gas-discharge devices

These devices incorporate a triode electrode structure, the visible indication being given by a plasma developed at the anode structure, which may be segmental or in the form of a symbol outline. In another form, a diode structure is used, and the whole panel may be illuminated. Largely obsolete, these devices require high voltage drive.

6.10.6 Display drivers

Multi-element displays may be driven continuously or in time-division multiplex. The latter method requires more complex drive circuits but offers a useful power saving and reduces device temperature considerably, thus aiding reliability. Multiplex operation often requires device currents to be increased to preserve display brightness, which affects the choice of driver device. A large range of dedicated integrated circuits is available for these applications, which is a considerable advantage as discrete driver circuits often require very large numbers of components.

6.11 Charge-transfer devices

Developed from mos transistor technology, these devices are of several kinds. In each case, charge is stored in a series of capacitor elements having metal gate electrodes insulated from a semiconductor substrate by an oxide layer, and is transferred from one capacitor to another by switching the gate potentials.

In the surface-channel charge coupled device, the semiconductor plates of the capacitors are formed by the creation of a reverse biased inversion layer in the substrate. A high gate voltage is required for this. The device has the disadvantage that charge is lost due to surface trapping of carriers. While this is not very significant for a few elements and at low speeds, it limits many applications of the device. This loss does not occur in the buried-channel device, where a permanent channel is formed below the surface by diffusion. Charge losses at each transfer are considerably reduced by this means.

Bucket-brigade charge-transfer devices have individual isolated channels diffused into the substrate, and the gate electrodes each overlap two channels. The structure is equivalent to a chain of field effect transistors, each with a capacitive load. Charge packets propagate along the chain of devices as they are switched in turn by (usually) a two-phase clock. Charge transfer losses can be extremely low in these devices.

All of these variants can be used as analogue or digital shift-registers, and can handle analogue signals with a high dynamic range (40–60 dB). As such, they can be used as delay lines and in the realization of digital filters. The use of buried-channel devices as image sensors in video cameras is discussed in Part *3*.

6.12 Piezoelectric devices

The piezoelectric effect is the appearance of an electric charge on the surface of a crystal which is mechanically stressed. Conversely, the application of an electric field will produce mechanical strain, i.e. a change in dimensions. Most practical piezoelectric materials are ceramics, with the notable exceptions of quartz, Rochelle salt (potassium sodium dihydrogen tartrate) and ADP (ammonium dihydrogen phosphate).

Ceramic capacitors often show unwanted piezoelectric effects. Useful devices based on piezoelectric materials include solid-state ‘sounders’ (tuned transducers for audible signalling), ultrasonic transmitters and receivers, and accelerometers. Quartz crystals are very widely used in stable-frequency oscillators, at frequencies from about 1 kHz to 100 MHz.

6.13 Electromagnetic and thermal devices

6.13.1 Electromagnetic devices

The electromagnetic relay has been in use for a very long time, and there are still many applications where a relay offers a better solution than any available solid-state switching device, for example, when a number of mutually isolated circuits have to be controlled by one signal. New constructions, with reduced size, increased ratings and high reliability, are continually being developed.

Mechanical motion, either linear or rotary, may be obtained with solenoids, stepping motors or continuous motors. In each case, care is required to control inrush currents and voltage transients caused by the inductive impedance of the device. In the interests of electromagnetic compatibility, brushless (electronically commutated) motors should be preferred.

Moving coil loudspeakers are usually used for speech reproduction and audible signalling. It should be noted that the precise way in which the device is mounted in the equipment profoundly affects the quality of sound reproduction, and care, based on experience, is necessary to optimize the conditions.

6.13.2 Thermal devices

Thermostatic switches are increasingly used as protective devices to minimize the consequences of overheating in equipment, due to either internal or external fault conditions. They are obtainable as encapsulated devices in a range of temperature ratings, with normally closed and normally open types available.

Thermal fuses are also widely used for the protection of transformers and power semiconductors. They are available in auto-resetting and non-resettable types. Fuse-links are of major importance in the design of safe, reliable equipment, and in the minimization of consequential damage arising from a fault. Incorrect design, leading to the destruction of a single printed circuit board, may involve costs of astronomical proportions if a broadcast programme or a computer system is thereby compromised.

Fuses are available in a wide variety of types, with different fusing current/time characteristics, and different maximum prospective current ratings. It is most important to take note of these when selecting the appropriate type of fuse for a given application. For miniature cartridge fuses, current/time characteristics are denoted by code letters TT, T, M, F and FF (in order of decreasing arcing time), while prospective current ratings are divided into 'low rupturing capacity' and 'high rupturing capacity' types. The former usually have glass encapsulation, and may explode if called upon to break a very large fault current, while the latter have ceramic encapsulation and very much higher maximum breaking current ratings.

6.14 General considerations in selecting active and other devices

For all types of component, the manufacturers' data sheets for appropriate devices give a good practical guide to the characteristics which should be taken into account, although the relative importance of the various characteristics naturally varies with the application. It is useful to refer to data sheets published by such manufacturers as Philips, who regularly issue comprehensive data sheets, using standard terminology. There may also be standard detail specifications in IEC, CECC, BS 9000 or other published standards for components of assessed quality, and in JEDEC and Pro-Electron registrations. It should be noted that if a characteristic is not specified in the data (i.e. numerical values are not given), its value(s) may not be controlled, or may be changed without notice, so that a large variation between samples may occur.

Manufacturers and distributors vary greatly in the type and quality of applications support services offered. This should be taken into account when choosing a device and supplier. Careful account should also be taken of cost: it may be less costly to use an 'industry standard' device of higher than necessary performance rather than an uncommon device which just meets the requirements. For new designs, 'current types' should always be used, rather than 'maintenance types'.

B L Smith
Chief Technical Writer, Thomson-CSF

7 Thermionics, Power Grid and Linear Beam Tubes

The first requirement for working in the microwave domain is a microwave source. Only when quasi-monochromatic sources began to appear in the 1930s, did microwave operations become possible.

7.1 Thermionic tubes

These sources were *microwave tubes*, and they are still with us in spite of the appearance in the 1960s of solid-state microwave sources. Tubes are by far the most powerful sources, especially at the highest frequencies of the microwave spectrum (up to 1 THz). If only for this reason, microwave tubes are here to stay for a long time.

7.1.1 Common principles

7.1.1.1 Difficulties at high frequencies

The earliest radio frequency tubes were triodes and multi-grid tubes. At low frequencies, the principle of operation of these tubes is quite simple. The rf signal is applied to the grid facing the cathode, which is a source of electrons sensitive to the applied electric field, connected to the negative pole of a power supply. As the field changes, the current emitted by the cathode changes in proportion. The electron flow, crossing the grid, reaches an electrode connected via the impedance of the useful load to the positive pole of the power supply. This impedance thus conducts the electronic current and develops an rf voltage at its terminals. This voltage is usually much larger than the one applied on the grid, so that the tube presents a large gain (15–20 dB).

However, as the frequency of operation is increased, at frequencies of the order of 100 MHz, the behaviour of the amplifier begins to deteriorate. The deterioration has two main origins:

- *Parasitic impedances*, which are negligible at lower frequencies, become important. These are the reactances due to the inter-electrode capacitances as well as the stray capacitances between connections. Even more important are the inductances of the connections in series between circuits and electrodes. Furthermore, these connections, whose lengths may be comparable to a quarter of free space wavelength, begin to radiate. Thus, amplifiers and oscillators become extremely difficult to adjust. Parasitic oscillations are often observed. Gain and efficiency are severely affected.
- *Electron transit times* between electrodes, especially between cathode and modulating grid, become of the same order of magnitude as the rf period. These transit times are simply due to the fact that electrons are massive particles and obey the laws of classical (or in the case of high energies, relativistic) dynamics. While electrons move in the inter-electrode spacing, they are submitted to an rf field which varies appreciably during their transit. If the frequency is sufficiently high, the field may even reverse itself during the transit time. If this happens in the cathode/grid region, some of the electrons will be slowed down and then reflected towards the cathode, thus reducing the net current extracted, while others will oscillate. In both cases, a large amount of rf source energy is absorbed by the electrons. Similar phenomena, though less pronounced because the electrons are faster and transit times smaller, take place in the grid/anode region. This results in a sharp drop in gain, power and efficiency.

A third source of deterioration can be mentioned. *Ohmic losses*, due to skin effect, increase as the square root of the frequency. This factor becomes really critical at higher microwave frequencies.

7.1.1.2 Solutions

To solve the problems summarized in section *7.1.1.1*, a few principles may be followed, which are basic to the concept of microwave tubes.

7.1.1.2.1 Integration of microwave circuits

Microwave circuits become an integral part of the tube. They are usually completely included inside the vacuum envelope. Active electrodes are part of the circuits which are no longer made of lumped elements (self inductances, capacitors, resistors, transformers, etc.). They are instead distributed, being of resonant cavity or waveguide type. They are usually completely shielded, so that they do not radiate. They are connected to external sources and loads by means of coaxial cables and waveguides which also do not radiate. All stray capitances and inductances are reduced to a minimum and become part of the microwave circuit.

These circuits show some improved qualities compared with their low frequency, lumped element counterparts. They have no radiation, lower losses (intrinsic Qs of the order of 10 000 are common at S-band, whereas they seldom reach 1000 for lumped element circuits at low frequencies) and constant geometry resulting in constant rf properties. In fact, these advantages are such that, for high power transmitters, the tendency is now to replace lumped element circuits at as low a frequency as possible (30 MHz) by cavity-type resonators integrated with the tetrode. This trend is limited only by the bulk of the resonators which becomes considerable at low frequencies.

7.1.1.2.2 Reduction of transit time

An obvious step to alleviate the problems caused by transit time is to reduce it as much as possible by decreasing the inter-electrode spacing. For instance, by reducing the cathode/grid spacing down to 60 μm, triodes have been operated up to C-band (6 GHz). This approach, however, is severely limited by a number of factors:

- It is very difficult to obtain and to maintain such short distances over a wide area, especially since the cathode is at a temperature of 700–800°C (for an oxide coated cathode).
- The cathode/grid capacitance is inversely proportional to this distance. This limits the frequency of the input circuit.
- The tube becomes microphonic due to capacitance variations under mechanical vibrations.

Because of these limitations, grid modulated tubes (triode, tetrode, klystrode) operate normally in the range of 0–1 GHz. A few tubes, working mostly under pulsed conditions, reach 3 GHz.

7.1.1.2.3 Use of transit time

By far the most effective way to counteract the negative aspects of transit time is to make use of its positive consequences. Common features of tubes whose operation is based on the use of transit time (i.e. klystrons, travelling wave tubes and crossed field tubes) and which constitute the overwhelming majority of microwave tubes, are:

- Electron beams are launched unmodulated and accelerated until they reach a constant average velocity.
- At this velocity, the electrons drift for several rf periods (e.g. klystron and magnetron) or even several tens of periods (travelling wave tube and crossed field amplifier) while interacting with the microwave circuits.
- The general mechanism for intensity modulating the beam is velocity modulation. In the klystron, the beam passes a narrow gap, which is part of a resonant cavity, across which a longitudinal rf electric field is present. Depending on the phase at which they cross the gap, electrons are either accelerated or decelerated in a periodic fashion. Drifting in a field free region, fast electrons catch up the slower ones, forming periodic electron bunches which are the major part of the current modulation; a much smaller part originates directly in the velocity modulation itself. In travelling wave interaction, the beam is accompanied by a quasi-periodic field pattern which, in a frame of reference moving with the beam, is almost static. Here again, some electrons are accelerated while others are decelerated and bunches appear.
- The rf energy exchange takes place in vacuum, without any impact of the electrons on the rf structure. In the klystron, the bunches cross the gap of a final cavity resonator and are slowed down by the field developed across the gap. Since energy is conserved, this decrease in the kinetic energy of the beam is transformed into rf energy in the cavity. Similarly, in travelling wave tube interaction, the equilibrium position of the bunches is such that they are submitted to a retarding electric field, so that they are continuously slowed down and give up energy to the rf circuit all along the interaction space.
- After interaction has taken place, the spent beam is usually collected on an independent electrode, the *collector*, having good thermal dissipation properties. Only in crossed field tubes such as the magnetron or crossed field amplifier does the beam, due to the constraints imposed by the geometry of the tube, eventually impinge the microwave circuit which is used as a collector.

7.1.2 Microwave circuits for electron tubes

As was seen in section *7.1.1*, the microwave circuits used in microwave tubes are cavity resonators and periodic slow wave structures. Their properties can be derived from the application of Maxwell's equations subject to the boundary conditions imposed by their geometry. Demonstrations of their properties can be found in references 2–4.

7.1.2.1 Resonant cavities

Any empty volume surrounded by a good conducting material (usually metallic) can be considered as a resonant cavity. If the conductivity is large enough for the losses to be considered negligible, the cavity, when excited by a microwave source, will exhibit electromagnetic fields only at discrete frequencies which form an infinite spectrum. These frequencies are determined by the solution of the following system of equations:

$$\Delta\vec{V} + k^2\vec{V} = O \qquad (7.1)$$

$$div\vec{V} = O \qquad (7.2)$$

where Δ is the Laplacian operator
$\frac{\delta^2}{\delta x^2} + \frac{\delta^2}{\delta y^2} + \frac{\delta^2}{\delta z^2}$
$k = \omega/c$ and $\omega = 2\pi f$, the angular frequency.
$\vec{V}$ is a vector which can be either the vector potential $\vec{A}$, the electric field $\vec{E}$ or the magnetic field $\vec{H}$. This system is associated with the boundary conditions which are:

- zero tangential field at the boundary for $\vec{A}$ or $\vec{E}$: $\vec{A}\times\vec{n}= 0$ or $\vec{E}\times n-0$.
- zero normal field at the boundary for $\vec{H}$:$\vec{H}.\vec{n}=0$, $\vec{n}$ being a unit vector normal to the boundary.

The system identified in equations (7.1) and (7.2), subject to the boundary conditions, has solutions only for a set of discrete real values of k which are its proper or resonant values k_n. For each k_n, the field configuration, be it electric or magnetic, can be computed as a solution of these equations. These solutions are called *resonant modes* and designated by the mode number so that peak field values at frequency $\omega_n = k_n c$ are $\vec{E}_n$ and $\vec{H}_n$.

These fields form a complete orthogonal set. It can be shown that

$$\begin{aligned} &\int_v \vec{E}_m\cdot\vec{E}_n \, dv = 0 \quad \text{when} \quad m \neq n \\ &\int_v \vec{E}_m\cdot\vec{E}_n \, dv \neq 0 \quad \text{when} \quad m = n \end{aligned} \qquad (7.3)$$

Most cavities in practical use have at least an axis of symmetry. With respect to this axis Oz, one can define transverse electric (TE) and transverse magnetic (TM) modes. The mode number is then defined by the prefix TE or TM followed by three numbers. For rectangular cavities, these are the number of variations of the field (quasi-half waves) in each direction Ox,

Oy and Oz respectively. In the case of cylindrical coordinates, the first number refers to the variations along r, the second in the θ direction, the third in the z direction.

7.1.3 Common technology: cathodes

The cathode is the source of electrons in every electron tube. The current density of electron emission from the cathode ranges from milliamperes to ten amperes per square centimetre of cathode area. Three mechanisms for emission of electrons from cathodes are usually used. These are:

- thermionic emission
- secondary emission
- field emission

Most microwave tubes, such as klystrons and travelling wave tubes, employ only thermionic emission; in the power grid tubes both thermionic and secondary electron emission are used.

No real cathodes meet the ideal characteristics. For instance, the impregnated cathode which consists of porous tungsten and magnesium impregnated with barium calcium aluminates must be heated to a temperature in the vicinity of 1050°C to produce an appreciable amount of electron emission. At that temperature, the current density is limited to the order of 1 A/cm². Because of the necessity for high temperature, the key constituents of the cathode responsible for the emission evaporate and therefore are depleted from the surface down to a region where the internal pressure is not high enough to allow the active element to migrate towards the surface.

7.1.3.1 Thermionic emission

Electron emission from the solid results when electrons in the solid have sufficient energy directed towards the surface to overcome the potential barrier (work function) to escape from the solid into vacuum. The thermionic electron current can be predicted by using the density of electron energy states and the probability of their occupation.

When the potential of the anode placed in front of the cathode becomes less negative, the number of electrons entering the space between cathode and anode increases, and finally a negative space charge of detectable density is formed in front of the cathode. This space charge causes an increase in potential in front of the cathode, adding to the potential barrier corresponding to the work function. Only faster electrons can overcome the barrier, while the slower ones will overcome the potential barrier but will have to return to the cathode after having penetrated some distance into the vacuum. At higher anode voltage, the potential barrier due to space charge disappears, and all electrons emerging from the cathode surface reach the anode (the saturation range). The equation for the saturated current density formulated by Schottky is:

$$J_s = J_o \exp \left(\frac{4.4}{T} \sqrt{\frac{V}{d}}\right) \qquad (7.4)$$

with $$J_o = AT^2 \exp \left(\frac{e}{\kappa} \frac{\phi}{T}\right)$$

where V is the applied voltage (V),
d the distance between anode and cathode (cm),
A a universal constant, value 120 cm² T^{-2},
e electron charge (C),
ϕ work function (V),
κ Boltzmann constant (WS^{-1} Kexp)
T temperature (K).

The most significant aspect of equation (7.4) is the exponential variation of the current density with the work function and the reciprocal of the temperature (see *Figure 7.1*).

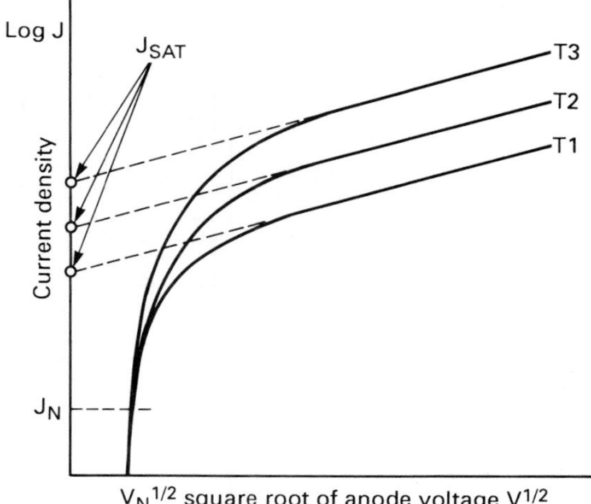

Figure 7.1 Schottky plots at three different temperatures for a planar diode, defining the zero-field saturated current J_{SAT} and showing the nominal current density, J_N, at the operating voltage, V_N

7.1.3.2 Thermionic cathodes

The first thermionic cathodes used in quantity were those employed in the early radio tubes. They consisted of pure tungsten or carburized thoriated tungsten filaments. The latter was considered to be the first type of dispenser cathode due to the fact that the thorium was dispensed by diffusion through the bulk of the wire and formed a monolayer on the emitting surface. They were directly heated cathodes, i.e. the filament responsible for the electron emission was heated by passing a current through it. The operating temperatures ranged from 2200 K for pure tungsten to 2000 K for carburized thoriated tungsten.

The major progress was the introduction of oxide cathodes by Wehnelt in 1904. He observed that a platinum filament covered with alkaline earth material was emitting electrons in vacuum at temperatures in the range of 800–1000°C. The improvement was the use of a nickel base material in various forms, coated with a mixture of barium, calcium and strontium oxides. These were used for more than 60 years at current densities of a few to a hundred of milliamperes per square centimetre. The requirement of higher current densities and cathodes withstanding more severe environment led to impregnated cathodes, discovered at Philips by Lemmens and Looges in the early 1950s.

They used a porous tungsten matrix in which was impregnated a barium aluminate having the eutectic composition $5BaO, 2Al_2O_3$. Levi improved the quality by introducing calcium oxide into the eutectic, and the impregnant became $5BaO, 3CaO, 2Al_2O_3$ (known as the *B type*). The electron emission was enhanced by a factor of 5. Some competitors proposed other compositions such as $4BaO, 1CaO, 1Al_2O_3$ (the *S type*) or $6BaO, 1CaO, 2Al_2O_3$ (*Brodie's composition*).

Current densities obtained from such impregnants range from 0.1 to 1 A/cm² when operated from 1000 to 1100°C. The improvement in current densities from B, S or the third variety of impregnant was obtained with the *M type* cathodes.

Basically the M type is a B or S type cathode covered with a film of several thousand angstroms of one of the metals of the platinum group (Os, Ir, Re). The effect of the thin film is to reduce the operating temperature of the cathode by approximately 80°C for the same electron emission density as obtained with B or S cathodes. Unfortunately, if higher current densities

Origin	Nature of the reservoir	Impregnated cathodes	Designation
1950 Philips Co Lemmens Jensens Loojes	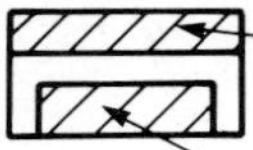 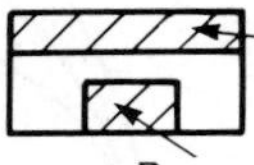 Porous W Pressed Ba Ca Sr carbonates Porous W Pressed Ba Ca Sr carbonates + W power		L
1955 Philips Levi		 A – B W Impregnated with 5-0-2, 5-3-2, 4-1-1	A B S
1955 Siemens from Katz	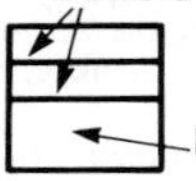 Double porous W layer Ba Ca aluminates + W powder		
1966 Philips Zalm		 Os, Ir B type	M
1976 Varian		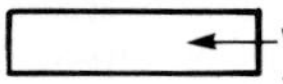 W + Ir impregnated 6 – 1 – 2	M M
1979 TH – CSF		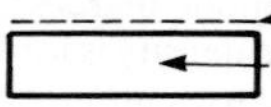 Os W + Os impregnated 5 – 3 – 2, 4 – 1 – 1	C M M
1979 Telefunken		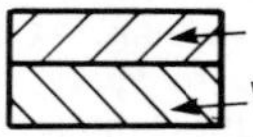 Co-pressed W and W + aluminates	
1980 TH – CSF	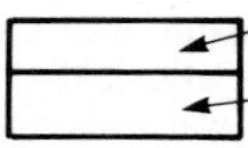 Porous W Impregnated cathode		S P

Table 7.1 Impregnated cathode types

are required, the increase in temperature is sufficient to cause the diffusion of the thin layer into the tungsten porous body and may cause pore clogging.

A novel type of porous matrix was then developed. It consisted of mixing and sintering together tungsten powder and powder of a metal of the platinum group in weight concentration typically 50:50 for tungsten osmium and 8:20 for tungsten iridium.

MM in *Table 7.1* refers to these *mixed matrix* types and CMM to *coated mixed matrix* cathodes. The latter may have a body of tungsten osmium or tungsten iridium coated with a thin film of osmium.

The numbers 5-0-2, 5-3-2, 4-1-1 and 6-1-2 represent the chemical proportions of barium, calcium and aluminium compounds in A type, B type and S type cathodes respectively.

7.1.3.3 Impregnated cathode operation

The open pores on the surface of the tungsten porous matrix have the shape of slots rather than round holes. When the cathode is heated to the operating temperature, chemical reactions take place in the matrix between the barium and calcium aluminates and the tungsten, and free barium is generated that migrates towards the surface and spreads on it to form an almost complete monolayer (*Figure 7.2*(a)). As the cathode is used at nominal temperature, the slot type pores tend to get smaller and smaller and wind up as separate small round pores. This is due to two cumulative phenomena: the chemical reaction which leaves residues and the thermal reconstruction of the surface.

Barium which was dispensed from very near the surface at the beginning of life comes, after a few thousand hours of operation, from deeper and deeper regions of the matrix, and comes in smaller and smaller quantities due to a drop of internal pressure. The emission becomes more and more patchy, and the work function distribution varies from a sharp distribution having a σ of 0.075 to a broader distribution where the σ is 0.25.

7.1.3.4 Life considerations

In most applications long life is required, either to minimize the replacement costs (e.g. a ground station) or limit redundancy cost (satellite applications). With B and S type cathodes set at 0.5–1 A/cm^2 and operating temperatures of 950–1000°C, life-

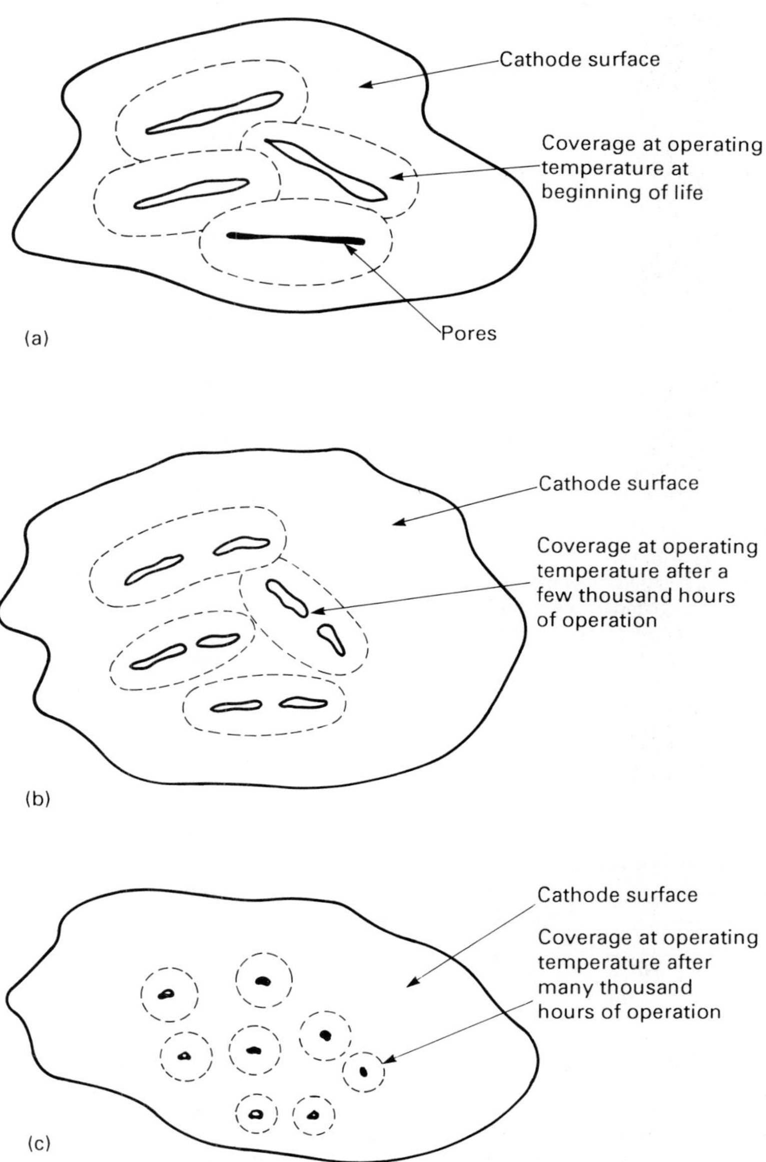

Figure 7.2 Evolution of impregnated cathode pores over lifetime

times of 3×10^5–10^6 hours can be expected. With M and MM type cathodes, higher current densities can be expected (1–3 A/cm^2 for lifetimes exceeding 10^5 h). CMM type cathodes allow operation at 4–6 A/cm^2 with the same life expectancy and temperature in the range of 1000–1050°C. *Figure 7.3* charts the life against nominal current densities extrapolated from experimental values at shorter lives.

One or more electrons in the material gain enough energy to be emitted.

The number of secondary electrons emitted from a surface depends mostly on the nature of the surface, and also on the energy and the striking angle of the impinging electrons.

The average ratio, δ, of secondary electrons emitted to the number of primary electrons producing them varies from less than one up to eight or ten. $\delta = i_s/i_p$. *Figure 7.4* shows values of this ratio for some common metals. The characteristics of secondary emission are known from experiments; however, the process of emission is rather complex and is difficult to deal with theoretically. The number of secondary electrons is low at low primary energy (which is easily understandable) and also very low at high primary energies. The reason for this is that the high energy primary electrons penetrate very deeply into the material, where they excite electrons which have a very small probability of reaching the surface to escape.

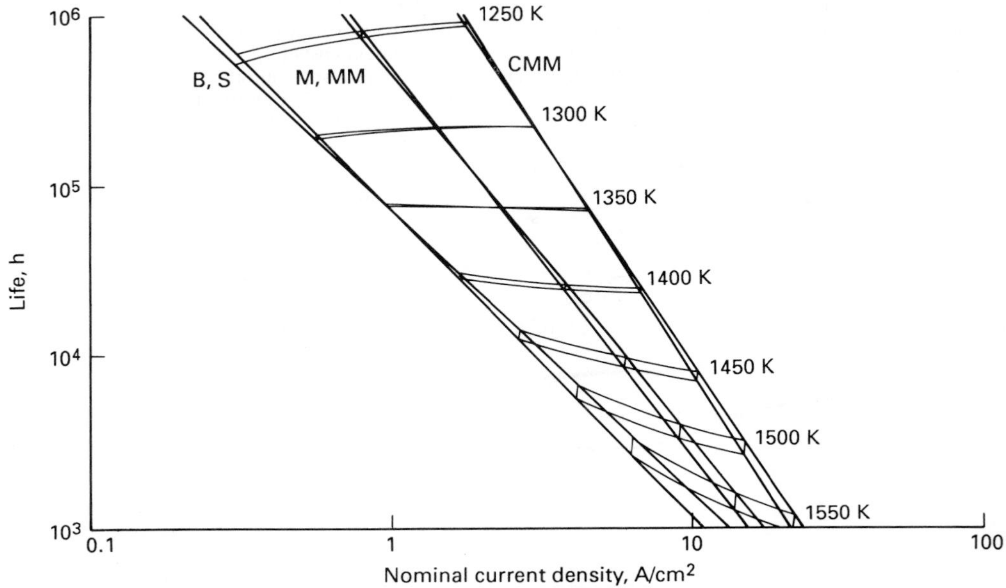

Figure 7.3 Extrapolated cathode life *versus* current density

7.1.3.5 Secondary emission

Another form of electron emission that plays an important role, especially in power grid tubes, is secondary emission. This occurs when a surface is bombarded by electrons or ions of appreciable kinetic energy.

Upon striking the material surface, a primary electron shares its kinetic energy with other particles in the immediate vicinity.

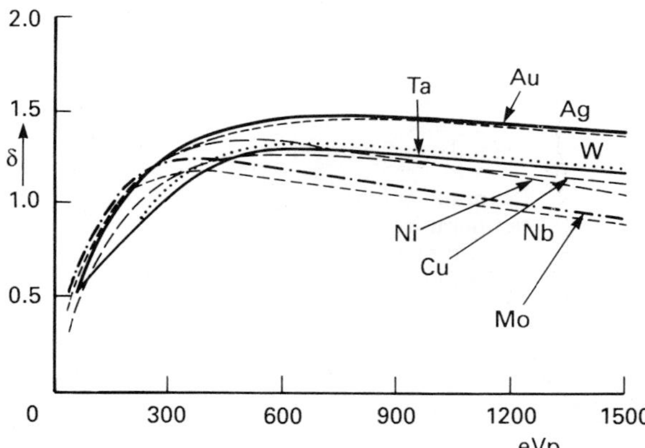

Figure 7.4 Secondary electron emission coefficient for high work function metals (after Warnecke)

7.2 Power grid tubes

7.2.1 Vacuum diodes

Even though the use of vacuum diodes has declined considerably, a knowledge of their properties aids understanding of the principles of operation of power grid tubes.

A vacuum diode consists of two electrodes in a vacuum: a thermoelectronic *cathode* and an *anode*. When the cathode temperature is high enough, it can emit electrons from the surface. The anode is brought to a positive voltage, V_p, with respect to the cathode, and thus attracts electrons emitted by the cathode, giving rise to a current, I_p, between the cathode and anode. The curve of I_p versus V_p is the *diode characteristic*.

For a given cathode temperature T, I_p increases with V_p up to a maximum current I_m. For a higher cathode temperature T the characteristic curve is the same up to I_m, but continues to increase to a higher maximum current $I_{m'}$.

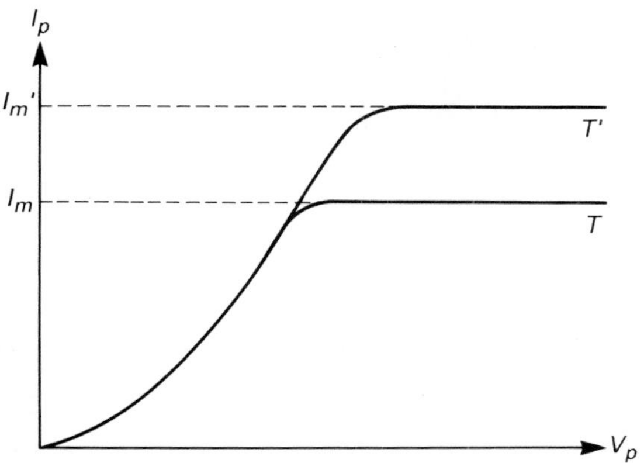

Figure 7.5 Diode characteristic curves

The maximum current I_m or $I_{m'}$ is due to saturation of the cathode. The number of electrons that can be furnished by the cathode is limited for a given temperature, thus limiting the current to a maximum which depends on the temperature. Note that the diode characteristic has the same form for any value of cathode temperature (Figure 7.5).

The diode characteristic is expressed in its most simple form by the Child-Langmuir relation:

$$I = KV^{3/2} \tag{7.5}$$

where K is called the *diode perveance*.

The anode thus attracts a given quantity of electrons, and any excess electrons stay in the vicinity of the cathode to form a *space charge*, creating a minimum potential near the cathode.

This space charge, which depends on the cathode temperature T and the voltage V, regulates the electrons. Those electrons having an initial velocity sufficient to penetrate the space charge region will be stopped by the space charge and will be confined to the cathode region.

7.2.2 The triode

A third electrode, in the form of a grid between the cathode and the anode of a diode, was introduced in 1907 by Lee de Forest.

The potential at the surface of the cathode is the superposition of the anode potential and the grid potential. When this potential is negative, there is no emission. This may be achieved by a sufficiently negative grid potential.

When the grid is less negative, certain regions of the cathode will 'see' a positive potential and be able to emit electrons which will be attracted to the anode. By further decreasing the negative potential of the grid, the area of the emitting zones increases. Beyond a certain point, all of the cathode surface will emit. When the grid potential becomes positive, some of the electrons emitted will be collected by the grid, giving rise to a grid current.

With small variations of grid voltage, the anode current can be controlled over a wide range from zero to maximum. This is the basic phenomenon which allows a large anode signal to be obtained from a small signal applied to the grid through appropriate circuits between the electrodes.

7.2.2.1 Characteristic curve sets

For each set of voltages V_a and V_g, there will be a corresponding anode current and grid current. Three types of curves can be drawn:

- currents as a function of anode voltage V_a, for different values of grid voltage V_g, called the *Kellogg diagram* (*Figure 7.7*)
- *currents as a function of grid voltage, for different values of anode voltage; this type of curve set is rarely used (Figure 7.8*)
- grid voltage versus anode voltage, for different constant current values; this is the most commonly used set of curves (*Figure 7.9*)

From these curves, the characteristic coefficients of the tube can be defined for any operating point. From the three principal variables I_A, V_A and V_G, three parameters can be defined supposing one of the variables constant:

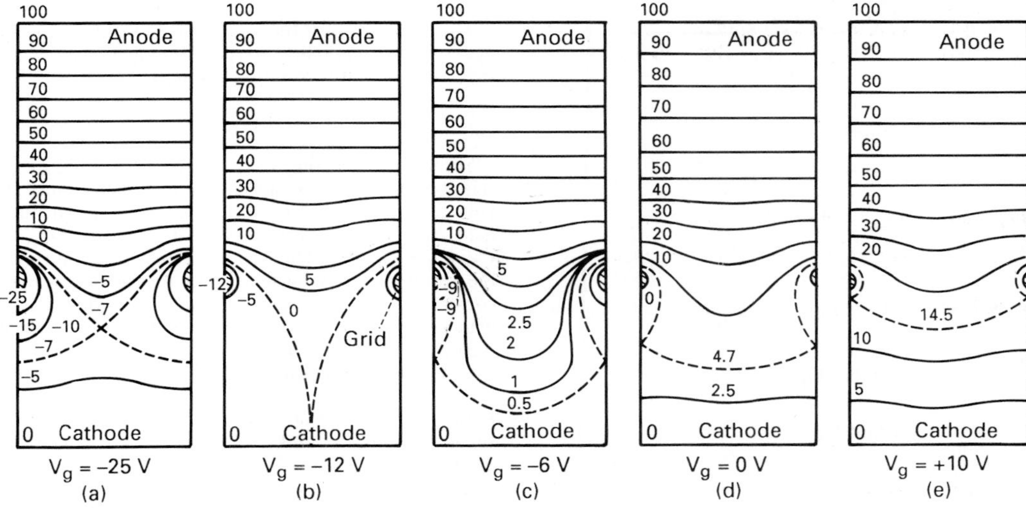

Figure 7.6 Equipotential curves

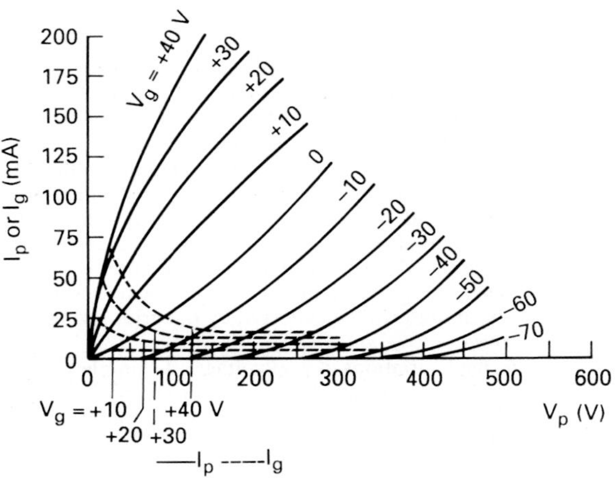

Figure 7.7 $I_p - V_p$ network: Kellogg diagram

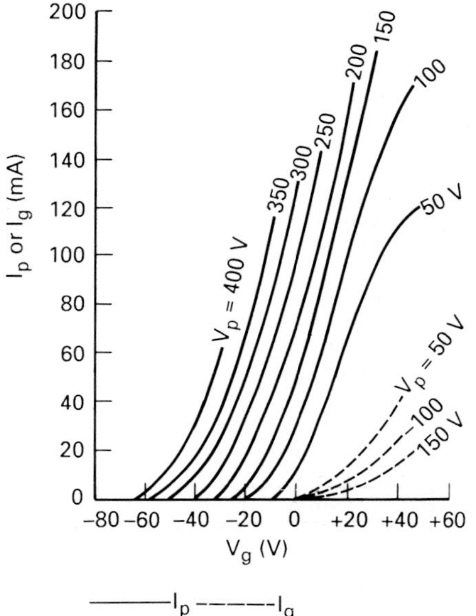

Figure 7.8 $I_p - V_g$ network

The amplification, for I_A = constant:

$$\mu = \frac{dV_A}{dV_G}$$

This unitless coefficient gives the voltage amplification of the tube.

The transconduction or tube slope:

$$s = \frac{dI_A}{dV_g} \quad \text{for constant } V_A$$

This parameter is usually given in milliamperes per volt.

The internal resistance of the tube:

$$R_i = \frac{dV_A}{dI_A} \quad \text{for constant } V_G\ (\Omega)$$

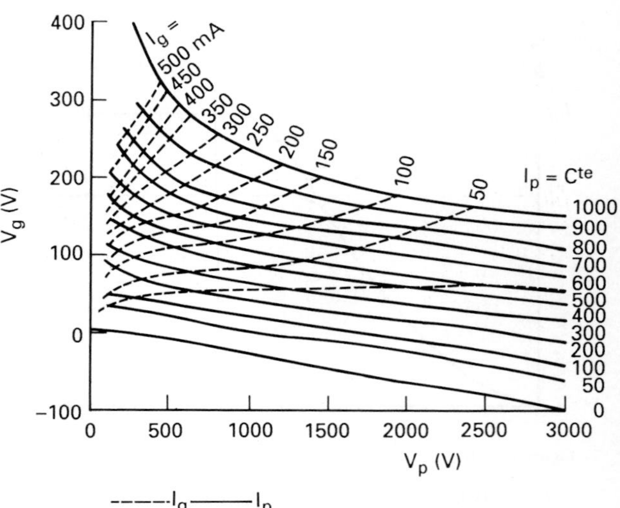

Figure 7.9 $V_p - V_g$ network

Between the three parameters, we have the relation $\mu = sR_i$. At each point on the curves of each of these sets, a linear relation between V_A, V_G and I_A can be obtained by considering the tangent of the curve:

$$R_i = \mu V_G + V_A \quad or \quad I_A = sV_G + \frac{V_A}{R_i} \qquad (7.6)$$

It is often useful to relate the three parameters defined above to other electrical quantities or physical dimensions of the tube electrodes.

If there are relatively few electrons emitted, and so a negligible space charge, there is a simple relation between the charge Q_{KG} on the cathode, the capacitance C_{KG} between grid and cathode, and the potential difference V_G:

$$Q_{KG} = C_{KG} V_G$$

The charge Q_{KG} is the integral of the surface charge density σ_{KG} over the cathode surface. The electric field E_{KG} at the surface of the cathode is thus given by:

$$E_{KG} = 4\pi\sigma_{KG}$$

Similarly, neglecting space charge, $Q_{KA} = C_{KA} V_A$ gives the relation between the anode/cathode capacitance C_{KA} and voltage V_A, giving rise to an electric field E_{AK}. The total electric field $E_{KA} + E_{KG}$ varies as the sum $Q_{KG} + Q_{KA}$ and so as $C_{KG}V_G + C_{KA}V_A$. The current between cathode and anode is the result of this total field; thus the current is constant for constant $C_{KG}V_G + C_{KA}V_A$. When V_G varies, V_A must vary with the ratio C_{KG}/C_{KA}. The definition of the amplification factor can be written: $\mu = C_{KG}/C_{KA}$.

It should be noted that:

- the capacitances are those corresponding to the electrode surface geometry, not including stray capacitance due to electrical connections, for example
- C_{KA} is less than C_{KG}, because the distance between cathode and anode is greater than that between cathode and grid, and also the grid acts as an electrostatic screen. Therefore μ is always greater than unity

7.2.2.2 Calculation of amplification factor

The amplification factor of a triode can be related to the dimensions of the various electrodes.

The principle of the calculation is to express the potential created by the charges on the different electrodes as a function of position for any point. From this result, the field at the surface of the cathode is calculated, and as before, one can calculate the amplification factor.

Note, however, that these calculations can only be solved in closed form by using approximations. Thus the values obtained from these formulas are only approximate.

For example, the calculation for a cylindrical triode gives:

$$\mu = -\frac{N\ln(r_a/r_g)}{\ln[2\sin(Np/2r_g)]}$$

In the case of a planar triode:

$$\mu = \frac{2rd_{gp}}{r\ln[2\sin(rp/a)]}$$

7.2.2.3 Operating class

We have seen that the triode characteristics can be represented, to a good approximation, by the equation:

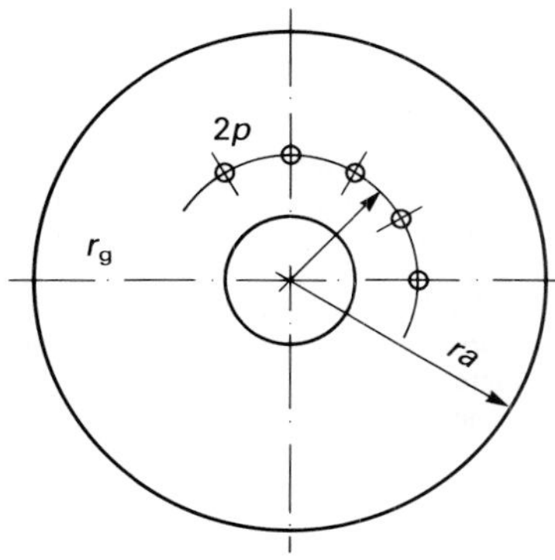

Figure 7.10 Configuration of a cylindrical triode. The grid consists of N bars, each of diameter $2p$

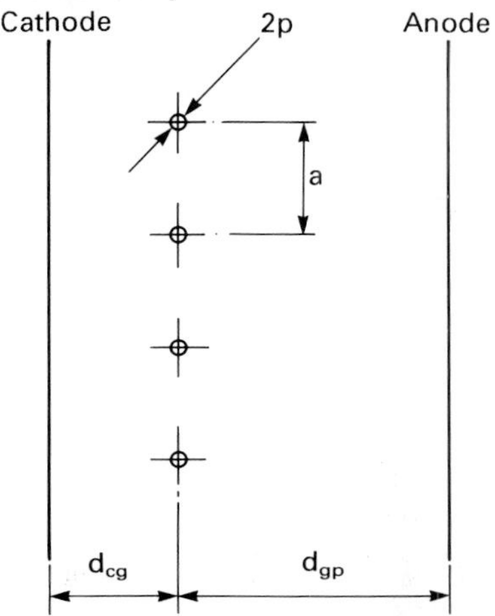

Figure 7.11 Configuration of a planar triode

$$R_i i_p(t) = \mu v_g(t) + v_p(t)$$

where i_p, v_g and v_p are the values of plate current, grid voltage and plate voltage, respectively, as functions of time t.

Generally, an operating point is chosen, at some constant values $-V_{G0}$, V_{p0}. For a grid voltage signal, applied between the grid and the cathode, which varies as $V_G \cos \omega t$, a voltage $-V_p \cos \omega t$ is observed on the anode. The phase reversal between the output circuit and the input circuit gives rise to the sign change. We can write:

$$v_g(t) = -V_{G0} + V_G \cos \omega t$$

$$v_p(t) = +V_{p0} - V_p \cos \omega t.$$

The anode circuit is loaded by an impedance R_p such that $V_p = R_p I_p$ where I_p is the amplitude of the fundamental anode current. Defining $V_0 = V_{p0} - V_{G0}$, gives, for a certain angle θ_0 where $i_p \to 0$:

$$\cos \theta_o = \frac{V_o}{V_G - R_p I_p} \qquad (7.7)$$

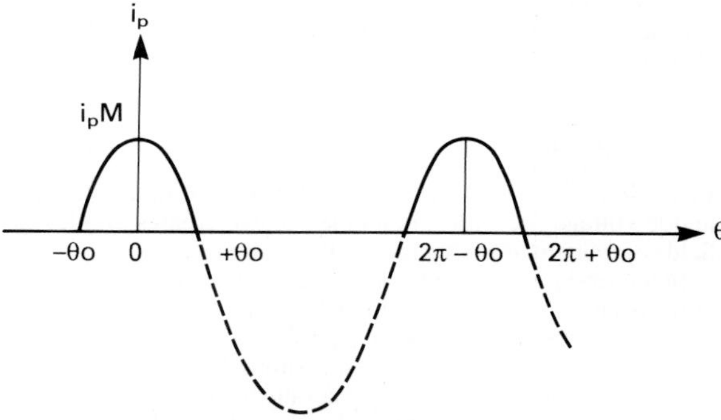

Figure 7.12 Anode current in operation

It is seen that $\theta = \omega t$ can take values between $\pm\theta_0$, i.e. $-\theta_0 < \theta < \theta_0$. θ_0 is thus a critical angle.

The steady-state component is given by I_{p0}. The ratio $I_0/I_{p0} = r$ is important because it determines the conversion efficiency of the tube and its circuit:

$$r = \frac{I_p}{I_{po}} = \frac{\theta_o - \sin\theta_o \cos\theta_o}{\sin\theta_o - \theta_o \cos\theta_o} \qquad (7.8)$$

The efficiency η is defined as the ratio of the effective power out $V_p I_p/2$ for an applied power $V_{p0} I_{p0}$ which gives:

$$\eta = \frac{1}{2}\frac{V I_p}{V_{po} I_{po}} = \frac{1}{2}\frac{V_p}{V_{po}} r$$

Another interesting parameter which can be derived from these formulas is the maximum current which the cathode must deliver. The current is maximum for $t = 0$:

$$I_{pmax} = \frac{\mu V_G - R_p I_p}{R_i}(1 - \cos\theta_o)$$

$$= \frac{\pi(1 - \cos\theta_o) I_p}{\theta_o - \sin\theta_o \cos\theta_o}$$

We then define the parameter M:

$$M = \frac{I_{pmax}}{I_p} = \frac{\pi(1 - \cos\theta_o)}{\theta_o - \sin\theta_o \cos\theta_o} \qquad (7.9)$$

which gives:

$$\frac{I_{pmax}}{I_{po}} = \frac{I_{pmax}}{I_p}\frac{I_p}{I_{po}} = Mr \qquad (7.10)$$

The different classes of tube operation can be defined in terms of θ_0:

Class A : $\theta_0 = 180°$, $r = 1$, $M = 2$, anode current always present

Class B : $\theta_0 = 90°$, $r = \pi/2$, $M = 2$, anode current half the time

Class C : in general one considers $\theta_0 = 60°$, $r = 1.7936$, $M = 2.5575$

We can thus conclude that class C operation is interesting because it offers the best efficiency; however, the product $M.r$ is very large, which requires high peak current from the cathode.

Class A operation gives lower efficiency (< 50 per cent), but the peak current is small. Class B represents a compromise which is often acceptable. Using these same relations between the different parameters, one can use Fourier analysis to calculate the different harmonics created by pulses of anode current.

7.2.3 Tetrodes

In a triode, the anode current depends substantially on the anode voltage, because the anode voltage creates an electric field component at the surface of the cathode.

In a tetrode, a *screen grid* is added between the control grid and the anode in order to diminish this effect. By creating an electrostatic 'screen' between the anode and cathode, the screen grid virtually eliminates the anode field at the cathode surface. This additional grid is generally held at a fixed positive potential with respect to the cathode, and strongly assists the extraction of electrons.

A theory analogous to that elaborated for triodes can be constructed from electrostatic theory, leading to similar equations:

$$i_p = S\left(V_G + \frac{V_{G2}}{\mu_{G2}} + \frac{V_p}{\mu_p}\right) \qquad (7.11)$$

One can calculate μ_{G2} and μ_p by formulas similar to those for triodes. In practice, μ_p has a value of several hundred, or even several thousand, which tends to confirm that i_p is virtually independent of V_p.

The same classes of operation are defined as for the triodes.

7.2.3.1 Secondary emission phenomena in tetrodes

Consider a tetrode with electrode potentials V_A for the anode, V_{G1} for the control grid and V_{G2} for the screen grid. The cathode current is then shared between the three electrodes (unless V_{G1} is negative).

Furthermore, suppose that V_{G1} and V_{G2} are fixed while V_A decreases from a value much greater than V_{G2}. Then as V_A decreases, I_A decreases first and I_{G2} increases. If there is secondary emission of electrons from the screen, these will be attracted by the anode if $V_A > V_{G2}$, causing an increase of I_A and decreasing I_{G2}. (This can even become negative if the secondary emission coefficient is greater than 1.)

On the other hand, when $V_A < V_{G2}$, the secondary electrons emitted from the screen grid cannot go towards the anode, and thus return towards the screen grid. Similarly, the secondary electrons emitted by the anode may be captured by the screen grid causing a decrease of anode current and an increase of screen grid current.

Practically speaking, this results in anomalies in the variations of the anode current as shown in the tetrode characteristics in *Figure 7.13*.

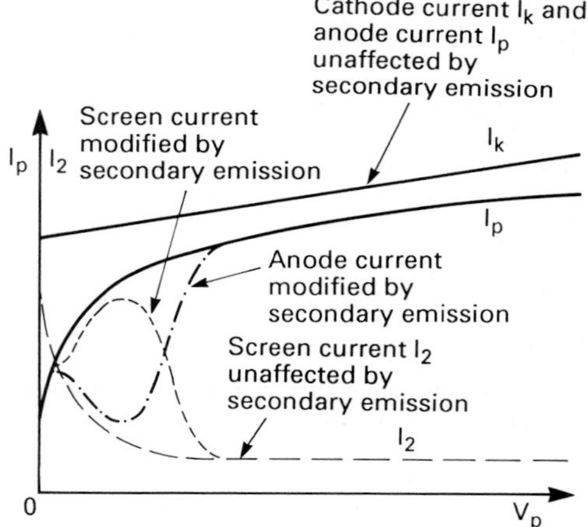

Figure 7.13 Tetrode characteristics

It is necessary for V_{G2} always to be less than V_A. To overcome this drawback, a third, *suppressor*, grid is added between the screen grid and the plate, and maintained at or near the cathode potential. This potential barrier pushes the secondary electrons back towards the anode. *Figure 7.14* shows the potential variations in such a tube, called a *pentode*. As shown by the characteristic curves, it is possible to operate the tube in the region where $V_p < V_{G2}$.

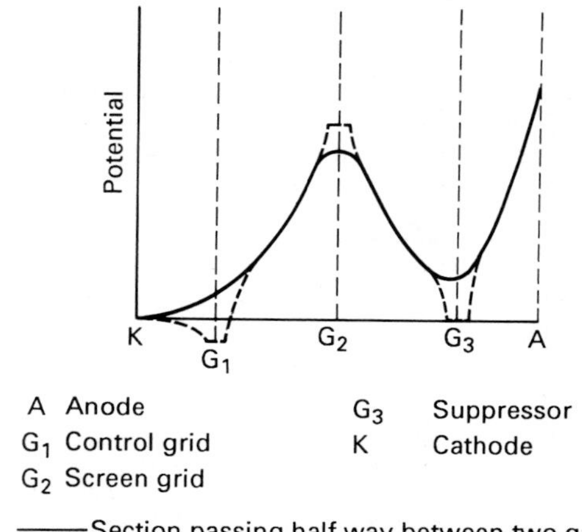

Figure 7.14 Potential variations in a pentode

This solution has been mostly used in receiving tubes. In power tubes, on the other hand, the presence of another grid complicates the technology. Also it is less necessary because the screen grid voltages are such that the secondary emission phenomena are weaker.

7.2.4 Power grid tubes

7.2.4.1 High frequency operation

Increasing the operating frequency of a tube eventually compromises certain performance characteristics, e.g. gain or efficiency. Several effects come into play:

- *Inter-electrode capacitance*, negligible at low frequencies when compared to the output circuit itself, becomes more and more important with increasing frequency. In the case of high frequencies (50 MHz–1000 MHz), it is common practice to use these inter-electrode capacitances as capacitances in the oscillating circuit itself.
- *Electrode inductance* can lead to considerable phase differences between the voltages appearing on the various electrodes inside the tube. To resolve this, most tubes use a coaxial geometry.
- *Skin effects*. Eddy currents are induced in a very shallow layer at the surface of conductors exposed to high frequency electromagnetic radiation. Appropriate materials and electrode geometry must be used to minimize such effects.
- *Electron transit time*. For low frequencies, electron transit time can be considered for all practical purposes to be instantaneous (i.e. negligible). The electron velocity v, after acceleration by a potential V, is given by $v = (2e/m)^{1/2}V^{1/2}$. The transit time can be calculated, for simple geometry, from the inter-electrode spacing. In general, the values are between 10^{-11} and 10^{-7} seconds. For tubes operating with much longer characteristic times, electron transit between electrodes can be considered to be instantaneous.

If the electron transit time is not instantaneous, other phenomena may arise. Consider the case of an alternating signal voltage V_g on the grid, which creates an electric field at the surface of the cathode and starts a current flowing. As the electrons advance, the voltage V_g will continue to vary, and could perhaps even change polarity, pushing the electrons back towards the cathode, heating it even more upon impact.

One effect of a finite electron transit time is to introduce a phase difference between the voltage and the current at the level of the grid. A practical consequence is that a smaller average current flows between the different electrodes. In order to recover the current lost in this manner, which increases with increasing frequency, the accelerating voltage must also be increased, leading to a decrease of both the gain and the efficiency of the tube. Beyond some frequency, the tube is no longer useable.

To increase the maximum frequency at which these tubes can be used, the inter-electrode spacing must be reduced. In this way, triodes can be constructed for operation at frequencies of several gigahertz.

7.2.4.2 Grid tube linearity

Amplifiers using grid tubes are sometimes required to transmit more than one signal at a time. If the tube characteristics were strictly linear, this would be no problem. Unfortunately, the linear relationship given previously is only a good approximation. Certain areas of the curve networks, particularly for low current or very high current values, are better represented by a polynomial expansion such as:

$$I_a = \sum_{j=1} A_j \left(V_g + \frac{V_p}{\mu} \right)^j$$

If two sinusoidal signals of frequencies f_1 and f_2 are amplified simultaneously, the output signal will contain components of the form $mf_1 \pm nf_2$. Although many of the spurious signals can be reduced or eliminated using a band-pass filter, there will still be those within the amplification band, e.g. between f_1 and f_2, that can lead to amplification anomalies at the output.

The calculation of the amplitude of the spurious signals introduced in multi-carrier operation is now relatively accurate using modern computer methods. The characteristic curve networks are measured and input to the computer, which then performs a harmonic analysis using Fourier methods to calculate the spurious amplitudes.

7.2.4.3 Grid tube technology

7.2.4.3.1 Cathodes

The two types of cathode in common use are oxide cathodes and thoriated tungsten cathodes.

Oxide cathodes are generally heated indirectly. Their main advantage is a continuous emission surface (see *Figure 7.15*). Their average current is of the order of 200 mA/cm². This makes them particularly useful in pulsed tubes with small duty cycles (of the order of 1 per cent or less), where they can deliver several amperes per square centimetre during the pulse, thus allowing high peak power levels to be obtained.

Figure 7.15 Oxide cathode (Thomson-CSF)

One drawback is that they are rather sensitive to ion bombardment. Any residual gases in the tube when ionized by the electron flow will be attracted to the cathode surface, resulting in 'cathode poisoning' and decrease of cathode emission. This effect is aggravated by higher tube operating voltages, which lead to greater ionization.

Thoriated tungsten cathodes are generally made of thin wires, which can be arranged on the surface of an imaginary cylinder as in *Figure 7.16*. To compensate for the effects of thermal dilation when the cathode is heated, the geometry of the array of individual wires may be obtained by supporting springs. Alternatively, the cathode may be composed of two equally spaced helical windings symmetrically orientated on the cylindrical surface. Such cathodes are directly heated and operate at about 2000 K. Saturation currents of the order of 3 A/cm² can be obtained, making such cathodes useful for very high power tubes (e.g. 1 kW–1 MW). In addition, they are resistant to ion bombardment, allowing the use of high accelerating voltages.

The most important drawback is the very high operating temperature, which requires a lot of heater power.

Figure 7.16 Thoriated tungsten cathode (Thomson-CSF)

Figure 7.17 Metallic grid (Thomson-CSF)

7.2.4.3.2 Grids

The grid electrode is the most difficult electrode to make, as it must satisfy many criteria, some of which are mutually exclusive, leading to a design compromise:

- The grid should have a geometrical form as perfect as possible, to ensure that the cathode-to-grid spacing is as accurate as possible in spite of the very small distance.
- The electrical field created by the grid at the surface of the cathode must be as constant as possible over the whole surface, requiring numerous small wires very evenly spaced.
- The grid should be transparent enough to avoid intercepting too much cathode current.
- To avoid thermal emission, the grid should have as low a temperature as possible when in operation. The thermal emission is the result of cathode evaporation of emissive material which is deposited on the grid, which then becomes a spurious emitter when the grid temperature is sufficiently high. The best way to lower grid temperatures is to use thermally 'black' materials for grid construction.
- If possible, the grid material should have a low coefficient of secondary emission.
- As the grid is required to conduct high frequency currents, it should be made of material of high electrical conductivity.
- The material should also be refractory, i.e. have a low vapour content under nominal operating conditions.

The two types of grid material commonly in use are metallic or graphite.

Metallic grids are commonly made of molybdenum, tantalium, niobium or tungsten. Unfortunately, all these materials have troublesome tendencies towards both thermal emission and secondary emission.

To reduce secondary emission, the grid surface may be 'blackened' by coating with powders such as zirconium, tantalum carbide or graphite. To reduce thermal emission, gold or platinum plating may be used. The gold or platinum combines with the emissive material deposited on the grid, increasing the work function and thus decreasing the grid thermal emission.

Graphite grids offer a low coefficient of secondary emission, as well as good thermal radiation properties. Ordinary graphite, however, is very fragile and difficult to machine. Pyrolytic graphite, on the other hand, presents in addition to the above qualities, improved thermal and electrical conductivity, together with excellent mechanical properties.

Grid blanks are obtained in the desired form by vacuum deposition of graphite obtained by high temperature "cracking' of hydrocarbons (at about 2000°C). The grid is then machined either by laser or by sandblasting. This type of electrode has led to considerable increase in the reliability and performance obtainable from power grid tubes.

Figure 7.18 Pyrolytic graphite grid (Thomson-CSF)

7.2.4.3.3 Anodes

Anodes are required to dissipate most of the power which is not supplied to the output circuit of the power grid tube. As the efficiency of the tube is gerally quite a bit less than 100 per cent, considerable power must be evacuated from the anode towards a cooling circuit to keep its temperature within acceptable limits. Furthermore, because the anode is at a high voltage during operation, the coolant must be electrically insulating.

Several types of cooling may be used, depending on the power to be evacuated and the environmental conditions of tube use.

Radiation cooling is the simplest: the anode radiates waste heat. As the temperature may reach several hundred degrees, this type of cooling may not be used when the anode is under vacuum, e.g. in a transparent (glass) vacuum envelope. The anode will be made of a refractory material such as nickel, tantalium or molybdenum, whose surface will be blackened to increase the radiation efficiency. Graphite is also used successfully.

Conduction cooling can be used for low power dissipation by strong mechanical contact between the anode and an external heat sink. If electrical insulation is required, the thermal conductor to the heat sink may be made of berylium oxide.

Forced air cooling can be used for power levels up to about 30 kW. In order to improve the thermal exchange with the moving air, cooling fins are welded to the anode to increase the surface area (see *Figure 7.19*). This method, simple in principle, becomes more troublesome at higher power levels as the air flow rates become large (several tens of cubic metres per minute). Fans for such air flow rates are noisy, bulky, power hungry and vibrating, which can lead to equipment reliability problems in practical installations.

Figure 7.19 Anode with cooling fins (Thomson-CSF)

Water cooling allows the dissipation of higher power levels by immersion of the anode in water. As the water heats up, it is replaced by an incoming cool water flow, and the heated water is pumped to a thermal exchanger (water/water or water/air) to be cooled and reintroduced into the circuit.

To avoid the formation of thermally insulating deposits on the anode, the water must be distilled. In order to provide electrical insulation of the anode, all water connections in the vicinity of the tube are of insulating materials.

Water flow rates are commonly of the order of a litre per minute per kilowatt to be dissipated. The maximum water temperature at the output is about 50°C for acceptable cooling efficiency. At higher limits, the anode locally heats a thin film of water at its surface, creating vapour bubbles which cover the anode surface and keep it from contacting the water. The anode cooling no longer works and the temperature rises sharply, which could lead to a critical situation. This is the phenomenon of *calefaction*.

Water vaporization cooling uses the latent heat of vaporization to perform heat transfer. The diagram of Nukiyama (*Figure 7.20*) represents the heat flow across a boundary surface at a uniform temperature as a function of the temperature of the surface. When the heat flow is 10 W/cm^2, the surface is at 110° and we have steady-state boiling. Further heating to point M corresponds to 125° and 135 W/cm^2. Continuing through an instable zone to the point L (Leydenfrost point), situated at about 30 W/cm^2 and 240°C, leads to thin-film vaporization. As seen in the figure, the next stable point after point M is the point Q situated at 1100°.

The Leydenfrost point is the temperature at which there is minimum heat transfer, corresponding to the temperature at which there begins to be a thin film vaporization process at the surface of the metallic anode immersed in a fast-moving water flow.

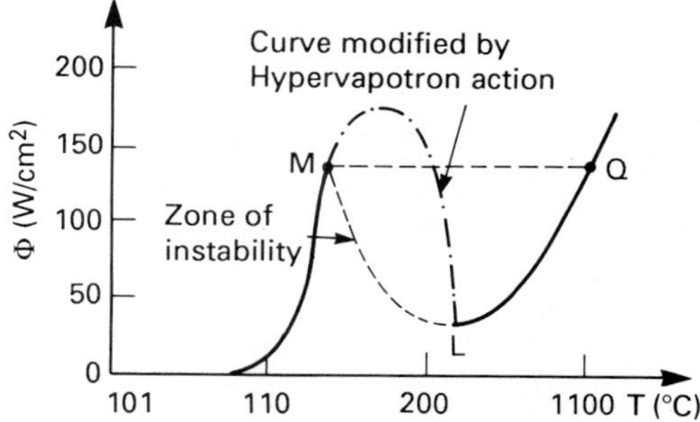

Figure 7.20 Diagram of Nukiyama. The scale of the abscissa is logarithmic

Figure 7.21 Vapotron (Thomson-CSF)

A Vapotron-cooled anode (*Figure 7.21*) is covered on its external surface by projections which create a stable thermal gradient about the critical point. The extreme end of the projection is always cooler than the area where the vaporization occurs. Under these conditions, a heat flow of 150–200 W/cm² can be obtained. If, instead of projections, the anode is covered by narrow grooves, the temperature gradient extends from the Leydenfrost point at the bottom of the grooves, to the point C temperature at the other end.

With this type of system, heat flow up to 300 W/cm² can be obtained. Practically, the anode is immersed in water, and the vapour created is captured, condensed and returned by gravity to the system water reservoir.

Hypervapotron cooling builds upon the Vapotron cooling concept to obtain even greater cooling capacity. During operation, the grooves of the thin groove Vapotron are periodically filled with vapour which escapes and is replaced with water. The idea of the Hypervapotron is to immediately recondense the vapour with a fast, turbulent flow of cold water. Outside the tube, the cooling system is then similar to a simple water cooling circuit, except that the output water temperature may reach 90° or even 100°. On the other hand, this type of cooling allows evacuation of up to 2 kW/cm² of anode surface.

Figure 7.22 Hypervapotron (Thomson-CSF)

7.3 Linear beam tubes

Linear beam tubes are the most versatile devices used for generation and amplification of energy at microwave frequencies. The usual forms are klystrons and travelling wave tubes (twts).

To produce microwave power, a high density electron beam is extracted from a cathode and accelerated by dc voltages to relatively high velocities. This accelerated beam must be controlled to have well-defined trajectories by a combination of focusing electrodes and magnetic field. The magnetic field confines the electron flow to a relatively narrow, straight channel, so that it can interact with suitable electromagnetic circuits along its trajectory and to prevent interception of the beam by these circuits.

The steady state trajectories are then modified by modulating the electron velocity through the interaction of the electrons with time varying electromagnetic fields produced by some particular circuit geometry, i.e. a cavity resonator as in a klystron, a series of cavities in coupled cavity twts, or a modified waveguide such as a helix.

The modulation of the velocity changes the successive electron trajectories as a function of the entrance phase into the modulating field. The initially uniform beam becomes non-uniform as the accelerated electrons tend to overtake the decelerated ones resulting in a time varying current density (*bunching*).

The time varying current passing through the electromagnetic fields associated with a circuit (cavity, coupled cavities, helix, etc.) transfers power to the field, the beam kinetic energy being transformed into electromagnetic energy in the circuit, which is then delivered to some transmission system.

Finally, the remaining kinetic energy of the electrons is converted into heat in a collecting electrode or *collector*. It is possible, at least in principle, to operate the collector at a depressed potential, i.e. a potential relative to the cathode lower than the potential of the main interaction region of the device. The electrons will then strike the collector electrode with lower kinetic energy. There is a saving in power at the cost of an extra power supply.

Linear beam tubes differ one from another principally in the characteristics of their interaction circuits.

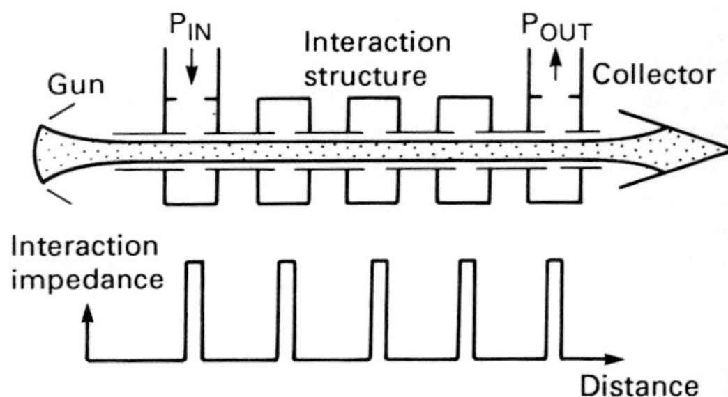

Figure 7.23 Klystron

In a *klystron* (*Figure 7.23*), strong interaction takes place in a small number of cavity gaps which initially modulate the beam and, in the output cavity, extract the energy from the very strongly bunched beam. Intermediate cavities, generally not being loaded by external coupling, will show a relatively high Q, resulting in a high coupling impedance and large power gain. On the other hand, the klystron is necessarily a fairly narrow bandwidth device.

In *travelling wave tubes* (*Figure 7.24*), the energy propagates along a slow wave structure (helix or coupled cavities) which presents to the electron beam a uniform coupling impedance, which however is much lower than the impedance of an unloaded cavity resonator. The bandwidth will be broader, but it is quite obvious that synchronism between the electron beam and the travelling wave is required. This implies constant voltage operation, generally demanding expensive well regulated high voltage power supplies. Klystrons are much more flexible and operate from simpler power supplies.

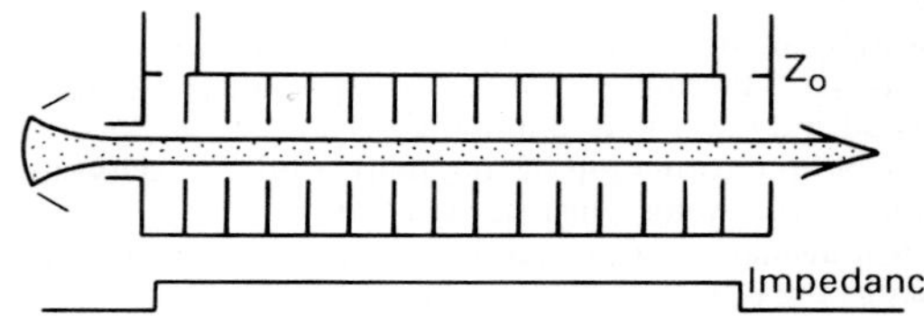

Figure 7.24 Travelling wave tube

The three major functions in linear beam tubes, *beam generation*, *interaction* and *dissipation*, are accomplished in three regions that are sufficiently well separated to allow optimization of each almost independently of the others.

7.3.1 Electron beams for linear beam tubes

Linear beam tubes operate at dc beam voltages and currents ranging from one or two kilovolts and a few tens of milliamperes for low power, low frequency travelling wave tubes to more than 300 kV and several hundred amperes for multimegawatt klystrons.

All linear beam tubes make use of electron guns operated under space charge limited cathode electrode emission and therefore obey the Child-Langmuir law: $I_0/V_0^{3/2}$ = a constant = φ perveance, where V_0 and I_0 refer to beam voltage and current.

The design of modern electron guns requires the solution of the equations of electron motion taking space charge and magnetic field into account.

7.3.1.2 Beam control

Some beam tubes such as klystrons do not require an accurate setting of the operating voltage, and the power output can be adjusted simply by changing the cathode to body voltage in a diode electron gun. In beam tubes such as twts requiring synchronism between the electron beam and the wave propagating in the structure, the beam voltage is imposed, and therefore adjustment of gain or power output requires control of the beam current. This is done by means of a control electrode which can be a modulating anode, a focus electrode, an intercepting grid, or a non-intercepting 'shadow' grid.

7.3.2 Beam focusing

As linear beam tubes must use long, thin electron beams, beam spreading due to space change forces becomes of prime importance. The universal beam spread curve in *Figure 7.26*

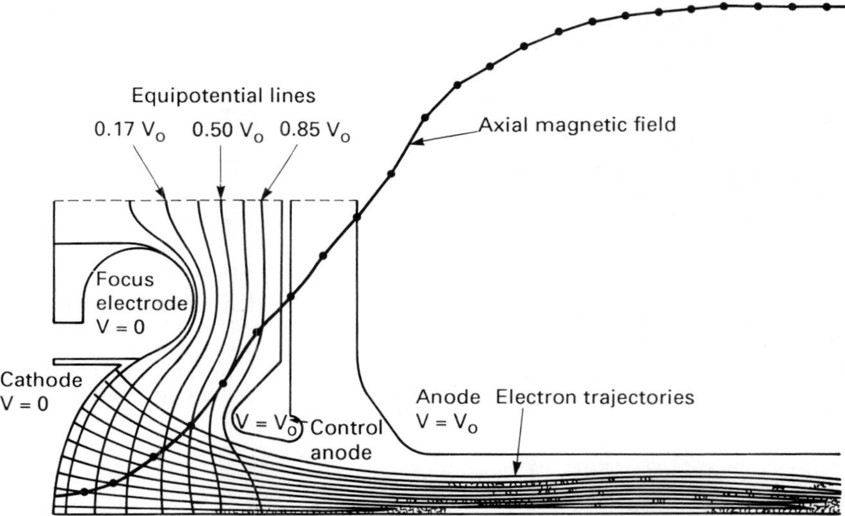

Figure 7.25 Pierce type electron gun

The *Pierce type* electron gun using a spherical cathode is by far the most frequently used (see *Figure 7.25*). Its design is relatively simple. The limiting parameters are perveance and area convergence. The great majority of existing tubes operate with beam microperveance ranging from 0.3 to 2.0, and area convergence from 10 to 70. Exceptionally, convergence of 100 at microperveance 1.0 [5] and convergence 30 with a hollow beam of microperveance 5.5 [6] have been achieved.

7.3.1.1 Cathodes

Two types of cathode are used in linear beam tubes, the oxide coated cathode and the dispenser type cathode.

The use of *oxide coated cathodes* is limited to low power devices where the cathode current density can be kept below 300 mA/cm², and below 200 mA/cm² for long life devices. They are still often used in high power pulsed devices such as klystrons even at multimegawatt level corresponding to peak cathode current density in excess of 5 A/cm², provided that the rms value is kept lower than 300 mA/cm².

Dispenser cathodes as described in section *7.1.3* offer a much better resistance to poisoning and poor vacuum than oxide cathodes. They can operate at much higher direct current density. Satisfactory operation exceeding ten years has been demonstrated on twts operating on geosynchronous satellites with B or S type cathodes working at 1 A/cm².

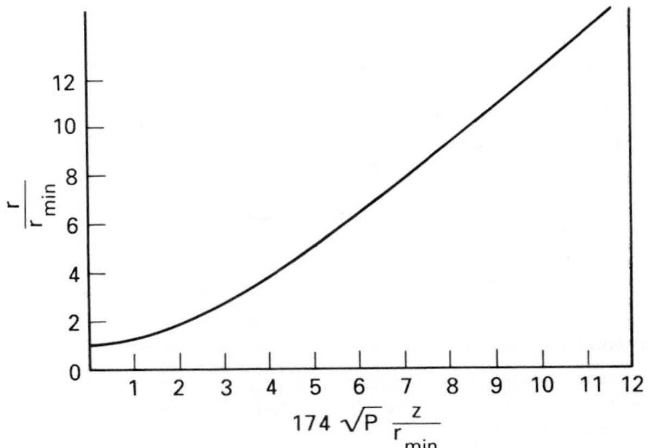

Figure 7.26 Universal beam spread curve

shows that, in a typical one microperv beam, space charge effects are already significant after a drift distance of only one minimum beam diameter, and beam diameter has doubled after only five minimum beam diameters drift. As a beam length to diameter ratio in excess of 100 is not unusual, a focusing structure is clearly necessary. The focusing structure can use a uniform axial magnetic field or periodic magnetic fields.

7.3.2.1 Focusing with uniform magnetic field

A beam immersed in an axial magnetic field is in equilibrium if a continuous balance exists between space charge, magnetic and centrifugal forces.

The magnetic focusing field is commonly obtained by use of a solenoid terminated at both ends by iron pole pieces with a centre hole for passing the electron beam. The solenoid is generally shielded with iron on its outside diameter. The coils are bulky and heavy, being several hundred kilograms for an S-band (3 GHz) multimegawatt klystron. Liquid cooling is generally required.

A number of tubes needing only a short interaction region, such as klystrons, may be focused with permanent magnets. For example, a klystron delivering 1 kW at 4.4–5.0 GHz is focused by a 20 kg permanent magnet.

7.3.2.2 Periodic permanent magnet focusing

A substantial reduction in system weight can be achieved by using periodic permanent magnet (ppm) focusing.

The electron trajectory in a magnetic field B is given by:

$$\frac{d^2r}{dz^2} + \frac{reB^2}{8mV_o} = O \ (e>o)$$

This equation shows that B appears only in its squared value, i.e. the electron trajectory is independent of the polarity of the magnetic field. An alternating field between $+B_0$ and $-B_0$ as in *Figure 7.27*(b) will focus the beam exactly as the continuous field B_0 in (a).

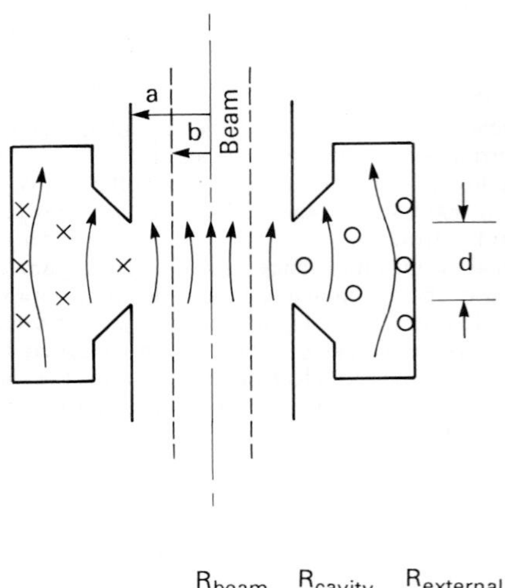

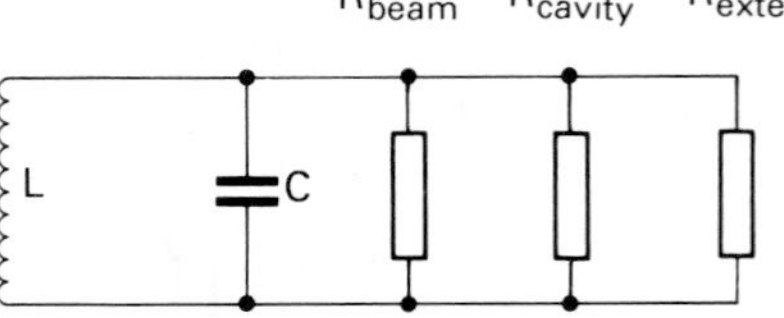

Figure 7.28 High power klystron cavity and its equivalent circuit

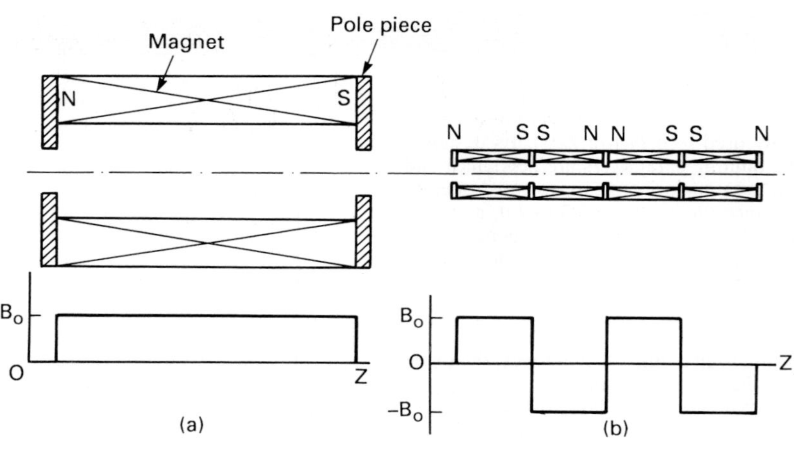

Figure 7.27 Permanent magnet focusing using (a) a continuous field, (b) an alternating field

PPM focusing is used for the great majority of helix twts. The saving in weight using ppm rather than uniform field focusing is one to two orders of magnitude.

7.3.3 Klystrons

Klystron amplifiers are extensively used as final amplifiers in microwave transmitters. Typical applications include vhf television transmitters up to 50 kW, tropospheric communication systems at L-band, S-band and C-band, with power from 1 to 20 kW, and earth to satellite links at C-band, X-band and Ku-band with power from 1 to 10 kW.

Multi-cavity klystron amplifiers are also used extensively in pulsed radar transmitters operating from uhf through X-band. Large ground based tri-dimensional radars operate at S-band with peak power output up to 30 MW, and several tens of kilowatts of mean power.

7.3.3.1 Velocity modulation

The velocity of electrons in the beam is periodically changed by the rf field in the input cavity causing bunches to form as accelerated electrons overtake decelerated ones. However, mutual repulsion forces between electrons tend to impede the rate at which fast electrons overtake slow electrons.

7.3.3.2 The cavity resonator

The cavity resonator is the basic circuit element in a klystron. It is designed to concentrate the electric field in the region of coupling with the beam (the *gap*). This gap must be made fairly short. A typical cavity has the shape shown in *Figure 7.28*.The shunt impedance of the cavity is:

$$Z = \frac{R}{1 + jQ\,[\omega^2-\omega_o^2)/(\omega_1\omega)]} \tag{7.12}$$

where ω_0 is the resonant frequency $1/\sqrt{LC}$. The most important parameter of a klystron cavity is its characteristic impedance defined by:

$$Z_o = \frac{R_{shunt}}{Q} = \frac{1}{\omega_o C}$$

The value of the capitance C is very close to the low frequency capacitance of the gap.

The interaction between beam and cavity is characterized by:

- the *beam coupling coefficient M*, which is the ratio of the rf current induced in the cavity gap to the current carried by the beam at the entrance of the gap
- the *beam loading resistance* R_b, which expresses the fact that, for a finite transit time of the electrons in the rf gap, the velocity modulation of the beam takes some energy away from the resonator.

7.3.3.3 Gain and bandwidth

The multi-cavity klystron is capable of providing extremely high gain but is generally considered as a narrow band device. It is however possible to trade gain for increased bandwidth by *stagger tuning* cavities. The resonances of cavities are distributed across the bandwidth in a manner similar to that of a distributed amplifier.

The mathematical analysis of the gain of a klystron incorporating a single intermediate cavity is relatively easy. However it becomes more and more complex as the number of cavities is increased.

The gain bandwidth product of a klystron is proportional to R/Q, so the gap capacitance C should be minimized for maximum bandwidth. However care should be taken not to decrease C by too large an increase of the gap length resulting in a poor coupling coefficient. The gain bandwidth product of a klystron is also proportional to the dc beam conductance:

$$\frac{\Delta f}{f} \propto \frac{R/Q}{R_o} = \frac{R}{Q} P_o^{1/5}k^{4/5} \tag{7.13}$$

where R_0 is the beam dc impedance, P_0 the dc beam power and k the perveance.

7.3.3.4 Power and efficiency

The small signal computation cannot predict the klystron behaviour when driven to saturation. The modulated beam can be considered as a constant current generator delivering its energy in the shunt impedance of the output cavity. There is an optimum value, for if the load resistance is too high the voltage generated across the output gap will exceed the accelerating beam voltage V_0, and electrons will be reflected resulting in a power loss.

Maximum efficiency is achieved when the maximum rf current component is obtained in the beam with minimum beam velocity spread.

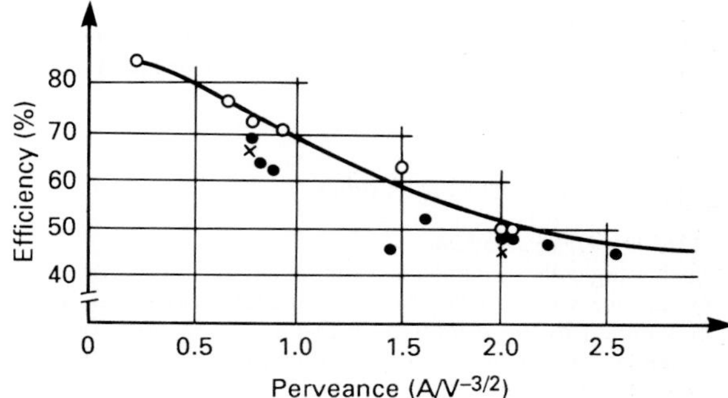

Figure 7.29 Klystron efficiency versus perveance

It has been shown that the maximum achievable efficiency is a function of beam perveance. At high perveance, or high beam current density, space charge repulsion forces become greater, thus limiting beam bunching. The curve of *Figure 7.29* shows computed and experimental efficiency versus beam perveance. As can be seen from equation (7.13), bandwidth is almost proportional to perveance, and so there must be a trade off between efficiency and bandwidth.

7.3.4 Travelling wave tubes

The travelling wave tube (twt) and the multi-cavity klystron amplifier have many features in common: electron gun, electron beam, necessity of a focusing system and collector. They

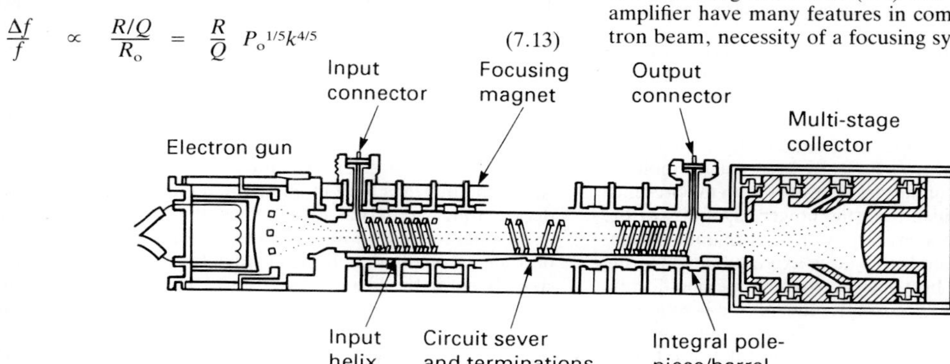

Figure 7.30 Helix travelling wave tube

differ mostly by their rf circuits, and by the mechanism for converting the kinetic energy of the beam electrons into microwave energy.

The rf circuits used in twts are spaced periodically along the tube axis. Such periodic circuits produce important reductions of the velocity of the signals transmitted. They are often referred to as *periodic slow wave structures* or *delay lines*.

The rf field inside the rf circuit must satisfy Maxwell's equations. Also it must satisfy a set of boundary conditions on the rf structure. For a periodic structure, these conditions have specific implications, described by Floquet's theorem.

Floquet's theorum states that for an infinite, lossless structure with periodic length p, propagating a wave of frequency ω in the z direction, the fields, electric or magnetic, at point x, y, z are the same as at point $x, y, z + np$ to within a phase shift, i.e.:

$$\vec{E}(x,y,z+np) = \vec{E}(x,y,z)e^{-jn\phi}$$

n being any positive or negative integer. ϕ is the phase shift between adjacent cells, which depends on the frequency of the signal.

7.3.4.1 *Helix travelling wave tubes*

7.3.4.1.1 *Helix delay line*

Technologically, the helix is a simple circuit; it is made from a metal wire, wound helically around the tube's axis. This helix is maintained in the cylindrical metal envelope of the tube by means of three dielectric support rods (see *Figure 7.31*).

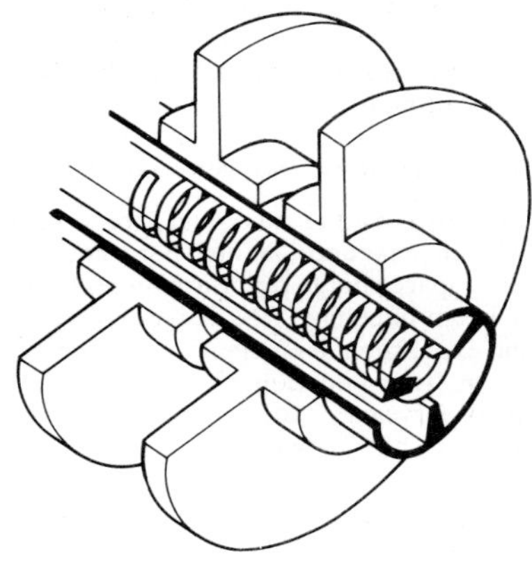

Figure 7.31 Helix with support rods, vacuum envelope and pole pieces

The electrical properties of the helix are shown on the Brillouin diagram (*Figure 7.32*). It can be seen that the fundamental component has a nearly constant velocity over a considerable bandwidth.

This leads to twts having operating bandwidths from one to nearly three octaves, and to a basically low cost technology.

The only disadvantage of the circuit results from the presence of dielectric supports, which limit the power capability of the structure. Therefore, the average power performance obtainable from such tubes depends to a large extent on engineering innovations, allowing the best designs to overcome significantly the basic thermal problem, and to offer the many advantages of the helix at the medium power levels. The maximum average power level is obtained with the very efficient brazed helix technology.

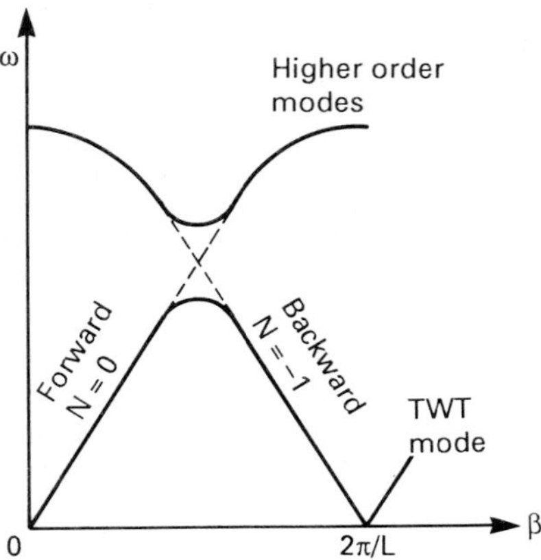

Figure 7.32 Brillouin diagram for the helix

7.3.4.1.2 *Travelling wave interaction*

In the travelling wave tubes, the interacting field (fundamental component for the helix twt) travels along the axis with a velocity slightly higher than that of the injected electron beam. Accelerating and decelerating forces are exerted by the field on the electrons, depending on their position relative to the wave.

Figure 7.33 shows how electrons seeing a field in the opposite direction to their speed are accelerated by the field, while those electrons seeing a field in the same direction as their speed are decelerated.

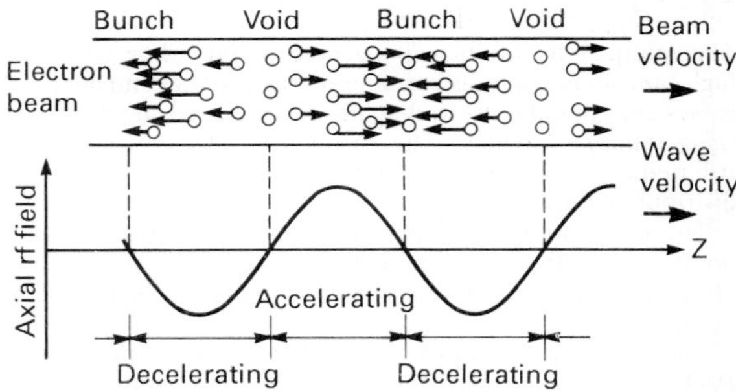

Figure 7.33 Electron bunching

The consequence is that accelerated electrons advance on the average beam, while decelerated electrons are slightly delayed. This velocity modulation gives rise to space charge density modulation; accelerated electrons form bunches with decelerated ones in front of them. At the same time, decelerated electrons increase the distance which separates them from accelerated ones in front of them, creating voids of charge. Bunches and voids travel at the beam average velocity along the axis.

Because there is an excess velocity of the beam over the field wave, the bunches tend to enter the regions where the field is decelerating, and voids those where the field is accelerating.

In other words, more electrons become decelerated than become accelerated, and the average kinetic energy decreases steadily along the beam. The energy gained from the beam is transferred to the source of the forces acting on the electrons, i.e. to the travelling field.

As the increasingly bunched beam proceeds along the tube and sees an increasingly intense field, as a result of the continuous energy transfer from beam to wave, the average kinetic energy of the beam decreases, reducing more and more the excess speed of the beam over the wave.

When that excess speed becomes zero, the process of energy transfer reverses, and the power of the wave reaches a maximum, or *saturated*, level.

If the interaction process is continued farther, then the rf power diminishes, the energy transfer being now from the wave to the beam, and the tube is in a so-called *over-saturated* state.

The geometrical point at which saturation occurs is normally the location of the rf output connector.

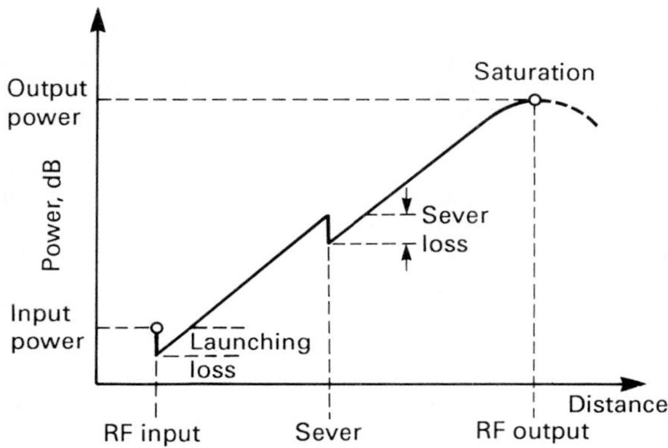

Figure 7.34 Variation of power with distance along the helix

The variation of the rf power along the axis is shown in *Figure 7.34*. It is seen that the gain per unit length is constant over most of the length. Over this distance, the tube operates as a linear amplifier, and interaction generates low distortion. It is only in the last portion of the delay line, near the rf output, that the gain per unit length decreases, becoming zero at the saturation point. This indicates a strong non-linearity of interaction, giving rise to distortion.

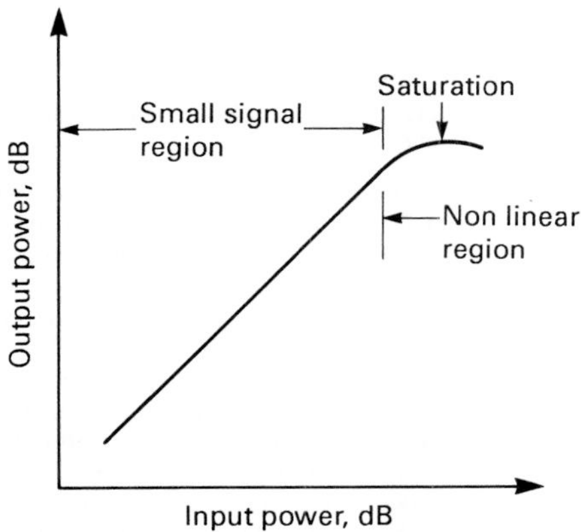

Figure 7.35 Output power versus input power in a helix travelling wave tube

Figure 7.35 shows the variation of rf output, when the input rf level is varied from very small values up to the nominal value producing saturation at the output end. The curve shows a pure linear distortion-free amplification, up to a level approximately 10 dB below the nominal. In that region, the tube is said to operate in *small signal*, or *low distortion*, mode. At higher levels, the non-linearity appears clearly when the curve departs from the straight slope. The tube is operating near saturation, with distortion, but in its maximum power and maximum efficiency mode.

7.3.4.1.3 Small signal gain

The small signal gain is given by:

$$G = A + 47.8\,CN$$

where G is the small signal gain in dB, A is a loss factor of 10–16 dB approximately, C is the gain parameter, and N is the number of electronic wavelengths along the beam.

C is defined by:

$$C = \left(K_o \frac{I_o}{4V_o} \right)^{1/3}$$

where I_0 is the beam current, V_0 is the beam voltage, and K_0 is the coupling impedance.

7.3.4.1.4 Stability

Like any amplifier, the helix twt is exposed to self-excited oscillations if the gain is too high and if a feedback mechanism exists.

The first basic oscillation mechanism is due to the successive reflections of the amplified energy on the output mismatch, and, after inverse travel through the entire tube, on the input mismatch, thus closing the instability loop. To prevent this instability, the helix is severed to form at least two physically independent sections, each terminated at one end on an rf connector and at the other end on a very good internal rf load made of carbon deposited on the dielectric helix supports.

The second feedback mechanism is due to the n = -1 space harmonic of the helix, which feeds the energy from output to input while interacting with the beam. This *backward wave interaction* becomes especially harmful for beam voltages greater than 7 kV. Various means are used by twt manufacturers to discourage this oscillation. The best designs allow modern tubes to be operated at a beam voltage higher than 15 kV and thus to reach respectable rf power levels.

7.3.4.1.5 Power output and efficiency

The electronic efficiency is related to the gain parameter by:

$$\eta_e = 2C = 2 \left(K_o \frac{I_o}{4V_o} \right)^{1/3}$$

On practical tubes, the electronic efficiency can reach values up to approximately 20 per cent. The overall efficiency can reach values well above this, by the use of *depressed collectors*, i.e. collectors whose voltage is negative with respect to the rf circuit voltage.

This slows down the electron beam entering the collectors, and allows the recovery of a large proportion of the kinetic energy remaining in the beam after interaction. Overall efficiencies up to 40 per cent and exceptionally up to 55 per cent are thus possible.

7.3.4.1.6 Applications

Helix travelling wave tubes are used in many application areas:

- telecommunications, including troposcatter, surface-to-satellite and satellite-to-earth transmitters
- electronic countermeasures (jamming)
- radar
- laboratory amplifiers

The telecommunication applications take advantage of the excellent fine grain characteristics (i.e. amplification characteristics in which noise and similar unwanted perturbations are of high-frequency and produce only small-area fine-grained disturbances to the signal) of the helix twt in narrow-band use (due to its natural broad band), of its low cost and long life demonstrated on many existing systems.

The helix twt, due to its extreme broad band, is commonly used as a laboratory amplifier covering all standard bandwidths.

7.3.4.2 Efficiency improvements

Efficiency optimization is of prime interest in systems such as airborne equipment and particularly space systems where power budget is a major parameter.

A small improvement in efficiency of a twt for a space application produces substantial benefits. Fewer solar cells are required as well as fewer batteries and smaller power supplies. The reduced power dissipation makes the thermal balance problem easier. DC input power represents the basic parameter in almost all communication satellites from which all other design considerations are determined. Consequently, considerable efforts have been made to increase practical efficiency of space twts. This has been done in two directions, one by increasing interaction efficiency and the other by introducing multi-stage depressed collectors.

7.3.4.2.1 Interaction efficiency improvement

The gain of a travelling wave tube is sensitive to the beam velocity. In the design of a tube, the beam voltage and the circuit (helix or coupled cavities) period are matched to give a flat gain characteristic against frequency. This match, however, will probably not result in the optimum power transfer from the beam to the circuit. In fact, if we consider the interaction between the electron beam and a uniform periodic structure, when energy is transferred from the beam to the electromagnetic wave in the structure, the loss of beam kinetic energy in favour of the rf field results in a reduced beam velocity and loss of synchronism in the tube region where the interaction is the greatest.

By reducing beam voltage, i.e. by resynchronizing beam and wave in the large signal region of the rf structure, efficiency is improved, but maximum efficiency and maximum gain are not obtained for the same beam voltage. This is an undesirable situation for the system performance, and the solution is to gradually reduce the circuit wave velocity in the large signal region by winding helices with variable pitch or progressively reducing the distance between successive gaps in coupled cavity twts.

Optimization of the tapered structure has been made using large signal computed codes. The results can be quite spectacular. For example, a helix twt operating at 12 GHz at a power output of 20 W has a beam efficiency of 13 per cent with a constant pitch helix. The same tube using an optimized tapered helix will exhibit 23 per cent beam efficiency without degradation of other parameters such as linearity or gain flatness.

7.3.4.2.2 Depressed collectors

One of the main advantages of linear beam tubes, compared for instance with crossed field devices, is the separation of the tube's regions, where the beam is formed, where rf interaction takes place, and where the spent beam is collected. It is then generally possible to implement some degree of sophistication in the design of the collector region when this is desirable.

In the majority of beam tubes, only a relatively small fraction of the beam energy is converted into rf power. Except for a few very high efficiency klystrons, more than half of the supplied energy remains in the beam at the exit of the interaction region. If all electrons were uniformly decelerated in the rf interaction, it could be possible to collect all of them on a collecting electrode set at a voltage just corresponding to the fraction of energy converted to rf. For instance, an ideal twt showing 20 per cent beam efficiency could operate at a cathode-to-collector voltage set to 20 per cent of the beam voltage. The net result would be a perfect 100 per cent efficiency.

Unfortunately the interaction process in linear beam tubes does not result in a uniform electron deceleration. The large signal computer codes give the distribution of electron velocities at the output of the rf interaction region. *Figure 7.36* shows the beam energy spread computed for two tube types: a high efficiency high power klystron operating at 70 per cent interaction efficiency at uhf and a helix twt operating at 23 per cent beam efficiency at X-band. It can be seen that a fraction of the beam out of phase with the rf has gained rather than lost energy in the interaction process. The energy spread is very different for a high efficiency klystron, where some electrons have lost all their energy. Any attempt to collect the beam at reduced voltage will result in a reflection of electrons towards the rf structure. If the collector were set to 80 per cent of the beam voltage, $V/V_0 = 0.2$, approximately 30 per cent of the beam would be reflected, which would cause very serious problems.

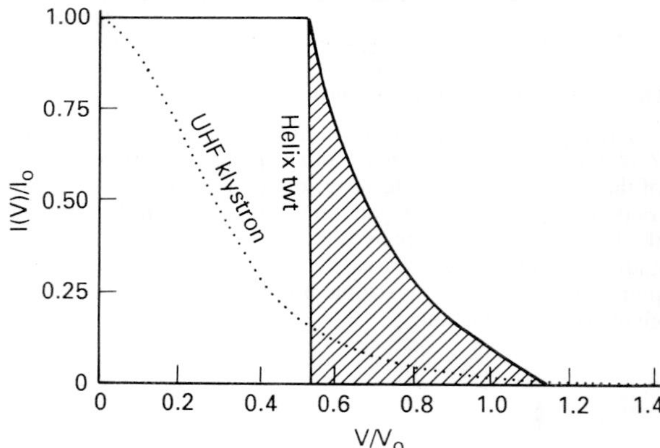

Figure 7.36 Beam energy spread for a high efficiency klystron and a helix travelling wave tube

In the travelling wave tube, no electron has lost more than 43 per cent of its energy. It would then be theoretically possible to collect the total of the beam at 43 per cent of the beam voltage, resulting in an improvement of the efficiency.

In a practical case, the electrons cannot 'land' on the collector at strictly zero velocity as space charge effects would cause their deflection. Some voltage margin must be allowed, but from *Figure 7.36* we can see that if we can collect all of the beam at 45 per cent of its accelerating voltage, we could still collect one third of that beam at one half of that voltage, i.e. 22.5 per cent.

Many twts have been built making use of a two-stage depressed collector. When the tube is operated at full output power, approximately two thirds of the beam current are collected on the first collector, at the highest voltage with respect to the cathode, and one third on the second collector. In the absence of rf power, practically all the beam is collected on the low voltage electrode. The power drained by the tube is therefore reduced by a half at no rf drive and one sixth at full power output.

More collector stages can be implemented at the cost of a more complex power supply. *Figure 7.37* shows a cross-section of a three stage depressed collector as used in a medium power twt (50–100 W) at Ku-band (12 GHz) designed for direct or semi-direct high definition television broadcasts from a geo-synchronous satellite. This tube has an interaction efficiency of 23 per cent and a very remarkable nearly 60 per cent efficiency when operated with the three stage depressed collector.

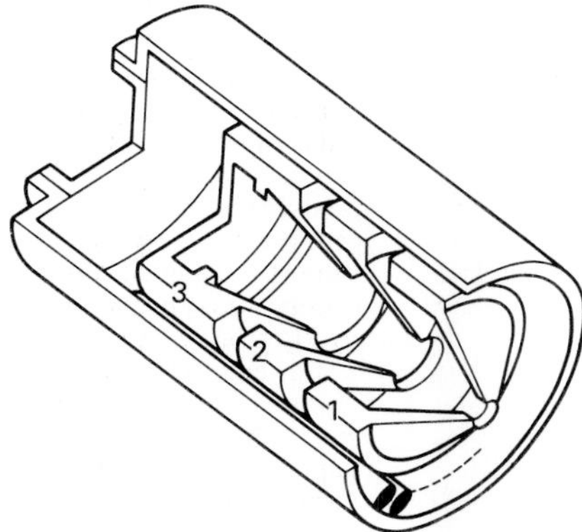

Figure 7.37 Three stage depressed collector

All electrons entering the collector structure have enough energy to enter the space between the first and second collector. Some electrons have lost too much kinetic energy to overcome the retarding field between these collectors and are reflected and collected on the back of the first electrode.

The same process takes place between the second and third electrodes. The electrons are sorted between the three electrodes according to their velocities. In this design, collecting electrodes are press-fitted against ceramic rods inside the vacuum envelope. The ceramic rods insure both electrical insulation between electrodes and heat transfer to the vacuum envelope which in turn must be cooled by convection, conduction or radiation.

Acknowledgements

This text was prepared with the aid of several engineers at Thomson Tubes Electroniques: B. Epstein, A. Schroff, P. Gerlach, R. Metivier, and their assistance is gratefully acknowledged.

References

1 GILMOUR, A S, Jr, *Microwave tubes*, Artech House (1986)
2 RAMO, S, WHINNERY, J R and van DUZER, T, *Fields and waves in communication electronics*, John Wiley & Sons (1965)
3 SLATER, J C, *Microwave electronics*, Van Nostrand (1950)
4 STRATTON, J A, *Electromagnetic theory*, McGraw-Hill (1941)
5 STRAPANS, A, McCUNE, E W and RUETZ, J A, 'High-power linear-beam tubes', *Proc IEEE*, **61**, March 1973, pp 299–330
6 McCUNE, E, 'A 250 kW cw X-band klystron', *IEEE Int Electron Devices Meet*, Washington (1967)

Acknowledgements

References

Part 3 Fundamentals of Colour Television

P G J Barten
Barten Consultancy

8 Electron Optics in Cathode Ray Tubes

The heart of a cathode ray tube is the electron gun in which one or more electron beams are generated. The beams are intensity modulated by the video signal and deflected by the magnetic field of a deflection unit at the outside of the tube in order to scan the display screen. This screen is covered by fluorescent material (*phosphor*) which converts a part of the energy of the impinging electrons to light of the desired colour. In this way, a picture is generated corresponding to the information contained in the video signal. To preserve the emission of electrons and their passage to the screen, the tube is evacuated to a pressure of 10^{-6} N/m^2.

The sharpness of the picture depends to a large extent on the spot size of the electron beam. Therefore the electron optical design of the electron gun is very important for the resolution quality of the tube.

8.1 Beam forming in an electron gun

In an electron gun, the electrons are generated by a flat thermionic cathode. After leaving the cathode, the electrons are slightly accelerated and focused in the *crossover* at a short distance from the cathode.

The beam diverging from the crossover passes first through an electrostatic prefocus lens, which narrows the divergence angle of the beam, and then through the main lens, which focuses the beam in a spot on the screen. This spot is in principle an image of the crossover (*Figure 8.1*).

During their passage through the prefocusing lens and the main lens, the electrons are accelerated to a velocity corresponding to a voltage in the range of 18–30 kV. When the electrons have left the gun, the voltage, and therefore also the electron velocity, is constant. In the first part of this equipotential space, the magnetic deflection is applied.

The final spot size is determined by the spread of the thermal velocities of the electrons, the spherical aberration of the electron optical system, the Coulomb repulsion between the electrons, and the defocusing caused by the deflection field. These effects are treated in more detail in the following sections.

In a colour picture tube, three beams are generally used, their intensity being modulated by the red, green and blue video signals. The beams can be arranged in a triangle (*delta configuration*) (*Figure 8.2*) or in-line in a horizontal plane.

Nowadays, only the latter configuration is used. The beams are usually generated in an integrated gun, which has separate

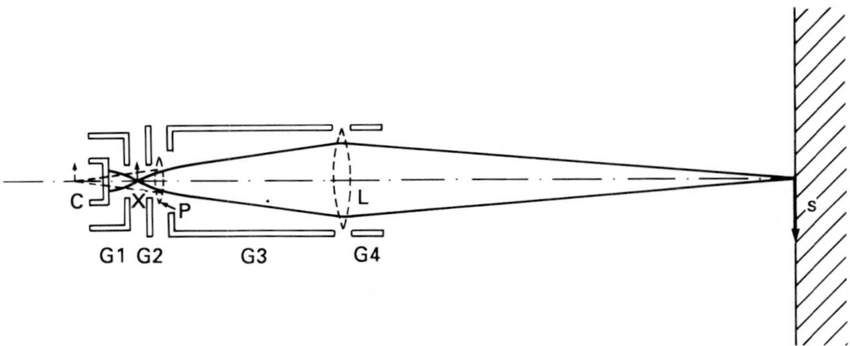

Figure 8.1 Principle of spot forming in a crt. After the electrons have left the cathode, C, the electron beam forms a crossover, X, at a short distance from the cathode. This crossover is imaged on the screen, S, by a main lens, L, after the opening angle of the beam has been narrowed by a prefocus lens, P,. The lens fields are formed by the configuration of the electrodes, G1, G2, G3 and G4

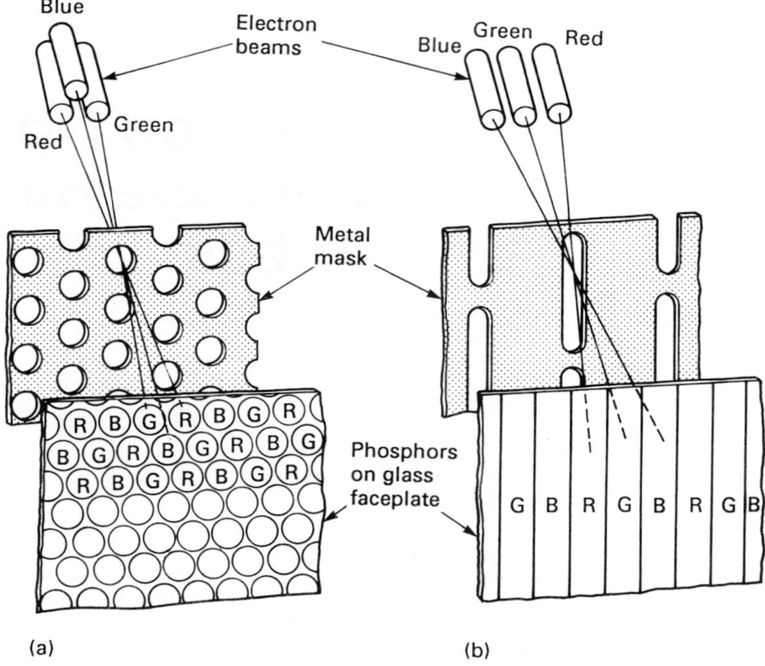

Figure 8.2 Delta configuration (*left*) and in-line configuration (*right*) of the electron beams in a colour crt

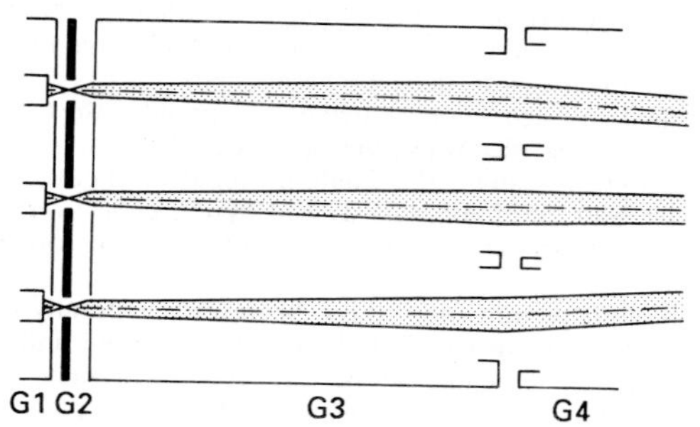

Figure 8.3 Integrated gun for a colour crt. The three beams are generated by separate cathodes, but guided and focused by common electrodes with separate apertures for each of the beams

cathodes, but common electrodes with separate apertures or the beam forming and focusing for each of the three beams (*Figure 8.3*).

Only in the gun of the Trinitron tube[1] is a common main lens used. In this gun, the three beams cross the tube axis in the centre of the lens. See *Figure 8.4*.

The electron optical beam forming in an integrated colour gun is not essentially different from that in a single beam gun, as used for monochrome tubes and projection tubes, and is not therefore separately treated here. The principles of the colour selection are considered in section *9*.

8.2 Electron emission of the cathode

The electron beam in a cathode ray tube usually originates from an indirectly heated cathode consisting of a nickel cap covered with a thin layer of a mixture of barium oxide, strontium oxide and calcium oxide, known as an *oxide cathode*. An internal

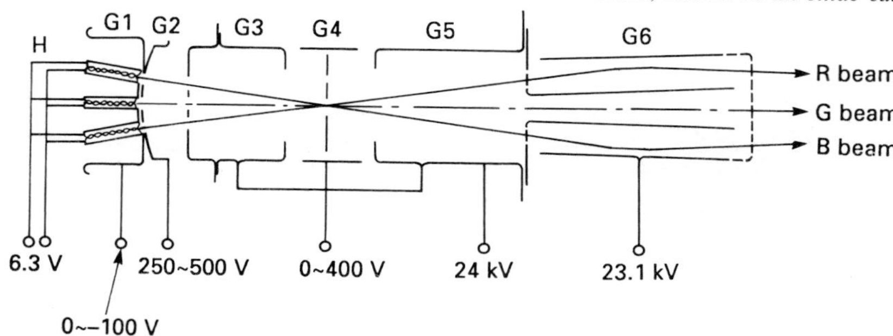

Figure 8.4 Cross-section of a Trinitron gun. This is an integrated gun with a common main lens for the three beams. The beams cross the axis in the centre of this lens

filament heats the cap to a temperature of about 810°C. At this temperature, sufficient free electrons can leave the surface, because the work function of the oxidic surface has been decreased substantially by the adsorption of free barium supplied by a reaction between barium oxide and reducing activators in the nickel cap (*Figure 8.5*).

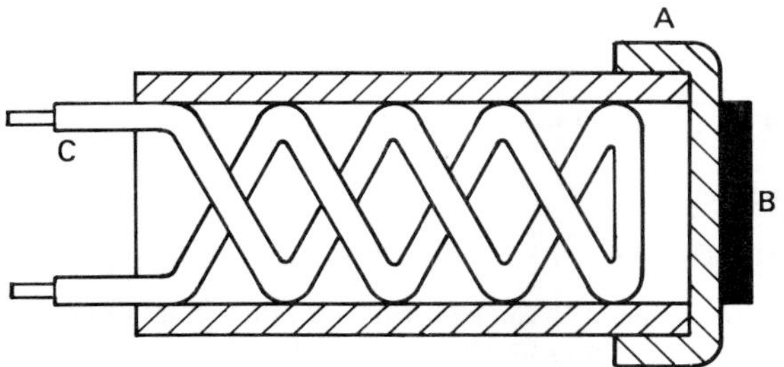

Figure 8.5 Oxide cathode consisting of a nickel cap, A, covered with a thin layer of barium, strontium and calcium oxide, B, and heated by an internal filament, C

If the emitted electrons are not removed from the neighbourhood of the cathode by a strong electric field, they will form a cloud of negative space charge which will repel further emitted electrons back to the cathode. With an accelerating electric field in front of the cathode, electrons can be drawn from this space charge cloud. The intensity of the beam can be modulated by varying the strength of this field. This type of emission is called *space charge limited emission*.

At high field strengths the emission saturates. For oxide cathodes, the current density at saturation is about 20 A/cm^2. However, the normally used current density is much lower. In view of the lifetime of the cathode, the allowable long term average current density of oxide cathodes is limited to about 1.4 A/cm^2 in the centre of the cathode. At this load the lifetime of the cathode is about 10 000–15 000 h before the current decreases to half its original value.

In cases where higher current densities are required, *impregnated cathodes* are used. In the impregnated cathode, a tungsten sponge impregnated with barium-calcium aluminate is used. The operating temperature of this cathode is about 1050°C, or about 240°C higher than that of an oxide cathode. At this temperature the barium-calcium aluminate reacts slowly with the tungsten, releasing free barium. Impregnated cathodes are more expensive than oxide cathodes, but have a longer life and can be used up to about 5 A/cm^2 at long term average conditions.

In a flat diode where an anode is placed directly opposite to the cathode, the space charge limited current is given by Child's law:

$$j = A\,\frac{V^{3/2}}{d^2} \tag{8.1}$$

where j = current density,
V = voltage difference over the diode,
d = distance between cathode and anode,

and A is a physical constant given by:

$$A = \frac{4}{9}\,\epsilon_0\sqrt{2e/m} = 2.334\cdot10^{-6}\ \text{ampere/volt}^{3/2} \tag{8.2}$$

where ϵ_0 = dielectric constant of vacuum,
e = electron charge, and
m = electron mass

This relation can be derived, assuming that the initial velocities of the electrons perpendicular to the cathode are negligible. If the quotient V/d is replaced by the field strength, E, on the cathode, equation (8.1) can also be used for the calculation of the emission current in an electron gun. In this way one obtains:

$$j = A\,\frac{E^{3/2}}{d^{1/2}} \tag{8.3}$$

where E is the value of the absolute electrical field strength at the cathode surface in the absence of space charge. The quantity d is the *equivalent diode distance*.

In a rotationally symmetrical gun, the field strength,E, may be assumed to decrease parabolically with the distance r from the axis:

$$E(r) = E_0\,(1-r^2/r_0^2) \tag{8.4}$$

where E_0 = field strength in the centre of the cathode, and
r_0 = radius of the emitting area

The total beam current, i, can now be calculated with the aid of the relation:

$$i = \int_0^{r_0} j(r)2\pi r dr \tag{8.5}$$

This yields:

$$i = \frac{2}{5}\,\pi A\,\frac{E_0^{3/2}r_0^2}{d^{1/2}} \tag{8.6}$$

where, according to Ploke,[2] the equivalent diode distance has been assumed to be independent of r. With the aid of this expression, the beam current can be calculated if E_0, r_0 and d are known.

8.3 Drive characteristic

The electrode configuration in the neighbourhood of the cathode basically consists of two flat plates with a circular aperture (*Figure 8.6*).

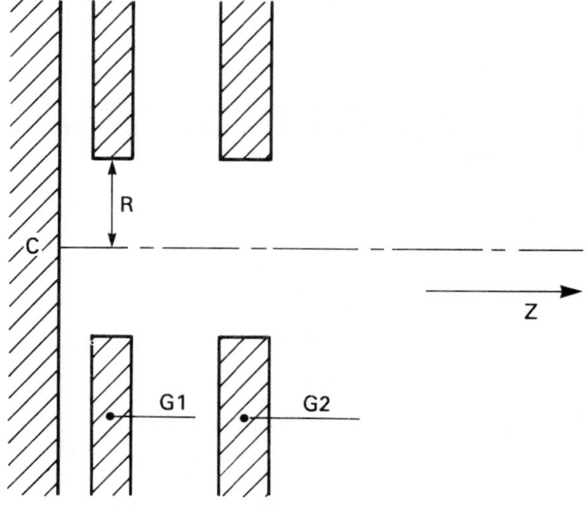

Figure 8.6 Electrode configuration in the triode part of an electron gun: C is cathode, G1 is first grid and G2 is second grid

For historical reasons, dating back to the time of radio valves, these electrodes and also the following electrodes are called *grids*, though there is no resemblance to a mechanical grid structure. The unit consisting of cathode and first two grids is called the *triode*, also for historical reasons. The second grid is kept at a voltage which is a few hundred volts positive with respect to the cathode. The electric field due to this voltage draws electrons from the space charge cloud in front of the cathode. This action is partly counteracted by the first grid which is at a negative potential with respect to the cathode. The beam current can be modulated by varying the voltage on the first grid, a process known as *grid drive*.

However, nowadays the beam current is generally modulated by varying the voltage on the cathode with first grid earthed. This is called *cathode drive*. In this case a smaller voltage variation is needed to vary the beam current. The other grids are kept at fixed voltages with respect to the (earthed) first grid. For the calculation of the current/voltage characteristic of the gun, equation (8.6) can be used, after expressing the quantities E_0, r_0 and d in known gun parameters.

Let us first take E_0, the field strength in the centre of the cathode. For cathode drive, E_0 can be described by the following linear relation:

$$E_0 = a(-V_c + D_2V_{g2} + D_3V_{g3}) \tag{8.7}$$

where a, D_2 and D_3 are constants depending on the geometry of the gun, and V_c, V_{g2} and V_{g3} are the voltages on the cathode, the second grid and the third grid respectively. The effect of further electrodes may be neglected. The constants D_2 and D_3 are the *Durchgriff* or penetration factors of the second and third grid respectively.

If the voltage on the cathode is increased, the value of E_0 and also the beam current will decrease. According to equation (8.6), the beam current is zero at $E_0 = 0$. The value of V_c for this situation is the *cut-off voltage* and indicated by V_{co}. Substituting $E_0 = 0$ and $V_c = V_{co}$ in equation (8.7) gives:

$$V_{co} = D_2V_{g2} + D_3V_{g3} \tag{8.8}$$

The cut-off voltage is usually chosen in the range 100–200 V. According to equation (8.8) it can be adjusted to the desired value by the choice of the voltage on the second grid. The effect of the voltage on the third grid is only a few volts. Equation (8.8) can now be used to simplify equation (8.7). This yields:

$$E_0 = a(V_{co}\text{-}V_c) \tag{8.9}$$

The quantity a in this expression is mainly determined by the geometry in the immediate neighbourhood of the cathode. A good approximation formula for a is:

$$a = \frac{w}{R}\,[1 + 0.627(s_{01}/R)^2 - 0.115(s_{01}/R)^4 + 0.030(s_{01}/R)^6]^{-1} \tag{8.10}$$

where w = 1.326...(electron optical constant),
R = radius of grid one aperture, and
s_{01} = distance between cathode and grid one

As s_{01}/R is usually about 0.2, the effect of the second and higher terms in this formula is only small.

The next quantity to be considered is the radius of the emitting area r_0. At this radius, the field strength in the absence of space charge is zero. From the potential field in the neighbourhood of the cathode, the following approximation formula can be derived:

$$r_0 = R\sqrt{\frac{V_{co}-V_c}{V_{co}+bV_c}} \tag{8.11}$$

where:

$$b = 0.94[1 + 2.07(s_{01}/R)^2 + 1.23(s_{01}/R)]^{-1} \tag{8.12}$$

The value of the constant b varies only slightly for different gun designs. For $s_{01}/R = 0.2$, b = 0.87.

The third quantity to be determined is the equivalent diode distance, d. For this quantity, the following empirical relation can be used:

$$d = \frac{a^2R^2}{w}\,(1-D)r_0 \tag{8.13}$$

where:

$$D = D_2 + D_3 \tag{8.14}$$

The value of D can be calculated with the aid of equation (8.8) by measuring the change of cut-off voltage at a variation of V_{g2} and V_{g3}. D can also be determined from the gun geometry with the aid of the following approximation formula:

$$1/D = 1 + \frac{W}{2}\,\frac{s_{12}}{R}\,\frac{\exp\{2w(s_{01}+t_1)/R\}}{1+(s_{01}/R)^2} \tag{8.15}$$

where t_1 = thickness of grid one, and
s_{12} = distance between grid one and grid two

This formula holds for equal aperture diameters in grid one and grid two.

Substituting the relations (8.9), (8.11) and (8.13) for E_0, r_0 and d respectively in equation (8.6) yields the following formula for the current/voltage characteristic:

$$i = K\,\frac{(V_{co}-V_c)^{9/4}}{(V_{co}+bVc)^{3/4}} \tag{8.16}$$

where:

$$K = \frac{2}{5}\pi A\sqrt{\frac{waR}{1-D}} \tag{8.17}$$

The maximum current to be used is the current at which $V_c = V_{g1} = 0$. At negative values of V_c, a current would flow to the first grid. For good operation of the tube, this situation must be avoided. Substitution of $V_c = 0$ in equation (8.16) gives:

$$i_{max} = KV_{co}^{3/2} \tag{8.18}$$

The proportionality constant, K, given by equation (8.17) is the *perveance* of the gun. Its numerical value can be determined by substituting equations (8.2) and (8.10) in equation (8.17). This yields:

$$K = \frac{3.889\ \mu A/V^{3/2}}{\sqrt{(1-D)\,\{1 + 0.627(s_{01}/R)^2 - 0.115(s_{01}/R)^4 + 0.030(s_{01}/R)^6\}}} \tag{8.19}$$

From this formula, it follows that K does not vary much for usual gun designs. It generally amounts to about 4 or 4.5 $\mu A/V^{3/2}$. This means that i_{max} is about 4 or 4.5 mA at a cut-off voltage of 100 V.

For the description of the current/voltage characteristic, the drive voltage V_d is often used:

$$V_d = V_{co} - V_c \tag{8.20}$$

Substitution in equation (8.15) gives:

$$i = \frac{KV_d^{9/4}}{\{(1+b)V_{co}-V_d\}^{3/4}} \tag{8.21}$$

The current/voltage characteristic is sometimes also expressed in the form of an exponential relation, where the exponent is called *gamma:*

$$i = \text{const. } V_d{}^{\gamma} \qquad (8.22)$$

This relation can strictly be valid only in a part of the characteristic. With the aid of equation (8.21) one finds $\gamma = 2.25$ for low currents and $\gamma = 2.25 + 0.75b \approx 2.9$ for high currents. For the total characteristic, an average gamma value of 2.5 can be used. To take equation (8.22) into account, the drive voltage of a crt is usually gamma corrected in the reverse direction.

The cut-off voltage V_{co} used in the foregoing equations is the electron optical cut-off voltage defined by the condition $E_0 = 0$ (the *Laplace cut-off voltage*). For this condition the beam current is zero, if the effects of initial velocities of the electrons are neglected. The Laplace cut-off voltage corresponds very well with the raster cut-off voltage, which is the voltage at which an evenly lit raster starts to become absolutely dark. Measurement of the raster cut-off voltage, however, is usually not very accurate. Therefore, the cut-off voltage is often determined by measuring the voltage at which the focused spot of a non-deflected beam disappears. To avoid the effect of ambient light, the measurement is carried out in the dark. The voltage found in this way is called the *spot cut-off voltage*, V_{cos}.

However, the spot cut-off voltage appears to be about 6–12 V larger than the Laplace cut-off voltage. This is because, under the extreme conditions of this measurement, the initial velocities of the electrons perpendicular to the cathode surface may not be neglected. Light of a few electrons that are able to leave the cathode against a repelling field is still visible. The difference from the Laplace cut-off voltage can be described by the following empirical relation:

$$\Delta V_{co} = 1.7\, V_{cos}^{1/3} \qquad (8.23)$$

This relation can be used to calculate V_{c0} from the measured spot cut-off voltage V_{cos}:

$$V_{c0} = V_{cos} - \Delta V_{cos} \qquad (8.24)$$

In practice, the exact value of the constant in equation (8.23) will depend on the ‘darkness’ of the measuring conditions.

8.4 Cathode load

The lifetime of a crt is generally determined by the life of the cathode. It is usually expressed as the number of hours before the emission current drops to 50 per cent of its original value. Apart from the vacuum in the tube and the cathode temperature, the lifetime of a cathode depends on the current density of the emission current (the *cathode load*). The closer the cathode load is to the saturation current density of the cathode, the shorter the life will be. The cathode load can be calculated with the aid of the formulas given in the two foregoing sections.

Inserting equation (8.4) in equation (8.3) gives:

$$j(r) = \frac{AE_0^{3/2}}{d^{1/2}} (1 - r^2/r_0^2)^{3/2} \qquad (8.25)$$

The current density has a maximum value in the centre of the cathode. This amounts to:

$$j_o = \frac{AE_0^{3/2}}{d^{1/2}} \qquad (8.26)$$

If this formula is compared with equation (8.6) one obtains:

$$j_o = \frac{5}{2} \frac{i}{\pi r_0^2} \qquad (8.27)$$

So, under the assumptions made in section *8.2*, the cathode load at the centre is 2.5 times larger than the average cathode load. Inserting equation (8.11) into this formula and using equations (8.16) and (8.20) yields:

$$j_o = \frac{5}{2} \frac{K^{4/3}}{\pi R^2} \frac{V_d^2}{i^{1/3}} \qquad (8.28)$$

This equation can be used for the calculation of the cathode load at a given beam current, if the drive voltage is given. The variation of the cathode load with beam current can be found by eliminating the drive voltage with the aid of equation (8.22):

$$j_o = \text{const. } i^{(2/\gamma - 1/3)} \qquad (8.29)$$

If the values of γ given in section *8.3* are used, one finds that the cathode load is nearly proportional to the square root of the beam current.

As we shall see, a small radius of the emitting cathode area is favourable for obtaining a small spot rise on the screen. For a given drive voltage condition, the value of r_0 can be adapted by the choice of the radius of the aperture in grid one, as can be seen from equation (8.11). To keep the emitting area small, the cathode load has to be chosen as close as possible to the allowable limit. As was mentioned in section *8.2* for oxide cathodes, this limit is about 1.4 A/cm², and for impregnated cathodes about 5 A/cm² at long term average beam current.

8.5 Electron optical image forming

Before going further into the electron optics of a crt gun, some general properties of electron optical image forming must first be treated.

For the velocity v of an electron, the following equation holds:

$$\frac{1}{2} mv^2 = \frac{1}{2} mv_0^2 + e\phi \qquad (8.30)$$

where v_0 = electron velocity on leaving the cathode, and
ϕ = potential in the electric field with respect to the cathode

So the velocity of the electron anywhere in the system is given by:

$$v = \sqrt{\frac{2e(\phi + \epsilon)}{m}} \qquad (8.31)$$

where:

$$\epsilon = \tfrac{1}{2} mv_0^2/e \qquad (8.32)$$

If the electron travels from a domain with a potential ϕ_1 to a domain with a potential ϕ_2, separated from each other by an electrically conductive double gauze, the electron will change its direction of movement in such a way that the velocity component parallel to the separation plane remains the same (*Figure 8.7*).

This means:

$$\sqrt{\phi_1 + \epsilon}\, \sin\alpha_1 = \sqrt{\phi_2 + \epsilon}\, \sin\alpha_2 \qquad (8.33)$$

where α_1 and α_2 are the angles of the electron trajectory with

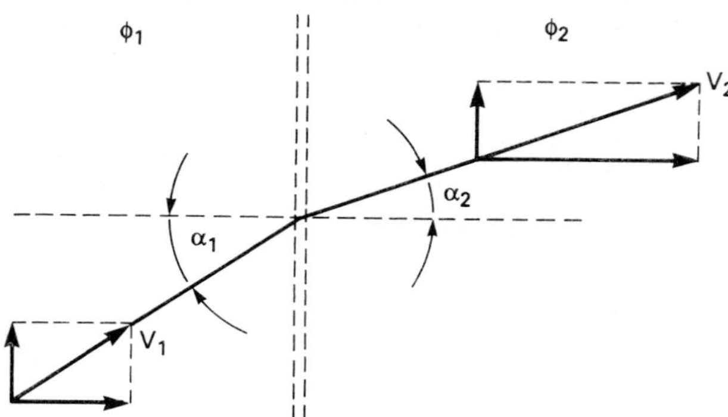

Figure 8.7 Refraction of an electron travelling from a domain with electric potential ϕ_1 to a domain with electric potential ϕ_2. The velocity component v sin α, parallel to the separation plane, remains the same

the normal to the separation plane. This formula is clearly analogous to Snell's law for the refraction of light at the transition from a medium with refractive index n_1 to a medium with refractive index n_2:

$$n_1 \sin\alpha_1 = n_2 \sin\alpha_2 \tag{8.34}$$

Apparently the quantity $\sqrt{(\phi + \epsilon)}$ fulfils the same role in electron optics as the refractive index n in optics. In practice, ϵ can often be neglected with respect to ϕ. A practical difference between electron optics and light optics is that, in the space between the electrodes, the potential ϕ varies continuously instead of in discrete steps. This means that the refraction can be considered to take place in an infinite number of small steps. At each step an equipotential plane acts as a separation plane.

The electrons are accelerated or retarded by the field strength perpendicular to these planes. In this way, lenses are formed between the electrodes of an electron optical system analogously to optical lenses. The strengths of these lenses can be continuously varied by varying the voltages on the electrodes (*Figure 8.8*).

In light optics, the sine law of Abbé plays an important role in image forming. It says that, for an ideal imaging system, the following relation holds between object and image:

$$r_1\, n_1 \sin\alpha_1 = r_2\, n_2 \sin\alpha_2 \tag{8.35}$$

where r_1 and r_2 are the sizes of object and image respectively, n_1 and n_2 are the refractive indices in object space and image space, and α_1 and α_2 are the angles between an imaging ray and the optical axis in object space and image space.

In electron optics this relation takes the form:

$$r1\, \sqrt{\phi_1 + \epsilon}\, \sin\alpha_1 = r2\, \sqrt{\phi_2 + \epsilon}\, \sin\alpha_2 \tag{8.36}$$

This is Helmholtz-Lagrange's law. In light optics, the refractive indices in object and image space are often the same. In electron optics, there is generally a large difference between the refractive indices in these spaces, much larger than would be possible in an optical system.

The validity of equation (8.36) is not restricted to object and image planes, but can also be extended to what are known as *supplementary* planes. For supplementary planes, each point in one plane corresponds to a direction in the other and vice versa, whereas for object and image planes, each point in one plane corresponds to a point in the other.

For object and image planes of an optical imaging system, the ratio r_2/r_1 is the magnification, M. From equation (8.36), the following relation between magnification and beam divergence angle at object and image planes can be derived:

$$M \sin\alpha_2/\sin\alpha_1 = \sqrt{(\phi_1+\epsilon)/(\phi_2+\epsilon)} \tag{8.37}$$

where α1 and α2 are the divergence semi-angles of the beam at the object and image sides of the system respectively. This equation states that the product of the linear magnification, M, and the angular magnification, $\sin\alpha_2/\sin\alpha_1$, is constant and that the value of the constant is determined by the ratio of the voltage at the object and image sides of the system. From this relation can be derived:

$$M = \sqrt{\frac{\phi_1}{\phi_2}}\;\frac{s_2}{s_1} \tag{8.38}$$

where s_1 and s_2 are the object and the image distances respectively, and ϵ has been neglected with respect to ϕ. Analogously to a light optical system, the following lens formula holds for these distances:

$$\frac{\sqrt{\phi_1}}{s_1} + \frac{\sqrt{\phi_2}}{s_2} = \frac{\sqrt{\phi_1}}{f_1} = \frac{\sqrt{\phi_1}}{f_2} \tag{8.39}$$

where f_1 and f_2 are the focal lengths at the object and image sides of the lens, respectively. All distances are defined with respect to what are known as the *principal planes* of the system.

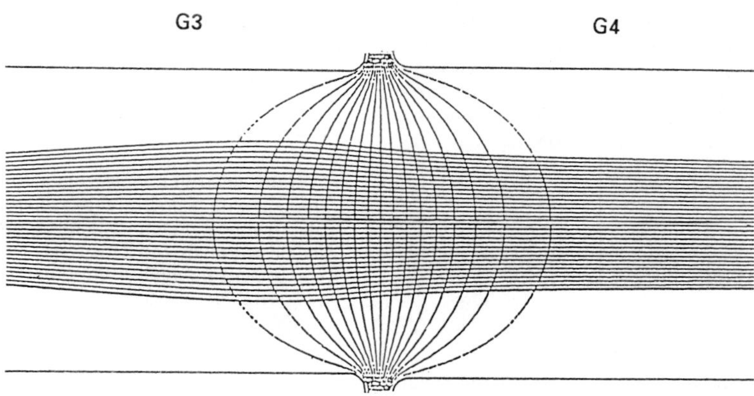

Figure 8.8 Electron trajectories and cross-sections of equipotential planes in the field of a uni-diameter electron lens

8.6 Forming the crossover

After leaving the cathode, the electron beam of an electron gun is focused in the crossover. This crossover serves as the object for the imaging system and is therefore of great importance for the beam forming in the gun.

The forming of the crossover is schematically indicated in *Figure 8.9*. Because of transverse thermal velocities, the electrons leave the cathode at various angles. They are first accelerated in a direction perpendicular to the cathode by the field in the immediate neighbourhood of the cathode. Because of this acceleration, the axial velocity of the electrons becomes large compared with the transverse velocity. The trajectories, therefore, become nearly parallel to the axis. A little farther away from the cathode, the trajectories are bent towards the axis by the shape of the electric field between the first and second grid. In this way, the narrowest cross-section is formed in the plane AB. In the neighbourhood of this plane, the

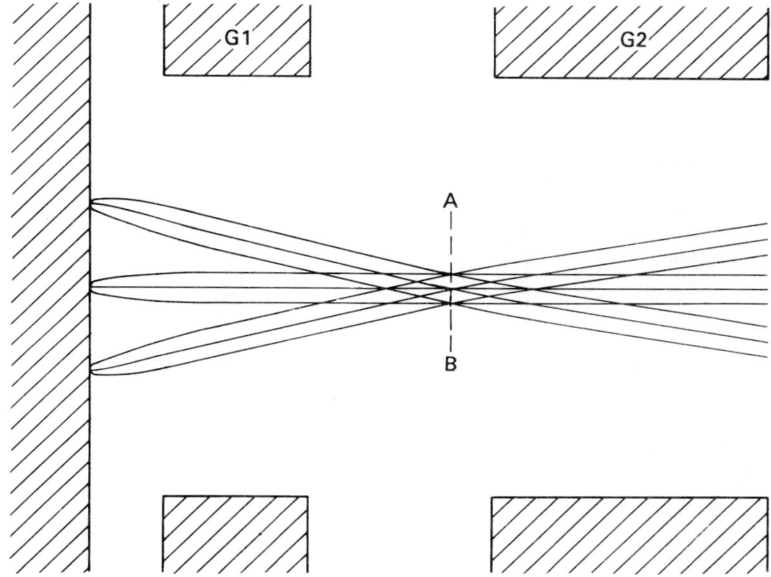

Figure 8.9 Forming of the crossover in the triode part of an electron gun. The electron beam has its narrowest cross-section in the plane AB

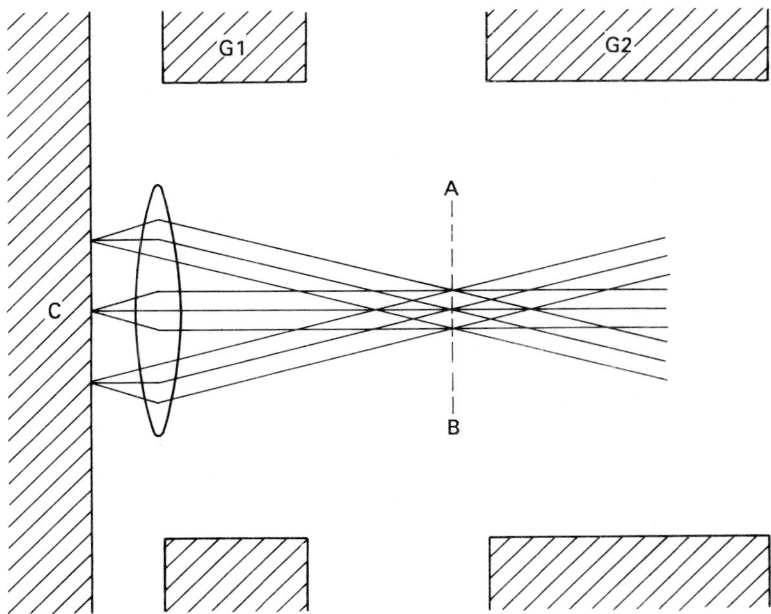

Figure 8.10 Optical analogue of the lens action in the triode part of an electron gun

trajectories cross the axis.

Under the idealized conditions assumed here, all electrons that leave the cathode in the same direction are focused in a single point of the crossover plane AB. This is further clarified in *Figure 8.10*, where an optical analogue of the lens action in the cathode area of the gun is given. The lens action is represented by a single optical lens at a short distance from the cathode. This lens is called the *cathode* lens. The cathode surface forms the object focal plane of the cathode lens and the crossover the image focal plane. As the voltages and therefore also the refractive indices at both sides of the lens are different, the object and image focal distances of this lens are also different.

8.7 Thermal spot size

As directions on the cathode are converted into points in the crossover and points on the cathode are converted into directions in the crossover, cathode plane and crossover plane are supplementary planes in the sense mentioned in section *8.5*. This makes it possible to apply the Helmholtz-Lagrange law to the beam in these planes. In this way, a prediction of the size and intensity distribution of the crossover can be obtained.

According to Maxwell, the distribution of the transverse velocities of the electrons leaving a cathode, situated in the XY plane, is given by:

$$dn = \text{const.} \exp\{-\frac{m}{2kT}(v_x^2+v_y^2)\}\, dv_x dv_y \tag{8.40}$$

where v_x = velocity in X direction,
v_y = velocity in Y direction,
k = Boltzmann's constant, and
T = absolute temperature of the cathode

The constant preceding the exponential expression takes account of the integration over the velocities in the Z direction and of the integration over the total emitting area. If ideal linear imaging is assumed, and the effect of the spread of velocities in the Z direction is neglected, all electrons with initial velocity components v_x and v_y are imaged in a single point of the crossover plane, with coordinates x and y given by:

$$x = cv_x \text{ and } y = cv_y \tag{8.41}$$

where c is a constant. Replacing v_x and v_y in equation (8.40) by x and y with the aid of this expression gives:

$$dn = \text{const.} \exp\{-\frac{m}{2kT}(x^2+y^2)/c^2\}\, dxdy \tag{8.42}$$

with a different constant preceding the exponential expression. The current density in the crossover is given by $j = edn/dxdy$. If, further, $r = \sqrt{(x^2+y^2)}$ is introduced, one obtains the following expression for the current density distribution in the crossover:

$$j(r) = \text{const.} \exp(-\frac{m}{2kT} r^2/c^2) \tag{8.43}$$

This can be written in the form:

$$j(r) = j_0 \exp(-r^2/\rho^2) \tag{8.44}$$

where j_0 is the current density in the centre of the crossover, and ρ is given by:

$$\rho = c\sqrt{\frac{2kT}{m}} \tag{8.45}$$

From equations (8.43) and (8.44), it follows that the current distribution in the crossover is Gaussian. The size of this distribution is characterized by the value $r = \rho$ where the current density has decreased to 1/e of its value in the centre. From equations (8.41)–(8.45), it follows that the electrons at this distance from the axis have left the cathode with a transverse velocity:

$$\sqrt{v_x^2+v_y^2} = \frac{\sqrt{x^2+y^2}}{c} = \frac{\rho}{c} = \sqrt{\frac{2kT}{m}} \tag{8.46}$$

This can be written in the form:

$$v_0 \sin\alpha_c = \sqrt{\frac{2kT}{m}} \tag{8.47}$$

where v_0 = total initial velocity of the electrons, and
α_c = angle of the initial velocity with the z axis

The initial velocity v_0 can also be expressed in the voltage ϵ defined by equation (8.32). This yields:

$$\sqrt{\epsilon}\sin\alpha_c = \sqrt{\frac{kT}{e}} \tag{8.48}$$

Electrons with this initial velocity and exit angle are imaged in the crossover at a distance ρ from the centre. Applying the Helmholtz-Lagrange law (equation (8.36)) to the data in cathode and crossover plane gives:

$$r_0\sqrt{\epsilon}\sin\alpha_c = \rho\sqrt{(\phi_{cr}+\epsilon)}\sin\alpha_{cr} \tag{8.49}$$

where r_0 = radius of the emitting area on the cathode,
ϕ_{cr} = potential in crossover with respect to cathode, and
α_{cr} = semi-angle of the beam in the crossover

Substituting from equation (8.48) gives:

$$\rho = \frac{r_o}{\sin\alpha_{cr}}\sqrt{\frac{kT}{e(\phi_{cr}+\epsilon)}} \tag{8.50}$$

where ϵ usually is negligible compared with ϕ_{cr}.

Equations (8.44) and (8.50) are valid only under ideal imaging conditions. As far as these conditions also apply to the image of the crossover on the screen, the spot on the screen also has a Gaussian distribution. Analogously to equation (8.44), the current density on the screen is given by:

$$j(r) = j_0 \exp(-r^2/\rho^2) \tag{8.51}$$

where now j_0 = current density in the centre of the spot on the screen, and
ρ = distance where the current density on the screen has decreased to 1/e of its value in the centre

The value of ρ in this expression can be determined without detailed information of the lens system of the gun, by again using the Helmholtz-Lagrange law. Applying this on the cathode and screen plane gives:

$$r_0\sqrt{\epsilon}\sin\alpha_c = \rho\sqrt{(V_s-V_c+\epsilon)}\sin\alpha_s \tag{8.52}$$

where V_s = screen voltage, and
α_s = semi-angle of the beam at the screen

Substituting from equation (8.48), and disregarding ϵ on the right-hand side of equation (8.52) yields:

$$\rho = \frac{r_0}{\sin\alpha_s} \sqrt{\frac{kT}{e(V_s - V_c)}} \quad (8.53)$$

The semi-angle of the beam at the screen side of the gun can be expressed in the beam diameter Φ_L in the main lens and the image distance s_2 between main lens and screen. To a close approximation:

$$\sin\alpha_s = \frac{\Phi_l}{2s_2} \quad (8.54)$$

The spot size on the screen is usually defined by the width where the current density has decreased to $(1/e)^3 \approx 5$ per cent of its value in the centre (see *Figure 8.11*). This size corresponds approximately to the diameter observed by the eye with the aid of a microscope[3].

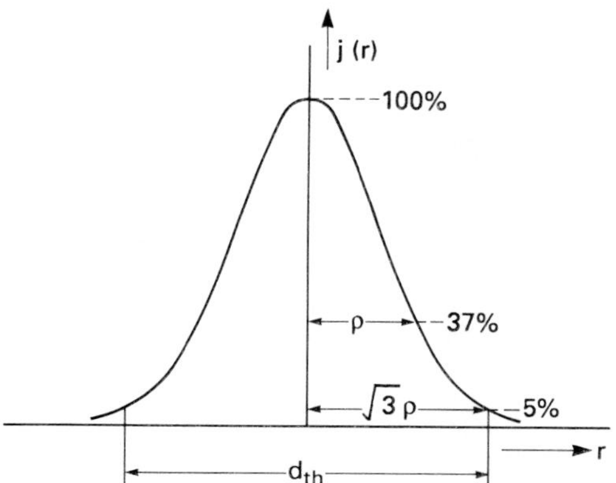

Figure 8.11 Current density distribution of a Gaussian spot

From equation (8.51), it follows that this size is given by:

$$2r = 2\sqrt{3}\rho \quad (8.55)$$

As the spot size defined by this equation is determined only by thermal velocities of the electrons, it is called the *thermal spot size* d_{th}. Using equations (8.53) and (8.54), one obtains:

$$d_{th} = 4\sqrt{3}\,\frac{s_2 r_0}{\Phi_L} \sqrt{\frac{kT}{e(V_s - V_c)}} \quad (8.56)$$

For an oxide cathode with T = 1083 K. the value of kT/e = 0.0933 V. So for this cathode:

$$d_{th} = 2.12 \frac{r_0 s_2}{\Phi_l \sqrt{V_s - V_c}} \quad (8.57)$$

if V_s and V_c are expressed in volts.

The beam diameter in the main lens is generally almost proportional to the radius of the emitting area on the cathode. Furthermore, as the variation of the cathode voltage V_c can be neglected with respect to the screen voltage V_s, the thermal spot size is nearly independent of beam current. In designing a gun, the thermal spot size can be reduced only by increasing ϕ_L or reducing r_0. The first possibility is generally limited by spherical aberration of the main lens and deflection defocusing. The second possibility is limited by the maximum allowable value of the cathode load (section *8.4*).

Because of aberrations of the lens system and space charge repulsion between the electrons, the real spot size on the screen will generally be larger than the above derived thermal spot size. These other contributions to the spot size are considered in the following sections. They distort the Gaussian current density distribution. At currents below 100 μA, their effect is generally small. So, for currents up to this level, the spot size is approximately equal to the thermal spot size, and the current density distribution is nearly Gaussian. This is the case in monochrome high resolution tubes, for instance, which are usually designed for current levels of about 100 μA.

8.8 Cathode lens

The description of the cathode lens given in section *8.6* is an idealized one. In reality, it shows aberrations like every other lens. As the image is located on the optical axis, only spherical aberration has to be taken into account. This also holds for the other lenses of the gun system. Because of this aberration, trajectories starting normal to the cathode do not intersect the axis in a single point, but form a *disk of least confusion* as shown in *Figure 8.12*. Before going into further details of the spherical aberration of the cathode lens, some general properties of this lens field will be discussed. By considering these properties, the beam angle in the crossover can be predicted without detailed knowledge of the electrode configuration of the triode.

In a rotationally symmetrical gun with a flat cathode, the voltage distribution in the immediate neighbourhood of the cathode can be described by the following series expansion:

$$\phi(r,z) = E_0 r_0 \left(\frac{z}{r_0}\right) \left\{1 - \left(\frac{r}{r_0}\right)^2 + \frac{2}{3}\left(\frac{z}{r_0}\right)^2\right\} + \ldots \quad (8.58)$$

where $\phi(r,z)$ is the voltage with respect to the cathode, and z is the coordinate along the axis of the triode. This equation holds for the situation without space charge and is valid only to the extent that higher order terms may be neglected. From the formula, it can be seen that the distances in this distribution are scaling with r_0 and the voltages with $E_0 r_0$.

If two field configurations with the same value of r_0 but different value of E_0 are compared, the trajectories in the immediate neighbourhood of the cathode will be the same. This follows from Snell's law given by equation (8.33), if ϵ is negligible compared with ϕ (the value of ϵ is of the order of 0.1 V). For situations with different r_0, the field dimensions and also the trajectories scale with r_0.

This is illustrated by *Figure 8.13*, where the trajectories in the triode of a gun are shown for two different beam currents (0.2 and 2 mA). From the uniformity of the trajectories, the following general properties can be derived:

$$\sin\alpha_{cr} \approx 0.2 \quad (8.59)$$

and

$$\phi_{cr} \approx 9E_0 r_0 \quad (8.60)$$

In practice some deviation from these values may occur. However, the following relation of the given quantities appears to have a more general validity and will therefore be used in further calculations:

$$\sin\alpha_{cr} \sqrt{\phi_{cr}} = 0.6 \sqrt{(E_0 r_0)} \quad (8.61)$$

With the aid of equations (8.9), (8.11) and (8.16), this relation can also be expressed as:

$$\sin\alpha_{cr} \sqrt{\phi_{cr}} = 0.6 \sqrt{(aR)}\,(i/K)^{1/3} \quad (8.62)$$

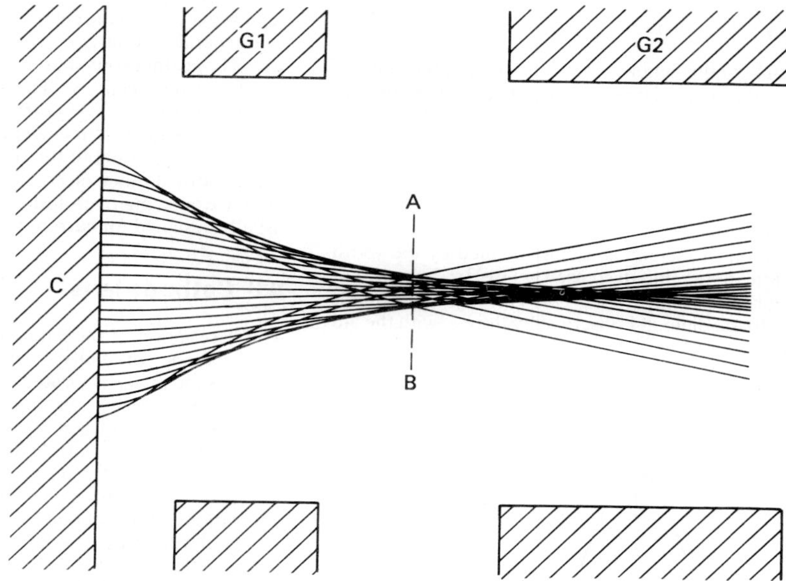

Figure 8.12 Spherical aberration of electron rays in the crossover

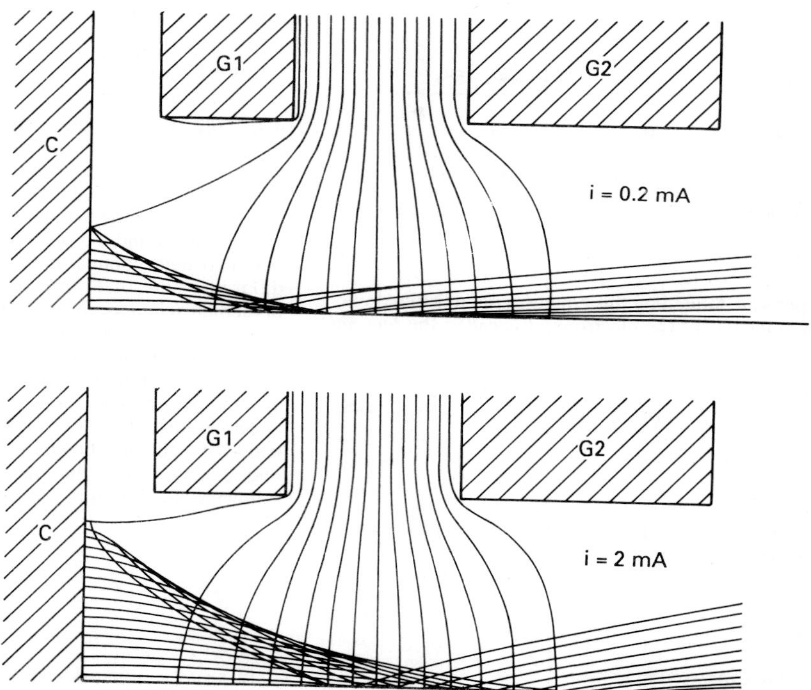

Figure 8.13 Half cross-section of beam trajectories and equipotential planes in the triode part of a gun at two different current levels. The trajectories of both plots roughly scale with the radius of the emitting area on the cathode

This equation will later be used for the calculation of the beam diameter in the main lens. It can also be used together with equation (8.50) to calculate the thermal size of the crossover.

Because of the uniformity of the trajectories in the cathode area, it could also be expected that the size of the disc of least confusion caused by spherical aberration of the cathode lens would scale with the radius of the emitting area of the cathode. In reality, however, the size of this disc does not scale with r_0 but with r_0^3. This is due to higher order terms in the series expansion of the potential distribution. The effect of spherical aberration of the cathode lens appears to be described by:

$$d_{sac} = 0.2\,\frac{r_0^3}{R^2} \tag{8.63}$$

where d_{sac} is the diameter of the disc of least confusion of the crossover. The numerical constant in this equation holds for the normal situation of a flat cathode with a circular aperture in the first grid. It does not depend on the further configuration in the triode.

8.9 Prefocus lens

Before the crossover is imaged on the screen by the main lens, the divergence angle of the beam is usually reduced by the field between the second and third grid. This action is called *prefocusing*, and the field configuration that performs the prefocusing is the *prefocus lens*. The prefocus lens forms a virtual image of the crossover behind the cathode (see *Figure 8.14*). This image is called *virtual crossover*. The virtual crossover is generally larger than the real crossover.

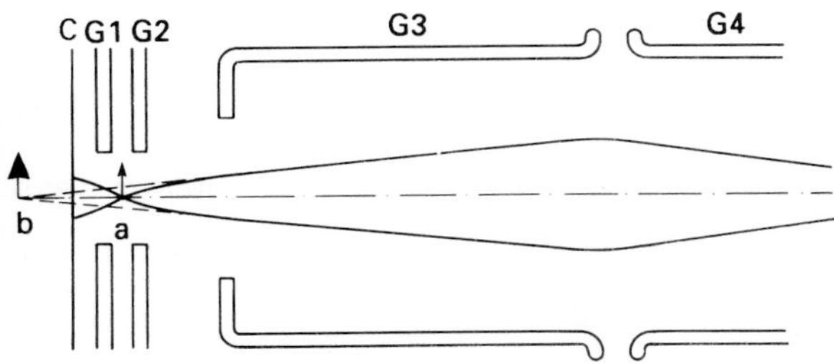

Figure 8.14 Forming of a virtual image (b) of the crossover (a) by the prefocusing field between G2 and G3. This image is called *virtual crossover* and is the object of the main lens formed by the field between G3 and G4, which images the virtual crossover on the screen

The magnification from crossover to virtual crossover is determined by the magnification factor M_P. For usual guns, the value of M_P is between 1 and 2, mostly in the neighbourhood of 1.3. As the axial position as well as the voltage of the crossover varies with beam current, the value of M_P also varies with beam current. In most cases, however, this variation can be neglected.

After leaving the prefocus lens, the beam enters the main lens. For the semi-angle α_L of the beam at the object side of the main lens, application of equation (8.37) gives:

$$\sin\alpha_l = \frac{\sin\alpha_{cr}}{M_p}\sqrt{\frac{\emptyset_{cr}}{V_{g3} - V_c}} \tag{8.64}$$

For the sake of simplicity, it is assumed here that the prefocusing occurs only between the second and third grid and that the voltage at the object side of the main lens is equal to the voltage of the third grid. In some types of guns, an additional prefocusing takes place after the beam has entered the third grid. In that case, the same formula can be used, but for M_P the total prefocusing magnification factor has to be taken and V_{g3} has to be replaced by the voltage at the object side of the main lens.

Using equation (8.62), equation (8.64) can be written in the form:

$$\sin\alpha_L = \frac{0.6}{M_p}\sqrt{\frac{aR}{V_{g3}-V_c}}\left(\frac{i}{K}\right)^{1/3} \tag{8.65}$$

From equation (8.65), it can be seen that the divergence angle of the beam entering the main lens varies with the cube root of the beam current. Only if M_P varies with beam current may a departure from this rule be expected.

The choice of the right beam angle at the object side of the main lens plays an important role in the design of an electron gun. The angle is determined by the choice of the strength of the prefocus lens. However, an optimal choice is generally possible for only one beam current. (See section *8.12*.)

Like other lenses, the prefocus lens also causes spherical aberration. The effect of spherical aberration is generally proportional to the third power of the beam diameter in a lens. The beam diameter in the prefocus lens is approximately proportional to the radius of the emitting area on the cathode. Therefore, the spherical aberration effect of the prefocus lens can roughly be described by:

$$d_{sap} = c_{sap}\,M_p\,\frac{r_0^3}{R^2} \tag{8.66}$$

where d_{sap} is the diameter of the disc of least confusion of the virtual crossover due to the spherical aberration of the prefocus lens, and C_{sap} is a constant which has been made dimensionless by the introduction of the factor R^{-2} in the formula. As the contribution of the prefocus lens to the total spherical aberration of the system is usually small compared with the contribution of the main lens, this rough description is generally sufficient. The spherical aberration constant in the formula usually has a value between 0.2 and 0.6. In special cases (see section *8.12*), the value can also be negative.

8.10 Main lens

The main lens images the virtual crossover on the phosphor screen. The simplest type of main lens consists of two electrodes with a small gap in between. It needs two voltages and is therefore known as a *bi-potential* lens. The focusing action of this lens is determined by the voltage ratio η defined by:

$$\eta = \frac{V_1 - V_c}{V_2 - V_c} \tag{8.67}$$

where V_1 and V_2 are the voltages on the first and second electrode, respectively. The first electrode is often the third grid of the gun. In this case $V_1 = V_{g3}$. The second electrode is generally at screen potential, so generally $V_2 = V_s$. The voltage V_1 is the *focus voltage*. The focus voltage is usually between 20 and 40 per cent of the screen voltage. If it is at the low end of this range, the gun is called a *lo-bi* gun and if it is at the high end, it is a *hi-bi* gun.

The magnification M_L of the main lens is determined by its focal length. From equations (8.38) and (8.39) it follows that:

$$M_L = \frac{s_2}{f_2} - 1 \tag{8.68}$$

as in light optics. As the lens is not a 'thin' lens, object and image distance must be taken with respect to the principal planes of the lens. These are situated at the low voltage side of the gap between the two electrodes, at distances h_1 and h_2 respectively from the centre of the gap. So, if the distance from the centre of this gap to the screen is L, the image distance is:

$$s_2 = L + h_2 \tag{8.69}$$

The most basic type of bi-potential lens is the *uni-diameter lens* consisting of two coaxial cylinders of equal diameter (see *Figures 8.8* and *8.14*). For this lens the following approximation formulas hold:

$$f_2 = 1.626 \exp(4.40\eta)\, D_L \tag{8.70}$$

$$h_1 = 0.639 \exp(2.315\eta)\, D_L \tag{8.71}$$

$$h_2 = 1.311 \exp(2.83\eta^2)\, D_L \tag{8.72}$$

The quantity D_L in these equations is given by:

$$D_L = \{1 + 0.5(s/D)^2\}\, D \tag{8.73}$$

where D = internal diameter of the cylinders, and
s = gap width between the cylinders

The same formulas can be used for the main lens of an integrated colour gun consisting of two plates with three holes (one for each beam) facing each other. See *Figure 8.15*. In this case, D is the diameter of the apertures and s the distance between the plates.

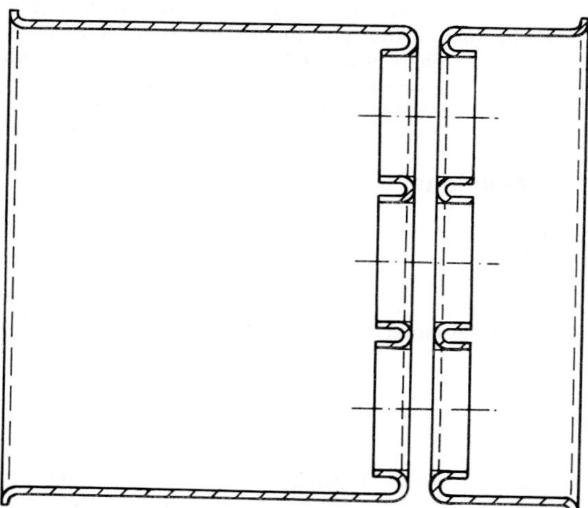

Figure 8.15 Main lens configuration of an integrated gun

Formula (8.70) can also be applied to other types of bi-potential lenses by adaptation of the value of D_L. The quantity D_L obtained in this way is called the *equivalent lens diameter*. It is defined as the diameter of a uni-diameter cylinder lens with an infinitesimally small gap and the same focal length. The equivalent lens diameter is a very useful quantity for the comparison of different lens designs.

Because of spherical aberration, a point shaped object on the axis of the lens is not imaged in a single point on the phosphor screen, but in a disc of least confusion, where the beam attains its narrowest cross-section. This situation is similar to that shown in *Figure 8.12* for the crossover. For the diameter d_{sal} of this disc, the following general relation can be derived:

$$d_{sa1} = \text{const.}\ (M_L+1)\Phi_L^3/D^2 \tag{8.74}$$

where D is an arbitrary geometric dimension of the lens. The value of the constant depends on the type of lens. For a uni-diameter cylinder lens this equation takes the form:

$$d_{sal} = \frac{0.143}{\sqrt{\eta}}\,(M_L+1)\Phi_L^3/D_L^2 \tag{8.75}$$

This formula also holds in good approximation for other types of bi-potential lenses, if the equivalent lens diameter defined by equation (8.70) is used for D_L.

In these formulas, Φ_L is the beam diameter in the principal planes of the lens. The value of Φ_L depends on the diverging angle of the beam at the object side of the lens. To a close approximation:

$$\Phi_L = 2s_1 \sin\alpha_L \tag{8.76}$$

The object distance s_1 in this formula can be expressed in terms of the distance L between lens and screen by using equations (8.38) and (8.69):

$$s_1 = \frac{L+h_2}{M_L}\sqrt{\frac{V_{g3}-V_c}{V_s-V_c}} \tag{8.77}$$

If further use is made of equation (8.65) for $\sin\alpha_L$, one obtains:

$$\Phi_L = 1.2\,\frac{(L+h_2)}{M_P M_l}\sqrt{\frac{aR}{V_s-V_c}}\ \left(\frac{i}{k}\right)^{1/3} \tag{8.78}$$

For a required gun design, nearly everything in this relation is fixed except for the values of M_P and M_L.

From equations (8.74) and (8.75) it follows that the spherical aberration decreases quadratically with increasing lens diameter. An increase of this diameter is, however, limited by the neck size of the tube. This must be as small as possible to restrict the amount of deflection power required. Within this mechanical limit, tube makers are trying to make lenses with the largest possible equivalent lens diameters. Examples are the OLF (*overlapping field*) lens of Matsuchita, the LAT (*large aperture thick metal*) lens of Toshiba, the XL (*expanded lens diameter*) lens of RCA, the EA (*elliptical aperture*) lens of Hitachi and the Polygon lens of Philips.

Lenses with larger effective diameters are characterized by a more gradual increase of the potential on the axis of the lens. This gradual increase can also be obtained by using extra electrodes on intermediate voltages, at the cost of extra connection pins or of internal voltage dividers with potential stability problems. This last solution can take the form of a spiral of high ohmic resistive material applied to the inside of a glass cylinder, acting simultaneously as a lens electrode and as a voltage divider (*spiral lens*).

A large effective lens diameter can also be realized by using magnetic focusing instead of electrostatic, with the aid of focus coils outside the neck. By this means, the neck size forms no restriction for the dimensions of the focusing field. This solution is often applied in projection tubes, at the cost of extra circuitry to generate a stable focusing current.

An increase of equivalent lens diameter generally entails a reduction of main lens magnification, as can be seen from equations (8.68) and (8.70), and consequently also an increase of gun length, as can be seen from equation (8.77). From equations (8.74) and (8.75), it follows that a reduction of lens magnification causes a further reduction of spherical aberration, if the beam diameter can be kept the same. According to equation (8.78), this is only possible if the total magnification $M = M_P M_L$ is kept the same. This means that the magnification of the prefocus lens must be increased, otherwise the decrease of spherical aberration due to an increase of the equivalent lens diameter could even be nullified by the increase of the beam diameter in the lens. Beam diameter and magnification also affect other contributions to the spot size (see section *8.12*).

In the case of a bi-potential lens, the spherical aberration can also be decreased by increasing the voltage ratio η, as can be seen from equation (8.75). This was the reason for the introduction of the hi-bi lens. Just like an increase of equivalent lens diameter, an increase of focus voltage causes a reduction of lens magnification and an increase of gun length. The beam diameter increases as well, unless the total magnification is kept constant by an increase of the prefocus magnification.

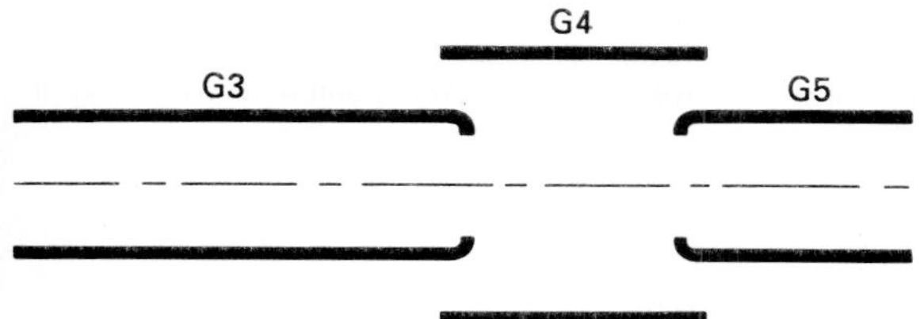

Figure 8.16 Electrode configuration of a uni-potential (UPF) lens used in a monochrome tube

In monochrome tubes, a uni-potential (*UPF*) lens is generally used. Such a lens consists of three coaxial cylindrical electrodes (see *Figure 8.16*). The outer electrodes are at the same voltage as the screen, and the central electrode is earthed or connected to an adjustable voltage close to earth. The spherical aberration behaviour of this lens is not so good as that of a bi-potential lens with the same mechanical diameter, but the gun is shorter and the generation and connection of a focus voltage of some kilovolts is obviated. The uni-potential lens can also be designed for the use of a much higher voltage on the central electrode. In that case, the spherical aberration is about equal to that of a bi-potential lens with the same focus voltage. This last type is called a high uni-potential (*HI-UPF*) lens. It is used by Sony in the Trinitron gun and by Hitachi in their projection tubes and also in an integrated version in their colour tubes.

In some types of guns, the number of lens electrodes is even extended to four. In these guns, an accelerating bi-potential lens is preceded by a high or low uni-potential lens. These lenses are respectively *hi-uni-bi* and *low-uni-bi* lenses (see *Figure 8.17*).

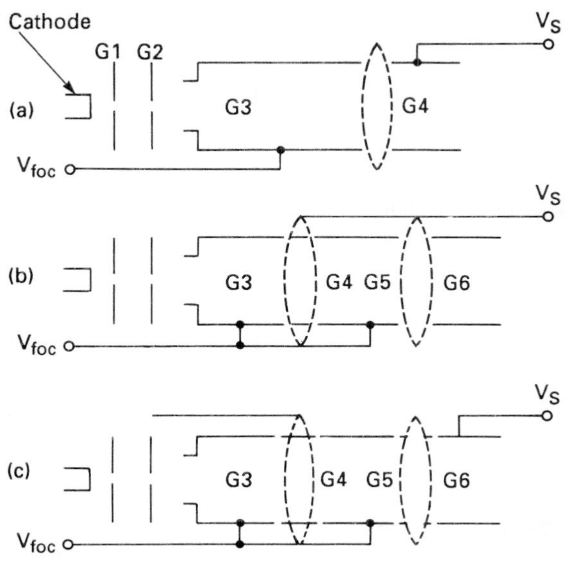

Figure 8.17 Electrode configuration of (a) a bi-potential lens, (b) a hi-uni-bi lens, and (c) a low-uni-bi lens

In high-uni-bi lenses, the central electrode of the UPF lens is connected to screen potential; in low-uni-bi lenses, it is usually connected to the voltage of the second grid. In the latter, there is less chance of high voltage arcing. In both cases, the uni-potential lens acts as a prefocus lens for the bi-potential main lens, so that a weaker main lens can be used without too large a beam diameter. In this way, the spherical aberration of the main lens can be reduced. However, the extra prefocus lens also adds spherical aberration to the system. It is therefore questionable if much is gained by such a lens configuration. Anyhow, such a system can be useful when strong prefocusing is required.

8.11 Effect of space charge on spot size

The Coulomb repulsion between the electrons in a beam can generally not be neglected. This *space charge effect* consists of a diverging action on the electron trajectories of the beam. The divergence of the trajectories inside the gun slightly modifies the effective strength of the electron lenses with little change to the spot size on the screen, but the divergence of the trajectories outside the gun, between gun and screen, sometimes considerably increases the spot size. Defocusing of the beam is usually prevented by adaptation of the focus voltage, but the focused spot is larger than it would have been in the absence of space charge.

For the simplified case of a laminar beam with a radius r, the radial field strength E_r at the edge of the beam is:

$$E_r = \frac{i}{2\pi\epsilon_0 rv} \tag{8.79}$$

The trajectory equation for the electrons at the edge of the beam is given by:

$$m\frac{d^2r}{dt^2} = eE_r \tag{8.80}$$

Double integration of this equation over the distance L between the centre of the main lens and the screen gives:

$$\Delta r = \frac{eiL^2}{4\pi\epsilon_0 rmv^3} \tag{8.81}$$

if it is assumed that Δr is small compared to r. The spot size increase due to space charge is $d_{sc} = 2\Delta r$. Substituting v with the aid of equation (8.31) and regarding ϵ as negligible compared with V_s, one obtains:

$$d_{sc} = p\frac{iL^2}{\Phi(V_s-V_c)^{3/2}} \tag{8.82}$$

where $\Phi = 2r$ is the beam diameter, and p is a physical constant given by:

$$p = (2^4\pi p\epsilon_0\sqrt{2e/m})^{-1} = 30.30\times10^3\ V^{3/2}/A \tag{8.83}$$

Equation (8.82) gives a good approximation for the spot size increase due to space charge in the case when the beam diameter is nearly constant from gun to screen. In that case, Φ has to be taken equal to the beam diameter Φ_L in the lens. In general, the beam diameter varies from the value Φ_L in the lens to the value d of the spot size on the screen. If a linear variation between these values is assumed, the following modification of equation (8.82) can be derived:

$$d_{sc} = 2p \frac{iL^2}{\Phi_L (V_s - V_c)^{3/2}} \frac{d/\Phi_L \{\ln(d/\Phi_L)-1\}+1}{(d/\Phi_L - 1)^2} \tag{8.84}$$

This formula gives a good approximation of the spot size increase in more general cases. It should be remarked that the spot size d in this equation is the total spot size, including the increase due to space charge. This means that this equation must be solved iteratively. For $d = \Phi_L$ the formula reduces to equation (8.82). More elaborate treatments have been published by Schwartz[4] and Hollway.[5]

8.12 Spot size in screen centre

The total spot size in the centre of the screen is determined by the contributions of spherical aberration, thermal velocities and space charge (*Figure 8.18*).

As the effect of spherical aberration is determined to a large extent by the rays at the edge of the beam, and outer rays in the cathode lens generally remain outer rays in prefocus lens and main lens, it may be assumed that the spherical aberration of these lenses add linearly. In this way, the diameter of the disc of least confusion on the screen due to spherical aberration becomes, after multiplication of the contributions of cathode lens and prefocus lens by the magnification to the screen:

$$d_{sa} = M_P M_L d_{sac} + M_L d_{sap} + d_{sal} \tag{8.85}$$

where d_{sac}, d_{sap} and d_{sal} are given by equations (8.63), (8.66) and (8.75), respectively.

Recently, electron guns have been introduced where the prefocusing is so designed that inner and outer rays in the cathode lens and the main lens are interchanged. In this way the spherical aberration of the cathode lens partly cancels that of the main lens. This is called an ART (*aberration reducing triode*) gun[6]. For such a gun, equation (8.85) can still be used. In that case, the ART effect can be taken into account by assigning a negative value of the order of -0.2 to the constant c_{sap} in formula (8.66).

The combined effect of spherical aberration and thermal velocities on the spot size on the screen can be found by a convolution of the Gaussian intensity distribution due to the thermal velocities with the intensity distribution of the disk of least confusion due to spherical aberration. The 5 per cent width of the resulting distribution is given by the following approximation formula:[3]

$$d = \sqrt{d_{sa}^2 + (0.3d_{th})^2} + 0.7d_{th} \tag{8.86}$$

To take account of the additional effect of Coulomb repulsion, the charge of the beam can be considered to be located on the axis of the beam. This causes the beam to expand in a direction away from the axis in every point of its cross-section.

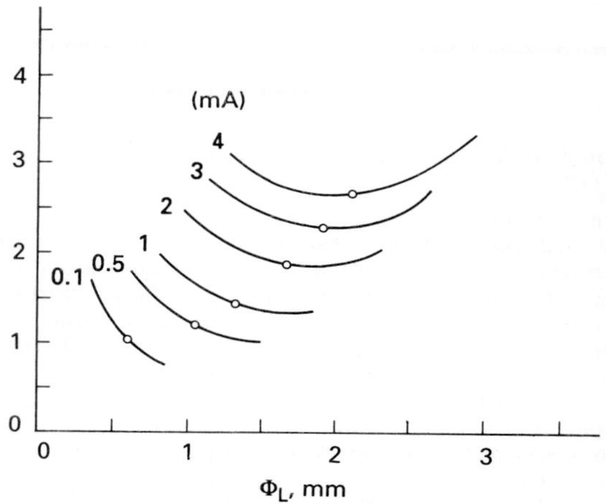

Figure 8.19 Calculated spot size at various beam currents as a function of the beam diameter in the main lens. The open circles correspond to an actual gun design

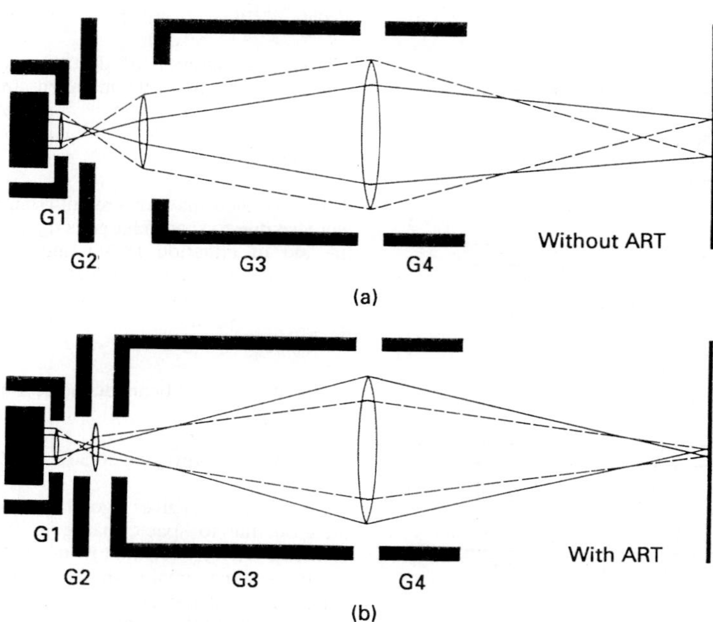

Figure 8.18 Effect of spherical aberration (a) in a conventional gun without ART (*aberration reducing triode*) and (b) with ART

Therefore, the space charge effect on the intensity distribution of the spot cannot be obtained by a convolution. Instead, the value of d_{sc} given in the foregoing section has simply to be added linearly to the other contributions. In this way the total spot size in the centre of the screen becomes:

$$d = \sqrt{d_{sa}^2 + (0.3d_{th})^2} + 0.7d_{th} + d_{sc} \quad (8.87)$$

where d_{sa}, d_{th} and d_{sc} are given by equations (8.85), (8.56) and (8.84), respectively.

The contribution of the spherical aberration of the main lens to the total spot size increases with the third power of the beam diameter in the lens. All other contributions are inversely proportional to this diameter. Therefore, to a close approximation, equation (8.87) can be written in the form:

$$d = \frac{A}{\Phi_L} + B\,\Phi_L^3 \quad (8.88)$$

where A and B are constants depending on the beam current and some other gun data. As an example, a graph of d as a function of Φ_L is given in *Figure 8.19* with the beam current i as parameter.

During the design of the gun, the prefocus magnification will be chosen so that the spot size is minimum. The minimum can be calculated by differentiating equation (8.88) with respect to Φ_L. This gives, for the value of the beam diameter at the minimum:

$$\Phi_L = \{A/(3B)\}^{1/4} \quad (8.89)$$

and for the minimum value of the spot:

$$d_{min} = 4(A/3)^{3/4}B^{1/4} \quad (8.90)$$

At this minimum, the contribution of the spherical aberration of the main lens to the spot size is:

$$d_{sa1} = (A/3)^{4/4}B^{1/4} \quad (8.91)$$

So, at the minimum, the contribution of the spherical aberration of the main lens amounts to one-quarter of the total spot size. From equation (8.90), it can further be seen that a reduction of spherical aberration of the main lens by a factor n reduces the minimum spot size by a factor of only $n^{1/4}$.

When degaussing a gun an optimum choice for the value of the beam diameter in the main lens is generally possible at only one beam current. This choice also determines the values of the beam diameter at other currents, as follows from equation (8.78). At a lower beam current, Φ_L is smaller than the optimum value for that current, whereas at a higher beam current, Φ_L is larger. The current at which the gun is optimized must therefore be carefully chosen in the range of currents to be used.

8.13 Deflection defocusing

The magnetic deflection field not only deflects the beam to the various points on the screen, but also exerts a lens action which defocuses and distorts the spot outside the centre of the screen. The lens defocusing consists basically of a rotationally symmetrical component and an astigmatic component. The rotationally symmetrical component brings the beam to a focus before it reaches the screen. Because of this, the spot is overfocused on the screen. This means that the spot shows a bright core, surrounded by a halo. The astigmatic component increases the overfocusing in one direction and decreases it in the perpendicular direction. Both effects increase quadratically with deflection.

If the astigmatic component is small, as is the case in monochrome data display tubes and projection tubes, the deflection defocusing can be corrected by weakening the main lens with the aid of a voltage that varies quadratically with deflection. This correction is called *dynamic focus*.

In colour tubes, three beams must be deflected at the same time. In a tube with a delta gun arrangement, corrections of the directions of the beams are needed to keep the beams converged (i.e. coinciding on the screen) during deflection. These corrections have to vary with deflection and are therefore rather complicated. They are called *dynamic convergence* corrections.

In an in-line colour tube with a self-converging deflection yoke, such corrections are not needed. The deflection field is so designed that the beams are automatically converged over the entire screen when they are converged in the centre. The convergence in the centre is obtained by a system of permanent magnetic correction rings on the neck of the tube or a small magnetic correction ring inside the tube that corrects for small deviations of the direction of the beams.

The astigmatic lens action of the self-converging deflection field cancels the overfocusing in the horizontal cross-section of the spot at the cost of a larger overfocusing in the vertical cross-section. In this way, the horizontal cross-section of the spot is automatically focused during deflection. However its size is not constant, but increases considerably from the centre of the screen to the edge: by 40 per cent in the case of a 90° tube and 100 per cent in the case of a 110° tube.

The vertical overfocusing of the deflected beams in such a system can be partly corrected by the use of a gun with astigmatic beams.[6,7,8,9] On entering the deflection field, the vertical cross-section of these beams is slightly underfocused (and often also reduced in size). This results in a reduction of vertical overfocusing during deflection, at the cost of a slightly larger vertical spot for the undeflected beam in the central part of the screen. The required astigmatism of the beams is usually obtained by means of a horizontal slit at the grid 3 side of grid 2, or a vertical slit at the grid 2 side of grid 1.

Vertical overfocusing can be completely corrected by using an additional quadrupole lens in the focus electrode of the main lens[10] (see *Figure 8.20*). The quadrupole lens divides the focus electrode into two parts. The first part is connected to the focus voltage required for the centre of the screen, and the second part is connected to a voltage that varies quadratically with deflection. This voltage changes the strength of the quadrupole lens and the main lens simultaneously. In this way, the astigmatic as well as the rotational symmetric lens action of the deflection field can be corrected. The system is known as DAF (*dynamic astigmatism and focus*).

Though a dynamic quadrupole system can correct vertical overfocusing of the spot in a self-converging system, it does not reduce horizontal spot growth during deflection. This spot growth in horizontal direction together with a spot size reduction in vertical direction is due to a distortion caused by the deflection field. Horizontal spot growth can be reduced only by using a non-self-converging deflection field, as in delta gun systems, but at the cost of having to apply dynamic convergence corrections.[11,12]

8.14 Resolution

The resolution of a crt is mainly determined by the spot size. The relation between spot size and resolution can be studied by a Fourier analysis of the effect of the spot size on the modulation depth of the various wave components of a

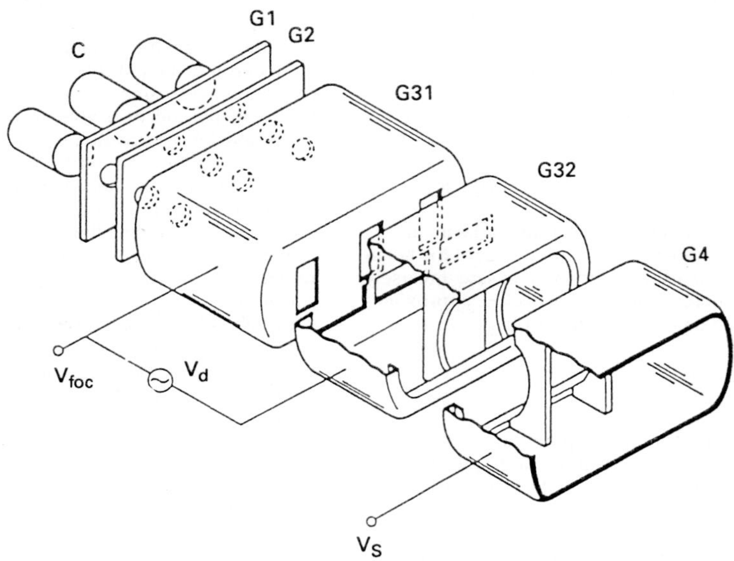

Figure 8.20 Gun with dynamic quadrupole lenses for simultaneous correction of spot astigmatism and focus at deflection. Quadruple fields of varying strength are generated in the gap between G31 and G32 by varying the voltage between these electrodes

luminance pattern to be displayed. The wave components are characterized by their spatial frequency, which is the inverse of the wavelength. The capability of the spot to reproduce these components is expressed by the modulation transfer function (*mtf*). The mtf gives the factor by which the modulation depth of an original wave component is reduced in the reproduction pattern as a function of the spatial frequency. It can be obtained by a Fourier transform of the intensity distribution of the spot. For a symmetrical distribution, the mtf is given by:

$$M(v) = \frac{\int_{-\infty}^{+\infty} l(x)\cos(2\pi vx)\,dx}{\int_{-\infty}^{+\infty} l(x)\,dx} \tag{8.92}$$

In this equation, v is the spatial frequency and l(x) the line-spread function defined by:

$$l(x) = \int_{-\infty}^{+\infty} j(x,y)\,dy \tag{8.93}$$

where j(x,y) is the point-spread function or intensity distribution of the spot.

The intensity distribution of a Gaussian spot with a 5 per cent width d is given by:

$$j(r) = j_0 \exp(-12r^2/d^2) \tag{8.94}$$

as follows from equations (8.51) and (8.55). Application of equations (8.92) and (8.93) yields:

$$M(v) = \exp(-\pi^2 d^2 v^2/12) \tag{8.95}$$

A graph of this function is given in *Figure 8.21*. To a good approximation, this formula also holds for a non-Gaussian spot with the same 5 per cent width.[13]

For a quick reference of resolution capacity, the spatial frequency where the mtf has decreased to the value 0.5 can be used. From equation (8.94) it follows that this spatial frequency is given by:

$$v_{0.5} = \frac{2\sqrt{3\ln 2}}{\pi d} \tag{8.96}$$

In practice, much higher spatial frequencies will also be resolved, because the eye can still easily detect wave components with a modulation depth of 10 per cent or less. The relation between resolution and visual sharpness has been dealt with elsewhere.[13]

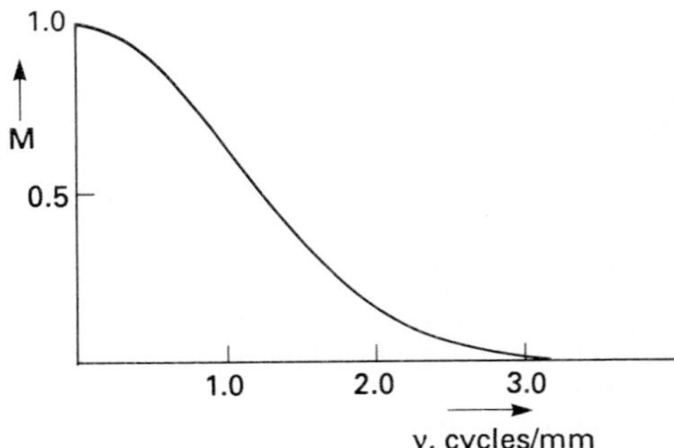

Figure 8.21 Modulation transfer function (mtf) of a spot with a Gaussian intensity distribution

References

1 YOSHIDA, S, OHKOSHI, A and MIYAOKA, S, 'The Trinitron—a new color tube', *IEEE Trans on Broadcast and TV Receivers*, **BTR-14**, 2, 19–27 (July 1968)
2 PLOKE, M, 'Elementare Theorie der Elektronenstrahlerzeugung mit Triodensystemen', *Zeitschrift für angewandte Physik*, **3**, 441–449 (1951) and **4**, 1–22 (1952) (in German)
3 BARTEN, P G J, 'Spot size and current density distribution of CRTs', *Proc SID*, **25**, 3 (1984)
4 SCHWARTZ, J W, 'Space-charge limitation on the focus of electron beams', *RCA Rev*, **18**, 3–11 (March 1957)
5 HOLLWAY, D L, 'The space-charge controlled focus of an electron beam', *Jour Brit IRE*, **24**, 209–211 (Sept 1962)
6 GERRITSEN, J and BARTEN, P G J, 'An electron gun design for flat square 110° color picture tubes', *Proc SID*, **28**, 1, 15–19 (1987)

7 BARTEN, P G J, 'The 20AX system and picture tube', *IEEE Trans on Broadcast and TV Receivers*, **BTR-20**, 4, 286–292 (Nov 1974)
8 BARTEN, P G J and KAASHOEK, J, '30AX self-aligning 110° in-line color TV display', *IEEE Trans on Consum Electron*, **CE-24**, 3, 481–487 (Aug 1978)
9 CHEN, H Y and HUGHES, R H, 'A high performance color CRT gun with an asymmetrical beam forming region', *IEEE Trans on Consum Electron*, **CE-26**, 3, 459–465 (Aug 1980)
10 ASKIZAKI, S, SUZUKI, H and SUGAWARA, K, 'In-line gun with dynamic astigmatism and focus correction', *Proc SID*, **29**, 1, 33–39 (1988)
11 GERRITSEN, J and SLUYTERMAN, A A S, 'A new picture tube system with homogeneous spot performance', *Proc Japan Display '89*, 458–461 (Oct 1989)
12 SUZUKI, H, MITSUDA, K, MURANISHI, H, IWASAKA, K and ASHIZAKI, S, '27V-in. flat-square high-resolution color CRT for graphic displays', *Proc Japan Display '89*, 554–557 (Oct 1989)
13 BARTEN, P G J, 'The SQRI method: A new method for the evaluation of visible resolution on a display', *Proc SID*, **28**, 3, 253–262 (1987)

Bibliography

KLEMPERER, O, *Electron Optics*, University Press, Cambridge (1939 and 1953)
COSSLETT, V E, *Introduction to Electron Optics*, University Press, Oxford (1946 and 1950)
JACOB, L *An Introduction to Electron Optics*, Methuen, London, John Wiley, New York (1951)
STURROCK, P A, *Static and Dynamic Electron Optics*, Cambridge University Press, Cambridge (1955)
GRIVET, P, *Electron Optics*, Pergamon Press, Oxford and New York (1965 and 1972)
SEPTIER, P, et al., *Focussing of Charged Particles*, **1** and **2**, Academic Press, New York and London (1968)
EL-KAREH, A B, and EL-KAREH, J C J, *Electron Beams, Lenses and Optics*, **1** and **2**, Academic Press, New York and London (1970)
SZILAGYI, M, *Electron and Ion Optics*, Plenum Press, New York and London (1988)
HAWKES, P W, and KASPER, E, *Principles of Electron Optics*, **1** and **2**, Academic Press, New York, London (1989)

R G Hunt D Sc, FRPS, FRSA, MRTS
Professor of Physiological Optics, The City University

9 Colour Displays and Colorimetry

9.1 Types of colour display

The display device in a colour television system has to be capable of receiving the red, green and blue picture signals and using them to produce the appropriate amounts of red, green and blue light at each point of the picture. The most widely used display devices are of three main types: shadow-mask tubes, Trinitron type tubes, and triple projection devices.

9.1.1 Shadow-mask tubes

The principle of the shadow-mask tube is illustrated in *Figure 9.1*. The red, green and blue picture signals are applied to the electron guns marked R, G and B, respectively, and all three electron beams from the guns scan the phosphor screen together. However, the screen consists of triads of dots of three different phosphors and, between the guns and the screen, a metal plate with holes in it (the shadow-mask) which ensures that the electron beam from gun R lands only on phosphor dots that produce red light, that from gun G only on dots producing green light, and that from gun B only on dots producing blue light.

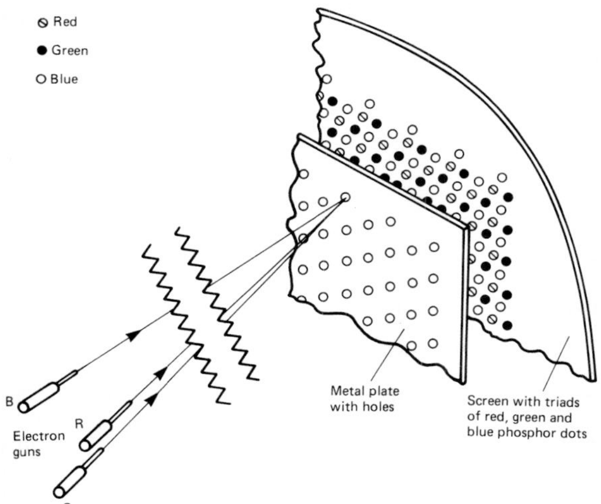

Figure 9.1 Principle of the shadow-mask tube

The rows of dots do not have to be aligned with the lines of the picture, but moiré patterns caused by beats between the line structure and the dot pattern arise at certain angles. As these are worst at $\pm30°$ and negligible at $0°$ it is arranged for the lines of the picture and the lines of the dots to be more or less parallel. For 525-line displays, the shadow-mask usually has about 357 000 holes, which provide about 520 lines of holes with about 690 holes in each line; hence the maximum definition of the tube amounts to about 345 picture-point pairs along a line and 260 picture-point pairs vertically.

If the three electron beams were small enough to irradiate, on the average, not more than one line of holes and its associated triad of phosphor dots, then the tube would not restrict the definition much in a 525-line system. However this would be a rather critical condition in which to operate, and each electron beam normally irradiates more than one line of these holes and their associated triads of dots. There is therefore some theoretical loss of definition, but other factors, such as interlacing, may make the loss unimportant in practice[1].

For 625-line displays, the tubes usually have about 440 000 holes providing about 575 lines of holes, with about 770 holes in each line. In a 56 cm tube, the distance between adjacent dots is usually only about 0.4 mm, so that very great accuracy is required in constructing these tubes[2].

Shadow-mask tubes are often used in colour video display units (VDUs) for viewing data generated by computers. In this case the viewing distance is usually only about 0.5–1 m, instead of about 2–3 m typical for viewing normal pictorial television. It is therefore necessary to use tubes having finer dot structures in VDUs. The size of the dot structure is usually quoted as the *triad pitch*; by this is meant the distance, p, between adjacent holes in the mask. Adjacent rows of phosphor dots are then separated by $p/2$ (see *Figure 9.2*), and the distance between adjacent phosphor dots is $p/\sqrt{3}$. Thus, in the case where (for typical pictorial television) adjacent phosphor dots are separated by 0.4 mm, the triad pitch is given by:

$$p = 0.4 \times \sqrt{3}$$

which is equal to about 0.7 mm. For VDUs, triad pitches of

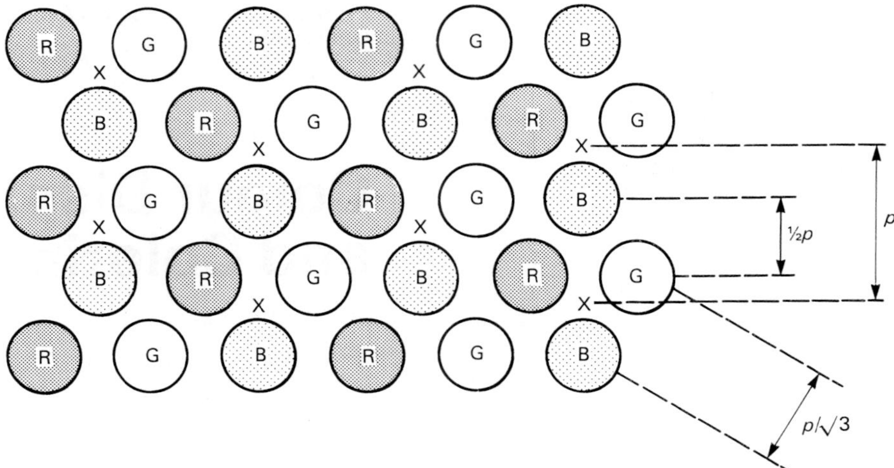

Figure 9.2 Arrangement of phosphor dots in shadow-mask tubes. Positions of shadow-mask holes are indicated by crosses. The triad pitch is p

about 0.3 mm (or sometimes about 0.2 mm) are usually used. In each vertical triad pitch there are two lines of holes.

The three beams of a VDU must be very accurately registered over all the display area, because mis-registration is very noticeable when small symbols are displayed, especially, as is often the case, against a black background. For this reason, special registration adjustments are usually provided in VDUs and registration to within a half, or a third, of a television line width is usually desirable.

The nominal spot size in shadow-mask tubes may be regarded as corresponding to the diameter where the luminance is half the maximum, when all three guns are firing. In VDUs, this spot size is usually about twice the triad pitch. It cannot be smaller than this, because smaller spot sizes result in the spot having variable colour when writing small symbols and make it difficult for the eye to locate the centre of a spot or a line. Thus, in the case of a triad pitch of 0.3 mm, the nominal spot size would be about 0.6 mm.

When using VDUs it is not normally necessary to use the line standards adopted for broadcast television. For a display tube height of 280 mm (typical of tubes having a diagonal of 48 cm or 19 in), a spot size of 0.6 mm corresponds to 280/0.6 = 467 lines. However, the use of more lines than this is common, and as many as 1000 are sometimes used. The excess lines are useful in reducing the incidence of spurious patterns (aliasing) and of jagged edges to lines that should be smooth.

For pictorial television, spot sizes may be similar in diameter to the triad pitch, because small symbols are not often displayed. Thus, for a display height of 325 mm (typical of tubes having a diagonal of 56 cm or 22 in), a triad pitch and nominal spot size of 0.7 mm corresponds to 325/0.7 = 464 lines. The use of more lines than this (525 or 625) in practice, again reduces the incidence of aliasing and jagged edges.

9.1.2 Trinitron type tubes

A three-gun tube in which the phosphors are laid down in stripes, instead of in dots, is the Trinitron. In this tube, the three electron guns lie in the same horizontal plane, and a metal plate with vertical slots in it is positioned so that the electrons from one gun can reach only vertical stripes of a phosphor that produces red light, those from another gun only the stripes that produce green light, and those from the third gun only the stripes that produce blue light (see *Figure 9.3*).

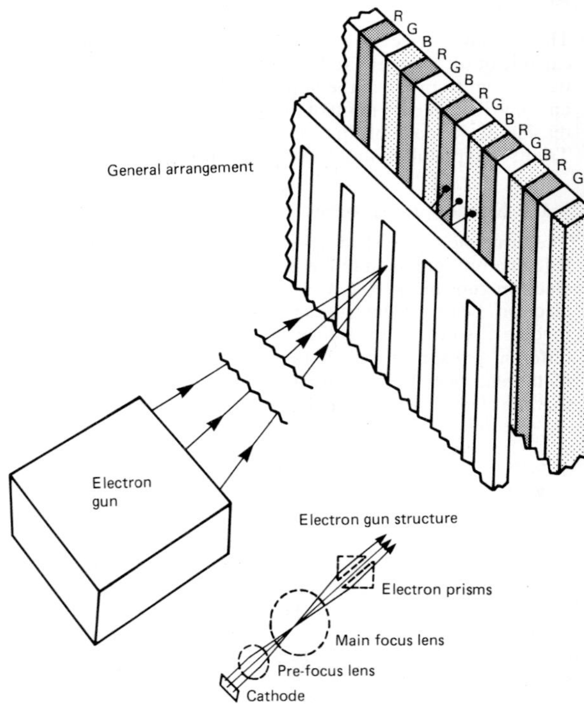

Figure 9.3 Principle of the Trinitron tube

This tube has certain advantages over the shadow-mask tube:

- Deflection of the three electron beams is easier because the gun construction enables the neck of the tube to be smaller.
- The displayed picture emits twice as much light per unit area because, for the same spot size, the beam current can be increased by a factor of 1.5 times, and the stripes of phosphor cover 1.33 times as much area of the tube faceplate.
- Vertical resolution is not affected by the screen structure so

that there is no moiré pattern or loss of vertical resolution by the screen.

- Adjusting the convergence to obtain registration of the three images is easier because the three beams are in a single plane.

The triads of phosphor stripes may be up to about half a millimetre wide, giving about 600 triads in a tube of 300 mm width. For equal horizontal and vertical definition the luminance signal should be able to resolve about 350 cycles per line (e.g. 525 × ½ × 4/3 black/white pairs in a system having 525 actual picture lines). The number of triads of vertical lines required is therefore ideally not less than about 700, but, as in the shadow-mask tube, smaller numbers can be used without too much apparent loss of definition because the actual visual appearance is complicated by interlacing and various other factors[1].

The Trinitron tends to be used for smaller displays than the shadow-mask tube.

9.1.3 Self-converging tubes

In Trinitron and conventional shadow-mask tubes, it is necessary to provide dynamic convergence correction. This is required because stronger magnetic fields are needed to bring the three electron beams into coincidence around the centre of the picture, than those required for the corners, which are farther away and therefore have longer electron paths. As the three electron beams scan the picture, the amount of convergence is therefore adjusted dynamically according to their position in the scan.

In the *precision in-line* tube[3], the three electron guns are arranged parallel to one another in the same horizontal plane, as in the Trinitron tube. However, they do not have dynamic convergence correction, but a special deflection coil is accurately cemented to the neck of the tube. This coil is designed to converge the three electron beams on to the shadow-mask at all positions in the picture. Such a coil can be made to do this only for horizontal or for vertical fans of electron beams; in this case the horizontal fans are converged, and the vertical fans converge before the mask is reached and then separate out into short vertical lines. However, by making the shadow-mask with vertical slots, instead of holes, the efficiency with which it allows the electrons through is about 16 per cent, which is similar to that of a conventional shadow-mask tube (although less than the 20 per cent of a Trinitron tube).

After passing through the slots, the electrons hit the red, green and blue phosphors, which are laid down in stripes as in the Trinitron tube. By making the slots in the shadow-mask discontinuous, the mask is sufficiently rigid to enable it to be made with a spherical profile, as in the conventional shadow-mask tube, rather than cylindrical as in the Trinitron tube. The stripes of phosphor are 0.27 mm (0.0108 in) wide, so that each colour is repeated every 0.81 mm (0.0324 in). The geometry of the phosphor stripes, the slots and the electron guns, is arranged to result in the electrons from each gun landing on phosphor of only one colour. The electron guns are mounted 5.08 mm (0.200 in) apart from one another. The precision in-line tube is particularly suitable for small and medium picture sizes; it can also be made using dots rather than short vertical lines.

9.1.4 Triple projection devices

When projection television devices are being used, it is possible to use the triple projection principle by having three projection television tubes arranged so that they throw red, green and blue images onto a single reflecting screen[4]. The problems of registration of the three images have to be overcome, but when the final display is wanted in projected form, the method is usually the best to adopt. Triple projection is used for the display of colour television pictures to large audiences, and sometimes also for displaying the terrain in simulators used for training the crews of aircraft and ships.

9.1.5 The luminance of reproduced white

The introduction of rare-earth red phosphors enabled whites of luminance about 50 cd/m² to be attained. If the screen has the same luminance at all angles of viewing, this corresponds to the emission of about 50π, or 160, lm/m². For a screen area of 0.15 m², the total emission is therefore about 24 lm. At 3 lm/W this would require about 8 W, or an effective beam current of about 0.3 mA at 25 kV anode voltage.

Subsequent improvements to the phosphors, such as the use of europium activated yttrium oxysulphide for the red, and copper activated zinc cadmium sulphide for the green, made whites of about 85 cd/m² attainable; slight modification to the chromaticities used, and improved screening techniques, have further increased the luminance to about 120 cd/m² in modern tubes[2]. By filling the interstices between phosphor dots with a black absorbing material, it is possible to increase the transmission of the faceplate of the tube (which is normally grey to reduce the effects of ambient illumination), and this further increases the luminance.

High luminance in the display is desirable both because colourfulness increases with luminance, and because a given level of ambient illumination will be less harmful. Flicker caused by the field frequency, however, becomes more noticeable as the luminance rises.

9.2 Colorimetric principles

9.2.1 Trichromatic matching

Colorimetry[5] is based on the experimental fact that observers can match colours with additive mixtures of three reference colour stimuli, normally a red, a green and a blue. This is possible because, in colour vision, the retina of the human eye transduces the incident radiant power of the light to electrical signals by means of only three spectrally different types of receptor, known as *cones*. There is a fourth spectrally different type of receptor, known as a *rod*, but the rods give only monochromatic vision at low levels of illumination. At levels

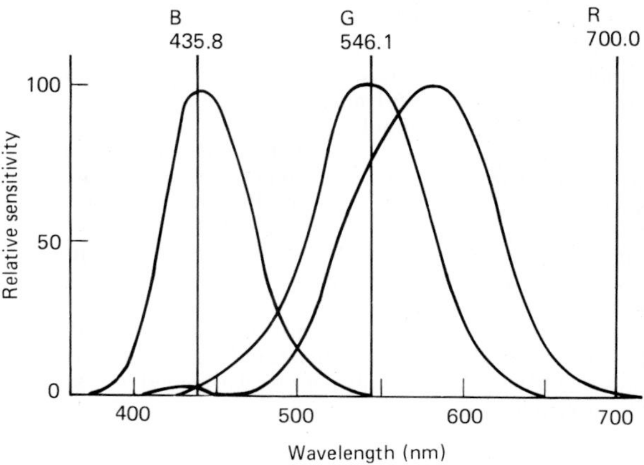

Figure 9.4 Probable spectral sensitivity curves of the cones of the eye, with the wavelengths of the CIE RGB primaries

high enough for colour perception to be operating effectively, it can be assumed, for the purpose of practical colorimetry, that the rods are inoperative. In *Figure 9.4*, spectral sensitivity curves typical of those believed to be characteristic of the cones are shown.

It is clear that one type of cone has a peak sensitivity at about 580 nm, another at about 540 nm, and the third at about 440 nm. Hence, red light stimulates mainly the first type of cone, green light mainly the second type, and blue light mainly the third type. Therefore, if beams of red, green and blue light can be varied in their amounts, and additively mixed together, the combination can be made to produce a very wide range of excitations of the three different types of cone. By adjusting the amounts until the cone excitations are the same as those produced by another colour stimulus, a match can be made. The amounts of the red, green and blue needed to make the match can then serve as a measure of the colour of the other stimulus, and these amounts are known as *tristimulus* values.

If yet another colour stimulus, having a different spectral radiant power distribution, was also matched, different tristimulus values would indicate that the colour looked different, while identical tristimulus values would indicate that it looked the same. Colours having identical tristimulus values but different spectral radiant power distributions are called *metamers*, and the phenomenon *metamerism*. The greater the difference in spectral radiant power distribution between two matching colours, the greater is said to be the degree of metamerism.

9.2.2 The CIE 1931 standard colorimetric observer

For tristimulus values to provide a satisfactory basis for the measurement of colour, various elements of the system must be standardized.

- If the colours of the red, green and blue reference-colour stimuli are changed, even slightly, the tristimulus values for a given stimulus being matched will also change.
- Even if observers having abnormal colour vision (colour blind observers) are excluded, individual observers differ slightly from one another in their tristimulus values for a match — a phenomenon often referred to as *observer metamerism*.
- The angular size of the field of view affects the colour match.

The CIE (Commission Internationale de l'Eclairage) has therefore defined a standard set of reference-colour stimuli, and a standard set of tristimulus values for them to match all the wavelengths of the visible spectrum. These data constitute the CIE 1931 standard colorimetric observer. The reference-colour stimuli are monochromatic radiations of wavelength 700 nm for the red stimulus (R), 546.1 nm for the green stimulus (G) and 435.8 nm for the blue stimulus (B).

If a typical white colour is matched, and the amounts of red, green and blue are measured in photometric units, such as lumens or candelas per square metre, it is found that, with any reasonably typical set of red, green and blue reference-colour stimuli, there is a great imbalance in the three amounts, the amount of green being the greatest, and the amount of blue being much smaller.

Thus, with the three CIE reference-colour stimuli, RGB, it is found that 5.6508 lm of the *equi-energy illuminant* (a hypothetical white having equal energy per unit wavelength throughout the spectrum) is matched by:

1.0000 lm of R,
4.5907 lm of G,
0.0601 lm of B.

Because white is a colour that is not biassed towards either red, green or blue, it is desirable in a colorimetric system for whites to be matched by roughly equal amounts of the three reference-colour stimuli. This is achieved by using units of different photometric magnitudes for each of the three reference colour stimuli. In the above case, 1.000 lm is still used for R, but 4.5907 lm is used for G, and 0.0601 lm is used for B. Then 5.6508 lm of the equi-energy illuminant would be matched by:

1.0000 lm of R,
1.0000 new unit of G,
1.0000 new unit of B.

Other whites, of slightly different colours, would then be matched by amounts of R, G and B that, although not exactly equal to one another, would be not very greatly different.

Using these new units for the CIE reference-colour stimuli, RGB, the amounts of them required to match a constant amount of power (per unit wavelength) of each wavelength of the visible spectrum are shown in *Figure 9.5*.

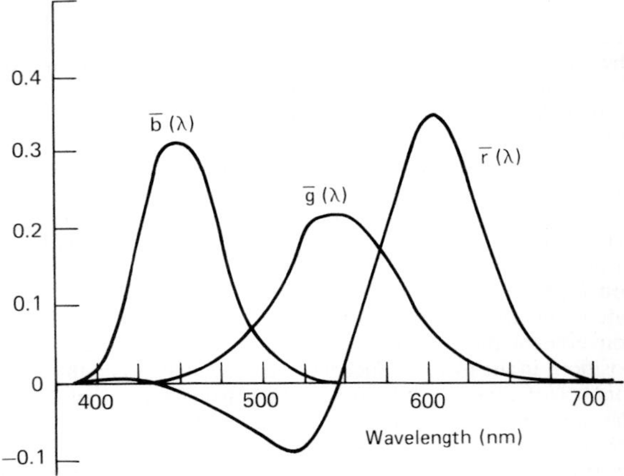

Figure 9.5 Colour-matching function for the CIE RGB primaries

It is clear from this set of curves that, at some wavelengths, one of the three amounts is negative. Some colours cannot be matched by an additive mixture of the three reference-colour stimuli. This is most obviously the case for the blue-green part of the spectrum around 500 nm. The reason for this can be seen by referring again to *Figure 9.4*. It is clear that the R stimulus will excite only the cones whose peak sensitivity is at around 580 nm (the 580 cones), and the B stimulus will excite mainly the 440 cones, but the G stimulus, although exciting the 540 cones most strongly, also excites the 580 cones to a considerable extent. At 500 nm, the cone excitations are approximately in the ratios of (1 of 580):(2 of 540):(1 of 440). But the G stimulus excites the cones in the ratio of about (1-1/2 of 580): (2 of 540). This means that, when trying to match the blue-green of 500 nm, even without any R present, the G stimulus produces too high a ratio of 580 cone to 540 cone stimulation.

The only way to make a match is therefore to add some R stimulus to the 500 nm colour, so that the combination then produces a higher ratio of 580 cone response to 540 cone response. By adding just the right amount of R stimulus to the 500 nm colour, it is found that a match can then be made by adjusting the amounts of the G and B stimuli appropriately. When this is done, the amount of R in the match is counted as negative. This problem of unmatchable colours occurs (although to different extents) with all sets of reference-colour stimuli, and is caused by the degree of overlap of the three curves of *Figure 9.4*.

The curves of *Figure 9.5* are known as colour-matching functions and are of great importance in colorimetry. They are denoted by symbols of the type $\overline{r}(\lambda)$, $\overline{g}(\lambda)$ and $\overline{b}(\lambda)$, and they enable tristimulus values to be calculated from spectral radiant power distributions (see section *9.2.4*).

9.2.3 CIE standard illuminants

The CIE defines a *source* as a physical emitter of light, such as a lamp or the sun and sky. The term *illuminant* refers to a specific spectral radiant power distribution, not necessarily provided directly by a source, and not necessarily realizable.

The following standard illuminants have been defined by the CIE for colorimetric purposes:

Illuminant A	representing light from a Planckian radiator at 2856 K
Illuminant C	representing average daylight with a correlated colour temperature of approximately 6774 K
Illuminant D_{65}	representing a phase of daylight with a correlated colour temperature of approximately 6504 K

A *Planckian radiator* is an illuminant whose spectral radiant power distribution is in accordance with Planck's radiation law, the nature of the radiation depending only on the temperature, usually expressed in kelvins.

An artificial light source of the incandescent type (e.g. a tungsten filament lamp) usually emits light than can be closely matched in colour by that from a Planckian radiator at a particular temperature, and this is said to be the *colour temperature* of the source. The spectral radiant power distributions of incandescent sources are also usually similar to those of Planckian radiators. However, the spectral radiant power distributions of discharge lamps (including fluorescent lamps) and of daylight are considerably different from those of Planckian radiators, and it is frequently impossible to achieve a close colour match with the light from a Planckian radiator at any temperature. In such cases, the light source can be given a *correlated colour temperature*, which is the temperature of the Planckian radiator yielding the nearest possible colour match.

Some correlated colour temperatures typical of sources often met with in practice are as follows:

North sky light	7500 K
Average daylight	6500 K
Xenon (arc or flash)	6000 K
Sunlight plus skylight	5500 K
Fluorescent lamps	3000–6500 K
Studio tungsten lamps	3200 K
Floodlights	3000 K
Domestic tungsten lamps	2800–2900 K
Sunlight at sunset	2000 K
Candle flame	1800 K

Spectral radiant power distributions of CIE standard illuminants A and D_{65} are tabulated at 5 nm intervals in *Table 9.1*.

λ (nm)	A	d_{65}
380	9.80	49.98
385	10.90	52.31
390	12.09	54.65
395	13.35	68.70
400	14.71	82.75
405	16.15	87.12
410	17.68	91.49
415	19.29	92.46
420	20.99	93.43
425	22.79	90.06
430	24.67	86.68
435	26.64	95.77
440	28.70	104.86
445	30.85	110.94
450	33.09	117.01
455	35.41	117.41
460	37.81	117.81
465	40.30	116.34
470	42.87	114.86
475	45.52	115.39
480	48.24	115.92
485	51.04	112.37
490	53.91	108.81
495	56.85	109.08
500	59.86	109.35
505	62.93	108.58
510	66.06	107.80
515	69.25	106.30
520	72.50	104.79
525	75.79	106.24
530	79.13	107.69
535	82.52	106.05
540	85.95	104.41
545	89.41	104.23
550	92.91	104.05
555	96.44	102.02
560	100.00	100.00
565	103.58	98.17
570	107.18	96.33
575	110.80	96.06
580	114.44	95.79
585	118.08	92.24
590	121.73	88.69
595	125.39	89.35
600	129.04	90.01
605	132.70	89.80
610	136.35	89.60
615	139.99	88.65
620	143.62	87.70
625	147.24	85.49
630	150.84	83.29
635	154.42	83.49
640	157.98	83.70
645	161.52	81.86
650	165.03	80.03
655	168.51	80.12
660	171.96	80.21
665	175.38	81.25
670	178.77	82.28
675	182.12	80.28
680	185.43	78.28

λ (nm)	A	D_{65}
685	188.70	74.00
690	191.93	69.72
695	195.12	70.67
700	198.26	71.61
705	201.36	72.98
710	204.41	74.35
715	207.41	67.98
720	210.36	61.60
725	213.27	65.74
730	216.12	69.89
735	218.92	72.49
740	221.67	75.09
745	224.36	69.34
750	227.00	63.59
755	229.59	55.01
760	232.12	46.42
765	234.59	56.61
770	237.01	66.81
775	239.37	65.09
780	241.68	63.38

Table 9.1 The spectral power distributions of CIE standard illuminants A and D_{65}

9.2.4 The XYZ system of colour specification

The presence of negative tristimulus values in red, green and blue systems of colorimetry has led the CIE to adopt a system in which a new set of tristimulus values, XYZ, are obtained from RGB by using the following equations:

$$X = 0.490\,00\,R + 0.310\,00\,G + 0.200\,00\,B$$
$$Y = 0.176\,97\,R + 0.812\,40\,G + 0.010\,63\,B$$
$$Z = 0.000\,00\,R + 0.010\,00\,G + 0.990\,00\,B$$

This simple transformation was carefully designed so that all colour stimuli would have all positive values of X, Y and Z. It was also designed so that the coefficients in the equation for Y, that is 0.176 97, 0.812 40, and 0.010 63, are in the same ratios as 1.0000, 4.5907, and 0.0601, the photometric values of the units used for expressing the amounts of R, G and B. This means that the tristimulus value, Y, is proportional to the luminance, and hence the ratio of the values of Y for any two colours, Y_1 and Y_2, is the same as the ratio of their luminances, L_1 and L_2. Hence:

$$\frac{Y_1}{Y_2} = \frac{L_1}{L_2}$$

The transformation was also designed so that, for the equi-energy illuminant, the values of X, Y and Z are equal to one another. The above equations can also be used to transform the colour-matching functions to the XYZ system, thus:

$$\bar{x}(\lambda) = 0.490\,00\,\bar{r}(\lambda) + 0.310\,00\,\bar{g}(\lambda) + 0.200\,00\,\bar{b}(\lambda)$$
$$\bar{y}(\lambda) = 0.176\,97\,\bar{r}(\lambda) + 0.812\,40\,\bar{g}(\lambda) + 0.010\,63\,\bar{b}(\lambda)$$
$$\bar{z}(\lambda) = 0.000\,00\,\bar{r}(\lambda) + 0.010\,00\,\bar{g}(\lambda) + 0.990\,00\,\bar{b}(\lambda)$$

These colour-matching functions, $\bar{x}(\lambda)$, $\bar{y}(\lambda)$ and $\bar{z}(\lambda)$, are shown in *Figure 9.6*; they are the most important spectral functions in colorimetry. They enable tristimulus values, XYZ, to be calculated directly from spectral radiant power data. If the radiant powers at wavelengths 1, 2, 3, ..., are $P_1, P_2, P_3, \ldots$, and the values of these colour-matching functions are $\bar{x}_1, \bar{x}_2, \bar{x}_3, \ldots$, $\bar{y}_1, \bar{y}_2, \bar{y}_3, \ldots$, and $\bar{z}_1, \bar{z}_2, \bar{z}_3, \ldots$, at the same wavelengths, then the tristimulus values are given by:

$$X = k(P_1\bar{x}_1 + P_2\bar{x}_2 + P_3\bar{x}_3 + \ldots)$$
$$Y = k(P_1\bar{y}_1 + P_2\bar{y}_2 + P_3\bar{y}_3 + \ldots)$$
$$Z = k(P_1\bar{z}_1 + P_2\bar{z}_2 + P_3\bar{z}_3 + \ldots)$$

For reflecting and transmitting samples, the constant k is usually chosen so that X, Y and Z are all equal to 100 for the perfect reflecting or transmitting diffuser (which reflects or transmits all the light at every wavelength) when similarly illuminated. The values of Y then usually give the luminance factor, reflectance factor, reflectance, or transmittance, in all cases as a percentage.

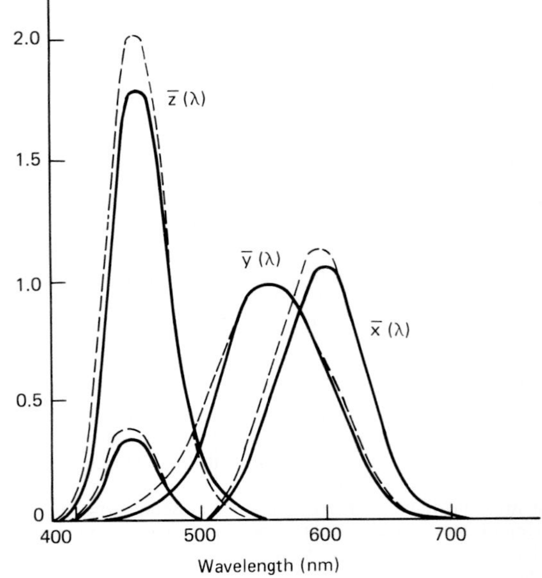

Figure 9.6 Colour-matching functions for the CIE XYZ primaries. Full lines for the 1931 (2°) observer; broken lines for the 1964 (10°) observer

For self-luminous sources, k can be chosen so that Y = 100 when they are used to illuminate the perfect reflecting or transmitting diffuser. For self-luminous objects, such as typical television displays, k can be chosen so that Y = 100 for a suitably chosen reference white in the scene considered. In all these cases, the absolute photometric level can be indicated by quoting the luminous flux, luminous intensity, illuminance, luminance, luminous exitance or light exposure, as appropriate, in addition to the tristimulus values, X, Y and Z. However, if k is set equal to 683, and P(λ) is the spectral radiometric quantity corresponding to the photometric measure required, then this will be given directly by the Y tristimulus value. The symbols X_a, Y_a and Z_a, can be used for such absolute tristimulus values to distinguish them from the usual relative tristimulus values X, Y and Z.

When calculating tristimulus values XYZ, or $X_aY_aZ_a$, the summations are usually carried out at 5 nm intervals throughout the visible spectrum, but intervals of 1, 10 or 20 nm may sometimes be used instead. A table of values of $\bar{x}(\lambda)$, $\bar{y}(\lambda)$ and $\bar{z}(\lambda)$, at 5 nm intervals is given in *Table 9.2*.

9.2.4.1 The CIE 1964 supplementary standard colorimetric observer

As mentioned earlier, colour matches are affected by the angular subtense of the field of observation, and the CIE has a set of colour-matching functions, $\bar{x}_{10}(\lambda)$, $\bar{y}_{10}(\lambda)$ and $\bar{z}_{10}(\lambda)$, for fields of view in excess of about 4° (see *Figure 9.6*); they constitute the colour-matching properties of the CIE 1964 supplementary standard colorimetric observer. In television applications, the areas of colours of interest in displays usually have angular subtenses less than 4° and the CIE 1931 standard colorimetric observer is therefore the appropriate one to use.

λ (nm)	$\bar{x}(\lambda)$	$\bar{y}(\lambda)$	$\bar{z}(\lambda)$
380	0.0014	0.0000	0.0065
385	0.0022	0.0001	0.0105
390	0.0042	0.0001	0.0201
395	0.0076	0.0002	0.0362
400	0.0143	0.0004	0.0679
405	0.0232	0.0006	0.1102
410	0.0435	0.0012	0.2074
415	0.0776	0.0022	0.3713
420	0.1344	0.0040	0.6456
425	0.2148	0.0073	1.0391
430	0.2839	0.0116	1.3856
435	0.3285	0.0168	1.6230
440	0.3483	0.0230	1.7471
445	0.3481	0.0298	1.7826
450	0.3362	0.0380	1.7721
455	0.3187	0.0480	1.7441
460	0.2908	0.0600	1.6692
465	0.2511	0.0739	1.5281
470	0.1954	0.0910	1.2876
475	0.1421	0.1126	1.0419
480	0.0956	0.1390	0.8130
485	0.0580	0.1693	0.6162
490	0.0320	0.2080	0.4652
495	0.0147	0.2586	0.3533
500	0.0049	0.3230	0.2720
505	0.0024	0.4073	0.2123
510	0.0093	0.5030	0.1582
515	0.0291	0.6082	0.1117
520	0.0633	0.7100	0.0782
525	0.1096	0.7932	0.0573
530	0.1655	0.8620	0.0422
535	0.2257	0.9149	0.0298
540	0.2904	0.9540	0.0203
545	0.3597	0.9803	0.0134
550	0.4334	0.9950	0.0087
555	0.5121	1.0000	0.0057
560	0.5945	0.9950	0.0039
565	0.6784	0.9786	0.0027
570	0.7621	0.9520	0.0021
575	0.8425	0.9154	0.0018
580	0.9163	0.8700	0.0017
585	0.9786	0.8163	0.0014
590	1.0263	0.7570	0.0011
595	1.0567	0.6949	0.0010
600	1.0622	0.6310	0.0008
605	1.0456	0.5668	0.0006
610	1.0026	0.5030	0.0003
615	0.9384	0.4412	0.0002
620	0.8544	0.3810	0.0002
625	0.7514	0.3210	0.0001
630	0.6424	0.2650	0.0000
635	0.5419	0.2170	0.0000
640	0.4479	0.1750	0.0000
645	0.3608	0.1382	0.0000
650	0.2835	0.1070	0.0000
655	0.2187	0.0816	0.0000
660	0.1649	0.0610	0.0000
665	0.1212	0.0446	0.0000
670	0.0874	0.0320	0.0000
675	0.0636	0.0232	0.0000
680	0.0468	0.0170	0.0000
685	0.0329	0.0119	0.0000
690	0.0227	0.0082	0.0000
695	0.0158	0.0057	0.0000
700	0.0114	0.0041	0.0000
705	0.0081	0.0029	0.0000
710	0.0058	0.0021	0.0000
715	0.0041	0.0015	0.0000
720	0.0029	0.0010	0.0000
725	0.0020	0.0007	0.0000
730	0.0014	0.0005	0.0000
735	0.0010	0.0004	0.0000
740	0.0007	0.0002	0.0000
745	0.0005	0.0002	0.0000
750	0.0003	0.0001	0.0000
755	0.0002	0.0001	0.0000
760	0.0002	0.0001	0.0000
765	0.0001	0.0000	0.0000
770	0.0001	0.0000	0.0000
775	0.0001	0.0000	0.0000
780	0.0000	0.0000	0.0000

Table 9.2 The CIE colour-matching functions $\bar{x}(\lambda)$, $\bar{y}(\lambda)$ and $\bar{z}(\lambda)$

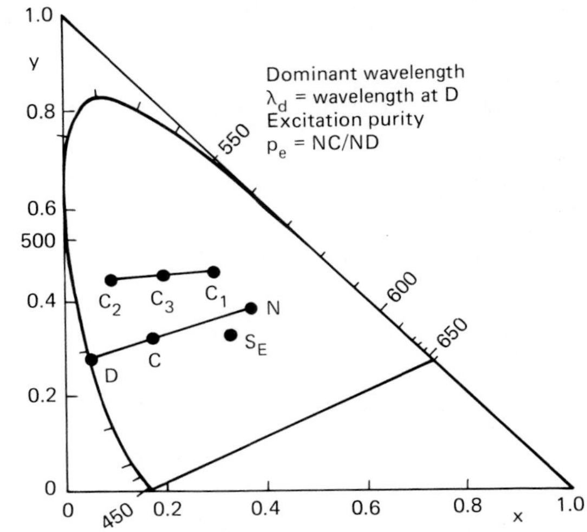

Figure 9.7 x,y chromaticity diagram

9.2.4.2 Chromaticity coordinates

Important colour properties are related to the relative magnitudes of tristimulus values. It is therefore useful to calculate chromaticity coordinates, and this can be done as follows:

$$x = X/(X + Y + Z)$$

$y = Y/(X + Y + Z)$
$z = Z/(X + Y + Z)$

Since $x + y + z = 1$, if x and y are known, z can be deduced from $z = 1 - x - y$. It is, therefore, customary to plot, in two-dimensional diagrams, y against x, as shown in *Figure 9.7*. These diagrams are called *chromaticity diagrams*, and provide useful 'maps' of colours. In *Figure 9.7*, the curved line represents the colours of the spectrum, and the area bounded by this curve and the straight line joining its two ends represents the complete gamut of all real colours.

If, in a chromaticity diagram, a colour C_1 plots at x_1y_1, and another colour C_2 plots at x_2y_2,, then the position of C_3, the colour formed by the additive mixture of C_1 and C_2, is such that it lies on the straight line joining the points x_1y_1 and x_2y_2, as shown in *Figure 9.7*. The point representing C_3 divides the line joining the points representing C_1 and C_2 in the ratio such that:

$$\frac{C_1C_3}{C_2C_3} = \frac{L_2/y_2}{L_1/y_1}$$

where L_1 and L_2 are the luminances of C_1 and C_2 respectively.

9.2.4.3 Dominant wavelength and excitation purity

In *Figure 9.7* are illustrated the derivations of two measures that correlate more closely with perceptual attributes of colours than tristimulus values or chromaticity coordinates. The point C represents the chromaticity of the colour considered; the point N represents that of a suitably chosen reference white or grey (usually the chromaticity of the illuminant, but this is normally different from the equi-energy illuminant, S_E); and the point D lies on the spectral locus intersected by the line NC produced. The wavelength corresponding to the point D is then termed the *dominant wavelength*, λ_d (if the point D lies on the line joining the two ends of the spectrum, then it is produced in the other direction to give the complementary wavelength, λ_c).

Dominant wavelength provides a measure that correlates approximately with the hue of the colour. The ratio NC:ND is termed the excitation purity, p_e, and correlates approximately with saturation (colourfulness judged in proportion to brightness, see *Table 9.3*).

Colour attributes	Correlates
Hue. Denotes whether the colour appears reddish, yellowish, greenish or bluish.	Dominant wavelength, λ_d. *CIE 1976 hue-angle, h_{uv} or h_{ab}.
Brightness. Denotes the extent to which the colour appears to be emitting or reflecting more or less light.	Luminance, L.
†*Colourfulness*. Denotes the extent to which the colour appears to exhibit a hue.	Not yet available.
†*Saturation*. Denotes colourfulness judged in proportion to brightness.	Excitation purity, P_e. *CIE 1976 saturation, s_{uv}.
Lightness. Denotes brightness judged relative to the brightness of a similarly illuminated area that appears to be white.	Luminance factor, L/L_n. *CIE 1976 lightness, L^*.
†*Chroma*. Denotes colourfulness judged as a proportion of the brightness of a similarly illuminated area that appears to be white.	*CIE 1976 chroma, C^*_{uv} or C^*_{ab}.

*These correlates are approximately uniform with the attribute.

†Saturation and chroma are both relative colourfulnesses. In a series of colours of constant chromaticity but reducing luminance factor (a *shadow series*), the saturation remains constant (because the falling colourfulness is judged relative to the falling brightness of the samples), but the chroma reduces (because the falling colourfulness is judged relative to the constant brightness of the reference white).

Table 9.3 Colour attributes and the correlates

For reflecting or transmitting samples, *luminance factor* can also be evaluated. Luminance factor is equal to L/L_n where L is the luminance of the sample, and L_n that of the reference white. Luminance factor correlates with lightness. Hence, dominant wavelength, excitation purity and luminance factor, provide correlates of hue, saturation and lightness, respectively.

9.2.5 Approximately uniform colour systems

9.2.5.1 Uniform chromaticity coordinates u',v'

Chromaticity diagrams are very useful in colorimetry, but, although the CIE x,y diagram has been widely used in the past, it does suffer from one important disadvantage: the colours in it are not uniformly distributed.

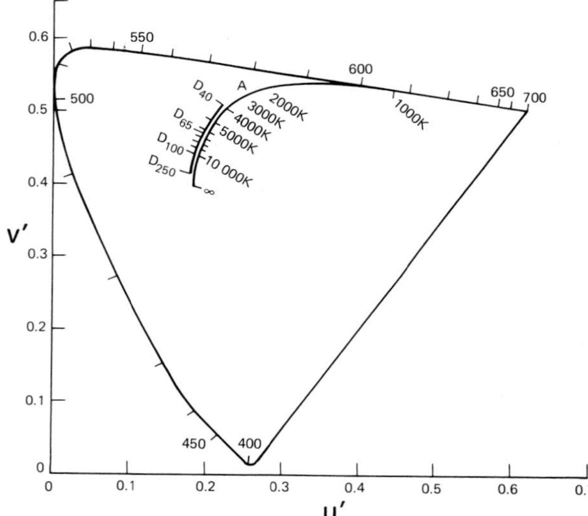

Figure 9.8 u', v' chromaticity diagram, with locus for planckian radiators having colour temperatures from 1000 K to infinity, and for daylight illuminants having correlated colour temperatures from 4000 K(D_{50}) to 25 000 K(D_{250})

In *Figure 9.8* is shown a chromaticity diagram in which are plotted:

$$u' = \frac{4X}{X + 15Y + 3Z} = \frac{4x}{-2x + 12y + 3}$$

$$v' = \frac{9Y}{X + 15Y + 3Z} = \frac{9y}{-2x + 12y + 3}$$

In this chromaticity diagram the colours are more nearly uniformly distributed. It is known as the CIE 1976 uniform-chromacity scale diagram, or the CIE 1976 UCS diagram, often referred to as the *u',v' diagram*. (In 1960 the CIE introduced a similar diagram in which u and v were plotted, where u = u' and $v = \frac{2}{3}v'$; this u,v diagram has now been superseded by the u',v'

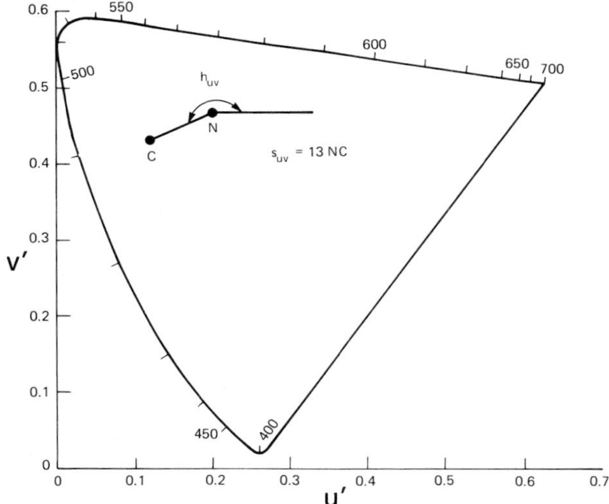

Figure 9.9 The derivation of hue-angle, h_{uv}, and saturation, s_{uv}, in the u′,v′ diagram

diagram.) The u′,v′ diagram is very useful for representing the additive mixtures of the light emitted by phosphor primaries in television displays.

9.2.5.2 *CIE 1976 hue-angle and CIE 1976 saturation*

The u′,v′ diagram can provide better correlates of hue and saturation than dominant wavelength and excitation purity. In *Figure 9.9*, the point C represents the chromaticity of the colour considered, and the point N that of a suitably chosen reference white or grey. The angle between a line from N horizontally to the right and the line NC is the CIE 1976 hue-angle, h_{uv}, and correlates with perceived hue better than dominant wavelength. The distance NC (when multiplied by 13) is the CIE 1976 saturation, s_{uv}, and correlates with perceived saturation better than excitation purity.

These better correlates arise in part from the better uniformity of colours in the u′,v′ diagram, as compared with the x,y diagram, and in part from the different type of formulation of h_{uv} and s_{uv}.

9.2.5.3 *Uniform colour spaces*

Uniform chromaticity diagrams, like all other chromaticity diagrams, only represent proportions of tristimulus values, not their actual values. They therefore only represent uniformly the magnitudes of colour differences for stimuli all having the same luminance.

In general, when two colours differ, they will not necessarily have the same luminance. Colour differences therefore have to be evaluated in three-dimensional colour space, rather than on a two-dimensional chromaticity diagram. The CIE has developed two such spaces: the CIE 1976 ($L^*u^*v^*$) colour space, also called the CIELUV colour space, and the CIE 1976 ($L^*a^*b^*$) colour space, also called the CIELAB colour space. The CIELUV space is more directly applicable to television, since it incorporates the u′,v′ chromaticity diagram already described. It is illustrated in *Figure 9.10*.

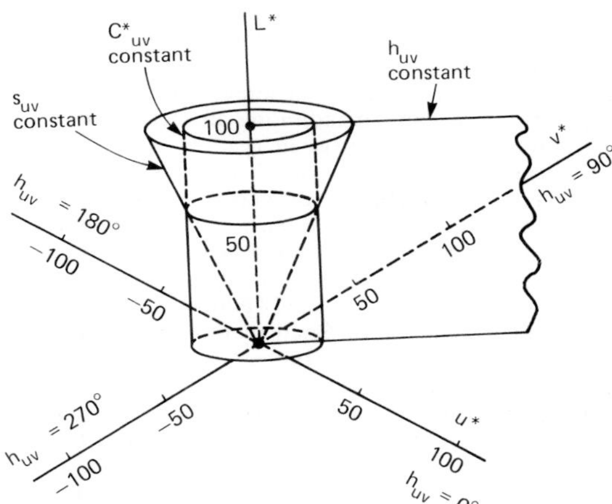

Figure 9.10 CIELUV colour space

The CIELUV space is produced by plotting, along rectangular coordinates, the quantities, L^*, u^* and v^*, defined as follows:

$$L^* = 116(Y/Y_n)^{1/3} - 16$$
$$u^* = 13L^*(u' - u_n')$$
$$v^* = 13L^*(v' - v_n')$$

where Y, u′ and v′, refer to the colour considered, and Y_n, u_n and v_n', refer to a suitably chosen reference white. (If Y/Y_n is less than 0.008 856, then L^* is evaluated as 903.3 Y/Y_n, instead of by the formula for L^* given above.

The total difference between two colours whose differences in L^*, u^*, and v^* are ΔL^*, Δu^*, and Δv^*, respectively, is then evaluated as:

$$\Delta E^*_{uv} = \{(\Delta L\}^*)^2 + (\Delta u^*)^2 + (\Delta v^*)^2\}^{1/2}$$

In this $L^*u^*v^*$ system, approximate correlates of perceptually important colour attributes, as shown in *Figure 9.10*, may be calculated as follows:

CIE 1976 lightness:

$$L^* = 116(Y/Y_n)^{1/3} - 16$$

where Y/Y_n must not be less than 0.008 856.

CIE 1976 u,v saturation:

$$s_{uv} = 13\{(u' - u\}_n')^2 + (v' - v_n')^2\}^{1/2}$$

CIE 1976 u,v chroma:

$$C^*_{uv} = (u^{*2} + v^{*2})^{1/2} = s_{uv}L^*$$

CIE 1976 u,v hue-angle:

$$h_{uv} = \arctan \{(v' - v\}_n')/(u' - u_n')\}$$
$$= \arctan (v^*/u^*)$$

CIE 1976 u,v hue-difference:

$$\Delta H^*_{uv} = \{(\Delta E\}^*_{uv})^2 - (\Delta L^*)^2 - (\Delta C^*_{uv})^2\}^{1/2}$$

h_{uv} lies between 0° and 90° if v^* and u^* are both positive, between 90° and 180° if v^* is positive and u^* is negative, between 180° and 270° if v^* and u^* are both negative, and between 270°

and 360° if v^* is negative and u^* is positive. CIE 1976 u,v hue-difference is introduced so that a colour difference ΔE^* can be broken up into components ΔL^*, ΔC^* and ΔH^*, whose squares add up to the square of ΔE^*. The hue-difference, ΔH^*_{uv}, is to be regarded as positive if indicating an increase in h_{uv} and negative if indicating a decrease in h_{uv}.

CIE 1976 u,v chroma, C^*_{uv}, has been designed to correlate with perceived chroma. This is the perceptual attribute defined as colourfulness judged as a proportion of the brightness of a similarly illuminated area that appears white or highly transmitting. It is equal to the product $s_{uv} L^*$, and the multiplication of the correlate of saturation, s_{uv}, by L^* allows for the fact that, for a given difference in the chromaticity of a colour from that of the reference white, its colourfulness decreases as the luminance factor is reduced. By being based on relative tristimulus values (X, Y, Z) and not on absolute tristimulus values (X_a, Y_a, Z_a), this measure C^*_{uv} does not change as the level of illumination is changed.

Thus, an orange and a brown may have the same chromaticities, and therefore the same values of s_{uv}, and the same saturation. But the lower value of L^* for the brown will result in it having a lower C^*_{uv}, and it appears of lower chroma. If the illuminance level is changed, the values of C^*_{uv} will not change, and this represents the fact that the perceived chromas of the orange and brown remain fairly constant over a wide range of illuminances. At lower illuminances both the orange and the brown will look less colourful than at higher illuminances; they will also look less bright at the lower illuminances.

Under specified viewing conditions, luminance can usually provide an approximate correlate with brightness, but does not provide a perceptually uniform scale. However, there is at present no agreed measure that provides a correlate for colourfulness (see *Table 9.3*).

The CIELAB system, which was designed to be similar to certain systems used widely in the colorant industries, is similar to the CIELUV system, but has no associated chromaticity diagram and no correlate of saturation. The CIELAB space is produced by plotting along rectangular coordinates the quantities, L^*, a^* and b^*, defined as follows:

$$L^* = 116(Y/Y_n)^{1/3} - 16$$

where $Y/Y_n \geq 0.008\,856$.

$$a^* = 500\{(X/X\}_n)^{1/3} - (Y/Y_n)^{1/3}\}$$

where X/X_n, Y/Y_n and $Z/Z_n \geq 0.008\,856$.

$$b^* = 200\{(Y/Y\}_n)^{1/3} - (Z/Z_n)^{1/3}\}$$

where X/X_n, Y/Y_n and $Z/Z_n \geq 0.008\,856$. X, Y and Z refer to the colour considered, and X_n, Y_n and Z_n refer to a suitably chosen reference white. Colour differences in this system are evaluated as:

$$\Delta E^*_{ab} = \{(\Delta L\}^*)^2 + (\Delta a^*)^2 + (\Delta b^*)^2\}^{1/2}$$

Approximate correlates of lightness, chroma, and hue in this system are calculated as follows:

CIE 1976 lightness:

$$L^* = 116(Y/Y_n)^{1/3} - 16$$

where $Y/Y_n \geq 0.008\,856$.

CIE 1976 a,b chroma:

$$C^*_{ab} = (a^{*2} + b^{*2})^{1/2}$$

CIE 1976 a,b hue-angle:

$$h_{ab} = \arctan(b^*/a^*)$$

CIE 1976 a,b hue-difference:

$$\Delta H^* = \{(\Delta E\}^*_{ab})^2 - (\Delta L^*)^2 - (\Delta C^*_{ab})^2\}^{1/2}$$

These spaces are intended to apply to comparisons of differences between reflecting object colours of the same size and shape, viewed in identical white to middle-grey surroundings, by an observer photopically adapted to a field not too different from that of average daylight. They are not necessarily applicable to self-luminous displays such as are used in television, without appropriate modifications.

A summary of colour attributes and their correlates is given in *Table 9.3*.

9.3 Chromaticities of display phosphors

9.3.1 Introduction

The choice of the colours emitted by reproduction phosphors is important because it affects:

- the gamut of colours that can be reproduced,
- the spectral sensitivities that are optimum for the three colour channels of the camera,
- the maximum luminance attainable on the display.

The phosphors are usually chosen to strike the best compromise between these three factors.

9.3.2 NTSC phosphors

When the NTSC (National Television Systems Committee) system was originally set up in the USA in 1953, the best phosphors then available were such as to produce primary colours having the following chromaticities:

	x	y	u′	v′
Red	0.67	0.33	0.477	0.528
Green	0.21	0.71	0.076	0.576
Blue	0.14	0.08	0.152	0.195
Illuminant C	0.3101	0.3162	0.2009	0.4610

The system used illuminant C as the reference white.

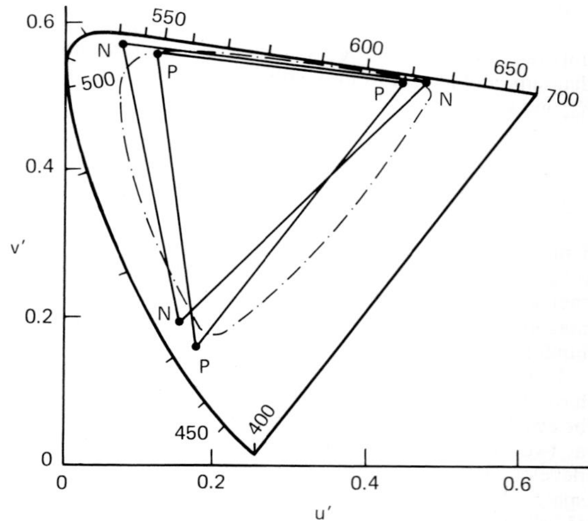

Figure 9.11 Chromaticity gamuts for NTSC phosphors (N), EBU phosphors (P) and real colours (broken line curve), in the u′,v′ diagram

The positions of these primaries in the u′,v′ diagram are shown in *Figure 9.11* by the points marked N. The triangle connecting the N points represents the gamut of chromaticities that can be reproduced. It is clear that there is a region of colours near the blue-green part of the spectrum, 470–530 nm, which lie outside the triangle and therefore cannot ever be displayed. There is also an even larger area between the edge of the triangle and the line joining the two ends of the spectrum. This area represents red, magenta, purple and violet colours that cannot be displayed. Although these two areas comprise quite a large proportion of the total gamut of real colours, the gamut of typical surface colours is considerably smaller, as shown by the broken line in *Figure 9.11*, and most of that area is covered by the triangle.

9.3.3 EBU phosphors

Since 1953, a much wider range of phosphors has become available, and the EBU (European Broadcast Union) has adopted the following set of chromaticities to represent typical phosphors that are used currently[6]:

	x	y	u′	v′
Red	0.64	0.33	0.451	0.523
Green	0.29	0.60	0.121	0.561
Blue	0.15	0.06	0.175	0.157
Illuminant D_{65}	0.3127	0.3290	0.1978	0.4683

This system uses illuminant D_{65} as the reference white.

The position of these primaries is also shown in the u′,v′ diagram in *Figure 9.11* by the points marked P. The corresponding triangle shows that the displayed gamut is even more restricted for blue-green colours. However these phosphors are capable of giving pictures of much higher luminance, and this increases the colourfulness of the displayed colours sufficiently to offset the loss of saturation of the blue-greens.

The tolerances for the EBU phosphors are such that their chromaticities should lie somewhere between the following four points specified for each colour:

red		green		blue	
u′	v′	u′	v′	u′	v′
0.441	0.530	0.115	0.562	0.157	0.159
0.441	0.520	0.119	0.570	0.174	0.170
0.461	0.518	0.128	0.560	0.183	0.154
0.461	0.526	0.124	0.552	0.176	0.146

9.3.4 Camera spectral sensitivities

For any set of primaries, there will be a corresponding set of colour-matching functions showing the amounts of the primaries needed to match each wavelength of the spectrum. This set of colour-matching functions then shows what the spectral sensitivities of the three colour channels of the camera should be. The two sets of colour-matching functions for the NTSC and EBU primaries are shown in *Figure 9.12*. Both sets have negative portions, which are slightly more pronounced in the case of the EBU phosphors because, in the case of the green, the EBU phosphor is more yellow than the NTSC phosphor.

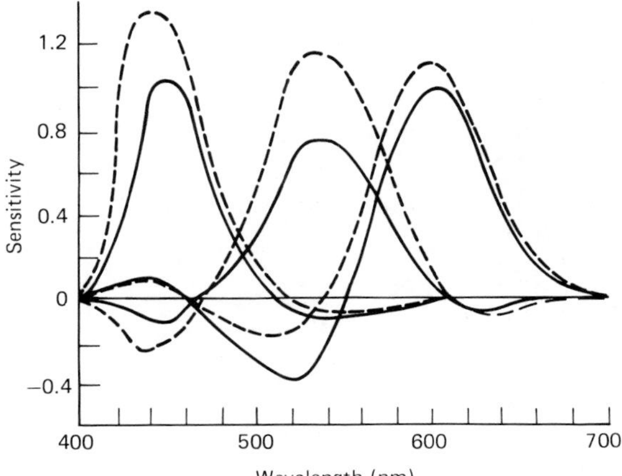

Figure 9.12 Colour-matching functions for the NTSC phospors (broken lines) and the EBU phosphors (full lines)

9.3.5 Matrixing

It is possible to realize negative portions in camera sensitivity curves by the technique of *matrixing*. Modified red, green and blue signals $R_mG_mB_m$ are obtained from the camera tube signals RGB by a circuit whose algebraic equivalent is a matrix:

$$R_m = + 1.14\,R - 0.18\,G + 0.04\,B$$
$$G_m = - 0.06\,R + 1.23\,G - 0.17\,B$$
$$B_m = - 0.03\,R + 0.02\,G + 1.01\,B$$

The numerical coefficients in the above set of equations are only given as an example of typical values that may be used.

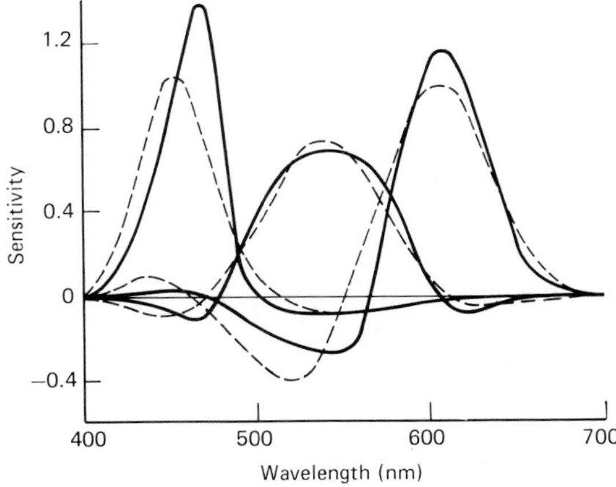

Figure 9.13 Typical matrixed camera sensitivities (full lines), and colour-matching functions for EBU phosphors (broken lines)

In *Figure 9.13* are shown the spectral sensitivities corresponding to the matrixed signals of a typical camera, together with the colour-matching functions for the EBU phosphors. The two sets of curves are only roughly similar, but the matrixing step usually produces very significant improvements in colour reproduction. What matrixing can do is to give correct colorimetric reproduction within the phosphor gamut; colours lying outside the gamut still cannot be displayed, and usually move to the edges of the triangle.

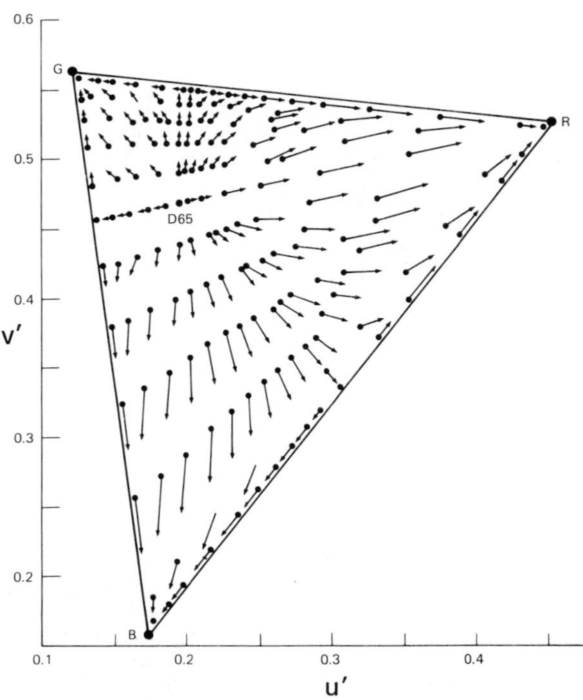

Figure 9.14 Arrow heads show how the chromaticities of the dots are distrorted by an increase in system gamma of 1.27

9.3.6 Effect of gamma correction on colour reproduction

If a system is designed to give correct colour reproduction within the phosphor gamut for an overall system gamma of 1, and is then used with a display device that results in an overall gamma of 1.27, then the distortions shown in *Figure 9.14* are produced.

In practice, this can occur because the transmitted signals are down-gammaed by the power of 1/2.2, and typical receivers operate at gammas of about 2.8.

References

1 JESTY, L C, *Proc IEE*, **105B**, 425 (1958)
2 WRIGHT, W W, *Jour RTS*, **13**, 221 (1971)
3 NEATE, J, *Television*, **23**, 344 (1973)
4 FEDERMAN, F and POMICTER, D, *Jour RTS*, **16**, vii (May-June 1977)
5 *IBA Technical Review* 22, Light and colour principles (1984)
6 BREMA, *Radio and Electronic Engnr*, **38**, 201 (1969)

Bibliography

HUNT, R W G, *The Reproduction of Colour*, 4th Ed., Fountain Press, London (1987)

S J Lent C Eng, MIEE
Engineering Research Department, BBC

10 Pickup Tubes and Solid-state Cameras

Prior to the start of colour television broadcasting, cameras used mainly orthicon type[1] pickup tubes. These were quite large, fairly complex, and had other disadvantages. Signal production was based on *photoemissive* targets which emit electrons in proportion to the brightness of the image focused on them. Without electron multiplication or secondary emission techniques, the sensitivity was restricted because the energy required to emit photoelectrons came from the incident light. Only a fraction of the theoretical limit (approximately 500 μA/lm) was approached in practice.

In the 1930s, work started on the development of target materials with *photoconductive* properties, in which the electrical resistance changes with exposure to light. This means that a current flowing through the target from an external source is controlled by incident illumination, so that very high photo-efficiencies (theoretically in excess of 10 000 μA/lm) can be obtained. This is similar to the manner in which very small changes of gate charge in a field effect transistor control the correspondingly much greater changes of source/drain current.

The first tube to be developed with a photoconductive target was the vidicon[2], which although suitable in many respects for television use, was characterized by image retention and poor dynamic response (the ability to follow rapid changes in illumination) due to photoconductive lag. Eventually these problems were overcome, and in 1964 the plumbicon tube[3] appeared with an improved photoconductive target. This offered outstanding advantages over previous types, such as high sensitivity, fast dynamic response, good resolution, simplicity, robustness and relatively small size. These features, amongst others, were ideally suited to its application in all fields of television and particularly for public service colour television where it has been the main type of pickup tube in use prior to the introduction of solid-state sensors in cameras.

The first colour cameras based on solid-state sensors appeared in 1979 following the introduction of ccd (charge coupled device) principles[4,5] in 1970. The cameras employed a single sensor[6] with integrated colour stripe filters and were intended for the consumer market to supplement the increasing use of home vtrs. Since that time, ccd imagers have improved very considerably and reached a stage where cameras employing three ccds in place of pickup tubes are being used for some applications in the colour television broadcast field. With the continued development of solid-state sensors, it is highly likely that cameras of broadcast standard, using only one sensor, will be introduced in the fairly near future.

In comparison with conventional pickup tubes, ccd imagers have many significant advantages. They are much smaller, and provide lighter tube image formats with the absence of registration adjustments, geometric distortion, capacitive lag or burn-in effects. They also provide instant operation (no warm-up time), have low power consumption, are robust and have a long life potential. The resolution of the ccd is limited by aliasing effects due to the sensors' structure, but this is likely to improve in future developments by using pre-filtering techniques and by increasing the number of sensor elements in the imager.

10.1 Photoconductive tubes: principle of operation

Figure 10.1 is a schematic diagram of a photoconductive tube with its target, electron gun and associated magnetic coils. Electrons emitted from the indirectly heated cathode pass through control grid 1 and form a beam which is accelerated by grid 2 into the cylindrical electrode grid 3. The beam continues

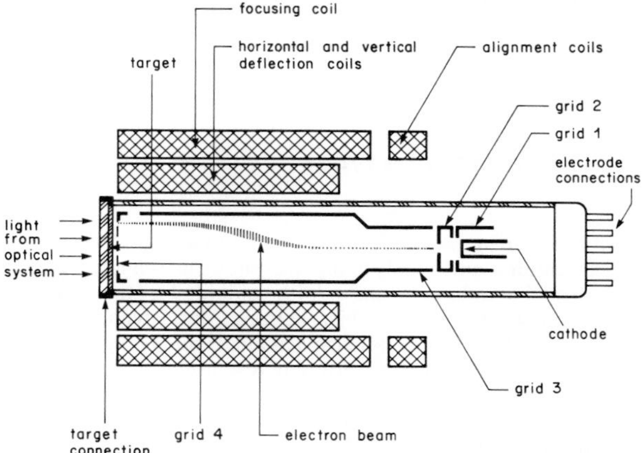

Figure 10.1 Typical electrode and coil arrangement of a photoconductive camera tube

on through grid 4 to reach the target at a relatively low velocity. Grid 4 is a fine mesh construction and provides a uniform decelerating field in front of the target.

Typical operating electrode voltages are:

cathode	0 V
grid 1	−30 – −100 V (without blanking pulses)
grid 3	300 V
grid 4	675 V
target	40 V

The beam is magnetically focused onto the target by means of the axial field from a focus coil, in combination with a suitable voltage on grid 3. (Fine focus can be set by varying either this voltage or the magnetic field.) A transverse magnetic field from two alignment coils provides adjustment to ensure that the beam lands normal to the target surface. Horizontal and vertical beam scanning is magnetic and provided by the deflection coils.

The target, which is basically a thin photoconductive layer deposited on the inside of the tube faceplate, can be described as consisting of a very large number of picture elements, effectively produced by the scanning beam and the target structure. Each of these elements consists of a small capacitor shunted by a light sensitive resistor. All the elements are connected at one end, by a transparent common conducting film between the faceplate and the target, to the external signal electrode which is held at about 40 V through a load resistance.

If the target is completely unlit, some of the higher energy electrons from the beam are attracted to the target until it reaches a potential close to that of the cathode. This results in a potential difference across each picture element, from the conducting film to the beam side of the target layer. Without illumination, the target material thus acts as an insulator allowing only a very small fraction of the charge on each element to leak away. This charge loss is restored during each new beam scan, which produces a so-called *dark current* flowing through the target and load resistance.

When the target is illuminated, the shunt resistance in each element is reduced by photoconductivity, and further charge leaks away by an amount dependent on the light incident on each element. Thus a charge pattern is built up on the beam side of the target layer, corresponding to the optical image focused on the faceplate. When the beam scans the target and re-establishes it to cathode potential, a capacitive current flows through the target via the load resistance, producing a signal voltage (negative polarity for highlights), which is continuously representative of the optical image.

10.2 Camera tube types

Tubes are manufactured for different applications, to various specifications, and mainly in the sizes shows in *Table 10.1*. They are available in basic forms or with combined features such as act (anti-comet tail) and bias lighting, dbc (dynamic beam control) with loc (low output capacitance) and bias lighting (see section *10.2.2*), and with different resolution characteristics. They are also graded for optimum performance in any of the luminance, red, green or blue camera channels. A typical tube is shown in *Figure 10.2*. Details of the various types and applications are to be found in the manufacturers' handbooks.

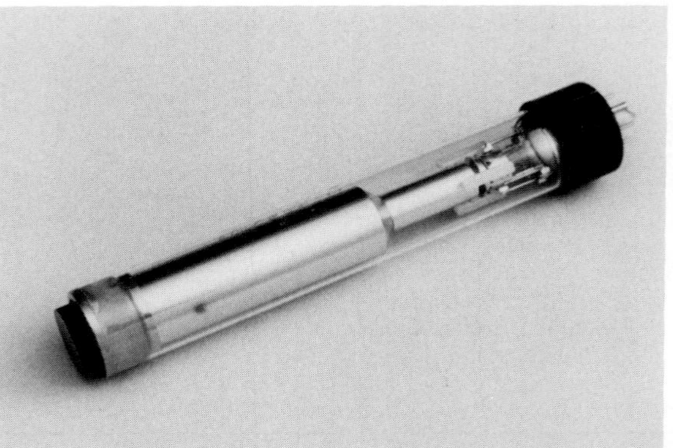

Figure 10.2 A typical lead oxide tube

Diameter (nominal)	Length (approx)	Target image area (quality rectangle)
30	205	12.8 × 17.1
25	160	9.6 × 12.8
18	105	6.6 × 8.8
14	73	4.8 × 6.4

Table 10.1 Camera tube dimensions (in mm)

Photoconductive tubes fall into two general categories: those suitable for broadcast quality television cameras[7], both monochrome and colour, and the remainder which are used for closed circuit television, industrial, military and other applications. The former have to meet stringent requirements for broadcast colour television, including high sensitivity, linear response to illumination, good resolution, fast dynamic response (low lag), low dark current, minimal shading, low optical flare and the ability to handle excessive highlights. Other factors of importance are long life, resistance to image 'burn-ins', freedom from blemishes and the ability to operate over a wide range of temperatures.

Many of these characteristics are inherent properties of the photoconductive materials used in the targets, and there are currently two main types of tube in use with target materials that provide these features to a satisfactory extent. These are described in later sections, with other specialized types of tubes.

10.3 Tubes for broadcast television

10.3.1 Plumbicon, leddicon, vistacon and saticon

With the exception of the saticon[8] all these tubes[9,10] have lead oxide targets and the same associated basic performance characteristics.

A *plumbicon* target[11] consists of the optically flat tube faceplate with a transparent conducting film of stannic oxide (SnO_2) on the inside, connected to an external signal electrode. Next is a photoconductive layer of lead monoxide deposited on the conductive film. The inner or scanned surface of the lead monoxide is doped to form a p-type semiconductor, and that close to the faceplate is doped to form an n-type semiconductor. This is analogous to a reverse biased p-i-n diode which allows current to flow in one direction when it is illuminated, but otherwise is an insulator with an extremely small reverse current, which is the dark current. The variation in dark current over the target area is very small, providing a good black level uniformity, which is an essential requirement in colour television.

The *saticon* was introduced more recently than the plumbicon but is now well established in many different forms for various applications, and is used extensively in the broadcast field. Like the lead oxide tube, the faceplate has a similar transparent conducting film of stannic oxide connected to the signal electrode. The photoconductive target has several graduated layers, with different combinations of selenium, tellurium and arsenic. On the scanned side of the target a thin porous layer of antimony trisulphide (SbS_3) is deposited to minimize any secondary emission effects produced by the scanning beam. As in the lead oxide target, dark current is very small and the effects very uniform over the scanned area, but the tube is capable of greater resolution than lead oxide, at the expense of lag performance.

10.3.2 Highlight operation, act and hop tubes

The illumination transfer characteristic for lead oxide and saticon tubes is almost linear, up to a maximum signal level depending on the maximum beam current. In order to avoid excessive lag and loss of resolution, the beam current is typically set to a level sufficient to stabilize twice the normal expected peak signal current. At extremely high light levels, the beam current is thus insufficient to replace the charge fully, and the target becomes unstabilized with *blooming* accompanied by loss of detail in the highlight areas. As several beam scans are required to re-establish stabilization in these areas, this condition leads to *comet tails* which follow moving highlights.

In order to reduce this light overload effect, lead oxide tubes have been developed, known as anti-comet tail (act) plumbicons,[12] and highlight overload protection (hop) leddicons.[10] Both types are equipped with special electron guns and operate on the same principal rise on the target elements.

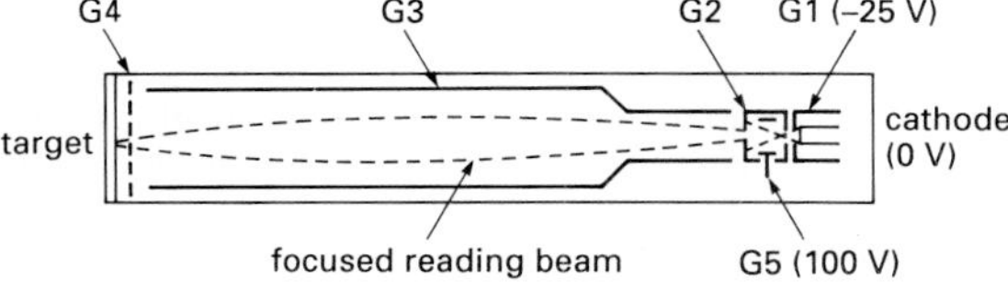

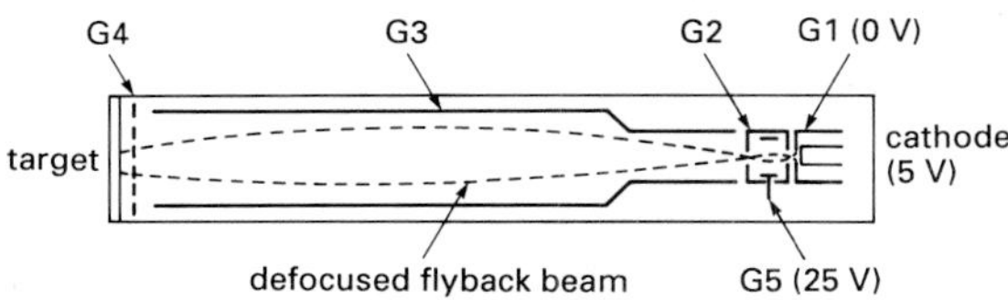

Figure 10.3 Diagram of a hop tube in the readout mode (*top*) and the flyback mode (*bottom*)). Grids G2, G3 and G4 are at normal tube potentials

The operation of an hop tube is shown in *Figure 10.3*. During the line flyback period, the control grid G1 is pulsed positively to increase the beam current, while the auxiliary grid G5 is pulsed negatively to produce a large diameter defocused beam, and the cathode is pulsed to a positive potential. Thus during line flyback the defocused beam, of about 100 μA, scans the target and recharges the extreme highlight areas, bringing them to the same positive potential as the cathode. Target elements with picture information below this potential are not affected until swept by the next normal scanning beam, which should not encounter any excessively charged areas. Thus a 'knee' followed by a saturation level is introduced into the light transfer characteristic of the tube. In this way, highlights of more than five lens stops in excess of normal beam capability can be handled without blooming, comet tail effects, or picture distortion.

10.3.3 Diode gun tube and dynamic beam control

In a conventional triode gun as described previously, the control grid G1 (see *Figure 10.1*) and the second grid G2 converge the electron beam to a point between the grids and the target. The resultant high density, and greater interaction of the electrons in the converged region, creates an increased electron energy spread within the beam. This increases the beam resistance and consequently the target capacitive lag.

The *diode gun tube*[13] is a development in which a triode gun has the control grid G1 biased positively with respect to the cathode, and thus operates in a diode mode with the grid drawing an appreciable current. This reduces the beam convergence, and results in a lower electron energy spread. The beam resistance is thus lower, resulting in reduced target lag. Thus a thinner target layer can be used for the same amount of lag as would be experienced with a conventional tube, but with a consequent increase in resolution, a particularly useful feature in smaller tubes.

A much larger beam reserve is available with the diode gun configuration and enables a system of *dynamic beam control* (dbc)[14] to be used to handle excessive highlights in plumbicons. A similar arrangement used with saticons is known as *automatic beam optimization* (abo).[15] In both methods, the return beam current is measured by comparing an amplified version of the signal voltage with a voltage derived from the beam current. This difference signal is arranged to control the beam current, so that it automatically increases in areas of the target where excessive highlights occur. Typically, these systems handle a light range equivalent to about four lens stops, somewhat less than that supplied by act or hop tubes, but adequate for small camera applications.

10.3.4 Low output capacitance tubes

The signal/noise ratio in a camera is determined mainly by the input stages of the camera pre-amplifier. However the limiting performance is dependent on the effective output capacitance of the camera tube (in the scanning yoke assembly), which is in parallel with the head amplifier input. Any reduction in this capacitance will improve the signal/noise ratio and the sensitivity of the camera.

In the loc tube[13,16] lower output capacitance is achieved by reducing the size of the transparent conducting film on the target, so that it is just larger than the scanned area, and replacing the external target contact ring around the faceplate with a smaller single contact point. Further improvements are obtained by using the loc tube in a special scanning yoke that contains the first fet stage of the video pre-amplifier, with only a very short connection to the tube target.

10.3.5 Tubes with mixed focus and deflection fields

Instead of the conventional all-magnetic system of focus and deflection[17] shown in *Figure 10.1*, electrostatic focus and magnetic deflection has been available for some time, particularly in the smaller sizes. As no focusing coils are required, the savings in power, size and weight are significant in the design of the smaller hand-held cameras. A further important advantage is that there are reduced interaction effects between focusing and deflection. The picture can be rotated, magnified or reduced in size by means of the focus voltage.

More recently, the so-called mixed field (ms) tubes incorporating the *deflectron* principle[18] have appeared in the range of saticon developments, where the focus is magnetic and the deflection is electrostatic. In this type of tube, the scanning electrodes are deposited in a particular pattern on the interior of the glass walls, between the electron gun and the target, so that only a relatively compact focus coil is required. In addition to the considerable economies obtained in power, space and weight, improvements in resolution, particularly in its uniformity over the target area, are also provided. For example, the resolution capability of a compact 25 mm dis (diode-gun, impregnated-cathode saticon) tube[19] developed for high definition television, is in the region of 40 per cent at 800 lines. In this tube, the gun has been specially developed to achieve very high resolution compatible with low beam discharge lag, and has a large beam current requirement typically up to 1.5 μA. In order to withstand the high electron emission density, the cathode is 'reinforced' with barium.

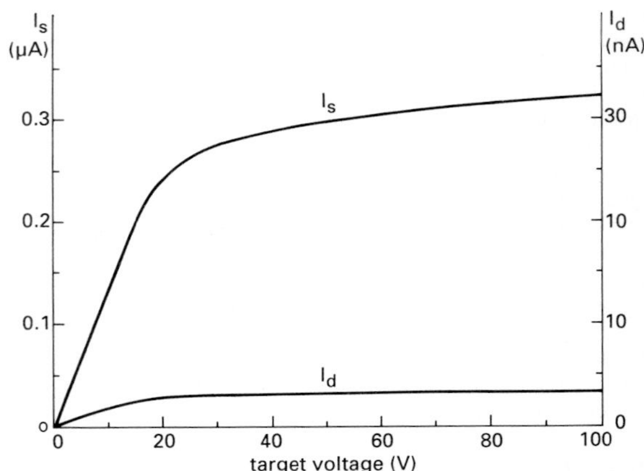

Figure 10.4 Signal current I_s, and dark current I_d, characteristics of a lead oxide tube, with constant level of illumination and varying target voltage

10.4 Performance characteristics of broadcast standard tubes

10.4.1 Sensitivity

Sensitivity comparisons between different camera tubes need to relate to conditions in which the same angle of view and the same depth of focus are required.

The aperture F (i.e. the f/number) of a lens is defined by:

$$F = f/d$$

where f is the focal length and d is the entrance pupil (effective lens diameter).

Cameras using different size tubes, all focused onto an object at the same distance, will have the same angle of view if the focal lengths of the lenses are proportional to the corresponding target image diagonals, and will have the same depth of field if the lenses have the same entrance pupil dimensions. These parameters[11] hold when the lens apertures are proportional to the target diagonals. Under these conditions, tubes (i.e. targets) of equal sensitivity will produce equal signal currents. Some typical lens apertures used for sensitivity tests and providing approximately equal depth of focus conditions for the three main tube sizes are shown in *Table 10.2*.

Tube diameter (mm)	Target diameter (mm)	Aperture (f/number)
30	21.4	4.0
25	16.0	3.0
18	11.0	2.0

Table 10.2 Aperture for same depth of focus with different tube sizes

The sensitivities of lead oxide and saticon tubes vary with target voltage, up to a point where, for constant illumination, both signal current and dark current increase by only relatively small amounts with further increases in the target voltage. This is shown in a graph (*Figure 10.4*) of the relevant characteristics typical for a lead oxide tube. It will be seen that these are very similar to those typical of a diode.

Normally the tube manufacturers recommend an operating target voltage above the 'knee' voltage, so that the sensitivity is stable, and other target characteristics provide an optimum performance.

The light transfer characteristic can be represented mathematically by:

$$I_s \propto E\gamma$$

where I_s represents signal current and E is illumination.

Thus γ is the slope of the transfer characteristic when plotted on logarithmic scales. For both types of tube this is a straight line, showing that γ is constant with a value between 0.9 and 1 for signal currents up to about 1 μA. In this region, the sensitivity can be expressed in μA/lx or μA/lm, without specifying the illumination level at which the sensitivity was measured. However, the illuminant must be specified, and normally it is assumed to be incandescent light at a colour temperature of 2856 K (see section *10.4.2*). The linear light transfer characteristic is a very desirable property of tubes for use in colour television cameras, as this feature greatly assists in the process of obtaining accuracy in colour reproduction, over a wide range of lighting contrast conditions.

The luminous sensitivity S_l of a camera tube is normally defined as the average signal current I_s generated by unit luminous flux falling uniformly on the scanned area A of its target. If L is the illuminance of the scanned area in lumens per square metre then:

$$S_l = \frac{Is}{AL} \quad \mu A/lm$$

and the typical luminous sensitivity of a lead oxide or saticon tube is approximately 350 μA/lm.

A better indication of the peak signal current I_p likely to occur when scanning (allowing for line blanking B as a percentage of the total line period) is obtained from:

$$I_p = \frac{100}{100-B} . I_s = pI_s$$

where p = 1.3 in the CCIR system.

In a monochrome camera, the target illumination L is related to the scene illumination N by:

$$L = N . \frac{rt}{4F^2(1+m)^2}$$

where r is the average scene reflectivity, t is the lens transmission factor, F is the lens aperture (f/number), and m is the linear magnification from scene to target.

A similar relationship holds for the red, green and blue channels of a colour camera, but additional complexity arises due to the allowances that must be made for the optical colour analysis components (see section *10.4.3.*).

10.4.2 Spectral response

The spectral response characteristics of a camera tube show how the radiant sensitivity S_r of a camera tube varies with wavelength. The radiant sensitivity is the average signal current produced by a tube per unit radiant energy falling uniformly on the scanned area of the target. Radiant energy is often expressed as mA/W or μA/μW, and at a particular wavelength λ is related to the luminous sensitivity S_l by:

$$S_r(\lambda) = 0.68\ V(\lambda)\ S_l(\lambda)$$

where $V(\lambda)$ is the normalized spectral sensitivity of the eye at wavelength λ. $V(\lambda)$ is an internationally agreed function normalized to a peak value of unity at a wavelength of 555 nm.

Lead oxide and saticon tubes have similar spectral response characteristics with their peak responses occurring at 500 nm and 530 nm respectively. In the red region, the response of the saticon continues to about 730 nm, somewhat beyond that for lead oxide. However an extended version of the latter, in which the target material includes a sulphur component, has a response extending into the near infrared. These characteristics are shown, and can be compared with those of other tubes, in *Figure 10.5*. This illustration also shows the curve for unit quantum efficiency, i.e. when each photon releases one electron which effectively becomes part of the signal current.

10.4.3 Colour sensitivity

A particular method[17] often used by tube manufacturers for specifying the sensitivity of tubes refers to luminance for a monochrome camera, or red, green and blue for the separate channels of a colour camera. The tube is lit by an illuminant with a colour temperature of 2856 K, and for a colour camera an appropriate, closely specified colour filter is inserted in the light path. The signal current obtained is a measure of the colour sensitivity, and is expressed in microamperes per lumen of white light before the filter. Some typical sensitivities obtained in this way and based on manufacturers' figures are shown in *Table 10.3*.

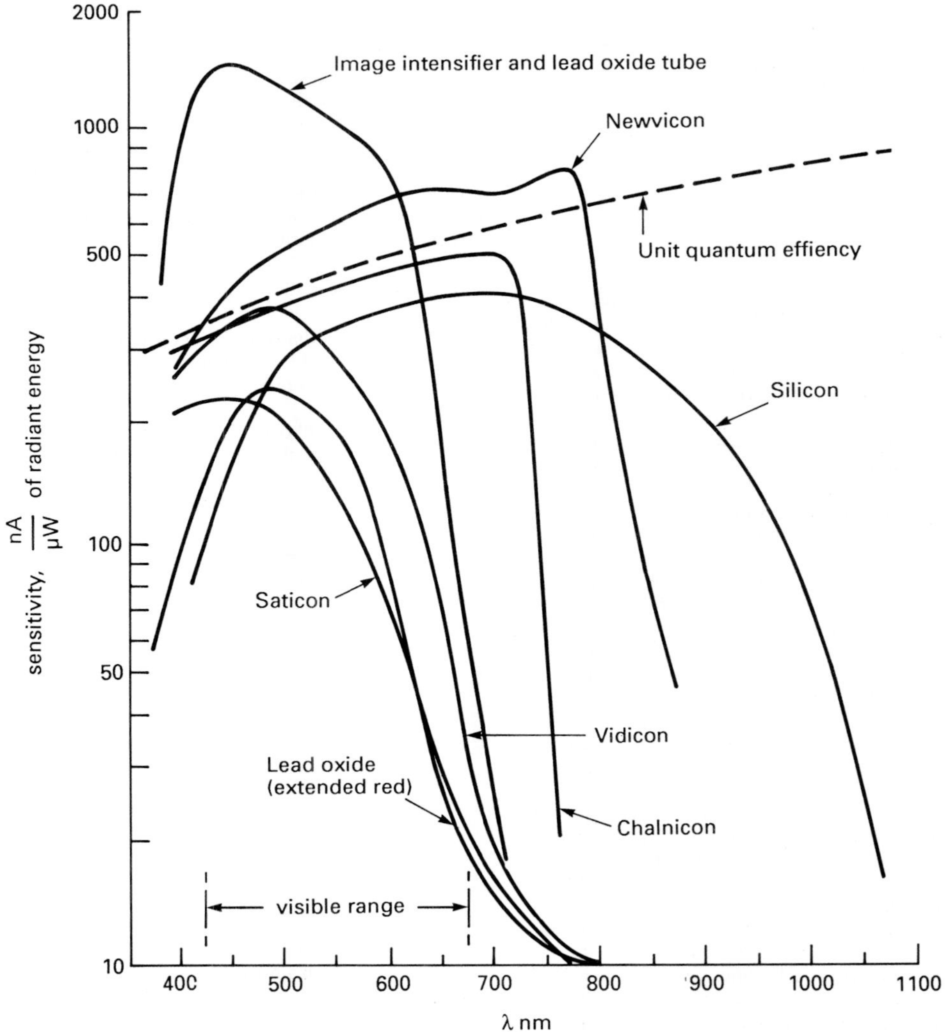

Figure 10.5 Spectral responses of photsensitive targets

	Lead oxide	Saticon
Luminance	375	350
Red	80	120
Extended red tube	115	
Green	165	150
Blue	38	80

Table 10.3 Typical colour sensitivities of tubes in μA/lm (with white light at 2856 K before the colour filter)

Table 10.4 shows some working beam currents and signal currents, corresponding to peak white, typical of a 25 mm diameter tube. For optimum lag and resolution, beam current is normally set to stabilize twice the signal current expected for peak white. The corresponding currents in larger and smaller tubes are higher and lower respectively than those shown.

	Signal current	Beam current
Luminance	200	400
Red	100	200
Green	200	400
Blue	100	200

Table 10.4 Typical signal (white) and beam currents (in nA) for a 25 mm tube)

10.4.4 Resolution

The resolution performance of a photoconductive tube depends mainly on the electron gun design, the thickness of the target layer and the target material. In a thick target layer, light is dispersed leading to loss of resolution. Differential absorption can take place in the layer; in a lead oxide target, red light is absorbed less than green or blue, so that red penetrates the layer to a greater depth and suffers greater scattering with further loss of resolution. This can result in a resolution difference between red and blue at 5 MHz of more than 10 per cent. A reduction in the target layer thickness can produce an improvement in resolution but increases other problems such as lag.

Tube resolution is expressed in terms of the modulation depth of a square-wave signal as measured on a waveform monitor. The signal corresponds to a test pattern of black and white bars of equal width (vertically positioned for horizontal resolution and horizontally for vertical resolution) on a chart illuminated by a 2856 K tungsten light; a 50 mm lens with an aperture of f/5.6 is normally specified. The spatial frequency of the pattern is normally specified in terms of the equivalent video frequency for the scanning standard used. It can also be specified in terms of TV lines where, for a square-wave resolution pattern in PAL System I, with a 4:3 aspect ratio and an active line scan duration of 52 μs (after line blanking), F=4N/2×3×52) MHz, where N is the total number of pattern lines (both black and white) per picture height (assuming the pattern lines to be horizontal) and F is the frequency of the video signal producing the same spatial frequency (assuming the pattern to be vertical).

In this example N = 78F, so 400 TV lines give a frequency of 5.13 MHz.

At 400 TV lines, the horizontal resolution of a typical high quality 25 mm tube intended for use in a green channel would be in the region of 50–60 per cent.

10.4.5 Aperture distortion

In a camera tube, the resolution is determined by the shape and cross-sectional area of the scanning beam where it lands on the target, and by the target structure. Normally the beam spot is significantly larger than the target elements and is more important in terms of limiting resolution. The effect of a finite beam cross-section is to produce aperture distortion resulting in high frequency loss in the camera output signal, thus impairing the resolution of the reproduced image. (A similar loss takes place in the display tube due to the scanning beam spot size.)

If the scanning beam had an infinitely small cross-sectional area, the corresponding output signal from the camera tube would rise instantaneously as the beam passed over an abrupt vertical black/white image transition on the target. In practice, the beam has significant size and will take a finite time to cross the transition. At the point where half the beam spot covers black and the other half covers white, the tube output will be intermediate between that corresponding to black, and that to white. The tube output therefore, represents an integral of the amount of light that has fallen on the particular area covered by the beam spot since it was last scanned.

Photoconductive tubes have a relatively low velocity beam so that its actual landing position and spot shape can, to some extent, be influenced by the target charge on the intended landing area. These are *beam pulling* and *self-sharpening* effects, which in the latter case means that the spot size decreases with increased charge. Thus the profile of the spot or beam aperture is difficult to define accurately. In the ideal case, if the beam cross-sectional area is constant and assumed to be rectangular, and the target image consists of a vertical bar pattern with the contrast changing sinusoidally in the line direction, then quantitatively for different pattern frequencies:

$$\text{relative response} = 20 \log_{10} \frac{\sin\Omega}{\Omega} \text{d B}$$

where $\theta = \pi lf/v$ radians (directly proportional to frequency if l and v are fixed) and l is the dimension of the beam in scanning direction, f is the frequency of sine-wave pattern, and v is the scanning velocity.

The curve of the above relative response expression is plotted in *Figure 10.6*. As the frequency is increased the response

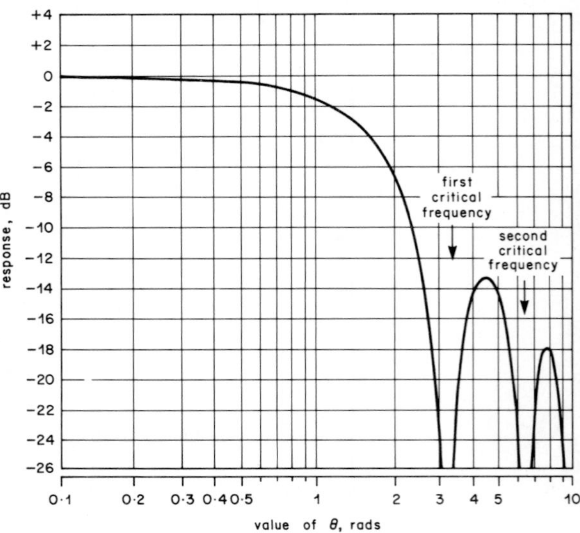

Figure 10.6 Curve of $20 \log_{10} \frac{\sin\Omega}{\Omega}$ dB

decreases, falling to zero when $\theta = \pi$ rad. This occurs when $\pi lf/v = \pi$ or $f = v/l$. This frequency is known as the *first critical frequency* and for a given scanning speed is inversely proportional to l.

10.4.6 Lag

When the illumination on a tube target changes rapidly, there is a delay, or *lag*, in the corresponding change in signal current. In a photoconductive tube with target stabilization achieved by means of a low velocity scanning beam, there are two types of lag:

- *photoconductive lag*, which is mainly a function of the target material,
- *capacitive lag*, which is determined by the way in which the electron beam discharges the target layer.

Photoconductive lag in broadcast standard tubes can, for most practical purposes, be considered to be negligible so that the speed of response to changes in illumination, or the dynamic resolution, is almost exclusively determined by capacitive lag. This is normally specified in terms of *build-up lag* for transitions from dark to light, and *decay lag* for transitions from light to dark. These effects are most noticeable when the illumination on the target is weak, as when a camera is viewing a 'low-key' scene.

The mechanism of build-up lag is basically as follows. When the target is unlit, with only dark current flowing and close to cathode potential, most of the beam electrons can land giving the target a slightly negative potential with respect to the cathode. If low level illumination suddenly reaches the target, photoconduction causes the potential on the scan side of the layer to rise positively by a small amount, although remaining slightly negative with respect to the cathode. An increasing number of lower energy beam electrons will then reach the target in the first few scans until the beam current landing balances the signal current plus the dark current, and target stabilization is reached at the new potential. During this period, it is said that the beam acceptance increases or, alternatively, that the beam resistance decreases.

Decay lag can be explained by considering the illumination to be suddenly reduced. Most of the layer charge is then removed in the next few scans. As the layer voltage falls however, beam electrons reach the target at a decreasing rate, until again only higher energy electrons land. This extends the period required to reach stabilization, when the beam current just balances dark current. Now it is said that the beam acceptance decreases or, alternatively, the beam resistance increases.

Build-up lag is normally measured[17] after the target has been unlit for at least 10 s. It is expressed as the percentage ratio of the intermediate current to the final current, at 60 ms and 200 ms intervals after the illumination has been restored.

Decay lag is normally measured after the target has had at least 5 s of illumination. It is expressed as the percentage ratio of the residual signal current to the initial signal current, at 60 ms and 200 ms after the illumination is removed.

The measurements are normally carried out with:

- low key conditions (low illumination), with peak signal currents typically 20 nA and 40 nA,
- high key conditions (high illumination), with peak signal currents typically 100 nA, 150 nA or 300 nA, depending on the tube type and working signal currents.

A light source with a colour temperature of 2856 K is normally used. If the tube is to operate in the red, green or blue colour channel of a camera, an appropriate colour filter is inserted in the light path (see section *10.4.3*).

10.4.6.1 Lag reduction

The layer capacitance of the target and the beam resistance are mainly responsible for capacitive lag in the target stabilization process. Beam resistance is to a large extent a function of the electron gun, and new guns[8] have been introduced with designs that minimize the energy spread of electrons in the beam. A thicker target reduces the effective layer capacitance and thus increases the layer potentials produced by the charge image, resulting in improved beam acceptance. However, a thicker target would also result in lower resolution due to increased light dispersion in the layer.

A relatively simple and successful method of reducing lag is to artificially induce additional dark current, so preventing the target elements from becoming more negative than is consistent with good beam acceptance. This is achieved by flooding the target with a very small amount of *bias light* which induces a few nanoamperes of additional dark current. The resultant small pedestal signal produced is subsequently removed in the signal processing stages. In *Figures 10.7* and *10.8* the typical lag curves are shown for one type of lead oxide tube. Characteristics for saticon type tubes, including the effects of bias lighting, are broadly similar, with the main differences occurring in the 'tails' of the decay lag curves.

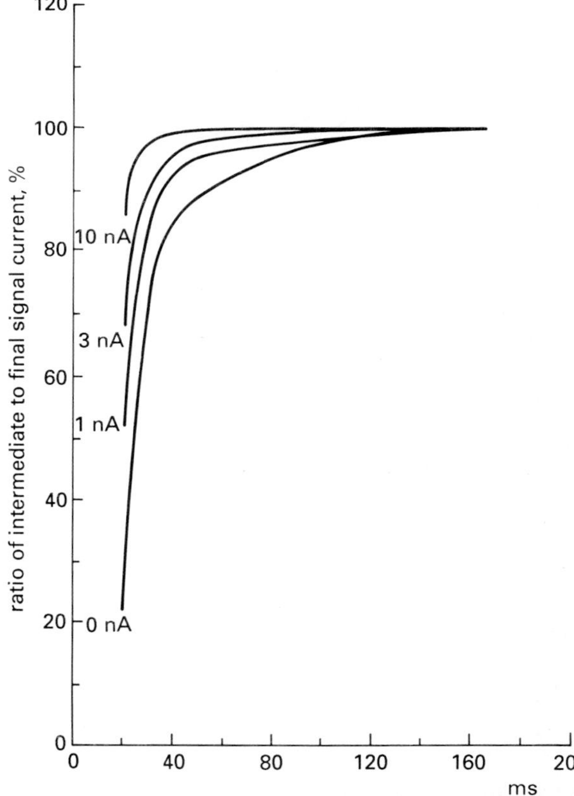

Figure 10.7 Typical build-up lag characterisitics of a lead oxide tube showing the effect of bias light. Signal current/beam current = 20/200 nA

Bias lighting can be introduced directly onto the tube faceplate via the camera optics, or the prism block in a colour camera. Differential lag control can then be achieved by adjusting the colour of the light by means of filters. An alternative method[12] is by conducting light internally via a lightpipe from a source within, or close to, the tube base onto the scan side of the target as shown diagrammatically in *Figure*

10.9. With this method, lag can be controlled by adjustment of the current in the light sources.

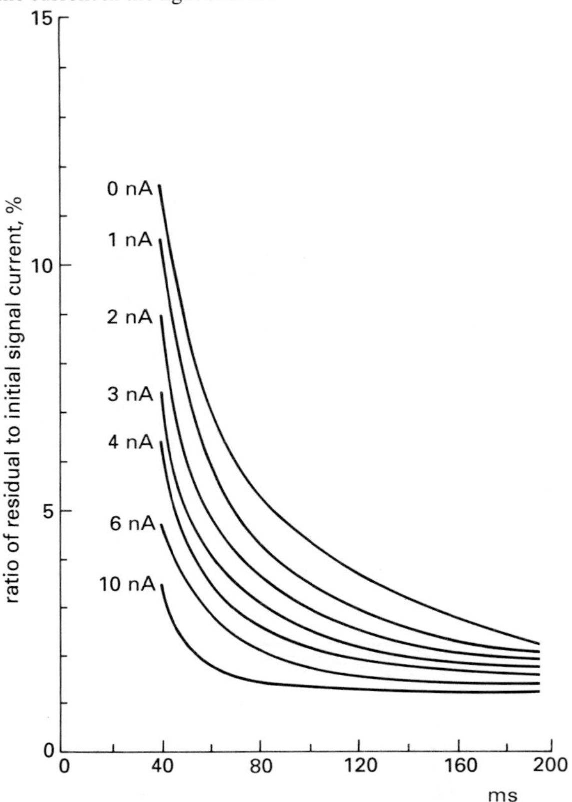

Figure 10.8 Typical decay-lag characterisitics of a lead oxide tube, showing the effect of bias light. Signal current/beam current = 20/200 nA

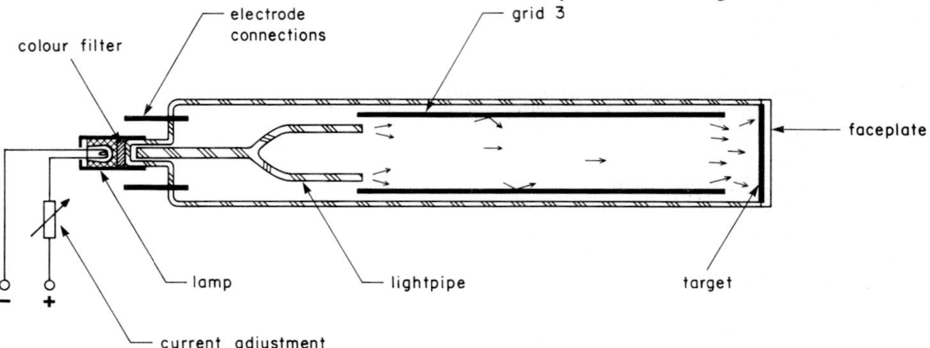

Figure 10.9 Tube with bias light injection

10.4.7 Image retention

Image retention, or *burn-in*, can be produced by prolonged exposure of the tube target to a very bright stationary image. Neither lead oxide nor saticon tubes are prone to this effect, provided the recommended applied target voltages are used and the target is not exposed in the absence of the scanning beam. If burn-in does occur, it can often be removed in a fairly short period when the beam is restored. However the manufacturers' recommendations should be consulted for the correct procedure.

10.4.8 Stray light and flare

Some of the diffuse light not absorbed by the tube target can be deflected back onto the faceplate where it may produce halation or flare. Lead oxide targets have a greater reflectivity in this respect than saticon targets, particularly to red light. In order to reduce this effect, lead oxide tubes normally have an anti-halation disc of 3–6 mm thickness cemented to the faceplate. Incident light reflected from the target passes back through the disc, is reflected from its outer face instead of the faceplate, and is absorbed by the blackened walls of the disc. Further flare reduction can be obtained if the outer surface of the disc is masked to leave an aperture slightly larger than the image area on the target.

10.5 Tubes for non-broadcast colour television

These are tubes intended for use where mainly non-broadcast standards are acceptable, such as closed circuit, industrial, military and consumer colour television. In many cases, lower grades of broadcast quality tubes are used, although high quality tubes have been developed for special applications, such as use with image-intensifiers in medical and scientific work.

10.5.1 Filter integrated colour tubes

In these, the optical colour analysis and signal generation is accomplished in a single tube[21,22] by the use of a set of fine stripe colour filters interposed in the optical path between the faceplate and the target. The video output from the tube consists of a baseband or luminance signal and a subcarrier, or carriers, with sidebands resulting from the electron beam scanning the spatial image of the stripe filters, superimposed on the image of the object scene. The output signal is then electronically decoded to recover the colour information in the sidebands of the subcarrier(s), which is then matrixed with the low frequency luminance component, to form conventional red, green and blue colour separation signals.

Various systems exist using either two or three stripe filter colours, and with differing arrangements of stripe orientation, using specially developed vidicon or saticon tubes. The resultant horizontal resolution is somewhat reduced compared with conventional three-tube colour cameras, and strong low frequency beat patterns can result from interaction between the filter stripes and detail in the picture. These patterns, however, can be reduced by the use of a spatial pre-filter in the optical path to the tube which attenuates the finer picture detail.

10.5.2 Vidicon

The vidicon[2] has a photoconductive target of antimony trisulphide (Sb_2S_3), built up in a number of sub-layers. Its properties are dependent on the composition of the sub-layers which are varied as appropriate for the tube's intended application. The light transfer characteristics are non-linear and, like sensitivity, are dependent upon the applied target voltage. A thin target provides good resolution, but suffers from high inherent photoconductive lag and is prone to burn-in. Dark current is relatively high and is very dependent on target voltage and temperature. It is also non-uniform to an extent that gives rise to noticeable background shading in the picture. Most of these features render the tube unsuitable for anything other than lower grade colour television applications.

10.5.3 Chalnicon

The chalnicon[23,24] has a cadmium selenide (CdSe) target and takes its name from the collective term *chalcogenide*, for selenides and sulphides. It has a very high sensitivity with a quantum efficiency approaching unity over the visible spectral range and extending to about 700 nm, as shown in *Figure 10.4*. It has a good performance in most other respects but has excessive lag (mainly capacitive) which would be unacceptable for broadcasting.

10.5.4 Newvicon

The newvicon[25] has a target composed of cadmium and zinc tellurides (ZnCdTe). It provides exceptionally high sensitivity with a quantum efficiency greater than unity over most of the visible spectrum (*Figure 10.4*) and extending into the infrared region to about 800 nm. Other notable good features are high resolution and minimal 'blooming' effects with very bright highlights. However, excessive photoconductive lag and dark current probably limit the applications to low light work, such as in surveillance.

10.5.5 Silicon diode tubes

These have targets effectively composed of silicon diode arrays[26]. They provide high resolution, high sensitivity with an extended red response, and can have good lag performances. However the dark current is excessive, temperature dependent, and prone to be very non-uniform over the target area.

10.5.6 Image intensified tube

This consists of a compact image intensifier coupled to a specially constructed photoconductive tube.[27] In this combination, an optical image is focused onto the photo cathode of the image intensifier, which emits electrons corresponding in density to the brightness of each point of the image. The electrons are accelerated to hit a small phosphor screen at the other end of the intensifier, which displays a focused electron image to the tube through a fibre-optic coupler. The image is brighter than that falling on the photo cathode, and the phosphor colour is chosen to match the spectral sensitivity of the tube. The overall sensitivity of this combination, compared with that of the tube, is increased by a factor of at least five. The picture quality is somewhat lower than normal broadcast requirements, and under very weak lighting conditions can be characterized by photon noise. This system probably has most applications in the scientific and medical fields.

10.6 Charge coupled devices

A ccd is a semiconductor storage array[28] consisting of regularly arranged mos capacitors with a common substrate, the separation between adjacent capacitors being small enough to enable charge interaction to take place. CCDs can be used for high density digital storage, analogue storage or delay, and also as photoelectric sensors.

10.6.1 Operation

An mos capacitor is shown in *Figure 10.10* consisting of a deposited metal electrode, insulated by a film of silicon dioxide, on a p-type silicon substrate. When a positive voltage is applied to the electrode a depletion region is created below the silicon dioxide/substrate interface, forming a 'potential well' with a depth depending on the applied electrode voltage. Any free electrons in the immediate region will be attracted and flow into this well.

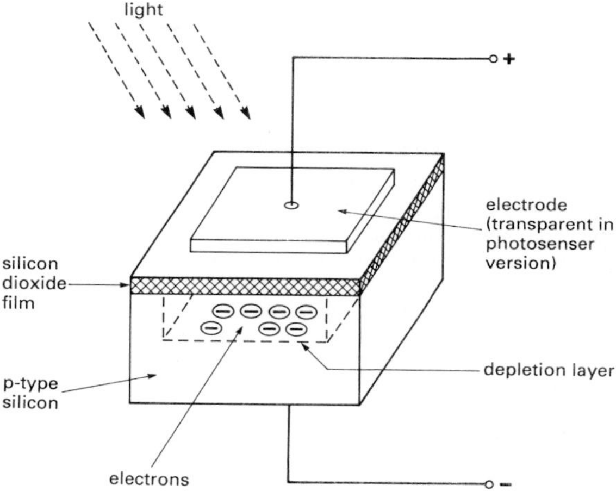

Figure 10.10 MOS capacitor and photo-sensor

The operation of a ccd consisting of a chain of mos capacitors joined by a common substrate is shown in *Figure 10.11*. If an electron charge is stored in the well under electrode P1 held at a positive voltage, and the same voltage is applied to electrode P2, the well under P1 with some of the charge enlarges and extends under P2. If the voltage on P1 is then reduced, the charge flow increases to the P2 well and the transference is finished when the voltage on P1 equals zero. Thus by applying suitably shaped and phased trains of *transfer clock pulses* to the electrodes, an electron charge can be moved along the ccd which then functions as an analogue shift register.

10.6.2 CCD imager

Light striking the surface of an mos capacitor generates an electric charge, proportional to the light intensity, within the semiconductor. The charge is stored in a potential well created in the substrate and can be moved as required, by means of the ccd transfer process described in section *10.6.1*. In order to improve the photoelectric conversion efficiency, the construction of the capacitors intended to operate as light sensors is modified. With front illumination, transparent electrodes can be made of polysilicon, or stannic oxide which has a more uniform spectral response. Illumination from the rear of the sensor requires the substrate to be very thin, which can introduce other problems.

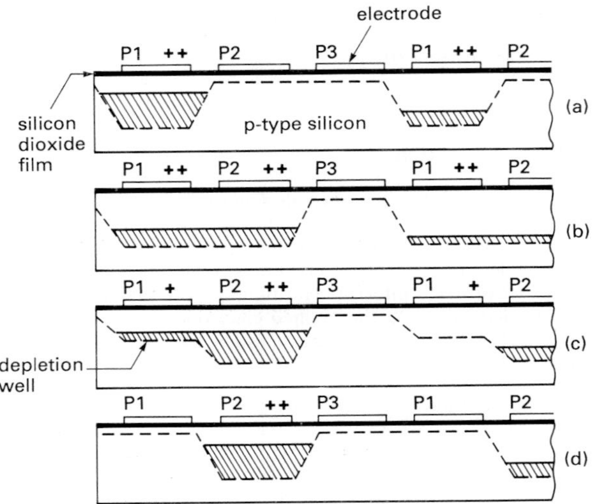

Figure 10.11 CCD operation, showing electric charge transfer

A ccd imager consists of an array of mos photosensors[29,30] suitably combined with ccds to form an area imaging device. This senses a whole picture focused onto it by the optical system, and the resultant charge pattern produced in the sensors is moved out from the array and into a series of ccd shift registers, to the output of the device. After suitable filtering, the resultant signal is passed onto the conventional video processing circuits. This is analogous to a pickup tube target, with the scanning provided by clock pulses to the ccds and the video output appearing at a single output terminal. Two main types of imager used in broadcast quality cameras are *frame transfer* and *interline transfer*. These differ only in the way in which sensor charges are moved around and out of the device.

In the frame transfer imager[31] (see *Figure 10.12*), the upper half consists of an array of sensors forming the light sensitive image area. The lower half of the device forms the storage/readout area which is masked to exclude any light. During vertical blanking, the sensor charges corresponding to the picture information are clocked out rapidly down into the frame storage area, which is then emptied line-by-line during line blanking into the horizontal (read out) shift register. The contents of this are then clocked to the output terminal during the active picture line. The sensors are still responsive to light while acting as shift registers during the vertical blanking period, so the transferred signal can be contaminated with unwanted information. This can result in some vertical blurring of the image, which can be prevented by the use of a synchronized mechanical shutter to mask light from the sensors during the transfer period.

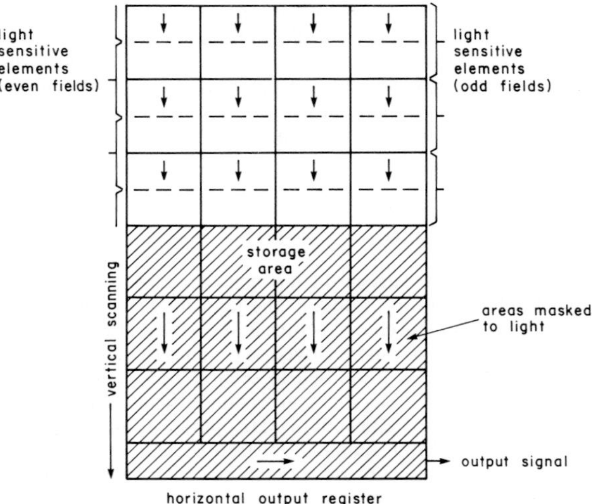

Figure 10.12 Principle of operation of a frame transfer imager

Figure 10.13 A frame transfer imager. The light sensitive region appears as the dark area in the centre of the imager

A disadvantage of this type of imager, an example of which is shown in *Figure 10.13*, is that it is required to be at least twice as large as the optical image. However, since the light sensitive area is larger and contains only sensor cells, sensitivity per unit area is high compared with interline structures.

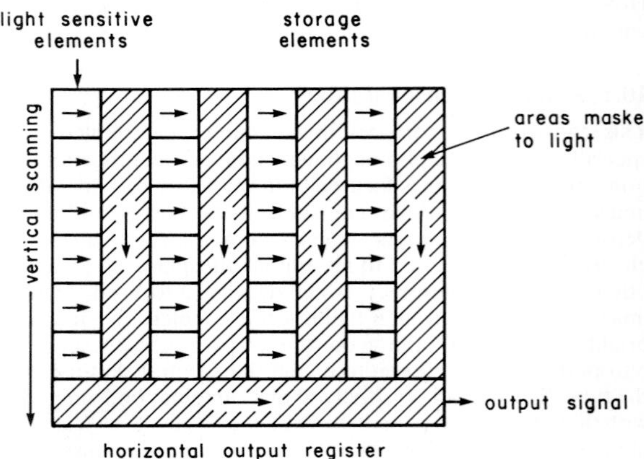

Figure 10.14 Principle of operation of a interline transfer imager

In the interline transfer imager[32] shown in *Figure 10.14*, the array consists of interleaved columns of sensors and vertical shift registers, feeding into a horizontal (read out) shift register. All the shift registers are masked from light. During field blanking, the sensor charges representing the picture information are moved sideways into the vertical shift registers. These charges are then clocked downwards, line-by-line during line blanking, into the horizontal register. The contents of this are then read out to the output terminal during the active line period, as in the frame transfer device.

This type of imager has an advantage[33] in that its total size is only slightly larger than the optical image area. However this region which contains the sensors is partly occupied by shift registers further reducing the sensitivity per unit area compared with the frame transfer device.

It should be noted that the imager diagrams shown in *Figures 10.12* and *10.13* are very idealized. Additional practical features would include control electrodes, channel stop diffusions to separate elements, overflow drains to absorb surplus charge, and aperture masks, all of which limit the effective size of each sensor element.

10.6.3 CCD imager resolution

The resolution of an imager is determined by the number of sensor elements it has in both the horizontal and the vertical directions of the array. The greater the number of elements, the finer will be the detail that the imager can resolve. It is shown in section *10.4.4* that, for a camera tube, the horizontal resolution is mainly determined by the velocity, v, and the cross-sectional area of the scanning beam. If it is assumed that the cross-section is rectangular, with length l, then the first critical frequency F_c (resolution falls to zero) = v/l.

The imager array consisting of discrete sensor elements constitutes an optical sampling system[34] where sampling frequency F_s in megahertz equals (the number of horizontal elements)/(active line time in microseconds).

According to sampling theory, the resolution of the imager will fall to zero at F_s and there is a correspondence between F_s and the critical frequency F_c for camera tube resolution, if the equivalents for beam velocity and length of the beam cross-sectional area are derived from the sensor array parameters. The resolution response of an imager will thus approximately follow the curve shown for sin θ/θ in *Figure 10.6*. With sampling, the maximum frequency that can be produced faithfully occurs at $F_s/2$, the Nyquist limit. Above this the required data signal becomes corrupted with spurious aliasing components.

In the vertical direction, there is a similar sampling action and the spatial frequency response in this direction also follows the sin θ/θ shape. The vertical sampling frequency corresponds to the number of active lines making up the picture, and above the corresponding Nyquist limit, as in the horizontal direction, the required information will have additional vertical aliasing signals.

In frame transfer imagers, all the lines in one picture are read into the shift registers after one field, and each sensor element has a light integrating period of only one field. Line interlace, required by most broadcast systems, is achieved by varying the potentials on the electrodes controlling the photosensitive area. This causes two interlaced sets of lines, vertically displaced from each other on the imager by one half-line space, to be read out on alternate fields, as shown in *Figure 10.11*.

With interline transfer imagers, correctly interlaced fields (sets of lines displaced vertically by one line space), can be read out. Compared with the field transfer imager, the vertical resolution is doubled for the same number of element rows, and the charge integration time for each sensor element is also doubled. However, with television systems operating at 50 and 60 fields/second, two field integration times produce considerable blur on image movement. In order to keep this to an acceptable level, two adjacent lines can be read out together thus completely discharging the image area every field period. Interlace is maintained by reading out and combining together the signals from lines 1 and 2, 3 and 4, 5 and 6, etc. on odd fields, then lines 2 and 3, 4 and 5, 6 and 7, etc. on even fields. This results in a reduction of vertical resolution which then becomes similar to that of a frame transfer imager with the same number of vertical elements.

For a typical interline transfer imager with 500 horizontal by 582 vertical sensor elements, a 52 μs active line time, 576 active lines per picture and 50 fields per second, then:

- horizontal sampling rate = $500/(52 \times 10^{-6}) = 9.62$ MHz
- horizontal Nyquist limit = 9.62/2 = 4.81 MHz
- vertical sampling rate = 576 cycles/picture height, equivalent to a horizontal sampling frequency of 14.8 MHz (see section *10.4.4*).
- vertical Nyquist limit = 288 cycles/picture height, equivalent to a horizontal sampling frequency of 7.4 MHz.

10.6.4 Reduction of alias signals

The apertures of imager sensor elements are very well defined, being set by masks in the manufacturing process. This is in contrast to camera tubes where the aperture of the scanning beam is much less distinct. Thus one significant disadvantage of cameras using ccd imagers is the increased visibility of moiré patterns, particularly in the vertical direction, due to aliasing. These effects can be reduced in both directions, by including a low-pass optical pre-filter in the camera lens system to exclude from the imager any spatial frequency components which will produce aliasing. One type of filter comprises a diffraction pattern etched into a glass plate; a number of alternative methods have been proposed.

The spatial frequency response of such a pre-filter is chosen as a compromise between alias signal reduction and providing an acceptable resolution performance, in both directions. Clearly, with an increased number of sensor elements on the imager, the compromise becomes less critical.

10.6.5 Spatial offset

This is a technique[35] used in cameras employing three imagers, to enhance the horizontal resolution and to reduce aliasing effects. It consists of physically off-setting the horizontal registration of the green imager, relative to the red and blue imagers, by an amount equal to half the horizontal sensor element spacing. The clock pulses driving the green imager are correspondingly delayed by an equivalent amount.

The luminance signal from the camera, which is a combination of red, green and blue signals, is then equivalent to a signal which has been derived with twice the actual sampling frequency. Thus the luminance resolution is increased and the pre-filtering frequency can be raised. Any alias signals contained in the individual colour signals will, to some extent, cancel each other out in the camera output, as the green alias signals will be phase inverted with respect to those of red and blue. This results from the 180° phase shift introduced to the green sampling carrier by the spatial offset of the green imager.

10.6.6 CCD imager sensitivity and spectral response

The signal/illumination transfer characteristic of the imager has a gamma of unity, like the camera tube, up to a saturation level. The spectral sensitivity performance is similar to that shown for silicon in *Figures 10.7* and *10.8* but with the difference that the imager response is sharply reduced above 600 nm, to minimize unwanted infrared sensitivity.

In contrast with camera tubes, the signal output of the imager is derived as a voltage instead of a current, and the main noise source is within the imager rather than in the amplifier following a camera tube. Thus, when measuring the relative sensitivities of imagers and tubes it is more convenient to compare the performances of two similar cameras, one using tubes and the other using ccds. The latest broadcast standard cameras using tubes and corresponding cameras with recently developed imagers have comparable sensitivities.

10.6.7 Imager highlight operation

Extreme highlights generate excessive charge which eventually produces a charge overflow. In recent imager designs this is absorbed by *overflow drains* which are interleaved with the ccds; thus charge spread is contained and highlight 'blooming' and 'stick' effects associated with tubes are prevented. Overflow drains can absorb more than ten times the normal light saturation level.

Highlights in excess of about ten times saturation level can, under some circumstances, produce 'smear' effects with interline transfer imagers. These appear as reddish vertical lines extending from the overload site, to both top and bottom of the picture, and are comparable to some common lens-produced specular effects. It is a breakdown condition produced by long wavelength light penetrating deeper into the silicon structure, and generating charge beneath the potential wells. Most of this is collected by the drains, but some directly enters the vertical shift registers and corrupts the required image charges.

10.6.8 Colour filter integrated imagers

These are similar in principle to tubes with combined colour filters, used in single tube colour cameras, and described in section *10.5*. The imagers incorporate fine stripe or mosaic type filters of two or three colours in the light path to the sensor. At present they are used in the so-called *single ccd* cameras[36] intended mainly for the domestic market. The colour filter stripes, or *mosaics*[37], are aligned with the sensor pattern in a particular way to simplify decoding.

Usually the signal clocked out of the imager consists of a baseband luminance signal, and subcarrier(s) with sidebands resulting from the sampling actions of the sensor array and the colour filter patterns. The red, green and blue separation signals are then recovered by a decoding process, as previously described for filter integrated camera tubes. Horizontally, the effective sampling frequency for the luminance signal can be the same as for a conventional imager. The sampling frequencies for the colour signals are much lower, resulting in lower resolution, due to the wider spacing between samples of the same colour.

Aliasing effects can be produced by the different sampling rates and can be reduced with optical pre-filtering. This requires a relatively low cut-off frequency to suppress aliasing due to the colour filter sampling, and this limits the luminance resolution.

10.6.9 Electronic shutter

The electronic shutter is a feature incorporated in some recently developed interline transfer imagers. It enables the light integration or 'exposure' time of the sensors to be varied from, say, 20 ms (one field) down to 1 ms or less. Thus, fast moving images can be reproduced with minimal movement blur, but at the expense of a proportional loss of camera sensitivity. This function is achieved by the inclusion in the imager interline transfer areas of additional drains, which can be controlled to waste some of the image charge.

References

1 McGEE, J D, 'A review of some television pick-up tubes', *Jour IEE*, **97**, 50 (1950)

2 WEINER, FORGUE and GOODRICH, 'The vidicon photoconductive camera tube', *Electronics* **23**, 5 (1950)

3 DE HAAN, E F, VAN DEN DRIFT, A and SCHAMPERS, P P M, 'The plumbicon, a new television camera tube', *Philips Technical Review* **25**, 133 (1963, 1964)

4 SEQUIN, C H and TOMPSETT, 'Charge transfer devices', Academic Press, New York (1975)

5 BARBE, D F and CAMPANA, S B, 'Imaging arrays using the charge coupled concept', *Advances in imaging pick-up and display*, Vol 3, Academic Press, New York (1977)

6 DILLON, P L P et al, 'Colour imaging system using a single ccd array', *IEEE Transactions on Electron Devices*, ED-25, **2** (1978)

7 LENT, S J, 'A review of image sensors for colour television cameras', *International Broadcast Engineer* **11**, 170 (1980)

8 NEUHAUSER, R G, 'The saticon colour television camera tube', *SMPTE Journal* **87** (1978)

9 LEVITT, R S, 'Performance and capabilities of new plumbicon TV camera pick-up tubes', *SMPTE Journal* **79** (1970)

10 TURK, W, 'The leddicon TV camera tube', *International Broadcast Engineer* **11**, 170 (1980)

11 'Plumbicon — The high performance TV camera tube for closed circuit television', Industrial Electronics Division, Mullard Ltd (1968)

12 SCHUT, T J and WEIJLAND, W P, '30 mm plumbicon camera tubes with fibre-optic faceplate, anti-comet tail gun and lightpipe', *Mullard Technical Communication* **11**, 109 (1971)

13 SCHUT, T J, 'Developments in television camera tubes', *International Broadcast Engineer* **11**, 170 (1980)

14 LOHNES, W, 'Highlight handling with diode-gun plumbicon tubes', International Broadcasting Convention (1980)

15 MOCHIZUKI, T and OHNISHI, K, 'Circuit operation and analysis of automatic beam optimiser (ABO)', *NHK Technical Monograph*, **35** (1986)

16 NEUHAUSER, R J, 'The RCA range of broadcast camera tubes', *International Broadcasting Engineer* **11**, 170 (1980)

17 'Plumbicon camera tubes and accessories', Technical Handbook Book 2, *Valves and tubes* Part 2a, Mullard Ltd (1985)

18 KURASHIGE, M et al, '2/3 inch magnetic focus electrostatic-deflection (MS) camera tube and deflection driver', *NHK Technical Monograph*, **35** (1986)

19 ISOZAKI, Y et al, '1-inch saticon for high definition colour television cameras'. *IEEE Transaction on Electron Devices*, ED-28, No.12 (1981)

20 VAN DE POLDER, L J, 'Target stabilization effects in television pick-up tubes', *Philips Research Reports*, **22**, 2 (1967)

21 KUBOTA, Y and KAKAZAKI, T, 'An eng camera using a single pick-up tube', International Broadcasting Corporation, IEE Conf Pub 181 (1980)

22 PRITCHARD, D H, 'Stripe-colour encoded single tube colour television camera systems', *RCA Review*, **34** (1973)

23 YOSHIDA, O, 'Chalnicon, a new camera tube for colour TV use', *Japan Electronic Engineering* (October 1972)

24 YOSHIDA, O, 'Recent chalnicon developments', 7th Symposium on Photoelectronic Image Devices 1978

25 'High-sensitivity photoconductive TV camera tube', *Japan Electronic Engineering* (January 1974)

26 SMITH, G E, 'The silicon-diode array camera tube', Bell Telephone Laboratories Proceedings of the 1970 Solid State Circuit Conference

27 ALLER, VAN G and SCHUTT, Th G, 'Combined image intensifier and plumbicon tube for studio colour cameras', *Philips Electronic Applications Bulletin* **32**, 3 (1973)

28 SHELDON, I, 'An introduction to ccd technology', Sony Broadcast Ltd, Basingstoke

29 ALUN, J, 'CCD imaging array technology', *Electronic Product Design* (October 1985)

30 BURT, D J, 'Development of ccd area imager sensors for 625-line television applications', *Radio and Electronic Engnr*, **50**, 5 (1980)

31 HURST, R N, 'The frame-transfer approach to charge coupled devices', Fact Sheet-RCA Broadcast Systems, USA.

32 HOAGLAND, K A, 'Television applications of interline transfer ccd arrays', NASA/JPL Conference on ccd technology and applications, Washington (1976)

33 'Frame transfer, X/Y and interline image sensors, — How do they compare?' *Philips Tech Pub* 170, Electronic Components and Materials (1985)

34 NORDBRYHN, A, 'The dynamic sampling effect with ccd imagers', Applications of Electronic Imaging Systems, *SPIE*, **143** (1978)

35 HOAGLAND, K A, 'Image shift resolution enhancement techniques for ccd imagers', *SID Digest* (1982)

36 ACKI, M et al, '$^2/_3$ inch format mos single chip color imager', *IEEE Transaction on Electron Devices* (1981)

37 TAKEMURA, Y and OOI, K, 'New frequency interleaving ccd colour television camera', *IEEE Transactions on Consumer Electronics*, CE-28, 4 (1982)

C K P Clarke
Senior Engineer, BBC Research Department

11 Colour Encoding and Decoding Systems

11.1 Introduction

The signals produced by colour television cameras are in the form of red, green and blue (RGB) colour signals. Cathode ray tube colour displays also operate with RGB. Although RGB signals can be of high quality, the three channels required represent an inefficient use of bandwidth and circuitry. Furthermore, impairments can arise if the three channels are not accurately matched. More efficient and rugged methods of colour encoding are therefore required at other points in the signal chain, such as for studio processing, recording, distribution and broadcast emission.

Several additional methods of colour encoding have been developed. Some produce composite signals, such as PAL, NTSC and SECAM, which retain compatibility with monochrome receivers. Such systems have been in use for many years. Now, however, component coding methods such as 4:2:2 digital components and MAC are being introduced. These methods sacrifice direct compatibility for advantages such as improved signal processing and picture quality.

Compatibility between systems is increasingly a problem in television broadcasting, especially for the new media: satellites, wide-band cable and new terrestrial services. This is particularly so in the case of the colour systems now envisaged. Here, all the main systems are presented in common terms, making the similarities and differences more apparent.

11.2 Colour signal relationships

All the colour systems described here are based on encoding of a luminance signal and two colour-difference signals, instead of the red, green and blue colour separation signals.

11.2.1 Gamma

The methods use signals pre-corrected for the assumed gamma of the crt display. Thus, the gamma-corrected colour separation signals are denoted R′, G′ and B′, with the prime (′) signifying that pre-correction has been applied. PAL, SECAM and MAC/packet systems are generally matched to a display gamma of 2.8, while a gamma of 2.2 is assumed for the NTSC system. The 4:2:2 digital components system uses signals appropriate for the associated composite colour standard, hence generally 2.2 for the 525-lines, 60 fields/second scanning standard and 2.8 for the 625-lines, 50 fields/second standard.

11.2.2 Luminance and colour-difference equations

The use of luminance Y and the two colour-difference signals B-Y and R-Y provides improved compatibility with monochrome systems by extracting the common luminance content of the colour separation signals. Also, the bandwidth of the colour-difference signals can be reduced, resulting in improved coding efficiency. In all the encoding systems described here, the luminance signal is defined as:

$$Y' = 0.299R' + 0.587G' + 0.114B' \quad (11.1)$$

based on the NTSC signal primary colour chromaticities[1]. The colour-difference signal relationships derived from this are:

$$B'-Y' = -0.299R' - 0.587G' + 0.886B' \quad (11.2)$$

and

$$R'-Y' = 0.701R' - 0.587G' - 0.114B' \quad (11.3)$$

At the display, the colour separation signals can be regained by applying the inverse relationships:

$$R' = Y' + (R'-Y') \quad (11.4)$$

$$G' = Y' - 0.194(B'-Y') - 0.509(R'-Y') \quad (11.5)$$

$$B' = Y' + (B'-Y') \quad (11.6)$$

It should be noted that the analysis characteristics used in cameras for $^{625}/_{50}$ colour signals are based on different primary colour chromaticities from those chosen for the NTSC system (on which equation (11.1) is based). These match the chromaticities of present-day display phosphors more accurately. Retention of the coefficients in equation (11.1) results only in slight grey-scale inaccuracies in the compatible monochrome picture[2]. In the NTSC system, it is assumed that correction is applied in the display circuitry of the receiver.

11.2.3 Constant luminance coding

In principle, colour coding using a luminance signal and two colour-difference signals produces a system in which distortion or perturbation of the colour-difference signals leaves the displayed luminance unaffected. Systems that maintain this principle are termed *constant luminance*.

The luminance signal of equation (11.1) is synthesized by matrixing the gamma-corrected colour separation signals, and consequently the luminance produced is not equivalent to that of a monochrome system. As a result, the compatibility of the colour signal reproduced on monochrome receivers is adversely affected, so making areas containing highly saturated colours darker than they should be. In modulated systems, however, a further factor influencing monochrome compatibility is the presence of the colour subcarrier. In this case, the effect of the tube brightness non-linearity on the subcarrier signals is to produce an additional average brightness component in highly coloured areas. To a first approximation, this offsets the losses resulting from matrixing non-linear signals.

Although the luminance component of a colour signal is not identical to a monochrome signal, the correct luminance can be reproduced by the colour display. This can be explained by considering that a correction for the inaccurate luminance signal is carried as part of the colour-difference signals. However, when the colour-difference signals are limited to a narrower bandwidth than the luminance, the high frequency content of the correcting component is lost. This results in a failure of constant luminance on colour transitions. Even so, the presence on colour transitions of subcarrier signals not removed by the notch filter tends to offset this effect, as explained above for the case of the compatible monochrome signal.

A closer equivalent to the monochrome signal could be obtained by matrixing linear colour separation signals to produce luminance. This would avoid the failure of constant luminance on colour transitions and would improve the compatibility of the monochrome signal. Such techniques are currently being considered for high definition television systems. Nevertheless, because linear signals are much more susceptible to noise, gamma correction or some form of non-linear pre-emphasis is still necessary to obtain satisfactory noise performance over the transmission path.

11.3 Composite colour systems

Broadcast composite colour signals are based on three systems of colour encoding: PAL, NTSC and SECAM. The three systems have the similarity that each consists of a broad-band luminance signal with the higher frequencies sharing the upper part of the band with chrominance components modulated onto subcarriers. *Figure 11.1* shows a typical distribution of luminance and chrominance frequency components for a PAL signal.

The composite approach provides a form of signal which is directly compatible with monochrome receivers. Many features of the colour signal have been chosen to optimize the compatibility of the systems with monochrome operation, particularly the visibility of the subcarriers on the monochrome picture.

The following sections describe the principal features of the three systems, highlighting their individual differences. While PAL and NTSC use amplitude modulation for the colour-difference signals, PAL with an additional offset, SECAM uses frequency modulation. Each of the systems is capable of providing high quality pictures and each has its own strengths and weaknesses. Although a development of the earlier NTSC system, the PAL system is simpler in some respects and is therefore described first.

11.3.1 PAL

11.3.1.1 Development

The distinguishing feature of the PAL colour system developed by Bruch[3] is that it overcomes the inherent sensitivity of suppressed carrier amplitude modulation to differential phase distortion. This is achieved by encoding the two colour-difference signals, U and V, with the phase of the V subcarrier reversed on alternate television lines, thus leading to the name *Phase Alternation Line*. The PAL system was developed primarily for the 625-lines, 50 fields/second scanning standard, first used in Europe.

11.3.1.2 Colour subcarrier frequency

The colour subcarrier frequency f_{SC} used for system I PAL signals is 4.433 618 75 MHz ±1 Hz. Systems B, D, G and H use the same frequency, but with a wider tolerance of ±5 Hz.

The subcarrier frequency and the line frequency (f_H) are linked by the relationship:

$$f_{SC} = \left(283\tfrac{3}{4} + \frac{1}{625}\right) f_H \qquad (11.7)$$

$$= \frac{709\ 379}{2500} f_H$$

Therefore the subcarrier phase at a point in the picture is subject to a cycle that repeats every eight field periods (2500 lines). PAL signals in which the specified relationship is not maintained are termed *non-mathematical*. The relationship of equation (11.7) was chosen to minimize the subcarrier visibility by providing the maximum phase offset from line to line and from picture to picture, subject to the constraint that the

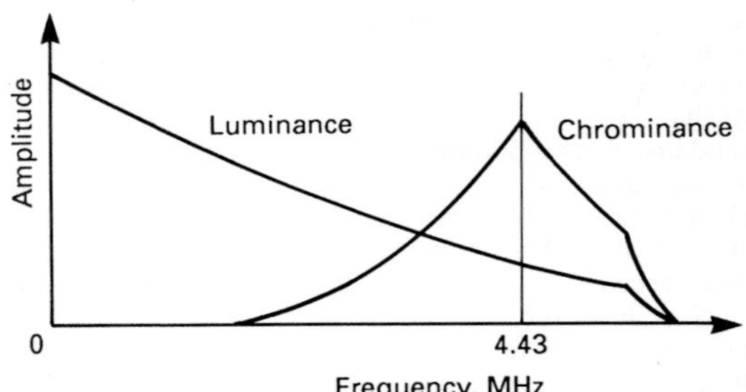

Figure 11.1 Positions of the main frequency components in a composite PAL signal

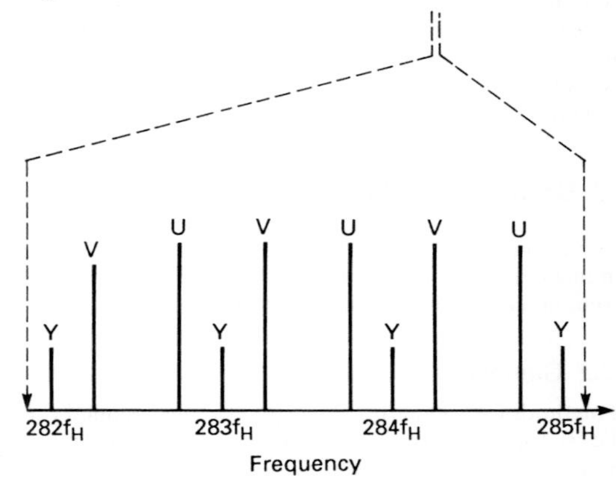

Figure 11.2 The interleaved structure of the high-frequency region of a composite PAL signal, showing luminance (Y) and chrominance (U and V) components

alternate-line phase inversion of the V signal has already introduced a half-line frequency offset between the two subcarriers. Thus, in the high frequency region of the PAL signal, the main components of the Y, U and V signals are interleaved as shown in *Figure 11.2*.

11.3.1.3 Colour-difference signal weighting

The weighted colour-difference signals, U and V, are given by the relationships:

$$U' = 0.493(B'-Y') \qquad (11.8)$$

and

$$V' = 0.877(R'-Y') \qquad (11.9)$$

The weighting factors have been chosen to limit the amplitude of the colour-difference signal excursions outside the black-to-white range to one third of that range at each end. The inverse relationships required in a decoder are:

$$B'-Y' = 2.028\,U' \qquad (11.10)$$

and

$$R'-Y' = 1.140\,V' \qquad (11.11)$$

11.3.1.4 Colour-difference signal filters

The PAL colour-difference signals are filtered with a low-pass characteristic approximating to the Gaussian template shown in *Figure 11.3* to provide optimum compatibility with monochrome receivers. This wide-band, slow roll-off characteristic is needed to minimize the disturbance visible at the edges of coloured objects because a monochrome receiver includes no further filtering. A narrower, sharper-cut characteristic would emphasize the subcarrier signal at these edges, widening the transitions and introducing ringing.

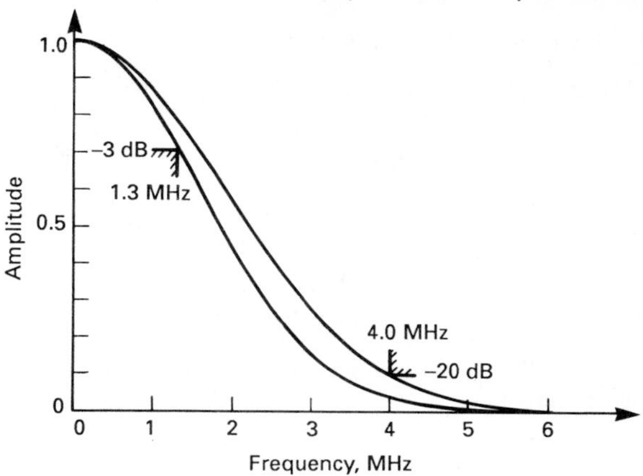

Figure 11.3 Gaussian low-pass filter characteristics for use in PAL coders showing the limits specified for systems B, D, G, H and I

11.3.1.5 Chrominance modulation

The PAL system uses double sideband suppressed-carrier amplitude modulation of two subcarriers in phase quadrature to carry the two weighted colour-difference signals U and V. Suppressed-carrier modulation improves compatibility with monochrome receivers because large amplitudes of subcarrier occur only in areas of highly saturated colour. Because the subcarriers are orthogonal, the U and V signals can be separated perfectly from each other. However, if the modulated chrominance signal is distorted, either through asymmetrical attenuation of the sidebands or by differential phase distortion, the orthogonality is degraded, resulting in crosstalk between the U and V signals.

The alternate-line switching of PAL protects against crosstalk by providing a frequency offset between the U and V subcarriers in addition to the phase offset. Thus, when decoded, any crosstalk components appear modulated onto the alternate-line carrier frequency, in plain coloured areas producing the moving pattern known as *Hanover bars*. This pattern

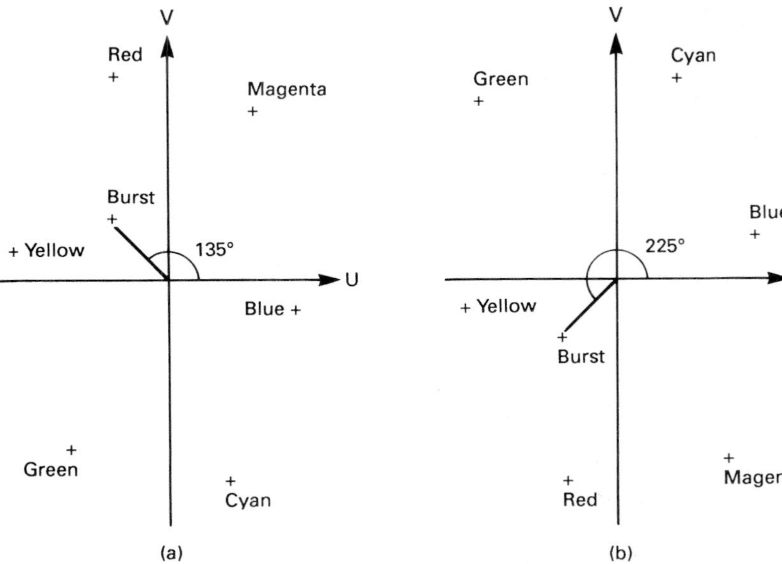

Figure 11.4 Vector diagrams showing the positions of the primary colours and their complements (a) on lines with the burst at 135° (line n) and (b) on lines with the burst at 225° (line $n+1$)

can be suppressed at the decoder by a comb filter averaging equal contributions from switched and unswitched lines.

The PAL chrominance signal can be represented mathematically as a function of time, t, by the expression:

$$U'\sin\omega t \pm V'\cos\omega t \qquad (11.12)$$

where $\omega = 2\pi f_{SC}$. The sign of the V component is positive on line *n* and negative on line *n*+1 (see section *11.3.1.7*).

As an alternative to the rectangular modulation axes, U and V, the modulated chrominance signal can be regarded as a single subcarrier, the amplitude and phase of which are modulated by the saturation and hue, respectively, of the colour represented. Thus the peak-to-peak chrominance amplitude 2S is given by:

$$2S = 2\sqrt{(U'^2 + V'^2)} \qquad (11.13)$$

and the angle α relative to the reference phase (the +U axis) is given by:

$$\alpha = \pm\tan^{-1}(V'/U') \qquad (11.14)$$

on alternate lines. Vector representations of the primary colours and their complements are shown in *Figure 11.4* for the two senses of the V-axis switch.

The modulation process is shown in spectral terms in *Figure 11.5*. In this figure, (a) represents the baseband spectrum of a full bandwidth colour-difference signal, the high frequency components of which are attenuated by Gaussian low-pass filtering to produce the spectrum of (b). When convolved with the line spectrum of the subcarrier signal, (c), this produces the modulated chrominance signal spectrum shown in (d).

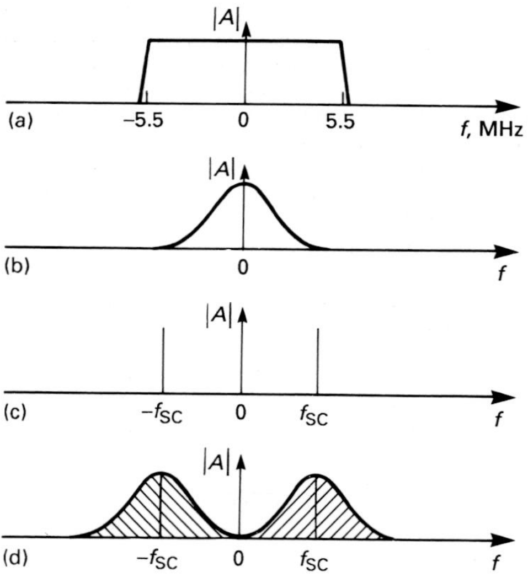

Figure 11.5 Frequency spectra in PAL chrominance modulation: (a) the baseband colour-difference signal, (b) the Gaussian filtered colour-difference signal, (c) the subcarrier sine-wave, (d) the modulated chrominance spectrum produced by convolving (b) and (c)

11.3.1.6 PAL encoding

The main processes of a PAL coder are shown in *Figure 11.6*. Gamma-corrected colour separation signals are converted to YUV form by a matrix combining the relationships of equations (11.1)–(11.3) with those of equations (11.8) and (11.9). The chrominance signal is formed by a pair of product modulators which multiply the orthogonally phased subcarrier waveforms by the low-pass filtered baseband U and V signals. The modulated subcarrier signals are added to a delayed version of the luminance signal Y, timed to compensate for delays in the chrominance circuitry, producing a composite PAL signal at the output.

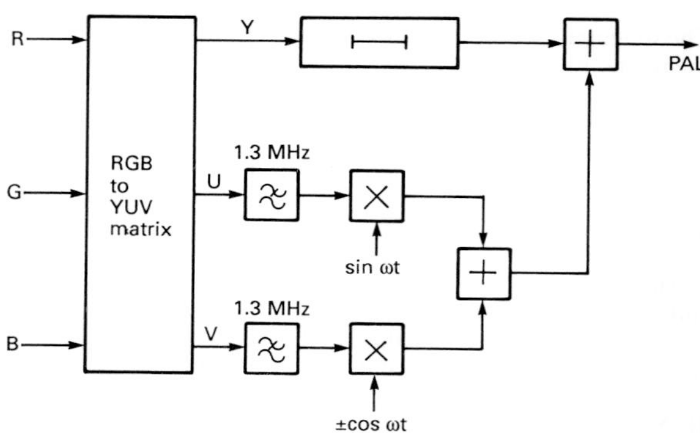

Figure 11.6 The main processes of a PAL encoder

In practice, PAL coders contain no band-limiting low-pass filter at their outputs. Because of this, the upper chrominance sidebands, which extend well beyond normal luminance frequencies, may sometimes be retained up to the broadcast transmitter, at which point the portion of the upper sideband above the nominal luminance bandwidth (5.5 MHz for system I) is removed. The partial loss of the upper sideband causes ringing and desaturation at sharp chrominance transitions. The resulting impairments to the picture are usually only noticeable in systems B, G and H.

11.3.1.7 Colour synchronization

Information about the subcarrier reference phase and the V-axis switch sense is transmitted with the signal by means of a colour synchronizing burst situated on the back porch of the video waveform, following the line synchronizing pulse. The signal timings and amplitudes of the colour reference burst are shown in *Figure 11.7*.

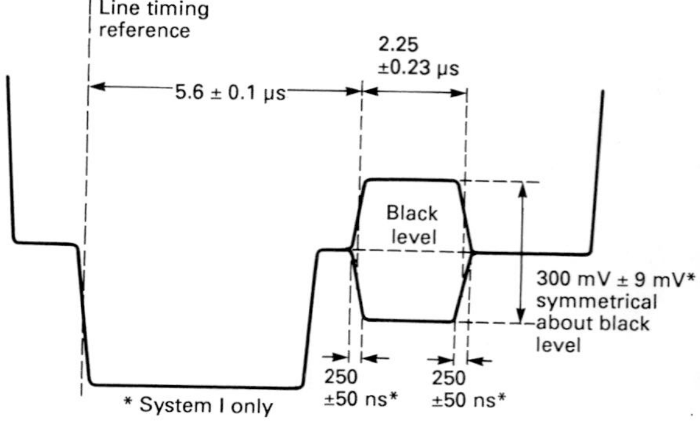

Figure 11.7 Waveform amplitudes and timings for the PAL colour burst in systems B, D, G, H and I. Items marked with an asterisk apply only to system I

The burst phase is 135° on lines where the V-signal is in the true sense (known as line *n*), and 225° on lines where the V-signal is inverted (line *n*+1). As a complete picture contains an odd number of lines, the V-switch sense changes from one

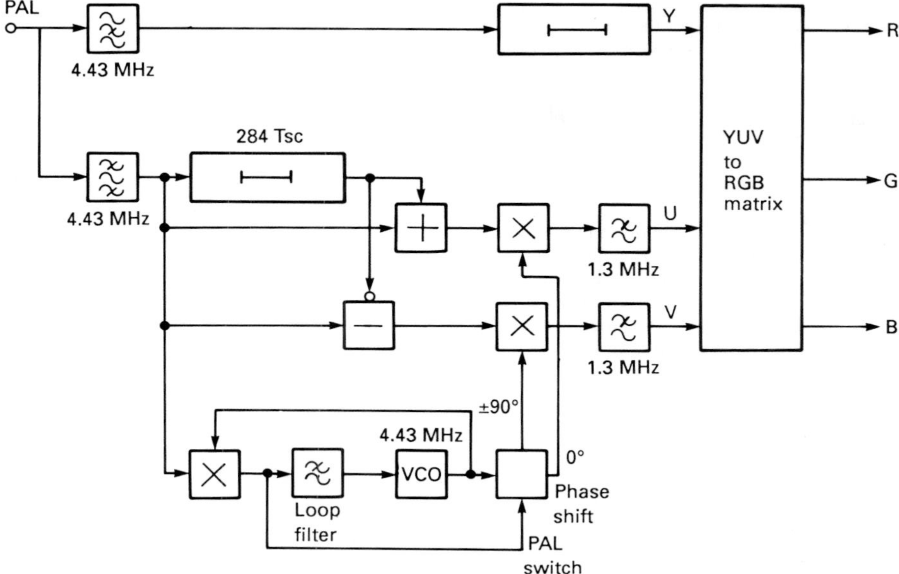

Figure 11.8 The main processes of a conventional delay-line PAL decoder

picture to the next, being positive on line 1 of the first field and negative on line 1 of the third field. In the eight field cycle of the PAL signal, the first four fields are identified by having a subcarrier reference phase ϕ in the range $-90° \leq \phi < 90°$ at the beginning of the first field. To facilitate editing, the EBU[4] has defined the value of ϕ at the beginning of the first field to be $0° \pm 20°$.

Colour bursts are omitted from nine consecutive lines during each field blanking interval, in a sequence that repeats every four fields. The lines with no burst are detailed in *Table 11.1*.

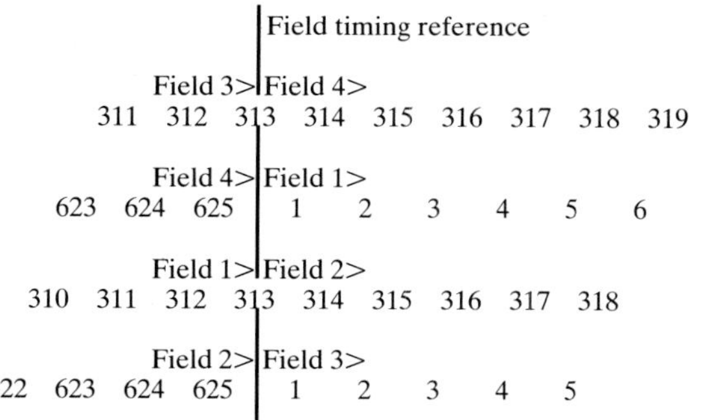

Field timing reference

Field 3>|Field 4>
311 312 313 314 315 316 317 318 319

Field 4>|Field 1>
623 624 625 | 1 2 3 4 5 6

Field 1>|Field 2>
310 311 312 313 314 315 316 317 318

Field 2>|Field 3>
622 623 624 625 | 1 2 3 4 5

Table 11.1 Lines with no burst in the PAL signal

11.3.1.8 PAL decoding

The main processes of a PAL decoder, shown in *Figure 11.8*, consist of separating the luminance and chrominance signals and demodulating the chrominance to retrieve the colour-difference signals U and V. Then, for display, the Y, U and V signals are converted to RGB by a matrix circuit embodying the relationships of equations (11.10) and (11.11), combined with those of equations (11.4)–(11.6).

In a conventional PAL decoder, as normally used in a domestic receiver, the modulated chrominance signals are first separated from the low frequency luminance by a Gaussian bandpass filter centred on the subcarrier frequency. The subcarrier signals are then comb filtered by averaging across a delay of duration 283½ or 284 cycles of subcarrier, as shown in *Figure 11.8*. With a 284 cycle delay, the V subcarrier signals are in anti-phase across the delay and so cancel at the output of the adder, while the U subcarrier signals are co-phased and are added. Similarly, the subtractor cancels the U components to leave the V signal. In each case, the averaging process suppresses any U–V crosstalk components (*Hanover bars*) caused by distortion of the quadrature subcarrier signals. Product demodulators fed with appropriately phased subcarrier signals are used to regain the baseband U and V signals, with subsequent low-pass filters to suppress the twice subcarrier components. The Gaussian bandpass filter and the post-demodulation low-pass filters combine to provide somewhat less chrominance bandwidth than the PAL signal contains. Although this sacrifices some chrominance resolution, it results in less cross-colour, caused by luminance signals being demodulated as chrominance.

The luminance signal is obtained by suppressing the main subcarrier frequency components with a simple notch filter, usually having a maximum attenuation at about 20 dB. The luminance path includes a delay to compensate for the chrominance processing circuits, so that the signals remain aligned horizontally. However, no attempt is made to compensate for the vertical delay of the chrominance comb filter. This introduces a mean delay of half a line relative to the luminance, so that coloured objects are displaced down the picture by this amount.

Quadrature subcarrier signals for the demodulators are generated by a voltage-controlled crystal oscillator locked to the reference phase of the incoming colour bursts. A method frequently used is to allow the alternating bursts at 135° and 225° to pull the oscillator on alternate lines. With a relatively long loop time constant (usually between 30 and 100 line periods), the oscillator takes up the mean phase of 180° relative to the +U reference axis. The error signal of the phase detector then provides the alternating V-switch square-wave signal directly.

In difficult reception conditions, the received signal may have significant variations of gain across the video band which could lead to incorrect colour saturation in the displayed picture. To reduce this effect, receivers incorporate automatic colour control circuitry to adjust the gain of the colour-difference amplifiers in response to the received amplitude of the colour burst.

11.3.1.9 System performance

The phase alternation property of the PAL signal substantially overcomes the susceptibility of suppressed carrier amplitude modulation to differential phase distortion. With severe distortion, however, PAL averaging causes a desaturation of the resulting picture. Also, the PAL switching halves the vertical chrominance resolution obtainable from the system. Another slight penalty is the subcarrier visibility that results from the subcarrier offsets of ¼ and ¾ line frequency.

11.3.1.10 PAL system variations

The baseband PAL coding parameters of the B, D, G, H and I standards are substantially the same. Some differences do arise, however, as a result of the constraints of the radiated signal parameters shown in *Table 11.2*.

PAL system	*Channel spacing (MHz)*	*Sound spacing (MHz)*	*Luminance bandwidth (MHz)*	*Vestigial sideband (MHz)*	*Chrominance sideband (MHz)*
B	7	5.5	5	0.75	0.57
D	8	6.5	6	0.75	1.57
G	8	5.5	5	0.75	0.57
H	8	5.5	5	1.25	0.57
I	8	5.9996	5.5	1.25	1.07
M	6	4.5	4.2	0.75	0.62
N	6	4.5	4.2	0.75	0.62

Table 11.2 Radiated signal parameters of PAL systems

In system B/PAL, intended for 7 MHz channels, the upper chrominance sideband is constrained to 570 kHz by the 5 MHz nominal signal bandwidth. System G/PAL, although intended for 8 MHz channels, maintains the same frequency allocations within the channel as B/PAL for compatibility. Systems B/ and G/PAL are widely used in Western Europe, Scandinavia, Australia, New Zealand and in parts of Africa. System H/PAL, used in Belgium, retains the 5 MHz bandwidth in the main sideband, but uses some of the extra bandwidth of the 8 MHz channel to extend the width of the vestigial sideband. On the other hand, system D/PAL, used in China, extends the signal bandwidth to 6 MHz, which provides capacity for the full 1.3 MHz of the upper chrominance sideband. System I/PAL, used principally in the British Isles, Hong Kong and South Africa, combines the two features, both extending the vestigial sideband and providing 5.5 MHz in the main sideband. Thus the upper chrominance sideband extends to 1.07 MHz.

Variants of PAL for 6 MHz channels incorporate greater differences. System N/PAL, used in Argentina, has a subcarrier frequency of 3.582 056 25 MHz to match the 4.2 MHz bandwidth. This conforms to the relationship:

$$f_{sc} = (229\tfrac{1}{4} + \frac{1}{625})\, f_H \qquad (11.15)$$

System M/PAL, used in Brazil, is a 525-lines, 60 fields per second version of PAL for a 4.2 MHz signal bandwidth. M/PAL has a subcarrier frequency of 3.575 611 49 MHz given by the relationship:

$$f_{sc} = 227\tfrac{1}{4} f_H = \frac{909}{4}\, f_u \qquad (11.16)$$

This system ignores the picture rate component of the subcarrier frequency, making the subcarrier more obtrusive on monochrome receivers.

11.3.2 NTSC

11.3.2.1 Development

The NTSC system, adopted for use in the USA in 1953, takes its name from the National Television System Committee[5] convened to recommend its parameters. It is based on the system M monochrome television standard, having 525 lines per picture and 60 fields/second, with a 4.2 MHz bandwidth and designed to occupy channels at a 6 MHz spacing. The relatively low horizontal resolution of this system placed severe constraints on the development of a compatible chrominance signal.

Colour information is transmitted using suppressed carrier amplitude modulation of two orthogonally phased subcarriers to carry two weighted colour-difference signals, known as I and Q (*in-phase* and *quadrature*).

11.3.2.2 Colour subcarrier frequency

The NTSC subcarrier frequency of 3.579 545 MHz ±10 Hz was chosen to fall in the upper part of the 4.2 MHz luminance band to reduce its visibility on monochrome receivers. It was also chosen to have a half-line frequency offset so that peaks and troughs of adjacent lines would provide a degree of visual cancellation. Accordingly, the subcarrier frequency is related to the line frequency by the following relationship:

$$f_{sc} = (229\tfrac{1}{4} = \frac{455}{2}\, f_H \qquad (11.17)$$

This causes the main luminance and chrominance components in the high-frequency part of the video signal spectrum to be interleaved as shown in *Figure 11.9*. The subcarrier phase at a point in the picture repeats every four fields.

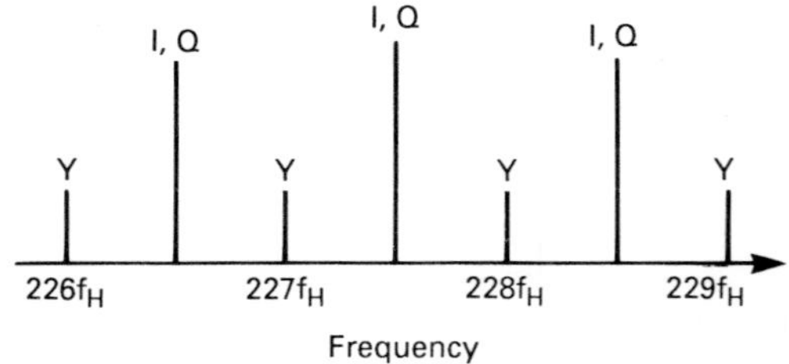

Figure 11.9 Positions of the main luminance (Y) and chrominance (I, Q) components in the high-frequency part of an NTSC signal

A further factor in the choice of colour subcarrier frequency was that any beat between the system M sound carrier (mean frequency 4.5 MHz) and the subcarrier would result in a frequency also having a half-line offset (approximately 920 kHz). This choice of subcarrier produced small changes in the nominal line and field frequencies to 15 734.264 Hz and 59.94 Hz, respectively, for colour operation, although these values remain within the original tolerances of system M.

11.3.2.3 Colour-difference signals

The NTSC system is based on the principle that significantly lower resolution is acceptable for some colours, notably for blue/magenta and yellow/green. So, instead of using the B-Y and R-Y signals directly, the signals are first weighted as in PAL, to limit their maximum excursions, and then phase rotated by 33°. Thus, the I and Q colour-difference signals, used to modulate the two orthogonal subcarriers, are given by:

$$I' = -U'\sin 33° + V'\cos 33° \quad (11.18)$$

and

$$Q' = U'\cos 33° + V'\sin 33° \quad (11.19)$$

with U and V given by equations (11.8) and (11.9). As can be seen from *Figure 11.10*, this rotation aligns the Q axis with the regions of colour least affected by a loss of resolution, so allowing the signal to be encoded with a low bandwidth.

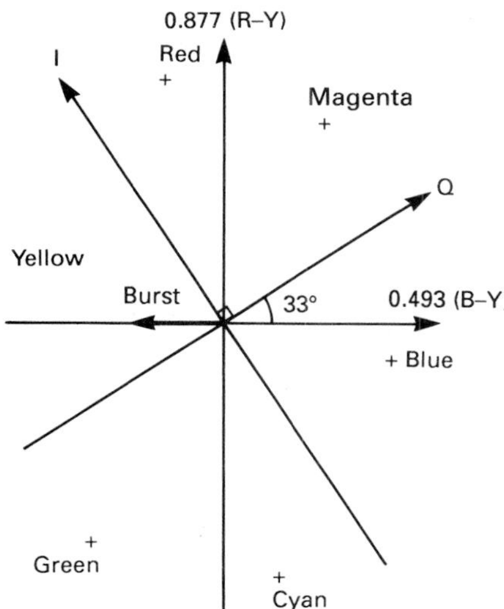

Figure 11.10 Positions of the I and Q modulation axes of the NTSC system relative to the weighted colour-difference axes (U and V). The positions of the colour primaries and their complements and the NTSC colour burst are also shown

In terms of direct relationships to B-Y and R-Y, the I and Q signals are given by:

$$I' = -0.269(B'-Y') + 0.736(R'-Y') \quad (11.20)$$

and

$$Q' = 0.413(B'-Y') + 0.478(R'-Y') \quad (11.21)$$

The inverse relationships are:

$$B'-Y' = -1.105I' + 1.701Q' \quad (11.22)$$

$$R'-Y' = 0.956I' + 0.621Q' \quad (11.23)$$

11.3.2.4 I and Q filters

The wide-band I signal uses a slow roll-off characteristic (*Figure 11.11*(a)) very similar to that used for the U and V signals in PAL. The Q signal, however, is limited to about 0.5 MHz bandwidth by the template of *Figure 11.11*(b).

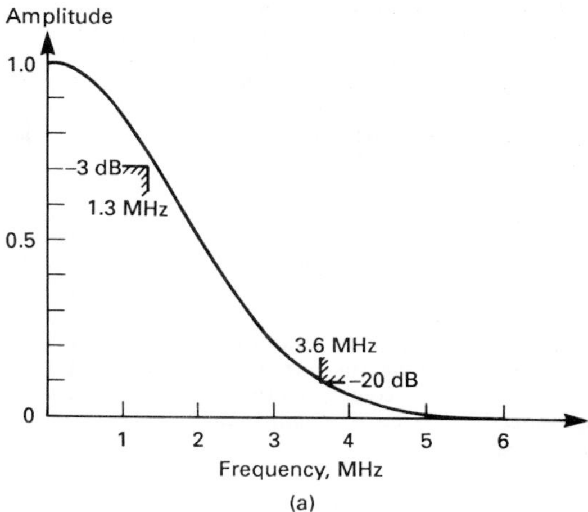

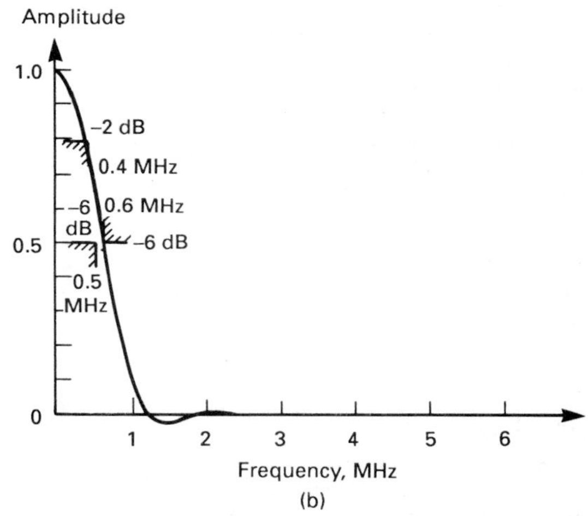

Figure 11.11 Limits for colour-difference low-pass filters for NTSC coders, showing some appropriate characteristics (a) for the I channel and (b) for the Q channel

11.3.2.5 Chrominance modulation

The NTSC modulated chrominance signal can be represented mathematically by the expression:

$$Q'\sin\omega t + I'\cos\omega t \quad (11.24)$$

where $\omega = 2\pi f_{SC}$.

The peak-to-peak chrominance amplitude 2S is given by:

$$2S = 2\sqrt{(I'^2 + Q'^2)} \quad (11.25)$$

As the signal bandwidth of the NTSC system is only 4.2 MHz, the spectrum space available for the modulated chrominance components is relatively limited. In particular, making the subcarrier a fine, high frequency pattern to reduce its visibility results in a significant loss of the upper chrominance sidebands. If both colour-difference components were wide-band, the asymmetry would cause crosstalk between the two signals. However, as the Q signal is low bandwidth, it remains a double-sideband signal and causes no crosstalk. Part of the upper sideband of the wide-band I signal, on the other hand, is lost. When demodulated, some I components do cross into the Q signal, but at higher frequencies than the true Q components. Therefore, the I–Q crosstalk components can be removed by low-pass filtering the demodulated Q signal.

11.3.2.6 NTSC encoding

The main processes of an NTSC encoder are shown in *Figure 11.12*. RGB signals are matrixed to YIQ form using the relationships of equations (11.1)–(11.3), combined with those of equations (11.20) and (11.21). The I and Q signals are filtered according to the templates of *Figure 11.11* and modulated onto orthogonal subcarriers. The modulated chrominance signals are combined with the luminance signal, appropriately delayed to compensate for the chrominance processing.

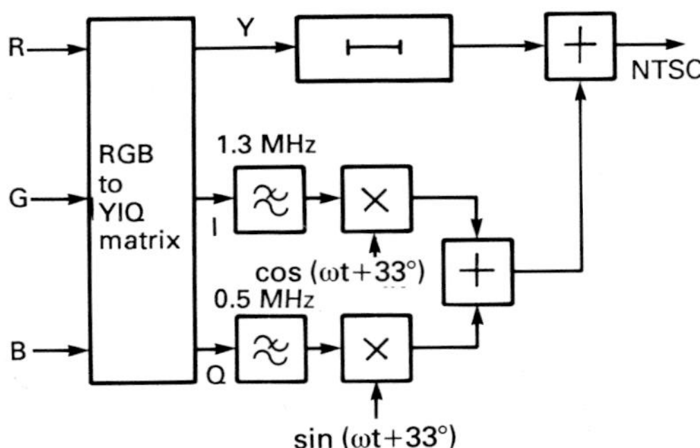

Figure 11.12 The main processes of an NTSC encoder

11.3.2.7 Colour synchronization

Subcarrier reference phase information is carried in the composite NTSC signal by a colour burst occupying a position on the back porch, following the line synchronizing pulse, as shown in *Figure 11.13*. The phase of the burst is 180° relative to the B-Y reference axis (see *Figure 11.10*). Colour bursts are omitted from those lines in the field interval starting with an equalizing pulse or a broad pulse.

The NTSC reference subcarrier and the line timing reference point (the half amplitude point of the falling edge of the line synchronizing pulse) are not only related in frequency by

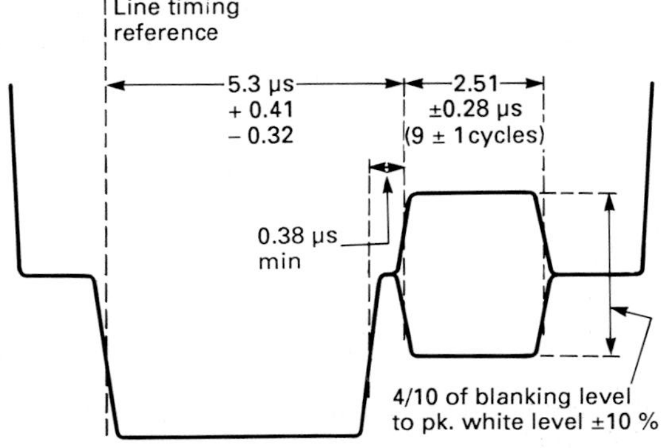

Figure 11.13 Waveform amplitudes and timings for the NTSC colour burst

equation (11.17), but are also related in phase. The line timing reference point is defined as being coincident with zero crossings of the reference subcarrier.

The first field in the four field sequence of an NTSC signal is defined by two factors. The first equalizing pulse of the field interval must:

- occur at a line pulse position,
- coincide with a negative-going zero crossing of the reference subcarrier.

11.3.2.8 NTSC decoding

The main processes of a simple NTSC decoder are shown in *Figure 11.14*. The colour-difference signals are separated by product demodulators, with the low-bandwidth Q filter used to remove crosstalk resulting from the asymmetrical sideband I signals. A 3.58 MHz notch filter is used to remove the main subcarrier components from the luminance signal. Y, I and Q signals are matrixed to RGB according to the relationships of equations (11.22), (11.23) and (11.4)–(11.6).

The locally generated subcarrier reference, sin ωt, is synchronized to the incoming colour burst waveform, sin (ωt + 180°). Phase-shifted versions of the reference phase are fed to the product demodulators.

With NTSC signals, comb filters based on line delays are frequently used in decoders to improve luminance resolution and to reduce cross-colour impairments.[6]

11.3.2.9 System performance

The NTSC signal is particularly susceptible to differential phase distortion. This causes high frequencies at one average level to be delayed more than those at another level, so altering the phase relationship between the burst and the chrominance signals of the active-line video. This results in hue errors in the final picture. Also, sideband asymmetry causes hue errors on colour transitions. This makes the NTSC signal much less robust than PAL or SECAM.

11.3.2.10 Countries using system M/NTSC

The NTSC system is used in North America, most of Central America and the Caribbean, and much of South America with the notable exceptions of Argentina and Brazil. The system is also used in Japan, South Korea and Burma.

The radiated signal is substantially similar in all cases, using a 0.75 MHz vestigial sideband and a 4.2 MHz main sideband within a 6 MHz channel.

11.3.3 SECAM

11.3.3.1 Development

The SECAM system, invented by De France[7], was developed in Europe for 625/50 scanning, like PAL, as a means of overcoming the susceptibility of the NTSC system to differential phase distortion. Instead of relying on two subcarriers remaining orthogonal, the colour-difference signals are transmitted separately on alternate lines, so that the signal from the previous line has to be held over in the decoder by a line memory. This led to the name *Sequential Colour with Memory*.

Early versions of SECAM used amplitude modulation in common with PAL and NTSC, but this proved unsatisfactory. This was because the substantial differences in the signals on alternate lines resulted in poor compatibility for monochrome receivers. The use of frequency modulation gave some improvement, but considerable optimization of the signal parameters was needed to achieve acceptable compatibility.

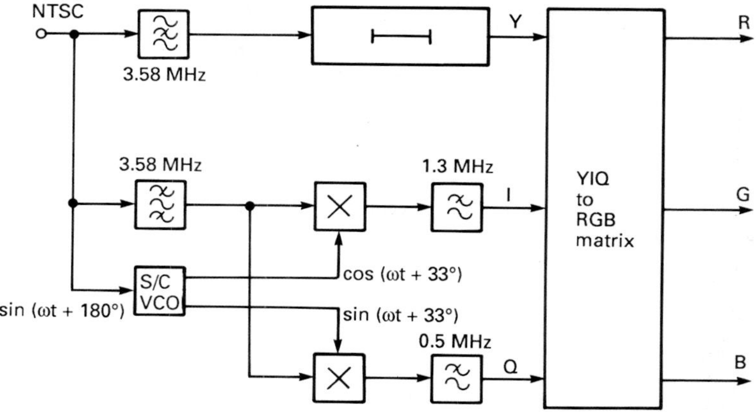

Figure 11.14 The main processes of an NTSC decoder

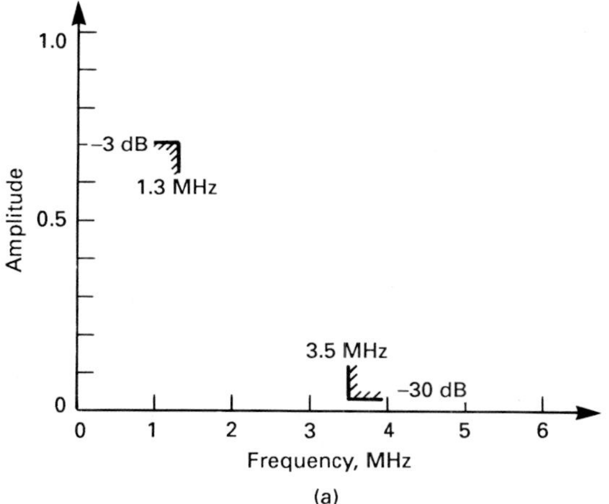

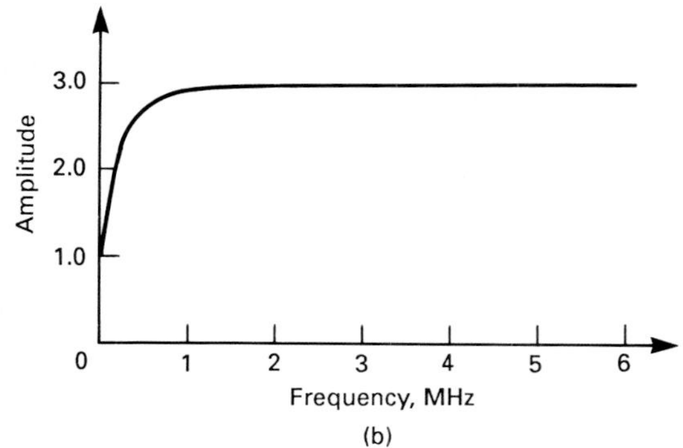

Figure 11.15 Colour-difference signal filtering in the SECAM system: (a) amplitude limits of the $\mathbf{D_R}$ and $\mathbf{D_B}$ low-pass filters, (b) the video pre-emphasis characteristic applied to $\mathbf{D_R}$ and $\mathbf{D_B}$

11.3.3.2 Colour subcarriers

The subcarriers on alternate lines use different undeviated frequencies:

$$f_{0R} = 4.406\ 250 \text{ MHz} \pm 2 \text{ kHz} \tag{11.26}$$

and

$$f_{0B} = 4.250\ 000 \text{ MHz} \pm 2 \text{ kHz} \tag{11.27}$$

These frequencies have the following relationships to the 625/50 line frequency:

$$f_{0R} = 282f_H \tag{11.28}$$

and

$$f_{0B} = 272f_H \tag{11.29}$$

Unlike PAL and NTSC, the fm colour signals are present even in monochrome areas of the picture, thus aggravating the problem of compatibility. Although the use of different line-locked frequencies on alternate lines reduces the chrominance visibility, this is combined with a complex pattern of phase reversals with two components. There is an inversion from one field to the next, and this is combined with a further pattern, either:

0°, 0°, 180°, 0°, 0°,

or

0°, 0°, 0°, 180°, 180°, 180°,

thus resulting in a twelve field sequence[8] for the SECAM signal.

11.3.3.3 Colour-difference signal processing

As with PAL and NTSC, the R-Y and B-Y signals are weighted before modulation, producing signals known as D_R and D_B, given by:

$$D'_R = -1.902(R'-Y') \tag{11.30}$$

and

$$D'_B = 1.505(B'-Y') \tag{11.31}$$

Also, the colour-difference signals are low-pass filtered with a characteristic matching the limits shown in *Figure 11.15*(a). This is very similar to PAL colour-difference filtering, but with a slightly sharper cut-off.

In addition, video pre-emphasis is applied according to:

$$F(f) = \frac{1 + j(f/f_1)}{1 + j(f/3f_1)} \tag{11.32}$$

where $f_1 = 85$ kHz.

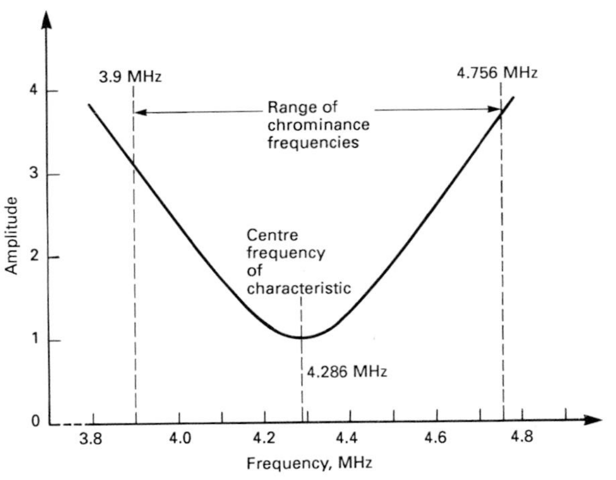

Figure 11.16 The SECAM rf pre-emphasis characteristic

chrominance signals to the range 3.900–4.756 MHz. The modulated chrominance signal is then filtered with the rf pre-emphasis characteristic of *Figure 11.16*, which further reduces the visibility of the subcarrier near the rest frequencies. This is given by the relationship:

$$G = M_0 \frac{1 + j16\,F}{1 + j1.26\,F} \tag{11.33}$$

where $F = f/f_0 - f_0/f$ and $f_0 = 4.286$ MHz.

M_0 sets the minimum subcarrier amplitude at f_0 to be 161 $mV_{p\text{-}p}$ in a standard level signal. Thus the chrominance components are represented mathematically by the expressions:

$$G\cos 2\pi(f_{OR} + \Delta f_{OR}\textstyle\int_0^t DR^*dt) \tag{11.34}$$

or

$$G\cos 2\pi(f_{OR} + \Delta f_{OR}\textstyle\int_0^t DB^*dt) \tag{11.35}$$

on alternate lines.

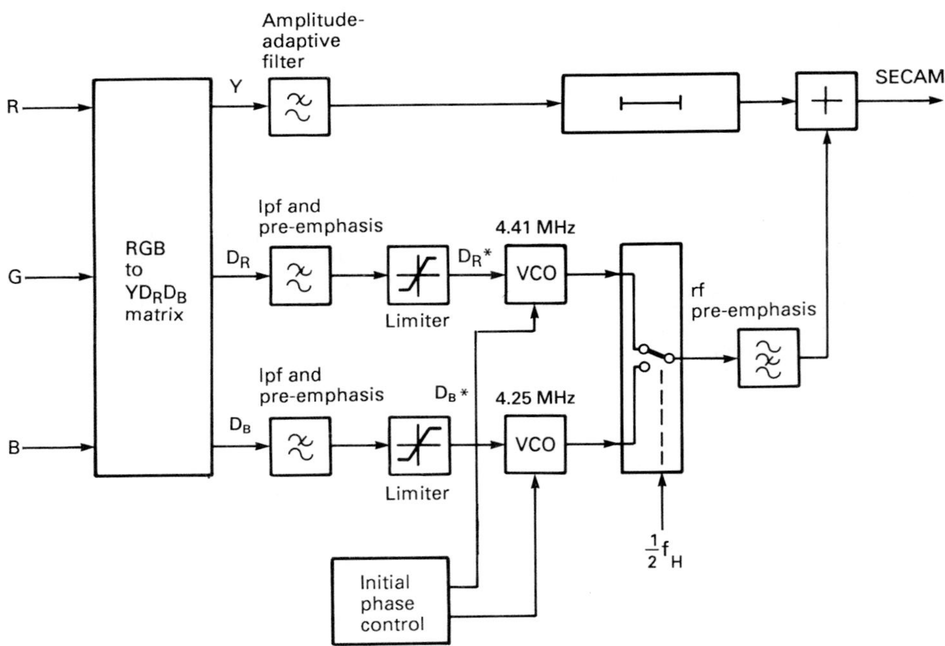

Figure 11.17 The main processes of a SECAM encoder

This produces a lift of about 9.5 dB above 750 kHz as shown in *Figure 11.15*(b). The pre-emphasis allows a relatively low amplitude of subcarrier to be used in plain coloured areas, thus further assisting monochrome compatibility.

11.3.3.4 Frequency modulation

The weighted, filtered and pre-emphasized colour-difference signals, D_R^* and D_B^* are used to frequency modulate the subcarriers of section *11.3.3.2*, producing nominal deviations of 280±9 kHz for D_R^* and 230±7 kHz for D_B^*, corresponding to the saturations of 75 per cent colour bars[9]. Limiters prevent deviations greater than +350 or -506 kHz for D_R^*, and +506 or -350 kHz for D_B^*. This restricts the risetime of fast, large amplitude colour transitions. The asymmetry, combined with the offset between the two subcarriers, limits the modulated

11.3.3.5 SECAM encoding

The main features of a SECAM encoder are shown in *Figure 11.17*. After matrixing to Y, D_R, D_B form, the colour-difference signals are filtered with a combined low-pass and pre-emphasis characteristic. Limiters then restrict the amplitude range of the resulting signals, before they are applied to voltage controlled oscillators for the two subcarrier frequencies. The initial oscillator phase is selected by a sequence generator according to one of the relationships in section *11.3.3.2*. Each modulated colour-difference signal is then selected on alternate lines and weighted by the rf pre-emphasis characteristic before being added to an appropriately delayed luminance signal.

Because of the fm capture effect, cross-colour in the SECAM system becomes a problem only if there is a significant

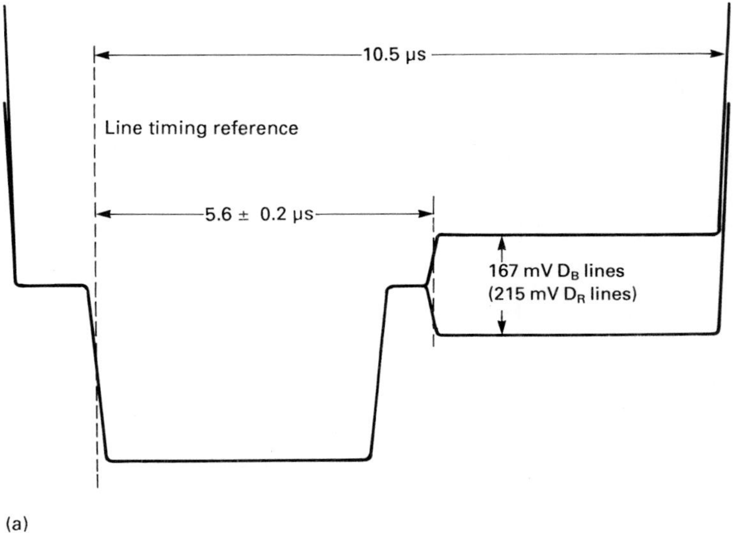

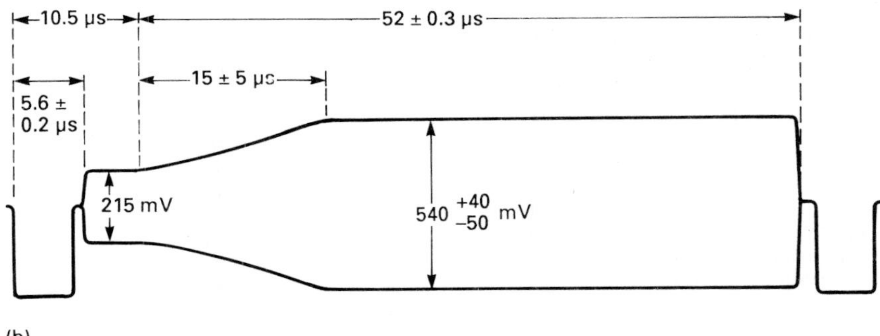

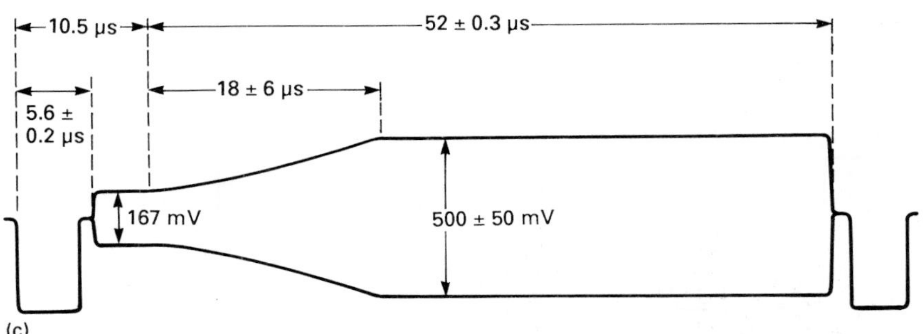

Figure 11.18 Colour synchronizing waveforms in the SECAM signal: (a) line synchronizing signal (D_B line), (b) D_R field synchronizing signal, (c) D_B field synchronizing signal

amplitude of luminance. For this reason, SECAM coders sometimes incorporate level-sensitive filters to retain some high-frequency luminance, while limiting the larger amplitude components that might produce cross-colour.

11.3.3.6 Colour synchronization

The SECAM signal provides for two methods of synchronizing the bistable switching sequence in the coder and decoder. First, the subcarrier signals on each line begin before the active-line period as shown in *Figure 11.18*. From this, the differing frequencies of the two carriers can be sensed on a line-by-line basis.

Alternatively, the field intervals contain a sequence of nine lines of identification signals, occupying lines 7–15 on the first and third fields and lines 320–328 on the second and fourth fields. Line 7 contains D_R^* signals on the first field and D_B^* signals on the third, while line 320 contains D_B^* signals on the second field and D_R^* signals on the fourth.

The large deviation and long duration of the field identification signals provide a very rugged means of regenerating the colour sequence in the decoder, whereas the line reference signals are short and have a much smaller frequency difference. However, because the field blanking lines represent a valuable resource for other uses, line reference synchronization is being encouraged in preparation for suppression of the field identification signals.

11.3.3.7 SECAM decoding

Figure 11.19 shows the main units of a SECAM decoder. The chrominance is filtered with a bell-shaped bandpass characteristic to remove the rf pre-emphasis and then applied to a line delay. The delayed and undelayed signals are switched into the D_R and D_B discriminators on alternate lines, so that each receives lines of the appropriate signal continuously. After demodulation, the low-pass filter provides video de-emphasis.

The luminance channel includes a filter with two notches at the rest subcarrier positions, thus providing good suppression of the subcarrier in monochrome areas. The luminance signal is delayed to match the processing delays of the chrominance circuits before being matrixed to RGB using the relationships in equations (11.4)–(11.6), combined with the inverse colour-difference weighting factors for D_R and D_B signals, given by:

$$R'\text{-}Y' = -0.526\ D_R \tag{11.36}$$
$$B'\text{-}Y' = 0.664\ D_B \tag{11.37}$$

11.3.3.8 System performance

The resistance of the SECAM system to differential gain and differential phase distortions makes it particularly advantageous for poor quality transmission links and magnetic recording. Also, the modulation parameters have been optimized to balance chrominance noise performance against subcarrier visibility to obtain reasonably satisfactory monochrome compatibility. However, the main disadvantage of the SECAM system is the non-linearity of the fm signal. This prevents mixing SECAM signals without separating and demodulating the chrominance. Also, the combined effect of luminance filtering in the coder and the decoder results in rather less high-frequency resolution than for PAL. The fm chrominance signals prevent the use of comb filtering in the decoder to improve luminance resolution.

11.3.3.9 System variations

SECAM encoding is used in virtually the same form in several broadcast signal standards. System B/G SECAM, used mostly in North Africa, the Mediterranean area and the Middle East, has only 5 MHz bandwidth. System D/K SECAM, used mostly in Eastern Europe, is similar, but provides 6 MHz bandwidth. System K1 also has 6 MHz bandwidth, but with the vestigial sideband extended to 1.25 MHz. This standard is planned for much of Africa. SECAM L, used mainly in France, is similar to system K1, although it differs from all the other standards through its use of positive vision modulation and am sound.

11.4 Component colour systems

While the composite colour signals described in the previous section were developed for compatible transmission through existing monochrome channels, the development of component systems is less constrained. With component systems, the extent of compatibility is limited to the need to co-exist with current studio and broadcast standards. In practical terms, this depends on maintaining reasonably straightforward conversion processes between the two types of system.

Although several component coding systems have been developed, only two forms have achieved widespread international recognition, the *4:2:2 digital components system* conforming to CCIR Recommendation 601[10] and time-division-based *multiplexed analogue components (MAC)*[11]. Both these systems achieve improved picture quality over composite

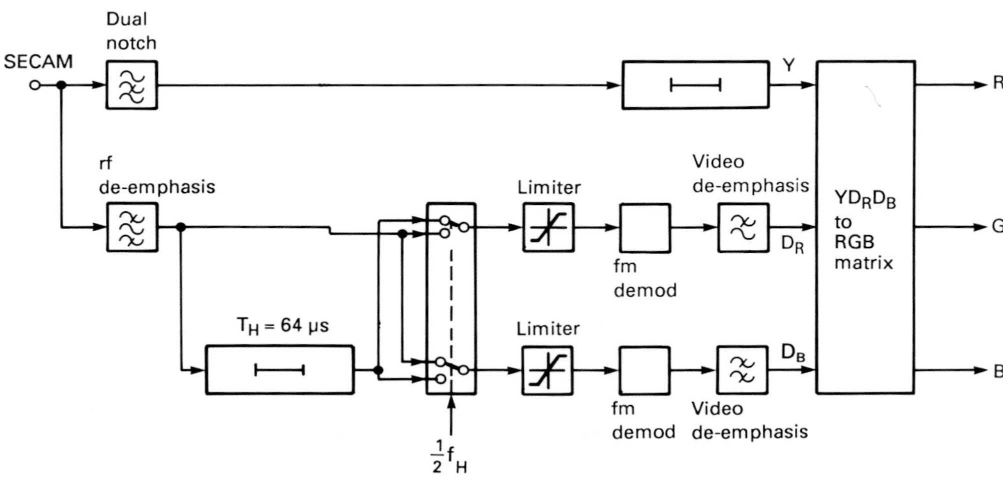

Figure 11.19 The main processes of a SECAM decoder

signal coding through encoding the luminance and colour-difference signals in a form that allows easy separation without cross-effects. However, the transmission bandwidths required for the signals are greater.

11.4.1 4:2:2 digital components

As well as providing better signal quality, the freedom from cross-effects of component systems has considerable advantages for digital signal processing, such as for 'special effects'. The lack of a colour subcarrier eliminates the complex phase relationships which cause a multi-field sequence in composite systems, thus allowing the signals to be stored or combined more simply. The freedom from distortion of digital signal processing and the availability of arithmetic and storage devices from computer technology has allowed very complex production techniques to become commonplace, which could never have been contemplated with analogue circuitry.

11.4.1.1 Background

The parameters of the 4:2:2 components standard were determined as the result of lengthy discussions between a large number of broadcasters,[12] primarily under the auspices of the European Broadcasting Union and the Society of Motion Picture and Television Engineers. Preliminary investigations were constrained by the need to limit the total data rate of a digital component standard. In addition to the general disadvantage of requiring higher speed logic for greater data rates, there were two specific limitations, resulting from the capacity of telecommunications links and the limit of recordable data rates at that time, each of which provided 140 Mbit/s with blanking removal or 160 Mbit/s gross. This allowed capacity for sampling rates of 12 MHz for luminance and 4 MHz for the two colour-difference signals with each sample quantized to eight bits.

However, in April 1980, demonstrations at the BBC Designs Department in London showed that, although the basic picture quality provided by this system was acceptable, the 4 MHz colour sampling was inadequate for high-quality colour-matting[13]. Improvements in recording techniques and the development of suitably transparent bit-rate reduction techniques for transmission links allowed higher sampling rates to be considered. As a result, early in 1981, sampling frequencies of 13.5:6.75:6.75 MHz were chosen after further tests at IBA, Winchester and at the SMPTE Winter Conference in San Francisco[12,14]. This formed the basis of the compatible dual standards for 525-line and 625-line scanning, subsequently known as 4:2:2 digital components.

The 4:2:2 terminology reflects the assumption in Recommendation 601 that a family of compatible standards could be used, provided that conversion to 4:2:2 sampling is straightforward. In practice, this restricts the choice to sampling frequencies with simple relationships to 4:2:2, such as 4:4:4, 4:1:1 or 2:1:1.

11.4.1.2 Sampling structure

The digital coding standards of Recommendation 601 are based

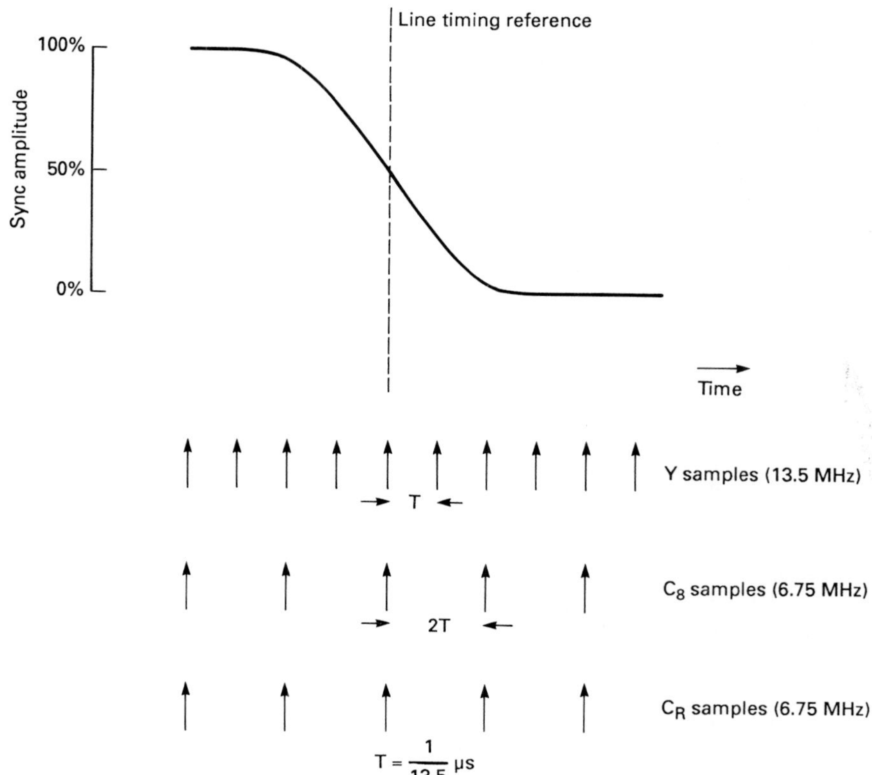

Figure 11.20 Positions of Y, C_B and C_R samples relative to the analogue line timing reference point

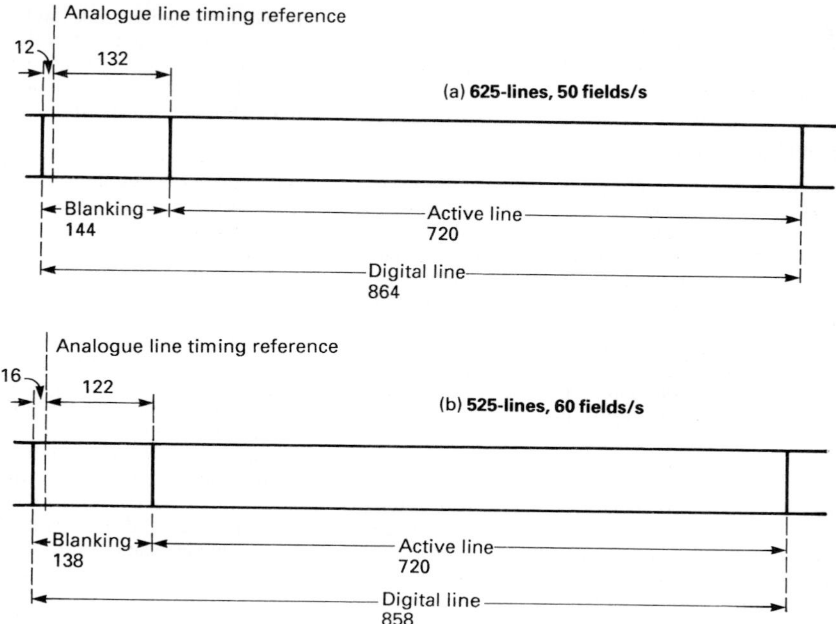

Figure 11.21 Number of 13.5 MHz samples in each section of the digital line for 625/50 and 525/60 scanning standards

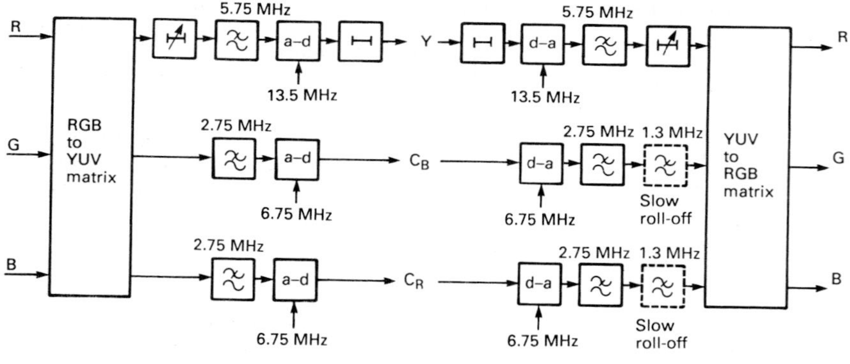

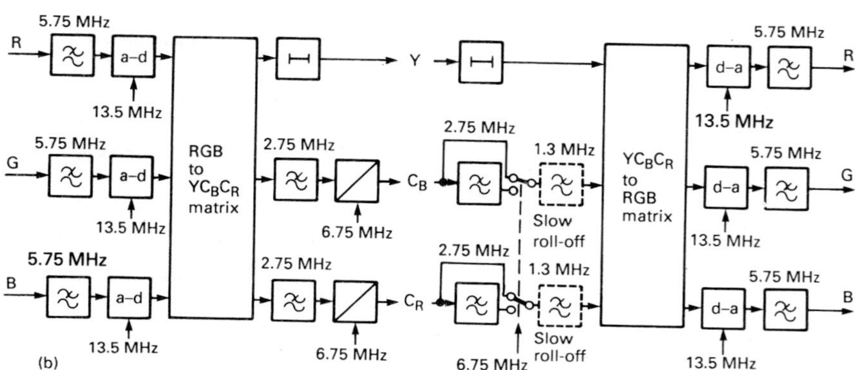

Figure 11.22 Alternative methods of conversion between analogue RGB and digital YC_BC_R signals: (a) with the C_B and C_R signals sampled directly at 6.75 MHz, (b) with digital encoding of the RGB signals

on line-locked sampling. This produces a static orthogonal sampling grid in which samples on the current line fall directly beneath those on previous lines and fields, and exactly overlay samples on the previous picture. This orthogonal sampling structure has many advantages for signal processing, including the simplification of filters and repetitive control waveforms.

The sampling is not only locked in frequency, but also in phase so that one sample is coincident with the line timing reference point (the half-amplitude point of the falling edge of the line synchronizing pulse) as shown in *Figure 11.20*. This feature is also applied to the lower rate colour-difference signal samples, known as C_B and C_R. This ensures that different sources produce samples nominally at the same positions in the picture.

The fundamental sampling frequency of 13.5 MHz gives a measure of commonality between 625/50 and 525/60 systems because this frequency has the property of being an exact harmonic of the line rate on both scanning standards. This produces 864 or 858 luminance samples and 432 or 429 colour-difference samples per line, respectively. When the 13.5 MHz sampling frequency was chosen, the problem of the ninth and eighteenth harmonics causing interference to the international distress frequencies of 121.5 MHz and 243 MHz was not considered. It is believed, however, that interference can be avoided by careful equipment design.[15]

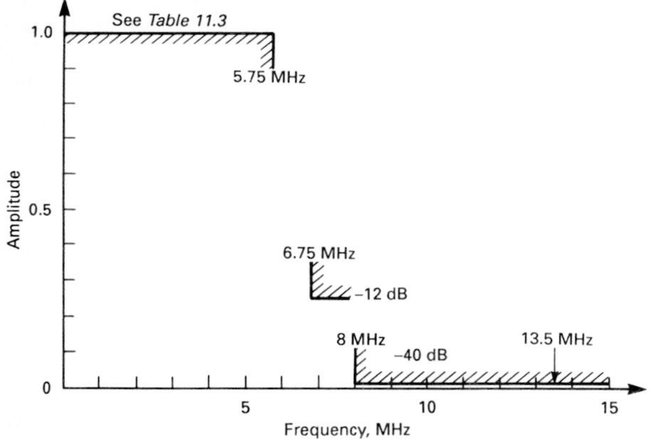

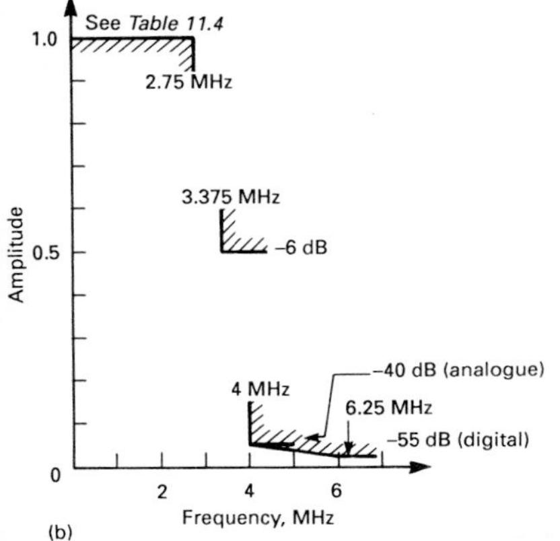

Figure 11.23 Amplitude limits (a) for luminance and (b) for colour-difference filters for 4:2:2 component signals. When used for analogue post-sampling conversions, it is assumed that sin x/x correction is included to give, overall, the passband response shown. In (b), a greater stopband attenuation is specified for digital filters because no sin x/x attenuation occurs

The different formats of the digital line for the two standards are shown in *Figure 11.21*. Each digital line consists of a blanking period followed by an active-line period. The analogue line timing reference falls in the blanking period, a few samples after the beginning of the digital line. The active-line period of 720 samples for luminance and 360 samples for each colour-difference signal is common to both standards. This is sufficiently long to accommodate the full range of analogue blanking tolerances on the two standards. As a result, analogue blanking need only be applied once, preferably in picture monitors or at the conversion to composite signals for broadcast transmission. Samples at the beginning and end of the blanking period are set aside for synchronizing codes, details of which are given in section *11.4.1.7*.

11.4.1.3 Conversion methods

Two approaches can be used for the conversions between the analogue RGB signals produced by cameras and accepted by displays, and the digital YC_BC_R signals used for processing. These are shown in *Figure 11.22*. In that figure, (a) shows digital conversion of YC_BC_R signals with the colour-difference signals sampled directly at 6.75 MHz. While providing a relatively simple implementation, this arrangement depends on the accuracy of setting the individual conversion circuits for its colour balance stability. Also the widely differing delays in the analogue luminance and colour-difference signal filters need to be accurately compensated. In contrast, converting the RGB signals as shown in *Figure 11.22*(b) produces matched delays in the analogue circuitry and includes the gain factors digitally, although the digital matrixing results in a greater degree of complication.

Both methods can include slow roll-off filters in the conversions for display or for composite coding. This avoids the ringing impairments that would otherwise occur due to the sharp-cut colour-difference low-pass filters. If the slow roll-off filters were always used, there would be a rapid loss of resolution when passing through a series of conversion processes.

11.4.1.4 Luminance filters

The use of sampling requires the bandwidth of the digital signal to be specified accurately. Also, in circumstances where mixed analogue and digital working may occur, the signals may pass through many conversion processes in cascade. For these reasons, the filters used in the conversion processes have to adhere to very tight tolerances, particularly in the passband region, to avoid a build-up of impairments. It is assumed that the post-conversion low-pass filters include sin x/x correction.

The template for luminance filter characteristics is shown in *Figure 11.23*(a). The amplitude characteristic is essentially flat to 5.75 MHz, with attenuations of at lease 12 dB at 6.75 MHz

Frequency range	*Design limits*	*Practical limits*
Ripple:		
1kHz – 1MHz	±0.005dB	increasing from ±0.01 to ±0.025dB
1MHz – 5.5MHz	±0.005dB	±0.025dB
5.5MHz –5.75MHz	±0.005dB	±0.05dB
Group delay:		
1kHz – 5.75MHz	0 increasing to ±2ns	±1ns increasing to ±3ns

Table 11.3 Luminance filter ripple and group delay tolerances

and 40 dB for frequencies of 8 MHz and above. Details of the passband ripple and group delay specifications are summarized in *Table 11.3*. The same template applies for RGB filters used in the conversion arrangements of *Figure 11.22*(b).

Frequency range	*Design limits*	*Practical limits*
Ripple:		
1kHz – 1MHz	±0.01dB	increasing from ±0.01 to ±0.05dB
1MHz – 2.75MHz	±0.01dB	±0.05dB
Group delay:		
1kHz – 2.75MHz	increasing from 0 to ±4ns	increasing from ±2 to ±6ns
2.75MHz – f_{-3dB}		±12ns

Note: The use of transversal digital filters ensures zero group delay distortion.

Table 11.4 Colour-difference filter ripple and group delay tolerances

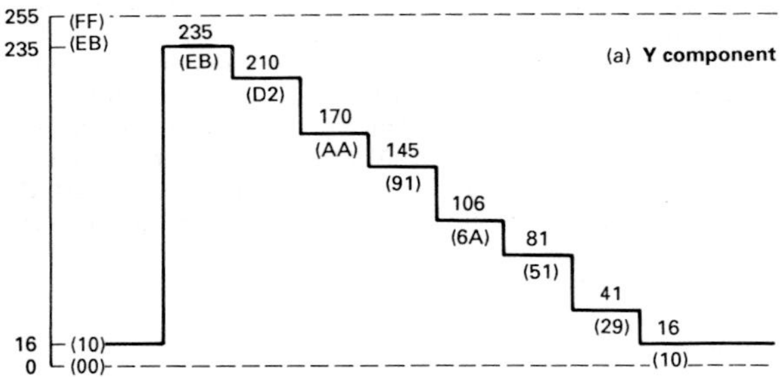

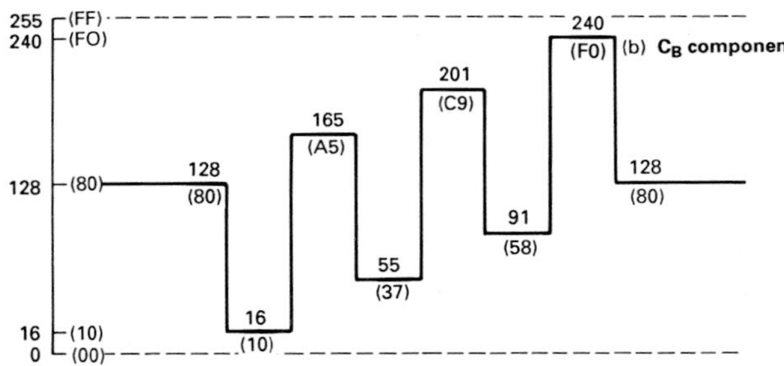

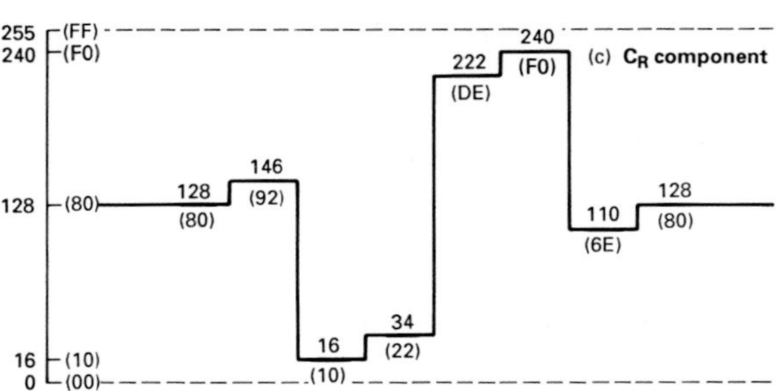

Figure 11.24 Coding levels in a 4:2:2 digital component 100 per cent colour bars signal. Numbers in parentheses show the codes in hexadecimal form

11.4.1.5 Colour-difference filters

The template for colour-difference filters is similar to a scaled version of the luminance filter, with the response essentially flat to 2.75 MHz as shown in *Figure 11.23*(b). However, as aliasing is less noticeable in the colour-difference signals, the attenuation at the half sampling frequency is relaxed to 6 dB. For digital filters as would be used in the arrangement of *Figure 11.22*(b), the attenuation requirement in the stopband is increased to 55 dB as shown because of the lack of attenuation from sin x/x sampling loss. For digital filters there is a particular advantage in using a skew-symmetric response passing through the -6 dB point at the half sampling frequency as this halves the number of non-zero coefficients in the filter, thus reducing the number of taps. Passband ripple and group delay tolerances are shown in *Table 11.4*.

The slow roll-off response used at the final conversion to analogue for viewing follows the form of a Gaussian characteristic such as that of the PAL coder response shown in *Figure 11.3*. With outputs to composite encoders, the normal low-pass filters of the encoder will fulfil this function.

11.4.1.6 Coding ranges

The eight-bit coding of Recommendation 601 corresponds to a range of 256 codes, referred to as 0–255. Levels 0 and 255 are reserved for synchronizing information (detailed in section *11.4.1.7*), leaving levels 1–254 for signal values. For luminance, a standard level signal extends from black at 16 to white at 235. Thus a small margin for overshoots is retained at each end of the coding range. The colour-difference signals occupy a symmetrical range about a zero chrominance level of 128, thus extending down to level 16 and up to level 240. The coding levels corresponding to a 100 per cent colour bars waveform[9] are shown in *Figure 11.24* for the Y, C_B and $C_{R\}}$ signals.

Additional gain factors have to be used in the conversion to digital form to normalize the range of the B-Y and R-Y signals, given by:

$$C'_B = 0.564(B'\text{-}Y') \tag{11.38}$$

and

$$C'_R = 0.713(R'\text{-}Y') \tag{11.39}$$

In digital conversions of the form shown in *Figure 11.22*(b), it is also necessary to take account of the factor 224/219 to allow for the different quantizing ranges of the luminance and colour-difference signals.

line numbers	F	V	H(EAV)	H(SAV)
625/50				
1-22	0	1	1	0
23-310	0	0	1	0
311-312	0	1	1	0
313-335	1	1	1	0
336-623	1	0	1	0
624-625	1	1	1	0
*525/60**				
1-3	1	1	1	0
4-9	0	1	1	0
10-263	0	0	1	0
264-265	0	1	1	0
266-272	1	1	1	0
273-525	1	0	1	0

* Line numbers are in accordance with current engineering practice for 525/60 in which numbering starts from the first line of equalizing pulses, instead of the first line of broad pulses as had previously been used.

Table 11.5 Timing reference indentification in codes

Binary								Hexadecimal
X				Y				XY
1	F	V	H	P_3	P_2	P_1	P_0	
1	0	0	0	0	0	0	0	80
1	0	0	1	1	1	0	1	9D
1	0	1	0	1	0	1	1	AB
1	0	1	1	0	1	1	0	B6
1	1	0	0	0	1	1	1	C7
1	1	0	1	1	0	1	0	DA
1	1	1	0	1	1	0	0	EC
1	1	1	1	0	0	0	1	F1

Table 11.6 Parity values for timing reference codes

11.4.1.7 Synchronizing codes

The positions of the active-line samples are marked by *end of active video* (EAV) and *start of active video* (SAV) codes, which occupy the first and last sample positions respectively in the digital blanking period. Originally defined as a four word sequence in a 27 MHz multiplexed signal, the format of the codes in the corresponding separate component Y, C_B and C_R signals is shown in *Figure 11.25*.

The preamble codes, often referred to by their hexadecimal values FF, 00 and 00, mark the position of the line label word known as XY. In binary form, X has values 1FVH, with F giving odd/even field information, V denoting vertical blanking and H differentiating between EAV and SAV codes. The values of F, V and H are defined in *Table 11.5*. The Y portion of the XY word provides a four-bit parity check word $P_3P_2P_1P_0$ for the X data, as defined in *Table 11.6*.

The beginning of the EAV code marks the beginning of the digital line period, which starts slightly before the corresponding analogue line. The start of the digital field occurs at the start of the digital line which includes the start of the corresponding analogue field. The half lines of active picture in the analogue signal are accommodated as full lines in the digital signal.

Unused samples in the digital blanking intervals are set to blanking level, i.e. 16 (10 hex) for Y, and 128 (80 hex) for C_B and C_R.

11.4.1.8 Parallel interface

In the parallel interface format, the samples of the 4:2:2 Y, C_B and C_R signals are formed into a 27 MHz multiplex in the order C_B, Y, C_R, Y. In the multiplex sequence, the samples C_B, Y, C_R are co-sited (all correspond to the same point on the picture). The EAV and SAV codes thus form the four word sequence FF, 00, 00, XY in the multiplexed signal.

The eight bit samples and a 27 MHz clock signal are conveyed down nine pairs in a multi-way cable with 25 way D type male pin connectors at each end. The individual bits are labelled DATA 0–7, with DATA 7 being the most significant bit. The pin allocations for the individual signals are listed in *Table 11.7*. Equipment inputs and outputs both use female sockets.

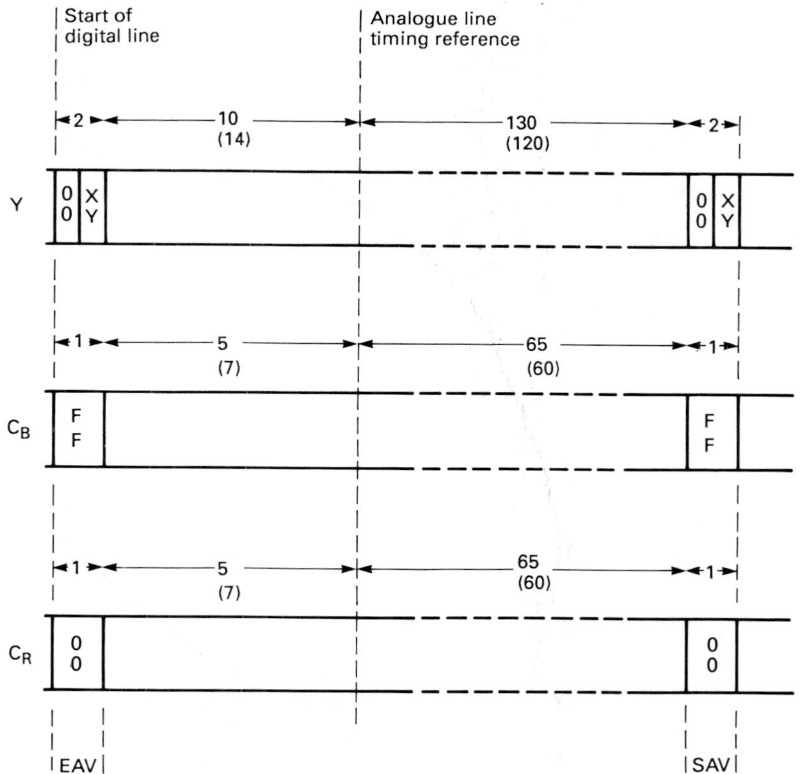

Figure 11.25 Positions of the EAV and SAV codewords in Y, C_B and C_R digital signals for 625/50 scanning. Values in parentheses apply for 525/60 scanning

1	Clock A	14	Clock B
2	System ground	15	System ground
3	Data 7A	16	Data 7B
4	Data 6A	17	Data 6B
5	Data 5A	18	Data 5B
6	Data 4A	19	Data 4B
7	Data 3A	20	Data 3B
8	Data 2A	21	Data 2B
9	Data 1A	22	Data 1B
10	Data 0A	23	Data 0B
11	Spare A-A	24	Spare A-B
12	Spare B-A	25	Spare B-B
13	Cable shield		

The notation A and B is used to denote the two terminals of a balanced pair. For a logic 0, A is negative with respect to B, and, for a logic 1, A is positive with respect to B.

Table 11.7 Pin connections in the 25-way parallel interface

Signals on the parallel interface use logic levels compatible with ecl (emitter-coupled logic) balanced drivers and receivers, and the nrz (non-return to zero) format allows for transmission over distances of 50 m unequalized or 200 m with equalization. The rising edge of the essentially square, 27 MHz clock waveform is the active edge and is timed to occur in the middle of the bit-cell of the data waveforms at the sending end.

11.4.1.9 Serial interface

Conversion to the serial interface format allows the signals of the 27 MHz multiplex to be conveyed down a single bearer, either coaxial cable or optical fibre. This is achieved by converting the eight bit samples to a nine bit format according to an adaptive mapped encoding procedure. The nine bit values are transmitted as nrz data, least significant bit first, to produce a 243 Mbit/s data stream. The eight to nine bit conversion process includes extra coding for clock recovery and word synchronization[16].

For coaxial connections, 75 ohm cables with BNC connectors are used. A specification for optical fibre interfaces has yet to be agreed.

11.4.2 MAC

Analogue component systems have received attention recently in three main areas of broadcasting:

- for broadcast emission, particularly by satellite,
- for low-cost professional video tape recorders,
- as a possible low-cost alternative to digital components in studios.

The possibility of using analogue components for cable distribution has also been examined.

Frequency-division multiplexing was favoured at one time for low-cost studio processing[17] because of its similarity to current composite technology. Thus, a system with a 9 MHz out-of-band subcarrier could avoid the band sharing of PAL and provide similar bandwidths to 4:2:2 components.

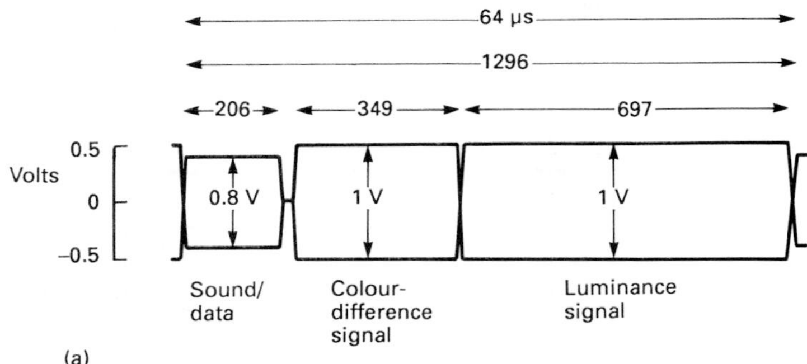

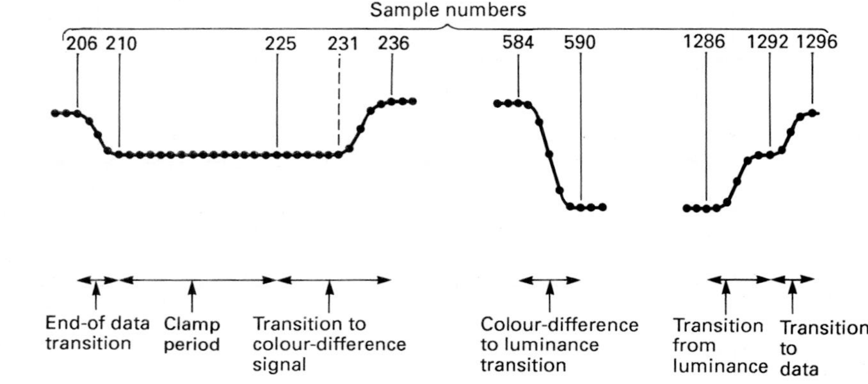

Figure 11.26 (a) Duration of the line segments and (b) details of the signal transitions in D-MAC/packet signals. In D2-MAC, the data period extends to 209 samples. C-MAC contains no baseband signal in the data period because the data signal is added by multiplexing at rf

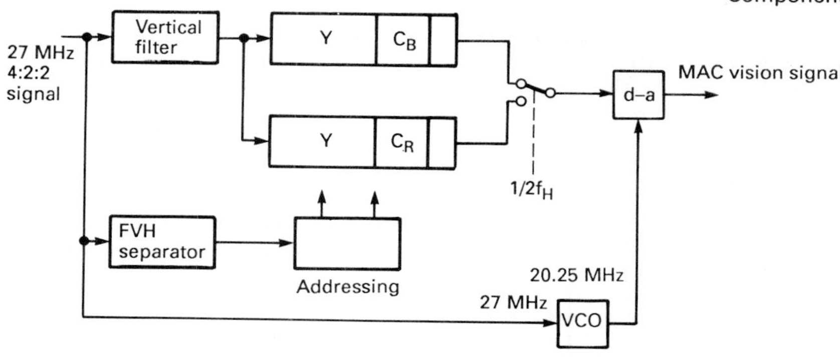

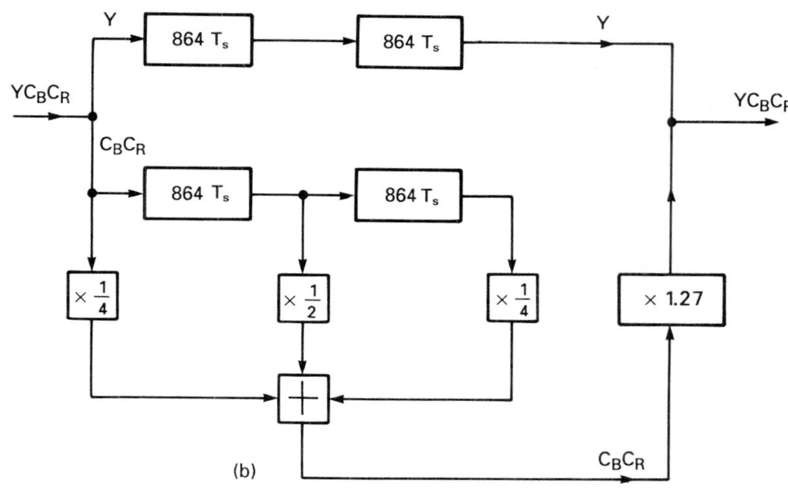

Figure 11.27 MAC vision signal encoding: (a) assembly of the multiplexed signal from 4:2:2 digital components, (b) an example of vertical filtering including scaling

Time-division multiplexed analogue component signals, generally referred to as MAC, time-compress the luminance and colour-difference waveforms, so that both can be accommodated on the same line. The parameters of these systems vary mainly in the compression factors used for luminance and chrominance, the synchronizing and clamping arrangements, and the overall bandwidth requirements of the multiplexed signal. In most systems, the two colour-difference signals are transmitted alternately on consecutive lines. Time-division multiplexed systems of this form have proved particularly suitable for recording, and for broadcast and cable systems. This section concentrates on the MAC/packet system of coding.

11.4.2.1 Development of the MAC/packet family

Although time-multiplexed component systems had been tried before, the Independent Broadcasting Authority started the development of such a system for DBS (direct broadcasting satellite) applications[18] in 1981. The vision parameters used gave a noticeable improvement in picture quality over comparable composite systems, particularly eliminating cross-colour and cross-luminance effects. Several different variants were developed, mainly differing in their sound and data transmission formats, but also using different compression factors from those originally proposed. The main features of these systems are:

A-MAC: separate sound carrier
B-MAC: multi-level sound coding; baseband time-mpxing
C-MAC: 2–4 PSK sound modulation; rf time-mpxing
D-MAC: duo-binary sound coding; baseband time-mpxing
D2-MAC: as D-MAC, but half the coding rate
E-MAC: enhanced vision parameters (e.g. wide aspect)

Although the systems have been developed in the context of 625-line, 50-field scanning with 2:1 interlace, a number of features have been incorporated, designed to enhance their compatibility with the 4:2:2 digital components system. Development of the C, D and D2 systems, known collectively as MAC/packet systems, has been co-ordinated to enhance their common features. C-MAC is advantageous for fm satellite applications, while D and D2-MAC are more suited to vsb-am for broadcast and cable applications. The lower data rate of D2-MAC allows transmission in the standard 7 or 8 MHz channels used for PAL and SECAM.

11.4.2.2 MAC/packet multiplex format

Most lines of a MAC signal include three main parts: a sound/data burst, a time-compressed colour-difference signal and a time-compressed luminance signal. The line segment durations are shown in *Figure 11.26* in terms of 20.25 MHz clock periods. A code at the start of the sound/data burst on each line provides for line and frame synchronization. The extra sound and data capacity of the C and D systems is arranged so that sound and data for D2-MAC signals can be separated easily. A grey level and zero chrominance clamping reference follows the sound/data segment.

The field blanking intervals (lines 623–22 and 311–334) contain other line formats. Lines 312 and 623 are set aside for test signals, line 624 contains reference signals and line 625 contains data for service identification, frame synchronization and, by counting from the frame synchronizing word, for line synchronization. The remaining lines can be used for other purposes.

11.4.2.3 Luminance signal

The MAC/packet signals are based on 625/50/2:1 scanning with a gamma of 2.8 and generally with an aspect ratio of 4:3, although provision is made for increasing this to 5:3. The parameters are based on a clock frequency of 20.25 MHz ±2.5 parts in 10^7.

This has been chosen so that the signal can be encoded directly from digital 4:2:2 components. The luminance compression ratio of 3:2 therefore corresponds to 13.5 MHz samples being read at 20.25 MHz. Accordingly, the frequency content of the time-compressed luminance is a 3:2 scaled version of the digital luminance filter described in section *11.4.1.4*.

The luminance amplitude range is 1 V_{p-p}, with black level at -0.5 and white at 0.5 relative to the 0 V clamp level.

The active picture signal occupies lines 24–310 on the first field and lines 336–622 on the second field. The luminance signal is set to black on lines 23 and 335 to take account of chrominance vertical filtering in the decoder.

11.4.2.4 Colour-difference signals

A compression ratio of 3:1 is used for the colour-difference signals, corresponding to 6.75 MHz samples being read at 20.25 MHz. As with luminance, the compressed colour signals are filtered with a 3:1 scaled version of the colour-difference signal filters described in section *11.4.1.5*. It is assumed that the MAC receiver contains a slow roll-off filter similar to that in *Figure 11.3* to eliminate ringing on sharp transitions.

The colour-difference signals are carried on alternate lines, odd-numbered and even-numbered lines carrying the U and V components respectively, to produce a frame-locked pattern. The colour sequence is identified by counting from the frame sync. Colour signals are transmitted in the line prior to that

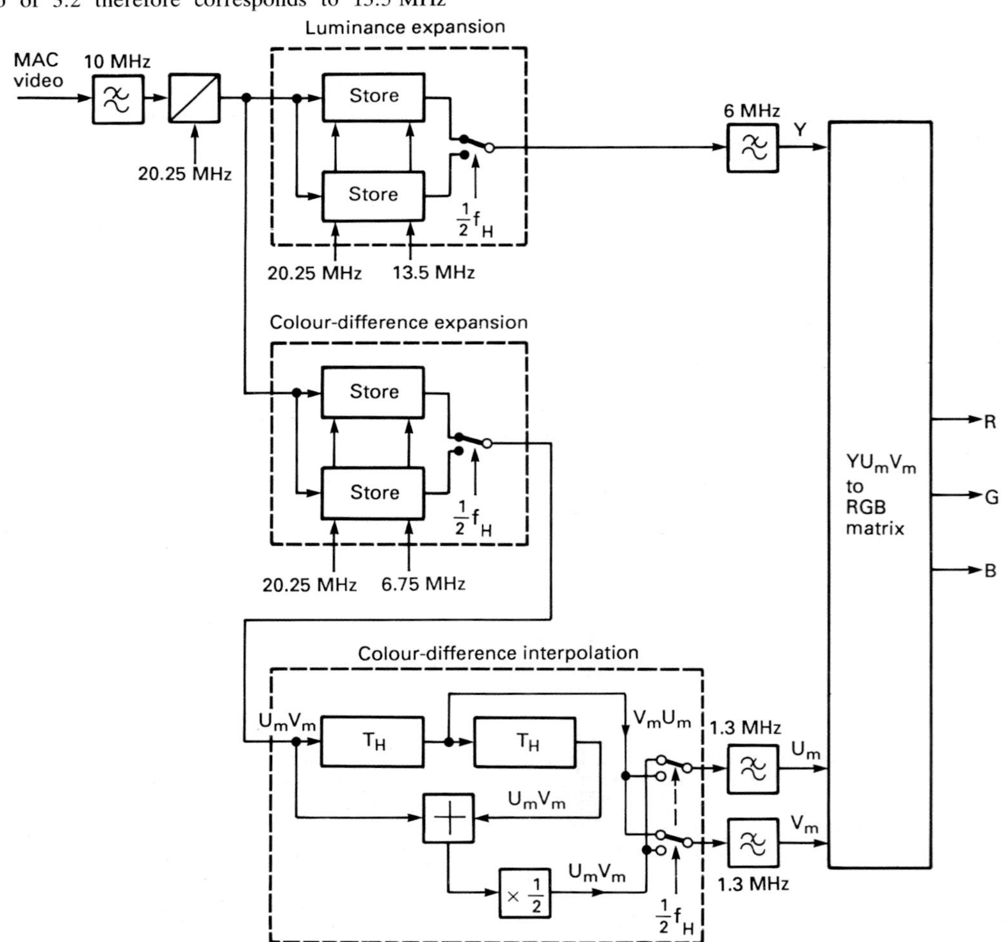

Figure 11.28 The main processes of a MAC vision signal decoder

containing spatially coincident luminance on the assumption that a ½:1:½ vertical filter will be used in the decoder to interpolate the missing chrominance information. Lines 23 and 335 contain colour-difference signals for interpolation.

The transmitted colour-difference signals are weighted according to the relationships:

$$U'_m = 0.733(B'-Y') \quad (11.40)$$

$$V'_m = 0.927(R'-Y') \quad (11.41)$$

With these weighting factors, the amplitude range of the colour-difference signals would extend to 1.3 V_{p-p}. While the transmission of 100 per cent saturated colours is not ruled out, these would exceed the normal 1 V_{p-p} amplitude limit of the multiplexed MAC signal. The inverse relationships, used in a decoder, are:

$$(B'-Y') = 1.364\,U'_m \quad (11.42)$$

$$(R'-Y') = 1.079\,V'_m \quad (11.43)$$

11.4.2.5 MAC encoding

The parameters of a MAC signal have been chosen to simplify encoding from a 4:2:2 digital components signal. The main features of the conversion process are shown in *Figure 11.27*(a). The chrominance components of the signal are first filtered vertically to reduce the amount of aliasing caused by the subsequent alternate line coding. *Figure 11.27*(b) shows an example of such a filter; more complicated filters can be used to optimize the retention of wanted frequencies and to suppress those that would cause aliasing.

The luminance signal is delayed by two line periods to take account of the vertical filtering delays of the coder and the decoder. The chrominance components are also scaled to occupy the appropriate coding range relative to the luminance signal. The factor 1.27 takes account of the 1.3 V_{p-p} range for 100 per cent saturation signals and the 219/224 coding range ratio of the 4:2:2 luminance and colour-difference signals.

The samples of the multiplexed YC_BC_R signal are then written into different parts of two storage registers, which have separate regions for the Y, C_B and C_R components. While the samples are being written into one register, the assembled samples of the previous line are read from the other register using a 20.25 MHz clock. A further region of the store can be used to multiplex digital data and sound signals for D and D2-MAC coding or to leave a time period for subsequent multiplexing of 2–4 PSK modulated data and sound signals in C-MAC coding.

When derived from composite signals, the lines used to form the MAC chrominance signal are chosen according to the following rule. The odd frames of the MAC signal are derived from the first and second fields of the composite signal, that is, for PAL, those in which the odd-numbered lines have the colour burst at 135°, or for SECAM, those in which the odd-numbered lines contain the D_R signal.

11.4.2.6 MAC decoding

The MAC decoder makes use of the conversion from 20.25 MHz to 13.5 and 6.75 MHz sampling rates to perform the 2:3 and 1:3 signal expansions for the luminance and colour-difference signals as shown in *Figure 11.28*. As in the coder, pairs of storage registers are used, one accepting the incoming samples at 20.25 MHz while samples are being read out from the other, either at 13.5 MHz for luminance, or at 6.75 MHz for the colour-difference signals. The writing and reading processes are interchanged on alternate lines. Vertical interpolation for the colour-difference signals substitutes the average of the adjacent line signals using the contributions of ½, 1, ½ from successive lines. The one-line period delay which this introduces in the colour signals is compensated by the additional one-line period delay included in the luminance signal during the encoding process.

Before matrixing, the sampled signals are low-pass filtered to remove the sampling frequency and provide sin x/x correction appropriate to the 13.5 and 6.75 MHz sampling rates. The overall effect of the colour-difference low-pass filters is to provide a slow roll-off characteristic, similar to the Gaussian filters shown in *Figure 11.3*. The Y, U_m and V_m signals are matrixed to produce RGB according to the relationships of equations (11.42) and (11.43), and (11.4)–(11.6).

Colour-difference cut point

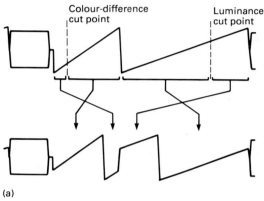

Luminance cut point

(a)

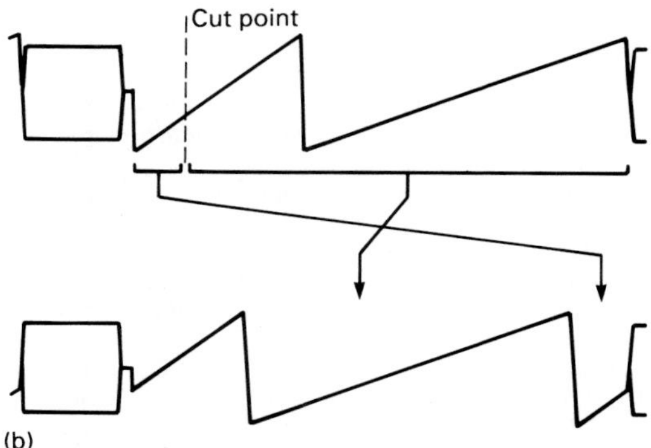

Figure 11.29 MAC picture signal scrambling using: (a) double-cut component rotation, (b) single-cut line rotation

11.4.2.7 Picture signal scrambling

The MAC/packet vision signal specification includes special provision for conditional access transmission[19]. Two methods can be used, known as double-cut component rotation and single-cut line rotation[11]. While the double-cut method is more secure, the single-cut method is more robust for vestigial sideband amplitude modulation applications.

In the *double-cut* method, 256 possible uniformly spaced cut-point positions are defined in both the luminance and colour-difference signal portions of the line. On each line, independent cut-point positions are chosen for the luminance and colour-difference components, and the signals are rotated within each component as indicated in *Figure 11.29*(a). The portions of the signal adjacent to the cut are repeated and the transitions to blanking level shaped to avoid distortion. Cross-fades are used between the original ends of the components to avoid revealing the cut-point location. The cut-point positions are chosen from a pseudo-random binary sequence, with a synchronized generator in the decoder used to identify the cut positions for descrambling.

In the *single-cut* method, only one cut-point is chosen on each line, from the 256 possible positions in the colour-difference signal portion. The colour-difference and luminance portions of the line are then both rotated about the single cut position as shown in *Figure 11.29*(b). Similar techniques to those of the double-cut method are used for avoiding distortion at the cut point and disguising the original blanking positions.

References

1 BINGLEY, F J, 'Colorimetry in color television - parts 1, 2 and 3', *Proc IRE* Part 1, **41**, pp 838–851 (1953), Parts 2 and 3, **42**, pp 48–58 (1954)

2 'Colorimetric standards in colour television', CCIR Report 476-1, XVIth Plenary assembly, Dubrovnik, XI-1, pp 42–43 (1986)

3 BRUCH, W, 'The PAL colour TV system — basic principles of modulation and demodulation', *NTZ Communications Journal* **3**, pp 255–268(1964)

4 'Timing relationship between the subcarrier reference and the line synchronizing pulses for PAL recordings', EBU Technical Statement No D 23-1984 (E), Technical Centre, Brussels (1984)

5 'Compatible color television', Federal Communications Commission Document No 53-1663 (1953)

6 KAISER, A 'Comb filter improvement with spurious chroma deletion', *SMPTE Journal* **86**, pp 1–5 (1977)

7 DE FRANCE, H, 'Le systéme de télévision en couleurs sequentiel simultané', *L'Onde Electr.* **38**, pp 479–483 (1958)

8 SABATIER, J and CHATEL, J, 'Qualité des signaux de télévision en bande de base', *Revue Radiodiffusion-télévision* **70**, pp 12--21 (1981)

9 'Nomenclature and description of colour bar signals', CCIR Recommendation 471-1, XVIth Plenary assembly, Dubrovnik, XI-1, pp 39–41 (1986)

10 'Encoding parameters of digital television for studios', CCIR Recommendation 601-1, XVIth Plenary assembly, Dubrovnik, XI-1, pp 319–328 (1986)

11 'Specification of the systems of the MAC/packet family', EBU Tech 3258-E, Technical Centre, Brussels (1986)

12 GUINET, Y, 'Evolution of the EBU's position in respect of the digital coding of television', *EBU Rev Tech* **187**, pp 111–117 (1981)

13 JONES, A H, 'Digital television standards' IBC80, *IEE Conf Publ* No 191, pp 79–82 (1980)

14 'A report of digital video demonstrations using component coding'. Special issue, *SMPTE Journal* **90**, pp 922–971 (1981)

15 'Interfaces for digital video signals in 525-line and 625-line television systems', CCIR Report 1088, XVIth Plenary assembly, Dubrovnik, XI-1, p 361 (1986)

16 'Interfaces for digital component video signals in 525-line and 625-line television systems', CCIR Recommendation 656, XVIth Plenary assembly, Dubrovnik, XI-1, p 356 (1986)

17 RICKARDS, A J and BROWN, I G, 'An intermediate coding system' IBC84, *IEE Conf Publ* No 240, pp 42–47 (1984)

18 LUCAS, K and WINDRAM, M D, 'Direct television broadcasts by satellite, the desirability of a new standard' IBA E&D Report No 116/81 (1981)

19 'General characteristics of a conditional-access broadcasting system' CCIR Report 1079, XVIth Plenary assembly, Dubrovnik, XI-1, pp 142–150 (1986)

Bibliography

Composite systems — general

'Characteristics of television systems', CCIR Report 624-3, XVIth Plenary assembly, Dubrovnik, XI-1, pp 1–33 (1986)

SIMS, H V, *Principles of PAL colour television and related systems*, Iliffe, London (1969)

CLARKE, C K P, 'Colour encoding and decoding techniques for line-locked sampled PAL and NTSC television signals' BBC Research Dept Report RD 1986/2

Colorimetry

HUNT, R W G, *The reproduction of colour in photography, printing and television*, 4ed, Fountain Press, Tolworth (1987)

PAL system

BRUCH, W, 'PAL a variant of the NTSC colour TV system — selected papers I and II', *Telefunken-Zeitung* (1965 and 1966)

'Specification of television standards for 625-line System I transmissions in the United Kingdom', (second impression). Radio Regulatory Division, Dept of Trade and Industry, London (1985)

NTSC System

FINK, D G (Editor), *Color television standards — Selected papers and records of the NTSC*, McGraw-Hill, New York (1955)

PRITCHARD, D H, 'U.S. color television fundamentals: a review', *SMPTE Journal* **86**, pp 819–828 (1977)

SECAM system

WEAVER, L E, 'The SECAM color television system', Tektronix, USA (1982)

4:2:2 digital components

'Encoding parameters of digital television for studios', CCIR Recommendation 601-1, XVIth Plenary assembly, Dubrovnik, XI-1, pp 319–328 (1986)

'Digital coding of colour television signals', CCIR Report 629-3, XVIth Plenary assembly, Dubrovnik, XI-1, pp 329–337 (1986)

'Interfaces for digital component video signals in 525-line and 625-line television systems', CCIR Recommendation 656, XVIth Plenary assembly, Dubrovnik, XI-1, pp 346–358 (1986)

DEVEREUX, V G , 'Performance of cascaded video PCM codecs', *EBU Rev Tech* **199**, pp 114–130 (1983)

STICKLER, M J, NASSE D and BRADSHAW, D J, 'The EBU bit-serial interface for 625-line digital video signals' *EBU Rev Tech* **212** pp 181–187 (1984)

MAC

'Specification of the systems of the MAC/packet family', EBU Tech 3258-E, Technical Centre, Brussels (1986)

'Specification du système D2-MAC/paquet' (The Blue Book) September 1985

SEWTER, J B and WOOD, D, 'The evolution of the vision system for the EBU DBS standard' IBC84, *IEE Conf Publ* No 240, pp 175–179 (1984)

Acknowledgment

The author wishes to thank his colleague Andrew Oliphant for reading the manuscript and the Director of Engineering of the British Broadcasting Corporation for permission to contribute this section.

Part 4 Broadcast Transmission

R S Roberts C Eng, FIEE, Sen MIEEE
Consultant Electronics Engineer

12 Radio Frequency Propagation

Radio frequencies form a small part of a wide range of types of energy transmission by means of electromagnetic fields.

All forms of electromagnetic radiation into space are characterized by four common factors: their speed of propagation in free space is the same (300×10^6 m/s), and they are subject, like light, to the laws of reflection, refraction and attenuation. The primary source of electromagnetic radiant energy is the sun. In this section we consider the small range of frequencies associated with broadcasting of television. The reception of those frequencies is dealt with in section *54*.

Clarke Maxwell indicated in 1864 that electromagnetic radiation could be established in space by electrical means. The physicist Hertz, in 1887, proved experimentally that electromagnetic fields could be set up in space and detected, as Maxwell had predicted. He also showed that such radiation obeyed the same laws as determine the behaviour of light. At a time when intercity and intercontinental communications were carried out by means of wire links, Marconi saw that this radiation provided the possibility for a 'wireless' means of communication. He put together a communication system consisting of a generator of radio frequency power, a radiation system, a receiving system and a detection system. His work culminated with his famous transatlantic digital signals in 1901 that launched radio communications.

12.1 Theoretical principles

12.1.1 Magnetic and electric fields

A direct current passing through a conductor establishes a

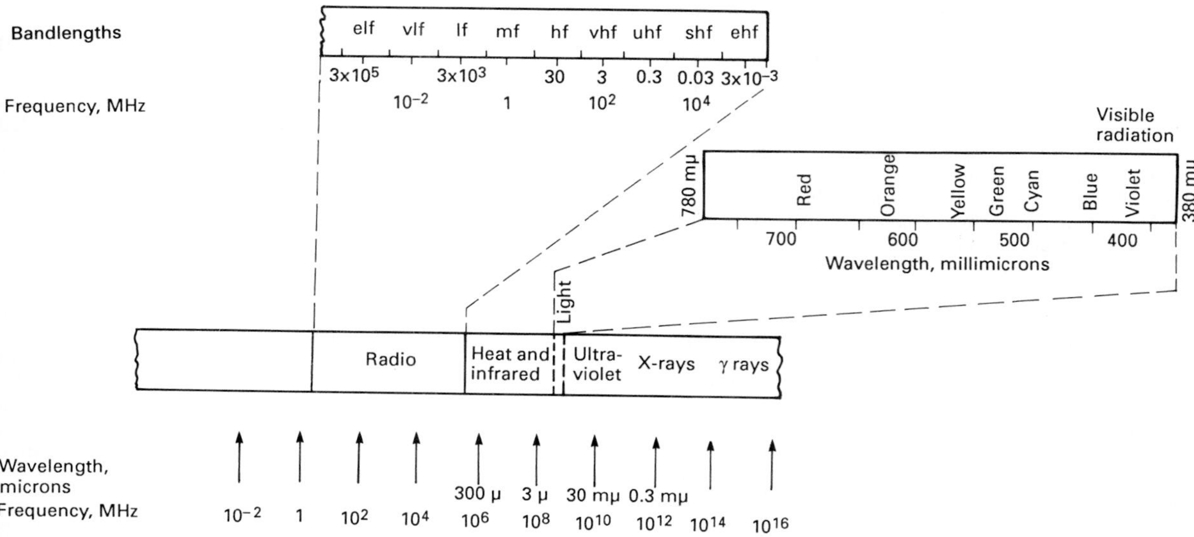

Figure 12.1 Television broadcasting takes place in the vhf and uhf bands. The European vhf bands are Band I (41–68 MHz) and Band III (174–216 MHz). The uhf bands are Band IV (470–582 MHz, UK channels 21–24) and Band V (614–854 MHz, UK channels 39–68). 11.7–12.5 GHz is allocated for television broadcasting via satellites

magnetic field around the conductor that extends into space to an infinite distance. However, the magnetic field strength falls at a rate inversely proportional to d^2, where d is the distance from the conductor. Thus, the static field strength falls rapidly to very low values over very short distances.

Energy has been taken from the supply to establish the field. The conductor has the property of inductance, and if the current flow is stopped, the magnetic field collapses with resulting self-induced voltage, the whole process returning energy to the supply.

Let us now consider the electric field. Hertz attached a pair of plates to a current-carrying conductor, identifying electric *poles* with a difference of potential between them, and an electric field in the space between them. The electric field has a potential gradient which is measured in volts per metre. A pair of plates 3 m apart with a potential difference of 60 V between them will produce a gradient of 20 V/m. A conductor lying in this field parallel to the flux lines will have a potential developed between its ends of a value proportional to its length, e.g. 0.5 m in the field will develop 10 V.

The static fields thus consist of rings of magnetic flux concentric with the conductor, and lines of electric flux near-parallel to the conductor, as shown in *Figure 12.2*.

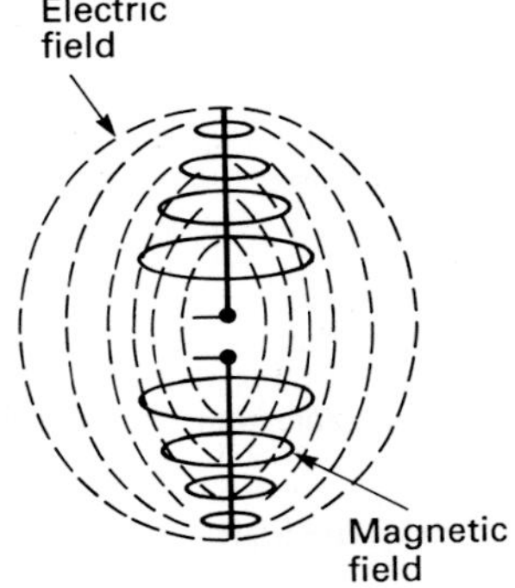

Figure 12.2 The magnetic and electric fields associated with an energized conductor

These local fields in close proximity to the conductor are termed *induction fields*. As indicated, the electric field can *induce* a voltage between the ends of a conductor and, as Faraday discovered in 1831, a relative movement between a conductor and the magnetic field will develop a voltage in the conductor.

Another conductor, positioned at a distance remote from that which is generating the fields, can be immersed in the fields, but will experience no effects if neither the conductor position nor the strength of the fields is changing. If the current is changing, a very different situation exists. The field intensities will be changing, and a changing voltage will be developed in a remote conductor.

12.1.2 Alternating fields

If the current is alternating, the induction field effects near the conductor will exist as with steady current, and their strength will vary as $1/d^2$. However, there will be an additional effect. The alternating fields will be sweeping away from the conductor with the speed of light, in the form of energy-carrying electromagnetic fields. At a point remote from the radiating conductor, another conductor can be erected parallel to the electric flux lines, and it will have an alternating current developed in it at the frequency of the field alternations. Power can be abstracted from the passing 'wave', because the current developed in the second conductor can be passed through a load. This is the principle established by Maxwell in 1864.

The radiated fields are thus a load on the energy source that is used to develop the fields. This load can be represented by an imaginary resistor having *radiation resistance* or *antenna impedance*. *Figure 12.3* shows, at a point remote from a radiating system, the field flux lines. They will be varying in phase, and are related to each other in quadrature in a plane at right-angles to the direction of propagation. The whole field system, advancing along the direction of propagation, is termed a *wavefront*.

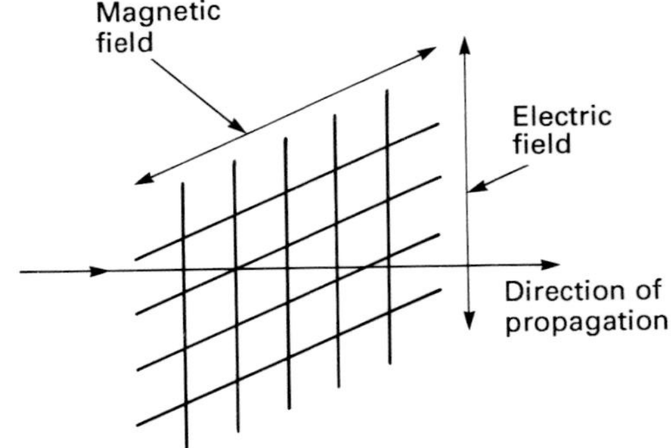

Figure 12.3 At a distance remote from the energized conductor, the radiated fields and direction of propagation are mutually orthogonal

12.1.3 Velocity and frequency

To sustain the concept of radiation of energy in the form of an electromagnetic wave in free space, the permittivity, e_o, of the medium through which propagation is taking place can be expressed in farads per metre as:

$$e_o = 4\pi \times 10^{-7} \text{ F/m} \tag{12.1}$$

The permeability, μ_o, in henries per metre is:

$$\mu_o = 10^{-9}/36\pi \text{ H/m} \tag{12.2}$$

The velocity of propagation through the medium is given by:

$$v = \frac{1}{\sqrt{e_o \mu_o}} = 3 \times 10^8 \text{ m/s} \tag{12.3}$$

The velocity given by equation (12.3) is often expressed as a constant c. It relates to all forms of electromagnetic radiated fields, i.e. radio, light, infrared, ultraviolet, etc. The constant c is not exactly 3.0×10^8. Experiments conducted over many years have yielded 2.997 925 as a more accurate value.

The travelling alternating fields will establish a *wavelength* in space. This is the distance between two corresponding points on consecutive repetitive cycles in the direction of propagation. The frequency and length of the wave for any electromagnetic radiation is related to c by $\lambda = c/f$, where λ is the wavelength in metres and f is the frequency in hertz.

The atmosphere surrounding the earth is not the vacuum of free space, and propagation through the atmosphere at altitudes up to 1000 km or so requires some modification to equation (12.3). If the atmosphere has relative values of permittivity e_r and permeability μ_r, the velocity of propagation becomes:

$$v = \frac{1}{\sqrt{\mu_r e_r}} \text{ m/s} \tag{12.4}$$

Fortunately, e_r and μ_r are generally near unity for most terrestrial communications and v approximately equals c, but for propagation along transmission lines (for example) the value for e_r will be very different from unity.

12.1.4 Impedance of space

The radiated energy is contained in the electrical and magnetic fields, and is given by:

$$\frac{\frac{1}{2}E^2 \times 10^{-9}}{36\pi} = \frac{\frac{1}{2}H^2 \times 4\pi}{10^{-7}} \qquad (12.5)$$

where E is the electrical field strength and H is the magnetic field strength.

Equation (12.5) provides the concept of impedance, given by:

$$E/H = 120\pi, \text{ or } 377 \text{ ohms} \qquad (12.6)$$

12.1.5 Radiated energy

From equations (12.5) and (12.6), the energy in each square metre of wavefront is given by $E^2/120\pi$ or $120\pi H^2$ W/m^2.

Consider an isotropic radiator at the centre of a sphere, radiating P watts uniformly in all directions. If the sphere has radius d its surface area is $4\pi d^2$, and the power flux per unit area is:

$$P_a = P/4\pi d^2 \text{ W/m}^2 \qquad (12.7)$$

From this and the energy formulae given above,

$$\frac{E^2}{120\pi} = \frac{P_a}{4\pi d^2}$$

$$E^2 = \frac{120\pi P_a}{4\pi d^2}$$

$$E = \frac{\sqrt{30p_a}}{d}$$

$$= \frac{5.5\sqrt{P_a}}{d} \text{ V/m} \qquad (12.8)$$

The most important conclusion from equation (12.8) is that the field strength of the radiated fields varies inversely with distance d from the radiator, whereas the local induction fields vary as $1/d^2$. The induction fields play no part in the radiation field but, as will be seen, they are very useful in antenna design.

12.2 Practical considerations

The outline of theoretical principles in section *12.1* concerns radiation in free space, and introduces the concept of an isotropic radiator. These principles require considerable modifications for practical implementation, because (a) no practical isotropic radiator can exist, and (b) simple terrestrial systems are not in free space. The radiation process cannot ignore the presence of the earth itself, and the gaseous atmosphere that surrounds it.

A step towards realism is the *Hertzian dipole*. This theoretical concept, based on the radiator used by Hertz, consists of a very short conductor, terminated with electrical charges at each end, along which flows a current of uniform density. The more practical antenna is the *half-wave dipole* shown in *Figure 12.4*.

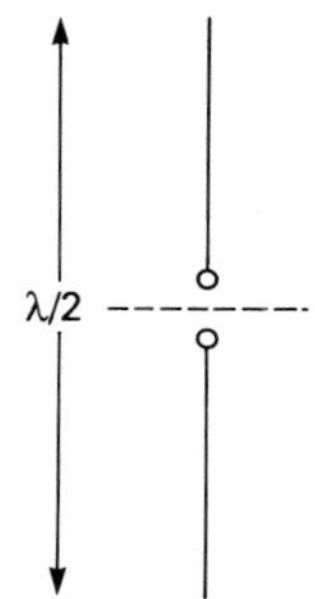

Figure 12.4 The λ/2 dipole. The antenna terminals are balanced with respect to ground

12.2.1 The rf spectrum

A wide range of frequencies, shown in *Figure 12.1*, is used for radio frequency communication. The boundaries at the extremes are ill-defined. Very low frequencies of a few hertz are used, and the upper limit extends to about 300 GHz, encroaching on the infrared end of the light spectrum. Today, the boundaries between radio and light are becoming difficult to separate.

For convenience, the rf spectrum is divided into a number of bands with international designations, as shown in *Table 12.1*.

Table 12.1 International frequency bands

Band	Frequency	Wavelength
vlf	< 30 kHz	> 10 000 m
lf	30-300 kHz	10 000 – 1000 m
mf	300-300 kHz	1000–100 m
hf	3-30 MHZ	100–10 m
vhf	30-300 MHz	10–1 m
uhf	300-3000 MHz	1–0.1 m
shf	3-30 GHz	10–1 cm
chf	30-300 GHz	10–1 mm

Frequencies above 1 GHz are often termed *microwave*

The '3–30' frequency classification is not arbitrary, but is derived from the differing modes of propagation. For instance, lf and vlf waves are propagated over the surface of the earth. The mf waves are 'ground waves' but, at times, use the ionized upper atmosphere. Propagation at hf is only possible using the ionosphere, and, at about 30 MHz, waves will be lost into space. Waves at vhf and uhf use the oldest method of electromagnetic communication; the principle is the same as the lighthouse. The transmitter antenna is situated at a high point, and the radiation is delivered to the terrain within visible range. The shf and ehf bands have wavelengths so short that another method of propagation can be used. The radiator can be situated at the focus of a parabolic reflector, and a narrow beam can be directed between the transmitter and receiving systems.

The television broadcast service commenced operations in the vhf band and later established a service in the uhf band. *Figure 12.5* shows a radiating system situated at the top of a high mast which, in turn, is sited on high ground. The radiator can direct the radiated fields towards the ground and thus illuminate a service area with a radius R to the distant horizon where the earth casts its massive shadow.

This method of broadcasting has many advantages. No fading is experienced (as may happen with lower frequencies), and field strengths are steady and unvarying. The useful range of the transmitter is determined by geography, and not by transmitter power. The only effect of power at vhf or uhf is to determine the signal/noise ratio at the receiver.

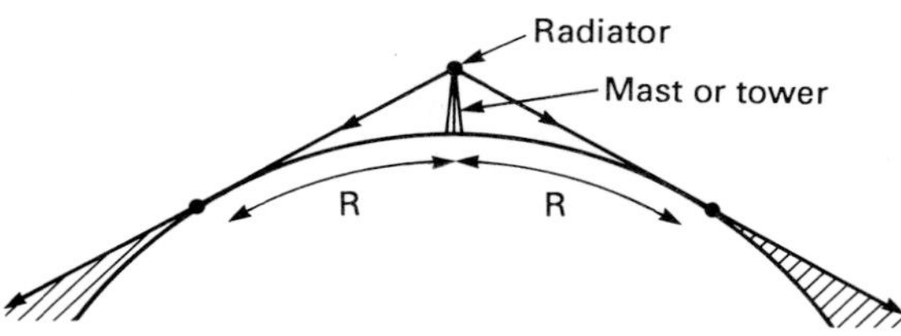

Figure 12.5 Range of a vhf transmitter

In addition to the band classifications shown in *Table 12.1*, there is another system of band classification for frequencies higher than 1 GHz, in which bands are designated by letters. These band classifications are of considerable interest now that satellites are being used for broadcast purposes.

Letter designations originated during World War 2, when early UK radar systems used wavelengths of about 10 cm and 3 cm, and it was useful to 'hide' their frequencies by the designations 'S' and 'X' respectively. Since the war, many radar and other applications of frequencies above 1 GHz have used a range of other letters for specific bands, and *Table 12.2* shows the IEEE Standard 521-1976, which was revised in 1984.

Table 12.2

Band	*Frequency Range (GHz)*
L	1 – 2
S	2 – 4
C	4 – 8
X	8 – 12
Ku	12 – 18
K	18 – 27
Ka	27 – 40
V	40 – 75
W	75 – 110
mm	110 – 300

Satellite broadcast operations differ from radar in that two frequencies are required for each channel, one for the 'up' path and another for the 'down' path. For example: a satellite link might be quoted as 'C 4/6', thus designating the two frequencies involved in Band C for up and down paths.

It is of note that the international allocations for DBS (Direct Broadcast from Satellites) span the X and Ku Bands.

12.2.2 The practical radiating element

Let us go back to Hertz and Marconi. Hertz used a conductor divided at the centre to allow energy to be fed to the system, as shown in *Figure 12.6*(a). Such a system is termed a *dipole*. Marconi, seeking a communication system with commercial possibilities, used a vertical conductor. *Figure 12.6*(b) shows an important feature of this form of radiation. The earth behaves as a reflecting surface so far as the radiating element is concerned, in the same manner as a mirror will show a second source of a radiator of light to an illuminant above the mirror. The Marconi modification of the Hertz conductor is used extensively, and all antenna elements are of either Hertz or Marconi form.

To determine the length of the radiating element, consider *Figure 12.7*. The dipole has a capacitance between its two halves, and the current flow in the conductor establishes inductance. To energize the dipole from a source at the centre, the capacitance must be fully charged, and it will then discharge through the inductance of the system in the manner of a tuned circuit. Clearly, a finite time is required to charge and discharge C, and the system needs to be tuned to the energizing source.

It so happens that a very thin conductor requires a quarter cycle of the resonant period for energy from the source to fully charge the capacitance in each half of the dipole. The discharge of the capacitances will require another quarter cycle to become complete, and the energy in the system will be in phase with the source at the input terminals. Each half of the system requires to be a quarter-wavelength long for this optimized performance.

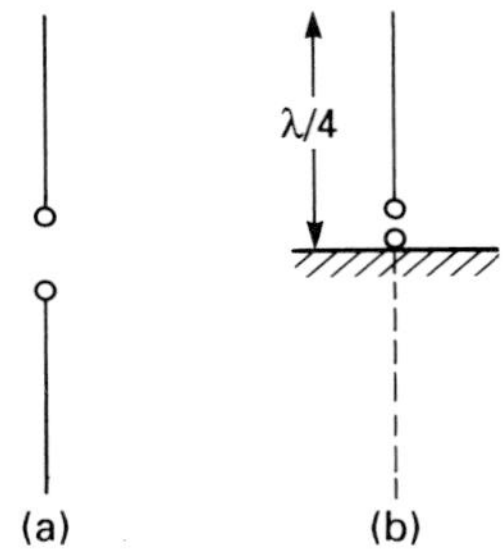

Figure 12.6 A λ/4 antenna, in which the ground plane and the image play a vital part

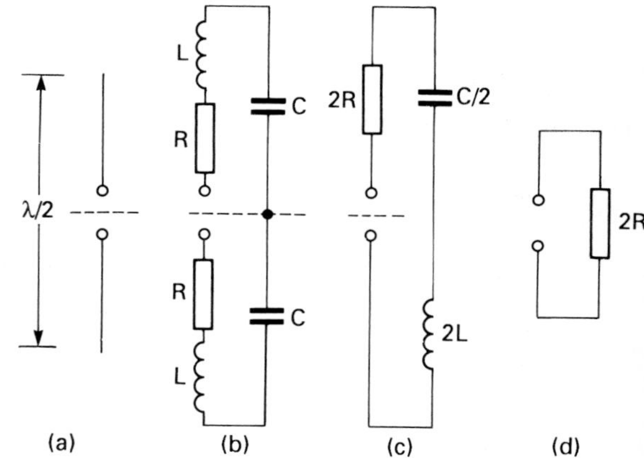

Figure 12.7 A half-wave dipole (a), and three equivalent circuits

In practice, the practical radiating system cannot consist of a very thin conductor; a practical dipole must be self-supporting. The dipole in practical form will have too great a capacitance, but this can be compensated for by reducing the inductance (i.e. the overall length). The practical dipole thus has an overall length less than a half-wavelength by a factor determined by the ratio of operating wavelength to the conductor diameter. The Marconi *monopole*, similarly, requires to have a height less than a quarter-wave. *Figure 12.7* shows the balanced dipole at (a), and at (b) the capacitances and inductances are shown as discrete component values; (b) also shows the addition of R, representing the radiation resistance. The equivalent circuit of the resonant radiator is in (c) and (d) and, considering the system as a tuned system, it is seen that the resonant frequency for the dipole is given by:

$$\omega = \frac{1}{\sqrt{[(C/2) \times 2L]}}$$

and, for the Marconi monopole, the frequency will have the

same value, i.e.:

$$\omega = \frac{1}{\sqrt{(LC)}}$$

12.2.3 Antenna impedance

The tuned radiator has an impedance which, at resonance, is resistive and is in the region of 73 ohms. This value is derived from an integration of the radiated power over the whole inside surface of the sphere, considered earlier in section *12.1.5*. Note that the impedance of the Marconi monopole will be half of the dipole value. In practice, there are several factors that can affect the value of antenna impedance.

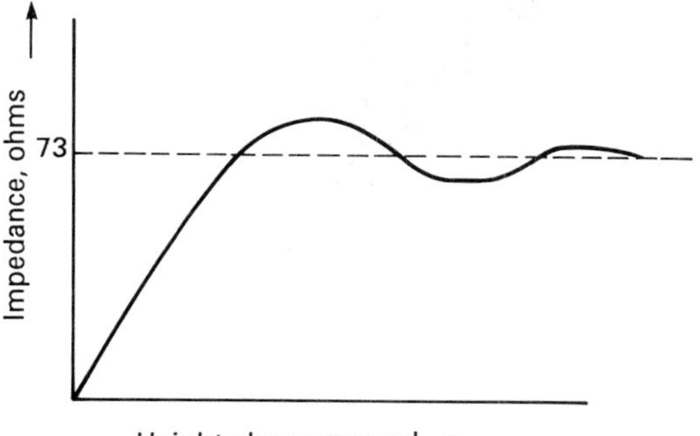

Figure 12.8 Variation of dipole impedance with its elevation

The presence of the ground cannot be ignored, and a dipole impedance will vary with its height above ground (see *Figure 12.8*). Precise values are not given because they will depend on the ratio of diameter:wavelength. When this ratio is small, the impedance may not be affected very much, but at uhf the diameter may be an appreciable fraction of a wavelength, and this could result in an impedance lower in value than that for a very thin conductor. Another factor that can affect an antenna impedance is the presence of other conductors or antenna elements in the near vicinity of the radiator. The effect is, generally, to reduce the radiator impedance.

12.2.4 The folded monopole and dipole

Folded versions of the half-wave dipole and the quarter-wave monopole antennas provide a useful means for changing the basic impedance. *Figure 12.9*(a) shows a λ/4 Marconi or monopole system; (b) shows it in more diagrammatic form with an input voltage *V*. In practice, the spacing between the two halves of the loop is a very small fraction of the operating wavelength. At (c) is shown the generator *V* replaced with three generators, each with a magnitude *V*/2, and so phased as to produce the same effect as at (b). Generators 1 and 2 are in phase and provide the voltage *V* of (b). Generators 2 and 3 are in phase round the loop, but this loop is a λ/4 section, shorted at the top, thus presenting a high impedance at the generator end (see section *24*), and the resulting current is near zero.

Generator 1 alone feeds, in parallel, the two limbs of the antenna, the current being *V*/2 divided by Z_1, where XZ_1 is the impedance of a λ/4 antenna. This current divides equally between the two elements so that the current in the left-hand branch is half of the total, i.e. *V*/4 divided by Z_1. If Z_1 is 37.5 ohms for a single monopole element, the impedance of the driven element is given by:

$$Z = V/I = 4 \times Z_1$$

This is approximately 150 ohms.

The folded antenna thus introduces a multiplying factor of four. The folded dipole has an input impedance of 300 ohms. The multiplying factor may be varied by making the diameters of each half of the fold unequal. If the diameter of the undriven half is larger than the diameter of the driven half, the factor is larger than four. If the driven half has a larger diameter than the undriven half, the multiplying factor will be less than four.

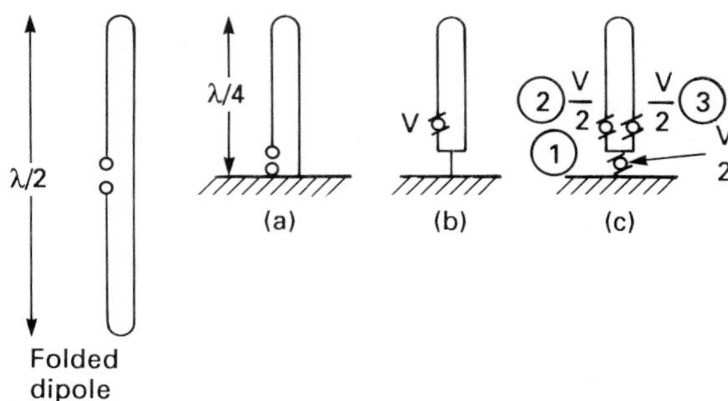

Figure 12.9 A monopole with its equivalent circuit

Another useful feature of the folded antenna is a wider bandwidth than that provided by the single resonant dipole or monopole. Any tuned antenna is essentially a single frequency device. A dipole, driven by a source lower in frequency than that for which it is designed, will have a reactive input impedance of the form $R - jX$. In folded form, another property comes into operation. Each half of a folded dipole consists of a λ/4 section, but at a lower frequency of operation the section becomes effectively less than λ/4. The section will now present an inductive reactance, and this can compensate in some degree for the capacitive reactance at the antenna input. Similar compensation is provided for operation at a frequency above the designed frequency of the antenna.

12.2.5 Directivity, the polar diagram and antenna gain

We have considered the concept of radiation from an isotropic source at the centre of a sphere, and the effects of such radiation at the surface of the sphere. The field strength would be the same at any point on the sphere surface. It can be seen from *Figure 12.2* that a practical dipole cannot radiate in this manner.

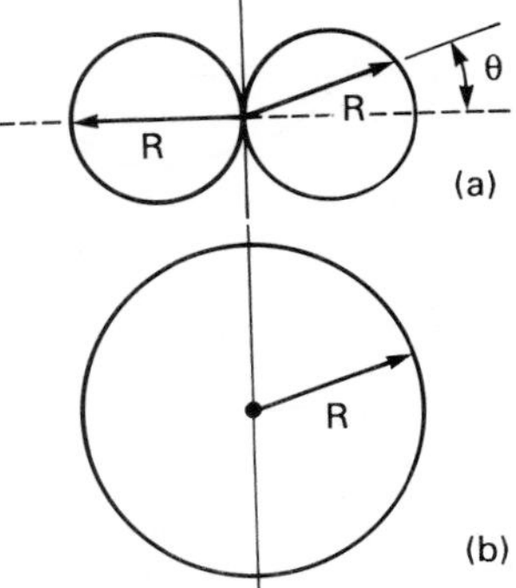

Figure 12.10 Relative values of field strength of a dipole along the direction *R*. (a) is elevation, (b) is plan view

The intensity of the magnetic field is at a maximum along a plane normal to the axis of the dipole through its centre. At the ends of the dipole, the current is zero, no magnetic field can exist, and radiation from the ends must be zero. Thus, radiation

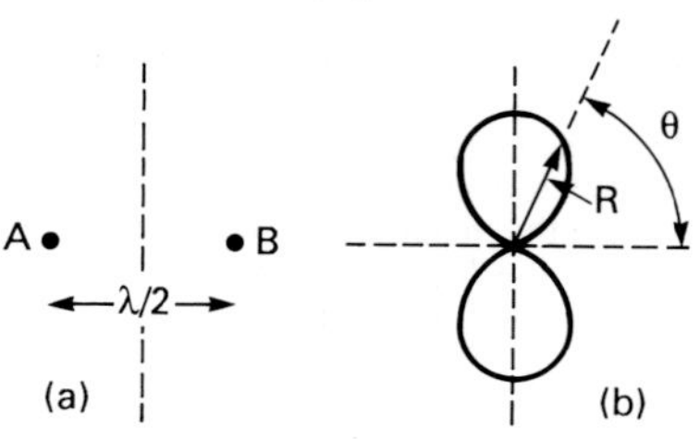

Figure 12.11 Two vertical radiators, viewed in plan (a), with their currents in phase, have the polar diagram (b)

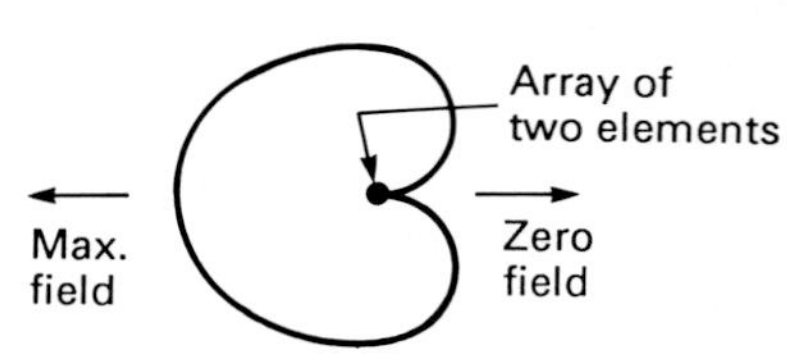

Figure 12.12 The polar diagram provided by the two-element array of *Figure 12.11* when one element is directly driven from a power source, and the other is energized from the induction field of the driven element

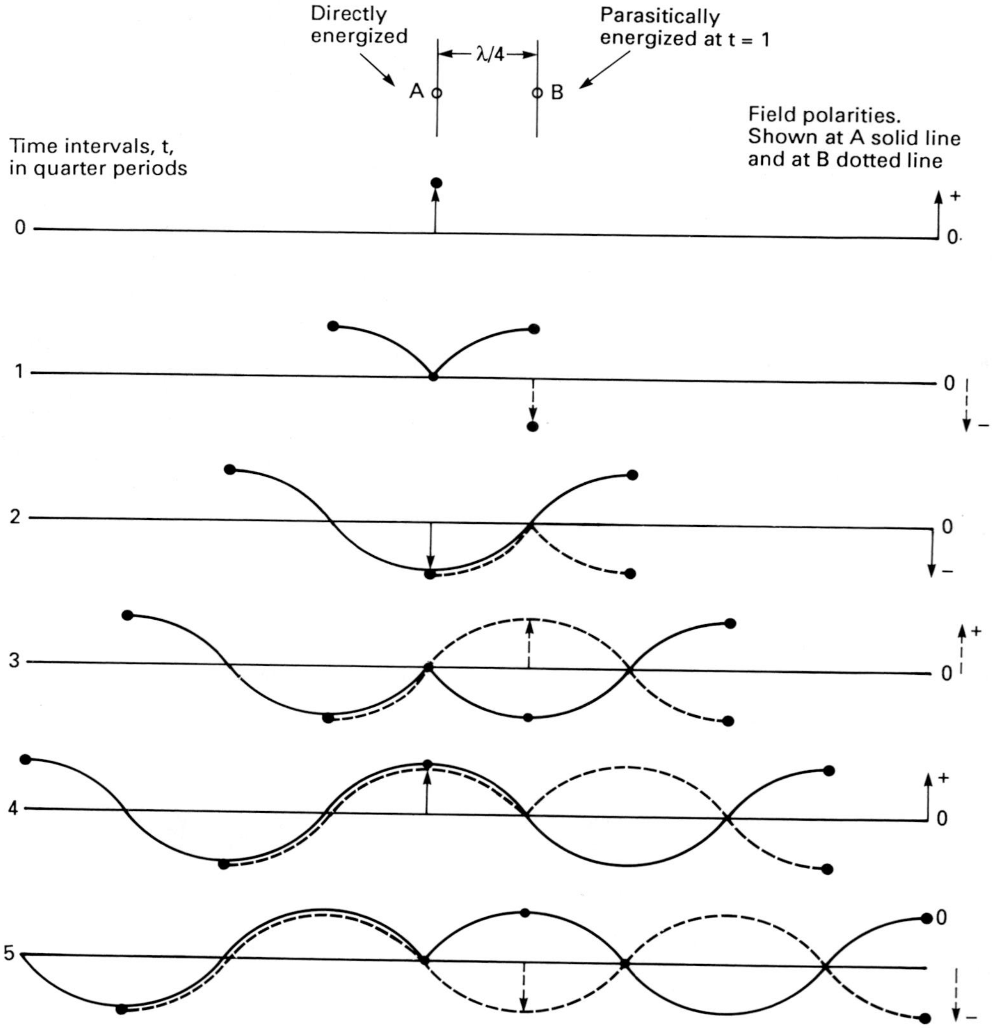

Figure 12.13 A cycle commences at $t = 0$, with the field at A having a maximum positive value. This field radiates uniformly in all directions and reaches B at $t = 1$, energizing B to a maximum negative value. Both radiators are now radiating uniformly in all directions as shown at $t = 2$. From this time onwards, the combined fields cancel in the direction A to B, but are in phase in the direction B to A

from a dipole could be depicted by *Figure 12.10*(a) which shows a *polar diagram* in which the dipole axis lies along the solid line, and the length R depicts the relative field strength at any angle ø, compared with the maximum value at 0°. The variation for a dipole is approximately as cos θ. Looking down in plan on the dipole, radiation is uniform in all directions, as shown in (b). The polar diagram for this case would be a circle.

In a practical case, using a dipole as a radiator, a surface of uniform radiation field strength would be represented three-dimensionally by a torus instead of a sphere. The dotted line in *Figure 12.10*(a) represents the ground plane that establishes the monopole.

The radiant energy from an isotropic source and from a dipole would be the same for the same power fed to each, but the three-dimensional distribution of this energy is very different. The field strength maximum for the dipole would be greater in value than for the isotropic case. The increase would be about 1.64 times or 2.16 dB over the isotropic. Equation (12.8) using a practical dipole should now read:

$$E = \frac{5.5\sqrt{1.64}\sqrt{P_a}}{d} = \frac{7\sqrt{P_a}}{d}\ \text{V/m}$$

Increases of this type are often referred to as *antenna gain*. In this case the gain is with respect to an isotropic system, but other antenna systems may refer to their gain in terms of a single dipole.

Any vertical radiator will have a circular polar diagram when viewed from above. The half-wave dipole can be used horizontally, in which case the radiation pattern, viewed from above, will be the *figure-of-eight* shown in *Figure 12.10*. It is now highly directive with two directions where maximum field strength is experienced, and two directions of zero field strength.

Most modern broadcast antennas require better directivity than can be provided by a simple dipole system. Unidirectivity is required to some extent, i.e. a single maximum (or minimum) field strength directed into specific directions, either for obtaining the best possible field strength in the maximum direction, or for reducing possible interference to other services in a zero direction. This can be achieved by the use of more than one antenna element as a radiating system or *array*.If two vertical radiators (A and B in *Figure 12.11*) are each fed with the same power with their currents in phase, and spaced $\lambda/2$ apart, they will each radiate uniformly in all directions, but their fields will cancel in the direction AB and add in the direction at right-angles. These field interactions will result in the overall radiation pattern for the array as shown in *Figure 12.11*(b). By varying the spacing, the current magnitudes and their phase relationship, it is possible to obtain an infinite variety of polar diagrams with maxima and minima in various angular positions.

One particular combination has a special significance. If the spacing is $\lambda/4$ and the currents have a phase difference of 90°, *Figure 12.12* indicates how the fields will add in one direction but will cancel in the other direction. This array is unique in that there is only one maximum and only one zero. The array is unidirectional, and a further change of 180° in either of the current phases will turn the diagram through 180°. There would be some practical difficulty in supplying the two radiators with the correct phase but, fortunately, a very simple system can be used to achieve the required result.The induction fields that exist in close proximity to a radiating antenna element normally take power from the source and return it to the source as the fields vary through each cycle. The average power per cycle is zero. If only one of the two elements shown in *Figure 12.11* is energized from a source, the second element can be energized by the induction fields. Such an element would be termed *parasitic*.*Figure 12.13* shows how the fields between the two elements, one energized and the other parasitic, will interact. The two elements are shown $\lambda/4$ (= 90°) apart, and the driven element A is assumed to start with a maximum positive current at $t = 0$. At $t = 1$ the field maximum will have travelled through space by a distance $\lambda/4$ all round the element A. The solid lines in successive quarter-cycle time intervals will show the field magnitudes in the space surrounding the element A. The broken lines show the field contribution by the parasitic element B. At $t = 1$ the radiated maximum field from A reaches B, and induces a maximum negative current in B (*Lenz' law*), which then commences to radiate in the same manner as A. The B field variations with time are seen to assist the A field radiating in the direction B to A, and cancel in the direction A to B, thus producing the polar diagram of *Figure 12.12*.

Further properties of an antenna radiating system are discussed in section *54*.

I M Waters C Eng, MIEE
Product Manager Transmitters, Varian TVT Ltd

13 Television Transmitters

13.1 Specifications

Terrestrial television transmitters are fed with baseband video and audio signals and deliver vision and sound modulated radio frequency power to an antenna. Transposers which receive off-air signals and transpose them without demodulation to another frequency for retransmission are covered in section *14*.

The characteristics of television broadcasting standards B, D, G, I, K, K′, M and N defined by the CCIR and used in different parts of the world are given in section *1.9*. The aspects of these which determine the design of a transmitter are:

- carrier frequency band
- vision bandwidth
- vision vestigial sideband characteristic
- vision modulation levels for black and peak white
- vision group delay characteristic
- number of sound carriers and modulation mono/stereo
- vision to sound carrier(s) frequency spacing
- vision to sound power ratio(s)
- sound pre-emphasis and modulation deviation.

Television transmitters operate in three frequency bands. Allowing for regional and national variations these are:

- Band I, vhf low band, 47–88 MHz
- Band III, vhf high band, 174–233 MHz
- uhf, 470–860 MHz

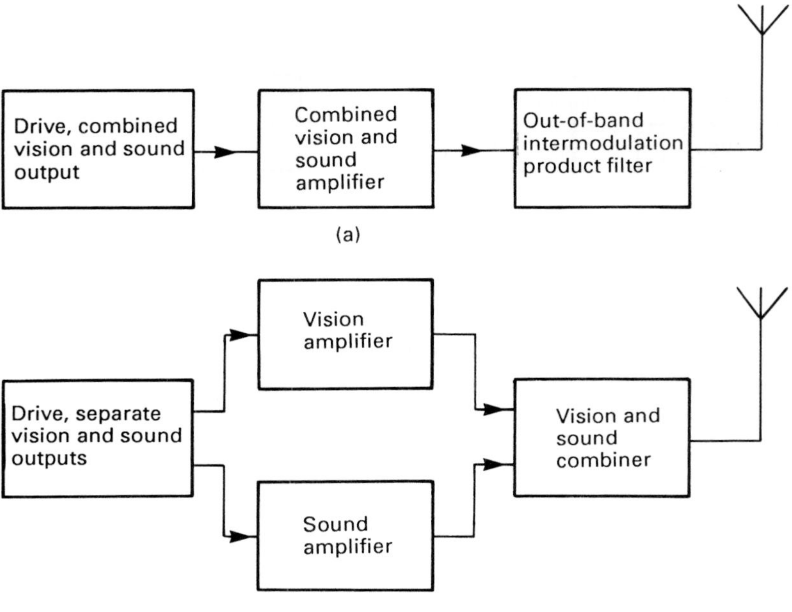

Figure 13.1 Basic types of television transmitter: (a) combined amplification, (b) separate amplification

Transmitters in most countries employ a vestigial sideband negative amplitude modulated vision carrier designated A5C, and one or two frequency modulated sound carriers designated F3E. The sound carrier(s) are higher in frequency than the vision. Provided that the vision transmitter has adequate linearity, differential gain and differential phase performance, the type of colour transmission (PAL, NTSC or SECAM) is not of prime significance.

Vision ouput power, defined as the instantaneous continuous wave (cw) power during the sync period, may be from 1 W to 60 kW for a single transmitter. Transmitters are operated in parallel for powers of up to 240 kW. When one sound carrier is used, its power is normally 10 dB below that of vision. With two carriers, the first is 10–13 dB and the second 20 dB below the peak sync vision power.

13.2 Basic transmitter types

Two basic types of television transmitter, one using combined and the other separate vision and sound amplification, are shown in *Figure 13.1*. In the first, modulated vision and sound signals are combined at low level and amplified together by a single amplifier chain. In the second, vision and sound are amplified separately and combined at high level.

Combined amplification has the advantage that, as only one amplifier is used, the transmitter is simpler, smaller and less expensive.

However, when the voltage representing 10 per cent sound power is added vectorially to the voltage representing 100 per cent peak sync vision power, the peak instantaneous power required is 173 per cent of that required for vision only. The linearity of the transfer characteristic is also important. Any non-linearity will lead to the generation of both in-band and out-of-band intermodulation products (ips) between the vision carrier, the colour subcarrier and the sound carrier(s). When two sound carriers are employed, ips will also be generated between them.

As non-linearity increases toward the extremities of the transfer characteristic it may be necessary, depending on the pre-correction available, to restrict the amount used.

In practice, it is normal to employ a final amplifier stage with a power handling capability 2.0–2.5 times that required for vision only. As this is normally operated in class A, with constant supply current regardless of modulation, efficiency is low and electricity consumption and cost of ownership high.

A compromise is necessary between ips and efficiency. In-band products become just visible when they are 40 dB below peak sync for SECAM, 48 dB below for NTSC or 52 dB below for PAL. The level that can be accepted depends on transmitter application. A PAL transmitter feeding transposers, that will further degrade performance, requires a level of -60 dB. For one at the end of the chain, this may be relaxed to about -52 dB.

It is also generally necessary to filter the transmitter output to suppress out-of-band products to a level of about -60 dB to avoid interference with services in adjacent channels.

As performance depends on critical non-linearity pre-correction, a combined amplification transmitter is more difficult to adjust and maintain.

The use of combined amplification is generally restricted to 10 or 20 kW peak sync by the availability of tubes of adequate capability and efficiency. The development of systems using two sound carriers, giving rise to more intermodulation products, has made combined amplification less attractive.

While *separate amplification* transmitters are larger and more expensive, each amplifier chain can be designed economically for the power required. Depending on the amplifier devices used, vision amplifiers can be operated in class A, AB or B and sound amplifiers handling one carrier in class AB or C so giving a significantly better efficiency.

The isolation provided by the high power combiner is such that there is virtually no interaction between the vision and sound amplifiers, and ip problems between them do not normally occur.

Adjustment is comparatively simple and long term stability good.

However, when two sound carriers are used, difficulties in combining the outputs of two amplifiers with a small frequency separation leads to a requirement to amplify the two carriers in

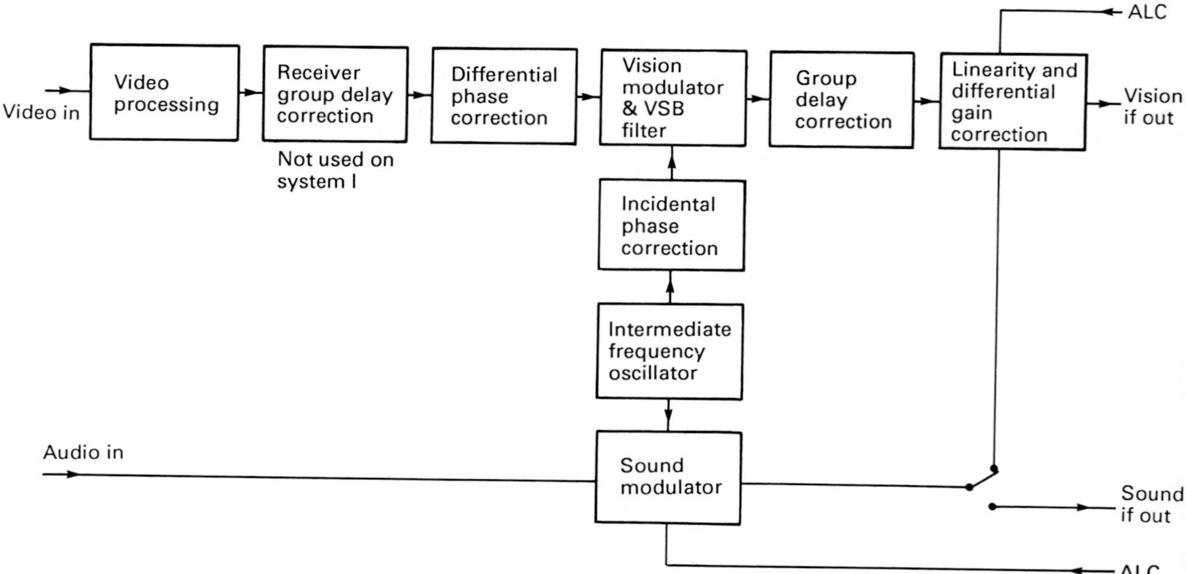

Figure 13.2 Typical television modulator

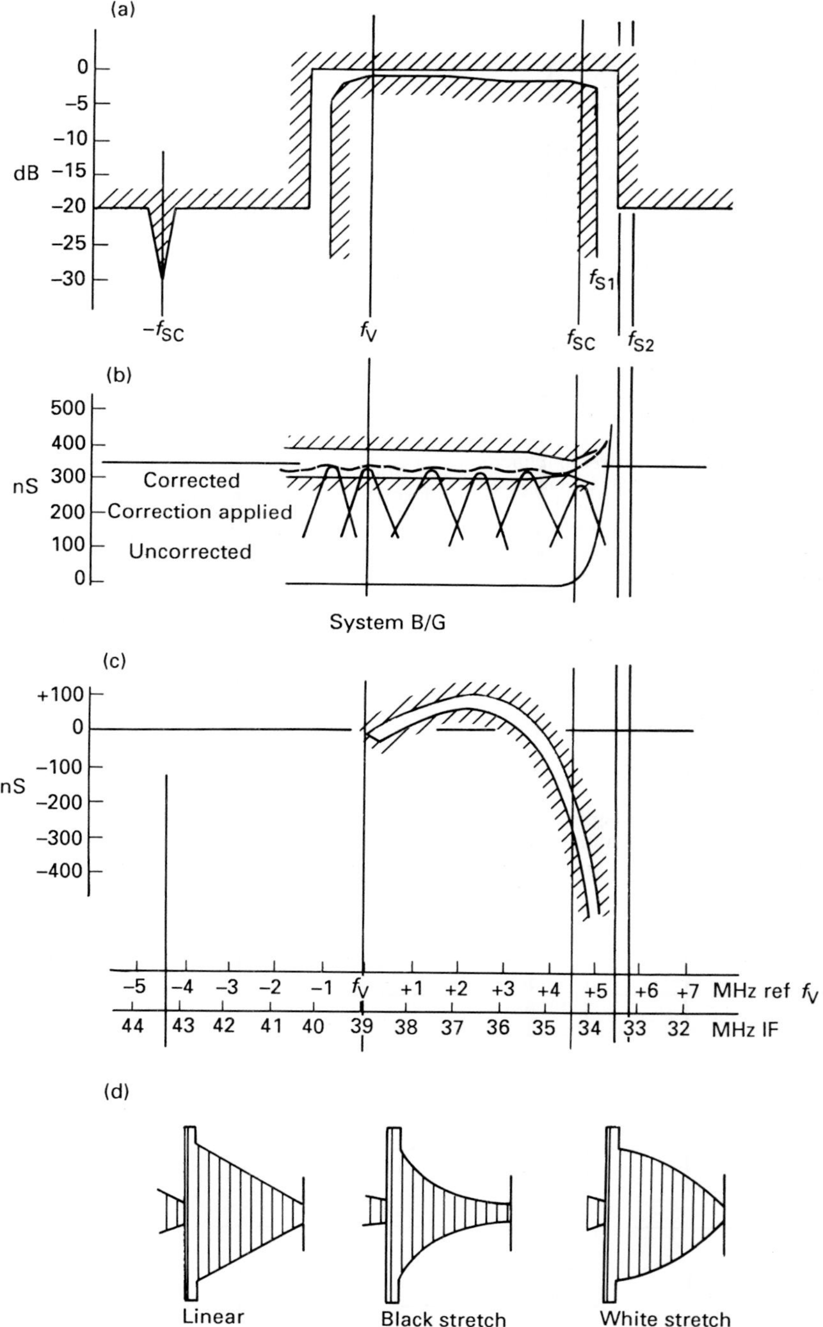

Figure 13.3 Pre-corrections: (a) amplitude/frequency response limits, (b) linear group delay response limits, (c) typical receiver group delay pre-correction limits, (d) linearity pre-correction applied to if envelope

a single sound output stage. When this is done, the stage requires to be operated in a more linear class AB mode, with sound non-linearity pre-correction provided in the drive to reduce ips between the carriers to an acceptable level.

13.3 Drive systems

All modern television transmitters employ a solid-state low level intermediate frequency (if) modulated drive or exciter system. This consists of a modulator, an up-converter and a power supply.

The modulator produces modulated vision and sound signals at intermediate frequency. Frequency response shaping and pre-correction is applied as required by the broadcasting standard used (B, G, K, etc.) and the needs of subsequent amplifier stages with mono or stereo sound. The up-converter transposes this output to the required vhf or uhf carrier frequencies.

13.3.1 Pre-correction

A modulator introduces pre-corrections adjusted to be equal and opposite to those encountered in subsequent stages to maintain the transmitter output within close specification limits.

By enabling the use of a larger part of the power amplifier's transfer characteristic, transmitter efficiency is increased significantly.

A vision signal is double sideband up to the mirrored vestige cut-off frequency and single sideband above it. Pre-corrections, such as those for receiver group delay and differential phase, relating to the single sideband region may be applied to the video signal prior to modulation. Pre-corrections such as linear group delay, which require independent correction over the whole pass-band, both below and above the carrier, are applied to the modulated if envelope.

13.3.2 Modulator

The block diagram of a typical basic modulator is shown in *Figure 13.2*.

The video signal is processed typically by clamping at black level without distorting the colour burst, regenerating the sync and peak white limiting the luminance without distorting the colour.

For standards other than I, group delay errors in receivers have been standardized so that pre-correction can be applied more economically once in the transmitter. Curves vary between countries, but are typically as shown in *Figure 13.3*(c).

Differential phase distortion (the variation of the colour subcarrier phase with modulation level) occurring in amplifier stages is pre-corrected prior to modulation.

The video signal modulates an if carrier. The double sideband modulated if is then passed through a filter, normally a surface acoustic wave (saw) type, to give the vsb amplitude/frequency characteristic required by the broadcasting standard used. The amplitude/frequency response is typically as shown in *Figure 13.3*(a).

The saw filter is phase linear and requires no group delay correction. Correction is provided for group delay distortion caused by vision pass-band filtering in the vision/sound combiner. This may be either at if in combined amplification transmitters or at high level in separate amplification types. Typically six discrete correctors, as illustrated in *Figure 13.3*(b), are adjusted to equalize group delay across the vision pass-band to compensate for phase distortion introduced by the combiner at the highest video frequencies.

The residual error, with receiver pre-correction disconnected, is typically less than 50 ns or about half a picture element.

Non-linearity of transfer characteristics which takes differing forms with tetrodes, klystrons and solid-state amplifiers, is pre-corrected so that the amplifier is used most efficiently consistent with a linear output. Several correctors operating at different modulation levels are provided to enable pre-corrections of the forms shown in *Figure 13.3*(d) to be introduced. As differential gain, the variation of gain at colour subcarrier frequency with modulation level, is not necessarily corrected optimally by the linearity correctors, additional differential gain correctors operating in the colour subcarrier band of frequencies only are often provided.

In combined amplification transmitters, the vision and sound signals are combined at if prior to linearity pre-correction. Automatic level control (alc), which maintains the transmitter peak sync output constant, is provided using a reference derived from a coupler at the transmitter output.

The if carrier of 38.9 MHz for standards B, G, I, K and K′, or 45.75 MHz for M and N, is generated by a crystal oscillator.

Incidental phase (the variation of vision carrier phase with

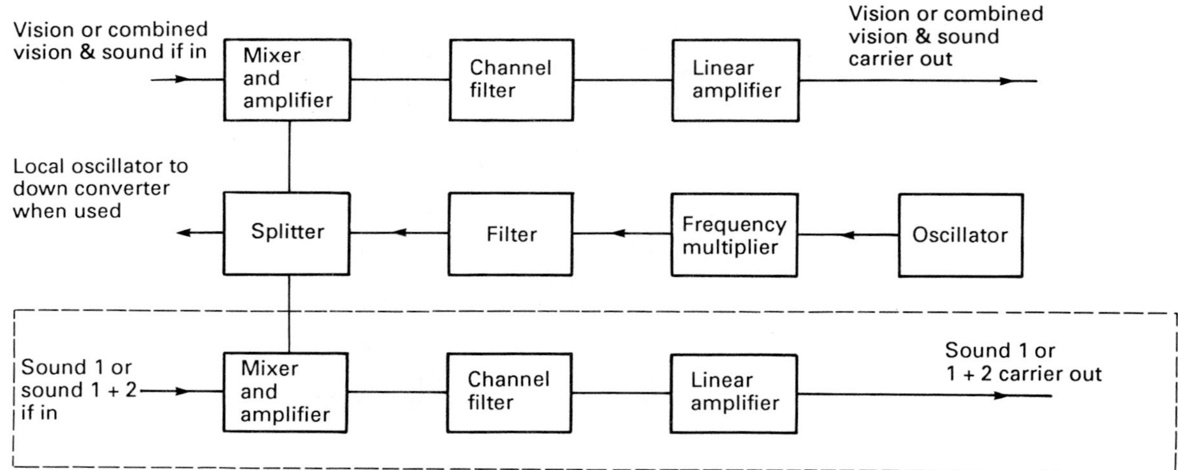

Figure 13.4 Up-converter

modulation level) which occurs in amplifier stages is corrected by phase modulating the if carrier fed to the vision modulator.

The audio signal frequency modulates an if carrier locked to the vision and separated from it by the vision/sound carrier spacing of the broadcasting standard used. The deviation and pre-emphasis employed are also in accordance with this standard. Sound alc is provided.

In combined amplification transmitters the sound is combined with the vision in the modulator; in separate amplification types it is fed separately to the up-converter.

The vision and sound outputs of the modulator are at a typical level of 1 mW.

13.3.3 Up-converter

The up-converter contains two identical chains. Each comprises a double balanced mixer, an amplifier, a channel filter and a further amplifier. These convert the if outputs of the modulator to the radiated carrier frequencies (*Figure 13.4*).

The local oscillator signal is generated by a synthesizer or a temperature stabilized crystal oscillator with multiplier and filter. It is split to feed both mixers.

To facilitate the suppression of unwanted mixing products, the local oscillator is placed higher in frequency than the carriers.

The vision sidebands produced in the modulator must therefore be inverted laterally and the sound carrier placed lower in frequency than the vision. The output of a mixer is typically at a level of 1.0 W.

In combined amplification transmitters the sound up-converter chain is omitted.

13.3.4 Precision frequency control and off-set

Service planning normally places transmitters using the same channel sufficiently far apart to avoid co-channel interference. However the need to fit in more services and/or abnormal propagation can cause interference in fringe areas.

This may be reduced if the carrier frequencies of both transmitters are precisely controlled, to be within ±2 Hz, and mutually off-set by a multiple of one-third of the horizontal scanning frequency normally ±5/3 H. When this is done the two sets of received vision sidebands interleave, and the observed interference patterning is reduced by as much as 16 dB.

When precision off-set is used, the local oscillator in the up-converter is augmented by a synthesizer which locks the carriers to a high precision external source, such as a rubidium oscillator, and provides the off-set.

13.4 Multi-channel/stereo sound

Two methods are employed to add multi-channel/stereo sound to a television transmission:

- the additional information is carried on one or more subcarriers superimposed on the sound carrier
- a second sound carrier is added

The characteristics of three systems used in various parts of the world are summarized in sections *13.4.1–13.4.3*.

13.4.1 The BTSC system

The BTSC (Broadcast TV System Committee) system was developed in the USA and has been adopted in some standard M countries. The original mono sound channel remains unchanged, with additional information carried by one or more subcarriers added to the mono baseband signal.

Subcarrier 1, with a frequency of 31.468 kHz (2 H), carries extra stereo information using the Zenith pilot tone system and dbx companding. (DBX companding is specified as part of the system to improve the s/n ratio.) The subcarrier is amplitude double sideband suppressed carrier (am-dsb-sc) modulated with a L-R signal which in the receiver is dematrixed with the mono L+R to give left and right audio. A cw pilot tone of 15.734 kHz (1 H) is also transmitted.

Subcarrier 2, with a centre frequency of 78.67 kHz (5 H), is frequency modulated to carry a dbx companded 10 kHz bandwidth second audio programme (sap).

Subcarrier 3, with a frequency of 102.271 kHz (6.5 H), is used for professional (pro) non-broadcast purposes. It is either narrow-band frequency modulated to carry a 3.4 kHz speech communication channel, or frequency shift keyed (fsk) to provide a 1.5 kHz data channel.

The total sound carrier, when fully loaded with subcarriers, has a deviation of ±73 kHz and an overall bandwidth of 400 kHz.

13.4.2 The IRT system

The IRT (Institut für Rundfunktechnik) system was developed in West Germany and adopted by some standard B/G countries. The original sound channel remains unchanged except that its power is reduced from -10 to -13 dB reference peak sync. It carries either mono or (L+R)/2 for stereo.

A second carrier, with a centre frequency 242.8175 kHz (15.5 H) higher than the original and a power -20 dB reference peak sync, is frequency modulated with a maximum deviation of ±50 kHz to carry either the stereo R signal, or another separate mono sound programme.

Identification control tones of 117.5 Hz (H/155) or 274.1 Hz (H/57) 50 per cent amplitude modulate a subcarrier with a frequency of 54.6875 kHz (3.5 H) superimposed on the second carrier and used to switch receivers for stereo or two channel reception respectively. Both carriers and the control tones are locked to the horizontal scan frequency (H).

13.4.3 The United Kingdom system (NICAM 728)

The UK system developed by the BBC is applicable to standard I, with a version adapted to standard B/G. The original sound channel remains unchanged except that the power is reduced to be -10 dB reference peak sync.

A second carrier, with a centre frequency 552 kHz above the original (nine times 728 kHz above the vision carrier) and a power -20 dB reference peak sync, is digitally modulated using differential quadrature phase shift keying (qpsk). Modulation is by a 14/10 bit near instantaneously companded 728 kbit/s data stream which carries stereo or two separate sound programmes with very high quality.

13.4.4 Implementation

In subcarrier systems, the composite baseband signal is generated by a coder which is usually part of the programme input equipment (pie). Provided that the sound channel has adequate deviation and bandwidth, multi-channel/stereo sound has little effect on the design of a transmitter.

Two carrier systems complicate the design of a transmitter considerably. It is necessary to generate and modulate the second carrier, combine it with the first, amplify both together without significant intermodulation, control and measure the power of both and combine them with the vision. For the IRT system, all three carriers and the ident tones have to be locked to the horizontal scan frequency.

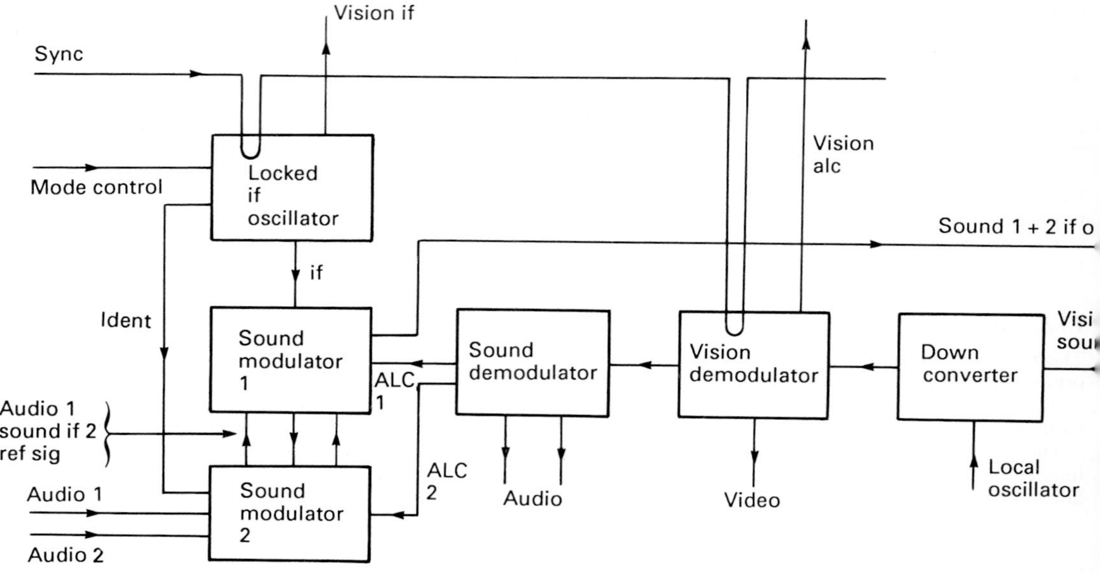

Figure 13.5 Adaptation of modulator shown in *Figure 13.2* for two carrier sound

Figure 13.5 shows the adaptation of the modulator in *Figure 13.2* for two carrier (IRT) sound. The crystal if oscillator is replaced by an oscillator that employs a phase locked loop to lock the 38.9 MHz if to the 15.625 kHz horizontal scan frequency. It also generates the ident signals.

The if feeds the sound modulators 1 and 2 where further phase locked loops are employed to produce the two sound carrier frequencies, which are frequency modulated by audio signals as required by the type of sound transmission in use. Ident signals are added to the second carrier. The output of modulator 2 is fed to modulator 1 where the two carriers are combined. Non-linearity in the sound amplifier chain is pre-corrected to minimize the generation of ips, and pre-correction is applied for fm to am conversion caused by a single notch vision/sound combiner if one is used.

In order to control and measure the power of the two carriers individually, a demodulation system as shown in *Figure 13.5* may be employed. A sample of the transmitter output is down-converted to if using a feed from the local oscillator. A synchronous detector demodulates video to feed a level detector for vision alc and power metering. A video output may be provided. An intercarrier output from the synchronous detector is converted to a low frequency to enable the two sound carriers to be separated and rectified to provide individual alc, power metering, deviation measurement and audio.

For digital systems, a digitally modulated carrier with an overall bandwidth of 700 kHz at the second sound frequency is combined with the original carrier, pre-corrected and combined with the vision in a similar manner.

The implications of stereo sound for vision and sound combiners are explained in section *13.8*.

13.5 Tetrode transmitters

Tetrode transmitters with powers from 1 to 50 kW are employed in Bands I and III and with powers from 1 to 20 kW at uhf. Although higher power types have been employed, constraints imposed by the availability of suitable tubes and efficiency generally limit the use of combined amplification to 5 or 10 kW (*Figure 13.6*).

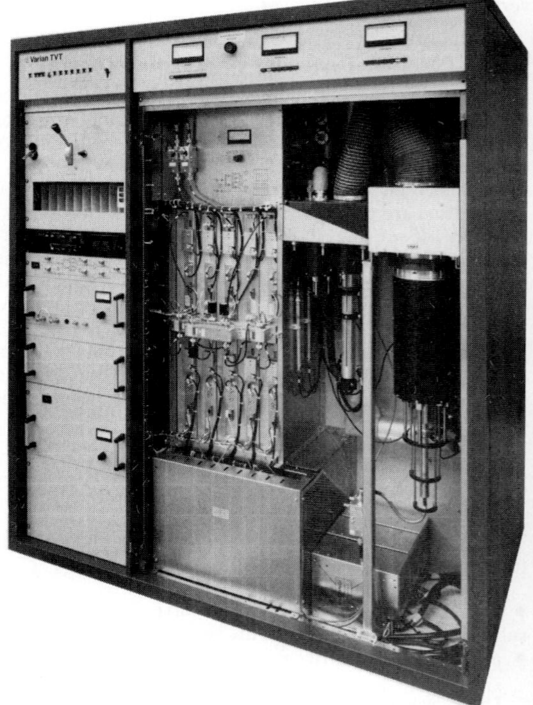

Figure 13.6 10 kW uhf two tetrode television transmitter (Varian TVT Ltd)

13.5.1 Combined amplification types

The block diagram of a typical 5 kW combined amplification tetrode transmitter is shown in *Figure 13.7*. A drive system

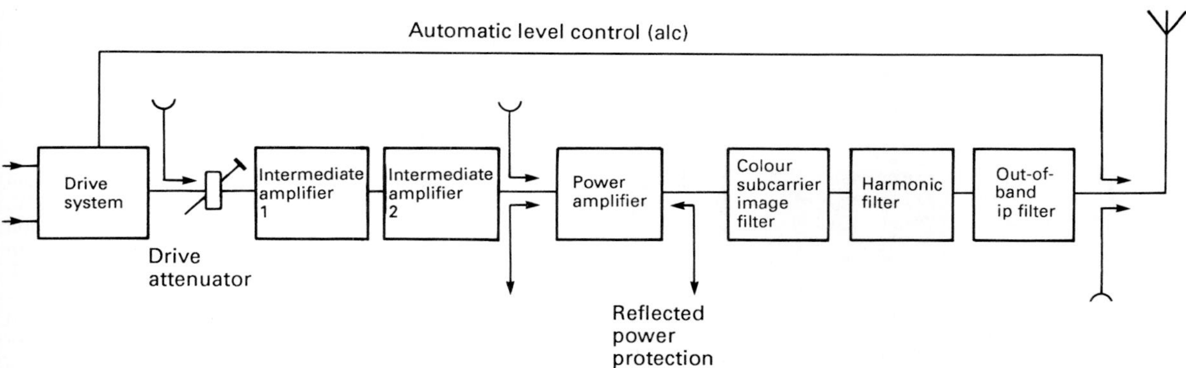

Figure 13.7 Combined amplification transmitter

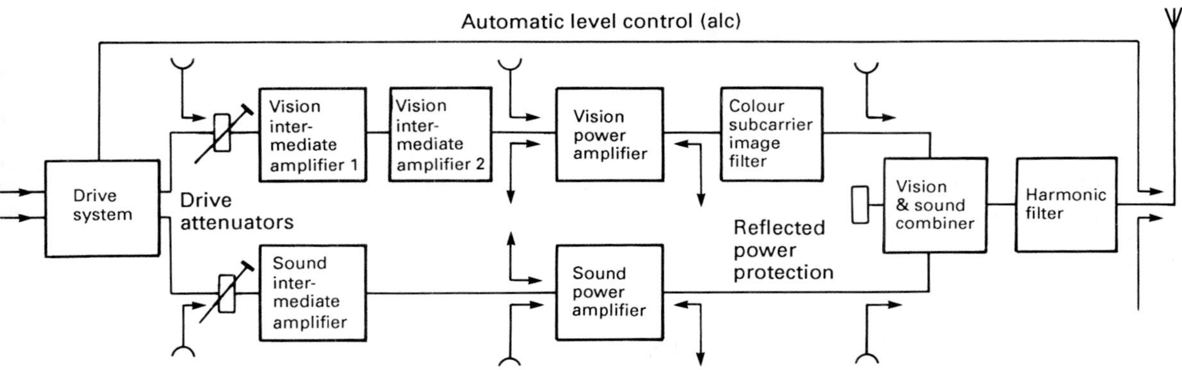

Figure 13.8 Separate amplification transmitter

providing a combined vision and sound output of 1 W is employed. As the power gain of the final tetrode power amplifier is 15–16 dB, an intermediate or drive amplifier with a gain of 22–23 dB and an output of about 200 W (peak sync) is required. This may be a solid-state amplifier driving another tetrode, or completely solid-state. The need for good linearity to minimize the generation of ips requires that all amplifier stages are operated in class A.

A variable attenuator at the input to the intermediate amplifier adjusts the drive level to take up variations in gain over the frequency range, while a directional coupler at the output of the transmitter feeds a signal back to the drive to control the alc circuits and keep the output constant.

Forward directional couplers provided at various points are used for test purposes, while a reverse coupler at the output of the final amplifier feeds the control logic to shut down the transmitter if power reflected from the load becomes excessive.

Power supplies are conventional. The bias on grid 1 of the tetrode final amplifier of approximately -100 V and the supply to grid 2 of approximately 800 V require to be well regulated, but the anode supply of 5–6 kV is less critical. To obtain the full life from a tube with thoriated tungsten cathode, the heater voltage must be held within ±1 per cent. Up to 5 kW forced air cooling is normally employed for the intermediate and power amplifiers, but at higher powers, the anode of the output tetrode requires to be water or vapour condensation cooled and grid 2 water cooled.

The output of the transmitter is filtered to attenuate the colour image at $f_V - f_{SC}$, harmonics at multiples of the carrier frequencies and out-of-band products caused mainly by intermodulation between f_V and f_{S1}/f_{S2}, which appear both above and below the channel. f_V is the vision carrier frequency, f_{SC} the colour subcarrier frequency, and f_{S1} and f_{S2} the frequencies of the first and second sound carriers.

For powers up to 1 kW, these functions can be performed by a single band-pass filter. At higher powers, where this is impractical, harmonics are attenuated by a low-pass filter, while stub notch filters are tuned to attenuate the colour image and the most significant ips.

13.5.2 Separate amplification types

Separate amplification tetrode transmitters are generally used at higher powers, up to 50 kW at vhf and 20 kW at uhf. The block diagram of a 10 kW equipment is shown in *Figure 13.8*.

A drive providing separate vision and sound outputs is used. The vision amplifier chain is similar to that of the combined type, except that the intermediate amplifier must produce 400 W ps (*peak sync*) of drive, and the power amplifier will deliver 10 kW in vision only service. As linearity requirements are less stringent, the stages can be operated in class AB with improved efficiency.

The sound amplifier chain may be a solid-state amplifier driving a smaller tetrode or all solid-state. If only one sound carrier is used, the stages may be operated in class B or C. If two carriers are used, the need to minimize intermodulation between them requires good linearity and operation in class A or AB.

The vision output is filtered to attenuate products at colour image frequency by a notch filter before the two signals are combined in a vision and sound combiner (see section *13.8*).

Figure 13.9 60 kW uhf two klystron television transmitter (Varian TVT Ltd)

The colour image filter may optionally be included in the combiner. The output is finally passed via a low-pass filter to attenuate harmonics.

13.6 Klystron transmitters

Klystron transmitters, used at uhf only, have basic powers from 10 to 60 kW ps and are used in combinations up to 240 kW. Units up to 30 kW may use a single tube with combined vision and sound amplification, otherwise separate amplification with usually two identical tubes for vision and sound is employed.

If klystrons are substituted for tetrodes in the output stages and circulators added at the tube inputs to absorb reflected power, the block diagrams in *Figure 13.7* and *Figure 13.8* are relevant also to klystron transmitters. As the gain of a klystron is typically 30 dB, a smaller solid-state intermediate amplifier is used.

13.6.1 The klystron tube

The klystron is an electron transit time tube which consists essentially of three parts: a gun that generates an electron beam, a radio frequency interactive section providing power amplification, and a collector to dissipate the used beam.

Tubes commonly have four or five resonant cavities either *integral* within the vacuum containment or *external*, clamped to the outside of the tube and coupling with the beam via ceramic

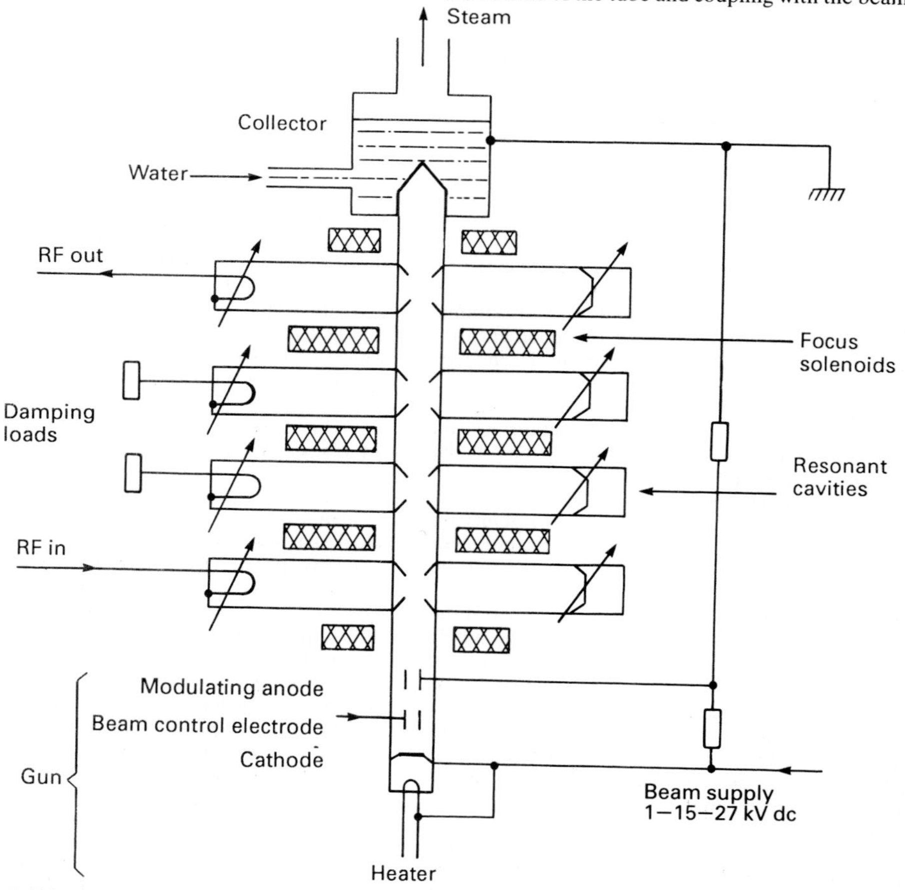

Figure 13.10 Vapour cooled klystron tube with four external cavities

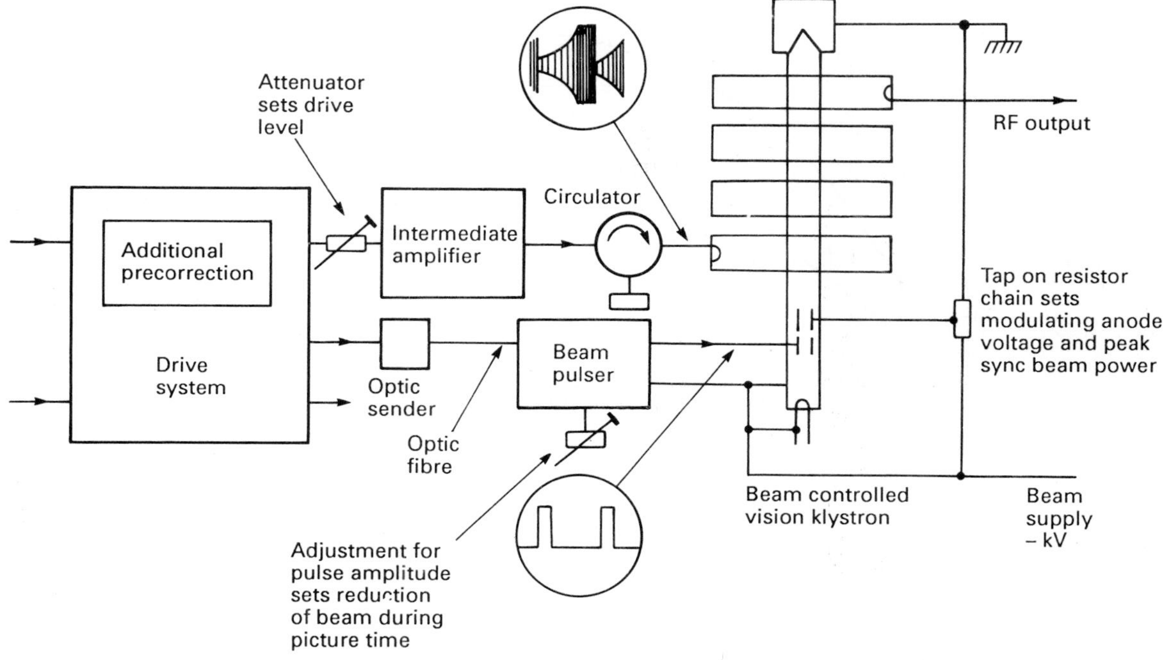

Figure 13.11 Adaptation of vision chain of separate amplification klystron transmitter for beam control

windows. Both types have merits and disadvantages. Integral types are predominant in the USA and externals in Europe.

The beam generated by the gun is accelerated by a dc potential of typically 15–27 kV. Its power is controlled by the voltage on the modulating anode. It is focused by an axial magnetic field produced by a solenoid and passes through the drift tube to the collector (*Figure 13.10*).

Drive power of typically 1–50 W, depending on type, is applied via a coupling loop in the first cavity where it produces an oscillatory field across the first interactive gap. This accelerates and decelerates the electrons in the beam, causing them to bunch. This bunching is enhanced by the intermediate cavities, while output power is extracted by a loop from the final cavity.

The used beam is dissipated in the collector, which is cooled by air, vapour or water. The body may be water or air cooled and the gun air cooled.

The power gain of a four cavity klystron is typically 30–40 dB. The transfer characteristic is non-linear, with gain reducing towards maximum power, so requiring considerable pre-correction. The cavities are tuned and the intermediate ones damped to give the bandwidth, of about 8 MHz, necessary for TV service.

13.6.2 Beam pulsing

While efficiency is important for all transmitters to reduce electricity consumption, it is particularly so for klystrons used in higher power uhf types where electricity cost is usually the largest single item in the cost of ownership.

A klystron vision power amplifier operates basically in class A with the beam current constant regardless of modulation. Improvements in tube electron optic design and in methods of tuning have increased efficiency over the 30 years in which klystrons have been used. A basic (unpulsed) figure of merit (fom) or efficiency, defined as peak sync power divided by beam power, of 42–55 per cent is now typical.

This basic figure of merit is normally improved by reducing the beam power by up to 35 per cent during picture time to a level adequate for providing black level power. It is only increased to full value for the approximately 8 per cent of the time during which peak sync power has to be radiated. This reduction is achieved by applying sync pulses either to the modulating anode or to a beam control electrode in the gun.

Figure 13.11 shows how the vision chain of the separate amplification klystron transmitter in *Figure 13.8* may be adapted for beam control by the latter method. A vision tube with a basic figure of merit of, say, 43 per cent, measured at the tube output, can be increased to 66 per cent and the overall transmitter electricity consumption reduced by 26 per cent compared with the non-pulsed tube.

Sync, derived from video in the drive, is fed to a 'pulser' which applies pulses of up to 1000 V peak amplitude between the beam control electrode and the cathode. As the pulser operates at cathode potential, the connection to it is via an optic fibre providing high insulation. The pulse polarity is such that beam current is reduced during picture time by an amount proportional to pulse amplitude, but allowed to rise during sync time to its maximum value, preset by the voltage applied to the modulating anode.

The transfer characteristic, showing drive and output waveforms for a vision klystron, with both normal and pulsed operation, is illustrated in *Figure 13.12*.

The following points should be noted:

- As black level is brought near to saturation, precise linearity pre-correction in this region has to be provided in the drive.
- Sync is removed from the modulated rf drive applied to cavity 1. The radiated sync is provided by the pulser.
- A phase shift, of approximately 40°, introduced into the vision carrier when the beam power is changed, is corrected by modulating the incidental phase pre-corrector in the drive.
- A reduced beam couples less effectively with the cavities to give less gain necessitating a higher drive power.

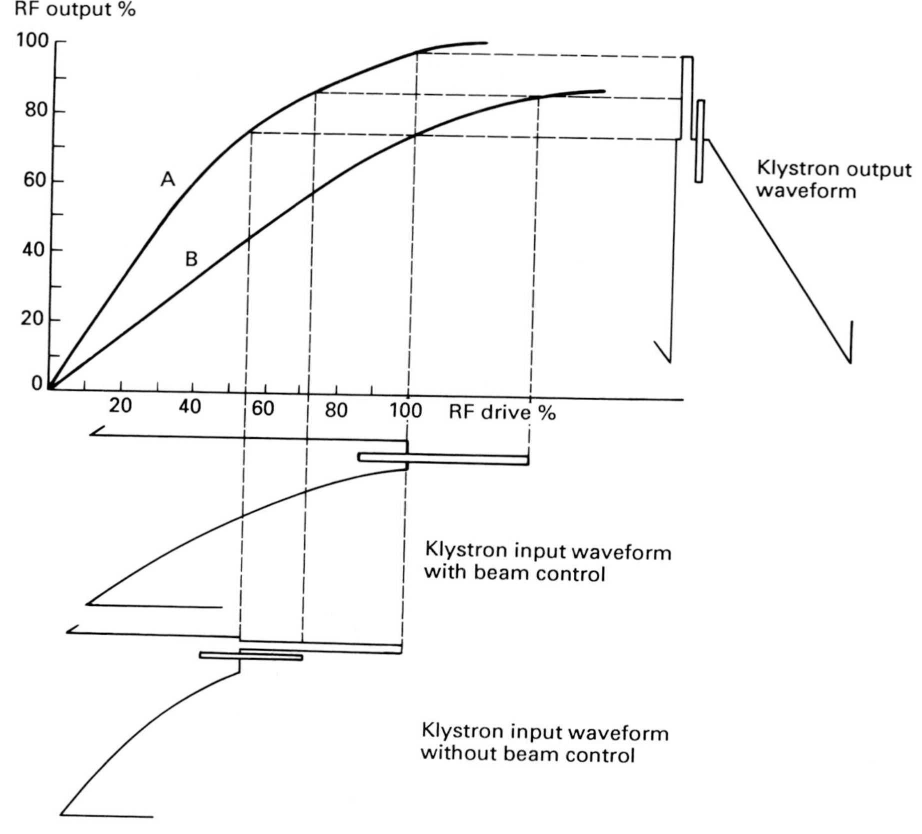

Figure 13.12 Transfer characteristics of vision klystron amplifier, with and without beam control. A is the transfer characteristic with full beam power, B is the transfer characteristic with beam reduced. Note that the output sync is provided by beam pulser

Figure 13.13 22 kW solid-state vhf television transmitter (Larcan Communications Equipment Inc)

13.6.3 Higher efficiency klystron derived tubes

The trend to higher powers together with increasing energy costs has prompted the development of two klystron derived tubes. There are the Klystrode tube and the multi-stage depressed collector (msdc) klystron. Either of these enables transmitter electricity consumption to be reduced by about a further 40-50 per cent compared with the most efficient pulsed klystrons.

The *Klystrode* has a cathode and grid like a tetrode, with a drift tube and output cavity like a klystron. As it operates in class B, the beam current varies with modulation to give a figure of merit, when transmitting an average picture, of 123 per cent. As the gain is approximately 22 dB, the drive required falls between that of a klystron and a tetrode.

The *msdc klystron* is identical to a modern pulsed klystron, as described in sections *13.6.1* and *13.6.2*, up to the end of the rf interactive section. The spent beam then enters a collector assembly where electron groups having differing exit velocities, due to the amplitude modulation of the television signal, are collected selectively by four collector stages at appropriate potentials. This results in a much lower spent beam power loss and the generation of less waste heat. A figure of merit of 130 or greater is achieved.

13.7 Solid-state transmitters

The radio frequency power available from solid-state devices has increased significantly in recent years. Fully solid-state

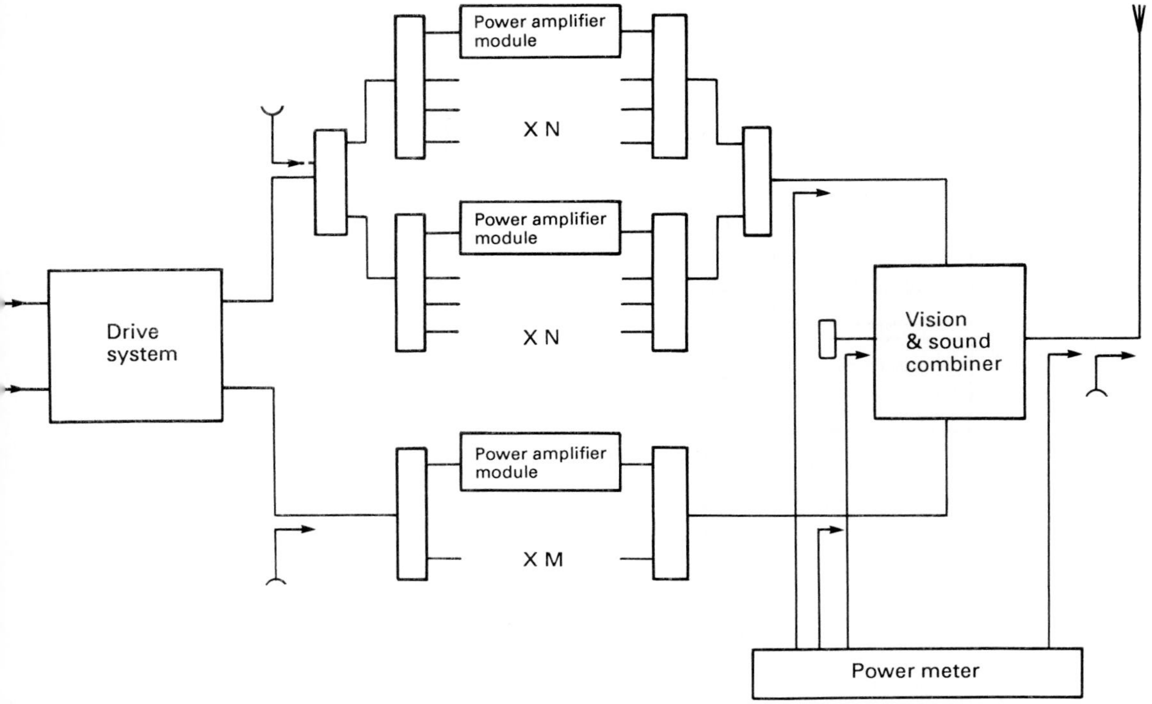

Figure 13.14 Typical vhf or uhf solid-state transmitter. The number of vision power amplifier modules N and sound modules M is determined by the output of the transmitter

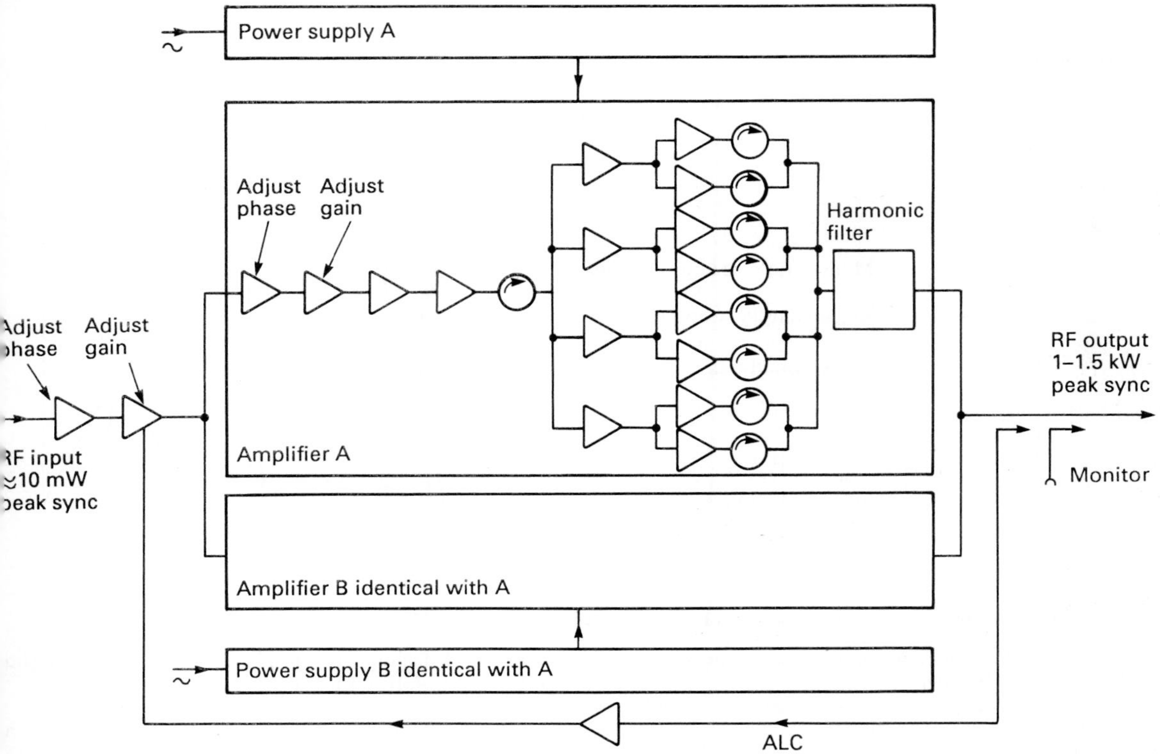

Figure 13.15 Typical individual solid-state power module

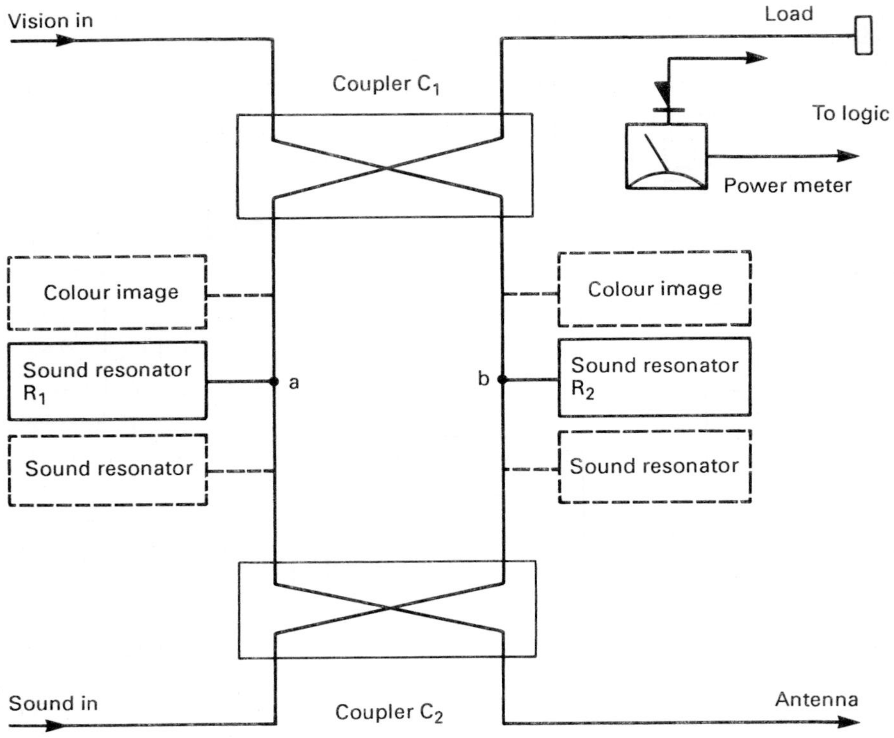

Figure 13.16 Vision and sound combiner

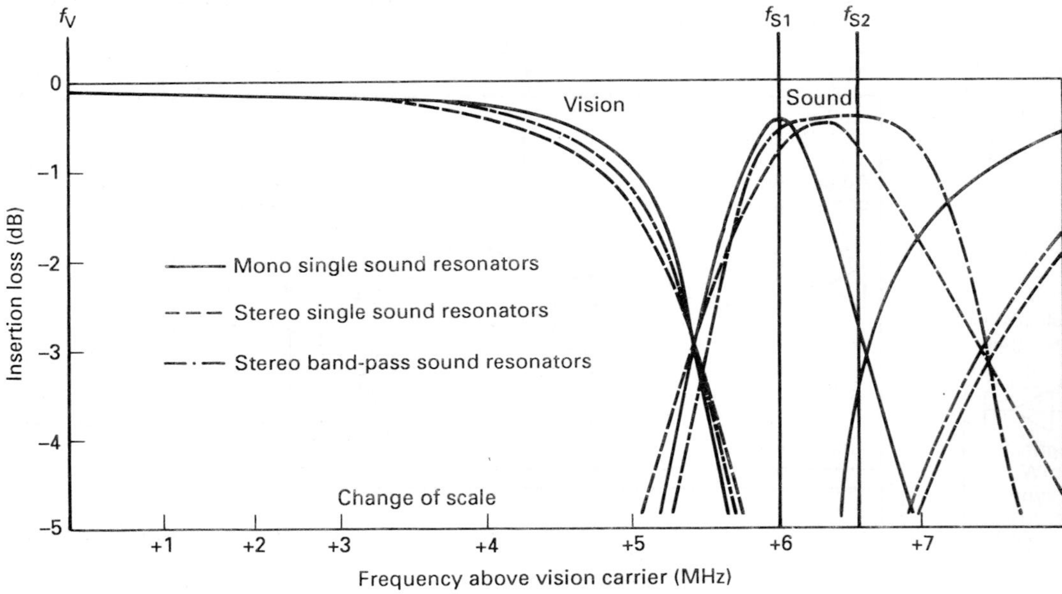

Figure 13.17 Amplitude/frequency response of typical vision and sound combiners, Standard I

television transmitters are now available with outputs up to 30–40 kW in Bands I, III and uhf.

For efficiency, cooling and size reasons, combined amplification requiring linear class A amplifiers is employed only up to a few hundred watts. Separate amplification is used for higher powers.

The block diagram of a typical vhf or uhf transmitter is sho in *Figure 13.14*. It consists of:

- a drive system with separate vision and sound inputs wh may be duplicated for extra reliability
- a vision power amplifier comprising a number of identi

modules operated in parallel to provide the required output power; each consists of an rf amplifier with a gain of about 40 dB and an individual output of 1–1.5 kW together with a dedicated power supply
- a sound power amplifier employing a lesser number of similar modules
- power splitters and combiners; to minimize the effect of failures, both amplifiers are constructed from two identical parts in parallel
- a vision and sound combiner

The block diagram of an individual power amplifier module, which also consists of two identical parts in parallel, is given in *Figure 13.15*.

Each half is provided with individual adjustments for gain and phase to optimize the summing of power at the module output. Each full module is likewise provided with adjustments for gain and phase to optimize the summing of power in the final combiners.

The solid-state devices used are either *bipolar* or *fet* with an individual gain in the region of 7–10 dB and an individual output of 70–150 W. They are operated in class AB for best efficiency, ease of cooling and device life.

The modules are broad-band over a group of channels, if not over the whole operating frequency band. They require no tuning on installation. The output devices are paralleled using ferrite circulators which provide individual protection against excessive reflected power. Individual harmonic filters and protection against excessive temperature are also included.

If an individual module fails, the transmitter will remain on-air with some reduction in power. The reduction is less for higher power transmitters with more paralleled modules. Means are provided to replace a defective module while on-air without programme interruption.

The power supplies are either *switch mode* or *switch regulator* types. They provide protection for the rf modules against excessive voltage or current. Protection against mains supply transients is also included.

Solid-state transmitters are air cooled either by dedicated fans for each module assembly, or by external fans which are usually duplicated. The cooling air is either ducted over heat sink fins attached to the modules, or directed by jets individually towards the transistors.

13.8 Vision and sound combiners

A vision and sound combiner is a fundamental part of all separate amplification television transmitters. It combines vision and sound to the antenna with minimum loss and negligible interaction between the power amplifier stages.

Combiners are constructed from coaxial components up to 50 kW peak sync and from waveguide for higher powers. The arrangement is shown in *Figure 13.16*.

It consists of a hybrid ring containing two 3 dB couplers interconnected by short transmission lines across which two or more high-Q resonators are connected. Combiners may be either vision or sound reflecting; the latter will be described.

The vision input is split into two equal parts by coupler C_1. These bypass the resonators R_1 and R_2, which are anti-resonant at vision frequencies. Their phase and amplitude are such that they recombine in coupler C_2 and are fed to the antenna. The sound input is split by coupler C_2. Resonators R_1 and R_2, resonant at sound frequency, introduce effective short-circuits at points a and b which reflect the sound back to coupler C_2. Again the amplitude and phase is such that they recombine in C_2 and are fed to the antenna. Any sound power not reflected combines in C_1 and passes to the load.

A probe monitors the load power as an indication of resonator tuning and feeds a signal to the transmitter logic to shut it down if this is excessive.

The amplitude/frequency responses of typical combiners for mono or stereo sound with single or band-pass resonators are shown in *Figure 13.17*. The sound resonators also shape the upper edge of the vision pass-band.

With BTSC stereo the sound carrier bandwidth, when fully loaded with stereo, pro and sap is 400 kHz. Ideally band-pass resonators should be used, but experience shows acceptable results are obtained by appropriately tuned single resonators.

For IRT and NICAM 728 stereo, using two carriers, band-pass sound resonators should also ideally be used. Again experience shows that adequate quality is obtained if single resonators are tuned to a frequency between the two carriers and pre-correction is provided in the drive for the fm to am conversion caused by the carriers deviating on the sides of the resonator response curves.

It is also good practice to include, within the combiner ring, the colour subcarrier image filters which suppress out-of-band radiation at f_V-f_{SC}. Energy at this frequency is then reflected to the load and not back to the vision power amplifier.

13.9 Control logic and safety

All transmitters incorporate a logic system which:

- controls the start-up sequence to on-air, applying power supplies, cooling, etc., in the correct order and proving the presence of each before applying the next
- maintains surveillance over the functioning of the transmitter when on-air, detects the onset of any fault and takes automatic action to shut down and protect the equipment
- interfaces, when required, with equipment to control the start/stop of the transmitter automatically or from a remote location
- interfaces, when required, with equipment conveying normal and fault status and analogue telemetry information to a remote location

In klystron and tetrode transmitters, although the power supplies, cooling, etc., take differing forms, the control logic is similar in principle. Little control logic is required in solid-state transmitters.

13.9.1 Safety

All transmitters must comply with the international safety standard defined in IEC 215. This requires, among other things, the provision of a positive mechanical interlock between the isolation of incoming supplies, the earthing of high voltage circuits and access to those parts of the equipment employing high voltages. The contacts of the earthing switch must be visible to the operator.

13.10 Transmitter specifications

The following typical specification is applicable to tetrode, klystron or solid-state transmitters. It lists performance parameters essentially independent of the operating standard in use.

13.10.1 Vision

- *Luminance linearity*: better than 0.9 (0.9 dB) from 10 to 75 per cent of peak sync. It is measured by modulating the transmitter with a ten step grey scale, and using a waveform monitor to compare the largest and smallest steps of the differentiated demodulated output.

- *Differential gain*: better than 0.95 (0.44 dB) from 10 to 75 per cent of peak sync. It is measured by modulating with a ten step grey scale with 20 per cent superimposed subcarrier, and using a waveform monitor with high-pass filter to compare the amplitude of subcarrier on any step with that on the black level step.
- *Differential phase*: better than ±3° from 10 to 75 per cent of peak sync. It is measured by modulating with the differential gain test signal, and using a vectorscope to measure the difference in phase between the colour burst and any other modulation level.
- *Incidental phase modulation* of vision carrier: less than 4°. It is measured by modulating with a ten step grey scale, and using a calibrated waveform monitor, with the vertical input fed from the video output and the horizontal input from the quadrature output of a synchronous detector, both via low-pass filters, observing the change of phase with modulation.
- *LF noise* (<1 kHz): better than 50 dB_{p-p} below peak picture amplitude.
- *HF noise* (>1 kHz): (unweighted) better than 60 dB_{rms} below peak picture amplitude. Noise is measured with a noise meter and appropriate filters to observe the demodulated output.
- *Low frequency response*: less than 2 per cent tilt on a 50/60 Hz square-wave. It is measured using a waveform monitor to observe the demodulated output.
- *Black level* and *peak sync* output stability: less than 2 per cent variation. These are measured by modulating with a signal alternating between black and peak white, and using a dc coupled waveform monitor to observe the demodulated output.

Amplitude/frequency response and group delay are standard related parameters measured with a sideband analyzer and a group delay measuring set respectively.

13.10.2 Sound

- *Audio frequency response* (mono): better than ±0.5 dB from 30 Hz to 15 kHz relative to the pre-emphasis curve employed. This is measured by modulating with a variable audio frequency and measuring the level of the demodulated output.
- *Harmonic distortion*: better than 0.5 per cent from 30 Hz to 15 kHz at up to ±50 kHz deviation. It is measured with a noise and distortion measuring set.
- *FM noise*: better than 66 dB_{rms} for standards using ±50 kHz deviation, or 60 dB for standards using ±25 kHz, with reference to normal deviation with de-emphasis. This is measured with a noise and distortion measuring set.
- *AM noise*: better than 50 dB_{rms} without modulation below rectified carrier. This is measured by observing rectified output on an oscilloscope and comparing the ac to dc component.
- *Synchronous am*: less than 1 per cent at ±50 kHz deviation. It is measured by observing the amplitude of the demodulated carrier with and without modulation.

Deviation and pre-emphasis are standard related parameters. Deviation is measured with a deviation meter, often built into the transmitter. Pre-emphasis is measured by modulating with a variable audio frequency and measuring the level of the demodulated output with and without the pre-emphasis network in circuit.

13.10.3 General

- *Carrier frequency stability* vision or sound: better than ±200 Hz per month when not using precision control. It is measured using a frequency counter.
- *Cabinet radiation*: more than 60 dB below the field that would be produced if the transmitter peak sync power were to be applied to a dipole at the centre of the cabinet. It is measured using a field strength meter.
- *Acoustic noise*: better than 65 dBA measured 1 m in front of the transmitter and 1 m from the floor.

13.11 Reserve systems

Most television transmitters are provided with reserve facilities to enable the service to continue, perhaps at reduced power, if a failure occurs. There are four main systems: passive, active (or parallel), multiplex and integral modular reserve.

13.11.1 Passive reserve

Passive reserve systems employ two complete transmitters, one in operation, the other in standby (*Figure 13.18*). If the operating transmitter fails, its logic causes it to shut down. A changeover switch, usually operated automatically, connects the reserve transmitter to the antenna and the failed transmitter to a test load for servicing. The reserve transmitter logic then brings it on-air.

The reserve may have a similar or lower power than the main.

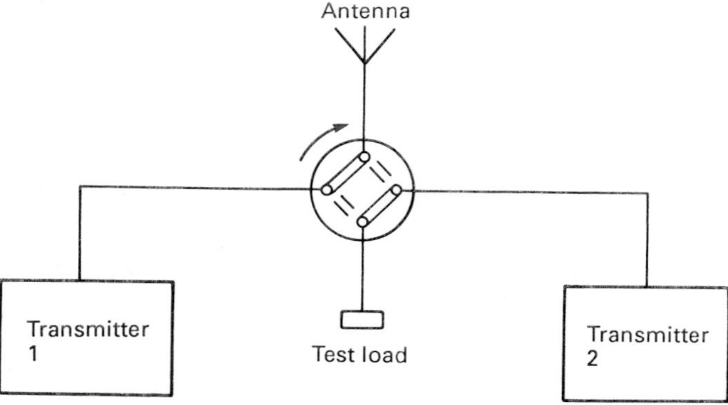

Figure 13.18 Passive reserve

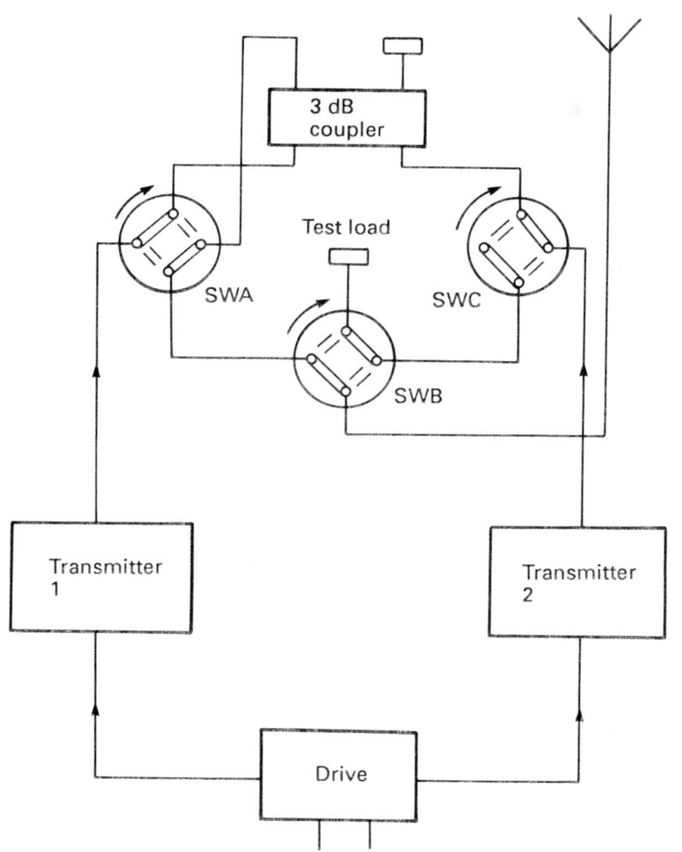

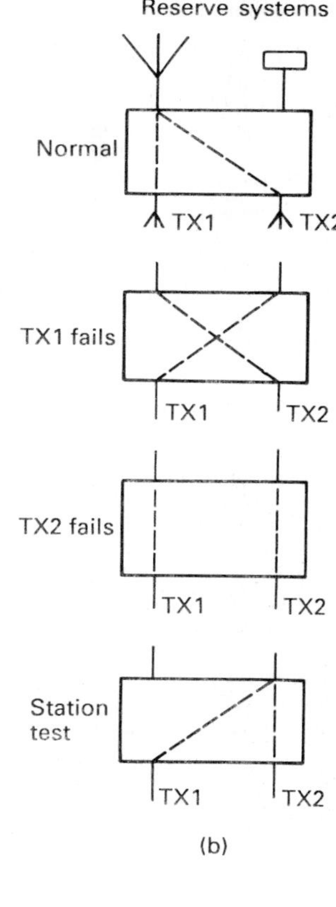

Figure 13.19 Parallel operation of transmitters: (a) using a 3 dB coupler with bypass switches, (b) using a switchless combiner. In (a) the switches are shown as in normal operation. If transmitter 1 fails, SWA, SWB and SWC operate; if transmitter 2 fails, SWA and SWC operate. For station test, SWB operates.

If similar, transmission is resumed at full power after a break of a few seconds.

It is normal to keep reserve tube transmitters operational with their heaters run partially at 'black heat'. Under this condition the tube is ready to deliver power instantly on the application of full heater and ht volts. Standby hours at black heat are not counted in tube warranty calculations.

13.11.2 Parallel reserve

Parallel or active reserve systems, shown in *Figure 13.19*, are used when the output required is greater than can be obtained conveniently from a single transmitter and/or no break in the service can be allowed.

The vision and sound outputs of a drive system, which is usually duplicated for added reliability, are split and fed to two identical transmitters. The transmitter outputs are then combined in a paralleling diplexer to feed the antenna. Means are provided to adjust the drive amplitude and phase for optimal combining.

If either transmitter fails, the service continues uninterrupted, but as half the power of the remaining transmitter will be radiated and half dissipated in the diplexer ballast load, the effective power will be 6 dB below normal. In some cases this is not significant, and the station is left in this state until servicing can be carried out.

The paralleling diplexer can be either a 3 dB coupler or a switchless combiner. When a 3 dB coupler is used, coaxial switches or manual U-links may be provided. These are operated during a convenient programme break to switch the remaining transmitter to the antenna directly, so giving a -3 dB reserve, until servicing is done. With a switchless combiner, phasing arrangements, which may be operated on-air, enable any of the configurations shown in *Figure 13.19* to be employed.

13.11.3 Multiplex reserve

Multiplex reserve (mpx) is provided at two levels: manual or automatic.

Manual operation requires interconnections to be provided within the transmitters so that, if either output stage fails, vision and sound can be combined in the drive and amplified by the stage remaining operational. This is then connected to the antenna directly bypassing the vision and sound combiner.

Automatic operation, used mainly at uhf, employs a main drive with separate outputs to feed two separate klystron amplifiers each with its own power supply and cooling. These amplify vision and sound which are then combined and fed to the antenna. If either amplifier fails, this is detected by the logic which:

- closes down the failed amplifier
- selects a standby drive with a combined vision and sound output which is fed to the amplifier remaining serviceable
- operates coaxial switches to connect its output to the antenna directly and that of the failed amplifier to load

If the main drive fails, the system also continues to operate in the multiplex mode. For economy reasons, the beam power of the amplifier normally used for sound is reduced, but increased automatically if it has to handle a combined signal.

13.11.3.1 Reserve power

The reserve power available from a multiplex transmitter depends on the level of in-band and out-of-band intermodulation products (ips) that can be tolerated during short term emergency use. With optimal pre-correction, the maximum reserve is about 3 dB below normal. The sound power is sometimes reduced to permit a higher vision output for a given ip level.

13.11.4 Integral modular reserve

Integral modular reserve (imr) is inherent in solid-state transmitters constructed from a number of essentially independent parallel amplifier/power supply modules. If a single module fails, the transmitter remains on-air, but the power will be reduced by an amount depending on the number of modules used.

The reduction is therefore less for higher power types. If the drive is duplicated with automatic changeover, and it is accepted that the final power combiners and the vision/sound combiner, being passive, are not likely to fail, then a single solid-state transmitter may be considered adequate for reliable broadcast service. A conventional reserve is then not needed.

If however the broadcaster operates an extended or 24 hour schedule, it is not possible to test the performance of the transmitter. In such cases, two completely separate solid-state transmitters are employed in a conventional parallel arrangement. This may use a switchless combiner to enable either transmitter to be transferred to load for test while the other maintains the service at half power and without any interruption.

13.12 Programme combiners

Television transmitters for various services are often co-sited and feed antennas on the same tower. When the transmitters are in the same frequency band but with an appropriate frequency separation, the outputs of at least four transmitters may be combined to feed a broad-band antenna (*Figure 13.20*).

Such combiners consist of assemblies of high power broad-band filters each transmitting one band of frequencies, typically 8 MHz wide, while rejecting the others.

For power handling and reliability reasons, antennas are often constructed from two halves each fed by an individual transmission line. In such cases, the output of each transmitter is split to feed two programme combiners each in turn connected to half the antenna.

13.13 High power multi-parallel systems

Multi-parallel systems are employed when an output power is required greater than that obtainable from two transmitters operating in parallel. Systems constructed from combinations of 60 kW klystron power amplifiers are employed, mainly in the USA, with outputs of 120, 180 and 240 kW.

These are used with antennas that have broad vertical radiation patterns and hence low gain. This combination gives the maximum permitted effective radiated power (erp), of up to 5 MW, toward the horizon and at the same time a high signal strength close to the station.

The block diagram of a typical six klystron 240 kW system is shown in *Figure 13.21*. The use of main and standby drives and a switchless final power combiner enables a -3 dB reserve power to be available if a failure occurs.

13.14 Programme input and monitoring equipment

All television transmitting installations include some programme input equipment (pie) and monitoring equipment. Their scope varies widely with station size and user philosophy. Some or all of the following may be used.

13.14.1 Programme input equipment

- switching vision and sound between main and reserve programme feeds
- correction of distorted incoming signals
- generation of local test signals for examination of vision

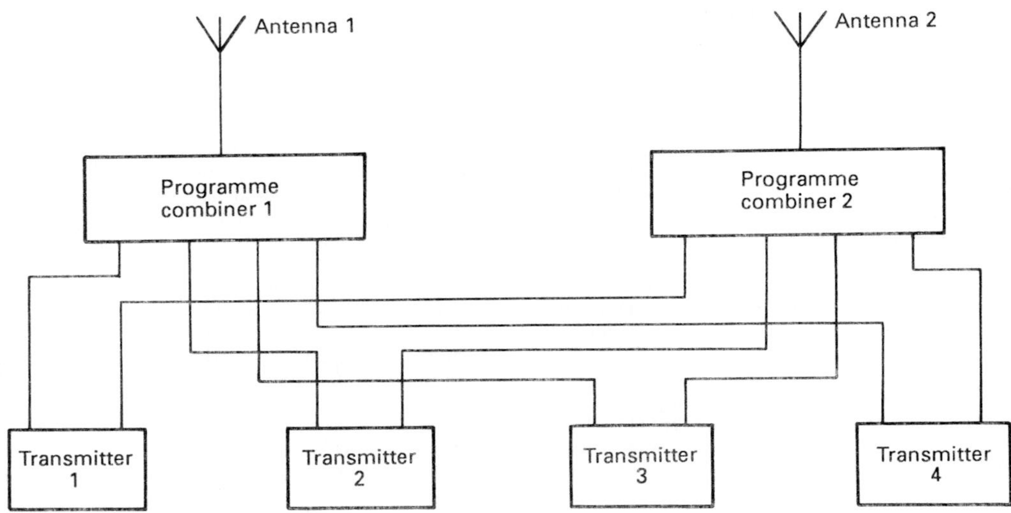

Figure 13.20 Programme combiners

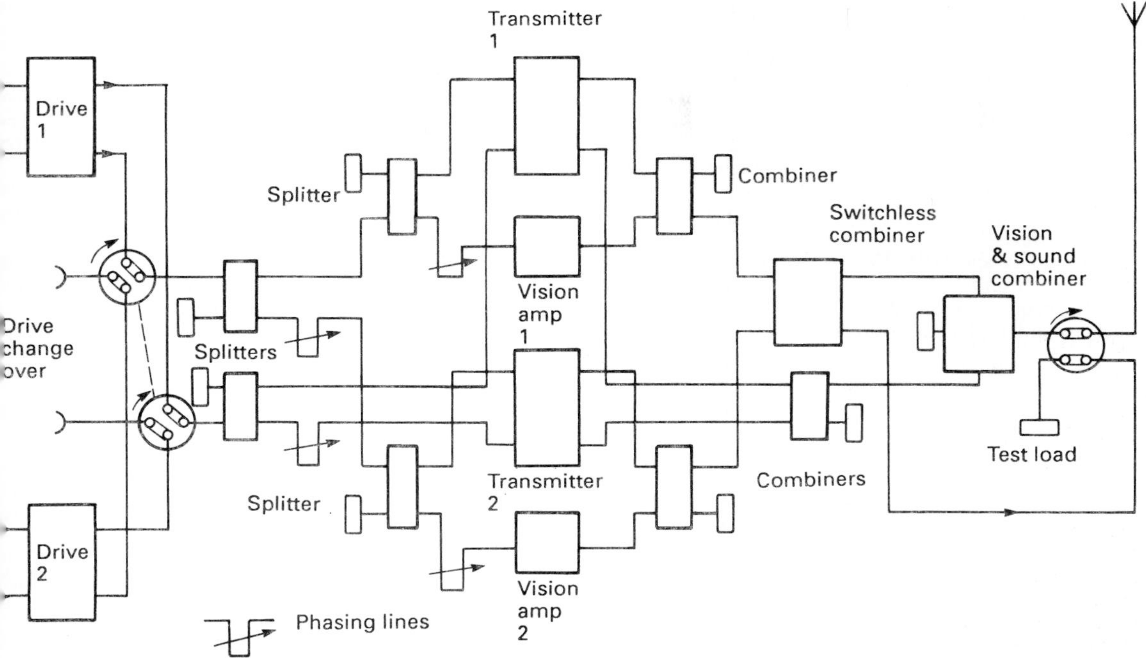

ure 13.21 Typical six klystron 240 kW uhf television transmitter system. The switchless combiner configurations are: normal A + B → C, 240 kW; smitter 1/amplifier 1 fails B → C, A → D, 120 kW; transmitter 2/amplifier 2 fails A → C, B → D 120 kW

ndwidth, luminance linearity, differential gain, differential ase, noise, etc.
decoding/transcoding sound delivered by sound in sync
limiting of audio signal (vision limiting usually part of vision dulator)
generation of composite baseband signal for multi-channel/ reo sound
deletion and re-insertion of vertical interval test signals (vits)
regeneration of teletext signals

.14.2 Monitoring

measuring the output power of the transmitter when dissi- ted in a test load provided with calorimetric or calibrated obe means of measurement
demodulation of a sample of the transmitter output, from a ectional coupler in the output transmission line, by a ecision Nyquist demodulator to give high quality video and dio
monitoring picture and video waveform in amplitude and ase
monitoring of audio by level meter and loudspeaker
measuring of audio frequency response, harmonic distortion d noise
monitoring of transmitter video performance by continuous alysis of vits, providing status information controlling auto- atic selection of reserve facilities if quality is degraded below a edetermined level
provision of remote control and status/analogue telemetry erfaces to a remote control or maintenance base

bliography

gineering Handbook, 7th ed, Dept of Science and Technology, National Association of Broadcasters, Washington

HARWOOD, A J, 'Technical constraints in service area planning', *IBA Tech Rev* No 7 (1976)

Specification of Television Standards for 625-line, System I, Transmissions in the United Kingdom, Dept of Trade and Industry, Radio Regulatory Division, London (1984)

Safety Requirements for Radio Transmitting Equipment, Publication 215, 3rd ed, International Electrotechnical Commission, Geneva (1987)

'Recommendations and reports of the CCIR', *X Part 1*, Broadcasting Service (Sound), Report 795, XVth Plenary Assembly, Geneva (1982) (Refers to IRT multi-channel/ stereo sound system)

'Multi-channel television sound transmissions and audio processing requirements for the BTS system', *OTS Bulletin* 60, Office of Science and Technology, Federal Communications Commission, Washington (1984)

Specification of a Standard for UK Stereo with Television Transmissions, Revision 4, BBC Engineering (1986)

WELLBELOVED, R and KYFFIN, M N, 'The transmission of dual channel sound', International Broadcasting Convention 1988, Conference Publication No 293

External Cavity Klystrons for UHF TV Transmitters, Technical Services Dept, Varian Microwave Tube Division, Palo Alto, California

SYMONS, R S, 'Klystrons for UHF television', *Proc IEEE*, **70**, No 11 (1982)

BENNETT, A J, 'New modulation techniques for increased efficiency in UHF TV transmitters', *Proc IEEE*, **70**, No 11 (1982)

BADGER, G M, 'Klystrode technical performance for modern high efficiency UHF TV transmitters', 11th International Broadcasting Convention (1986)

SHRADER, M B, 'Klystrode technology up-date', National Association of Broadcasters, 42nd Annual Broadcast Engineering Conference, Las Vegas (1988)

OSTROFF, N S, WHITESIDE, A H, SEE, A and KIESEL, R C, '120 kW Klystrode transmitter for full broadcast service', National Association of Broadcasters, 42nd Annual Broadcast Engineering Conference, Las Vegas (1988)

McCUNE, E, 'Final Report, multi depressed collector klystron project', National Association of Broadcasters, 42nd Annual Broadcast Engineering Conference, Las Vegas (1988)

IKEGAMI, T, 'A newly designed solid state television transmitter', Colloquium on Advances in Solid State Techniques for Broadcast Transmitters, IEE Professional Group E14, Digest No 1988/16 (1988)

GERRARD, J A and DRURY, D, 'Design philosophy for solid state transmitters', Colloquium on Advances in Solid State Techniques for Broadcast Transmitters, IEE Professional Group E14, Digest No 1988/16 (1988)

MIKI, N, 'Technologies for solid state television transmitte Colloquium on Advances in Solid State Techniques Broadcast Transmitters, IEE Professional Group E Digest No 1988/16 (1988)

CORBEL, P, 'New developments in transistorised televisi transmitters', Colloquium on Advances in Solid State Te niques for Broadcast Transmitters, IEE Professional Gro E14, Digest No 1988/16 (1988)

TEW, B G, 'System implications of high power multi-ampli UHF television transmitters', 12th International Broadca ing Convention, Conference Publication No 293 (1988)

VAUGHAN, T J, 'RF components for high power and su power UHF TV transmitting systems', 12th Internatio Broadcasting Convention, Conference Publication No (1988)

P Kemble C Eng, MIEE, B Sc
Principal Engineer, IBA

14 Transposers

In an ideal situation, all potential viewers would be provided with television programmes by means of a central high power transmitter. This would be modulated with audio and video signals from the studio, connected by means of a landline. Such an arrangement would represent the most economical means of distribution in terms of *cost per viewer*.

Unfortunately, such a simple situation is possible only at a limited number of locations. Intervening hills, buildings and even trees attenuate direct reception of a high power transmitter, particularly at uhf, causing a shadow in its coverage. Substantial numbers of people can be affected.

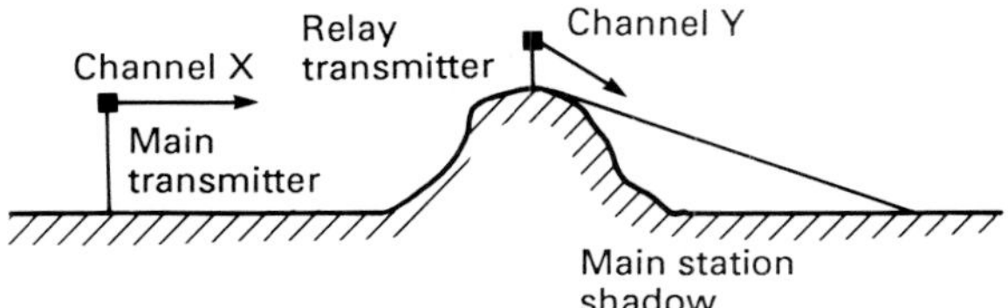

Figure 14.1 Relay on hill transposes channel X to channel Y to fill in shadow behind hill

Rather than provide landlines and expensive modulators at every transmitter site, a more satisfactory solution is to provide lower power relay transmitters to fill in the shadows. The principle is shown in *Figure 14.1*. The relay is sited at a carefully chosen location which can both see the target area to be served and also receive good signals from the main transmitter. Despite the shadow, in most cases there will actually be some residual signal in the target area from the principal station, so if the relay station were to transmit on the same channel, there would be unacceptable co-channel interference to viewers of the relay and of the main station. The equipment at the relay site therefore changes the frequency (or *transposes*) the incoming signal from the parent transmitter on channel X and re-transmits on a clear channel (channel Y). In some areas it may be necessary to build a chain of relay stations, each serving its own target area and providing a signal feed onwards to the next. In that case several different channels will be used by the various stations.

Although this technique is simple enough in concept, in countries like the UK (which has around 900 relay sites as well as 50 high power main sites) very complex planning is necessary. This is because each site transmits four different pro-

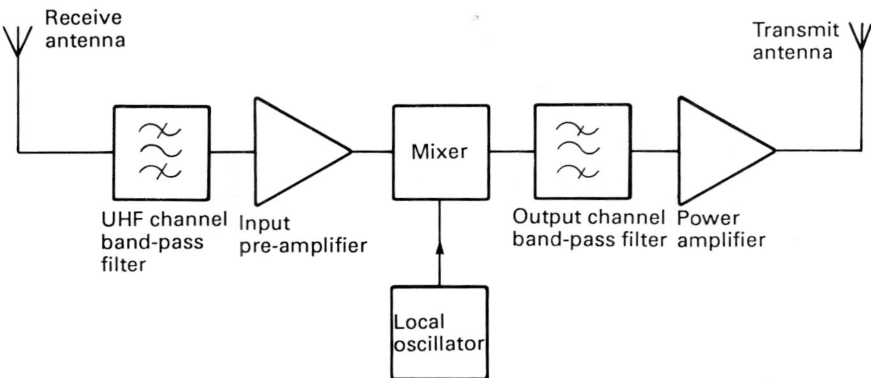

Figure 14.2 Single conversion transposer

grammes (possibly five in due course), so groups of channels have to be transposed. As the network expands it becomes increasingly difficult to find four clear channels for each site to use. In many cases no completely clear transmit channel can be found, and the service area is limited by co-channel interference from a remote station.

14.1 Transposer configurations

14.1.1 Single conversion

The block diagram of a very simple transposer is shown in *Figure 14.2*. After some uhf band-pass filtering and pre-amplification, the incoming signal is mixed with a local oscillator to generate the desired output channel directly. For example, if the input is channel 30 (543.25 MHz vision) and the desired output is channel 60 (783.25 MHz vision), then a local oscillator at the difference frequency of 240 MHz will produce the required output (as well as 303.25 MHz, which can be removed by filtering).

Problems can arise with this simple technique which limit its usefulness. It would be impracticable, for example, to receive channel 22 and transmit channel 52. The local oscillator would again have to be at 240 MHz, but this time its second harmonic falls in the receive channel and is likely to produce unacceptable interference patterning. Such relationships, and those due to harmonics of the local oscillator, can result in a large percentage of transpositions having to be avoided. While this probably will not matter in countries having a small requirement, and the technique is in use, it would be an unacceptable restriction in the UK. A further problem in a crowded spectrum is that the uhf filtering will not be able to provide much rejection of any carriers on the adjacent channels, which will therefore become unwanted re-radiated signals.

14.1.2 Double conversion

The more commonly used transposition technique is shown in *Figure 14.3*. This is a double conversion process, with a large part of the transposer operating at a standard intermediate frequency (typically 38.9 MHz for vision and, in system I, 32.9 MHz for sound).

As in single conversion, there is a band-pass input filter, A, which prevents strong neighbouring transmissions, perhaps from the same site, overloading the input amplifier. It also helps to improve the signal/noise ratio by preventing noise on the image frequency from being mixed down and added to the if. (Other techniques for this are mentioned below.)

A low noise high gain pre-amplifier, B, enables the transposer to achieve a good noise figure, typically 8 dB, and this feeds the input mixer where conversion to if occurs.

The mixer, C, may be a simple diode bridge, or a slightly more elaborate arrangement as shown in *Figure 14.4*. This is an image rejection type.

Both the wanted input signal, f_V, and the image frequency, f_{IM}, are divided into two paths by the first quadrature splitter. Thus the input to the top mixer is: $f_{IM}<90$ and $f_V<90$ plus the local oscillator. The usual sum and difference products result as indicated in *Figure 14.4*. The inputs and products are similar for the bottom mixer, except for the phase difference. It will be

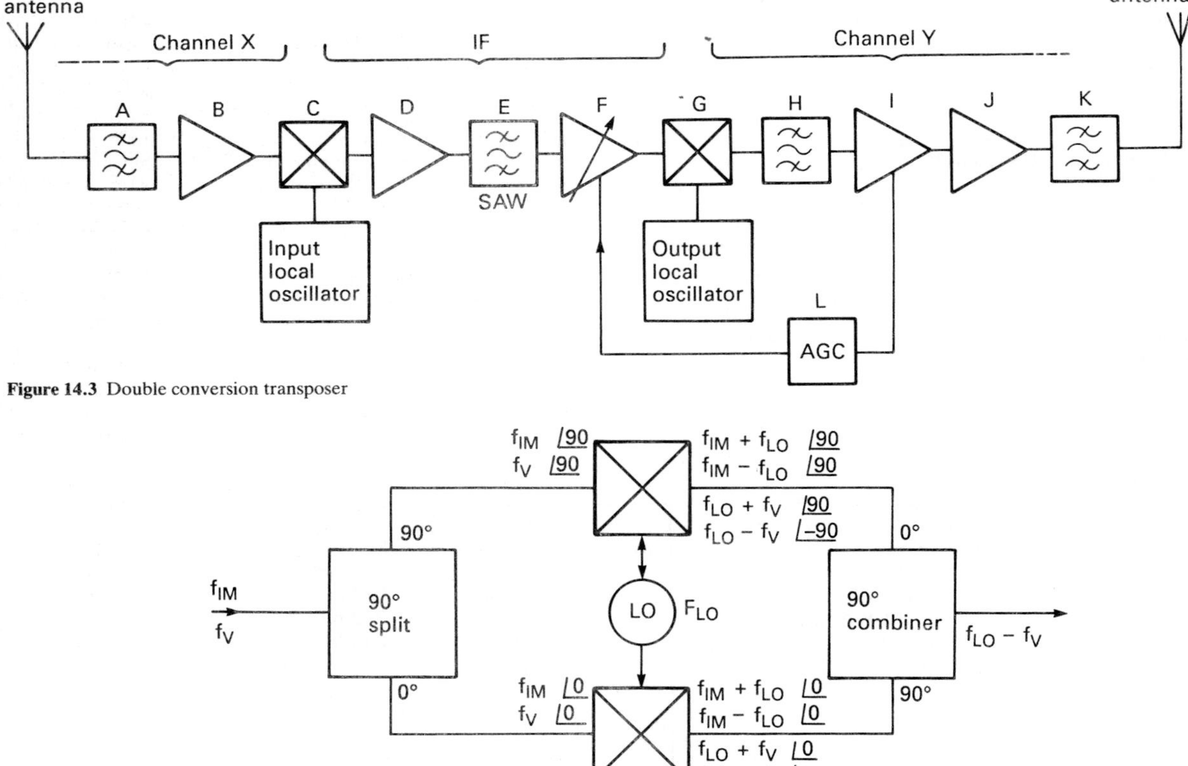

Figure 14.3 Double conversion transposer

Figure 14.4 Image cancelling mixer

seen that the phase of the various products is such that all but the wanted products cancel out in the final quadrature coupler.

This technique helps to reduce cost in volume production because it allows the input image filter to be eliminated; this has to be tuned to the required channel and is a relatively expensive manual operation.

Most of the gain in the transposer is at if where stability is easier to achieve, in fixed amplifier D and variable amplifier F (*Figure 14.3*).

One problem quoted for the direct conversion transposer was the rather poor frequency response of the uhf filter, A, which allows re-radiation of lower adjacent sound and upper adjacent vision carriers. This is undesirable for two reasons. First, if the level is sufficiently high the extra carrier may cause intermodulation problems in the power amplifier of the transposer itself. This will degrade the viewer's picture. Second, the adjacent carrier may itself cause a visible pattern for the viewer, or may interfere with some other service.

The if filter, E, in *Figure 14.3* is to eliminate these problems by providing a carefully tailored band-pass response. This is now easily achieved by means of a *saw* (surface acoustic wave) type device which can be designed for this frequency range.

It is readily possible to provide notches at (f_V-2) MHz and (f_V+8) MHz of at least -50 dB with respect to vision, so adjacent channel interference is now virtually a non-problem. A typical saw frequency response is shown in *Figure 14.5*.

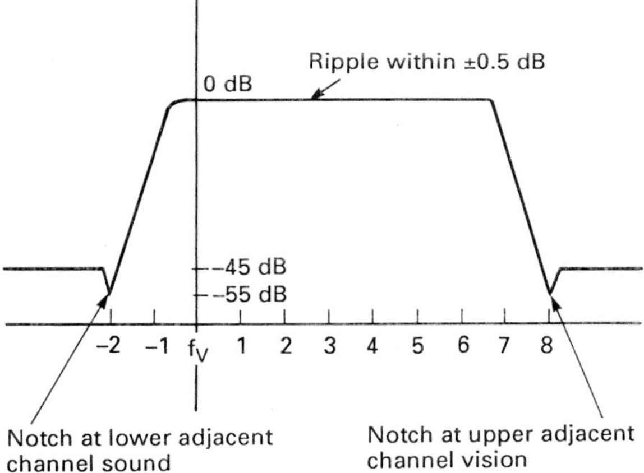

Figure 14.5 SAW filter response

The only problem likely to remain is in the exceptional situation where the aci is so high, even after receive aerial azimuth and polarization discrimination, that the transposer front end is still overloaded. Levels of aci which equal the field strength of the wanted signal should be acceptable.

While saw filters have the desirable feature of predictably providing the sharp cut-off and good frequency response needed with no manual adjustment, the physical length of the device needed to achieve it results in an attenuation of the wanted signal of typically 30 dB. (As the name implies, the signal propagates in the device as a surface wave between two electromechanical transducers.)

SAW filters have quite a large temperature coefficient and so require some simple temperature control to keep the notches on frequency. Amplifier D (*Figure 14.3*) is provided to make up the lost gain, but since the input power to be applied to the saw filter is limited to about +10 dBm, and allowance must be made for maximum input signals, pre-amp gain, etc., it will not be possible to make up all the lost gain at that point.

Introduction of the saw filter does therefore marginally degrade the transposer noise figure. One other advantage it does bring is the elimination of the need for group delay correction; this is necessary when LC filters are used to provide channel shaping because of their phase characteristics. SAW filters can have their frequency response and group delay response controlled almost independently, and can have virtually constant group delay despite the deep notches.

Finally, the signal is mixed back up to the desired output channel, before being amplified to the appropriate power.

Like the rest of the system, the output amplifiers I and J (*Figure 14.3*) have to handle both vision and sound carriers. At low signal levels this presents no particular problem, but with increasing power the amplifier must remain sufficiently linear to minimize the generation of distortion products. These fall into two categories, those inside the transmission channel which would affect the received picture quality and those outside the channel which are liable to interfere with other services. These distortions are discussed in more detail in section *14.3.2*.

Band-pass filter H removes local oscillator feed through and the unwanted mixing product, to prevent these reaching the output amplifier.

Finally output filter K prevents out-of-channel radiation, and may include notches to give better attenuation of the principal unwanted products at (f_V-6) MHz and (f_V+12) MHz which lie in adjacent TV channels.

As well as the direct signal path there are other functions that need to be provided. Because the transposer is receiving off-air signals, it is normal for some signal level variation to occur. Over sea paths, fading of 40 dB is not unusual. The transposer must include *automatic gain control* (agc) to hold the output power constant.

For this agc, it is quite common practice to monitor the power at the final output of the equipment. However this is not the ideal position for the detector. As the gain of the output amplifier begins to fall, due to either the ageing of a valve (tube) or loss of output transistors, the drive level will be increased to try to compensate. The effect will be to drive the output into limiting and produce more and more distortion. The viewer's picture will rapidly become unacceptable. Had the power simply been allowed to decline with the falling amplifier gain, then the service would have been satisfactory for much longer. To counteract input fading, it is preferable to keep just the if level constant. If necessary, any small variation in output power due to amplifier temperature changes or progressive failure can be minimized by means of a 2–3 dB range auxiliary agc loop round the output stages alone.

The performance of the main agc would typically hold the vision carrier constant within 1 dB over a 40 dB range of input signals. Below about -60 dBm input level, the signal will become too noisy to provide a satisfactory picture, so the transposer output can be allowed to fall or be switched off entirely. The maximum signal level is unlikely to reach -20 dBm. In fact, transposers are generally designed to be provided with nominally -40 dBm input. The input is more likely to suffer deep fades, due to aircraft flutter or in some cases propagation across tidal water, than it is likely to be significantly enhanced, so the agc window is offset in this way.

In the case of systems provided with an automatic changeover to a passive reserve, it is desirable to ensure that the agc thresholds of the A and B sides are set such that B decides it has lost its input before A. Unless this is true, deep fades may cause the A output power to begin to fall while B is still reporting signal received. This will cause equipment changeover when there is actually no fault.

Proper design of the agc also requires careful consideration of two other points:

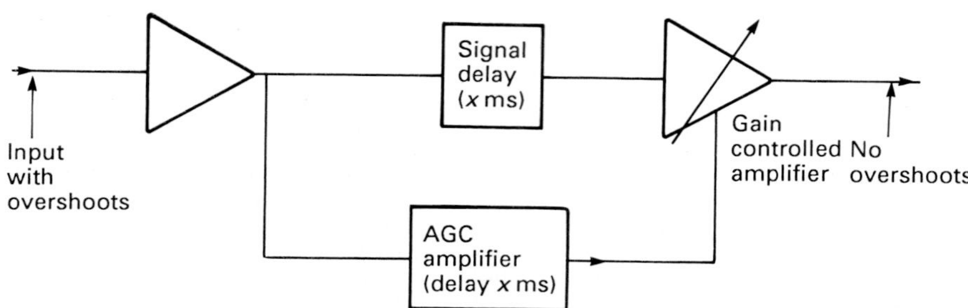

Figure 14.6 Feed-forward gain control

- It must respond sufficiently quickly to cope with fast flutter, but not so fast that it distorts sync pulses as if they are unwanted power variations.
- There must not be excessive power overshoots in the rf drive, particularly if the transposer is followed by a power amplifier.

Excessive power overshoots may cause a valve to trip or transistors to fail. Overshoots could arise when the parent station first comes on-air each day, since without an input signal the agc gain will have been maximum. A fast overpower limiter is one partially effective solution, but it is better to avoid the overshoot altogether by a *feed forward* technique as indicated in *Figure 14.6*.

A power detector controls a downstream attenuator with appropriate delay in the signal path to allow time for the attenuator to operate before the overpower arrives. If a limiter is used, it must not be inside the agc loop itself, or oscillation will occur while the overdrive input prevails.

The other major function provided within the transposer is the local oscillator arrangement. There have been a number of different "standard' intermediate frequencies used by transposers, e.g. 31.25 MHz, 32.7 MHz, 38.9 MHz, 39.5 MHz and 40.75 MHz. Currently 38.9 MHz for vision and 32.9 MHz for sound is popular, therefore the local oscillator is 38.9 MHz above vision carrier. The choice is largely for compatibility of test equipment with if modulated high power transmitters which also commonly use that frequency.

The output frequency stability of the transposer must be very good in order to obtain the benefits of offset working for subjective improvement of co-channel interference. The output depends on the incoming frequency and that of the two local oscillators. In a tandem chain of transposers the final frequency has the possibility of a build-up of numerous errors. The longest chain in the UK has a main station feeding five relays.

Although it is possible to correct automatically for any error on the incoming signal, most transposers do not do this, and for practical reasons the stability of each local oscillator is within 250 Hz over three months.

Much equipment simply uses two independent vhf crystal oscillators multiplied up to the required frequency. The crystals are operated at their temperature inversion point for optimum stability, but their setting up time together with the lead time taken to supply an aged crystal on the required frequency means that it is a time consuming process during manufacture or should it be required to change channel.

The same applies to spares. It will be necessary to hold at least two spare crystals for each frequency, since when a spare finally comes to be used there is a danger that it will not pull onto frequency and will have to be rejected. Thus, although crystal oscillators are capable of providing excellent stability and phase noise, there are also disadvantages.

With the availability of low power integrated circuits, more complex local oscillator systems using synthesizers are now common. These use a low frequency (often 1 or 5 MHz) crystal oscillator as the basic reference, and both input and output local oscillators are phase locked to it through a chain of dividers. Any drift of the reference frequency will cause the two local oscillators to move in the same direction, so the error on the output will be minimized.

If it were only necessary to operate on exact channels the synthesizer would be much simpler, since the frequency increments would always be 8 MHz, and this could be the reference and comparison frequency. In practice it is also necessary to operate on offsets of a fraction of line frequency (+ $^5/_3$ line, zero and -$^5/_3$ line are used in UK) so now the comparison frequency has to be low enough to allow that increment.

With the final frequency being divided down more and more, its random phase noise increases, and careful design is necessary to keep it acceptable. Phase noise on the local oscillators is transferred to the output carriers and can potentially cause problems on fm sound reception if the receiver does not use intercarrier sound recovery, and also on pictures received on true synchronous receivers where phase noise is converted into low frequency am noise and degrades the video s/n ratio.

Although synthesizers are not without their own difficulties, they do have the big advantage that a standard unit can be used in any equipment, no matter what the channels. Fundamentally only a standard crystal is required and channel selection is made simply by means of switches.

14.1.3 Active deflectors

It was mentioned initially that in nearly all cases the transmit and receive frequencies have to be different in order to prevent mutual interference in the service areas of the parent and its relay. However, in some locations where the topology gives excellent screening, it is possible to retransmit the incoming frequency.

This system is known as an *active deflector*. The gain required of the system is easily established if the received signal level and

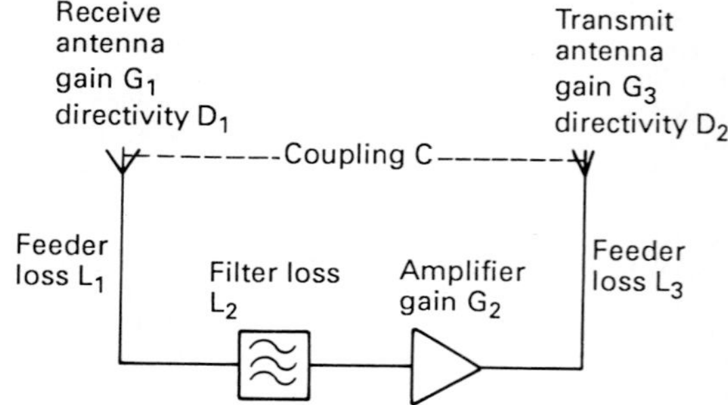

Figure 14.7 Active deflector

e necessary erp are known. Referring to *Figure 14.7*, the gain made up as follows:

$$\text{overall gain} = G_1 - L_1 - L_2 + G_2 - L_3 + G_3$$

There is an additional requirement that the loop gain must be ss than unity at all frequencies to prevent oscillation; i.e.:

$$C > (G_1 - D_1) - L_1 - L_2 + G_2 - L_3 + (G_3 - D_2)$$

A band-pass filter can help to meet the requirement out-of-annel, but antenna isolation and pickup between cables has be carefully established in-band. Active deflectors are mmonly used by "self-help' schemes where the target is too nall for it to be viable as a site to be constructed by the oadcasters. The installation sometimes does not even use an nplifier, just high gain reception and transmission antennas ther side of a hill top to "bend' the signal into an isolated lley. In all cases the scheme must be officially licensed, to able a check that the re-radiated signal will not cause terference to a service elsewhere.

4.2 Design philosophy

transposer for a single service has a block diagram as shown *Figure 14.3*. For several services, costs can be reduced by oviding as much common equipment as possible. This has en made easier by the continued trend to solid-state and iniaturization. It is normal to use not only common buildings d electricity supplies, but also common antennas for recep-n with splitting filters used to separate the signal for each rvice and a combining unit to combine the outputs for a mmon transmitting antenna. A station block diagram is own in *Figure 14.8*.

As well as the obvious cost saving, a common broad-band ansmission antenna helps to ensure similar coverage for each rvice. In fact, it is common practice for a station output to be lit to drive an antenna that is in two halves. This arrangement means that if the two antennas are similar but diverge from the ideal impedance in the same way, the result of the reflections will tend to be dumped in the load of the output splitter. This provides a better match for the amplifier. It also means that if one half antenna develops a fault, the station can continue to transmit on the other half.

For the lower powers (2 W and 10 W) it is possible to fit several transposers into one rack, and then even dc power supplies and perhaps the common reference oscillator for the local oscillator synthesizers can be shared. Some redundancy is needed to ensure that a single fault does not result in total loss of output.

Design of the station will depend upon the circumstances of the operator. This includes factors such as the reliability of the service required, the accessibility of the sites, the availability and skill of maintenance personnel, the total number of equipments which are in the network, and the level of spares backup.

Where small, remote populations are to be served, programme outages of some hours duration may be acceptable. This would certainly not be true for a large population centre. In the latter case, improved reliability can be achieved by providing reserve equipment (either passive reserve with automatic changeover, or simultaneous parallel operation of two similar amplifiers). In fact, loss of public electricity supply is often the major reason for loss of transmission, especially in the more remote areas, and it may even be considered worthwhile to provide two incoming supply feeds or a standby generator on site.

Early transposers used a thermionic valve for the power amplifier. The programme output would be dependent on a single device which is liable to catastrophic failure. It is therefore normal in valved equipment to provide a duplicate transposer to take service automatically on loss of the main (see *Figure 14.9*). Alternatively, two identical amplifiers could be used in parallel to generate the required power. This has the advantage of avoiding the short loss of output during the warm-up and changeover sequence. On loss of one half the output,

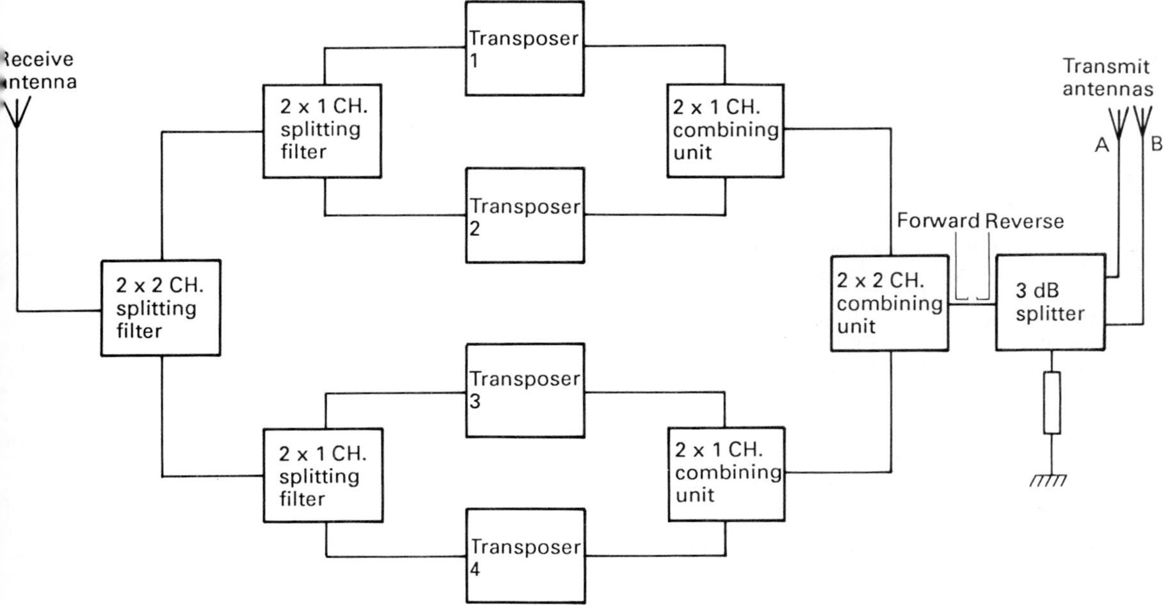

gure 14.8 Typical station block diagram

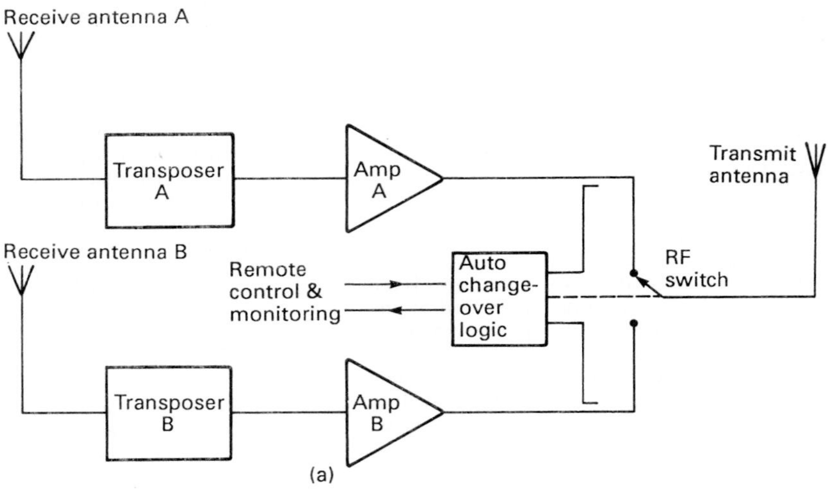

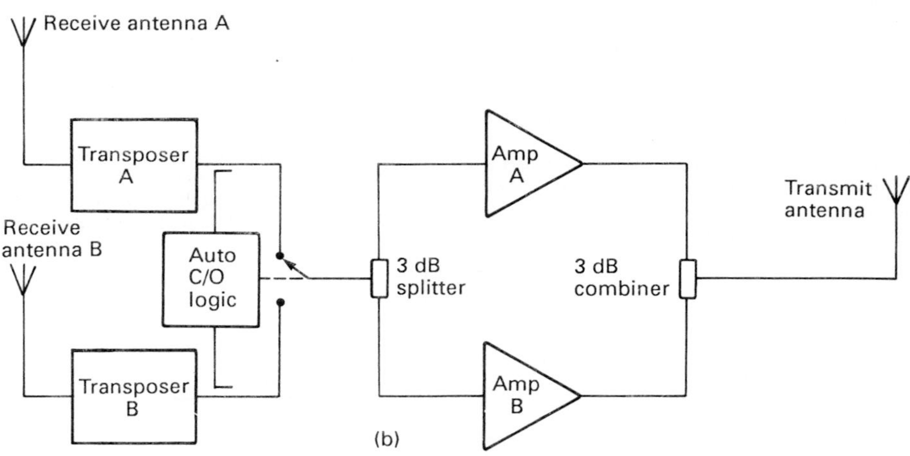

Figure 14.9 Provision of a duplicate transposer: (a) full passive reserve, (b) parallel or active reserve

power will reduce by 6 dB until the maintenance personnel can fix the fault or reconfigure the working half direct to antenna to be only 3 dB below normal.

Once standby equipment has been provided, it also becomes necessary to provide telemetry for remote monitoring. Otherwise the station will still be off the air before anyone realizes that something is amiss, and now both the main and the reserve equipment have to be repaired. Nothing is gained by the provision of a reserve without remote monitoring, except for extending the time between total station failures.

It is usual to draw up a simple algorithm along the lines shown in *Figure 14.10*, which can be used for determining whether telemetry and/or reserve equipment is necessary. Although there will always be exceptions (e.g. the station may justify a standby simply because it is often inaccessible in severe winters), most cases are simply analyzed in this way. A decision is required about providing telemetry at the parent station because if the dependent station is reported off-air, it could be because its parent has failed. If the parent is not too far away will be easier simply to visit both sites. If the parent is in remote location it would be better to provide telemetry remove the need for a possibly unnecessary visit.

Early uhf transposers used a thermionic valve even at the 10 W level, but *transistors* began to replace them in the early 1970s. Four BLX98 devices in parallel was one arrangement f 10 W, configured between couplers as shown in *Figure 14.1* This arrangement is still commonly used with other devices a at higher levels.

It allows a good return loss for input and output, and the amplifier is relatively independent of the parameters of particular device. It provides protection against a single failu in the signal path causing a total loss of output, and this can taken a step further by distributing the power supply in the w shown.

The advantages offered by transistors have meant that th have been used ever since at increasing power levels as suitab evices became available. The reasons for using them include:

reduced component stress and safety; high voltage power ıpplies are not required
long life; no expensive consumable valves every few thouınd hours
excellent programme reliability; loss of one or two amongst ıultiple parallel devices gives negligible loss of output power
good stability of performance; multiple devices between ›uplers mean that equipment performance is less critically ependent upon the characteristic of a single device
inherent good return loss of modules by combining and ›litting in 3 dB couplers

Since relay sites are usually unattended and frequently in remote locations, stability, reliability and ease of maintenance are particularly valuable features. With solid-state amplifiers, it has also become the norm to design them as plug-in broad-band modules, covering the whole of Bands 4 and 5 without on-site adjustment. Channel conscious filters, which are passive devices and therefore less likely to fail, are mounted separately in the rack.

The time spent on site to repair a fault is reduced to identifying the faulty module by means of inbuilt metering and then substituting a pre-tested broad-band spare. The actual repair of the fault is performed back at base.

An associated benefit is that the cost of building and maintaining the site itself is greatly reduced. The working area needed is much less, and because personnel will only be present

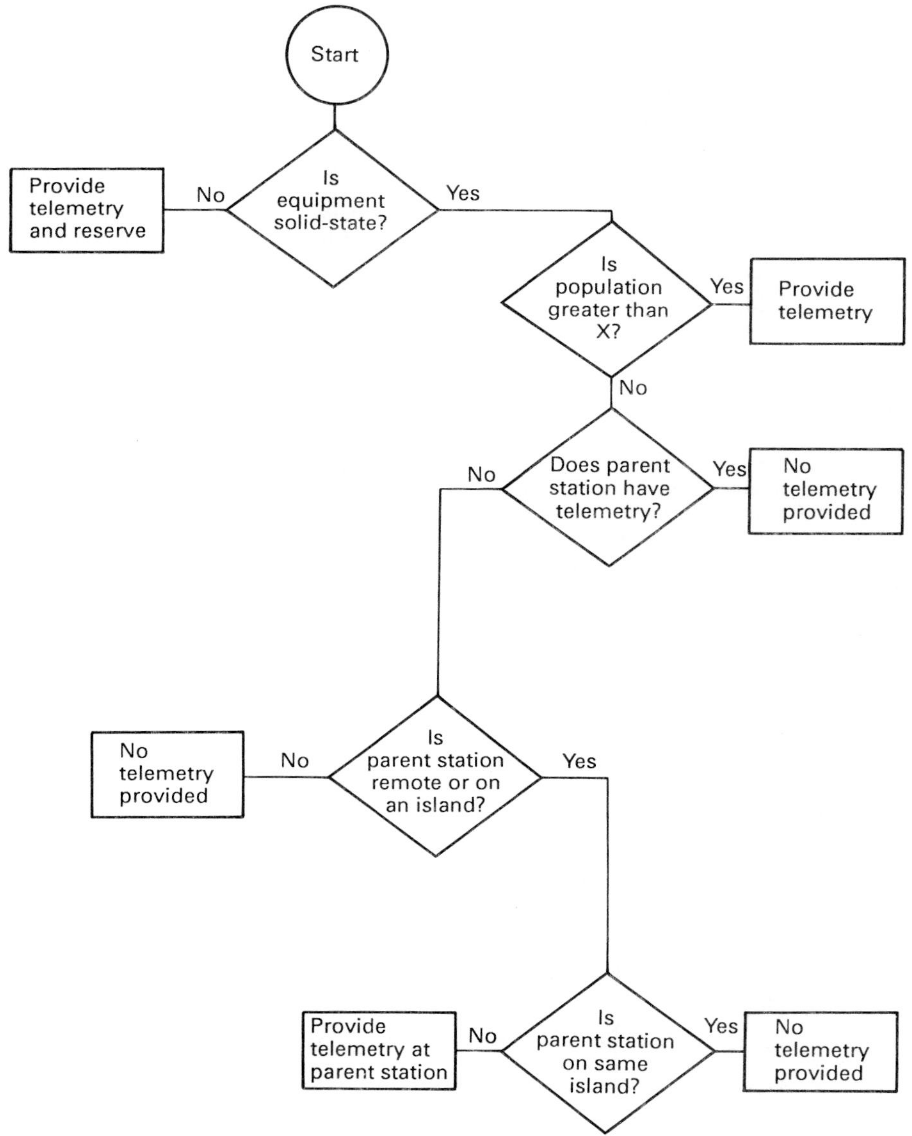

ȝure 14.10 Possible algorithm to decide need for telemetry

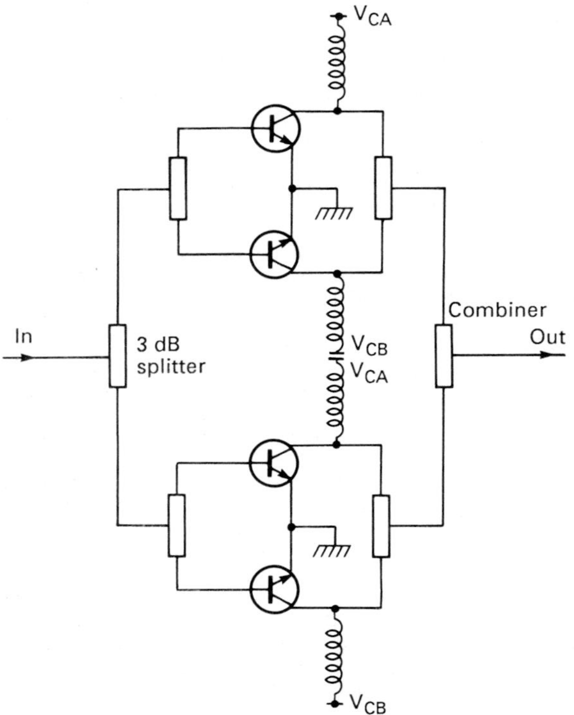

Figure 14.11 Typical broad-band amplifier configuration

for a short time, domestic facilities like kitchens and toilets are not necessary.

Such economies have allowed relays to be provided for communities far smaller than would otherwise have been possible. In the UK, where uhf coverage for the four services already exceeds 99.5 per cent, relays are still being constructed to serve targets as low as 200 people. Relays for so few people (and even for much larger groups) are in no way commercially viable in terms of any increased advertising revenue generated, since the percentage audience increase is so small. They are provided simply as a public service responsibility by t broadcaster.

This highlights why all aspects of the station design have to carefully considered to ensure that expenditure is reasonab while achieving the desired standard. Even so, the capital cc of a station to provide four uhf services to 200 people is mo than £100 per viewer.

14.3 Transposer performance

The technical characteristics required of a transposer w depend on its application. In countries where the broadcast may be a private business and only one or two transposers a required for the community, which is remote from the nex then the specification will be just sufficient to meet minimu requirements of the licensing authority. Tighter specificatio mean higher cost.

14.3.1 Frequency stability

In Europe, the high population density and the need f multiple services in many countries demands much mo stringent requirements for the transmission equipment. Th fundamental requirement is to minimize interference to ar from other stations. Apart from the actual frequency planni that this entails, it necessitates a high degree of frequen stability. To enable the maximum number of stations to b built, with a limited amount of spectrum, the same channels a used many times in different parts of the country. So that th geographical distance between re-use is as short as possible, us is made of the subjective improvement gained by controllir the relative frequency within small limits.

CCIR Recommendations (Document 11/1028-E, Dubrovni 1986) show that if the interfering frequency is random, then must be much lower in amplitude to be imperceptible tha when the two signals are related to each other in certain define ways. Then the interference can be at a higher amplitude for th same subjective effect. This allows more stations to be packe in.

The effect is due to the structure of the pattern that produced on the screen. As the relative frequency of th interfering signal increases, a cyclical change in the visibility the dot pattern occurs every 50 Hz. Superimposed on this

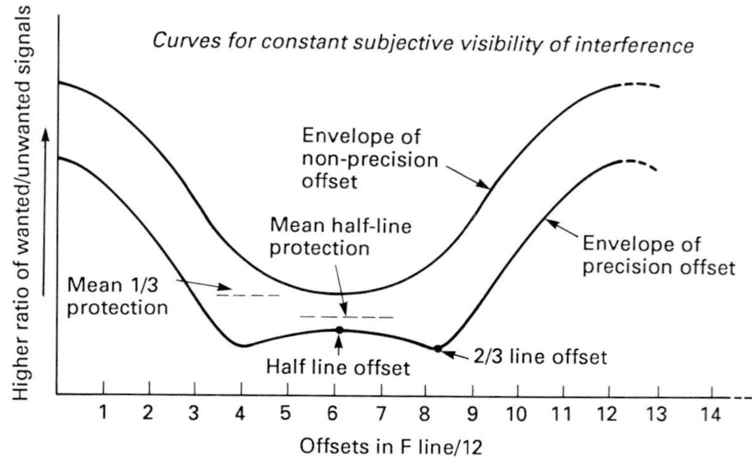

Figure 14.12 Change in protection ratio with different offsets. 25 Hz change in frequency difference moves operation from one envelope to the othe

another cyclical change based on line frequency. It will be seen from *Figure 14.12* that the optimum condition occurs when the frequency separation is precisely $^1/_3$, $^2/_3$, $^4/_3$, $^5/_3$, etc., of the line frequency. Since the frequency cannot (usually) be guaranteed to be within the one or two hertz necessary for the optimum condition, it has to be accepted that the protection will on average lie midway between the precision-best and precision-worst limits. This is nevertheless a distinct improvement compared with zero offset.In the UK it is normal to use this point on the curve (actually $^5/_3$ line offset), since it allows a group of three transmitters to be protected against each other. One operates at -$^5/_3$, one at zero offset and one at +$^5/_3$.

It should be noted that, unless true precision offset is in use, the mean $^1/_2$ line offset condition is better than the mean $^1/_3$ offset case. Therefore, exceptionally, in particularly troublesome cases of interference, $^1/_2$ line offset is used. It is not used normally because it only allows a pair of stations to be protected instead of three and would therefore reduce the total number of sites possible in a country.

In 625-line non-precision working with zero line offset, 61 dB protection is required for limit of perceptibility, whereas precision frequency control on $^5/_3$ line offset requires only 36 dB.

Extensive networks do not often use precision offset because of the complexity and expense of implementing it. Many transposers simply use an ovened crystal for each local oscillator, but even those that already use frequency synthesis from a low frequency reference crystal would also need to be modified so that they can be locked to a new reference of adequate stability.

An alternative to an on-site source would be to derive the standard from a central reference over-air. This might be:

- the 198 kHz BBC transmission
- obtained by demodulating the vision carrier to extract line frequency which has been derived from an atomic reference in the studio
- obtained by demodulating a reference burst which has been added to a spare line in the vertical blanking interval

14.3.2 Intermodulation

The other possible source of interference to others is the presence of unwanted frequency components in the output. They might be spurious signals unrelated to the vision or sound and caused by unexpected oscillation resulting from high gain and poor screening at frequencies well away from the intended output. This type of problem, once recognized, is easily cured.

Another form of unwanted output is due to the inherent distortion of the power amplifier. As described above, it has been normal to process both vision and sound in a common path, simply because of the difficulty of splitting them. In an ideal amplifier this would be no problem. In practice no system has a perfectly linear transfer characteristic, and the result is an output containing mixing products, principally between the three major input frequency components: vision carrier (f_V), sound carrier (f_S) and chrominance subcarrier (f_{SC}).

If the non-linear transfer characteristic is represented by a polynomial containing terms up to the third (i.e. $y = A + Bx + Cx^2$), then by substituting for x the input comprising the three carriers mentioned above, the expansion shows that, apart from the amplified wanted signals, there are:

second order terms which produce:

- the second harmonic of each input carrier (e.g. $2f_V$)
- sum and difference signals from the input carriers taken in pairs (e.g. f_V+f_S and f_V-f_S)

(All the above terms occur well away from the wanted frequencies.)

third order terms which produce:

- signals which appear at the same frequency as the input carriers and therefore change their characteristics (resulting in cross-modulation from one carrier to another)
- third harmonics of the inputs (e.g. $3f_V$)
- signals which fall inside and close to the channel (e.g. $2f_V-f_{SC}$)

It will be seen that the third order products are the most difficult to deal with, because many are actually within the wanted channel and so cannot be removed by filtering. In the UK system I, the dominant terms occur 1.57 MHz on either side of the vision carrier ($f_V \pm (f_S - f_{SC})$) and can produce an annoying pattern especially in areas of saturated red, since the chrominance and luminance levels are both high with this modulation. There is also a term that causes vision modulation to occur on the sound carrier and in severe cases can degrade sound demodulation.

Since filters will not help, one recourse is to reduce the power output of the amplifier so that it is operating on a more linear part of its characteristic and the effects are minimized. This technique is used in the smaller transposers of up to about 10 W output. It is feasible because semiconductors have been available for many years such that just four devices in class A will do the job. Thus despite the low efficiency, the design is simple and removal of the dissipated heat is not a problem.

With increasing power output requirements, it is no longer economic to underrun the amplifier, and a new technique was developed. The concept is simply to add a low level stage with the inverse transfer characteristic to the output amplifier, so that the overall characteristic becomes a straight line.

Effectively a linear system has been produced, which therefore has no distortion. Of course, it is not quite that easy, but nevertheless open-loop systems can easily achieve a long term improvement of at least 6 dB in the level of the *intermodulation products*, or alternatively an amplifier can be used at 3 dB more output power than before pre-correction. This technique is a cost effective way of producing more power, since a low level stage replaces the need for extra power amplifier stages — which become increasingly more difficult and expensive to add. It is not possible to just keep adding more devices in parallel, because heat disposal becomes difficult and the extra power generated is largely wasted in the longer cables and more numerous couplers. The law of diminishing returns applies.

A popular version of *linearity corrector* is shown in simple form in *Figure 14.13*. The input signal is split through two main paths. In the top path it is applied to amplifier B which is deliberately driven hard to produce distortion. In parallel it also passes through identical amplifier C, but is first attenuated (E) so that amplifier C does not distort. The attenuators E and F ensure that the gain of each path in the upper arm is identical so that when they are recombined, the original carriers cancel, leaving just the distortion. The distortion is then added to the main signal arriving through amplifier D. By adjusting the amplitude (A) and the phase (G) of the distortion, it can be made to largely cancel out the distortion occurring in the final amplifier. The main problem in designing such a pre-corrector is in choosing a low level amplifier which has as near possible the same characteristic as the high power output stage.

14.4 System performance

Although there are several parameters which must be controlled in a chain of transposers to ensure acceptable pictures at the output, a fundamental one is the *signal/noise ratio*. The

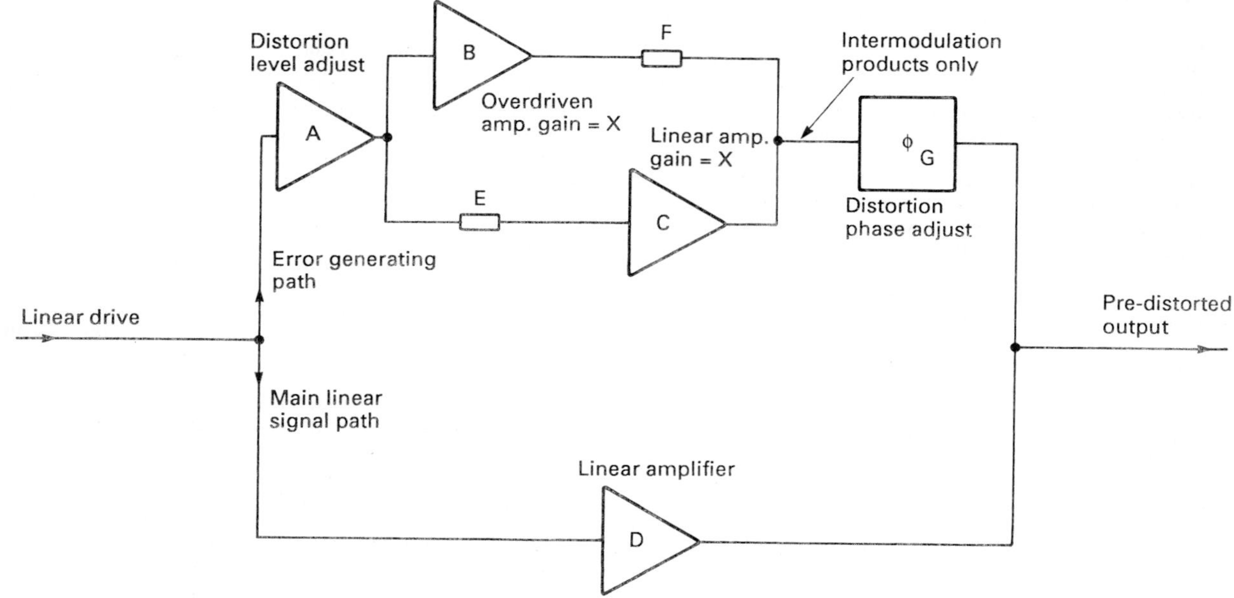

Figure 14.13 Linearity pre-corrector

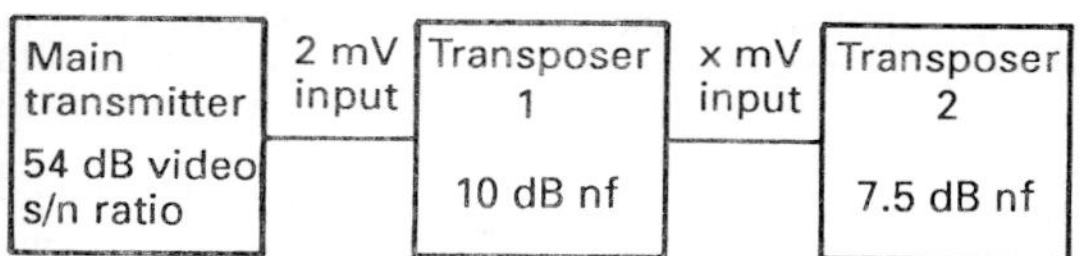

Figure 14.14 Transmitter and two transposers

following example of its calculation refers to *Figure 14.14*. It is of direct interest in establishing what minimum field strengths are acceptable, complexity of antenna arrays and consequently size and strength of the support structure.

Consider the chain shown in *Figure 14.14*. The main station has a video s/n ratio of 54 dB and the first transposer a noise figure of 10 dB and a terminated input signal of 2 mV. The second transposer has a noise figure of 7.5 dB, and it is desired to know what signal level is required to achieve an output video s/n ratio of 40 dB.

For system I, a signal level of 1 mV through a device having a noise figure of 8 dB results in a video s/n ratio of 44 dB. Different signal levels or noise figures change the resulting video noise proportionately. Therefore the video s/n resulting from the first transposer, with the values given, will be 48 dB if the input video had been noise free. As the output of the main transmitter only had 54 dB s/n, the combined effect is found by adding the noise powers:

$$10^{-a/10} = 10^{-b/10} + 10^{-c/10}$$

where a is the overall s/n dB, b is the parent s/n dB, and c is the transposer s/n dB.

This gives -47 dB s/n video as the output of the first transposer.

The above formula can now be used to establish what video s/n ratio the second transposer must produce on its own, if fed with clean video, so that when *actually* fed with -47 dB s/n video the result will be -40 dB. This turns out to be -41 dB. Since the transposer has a noise figure of 7.5 dB, an input signal level 3.5 dB below 1 mV, or 0.67 mV, will produce that result. In this way it is possible to design a receive antenna system for the second relay to provide a particular quality of output, if the available field strength is measured as part of an initial site test.

In a 50 ohm system, the terminated volts V_t available from a dipole in field strength E V/m is given by:

$$V_t = 0.13E\lambda$$

where λ is the wavelength.

The procedure described can be applied to any number of tandem relays to determine if the final output will be satisfactory.

Future trends in transposer design will almost certainly be towards achieving ever higher power with solid-state amplifiers. This in turn depends upon suitable devices becoming available. To a large extent, research and development in semiconductors follows from military projects where large budgets are available, rather than directly from requirements of broadcasters.

14.5 Future developments

Two factors are likely to influence the design of new equipment in the next few years. The first is the desire to achieve ever higher powers with transistors. The second is the introduction of dual carrier sound. The latter could be used for data transmission, but is more likely to be used to allow stereo sound or, in some countries, an alternative language.

14.5.1 Dual carrier sound

The second carrier will be added at 6.552 MHz above vision and at a level of -20 dB. It will be modulated by a four-phase qpsk digital signal with a bit rate of 728 kHz. The frequency of the carrier was selected to ensure maximum compatibility with the existing system, to minimize its visibility on the picture and to reduce its potential interference with the upper adjacent nnel. The level of the carrier, plus the associated reduction amplitude of the conventional sound carrier from -7 dB to dB was also chosen to minimize the generation of new ermodulation products.

The principal ip will occur at 0.552 MHz above vision and can entially cause an annoying low frequency pattern on the ture. It is to be noted that ips due to the traditional vision, our subcarrier and fm sounds are worst in areas of saturated , where the amplitudes are high. This condition arises atively infrequently and usually only in parts of the picture. the other hand the two sound carriers will always be present d so the new 0.552 MHz ip will appear all over the picture d become worse as the luminance level increases.

The information on the new carrier actually undergoes a udo-random scrambling process before modulation to prove its ruggedness, so even with no audio input the energy l be dispersed. This helps slightly in improving the subjective ibility of the pattern, but there is still a significant amount of ergy at the carrier frequency.

t is interesting to note that, because of the frequencies osen, the 0.552 MHz is actually offset from vision by close to multiple of one-third line frequency, which also helps to luce its visibility. Nevertheless the introduction of stereo und will be a new test on the linearity of all common plification transposers which have to handle it, and it may ult in more frequent maintenance visits to keep the equip- nt in adequate alignment.

.5.2 Separate amplification

e other advance is the introduction of separate amplification vision and sound carriers, together with the use of class AB. parate amplification has always been used in high power main stations to avoid intermodulation between vision and sound, but the video and audio signals are separately available, are separately modulated, and are therefore easily kept apart.

Until recently, it has not been practical to split vision and sound rf components when receiving them from the parent station off-air, hence the use of common amplification and the need for the linearity of class A. With the frequency responses that can be achieved with saw filters, the situation has changed. The complete signal is applied simultaneously to two such filters at if. The output from one is the vision component, while the other passes only the sound. Subsequently they can be amplified independently, avoiding intermodulation,

Although good linearity is always necessary for an amplitude modulated system, it is now less critical and class AB can be used. This means that the efficiency of the amplifier is increased and the all-important heat dissipation reduced, both in the rf amplifier and in the power supplies because of the lower average current drawn.

The availability of transistors rated at 50 W at 1 dB compression in class AB compares with the 8 W (uncorrected) available from devices used in common amplification class A. This change now enables 1 kW transposers to be solid-state, whereas for several years 200 W was the state of the art.

A 1 kW equipment could use 32 50 W transistors in the vision amplifier, allowing for worst case gains and output combining and filtering losses. This is the same number of devices as are now used to generate 200 W in class A. There would also be a sound amplifier, probably still in class A because of the need to handle two sound carriers, but the complete system would not be significantly more complicated to achieve a five-fold increase in output. It is therefore to be expected that its reliability would be similarly good.

P A Crozier-Cole
Head of Telemetry and Automation Section, IBA

15

Remote Supervision of Unattended Transmitters

many years, television transmitters have been designed for ttended operation. The move away from attended opera- began in the early 1960s when medium powered vhf TV nsposers were introduced at relay stations. At first, these ipments were quadruplicated to ensure adequate avail- ity of service, such were the doubts about this mode of ration. However, it soon became apparent that duplicated ve equipment in the programme chain gave ample con- ity of service. In fact, the control system for four sets of nsposers proved to be a liability in itself, and the experience ned with it underlined the need for accurate and consistent t detection coupled with changeover logic which needed to at least as reliable as the transposers themselves.

Vith the introduction of high power uhf transmitters for the -line colour services in the UK in the mid 1960s, unattended ration was an essential requirement, since no broadcaster ld have contemplated the number of operational staff who ght otherwise have been required to run a national network. The basic control requirement was the need to turn the nsmitter on and off each day. Although the life of thermionic ices is probably shortened by constant switching on and off, power consumption of large transmitters is not insignificant generally outweighs valve (tube) or klystron replacement ts. Once remote controls have been provided, there is a responding need to know whether they have taken effect. In eral, there will be a pair of indications for every control ction supplied. Thus the need for a two-way supervisory tem emerges: in an *outward* direction from a control centre remote transmitting stations, and in an *inward* direction rying indications and other information. It is then a straight- ward matter to handle much other supervisory information the form of alarms, *on/off* indications and analogue ntities.

Jsually, the control regime adopted is a combination of note and automatic control. *Automatic control* ranges from f-resetting overload protection circuits in the transmitter to re recently introduced feedback loops which adjust video -correction of the television waveform. *Remote control* can regarded as a back-up to automatic control. Ideally, a station eft in the *automatic mode*, so that if no contact with the tion is possible, it largely looks after itself, exercising the ernal switching options that seem most appropriate. At sent, all the functions can be expressed relatively simply, but in future it is quite possible that knowledge-based or 'expert' systems will be able to take the control techniques a step further.

It is normal practice to duplicate all items of active equipment in the programme path at a transmitting station, to ensure that no single fault interrupts the service. A broadcaster who may have started operations with single-ended equipment is proba- bly going to invest in a second transmitter and a changeover arrangement before considering a supervisory system. However, once a control system is in place, it is possible to cater for some double fault conditions, although of course the number of possible combinations making up double faults no doubt exceeds the capability of practical control logic, even software based, and it is not normally considered to be either necessary or worthwhile.

An automatic system can make simple changeover decisions, based on the inputs it receives from signal detectors along the programme path, and in the absence of outside intervention it will execute the changeover.

The operator in his distant control room may have more information available to him which leads him to override the automatic arrangements. He may wish to inhibit a changeover, reset a locked-out overload circuit or force a changeover that should have taken place. He needs to be able to switch the station to a *remote control* mode and then execute the function. The operator therefore needs clear, unambiguous alarms and indications to help him judge what is necessary, a robust remote signalling method that will send only the wanted controls and not be corrupted by data errors, and finally a set of indications to confirm to him that the control has taken effect.

It is important to bear in mind that nowadays it is very unusual to be able to monitor the off-air signals from all the TV stations supervised from one point. It will normally be possible to monitor the nearest or most powerful, which may well be co- sited with an area control room, but in general the main reliance is placed on telemetry from a whole group of stations.

The control room operator will probably have a wide variety of station types to contend with. Most networks contain equipment of different ages, and they will also have equipment ranging from high powered transmitters with pairs of klystrons to low powered solid-state relay stations. Much can be done to assist the operator in his task and the design of suitable control rooms is discussed in section *15.5*.

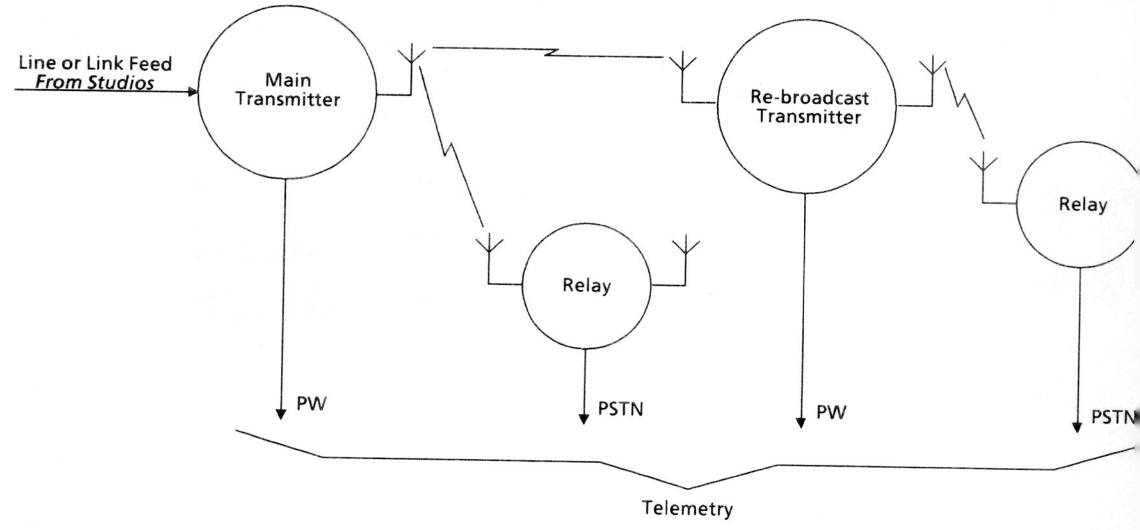

Figure 15.1 Representative transmission network

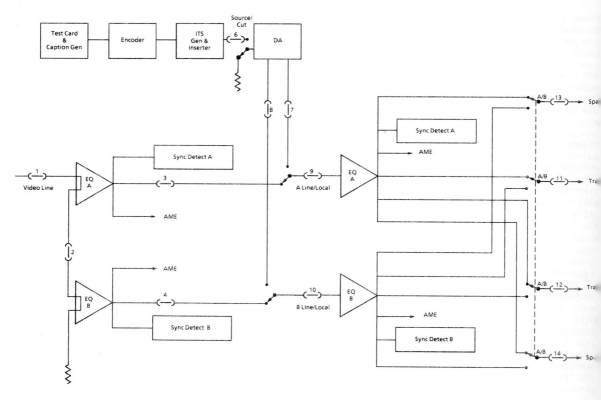

Figure 15.2 Programme injection point video PIE

15.1 TV transmitter network

A representative TV transmitter network is shown in *Figure 15.1*. A main transmitter is shown, fed by a line or microwave link from a studio or *Network Switching Centre* (NSC), which covers the main service area and acts as the parent or source a larger group of transmitters and relay stations.

Where a good unobstructed path exists, it is often m economic to feed secondary transmitters from high quality broadcast receivers, rather than rent further circuits fr telecommunications service providers. In practice, both methods are used, according to the circumstances.

It can be seen that some stations are crucial to the whole area served, while others may be of only relatively minor importance, serving a small population. The stations can therefore be arranged in a hierarchy according to factors such as:

- primary population coverage
- dependent population coverage
- percentage of area served
- position in a chain of stations

The level of remote control and supervision can then be matched to the operational needs for each type of station. A further factor to take into account is the accessibility of a station to a maintenance team. If travel from a maintenance base to a station is lengthy or difficult, telemetred information to a team manager can assist him in judging his priorities.

From the above factors, a broadcasting organization can decide what fault reporting times are tolerable, and hence decide how to structure the telemetry network. Crucial stations probably justify near instantaneous fault reporting, over *private wires* (PW) or dedicated radio links. Less important stations can be supervised over a *public switched telephone network* (PSTN), giving an access time of a few minutes, allowing time for retries if the first dialling attempt does not succeed. More recently, in the UK, the use of very low power narrow-band vhf or uhf unregulated channels has been authorized, giving an alternative low cost telemetry route over distances up to 40 km (25 miles).

15.2 Main station systems design

A representative transmitting station is described here including the scope of possible monitoring, supervisory and control systems.

Figure 15.2 shows the *programme input equipment* (PIE) for the vision signals. This connects an incoming programme feed to a pair of transmitters. Every active item of equipment in the programme path is duplicated, even though in many cases only a single line feed is available. In fact, the single line feed is probably backed up by reserve circuits and other duplicated amplifiers in the path.

Two basic switching functions are shown, a *line/local* changeover and an *A/B* changeover. In normal operation, the station would be line fed, but in the absence of incoming programme (detected from the presence of composite line and frame sync pulses), the logic governing the PIE switching selects the local side chain of signal sources. The side chain can produce a test card or a variety of apology captions to suit the circumstances of the interruption to programme.

Incoming signals will normally include the *insertion test signal* (ITS) in the vertical blanking interval, which is used for automation correction and monitoring purposes. An ITS generator is therefore needed in the local side chain when the incoming signal is absent. The A/B changeover merely ensures that a failure in one of the equalizers does not interrupt the service. The first pair of equalizing amplifiers makes up for cable losses between the incoming circuit termination and the PIE bay in another part of the building. The second pair of equalizers pre-corrects for the losses in the cables to the transmitters. In this way, it can be ensured that a jackfield at the heart of the PIE is at a flat equalized point in the system, and cross patching of sources and destinations for test purposes can be carried out with impunity.

A pair of transmitters is almost always used, either operating in parallel where sheer rf power or continuity of service is the paramount consideration, or operating as a main and reserve transmitter, often of a lower power, where the finite break in transmission can be tolerated. In either case, twin video and audio feeds are required from the PIE.

The corresponding arrangements for the sound signals are shown in *Figure 15.3*. In this case, two incoming sound lines are

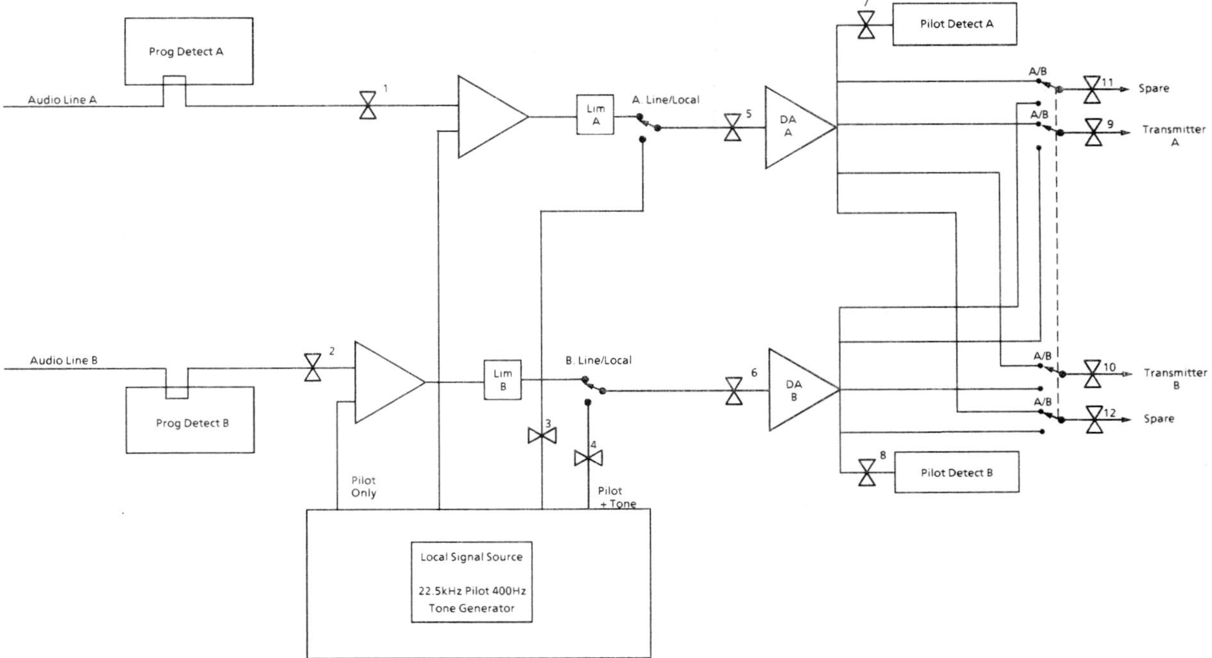

Figure 15.3 Programme injection point audio PIE

shown, and line failures are identified by *programme break detectors*. Simple forms of these can never be very satisfactory devices because their operation has to be an arbitrary compromise between sensitivity and operating time. They must not give spurious alarms during long silences but must respond to genuine line faults reasonably quickly. Unfortunately, it is not feasible to check line continuity with a super-audio pilot tone as in the rest of the system, because the frequency response of 'music' lines is deliberately rolled off above 10 kHz for other reasons.

Methods have been developed to compare samples of the audio signal at both ends of a circuit and to generate alarms if they differ. One method adds inaudible vlf tones to the sound signal, and another inserts data bits into the ITS. These are both relatively expensive solutions to a minor problem, and the need for them is being overtaken by the widening use of sound-in-sync distribution.

From the PIE onwards, it is useful to be able to check the continuity of the sound chain by pilot tone (22.5 kHz). This can be detected at the PIE output to operate the A/B changeover and also at the transmitter output to check that the modulators are operating.

Limiters are normally included in the PIE at the transmitting station to prevent over-deviation of the sound carrier if incoming levels are too high, although the law which such limiters should follow and their subjective effect is the subject of seemingly endless debate. In practice, they operate rarely in a well managed network and are only included as a safeguard.

The introduction of NICAM stereo sound to television has taken advantage of the development of dual channel sound-in-sync (dsis) which will displace conventional sound circuits. *Figure 15.4* shows an arrangement currently in use which fits between the video and audio schemes of *Figures 15.2* and *15.3*. The PIE now produces a third pair of outputs to the transmitter: the quadrature phased shift keyed (qpsk) 6.552 dual channel sound (DCS) subcarrier, again with its own A/B changeover. Signal processing in the DCS chain includes a NICAM reframer which enssures continuity of the bit stream to the modulator, even if there are asynchronous source switches farther back in the chain.

The mono signal for the original fm sound carrier is derived from the incoming dual channel signal, either as a summation of left and right if it is stereo, or from only one of the channels if it is a dual language service.

In an actual transmission network, there can be a considerable variety of programme feed arrangements. A line-fed station has been described above, but another major type is the re-broadcast station where the line equalizers and local side chain are simply replaced by a pair of re-broadcast link (RBL) receivers. Some stations are fed by privately owned microwave links, which from a system point of view resemble RBL stations. As network operation develops, opt-out facilities are being provided where minor stations are detached from the main network for local programmes for part of the day. In this case, a new line has to be provided, but the side chain is replaced by an RBL receiver as a reserve feed from the main network.

From an interfacing and supervisory point of view, the transmitter subsystems themselves differ considerably from one another in architectural detail. While the basic split is

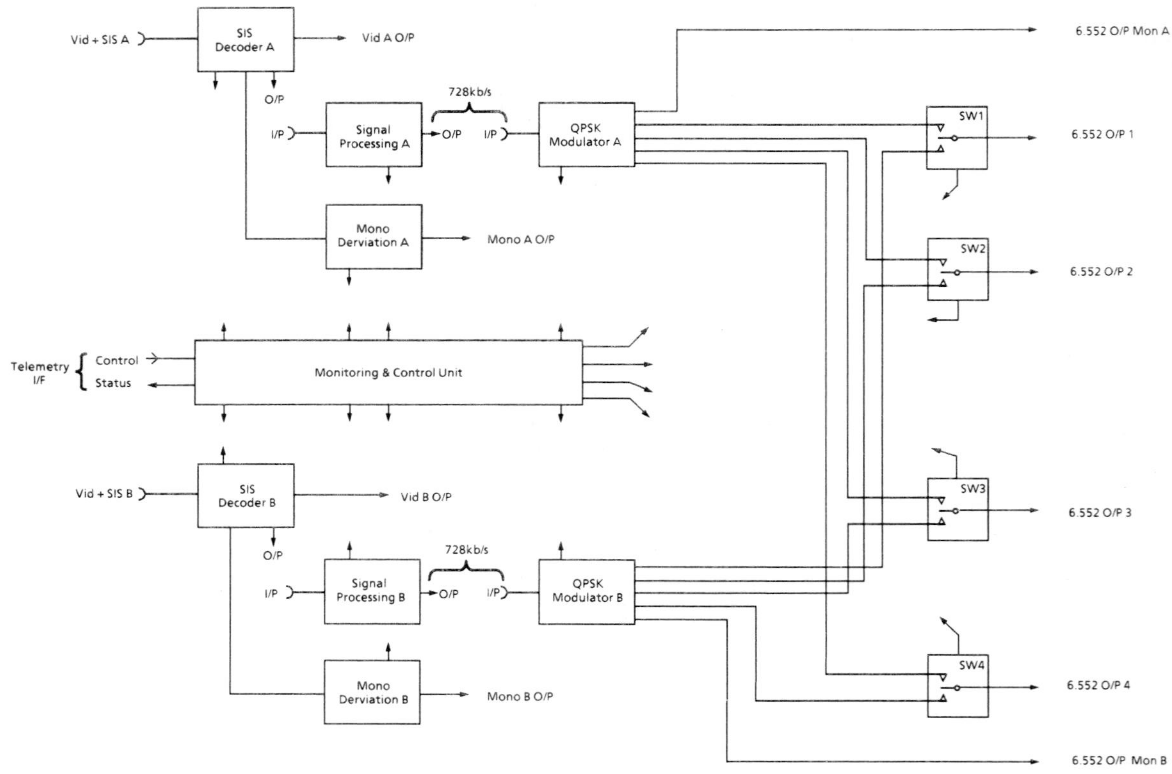

Figure 15.4 Dual channel sound PIE

between parallel or *main and reserve* operation, different manufacturers offer different drive, modulation and phasing arrangements, having different mimic diagrams and control options.

This all points to the need for a judicious balance between standardized interfaces and ample flexibility in a supervisory system, so that it is not too expensive and effort consuming to implement, but does not constrain the system design of the complete station.

Before considering interfacing in more detail, followed by telemetry and control rooms themselves, we need to look at automatic monitoring systems.

15.3 Automatic on-site monitoring

Reference has been made in section *15.2* to the need to monitor the output of transmitting stations, and that this can sometimes be done directly off-air at a control station. Monitoring can be subdivided into two categories: subjective monitoring of picture content and production standards, and the strictly technical performance of the transmission network.

Technical performance monitoring can be based largely on measuring the distortions of the parameters explored by the ITS waveform in the vertical blanking interval. The parameters measured include:

- sync pulse amplitude
- bar amplitude
- luminance non-linearity
- differential gain
- differential phase
- 2T pulse K
- chrominance/luminance gain inequality
- chrominance/luminance delay inequality
- lf error
- bar tilt
- chrominance/luminance crosstalk
- noise

Noise is in fact measured from a 'quiet' line in the vertical blanking interval reserved for the purpose. The working definitions of these measurements are given in section *15.3.1*.

ITS distortions can be measured by automatic monitoring equipment (AME) located at the station, sampling the signal at the line inputs, the transmitter inputs and the output of a high quality demodulator, itself switched to either transmitter output or the combined output of the station to the antenna feeder.

Figures 15.5 and *15.6* show two typical switching schemes for AMEs at RBL fed main and reserve transmitters and line fed parallel transmitters respectively.

An AME can be divided into two parts, the ITS analyzer which quantifies the measured distortions and presents the results in numerical form as percentages of bar amplitude or dB, etc., and a supervisor section. The latter can contain reference limits against which to compare sets as *caution* and *urgent* limits.

The *caution limits* would be set to correspond to maintenance limits, within which the network should normally be operated. In other words, when caution limits are passed, transmitter pre-corrections probably need to be re-adjusted on the next maintenance visit.

The *urgent* limits can be set to correspond to gross distor-

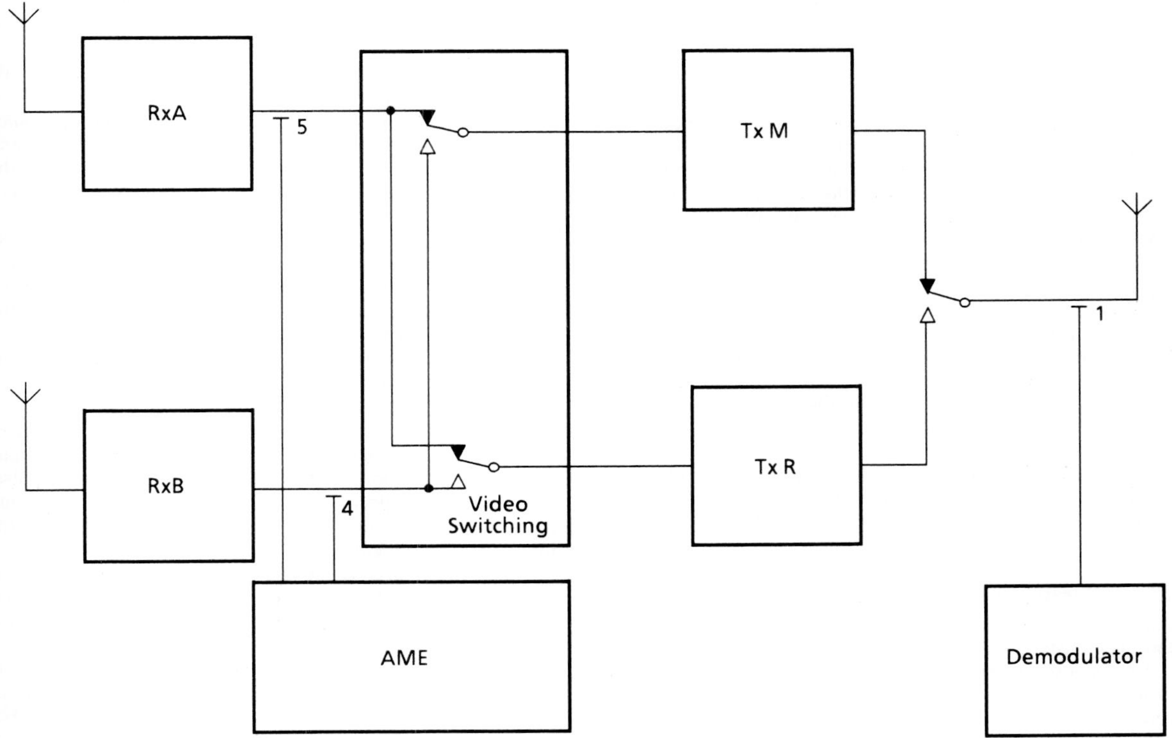

Figure 15.5 RBL fed main/reserve transmitters

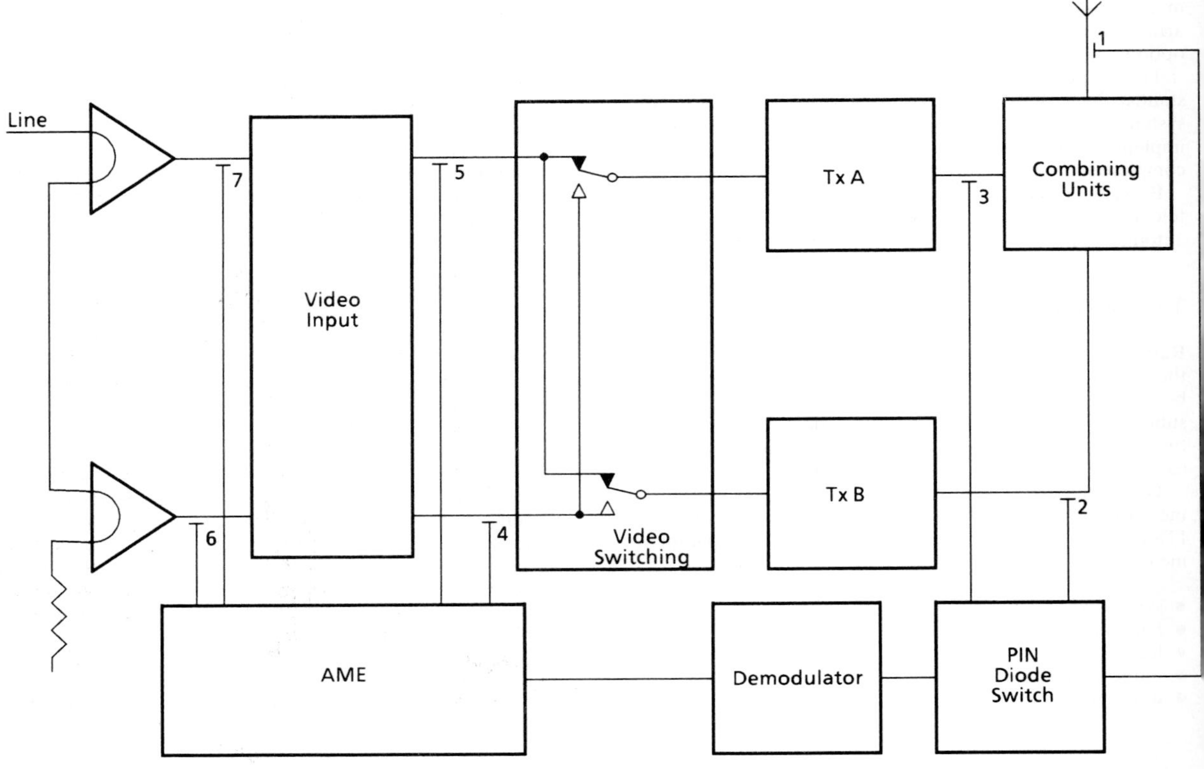

Figure 15.6 Line fed parallel tranmitters

tions, beyond which the offending item of equipment should not be in use. This might amount to switching off one of a parallel pair of transmitters or changing over the transmitters or RBL receivers in use. Thus the on-site AME gives the option of closing the control loop at the station itself without the need for outside intervention. Not all broadcasters are happy with this degree of on-site automation however, and prefer to use the AME as an indicator rather than an executive tool.

Before switching decisions can be taken on site, account has to be taken of the status of the station, i.e. whether alternative equipment is available, whether the distant control room has enabled automatic switching, etc. *Figure 15.7* shows the way in which AME data can be processed before reaching an executive recommendation.

The AME supervisor also contains the scanning sequences of the ports from which samples are taken. A simple sequence normally samples each port briefly, but if limits are passed then more specific search routines are entered to localize the faulty equipment. This places a premium on the measuring time needed by the equipment, but which is perhaps less important if executive control is not required.

15.3.1 Parameter measurement definitions

Parameters measured for monitoring technical performance have been listed in section *15.3*.

Luminance bar amplitude is determined by measuring the difference between the level at the mid-point of the bar and a reference point at blanking level. The result is expressed as a percentage of the nominal value of 700 mV (i.e. 700 mV = 100 per cent). The position of the blanking level reference is determined by a clamp which operates for nominally 630 ns commencing approximately 8.4 μs after the half-amplitude point of the leading edge of each line synchronizing pulse.

Sync amplitude is determined by measuring the difference between the mid-point of the last broad pulse of each field and black level. The result is expressed as a percentage of the nominal sync amplitude of 300 mV (i.e. 300 mV = 100 per cent).

2T pulse bar ratio is determined by measuring the ratio of the 2T pulse with respect to bar amplitude and expressing this as a percentage. The peak amplitude of the 2T pulse is measured with reference to the same blanking level point as for the luminance bar measurement.

Chrominance/luminance gain is determined by measuring the positive peak amplitude of the chrominance bar and expressing this as a percentage of the mid-point of the luminance bar.

Chrominance/luminance delay is determined by measuring the time difference in nanoseconds between the luminance and chrominance components of the composite sine-squared pulse. This difference is positive if the symmetry axis of the chrominance component is delayed relative to the symmetry axis of the luminance component.

Luminance non-linearity is determined by measuring the largest and smallest step amplitudes of the staircase waveform signal and expressing the difference as a percentage of the largest.

Signal/noise ratio. The noise bandwidth is limited at both high and low frequencies by filters. The low frequency is limited by a first order network having its corner frequency at 100 kHz. The high frequency is limited by the CCIR (Recommendation 567) low-pass filter. The result is expressed in dB relative to 700 mV.

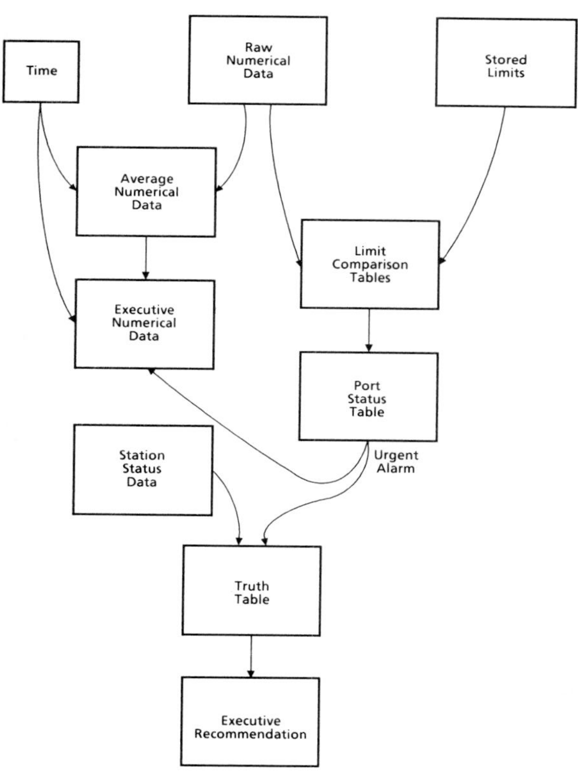

Figure 15.7 AME data processing

Chrominance/luminance crosstalk is determined by measuring the amplitudes of two points within the chrominance bar after suppression of the subcarrier and relating these to the amplitude of a third point after the chrominance bar.

LF error is determined by extracting the fluctuations of blanking level measured at a point 9 μs $\pm$300 ns after the half amplitude of the leading edge of each line sync pulse. The resulting waveform is frequency weighted. The peak-to-peak of the output from the weighting network is measured and the result expressed in millivolts.

Bar tilt is determined by measuring the difference between the level of the luminance bar 1 μs after the half-amplitude point of its leading edge and the level 9 μs after the half-amplitude point of its leading edge and expressing it as a percentage of the luminance bar amplitude.

2T pulse K rating. The base of the 2T pulse may be compared with an electronically generated mask which is similar to the time weighted oscilloscope graticule recommended by the CCIR for manual measurement on 625-line systems. The mask should be continuously variable between 0 and 10 per cent K. It should automatically adjust to the amplitude of the highest overshoot or undershoot before or after the 2T pulse within a time interval restricted to -0.8–-0.2 μs before the centre of the pulse and +0.2–+1.2 μs after the centre of the pulse.

Differential gain is expressed by two values $+x$ per cent and $-y$ per cent and is determined by evaluating the amplitude modulation of the colour subcarrier superimposed on the staircase. The value is determined by implementation of one of the following equations:

Either:

$$100 \quad \left(\frac{A_{max}}{A_o} - 1\right)\% \quad = x$$

or:

$$100 \quad \left(1 - \frac{A_{min}}{A_o}\right)\% \quad = y$$

The greater of x or y is displayed. A_o is the amplitude of the subcarrier on the black step.

Differential phase is expressed by two values $+x^o$ and $-y^o$ and is determined by evaluating the phase modulation of the colour subcarrier superimposed on the staircase.

15.4 Station controller concept

We can now draw together all parts of a complete transmitting station into a system which has to be controlled and then interfaced to the outside world for supervisory purposes (*Figure 15.8*).

The transmitters usually have their own controller, because automatic transmitter changeover is governed by fault detectors, protection circuits, timers and interlocks, best left to the transmitter manufacturer to offer as a working system. With processor based transmitter controllers, and a station controller that is also processor based, it is possible to have serial data interfacing to the transmitters, i.e. an RS 232 interface with the detailed functions being consigned to a software specification. This approach has greatly decreased installation time on site, once a basic system has first been 'de-bugged'.

Then there is the logic controlling the PIE itself. In some cases, this can be separate logic unaffected by failures elsewhere in the system, or being fairly trivial, it can be merged into the function of the station controller.

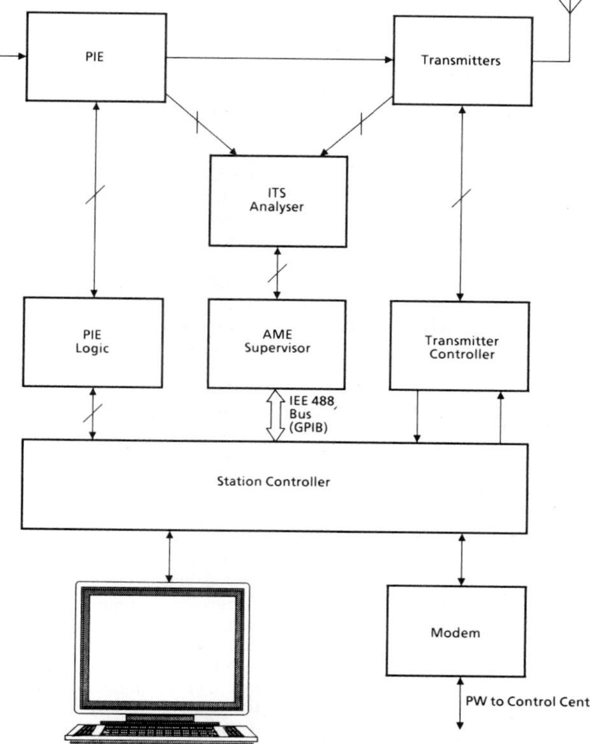

Figure 15.8 Transmitting station control system

Thirdly, the AME supervisor needs to work together with the station controller, either passing on numerical data for transmission outside the station or recommending executive controls based on limit transgressions. The *general purposes instrumentation bus* (GPIB), i.e. the IEC 652-1 or IEEE 488 bus, has been used for this interface, to handle the volume of numerical data, but the trend is to a serial data transfer at an appropriate rate.

To avoid the need for rigid hardware display and control panels, VDUs and keyboards are now used, fed by another Standard RS232 9600 baud serial data link. Finally, there is a serial port to a modem and away to the distant control centre by private line.

To provide the flexibility required in a station controller, it is important that the software should be effectively modularized. This means that each module has to perform a specific function quite independently of any other part of the program. It means that the interface between a module and the rest of the program must not be affected if the module itself is re-written. On the positive side, it means that modules can be documented and tested independently of the others, and fault finding and correction in operation becomes easier and more systematic. The functions of such modules will include telemetry, PIE control, station control and display interface.

The provision of serial links between major parts of the system allows the use of inexpensive fibre-optic connections to break up some of the earth loops which otherwise exist at a transmitting station and which in the past have compounded the damage done by lightning strikes (see section *16.3.8*). Similarly, direct logic connections from housekeeping functions such as door switches and dehydrator pressure switches are best made via opto-isolators.

Control no.	*Mnemonic*	*Parallel or Passive Reserve*	*Description*
1 2	TXA RESET TXB RESET	P and P R	Applies a lockout reset to the appropriate transmitter.
3	TXC START	P and P R	sets the SYSTEM START bit to initiate a system run-up, provided AUTO START is clear.
4	TXC STOP	P and P R	Clears the SYSTEM START bit to initiate a system rundown with immediate removal of ht provided AUTO START is clear.
5	TXC AUTO	P and P R	Sets the AUTO START bit so that SYSTEM START is affected by the VIDEO REC A/B bits
6	TXC NOAUTO	P and P R	Clears the AUTO START bit so that SYSTEM START is not affected by the VIDEO REC A/B bits.
7	TXC END	P and P R	Prematurely ends the Initial Fault delay regardless of the length of time it has run, assuming the delay has been started. Sets INIT DEL UP.
8 10	TXA STOP TXB STOP	P	Causes the appropriate transmitter to stop, removing ht immediately but remaining with HEATERS ON for a period of ten minutes. If the transmitter is already off, for some other reason, this control will prevent it from switching on. Sets the MANUAL STOP bit.
9 11	TXA NOSTOP TXB NOSTOP	P	Cancels controls 8 or 10 as appropriate, permitting SYSTEM START to control the Run-up and Rundown of the transmitter. Clears the MANUAL STOP bit.
12	TXC S/D	P	Enables the automatic shutdown function so that one transmitter may be shutdown under certain conditions. Sets the AUTO S/D SET bit.
13	TXC NOS/D	P	Disables the automatic shutdown function so that no transmitter may be shut down as a result of a VIDEO COMP fault. If a transmitter is already shut down in this manner, its shutdown bit will be allowed to run up. Clears the AUTO S/D SET bit.
14	TXC C/O	P	Enables the automatic change-over function of the drives. Sets the AUTO C/O SET bit.
15	TXC NOC/C	P	Disables the automatic change-over function of the drives, Clears the AUTO C/O SET bit.
16 17	TXC PREFA TXC PREFB	P	Selects the preference of the appropriate drives.
18 19			
20 21			
22 24	TXA START TXB START	P R	Instructs the appropriate transmitter to run up to ht. If the transmitter is already on, this control will hold it on regardless of the state of SYSTEM START. Sets the MANUAL START bit.
23 25	TXA NOSTART TXB NOSTART	P R	Cancels controls 22 and 24 shutting down a transmitter unless of course it is receiving SYSTEM START. The transmitter being shut down will lose ht immediately and remain on BLACK HEAT assuming SYSTEM START is set, otherwise it will switch right off. Clears the MANUAL START bit.
26 27	TXA SELECT TXB SELECT	P R	Instructs the feeder switch to connect the appropriate transmitter to the antenna, regardless of whether or not it is switched on and whether or not SYSTEM START is set.
28	TXC C/O	P R	Sets the AUTO C/O Sets the AUTO C/O SET bit and enables the automatic change-over function, permitting an automatic change-over under certain fault conditions.
29	TXC NOC/O	P R	Clears the AUTO C/O SET bit and inhibits the system from carrying out an auromatic change-over.

Control no.	*Mnemonic*	*Parallel or Passive Reserve*	*Description*
30	TXC COND	P R	Sets the COND C/O SET bit to restrict the type of fault which will initiate an automatic change-over, assuming AUTO C/O SET is set. Only VISION CARR and/or SOUND CARR clear will initiate a change-over.
31	TXC NOCOND	P R	Clears the COND C/O SET bit and cancels control 30, permitting an automatic change-over to be initiated by VISION CARR, SOUND CARR, VIDEO COMP or AUDIO COMP, assuming AUTO C/O SET is set.
32	TXC AME	P R	Sets the AME ACTION bit and initiates a change-over subject to the presence of SYSTEM START, the expiry of INIT DEL UP, AUTO C/O SET being set and COND C/O SET clear. This initiation causes the Reserve transmitter to run up and only when this is READY FOR HT will the feeder switch be changed over. After change-over, the COND C/O SET bit will be set. If AME ACTION is already set, this control has no effect.
33	TXC NOAME	P R	Clears the AME ACTION bit. If the transmitters have already changed over, this control has no other effect. If the transmitters have not already changed over, this control cancels any change-over initiation and runs down the reserve transmitter which remains on BLACK HEAT assuming SYSTEM START is set.
34 35	TXA READ TXB READ	P and P R	Requests a print out of the status of the appropriate transmitter.
36	TXC READ	P and P R	Requests a print out of the status of the transmitter system.
37 38	TXA STATUS TXB STATUS	P and P R	Requests a print out of the status of the appropriate transmitter in a string of 8-bit data words.
39	TXC STATUS	P and P R	Requests a print out of the status of the transmitter system in a string of 8-bit data words.
40	I/O nn	P and P R	Requests a print out of the data at Port nn.
41	PRINT nnnn	P and P R	Requests a print out of the data in memory location nnnn.
42	PRINT nnnn oooo	P and P R	Requests a print out of the data in memory locations nnnn to oooo.
43	TEST RAM	P and P R	Checks the read/write memory and prints out the addresses of any faulty memory locations together with the data written into and read from the location.
44	TEST ROM	P and P R	Checks the read-only memory for correct data sum and prints out the expected and measured counts for each ROM chip.

Table 15.1 Transmitter remote controls

Indicator no.	*Indication*	*Description*
1	GO HOME	The transmitter or transmitter system is in a state to allow the operator to Go Home. It indicates that all important isolators are on and that the switching frame switches or links are in the appropriate positions.
2	COOLING ON	The full transmitter cooling is on, which is the first event after switch on and last event before switch off.
3	HEATERS ON	The full heater volts are applied to the klystron.
4	BLACK HEAT	Reduced heater volts are applied to the klystron.
5	READY FOR HT	The transmitter is ready for immediate application of ht, when focus current and full heater volts are present and the heater delay has expired. This indication should remain true even when ht is switched on.
6	HT ON	The ht is switched on.
7	VISION CARR	Vision carrier is present.
8	SOUND CARR	Sound carrier is present.
9	VIDEO COMP	Video modulation is present on the transmitter output or absent from both input and output.
10	AUDIO COMP	Audio pilot tone is present on the modulator output or absent from input and output.
11	LOCKOUT	The transmitter ht is locked out.
12	COOLING FIT	A group indication to give an alarm when water tanks are low in level, heat exchangers are venting steam etc. For the transmitter system this would include the test load and diplexer load system.
13	MANUAL STOP	The transmitter has received a remote manual stop control.
14	MANUAL START	The transmitter has received a remote manual start control.
15	FIRE	The transmitter heat sensor is indicating an excess termperature.
16	AUTO S/D SET	Automatic shutdown of one transmitter is enabled.
17	AUTO C/O SET	Automatic change-over of the drives (parallel) or transmitters (passive reserve) is enabled.
18	AUTO START	The SYSTEM START is affected by the presence of VIDEO REC A/B.
19	VIDEO REC A/B	The appropriate transmitter is detecting incoming field syncs.

Indicator no.	Indication	Description
20	SYSTEM START	The system is being instructed to run up.
21	INIT DEL UP	The initial delay which inhibits auto shutdown (parallel) or auto change-over (passive reserve) has expired.
22	VIS IN USE A/B	The appropriate vision drive is in use.
23	SND IN USE A/B	The appropriate sound drive is in use.
24	L.O. IN USE A/B	The appropriate local oscillator is in use.
25	PREF DRIVE A/B	Indicates which drives are preferred.
26		
27		
28	SHUTDOWN A/B	The appropriate transmitter is shut down due to a VIDEO COMP fault.
29	PHASE FLT V/S	The vision or sound transmitter phase error exceeds 20°.
30	PSU FAULT	One of the duplicate power supplies for the transmitter controller or for any other unit in the system, is faulty.
31	COND C/O SET	The conditional change-over is set, restricting the initiation of an automatic change-over to VISION CARR and/or SOUND CARR being clear, assuming AUTO C/O SET is set.
32	A/B TO ANTENNA	The appropriate transmitter is switched to antenna.
33	AME ACTION	The system has received a control to initiate an auto change-over resulting from remote automatic monitoring equipment.
34	A TO ANTENNA	Indicate which passive reserve transmitter was to antenna prior to the last automatic change-over, and its status immediately before the change-over switch is instructed to operate.
35	B TO ANTENNA	
36	HT ON	
37	VISION CARR	
38	SOUND CARR	
39	VIDEO COMP	
40	AUDIO COMP	
41	TX A&B TO AE	The feeder switching frame is routing both transmitters to aerial via the diplexer and splitter.
42	LOCAL	The appropriate transmitter has been switched to local control.
43	TX CONTR FLT	The controller watch-dog timer has expired, following a fault in the program execution.

Table 15.2 Transmitter remote indications

Tables *15.1* and *15.2* give an idea of the scope of remote controls, alarms and indications. They give a schedule of transmitter subsystem functions used on the UK Channel Four network. Table *15.3* is a similar list, in the form of GPIB nmemonics, for the AME subsystem.

15.4.1 Telemetry

Industrial telemetry has developed steadily over the past 30 years and conventionally uses frequency shift keyed (fsk) tones in the frequency pass-band of speech telephone circuits, i.e. in the band 300–3 000 Hz. Telemetry data can thus be transmitted over rented lines (PWs), public switched telephone network (PSTN) connections set up by automatic dialling and answering units, or equivalent radio telephony (RT) circuits. In practice, all three are used, depending upon geographical and economic factors.

Function	Mnemonic	Message format From AME	Message format From SC
Interaction Handshake Mnemonics			
ACKNOWLEDGE	IA----	A	A
NOT ACKNOWLEDGE	IN----	A	A
ILLEGAL PORT DATA ACCESS	IDP---	A	–
ILLEGAL COMMAND (AME LOCAL)	ICL---	A	–
POWER ON ACKNOWLEDGE	IP----	–	A
Commands Mnemonics			
RESTART MAIN PORT SEQUENCE	CRS---	A	A
DRIVE CHANGEOVER REPORT	CRD---	A	A
VIDEO CHANGEOVER REPORT	CRV---	A	A
INHIBIT TX EXEC REC	CETI--	A	A
RESET TX EXEC INHIBITION	CETR--	A	A
INHIBIT VIDEO EXEC REC	CEVI--	A	A
RESET VIDEO EXEC INHIBIT	CEVR--	A	A
INHIBIT ALARMS	CAI---	A	A
RESET ALARMS INHIBITION	CAR---	A	A
EXTERNALLY SWITCH TO PORT y	CSP-Oy*	A	A
SWITCH TO AUTO	CSA---	A	A
SET CLOCK	CC----	B	B
SET TIMED DATA COLLECTION			
(ALL PORTS)	CTA---	A	A
(PORT y)	CT-Oy*	A	A
CANCEL TIMED DATA COLLECTION	CTS---	B	B
GPIB LINE TEST	CL----	D	D
Station Status Mnemonics			
STATION STATUS	S-----	C	C
Executive Recommendation Mnemonics			
MAIN TO RES. TX CHANGEOVER	ETC---	H	A
TX A OFF	ETA---	H	A
TX B OFF	ETB---	H	A
DRIVE A TO B CHANGEOVER	EDC---	H	A
SWITCH TO VIDEO A	EVA---	H	A
SWITCH TO VIDEO B	EVB---	H	A
EXEC REC ACKNOWLEDGE	EA----	H	A
Alarms Mnemonics			
ALARMS	A-----	I	A

*Note: y = 1 to 8

Table 15.3 AME GPIB mnemonics

15.4.1.1 PW telemetry

For a network of prime main stations, rented private circuits are often used, and there is some economy in grouping them as a *multipoint* in which all outstations can hear messages from the master station and the master station can hear individual outstations. The outstations themselves are not intended to hear one another, although this can happen if the impedance matching at the line junctions is incorrect.

Most telemetry systems are governed by the master station which interrogates each station in turn, using its unique digital address, and looking for any change-of-state bits or characters. If there has been a change of state, then it will interrogate the outstation in full by demanding a series of reply words carrying the status details. Most systems will seek a full reply on a regular basis anyway, even if changes of state are not flagged, but the short interrogation contributes to a quick response to genuine alarms. In a typical system running at the relatively low data speed of 200 bauds, it will take 10 s to scan a network of 20 stations for change-of-state bits, while a full update from each station can be achieved every two minutes unless interrupted by higher priority tasks.

The most obvious higher priority task is the sending of controls. A typical system will send a control word, which will be protected by a parity check and rejected if wrong, have it echoed back correctly by the outstation and then follow with an 'executive' word. In the system referred to above, this can be carried out within half a second if there are no data faults. Much higher data rates are now reasonable, giving corresponding gains in system response time.

In the interrogation sequences, outstations are addressed one at a time, data transmission only takes place in one direction at a time (half duplex operation), and finite times are allowed for responses. If no response is forthcoming, there will be one or two more interrogations before a fault is flagged. Telemetry data differ in this respect from the type of data communications between computers or terminals in a network. In the latter, data exchange is normally intermittent, but with long strings of data corresponding to keyboard entry or the contents of one page of display on a VDU. This makes it hard to share the same circuits for telemetry and computer data, but if the two can eventually be reconciled, e.g. by using intelligent statistical multiplexers at high data rates, the advantages for operators such as broadcasters will be considerable.

15.4.1.2 PSTN telemetry

Sending telemetry data over the PSTN brings with it a number of unique problems, but the outstanding advantage is of course the relatively low cost of line connections which are required only intermittently and for very short periods.

A PSTN telemetry system is normally quiescent but would operate in the following circumstances:

- when the control centre initiates a call for up-dating or control purposes
- when an outstation makes a daily routine call, possibly triggered by receipt of a parent transmission
- when an outstation makes a genuine alarm call

Because PSTN lines may well be busy, it is permissible for unattended stations to make several call attempts before giving up, and at the control station the incoming lines can be arranged as a hunting group. This means that outstations have to dial only one number, and the PSTN exchange automatically passes incoming calls along the group, up to the limit of the number of lines allocated.

In PSTN telemetry it is important to attach the time of an event to the data at the outstation when it happens, so that the uncertainty introduced by call delays does not confuse the subsequent interpretation of a sequence of events. In fact, it is now the practice to provide a real time clock at all outstations for this purpose, which can be remotely synchronized with the control station from time to time.

In both PW and PSTN telemetry, it is also useful for the outstation to maintain a short log of timed events, so that if telemetry is lost for some reason and then restored, it is possible to reassemble the log held at the control station with events in the correct sequence.

15.5 Control centres

Complete networks can be operated and supervised from control or operations centres, which often form a communications hub within a broadcasting organization.

The functions of an operations centre may be categorized as supervision, control and communications.

15.5.1 Supervision

The most obvious aspect of supervision is the receipt, logging and display of information gathered by telemetry (sections *15.4.1*and *15.4.2*). Ideally, the job of a control centre will be a passive one, in which the satisfactory operation of an automated network is merely observed in action.

Secondly, off-air monitoring of convenient picture sources can be provided, and if full attention is paid to factors such as interference-free reception, comfortable viewing conditions, top quality picture monitors, correct viewing distances and background lighting at the correction colour temperature, then it will be possible to make expert judgements of subjective picture quality in terms of grades and studio defined criteria.

The third aspect of supervision is that of sound quality monitoring. Again, attention to detail such as the choice of studio quality loudspeakers, the absence of background disturbance, and the correct reverberation times in the monitoring room, makes it possible to give valid sound quality judgements if required.

Current trends are however going away from this regulatory approach to quality monitoring away from studios, and sound monitoring in particular is made very difficult by the bustle of other activities in a control room and the sheer number of sources to be monitored. If sound continuity is the usual prime consideration, then visual indicators, either PPM meters or bar graph displays within the associated picture, can be used. The introduction of dual channel and stereo, albeit of near cd quality, has exacerbated all these problems recently.

Teletext also sometimes demands monitoring. Again, a distinction can be made between pure 'technical monitoring' and subjective monitoring. Technically, it is feasible to provide instruments measuring data quality in terms of eye height, error rates, etc., but good looking data can of course be carrying wrong or no information at all, while fully satisfying the technical criteria. With the increasing importance of Subcarrier User Groups (SUGs) using teletext services, the monitoring requirement is an editorial one, checking that correctly identified text has been broadcast in the right time slot. At present, this can only be seen as an effort intensive activity, not lending itself to automatic methods.

15.5.2 Control

Although most networks are intended to operate automatically with the minimum of operator intervention, many occasions arise when the application of remote control can restore or improve the service, without the need to dispatch a maintenance team on an unscheduled visit. Unexplained or chance overload trips can be reset remotely, a drive transmitter giving out patterning can be switched out, and locally generated apologies can be selected according to the circumstances.

Remote controls need to be totally secure and reliable with no risk of corruption from small data errors, and indications are needed to show unambiguously that controls have taken effect. This is where it is necessary to have a two way supervisory circuit, rather than an outgoing one only, because controls can be echoed before execution and indications returned from stations outside the practicable off-air monitoring area.

The variety of detailed systems being controlled will be large, and not easy for the operator to remember in an urgent situation. One way to ease the work load is to provide comprehensive mimic diagrams of each system, which can be broken down into sub-pages showing just one aspect of the station, e.g. the vision PIE, and also showing the control options available. From well designed and presented mimics, using colour sparingly but meaningfully, the operator should be able to see quickly and easily what control is called for and what its effect will be.

The VDU and keyboard provided with the supervisory system can then be used for many other purposes, including:

- an index, listing all stations in alphabetical order with station numbers
- a list of stations not equipped with telemetry but recording their status as reported in other ways
- station details, carrying fixed information in a uniform format
- a scratch pad, for messages between operational shifts
- mobile maintenance team locations, either at base or in action on site
- displays of programme circuit performance
- categorized subjective picture faults
- diagnostic displays of the supervisory computer itself, perhaps as a dummy outstation, showing which side is in use, the state of the reserve processor and various peripherals
- preset controls which can be set up in advance
- schedules of which outstations are looked after by which maintenance bases
- AME limits for programme circuits and transmitter performance

Earlier control rooms were provided with hard wired display panels giving a mimic diagram of the network itself, with the status of each station represented by red, amber and green leds signifying:

green = at full power with no faults
amber = with a minor fault, including reduced power
red = off the air

This gives an operator or a supervisor an instant overview of the region under control. By adding touch contacts to each station area, very quick access to the appropriate VDU header page can be provided. A more up-to-date approach is to provide all the facilities in software, using the proprietary 'windowing' techniques which are now available in industrial systems.

In the design of a supervisory and control system, decisions have to be taken on the priority of the various tasks, given that an economically sized processor cannot provide an apparently instant response to all the demands made upon it. The priorities chosen in a typical network are:

- preset transmission controls
- keyboard controls
- change-of-state scanning of outstations

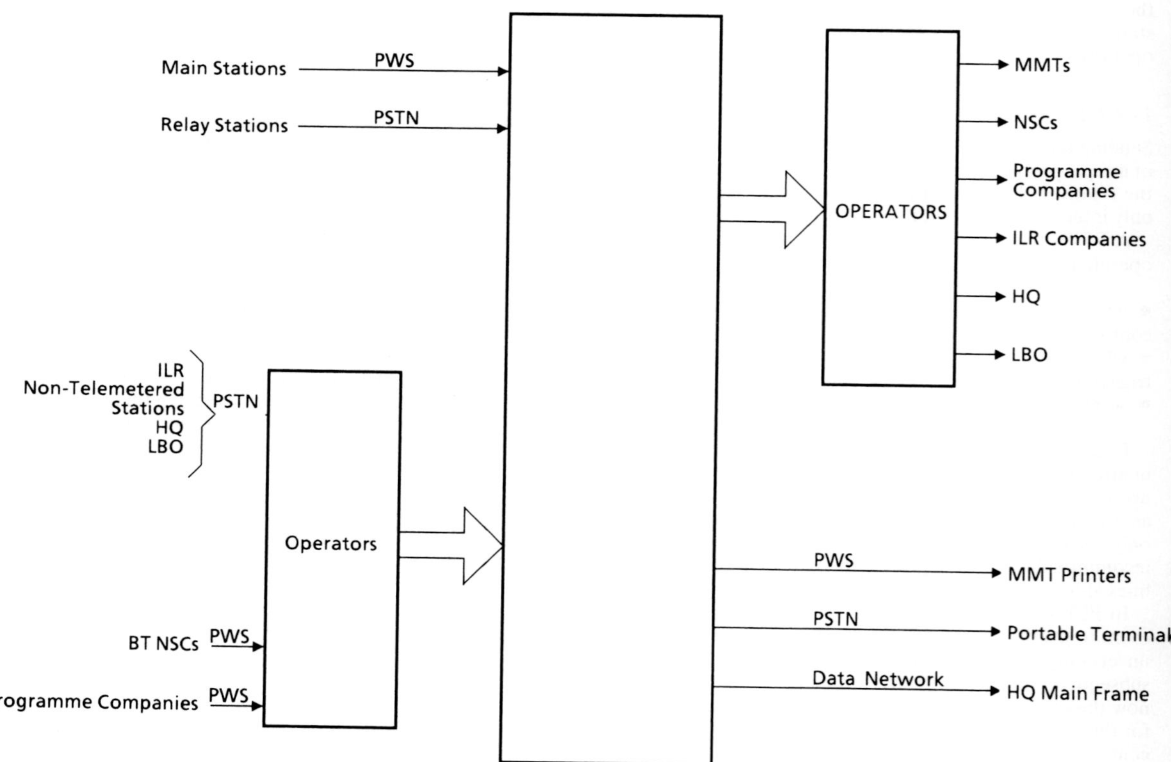

Figure 15.9 Operations centre information flow

- full scanning of particular stations as revealed by previous item
- AME sub-routines driven by the control centre
- routine full scanning of all outstations

15.5.3 Communications and the future

This is the final and possibly the most important function of a control centre. Having amassed a great deal of operational information, a broadcaster needs to think through what is to be done with it, who needs it, where it needs to be sent and ways in which the system may need to be changed as operational requirements change.

Figure 15.9 shows the flow of information in and out of a control centre. Much information will arrive by means other than by telemetry, needing to be sifted, prioritized and passed on. In the short term, reported faults need to be repaired by maintenance teams, who are therefore equipped with intelligent terminals which pick up the events at their own stations from a data network radiating from the centre.

Parties outside the organization may all need punctual and accurate reports of faults as they arise and the expected time to repair. It is therefore justifiable to use additional private circuits for near instantaneous speech communication to their corresponding operations rooms.

In the longer term, planning engineers may wish to examine the track record of particular parts of the system when specifying the next generation of equipment, or maintenance management may wish to compare the performance of similar equipment from different manufacturers. There is therefore a need to compile performance statistics from telemetered information. The current approach is to regard this as management information and to separate it from day-to-day operational information. However, there has to be an information interface between a supervisory system and the management information system of the complete organization.

In some organizations, data networks are now being put in place to encourage this integration of two systems, sharing it with inter-linked private telephone exchanges, office automation services and electronic mail. Such systems are likely to be in a constant state of evolution as the state of the art, and the users' perceptions of the possible, continue to leap-frog one another.

J P Whiting M Sc, C Eng, FIEE
Head of Power Systems, IBA

16 Transmitter Power System Equipment

16.1 Electricity supplies

16.1.1 Generation, transmission and distribution

The generation, transmission and distribution of electrical energy is undertaken internationally at different voltage levels, but a typical system is illustrated in *Figure 16.1*.

From the primary substations the high voltage system, usually between 10 and 13 kV, distributes electricity to the majority of larger consumer loads and also to the urban and rural secondary substations. At the secondary substations, it is transformed to between 380 V and 450 V three-phase and 200 V to 260 V single-phase for use by the remaining smaller consumer loads. Typical underground and overhead hv distribution systems are shown in *Figures 16.2* and *16.3*. Most transmitter stations operate from these voltage levels and will be fed from either underground or overhead systems, or a combination of both.

The cost of an underground supply system can be anything up to 20 times that of an overhead supply system, and therefore much of the transmission and rural distribution is by overhead line conductors.

16.1.2 Power system faults

A short-circuit fault on the system, either between phases or between phase and earth, will result in a fault current flowing to the fault many times greater than the load current. The equipment needs to be able to withstand the increased electromagnetic forces and thermal stresses caused by the fault current, particularly during the first few cycles.

For these reasons, it is necessary to calculate the maximum possible fault current in order to specify the equipment withstand rating. The most onerous fault condition will invariably be a three-phase fault, and this fault configuration conveniently reduces to a three-phase star connection. Consequently, circuit fault calculations can be treated as for a single-phase circuit.

The fault current, I_f, is given by:

$$I_f = \frac{V_b}{Z_t}$$

where Z_t is the total impedance from the source to the point of the fault, and V_b is the selected voltage, to which all impedances have to be referred. The phase fault current, I_f, has to be multiplied by 3 V_{ph} to obtain the MVA rating for a given phase voltage V_{ph}.

Usually, the impedance of the transformer is given as a percentage value (Z_p), based on its rated MVA, and this value has to be converted to its actual value (Z_a).

A worked example to illustrate the principles of fault calculations follows in section *16.1.2.1*. The topic is dealt with in greater detail in Reference 4.

16.1.2.1 Fault calculation

An electrical load is fed from an 11 000 V/440 V power transformer via two cables in parallel. Calculate the fault current in kiloamps at the load if:

1 the electricity supply fault level is 150 MVA
2 the power transformer rating is 1000 kVA and has a percentage resistance of 4.5 per cent
3 one cable impedance is 0.1 ohms resistance and 0.02 ohms reactance, the other is 0.056 ohms resistance and 0.014 ohms reactance

The procedure is to: (a) change the fault level of 150 MVA into a source impedance; (b) change the percentage reactance of the transformer into an ohmic reactance; (c) convert the two parallel impedances of the cables into a combined impedance. Finally, (d), add these three phase impedances together and divide into the phase voltage to obtain the phase fault current.

(a) The source fault rating (S_s) is 150 MVA, and its impedance, i.e. the impedance of the source, can be assumed to be reactive. It is:

$$S_s = 3V_pI_f$$

where $I_f = V_pZ_s$.

Substituting, $Z_s = j3\ V_p{}^2/S_s = j3\ (440/\surd 3)^2/150 = j0.0013$ ohms

(b) The transformer impedance is given in percentage terms, Z_{tp}, related to full load rating, S_t. This is usually the case and is expressed as follows:

$$Z_{tp} = I_f\ Z_t/V_{ph}$$

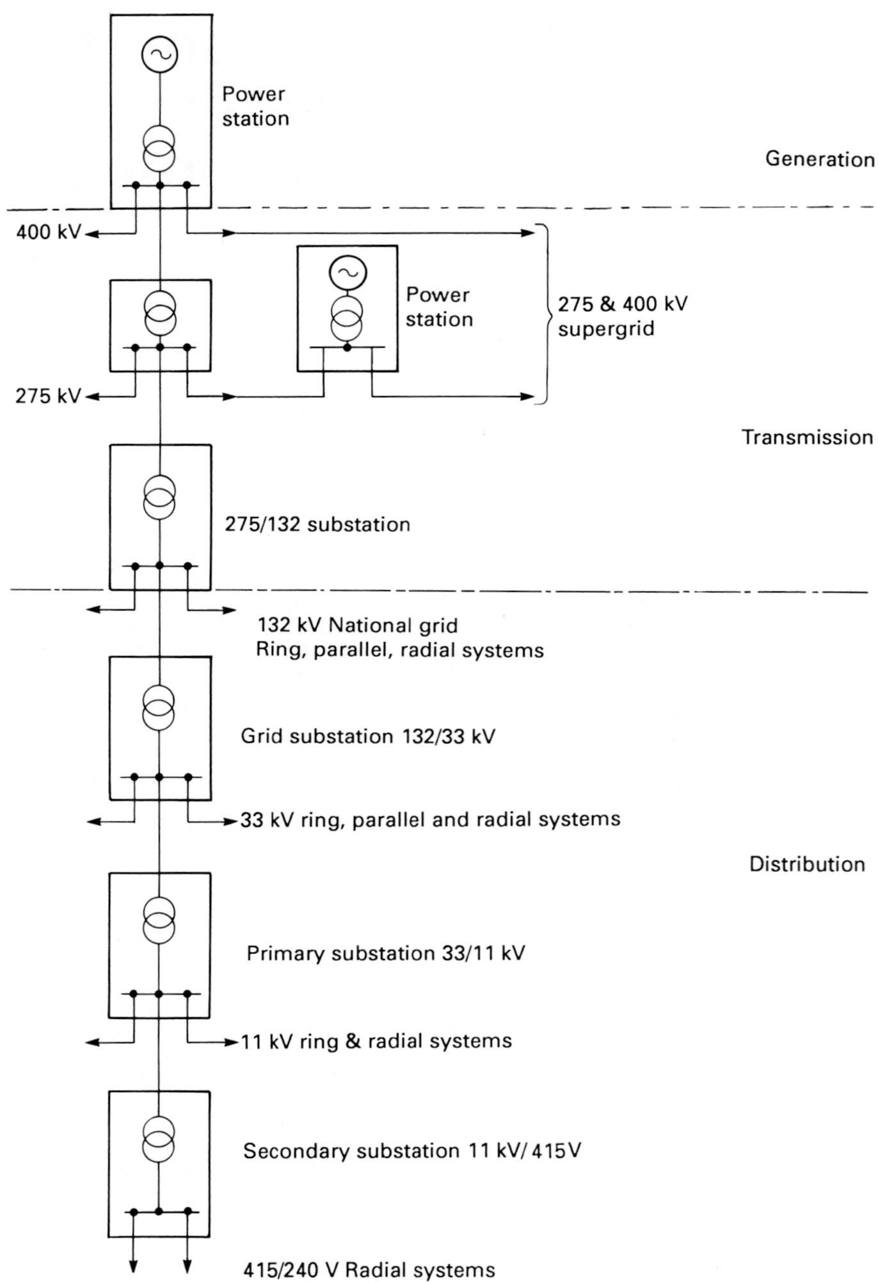

Figure 16.1 General configuration of the generation, transmission and distribution of electricity in the UK

$= j0.013 + j0.0087 + 0.036 + j0.0083$

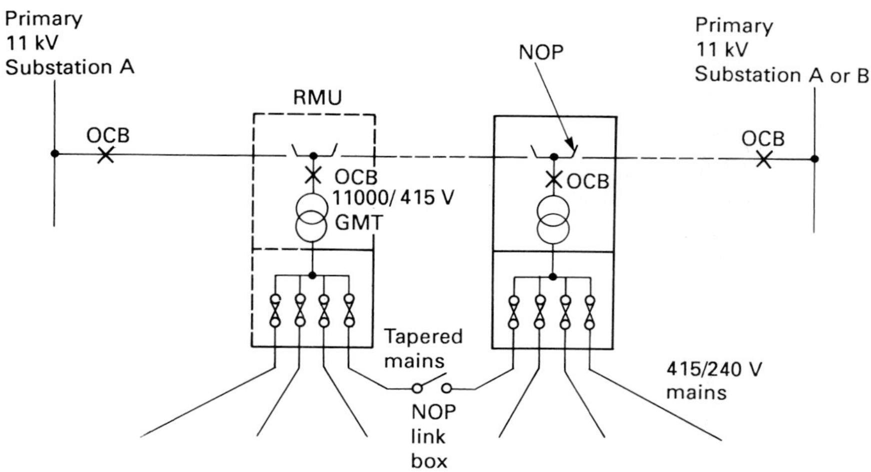

Figure 16.2 Typical 11 kV distribution underground system in the UK. OCB = oil circuit breaker; NOP = normally open point; RMU = ring main unit; GMT = ground mounted transformer

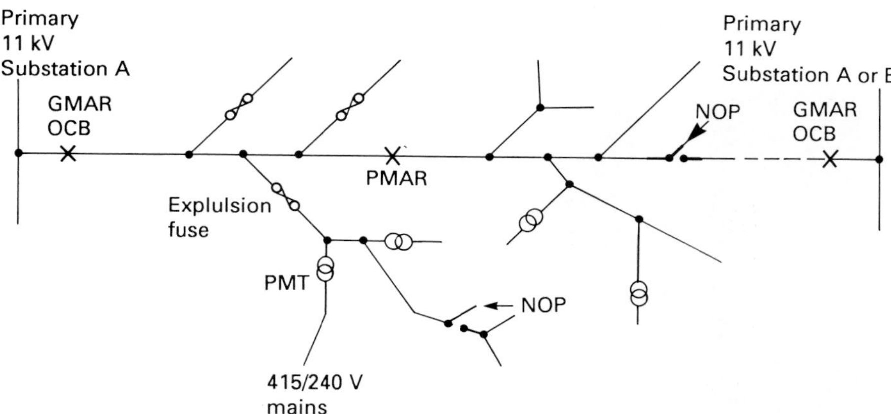

Figure 16.3 Typical 11 kV distribution overhead system in the UK. GMAR = ground mounted auto-reclose oil circuit-breaker; PMT = pole mounted transformer; PMAR = pole mounted auto-reclose circuit-breaker

Also $S_t = 3V_{ph} I_f$

Substituting, $Z_{tp} = S_t/3\ V_{ph} \times Z_t/V_{ph}$.

Therefore, $Z_t = 3\ V_{ph}{}^2/S_t = 0.045 \times 3 \times (440/\surd 3)^2/1 \times 10 = j0.0087$ ohms

(c) The cables in parallel have the following impedances:

$$0.1 + j0.02 = 0.102\ /\text{-}11.31°$$

and

$$0.056 + j0.014 = 0.058/\text{-}14.04°$$

Using the product/sum equation for parallel impedances, the combined impedances of the two cables in parallel are:

$$Z_c = \frac{0.102/-11.31° \times 0.058/-14.04°}{0.156 + j0.034} = 0.036 + j0.0083 \text{ ohms}$$

(d) The total phase impedance is therefore:

$$Z_{ph} = Z_s + Z_t + Z_c$$

$$= j0.036 + j0.0183 = 0.0404/\text{-}26.95°$$

$I_f = V_{ph}Z_{ph} = 440/\surd 3/0.0404/\text{-}26.95° = 6.67/\text{-}26.95°\text{kA}$

16.1.3 Supply reliability

In practice, the fault is located and isolated and other consumers restored by switching operations, leaving only those consumers directly connected to the faulty section off supplies until the fault is repaired. Restoration of supplies is therefore within 'switching times' for the majority of consumers.

Rural networks are fed by overhead systems and are therefore subjected to faults of a transient nature, such as lightning, ice and windborne objects. Often these are non-damage faults, and so the controlling circuit-breaker is arranged to reclose automatically. A typical sequence is shown in *Figure 16.4*. Transmitter supplies may be subject to such sequenced interruptions of supply.

Urban systems are less likely to fault because they are predominately underground, but when a fault does occur, for

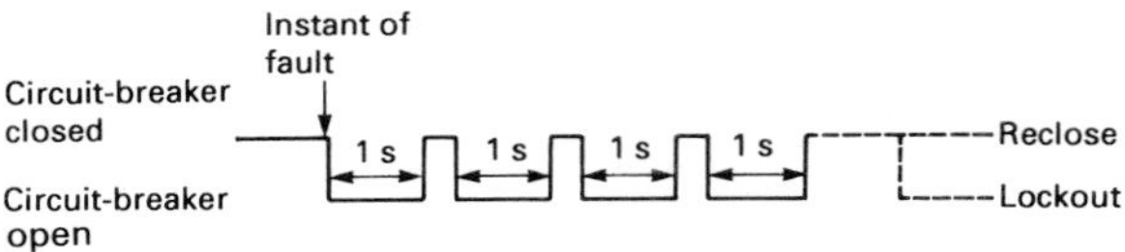

Figure 16.4 Pole mounted auto-reclose sequence

example, due to a mechanical digger or land subsidence, it is usually of a persistent nature and auto-reclose techniques are inappropriate.

So rural systems are likely to have more interruptions of short duration, while urban systems are likely to have fewer interruptions of longer duration.

Reliability can be designed into the supply circuit, but its improvement may not always be financially viable, unless the consumer is willing to contribute towards the cost.

16.1.4 Metering

The metering point of the supply is usually where the legal and financial responsibility changes from the Electricity Utility Board to the consumer. When the metering current transformers are located in the hv switchgear, the consumer owns, and therefore has to operate and maintain, the hv switchgear, the step-down power transformers and the interconnecting cabling.

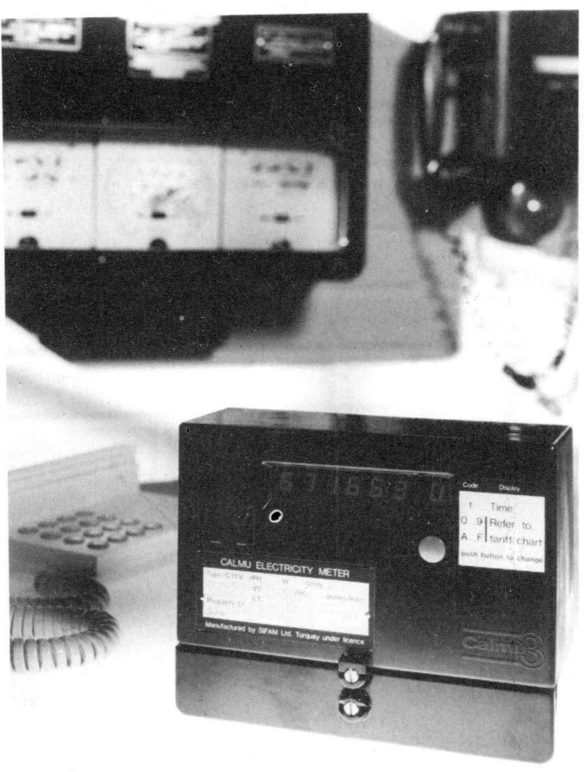

Figure 16.5 Three-phase solid-state meter (CALMU) with its former electromechanical equivalent in the background — note the reduction in size

In the case of small supplies, up to about 50 kVA, metering is connected directly into the mains cables. For larger supplies, the metering is operated from IEC 185 Class 1 (BS 3938) current transformers of appropriate ratio. If the supply is above 415 V, voltage transformers to IEC 186 (BS 3941)are used to reduce the metering potential to 110 V. (See also section *16.2.2.5*.)

The type of metering equipment required depends on the tariff structure. However, while a utility must charge for the energy taken, the charge must be in proportion to the current taken, because the equipment provided by the utility must be capable of supplying that current. A consumer with a poor power factor would take a larger current from the supply than a consumer with a good power factor.

Most meters are of the *induction* type. An induction meter reads the product of the current passing through the meter, the voltage applied to it, and the cosine of the angle between them, i.e. it measures power and is therefore a kWh meter. It is used extensively for metering purposes. However, while it is a relatively cheap instrument, it does not take the consumer power factor into consideration, and therefore some compensation has to be applied to remedy this limitation. The factor is given in appropriate tariffs. The same instrument reads reactive kVAr when a phase shift of 90° is introduced into the voltage circuit.

The *kVA meter* is the other instrument widely used. This is a combination of a kW meter and a kVAr meter, both acting on one disc.

Some tariffs are based on the kVA meter and some on the kW and kVAr meters. The tariff structure therefore needs to reflect this difference in metering policy.

The introduction of solid-state electronic technology to metering has completely revolutionized metering techniques. The systems offer higher accuracy and record any combination of kW, kVA, kVAr, kWh, kVAh, kVArh, maximum demand values and power factor. Moreover, the metering will cater for time of day spot pricing and will compute energy charges.

A significant feature is that the new systems are *interactive*, inasmuch as the meter can be controlled remotely, either by radio teleswitching or by mains borne signalling, to switch loads such as, for example, domestic heating and hot water loads. Intelligent metering can transmit the energy data via telephone links, radio or mains carrier to a remote centre for billing purposes and energy management statistics.

Further information on metering can be obtained from Reference 4.

16.1.5 Tariff structures

The cost of the electricity consumed at a transmitter station is a major revenue item.

Supplies are mostly charged on a maximum demand tariff. These vary in structure and in unit costs. A typical structure is shown in Table *16.1*.

16.2 Power equipment

16.2.1 Switchgear

Switchgear and its associated protection equipment is used to control and distribute electrical energy in a safe manner. The term *switchgear* includes circuit-breakers, switches, isolators, combination of switch and fuse units, busbars, protective relays and fuses.

These items are usually combined together with a busbar

Maximum demand tariff comprises:	*Determined by:*	*Possible variations*
Fixed charge	Tariff fixed value	None
Availability charge	Mutually agreed value	Should be kept as low as possible by keeping the figure under review
Maximum demand charge which invariably includes a power factor improvement clause	Metered value (power factor derived)	Should be kept low by improving load factor Power factor improved to 'best value
Units KWH	Metered	Advantage should be taken, if possible, of any differential rates, i.e. use of energy during off-peak periods
Off-peak units	Time metered	
Fuel adjustment clause	Tariff fixed	None

Table 16.1 Commercial/industrial tariff structure

system to form a switchboard. Switchboards which are assembled and tested at the manufacturer's works and delivered to site as composite units, are referred to as *factory built assemblies* (FBAs) and comply with IEC 439 (BS 5486).

Under this standard, provision is made for four classes of design (Forms 1–4) and cater for increasing standards of insulation and circuit segregation between incoming and outgoing circuits. For a variety of reasons, including transport to site, switchgear may have to be erected on site. These units are known as *custom built assemblies* (CBAs).

Figure 16.7 415 V auxiliaries board

Figure 16.6 11 kV oil circuit-breaker switchboard

16.2.1.1 Arc interruption

The action of opening an ac power circuit invariably produces an arc at the contact tips. Where the fault power factor is relatively high, i.e. between 0.8 and 1, as when interrupting load currents, arc interruption is not so difficult. However, under low power factor short-circuit fault conditions, the voltage across the contact gap at a current zero will be near its maximum and will therefore attempt to restrike the arc (see *Figure 16.8*).

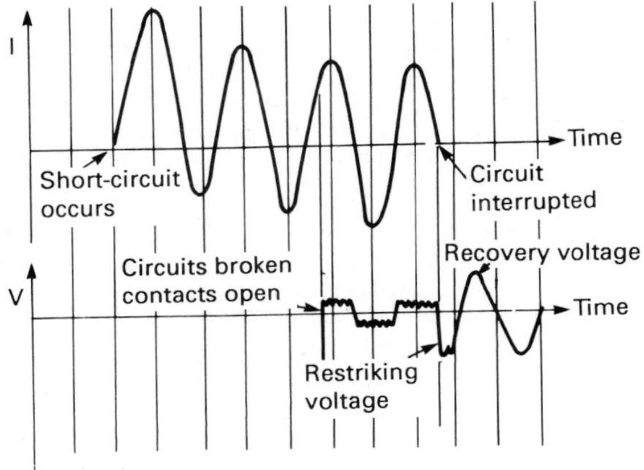

Figure 16.8 Circuit interruption

To interrupt the arc requires the highly ionized gaseous path between the opening contacts to be de-ionized. The dielectric strength between the contacts needs to increase sufficiently to be able to withstand the rising voltage, referred to as a *restriking voltage*, impressed across the gap. This is achieved by the use of an arc pot (see sections *16.2.1.3* and *16.2.1.4*).

16.2.1.2 Operating mechanism

Magnetic forces, proportional to the square of the current, produce mechanical stresses which are particularly high under fault conditions. *Figure 16.9* illustrates the direction of the forces when closing or opening a circuit-breaker. The arc is forced outwards, and the moving contacts are forced downwards, and this force assists the speed of the break or resists the closing action.

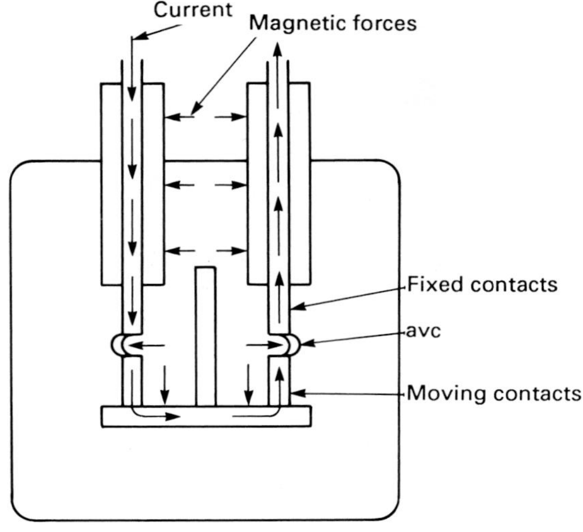

Figure 16.9 Electromagnetic forces in a circuit-breaker due to short-circuit currents

The fault closing capacity of the circuit-breaker needs to be higher than its breaking capacity, because it needs to be able to withstand the higher currents present during the first few cycles.

The operating mechanism provides the important role of closing and tripping the circuit-breaker both manually and automatically and under load and fault conditions. Closing is achieved using a large solenoid or spring-charged release coil, and tripping is afforded by a small shunt trip coil.

The closing mechanism must provide sufficient energy to accelerate it against friction, springs and electromagnetic forces. It must latch into the closed position without noticeable bounce on the contacts.

There are three categories of closing mechanisms, and the type selected depends upon many factors of which cost is perhaps the most important. They are solenoid operation, spring assisted and manual dependent.

Manual mechanisms, while still in service, are not now considered safe because the closing force is dependent on the operator. *Spring assisted mechanisms* are the most common and available as hand charged, hand wound or motor wound spring. *Solenoid mechanisms* are the most expensive. They require a high capacity battery to provide the necessary energy for closing. They have the advantage that they are immediately ready for a second reclosure if necessary, while a motor wound spring takes some seconds to recharge its springs.

The opening energy has to be instantly available and independent of the normal power supply, so the shunt trip coil is usually supplied from a battery source. If trip initiation occurs during the closing operation, the circuit-breaker must trip immediately.

16.2.1.3 HV oil circuit-breakers

The operational requirement of a circuit-breaker, as laid down in IEC 694 (BS 162, BS 6581), is that in addition to its normal rated close and trip duty, it should be able to make onto and, in conjunction with its protection, to break its rated fault current.

Most modern oil circuit-breakers are fitted with an *explosion pot*. The arc heat energy decomposes the oil to liberate a mixture of gases which exert pressure on the oil, and, by careful design of the arc pot, cool oil is forced across the arc path (see *Figure 16.10*).

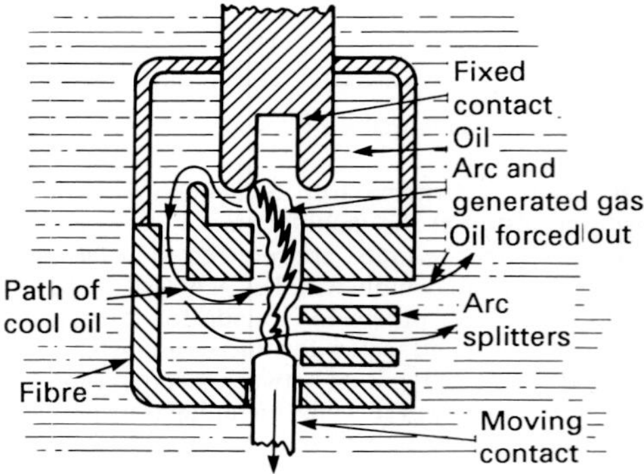

Figure 16.10 Explosion pot of a cross-jet oil circuit-breaker

16.2.1.4 LV air circuit-breakers

Air circuit-breakers should comply with IEC 157 (BS 4752). Arc interruption is achieved by extending the arc path across splitter plates (*Figure 16.11*). The electromagnetic effect of the current loop causes the arc to rise between the splitter plates into the arc shutes. The resistance of the extending arc brings the voltage across the contacts more into phase with the current and so assists in the arc interruption process.

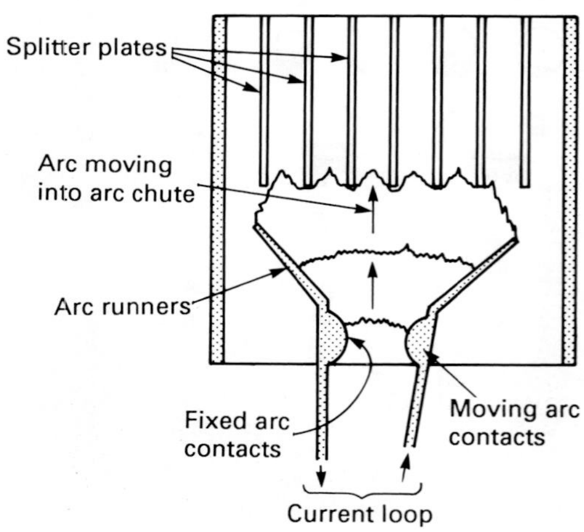

Figure 16.11 Chute-type air-break circuit-breaker

16.2.1.5 Miniature circuit-breakers and moulded case circuit-breakers

Miniature circuit-breakers (mcbs), and moulded case circuit-breakers (mccbs), should comply with IEC 292-1 (BS 4941), and IEC 158-1 (BS 5424) respectively. Both have a similar arc interruption process to that of the air circuit-breaker. On certain designs, circuit interruption within the first quarter of a cycle produces a current limiting or 'cut-off' effect similar to that exhibited by the hrc fuse (see section *16.2.2.3.3*).

16.2.1.6 Fuse-switches

The operational requirements for fuse-switches, as laid down in IEC 408 (BS 5419), are that, in addition to their normal rated opening and closing duty at a specified power factor, they should be able to make onto and withstand their rated fault current until their fuses interrupt the circuit.

16.2.1.7 Switches

In addition to their normal rated opening and closing duty at a specified power factor, the operational requirement for switches, as laid down in IEC 265 (BS 5463), is that they should be able to make onto and withstand their rated fault current until the fault current is cleared by the system protection.

16.2.1.8 Isolators

Isolators need only to carry their rated current, to open and close negligible, i.e. no-load, current and to carry their rated fault current for a specified duration. These requirements are laid down in IEC 265 (BS 5463).

16.2.2 Protection equipment

Protection is provided to detect a fault quickly and initiate rapid isolation of the fault to limit the energy 'let-through' so as to reduce the physical risk to personnel and restrict damage to the equipment. A secondary, but important, requirement of the protection equipment is that it should indicate or 'flag' its operation so that the location and the type of fault may be analysed.

16.2.2.1 Discrimination, sensitivity, stability and protection zone

The protection, shown in *Figure 16.12*, should be so adjusted that in the event of a fault occurring at F1, circuit-breaker CB1, the 'minor' circuit-breaker, should discriminate with CB2, the 'major' circuit-breaker, to isolate the fault so that the supplies to load A are unaffected. Therefore, the protection on CB1 has to be more sensitive than that on CB2, and the protection on CB2 needs to remain stable.The total operating time of a circuit-breaker comprises:

1 the protection operating time
2 the trip mechanism operating time
3 the arc interruption time

Each of these have tolerances that need to be catered for. The total operating time of the minor circuit-breaker (1 + 2 + 3) must be less than the protection operating time (1) of the major circuit-breaker. The difference between these, including the tolerances, is the *discriminating time* and is usually considered to be between 0.35 and 0.5 s.

The *zone* of protection for CB2 will be between CB2 and CB1.

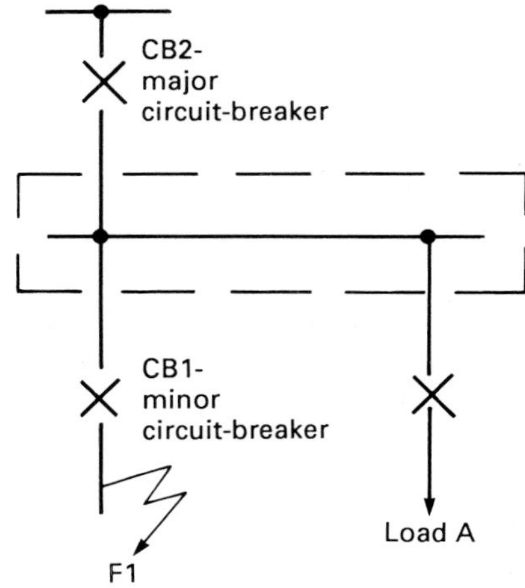

Figure 16.12 Protection discrimination, stability and sensitivity

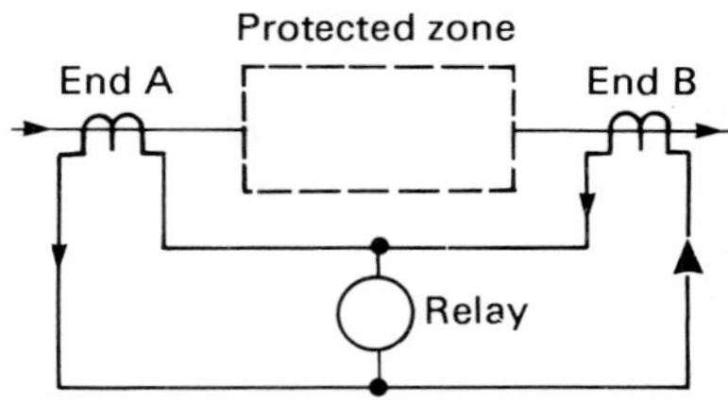

Figure 16.13 Circulating current system. Under through load or external fault conditions, the relay should not operate

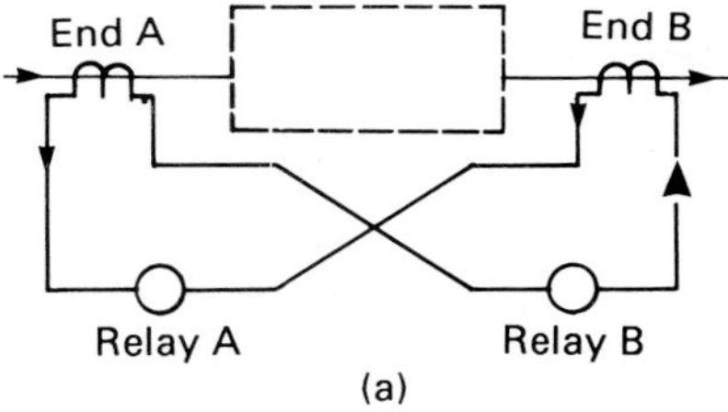

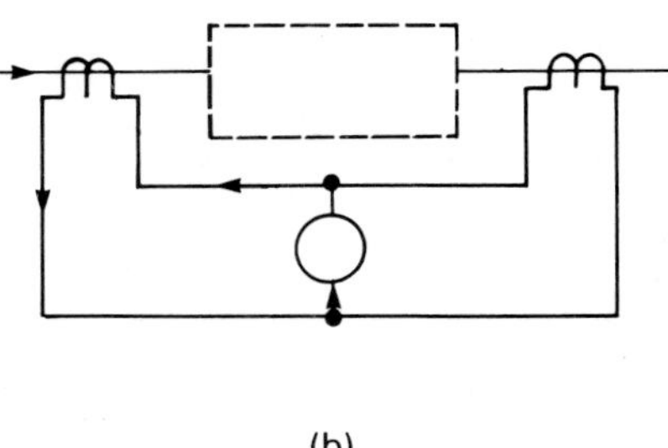

Figure 16.14 (a) balanced voltage unit protection system; (b) circulating current system. Under internal fault conditions, the relay should operate

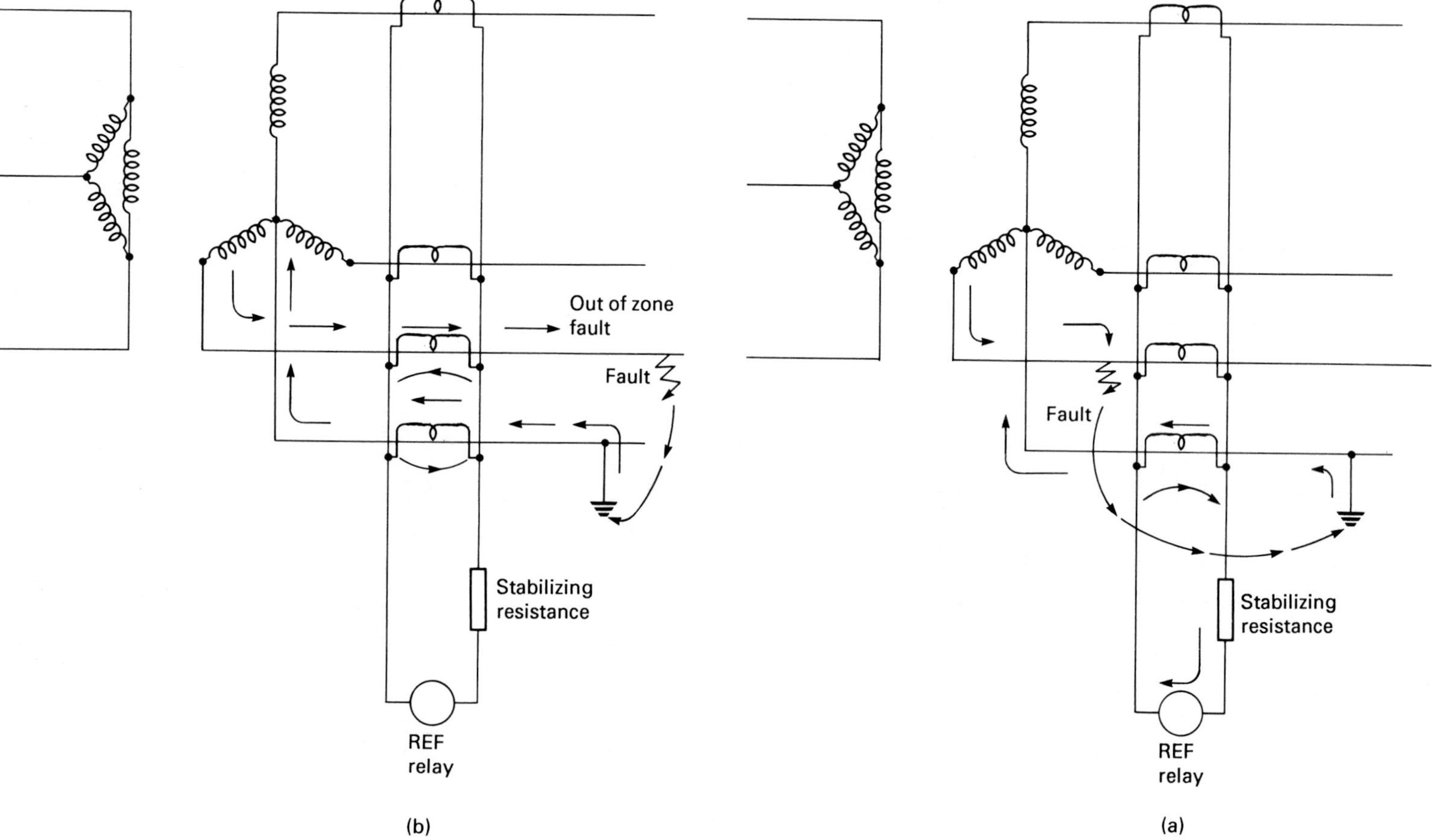

Figure 16.15 Restricted earth-fault protection applied to a four-wire system using four current transformers: (a) the system is stable for an external

Fault calculations are necessary to achieve these objectives (illustrated in section *16.1.2.1*) and for more complex networks a program to predict the relay settings has been written[1].

16.2.2.2 Unit protection

Unit protection (*Figure 16.13*) operates on the principle that current entering the protected zone must be equal to that leaving it. These two quantities are compared, and the difference is fed to a relay which is set to operate at a sensitive value.

Under normal load conditions, the difference amount will be small, but should a fault occur in the protected zone it will increase and the instantaneous relay will operate to initiate the isolation of that zone. Should a fault occur beyond the zone, then the two quantities, although greatly increased, should continue to balance and the protection remain stable. Unit protection is high speed in operation and has low sensitivity, but care has to be taken to ensure stability during through fault conditions.

This form of protection is expensive and therefore is used extensively only on the higher voltage systems. A second and similar unit protection system operates on the balanced voltage principle and is shown in *Figure 16.14*. At broadcast stations, they are occasionally used to protect the power transformers.

Restricted earth-fault protection is another unit form which balances current transformers rather than sets current transformers as in the systems described in section *16.2.2.2*. The REF relay (*Figure 16.15*) is an instantaneous attracted armature type, and the system illustrates the operation of the protection for a fault in-zone and out-of-zone.

16.2.2.3 Non-unit protection

Non-unit protection schemes include all forms of overcurrent and earth-fault protection and are used extensively. The characteristics of the more commonly used types only are described.

16.2.2.3.1 Inverse definite minimum time relay

The inverse definite minimum time (IDMT) relay has an induction disc upon which a torque is exerted by the interaction of fluxes produced by the relay operating current in such a manner as to provide the various relay characteristics, to IEC 255 (BS 142), as shown in *Figure 16.16*.

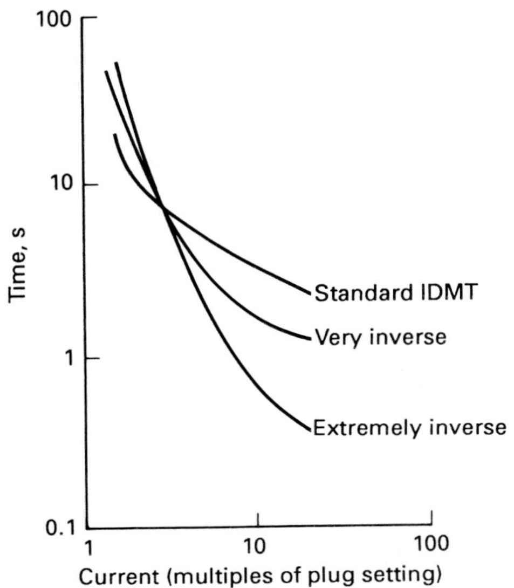

Figure 16.16 Overcurrent relay characteristics

The IDMT relay is provided with two adjustments. The *time setting multiple* (TSM) calibrated from 0 to 1, is a means to adjust the travel distance of the contact attached to the disc and therefore the operating time.

The other adjustment is the *plug setting* (PS), which provides seven steps of percentage current sensitivity settings at which the relay disc will start to rotate.

Variation of the PS setting has the effect of moving the characteristic horizontally on the current/time graph, and vertical adjustment is obtained by varying the TSM setting. It will be observed, therefore, that, within the range of its adjustments, a whole variety of relay characteristic positions may be achieved.

The relay characteristic most suitable for grading with fuses is the *extremely inverse* because it is similar to that of the fuse.

Figure 16.17 shows an IDMT relay grading with a fuse at a transmitter station supplied at 11 kV.

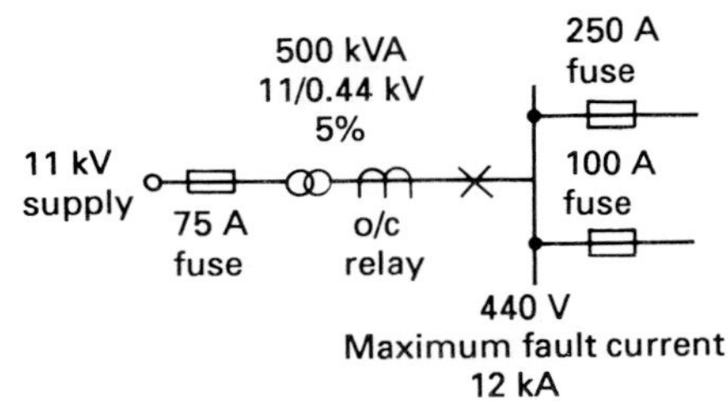

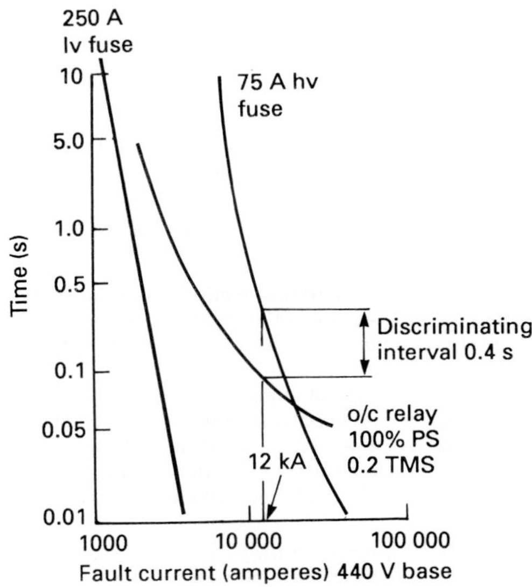

Figure 16.17 Protection grading at a typical transmitting station

In order to include the different curves on one graph, it is necessary to refer all the current values to a common base voltage which, in the example, is 440 V. The curve for the IDMT relay and the 75 A high-voltage fuse clearly shows a discriminating interval of 0.4 s at the maximum fault level of 12 kA. There is a generous margin between the 250 A lv fuse and the relay, and this is brought about by the effect of the fuse cut-off characteristic. Faults on the lv distribution system will be cleared by the appropriate circuit fuse leaving the other lv circuits at the lv switchboard unaffected.

16.2.2.3.2 Thermal and magnetic protection

Thermal and magnetic devices are used extensively because of their comparative low cost and simplicity.

A *thermal* type operates with current passing through a bimetallic element which is arranged to trigger a spring trip mechanism. The thermal device is inaccurate because its operating time is influenced by the heating effect of the load current passing through it prior to the fault and also by its ambient enclosure temperature. The more sophisticated designs incorporate compensation for these effects.

A *magnetic* type, which may be time delayed or instantaneous in operation, is dependent upon the attractive force exerted on a plunger in the magnetic field of a coil carrying the fault current. The time delay feature is achieved by the movement of a piston in a dashpot containing constant viscosity silicon fluid.

Both thermal and magnetic protection are integral parts of the moulded case and miniature circuit-breakers. The thermal device provides the protection on low values of overcurrent, and the magnetic device provides the fast acting short-circuit protection (*Figure 16.18*).

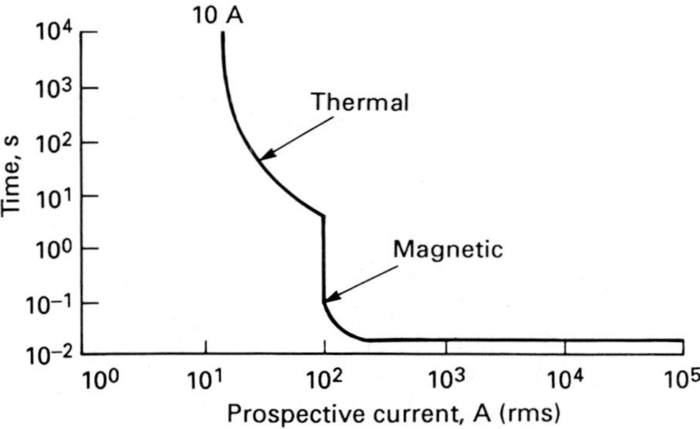

Figure 16.18 Time/current characteristics for Type 3 miniature circuit-breaker to BS 3871

The mcb overcurrent setting is fixed at some multiple integer, e.g. Type 3 is set between 4 and 7 times its rated current, and the withstand short-circuit rating varies up to a maximum of 16 kA. The mcb is a sealed unit and so 'tamperproof'.

The mccb has a range of adjustments for both the overload and the instantaneous overcurrent settings and has short-circuit withstand ratings of up to approximately 50 kA. Optional features include solenoid closing, remote tripping, interchangeable protection modules and plug-in circuit-breakers.

16.2.2.3.3 High rupturing cartridge fuses

The rewireable fuse has been superseded by the high rupturing capacity (hrc) fuse because it has a definable short-circuit interrupting rating and a non-deteriorating operating characteristic.

An hrc fuse, to IEC 439 (BS 88), comprises a ceramic body, containing specially designed fuse elements connected between the metal end caps. It is filled with pure granulated quartz.

The ratio of the minimum fusing current to its actual current rating is the *fusing factor* and has assigned values of P, Q1, Q2 or R, as listed in Table *16.2* and stipulated in IEC 439 (BS 88).

Class of fuse-links	*Fusing factor*	
	Exceeding	*Not exceeding*
P	1.00	1.25
Q1	1.25	1.5
Q2	1.5	1.75
R	1.75	2.5

Table 16.2 Fusing factors

It will be seen from *Figure 16.19* that the fault current is interrupted by the fuse before it reaches its peak value. This is referred to as the *cut-off* current and limits damage to the equipment by considerably restricting the thermal stresses and electromagnetic forces.

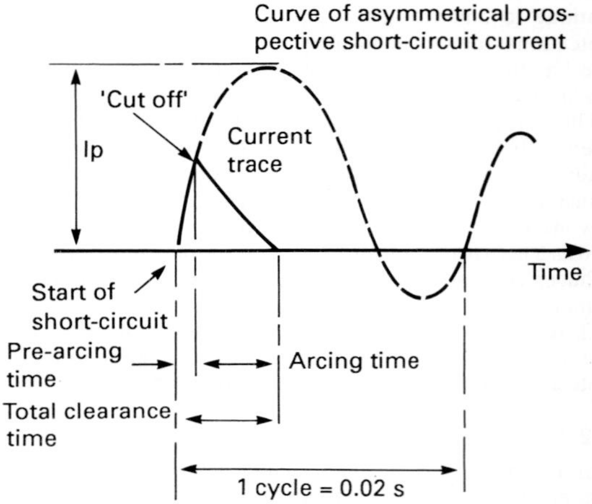

Figure 16.19 Cut-off feature of hrc fuses

Where the system fault level exceeds that of the rating of the equipment, the hrc fuse must have a cut-off current that is less than the withstand capacity of the equipment. The cut-off currents for a range of fuses are given in *Figure 16.20*. For example, a 60 A fuse subjected to a fault current of 50 kA rms would limit the cut-off current peak to approximately 7 kA.

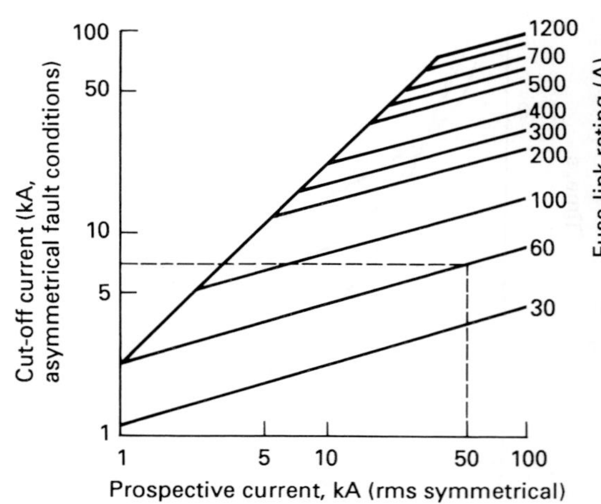

Figure 16.20 Cut-off current characteristics for hrc fuse links

For overcurrents involving fuse operating times less than 10 ms, it is necessary to base discrimination on the *let-through* energy which is expressed in terms of I^2t. The total I^2t value for the minor fuse should be less than the pre-arcing I^2t value for the major fuse. The I^2t values for a typical range of fuses are shown in *Figure 16.21*.

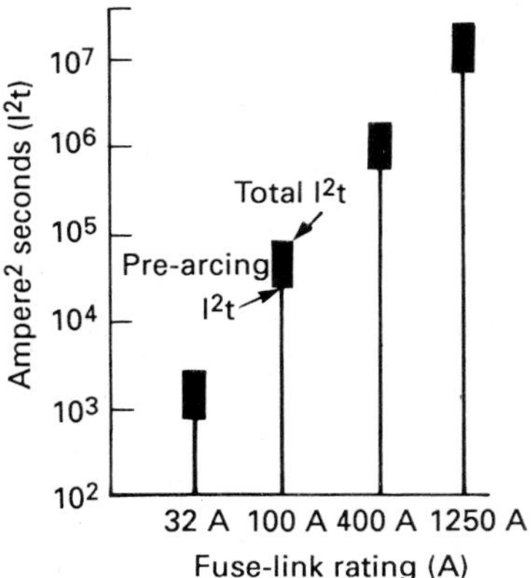

Figure 16.21 I^2t characteristics for hrc fuses

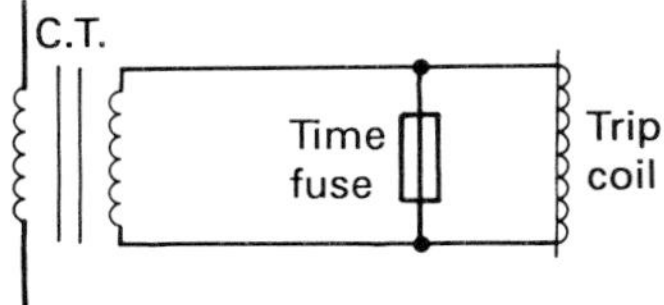

Figure 16.22 Current-transformer operated direct-acting trip coil

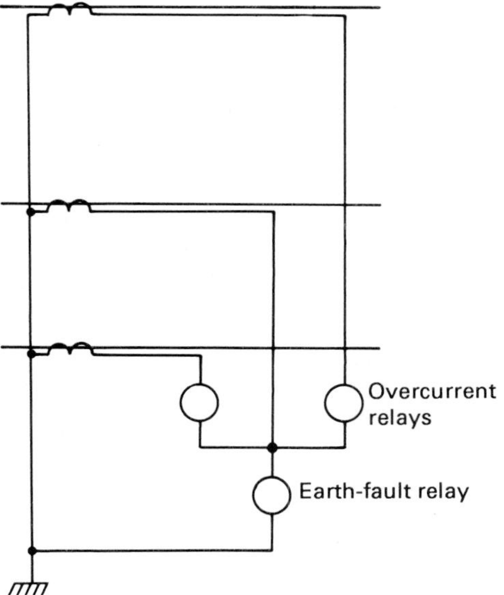

Figure 16.23 HV restricted earth-fault protection

The use of an hrc *time fuse* connected in parallel with the circuit-breaker trip coil provides the hv circuit-breaker with a time characteristic similar to that of the hrc fuse. The circuit-breaker trip coil is energized only after the fuse has operated (*Figure 16.22*).

The hv fuse differs from the lv fuse in that it does not have the ability to operate at low fault currents without the elements 'burning back' until an arc is struck between its end caps to create an explosion. To avoid this disaster, current is diverted through an ignition wire which detonates to operate a striker pin. This in turn triggers the operation of a circuit-breaker or contactor to isolate all three phases.

16.2.2.4 Earth-fault protection

The detection of a fault between a phase conductor and earth is achieved by connecting an appropriate relay in the residual circuit (*Figure 16.23*).

The calculation of asymmetrical fault currents on three-phase systems involves the use of symmetrical components. The appreciation of this mathematical tool is not difficult but is outside the scope of this text[2].

Relays used for this purpose can be either an *inverse definite minimum time* type, and set to discriminate in a manner identical to that of the overcurrent IDMT protection, or an *attracted armature instantaneous* type.

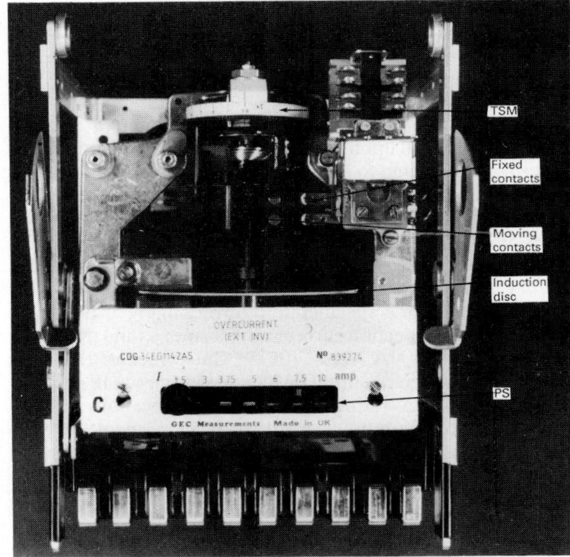

Figure 16.24 Inverse definite minimum time relay

Low current applications often use a core balance type of current transformer, where the phase and neutral conductors are passed through the core as shown in *Figure 16.25*. Under normal healthy conditions, the sum of the magnetic fluxes in the transformer will be zero, but should an earth-fault occur, the balance will be upset, and the residual flux will operate the relay.

The current operated *residual connected device* (rcd), is one form of this protection, and operates when the out of balance current exceeds the tripping sensitivity of the relay (these range between 10 and 500 mA).

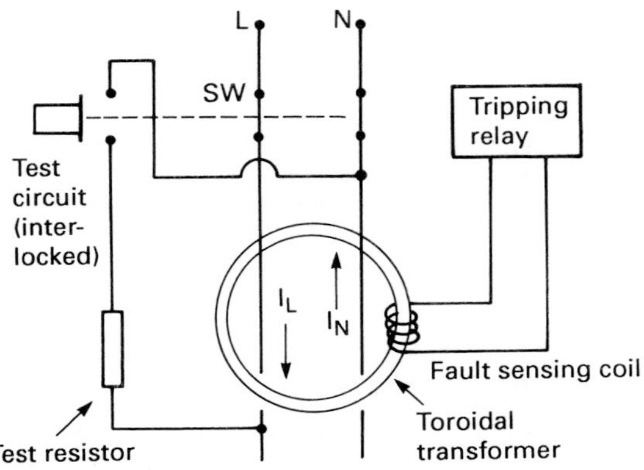

Figure 16.25 A residual connected device

16.2.2.5 Current and voltage transformers

For a relay to be operationally discriminative, it must be provided with accurate quantities of current and voltage.

16.2.2.5.1 Current transformers

A protection current transformer differs from a metering current transformer in that the former must retain its ratio for very high fault currents, whereas the latter must:

- maintain defined accuracy at load currents
- saturate at high fault currents to protect the metering from excessive voltages

The ability of a protection current transformer to fulfil its function will depend upon its design and on the load impedance (or *burden*) connected to it. A current transformer needs to be accurate in both its ratio and phase displacement.

Protection duty current transformers are listed as accuracy classes 5P or 10P. The figure defines the maximum composite error in percentage at a specified overload value, called the *accuracy limit factor*. 'P' indicates a protection duty. The accuracy limit factor is specified by a further figure 5, 10, 15, 20 or 30. A typical protection current transformer would therefore be defined as 10 VA class 10PI5 which means that the transformer has a rated output of 10 VA of accuracy 10 per cent at 15 times its VA rating. For certain applications, where current transformer outputs need to balance under through fault conditions, as with unit and restricted earth-fault protection, the class description is not sufficient and the specification needs to be more explicit. It is referred to as a Class X current transformer.

The accuracy class of metering current transformers is listed in IEC 185 (BS 3938) as 0.1, 0.2, 0.5, 1, 3 or 5. Each class has a defined ratio and phase displacement error at rated current and frequency. For example, a typical metering current transformer would be defined as 15 VA Class 0.5.

A current transformer should never be operated with its secondary winding on open circuit, because the primary current becomes, in effect, the magnetizing current and so the induced secondary voltage will be very high. The voltage can be dangerous to personnel and can cause the failure of the transformer insulation.

16.2.2.5.2 Voltage transformers

A voltage transformer, IEC 186 (BS 3947), is required to transform the system high voltages to the operating voltages of the relay, usually 110 V and 55 V to earth. Voltage transformers are similar to small power transformers, but are designed to maintain ratio and phase errors within very close limits for a specified range of secondary burdens.

16.2.3 Power transformers

Perhaps the greatest single reason for the adoption of alternating current as the mode for transmission and distribution of electrical energy has been the transformer. Using a transformer, voltages can be changed from one value to another in a convenient manner with high efficiency. If required, using static devices, various dc voltage outputs can readily be obtained. Power transformers should conform to IEC 76 (BS 171).

16.2.3.1 Voltage regulation

The approximate equivalent circuit of a transformer together with its phasor diagram is shown in *Figure 16.26*. The regulation of a transformer is the change in output voltage from V_1 to V_2 caused by the load current increasing, and is a function of the internal resistance and reactance values of the transformer.

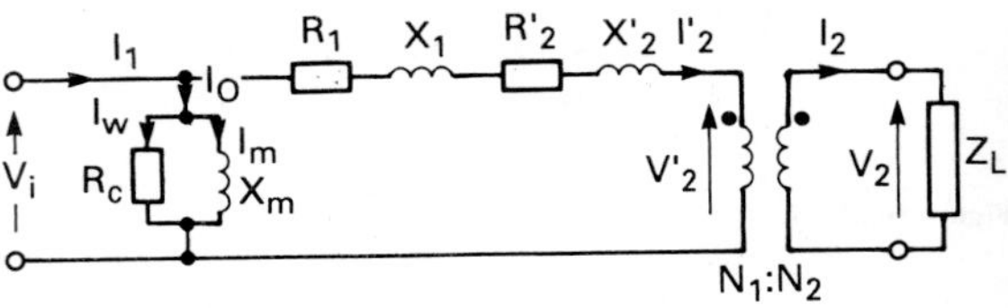

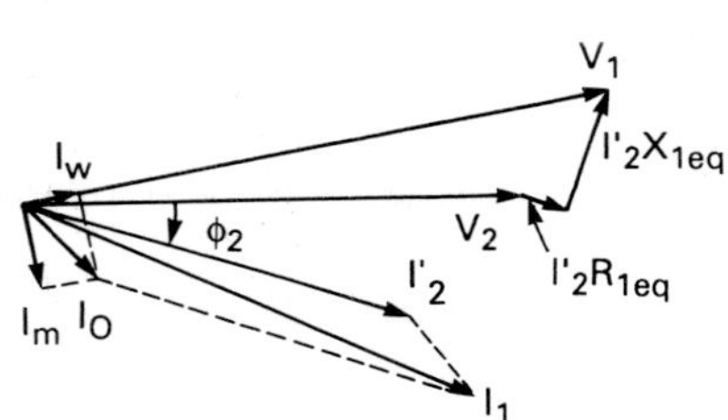

Figure 16.26 Transformer equivalent circuit

16.2.3.2 Transformer loss and efficiency

Transformer losses may be divided into two categories: those which vary with load current, referred to as *copper losses*, and those which are practically constant at all loads, referred to as the *iron losses*. The iron losses can be measured with the transformer on no-load and comprise hysteresis losses and eddy-current losses.

The power loss due to *hysteresis* for a given transformer is proportional to the supply frequency. The *eddy current* loss, due to circulating currents set up in the core material laminations is, for a given transformer, proportional to the square of the supply frequency. Because of the non-linear nature of the transformer core, harmonics, predominantly triple-n harmonics, are generated into the power source system.

The transformer efficiency is given by the equation:

efficiency = output power/input power
= (input power – losses)/input power
= 1 – {(copper losses + iron losses)/input power}

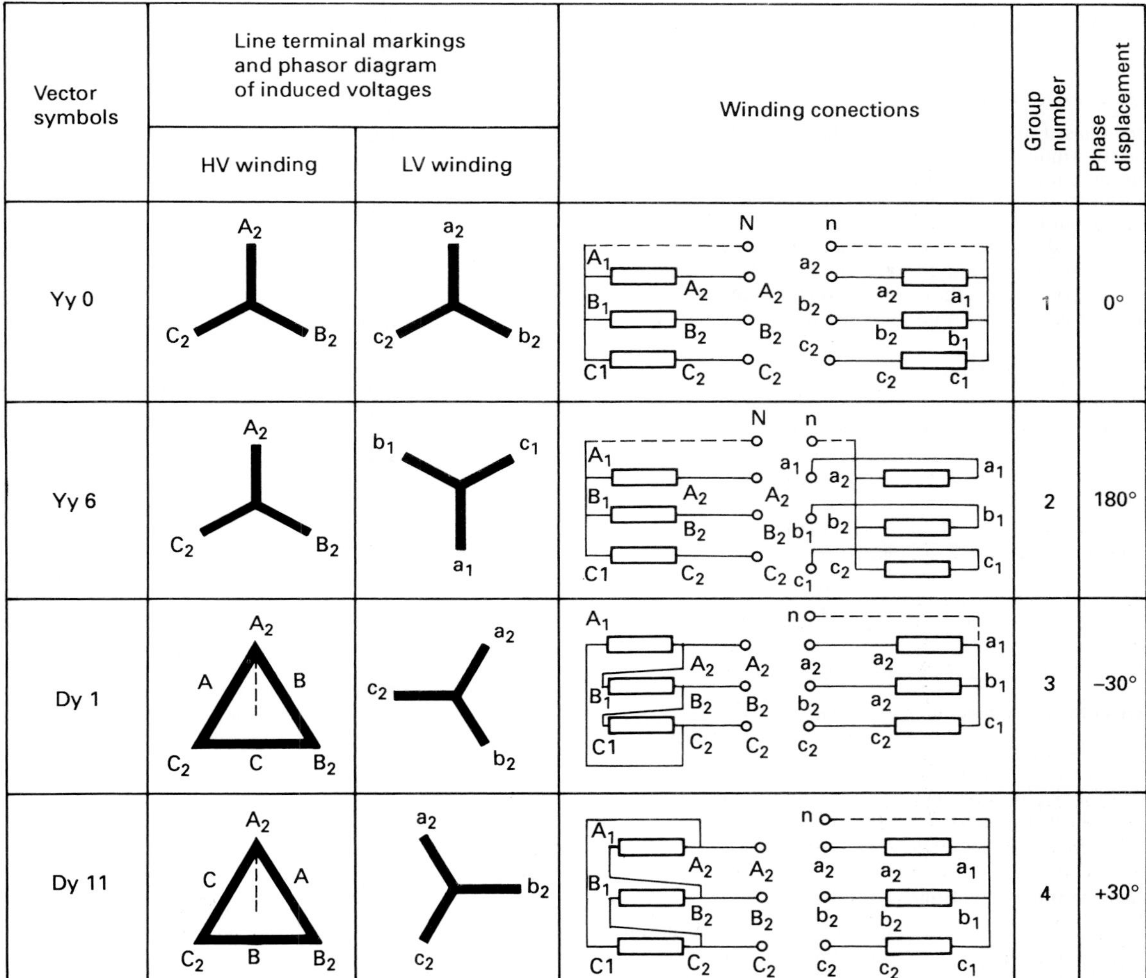

Figure 16.27 Three-phase transformer winding connections

16.2.3.3 Transformer cooling

The heat produced by the losses must be removed to prevent damage to the winding insulation. Various methods of cooling are listed in Table *16.3*.

Type of transformer	*Oil circulation*	*Cooling method*	*Abbreviation*	
			Dry	*Mineral oil*
Dry	—	Natural	AN	—
	—	Blast	AB	—
Oil immersed	Natural thermal head only	Air natural	—	ON
		Air blast	—	OB
		Water	—	OW
	Forced by pump	Ait natural	—	OFN
		Air blast	—	OFB
		Water	—	OFW

Table 16.3 Classification of methods of cooling

Insulating material allows for a temperature rise of up to 150°C, so that the use of dry type transformers is extensive. Resin cast transformers are impervious to moisture, and resistant to mechanical damage and thermal stress. They have a high impulse voltage withstand and need little attention. They are widely used indoors where the oil-filled type would present an unacceptable fire risk and would need to be provided with means to contain and limit the spread of leaking oil.

A conservator tank can be connected above an oil-filled transformer and partly filled with oil. This enables the transformer to remain completely full by allowing for the changes in the volume of the oil to take place under varying load conditions. The comparatively small conservator tank reduces the surface area of the oil in contact with the air and so reduces oxidation of the oil and corrosive effects.

A Buchholz relay can be mounted in the connecting pipework between the main transformer tank and the conservator tank. It comprises two floats each having a set of contacts. One indicates an alarm by responding to slow moving gas generated in the oil by an incipient fault. The other will be operated by the turbulence of a surge of oil and gas caused by a serious fault and will initiate a trip operation.

16.2.3.4 Transformer connections

Three-phase transformers, used for larger supplies, have three separate windings on the primary and secondary and can be connected in a number of configurations. The more common connections are shown in *Figure 16.27*. The most common arrangement for a step-down transformer is for the primary to be connected in delta, at 11 kV, and the secondary in star, at 415 V. This is referred to as a Dy 1 or Dy 11 connection depending on whether the connections give a secondary voltage lagging or leading the primary reference voltage respectively.

16.2.3.5 Auto-transformers

Another winding arrangement is that of the auto-transformer. The input voltage is applied to the primary winding and the output voltage derived from a tapping on the same winding (*Figure 16.28*).

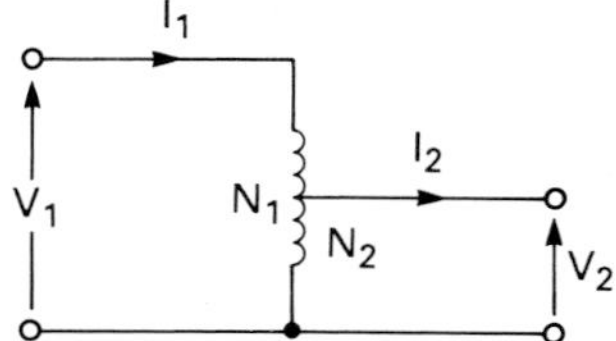

Figure 16.28 Auto-transformer (step down)

The advantage of such an arrangement is that the transformer uses less copper, and there are therefore less copper losses and a higher efficiency. It is lower in cost, but suffers from the disadvantage that if the secondary faults on open circuit, the output voltage changes to the input voltage.

16.2.4 Automatic voltage regulators

It has been the practice to confine the voltage at the transmitter to a ±0.5 per cent window using an automatic voltage regulator (AVR) from a source which can vary between +6 and -10 per cent.

There are various types of voltage regulator, and these include solid-state, variable transformer and moving coil.

16.2.4.1 Solid-state AVR

The connections of a solid-state automatic voltage regulator are shown in *Figure 16.29*. Two transducers, TD1 and TD2, are connected across a proportion of the output winding of the isolation transformer, T1. The output is taken from the common point of the two transducers, whose inductance is varied by the control winding current, so that the output voltage will range between V_1 and V_2.

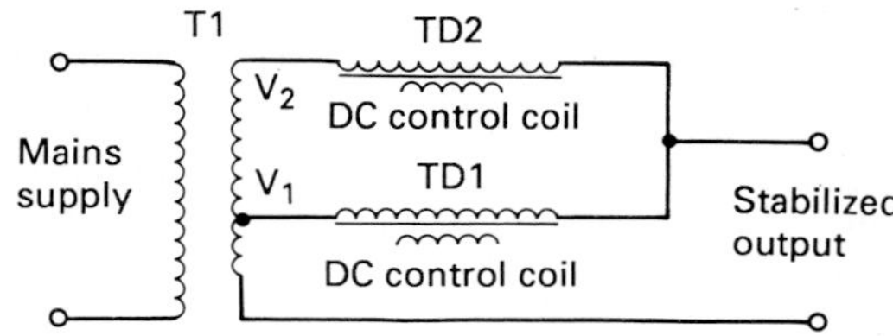

Figure 16.29 Solid-state automatic voltage regulator

A transistor operational amplifier compares the voltage across the load with a highly stable reference voltage. Any voltage error is amplified and directly coupled to the control winding of the transductors to modify the output voltage.

This type of regulator has no mechanical nor conductive moving parts and is more accurate than an electromechanical one. The transductor is not susceptible to damage from short-circuits or voltage surges, and the amplifier operates at low voltage and current levels. However, harmonic filters are necessary to reduce distortion, and the level of distortion is affected by only moderate input voltage variations.

A thyristor controlled regulator is another type of solid-state voltage regulator. The thyristors are connected in inverse parallel as a bridge so that the current will be controllable in either direction by applying the appropriate control signal to the gate connections. Two methods of gate control are possible. *Phase control* regulation is achieved by 'point-on-wave' switching. *Burst firing* regulation is achieved by conducting in bursts of full half cycles interspersed with non-conducting periods. Phase control regulators have limited rating because the harmonic generation causes distortion to the supply.

16.2.4.2 Variable transformer AVR

A variable transformer regulator (*Figure 16.30*) has a transistorized servo-amplifier to monitor the output voltage and control the operation of a geared reversible motor. The motor drives the moving arm of the variable transformer, the output of which supplies the primary winding of a fixed ratio auxiliary transformer. The secondary winding of the auxiliary transformer is connected in series between the supply and the load, and this affords the buck-boost action to control the output voltage.

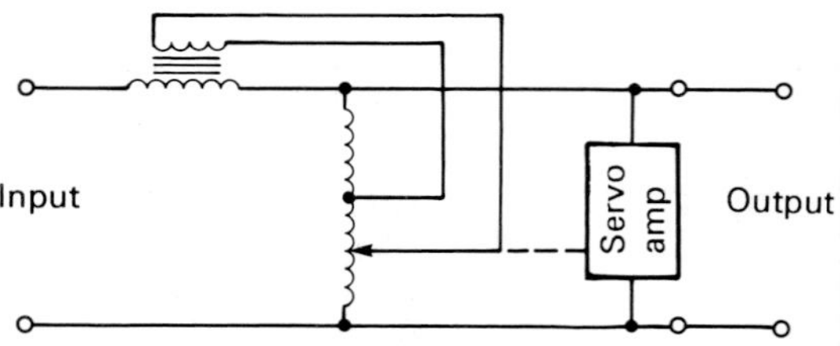

Figure 16.30 Variable transformer automatic voltage regulator

The advantage of this type of regulator is that it does not generate any significant harmonic distortion. The disadvantages are that it has mechanical and conductive moving parts.

16.2.4.3 Moving coil AVR

A moving coil type of regulator comprises a primary winding consisting of two coils wound on the upper and lower halves of a magnetic core over which an isolated short-circuited coil can move up and down. The division of voltage between the two primary coils is determined by their relative impedance.

With the moving coil in the position *x* (*Figure 16.31*(a)), the impedance of coil *x* will be small, and so the greater part of the voltage will appear across coil y. When the moving coil is in the position y, the greater part of the voltage will now appear across coil x.

Two additional coils are connected so as to inject a buck-boost voltage in series with the line. Any desired values of buck-boost can be provided by choosing a suitable number of turns for the two coils r and s.

The moving coil is driven by an induction disc motor, or, on the larger sizes, an induction motor, controlled by a voltage measuring relay. The advantages of a moving coil regulator are

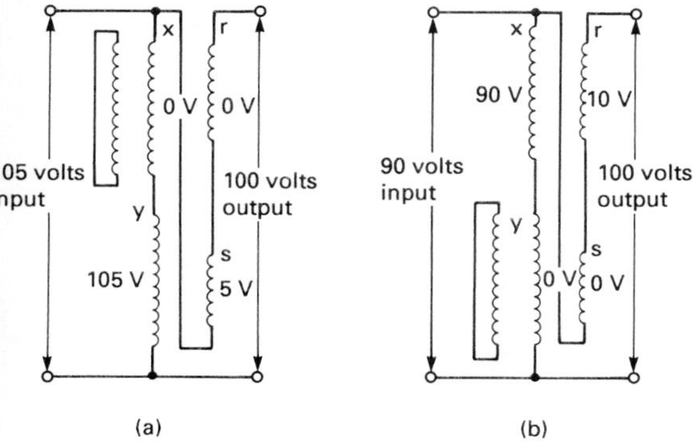

Figure 16.31 Moving coil automatic voltage regulator

that it is very robust, does not introduce significant harmonic distortion and has no conductive moving parts. The disadvantage is that it does have mechanical moving parts.

16.2.5 Motors

The operation of electric motors are achieved by the interaction of electromagnetic field systems arranged in such a manner as to cause a rotational torque. The different winding arrangements distinguish one form of motor from another. Further details are available in Reference 4.

16.2.5.1 Direct current motor

A dc motor derives its rotational torque from the interaction of electromagnetic fluxes produced by the fixed stator field winding and the rotor winding. A commutator reverses the current direction in the rotor conductor as it moves from the influence of one pole to another, so that the same rotational direction is maintained. The operating characteristics of speed and torque are shown in *Figure 16.32*.

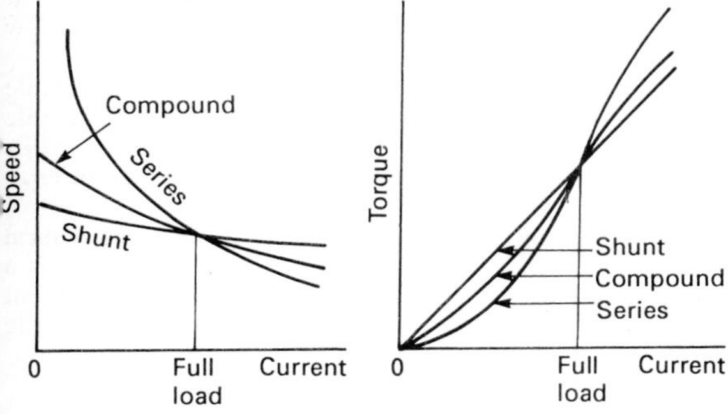

Figure 16.32 Speed and torque characteristics of a dc motor

16.2.5.2 Alternating current motor

The ac induction motor is used extensively in both single-phase and three-phase forms.

16.2.5.2.1 Three-phase induction motor

A three-phase induction motor has a primary winding, usually the stator, which produces a rotating electromagnetic field. The speed of rotation, n, is proportional to the frequency of the supply, f, and inversely proportional to the number of pairs of poles, p.

The rotor has an evenly distributed arrangement of conductors that forms a closed circuit, through which currents can flow.

At start, the synchronously rotating field of the stator cuts the rotor windings to induce a high rotor current, the field of which interacts with the stator rotating field and produces a rotational torque to accelerate the rotor in the same direction as the rotating field. As the rotor accelerates, the rate at which its conductors are cut by the synchronously rotating field decreases, and therefore so does the rotor induced voltage, current and accelerating torque.

The rotor cannot run at synchronous speed, since the rotor current is induced only where there is relative speed between the rotor conductors and the synchronously rotating field. The difference between the synchronous speed and rotor speed is known as the *slip*.

As the mechanical load on the rotor increases, the rotor slows down, the slip increases, the induced rotor emf and current increase so that a greater torque is developed. This process continues until the *pull-out* torque of the machine is reached and the motor stalls. The operating characteristics are shown in *Figure 16.33*.

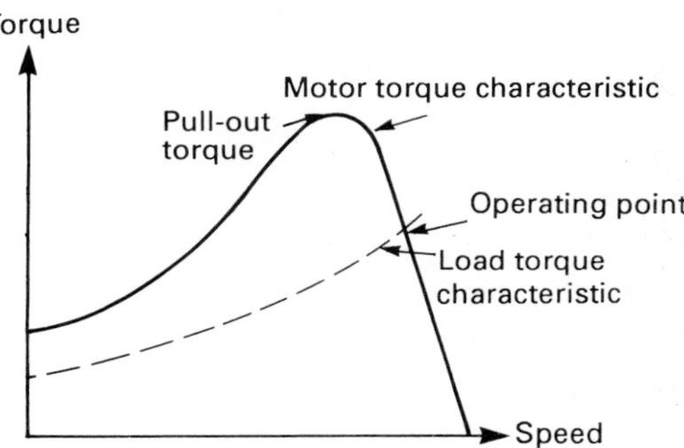

Figure 16.33 Induction motor torque characteristics

Three-phase induction motors are available with either a squirrel cage or slip ring rotor winding.

A *squirrel cage rotor* comprises solid conductors connected together at each end of the rotor. This robust design enables it to be built for reliable service at a relatively low cost. It is therefore a general purpose motor. Because of the high starting current when switched direct on-line, a squirrel cage motor may cause a dip in the supply voltage, and this may adversely affect other equipment. The starting current can be limited by *star/delta* or *auto-transformer* methods.

The star/delta method connects the windings initially in star so that the line voltage divided by $\sqrt{3}$, that is approximately 58 per cent of the line voltage, is applied to each winding. When the motor has run up to speed, a changeover switch is operated to connect the windings in delta so that the full line voltage is applied to each winding.

The starting current is limited in the auto-transformer method by connecting the motor to the auto-transformer tappings, which are adjusted to reduce the applied voltage to about 50–80 per cent supply voltage as determined by the extent to which the starting current needs to be limited.

A *slip ring rotor* has conductors wound in slots, the ends of which are brought out to slip rings. An external resistance is connected via the slip rings to limit the starting current. Its value is slowly reduced as the speed of the motor increases until the slip rings are shorted out.

16.2.5.2.2 Single-phase induction motor
The single-phase winding on the stator does not produce a rotating but a pulsating field. Consequently the rotor, which is usually squirrel cage connected, will not start. A pulsating field can be considered to comprise two fields of equal magnitude rotating in opposite directions. So if the rotor is 'artificially' started in one direction or the other, it will continue to rotate in that direction. This initial start is effected by adding a start winding, electrically a few degrees out-of-phase with the main stator winding. The field moves from the start winding to the main winding, and so the rotor receives a pulse start torque. The start winding is switched out of circuit when the rotor is up to speed. The single-phase induction motor displays similar characteristics to those of the three-phase induction motor.

Another type of single-phase motor has a commutator fitted to its rotor to give a pole and armature system identical to that of a series dc motor. Such a dc motor will produce unidirectional rotation on ac because its field and armature currents change direction simultaneously. It is, therefore, termed a *universal motor* and is commonly used in small sizes.

16.2.5.2.3 Motor control and protection
The control equipment for motors should provide safe and efficient operation of the motor and its associated mechanical drive. This covers the aspects of starting, stopping, reversing and speed control, tripping the motor automatically under abnormal and fault conditions, and the isolation of the motor and control gear for maintenance purposes.

In general, motor protection is required against the conditions of overload, open circuit, low voltage, overcurrent and earth-fault.

To allow for both starting current and short duration peak loads, the overload trip device incorporates an inherent time delay using a dashpot or bimetallic strip. To cater for the thermal loading conditions of the motor, the bimetallic element can be designed to have similar heating and cooling characteristics to those of the motor it protects.

Thermistor trip devices are sometimes built into the structure of the motor at its various hot spots. In this way, the heating of the trip device is directly related to the actual motor hot spot temperature rather than a simulation of the temperature in terms of the motor current.

Once running, a three-phase motor may continue to run as a single-phase motor after losing one phase of the supply. It will take a higher current than normal, and this may cause a winding to overheat. Therefore, as a safeguard, a relay is included in the protective gear to trip the motor control switch.

A motor contactor opens automatically if the supply voltage falls below its *drop-out* value. BS 5424 and IEC 947-4 set this value at 60 per cent. Similarly, mechanically latched devices, such as a circuit-breaker, can be released by the plunger of a solenoid dropping if the supply voltage falls below a certain value.

16.2.6 Cables

For fuller information see Reference 4.

16.2.6.1 Cable materials

The general construction of a cable is shown in *Figure 16.34*.

16.2.6.1.1 Conductors
Copper is extensively used as a conductor, but aluminium is a competitive alternative. The conductivity of aluminium is approximately one third that of copper, but by using solid conductors its overall size is brought back to approximately the same size as that of stranded copper.

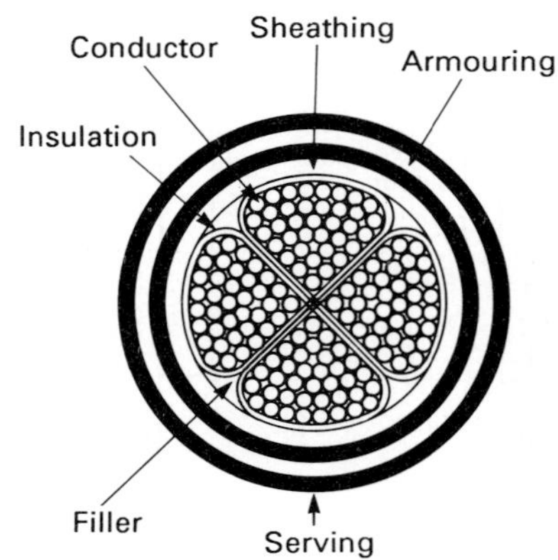

Figure 16.34 Typical cable construction

16.2.6.1.2 Insulation
Oil impregnated paper insulation to IEC 55 (BS 6480) will withstand excessive temperature changes, and the lead sheath, used to prevent ingress of moisture, forms a good continuous earth return conductor. The stringent jointing techniques required by paper insulation, its low mechanical strength, its low thermal capacity, and its hygroscopic nature have led to the introduction of thermoplastic alternatives.

At high temperatures, pvc becomes soft; at low temperatures, it becomes rather brittle. Consequently a compromise is necessary, and three types emerge.

General purpose grade has excellent resistance to oils and chemicals. It is self extinguishing when ignited and the source of flame removed. However, when pulled into conduit there is a tendency for it to bind with other conductors.

Hard grade is formulated with less plasticizer and shows less tendency to soften at high temperatures. Its mechanical properties are preferable. Its improved performance at high temperatures is obtained at the expense of its performance at low temperatures, displaying a cold shatter characteristic below 10°C.

Heat resisting grade is obtained by replacing the volatile plasticizer in the two previous types by relatively non-volatile plasticizer. This produces a type of pvc cable that retains its flexibility when subjected for long periods to temperatures as high as 85°C.

Cross-linked polyethylene (xlpe) has high tensile mechanical strength, good chemical resistance and heat stability. It is a superior dielectric to pvc, is physically smaller, has excellent resistance to abrasion, is inflammable, but exudes very little smoke and corrosive gases.

16.2.6.1.3 Sheathing
Cable sheathing is used to provide a water barrier and in some cases mechanical protection and earthing.

The excellent corrosion resistance properties of lead sheathing are particularly important. The fatigue resistance of pure lead and its mechanical strength can be improved by the addition of small proportions of certain alloys.

The use of an alternative to lead has always been investigated because of its weight, and aluminium is an obvious choice. Aluminium extrusion produces a sheath less flexible than lead, but a corrugated construction improves its flexibility. Care is necessary in service to prevent sheath corrosion, as aluminium occupies a high place in the galvanic series.

16.2.6.1.4 Armouring

Cable armouring is used to protect against mechanical damage. Steel tape armour provides good protection against such damage, but steel wire armour is preferable where additional longitudinal stresses may occur during installation. Wire armour can be supplied with tinned copper wires to lower the overall resistance of the armouring for earth continuity purposes.

Single-core cables for ac systems do not usually have armour because of the induced losses in the armouring. Where armour is necessary, non-magnetic tapes or wire must be used.

Aluminium type armouring has the advantage that it is lighter in weight and it affords a higher conductivity. However, care needs to be taken before installation that the pvc oversheath is not damaged, as this would allow ingress of moisture which could lead to corrosion and ultimately the loss of the earth return path.

16.2.6.1.5 Serving

The life of a cable may depend upon the degree of overall protection against chemical corrosion, electrolytic action and mechanical damage. Bituminized paper tapes are applied immediately over lead sheath cables followed by cotton or hessian tapes.

For unarmoured cables, pvc serving is the usual form of protection, and for armoured cables, an extruded pvc serving may be necessary in highly contaminated soil.

16.2.6.2 11 kV cables

Three core, paper insulated, oil impregnated, belted construction, lead alloy sheath and galvanized steel wire armouring (pilcswa) cable, IEC 55 (BS 6480 Part 1), has been used extensively. The conductors can be either stranded copper or solid aluminium (sac) and the armouring can be steel tape (sta).

The introduction of aluminium sheathed cable, IEC 55 (BS 6480 Part 2), as an alternative produces a cheaper cable. It is less flexible but is sufficiently robust to dispense with armouring. The sheath can be of either the straight or corrugated form and, because of its low weight and easy jointing, its installation costs tend to be low.

PVC and cross-linked polyethylene insulated cables are being used more extensively.

16.2.6.3 Medium voltage cables

PVC insulated, single wire armoured, pvc sheathed (pvc swa pvc) cables, to IEC 502 (BS 6346), are being superseded by xlpe swa pvc cable to IEC 502 (BS 5467). The jointing procedures are similar to those of pvc.

16.2.6.4 Wiring cables

The traditional rubber insulated cable, to IEC 227 (BS 6007), has been superseded by the considerably smaller equivalent pvc cable to IEC 227 (BS 6004). With insulation up to 500 V, it can be obtained either in single solid core or flat twin core pvc insulated, pvc sheathed with bare copper earth continuity conductor. The pvc insulation is designed to have a higher tensile strength, higher resistance to deformation, and a higher insulation resistance than the sheath. The sheath compounds are usually formulated to provide good abrasion resistance and yet have easy tear properties to facilitate stripping at terminations.

16.2.6.5 Mineral insulated cables

Mineral insulated cable comprises a solid conductor embedded in densely compacted magnesium oxide and contained in an extruded aluminium or copper tube which forms the sheath of the cable. The compressed mineral oxide will withstand high temperatures, and the metal sheath is a good conductor of heat, so the cable has high current ratings and the ability to withstand some exposure to fire. Due to the hygroscopic nature of the insulation, it is necessary for the ends to be carefully sealed. Where additional protection is desired for corrosive environments, the cable can be supplied with an overall pvc sheath. The cable does have a low impulse strength, and if voltage spikes are likely to be present in the electrical supply, it may be necessary to install surge absorbers.

16.2.6.6 Cable selection

There are three parameters to be taken into consideration when selecting a cable size:

- thermal rating
- voltage drop
- short-circuit rating

The operating temperature of a cable should be less than the thermal rating of its insulation. It is conditioned by the environment, e.g. whether it is installed above or below ground and the proximity to other current carrying cables. IEC 364 and IEE Regulations apply factors for the cable installation environment in order to assess its current rating.

The voltage drop requirements are that, from the origin of the circuit to any point in the circuit they should not exceed 2.5 per cent of the nominal voltage at the designed current, disregarding starting currents. However, the voltage drop of cable installations above 1 kV rating is usually not significant.

Subsequent to a system fault, large currents set up thermal stresses and electromagnetic forces in the cable. The adiabatic relationship is expressed by:

$$I = Sk/\sqrt{t}$$

where S is the cross-sectional area in mm^2
I is the rms fault current
t is the operating duration of the fault current
k is a factor related to the materials of the cable

A graph showing a typical cable adiabatic characteristic is shown in *Figure 16.35*, and, provided the protection operates below this, the cable is adequately protected.

The electromagnetic forces create a cable 'bursting' effect within multicore cables. A typical characteristic is also included on the graph.

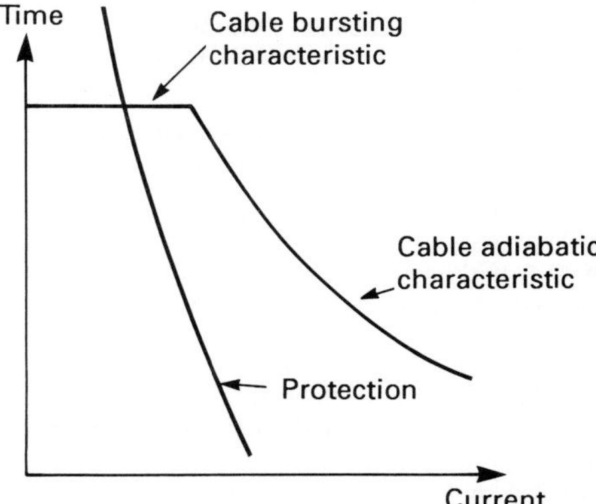

Figure 16.35 Cable adiabatic characteristic

16.3 Transmitter installations

16.3.1 Transmitter power distribution system

The main uhf transmitting stations are provided with duplicate hv supplies from a primary substation (see section *16.1*) fed either radially or from a ring system, or radially from different primary substations. Various arrangements are shown in *Figure 16.36*. At the Type 2 transmitting stations, the lv bus-section circuit-breaker is normally open so that the transformers share the load. At the Type 3 transmitting stations, where auto-changeover facilities are provided, one transformer is on standby.

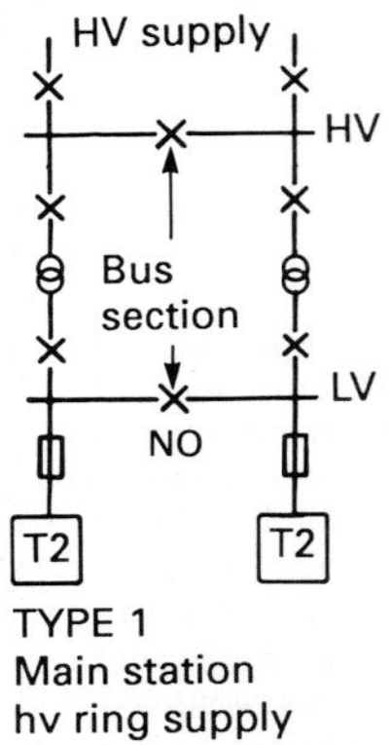

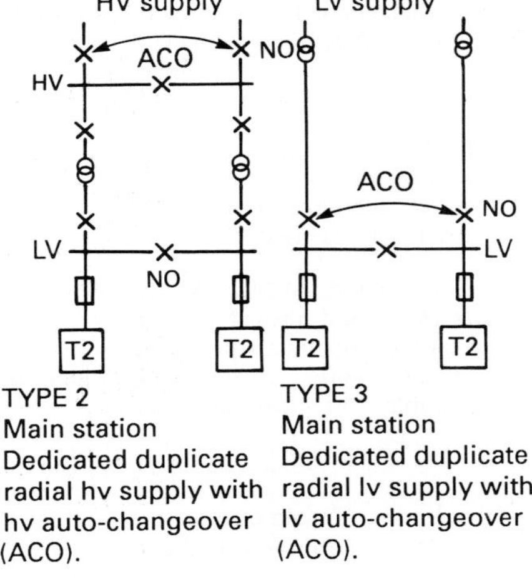

Figure 16.36 Station types showing power supply arrangements. NO indicates *normally open*

In the instances where the main station is fed from a ring system Type 1 station, a loss of supply from one section of the ring main would leave the supplies to the site unaffected. No auto-changeover facilities are therefore needed.

A simplified block diagram of the control circuit for the automatic changeover is shown in *Figure 16.37*. An under-voltage relay monitors each of the incoming supplies. In the event of the loss of the selected supply, a delay is initiated. If the supply is not restored during this period, the auto-changeover is initiated provided the alternate supply is available. Subsequent to a successful changeover, the system will not revert to the selected supply when it is restored. This feature prevents

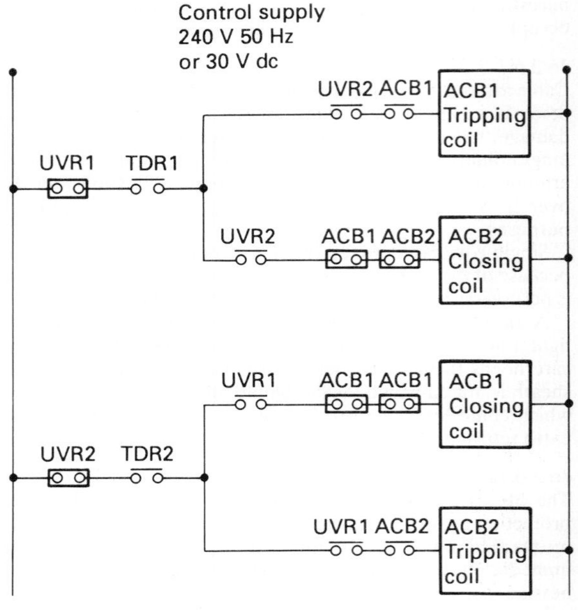

Figure 16.37 Automatic changeover control scheme. TDR = time delay relay supply; UVR = undervoltage relay supply; ACB =automatic circuit-breaker

unnecessary auto-changeover sequences and consequential interruptions to the programme transmissions.

Remote indication to a regional control centre over a telemetry system provides data as to the method of feeding the station and the availability of the main supplies.

The lv switchboard provides supplies to the station heating and lighting equipment as well as to the transmitters and their auxiliaries. The complete electrical installation must comply with IEC 364, *Electrical Installation of Buildings*, in matters of electric shock, fire, burns and injury from mechanical movement of electrically actuated equipment[12]. The standard qualifies requirements in the matters of safety design, protection against direct and indirect electric shock and against thermal and overcurrent effects. It lays down minimum requirements for electrical isolation and switching and rules for the selection, erection, inspection and testing of electrical equipment.

16.3.2 Supply reliability and availability

The number of supply interruptions averaged out over a number of years is referred to as supply *reliability*. *Availability*, on the other hand, relates to the average duration of such events. Reference 3 deals with the topic of reliability in greater detail.

16.3.3 Standby generation

Standby generation is provided at transmitter sites when improved security of the power supplies is required. Should the main supply be interrupted or deviate outside predetermined limits, the plant will start automatically and take on load.

16.3.3.1 Engine performance

The vast majority of standby generators are diesel driven. Air is taken into the cylinder through the injection valves and compressed. Fuel is introduced by the injector at the top of the compression stroke where it mixes with the air in the cylinder. The engines can be two-stroke or four-stroke cycle and are driven at synchronous speed (i.e. 1500 rev/min for a two-pole alternator).

The engine speed is controlled by the governor, which can be of mechanical, hydraulic or electronic design. The accuracy of the governor is defined in ISO 3046 (BS 5514, Part 4). The requirements are for the steady-state response not to exceed a ±1 per cent tolerance of the rated frequency, and the dynamic response to a sudden load change not to exceed 15 per cent frequency deviation and to return to a steady-state condition within 15 s and within 5 per cent frequency tolerance.

The power rating of the engine is important and is defined in ISO 3046 (BS 5514, Part 10). *Overload power* is defined as the ability to deliver an additional overload capacity of 10 per cent for one hour in any 12 consecutive hours of continuous running without detriment. The power rating is very much affected by altitude.

An important criterion of engine design is its ability to accept a proportion of full load from an initial cold start. This will, to a large extent, depend upon the type of aspiration of the engine and the inertia of the set.

The energy in the exhaust gases can be used to drive a turbocharger to compress more air into the cylinder and hence to produce increased power output from the engine. Further additional power can be obtained by using a two-stage turbocharger. The use of the turbocharger does mean that the engine can only accept approximately 80 per cent of its rating in one step from a cold start.

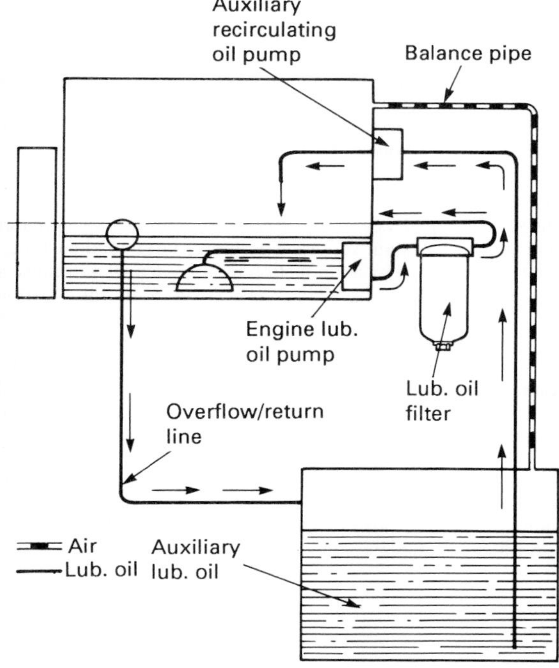

Figure 16.38 Diesel lubricating system

16.3.3.1.1 Lubrication

Predominantly, two oil circulating systems are used: a *dry sump* system, in which the sump oil is pumped to an exterior reservoir tank from which the oil is pumped at a maintained pressure into the engine, and a *wet sump* arrangement (*Figure 16.38*) which uses the sump as the reservoir.

Probably the two most onerous operating conditions are the *start* condition where the cylinder wear is greatest, and the *extended run* situation where the lubricating oil may become contaminated. The former condition is partly solved by pre-priming systems, and the latter by ensuring that adequate quantities of lubricating oil are available.

16.3.3.1.2 Cooling and ventilation

A considerable volume of air is required by a diesel for cooling and combustion, bearing in mind that 10 per cent of a diesel generator set rating has to be dissipated. The performance of the engine depends upon the quality and quantity of the air supplied for combustion. Diesels can be either air cooled, up to about 25 kW, or water cooled, where the radiator can be mounted directly onto or remote from the engine, and a heat exchanger or cooling tower might be included.

Most air inlet and outlet vents to a diesel engine room are of the automatic louvre type, so that it is kept in dry and warm conditions during non-operational periods.

16.3.3.1.3 Exhaust systems

An internal combustion engine will operate more efficiently when its exhaust gases are discharged with the designed level of back pressure. The aim therefore is to keep the exhaust system route as short as possible and with the least number of bends. A single silencer should be installed with a tailpipe of specified length. Where silencers are installed in tandem, the first silencer should be located as close to the engine as possible. High frequency noise reduction requires an absorptive design; low frequency noise reduction requires a reactive design. Wide frequency noise reduction requires both or a combination unit.

16.3.3.1.4 Mechanical protection

The protection of an engine comprises at least a low lubricating oil pressure and high engine coolant temperature.

16.3.3.1.5 Noise levels

The noise level within a few metres of an engine can be in the region of 100 dBA. If the set is housed in a brick building, external noise level is reduced to about 75 dBA. Any further reduction would need acoustic treatment, and its implementation becomes relatively costly.

16.3.3.2 Alternator

An alternator, three-phase or single-phase, is designed to the appropriate sections of IEC 34-1 and IEC 34-13 (BS 4999 and BS 5000). The continuous rated temperature should not exceed the particular class of insulation given in IEC 85 (BS 2757). The rating of an alternator, usually given in kilowatts or kilovolt-amps at an assumed power factor of 0.8 lagging, should not be less than that of the engine and should include the overload conditions referred to in section *16.3.3.1*.

16.3.3.2.1 Excitation system

The alternator field winding on the rotor is usually of the salient pole type and is supplied with dc via diodes mounted on the rotor from the rotor ac exciter winding (*Figure 16.39*). The stator dc exciter winding is controlled by the dc output from the AVR to control the output voltage of the alternator.

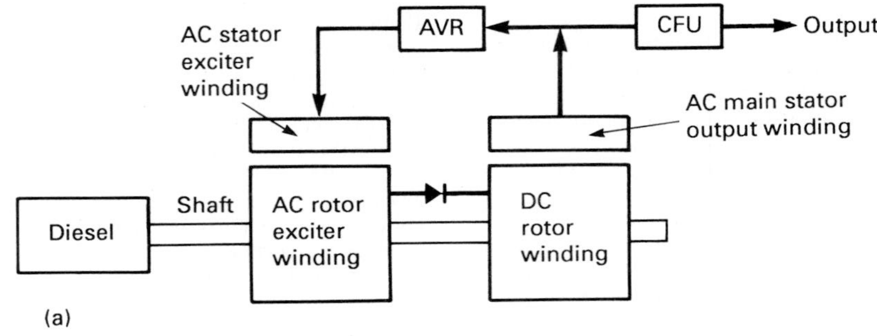

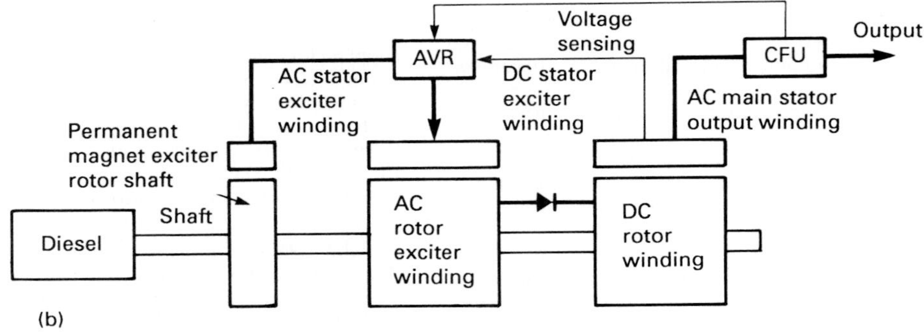

Figure 16.39 Excitation: (a) self excited generator, (b) separately excited generator. AVR = automatic voltage regulator; CFU = current forcing unit

The AVR source can be taken from either the alternator output (i.e. *self excited*) or from a rotor mounted permanent magnet pilot exciter (i.e. *separately excited*). The latter system provides a purely sinusoidal supply to the AVR, whereas the former may have a distorted supply because of harmonics taken by the load. This may affect the designed performance of the AVR. The separately excited system also affords a faster voltage rise time on start-up because it does not rely on the alternator residual voltage.

A further feature of the excitation system is a current compounding circuit to provide additional excitation proportional to the load current, giving a much faster response to load changes such as motor starting currents. Current compounding under fault conditions will ensure that an output of 300 per cent of the rated full load is maintained for a period of up to 10 s (see *Figure 16.40*).

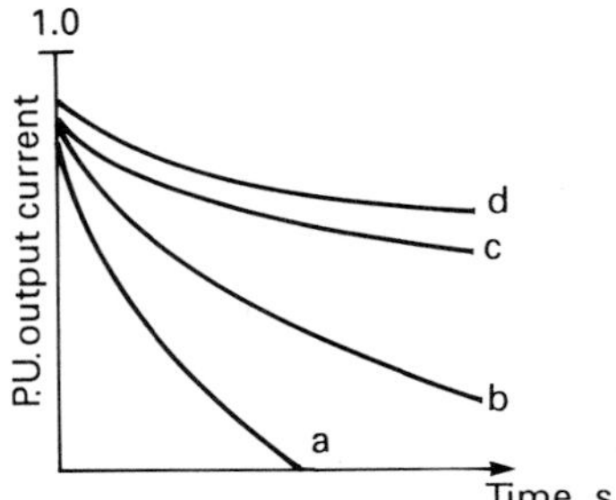

Figure 16.40 Response characteristics for various excitation systems: (a) self excited, (b) separately excited, (c) self excited with current forcing unit, (d) separately excited with AVR

16.3.3.2.2 Alternator protection

Under fault conditions, provision must be made to trip the main circuit-breaker, suppress the excitation system, and shut down the fuel to the prime mover.

The protection of the alternator will have some form of restricted earth-fault to cater for a stator main winding fault to earth together with three-phase overcurrent protection, as recommended in the ESI Engineering Recommendation G59 *Recommendation for the connection of private generating plant to the Electricity Boards' distribution system.*

16.3.3.2.3 Load

The source impedance of the alternator is considerably greater than that of the main supply. Therefore, large load currents, like those imposed by an induction motor starting, will cause a substantial voltage dip. This may affect both the starting performance of the diesel engine and the performance of other voltage sensitive loads connected to the alternator system.

An often forgotten aspect is the harmonic content of the connected loads. Odd harmonics, triple-n in particular, will tend to 'flatten' the waveform, and this may be seen as an undervoltage condition by other voltage sensitive equipment. *Figure 16.41* illustrates this.

The switchgear interlocking must ensure that the generator supply cannot be paralleled with the mains supply. The supply company must be notified of any standby generation installation which also has a mains supply to the site.

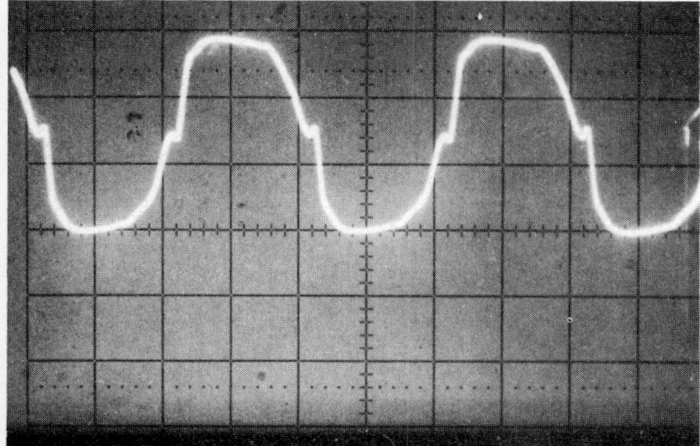

Figure 16.41 Standby generator output voltage waveform with harmonic load

16.3.3.2.4 Parallel operation

Parallel operation of private generation with the public supply system at high voltage must comply with the ESI Engineering Recommendation G59.

Before parallel operation, with either the supply system or with other generation plant, the alternator output voltage and frequency have to be adjusted to match the 'running' system.

The sharing of the power and the reactive power is a function of the governor controlling the engine and the AVR controlling the alternator, respectively.

More detailed information on the control and protection of parallel generation may be found in other publications[4].

16.3.4 Uninterruptible power supplies

Where short supply interruptions of approximately 10–15 s are acceptable, standby generation can be used. For application where an interruption however short cannot be tolerated, an uninterruptible power supply (UPS) has to be used.

UPS systems divide into either *rotary* or *static*. *Figure 16.42* shows a typical system.

The rectifier fulfils two functions: power to the inverter and to the battery charger. It usually comprises a thyristor or transistor bridge circuit with smoothing on its output and filter circuits on its input. The inverter comprises one of the following:

- switching thyristors and/or transistors to switch the dc input in blocks of variable duration to give, after smoothing, a sinusoidal waveform; filter circuits are necessary to reduce the harmonic content of the output
- a dc motor driven generator which affords galvanic isolation between input and output circuits

The battery system usually floats across the dc link, although in some systems it may be separately charged and connected to the dc link when the mains supply fails. On an interruption or 'brown-out' of the supply, the critical load is maintained from the battery via the inverter for anything up to approximately one hour. Should autonomous operation be required for longer than the battery capacity, then standby generation is necessary.

Should the UPS fail, the by-pass switch, which is usually a static device, will change to the main supply. In the event of a fault on the UPS output, the by-pass switch will provide a low-impedance source to ensure the operation of the protection. A manually operated by-pass switch can be used for maintenance purposes.

The rotary system is less efficient, and its transient time response is slower than that of the static UPS. The rotary system fulfils the same function as the static inverter. The dc motor drives the alternator and operates from the rectifier under healthy supply conditions and from the battery during loss of supply. The output frequency is controlled by feedback to the dc drive, the voltage via the alternator field circuit. A rotary UPS has other more sophisticated motor-generator arrangements than the one described.

Many critical loads comprise switched-mode power supplies (SMPS) which generate a 'peaky' waveform. It is customary to take into account the peak current, (or *crest* factor) rather than the average current to arrive at the appropriate UPS rating. For this kind of load, the rotary systems require only the alternator rating to be increased; the static system will require derating. A rotary system also affords complete electrical, galvanic isolation from the supply. A static system has an advantage in weight and space requirements, and the cost per kVA installed tends to be less.

The selection of either a rotary or a static UPS will depend on the characteristics of both the critical load and the supply. The relative life cycle costs should also be taken into account. Other technical matters are dealt with in greater detail elsewhere[5].

16.3.5 Alternative sources

In the wake of the energy crisis, there followed considerable research into, and development of, alternate energy sources. More recently, alternate sources are being considered for low-power remote transmitter sites where the cost of providing mains supply is becoming prohibitive.

16.3.5.1 Solar power

A photovoltaic cell converts solar radiation into electrical energy. The intensity of the sunlight, i.e. the insolation level, is

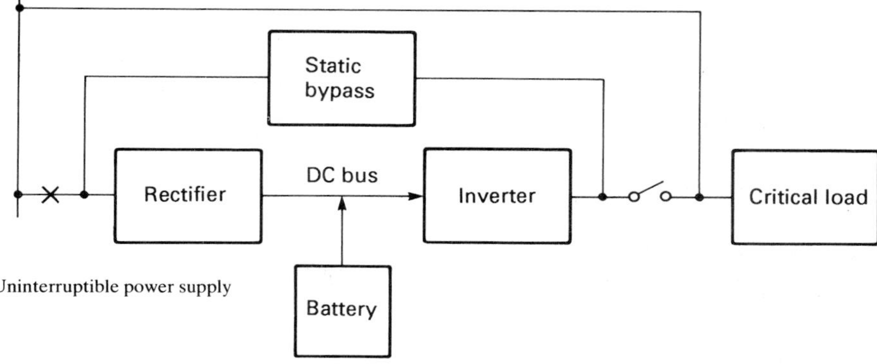

Figure 16.42 Uninterruptible power supply

about 1350 W/m² at the distance of the earth from the sun. The maximum power that can be delivered to an external load is typically up to 150 W/m² of solar panel area with up to 15 per cent efficiency and 0.5 W per cell. A typical voltage/current characteristic is given in *Figure 16.43*.

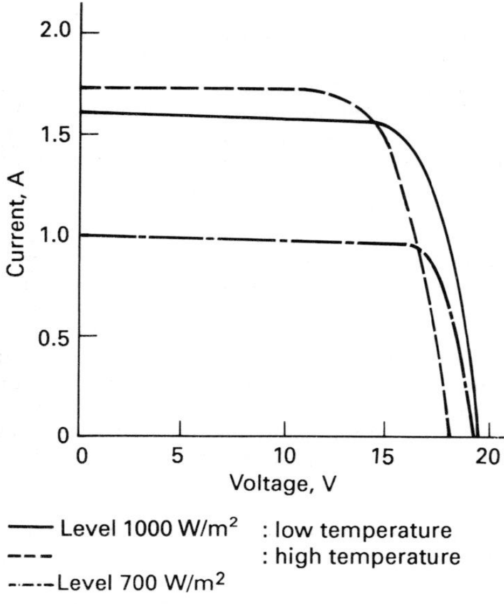

Figure 16.43 Voltage/current characteristics of solar power module

Modules of cells connected in series or parallel can give the required voltage and current outputs. Solar energy variations during the day/night cycle and throughout the year reduce the average insolation level. A battery storage system is therefore necessary and determines the operating voltage of the solar cell. A charge regulator is incorporated to protect the battery from overcharging. A diode is included to prevent reverse current and the discharging of the battery. A typical arrangement is shown in *Figure 16.44*.

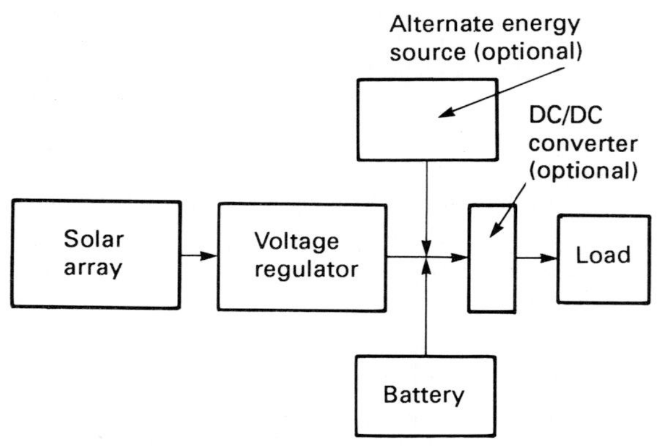

Figure 16.44 Solar generator systems

Research is being conducted into a photo-electro-chemical cell. This uses light energy to drive a chemical reaction which has a capacity to store energy during cloudy or night periods.

16.3.5.2 Wind power

The power available to a wind generator is directly proportional to the area swept out by the blades and to the cube of the wind velocity. In a strong wind of 10 m/s, the power output is about 250 W/m² of the swept area. Power generation is affected by the extreme variations in the strength and speed of the wind.

Horizontal axis propeller designs operate at constant speed using blade pitch control with a synchronous generator, while others operate at variable speeds and drive an induction generator. The blades are free to rotate about a vertical axis to keep them at right-angles to the wind direction. The main shaft, gearbox and generator are mounted at the tower top.

Vertical axis designs do not need to rotate into the wind, and the gearbox and generator are located at ground level. The torque and electrical output varies cyclically over each revolution.

The satisfactory application of the wind generator is limited to the selection of a suitable site location and to offsetting the high maintenance costs.

16.3.5.3 Thermoelectric generators

Heat produced by a gas or oil fired burner is directed onto one side of a thermoelectric energy converter based on the Seebeck[4] principle (*Figure 16.45*). The other side of the convertor is cooled by natural air convection, and the resulting temperature differential causes dc power to be produced up to approximately 1 kW at 12, 24 or 48 V.

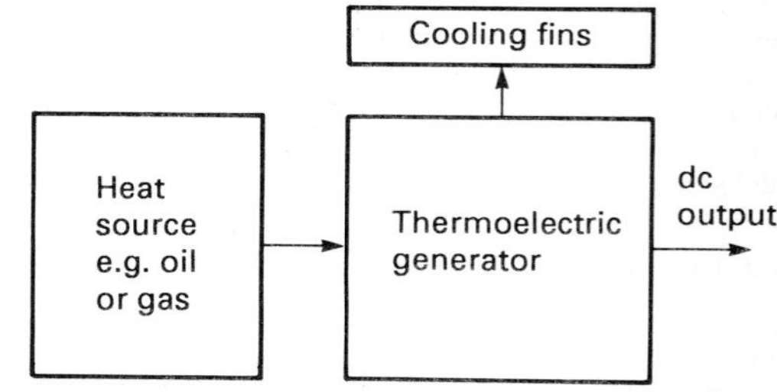

Figure 16.45 Thermoelectric generator

The generated heat can be used for space heating. As there are no moving parts, the thermoelectric generator is silent, has a high reliability of some 9000 hours MTBF (mean time between failures), and a life cycle of around 20 years. Preventive maintenance is required annually. If necessary, it can be run in continuous mode so that a dc battery system installation need not be necessary.

16.3.5.4 Fuel cells

A fuel cell converts the chemical energy in a fuel directly into electrical energy by an electrochemical process. The cell comprises two electrodes, separated by an electrolyte, which are connected to an external circuit. As fuel is supplied to one electrode and an oxidant to the other, the cell develops a dc voltage across its electrodes. Cells are connected in series to provide the required voltage[6].

16.3.6 Power factor correction

The apparent power taken by an inductive or capacitive load will be greater than the power actually converted into useful energy. The ratio of the two is the power factor:

$$\text{power factor} = \text{useful power (W)} / \text{apparent power (VA)}$$

By keeping the power factor as near to unity as possible, the voltage regulation of any distribution system will be improved, the system losses will be reduced, and any tariff penalties will be avoided.

It is usual to improve the power factor of an installation by connecting capacitors in parallel with the system. Capacitors are reliable, efficient and have few maintenance requirements.

It is preferable for the capacitor to be installed as near as possible to the connected loads, but due to the diversity in the locations of the load, it will often be more economic to use bulk correction capacitors at the source switched in steps as the power factor of the loads varies.

16.3.7 Earthing

The earthing of electricity supply systems is usually governed by the Electrical Utility. It requires system earthing so as to restrict the potential on each conductor to a value appropriate to the insulation level, and to ensure, for safety reasons, efficient and fast operation of the protection in the event of an earth fault.

The consumer is legally responsible for providing an earth to comply with the requirements. However, the Electricity Utility will normally offer an earth terminal connection, provided an indemnity is agreed by the consumer absolving the Utility from the responsibility of any consequences in the event of the loss of that earth.

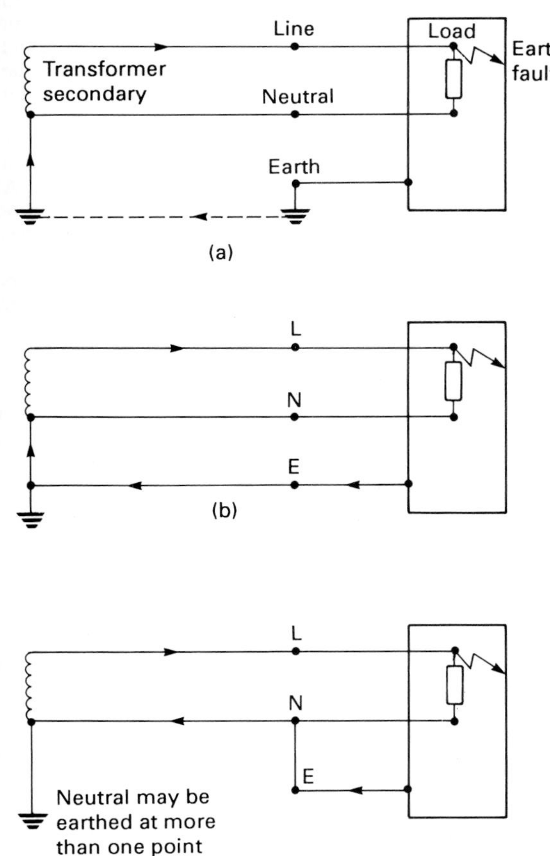

Figure 16.46 Earth loop impedance paths: (a) earth return through soil (IEC 364 TT system), (b) system with earth conductor (IEC 364 TN-S system), (c) protective multiple earthing (IEC 364 TN-C-S system)

Some Electricity Utilities are implementing a *protective multiple earthing* (pme) policy which utilizes a combined neutral and earth conductor. In this way an earth terminal can be provided at a much reduced cost. The neutral is connected to the earth electrode system at or near the supply transformer, at other intermediate points, and at the remote end of the pme system.

Some earthing systems are shown in *Figure 16.46*. The earth-fault current will be limited by the impedance of the earth-fault loop path, and this comprises the supply transformer secondary phase winding, the line conductor to the point of the fault, the earth protective conductor of the consumer's installation, and the return path to the transformer neutral.

The IEC and the *IEE Wiring Regulations* allow overcurrent devices, i.e. hrc fuses or miniature circuit-breakers, to protect for earth-faults provided that, with a fault of negligible impedance, the earth-fault impedance is sufficiently low to produce a current that will cause the overcurrent protection to clear the fault within a stipulated time.

Where the earth-loop impedance is too high to operate such overcurrent protection, a current operated rcd (section *16.2.2.4*) must be used. To avoid the risk of serious shock, the maximum voltage which may be sustained on exposed metalwork should not exceed 50 V. The effects of current on the human body as given in IEC 479 are shown in *Figure 16.47* together with a superimposed characteristic of a 30 mA rcd.

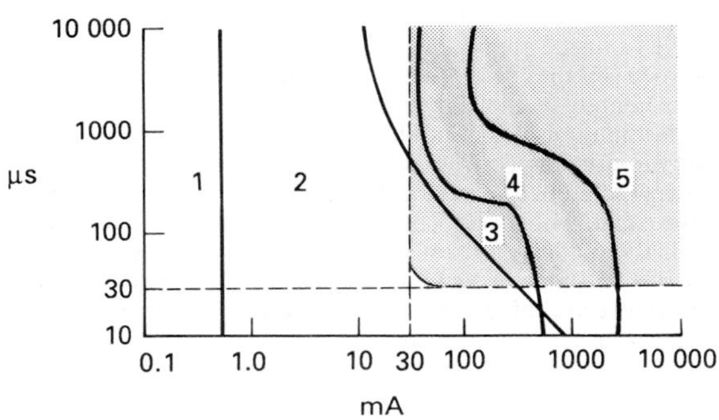

Figure 16.47 The effects of current on the human body (from IEC 479). In area 1, there is usually no reaction effect; in area 2, there is usually no patho-physiologically dangerous effect; in area 3, there is usually no danger of fibrillation; in area 4, fibrillation is possible (up to 50 per cent probability; in area 5, there is fibrillation danger (more than 50 per cent probability); in the shaded area, protection is afforded by a 30 mA current operated rcd at 30 ms

The sensitivities of the rcd range from 10 to 500 mA, but they do not afford a cut-off current characteristic. However, the overall tripping time is within 30 ms, which would prevent fibrillation of the heart.

16.3.7.1 Earth electrode system

The resistance of the connection to true earth potential of an electrode system depends upon the resistivity of the surrounding soil. The composition of the ground will give a general indication of its resistivity. For example, marshy ground will be between 2 and 3.5 ohms per metre and rock will be greater than 100 ohms per metre.

To obtain a low overall resistance, the current density in the soil in contact with the electrode should be as low as possible.

This can be achieved economically by having one electrode dimension large in comparison with the other two, such as a rod or strip. The curves in *Figure 16.48* give an indication of the resistance of plate, rod and strip electrodes installed in homogenous ground which has a soil resistivity of 100 ohm-metres.

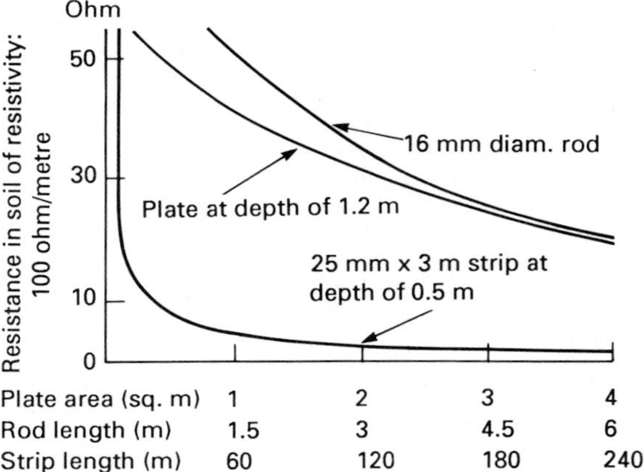

Figure 16.48 Earth electrode resistance

If one plate or rod of reasonable dimensions does not give the required minimum value of resistance, additional units can be connected in parallel. Provided each plate or rod in a nest is installed outside the main resistance area of any other earth electrode in the nest, the combined resistance is reduced approximately by the inverse ratio of the number installed.

Telecommunications and high frequency equipment circuits often require a 'noise free' reference earth. It is important that the earth reference points for the various parts of the system are at the same potential. The connections to the main earth terminal should not form an earth-loop path for earth current to circulate and induce signals into the system. Sometimes it is necessary to provide a separate earth insulated from the general earth and connected direct to the common earth terminal. This topic is dealt with in greater detail elsewhere[7].

16.3.8 Lightning

Lightning is an electrical discharge created by an atmospheric condition that disturbs the normal electric field balance between the earth and the ionosphere.

16.3.8.1 The nature of lightning

It is generally accepted[8,9] that there are four main types of lightning, i.e. positive and negative downward flashes and positive and negative upward flashes. Of these, the negative downward flash is the most common. Accordingly, only this type will be described (*Figure 16.49*).

The earth carries a positive charge, and the ionosphere, up to 50 km from the earth's surface, carries a negative charge. A negative downward lightning discharge commences when a negative potential leader advances towards the earth in steps of about 50 m seeking the path of greatest conductance.

The high potential difference between the charged cloud and earthed objects produces upward streamers. These currents can amount to a few microamperes and can occur for long periods without a lightning flash occurring.

The establishment of a complete path between the leader and the streamer produces a return stroke of current from the earth to the cloud, the magnitude of which can be between 25 A and 200 kA. The time taken for the stepped leader to reach the ground is about 200 ms, and the return stroke lasts for about 100 μs.

The return stroke can be regarded as a high frequency discharge from a constant current source which can reach a peak value in 1.2 μs and decays to half value in about 50 μs. A generalized waveshape of a return stroke is shown in *Figure 16.50*. The slope of the rise of the current generates frequencies of the order of 500 kHz.

It is common for the lightning flash to produce further leaders and return strokes over the same ionized path. The leader stroke does not in this case involve the stepped leader process and is therefore much faster reaching the earth.

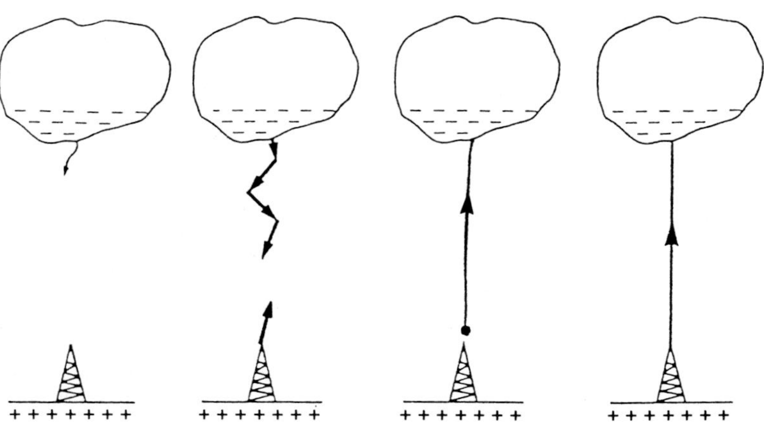

Figure 16.49 Lightning flash development

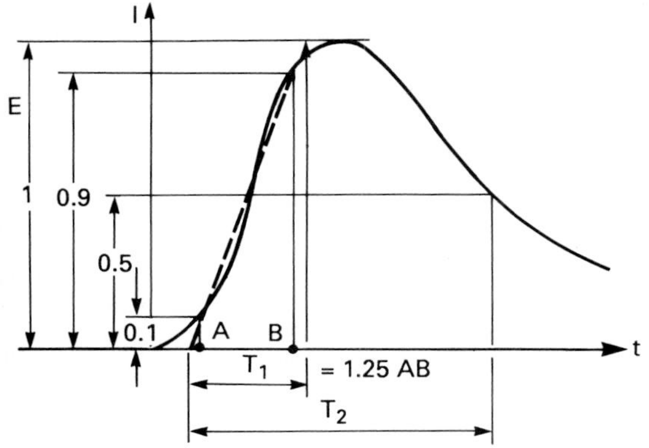

Figure 16.50 Generalized waveshape of return stroke lightning current. T_1 is typically 1.5 μs, T_2 typically 50 μs

16.3.8.2 Lightning protection system

If a structure is struck by lightning, the function of its lightning protection systems is to provide an adequate path to earth which will be able to discharge a strike of average severity.

A lightning conductor can provide protection against a direct lightning strike to a structure by attracting the strike to itself. In the simple case of a single vertical conductor, the zone of protection can be regarded as a cone with its apex at the highest point of the conductor and a base having a radius equal to the height[10].

At the majority of transmitting stations, the area of the site buildings is within the protective zone of the mast or tower, and direct lightning flashes to other than the antenna structure are unlikely and unknown.

A lightning protection system, as defined in BS 6651, comprises the air termination, the down conductor, and the earth termination.

16.3.8.2.1 Air termination

The function of the air termination is to create a point of strike for downward flashes, or, more likely, a point of discharge for upward flashes. It usually comprises a 1 m vertical rod and affords partial protection for the antennas.

16.3.8.2.2 Down conductor

Where the support for the antennas is a metal tower or mast, a down conductor is unnecessary, but where the tower is of concrete construction, it is important to use the steel reinforcement and to ensure that it is continuous. It should also be supplemented with down conductors and adequately bonded at vertical intervals in a manner recommended by BS 6651.

16.3.8.2.3 Earth termination

Should the antenna support structure receive a direct strike, the lightning current will disperse through the steel structure, or the concrete reinforcement, down to earth. As the current is discharged, it produces a voltage across the surge impedance of the earth electrode system which will momentarily raise the site potential to a high value above true earth.

Provided the bonding is correct, a low surge impedance for the earth electrode system would seem unnecessary, because the whole earth system will momentarily rise in potential. However, where other utility services, e.g. electricity supplies and telecommunications, enter the site, a remote earth is consequently introduced, so that longitudinal voltages are liable to cause insulation failure and high discharge currents.

The earth resistance, therefore, must be kept low and preferably below 10 ohms. The use of reinforced concrete as part of the earth electrode system has good effect, and semiconducting concrete should also be considered.

Adequate and frequent *bonding* should ensure that the installation will produce an equipotential cage. This should prevent side-flashing and electric shock to any person in contact with different parts of the installation.

Voltages on the surface of the ground in the vicinity of earth electrodes must be restricted to safe values. This can be achieved by using electrodes that form a ring around the area to be protected. The electrode should be buried deep enough to reduce the surface potential.

16.3.8.3 Effects of lightning

The magnetic fields produced by the lightning current can cause mechanical forces, so that sharp bends will tend to straighten, small loops will try to expand, and parallel conductors, with current flowing in the same direction, will tend to move towards each other.

The response of an earth electrode system relative to a lightning stroke depends upon its surge impedance. Reflections of the impulse wave will occur from the mismatch impedance points of the earthing system.

The duration of a series of lightning pulses, and consequently the adiabatic effect on the lightning protection system, is usually within the capability of the size of the system which is designed to cater for the electromagnetic forces.

The rapid change in the current rate will induce voltages in other metallic parts of the installation and unscreened conductors of the electrical circuits. Secondary discharge protection is required as discussed in section *16.3.8.4*.

16.3.8.4 Insulation coordination

Insulation coordination is necessary to avoid the breakdown of system equipment. It relates the designed impulse strength of the equipment to the parameters of the transient overvoltages to which the equipment is likely to be subjected.

The transient may be due to any system disturbances such as the effects of lightning, switching operations or fault events.

To avoid the installation impulse strength being exceeded, the energy in the transient must be either safely absorbed, i.e. *voltage clamped*, or diverted to earth, i.e. *crowbarred*.

16.3.8.4.1 Voltage clamping devices

Voltage clamping devices utilize filter circuits that have the effect of changing the transient from a high peak value for a short time to a sufficiently lower peak value for a longer time. These devices have a non-linear impedance that is dependent upon either current flowing through or the voltage applied across the device.

Selenium cells are based on rectifier application technology, but the extent of application has diminished in favour of more advanced materials.

Zener diodes are silicon rectifier technology and have very effective clamping characteristics. They are available in voltage ratings down to a few volts but have limited energy dissipation capability.

Varistors use a variable resistance material, such as silicon carbide or metal oxides.

The relationship between the current, I, through the device and the voltage, V, across its terminals is given approximately by:

$$I = kV^x$$

where x represents the non-linear characteristic of conduction. With higher values of x, the clamping is more effective.

The first varistors used silicon carbide which has a low x value. It therefore has a wide application in high voltage surge resistors, but it needs a series gap to stop the power follow-through current when the voltage returns to the normal operating value.

More recently, metal oxide varistors have been developed using zinc oxide which gives x values of up to 30.

The quiescent power consumed by the varistors must also be acceptable to other circuit constraints. High x value devices consume low power at the rated operation voltage, but a small increase in the voltage can cause a large increase in the consumption. The characteristics of different types of transient suppressor devices are shown in *Figure 16.51*.

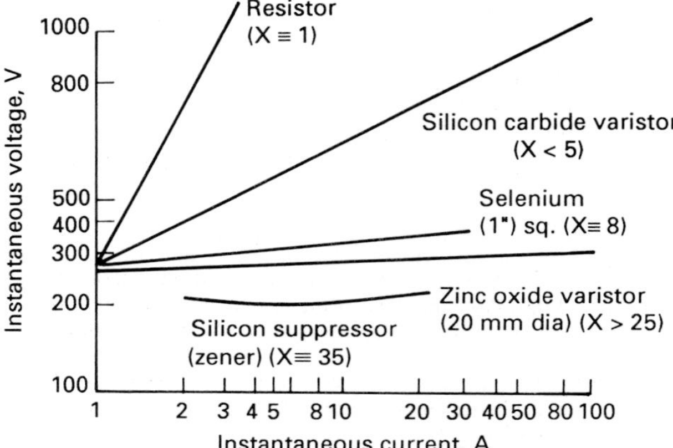

Figure 16.51 Characteristics of transient suppressors

16.3.8.4.2 Crowbar devices

The devices designed to divert the current to earth by a short-circuiting action operate on the principle of a change from high to low impedance by a switching process which can be inherent in the device, such as a spark gap, or can be triggered by a sensing device, as in the case of thyristor operation.

Gas discharge devices are particularly used in the protection of communications circuits where there is no problem of power follow-through current. The devices comprise single and triple metallic electrodes within a hermetically sealed tube containing gas at a reduced pressure, to allow a wider gap spacing. The triple unit enables both the line and the return wires to be connected to earth within the common tube instead of using two separate single gas protectors. The spark over from each line, therefore, occurs at the same instant, and this prevents transverse voltages in the circuit.

16.3.9 Battery equipment

Batteries are used for many different applications including:

- engine starting
- maintaining supplies for telemetry equipment
- back-up for solar and wind power systems
- switchgear tripping and control
- uninterruptible power supplies
- emergency lighting
- fire alarm systems
- hand-held portable equipment

16.3.9.1 Secondary cells

The secondary batteries most widely used in power system installations are the lead/acid and the nickel-cadmium types.

16.3.9.1.1 Lead/acid cells

Lead/acid batteries are manufactured in various forms in sizes up to about 1000 Ah. A lead/acid cell comprises two lead plates in an electrolyte of dilute sulphuric acid. The open-circuit voltage of a fully charged cell is about 2.1 V and this falls to about 1.8 V when the cell is discharged.

The cells may be charged by either constant voltage or constant current methods, but the former is more usual. The charger output voltage is set to about 2.3 V per cell where the charger is permanently connected, i.e. float charging. The specific gravity of the electrolyte varies with the state of the charge reaching a value of 1.22–1.27 when fully charged and falling to about 1.18 when discharged.

A high performance Planté cell has pure lead plates, is ideally suited to standby applications, and will tolerate the level of overcharge when permanently connected to a constant voltage charger. It has an operating life of about 20 years when properly maintained.

Tubular plate cells are used for stationary and mobile applications. They will withstand a large number of charge and discharge cycles, and on standby float charge duty they have a life of about 10 years. Their high rate discharge performance is not very good.

An automotive battery will produce around 3 kW of power for short periods and, provided it is not subject to overcharging or long periods of undercharge, it will have a useful life of 3–5 years. The plate construction, while being rugged and comparatively cheap, is very susceptible to low levels of continuous overcharge and also causes these cells to develop a high self discharge rate.

The electrodes in the sealed lead/acid type of cell are of a wound construction and made from pure lead. The most significant feature of this design is the means by which, in the overcharge condition, the gases produced by electrolysis are recombined instead of being released to the atmosphere as they are in the conventional vented cell. The useful life of this cell is considered to be approximately 10 years.

The main characteristics of the various types of lead/acid cells are given in *Table 16.4*. Typical charge and discharge characteristics are given in *Figure 16.52* for high performance Planté cells.

	General characteristics			
Type of cell	*Float voltage per cell (to maintain full charge), volts*	*Life expectancy on float voltage, years*	*Life expectancy* charge/discharge cycling performance, cycles*	*High rate performance*
Planté	2.25	20	500	Medium
Tubular plate	2.25	10-12	800	Low
Flat plate	2.25	10-12	800	Medium
Automotive	2.25	1½	120	High

* Based on discharge to 50 per cent of nominal 10h capacity followed by complete recharge

Table 16.4 General characteristics of lead/acid cells

16.3.9.1.2 Nickel-cadmium cells

The nickel-cadmium cell comprises nickel-cadmium positive and negative plates in an electrolyte of about 30 per cent potassium hydroxide and distilled water. The open-circuit voltage for a fully charged nickel-cadmium cell is about 1.4 V, and this falls to about 1.1 V per cell when the cell is discharged. Nickel-cadmium cells may be charged by either constant current or constant voltage methods. However, in the case of

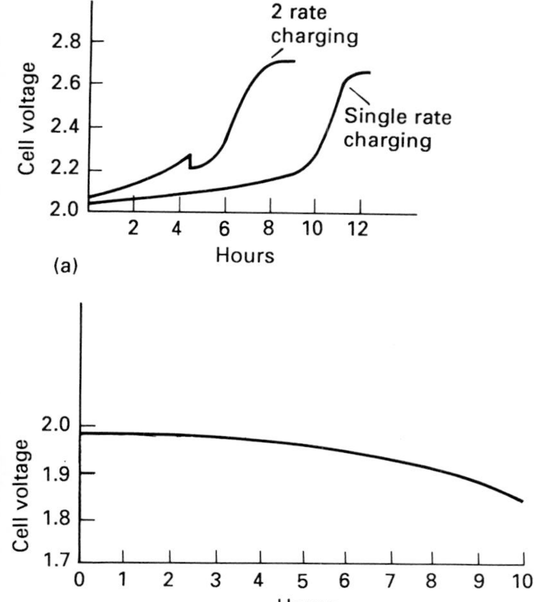

Figure 16.52 Characteristics of high performance Planté cells: (a) typical recharge and (b) discharge characteristics

sealed cells, the constant current method is invariably used to avoid overcharging.

Constant voltage charging for the vented type is usually set to 1.45 V per cell, for 'float' operation. Unlike lead/acid batteries, the specific gravity of the electrolyte does not give an indication of the state of charge.

The modern vented nickel-cadmium cell employs a pocket plate construction which is extremely rugged. It has a very long life, exceeding 20 years, and will withstand considerable over and undercharging without sustaining permanent damage. It is also capable of providing high rate discharge and will tolerate very high charge rates.

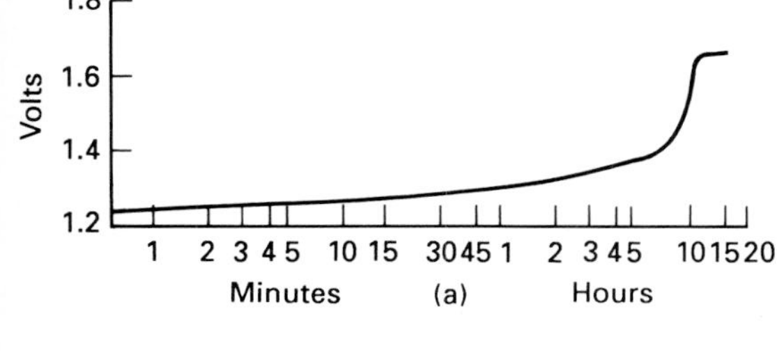

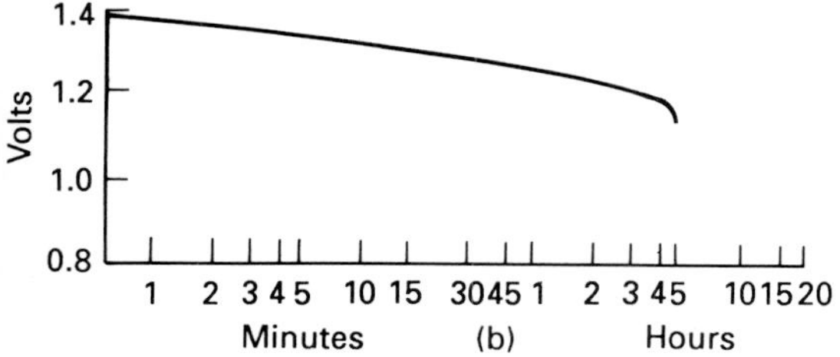

Figure 16.53 Characteristics of nickel-cadmium cells at 25°C: (a) constant current charge characteristics, (b) discharge characteristics. Nominal discharge voltage is 1.2 V per cell

As with the sealed lead/acid cell, the essential feature of the design of the sealed nickel-cadmium cell is to achieve recombination of the gaseous products and so sustain the volume of electrolyte. The useful life of the cell is usually about five years.

The disadvantage with all nickel-cadmium cells, but more pronounced with the sealed type, is the depression in voltage that can occur when the cells are subjected to repetitive levels of partial discharge followed by recharge. Typical charge and discharge characteristics are given in *Figure 16.53*.

16.3.9.2 Battery chargers

Modern constant voltage and constant current battery chargers are usually completely solid-state using either a transistor or a thyristor controlled unit to regulate the output. In addition to voltage control, the chargers are current limited to prevent too high a charge rate when the battery is in a low discharge condition. Additional features such as a boost charge rate, under and overvoltage, and earth fault alarms are usually incorporated. More recently, and for several reasons, emphasis has been placed on the use of battery charge condition monitors[11]. Additional smoothing is often required for charges associated with telecommunications equipment.

16.3.10 Fire alarm and protection systems

Fire and the production of smoke can kill or maim by asphyxiation, irradiation, poisoning or burning. The lack of visibility in smoke filled rooms is also a major threat to human life.

The essence of a successful fire protection system is to minimize the delay between the start of the fire and the action taken to combat it.

The requirement of a fire alarm system is to:

- raise an alarm and indicate a means of escape
- limit the damage to property by activating fire fighting equipment

The protection of property by fire detection, alarm and extinguishing systems may be financially advantageous because of the reduced insurance costs. However, in certain premises it is a legal requirement to provide an effective means to warn against the outbreak of fire.

Fire protection systems divide into three categories: fire detection and alarm systems, portable equipment and automatic extinguishing systems, and emergency lighting.

The use of automatic equipment is necessary to safeguard property that is unattended.

16.3.10.1 Fire detection and alarm systems

The alarm system should be designed and installed in accordance with CEN/TC72-EN54 (BS 5839 Part 1, 1988), which covers such matters as the division of the premises into suitable zones.

Conventional systems operate using *on* and *off* detectors located in zones. The advent of intelligent systems using analogue detectors with individual identity and 'watchdog' monitoring will afford rapid and high integrity fire detection.

Each zone system comprises fire detectors and manual call points connected into control and indicator equipment which is fed from a small uninterruptable power supply, and actuates audible and visible warnings. The wiring should be kept entirely separate from all other wiring services. PVC insulated cable should be protected by conduit or trunking, or mics cable (section *16.2.6.5.*) should be used.

Fire detectors are designed to respond to one or more of the three characteristics of fire, i.e. heat, smoke or flame, and these are summarized in Table *16.5*.

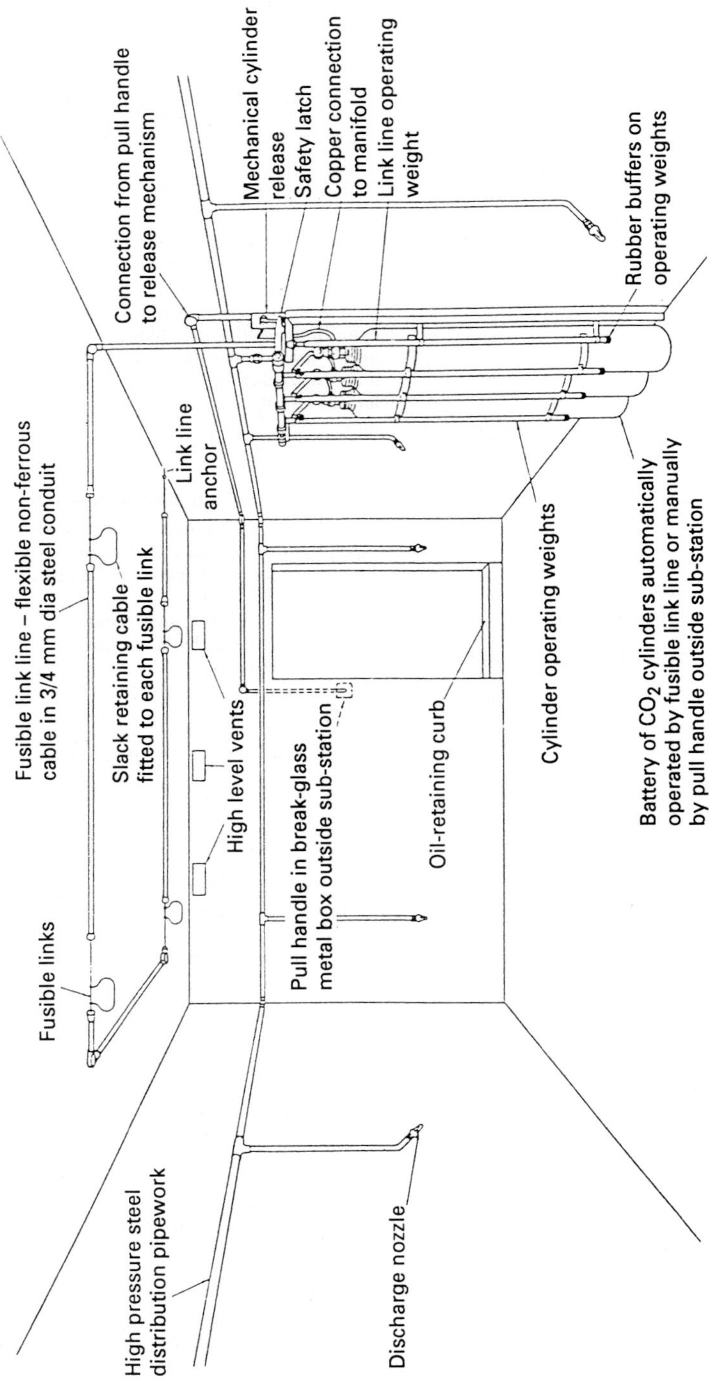

Figure 16.54 Typical arrangement of carbon dioxide fire extinguishing installation

Detector	Type	Fire category	Best application	Integrity
Heat	a) Fixed temperature elements b) Rate of temperature rise elements	A,B,C	for fast heat take off	good
Smoke	a) Ionization detectors b) Optical detectors	A,B,C	for slow heat take off	care in dusty environment
Flame	a) Infrared b) Ultraviolet	B	flammable environment	care in location

Table 16.5 Fire detection

In some situations, a combination of more than one type of detection may be preferable. Detectors should comply with CEN/TC72-EN54 (BS 5445) for industrial applications.

16.3.10.2 Extinguishing systems

16.3.10.2.1 Portable equipment

Portable fire-fighting equipment, such as fire extinguishers and fire blankets, complement an automatic fire alarm system. It is important that the appropriate extinguishing agent be used.

16.3.10.2.2 Automatic extinguishing systems

The fire detection system initiates an extinguishing system to release an appropriate agent to flood the protected zone.

The main extinguishing agents used at installations containing electrical equipment are carbon dioxide (CO_2), Halon 1211 or Halon 1301.

Carbon dioxide is a dry non-corrosive gas which does not conduct electricity. Its extinguishing effect is obtained by diluting the atmosphere to a point where the oxygen content is no longer sufficient to support combustion.

A typical layout of a carbon dioxide installation, for an hv switch room, is shown in *Figure 16.54*. Carbon dioxide is stored in liquid form under pressure in steel cylinders. When released, it is rapidly discharged as a gas which appears as a white mist. This is due to the mixture being frozen by the extremely low temperature of the gas and the presence of finely divided particles of solid carbon dioxide dry ice.

Carbon dioxide is not poisonous but, at concentrations above 5 per cent volume, judgement becomes impaired, and at concentrations above 10 per cent death by asphyxiation may occur. However, concentrations of up to 50 per cent are required to deal effectively with a fire, so facilities must be provided to lock-off the automatic system before entering the protected zone.

There is an increasing use of vapourizing liquids which are halogenated derivatives of simple hydrocarbons, the halogens being fluorine, chlorine and bromine.

The types commercially available are bromochlorodifluor methane, BCF, known as Halon 1211 and bromotrifluoro methane, BTM, known as Halon 1301. Halon gas needs a concentration of volume of about 5 per cent. It does not wet nor leave a residue, and it can be effective in three main classes of fire as defined in ISO 8421 (BS 4422):

Class A combustible material
Class B flammable liquids
Class C fires including live electrical circuits

Since a low concentration is required to extinguish most fires, it has a low degree of inhalation and is regarded as a safe agent for human contact. It is more expensive than carbon dioxide.

16.3.10.3 Emergency lighting

There are two options offered in BS 5226 Pt 1 for the design of emergency lighting, i.e. the defined and undefined escape routes.

A *defined* escape route can be up to 2 m wide, with the centre line illuminated to a minimum of 0.2 lux, while 50 per cent of the route width should be lit to a minimum of 0.1 lux.

An *undefined* escape route covers open areas, and the horizontal illuminance over the whole area should not be less than 1 lux. This latter system of escape route is more common because of the difficulty of keeping a defined route unobstructed and limiting the use of the site for changing circumstances.

References

1 WHITING, J P, 'Computer prediction of IDMT relay settings and performance for interconnected power systems', *Proc IEE*, **130**, 139–147 (1983)
2 *Protective Relays Applications Guide*, GEC Measurements, Pubn G-1011A
3 WHITING, J P, 'Reliability of power supplies and systems at broadcast transmission stations', CIRED Brighton, *10th Intern Conf Electricity Distribution*, IEE Pub. 305, Part 5 (1989)
4 LAUGHTON, M A and SAY, M G, *Electrical Engineer's Reference Book*, Butterworths (1985)
5 WHITING, J P, 'UPS systems for satellite broadcasting and computer installations', *Frost & Sullivan Conf Uninterruptible Power Supplies*, London (1989)
6 McDOUGALL, A, *Fuel cells*, Macmillan (1976)
7 WHITING, J P, 'The design of lighting protection systems at broadcast stations', *ERA Lightning Protection Seminar* (1987)
8 ELECTRA, 'Lightning parameters for engineering application', Cigre Study Committee 33
9 GOLDE, R H, *Lightning: Physics of Lightning, Vol 1*, Academic Press
10 GOLDE, R H, *Lightning: Lightning Protection, Vol 2*, Academic Press
11 WHITING, J P, 'A low-cost battery state-of-charge monitor', *ERA Conf Power Sources & Supplies*, Part 2, London (1987)
12 JENKINS, B D, *Commentary on the 15th Edition of the IEE Wiring Regulations*, Peter Peregrinus

Bibliography

UK Legislation

Health and Safety at Work Act 1974
Electricity at Work Regulations 1989
Electricity Supply Regulations 1988
Energy Act 1983
IEE Wiring Regulations, 15th Edn (1981)
Fire Precautions Act 1972

Power plant

SAY, M G, *Performance and Design of AC Machines*, Pitman (1958)

Central Electricity Generating Board, *Modern Power Station Practice*, **4** (1971)

SAY, M G, *Introduction to the Unified Theory of Electromagnetic Machines*, Pitman (1971)

FITZGERALD, A E, KINGSLEY, C Jr and UMANS, S D, *Electric Machinery*, McGraw-Hill (1983)

Power systems

ELGARD, O I, *Control Systems Theory*, McGraw-Hill (1967)

GUILE, A E and PATTERSONS, W, *Electrical Power Systems*, **1** and **2**, Pergamon Press (1977)

Power cables

BUCKINGHAM, G S, 'Short-circuit ratings for mains cables', *Proc IEE*, 108(A) (1961)

GOSLAND, L and PARR, R G, 'A basis for short-circuit ratings for paper insulated cables up to 11 kV', *Proc IEE* 108(A) (1961)

PARR, R G, 'Bursting currents of 11 kV 3-core screened cables (paper-insulated lead-sheathed)', *ERA Report F/T* 202 (1962)

Power system protection

WRIGHT, A and NEWBERRY, P G, *Electric Fuses*, IEE Power Engineering Series No 2, Peter Peregrinus

Developments in Power System Protection, IEE Conf Publ No 185, Peter Peregrinus

Power System Protection, Edited by the Electricity Council

Protective Relays — Application Guide, GEC Measurements

Power system harmonics

Limits for Harmonics in the UK Electricity Supply System, Engineering Recommendation G5/3, London (1976)

D J Bradshaw BSc(Eng), AMIEE
Design and Equipment Department, BBC

17 Diagnostics in Computer Controlled Equipment

Increasingly, traditional hard-wired logic systems employing large numbers of circuit elements are being replaced by software controlled systems. These permit a wide range of sophisticated control schemes to be implemented without requiring vast amounts of electronics but, as far as maintenance staff are concerned, they bring with them certain potential problems, among which are:

- The 'invisible' nature of software; examination of a circuit diagram is not sufficient to discover the way in which a software controlled module will function.
- The reliability of software equipment; this is increased through the reduced number of components required which leads, in turn, to lack of familiarity with the equipment on the part of maintenance staff.

Conventional fault location techniques based, for example, on the use of an oscilloscope to trace a signal through an equipment, are of very limited use in a software controlled system. Fortunately, however, it is possible to utilize the processing power of the computer to assist in the fault detection and diagnosis operation. The use of diagnostics programmes permits faults to be diagnosed to module level for first-line maintenance (on-site module replacement) and, possibly, to component level for second-line maintenance (module repair).

17.1 Computer systems

Computer systems comprise the following main elements:

- The *central processor* (CPU) which executes the sequence of instructions comprising the control programme (in most control systems, the CPU will be based on a microprocessor).

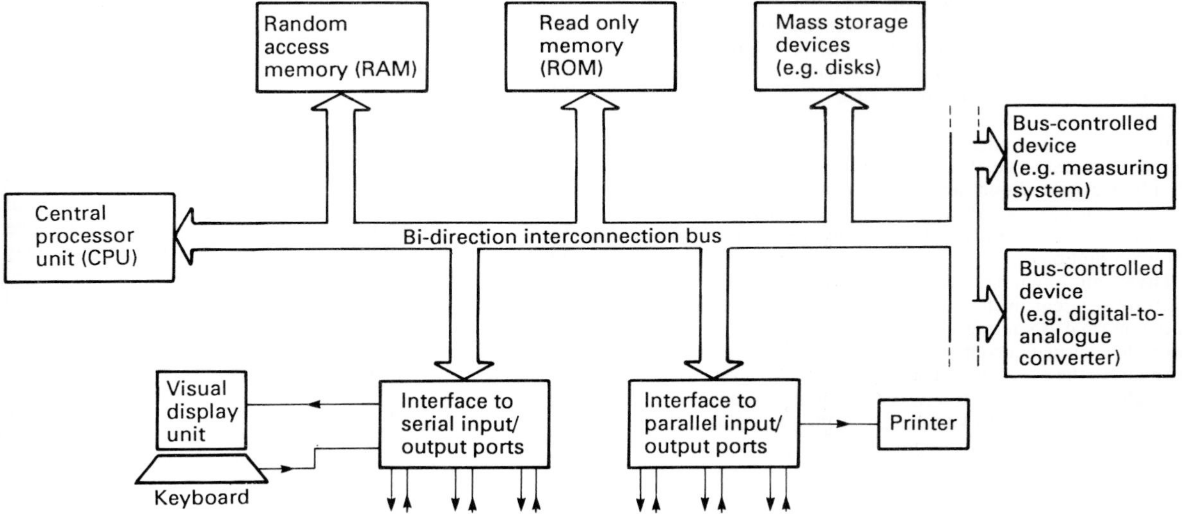

Figure 17.1 Schematic of a computer controlled system

- The *store for the instructions*, usually held in non-volatile read only memory, ROM.
- The *store for other data* generated during the execution of the programme and usually comprising random access memory, RAM.
- The means for the computer to *communicate* with the outside world.

Figure 17.1 shows the main parts of a computer controlled system. The CPU communicates with the instruction and data stores and with the interfaces to the equipment under control via a multi-way bus which, in many systems, carries time division multiplexed addresses, data and programme instructions.

17.2 Types of fault

There are three main types of fault that can affect computer controlled equipment:

- malfunctions of components or connections,
- loss or corruption of data stored in semiconductor memory,
- software faults.

Malfunctions of components or connections, such as a failed integrated circuit or a short-circuited mother-board connection, can be intermittent in their occurrence when devices are operated close to their limits or the equipment is subject to temperature or supply voltage variations. Faults of this kind are generally irreversible and require maintenance effort to correct them: they are usually referred to as *hard* faults.

Loss or corruption of data stored in semiconductor memory could be corruption of RAM contents, spurious modification of a CPU's internal registers or status bits (such as interrupt enable/disable bits), or change in the mode of operation of a programmable peripheral device. This type of failure, known as a *soft* failure, can be caused by a momentary disturbance on a power supply line or by radio frequency interference but is essentially reversible in that normal operation can usually be restored by resetting and re-initializing the system. However, the consequence of such a failure can be just as serious as a component failure in that a programme misoperation will probably result.

This type of failure is not unique to computer controlled systems; it is possible in almost any system using storage elements (flip-flops or latches) and was frequently overlooked in hard-wired systems, with the occasionally serious result of a 'latch-up' condition. Good design and installation practice should minimize the likelihood of errors of this type occurring, but it is difficult — if not impossible — to eliminate them completely.

Whilst software faults are not faults in the strict sense, they can result in just as great a malfunction when the necessary set of unforeseen conditions arises. The avoidance of faults of this kind demands that great care be taken during the programme writing phase, followed by extensive and rigorous testing for all known conditions.

There can, of course, be development errors present in the software resident in the control system. Whilst not apparent under normal operating conditions, they reveal themselves later when particular combinations of events occur or if changes are made to the use of the equipment. It is not generally feasible for such faults to be rectified by other than the original supplier because of the detailed knowledge of the programme that is required.

17.3 Tasks for diagnostics

In dealing with faults of the types outlined in section *17.2*, diagnostics routines have a number of tasks to perform, among which are:

- When power is first applied to the system (or re-applied following a failure of the supply), they must verify that all the parts of the system are present and capable of correct operation. This will verify that the system is correctly installed and ready for service.
- During normal operation, they must perform continuous checks on the system to verify that no part has been disconnected, that memory is functioning correctly, and that the programme store (usually EPROM, which has a finite life) is uncorrupted. The objective is to detect a latent fault condition before it can cause a malfunction and possible damage.
- When a failure is detected, they must provide information to permit the control system to revert to a *fail-safe* condition, report the fault and display diagnostics information so that the defective module can be replaced and operation resumed.
- When a module is to be repaired, they must provide assistance in diagnosing the fault down to component level so that a rapid repair can be made and the module returned to the stock of spares, so reducing the number of spares that are required.

17.4 Types of diagnostics

The tasks outlined in section *17.3* are performed by two main classes of fault diagnosis technique which can be categorized as *closed loop* and *open loop* diagnostics. Each of these has a role to play in the location and elimination of a fault in a computer controlled system.

17.4.1 Closed loop diagnostics

Closed loop diagnostics are those where the computer both applies the test stimulus and monitors its result. Most control computers spend a substantial amount of time effectively idle, waiting for an external event to occur to which they must respond. This idle time can be utilized for running a diagnostics programme on a near-continuous basis as a background task to the main programme. Diagnostics of this form are obviously well suited to use in unattended systems where they can be used to perform continuous monitoring of the system and so provide early detection of a potential fault situation. Clearly, the software for closed loop diagnostics routines running as a background task needs to be fully integrated with the control system software, resident in ROM or on the system disk, if fitted.

Tests have to be performed on all the component parts of the system to determine whether or not they are performing correctly, but care needs to be taken to prevent the diagnostics programme from interfering with the normal operation of the main programme. This consideration may limit the extent of the tests, particularly on RAM where the contents must not be changed as a result of the test and where the time available may not permit exhaustive testing. In general, closed loop diagnostics programmes running as a background task in the control system give an early indication of a latent fault condition without identifying the precise nature of the fault; for this, the more searching tests possible with open loop diagnostics are better suited.

The diagnostics programme that should run at power-up is a special case of closed loop diagnostics. Since the system is not yet operational, the diagnostics can perform detailed tests without the constraints of time and the need to avoid interfering with the normal operation of the system.

A frequent limitation of closed loop diagnostics is that, unless the diagnostics requirement has been taken fully into account during the design of the system, it may be impossible for the peripheral devices to be tested. This can be overcome if the peripherals themselves contain intelligence and can respond to an interrogation from the background diagnostics programme with some kind of status message; at least this will indicate that they are connected, set to the correct message format and interface protocol (e.g. correct data transmission speed and number of bits in the data word).

17.4.2 Open loop diagnostics

Open loop diagnostics are those which involve both the computer and an operator. Either the computer generates and applies a stimulus signal but an operator observes and interprets its effects, or the operator applies a stimulus signal to which the computer responds. For example, in a diagnostics mode a computer could generate a repetitive signal (e.g. a square-wave signal) that can be traced by an operator using an oscilloscope, or the computer illuminates a lamp in response to the operator pushing a switch. It is in the area of input/output ports and peripheral devices that the open loop tests are particularly useful. Frequently it is not possible for a closed loop to be formed that extends through the input/output ports to permit testing of the peripheral circuits and devices by the computer, and in these cases the assistance of an operator is required.

Whilst open loop diagnostics are unsuitable for fault detection in unattended transmission equipment, since they require the presence of an operator and generally an equipment has to be removed from service for the tests to be performed, their value is realized when the faulty equipment has to be repaired. Then the diagnostics, if carefully designed, can contribute to the rapid identification of the precise nature of the fault and the early return of the equipment to service.

The waveform appearing within a computer controlled system (particularly on the interconnecting signal bus) can be very complex, and the usefulness of simple test equipment such as oscilloscopes is limited. It is, however, desirable for the principles followed for fault diagnosis in software controlled equipment to be similar to those used on conventional equipment, primarily signal tracing. This involves the operator tracing a signal through an equipment until its form no longer corresponds with that expected. Fortunately, at least one technique exists which permits this method of fault diagnosis, as described in the following section.

17.4.2.1 Signature analysis

A signal tracing technique developed to deal with the problems of signal tracing in digital systems and known as *signature analysis* is very suitable for open loop testing in modest computer based systems. It is particularly useful on multiplexed address and data lines where no other technique is practicable for field servicing.

If a piece of logic circuitry is stimulated repeatedly with a fixed data pattern, the circuitry will exhibit a repeated definable data pattern at its output. The presence of this expected pattern in response to the known stimulus is an indication of correct circuit operation. However, so that it can be compared with the correct pattern, some means is required to store and display the output signal. Long data streams are inconvenient and a compressed form would be more useful. It is the function of a signature analyser to compress a data stream and display a unique representation of it, and this is achieved as follows.

Data appearing at a given circuit node in response to a defined input stimulus sequence is sampled, for a period governed by start and stop signals, by clocking it into a feedback shift register.

The contents of the shift register at the end of the sampling period are displayed as a four-character display, and this characterizes the behaviour of that node.

In practice, a 16 bit shift register is used, the parallel outputs of which are used to drive a four-digit hexadecimal display, these four digits being the *signature* of the node. Using a 16 bit register enables up to 2^{16}-1 (65 535) different signatures to be identified.

The essential elements of a signature analyser are illustrated in *Figure 17.2*.

The stimulus sequence must be designed to cause at least one change of state at all nodes to be tested, or several sequences will be required. It is necessary to specify the points in the circuit from which the start and stop signals for the sampling period are to be obtained, and the clock phase to be used. Also a logical test procedure may require special test sockets, switches or links to break feedback loops and modify the operation of parts of the circuit under test. Consequently, the use of signature analysis as a diagnosis technique should be designed-in from the outset. (The technique can, of course, be applied to production testing as well as to fault diagnosis in the field.)

The equipment documentation should indicate the expected signatures on the circuit diagram, and the maintenance information should give details of the switch settings and links that may be required.

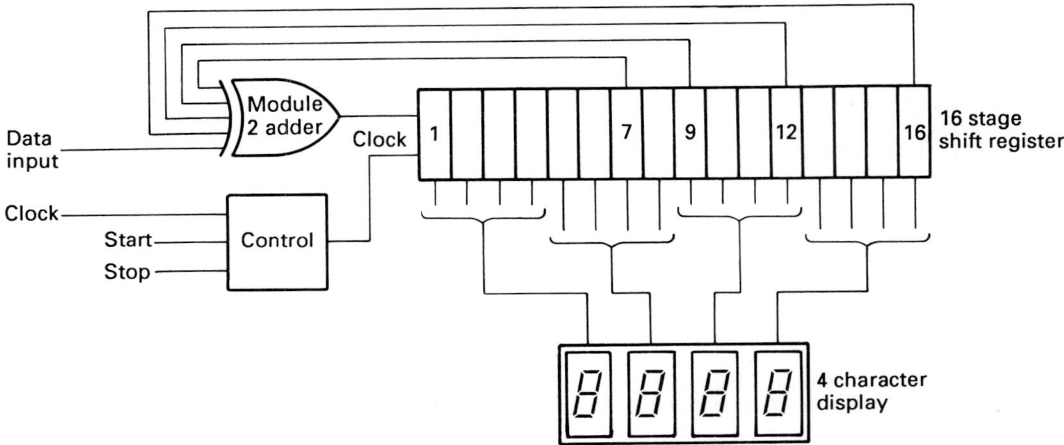

Figure 17.2 Schematic of a signature analyser

The generation by the computer of the test stimuli for signature analysis is one of the most versatile, yet powerful, forms of open loop diagnostics, particularly for fault-finding at component level.

17.5 Response to detecting a fault

In general, all that can be done on detecting a *hard* fault in the control system is for normal operation to be suspended, all outputs to be set to a safe condition, a report to be made to the control centre that a fault has occurred and to present as much diagnostic information as possible to the maintenance staff, ideally at least identifying the faulty module so that it can be replaced and service restored with the minimum delay.

In the event of a *soft* failure being detected, it may still be desirable for the failure to be reported, but what is primarily required is that the control system be reset and re-initialized to restore normal operation. This process can cause significant interruption to the control system behaviour, due to the loss of data stored in RAM, and special provision may be necessary to report this condition.

At this point the difference between a hard failure and a soft failure becomes apparent. A hard failure will continue to be present even after a system restart, whereas a soft failure should be cleared by this operation. In general, therefore, when a fault is detected the control system should report it, display diagnostic information and attempt to restart the system by applying a reset.

17.6 Watchdog timer

Detection of the occurrence of a soft failure usually means detection of a halted programme, and this can be achieved by the use of a hardware *watchdog* timer. The use of such a timer is essential for the reliable operation of unattended microcomputer control systems. It enables the control system to attempt to continue to operate correctly in the presence of random errors caused, for example, by extraneous interference such as mains-borne voltage spikes. More importantly, perhaps, it helps to prevent the control system from functioning in an incorrect, possibly hazardous manner.

A watchdog timer generally comprises a simple timing circuit (e.g. a monostable — not a device dependent on the CPU clock which would fail to generate an alarm if the clock should stop) that can be reset during normal programme operation and which, if not reset, will apply a restart signal to the CPU. The timer should be resident in the CPU module to prevent its disconnection. Resetting the timer can be achieved, for example, by addressing it through an I/O port.

The software to be executed must reset the timer sufficiently often to prevent a CPU restart. If a hardware fault or extraneous interference (e.g. EPROM or voltage dip, respectively) should cause the CPU to halt or otherwise malfunction, then a restart signal will be generated and applied to the CPU.

Following the restart, the CPU may execute a diagnostics routine, but the software should be able to differentiate between the power-up situation (when RAM contents are random data) and a restart (when the RAM may contain essential data the loss of which would be unacceptable). Since RAM data may have been corrupted, it should be verified after a restart, using some form of error detection/correction strategy.

The watchdog timer can be used by the CPU itself to force a restart if a background diagnostics routine detects a fault which can be cleared by a system restart.

The watchdog timer can be connected to external circuitry so that the frequency of restarts can be recorded or to enable an alarm to be given if, say, the watchdog timer is generating a near-continuous restart signal because of a system hardware fault affecting programme operation.

If a watchdog timer is incorporated, it may have to be disabled when interactive diagnostics are invoked, to prevent any restarts interfering with their use.

The timer reset should not depend on a complex lsi device because of the indeterminate effects of a spurious impulse, and a simple counter/timer realized in ttl or cmos is probably more reliable. Similarly, the timer reset signal itself should be derived in such a way as to minimize the likelihood of a spurious reset being generated during a controller malfunction.

17.7 Implementation of diagnostics

The diagnostics routines for the closed loop diagnostics that run at power-on, in response to a reset from the watchdog timer or continuously in the background, must of course be fully integrated with the main control programme.

Because of the way in which programme instructions and data are carried on the interconnecting signal bus, any short-circuit between bus lines or between a bus line and earth will have a serious effect on the operation of the computer. This type of fault is not uncommon, particularly on printed circuit mother-boards or memory cards with very closely spaced printed conductors. Consequently, the hardware should be so designed that short-circuits on the mother-board do not prevent the CPU from executing its own self-test programme, and faults on other plug-in modules must not prevent the CPU from determining on which module the fault exists. To achieve this, the following criteria can be applied:

- The ROM containing the self-test programme, any RAM that this programme needs for use as a temporary store, and the output and display devices required to report the diagnostic information must be resident on the CPU module itself, along with the watchdog timer and system-reset circuitry.
- All signal busses and control signals on the CPU module must be adequately buffered from the connection to other modules via the mother-board.
- During self-test, the CPU must be able to isolate itself from any inputs from the mother-board which would inhibit proper CPU operation.
- All plug-in modules must have sufficient buffering to isolate their internal circuitry from the mother-board connection, so that a short-circuit between two address lines on a RAM module will be detected as a RAM module fault and not as a mother-board fault.

Ideally, the open loop diagnostics programmes should be integrated with the system software as this generally results in the shortest time to diagnose a fault and is usually the most economic implementation. The use of integrated diagnostics does not disturb the system as much as the connection of an external test instrument or diagnostics module and so is less likely to create conditions different from those in which the fault occurred.

If it is possible for the open loop diagnostics to be made interactive, fault location to component level is facilitated still further. In this case, a terminal of some kind (e.g. a visual display terminal or a keyboard printer) can be used to select programmes designed to test specific modules or assemblies of components and to display the results of the tests. The fault location process can be simplified even further if the interactive diagnostics are menu driven and incorporate a *help* facility to overcome any lack of familiarity with the equipment. This type of assistance is really only practical for large systems or those to which the more detailed diagnostics software can be added from a floppy diskette or emulator, for example.

If memory size limitations prevent the incorporation of open loop diagnostics routines with the control programme, and the computer uses a microprocessor, it is frequently possible for diagnostics to be added by using the emulation technique described in the following section.

17.8 Emulation

This technique allows the operation of a microprocessor in the equipment under test to be emulated by an external test instrument. The instrument is connected by a multi-way cable to the socket in which the microprocessor would be fitted normally (see *Figure 17.3*). Control of the equipment under test can now be exercised from the external emulator, and diagnostic routines can be run on the emulator that would be too large to be integrated with the system software.

Using an emulator, a detailed analysis of the operation of the system can be made. Break points can be specified which cause the programme to halt at predetermined points and allow the memory contents to be examined and modified as required. Alternatively, the programme can be *single-stepped*, allowing the operation of the programme to be followed in detail.

Using in-circuit emulation, the diagnostics can be made highly interactive without having to provide interfaces to terminals, etc., in the equipment to be tested. The technique does, however, require easy physical access to the microprocessor device.

Until recently, emulators were used primarily as tools for the development of software, particularly during the hardware/software integration phase. As such they were not ideal for fault diagnosis work. Recently, however, instruments suitable for use in the field and designed specifically for maintenance purposes have begun to appear. These enable stimulus sequences for signature analysis to be applied via the microprocessor socket and incorporate a test probe for collecting signatures. They also permit the response of a known good module to be measured and stored for later comparison with a suspect module.

17.9 Diagnostic tests

17.9.1 Read only memory

Since EPROM devices have a finite life and are the least reliable devices in most computer based control systems, a test of EPROM and ROM in the closed loop power-up and re-start diagnostics is essential.

Fortunately, read only memories are relatively easy to test. Their contents are fixed and, generally, it is possible to compare these with a reference at full system operating speed and to have confidence in the result.

Several techniques exist for ROM testing, most of which are based on the derivation of a check sum from the contents of the ROM. For example:

- A sum of the entire contents of an EPROM or ROM device. This sum is compared with the check sum held in a data table forming part of the diagnostics.
- Use of a word stored in a device and chosen so that the check sum is all ones or zeros. An all-ones check sum is preferred, based on the failure modes of ROM devices and failure of the whole memory — many address decoding faults, in which blocks of memory would be read twice, give even parity, as would drive or decode circuit faults.

The comparison of device check sum with a reference value stored in a data table has the merit of verifying not only that the device is functional but also that the correct programmed device has been installed.

A variant of the simple check sum is the skew check sum, which can detect multiple word errors. Here the check sum calculation is performed, for example, between bit n of word m and bit $n+1$ of word $m+1$.

17.9.2 Random access memory

A random access memory device is subject to a number of possible failure modes:

- faulty address decoding,
- individual cells stuck at zero or one,
- data-dependent faults,
- degradation of access times,
- saturation in sense amplifiers,
- refresh problems in dynamic memories.

Full testing of RAMs is extremely time consuming. For example, an exhaustive self-test for data-dependent faults in a single 16 k RAM would take several hours. Consequently, only limited testing can be envisaged.

Where a RAM test is part of the power-up diagnostics (or if it is permissible for the data held in RAM to be overwritten), one test method is to load every RAM location with a bit pattern related (by means of an algorithm) to the address of that location. The RAM locations are then tested to check that they

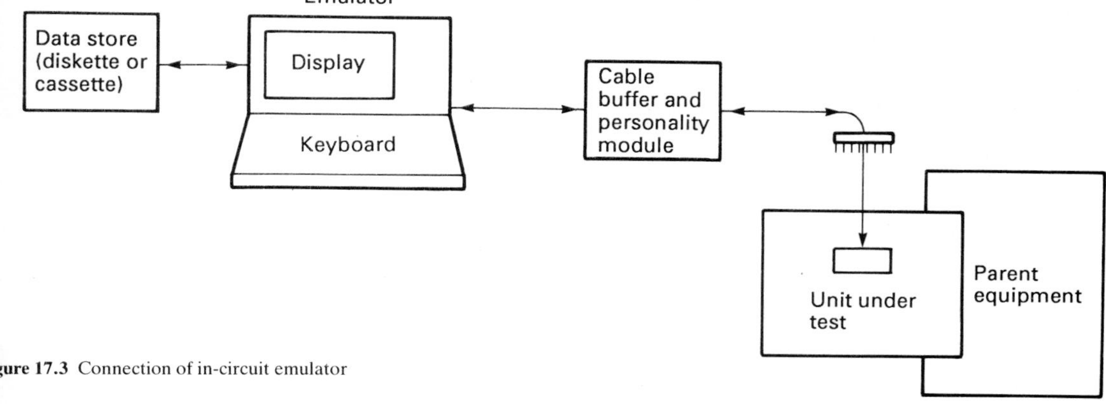

Figure 17.3 Connection of in-circuit emulator

each contain the correct bit pattern. This test can reveal address decoding errors as well as defective RAM devices.

When RAM testing is performed as a background task, care must be taken not to interfere with the performance of the system. It is necessary to save the contents of the RAM locations under test and to prevent those locations from being accessed by the control program until the test is completed and the contents have been restored. This can be achieved by disabling all interrupts during the test, which must now take only a short time to perform. Consequently, it will be necessary to test only a limited amount of RAM at a time and only simple tests, such as loading locations with fixed data, will be possible.

When data is normally held in non-volatile RAM and background testing is employed, the fact that a RAM test is in progress should be stored in a non-volatile register to prevent false data being used in the event of a power-supply disturbance causing a restart while the RAM still contains the test data.

It is possible for extraneous interference to corrupt the contents of RAM, a fault which may go undetected with simple functional testing. Test data can be loaded into a number of RAM locations and tested as part of the background diagnostics. Also, if security of data is of vital importance, then error detection and correction techniques should be employed.

17.9.3 Input/output tests

Input/output testing is an area where operator involvement is usually a requirement. For complete testing, the operator has either to create the *loop-back* for automatic testing or to become part of the loop himself.

17.9.3.1 Parallel input/output

Many programmable peripheral interface and controller devices have a few registers that can be accessed either via ports or as memory. Some of these are like a small RAM and can b tested as such.

Limited testing of parallel input/output ports can be do through the interactive diagnostics. The operator causes th output port to change state either as a response to an extern event (e.g. a switch closure), or under the control of th operator through a terminal.

17.9.3.2 Serial input/output

Serial input/output is usually achieved with universal sy chronous or asynchronous receiver/transmitters (usarts). Ty ically, a serial interface would be based on the V.24[2] (RS23 C[3]) or V.28[4] specification and would employ a usart integrate circuit device.

Some designs of usart have a facility for internally connectir the transmit-data port to the receive-data port under softwa control for diagnostics. If a usart not having this facility is use then a loop-back connector can be fitted or a display termin connected. Using a loop-back connector at the remo peripheral location enables the line connection circuits to b tested.

References

1 *Application articles on signature analysis*, Hewlett-Packa publication 02-5952-7542

2 *List of definitions for interchange circuits between da terminals and data circuit-terminating equipment*, CCIT Recommendation V.24

3 *Interface between data terminal equipment and data com munication equipment employing serial binary data inte change*, EIA Standard RS232-C

4 *Electrical characteristics for unbalanced double-current inte change circuits*, CCITT Recommendation V.28

G W Wiskin B Sc, C Eng, MICE, MI Struct E
Architectural and Civil Engineering Dept, BBC
R G Manton B Sc(Eng), PhD, C Eng, MIEE
Transmission Engineering Dept, BBC

18 Masts, Towers and Antennas

18.1 Civil engineering construction

A proposal for a broadcasting or relay station will include a knowledge of suitable antennas and at least an appropriate location. In order to establish if such a proposition is viable a budgetary estimate of cost to meet a target completion date will be required. Preliminary information from many sources will have to be collated to produce the necessary financial, programming and planning details.

18.1.1 Preliminary research

18.1.1.1 Aims

The location of the site must be established at an early stage as site costs and planning considerations can vary considerably from site to site. Some concept of the sort of structure necessary to support the proposed antenna loading must have been established prior to visiting the proposed sites.

18.1.1.2 Maps

1:50 000 survey maps provide sufficient topographical information to:

- estimate the height of all the antennas that are required,
- assess the length and slope of any access track that is to be established,
- show possible traffic restrictions between the nearest main road and the site.

Larger scale maps will be necessary to assess the size and shape of the plot where a new structure, building or access are to be constructed. If a structure already exists on the site, the owners will probably be able to supply a site plan showing all the features.

18.1.1.3 Site access

However detailed the preliminary research has been, it is imperative to walk the site. The access route or routes from the nearest main road must be checked for low bridges, weight restrictions, tight corners, steep hills, etc.

The type of vehicles requiring access to the site will depend on the construction and operation. Construction traffic will typically include ready-mixed concrete trucks during foundation work. Not only are these vehicles heavy, but they are nearly 5 m high. Lorries will normally be used to supply steelwork, bricks, etc. Earth-moving plant can easily traverse most terrain, but slowly. If it is travelling a distance, it will probably be transported on a low loader to an unloading point as close as possible to the site. Cranes may be needed to lift steelwork or antennas.

Operational traffic will be lighter, perhaps cars for all-weather surfaced roads or four-wheel drive vehicles for access tracks. However, operational maintenance traffic will require access at any time of the day or night throughout the year. Consideration must therefore be given to surface finishes that may be affected by the weather. Conditions, such as flooding rivers, fog or ice and exposed areas liable to high winds or drifting snow all cause problems.

18.1.1.4 Local knowledge

It is extremely helpful to the project if those undertaking surveys can meet the local landowners and tenants, to ensure that boundary marks and access routes are acceptable to all. In particular where wayleaves or shared access are necessary, any practical restrictions must be determined. If good relationships are established with local people early in the project, later problems can be sorted out amicably by the original protagonists. However, if early grievances are not resolved, they will escalate, possibly causing delays to contractors and thus additional expense.

Local information can be useful in assessing the route for access tracks, the ground conditions and underground service runs. It may even indicate temporary storage areas, or construction access outside the site boundary in cases where the chosen site is large enough only for the final works. If the site is restricted for space or adjacent to occupied premises, the safety aspects of construction and maintenance may be more difficult to satisfy.

18.1.2 Surveys

18.1.2.1 Topographical survey

A detailed topographical survey of the site should be undertaken either to check values and update the existing site plan or to produce a new one. The resulting plan will enable the planners and engineers, not having visited the site, to contribute to the work. The plan should be drawn to a scale not smaller than 1:500, so that details such as boundaries, buildings, other structures, overhead and underground services and street furniture can be clearly marked. Contours or spot heights and coordinates of control points should also be given.

18.1.2.2 Soil survey

A soil survey may be necessary for some sites. This could be limited to establishing the depth of peat that overlays the rock across the site using trial pits and probes. At the other end of the scale, it could involve a substantial number of boreholes, with the appropriate sampling and laboratory tests. These may be necessary to establish the bearing capacity and shear strength of the different strata underlying the site. The water table will be of interest during foundation construction, particularly if saline and tidal. Parameters such as the need to use sulphate resisting cement in the concrete, or whether to design foundations as submerged, will also be obtained from the survey data.

18.1.3 Programming and approvals

18.1.3.1 Programme

At the start of a project it will be necessary to obtain information from several sources on duration of work items and their interaction with others. The following items should be evaluated in programming the construction of a new station:

- project brief including budgetary cost,
- financial approval,
- preliminary/sketch designs,
- site acquisition and planning approval,
- finalized requirements and let contracts,
- construction of access track,
- design, supply and construction of transmitter building,
- design, supply and erection of antenna support structure,
- fencing and landscaping,
- supply and installation of transmitter equipment,
- supply and erection of antenna,
- testing.

Where existing structures can be utilized, these items will not all be necessary. All programming constraints must be satisfied. Typically, arrangements need to be made so that in the interests of safety, work does not take place under the antenna support structure whilst men are working aloft.

18.1.3.2 Approvals

The local planning authority will generally require to see the site plan (1:200 scale) and an elevation of the proposed structure showing the antenna configuration intended. Specifications and calculations may also be required to obtain local building approvals.

It must therefore be clearly established what type, size, height and bearing is planned for each antenna. However, structures will undoubtedly be designed to carry a number of speculative antennas to allow for future expansion and/or allow other users to share the facilities. Such antennas may be excluded from the original planning application, but the structural engineer will still require the data as part of his design brief. Before a type of structure can even be selected, the designer will require the types and sizes of feeders to each antenna, the required means of access up the structure and the design wind speed.

18.1.4 Structures

18.1.4.1 General

The structure may be a simple wood pole or a tall guyed mast, but the principles in selection will remain the same. They are that the structure shall:

- be strong enough to withstand (a) the maximum design wind speed, with the specified antenna loading, and (b) the specified wind speed and icing conditions with that antenna loading,
- be stiff enough to limit the deflection of each antenna to less than that specified at its operational wind speed,
- be safe to be climbed by those staff trained to do so,
- be constructed within the budget and time scales allocated,
- be maintainable for its intended life-span,
- not impose unacceptable environmental or physical conditions on the locality.

18.1.4.2 Poles

Cylindrical poles of wood, steel or aluminium or welded lattice poles can support light antennas up to a maximum height of approximately 17 m in low wind speed areas. They rely on their bases being buried at a sufficient depth in compacted ground (or concrete) to stabilize them. Access is normally from a removable ladder, with step bolts over the top section. Antenna installation and maintenance work should therefore only be undertaken by fit trained personnel.

Figure 18.1 A 17 m timber pole

18.1.4.3 Towers

Self-supporting towers can vary in height from 10 m to 300 m. The ratio of the tower height to the base width of the section under consideration should not exceed 8:1 over the top 40 m whilst carrying omnidirectional uhf antennas and microwave dishes. For all other structures the ratio should not exceed 10:1. However, for every additional 60 m of tower height the ratios should be reduced by 1.

The face width of the structure should be no less than:

- 0.4 m where external access only is provided,
- 1.2 m for a square tower with an internal ladder,
- 1.45 m for a triangular tower with an internal ladder.

Moreover, the face width of the structure should be no less than half the width of any ancillary supported at that level.

The base width of the structure should be as large as possible to minimize the foundation forces, but not so large that the exposed face area of the tower increases too much. This optimization may be achieved with sloping leg members. However, it is recommended that the point of intersection of the projected lines of the legs is higher than 60 per cent of the tower height.

Figure 18.2 A 45 m tower

A bend line occurs at the point where the tower legs change slope. Several bend lines will occur in 'eiffelized' structures. A large horizontal force is developed at these bend lines, and there are usually local moments resulting from the location of the bend just above or below the member joint node. To cater for these circumstances, a horizontal member must be provided at this level across the full width of the tower.

The shape of the main bracing members will depend on the *height:width ratio* (h/w) of the panel and its size. For the top narrow parallel sided panels of $h/w>1.5$, 'Z' shapes may be used. 'X' shapes are common where $0.7<h/w<1.5$. However, 'K' shapes are used for large panels where $0.5<h/w<1.0$ and often where access is required through a tower face, e.g. from an internal ladder to an external platform.

The main members so described form the main structural frame and will be designed in accordance with the specified standard[1,2,3] to resist forces due to wind and gravity loads on the structure and its ancillaries. However, secondary members may be needed to restrain the main members against buckling. The secondary members must be capable of providing such restraint, and have adequate strength to support ice, platforms or ancillaries that load them.

Auxiliary members, including antenna mounting and platform steelwork, will be designed to support their local loads plus ice where applicable.

The tower legs are usually supported on individual foundations.

The critical design criteria however are rarely bearing or shear, but uplift and overturning. So a foundation bearing directly onto sound rock which would satisfy any building, may not be adequate unless satisfactorily anchored to the rock. A cone of soil above the foundation pad may be utilized to provide the dead weight. However, where the ground area required to resist the uplift forces at one leg impinges on a similar area around another leg, a combined foundation should be considered. For small structures this takes the form of a raft foundation.

Figure 18.3 A 225 m mast with replacement under construction

18.1.4.4 Masts

Guyed masts vary in height from 10 m to some of the world's tallest structures. However, the following parameters will provide a reasonable structural profile.

The mast column will be supported at various levels by sets of tensioned stays. The ratio of the height between stay levels and the face width of the column should not exceed 40:1. The face width of the column should conform to those parameters given for towers. The bracings shapes and number designs will also conform to the parameters given in section *18.1.4.3*.

The normal stay arrangements are for three stay lanes 120° apart for triangular mast columns and four stay lanes 90° apart for square mast columns. These stays will be anchored to foundations so that the vertical angle between the stay and the ground plane is between 30° and 60°. To minimize costs, these foundations may be arranged to support several levels of stays.

The stay anchors are usually blocks of concrete of sufficient weight to resist the uplift forces, and sufficient width and depth to resist the sliding and overturning forces. For light or temporary structures, ground anchors can be employed, and rock anchors can provide a satisfactory anchorage solution for large structures on suitable underlying rock.

The column base is usually tapered where possible to form a structural pin. This largely reduces the foundation design condition to one of direct bearing. The vertical load is mainly due to the thrust in the column exerted by the tensioned stays. However, where a fixed based mast is installed, the foundation designer will have to consider much larger shear and overturning forces.

Figure 18.4 Roof mounted poles

18.1.4.5 Roof-mounted structures

Roof-mounted structures are potentially the easiest and cheapest to utilize. However, they have the reputation of being time-consuming projects and prone to problems.

Access to the pole, mast or tower site will be either up the face of the building or via internal lifts and stairs. Short, light members and protection to finishes will be major requirements. Protection to roof finishes will be particularly important to prevent water penetration into the building. Wherever possible, it is wise to avoid puncturing the existing finishes, even to provide anchorages.

It is important that the antenna support structure loads are transmitted directly to the building frame. The assumption that the building can transmit the loads even short distances onto its structural frame can lead to local failures.

The high intensity short duration loads experienced by these structures are known to have caused damage where only normal building design standards have been applied, so antenna support factors of safety should be maintained.

18.1.4.6 Existing structures

Where it is proposed to utilize an existing structure, the antenna and feeder type, their location and the method of attachment should be agreed and approved by the owner prior to installation. This may take additional time, but can prevent unnecessary costs due to misunderstandings and unacceptable details.

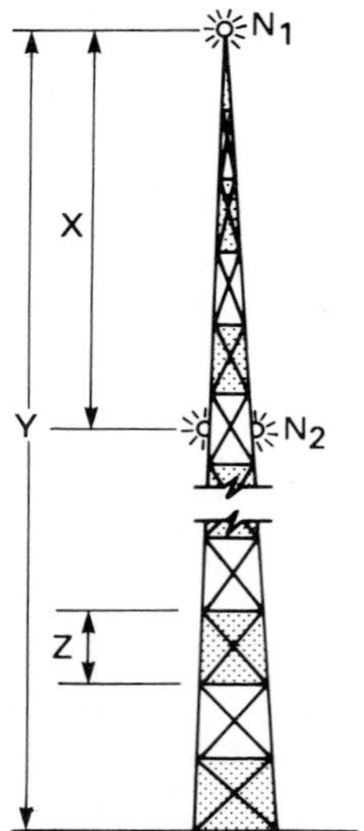

$$\text{Number of lights} = \frac{\text{Y (in metres)}}{45}$$

$$\text{Light spacing} = X = \frac{Y}{N} \leqslant 45\text{m}$$

$$\text{Band spacing} = Z = \frac{Y}{7\ (9,11,\text{etc})} \leqslant 30\text{ m}$$

Figure 18.5 Marking and lighting a tall structure

18.1.5 Operational considerations

18.1.5.1 Access

Once the type and size of structure has been ascertained, it is easy to provide something to meet the initial requirements and lose sight of its reason for existence. Antennas are to be attached to the structure, so someone will be required to install them and service them. Depending on the height and location of the structure, aircraft warning lights may be needed. If so, they too will need servicing. The designer's brief should indicate to which parts of the structure access is required, what type of person will require to use that access and how often.

The simplest access will be to climb the face of the structure. This can utilize step bolts, an external ladder or face bracing, with or without ladder step bracing. This form of access is suitable only for regular climbers who have demonstrated their ability and fitness. Even so, they must be provided with suitable safety equipment and trained to use it. If a fail safe fall arrester device is provided alongside the ladder, other properly equipped, trained, fit personnel could also ascend safely.

Since climbing is tiring, especially in cold windy weather, a vertical or near vertical cat ladder is normally provided for the full height of the structure, with rest platforms at 10 m intervals. These rest platforms are normally of sufficient size for only one person to stand or sit. They can also be used as passing places. Working platforms, with flooring, handrails and toe boards may be specified, typically at antenna mountings, aircraft warning lights or ladder changeover levels. In some cases, these platforms will provide access from the inside to the outside of the structure. Detailing these areas to maintain acceptable unobstructed walkways can prove difficult, unless incorporated in the original design.

Where occasional climbers require access to parts of the structure, ladders with safety hoops and/or lifts and platforms that satisfy recognized safety standards[4] should be provided. Temporary mechanical access for antenna erection and maintenance of the structure will be necessary on larger structures. Appropriate locations may therefore be designated as safe rigging points, and possibly have a lifting jib built in. All rigging equipment, chairs, winches and lifts will have to be tested, checked by a competent person, recorded and certificated in accordance with the local/national safety requirements[4].

18.1.5.2 Maintenance

Maintenance work on the *structure* (as opposed to the antenna) is essentially required to ensure that it will be able to perform its function safely for its intended life span. Regular checks on the condition of the surface are necessary to detect any corrosion or damage. Repairs can then be initiated as required.

The designer can, at the risk of increasing the capital cost, minimize the amount of maintenance that will be necessary. He can ensure that the materials supplied and applied will last as long as reasonably possible. Steelwork would be galvanized and possibly painted for additional corrosion protection as well as for aircraft warning bandings. Mast stays would be made up of galvanized steel wire strands bedded in an impervious material, or be of a non-corrosive material.

The stays will need retensioning to maintain the correct initial tensions for stability and control of possible stay oscillations. The regularity of retensioning will depend on the age of the stays and their construction. Typically prestressed steel wire rope will bed down over an initial period of 12 months, and once adjusted should then perform without attention for another five years. Non-metallic ropes generally perform less well and need more regular attention.

18.1.5.3 Radio frequency hazards

Whether the work be inspection, repair, painting or greasing, it is likely that someone will need to work in the vicinity of operating antennas. The radiation field must therefore be checked with a hazard meter to ensure that it is within the permissible limits[5]. People may not remain in areas where such limits are exceeded (see *Table 18.1*), so the transmission from the offending antenna will have to either operate at reduced power or be shut down. Where reduced power or shutdowns are essential, maintenance will have to be programmed to take place at times convenient to the broadcasters' schedule. This may mean inefficient use of time on the structure, and increase the cost.

	Root mean square values		
Frequency, GHz	*Electric field strength V/m*	*Magnetic field strength A/m*	*Power density W/m²*
0.03 – 0.4	61.4	0.163	10
0.4 – 2	$97.1\sqrt{f}$	$0.258\sqrt{f}$	25f
2 – 300	137	0.364	50

Table 18.1 Reference levels for continuous exposure to electromagnetic fields

Sometimes an antenna will radiate a high rf field where shutdowns or reduced power working are difficult to organize, e.g. an antenna broadcasting 24-hour television. In this case, to ensure continuity of service a reserve antenna operating from the same structure or an adjacent one should be considered. This will increase the total loading on the structure, so is best considered and allowed for at the design stage. Alternatively, an existing structure may be re-analyzed and possibly strengthened to accommodate additional antennas as requirements change.

18.1.5.4 Security

All sites should be so protected that the general public, behaving reasonably, will not cause damage to the installation, neither will they be harmed by it. However, children and animals will not always behave reasonably so they must be prevented from causing trouble.

The outermost line of security will be the site boundary fence. Depending on the type of fence and type of station, a local security fence may be required immediately around the installation, suitably posted with notices warning of the appropriate hazards, i.e. high voltages and non-ionizing radiation. If this fencing is climbable, and the site is accessible, some anti-climbing guard will be required on the structure to prevent unauthorized ascent. Suitable protection is required to prevent cows eating stay grease and rubbing undone unsecured rigging screws. Rodents and camels have been known to eat insulated cable. Where vandalism or terrorism is a problem, it may prove necessary to provide security guards or electronic surveillance equipment.

To ensure that access is limited only to authorized persons, a pass system should be used for attended sites, and keys issued to appropriate people for unattended sites.

18.1.6 Structural design

18.1.6.1 Loading

The main loading on the structure will be the wind force exerted against the structural frame and all the ancillary ladders, platforms, antennas and feeders attached to it. National standards[1,2,3] give guidance to obtain the appropriate design wind speed for the terrain encountered within that country. These structures are normally designed using the mean hourly wind speed likely to be exceeded at least once in 50 years in that topography. The associate gust duration will vary between 1 and 16 s depending on the size of the structure.

The critical wind directions for a *square structure* are:

- directing onto a face, which normally gives maximum bracing forces,
- into a corner, which normally gives maximum leg forces and foundation loads.

For a *triangular structure* the above two directions will give the maximum and minimum leg loads, but an additional direction is:

- in the plane of the face, giving critical bracing forces under some conditions.

These wind directions will provide the maximum and minimum stay tensions for guyed masts.

When ice loads are considered, the effective wind area of all items will be increased. In some cases, it may not be sufficient to allow for a thickness of radial ice around members. Under extreme conditions, masts in particular are known to ice across the full face of the structure when closely spaced feeders are attached to, or adjacent to, that face. On most structures, severe icing is unlikely to occur at the design wind speed, so a more appropriate wind speed is selected. It is therefore unusual for the icing condition to be critical for towers at low altitudes in temperate climates like that in the UK.

For masts taller than 100 m, a series of load cases should be considered to assess the susceptibility of the mast to local gust loads. These can be simulated by alternately loading and unloading adjacent spans as recommended in IASS[6]. Similarly, masts prone to icing should be analyzed with ice on stays on one side only.

Mast stay tensions should be assessed for performance of the slackest stay, as this could lead to an instability condition. The slackest conditions should be outside the working range of wind speeds, and can be assessed by analyzing results at varying wind speeds, say 0.7 × design wind, design wind and 1.2 × design wind.

Masts designed with initial stay tensions in the region of 10 per cent of the ultimate breaking load of the stay rarely exhibit instability problems including those caused by stay galloping.

18.1.6.2 Factors of safety

Factors of safety at the design wind speed condition are usually taken as:

3	for stays (ropes and fittings),
1.7	for buckling of steelwork,
2	for foundations using soil properties,
1.5	for foundations using only dead weight of the block.

Factors of safety at a survival wind speed condition (1.2 × design wind) are usually taken as:

1.5	for stays,
1.15	for buckling of steelwork,
1.2	for foundations.

Where this condition is satisfied, the factors of safety at the design wind speed can be relaxed.

18.2 Electrical design of antenna systems

The provision of an antenna system for the transmission of terrestrial television signals involves the exercise of many engineering disciplines from economics through transmission line theory and vector arithmetic to metallurgy. The design of the antenna cannot be considered in isolation because, for a given service coverage, its physical shape and hence cost are inextricably bound up with the economics of transmitters, feeders and support structures. It is unlikely, therefore, that an ideal antenna for a given purpose can be bought 'off the shelf'. The need to propagate adequate signals to a given service area at lowest cost and at the same time avoid causing interference to viewers in neighbouring areas, results in many antennas having to be designed for specific sites.

18.2.1 Definition and design philosophy

The antenna system comprises all the equipment that is necessary to carry radio frequency energy from the transmitter(s) and to propagate it into space. In addition to the antenna itself with its feeders, the largest installations may include diplexers for combining the output power of transmitters operating on the same carrier frequency, and *channel combiners* for combining the signals of several channels into a single antenna. An antenna system which serves a large population may be split into sections to give reliability of service.

The antenna itself may be fed in two halves, one above the other, through separate main feeders. Transmitters may consist of pairs of amplifiers with common drives where, in the event of failure of one transmitter or half-antenna, the service can continue to run with a reduction in *effective radiated power* (erp) of 6 dB over most of the service area. There is a need to diplex and split the outputs of two amplifiers rather than feed them directly to separate half-antennas. The reason for this is considered in section *18.2.12*. Low power stations serve smaller populations and are more easily engineered to have higher factors of safety. Usually, therefore, they have single sound and vision transmitters per channel with a single antenna and feeder. Schematics of typical high and low power antenna systems are shown in *Figure 18.6*.

18.2.2 Types of antennas

The following paragraphs describe antennas for the uhf bands 470–860 MHz (wavelengths 0.64–0.35 m). The principles for vhf (30–300 MHz) remain the same, although some of the components are larger because of the increased wavelength.

18.2.2.1 High power antennas

The elementary radiating part of a high-power uhf antenna is a panel of radiating elements, typically about 0.5 m wide and four wavelengths high. The panel may support eight horizontal dipoles or four vertical slots in front of a reflecting screen; either system provides horizontal polarization. The rear of the panel usually has a single input connector and houses a system of branch strip-line feeders and matching sections behind a rear cover. Up to eight panels may be required in a single tier, depending on restrictions imposed by any support structure, to obtain a near-omnidirectional *horizontal radiation pattern* (hrp) or whatever directional hrp is called for by the service planning engineer. The panels will be fed by a suitable feeder harness to give the appropriate feed currents.

The same horizontal pattern of panels and feeders will be repeated in a number of tiers in each half-antenna to give the required antenna gain. There are, however, important considerations to be taken into account in the design of the feeder system which takes power from the main feeders to the individual tiers. These are described in section *18.2.5*.

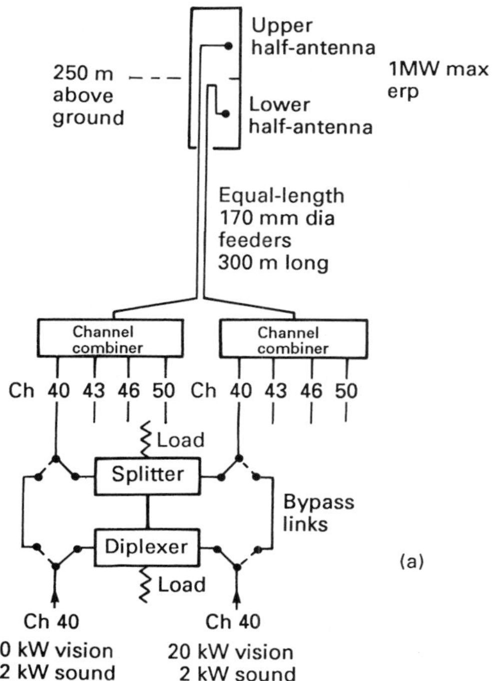

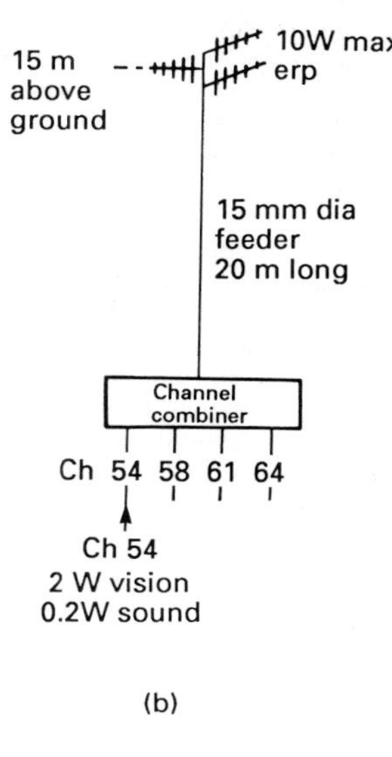

Figure 18.6 Complete transmitting antenna systems: (a) high power, (b) low power

Figure 18.7 Interior view of a high power uhf antenna (Alan Dick & Co Ltd)

Panels for a high power station may be arranged around the faces of a square or triangle with interior access for a maintenance engineer. The level of radiation inside should ideally be restricted so that the maintenance engineer can climb through the antenna without hazard when it is powered.

Usually all the panels will be housed inside a cylinder of glass-reinforced plastic (grp) to reduce wind loading, to prolong the antenna's life and to make the environment more convenient for maintenance. The panels may be fixed to the sides of a steel lattice mast, in which case the grp cylinder may be a relatively thin shell. Alternatively, the grp cylinder may be thick enough to be load-bearing and may support the panels as a cantilever without the benefit of a load-bearing spine. In either case, the dimensions across the antenna seen in plan must not exceed a wavelength or two, otherwise it will not be possible to maintain a reasonable hrp over the bandwidth required for two or more channels.

Whatever the type of antenna, it is important to minimize bimetallic corrosion by complying with a standard code of practice for materials in contact, e.g. BS PD5484:1979.

18.2.2.2 Low power antennas

Vertical polarization is used for most UK low power relay stations to minimize interference to reception in main station areas. (This is not a common practice elsewhere.) Antennas giving almost omnidirectional hrps may consist of up to 16 vertical dipoles stacked collinearly off the side of a single metallic support pole. The largest antennas may be housed inside a grp cylinder 0.4–0.9 m in diameter; the antenna would then be lowered through the tower for maintenance purposes. Alternatively, more complicated hrps may be produced by combinations of log-periodic antennas fed with specified currents and pointing in appropriate directions.

18.2.3 Design procedure

Primary data for the design of an antenna system will be generated by the service planning engineer who will state the location, the height on the support structure and the erp required in all directions of azimuth (and possibly at some angles of depression) from the antenna. This will be given in the form of a template (see *Figure 18.8*) stating the minimum erp required in some directions and the maximum erp that can be permitted in other directions. (The *erp* is the power which would have to be fed into a suitably oriented half-wave dipole in the same position as the transmitting antenna in order to produce the required field in the specified direction. For negative video modulation with positive synchronizing pulses, the transmitter power and erps are expressed in terms of the power at the peak of the synchronizing pulses (kW sync).)

The antenna designer starts by arranging radiating elements (panels or log periodics) so that they produce an hrp to fit the required template on all the required frequencies. Initial work can be carried out by computation but, for all but the simplest directional hrps, the hrp has to be proved by measurement. It is necessary for the designer to construct at least one tier of the antenna on a turntable so that its hrp can be measured by rotation.

The measuring antenna for this purpose has to be at a suitable distance and height above ground so that reflections from other objects and the ground itself do not interfere and so that the field from the measuring antenna, used as a transmitter, illuminates the measured antenna uniformly. It is reasonable to specify that the max:min ratio of a four-channel antenna omnidirectional hrp should not exceed 5 dB in any channel. Once the hrp has been established, the ratio of maximum power gain to mean power gain can be determined by dividing the area of the circle circumscribing the hrp diagram (plotted radially in linear voltage) by the area of the hrp diagram itself. Following

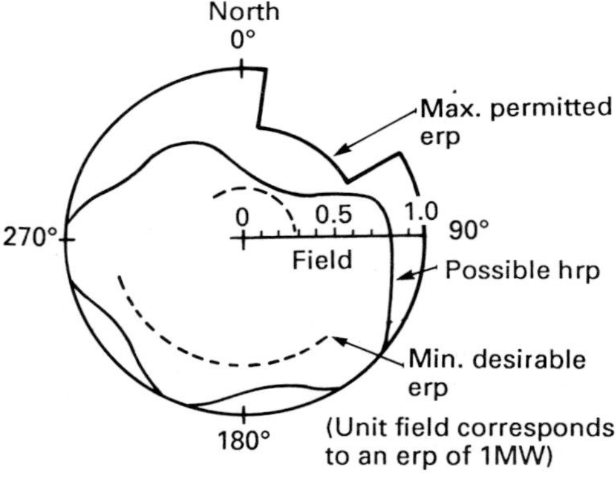

Figure 18.8 Template for horizontal radiation pattern

this, the designer can calculate the required mean intrinsic antenna gain, G, from a knowledge of the transmitter power available and various assumed losses in the system as shown in *Table 18.2*. In this example, G = 15.8 dB.

Transmitter peak sync power	40kW	
Max erp required	1 000kW	
System max gain required		+14dB
Network and combiner loss	−1.2dB	
Main feeder loss	−1.5dB	
Distribution feeder loss	−0.6dB	
Beam tilting and null-fill loss (see section 18.7.5.1)	−0.5dB	
Antenna max/mean gain	+2.0dB	
Antenna mean intrinsic gain	+G dB	
System max gain available		G−1.8dB

Table 18.2 An example of the summation of losses and gains in a high power transmission system

18.2.4 Antenna gain

The mean intrinsic power gain of an omnidirectional stack of dipoles in a panel-type antenna is determined by the vertical length of the antenna measured in wavelengths. It is given approximately by:

$$10 \log_{10} (1.2 \text{ antenna length/wavelength}) \text{ dB}$$

In the example given in *Table 18.2*, the appropriate intrinsic gain can be provided by a 32-wavelength antenna, which can conveniently consist of eight tiers of four-wavelength panels.

The maximum intrinsic gain of an array of log-periodic antennas has to be calculated from a knowledge of the maximum gain of a single antenna. If there is more than one antenna in a single tier, the maximum power gain of the single antenna will be reduced in the ratio of the areas of the array hrp to that of the single antenna. The power gain can be increased by a factor approximately equal to the number of tiers.

18.2.5 Vertical radiation pattern

The vrp of a vertical stack of identical radiating elements, regularly spaced and fed with equal co-phased currents, consists of a main lobe in the horizontal plane and a number of subsidiary lobes above and below the horizontal which are separated by nulls. The first step in adapting this for broadcasting is to tilt the main lobe downwards towards the edge of the service area to avoid wasting power. The angle of tilt, θ_T, will typically be about 0.5°, and this can be achieved by feeding the lower tiers through progressively longer feeders so that the phase difference between adjacent tiers is about $360D \sin\theta_T°$, where D is the number of wavelengths between tiers. The resulting vrp is shown in *Figure 18.9*(a). The next step is to modify the phases in such a way that an adequate signal (which does not vary too much from one frequency to another) is provided at all angles of declination θ down to about 25°.

18.2.5.1 Specification

A specification that can reasonably be applied to *null filling* is that no amplitude, E_θ, in the vrp should be less than 50 per cent of the locus of the envelope of the maxima of the subsidiary lobes of the unfilled vrp, i.e.:

$$E_\theta \geqslant E_{max}/2\pi A \sin(\theta - \theta_T) \text{ for } 90°/\pi A<\theta<15°$$

where θ is the angle of declination from the horizontal, E_{max} is the maximum amplitude of the main beam, and A is the length of the antenna in wavelengths.

A further constraint that helps to ensure uniformity of field is that the ratio of any adjacent maximum to minimum in the erp should not exceed 6 dB. *Figure 18.9*(b) shows the vrp of an antenna consisting of eight tiers of four-wavelength panels which has been successfully filled to 9° of declination by phase perturbations. In practice, the antenna designer has to resort to other methods, such as filling the nulls of individual panels, in order to fill nulls at greater angles. Methods of filling nulls have been described by Hill[7]. The action of filling nulls and tilting the beam of an antenna results in a reduction of antenna gain which is referred to as *beam tilt and null fill loss* (see *Table 18.2)*.

18.2.5.2 Maximum aperture

Because the 3 dB beamwidth of an antenna, in degrees, is given approximately by 50/(antenna length in wavelengths), the limiting length of antenna that can be used in practice is about 40 wavelengths, this being determined by the structural stability of mast plus antenna that can reasonably be achieved. Under

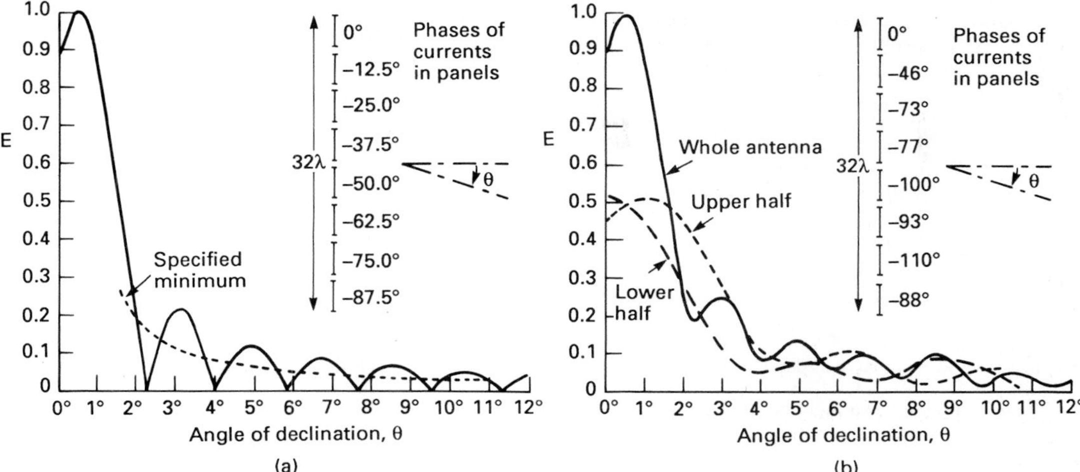

Figure 18.9 Vertical radiation patterns of eight tiers of four-wavelength panels: (a) beam tilted, not null filled (loss = 0.02 dB): (b) beam tilted, null filled (loss = 0.45 dB)

the worst environmental conditions, the antenna must not be allowed to depart from the vertical by more than 20/(antenna length in wavelengths) degrees.

18.2.5.3 Computation

The vrp of a high-power uhf or vhf antenna cannot be measured directly in the same way as an hrp. This is because of the physical size of the whole antenna and the large unobstructed area that would be required for measurement. In practice, the feedpoints on each tier of panels are fitted with directional couplers or voltage probes so that the relative levels of applied voltages can be measured. The vrp, E_θ, can then be computed from the formula:

$$E_\theta = \left| \Sigma_{n=1}^{N} V_n(\Theta) C_n \exp j \left\{ \varnothing_n - (2\pi d_n \sin\Theta) \right\} \right|$$

where N is the number of tiers, $V_n(\theta)$ is the vrp of the nth tier, C_n is the amplitude of current in the nth tier, θ_n is the phase of current in the nth tier, and d_n is the distance in wavelengths of the centre of the nth tier below the reference plane.

The intrinsic gain, G_d, of the antenna relative to that of a dipole is:

$$\frac{1}{1.64} \int_{-\pi/2}^{+\pi/2} \frac{E_\Theta^2}{E_{\Theta max}^2} \cos\Theta \, d\theta$$

The beam tilt and null fill loss is:

$$10 \log_{10} \frac{E_{\Theta max}^2}{N \Sigma c_n^2} \text{ assuming } V_n(\Theta)_{max} = 1$$

VRPs of antennas consisting of dipoles or log-periodic antennas may also be beam tilted and null filled by various means, if necessary (*Figure 18.9*).

18.2.6 Feeders

All vhf and uhf antennas are constructed so that they can be fed by coaxial (unbalanced) feeders. Feeders vary in size from 170 mm diameter main feeders, which are needed to carry large amounts of energy over large distances with a minimum of attenuation, to 15 mm diameter feeders, which may be used to feed the elements of a low power antenna. Most antennas and transmitters are standardized to an impedance of 50 ohms but, where 170 mm diameter feeders have to be used to minimize attenuation above 850 MHz, it may be advisable to use 60 or 75 ohm feeders with suitable transformers. The purpose of this is to extend the usable frequency range before the TEM energy breaks up into waveguide modes.

For convenience of installation and to keep down the cost of maintenance, most feeders used in the UK are of the continuous semiflexible type with corrugated conductors; outer conductors are protected by a sheath of black pvc. The largest feeders, and those that carry the highest powers, are mainly air-spaced with helices or spacers of dielectric to support the inner conductor. These feeders need a pressurized supply of dry air or inert gas (about 100 mb) to prevent the ingress of moisture. Smaller feeders often have a foam dielectric to completely fill and hermetically seal the space between the inner and outer conductors.

18.2.6.1 Feeder ratings

Table 18.3 shows the mechanical and electrical properties of some of the semiflexible coaxial feeders that are available. The power rating and attenuation are proportional to 1/√frequency and √frequency respectively. The power rating also depends on the ambient temperature and standing wave ratio. Reference should be made to the manufacturers' catalogues for details of this. The peak rf voltage and mean power that should be taken into account for multichannel antenna systems depend on transmission standards.

In the UK, the sound carrier is frequency modulated and is one-tenth of the vision sync pulse power. The greatest average power occurs in a black field, where the rms carrier voltage for the picture content rises to 76 per cent of the rms voltage of the sync pulses. It may then be shown that the equivalent mean power for both sound and vision is 0.71 sync power (the corresponding mean power for an average picture is only 0.53 sync power). Mean powers for each channel must be added.

The peak rf voltage for UK standards is 1.86 times the sum of the rms voltages in the sync pulses of each channel. Before choosing a suitable feeder, the above figures should be multiplied by any safety factors that are thought to be necessary. Typically these are 1.5 for mean power and 2.0 for voltage but they will depend on what system of feeder protection is employed (see section *18.2.9*). The mean power

Diameter over sheath, mm	*Dielectric**	*Velocity factor,* %	*Max peak rf volts,* kV	*Mean power at 600MHz,* kW	*Attenuation at 600MHz,* dB/100m	*Min bending radius,* mm
170	A+H	97	17	53	0.49	1200
120	A+H	97	12.5	26	0.72	500
90	A+H	96	9.7	17	0.93	380
50	A+H	95	5.2	6.1	1.61	180
50	F	88	5.6	4.6	1.96	300
30	A+H	93	2.7	2.7	2.98	100
30	F	88	3.0	2.5	3.10	120
15	F	88	1.6	1.1	5.48	70

* Dielectric : A+H = air+helix; F = foam

Table 18.3 Semiflexible 50-ohm coaxial feeders (Radio Frquency Systems, Hannover. Division of Kabelmetal Electro GmbH)

and peak voltage, expressed as a function of vision sync pulse power, vary from one country to another as the transmission standards change.

18.2.6.2 Feeder uniformity

Main feeders, in particular, must have a uniform impedance if the transmitted picture quality is not to be marred by multiple images. To ensure this, the input voltage reflection coefficient should not exceed 5 per cent at any frequency within the video band of each channel when the feeder is terminated by a resistor equal to its average characteristic impedance. In addition, the voltage reflection coefficient of a 0.1 μs sine-squared pulse at vision carrier frequency or colour subcarrier frequency should not exceed 1½ per cent.

18.2.7 Lightning protection

Any antenna which forms the topmost part of a mast should be equipped with 1 m lightning protection spikes electrically bonded to the mast structure. In addition, the outer conductors of all coaxial feeders inside a mast should be electrically bonded to the mast at the antenna and where they leave the mast at ground level.

18.2.8 Antenna impedance

If the impedance of an antenna and its distribution network at the top of the mast is not sufficiently well matched to the characteristic impedance of the main feeder, a fraction of the signal applied to the antenna will be reflected back to the transmitter via the channel combiners and diplexers. In general, the output stage of the transmitter will not absorb this signal, and a large percentage of it will be returned to the antenna. Here it will be transmitted, but delayed in time and attenuated by two traversals through the main feeder and equipment at ground level. The reflected signal may then be seen by viewers as a delayed image or ghost.

For an antenna with a feeder run longer than 50 m, acceptable limits for the levels of delayed signal, expressed as percentages of the primary signal amplitudes, are obtained if values lie below one of the two alternative lines shown in *Figure 18.10*. For the reflection coefficient of the antenna itself, these figures may be relaxed by twice the feeder attenuation and 2.5 dB for loss in equipment at ground level. *Figure 18.10* applies to the UK transmission standard, where the video band extends 5.5 MHz above the vision carrier and 1.25 MHz below it. Similar lines can be drawn for the different video bands used in other countries.

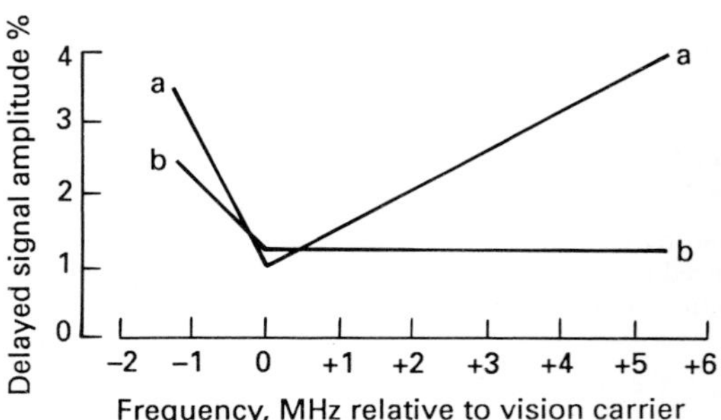

Figure 18.10 Impedance specification : alternative maximum permissable levels of delayed signal

f the antenna is divided into halves it is usual to permit three times the amplitude of the normally specified delayed image to be radiated when only a half-antenna is being used under emergency conditions. For antennas with shorter main feeders, a given delayed image will be less visible, therefore a relaxed specification can be applied. For example, the reflection coefficients in *Figure 18.10* can also be increased by a factor of 3 if the feeder is less than 20 m long.

18.2.9 Directional couplers

Directional couplers may be used at various points in an antenna system to monitor the forward and reverse flows of power. They are particularly useful if situated at the upper or lower ends of the main feeders, where a change in the ratio of reverse to forward power may indicate a fault in the antenna. For this purpose, the directivity of the reverse coupler needs to be about 40 dB. The output of the couplers can, if necessary, be made to provide executive control over the transmitter output power if the reverse power reaches a predetermined value.

18.2.10 Hybrid or diplexer

A directional coupler, which splits power equally between its direct output and its coupled line (*Figure 18.11*), constitutes one form of hybrid or diplexer. This form of hybrid has a phase difference of -90° between its two output ports.

There are other forms of coaxial hybrid, where the power is split equally, but is either co-phased or 180° out of phase, depending on which input port is used. Hybrids are used to diplex the power of two transmitters, to split power equally between two loads, to provide quadrature phase feeds in phase-rotating systems, or to form parts of channel combiners.

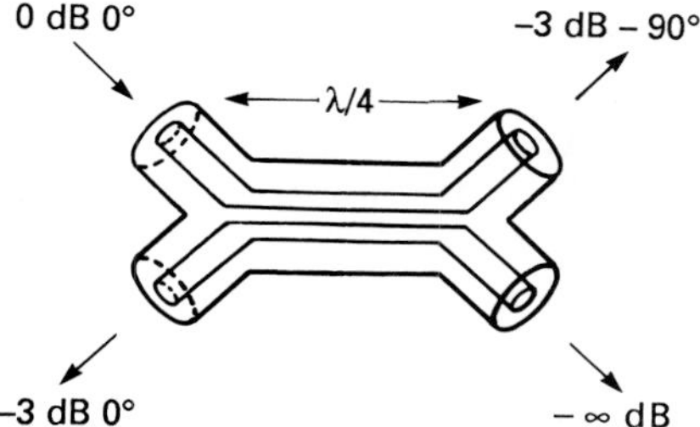

Figure 18.11 '3 dB coupler' hybrid

18.2.11 Channel combiners

Channel combiners are required at ground level when more than one channel is fed into a single antenna; it is usual to provide duplicate combiners when the antenna is fed by two main feeders. The main principles of all combiners for television are the same. For example, the input ports for the several frequencies must be isolated from each other by at least 30 dB to keep intermodulation to an acceptable level. The input voltage reflection coefficient must be maintained at a low level, over as wide a band of frequencies as possible, by means of a suitably connected absorber load. This load will help to ensure the stability of transmitters and provide a sink for spurious frequencies and remaining intermodulation products.

The principal elements of channel combiners and also of sound/vision combiners are hybrids and resonators. The two combiner configurations are shown in *Figure 18.12*. The Rotamode described by Hutchinson[8] can be arranged to have one narrow band input port and one wide band input port, as would be required for a sound/vision combiner. Alternatively, it can be used to combine channels in series as in *Figure 18.12*(a). The three combiners shown in (b) consist of hybrids with transmission line resonators. They have equal bandwidth ports with alternate stop and pass frequencies; such combiners can be used to add channels in parallel. The subject of channel combiners in general has been described by Manton[9].

18.2.12 The overall system

A schematic for a complete antenna system with duplicate main feeders and pairs of transmitters is shown in *Figure 18.6*(a). At first sight it is not obvious why separate transmitters should not feed separate half-antennas independently, without the complication of diplexing output powers and then re-splitting them to feed half-antennas. This becomes clearer on examination of *Figure 18.9*(b). It will then be seen that the vrp of the whole antenna in the vicinity of alternate minima is formed by the subtraction of the vrp of one half-antenna from that of the other. Consequently any slight differences between the modulation characteristics of signals fed to the two half-antennas will be exaggerated in areas covered by these minima.

The effect of using a transmitter diplexer is to provide a sink load for modulation differences and to provide a unified output with average modulation characteristics that can be split between two half-antennas. (It follows that, if one transmitter fails, the remaining transmitter will continue to operate but will transfer half of its power to the sink load and the other half to the antenna.)

The use of a hybrid as a splitter transformer ensures that any delayed images, caused as a result of voltage reflections from the antenna or feeders, are not exaggerated in areas served by minima of the vrp.

Where a pair of transmitters are fed by a common drive it is possible to increase the reflection loss at ground level, and hence reduce the level of delayed images, by delaying the signal feed to one transmitter by 90° of phase and delaying the signal from the output of the other transmitter by a similar amount. The path lengths for the reflected signals then differ by 180° and, provided that the output stages of the transmitters are identical (and are both operating), the reflected signals will be entirely absorbed in the sink load of the diplexing hybrid. (The splitting hybrid is still necessary.)

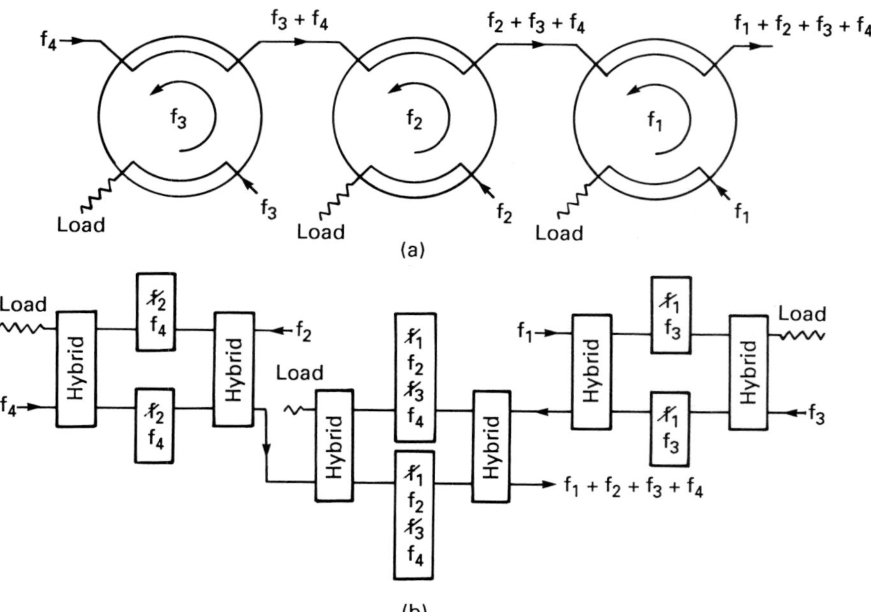

Figure 18.12 Four-channel combiners: (a) Rotamode rotating mode resonator combiners, (b) hybrids with transmission line resonator combiner

References

1 BS 8100: 1986, *Lattice towers and masts*, British Standards Institution
2 EIA Standard RS-222-D, *Structural standards for steel antenna towers and antenna supporting structures*, Electronics Industries Association
3 CAN/CSA-S37-M86, *Antennas, towers and antenna-supporting structures*, Canadian Standards Association (September 1986)
4 *The Construction Regulations 1961 and 1966*, HMSO
5 *Advice on the protection of workers and members of the public from the possible hazards of electric and magnetic fields with frequencies below 300 GHz*, National Radiological Protection Board (May 1986)
6 *Recommendations for guyed masts*. Working group No 4, International Associated for Shell and Spatial Structures
7 HILL, P C J, 'Methods of shaping vertical radiation patterns of vhf and uhf transmitting aerials', *Proc IEE*, **116**, No 8, 1325 (1969)
8 HUTCHINSON, R, 'Rotamode filter networks', *Communications and Broadcasting* (Journal of Marconi Communication Systems Ltd), **6**, No 2, 15 (1981)
9 MANTON, R G , 'Channel combiners for radio-frequency transmitters', *JIERE*, **55**, No 10, 335 (1985)

J H Causebrook B Sc, PhD, MIEE, C Eng, AMIOP
Service Area Planning Section, IBA

19 Service Area Planning

The steps in the process of planning coverages for broadcast networks are:

1 Recognize the area requiring a service.
2 Study the topography and population distribution.
3 Choose a site and antenna height to serve the area, and satisfy other criteria of access, ownership and environment.
4 Choose channels, frequency offsets and polarization, from a knowledge of administrative constraints and other uses of the spectrum.
5 Choose the power and antenna radiation pattern, to serve the area but avoid interference.
6 Consider the way in which the transmitter will be fed with programme, preferably off-air from another transmitter in the same frequency band.
7 Check the validity of the above plans by conducting a site test with a portable transmission system and making reception tests in the area.
8 Process the plans so that they receive the necessary national and international approval, following the rules of the International Radio Consultative Committee (CCIR).
9 See that the major objectives are achieved by making a field strength survey when the station is built.

The propagation information given in the following paragraphs is not restricted to television, because it is independent of the modulation. However, the very large bandwidth required for television means that it is pointless to consider wavelengths below 30 MHz for this purpose.

19.1 Basic theory

The fundamental theory of propagation is embodied in Maxwell's equations, but these are difficult to remember, and even more difficult to solve for a practical problem. For service planning purposes, it is rarely necessary to resort to this depth of theory.

Thus, instead of solving for the vector field of both electrical and magnetic components, it is nearly always sufficient to consider the field as a single scalar, which will be designated here by ϕ. The problem to be solved can now be expressed in a wave equation:

$$\nabla^2 \emptyset + k^2\phi = -\delta \qquad (19.1)$$

where ∇^2 = an operator that produces the second differential, with respect to space coordinates, of the function which follows it, i.e. ϕ.
δ = the current density of the source, i.e. the currents in the antenna described as if they existed at a point.
k = 2π divided by the wavelength.

In any particular case, this equation must be solved from a knowledge of the boundary conditions which must satisfy:

$$\delta\phi/\delta n = jk\phi Z \qquad (19.2)$$

where Z is the surface impedance and n is the normal to the surface.

If a value has been obtained for ϕ, it is possible to obtain the magnetic field from:

$$H = jk\phi \qquad (19.3)$$

The electric field can then be obtained from:

$$E = H120\pi \qquad (19.4)$$

where 120π is known as the *impedance of free space*.

The quantities E and H are vectors at right-angles to the direction of propagation and at right-angles to each other. On a boundary surface, there is also a radial electric field which is given by:

$$E_r = E\,Z \qquad (19.5)$$

A simpler method of making field strength predictions is via prescriptive techniques, many of which are given in this section. These prescriptions usually require the calculation of an attenuation factor, A, which may be applied to a reference. Different forms of propagation, and historical convention, mean that several references have to be considered:

- *below 30 MHz*, it is the field strength that would be obtained over a perfectly conducting flat plane
- *between 30 MHz and 3 GHz*, it is the free space field strength
- *above 3 GHz*, it is the free space loss

In the first two cases, applying the attenuation to the reference gives the electric field strength, E, but for the last case it is a path loss.

19.2 Planning for frequencies between 30 and 3000 MHz

The reference level for these frequencies is usually taken to be the *free space field* (fsf) strength. This is based on the transmitter radiating from a half-wavelength dipole and nothing existing in the space to the reception point. If this antenna radiated equally in all directions, the power per unit area at a given distance would be the transmitter power divided by the surface area of a sphere of radius equal to that distance. However, in its direction of maximum radiation, the half-wavelength dipole radiates 1.64 times this power.

The power passing through an area of 1 m^2 can also be equated to the vector product of the electric and magnetic field strengths, and the magnetic field can be equated to the electric using equation (19.4).

These two ways of deriving the power may now be equated and rearranged to be explicit in the electric field strength. This leads to an expression for the free space field strength in decibels relative to 1 μV/m (dBμ) of:

$$FSF = 107 + 10 \log P - 20 \log D \qquad (19.6)$$

where P is the power in kilowatts and D is the distance in kilometres.

This fsf is modified by the earth's surface, buildings, trees and the atmosphere. The irregular terrain can be stylized into shapes which have simple geometries. Classic calculations can then be used to calculate an attenuation over these shapes.

Hills obstructing a path can be stylized into 'knife-edges' and the attenuation determined by a diffraction calculation first devised by Fresnel. This uses a parameter:

$$v = h \sqrt{kd/(\pi ab)} \qquad (19.7)$$

where h, a and b are defined in *Figure 19.1*. This parameter is the limit of integration in Fresnel's diffraction formula, but it is possible to use a graph of the type shown in *Figure 19.2* to determine the required attenuation.

If many edges exist on the path between transmitter and receiver, the rigorous solution requires a multiple integral, which is very hard to solve. Therefore, several people have devised simple approximate solutions to the problem. A method which gives good results is attributable to Deygout[1]. It depends upon taking each edge in turn, as if it alone existed on the path, and discovering which one gives the greatest attenuation.

The selected edge is then used to divide the path in two, and the process is repeated on the two halves, assuming that the selected edge is now a terminal. This produces two further attenuations from two edges which may again be used to divide the path. When the process is complete, the total attenuation attributed to the path is the sum of the primary value together with those derived from the divided portions. This process is illustrated in *Figure 19.3*.

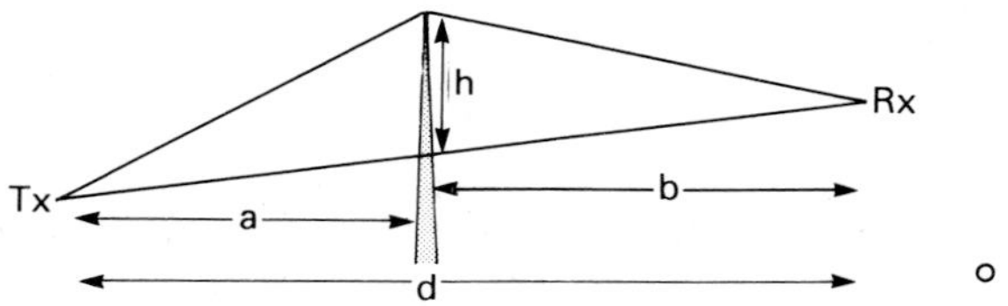

Figure 19.1 knife-edge diffraction

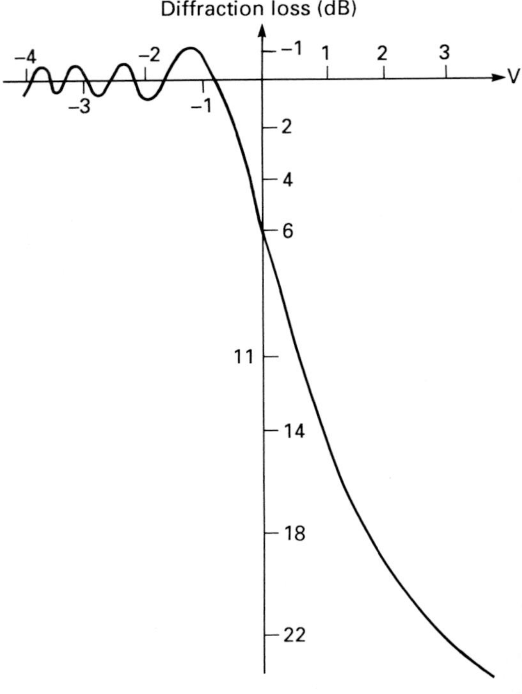

Figure 19.2 Fresnel diffraction loss

The simple knife-edge stylization underestimates the real attenuation on nearly all paths. This shortfall may be overcome by approximating the profile to various forms of surface with a simple geometry, e.g. spheres, cylinders or wedge shapes. These surface calculations often overestimate the attenuation. Thus, a reasonable approximation to the actual value is an interpolation between the knife-edge and a surface calculation. However, in the case of a sea path, a simple spherical loss calculation should be made.

Buildings and trees also obstruct the path to create a further attenuation, called the *clutter loss*. In this case, it is even more difficult to make definitive calculations, because of the highly irregular nature of these objects. Attempts have been made to devise sophisticated predictions, but for most purposes it is better to accept some simple statistical results as follows:

	dB vhf	dB uhf
Urban	8	12
Suburban	6	8
Wooded	4	8
Rural	2	4
Rx foreground slope:	change above by:	
facing Tx	-2	-4
away from Tx	4	4

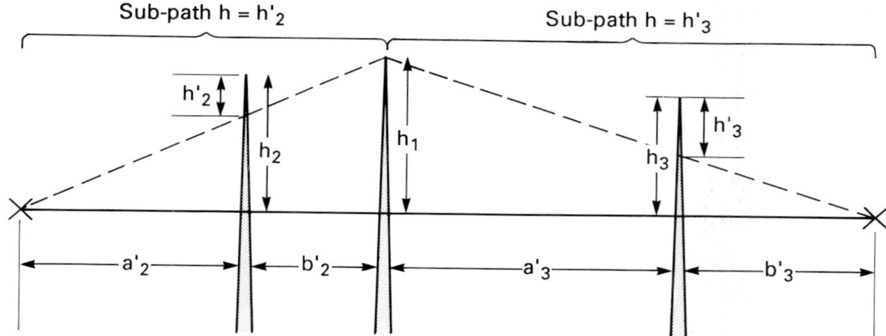

Figure 19.3 The Deygout construction

On a path that is fully line-of-sight, the field strength is modified by a ground reflected signal. Calculation is again made difficult by the irregular nature of the surface. If there is sufficient clearance of the ground so that the v parameter of equation (19.7) is less than -0.78, and the ground is sufficiently flat to be able to ascribe a simple reflection point, attenuation may be calculated from:

$$A = -10 \log \{1 + r^2 - 2r \cos (k\delta - \beta + \pi)\} \quad (19.8)$$

where r is the reflection coefficient, δ is the path difference between direct and reflected rays, and β is the phase change on reflection.

Typical values of r and β are shown in *Figure 19.4* for vertical and horizontal polarizations. The dip which occurs in the value of r for vertical polarization is the *pseudo Brewster angle*. In most practical cases, the grazing angle is small, which allows a simplification, particularly for horizontal polarization, of:

$$r = 1, \beta = \pi, \delta = 2st/d \quad (19.9)$$

where s and t are the heights of the receiving and transmitting terminals above the reflecting plane.

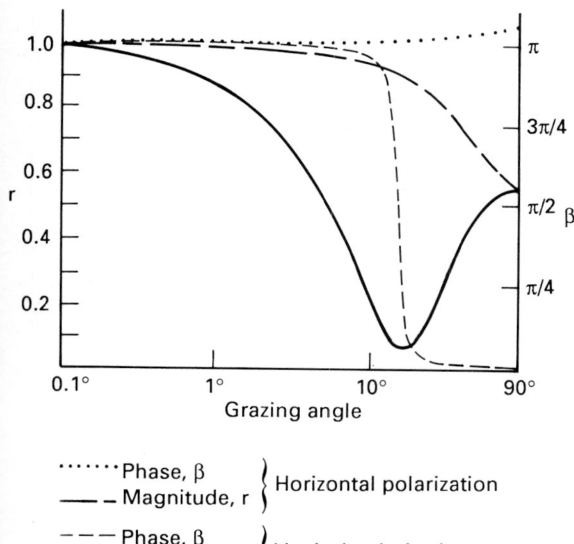

Figure 19.4 Typical values of reflection coefficient

When these values are substituted into equation (19.8), we get:

$$A = -10 \log \{4\sin^2 kst/d\} \quad (19.10)$$

This calculation is often insufficient because of the irregularity of the ground. The problems caused by reflections occurring on a real terrain are still very hard to predict. Probably the best hope for the future will be found in a method using a finite element technique.

Solutions require a very accurate description of the terrain, because results depend on a vector product of signals of small wavelength. Thus, a small error in the derivation of differences in path length results in large errors of field strength.

Vertical objects on the terrain often cause reflections which arrive at the receiver with a long difference in path from the direct wave. These objects are usually man-made, but can be cliff-like features. The problem caused is predominantly that of modulation distortion. For TV, this often manifests itself as a 'ghost' image.

It is often possible to predict the strength of the reflected signal by considering the pattern of radiation which occurs from a rectangular aperture. In this case, the beam in the direction of specular reflection is broadened by diffraction effects to produce a radiation pattern in the form of a 'sinc' function (sinc $\times$ = (sin x)/x).

The strength of the reflected signal is directly proportional to the incident field strength and the width and length of the object, and inversely proportional to the wavelength and distance from the object to the receiver.

The reflected signals are less tolerable if they are caused by moving objects such as cranes, ships, planes, vehicles and wind turbines. The latter are becoming more numerous and larger, thus representing a serious new threat to broadcast reception[2]. The degradation is serious for reception on the opposite side of the object from the transmitter. This is the *forward scatter* region. The calculation system that may be employed in this region is based on Babinet's principle, which is illustrated by *Figure 19.5*. This principle is a form of superposition theory which can be found in many texts on optics and electromagnetism.

The field strength is also modified by the existence of the troposphere, especially over long distances and for co-channel interference. This modification is because the refractive index of the atmosphere varies in space and time. The most obvious phenomenon is that of reducing refractive index with height, which bends the wave back towards earth.

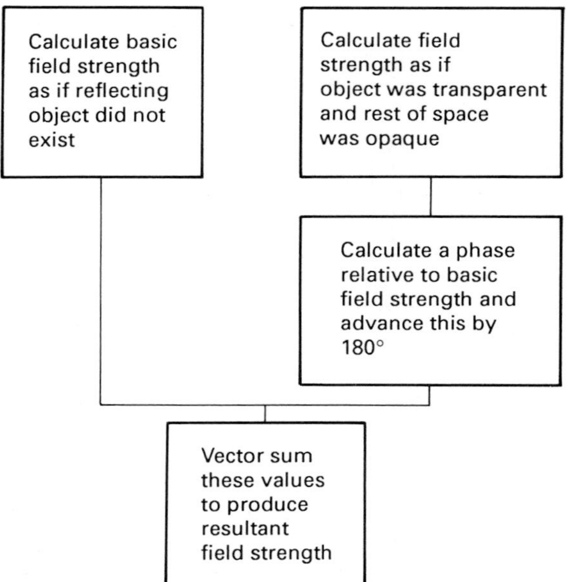

Figure 19.5 Babinet's principle

The simple way of allowing for this is to ascribe an earth radius which is larger than the physical value. For average propagation conditions, this value is taken to be equal to 4/3 times the true radius, but for small percentages of the time the effective radius can be much larger than this. In fact, the wave can bend with a radius much larger than that of the earth, which leads to the condition called *ducting*. This occurs mostly over a sea path and is a serious source of co-channel interference. At uhf over the English Channel and the North Sea, it occurs for about 5 per cent of the time, and over warm seas for much larger percentage of the time.

In addition to the simple reduction with height, there are more irregular changes of refractive index in the atmosphere. These can take the form of rapid layer changes or changes of a blob-like nature. The former reflect the wave, and the latter cause scatter propagation. Both tend to increase the field at distant points and again cause co-channel interference. The strength of the scatter signal is dependent upon the angle between the receiver and transmitter horizon lines, among other factors.

The frequencies above 30 MHz are not generally prone to interference by reflections from the ionosphere, but it is possible for some reflections to occur up to about 100 MHz from the E-layer when the ion density is enhanced by a mechanism called *sporadic E*. This occasionally gave severe picture degradation in the UK from transmitters in central Europe (about 1500 km away), when these lower frequencies were used for TV broadcasting.

The calculations given here are too lengthy and complicated to be completed by hand for the planning of a broadcast network, so extensive use is made of computers. They are particularly necessary for interference calculation and hence frequency planning. In the UK, a large terrain data base is used, and a data base containing many of the parameters of transmitters in NW Europe. Thus, if a new transmitter is required, it is possible to insert the details and obtain the impact it will cause to existing services and the impact they will have upon it. This is done by profiles being automatically produced from the terrain data base followed by the calculations given above. These same types of calculation also provide a computer drawn coverage area.

19.3 Planning for frequencies above 3 GHz

The reference level for these frequencies is the *free space loss* (fsl); this should not be confused with the free space field of section *19.2*. The fsl is defined as the power transmitted isotropically, divided by the power received on an isotropic

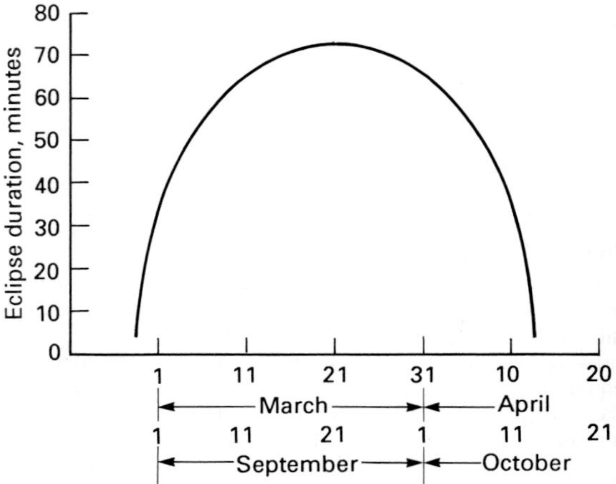

Figure 19.6 Eclipse duration

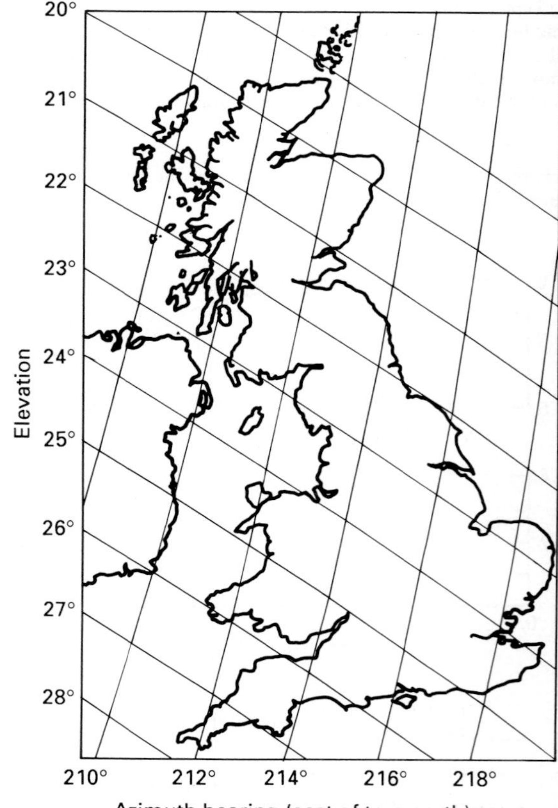

Figure 19.7 Receiving antenna pointing for satellite at 31°W

antenna. Such an antenna is only a theoretical concept, but if it existed it would have a capture area of wavelength squared divided by 4π. This results in the value of fsl in decibels being given by:

$$FSL = 20 \log 2kd \qquad (19.11)$$

A direct broadcast satellite (dbs) needs to be in an orbit about 36 000 km above the equator. Power is obtained from solar cells which will be eclipsed by the earth's shadow to an amount shown by *Figure 19.6*. This eclipse time is put outside normal viewing hours by the satellite being about 30° W of the main target. For the UK, this makes the transmitter to receiver distance, d, about 39 000 km.

The frequencies allocated for satellite broadcasting are around 12 GHz. Applying these factors to equation (19.11) gives a value of fsl of 206 dB. The receiving antenna pointing for such a satellite can be given by a chart of the type shown in *Figure 19.7*. It can be seen that even for the north of the UK, the elevation angle is such that many terrain features should be cleared by the signal. However, there will be some locations that will be obscured by hills, buildings or trees. A suitable location will exist, for a satellite at 31° W, if the sun falls on the chosen location for the antenna at 3 pm in early March or mid October.

At microwave frequencies, diffraction losses are so large that most paths are designed to have a good clearance. To quantify the amount of clearance, it is helpful to consider the concept of Fresnel zones, in which the space between the terminals is divided into ellipsoids. The *N*th zone is defined by the fact that the distance from the transmitter to its surface and then to the receiver is *N* half-wavelengths greater than the direct distance from transmitter to receiver. If a knife-edge ends at the lower boundary of the first Fresnel zone, it has a v parameter of -1.414, but for the diffraction loss to equal zero, v = -0.78, and a reflecting surface will cause a reduction from the free space value for all v > -0.82.

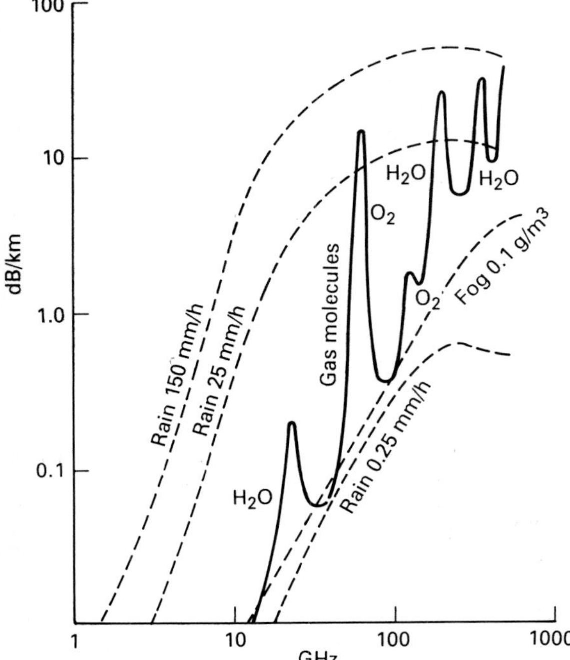

Figure 19.8 Absorption through the atmosphere

Thus, the microwave engineer usually allows objects to penetrate into the first Fresnel zone to the point where v = -0.82. Loss from free space is then dependent only upon atmospheric absorption and scattering out of the beam. This is caused by rain, fog, water vapour and other gas molecules.

Although the losses tend to increase with frequency, this is not the rule, because losses sharply increase at the resonant frequency of molecules, as can be seen in *Figure 19.8*. These attenuations have the obvious disadvantage of reducing coverage, but they do allow frequent co-channel re-use without causing interference.

The transmitter power multiplied by the gain of the transmit antenna relative to an isotropic source provides the *effective isotropic radiated power* (eirp). For a direct broadcast satellite, this should be about 60 dBW. The received carrier power is then given by:

$$C(dBW) = eirp(dBW) - fsl + GI - A \qquad (19.12)$$

where GI is the receiving antenna gain (dB) relative to isotropic reception (e.g. a 0.6 m dish at 12 GHz gives 35 dB), and A (dB) is the attenuation which may be caused by diffraction, rain or atmospheric absorption. For satellite reception in Europe, very short periods of heavy rainfall can increase A by 6 dB.

In addition to satellite broadcasting, microwave frequencies are beginning to be used for terrestrial broadcast services. The frequencies which are being considered are near 2.5 GHz, 12 GHz and 30 GHz. These proposals have been variously called *multipoint video distribution system* (MVDS), *multipoint microwave distribution system* (MMDS) or *millimetre-wave multichannel multipoint video distribution services* (M^3VDS).

With such allocations, there could be as many as 12 extra programmes on 2.5 or 12.5 GHz or even 25 programmes at 40 GHz, covering a large percentage of the population. These services might use fm vision modulation instead of the am system at uhf.

19.4 Modulation systems

Amplitude, frequency, or some form of pulse modulation, may be used to put a signal onto a carrier. This is relevant to the service planner because it influences signal/noise ratio, bandwidth requirement, and factors such as ghost visibility and susceptibility to interference.

An important factor in the noise performance of a system may be called the *demodulation improvement factor*, D. This is the relationship between the carrier/noise ratio and that of the signal/noise ratio. The carrier/noise refers to any point in the receiver which comes after significant sources of noise and before detection and significant filtering. In some systems, it may be unclear what is meant by carrier power, e.g. in TV system I the carrier is partially suppressed so it is defined as the mean power during the synchronizing pulse (*peak sync power*). Signal/noise may be quoted for unweighted or weighted measurement. The unweighted definition is used here.

The value of D for several systems is:

luminance of TV, system I	-8
amplitude modulated sound	20 log m
fm without pre-emphasis	20 log r + 5
fm sound for TV, system I	21.5

where m is the depth of modulation index, and r is the ratio of maximum deviation frequency to baseband frequency.

The bandwidth B in dB/Hz, which should be applied when calculating output noise power, is that of the baseband.

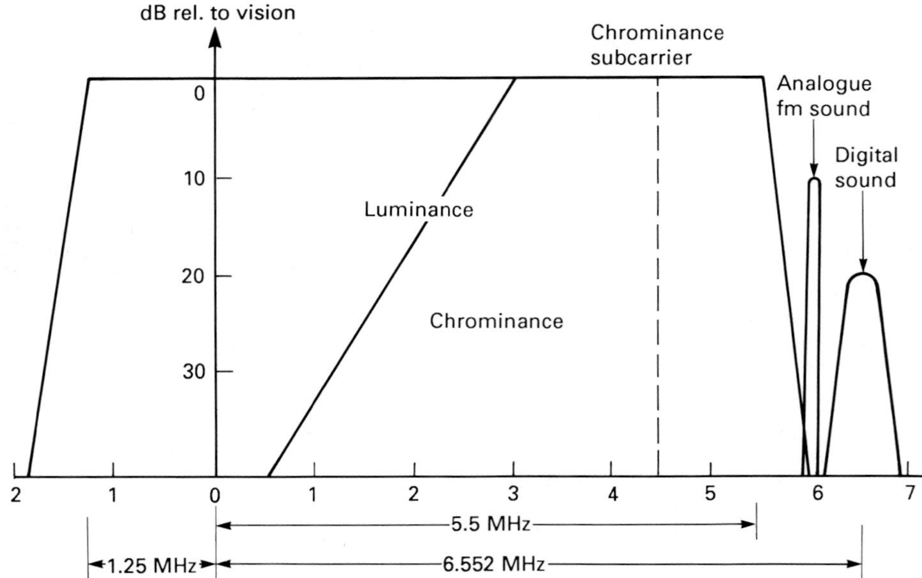

Figure 19.9 Spectrum of TV system I with digital sound

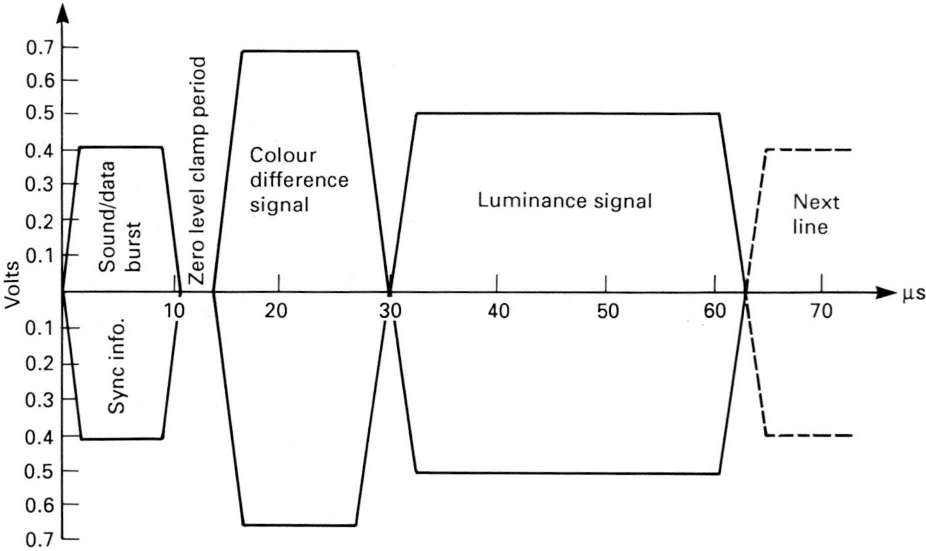

Figure 19.10 D-MAC picture line with sound/data burst

The rf bandwidth of an fm signal is twice the sum of the maximum deviation frequency and the maximum baseband frequency.

The colour subcarrier for TV system I is amplitude modulated on a subcarrier which is separated from the luminance carrier by 4.433 618 75 MHz, but still inside the bandwidth of the luminance signal. This results in a degradation called *cross-colour*. The sound is separated from the luminance carrier frequency by 6 MHz and is fm modulated with a deviation of 50 kHz. The digital stereo sound, NICAM 728, is spaced from the luminance carrier by 6.552 MHz and has a bandwidth of 728 kHz. These relationships are shown in *Figure 19.9*.

The problem of cross-colour is overcome in the MAC system where the separation of luminance/chrominance/sound/data is by time division, and each of the signals is time compressed. The arrangement is such that onset of noise is approximately equal for each of the sets of information. The MAC system can be amplitude or frequency modulated, but for satellites, fm must be used because the power required is 24 dB less. In this case, the deviation is 13 MHz/V, which makes the total bandwidth 27 MHz (*Figure 19.10*).

In an fm system, a reflected signal is liable to produce a ghost signal that is more smeared, and hence less objectionable, than that of an am system.

The protection ratios needed between TV systems using fm modulation have not yet been fully defined, but they are liable to lie between those for am using non-precision offset and precision offset. Thus, despite the fact that fm modulation requires greater bandwidth, its spectral efficiency is not necessarily worse because the frequency re-use distance is reduced. In any case, the power efficiency of fm is very much greater. This is very important for satellite transmissions where power is obtained from solar cells.

19.5 Received signal

A half-wavelength dipole has an impedance of 73.2 ohms, and it generates an open circuit voltage of the field strength multiplied by the wavelength upon π. Allowing for an antenna gain, G dB, and a cable loss, L dB, the carrier power in decibels relative to 1 W is given by:

$$C = 20 \log E/k + G - L - 10 \log 73.2 \qquad (19.13)$$

Allowance should be made for a system of other impedance, R ohms. Then the voltage, in decibels, across the input terminals of the receiver is given by:

$$V = 20 \log E/k + G - L + 10 \log R/73.2 \qquad (19.14)$$

The noise power is given by:

$$N = F + B - 204 \qquad (19.15)$$

where F = $10 \log \{\alpha T/290 + (1-\alpha) + (f-1)\}$ (antenna + feeder + receiver noise contributions)
α = antilog (-L/10)
T = antenna temperature
f = receiver system noise factor

The antenna temperature arrives from three major sources: galactic (significant below 500 Hz), gaseous (above 10 GHz) and the earth's surface. The values of F are given in *Figure 19.11*, for no feeder loss.

From the above, the wanted field strength (wfs) in decibels relative to 1 μV/m is given by:

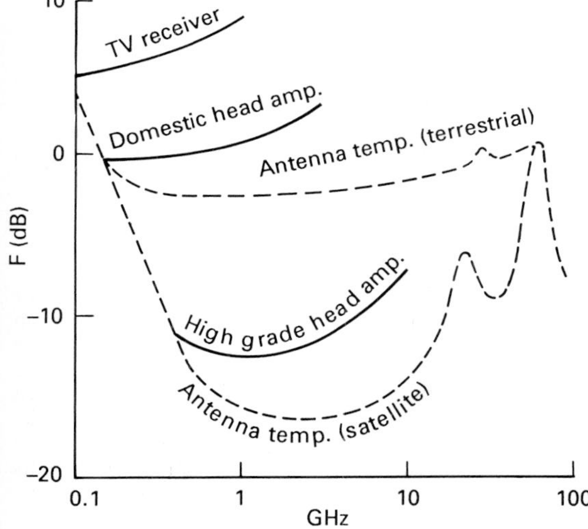

Figure 19.11 Contributions to noise power

$$WFS = S - D + F + B + 20 \log k - G + L - 65.4 \qquad (19.16)$$

where S dB is the signal/noise ratio required for adequate reception. This is about 33 dB for vision and 53 dB for sound broadcasting.

Table *19.1* shows the wanted field strengths, for TV to overcome noise in Bands 4 and 5 and is divided into the cases of average and good domestic receiving locations.

	Band	*G*	*L*	*F*	*WFS*
Average receiver	5	10	4	8	70
	4	10	3	7	64
Good receiver	5	13	0	0	55 (but 58 used)
	4	13	0	0	51 (but 52 used)

Table 19.1 Field strengths to overcome noise in bands 4 and 5

In addition to adequate field strength, the picture and sound quality must be reasonably free from interference and multipath impairment. This quality is assessed using a five point impairment scale, as follows:

1 very annoying
2 annoying
3 slightly annoying
4 perceptible but not annoying
5 imperceptible

The main types of interference are *co-channel* (cci), *local oscillator* (loi), *adjacent* (aci) and *image*. For TV, cci appears as light and dark bands on the screen or sometimes a ghost image of the unwanted picture. The bands are nearly horizontal straight lines, often moving at a constant rate. LOI consists of a set of wavy lines or of a moiré type pattern. ACI is of two types: firstly from the lower adjacent fm sound creating a moiré like pattern shimmering with the sound modulation, secondly from the upper adjacent with near vertical lines.

The acceptibility of cci depends not only on the relative levels of the wanted and unwanted signals but also on small frequency differences between the two. The interference bands can be analyzed into two components: the number of oscillations across the screen, *a*, and the number of oscillations down the screen, *b*. The offset frequency is then given by (15 625*a* $\pm$50*b*).

Vision carrier frequencies usually differ from the nominal by 0, 5/3 or -5/3 times line frequency, with a tolerance of $\pm$500 Hz. Thus, *a* has values of 0, 2 or 3 and *b* of between 0 and 20 for zero offset or between 80 and 120 for 5/3 and 10/3 line offsets.

The most annoying pattern is slow moving with about 10 bands. The pattern is stationary when the offset differs from the nearest multiple of line frequency by an exact multiple of field frequency. At an offset which differs from the nearest line frequency harmonic by an odd multiple of half the field frequency, a flickering pattern occurs which is least annoying. To keep the patterns in the least visible condition, a precision of $\pm$1 Hz is required.

These facts translate to the protection ratio curves given in *Figure 19.12* for continuous interference. For interference enhanced by tropospheric effects, which last for small percentages of the time, the protection ratio is increased by about 10 dB.

At a TV site which re-broadcasts a signal picked up from an existing transmitter, assessment of acceptability must meet more stringent criteria than those at a domestic location. In the UK, two types of site are recognized, dependent on the importance of the new transmitter. The most stringent of these standards is called P and the other is Q. The required criteria are laid out in Table *19.2*.

If the interfering signal is from a different direction from the wanted signal, it is necessary to apply the protection provided

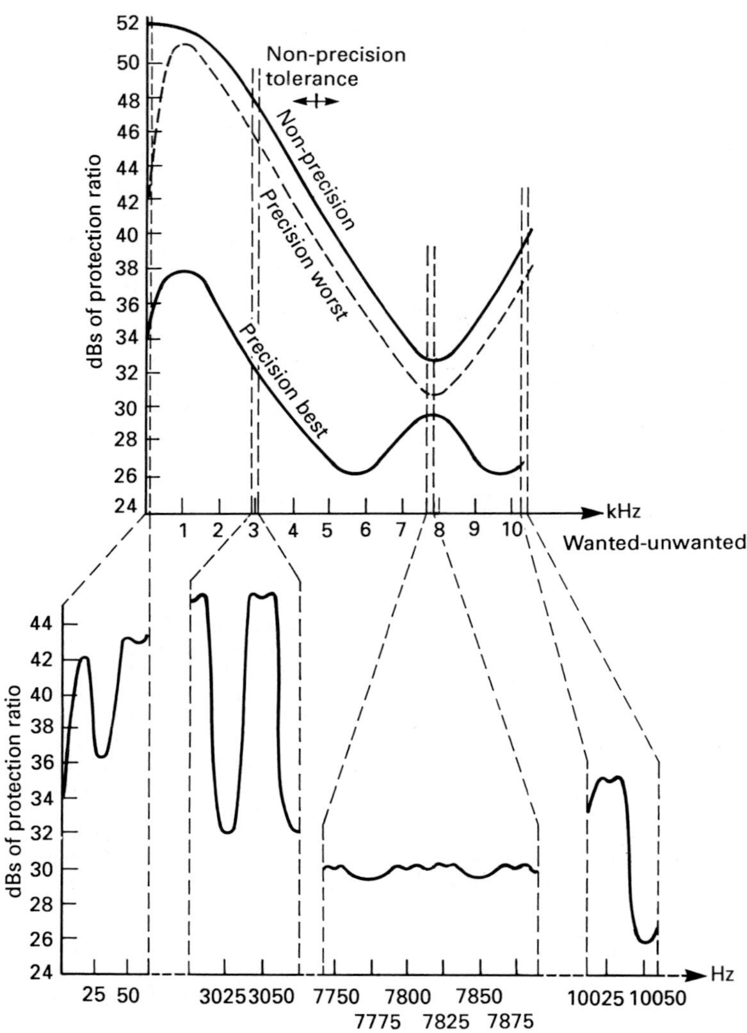

Figure 19.12 Protection ratios for continuous interference. The lower set of curves are expansions of parts of the upper graph

			P (dB)	*Q* (dB)
Signal/noise ratio			39	33
Ghost delay $< 1\mu s$			28	24
Ghost delay $> 1\mu s$			31	28
Chrominance/luminance gain inequality			+2/−4	+2/−6
Sound/vision carrier ratio			+2/−4	+2/−6
Intermodulation products			48	40
CCI protection ratios:	%time	offset		
	50	zero	55	50
	50	±5/3	45	40
	1	zero	45	40
	1	±5/3	30	30

Table 19.2 P and Q criteria

by the receiving antenna. In the case of domestic reception, it is usual to apply a simplified pattern based on what should be obtained from a reasonable antenna.

Future developments in receiver technology will bring possible changes in service planning parameters. These could well reduce some forms of degradation such as interference and multipath images. There could also be more radical changes such as wider screens, higher definition, and spectral compression techniques, which might involve considerable changes in the planning parameters.

19.6 Measurement technique

A service area planning team wishing to measure either vhf or uhf usually has a vehicle fitted with a pneumatic mast that can raise an antenna to 10 m. This is a typical height at which an antenna would be placed on the usual two storey domestic dwelling. The antenna on the mast should have a performance which is close to that used by the public, but often a log periodic type is installed because a wide range of frequencies may be measured. It is convenient if the antenna is mounted so that its polarization can be changed automatically from within the vehicle.

A head amplifier can be fitted in such a way that it may be switched out for accurate measurement, but switched in to assess weak signals. The cable must be robust and all connections should be well made to avoid losses and mis-matches.

The measuring receiver should work off vehicle voltage and be compact. For most purposes, an accuracy of ±1 dB is sufficient. It should have a wide dynamic range and cover all the frequencies required. If the receiver is designed to cover different types of services, it should have a system for changing the bandwidth, e.g. 300 kHz for terrestrial TV. To enable a direct reading of field strength, it requires the capacity to store antenna and cable calibration factors and combine these with the measured voltage. This process is an application of equation (19.14).

Modern equipment should have an output capable of being fed to a computer to facilitate the processing and logging of results. During a survey, it is often desirable to determine many different channels at each location. Thus it is convenient to be able to store information on each of these so that they may be called up sequentially by a single pushbutton operation.

As well as the numerical determination of field strength, there needs to be a system to determine quality of signal. For TV, this is a monitor of sufficient definition to see clearly any degradation that may be perceived in the home. It should be capable of providing teletext, because degradation of these signals will occur with multipath signals of shorter delay time than those which can be observed as a ghost image on the TV screen.

For fm sound, it is possible to use the fact that multipath signals result in a change of field strength, which is highly dependent upon frequency. Some measuring receivers contain an am/fm display, which may be read to determine relative magnitude, number and delay times of the multipath signals. The vehicle should also be fitted with a high quality receiver for subjective assessment of the sound quality. Co-channel interference may often be perceived as noise, until the unwanted signal becomes the stronger when it will be heard with the original signal causing noise.

The variation of signals in space is such that it is desirable to make small movements of the antenna about each point to assess the magnitude of this problem. These may be done by changes of height or by small lateral movements. However, the latter should be made with great care, because serious accidents have occurred by hitting power lines or other overhead objects.

19.7 Site testing

Confidence in the ability to predict field strength is not always sufficient to build expensive stations purely on the evidence this provides. One of the problems of prediction is the lack of accurate terrain data, particularly about trees and buildings. Thus, it is often desirable to carry out site tests. These may be done with a trailer mounted pneumatic mast, which can raise the antenna to about 35 m. For larger stations, more elaborate methods are needed, such as raising the antenna on a balloon. The power and antenna radiation pattern need not be similar to the final values envisaged, because scaling may be applied.

When the test transmitter is established, a team will carry out measurements in the way described above. The measurements may be made to 'blanket' the area under consideration, but it is often more economic to measure only in areas which prediction and skilful map reading have shown to be relevant for the production of coverage area contours.

During the course of a site test, it is often necessary to make assessments of incoming signals to find an adequate programme feed for the new transmitter. This involves field strength determination, picture grading , and a measurement of luminance relative to both chrominance and sound.

References

1 DEYGOUT, J, 'Multiple knife-edge diffraction of microwaves', *IEEE Trans. Antennas & Propagn.*, **AP-14**, 4, 480–489 (1966)

2 CAUSEBROOK, J H, 'Distortion of radio wave signals by wind turbines', *JIERE*, Supplement to **58** No 6, Sept–Dec 1988

Part 5
Distribution of Broad-band Signals

R Wilson BSc
Continental Microwave

20 Microwave Radio Relay Systems

20.1 Types of microwave link

Different types of microwave link system are available for:

- electronic news gathering
- outside broadcast
- fixed radio relay

The first two are covered in greater detail in section *42*.

20.1.1 ENG links

Equipment intended for electronic news gathering is compact, lightweight and capable of quick and easy assembly on location. Transmitters and their associated antennas (often omnidirectional) are generally mounted together on cameras or back-packs. Receivers are normally tripod mounted with directional antennas.

Circular polarization is usual in ENG applications to reduce the effect of reflections from buildings, trees, etc., in the vicinity.

Transmitters and receivers are frequency agile with 2–10 rf channels (depending on the size of the units), and typically operate in the 2 or 2.5 GHz bands, though equipment is also available at higher frequencies. ENG links normally convey one video and up to two audio signals.

20.1.2 OB links

Outside broadcast links, though portable, are also very rugged and offer high levels of performance over long paths. Transmitters and receivers comprise control and head units. Control units contain power supplies, modulators or demodulators and are often rack-mounted in OB vehicles. Head units include if and rf modules and are nearly always mounted on heavy duty tripods with 0.6 or 1.2 m parabolic antennas.

Synthesized local oscillators provide frequency agility over a relatively wide frequency range (approximately 500-600 MHz), and equipment is available to cover bands between 2 GHz and 24 GHz. OB links will carry one video and up to four audio channels (plus continuity pilot on occasion).

In addition, tripod panning brackets and rf multiplex allow several head units to be connected to a common antenna thereby offering dual (*duplicated*) or bidirectional (*duplex*) operation or both.

20.1.3 Radio relay links

Microwave radio relay links comprise 'slim-line' or 19 inch (48¼ cm) rack-mounted equipment and are used generally to transmit video and audio signals from a studio to a broadcast transmitter or from one regional centre to another. A system typically comprises the following items:

radio equipment

- rack-mounted transmitters and receivers
- waveguide or coaxial branching (filters and circulators to multiplex transmitters and receivers to a common antenna)
- waveguide or coaxial feeder and accessories between radio equipment and the antenna
- parabolic antennas with diameters ranging from 0.6 m to 3.7 m
- interface steelwork between antenna and tower
- service channels (omnibus order wire, express order wire and other telephone channels above the video)
- supervisory equipment
- power equipment (solar cells, diesel generators, chargers and batteries)

civil works

- antenna support structure (tower or building wall or roof)
- radio equipment room or cabin
- air conditioning

20.2 Microwave radio relay systems

20.2.1 Transmitters

Three types of microwave transmitter are in general use; those in which:

- the video signal directly modulates the rf carrier (*direct modulation*)
- the video signal modulates an intermediate frequency (if) which is then up-converted to rf (*single conversion*).

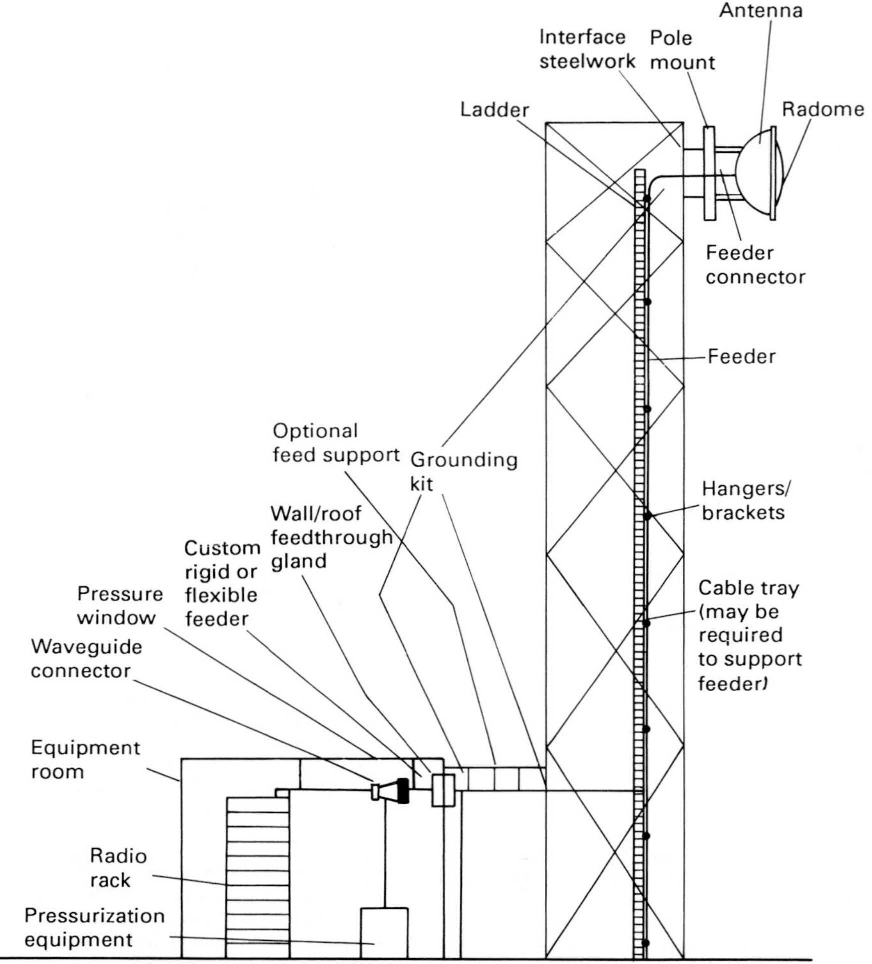

Figure 20.1 Radio relay link terminal

- as above but with two intermediate frequencies (*double conversion*)

20.2.1.1 Direct modulation

In a transmitter employing direct modulation, the video signal is amplified, pre-emphasized and filtered (5.5 MHz cut-off for 625-line PAL systems) to remove noise or spurious tones above video. Audio signals (up to four) are each applied to an audio modulator, which frequency modulates a subcarrier between 5.8 and 8.59 MHz. The video signal and audio subcarriers are then combined into a composite signal, which frequency modulates a voltage controlled oscillator operating at the final rf frequency. In some systems, modulation takes place at a lower frequency, which is then multiplied up to the final frequency. An rf amplifier and band-pass filter usually follow.

20.2.1.2 Single conversion

Most microwave transmitters used in radio relay applications employ an if at 70 MHz.

Again, the video signal is amplified, pre-emphasized, filtered and combined with audio subcarriers before being fed to a voltage controlled oscillator, centred on 70 MHz rather than rf.

The modulator is followed by an up-converter comprising a mixer and local oscillator (tuned 70 MHz below the required final transmitter frequency). The output from the modulator, which is centred on 70 MHz, is thereby translated to rf. Three significant frequencies are present at the output of the mixer, only one of which is required:

- local oscillator (F_o)
- lower sideband of the mixing process (F_o – 70 MHz)
- upper sideband of the mixing process (F_o + 70 MHz)

A band-pass or high-pass filter selects only the upper sideband which is then passed to a solid-state amplifier (using bipolar transistors below 2.5 GHz or GaAs fets above).

20.2.1.3 Double conversion

Instead of directly up-converting from 70 MHz to rf, double conversion transmitters employ a second rf (usually between 500 MHz and 1 GHz). The signal is up-converted from 70 MHz to a frequency in the region of 800 MHz and is then band-pass filtered and amplified. It is mixed with another local oscillator signal tuned 800 MHz below the final frequency. The output

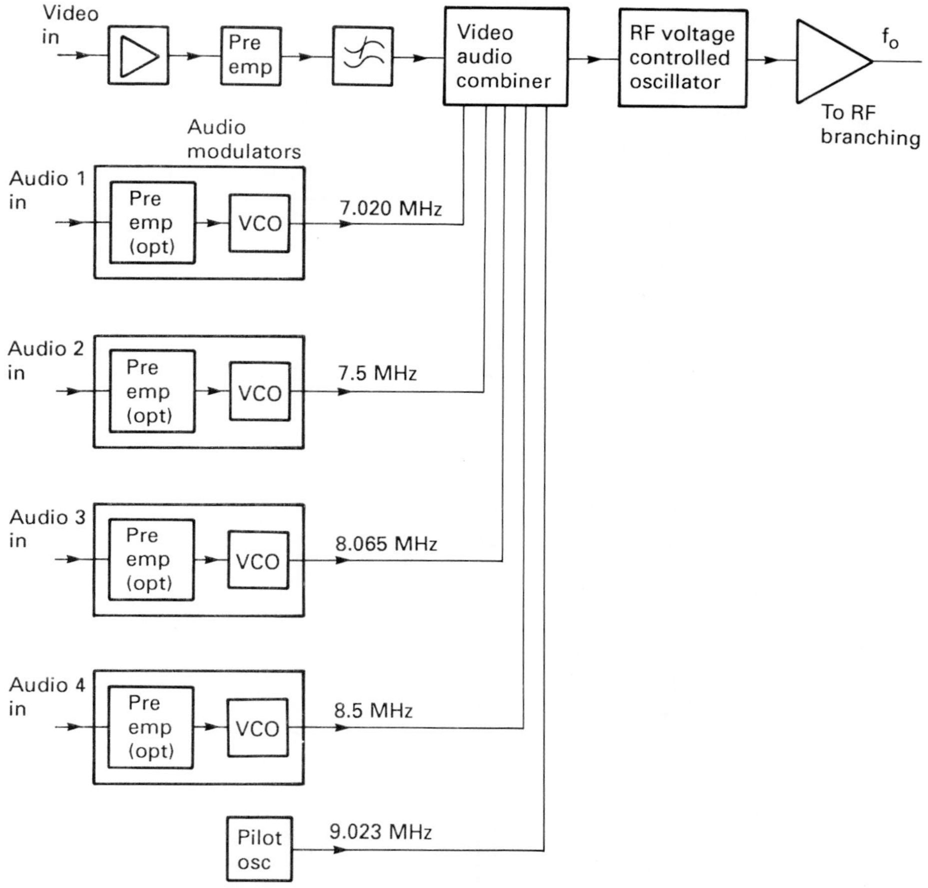

Figure 20.2 Microwave transmitter block schematic (direct modulation)

signal is passed through a relatively broad-band filter (approximately 500 Hz wide) and amplified again.

In all cases, the signal is fed either directly to the antenna port (usually at the rack top) or through rf branching filters and circulators, if a number of transmitters and/or receivers are multiplexed together.

The advantage of double conversion is that a transmitter does not require an integral channel filter and therefore can be tuned to any frequency within a designated frequency band. This is a significant consideration with regard to spares holdings.

20.2.2 Receivers

Microwave receivers for radio relay applications are usually *single conversion*. Double conversion is generally reserved for ENG or outside broadcast systems.

20.2.2.1 Single conversion

The rf signal enters the system through a band-pass filter, approximately 30 MHz wide, which is normally located in the rf branching.

The signal is then passed either through an optional low noise amplifier (lna) or directly to the shf mixer where it is mixed with a local oscillator signal tuned to 70 MHz below the incoming signal. The resulting if (centred on 70 MHz) is applied to a fixed gain pre-amplifier, is band-pass filtered, and is then equalized to minimize group-delay and level variations across the band. Further amplification stages with automatic gain control (agc) follow, to provide a constant output of 0.5 V_{rms} for application to the if demodulator.

The signal is limited (*clipped*) to remove any amplitude modulation introduced by noise and distortion in the if and rf stages, and the resulting square-wave is amplified and then low-pass filtered to remove harmonics. Further amplification follows before the signal is applied to the fm discriminator, which converts frequency modulation into amplitude modulation. Most types of discriminator used in microwave links comprise two tuned circuits operating in parallel, one tuned approximately 15 MHz above the if and the other tuned 15 MHz below. Diodes then rectify the signals emerging from both tuned circuits.

Following demodulation, the composite signal (comprising video and audio subcarriers) is split between two outputs. One, destined to be the video signal, is fed through a low-pass filter (5.5 MHz cut-off for 625-line PAL systems) to remove the audio subcarriers and is then de-emphasized and amplified to 1 $V_{p\text{-}p}$.

The second, destined to be the audio, is split four ways, to

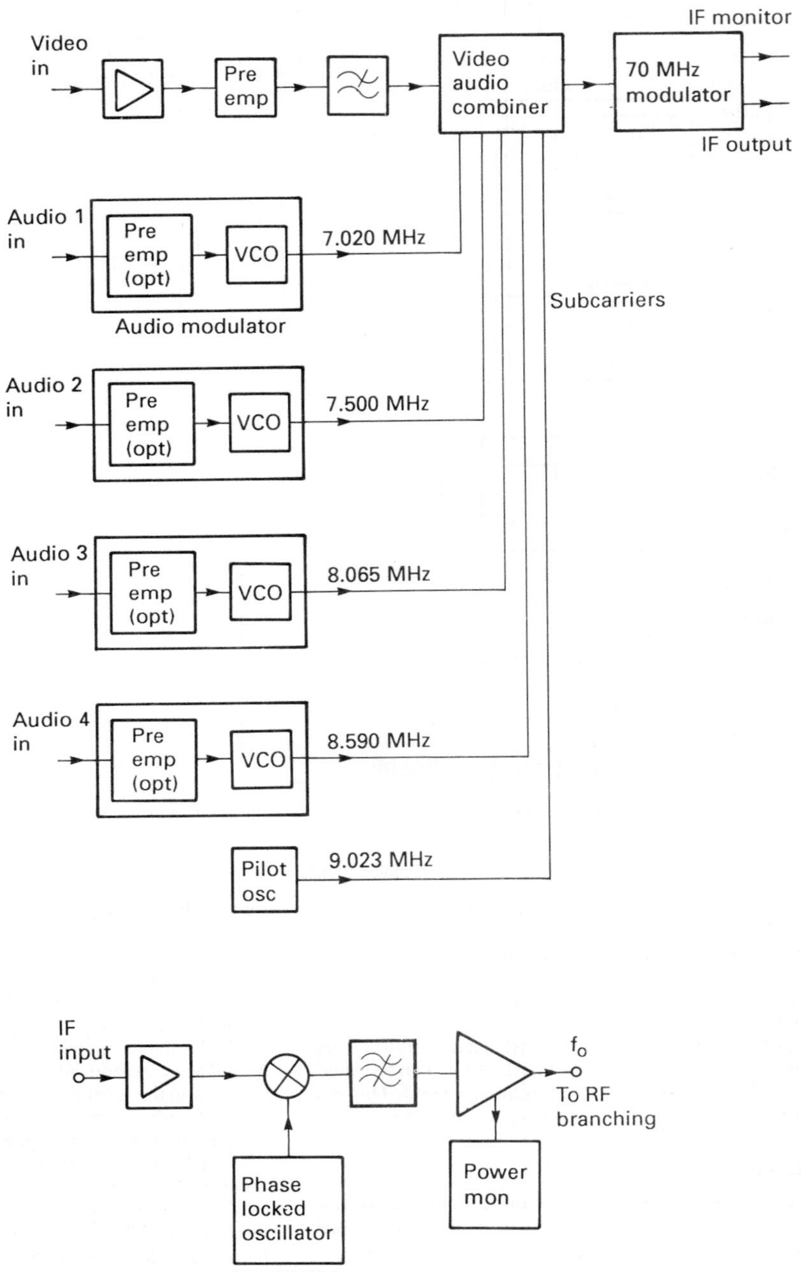

Figure 20.3 Microwave transmitter block schematic (single conversion)

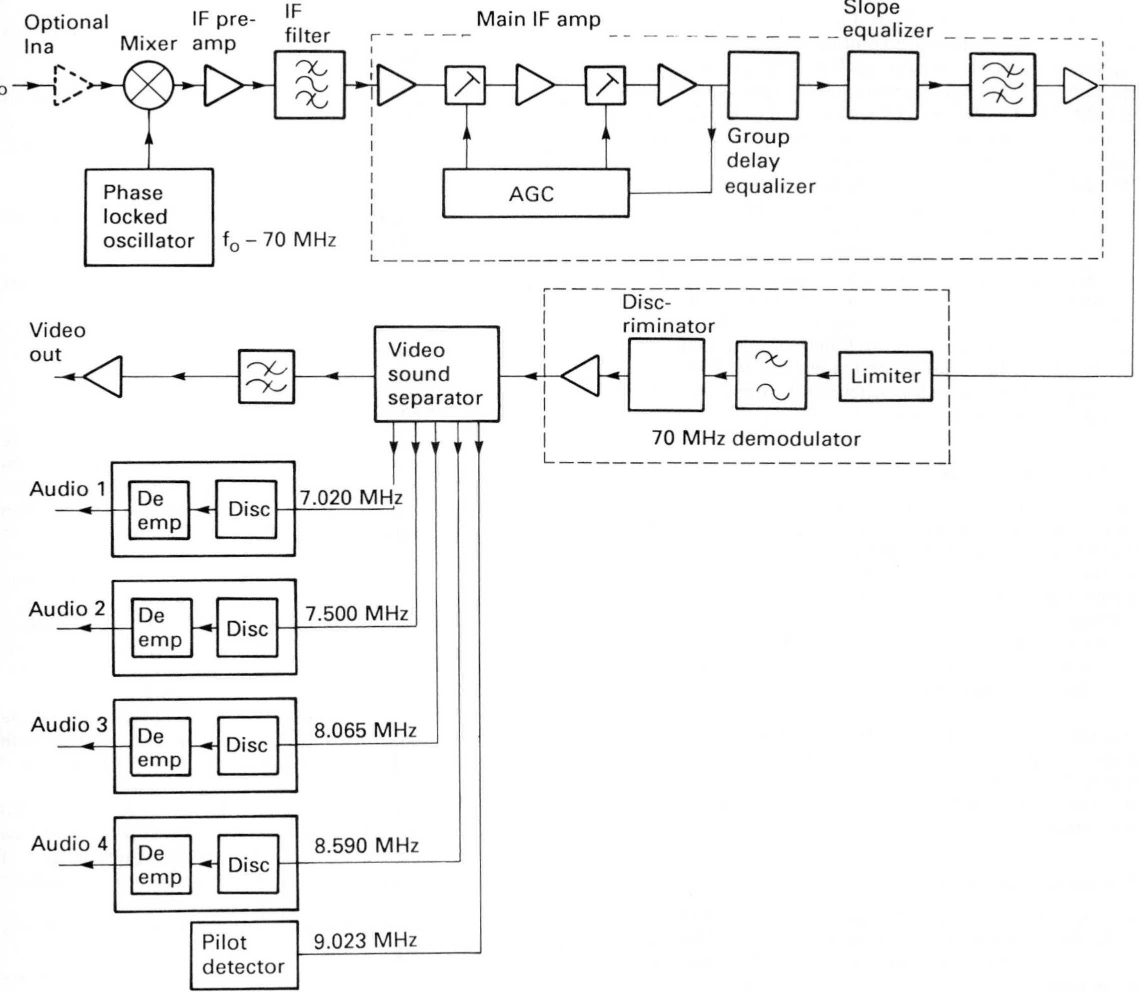

Figure 20.4 Microwave receiver block schematic (single conversion)

each of the four audio demodulators. Each audio demodulator includes a band-pass filter, tuned to one of the subcarrier frequencies, followed by a limiter and fm discriminator, which converts the frequency modulated on subcarrier into amplitude modulation which is then rectified. The audio signal is then filtered and amplified.

20.2.3 Service channel equipment

In a complex microwave network, it is usual to include speech and data channels, above the video and audio programme signals, for the transmission of important network management information.

20.2.3.1 Order-wire channels

Order-wire channels allow voice communication between sites on two-way systems. A single channel vhf or uhf radio link is sometimes used where the main microwave system is one-way only.

During installation, commissioning and maintenance activities, it is very convenient if engineers working on the system can communicate between sites. An *omnibus order-wire channel* is normally included to provide this facility. The equipment comprises an order-wire panel with telephone handset, loudspeaker (or buzzer) for calling, 3.4 kHz low-pass filters and amplifiers for two-way voice communication. A multi-way bridge is used at repeater and branching stations to allow omnibus transmission and reception from two or more directions simultaneously.

Omnibus calling, by voice (via the loudspeaker) or buzzer, alerts all other stations simultaneously. More sophisticated omnibus order-wires employ selective calling, where a keypad is used to call a particular station which has its own unique identity code.

An *express order-wire channel* provides direct voice communication between two sites only, between a network control centre and a route switching station, or between two regional control centres. Express order-wire channels are always separate from omnibus channels.

It is not unusual for a large microwave network to employ several express order-wire channels between important sites.

20.2.3.2 Telephone channels

The service channel is often required to carry an additional 12 or 24 voice channels. These are combined together in equipment called *frequency division multiplex* (fdm) in 4 kHz channels; hence 24 channels occupy spectrum between 12 kHz and 108 kHz. The bandwidth of each channel is 3.1 kHz, and terminations are either two-wire or four-wire with E & M (earth and mark) signalling (ringing information).

E & M signalling is a facility provided by telephony multiplex and some order-wire equipment to convey on-hook and off-hook signals over a standard voice channel. As the bandwidth of a standard channel is from 300 Hz to 3.4 kHz, it is not possible to send this information as low frequency pulses. In the case of analogue multiplex, on-hook/off-hook signals are transmitted as the presence or absence of a 3825 Hz tone. A 3.4 kHz low-pass filter prevents the tone from being heard in the audio band. At the receiving end, a narrow 3825 Hz band-pass filter followed by a detector switches a dc supply on and off to reconstitute the original signal.

20.2.3.3 Supervisory equipment

It is essential for a network manager to know whether all the equipment is fully operational in a large microwave link system. As many sites are unattended, supervisory equipment is used to transmit the status of radio, multiplex, power, pressurization and various other site alarm conditions through a telephony audio channel. Typically, 16 alarm states are *time division multiplexed* (tdm) into a composite signal which is then applied to a frequency shift keying (fsk) modulator centred on approximately 2 kHz. Some transmitters cater for many more than 16 inputs.

The signals are demodulated at the control or master station to provide a display of the alarms at each remote site. The display may simply comprise a panel of leds, or it may be based on a microcomputer which will show alarm states on a VDU or as a paper print-out if required.

20.2.4 Feeder equipment

Transmitters and receivers are multiplexed together with coaxial cable or rigid, rectangular waveguide. The radio output, often called the *antenna port*, generally comprises a coaxial connector for equipment up to 3 GHz and a square or rectangular waveguide flange for higher frequencies.

A coaxial cable or elliptical waveguide feeder is then used between the radio output port and the antenna.

At frequencies below 3 GHz, it is normal to use cable with either air or foam dielectric. Cable diameters range from quarter-inch to 1⅝ inches (6.35 mm–4.13 cm), and insertion loss decreases with increasing diameter. It is also lower for air dielectric cables. However, as pressurization equipment is required with air dielectric cables, they are not generally used nowadays.

Corrugated, elliptical waveguide is used at frequencies above 3 GHz. This is relatively low loss and is sufficiently flexible to accommodate the inevitable bends in a typical feeder run.

Connectors at each end of the waveguide incorporate transformers which match it to the rectangular waveguide used in the radio branching and antenna feed assembly. A different size of waveguide is required for each frequency band of operation.

Accessories for feeder systems typically comprise:

- connector at each end
- wall or roof feed-through gland to provide a tidy and weatherproof entry into a radio equipment room
- grounding kit (usually three) to ground the feeder at the antenna, at the base of the tower and at the point of entry to the equipment building
- flexible 'tails' for use with relatively rigid, large radius cables or waveguide which cannot be connected to the radio and antenna easily
- hangers and brackets to secure the feeder to the tower and to building walls or ceilings; cable trays or ladders may be required to support the feeder on the tower, in the equipment room or in between

Pressurization equipment is needed for air-filled cable and waveguide systems comprising:

- automatic dehydrator and pump to maintain an excess pressure of 5–10 psi (0.35–0.70 kgf/cm^2)
- manifold (if there are two or more feeders)
- pressure window (for elliptical waveguide only)

20.2.5 Antennas

Circular, parabolic antennas are generally used in microwave radio relay systems. They are economic to manufacture and provide high gain, good side-lobe suppression and high front/back ratio. Reflectors are normally spun from solid aluminium sheets, though grid structures, which are lighter and offer less wind resistance, are often employed at frequencies below 2.5 GHz. Bandwidth is typically 10 per cent of the operating frequency.

Antennas are available in standard, low vswr (*voltage standing wave ratio*) and high performance versions, and feeds may be single or dual polarized (horizontal, vertical or both). Circular polarization is rarely used in fixed link systems.

High performance versions use a side shield to improve the front/back ratio and side-lobe suppression. Sizes range from 0.6 m up to 3.7 m diameter, although the most commonly used are between 1.8 and 2.4 m.

Nearly all antennas are equipped with a panning frame which mounts onto a vertical 115 mm diameter steel pole. Interface steelwork secures the pole to the tower. Alternatively, if antennas are deployed on roof tops or building walls then other types of interface steelwork will be required.

Antennas intended for use in very high winds employ diamond mounts (four fixed positions) rather than poles.

Radomes which fit across the front of the antenna are optional on all but high performance antennas where they are integral. They are used to prevent ice build-up around the feed and to reduce wind loading.

20.3 Fixed link configurations

The most commonly used configurations are unidirectional, bidirectional, unduplicated and duplicated (hot-standby, twin path and N+1).

20.3.1 Unduplicated, unidirectional configuration

Transmitters and receivers are connected directly to the antenna port via a waveguide or coaxial channel filter. Unduplicated equipment is economic but, in the event of an equipment failure, video and audio traffic will inevitably be lost.

20.3.2 Unduplicated, multichannel, unidirectional configuration

Several transmitters and receivers are multiplexed to a common antenna by means of rf branching.

RF branching comprises an arrangement of channel filters and ferrite circulators as shown in *Figure 20.6*(a). The signal

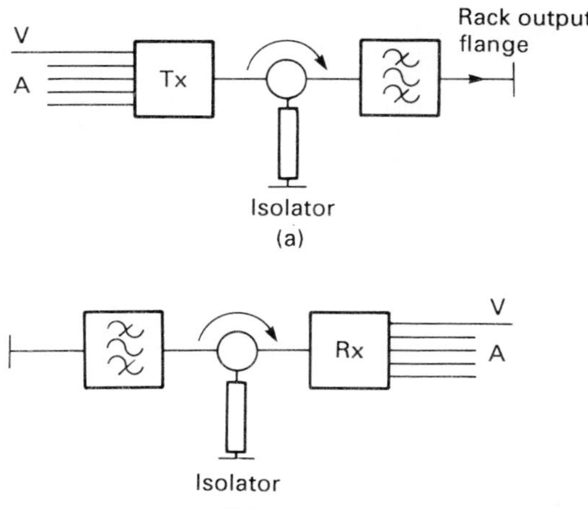

Figure 20.5 Unduplicated unidirectional configuration: (a) transmitter, (b) receiver

from transmitter A is directed to the antenna port via the channel filter and circulator. The output of transmitter B, which is at a different frequency from transmitter A, is directed by the circulator to the channel filter of transmitter A from which it is reflected back towards the circulator. Both signals then emerge (combined) from the same port of the circulator.

Frequency spacing between transmitters is dependent on the bandwidth and roll-off characteristics of the channel filters and is typically 49 or 56 MHz for television transmission.

It is, of course, possible to multiplex more than two transmitters in this manner. The penalty is the additional loss introduced each time the signal is 'bounced' off an adjacent filter and the number of passes through circulators.

Signals are separated in the same way at the receiver.

20.3.3 Bidirectional configuration

RF multiplex can also be configured to connect transmitter(s) and receiver(s) to a common antenna. This provides bidirectional (*duplex*) configuration.

20.3.4 Duplication (protection)

Equipment can be duplicated to prevent loss of traffic. Two transmitters and two receivers carry the same video and audio signals and, in the event of a failure, an automatic changeover switch selects the operational equipment.

There are various ways in which transmitters and receivers can be multiplexed together, and these lead to the arrangements defined in sections *20.3.4.1–20.3.4.3*.

20.3.4.1 Hot-standby

Video and audio signals are split equally between two transmitters operating on the same frequency. Transmitter output signals are fed to a switch which can be either coaxial or waveguide.

Coaxial relay and rotary waveguide switches have relatively low insertion loss (<0.5 dB) but operate quite slowly (typical transfer times are in the region of 20–30 ms). PIN diode switches, however, change over in just a few microseconds but are more lossy (typically 2 dB). Changeover switches can have three or four ports.

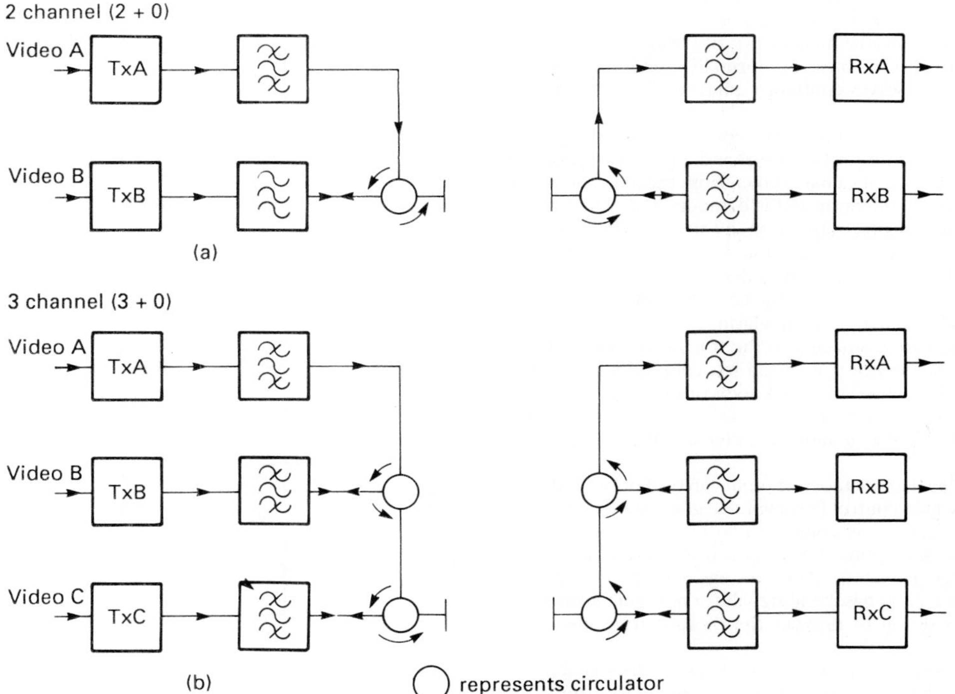

Figure 20.6 Unduplicated multichannel unidirectional configuration: (a) two channels (2 + 0), (b) three channels (3 + 0)

A *three port* switch reflects power from the unused transmitter back into an isolator fitted to the output of the transmitter. The isolator must be capable of accepting the full power of the transmitter continuously. A *four port* switch directs the power of the unused transmitter into a load fitted to the switch.

Monitoring circuits in the transmitter check for correct operation of various modules and raise an alarm if a fault is detected. An alarm causes the changeover switch to select the other transmitter.

20.3.4.2 Twin path

A twin path system does not employ transmitter switching. Both transmitters operate simultaneously on two separate frequencies, and their outputs are combined with rf branching to feed one antenna. Receiver signals are split in a similar way.

Both receiver outputs are connected to a video/audio switch which selects the better channel.

Twin path duplication is less efficient spectrally as two bearer

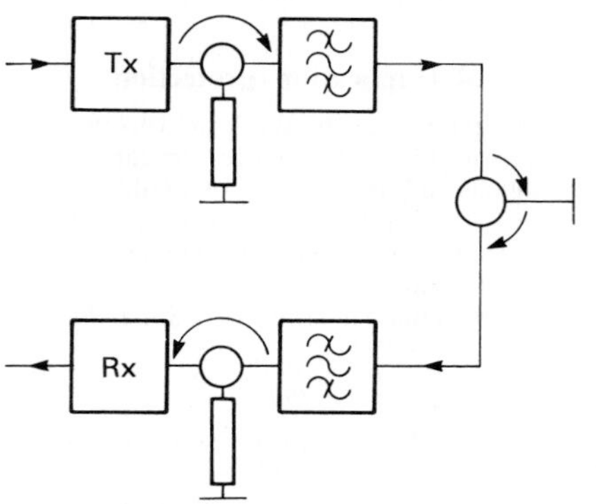

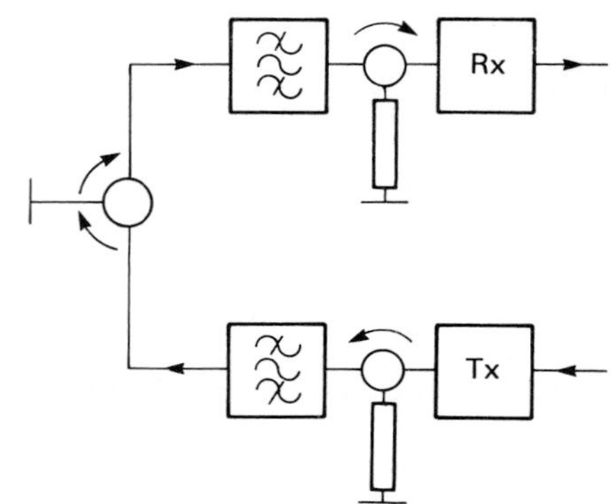

Figure 20.7 Bidirectional configuration

At the receiving end of the link, the incoming rf signal is split in a waveguide or coaxial hybrid circuit and fed equally to both receivers. Switching takes place at baseband if monitoring circuits detect a fault in the operational receiver.

An equal split at rf between the two receivers results in a 3 dB loss in both signal paths. An unequal split, 10:1 or 100:1 (10 dB or 20 dB), reduces the loss into one receiver at the expense of the other. Receiver switching is then biased to the channel with the least attenuation (the main or priority channel).

Transmitter performance is normally monitored simultaneously in two ways: continuity pilot level, and rf power output.

An 8.500 MHz or 9.023 MHz continuity pilot signal is injected into the transmitter baseband prior to any active circuits. The 70 MHz if signal is taken from the monitor output of the frequency modulator and is demodulated. The level of the resulting pilot signal output is monitored. If it falls by more than a predetermined value (usually 3 dB), an alarm is activated (led or dry contacts or both).

RF power is monitored after the final stage of amplification. Again a 3 dB fall will activate an alarm.

These alarms are connected to the transmitter switch logic.

Receiver performance is assessed by: continuity pilot level, and received signal level or signal/noise ratio.

Continuity pilot level is measured close to the receiver output and, if a fall of 3 dB or more is detected, then an alarm is activated.

The signal level at the receiver input can be determined from the automatic gain control (agc) voltage generated in the main if amplifier. This agc is compared with a reference voltage which is set to correspond to the agc value at threshold.

An alarm is activated when the agc voltage crosses this threshold value. IF or baseband signals (or both) are also muted under this condition to prevent the onward transmission of noise.

The noise level in the baseband close to the pilot signal is also used as a switching criterion. A noise detector measures the level and raises an alarm if it exceeds the threshold value.

All these alarms are connected to the receiver switch logic.

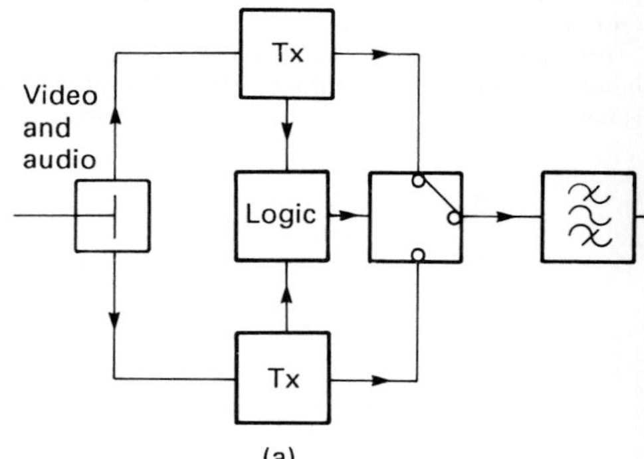

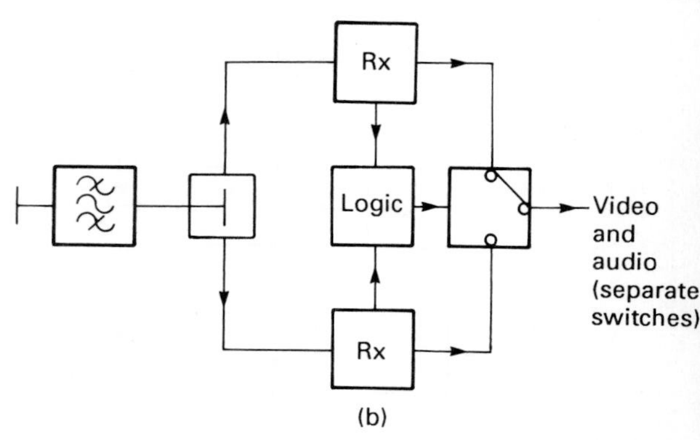

Figure 20.8 Hot-standby configuration: (a) transmitter, (b) receiver

frequencies are needed. However, the need for transmitter logic and switching equipment makes the hot-standby approach more costly.

20.3.4.3 N+1 protection

Hot-standby and twin path configurations are most suitable for protecting a single microwave channel. In a system carrying two or more independent channels, duplication of each is both costly and unnecessary. Instead, the normal procedure is to provide one additional rf channel to protect several (N) others.

Detection of a fault at the receiver causes the receiver switching logic to send an alarm signal to the transmitter end of the link via a return channel.

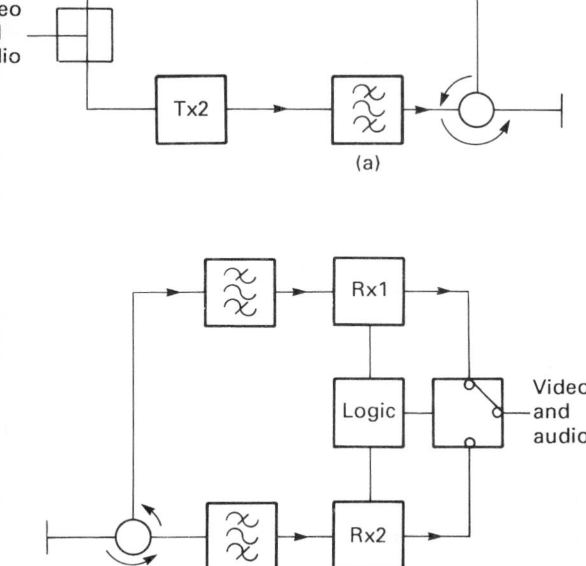

Figure 20.9 Twin-path configuration: (a) transmitter, (b) receiver

This alarm signal causes the transmitter switching logic to switch the affected signal from the failed channel to the protection channel.

The return channel required for N+1 operation may be a broad-band microwave link, a narrow-band vhf/uhf radio link or even a standard telephone line.

Confirmation that switching has taken place at the transmitter is then sent to the receive end, and the switch associated with the failed channel selects the protection channel at the receiver.

20.4 System planning

Fixed microwave links are generally used in the television industry for the transmission of video and audio signals between studios and broadcast transmitters and between regional centres.

Studios and broadcast transmitter sites are chosen to suit their prime functions and not for reasons of microwave link propagation. Therefore the planning engineer must choose an economic route commensurate with the required performance.

For relatively short paths with tall broadcast transmitter masts, line-of-sight propagation can be often achieved with just one hop. However, where the separation between sites is greater, or where obstructions intrude into a line-of-sight path, then two or more hops are required in a tandem connection. Intermediate stations are known as *repeaters*.

The tasks involved in planning a microwave link are:

- Prepare a path profile between sites and determine antenna heights to meet clearance criteria defined below. If it is obvious at the outset that a direct link between studio and transmitter is not possible, then proceed to the next step.
- Choose likely repeater site(s) from a detailed map, or by visual inspection of the area, looking for suitably elevated positions with road access which provide suitable path lengths. If possible, avoid positioning repeater stations on hill tops away from roads; access costs are likely to be very significant during construction and maintenance. If a suitable station already exists which another user is prepared to share, then a considerable saving in civil works costs will obviously result.
- Prepare path profiles between likely repeater stations and determine antenna heights to meet clearance criteria. A site and route survey will always be required to confirm information derived from maps.
- Choose routes that offer a reasonable compromise between hop length and antenna height.
- Perform calculations to determine transmitter output power, receiver noise figure and antenna size needed to provide a suitable fade margin and s/n ratio.
- Specify building and tower requirements.

20.4.1 Path profiles

The first task in assessing the suitability of a route is to determine clearance between the microwave beam and potential obstructions along the route. The terrain is analyzed by plotting ground height contours taken from a topographical map, with a scale of 1:50 000 or less, onto specially prepared graph paper with curved lines representing the earth's curved surface. The microwave path is represented by a straight line drawn between the two antennas.

Atmospheric refraction causes ray banding at microwave frequencies, and it is normal to assume an effective earth radius (k) 4/3 greater than actual. In practice, this means that the microwave beam is bent slightly downwards by the gradual change in the atmosphere's refractive index with height. The microwave signal can therefore be received some way beyond the optical horizon.

Changes in atmospheric refraction can occasionally cause the value of k to drop below 2/3. This prevents the microwave signal from being received at the optical horizon.

It is often more convenient to plot ground height on rectilinear graph paper and add the earth bulge for two values of k: 4/3 and 2/3.

Earth bulge (EB) may be expressed in metres as:

$$EB = \frac{d_1 d_2}{12.8k}$$

where d_1 and d_2 are the distances from a particular point on the path to each end in kilometres.

20.4.2 Fresnel zone

The next step is to calculate the radius of the first Fresnel zone, which is defined as the locus of all points surrounding a radio beam from which reflected rays would have a path length one half-wavelength greater than the direct ray.

The value in metres of the first Fresnel zone radius is given by:

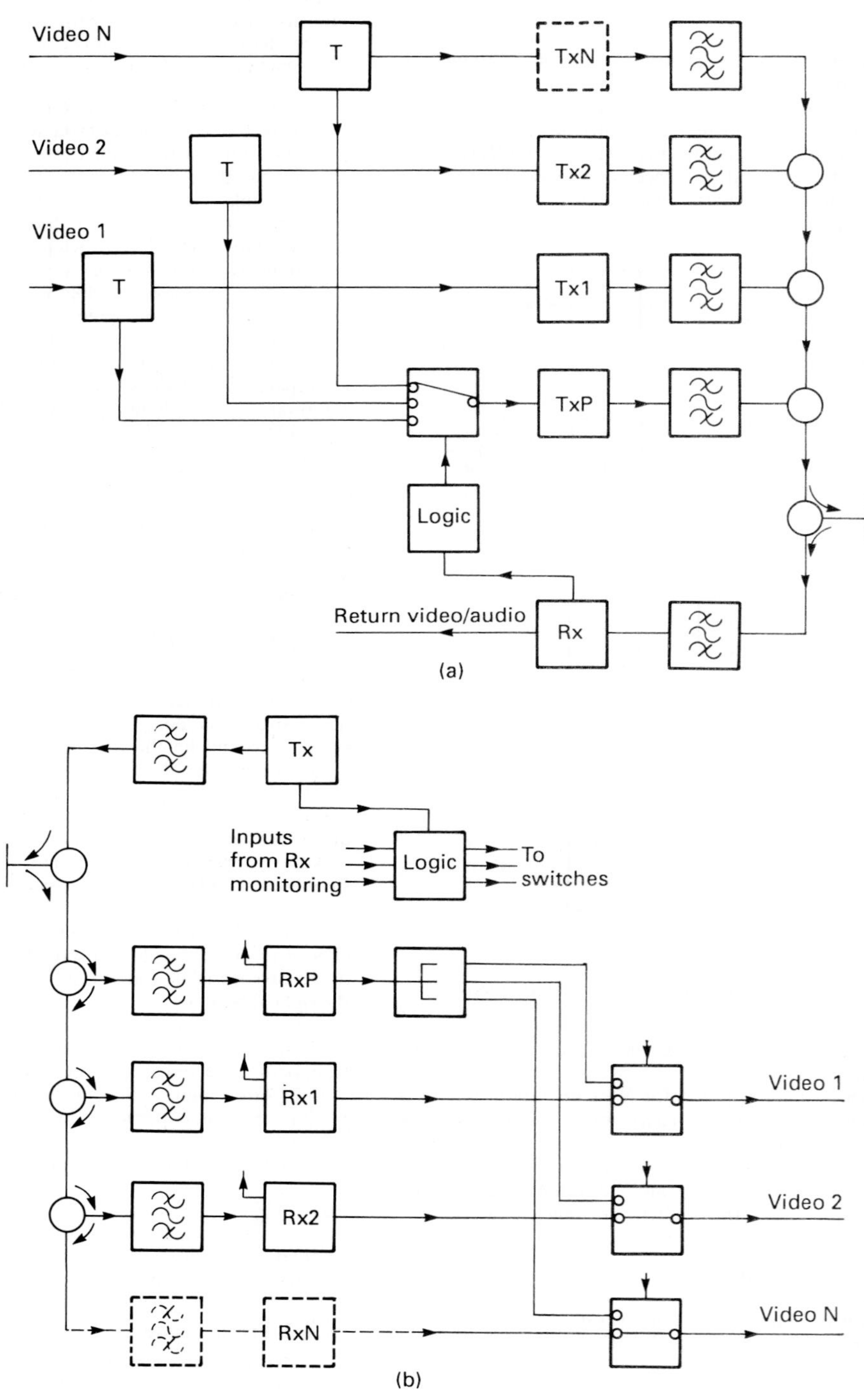

Figure 20.10 N+1 terminals: (a) transmitter, (b) receiver

$$R_{FZ} = 17.3 \sqrt{\frac{d_1 d_2}{F(d_1+d_2)}}$$

where d_1 and d_2 are the distance to each end of the path in kilometres, and f is the frequency in gigahertz.

Figure 20.12 shows that free space propagation loss corresponds to transmission of 60 per cent of the radius of the first Fresnel zone.

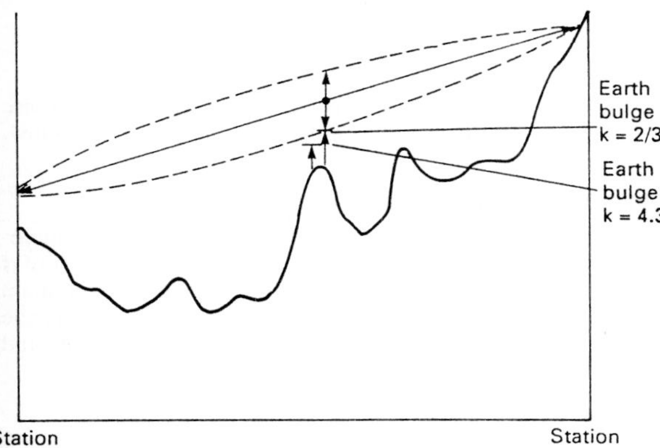

Figure 20.11 Path profile on rectilinear graph paper

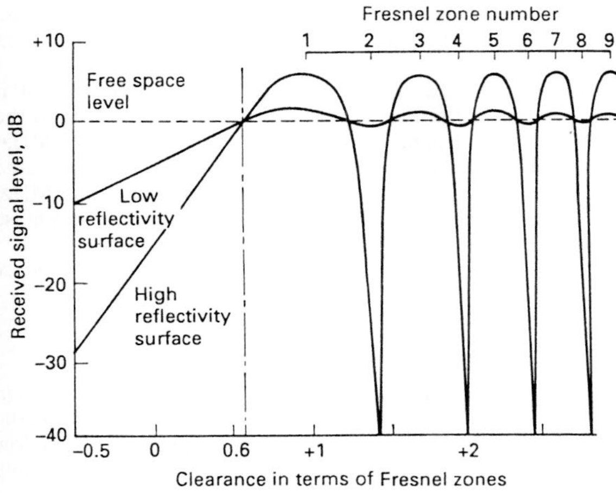

Figure 20.12 Received signal level versus clearance

20.4.3 Clearance criteria

Two clearance criteria are commonly applied to microwave links:

- k = 4/3: clearance = $0.6R_{FZ}$ + EB
- k = 2/3: clearance = $0.3R_{FZ}$ + EB

The second requirement is generally used for longer paths where the earth bulge is greater and where atmospheric effects are more significant. The choice of $0.3R_{FZ}$ is a compromise, but values of k as low as 2/3 occur only for a small percentage of the time and therefore it is normally accepted.

Path loss under normal, unfaded conditions should be close to the predicted value if these conditions are met.

20.4.4 Received signal level

The power of the signal at the microwave receiver input is dependent on the following factors:

- branching losses at both ends of the link (dB) (as stated by the radio equipment manufacturer)
- waveguide or cable feeder losses at both ends of the link (dB) (as stated by the feeder equipment manufacturer)
- propagation losses
- antenna gains (dB) (as stated by the antenna manufacturer)

Propagation loss in decibels under normal conditions is $20 \log(4\pi d/\lambda)$, where d is the path length and λ is the wavelength (both in metres). It may alternatively be expressed as $20 \log(4\pi df/c)$, where f is the frequency (Hz) and c is the velocity of light (3×10^8 m/s).

Therefore, the received signal level in dBW is equal to transmitted power minus branching losses minus feeder losses minus propagation loss plus antenna gains.

A typical calculation is shown in *Table 20.1*.

Equipment type	VFL	70	
Frequency	GHz	7.5	
Path length	km	55.0	
Path loss	dB	144.8	
Transmitter power	dBW	2.0	Stated by manufacturer
Transmitter losses			
Branching loss	dB	1.5	Stated by manufacturer
Other losses	dB	0.5	Connectors etc
Waveguide loss	dB	2.5	Length in m × loss/m
Total TX loss	dB	4.5	
TX antenna parameters			
Diameter	m	1.8	
Gain	dB	40.8	Stated by manufacturer
Radiated power	dBW	38.3	
Received power	dBW	−106.5	
RX antenna parameters			
Diameter	m	2.4	
Gain	dB	43.3	Stated by manufacturer
Receiver losses			
Waveguide loss	dB	3.5	Length in m × loss/m
Other losses	dB	0.5	Connectors etc
Branching loss	dB	1.5	Stated by manufacturer
Total RX losses	dB	5.5	
Signal at RX input	dBW	−68.7	
Receiver threshold (video)	dBW	−106.7	Item 15 + 12dB
Fade margin	dB	38.1	Item 12 − item 13
KTBF noise	dBW	−118.7	
where:			
KT	dBW	−203.5	
B	MHz	30	
F	dB	10	
Carrier/noise	dB	50.1	Item 12 − item 15
Video noise bandwidth	MHz	4	Standard for 625 line PAL
FM deviation (pk)	MHz	4	Standard CCIR
FM improvement	dB	2.8	
Bandwidth improvement	dB	7.8	
Emphasis improvement	dB	2	
Signal/noise UWTD (fade dependent only)	dB	62.7	
Fixed noise contrib (unweighted)	dB	67.0	Fixed noise generated by amplifiers, oscillators, etc

Total signal/noise (unweighted)	dB	61.3	
Weighting improvement	dB	10.9	
Fixed noise contrib (weighted)	db	71.0	Fixed noise generated by amplifiers, oscillators, etc
Total signal/noise (weighted)	dB	69.1	
Availability estimate assuming multipath (Rayleigh) fading			
Path roughness 4 for smooth terrain inc over water 1 for average terrain with some roughness 0.25 for mountainous, very rough terrain			
Parth roughness value (A)		1.0	
Conversion from worst month probability to annual (B): 0.5 for large lakes or dimilar hot, humid areas 0.25 for large inland areas 0.125 for mountainous or very dry areas			
Conversion from worst month probability to annual value (B)		0.25	
Estimated availability		99.9971	

Derived using the following expression to account for multipath (Rayleigh fading)

30logd+10log(6abf)−fm−70

where			
	D	=	Path length (km)
	A	=	Path roughness coefficient
	B	=	Climate coefficient
	F	=	Frequency (GHz)
	FM	=	Fade margin (dB)

Table 20.1 System performance example for an unduplicated 55 km path at 7 GHz

20.4.5 Fade margin

One of the most important parameters of a radio link is the fade margin. This is defined as the difference between the *normal* or *unfaded received signal level* and the *lowest received signal level* which will provide an acceptable output from the receiver expressed in decibels.

The lowest received signal level which provides an acceptable output is defined as the receiver threshold for which there are two commonly used definitions:

- the rf signal level at the receiver input below which video or audio s/n ratio is unacceptable
- the rf signal level at the receiver input below which fm improvement ceases. This is normally referred to as the *threshold of fm improvement*, and in video systems it is approximately 10-12 dB above the receiver noise floor.

Receiver noise floor is given by KTBF, where K is Boltzmann's constant, T is the noise temperature of the receiver in degrees K, B is the noise bandwidth of the receiver usually taken as the 3 dB points of the if filter (Hz), and F is the noise figure at the receiver input (dB).

For the purposes of performing calculations, the following formula is easier to use:

$$\text{Noise (dBW)} = -203.5 + 10 \log B(\text{Hz}) + F(\text{dB})$$

The *fade margin* determines the availability of the link in the presence of fading; the greater the fade margin, the lower the link outage time. The relationship between availability and fade margin is given in section *20.5.1*.

20.4.6 Carrier/noise ratio

The carrier/noise ratio (c/n) is the ratio of the received signal (carrier) level to the noise power at the receiver input, expressed in dBm or dBW, i.e.:

$$c/n = \text{received carrier level (dBW)} - \text{KTBF (dBW)}$$

Video signal/noise ratio (s/n) is derived from the carrier/noise ratio taking into account the improvement factors listed below.

20.4.7 FM and bandwidth improvement factors

Frequency modulation (fm) was first introduced into microwave links to avoid problems associated with poor linearity of rf devices, which would have resulted in severe distortion in am systems. FM also brought a useful feature, which was the increase in video signal/noise ratio resulting from fm and bandwidth improvement factors.

FM improvement, in decibels, is defined as:

$$20 \log \frac{f_f}{F_m}$$

where d_f is the peak deviation (MHz), and F_m is the highest frequency (5.5 MHz for 625-line PAL).

Bandwidth improvement, in decibels, is defined as:

$$10 \log \frac{B}{b}$$

where B is the receiver bandwidth (to 3 dB points in MHz), and b is the video filter cut-off frequency (to the 3 dB points in MHz).

There is bandwidth improvement because the rf and if bandwidths of the receiver are considerably greater than the video bandwidth at baseband, i.e. 30 MHz instead of 5.5 MHz. Noise on both sides of the carrier (±15 MHz) is demodulated into the baseband and therefore extends out to 15 MHz. However, as the baseband is filtered above 5.5 MHz, only noise within ±5.5 MHz of the carrier degrades the signal.

FM and bandwidth improvement are the *capture noise quieting effect* associated with fm transmission.

20.4.8 Pre-emphasis improvement

Pre-emphasis in television microwave links reduces the level of large amplitude, asymmetric, low frequency components in the video signal by 14 dB relative to high frequencies. These low frequency components would cause problems in maintaining the mean frequency of the modulator and would modulate the colour subcarrier as a result of non-linearity in the baseband. This effect is manifested as differential phase and gain distortion, chrominance and luminance inequalities and crosstalk.

The pre-emphasis characteristic is defined in CCIR Rec. 405-1. The pre-emphasis factor is 2.0 dB for triangular fm noise (system I).

20.4.9 Weighting

As the eye does not perceive the effect of noise equally at all frequencies, measurements of the video s/n ratio are taken through a luminance weighting network as defined in CCIR

Report 637-2. The weighting improvement is 10.9 dB for triangular fm noise (system I). (CCIR Report 637-2 includes further details on emphasis and weighting factors for system I and other CCIR systems.)

The overall signal/noise ratio for luminance (unweighted) is:

$$s/n = c/n + (fm + bw + \text{pre-emphasis improvements})$$

The s/n ratio for luminance (weighted) is given by:

$$s/n = c/n + (fm + bw + \text{pre-emphasis} + \text{weighting improvements}).$$

20.5 Additional losses

20.5.1 Availability

The propagation loss calculated in section *20.4.4* took account only of fixed losses. However, the signal power arriving at the receiver in a real radio system can vary significantly and can fade well below the anticipated value as atmospheric conditions change.

Multipath fading over a microwave link is due to the signal arriving at the receive antenna from a number of different routes such as:

- the direct path between antennas
- ground and building reflections
- refraction in the atmosphere

Sometimes the signals add in phase, and the resultant is stronger than anticipated. At other times the signals cancel, and the resultant is weaker.

For short periods, almost complete cancellation can take place; the signal level then falls below the receiver threshold and transmission is lost. Fortunately, severe fading of this magnitude is relatively uncommon and occurs for only short periods.

Multipath fading is dependent on the following factors:

- ground roughness (smooth surfaces cause pronounced reflections)
- climate (causes changes in refractive index of the atmosphere which result in rays taking different routes to the receiver)
- path length (longer paths increase the opportunity for these events to take place)
- frequency (small delays represent a larger proportion of a wavelength)

Vigants derived the following expression to predict signal availability in the presence of Rayleigh fading:

$$\text{Availability (\%)} = 100(1\text{-}10^{\,30 \log d + 10 \log (6ABf)) - F - 70})$$

where
d = path length (km),
A = path roughness,
B = climatic conditions,
f = frequency (GHz),
F = fade margin (dB).

The path roughness (A) is 4 for smooth terrain including over water, 1 for average terrain, or 0.125 for mountainous or very dry areas.

The climatic conditions (B) are 0.5 for hot, humid areas, 0.25 for large inland areas, or 0.125 for dry inland areas.

The fade margin can alternatively be calculated on the basis of a given availability by rearranging the expression:

$$\text{Fade margin} = 30 \log d + 10 \log (6ABf) - 10 \log (1\text{-}\frac{A}{100}) - 70$$

where A = required availability (%).

20.5.2 Diffraction losses

There are occasions when it is impractical to provide sufficient antenna height to meet the clearance criteria stated in section *20.4.3*, and plans for that particular route may have to be abandoned. However, it may be possible to tolerate the additional loss introduced as the signal is diffracted over the horizon. *Figure 20.12* illustrates loss versus penetration into the Fresnel zone.

Transmission of 60 per cent of the radius of the first Fresnel zone is represented as 0 dB as expected. Additional loss is determined by the physical characteristics of the obstructing surface, i.e. whether it is smooth and rounded or sharp edged. Expressions for calculating diffraction loss are as follows:

$$\text{Diffraction loss} = A(X_0) - B(X_1) - B(X_2) - 20.5$$

where
$X_0 = d_0 \times B_0$
$X_1 = d_1 \times B_0$
$X_2 = d_2 \times B_0$
and
$B_0 = 670(f/A_e{}^2)[1/3]$
$d_1 = (2A_eH_1)[1/2]$
$d_2 = (2A_eH_2)[1/2]$
where
A_e = effective earth radius (km)
H_1 and H_2 = antenna heights (km)
f = frequency (MHz)
and
$A(X_0) = 0.057\,510\,4 \times X_0 - 10 \log X + 2.066$
$B(X_1) = 0.057\,510\,4 \times X_1 - 10 \log X_1 + 0.71$
$B(X_2) = 0.057\,510\,4 \times X_2 - 10 \log X_2 + 0.71$

2.5.3 Rainfall

At frequencies above 10GHz rainfall can cause severe attenuation. At 23GHz, it can be the dominant factor assessing fade margin requirements in many parts of the world.

As an example, attenuation (per kilometre) caused by a rainfall rate of 50mm/h at GHz is 1.3dB, while at 23GHz it is 6.3 dB. Path lengths at 23GHz are usually limited to 10km or so.

2.5.4 Gases

Attenuation from atmospheric gases is negligible at most frequencies used in most radio relay applications. Attenuation (per kilometre) is below 0.01dB at 10GHz and 0.3dB at 23GHz.

The effect of rainfall and gases is covered extensively in CCIR Volume V.

20.6 Improving availability

Signal availability can be improved by using space or frequency diversity at the receiver.

20.6.1 Space diversity

Space diversity requires two antennas each with its own receiver. The ideal spacing between the antennas is determined by several factors such as antenna height, path length, frequency and whether diversity is required to counter the effects

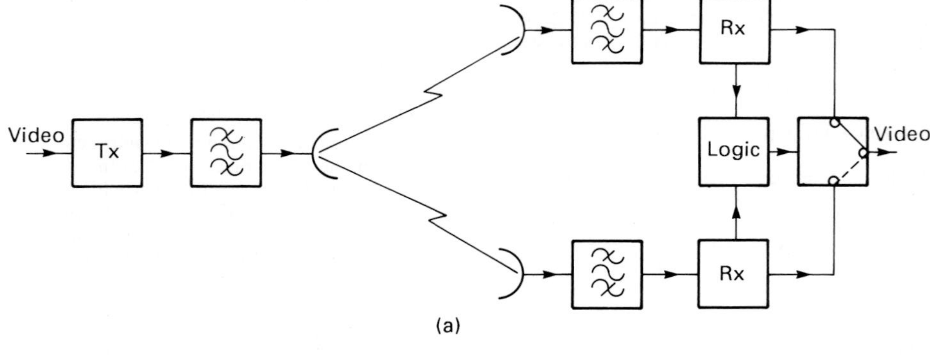

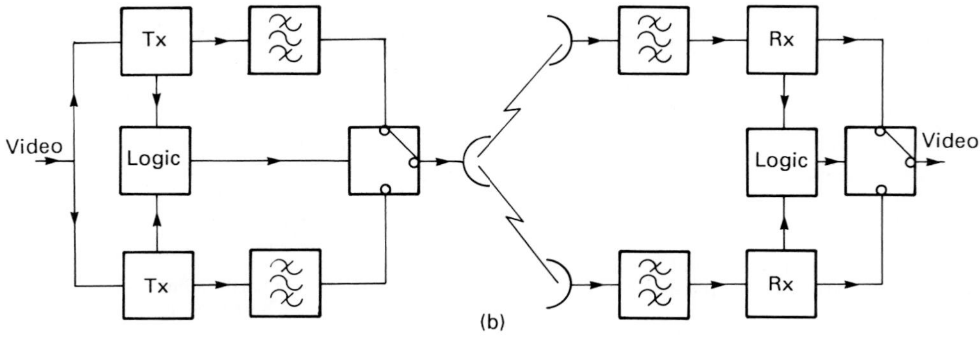

Figure 20.13 Space diversity configuration: (a) transmitter unduplicated, (b) transmitter duplicated

of specular reflection from very smooth surfaces such as water, or whether it is to counter atmospheric effects.

20.6.1.1 Reflections from water

When a microwave signal passes over water it is possible, at certain times, for the reflected and direct rays to cancel at the receiving antenna. If a second antenna is placed above or below the original at such a distance that the path length of the reflected ray is increased or decreased by one half wavelength, the rays will add instead of cancelling.

Monitoring circuits in the receivers detect changes in received signal level and switch from one receiver to the other as the signal improves or degrades.

To calculate antenna separation, it is first necessary to establish the point of reflection along the path by solving the following equation for n:

$$\frac{2n^2-3n+1}{12.8\ k} = \frac{H_b}{D^2} - \frac{(1-n)\ H_a}{nD^2}$$

where

H_a = height of antenna at site A above mean sea level (m)
H_b = height of antenna at site B above mean sea level (m)
D = path length along earth's surface (= $D_a + D_b$)
n = ratio of distance from reflection point to site A (D_a/D)
k = effective earth radius factor (normally 4/3)

The next step is to calculate H_{ra} and H_{rb} where:

H_{ra} = height of tangential plane above mean sea level at site A = $(nD)^2/17kR_0$
H_{rb} = height of tengential plane above mean sea level at site B = $(1-n)^2D^2/17kR_0$

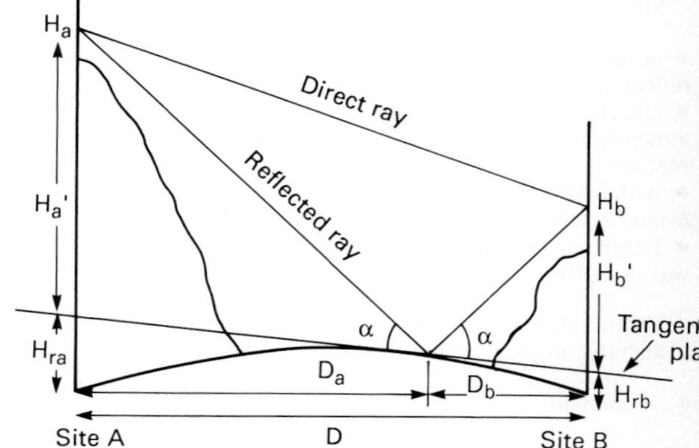

Figure 20.14 Diagram of ray geometry

R_0 = true earth radius in km, i.e. 6390

Then

H_a' = height of antenna in metres above the tangential plane at site A = $H_a - H_{ra}$
H_b' = height of antenna in metres above the tangential plane at site B = $H_b - H_{rb}$

Then antenna spacing in metres is:

$$s = \frac{3D}{4fH} \times 10^5$$

where

D = path length (km)
f = frequency (MHz)
H = height of transmitting antenna above tangential plane, i.e. either H_a' or H_b' depending on which end is the transmitter site.

20.6.1.2 Protection against atmospheric effects

When space diversity is used to protect against Rayleigh fading or ducting (where a ray is trapped in an atmospheric layer), antennas should be spaced as widely as possible. Generally, tower height sets a practical limit on the spacing, but a minimum of 150 wavelengths should be the target.

CCIR Report 376-4 includes a nomogram showing the improvement from space diversity.

20.6.2 Frequency diversity

The disadvantage of space diversity is that it doubles the cost of the antenna and feeder equipment at the receiving terminal and may require a stronger tower.

Frequency diversity is a similar configuration to twin-path but with the frequencies more widely separated, i.e. approximately 5–10 per cent.

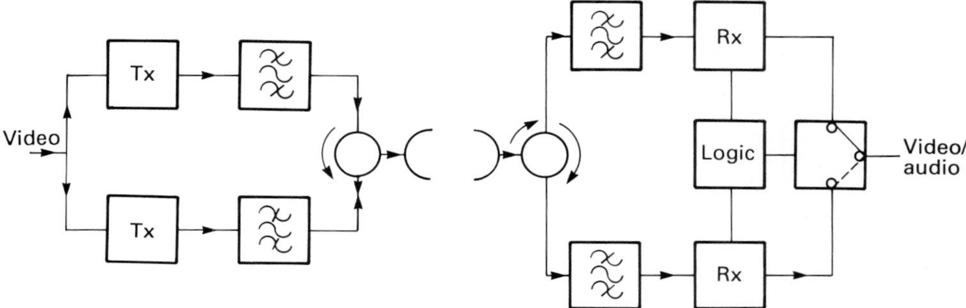

Figure 20.15 Frequency diversity configuration

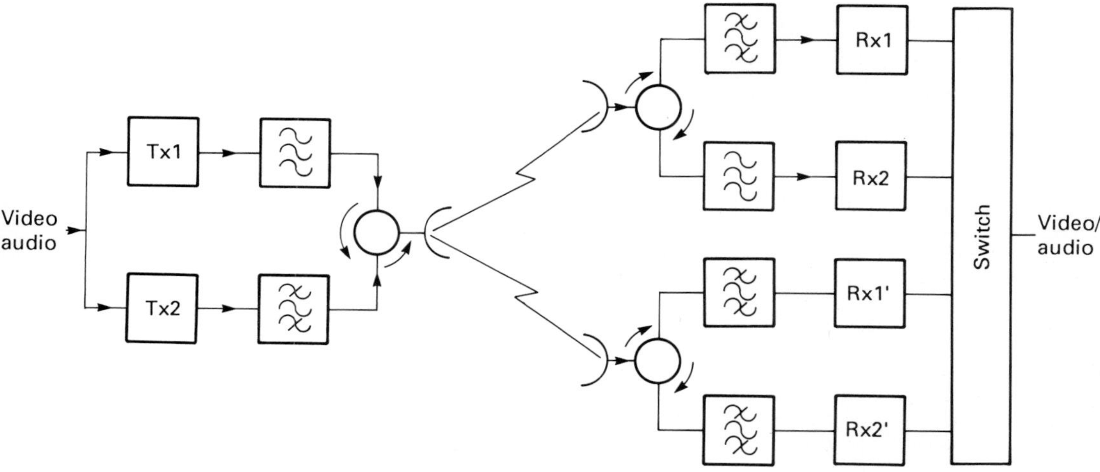

Figure 20.16 Space and frequency diversity configuration

B Flowers MRTS
Head of Eurovision Control Centre, European Broadcasting Union

21 Intercity Links and Switching Centres

21.1 Historical development

When the British Broadcasting Corporation (BBC) started public television broadcasting in November 1936, the studio and 17 kW vhf transmitter (34 kW effective radiated power) were located together at Alexandra Palace, situated on a hill in north London. This avoided the problem of how to convey the 405-line television signal, with a bandwidth of about 3 MHz, from the studio to a distant transmitter.

By May 1937, the BBC was able to provide outside broadcast coverage of the coronation of King George VI, using an equalized coaxial cable from Hyde Park Corner to Alexandra Palace, a distance of 13 km.

In June 1946, when the BBC commenced its post-war television service, coaxial cable was still the best means of conveying signals between cities. Indeed one of the BBC's test transmissions, intended to enable television receiver manufacturers to align their products, was a film showing how the General Post Office (GPO) laid a twin tube coaxial cable, with repeaters, from London to Birmingham. The repeaters provided compensation for the cable losses, primarily high frequency attenuation, thereby ensuring good quality and a reasonable signal/noise ratio of the demodulated signal at Sutton Coldfield, the Birmingham area transmitter. The video signal was transmitted as amplitude modulation of a 6.12 MHz carrier, with partially suppressed upper sideband, utilizing a bandwidth of 3–7 MHz. Coax tubes were subsequently installed from Birmingham to Manchester, and from London to Cardiff, utilizing amplitude modulation of a 1 MHz carrier, with partially suppressed lower sideband, requiring a bandwidth of 500 kHz–4.5 MHz.

The only alternative means of conveying a television signal several kilometres was by using the same amplitude modulation of a vhf transmitter as was utilized by the main public transmitters. This system was employed for outside broadcasts in the late 40s.

Then in 1952 intercity links entered a new era with the introduction of shf links (links at *super high frequency*, i.e. 3–30 GHz). The development of radar (*radio detection and ranging*), and the invention of the travelling wave tube by Dr R Kompfner at the Clarendon Laboratory, Oxford University, during World War II, gave microwave engineers the required technology to transmit and receive narrow beams of fm modulated carriers with frequencies of a few gigahertz. In fact we still denote the various frequency bands in this part of the radio spectrum with the letters adopted by radar development engineers, as follows:

P	420–450 MHz
L	1–2 GHz
S	2–4 GHz
C	4–8 GHz
X	8–12 GHz
Ku	12–18 GHz
K	18–27 GHz
Ka	27–40 kHz

The official ITU frequency bands, as defined by the ITU Radio Regulations, Article 2, are listed in Table *21.1*. The relationship between them and the letter codes is indicated in *Figure 21.1*.

The big advantage of shf links over coaxial cable distribution is that the shf link towers require only small areas of land every 30 km or so, whereas coaxial cables must be laid in suitable underground ducts.

Nowadays, coaxial cables are used mainly for local connections between broadcasters' facilities within cities, utilizing amplitude modulation and/or frequency modulation of an rf carrier. In the USA, it is common practice to add one or two audio channels on fm subcarriers above the baseband video signal.

Coaxial cables are also widely employed for the distribution of cable TV to the general public, utilizing rf carriers modulated in the same way as the broadcasters' transmitters. In Belgium, for example, 85 per cent of homes are connected to cable distribution of about 16 TV channels.

In the 1950s, shf links were built throughout Europe and North America, forming the contribution and distribution networks for the national broadcasters. Moreover, the early 50s saw the first experimental international transmissions, starting with a 405-line transmission from Calais to London in 1950. This was followed by programme exchanges between France and England in 1952, using an optical converter to change between the French 819-line standard and the British 405-line standard.

Band number	*Symbols*	*Frequency range (lower limit exclusive, upper limit inclusive)*	*Corresponding metric subdivision*	*Metric abbreviations for the bands*
4	VLF	3 – 30 kHz	Myriametric waves	B.Mam
5	LF	30 – 300 kHz	Kilometric waves	B.km
6	MF	300 – 3 000 kHz	Hectometric waves	B.hm
7	HF	3 – 30 MHz	Decametric waves	B.dam
8	VHF	30 – 300 MHz	Metric waves	B.m
9	UHF	300 – 3 000 MHz	Decimetric waves	B.dm
10	SHF	3 – 30 GHz	Centimetric waves	B.cm
11	EHF	30 – 300 GHz	Millimetric waves	B.mm
12		300 – 3 000 GHz	Decimillimetric waves	

Note Band number N extends from 0.3×10^{N} Hz to 3×10^{N} Hz.

Table 21.1 ITU frequency bands

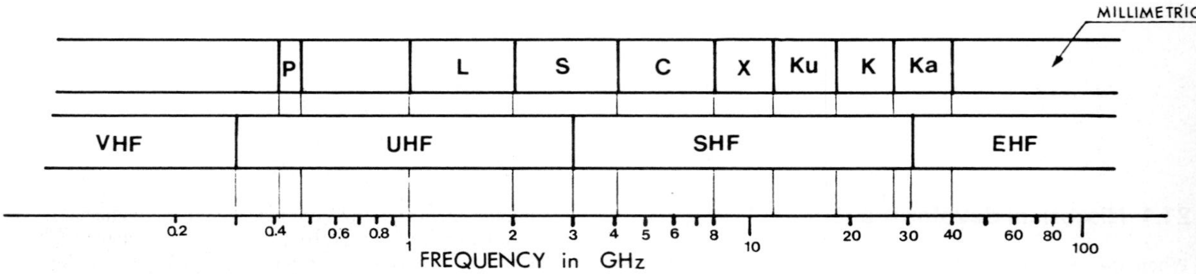

Figure 21.1 Relationship between letter codes and frequency bands

On 2 June 1953, the coronation of a British monarch once again provided the impetus for progress. TV coverage of this event was relayed live from London to France, the Netherlands and Germany, thereby arousing great interest in the new concept of international television.

The birth of Eurovision is considered to be 6 June 1954, which was the start of a series of international exchanges between eight European countries, i.e. Belgium, Denmark, France, Germany, Italy, Netherlands, Switzerland and the United Kingdom, coordinated from Lille in France, and known as the *Lille experiment*. There was a total of 55 Eurovision transmissions in 1954, since when there has been a steady increase to about 30 000 Eurovision transmissions in 1988. The word *Eurovision* was invented in the early 1950s by George Campi, a journalist working as TV critic for the London *Evening Standard* newspaper. Meanwhile the shf network, available for international television in the European Broadcasting Area, increased from 10 000 km in 1955 to nearly 300 000 km by 1985, when the Eurovision network was augmented by the addition of two transponders on Eutelsat I-F2 (flight 2 of Eutelsat series I).

21.1.1 The BBC transatlantic slow scan system

Before considering the development of satellite links, a description of the first system to be used for regular transatlantic television transmission is of interest. By the late 50s, it took about ten hours to send film across the Atlantic Ocean by air freight. Very important news film was sometimes flown across in about five hours, on board a Hustler jet of the United States Air Force.

At this time there were already several transatlantic telephone cables in operation (TAT cables), so BBC engineers built a slow scan television system to transmit pictures across the Atlantic, between the BBC Television News Centre at Alexandra Palace, London, and NBC New York or CBC Montreal.

These long distance music circuits had an audio bandwidth of about 6 kHz, but after equalization a usable bandwidth of only 5 kHz was available. The bandwidth of the BBC 405-line system was 3.5 MHz, i.e. about 700 times greater than that of the audio circuit. Therefore a slow scan system would normally take about 700 s to transmit 1 s of pictures, which was clearly unacceptable. Several tricks were performed to reduce this ratio to 100:1.

The system was essentially a slow scan telecine transmitting to a slow scan film telerecorder. A reduction of 2:1 was achieved by sending only every second film frame, and printing it on two successive frames at the receiving end. A further reduction of 2:1 involved sending only 200 lines without interlace, with a consequent reduction of vertical resolution. A similar reduction of horizontal resolution produced a slow scan signal with a bandwidth of about 5 kHz. Long persistence cathode ray tubes, as used in radar displays, enabled the operators to monitor the pictures at both the sending and receiving ends.

This system was used for transatlantic transmission of short news items from 1959 to 1962, and although the resultant pictures were of mediocre quality, with rather jerky movement portrayal, they were acceptable for urgent news material.

21.1.2 Communication satellites

In July 1962, the first live high quality transatlantic television transmissions took place via the *Telstar* satellite. This was a low-orbit satellite, which circled the earth once every 90 minutes, with only 20 minute windows of availability. Earth stations were built at Andover, Maine, USA, at Pleumeu Boudou, Brittany, France, and Goonhilly Downs, Cornwall, UK, with the ability to track the satellite. The Andover and Pleumeu Boudou earth stations utilized very large horn antennas, with spherical radomes for protection against the elements, whereas the GPO antenna at Goonhilly Downs used a large parabolic dish, costing less than half as much as a horn antenna.

Horn antennas are less susceptible to man-made interference, but in practice this is not a problem, and nearly all subsequent satellite earth station antennas were designed with parabolic reflectors. Like terrestrial links, these satellite links utilize frequency modulation, which has good inherent immunity to interference. In practice, the most common cause of interference on satellite links is crosstalk from an adjacent channel in the satellite transponder, especially when one transponder is shared by two television channels.

Figure 21.2 BRT/RTBF shf links tower at the Belgian television centre, Brussels (BRT)

In 1965, *Early Bird*, the first geostationary communications satellite, entered service for transatlantic television, heralding a new era, and leading to the birth of *Intelsat*, an international consortium of more than a hundred PTTs, which provides intercontinental telephony, data and television circuits between the member countries, with the deployment of geostationary communication satellites above the Atlantic, Pacific and Indian Oceans.

21.1.3 Eurovision development

The technical coordination of international television in the European Broadcasting Area was undertaken by the Technical Centre of the European Broadcasting Union (EBU) in Brussels, Belgium. From January 1956, a coordination centre was established on the sixth floor of the Palais de Justice in Brussels. This location was chosen because the Belgian shf links were installed in the dome of the Palais de Justice when Belgian television started in 1953. In 1979, the Eurovision Control Centre (EVC) was transferred to the Belgian Television Centre at Boulevard Reyers on the east side of Brussels, where an 89 m shf links tower was built.The Belgian television services own and operate their own shf links, as do the broadcasting services of France, Spain, Portugal, Italy, Yugoslavia and most of the North African countries.

In contrast, the northern and eastern countries of Europe have television shf links which are owned and operated by their respective PTTs.

Figure 21.3 Parabolic dish and horn antennas on the BRT/RTBF shf links tower (BRT)

In the late 1960s, shf links were established across the Mediterranean Sea, some of which depended on tropospheric scatter to bridge the distances involved. Even normal shf links function beyond direct line of sight distance, due to refraction by the earth's atmosphere, the density of which decreases with increasing altitude. The resultant extension of range, during normal weather conditions, can be considered as equivalent to straight line propagation on an earth whose radius has been increased by a factor of $4/3$.

Following the introduction of colour television in Europe from 1967 onwards, the performance of the shf links and switching centres had to be improved. PAL and SECAM, unlike NTSC, are relatively immune to differential phase distortion, but cumulative high frequency attenuation proved to be a problem in Europe on long international circuits. Of course it could be equalized at the receiving end, but nevertheless it produced a deterioration of the chrominance signal/noise ratio.

The introduction of insertion test signals (ITS) on lines 17, 18, 330 and 331 in the frame suppression period, enabled the performance of the circuits to be monitored at all times and improved when necessary. The test signals concerned were defined by CCIR Recommendation 473-1 for 625-line signals, which became the universal line standard in Europe, with the

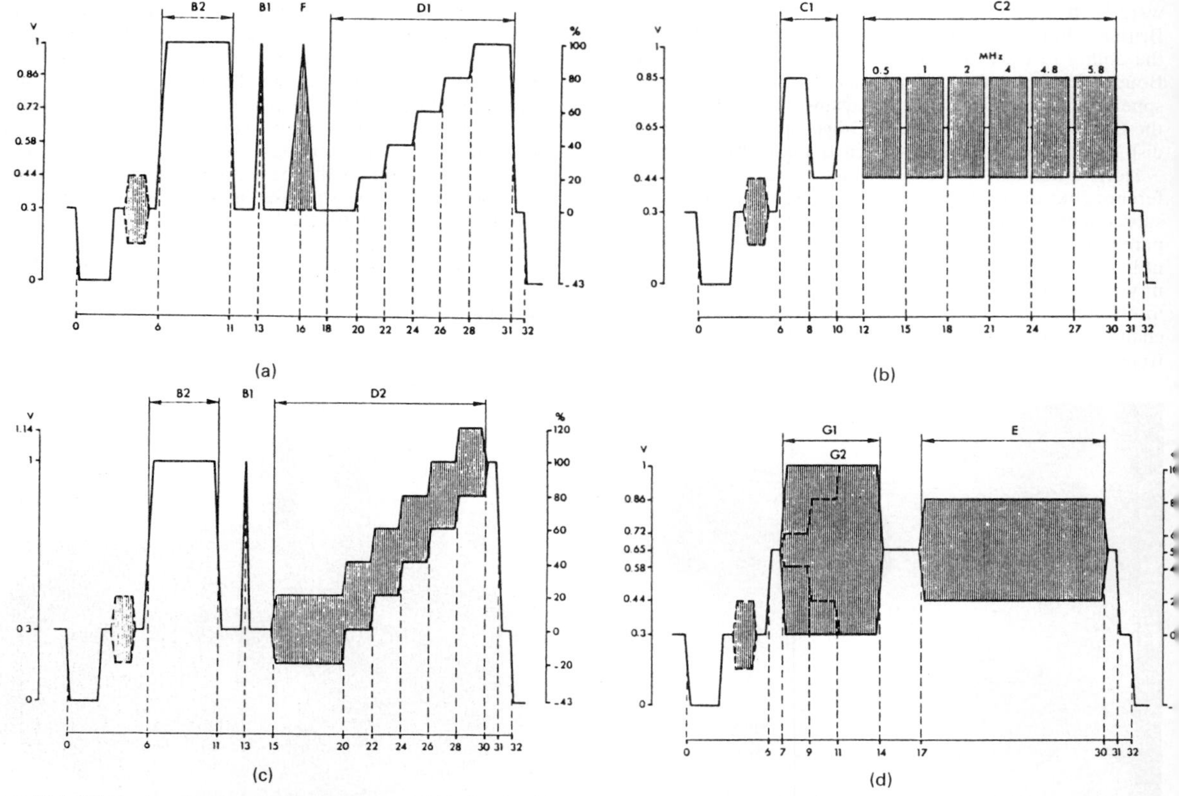

Figure 21.4 625-line insertion test signals to be inserted (a) in line 17, (b) in line 18, (c) in line 330, (d) in line 331. B1: 2T pulse, B2: luminance bar, C1: two-level grey pulse, C2: 'multi-burst' signal, D1: five-riser 'staircase' signal with superposed chrominance subcarrier, E: composite mid-grey bar, F: 20T composite pulse, G1: composite luminance bar, G2: three-level; composite luminance bar

introduction of colour TV in the United Kingdom and France in 1967.

All other European countries had adopted the 625-line system from the start of their black and white services, with the exception of French speaking Belgian television and Luxembourg television, which both employed the French 819-line system initially.

21.1.4 USA development

In North America, insertion test signals are known as *vertical interval test signals* (VITS). They are inserted on line 17 of both odd and even fields (*Figures 21.4* and *21.5*).

The three big American commercial networks, ABC, CBS and NBC, formerly distributed their programmes around the USA by means of shf links, provided by AT&T long lines division. These networks went around the states in a clockwise direction, distributing their programmes from New York or Washington to their affiliates. This 'round robin' arrangement enabled regional centres to break the loop and send programmes back to New York when required.

In the late 1970s, the Public Broadcasting Service (PBS) adopted the new technology of domestic satellite (DOMSAT) distribution to deliver its programmes to regional broadcasters all over the USA, using a Westar C-band satellite. The programmes were sent on four transponders at hourly intervals, to take account of the four time zones across the USA.

The three other networks followed suit in the early 1980s, using several satellite channels to distribute programmes from New York to the other states, and other satellite channels to bring contributions from regional studios back to New York. NBC New York, for example, installed a system to control remotely the regional up-leg transmitters by sending coded instructions in digital format, known as the *satellite network management system*. This utilizes *single carrier per channel* (scpc) to distribute the remote control data by satellite. The unit which interprets these instructions and activates the regional up-leg transmitters is known as the *pup* (portable uplink package).

NBC currently leases seven transponders on RCA Ku-band satellites. ABC and CBS lease C-band satellite transponders, which are comparatively immune to signal path attenuation in the presence of heavy rain storms. On the other hand, the Ku-band transmit and receive dishes are smaller than C-band dishes, so they can be located at the broadcasters' premises, whereas the larger C-band dishes are used in conjunction with shf links to distribute programmes to broadcasters in a given area. ABC and CBS are now implementing *satellite news gathering* (SNG), using Ku-band satellite transponders.

21.1.5 Audio transmission

In the 1950s and 1960s, the sound component of international television transmissions was sent via separate PTT wide-band audio circuits. From January 1962, the EBU leased a network of wide-band audio circuits, centred on Brussels, for this purpose. In addition, for live multilateral transmissions, the receiving broadcasting services order their own commentary circuits from the origin, to mix with the 'international sound', or 'ambience', at the point of reception. For single destination

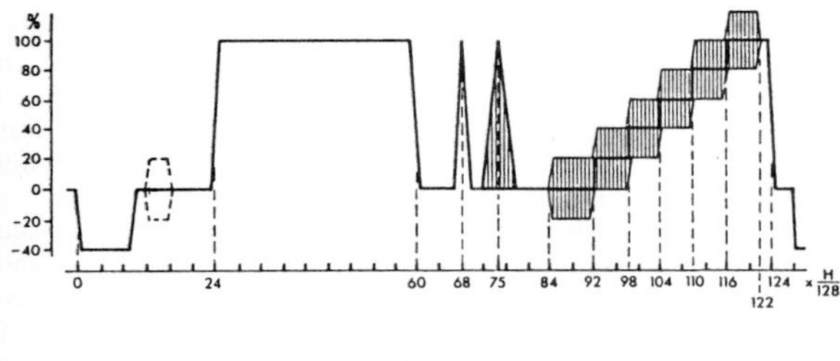

(a)

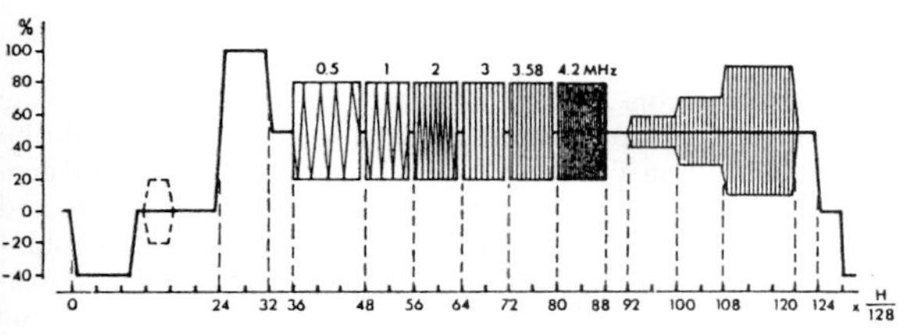

(b)

Figure 21.5 525-line vertical interval test signals to be inserted (a) in line 17 of the first field, (b) in line 17 of the second field

transmissions, the sound channel provided by the EBU network is available for the complete audio signal.

The separate routeing of audio signals applied also to Intelsat circuits in the early days, and loss of audio was a regular occurrence.

A geostationary satellite must be positioned at an altitude of about 35 900 km above the equator, giving an orbital period of 24 h. Hence, provided the satellite is travelling from west to east, it remains almost stationary as seen from the earth. The total length of an up-leg and down-leg satellite circuit is about 75 000 km, depending on the location of the earth stations. Since the speed of light is 300 000 km/s, this gives a transit time of 250 ms. Consequently, loss of video/audio synchronization can occur if the signals follow paths with a different number of satellite hops. Solid-state audio delay units are now available to restore synchronism.

By the early 1970s, BBC engineers had developed a digital audio transmission system for television sound, known as *sound-in-sync* (sis). This system samples the audio signal at twice line frequency, e.g. 31.25 kHz for 625-line systems, using 10 bits per sample in the analogue/digital conversion. The resulting 20 bits, for two samples per television line, plus one start bit, are inserted in the line-sync period of the television waveform. An active compander system, employing an amplitude modulated pilot tone locked to line frequency, improves the overall performance to be equivalent to a simple 12 bits per sample coding system, with a bandwidth of 25–14 000 Hz.

This development was welcomed by the EBU as the solution to most of its operational audio problems, and in 1974 the Eurovision wide-band audio network was replaced by the sound-in-sync system. Since 1974, the international sound signal has been subject to far fewer breakdowns, as it automatically accompanies the video signal throughout the Eurovision network.

Moreover, the high quality of the audio signal (0.3 per cent total harmonic distortion, signal/noise of about 60 dB weighted) is maintained regardless of distance and cumulative distortion of the analogue video signal. Naturally, if this distortion is excessive, the sound-in-sync decoders cannot decode properly, producing characteristic crackling on the audio output signal. In this situation, it is customary to decode and recode the audio en route, thereby incurring a slight increase in quantizing noise. SIS decoders restore the synchronizing waveform to its correct form, but a simpler pulse blanking unit can be used to provide a sync which is non-standard, but quite adequate for vision monitoring purposes. Some modern picture monitors are sis compatible.

In the USA, sis was not adopted for the simple reason that the AT&T long lines tariff included the provision of one or two audio channels with the video circuit at no extra cost. However, the Canadian carrier Bell Canada adopted the 525-line version of sis on its terrestrial circuits. PBS sends audio as digital modulation of a 5.8 MHz subcarrier, known as the *DATE* system (Digital Audio Transmission Equipment).

Most European national networks utilize the audio channels that are built into terrestrial shf links. There are normally four such audio channels available on fm subcarriers above the video baseband signal. Spectrum analyser displays of this system are shown in *Figures 21.6* and *21.7*.

These audio channels are not normally suitable for international transmission due to incompatibility between national

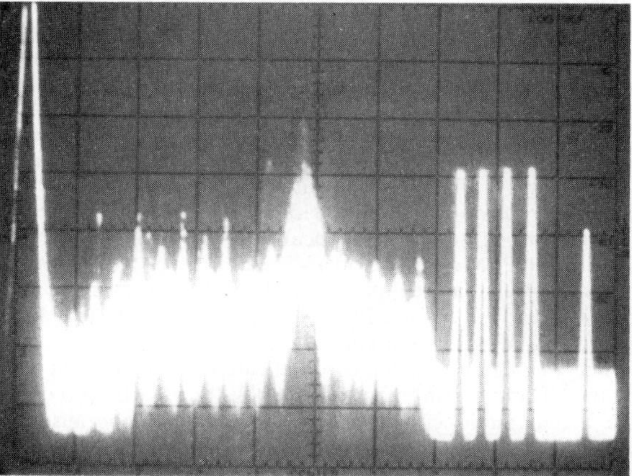

Figure 21.6 Baseband spectrum (0 – 10 MHz) of a PAL test pattern with four fm audio channel subcarriers at 7, 7.36, 7.74 and 8.14 MHz, plus a continuity pilot at 9.023 MHz (BRT)

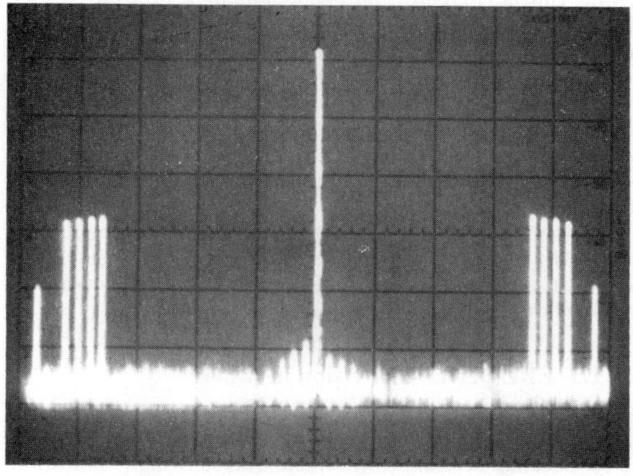

Figure 21.7 IF spectrum (70 MHz +– 10 MHz) of Figure 21.6 carriers. Unmodulated carriers are shown for clarity. (BRT)

systems. Moreover, at baseband switching centres, the audio and video are separated, with the possibility of losing the audio signal. However, the international link from Zagreb, Yugoslavia, to Ankara, Turkey, utilized this sound-on-carrier system for about ten years after the introduction of sound-in-sync on the Eurovision network, because the circuit transits Bulgaria, where the national broadcaster, Bulgarian Television (BT), is a member of the OIRT and Intervision, rather than the EBU and Eurovision. Nowadays, several Intervision services utilize sis, as does the international vision circuit Zagreb — Ankara. Intelsat satellite circuits do not carry sis but they do associate one or two audio channels with every vision circuit, thereby avoiding the problems of lost audio and loss of video/audio synchronization which occurred in the 1960s.

21.1.6 Audio levels

Most European broadcasters monitor audio levels with a PPM (*peak programme meter*) whereas American broadcasters have traditionally used a VU meter. Recently, some new meters have become available in the USA which combine both systems, taking advantage of led displays to indicate mean and peak values of modulation.

The EBU PPM has an integration time of 10 ±2 ms, whereas a VU meter has an integration time of about 300 ms. Therefore, if a PPM and a VU meter are aligned to indicate 0 db when fed with a 1 kHz tone at 0 dBm, when the tone is replaced by speech or music, the PPM will indicate peaks up to 10 dB higher than the VU meter, depending on the exact nature of the modulation. At a zero reference point, 0 dBm tone corresponds to 0 dBm0, which is normal line-up tone reference level (EBU 'test' level). Peak permitted level, as defined by the CCITT in Recommendations J13 and J14, is 9 dB above this 0 dBm0 line-up tone. Hence peaks are normally set to reach +9 dB on a PPM and 0 dB on a VU meter. Even a PPM reads 1 or 2 dB lower than real peaks, so the BBC, for example, sets peaks to read +8 dB on a PPM. The EBU sets peaks to +4 dB, leaving a 5 dB safety margin.

The VU meter situation is complicated by the fact that 0 dB on the VU meter can correspond to +8 dB, +4 dB, or 0 dB on the actual audio circuit. This situation arose because the original VU meter was a simple passive meter preceded by a rectifier, taking power from the audio circuit. To avoid loading the circuit asymmetrically, a pad was used to isolate the VU meter from the line. With 0 VU corresponding to +8 dBm on the line for continuous tone, modulation peaks can reach +18 dBm on the line, which defines the required headroom. Using this alignment, frequency modulation with hf pre-emphasis may cause overmodulation on shf link and satellite audio circuits, so American broadcasters now send 0 VU at +4 dBm or 0 dBm to line. In the European context, 0 dBm0 tone may also be sent to line at various absolute levels, +6 dBm being a common standard. This just depends on the agreed levels between the broadcaster and the local PTT. For sound-in-sync, 0 dBm0 corresponds to 0 dBm, at the input to the coder.

As a general rule, provided 0 dBm0 tone is set to read 0 dB (i.e. 100 per cent) on a VU meter, and 0 dB (i.e. EBU 'test' level) on a PPM, the two systems should concur when peak modulation levels are set in the normal way.

21.1.7 EBU permanent vision network

By 1967, Eurovision traffic had increased to over 2000 transmissions a year, so that the leasing of a permanent network of shf links became economically viable. As a rough indication, leasing a vision circuit is worthwhile if the average utilization exceeds 1000 minutes a month. The Permanent Vision Network (PNV) grew to a total of about 18 000 km by 1985, when the introduction of two leased satellite transponders led to a 30 per cent reduction in the total length of leased terrestrial circuits.

By 1987, the density of television traffic from the USA to continental European broadcasters justified the lease of a transatlantic television channel (TTC) by the EBU. This was established in December 1987, just in time for the Reagan/Gorbachev summit meeting in Washington. It comprises a terrestrial shf link from Washington to New York, and a leased transatlantic satellite link from New York, via Staten Island Teleport, to Paris, Frankfurt and Rome. The British broadcasters prefer to retain their association with *Brightstar*, a two way transatlantic satellite link leased by Visnews, with its European gateway in London, and USA gateways in New York and Washington (*Figure 21.8*).

21.1.8 Optical fibres

To complete this historical survey of intercity links, the development of optical fibre transmission systems should be considered. Long distance fibre-optic circuits utilize digital pcm transmission, thereby preserving a good signal/noise ratio and avoiding cumulative distortion. Fortunately, a universal standard has been agreed for digital sampling of 525-line and 625-line television signals, namely the 4.2.2 system. A basic requirement of all analogue/digital conversions is that the sampling rate must be at least twice the bandwidth of the signal to be sampled.

A colour television signal is composed of three components, namely luminance (Y) plus two colour-difference signals (U) and (V) (i.e. weighted values of R-Y and B-Y). Since the human eye has a greater acuity for black/white (luminance) information than colour (chrominance) information, the U and V signals do not require such high definition as the Y signal. Hence Y is sampled at 13.5 MHz, whereas U and V are sampled at 6.75 MHz. To keep quantizing noise to an acceptable level, at least 8 bits per sample are required. (For studio quality, 10 bits per sample are used within the broadcaster's premises, where very high bit rates can be handled locally without difficulty.) Therefore straight digital coding would require at least:

$$(13.5 + 6.75 + 6.75)8 = 216 \text{ Mbit/s}$$

without including parity bits for error detection. Fortunately, various bit rate reduction techniques are available, which can reduce the bit rate considerably. Also, for example, a number of systems have been developed to reduce the net bit rate to 140 Mbit/s (IRT, IBA).

The bit rate can be reduced even further, to, say, 34 Mbit/s, but with present techniques, while this produces a vision signal that may be of adequate quality for transmission to the public, questions remain as to whether it will be entirely adequate for production techniques such as colour-matte. In contrast, bit rates of 2 Mbit/s for 625-line signals and 1.5 Mbit/s for 525-line signals, are used for video conference services, where the resultant blurring of movement is tolerated.

Fibre-optic circuits have already made a big impact on local connections within cities, from the broadcaster to the local PTT and vice versa. These circuits employ frequency modulation of a carrier, with one or two subcarriers for audio channels. The video signal remains in composite form, unlike the component coding of digital fibre-optic systems. A typical frequency for the carrier is 32 MHz with a peak deviation of ±14 MHz. 70 MHz is also used, which has the advantage that standard if equipment can be utilized.

These analogue fibre-optic systems provide a very high quality of transmission for a distance up to about 60 km in the

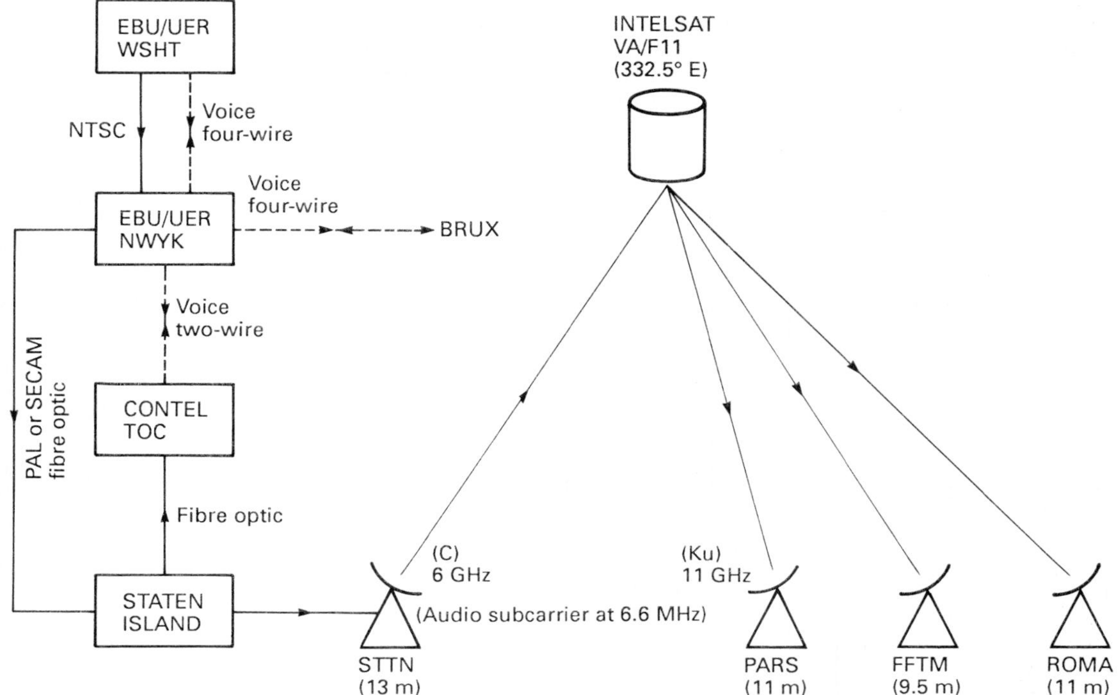

Figure 21.8 EBU/TTC (Transatlantic Television Channel)

case of laser modulators. LED modulators can be used for shorter distances of a few kilometres.

Composite digital transmission is another alternative, whereby a composite NTSC, PAL or SECAM signal is sampled at three times colour subcarrier frequency, usually with 9 bits per sample, since the maximum signal amplitude is higher than that of component signals.

Optical fibres have considerable advantages over coaxial cables. They are completely immune to rf interference, so they can be installed alongside electric railways, for example. A cable containing eight optical fibres has a diameter of about 20 mm. Moreover, a single optical fibre can transmit over 1 Gbit/s. A second light source of a different wavelength can double this capacity.

The European Economic Community has established a project called *RACE* (research and development in advanced communications – technologies in Europe) to promote high capacity data links between the twelve countries of the EEC. European industry, broadcasters, PTTs, and the EBU are involved in this project, which will eventually provide intercity optical fibre circuits, capable of carrying high definition television and many other data signals.

21.2 Eurovision links

Figure 21.9 shows in a simplified form how international terrestrial and satellite circuits are coordinated in the European Broadcasting Area. Normally, the ITC is the national PTT switching centre, and the CNCT is the national broadcaster's international control room, which acts as an interface between the broadcaster and the EVC. The CNCTs are connected to the EVC by four-wire *technical coordination circuits*, which are connected in (*n*-1) conference mode.

(n-1) conference mode describes a system to interconnect a number (*n*) of four-wire communication circuits in such a way that each remote terminal receives a mix of the modulation from all other terminals, but excluding outgoing modulation from the terminal in question. This arrangement enables any number of broadcasters, in different locations, to talk together as if they were sitting around a conference table. A given terminal must not receive its own output back from the central conference unit, otherwise oscillation may occur when the microphone is opened, due to the Larsen effect. The alternative of cutting the loudspeaker completely when the *speak* key is pressed is unacceptable, because in this case the speaker cannot hear other people trying to intervene. In practice, the loudspeaker level should be automatically attenuated by about 15 dB, when the microphone *speak* key is pressed. This system is known as *autodim*.

The main communications system for the new Eurovision Control Centre in Geneva will be based on a 64-channel (*n*-1) conference system. Each four-wire channel will be selectable to one of eight independent conference groups, which will be created by software control of the 64-channel (*n*-1) conference matrix X-points, giving maximum flexibility.

A similar network of *programme coordination circuits* (pccs) enables EBU member services to discuss programme matters, primarily concerning the four daily Eurovision News exchanges (EVNs).

Mobile earth stations provide up-legs for special events when appropriate. For example, two mobile earth stations were flown to Reykjavik for television coverage of the Reagan/Gorbachev summit meeting in October 1986.

The EBU leased transponders in Eutelsat I-F2 have an rf bandwidth of 80 MHz, which is wide enough to accommodate two television channels per transponder. Hence, from 1987 an additional television channel was created in one leased transponder, and a pcm stereophonic *Euroradio* channel was created in the other leased transponder.

This third television channel is foreseen for use with mobile

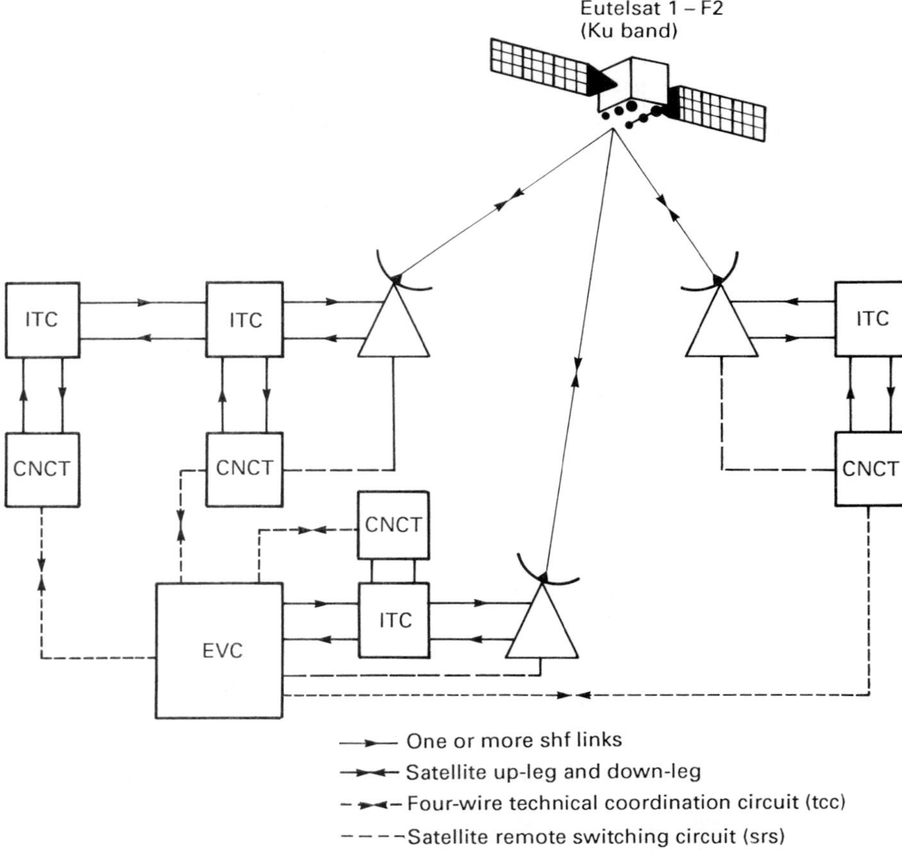

Figure 21.9 Eurovision links. ITC: International Television Centre, CNCT: National Technical Control Centre, EVC: Eurovision Control Centre

Figure 21.10 Mobile earth stations at Reykjavik, October 1986 (EBU)

Figure 21.11 Eutelsat I, flight 2 (British Aerospace)

earth station up-legs, and the Euroradio channel is foreseen for the distribution of EBU concerts and other stereophonic radio transmissions.

The situation in mid 1989 was that efforts were being made to squeeze two television signals plus one Euroradio channel into each transponder, giving a total of four television channels and two Euroradio channels in the two leased transponders.

The Euroradio digital stereo signal utilizes 1.024 Mbit/s, with a nominal up-leg transmit power of 70 dBW, whereas the fm television channels occupy 36 MHz of rf bandwidth, with a

Western Spotbeam
Eastern Spotbeam
Atlantic Spotbeam
Eurobeam

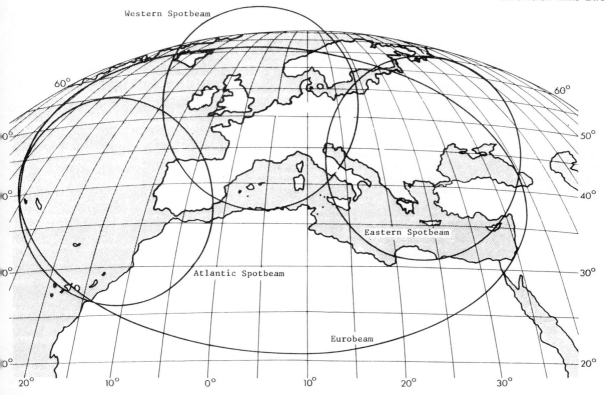

Figure 21.12 Footprint of Eutelsat I-F2 Eurobeam

Figure 21.13 Danmarks Radio satellite earth station for use with Eutelsat I-F2 (7.6 m dish) (Hanne Boock)

nominal up-leg transmit power of 76 dBW. All these signals utilize the satellite's Eurobeam antenna, whose footprint covers the whole of the European Broadcasting Area, as defined by the ITU.

Satisfactory reception of the television signals requires an earth station antenna of at least 7–8 m diameter, whereas the more robust digital stereophonic radio signals can be received with a dish of only 3 m diameter. Consequently Euroradio reception can be easily installed at the broadcaster's premises.

The satellite earth stations utilizing Eutelsat I-F2 for Eurovision are listed in Table *21.2*.

In fact, Eurovision traffic was transferred from Eutelsat I-F2 to Eutelsat I-F1 in December 1989, with another transfer to Eutelsat I-F5 foreseen in December 1990. From 1993, EBU will lease four wideband (72 MHz) transponders in a second generation, Eutelsat II, satellite, thereby doubling their leased space-segment capacity.

Eutelsat standard earth stations

Country	*Location*	*Transmit channels*	*Receive channels*
Belgium	Lessive	1	2
Sweden	Aagesta	1	2
Austria	Aflenz	1	
France	Bercenay	1	1
Germany	Usingen	1	1
Italy	Fucino	1	1
Spain	Guadalajara	1	1
Switzerland	Leuk	1	1
United Kingdom	Madley	1	1
Cyprus	Makarios	—	1
Portugal	Sintra	—	1
Turkey	Ankara	—	1

Earth stations at broadcasters' premises

Country	*Location*	*Transmit channels*	Receive channels
(In service)			
Algeria	Bouzareah	1	2
Denmark	Copenhagen	1	2
Egypt	Cairo	—	2
Germany	Hamburg	—	2
Iceland	Reyykjavik*	—	2
Israel	Jerusalem	—	2
Jordan	Amman	—	2
Netherlands	Hilversum	—	2
Norway	Nittedael	1	2
Tunisia	Tunis	1	2
Yugoslavia	Zagreb	1	2
(Planned for later)			
Finland		1	2
Greece		1	2
Ireland		1	2
Libya		1	2
Luxemburg		1	2
Monaco		1	2
Morocco		1	2

*A 5.5m dish at Skyggnir, the Icelandic PTT earth station site, is being used temporarily.

Table 21.2 Satellite earth stations utilizing Eutelsat I-F2 for Eurovision

21.2.1 Quality control

Insertion test signals (ITS) can be measured manually, by means of a specialized oscilloscope or television waveform monitor. In an operational environment, manual measurement is too time consuming, so automatic ITS measurement units are installed in most CNCTs, ITCs and satellite earth stations. There are numerous different lists of tolerances for the various parameters, each applicable to different sections of a television network. The target values adopted by the EBU for the two principal elements of the Eurovision network are listed in *Table 21.3*.

	Tolerance T	S
Luminance bar amplitude error	±12%	±6%
Baseline distortion	± 3.5%	± 4%
2T pulse/bar rtio	± 9%	±18%
Peak differential gain	± 9%	±13%
Chrominance/luminance gain inequality	±11%	±13%
Signal/noise ratio (weighted)	54dB	51dB
Bar slope	± 3.5%	± 4%
Line-time non-linearity	*	12%
Chrominance/luminance delay inequality	±100ns	±90ns
Chrominance/luminance intermodulation	*	*
Peak differential phase	±5°	±6°
Transient sync compression	*	*

* not specified

T: a terrestrial international connection, (CNCT-ITC-ITC-CNCT)
S: an international connection via Eutelsat (CNCT-ITC-Eutelsat-ITC-CNCT)

Table 21.3 Tolerances for terrestrial (T) and Eutelsat (S) international connections

Weekly tests are made to check the performance of all EBU leased terrestrial and satellite circuits. Occasional circuits, and temporary lease circuits, are also tested when necessary, in particular just prior to major events. However, the big advantage of ITS, compared with full-field test signals, is that measurements can be carried out during programme transmission, whenever there is doubt about the performance of a circuit.

21.2.1.1 Luminance bar amplitude error

The luminance bar amplitude is defined as the difference in level between the mid-point of the white bar on line 17 and a reference point 4 μs before the first riser of the staircase. The percentage error is the difference between the bar amplitude and 0.7 V, being positive if the bar amplitude is greater than 0.7 V and negative if the bar amplitude is less than 0.7 V.

Overmodulation of shf links produces non-linear distortion, whereas a low modulation level incurs the penalty of a poor signal/noise ratio.

21.2.1.2 Baseline distortion

This is defined as the difference in level between a point 400 ns after the half-amplitude point of the trailing edge of the luminance bar, and a reference point 4 μs before the first riser of the staircase on line 17, expressed as a percentage of the luminance bar amplitude. The sign is positive if the level immediately after the luminance bar is higher than the level at the reference point.

Baseline distortion is indicative of low frequency distortion, which produces a characteristic streaking impairment of the picture. It may also cause crosstalk from sis modulation on the picture. Some types of scrambling are particularly adversely affected by baseline distortion.

21.2.1.3 2T pulse/bar error

This is defined as the difference between the 2T pulse amplitude and the bar amplitude, expressed as a percentage of the bar

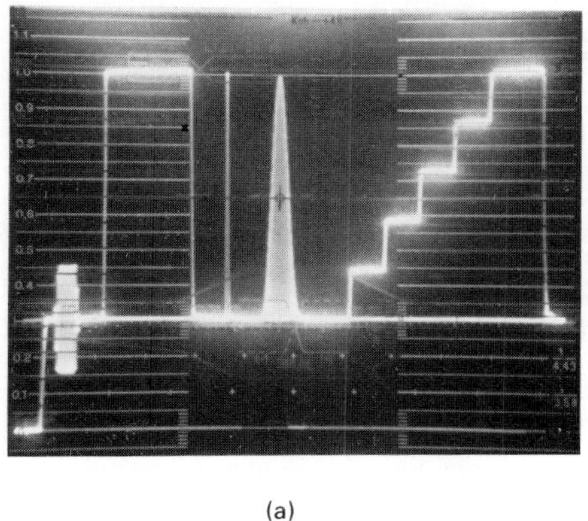

(a)

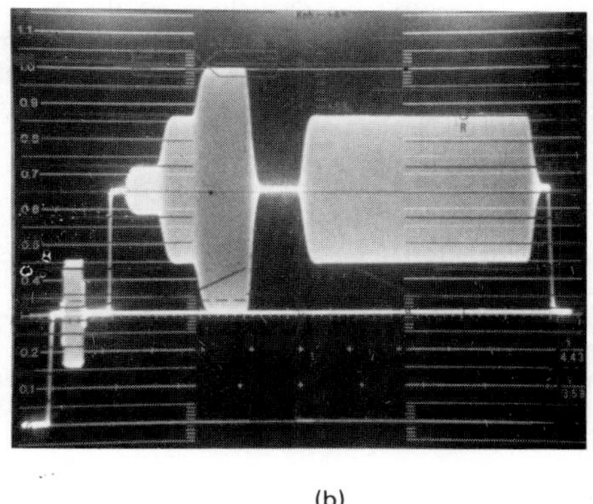

(b)

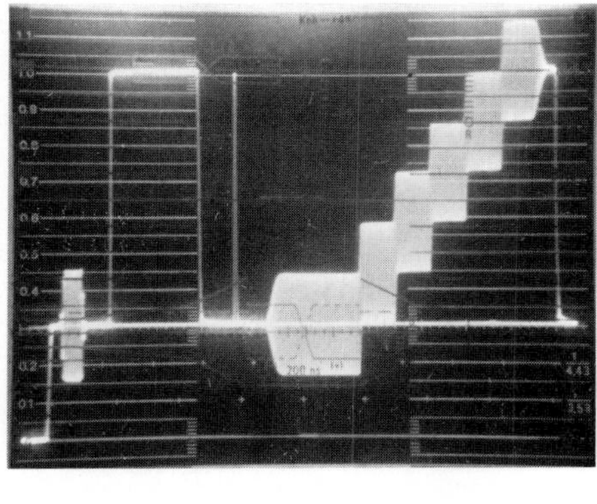

(c)

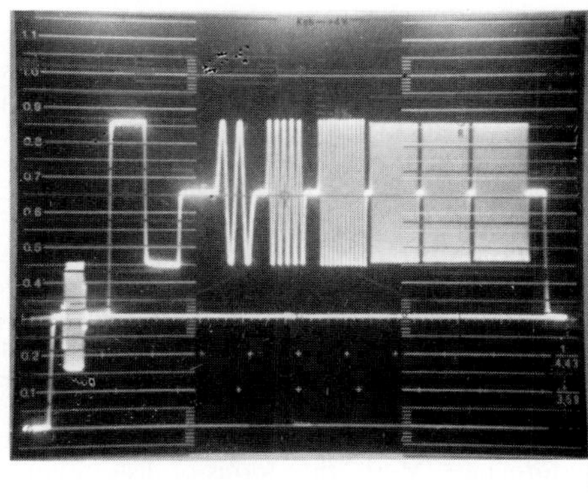

(d)

Figure 21.14 Insertion test signals: (a) line 17, (b) line 18, (c) line 330, (d) line 331 (EBU)

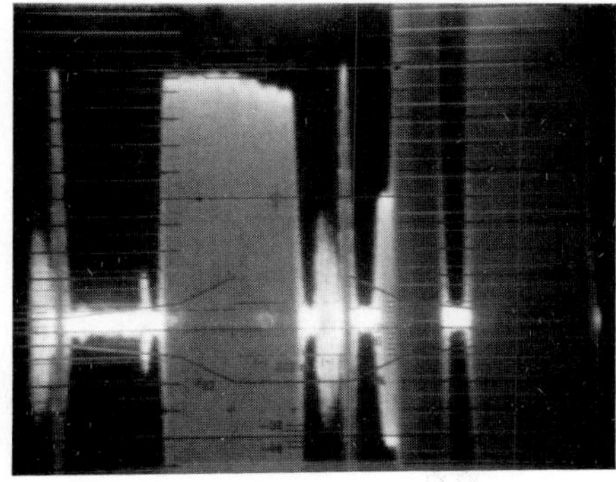

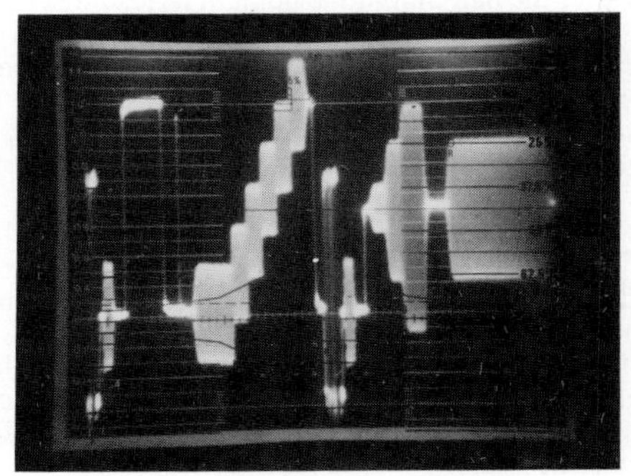

Figure 21.15 Line 330 via chrominance filter showing 7 per cent differential gain (EBU)

amplitude. The sign is positive if the 2T pulse amplitude is higher than the bar amplitude, and negative if lower.

The 2T pulse/bar ratio is sensitive to the medium and high frequency response of a circuit. A low 2T pulse/bar ratio implies reduced horizontal resolution. Overshoot of the 2T pulse trailing edge and ringing produce contouring around vertical or near vertical outlines in the picture. SIS utilizes a group of 2T pulses, so severe 2T distortion can cause sis decoders to malfunction.

21.2.1.4 Peak differential gain

Peak differential gain is defined as the maximum difference in amplitude of the colour subcarrier, superimposed on the staircase in line 330, from its amplitude at black level, expressed as a percentage of its amplitude at black level.

Differential gain causes luminance dependent saturation errors on PAL and NTSC pictures, whereas SECAM is relatively immune to differential gain.

21.2.1.5 Chrominance/luminance gain inequality

This is defined as the difference between the peak-to-peak amplitude of the full amplitude burst of the chrominance signal on line 331, and the bar amplitude on line 176, expressed as a percentage of the bar amplitude. The sign is positive if the chrominance burst amplitude is greater than the luminance bar amplitude, and negative if less.

Chrominance/luminance gain inequality causes colour saturation errors on PAL and NTSC, whereas SECAM is relatively unaffected. However, a very low chrominance level produces noisy colour in SECAM pictures.

21.2.1.6 Signal/noise ratio

Signal/noise ratio is the ratio in decibels of the luminance bar amplitude to the rms value of the noise measured over a bandwidth 10 kHz–5 MHz. In practice, noise is measured on line 22, using a high-pass filter of 200 kHz, to avoid measurement errors due to the presence of residual waveform distortion during the measurement period.

Signal/noise ratio is the most fundamental measurement of a circuit's performance. Most other forms of distortion can be corrected by equalization, but a poor signal/noise ratio is difficult to improve. However, field store devices can achieve an improvement of several decibels, except where movement occurs, by taking advantage of the repetitive nature of a television signal, as opposed to the random nature of noise.

21.2.1.7 Bar slope

Bar slope is defined as the difference between the level of the bar 1 μs before the half-amplitude point of its trailing edge, and the level 1 μs after the half-amplitude point of its leading edge, expressed as a percentage of the bar amplitude. The sign is positive if the trailing edge is higher than the leading edge, and negative if lower.

Bar slope is another measurement of low frequency distortion, similar to base-line distortion.

21.2.1.8 Line-time luminance non-linearity

This is the difference between the amplitudes of the largest and smallest risers of the luminance staircase, expressed as a percentage of the largest. It can be measured by passing the signal through a differentiating filter (a notch filter centred around 1.1 MHz), then comparing the amplitudes of the differential staircase risers.

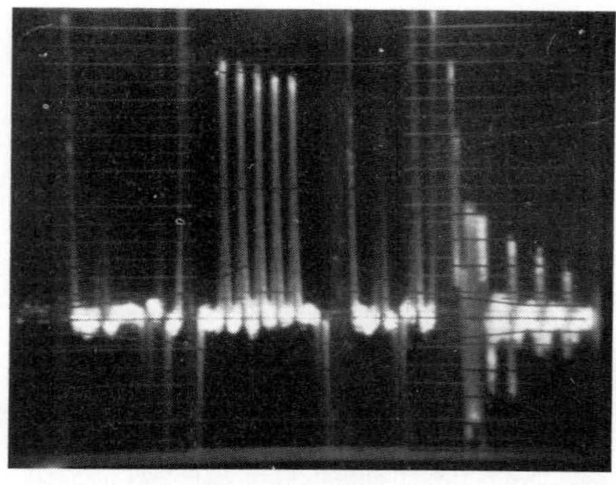

Figure 21.16 Line 17 via differentiating filter showing 5 per cent line-time non-linearity (EBU)

Luminance non-linearity produces saturation errors for NTSC, PAL and SECAM signals. Black or white crushing are severe forms of luminance non-linearity.

21.2.1.9 Chrominance/luminance delay inequality

This is defined as the lag (in nanoseconds) of chrominance with respect to luminance, as seen in the 20T pulse of line 17. If the chrominance lags the luminance, the delay inequality is positive, and if the chrominance leads the luminance, the delay inequality is negative.

Chrominance/luminance delay inequality produces a lateral displacement of colour in the picture. The classic example is

(a)

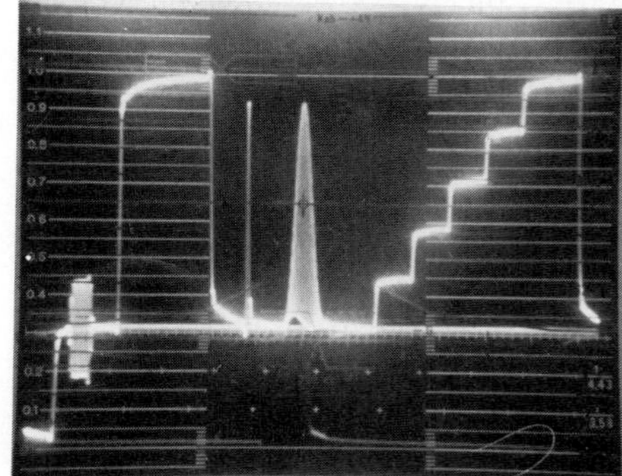

(b)

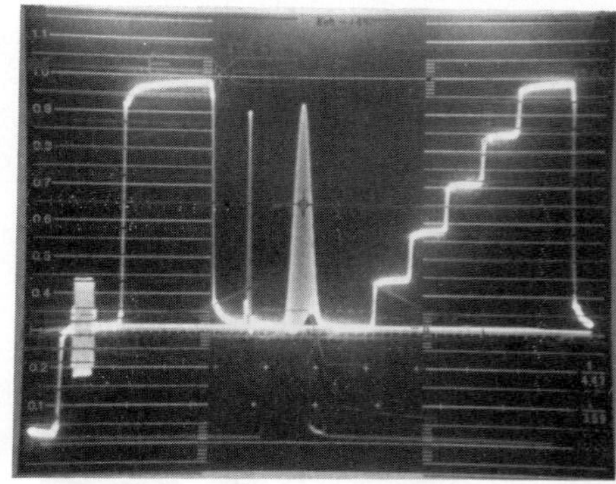

Figure 21.17 Insertion test signal, line 17, showing (a) chrominance/luminance lag of 88ns, 7% bar slope, and −9% 2T pulse/bar ratio, and (b) chrominance/luminance lead of 96 ns, 7% bar slope, and −9% 2T pulse/bar ratio (EBU)

when footballers' coloured shirts precede or follow them around the pitch.

21.2.1.10 Chrominance/luminance intermodulation

This is the change in level of the luminance component of the full amplitude chrominance burst on line 331, expressed as a percentage of the bar amplitude on line 17. The sign is positive for upward displacement of the luminance component, and negative for downward displacement.

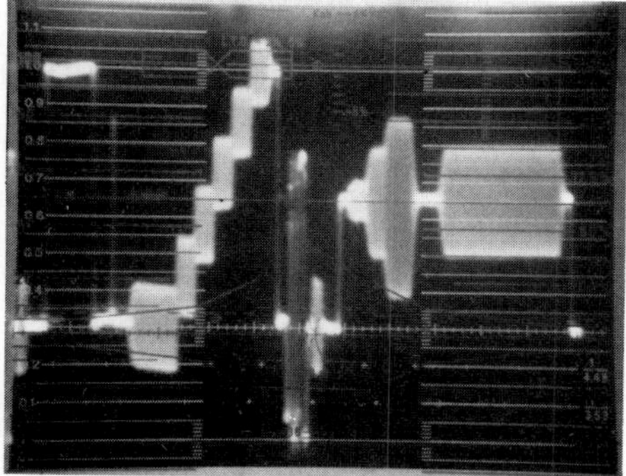

(a)

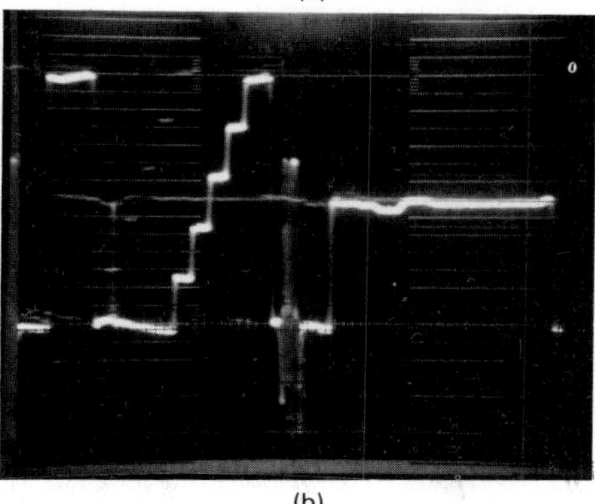

(b)

Figure 21.18 Line 330 and 331 (a) showing downwards displacement of chrominance, i.e. −3 per cent chrominance/luminance intermodulation, (b) via low-pass filter showing −3 per cent chrominance/luminance intermodulation (EBU)

Chrominance/luminance intermodulation causes saturation errors by changing luminance levels in areas of high saturation.

21.2.1.11 Peak differential phase

Peak differential phase is defined as the maximum error in subcarrier phase as the luminance level varies from black level to peak white. It is measured by comparing the phase of the colour subcarrier on the staircase of line 330 with the phase of the subcarrier at black level.

Differential phase causes colour changes in NTSC pictures. PAL is relatively immune, although large values of differential phase produce slight colour desaturation. SECAM is virtually immune to differential phase.

21.2.1.12 Transient sync compression

This is defined as the largest change in synchronizing pulse amplitude, expressed as a percentage of black level synchronizing pulse amplitude, following an abrupt 100 per cent change in average picture level from black to peak white.

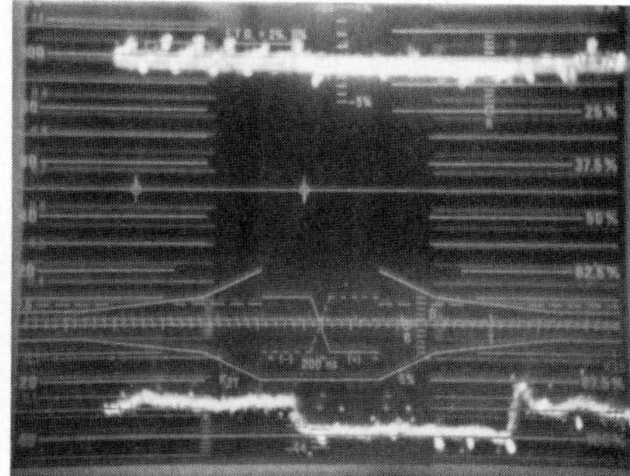

Figure 21.19 Slow-scan display of sync amplitude showing transient and steady-state compression folowing change in apl from black to white (EBU)

Transient sync compression is sometimes a serious problem on very long terrestrial circuits, causing temporary loss of synchronization following abrupt changes in apl.

21.2.2 Automatic video equalizers

The presence of ITS on international signals permits automatic equalization of certain parameters. Automatic video equalizers are installed at certain key points on the Eurovision network, i.e. Paris, Brussels, Copenhagen, Milan and Vienna. Vienna is the normal point of reception for news items from Intervision.

The Intervision video signals are equalized before being pre-recorded by Austrian television (ORF) for subsequent transmissiion in a Eurovision news exchange.

Intercontinental news items are normally pre-recorded at the EVC on C-format vtrs, prior to distribution in one of the daily EVNs. Since the received signals are sometimes of dubious quality and stability, a field-store synchronizer, preceded by an automatic equalizer, serves to stabilize and equalize the incoming signals.

21.2.3 Source identification

On the Eurovision network, lines 16 and 329 carry *insertion data signals* (IDS), which are digitally coded source identification. IDS decoders are installed in ITCs and CNCTs to insert the decoded IDS identification in a corner of the displayed picture. The identification is of the form:

LNDN
BBC1

and the digit can be varied from 0 to 7, so different feeds from the same originating broadcasting service can be unambiguously identified. This is very useful when there are two similar programmes, such as simultaneous football matches, sent from the same CNCT to the Eurovision network.

21.3 Switching centres

The role of a television switching centre is to connect incoming video and/or audio signals to one or more outgoing circuits at the appropriate time. This is usually done at baseband, since local sources and destinations are baseband signals. Moreover, the baseband signals can be processed in various ways, whereas if signals cannot. Nevertheless, if switching is employed in some cases, where simple transit connections are required.

Nowadays, some switching centres are partially automated, with computers to memorize the daily switching schedule and implement the required interconnections. The possibility of manual intervention is essential however, to ensure realtime flexibility.

21.3.1 Switching matrix design

A modern switching matrix, or routeing switcher, normally caters for one video plus two audio channels. Two channel audio for television broadcasting is already available in Germany and the Netherlands, for example, to be followed soon by the broadcasters of several other countries. The BBC and the IBA have developed two-channel sound-in-sync systems to distribute these signals, utilizing quaternary coding instead of binary coding to double the capacity of the system. Reed relays are commonly used to provide the audio switching connections, whereas video switching is based on solid-state circuitry, where diodes are biassed on to provide through connection (see *Figure 21.20*). Alternatively c-mos integrated circuits are now available for both video and audio switching, which connect any of eight inputs to an output when biassed accordingly.

Precautions are necessary to ensure that there is no crosstalk between channels, especially for the high frequency component of the video. The overall performance of the switcher must ensure that all forms of distortion are negligible, for both video and audio signals. Sufficient headroom must be provided for maximum permissable modulation levels. Video switchers can usually be synchronized to a local sync reference, to ensure that switching occurs during the field blanking interval for synchronous sources.

Since a given input channel can feed any number of output channels, the basic information required to control a video/audio switcher is the appropriate input channel that is to be connected to a given output channel. Each output channel has a dedicated selector, which may be a row of pushbuttons or a keypad with an led alphanumeric display.

Most modern selectors send their switching instructions as serial data (RS 232), which includes an address to identify the output concerned and an address to identify the required input. The selectors usually communicate with the switcher via a common data line, which may be a coaxial cable for serial data, or alternatively a parallel data-bus.

The most sophisticated selectors have a preselection facility to reduce the risk of switching errors. Moreover, some selectors can be programmed to control any output channel or channels, giving complete flexibility. It is common practice to provide a list of interconnections on a VDU for large switching matrices.

21.3.2 Network switching

The precise nature of CNCT and ITC switching arrangements varies considerably from country to country, although they are based on the configuration shown in *Figure 21.9*. In general, the PTTs do not carry out *hot switching* (switching on cue), this being the broadcasters' responsibility. The PTTs or RPOAs (*Recognized Private Operating Agencies*) provide interconnections at specified times, in accordance with bookings made by the broadcasters.

The booked periods include a preparatory period of five or ten minutes, for live transmissions. When a transmission is recorded at both the origin and destination(s), no preparatory period is foreseen, but in the case of sequential unilateral transmission from a given origin to various destinations, a five minute switching gap is foreseen between transmissions. This gap is an invaluable safety measure to absorb overruns, which may be due to technical problems or programme problems delaying the start of a transmission. These preparatory period rules have been formulated by the EBU, but broadcasters in other areas of the world follow similar procedures.

In the case of permanently leased circuits, the PTTs are not required to carry out preliminary line-up procedures, whereas for occasional circuit bookings, a 30 minute line-up period is foreseen, as defined by CCITT Recommendations N54, N55 and N62. For Intelsat bookings, line-up procedures are defined in the SSOG (*Satellite Systems Operations Guide*) published by Intelsat. In addition, STOC-TV (*Satellite Technical and Operational Committee*) has published the *Gold Book*, which gives guidance for transmissions between North America and other parts of the world.

In some cases, the CNCT remotely controls the ITC switching matrix, thereby reducing the number of national extension circuits required to carry out hot switching. This applies to Hamburg, Frankfurt and Vienna, for example. In countries where the ITCs belong to the national broadcasters, a much

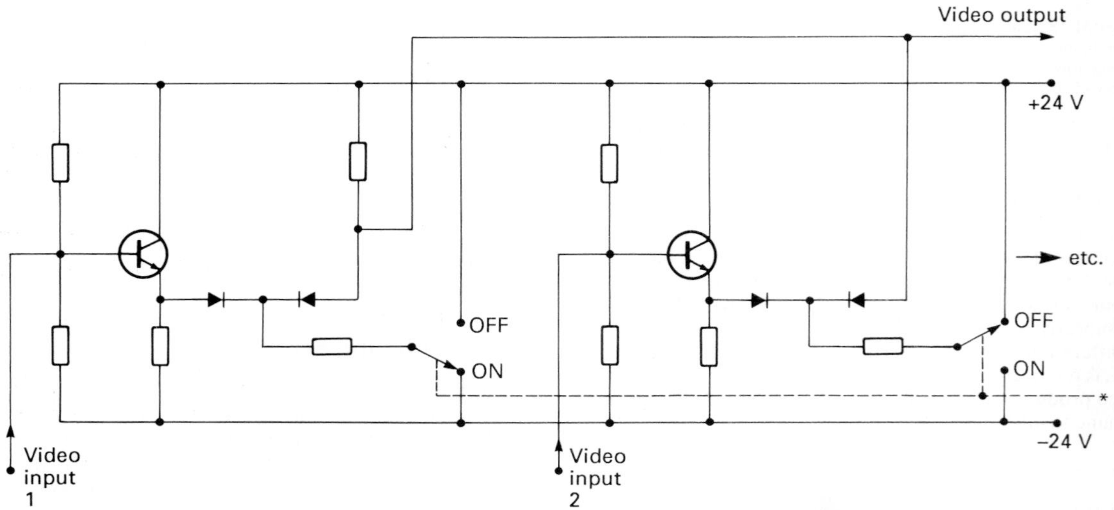

Figure 21.20 Basic principle of diode switcher for video

closer cooperation between CNCT and ITC is possible, such that some switching is done by ITCs and some by the CNCTs. In fact, in some countries, where the links belong to the broadcasters, the ITC and CNCT are virtually the same place, as for example in RAI Milan and RAI Rome. On the other hand, at Albis in Switzerland, the Swiss PTT carries out programme switching for the Swiss broadcasting services and for the EBU, in collaboration with the Eurovision control centre in Brussels.

Incidentally, it is expected to transfer the EBU Technical Centre and the Eurovision Control Centre from Brussels to Geneva in 1989 and 1991 respectively, thereby bringing the EBU technical, programme and legal services together.

21.3.3 Remote Network Switching

In 1970, a system to control remotely the switching matrices at Lille and Frankfurt from the EVC in Brussels was implemented, known as RNS-1. After two or three years, it was no longer required at Lille, when separate permanent circuits were leased each way between Brussels and London, and Brussels and Paris. However, the Frankfurt RNS-1 gave twelve years of reliable service, until it was replaced in 1982 by RNS-3, whereby *ARD Sternpunkt* remotely controlled the DBP switching matrix at Frankfurt.

RNS-2 was a computerized project, which was expected to control remotely nine ITCs in the central European area from EVC Brussels. To facilitate the selection process, circuits were selected, as opposed to actual matrix cross-points, and the resulting network configurations were stored in pre-selection memories. This approach was fine for circuits with dedicated input and output connections at the ITC switching matrices, but it could not cope very well with occasional circuits, which do not necessarily have dedicated connections to the matrices.

RNS-2 was also delayed by numerous faults. By the time it was ready for service in 1980, the volume of traffic on the Eurovision network had increased to the point where the capacity of the system was inadequate, so it was abandoned and replaced by the RNS-3 system. This gives remote control of the ITCs to the local CNCTs, whose staff are fully aware of all relevant details of the local situation.

Satellite earth stations, working with the EBU leased transponders of Eutelsat I-F2, are also remotely controlled by the corresponding CNCTs. NBC's remotely controlled *skypath* system is described in section *21.1.4*.

21.3.4 Romainville

The French ITC at Romainville, on the east side of Paris, is one of the biggest and most up-to-date switching centres in the world, so it is a good example to illustrate how a modern switching centre functions.

It has two basic responsibilities:

- to distribute the programme outputs of the national broadcasters to their respective transmitters
- to establish contribution circuits for these broadcasters to receive programme contributions from elsewhere in France or abroad

Romainville routes video and audio signals for seven television production houses in France, i.e. TF1, A2F, FR3, Canal Plus, LA5, M6 and TV5, which is a francophone co-production service with the French-speaking Belgian, Swiss and Canadian television services, distributed via the Eutelsat I-F2 satellite (see *Figure 21.21*). National and local radio programmes are also routed via Romainville, using audio channels of the shf links.

Figure 21.21 Romainville shf links tower and switching centre (R Gregoire, SYGMA)

The average number of transmissions handled is 150 a day, rising to 300 a day during busy periods. Routeing of the signals is achieved by means of five switching matrices:

- if matrix for incoming signals (50 × 60)
- if matrix for outgoing signals (50 × 60)
- video matrix
- audio matrix (the video and audio matrices constitute a television matrix (80 × 80))
- radio matrix (80 × 80)

These matrices are interconnected as shown in *Figure 21.22*. They can be operated either in a computer assisted mode, according to a programmed schedule of switching requirements, or in manual mode. The latter system allows a step-by-step selection of interconnections, but shows the operator which transmissions will be interrupted by his or her actions.

A bank of 75 monitors enables the operators to monitor their selections and pre-selections. Certain monitors and their associated waveform monitors are dedicated to quality control. High quality audio monitoring is also available. The switching centre is connected to the outside world via nine parabolic dishes, numerous coaxial cables for local connections, and two satellite receive only dishes at ground level for the Eutelsat I-F1 and Telecom 1 satellites. The tower has a height of 108 m, and the diameters of the platforms range from 26.3 m to 43.4 m. The platforms were constructed at ground level, then slowly hoisted into position by means of steel cables. Romainville tower entered service in 1986.

21.4 Summary

The first intercity television links in the 1940s were coaxial tubes

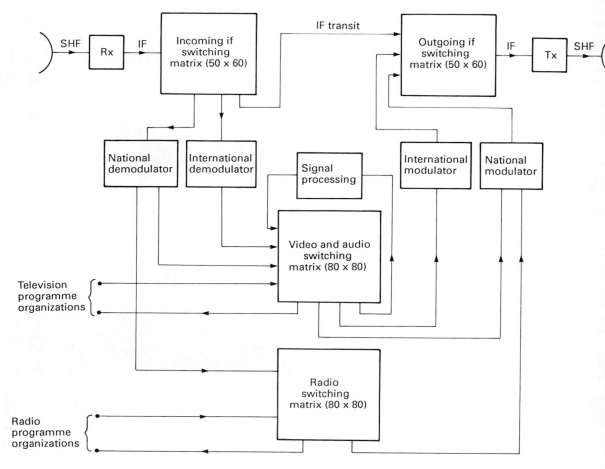

Figure 21.22 Romainville, Paris, switching centre schematic diagram

Figure 21.23 TVRO dish for reception of TV5 from Eutelsat I-F2 at RTBF, Brussels (EBU)

carrying amplitude modulated carriers. Then, in the 50s, shf links were constructed, which greatly facilitated transmission of television signals, using frequency modulated microwave carriers. In the 1960s, communication satellites introduced intercontinental television transmission, and by the 70s they were also extensively used for domestic purposes in North America. Eurovision adopted leased satellite transponders, in collaboration with Eutelsat, from the end of 1984, and Intervision used Intersputnik satellites.

Geostationary satellites enable large countries, such as the USA, Canada, India, China and Australia, to distribute television programmes throughout their vast areas with relative ease. Satellites also provide a very flexible means of news gathering, particularly from remote areas. In Asia, the ABU organizes daily news exchanges via Intelsat.

During the 1980s optical fibres were introduced for local circuits, and in the 1990s they will also be used for long distance circuits, utilizing pulse code modulation of component signals. It remains to be seen to what extent optical fibre circuits will supplement shf links and satellite circuits for television transmission.

In many ways, geostationary communication satellites are ideal for point-to-point television transmission, especially when the earth stations are located at the broadcasters' premises, thereby minimizing costs. However, optical fibre circuits, with

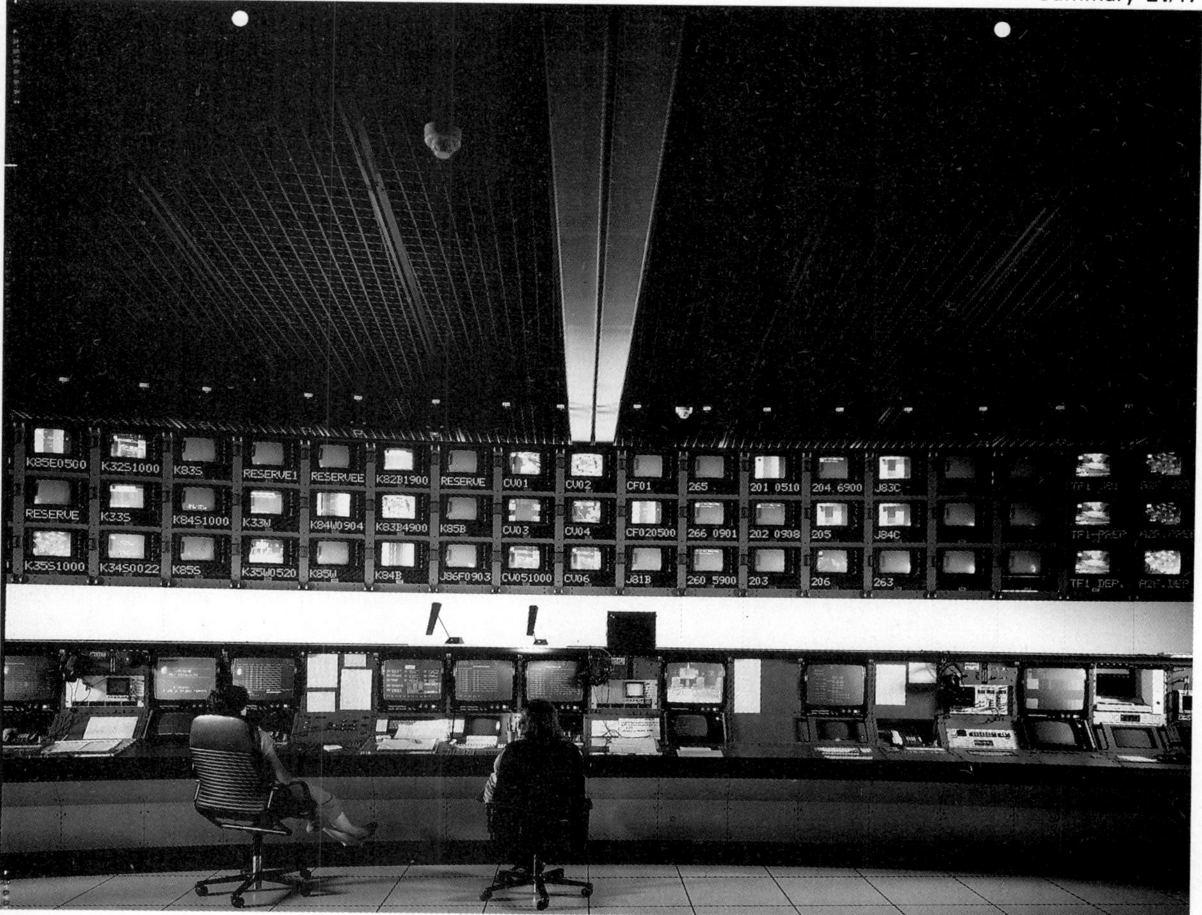

Figure 21.24 Romainville switching centre operations room (A Nogues, SYGMA)

their very high bit rate capacity, will probably play a prominent role in the transmission of digital television signals in the future.

Bibliography

CASTELLI, E, GALLAGHER, J C, PULLING, M and STUCCHI, N, *Radio-relays for television*, EBU Technical Monograph No 3111

MATHIEU, M, *Télécommunications par faisceau hertzien*, CNET, Paris (1979)

EVANS, B G, *Satellite communications systems*, IEE (1987)

Reference data for radio engineers, ITT

WEAVER, L E, *Video measurement and the correction of video circuits*, EBU Technical Monograph No 3116

WOOD, C B B, PADEL, S H and RAINGER, P, 'Cablefilm Equipment', *Jour SMPTE*, **70**, No 7, p 494 (1981)

WATSON, S N, 'Cablefilm Equipment', *Jour Brit Radio Engineers*, **20**, No 10, p 759 (1960)

Performance specification of equipment for EBU insertion signals (625-line television systems), EBU Tech 3209

The EBU standard PPM for the control of international transmissions, EBU Tech 3205

BLEAZARD, G B, *Introducing satellite communications*, NCC (1985)

MARAL, G and BOUSQUET, M, *Satellite communications systems*, Wiley (1986)

SANDBANK, C, *Digital television*, Wiley (1990)

J L E Baldwin B Sc, M Inst P, FRTS, FSMPTE
Consultant

22 Television Standards Converters

Television standards conversion is the process of accepting pictures on one television standard with one particular number of lines per picture and of fields per second and, from this input, producing output pictures on another television standard with a different number of lines per picture and/or fields per second. Typical conversions would be from 525/59.94 to 625/50 or vice versa.

For such conversions, there is also typically a change of the colour system, from NTSC to PAL or SECAM in the case of 525/59.94 to 625/50 conversion, but it should be noted that where there is no change in the number of lines, or in the fields per second, then the operation is called not standards conversion but *transcoding*.

As the spatial and temporal resolution of picture sources has increased over recent years, the normal Nyquist limits are to some extent being exceeded. When the eyes are tracking movement, the unwanted alias components are largely filtered out by the characteristics of the eye, so these resolution increases result in subjective improvements in picture quality. However, standards converters usually do not have this tracking filter characteristic, so there are compromises which have to be made, and these are subjective in nature.

In some ways, standards conversion is almost an art form. Due to the alias components, the input pictures need some interpretation to achieve near to optimum processing, so that the wanted signal may be derived, without noticeable artefacts. It is a complex subject and this section indicates some of the more significant problems and attempts that have been made to find solutions. The results of conversion can be very satisfactory but do depend on how critical are the input pictures.

It is normally assumed that an ideal conversion would result in output pictures indistinguishable from those produced by a camera operating on the output standard. However, it should be realized that this cannot even be achieved theoretically when converting from a lower resolution standard to one of higher resolution.

22.1 Background

22.1.1 Optical standards conversion

The principle is essentially simple in concept. It uses a camera, operating on the output standard, to look at a high quality display of the input pictures. Provided input and output field frequencies are the same, the results are tolerable. However, having an extra display and camera in the signal path causes problems to accumulate, such as noise, lag, flare and gamma.

Also, there are more fundamental problems with beats between the two line structures, which become more significant as the vertical resolution increases. This type of conversion was used in the early days of Eurovision, and it was reasonably satisfactory, if allowance were made for the novelty and topicality of the programmes.

A further problem arises when the input and output field frequencies differ, as when converting from a standard with a 60 Hz field frequency to one using 50 Hz. In such a case, there will be occasions, every six input fields, where there will be light resulting from two display scans falling on the camera target between one output scan and the next. Since camera tubes integrate the light from one output scan to the next, there will be a brightness modulation of the output at 10 Hz, the beat frequency between the two field frequencies. A similar problem occurs for a conversion in the reverse direction, due in this case to there being no input field scan between one output scan and the next.

The extent of the brightness modulation is a function of the decay characteristic of the phosphors used in the display. This would need to be very long to decrease the brightness modulation to acceptable limits, but has to be relatively short to prevent objectionable smearing on fast movement. No satisfactory compromise between these two effects appears to be possible. However, ingenuity largely overcame these conflicting requirements by adding a reference white pulse, just outside the active line time, and using the amplitude of this pulse recovered by the camera tube to modulate the video gain of the camera, so that the undesirable brightness modulation of the picture was removed.

Using this technique and many other detailed improvements, Bosch Fernseh produced an optical converter that was widely used and generally gave satisfactory results.

22.1.2 Non-optical field-rate standards conversion

A British patent applied for in June 1959 described how magnetic tape recording could be used to duplicate, or delete, a chosen fraction of the number of lines and some other fraction of the number of fields. One case considered was the conversion of television pictures from 525/60 to 405/50. It appears to be the first proposal for a non-optical television standards converter. The ideas were not proceeded with, as special head wheels would have been required, and the acceptability of the results from such a machine were somewhat doubtful.

Early Bird was launched in 1965, and this noticeably increased the pressure to develop a field-rate standards converter. The first field-rate converter without an optical image was produced by the BBC, using an array of quartz acoustic delays, but this did not convert the number of lines, so there were 50 lines blank at the top and at the bottom of the screen, in conversions to 625 lines. The active line time was shrunk to make the aspect ratio correct by sampling the active line at, it is believed, 576 equally spaced points and storing the samples in individual capacitors. These capacitor stored samples were then read out at a higher rate to compress the active line time appropriately.

The conversion, although obviously imperfect, was considered to be better than the alternative optical approach available at the time, but some viewers complained that they had been "robbed' of the information in the border surrounding the picture. This converter operated on a precise 6:5 ratio between the field frequencies, and of course the NTSC field frequency is 1 part in 1001 lower than 60 Hz. For tape inputs, this could be overcome by playing the tape that little bit faster.

The next noteworthy improvement was in the BBC *advanced field store standards converter* which used an array of electrical delay lines as well as the array of quartz acoustic delay lines, so that the input signals could be made to arrive at precisely the correct time for processing to form a normal 625-line signal.

To overcome gain variations between different combinations of delay lines, frequency rather than amplitude modulation was used for the signal passing through them. So that the same interpolation could be used for chrominance as for the luminance signals, the chrominance subcarrier was shifted to a multiple of line frequency. Mixing of frequency modulated signals cannot be achieved by conventional means, so interpolation was difficult. However, a 50:50 interpolation was provided by a system which in principle is similar to multiplying the two signals together, selecting the frequency band which contains the sum signal from this process and dividing this signal frequency by two.

The real breakthrough in standards conversion came when it was possible to use digital techniques. The cost and size of high speed digital storage was the main deciding factor, although the speed of digital processing before 1971 would have presented major problems, which could be overcome only with greater complexity.

In 1969 the cost of high speed storage was about 50 pence per bit (about $1.20), but it then dropped with the introduction of mos devices to about 9 pence per bit and a package could store 192 bits rather than 2. More important was that a number of manufacturers were planning to introduce chips which could store 1024 bits, and they were forecasting that the introductory price would be about 1.6 pence per bit falling in production to perhaps 0.4 pence (about 1 cent). Some manufacturers even forecast that the production costs in the long term could fall to a few millipence per bit.

The BBC and the IBA developed digital line standards converters to operate between 625 and 405 lines and both of these were demonstrated to the Technical Committee of the EBU in 1971. The BBC converter used linear interpolation between two lines, whereas the IBA converter could interpolate using four lines of information.

As the cost of the integrated circuits for two fields of storage had dropped to about £15 000 ($36 000), the IBA decided to proceed with the development of a model to prove the feasibility of digital field-rate conversion.

Pictures through DICE (*Digital Intercontinental Conversion Equipment*) were first shown at the 1972 International Broadcasting Convention. They created considerable interest, and the IBA was asked to arrange comparative tests and demonstrations with the BBC *Analogue Electronic Converter* and a Bosch Fernseh *Optical Converter*.

The signal origination and routeing were chosen so that there would be no intrinsic advantage to any converter.

As a result of these demonstrations the IBA were asked if it would allow the digital converter to be used operationally on all the unilaterals passing from the United States to Europe on the occasion of the presidential election.

Figure 22.1 Installation at ITN showing the experimental DICE 525 to 625 converter in the middle; the two racks at the extreme left are two of the seven racks of an analogue field rate converter (ITN)

Thus the experimental 525 to 625 DICE, developed only to prove feasibility, was first used operationally in November 1972, on the occasion of the re-election of President Nixon, when it was used on every unilateral passing to Europe. Essentially overnight, the quality of conversion made by DICE established internationally that digital processing of pictures was not only possible, but the quality achieved was noticeably better than that achievable with comparable analogue equipment.

The IBA had intended to strip the converter down to complete the development, and to rebuild it in a bi-directional form so that it would also work in the 625 to 525 direction. Instead it was installed at Independent Television News, where it was used during the ensuing years, not only for live satellite conversions but also for a much greater number of tape conversions. The development continued, resulting in a more advanced bi-directional DICE.

As the cost and size of digital storage continued to decrease, it became practicable to use more fields in the interpolation and so achieve better vertical resolution on static pictures. The BBC ACE converter used four field storage and fixed interpolation and more recently AVS have introduced a converter ADAC, again with four fields of storage but with interpolation which adapts depending on motion.

With high definition television, a further refinement has been introduced, initially by NHK, which determines the velocity of movement and compensates for this velocity.

22.2 Movement portrayal

22.2.1 Movement effects on film

The most significant problems of standards conversion arise during movement. An analogy that facilitates explanation lies in cinematographic techniques. The use of 24 frames a second in the camera, but projecting each film frame twice to give a flicker frequency of 48 Hz, has been the standard of the movie industry for some 60 years.

For film cameras that use a shutter rotating at 24 rev/s, the shutter is sectored so that the light from the scene is allowed to form an image on the film during about a 180° rotation of the shutter. During the other 180°, when the shutter obscures the light, the film is pulled down to the position for the next exposure. The angle of shutter opening can be changed, but 180° is normal.

Consider a white object travelling against a black background, such as a white vertical bar moving horizontally. In *Figure 22.2*, the plane of the chequerboard represents zero light, and the *movement per frame* shows the distance the image moves from one film frame to the next. The position of the image of the bar on the film frame moves progressively to the right, being blanked out by the shutter while the film is pulled down.

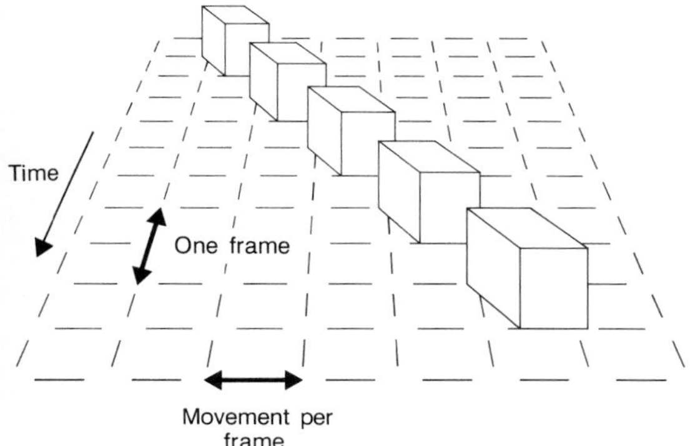

Figure 22.2 A representation of a shuttered image, moving to the right, of a white bar against a black background

Due to the motion of the object during the exposure of a single film frame, the image recorded by the film will be blurred (see *Figure 22.3*). This blurring is shown in greater detail in *Figure 22.4*. Here the motion of the white bar, progressively moving to the right, is as though the bar had been illuminated by a strobe lamp flashing every 2 ms. Successive traces have been displaced downwards. The extent of the blur can be seen to equal two-thirds of the width of the bar; since the shutter was open for 1/48 s, it can be deduced that the speed of bar was 32 bar widths a second.

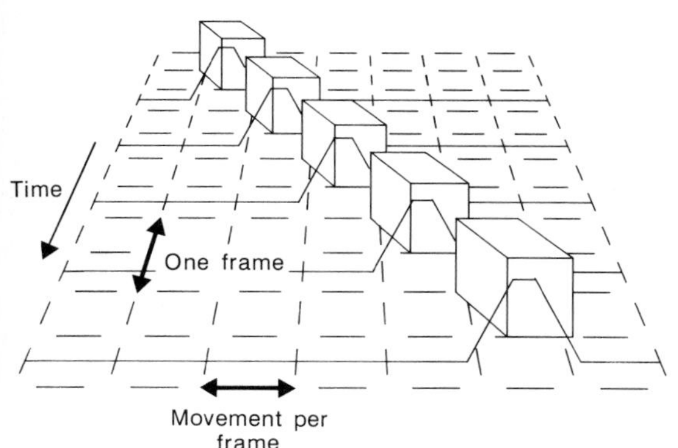

Figure 22.3 The same view of the shuttered image shown in *Figure 22.2* with the superimposed indication of film exposure

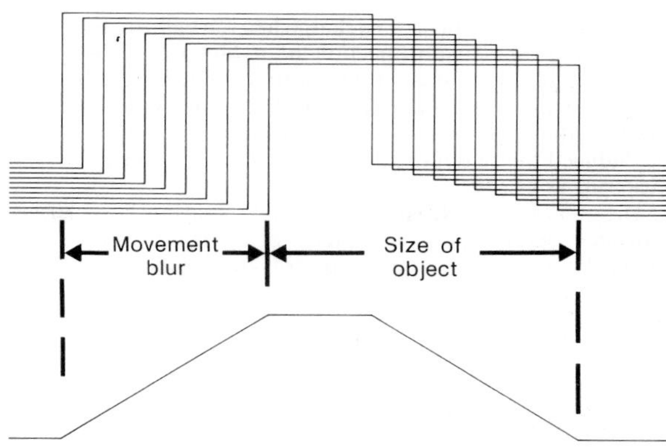

Figure 22.4 The blurring of the image on one film frame

For a cinema screen of conventional brightness, a flicker frequency of 48 Hz is about the lowest that would be acceptable and this is used almost invariably. The film projector has a shutter with two 90° apertures, which interrupts the light twice per revolution. During only one of these interruptions is the film pulled down to the next film frame, so each picture on the film is projected twice before moving to the next picture on the film. This is a form of standards conversion.

In *Figure 22.5*, each blurred film picture is shown projected twice before progressing to the next film picture. The representation shown here is appropriate for the case when the eyes are not following the motion. However, it is necessary to cancel the average motion to understand what the appearance will be when the eyes are following the motion. This can be seen by altering the viewpoint, as in *Figure 22.6*.

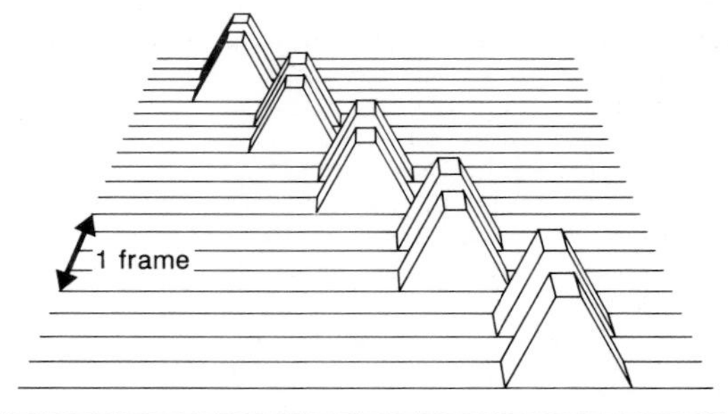

Figure 22.5 The perceived motion of the bar if the eyes do not follow the movement

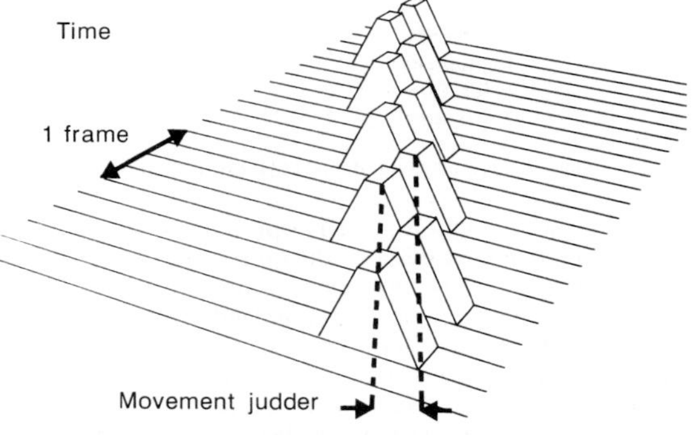

Figure 22.6 A change of viewpoint for *Figure 22.5* gives a good indication of the perceived motion obtained when the eyes are tracking the movement of the bar

In *Figure 22.6* it can be seen that, when the eyes are following the motion, the first projection of a film frame is always advanced in the direction of motion whereas the second projection is always retarded. The first projections fall along one line, the second projections falling on a different line. It can be deduced that a double image will be seen with each image flickering at 24 Hz.

The smoothness of the movement is also impaired by movement judder also at a frequency of 24 Hz. The conclusions drawn from this diagrammatic portrayal correspond well with the subjective impression gained by observing the motion portrayal achieved in the cinema.

22.2.2 Movement portrayal of early non-optical converters

The method used in section *22.2.1* for considering the motion portrayal of film is equally applicable to electronic standards conversion. Just as the film accumulates exposure while the shutter is open, so the elementary areas of the camera tube accumulate charge between the times when it is read by the scanning beam. Essentially all the accumulated charge is read out on each vertical scan, so the effective shutter duration is $^1/_{60}$ s for 525-line/60 fields a second cameras.

The illustrations use precisely the same scale of actual time and of actual movement as was used for film, but as a 525/60 frame lasts $^1/_{30}$ s, compared to $^1/_{24}$ s for motion picture film, the pitch of the chequerboard has changed. There is no change of size of movement or of time: only a change in the units in which it is measured. *Figure 22.7* represents the same bar, moving at the same speed from left to right. The speed of the bar is still 32 widths a second. The lines dividing the moving bar show the boundaries between which charge is accumulated by the elements of the camera tube. There is less blurring than with film because the effective shutter time is $^1/_{60}$ s, rather than $^1/_{48}$ s.

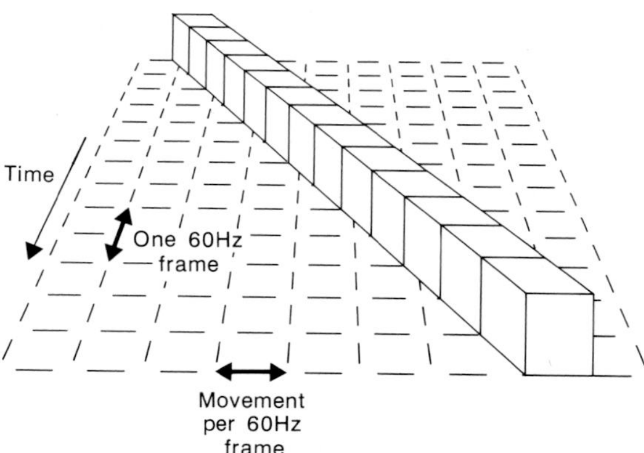

Figure 22.7 The motion portrayal of a 60 Hz television camera

The charge accumulations, when displayed during individual fields, are shown in *Figure 22.8* in the same way as was used for the projection of film. The movement seems fairly smooth, as it should be, but this can be better judged by changing the viewpoint to obtain the effect of the eyes tracking the motion.

As can be seen from *Figure 22.9*, the successive images all fall nicely on a straight line. This figure is incorrect in one very important respect in that it shows the display as having no decay in the time from one field to the next. This is obviously incorrect for conventional tube displays, although it is substantially true for Eidophors and may be true for solid-state displays.

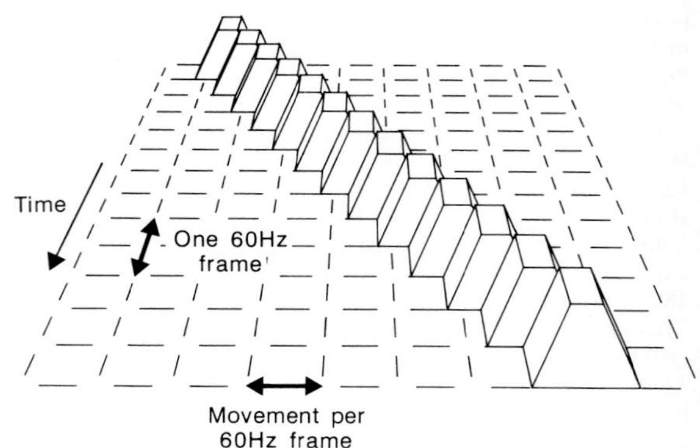

Figure 22.8 The motion portrayal of a displayed 60 Hz television standard if the eyes are not tracking the movement of the bar

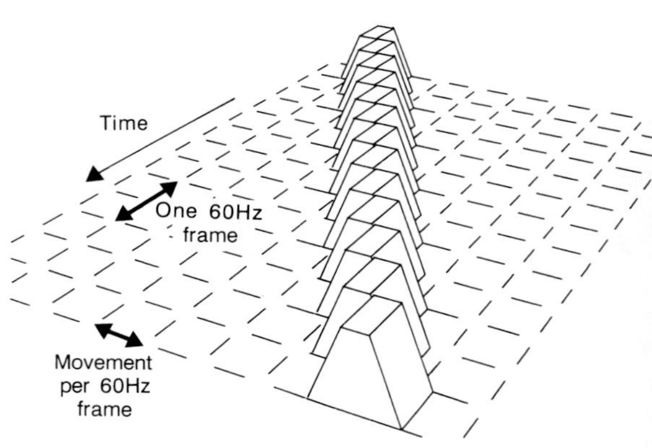

Figure 22.9 The motion portrayal of a displaced 60 Hz television standard when the eyes are tracking the movement of the bar

The effect of the phosphor decay is shown in *Figure 22.10*. It can be seen that it is not in conflict with *Figure 22.9*, as far as smoothness of movement is concerned, but the additional blurring, resulting from the eyes tracking the average motion, is much less. The initial decay of light output is very rapid, the majority of the light output being in the first 10 per cent of a field duration from the time of excitation. The initial brightness,

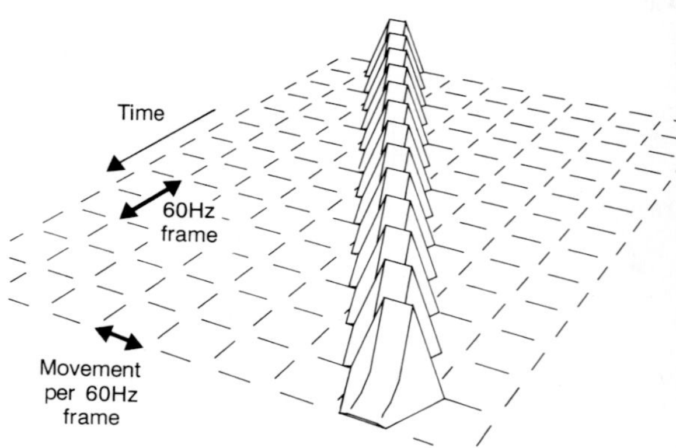

Figure 22.10 The motion portrayal of a 60 Hz standard, when the eyes track the motion, allowing for the decay of the phosphor

before the decay, is the determining factor. In subsequent illustrations only the initial brightness will be shown.

22.3 Judder in standards converters

If every sixth field were omitted, there would be a judder of about two-thirds of that for motion picture film but at a frequency of 10 Hz instead of 24 Hz (see *Figure 22.11*). Both of these frequencies are readily visible at normal brightness levels but the annoyance at 10 Hz would be greater for the same amplitude. Overall, the results might be about comparable to motion picture film.

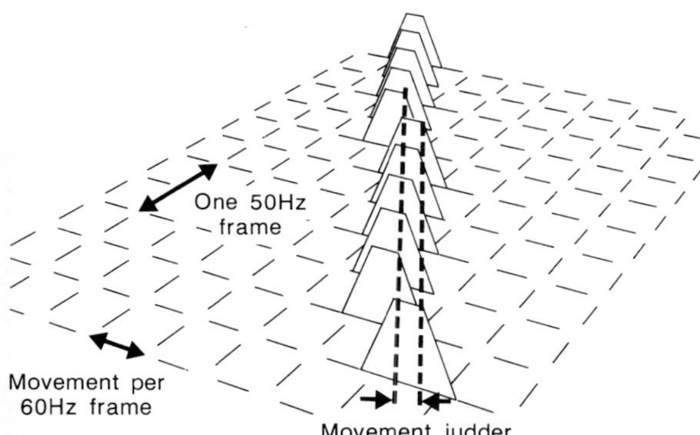

Figure 22.11 Motion performance of a 60/50 Hz field rate converter obtained by omitting every sixth field

It could be argued that motion picture film is widely accepted, so this level of performance should be equally acceptable. However, the rate of panning a camera when shooting a film is very carefully limited as a function of focal length, so that movement judder is not too distracting under this most critical condition. Television programmes are produced without this constraint, so a programme may contain rapid movement which only causes movement blurring on the original standard without judder. The rate of panning may, quite reasonably, be so high that it would cause unpleasant judder after such a conversion.

Such judder would only occur during movement, and the panning rate has a considerable effect on the subjective impression. A pan that takes a stationary object 10 s to pass from one side of the screen to the other will last 600 input fields, and the peak-to-peak amplitude of the judder will be four-fifths of a six-hundredth of the screen width — sufficiently small that it is not disturbing even if it is visible. At higher speeds, greater than a screen width a second, the eye is most unlikely to follow the average motion, so the judder becomes much less noticeable. This leaves a speed range between a tenth and one screen width a second where judder may be important.

Making an allowance for the incidence of movement judder, it seems that in this respect such a converter would have been considered acceptable, in the absence of anything better, at least for news items.

The BBC advanced field store standards converter used 50:50 interpolation on two successive output fields of the five output field sequence, the remaining output fields using no interpolation. The resulting movement performance is shown in *Figure 22.12;* the movement judder has been decreased to about a half of that which would be obtained without interpolation. The picture quality was undoubtedly improved compared to earlier realized converters and was claimed to be good enough for entertainment programmes, as well as news.

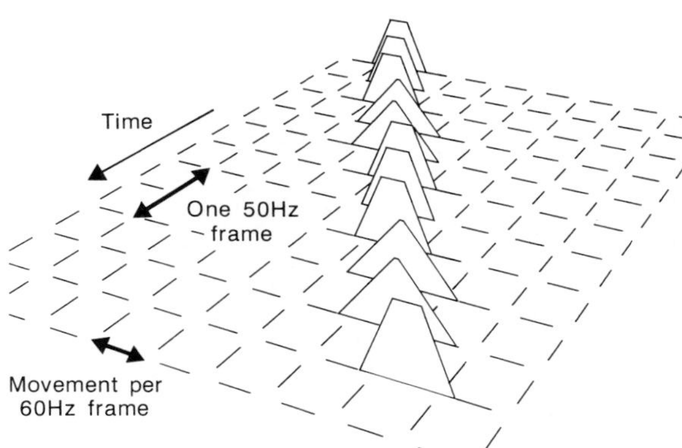

Figure 22.12 Motion performance of a 60/50 Hz field rate converter using 50:50 interpolation on two successive output fields

The philosophy adopted for DICE was that each field had to be treated as a complete picture since the cost of field stores was so high in 1972 that only two composite field stores could be considered. Simple linear interpolation between fields was the best that could be achieved, given the constraint of just two field stores. *Figure 22.13* shows what could be achieved in theory with such a two field interpolation. It seems that there should be no movement judder, but there is a noticeable variation of movement blurring which is inevitable with linear interpolation between two fields; this variation is in the ratio 2:1.

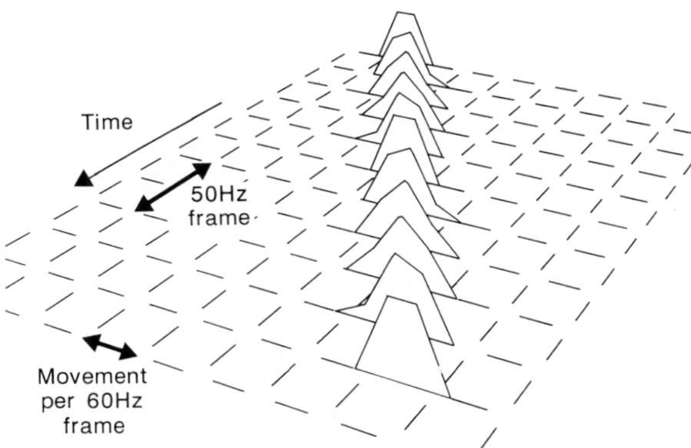

Figure 22.13 Theoretically achievable performance of a 60/50 HZ field rate converter using two field stores

The three-dimensional representations used give a good general impression, but the precise variation of response is not clear. The perspective that made them more understandable, prevents them being used for comparing detail. By moving the viewpoint to infinity and magnifying the image to compensate, *Figure 22.14* results. This again is viewed from a direction corresponding to the eye tracking the motion.

The sequence of output fields starts at the top and moves progressively downwards until the sequence repeats. By lowering the viewpoint, until the first and last fields of *Figure 22.14* are superimposed, *Figure 22.15* is obtained but the vertical scale has also been expanded. These two illustrations need to be used together to compare the output on successive fields, *Figure 22.14* being used more as an index to identify the order of the fields in *Figure 22.15*, and also to confirm the shape of the response on individual fields.

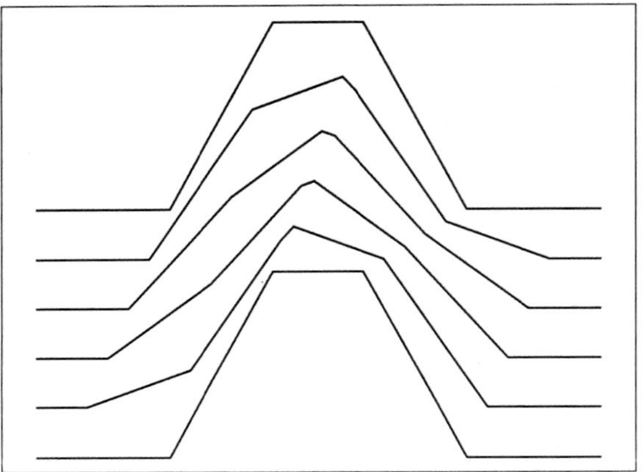

Figure 22.14 View from infinity; less comprehensive but greater accuracy

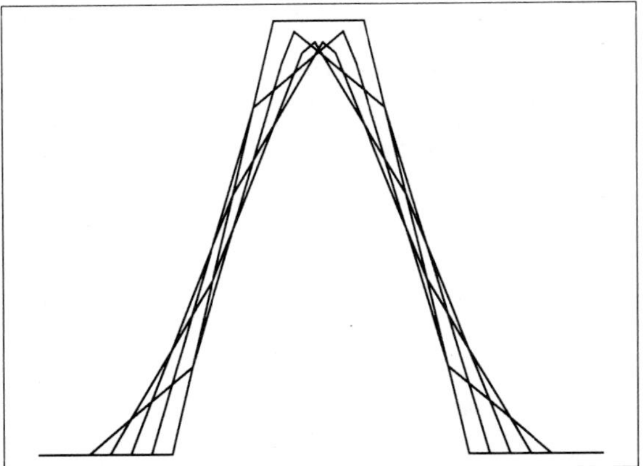

Figure 22.15 Infinity view; more confusing but better for comparison

Figure 22.15 has particular importance in showing the variations of the signal on successive fields. For speed of movement greater than half a bar width per input field, the output amplitude on the interpolated fields never reaches that obtained on the non-interpolated fields. In this case, the speed of movement was 32 bar widths a second or $^{32}/_{60}$ of a bar width per input field. For slower rates of movement, the peak signal would be independent of the interpolation.

The areas under the curve are the same on each and every output field. At first sight, this meets the requirements necessary to prevent the display flickering at the difference frequency between the two field rates. However, the effect of the gamma of the display under large signal conditions has been ignored; a certain loss of brightness arising from a reduction in the area near to white will not be fully compensated by an equal increase in the area near to black. If the depth of modulation be small then interpolation works relatively well, but this must be expected to degrade for higher depths of modulation if the interpolation is operating on gamma corrected signals.

22.4 Interpolation

For interpolation to work, it is necessary for there to be continuity of form. The interpolation can only be expected to be satisfactory to the extent that this exists. For example, it is not possible to interpolate the information in one field of any conventional composite signal with that in an adjacent field since the phase of the subcarrier would be different.

Even for an interlaced monochrome signal, simple linear temporal interpolation between corresponding lines on two consecutive fields cannot be achieved because in principle the lines do not correspond.

Such temporal interpolation cannot be purely temporal since interlace arranges that the lines on odd and even fields are vertically displaced on the picture; an attempt at simple linear temporal interpolation would therefore also result in unwanted vertical interpolation.

It is often necessary to change the natural form of the signal to render it suitable for interpolation; however, sometimes it is possible to arrange a system of interpolation which makes special allowance for the nature of the signal.

22.4.1 Number of fields for movement interpolation

Movement, or more correctly *temporal*, interpolation requires a minimum of two fields of information to be available simultaneously. Where two field temporal interpolation is used, it is necessary to make allowance for the interlace. The temporal interpolation used must either remain constant for an output field or change smoothly and progressively through the field; in particular, it must not change significantly from one line to the next as this would cause a sideways step on the vertical edges of objects moving horizontally.

To prevent vertical displacements occurring in sympathy with the temporal interpolation, it is necessary to precede this temporal interpolation by a vertical interpolation differing for the two fields to make the effective position of the line structure identical for the two fields. If this were not done, the converted picture would vibrate up and down in sympathy with the temporal interpolation and would be most noticeable when there was little movement in the input pictures. In DICE, this vertical interpolation was organized so that it was always producing information appropriate for the nearest line, of a hypothetical 525-line progressive raster, to the vertical position which successive output lines would take on display.

If the input pictures were static no temporal interpolation would be required, and information from two successive input fields could be used in the formation of each output field. This would remove the artefacts of interlace on the input standard for static pictures and, since there would be twice the number of samples or lines in the vertical direction, the theoretical vertical resolution could be doubled compared to a converter which has, at times, to produce output pictures from the lines of one input field.

It would, of course, be possible to use information equally from both fields in those parts of the picture which were static and to use appropriate interpolation between them in parts where movement was occurring. However, it must be realized that the variation of the vertical resolution between one mode and the other could itself represent a worse artefact than that resulting from a continuous impaired resolution.

If three fields of storage are used, then the interpolation can be arranged so that the sum of the coefficients of the two outer fields is constant and equal to the fixed coefficient used for the centre field. This ensures that information from odd and even fields make equal contribution, so the vertical resolution can be that which could be achieved from a sequential source, at least for static pictures. The converter would not have to adapt to achieve this vertical resolution, it could be in this mode continuously.

However, the fixed central coefficient limits the choice of the interpolation function. The temporal aperture would be substantially constant which is good, but it is equal to the maximum which would occur for two field interpolation. Probably the reason why it has not been explored further is that the price for a field of storage dropped quite rapidly, from that at which only two field stores could be considered to a price where four field stores could be used.

If information from four fields is simultaneously available, the condition to have an equal contribution from odd fields and even fields is less constraining than it was for three field interpolation. As there will always be a pair of odd fields and a pair of even fields, the distribution of coefficients between the fields of a pair is without constraint. There is also the undeniable initial attraction that an interpolation function of sine-squared shape and having a half amplitude duration equal to two fields will meet the requirement of equal contributions from odd and from even fields (see curve *a* of *Figure 22.16*). It seems therefore to be an excellent choice.

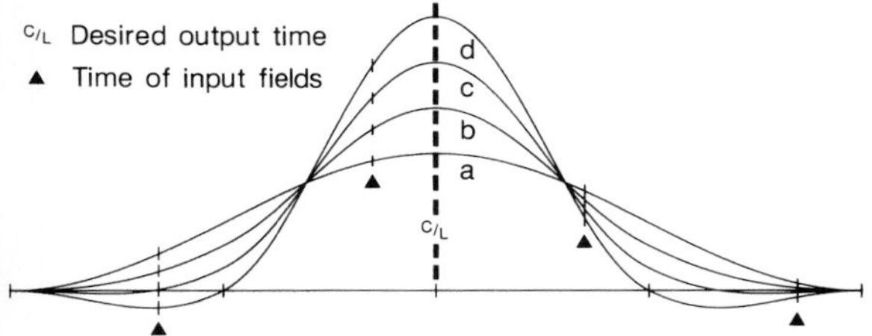

Figure 22.16 Four field interpolation functions with various compromises between vertical and temporal resolutions

However, it must also be realized that unless the vertical interpolation also takes equal contributions from odd and from even picture lines, then some lines could come predominantly from field two, whereas others could come predominantly from field three, of the four fields that are stored at any time. This could cause problems on movement.

Unfortunately, an interpolation function of half amplitude duration equal to two fields in duration does cause visible smearing on movement at certain speeds, so the obviously apparent benefits of four field interpolation are not fully realizable in practice.

Figure 22.17 An adaptive four field interpolation converter (Applied Video Systems Ltd)

One approach is to adopt temporal/vertical interpolation, which is a fixed compromise between vertical resolution and temporal smearing (such as curve *b* of *Figure 22.16*); this is the approach used in the BBC *ACE* converter design.

A more recent approach is to make the system adaptive, dependent on the picture temporal variation, so that high vertical resolution is available on static pictures. This is sacrificed when motion occurs so that different temporal interpolation is used which results in less blurring. *Figure 22.16* shows four different compromises between vertical and temporal resolution. If only two modes were available, the switching between modes might be disturbing, so in the AVS *ADAC* converter (*Figure 22.17*) more modes are provided which are graduated and selected depending on the temporal variation.

22.4.2 Interpolation functions

Conventionally, it is assumed that the interpolation function chosen also specifies the frequency response of the interpolation. Unfortunately, this is not the whole truth. In addition, there is a strong implication from the usage of the phrase *the frequency response of the interpolation function* which can be misleading.

In more normal usage, the process of interpolation is used to derive the value at many different positions between one sample point and the next, and all the parts of the interpolation function are used between one sample and the next. It may well happen that the *instantaneous frequency response* may differ from one interpolation position to another, but the average frequency response of the interpolation will be that which can be derived from the interpolation function. By *instantaneous frequency response* is meant the frequency response that would arise if the coefficients used for that position were used all the time.

The different interpretations of frequency response may be seen more clearly in connection with *Figure 22.18*. The three curves show different interpolation functions, and each would have a frequency response differing from the others. Each of the three interpolation functions has a half-amplitude width equal to the spacing between samples. Suppose the purpose of the application is to alter the timing of a programme signal with respect to the samples but to leave the sampling rate unchanged.

Consider two cases. The first will be when zero retiming is necessary, and here only the one coefficient B of 100 per cent

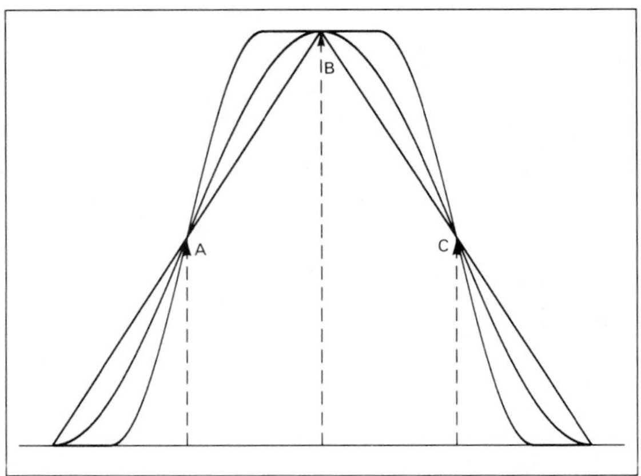

Figure 22.18 Three interpolation functions with different frequency responses

would be used. In this case, the frequency response will be completely flat. The second case is where the required retiming is the maximum of precisely half the sample spacing. In this case, the two coefficients A and C would be used each of 50 per cent. The frequency response will be cosinusoidal falling to zero at half the sampling frequency and continuing to fall until it reaches a gain of -1 at the sampling frequency.

So there are three interpolation functions with three different apparent frequency responses each of which can result in a frequency response which is either flat or drooping cosinusoidally depending on the required timing shift.

In the context of standards conversion, the desirable feature of interpolation is that it should produce a signal appropriate for the position where it will be displayed. To take a specific case, let us consider vertical interpolation and assume that the television signals use progressive scanning. In theory, it is reasonable to define the wanted output signal, on a particular line, as being the most positive excursion of a cosine wave of a particular amplitude and vertical frequency.

It is possible to measure vertical frequency in many different ways. One that is frequently used is *cycles per picture height*. However, for this purpose it is more convenient to measure it in *degrees per line*, and the lines that will be used are those of the input before interpolation. If the span of the interpolation function is an odd number of lines, since interpolation is being considered rather than extrapolation, the output line will have an offset with respect to the central line of between plus and minus half a line. For an even number of lines, the offset will have the same range about a point midway between the input lines.

For each value of offset it is possible to calculate the phase of the cosine signal appearing at each line being used by the interpolation, and the output of the interpolator will be the algebraic sum of the products of each signal and its appropriate coefficient. Of course, at this stage the coefficients are not known. The real part of the output of the interpolator should be equal to the defined wanted output signal, and the imaginary component should be zero.

Each definition of a vertical frequency at which a specified real response will be obtained provides one equation containing the coefficients, and each definition of a frequency at which the imaginary response will be zero also provides an equation. It is therefore just a matter of defining an appropriate number of frequencies to obtain the coefficients for one offset. Using the same frequencies for all offsets defines the interpolation function.

Using this method, the real and imaginary frequency responses are shown in *Figure 22.19* for a two line interpolator. The real part has been defined at zero frequency A to have a response of 1. The imaginary part has been defined to have a zero in its response at point F. The zero of the imaginary response at B arises as a result of it being the sum of sine terms, which are all zero at zero frequency. Nine different offsets were used in these figures, from $-\frac{1}{2}$ to $+\frac{1}{2}$ in steps of $\frac{1}{8}$. Some of these curves precisely overlap other curves. The real part of the response has the inevitable variation between being flat and being cosinusoidal; that is inevitable with two line interpolation.

Figure 22.20 shows the case for three line interpolation, the point E being arbitrarily chosen to have the same frequency as F but to have a somewhat reduced real response compared to that of point A.

Figure 22.21 is the four line interpolation case where an additional point D has been defined for the imaginary response, which has been markedly improved. Point D has been chosen to be at half the frequency of F.

Figure 22.22 shows the five line interpolation case, the additional point being C.

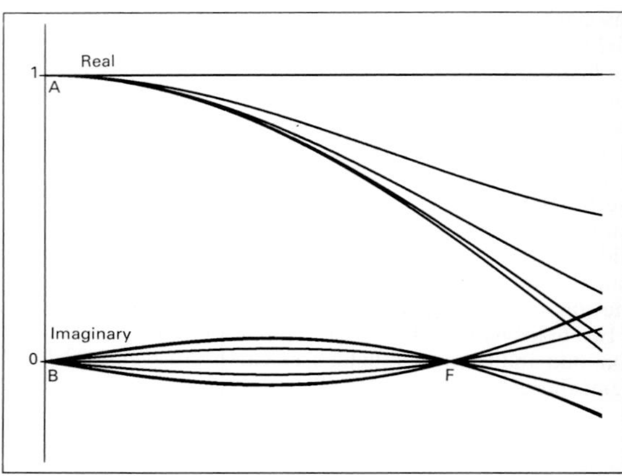

Figure 22.19 Real and imaginary response variation for a two-tap interpolator. See text for significance of symbols

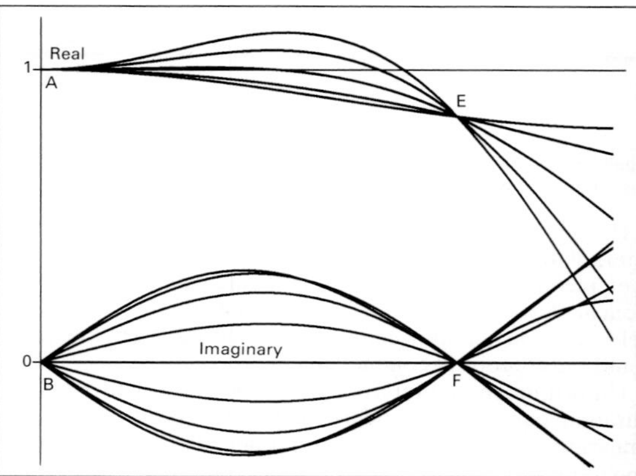

Figure 22.20 Real and imaginary response variation for a three-tap interpolator

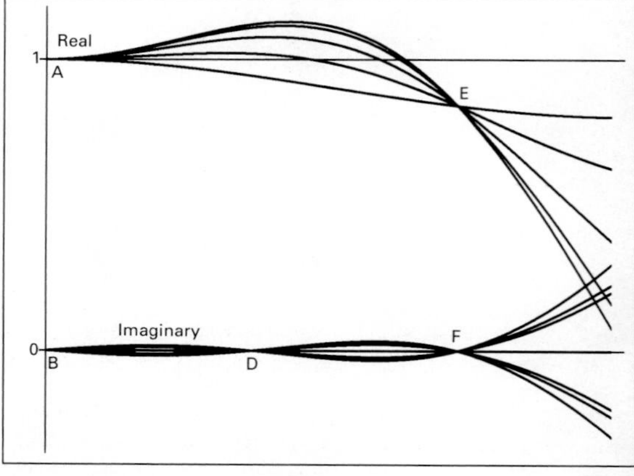

Figure 22.21 Real and imaginary response variation for a four-tap interpolator

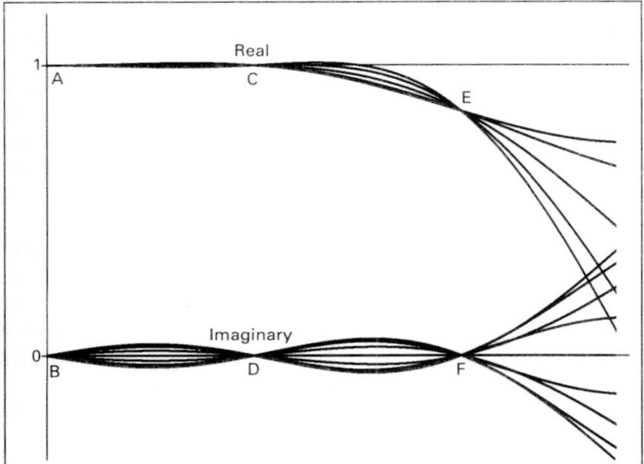

Figure 22.22 Real and imaginary response variation for a five-tap interpolator

To illustrate the changes as the number of coefficients is increased, the frequencies of the points have been held constant and either used or left unused. The choice of frequencies is noticeably unfair to the three line case of *Figure 22.20*. It is reasonable to assume that a given broadening of the envelope is equally significant for the real and the imaginary components, provided the frequency is the same. However, the significance of a given broadening of the envelope decreases rapidly with increasing frequency.

Although, as shown in *Figure 22.22*, the five line case looks good, experience shows that the improvements that could be made to the resolution by tolerating a greater broadening of the real response is likely to be justified subjectively.

The advantage of this approach is that it shows the dynamic variation of frequency response that results from the interpolation function in a way that is both relevant and easily understandable, but some experience is needed to understand the sensitivity of the display. If the real response is kept equal to unity it is perhaps of interest that, as the number of coefficients approaches infinity, so the interpolation function approaches a (sin x)/x function.

22.5 Future of standards conversion

It is debatable if the present quality of the best interpolating standards converters is satisfactory. If pictures are selected to be very critical, then undoubtedly the impairments arising from conversion are also very obvious. However, in this situation ordinary television scanning can result in obvious impairments; probably the worst of these would arise from the effects of interlace.

It is the frequency of occurrence of conversion impairments, as well as the possible magnitude of the impairment, which are important. If all programmes had to be converted, then the quality of conversions would have to be far higher than it would need to be if only 5 per cent had to be converted. Undoubtedly, if a news item is very newsworthy, it will be shown even if the quality is substandard. Amateur 8 mm film has been used at times when it is the only record of a news item. The quality needed for a programme can often depend on the importance of that programme to the prestige of the broadcaster. Generally, an impairment arising in the conversion from a foreign standard will not affect a programme that is considered to be a prestige product.

There is little evidence that programmes that are of adequate quality in their original standard become inadequate when passed through a high quality converter. Unfortunately, the majority of converters are not of such high quality.

22.5.1 CCD picture sources

The characteristics of ccd cameras differ from ordinary tube cameras particularly in the temporal aperture and to a certain extent in terms of vertical resolution. The difference in vertical resolution would tend to result in a somewhat higher level of vertical aliasing which would tend to make standards conversion a little more difficult, but since this would also increase the problems of interlace it seems probably that the amount of vertical aperture correction would be decreased so leaving the problems of standards conversion substantially unchanged.

The reduced temporal aperture of ccd cameras used for slow motion would have an adverse effect on converters in that it will increase movement judder. However, this will also have an adverse effect on normal viewing so again it will not appear normally in the signals applied to standards converters. There will remain a small reduction in temporal aperture under normal conditions due to the time taken for charge transfer in ccd cameras; this will result in a small deterioration of standards conversion quality.

22.5.2 HDTV

A 1250-line picture viewed at three times picture height and a 625-line picture viewed at six times height result in the same angular resolution and number of picture lines per degree, so the problems of conversion between high definition standards will be comparable to the technical problems of conversion between conventional definition standards. Of course the size of the field stores will be about five times larger, and the speed of the mathematical operations will have to increase proportionately, but these do not represent a significant theoretical problem.

22.5.3 Motion compensation

The next method of standards conversion is likely to measure the velocity of movement and to compensate for this motion in the interpolation.

Figure 22.23 shows at the left the same parts of three successive fields, the first being at the top and the last at the bottom. At the right, superimposed on the axes of vertical and horizontal velocity, is a straight sloping line which is the locus of

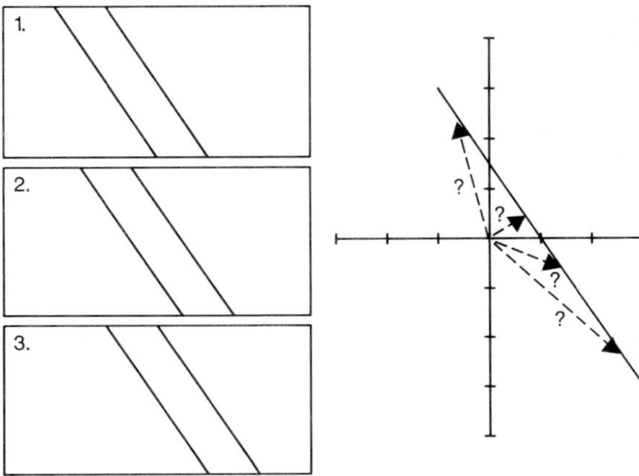

Figure 22.23 Successive simple input pictures and undefined motion vectors

all possible motion vectors that can describe the motion on the left. The scale of this graph, if considered as motion per frame, is the same as the part pictures. Obviously, no unique vector can be determined, but this is not significant since any vector from the origin finishing on the line does comply with all the available information.

Figure 22.24 shows the same three successive fields, but additional details *a*, *b* and *c* are shown in each part picture. Now there appears to be a conflict between three quite different vectors shown at the right. It may seem at first that the three part pictures cannot represent actual pictures from a scene, but the region between the two sloping lines could be an image of, for example, the sloping windscreen pillar of a car, *a* could be a shadow cast by a vertical object, *b* could be a reflection in the shiny paint and *c* could be a mark on the paint.

If the shadow *a* is looked at more closely, it can be seen, from the variation of its width, that it is not moving vertically but is completely static. It appeared to move because the shadow was visible only where it fell on the windscreen pillar.

Of the motion vectors shown, vector *c* should be used for object *c*, *b* for object *b*, but for object *a* the upper and lower edges are moving with the velocity vector *a* but the body of the shadow is stationary. Only vector *c* is really appropriate for the lines representing the windscreen pillar, although *a*, *b* or *c* would all be satisfactory for the picture as drawn; if some texture was visible on the paint of the pillar, then only *c* would be appropriate. It should be realized that texture can be quite

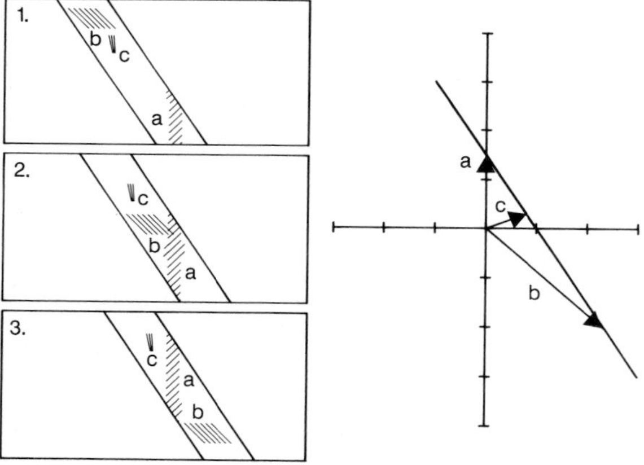

Figure 22.24 More complex input pictures and their apparent motion vectors

clearly visible, even when it is smaller than the noise level. If object *c* were missing, would vector *c* still be determinable from texture?

If a standards converter relied on motion vectors, then these must be totally reliable; that is not achievable. What may be achievable is to have a converter that relies on interpolation but which uses motion vectors only when those motion vectors are known to be reliable. Fortunately the motion vectors tend to be most reliable when the use of vectors would be most beneficial.

Even if motion vectors worked perfectly for motion, it does not mean that conversion would be perfect. Temporal effects certainly occur with motion, but they also occur independently of motion. A simple but extreme case would be a strobe light on a vehicle. It may be possible to use motion compensation for the vehicle but not for the strobe light. There would then be a real possibility that the strobe light would appear to wander slightly with respect to the vehicle.

The leaves of a tree may be fluttering in the wind. As they flutter, the camera may sometimes be seeing the sunny side, sometimes the shadow. There would be large temporal effects, not directly attributable to the translational movement of the leaf, but at times these could result in significant erroneous motion vectors.

The use of motion vectors is potentially dangerous in that a wrong estimation can cause unbelievable effects. For example, a rigid object might suddenly deform when a shadow passed over it or if it were partly obscured by another object.

There are many alternative proposals for determining motion vectors. The best system may not be the one that is accurate for the highest proportion of pictures, but the one which reliably warns that the results may be in error, so that motion compensation of the feature would be abandoned, rather than risk using erroneous motion compensation.

At the boundary of a moving foreground object, on the input pictures, there will be overlap areas where the blurred edge of the moving object overlaps and partially obscures stationary background objects. This will cause problems with the movement vector evaluation, but in addition, these areas must not be used in the movement compensation because they inevitably contain information that is moving at a different speed to the evaluated vector.

The quality of conversions on critical material is much improved on 20 years ago. Some people have claimed that motion compensation will eliminate conversion impairments; this is not true. However, motion compensation used with care would be beneficial on some critical pictures, without causing too much degradation on other material. This may, on average, provide a useful improvement to the quality of conversions.

K Davison Manager Communications, Thames Television

G A Johnson B Sc, C Eng, MIEE
Deputy Head Engineering Services, ITV Association

23 Satellite Distribution

The following sections describe features of satellite communications that are of particular interest to television broadcasters and facility companies. They consider the availability of fixed satellite services (FSS) for point-to-point and point-to-multipoint television transmissions and indicate limitations that the use of communications satellites may impose on television production techniques. As national and international restrictions on access to FSS are eased, the conditions for private uplinking by the user are discussed and the application of a power budget is described.

23.1 Background

The first satellite to be equipped with an active transponder (receiving and transmitting equipments) for television transmissions was *Telstar 1* launched by the United States in 1962. It flew in an elliptical orbit between 914 and 5436 km (568 and 3378 miles) above the earth's surface. Each orbit took three hours and contact between the east coast of America and ground stations near the west coasts of England and France was limited to periods of 20 minutes which did not always suit the broadcaster's needs. Extremely large, expensive, complex steerable antennas were required at the ground stations to track the satellite as it moved across the sky.

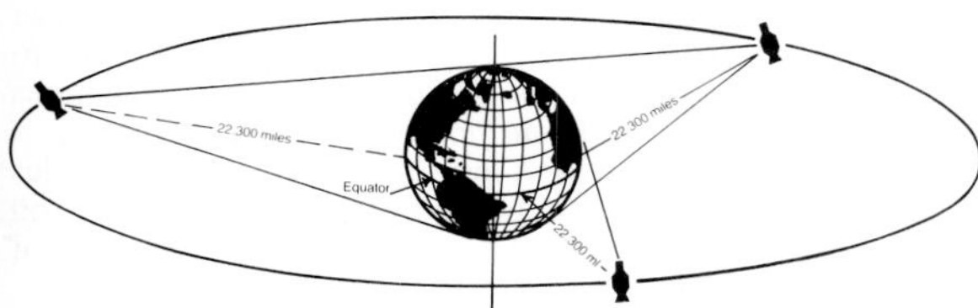

Figure 23.1 Three geostationary satellites at an altitude of 35 800 km above the equator can provide virtually complete coverage of the earth's surface

A method to provide almost continuous coverage had been suggested by Arthur C Clarke in 1945. Three satellites would act as repeaters in the sky and be equally spaced in the equatorial plane in a circular orbit at a critical altitude of 35 787 km (22 237 miles). The direction of rotation would be the same as that of the earth; at this altitude the time taken for one orbit would be precisely one sidereal earth day. Consequently, each satellite would be fixed above a particular location on the equator and appear stationary to anyone on earth. This is the only orbit to allow fixed antennas to be deployed on the satellites and on the ground. The first such *geostationary* satellite designed to handle television signals was launched in August 1964.

23.1.1 Transponders

Like the early active satellites, modern designs are equipped with transponders to handle the signals. A *transponder* comprises a low noise receiver, a frequency converter and a high power amplifier . Many recent satellites are equipped with matrices to allow transponders to be switched by ground command between different receiving and transmitting antennas. Flexible connectivity is considered in more detail in section *23.4.5*.

Until recently, the transponder high power amplifier (hpa) has invariably been implemented as a travelling wave tube amplifier (twta). In new designs having hpa outputs up to 10 W, the twta is being challenged by the more efficient solid-state power amplifier (sspa) which also offers a more nearly linear transfer characteristic.

Many transponders on multi-purpose communications satellites are available for long term and occasional television leases.

23.1.2 Footprints

The service area associated with a downlink is usually described in terms of a footprint map of the appropriate part of the earth's surface marked with contours representing either transmitted equivalent isotropic radiated power, eirp (dB(W)) or received power flux density (dB(W/m^2)). Transponders are normally operated with *backoff*, or power reduction, from these maximum possible levels which usually correspond to hpa saturation (in the region of the knee in its transfer characteristic). The user should always check with the satellite operator the backoff applying to a particular transponder. The point near the centre of the footprint at which the flux density is greatest is known as the *bore-sight*.

The shape and area of the footprint is determined by the feeder arrangements to the transmitter antenna and its shape and beamwidth. The footprint also indicates the approximate area over which an uplink may successfully access a transponder on the satellite.

23.2 Satellite operators

Telecommunications administrations were the main driving forces behind non-military exploitation of communications satellites. They saw satellites as an economic way to provide extra capacity for mainly international telephony traffic. Of most interest to major broadcasters are the international communications satellites which cover the Atlantic, Indian Ocean and Pacific regions. The world map in *Figure 23.2* shows the global beams of satellites located at these positions.

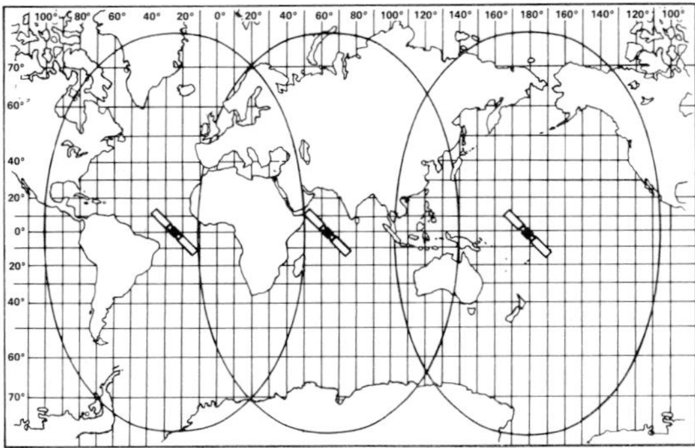

Figure 23.2 The global beams of equatorial satellites over the Atlantic, Indian Ocean and Pacific

A large percentage of the world's most heavily populated areas are able to communicate with each other by using a single satellite hop. This is cheaper than multiple hops, and the amount of delay and other impairments that the signal is subjected to is kept to a minimum.

23.2.1 Intelsat

In 1964, a number of countries agreed to establish a satellite telecommunications body to serve the whole world. In 1971, this agreement saw expression as *Intelsat*, the International Telecommunications Satellite Organization. For many years, this consortium held the monopoly for the provision of international public and private satellite services. Its membership numbers more than 110 countries which are represented in most cases by the national telecommunications operator (usually the PTT). It has added satellites to its system to meet the increasing demand for international circuits. The television user can now book occasional circuits on eight Atlantic Ocean, three Indian Ocean and two Pacific Ocean international communications satellites operated by Intelsat. Its revenue from television is about 6 per cent of the total income from communications satellite traffic.

Intelsat world-wide coverage is supplemented by several national domestic satellites and regional systems such as those provided by the *Eutelsat* consortium of European PTTs and *Intersputnik* in the USSR.

23.2.2 North American domestic communications satellites

The USA is served by many *domsats* which fill the gap in world coverage between the Pacific and Atlantic satellite systems. The domsats meet the demand for a very high level of internal traffic within North America and provide capacity to extend Pacific and Atlantic traffic across the American continent. There are 14 domsats to carry occasional traffic split between the 6/4 GHz and 14/11 GHz bands. (The notation 6/4 GHz indicates an uplink frequency of 6 GHz, downlink at 4 GHz.) Most US domsats provide *conus* (the 48 contiguous states of the USA) footprints and some are extended to cover Alaska, Hawaii and Puerto Rico. The Canadian *Anik* satellites (C2 at 110°W and C3 at 117.5°W) carry 14/11 GHz band occasional traffic to Canada and parts of the USA. The Mexican *Morelos* F1 and F2 satellites at 113.5°W and 116.5°W provide occasional 6/4 GHz circuits to central America. The occupancy of most American domsats varies from month to month; in addition to the leased channels, a large amount of occasional traffic is also carried from transportable ground stations.

The North American domsat system is very responsive to market changes. An organization which proved helpful one year may not exist the next, and a broadcaster must either maintain a regular and continuous contact with the many carrier companies involved or avoid these complications by booking with a satellite servicing company such as *BrightStar*.

23.2.3 Eutelsat

Eutelsat provides a 14/11 GHz service in Ku-band with a footprint covering North Africa, Western Europe and the European broadcasting region. The European Broadcasting Union leases some transponders on a long term basis to augment its permanent terrestrial TV network. Occasional circuits may be booked on some Eutelsat satellites.

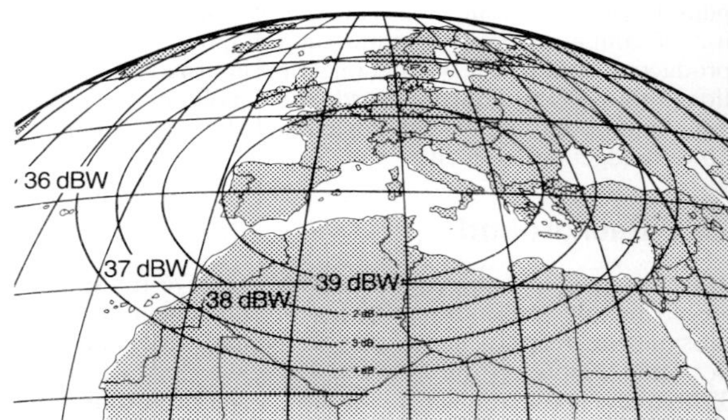

Figure 23.3 The Eutelsat footprint

23.2.4 Intersputnik

The *Gorizont* satellites are part of the USSR's international geostationary communications satellite system, Intersputnik, which links the former Warsaw pact countries and others such as Cuba and North Korea. Currently, regular access to Intersputnik is not directly possible from most Western nations but can be achieved via an *Intervision* country having a connection to the *Eurovision* network. Intersputnik provides transponders for occasional traffic in the 6/4 GHz band, and like other geostationary satellites it cannot deliver signals of adequate strength to near polar latitudes.

In order to provide a service to the significant populations in areas which lie at about 70°N, the USSR developed the *Molniya* system which carries communications satellites in an elliptical orbit inclined at 63° to the earth's equatorial plane. As a satellite in this system travels into space, it appears to hover for a period of about six hours (see *Figure 21.4*). By switching signals at appropriate times between the four satellites distributed along the orbit, the *Molniya* system provides continuous coverage of its service area.

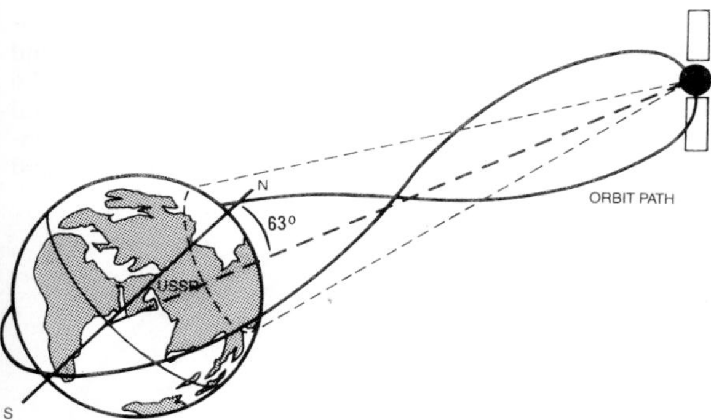

Figure 23.4 Elliptical orbit of Moliya satellites

23.3 Satellite applications

A broadcaster may wish to use a communications satellite for one or more of the following major applications:

- distribution of programme streams or networks to cable networks or affiliated stations
- scheduled or occasional contributions from remote points to a network studio centre
- totally unscheduled contributions of ENG material to a studio centre (often referred to as SNG — satellite news gathering)

The first application would normally be served by a permanently leased transponder and the third application by occasional bookings.

23.3.1 Regulatory considerations

In countries where the telecommunications services have not been deregulated, the user is obliged to make occasional bookings or long term leases with the public telecommunications operator authorized by the government. This is often the government controlled PTT. Even in deregulated countries, the user may be limited to booking through two or three approved companies. Where governments and satellite operators permit a broadcaster to operate its own transmitting ground station, a stringent technical performance specification must be adhered to. This is to protect other users of the satellite and prevent interference to other satellites in the same cluster or in adjacent 'slots', and to terrestrial services.

23.4 Satellite management

23.4.1 Station keeping and the TT&C control centre

At the end of its launch phase, a satellite is manoeuvred to its allocated station (or *slot*) in the geostationary orbit. If the influence of micro-meteorite impact, magnetic and gravitational fields and the pressure of solar radiation on the structure of the satellite were negligible it would remain at its station in the 'Clarke' orbit indefinitely without expending any further energy. In practice these effects are not insignificant, and they cause the satellite to change its orientation and gradually to drift off station and out of geostationary orbit.

A liquid propellant is carried which can be released in appropriate volumes and direction by thruster jets to compensate for orientation and positional errors. The quantity of propellant carried is limited by payload considerations. When this is spent, perhaps after seven years or more, the satellite will drift to a new non-geostationary orbit and its useful life will be over. Control of the release of propellant is determined in part by an on-board automatic control system and in part by instructions from the *telemetry, tracking and command* (TT&C) ground station.

The TT&C system on board the satellite makes environmental measurements and monitors the performance of many aspects of its operation. The resultant status data signals are sent digitally to the TT&C ground station which continually computes the satellite's precise location. Control signals are generated when necessary to keep station better than ±0.1° in N-S and E-W directions and to maintain beam pointing accuracy to ±0.1°. These and other command signals are uplinked to the satellite in digital form.

During the launch phase the satellite's position is varying greatly so an omnidirectional antenna is deployed for the telemetry and command beacons. These operate on similar frequencies which are chosen to be between about 130 MHz and 4 GHz (where atmospheric absorption is lowest) during this transient phase. Under normal orbiting conditions, the carrier frequencies are within the band allocated for the main transponders and employ a directional antenna aimed at the TT&C control centre.

23.4.2 Powering of satellite services

Some satellites are powered by small nuclear reactors, but those used by the broadcasters are powered by solar cells. End-of-life power demand for a large satellite in the FSS may be more than 2 kW. Rechargable Ni-Cd batteries provide a temporary source of power for short twice-yearly periods of eclipse when the earth shades the satellite from the sun. The battery supply powers essential services during the eclipse, but on most satellites it is not adequate to provide the high total power required by the communications transmitters.

If the satellite is placed in the geostationary orbit at the same longitude as a country which is receiving its signals, the eclipse will occur around midnight, varying in duration from zero at 22 days on either side of each equinox to 72 minutes at the spring or autumn equinox itself. By choosing a parking position for the satellite to the west of the country being served, the time when the eclipse occurs can be delayed by 4 minutes for each longitudinal degree shifted.

The early hours of the morning are, statistically, periods of low traffic density, and so this strategy is employed with domestic satellites to minimize the effect of the eclipse. Clearly the timing of the eclipse can not be optimized in this way for a satellite serving two continents or a large international region.

Two other undesirable, but predictable, natural phenomena disturb the continuity or quality of the signal. The most important of these occurs when the sun aligns with the satellite in the beam of the receiving antenna. When this happens, for periods of about 10 minutes on five consecutive days twice a year, the ground station receives an enormous increase of solar noise. The effect is either an interruption (*outage*) or severe degradation of the signal.

Less serious is the deterioration of transmission performance brought about by the eclipse of the satellite by the moon. In the decade to the end of the century, a satellite at 19°W will experience 22 instances of moon solar eclipses. The level of shadowing will vary from 1 to 100 per cent and the duration from 1 to 64 minutes. On 17 January 1999, solar energy will be totally blocked for 51.4 minutes, starting at 12.03 GMT.

23.4.3 Coverage

In principle, it is possible to provide virtually complete coverage to the major populated areas with three satellites equidistantly spaced around the equator as shown in *Figure 23.1*. Such a scheme of three satellites provided the possibility of introducing worldwide communications, including low-density traffic routes to and from remote areas. The first geostationary satellites employed broad-beam antennas which covered all of the earth's surface visible from satellite. This amounted to about 42 per cent of the total surface area.

Two major drawbacks were evident; one was the need for large receiving antennas to pick up the weak signals from the satellites, and the other was the inevitable waste of energy over the oceans and deserts. With the later introduction of spot beam antennas, the effective radiated power was increased and directed onto areas of high population.

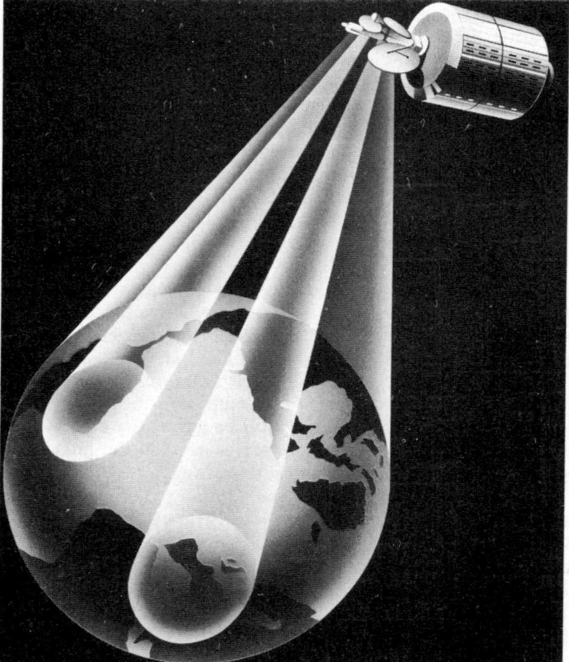

Figure 23.5 The use of directional antennas enables energy to be concentrated where needed

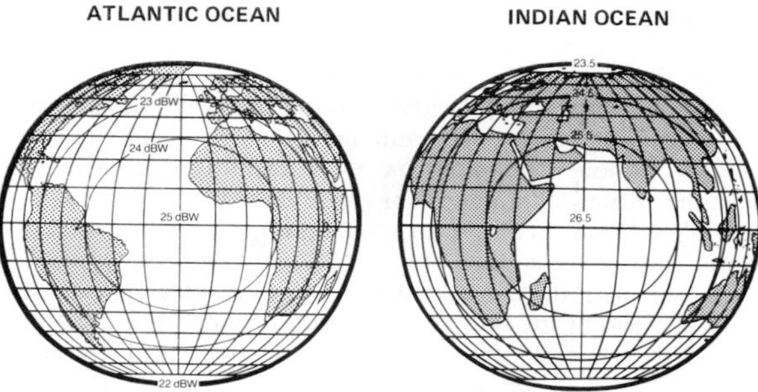

Figure 23.6 Two examples of 4 GHz band footprints

In the 4 GHz band, the beam bore-sight eirp (*equivalent isotropically radiated power*) for a typical full bandwidth transponder ranges from 26.5 dB(W) for global coverage (42.4 per cent of earth's surface, 17.5° × 17.5° beamwidth), through 28 dB(W) for hemisphere coverage (20 per cent) to 31 dB(W) for zonal coverage (10 per cent, 14° × 5°). Typical figures for spot beams (beamwidths less than 3° × 2°) in the 11 GHz band are 44.5 dB(W) (inner) and 41.5 dB(W) (outer). *Figure 23.6* illustrates two examples of 4 GHz band global footprints, and *Figure 23.7* shows how the radiated power has been concentrated into two 'hemisphere' beams onto the American and African/European land masses.

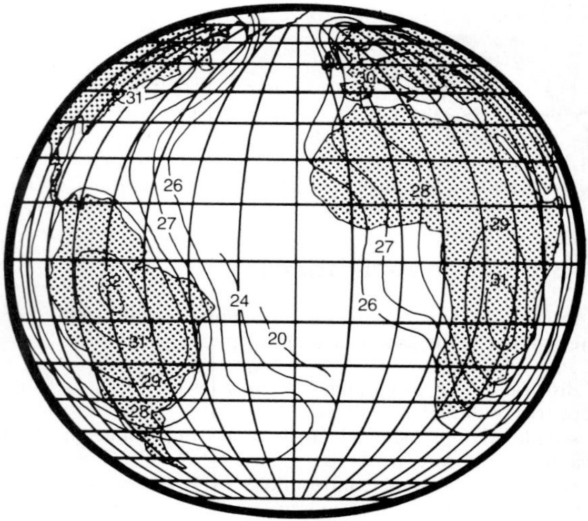

Figure 23.7 Concentration of radiated power onto land masses

23.4.4 Frequency re-use

The current range of communications satellites is equipped with multi-beam antenna (mba) systems which allow several beams to be transmitted simultaneously. A satellite may be equipped with broad global and steerable directional spot beams operating on the same frequencies and possibly using opposite linear or circular polarizations. By these means, a particular frequency may be re-used by two, four or even six beams.

23.4.4.1 Frequency bands for FSS

Until the start of the 1980s, nearly all fixed satellite services operated in the 6/4 GHz region of C-band. When satellite/earth communication started, this part of the spectrum had several advantages over higher frequency bands. The technology was available to provide transponder output power of 5 W or more; attenuation caused by the atmosphere, rain or other weather conditions is low. The most important disadvantage is the sharing of the 4 GHz and 6 GHz bands with terrestrial (radio relay) services. There is a great potential for mutual interference. The following analysis highlights one argument to support a move to higher frequencies.

The eirp is determined from the antenna gain which is related to the beamwidth for the coverage required.

The gain of the antenna having an effective area A is given by:

$$G_t = \frac{4\pi A}{\lambda^2} = \frac{\pi^2 D^2 \eta}{\lambda^2}$$

where η is efficiency (in practice between 50 and 80 per cent). For a circular aperture of diameter D the following approximate relationship holds for beamwidth b:

$$b = \frac{75\lambda}{D} \text{ degrees}$$

where λ is wavelength.

From these relationships it follows that an antenna of a given aperture provides 10 log $(11/4)^2 = 8.8$ dB more gain at 11 GHz than at 4 GHz. The beamwidth at 11 GHz is $4/11 = 0.36$ times that at 4 GHz.

It follows that 14/11 GHz operation permits smaller dishes to be used, and narrower spot beams can be produced. Because 14/11 GHz is less prone to interference, ground stations will operate satisfactorily in urban areas and are permitted to operate at higher power levels. These considerations have led most new communications satellites to be equipped to operate in both 6/4 GHz and 14/11 (or 14/12) GHz bands, and many satellites which have yet to be launched will carry only 14/11 (or 14/12) GHz transponders. *Cross-polar discrimination* (xpd), which is the key to successful frequency re-use based on either orthogonal or oppositely sensed circular polarizations, can be severely degraded by atmospheric conditions especially at frequencies above 10 GHz.

23.4.5 Flexible connectivity

Each receiving antenna is permanently connected to its own low noise amplifier (lna) in most satellites. The lna output signals are grouped together for down-conversion. Following conversion, in modern designs the signals can be switched in a matrix to the hpas. The hpas in turn are connected to the transmit antennas through a router. This high level of transponder and antenna flexibility allows the satellite to be reconfigured by commands from the TT&C centre in order to bring a spare transponder into service or to change the coverage to account for changed traffic conditions.

This flexible connectivity can be extended on some satellites to allow a 14 GHz uplink to be routed to a 4 GHz downlink (or a 6 GHz uplink to a 11 GHz downlink). This type of connection is known as *cross-strapping* and may be used to advantage by the operator of a private 14/11 GHz ground station to communicate with a major distant 6/4 GHz ground station. A transponder attracts a higher tariff when it is cross-strapped.

23.5 Point-to-point connections for television

The component parts of a television service are shown in *Figure 23.8*. They are:

- the terrestrial extension circuits from the studio or outside broadcast location to the ground station; these may be divided into the local circuit (or loop) from the studio to the telecommunications operator (the PTT) and the main link from PTT to the satellite ground station
- the uplink (or up-leg) from that particular ground station to the satellite
- the transponder
- the downlink (or down-leg) to the ground station (or stations) scheduled (or cleared) to receive these signals
- the circuits from the receiving ground station to the final destination, usually the TV station or studio centre.

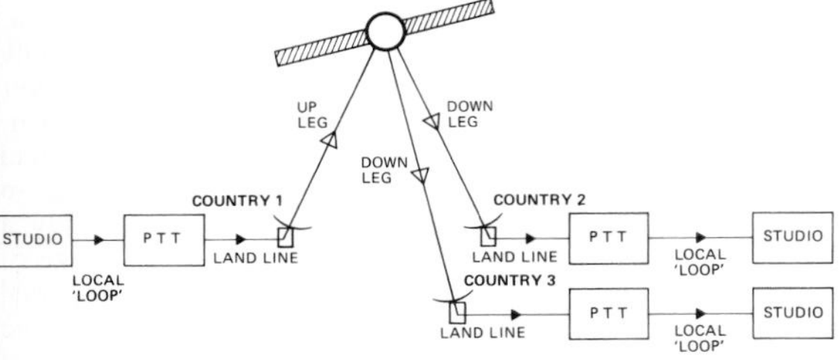

Figure 23.8 Point-to-point connections for a television service

The terms *up-leg* and *down-leg* describe the microwave links to and from the satellite. The term *space segment* is used to describe the up-leg and down-leg paths together and often includes the transponder in the satellite.

The normal access point for a satellite link is the *gateway*. The fees for space segment circuits are charged between gateways and include the cost of any terrestrial circuits between gateways and ground stations. Any circuits needed between the studio or television centre and the gateway must be paid for additionally. In the USA and Canada the gateway is located at the ground station, but in many other countries the gateway is the capital or other major city. Additional fees for the studio/gateway connection add to the total cost and may complicate the booking arrangements.

Where a broadcaster in Europe receives a signal which originates in, for example, New Zealand there will be two space segments, i.e. two up-legs and two down-legs. If the route is properly planned, one midpoint ground station will be selected to access both satellites, i.e. Hong Kong in the example given in *Figure 23.9*.

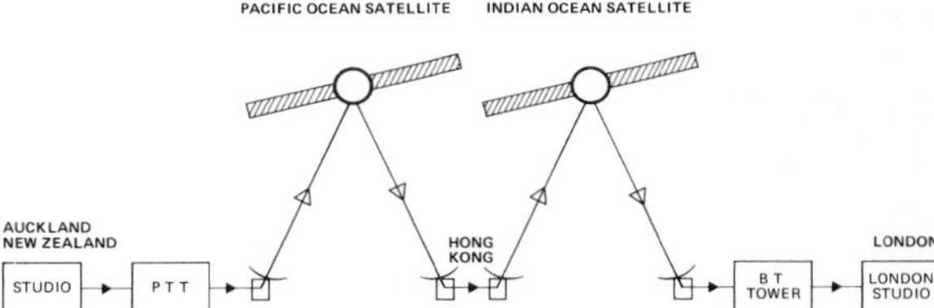

Figure 23.9 Point-to-point connections involving two space segments

23.5.1 Leasing of transponders

Transponders and the associated links may be leased either for *occasional use* periods as required or *continuously* (three months and longer). Depending on the satellite, bookings may be made with the telecommunications administration (such as *British Telecom International* in the UK) or other national signatory to Intelsat (such as *Comsat*, the Communications Satellite Corporation, in the USA) or with a private organization which provides a satellite booking service (such as *BrightStar*). Organizations providing booking services hold permanent leases on some international satellites and many domsats.

Where a broadcaster identifies a need for a prolonged transponder lease, it may be more economical to arrange a lease for an extended period directly with the satellite operator. Intelsat, for example, provides pre-emptible services, and each signatory participating in a lease may operate as many receiving ground stations as desired at no extra cost. There is, however, an additional charge for each signatory (in excess of two) participating in the lease. The participating signatories may designate one or more ground stations to transmit TV services.

The cost of a lease is determined by the type of satellite, the type of transponder, the type of connectivity, the type of video channel and the bandwidth. The service can be provided on any satellite where capacity exists that meets the requirements of the user. Normally this service is not carried on a primary or major path satellite.

At the start of 1990 the following transponders were available for occasional use:

Flight	*Position*	*Beam*	*Band*	*Path*	*Available television channels*	*Channel rf bandwidth*
INTELSAT SERVICES						
Atlantic Ocean Region (AOR)						
VA F10	24.5°W	global		Primary	two	20 MHz
V F4	34.5°W	global	C, Ku	Major 1	two	20 MHz
V F6	18.5°W	global		Major 2	two	36 MHz
VA F11	27.5°W			BrightStar	one	
V F2	1.0°W			Spare 359°	three max	
VA F11	27.5°W	W hemi		Main reserve	(by special	20 MHz
		W hemi	Ku	332.5°	arrangement)	
V F13	53°W		Ku	Spare 307°	(by special arrangement)	
Indian Ocean Region (IOR)						
VB F15	60°E	global	C	Primary	two	20 MHz
V F5	63°E	global	C, Ku	Major 1	two	18 MHz
V F7	66°E		Ku	Spare 66°	(by special arrangement)	
Pacific Ocean Region (POR)						
V F1	174°E			Primary	two	
V F8	179°E	W hemi	C	Major 1	two	20 MHz
OTHER SERVICES						
AOR						
France						
Telecom F1	8°W	Semi-global	C		one	20 MHz
		Euro	Ku		one	36 MHz
USSR						
Gorizont-12 (Statsionar 11)	11°W	C		(via Intersputnik)	34 MHz	
Gorizont-15 (Statsionar 4)	14°W		C, Ku		(via Intersputnik)	34 MHz
USA						
PanAmSat F1	45°W		Ku		one	
IOR						
European PTTs						
Eutelsat-I F2	7°E	W spot	Ku	Primary	one	72 MHz
Eutelsat-I F5	10°E					
Eutelsat-I F4	13°E					
Eutelsat-I F1	16°E	Euro (up) W Spot (down)	Ku		one	
France						
Telecom F2	5°W	Euro	Ku		one	36 MHz
Arab League						
Arabsat-1A F1	19°E		C		one	
Arabsat-1B F2	26°E					
USSR						
Gorizont-11	53°E				(via	
Gorizont-3	40°E		C		Intersputnik)	
POR						
Australia						
Aussat A1 Cen Aus/ North East/ National B	106°E		Ku		one	45 MHz
Aussat A2 South East/ National A	156°E		Ku		one	45 MHz
Aussat A3 South East/ National A	164°E		Ku	Spare	one	45 MHz
Japan						
CS-2A	132°E		C			
			Ka			
CS-2B	136°E		C			
			Ka			
BS-2A	110°E		Ku			
	BS-2B		Ku			
Indonesia						
Palapa B1	108°E		C	S.E. Asia Primary	one	36 MHz

Table 23.1 Sample of transponders available for occasional use

23.5.2 Half transponder operation

Analogue television signals are transmitted to and from the satellite in frequency modulated form. A *full transponder* has a typical bandwidth of 36 MHz and some are split to allow two television channels of 18/20 MHz bandwidth to be carried. Not surprisingly, a *half transponder* is less costly to lease than a full transponder, but there is a disadvantage when two television signals share the same transponder. A decrease of power (or *backoff*) of about 2.5 dB becomes necessary to reduce to an acceptable level intermodulation distortion introduced by twta non-linearity. This reduced transponder power is then shared equally between the two signals (a further decrease of 3 dB). The deviation, too, is reduced in this mode of operation. It is shown later that the carrier/noise ratio is however improved by 2.6 dB because of the smaller noise bandwidth of a 20 MHz half transponder. This becomes very significant when operating at marginal values of carrier/noise ratio.

The video signal deviates the fm carrier of a full transponder by 13.5 MHz/V (6.75 MHz peak deviation) on many communications satellite services. Eutelsat requires a 1Y peak-to-peak video signal to deviate a video carrier by 25 MHz for a full transponder and by 19 MHz for a half transpander. Pre-emphasis as described in CCIR Recommendation 405 is applied at the uplink ground station, and a compensating de-emphasis is employed at the downlink ground station to improve signal/noise performance (by about 2 dB for the luminance component of a 625/50 signal).

23.5.3 Protection of other services

An *energy dispersal* waveform is usually added to the video signal before modulation. This ensures that interference that could otherwise be introduced into satellite or terrestrial multi-channel telephony systems is spread over many channels. A symmetrical triangular waveform locked to the television field synchronizing pulse is preferred. Its frequency is usually one half or one quarter of the field frequency and is phased so that its discontinuity of slope occurs outside the active picture area.

The dispersal waveform has an amplitude which is typically 5–10 per cent of the video signal. The ground station receiver must remove the picture flicker introduced by the waveform. Several methods are possible. Two stages of black level clamping of the demodulated signal will reduce the flicker to an imperceptible level. An alternative approach is to apply a narrow-band frequency feedback technique to the if stage of the receiver. This is suitable only if a very low dispersal frequency is employed (not greater than one twentieth of the field frequency). The method also has the advantage of providing threshold extension which allows pictures of acceptable quality to be received under lower values of carrier/noise ratio than would be possible with a conventional frequency discriminator.

23.5.4 Sound signals

In the case of NTSC, PAL and SECAM signals, a transponder usually carries both video and associated sound signals (monophonic, stereophonic or multiple channel). The sound signals (which may be processed by a proprietary noise reduction companding system such as Wegener Panda) are carried in the range 5.4–8.0 MHz (relative to the vision carrier frequency). Monophonic signals may alternatively be carried by the sound-in-sync method (Varian TVT/BBC system), and sound-in-sync variants may also be used for dual channel services (RE Instruments/IBA, Varian/BBC, Vistek/ITV Association or BTS). In the case of multiplexed analogue component signals (B-MAC, C-MAC, D-MAC and D2-MAC), the sound signals are in digital form (adaptive delta modulation or MAC/packet for B-MAC, and MAC/packet for the other MAC signals) and are time division multiplexed with the analogue video components.

23.5.5 Interaction between sound and video signals

The signal/noise ratio of the video signal is slightly reduced through the deviation of the main carrier by a sound subcarrier. For the converse reason, the signal/noise ratio of the sound signal depends slightly on picture content. A further small picture impairment is caused by intermodulation products. These unwanted effects may become significant where two or more subcarriers are provided. Unwanted interaction does not occur with tdm sound signals (sound-in-sync or MAC/packet).

23.6 Factors affecting programme production

23.6.1 Transmission delays

The transmission delay, t, between a ground station and the satellite is given by:

$$t = \frac{d}{c}$$

$$= \frac{1}{c} \left(R_e^2 + R^2 - 2 R_e R \cos \phi\right)^{1/2}$$

where

c = velocity of light
d = distance between satellite and ground station (*slant range*)
R_e = radius of the earth
h = altitude of the satellite
R = $R_e + h$
ϕ = angle subtended at the centre of the earth between the directions of the satellite and the ground station

Thus the transmission delay, t_d, for a single hop between ground stations lies between the limits:

$$\frac{h}{c} \leq t_d \leq \frac{2h}{c} \left(1 + \frac{2R_e}{h}\right)^{1/2}$$

For a geostationary satellite, this delay is in the range 240–280 ms. To this must be added the delay introduced by the terrestrial extensions to the broadcaster's premises.

23.6.2 Sound signal

Sound signals must always follow the route of the video signal with which they are associated as closely as possible, since the observer can tolerate only a small differential audio/video delay. For many types of television programme material, the absolute delay introduced by the point-to-point transmission path is unimportant. A simple playout of a complete package of sports or news items falls into this category. Problems arise, however, when a two-way discussion or interview is carried out over a satellite link since a question takes at least one quarter-second to reach the ear of a participant at the distant end and a further quarter-second for the answer to be received by the questioner. Using two satellites in cascade doubles the delay. This is not an easy problem to overcome for a fast moving two-way interview or discussion. It is sometimes possible to book sound circuits that are routed by undersea cable in one direction to reduce the delay, but few of these circuits are available.

The audio bandwidth on a satellite is 15 kHz per channel (optionally 7.5 kHz on some), and the limiting factor of circuit quality is the performance of the landline from studio to ground station (*backhaul*). This is particularly important in the USA where generally a terrestrial programme audio circuit may have a bandwidth of only 5 kHz which is adequate for speech but not for music.

Where a two-way interview is required, it is essential that only *clean reverse sound* (or *mixed-minus sound*) is fed to the participants. If any trace of the distant end voice or sound is relayed back it is heard as a disturbing echo which makes conversation almost impossible. Great care is necessary in adjusting audio levels to ensure that the reverse feed stays clean. The use of deaf aids (ear-pieces) in participants' ears is preferred, since it is much more likely to allow a successful conversation than if foldback loudspeakers were used in the studio.

23.6.3 Studio control of signal levels

23.6.3.1 Video

On most 6/4 GHz Intelsat services, each occasional television channel occupies a half transponder of 20 MHz bandwidth. This arrangement handles 525 NTSC television signals rather better than 625 PAL which can suffer from visible distortions and adjacent channel crosstalk, e.g. colour bar breakthrough. To reduce this problem, Intelsat insists that only the 75 per cent version of the colour bars test signal is allowed to be sent over the system and that it is limited to less than 30 s duration. This is usually easy enough to arrange in master control rooms, but care should also be taken that 100 per cent colour bars from the national network or from videotape machines are never presented to the satellite connection. High level video signals that can be produced by vtrs operating in the *fast rewind* mode must never be allowed to be passed over a satellite path.

High level signals with fast rise and fall times, such as white captions, can severely test the system since pre-emphasis increases the deviation of the fm carrier at high video frequencies. Under these conditions, the fm carrier may instantaneously exceed the available radio frequency bandwidth. This generates truncation noise which can manifest itself as picture tearing on highlights and streaking following sharp transients.

23.6.3.2 Sound

Confusion can result from the differing line-up tone levels and metering instruments used around the world. Major European broadcasters use a peak programme meter (PPM) with 0 dB*u* as the line-up level (u = 0.775 V), which is 8 dB less than the peak audio level. Many other countries use a volume unit (VU) meter with line-up at +8 dB*u* (i.e. at peak level).

It is important that a satellite booking includes adequate time to make measurements and adjustments to correct for errors in video and sound levels. It is good practice to start a booking early enough to play out programme material during the pre-transmission period over the satellite system so that variations in monitoring systems can be corrected and action can be taken to minimize distortion of the audio signal. This is often manifested as heavy sibilance of voices which can be improved by reducing the *send* level.

23.7 Transportable ground stations

It is technically possible to transmit directly to communications satellites from transportable ground stations. In the USA this has been possible since 1976 in the 6/4 GHz band by using domestic communications satellites. The ground stations for this band are rather large (up to 10 m, or 30–33 ft diameter) and are transported by articulated trucks. Satellite servicing vendors such as Netcom Enterprises, Videostar Connections Inc and BrightStar Communications Ltd can provide up-leg ground stations and can negotiate the necessary space segment on US domsats on behalf of the broadcaster.

23.7.1 Privately operated uplinks

More recently it has become possible to rent transponder time in the Ku-band on the Intelsat system (14 GHz is used for the uplink and 11 GHz for the downlink). At these frequencies, considerably smaller ground stations can be used. These use typically 3 m (10 ft) diameter antennas which are generally trailer mounted and designed to be containerized for long distance shipment.

Figure 23.10 A VSAT terminal (Marconi)

To provide a successful uplink for an event covered as an outside broadcast from a remote site, the ground station must be robust, reliable and highly mobile, easy to set up quickly and designed with adjustable eirp of an adequate rating to allow operation from any point within the satellite footprint. Improved side-lobe performance may be necessary in order to keep interference low to other satellites in the vicinity. Offset antennas can be particularly effective in allowing near and far angle side-lobe levels to be reduced.

The station must also incorporate downlink receiving equipment to allow the satellite to be acquired and communication to be set up with the TT&C centre, the receiving ground station and the receiving television station. The receiver will also be used to check that another user is not already accessing the transponder. Satellite operators require ground stations to meet transmit and general technical specifications as a condition for access to their transponders.

The mobile ground station must be equipped to measure the eirp of its transmitter, its frequency and frequency deviation (including the deviation associated with the dispersal waveform). Units like these are operated both by telecommunications administrations (e.g. British Telecom International (UK), Telespazio (Italy), Direction Generale des Télécommunications (France)), and by broadcasters (e.g. BBC, TDF, CBS). BTI, for example, offers a range of mobile units equipped with 5 m diameter dishes with eirps between 75 and 78.5 dB(W) down to one unit fitted with a 1.5 m square antenna and having an eirp of 67.5 dB(W).

The need to respond swiftly to news stories wherever they might occur in the world led Independent Television News (UK) to commission a design study for an even smaller ground station able to be packed in flight cases. This 14/11 GHz VSAT (*very small aperture terminal*) is basically simple in design and has the lowest possible power consumption. It uses a 2.1 × 1 m (7 × 3.3 ft) elliptical section Gregorian offset antenna which is small enough to be transported by passenger aircraft as accompanied baggage and is normally deployed in panel trucks or similar rental vehicles. This antenna design is marketed as a component of the Newshawk system.

23.8 Carrier/noise derivation

23.8.1 Calculation of received power

The power flux density P_d of a signal received at a distance d from a transmitter of eirp P_e is given by:

$$P_d = \frac{P_e}{4\pi^2 d^2} \ (\mathrm{Wm^{-2}})$$

If the receiving antenna has an aperture A then:

$$P_r = P_d A \ (\mathrm{W})$$
$$= \frac{P_e A}{4\pi^2 d^2} \ (\mathrm{W})$$

Where P_r = power received at the terminals of the aerial. Now the gain if the receiver antenna G_r at wavelength λ is:

$$G_r = \frac{4\pi A}{\pi^2}$$

$$\text{Thus, } P_r = G_r P_e \left(\frac{\lambda}{4\pi d}\right)^2$$

In terms of dB(W) this relationship may be written:

$$P_r = P_e + G_r - 10 \log \left(\frac{4\pi d}{\lambda}\right)^2 \ \mathrm{dB(W)}$$

The right-hand term is the path loss L_s (or *free space attenuation*) which is seen to be dependent on wavelength (and therefore frequency). If the slant range d is not known, it may be calculated from the relationship:

$$d^2 = R_e^2 + (R_e + h)^2 - 2R_e (2R_e + h) \cos\theta \cos\lambda$$

where R_e = radius of earth (6.278 × 10^6 m)
h = altitude of satellite (35.787 × 10^6 m)
θ = latitude of receiving point
λ = difference in longitude between satellite and receiving point

The value of d will lie between 35 787 km, where the ground

station is directly below the satellite on the equator, and 41 679 km ifthe ground station is on the satellite's horizon. Table *23.2* shows the variation of L_s with frequency with slant range.

Frequency (GHz)	*Free space loss L_s* (dB)		
	d = 35 787 km (*zenith*)	*d 39 000 km* (*typical*)	*d = 41 679 km* (*horizon*)
4	195.6	196.3	196.9
6	199.0	199.8	201.4
11	204.4	205.1	205.7
12	205.1	205.8	206.4
14	206.5	207.2	207.9

Table 23.2 Variation of free space loss with frequency

In practice, loss due to the antenna direction pointing error L_d must be accounted for, and atmospheric loss L_a and rain/precipitation loss L_p loss must be added to the path loss. At 4–6 GHz, L_p is very small, but in the 11 and 14 GHz bands it can be as high as 10 dB for short periods during very heavy rain. L_a depends upon the elevation angle of the satellite at the ground station, and is in the range 0.5 – 1.5 dB for angles between 90° and 5°. Uplinking at low angles of elevation is subjected to regulatory restrictions in order to minimize interference with terrestrial services.

The fuller expression for received carrier power is:

$$P_r = P_e + G_r - (L_s + L_a + L_p) - L_d \text{ dB(W)}$$

23.8.2 Calculation of receiver noise

The noise performance of the receving system is determined by referring all significant noise contributions to the antenna terminals. Noise figures (N) of individual elements (first amplifier, second amplifier, second amplifier, mixer, etc.) are each converted to noise temperatures (T) according to the relationship:

$$T = T_o (N—1)$$

where T_o = room temperature (290 K).

The cable connecting the antenna to the first amplifier will introduce a loss (l_c), thereby making a noise contribution $1/l_c$ to the system. The corresponding noise temperatire for the cable is:

$$T_c = 290. \frac{1 - l_c}{l_c} \text{ (kelvin)}$$

Suffixes c, 1, 2, and m refer to the cable,the first and second amplifier stages and the mixer. The noise temperature of each element is referred to the antenna terminals, and so the system noise temperature is:

$$T_s = T_a + T_c + \frac{1}{l_c} . t_1 + \frac{1}{l_c} . \frac{1}{G_1} . T_2 + \frac{1}{l_c} . \frac{1}{G_1} . \frac{1}{G_2} . T_m$$

where T_a = antenna noise temperature (clear sky).

In a ground station receiving system the noise temperature T_B of the low noise block (lnb) is usually quoted, and so the relationship may be rewritten:

$$T_s = T_a + T_c + \frac{1}{l_c} T_B$$

Noise power P_n in a system noise bandwidth B_n is given by:

$$P_n = K T_s B_n \text{ (W)}$$

where K is Boltzmann's constant ($1.38 \times 10—^{23}$ $WK^{-1}Hz^{-1}$). 10log k = −228.6 dB($WK^{-1}Hz^{-1}$). In terms of dB(W):

$$\begin{aligned} P_n &= 10 \log(k\, T_s\, B_n) \\ &= 10 \log T_s + 10\log B_n + 10\log k \\ &= 10 \log T_s + 10\log B_n - 228.6 \end{aligned}$$

23.8.2.1 Carrier/noise ratio

Since the carrier/noise power ratio is P_s/P_n it, may be described in terms of dB (W) as:

$$P_s - P_n = P_e + G_r - 10\log T_s + 228.6 - 10\log B_n - L_s - (L_a + L_p) - L_d \text{ (dB)}$$

This relationship, usually written as *c/n ratio*, may be used directly if each individual term is known or can be estimated. Very often a figure of merit G/T (*gain/noise temperature ratio*), equivalent to G_r (dB) − 10 log T_s (dB(K^{-1})), is quoted for a receiving system. This allows the previous equation to be rewritten:

$$(c/n)_{dB} = P_e + (G/T_s) \text{ dB}^{k-1} + 228.6 - 10\log B_n - L_s - L_a - L_p - L_d$$

Using this relationship, it is instructive to compare the values of c/n for full and half transponders with 36MHz and 20MHz bandwidths respectively.

For the half transponder, $10\log B_n$ = 73.0 dB($Hz-^{1}$), whereas for the full transponder $10\log B_n$ = 75.6 dB ($Hz-^{1}$). Thus operating with a half transponder improves the c/n ratio by about 2.6 dB. The same level of improvement will be achieved if the signal from a full transponder is restricted to a 20 MHz pass-band within the receiver. In this case, however, other picture impairement will be introduced.

It was shown earlier that space loss (L_s) is 8.8 dB greater at 11 GHz than at 4 GHz. This allows the c/n figures for half transponders at 4 GHz and 11 GHz (where the slant range is 39 000 km) to be compared:

$$\begin{aligned} (c/n)_{4GHz} &= P_e + G/T_s - 40.7 - L_a - L_d \text{ (dB)} \\ (c/n)_{11GHz} &= P_e + G/T_s - 49.5 - L_a - L_d - L_p \text{ (dB)} \end{aligned}$$

23.8.2.2 Signal/noise ratio

The signal/noise performance of satellite link is of major interest to the user. Over a range of c/n values, s/n increases in line with c/n reaching a maximum limiting value determined by intermodulation noise. There is a critical *threshold* value of c/n below which s/n reduces much more rapidly than c/n. For a conventional design of fm demodulator, threshold c/n is around 10 — 12 dB.

Pictures received at the fm threshold exhibit impairment, mainly black and white noise spikes (of severalmicroseconds duration) often referred to as *sparklies*. Below threshold, picture quality is unsuitable for programme purposes. To be clear of threshold noise, a satellite service has a target c/n ratio of at least 14 dB. The following expression applies in the linear region above threshold:

$$s/n = (c/n).\, r.\, \frac{f_d}{f_v}\, 2.\, \frac{3B}{f_v}$$

which in terms of dB is:

$$s/n = c/n + 20\log r + 10\log 3 + 20\log \frac{f_d}{f_v} + 10\log \frac{B}{f_v}$$

where f_d = peak-to-peak deviation

f_v = maximum baseband video frequency
B = radio frequency bandwidth
r = $(f_d/f_l) - 1$
f_l = peak-to-peak deviation corresponding to nominal luminance signal

The effect of pre-emphasis of the baseband video signal (and the consequent de-emphasis) is to improve the unweighted luminance s/n ratio of a PAL signal by about 2 dB.

The Intelsat satellite system operations guide gives the following values for s/n ratio (unified weighted) which apply on one hop between permanent ground stations:

	525/60	625/50
Full transponder	53.3 dB	50.1 dB
Half transponder	48.7 dB	47.1 dB

The unified weighting factor for PAL system I is 11 dB. A signal/weighted noise ratio of 47 dB corresponds to a picture grade of about 4.5 on the CCIR five point which is equivalent to a picture quality between good and excellent.

The previous equation, in the case of a 625/50 PAL signal, provides the following results:

	$(s/n)_{uw}$ (dB)	$(s/n)_w$ (dB)
Full transponder	c/n + 19.8	c/n + 28.8
Half transponder	c/n + 17.2	c/n + 26.2

At the lowest acceptable value of c/n (=14 dB) which might apply at difficult outside broadcast locations with a transportable ground station, the signal/weighted noise ratios become 42.8 dB (full transponder) and 40.2 dB (half transponder) corresponding to picture grades of 4 (good) and 3.8.

23.8.2.3 Power budgets

As the following example shows, a power budget calculation allows the user to determine whether a proposed link will produce a signal of broadcast quality.

The following figures are representative of a 14/11 GHz single hop link from a private uplink at an outside broadcast event to a major receiving ground station. The slant range for each ground station to the satellite is taken as 39 000 km.

The following values are used: uplink frequency 14 GHz, power 400 W, antenna diameter 1.8 m, aperture efficiency 65 per cent.

Satellite G-over-T (G/T) = 3 dB (K^{-1}); transponder gain = 111.1 dB; downlink frequency = 11 GHz; power = 10W; receiving ground station G/T = 30 dB(K^{-1}).

			Uplink		Downlink	
(a)	*Calculation of received power*					
P_t	TX power	dB(W)	26.0		10.0	
L_t	TX system loss	dB	−0.7		−1.0	
fiG$_t$	TX antenna gain	dB	45.4		35.0	
P_e	TX eirp	dB(W)		70.7		44.0
L_a	Atm. loss (up)	dB		−0.7		n/a
L_s	Free space loss	dB		−207.2		−205.1
L_a	Atm. loss (down)	dB		n/a		−0.5
G_r	RX antenna gain	dB		37.1		56.6
L_r	RX system loss	dB		−1.0		−1.0
P_r	Received power	dB(W)		−101.1		−106.0
(b)	*Calcualtion of receiver noise power*					
T_s	System noise temp	K	2000		460	
T_s	System noise temp	dB(K)		33.0		26.6
B_n	RX noise bandwidth	MHz	20		20	
B_n	RX noise bandwidth	dB(Hz)		73.0		73.0
k	Boltzmann's const.	dB($Wk^{-1}Hz^{-1}$)		−228.6		−228.6
P_n	RX noise power	dB(W)		−122.6		−129.0
(c)	*Calculation of received carrier/noise ratio*					
	$10\log(c/n) = P_r - P_n$	dB		21.5		23.0

The downlink calculation assumed a noise-free signal within the satellite. In practice, noise contributions from the uplink, the satellite receiver and transponder intermodulation products must be allowed for. The true down-link carrier/noise ratio is related to these contributions thus:

$$(c/n)_d^{-1} = (c/n)_u^{-1} + (c/n)_i^{-1} + (c/n)_r^{-1}$$

where d,u,i,r refer respectively to the downlink, uplink, intermodulation and the receiving ground station.

Note that this relationship is between natural numbers (i.e. not in decibels). Using figures from the example above:

$$(c/n)_{d-1} = (141.3)_{-1} + (c/n)_{i-1} + (199.5)_{-1}$$
$$(c/n)_{d-1} = 0.012 + (c/n)_{i-1}$$

If intermodulation noise could be ignored:

$(c/n)_{d-1}$ = 82.7 times = 19.2 dB
If $(c/n)_i$ = 30 dB (=1000 times):
$(c/n)_d$ = 76.4 times = 18.8 dB

From these calculations we can see that the influence of noise in the uplink has worsened the true downlink carrier/noise figure by about 3.8 dB and that intermodulation noise has worsened it by about 0.4 dB. If the uplink carrier/noise is increased by 4.6 dB, the true downlink carrier/noise will improve by 2.0 dB.

Returning to the figures in the example, we can see that the uplink margin above carrier/noise threshold is about 11 dB and the correseponding downlink figure is about 8 dB (=18.8 − 11 dB). These show that an acceptable signal will be received even if heavy rain on either the uplink or the downlink should cause a significant deterioration in reception conditions.

A further stage of calculation is needed to determine whether the looped-back video signal at the OB ground station will be of adequate quality for setting-up and monitoring purposes. At 11 GHz the receive antenna gain will be 43.3 dB, and repeating the power budget calculation shows that its received power P_r −119.3 dB(W). A modern low noise block will have a noise temperature T_B of, perhaps, 170 K (corresponding to a noise factor of 2 dB). An assumed antenna coupling/cable loss of 0.5 dB (a power loss of 12.2 per cent) degades T_B to 191.0 K. Adding a clear sky noise temperature of 79 K produces a system noise temperature of 270 K (i.e. 24.3 dB). The received noise power P_n is therefore 24.3 + 73.0 − 228.6 = −131.3 dB(W) which provides a c/n figure of −119.3 −(−131.3) = 12.0 dB.

Allowing for $(c/n)_i$ = 30 dB produces a true downlink carrier/noise figure of 11.5 dB (a worsening of only 0.5 dB when compared with the value just derived). This would be just acceptable for measurement purposes, but a 625/50 PAL signal would exhibit some sparklies since it is very close to the threshold value. It would be very desirable, in this case, to employ a receiver with threshold extension to provide a margin of, say, 5 dB to allow for bad weather conditions.

23.9 Future developments

Reliable higher power transponders are being developed. Utilizing more linear amplifiers of greater efficiency, additional backoff associated with half transponder usage will be reduced. Higher power will allow a further fall in the size of ground stations. The application of uplink power control (upc) will become more usual. With upc, the ground station power is adjusted automatically to compensate for temporary signal fades caused by precipitation losses. Additional on-board processing will allow greater transponder flexibility. Switchable receiver attenuators will accommodate a range of ground station powers. New designs will permit the power output to be backed-off when uplink propagation conditions allow. Under ground control, transponder-to-antenna connections will be reconfigured as required to tailor footprints for special requirements. Satellites will increasingly be equipped with adequate back-up power to maintain the operation and performance of all transponders during periods of eclipse.

Services will become available using frequencies in the region of 30/20 GHz which will allow even smaller ground stations and narrower beams under certain conditions. Satellite spacing in the Clarke orbit can be reduced at these frequencies allowing more satellites to be accommodated. The expected life of communications satellites will increase as new methods of station keeping, such as electrically powered ion thruster jets, are introduced.

Satellite-to-satellite links will be introduced allowing greater distances to be linked with one space segment.

The major ground station development will be lnas designed with new simple devices and having noise temperatures less than 70 K (noise figure less than 0.94 dB).

R S Roberts C Eng, FIEE, Sen MIEEE
Consultant Electronics Engineer

24 Coaxial Cable and Optical Fibres

24.1 Cable transmission

The use of cable for distribution of broadcast signals is not new. Provision of a central reception point for radio sound programmes and a distribution network dates from 1924 in the UK. Modest networks distributed television programmes before 1939. More than 13 percent of television viewers in the UK receive their programmes by means of cable, with much higher percentages in other parts of Europe, Canada and the USA. Existing UK networks must conform to various technical requirements, and must be licensed. Various BSI Specifications and other documents exist (see Bibliography).

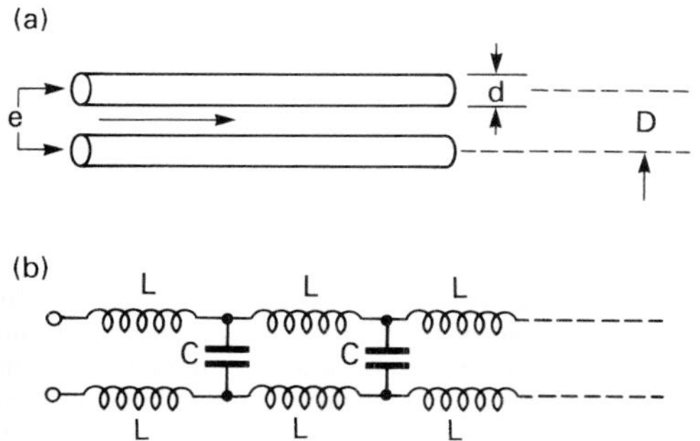

Figure 24.1 Balanced transmission line (a) and its equivalent circuit (b). Its characteristic impedance is double that of the coaxial line in *Figure 24.2*

Cable distribution systems must conform to the broad principles governing the transmission of energy from a source to a load. A cable distribution system involves three main parts:

- the *head end* assembles and combines the signals to be distributed, and feeds them to
- the *network*, from which they are tapped off to final cables to supply
- the *subscriber outlets* in individual premises

Network requirements are that:

- losses must be as low as possible
- no radiation should take place that could cause interference to other services remote from the network
- the entire system must be screened from possible interference by the fields radiated by other services

24.1.1 The transmission line

The pair of conductors shown in *Figure 24.1*(a) constitute a transmission line, connected to an rf source e, the energy progressing along the line as shown towards the termination at the end of the line.

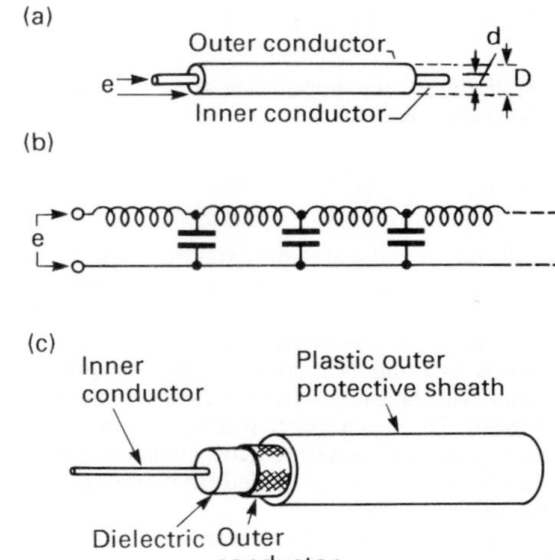

Figure 24.2 The two-conductor transmission line in coaxial form (a), its electrical equivalent (b) and its physical construction (c)

There is a capacitance C between the two conductors and, if the line is termination by a short-circuit, an inductive loop L is formed. If the line length is doubled, the capacitance is doubled and the inductance of the loop is doubled, but the ratio L/C remains constant. The line may be considered to consist of a succession of unit lengths, each with capacitance and inductance as shown in *Figure 24.1*(b). The line is *balanced*, i.e. neither input terminal is at earth potential.

The currents flowing in the two conductors will be in opposite directions, and their magnetic fields will therefore tend to cancel. Radiation of any energy from the line would be a loss of energy from the system, and the field-cancellation effect helps to minimize losses from the line.

The most widely used form of the transmission line is the *coaxial* construction shown in *Figure 24.2*. The two-conductor line now takes the form of an outer conductor enclosing an inner, axial conductor.

Balanced lines are not used in cable distribution networks, and we will be concerned only with the unbalanced coaxial line of *Figure 24.2*.

The *unbalanced* line has an immense practical advantage over the balanced line in that the outer conductor can be at earth potential, thus fully screening the inner conductor. This feature prevents radiation from the line, and prevents interfering signals from being received by the line inner conductor.

24.1.2 Impedance

If the conductor resistance is low enough to ignore, and the shunt resistance provided by the insulation is high enough to ignore, the impedance 'seen' by the source in *Figure 24.2*(a) is given by:

$$Z_o = \sqrt{(L/C)} \text{ ohms} \quad (24.1)$$

where L is the inductance per unit length, and Z_o is the *characteristic impedance* of the line.

The dimensions and spacing of the conductors, together with the nature of the insulation between them, will determine the values of the inductance and capacitance and so the value of Z_o. The impedance can be expressed in terms of the dimensions of the outer conductor D and the inner conductor d (*Figure 24.2*). The permeability μ and permittivity ϵ of the dielectric must be taken into account, and equation 24.1 becomes:

$$Z_0 = 138 \log_{10} \frac{D}{d} \sqrt{\left(\frac{\mu}{\varepsilon}\right)} \text{ ohms} \quad (24.2)$$

For completeness, the impedance for the balanced line of *Figure 24.1*(a) is given by:

$$Z_0 = 276 \log_{10} \frac{D}{d} \sqrt{\frac{\mu}{i}} \text{ ohms} \quad (24.2)$$

For air spacing, μ and ϵ are approximately equal to 1, and this part of equation 24.2 can be ignored. For other insulating materials, such as are used in practical cables, μ will probably be near 1, but ϵ will have higher values.

Equations 24.1 and 24.2 do not include frequency; a transmission line will transmit all frequencies from infinitely high to direct current. The signal source will 'see' an impedance Z_o, but as this impedance consists of reactive elements, the impedance will absorb no power, provided that the resistance and dielectric losses are zero.

The characteristic impedance can have any value, determined by the ratio D/d, but in practice, manufacturing limitations exist. The clearance between the inner and outer conductors cannot be too small, and the inner conductor diameter cannot be reduced without increasing resistance losses. These considerations limit the available range of impedance for coaxial lines to about 20–200 ohms. (For a balanced line the range is about 150–800 ohms.)

The value of cable impedance used for distribution networks has been standardized in the UK at 75 ohms, but 50 ohms has been used elsewhere in Europe. With air spacing, a ratio D/d of 3.6 gives an impedance of 75 ohms, but the practical cable using an insulator for spacing will have a different ratio to provide a Z_o of 75 ohms, as defined by equation 24.2.

24.1.3 Cable losses

Any cable will have some conductor resistance and insulator loss. This loss would be a minimum if the dielectric between the conductors were air and the conducting surfaces had a large area. Unfortunately, some insulation is required between the inner and outer conductors to ensure that the inner is accurately centred in the outer, and a large surface area for a given impedance requires a large overall diameter for the cable.

Power loss from a balanced cable has been mentioned, where energy radiated represents a 'radiation resistance' loss that would add to the total losses by the cable. Some coaxial cables use a woven braid for the outer conductor, and some energy leakage can take place through the meshes of the weave. This may not be serious if the power levels are low but, where the power levels on the cable are large, some regulations require a double-woven cable or a solid sheath to be used.

Cable losses are usually expressed in decibels per unit length, and they increase as frequency is raised in a complicated manner, mainly due to the increase in *skin resistance* of the conductors. The increase in decibel loss is approximately proportional to $\sqrt{f}$. Thus, for example, a length of cable that has a loss of, say, 2 dB at 50 MHz, will have a loss of the order of 4 dB at 200 MHz and 8 dB at 800 MHz.

24.1.4 Matching and termination

For maximum transfer of power from a source e into a load R_L consider the example shown in *Figure 24.3*. Assuming a value of 75 ohms for R_s and 10 V for e, the resulting circuit voltages and currents are shown in *Figure 24.4*.

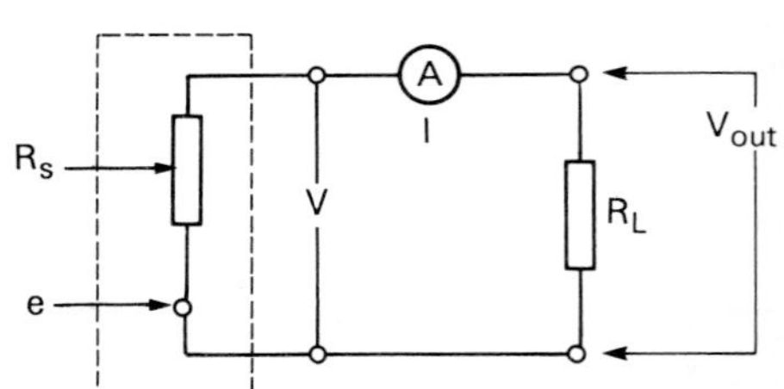

Figure 24.3 A power source connected to a load

It is seen that, for maximum transfer of power from the source to the load, the load impedance must equal in value the internal impedance of the source. This is termed a *matched* condition and, in a simple case where a source feeds a line and the far end of the line feeds the input terminals of a receiver, is a two-fold process: the source feeds its power into the line, and the line feeds the receiver at its input. The line must match the source, and the receiver input impedance must match the line. For the usual UK network, the source output impedance will be 75 ohms, the network impedance will be 75 ohms and the receiver input impedance will be 75 ohms.

Figure 24.4 shows that the matched condition is not sharply defined, and it extends over a broad band. In fact, if the load is half or twice the correct value required for matching, the power in the load changes by only 0.5 dB, but mis-matching produces other effects, and must be avoided as far as possible.

A line with a source of power e at one end, and a terminating load Z at the other, is depicted in *Figure 24.5*. Consider a transmission of rf energy advancing along the line from e to Z. If the line is terminated with an open circuit (an infinitely high impedance) the energy cannot be dissipated when it reaches the end of the line, and it is returned (reflected) back along the line

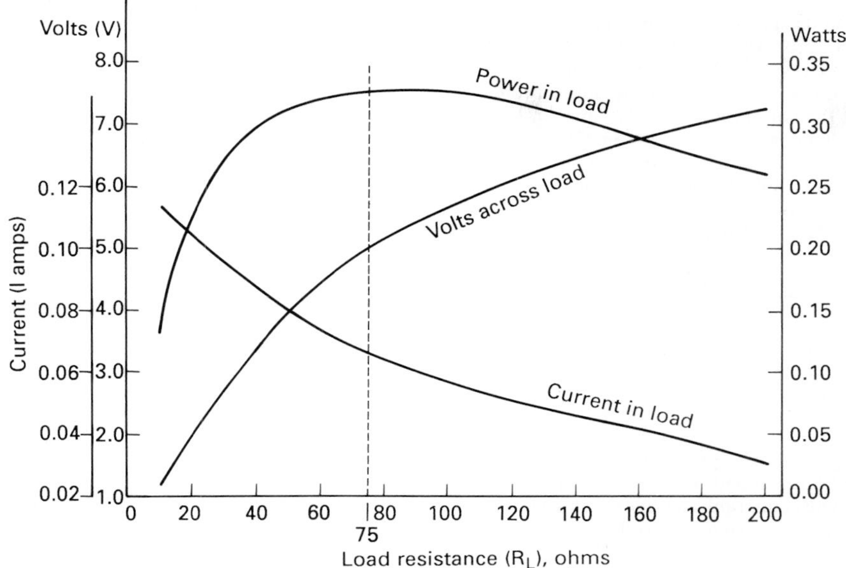

Figure 24.4 The current, voltage and power relationships in *Figure 24.3*, as the load is varied

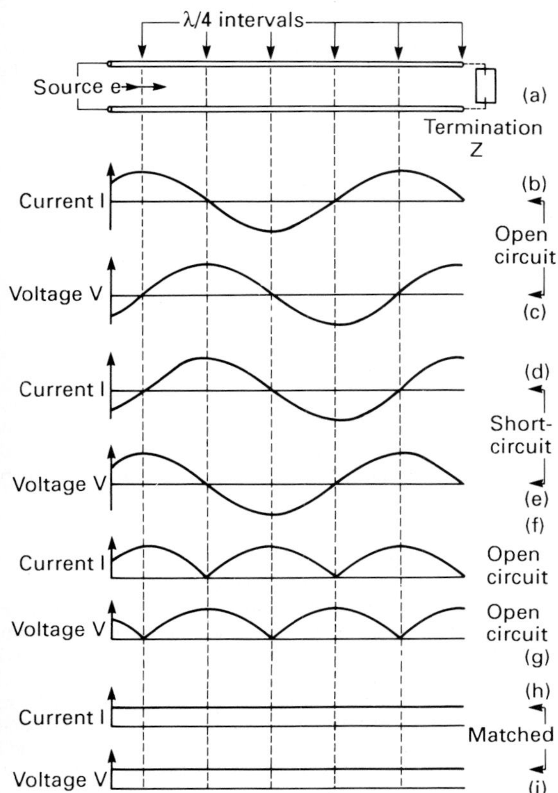

Figure 24.5 A long line fed from an rf source showing current and voltage waveforms along it for various termination conditions (see text)

towards the source. No current can flow in the termination (b), and the voltage can be a maximum (c). The energy advancing along the line towards the termination will now experience interference with that reflected from the termination, this effect producing positions along the line where additive and subtractive values of current or voltage give rise to 'stationary' patterns or *standing waves*.

In *Figure 24.5*, (d) and (e) show similar standing wave effects for a termination consisting of a short-circuit. In this case, no voltage can exist at the termination, but current can have a maximum value. All four standing wave patterns show that the line now has a distribution of voltage and current along its length that is not uniform.

A voltmeter connected across the line between the two conductors will indicate maximum, minimum or intermediate values, depending on its position along the line. The currents and voltages in adjacent half wavelength sections of the standing wave will be in phase opposition.

If we now examine the open circuit line for amplitude only (e.g. by sliding a voltmeter along it), we obtain the variations shown in (f) and (g). The connection of a load to the line would have to be at a point where the voltage and current have suitable values, and the system would then function on one frequency only. This is not very practical, but the example serves to show that standing waves must be eliminated as far as possible.

This can be done in two ways. One is a theoretical concept only, and simply recognizes that a line of infinite length would absorb all the energy that is sent along the line, and would not reflect any back to the source. The second, practical method is to note that a termination with a finite value of impedance somewhere between zero and infinity can simulate a line of infinite length and absorb all the energy it receives without any reflection back to the source. Such a termination would have the same impedance value as the characteristic impedance Z_o of the cable, and would equal $\sqrt{(L/C)}$ ohms.

The resulting current and voltage along the line would then be as shown in (h) and (i), except that in the practical case, where line losses will exist, there would be a downward slope of the current and voltage values as the source energy moves away along the line.

24.1.5 The cable network

The head end is shown in its simplest form in *Figure 24.6*. A number of sources, S_1, S_2, etc., are assembled in a combining unit and, although amplifiers are shown for each source, the object is to process the source signals so that the output from the combining unit is at the correct level for the input to the main trunk amplifier. This may entail attenuating some sources, or changing their frequency to position them in a more favourable place in the band available for distribution.

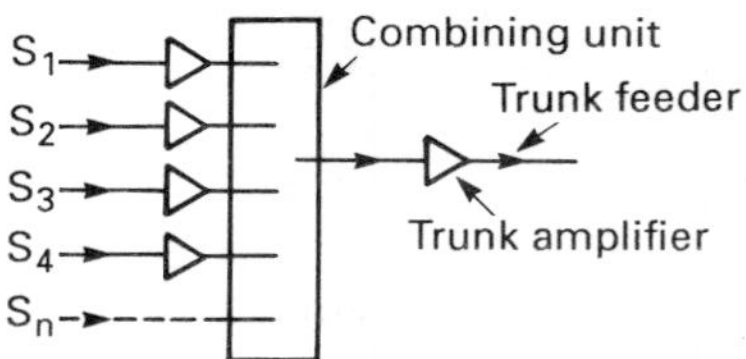

Figure 24.6 Head and essentials

The head end is usually sited in a position where antenna systems for off-air reception of broadcast signals can provide the best possible signals. The main trunk amplifier is a high level amplifier that feeds the combined signals into the main trunk cable that links the head end to the main centre for distribution.

Figure 24.7 shows the principles used in the distribution network, which subdivides into spur, distribution and subscriber feeders to the individual outlets. Each network is unique and there are many variants of the basic scheme. For example, the head end may be sited near the centre of an area to be served, and several trunks may be used radially to cover the area.

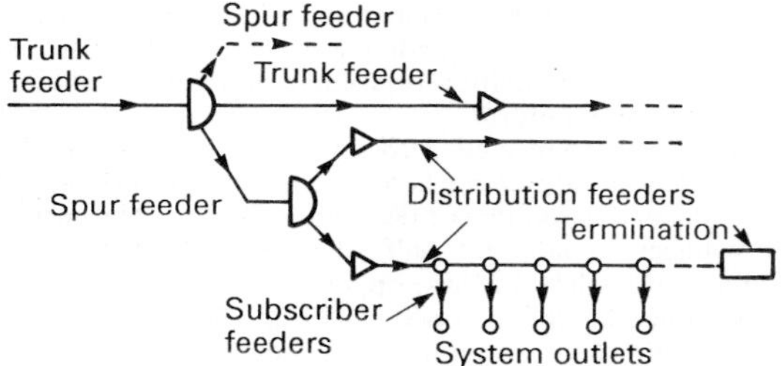

Figure 24.7 The network supplied by a high level, main trunk feeder

Another feature concerns the many amplifiers on the network and their power supply. Locally derived mains power may be difficult to provide at some remote amplifiers and, in these cases, it is possible to feed power along the network cable as shown in *Figure 24.8*.

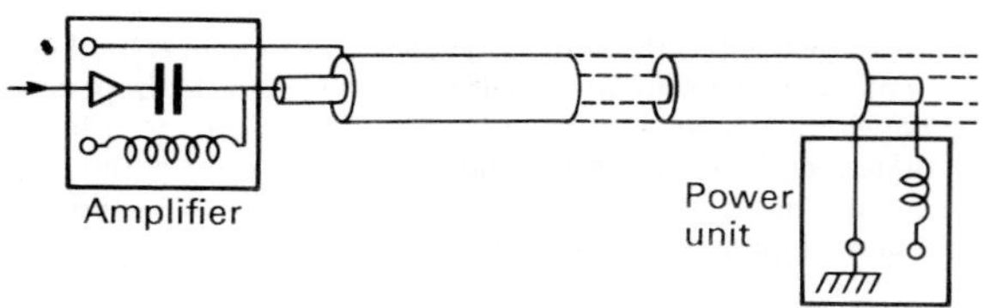

Figure 24.8 A method of feeding a power supply along a signal feeder

Regulations require that the subscriber outlet sockets must be isolated from the network by capacitors or transformers so that if a fault develops in a receiver of a type that renders the receiver input socket live to high voltages or mains supply voltages, the network itself does not become dangerously live.

A further isolation of about 20 dB is required between any outlet socket and the network, as explained in section *24.1.6*.

We have seen in section *24.1.3* that cable losses vary as $\sqrt{f}$, and thus a wide-band source of signals will experience a greater attenuation at the hf end of the band than at the lf end (see *Figure 24.9*). Equalizing units can be inserted into the network at any point, but it is convenient to equalize at the amplifiers which are being used to restore the attenuated signals to their original launch value.

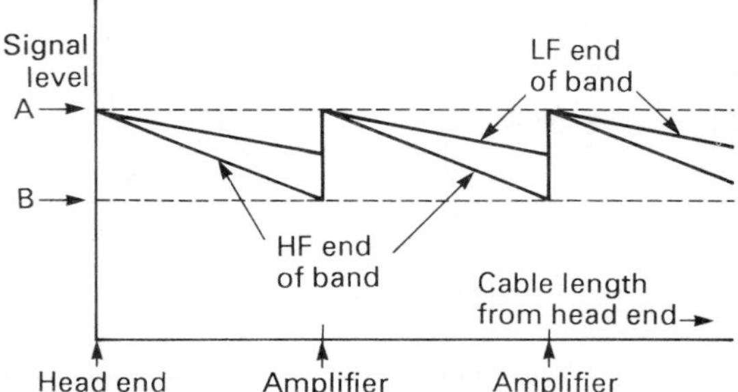

Figure 24.9 Cable attenuation and network amplifiers

Figures 24.6 and *24.7* show combining and splitting units and T connections between subscriber feeders and distribution feeders. All of these must have input and output impedances that match the system impedance, and must not introduce any undue loss. A principle used for splitting from one cable into two is shown in *Figure 24.10* where the system's 75 ohm impedance is maintained on each line.

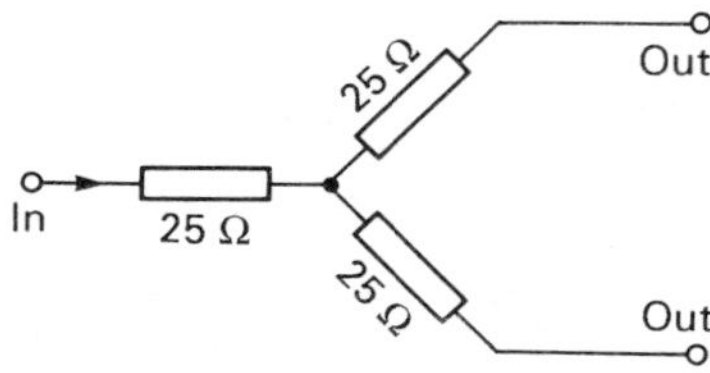

Figure 24.10 Basic principle for dividing a cable signal source to feed two cables while maintaining matching

A splitter can operate with its function reversed as a combiner. However, such a system that uses resistors would absorb some power and, in practice, a transformer would be used to effect the same result. A directional coupler can also be used for combining or splitting.

24.1.6 Cable amplifiers and signal levels

The most significant aspect of cable operation is the amplifiers that are used on the network. *Figure 24.9* shows how signals are launched at a level A. As they progress along the cable, some energy is lost due to cable attenuation, and a minimum level B is reached. At this point an amplifier is used to raise the level back to A. The associated circuitry includes that necessary to rebalance the levels of high and low frequency signals in the band.

Four basic factors that determine the values of levels A and B are:

- Any outlet socket used for the connection of a standard television receiver to the system must supply any of the distributed television channels at a minimum level of 1–2 mV, if the signal/noise ratio of the receiver is to be acceptably high. The receiver input voltage must not exceed 5–10 mV if receiver overload effects are to be avoided. Any broadcast fm sound channels being distributed must provide signals at a minimum of 0.5 mV and a maximum of 5 mV.

- Cable standards require that isolation of about 20 dB must be provided between any outlet socket and the network. This ensures that, if a fault develops in a receiver of a type that might feed into the network and affect other receivers (e.g. instability or oscillation), a minimum isolation of about 40 dB exists between any two receivers.
- Many cables use a woven conducting braid for the outer conductor, and so the screening is not complete. A double-woven braid is often used to increase the screening efficiency, but some radiation takes place from any braided cable. It is important that signal levels on the network are not so high that radiation constitutes an interfering signal to other services. A cable with a solid tubular outer conductor is the most satisfactory and is widely used for trunk feeders where the signal levels are at their maximum. The screening of the cable from interfering fields from other services is particularly important because, if an interfering signal finds its way into one of the amplifiers on the system, it may not be eliminated by any subsequent filtering.
- A large network requires some form of *automatic gain control* (agc). Cable attenuation is temperature sensitive. Copper has a positive temperature coefficient of resistance, and the network attenuation due to resistance will increase during hot weather, becoming lower in a colder ambient temperature. The agc control signal is usually derived from one or more pilot signals, distributed from the head end along the signal channels. The control is applied to a few selected amplifiers on the network.

If a required broadcast signal level of 2 mV is required at the remote outlet on a distribution feeder, a level of 20 mV is required on the distribution feeder at this point. If the feeder loss is, say, 6 dB, the level at the distribution amplifier output must be 40 mV.

The input to the distribution amplifier will be attenuated by splitting. The input power to the splitter will divide between the two output paths, each of which will be −3 dB on the input level. The output from the trunk splitter will need to be 80 mV and the level on the trunk feeder in the region of 100 mV. The head end launch level needs to be higher than 100 mV to allow for the trunk cable attenuation and to provide an ample design margin.

The maximum launch level A in *Figure 24.9* is determined by:

- signal/noise ratio
- signal level
- amplifier distortion effects

The s/n ratio will be determined by the network noise which will be amplified along with the signals, and the noise generated by the amplifiers.

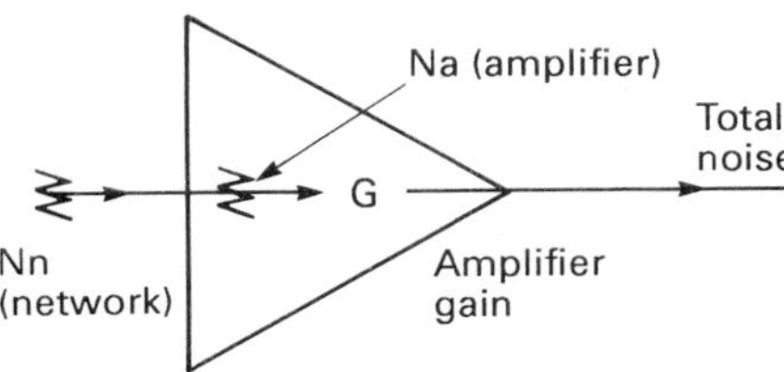

Figure 24.11 Network and amplifier noise

Any resistor generates noise which is given by:

$$E = \sqrt{(4kTBR)}$$

where k is Boltzmann's constant, T the temperature (kelvin), B the bandwidth and R the resistance.

The network resistance is 75 ohms and, for a bandwidth of 5.5 MHz, E becomes about 2.5 μV which, matched into 75 ohms, becomes a 1.25 μV noise level at an amplifier input.

Figure 24.11 shows the two sources of noise, and their relationship to an amplifier. The contribution by the amplifier gives it a *noise factor* (nf) given by:

$$\text{nf} = \frac{G(N_n + N_a)}{GN_n} = 1 + \frac{N_a}{N_N} \qquad (24.4)$$

With no signals input to the system, the output at any outlet socket will consist of noise. With a signal input to the system, the signal level at the output sockets needs to be high enough to provide an acceptable s/n ratio. A minimum value of this ratio is usually taken as about 45 dB.

There are several constraints that prevent the s/n ratio being as high as we would choose to make it. One has been mentioned: signal levels must not be so high that possible radiation from the system becomes a source of serious interference to other services. Interference can be set up at nearby receivers that are not connected to the network, but receive their programmes off-air, using their own antenna system. The most serious design consideration for the network is the degree of non-linearity in the operation of the network amplifiers.

All amplifiers are non-linear in their operation to some extent. A single input to an amplifier will, if the amplifier has a linear input/output transfer characteristic, be reproduced at the output as an amplified and faithful copy of the input waveform. Any non-linearity will result in waveform distortion and consequent harmonic generation.

If two or more signals are supplied to the input of a broadband linear amplifier, the signals will be amplified in a distortionless fashion with no mutual interference effects. However, if non-linear operation takes place in the amplifier, mutual interference effects arise between the signals, and a number of spurious signals are generated.

Consider an input consisting of two signals:

$$E = A \cos \omega_1 t + B \cos\omega_2 t \qquad (24.5)$$

applied to an amplifier having a gain G. The output due to non-linearity may be expressed as:

$$E_{out} = G_1 E_{in} + G_2 E_{in}^2 + G_3 E_{in}^3 \ldots \qquad (24.6)$$

G_1, G_2, G_3 etc. do not have the same meaning as the gain G, and they will have different values from G, the extent of the difference being determined by the levels A and B, and the non-linear law of the amplifying device when it is being driven under these particular conditions.

The output will consist of:

First order	$G_{1in} = G_{1A} \cos \omega_{1t} + G_{1B} \cos \omega_{2t}$
Second order	These will include a dc component, second harmonics of ω_1 and ω_2 and other even harmonics, and difference frequencies with intermodulation between the two signals.
Third order	These will include third and odd harmonics, original frequencies with cross modulation, and difference frequencies with intermodulation.

If the input signal consists of three frequencies (as in a single television channel, with three carrier frequencies), a further set of output signals is generated. To handle two television channels (i.e. six carrier frequencies), the interaction effects between channels become more serious and, of course, even more severe as more channels are added. If an amplifier performance with respect to spurious frequencies and intermodulation is decided for a given number of channels, the addition of further channels will require the amplifier gain to be reduced. The amplifier output will be divided in a random statistical fashion, but a rule-of-thumb suggests that the reduction in gain should be about 3 dB for each doubling of channels.

We have seen that noise is increased as amplifiers are cascaded in a network and, to establish a high signal/noise ratio, the signal input to an amplifier should be as large as possible. To keep spurious frequency levels and effects as low as possible, the signal inputs to an amplifier must be kept as low as possible. These two conflicting requirements for amplifier performance require that a compromise must be accepted for amplifier gain, input signal levels and number of channels that will provide an s/n ratio with a minimum value of about 45 dB at each outlet, for each channel that is being distributed.

24.1.7 Talkback

Modern cable systems use wide-band amplifiers that operate over a frequency range of 40–400 MHz or 40–850 MHz, although the recent removal of broadcasting from Band 1 (40–68 MHz) has tended to make the lowest frequency for a distribution system around 75 MHz. In either case, a wide range of frequencies, from dc up to the lowest distributed frequency, is available for a subscriber to 'talk back' to the head end.

The network attenuation is reduced, and a few low-power amplifiers would be required, e.g. a cable attenuation figure at 40 MHz is halved at 10 MHz and reduced to a quarter at 2.5 MHz. Relatively simple filters will ensure that the forward and backward signal information will not interfere with each other.

24.2 Optical fibre transmission

Development of transmission systems using light waves within optically transparent fibres may be considered to have commenced in the 1960s, when silica glass fibres were produced with an attenuation in the region of 1000 dB/km. A decade later, attenuation had been reduced to about 20 dB/km. By 1989, fibre cables were available off the shelf with attenuation of about 0.5 dB/km, with several manufacturers producing cable with much lower values of attenuation.

Light waves and radio waves differ only in their wavelength. Radio waves may be propagated along confining tubes in waveguides, and be guided in the ionosphere (as in 3–30 MHz band radiation) through refraction and reflection.

Visible light spans a frequency range from about 385–790 THz (see *Figure 12.1*) bordered by invisible bands in the infrared and ultraviolet. It is now possible, using lasers, to generate a light beam of a single, discrete wavelength.

Such a light source can be a *carrier wave* which can be modulated by any of the usual methods. The carrier frequency is so high that the bandwidth resulting from carrier modulation can be very wide and still be only a small fraction of the carrier frequency. As a consequence, the effects shown in *Figure 24.9*, although present to some degree, will show negligible differences between the low and higher frequencies resulting from signal modulation.

Modulation can be by analogue or digital means, but most fibre systems use digital modulation, with all the familiar advantages of digital transmission, such as operation at a constant amplitude and the many methods that can be used for modulation.

24.2.1 The fibre transmission line

A transmission fibre can be of any material, solid, liquid or gas, that conducts light with minimum attenuation. Of all the possible transmission media, glass offers the most advantages. Glass can be drawn into fibres relatively easily, and has a mechanical robustness that is not readily available with many other substances, such as plastics.

Figure 24.12 shows a glass fibre core surrounded by a glass sheath or *cladding*. The refractive index of the core is higher than that of the cladding, and the transmission paths are confined to the core.

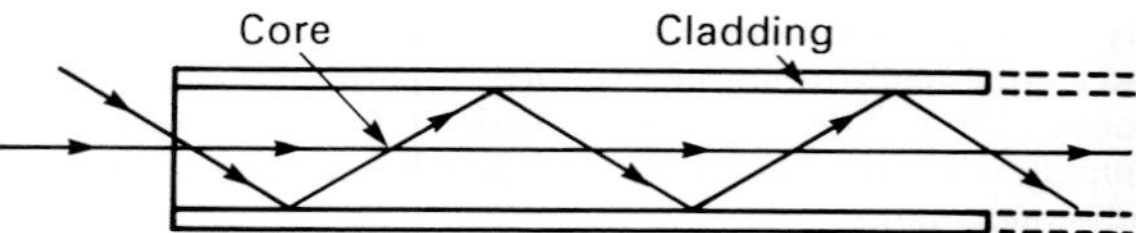

Figure 24.12 Transmission along a multimode fibre cable

A limitation is that light rays entering the core at an angle will take a longer time for transmission along the fibre than those parallel to the axis. Each bit of a digital pulse signal contains many component frequencies, and these must be retained if the bit is to have short rise and fall transmission times. The late arrival of some of the component frequencies will have the effect of 'broadening' the signal bit and thus reducing the possible bit rate (an effect known as *dispersion*).

This transmission is termed *multimode*, and the pulse broadening effect sets a limit to the transmission bit rate that can be used for a given length of cable. An associated effect is *attenuation*, the combined effect being termed *modal dispersion*.

There are two possible solutions to the problem of dispersion. One is to fabricate the fibre so that its refractive index varies from a high value at the centre of the core, where the velocity of propagation is lower, to a low value at the outer diameter where the velocity is comparatively high. The changes in index values are obtained either in steps or in a continuous manner.

The second method is to reduce the core diameter to a value comparable with the wavelength of the light source, typically 3–10 μm. Such a core can support only one mode of propagation, and is termed *monomode*. Two practical difficulties exist with the use of monomode cable. The light source must be a laser, as a beam is needed small enough in diameter to couple into the fibre core with reasonable efficiency. With such a concentration of the light source, the laser power must be kept low to avoid excessive temperature rise of the core at the point of beam injection.

24.2.2 Attenuation and dispersion

Any medium used for light transmission will introduce some attenuation, i.e. the *transparency* can vary. *Figure 24.13* shows how a typical silica fibre may attenuate light sources of different wavelengths. The broken line shows the *Rayleigh scatter* or *Rayleigh dispersion*, but the solid curve is representative of a cable that might be used today. The solid curve shows the combined effects of Rayleigh scatter and other effects.

The Rayleigh curve is the result of imperfections in the glass such as cracks, bubbles and, surprisingly, water ions that result in scatter and absorption with attenuation peaks at 1.25 and 1.39 μm. The minute imperfections have dimensions comparable with the wavelength of the light being transmitted, and the attenuation decreases in proportion to $1/\lambda^4$.

Optical materials also experience *chromatic dispersion*, caused by the fact that the refractive index varies for different wavelengths. A light pulse includes various wavelengths, and no source is perfectly monochromatic, so that the various wavelengths in the light pulse travel along the fibre at different speeds.

At any particular wavelength, dispersion is measured as the amount of delay in picoseconds per kilometre, per nanometre difference in wavelength. At 850 μm, for example, chromatic dispersion in silica fibre is typically about 90 ps/km/nm. This figure falls with increasing wavelength to a minimum at about 1.31 μm.

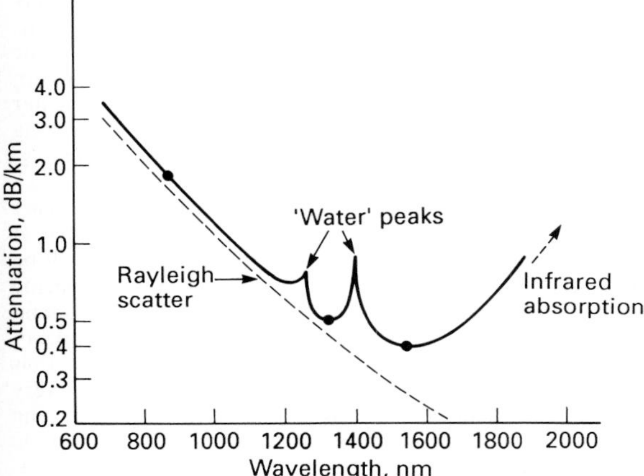

Figure 24.13 Example of the attenuation of a silicon fibre cable

Towards the long wave end of the curve, the falling attenuation is met by a rising infrared absorption loss due to molecular resonance with silicon oxide, and similar causes. The combined falling and rising attenuations as wavelength is increased result in the minimum around 1.55 μm.

24.2.3 Communication systems

A communication system requires a fibre cable with low attenuation at clearly defined wavelengths. It then requires a light source generator that will operate at maximum efficiency at wavelengths of low attenuation. At the end of the cable, a photo detector is required that functions at the same wavelength as the light source, with peak efficiency.

Historically, the early systems had difficulty in equating the need to have light sources and detectors each operating at the same wavelength with maximum efficiency at a wavelength suitable for transmission with minimum loss. 850 μm was the general choice, as this provided the necessary compromise, using available light sources and detectors. More recently, the lower cable attenuations at 1.31 μm and 1.55 μm have become useable through the development of appropriate semiconductor devices. 850 μm is often referred to as *short wave*, with 1.31 and 1.55 μm being termed *long wave*.

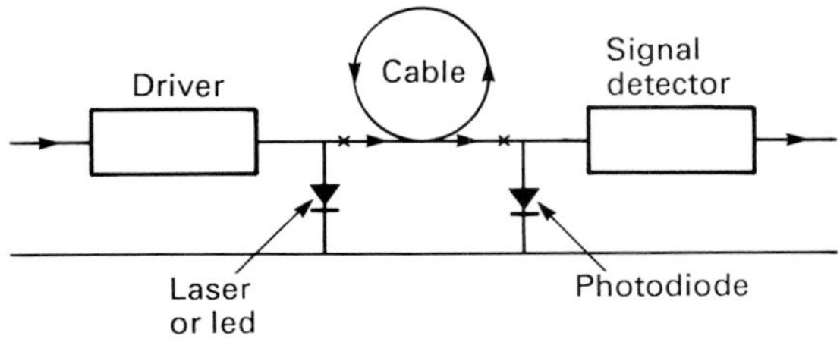

Figure 24.14 A basic fibre system

The essentials of a basic system are shown in *Figure 24.14*. The light source can be an led, where the light output is proportional to diode current, and where linear operation is required, as might be the case for some analogue signals.

A laser can provide more light output, and so provide a better signal/noise ratio, but its operation is non-linear and is a good choice for a digital system. The non-linearity results from its behaving as a diode up to a threshold where laser action starts, with a massive increase in output.

While coupling into a monomode fibre is carried out more efficiently with the small diameter light beam of a laser, the concentration of power that results may mean that the available laser output may need to be reduced to limit the temperature rise of the glass at the point of power injection. In practice, a trade-off has to be made between the low coupling loss due to the narrow beam, and the power loss that might have to be made to limit the temperature. As an example of laser operation, one particular make, operating at 850 μm, has its threshold current at about 50 mA, and can launch more than 1 mW into the fibre.

Detectors were usually reverse-biassed p-n photodiodes, which rely on the production of electron holes by photons. The *quantum efficiency* is the ratio of the number of electron holes to the number of photons. For example, 100 photons creating 95 holes would give the device a quantum efficiency of 95 per cent. The quantum efficiencies of silicon and germanium vary with frequency; as a result, silicon detectors are more efficient at 850 μm, while germanium would be used in long wave systems.

24.2.4 Splicing and connecting

Fibre cables are mechanically quite rugged, but they are sensitive to crushing, longitudinal strain and internal stresses caused by sharp bends. Temperature changes can change their characteristics. For cable laying, the fibre cables are often housed loosely in a semi-flexible tube. This outer tube can take all the tension stresses involved in drawing through ducts, and the tube is tough enough to prevent any attempt to bend it round a small radius.

Splicing posed many early problems but there are now many designs for use in the field. The procedure is to diamond-cut, cleave and polish the cut ends. These are held in a jig that butts the ends, which are then welded with an oxy-hydrogen micro torch. With care, such a joint can achieve a loss lower than 0.1 dB. Clearly, splicing losses must be as low as possible because they add to the general system loss.

There are four possible causes for splicing loss:

- *separation* of the cable ends produces negligible attenuation for separation up to one-tenth of the core diameter, but 1 dB for a separation of half a diameter
- an *angle* at the butt of 2° can produce 0.2 dB loss
- a *shift* of the butt ends up to one-tenth of the diameter introduces 0.3 dB
- if the *diameters* of the butt ends differ by 10 per cent, the loss can be 0.5 dB

Plug-and-socket or other forms of *connectors* pose similar problems to those of splicing, but in a more acute form due to the need for mechanical accuracy that will ensure that the losses are low but, above all, are repeatable. One form uses lenses at the 'plug' and at the 'socket' so that the actual light path is made or broken with very little loss, and is repeatable.

24.2.5 Associated equipment

Power dividers may consist of a totally integrated device of the form shown in *Figure 24.15* which is a photolitho optical waveguide that splits a light beam into two paths.

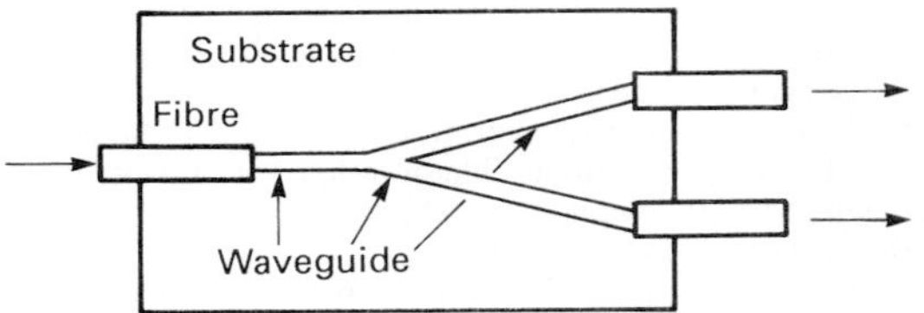

Figure 24.15 A waveguide splitter. The guide sections are photolitho deposited

Any splitter can be used as a combiner, i.e. two beam sources can be combined into a single path by reversing the direction of the paths.

Another type of divider (or combiner) is shown in *Figure 24.16*, in which two fibres have ground faces in contact.

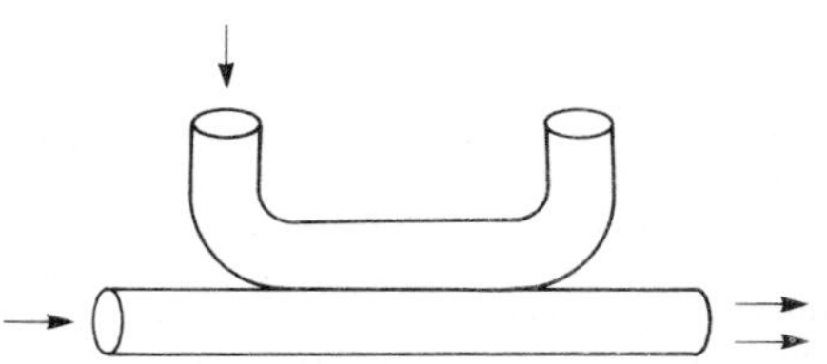

Figure 24.16 A method for combining two signals into a common output, similar to the coupler often used with coaxial lines

An interesting variation of *Figure 24.15* has metallic electrodes adjacent to each of the optical waveguides. The refractive index in each guide can be changed by the application of an electric field across the guides. The device can thus be used as a switch whereby a choice of either of the two output paths may be made.

A further application of this device is to increase the signalling speed. Digital modulation of a laser diode in *on/off* fashion requires a relatively long time and, at about 500 Mbit and above, the laser can no longer respond accurately. The device used as a switch can permit the diode to remain *on* continuously, the bit signals switching the light beam from the transmission path to a sink, with almost no constraints on switching speed.

A more recent development has seen the appearance of light-frequency generators which have a high order of frequency stability, suitable for use as the local oscillator of a superheterodyne receiver. The receiver changes the frequency of the detected signal to an intermediate frequency of about 1–2 GHz for subsequent processing. The usual laser is not suitable for superhet purposes, the bandwidth being too wide and the frequency stability quite inadequate. Viable oscillators have now been produced, and receivers can now provide a very useful gain of 20 dB or so over the use of a simple diode detector.

It has been seen that basic path loss can be very low. It is now possible, with that extra 20 dB of gain, to obtain very long paths (e.g. 100–200 km) without the use of repeater stages, and it is probable that all future fibre systems will use superhet receivers.

24.3 Future developments

It is probable that coaxial copper systems will not receive any spectacular changes in design or technique, but these systems will always find a place in 'short-haul' networks for economic reasons. Repeaters are amplifiers of relatively simple design, compared with those for a fibre optic system where a repeater entails demodulating the signal to baseband, and remodulating a new carrier — a relatively expensive operation. Fortunately, fibre systems will use very few, if any, repeaters.

Fibre systems are capable of very low system loss, but many lines of research are proceeding, each of which is aimed at reducing attenuation to the point where no repeaters will be required. *Figure 24.13* shows the general trend of reducing attenuation as wavelength is increased, with infrared attenuation predominating over Rayleigh scatter as the main source of loss for wavelengths beyond about 1.5 μm. Other materials than silica are being investigated, some of which can have very low losses when operated at wavelengths in excess of 2 μm. The new glasses are metal oxides and halides, and calcogenide glasses, all with very optimistic theoretical figures for attenuation, such as 3×10^{-4} dB/km at wavelengths in the range 2–12 μm.

Improvements in s/n ratios are the object of investigations into the use of higher powered sources, and reductions in the noise generated by lasers.

Wavelength multiplexing, using two or more sources, is being developed, using dichroic filters or gratings to extract a required source from a wide-band carrier of several channels.

Bibliography

BS 415 *Safety requirements for mains-connected apparatus*

BS 5425 *Coaxial cable for wideband distribution systems* (1986)

BS 6330 *Code of practice for reception of sound and television broadcasting* (1983)

BS 6513 *Wideband cabled distribution systems* (1984)

BS 6558 *Optical fibres and cables* (1985)

IEC Pubn 96-1 *Radio-frequency cables* (1986)

IEC Pubn 728 *Cabled distribution systems* (1982)

IEC Pubn 794 *Optical fibre cables* (1984)

A F Reekie AMIEE
Formerly Senior Engineer, European Broadcasting Union

25 Tree-and-branch Wired Networks

The arrangement most commonly used for cable distribution of sound and television programmes is that known as *tree-and-branch*, in which the signals to be distributed are brought together at the head-end and conveyed in frequency multiplex to the individual subscribers by means of a network of cables configured in a hierarchical structure resembling that linking the roots of a tree to the individual leaves. This analogy is continued in the naming of certain parts of the network, so that the cables fed directly from the head-end are *trunks* and these in turn generally feed various *branch* cables. The outlets in the individual subscribers' homes are fed by *subscribers' feeders*, each connected at a junction box (*tap*) on a branch cable. In principle, the full frequency multiplex is delivered at every subscriber's outlet, so that the programme that is may be selected by the operation of the receiver's own tuner.

The evolution of the design of such networks has been determined largely by the development of 'over the air' broadcasting as, notwithstanding the relatively recent introduction of receivers intended to be particularly suitable for connection to cable, all distribution networks have to deliver signals that correspond in all significant aspects to those broadcast by terrestrial transmitters in the region concerned. The aim of good cable network design is to ensure that a signal satisfying the technical quality requirements is delivered to all subscribers at the lowest cost. In practice, considerable skill is needed to balance the various factors that determine performance and cost, while complying with the numerous constraints, both natural and man-made, that apply to each installation.

Because these constraints vary so much from one cable network to another, this section is inevitably limited to a discussion of the general principles, rather than the features that are characteristic of an individual cable network. Although typical values are given where appropriate, it should be appreciated that they do not necessarily apply in all cases.

Eight different colour television systems are used for terrestrial broadcasting in various parts of the world. They are described in CCIR Report 624 and identified by certain letters of the alphabet (see Table *25.1*). Most of the 45 million subscribers to cable networks in North America, where the 525-line system M-NTSC is universal, are supplied through tree-and-branch networks. This configuration is also predominant in Western Europe, but most of the 15 million subscribers there are supplied with system B-PAL signals.

As the situation in North America has been described by Coll and Hancock[1], the following survey refers mainly to the case of the distribution of system B 625-line, vestigial sideband, amplitude modulated, vision signals, the associated data signals and frequency modulated sound signals. However, as the various 625-line systems have much in common, the following description may be assumed to apply to all of them, except where particular systems are specifically mentioned. Direct satellite broadcasting using MAC/packet signals began in 1989, and a discussion of the problems involved in the distribution of these signals by cable is given in section *25.9*.

25.1 Economic factors

The technical characteristics of tree-and-branch cable networks can be usefully discussed only in the context of the economic aspects, because the construction of any cable network represents a substantial investment on which it is necessary to obtain a satisfactory return, and the operating costs must also be covered. Because the investment is largely independent of the proportion of potential subscribers who do in fact become subscribers (the *penetration*), whereas the main source of revenue comes from the initial *connection charge* and the fees paid periodically by the subscribers, it is essential to achieve as high a penetration as possible. In fact, the initial connection charge does not usually form a large part of the revenue, because in many cases it is reduced or waived when cable service first becomes available, as an incentive to potential new subscribers, and it acts mainly as a deterrent against successive disconnections and re-connections (*churn*).

In most cases, installation and operation of a cable network is undertaken in accordance with the terms of a *franchise* or licence issued by either the local authority or a specialized Government agency, or both. An exclusive authorization to install and operate the cables over and under the public streets within a specified area during a limited period is granted in exchange for certain constraints and payments. The terms of such franchises are usually critical for the economic viability of the network.

The main selling point of cable distribution is, of course, the programmes carried. The availability of a suitably attractive

System	*Lines/picture*	*Field/s*	*Channel width,* MHz	*Sound-vision spacing* MHz
M	525	60	6.0	4.5
N	625	50	6.0	4.5
B	625	50	7.0	5.5
G	625	50	8.0	5.5
I	625	50	8.0	6.0
D,K,L	625	50	8.0	6.5

Table 25.1 Characteristics of various television systems

6 MHz channels (system M)				*7 MHz channels (system B)*			
channel ident.	*lower edge* MHz	*vision carrier,* MHz	*upper edge,* MHz	*channel ident.*	*lower edge,* MHz	*vision carrier,* MHz	*upper edge,* MHz
2	54	55.25	60	2	47	48.25	54
3	60	61.25	66	3	54	55.25	61
4	66	67.25	72	4	61	62.25	68
5	76	77.25	82	C	81	82.25	88
6	82	83.25	88	81 S02	111	112.25	118
A	120	121.25	126	82 S03	118	119.25	125
B	126	127.25	132	38 S04	125	126.25	132
C	132	133.25	138	84 S05	132	133.25	139
D	138	139.25	144	85 S06	139	140.25	146
E	144	145.25	150	86 S07	146	147.25	153
F	150	151.25	156	87 S08	153	154.25	160
G	156	157.25	162	88 S09	160	161.25	167
H	162	163.25	168	89 S10	167	168.25	174
I	168	169.25	174	5	174	175.25	181
7	174	175.25	180	6	181	182.25	188
8	180	181.25	186	7	188	189.25	195
9	186	187.25	192	8	195	196.25	202
10	192	192.25	198	9	202	203.25	209
11	198	199.25	204	10	209	210.25	216
12	204	205.25	210	11	216	217.25	223
13	210	211.25	216	12	223	224.25	230
J	216	217.25	222	90 S11	230	231.25	237
K	222	223.25	228	91 S12	237	238.25	244
L	228	229.25	234	92 S13	244	245.25	251
M	234	235.25	240	93 S14	251	252.25	258
N	240	241.25	246	94 S15	258	259.25	265
O	246	247.25	252	95 S16	265	266.25	272
P	252	253.25	258	96 S17	272	273.25	279
Q	258	259.25	264	97 S18	279	280.25	286
R	264	265.25	270	98 S19	286	289.25	293
S	270	271.25	276	99 S20	293	294.25	300
T	276	277.25	282				
U	282	283.25	288				
V	288	289.25	294				
W	294	295.25	300				

Table 25.2 Cable channel identification (systems M and B)

selection of programmes exclusive to cable at the head end is the key to achieving high penetration. The most favourable conditions occur in relatively small countries like Belgium where the cable networks have been able to satisfy subscribers not only by distributing the television programmes broadcast in the neighbouring countries, but also, where necessary, by converting these signals into the local transmission standards, thereby allowing less expensive single-standard receivers to be used. In this situation, an average penetration of more than 80 per cent has been achieved. On the other hand, in countries where a wide range of attractive signals can already be received by means of the use of individual antennas, i.e. where virtually all new cable networks are located, existing television viewers will not be induced to become subscribers unless the cable is perceived as offering better value for their money than individual reception, and in these conditions penetration of more than 30 per cent is considered to be high.

The relatively rapid growth of cable distribution in North America has undoubtedly been due largely to the structure of broadcasting in the United States and Canada, where first the output of the numerous independent terrestrial stations and later the availability of programme services exclusive to cable and supplied by transmissions in the fixed satellite service (FSS) provided cable operators with what they needed to attract subscribers. Although small cable networks have been in operation for many years in areas of Western Europe where over-the-air reception of terrestrial broadcasts is unsatisfactory, the present period of rapid expansion of broad-band cable networks in major cities did not begin until the introduction of trans-national programme services transmitted in the FSS and specifically intended for distribution by cable, starting with *Sky Channel* in April 1982. It should be noted that, although these new programme services have had to operate unprofitably for a long time, and some of the pioneering ones have fallen by the wayside, there is no lack of demand for the available satellite facilities, as the expected growth of the potential audience should in due course produce a corresponding increase in revenue without significant extra costs.

According to surveys of subscribers and potential subscribers to cable networks, most people become subscribers not only to have access to certain programme services but also to obtain them with a high technical quality. Although individual subscribers usually confine their viewing to a small sub-set of the available programme services, they consider it to be advantageous to have a wide range of choice, presumably because they expect that they will then be more likely to be able to find something worth watching at all times. Provided the programme signals can be obtained at a reasonable cost at the head-end, and the network is not already at one of the 'capacity thresholds', the marginal cost of distributing one more programme signal is relatively low. However, once a programme has been included in the selection distributed, it cannot suddenly be removed without protests from subscribers. The number of programme services available for distribution by cable has been increasing rapidly, and it is very desirable to ensure that spare capacity is available to accommodate additional programmes in future.

The capital investment required for the installation of a cable network obviously depends on the characteristics of the network, notably its size, the number of programmes to be carried, and whether the cables are buried underground, attached to the buildings or suspended from poles. The effect of variations in size can be eliminated by considering the amount invested per subscriber's outlet; in 1988 and 1989, figures in the range £125–£400 with an average of £175 were quoted in large western European countries. The corresponding monthly subscriptions (included value added tax) for connection to a network supplying about 15 television programmes ranged from about £5 in Belgium (with penetration of about 80 per cent) to £15 in France (with penetration of about 10 per cent)[2].

A particular economic advantage of the cable network is that it enables all subscribers to benefit from the savings obtainable by performing certain technical operations on the signals of each programme at the head-end, instead of individually at each receiver. The introduction of direct satellite broadcasting (dbs) provides a valuable opportunity for putting this into practice, because potential cable subscribers can avoid the expense of installing individual antennas for dbs, and at least reduce the cost of adapting their existing receivers for dbs, simply by becoming cable subscribers, provided that the cable operator has installed the corresponding equipment at the head-end.

25.2 Standards

It is, of course, essential at all times to comply with all the numerous laws, regulations and standards applicable to the installation and operation of cable distribution networks. We cannot consider in detail the non-technical aspects here, and we can only draw attention to the technical requirements regarding electrical safety, equipment approval, wiring practice, etc., which vary so extensively from one country to another. Details of the regulations that apply only to cable networks can be obtained from the national telecommunications administration and the cable regulatory authority (if a separate body). Internationally accepted current practice and standard methods of measurement are specified in IEC Publication 728-1[3]. As it is obviously impossible to take account of all the different national technical requirements in preparing a general survey, it should be noted that the following description does not necessarily apply in all practical cases.

25.3 Receiver characteristics

Because the objective of every distribution network is the supply of signals corresponding to the requirements of the receivers to be connected at the subscribers' outlets, it is appropriate to consider these characteristics first. Unfortunately, these characteristics are not completely standardized, even though there has been a trend towards uniformity in recent years. So far as the rf input is concerned, the single nominally 75 ohm IEC standard female coaxial connector is now virtually universal on system B-PAL receivers. Modern receivers for system M-NTSC are equipped with a type F female coaxial connector, and those for system L-SECAM have the French standard connector, which is similar to the IEC standard but slightly smaller.

In many cases, the receiver chassis may be connected to one side of the mains electricity supply, and so the rf connector must be sufficiently isolated to avoid any risk of electric shock. It appears to be difficult to achieve this requirement simultaneously with that for constant impedance and protection against leakage.

Where the receiver is isolated from the mains electricity supply, it is likely to be equipped with several other connectors, such as video and audio inputs and outputs. In place of the various proprietary designs which tend to come and go, the internationally standardized 21-pin *Euro A/V* connector (also known as *SCART* and *Peritelevision*, is becoming more common, as it is mandatory in some European countries. In North America, independent luminance and chrominance inputs intended for NTSC signals from the new 'super' video cassette recorders, etc., are provided on some receivers, and some manufacturers envisage incorporating two rf input connectors with an internal *A/B switch* to allow the user to select signals obtained from an individual antenna as well as those supplied by the cable network.

The availability of access to a composite video input in this way may provide a convenient point for the connection of descrambling equipment when this technique is used for controlling access to particular programmes[4].

The tuning range of older receivers usually corresponds to the bands allocated for broadcasting in the area where they were bought, and it may also be difficult to tune them to channels offset from those used for broadcasting, at least in the vhf band. Many modern receivers, on the other hand, can be tuned freely, not only to any frequency in the broadcasting bands but also in much of the vhf range between these bands, and above them up to 300 MHz.

Unfortunately, older designs of receiver tend to be relatively intolerant of the simultaneous presence of numerous rf signals, particularly in adjacent channels to the wanted signal. Another receiver characteristic that must be taken into account is the possible presence of leakage from the local oscillator at the rf connector, at a frequency which in system B is generally 38.9 MHz above the vision carrier frequency of the signal to which the receiver is tuned.

25.4 Characteristics of signals at subscriber's outlets

The essential task of the cable network is to deliver signals at the subscriber's outlet that, as far as possible, enable the subscriber's receiver to deliver the optimum picture quality. The main aspects involved are:

- relative and absolute signal amplitudes
- freedom from interfering signals of various types
- freedom from echoes
- channelling plan

Values for all except the last of these are specified in IEC Publication 728-1. Current practice in devising channelling plans is described in section *25.5*. It is obviously important to ensure that the signal throughout each channel used for distribution has a substantially flat amplitude/frequency response, notwithstanding the 'tilt' inevitably introduced by the cable's attenuation characteristics and compensated by those of the amplifiers used. What may be less obvious is the need to keep the absolute signal amplitudes within tight limits, despite the fact that most television receivers have effective automatic gain control (agc) circuits.

The maximum and minimum values for carriers specified in IEC Publication 728-1 (83 and 57 dBμV respectively for television channels in the range 30–300 MHz, and 80 and 47 dBμV respectively for stereophonic radio signals in Band II) are not intended to represent a safe operational range, but extremes which must never be exceeded. The tuners of many receivers are very susceptible to overload in the presence of numerous signals, and an attenuator of about 10 dB often provides an effective remedy if the picture quality on a receiver is disappointing when it is fed from a broad-band cable for the first time.

Because the attenuation of the cable varies with changes in ambient temperature, it is customary to ensure that the overall gain of the distribution network remains constant by the use of amplifiers equipped with agc circuits which operate by reference to one or more unmodulated pilot signals.

When a cable network is introduced in an area for the first time, some subscribers may have receivers with characteristics that are only just adequate for off-air reception. In the past, when a cable network started operation it often delivered signals in alternate channels in the vhf broadcasting band only. However, this corresponds to the distribution of a maximum of six programme services, which would now be regarded as insufficiently attractive in the absence of special circumstances, such as standards conversion permitting single standard receivers to be used.

In order to distribute additional programme services, while allowing the older receivers to remain in use, the obvious solution was to deliver them as signals in the uhf band, which also solved the problem in areas such as the United Kingdom where the vhf band is no longer used for broadcasting. Because the relatively high attenuation at uhf makes it impractical to distribute uhf signals as such throughout the network, transposition must be performed from the distribution frequency to uhf near to each subscriber.

Each of the three possible arrangements for transposition has its own advantages and disadvantages. It should be noted that more than one arrangement may be used in a single network if the signals at the distribution frequency are made available at each subscriber's premises.

25.4.1 Transposition arrangements

The transposition to uhf can be performed either at the end of the distribution network or in each subscriber's premises. In the latter case, either block conversion or selective conversion may be used, whereas in the former case only block conversion is possible.

The main advantage of the use of *selective converters*, which change the frequency of whichever input signal the subscriber has selected to a fixed vhf or uhf channel within the receiver's tuning range, is that they attenuate all unwanted signals before the receiver's rf input connector, and so allow adjacent-channel transmissions to be received satisfactorily by receivers that could not otherwise do so. However, they tend to be relatively expensive and inconvenient to use, notably in the case of receivers equipped with remote control facilities as these can no longer be used to select the programme service.

The use of *block converters* may be relatively inexpensive, especially where a single converter serves several subscribers, but it does not by itself allow adjacent-channel operation to be introduced. Unless prevented by the local regulations, or the wish to use the converter in connection with access control, most cable network operators prefer to provide selective converters temporarily to those subscribers whose receivers are insufficiently selective, as a means of introducing adjacent-channel transmission. It should be noted that, in West Germany, in the case of channels outside the bands allocated for broadcasting, only selective converters may be used that are equipped with a device to prevent operation in the absence of a television signal.

25.5 Channelling plans

The constraint of compatibility with existing broadcast transmissions is an essential factor in the design and operation of tree-and-branch cable networks, and this is particularly important in the case of the channelling plan. Almost without exception, terrestrial television broadcasting operates with contiguous channels grouped within the frequency bands allocated for this purpose. The width of each channel depends on the television system: 6 MHz with the 525-line 60 fields/s system and 7 or 8 MHz with the various 625/50 systems (see Table *25.1*). The first tree-and-branch cable networks followed the existing channelling arrangements directly, leaving at least one unused channel between those occupied by programme signals.

The selectivity of receivers intended for off-air operation was more than sufficient under these conditions. If there were any broadcast signals in the same band which were likely to be received locally, care was taken to ensure that the channels concerned were not used within the cable network, in order to prevent possible interference due to leakage into, or out of, the network. In almost all cases, the use of alternate channels also ensured that the local oscillator frequency tuned to one of the cable programmes would fall into a vacant channel (or outside the band), and thus there was no risk of interference from this source.

The number of programmes that could be distributed this way in the vhf bands was, of course, limited to about half the number of vhf channels available for broadcasting, i.e. about six. When the number of programmes to be distributed exceeded this number, there were various possible solutions, but a change in the channelling plan was not considered to be an option. This was no doubt because the loss of compatibility in general, and of the above mentioned benefits in particular, was thought to be too serious. At most, in certain networks, a channel at the edge of the band was offset from the nominal frequency by about 1 MHz, to allow both it and the next channel within the band to be used to carry programmes without putting too stringent demands on the receivers' adjacent-channel selectivity.

In many networks, the next stage was to introduce operation at frequencies lying between Band II and Band III (i.e. in the range 108–174 MHz), because the pass-band of the existing amplifiers already included these frequencies. In order to ensure that the same pattern of relationships between the frequencies of the signals in one channel to those in another applies in the case of the new channels, they are defined by extrapolation of the channelling plan already adopted for broadcasting within Band III. In other words, the additional frequency band is divided into contiguous channels having the same width: 6 MHz for systems M and N, 7 MHz for system B, and 8 MHz for the other systems.

Not all of these extra channels are equally suitable for use, notably because of the risk of leakage into, and/or out of, the cable. The lower part of this frequency range is allocated to various aviation radio services, including the emergency frequency of 121.5 MHz, while the higher part is occupied by various fixed and land mobile radio services. In both cases, there is a significant potential for mutual interference, and therefore it is necessary to check that no problems arise in practice whenever operation in these 'mid-band' channels is required — and before they are actually used for distributing programmes. Similar precautions should also be taken when frequencies above Band III are used.

25.5.1 Channel assignment planning

In principle, it is preferable to distribute programme signals in channels different from those in which they are received at the head end of the cable distribution network. In particular, this provides an easy method of distinguishing between direct reception and leakage from the distribution network. The preparation of an assignment plan is more of an art than an exact science, because both technical and non-technical factors must be taken into account in determining which programmes are distributed in which channels.

With regard to the technical factors, it is necessary in particular to comply with the constraints that apply to transposition from the channel of reception to the channel of distribution (generally due to the fact that with certain pairs of channels there is a risk of interference caused by the presence of a harmonic of the transposer's local oscillator frequency within the output channel). Details of these 'forbidden combinations' are given in the documentation supplied by the manufacturers of transposer equipment.

Another important point is the fact that the deterioration of the technical quality of the distributed signal occurring along the distribution network (and in the receiver) is greater in the higher numbered channels than in the lower numbered ones. When any other aspects specific to the particular network concerned, such as the proportion of receivers that can be tuned to channels outside the broadcasting bands, are taken into account, the available channels can be placed in a *technical order of merit*.

Obviously, every subscriber (and potential subscriber) to the cable network has their own general order of preference for access to the various available programmes. If they could have access to only one of the programmes carried on the cable, the programme they would then choose can be considered to have the greatest value to them. Similarly, by analysis of the choices that would be made in the case of larger numbers of available programmes, an *order of preference* for the ensemble of available programmes can be established. Although these orders of preference must also vary from time to time, the aggregate 'hierarchy' of these preferences is stable enough to be taken into account in the channel assignment plan.

In practice, the most satisfactory arrangement appears to be one in which this aggregate order of preference corresponds to the technical order of merit, so that the most popular programmes are carried in the channels having the highest technical performance. In most, if not all, cases, this means that the programmes intended to have a broad appeal to audiences in the area served by the distribution network are carried in the 'best' channels, while programmes addressed to more specialist audiences (linguistic minorities, participants in formal education, etc.) are carried in the others.

After the channels available in the frequency band between Bands II and III have been occupied, it is also possible in some areas to add a couple of channels between Bands I and II (68–87.5 MHz). Unless the local receivers can tune to them already, however, it is generally considered to be preferable to start adding channels above Band III. This requires amplifiers having a wider pass-band than those suitable only for signals lying in Bands I and III and the frequencies between them.

On the basis of the constraints applying to amplifier design as well as receiver tuner characteristics, an upper limit of about 300 MHz is generally the next step, giving ten channels 7 MHz wide above the top of Band III in the case of system B.

In order to enable the various channels to be identified conveniently, they are numbered or given letters. Unfortunately, there are several different ways in which this can be done, and several different systems are in use. The most important frequencies for some of these identification systems are given in Table *25.2*.

25.6 Networks

25.6.1 The distribution network

The part of the network that has to be installed to supply the signals to an individual subscriber consists of the subscriber's feeder and the subscriber's outlet. Where Band II radio signals are distributed by the network, the subscriber's outlet normally includes two coaxial connectors, and a simple filter is used to provide at least sufficient isolation between them for no noticeable impairment to television reception to occur if the radio connector is left unterminated.

The isolation is not normally sufficient to protect television reception against interference due to leakage signals from the local oscillator of a radio set tuned for operation in Band II, or vice versa. This is presumably because subscribers are willing to take the responsibility for tuning their radio and television receivers in a compatible way when they wish to use them simultaneously.

In order to prevent inadvertent mis-connections, the television signal should preferably by supplied by a male coaxial connector and the Band II radio signal by a female. As cable subscribers tend to have a greater interest in audiovisual reception than non-subscribers, it is likely that many of them will have more than one television receiver. The design of the standard subscriber's feeder is such that it is not intended to be suitable for supplying signals to two television receivers, however. Although some subscribers may be willing to tolerate the impairment that they find when they install a simple matched splitter for this purpose, cable operators normally try to discourage this practice.

The subscriber's feeders of tree-and-branch cable networks universally consist of coaxial cable, and it appears to be unlikely that an alternative means of transmission will be adopted in the near future. The reason for this lies mainly in the fact that almost all receivers are fitted with coaxial input connectors, and these feeders are numerous and short. It is also important to ensure that they can easily be connected and disconnected to the taps on the branch cables, in accordance with commercial requirements. Although in the past some cable operators used individual 'clip-on' taps designed to make contact with the branch cable at any convenient point, these were found to be insufficiently reliable in practice, and the use of multi-way junction taps has become virtually universal.

The distributed signal is attenuated by these passive components, which also provide the required degree of isolation between different feeders. Manufacturers provide a range of taps having different amounts of attenuation, in order to take account of the reduction in signal level occurring along the branch cable on which they are to be installed. The attenuation available typically ranges from 6 to 35 dB in 5 dB steps. Incidentally, all unused taps should be terminated with 75 ohm resistors in order to maintain the designed loading and impedance conditions.

25.6.2 The trunk network

For many years, the trunk distribution network also consisted of a network of coaxial cables, but recently the development of optical technology has reached the point at which the relatively low cost of optical fibres makes this the most economical solution for transmissions over long distances, despite the cost of the terminal equipment required[5]. Where coaxial cable is used, the design of the trunk distribution network is based on the principle that the attenuation of an electromagnetic signal due to transmission through a homogenous cable is directly proportional to the distance and inversely proportional to the frequency of the signal. To be able to deliver a signal of the required amplitude at the end of a long distribution network, it is therefore necessary to amplify it at intermediate points.

When several signals at different frequencies are present simultaneously, the amount of amplification applied to restore the initial amplitude depends inversely on the frequency. The distortion inevitably introduced by practical amplifiers depends largely on the amplitude of the output signal, and is cumulative when the signal passes through one amplifier after another. The task of the designer is then to devise the most economic arrangement of cable and amplifiers, taking account of the maximum length of the network, the characteristics of the broad-band amplifiers to be used, the cable available, the range of frequencies to be distributed, and the required amplitude and signal/noise ratio of the signal at the output of the network. A computer-aided design technique for this calculation has been described by Niels and Tollestrup[6].

The characteristics of cables and amplifiers should, of course, be obtained from the documentation published by manufacturers. Typical values of the attenuation of cables at 300 MHz range from 2 to 3 dB/100 m for trunk cables and from 5 to 9 dB/100 m for subscribers' feeders[7]. In modern cable networks operating at up to 300 MHz, the average amplifier spacing is about 500 m in the trunk section and 250 m in the branches. With a maximum of about 20 amplifiers in cascade, this provides coverage within a radius of about 5 km from the head-end[8].

One advantage of the use of coaxial cables is that they automatically provide a means of transmitting power to amplifiers along the trunk network, which can thus be located wherever is most appropriate, regardless of the availability of mains power supplies. It is not generally considered necessary to make special arrangements to ensure that distribution amplifiers remain operating in the event of a failure of the local mains supply, because such a failure would also prevent operation of most, if not all, the television receivers fed by the network downstream from the amplifier. However, it may be worthwhile providing emergency power supplies for at least some of the head-end equipment and trunk amplifiers, especially in cases where a local power failure is unlikely to affect the whole of the area served by the network.

25.7 The head-end

Even the best designed cable network introduces a small degree of impairment to all the signals distributed, and so it is very important to ensure that the input signals are of the highest possible technical quality. For this reason, the most appropriate location for the head-end is one where such reception can conveniently be achieved, and which is also convenient for the design of the distribution network, i.e. it should ideally be both relatively central and relatively high. In many urban areas, the top of a tall building provides a suitable location, not only for direct reception in the vhf and uhf bands, but also for the terminals of the microwave links bringing in signals from broadcasts transmitted too far away for direct reception.

In several European countries, the national telecommunications administration operates its own network of such microwave links to supply the head-ends of cable networks with the television programmes broadcast in neighbouring countries[9]. This avoids the unnecessary multiplication of such links that would occur if each network had to install its own microwave equipment for this purpose. Reception of programmes intended exclusively for cable distribution by means of transmissions in the fixed satellite service in the 11 GHz band is also possible at such locations, although care must be taken to provide protection of the parabolic antenna against possible interfering terrestrial signals, and the risk of damage in strong winds. As well as these facilities for the reception of externally produced signals and for processing them into a form appropriate for distribution, the head end may also contain those required for local origination, when this is permitted by the competent authority[10].

Depending on the operating priorities, the locally originated signals may be obtained from:

- test pattern generators
- caption generators (channel assignments, fault warnings, etc.)
- equipment for generating, and possibly displaying, a sequence of teletext pages (local announcements, advertisements, the 'rolling page' from a broadcast teletext service, etc.)
- automated local time and weather equipment
- a generator of a scrolling display of news text obtained from a specialist agency
- a 'mosaic' generator, producing a montage of electronically reduced displays of 16 distributed programmes on a single screen
- playback equipment for video recordings on various formats
- electronic cameras in a local studio
- electronic cameras at a remote location (possibly via 'upstream' transmission facilities provided in the distribution network)

The head-end control room must, of course, be equipped with the corresponding facilities for individually processing, monitoring and switching all the distributed signals, and for dealing with enquiries from subscribers when the main office is closed. In principle, the whole distribution network of a tree-and-branch network is *transparent*, in the sense that the only points in it at which the signals are treated individually are the head-end and the selective converters, if any.

25.8 Distribution of sound and data signals

Attention should be paid to the following points in order to ensure that the network will distribute sound and data signals satisfactorily.

25.8.1 Sound accompanying television

The main monophonic sound signal in each television channel is transmitted by means of a frequency-modulated carrier (am in the case of system L) at a frequency several megahertz above the associated vision carrier, as shown for each system in Table *25.1*. When broadcast, the sound carrier is already some 10 dB below the vision carrier, but as satisfactory reception of monophonic signals can easily be achieved at even lower relative amplitudes it is customary to attenuate them further by up to 10 dB, in order to minimize adjacent-channel interference. The introduction of stereophonic or two-channel sound transmissions systems in certain countries has required a number of minor changes in the channel processing equipment at the head-end of the cable networks carrying the signals concerned. These signals are also more susceptible than monophonic sound to impairment due to the presence of short-delay echoes in the distribution network.

In the *two carrier* system, which involves the transmission of a second fm sound carrier at 5.742 MHz above the vision carrier in system B, care must be taken to ensure that the pass-band of the channel filters at the head-end extends to beyond this frequency. Similar attention to the pass-band of the channel filters is also required in the case of the digital system involving the transmission of a 728 kbit/s dqpsk (digital quadrature phase shift keying) signal[11]. When this system is used in a 7 MHz channel, a minor modification of the vestigial sideband filter in the upper adjacent channel is required in order to prevent mutual interference. Where the additional sound channel is transmitted by means of the modulation of the main sound carrier with a frequency multiplex signal, such as the MTS system in North America, it is necessary only to ensure that the sidebands around that main sound carrier are not attenuated.

25.8.2 FM radio

Although tree-and-branch cable distribution networks are generally intended mainly to convey television signals, the marginal cost of providing frequency modulated radio signals in Band II (87.5–108 MHz) is so low that this is almost universally done, at least where television signals in both Band I and Band III are carried. Monophonic signals put no more demands on network performance than those conveying fm television sound, but satisfactory reception of the familiar pilot-tone stereophonic radio signals requires a relatively high signal/noise ratio and linearity.

The use of additional subcarriers within the audio multiplex of some fm radio transmissions, such as those carrying data (e.g. the radio data system RDS)[12], increases the risk of audible intermodulation, even if these additional services are not used by subscribers. According to IEC Publication 728-1, the amplitude of monophonic fm carriers at the subscriber's outlet should lie in the range 37–80 dBμV, and stereophonic carriers should be in the range 47–80 dBμV. Furthermore, the relative amplitudes of any two adjacent carriers should not exceed 8 dB (6 dB if their nominal frequencies are less than 600 kHz apart). In practice, the amplitude of the biggest fm radio signals is kept about 10 dB below the amplitude of the vision carriers of television signals in Band III.

Where all the fm signals to be distributed are received from co-sited transmitters having similar radiated power, it may be sufficient to feed the input to the distribution network from a directional antenna through a band-pass filter. The 'capture effect' of fm reception provides adequate protection against leakage into or out of the cable if this is kept within the limits required for television.

In many cases, however, individual processing of each fm radio signal, including agc and transposition to another frequency, is necessary to satisfy the most critical subscribers. A minimum nominal frequency separation of 400 kHz between adjacent fm carriers is specified by IEC Publication 728-1 for high fidelity transmission. All nominal carrier frequencies should be integral multiples of 100 kHz, in accordance with broadcasting practice. Special care must be taken at the head end to avoid multipath reception, to which some receivers are particularly sensitive.

25.8.3 Digital radio multiplex for dbs

In the context of the preparations for the introduction of direct satellite broadcasting in West Germany, there is broadcast a 20.48 Mbit/s digital multiplex conveying 16 high quality stereophonic programmes within one of the five channels assigned in the WARC-BS 1977 Plan for dbs[13]. To enable subscribers to cable networks to receive this multiplex signal, it is proposed to distribute it within the range 111–125 MHz; the specifications for the corresponding receivers require provision to be made for such an input, as well as one in the range 950–1350 MHz, obtained from the dbs outdoor unit. Unfortunately, the introduction of this new transmission system has had to be deferred because the solar panels of the first German direct broadcasting satellite, TV-SAT, failed to deploy correctly when it was placed in orbit in October 1987.

25.8.4 Data signals

Data signals are already being broadcast operationally in multiplex with vision, notably for the teletext service, and experimental broadcasts of several new types of data transmission have already begun:

- 728 kbit/s dqpsk, for two television sound channels (see section *25.8.1*)
- radio data (rds) in multiplex with fm radio signals (see section *25.8.2*)
- 20.48 Mbit/s for 16 stereo radio programmes (see section *25.8.3*)
- 10.125 Mbit/s duobinary data from D2-MAC/packet dbs (see section *25.9*)
- 20.25 Mbit/s data from C-MAC and D-MAC/packet dbs (see section *25.9*)

It is clear from the experience obtained with the transmission of teletext signals (nrz binary data at approximately 5.7, 6.2 and 6.9 Mbit/s for Telidon, Antiope and UK teletext respectively) that satisfactory television transmission does not necessarily guarantee a sufficiently low bit-error rate for data reception. In fact, the presence of one or more echoes with a delay corresponding to an integral multiple of the period of the data signal is particularly harmful. The criteria for tolerable echo characteristics given in IEC Publication 728-1 apply only to the subjective television picture quality, and more stringent values appropriate for the various types of data signal are being studied.

25.9 Distribution of MAC/packet signals

In accordance with a directive of the European Community, only members of the MAC/packet family may be used for a direct broadcasting service within the 12 member-states of the European Community, and these systems have already been adopted for certain other satellite transmissions. As explained in section *29*, the MAC/packet systems specified for this application employ a time-division multiplex technique to transmit the time-compressed luminance and chrominance components of a 625/50 colour video signal and digital data conveying the associated sound and auxiliary signals. In order to minimize the satellite power requirements, the dbs transmissions employ frequency modulation, each programme occupying a channel 27 MHz wide in the band 11.7–12.5 GHz. Taking into account the use of orthogonal polarization, different orbital positions and the directivity of receiving antennas, the 1977 Geneva Plan for dbs provides for a total of 40 such channels within the band with their nominal centre frequencies spaced at intervals of 19.18 MHz (see Tables *28.1* and *28.2*).

Even after block transposition to what is becoming known as the *first intermediate frequency* range of 950–1750 MHz, at which they will normally be input to the receiver in the case of individual reception and small collective antenna installations, these signals are not directly suitable for transmission through cable distribution networks, particularly because they occupy such large channels. Indeed, the desire to be able to distribute MAC/packet signals within the channels of existing cable networks has been largely responsible for the development of the D2-MAC/packet system. Of course, the obvious solution to this problem is the conversion of the MAC/packet signals to whichever of the systems listed in Table *25.1* is already being used locally, so that subscribers will be able to watch the new programmes with their existing receivers. However, while satisfying the desires of some subscribers, this would undoubtedly frustrate others, in particular by depriving them of the main advantages of the MAC/packet system: the improvement in picture quality and the provision of additional sound and data services.

The obvious solution is distribution in the form of a vestigial-sideband amplitude modulated (vsb/am) signal, similar to those of the existing terrestrial television systems. In this case the bandwidth must be sufficient for transmission of both the vision components and the digital data multiplex[14, 15, 16]. In all MAC systems, the baseband video bandwidth corresponds in principle to 0.75 times the standard digital video sampling frequency defined in CCIR Rec. 601, i.e. $0.75 \times 13.5 = 10.125$ MHz, whereas a bandwidth twice as great is required for transmission of the binary data multiplex of the C-MAC/packet system, which has an instantaneous bit-rate of 20.25 Mbit/s. It is not possible to reduce the bandwidth of the digital multiplex to match that of the video components by means of a low-pass filter because this would result in an unacceptable increase in the bit-error rate. However, in a transmission network having characteristics suitable for analogue video signals, the bandwidth required for the digital multiplex could be reduced by adopting a coding technique in which more than two logic levels are defined.

In particular, the use of duobinary coding (see section *29*) would halve the required bandwidth to 10.125 MHz without the need for additional constraints on the maximum relative amplitude of echoes. This corresponds to what is known as the *D-MAC/packet system*. Similarly, if the original instantaneous bit rate of the data multiplex was halved to 10.125 Mbit/s, the duobinary signal would require only 5.0625 MHz and the bandwidth of the video components could be reduced to match, with only a loss of horizontal resolution, by means of a low-pass filter. This corresponds to the case of the *D2-MAC/packet system*, which has already been shown experimentally in existing cable networks in Switzerland to permit satisfactory transmission and reception within a 7 MHz channel adjacent to similar channels conveying PAL-B signals. In fact, the specifications for the D-MAC/packet system make provision for the subdivision of the data multiplex into halves such that either of them may easily be converted for transmission as a D2-MAC/packet signal.

The vsb/am standards for terrestrial television broadcasting, defined in CCIR Report 624, require the vestigial (lower) sideband to be transmitted without attenuation in the neighbourhood of the carrier, so that the full Nyquist filtering must be performed in the receiver. To ensure full compatibility at the receiver, cable distribution networks comply with these standards for fdm systems, although this requires the use of wider channels than operation with half-Nyquist filtering. If half the Nyquist filtering is performed at the input to the distribution network and half at the receiver (so that the signal distributed at the carrier frequency is attenuated to 0.707 of its full value), it should be possible to obtain satisfactory reception of vsb/am D-MAC/packet signals with adjacent-channel operation in channels 10.5 MHz wide (which corresponds to 1.5×7 MHz). Unfortunately, such operation would be satisfactory only if all equipment involved complied with stringent tolerances, which would be difficult to achieve in practice.

Studies have indicated that operation with full Nyquist filtering at the input to the distribution network would be optimum from the point of view of network design, but its wider implications have not yet been examined. Taking into account the need to ensure that cable subscribers will be able to enjoy all the advantages of the broadcast MAC/packet signals that will be available to users of individual antennas, the European broadcasting organizations represented by the EBU have proposed that 12 MHz channelling should be adopted for the distribution of all vsb/am members of the MAC/packet family. This would also provide some spare capacity for possible further compatible enhancements in the context of the evolution towards high definition television (HDTV).

Provision for temporary national options with channel spacing of 7 or 8 MHz is also envisaged. In either case, it is likely that frequencies in the range 300–450 MHz, which have not so far been used in European cable networks, will be preferred for the distribution of vsb/am MAC/packet signals[17]. CENELEC Technical Committee 106 is providing a forum for the formation of a consensus on this aspect.

References

1 COLL, D C and HANCOCK, K E, 'A review of cable television: the urban distribution of broad-band visual signals', *Proc IEEE*, **73**, April, 773–787 (1985)

2 GENSOLLEN, M, and VOLLE, M, 'Evaluating the cost of a cable TV network and its profitability', *Symposium Record*, CATV Montreux, 466–473 (1985)

3 'Cabled distribution systems for television signals, 30–1000 MHz', *IEC Publ 728-1*, 2nd edn, Geneva (1986)

4 CICIORA, W S, 'The long road to the EIA multiport', *Symposium Record*, Montreux (joint vol.), 157–164 (1987)

5 CELLMER, J, et al, 'Hybrid network: a new design for broadband networks', *Symposium Record*, CATV Montreux, 156–175 (1987)

6 NIELS, C and TOLLESTRUP, M, 'CADS: computer-aided design of TV', *Symposium Record*, CATV Montreux, 133–143 (1985)

7 THONNESSEN, G and ZAMZOW, P, 'Low-loss coaxial cables for CATV networks in Europe', *Symposium Record*, CATV Montreux, 229–249 (1987)

8 HORN, U, 'Demands on a 440 MHz CATV system — possibilities of realisation', *Symposium Record*, CATV Montreux, 71–84 (1985)

9 STILKERICH, H, 'The CATV interconnection network of the Deutsche Bundespost', *Symposium Record*, CATV Montreux, 138–155 (1987)

10 LAYCOCK, D, 'The Your Channel experience'. *Symposium Record*, CATV Montreux, 357–365 (1987)

11 ELY, S R, 'Progress and international aspects of digital stereo sound for terrestrial television', *IBC '86 Conference Publ*, 138–143 (1986)

12 WRIGHT, D T and EDWARDSON, S M, 'Review of broadcast radio-data systems', *IBC '86 Conference Publ*, 85–88 (1986)

13 TREYTL, P, 'Digital sound service for direct broadcasting satellites', *DFVLR*, Munich (1983)

14 O'NEILL, H J and AVON, P A, 'The distribution of C-MAC in cable systems', *IBC '86 Conference Publ*, 310–316 (1986)

15 QUINTON, K C, 'Carriage of MAC/packet signals in MATV and CATV networks', *Symposium Record*, Montreux (joint vol), 85–88 (1987)

16 SCHLOGL, H, 'Head-end components for satellite reception in individual and CATV/MATV systems', *Symposium Record*, Montreux (joint vol), 89–130 (1987)

17 AFSAR, E, 'Technical options for increasing the frequency range and the channel capacity of existing CATV networks', *Symposium Record*, CATV Montreux, 183–195 (1987)

K C Quinton MBE, B Sc, F Eng, FIEE, FRTS
Formerly Director of Research, British Cable Services Ltd

26 Switched-star Networks

Most cable TV systems in service are of the tree-and-branch type described in section *25*. They have developed as community antenna systems (CATV), implying that their prime function is the relay of broadcast programmes, and they were designed to deliver signals in the manner for which domestic TV (and radio) receivers have been designed. Thus coaxial cables carry signals in frequency division multiplex (fdm) and channel bandwidths suit the local broadcast standards. Multiple signals arrive in the subscriber's home arranged in fdm.

Switched-star networks are mostly confined to installations in the UK and the Netherlands and even here they are in a minority. They offer many advantages over tree-and-branch systems (see section *26.3*), but their initial cost is noticeably higher. Unless the secondary distribution cables are laid underground they can be unsightly. The equipment in the switching point nodes, which interface the primary and secondary distribution networks, can be inappropriate for mounting on poles or the outside faces of property.

This section deals with the technology and service aspects of switched-star networks. Section *25* describes system performance requirements such as choice and equipping for the head end installation, the permissible limits for linear and non-linear distortions in the transmission path, signal delivery levels and safety for the domestic terminals, etc., as these are common to both types of topology. Indeed, national and international specifications for broad-band cable systems do not differentiate between the alternatives. A choice in favour of switched-star will usually result from either a preference given by the licensing authority or an operator's wish to obtain maximum 'future proofing' or perhaps lower annual costs. Switched-stars may also be preferred if advanced interactive services are envisaged or integration with telecommunications is required.

26.1 Origins

In the majority of UK cable networks over the period 1952–1975, the principle of using a single coaxial cable for distribution was not followed because TV was added at hf to existing radio relay networks using twisted pairs and space-division multiplex. Those systems lost popularity as the number of TV programmes increased.

The first switched-star networks were a natural development. Following the principle that the primary, or trunk, cabling comprises less then 20 per cent of all cable purchased, then, by using small, low bandwidth and therefore cheap cables for subscribers' feeders, high programme capacity could be achieved provided programme selection was done at the interface of primary and secondary networks[1]. Small experimental switched-star systems, based on twisted pair subscribers' cables, entered public service in the Netherlands at Arnhem, Soest and Sluis. These were soon to be displaced by substituting coaxial cables throughout, using frequency division multiplex on the trunks, tuners in place of on/off circuits in the switching points (sps) and removing the hf-vhf domestic converters but retaining remote programme selection by numeric keypad.

26.2 System description

A switched-star system comprises a head-end and a primary (or trunk) network leading to switching points from which discrete subscribers' feeders lead to each dwelling. While there are variations, described in section *26.5*, the trunk is normally engineered as for a tree-and-branch system, i.e. the switching points are tapped on to a branching network of one or more coaxial cables carrying TV signals in fdm (*Figure 26.1*).

Within each switching point there is one or more dedicated switching circuit for each subscriber's outlet. In this context, *outlet* is defined as single programme delivery; if two or more selected programmes are required simultaneously in the dwelling then there must be an appropriate number of switches even if, for vcr plus TV, two switches serve one physical outlet socket. The operation of a switching point can be visualized by regarding each switch as a TV tuner remotely controlled by a subscriber to the service, the output frequency being fixed irrespective of the programme chosen. The cable operator has an overriding control over each switch, thereby enabling conditional access, or, if required, he can remove service by inhibiting all selections.

Figure 26.2 shows the difference between branching and star secondary networks. The latter has bundles of cables, one per dwelling. Current switching points serve from 16 to 600 outlets. Generally, a large unit will have a local power input, contain trunk amplification, and also have substantial computing and

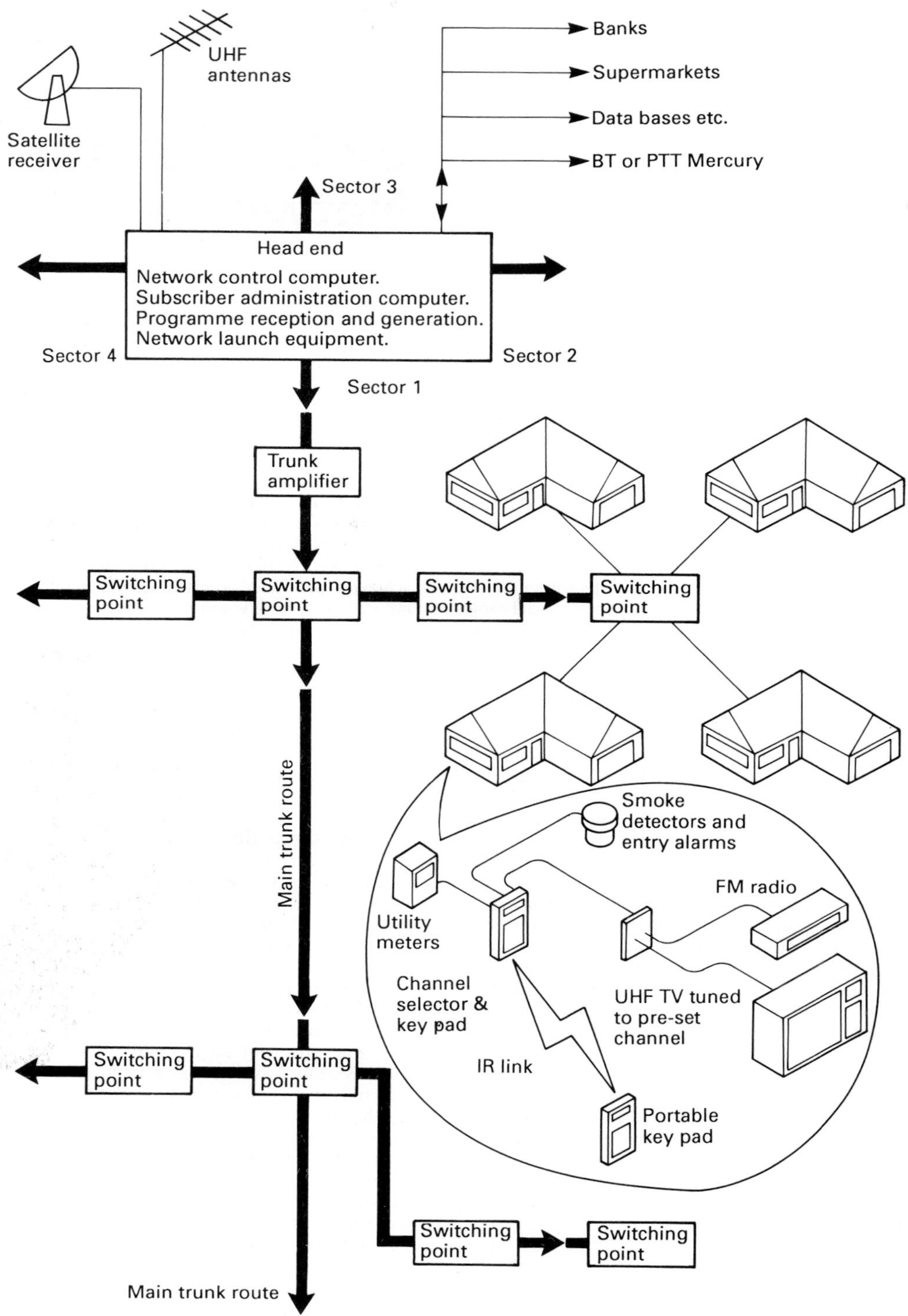

Figure 26.1 Star network layout

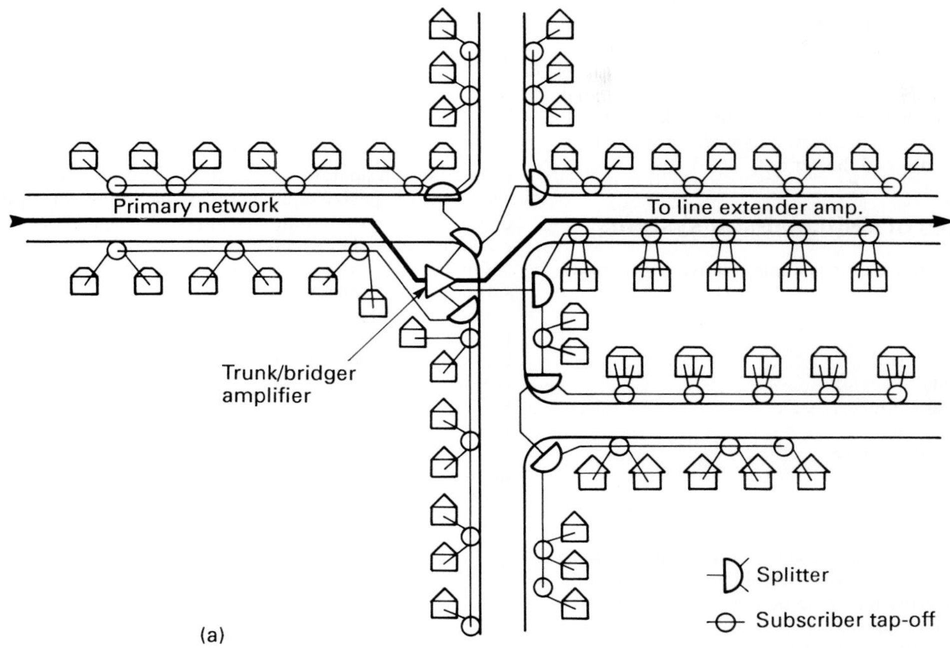

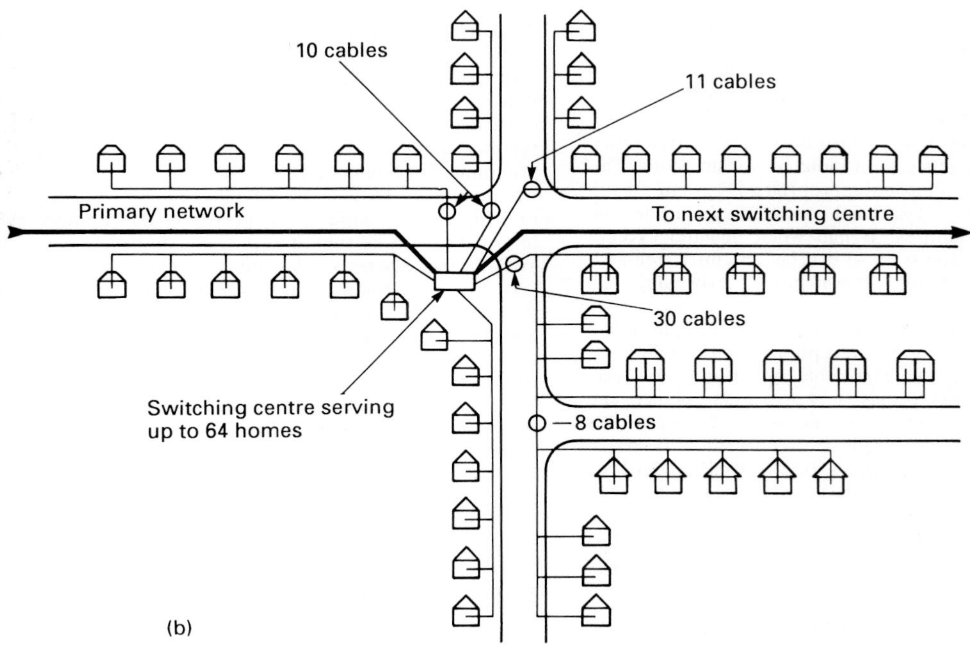

Figure 26.2 Distribution options: (a) shows a branching secondary network, and (b) a star secondary network (IEE)

data communication equipment. A small sp is likely to take power from the trunk coaxial and have minimal local intelligence. It has the advantage that the fewer outlet cables will be shorter, reducing the average cable cost per dwelling. See also references 2 and 3.

When a subscriber's feeder carries more than one TV signal these are transmitted in fdm. The subscriber's keypad will control two 'outlets' if one is dedicated to a vcr, in which case the TV will be tuned to f_1 and the vcr to f_2.

26.3 Advantages of switched-star systems

The advantages of switched-star systems relative to tree-and-branch topography are:

- flexible capacity
- conditional access control
- pay-per-view facility
- privacy of communication
- fibre-optic transmission
- protection against interference

26.3.1 Programme capacity

A cable system to serve, say, 100 000 dwellings or more requires a major investment, with completion taking several years and subscriber growth lagging behind construction progress. An operator may have to look beyond ten years from construction start for satisfactory profitability, particularly if all cable is ducted underground. It is extremely difficult to make an accurate forecast of required programme capacity 10–20 years ahead.

The *switched-star* system has the merit that capacity can be augmented indefinitely by adding to the trunk cabling and with minor additions in the switching points. The subscribers' feeders and domestic equipments would be undisturbed. It is the cost of these feeders and of their installation which dominate construction cost.

A *tree-and-branch* system operating up to 450 MHz, a reasonable limit for a 100 000 dwelling system, has an estimated maximum capacity of about 32 channels 8 MHz in width after taking into account regulatory constraints on permitted radiation and likely interference from radiated transmissions. This estimate is for the UK, using appropriate DT1 and British Standard specifications[4,5]; national conditions should be considered for other countries. If and when demand exceeds capacity, it becomes necessary to overlay an entire tree-and-branch network and provide A-B switches in the dwellings — a very costly operation.

This subject of capacity should be viewed in the light of two developments. Firstly, direct reception from satellites is expected to grow rapidly. The cable operator is in direct competition with this new provision of programme choice, and although he can offer a more economic means of reception his previous advantage of choice will be diminished. If he transcodes higher quality satellite signals to terrestrial standards he can offer further economies, but at a later date he might have to distribute the same programme simultaneously to the enhanced (satellite) standard to avoid disconnections in favour of direct reception. There would then be a significant increase in cable bandwidth requirement.

Secondly, it will be seen from section *27* that developments in interactive services are forecast. These essentially non-entertainment services will require channel capacity on the cable; if this is not available, the entertainment choice will need to be restricted or the new potential sources of revenue will have to be turned away.

26.3.2 Conditional access control

Modern cable systems usually offer a basic service with additional programmes available for extra subscriptions. These might be available in tiers or on an 'a la carte' basis. Methods used for withholding or releasing these are termed *conditional access* controls. In a switched-star system, the operator addresses memories in the switching points, and each time a subscriber selects a programme reference is made to memory prior to switch operation to permit programme release or not. There is no special equipment in the home, and piracy of unauthorized programmes is impossible since these cannot reach the subscriber's feeder.

On tree-and-branch systems, all programmes reach every subscriber, so for those subject to conditional access it is necessary to adopt scrambling and for appropriate subscribers to have descramblers to gain access. These operations inevitably produce some degradation in picture or sound quality although this may not be discernable to the average viewer. Scrambling on cable systems appears not to be standardized; vendors of equipment use differing methods, usually protected by copyright. Once a cable operator has adopted a particular method he will be constrained in his sources of descramblers, which is not in his best interests.

It should also be noted that any change in TV waveform, e.g. a new satellite broadcast standard, could result in a second descrambler being required in the home.

Provision of conditional access control for tree-and-branch systems comparable with that of switched-star systems, so that the operator can make adjustments to programme release by command signals on the cables, requires domestic addressable descramblers. These are expensive items physically inaccessible to the operator without the subscriber's permission. Subscriber removals to another district are a source of losses.

26.3.3 Pay-per-view

A switched-star system embodies communication from subscribers to switching points for programme selection. An operator can transmit to switching point memories, information concerning creditworthiness of subscribers. The memories can also recognize personal identification numbers (PINs). It is therefore possible to accept or reject a request for a particular programme of limited duration from a person allocated a PIN without the need to communicate the request to the head end in real time. Thus impulse pay-per-view is facilitated. Billing information can be routed to the head end when convenient, avoiding message peaks. A switching point can also be made to take advance bookings, i.e. it could release a selection for a given period. It can also be equipped with text generation for acknowledgements, etc.

In a tree-and-branch system, there is no equivalent distributed intelligence, so pay-per-view requests are made either by telephone or by upstream messages, both of which are subject to significant delays. In the second case, the main cause of delay is the need to segment the network with switching at suitable nodes to avoid intolerable accumulation of noise and interference accompanying upstream signals and to connect segments in sequence[6].

26.3.4 Privacy of upstream communication

The discrete connection between a subscriber and switching point ensures that no other subscriber can eavesdrop on his messages. This is of particular importance in some interactive operations, e.g. telebanking. In a tree-and-branch system, neighbours are connected to a common cable.

26.3.5 Fibre-optic transmission

The advantages of optical fibres are their total immunity to electromagnetic interference and storm damage to equipment, and they provide isolation to power voltages.

TV receivers require picture signals to be delivered in analogue form, and this is unlikely to change in the foreseeable future. In the following description references to am and fm apply to the electrical signals at the optoelectronic terminals; the light modulation is always one of varying intensity. Up until 1988, multi-channel analogue transmission on fibre was constrained to the use of fm signals because of poor laser linearity. That year, two breakthroughs were recorded: the distributed feedback laser which gave much improved linearity, and the Mach-Zehnder external optical modulator which, although non-linear, can be accurately compensated[7]. In 1989, up to 40 vsb/am TV signals could be carried successfully on one fibre using these pre-production devices. However, the device costs, even those predicted for quantity production, are likely to restrict their use to primary network service.

The permissible attenuation between optical sender and receiver falls as the number of TV signals is increased, and it is much less for am than for fm. Reference 7 suggests that for 18 TV channels using vsb/am, the attenuation limit is about 10 dB. While this is sufficient for 20 km fibre runs or more, there is clearly inadequate power available for splitting to serve several receivers.

One is led to the conclusion that the delivery of many programmes using fibre into the home, following tree-and-branch topology, is impracticable.

A switched-star system reduces these problems because the number of channels to be transmitted simultaneously in the subscribers' feeders is very small and the sending power is not shared between dwellings. Further, the low bandwidth required of the senders makes the use of inexpensive leds a practical proposition. LEDs tend to be more linear than lasers, but an interesting experiment with a laser from a CD player, carrying few TV signals, has been reported[8]. When a switched system with optical final distribution is designed, a decision to use fibres in the trunk circuits will affect whether am or fm is used. The only way that *am* can be used here is by adopting a star layout to avoid intermediate amplification between head end and sps. The remaining options are *fm* or *digital*, as the optical loss budget then permits tapping the sps on the trunks. In the first case it can be economic to continue with fm on subscribers' feeders; in the second, demultiplexing and DAC will be required in the switching point. Purely optical switches for sps appear to be inappropriate because selective circuits are essential sp components, although one could conceive one fibre per programme in the trunks and an optical crosspoint matrix. A final alternative is to follow a telecommunications principle of digital tdm in trunks and switches within sps which gate out the time slots allocated to a particular programme.

In countries where overhead cabling is common, e.g. USA and Japan, the bulk of cables leaving sps may be unacceptable unless optical fibres are adopted.

26.3.6 Protection against interference

A cable system is most vulnerable to ingress of interference where signal levels are lowest. This is always on the few decimetres of cable leading into homes. Levels are higher at amplifier inputs because of accumulation of amplifier noise. The major interference sources are high power broadcast transmissions and mobile radio telephony transmitters.

To combat the first, an intercarrier beat can be removed by locking a cable carrier to that broadcast, but this is likely to conflict with a requirement on cable systems to use harmonically related cable carriers to minimize intermodulation distortion produced in equipment. Dealing with the second is a matter of screening and it is costly to increase this on the high proportion of cable which is close to subscribers. A statistical chance can be taken, e.g. to allow for noticeable interference for the small percentage of time when a mobile transmitter might be close to a dwelling.

Also, the domestic terminal equipment, the TV set or to a less extent the set-top converter, can be vulnerable to longitudinal interference signals.

These are serious problems for tree-and-branch systems where the stacking of 30 or more carriers gives little opportunity for avoiding action. In the switched-star case, where only two or three vision carriers occur on subscribers' feeders, and these can be at vhf or uhf, there are no problems in finding 'clear' TV channels.

26.4 Disadvantages of switched-star systems

As has been said, the main disadvantage of a switched-star system compared with tree-and-branch construction is the overall higher cost and the higher proportion of this that is independent of subscriber penetration. The computing and data communications items in switching points together with the enclosing street cabinets involve significant initial investments. Subscribers' feeders consume considerably more cable than in the branching case. Whether they are installed as required or, as in telephony practice, laid at the outset and routed within dwellings when required, is an operator's choice. The sp switches together with their domestic keypads will probably be of comparable cost to the branching equivalents of line tap plus addressible set-top. However, in some designs, several switches are combined within one module to give economies of space and cost. This leads to under-utilization, at least temporarily, compared with adoption of individual switches.

If underground network construction is mandatory, then trenching and ducting costs are generally independent of system choice. Overhead wiring, whether on poles or property, is always much easier to accommodate for branching configurations as the number of cables is only one or two. The flexibility is also much better. For example, if a large dwelling is converted or replaced by several small dwellings, then on a branching network additional taps and drop cables might suffice. On a star layout new long cables have to be provided and, if the local sp is fully utilized, an extra sp (possibly co-sited) might be needed.

Considering trends in *capital costs*, tree-and-branch technology is mature, and equipment items are highly developed and mass produced. Cost reductions are more likely with the embryonic switched-stars. Probably, the switching costs will fall below those of addressable descramblers. On *running costs*, maintenance and amortization charges favour switched-stars, largely because complexity of domestic items is avoided.

26.5 Practical switched-star system

Switched-star systems may be classified by the switching method used in the switching points.

26.5.1 Baseband switching

An experimental system was commissioned in the mid-1970s in Japan[9]. 300 subscribers were each connected by two optical fibres to one centre. The main purpose was to explore public interest in information and educational services from a 24

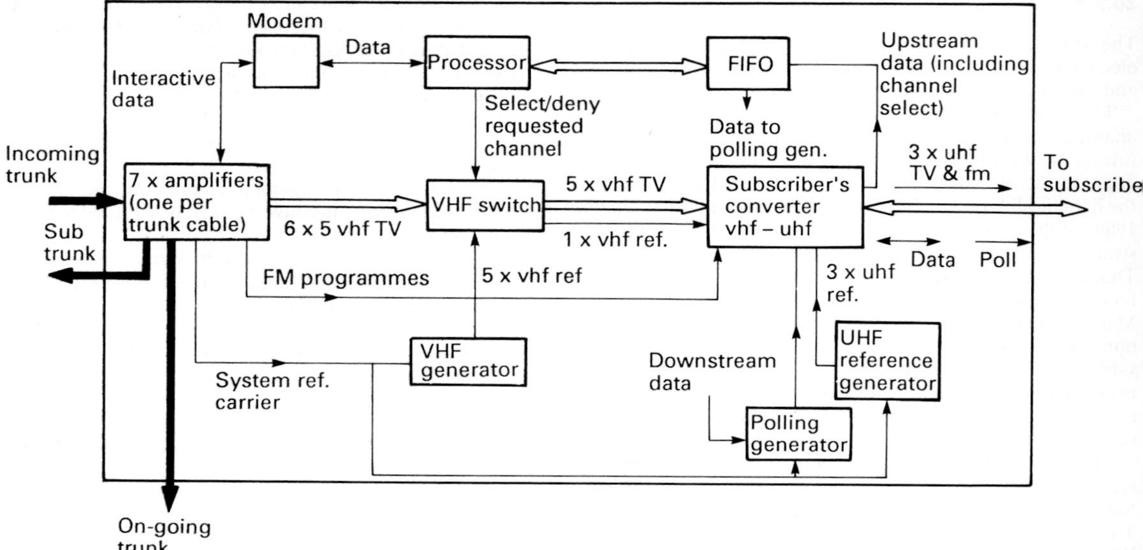

Figure 26.3 Switchpoint (British Cable Services Ltd)

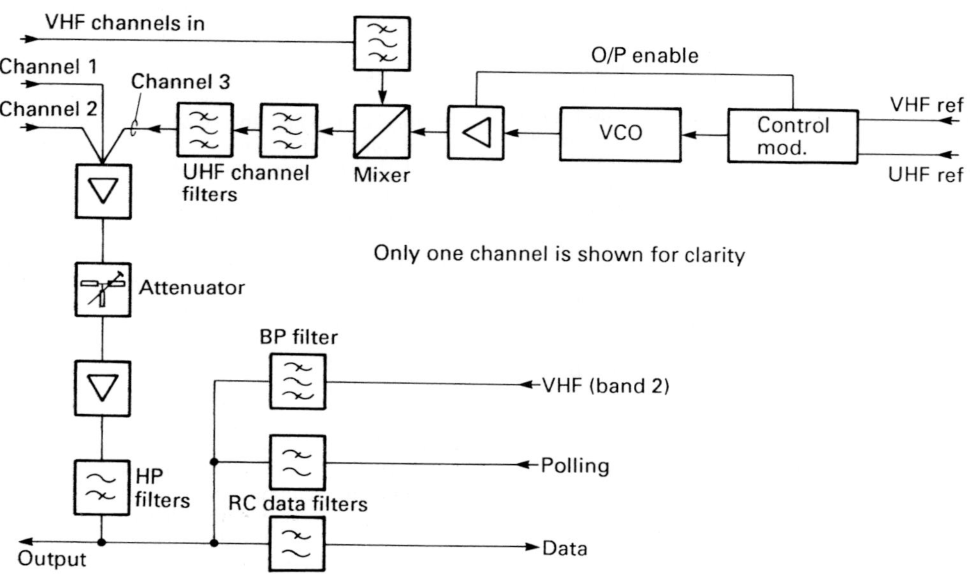

Figure 26.4 Switching point subscriber converter (British Cable Service Ltd)

channel menu; the network was closed down after the required information was obtained.

A system embodying optical fibre trunks in a branching layout where each channel is reduced to baseband in sps was described in 1984[10]. The number of sp receivers is set by the number of trunk channels; in this particular design, three fibres each carrying up to four TV channels are dedicated to each sp from the head end to give a video library service so there are 12 extra receivers. Receivers are followed by semiconductor switching matrices using custom vlsi (very large scale integration) to provide up to 48 inputs routeing to 16 outlets. The 'library' trunk fibres do not require signal amplification en route. SP design suits up to 300 subscribers.

A system was described in 1985[11] in which one trunk coaxial followed the tree-and-branch principle of all channels carried in fdm and in which each channel was reduced to baseband in sps. Fixed tuned receivers were again followed by one vhf modulator per outlet, connecting by semiconductor crosspoints, 16 modulators being on one board. SP design was to suit 100 dwellings.

26.5.2 HF switching

The experimental systems mentioned in section *26.1*, in the Netherlands, used one coaxial per channel in the trunks, i.e. space division multiplex, with vision on an 8.86 MHz carrier and sound at 3.36 MHz[12]. Switches were semiconductor with a bandwidth exceeding 12 MHz, the signals being passed to subscribers at hf followed by converters to vhf in the homes.

Another similar system was used experimentally in Italy[13] but the vision carrier was 13.2 MHz.

In both of the above systems, subscriber connections were by individual twisted pairs.

26.5.3 VHF switching

Apart from the Thorn-EMI switched system[11], systems described so far require in each sp a total of crosspoints equal to the number of trunk channels multiplied by the number of outlets. Any subsequent programme augmentation beyond the initial design capacity will require major sp changes. To avoid these, the most popular approach is to switch by using frequency-agile converters with fixed output frequencies, one converter being required for each outlet.

At least four designs are in service that use a single coaxial trunk carrying signals in frequency division multiplex with tuners in sps remotely controlled by subscribers. They are probably the cheapest switched-star systems and have been designed for various sp sizes. The largest contain the most computing and memory power and some have local text and picture generation. A further advantage is that extensions to, or conversion of, a tree-and-branch system are possible because the trunk network is common. In the Netherlands, conversion of non-switched *mini-hub* networks is taking place.

In Germany, current trends are to use star points as locations for scramblers[14] to give conditional access control. Initially, vhf switching is confined to injecting scrambling signals, but it is anticipated that true switching to outlets will be introduced when programme requirements demand two coaxials in the trunk network.

One system combines vhf matrix switching, between several trunk coaxials, with frequency-agile converters[15]. The basis for this idea is that, sooner or later, more than one trunk coaxial will be required to provide channel capacity so crosspoint switching is designed in. Initially the trunk operates up to about 200 MHz, carrying five TV channels widely disposed on each of six coaxials. Increasing capacity beyond 30 channels is then a matter of increasing the trunk planning frequency and/or reducing the trunk channel spacing.

26.5.4 Circuit configurations

A brief description of the British Cable Services *System 8* switchpoint and domestic items serves to illustrate switched-star design[16]. Others may be simpler.

Figure 26.3 shows the trunk cabling entering and leaving on the left, comprising six coaxials each carrying five TV channels at vhf with 19 MHz spacings and one coaxial carrying Band II fm plus a reference carrier. The amplifiers feed busbars making back-plane connection to six crosspoints at each outlet. Also within the vhf switch is a set of five crosspoints per outlet switching vhf reference frequencies. When the subscriber selects a programme, by two keypad buttons, the two bytes are polled and reach the processor which, if enabled for that programme, selects the reference frequency and coaxial accordingly. The subscribers' cable can carry up to three independently selected programmes, each having a uhf carrier. When the subscribers' converter is plugged in, it will be preset to take one uhf reference. The converter local oscillator is controlled by a phased-locked loop (pll) referred to the sum of vhf and uhf reference inputs so it tunes to the required channel and delivers an output which, because of the precision of the incoming programme frequencies, will be frequency locked to other outputs on that uhf channel leaving the switchpoint. This minimizes any crosstalk effects, making them similar to crossmodulation for which the BS 6513 limit is $46 + 10 \log_{10}(N - 1)$ in decibels where N is the total number of programmes. To reduce unwanted signals, the local oscillator frequency is divided by 256 before entering the pll comparator, and the vhf and uhf references are less than 5 MHz.

The switchpoint converter for one outlet is shown in *Figure 26.4*. The other two selected TV programmes and Band II inputs are added at the top left. The uhf channel filters ensure that other programmes do not reach the subscriber, the mixer being presented with many signals. The three TV signals pass through an amplifier with output sufficient to provide more than 1 mV at the domestic outlet through 250 m of cable.

An inexpensive domestic terminal is illustrated in *Figure 26.5*. It assumes one outlet socket in the living room and another elsewhere. The upper channel selector is identical to the lower one shown in full.

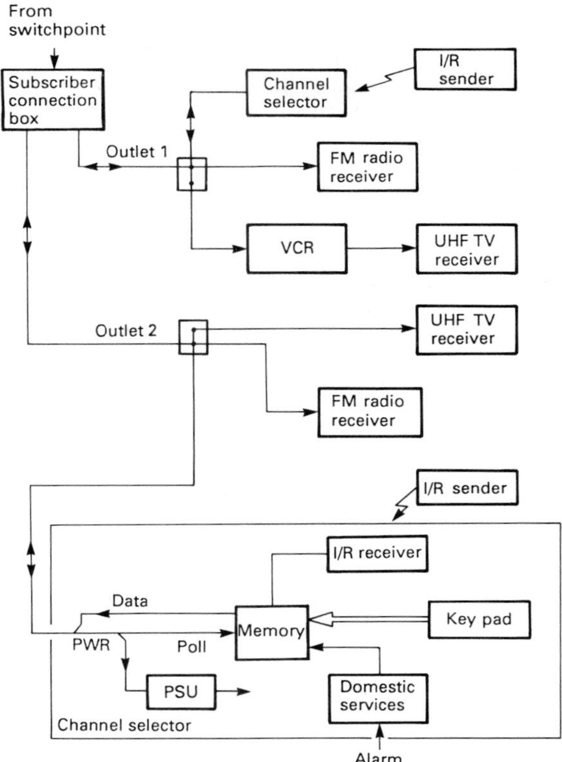

Figure 26.5 Subscriber installation (British Cable Services Ltd)

26.5.5 Further features of practical systems

In general, fm radio programmes are not switched but delivered continuously in frequency division multiplex. The TV signal from each switch comprises only one programme for delivery to the subscriber so there is significant unused bandwidth on the feeder.

Some operators are taking advantage of this to give continuously up to eight TV programmes in fdm. These would form a 'basic tier'. There are two advantages in this: firstly, revenue is obtained from families not wishing to subscribe for more programmes and the cost of an sp switch is avoided, and secondly, one programme can be viewed while another is recorded provided not more than one of these is outside the basic tier. This saves a second sp switch. The disadvantage is the increased likelihood of subscriber confusion where some programmes are obtained by set tuning and others by cable keypad.

The non-linearity of switches, whether crosspoints or converters, is significant and has to be taken into account in the overall distortion build-up.

References

1 UK Patent No. 1 158 918 (1966)
2 TYLER, P, 'Multistar configurations for wideband local networks', *Cable TV Eng*, **12**, No. 7 (1984)
3 BARDUREK, G R, 'Node sizes and switch sizes for star-type distribution systems', *Cable TV Eng*, **12**, No.10 (1985)
4 Dept of Trade and Industry (UK) Specifications MPT 1510 and MPT 1520
5 British Standard No. 6513 Part 3 Appendix
6 ALLORA-ABBONDI, G, 'Interactive communications networks', *Int TV Symposium*, Montreux (1985)
7 CHIDDIX and PANGRAC, 'Fiber backbone — Multi-channel am video trunking', *Int TV Symposium*, Montreux (1989)
8 TAN, PIKAAR and NIJNUIS, 'CATV distribution over a fibre optic local network', *Int TV Symposium*, Montreux (1989)
9 KAWAHATA, M, 'Development of optical information transmission system in Higashi Ikoma', *3rd Euro Electro-Optics Conference*, Geneva (1976)
10 RITCHIE, W K, 'The British Telecom switched-star cable TV network', *British Telecom Tech Jour*, **2** No. 4
11 ETHERIDGE, A J, 'The Thorn-EMI switched cable TV system', *Int TV Symposium*, Montreux (1985)
12 GORGINI, E J, 'Solid state switched cable TV system', *Proc Eurocon '77*, Venice, **1**
13 BONAVOGLIA, L, 'A CATV system with backward channel', *Munchner Kneis Symposium*, Springer-Verlag (April 1977)
14 KOCHAN, R, 'Extendable Pay-TV systems', *Int TV Symposium*, Montreux (1987)
15 QUINTON, K C, 'A high capacity cable TV proposed for the UK market', *Cable Television Eng*, **12**, No. 2 (1983)
16 SMART, D H, 'Innovative aspects of a switched-star cabled TV distribution system', *NCTA Annual Convention*, Las Vegas (1987)

K C Quinton MBE, B Sc, F Eng, FIEE, FRTS
Formerly Director of Research, British Cable Services Ltd

27

Interactive Applications of Multi-channel Cable Systems

nteraction may be regarded as a process of reciprocal action. he newspaper and broadcast industries have long been nvolved in interactive operations. An article in a newspaper night cause a letter to the editor; an advertisement seeking the eturn of a coupon establishes a connection between vendor nd potential purchaser. With commercial TV advertising, a emand can be stimulated.

A cable system, unlike a telecommunications network, has he broadcast feature of linking a single source to multiple utlets. Its TV capacity is high, and it can offer more services han commercial TV broadcasting. In addition it can provide or dialogue between provider and customer. It can also offer ocalized services.

These possibilities were first recognized in the UK in 1974[1] here a study programme was proposed. At that time UK cable ystems were constrained to the relay of broadcast programmes. The impetus for change stemmed from the perceived rowth of an information technology industry and a recommendation that the control of cable systems should be transerred to the Department of Industry[2]. A list of potential cable ervices was produced in 1982[3].

Concurrently, various experiments in interactivity were eing made on cable systems in the USA.

Generally, interactive operations on cable are still at an early tage of development. Potential user companies wish to have a ubstantial audience capable of responding, yet cable operators re reluctant to invest in the additional facilities without ssurance of an adequate return. Predominantly, this investment will be in the provision of communication channels upstream' from domestic and business terminals and in quipment to interface with service providers. Many of the ources of information on interactive services describe proosals with technical solutions, but the popularity or viability re yet to be proved.

27.1 Background

n the 1970s, several experiments were conducted to explore he potential of cable delivery combined with upstream communication from the home. Most were either wholly or partially ponsored by public institutions. Interactive video installations lso found a place in universities. A useful reference is the record of a meeting of the Munchner Kreis in 1977[4]. This contains:

- descriptions of educational projects sponsored by the National Science Foundation, USA
- developmental models to suit pay-per-view, closed user group education and monitoring of meters and alarms, contributed by Michigan State University
- report on adult education experiments in Spartanburg, California
- a contribution from the Mitre Corporation illustrating the background to the 1972 FCC Operational Rules and Standards requiring new cable systems to provide community services and two-way capability
- system design information on the Orlando cable network to carry data or pictures upstream
- experience with college courses by cable in Oregon
- technical details and experience with information services on the experimental fibre-optic system in Osaka and the coaxial system in Tama, Tokyo
- two-way data communications on CATV using multiple concentrators in experiments by CCETT in Rennes
- proposals for education and consultation using CATV with the telephone, contributed from Delft

One of the most notable experiments of the 1970s was with Warner Cable's *Qube* system in Columbus, Ohio. This featured polling, parental control of access, home security services and pay-per-view. It also had continuous audience monitoring for all channels, which revealed the low interest in broadcast children's TV programmes and prompted initiation of the premium channel *Nickelodeon*. The success of this is attributed to the absence of advertisements directed to children (so parents did not switch off).

27.2 Potential interactive services

Services potentially available on cable systems were listed in a UK Cabinet Office report[3]. They are:

- terrestrial TV and radio channels (from inside or outside the country)

- satellite TV and radio broadcasts (also originating at home or abroad)
- subscription TV (films, sport, arts, etc.)
- specialized subject channels (e.g. news, education, religious programmes, health)
- specialized audience channels (e.g. for particular ethnic groups, for different age groups — children, aged, etc., for people with impaired hearing)
- local channels (e.g. local and national government, information, 'what's on', consumer information, programmes by and about community groups)
- other services (e.g. fire and burglar alarms, control of heating systems, remote meter reading, shopping, banking, betting, etc. from home, opinion polling, video games, electronic mail/messaging, interactive computer-assisted learning, general videotex information, software supply to home computers, access to national and international communications, wide-band business communications)

The dependence of these services on technical provisions are outlined in Table *27.1*. The technical provisions shown in that table are the communication requirements from the home or business back towards the service provider. In many operations in the USA, use has been made of the telephone connection, by voice or data (via a modem). The 800-code addressing can remove a call charge. Also sending equipment can be provided which automatically identifies the caller and, for data signals, the key operations can be simplified if service provider instructions (by pictorial text or audio) are followed.

Service	Provision
ENTERTAINMENT	
Basic tier	
5 terrestrial broadcasts, BBC and ITV	
4 cable-orientated programmes	
1 or more non-premium satellite programmes	
Premium programmes	
1 cable-originated subscription TV	(a)
1 sport	(a)(b)(d)
1 arts	(a)(c)
1 or more premium satellite programmes	(a)
NON-ENTERTAINMENT, TV BASED	
Basic	
programme guide	
cabletext	
educational	(b)(d)
local/national government	(e)
Premium	
news	(a)
financial news	(a)
Leased	
Basic	
shopping, banking, insurance,	(b)
building societies, betting, etc	(b)
private broadcasting	(a)
Premium	
home study courses	(d)
NON-ENTERTAINMENT, DATA BASED (leased or premium)	
security, meter reading, audience measurement, funds transfer, downloaded computer software, electronic mail	(f)

Notes : Associated technical provisions

(a) Control of access to TV channels:
Premium services and subscription TV
Private broadcasting
(b) Upstream communication by numeric keypad
(c) Pay-per-view access control and billing
(d) Upstream communication by alphanumeric keypad
(e) Polling
(f) Data transmission: Upstream
Two-way

Table 27.1 Dependence of cable services on technical provisions
Reproduced from 'Local Telecommunications 2' (IEE) with permission

The major drawbacks of telephone use are the tempora blocking of normal telephone calls and time delays due to busy provider's switchboard. However, the telephone offe the advantage of realtime dialogue, and not all of the intera tive cable systems have this facility.

In Table *27.1* upstream communication is assumed to t through the cable network.

It is most important, whenever an interactive service contemplated, to decide the maximum permissible transmi sion delay for messages from a subscriber to a service provid upstream as this will determine an important cable syste parameter.

27.3 Pay-per-view

The alternatives to the established methods of funding pr grammes (i.e. licence fees and advertising) are subscriptio (usually on a monthly basis) and pay-per-view (ppv) in which payment is made for a particular event, such as a major sportir event or a cultural performance, e.g. opera or ballet. The fil industry is supportive of ppv because returns are proportion to audience, as with cinema presentations. A 1985 opinion pc in France showed that 41 per cent of the public favoured pp against 14 per cent for subscription TV[5].

To operate a ppv arrangement there are two alternative Firstly, the event can be preceded by a 'trailer' lasting, say, 3 minutes during which orders can be accepted by keypa booking while the subscriber is viewing that channel. Th message can be stored in the switching point (sp) in a switche star cable system. If the subscriber is creditworthy (informatio held in the sp) the switch will remain operative when the hea end computer signals the event start. Other subscribe switches will be inhibited on the channel at that time. Th billing information will be routed to the head end whe convenient. On tree-and-branch systems, either the set-top un will function as described for an sp or the message will pass the head end. In this case, it must carry both the subscriber identification and the event number as the head end is obliviou of the channel being viewed (unless there is only one pp channel). The event material will be scrambled after a fe minutes and the head end, having checked creditworthines will have addressed appropriate descramblers to release th programme. It will also originate billing information.

American experience shows that purchase of ppv events o impulse is about ten times more successful than ordering via th telephone[7].

Secondly, provision can be made for advance bookings. I this case the event must have an identification number, and will be immaterial which channel any subscriber is viewin while he places his order. Clearly the computer memor systems have to be more extensive for advance booking Echoing of the order is advisable, using a text respons individually addressed, particularly for advance booking otherwise failure to provide the event will be a source c dissatisfaction, it being unclear whether the cause is an error i ordering, inadequate credit rating or system malfunction.

Inclusion of a personal identification number (PIN) whe placing the order is recommended, to stop ordering by childre babysitters, etc.

27.4 Private broadcasting

Private broadcasting is also referred to as a *closed user grou* operation. The conditional access arrangements embodied i cable systems for providing premium services can be used to restrict the audience according to the wishes of a service provider. Examples are informative pictures to members of a police force, update courses for staff of a citizen's advice bureau and pharmaceutical companies' information to medical practitioners. The list of 'permitted' addresses is entered at the head end and then, during the event, all access is inhibited except for those addresses. Interactivity can comprise:

- entry of a PIN by the individual concerned at that address to gain access
- communication of questions from recipients using a keypad

27.5 Interactive education

Possibly the earliest report of distance learning with an interactive facility is by Professor Bordewijk of Delft University[8]. His arrangement combined a *scribophone*, a device for sending or receiving handwriting from a pad via a telephone pair, with a TV camera. Tutor and students each had a scribophone but one camera was used by the tutor. By this means a lecture could be given using an 'electronic blackboard', and students could take part using telephone connections in conference configuration. Clearly, alphanumeric keypads for the students, with their signals routed either by telephone pair or on the cable network, would be an alternative to scribophone use[9].

In Westminster, UK, where the cable system embodies a video library and each subscriber's keypad has controls for playing a video disc, educational facilities are available. These range from programmes available to schools to practical sessions for householders, e.g. maintenance of domestic appliances, cars, etc. Interactivity is restricted to disc choice, rewind and freeze frame[10].

Access to data bases complements studies and can be provided either by videotex-teletext conversion within the cable system or by routeing data to domestic personal computers (see section *27.11*). There are various applications in university colleges[4].

27.6 Video libraries

Provision of a programme, on demand, exclusive to one home requires the availability of a discrete connection between head end and subscriber. The only practical way to achieve this is by following telephone principles of switching. In the case of cable, the number of lines leaving the head end can be constrained by the use of further switching in switching points in switched-star systems. In Westminster[10] each sp serves 300 subscribers, and a maximum of 12 can have access to the library at any one time. This is achieved by a star network of optical fibres to the sps. Advantages are an acceptable bulk of fibres at the head end (333 would serve 100 000 subscribers at 12 TV channels per fibre or 1000 at four channels per fibre as currently in Westminster) and no amplification requirements along the way.

In France, experiments have been conducted[11] with video library services ranging from database text through to movies from video tapes. Previous experience using data was gained from Minitel operation. In Biarritz, 15 subscribers share two channels directly accessing the library[12].

In the USA, a World Video Library service offers ten programmes each week[13], and to accommodate requests for different start times more than ten channels on the cable network are dedicated to this. Subscribers have special converter boxes and pay around $3 for their choice.

Even with few connections dedicated to library access, transmission provision is very costly, as is the central hardware, software and labour involvement. There is little evidence of commercial viability.

Reports from the USA[14] and the UK show that a much less ambitious library service can successfully be operated on one channel available to all. Typically, the library holds 'pop videos', usually much shorter than movies, and subscribers signal their choice from a catalogue. The title then enters a queue which an operator ('video disc jockey') follows for the playing sequence. Times when the items will be shown are provided by subtitle-style text announcements. Popular with predominantly young people throughout 24 hours, this service is referred to as a *video juke box*.

Essential to any video library is a catalogue, presented visually. Readers familiar with accessing videotex information sources by successive menus will realize that for comprehensive library facilities, creating and maintaining a catalogue can be very expensive. In Montreal, a library of about 5500 items is available but the videotape loading is done manually[15].

27.7 Opinion polling

Any cable system with basic upstream communication links should be able to provide for votes to be gathered from viewers. An example of the simplest arrangement would be a programme including a talent show. The announcer might say: *If your vote is for Mr X press your button now*. The upstream data, one bit per vote, should reach the head end very quickly, even on a large tree-and-branch system, and be totalled, so that the next vote can be taken. Another arrangement permits voting on a scale 0–9 using a numeric keypad. Votes for up to ten contestants can then be gathered simultaneously. In both cases, the polling will only be meaningful if steps are taken to inhibit multiple key signals. In one switched-star system this is achieved by accepting only the last entry prior to the *event finished* signal.

Polling results cannot be taken too seriously because there is no knowledge of the number of viewers casting one vote; also the voter might be a very young child. Nevertheless, audience participation adds to programme interest, and a sense of involvement can be helpful for such as the housebound.

27.8 Interactive teleshopping

This is considered by many to be the interactive service with the greatest potential. It is also the one suffering from the greatest inertia because those most likely to be involved, i.e. international stores groups, have a huge investment in shops each dealing with thousands of customers while cable systems with interactive capability are in their infancy so do not, so far, offer a comparable customer base.

27.8.1 Tele-offering

This is the simplest service associated with shopping. A vendor spends, say, 30 minutes demonstrating various goods or services on a channel available to all subscribers. The offerings are numbered using subtitles, and subscribers can use numeric keypads to ask for further information. This is equivalent to returning a coupon in a newspaper or magazine. No orders are placed, there is no need for a PIN, and the cable operator routes the subscriber's details and item number to the vendor when convenient.

27.8.2 Impulse buying

Ability to order on impulse has similar advantages to those of impulse pay-per-view. Presentations are made as for tele-offering, but orders can be placed. Normally, this will require the subscriber to enter a PIN number or credit card number. The upstream capability of the cable system determines the method of communication. One tree-and-branch method is to store an order in the set-top unit until it is polled from the head end. By insisting that the *order* button is pressed during the time that the required item is offered, the message sent up the cable identifies the item by a time code generated in the set-top unit.

In a switched-star system featuring character generators in the sp with presentation in subtitle mode, a subscriber can order in a more leisurely fashion by referring to the item number. Using a keypad they can enter quantity, colour code, etc., with character echoing to check the entry for errors. They would also have a *cancel* key. Order details are polled from the switchpoint.

27.8.3 Text shopping services

Some experience has been gained using videotex, notably in France where, at the end of 1988, four million *Minitel* terminals, about one for every ten households, were connected to telephone pairs. The four main tele-supermarkets in Paris serve more than 15 000 customers by this means with groceries, achieving 100 million francs annual sales with home deliveries within 24 hours of ordering[16]. Clearly, with videotex-teletext conversion and upstream keypad communication provided on cable systems, cable subscribers could have a similar facility. This arrangement was described in 1989[17].

Shopping arrangements described in sections *27.8.1* and *27.8.2* do not require an ability for the subscriber to receive vendors' lists on demand, but this is essential for shopping such as grocery, where the number of items on offer might be several thousands. The application of teletext for downstream list delivery, and for any individual messages, is achieved by allocating a three-digit teletext page number with a four-digit sub-page number to each cable subscriber. The teletext receiver is set to receive this, and the message for the subscriber remains until replaced although it is transmitted only once in two consecutive TV fields.[18] This method of communication is equally applicable to tree-and-branch and to switched-star networks.

In Zaltbommel, in the Netherlands, an experimental system[19] provides information and shopping services. In this, downstream messages reach switching points via full-field teletext and are then converted to analogue TV pictures and applied to the appropriate subscriber's cable.

27.8.4 Still pictures in sequence

Still pictures used for advertising can be delivered by either digital or analogue transmission, the latter being preferred to serve standard TV receivers. The general principle is to provide product information with descriptive text, each item being displayed for a few seconds. In Britain, the application so far has been most successful in selling items typically found in local newspapers in the classified advertisements, e.g. secondhand domestic hardware, cars and houses. Theatre programmes and other local activities also benefit from pictorial presentation.

An advantage of electronic advertising of an individual item is that the advertisement can be removed as soon as a sale is made, removing the frustration of both vendor and potential purchaser thereafter which is often experienced when using newspapers.

At the cable system head-end, the advertisement will be stored digitally, giving the advantage of adjustable picture dimensions, blanking and text insertion using a keyboard. Pictures can be originated from colour photographs or by an 'electronic paintbox'. Available hardware and software provide full broadcast picture quality; British Telecom's *Photo Videotex* and Diverse Picture Systems' equipments both yield a resolution in excess of 400 000 pixels (see section *36*).

27.8.5 Still pictures on demand

This is an equivalent of a mail order catalogue, through which a cable subscriber can browse or, using an index, seek a particular product (or service). An advantage is the ability to update prices, availability and other information immediately without waiting for the next mail shot.

A method in use in the Netherlands and elsewhere is to store the catalogue as in section *27.8.4* but to carry the pages in a dedicated cable channel in analogue form as a sequence of discrete TV fields or frames. Following on-screen instructions and using a keypad, the subscriber selects a page which is then entered into a local store and reproduced continuously as a still frame for as long as required. This apparatus is known as a *frame grabber*. It may be in a switching point or could be in the home.

An interactive system will provide for dialogue with the vendor, including ordering, and the downstream message may be carried by teletext, individually addressed, using text insertion.

It is, of course, possible to extend the service by providing an in-depth picture sequence, e.g. allowing an estate agent to show a client various rooms in a house on demand or a package tour operator to show hotel details pictorially if required.

27.9 Telebanking

The prime purpose of cable TV systems is to serve domestic customers, and they can provide communication between the home and clearing banks. Business telebanking is likely to be conducted by two-way encrypted data signals.

A system requirement is provision of secure (private) connections, and in the downstream direction four possibilities are:

1 data encryption
2 closing of a private broadcasting ability to one address at any one time
3 adaptation of a video library service
4 downstream, individually addressed, teletext data

Option 1 requires special domestic equipment, with a unique code for each home, so this is unattractive. Option 2 is feasible; indeed, with a complete channel available which will support the teletext rate of 6.9 Mb/s, a message can be sent in a very short time. However, the normal provisions for closed user group services are not engineered for the very rapid access re-routeing requirement necessary if several customers are on line. For option 3, it is doubtful whether many cable systems will provide for a video library due to the high cost of extra trunk circuits and plant.

A solution to the problem, which gives good privacy, is to adopt the teletext message principle described in section *27.8.3* for teleshopping. The privacy stems from the fact that an eavesdropper would need to know the teletext page number of the recipient and be receptive to that page during the two fields that the message is sent (it is only sent once). Several clearing banks have accepted this method on the basis that the information displayed does not need to identify the intended recipient. In switched-star systems,a private message can be produced in analogue form in switching points; the data input can be blocked there so complete privacy is then assured.

Although the teletext message route through to subscribers is equally applicable to tree-and-branch and switched-star networks, upstream messages cannot be regarded as private on tree-and-branch networks because the isolation between subscribers on one feeder may be as low as 30 dB. Eavesdropping is thus practicable.

27.10 Telebetting

To operate a telebetting service on a cable TV system, a critical requirement is either a very fast upstream message capability or a message store arrangement which records the exact time that the bet was placed. Even so, the bookmaker will not tolerate much delay as placements can alter the odds. Upstream traffic will include very high peaks, and the system must handle these with minimum delay.

It should be presumed that the event is shown as live pictures, preferably with inserted text to communicate with individual participants echoing their entries so that errors can be corrected. It will normally be necessary for anyone placing a bet to have an account with the bookmaker providing the service, to ensure creditworthiness. Use of a PIN will also be essential to identify the individual.

27.11 Miscellaneous services

These are services referred to in section *27.1* and Table *27.1* that have not been itemized in the subsequent sections. It should be possible to identify the technical arrangements that are described which can be used to facilitate the outstanding services.

Concentration has been on the use of a keypad and domestic TV receiver in the home. Clearly, if a personal computer is introduced, horizons are widened. For example, a cable operator could have a library of computer programs at the head end and make these available to individuals using a pay-per-view access facility. The preferred method of data delivery is by teletext, but until receivers become available which embody teletext-to-videotex conversion the data might have to be sent at a slow rate. Receivers are available that produce hard copy from teletext. Another possibility is the provision of an electronic mailbox at a cable head-end. While incoming mail could be delivered to the TV screen, it would be best to use a word processor for originating mail. Local advertisers could use the cable mailbox to advantage.

Two-way low speed data, whether interrogating alarms or utility meters, is interactive. These services exploit the point-to-multipoint feature of a cable network and most likely will use data addressing for premises identification.

References

1 'Report on cable communication systems by the National Electronics Council, *National Electronics Rev*, **10** 1, pp 12–14 (1974)
2 *Information technology*, Report by Cabinet Office Advisory Council for R & D, HMSO (1980)
3 *Cable systems*, Report by Cabinet Office IT Advisory Panel, HMSO (1982)
4 KAISER, MARKO and WITTE, 'Two-way cable television', *Munchner Kreis Symposium* (April 1977)
5 FLICHY and MICHON, 'Montpellier multiservice experiment', *Intern TV Symposium*, Montreux (1987)
6 BRUGLIERA, V, 'Pay-per-view — Déjá vu?', *Intern TV Symposium*, Montreux (1987)
7 ALDEN, P J, 'The American experience with interactive services', *Intern TV Symposium*, Montreux (1985)
8 BORDEWIJK, J L, 'On the marriage of telephone and television', *IEEE Trans on Communications*, *Com-32*, No 1 (Jan 1975)
9 KEGAL and BONS, 'On the digital processing and transmission of handwriting and sketching', *Eurocon*, Venice (1977)
10 RITCHIE and SEACOMBE, 'The Westminster multiservice cable-TV network — Experience and future developments', *Intern TV Symposium*, Montreux (1987)
11 DUPUIS and PONCIN, 'Experimentation des services de télévideotheque', *Intern TV Symposium*, Montreux (1985)
12 LACOTT, CAYLA, MAHIER, LAMICHE and VEYRES, 'Video libraries and interactive video systems: Hardware and software', *Intern TV Symposium*, Montreux (1985)
13 'W V L offers two way PPV', *Cablevision*, (Jan 22 1984)
14 'C V J offers interactive juke-box via cable', *The Videodisc Monitor*, **4**, No 4, (April 1985)
15 'Cable Canada: Cablodistribution', *Cable and satellite Europe (GB)*, (Sept 1985)
16 McCLAY, A, 'Focus homeshopping', Comité International des Entreprise á Succursales (Paris)
17 ROBERTS and VINCETT, 'Videotex and interactive services on tree and branch cable TV networks', *Intern TV Symposium*, Montreux (1989)
18 QUINTON, K C, 'A high capacity cable TV proposal for the UK market', *Cable Television Eng*, **12**, No 2 (Feb 1983)
19 HOFMAN and KRIJGER, 'The provision of interactive services in a CATV distribution network — The Zaltbommel experiment', *Intern TV Symposium*, Montreux (1987)

Part 6 Direct Broadcasting by Satellite

D Wood
European Broadcasting Union

28 DBS Systems: Planning and Fundamentals

Satellites have been used for many years in the transmission of signals from one terrestrial point to another. Services of this kind are known as *fixed satellite services* (fss) and operate at relatively low powers. The equipment required to receive signals from the satellite is usually relatively large and expensive, although 'quasi-broadcasting' in fss bands may be practical. Fixed service satellites are used for relaying television programmes from one country to another, e.g. via the Eurovision network. For *direct broadcasting by satellite* (dbs), the satellite transmitted signals are at higher power levels so that the signal can be received by a relatively small receiving antenna connected to the television set in the home. The term 'dbs' is more widely used than the formal ITU term, *broadcasting satellite service* (bss).

A number of European countries have plans for the introduction of dbs services. In a dbs system, programmes from a ground station are transmitted to the satellite in a fixed satellite service band. Bands specifically intended for this purpose (*feeder-links*) are used. The ground station, or possibly a separate station, also transmits signals for the control and monitoring of the satellite. The satellite converts the programme carrying signals from the ground station (the *uplink*) to a frequency in the band allocated to dbs and retransmits them in a beam aimed towards the service area.

Two types of reception are possible and are defined by the ITU *Radio Regulations*: individual reception and community reception. For *individual reception*, the viewer needs two items of equipment. The first is an outdoor unit, which must be mounted where there is a clear view of the satellite. This unit is likely to consist of a parabolic antenna (or dish) between 30 and 90 cm in diameter, or possibly a flat panel, together with the associated system to convert the signal to a lower frequency suitable for transmission, by coaxial cable, to the second unit, the indoor unit. This is likely to consist of a channel selector mounted on or near the television receiver, and eventually such systems may be included in the receiver itself.

For *community reception*, the signal from the satellite is received by a relatively large antenna, possibly a 2–3 m dish, down converted and distributed by cable to a number of receivers, e.g. in a block of flats. With modulation conversion, it is possible to use conventional cable networks for distribution over a wide area.

DBS differs from conventional terrestrial broadcasting in that full national coverage can be achieved by a single transmitter on a satellite. With individual reception, some viewers may not be able to receive the transmissions because they do not have an unobstructed view of the satellite. This problem is more likely to occur in built-up areas with the obstruction taking the form of a neighbourhood building or other structure. Geographical features can also obscure the view of the satellite. As examples, the angles of elevation required in the UK to achieve a boresight with a geostationary satellite at 31°W, the orbital position allocated to the UK, would be from about 20° in the north of Scotland to 28° at Land's End in the south west of England.

Some years ago, the BBC carried out a survey to estimate the proportion of houses which might be unable to receive transmissions from a UK direct broadcasting satellite using an individual antenna. At 3 pm on or about October 13, the sun appears to an observer in the UK to be in the same position in the sky as that which a satellite in the orbital position assigned to the UK would occupy. A sample of several hundred householders was asked to observe which parts of their premises were at sunlight. The conclusions of this survey was that nearly all householders should be able to find at least one position for an individual receiving antenna on their premises.

Frequency modulation or another constant amplitude modulation method will be used for dbs.

The receiver will probably be a double superhet. The first stage will be mounted on the receiving antenna and can employ a Gunn oscillator and a tuned cavity at a fixed frequency below the signal frequency, and give a first intermediate frequency in the range 900–1300 MHz. This signal will arrive at the indoor unit and be decoded.

28.1 Geostationary orbit

Not all orbits are suitable for satellite broadcasting. Generally, such systems are based on the unique case of the geostationary orbit. This is a circular equatorial orbit, in which the period of revolution of the satellite is equal to the period of rotation of the earth, and the direction of movement of the satellite is in the direction of the rotation of the earth.

Such a satellite appears to be stationary in the sky when viewed from any point on the earth and is fixed at the zenith of a given point on the equator, the longitude of which is that of the satellite. This longitude describes the position of a satellite on the geostationary orbit. The great advantage of the geostationary orbit is that it enables a fixed receiving antenna to be used on the ground. This is so important that almost all the systems studied in connection with satellite broadcasting are based on geostationary satellites.

The characteristics of the ideal geostationary orbit are:

Period, t:	86 164.091 s = 23 h 56 min 4.091 s
Equatorial radius of the earth, r:	6378.16 km
Altitude, h:	35 786.04 km
Radius of the orbit, r + h:	42 164.20 km
Speed of the satellite:	3.074 662 km/s
Length of 1° arc:	735.904 km

28.2 Satellite geometry

The geometry of a geostationary satellite can be seen from *Figures 28.1* and *28.2*. The position of the satellite S is defined by its longitude. The latitude is determined by the geostationary orbit. The receiving point P is taken to be at a longitude λ, and a latitude φ.

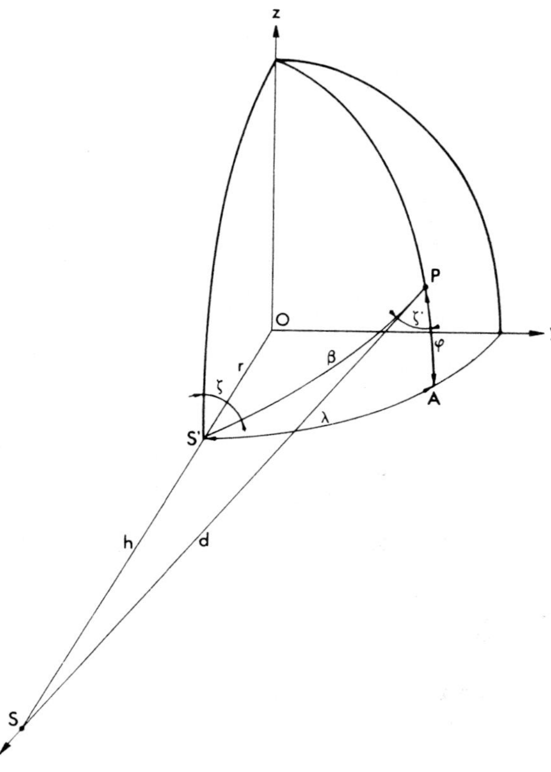

Figure 28.1 Evaluation of the angles at which a point P on the earth is seen from the point S′ on the earth's surface situated vertically below a geostationary satellite, S (EBU)

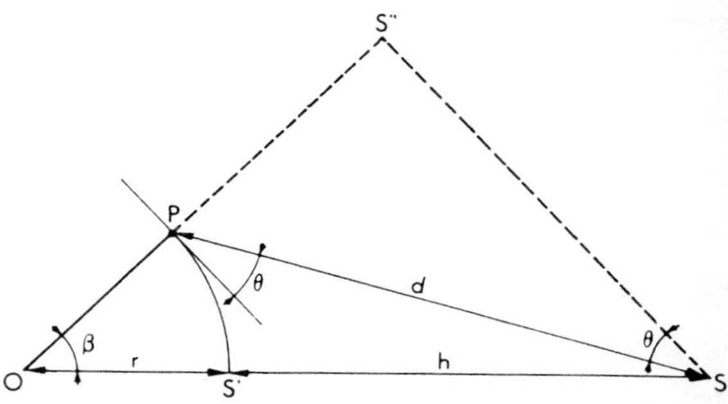

Figure 28.2 Evaluation of θ, the angle of elevation of the satellite S (EBU)

Taking the difference in longitude to be $\lambda = \lambda_1 - \lambda_S$, it can be shown that:

azimuth $\zeta'^{\circ} = \tan^{-1}(\tan\lambda/\sin\varphi)$,
elevation $\theta^{\circ} = \tan^{-1}(\cos\beta - \delta)/\sin\beta$

where $\delta = r/(r+h) = 0.151\ 269$ and $\beta = \cos^{-1}(\cos\varphi\cos\lambda)$, and distance d km $= 35\ 786\ \sqrt{(1 + 0.419\ 99(1-\cos\beta))}$.

28.3 Satellite coverage

Calculation of real coverage areas requires a computer study. A simplified picture for a circular beam can be obtained as follows. The convention is adopted (see *Figure 28.3*) that the coverage extends to a distance from the centre of the coverage area (boresight from satellite antenna, S) corresponding to a semi-beamwidth angle δ.

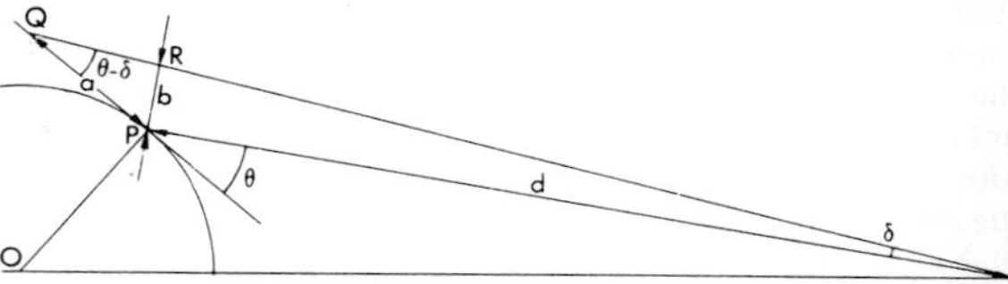

Figure 28.3 Geometrical representation of the plane containing the centre of the earth, the satellite and a point P on the earth (EBU)

In general, the coverage area will be elliptical. The semi-major axis, a, can then be shown to be approximately $a = b/\sin(\theta - \delta)$ where the coverage radius in the transverse direction, b, the semi-minor axis, can be calculated as $b = d\tan\delta$.

28.4 Propagation

An isotropic point source radiating a power *P* watts will give rise to a power flux density, pfd, at distance *r* metres defined by:

$$\text{pfd} = P/4\pi r^2\ \text{W/m}^2.$$

The power radiated is realized by using a transmitter of power P_t together with an antenna of gain G_t in the relevant direction. The product $P_t \times G_t$ is called the *equivalent isotropic radiated power* (eirp), and is in the range of about 62 dBW for direct broadcast satellites. The power available from the receiving antenna, P_r, is pfd × antenna aperture.

Antenna aperture is related to receiving antenna gain, G_r, by:

$$\text{antenna aperture} = G_r\lambda^2/4\pi$$

where λ = wavelength.

Combining the above and substituting P_t and G_t in place of P, gives:

$$P_r = P_t G_t G_r \lambda^2/(4\pi r)^2$$

For a geostationary orbit at 12 GHz, the free space attenuation $\lambda^2/(4\pi r)^2$ is about 206 dB in Europe. Directly below the satellite it is 205.1 dB increasing to 206.4 dB at the horizon.

Thus, for example, in the UK, P_r dBW = P_t dBW + (G_t + G_r-206) dB.

28.5 Eclipse of a geostationary satellite

When certain conditions coincide, the sun, the earth and the satellite lie on a straight line and the satellite passes into the shadow of the earth. From the point of view of an observer on the satellite, the sun is eclipsed. An observer on the earth will see an eclipse of the satellite. Normally, broadcasting satellites will not be able to transmit programmes during the period of the eclipse. The weight of batteries required to replace the solar power would be prohibitive.

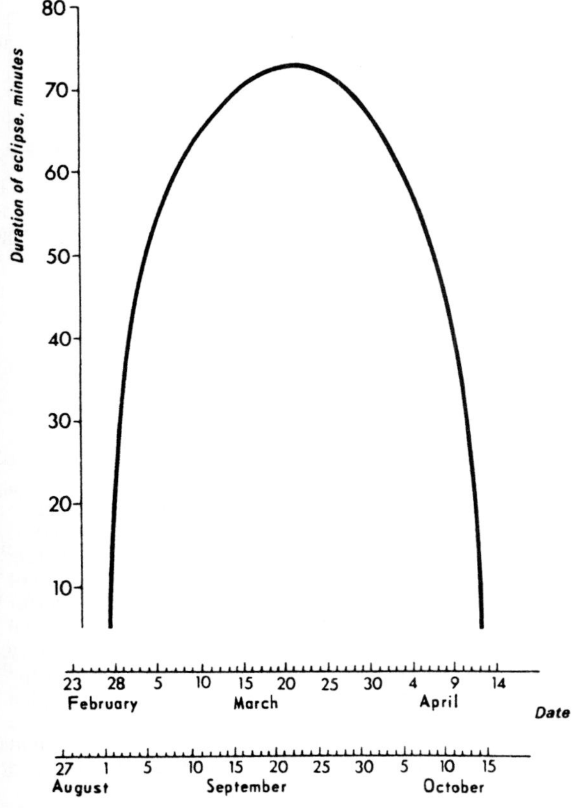

Figure 28.4 Real duration of an eclipse as a function of calendar date (EBU)

Figure 28.4 shows the duration of the eclipse as a function of calendar day. The maximum duration of the eclipse is 72 minutes, occurring on March 21 and September 23. The duration of the eclipse becomes zero 22 days either side of the equinox, and an eclipse will occur therefore during the two following periods: in spring from February 27 to April 12, and in autumn from September 1 to October 15.

The duration of the eclipse is longer than one hour for a total of 52 days in a year (March 8–April 3, and September 9–October 6).

The mid-point of the eclipse always occurs at midnight true solar time at the longitude of the satellite. The eclipse can be advanced or retarded by moving the satellite towards the east or west respectively. A change of position of 1° in longitude corresponds to an alteration of 4 minutes.

28.6 Signal/noise ratio and picture quality

In order to define service requirements in terms of signal/noise ratio, a specification is used based on a degree of subjective quality that is recognized as being not very good: this is normally graded 3.5 on the CCIR 5 point scale for conventional television. This corresponds to an impairment between *perceptible but not annoying* and *slightly annoying*. For a 5 MHz bandwidth PAL/SECAM picture, this implies a signal/noise ratio of about 33 dB unweighted. This is seen as the minimum value and it is considered that it should be achieved at the limit of the service area, during 99 per cent of the least favourable month (from the point of view of propagation).

In addition to this, a special specification for the carrier/noise ratio must be given because of the fm threshold. Although the quality above is specified for 99 per cent of the least favourable month, it would be unacceptable if the fm threshold were reached during 1 per cent of the least favourable month. Consequently, the carrier/noise ratio is specified to lie above the threshold during 99.9 per cent of the least favourable month.

The two specifications suggested for planning associated with PAL/SECAM are:

c/n ⩾ 10 dB (99.9 per cent of the least favourable or worst month)
c/n ⩾ 14 db (99 per cent of the least favourable or worst month).

28.7 Protection ratios

In choosing the subjective quality by which the protection ratios are to be defined, it is not possible to proceed as in the case for noise and accept a minimum quality that will be near mid-opinion. In a satellite system, interference may well have a relatively continuous effect without significant variation. Future developments in receiving systems are more likely to give an improvement in the noise rather than in interference. For interference, therefore, CCIR impairment grade 4.5 was used for the WARC 1977 Plan (see section *28.9*). The results were based on a large number of subjective assessments using PAL/SECAM. They indicate that for co-channel interference (0 MHz channel spacing), the protection ratio required to give grade 4.5 is just greater than 30 dB. The protection ratio required to give grade 4.5 with a 19 MHz channel spacing (adjacent channel interference in the WARC 77 Plan) is 14 dB. In order to permit small variations of power, the protection ratios adopted for planning were 31 dB cci and 15 dB aci. Tests since made with the MAC/packet family (see section *29*) indicated that these ratios would still give adequate protection.

28.8 Receiver sensitivity

It is usual practice to specify the overall performance of the receiving system in terms of its G/T, where G is the antenna gain and T is the effective noise temperature.

$$G/T = \frac{\alpha\beta G_m}{\alpha T + (1-\alpha)T_0 + (n-1)T_o}$$

Channel group	Frequency, GHz	Orbit position	Country	Polarization R-hand or L-hand
1	11.72748	19°W	France	R
5	11.80420	37°W	San Marino	R
9	11.88092	5°E	Turkey	R
13	11.95764			
17	12.03436			
2	11.74666	13°W	Ireland	R
6	11.82338	19°W	W. Germany	L
10	11.90010			
14	11.97682			
18	12.05354			
3	11.76584	37°W	Liechtenstein	R
7	11.84256	31°W	Portugal	L
11	11.91928	19°W	Luxembourg	R
15	11.99600	5°E	Greece	R
19	12.07272			
4	11.78502	37°W	Andorra	L
8	11.86174	31°W	United Kingdom	R
12	11.93846	19°W	Austria	L
16	12.01518			
20	12.09190			

Further assignments, all on orbit position 5°E, are as follows:

Channels	Country	Polarization
2, 6, 10	Finland	L
14, 18	Norway	L
4, 8	Sweden	L
12, 16, 20	Denmark	L

Table 28.1 The WARC-BS 1977 Plan: lower half 11.7 – 12.1 GHz

where α = total coupling losses
β = total of antenna and installation ageing losses
G_m = maximum isotropic gain of antenna
T = effective temperature of the antenna (~150 K)
T_o = noise factor reference temperature (290 K)
n = overall noise factor in the receiver, referred to the receiver input.

The specification proposed at the time of the WARC-BS 77 Conference (see section *28.9*) was a G/T of 6 dB/K. This could be achieved with a 0.9 m diameter antenna and a noise factor of 8 dB, with a margin for coupling loss, pointing loss and ageing loss.

However, since the conference, achievable noise factors have fallen to about 4 dB or less. This means that the same G/T can be achieved with a smaller antenna, or that a given service may now be receivable over a wider area (subject to interference constraints), or indeed that the quality obtained at the edge of the service area will be better than grade 3.5, which may be useful for HDTV.

28.9 The WARC-BS 1977 plan for 12 GHz dbs in Regions 1 and 3

The basis for satellite broadcasting in the 12 GHz band in Europe was set at a World Administrative Radio Conference in 1977. The conference formulated plans for future satellite broadcasting services in the frequency band 11.7–12.5 GHz. The outcome was a world agreement for satellite broadcasting with an associated plan covering ITU Regions 1 and 3 (most of the world except North and South America). (A similar plan for ITU Region 2, the Americas, was established by a Regional Administrative Radio Conference in 1983.) The date of entry into force of the 'Final Acts' was January 1, 1979, and the duration of the validity of the provisions in the associated plan was at least 15 years from the date of entry into force of the Final Acts, i.e. until 1994. If, before that date, no competent conference has made provisions to replace the plan and agreement, they will continue in force.

Region 1 (which includes Europe) has the use of 40 channels (11.7–12.5 GHz), and Region 3 uses 24 channels (11.7–12.2 GHz).

Guard bands are provided of approximately 14 MHz at the lower extremity and 11 MHz at the upper extremity of the band. The orbital positions are arranged generally on a regular spacing of 6° with a discontinuity at around 34°E. The United Kingdom has been assigned five channels for an orbital position at 31°W, and most other European countries have a similar number of assignments.

The technical criteria used in Region 1 in formulating the plan were based on the assumption that a conventional 625/50 PAL or SECAM colour television signal would be used with a

Channel group	Frequency, GHz	Orbit position	Country	Polarization R-hand or L-hand
21	12.11108	37°W	Monaco	R
25	12.18780	31°W	Iceland 1	L
29	12.26452	19°W	Belgium	R
33	12.34124	5°E	Cyprus	R
37	12.41796			
22	12.13026	19°W	Switzerland	L
26	12.20698			
30	12.28370			
34	12.36042			
38	12.43714			
23	12.14944	37°W	Vatican	R
27	12.22616	31°W	Spain	L
31	12.30288	19°W	Netherlands	R
35	12.37960	5°E	Iceland 2	R
39	12.45632			
24	12.16862	19°W	Italy	L
28	12.24534			
32	12.32206			
36	12.39878			
40	12.47550			

Further assignments, all on orbit position 5°E, are as follows:

34	Sweden	L
38	Norway	L
24, 36	Nordic 1	L
22, 26	Nordic 2	L
28, 32	Nordic 3	L
30, 40	Nordic 4	L

Table 28.2 The WARC-BS 1977 Plan: upper half 12.1 – 12.5 GHz

frequency modulated sound subcarrier. This frequency modulates the main carrier with a deviation of 13.5 MHz/V, which leads to an overall bandwidth of about 27 MHz. The channel spacing is 19.1 MHz, so there is an overlap between channels.

It was agreed that the planned power flux density to be provided by the broadcasting satellites would be -103 dBW/m^2 at the edge of the service area. It was also agreed that an energy dispersal signal of 600 kHz would be applied.

Particular features of the plan are that in Europe, eight countries in western Europe share a common orbital position, and six countries in eastern Europe share a common orbital position. The Nordic countries (Denmark, Finland, Norway and Sweden) arranged to share in a common supranational beam, having two channels allocated to each country concerned. The number of channels in each national beam is correspondingly reduced to three. Supranational beams are also provided for a number of Arabic countries.

The aim of the plan was to select frequencies and orbital positions so as to minimize interference between the various assignments in the plan. The elements used included receiver selectivity, receiving and transmitting antenna directivity and polarization.

28.9.1 Receiver selectivity

The if response of the receiver largely determines the receiver selectivity. Only co-channel and adjacent channel interference levels were taken into account. It was assumed that receiver selectivity would allow the level of signals beyond the first adjacent channel to be ignored.

To assess the quality of the plan, each assignment is judged by a single criterion called the *equivalent protection margin*. This is a measure of the effective sum of the co-channel and adjacent channel interference signals expressed in terms of a single equivalent co-channel interference. The equivalent protection margin is numerically the value in decibels by which the ratio of wanted signal to total interference exceeds the agreed co-channel protection ratio.

28.9.2 Directivity of the receiving antenna

Signals reaching the receiving antennas outside the main beam are considered as being reduced in level by the amount indicated by the reference pattern agreed by the conference. Since the receiving antenna is assumed to be directed towards the orbital position of the wanted satellite, the effect of receiving antenna directivity is dependent upon the orbital separation between the wanted and interfering satellite.

28.9.3 Directivity of the transmitting antenna

Outside the service area it is assumed that the transmitted signal falls off in accordance with the conference agreed reference pattern for the transmitting antenna.

Channels	Orbit positions −31°		−25°		−19°	
	R	L	R	L	R	L
1, 5, 9, 13, 17				LBX	F	
2, 6, 10, 14, 18	IRL	GNP	ALH	TGO	ZA2	D
3, 7, 11, 15, 19	LBR	POR + AZR		LBZ	LUX	BEN
4, 8, 12, 16, 20	G	CPV	ALI		ZA1	AUT
21, 25, 29, 33, 37	HVO	ISL		MRC	BEL	NMB
22, 26, 30, 34, 38		CTI	TUN		NIG	SUI
23, 27, 31, 35, 39	SRL	E + CNR	GHA		HOL	GNE
24, 28, 32, 36, 40				NGR		I

ALH, ALI — Algeria beams
AUT — Austria
AZR — Azores
BEL — Belgium
BEN — Benin
CNR — Canary Islands
CPV — Cap Verde
CTI — Ivory Coast
D — Federal Republic of Germany
E — Spain
F — France
G — Great Britain
GHA — Ghana
GNE — Equatorial Guinea
GNP — Guinea-Bissou
HOL — The Netherlands
HVO — Burkina Faso (formerly Upper Volta)
I — Italy
IRL — Ireland
ISL — Iceland
LBR — Liberia
LBX, LBZ — Libya
LUX — Luxembourg
MRC — Morocco
NGR — Niger
NIG — Nigeria
NMB — Namibia
POR — Portugal
SRL — Sierra Leone
SUI — Switzerland
TGO — Togo
TUN — Tunisia
ZAI 1, ZAI 2 — Zaire beams
R — right-hand polarization
L — left-hand polarization

Table 28.3 Allocations associated with orbit positions at −31°, −25° and −19°

28.9.4 Polarization discrimination

Circular polarization is used, to assist frequency re-use, and interference is reduced by ensuring that the interfering signal has an opposite polarization from the wanted signal. The cross-polar patterns of the transmitting and receiving antennas are shown in *Figures 28.5* and *28.6*.

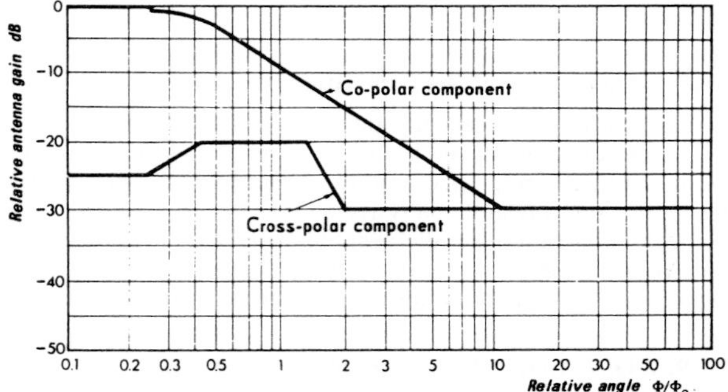

Figure 28.5 Co-polar and cross-polar diagrams of the antenna for individual reception (EBU)

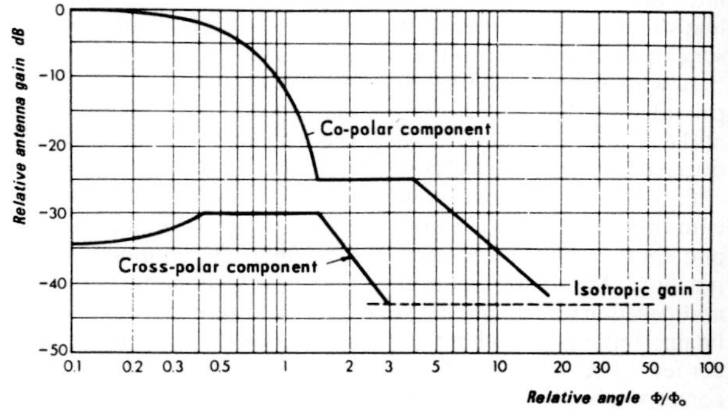

Figure 28.6 Co-polar and cross-polar diagrams of the transmitting antenna (EBU)

Countries are not of a regular size and shape, so the transmitting beam which is normally elliptical and which must serve the whole of the country's territory must also inevitably cover some parts of neighbouring countries. In addition, it is necessary to allow a margin of 0.1° in each direction to cover satellite pointing errors. The WARC 77 Plan was based on the use of so-called polygon points. These points form the corners of a convex polygon which encloses the territory concerned.

From these, the computer is able to calculate the minimum elliptical beam required. The minimum practical transmitting beam is assumed to be a circular one whose diameter subtends an angle of 0.6° at the satellite. This is larger than required to give coverage of small countries and consequently, in such cases, the radiated power needs to be reduced to a value such that the agreed value of power flux density at the limit of the service area is provided and that protection from interference is not provided outside the polygon defining the country concerned.

The following groupings were realized in the plan:

at 19°W: France, West Germany, Belgium, the Netherlands, Luxembourg, Italy, Switzerland, Austria
at 1°W: Poland, East Germany, Czechoslovakia, Hungary, Romania, Bulgaria
at 5°E: Norway, Sweden, Finland, Denmark
at 25°W: Libya, Tunisia, Algeria, Morocco.

Bibliography

MERTENS, H, 'Satellite broadcasting: design and planning of 12 GHz systems', *EBU Publication Tech* 3220-E (March 1976 (re-edited in 1981))

Satellites for Broadcasting, IBA Technical Review 11 (March 1979 (revised edition))

Broadcasting Satellite Systems, CCIR Publication (1983)

Broadcasting Satellite Service (Sound and Television), CCIR Recommendations and Reports of the CCIR, **10** and **11**, Part 2 (1986)

D Wood
European Broadcasting Union

29 DBS Transmission Systems

The World Administrative Radio Conference WARC-BS 1977 established the planning for broadcasting satellite services (bss) in the frequency bands 11.7–12.2 GHz for ITU Region 3 (Australasia and Asia) and 11.7–12.5 GHz for Region 1, (Europe, Africa and the part of the Soviet Union in Asia). See section *28.9*.

The WARC-BS 1977 planning was based on the quality requirements of conventional composite television systems (PAL, SECAM and NTSC, with analogue sound). However the plan allows other systems to be used, provided they do not cause more interference than the planned conventional systems. There are certain stated limits which must not be exceeded (cci 31 dB, aci 15 dB) for the just perceptible threshold of interference (CCIR grade 4.5).

At the present time, it is unlikely that PAL or SECAM will be used in Europe because a new transmission system (the MAC/packet family) is the subject of a directive by the European Community to its member states. Also, in the Federal Republic of Germany there are plans for a digital sound broadcasting system (dsr) which carries 16 stereophonic signals in one dbs broadcasting channel.

In Japan, on the other hand, satellite broadcasting began in 1984 with an NTSC vision system coupled with a new digital sound subcarrier system. Subsequently, an HDTV broadcasting system called MUSE which is capable of being used in a 27 MHz dbs channel has been developed by the Japanese Broadcasting Corporation (NHK).

In Europe, a further member of the MAC/packet family, HD-MAC, is being developed for 27 MHz channels. 12 GHz dbs service operations in the 50 Hz world will have the option of beginning a new service directly in HDTV with HD-MAC or of continuing an already existing MAC/packet service with HD-MAC.

Wideband analogue and digital HDTV emission formats which may be candidates for dbs use in the 22 GHz band, should it become available, are at an earlier stage of development in several parts of the world[1].

Studies are also being made in Europe with a digital sound satellite broadcasting format *ads* (advanced digital system) for use in the range 0.5–3 GHz, should this be possible. The service audience could include automobiles[2].

In 1983, the Region 2 (the Americas) dbs channels were the subject of a planning conference (RARC-BS 1983), and a plan was established using 24 MHz wide channels. Once again the plan was based on conventional composite signals, but with agreement that alternative systems could be used provided that certain interference limits were not exceeded. At this planning conference there was considerable interest in the use of MAC-type systems for dbs, although no dbs services were then planned in North America, largely because the economic balance was not considered to be sufficiently favourable.

The worldwide pattern of dbs broadcasting standards is therefore not simple, and it is possible to indicate here only some of the key features of certain systems. Particular emphasis is placed on the MAC/packet family because of its importance. Some historical background is necessary to understand why the systems are as they are.

29.1 MAC/packetfamily

29.1.1 General principles

In 1981, agreement was concluded in the CCIR on a digital television studio standard (CCIR Rec. 601) based on the use of separate YUV components, rather than a composite structure like PAL, SECAM or NTSC[3]. One of the parameter values to be established was the bandwidth of the components. There was evidence to suggest that increasing the luminance resolution beyond 4.5 MHz in a 625/50 system produces very little return in terms of visible improvement in picture quality at viewing distances of 6 H or more. Equally there was evidence that a colour-difference bandwidth of about one third of the luminance bandwidth provides a good subjective quality balance. However there were other arguments which led to the choice of rather higher bandwidths for Rec. 601. The main one (at the time) was that allowance should be made, in terms of a bandwidth margin, for the requirements of picture processing systems in studios, such as colour-matte. This led to the choice of about 5.5–6 MHz luminance bandwidth, and colour-difference bandwidths of half that value.

To support the higher luminance bandwidth it was also suggested in Europe that such bandwidths, if broadcast and shown on domestic screens of moderate size, could provide a quality bridge between conventional composite television quality and a future high definition television system. This quality could well be difficult to distinguish from HDTV itself on moderate-sized screens, and therefore it could be an attractive possibility for broadcasters and consumers in the short term, until low cost, lightweight, very large HDTV displays are available for the consumer market.

For Europeans, the next available broadcast band is the 12 GHz dbs band, and after agreement on CCIR Rec. 601, the IBA in the UK devised a method of broadcasting the Rec. 601 quality in the channels allocated by the WARC-BS 1977. They used a technique that is not new in concept but which was only recently practically implementable in receivers, and which is termed the MAC system (*Multiplexed Analogue Components*)[4]. Essentially the system is a time multiplex of luminance and line-sequential colour-difference signals which have been time compressed so that both fit together, next to each other, in the active line time, rather than a frequency division multiplex as in conventional composite systems. An evaluation of alternative 12 GHz dbs signal formats which included digital vision systems, MAC systems, and improved composite systems, led in 1983 to the MAC system being chosen[5]. It was thought to permit future expansion to a greater extent than improved composite systems, and it offered higher picture quality than the digital bit-rate reduction systems practical for the 27 MHz rf bandwidth available in the WARC-BS 1977 plan.

Studies were made in parallel on digital sound and data coding systems to accompany vision for dbs; the classification system used for them is:

A systems: frequency multiplexing, with the picture signal, of a digitally modulated sound and data subcarrier

B systems: baseband time multiplexing, with the picture signal, of a full response digitally coded sound and data signal

C systems: rf time multiplexing of a sound and data system with the picture signal

D systems: baseband time multiplex of the picture and a sound and data system with partial-response coding (more specifically, duobinary coding)

The type of system which is most efficient in terms of available audio channels and failure characteristic for the available rf bandwidth, assuming that the line blanking period is used for a digital burst, is the *C system*, and this system was originally proposed (using 2-4psk) by the EBU for dbs in conjunction with the MAC vision system.

The C-MAC system has (for a 27 MHz bandwidth dbs system) a data capacity of about 3 Mbits/s, and this can be used for various digital audio or data services. Before the transmitter, a multiplex combines the various digital signals into a single bit stream, and at the receiver a demultiplexer separates the signals so that the user can choose the service he wants.

The main criteria by which a method of multiplexing can be judged are flexibility and efficiency. The *efficiency* is the ratio of useful data to labelling data in the bit stream. A balance must be struck between the two, of the highest efficiency possible, commensurate with the required *flexibility* and with practical receiver design. Two classes of systems were studied, the *packet* system, in which consecutive packets of data are sent, comprising a label or header, followed by bytes of useful data, and the *structure map* or *fixed format* system, in which the packet routeing information is sent more slowly and less often. Both systems were proved to work, and for reasons partly technical and partly non-technical the packet system (originally proposed by the CCETT, France) was chosen. The name thus given to the original EBU-proposed dbs system in Spring 1983 was the *C-MAC/packet* system.

The EBU C-MAC/packet system had initially a mixed reception among receiver manufacturers. Although the C system is the most efficient modulation method for satellite broadcasting, it was argued to be incompatible with cable systems and inconvenient for other transport media such as vcrs. In response to this, studies were made to develop an am version of the system, which could be used on cable networks. Two types of cable network channel were considered, one with an available bandwidth of about 7 MHz and the other with an available bandwidth of about 11–12 MHz. For the latter, the *D-MAC/packet* format was developed. In this system, the sound and data signals in the horizontal blanking are duobinary coded and amplitude modulate the carrier along with the vision signal. The duobinary system was proposed rather than quaternary (four-level) coding, because it is less sensitive to transmission impairments (such as lf gain irregularities) but is, on the other hand, more efficient than a simple binary (two-level) coding scheme. It was considered to provide the most efficient coding method consistent with the kinds of transmission line impairments likely to occur in cable networks.

Following the development of the D-MAC/packet system in Europe, it was necessary to derive from it a second format for the circumstances where the system basebandwidth could not be extended beyond about 6 MHz. This is the case for many current cable systems in Europe now using conventional PAL and SECAM. Consequently a system with half the bit-rate of the D system was developed, the *D2-MAC/packet* system. The format of the C and D systems was adjusted such that the data could very easily be divided into two parts. In fact, the data burst for the C and D systems can be considered as two side-by-side D2 data blocks, which are time compressed so that they both fit into the time period of the horizontal blanking.

Following this, there was pressure by European receiver manufacturers to develop a broadcast fm version of the D2-MAC/packet system for use in France and Germany, so that one single baseband system could be used throughout all parts of the broadcast chain; subsequently, a broadcast version of the D-MAC/packet system was developed following the decision in the UK to adopt this system to minimize receiver costs. The family has currently three members, the C-MAC/packet system for dbs and the D-MAC/packet and the D2-MAC/packet systems for cable networks or dbs[6,7]. These systems are now the subject of an EEC directive and CCIR Recommendation 650[1,8].

The MAC/packet time multiplex retains a 625-line raster with 50 fields (or 25 frames) per second. In these frames, we have in principle the option of inserting, as required, any analogue or digital component we wish, with their positions being signalled by *time multiplex control* data carried in line 625. The result, in terms of services, is the possibility of transmitting in any given channel a variable and evolutionary combination of moving pictures, still pictures, graphics, sound signals of a variety of quality and any form of data. The packet multiplexing system allows the coexistence of several digital channels without restrictions as regards their individual bit-rates, their repetition rates or their timing. Nevertheless, there are certain configurations which are now well defined.

29.1.2 Vision coding

In all members of the MAC/packet family (C, D and D2), the picture is coded using MAC in exactly the same manner. In principle, the quality is the same and there is the same scope for future extensions. However, the filtering involved in the am/vsb modulation to restrain the D2 signal within the confines of a 7 or 8 MHz channel results in a loss of horizontal definition compared to the C or D system. Apart from this, all discussion of the relative merits of the three systems is concerned with the digital sound and data part.

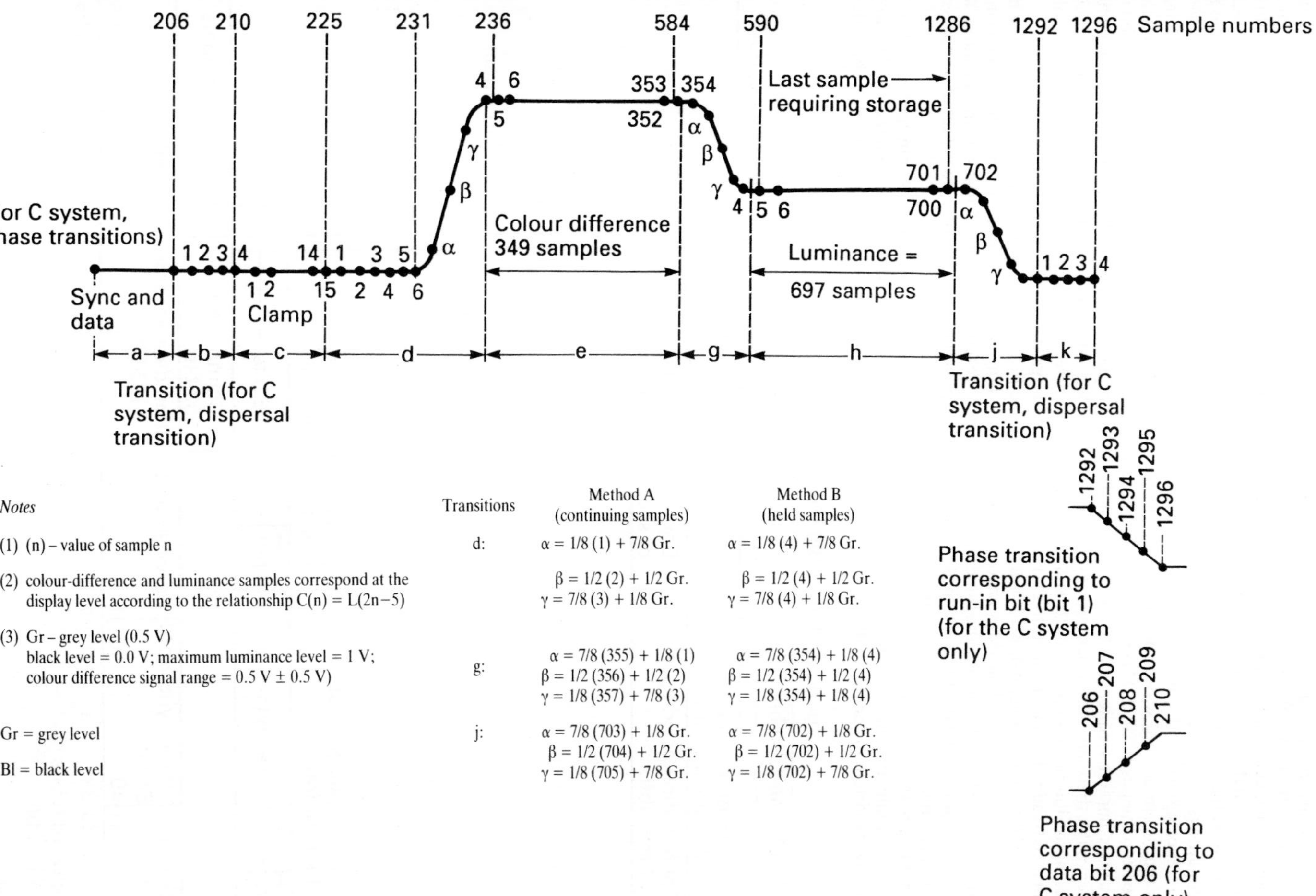

Notes

(1) (n) – value of sample n

(2) colour-difference and luminance samples correspond at the display level according to the relationship C(n) = L(2n−5)

(3) Gr – grey level (0.5 V)
black level = 0.0 V; maximum luminance level = 1 V;
colour difference signal range = 0.5 V ± 0.5 V)

Gr = grey level

Bl = black level

Transitions	Method A (continuing samples)	Method B (held samples)
d:	α = 1/8 (1) + 7/8 Gr.	α = 1/8 (4) + 7/8 Gr.
	β = 1/2 (2) + 1/2 Gr.	β = 1/2 (4) + 1/2 Gr.
	γ = 7/8 (3) + 1/8 Gr.	γ = 7/8 (4) + 1/8 Gr.
g:	α = 7/8 (355) + 1/8 (1)	α = 7/8 (354) + 1/8 (4)
	β = 1/2 (356) + 1/2 (2)	β = 1/2 (354) + 1/2 (4)
	γ = 1/8 (357) + 7/8 (3)	γ = 1/8 (354) + 1/8 (4)
j:	α = 7/8 (703) + 1/8 Gr.	α = 7/8 (702) + 1/8 Gr.
	β = 1/2 (704) + 1/2 Gr.	β = 1/2 (702) + 1/2 Gr.
	γ = 1/8 (705) + 7/8 Gr.	γ = 1/8 (702) + 7/8 Gr.

Figure 29.1 Representation of the waveform (unscrambled) of the transmitter modulating signal (before pre-emphasis) (EBU)

The principle of multiplexed analogue coding for the luminance and colour-difference components is that these components are coded separately and transmitted in turn within the duration of part of one scanning line. To make the information fit into the available time slot, the components have to be compressed in the time domain (and there has to be corresponding temporal expansion in the receiver). These operations can be achieved by analogue/digital conversion and then loading and reading a line store at the appropriate speeds.

For the normal video conditions, the compression ratios for luminance and colour-difference signals are 3:2 and 3:1 respectively. *Figure 29.1* shows the recommended waveform of one picture line. The basic clock frequency is 20.25 MHz which is exactly 3/2 times the luminance sampling frequency in the 4:2:2 studio standard (CCIR Rec. 601); there are 1296 clock periods per line. The luminance compression ratio is equal to 3:2, and the non-compressed luminance bandwidth is equivalent to that of the studio standard (5.75 MHz) for a channel bandwidth of 8–9 MHz. The compression ratio of the colour-difference components is twice that of the luminance, thus giving the same bandwidths as in the Rec.601 studio standard for a channel bandwidth of 8–9 MHz. Each line of picture information contains 697 luminance sampling periods (about 34 μs in total) and 349 colour-difference sampling periods (about 17 μs). Weighted R-Y and B-Y are transmitted line-sequentially.

After expansion, the colour-difference bandwidth can be up to 3 MHz, although, to reduce the noise bandwidth, it may be advantageous to reduce this to about 2 MHz in the receiver.

Subjective tests on certain critical still pictures coded in C-MAC/packet with the suggested pre-emphasis network have shown that after passing through a satellite simulator with a high carrier/noise ratio (about 20 dB), the mean quality is about 0.2 CCIR quality grade lower than the value reached by the digital 4:2:2 studio standard itself[9]. Furthermore, the first threshold effects are seen at a c/n ratio of about 10 dB with a normal demodulator and 8 dB with a threshold-extension demodulator. The total failure point, corresponding to quality grade 1.5, is reached at a c/n ratio of 2–3 dB (it is reckoned that at this level of quality a viewer will cease to watch the picture, regardless of the inherent interest of the programme). This is about 1–2 dB after (and therefore better than) an equivalent PAL/SECAM signal.

To allow for future receiver developments, however, which might include noise reduction systems and edge enhancement built into the receiver, the MAC/packet specification also allows for the possible future use of a second set of compression ratios, 5:1 and 1.25:1.

The colour-difference component is transmitted before the luminance component, because low frequency distortion would be most visible in the colour-difference signals. The latter is

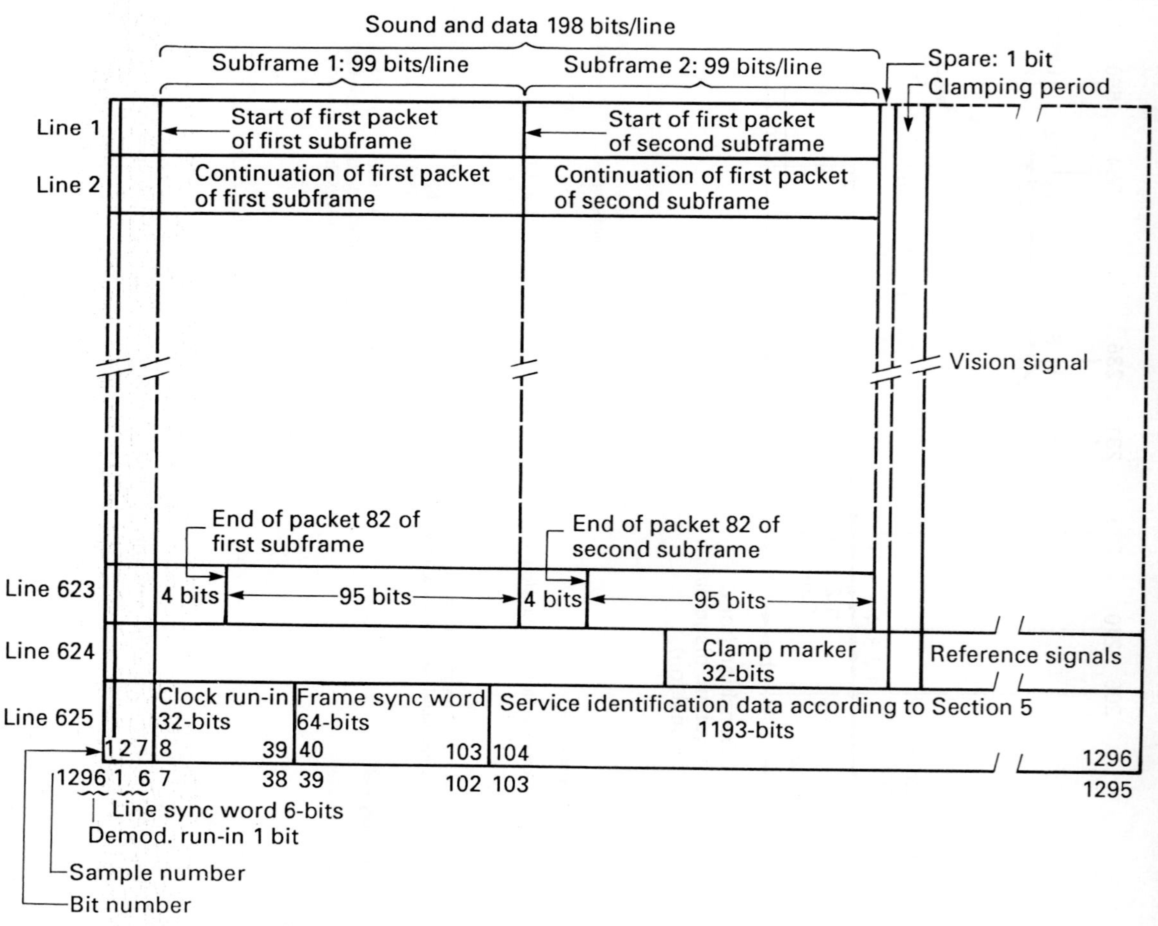

Figure 29.2 C– and D–MAC/packet systems transmission multiplex structure (not to scale) (EBU)

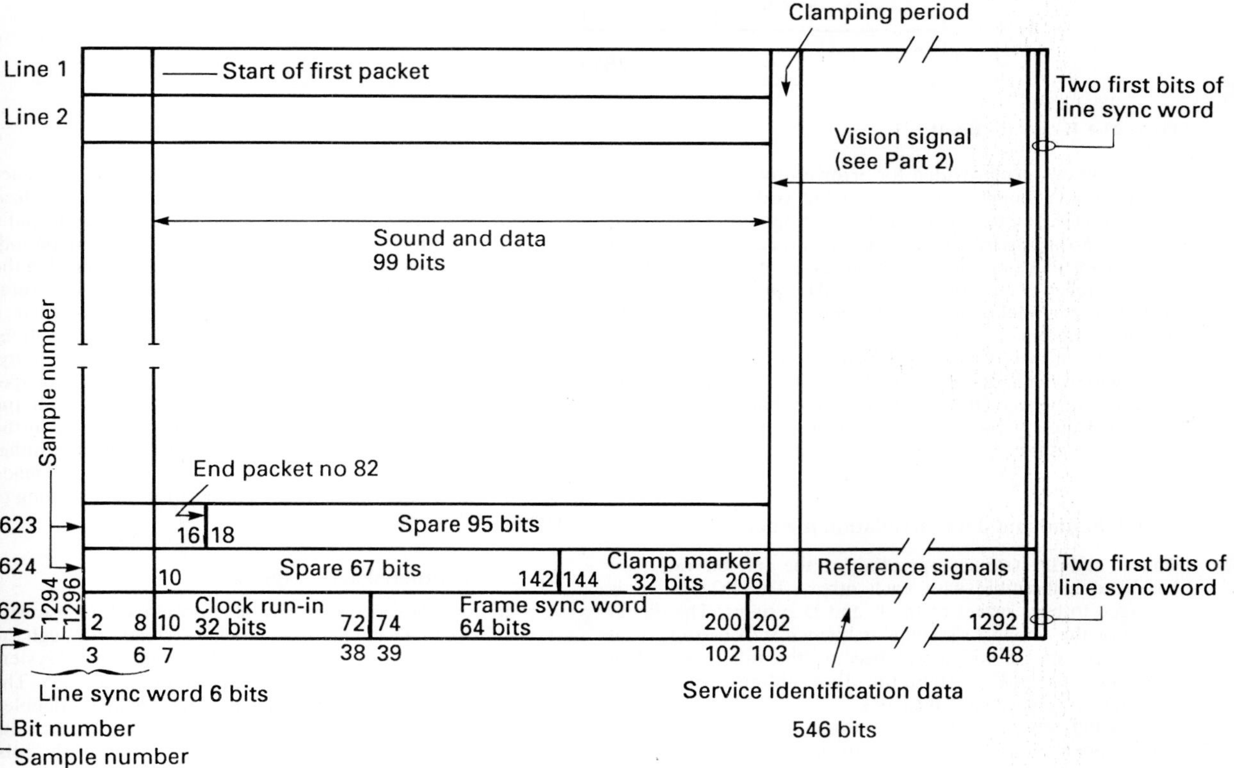

Figure 29.3 D2–MAC/packet system transmission multiplex structure (not to scale). Data bits occur on even-numbered sample points only (EBU)

therefore closest to the clamping reference at the start of the line.

The MAC/packet system is designed to have broadly matched vertical and horizontal definition. By dropping alternate lines of the colour-difference signals, the potential vertical resolution of the colour-difference information is halved with respect to luminance. The maximum amount of vertical detail possible in a picture is usually expressed in terms of *cycles/picture height* (c/ph). The cycle is the discernible pair of apparent stripes, one light and one dark. The Nyquist limit for the colour-difference signals is thus 71.87 c/ph and this corresponds to an equivalent frequency of 1.85 MHz. To reduce the potential alias components which would be produced by a source signal with components beyond this, the colour-difference signals must be vertically pre-filtered. The available resolution depends on the type of filter used. If a simple line averaging filter (1.1 filter) were used, it would cause a loss of 3 dB at 1.85 MHz, while with a 1,2,1 filter, the loss would be 6 dB. The total response however also depends on the type of post-filtering in the receiver. The current EBU recommendation is for a seven-tap pre-filter and a three-tap post-filter in the receiver.

For the MAC system, the use of large amounts of pre- and de-emphasis to reduce distortion on subcarriers (as PAL) is not necessary. Because of the triangular nature of fm noise and the spectral content of vision signals, however, a modest amount of high frequency emphasis is applied to give improved noise and interference performance taking account of the requirements of the WARC plan.

The transfer characteristic is:

$$H(f) = 0.7071 \cdot \frac{1 + jf/0.84}{1 + jf/1.50}$$

An optional non-linear characteristic is also under study.

29.1.3 Sound coding

Sound channels are digitally coded. The sampling frequency used is 32 kHz (or 16 kHz for sound signals for which a reduced audio bandwidth is acceptable). The resolution is that corresponding to 14 bits per sample in linear coding, although bit-rate reduction can be used in the form of near-instantaneous companding (the system commonly known as *nicam*) from 14 to 10 bits per sample. A 32 kHz sampling frequency and 14 bit/s sample allows an audio bandwidth of 15 kHz and an effective weighted signal/noise ratio of about 66 dB. The companding system in addition introduces programme modulated noise, but this is audible only on critical material.

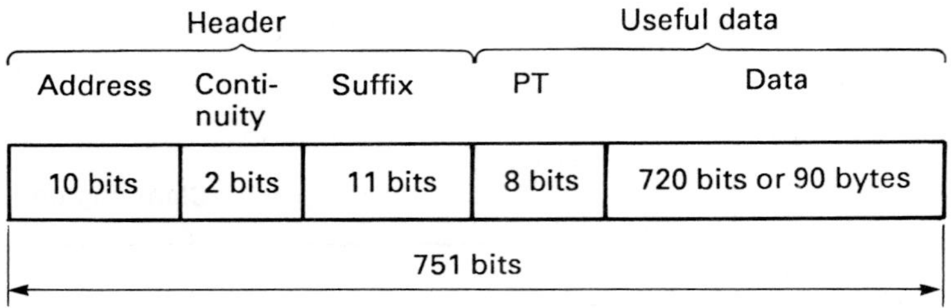

Figure 29.4 Packet structure (EBU)

Two options are provided for error-protection: either one parity bit per sample, or a Hamming code-forward error correction (fec) system covering the most significant bits of the samples. *Forward error correction* is the process of adding to a data stream additional bits whose value is related to the wanted bits. The receiver can use the additional bits to detect if there is an error in the wanted bits and, in some circumstances, correct it. The different configurations in the MAC/packet sound coding schemes allow broadcasters to select, on a case by case basis, within the limits of the total capacity available, the best compromise between the number of sound channels broadcast, their final quality and their failure characteristics (the latter affects the outer limits of the coverage area).

29.1.4 Sound and data modulation method

The C and D systems, operating at the full bit rate, are baseband compatible with each other. The D2 system is arranged to be a subset of the C and D systems. The digital frame of the C and D systems is divided into two identical subframes, one of which will become the full frame when transcoding to the D2 system; the other subframe will be lost, together with the services it carries.

The configuration adopted is known in *Figure 29.2* for the C and D systems. It is seen that the information (as distinct from the control data) is carried in lines 1–623; lines 624 and 625 are reserved for other purposes. In each of the first 623 lines, the first bit is a run-in bit used in differential demodulation for the C system. It is followed by a line synchronization word of 6 bits and then by two series of 99 information bits constituting the two subframes; the last bit of the burst (bit 206) is not yet allocated.

Figure 29.3 shows the corresponding structure for the D2 system. The line synchronization is identical, but the useful information is reduced to 99 bits per line.

The packet structure is identical in the three systems (see *Figure 29.4*). Each packet contains a 10-bit address which is recognized by the receiver and used to steer the data to the appropriate decoder (sound, teletext, etc.). The address length is sufficient for 1024 different types of service. The next two bits form a continuity index which can be used to detect, and thus to conceal, packet loss caused by incorrect recognition of the address owing to transmission errors. An 11-bit suffix then provides strong protection for the address through the use of a cyclic Golay code (23,12). The Golay code involves the addition of 11 further bits (calculated from the information bits) to 12 information bits. The receiver can use the additional bits to correct and detect up to three random errors in the information bits. In the packet structure, a so-called pt (*packet type*) byte indicates whether the subsequent data are samples of a digital sound channel or *interpretation blocks*, which serve to indicate in coded form the nature of the sound (mono, stereo, etc.). Out of a total of 751 packet bits, the useful data occupy 720 bits or 90 bytes.

Line 624 is to be used for signals such as level referenc signals (black, white, etc.), signals allowing adaptive equaliza tion for the picture and data in the presence of echoes, and special word which indicates the position of the clamp period this being necessary because the timing may be changed in th future if the data burst is shortened to accommodate picture with wider aspect ratio, etc. Line 625 contains, in particular, frame synchronizing word and data which indicate the position of all the elements of the tdm multiplex, to permit flexibility

Other service identification data are transmitted in a spe cialized channel of the packet multiplex (packets with th address 0); this gives a list of all the services available in th channel and the information needed for the automatic config uration of the receiver and its decoders to suit the choice mad by the user. The packet multiplex also carries data relating t conditional access.

29.1.5 Conditional access system

A conditional access system has been developed for use wit the family of C, D and D2-MAC/packet systems, although th service operator is free to choose the key management syster of his choice. The principles are outlined in *Figure 29.5*. Th different components (picture, sound, data) are first scramble by a pseudo-random binary sequence (prbs) from the scram bling sequence generator. The point in the cycle of the prb generator is defined by a *control word*. The 'local' control wor is a fixed binary word also built into the receiver and can be use when there is to be free access to the programme. The receive can descramble the signal without needing other decryptio data. Where access is to be controlled or not free, a variabl control word is encrypted by a two tier *authorization key* (basi and supplementary) which, in turn, is encrypted by a *distribu tion key*. To decrypt the encrypted authorization key, the use uses his personal distribution key together with the encrypte authorization key (the *entitlement management message*) whic may be transmitted with the broadcast signal or sent by som other physical means. To obtain the control word, the use needs the authorization key and the encrypted version of th control word. The signal can then be descrambled by a pseudo random binary sequence identical to that used for scrambling

The entitlement checking messages and entitlement manage ment messages (the encrypted version of the control word an the encrypted authorization key) can be both transmitted in th packet multiplex, or the emm can be conveyed to the subscribe by other means (e.g. post), depending on the type of ke management system (the *access control system*) adopted.

The MAC vision signal scrambling is done by cutting an rotating active picture line segments. It is possible to define, b the pseudo-random binary sequence, either a cut point in th luminance and in the colour difference (the *double-cut compo nent rotation process*) or a single cut point in the colou difference (the *single-cut line rotation process*). The first proces offers the greatest security but is somewhat more sensitive t line-tilt distortion and amplitude/frequency response distortio

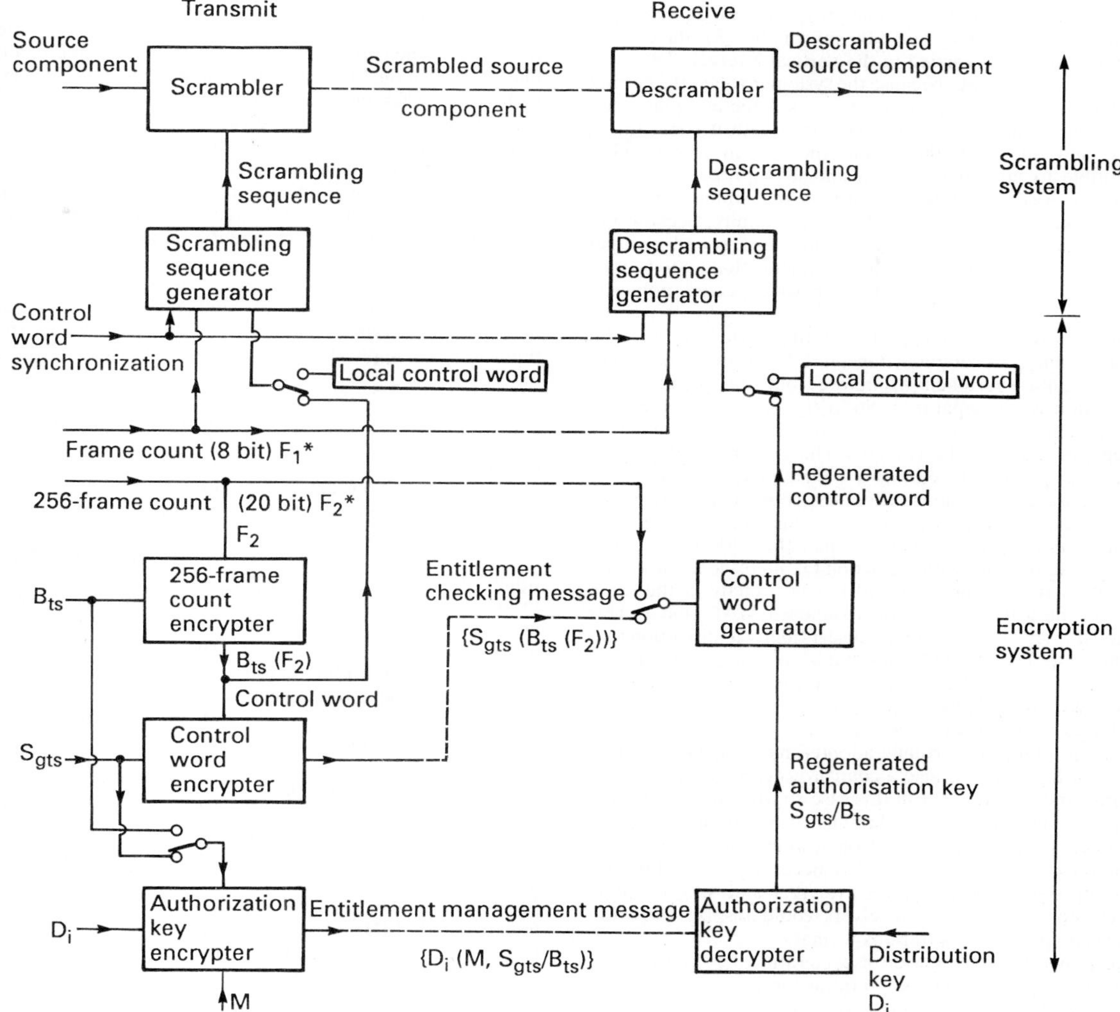

Figure 29.5 Generalized block diagram of the conditional access system. D = distribution key, i = individual or small group subscript, B = basic authorization key, t = time subscript, S = supplementary authorization key, g= user group subscript, s = service subscript, M = customer messages, F_1 = 8-bit frame count, F_2 = 20-bit 256-frame count, F_2* is the 20 msbs of a 28-bit frame count for which F_1 is the eight lsbs (EBU)

which can occur in cable networks and in the rf stages of receivers. The overall tolerance required for the line tilt is about 0.3–5 per cent depending on the masking effect of any noise in the picture.

The digital signal is scrambled by modulo-2 addition of a pseudo-random binary number.

29.1.6 HD-MAC

The broadcast transmission of further extended quality MAC pictures may be envisaged with pictures originated from, for example, high-definition video sources or 35 mm film. This requires the addition to the MAC system of a series of signal processes at both the sending and the receiving ends of the chain, intended to improve definition and to provide a wider aspect ratio. It is possible in normal MAC to transmit pictures having an aspect ratio of either 5.33:3 or 4:3. For HD-MAC a 5.33:3 aspect ratio will be used.

The C, D and D2 systems have the same theoretical scope for picture enhancement if they are transmitted in wideband channels (such as a satellite channel). However, if the D2 system is used on cable networks in 7 MHz channels and filtering is introduced during the am/vsb modulation, this sets an unavoidable limit to the horizontal definition.

There has been from the beginning of the development of MAC much interest in ways of making wide aspect ratio pictures available, in a compatible way, for future generations of receivers. The methods studied fall into three categories.

The first method may be called the *cut component method* (suggested by the IBA, Britain). Here the luminance and colour signals for a 4:3 picture are transmitted exactly as in the normal service configuration[10]. The burst length however, which is currently about 10 μs in the C-MAC/packet or D-MAC/packet systems, is reduced to about 4 μs and the spare space is used to carry luminance picture edge information. This edge information can be stitched on (added) in a wide screen receiver to the main luminance information to give a wider aspect ratio. The aspect ratio could then be about 4.7:3, or even 5:3 if an adjustment to the compression ratio of the edge information were made. The edge information for the colour-difference signals is carried in the vertical interval. The term 'cut component' applies because the system involves the cutting up of an original 5:3 picture into a 4:3 picture plus edge information. This cutting process brings with it some draw-backs, and one of them is that the system is somewhat susceptible to line-tilt distortion and lf amplitude/frequency distortion. Also, the system is not very well suited to the D2-MAC/packet variant of the MAC/packet family, because the shorter burst length would mean only one sound channel picture is available (and there are arguments that up to four or five sound channels are useful for wide screen domestic viewing).

The second method (suggested by the RAI, Italy) may be called a *continuous component method*. Here, the burst length is once again shortened as before, but the picture edge information is not separated. So in the active line there are luminance and colour-difference components which are longer in length than the standard lengths. These can be used to give a 4.7:3 or 5:3 picture. To extract a 4:3 sub-picture, only part of the luminance and colour-difference components are used. These would not necessarily be in the same place continuously and the sub-picture may need to pan the wide aspect ratio picture in accordance with the action taking place in the scene. The sub-picture elements would not be exactly in the same place on the active line as for a standard MAC signal. This method does not suffer from the susceptibility to distortion that the cut component method does, but is, in common with it, not well suited to the D2-MAC/packet member, because the shorter burst length means that only one sound channel is possible to accompany the picture.

The third and now virtually adopted method (proposed by the CCETT, France) is also a *continuous component method*, but the burst length is not in this case shortened. The normal active line time is used to carry luminance and colour-difference components which are derived from a wide aspect ratio signal. To obtain a 4:3 sub-picture it will be necessary to use different decompression ratios in the receiver (2:1 and 4:1), or alternatively conventional 4:3 receivers could display the wide aspect ratio pictures in letter box format.

Using this method, a wide aspect ratio of 5.33:3 gives a convenient set of decompression ratios for the 4:3 sub-picture. This concept is the one included in the HD-MAC system currently being developed.

The horizontal definition obtained with the MAC system is related to the clock frequency used for processing the picture. With the normal frequency of 20.25 MHz, the video bandwidth is about 8.5–9 MHz. However, if there are HDTV programme sources available the effective video bandwidth can be increased. In a 27 MHz WARC-BS 77 satellite channel the bandwidth could probably be extended without difficulty to 11–12 MHz without creating problems of interference with the adjacent channels, and the non-compressed luminance base-bandwidth could then be 7–8 MHz, assuming the compression ratio is maintained as 3:2.

Spectrum folding techniques can be employed to give effectively about 9 MHz uncompressed basebandwidth.

The HD-MAC system being developed uses a combination of diagonal filtering and an optional four-field sequence to achieve a 4:1 bandwidth reduction starting from a 1250/50 source. A technique known as *line sample shuffling* is used to arrange the samples in a 625/50 raster so that compatible reception is possible. The intention is to transmit about 1 Mbit/s of control data (DATV) to inform the HDTV receiver whether the picture should be built up over 2 or 4 fields for each sub-block of the picture and for motion compensation[1].

29.2 B-MAC system

In North America a Canadian company, Digital Video Systems (now part of Scientific Atlanta), developed 525-line and 625-line versions of the MAC system, similar to the EBU system but with some differences in approach, termed B-MAC[11]. The B-MAC system uses a B-type (as defined in section *29.1.1*) sound and data modulation system, but has a rather simpler type of multiplex structure than the MAC/packet system. Sound data in the horizontal blanking interval can be optionally either binary or quaternary coded. The B-MAC system, in addition, uses adaptive delta modulation for the sound coding rather than linear coding or near-instantaneous companding as in the EBU system. The 625/50 version is used in Australia for a quasi-broadcast medium power satellite feed to the Australian outback. The 525/60 version has been relatively successful in point-to-point applications. Some characteristics of the system[1,8] are:

	B-MAC (625)	B-MAC (525)
modulation frame rate:	25	29.97
no. of sample periods/line:	1365	1365
nominal clock frequency, MHz:	21.328	21.477
nominal vision bandwidth, MHz:	7.5	6.3
instantaneous bit-rate: (quaternary/binary) Mbits/s	14.22/7.11	14.32/7.16
aspect ratio:	4:3 or 16:9	4:3 or 16:9
no. of sample periods for luminance:	750	750
compression ratio for luminance:	3:2	3:2
luminance sampling frequency, MHz:	14.219	14.318
uncompressed luminance bandwidth, MHz:	5	4.2
colour-difference vertical pre-filter coefficients:	0.25,0.5,0.25	0.25,0.5,0.25
useful data burst length (q/b) bits/line:	102/51	102/51
no. of audio adm channels (quaternary/binary):	6/3	6/3
fm bandwidth, MHz:	24	24
frequency deviation, MHz/V:	16.5	17.5

29.3 Digital sound radio system

Somewhat before the EBU studies of MAC systems, a group of organizations in West Germany had been studying alternative methods of providing a number of digital audio services in one WARC-BS 1977 channel. The system developed has a 4psk continuous multiplex with a data rate of 20.48 Mbit/s, which allows up to 16 digital stereophonic sound channels. It is also possible to have a similar all digital sound system in the framework of the MAC/packet family, but unfortunately the two approaches are not compatible.

The digital sound radio (dsr) system was developed in the Federal Republic of Germany by a number of laboratories associated with the national administration, broadcasting organizations and others (IRT, DFVLR, TU Hanover, FTZ, AEG Telefunken)[12]. The objective is to provide a multi-channel sound only service via one (or more) of the available WARC-BS 1977 channels.

The design is such that if the transmission power planned for WARC-BS 1977 is used, the service is likely to be receivable in West Germany itself on a dish as small as 30 cm diameter. The potential coverage of the system with larger dishes is considerable, probably covering much of mainland Europe.

The system format is a relatively straightforward fixed format multiplex. The audio coding for the system is based on 32 kHz sampling frequency. This allows an audio bandwidth of up to about 15 kHz. Linear coding is used, and the system is based on 14 bits/sample, but a floating point system is used in conjunction with a scale factor, which means that the resolution is equivalent to 16 bits/sample. There is sufficient data capacity available to extend the system even further (up to 21 bits/sample resolution) if valuable.

32 samples, one from each of the 32 channels, are collected in two partial frames of equal length. These partial frames are used to generate a qpsk system. Each frame is used to modulate one of the two 2psk axes of the carrier to create the qpsk. At the beginning of each partial frame there is an 11-bit sync word followed by a special service bit and four 77-bit blocks which contain the audio samples of two stereo channels each. The structure is linked to the provision of BCH error protection . Bose-Chaudhuri-Hocquenghem codes (BCH) are general purpose sets of error-correcting codes which can be constructed to any length with various error-correcting capacities. The 63-bit BCH code word used in this system consists of 44 information bits and 19 check bits. The 11 msbs (*most significant bits*) of the audio samples of the left and right channels of two services are put together to form 44 information bits. Also the 77-bit block contains two additional information bits, one for each stereo channel. At the end of the 77-bit block, the remaining lsbs (*least significant bits*) of the four audio samples are transmitted without protection.

The dsr system has a 3-bit scale-factor word which defines into which of eight ranges the highest of a block of 64 consecutive audio samples falls. This is used to reduce the effects of undetected bit errors in the 11 protected msbs and reduce the impairment caused by bit errors in the 3 lsbs. With the addition of this technique, acceptable sound quality is said to be maintained to bit error ratios of $1{:}10^{-2}$.

Using the scale factor in the transmission format makes it possible to transmit additional information. Because the range is known, part of the sample word becomes redundant and can be occupied by bits giving information on further resolution. For signals 12 dB or more below limiting level, 16 bit resolution is possible. Therefore quantization noise is reduced (theoretically by about 12 dB) for signals below this level. The scale factor is heavily protected by a BCH code.

In the qspk modulation, the half or partial frames are processed into an rf signal with 20.48 Mbit/s bit-rate.

Outline data are:

number of audio channels:	16 stereo/32 mono
audio frequency range:	15 Hz–15 kHz
source coding:	pcm 16-14 bit floating point complementary binary offset
sampling frequency:	32 kHz
transmission frame:	2 parallel data streams each with 8 stereo channels
frame repetition frequency:	32 kHz
transmission bit-rate:	20.48 Mbit/s
symbol rate:	10.24 Mbaud
fec coding:	BCH 44/63
fec encoding block:	4 × 11 msbs from 2 stereo (or 4 mono) channels
modulation:	4psk

With each stereo programme, additional information on the broadcasting stations, such as station name, programme type and speech music identification, is available in a subframe derived from the information bit.

29.4 Japanese A-type NTSC system and MUSE 9 system

In Japan, a decision on the standard to be used for the first satellite broadcasting services had to be taken in 1982, because the services were due to start in 1984. After discussion at a national level, an A-type NTSC[14] system was chosen (see section *29.1.1*). The NHK system optionally allows either linear sound coding or near instantaneous companded sound coding.

Characteristics include:

symbol rate:	2.048 Mbaud
bandwidth of data:	1.2 MHz
sampling frequency of audio:	32 kHz or 48 kHz
method of coding:	near-instantaneous 14–10 bit/s or 16 bit/s linear coding
fec:	63, 56 BCH
no. of audio channels:	4 (15 kHz) or 2 (20 kHz)
data modulation method:	4psk
subcarrier frequency:	5.7272 MHz

In 1984 the MUSE 9 HDTV broadcasting system was developed [13]; this is an 1125/60 component coded system which uses diagonal filtering and temporal sub-sampling to achieve a 4:1 reduction in signal bandwidth. The resulting signal has an 8.5 MHz basebandwidth and is a candidate for either the WARC-BS 1977 or RARC-BS 1983 channels.

References

1 *Conclusions of the Interim Meetings of Study Groups 10 (Broadcasting Service — Sound) and 11 (Broadcasting Service — Television)*, CCIR, Geneva (November 1987)
Part 2 (Study Group 10) Part 2 (Study Group 11) Broadcasting — Satellite Service (Sound and television), CCIR, Geneva (November 1987)

2 *Advanced Digital Techniques for UHF Satellite Sound Broadcasting*, EBU Technical Centre, Brussels (August 1988)

3 *Recommendations and Reports of the CCIR Broadcasting Service (Television)*, **XI, Part 1**, CCIR XVI Plenary Assembly, Dubrovnik (1986)

4 LUCAS, K and WINDRAM, M, 'Direct television broadcasts by satellite — desirability of a new transmission standard', *IBA Experimental and Development Report 116/81* (1981)

5 MERTENS, H and WOOD, D, 'The C-MAC/packet system for direct satellite television', *EBU Review - Technical No 200* (August 1983)

6 *Specification of the systems of the MAC/packet family, Tech 3258 E (English), Tech 3258 F (French)*, EBU Technical Centre, Brussels (October 1986)

7 *Specification du Système D2-MAC/Paquets*, Télédiffusion de France (September 1985)

8 *Recommendations and Reports of the CCIR Broadcasting Satellite Service (Sound and Television)*, **X** and **XI Part 2**, XVI Plenary Assembly, Dubrovnik (1986)

9 SEWTER, J B and WOOD, D, 'The evolution of the vision signal for the EBU DBS standard', *Proc IBC 1984* (Published by the IEE)

10 'Compatible higher-definition television', *IBA Tech Rev*, No 21 (November 1983)

11 LOWRY, J, 'B-MAC — An optimum format for satellite television transmission', *SMPTE Journal* (November 1984)

12 'Digital sound service for direct broadcasting satellites' (in English), Federal Ministry of Research and Technology (BMFT), Federal Republic of Germany

13 NINOMIYA, Y, 'The concept of the MUSE system and its protocol', NHK Laboratories Note, 348 (July 1987)
14 *A Modulation System of Television Picture and Sound Signals for Satellite Broadcasting at 12 GHz*, CCIR Study Period 1982-86, Doc 10-11S/6

Y Imahori
Chief Engineer, NHK

30 Uplink Terminals

30.1 System design

30.1.1 Types and functions

Direct broadcasting by satellite needs *programme transmitting stations* to send broadcast programme signals towards the satellite, and *satellite control stations* for the satellite's mission equipment and bus equipment.

These two types of stations may be integrated in a single system, or constructed as independent systems or in hybrid forms. The following sections will consider a main earth station that has the function of programme transmission and satellite transponder control, and other earth stations that operate principally as programme transmitting stations.

30.1.2 Specifications

The requirements for uplink terminals depend on their function, on the performance specifications and on the transmission criteria for the broadcasting satellite, as well as on the role they are required to play in the direct satellite broadcasting system.

The functions required of each uplink terminal depend on the position which the uplink terminal system is to occupy, the area in which it is to be placed, and whether the system is to be mobile or not.

The transmission conditions of each uplink terminal (such as frequency, number of channels, transmitting power and polarization) are determined by considering the input/output parameters, the transmission system of the broadcasting satellite, its availability, the prevailing meteorological conditions, and the necessity to comply with the technical standards of WARC-BS (see Table *30.1*).

A suitable location will:

- be free from mutual interference (intermodulation, crosstalk) with other communication systems
- have good meteorological conditions (acceptable limits of rain, wind, snow) and ground conditions (earthquake free)
- have a clear vision of the stationary satellites and be unaffected by aircraft, etc.
- have access to stable electric power and stable video and audio transmitting links (terrestrial)

Furthermore, in order to ensure continuity of service, there need to be available stand-by terrestrial systems, which can provide back-up when the main earth station is not functioning or in the event of disaster, heavy rain, etc.

Thus, in designing uplink terminals, it is necessary to take into consideration the whole satellite broadcasting system and the functions, performance and economy of the terminals. *Figure 30.1* shows a satellite broadcasting system. The various types of station are described in section *30.2*.

30.1.3 Transmission systems

Various methods can be used to transmit video and audio signals via satellite. In Europe, the C-MAC and D-MAC systems are used, and in Japan, the NTSC (pcm for audio) system. Additionally, the MUSE and other systems are used experimentally for high definition TV. MUSE uses *Multiple sub-Nyquist sampling encoding* (see section *29.4*). Table *30.2* gives typical transmission system parameters for satellite broadcasting in Japan.

For the video signals, the main carrier is frequency-modulated with a signal in the 4.5 MHz band. For the audio signals, the subcarrier of 5.727 272 MHz is q-dpsk modulated with a pcm signal which is sampled at 32 kHz (A-mode) or 48 kHz (B-mode), and it is superimposed on the upper end of the video band. A data channel is provided in the audio format for application to teletext, facsimile and other forms of broadcasting.

30.1.4 Link budget

FM transmission is used in satellite broadcasting.

The direct broadcasting system is required to secure an overall carrier/noise ratio of about 14 dB (in accordance with WARC-BS). In general, at a c/n ratio below 10 dB, although it depends on the fm demodulation circuit, the threshold noise appears as a spot on the screen of the crt. It is therefore necessary to ensure that the c/n ratio is above this value by a suitable margin.

In the case of the uplink terminal, it is normally necessary to provide sufficient c/n to prevent the c/n of the uplink from affecting the c/n of the downlink. It is also necessary to make provision to avoid adverse effects of rain, snow or other meteorological conditions.

ITU: Region 1 (11.7 ~ 12.5 GHz)
Region 2 (12.2 ~ 12.7 GHz)
Region 3 (11.7 ~ 12.2 GHz)

Item	WARC-BS Region 1, 3	RARC SAT-83 Region 2
Satellite orbit separation	6°	–
Satellite station keeping accuracy	±0.1° (E–W) (N–S)	±0.1° (E–W) ±0.1° (N–S)
Satellite antenna direction,		±0.1°
pointing accuracy	±0.1°	
beam rotation	±2°	±1°
Transmitting wave polarization	Circularly polarized	Circularly polarized
Minimum elevation angle of receiving antenna	20°	20°
Transmission signal bandwidth	27 MHz	24 MHz
Power flux density in service area		
individual reception	−103 dBW/m²	−107 dBW/m²
community reception	−111 dBW/m²	–
Diameter of receiving antenna	0.9m	1.0m
G/T individual reception	6 dB/k	10 dB/k
community reception	14 dB/k	–
Carrier/noise ratio in service area	14 dB	14 dB
Power flux density for interference protection of terrestrial station	−125 dBW/m²/4 kHz	–
Energy dispersal frequency	22dB (600 kHz_{p-p} deviation)	–
Interference protection ratio for satellite to satellite	31 dB (with co-channel) 15 dB (with adjacent channel)	28 dB (with co-channel) 13.6 dB (with adjacent channel)

Table 30.1 Technical standard of WARC-BS

In general, a c/n ratio of about 20–24 dB produces no problem in respect of picture quality.

For the main earth station, which is the key facility of the satellite broadcasting system, the c/n is set with a margin of about 10 dB. Normally, for fair weather, the c/n is set about 30 dB; for rainy weather, power must be increased. In the worst conditions it is necessary to obtain back-up from other stations. Table *30.3* shows uplink parameters for a main earth station, sub earth station and transportable earth stations of the BS-2 satellite broadcasting system.

30.2 Earth stations

30.2.1 Main earth station

The main earth station is the key facility for transmitting television programmes to the satellite; it coordinates and monitors the satellite and all the earth stations for satellite broadcasting. Its functions are:

- transmission and reception of satellite broadcast programmes
- control of satellite transponder
- operation of the order-wire link
- monitoring of the receiving condition of the satellite broadcast signals over the service area by the data from the monitoring earth stations

The main earth station for the Japanese broadcasting satellite BS-2 is described in the following paragraphs.

It consists of two antennas for two satellite (BS-2a, BS-2b) operation with 14 GHz transmitters, 12 GHz receivers, command transmitters, telemetry signal receivers, sets of order-wire equipment, computer systems, and control and monitoring equipment. The major characteristics of the station are listed in Table *30.4*, and a block diagram is given in *Figure 30.4*.

30.2.1.1 Antenna equipment

For efficient transmission to the satellite, and for high sensitivity reception of very small signals from the satellite, large diameter antennas are advantageous, but limitations will be imposed by the effect on the building structure, strength, manufacturing costs, etc. Antennas of 5 and 8 metres diameter have been used which provide an uplink c/n in fine weather of not less than 30 dB and a rainy weather margin of 6–10 dB.

Design of the antenna equipment must provide adequate mechanical strength to withstand earthquakes and strong winds. The antenna equipment is fed with two TV signals, one command signal and two order-wire link signals.

The satellite tracking system is high precision and automatic, making use of the fact that the higher modes detected in the received telemetry signal from the satellite become zero when the system faces towards the front of the satellite.

Characteristics of the main earth station antenna are listed in Table *30.5*.

30.2.1.2 TV signal transmit/receive

This equipment consists of a baseband unit, a modulator/

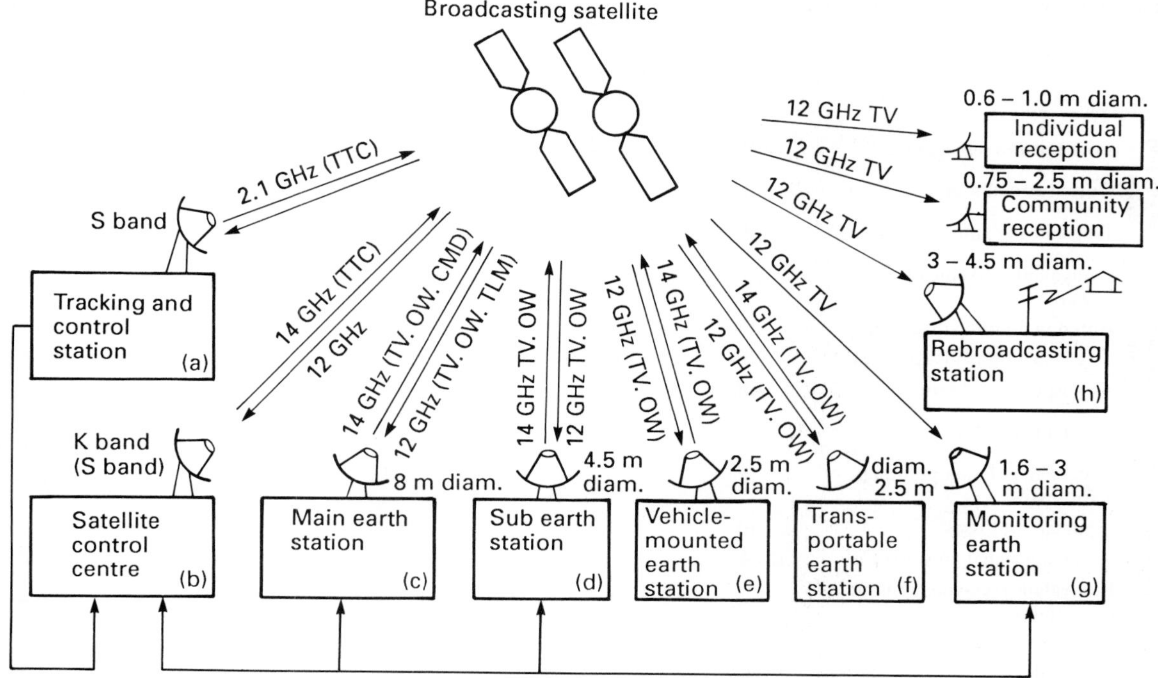

Figure 30.1 Satellite broadcasting system. (a) and (b) have the functions of control stations; (c) is a main earth station and has some of the functions of the programme broadcasting station and the control station; (d), (e) and (f) are programme transmitting stations

Table 30.2 Transmission system parameters of television picture and sound signals for BS-2 broadcasting

Television picture	*Sound*
Television system: 525-line NTSC system	Transmission mode: A/B
Maximum video frequency: 4.5 MHz (see *Figure 30.2*)	Sound signal bandwidth: 15–20 kHz
Type of modulation: fm	Sampling frequency: 32–48 kHz
Frequency deviation of carrier: 17 MHz_{p-p}	Quantizing and compounding: 14/10 bit/16 bit
Modulation polarity: positive	Number of frame bit: 2.048 Mb/s ± 10 b/s
Pre-emphasis: CCIR Rec. 405	Number of sound channel: 4/2
Frequency deviation of carrier by energy dispersal signal, symmetrically triangular, with frequency of 15 Hz: 600 kHz_{p-p}	Subcarrier frequency: 5.727272 MHz ± 16 Hz Frequency deviation of main carrier by the subcarrier: ±3.25 MHz $^{+10\%}_{-5\%}$
RF bandwidth: 27 MHz	Modulation method of subcarrier: Q-DPSK (see note)

Note: Q-DPSK = Quadrature differential phase shift keying

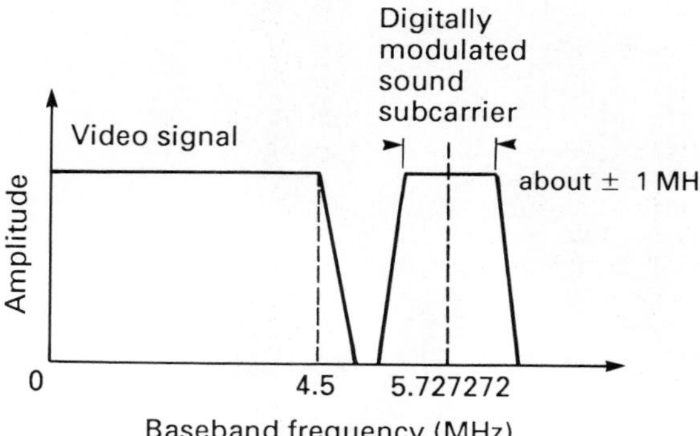

Figure 30.2 Spectrum of television picture and sound/data signals

demodulator unit, a frequency converter unit, a high power amplifier unit, a low noise receiver unit, a switching unit and a control supervisory console.

The high power amplifier employs a klystron or travelling wave tube amplifier (TWTA) and should be able to provide a 300 W–2 kW output. It should be installed near the antenna to minimize the feeder loss and the output circuit loss.

A redundancy system is provided to improve the reliability. It is possible to select the normal system or the redundancy system by remote control from the operation room.

30.2.1.3 Command transmitter and telemetry receiver

This equipment is provided to perform supervision of the operating conditions and control the *on/off* switching of the satellite transponder. Apart from the baseband unit which

Table 30.3 Uplink parameters of BS-2 satellite broadcasting

	System parameter		*Uplink (fair weather)* Main earth station	*Uplink (fair weather)* Sub earth station	*Uplink (fair weather)* Transportable earth station	*Downlink (rainy weather)*
Transmitting system	Transmitting signal power	dBW	(350W) 25.5	(500W) 27.0	(500W) 27.0	(100W) 20.0
	Output circuit losses	dB	4.2	1.2	1.0	2.3
	Transmitting antenna gain	dB	58.7 (8 mø)	54.0 (5 mø)	48.0 (2.5 mø)	39.0
	Equivalent isotropic radiation power	dBW	80.0	79.8	74.0	56.7
Propagating path	Antenna pointing error losses	dB	–	–	–	0.5
	Free space transmission losses	dB	– 207.1	– 207.1	– 207.1	– 205.6
	Atmosphere and rain losses	dB	1.0	1.0	1.0	2.0
	Receiving antenna gain	dB	37.0	37.0	37.0	37.6 (0.75m dia.)
Receiving system	Receiving antenna pointing error losses	dB	0.5	0.5	0.5	0
	Input circuit losses	dB	1.0	1.0	1.0	0
	Receiving signal power (c)	dBW	– 92.6	– 92.8	– 98.6	– 113.8
	Equivalent receiving system noise temperature	dBK	31.3	31.3	31.3	26.4
	Noise power density	dBW/Hz	– 197.3	– 197.3	– 197.3	– 202.2
	Bandwidth	dBHz	74.3	74.3	74.3	74.3
	Noise power (n)	dBW	– 123.0	– 123.0	– 123.0	– 127.9
Carrier/noise power ratio (c/n)		dB	30.5	30.2	24.4	14.1

Figure 30.3 The antenna of a main earth station (NHK)

performs command signal generation and telemetry signal decoding, etc., this equipment shares the TV signal transmit/receive equipment. The low noise amplifier unit, the down convertor unit and the signal demodulator unit in the telemetry reception part is used also by the TV signal transmit/receive equipment.

30.2.1.4 Supervisory control

The supervisory control equipment consists of a control console, a wall display, a command telemetry unit, an electronic computer and peripheral units, as follows:

Supervisory control console

- main earth station control console
- main earth station supervisory control
- command sending console
- telemetry supervisory console

Wall type display

- two air monitor units (for two channels)
- four monitor units (for two links)
- panel to indicate the operating conditions of the satellite and ground facilities
- alarm indicator panel

Electronic computer system

- central processing units (on-line, off-line, reception monitor)
- main memory unit (MOS 512 kbytes)
- auxiliary memory unit (magnetic disk magnetic tape)

The sub earth station backs up the main earth station, and undertakes transmission of emergency news programmes and others. It can also be used in the event of disaster, or of breakdown of the main earth station.

To be available in the event of a disaster, it needs to be installed as far away as possible from the main earth station. As it also has to function as the site diversity station, its distance from the main earth station is generally about 20–100 km. This is sufficient to make it available in most cases when the main station is suffering from rain attenuation.

A typical sub earth station consists of a 4.5 m diameter antenna, 14 GHz transmitters, 12 GHz receivers and sets of order-wire equipment.

Table 30.4 Major performance figures and specifications of main earth station for satellite BS-2

Parameter	Specification
Transmit overall	
TV	
TV video	
Input frequency	40 Hz – 4.5 MHz
Input level	$1\ V_{p-p}/75\Omega$
TV sound	
Input frequency	50 Hz – 15 kHz
Input level	0 dBm/600Ω
Modulation	
Sound	Quadriphase shift keying modulation
Video & sound	Frequency modulation
Intermediate frequency	140 MHz band
Transmit frequency	14 GHz band
Transmit output power	1.4 kW (max.)
Spurious level	< −46 dB
RF amptitude frequency response	Within ± 0.3 dB/centre frequency ± 8 MHz
RF group delay response	Within 2 nsec/centre frequency ± 8 MHz
Telemetry and command	
Transmit frequency	14.00028 GHz (BS–2a) 14.00371 GHz (BS–2b)
Transmit output power	500W (max.)
Spurious level	< −46 dB
Order-wire	
Input frequency	0.3 – 3.4 kHz
Modulation	Frequency modulation
Intermediate frequency	140 MHz band
Transmit frequency	14 GHz band
Transmit output power (at HPA output)	200W
Receive Overall	
TV	
Input frequency	12 GHz band
Input level	−80 – −60 dBm
Input vswr	< 1.25
System noise temperature	< 270K
IF	140 MHz band
Bandwidth	27 MHz
RF amplitude frequency response	Within ± 0.3 dB/140 MHz ± 8 MHz
RF group delay response	Within 2 nsec/140 MHz ± 8 MHz
Video	
Output frequency	40 Hz – 4.5 MHz
Output level	$1V_{p-p}/75\Omega$
Order-wire	
Input frequency	12 GHz band
IF	140 MHz band
Audio output frequency	0.3 – 3.4 kHz
Output level	0 dBm/600Ω

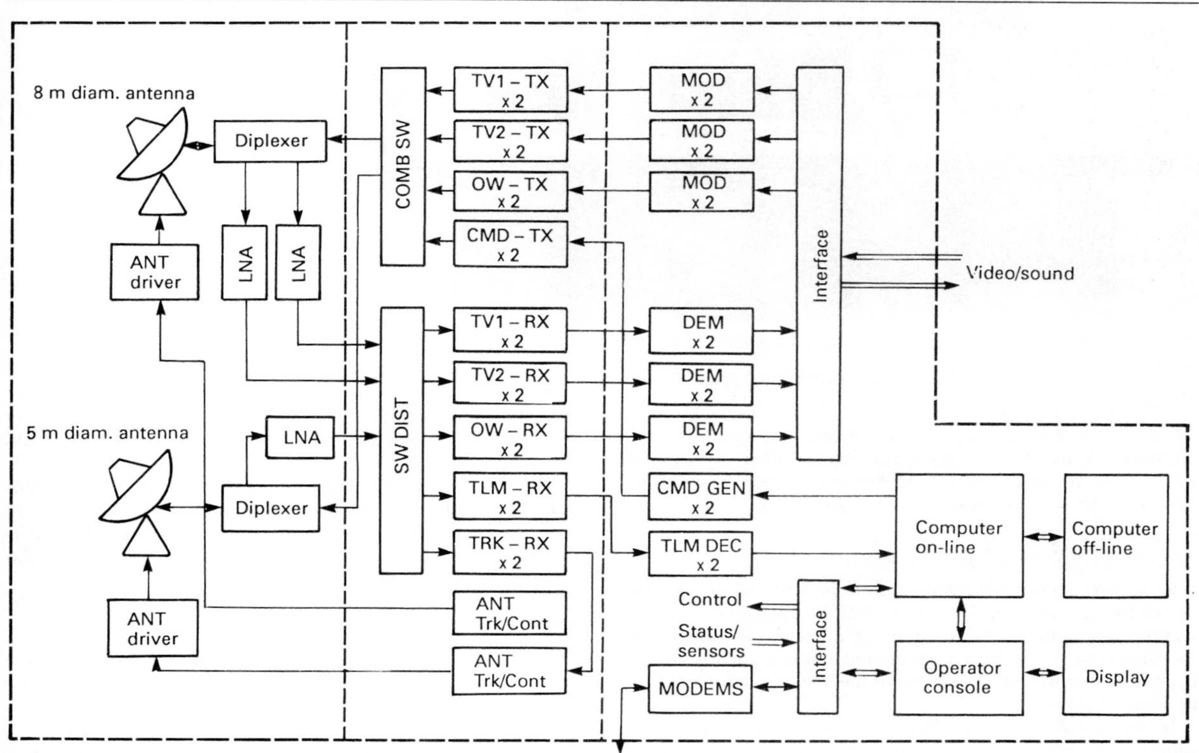

Figure 30.4 Block diagram of a main earth station

Table 30.5 Typical characteristics of earth station antenna

Type of antenna:	Cassegrain
Diameter of main reflector:	8m
Horn:	Conical corrugated horn
Tracking speed:	0.01°/sec
Auto-tracking:	Monopulse signal
Max. operable wind velocity:	60m/sec (peak)
Weight:	10t
Frequency:	Tx. 14.0 – 14.5 GHz Rx. 11.7 – 12.2 GHz
Gain (12G/14G):	58.7 dB/57.4 dB
VSWR:	Tx. < 1.4 Rx. < 1.25
Withstand rf power:	Mean 2.22 kW Peak 7.4 kW
Polarization:	Right hand (or left hand) circular polarization
Ellipticity:	< 0.9 dB

Figure 30.6 Sub earth station (NHK)

Figure 30.5 Broadcasting satellite operation centre room (NHK)

30.2.2 Vehicle mounted and transportable earth stations

These earth stations are able to access the satellite from anywhere in the broadcasting service area, and are utilized for news relaying in a disaster emergency or for programme relaying from outside the station at sports events, etc.

30.2.2.1 Vehicle mounted earth station

The vehicle mounted earth station is able to access the broadcasting station by securing a stable transmission path for lengthy programmes such as sports and events. Alternatively, the large vehicle on which it is mounted makes it quickly mobile to provide a short setting up time for the live relay of emergency news or local programmes.

It consists of a transmitter/receiver, an antenna and power supply equipment mounted on a vehicle. The antenna equipment can quickly be removed from its stored position to point to the broadcasting station. The antenna for vehicle mounted use is a Cassegrain of 2.5 m diameter.

A Cassegrain antenna is the most widely used system having dual reflectors. It has a hyperbolic subreflector that effectively creates a virtual focus between it and the hyperbolic main reflector.

The transmit/receive equipment can transmit one TV channel, receive two TV signal channels, and transmit and receive two order-wire channels (one channel per TV channel). It is capable of transmission and reception of two or four audio signal channels for each video signal channel.

The power supply equipment has a 25 kVA capacity to supply power to the transmit/receive equipment and also for lighting and other relaying equipment.

30.2.2.2 Transportable earth station

This is designed to be transported onto site by helicopter when land access is not available. The total weight (including the

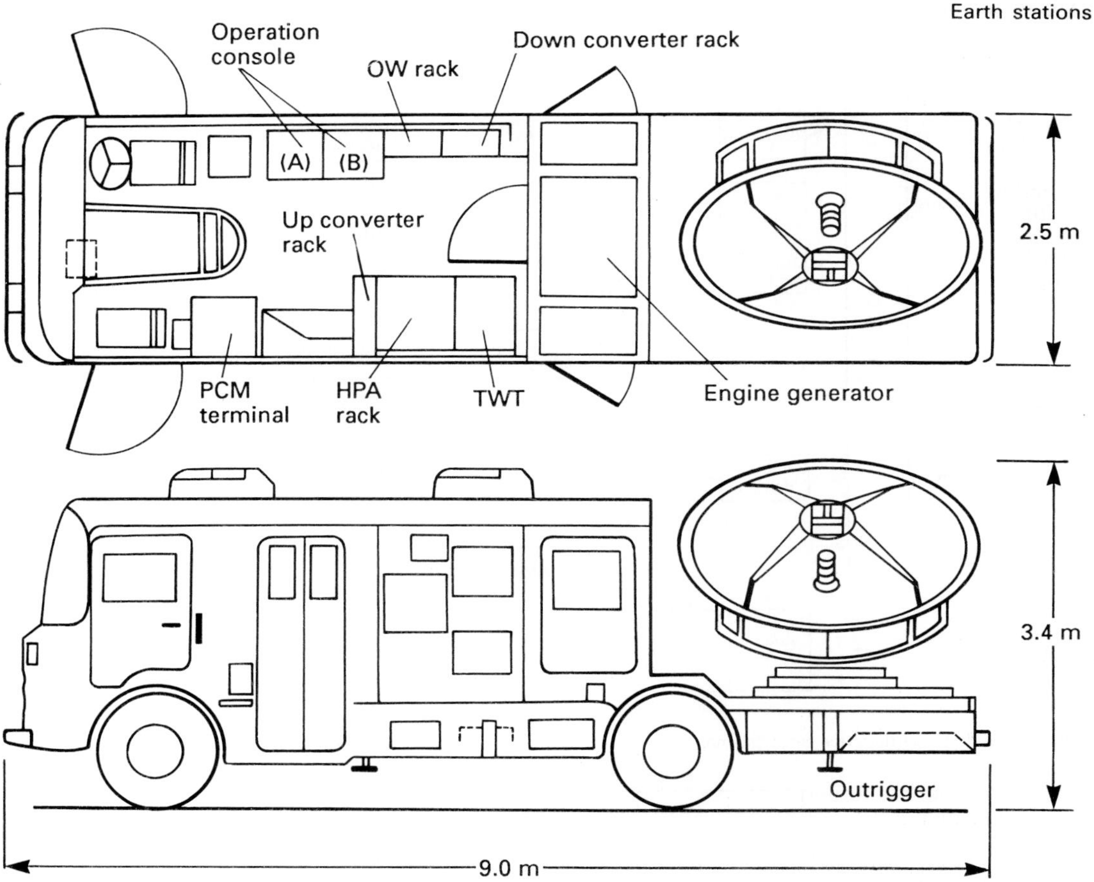

Operation console
OW rack
Down converter rack
(A) (B)
Up converter rack
PCM terminal
HPA rack
TWT
Engine generator
2.5 m
3.4 m
Outrigger
9.0 m

Figure 30.7 Configuration of a vehicle mounted earth station. Its total weight is about 10^3 kg

antenna, transmit/receive equipment and generator) does not exceed 750 kg, and the construction is compact and simple. Programmes can be relayed with a minimum of equipment.

A transportable earth station may consist of a 2.5 m diameter antenna, a 14 GHz 200–500 W output transmitter, a 12 GHz receiver, an order-wire unit, and a generator.

The antenna can be split into five parts for transportation, and the reflector is made of glass fibre reinforced plastics for light weight. The primary radiator employs a horn of corrugated structure to improve the aperture efficiency and to realize low side lobes of wide angle directivity.

The transmit/receive equipment consists of a baseband unit, a modulator, a frequency converter, a high power amplifier, a distributor, a low noise receiver, a demodulator, an order-wire unit, a monitor and a dummy load. It is capable of transmission and reception of one video channel and two audio channels. The high power amplifier has a 200–600 W output and employs a TWTA for amplification. Each component is designed to weigh below 50 kg for transportability.

The generator has a capacity of 4–6 kVA and supplies stable power to the transmitter/receiver. The engine and the generator are separate to facilitate transportation and weigh below 100 kg each.

30.2.3 Other earth stations

30.2.3.1 Interstations programme switching

An interstations switching control system provides smooth switching of programmes between stations. Television programme switching between transmitting earth stations is performed via satellite using uplink transmitter control. To achieve programme switching without signal overlap or break during the switchover, the propagation time from the earth station to the satellite and the time difference between the on-air station and the standby station to the satellite must be taken into account.

Figure 30.10 shows a programme switchover timing chart.

Figure 30.8 Transportable earth station on site (NHK)

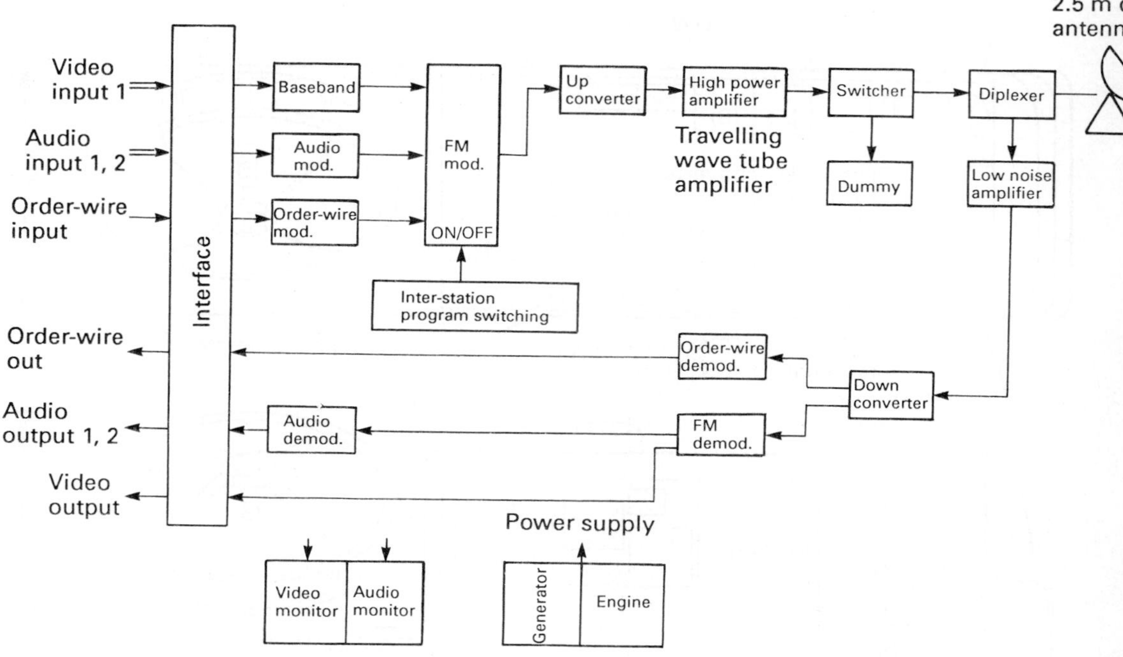

Figure 30.9 Transportable earth station block diagram

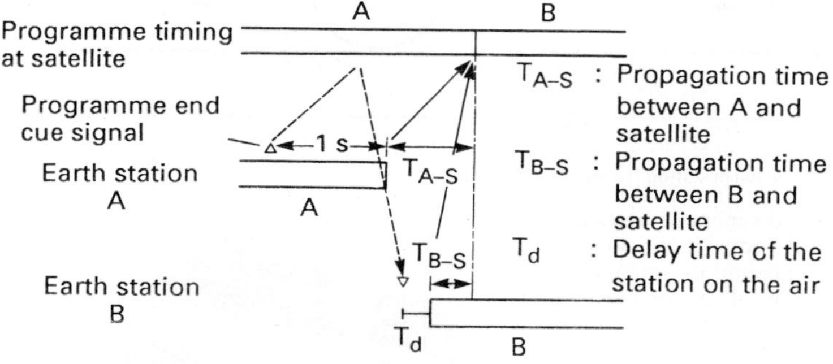

Figure 30.10 Fundamental programme switchover timing chart

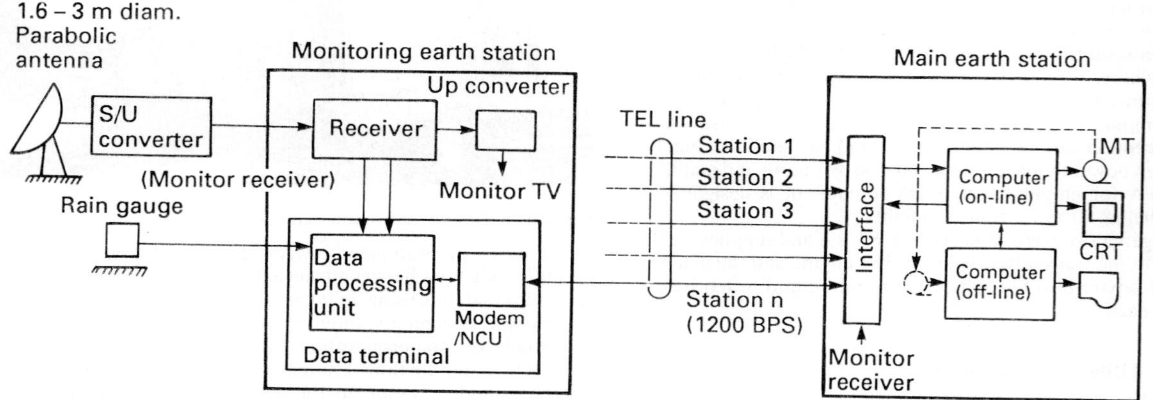

Figure 30.11 Monitoring earth station and data gathering/processing system

Figure 30.12 Rebroadcasting station on site (NHK)

The switching control system inserts a switchover control signal into the field blanking interval of the TV signal.

The switchover of programmes is achieved as follows, assuming that station A's programme is on-air via satellite while station B's programme is scheduled to follow it.

Station A will insert a programme switching cue signal (a *Q signal*) in the field blanking interval of its programme signal to notify station B that its programme is about to end. A specified time later it will turn off its transmitting carrier. Meanwhile, station B will turn on its carrier after an interval depending on the signal propagation time between station B and the satellite. This time is not dependent on the locality of station A.

30.2.3.2 Monitoring earth station

A monitoring earth station is used at several sites in the service area to receive and monitor the signal from the broadcasting satellite, and to supervise the condition of the broadcast service.

It consists of a 1.6–3.0 m diameter parabolic antenna, high stability receiving equipment to measure the received level, rain gauges, and received data processing equipment.

The received level data and rain data are sent via a terrestrial telephone line to a computer installed in the main earth station.

These data are processed and analysed for daily variation, seasonal variation and attenuation due to rainfall.

30.2.3.3 Rebroadcasting station

A rebroadcasting station receives and rebroadcasts dbs TV signals.

In the main service area, with the direct broadcasting system, it is possible to receive clear pictures using an antenna less than 0.6 m in diameter. But, at the fringe of the service area or inside a service area of low field intensities, a large diameter antenna is required which may be uneconomic for an individual. In these areas, therefore, the satellite broadcast signal is first received with a large diameter antenna and then converted into terrestrial signals for transmission to the service area.

A rebroadcasting station consists of a large diameter antenna (greater than 3 m), a set of receivers, demodulators, modulators, frequency converters, power amplifiers, a power supply and a station building.

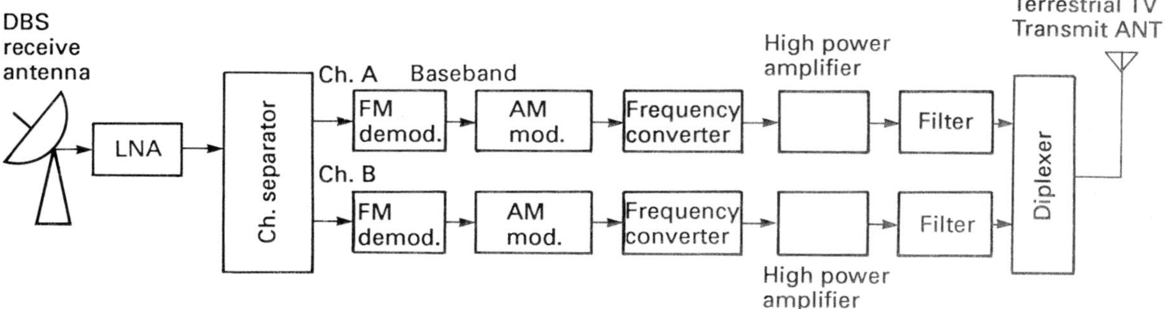

Figure 30.13 Rebroadcasting station block diagram

S Hirata
Senior Specialist, Toshiba Corp.

31 DBS Receivers

A dbs receiver is capable of receiving direct-to-home signals from the broadcasting satellite in the service area. As a piece of domestic equipment, its cost must not be excessive and its performance must be adequate.

As well as the features inherent in satellite broadcasting, dbs systems are expected to improve the quality of sound and vision and to realize diversified new broadcasting services.

The receiver is an important component of the complete dbs system. Efforts have been made to realize high performance receivers at reasonable cost.

Operational dbs systems using the 12 GHz band are currently in service or in preparation, and dbs receivers are on the market in some countries. This section describes the performance of the 12 GHz band dbs receiver as a consumer product.

31.1 DBS transmission systems

In conformity with the WARC-BS plan, satellite broadcasting systems have generally adopted frequency modulation for the rf signals. In spite of the weak signals inherent in satellite broadcasting, broad-band fm makes possible the reproduction of noise suppressing picture and sound quality. The baseband signal format has been evolved to fit satellite broadcasting transmission and to improve the quality of the signals and the service flexibility.

Three major dbs transmission systems are reported by the CCIR[1]: the MAC/packet family, the B-MAC system and the digital subcarrier/NTSC system.

The MAC/packet family and the B-MAC system adopt the principle of *time-division multiplexing* (tdm), which permits an improvement in picture quality. The sound/data signals are multiplexed at rf or baseband in the line blanking interval.

On the other hand, the digital subcarrier/NTSC system uses *frequency-division multiplexing* (fdm). The vision signals have almost the same parameters as those of a conventional M/NTSC system, for compatability with terrestrial television. The sound/data signals are frequency-multiplexed with the vision signal.

Figure 31.1 shows the multiplexing structures of each system. These systems apply digital techniques for sound and data transmission, in order to make full use of the capacity and the flexibility provided by the channel. The sound signal comprises the accompanying high quality sound signals and some possible additional independent sound signals. The data signal has the capability of transmitting digital data services such as teletext, facsimile and telesoftware, service identification and conditional access systems, etc.

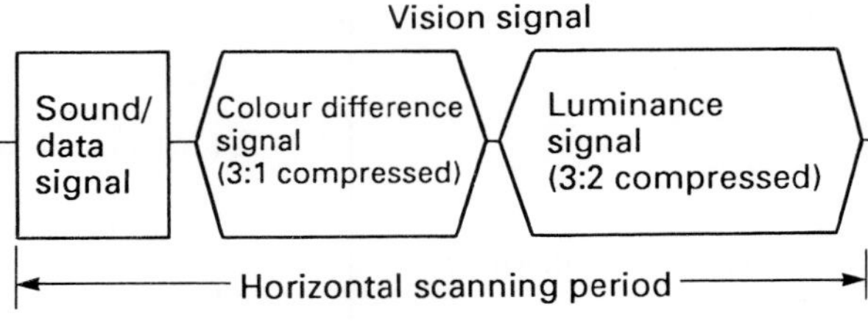

(a)

Digital sound/data
subcarrier (4-psk)
Deviation
Vision signal
4.5
5.73
Freq. (MHz)

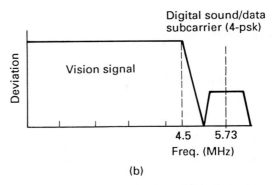

(b)

Figure 31.1 Multiplexing structures of the dbs systems: (a) MAC/packet family and B-MAC, (b) digital subcarrier/NTSC

The signal specifications for each system are shown in Table *30.1*. The principal variations between the receivers of each

Item / System	C-MAC	D2-MAC	B-MAC		Digital subcarrier/ NTSC
Country	EBU	W. Germany France	Australia	USA Canada	Japan
Nominal channel bandwidth (MHz)	27			24	27
Line/Field	625/50			525/60	
Vision coding/Modulation	MAC/fm			NTSC/fm	
Nominal sampling frequency (MHz)	13.5		14.22	14.32	—
Luminance [chrominance] compressed bandwidth (MHz)	5.6 [2.4]		5.0 [3.1]	4.2 [2.1]	4.5 [2.5/1.5]
Nominal transmitted baseband bandwidth (MHz)	8.4		7.5	6.3	6.3
Reference clock frequency (MHz)	20.25		21.328	21.477	—
Sound/data multiplexing principle	tdm* at rf	tdm* at baseband			fdm* (5.73 MHz)
Data coding/modulation	2–4 psk	Duobinary-fm	Quarternary/binary-fm		4-psk
Symbol rate (M baud)	20.25	10.125	7.11	7.16	2.048
Number of bit per symbol	1		2/1		1
Instantaneous bit rate (Mbit/s)	20.25	10.125	14.22/7.11	14.32/7.16	2.048
Mean data rate (Mbit/s)	3.08	1.54	1.59	1.60	2.048
Multiplex description	Flexible		Rigid		Flexible
Type of multiplex	Packet		Continuous		
Sound coding	14bit/sample lin* or 14/10 bit/sample nic*		Adaptive delta modulation		16 bit/sample lin* or 14/10 bit/ sample nic*
Protection (Error Correction)	1 parity bit/sample or 5 bit Hamming code/sample		2.33 bits/13 bit block		BCH (63, 56), SEC, DED*
Maximum number of audio channel	8	4	6/3		4 (15 KHz), or 2 (20 KHz)

*Note: tdm: Time-division multiplexing fdm: Frequency-division multiplexing
lin: Linear coding nic: Near instantaneous companding
SEC, DED: Single error correction, Double error detection

Table 31.1 Summary of signal specifications in dbs system

system relate to the way in which the baseband signals are processed.

31.2 Configuration

A typical dbs receiver system consists of four elements: an antenna, an outdoor unit, an interconnecting cable and an indoor unit (see *Figure 31.2*).

The receiving antenna is installed on or adjacent to the viewer's house or building and can be pointed in the direction of the incoming signals. The outdoor unit is directly attached to the receiving antenna. The indoor unit, which contains controls such as channel selection, is located close to the conventional TV receiver.

Generally, the 12 GHz dbs receiver system uses double frequency conversion to ease the problem of selectivity, image rejection and local oscillator radiation.

31.2.1 Receiving antenna

The incoming signals from the satellite are received by the antenna. The receiving antenna consists of a reflector, primary feed, polarizer and mount with aiming mechanism. A flat antenna will consist of an antenna board, built-in outdoor unit and mount with pointing mechanism.

31.2.2 Outdoor unit

The outdoor unit amplifies and converts the incoming signal (shf; 12 GHz band) received by the antenna into the first intermediate frequency (if) signal (uhf; 1 GHz band) for transmission through a coaxial interconnecting cable to the indoor unit. It consists of a low noise amplifier (lna), a first local oscillator, a mixer, and a first if amplifier.

31.2.3 Indoor unit

The indoor unit selects the desired channel from the first if signals, converts the signal into second if, and then demodulates the fm signal to the baseband signal. It consists of a second converter, a second if amplifier, an fm demodulator, a de-emphasis network, an energy dispersal signal rejection circuit, a baseband signal processor, a vhf/uhf remodulator, and a dc power supply. A baseband signal processor, a video signal processor and a sound/data signal processor are also provided.

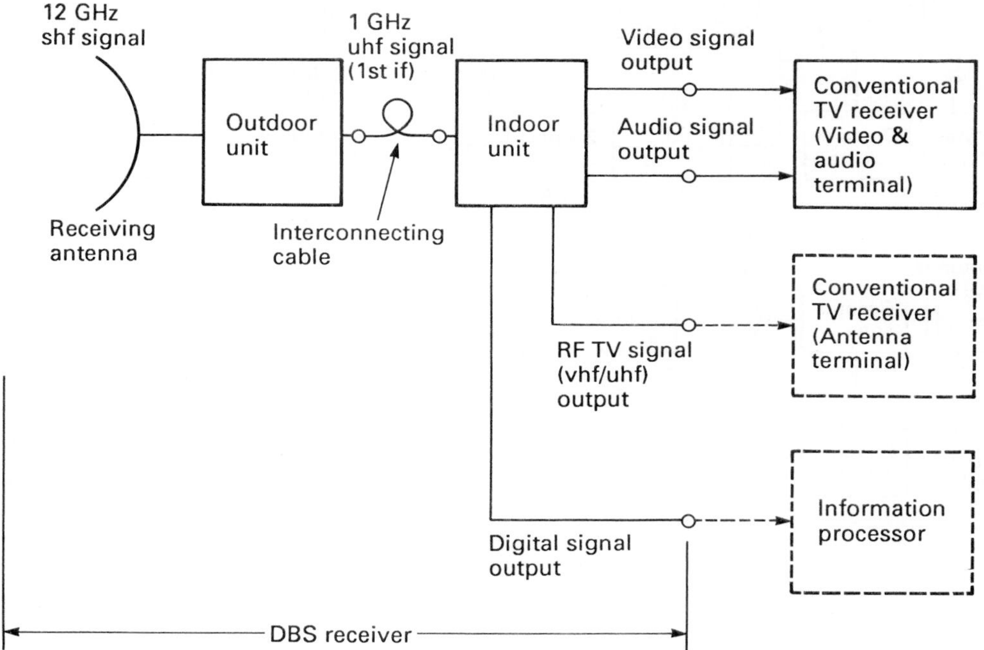

Figure 31.2 Basic configuration of dbs receiver

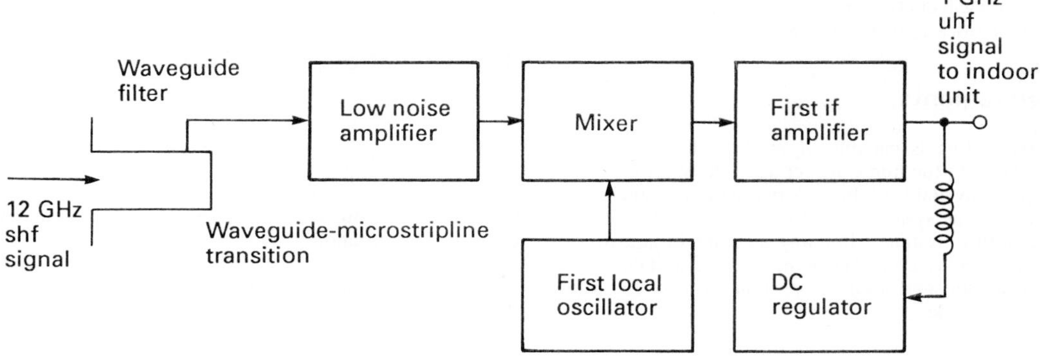

Figure 31.3 Outdoor unit

In the MAC/packet family and the B-MAC system, a MAC vision decoder is needed. For sound/data signal processing, the receiver needs a demodulator or decoder, a de-multiplexer and a digital sound decoder.

The indoor unit has various types of output interfaces, such as baseband video/audio signal outputs, rf TV signal output and digital signal output (bit-stream), etc. The indoor unit functions as an adaptor connected to the main receivers, such as the existing TV receiver or the information processor. It may be incorporated in the TV set.

31.3 Technical features

The satellite broadcasting systems have the following inherent technical features:

- *wide-area coverage*: a wide area can be covered simultaneously
- *high quality*: ghost free high quality picture and digitally modulated high quality sound
- *disaster immunity*: relaying is possible from anywhere via satellite
- *development potential*: introduction of new broadcasting systems is possible

In addition, the dbs receiver itself uses the following technologies:

- microwave technology for consumer use
- digital communication techniques
- digital audio and video signal processing
- data transmission and applications

Receivers are, of course, expected to comply with the technical requirements used in the planning of the WARC-BS 77 or RARC-SAT 83. For example, the recommended characteristics of home receiving equipment for satellite broadcasting in the 12 GHz band are shown in Table *31.2*.

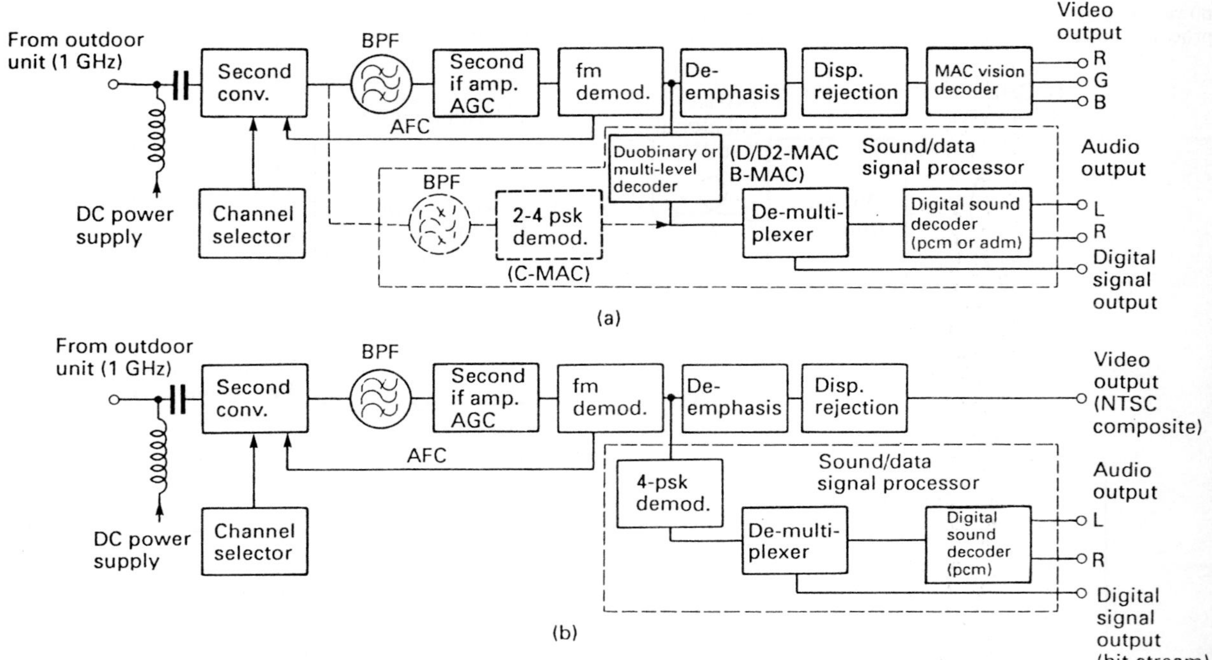

Figure 31.4 Indoor units:(a) MAC/packet family and B-MAC system, (b) digital subcarrier/NTSC system

The fact that the dbs receiver has been developed using these high technologies will enable it to play an important role as the centre of new broadcast receiving systems.

31.4 Requirements

Satellite broadcasting is capable of sending the same high quality TV signals to the home viewer as to the studio. These signals need to be available to the majority of the viewers in the service area, and the system should be harmonized with the existing and future systems. For any satellite broadcasting system, the receiver is a crucial element affecting as it does the system performance and cost. Important criteria for any receiver system include:

- low cost
- high performance
- unit interchangeability
- compatibility between dbs receivers and existing TV sets
- less spurious radiation
- protection against interference
- expandability to future satellite broadcasting systems
- ease of handling as a domestic appliance

Thus it is necessary to specify target performances for the receiving equipment and promote its standardization.

31.5 Operation and performance

31.5.1 Receiving antenna

Compared with the vhf signal, the shf (12 GHz) signal is extremely short in wavelength (about 2.5 cm) and may be captured by a parabolic dish.

The receiving antenna needs to have a gain that is sufficient relative to the required G/T (which indicates the sensitivity of the receiver—see section *31.5.1.1.2*) in the receiving area, and it should be available at a moderate cost, be easily installed, and retain a stable performance through its life.

Three types of receiving antennas are shown in *Figure 31.5*. The receiving antenna with a parabolic reflector (dish) provides excellent gain and directivity.

Figure 31.6 shows the construction of a parabolic receiving antenna. The outdoor unit is directly connected to the primary feed to minimize loss.

The signals reflected by the parabolic reflector focus at the top of a primary feed. The signals collected by the primary feed are introduced into the outdoor unit through a polarizer which converts the polarization of the signals from circular to linear.

Antenna	
Type and Gain	Type 60 75 90 100 120 (dB) 34.5 36.5 38.0 39.0 40.0
Output VSWR	< 1.3
Output port	WRJ-120 waveguide/BRJ-120
Outdoor unit	
Input port	WRJ-120 waveguide/BRJ-120 flange with waterproof
Noise figure	< 4 dB
Input signal level	− 80 dBm ± 10 dB/channel
Input vswr	< 2.5
Overall gain	48 ± dB
1st local oscillator frequency	10.678 GHz ± 1.5 MHz (−30° – 50°C)
leakage power	< −30 dBm
1st if	1.036 – 1.332 GHz
Indoor unit	
Input signal level	43 dBm +15 dB / −18 dB
Input vswr	< 2.5
2nd if	either 134.26 or 402.78 MHz
Leakage power of 2nd local oscillator	< −55 dBm
AFC gain	within ± 500 KHz against ± 2 MHz deviation
Received signal quality	
frequency response	video: +1 dB for 50 Hz 4.2 MHz, within −3 dB at 4.6 MHz sound: within +1 – −3 dB at 15 KHz for A mode within +1 – −3 dB at 20 KHz for B mode
linearity	video: Differential Gain < 5% Differential Phase <5°
bit error rate	$< 3 \times 10^{-4}$ at c/n = 9 dB, before error correction

(CCIR, Study Group, DOC, 10-11S/J-12, Japan)

Table 31.2 Characteristics of home receiving equipment

In the case of a flat antenna, the signals are received by several hundred small antenna elements on a board. Their currents are then collected and fed directly to the outdoor unit.

31.5.1.1 Antenna performance

Performance criteria for a dbs antenna are gain, efficiency, G/T, directivity and vswr.

31.5.1.1.1 Gain and efficiency

Antenna gain, G_r, is given by:

$$G_r = \eta \left(\frac{4\pi A}{\lambda^2} \right) \tag{31.0}$$

where η is efficiency, A is aperture area and λ is wavelength. The efficiency η is the product of six specific efficiencies:

$$\eta = \eta_m \cdot \eta_g \cdot \eta_r \cdot \eta_b \cdot \eta_i \cdot \eta_s$$

where η_m = efficiency due to reflector material
η_g = polarization efficiency
η_r = efficiency due to accuracy of reflector surface
η_b = efficiency due to blockage of radiated beam
η_i = illumination distortion
η_s = efficiency due to spillover at the surrounding area of reflector

The larger the aperture area is, the higher the gain. If the aperture area is constant, each efficiency must take a higher value in order to increase the gain of the antenna.

However, $\eta_i \cdot \eta_s$ has an optimum point, and its maximum value is about 0.8.[5] An effective surface accuracy better than 0.5 mm rms is desirable under all weather conditions. When the accuracy of the reflector surface is 0.5 mm rms, η_r is 0.94, so that $\eta_r \cdot \eta_i \cdot \eta_s$ is 0.75. Therefore, the upper limit of the antenna efficiency is considered to be approximately 75 per cent.

Using an offset parabolic reflector, antenna efficiency of more than 70 per cent has already been achieved in commercial antennas.

31.5.1.1.2 Overall performance of receiver and G/T

The *input carrier signal power*, C_i, of the outdoor unit is expressed by:

$$C_i = \left(\frac{W G_t}{L_t L_p} \right) \left(\frac{\lambda}{4\pi d} \right)^2 \frac{G_r}{L_r} \frac{1}{L} \text{ watts} \tag{31.1}$$

where W = the transmitting power of the broadcasting satellite
G_t = the transmitting antenna gain
L_t = the feeder losses between the transmitter and transmitting antenna
L_p = the pointing losses of the transmitting antenna
λ = the wavelength of the carrier signal
d = the distance between the satellite and the receiver
G_r = the receiving antenna gain
L_r = the feeder losses between the receiving antenna and the outdoor unit
L = the rain and atmospheric losses

The first term of the equation is the *equivalent isotropically radiated power* (eirp) expressed as follows:

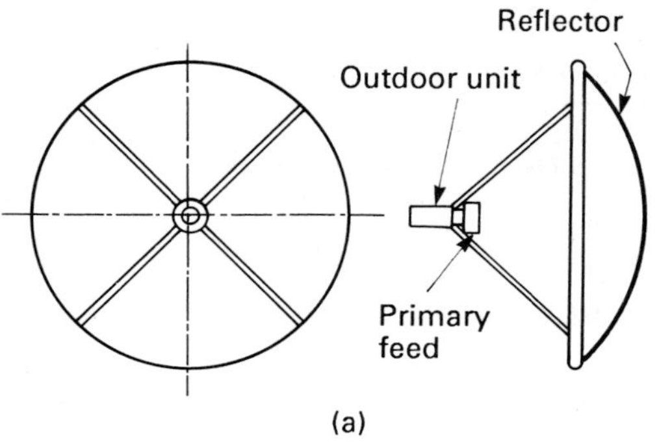

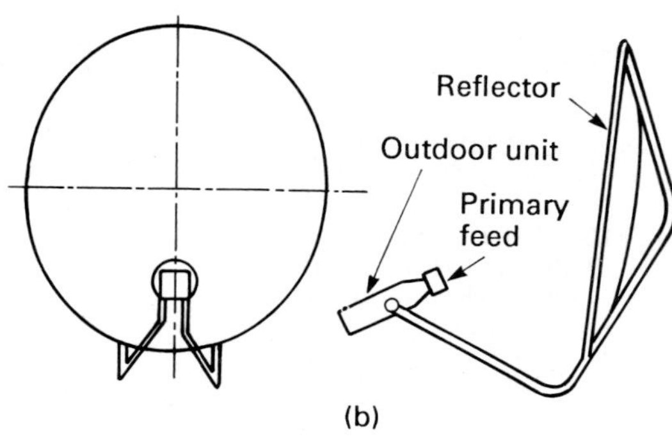

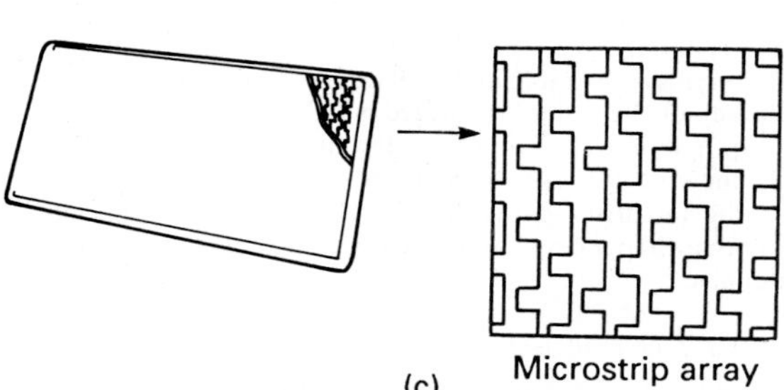

Figure 31.5 Types of receiving antenna: (a) centre feed parabolic antenna, (b) offset parabolic antenna, (c) flat antenna

$$\text{eirp} = \frac{W\,G_t}{L_t L_p} \text{ watts} \qquad (31.2)$$

The second term is the *free space path loss* (L_f):

$$L_f = \left(\frac{\lambda}{4\pi d}\right)^2 \qquad (31.3)$$

Therefore, when a receiving antenna with gain G_r is used, the *input carrier signal power* of the outdoor unit is expressed by:

$$C_i = \text{eirp}. \quad \frac{G_r}{L_f L_r L} \quad \text{watts} \qquad (31.4)$$

On the other hand, the input noise power of the outdoor unit includes the signal source noise (noise of antenna) and the noise generated in the receiver (converted value at the input port).

The *input noise power* of the receiver (outdoor unit) N_i is:

$$N_i = KT_aB + KT_0B(n-1) = KB[T_a+T_0(n-1)] \text{ watts} \qquad (31.5)$$

where
K = Boltzmann's constant, 1.38×10^{-23} W/Hz K
T_a = the noise temperature of the receiving antenna (K)
B = the bandwidth (Hz)
T_o = the reference temperature (K)
n = the overall noise figure of the receiver, expressed as a power ratio

The expression $T_a + T_o(n-1)$ is the *system noise temperature*, T_s.

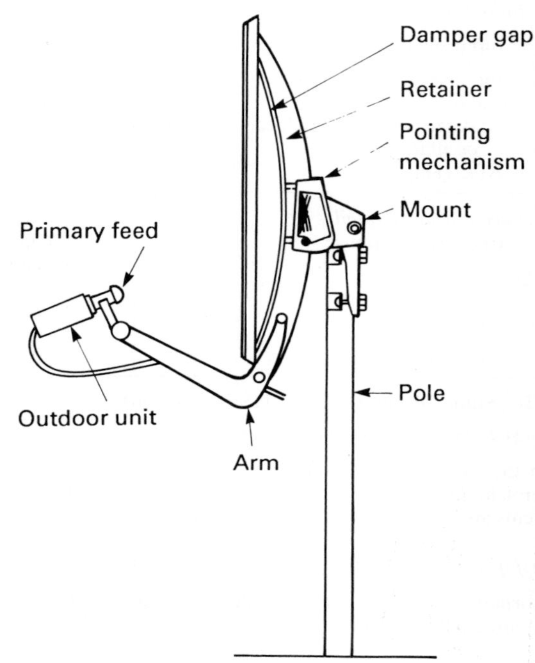

Figure 31.6 Construction of parabolic receiving antenna

The resultant carrier/noise ratio, C_i/N_i, at the input port of the receiver (outdoor unit) is calculated as follows:

$$C_i/N_i = 10 \log \frac{\text{eirp}}{KB[T_a + T_0(n-1)]L_f L_r L} \text{ dB} \qquad (31.6)$$

To attain excellent picture quality, a carrier/noise ratio of more then 14 dB is needed.

The *figure of merit* (G/T) is the receiving antenna gain (G_r) divided by the system noise temperature (T_s) a compound of the antenna and outdoor unit noise temperatures; it indicates the sensitivity of the receiver. From equations (31.4) and (31.5) the figure of merit is:

$$\frac{G_r}{T_3} = \frac{C_i}{N_i} \cdot \frac{L_f L_r L\,K\,B}{\text{eirp}} \qquad (31.7)$$

The figure of merit is defined in terms of the pointing error and the coupling loss by WARC-BS[3] as follows:

$$G/T = \frac{\alpha\beta G_r}{\alpha T_a + (1-\alpha)\,T_o + (n-1)\,T_o} \qquad (31.8)$$

where α = the total coupling losses, expressed as a power ratio,
β = the total losses due to the pointing error, polarization effect and ageing, expressed as a power ratio,
G_r = the effective gain of the receiving antenna, expressed as a power ratio and taking account of the method of feeding and the efficiency,
T_a = the effective temperature of the antenna,
T_o = the reference temperature (=290 K),
n = the overall noise figure of the receiver, expressed as a power ratio.

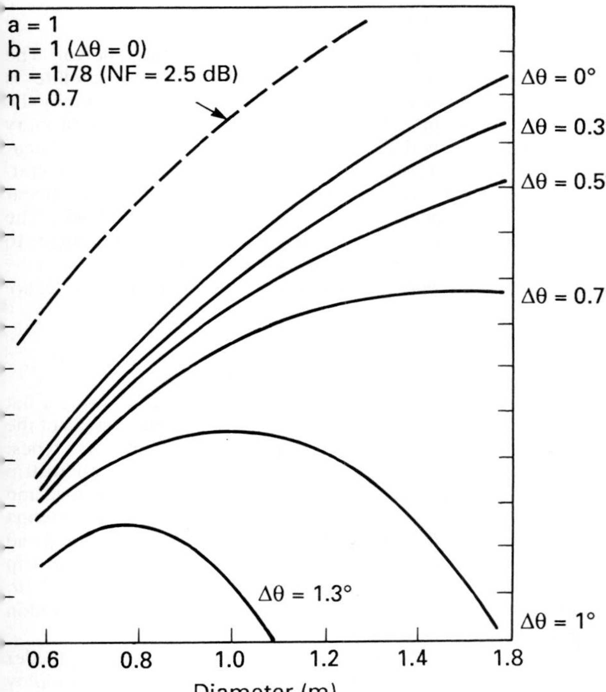

Figure 31.7 Relationship between antenna diameter and effective figure of merit. a = 0.9 b (considering losses due to overall pointing error $\Delta\Omega$), T_a = 150 K, T_o = 290 K, n = 2.51 (nf = 4 dB), f = 12 GHz, η = 0.6)

In WARC-BS planning, the value of the G/T is 6 dB/K for individual reception, and 14 dB/K for community reception.

However, an antenna for home use needs to be small and lightweight. An antenna with a relatively small diameter is now capable of providing the required figure of merit by virtue of its low noise figure and high efficiency. Even an antenna of 60–70 cm diameter provides a G/T between 10 and 12 dB/K.

At 12 GHz, the most popular shape of antenna for dbs reception is one with a conventional or offset parabolic reflector. The choice of diameter depends on the required figure of merit.

The relationship between the antenna diameter and the effective figure of merit is indicated in *Figure 31.7* (including the loss in gain introduced by the antenna pointing error).

31.5.1.1.3 Directivity

The directivity of the receiving antenna is expressed by the co-polar (same component as antenna polarization) and cross-polar (reverse component to antenna polarization) reference pattern. In order to avoid mutual interference between countries, WARC-BS has decided on the antenna reference pattern shown in *Figure 31.8*.

31.5.1.2 Reflector

The receiving antenna must provide not only high performance but also high reliability (including high resistance to wind), and ease of mass production and installation for domestic use.

Compared with a centre feed parabolic antenna, an offset parabolic antenna has the following advantages:

- There is no need to allow for a decrease in gain and degradation of the antenna pattern due to the scattering effect of the feed support and the blockage by the primary feed.
- It is possible to reduce the insertion loss by connecting the outdoor unit directly to the primary feed.
- Wind pressure is lower as it strikes a less curved surface. Adhesion of snow on the offset antenna is much less.
- It is easy to attach a cover to the outdoor unit to prevent temperature increases.

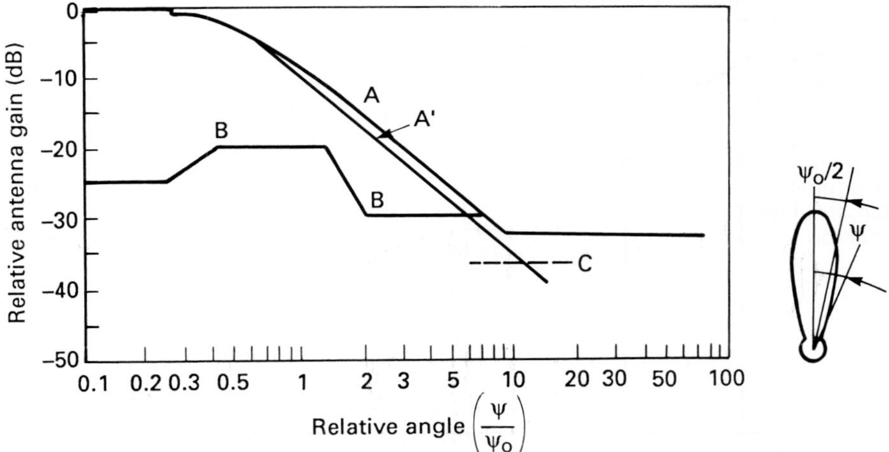

Figure 31.8 Co-polar and cross-polar reference patterns for receiving antennas (Regions 1, 3). Curve A: co-polar component for individual reception without side-lobe suppression; Curve A′: co-polar component for community reception without side-lobe suppression; Curve B: cross-polar component for both types of reception; Curve C: minus the on-axis gain (WARC-BS)

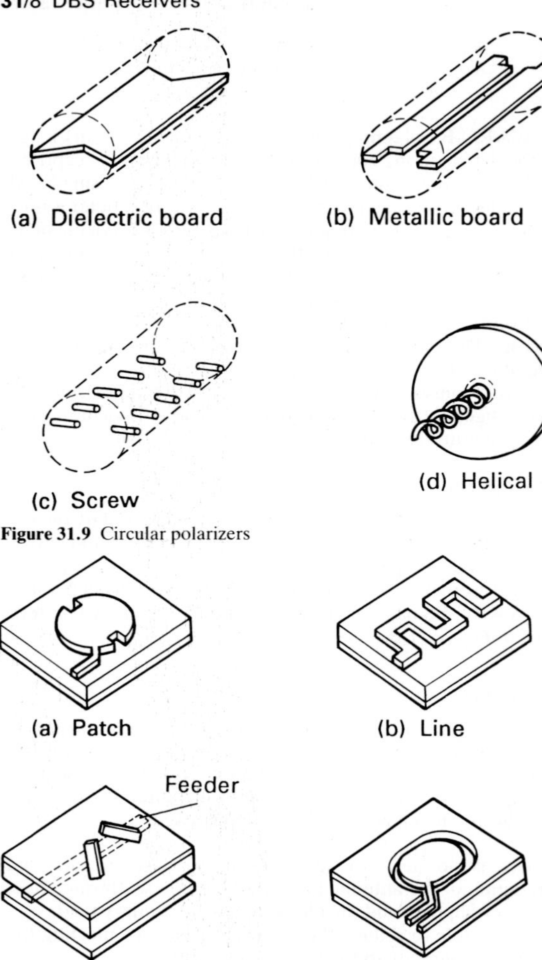

Figure 31.9 Circular polarizers

Figure 31.10 Patterns of microstrip antenna element

- It is easy to mass produce by pressing or plastics moulding techniques.

The reflector of the antenna is made of metal or reinforced plastics with conductive material such as embedded wire mesh, carbon fibre or evaporated aluminium.

The offset parabolic antenna has been proved to be able to offer high efficiency and reliability and be suitable for mass production.

31.5.1.3 Primary feed and polarizer

The primary feed is located at the focus of the reflector. The right-hand circular polarized signal from the satellite is reflected and changed to left-hand circular polarization by the reflector, then introduced into the primary feed. The primary feed with a circular polarizer converts the signal to linear polarization and feeds it to the input stages of the outdoor unit.

A circular polarized signal is considered to have two linear polarized components with a phase difference of 90°. The circular polarizer changes both linear polarized components to the same phase. The feed is usually horn type.

The various types of polarizers available are shown in *Figure 31.9*.

31.5.1.4 Flat antenna

A flat antenna receives signals from the satellite, using a flat board containing small built-in antenna elements instead of the reflector. Flat antennas are roughly divided into two types: *microstrip* and *slotted waveguide*. Recently, microstrip antennas using a printed circuit board have been developed and put on sale in Japan. In these, several hundred small antenna elements shaped on a board (antenna array) are combined and their currents are collected at the feed point. The microstrip element comes in various patterns as shown in *Figure 31.10*. The printed circuit board is made of a material such as teflon glass fibre.

The flat antenna using a simple microstrip has bigger feeder losses and narrower bandwidth, and so it is necessary to employ only the best material with a contrived antenna element pattern and a contrived construction of the board such as a multilayer construction in order to attain wideband and low-loss performances.

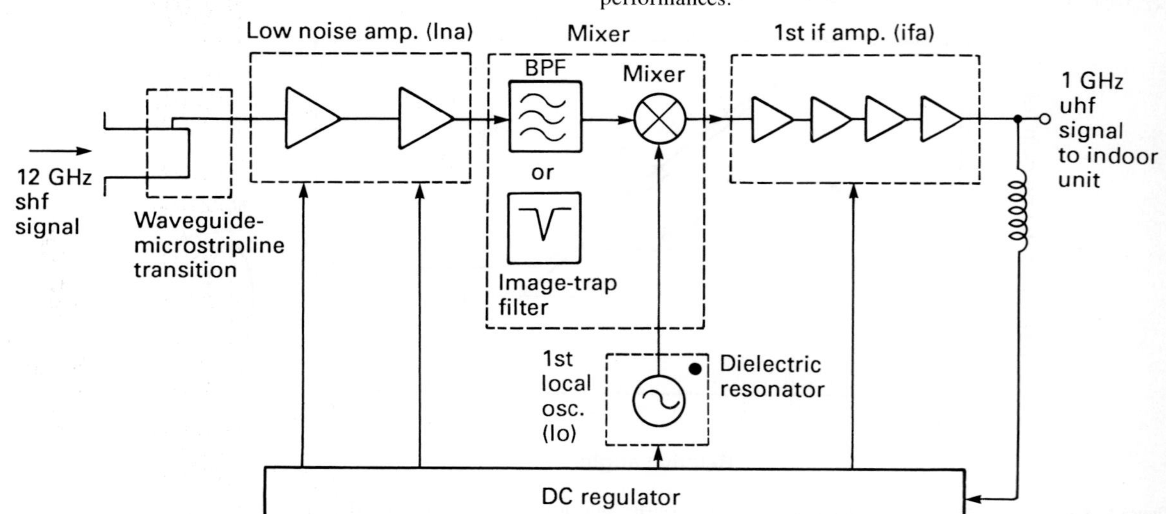

Figure 31.11 Outdoor unit (low noise converter)

The flat antenna is expected to be cheaper, easier to install and less obtrusive than the parabolic antenna. In addition, the flat antenna can incorporate an outdoor unit and has the capability of being pointed with electronic steering. However, the present flat antenna has the disadvantages of lower efficiency due to feeder losses and higher effective temperature of the antenna, particularly in a large unit. Further developments are necessary.

31.5.2 Low noise converter (outdoor unit)

The *low noise converter* (lnc) is one of the important components that determines the receiving signal quality of the dbs receiver. A block diagram is shown in *Figure 31.11*. The 12 GHz fm TV signals received are converted to the first if signals and then amplified in the uhf band. The lnc performs block-down conversion of the satellite TV signals to the first intermediate frequency range.

The signal collected by the antenna is introduced through the waveguide to the microstrip/line transition circuit. This circuit feeds the signals to the next low noise amplifier in microstrip/line mode. The lna is made of two to three stage gallium arsenide fets (mesfets). The output signal of the lna is mixed with the output of first local oscillator and converted to the first if signal.

A technology known as MIC (microwave integrated circuit) with low noise gallium arsenide fets and Schottky barrier diodes is mainly used to integrate the lnc circuits.

The output signal of the mixer is amplified by the first if stage. The dc power is supplied to the lnc by the indoor unit through the coaxial interconnecting cable.

To ensure reliability under severe environmental conditions, a hermetically sealed and waterproof construction is applied to the outdoor unit.

31.5.2.1 Waveguide microstrip/line transition

The signal collected by the feed device of the antenna is passed through the waveguide to the microstrip/line transition circuit, which is formed on the substrate.

Using the phenomenon of frequency cut-off in the waveguide, a high-pass filter is formed, which suppresses the image band. The first local oscillator frequency is usually allocated below the signal band. The image band is approximately 9–10 GHz, which may be used in a strong radar band. The outdoor unit with waveguide filter is effective for image suppression. Using this filter, image suppression of more than 50 dB is attained.

The waveguide microstrip/line transition circuit operates so

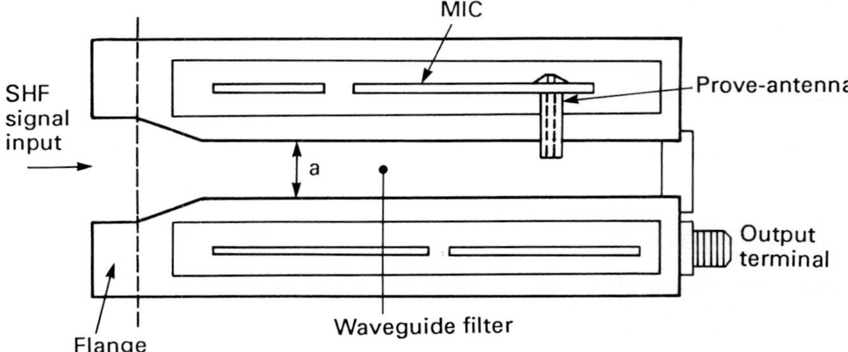

Figure 31.12 Principle of an outdoor unit

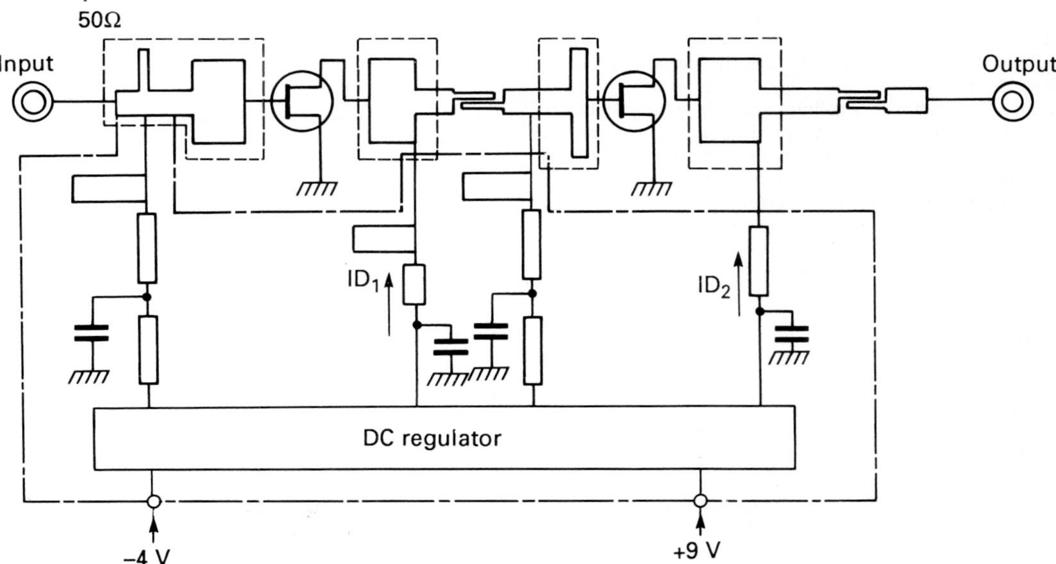

Figure 31.13 A low noise amplifier circuit. The broken line boxes are impedance matching circuits, designed to minimize the noise in the input and to maximize the gain in the output

as to transmit the signals in the waveguide to the microstrip/line of the low noise amplifier. This circuit may be formed by inserting a prove-antenna connected to the microstrip/line into the waveguide.

31.5.2.2 Low noise amplifier

The low noise amplifier is composed of a two to three stage amplifier using low noise devices such as gallium arsenide fets (see section *31.5.2.2.2*) or high electron mobility transistors (hemts — see section *31.5.2.2.3*) and is fabricated by MIC technologies. Sufficient gain must be provided by the lna to overcome losses in the transmission from it to the following receiver circuit and to override noise which originates after the lna.

31.5.2.2.1 Noise and gain of low noise converter

The key parameter for a low noise converter is the noise performance. In order to receive and amplify the weak signal from a satellite minimizing the noise generated in the receiver, a low noise figure (nf) and high gain are required.

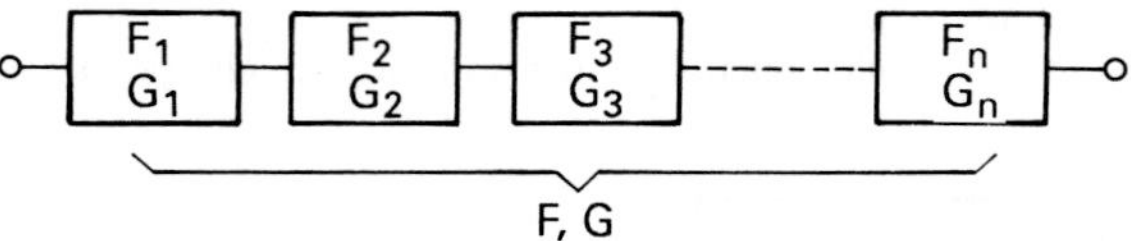

Figure 31.14 Low noise amplifier. F = noise figure, G = gain of amplifier

In the low noise amplifier shown in *Figure 31.14*, the overall nf is expressed by:

$$F = F_1 + \frac{F_2 - 1}{G_1} + \frac{F_3 - 1}{G_1 G_2} + \cdots + \frac{F_n - 1}{G_1 G_2 \ldots G_{n-1}} \quad (31.9)$$

Therefore, the overall nf is affected mainly by the noise figure of the first stage, and less by the noise figures of the succeeding stages depending on the gain of the preceding stage. Consequently, a high gain amplifier with low nf is required in the preceding stage.

In a practical lnc, two lossy components (a filter and a mixer) succeed the lna. Assuming the first stage nf to be low, and the filter and mixer nfs to be less than 0.1 dB, the lna gain should be more than 15 dB.

The gain associated with a gallium arsenide fet in practice is approximately 9 dB, and a two to three stage lna is necessary. The fet has 1.6 dB nf, and a typical noise factor for the lnc of 2.0 dB has been obtained. Table *31.3* shows an example of gain and nf allotment of a low noise converter.

	NF (dB)	Power gain
WG WM	0.1	—
LNA	1.7	18.0
MIX	5.0	− 5.0
IFA	2.5	35.0
Overall	2.01*	48.0

* Overall NF

$$NF = 0.1 + 10 \log \left(10^{1.7/10} + \frac{10^{(5+2.5)/10} - 1}{10^{18/10}} \right) = 2.01 (db)$$

Table 31.3 Typical gain and nf allotment of a low noise converter

The gain diagram is indispensable in designing a hig performance lnc. The lnc gain is determined after the consi eration of the noise factor of the indoor unit (which will n affect the receiver's nf) and the linearity of the lnc. Maximur interconnecting cable loss is about 15 dB for a 30 m cabl Assuming the noise factor of the indoor unit to be 12 dB, that the lnc to be 3 dB and the gain of the lnc to be 48 dB, then th total increase will be 0.02 dB. This is a negligible figure.

LNC gain linearity is usually measured by third orde intermodulation. The co-channel protection ratio is required t be 31 dB by WARC-BS. Assuming the intermodulation proc uct of the input signals to be 24 dB, the third order inte modulation measured at the output port should be less tha -55 dB. An overall gain of 48 dB is adequate for the lnc.

31.5.2.2.2 Gallium arsenide fet

The GaAs mesfet (gallium arsenide metal semiconductor fiel effect transistor) is an important device for the 12 Ghz ban lnc. The electron mobility of gallium arsenide is about six time that of silicon. The construction of a GaAs mesfet is simple bu it has high-frequency performance superior to a silicon transis tor in:

- decrease of the parasitic capitances by isolation effect using layer of semi-insulating GaAs
- decrease of the parasitic serial resistances and increase of th transconductance, g_m by a high electron mobility using a active layer of n-GaAs

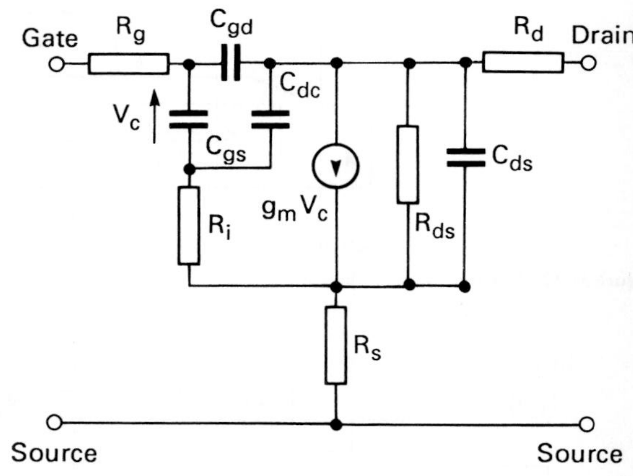

Figure 31.15 Equivalent circuit of GaAs fet

$$F_0 = 1 + 2\pi K_f C_{gs} \sqrt{\frac{R_g + R_s}{gm}} \quad (31.10)$$

where F_o is the optimal value of noise figure, K_f is the fitting factor and f the operating frequency

To achieve low noise performance, the epitaxial layer unde the gate must be etched into a recess, and the gate length mus be short. This requires the use of an ion-implantatio technique.

Characteristics of low noise GaAs fets in the shf band hav been greatly improved in recent years through the progress o semiconductor technologies.

A GaAs fet with a very low noise figure has become availabl at reasonable cost. Using low noise GaAs fets with a nois factor of 1.6 dB, a typical overall nf of about 1.9 dB has alread been achieved in the receivers on sale.

31.5.2.2.3 HEMT

In addition to the improvements of GaAs mesfets, an epoch-making new device, hemt (high electron mobility transistor), has recently been introduced into the lnc of the dbs receiver.

In 1978, it was discovered by Dingle and others of Bell Labs that the hetero-structure (junction of different semiconductors) of AlGaAs-GaAs has a high electron mobility. A hetero-structure has about double the electron mobility of a Schottky-barrier junction as is formed in the GaAs mesfet. The GaAs fet with a hetero-structure is a *hemt*.

The hemt is fabricated by a similar process to the GaAs fet. Because of its superior transconductance, the hemt has an extremely low noise performance.

With a hemt having a noise factor of 1.2 dB, a typical overall nf of about 1.5 dB has been achieved in receivers.

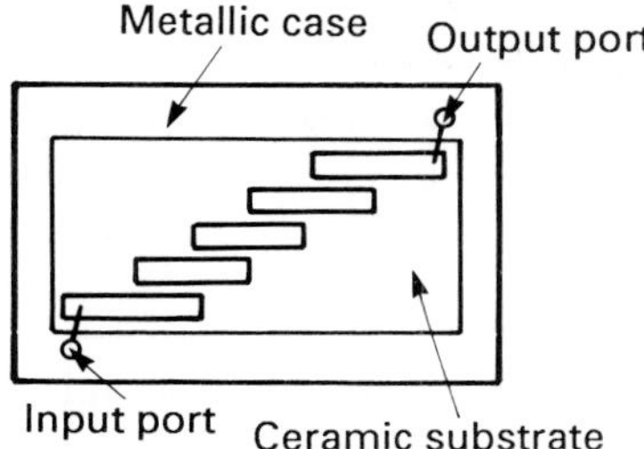

Figure 31.16 Construction of a band-pass filter

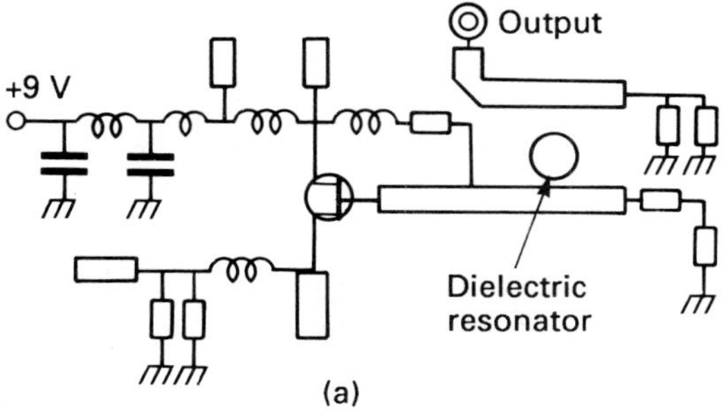

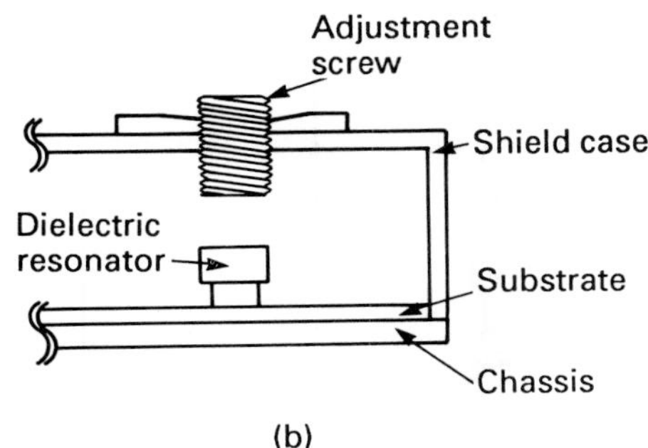

Figure 31.17 First local oscillator: (a) circuit, (b) adjustment mechanism

31.5.2.2.4 MIC

The receiver front end consisting of a low noise amplifier, local oscillator and mixer, is fabricated by the hybrid integrated circuit technology (called *microwave integrated circuit*) whereby the fets and the associated passive elements formed by microstrip/lines and stubs are mounted or formed on the same teflon or ceramic substrate. *Figures 31.13* and *31.17* show examples of the MIC. A more integrated version will eventually be possible, and microwave monolithic integrated circuits (e.g. active and passive components integrated with the same GaAs chip) may be foreseen.

31.5.2.2.5 Band-pass filter

A band-pass filter has two objectives: image suppression, and spurious radiation control. A typical filter is shown in *Figure 31.16*. The image suppression could be kept below 40 dB only by this filter. Instead of using the band-pass filter, an image-trap filter is used in the image recovery mixer (see section *31.5.24*) for image suppression.

31.5.2.3 First local oscillator

The first local oscillator is a simple GaAs fet oscillator stabilized by a dielectric resonator. The oscillator principle is usually either reflection or feedback oscillation. A GaAs fet used in the oscillator is sometimes a medium power unit. *Figure 31.17* shows an example of a first local oscillator circuit using a common source fet and adjustment mechanism.

Frequency drift of the local oscillator should be within the frequency space. A practical figure will be within 1 MHz, with the provision of afc (automatic frequency control) in the indoor unit. Using a dielectric resonator, stable oscillator frequency can be obtained in whatever severe thermal condition the outdoor unit may be placed.

31.5.2.4 Mixer

The 12 GHz TV signal is mixed with the first local oscillator signal, and converted to the first if signal (1 GHz). The mixer usually utilizes a balanced diode mixer, which effectively reduces local oscillator leakage. An alternative mixer is single ended with image recovery, which reduces insertion loss. The image-trap filter suppresses image signals and reflects image components generated by diode mixing. Image components are put into the diode mixer again and improve the conversion efficiency. *Figure 31.18* shows the image recovery mixer. Mixer diodes are GaAs Schottky barrier diodes. Silicon Schottky barrier diodes are also useful for the mixer.

31.5.2.5 First if amplifier

The first if amplifier amplifies 1 GHz uhf band signals, and drives the signals into the connecting cable. A significantly high output power stage is necessary in order to reduce the output stage distortions and/or intermodulations. The first if amplifier with or without gain adjustment and tilt-compensator consists of a four to five stage amplifier using a low noise high frequency transistor, and its gain is commonly 30–40 dB.

31.5.3 Indoor unit

The indoor unit has channel selection and fm demodulation functions (see section *31.3.4*). The video signals and the sound/data signals are derived by demodulation fm signals. In the C-MAC system, however, the sound/data signals are directly demodulated by a 2-4psk demodulator from rf signals. The derived sound/data signals are processed by demodulator and/or decoders to the baseband audio signals. The MAC vision signals are also decoded by a MAC vision decoder to baseband signals.

As the dbs signal format is different in each country, so the indoor unit has to be designed specifically for each.

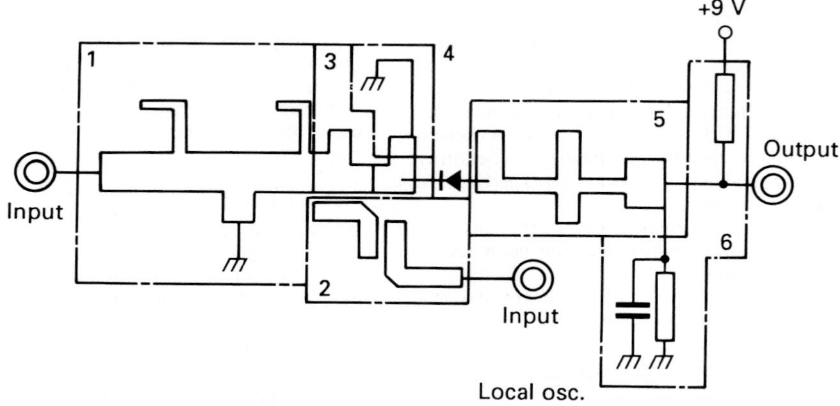

Figure 31.18 Mixer circuit: 1 image-trap filter, 2 band-pass filter, 3 impedance matching circuit, 4 short circuit for if signal, 5 low-pass filter, 6 bias circuit

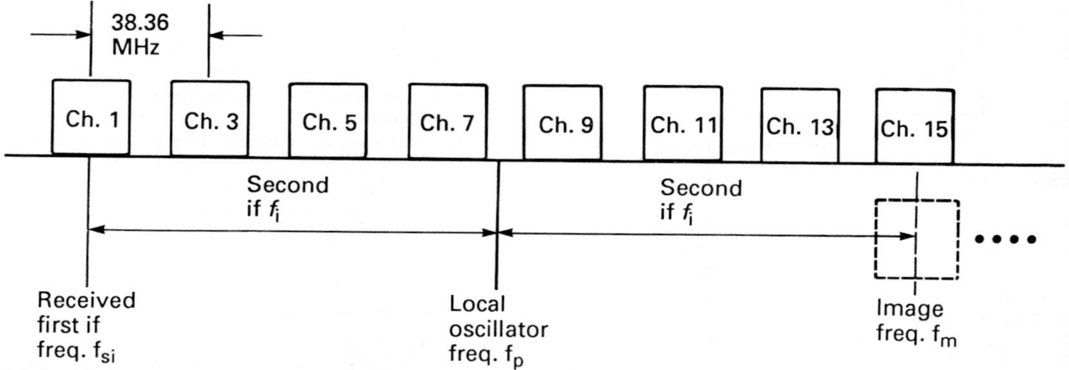

Figure 31.19 Relationship between second if and other frequencies

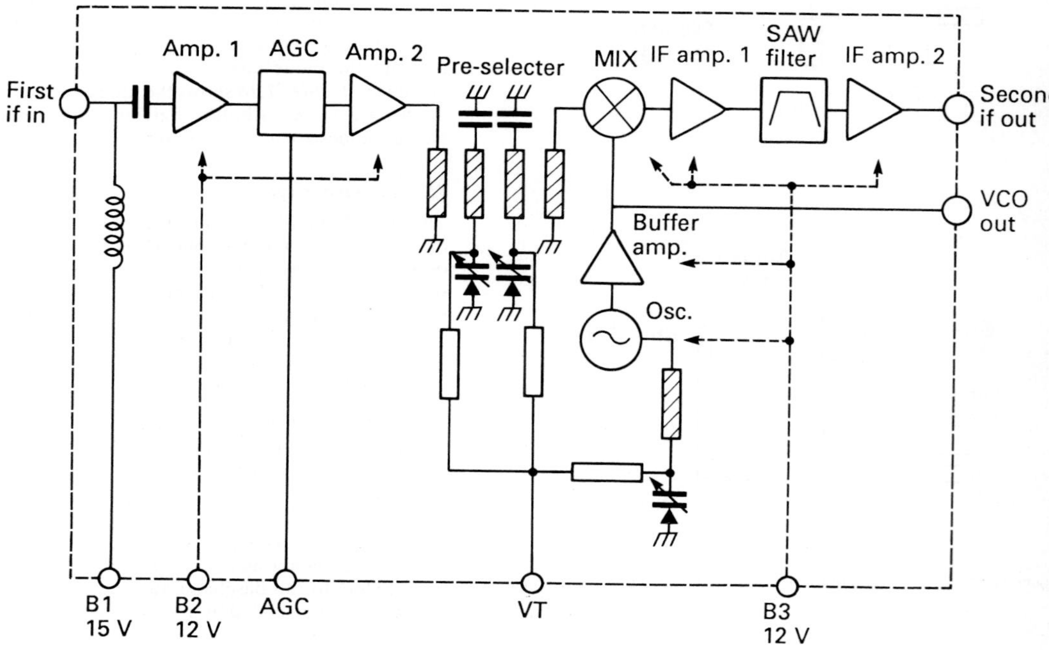

Figure 31.20 Configuration of second converter

31.5.3.1 Second intermediate frequency and bandwidth

The 12 GHz dbs receiver employs a double frequency conversion system.

The second if is freely selectable, but one must take into account high power terrestrial broadcasting transmitters, and interference from the second local oscillator radiation of other receivers. For community reception, a common outdoor unit is connected to multiple indoor units. When the selected value of the second if is smaller than the value of the total frequency bandwidth allocated to satellite broadcasting, local oscillator frequencies fall in part of the first if band. To keep this interference as low as possible by arranging the second local oscillator frequencies between any two adjacent channels allocated to that area, the following relationship is desirable for the second if:

$$f_i = F(n+\frac{1}{2})\ \text{MHz}$$

where F is 38.36 MHz for Regions 1 and 3 or 29.16 MHz for Region 2.

The relationship is valid when the selected frequency f_i is smaller than the value of the total frequency bandwidth in that area. However, if $2f_i$ is less than the total bandwidth, there is the possibility of image frequency interference, and insertion of a tunable band-pass filter may be advisable before the mixer.

The second intermediate frequency bandwidth is defined by WARC-BS to be 27 MHz (Regions 1 and 3) or 24 MHz (Region 2). Filter characteristics outside the band should be specified as co-channel protection ratio defined by WARC-BS.

31.5.3.2 Second converter

The second converter is one of the most important parts in an indoor unit, and is similar to the TV tuner, especially the uhf tuner.

It converts the uhf intermediate signal to a second if signal with a 27 MHz or 24 MHz bandwidth, set by its channel selection switch. The second converter consists of a first if amplifier with agc, a tunable preselector, a mixer, a local oscillator with buffer amplifier and a second if amplifier which includes a surface acoustic wave (saw) filter.

In order to control the gain of the second converter and minimize the intermodulation, an agc using PIN diodes is provided in the first if amplifier.

In its rf stage, a tunable preselector (band-pass filter) is preferable because of the reduction in image rejection and the reduction in the second local oscillator radiation and IM3. The tunable preselector, which is composed of a double tuned band-pass filter varicap diodes and controlled by the voltage tracking the second local voltage controlled oscillator, reduces image interferences effectively.

31.5.3.3 Second if amplifier

In the second if amplifier, filtering, gain control and limiter functions are provided.

The band-pass filter is used to suppress the out-of-band noise. The overall selectivity of the receiver is set by this filter with a bandwidth of 27 MHz or 24 MHz. The filter consists of a saw filter or a helical filter. The saw filter assures a high quality picture, with no need for adjustments, suppressing the out-of-band noise and giving superior phase delay characteristics. The helical filter is used in high second intermediate frequencies.

Gain control is provided for stabilizing the if signal level of the demodulator to obtain stable demodulation characteristics, and keep intermodulation distortion at low level.

The limiter suppresses undesirable am noise over a wide carrier frequency and improves the c/n threshold characteristics. In a phase locked loop (pll) demodulator, the limiter function is included.

31.5.3.4 FM demodulator

The fm demodulator converts the incoming frequency modulated signal to a baseband signal containing a video signal and digitally modulated sound/data signals. It can generate an afc signal.

The fm demodulator of the dbs receiver requires wide-band demodulation. There are two types of demodulators, the double-tuned discriminator and the phase locked loop demodulator shown in *Figure 31.21*. A pll type demodulator is most suitable for a dbs receiver because of its high performance. It has the advantage of s/n improvement in the picture under low c/n condition, known as *threshold extension*, simple adjustment and excellent linearity. Basically, it consists of a phase detector (or mixer), low-pass filter (or loop filter) and voltage controlled oscillator (vco). When the pll is locked to the incoming fm signal, the demodulated signal can be derived from the output of the loop filter. The performance of the pll is determined by the time-constant of the loop filter.

A pll fm demodulator improves the threshold c/n by about 2–3 dB.

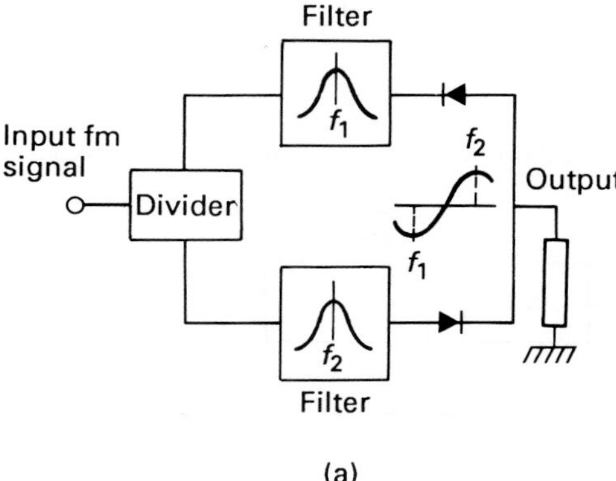

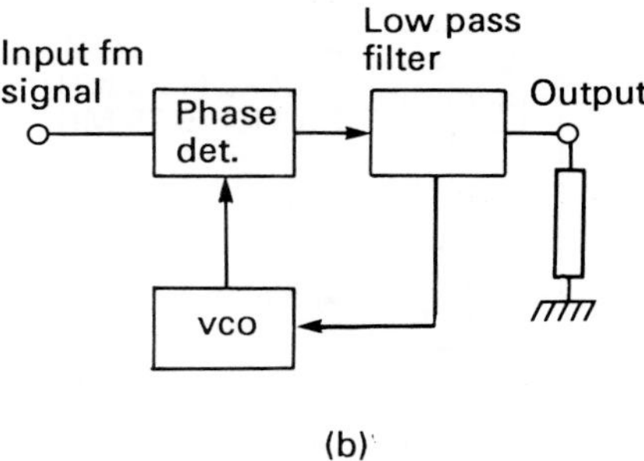

Figure 31.21 FM demodulator: (a) double-tuned discriminator, (b) phase locked loop demodulator

31.5.3.5 Energy dispersal signal rejection

The energy dispersal signal has a triangular waveform (a half, or a quarter, of the field frequency); it is added to the video signal in order to avoid concentration of the energy spectrum in fm transmission. The rejection circuit removes this triangular waveform; unless it is removed, there will be flicker interference on the picture. A high-speed peak clamping circuit or a pulse clamping circuit (*Figure 31.22*) is used.

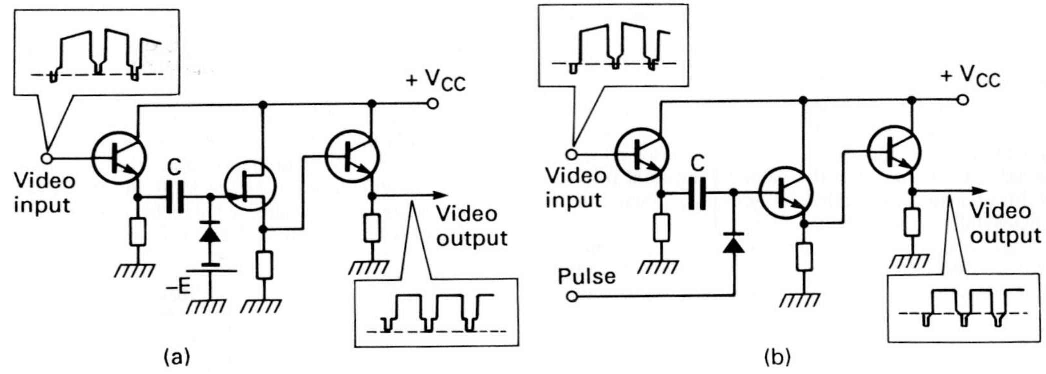

Figure 31.22 Examples of analogue energy dispersal rejection circuits: (a) peak clamp, (b) pulse clamp

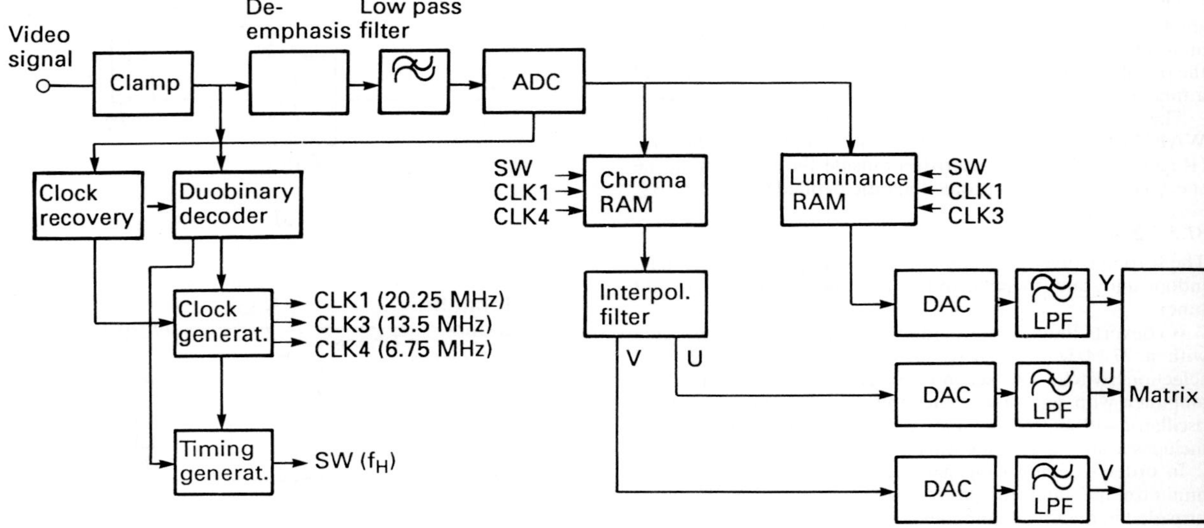

Figure 31.23 Notional diagram of a typical MAC vision decoder

31.5.3.6 De-emphasis

As the signal/noise ratio of the high-band video signal is degraded in fm transmission, the video signal is pre-emphasized in high-band, before transmission. So, the received video signal is de-emphasized by means of a network having a reverse characteristic to the pre-emphasis. The characteristic of the emphasis is defined by each transmission system.

31.5.3.7 Video signal processor

The digital subcarrier/NTSC system employs the conventional NTSC composite video format, so this system does not particularly require video signal processing in the dbs receiver. As the MAC/packet family employs time division multiplexing, it is necessary to decode the MAC vision signals into conventional video signals in the receiver. The MAC vision signal processor is described in the following section.

31.5.3.7.1 MAC vision decoder

The present vision standard specifications of the MAC/packet family is compatible in each system, so the circuit configuration of the MAC vision decoder is the same. *Figure 31.23* is a notional diagram of a MAC vision decoder. The output signals of the fm demodulator are applied to the MAC decoder through the clamp, de-emphasis, and low-pass filter.

On the other hand, clock signals and timing pulses are generated from the data burst signals, which are multiplexed during the line blanking interval.

The MAC vision signals are digitized by an ADC using sampling clock frequencies (20.25 MHz for C/D and D2-MAC and 21.477 MHz for B-MAC, 525-line systems). They are divided into two signals: one for the luminance and one for the colour-difference signal processing circuit (comprising line memories). The time expansion of the luminance signal, Y, is achieved by writing the digitized vision signals into a line memory at the above sampling frequency and reading them out at a frequency of 13.5 MHz (for C/D and D2-MAC) or 14.318 MHz (for B-MAC 525-line). This can be done by two line memories used alternately for read and write.

The colour-difference signals are sequentially transmitted within alternate lines as U and V. U is transmitted on odd active lines and V on even active lines. A chrominance signal is transmitted one line before the associated luminance signal.

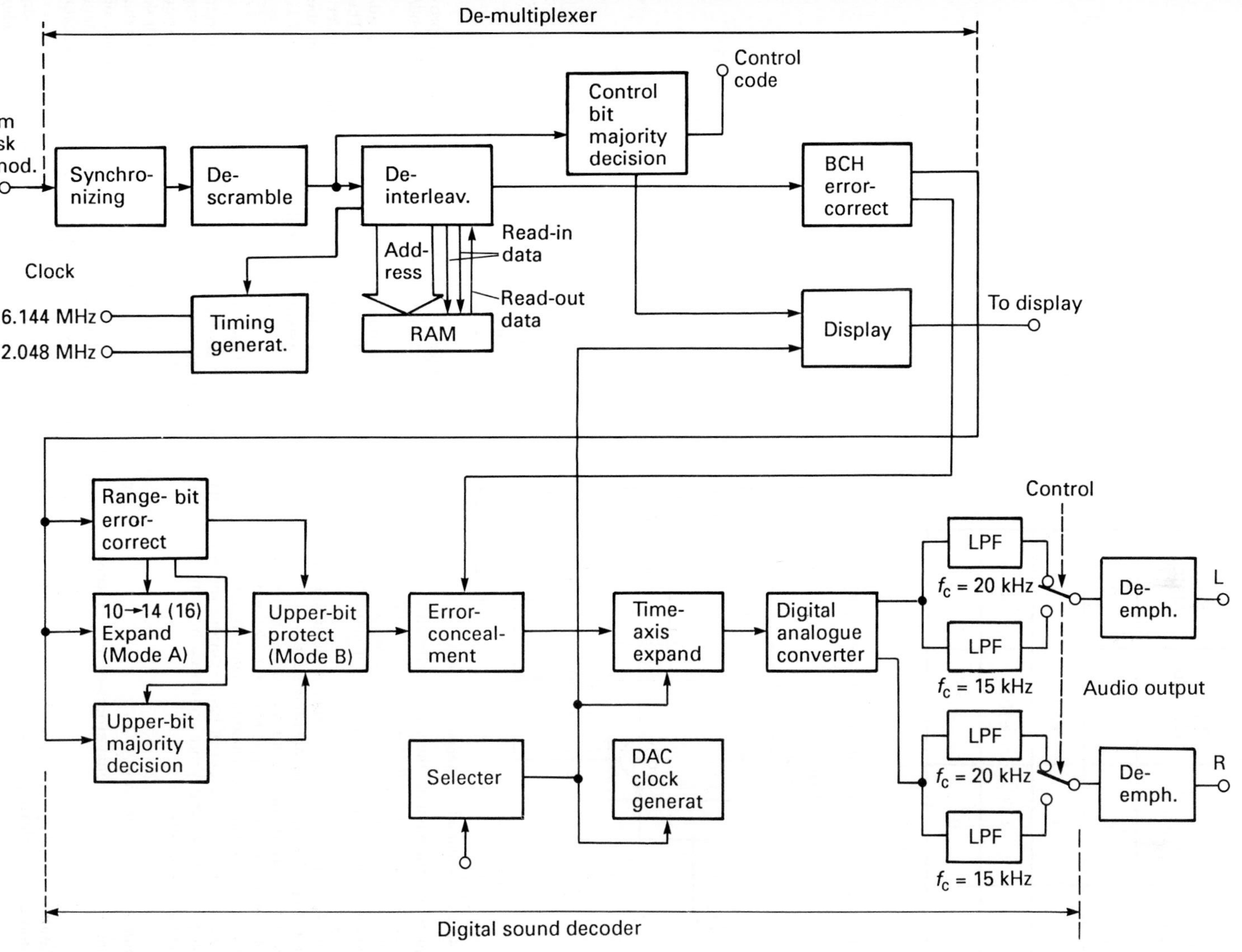

Figure 31.24 Sound/data signal processor (digital subcarrier/NTSC). LPF = low pass filter

Symbol		Phase-change	Rest state
y_n	x_n		
0	0	0°	0
0	1	+90°	1
1	1	±180°	2
1	0	–90°	3

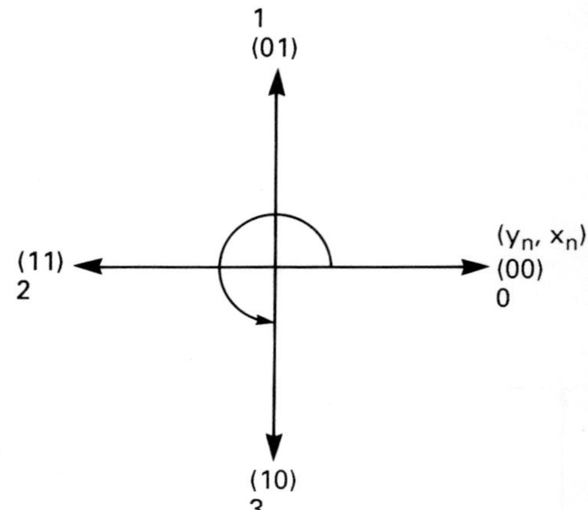

Figure 31.25 Transition rule of quadrature phase shift keying (4psk)

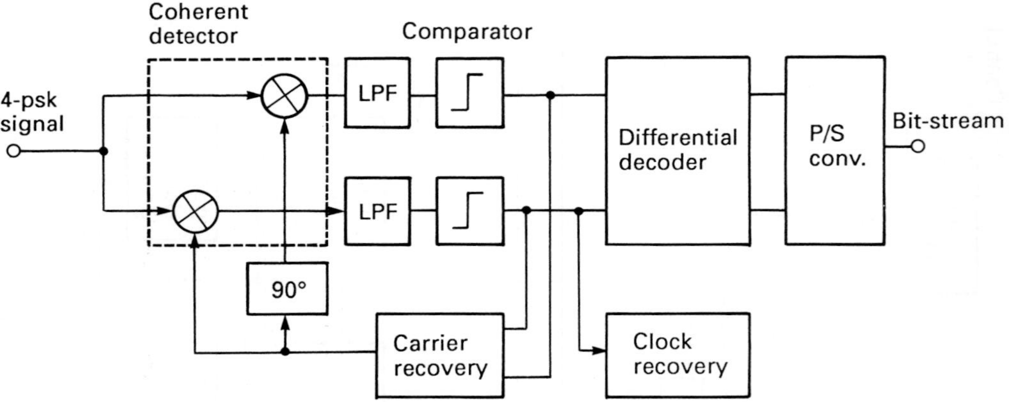

Figure 31.26 4psk demodulator (synchronous detection)

Due to sequential transmission, the colour-difference signal processing necessitates digital interpolating processing of two adjacent chroma lines. For interpolating processing, a 1:2:1 interpolation filter using a mean value of two adjacent chroma lines is employed.

The colour-difference signal processing is achieved in a similar manner to the luminance signal processing.

Y signals derived from the luminance signal processing circuit and U and V signals derived from the chrominance processing circuit are converted to analogue signals by a DAC, and are filtered to suppress the out-of-band spurious signals. Analogue signals of Y, U and V are converted by a matrix circuit to R, G and B signals.

31.5.3.8 Sound/data signal processor

The functions of a sound/data signal processor in the receiver comprise:

- demodulation or decoding of multiplexed sound/data signals
- transmission de-multiplexing of sound data and additional data
- digital sound decoding

For the dbs systems, sound signals are transmitted in coded forms together with the data signals and in the format of either continuous and rigid structure or flexible and packet structure.

In *de-multiplexing*, the data for particular services are processed and selected from the demodulated or decoded sound/data signals.

The modulation, multiplexing and coding forms of the sound/data signal are different in each transmission system. The sound/data signal processor (comprising demodulator/decoder, de-multiplexer and digital sound decoder) utilized in the receivers of each system are described in sections *31.5.3.8.1–31.5.3.8.3*.

31.5.3.8.1 Digital subcarrier/NTSC system

The sound/data signals are carried with a 5.73 MHz subcarrier (4psk modulated), which is frequency-multiplexed with the conventional NTSC vision signal. In the receiver, the subcarrier signal is separated by a 1 MHz band-pass and fed to a 4psk demodulator. The sound/data capacity (mean data rate) of the system is 2.048 Mbit/s. A continuous and rigid structure of sound/data multiplexing is employed. In the case of data transmission, however, the multiplexing employs the packet format in an additional data area.

In this system, the digital sound coding uses pulse code modulation. The system can provide four high quality audio channels (*mode A*) of 15 kHz bandwidth with near instantaneous 14/10 bit companding, or two very high quality audio channels (*mode B*) of 20 kHz bandwidth with linear 16 bit coding. *Figure 31.24* shows a sound/data signal processor of the digital subcarrier/NTSC system.

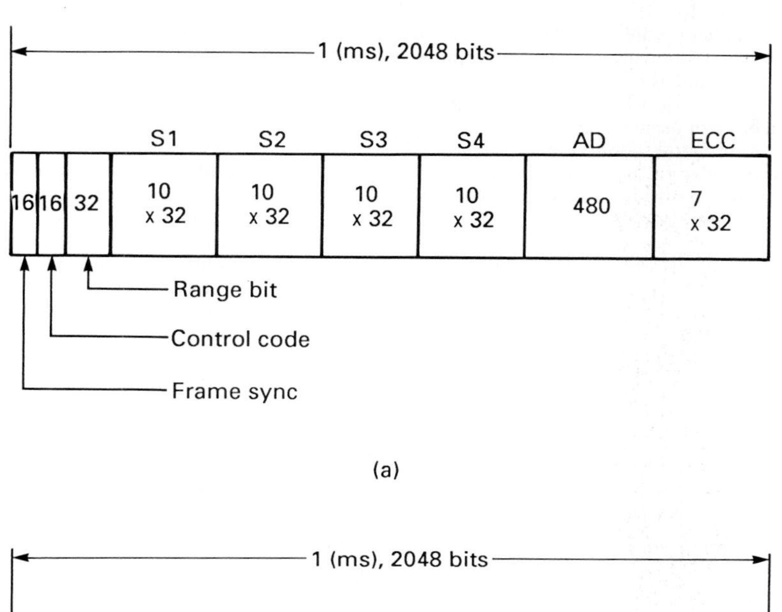

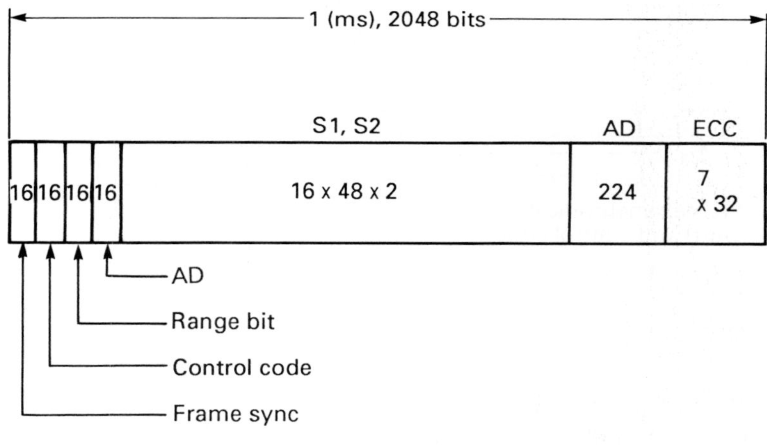

Figure 31.27 Sound/data frame structure (digital subcarrier/NTSC system): (a) mode A, (b) mode B; S_n = audio signal *n*, AD = additional data, ECC = error correction code

The 4psk (*quadrature phase shift keying*) is a four-state modulation method converting logic changes in signal pairs of bits (called *symbols*) of the bit-stream to the phase-change (rest states 90° apart) of the carrier signal by the rule shown in *Figure 31.25*. Because the phase/changes between one symbol and the next (differential logic) are transmitted, the receiver can easily recover them without detecting the absolute carrier phase.

For the 4psk demodulation, synchronous detection is mainly utilized as shown in *Figure 31.26*. In this method, the coherent detector, composed of two multipliers, performs a phase detection of an incoming 4psk signal by using two reference carriers, whose phases differ by 90°. Since the carrier of the incoming signal is suppressed, carrier recovery is first required. For synchronous detection, the carrier recovery is conducted by a phase locked loop circuit. To conduct the carrier recovery using pll, it is necessary for the output of the coherent detector to eliminate the rest-state phase shift components of the incoming carrier. For this purpose, methods such as Costas loop, remodulation and others are employed.

After the phase lock detection, the demodulated signals are discriminated and converted into two bit-streams by the comparator. A clock recovery circuit using a pll is also provided and regenerates a bit-synchronizing clock signal from the bitstream. The output of the 4psk demodulator is applied to a differential decoder having reverse logic to transmitted differential logic symbols. The output of the differential decoder, which has parallel two bit-streams, is applied to the parallel/serial converter and converted into a serial bit-stream.

The *de-multiplexer* in the receiver converts a bit-stream into a sequence of sound samples and data. It comprises such circuits as a frame sync circuit, a descrambler, a de-interleaver, a control code decoder and error correction circuit. (See *Figure 31.24*.) *Figure 31.27* shows a frame structure.

When the bit-stream is applied to the de-multiplexer, first a frame synchronization word of 16 bits is detected by a frame sync circuit from the bit-stream and is synchronized. The frame sync circuit has a flywheel effect so as not to mis-synchronize in the case of mis-detection or error occurrence.

In the descrambler, a pseudo-random binary sequence (prbs), which is previously added to randomize the transmission spectrum and to recover easily the bit-clock signals, is eliminated from the bit-stream by adding (modulo 2) the same prbs.

Descrambled sound/data signals are divided into control code decoder and de-interleaver. Error protection of control codes are performed by 8/15 majority decision of each 15 frames.

Interleaving is previously applied to the bit-stream in order to error-correct effectively, converting burst errors which occur in transmission to random errors. In this system, the 2048 bits of each frame are interleaved with a distance of 32 bits. At the RAM of the de-interleaver, bit read-in (transmission sequence) and read-out are performed with the sequence shown in *Figure 31.28*.

After the de-interleaving, error correction of BCH(63,56) is performed. The BCH(63,56) code makes one bit error correction and two bits error detection for one subframe of 63 bits.

BCH is from the inventors Bose, Chaudhuri and Hocquengham. (63,56) means that the total bit numbers of the block is 63 bits, and its information bit number is 56 bits. (The check code bit number is 7 bits.)

After the error correction, the sound-sample words are decoded by using sound coding law. Digital companding

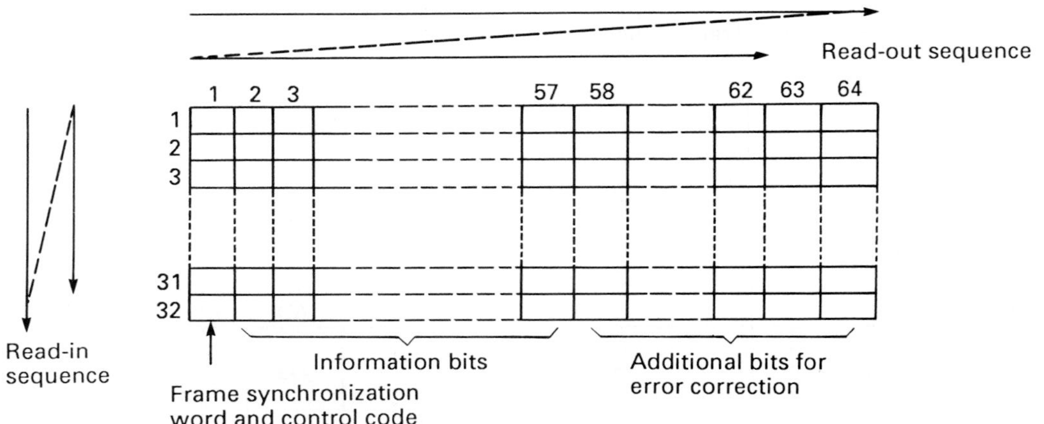

Figure 31.28 Bit interleaving sequence (digital subcarrier/NTSC system)

Range of largest magnitude "a"		Range number	Sound-sample bit number													
Positive	Negative		MSB 1	2	3	4	5	6	7	8	9	10	11	12	13	LSB 14
$a < 2^{13}$	$\lvert a\rvert \le 2^{13}$	0	* 0 1													
$a < 2^{12}$	$\lvert a\rvert \le 2^{12}$	1	* 0 1	0 1												
$a < 2^{11}$	$\lvert a\rvert \le 2^{11}$	2	* 0 1	0 1	0 1											
$a < 2^{10}$	$\lvert a\rvert \le 2^{10}$	3	* 0 1	0 1	0 1	0 1										
$a < 2^{9}$	$\lvert a\rvert \le 2^{9}$	4	* 0 1	0 1	0 1	0 1	0 1									

* Positive = 0
Negative = 1

Parts of the same bits as MSB bit

Bits used for companded code

Extracted bits

Figure 31.29 Near instantaneous companding

techniques are used to produce efficient coding formats for broadcast audio transmission.

In mode A of the digital subcarrier/NTSC system, *near instantaneous companding* is adopted in order to reduce the sound-sample word from 14 bits to 10 bits (as shown in *Figure 31.29*). Near instantaneous coding is done by using five coding ranges of compression to which each 32 successive samples in a period of 1 ms has been subjected. In the receiver, in order to reproduce original sound-sample words, range-codes are error corrected and sample words are expanded using range-codes and companding law.

After the error correction, an error concealment of a sound-sample word using error flags added in the error correction procedure is performed by the interpolation of two adjacent sound-samples. The error-concealed sound data are expanded along the time axis by DAC interfaces and are converted by a DAC to analogue audio signals.

31.5.3.8.2 MAC/packet family

The digital sound/data signals are multiplexed during the line blanking interval. In the receiver, the sound/data signals are derived and demodulated by a 2-4psk demodulator (in C-MAC), or decoded by a duobinary decoder (in D and D2-MAC). The data capacities (mean data rate) of each system are 3.08 Mbit/s in the C and D-MAC systems and 1.54 Mbit/s in the D2-MAC system, respectively. For the MAC/packet family, a data burst of each line is defined to contain 206 bits in the C and D-MAC systems and 105 bits in the D2-MAC system, respectively.

Using these bits, various forms of sound/data signal, synchronization word and control code are transmitted.

In the MAC/packet family, the digital sound coding adopts pcm. The systems are capable of providing eight high quality audio channels of 15 kHz bandwidth with near instantaneous 14/10 bit companding.

Figure 31.30 shows a sound/data signal processor of the MAC/packet family.

For the C-MAC/packet system, multiplexing is carried out at rf on the modulated signals. The sound/data signal is modulated using 2-4psk.

The 2-4psk is a modulation method converting logic 1 to a phase change of +90° and logic 0 to -90°. The phase-change of the carrier between adjacent bits is only 90° (*Figure 31.31*).

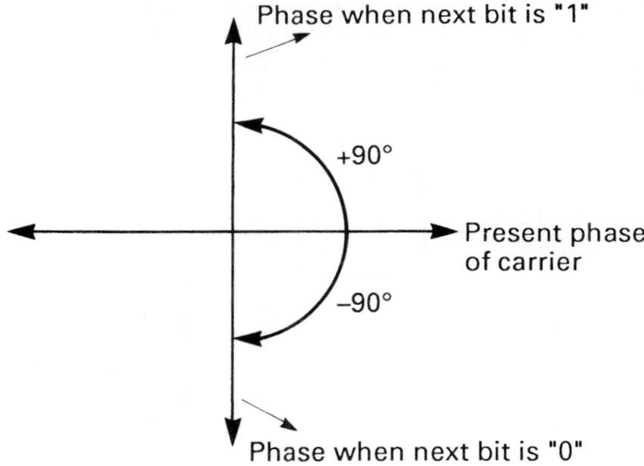

Figure 31.31 The rule of 2–4psk

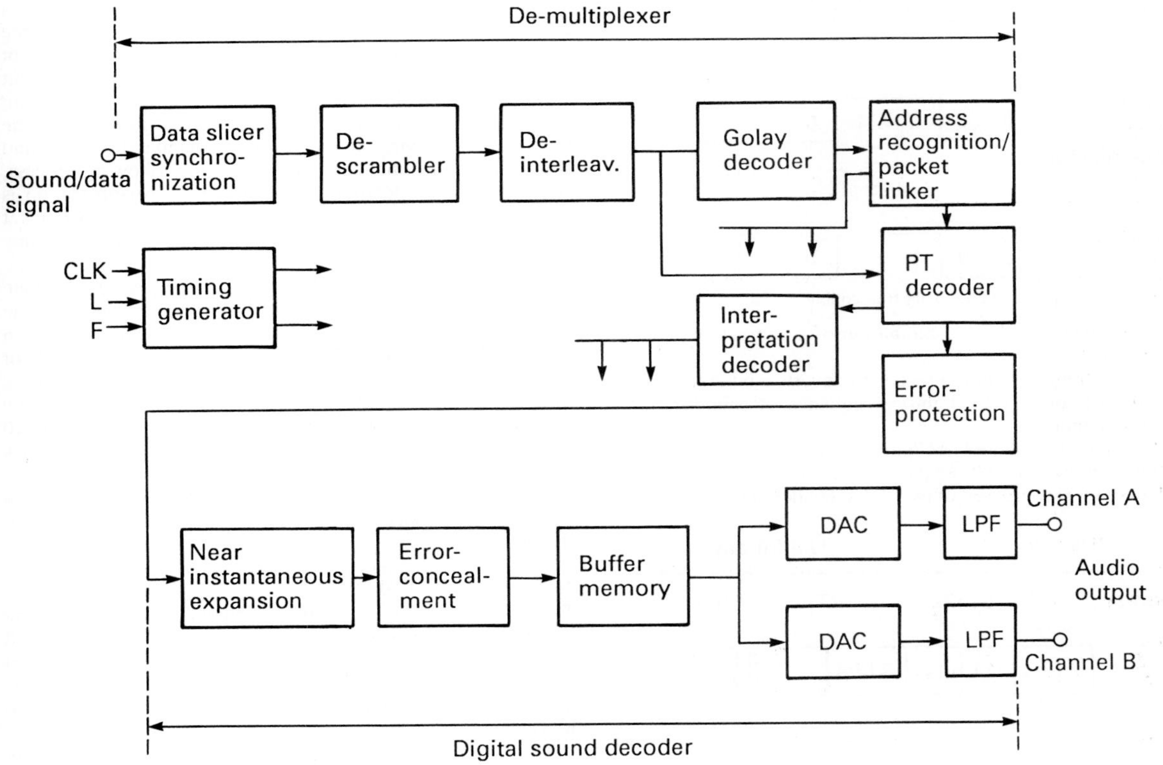

Figure 31.30 Sound/data signal processor (MAC/packet family)

The demodulation of the 2-4psk signal is accomplished by using methods such as differential demodulation and coherent demodulation.

The principle of differential demodulation is shown in *Figure 31.32*. The incoming signal through a band-pass is split into two paths. One is phase-shifted and delayed by one bit, then the two signals are multiplied together. The sound/data bit-stream can be derived from the result.

For the D and D2-MAC/packet systems, sound/data multiplexing is carried out at the baseband during the line blanking interval. The sound/data signal is coded in duobinary form. With this coding method, the signal has three characteristic levels where the extreme levels (+1,-1) represent logic 1 and the intermediate level represents logic 0. An example of duobinary transcoding is illustrated in *Figure 31.33*. By using duobinary coding, the sound/data signals can be transmitted with reduced bandwidth compared with nrz.

Selection of a particular service in the receiver is carried out by recognizing the address of the desired service carried in the packet header. A continuity index is used to assure the link between successive packets of the same service. The pt (*packet type*) code informs the sound decoder of the nature of the sound signal such as the coding law and audio bandwidth.

The de-multiplexer comprises a sync circuit, a de-scrambler, a de-interleaver, a Golay decoder, a pt decoder, an address recognition/packet linker and an error-protection decoder (see *Figure 31.30*).

The demodulated or decoded sound/data signals are applied to the de-multiplexer. Synchronization can be achieved either by *line sync word* (lsw) or by *frame sync word* (fsw). For sound/data synchronization, the start of the first packet is defined in line 1.

De-scrambling and de-interleaving are performed in the same way as in the digital subcarrier/NTSC system. In this system, the 751 bits of each packet are interleaved with a distance of 94 bits.

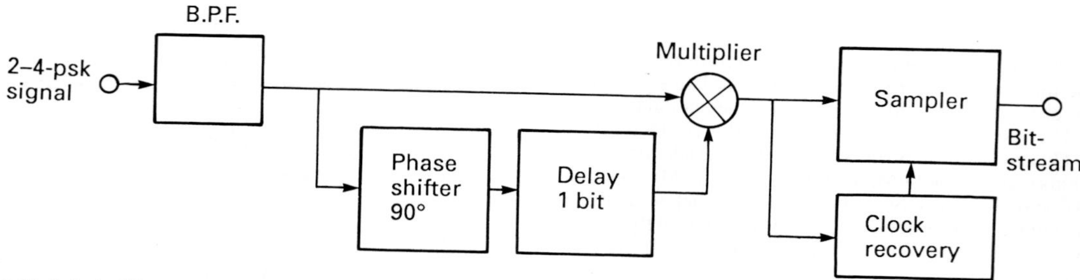

Figure 31.32 2–4psk differential demodulator

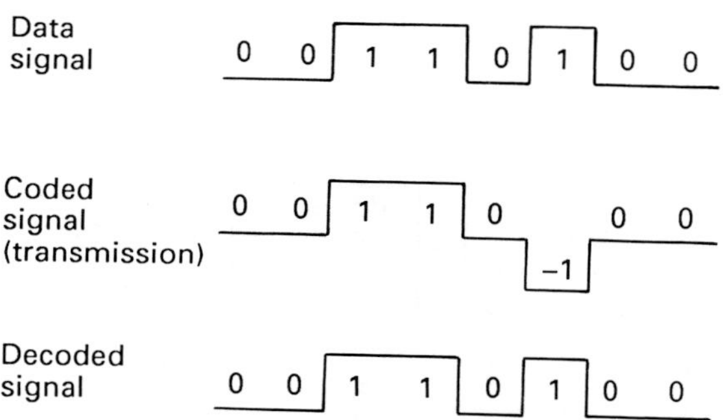

Figure 31.33 Example of duobinary transcoding

The decoder can be a combination of a comparator and a transcoding logic.

The *de-multiplexer* in the receiver converts a bit-stream and all types of sound packet into a sequence of sound-sample words and data.

The sound/data of the MAC/packet family are transmitted by packet multiplexing. All the packets have the same length and each of them is composed of two parts: header and useful data.

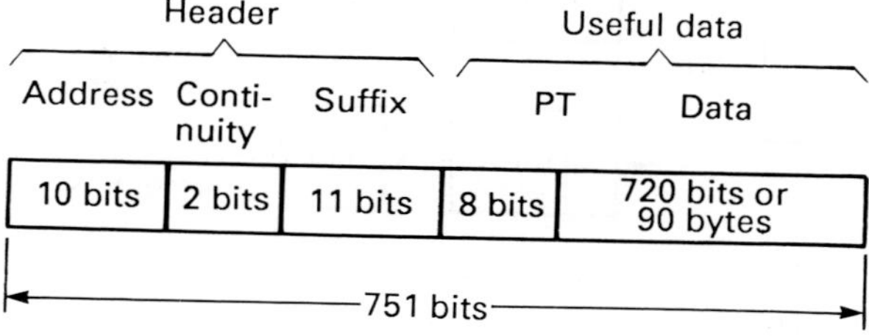

Figure 31.34 Packet structure

The de-interleaved signal is divided into two paths, via a Golay decoder and a pt decoder. At the Golay decoder, the header is protected by using Golay cyclic code. At the pt decoder, the pt code is also detected and decoded. The output of the Golay decoder is applied to the address recognition/packet linker. The header and the pt code recognize and decide the required service of the received sound/data signal and control the functional configuration of the following process. After address recognition/packet linking, error-protection is performed. The error-protection method of the MAC/packet family operates at two levels: 1 parity bit/sample and Hamming 5 bit/sample.

For transmission, *near instantaneous companding* is done with compression from 14 bits to 10 bits per sample in 32 sample (1 ms) blocks. All the samples in each block are coded with coding ranges of compression determined by the magnitude of the largest sample in the block, and a scale-factor code is formed to convey the degree of compression to the receiver.[6]

In the receiver, sound-sample words are expanded from 10 bits to 14 bits using the scale factor and the near instantaneous companding law.

After the error correction, sound-samples with uncorrectable errors are concealed by interpolating two adjacent sound-samples.

31.5.3.8.3 B-MAC system

The sound/data signals are multiplexed during the line blanking interval. In the receiver, the sound/data signals are derived and decoded by a multi-level decoder. The data capacity (mean data rate) is about 1.6 Mbit/s. A data burst of each line is defined to contain 100 bits (50 symbols), and a packet structure is not employed.

The B-MAC system adopts an *adaptive delta modulation* (adm)[8] as the sound coding method, and is capable of providing six high quality 15 kHz audio channels.

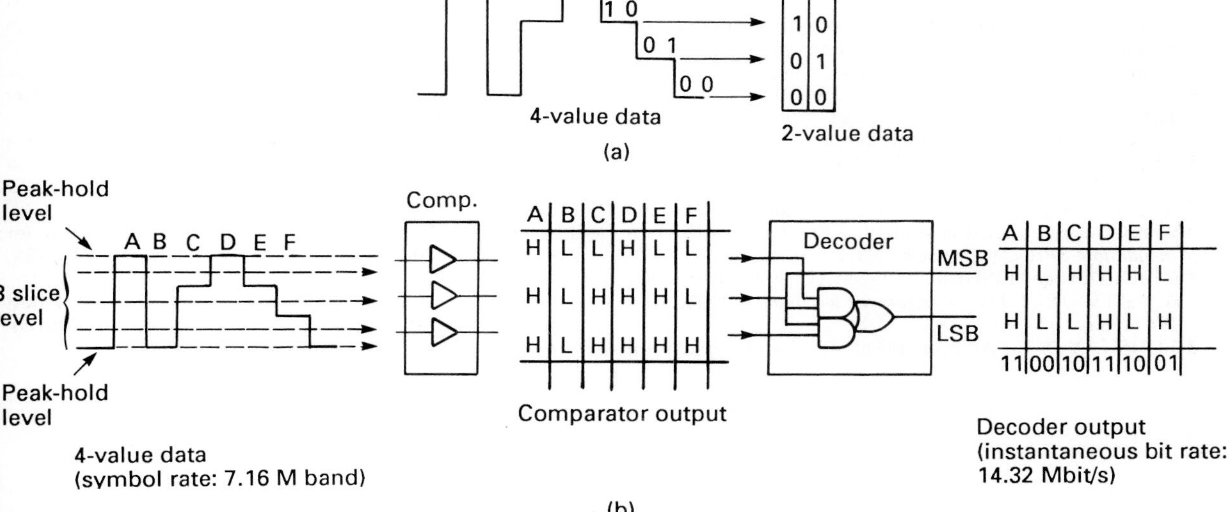

Figure 31.35 4 – 2 value transcoding: (a) transcoding rule, (b) operation

The sound/data signal is coded in *multi-level code*. Four level symbols are employed, each carrying 2 bits of information. Certain control data are transmitted using a two-level symbol which is a subset of the four-level symbols.

The decoder can use methods consisting of a peak-hold, comparators (or slicers) and a transcoding logic. *Figure 31.35* demonstrates the principle.

ADM is a digital sound coding technique based on delta modulation and has a quality at least as high as an equivalent pcm system. This system offers high quality audio performance at the relatively modest bit rate per channel of 220–350 kb/s.

31.6 IC development

In order to achieve high performance, mass production and reasonable cost, it is important to introduce integrated circuits into the dbs receiver.

Monolithic IC development steps for the dbs receiver have been undertaken in the following areas:

- *microwave*: The monolithic microwave IC approach is considered to be one of the better methods of producing cost-effective and reliable microwave devices with accurate configuration in large quantity
- *second if*: the monolithic analogue ICs make possible high performances, minimize components and avoid adjustments
- *baseband signal*: analogue and digital lsi technologies can realize complex, large-scale and high performance signal processors in simple configuration

At present, various types of IC for the dbs receiver have already been developed and some of them have been used in the receivers on the market.[7,9–13]

31.7 Re-transmission of satellite broadcasting

For maximum diffusion of satellite broadcasts, it is necessary to establish community reception systems or cable re-transmission systems to allow numbers of households to receive satellite broadcasting services.

For the dbs re-transmission, the D and D2-MAC/packet systems were developed to have media transparency. Duobinary data transmission makes it possible to convert the fm satellite channels to the vestigial-sideband amplitude-modulated cable channels with reduced channel spacing (of at least 10.5 MHz for D-MAC, or 7 MHz for D2-MAC).

Generally, the cable re-transmission systems take one of two approaches:

- transmit fm signals with wide bandwidth, without changing the rf modulation parameters of satellite broadcasting (suitable for a small-scale system)
- transmit signals by remodulating to am compatible with the terrestrial broadcasting system (suitable for a large-scale system)

31.8 Future expansion

The broadcasting satellite will not only supply television satellite broadcasting but will also serve as the channel for important new broadcasting media.

The expected developments include high quality multichannel digital sound broadcasting, conditional access broadcasting using scrambled signals. high definition TV, data broadcasting such as facsimile, teletext and still-picture broadcasting. To realize these broadcasting systems, additional reception equipment will be needed. It may be said that the satellite broadcasting systems have just moved from the phase of development of basic transmitting and receiving systems to the phase of development of new broadcasting systems using the satellite broadcasting media.

References

1 'Broadcasting-satellite service (sound and television)', *Recommendation and Reports of the CCIR*, **X** and **XI-Part 2**, Rep. 1074/Rep. 473-4, 68-99/190-211, Geneva (1986)

2 'Specification of transmission systems for the broadcasting-satellite service', *CCIR Special Publication* JIWP 10-11/3, Geneva (1986)

3 'Final acts on the world administrative radio conference for the planning of the broadcasting satellite service in frequency band 11.7–12.2 GHz (in Regions 2 and 3) and 11.7–12.5 GHz (in Region 1)', *ITU*, Geneva (1977)

4 HIRATA, S, 'DBS receiver design', Joint Session, *14th International TV symposium*, Montreux, 172 (1985)

5 STUTZMAN, W L and THIELE, G A, *Antenna Theory and Design*, John Wiley & Sons, New York, 433

6 'Specification for transmission of two-channel digital sound with terrestrial television system', *EBU Technical Recommendation*, SPB 424, Brussels (1987)

7 ZIBOLD, H and FISHER, T, 'D2-MAC: A new feature for digital TV', *IEEE Trans Consumer Electronics*, **CE-32**, No 3, 274 (1986)

8 FORSHAY, S E, 'An economical digital audio system for consumer delivery', *IEEE Trans Consumer Electronics*, **CE-31**, No 3, 269 (1985)

9 NEELEN, A H, 'An indoor unit for satellite TV', Broadcast Session, *15th International TV Symposium*, Montreux, 274 (1987)

10 HASEGAWA, K et al, 'Fully integrated fm demodulator circuits for satellite TV receivers', *IEEE Trans Consumer Electronics*, **CE-33**, No 2, 77 (1987)

11 DEHERY, Y F and DECLERCK, C, 'A chip set for a modular structure of D2-MAC/packet receiver', Broadcast Session, *15th International TV Symposium*, Montreux, 459 (1987)

12 RONNINGEN, L A, 'The Nordic ASICs and software for the C/D/D2-MAC/packet receivers', Session XX-DBS, *ICCR* (1987)

13 HIRATA, S, 'IC developments for dbs receivers', Latest Broadcasting Media and Equipment, Session 3-2, *International Broadcasting Symposium*, Tokyo (1985)

Part 7
TV Studios and Studio Equipment

Section 38 Film and Video Tape Editing
P. Audemars, London Weekend Television

38.1 Film editing
38.2 Video tape editing
38.3 Interrelationship between film and video editing
34.4 Future developments
35.5 The role of the editor

Section 39 Telecines
J. Millward, Rank Cintel Ltd

39.1 Telecine types
39.2 Film formats
39.3 Film transports
39.4 Sound reproduction
39.5 Optical and scanning systems
39.6 Colour response
39.7 Signal processing

R Stevens B Sc, C Eng, MIEE
Engineering Project Supervisor, Thames Television

32 Studio Planning and Requirements

32.1 Studio design

32.1.1 Types of studio centre

Television studios and studio centres are designed for many different reasons to do many different tasks. Most of the types fall into one of the following categories:

- *massive complex*, where studios which can produce all the various different programme types are located in one building,
- *large regional centre*, where there is a mixture of local production of programmes and programmes received from a network,
- *small regional centre*, where most of the programmes come from a network but local news and other programmes of local interest are made,
- *specialized centre*, where only programmes of a particular type are made, such as news or sport,
- *recording centre*, where drama and light entertainment programmes, or perhaps commercials, are recorded, the programmes being played out from video tape from another centre,
- *playout centre*, where programmes made outside and brought in either on video tape or by landline are put together in a transmission sequence; the complex may include some programme-making equipment — particularly for making promotional material,
- *non-broadcast studio*, where training videos can be made and which do not need to conform to broadcast quality standards.

32.1.2 Size of studios

Television studios are built in a variety of sizes, usually to suit the particular types of programming that they will produce. The

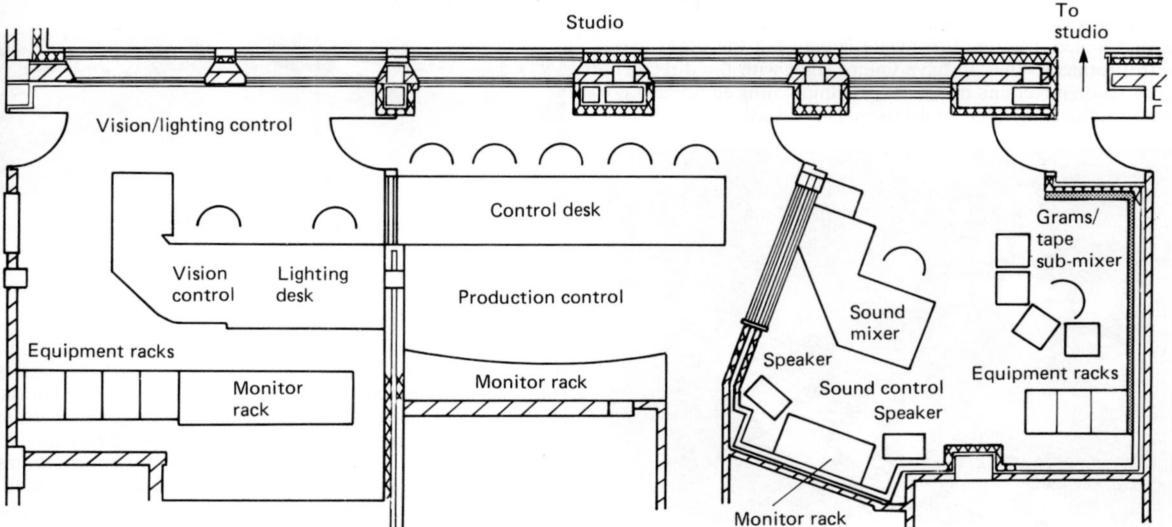

Figure 32.1 Plan view of studio control rooms

total useable space in a studio can be reduced by between 20 and 30 per cent if provision is made for a cyclorama curtain, a camera pull-back area (to give free access to the studio for scenery setting), audience seating and the requirements of fire regulations.

The range of studio sizes is as follows:

- *Large studio*. This is between 400 and 1000 m^2 and is used for light entertainment and drama programmes. The normal camera complement will be four or five but further cameras will probably be assigned to the studio for complex productions. Some of the cameras will be portable models to enable them to be manipulated into tight corners and to provide the greatest flexibility. Audience seating will also be provided, which can be folded up out of the way to accommodate large drama sets.
- *Medium studio*. This is between 200 and 400 m^2 and is used for smaller light entertainment (quiz) programmes, magazine and current affairs programmes. Again, seating for small audiences may be provided and there will normally be three or four cameras.
- *Small studio*. In size up to 200 m^2, this includes facilities for programmes such as news and sport where most of the programme material comes from outside the studio but studio space must be provided for interviews. Also included are single camera announcer (presentation) studios and remote controlled studios in parliament centres.

32.1.3 Layout of control rooms

The physical layout of a television studio and its associated control rooms depends on many factors and, on many occasions, compromises. Space for a new studio may be found as a part of an existing building (space may have been allocated when the building was originally erected), or an extension to an existing building, or in a new building that may have to be specially constructed. This may not be possible in the centres of major cities, and existing office-type buildings may have to be converted.

Associated with the studio there will normally be three control rooms, shown in *Figures 32.1–32.4*.

32.1.3.1 Production control room

The production control room is where the programme director sits and hence it is usual to locate it between the other two control rooms so that all have line of sight with the director.

The vision mixer and other vision manipulating equipment is located in this room and, whilst the personnel will vary with the requirements of different directors or different companies, the following operators are usually involved:

- director,
- vision mixer, perhaps with an assistant if there are complex production effects to be controlled,
- technical supervisor, responsible for the technical performance of the studio,
- production assistant,
- character generator operator,
- editorial/production staff depending on the nature of the programme.

Figure 32.2 Production control room (Thames Television)

32.1.3.2 Vision and lighting control room

The quality of the pictures is the joint responsibility of the lighting director and the vision control operator and hence it is important that they work together on creating the pictures the director wants. If there are a number of studios on the site it is advantageous to locate all the camera control units or camera base stations together in a set-up area (usually a part of the *central apparatus room*). However, if each studio is to be an isolated production unit, then all the camera controls will be in this area.

The operational staff will be:

- lighting console operator,
- lighting director,
- vision control engineer,
- make-up and wardrobe staff may also use this room as a monitoring point for their involvement in the programme.

Figure 32.3 Vision and lighting control room (Thames Television)

32.1.3.3 Sound control room

The sound mixing console is located in the sound control room, from where all the sound associated with the programme is controlled. There may be a sub-mixer for sound effects, music and any other pre-recorded inputs to the programme from records, tapes, compact disc or the digital sound storage media. It is here also that any telephoned interviews or contributions will be balanced and where communications will be controlled. This can be a most complex operation on some programmes, such as major sporting events where a number of outside broadcasts will be used.

Figure 32.4 Sound control room (Thames Television)

The operational staff will be:

- sound mixer,
- sound assistant,
- other sound staff as required depending on the complexity of the programme.

In many studios the control rooms are located on the first floor level alongside the studio itself to give a direct view down into the studio. This visual communication with the studio is found to be helpful but, with large sets, the view can be restricted. The limitations on the layout in some buildings prevent it altogether.

32.1.4 Studio requirements

To achieve many of the technical requirements of the studio itself requires close liaison between the architect and the acoustics and services consultants as well as the planning engineer. There are a number of major points to be considered.

32.1.4.1 Noise level

A television studio should have a *noise criterion* (nc) of 15–20, while the criterion for the sound control room, where a quality assessment of the programme sound is being carried out, could be relaxed to nc25. Unwanted noise in the studio is of three main types:

- structure born, perhaps from a neighbouring ventilation plant or lighting dimmer room, from noisy activities in the studio complex or from trains, traffic or aircraft passing close by,
- ventilation noise, caused by turbulence in the air ducting or by the air travelling at too high a velocity,
- noise caused by the operational staff or the technical equipment in the studio, perhaps by cooling fans.

32.1.4.2 Acoustic isolation

The usual way of solving the problem of acoustic isolation is to build the studio (and the sound control room) on the principle of a 'room within a room'. The studio has a double brick skin with the inner skin being laid on a separate concrete slab that has been isolated from the main building structure by springs or rubber pads. The acoustic properties are further enhanced by covering the walls with a 50 mm thick layer of mineral or rockwall over a 50–100 mm air gap. The reverberation time for an average studio used for drama should be 0.6–0.8 s. Smaller studios will have a shorter reverberation time of perhaps 0.3–0.6 s. If the studio is being used for musical recordings then a longer reverberation time of up to 1.4 s will be more pleasing.

32.1.4.3 Air conditioning

All the normal restrictions on air conditioning apply, but the velocity of the air must be low and the number of air changes per hour should be high. The system should have a quick response time to accommodate sudden changes in heat gain.

32.1.4.4 Lighting grid

The nature of the lighting suspension system will depend on the planned programme requirements of the studio. A saturation lighting rig will reduce the time required to set a lighting plot, but it may not be cost effective in a large studio. Whether the lights are to be suspended from motorized self-climbing hoists or from a grid and catwalk system, the suspension will be a part of the inner structure of the studio.

32.1.4.5 Floor finish

The important criteria for studio floors are that they provide a quiet and flat surface to permit the easy movement of cameras and are strong and durable to carry heavy weights of scenery and studio props. A 6 mm layer of linoleum over an asphalt base or the more expensive floated epoxy resin surfaces are used. They should be laid to a tolerance of 1 mm in about 2 m.

32.1.5 Studio planning

The studio planning engineer has to create a tool for the production and distribution of television programmes. However, there will be a number of constraints on him, some of which are listed here, but in no particular order:

- *Financial*. All have to work within a budget, and new equipment and buildings will be allocated much less than staff costs.
- *Programme requirements*. Identification is necessary of the particular types of programme which will be made in the studio. Even if their use cannot be foreseen at an early stage, it is well to equip studios with water and gas supplies, for example, for practical items.
- *Available space*. Account must be taken of the layout of technical areas, and these will interrelate to the facilities required and the operational techniques, staffing levels and overall technical performance of the installation.
- *Equipment choice*. Amongst considerations here are the likelihood of the equipment becoming obsolete during the useful life of the installation, the level of maintenance required, the need for new operational facilities to meet competition from other installations, and the required technical performance.
- *Operational techniques*. These will reflect the staffing levels required in the various areas and the tasks they will perform.
- *Future requirements*. Any installation should ideally be planned with room for additions or alterations to the equipment complement during the life of the installation so that new items can be added as they become available.

The major events of studio planning are shown in *Figure 32.5*.

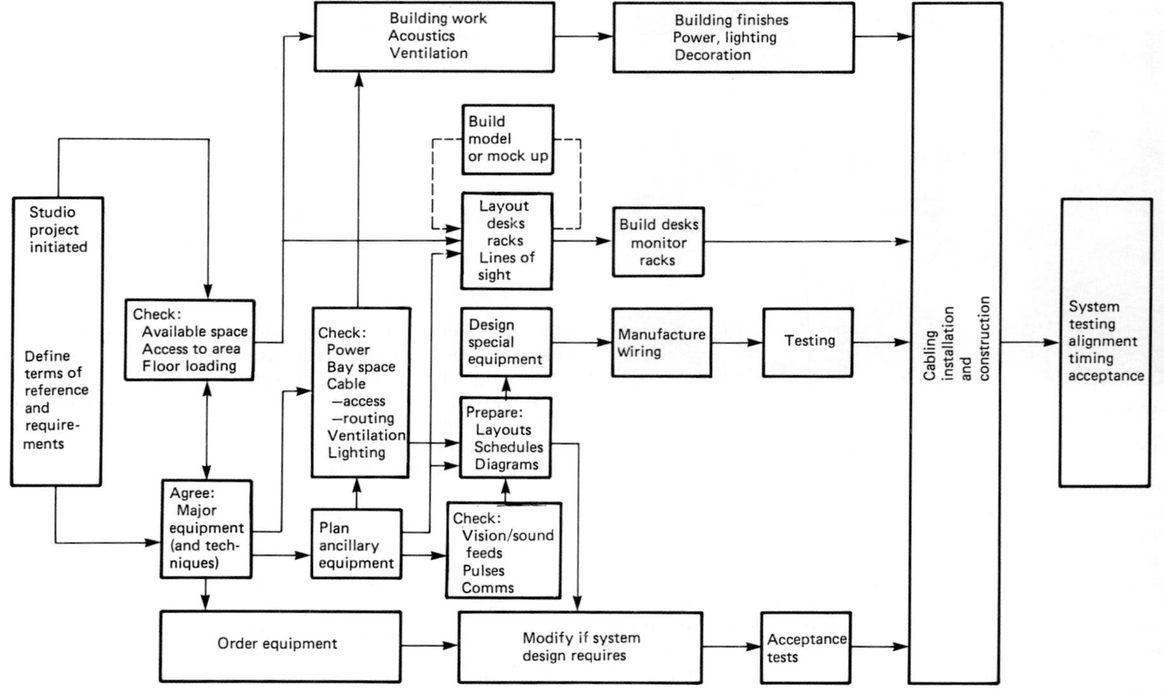

Figure 32.5 Project event block diagram

32.2 Basic video and audio facilities

32.2.1 Vision facilities

In order to obtain the most cost effective use of television studio equipment and to simplify system design, it has been the generally accepted technique to locate telecine and vtr machines centrally and assign their outputs to studios and the master control room (MCR) as required. The master control room and adjacent presentation area are the areas where central control over the television system in the station is provided and where programmes are switched to a local transmitter or to other outgoing lines. A typical facilities diagram is shown in *Figure 32.6*.

In recent years there has been a trend to copy film programmes onto video tape for transmission. The advantages are:

- If the film needs editing this can then be done without physically cutting the film by copying across the required portions.
- Videotape is becoming more economical, and a full length feature film can easily be stored on a single reel of tape. Reloading in the middle of a broadcast is therefore not required.
- The use of telecine machines is taken 'off line' in dedicated transfer suites and the number of machines can be reduced.

Centralizing the technical equipment in a *central apparatus room* (CAR) reduces the length of cable runs and leads to simpler equalization and timing problems and easier access for maintenance. The provision of air conditioning is also easier, and the equipment and cable operate in a more uniform environment. However, there are advantages in a dedication approach, where each studio is a completely separate and self-contained production unit with all the source equipment including vtrs located adjacent to the studio. This removes the need for assignment systems and makes it easier to accommodate new technical standards in the future without disrupting the complete studio complex.

The following is a list of sources to a studio vision mixer that may be dedicated to the studio or assigned to it:

- cameras,
- vtrs, including all the various formats,
- telecines,
- slide scanners,
- still stores,
- character generators,
- digital special effects,
- remote sources (e.g. from an outside broadcast).

32.2.2 Sound distribution

Outputs from sound sources are distributed on twin screened cable using a balanced signal on the two inner cores. Any interference induced onto the line will then be equal in both legs and will be removed by the common-mode rejection designed into the receiving equipment. An isolated and balanced signal is achieved by using an audio transformer but, with the increased high density of circuits in modern equipment, the balance about earth may be achieved electronically.

Jackfields are used in the audio distribution path to provide monitoring and simple rerouting facilities. Jack sockets may be connected in different ways, depending on the routing application, as shown in *Figure 32.7*.

The three routing applications are:

• *Normal* through circuit connection, where the circuit is broken by the connection of a jack plug into either socket. Usually 'listen' rows are connected in parallel with each row to provide monitoring facilities, but they take up valuable space.

• *Double normal* or *cross normal*, where the inner connections of each jack socket are wired across to the outer connections of the other socket. Inserting a plug into either socket provides monitoring facilities without breaking the circuit. A dummy plug must be inserted into the other socket to break the circuit away.

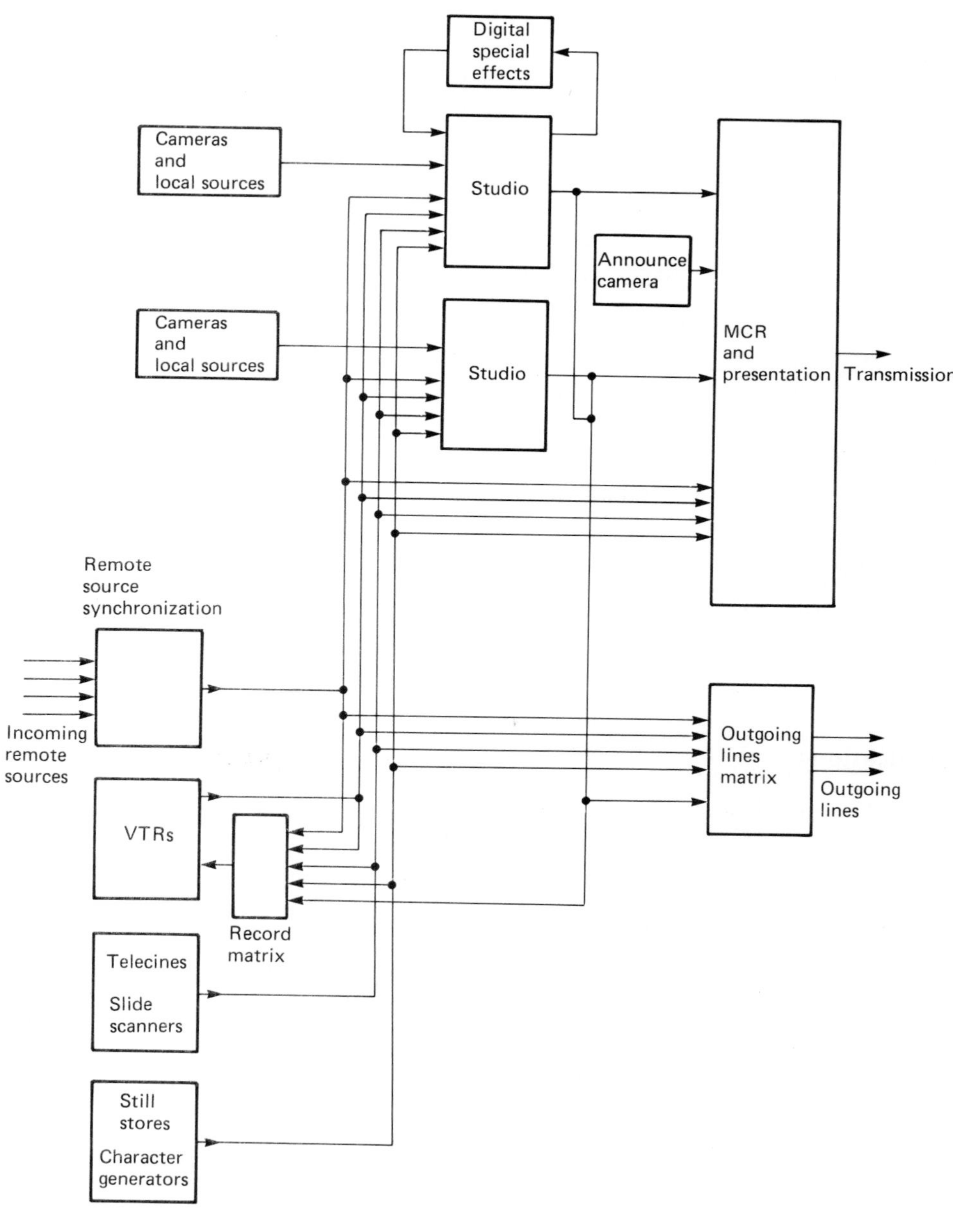

igure 32.6 Basic studio centre facilities diagram

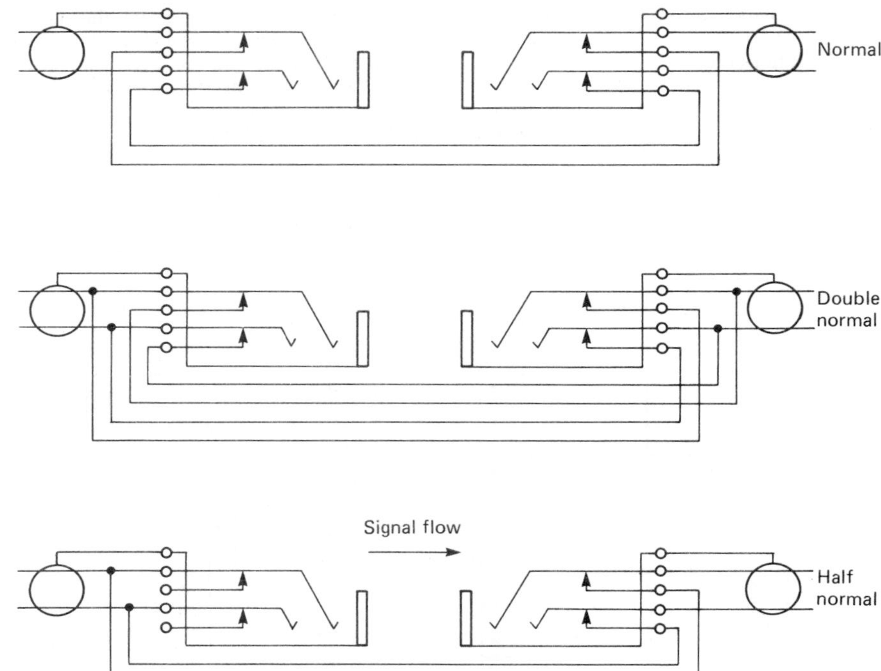

Figure 32.7 Normalling of audio jacks

- *Half normal*, where one jack is wired as in 'normal' and the other as in 'double normal'. Inserting a jack plug into one socket provides monitoring without breaking the circuit and inserting the plug into the other socket breaks the circuit away. This is most appropriately used where there is a defined direction of signal flow: monitoring can be achieved without breaking the circuit and yet, by connecting to the other socket, the circuit can easily be over-jacked in the event of an emergency.

32.3 Communications

Good communication links are vital between all those involved in a television production and, because different programmes may require different facilities or because the needs of the studio may develop during the economic life of the equipment, they should ideally be comprehensive, flexible and reliable.

There are two main networks of communication, which may require separate, but interconnected, systems. Firstly, those in the studio need to communicate locally, such as:

- on the studio floor:
  - cameramen,
  - floor manager,
  - boom operator,
  - lighting electricians,
  - lighting gantry,
  - rehearsal loudspeaker (muted when 'on air'),
  - programme presenters,
  - musical director,
- in the control rooms:
  - director } (using an 'open' microphone, heard at all times by everyone),
  - production assistant
  - sound control,
  - technical supervisor,
  - lighting console operator,
  - vision control.

Secondly, there are those in other areas who may be involved in the production and, in some systems, they can have their talkback assigned to the studio through the routing system. Such areas are:

- vtr,
- telecine,
- graphics and still stores,
- maintenance,
- MCR/presentation,
- other studios.

32.3.1 'All to all' system

In an 'all to all' system, each station is wired to all the others with audio pairs and sometimes with dc control and calling wires. Each talkback station has a microphone and send amplifier and a series of keys to switch the amplifier output to the relevant cable pair for the required destination. A receive amplifier is also included to drive a loudspeaker with the speech on the incoming line. The dc signalling level can provide call cues, but in simple systems this can be omitted and callers can identify themselves. The system is operationally simple and reliable as faults will generally only affect one station. However, much cabling is required and expansion of the system is not easy.

32.3.2 Centralized matrix

Rather than cable each station to every other, this system centralizes the switching and routing functions in a matrix to which each station can be connected by a four wire (two send, two return) cable and dc control circuits. The controls at each station will be the same as described in section *32.3.1*. Expansion of the system is possible by adding extra matrix cards, and special control functions can be added centrally. However a failure of the matrix causes a catastrophic failure of the whole system, so back-up power supplies are to be preferred.

32.3.3 Microprocessor control

Systems with a centralized matrix but making use of microprocessor control both centrally and in each station provide the most flexibility. The dc controlling wires are replaced by a data pair which, in some systems, also has the send and return audio multiplexed with it. The numbers of keys can be reduced and requirements programmed into the control system. It is thus easier to achieve control of open talkback and the conferencing of stations.

It is important that communication systems should have good audio quality. Microphones have agc amplifiers to prevent distortions in the system but too much control is often frowned on as it is important for those involved in the programme to hear, for example, that the director is becoming agitated! Howl round must be prevented either by dimming arrangements on loudspeakers or by careful positioning of microphone and loudspeakers. The system can be interfaced to radio talkback for those moving about the studio, using internationally agreed wavelengths.

32.4 Assignment systems

To make the most flexible and cost effective use of centralized items of equipment (mainly telecines, video tape recorders, character generators and still stores) an assignment system can be used. The outputs from these picture sources are assigned to the various studios and MCR/presentation as required. The following functions may be assigned:

1 forward vision,
2 forward programme audio (this may be a stereo signal or dual programme audio),
3 return vision and sound cue feeds,
4 communications,
5 timing pulses,
6 machine control.

The last item enables the source machine to be controlled from the studio. The EBU-SMPTE digital control interface standard allows a building block approach to the control of machines using distributed intelligence to transmit machine control functions in a standard form. For example, a standard control panel can control a number of different types of vtr machine.

In a small studio system with only a modest number of picture sources, the most economical method is to distribute sound and vision from all the machines directly to all destinations and assign only the required ancillary functions (3–6 above). However, this will increase the number of inputs to the vision mixer.

In a larger system, this would lead to unnecessarily large mixers and assignment is required. As a general rule, each studio should have a sufficient number of assigned lines to produce the most complex programme planned for the studio without the need to re-assign in the middle of the programme. The presentation studio should have sufficient assigned lines so that one or two complex programme junctions (perhaps with promotional material, commercials, announcements and local news headlines) can be presented without re-assignment.

Figure 32.8 shows the two types of assignment system: flexible and selective.

32.4.1 Flexible assignment

In a flexible assignment any source can be assigned to any input to any area. This normally requires a large matrix with an input dimension equal to the number of destinations. It can be controlled either centrally or from each group of destinations, so each studio can, for example, assign sources to itself. The system requires indicators at each destination to show which source has been assigned. These indicators can be controlled either from the assignment control system or by decoding source ident information that has been coded into, for example, a spare line in the field interval.

The system should have zoned power supplies so that a catastophic failure of the whole matrix cannot lose every assignment. An alternative is to break the matrix into smaller parts. There is some redundancy in the system as operators usually assign sources of a particular type to particular inputs on the mixer (e.g. vtrs always on inputs 6–8) and hence a number of the crosspoints may not by used.

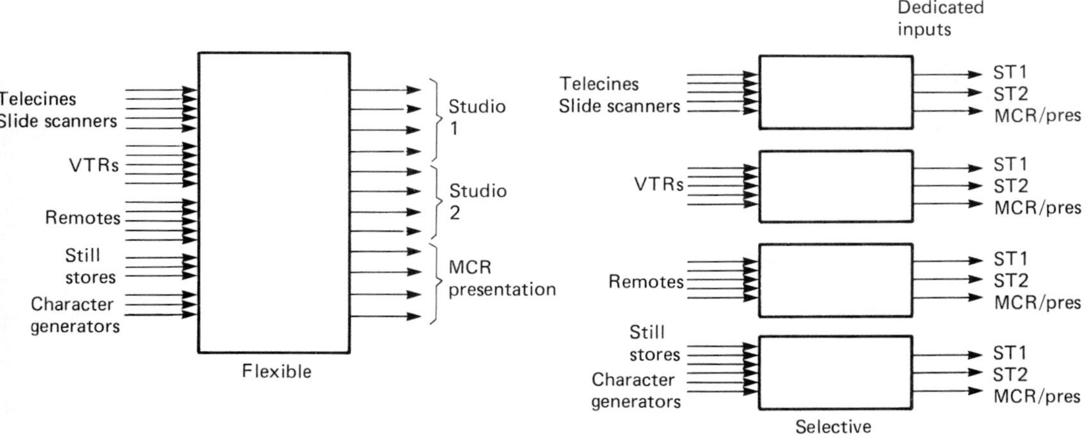

Figure 32.8 Flexible and selective assignment systems

32.4.2 Selective assignment

In selective assignment a discrete number of assigned lines is provided for each type of source to a particular area. This reduces the number of crosspoints and takes advantage of the dedicated inputs preferred by operators and outlined in section *32.4.1*. The matrix is controlled from the source rather than from the destination which is a further advantage because, if a machine becomes faulty, a new machine can be allocated locally and a re-assignment made down the same line to the studio. The studio need never know that a re-assignment has been made. The only disadvantage with this system is that it lacks total system flexibility, which may be required in certain studios for particularly complex programmes.

32.5 Cabling

32.5.1 Coaxial cable

Television systems using PAL, NTSC or SECAM coded signals are connected together almost universally by coaxial cable. The parameters of delay and attenuation of the cable affect the design of a studio system.

The cable used by most professional broadcasters has a solid plain copper inner conductor and an outer conductor of tinned copper wire braid. Like many other coaxial vision cables with a characteristic impedance of 75 ohms, it has a nominal velocity ratio of 0.666, which leads directly to the cable delays shown in *Table 32.1*.

Cable length	*Delay time*	*Degree of phase*	
m	ns	PAL	NTSC
1	5	7.99	6.45
2	10	15.97	12.89
3	15	23.96	19.34
4	20	31.94	25.79
5	25	39.93	32.23
6	30	47.91	38.68
7	35	55.90	45.13
8	40	63.89	51.57
9	45	71.87	58.02
10	50	79.86	64.46
20	100	159.72	128.93
30	150	239.57	193.39
50	250	399.29	322.32
100	500	798.58	644.64

Table 32.1 – Coaxial cable delays

32.5.2 Cable equalization

A PAL vision signal has a bandwidth of 5.5 MHz, and the response of a coaxial cable varies over this frequency range. At low frequencies the signal amplitude varies because of the variation in impedance, and at higher frequencies the rf losses (shown in *Figure 32.9*) come into play.

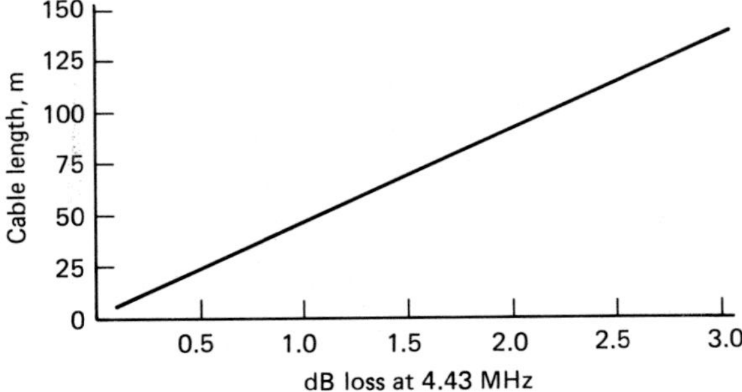

Figure 32.9 Attenuation of coaxial cable at subcarrier frequency

These losses must be equalized out and there are two basic devices for doing this:

- *Active equalizers*. These are amplifiers with a network in the feedback path which modify the gain of the amplifier to compensate for the cable losses, while maintaining the overall gain of the cable and amplifier at unity.
- *Passive correctors*. These complement the attenuation of the cable with further attenuation at different frequencies to give equal attenuation across the video band. They are usually designed to have a fixed total attenuation corresponding to the loss of the maximum length of cable that can be corrected, so that they can be followed by amplifiers with a standard gain setting.

In fixed studio installations, there are many advantages in using passive correctors. Each cable has its own corrector which, when properly set up, provides a path with a flat response for the signal. If the associated amplifier becomes faulty it can easily be changed with another with the same fixed gain without altering the equalization of the circuit in any way.

However, correct adjustment of passive cable correctors, whilst only carried out once, is a time consuming business; active correctors, with controls for gain and equalization, are much quicker to adjust. Versions have been manufactured where the adjustments are built into the rear plane of the amplifier housing and hence stay in circuit even when the amplifier card is changed, or where the correction network is on a plug-in card which can be easily removed from the amplifier and stay with the cable for which it is set up.

In a colour television system, the losses in all programme signal paths need to be equalized. Short cable lengths of only a few metres need not be equalized individually but, if there are a number of these in series, then the losses will build up. In these cases, it is best to try to locate equipment close together so that the signals do not have to make unnecessary journeys around the bays of equipment. If active correctors are to be used, it is always best to place the equalizer at the receiving end of the cable, as the amplifier is then not required to handle signal levels greater than the standard level. However, in some installations this may lead to unnecessarily large numbers of correctors.

Figure 32.10 shows how a vtr could be routed to an equalized signal distribution system in CAR where, by making cable lengths equal, the signals are flat at the mixing points. By allowing signals through the assignment system to be over-equalized, the number of equalizers can be reduced. The equal cable lengths also assist with timing the system.

32.6 Use of patch panels

Patch panels have removable links through which the vision signals are routed. Removal of a link allows the vision circuit to be immediately accessed at that point for test or emergency purposes. The decision whether to use patch panels and where to place them is always difficult. There are several considerations to be taken into account:

- Reliability is essential. If the studio centre is involved in live broadcasting, then consideration will have to be given during the system design to various emergency paths that can be brought into play if there are equipment failures. Ultimately access to picture sources and destinations on a patch panel will be the last defence to maintain signals on air in the event of serious failures.
- Patch panels provide useful monitoring points along the signal path for checking the presence and level of signals.

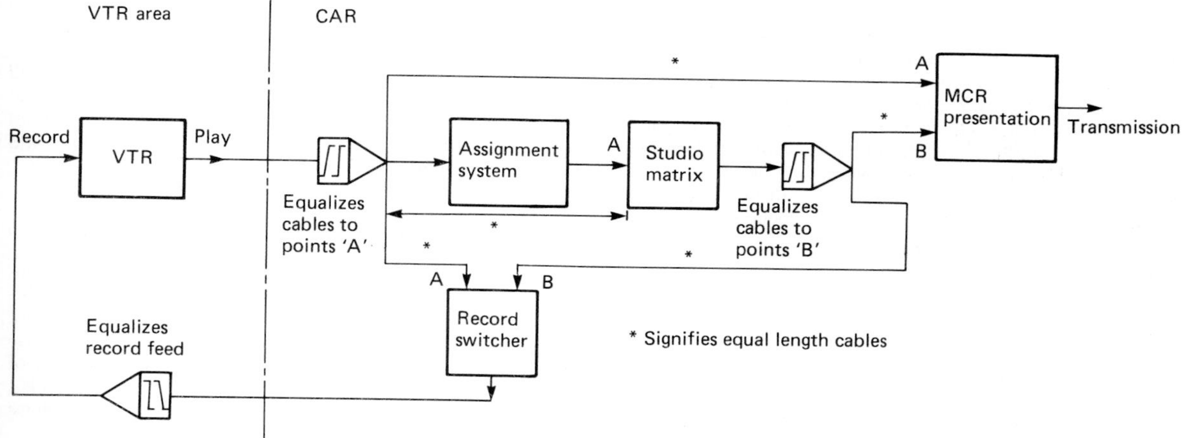

Figure 32.10 Equalized studio path distribution system

However, it may be that this function is better carried out by monitoring matrices at sensitive points along a signal path with the output signal fed to measuring equipment in a control room.

- A patch panel produces a small deviation in the characteristic impedance of the signal path and, if many are placed in series, they can add to other distortions and will contribute to a poor signal response.
- If 'over-patching' is going to occur, then all the signals at patch panels should, ideally, be flat. This restricts the places in the signal path where they can be inserted or adds to the complexity of equalization.

32.7 Television studio centre performance

In the UK, the Independent Broadcasting Authority is responsible for ensuring the high technical standards throughout the Independent Television network are maintained. Detailed codes of practice have been drawn up, which play an important role in the planning, equipping and operation of television studios. The codes specify limits to be realized on a day-to-day basis.

Amongst various different paths that can be measured through a studio complex, it is the studio path that is given here. The path comprises that part of the system that starts at the output of any source and is routed through the normal assignment paths, one studio and the normal studio output assignment path to end at the interface to vtr (for recording) or to master control (for transmitting through presentation). The path is shown in *Figure 32.10* and the performance figures are:

Signal levels

Signal level	0.7 V
Adjustment error	±0.2 dB
Signal level gain stability	±0.2 dB

Linear waveform distortion

2T pulse-to-bar ratio	½% K
2T pulse response	½% K
2T bar response	½% K
50 Hz square wave response	½% K
Chrominance/luminance gain inequality	±3%
Chrominance/luminance delay inequality	±20 ns

Non-linearity distortion

Luminance line time non-linearity	3%
Differential phase	±2°
Burst/chroma phase	±2°
Differential gain	±3%
Transient gain change, luminance	2%
Transient gain change, chrominance	2%
Transient gain change, sync	2%

Input/output impedance - return loss

Luminance	-30 dB
Chrominance	-30 dB
Low frequency	-30 dB

Noise

Weighted luminance (rms)	-64 dB
Weighted chrominance (rms)	-58 dB
Total low frequency random and periodic (p-p)	-45 dB
Low frequency random (p-p)	-52 dB
Interchannel crosstalk	-52 dB

32.8 Time code

Time code is an electronic means of identifying uniquely every frame of a video signal or every 1/25 s of an audio signal by a sequential digital code. It is useful in editing programmes, to identify particular frames and enable an edit system to find them, and in synchronizing audio and video machines together. To enable correctly colour framed edits to occur, time code is related to the PAL eight field sequence and the NTSC four field sequence. In NTSC, the standard is that field 1 of the four field sequence starts even numbered frames. In PAL, field 1 starts with frame 1 and every fourth frame after that.

Time code is the result of an internationally accepted standard agreed by the SMPTE and the EBU (in Tech 3097E). However, the systems have a major difference as the SMPTE code relates to 30 frames per second and the EBU code relates to 25 frames per second. In the 30 frame NTSC code a 'drop frame' bit indicates that two frames should be dropped each minute to more accurately represent the correct time as the frame rate is not exactly 30 frames/s.

Time code may be recorded on an audio track in its longitudinal (ltc) form or on two non-adjacent lines in the vertical interval of the vision signal in its vertical interval (vitc) form. A vitc should only be used in conjunction with ltc; it is a facility for providing a reliable readout from vtrs in slow and stop motion. Some time code readers can accept both types of code and use the most appropriate depending on the speed of the tape. Audio matrices can be used to route ltc but, if the time code is to retain its integrity at fast spooling speeds, the system bandwidth needs to be 150 kHz.

Included in the 80 bit ltc is:

the time of day:	6 segments of 4 bits each
the frame count:	2 segments of 4 bits each
sync word:	16 bits
space for user bits:	8 segments of 4 bits each

The user bits can be used to store any information, such as the programme title or a source identity code. VITC code has 90 bits. It is similar to the longitudinal code with the addition of synchronizing bits and, in place of the sync word, a cyclic redundancy check.

32.9 System synchronization

32.9.1 System timing

Encoded PAL, NTSC and SECAM vision signals reaching any mixing point in a television system (studio vision mixer, vtr edit mixer, presentation mixer, etc.) have to be synchronous in both frequency and phase before they can be safely mixed. Errors in line timing between the signals being mixed will cause picture shifts to occur; errors in subcarrier phase (in PAL and NTSC systems) will cause colour changes to occur.

Signals are synchronous (colour framed) when they are horizontally and vertically matched and on the same field. Generally accepted tolerances for the timing of these parameters in a PAL or NTSC system are:

line sync:	±12 ns
subcarrier:	±1.5°

For monochrome installation and for SECAM colour systems the tolerances can be relaxed to:

line sync:	±25 ns

Because of the modulation process in SECAM, the phase of the subcarrier is not relevant and it is only necessary to ensure the colour sequence information is in the correct phase.

32.9.2 Subcarrier-to-line phase relationship

When colour systems were first agreed, there was no fixed relationship between line phase and subcarrier phase. Early pulse distribution systems even distributed the line timing information on coded pulses separately from the subcarrier timing information. As there was a reference burst of subcarrier on each line, the phase relationship was not thought to be important.

However, with the increasing sophistication of equipment — mainly vtrs — and the increasing sophistication of editing techniques towards the end of the 1970s, the need for a fixed phase relationship between line and subcarrier phase was identified. It was found that when edits were made using recorded sequences with different line-to-subcarrier phase relationships, there were objectionable disturbances during playback.

PAL systems are divided into two fields by the 'odd' and 'even' fields, and then divided further into four by the alternating phase of the V signal which repeats over a four field sequence. There is a fixed relationship between line and subcarrier frequency:

$$f_{sc} = (284 - {}^{1}/_{4})f_h + {}^{1}/_{2}f_v$$

where f[,sc] is the subcarrier frequency, f_h is the line (horizontal) frequency and f_v is the field (vertical) frequency. From this it can be calculated that there are 283.7516 cycles of subcarrier on every line. Each line begins with almost exactly a 90° phase shift from the previous line. This has a repetitive pattern of eight fields giving rise to the PAL 'eight field sequence'. Fields 1 and 5, for example, are identical except for a subcarrier phase change of 180° between them.

Hence altering the subcarrier phase of a signal by 180° will move the signal in the eight field sequence by four fields. These changes can occur with time if the synchronizing pulse generator (SPG) does not have a stable subcarrier/line phase relationship(Sc/H) or if programmes are made in different locations with equipment driven from different SPGs that are each set up to a different phase relationship.

Because of the simpler relationship between subcarrier and line in NTSC systems, the sequence repeats after four fields. The phase is measured by extrapolating the reference burst at the start of line 10 backwards by 19 cycles to the leading edge of line sync and noting the phase of the subcarrier at the half amplitude point. The first field of the four field cycle is defined when the subcarrier is positive going at the half amplitude point. The subcarrier phase at this point, as defined by the EIA transmission standard RS 170A, is ±40° from its zero crossing point.

In PAL the phase is measured by extrapolation of the U component of the reference burst to the half amplitude point of the leading edge of line sync in line 1 of field 1. A tolerance of ±20° from the zero crossing has been recommended by the EBU (D23-1984).

32.9.3 Timing levels

In order to achieve *frequency* synchronism between picture sources, they must all be driven effectively with synchronizing pulses from the same generator.

In order to achieve *phase* synchronism the total delays (i.e. the total effective path lengths) from the SPG to the picture originating source and then on to the mixing point must be equal.

In a television complex with studios that are producing purely pre-recorded programmes, only one timing level is necessary: namely the inputs to the studio mixers. However, where a presentation function is also required, a further timing level at the input to the presentation mixer system is necessary. The studio timing level is earlier so that signals passing through the delay inherent in the studio system can still be timed with signals being used directly by MCR/presentation.

In simple systems the required timing can be achieved by the addition of switchable delays; these are switched in to delay the source to the presentation timing level and switched out to make the source earlier to the studio timing level. Ideally delays of this nature should be placed in the pulse feeds rather than the vision feeds so that they do not contribute to the losses in the signal path.

32.9.4 Delay lines

Lumped constant delay lines used for pulse delays do not need to perform to broadcast quality, but accuracy and stability of delay time is required otherwise the timing may drift outside the tolerance values.

However, there are occasions in television system design when timing delays cannot be put in the pulse feeds — for example when timing linear RGB feeds from camera channels into a mixer for chroma key — or when it is particularly convenient to put them in the vision feed. Recent technical advances in the manufacture of video delay lines have led to greatly improved performance and to their availability in dual-in-line packages with the necessary equalizing circuits built in to achieve full video response.

Typical performance figures of the Matthey zero loss video delay lines (with built in equalization and amplification) for various lengths of delay are given in *Table 32.2*.

Delay ns	*Amplitude ripple* dB	*Group delay ripple* ns	*2T pulse/bar ratio K-factor,%*
170	± 0.05	± 5	0.1
330	± 0.05	± 5	0.2
830	± 0.1	± 5	0.3
1330	± 0.15	± 7	0.4
1830	± 0.15	± 10	0.5

Table 32.2 Delay line performance figures

32.9.5 Synchronizing remote signals

Sources which are remote from a studio centre, such as feeds from an outside broadcast, are fed from separate synchronizing equipment and are therefore not in synchronism with the local pulses. There are three main ways of dealing with this situation: using genlock, reverse locking or digital synchronizers.

32.9.5.1 Genlock

To apply genlock, the remote signal is fed to the local SPG which locks to the subcarrier frequency and line and field timing information from the remote source. The line phase and subcarrier can then be adjusted for timing purposes. During this act of *genlocking*, miscounting techniques are employed to move the timing between that from the local oscillator to that from the remote signal, and the pulses exhibit non-standard timings for a few seconds. This has two drawbacks:

- Some equipment, in particular vtrs, are adversely affected in record and replay modes as they are mechanical devices and the capstan cannot follow the instantaneous changes to the pulses.
- Only a single remote source can be accommodated at any one time.

32.9.5.2 Reverse locking

For reverse locking, sync and subcarrier timing comparators at the local station derive an error signal between the local and remote pulse timings which is fed back to the remote SPG to correct the timing errors. The problems of genlock are overcome because the local pulses are not upset and the number of remote sources that can be accommodated is limited only by the number of comparators.

32.9.5.3 Digital synchronizers

Digital synchronizers store a remote signal in digital form and read it out in time with local pulses. Various techniques are used to store either a field or a frame of a fully coded PAL or NTSC signal or a signal decoded to its components. Care must be taken to delay the associated audio signal, otherwise, if the vision signal passes through a number of synchronizers, the differential delay will become objectionable, particularly as the sound will arrive before the vision.

32.9.6 Pulse distribution system parameters

The modern solution to pulse timing distribution makes use of the fact that all essential timing information is carried in the composite video itself. Timing information is normally distributed as a feed of mixed syncs and colour burst (known as *colour black* or *black and burst*). This signal contains all the necessary line, field and subcarrier timing information and is in a form that can be distributed through a normal vision signal chain. It is also a signal where the Sc/H phase information can be distributed.

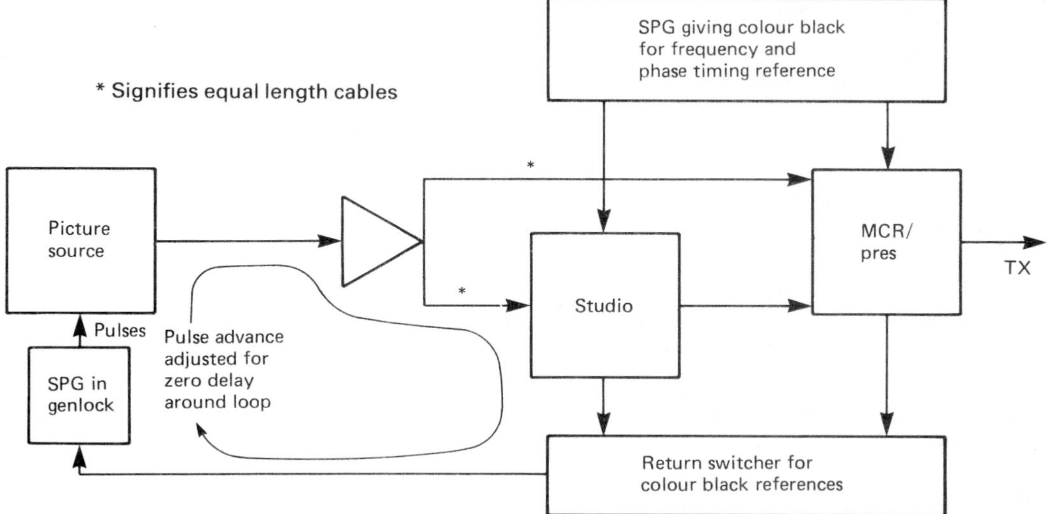

Figure 32.11 Principle of zero loop delay

Broadcast quality master sync pulse generators are equipped with very high quality oven-stabilized crystal-controlled subcarrier reference oscillators. The subcarrier signal is the timing reference, and all individual drive pulses are produced by separate circuits triggered from the subcarrier by the use of digital electronic techniques.

Discrete pulses can be regenerated from the colour black signal by use of an SPG operating in the genlock mode and locking to the feed of colour black. Many items of equipment are manufactured with built in SPGs and will accept colour black directly. To help identify the eight field sequence, a white pulse can be included on line 7 of field 1. Slave units can detect this pulse and synchronize to the eight field sequence.

The use of colour black helps fulfil the criteria for an ideal pulse system:

- It should not cause a major system failure if any part fails. If the central SPG (and hence the source of the colour black) or the distribution system fails then the individual source SPGs deslave and continue to generate pulses in local lock. The sources will no longer be in frequency synchronism and complex mixing effects cannot be achieved, but at least the sources continue to supply pictures — important for a 'live' programme. If the source SPG fails then only one source is lost.
- It should be easy to adjust and maintain the system timing tolerances. The use of the line and subcarrier phase controls on the SPG permits easy adjustment of timing.
- It should be easy to modify or extend to accommodate new equipment. Again, the use of colour black simplifies distribution of timing information and the system can also lock up to a remote source vision feed.

32.9.7 Zero loop delay

If the studio centre has been designed and built with equal path lengths from the assignment and distribution system to the mixing points, then the zero loop timing delay principle shown in *Figure 32.11* can be used.

The paths from the SPG to the source and from the source all the way to the mixing points are the same electrical length, and hence the source SPG can advance the phase of its output pulses to equal the delay around the pulse and video path. The assignment of reference colour black signal between the studio timing level and the presentation level moves the timing of the source in synchronism forwards or backwards in time.

32.9.8 Reverse locking systems

The Sc/H phase of a signal must be within the recommended tolerance window at the mixing point. Group delay in the distribution system can alter the phase and so can poor discipline if the subcarrier phase is adjusted without also adjusting the line sync timing.

Various items of test equipment are available which measure both the absolute Sc/H phase of a single signal and the relative Sc/H phase between two signals; one signal is taken as the standard and the other is measured with reference to it. Some reverse locking systems are available which, depending on their degree of sophistication, measure the subcarrier and/or line and field timing at the mixing point and route error correction signals back through the assignment system to the source SPG.

32.10 Analogue and digital component systems

32.10.1 Analogue systems

Each time a signal is coded to PAL or NTSC and then decoded — even using a comb line decoder — the signal quality is adversely affected by a loss of resolution and by PAL or NTSC 'footprint' errors. PAL and NTSC are a composite of the luminance and chrominance signals; crosstalk between the luminance and chrominance (cross colour) produces coloured effects in fine detail, and crosstalk between chrominance and luminance makes areas of saturated colour show up as a dot pattern on monochrome receivers. In SECAM the vertical chrominance resolution is reduced, but the use of frequency modulation prevents any subcarrier offset from reducing the visibility of the dot pattern.

Many items of studio equipment decode the vision signal to its luminance and chrominance components before processing the signal digitally, such as:

- synchronizers,
- digital production effects devices,
- slide stores,
- time base correctors.

Other items, such as character generators and electronic graphics devices, originate signals digitally in their component form. Of course, cameras and telecines are RGB component originating devices and colour monitors display component signals.

Component forms can be RGB (used in cameras, telecines, monitors, PAL or NTSC encoders) and YIQ (for NTSC systems) or YUV (Y, R–Y, B–Y) used in vtrs, mixers and synchronizers. Movement between forms can be achieved using a translator. The Y signal is unipolar and R–Y and B–Y are bipolar. Keeping signals in their component form prevents the signals being system conscious and avoids the deleterious effects of the coding process. Even if the signal is finally coded into PAL, NTSC or SECAM, noticeable improvements can be obtained in the edge crawl effects of captions and special effects, in the results from off-tape chroma key (colour separation overlay) and in the interface with digital equipment, such as that mentioned above.

Several different 'islands' can be built with component technology such as a studio, where all the source equipment is dedicated to it, transfer facilities for film to vtr and vtr edit suites. The output from one 'island' — usually a videotape — can be physically taken to the next process in the chain, e.g. from a studio to an edit suite. In these compact areas there are advantages in terms of reduced complexity and cost which can be made by using parallel connections over three or four cables for the component signals and syncs.

Colour black can still be used for timing control as that is the signal that much equipment still expects. Of course, subcarrier phase adjustment is not needed and timings can be relaxed to monochrome levels of, say, ±25 ns.

Some errors in component systems are more critical than they would be in composite coded systems, such as:

- differential dc level offset, to reduce noticeable black balance errors,
- channel level accuracy, to reduce white balance and saturation errors,
- interchannel timing accuracy; 40 ns between components is visible as colour fringeing, and so paths through equipment should be adjusted for equal delay; installation cables should be cut carefully to the same length.

Differential comparison of component signals is needed to test the system adequately to the required accuracy to remove colour casts.

While it is practical to have 'islands' using component technology, presentation and network switching functions will require access to many sources in the station. An assignment or routing switcher with access to studios, vtrs for programmes and commercials, outside broadcasts, graphics and character generators will still be required. To take advantage of components and to make system routing easier, distribution on a single wire is possible using time division multiplex techniques, such as the SMPTE S-MAC system. S-MAC compresses the luminance by 2:1 and the colour difference chrominance signals, based on a 50 per cent pedestal, by 4:1 onto a single channel. S-MAC requires a bandwidth of 11 MHz, and cables, routing switchers and distribution amplifiers must be able to accommodate this. If the bandwidth of the system becomes limited then the luminance and chrominance lose resolution 'gracefully'.

32.10.2 Digital systems

The component analogue video systems mentioned in section *32.10.1* suffer from distortions such as gain and black level variations, frequency response and group delay. Digital systems, however, are robust, flexible and reliable. Digital video tape recorders provide nearly transparent signal performance and allow the recorded signal to be used again, for example, as a further input to a digital vision mixer for the build up of complex re-entries and key effects. There are no variations in gain, no accumulative signal distortions and no variations in stability.

A digital component coding standard which does not resemble PAL, NTSC or SECAM has been developed. It is defined in CCIR Recommendation 601 and is known as the *4:2:2 digital production standard*.

13.5 MHz line locked sampling of the luminance signal allows orthogonal sampling patterns for both 625/50 and 525/60 systems. The colour difference channels are sampled at half this frequency (6.75 MHz). There are 720 samples/line for the luminance signal and 360 samples/line for each chrominance signal, with both 525 and 625 systems.

The EBU has defined a time-multiplexed, bit-parallel interface based on CCIR coding parameters in which eight bit video data words, derived from luminance and colour difference samples, are transferred at 27 million words per second (EBU Tech 3246-E). With signal equalization for distances over 30 m, the parallel interface works correctly and is cost effective up to about 200 m (i.e. up to the longest distances experienced in the average studio centre) on good quality cable. It then needs a repeater to correct timing skew. The cable is multicore (eight balanced signal pairs and the 27 MHz clock) which carries a multiplex of luminance and colour difference signals. To minimize differential delays, all cables must be of the same electrical length. Crosstalk between signal pairs is reduced by using balanced transmission on twisted pairs, and an overall screen crosstalk between cables. The connectors used are 25 way sub-miniature D-type with the functions shown in *Table 32.3*.

Contact	*Assignment*	*Contact*	*Assignment*
1	data 7	14	data 7 return
2	data 6	15	data 6 return
3	data 5	16	data 5 return
4	data 4	17	data 4 return
6	data 2	19	data 2 return
7	data 1	20	data 1 return
8	data 0	21	data 0 return
9	–	22	–
10	–	23	–
11	clock	24	clock return
12	system ground	25	system ground
13	chassis ground		

Table 32.3 Assignment of connector contacts for the parallel digital interface

Matrices working in the parallel format are physically large because they need to accommodate connectors and the necessary interconnections and to dissipate power.

For routes longer than 200 m, a serial link working at 243 Mbit/s may become cost effective. This requires a serializer and a deserializer at equipment inputs and outputs. Timing and equalization are less critical. Sources must be timed to within 18.5 ns, which is half the 27 MHz clock period. Delays through the system are greater than with analogue systems because of repeated latching in the line sending and receiving signals and in signal processing stages. The delay is in increments of one clock period at 27 MHz (37 ns).

Bibliography

PAL systems

THIRLWALL, C, 'PAL colour framing', *Professional Video*, **10**, No 4, 22 (1984)

Studios

THORPE, P R, 'A news studio for the 1980s', *International Broadcast Engineer*, **11**, No 173, 68 (1980)

BERRY, A, 'A new production centre for Open University broadcasting', *International Broadcast Engineer*, 26 (1982)

LEIGH SMITH, T, 'Limehouse goes on stream', *Professional Video*, **9**, No 12, 40 (1983)

LEIGH SMITH, T, 'Channel four at one', *Professional Video*, **10**, No 2, 30 (1983)

PARKER, P, 'Central's new East Midlands television centre', *International Broadcast Engineer*, **15**, No 195, 33 (1984)

REAY, D, 'HTV's new television complex at Culverhouse Cross', *International Broadcast Engineer*, **15**, No 197, 5 (1984)

DEVESON, J, 'A new television production centre for JPC Jordan', *Hardware International*, No 1, 12 (1986)

DEAVES, J G, 'An analogue component news centre', *Hardware International*, No 2, 12 (1986)

Component video

SCOTT, B G, 'Components in the studio centre — digital or analogue?', *IBC Conf Publication*, No 240, 24 (1984)

RICKARDS, A J, 'Components in the studio', *Television* (Journal of the Royal Television Society), **22**, Nos 5, 6, **23**, No 1 (1985–1986)

WEISS, S M, 'Practical considerations in implementing component video', *IBC Conf Publication*, No 268, 264 (1986)

BARRACLOUGH, J N and MAIN, A B, 'Experiences with digital video components', *IBC Conf Publication*, No 268, 280 (1986)

W H Klemmer
Broadcast Television Systems GmbH

33a Studio Cameras and Mountings – Cameras

The electronic colour camera is one of two important signal sources for the creation of television signals. The other is the telecine (see section *39*). It seems likely that eventually even the classic 35 mm film camera will be superseded by the electronic camera. The first sign of this is in the use of electronic cameras for high definition television (HDTV).

The range of professional electronic cameras can be split into the following categories:

- studio
- electronic cinematography (EC)
- electronic field production (EFP)
- electronic news gathering (ENG)
- industrial and other uses

This list shows the cameras in descending order, with respect to their picture quality and their design features. There are actually no real EFP cameras, so this application is covered by suitable adaptation of studio cameras or by high quality ENG cameras (*Figure 33.1*).

At the top of the list are EC and studio cameras, both claimed to have the highest possible picture quality within a chosen scanning mode. In addition, studio cameras are fully fitted out from an operational point of view, and are as far as possible automated. Manufacturers of modern studio camera systems offer not only large cameras, but also portable camera heads. These can be integrated into the system with full compatibility, with only small reductions in quality and features. Modern studio cameras exclusively use the three channels: red, green and blue.

Whether PAL, SECAM or NTSC is used, the difference does not affect the general system configuration. It only affects the encoder at the output of the camera system, and this will decrease in importance with the progressive introduction of component television. The system configuration described in section *33.1* applies also to HDTV cameras, even though they do not have any modulating encoders, as only component outputs are used.

For photoelectric conversion, studio cameras today still use camera tubes with image sizes of $^2/_3$ inch, 1 inch and $1^1/_4$ inches and with a plumbicon or saticon layer. In the 90s, a complete change, already introduced in the ENG/EFP area, will continue, i.e. the introduction of ccd area sensors. Once solutions have been found for the remaining problems affecting the introduction of ccd sensors into HDTV cameras, these sensors will eventually surpass camera tubes in all quality determining features. However, two disadvantages must be mentioned: less flexibility with respect to multi-standard camera operations, and more difficulty in correcting lens dependent colour registration errors.

A salient feature of modern studio cameras is the extensive automation of the working and tuning operations through the application of some microcomputer systems. This can provide, at the same time, rational data communication especially in multi-camera systems (*Figure 33.2*).

33.1 System structures

Every studio camera system includes a camera head, with zoom lens and viewfinder, which is connected through a camera cable to the camera control unit (CCU). The operational control panel (OCP) and, in the case of multi-camera operations, also a master control panel (MCP), are connected to this CCU as well. *Figure 33.3* shows a typical block diagram of a studio camera, illustrating one of the many possible arrangements of the cable interface.

33.1.1 Camera head

After optical colour separation in the prism, the signals for the three colour channels R, G and B are created by line scanning in the three camera tubes. After pre-amplification, this acquired signal current is then linearly and non-linearly processed in the head signal processor.

A microprocessor controlled correction system creates analogue correction voltages. These are firstly mixed into the scanning beam deflection unit (for correction of colour registration errors), and then additive and multiplicative errors are removed from the analogue video signal. The scanning beam intensity in the camera tubes is guided by the *dynamic beam control* (DBC) or *automatic beam control* (ABC) so that the scanning of high lights is possible.

33.1.2 Camera cable

There are various ways of transmitting the fundamental signals:

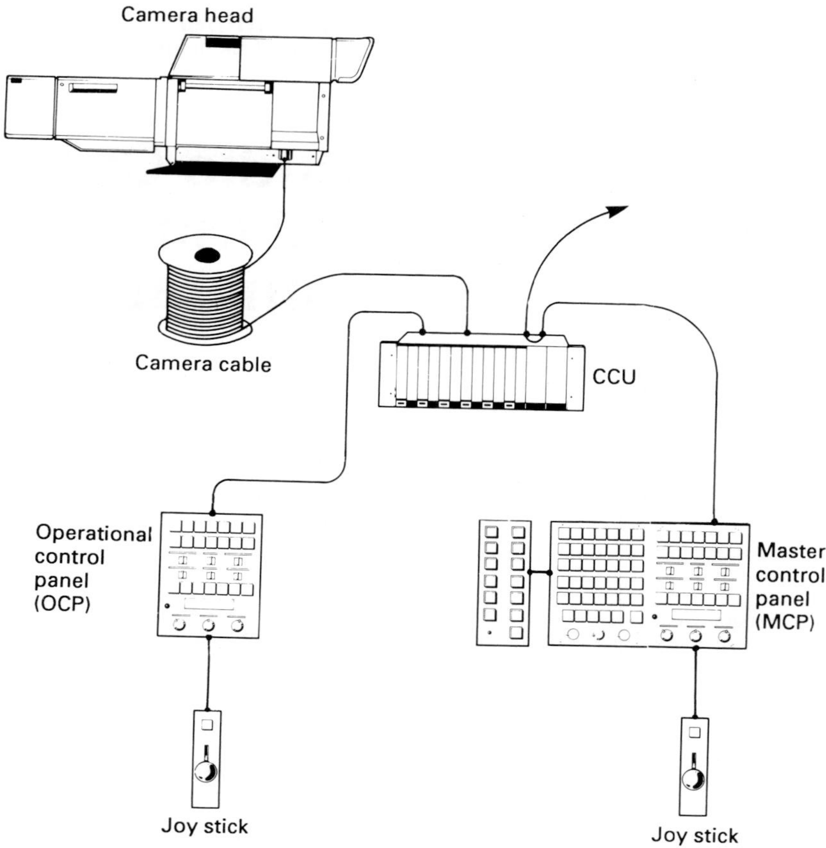

Figure 33.1 Components of a studio camera chain (Broadcast Television Systems GmbH)

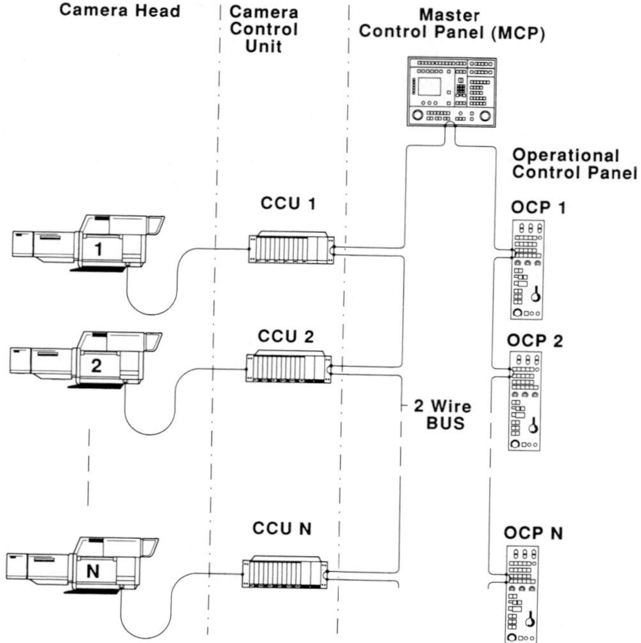

Figure 33.2 Multi-camera system (Broadcast Television Systems GmbH)

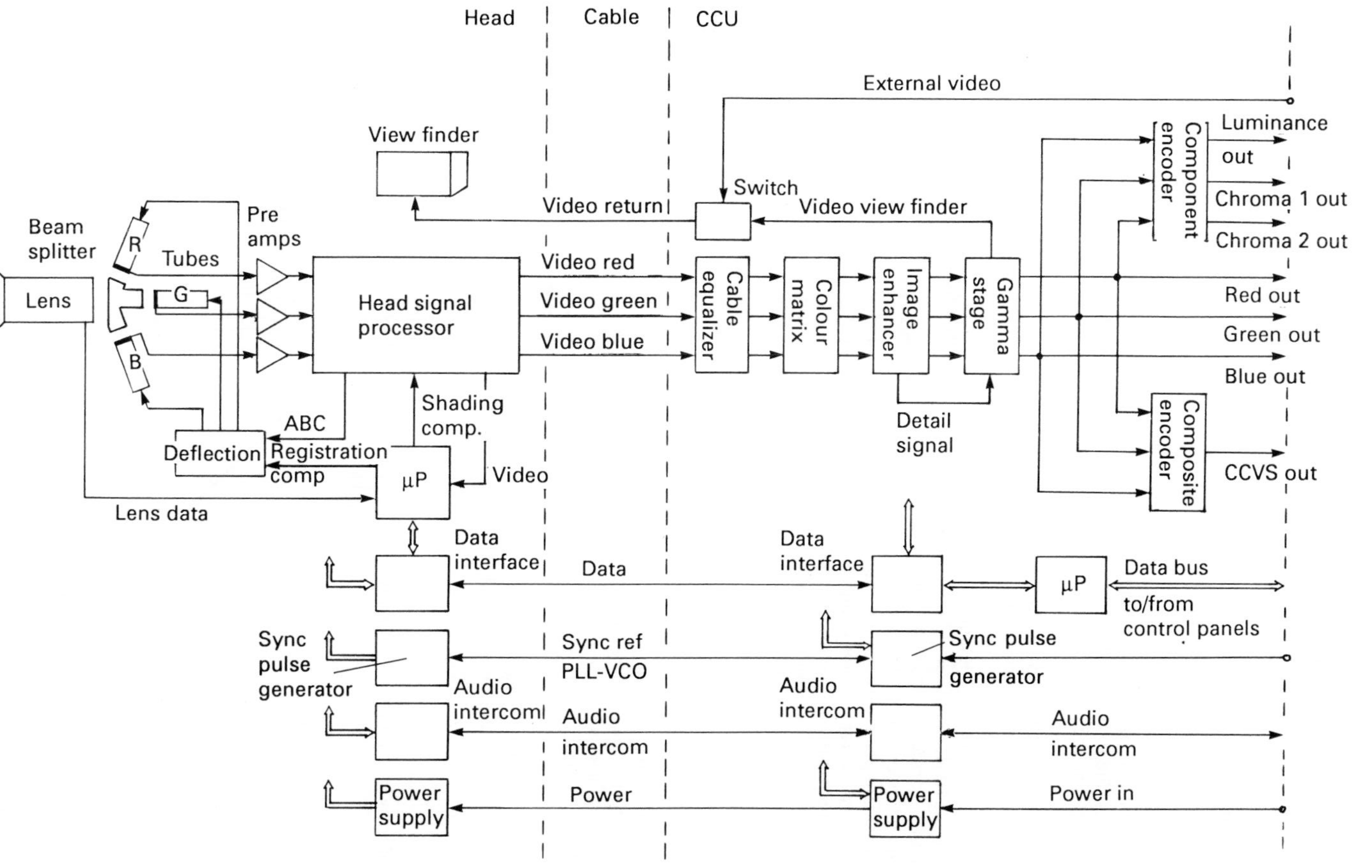

Figure 33.3 Block diagram of a studio camera including camera head, cable and CCU

- parallel transmission in base band via a multi-wire cable
- partly modulated transmission via a multi-wire cable
- time division multiplex transmission, i.e. data and impulses mixed in the blanking intervals
- fully modulated bidirectional transmission via a triax cable
- transmission in analogue or digital form via one or more optical fibres

33.1.3 Camera control unit

In theory, the units for the colour correction matrix, the frequency response correction and the gamma correction (see section *33.2.3*) should be arranged in the camera head to utilize fully the gamma stage for companding *before* the camera cable. In practice, this is not possible, due to the usual limitations of space and power dissipation. Hence the arrangement shown in *Figure 33.3* is usually used.

If signal transmission is via a multi-wire cable, the length dependent frequency response of the camera cable must then be adjusted.

After colorimetric correction and two-dimensional frequency response correction, the RGB signal is non-linearly pre-compensated in the gamma correction stage in order to compensate for the subsequent inverse non-linear transmission characteristic of the receiver. This pre-compensation stage has the advantageous and vital effect of transmission companding.

Following it, the signal output exists either as the component triplet RGB or YC_RC_B, or, after appropriate coding, as a composite video signal. The component triplets RGB or YC_RC_B enable transmission which is theoretically correct, at least if they are output in the full bandwidth and as long as appropriate measures have been undertaken in the camera and receiver.

Connected with the signal coded in a frequency division multiplex are various defects dependent on this principle. Modern studio cameras tend to use component outputs. They do so completely for HDTV applications.

33.1.4 Other units

Both the camera head and the CCU use the units data transceiver, synch pulse generator, audio intercom and power supply.

The *data transceivers* are not normally stand-alone units, but rather microprocessor interfaces in the head and CCU requiring a large amount of software. The data path from CCU to head carries the control and switching signals which originated at the control panel to the head. The data path from head to CCU, on the other hand, carries information on the camera head status and on eventual error diagnosis data.

The *pulse generators* in the head and CCU are connected via a phase locked loop (pll). The CCU clock generator is also coupled to an external clock via a further pll. High value microphones, which later provide hi-fi quality, can be connected to the camera head in modern studio camera systems.

For easy communication between operators, bidirectional intercom paths are planned.

The power supply used in the camera head is derived either from a high direct voltage (200–400 V) or from an alternating voltage with a high frequency. These supply voltages are then converted via dc/dc or ac/dc converters into the required dc voltages for the camera head.

33.2 System components

33.2.1 Optical block

The optical block is the unit for optical/electrical signal conversion, i.e. lenses, filter wheels, beam splitters and camera tubes. A detailed description of camera tubes and solid-state imagers can be found in section *10*.

The fundamental arrangement of the optical block is shown in *Figure 33.4*. The lens reproduces the scene to be observed on the photoconductive layers of the camera tubes via the beam splitters. Almost exclusively used are zoom lenses with focal length variations of 1:8 to 1:40 and f-numbers of 1:2.1 (image size 1¼ inch), 1:1.6 (image size 1 inch) and 1:1.2 (image size ⅔ inch). The three adjustable parameters of focal length, iris and focusing distance are usually altered via servo motors, which are built into the lens.

The operator at the camera head controls focal length and distance, whereas the aperture opening is controlled from the control panel or automatically, so that a constant image signal level is produced. The requirement for optimal, constant picture quality over the whole focal length range necessitates,

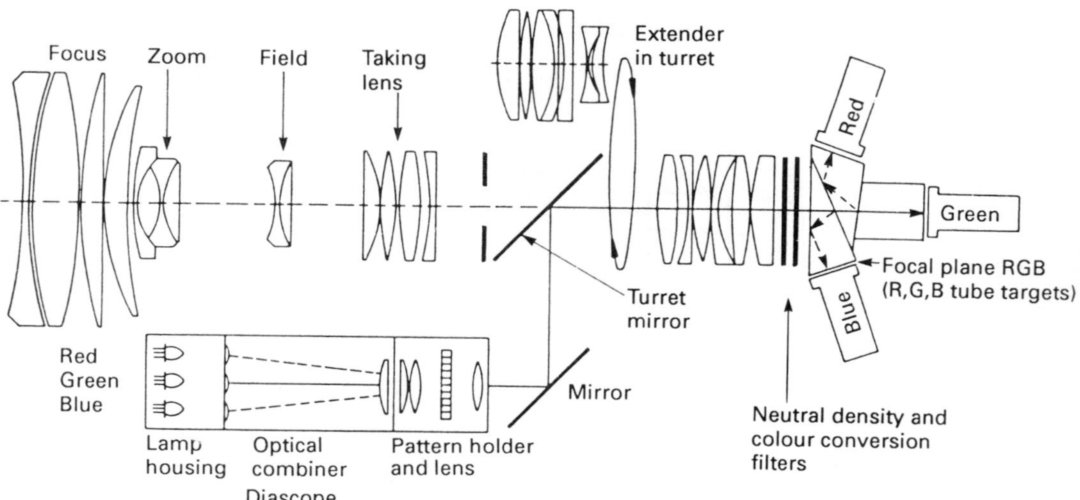

Figure 33.4 Optical block of an RGB camera: zoom lens, neutral density filter wheel, colour conversion filter wheel, beamsplitter and tube arrangement (Fujinon)

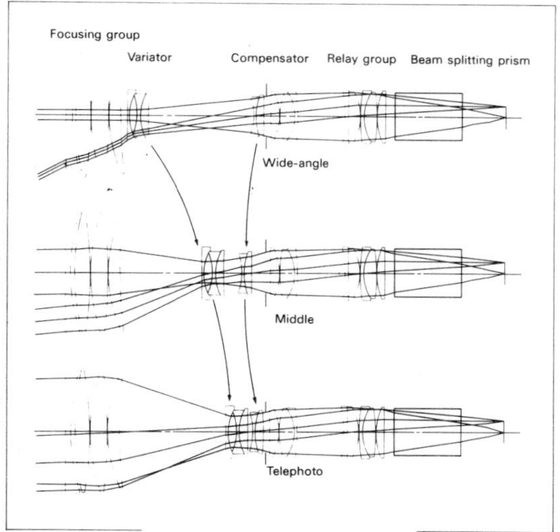

Figure 33.5 Optical path of a modern TV zoom lens shown for wide-angle, normal and telephoto operation (Canon)

for example, having the simultaneous complex shifting of several lens groups within the lens (*Figure 33.5*).

An important precondition for automatic camera alignment is that diascopes are built into the lens which are allowed to swing into the beam path. With the help of such a projected test slide, it is possible to adjust the parameters of image geometry, colour registration and white shading.

It can be recognized from the arrangement of the projector in the lens, that dynamic (i.e. setting dependent) chromatic errors of the main lens cannot be corrected at that point. For this reason, the lens gives actual values of the setting parameters or even correction data to the camera head, so that an electronic correction can take place there. Even so, the test projector simulates lens aberrations, which occur at a medium operating focal length or distance.

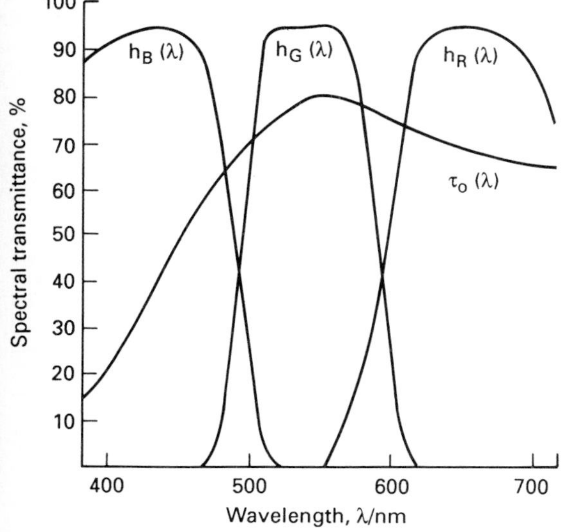

Figure 33.6 Spectral transmittance $\lambda_o(\lambda)$ of a lens ands spectral filtering function $h(\lambda)$ of a beam splitter

Important parameters determining quality of a studio lens are:

- resolution power and resolution distribution over the image field
- colour registration (lateral chromatic aberration)
- focusing preservation and tracking (longitudinal chromatic aberration)
- vignetting (brightness distribution over the image field)
- distortions of the image geometry

Section *33.4* deals with the electronic correction of lens errors, and section *33.3* with depth of field and sensitivity. *Figure 33.6* shows the spectral transmittance, denoted by $\tau_o(\lambda)$ of a studio lens in the wavelength range 380–780 nm, relevant to colour transmission.

Modern studio cameras contain two *filter wheels*, each with up to five filter plates. One filter wheel has neutral density filters to reduce light (manipulation of the depth of field). The second wheel is for colour conversion, i.e. adaption to various types of illumination. The depolarizing plate, also shown in *Figure 33.4*, is present in order to prevent colour falsifications on the dichroic layers in the beam splitter, due to eventual incoming linearly polarized light.

Three glass prisms, with dichroic layers at the connecting surfaces, form the *beam splitter*. The selective reflection at the dichroic layers is utilized to separate spatially the light currents (*Figure 33.7*). Correction filters are necessary at the three output boundary surfaces of the splitter. This is due to the transmission of partly unwanted wavelength ranges as a result of the use of only the wavelength selection for the dichroic layers. This leads to the typical spectral transmittance characteristics $h_B(\lambda)$, $h_G(\lambda)$ and $h_R(\lambda)$ of the complete splitter shown in *Figure 33.6*.

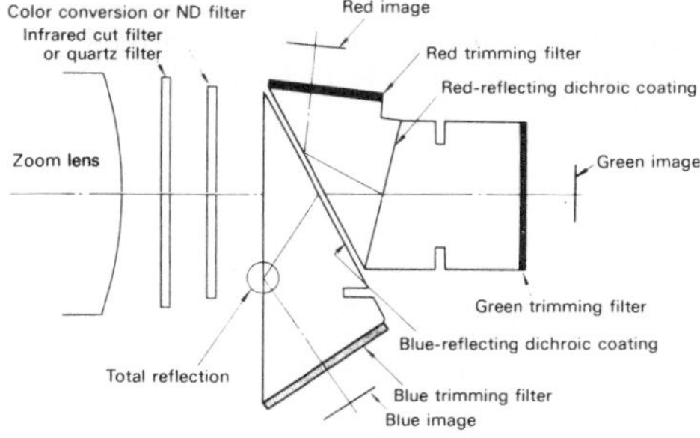

Figure 33.7 Colour separation optical system: beam splitter (Canon)

The three part beam splitter of *Figure 33.7* can be used for relative f-numbers up to f1.4. In general, for greater f-numbers of around f1.2, a four part beam splitter is required.

The *camera tube* is a crucial, quality determining, element in the electronic colour camera. The limits of all the quality determining parameters are set by camera sensors, i.e.:

- static resolution in horizontal and vertical directions
- dynamic resolution, i.e. temporal resolution and lag
- signal/noise ratio
- sensitivity (absolute and spectral sensitivity)
- signal dynamics
- stability of the scanning raster (colour registration)

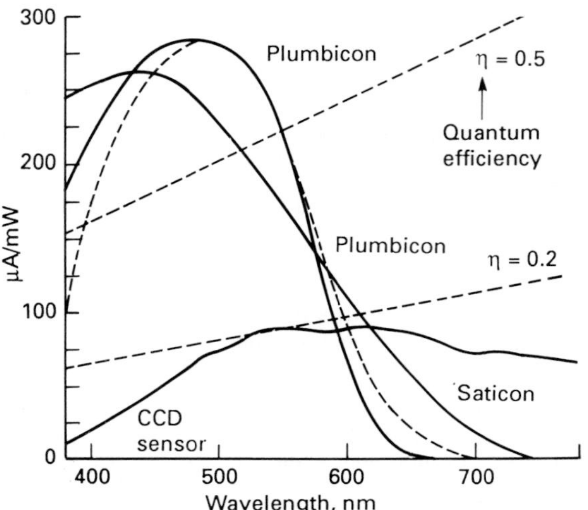

Figure 33.8 Absolute spectral sensitivity of different photoconductive layers

Tubes with plumbicon and saticon II or saticon III layers are used in studio cameras. *Figure 33.8* shows the spectral sensitivity distributions of these photoconductors.

The scanning raster sizes to scale, which have been produced for the image sizes 1¼ inch, 1 inch and ⅔ inch, as well as for aspect ratios 4:3 (standard television) and 16:9 (HDTV), are shown in *Figure 33.9*. The numerical values can be obtained from Table *33.1*.

In the past, camera tubes have had two coils for horizontal and vertical deflection, one coil for beam focusing, and possibly coils for beam alignment. With the development of electrostatic deflection, particularly in HDTV cameras, the deflection coils are not required. Instead, there are tube envelopes containing vapour deposited and laser cut deflection electrodes.

The concept of electrostatic deflection has fundamental advantages:

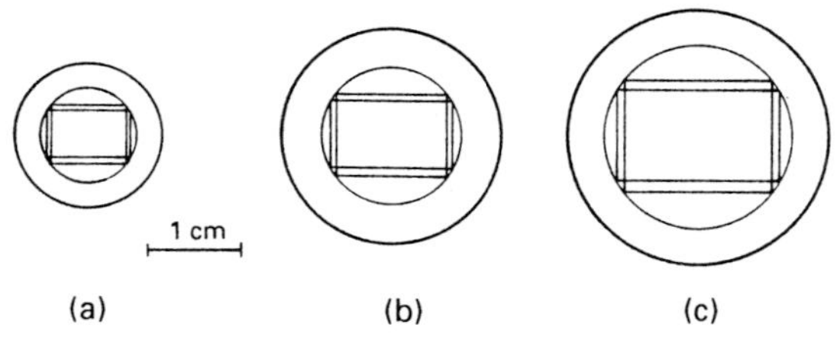

Figure 33.9 Scanning rasters for different image sizes, scaled: (a) ⅔ inch tube, (b) 1 inch tube, (c) 1¼ inch tube

- uniform resolution distribution over the whole picture
- less geometrical distortion
- possibility of higher deflection frequencies (HDTV)
- no thermal drift of the deflection unit with respect to the tube, i.e. stable raster registration

The photoelectric storage layer resists temporal changes in exposure. It takes several scanning periods (fields) for the charge of a picture element to be fully recombined. This effect is characterized by lag and depends on the structure and capacitance of the storage layer, and on the non-linear landing behaviour of the scanning electron beam, as well as on the signal level. Plumbicon layers are better here than saticon I and II layers.

To improve the behaviour of the lag, a bias light of approximately 3 per cent of the nominal light level is made to strike the camera tubes (*Figure 33.10*).

This bias light can either be injected from the front via the beam splitter, or it can be transported from behind via the tube envelope or special light conductor. It is necessary to be able to adjust separately the bias light in the three colour channels, in order to be able to set up uncoloured residual lags.

33.2.2 Video pre-processor

33.2.2.1 Pre-amplifier

At a spectral energy distribution $\phi(\lambda)$ in webers per nanometre, with respect to the spectral sensitivity functions of the previous

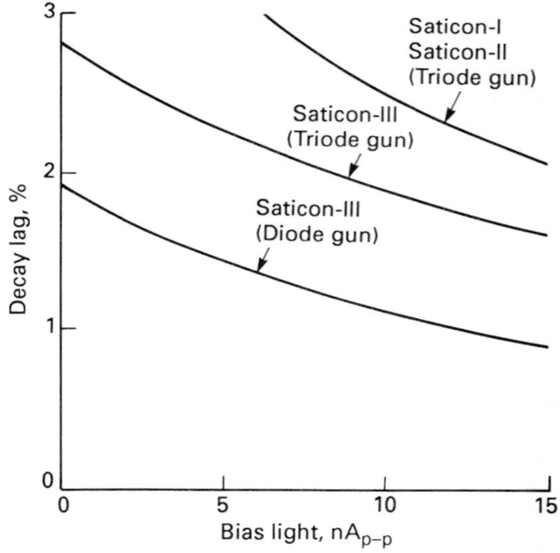

Figure 33.10 Decay lag characteristics for saticon tubes with several gun types, as a function of the bias light level. Scan size is 9.5 × 12.7 mm, signal current is 0.3 μA, beam current is 0.6 μA, and decay time is 3rd field (Hitachi)

Nominal diameter		Image diagonal	4 : 3		16 : 9	
Inch	mm	mm	Image size mm × mm	Image area mm²	Image size mm × mm	Image area mm²
⅔″	16.9	11	8.8 × 6.6	58.1	9.6 × 5.4	51.8
1″	25.4	16	12.8 × 9.6	122.9	14 × 7.8	109
1¼″	30	21.4	17.1 × 12.8	218.9	18.7 × 10.5	196.4

Table 33.1 Scanning raster dimensions

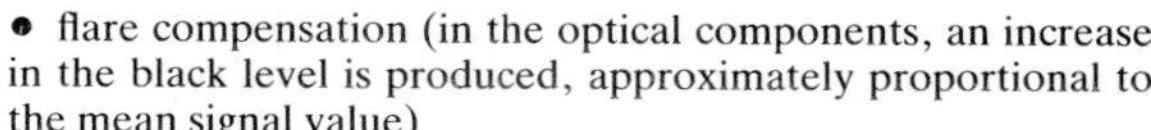

section, the following signal currents enter the lens:

$$I_R = k.\int \phi(\lambda).\tau_o(\lambda).h_R(\lambda).s(\lambda).d\lambda$$
$$I_G = k.\int \phi(\lambda).\tau_o(\lambda).h_G(\lambda).s(\lambda).d\lambda$$
$$I_B = k.\int \phi(\lambda).\tau_o(\lambda).h_B(\lambda).s(\lambda).d\lambda \quad (33.1)$$

Typical signal currents, at a studio illumination of 3200 K, are about 240 nA (R), 300 nA (G) and 160 nA (B) for picture white. One of the tasks of the video processor that follows is to create, for a picture white to be defined, the same signal level in the three channels via selective amplification.

The principal structure of the camera pre-amplifier is shown in *Figure 33.11*. The amplifier has the task of converting the above small signal currents into signal voltages of the order of magnitude of 0.3–0.7 V. For practical reasons, the transimpedance concept illustrated in *Figure 33.11* is exclusively used. The demand for an optimal signal/noise ratio results in the separation of the pre-amplifier into two sub-assemblies. The front part, consisting of a field effect transistor and a feedback resistor, is fixed directly to the top of the camera tube. It is possible to couple inductively the pre-amplifier via the so-called *Percival coil*, beside the direct coupling. The structure of the pre-amplifier determines the obtainable camera signal/noise ratio (see section *33.3.1*).

The functions of the processor usually residing in the camera head can be further split up as shown in sections *33.2.2.2–33.2.2.5*.

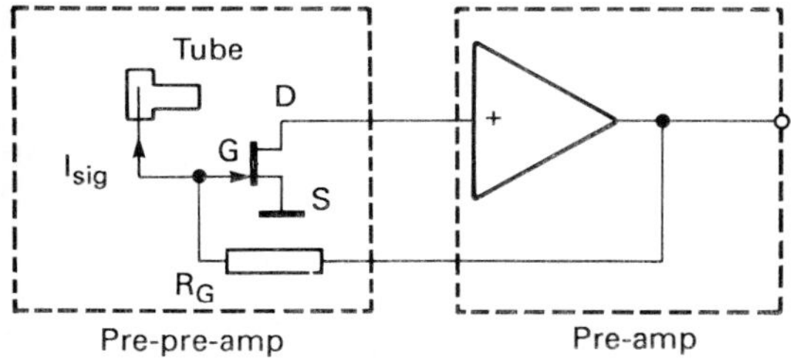

Figure 33.11 Schematic diagram of a camera pre-amplifier

33.2.2.2 Frequency compensation

- compensation of the frequency response of the pre-amplifier
- horizontal aperture correction (out-band correction, i.e. part compensation of the camera tube frequency response)

33.2.2.3 Additive correction and control

- bias light compensation (the bias light causes an unwanted increase in the black level)
- flare compensation (in the optical components, an increase in the black level is produced, approximately proportional to the mean signal value)
- black shading compensation (dynamic compensation of an eventual non-uniform image illumination by the bias light; this was achieved formerly by adding H, V sawteeth and parabolas, and now by free programmable signals from the correction computer)
- black level control (master black, i.e. colour neutral or colour selective, each obtained by the operator at the control panel)

33.2.2.4 Multiplicative correction and control

- gain selection (-6–+24 dB, for sensitivity matching)
- automatic white balance (matching of the red and blue gain with respect to the green)
- multiplicative shading compensation
- white level control (setting up of the colour balance via the operator at the control panel)

Multiplicative shading compensation involves dynamic compensation of level errors through lens vignetting, non-uniformity of the sensitivity of the photoconductive layer, and beam spacing modulation or velocity modulation formed by the

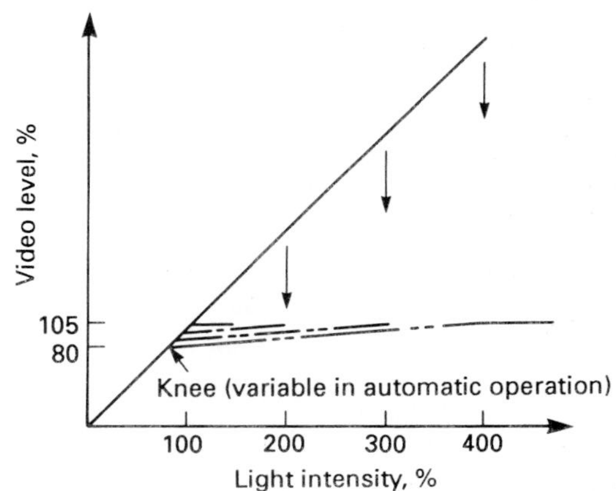

Figure 33.12 Transfer characteristic of a dynamic knee processor (Broadcast Television Systems GmbH)

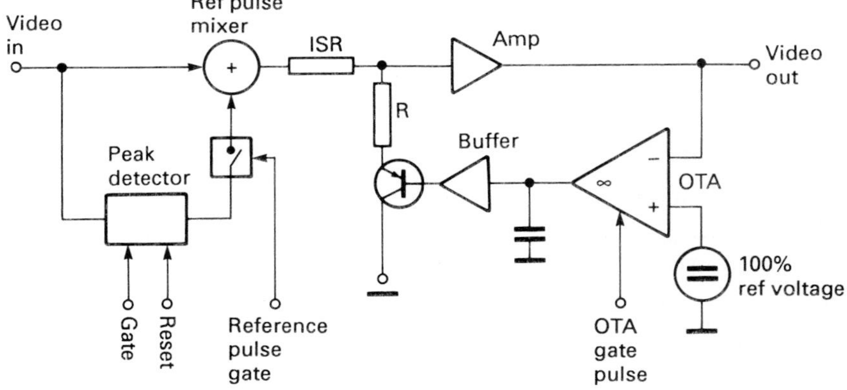

Figure 33.13 Schematic diagram of a dynamic knee processor

geometrical errors when scanning. Production of the correction signals is basically the same as for additive correction.

33.2.2.5 Non-linear control

- knee processor

The knee processor possesses a compressing transfer characteristic, through which the available contrast ratio of the camera can be increased by a factor of 4–6 with respect to conventional systems (*Figure 33.13*).

As *Figure 33.12* shows, the input signal values which exceed 100 per cent are compressed into the output range of approximately 80–100 per cent. In the case shown, this is dynamically produced, i.e. controlled from the peak value of the input signal, so that the maximum input value (assuming that it is greater than 100 per cent) is always transferred to 100 per cent at the output.

33.2.3 Main video processor

As can be deduced from *Figure 33.3*, the stages for the colour correction matrix, two-dimensional frequency response correction and gamma correction form the core of the main video processor.

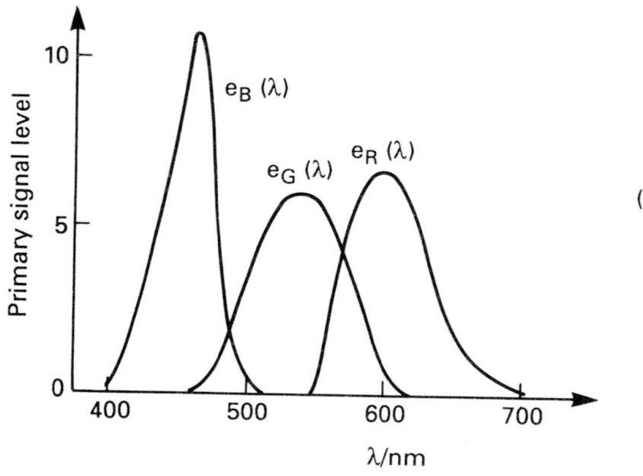

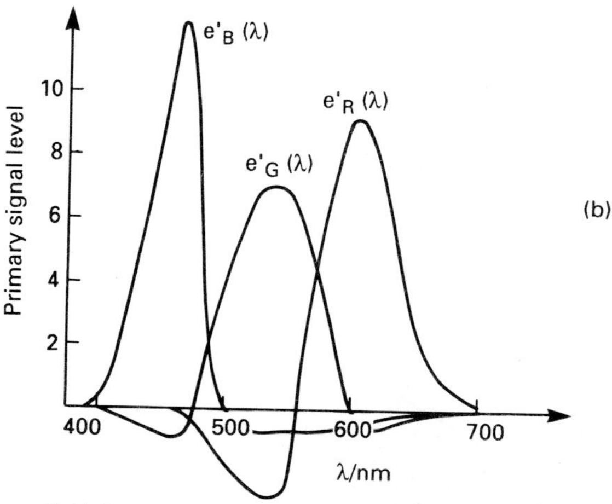

Figure 33.14 Spectral sensitivity of an RGB camera: (a) at input of the colour matrix, (b) at output of the colour matrix.
$e'_R(\lambda) = 1.45\, e_R(\lambda) - 0.45\, e_G(\lambda)$
$e'_G(\lambda) = -0.10\, e_R(\lambda) + 1.20\, e_G(\lambda) - 0.10\, e_B(\lambda)$
$e'_B(gl) = -0.05\, e_R(\lambda) - 0.10\, e_G(\lambda) + 1.15\, e_B(\lambda)$
(Lang[1])

33.2.3.1 Colour correction matrix

For a colorimetrically correct picture reproduction onto a screen, the camera's primary signals have to be matched to the screen phosphor. This occurs via a linear transformation, i.e. a weighted addition and subtraction of the input primary signals into an electronic matrix.

Figure 33.14 shows that the spectral sensitivities of the camera channels can be matched to a good approximation to the negative components of the spectral curves of the screen phosphor.

33.2.3.2 Frequency response correction (aperture and contour correction)

The lens and the camera tubes act as low-pass filters for the horizontal and vertical spatial frequencies. The two-dimensional frequency response of both components multiply, so that at high spatial frequencies the overall transfer characteristic falls rapidly. The behaviour is phase linear, but different in horizontal and vertical directions, as well as astigmatic in the diagonal direction.

Even though a camera (a system scanning line by line) needs a spatial pre-filtering at least in the vertical direction, the frequency response must be raised in the horizontal and vertical directions to give a satisfactory image.

After a thorough smoothing of the total frequency response (*aperture* correction), a peaking of the middle spatial frequencies (*contour* correction) is carried out. This ensures that the decreasing frequency response of the final reproduction system (monitor) is also simultaneously compensated for.

It is necessary to produce a two-dimensional phase linear high-pass filter. First of all, a detail signal is created and added to the main signal. The principle for the creation of this detail signal is the same for both spatial coordinates, i.e. displaced and weighted parts of the signal are subtracted in the x and y directions from this signal (*Figure 33.15*). A spatial discreet filtering then takes place, which is, as required, phase linear.

The creation of the circuit obviously takes place in the time domain, i.e. in the horizontal direction via delays with analogue or digital delay time elements, and in the vertical direction via line delays. The latter are conventionally achieved by ultrasonic delay lines (in the carrier frequency range) or digitally by line stores. The vertical displacement can naturally only occur in increments of one line, which implies a displacement of two spatial lines in interlace scanning.

In principle, the horizontal filter can be connected in series with the vertical filter, but a much more flexible arrangement is shown in *Figure 33.16*. The two-dimensional detail signal frequency response of this arrangement is shown as an example in *Figure 33.17*.

The raising of the frequency response naturally worsens the signal/noise ratio at high frequencies. To minimize this problem, a non-linear processing of the detail signal is carried out (*Coring technique*).

33.2.3.3 Gamma correction

The transfer characteristic of colour display tubes follows approximately a power law. Therefore, for a correct reproduction of colour, an inverse alignment has to take place at one point in the transmission system, i.e. the function $y = x^\gamma$ ($\gamma \approx 0.4$) must be realized. Traditionally and for companding reasons, this is carried out in the source equipment, i.e. the camera.

From a colorimetric point of view, gamma correction before the transmission channel can be a problem. The circuit shown in *Figure 33.18* uses the inversion of the fet transfer characteristic.

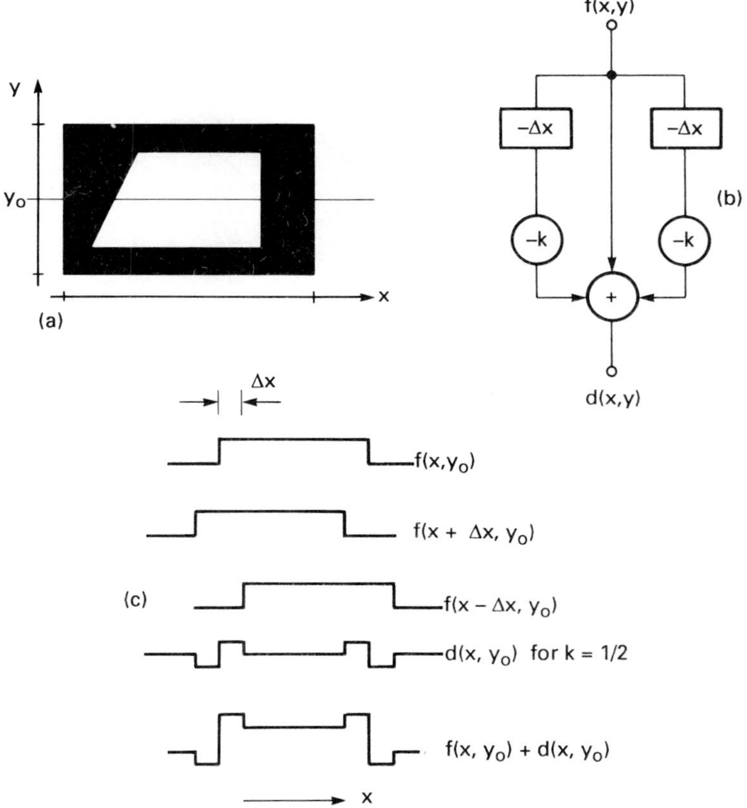

Figure 33.15 Generation of a horizontal detail signal for contour or aperture compensation: (a) image at the input f(x,y) of detail processor, (b) principle of a horizontal detail processor, (c) horizontal waveforms at the constant vertical position y_o

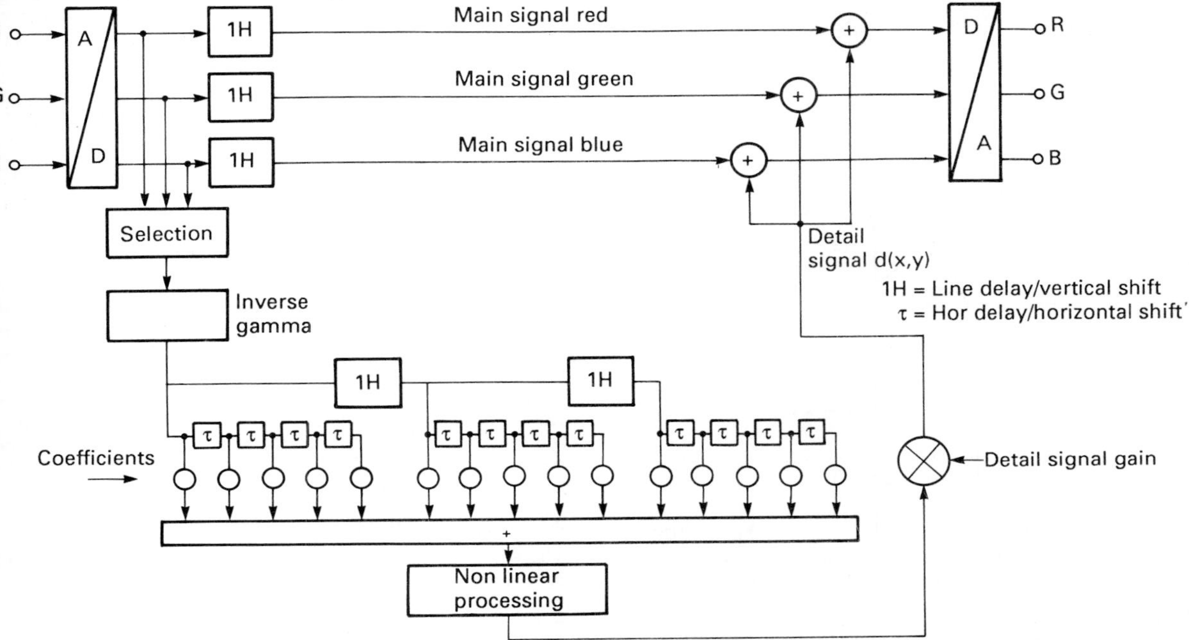

Figure 33.16 Schematic diagram of a two-dimensional digital image enhancer (Broadcast Television Systems GmbH)

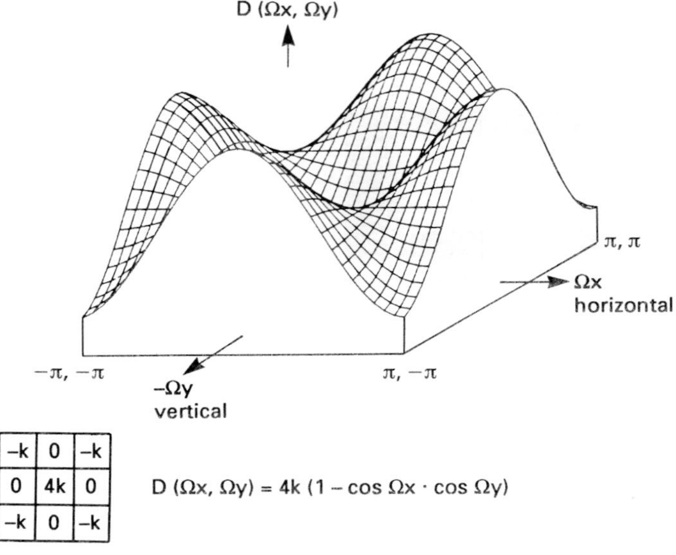

Figure 33.17 Two-dimensional frequency response of the detail signal d(x,y) in *Figure 33.16* but with a reduced (3 × 3) coefficient matrix (Broadcast Television Systems GmbH)

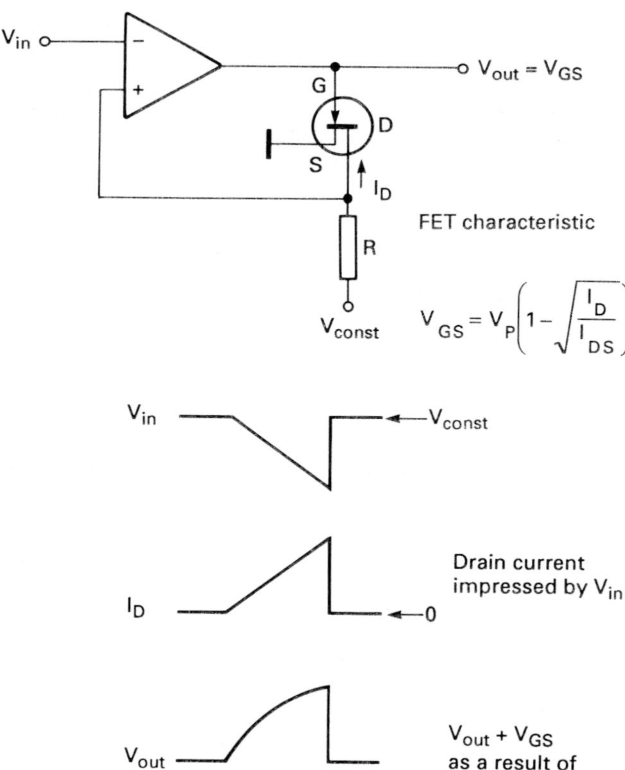

Figure 33.18 Schematic diagram of a gamma processor, based on the fet inversion principle

The corrections currently carried out for colour reproduction and gamma pre-compensation in the image source equipment are with respect to a particular standard reproduction device (a monitor) with given phosphors and transfer characteristics. In the process of defining and introducing new television systems, it is desirable to have a standard output interface for the source equipment and a standard input interface for the image reproduction equipment. Only then can a random coupling of the wide variety of source and display devices lead to optimal colour reproduction.

33.3 Operational characteristics

The characteristics of an electronic camera which determine its quality are:

- signal/noise ratio
- sensitivity
- static and dynamic resolution
- colour reproduction
- colour registration of the three scanning rasters

These characteristics interact and cannot be studied in isolation. For example, static and dynamic resolution are electronically improved at the cost of the s/n ratio, or a bad colour registration worsens the luminance resolution. An overall analysis is especially required for sensitivity and the s/n ratio. In selecting the nominal signal current for picture white, the limits for the s/n ratio and sensitivity are set at the same time.

33.3.1 Signal/noise ratio

The signal/noise ratio of an electronic camera with camera tubes is determined by:

- the nominal signal current
- internal noise of the first fet in the pre-amplifier
- effective capacitance at the amplifier input (i.e. the sum of

the target source capacitance, the fet input capacitance and unavoidable stray capacitances)

- noise due to the effective ohmic resistances at the amplifier input

The internal noise of the camera tube is negligible. Its effective value is

$$I_{rms} = \sqrt{2eI_{sig}B}$$

All the other noise sources are also negligible for a professional arrangement of the amplifier train.

A pre-amplifier with direct coupling to the camera tubes (*Figure 33.19*) has the following frequency response:

$$\frac{U}{RI} = \frac{1}{1+jf/f_g}$$

$$\text{where} \quad f_g = \frac{1}{2\pi RC_E/V_O}$$

i.e. a low-pass filter characteristic of the first order. This has to be compensated for either in the pre-amplifier itself or in one of the following stages. For this reason, the spectral noise power density has a curve that increases quadratically with frequency. This noise spectrum increase in the upper frequency range can be reduced by inductive coupling with a Percival coil as shown in *Figure 33.19*(b). This coil is arranged between the signal source and the noise source. By using the resonance peaking of L and C, a pre-emphasis for the signal current is introduced, without simultaneously increasing the noise.

For the analysis of the frequency response and the s/n ratio, for direct and inductive coupling, shown in the diagrams, the following definitions apply:

R: Feedback resistor of the pre-amplifier, or, in the case of the delivery of a target bias voltage, the parallel circuit of a feedback resistor and a bias voltage resistor (typical value: 1 M ohm).

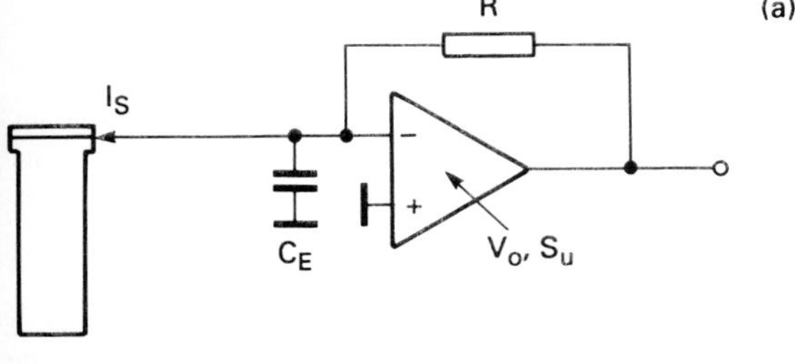

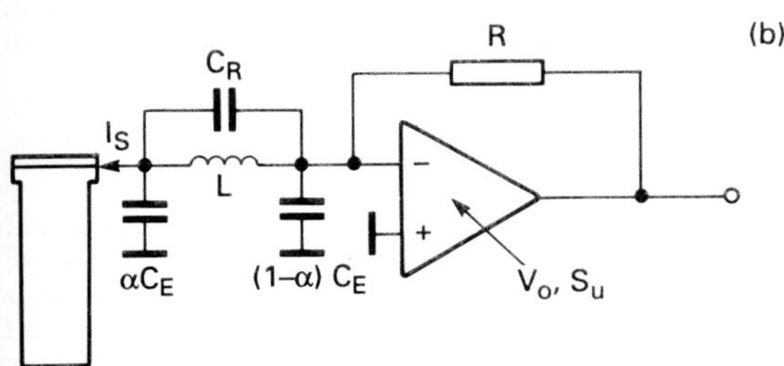

Figure 33.19 Components of a camera pre-amplifier, contributing to noise performance: (a) direct coupling, (b) inductive coupling of pre-amplifier and tube. Definitions of the quantities are given in the text

C_E: Total effective capacitance at the amplifier input. In the case of inductive coupling, this is split into $\alpha C_E + (1 - \alpha)C_E$ (typical value: 10 pF).

α: Distribution factor of C_E (typical value: 0.5).

L: Inductance of the Percival coil (value: 0–30 μH).

C_R: Resonant capacitance of the Percival coil (typical value: 1 pF).

S_U: Noise voltage density of the fet (typical value: 1 nV/√Hz). The noise characteristics of the fet are described here by its input noise voltage.

I_s: Nominal signal current for a 100 per cent video level in the green channel (typical value: 300 nA).

V_o: Open loop gain of the pre-amplifier. This is not relevant to the s/n ratio.

B: Bandwidth of the video signal, i.e. the bandwidth up to where the noise power density is integrated. B is 5 MHz for standard television and 20–30 MHz for HDTV.

The signal/noise ratio, dependent on bandwidth, B, is, for direct coupling, given by:

$$s/n\,(dB) = 10\lg \frac{I_s^2}{4kTB/R + BS_u^2/R^2 + {}^4\!/_3\pi^2 C_E^2 S_u^2 B^3}$$

Figure 33.20 shows the pre-amplifier frequency response (V_o = 300) over frequency, f, for various inductances, L. *Figure 33.21* shows the noise spectrum of the equivalent noise input current in pA/√Hz (independent of V_o), also for various inductances. *Figures 33.22* and *33.23* show, for direct coupling, the unweighted signal/noise ratio as a function of bandwidth, B, in MHz, with parameters S_U and C_E. The likewise unweighted s/n ratio for inductive coupling is shown in *Figures 33.24* and *33.25*, with parameters L and C_R. In *Figures 33.21–33.25*, as well as in the equation for s/n ratio, the frequency response correction has already been taken into consideration, i.e. a smooth overall characteristic is always assumed. The following values are used for the curves:

R	= 1 M ohm
C_E	= 10 pF
α	= 0.5
L	= 10 μH
C_R	= 1pF
S_U	= 0.9 nV/√Hz
I_s	= 300 nA

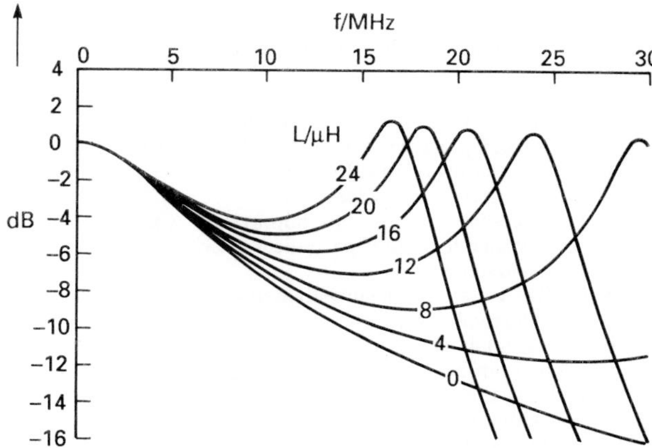

Figure 33.20 Frequency response of the pre-amplifier in *Figure 33.19*(b) for different values of the inductance, L

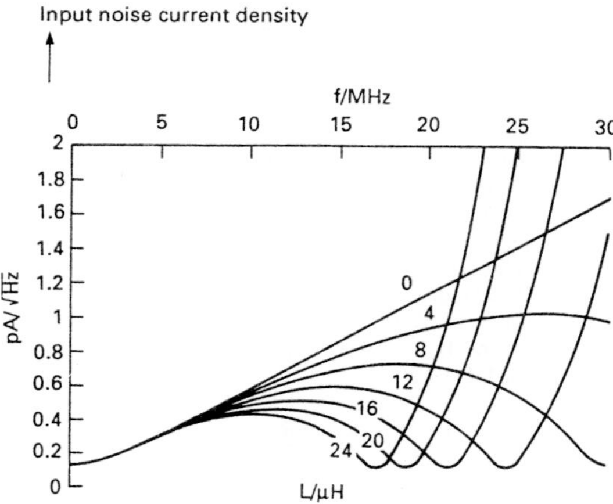

Figure 33.21 Equivalent input noise current densities in pA/√Hz of the pre-amplifier in *Figure 33.19* (b) for different inductances, L

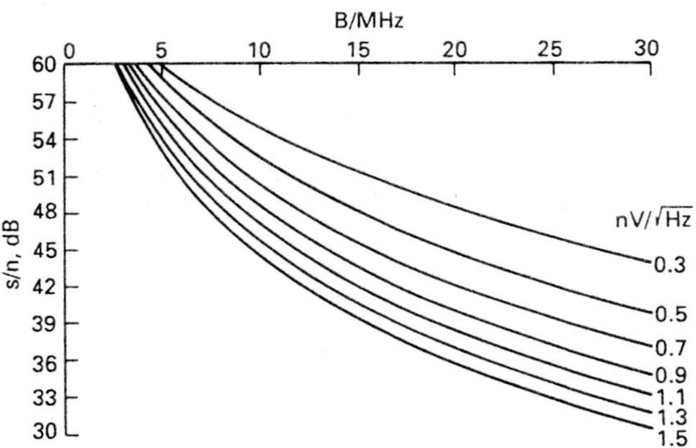

Figure 33.22 Unweighted s/n ratio as a function of the total bandwidth, B, for different noise levels of the fet in nV/√Hz

From the family of curves, the following primary results can be deduced:

- The noise voltage density, S_U, of the fet, as well as the input capacitance, C_E, have the greatest influence on the s/n ratio.
- With increasing bandwidth, B, the influence of all the values of the characteristic increases.
- The Percival coil has no importance when considering today's attainable input capacitances of about 10 pF in 5 MHz television systems.
- In HDTV systems, the unweighted s/n ratio, on introduction of the Percival coil, can be increased by about 4–6 dB. However, a laborious frequency response correction needs to be carried out at the same time (*Figure 33.20*). The s/n ratio gain, due to the coil, is in the upper frequency range. This is not so noticeable to the senses of a human, and the weighted s/n ratio gain is reduced to about 1–3 dB.
- The resonant frequency of the Percival coil must clearly lie above the system bandwidth.

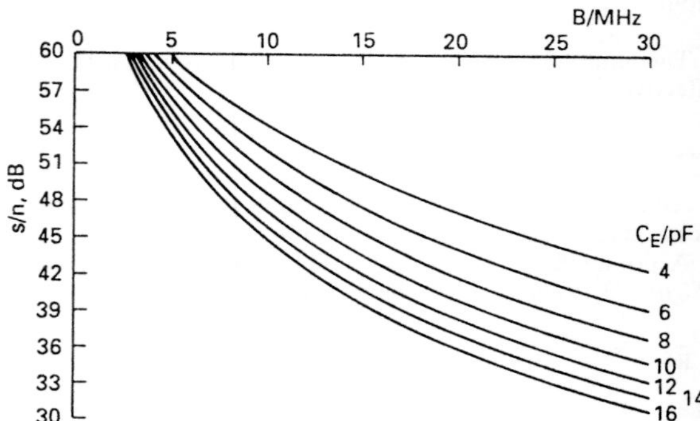

Figure 33.23 Unweighted s/n ratio as a function of the total bandwidth, B, for different input capacitances, C_E

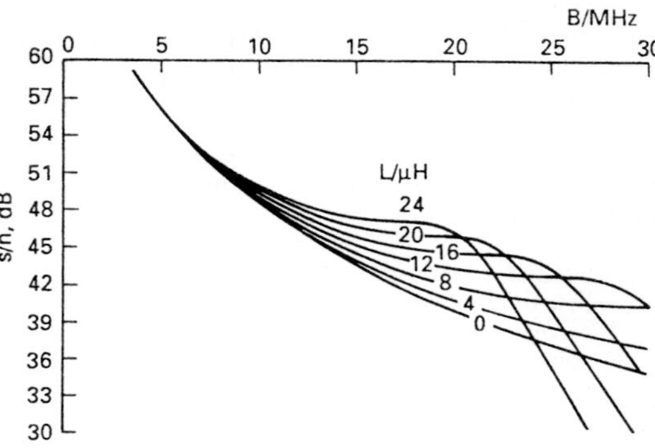

Figure 33.24 Unweighted s/n ratio as a function of the total bandwidth, B, for different inductances, L

33.3.2 Sensitivity

The sensitivity of an electronic camera is defined by the choice of lens and camera tubes. The sensitivity of a charge storing camera tube is, for example, defined as the quotient of the signal current, I_S, and the illumination on the tube target, E_{TG} (measured in lux). The tube is used in a TV system, with a total picture duration T_{tot}, and an active picture duration T_{act}. The camera tubes have an integral layer sensitivity S_i' (measured in amps per lumen) and a scanned target area A_{TG}. The signal current is calculated as follows:

$$I_s = S_i'.T_{tot}/T_{act}.A_{TG}.E_{TG} = S_i.A_{TG}.E_{TG}$$

The sensitivity of the tubes does *not* depend on the number of lines, on the frequency of image changes, or on the type of scanning (interlace or progressive). During scanning, whenever a stabilized state has been reached between the charging and discharging of the photoconductive layer, the charge which has flowed in via the photocurrent will also be completely removed again by the signal current. In other words, with a decrease in the integration time (less charge stored), a similar decrease in

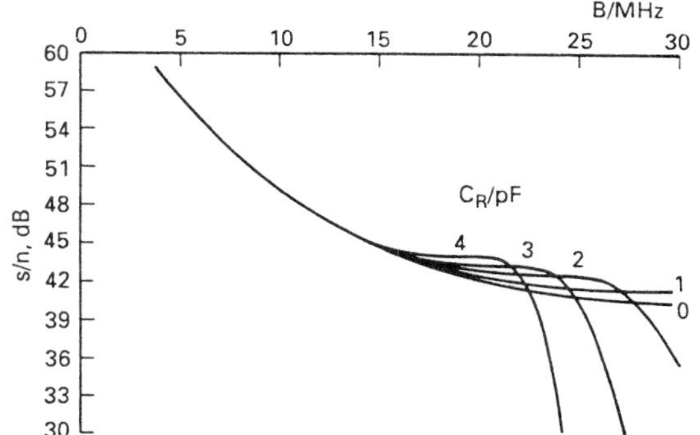

Figure 33.25 Unweighted s/n ratio as a function of the total bandwidth, B, for different resonance capacitances, C_R

the read-out time occurs (the charge is read out quicker). Hence the quotient, $I = \Delta Q/\Delta T$ remains constant.

The different television standards are only reflected by the signal current through the relationship between the total and active picture periods and through the scanned target area. Of course, an increase in the number of lines or the transition from interlace to progressive scanning leads to a decrease in the amplitude deviation of the target surface potential. Hence, with the same layer capacitance, an unchanged signal current is produced.

Camera tube manufacturers specify the sensitivity of the tubes by giving the three integral sensitivity values: $S_i(R)$, $S_i(G)$ and $S_i(B)$. The measurement of these values takes place in a scanning raster, with a given T_{tot}/T_{act}, at a given standard illuminance (e.g. 2856 K), and for a specified filter triplet, similar to that shown in *Figure 33.6*.

Typical values for saticon tubes are: $S_i(R) = 120\ \mu A/lm$, $S_i(G) = 150\ \mu A/lm$ and $S_i(B) = 80\ \mu A/lm$. Hence, the value of 150 $\mu A/lm$ implies that camera tubes with beam splitters connected in front as specified, at an irradiation of 10^{-3} lumen of white light of the specified light type, produce a signal current of 150 nA.

If the signal current for other filter curves and types of light is required, the integration produced from equation (33.1) must be carried out for a known spectral sensitivity distribution of the layer.

The relation between the scene illumination E_{SC}, and the target illumination E_{TG}, for a given medium lens transmission factor τ, scene remission ρ, and aperture number F, is given by:

$$E_{TG} = \tau\rho E_{SC}/4F^2$$

From this, the signal current can be deduced as a function of the scene illumination, i.e.:

$$I_S = S_i . A_{TG} . \tau\rho E_{SC}/4F^2$$

The relationship between illumination, E_{SC}, and the aperture number, F, is shown in *Figure 33.26* for the aspect ratios 4:3 and 16:9, as well as for the image sizes ⅔ inch, 1 inch and 1¼ inch. The aspect ratio 16:9 has, for the same target diagonals (due to the smaller area), 89 per cent of the sensitivity of 4:3 scanning.

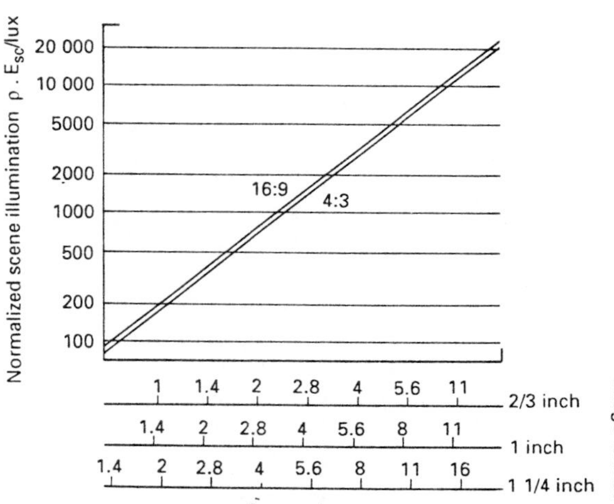

Figure 33.26 Sensitivity of a TV camera, for different f-numbers, image formats and aspect ratios. $I_s = 300$ nA, $S_i = 150\ \mu A/lm$, $\tau_{lens} = 0.8$

At first sight, according to *Figure 33.26*, the sensitivity of a camera appears to increase with an increase in the tube diagonals. In fact, the 1¼ inch camera, for the same aperture number, F, delivers a signal current which is a factor 1.8 times that of a 1 inch camera. The false deduction arises from the completely different reproduction geometry of the 1¼ inch camera. If two cameras with picture diagonals d_1 and d_2 are operated with the same layer sensitivity, aspect ratio, lens transmission factor, etc., then the ratio of their signal currents is:

$$I_{S2}/I_{S1} = (d_2/d_1)^2 . (F_1/F_2)^2$$

If the same visual angle and depth of field are required, the aperture numbers can be calculated as follows:

$$F_1/F_2 = d_1/d_2$$

If this is then substituted into the previous equation, the same signal currents are produced, i.e.

$$I_{S2}/I_{S1} = 1$$

Hence the sensitivity of the electronic camera is, for the same picture composition, independent of image size!

33.3.3 Depth of field

As a function of the object distance, a, the focal length, f, the aperture number, F, and the circle of confusion, h/N (image height divided by effective vertical resolution in lines), it is possible to differentiate between the near limit of depth, a_v, and the far limit of depth, a_H, i.e.:

$$a_v = \frac{a}{1+pa} \text{(a,}a_v \text{ normalized in metres)}$$

$$a_H = \frac{a}{1-pa} \text{(a,}a_H \text{ normalized in metres)}$$

where $$\frac{p}{F} = \frac{1000}{N}\ \frac{h}{mm}\ \frac{1}{(f/mm)^2} = \frac{4F\tan^2(\alpha_H/2)}{N/1000 . b/h . b/mm}$$

where h is the image height on the tube target, b is the image

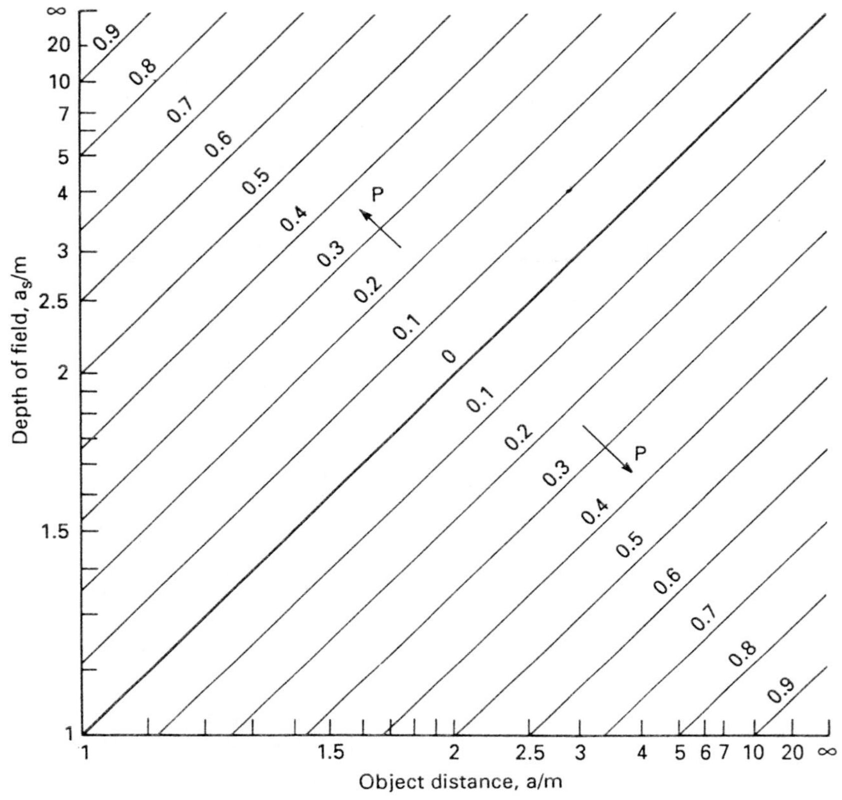

Figure 33.27 Calculation of depth of field as a function of the object distance and the parameter, p (see text)

size and α_H is the horizontal visual angle. With help from the normalized presentation in *Figure 33.27*, after calculation of p, the focusing range can be read off. An example might be:

aspect ratio:	b/h = 4:3
picture width (1 inch):	b = 12.8 mm
effective resolution:	N = 400
aperture number:	F = 2.0
horizontal visual angle:	$\alpha_H = 45°$
giving	p = 0.2

Figure 33.27 shows, for a setting distance of 2 m, a focusing range of approximately 1.4–3.3 m. As can be easily deduced, for the same focusing depth and visual angle in a HDTV camera, with N = 800 lines effective resolution and with an aspect ratio of 16:9, an aperture number of 5.6 would be required. This explains the focusing depth and focusing control problems in high resolution cameras.

33.4 Automation functions

The automation functions so far realized for modern studio cameras can be divided into basic set-up, pre-operational set-up and continuous automatic functions. The *basic set-up* is carried out the first time the system is put into operation and after a tube change, as well as periodically over large cycles. The *pre-operational set-up* includes algorithms, which optimize the camera state to the actual requirements of a scene. The *continuous automatic functions* are active during the actual picture operation. It is possible to realize the following individual functions.

33.4.1 Automatic basic set-up

Operational data for the camera tubes:

- mechanical back focus
- electrical focusing
- beam landing or beam alignment
- beam current
- beam current control (ABC)

Optimization of the scanning raster:

- mechanical rotation
- coarse registration
- geometry
- fine registration

Alignment of the video amplifier:

- flare compensation
- white shading
- black shading
- shape of the gamma correction

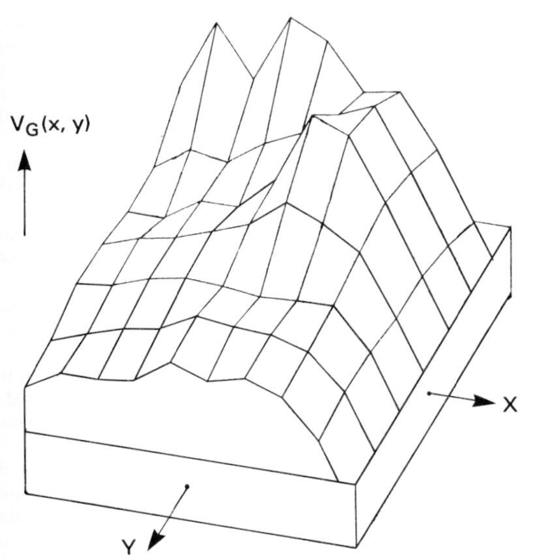

Figure 33.28 Typical waveform of a geometric correction signal as a function of spacial coordinates x and y

33.4.2 Automatic pre-operational set-up

- white balance
- black balance
- selection of conversion filters
- cable length dependent frequency response compensation

33.4.3 Continuous automatic functions

- iris control
- dynamic centring of the red and blue rasters
- dynamic lens error corrections

33.4.4 Error corrections

From the functions above, two important processes will be described in more detail as examples, namely, fine registration corrections of the scanning raster and dynamic lens error corrections.

33.4.4.1 Fine registration correction

The introduction of microcomputers into studio camera technology has led to conventional registration correction processes

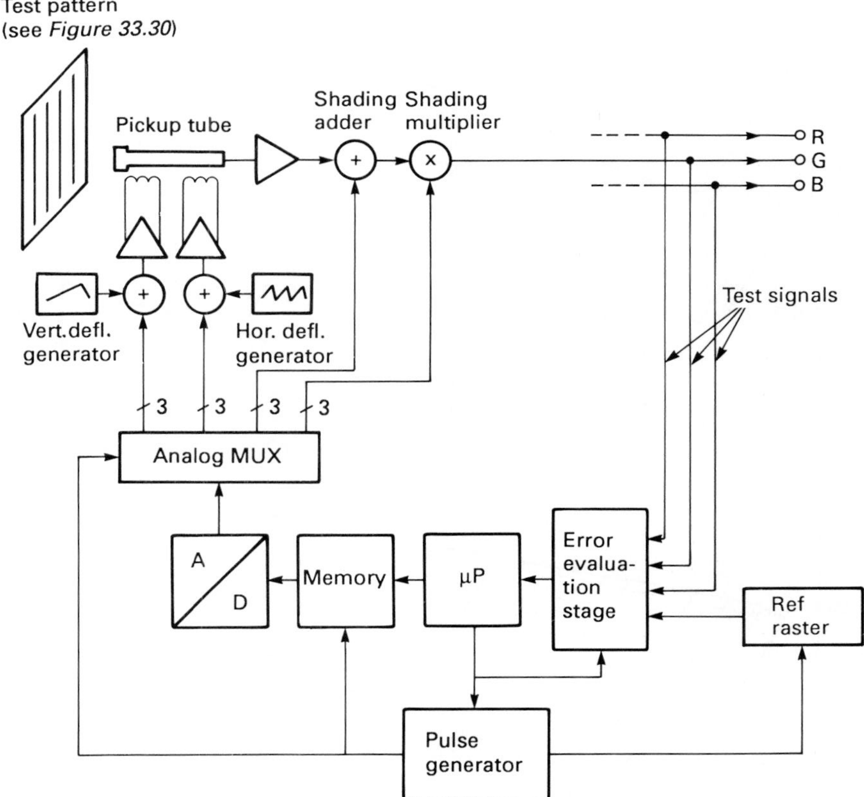

Figure 33.29 Block diagram of automatic registration and shading compensation, using a test chart as shown in *Figure 33.30* (Broadcast Television Systems GmbH)

Figure 33.30 Test slide to be used in a lens diascope for automatic registration and shading setup (Broadcast Television Systems GmbH)

being superseded by computer supported digital algorithms. In conventional processes, the horizontal and vertical signals of a given form (e.g. sawtooth or parabola) were mixed with the deflection currents. However, firstly an extensive balancing of numerous potentiometers was necessary, and secondly the obtainable quality especially in the border areas of the picture was limited due to a finite supply of periodic correction signals.

On the other hand, modern computer supported processes start by segmenting the image field, from which, by definition, an almost freely programmable correction signal can be created from the matrix of correction points containing $N_H \times N_V$ values. The flexibility of this signal is limited by the number of correction points and by the necessary horizontal and vertical interpolation processes between the correction points (*Figure 33.28*). A high number of correction points, e.g. 16, in the horizontal direction leads to a bandwidth of several 100 kHz for the correction signal.

Even though the horizontal correction signal is dependent on the vertical coordinates, and the vertical correction signal is dependent on the horizontal coordinates, two similar and (in the bandwidth) equivalent signals are obtained. The vertical deflection must also be able to transmit this additively mixed broad-band correction signal.

The correction points can now be found, either by menu driven manual tuning or by a fully automatic process. *Figure 33.29* shows the principle structure of the tuning system.

This, with the help of a *diascope* (test slide in the lens), measures and minimizes the registration errors. The test slide shown in *Figure 33.30*, as well as measuring two-dimensional registration errors (vertical depositing positions are also converted into temporal errors, due to the diagonal lines), also determines white shading errors above the wide white lines. The actual correction is via a control process, which determines the preliminary depositing position errors between, for example, red and green, and then iteratively changes the correction value until the remaining error is within the tolerance limits. The correction values found from correction point to correction point by this method are deposited in memory.

The signals interpolated from these values are always

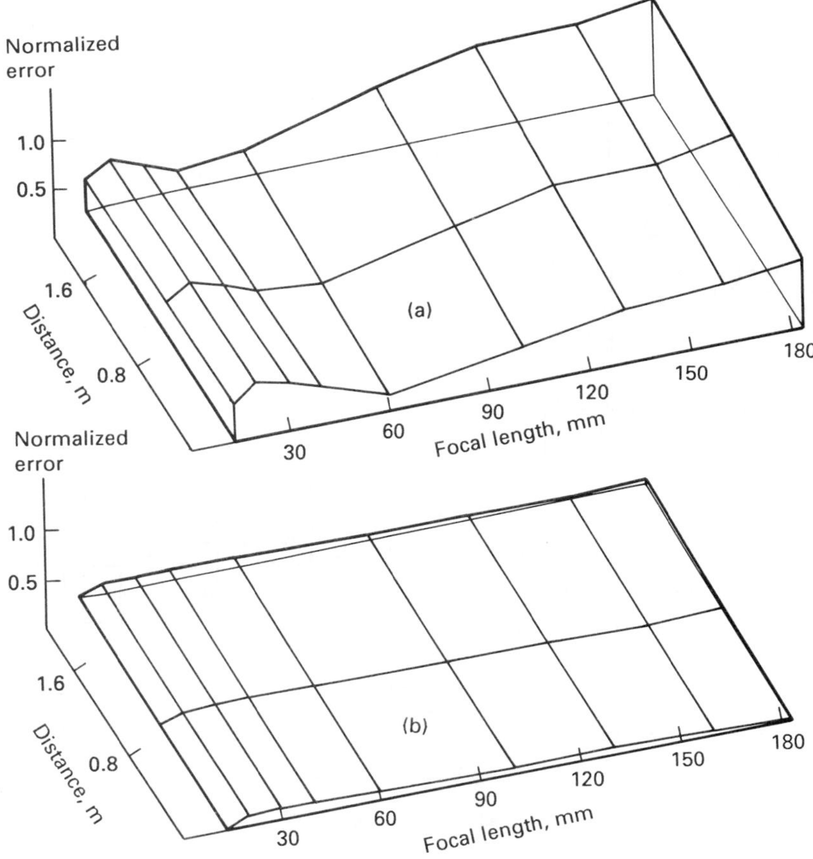

Figure 33.31 Typical lateral chromatic errors of a studio TV lens: (a) without dynamic lens error correction, (b) with dynamic lens error correction (Broadcast Television Systemms GmbH)

available during the actual camera operation. As already indicated, such a process can also sensibly be used for the correction of image geometry, and black and white shading.

33.4.4.2 Dynamic lens error corrections

Using a registration correction process as described above, no errors can be corrected which result from the three operational parameters focal length (f), setting distance (a) and aperture (F) of the lens. Of course, to a very good approximation, the error here is purely an error in the magnitude of the red and blue rasters with respect to the green one. Hence the following functions should be stored in the correction computer:

$$\begin{aligned}\text{H magnitude (R)} &= f_1\,(f, a, F)\\ \text{V magnitude (R)} &= f_2\,(f, a, F)\\ \text{H magnitude (B)} &= f_3\,(f, a, F)\\ \text{V magnitude (B)} &= f_4\,(f, a, F)\end{aligned}$$

The computer must also be given here the current actual values of f, a and F. The obtained correction values are then added to the deflection currents or voltages. *Figure 33.31* shows the typical errors dependent on the lens before and after correction.

Reference

1 LANG, H, *Farbmetrik u. Farben sehen*, R. Oldenburg Verlag, Munchen (1978)

Bibliography

PEARSON, D E, *Transmission and Display of Pictorial Information*, Pentech Press (1975)

KLEMMER, W, 'Multistandard HDTV camera', *Electronics & Wireless World*, 708–711

KLEMMER, W, 'Concept and realization of a HDTV studio camera', 14th Inter TV Symposium Montreux, Symp Record, Joint Session, 382–396

REIMERS, U, 'Origin and perceptibility of noise in a HDTV camera', *Frequenz 37*, 316–323 (1983)

KURASHIGE, M, '1 inch magnetic focus and electrostatic deflection compact saticon for HDTV', *NHK Laboratory Notes No 322* (Nov 1985)

KATO, S et al, 'Performance characteristics of improved pickup tube', *SMPTE Jour* (October 1983)

FRANKEN, A, 'Modulationsubertragungs-funktion einer Kamerarohre', *Fernseh u. Kino Tecknik*, 4 (1978)

KURASHIGE, M, 'Effect of self-sharpening in low-velocity electron beam scanning', *IEEE Trans Electron Devices, ED-25*, 10 (October 1982)

CANON, *TV Optics. The Canon Guide Book of Optics for Television Systems* (1986)

KLEMMER, W, 'The two dimensional resolution of pickup tubes in HDTV camera systems', *Image Technology*, 328–332 (July 1987)

P W Wayne
formerly Marketing Director, Vinten Broadcast Ltd

33b Studio Cameras and Mountings – Mounts

33.5 Positioning equipment

This is equipment that enables the operator to place the camera in the right place at the right height. It varies from simple tripods to pedestals and simple cranes. Section *33.6* deals with pan and tilt heads, and includes a look at different designs which have particular features or benefits.

33.5.1 Tripods

The simplest portable mounting for any camera is a tripod. There are dozens built to professional standards, but few achieve all the desirable characteristics, which are:

- lightness
- rigidity
- a good range of height adjustment
- simple height locking mechanisms
- a firm spreader or triangle
- useable on both smooth and uneven surfaces, and on soft as well as hard ground

Most tripods have spikes that can be covered with rubber feet. These may be included in the spreader, or form safety covers over spikes. All tripods are a compromise, but modern materials have enabled designers to make extremely rigid structures that are not too heavy to carry, and the newer of such tripods give a height range of 42–157 cm (16.5–61.75 inches), and collapse down to a package only 69 cm (27 inches) long.

When buying a tripod and spreader, check that:

- when fully extended on its spreader, the top platform resists horizontal rotation; such rotation is called *wind up*, and if it exists in any degree, it gives quite distressing effects on the start and finish of a pan
- the points or feet on the tripod fit very snugly into the ends of the spreader or points of the triangle
- the spreader has enough rigidity horizontally to ensure that the legs are splayed as near as possible to 120°
- vertical downward forces do not deflect the legs in any way
- all fittings and clamps are designed for ease of use with gloved hands

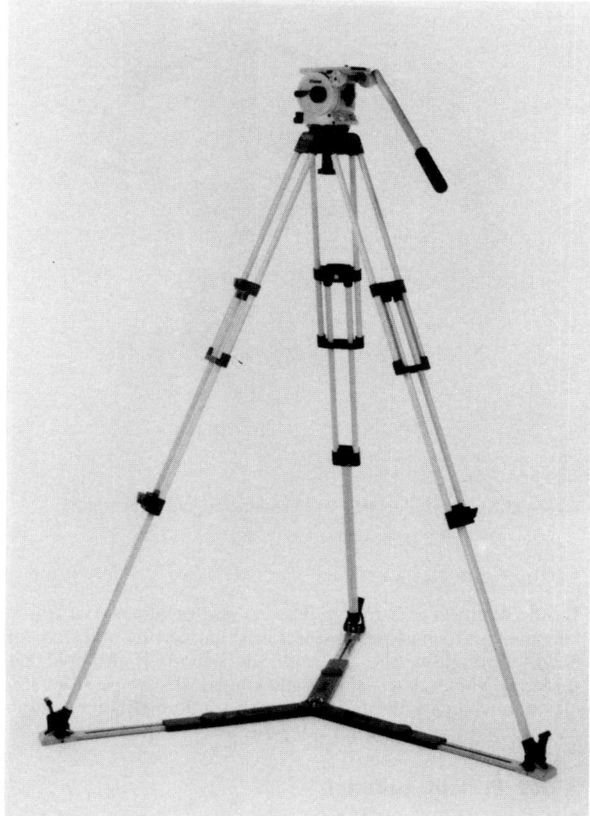

Figure 33.32 Vision two stage ENG tripod (Vinten)

A triangle is more rigid than a Y spreader, but is very difficult to fold up quickly, and much more difficult to use on uneven ground or a staircase, which requires using a flexible Y shaped spreader.

Most tripods in Europe use either a 100 mm or a 150 mm bowl. However, in the USA, and the film industry in particular, they use Mitchell and even Junior Mitchell fittings with clamps levelling the tripod by adjusting the legs.

33.5.1 Heavy duty tripods

Heavy duty tripods use four bolt fittings with the same Mitchell alternatives in the USA.

In order to make elevation changes, some heavy duty tripods are fitted with a simple crank elevation unit. These give a height adjustment of 61 cm (24 inches).

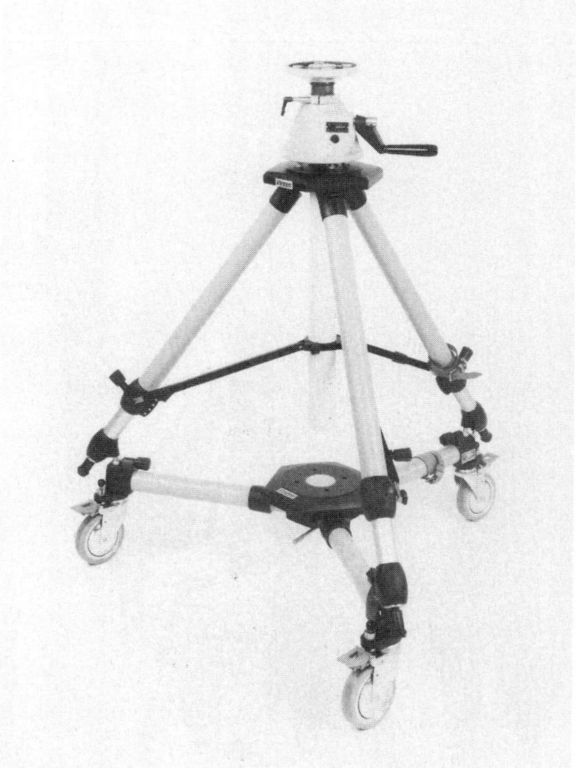

Figure 33.33 Heavy duty tripod, elevation unit and skid (Vinten)

33.5.1.2 Rolling base

By the addition of a rolling base (a *skid* or a *dolly*), a tripod becomes a moving camera platform that can be adjusted for height throughout the range of the tripod. If the skid has castoring wheels, some simple tracking moves are possible, but elevation changes are difficult in a continuous shooting studio environment, and certainly not possible on shot.

33.5.2 Portable pedestals

Small portable pedestals form the next category of mounting. They differ significantly from simple tripods and skids through the addition of an elevation unit. A telescopic column supports the weight of the camera, and pan and tilt head in poise. This column can be added to a tripod or specially constructed dolly wheels, and varies in sophistication from devices with simple castors to steerable systems as shown in *Figure 33.34*.

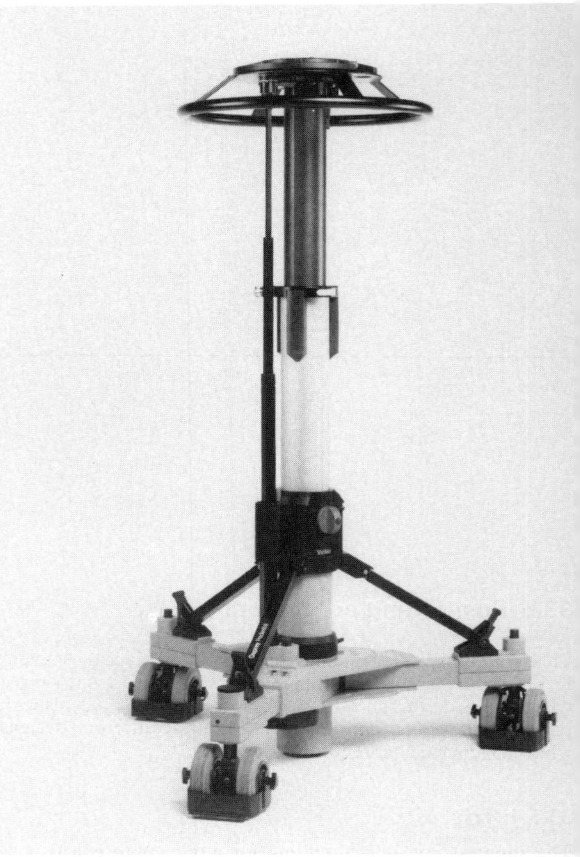

Figure 33.34 Osprey pedestal. This gives a height from a minimum of 65.5 cm (26 inches) to a maximum of 1.47 m (58 inches) (Vinten)

The most sophisticated of these can make steer and crab movements similar to big pedestals, but few can elevate 'on shot', and most have quite severe restrictions on the range of elevation possible.

Regardless of whether the balance system is hand cranked, hydraulic, pneumatic or spring, these low cost units have to be single stage, and the adjustable height range is therefore limited to about 420 mm (16 inches).

In a recent development, Vinten have brought out the Osprey pedestal, where the pneumatic elevation unit can be operated in a low and high position, each one giving a 16 inch height range and therefore an overall height of 32 inches (815 mm).

33.5.2.1 Studio pedestals

A studio television camera together with prompters and viewfinders can weight nearly 100 kg. The most satisfactory way of mounting it is to use a pedestal with crab and steer alternatives.

These are six wheeled devices on a triangular base of approximately 1 m side. They are designed to carry loads up to 150 kg or up to 70 kg. Both the lightweight and heavy duty pedestals can be single or three stage versions. The former have a 76–129.5 cm height range, and the three stage pedestals have a height range of 54.5–149.5 cm.

On all studio pedestals, steering and elevation is by means of a steering ring immediately below the pan and tilt head. When properly balanced, the cameraman should be able to raise and lower the camera on shot, and skilled operators will be able to move the pedestal almost as an extension of themselves. Tracking in and out, traversing left to right or any combination, makes it practical to follow a performer walking round a room.

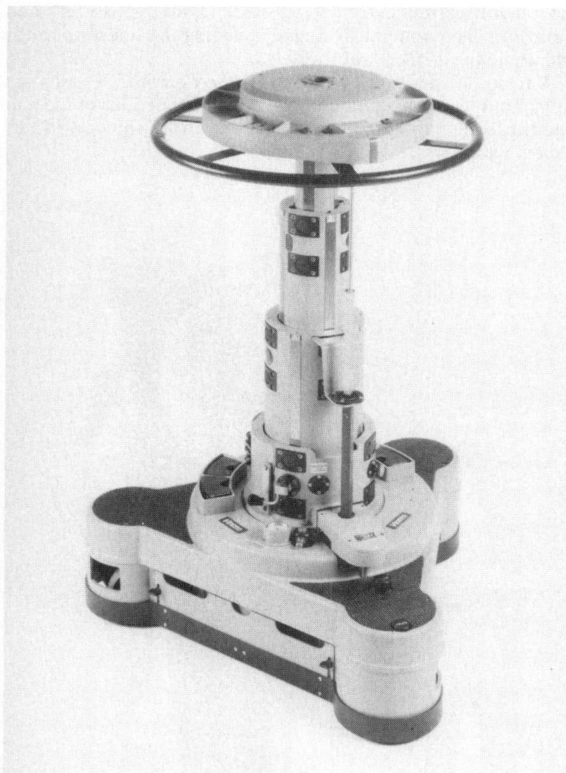

Figure 34.35 Fulmar pedestal (Vinten)

The secret lies in the crab/steer mechanism. When all three wheels are locked together in parallel, the pedestal is said to be *in crab*. Crabbing does not affect the angular orientation of the pedestal, and therefore the pan action of the head is not affected by pedestal movement.

The pedestal travels in the direction of the steer spot or pip on the steering ring. It may be surprising that this has proved to be the most versatile way of using pedestals. The steering position where two wheels are locked together and only one rotates is used only for negotiating awkward obstacles or turning the pedestal round on its base.

There are four types of pedestal:

- weight balanced
- spring balanced
- hydraulic operating against compressed gas
- pneumatic compressed gas only

Although other gases can be used, nitrogen, which is inert and inclined to be moisture free, is the perfect choice when it is available in a compressed form. Compressed air can be used, but the water that accumulates in the reservoir tanks has then to be emptied quite frequently.

33.5.3 Cranes

The Hollywood film industry created the ride-on camera crane, and we are all familiar with the picture of one of these huge devices and the director with his megaphone shouting instructions to the camera operator and his focus puller.

Some of these cranes were used in the early days of TV, and indeed some are still in use for the older TV studio cameras, although the advent of lightweight TV cameras is rendering this type of crane obsolescent. It gives a much higher point of view than is possible with a pedestal, whose maximum height must be limited to the height of the cameraman's arms fully extended.

The Nike and Tulip are examples of big cranes still in use.

A smaller two man crane such as the Kestrel can be used for high and low shots, but the boom cannot be moved (*swung*) from side to side. The reduction in crew is the main advantage when using the older heavier cameras.

Figure 33.36 Kestrel camera crane (Vinten)

The Kestrel crane shown in *Figure 33.36* is psupplied in either a manual or powered elevation form. The tracker steers and pushes the dolly and in the case of the manual version elevates the platform. Jib elevation controls are extended to the cameraman in the case of the powered version.

While it might seem hard for the tracker to have to elevate his cameraman, the job is made much easier by the jib arm balancing system, which is a pneumatic balancing ram adjusted to take the total load of camera, operator, etc. This makes both the motor drive and the effort required to manually raise and lower the jib very practicable.

33.5.3.1 Short crane arms

The advent of lightweight cameras has made practical short crane arms that are operated by standing cameramen only. The camera pan and tilt head is balanced, and this simple device gives a surprising height range from close to the floor to 1.8 m with even a short arm.

The crane arms can be mounted on almost any of the tripods and pedestals that have been described, to increase their versatility. Apart from height range, the most useful aspect of short cranes is being able to *track over* furniture and other obstacles.

Single camera operator cranes such as the 'mole' from Mole Richardson are also still in use, but becoming less popular as the same height can be obtained with a Merlin arm. The Merlin has the advantage of requiring only one operator instead of the usual crew of three or four for a mole crane.

The mole has a counterbalanced arm with a weight box at the rear end. This weight box is the control position for the arm swayer who positions the camera in space. The driver either walks along beside the crane or sits on a seat behind the arm swayer.

The Merlin arm and its counterpart in the USA, the Barber boom, is a comparatively new development which takes full advantage of the new lightweight cameras.

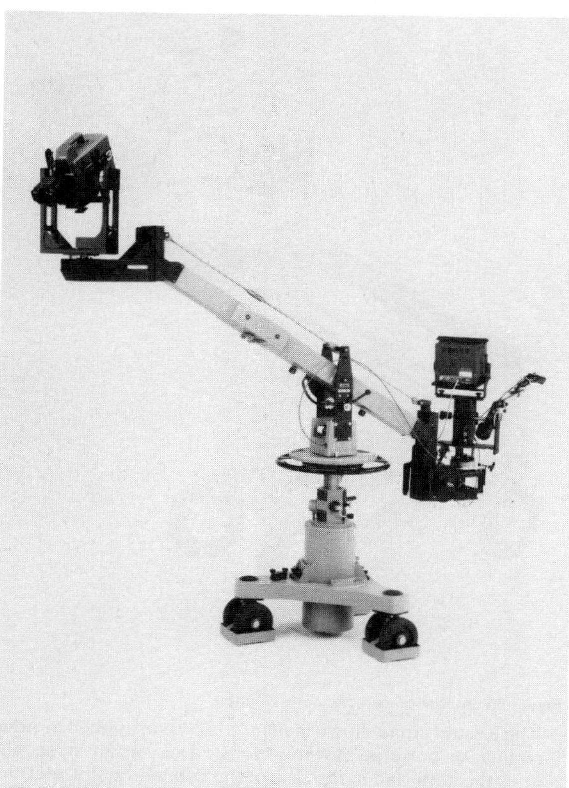

Figure 33.37 Merlin crane (Vinten)

The camera is slung in a cradle at one end of a boom arm and the operator stands at the other shorter end of the boom with pan, tilt, elevation, zoom and focus controls extended to him as well as the viewfinder.

33.6 Pan and tilt heads

A pan and tilt head is a device that enables the cameraman to point his camera at a subject, compose the picture well and, if necessary, follow the subject when it moves. It is occasionally necessary to simplify move left or right to see a greater field of view, but usually this movement is associated with following something.

The human eye is unable to pan smoothly unless it is following an object that is moving, and therefore a panning camera movement without something to follow seems unnatural. It is necessary to make camera movements as imperceptible to the viewer as possible.

Jerky movements, inaccurate movements and unnecessary movement are all extremely disturbing to the viewer, and professional equipment is designed to help the camera operator overcome these undesirable effects.

A tripod or pedestal, described in *Section 33.5*, carries the weight of the camera, but one of the prime functions of the pan and tilt head is to make the movement of the camera, in tilt, almost weightless.

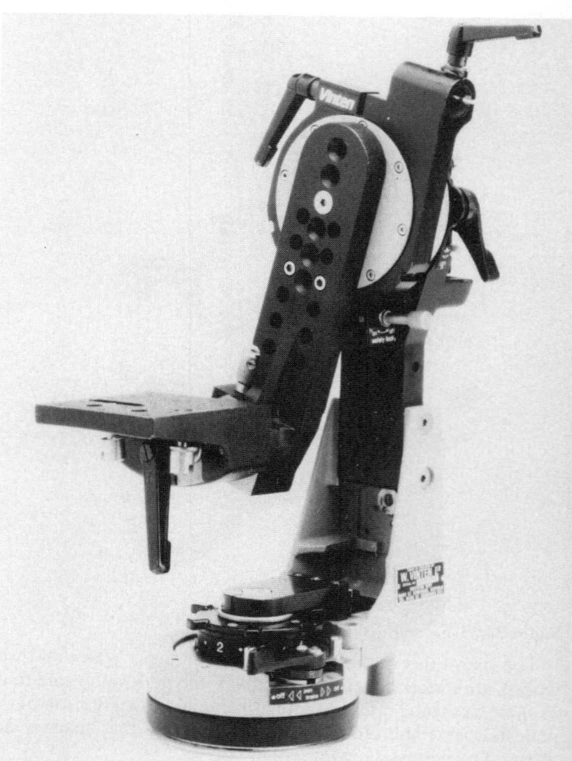

Figure 33.38 Swan Mk II post head (Vinten)

Various methods of achieving this weightless balance or counterbalance are used by different manufacturers. Since both pan and tilt are controlled by the single or double pan bars, the 'feel' of the movement in both axes has to be as near identical as possible. It is important to understand that balance and counterbalance do not impart 'feel' or 'drag' to the pan and tilt head. The solution to this problem is described in section *33.6.5*.

33.6.1 Post head

There are four basic balance methods used in pan and tilt heads, and the simplest of these is the post head. It consists of two identical swivels: one is under, or near, the centre of the horizontal camera; the other is mounted on a post attached to the previous swivel.

A correctly set up camera and lens can be balanced through its own centre of gravity, and therefore will remain where it is pointed until such time as the camera operator moves it to another position. As the two swivels are identical, the freedom of movement in both axes is the same.

Although the apparatus appears somewhat ungainly, the simplicity of the method of balancing has a lot to commend it, and post heads have earned high reputations and have dedicated users.

There are two disadvantages with post heads. The first is that, in order to carry heavy loads, they need to be massive and are only therefore practical for lightweight cameras. The second disadvantage is that access to internal circuitry, or mechanism of the camera, is restricted on one side by the post itself.

Some camera operators would say that, unless the post head is designed with the horizontal immediately beneath the centre of the lens axis, panning is an unnatural movement. Although this would appear to be so, tests have shown that, for most normal movement, being off axis by up to 15 cm is virtually undetectable.

All these disadvantages can be overcome by the second type of pan and tilt head, which is often referred to as a *gear* head and is one of the most popular mountings in the film industry.

33.6.2 Geared heads

A gear head consists of a quadrant of a circle. The camera is arranged to have its centre of gravity placed at the centre of that circle.

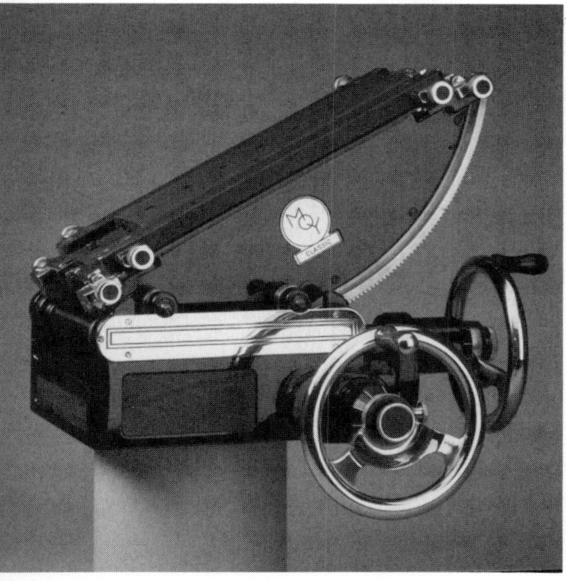

Figure 33.39 Moy classic geared (Ernest Moy Ltd)

This method achieves perfect balance, and movement is restricted only by the friction of the mechanism and the practical size of the quadrant. This type of head is particularly suited when the camera centre of balance alters significantly, as can be the case with film moving in a camera magazine, and when manipulating very heavy 35 mm or 70 mm film cameras.

The control of the camera movement using the gear wheels is a technique that needs practice, but the wide adoption for film cameras is an endorsement for the method. There are very few applications for this type of mounting in television operations although there is some suggestion that the move to high definition television may see the geared head finding some application in television production.

33.6.3 Cam heads

A gear head keeps the centre of gravity of the camera in the same position throughout the tilt range. Perfect balance can also be achieved by the use of a developed cam which allows greater tilt by letting the centre of gravity of the load move forwards and backwards without moving up and down.

The camera is, in effect, pivoting about its centre of gravity, and although its mass has to be shifted fore and aft a few centimetres, this requires very little effort. As with the two previous heads, the camera will remain at any tilt angle, and will only move to another one by the action of the camera operator.

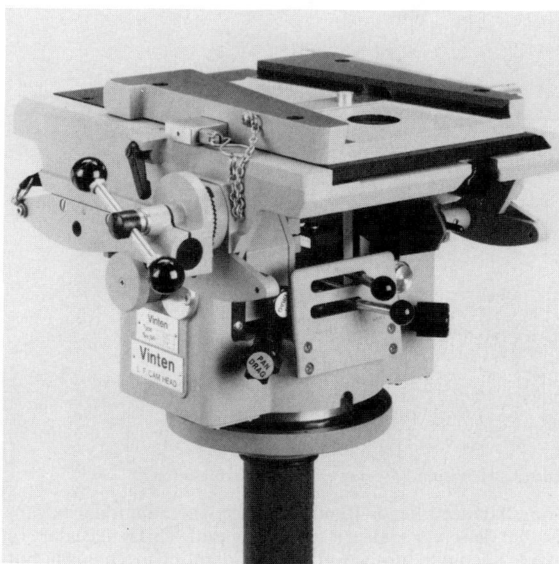

Figure 33.40 Mk 7A cam head (Vinten)

The disadvantage of this method of camera mounting is that, if different loads and different cameras are used, new cams have to be fitted for a normal studio operation using standard cameras and lenses. This is not a major disadvantage, and changing cams to accommodate prompters, etc., is made easier on the more modern cam heads which have external cams.

The preceding pan and tilt types were developed for film and TV cameras when most of these devices with their lenses weighed 13–18 kg (30–40 lb) or more.

33.6.4 Spring balanced pan and tilt heads

The post head became more practical when lightweight (hand-held) cameras came along, but it also became possible to use spring balance heads. Improvements in the use of springs have made it possible to design this type of head for up to 40 kg of load.

The key benefit of the spring balanced head is the ease of adjusting the balance when the camera load is changed. This requirement becomes more common in the field when interchange of, say, eyepieces and viewfinders may occur, or when batteries or recorders are added.

In this type of pan and tilt head, the weight of the camera is counterbalanced by some form of spring mechanism which may be either partially or fully variable. The amount of spring counterbalance needed depends on both the overall weight of the camera, and the height of its centre of gravity. Hence significant effect is felt when adding, for example, a viewfinder on top of the camera.

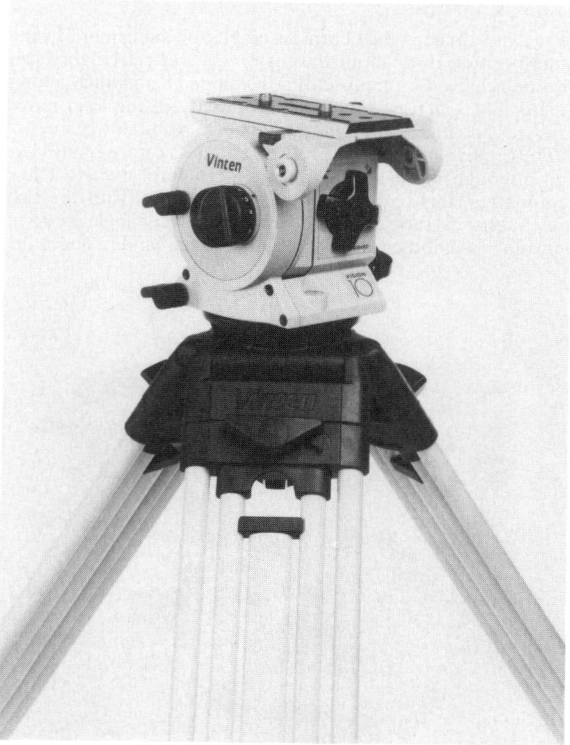

Figure 33.41 Vision 10 spring balanced head (Vinten)

Spring balanced heads frequently have the additional advantage that their mechanism is totally encased. This is necessary to provide adequate protection for the operating mechanism, but also means that they are particularly suited for operating in adverse climatic conditions on location.

They are also often compact and light, making them very easy to transport to and from location.

33.6.5 Drag mechanisms

As was mentioned in section *33.6*, the balance achieved by the pan and tilt head does not introduce drag to provide the right 'feel' for the camera, and indeed should not. However, particularly as cameras get lighter with the introduction of ccd technology, the inertia of the camera gets very small, potentially leading to shake and jerky movement as the cameraman starts or stops a movement.

To overcome this, it is usual to introduce artificial drag or inertia, which makes the job of the cameraman much easier when making smooth movements with the camera.

The lowest cost way of achieving this is to use a simple friction plate and a brake that can be applied or released depending on the speed of movement needed. The problem with this type of mechanism, particularly with lighter cameras, is that sticking can occur before the friction plate starts to move against the brake. This can lead to grabbing and hence jerking of the movement at start and stop.

A better method is to use fluid damping wherein a fluid is forced between friction plates, the pressure between which can be varied to increase or decrease the amount of drag. Precision construction is required for this type of mechanism, and hence it is more costly than non-lubricated friction, but the performance achieved is considerably better.

This type of fluid damping has benefits over simple hydraulic systems in that rapid moves cause hydrodynamic effects, enabling, for example, a whip pan to be performed.

33.7 System stability

In summary, the key requirement of the camera mounting is that it can provide a smooth movement of the camera as might be seen through the eye of an observer.

This quality of movement relies on a very stable platform, be it a tripod, a pedestal or a crane, and a balanced and controlled action through the mounting and the pan and tilt head.

While at the end of the day the mounting frequently costs far less than the camera on it, no cameraman should forget that his ability to shoot quality pictures relies more often on the quality of his mounting than anything else.

J Summers
formerly Lighting Director, BBC

34a Studio Lighting and Control – Lighting

34.1 The purpose of lighting

Artificial lighting is needed for four main purposes:

- To ensure the presence of sufficient light to obtain an intelligible and technically satisfactory reproduction of the original scene. The level required will depend on the sensitivity of the system used, and will be affected by the stop number of the lens, the presence of colour filters or diffusers and the use of additional equipment such as an artist prompting device.
- To keep the scene contrast within the limits acceptable to the TV system. Colour cameras currently in use handle a contrast range of 30:1, but a sunlit exterior scene may have an inherent contrast range of 500:1. Additional light is needed in the shadow areas to reduce this to an acceptable level.
- To give shape to objects and depth to a picture. A television image is two-dimensional. By using light and shade the illusion of the missing third dimension can be created.
- To interpret the emotional content of a scene. A lighting director has the opportunity to use creative skill to enhance the scene with a chosen style of lighting. It is possible to heighten the pictorial interest, or increase the emotional response of the viewer, by the subtle use of light and shadow.

34.2 Lighting sources

Light sources used to produce a photographic image fall into one of two categories: *hard* or *soft*. Hard sources produce a hard shadow; soft sources a soft one. The sun is an example of a hard source, and skylight on an overcast day an illustration of a soft. Soft sources are large in area relative to the distance they are from the scene. On a dull day the whole hemisphere of the sky is the source of light; as there is no one direction of light, no object can cast a shadow. A hard source is small in area relative to its distance from the scene. If an object is placed in the path of unobscured sunlight, it casts a sharply defined shadow.

34.2.1 Soft sources

A commercially produced soft source should have as large an area as is practicable and transmit an even light over its whole surface. Those that are smaller will cast soft, discernable shadows and are designed to be used in situations where space is limited. The lamps in soft sources are usually small, and therefore have dimpled reflectors situated behind them to scatter the light. In addition they are usually masked in some way to prevent any direct shadow-forming light leaving the luminaire.

Many matt, white reflecting materials can be used to provide an efficient soft source. If they are illuminated by one or more hard sources the reflected light will be soft. Cyclorama cloths, out of vision walls of a set, polystyrene sheets and card are a few examples of devices used in practice to obtain soft light. Large area soft sources become virtually shadowless and cannot be easily contained within limits. The principal use of soft light is to control the depth of shadow created by a hard light, whilst minimizing additional, conflicting shadows from itself.

34.2.2 Hard sources

34.2.2.1 Soft edged fresnel spotlight

This is a light source fitted with a lens, designed to focus light into an adjustable beam angle. When the lamp and reflector are moved away from the lens (*spotting*), the intensity of the beam is increased but coverage is reduced. If the luminaire is set for maximum beam angle it is said to be *fully flood*. It is fitted with *barn doors* (shadow casting blades) whose purpose is to control the shape of the light beam. The ability of the barn doors to control the beam reduces as the lamp is spotted. Colour filters, diffusers or *jellies* (light attenuators) may be fitted, as well as a *flag*. A flag is a blade similar to a barn door, but clamped to the lamp housing by a ball and socket arm. It is used as an adjustable shield to prevent stray light entering a camera lens and causing flare.

Spotlights are used to shape and texture objects by creating shade and shadow.

34.2.2.2 Profile spot

In optical design, a profile spot is similar to a photographic projector, but it is intended to project sharp images over shorter distances. These images can be produced from specially designed *gobos* in sheet steel, or from hand-made ones in aluminium foil.

Effects projectors give moving patterns, obtained from a continuous photographic image deposited on a plate glass disc, rotated by an electric motor. Another type rotates a circular metal gobo. If profile spots are suspended, care has to be taken with studio ventilation so that the effect does not sway.

34.2.2.3 Follow spot

A follow spot is a narrow angle spotlight that can be used over long throws, and panned to follow moving artists. It can be focused to give very sharp shadows, and cut off. Provision is made for beam shaping shutters, iris diaphragms with blackout discs, and colour frames.

34.2.3 Other sources

Some studios use dual source luminaires, which can be transformed fairly quickly from hard sources to soft ones, or vice versa. They can be dual wattage, which enables the light output to be altered whilst maintaining a constant colour temperature. They are most usefully employed as a saturated rig, where suspended luminaires are permanently installed over the whole studio area at a high density. Although they have disadvantages, such as heavy weight and compromised softlight size, they also have the advantages of speedy rigging, and flexibility in use.

A number of luminaires have been specially designed to light cycloramas, and they can be divided into two groups:

- *Top cyclorama units*. These can be floodlights used to light the top portion of the cyclorama as evenly as possible. An alternative design, employing a multi-curve reflector in each of four compartments, is capable of giving even illumination from top to bottom of the cyclorama. Each compartment is separately controlled so that, by fitting a different colour filter in each, many hues can be synthesized.
- *Ground row*. These floor units are frequently placed close to the bottom of the cyclorama, deliberately using the fall off effect of the light to create an artificial horizon. A unit may consist of four compartments in line, individually wired to facilitate colour mixing. One design of ground row unit overcomes the problem of the curve of the cyclorama by hingeing the rear of the compartments so that the unit can be adjusted to fit the bend.

There is a group of luminaires available that can be generally described as *disco lighting*. The range is large, and an exhaustive list could not be given as they are in an ever-changing market that responds to fashion. A common factor is that they are designed to be used in conjunction with smoke effect. The light sources themselves have a very low thermal inertia, so that they respond almost instantaneously to changes of input voltage.

Narrow angle or parallel beam sources are used to illuminate the smoke, creating beams of coloured light that can be made to pulse to music, or sequence in any predetermined manner. Some units are available with remotely controlled colour changing, and remote pan and tilt mechanisms. A large group of kinetic effects employs parallel beam sources mounted on a structure that can be rotated in either the horizontal or the vertical plane, or in both simultaneously, using variable speed motors.

There are units designed to fire a number of photographic flashguns in sequence, while others will give an effect similar to twinkling stars, by randomly firing a string of small discharge tubes.

Automated lighting systems are available that can remotely control all the operational functions of a luminaire. There is a spotlight unit using a metal halide arc, and a wash version with a tungsten source. Having its own microprocessor, each unit can store up to 1000 cues, controlling light hue, saturation, intensity, beam angle, beam edge and gobo patterns, together with pan and tilt movement. They can be linked to a control console to permit individual manual operation or pre-programmed cues, combining up to 1000 units.

Automated lighting has a clear application in disco and rock concerts, but because they are so versatile, they are a useful tool in any situation where remote control is needed.

34.2.4 Lighting console

Television is a photographic medium that gives instant pictures. To utilize this fact fully it is important to be able to make instant adjustments to the brightness of any luminaire, and electronically memorize the levels for future recall. This is the function of a *lighting console*. The combination of picture monitor and lighting console considerably extends the artistic choice of the lighting director, as any combination of luminaires may be selected at various dimmer levels, and the effect instantly seen. Changes to occur in vision may be set up, as either an instant or punch change or a fade across from one state to another. More subtle lighting changes, meant to go unnoticed by the viewer, may be made by using a group fader or by using manual adjustment.

34.3 Static portraiture

There are no rules in television lighting, only objectives. How they are achieved is up to each individual, and although custom and fashion may shape what is generally accepted, there is no reason for these to become a straightjacket worn by the lighting director. The following sections should be regarded as a guide rather than a rule book to the way that certain re-occurring situations can be dealt with.

34.3.1 Straight to camera

An arrangement of lights for a single sitting subject is shown in *Figure 34.1*. In this arrangement, **A** is the modelling, or

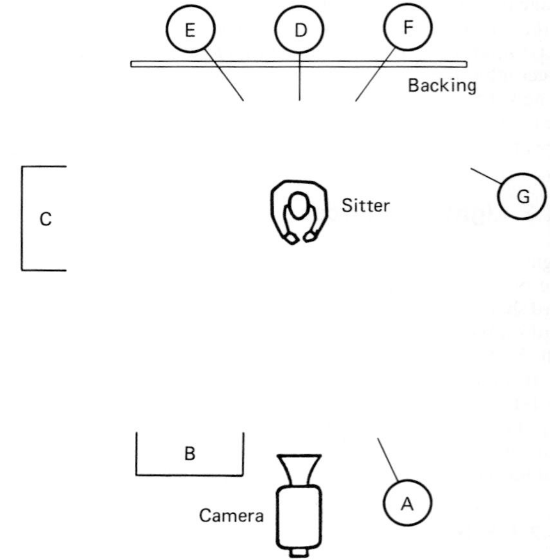

Figure 34.1 Lighting for a single static subject

keylight, and would normally be a fresnel spotlight. The light beam can be controlled by the barn doors to keep it off the backing. It can be placed either side of the camera. The shadow area of the face becomes greater as the keylight is raised in height, or the horizontal angle increased. The nose shadow should not be distractingly long, nor the eye sockets too dark. The position of the keylight is set to give roundness and depth to the face, probably within 0°–30° horizontal angle to the camera.

B is the fill light and will be a soft source. It is placed on the opposite side of the camera to the key, as close to it as possible, and set at the same height as the camera lens. It has a number of functions. It controls the depth of shadow created by the keylight on the front of the face, neck and clothes. It acts as an eye light, ensuring that if the sitter looks down, light will get into the eyes. If there are unflattering lines or wrinkles, this light will minimize them.

C is a soft source and is another fill light to control the shadow from the key on the side of the face. It is placed at eye height, and at 90° to the camera position, so that it does not cast a nose shadow.

D is a single backlight, a fresnel spot suspended over the top of the backing, directly behind the sitter. This gives specular highlights in the hair from the camera position, and adds depth by rimming the shoulders. The barn doors, or a flag, should be adjusted to keep any light off the camera lens. Sometimes studio space is limited, but the subject should not be placed too close to the backing, as this increases the angle of the backlight, and causes the background to be excessively sharp, and possibly distracting. A backlight of 30°–40° is ideal.

An alternative to single backlight is double backlight. Two spotlights (**E** and **F**) light the back of the sitters head at a horizontal angle of approximately 40° to the axis. Interesting specular highlights are now obtained on the sides of the face as well as the hair, defining the cheeks and jawline with more clarity. A highlight instead of a shadow is used to suggest the third dimension. Because backlight **E** lights the part of the face unlit by the keylight, the effect is stronger on this side of the face. Fill light **C** may not now be needed. Double backlight is often used on light entertainment productions, where the fuller effect is more suitable.

G is shown symbolically to light the backing. How the backing is actually lit will depend on its colour, luminance and texture. It was suggested earlier that the keylight is kept off the backing by the barn doors so that the backing can be lit separately. In this way, control of the backing may be obtained independently of the key. For maximum control of a lighting set-up, it is desirable to have only one function for each light source.

When all the luminaires have been set, they should be balanced, i.e. the final brightness of each should be determined by operating the lighting console. First the key is adjusted to obtain full exposure on the camera. Fill lights **B** and **C** must be used with discretion, as too high a level will destroy the modelling created by the keylight. The backlight and background levels chosen will be determined largely by the type of programme being made, and should be adjusted to suit its style. The picture monitor should be examined to see if there are signs of an unwanted colour shift due to excessive dimming of a light source. This is most likely to occur with fill lights, as very small quantities are needed. If necessary a smaller wattage luminaire should be used, or the light attenuated by means of a neutral density filter.

Consideration has been given to static portraiture involving one camera. Floor mounted lights can be used wherever practicable for this, for ease of adjustment. When more cameras are used, floor lights are no longer viable, as they impede camera movement, and there is more risk of lens flares. Suspended luminaires should be used, and because distances are increased to approximately four or five metres, higher wattage units must be employed.

34.3.2 Two-way interview

Lighting for a two-way interview is illustrated in *Figure 34.2*. Cameras 1 and 3 are positioned to take close-ups of participants while camera 2 takes a two shot. The two spotlights **A** and **B** each have two functions: to backlight the nearest head, and key the other. The barn doors are set to keep the light off the backing. Soft sources **C**, **D** and **E** are fill, and the background is lit with spotlights **F**, **G**, **H** and **J**.

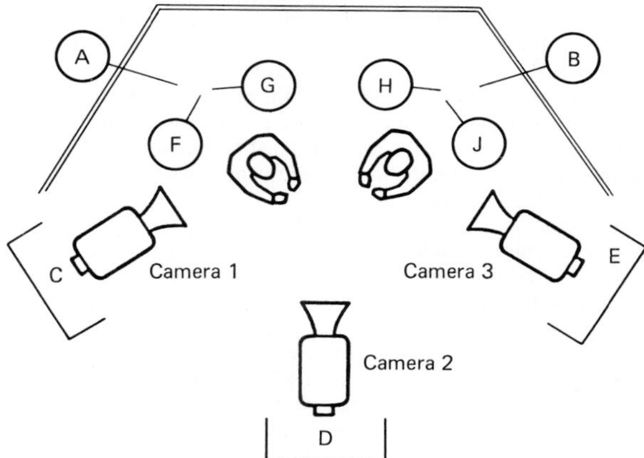

Figure 34.2 Basic lighting for a two-way interview

Although quick to rig, this lighting set-up has two disadvantages. The balanced level of **A** and **B** is adjusted to give keying levels on faces for correct exposure, and this may result in too much backlight on one or both participants. This problem can be reduced by using material called a 'jelly' in the bottom half of each of these spotlights, to attenuate the brightness of the backlight only. The other problem is that the positions of **A** and **B** are chosen for optimum backlight, and this may not give a satisfactory key angle. These difficulties are overcome by rigging separate backlights, the spotlights **K** and **L** shown in *Figure 34.3*.

These backlights are placed in the ideal position, directly in line with each cross camera lens, barn doored to light only the back of the participants. The keylights **A** and **B** may be moved to a more frontal position to give better modelling and smaller

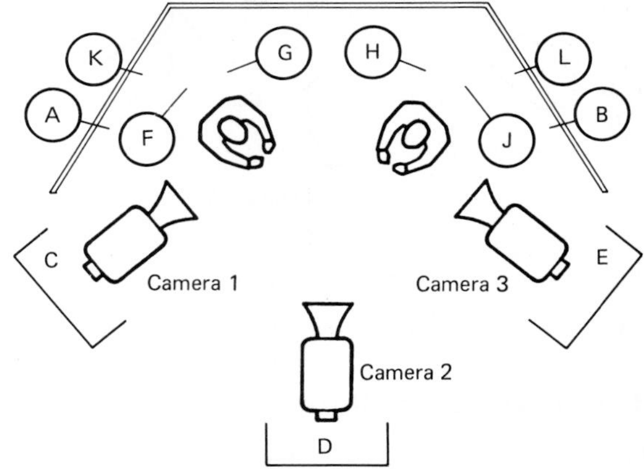

Figure 34.3 Improved lighting for a two-way interview with separate backlight

'nose shadow'. Their barn doors are set so that their light falls only on faces, and not on the backs of heads or backing. The set-up can be adapted for as many participants as necessary, so long as they are in two groups facing each other. The keylights are widened to cover each side and backlights added as necessary.

34.3.3 Three-way interview

The set-up for a three-way interview includes a central link person, who will introduce and close the programme on camera 2 (*Figure 34.4*). Camera 1 takes the camera right interviewee, camera 3 the left one. Cameras 1 and 3 will also take shots of the link person when turning to question either of the others. The difficulty is to obtain satisfactory lighting for this central position. If the plot shown in *Figure 34.3* is used, the result will be 'badger' lighting when seen on camera 2, with both sides of the face brightly lit by **A** and **B**, yet the eyes and the front of the face dark — a very unflattering picture. This position must be lit separately. The two keys **A** and **B** must be further barn doored so that they only light the faces of the interviewees, with no light falling on the link position.

Extreme care and accuracy of setting is needed for this operation, and it will help if the keys are placed more frontal to subjects, and brought slightly closer. These keylights are now so frontal that it is not advisable to use boom microphones for sound pickup, as boom shadows are very likely. A spotlight **M** is rigged directly over the central camera, to act as the link position keylight, barn doored closely so that it lights only this person. More accurate setting will be obtained using a lower powered unit closer to the subject. Double backlight (**N** and **P**) that is effective from all camera positions is useful.

Sufficient separation between the link position and the interviewees must be allowed in order to light this set-up satisfactorily.

Some quiz games take the same form as a three-way interview, but are on a larger scale. Instead of two interviewees, there are two teams, and a quiz master replaces the link person. The principles suggested can be adapted to lighting this situation.

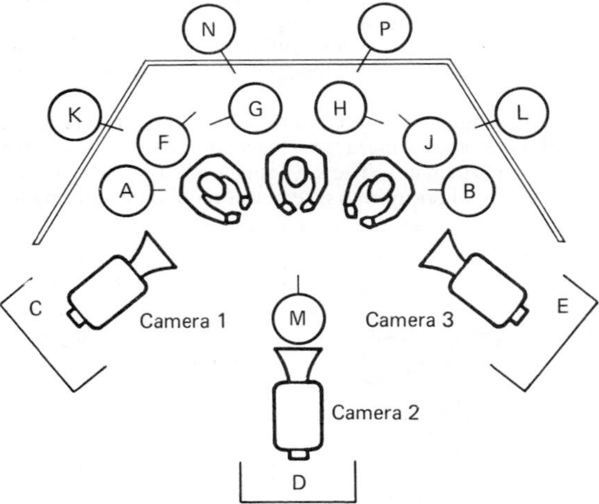

Figure 34.4 Lighting for a three-way interview or discussion

34.4 Moving portraiture

In circumstances where there is movement of both artists and cameras, as in a drama or a situation comedy production, the lighting requirements become more complex.

Many different treatments may be employed, but suggestions made here will continue to be based on applying portrait techniques. For simplicity the following diagrams do not show doors or windows. The effect of windows as light sources is dealt with later. It is assumed that a boom microphone will be used for sound pick-up, and therefore precautions must be taken to avoid boom shadows. All acting areas, known as *sets*, can be categorized as single sided, two sided or three sided.

34.4.1 Single sided sets

The single sided set is the simplest and most limited in production values. It will be used when economy of space or cost is desirable. The backing shape is shown diagramatically in *Figure 34.5*. It may be more complex, but essentially it consists of one wall of a building. It is apparent that if either the camera or the artist is given much movement, there is insufficient backing to prevent 'shoot off' as shown from camera position 'b'. If another artist is brought in, shots can only be taken in profile from the front, on a central axis, for the same reason.

Because the camera is restricted to frontal shooting, a keylight can be placed at a suitable angle to it to portray the desired dramatic mood. If it is positioned very frontally to the camera, it will give a high-key picture, and as it is moved to the side, more shadow will appear, and the effect will be low-key. Softlight is placed in the same relative positions as shown in *Figure 34.1* to control the depth of keylight shadow. Backlight is optional. If there is no window or other apparent light source behind the artist, the result may appear odd if backlight is used. Separate background lights should not be used unless very localized, as the artist may move close to the backing and distracting double shadows appear. It is better to use the keylight for this purpose, and plan in advance with the scenic designer a suitable tone for the backing.

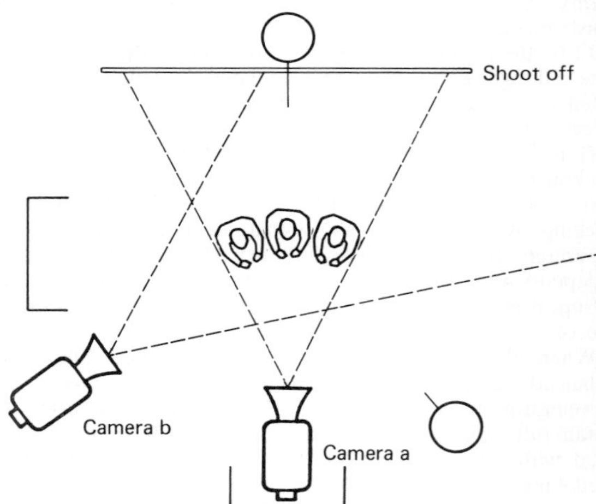

Figure 34.5 A single sided set with one moving camera

34.4.2 Two sided or L-shaped sets

These are sets consisting of two adjacent walls forming a corner. They give greater production flexibility, as now there is backing to enable cameras 1 and 2 (*Figure 34.6*) to take frontal shots of two or more artists facing each other. This is called *cross shooting*. A third camera takes a two shot.

Movement of artists is severely restricted. Action at the sides of the set will result in either shots with bad eye lines, or shoot off problems. If the action moves upstage, deeper into the corner, the cameras become cramped by the walls.

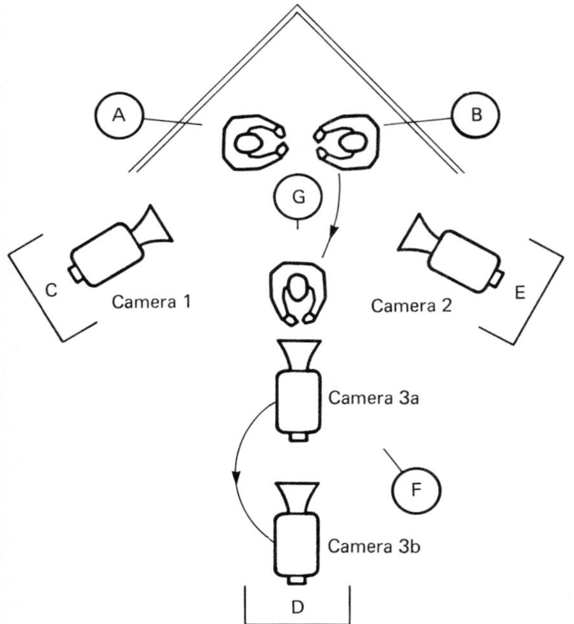

Figure 34.6 A two sided or L-shaped set

The lighting plot is based on that used for a static two way (shown in *Figure 34.2*) but separate background lights have not been used. Lights **A** and **B**, fitted with bottom half jellies, each have three functions; keylights, backlights and background lights. It was advised earlier to have only one function to each luminaire, but here lack of space forces otherwise. If there is no artist movement these functions can be separated by using additional luminaires. Soft fill lights **C**, **D** and **E** are placed behind the cameras, and rigged as low as practicable. Keylights have to be steep in these small sets; keeping the frontal softlight low in elevation helps to get light into the eyes. Bounce cloths, large white reflectors approximately 3.5 m long and 2.5 m wide, may be used as an alternative soft fill, but care is needed to position them so that they do not impede camera or boom movement.

There is another position that can be used for action, and that is downstage, shot on camera 3 in position 'b'. In effect this can be regarded as a single sided set, and lit as shown in *Figure 34.5*, using luminaires **C**, **D**, **F** and **G**. The keylight **F** must be barn doored so that it only lights this downstage area, or boom shadows may occur in the cross shooting part. Alternatively **F** can be controlled by using the lighting console, fading it up as the action approaches this position, and taking it out afterwards if the action returns upstage.

34.4.3 Three sided set

Three walled sets offer maximum production flexibility, allowing the possibility of considerable artist and camera movement, and shots with good eyelines. The lighting plot must be equally flexible to meet these production requirements. The principle suggested is to rig keylights for all areas appropriate for the planned action, then at any moment use only those immediately necessary. By using the lighting and console in this way, good modelling should result and boom shadows avoided. A high degree of skill is needed for this way of working, but the effort is rewarded by the quality of the final pictures.

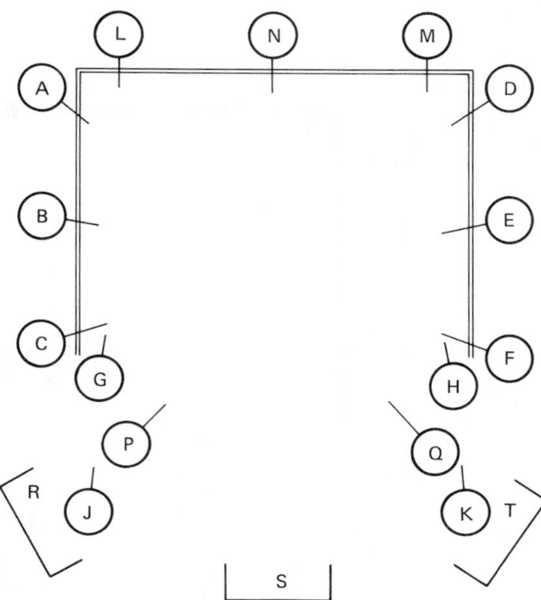

Figure 34.7 Lighting for a three sided set

The set-up is shown in *Figure 34.7*. **A**, **B** and **C** function as keylights, backlights and background lights, each covering the same area but from different angles. **D**, **E** and **F** perform the same functions from the other side, and all six luminaires have bottom half jellies fitted. At no time is more than one from each side in use, but they can be used in any combination. Choice depends on the mood to be created and the camera positions at the time. These six should offer suitable alternative keylights for the central area of the set.

When action approaches the side walls, these spotlights become unacceptably steep, and keylights **G** and **H** cover the upstage part, while **J** and **K** light the downstage area. These spotlights must light only the side portions, or doubling with the cross lights will occur. These side areas have **L** and **M** as backlights, and they should be fitted with bottom half jellies. **P** and **Q** are alternative frontal keylights, one of which should be used when the action in the middle is played out to a central camera. Backlight **N** covers this position. Softlights **R**, **S** and **T** are rigged at the front of the set, keeping their elevation low.

This plot will work well for productions employing continuous recording, and using several cameras. Care must be constantly exercised to ensure that the modelling created by a keylight is not destroyed by another, left up inadvertently. The aim should be to have only one keylight on the front of an artist at any time. The cross keys present a problem as they are also background lights, and it may become necessary on occasion to split these two functions, in order to fade out the key element. Where this is so, extra set lights are rigged.

34.5 Creative lighting

So far we have considered lighting only to achieve portraiture; no attempt has been made to create a feeling of realism in the pictures. Lighting may be modified to take this into account. Some productions will call for more emphasis on glamorous portrayal than on reality, while the converse may hold for others. Additional luminaires should be added to the plot to simulate the effect from light sources, such as windows, or artificial lights that appear in the set.

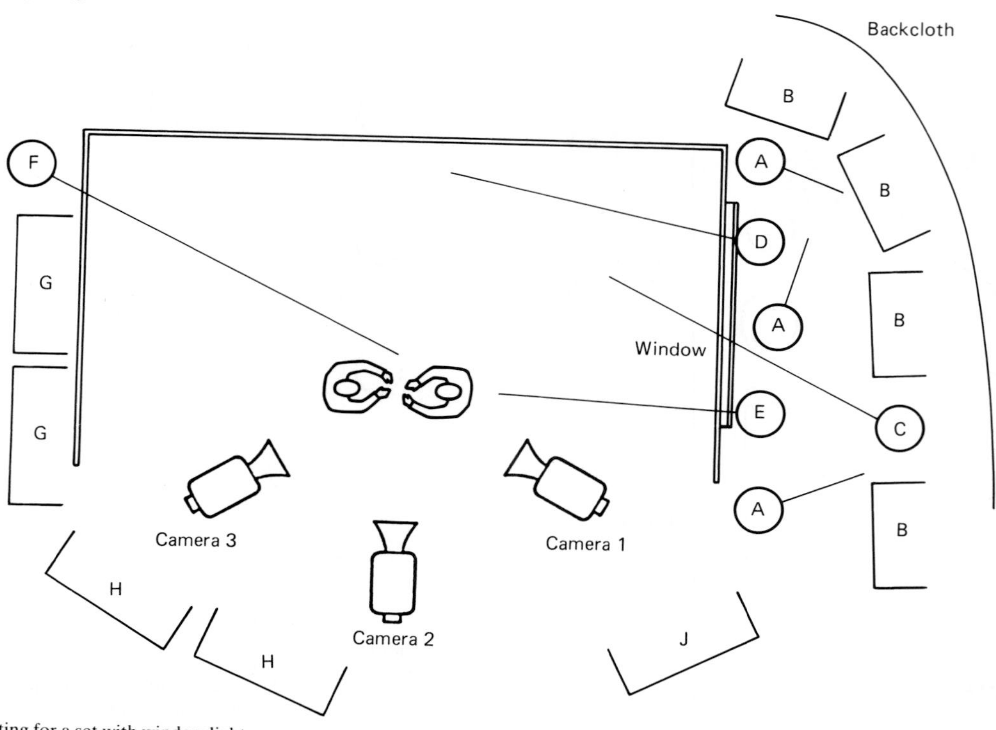

Figure 34.8 Lighting for a set with window light

34.5.1 Simulated daylight

Sets with windows, seen in daylight, require three groups of lighting: backcloth lighting, set lighting and artist lighting.

Backcloth lighting should be to a high brightness level in order to simulate an exterior appearance. It is most easily achieved using hard sources (fresnel spots **A** in *Figure 34.8*), but care must be taken to illuminate evenly. Space between the back of the set and the backcloth is often very limited, making these lights steep, and so ground-row units, **B**, can be added to supplement the level at the bottom of the backing.

Set lighting requires a powerful fresnel spot (**C**) positioned to simulate direct sunlight through the window. A pattern of light, modulated by the glazing bar shadows, will fall on furniture, walls and floor near the window, giving a realistic, sunny feel to the scene. By placing the light source downstage of the window, part of the back wall will receive some of the effect. Keeping it low in elevation will enable the effect to penetrate deeper into the set. If the window is small, it may be necessary to supplement this light on the back wall with a further fresnel spot (**D**), but care must be taken to avoid double shadows.

Artist lighting is achieved with a keylight (**E**) slightly downstage of the window, lighting the action and the opposite wall to a height of 1.75 m. It should not light the back wall or double shadowing will occur. A bottom half jelly should be fitted if the set is small. The key from the opposite direction (counter-key) should be well upstage (**F**), and if necessary a bottom half jelly fitted. It is barn doored off the back wall, and lights the action, but as little as possible should fall on the window wall. Lens flare can appear from camera 1 position, caused by spotlight **F**, so it may be necessary to use a flag on this luminaire. By counter keying from an upstage position, artists with their back to the window seen from camera 3 position will appear darker than those facing the window, seen on camera 1.

If more realism is required, soft sources **C** may be used instead of keylight **F**. Soft frontal fill **H** and **J** complete the plot. More soft light is rigged on the side farthest from the window (**H**), to give greater control of the shadow area created by upstage key **F**.

34.5.2 Artificial light sources

If a set has no specific light source visible, but is assumed to be lit conventionally by overhead and wall lighting, suitable pictures will be obtained by selecting keylights appropriate for camera and artist positions. An interior evening feeling will probably best be obtained by increasing the soft frontal light level, and so achieving a softer contrast in the pictures. Window walls can be brighter than shown in daylight scenes. If suitable to the design of the set, lit wall lights will reinforce the evening mood.

Sometimes a script will call for specific light sources in a scene, such as candle light, or firelight. Where this is so, the problems of keeping the light sources giving the effect out of camera shot must be solved at the planning stage. The effect of candle light can be simulated effectively by using a 150 W projector bulb, suitably mounted and protected, close to the candle, but hidden from camera view. A realistic firelight flicker can be obtained by feeding suitably placed floor spotlights to a sound-to-light modulator, a unit more often used on pop music programmes. Excellent pictures can result from using these effects, but all shots must be planned in detail for successful results.

Television lighting should, by its chosen style, always complement the production and create imperceptibly the appropriate mood for the viewer. The production style should be discussed and fully understood by all members of the production team, at the initial planning stage. Words are often inadequate to describe visual concepts, and books of paintings or photographs can often assist the exchange of ideas.

The suggested treatments might range through documentary realism, theatrical (or 'heightened') realism, a delicate high key ethereal approach, or any shade or combination of these. When a coordinated style has been agreed, the lighting director is able to design a lighting plot, and balance lighting during rehearsals, constantly guided by the mental images conceived at the planning stage.

34.5.3 Recent techniques

By definition, creative lighting cannot be taught, but it is possible by considering the latest techniques to develop a new idea or concept to solve a particular lighting problem. Some practitioners have abandoned the use of hard key portrait lighting, believing it to be too unreal. Except in areas close to windows admitting direct sunlight, rooms are softly lit by light reflected from walls. Some lighting directors have developed ways of reproducing this effect, obtaining modelling in faces solely by the use of soft light. In practice, soft light can be obtained in a number of ways:

- Soft sources approximately 1.25 m square, mounted on floor stands or rigged overhead, can be used either singly or in a group, to produce a high intensity soft source. Floor stands are best used on large sets, when the studio is working in the rehearse/record mode (ie breaking the production down to a series of short recordings by rehearsing part of a scene then immediately recording it). Because the production is broken down into a series of short takes, the floor stands should be less of an obstruction to camera movement.

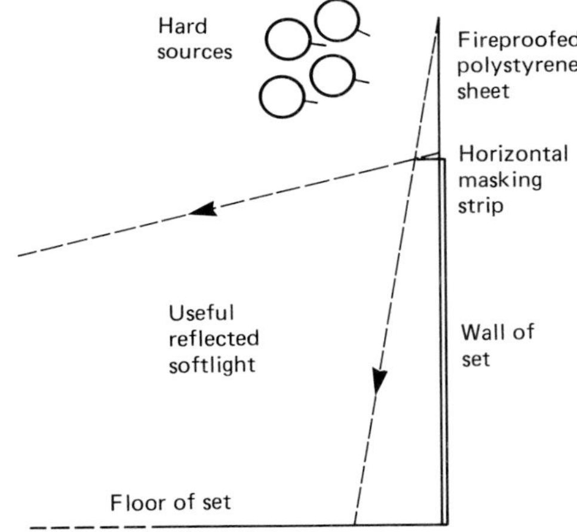

Figure 34.9 Reflected or 'bounce' lighting

- Sheets of fireproofed polystyrene 1.25 m by 2.5 m can be fixed horizontally to the top of set flattage. When these are lit with powerful hard sources, they reflect the light into the set, as shown in *Figure 34.9*. The light may be modified by fixing a reflecting surface to the polystyrene. If this has a metallic finish that is dimpled or perforated, a higher intensity of soft light is obtained. Some form of masking is necessary at the bottom of the reflectors, to prevent light streaking down the wall of the set below.
- Set walls that are light toned can be used as reflectors while they are out of vision. Pieces of white card can be hidden in sets, and used as reflectors when lit from above. Smaller pieces of polystyrene, 1.25 m square, fixed to lightweight stands and lit from above can be moved around more quickly and quietly than a soft luminaire.
- On some occasions it is necessary to reproduce the effect of very powerful soft source over a large area. This can be achieved by using a large number of hard sources from the same direction. There are pitfalls in this technique, as multiple shadows will be visible if the floor or backing is plain and unbroken. It is most likely to be successful when portraying a woodland exterior, where the studio floor is covered with turf and leaves, and the background is trees and bushes. By covering the whole set with hardlights at high density from the same direction, the effect of diffused sunshine can be obtained.

The soft key method of lighting gives a totally different look to the picture, when compared with that obtained from using hard key techniques. Many viewers claimed at one time that they could always spot the studio pictures. With the use of soft keying far greater realism has been introduced, and for certain drama productions this has proved to be an essential element. There will however always be a place for the hard key type of picture, and both styles are to be found in modern television studio lighting.

J Kelleher MIEE
Formerly Chief Engineer, Dynamic Technology Ltd

34b Studio Lighting and Control – Control

34.6 Lighting control system

The basic control console provides the following functions in the control of studio lamps, both individually and in groups:

- it switches them on and off
- it controls their brightness

34.6.1 On/off control

Because large individual lamp powers are used, typically from 1 kW up to 10 kW, at mains voltage, the actual on/off control of this power is indirect. The on/off switch itself operates remotely, either as an electric relay-type contactor or as an electronic semiconductor device such as a thyristor, to control the input of actual power to the individual lamp.

34.6.2 Brightness control

For the same reason, the control of brightness is undertaken indirectly. In this case, a small potentiometer, or *fader*, is provided adjacent to the on/off switch on the control panel. This supplies a reference brightness level, which controls the variable impedance placed in series with the source of ac power and the studio lamp itself.

The variable impedance in series with the lamp is called a lamp *dimmer*. It may be a variable inductance or an electronic device. In the former case, the control of the variable element is by a servo-controlled electric motor. The potential from the fader provides the reference input for setting the required brightness. The siting of the dimmers (as well as any associated controlling ancillaries) are in another area known as the *dimmer room*.

However, the variable impedance used is more usually an electronic device, i.e. a semiconductor thyristor dimmer. Here the fader potential is applied directly to the dimmer itself as the input brightness controlling signal. The basic lighting control module for an individual lamp takes the form shown in *Figure 34.9*.

34.6.3 Group control

When the control of more than one lamp is required, it will be necessary to replicate the individual lamp control module shown in *Figure 34.9* as many times as there are lamps in the studio.

In a television studio, this will involve hundreds of lamps. It will still only provide brightness levels for one setting in the studio and for one camera viewpoint. The practical requirement is for many more settings. Therefore, the multiplicity of individual lamp control modules, forming a single-setting lamp

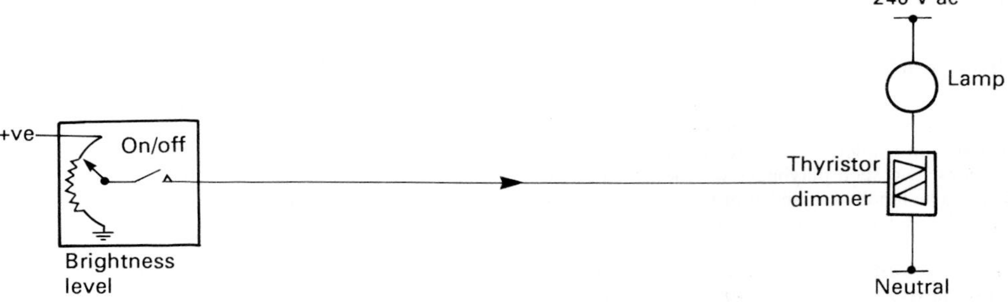

Figure 34.9 Basic individual lamp control module

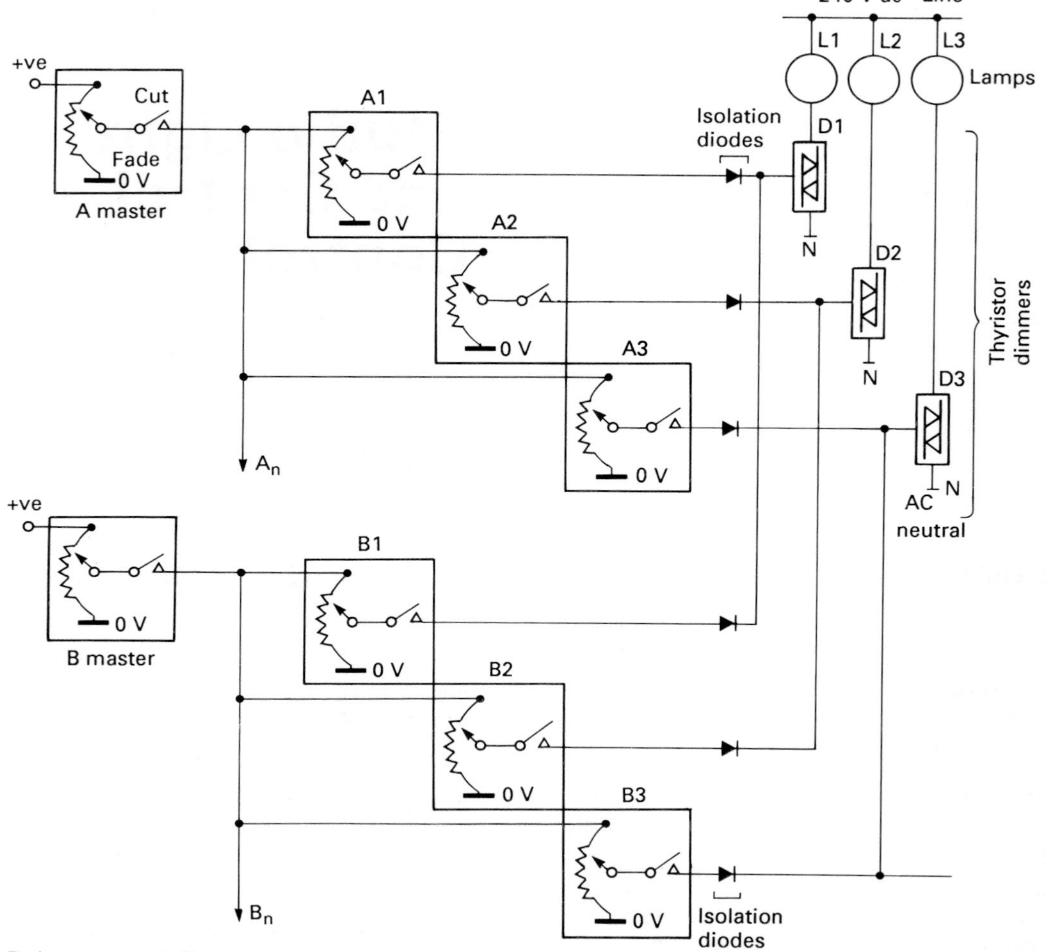

Figure 34.10 Basic group control of lamps

brightness panel, has to be duplicated, or even triplicated, to provide extra settings of lamp brightness for the various studio production lighting requirements.

A change between one group of settings and the next can be undertaken by the arrangement of *Figure 34.10*, which shows that the voltage supply to both sets of panel faders is by way of selector switches called *master cut* switches. By means of a separate group of *master faders*, of which two are shown, a more progressive control of the overall supply voltages is available to the fader panels controlling the studio lamps.

34.6.4 Playback panel

The arrangement of a master cut switch and fader controlling the overall voltage to a much larger group of individual lamp controlling switches and faders constitutes a *playback panel*.

Normally, at least two such playbacks will be provided, in which case they will be labelled A and B or just AB playbacks. A variant is for the A to become a master and B a preset playback.

The lighting plot, accordingly, is formed from the content of lamps within the playback. However, digital electronics has enabled an entirely new approach in the technique of lighting control systems.

34.6.5 Digital lighting control

Rapid development in lighting control techniques derived from four applications of digital technology:

- individual lamp control module
- operational store
- central processor or operational controller
- long term or archival store

34.7 Individual lamp control module

34.7.1 Selection of individual lamps

The lamp control module incorporates a numeric pad for selecting (or addressing) the individual lamp (*Figure 34.11*(a)). A three-digit number may be keyed, and is displayed (b). This indicates that the operation of the entire individual lamp module is devoted to the control of the condition of a lamp of that number. This includes the on/off switching of the lamp (c) and the adjustment of the brightness of the lamp (d), as well as monitoring (e) or measuring the existing brightness of the selected lamp.

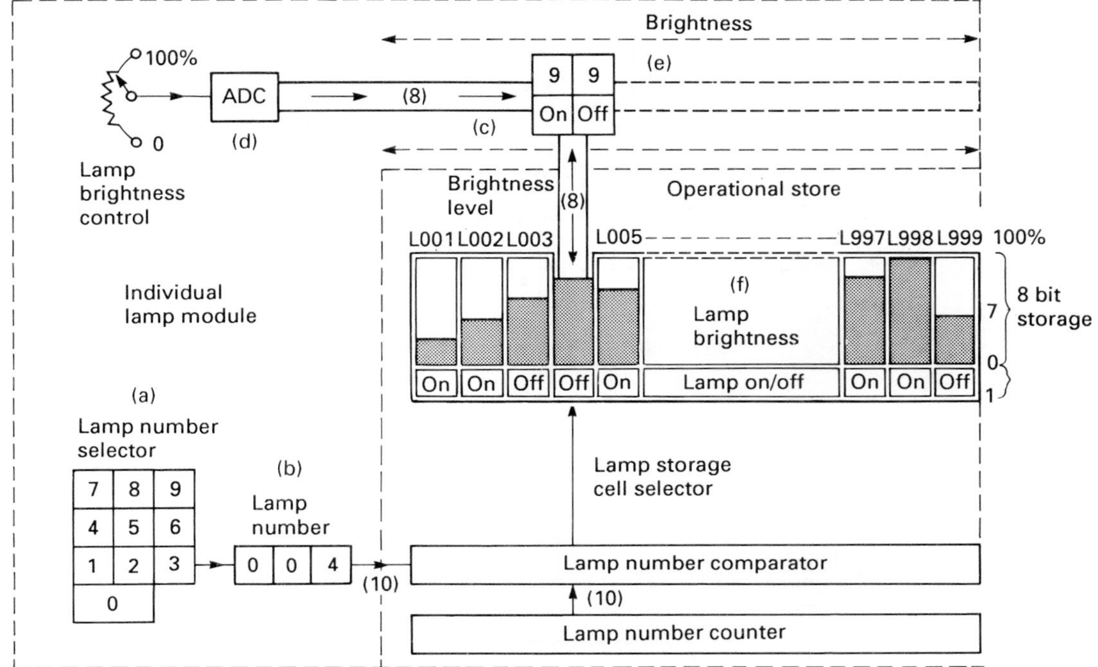

Figure 34.11 Individual lamp control module (analogue/digital)

The existing brightness of the selected lamp has been previously held in a lamp-condition holding register (f), which holds the brightness and on/off condition of every lamp in use in the studio. It is from such a register that the control information is drawn for the operation of the dimmers that control the condition of the studio lamps.

The individual lamp keypad address applies directly to a specific lamp within the context of this register, which is the basic module of the modern lighting control system.

34.7.2 Control of selected lamp brightness

In digital lighting control systems, it is usual to dispense with the actual quadrant fader/potentiometer combination and introduce a more digitally direct form of lamp brightness control. *Figure 34.12* shows a programmable up/down counter. The maximum count is usually 7 or 8 bits providing 128–256 brightness increments of level from *off* to full *on*. In practice, 1 per cent levels are commonly used. This allows the on/off control of the lamp to be combined into an 8 bit word associated with each selected lamp.

By programming the counter with an existing stored level of selected lamp brightness, the subsequent readjustment of the level becomes merely an adjust-of-count operation by means of the up/down impulses. These are generated by a suitably proportioned thumbwheel, which performs the function of the original fader/potentiometer in adjusting the lamp brightness. It is deeply embedded within the lighting control panel itself, so that a segment of the perimeter protrudes. In this way, the favoured association between the fader level and its progress across the quadrant-shaped profile is retained. This gives a helpful tactile indication of both the level and the increments of lamp brightness being introduced, without having to look down at it.

The individual lamp brightness parameter so created is held in the overall storage for the individual lamp conditions.

34.7.3 On/off bit control of lamps

The on/off bit is then used to inhibit/enable the 7 bit brightness signal to the actual DAC and thereby control the ultimate on/off condition of the studio lamp. The stored brightness level may be monitored separately by means of the unaffected brightness indicator.

34.8 Control of lamps in groups
34.8.1 Storage of lamp conditions

So far, consideration has centred on that aspect of lighting control requiring the selection and adjustment of individual lamps. As this proceeds, the result of each selection and adjustment is subsequently stored in a large capacity register capable of holding the resulting lamp data for all lamps finally selected. In practice, many of these registers will be involved as *holding* or *operational* stores for subsequent processing operations. It is for this reason that the various data-holding registers are termed *operational stores*.

34.8.2 Operational store

The store and its associated peripherals form the basic working module of the modern lighting control system. The complete module is shown in *Figure 34.13*. It consists of a register or digital store of the data for all selectable lamps.

The addressing is sequential and generated cyclically by the lamp address counter, as is conventional for digital storage systems. The number of address lines produced depends on the maximum number of lamps envisaged for the system. For example, up to 1000 lamps would require 10 address lines (2^{10} = 1024).

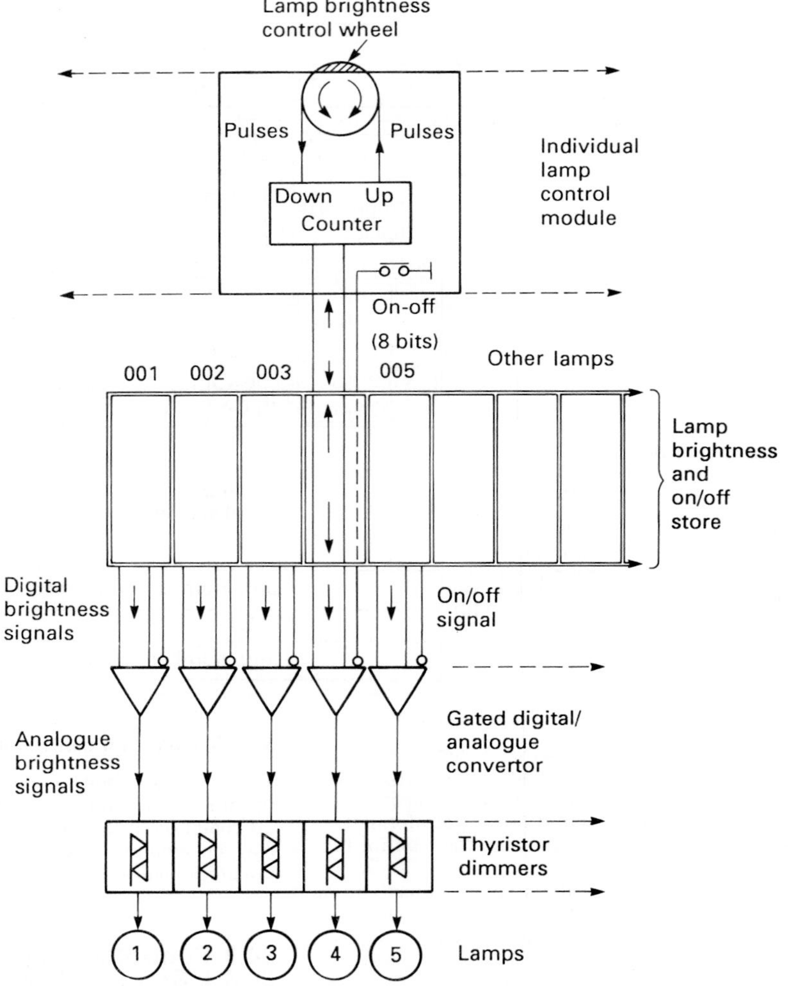

Figure 34.12 Individual lamp control module (digital/analogue)

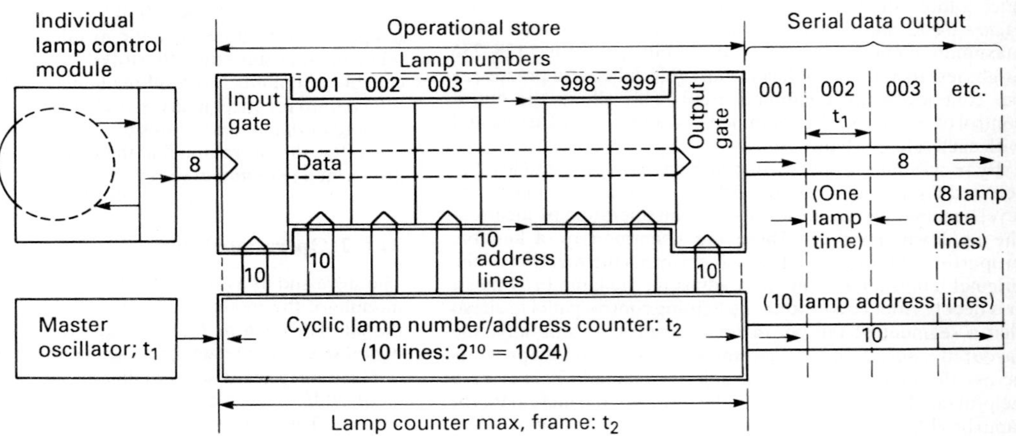

Figure 34.13 The operational store

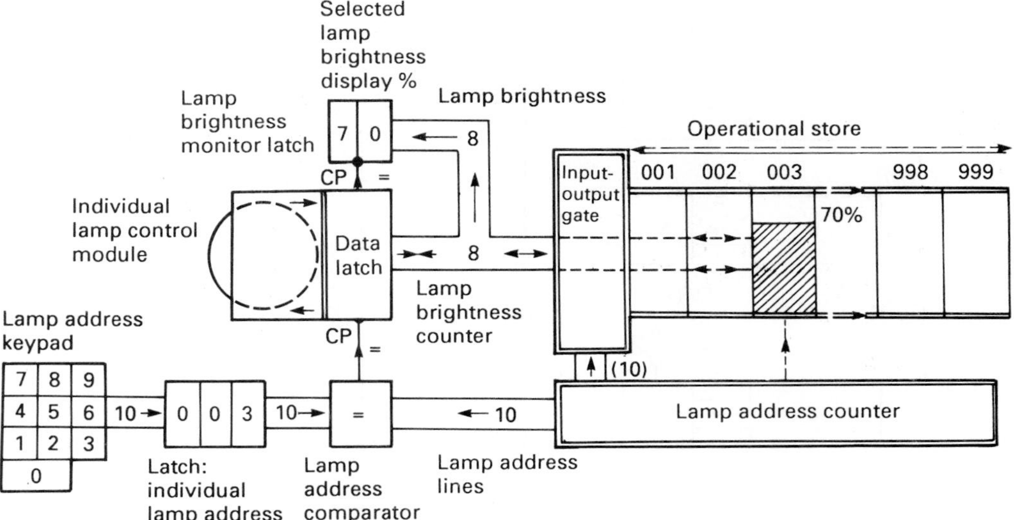

Figure 34.14 Operational store: individual lamp addressing

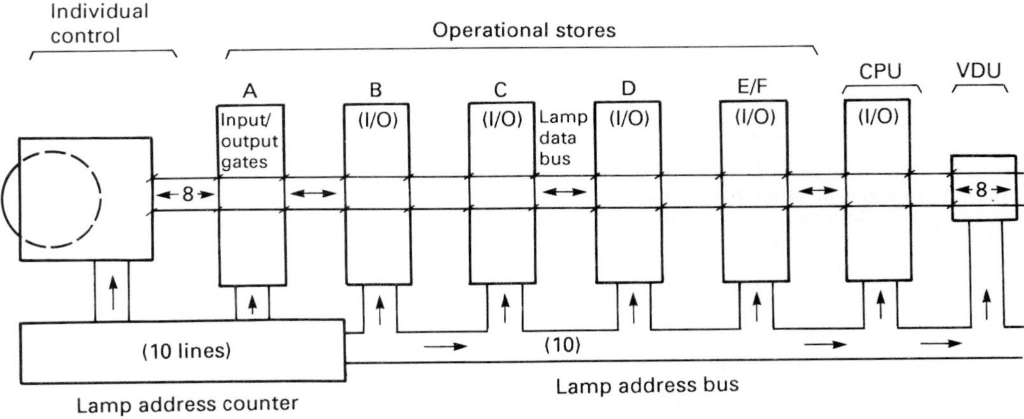

Figure 34.15 Operational stores: distribution of lamp brightness data bus stream

34.8.3 Individual lamp addressing

When the input to the store is from the individual lamp module, the keypad lamp number selected is compared with the sequence of the lamp address count until an equals pulse is produced when the two input numbers are the same. This pulse then allows the individual lamp data to be input and stored at that address using the input gate.

The output of lamp data from the store is in the form of a sequence of time-division periods of 8 bits data in parallel form, the length of each period being dependent on the master oscillator frequency feeding the counter.

Lamp data is normally processed through 8 way ribbon cable. The 8 bit parallel data are issued serially in time division multiplex, usually around 10–100 μs per lamp.

34.8.4 Individual lamp panel brightness monitor

The monitoring of the selected lamp brightness while being controlled from the fader at this panel, is achieved by using the same equals pulse to *latch* the particular lamp data (selected by the keypad from the lamp data stream) into the individual lamp brightness display. (See *Figure 34.14*).

34.8.5 Distribution of lamp brightness: data streams

The lamp data in the form of a time-division data bus stream is then distributed from the particular operational stores involved to other parts of the lighting control system, namely:

- to the individual lamp control module for control and monitoring
- to the more comprehensive information system provided by the visual display units (VDUs)
- to other operational stores
- to the central operational processor (CPU)

These distribution paths are illustrated in *Figure 34.15*.

34.8.6 Multiple operational stores

It is clear that the operational and technical functioning of modern lighting control systems depend on the ubiquitous

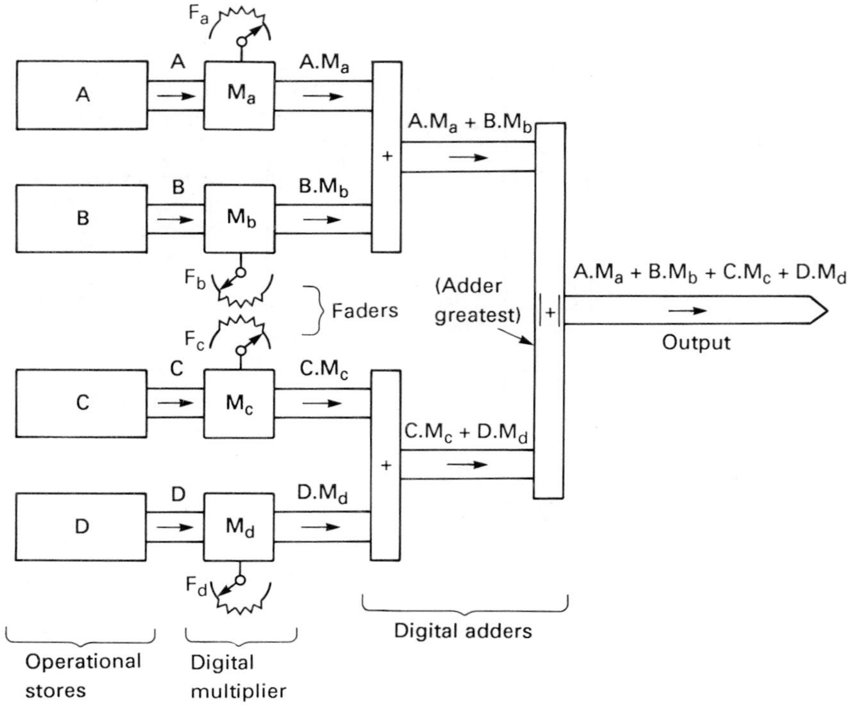

Figure 34.16 Operational stores in cross-fading

involvement of operational stores containing lamp brightness data from the various lamps involved. Sometimes the operational store may contain as little as one lamp; on other occasions it may include many hundreds of lamps at varying degrees of brightness, with some switched on and others switched off depending on the artistic and operational intent for that particular store.

34.8.7 Operational stores in cross-fading

During cross-fades, undertaken by the central processing unit, the lamp brightnesses in one store, here called the A store, are progressively reduced by multiplier M_a, while the lamp brightnesses in a second store, the B store, are progressively increased by multiplier M_b (see *Figure 34.16*). Simultaneously, another cross-fade would use a second pair of operational multipliers and stores associated in the same way, but labelled C and D.

34.8.8 Operational stores in lighting balancing and coloration

Another application using between four and six operational stores is illustrated in *Figure 34.17*. This meets the requirement encountered in stage and studio craft when the large sky-simulation curtain called the *cyclorama* is to be coloured as a general artistic background effect.

In *Figure 34.17*, the contents of the operational stores are:

A all the red lamps
B all the green lamps
C all the blue lamps
D all the white lamps
E particular modelling lamps for the artists
F other balance or setting lamps

By adjusting the output separately of A, B, C and D (via the central processor), the colouring of the backcloth can be made to match the hue of spectral colours, while the addition of white lights from D can influence the saturation or purity of colours separately, to produce precisely the required colour effect.

Independently, the adjustment of E and F achieve an artistic requirement for further lighting balance for the artists and settings within this coloured environment formed by the background.

34.8.9 Operational stores in summary control of the lighting

Another application involving two operational stores is when one, A, contains more lamps than the other, B (*Figure 34.18*).

The B store contains lamps that are to be switched off in A, by matching with similar lamp numbers in store A. This process is *minussing* (A-B). It enables lamps in A which match those in B to be switched off.

Alternatively, lamps in the B store can be added into the existing lamp content of the A store, a process called *plussing* (A+B).

34.8.10 Operational stores in large combinations

Operational stores are frequently encountered combined in pairs. In particular, the dual combinational-pair involving A/B and C/D forms the basis for automated and simultaneous cross-fades (see section *34.9.1*).

However, to meet the requirements for multiple plot inter-balancing, as when cycloramic coloration is involved, a more readily extendable grouping within an E/F combination is normally provided. These groupings are called *grouped sub-masters*.

34.8.11 Grouped sub-masters

Grouped sub-masters are shown displacing an original E/F pair combination in *Figure 34.19*. Normally, the sub-masters will be numbered — between 2 and 24, or even up to 30. Each will form an operational store, but with vestigial operational facilities for cut, manual fade and balance control, and store.

There will normally be no provision for automated fades and cross-fades.

34.9 Operational processing: the CPU

The function of the processor within the overall system is shown in *Figure 34.20*. The lighting control activities closely associated with the processor are:

- *progressive and transitional control:* fading and cross-fading,

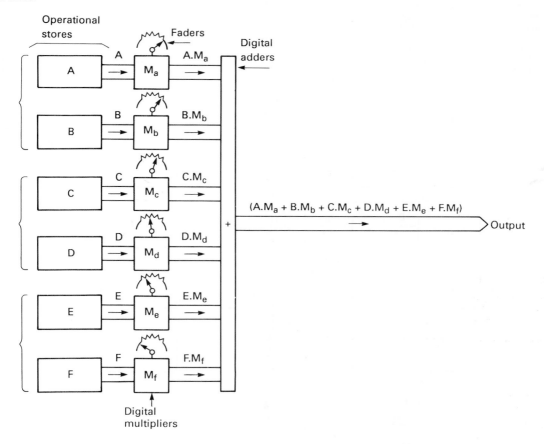

Figure 34.17 Operational stores in balancing

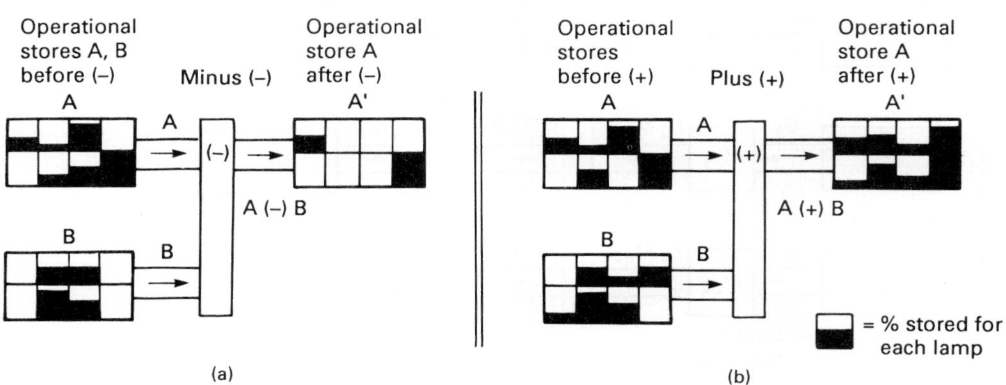

Figure 34.18 Operational stores in the arithmetic process of lighting: (a) minus (relative brightness of matched lamps is significant), (b) plus (here the relative brightness of matched lamps is significant – see plus protocol)

both manual and automatic; balancing of lighting from multiple operational stores

- *summary and sequential control:* cutting, sequencing, sorting, plussing and minussing, including sequence fading-related operations
- *special effects control:* lamp flashing and lamp chasing control
- *display indication control:* displaying of operationally required details using a visual display unit or crt monitor
- *main file control:* file amendment, loading and dumping with disk memory systems (see section *34.10*).

Each of the foregoing lighting control activities is now considered in the technical context of the type of processor control normally involved.

34.9.1 Gradual change in lighting control

34.9.1.1 Manual fading and cross-fading

Manual fading uses an 8 bit parallel multiplier. This is generally a static version because of the speed of operation required with lighting control processing (see *Figure 34.21*). The multiplicand input is the 8 bit time-division lamp data sequence which originates normally from a selected operational store (A in *Figure 34.21*).

The multiplier input is obtained independently by manual control of an up/down counter producing again 8 bits. The resultant 16 bit product is rounded off to 8 bits and fed to a digital adder as plot-product input A.

The B input to this adder is obtained from a similar arrangement, but from another operational store. In this way,

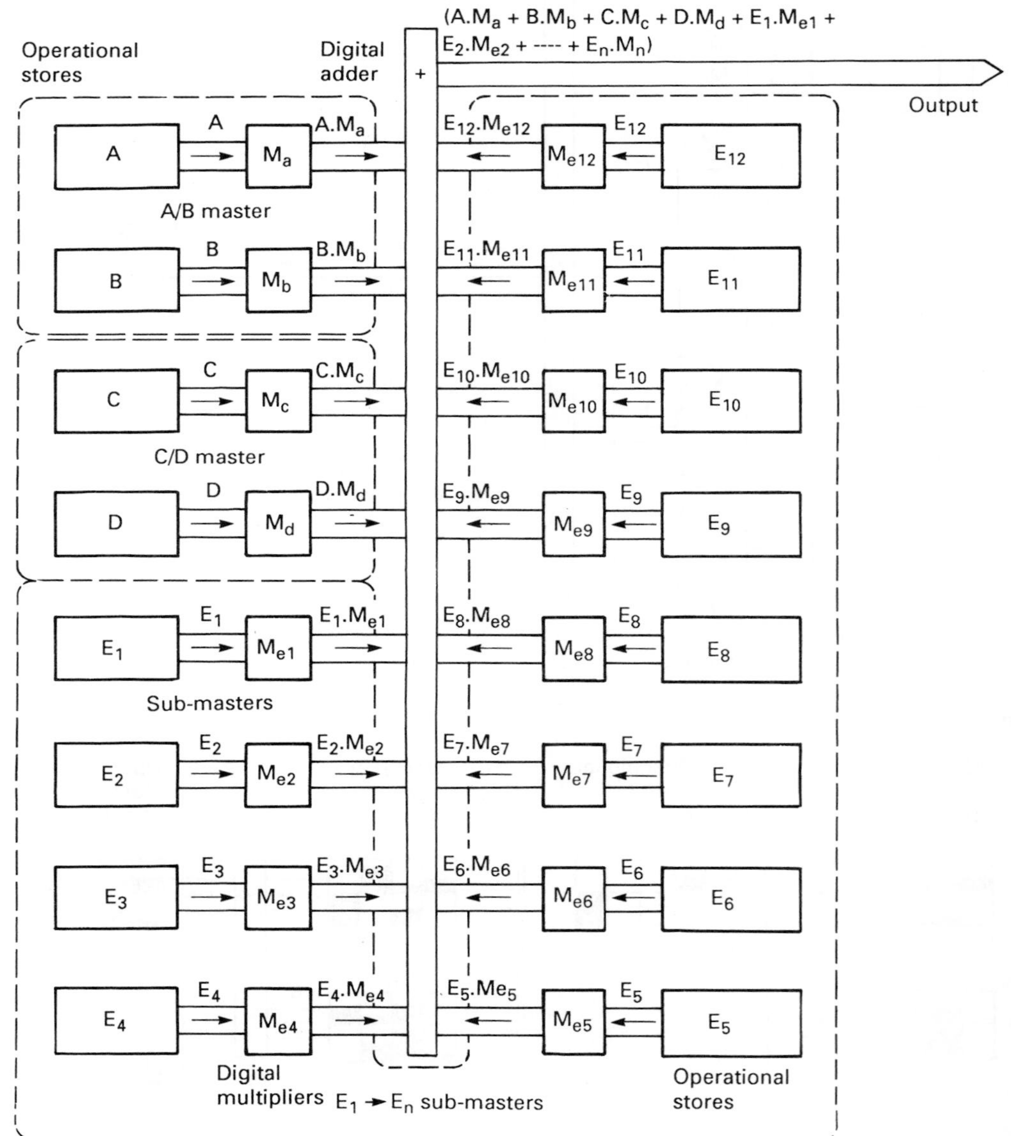

Figure 34.19 The sub-master playback, with AB and CD master playback

both the A and B lamp plots are combined by addition, following their separate multiplications by the multiplier stages. Algebraically, the combined outputs take the form: $AM_a + BM_b$, where M_a and M_b are multiplier factors between 0 and 1, controlled by manual faders F_a and F_b.

The fading-up operation consists of manually taking the fader from initially zero to maximum; fading-down being the reverse. Cross-fading is achieved by combining both operations at the adder outputs as shown in *Figure 34.21*.

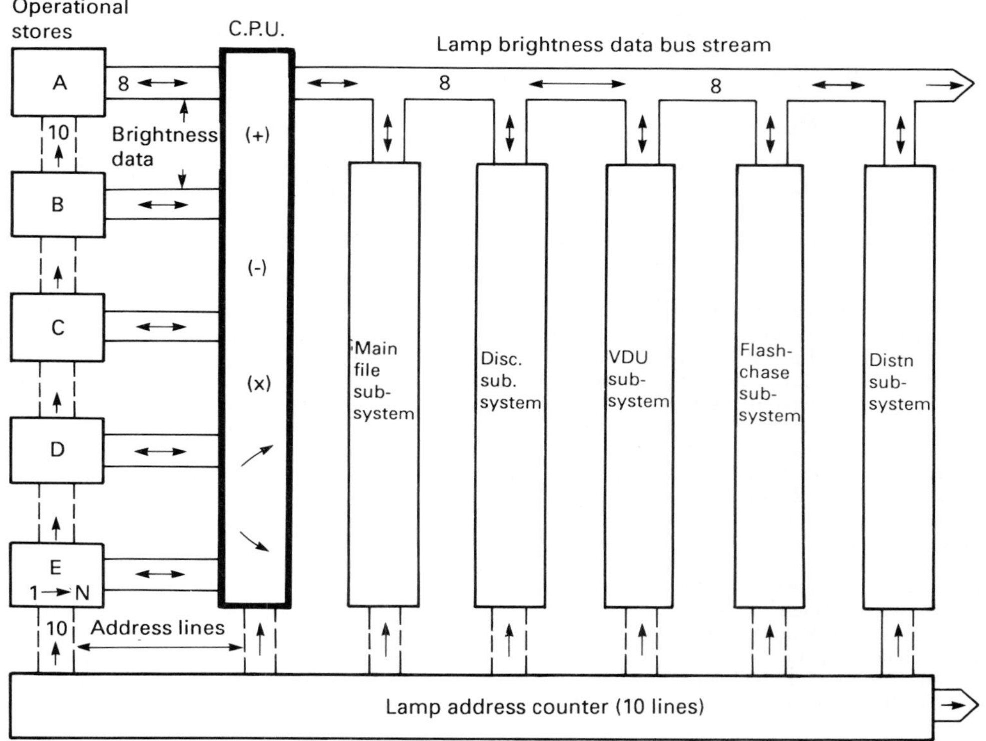

Figure 34.20 The central processor unit (CPU) in operational context

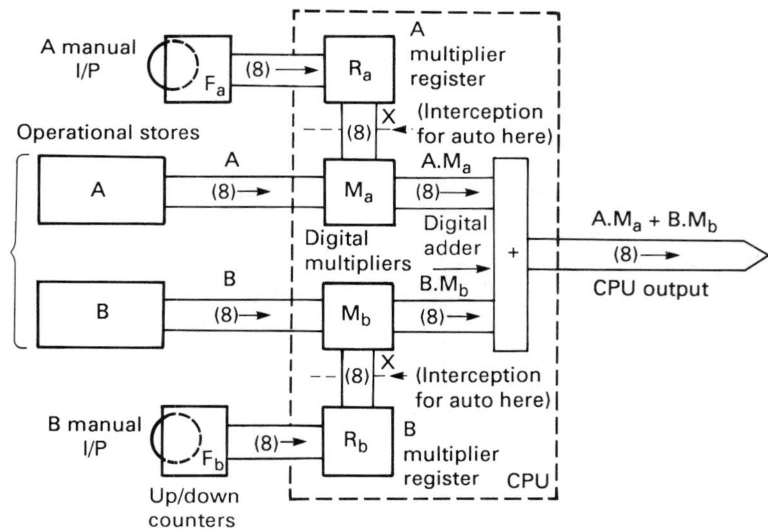

Figure 34.21 Manual cross-fades and the central processor unit

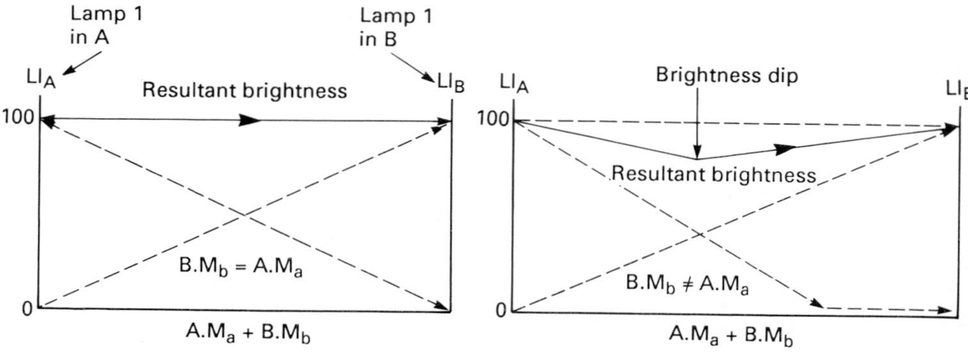

Figure 34.22 Manual cross-fade and brightness dip: L1A=L1B where L1A=L1 in A and L1B=L1 in B

34.9.1.2 Dipless cross-fading

When manually cross-fading between two lamp plots A and B, it is possible that the same lamps may exist in both lamp plots at the same brightness. The normal process of manually cross-fading described in section *34.9.1.1* would, nevertheless, introduce an unintentional dip in the resulting brightness unless special care is paid to controlling the up-fade progress with respect to the down-fade progress. This brightness dipping possibility between equal-brightness lamps is well illustrated by a histogram display of the effect in *Figure 34.22*.

Here, when equal-lamp brightness is involved between manual cross-fades on the same lamp in both lamp plots, a brightness dip may occur when the cross-fade rate profiles are asymmetrical. To overcome this possibility, the processor (as an option) may incorporate a dipless cross-fade facility. This consists of a comparator between both lamp numbers and lamp brightness data in A and B operational stores (see *Figure 34.23*). When equalities are located, these lamp numbers are bypassed around the multiplication process, remaining always at constant brightness.

There are occasions when this dipless option is not required, and it is then switched off operationally.

34.9.1.3 Balancing of lighting plots

This involves two or more operational stores with their associated faders and multipliers. Their outputs are combined by digital addition (*Figure 34.17*). In this operation the multiplier levels for balancing retain their intermediate positions.

34.9.1.4 Automated fading and cross-fading

The significant difference in automatic fading is that the path between the multiplier register, R_a, of *Figure 34.21* and the actual multiplier, M_a, is intercepted and a more complex and automated sub-system introduced to control the multiplication process. This is shown in *Figure 34.24*. The sub-system consists of a register, a binary rate multiplier and a digital up/down counter.

The *register*, (a) in *Figure 34.24*, stores the control word generated by the up/down counter, F_a. For automatic sequencing, this would be as (d), an array of similar registers with differing rates of fade. In the latter case, the selection of the appropriate rate register would form part of the automatic sequence process.

The *binary rate multiplier* (*Figure 34.24*(b)) is a pulse generator whose pulse rate can be altered by a control word presented to its rate control input. The input in this case is from the selected rate register. The output rate pulses from this generator are fed to a digital up/down counter that controls the multiplier input.

The *digital up/down counter* (*Figure 34.24*(c))has input from the binary rate multiplier, producing 8 bits for the multiplier input (e).

For cross-fades involving two operational stores, the sub-system is duplicated, and the two data product outputs are combined by a digital adder as with manual cross-fading. A histogram of complete automated cross-fades is shown in *Figure 34.29*. As with manual fading, the output is AM_a + BM_b.

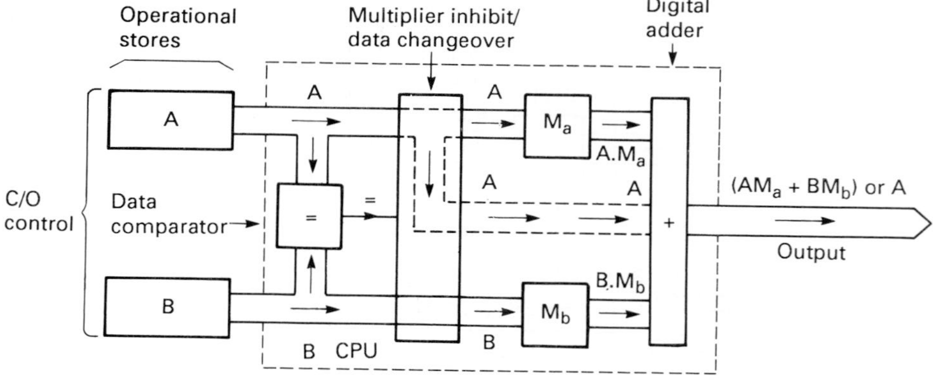

Figure 34.23 Manual cross-fade with dipless brightness option

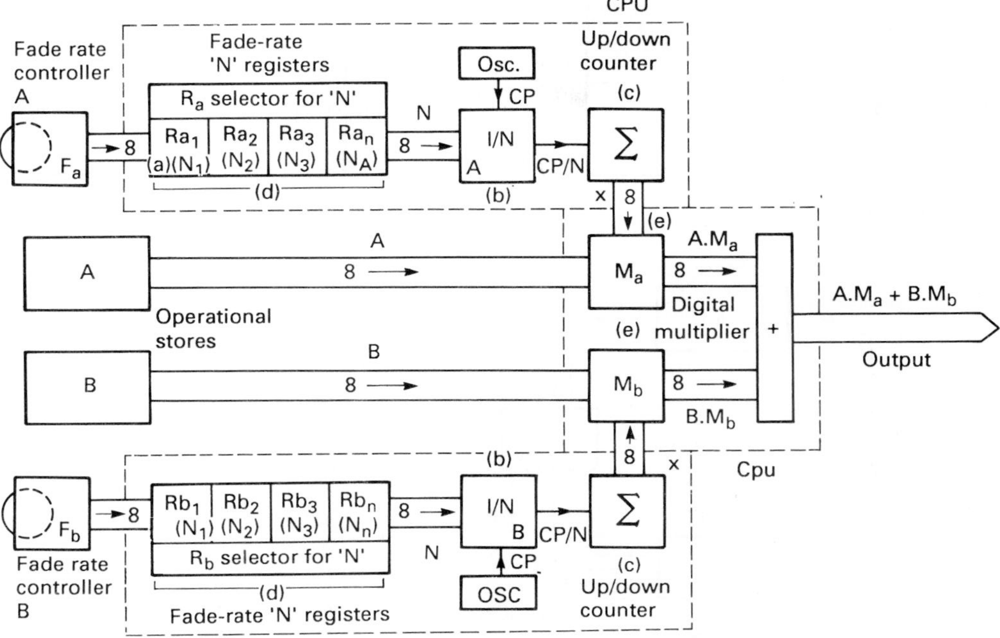

Figure 34.24 Automated cross-fades

34.9.1.5 Rationalized multiplication

By rearranging or rationalizing the algebra of the output expression, it is possible to improve the efficiency if not the elegance of the technical process in the automation of cross-fading.

When the cross-fade histogram is symmetrical, it follows that $M_a = M_b = M$ which becomes a single multiplier. Furthermore, by initially subtracting lamp brightness data in the A operational store from lamp brightness data in the B operational store, lamp by lamp in sequence, only one resultant signal need be fed to the multiplication process, while the second proceeds directly to the adder. Accordingly the algebraic equation (or algorithm) becomes:

$$M(B\text{-}A) + A = \text{output}$$

This is illustrated in *Figure 34.25*, and as a histogram in *Figure 34.26*.

34.9.1.6 Operational stores protocol

The involvement of more than one operational store in the automated cross-fade process requires a rigid and predetermined protocol to be observed in practice. This is mainly to avoid confusion in interpreting the outcome of automated processes.

Accordingly, it is convenient in some systems to use operational stores in dual combinations. These stores are allocated *master* and *slave* roles (also labelled sometimes as *output* and *preset*, or even *primary* and *secondary*).

34.9.1.6.1 Output/preset protocol

The use of dual operational stores in this form of protocol is illustrated in *Figure 34.27*, with the histogram in *Figure 34.28*.

The A store is the output normally controlling the lamps in the studio; B, not normally controlling studio lamps, is the preset or slave. Using the algorithm already described, the output following multiplication is M(B-A) + A as M varies between 0 and 1.

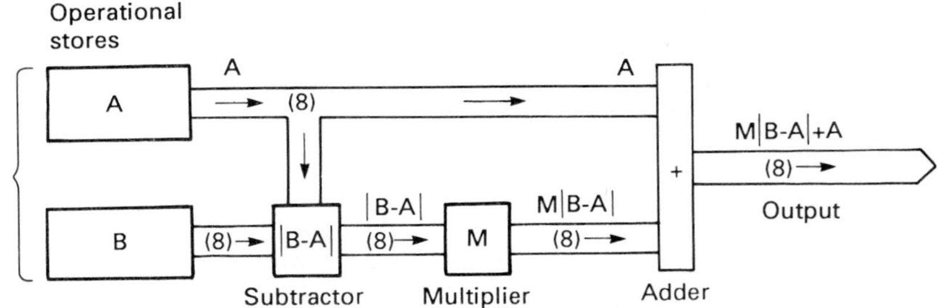

Figure 34.25 Rationalized multiplication

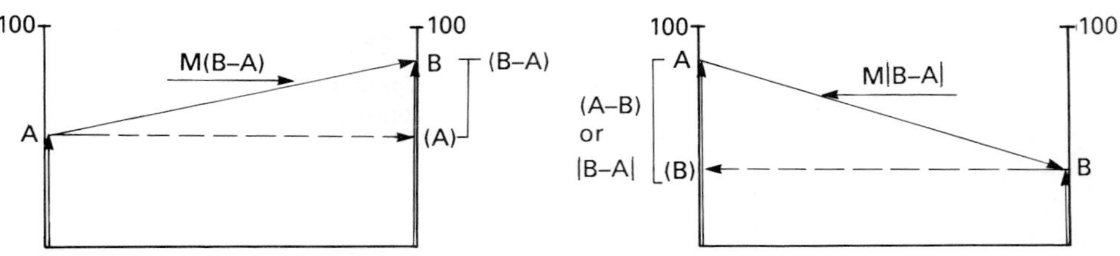

Figure 34.26 fade profile of M(B_A) + A

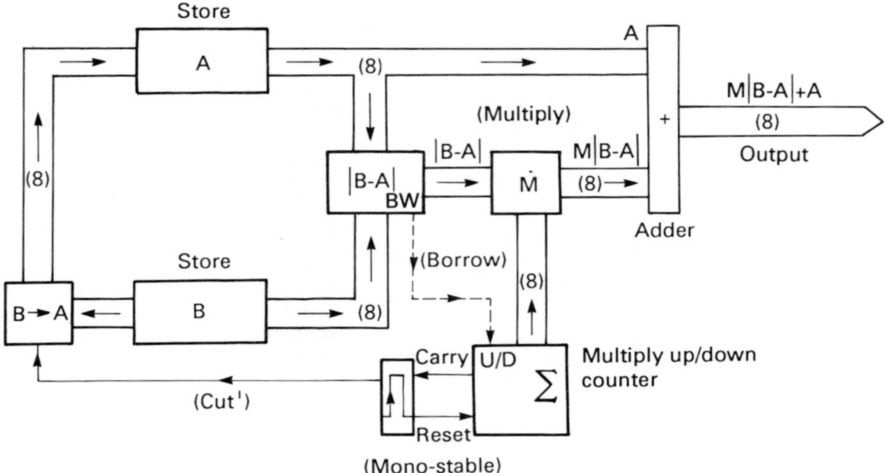

Figure 34.27 Output/preset protocol automated cross-fades. Note that *cut′* is an operational transfer of brightness data; (Cut′) is a simulated condition introduced automatically during cross-fade procedures

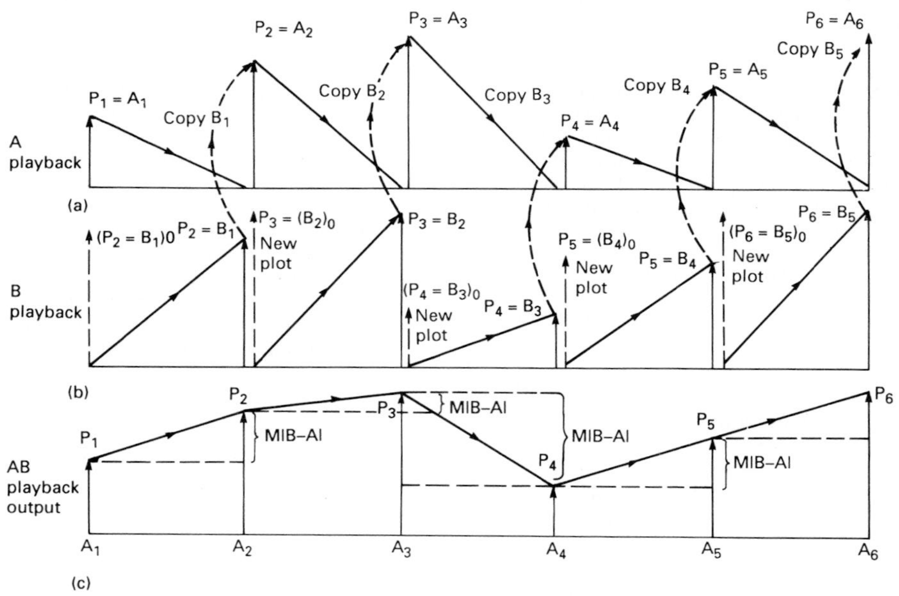

Figure 34.28 AB cross-fades with master/slave rationalizing protocol.
(a) 1. $P_1 - P_6$ are selected plots from main file. 2. $A_1 - A_6$ are progressive states within A playback. 3. At end of each cross-fade, the faded-up plot in B is copied into A. 4. Simultaneously, M is restored to 1; A is back in master control; B is faded down, ready for next plot, as slave.
(b) 1. $P_1 - P_6$ are selected plots from main file. 2. $B_1 - B_6$ are progressive states within B playback. 3. B playback in joint control during cross-fade time only. 4. At end of cross-fade, B plot is copied into A, and B is faded down simultaneously into slave role.
(c) AB output roles. Here A playback remains in virtual control during, and subsequent to, all cross-fades as a direct result of rationalization protocol

The lamp plot in A becomes faded-down and that in B becomes faded-up, under the transitional conditions of automated cross-fading. The protocol allows both A and B operational stores to control the studio lamps during the process of cross-fading only.

At the conclusion of the cross-fade (when B is in virtual control of the studio lamps) the protocol further requires the following immediate rationalizing of the situation which prevails:

- B lamp plot is copied into A store displacing the original A plot
- the multiplier input is returned to zero

The result is that the B lamp plot occupies the A store (now faded-up) as well as in the B store (now faded-down). The original A plot faded-down previously is cancelled. For a sequence of cross-fades, repeated selections of lamp plots from the main file would contribute further inputs to the B store.

34.9.1.7 Sequenced automated cross-fades

At the completion of each cross-fade, after the lamp plot from B has been copied into the now fully faded-up A store, the fully faded-down B store becomes available for further cross-fading in a continuous sequence. The new lamp plot, selected from the main file, is entered into the faded-down B store, and the cross-fade process repeated.

The histogram of *Figure 34.28* illustrates the progression of cross-faded lamp plots accordingly between A and B operational stores. By way of comparison, the histogram of *Figure 34.29* illustrates the *alternation* of control between A and B operational stores.

34.9.1.8 Simultaneous automated cross-fades

By providing a second dual operational store combination, e.g. C and D, a separate cross-fade may also control the studio lamps along with the progress of that from the A/B combination already described.

As an example, A/B and C/D simultaneous cross-fades are illustrated in *Figure 34.16*.

From *Figure 34.16* it will be seen that the A/B and C/D outputs are combined by means of a special form of digital adder called *adder greatest* which contains a comparator that modifies the addition process when two lamps, with the *same lamp number* but differing brightness, come together as inputs to this adder, from the A/B and C/D contributions.

The comparison process ensures that only the greatest brightness of the two identical lamp numbers is the output, rather than the simple addition of the two.

The combining of the outputs is illustrated in *Figure 34.30*. It will be seen that the *greater than* signal from the comparator controls the changeover gate to the greatest brightness lamp.

34.9.2 Summary change in lighting control

Lighting control operations include cutting, sequencing, sorting, plussing, minussing (including sequence plus and minus fades).

The involvement of the processor with these lighting control operations is described below.

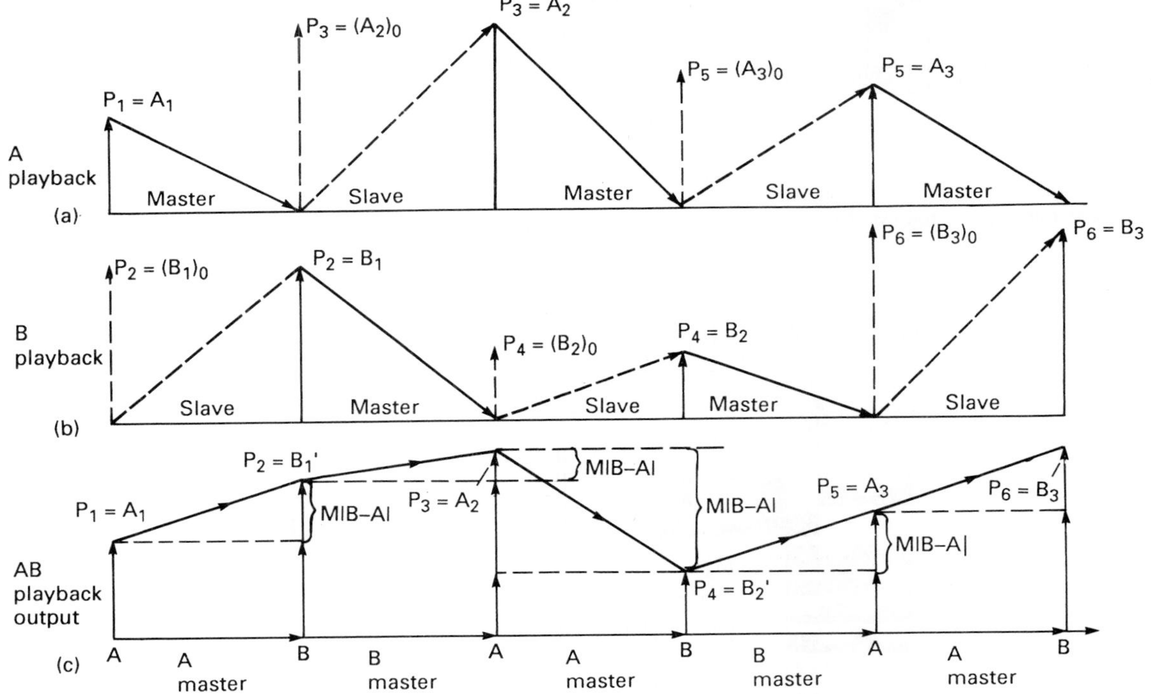

Figure 34.29 AB cross-fades without master/slave rationalization protocol.
(a) A input plots P_1, P_3, P_5 selected at main file. At the end of alternate AB cross-fades, A playback is faded down, losing all control of playback output. Accordingly, it adopts the role of a virtual slave, relinquishing its previoys master control (to B)
(b) B input plots P_2, P_4, P_6 selected at main file. At the end of alternate AB cross-fades, B playback is faded up, and is effectively in control of combined playback output. Accordingly, it adopts the role of a virtual master, relinquishing its previous slave (or 'blind') role (to A).
(c) indicates the virtual alternation of master/slave roles

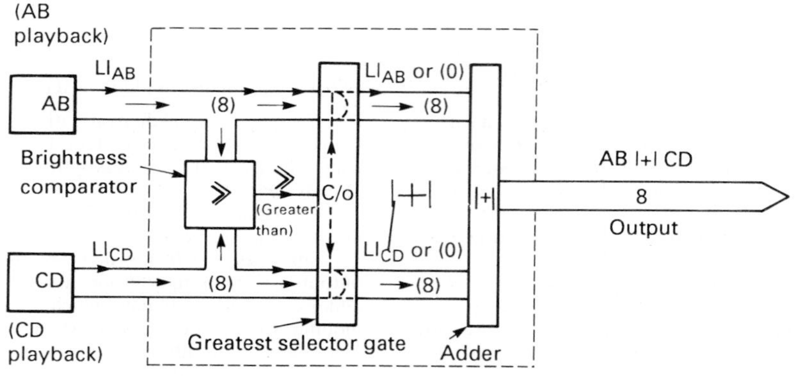

Figure 34.30 Adder greatest diagram. $L1_{AB}$ is lamp no 1 from AB playback; $L1_{CD}$ is lamp no 1 from CD playback

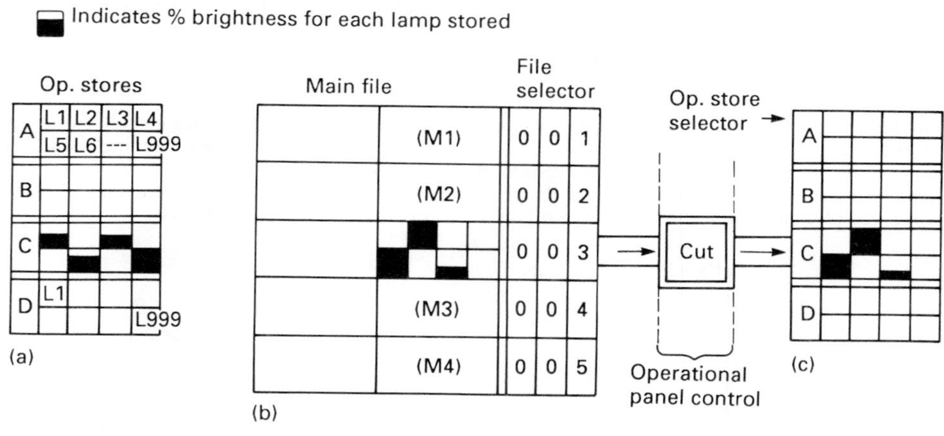

Figure 34.31 Cut: File to C store (M→C), (a) contents of C store before cut, (b) contents of selected file M3, (c) contents of C store after cut

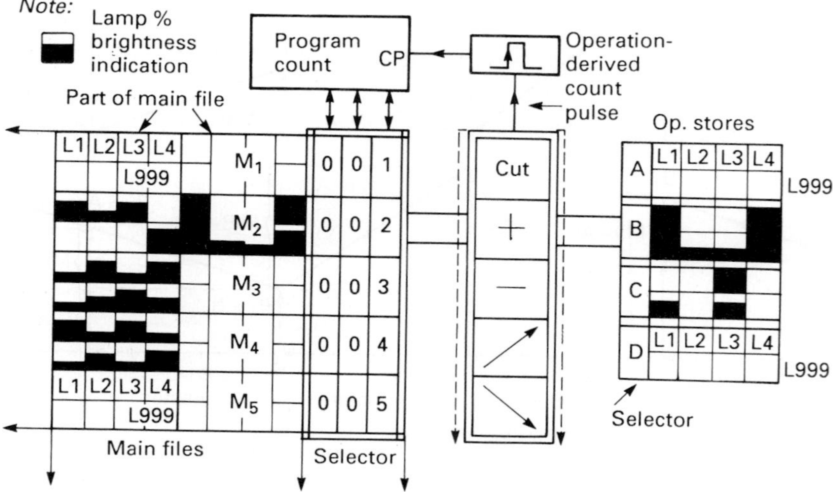

Figure 34.32 Sequence file address

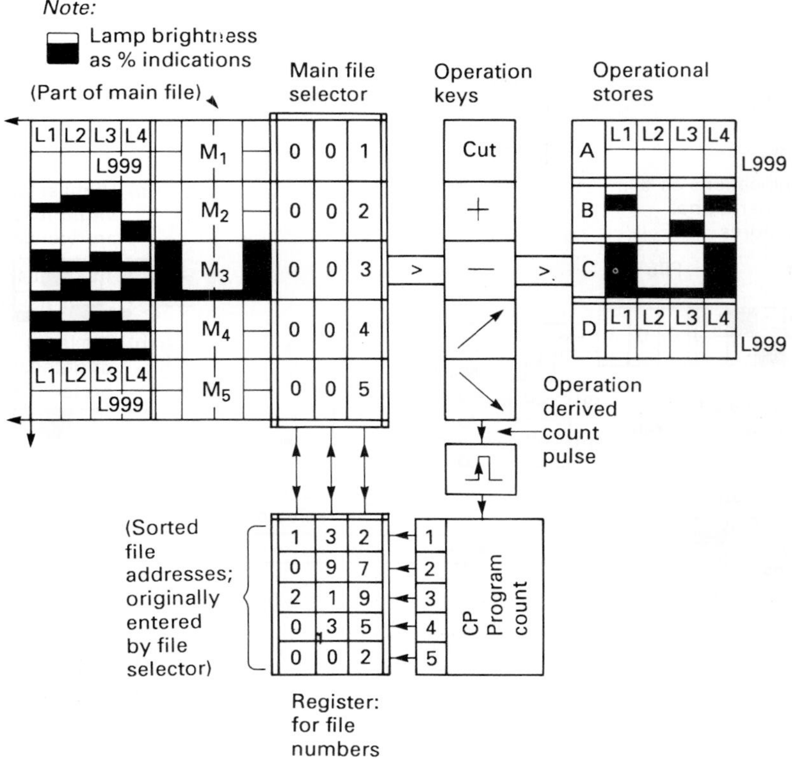

Figure 34.33 Sorting file addresses

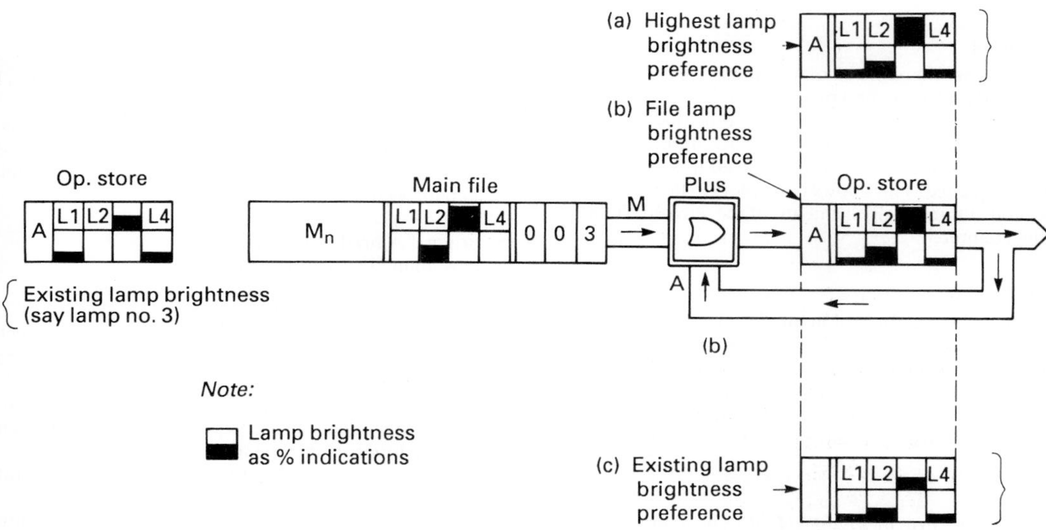

Figure 34.34 Plus (A+M). (a) The A store before plus (+), (b) the main file selection (M3)

34.9.2.1 Cutting

For cutting (M→A), the lamp plot is selected from the main file by the file selector (see *Figure 34.31*). The operational store is then allocated by the operational store selector, and finally the lamp plot is copied completely and instantly into that operational store by means of the *cut* button. The original lamp plot already existing in that allocated store is also instantly cancelled.

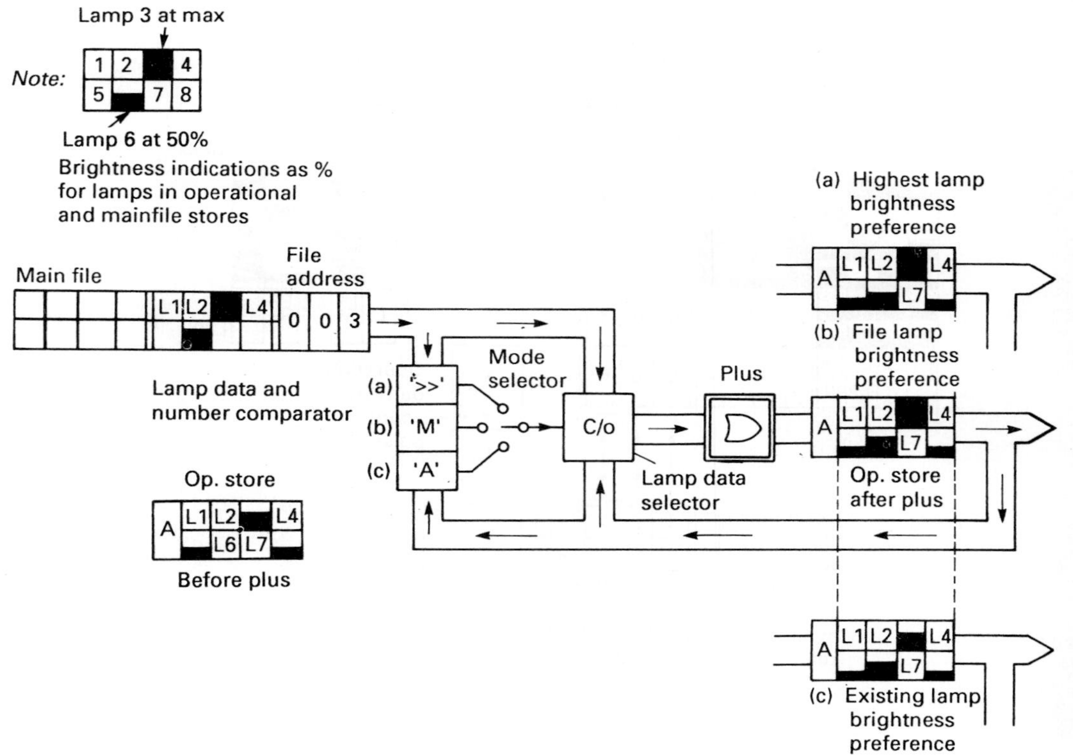

Figure 34.35 Selection of plus protocols

34.9.2.2 Sequencing

Sequencing applies to the control of the main file selection referred to above. By introducing a programmable counter into the file selector system (see section *34.10.4*), repeated operation of any operational control such as *cut*, *plus*, *minus*, *fade*, etc., allows the file selector to move on the file address one at a time in numeric sequence. This allows lamp plots to be recalled in a simple fashion automatically by means of the derived count pulse fed to the counter.

34.9.2.3 Sorting

Sorting is a more elaborate form of main file selection of lamp plots. It employs a *register* to store lamp plot numbers (which are not themselves in any numeric order) and allows them to be assembled into an ascending numeric series. This enables the lamp plot selection at the main file to be obtained sequentially and automatically as above by means of the derived pulse.

34.9.2.4 Plussing

Plussing (A+M) is shown in *Figure 34.34*. The lamp plots selected from the main file do not now displace the original lamp plots within the allocated operational store, but *augment* them.

If the original lamp plot in the A store already contains lamp numbers L1, L3, L5, L7 and L9, and the selected main file lamp plot contains lamp numbers L2, L4, L6 and L8, then plussing produces a final result in A of L1, L2, L3, L4, L5, L7, L8 and L9.

34.9.2.5 Operational protocol for plussing

It will be seen that if A store contains the lamps L1, L2, L3, L4 L5, L7 and L9, and the main file lamp plot selected contains L2 L3, L4, L6 and L8, then there will be a conflict of brightness fo the common lamps L2, L3 and L4, should these be different

This is reconciled by having the various plussing protocol pre-selectable as follows:

- the original A brightness prevails
- the new main file brightness prevails
- the greatest brightness prevails

Figure 34.35 shows the technical process of this pre-selectior for plussing. (b) and (c) show lamp number address compara tors between the main file and the A store signals. The resulting equals pulse generated by the contending lamps is accordingly pre-selected from either the a or b position to determine which of the two brightnesses shall be input to the A store.

Item (a) shows a further lamp brightness or data comparator This generates an inhibit/enable pulse. By this means, the changeover gate will only operate in favour of the brightness that generates the *greater than* pulse, as indicated, from the previously matched lamp numbers.

34.9.2.6 Sequence plussing for up-fades

This combines the features of:

- sequencing

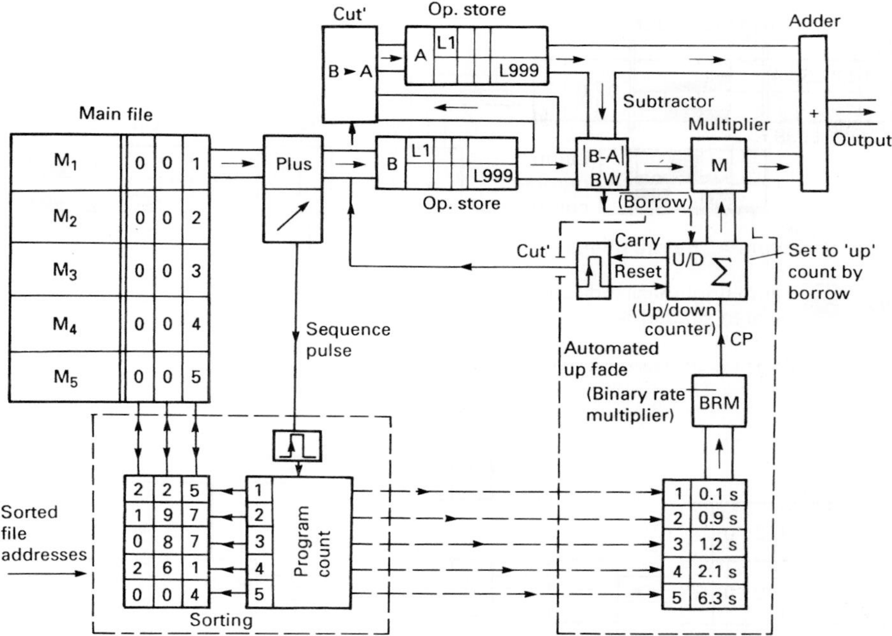

Figure 34.36 Sequence plussing for up-fades

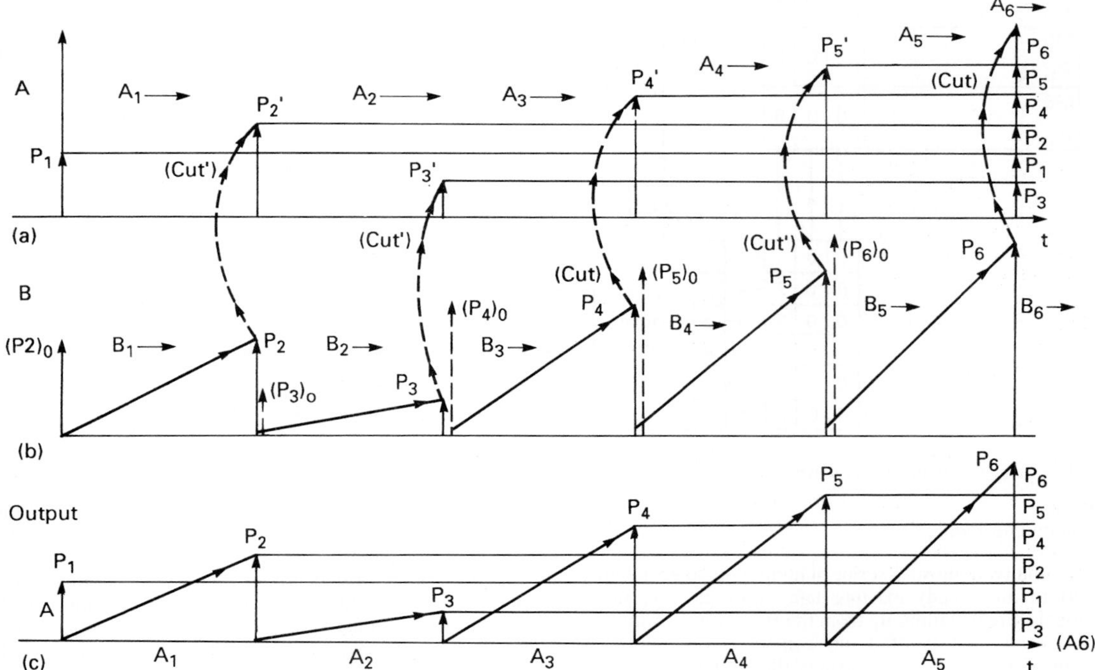

Figure 34.37 Sequence plussing for up-fades
(a) 1. $P_1 - P_6$ are selected plots from main file. 2. $A_1 - A_6$ are progressive states during sequence plussing for up-fades within the A playback. 3. At the end of each up-fade, the faded-up plot in B playback is copied (cut') into the A playback. 4. Simultaneously, the multiplier M is restored to 1. 5. A playback is then again in control.
(b) 1. $P_1 - P_5$ are selected plots from main file. 2. $B_1 - B_6$ are the progressive states within the B playback. 3. B playback in joint control during up-fade period only. 4. At end of up-fade, B plot is copied (cut') into A. 5. Simultaneously, M becomes 1, A assumes master role, with B available as slave.
(c) A playback retains master o/p role

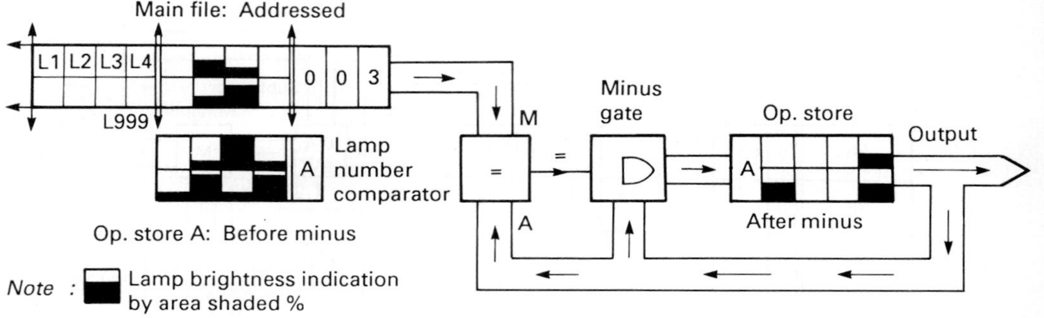

Figure 34.38 Minussing (A–M). Brightness differences between matched lamps are not of significance here

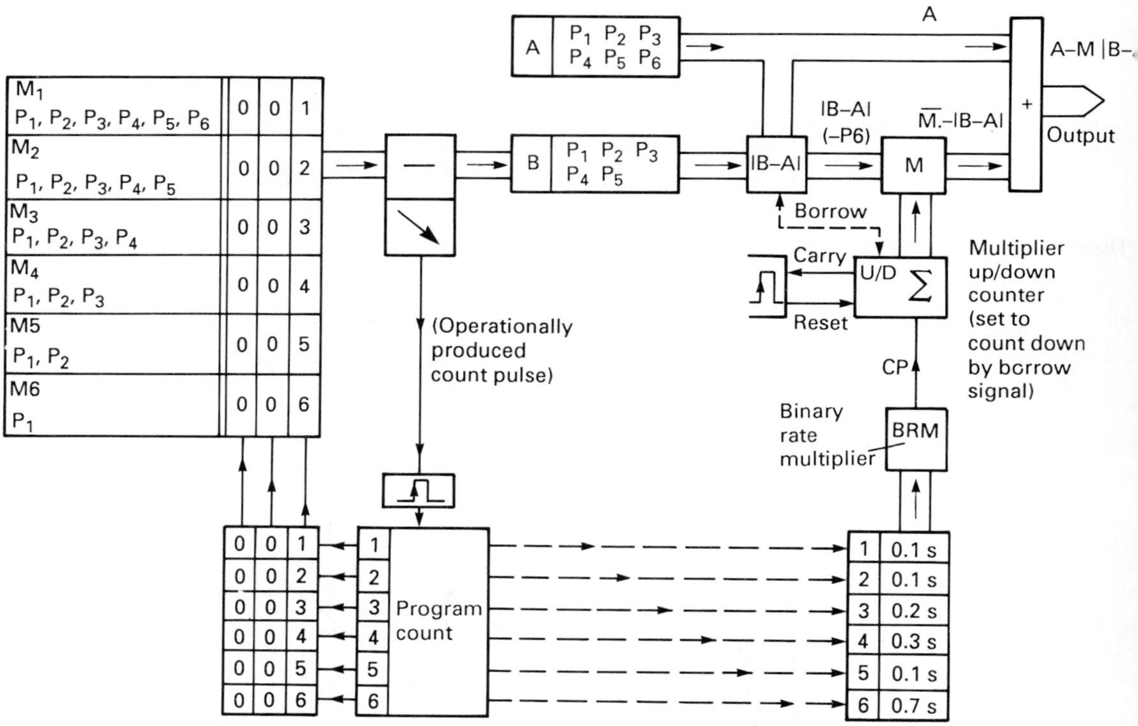

Figure 34.39 Sequence minussing for down-fades

- plussing
- automated up-fading

Operationally, it allows a series of lamp plots to be added-in (*plussed*) to an already existing lamp plot in, say, the A operational store by fading-up from the B store in a continuous sequence each time the fade-up control is operated. The histogram of *Figure 34.37* illustrates the summation process which finally fades all pre-selected plots from B into an all-embracing master lamp plot in the A operational store.

34.9.2.7 Minussing

In minussing (*Figure 34.38*), lamps are selectively switched off by means of the main file lamp plot whose lamp number contents match certain lamps within an existing lamp plot in the A store, which here is the operational store allocated. When all the lamps match, the result is a so-called "black-out' operation, and every lamp in the studio is switched off.

More usually, a selection of lamps is made in advance and stored in the main file. The actual matching of minussing lamps (within the main file lamp plot) to the operational store is by exclusive-or comparison which produces an equals signal, and gating out those lamps from the operational store involved, when recirculated through the minus gate.

34.9.2.8 Sequence minussing for down-fades

This facility combines the features of:

- sequencing
- minussing
- automated down-fading

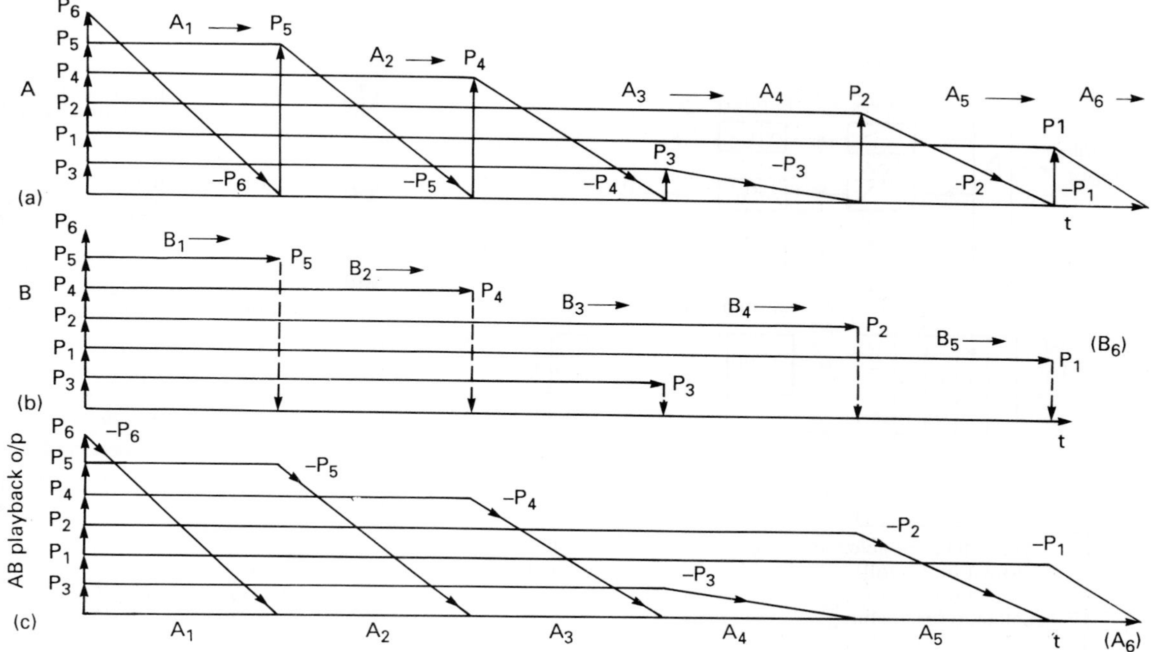

Figure 34.40 Sequence-minissing for down-fades
(a) 1. $P_1 - P_6$ are selected plots from main file. 2. $A_1 - A_6$ are progressive states during sequence-miniussing for down-fades in A. 3. At end of each down-fade of particular plot (selected by minus comparison between contents of $|B–A|$), the particular plot is finally withdrawn by the minus action. 4. Simultaneously, the multiplier is restored to 1.
(b) 1. $P_1 - P_6$ are selected plots from main file. 2. $B_1 - B_6$ are progressive states during sequence-minussing for down-fades in B playback. 3. Comparison between B and A decides plot in A to be faded down.
(c) Progression of selected plots in A playback for down-fading by the minussing comparison procedure

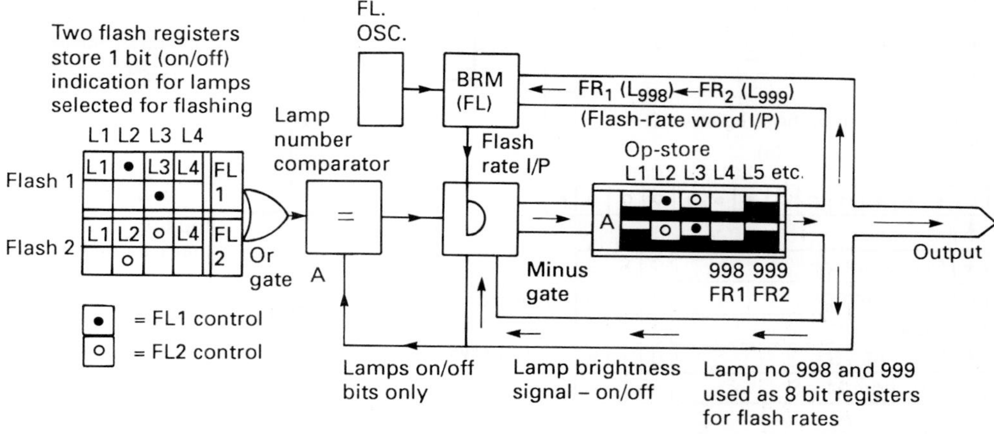

Figure 34.41 Lamp flashing (A–FL)

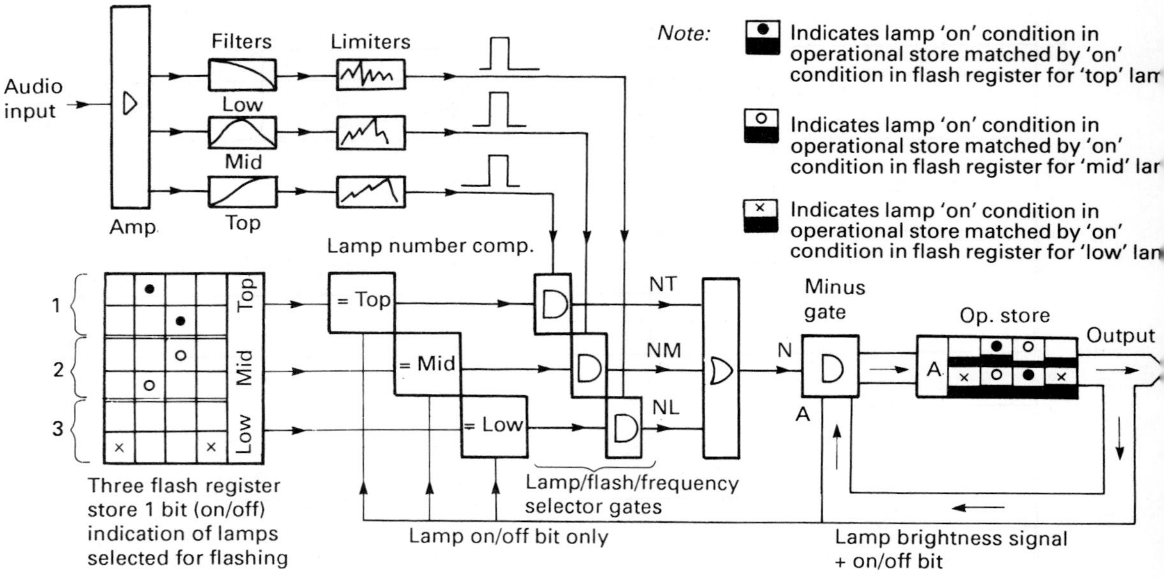

Figure 34.42 Lamp flashing: audio rate control (A–N)

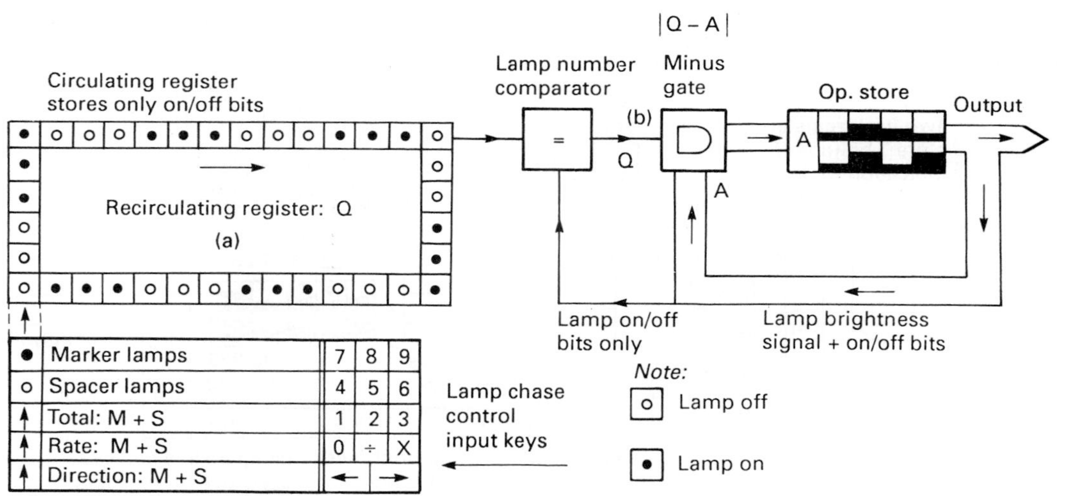

Figure 34.43 Lamp chasing (A–Q)

Operationally, by sequenced selection of the main file lamp plots, certain lamp numbers may be selectively faded-out, both progressively and in sequence, until all lamps are reduced to zero in brightness. The histogram illustrating this operation is shown in *Figure 34.40*. It is the inverse operation to the sequence plus up-fades described in section *34.9.2.6*.

34.9.3 Effects lighting control

34.9.3.1 Lamp flashing

This is an elaborated extension of the minussing process, by means of a sub-system shown in *Figure 34.41*. Two separate registers are provided, labelled *flash 1* and *flash 2* which allow a more dramatic form of lamp flashing where two sets of lamps alternate, i.e. (A-FL1) and (A-FL2).

These two registers are pre-loaded with the minussing lamps, and the selected operational store, A, contains the full complement of lamps. Accordingly, the matched lamps will be switched *on* and *off* at a rate decided by another pair of registers known as *flash-rate registers*. These will contain only 8 bits each for the purpose of flash rate control, and so they normally form part of each lighting plot, occupying two unused lamp addresses, e.g.lamp 998 and lamp 999. From here, the 8 bit word is fed to a binary rate multiplier, whose output rate controls the on/off intervals at the minus gate.

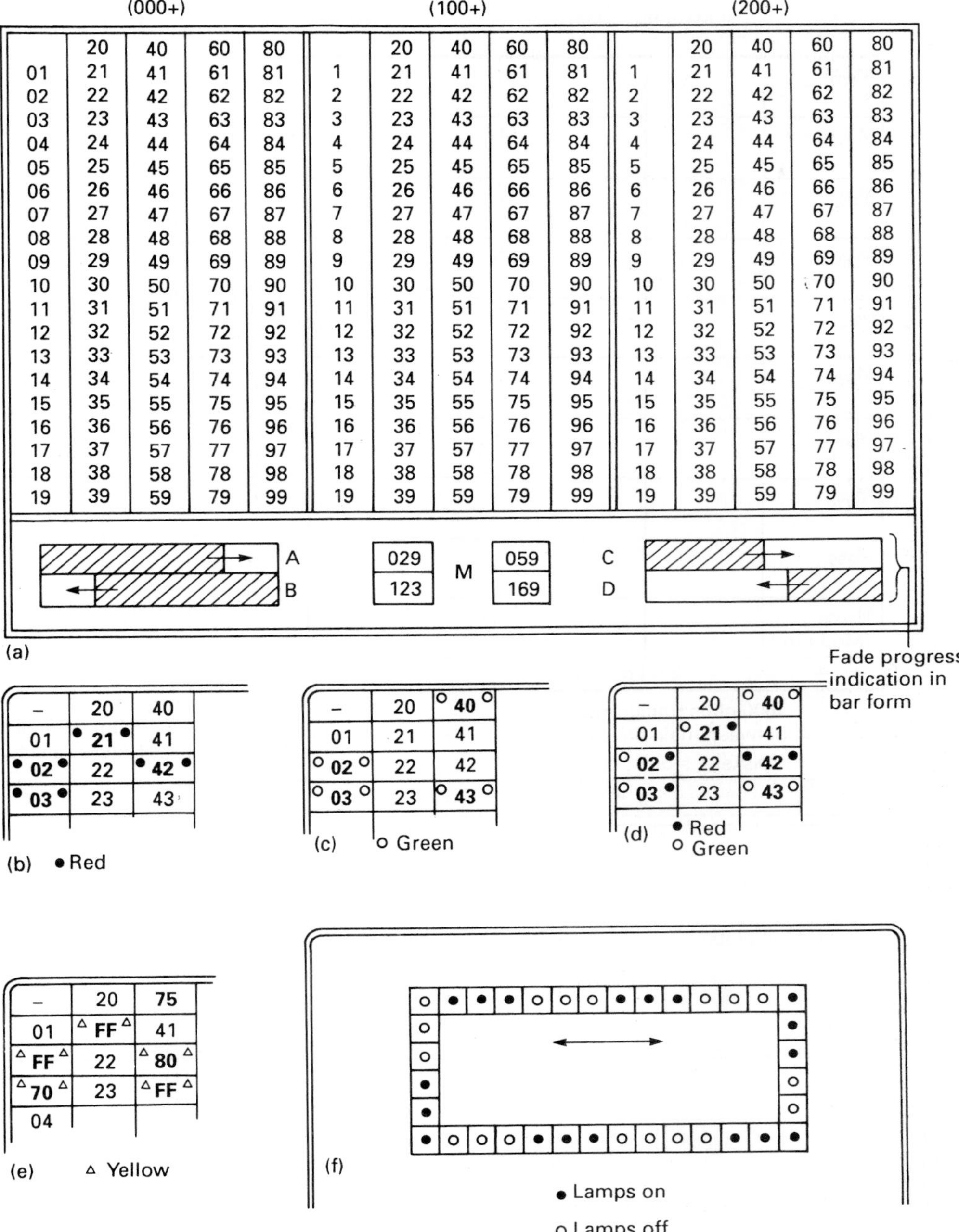

Figure 34.44 VDU format (a) display of lamp numbers and operational details, (b) studio output lamps switched on in red, (c) preset lamps to be switched on in green, (d) output and preset lamps combined in cross-fade, (e) display of lamp brightness displacing lamp numbers, (f) display of lamp chasing details

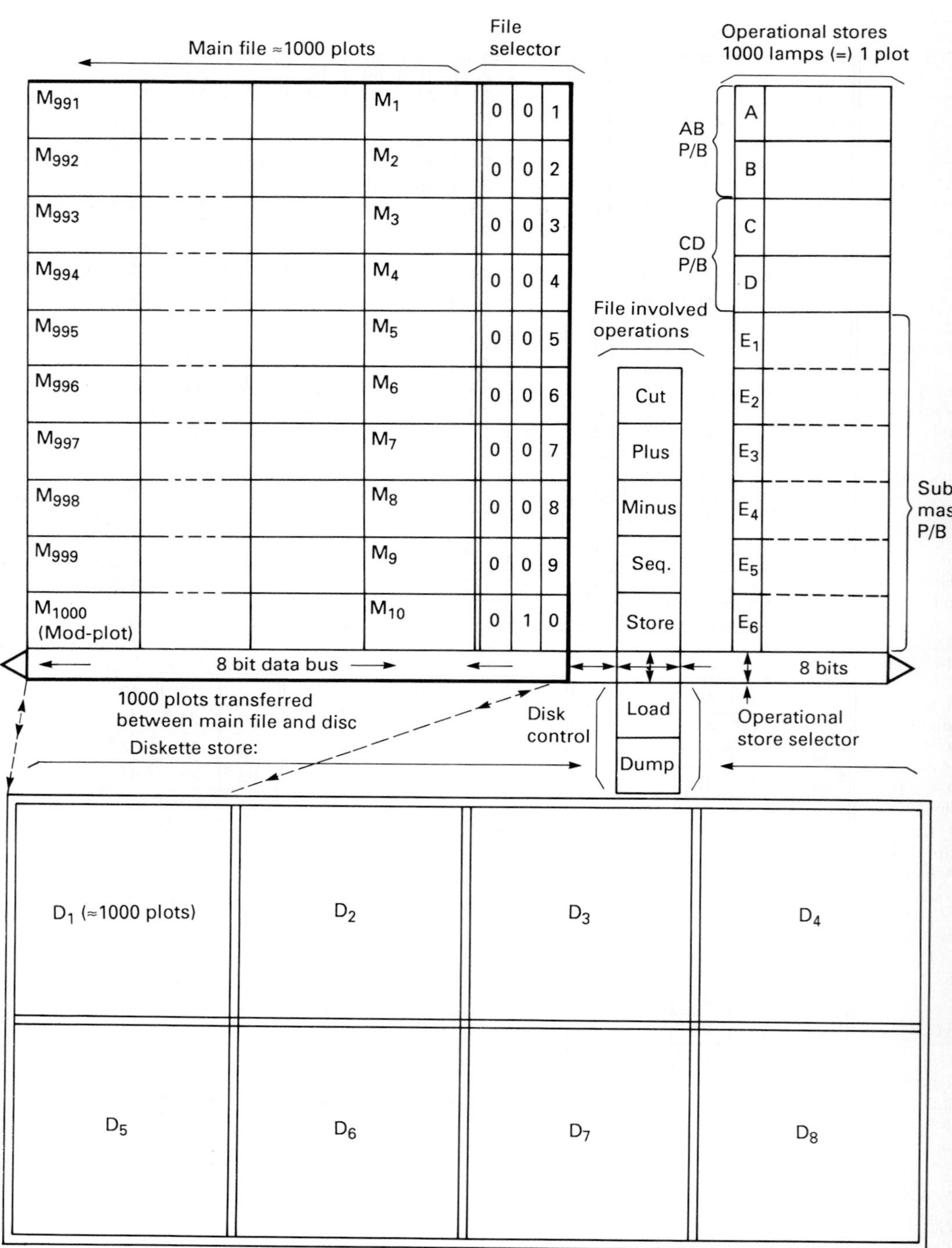

Figure 34.45 Main file in operational context with disk and operational stores

34.9.3.2 Audio control of lamp flashing

This is an adaptation of the lamp flashing facility. The difference is that the gating pulse, N, used at the minussing gate, is obtained by an external audio input signal (which is usually a musical signal). This audio signal is amplified and clipped to produce minus gating pulses, which are accordingly related to the rhythms of the musical input. By this means the lamps indicated in the flashing lamp, or minussing, register are gated at a rate decided by the pulses from the audio input.

By using bass, middle and treble filters in a musical input, the more rhythmic, bass form of flashing can be augmented selectively, by otherwise timed contributions from the outer parts of the musical spectrum, i.e. $(A\text{-}N_{bass}) + (A\text{-}N_{middle}) + (A\text{-}N_{treble})$.

34.9.3.3 Chain control of lamp flashing

This is an elaborate application of the minussing process. A chain of minussing pulses is involved with the selected operational store in the process (also referred to as *lamp chasing*).

In this case, the sub-system involved is complex and may well form an entirely separate facility external to the main lighting control system. However, the introduction of visual display units (or crt monitors) along with so-called "soft keys' has enabled modern systems to incorporate these control operations without occupying valuable main control panel space.

The control processing involved is shown in the sub-system drawing of *Figure 34.43* and consists essentially of a re-entrant circulation shift register, also called a *Johnson counter* (marked (a) in *Figure 34.43*) forming the basic chain — Q in this case. The output port of this shift register is used to control the minus gate of the minussing facility (b).

By means of a VDU mimic display of the circulating shift register contents, the following parameters are controlled by the soft keys provided:

- the length of the circulating shift register, i.e. the total number of lamps in the chain
- the length of the *on* lamp markers, i.e. the number of lamps switched on and circulating in the chain
- the length of the *off* lamp spacers, i.e. the number of lamps switched off and circulating in the chain
- the rate of progress of the shift of lamps around the circulating register forming the chain
- the direction of progress of the circulation

A typical mimic display of a chasing shift register and associated soft keys is shown in *Figure 34.44*(f).

The pictorial effect is a ring of lamps, some off, some on, continually circulating.

34.9.4 Visual display unit

The introduction of a VDU greatly enhances the scope of the lighting control system. Mimic diagrams can be used in association with soft keys to reduce the number of controls on the actual controlling panel, as well as to indicate more meaningfully the activity taking place or being controlled.

34.9.4.1 Types of display

The types of display available with the VDU, including colour displays in use, are:

- lists of all lamps within the system (usually in white display)
- lists of all lamps switched on in the various operational stores (usually in green display)
- lists of all lamps on in the studio (usually in red display)
- lists of dual-combinations of operational stores contributing to cross-fades (e.g. *output* in red numerals and *preset* in green numerals)
- *bar* indications of progress of cross-fades
- ancillary information such as main file selection
- textual indications or explanations of complex activities

34.9.4.2 Display formats

The most demanding of these are the separate requirements for the various lamp listings. Although up to 1000 lamps may be involved, it is impractical to try to display more than 300 of these as numerals at any one time on one page of display. All the lamp numbers can then be accommodated on two or three pages.

The display format of *Figure 34.44*(a) has been used satisfactorily for displaying up to 300 lamps on one page. The density of this presentation has been achieved by omitting the largely unnecessary hundreds digit, and relying on the spacial context of the particular number pairs to indicate the missing digit in each case.

When the listing has specifically to indicate lamps switched on, then the particular numbers, normally white, are changed to red if an output store is being checked (i.e. A or C), or to green if a preset store of a cross-fading dual-combination is being checked (B and D). See *Figure 34.44* (b) and (c).

Moreover, both contributions can be checked during the progress of a cross-fade, for instance, by having shared colour lamp numbers. In each case, one digit of the lamp numeral pairs is in red and one is in green (*Figure 34.44*(d)).

To display the brightness of all the lamps simultaneously, again numeral pairs are used from 00 to 99. 100, or maximum brightness, is indicated by the symbol FF. Here the brightness indication in yellow displaces normal lamp numerals in white (*Figure 34.44*(e)). For this, of course, further pages would be involved, labelled as *brightness pages*.

34.10 Main file control

34.10.1 The system

The main file is shown diagramatically in *Figure 34.45*.

The main file (or memory) of a lighting control system requires long-term retentivity, i.e. unlike the more short-term requirement for the operational stores, the main file system has to be a repository for the work so far undertaken. It is a long-term store, mainly for the lighting plots from the previously used operational stores. It also provides rapid access to diskette stored plots.

To protect the main file workload, it is usual to provide diskette-type long-term storage; this accommodates also the lighting plots of a more archival nature. The lighting plots of the shows expected to be regularly repeated are retained in this fashion.

The transfer of lighting plots to the main file from the diskette is the *loading* process; the transfer of the lighting plots to the diskette from the main file is the *dumping* process.

34.10.2 Storage capacity requirements

34.10.2.1 Operational store

As has been seen, the need is for a hundred or more increments of lamp brightness, or one binary word of 8 bits (termed a *byte*). Therefore, an operational capacity of 1 kilobyte (kb) is normal to provide for a single plot of up to 1000 lamps.

34.10.2.2 Main file

This may well have to provide long term storage for between 100 and 1000 lighting plots. Therefore, capacity may be

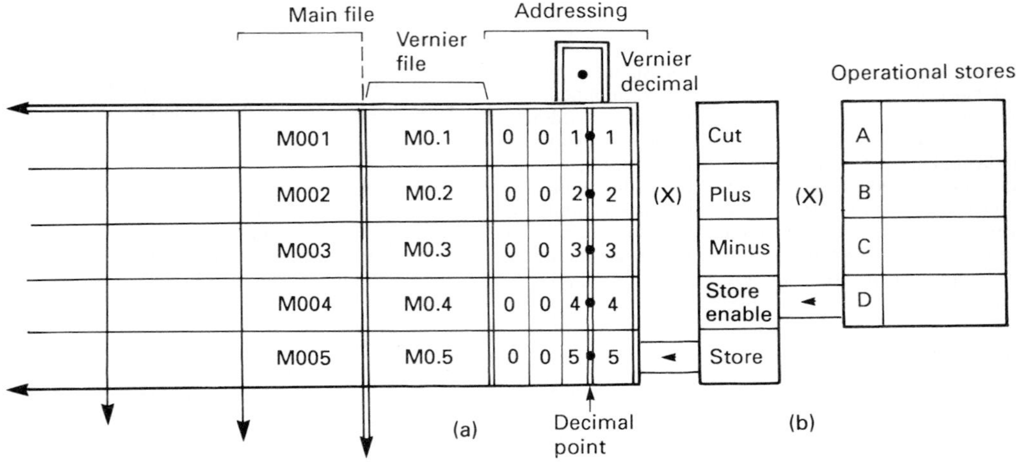

Figure 34.46 Main file: (a) Vernier decimal addressing; (b) storing

required for between 0.1 Mb and 1 Mb of operational data storage.

34.10.2.3 Central processor unit

The processing and computational software storage required in the CPU, normally in the form of so-called *firm-ware*, is of the order of 0.1 Mb.

34.10.2.4 Diskette

To hold the contents of from one to many main file loads, the capacity required of a diskette is between 1 Mb and 10 Mb.

34.10.3 Main file versus diskette

A main file provides instantaneous access to all the addressed lighting plots contained within it, while a diskette may sometimes require a few seconds to address and access the lighting plots it contains. This would not normally be operationally acceptable.

Despite this delay in accessing the lighting plots, the diskette has found further application in standby forms of lighting control, where it provides a secondary control facility for the control of the studio lamps. By this means, and using a separate and more basic control panel or console, the previously deposited lighting plots are separately accessed and processed into analogue voltages. They are then applied to the dimmers by separate means through isolation diodes (see *Figure 34.49*).

In this way, a production can be saved by the operation of the standby diskette facility where this is provided for emergencies.

34.10.4 Main file operations

34.10.4.1 Selection

Selection of the main file is by decimal key pad, similar to the selection of the studio lamp. However, there is an option available for introducing an intentional difference in the format for this latter selector, in order to prevent confusion of the operational requirement (i.e. whether it is a lamp or a plot that is being addressed). Certainly, the combining of both these operational selections at the one keypad is likely to create more opportunities for operational errors than having their different roles emphasized by their distinctive key formats.

34.10.4.2 Storing

For this operation, the operational store is selected, as is the main file address. The store operation is then enabled (for protection against inadvertent operation), followed by the pressing of the actual *store* button, to record the lighting plot at that file address (see *Figure 34.46(b)*).

34.10.4.3 Changing and summing

These activities form the bulk of the operations with the main file. The file is addressed at the main file address keypad and the desired operational store is selected. Then the *cut* button will load a lighting plot from the addressed file into that particular operational store, the *plus* button will add the selected plot to an existing plot within the selected operational store, or the *minus* button will switch off any lamps with duplicates in both the selected file and the particular operational store. (See *Figures 34.31*, *34.34* and *34.38*.)

34.10.4.4 Sequencing

When sequenced operations are involved, the file selection facility incorporates a programmable counter. This is stepped on, and so the file address is also stepped on, by one address at a time, on each occasion the *cut*, *plus*, *minus* or *fade* buttons are actioned. (See *Figure 34.32*).

34.10.4.5 Sorting

Normally, the lighting plots end up being stored at disjointed file addresses throughout the overall main file address assembly. However, when a sequence is needed, the comprehensive facility of sorting enables the required, but disjointed, file addresses to be collected within an array of registers into a numeric series. This allows a file addressing sequence to take place. (See *Figure 34.33*).

34.10.4.6 Vernier file numbering

By setting aside ten files called *verniers*, the file numbering facility may include decimal extensions to incorporate other files on a later occasion. This will accommodate file amendments and file interpolations. All that is required to interpose

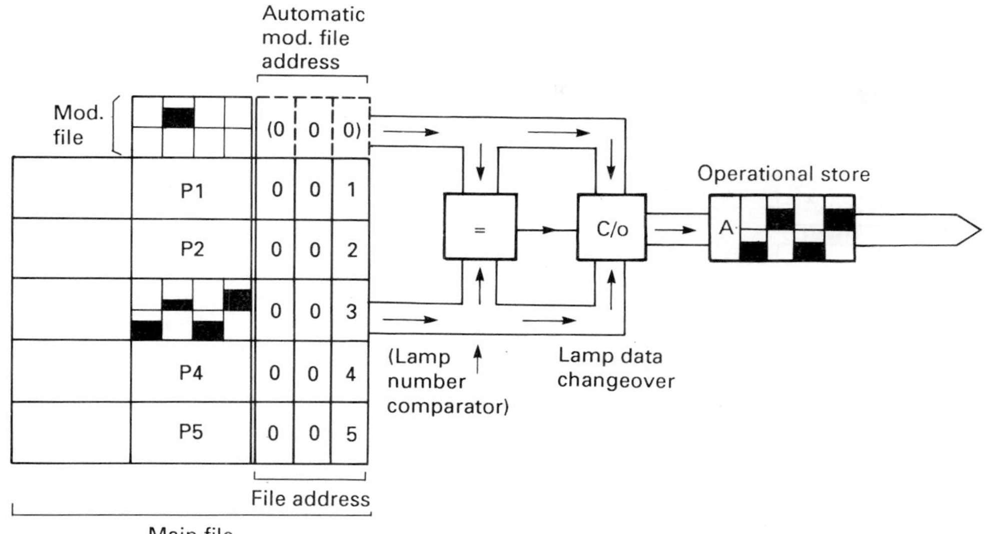

Figure 34.47 File data amendments (|MOD| A)

an amended file within the section occupied by a run of file numbers, is to add the vernier file to the already addressed original file number as a decimal addendum to that file number e.g. 199.1, 199.2, 201.1, etc. (See *Figure 34.37(a)*.)

34.10.4.7 Automated file amendments

It is possible to achieve automated file amendments by means of the so-called *modification* facility. This is tailored operationally for last minute, and unforeseen, changes resulting from alterations made late in rehearsals, or even during the final transmission.

Lamps requiring to be modified in brightness, and usually occurring repeatedly throughout many of the already recorded lighting plots within the main file, can be automatically amended to take the new brightnesses. The procedure is to store the lamps with the amended brightness levels into a specially set-aside file address; usually this is main file address number 000 or 999 (which are normally unused file addresses). Then the process outlined in *Figure 34.47* takes place.

When lamps within existing files have to be modified, a button marked *mod* is preset. Subsequently, the lamps within any further main file selections are subjected to a lamp by lamp comparison with the modified lamps stored in the mod file. If a lamp number match occurs, then the amended brightness from the mod file is automatically substituted for each matched lamp. When such amendments are no longer required, the mod facility is switched off.

34.11 Lighting control brightness distribution

34.11.1 Lamp brightness data bus stream

The lamp brightness data, processed by the CPU, emerge as an 8 bit parallel (but time-divided serially) data bus stream, Within it, the time period of t_1, measured in microseconds, and preset by the master oscillator of the lamp counter, is allocated to the brightness data for each lamp on the stream. The overall frame time, t_2 measured in milliseconds, is decided by the maximum number of lamps for which the system was designed. It will be seen that both these times achieve significance in the context of the sample and hold circuits in the process which follows.

34.11.2 Terminal processor

The studio lamp brightness data in the form above undergo two distinct conversion processes which prepare the serialized digital brightness data for ultimate application to the dimmer module in parallel form as a control voltage. When the terminal processor is situated adjacent to the CPU, the resulting control voltages have to be conveyed to the normally remote dimmer room by multiple multicore cables for application to the dimmer modules in parallel form. However, when the terminal processor is itself in the dimmer room, the lamp brightness serial data stream is channelled to it by means of the specially formatted data carrier systems, such as the DMX512 system. This is referred to as *remote multiplexed dimmer* control. It is in serial form and uses minimal inter-cabling as a result.

34.11.3 Brightness data demultiplexer

The brightness data bus, as shown in *Figure 34.48*, consists of 8 bit parallel data in time-divided serial form. It contains the complete brightness information for every lamp in the studio in 7 bit character form with an added on/off bit, normally incorporated as an eighth bit.

However, the dimmer control input requires the signal to be in analogue form, and furthermore, an exclusive analogue voltage input is required for each dimmer module. The analogue derived data, now in serial form, has also to be translated into parallel form for application to the dimmers.

34.11.4 Conversion of serial binary characters to analogue

The 7 bit binary data bus (i.e. without the on/off bit) is fed in its serial time-divided format to a single digital/analogue convertor. Accordingly, as the process evolves serially, the analogue form of the same time-divided data emerges, as shown in *Figure 34.48*(a). Each analogue sample occupies the same

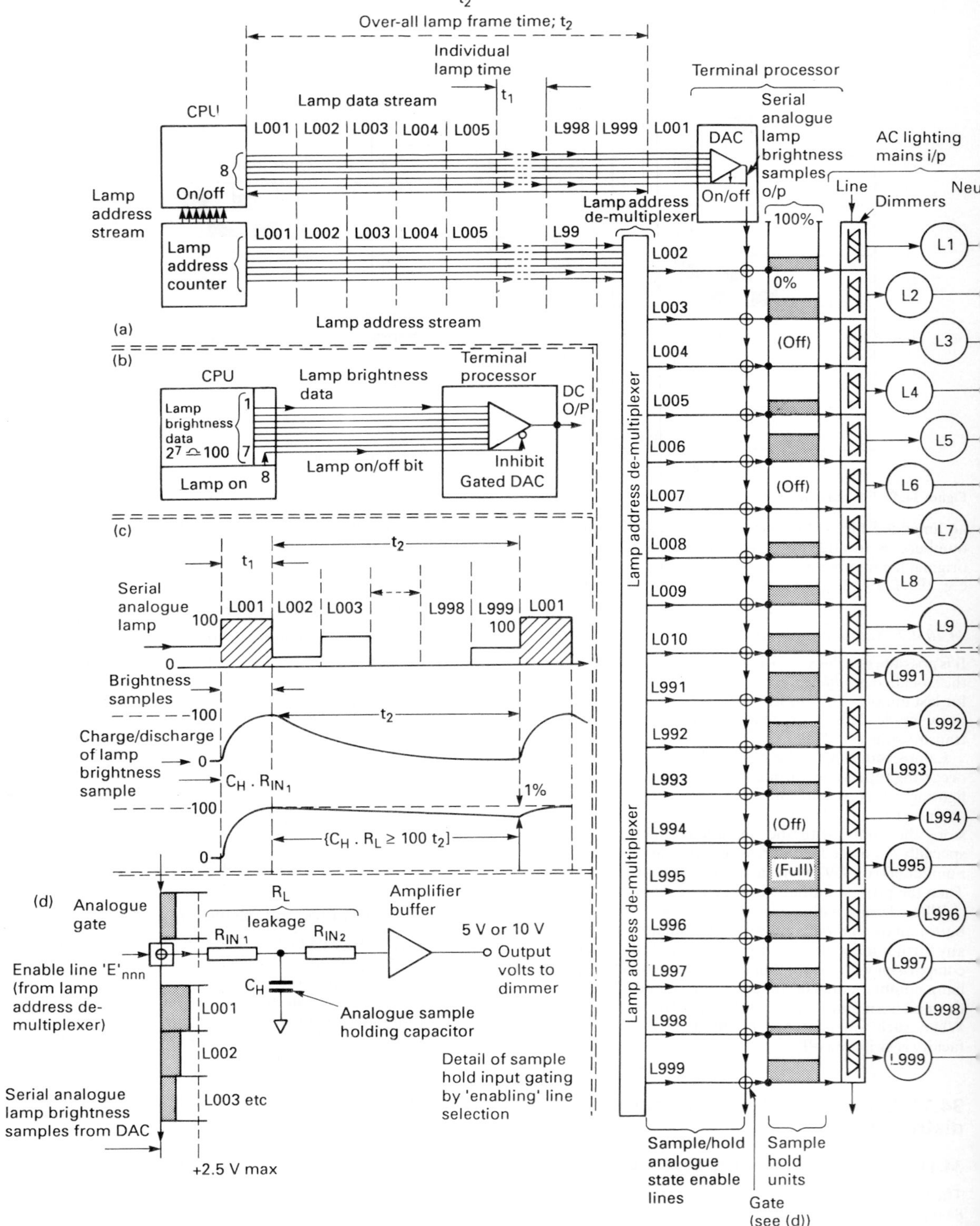

Figure 34.48 Lighting control data bus stream and terminal processor (a). Detail of the lamp on/off bit is shown in (b) and the sample and hold in (c) and (d)

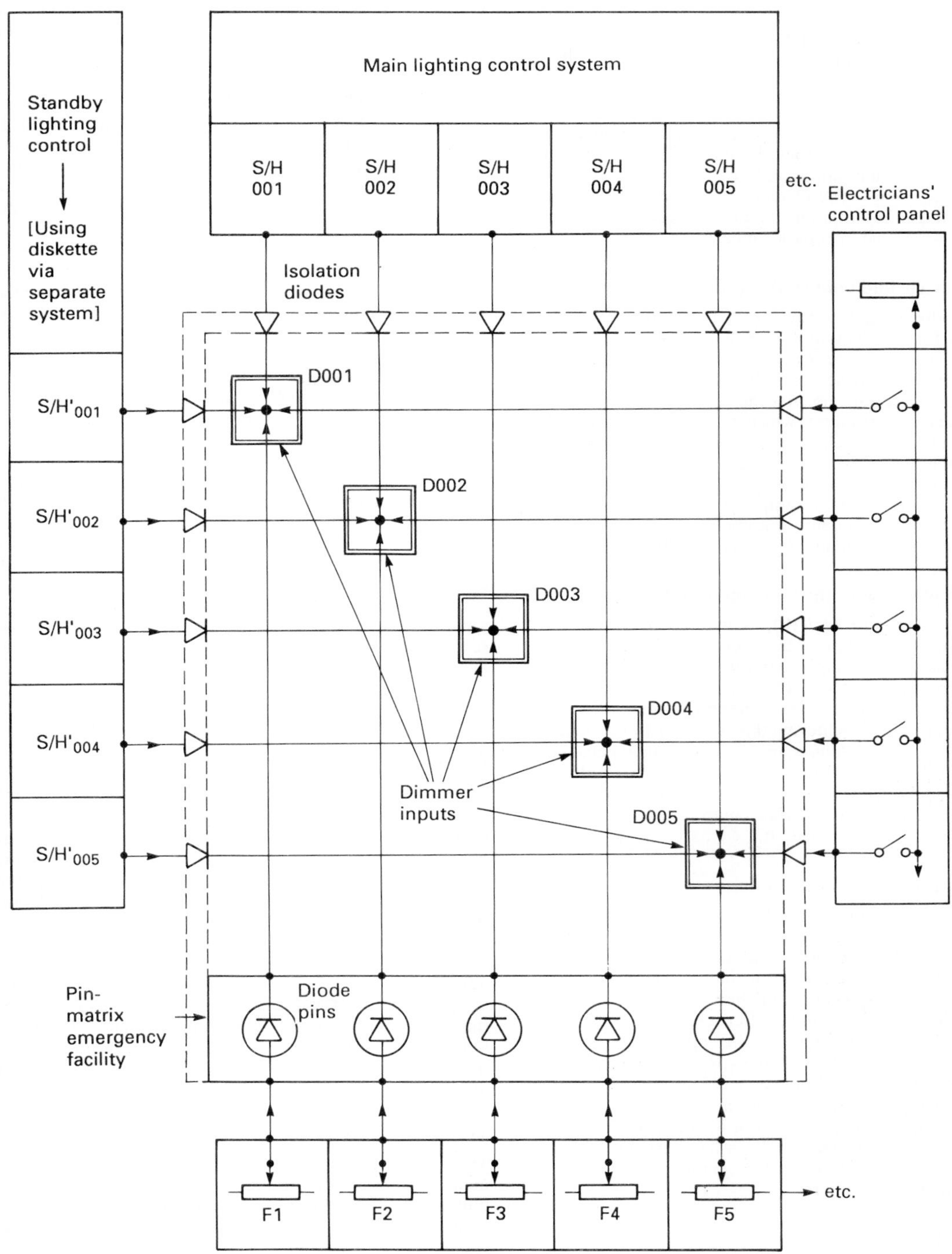

Figure 34.49 Isolation diodes

period of time as the 7 bit character from which it was derived. Usually, the amplitude of each voltage sample has a maximum value of about 2.5 V.

34.11.5 Conversion of serial analogue voltages to parallel

Conversion from serial analogue form into an array of separate voltages in parallel is by means of a bank of sample and hold units. There will be one unit for each dimmer module. The serial analogue signal is fed to each sample and hold unit, but only an "enabled' unit can accept any information from the serial analogue signal. The enabling pulse, which admits the relevant analogue information for the particular sample and hold unit, is generated by separate means.

34.11.6 Sample and hold enable pulse

The lamp address counter provides the required lamp address count feeds for the demultiplexing process shown in *Figure 34.48*. By this means, a series of time-shifted enable pulses is produced. Each of these, as well as being exclusive to a particular lamp address, also has the same time period as the sample of the analogue data to which it relates. Accordingly, by distributing the decoded enable pulses appropriately, the serial analogue time-division period is associated with an allocated sample and hold unit, and accordingly also a dimmer module.

In this way, all analogue samples within the serial signal become held, and subsequently buffered to provide the parallel form voltages for controlling the dimmer modules.

34.11.7 Application of the lamp on/off bit

The on/off bit (*Figure 34.48*(b)) is used to make the final decision (usually quite late in the overall distribution system) whether the associated lamp is to be *on* or *off* in the studio. This leaves the associated lamp brightness available in binary form, for monitoring purposes elsewhere in the system.

A convenient way of achieving this control is by introducing the on/off bit as an *inhibit* signal into the digital/analogue convertor itself, so inhibiting this device. As a result, the brightness signal conveyed serially to the sample and hold unit, and thereby to the dimmer module control input, is zero, and the associated lamp is switched off.

34.11.8 Sampling and cycle timings

There are two significant timings (see *Figure 34.48*(c)). Firstly, the time period of the sample pulse, t_1, is equivalent to the initial time period for the master oscillator used in the lamp address counter. Secondly, the cycle time t_2, which is the lamp counter frame time, spans the time between updating sample pulses for the same lamp number. Between these updates the leakage, R_L, of the hold charge in the capacitor, C_H, will be effective. Accordingly, $t_2 = 100\ C_H R_L$ is important as a technical design requirement, to avoid leakage of charge affecting the lamp brightness. Furthermore, t_1 is related to the time required by the sample capacitor to charge to maximum.

The buffer amplifier which isolates the capacitor will also raise the voltage level stored to a maximum of 5 or 10 V to control the thyristor dimmers.

34.11.9 Isolation diode

This diode forms an important final component in the distribution voltage feeds to the thyristor dimmers. It is shown in *Figure 34.49*. Its role is twofold:

- to isolate the delicate electronics of the lighting control system (via the sample and hold units) from the possibly destructive *blow back* voltages from the thyristor dimmer (240 V ac)
- to buffer the sample and hold control voltage (generated by the lighting control system) from other ancillary control units also contributing to the control of the dimmer

34.11.10 Other ancillary control units

The *pin matrix* is also an emergency lighting control facility (see *Figure 34.49*) and may control the studio lamps via access to the same dimmers by means of a *second* diode connection to the dimmer input without involving the main system. In this case of cathode to cathode diode connection, the highest voltage prevails to control the dimmer.

The selection and control of the lamps is greatly facilitated by the introduction of the so-called *pin matrix board*. This is associated with a separate bank of about ten analogue faders. The diodes are mounted within the pins. By inserting the pins into a line of lamp matrix holes of a separate panel associated with single faders, any number of studio lamps can be controlled by one or more of the faders.

The *electricians' panel* is another form of input diode connection to the same dimmer control point as used by the previous two control signals, i.e. the lighting control system and the emergency or pin matrix system. This connection allows the studio electricians to check out both the dimmers and the studio lamps separately, without involving the control system, which can be switched off.

Here the requirement is to switch on selected lamps at one fixed brightness setting, usually at maximum, by means of an array of switches.

The *standby lighting control system* (using diskette and separate system) incorporates a fourth set of diodes allowing separate and independent control of all the dimmers should the main system fail.

34.12 Lamps

34.12.1 Studio lamps

34.12.1.1 Incandescent filament lamps

These produce visible light as well as infrared energy as a result of the heating effect of the electric current flowing through the filament wire. Tungsten is particularly suitable for a filament material because of its high melting point and low evaporation rate at high temperatures. The filament wire is coiled to shorten the overall length and to reduce the thermal loss.

The proportion of the radiation from a filament which gives visible light increases sharply with temperature, so it is advantageous to operate the filament at the highest possible temperature. In a vacuum, this is limited by the evaporation of tungsten. The evaporation can be greatly reduced by filling the bulb with an inert gas, so the temperature can be raised without the life becoming too short, and the radiation efficiency is increased.

34.12.1.2 Tungsten halogen lamps

The conventional incandescent gas-filled lamp loses filament material by evaporation, much of which is deposited on the bulb wall. When halogen is added to the filling gas (and certain temperature and design conditions established), a reversible chemical reaction occurs between the tungsten and the halogen. Tungsten is evaporated from the incandescent filament, and some diffuses towards the bulb wall. Within a specified zone between the filament and the bulb wall, where the temperature conditions are favourable, the tungsten combines with the halogen. The tungsten halogen molecules diffuse towards the filament, where they dissociate, the tungsten being deposited back onto the filament, while the halogen is available for a further reaction cycle.

The improved efficiency and life of the tungsten halogen lamp over a conventional incandescent lamp does not in fact arise from the re-deposition of the tungsten onto the filament, but rather because the regenerative cycle prevents the accumulation of the tungsten on the bulb wall.

This allows the luminous output figure for the new lamp to be maintained throughout its life, so achieving up to 100 per cent lumen maintenance. The regenerative cycle also permits a radical change in the geometry and size of the lamp. This enables the lamp to operate at an increased gas pressure and hence increased density. The increase in gas density suppresses even further the evaporation of tungsten from the filament.

The options which are available as a result are higher efficiency or longer life. Furthermore, the reduction in size is quite significant. A tungsten halogen lamp has only one per cent of the volume of its conventional counterpart, and so may be more easily incorporated into optical systems.

The greatly improved efficiency is partly attributable to this very significant reduction in size, for now the bulb wall is made of silica and is very close to the filament itself. The clearance is so small that the gas convection currents cannot operate between the filament and the bulb wall to cause further heat losses from the filament. The remaining heat losses are from the filament conduction through the electrodes, and radiation through the silica wall.

The increase in efficiency of tungsten halogen lamps over conventional incandescent lamps is at least 50 per cent. The latter may produce 12 lumens per watt, whereas the tungsten halogen lamp will produce 15–35 lm/W.

34.12.2 Gas discharge lamps

34.12.2.1 Metal halide lamps

The few lines of visible radiation from high pressure mercury discharge lamps (MB) are shown in *Figure 34.51*(a). The absence of energy in other areas of the visible spectrum results in only moderately efficient luminosities of around 50 lm/W and extremely poor colour rendition indices of about 16–50 (out of the possible maximum of 100). It is this colour rendition aspect which is in most need of improvement when considering gas discharge lamps for use in television and film studios.

The method of improving the colour rendition of a mercury lamp, type MB, is to include more than one metal within the discharge tube, so that the emission lines occur over a wide range of the visible spectrum. However the use of other metals, as well as mercury, in the arc tube of a discharge lamp introduces problems, namely:

- The vapour pressure must be sufficiently high for these metals at the temperature of the arc tube wall to be excited into the discharge.
- The metals themselves must not react with the arc tube material or its electrodes at these temperatures.

The arc core temperature may reach 5700 K, so a silica body is used for the tube itself. The problems listed have been overcome by using halides of the metals rather than the metals themselves. The vapour pressure of the metal halide is generally higher than that of the metal itself; the reactivity in the case of alkali metals is less.

The iodide compound is used in practice. The compact source lamps, CID and CSI, may also use the chloride.

34.12.2.1.1 Halide cycle

When a metal halide lamp is first energized, the output spectrum is initially that due to mercury vapour, since the halides remain solidified on the relatively cool arc tube wall. As the arc tube wall temperature increases, the halides melt and vapourize. The vapour is carried into the hot region of the arc by diffusion and convection, and the temperature of the arc causes dissociation of the halide compound into halogen and metal atoms. The metal atoms are then excited in the high temperatures of the arc core and produce their characteristic spectral emission. The atoms continue to diffuse through the arc tube volume and, in the region of the relatively cool arc tube wall, metal and halogen atoms recombine in the form of the halide compound. This recombination process is particularly significant, in the case of the chemically active alkali metals, in preventing attack of the silica wall.

34.12.2.2 Compact source lamps

To achieve still higher source brightness, a lamp should have a short arc gap and a high electric field. This normally means that lamps operate at a higher pressure of mercury or other vapour (usually of several atmospheres). However, the fill pressure of a metal halide lamp when cold is less than atmospheric, so it is safe to handle. The construction of the CSI and CID lamps is very similar. The arc tube is of silica and filled with argon, mercury and the metal halides required. CSI tubes contain gallium iodide, thallium iodide and sodium iodide. CID tubes contain tin iodide and indium iodide.

34.12.2.2.1 Arc tube and jackets

The lamps can be used without the glass jacket to enclose the extremely hot silica arc tube itself. A 1 kW unjacketed lamp is shown in *Figure 34.50*.

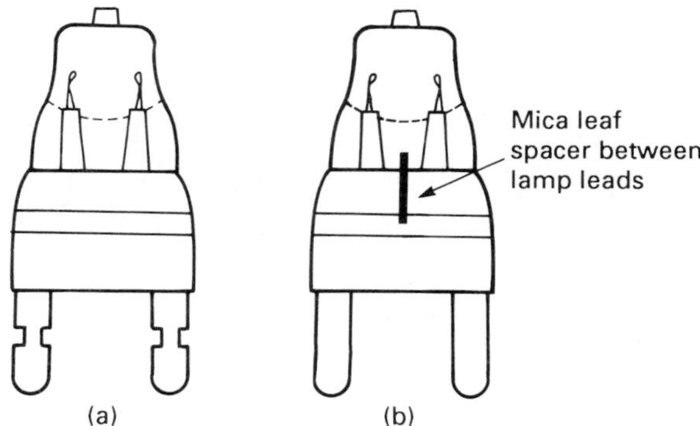

Figure 34.50 Gas discharge lamps, (a) CSI 1 kW, G22 bipin, 7 – 10 kV ignition pulse, (b) CID 1kW G38 bipin, 30 kV ignition pulse

The left-hand lamp (a) has a G22 bipin base and is designed for switch-on using ignition pulses of 7–10 kV, which means the lamp can only be switched on when the arc tube is cold. To achieve re-strike of the lamp while it is still hot requires an ignitor pulse of 30 kV.

To permit such a voltage to be used, without the possibility of arcing or tracking externally, the other lamp shown (b) is mounted on a G38 base. Here the arc tube base pinch is slotted and a mica leaf inserted between the lamp leads.

34.12.2.2.2 Characteristics

The halides are completely evaporated in the compact source lamps as a result of the higher power loading and consequently high arc tube temperatures. The emission spectra for both lamps are shown in *Figure 34.51*(b) and (c). They exhibit a combination of discrete lines and a continuum throughout the visible spectrum.

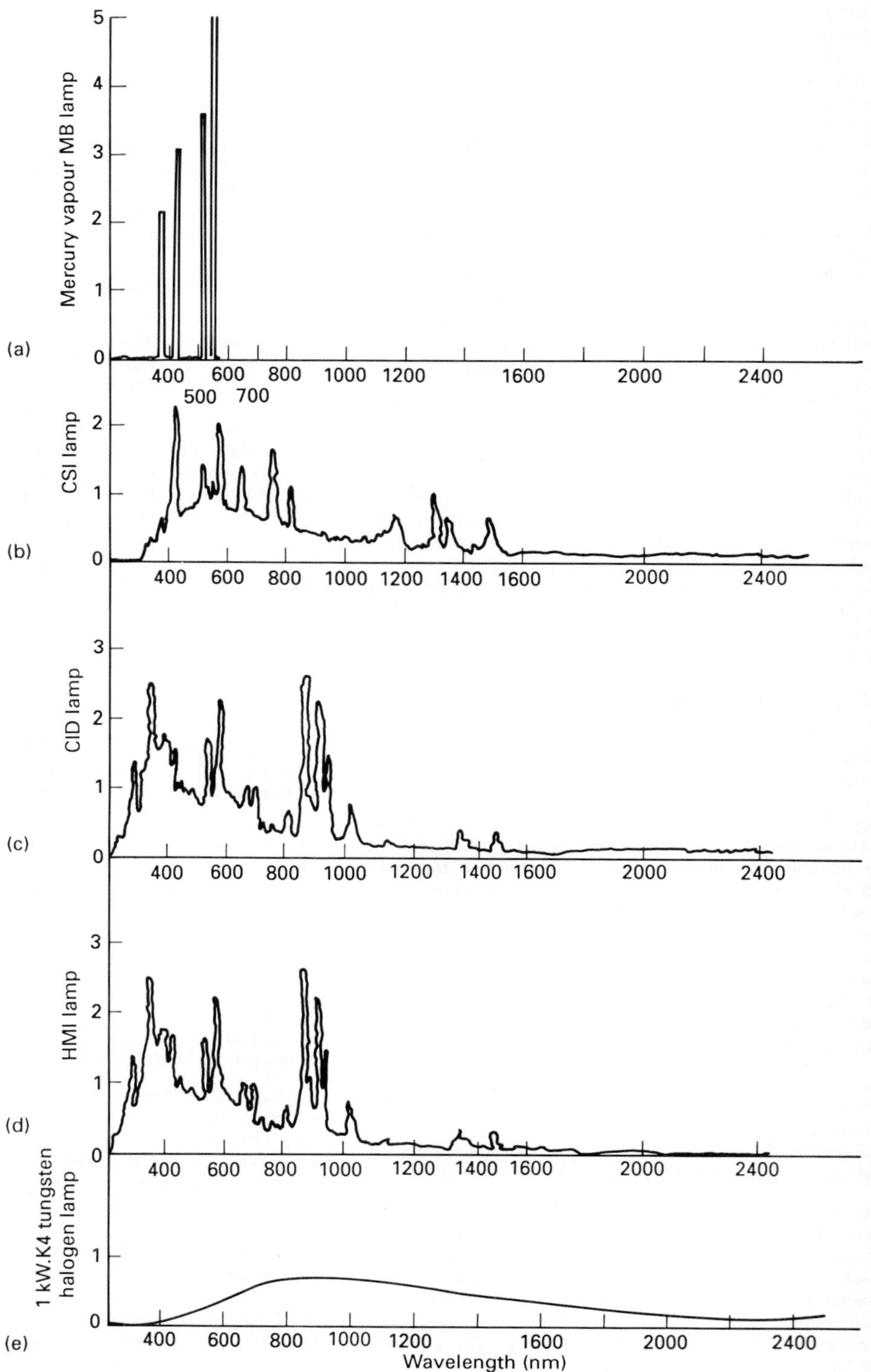

Figure 34.51 Lamp characteristics

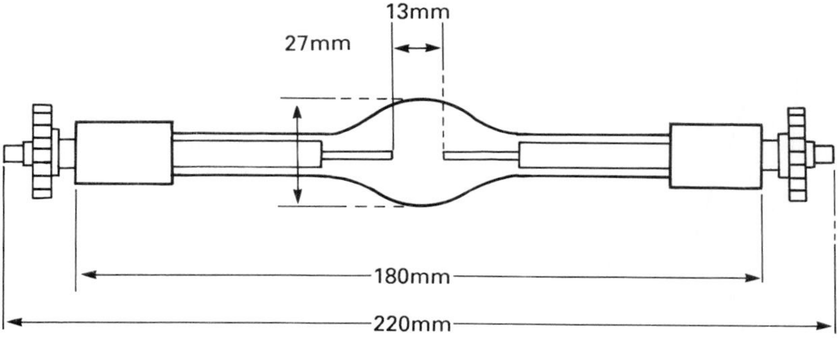

Figure 34.52 HMI 1200 W gas discharge lamp

The colour rendition index R_a is 80 or more for both types of lamp. Furthermore, the CID lamp (using tin halide) has a relatively low melting point and a high vapour pressure, which results in the spectrum being insensitive to power dissipation. This means the emission is unchanged for a mains voltage variation of up to 50 per cent (i.e. the colour temperature does not change).

34.12.2.3 Double ended lamps

A double ended HMI lamp (*Figure 34.52*) consists of a double ended, thick walled silica arc tube with axially mounted tungsten electrodes. The lamps are filled with mercury and argon, together with dysprosium iodide, thallium iodide and holmium iodide, giving a daylight spectrum, with a colour temperature of 5600 K (±400 K).

The high degree of spectral continuum results in a colour rendition index R_a greater than 90. The luminous efficiency is between 80 and 120 lm/W, and they produce typically 1100 cd/m^2. The double ended design facilitates hot restrike operation.

34.12.2.3.1 Circuits for metal halide lamps

The electrical characteristics of these lamps are very similar to corresponding mercury lamps, except that increased voltages are required for starting, run-up and restriking (following extinction). As with mercury lamps, the negative slope volt/amp characteristic requires a choke ballast. In particular, CSI, CID and HMI lamps have high filling pressures and require extra high voltages for cold starting, and even higher for hot restriking. This is usually achieved by generating from a spark gap a burst of high frequency pulse voltages every half-cycle of power. For this, the ignitor must be mounted very near the lamp. *Figure 34.53* shows a typical starter circuit.

34.12.3 Illumination characteristics of television lamps

34.12.3.1 Thermal radiation

When a body is heated to a high temperature, its constituent atoms become excited by numerous interactions, and energy is radiated in a continuous spectrum. This arises because the energy levels of the electrons in the solids are broadened to the point of merging into a continuous band. The extent of the energy levels of this continuum, plotted as packets of energy per wavelength as the temperature is increased, is shown in *Figure 34.54*.

The thermal radiation in *Figure 34.54* is from the purest source of non-selective energy: a *black body*. For other bodies, which are not so black, but more shiny and more reflective, such a curve would not be so continuous or calculable; they cannot therefore be used as bases for reference.

34.12.3.2 Thermal radiation peak

It will be seen from *Figure 34.54* that the packets of energy increase in value rapidly with increasing temperature. In fact, the wavelength of the maximum energy packet is inversely proportional to the temperature, T.

34.12.3.3 Colour temperature

As the temperature increases, the peak of the power radiation shifts from the red end of the spectrum, through the yellow, to the blue, and this agrees with the visual colour observation.

It is possible to calibrate this colour shift by colorimetric means, and to illustrate it on a chromaticity diagram. Such a diagram is the CIE 1970 version, which uses U and V coordinates. These chromaticity coordinates are obtained from the spectral power distribution curve of the black body radiator, and plotted for the various temperatures as a smooth

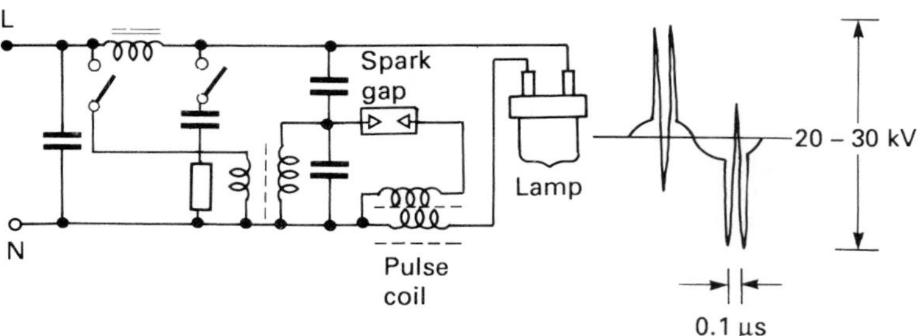

Figure 34.53 CSI Ignition circuit for hot restart

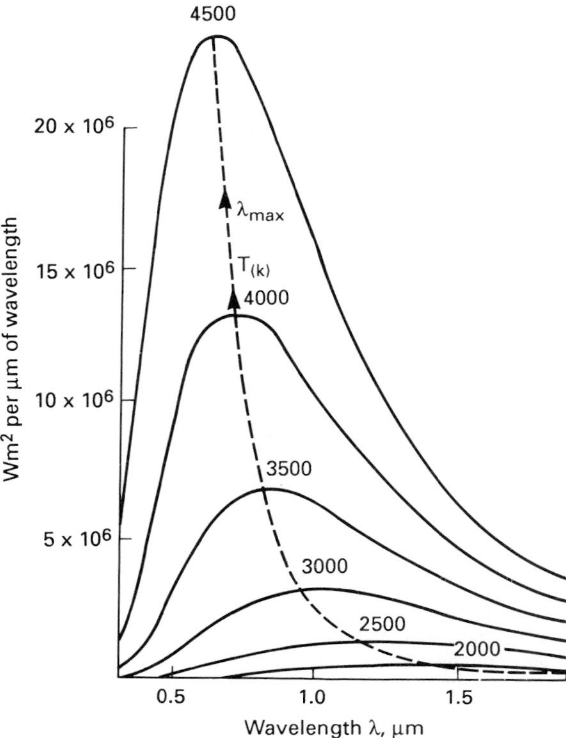

Figure 34.54 Black body radiation. The spectral power distribution curve for a black body radiator at various temperatures of incandescence

curve called the *black body temperature locus* (see *Figure 34.55* on opposite page).

The chromaticity measurement of a light source places it on a locus that identifies its colour temperature.

A light source that is not on this locus can nevertheless be quoted as a figure of *correlated colour temperature*. This is the temperature of the black body radiator whose perceived colour most closely resembles that of the source.

The recommmended method of determining the correlated colour temperature of a source, is to draw a line (the *ISO temperature line*) from the chromaticity of the source to cut the locus at right-angles (on the CIE 1970, U V chromaticity diagram). The ISO temperature line is shown in *Figure 34.56*.

34.12.3.4 *Standard illuminants*

The existence of the locus enables a set of standard illuminants to be established for reference and comparative purposes. These are indicated (*Figure 34.56*) as follows:

CIE Illuminant A: Colour temperature 2856 K. A standard for incandescent lamps.
CIE Illuminant B: Colour temperature 4874 K.
CIE Illuminant C: Colour temperature 6774 K.
CIE Illuminant D: Colour temperature 6500 K.

The Illuminant D is a daylight reference, approximately equivalent to the north sky.

34.12.3.5 *Colour rendition of lamps*

This is an important consideration in television operations where lamps, other than incandescent lamps, are used, i.e. gas discharge lamps. The latter do not have a continuum of radiated energy output, but large peaks of energy (or *lines*) between which little radiant energy is visible. Although they can be quoted as sources with chromaticity coordinates, e.g. with a correlated colour temperature, this information is insufficient without further details about the missing radiant energies. It results in a false representation of the colours from these lamps. However, an attempt has been made to introduce a *figure of merit*. This has had some acceptance in general lighting circles, and is called the *colour rendition index*, R_a.

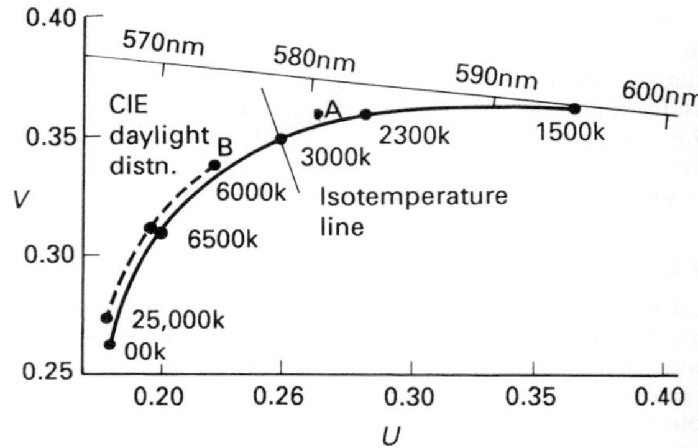

Figure 34.56 CIE (U, V) chromaticity diagram with black body locus and ISO temperature line, also locus of CIE daylight illuminants

The colour rendition index is derived from the colorimetric measurements made on selected Munsell chips, using a specified test lamp. The same measurements are made on the unknown lamp. The unknown lamp readings are then subtracted from the test lamp readings to establish the difference between the two sets. These colour differences are rms averaged and then labelled as ΔE. This is a single figure describing the overall error around the complete hue circle for the lamp in question for the colour samples used. In this form, however, it has not received a wide acceptance in television where a more usual unit is the *just noticeable difference* (JND). The JND approximately equals 4.6 times ΔE. Accordingly, the colour rendition index, R_a, which is a percentage scale, is introduced by subtracting this augmented colour difference error from 100, i.e. $R_a = 100 - 4.6\Delta E$.

The colour rendition indices for studio lamps are:

tungsten halogen: $R_a = 100$
high pressure mercury vapour: $R_a = 16$–48
metal halide discharge: $R_a = 70$–90

The metal halide lamps, such as CSI, CID and HMI, are considered to be satisfactory for outside broadcast television use.

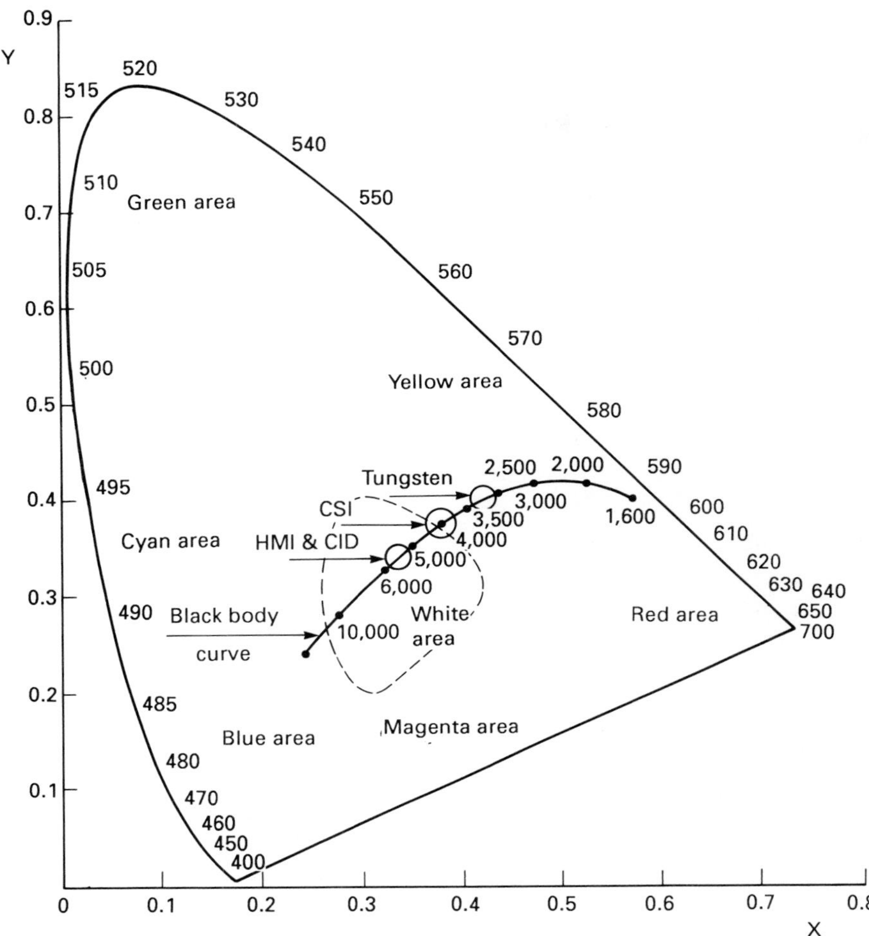

Figure 34.55 Colour temperature

D Bryan
Formerly Technical Director, Michael Cox Electronics Ltd

35 Mixers (Switchers) and Special Effects Generators

35.1 Definition of terms

35.1.1 Vision mixer

A vision mixer combines television pictures from several sources such as cameras, video recorders or telecine machines to produce a continuous programme. The pictures may be selected by instantaneous switch transitions in a time sequential manner or may be combined together on the screen in various ways, principally *mixing*, *wiping* or *keying*.

Variations of the basic studio mixer may be used for video tape editing or presentation, i.e. the linking of programmes, commercials, promotions and announcements fed to the transmitter.

32.1.2 Mixing (or dissolving)

Proportions of two or more picture signals are simply added together so that the two pictures appear to merge on the output screen. Very often this process is used to move on from picture A to picture B. In this case, the proportions of the two signals are complementary so that as the contribution of picture A changes from 100 per cent to zero, the contribution of picture B changes from zero to 100 per cent. This transition would conventionally be under the control of a manual fader lever arm (or *T-bar*) but it may also be electronically controlled at a preset rate.

To achieve certain visual effects, the proportions of the A and B pictures might not complement each other, being independently controlled by the two halves of a *split fader*. In this case, the equipment must protect against output overload and deviation from the standard levels of the synchronizing components of the output waveform. An extended form of this technique, possibly derived from the philosophy of audio mixing, is the 'knob a channel' mixer, in which each of perhaps eight inputs has its own independent fader lever arm.

35.1.3 Non-additive mixing

Non-additive mixing (*NAM*) combines two pictures electronically so that, at a given point on the screen, only the picture with the greater luminance at that point is visible. A fader lever arm may be used to effect a transition from picture A to picture B, or NAM may be offered with both signals contributing equally, the situation at the mid position of the fader lever arm if fitted.

NAM may be used to insert high luminance titles over lower luminance backgrounds or for creating ghostlike effects as the pictures appear combined and yet separate. A further use, where split fader lever arms are not provided for (additive) mixing, is to combine two pictures without reducing the intensity of either.

35.1.4 Wiping

The output picture is divided, spatially, between two input pictures, the shape of the division being an electronically generated geometric pattern. The pattern may vary from simple straight lines to complex shapes such as stars. This pattern may change size under fader lever arm or electronic control so that a transition from picture A to picture B is achieved. It is important to recognize that the input pictures themselves do not move or change size but only the boundary between them. This dividing line may be made diffuse or soft-edged and may be highlighted by an accompanying coloured border.

A simple, preset wipe such as a rectangle is sometimes known as a *split screen*. This is not normally used to effect a transition between sources and may be alternatively considered as a key.

35.1.5 Keying

The output picture is divided, spatially, between two input pictures, the division being controlled by some attribute of a video signal known as the *key* signal. The key signal may be one of the input picture signals (*internal* key or *self* key) or it may be a third signal (*external* key).

To achieve effects such as the insertion of titles or captions, the key signal amplitude is the attribute used to control the separation (*luminance* key). In this case, the video signal filling the caption may be the same as the key signal (usually white), or it may be an electronically synthesized colour signal. By generating key signals larger than the accompanying fill signals, borders around keyed captions may be achieved.

Chroma key, or CSO (*colour separation overlay*) in BBC terminology, uses the attribute of colour saturation of a particular hue to divide the screen between the input pictures.

One of the input pictures is normally also the key source. The scene to provide the key signal is normally set and lit to provide an area of high saturation of a particular hue. The chroma key system replaces this saturated area, e.g. a 'notice board' beside a news reader or a total background area in a drama studio, with the picture from the second picture source. In the example, this might consist of news material or a specially shot background picture.

As in wiping, the inserted picture size or geometry is not adjusted to fit the shape of the insert. If necessary this could be achieved by digital effects equipment, by specially shot recorded material or, if a live camera picture, by framing the infill shot appropriately.

It is, of course, necessary to ensure that no part of the chroma key scene, other than that to be replaced with the fill picture, contains any saturated colour of the chosen hue.

35.1.6 Types of switcher

Production switcher is the American term for a *studio vision mixer*.

Post production switcher is the American term for a *video tape editing mixer*. These mixers may either be specially designed for editing or they may be normal studio mixers interfaced to edit controllers.

Master control switcher is the American term for a *presentation mixer*. The British use it to refer to a routeing system associated with feeding station outputs to line.

Routeing switcher is an equipment which allows a number of input signals to be fed to a number of outputs. Each output destination can select only one input at a time but, normally, an input can be selected by more than one, indeed by all destinations if required. The video routeing switcher is often accompanied by similar systems for distributing programme audio, talkback audio, signalling tallies or cues, time code and possibly telephone connections. Sizes may vary from 2 inputs × 1 output upwards. 64 inputs × 64 outputs is not uncommon.

35.2 Routeing switchers

35.2.1 Matrices

Routeing switchers are represented schematically as XY matrices of input and output buses. At each bus intersection is a switch capable of joining input bus to output bus. These switches are knows as *crosspoint switches*. In early designs, they were electromechanical relays, but in modern designs they are electronic switches (see section *35.2.3*). The control system is arranged so that only one crosspoint per output bus may be turned on at a time. Inputs and outputs are normally buffered, inputs to provide a high input impedance or accurate termination over the video band, outputs to provide one or more 75 ohm outputs.

35.2.2 Performance

As a video signal may pass through many switching matrices between picture source and transmitter, it is essential that the performance of each matrix is as near perfect as possible. In addition to the common video performance parameters such as linear and non-linear distortions and noise, there are performance parameters peculiar to matrices, namely timing spread, switching disturbances, bus interaction and signal crosstalk.

35.2.2.1 Timing spread

It may be important, particularly in NTSC and PAL installations, that the timing of a video signal arriving at a destination, perhaps a mixer input, is constant to within one degree of subcarrier. This timing accuracy must be maintained when switching between inputs of a switching matrix feeding this destination. As this performance must be achievable on all outputs, the inputs of the matrix must be co-timed and the transition time from any input to any output should be identical. This is achieved by attention to the matrix topography and, ideally, by constant delays throughout crosspoint switches.

35.2.2.2 Switching disturbances

On some occasions, the result of a change of source selection made on a switching matrix is fed through following equipment to the transmitter. It is desirable that the resulting disturbance is minimal. A switch or cut between non-synchronous sources is bound to be visible, but a cut between synchronous pictures should not be objectionable. It is normal for the control system to make the change between sources during field blanking so that no effect is seen on the picture. This is best timed to occur between the end of field synchronizing pulses and the start of any insertion test or data signals.

Disturbances may also be caused by voltage spikes superimposed on the output signal, either by coupling of the switch controlling voltage onto the signal or by overlapping or gapping switch timing. Finally, disturbances may be caused by changes of dc level at the output which are not coped with satisfactorily by following black level clamp circuits. These changes may be caused by varying input to output offset voltages of crosspoint switches or they may be inherent in the design. Matrices may be dc coupled throughout, in which case differences of input dc voltages will cause steplike changes at the output, or they may be ac coupled in which case differences of picture content will cause the same effect.

The input buffers will ideally contain some form of dc restoration, usually diode dc restorers, as black level clamp circuits need clamp pulses, and switching matrices are normally capable of operating with non-synchronous inputs so a sync-separator and clamp pulse former would be needed for each input. Diode dc restorers will only operate satisfactorily with composite inputs.

35.2.2.3 Bus interaction

It is desirable that no changes of gain or timing occur to a signal selected on an output bus if that input is selected or de-selected on another output bus. This requires that the input buffer should have low output impedance and that the output buffer should have high input impedance. It is advantageous if the crosspoint switches themselves are buffers drawing minimal input current.

35.2.2.4 Signal crosstalk

It is important that minimum coupling exists between unselected inputs and outputs. This is particularly difficult in coded systems where there is high energy distributed around the subcarrier frequency. This is more liable to leak to adjacent circuits and may be demodulated to produce large area colour patterns depending on the relative subcarrier frequencies of wanted and unwanted signals. The main causes of signal crosstalk are incomplete isolation of crosspoint switches whether at low frequencies or due to shunt capacity at high frequencies, stray capacity to high impedance sections of circuitry, and shared earth current paths, accentuated at high frequencies by series inductance, particularly in the areas of line driving output stages.

35.2.3 Crosspoint switch design

The ideal crosspoint switch[1] should exhibit the following characteristics:

- small size
- transparent performance when *on*
- low dc offset
- low feed through when *off*
- fast switching action, ideally less than 100 ns
- low power dissipation
- ease of incorporation into large matrices

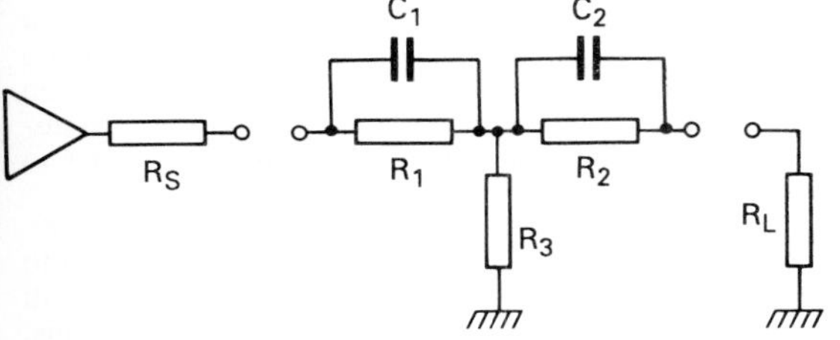

Figure 35.1 Equivalent circuit of unbuffered crosspoint switch

Figure 35.1 shows the equivalent circuit of an unbuffered crosspoint switch. R_1 and R_2 are low resistance when the switch is *on* and high when it is *off*. R_3 is low resistance when the switch is *off* and high when it is *on*. R_3 is provided to produce extra attenuation when the switch is *off*.

It is important that R_L is low to minimize capacitive crosstalk through stray capacitance such as C_3. R_L is, in fact, defined mainly by the output of an *on* crosspoint switch and so will be approximately equal to $R_S + R_1 + R_2$. It is therefore important that R_1 and R_2 are low when *on* so as to minimize R_L.

A practical realization of the unbuffered crosspoint switch is shown in *Figure 35.2*, using diodes as variable resistance elements.

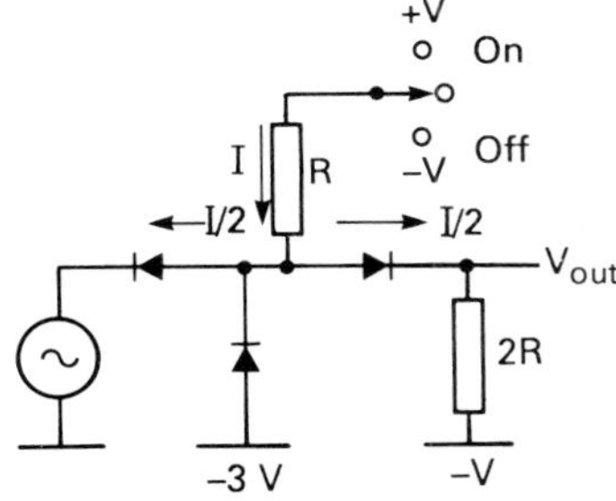

Figure 35.2 Practical unbuffered crosspoint switch

35.2.3.1 Buffered crosspoint switch

The circuit diagram of a buffered crosspoint switch is given in *Figure 35.3*. The three diodes of the diode switch are replaced by three bipolar transistors. S_3 is the control transistor. When it is turned *off*, S_1 and S_2 act as a pair of complementary emitter followers. When S_3 is *on*, its collector falls to the saturation voltage, turning off S_1 and S_2 and hence blocking the signal path. This circuit has several advantages:

- almost exactly unity gain
- low dc offset, temperature stable
- high attenuation when *off*
- low output impedance, not dependent on input bus impedance
- input bus buffered from output bus capacitance when *on*

The last feature is important as it means that the capacitive load of the output bus is not applied to the input bus. Without this feature, the capacitive loading on the input bus varies with the number of output buses selecting that input. This gives rise to changes of timing and high frequency response.

Crosspoint switches are often fabricated as thick film hybrid microcircuits to give consistent performance, small size and low assembly cost.

35.3 Vision mixer architecture

35.3.1 Switching matrix

With the exception of the 'knob a channel' type described briefly in section *35.1*, virtually all vision mixers consist of a switching matrix plus a number of mix/effects (m/e) systems. A mix/effects system may be as simple as a two input mixing amplifier, or it may have as many as four inputs which can be combined, together with internal synthetic colour sources, by

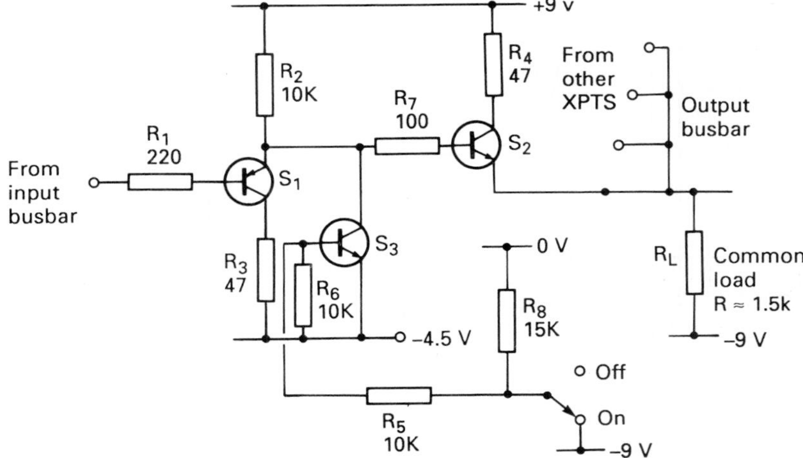

Figure 35.3 Buffered crosspoint switch

mixing, wiping and keying. The matrix is used to select, from the relatively large number of inputs, the signals to be fed to particular m/e system inputs. These selections will change during programme time, and so the performance of the switching matrix must not give rise to disturbances.

The number of matrix output buses feeding mix/effects system inputs, together with the number of matrix inputs, is sometimes used to give a rough indication of the size and power of the mixer. Other matrix outputs are often used for preview and utility purposes.

35.3.2 A/B mix/effects system

The simplest mix/effects system is one with two inputs, normally fed from two output buses of the switching matrix. The inputs and the buses feeding them are known as A and B, hence the name. The system is able to mix, wipe and possibly non-additively mix between inputs and to key one, usually B, into the other.

In early designs, the different effects would have been carried out in different amplifiers with a selector switch at the output. In modern designs, there would normally be only one set of processing electronics which would be controlled appropriately for the required effect.

35.3.3 Cascade systems

Because a combination of only two pictures on the screen simultaneously is artistically limiting, several A/B mix/effects systems may be combined in a cascade formation. The output of the highest m/e may be selected as the input to the next lowest and so on. When one m/e feeds another, the process is known as re-entry.

As an example, we could have a wipe between two inputs at M/E1 (*Figure 35.4*) forming the background for an artist, chroma keyed in at M/E2, with a caption keyed over everything at M/E3.

35.3.4 Timing

As video inputs to a mix/effects system must be identically timed, particularly for NTSC and PAL, delay lines are shown in *Figure 35.4* which compensate for the delays through the upstream mix/effects systems. These delays are typical of those found in most mixer architectures.

35.3.5 Transition type mixers

The cascade system cannot always achieve its apparent potential. Imagine, for instance, that in the earlier example, M/E1 has completed a wipe, leaving the chroma keyed artist in front of a single background picture with the overall title inserted at M/E3. If it is now required to mix the whole composite picture to a single source, this is not possible. It would have been possible, however, if the chroma key could have been transferred to M/E1 and the title key to M/E2 leaving M/E3 free to perform the mix.

A possible way round this difficulty is known as *multiple re-entry* where each mix/effects system is available as an input to every other. In other words, the cascade works in both directions, and the output point is selectable in some way. This, however, involves complex delay switching.

The problem was analyzed in detail by Central Dynamics Ltd of Montreal, Canada who, in 1976, introduced, as the CD 480 series, the transition type of mix/effects system which has subsequently been adopted by many other manufacturers[2].

In essence, this consists of a four input mix/effects system with a single lever for control. The inputs may be referred to as:

programme background
preset background
key level 1 (mid-ground)
key level 2 (foreground)

Transitions can be selected to take place at any of the three levels. Background transitions cause a change from *programme background* to *preset background*. Key transitions cause a change from *key-on* to *key-off* or vice versa depending on the initial conditions. The same type of transition, cut, mix or wipe, takes place at all levels.

To achieve the effect discussed in the earlier example, key 1 would have been *on* and set for *chroma key*. Key 2 would have been *on* and set for *title key*. Background wipes would have been achieved by selecting *background* transitions and wipes.

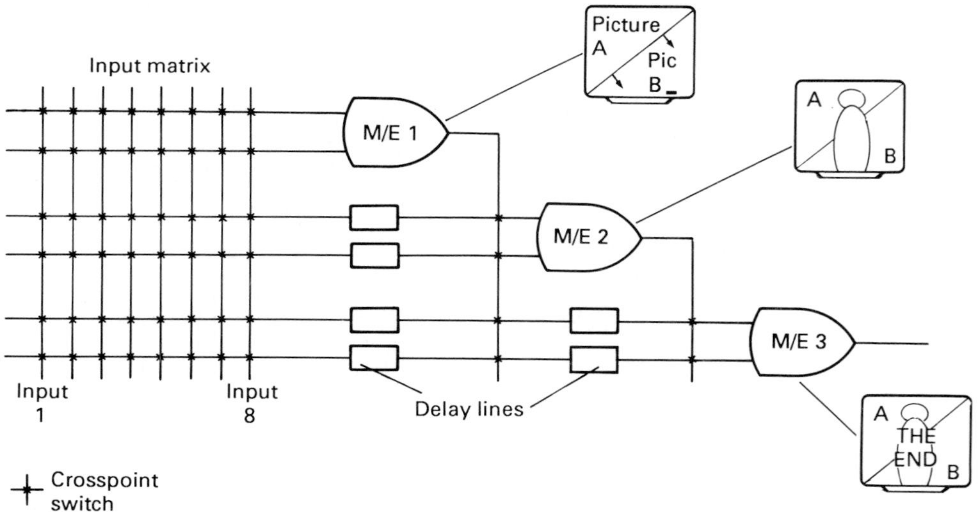

Figure 35.4 'Cascade mixer arrangement showing, in sequence, a wipe between two background pictures, artist superimposed by chroma key, and caption keyed over the M/E2 output

The final mix would have been achieved by selecting *background*, *key 1* and *key 2* transitions and *mix*.

Central Dynamics realized the transition effect by mixing or wiping between two cascaded A/B systems with three A/B mix/effects amplifiers in each. An important advantage of the system is that the cascade system not being transmitted can be fed to a preview monitor to show the result of the next transition.

Other manufacturers, including Michael Cox Electronics Ltd have used multi-parallel-input mix/effects amplifiers to achieve similar results. The apparent cascade effects are achieved by combination of the control signals rather than by physical cascading of the video amplifiers. This has the advantage of reducing the number of stages a signal may have to pass through and eliminating the need for compensating delays.

35.3.6 Cut row

Some mixers include a *cut row*, a matrix output bus which feeds all the main inputs plus the mix/effects outputs to the mixer output. This is considered desirable for live productions where instantaneous access to an input is desirable.

35.3.7 Flip-flop mixing

An extension of the cut row idea is to provide a pair of buses, *programme* and *preset*, with the ability to additively mix between them. The lever action is arranged so that, at the end of a mix from *programme* to *preset*, the sources selected on the two buses interchange so that the *programme* bus is again providing the main output. Thus the *programme* bus can also be used as a *cut row*. Flip-flop action is normally provided on the background buses of a transition type mix/effects system.

35.3.8 Downstream keyer and fade to black

The *downstream keyer* (dsk) is used to insert titles over the picture produced by the mix/effects systems and is normally the final operational part of the mixer. It is provided in this position as it is clear to the operator that it will produce an overall title irrespective of m/e operations. The final control may be an overall *fade to black*, often integrated electronically with the downstream keyer, but in some cases a separate unit, and in others part of an output processing amplifier.

35.4 Implications of colour system

In all systems, provided that the sources are identically timed, proportions of two or more luminance signals can be added and the resulting signal will represent the combined luminance.

NTSC and PAL have the advantage over SECAM that subcarrier phasors represent a linear mapping of the UV plane of colour space. Vectorial addition of the subcarrier components of two or more NTSC or PAL signals, assuming them to be of identical reference phase, will have the same effect as mixing light of equivalent chromaticities. It is therefore normal practice to mix NTSC and PAL in single channel mixers, i.e. luminance and subcarrier are not separated.

SECAM, having frequency modulated colour subcarrier, is at a disadvantage for mixing. The addition of subcarrier signals at two different frequencies does not yield an average frequency, which would be necessary for true colour mixing (ignoring the complicating effects of high frequency pre-emphasis). Furthermore, switching during picture time between two subcarrier frequencies, a process which might suggest the possibility of keying, is not acceptable. The instantaneous switch between two random sine-waves introduces a voltage step with inherent high frequency components. These high frequency components are demodulated as an interfering colour signal.

SECAM is normally mixed by partially demodulating to luminance and a sequential D_r/D_b signal, the latter being effectively baseband R-Y on one line and B-Y on the next. These signals add arithmetically to give correct colorimetric results. The mixer, however, must have two video channels, one capable of processing the bipolar D_r/D_b signal. It is important that differences in path length of the luminance and colour difference channels are equalized before re-coding to SECAM.

A disadvantage of this method is that there is no readily available method for cleanly separating the subcarrier from the luminance of SECAM signals. Comb filter methods do not work as, unlike PAL and NTSC, the luminance and chrominance spectra are not interleaved. It is normal to separate luminance by means of a low pass filter of around 3.1 MHz and to discard luminance above that frequency. The mixer is normally bypassed when not actually mixing, and entry and exit from the bypass mode are sometimes visible, if only due to the change in luminance bandwidth.

It has been shown[3] to be financially as well as technically advantageous in small SECAM installations, particularly with a majority of camera or telecine sources, to use a time channel RGB (or Y, R-Y, B-Y) mixer followed by a single SECAM coder. This approach may be made even more relevant with the advent of analogue component video tape recorders.

Analogue component video, RGB or Y, R-Y, B-Y can be processed in three parallel channels. It is an advantage to concentrate as much of the complexity as possible in the common control system and as little as possible in the video paths. It is important that the gains of the channels track accurately, otherwise colour errors will be introduced. It is also important that the timing and effective dynamic ranges of the control signals fed to the three channels are equal, otherwise colour effects may occur on, for example, the edges of wipes.

Time division multiplexed analogue component systems (S-MAC) have been proposed[4] which combine the advantages of analogue component video with single wire distribution. While mixing of these signals should be possible, it is thought unlikely that no keys and no more than the most basic wipes would be acceptable. Keying and wiping would involve the generation of a control system repeating itself three times at the different degrees of compression of the luminance and colour difference signals. The alternative method would be to de-multiplex to a common time scale, probably fully, but perhaps to the luminance time domain, and to process in a three channel system.

35.5 Mix/effects amplifiers

The term 'amplifier' is widely used, although it is not strictly appropriate here as there is normally no overall voltage or current gain. The basic requirement is for a voltage controlled attenuator, variable between unity gain and, ideally, infinite attenuation, in practice at least 60 dB over the video band. One such attenuator is provided for each input to the mix/effects system. The controlling voltage might be derived from a fader potentiometer for mixing, a wipe pattern generator for wiping, or from a key video signal for keying.

The A/B mix/effects system consists of two such attenuators while a typical multi-input version as used on a transition type mixer might consist of six. The linearity of the output versus control voltage is much more important in the latter case as there is complete freedom in the proportions of input signals mixed together and yet the output must sum to unity.

In a cascade of A/B mix/effects amplifiers, correct output is guaranteed if the sum of the two inputs applied to each pair of attenuators complement each other.

A further requirement is that, as the contribution from a given input is varied, no dc shift takes place. In other words, the signal should be attenuated from a full amplitude video signal down to a dc voltage equivalent to black level. This parameter is not critical for mixing, but rather for wiping and keying where the black level change would occur during picture time and could not be removed by subsequent black level clamping. To ensure that the signal black level reaching the attenuator is constant, a black level clamp will be employed.

A good test of the accuracy of the arithmetic of a mix/effects system is made by feeding a white signal into all the inputs, performing complex effects and looking for deviation from white at the output.

The device most widely used for achieving this voltage controlled attenuation is the double-balanced modulator integrated circuit such as the MC 1496 or TCA 240.

This can be configured in several ways. One common way is to use the two constant current tails, T_5 and T_6, to sink currents $I_b + I_s$ and I_b respectively, where I_b is a current equivalent to black level and I_s is signal current (zero for black). A small voltage applied to the bases of T_1 and T_3 (*Figure 35.5*) causes the current reaching R_L to change from $I_b + I_s$ when T_1 and T_3 are conducting, to I_b only when T_2 and T_4 are conducting.

A cascode arrangement is commonly employed to eradicate the effects on R_L of stray capacitance. A pair of such circuits would be adequate for an A/B mix/effects system.

Non-additive mixing is achieved, electronically, by feeding black level clamped video signals to the bases of a pair of emitter coupled n-p-n transistors.

The transistor with the higher base voltage, i.e. luminance, turns on, turning off the other and hence feeding its signal to the output. To achieve a sharp cut-off, the signals may be amplified to perhaps 3 V peak-to-peak at the transistor bases and subsequently attenuated.

35.6 Wipe pattern generation

The function of the wipe pattern generator is to produce contro

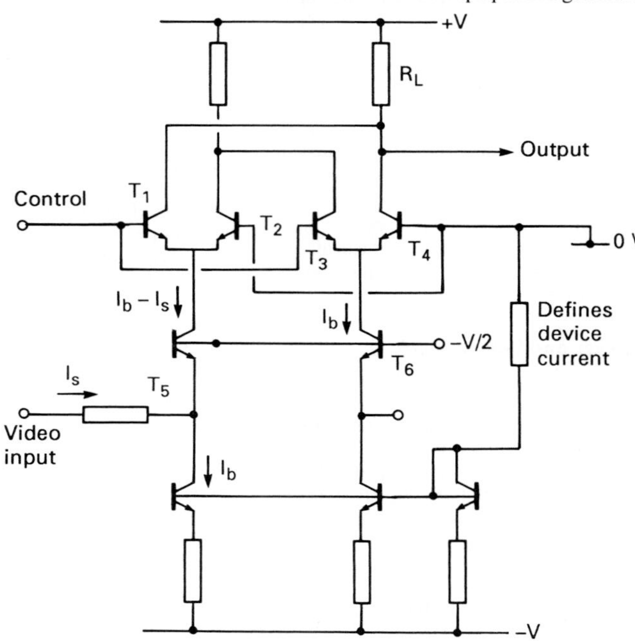

Figure 35.5 Simplified circuit of typical multiplier, e.g. MC 1496 or TCA 240

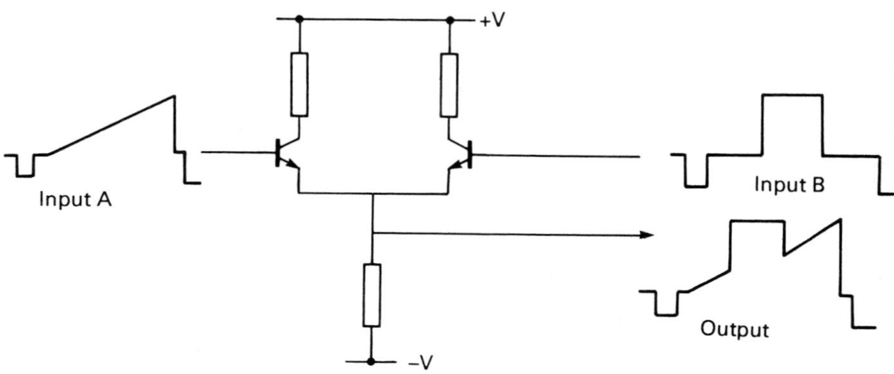

Figure 35.6 Simplified non-additive mixing circuit

signals which, when applied to a mix/effects system, will give rise to the moving, geometric divisions of the screen area known as *wipes*. Generation is divided into two parts: formation of waveforms defining the shape of the pattern, and from these producing movable edges with softness and border characteristics as required.

35.6.1 A simple wipe

The pattern waveforms consist of a voltage, dependent on the horizontal and vertical position X and Y on the television raster. It may be helpful to think in purely physical terms of X,Y and Z. As a simple example, consider an inclined plane sloping upwards from –1 at the left-hand side of the screen to +1 at the right-hand edge. In electrical terms, this is equivalent to a line rate sawtooth signal with no field rate component.

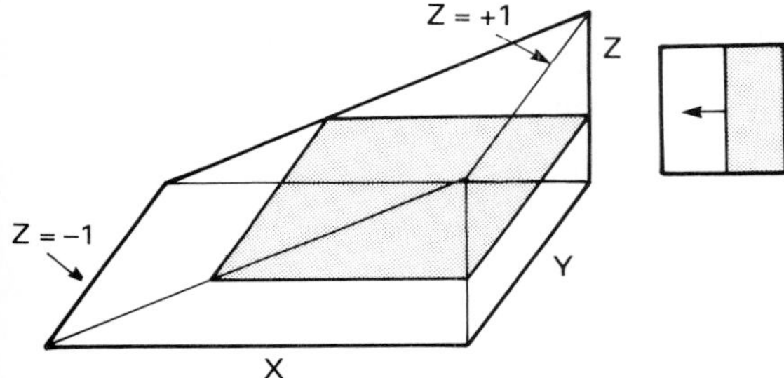

Figure 35.7 'Page' wipe generation

The waveform is converted into a mix/effects control signal by means of a level dependent slicing circuit whose threshold can be varied by means of a fader lever. This circuit will not be a high gain comparator but a circuit whose output, essentially a logic signal, changes state gradually as the input traverses a preset voltage range or window. The width of the window can be altered, the effect being to vary the width of the soft edge.

Borders are generated by the provision of two such level dependent circuits set to work at different thresholds. Simple logical combination of their output signals allows the derivation of a border control signal.

35.6.2 More complex patterns

More complex patterns are generated by modifying the three dimensional signal fed into the slicing circuit. Many patterns can be based on the two ramp signals sloping horizontally and vertically which themselves produce vertical and horizontal edges.

For example, performing a precision rectification (*Figure 35.8*) about the centre of one of these ramps produces a tentlike

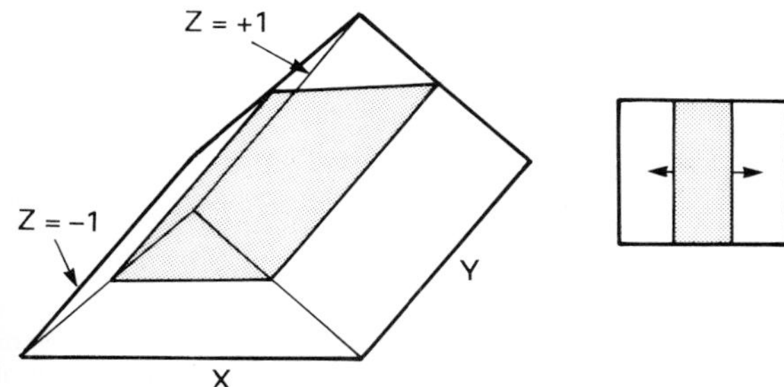

Figure 35.8 'barn door' wipe generation

waveform which, when sliced, gives the effect of wiping from two opposite sides of the screen to a line at the centre. Non-additive mixing of the two ramps, i.e. taking the larger at each point on the screen, creates a square corner insert as in *Figure 35.9*. Performing the same operation on two tentlike waveforms creates a central rectangular box.

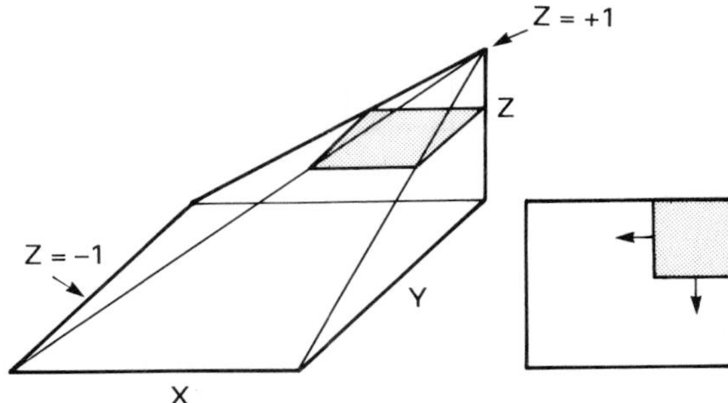

Figure 35.9 Corner wipe generation

35.6.3 Circle generation

The ramps can be manipulated, by squaring, adding and square rooting to generate a conical surface.

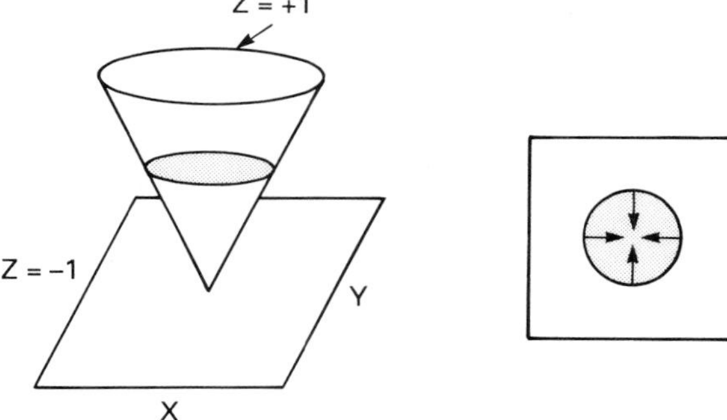

Figure 35.10 Circle wipe generation

The equation of a cone in polar coordinates, apex at the origin, is $h = kr$ where h is the height above the XY plane, k is a constant, and r is the radius.

By Pythagorus,

$$r = \sqrt{(X^2 + Y^2)}$$

Therefore,

$$h = k\sqrt{(X^2 + Y^2)}$$

This relationship is used to generate the conical voltage surface which, when sliced, will render a circular wipe pattern.

35.6.4 Rotation

If the basic ramps are added together in proportion, a diagonally sloping plane will be created. This on its own gives the diagonal wipe. If the proportions of horizontal and vertical ramps are varied, this diagonal will be seen to rotate. If a second mixture of the basic ramps is made whose axis of slope is 90° from the first, this new pair of ramps can be used as the basis of all the other patterns. Multiplication of the contributions of the two fixed ramps by sine and cosine functions will produce smooth rotation from a linear control (see *Figure 35.11*).

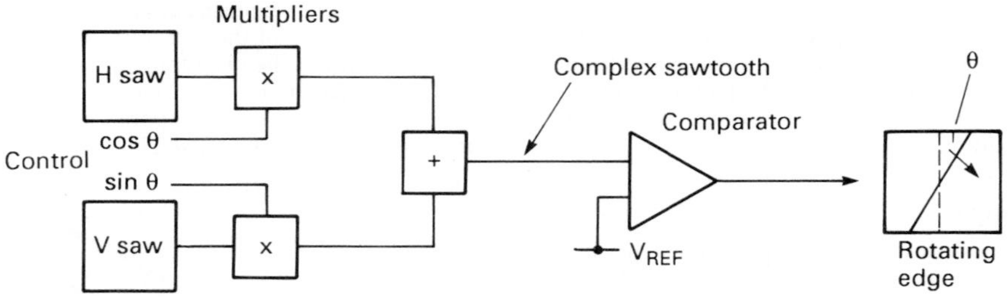

Figure 35.11 Principle of rotary wipes

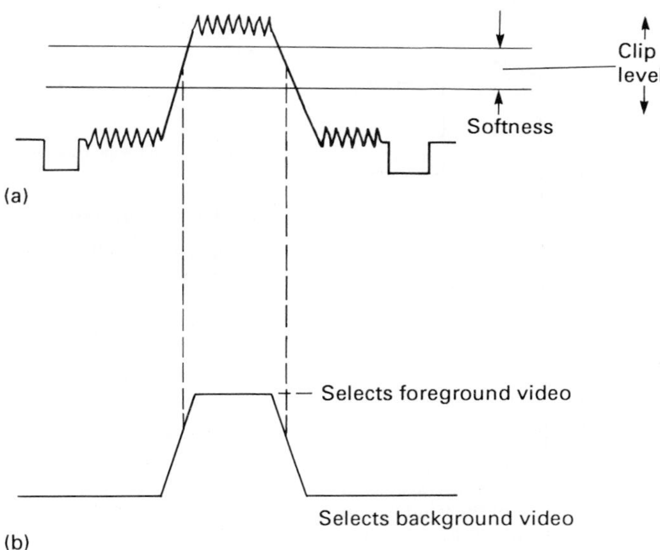

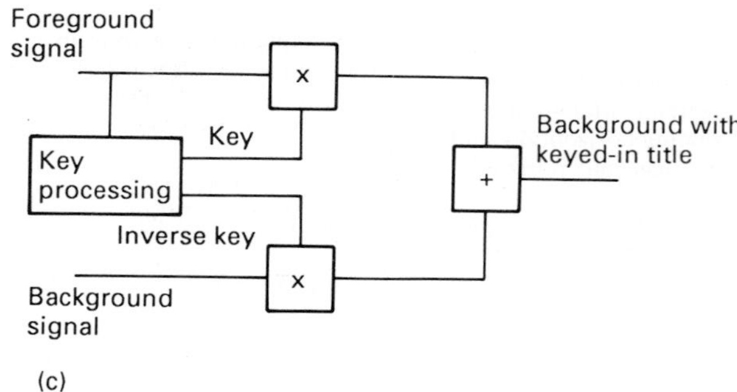

Figure 35.12 Luminance keying, (a) signal from caption camera, (b) processed key signal, (c) block diagram of system

35.6.5 Positioning

Certain of the patterns require to be positioned anywhere on the screen. This can be achieved by altering the initial voltage conditions of the X and Y ramp generators so that the zero crossing point is correspondingly moved. This movement is normally controlled by an XY potentiometric joystick.

35.7 Title keying

Titles or captions may originate from cameras, slides or character generators. In addition to the basic requirement of keying the title into a background, refinements include colouring and edging the characters.

35.7.1 Luminance keying

A control signal is derived, based on the amplitude or luminance of the key signal. A level dependent slicing circuit, similar to that used in wipe generation, is used. The output is essentially a logic signal, but, due to the finite window of the slicing circuit, one which is linear between its upper and lower limiting states. If the key signal has slow edges, these will be reflected in the shape of the edges fed to the mix/effects amplifiers, although some increase of slope would normally take place.

The level about which slicing takes place is known as the *clip level* and is normally a user control as is the *window width* or *key softness*. These parameters are adjusted to give the best visual results and are essential in the case of camera generated captions where noise and shading may be present on the caption signal.

35.7.2 Arithmetic errors

A traditional method for simple black and white caption insertion has been to multiply, in an m/e amplifier system, the caption by the control signal derived from itself, and to multiply the background signal by the inverse of this control signal. This might be expected to produce the effect of white lettering keyed into the background. Assume that the clip level or slicing point is set to the mid-point of the caption signal amplitude (the optimum case). At the mid-point of the edge rise-time, the amplitude of the control signal and its complement will both be half of their maximum value. The video contribution from the background signal will be 0.5 times the background signal amplitude. The contribution from the caption will be 0.5 times the caption signal amplitude, which is at this point 0.5 times maximum, or 0.25 times white luminance. If the background happens to be white, there will only be a signal of 0.75 times white luminance. In other words, there will be a grey rim around the edge of the caption where we might have expected an all white screen.

35.7.3 Matteing

A solution of the above problem is obtained by multiplying the background signal by the inverse of the control signal as before and then adding the caption signal to it, giving the correct arithmetic. This process has become known as *matteing*, a word ill-defined in video use being derived from film technology. (The term *colour matte generator* is used to describe a synthetic colour or colour field generator.)

To achieve satisfactory results, the caption signal needs to be free from noise and shading as these would simply add to the background. This requirement is better met by electronic character generators. Equally good results could be obtained by feeding not the caption signal but a totally white picture into the caption multiplier. This is a special case of colour filling of captions.

35.7.4 Colour filling

It is desirable for artistic reasons as well as for legibility to be able to colour inserted captions. This is achieved by feeding the video input of the caption multiplier with a synthetically generated colour signal (colour matte, background or field). This must conform to, and be phased with, the colour system in use. In PAL or NTSC, this signal is generated in a rudimentary coder with dc inputs controlling luminance and the colour modulators. In SECAM, a baseband generator, conforming to the D_r/D_b format, will be provided for the colour channel together with a separate luminance signal. For other colour vision mixing systems, suitable generators must be provided, e.g. RGB or Y, R-Y, B-Y.

35.7.5 Edge crawl

A problem, a manifestation of aliasing, exists, particularly in PAL and NTSC, due to the fact that subcarrier on the background signal and, more importantly, on a highly saturated colour fill signal can be controlled by very fast control signal edges derived from the luminance of caption signals. This is equivalent to modulating the subcarrier at much higher frequencies than the PAL or NTSC systems allow. The result is an almost instantaneous change from an area of high saturation to, say, black.

The subcarrier dots at this sharp edge are very visible and, due to the fractional line frequency offsets of the two systems, ragged. PAL suffers more than NTSC in that the picture frequency subcarrier offset causes the pattern to crawl down the screen. The effect is worsened if two saturated colours with non-aligned dot patterns are juxtaposed as caption and background. Also, on sloping edges which almost line up with the angle of the subcarrier dot pattern, the ragging takes on an objectionable low frequency form.

Band limiting of the control signal would produce an acceptable result as far as aliasing effects were concerned, but, as the luminance channel is also affected, the overall caption resolution would not be acceptable.

Two solutions to this problem suggest themselves: RGB or Y, R-Y, B-Y keying with correct band limiting of the colour difference signals, or the separation of the coded signals into luminance and chrominance and the use of band limiting, either of the chrominance control signal or of the modulated chrominance output.

35.7.6 Edging

To enhance the legibility of titles, they are often surrounded by a black, white or coloured edge. In general, this edge is obtained by using a key signal which is slightly larger than the caption. The background colour of the original caption will be visible in the edge area.

Character generators normally produce a separate key signal wider than the caption when an edge is required. For other sources, the vision mixer may have the means to extend the key signal both horizontally and vertically by means of delay elements (see *Figure 35.13*). To obtain the necessary delays of one or more television lines to extend the key signal vertically, the key signal might be digitized and RAMs or shift registers used as delay elements. The quality of the result will depend on the sampling frequency and the number of bits used. Older designs use glass delay lines.

The extension of the key signal involves shifting its centre downwards and to the right. If the original video is to be used to fill the key, it must also be shifted. This would involve digital techniques to delay the video quality fill signal by a television line or more. A more usual technique is to limit the fill to synthetic colour and to delay a gating signal only. Indeed the gating signal may be derived from signals delayed to widen the key.

An alternative, which allows the use of the original video as fill, uses digital delays of just short of one field to extend the key upwards and to the left.

35.8 Chroma key

Chroma key or *colour separation overlay* is the technique of replacing an area of saturated colour in a picture with part of another picture. A control signal is derived from the colour information in the first picture which, when applied to a mix/effects amplifier system, causes the two pictures to be combined appropriately.

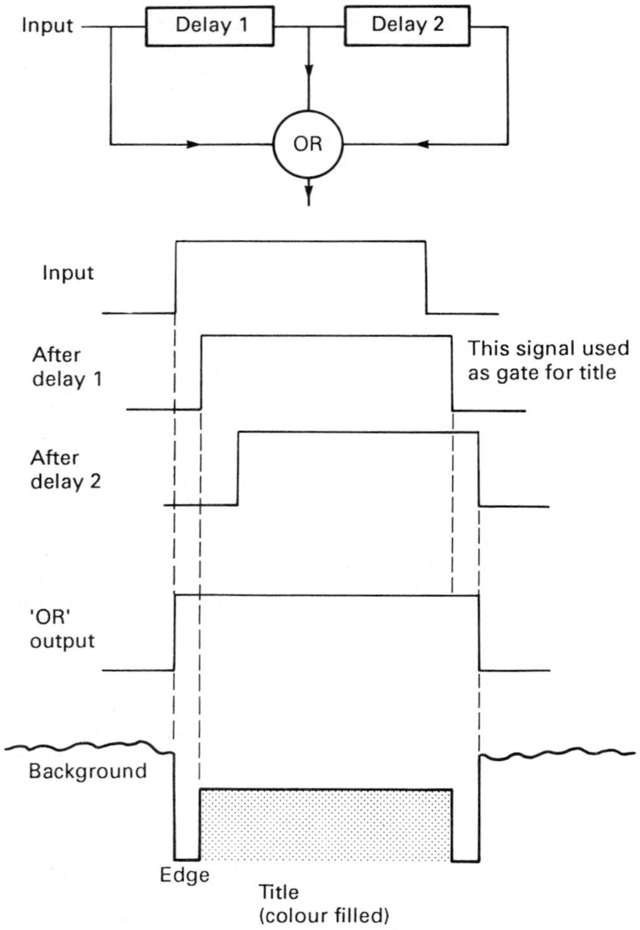

Figure 35.13 Title edge generation

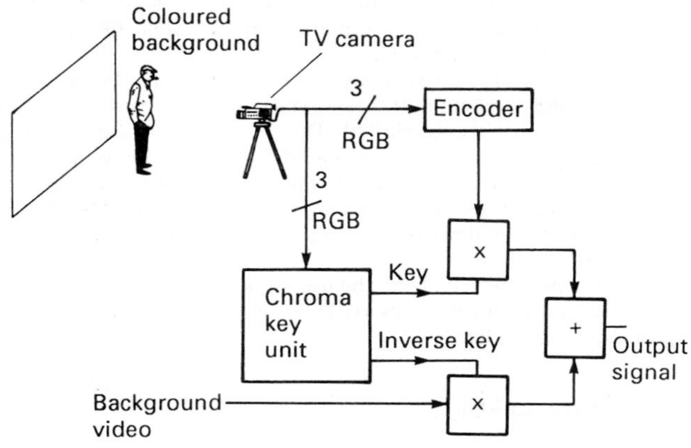

Figure 35.14 Chroma key

The best results are obtained when the control signal is derived from the original RGB separation signals, but it is possible to obtain reasonable results by decoding coded signals. The effects of edge crawl are a problem in chroma key as well as in title keying. Improvements can be achieved by carrying out the keying process in the component, RGB or Y, R-Y, B-Y, domain. Matteing techniques avoid rimming effects in the same way as described for title keying.

35.8.1 Key signal derivation

The simplest systems use a single colour difference signal, such as B-Y, derived either from a matrixing of R, G and B or by decoding a coded signal. This colour difference signal (*Figure 35.15*) is at a maximum for blue, zero for neutrals, and negative for complementary, yellow, hues. Only the positive polarity is used. The signal is sliced or clipped as in luminance keying, and a mix/effects amplifier control signal is produced.

If both B-Y and R-Y colour difference signals are derived and added together in proportions weighted as sine and cosine of a 360° control input, a colour difference signal of variable hue results. It is thus possible to provide a user control to select and optimize the keying colour.

35.8.2 More advanced key derivation

The simple colour difference signal described in section *35.8.1* appears as an inclined plane if its amplitude is plotted as height above a plane containing the RGB colour triangle. When sliced for control signal generation, this plane produces a 180° acceptance angle, or angle in which positive control signals exist, centred about the required hue. Adjustment of the hue control rotates the axis of slope of the plane.

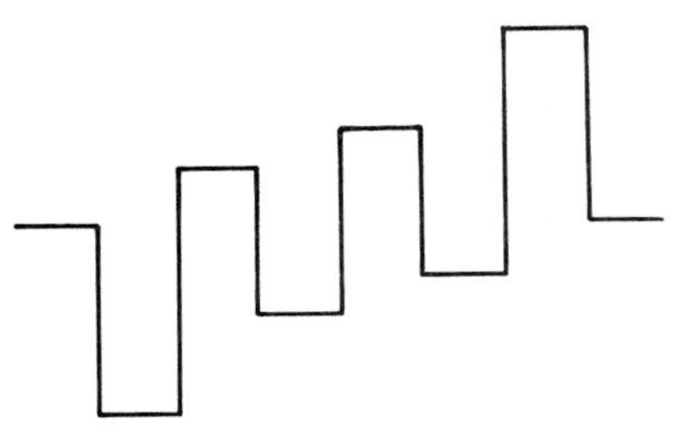

Figure 35.15 B-Y waveform for colour bars

Wright and Fletcher[5] described the non-linear combination of two such planes to reduce the effect of adjacent hues on the control signal. The form of their control signal was B-(R NAM G) where NAM signifies 'the greater of'. The acceptance angle in this case is 120°.

B Pethers of Michael Cox Electronics has developed this technique to produce continuous variations of both key colour and acceptance angle.

35.8.3 Key colour suppression

The object of key colour suppression, or *fringe elimination*, is to remove any traces of key colour from the chroma key camera picture. These traces might be caused by reflection from the coloured flat, light spill or imperfect keying around the foreground object. The principle is that a proportion of the key or control signal (colour difference derived) is subtracted from each of the RGB outputs of the camera. The proportions subtracted depend on the hue selected. The result is to reduce the values of R, G and B to the lowest of three for hues that contribute to keying. As this process is carried out with RGB signals, the fringe elimination equipment is fitted ahead of the chroma key camera's coder. Of course, the key signal, essential in producing this result, must be derived from RGB signals not processed in this way.

35.8.4 Matteing systems

An extension of key colour suppression is to use its self-cancelling properties to produce, from the chroma key camera, a signal in which the key coloured area is reduced to black. The key signal is used to multiply the fill picture in the normal way. The result is added to the key colour suppressed chroma key scene. It should be noted that key colour suppression only reduces R, G and B to the least of these, in key areas. A manual control is provided to subtract equal amounts of key signal from all three to reduce the output to black.

35.8.5 Shadows

To increase the realism of chroma keyed backgrounds, the facility to reproduce shadows thrown onto the coloured flat is often provided. The intensity of the signal produced by the chroma key camera is used to modulate the intensity of the inserted fill signal by multiplying the fill control signal.

35.9 Output processing amplifiers

It is sometimes necessary, at the output of a vision mixer, to provide an electronic system to ensure that the technical aspects of the signal conform to agreed standards, despite input irregularities.

Parameters to be monitored include:

- sync amplitude and timing
- chrominance amplitude and phasing
- luminance and chrominance peak excursions

For PAL and NTSC, methods exist for controlling sync and burst amplitude by measuring and topping up. However, the simplest and most effective methods employ either gated fading or sync and burst reinsertion.

Gated fading is a technique of gating mix/effects control signals with mixed blanking so that, during blanking time, a reference signal, such as internally generated coloured field, is selected.

Sync and burst reinsertion involves the use of an additional m/e type amplifier, controlled by a mixed blanking signal, which replaces the whole signal during blanking time with a reference sync and burst signal.

Both methods result in a stable output signal during blanking time irrespective of inputs or mixer settings. All sources must be timed into the mixer so that both horizontal timing and subcarrier phasing correspond with the output reference. Automatic timing correctors which adjust input timing are available.

Peak limiting is particularly important on 'knob a channel' mixers, and those with split faders for mixing, where overloads can easily be generated. It is not possible to limit the combined luminance and chrominance signals in a very satisfactory manner. They must be separated by filtering. The luminance must be limited at both black and white. The chrominance must be subjected to gain control which will be dependent on luminance as well as chrominance amplitude.

35.9.1 Fade to black

An overall fade to black facility is often incorporated into the output processing amplifier. This takes the form of a voltage variable attenuator, similar to a mix/effects amplifier, affecting the visible part of the signal while the synchronizing components are held stable in the normal way.

35.9.2 Non-synchronous operation

The mixer and processing amplifier may be required to deal with non-synchronous signals although not, of course, to mix them with other signals. The fade to black facility will still be required to operate. This will be achieved by synchronizing the processing amplifier to the non-synchronous video rather than to the station reference.

The degree of synchronizing may vary from generating a non-synchronous blanking signal, to allow gated fading to the input signal's own sync and burst, to the generation of a completely new sync and burst which can replace that of the non-synchronous signal.

35.10 Digital effects

A digital video effects system may be integrated with the vision mixer. The principal attribute of this equipment is its ability to move picture information about the screen area. This movement is achieved by digitizing the input picture information, storing the result in a RAM frame store and reading it out when required to take up its new position on the raster. The movement may be a simple horizontal or vertical shift, may involve decreasing or increasing size (*squeezing* or *zooming*) or may involve complex changes of geometry such as *rolling up*. The information must be stored in component form (usually Y, R-Y, B-Y to EBU/SMPTE format) because changing size would result in changes of subcarrier frequency if a composite format were used.

A key signal, equal to the new picture size, is provided so that shrunken pictures may be cleanly keyed into backgrounds by the vision mixer.

When size or geometry is changed, there is no longer a one for one relationship between input and output samples. Each output sample will be the result of the addition of contributions from a number of spatially adjacent input samples. For complex changes of geometry, the algorithm for computing the composition of a particular output sample will change over the picture area. The equipment must therefore be capable of very high speed real time arithmetic.

It is possible to process inputs and to route the resulting data to particular storage locations on the writing side of the frame store. In this way, it is possible to have, say, four quarter size pictures on the output screen at the same time using only one frame store.

35.11 Control

Modern designs use microcomputers to provide facilities such as memory when switched off, memory of settings which vary from input to input, and accurately timed transitions. A major improvement has been the introduction of serial control[6], removing the need for many multicore control cables. The control panel will carry many controls, both analogue and digital, as well as status indicators. The output of the analogue controls will be converted to digits, probably to 12 bit resolution, before being sent by serial link to the rack of processing electronics.

An important design requirement is that the data controlling the video should be updated during field blanking, otherwise effects will be seen on pictures.

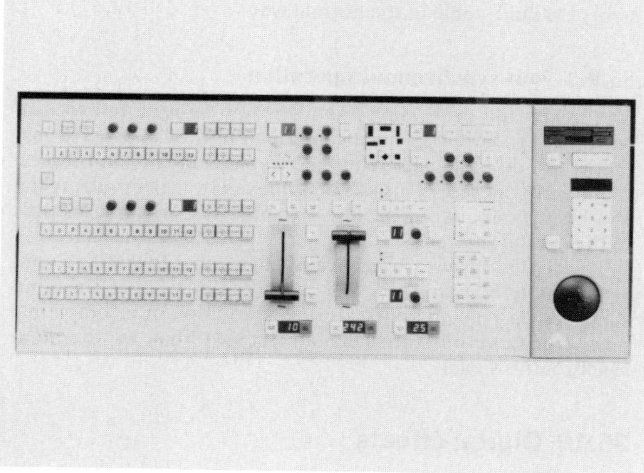

Figure 35.16 Arena vision mixer control panel (ACE)

35.11.1 Memory systems

With the large number of controls requiring fine adjustments, the production of fast sequences of complex effects is difficult. This limitation has been overcome by the introduction of memory systems capable of storing and reproducing complete sets of panel data. These can be linked together manually or automatically to create production sequences.

Additional features include smooth interpolation between two stored sets of analogue control values, and learning the trajectory of a manually controlled transition for subsequent re-creation.

Bulk storage facilities, such as floppy disks, may be provided for long term storage of the effects used in the production of a particular programme.

Early memory system designs were interposed between the control panel and rack electronics. Modern designs form an integral part of the computer control system.

35.11.2 Editor interfaces

Much of the creative work involved in making a programme may be done in the post-production or editing phase. Modern edit controllers are computers which control both the replay and record video tape recorders and the vision mixer on a frame accurate basis. EBU/SMPTE time code is recorded on a special track on the tapes to act as a time reference. The ability to repeat accurately a particular transition means that minute adjustments of timing may be made until the programme editor is happy, when the sequence will be recorded.

The simplest interface between edit controller and vision mixer is a trigger which starts an automatic transition, previously set up manually on the vision mixer's control panel. More comprehensive interfaces control more of the vision mixer's functions directly.

Early examples of this type were connected into the multicore cables joining the control panel to the rack. The edit controller generated control signals to replace those normally sent from the panel. Modern types use a serial link[6], joining the edit control computer directly to the vision mixer's control computer.

References

1 COX, M H, 'The switching and combining of signals', *Television Engineering: Broadcast, Cable and Satellite Part 2: Applications*, Royal Television Society

2 SKRYDSTRUP, O, 'Multi-function video processing in vision mixers', *Proc IBC 1976*, IEE Conf Pub No 145

3 COX, M H, *SECAM: The other European colour system*, Michael Cox Electronics Ltd (1980)

4 DALTON, C J and MALCHER, A T, 'Communications between analogue component production centres', *SMPTE Jour* (August 1988)

5 WRIGHT, R C H and FLETCHER, R.E, 'Development of operational techniques using CSO (Chromakey) and technological developments that could effect a change in production methods', *Proc IBC 1978*, IEE Conf Report No 166

6 JARRETT, P H, 'The E S bus remote control system — An introduction for prospective users', *EBU Review — Technical* (Dec 1987)

F M Remley Jr FSMPTE, MBKSTS
Technical Director Broadcasting Service,
University of Michigan

36 Computer Graphics and Animation

The hardware and software systems described here are designed to create and manipulate computer-generated video graphics images intended for use in television. Such images are created and stored in the memory of a suitable computer system by an artist. The stored data are then converted to video signals for incorporation into television productions

Modern computer-assisted graphics creation systems, though based on digital computer technology, are quite different from video systems designed for digital signal processing. Digital signal processing is described in Chapters 22 and 35 and is commonly employed in television system standards conversion, television recorder timebase error correction, video signal synchronization and manipulation, etc. In general, digital video signal processing apparatus is intended for real-time operation and thus must be designed around high speed analogue-to-digital conversion equipment, rapid access memory systems and high speed digital multipliers, digital adders and so on. Processing of the signal in parallel signal paths is often used in order to add speed to the system. Completely digital television systems have also been developed and are described in section 11.

36.1 Background

Video graphics creation, in the context of this chapter, is made possible by use of computer-based tools that assist an artist in the creative process. The use of computer-based drawing and design tools has revolutionized the preparation of graphics presentations for broadcasting and for training and other informational video uses. The roots of these applications lie in the discovery that equipment originally developed for computer aided design (CAD) in manufacturing industries could be adapted to the preparation of audiovisual materials like projector slides and overhead transparencies and, most relevant here, to television production use. As experience was gained, television users identified and demanded changes in hardware and software that led to improved systems for the specialized needs of video graphics artists. The relationship between CAD systems and video graphics systems remains a synergistic one, with a continuing two-way flow of information. Even so, relatively few system suppliers active in the industrial CAD market have chosen to enter the television production markets. In contrast, many manufacturers of traditional television apparatus have designed and developed computer-based equipment in order to improve the function and utility of the television production tools that they provide. It is such relatively specialized television digital graphics equipment that we will consider now.

Computer-based graphics systems have similar, but not necessarily identical, appeal both for industrial managers and for television producers. Most of these systems are capable of improving productivity, and so reducing personnel costs. They may also improve the accuracy and quality of work, since some kinds of projects can be checked by the computer system for errors, missing elements and so on. In the case of the graphic representation of physical objects, the use of computer-based systems permits three-dimensional image presentations with translations, rotations, textures, perspectives and lighting of the object images to be under total control of the system. Some kinds of image representation are possible with computer-based systems that would be impractical or impossible to achieve by manual means. Many kinds of animated image presentation are possible, with computer assistance speeding the preparation process.

36.2 Computer aided design

Industrial projects are often handicapped by slow and inaccurate preparation of drawings specifying design details for manufactured items, plans for building construction, etc. It has been amply demonstrated that the use of CAD equipment will allow a design department to produce drawings more rapidly. The usual industrial CAD facility is designed to replace as much as possible of the human labour involved in draughting with the more rapid and accurate work of a computer system. As an example, consider the preparation of the intricate designs for manufacturing complex integrated circuits. The photolithographic masks used in each manufacturing step for integrated circuits are all drawn by CAD systems controlling high accuracy plotting apparatus. It is unlikely that the semiconductor industry could survive if it depended solely on human draughting efforts because details of the design preparation and the process of checking the designs for errors are monumental tasks even with computer assistance.

36.3 Television graphics

In the case of television graphics, the requirements for computer assistance are somewhat different from those of industry. The aesthetic considerations almost always associated with television graphic production make it unrealistic to expect very much assistance in creative design from a computer system, let alone any overall creative decisions. In addition, television computer graphics systems are not usually applied in order to achieve large quantities of completed drawings. Quality not quantity usually rules the process. Television animation projects do require large numbers of similar drawings, as do many industrial and construction projects. But animation is not the main application of computer graphics in television.

Most television graphics systems will be under the control of a skilled graphics artist and that person will make basic project design decisions. Even a highly skilled video design artist can usually be helped in achieving desired visual effects by use of a suitable computer-based graphics system. Perspectives, texture renditions and lighting angles can be calculated and displayed at any time by the computer. This greatly speeds the design process, since the artist is freed from the routine effort so often associated with graphics preparation.

A simple example is in text information. Historically, an artist prepared such material by use of a pen or brush on artist's board, or perhaps by use of some form of manual printing process like letterpress or processed plastic film apparatus. The text was prepared for one television image at a time. For many years, however, it has been possible to prepare text displays using keyboard-operated electronic character generators that quickly and automatically provide a wide range of font sizes and designs, automatically centre or justify the lines of text, and allow instantaneous editing of text, colour, backgrounds, etc. These labour saving attributes are now incorporated into present-day graphics systems designed to generate a very broad range of images for television use.

36.4 Potential problems in graphics system designs

The use of computer-based tools can greatly assist an artist in the preparation of creative graphics displays for television applications. However, certain problems may remain in the graphic output of present-day systems. These problems are mostly technical in nature and should be taken into account when specifying computer graphics systems and when using such systems. The picture quality of computer-generated images continues to improve at a rapid pace, and since image quality is always among the most important of the criteria of system capability, any other drawbacks to system use can usually be tolerated.

36.4.1 Picture impairments

Aliasing is an impairment frequently observed in video images. The phenomenon often can be observed in the display of diagonal lines in television images and in this case results from the effects of the television scanning process. Here, however, we will consider only the type of aliasing that comes about as a result of computer-generated images being composed of a finite number of discrete samples of a (presumably) continuous analogue image that might exist in the physical world. The greater the number of digital samples present in the final digitized image, the less likely it is that aliasing will be a problem.

Aliasing is also known as 'exhibiting the jaggies', a phrase that is descriptive of the appearance of the phenomenon. When aliasing is present we observe that digitized images containing what should be smooth shape outlines exhibit more or less jagged edges. Lines lying nearly but not exactly parallel to the television raster scanning lines may show a step-shaped slope, rather than a uniform ramp.

Aliasing is annoying to the viewer because the eye is very sensitive to the characteristics of well defined, high-contrast edges that often occur in detailed images. This very fact allows sophisticated graphics systems to compensate, partially at least, for the sampled nature of digitized images by employing software-controlled anti-aliasing features. In this case, the computer purposely softens the edges of image transitions that are likely to exhibit aliasing. In colour systems this may be done by a combination of colour hue and saturation manipulation at boundary edges, coupled with digital spatial filtering.

Anti-aliasing systems that are limited only to the use of spatial filtering techniques, e.g. black and white graphics systems, tend to soften the image slightly. Anti-aliasing systems require adequate system memory, good software and some additional processing time. Most of them also impose a lower limit on coloured line width; lines that are finer than this limit are displayed as monochrome lines.

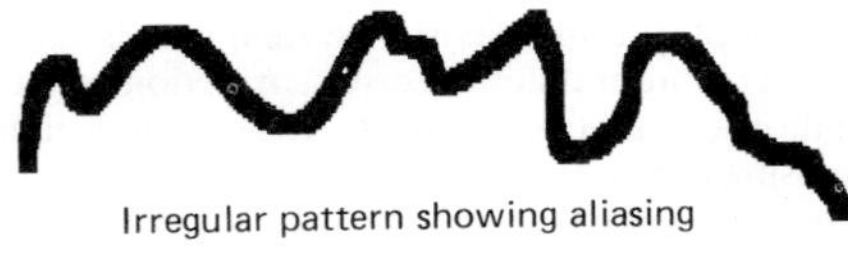

Figure 36.1 Examples of aliasing

36.4.2 Other picture defects

Computer-based graphics systems are known for image reproducibility and for well defined accuracy of presentation. These laudable characteristics can also contribute to an artificial appearance in some computer-generated video displays. In addition to aliasing defects, described above, such effects as *polygoning* and *motion jerking* can occur.

Polygoning is a problem primarily in three-dimensional graphics systems. Most such systems generate images exhibiting the appearance of solids by combining polygonal shapes

	Real-time operation	*Animation available*	*Can manipulate images*	*Colour images*	*Large palette of colours*	*Wide range of input tools*	*Precision drawings*
CAD system	No	Seldom	Usually	Often	Seldom	Yes	Yes
Digital video recorder	Yes	Yes	Video input signal	Yes	Original	No	No
Computer graphics system	No	Yes	Yes, in many ways Yes, 2 or 3 dimensions	Yes	Yes	Yes	Sometimes

Table 36.1 Comparison of computer based graphics systems

computed in three-dimensional space. Unsophisticated algorithms used in calculating the geometry may not allow computation of enough polygons to simulate accurately certain surface shapes.

Even if the resulting polygonal distortions are very small, they may lend an air of unreality to the image. Accordingly, a skilled artist will choose image shapes and surface textures that will minimize the visibility of residual polygon shape aberrations. It is also possible, of course, for a computer-generated image to appear unreal by virtue of its very perfection of shape and surface; real life is usually filled with minor imperfections that the eye is accustomed to seeing. A good artist will avoid this trap by applying details drawn from life experience to the design.

In the case of computer-generated animated images, motion defects may pose problems. These defects take the form of unrealistic image movements and are usually caused by memory or storage limitations, or by defective motion simulating algorithms. Such defects are most likely in complex, rapidly moving images. Motion defects can be solved in increasing the memory capacity, the computation algorithms or perhaps the overall computational speed of the system to accommodate the needs of the moving image.

36.4.3 Data exchange and system control

Most computer-generated video graphics systems have certain highly proprietary elements in their designs. This is true of both the hardware and the software, and may have unfortunate consequences for users.

Hardware incompatibility, especially signal incompatibility with existing video systems, may make installation of new or augmented graphics equipment very difficult. For example, a system based on Y and C analogue component video signals will require converting accessories to work with analogue or digital RGB signals or with digital composite video systems. As more facilities are designed to use video signals conforming to recognized analogue component standards and to CCIR Recommendation 601 digital video specifications, it is inevitable that compatibility will be incorporated in hardware designs and equipment interfaces. Until that time it will be important to understand clearly the system interface requirements of graphics systems before they are purchased.

It is also important to examine the control requirements of new graphics facilities designed for studio installation. The increasing popularity of the EBU/SMPTE ESbus standardized television equipment control interface allows the easy coordination of graphics generation systems with video recorders, production switching systems, character generators and other such equipment. Provisions for ESbus control are important for the satisfactory integration of new facilities and should be specified whenever possible.

Software compatibility could allow exchange between graphics systems of images in their most simple form. In other words, the commands necessary to generate an image can usually be stored by the system on a magnetic disk or other digital storage medium. The computer code thus stored can be used to recreate the original computer graphics image at any time, at least when used with the machine that generated the code. However, the details of arranging the stored data and the details of the command structure itself usually differ from machine to machine. In addition, no common graphics creating language has yet evolved. As a consequence of these facts, the exchange of image commands is only possible between equipment of the same manufacture. It is expected that standardization of commands and of data storage specifications will occur in the future and thus simplify the widespread use of new graphics systems.

36.4.4 Future improvements

As a general rule, most limitations presently imposed on computer-generated video graphics will fade in importance as both computer and video technologies advance. More digital memory will solve some kinds of problem. Faster processing speeds will solve others, especially when real-time motion effects are required. The cost of computer memory is dropping rapidly, and more powerful, smaller and less expensive computers are continually being introduced to the market. These developments, plus advances in image processing software and hardware will result in improved images produced by more easily acquired equipment. Standardization of control and interface hardware and of data and command software storage will make graphics systems even more useful production tools.

36.5 Hardware elements

Hardware designs applied to graphics images have historically used many common, easily obtainable computer system components. For example, similar computer central processor units can be used for general numerical computation, for preparing industrial drawings or for preparing graphics for television. In the present state of technological development, however, some of the limitations inherent in computer image processing have been solved by the design of special graphics computers (*graphics engines*) to replace slower or more memory-limited general-purpose computers, and thus increase the image quality or the speed of operation or both. Certain of these dedicated graphics systems, often having a high initial cost, are constructed around specialized central processing units using, for example, pipe-line or parallel-processing architectures in order to achieve high speed operation while generating either very detailed or very near to real-time animated images.

In computer graphics as in television, the time available for image formation and the final image resolution often contend for primary emphasis in equipment design. In the future it is expected that basic reliance will be placed on advanced general-purpose computers, probably equipped with packaged graphics

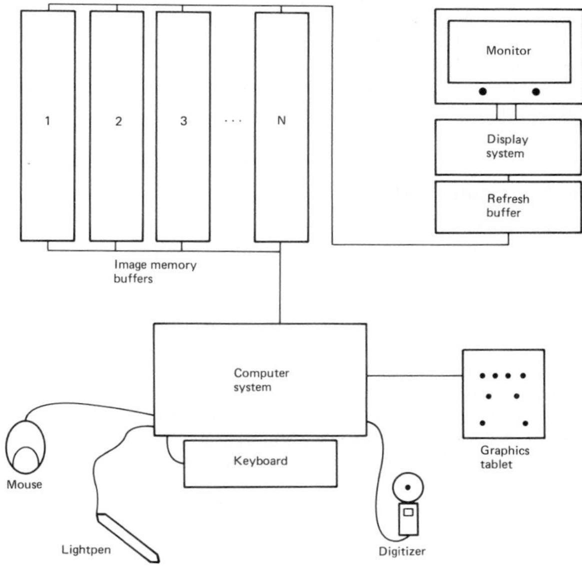

Figure 36.2 Elements of a computer graphics system

processing subsystems, rather than upon these special-purpose graphics computers. This will result in lower overall hardware costs.

36.5.1 Refresh buffers

Complex computer images require large amounts of computer memory for their creation. This memory can only be read, written to or refreshed at a specific rate determined by the processor clock designed into the computer. Image displays, especially those based on raster scanning, require rapid transfer of image data to the display device, and must maintain an image for operator viewing even while changes are being computed in the computer memory banks. The result of this situation may be that the bus-based computer memory system cannot keep up with the requirements of the display system. To solve this problem a subsystem of memory locations known as the *refresh buffer*, or more commonly but less precisely as the *frame buffer*, can be set aside. A refresh buffer usually stores a full frame of display information and serves only to refresh the display.

In the following discussion of the need for a refresh buffer, data values for 625-line television systems are indicated, with 525-line values shown in parentheses.

Weinstock[1] shows that for 625-line (525-line) television systems there are about 570 (480) active scanning lines after vertical blanking is removed. Given the 4:3 aspect ratio of the television image, he shows that about 760 (640) horizontal samples are needed for each line to achieve matched horizontal and vertical image resolution. The product of 760 by 570 (640 by 480) results in 433 200 (307 200) samples. Usually, each of these samples is represented in the graphics system image memory by two bytes (16 bits) of data. Two bytes per sample allow good accuracy in the specification of the colour of each pixel, and the value of each these two byte quantities makes up the final data sent to the display 25 (30) times each second. There thus is a calculated requirement for transmitting a 21.66 (18.43) Mb/s gross data rate to the display system. This significant data rate may exceed the transfer capabilities of the computer memory bus system. It can be drastically reduced by using a refresh buffer to store the display data before actual display refresh occurs. The refresh buffer can be reloaded as required with computed final display data at a rate perhaps only one-tenth that of the gross data requirements of the display. At the proper moment a rapid display updating is accomplished from the data stored in the buffer. Thus this system reduces the burden on the main central processing unit and random access memory (RAM) system of the computer system.

36.5.2 Frame buffers

Frame buffers, also called *frame stores*, are relatively commonplace in modern television systems. They are utilized for a variety of purposes in addition to their uses in computer graphic systems. These buffers consist of arrays of dedicated RAM integrated circuits, and they provide storage for one frame of video information that can be written to and read from at (usually) different clock rates. Television digital synchronizers and standards converters use such stores, together with digital video recorders and digital video effects systems of many types. The computer graphics application of frame stores is comparable to these other applications, and the refresh buffer concept is especially useful in the design of practical, low cost computer graphic systems. In addition, one or more frame buffers may be used as image memory for the data representation of each image pixel. The combination of a refresh buffer and one or more image frame buffers allows the graphics system to work at best efficiency, especially when long (e.g. two byte) words are associated with each pixel and require rapid manipulation.

36.6 User interface

A major difference in graphics system designs, and a factor that spells success or failure in the television field, is the so-called *user interface*. The user interface includes the physical objects manipulated by the artist in the creative process. In a traditional, non-computer graphics situation the user interface may include pens, drawing pencils, brushes, a drawing table and similar tools. In a computer aided graphics system the user interface may include items such as a graphics tablet (bitpad), mouse, trackball, light pen and keyboard together with the video monitor used to examine the results. Importantly, the user interface also includes the software-based creative tools provided for the operator. Of these, the software portion of the interface has usually proved to be the most difficult to define and to refine.

36.6.1 Interface software

In the early days of computer graphics it was assumed that the operator of a graphics system would be conversant with computer terminology, would be informed about the complexities and subtleties of a computer operating system such as UNIX or MS-DOS and might well know one or more high-level computer languages such as Fortran or C. Given the large number of computer-assisted graphics systems now used successfully in television production, it is clear that these earlier assumptions are invalid. Most successful contemporary systems use software interfaces based on the premise that the operator of the finished system will not be a computer expert but rather will be a creative artist. The success of computer graphics in television production is largely due to an improved human factor design of the user interface.

Often the interface software is considered to be proprietary, and the basic system design information (the *source code*) is not available to the end user. Potential purchasers of graphics systems are well advised to investigate this, since relatively simple changes to the system may be difficult and expensive unless contractual arrangements have been made to allow for modification of the source code. A degree of incompatibility between systems results from user interface differences, and graphics artists trained on one system may have some minor difficulties in adapting to another system.

A very common type of user interface is that developed by Xerox and introduced commercially some years ago in the Xerox Star computer workstation. The Star interface is very similar to that now used in the popular Apple Computer Macintosh microcomputer system, and has been adopted, with variations of all sorts, by many computer and workstation manufacturers. It almost always includes a mouse as a primary pointing tool, pull-down menus for user choices and icons to identify system components. *Icons* are stylized, consistent representations of such things as files, directories, subdirectories, disk drives and operating system options, and are displayed on the face of the terminal screen when choices are necessary. Choices are made by pressing the switch on a mouse ('clicking' the mouse) when the screen cursor is positioned on the desired icon or by touching the icon with the tip of a light pen. Machines using a graphics-based interface are usually well suited to graphics production, and this explains the frequent use of this type of user interface in television graphics systems.

All versions of the Star interface are supplemented with tools adapted to the design philosophy of the computer graphics system to which they are attached. In the case of systems intended to simplify the process of two-dimensional graphics preparation, the kind of system usually called a *paint* system (see section *36.6.2*), the user interface will provide a variety of paintbrush shapes and sizes, a means for applying colours to the images in a way similar to that of an air brush, a means for erasing mistakes, a means for 'filling' closed shape outlines with colour, a means for colour selection, usually in the form of a palette of colours and/or grey (luminance) values, and other useful tools for drawing straight and curved lines, geometric shapes and freeform shapes.

Figure 36.3 User interface with icons

36.6.2 Input tools

Television video graphics systems and CAD systems share a variety of user input devices.

Commonly, a computer terminal connected to a central processing unit (CPU) is used as the main element of the input system. This terminal may be a conventional crt display, an elaborate graphics-oriented technical workstation or a personal microcomputer operating as a controller for an internal or external graphics system. A keyboard is universally used to enter numeric data, to select choices from some kinds of on-screen menus, to generate text for labels, etc.

It is necessary to provide a work area upon which the creative artist will generate the television graphic. This work area is analogous to a blank canvas chosen by an artist when beginning an oil painting. In computer graphic systems, the work area is usually a computer defined imaginary space, part or all of which is displayed on a video monitor and made accessible to the artist by use of input tools. Several choices of input tool are possible, but in each case the result of using the tool influences an image displayed on the terminal monitor.

The device known as a *mouse* is a common form of input tool. A mouse usually takes the shape of a roughly rectangular box attached to the end of a flexible cable leading out of the terminal. It has on its top surface one or more momentary-contact pushbutton switches for sending commands to the computer system. The bottom surface of the mouse contains some sort of motion detector, often a rolling rubber ball driving optical movement encoders, designed so that when the mouse is guided manually over the surface of a horizontal work area the movement is converted into a corresponding pointer (cursor) movement on the video display. Thus, moving the mouse and pressing a control button may be used to make selections from menu items listed on the display screen, to draw lines on the display, to generate points and to delineate areas on the display, etc. The data supplied by the mouse itself are relative, not absolute, position data since the mouse may be lifted and placed anywhere on the work surface between input motions. These design factors make the mouse a very flexible input tool and the operation of the mouse soon becomes second nature to the artist.

The *trackball* is a close functional relative of the mouse. This device consists of a stand holding a partially recessed sphere. The sphere can be rotated by hand and optical or mechanical encoders convert the rotation into cursor movements on the display screen. The trackball also has one or more command switches and provides relative position information in a similar way to the mouse.

Many artists prefer an input device providing absolute rather than relative position data. Such a tool can take the form of a *graphics tablet* and a *stylus*. The graphics tablet, sometimes called a *bitpad*, is a work area that contains elements beneath its top surface to sense the position of a stylus as it is moved over the surface. This positional information is sent to the computer system for processing and display. The stylus/tablet combination permits the artist to make freehand sketches and to place the cursor precisely on the display image. The stylus is usually equipped with a nib that operates a momentary switch. Pressing the stylus against the surface of the tablet causes the nib to close a switch, thus commanding the computer system to store data about the exact point being touched. This selection process may identify image points, e.g. the beginning or end of a line, or it may identify a function selected from a menu of thousands printed on the tablet itself.

It is also possible to use a special point-selection device known as a *digitizer* as a data input tool. The digitizer is a movable device containing a viewing glass, usually magnified, and a reticule. It also has a command button. The digitizer is used to identify points located in a drawing or picture placed on the work area. Data identifying the digitizer location are sent to the computer when the switch is depressed. Thus, the outline of a drawn object, perhaps a map or other complex shape, is digitized into a series of points stored in the computer memory as image locations. Later, these outline data can be further processed by the graphics system and can be modified or incorporated into other images.

Figure 36.4 The Quantel Paintbox electronic painting and graphics system. The artist uses the electronic stylus, tablet, joystick and keyboard to control the various functions (Quantel Ltd)

Many CAD systems and some video graphics systems make use of the *light pen* for data input. This device is pen-shaped and contains a photosensor at its tip. When the pen is placed against the screen of the video display device the photosensor detects the position of the pen on the screen. Thus, the pen may be used to select items on the display screen or to draw effectively on the display screen in a way analogous to the use of a graphics tablet and stylus combination. The light pen also contains a finger operated switch for issuing commands to the system.

36.6.3 Display devices

Usually it is assumed that the final display for a graphics system will be a television monitor or receiver. However, some systems derived from CAD designs use vector display monitors rather than raster scan monitors during the graphics composition process. Vector display systems are nearly always designed for high resolution display of line images, in monochrome or at most a few colours. These lines may define all kinds of shapes and patterns, but they are produced by successive point-to-point vector deflection of the electron beam, rather than by television style scanning as in a raster scan system. Thus vector displays are not very useful for television graphics work and, in fact, may lead to graphics designs that are seriously deficient when converted from vector display, because some forms of aliasing may appear during the conversion process.

36.7 System classifications

One writer in the field[2] has identified several categories of graphics system, some of which are applicable to television needs. These are:

- high resolution photographic slide making systems,
- two-dimensional paint systems, bit-mapped design,
- two-dimensional paint systems, 'true colour' design,
- three-dimensional systems,
- hybrid 2D/3D systems,
- hybrid character-generator/paint systems,
- systems for animation use (e.g. 'tweening' systems).

The specification and purchase of computer-assisted video graphics systems is little different from the same process applied to other video equipment. However, it is very important to be well informed about the nomenclature used in describing these systems, since much of it derives from CAD and computer jargon and subtle meanings may not be clear to persons familiar with television jargon.

Many early computer graphics systems were used to generate output for display on a vector-deflection crt, or to provide photographic output or perhaps printed output. Partly as a result of this history, the resolution of computer graphics systems is often defined on the basis of the pixel, or picture element. Because there is no inherent aspect ratio in computer graphics systems, a square image is frequently assumed. Hence, resolution specifications of 512 × 512 pixels, or 1024 × 1024 pixels, or 2048 × 2048 pixels are very common in the computer graphics industry. On this basis, television systems require at least 512 × 512 pixel resolution to generate satisfactory images and, when the 3 × 4 aspect ratio is applied, the effective video resolution of some systems is 480 × 512 pixels. Lower values are unlikely to produce satisfactory results and higher values are becoming the rule as equipment costs continue to decrease.

The figures just cited are applicable to composite colour video signals. In fact, however, the latest video equipment is designed to use the sampling structure of CCIR Recommendation 601, the internationally agreed specification for studio component digital video signals. This specifies that the active portion (the unblanked portion) of each horizontal television line will contain 720 luminance component samples and 360 colour-difference component signal samples. The number of vertical samples is determined by the number of scanning lines of the television system in use — 525 or 625 samples. It is expected that most television computer graphics systems will sample in accordance with CCIR Recommendation 601 in the near future.

36.7.1 Photographic slide making systems

Computer-controlled graphics systems intended for slide making are characterized by emphasis on high resolution performance. Raster scan operation at 2048 × 2048, or even more, pixels per image is typical. Such systems make use of software specially optimized for visual graphic displays on large optical projection screens. Slide making systems sometimes are based on vector scanning techniques. The use of vector scanning makes the output graphic data independent of the recording system; the final slide resolution is largely a function of the film recorder used to prepare the slide.

These systems usually incorporate a proprietary photographic slide exposure camera and are relatively slow in displaying the final image. Slide exposure times of a few seconds to several minutes are typical, with the shorter times resulting from use of more memory and higher processing speeds in the computer system.

Equipment of this type is sold under the names Genigraphics, Dicomed, Matrix and AVL.

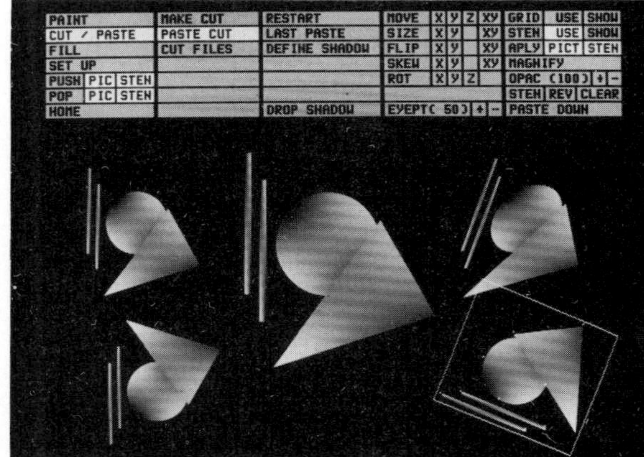

Figure 36.5 An Ampex AVA 3 user interface screen (Ampex)

36.7.2 Painting systems for television use

Television graphics painting systems are two-dimensional image creating devices, usually based on television raster-scan technology. In present television production facilities, the most common kind of graphics painting system is a bit-mapped system that uses a semiconductor memory display (frame) buffer. With this type of system, as many as 256 colours may be displayed in any one image out of a large palette of possible colours. Many designs of this type are based on 512 × 512 pixel resolution. A simple form of image animation is easily achieved with this kind of equipment. Known as *colour-cycling*, the process consists of rapidly changing specified image areas by writing new values into the display buffer, thus producing changes in the image and creating an effect of animated motion.

Variants of the basic painting system, known as *paintbox* or *true-colour* paint systems, feature sophisticated colouring techniques that permit a larger selection of colours from a very large palette of colours, and can accurately emulate air brush image-forming techniques and the application of transparent paints. Some designs are best suited to the production of still images, but with animation possible through frame duplication. This design approach is represented by the Ampex AVA system and the Quantel PaintBox.

Other, newer systems offer both colour cycling animation and carefully devised transparent paint effects. These are represented by the Aurora and Colorgraphics Artstar units, which are fully configured high-quality systems that operate on a stand-alone basis. Even newer alterations are finding their way to market. Some are lower cost systems based on the use of a powerful personal computer with a dedicated computer graphics board, such as the Truevision Targa frame buffer board and its associated software.

36.7.3 Three-dimensional systems

Three-dimensional graphics systems incorporate the evolutionary advances made in modern CAD graphics systems and add refinements especially suited to television production needs. The images produced by three-dimensional systems can be very impressive, but the investment in graphic artist effort in the image programming and creation functions can be significant.

Figure 36.6 PictureMaker three-dimensional system (Cubicomp Corp)

In general, a three-dimensional display starts with the creation of a two-dimensional image, or object outline, then adds a depth dimension by means of a wire-frame model. The wire frame is then covered with a computed surface, coloured, and moved and rotated in space. The rotation and movement involve very heavy computational demands on the system, especially if they must occur in real time. Accordingly, sophisticated three-dimensional systems are based on specialized graphics processor subsystems that are attached to powerful microcomputers, with the Cubicomp system as an example, or, at greater cost, systems may use special, dedicated computers designed to process image data at high speeds. The Bosch BGS 4000 is an example of the latter approach. Indeed, three-dimensional systems using special processors like the Pixar super minicomputer graphics engine or the Cray supercomputer are capable of impressive real-time generation and display of rapidly changing three-dimensional images, albeit at great expense. The results can be spectacular, and the realism of the images continues to improve as the software is revised to match advances in hardware throughput.

36.7.4 Other systems

Because the three-dimensional graphics systems described above are primarily designed to move simulated 3D objects through simulated 3D space, it may well be necessary to add a two-dimensional paint system to a three-dimensional object-oriented system in order to generate background and foreground images. This then creates a form of hybrid system that allows the artist easily to prepare the necessary foregrounds, backgrounds and other scenic visual elements. Systems such as this are readily available and can be assembled by a combination of 2D and 3D hardware and software.

Another form of hybrid system, a combined character-generator and paint system, can be very useful in the television studio. The Chyron IV is a well known example of this combination and provides very high quality text character presentation with accompanying paint facilities to permit elaborate image preparation. Many combinations and options are available that permit graphics displays of many kinds.

A specialized form of computer graphics device used in video animation work is called a *tweening* system. The function of this is to compute and draw the frames that occur between key frames of an animated production. The key frames are generated by an artist and the tweening system calculates the data necessary to draw the intermediate frames and thus create smooth animated motion.

References

1 WEINSTOCK, N, *Computer animation*, Addison-Wesley (1986)

2 BOUDINE, A, *Video Manager*, 21 (February 1987)

S Lowe Manager International Technical Training,
Ampex Ltd

37 Video Tape Recording

The frequency spectrum occupied by a video signal extends from 25 Hz or 30 Hz up to 5.5 MHz. This is equivalent to 18 octaves, with a maximum that is more than 350 times that of an audio recorder. It is essential to preserve the dc component, black level. Both the range and the dc requirement are considerations that demand a new approach to recording when compared with audio or data recording.

A third aspect of video that has to be accommodated is the high maximum frequency which, because of the colour subcarrier, is also the maximum energy part of the signal spectrum. The maximum frequency is proportional to the speed at which the recording head travels over the recording medium. Video recorders have head to tape speeds in the range of 200 inches (5.08 metres) a second for domestic machines to over 1000 inches (25.4 metres) a second for some professional machines.

The problems of recording video are eased by the nature of the television signal and the eye. The continuous visual scene is broken into segments, fields and frames, with non-picture periods amounting to 25 per cent. These breaks may be used to reduce the data rate of the recording channel. The relative lack of discrimination in the eye means that fine colour detail is not seen and so may be discarded by the low end recorders such as those used for home recording.

The requirement for a high relative speed for the video recording head places severe performance constraints on the transport system and the recording medium. The interface between the recording head and the magnetic medium must be close and even, while the speed of travel must be accurate and constant.

37.1 Frequency range

A typical audio recorder frequency specification would offer a range of 50 Hz–15 kHz. Between these frequencies, the replay level would be substantially flat within ±3 dB. To achieve this, a range of around eight octaves, considerable frequency equalization is required.

The replay output voltage of a magnetic system is proportional to:

- the record current, up to the point where the medium is fully saturated
- the signal frequency, up to the point where the recorded wavelength approaches the replay head gap width

The level of record current is variable and under the control of the circuit designer. The dependance on frequency is a fundamental of physics and can only be accommodated.

The replay head output voltage rises with signal frequency at 6 dB per octave (see *Figure 37.1*). Increasing the frequency causes the wavelengths to become shorter. As the wavelength on tape becomes comparable to the replay head gap length, there is a drop in induced emf. This can be seen in *Figure 37.2* where the gap length is equal to one wavelength. The resultant

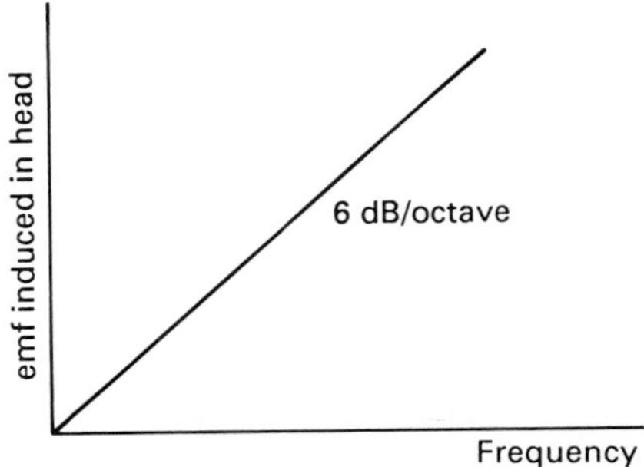

Figure 37.1 A graph of head output against frequency ignoring all loses and assuming a constant record for all frequencies

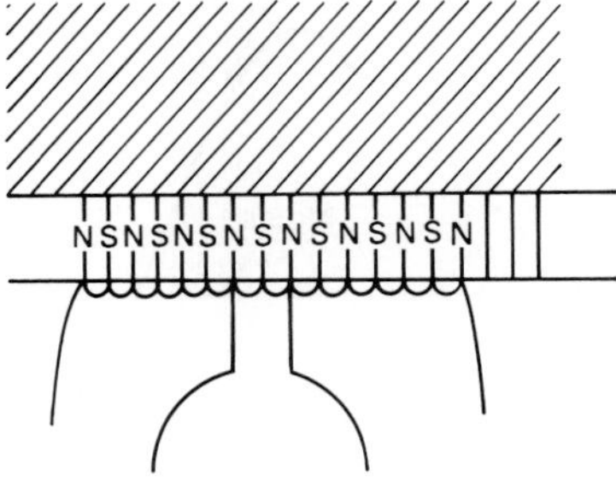

Figure 37.2 Flux distribution where wavelength equals gap length

flux through the head is zero, and as the tape moves it remains zero.

The frequency at which this occurs is the *extinction frequency* and is determined by the gap length and the tape speed. If the frequency is increased, so that the wavelength reduces still further, the output increases again. In audio and video recorders, only the output below the extinction frequency (f_{ext}) is used.

The response is similar to the function $y = \sin x/x$, and is shown in curve (a) of *Figure 37.3*. If this response is added to the rising frequency/output graph, then the characteristic shown in curve (b) is obtained. The frequency axis is normally logarithmic, and a somewhat distorted view of the response may be obtained.

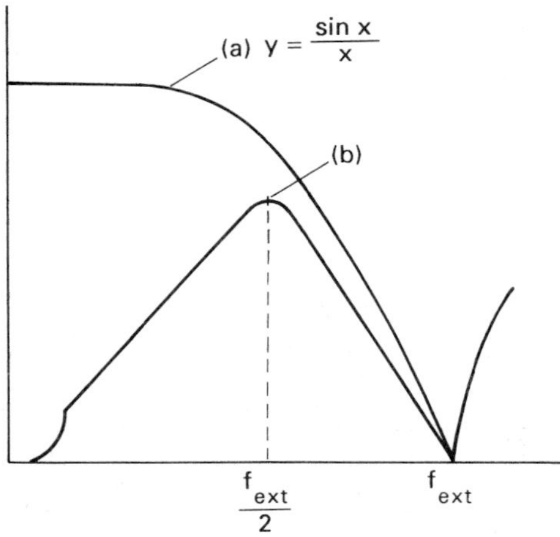

Figure 37.3 Response modified by finite gap length

Ignoring all other losses, such as eddy currents in the core, spacing loss from the separation between the head and the media or thickness loss from the depth of the magnetic medium, the peak of the response is at $f_{ext}/2$ indicating an equal passband above and below the peak. The logarithmic scale tends to compress the upper response.

37.1.1 Final response

The useable range is set by the ratio of *signal power* to *noise power*. The noise from the tape comes from unevenness of distribution of the magnetic particles in the medium, which is a mixture of binders and lubricants. For an audio system, a signal/noise level of 54 dB is acceptable. Special signal processing techniques, such as the pre-emphasise of low level input allows a significant improvement in signal/noise performance.

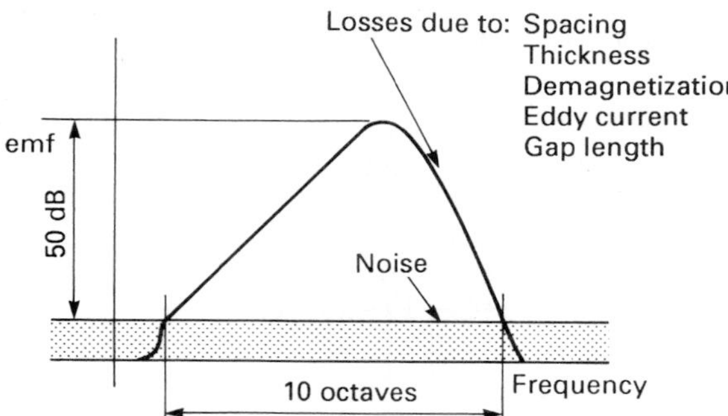

Figure 37.4 Final response

The final response is shown in *Figure 37.4*. For an acceptable signal/noise ratio, the system is limited to a dynamic range of about 50 dB. This gives a bandwidth of about ten octaves and is independent of tape speed since it represents a ratio between minimum and maximum frequencies. The upper frequency is limited by head to tape speed, gap length, tape resolution and tape finish. A bandwidth from 25 Hz to 12.8 kHz is equal to nine octaves, inadequate to record even the television line frequency of 15 kHz.

37.2 Modulation

37.2.1 Direct recording

Direct recording systems are those in which the composite video signal is applied direct to the input of the modulator. They require wide-band electronics and extremely stable transports, making the machines complex and subject to routine alignment and maintenance. They are not suitable for use in domestic or field recording situations.

Audio signals are recorded with amplitude modulation, and an increase in signal amplitude (*loudness*) results in an increase in flux amplitude. Video signals are recorded as a form of frequency modulation, and an increase in amplitude (*brightness*) results in an increase in frequency.

The carrier frequency is chosen to allow an acceptable head to tape speed and is usually between 6 and 10 MHz. The relationship between amplitude and frequency is specified for three well defined signal levels: *peak white*, *black level* and *sync tips*. The difference between the peak white frequency and the sync tip frequency is usually in the region of 1 MHz.

Frequency modulation used in vtrs differs in two respects from most other systems in common use:

- The centre frequency is very close to the highest modulating frequency.
- The modulation index (f_d/f_m) is lower than in most other systems.

The term *centre frequency*, when referred to video signals, is

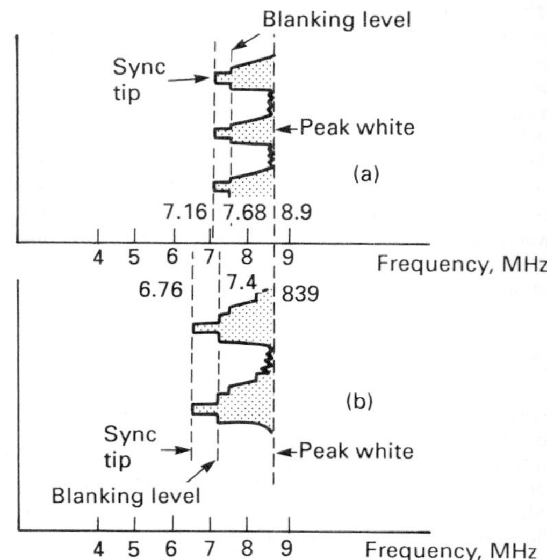

Figure 37.5 Frequencies used in (a) EBU C format and (b) EBU B format videotape recording systems

not easy to define. The signal does not have amplitude symmetry, and in the absence of a picture (*black level*) there are still the line and field synchronizing signals. It is convenient to regard the centre frequency as the modulator frequency at the mean level of the video, a low brightness picture.

Typical frequencies for the modulators in video recording are shown in Table *37.1*. They range from 3.0 to 10 MHz. The lowest centre frequency must therefore be higher than the highest video frequency plus the low frequency limit, set by the sync tip frequency.

Format	*Sync tip*	*Peak white*	*Chrominance*
C format	7.16 MHz	8.9 MHz	N/A
U-matic	3.8 MHz	5.4 MHz	685 kHz
Betacam – metal tape	6.8 MHz	8.8 MHz	5.433 MHz – 6.767 MHz
Betacam – oxide tape	4.4 MHz	6.4 MHz	5.433 MHz – 6.767 MHz
M II	4.9 MHz	7.0 MHz	

Table 37.1 Typical modulation frequencies for use in videotape recorders. These numbers may vary as the formats evolve

This will mean that the second and higher order lower sidebands will extend below the origin, and that the low harmonics of the upper sidebands will interweave with the carrier. This gives rise to moiré interference, so named because of the visual effect on the picture, which is a characteristic of vtr modulation systems.

The modulation index is low because of the high deviation and low modulating frequency. In a more usual system, e.g. fm radio, the deviation is likely to be less than 0.1 per cent of the carrier. In a vtr, it is nearer 5 per cent.

The maximum sinusoidal deviation that can occur would be for a video signal with an average mid-grey level and a frequency component with a peak value from peak white to black level. Such a situation is rare in monochrome signals, although black to white transitions contain large amplitudes of high frequency components. In colour signals, where the chrominance information is transmitted on a high frequency subcarrier, large amplitudes of high frequency are more common.

The worst case condition is 100 per cent amplitude, 100 per cent saturated bars. The peak-peak subcarrier amplitude will always be less than 0.9 of the peak-peak excursion from sync tip to peak white.

From the figures in Table *37.1*, a typical value for modulation index would be 0.54.

37.2.2 Choice of frequency modulation

The main reason for choosing fm for a vtr is to reduce the bandwidth to be recorded on tape. This is essential before the system can even begin to work, even in black and white. Instead of covering 18 octaves (25 Hz–5.5 MHz), the range is less than 1 octave for the frequency modulated signal, typically 7.16–8.90 MHz. The overall range, including the upper and lower sidebands is within 2 or 3 octaves (3–24 MHz).

There are other advantages in the use of fm:

- Variations in replayed signal amplitude caused by head to tape contact pressure changes do not affect the picture signal.
- An s/n ratio of 10 dB is adequate, allowing narrower tracks to be used than would be the case for am.
- The recovered picture signal is dependent upon the deviation and not on the absolute frequency. Changes in head to tape speed, such as between machines, are automatically compensated by the replay decoder.

The last point may be used to advantage where machines are upgraded to higher carrier frequencies as in U-matic and Betacam systems. Tapes may be interchanged, with restrictions on performance between systems.

37.2.3 Folded side bands

The composite colour signal has large amplitudes of high energy, particularly the subcarrier. These give rise to high energy sidebands in the modulated signal applied to the tape. The lower sidebands would fall below zero frequency but instead are reflected or 'folded'. For particular hues and saturations, the reflected second order side frequency will interfere with the first order lower sideband, and a beat between them will occur. This is decoded by the replay system as a signal frequency and is seen as a wavy vertical line pattern on the picture, more noticeable when a colour picture is viewed on a black and white screen.

The effect is similar to that seen when a fine pattern grating is placed over another similar one, and the name given is *moiré patterning*.

Moiré is a characteristic of a composite vtr signal chain. It is generated at the point where the modulation takes place and cannot subsequently be removed. Care is taken to choose frequencies for the carrier and modulation to minimize the effect.

37.2.4 Harmonic distortion

When a signal is passed through a non-linear circuit, harmonics are produced. A magnetic recording system is particularly prone to third harmonic due to the saturation process, lack of hf bias and the limiting action of the demodulator.

For a centre frequency of 6 MHz, the third harmonic would be 18 MHz. This is outside the fundamental pass-band, but the lower side frequencies of the harmonic may still appear within the pass-band of the system.

The modulation index increases with the order of the harmonic, and so there is more energy in the higher orders of the sidebands of the harmonics, i.e. the ones that appear in the pass-band. These lower side frequencies beat with the fundamental frequencies and also produced moiré effects.

37.2.5 Shelf working

As the carrier frequency is increased, there is an increase in the harmonic frequencies and in their side frequencies. This causes an increase in the moiré energy. As the frequency increases further, a point is reached where a side frequency no longer falls below zero or falls out of the pass-band. At this point there is a drop in total moiré, giving rise to the graph shown in *Figure 37.6*.

The carrier frequency is chosen to be just higher than the point at which there is a drop in the moiré. The shape of the

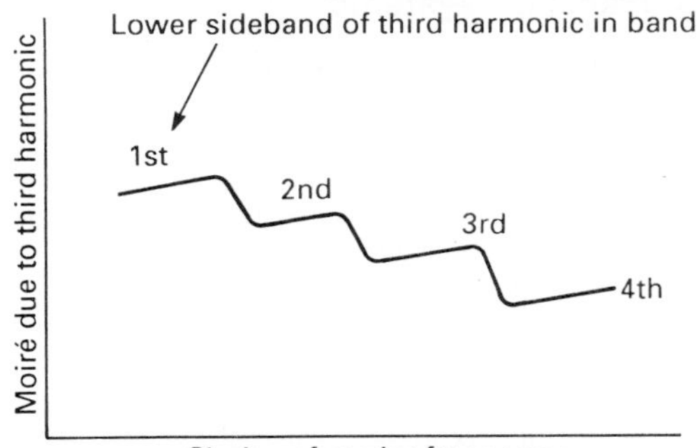

Figure 37.6 Shelf working

curve gives rise to the name *shelf working*. Each shelf has a band of frequencies associated with it resulting in the names *low band* and *high band* as applied to improved recording systems.

37.2.6 Colour-under or heterodyne recording

The bandwidth requirements and transport stability specifications make direct recording vtrs too complex and expensive for domestic use and too delicate for news or field recording. Lower bandwidth machines are acceptable for record/replay use where multiple generations are not required, e.g. in editing.

A reduction in bandwidth also reduces the tape consumption and hence the running cost of the system.

If the video channel bandwidth were simply to be reduced, to say 50 per cent, there would be a similar reduction in the resolution in the replayed picture. However this is not possible in a composite system where the colour subcarrier occupies a space in the spectrum close to the maximum signal frequency.

If the channel bandwidth is reduced, the colour subcarrier, and therefore the picture colour information, is lost. The adopted solution is to translate the high frequency subcarrier to a lower frequency before recording the video signal. This translation is effected by separating the chrominance part of the video from the luminance and then mixing it with a local oscillator. The lower sidebands are then filtered out and added back to the luminance signal.

A high-pass filter in the luminance signal path leaves a gap in the frequency spectrum below the lower sideband of the fm modulated video. The chrominance is therefore heterodyned and fitted *under* the luminance.

37.2.7 Component recording

The light entering the camera is broken into red, green and blue before being focused on the photoelectric transducers. These three signals are then combined to form the chrominance signal. A fourth signal, which may be provided from an additional photoelectric transducer or by combining the outputs of the red, green and blue channels, forms the luminance or black and white signal.

The first stage of combination of the RGB signals is to subtract the luminance, Y, signal from the red and blue signals. The resulting two signals, R-Y and B-Y, are the *component* or *colour difference* signals.

If these signals are applied to the modulator in the vtr, a reduction in bandwidth will reduce only the resolution but will not remove the colour. The penalty is that three channels have to be recorded instead of one. However, component recorders have advantages in tape economy, picture quality and robustness that make them valuable for mobile use.

37.3 Transport systems

37.3.1 High speed for video heads

Once the bandwidth of the video signal has been reduced to manageable proportions, the next requirement is a transport capable of giving the high head to tape speed required for the signal frequencies. These are in the region of 20 MHz when the upper sidebands are taken into account.

The speed required is set by:

- the length of the replay head gap
- the frequency to be recorded

The length of the record head gap is less important because the actual point of recording is just outside the trailing edge of the gap.

The replay head gap should be as small as possible to allow the highest frequencies to be recovered. However, as the gap length reduces, less flux is recovered from the tape into the magnetic circuit, and the output signal falls. Gap length is therefore the best compromise, and typical values range from 1 μm to 6 μm.

Head to tape speed is the wavelength of the highest frequency to be recorded divided by the gap length. For 15 MHz and a head gap of 6 μm, the head to tape speed would be 0.5 m/s.

For one hour of recording, a reel of tape 82 km (270 000 ft) long would be needed, travelling at 82 km/h (51 miles an hour). One early machine did use metal tape travelling at 20 miles/hour, but it had to be bolted to the floor, had four foot diameter reels for 20 minutes of recording and had to be operated behind closed doors. Not a practical system for the average home!

37.3.2 Segmented recording

One of the aspects of the video signal not present in other signals to be recorded, e.g. audio, is that it is packaged in lines, fields and frames. Each break offers an opportunity to stop recording and reset the system.

Early tape machines used two-inch wide tape across which video tracks were laid by recording heads mounted on a rotating drum. Each track contained 17 lines of the picture, which allowed some redundancy. The overlap time allowed the replay electronics to switch from the head just leaving the tape to the one just meeting the tape. The switch took place during the line synchronizing period.

The system, *quadruplex*, derived its name from the layout of the head drum which had a quartet of heads on it placed at 90° to each other. The recording and replay characteristics of each channel have to be identical, or horizontal bands are seen in the replayed picture. This 'banding' effect is the main signal deficiency of the quadruplex system which, nevertheless, occupied an unchallenged position as the worldwide standard for television programme exchange for 25 years.

37.3.3 Helical recording

The final blow to the quadruplex system was the arrival of the helical recording systems, particularly the non-segmented standards which were led by the A format, later to become the C format. The concept of video record and replay heads mounted on a rotating drum was maintained, but instead of a segment of the picture for each track, a complete field was recorded.

The signal system was similar to that of the quadruplex machines without the need for matching several channels for each picture.

Two main configurations are in common use, using a single video head or a double video head.

37.3.3.1 Single video head

The C format uses a drum which rotates at field rate, either 50 Hz or 60 Hz for the PAL or NTSC systems, on which is mounted a video head to record one active field.

The period when the head is moving from the end of one track to the beginning of the next is synchronized to be the field interval and is not recorded. The replay system regenerates the missing information, known as the *format drop out*.

An option is to add an extra video head to the drum to record the missing part of the signal, which amounts to 10 or 12 lines for PAL or NTSC. The replay system may then make use of this part of the signal to carry extra information, such as time code or programme data.

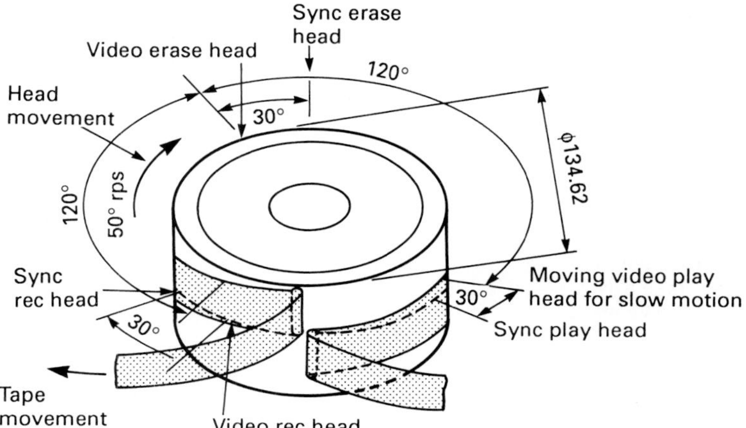

Figure 37.7 Typical helical scan tape transport and rotating head drum assembly, shows the professional broadcasting 1 inch C format. Most other formats limit the angle of wrap of the tape around the drum to 180°. To allow this, the drum has to carry two record/replay heads spaced 180° around the circumference. Each head records/replays a complete television field

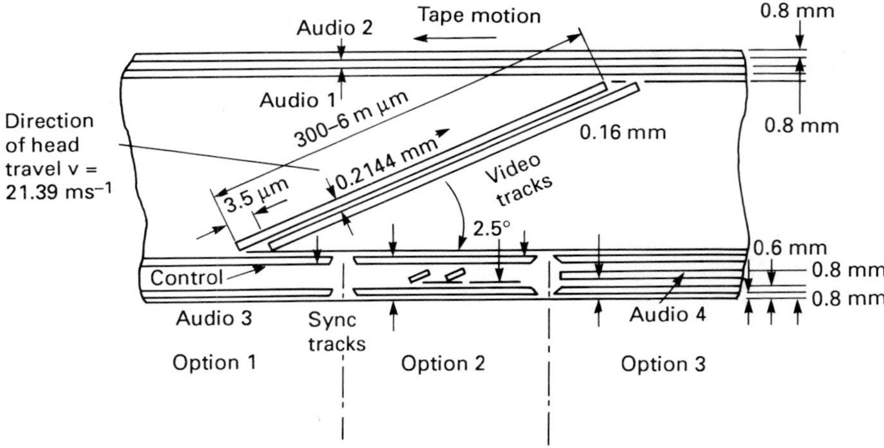

Figure 37.8 Track pattern for broadcast 1 inch C format (not to scale). The actual angles and the relative length to width are much greater. There is a guard band between the video tracks, and the video head gaps are perpendicular to the track

37.3.3.2 Double video head

Where the machine is to be used outside the broadcast environment, the extra cost of the circuit to replace the format drop out is too expensive. A drum with two heads mounted at 180° is used, and the replay electronics are switched between each field. There is a need to match the channels, but since the comparison has to be made sequentially rather than within one picture, it is less critical.

37.3.3.3 Segmented double video head

Formats which use one track per field demand long tracks that are difficult to follow on replay. The shorter tracks of the quadruplex are far more robust. Between the two extremes, there are the slant track formats of which the B format only remains in use.

Each track contains about one third of a field, which reduces the need to match channels and places less strain on the track following system.

37.3.4 Azimuth recording

To obtain the maximum high frequency response from a recording system, it is essential that the replay head gap is exactly parallel to the angle of the record head gap. This is referred to as the *azimuth angle;* an error in azimuth will reduce the high frequency response of a recorder. In the majority of applications, the gap is arranged to be perpendicular to the direction of travel of the head. For audio recorders, this is parallel to the tape edge; for video recorders, it is parallel to the track edge.

As the distance between the tracks is reduced, the replay head will begin to pick up information from the adjacent track. This is not very critical for an fm recording system, since a 2–3 dB difference in signal amplitude is enough to ensure that only the wanted signal is decoded. For an am signal, it is essential to keep cross-channel interference to a minimum (better than 50 dB if possible). Traditionally, this is done by leaving a gap (*guard band*) between one track and the next.

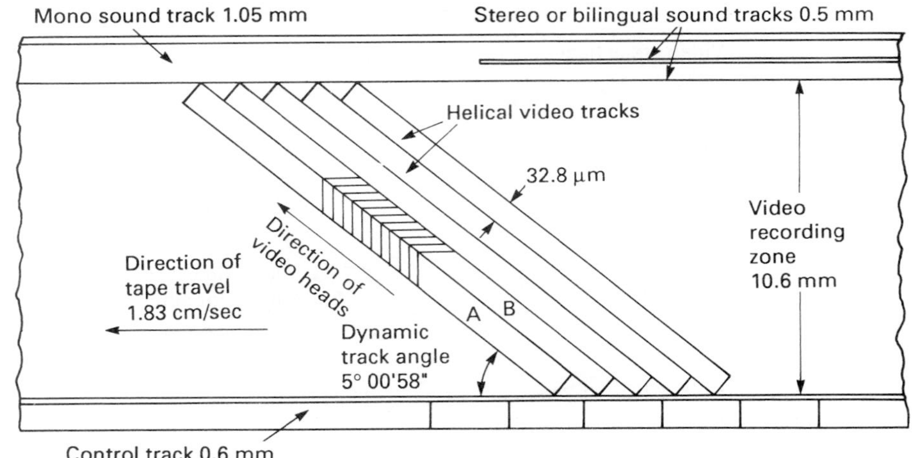

Figure 37.9 Practical track layout for domestic recorders using colour-under and azimuth offset. This is a Betamax layout where the tape is 12.7 mm (1/2 inch) wide, and one control pulse is recorded for every two video tracks

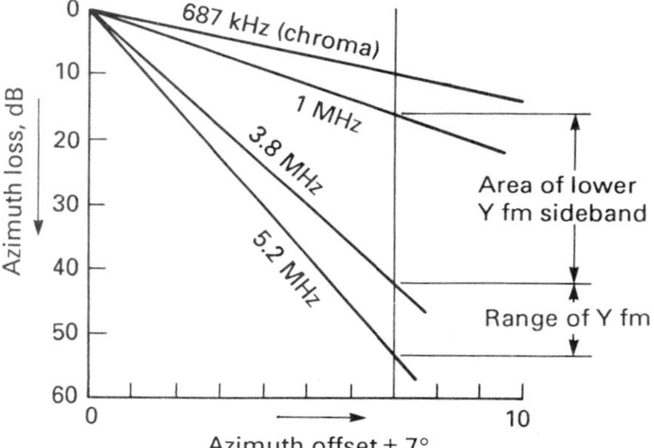

Figure 37.10 Effect of azimuth offset at different frequencies. To reduce tape consumption, the inter-track guard band is removed. To prevent crosstalk from one track to the next, the azimuth angle of the head gaps for field 1 and field 2 are offset from perpendicular. This prevents the high frequency of track B being read by the replay head of track A

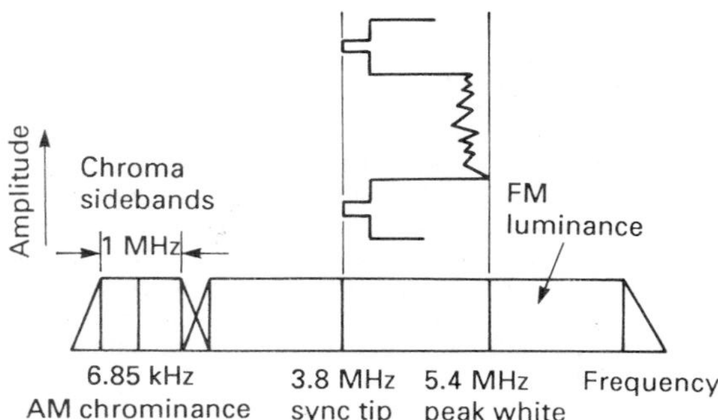

Figure 37.11 Colour-under spectrum showing how the frequencies of the colour information are changed in domestic and low end industrial machines, e.g. VHS, Betamax or U-Matic. This process allows the colour signal to be maintained on a restricted band width system

To achieve the most efficient use of tape, it is desirable to remove the guard bands but still retain the signal separation.

For a double head fm system, one head may be assigned to record the even fields while the other records the odd fields. The replay heads will be similarly controlled.

With short recorded wavelengths, such as are used for the luminance of the colour-under systems, azimuth angle is very critical. Just 5–10° of error between the record and replay heads will reduce the signal level by 40–50 dB for frequencies in the region of 3 MHz.

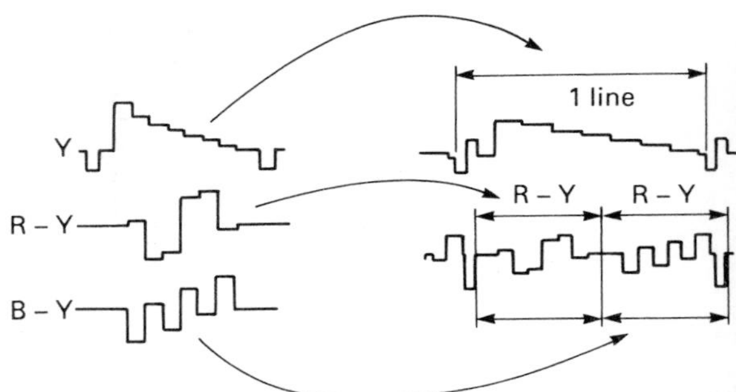

Figure 37.12 Betacam time division compressed video. To retain high quality but reduce the on-tape bandwidth component, recorders split the PAL or NTSC signal into luminance (Y) and colour difference (R-Y), (B-Y) signals. The colour difference signals require only half the bandwidth of the luminance and can therefore be compressed before being recorded one after the other on a single track. The luminance is recorded on a second track

In low cost and domestic vtrs, the odd fields are recorded with a head that is deliberately angled away from the perpendicular in one direction, while the even fields are recorded by a head angled in the opposite direction. In replay, each channel is recovered by a head with a suitable azimuth offset. Should the head stray onto the wrong track then the high frequency loss caused by the azimuth error will be more than adequate to suppress the unwanted signal. The guard bands are then not required. A separate method has to be used for the low frequency am colour signal in the colour-under vtrs.

37.4 Servo systems

All tape recorders are required to move the tape past the head at an even and accurate speed. In the case of a video recorder, it

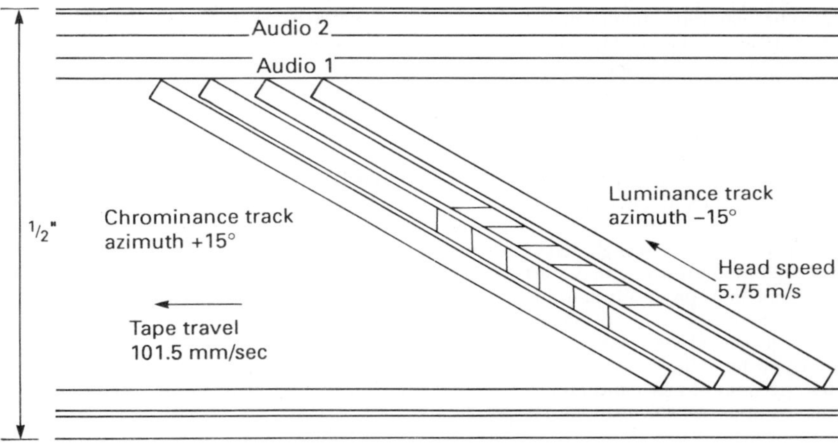

Figure 37.13 Typical component recording format using compressed colour difference signals and azimuth crosstalk elimination

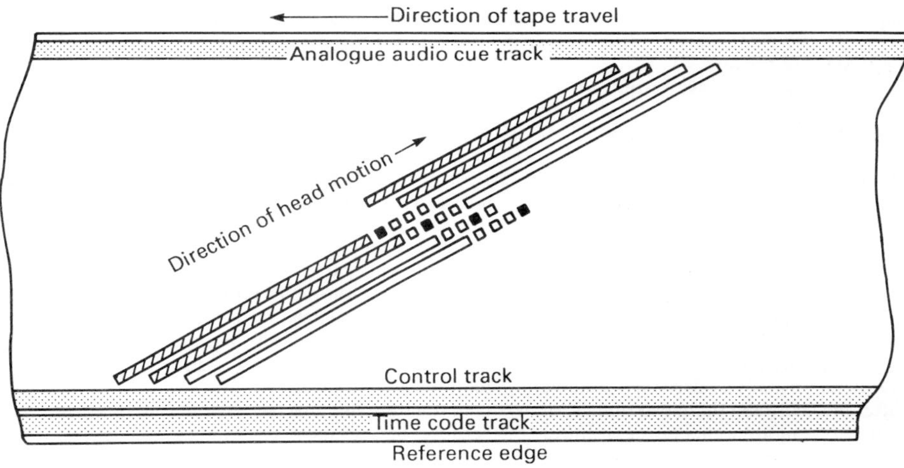

Figure 37.14 Component (D1) digital videotape recording format. Tape width = 19 mm, track width = 0.040 mm, track length = 170 mm

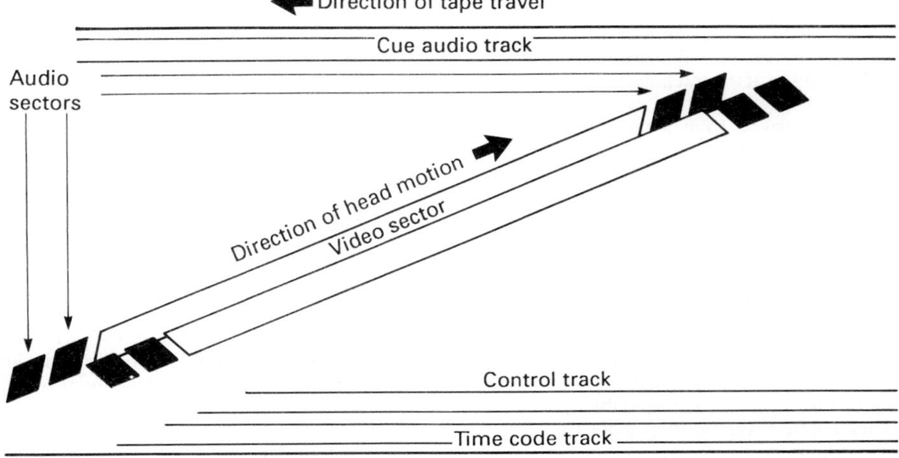

Figure 37.15 Composite (D2) digital videotape recording format

is also necessary to move the video head over the tape at a high speed. If a successful replay is to be achieved, some means of reproducing the movement of the tape and the head has to be found, preferably one which does not need an operator to control it.

The speed of the tape during the record process is set by the rotation speed of the capstan and its diameter. Provided that it has been made to an accurate dimension, and that the motor rotates at the correct and constant speed, the desired tape speed will be achieved. This is the basis of audio recorders and of the audio channels for a video recorder.

The video tracks on the other hand are laid at an angle across the tape, and it is therefore essential to know where they start and finish. Film solves a similar problem by physically marking the backing. Each frame has a sprocket hole, or holes, associated with it. When the film is projected, these holes place each frame in the correct position in front of the lenses and light source. Fine adjustment, to take account of mechanical tolerances, is made by adjusting the position at which the film is stopped for exposure.

In the same way, magnetic tape has a series of magnetic marker pulses on a longitudinal track similar and parallel to the audio tracks. This track is called the *control track*, and the pulses are used during replay to:

- regulate the longitudinal speed of the tape
- match the rotation of the video replay head drum to the position of the video tracks
- provide information as to which field in the PAL or NTSC sequence is being replayed

37.4.1 Speed servo

There are two speed servos in a video recorder: one for the capstan to control the speed of the tape, and the other for the drum to control the video head movement before the video tracks have been located.

The speed servo compares the frequency of the recovered pulses with a reference frequency, usually by deriving a suitable dc value for each. The difference between these two signals, or the *error signal*, is used to drive the motor either faster or slower as required.

37.4.2 Position servo

Once the video drum is rotating at the correct speed, it has to be regulated to bring the video replay head into contact with the start of each track as that track moves past the drum. This requires a measurement of the position of the track start and the head angular position.

The head angular position is derived from a tachometer attached to the drum drive. The pulses it generates are compared first with the video field pulse reference applied to the vtr. Once these are brought together, the capstan speed is regulated to bring the video track up to the video head as the new field and track starts.

37.4.3 Start up sequence

When the *play* or *record* command is issued, both the head drum and the capstan begin to rotate. The dominant control is a speed servo in each case. Maximum error exists, and the motors accelerate accordingly, limited only by the power available. At this time, the position errors will be swinging from a positive maximum to a negative maximum as the drum and capstan alternate between being ahead of, and behind, the desired position.

When the drum approaches the correct speed, the speed error signal will reduce and the rate at which the position error changes will also slow down. The position servo will finally take over and control the position of the video head relative to the reference. The speed is inevitably correct if the position is to remain correct.

37.5 Timebase correction

All colour vtrs require replay signal stability beyond what is mechanically possible. The allowable variation is of the order of a nanosecond (10^{-9}s) and may derive from the PAL or NTSC specifications.

For the PAL system, the colour subcarrier frequency to two decimal places is 4.43 MHz, and one cycle takes $1/4.43 \times 10^6 =$ 225.7 ns. The allowable phase error in the colour signal is $\pm 2°$. Therefore the timing error is roughly $(226 \text{ ns} \times 2)/360° =$ 1.2 ns.

To achieve this, the signal recovered from tape must be processed to remove the timing errors. The process is called *time base correction*, but is usually combined with other processing such as format drop out replacement, video amplitude, chrominance amplitude and phase and rf drop out concealment.

37.5.1 Timing error correction

The principle element of all time base correctors is a video store or delay. The off-tape video signal is put into the store at a rate governed by the rate of the off-tape signal; either the line sync pulses or the chrominance burst are used. The output from the store is controlled by a stable external reference, usually a station reference to which the vtr is locked for replay.

Signals which arrive too early are delayed, while those which arrive late pass through a reduced delay. To make it possible to advance some signals, all signals have to be delayed by an amount equal to half the normal maximum error. The vtr output must therefore be advanced with respect to the reference.

In quadruplex machines, the timing errors were small, because of the short tracks, but had large discontinuities, because of the head switches. The maximum error would never be more than a few microseconds, and so the store could be made with an analogue delay line, segments of which were switched in and out as required.

In helical machines, the timing errors may be large, especially where the recorder is portable, but there are no sudden jumps except during the field period when there is in any case plenty of time to recover. To provide enough delay to correct all likely errors, a delay of ten lines of video is needed. This is very difficult to achieve with analogue delay lines, as the output signal amplitude is dependent upon the amount of delay line that has been used.

The performance of the time base corrector (TBC) is dependent upon the accuracy with which the timing errors may be measured. The most readily available accurate reference is the colour burst at the start of each line. The zero crossing points are predictable from line to line and are directly related to the picture information.

The presence of noise will reduce the accuracy of the measurement, and therefore TBC performance is diminished for noisy recordings. To counteract this effect, some formats increase the amplitude of the recorded burst to improve the signal/noise ratio.

An additional requirement for the TBC of a helical vtr is the reconstruction of a proper PAL or NTSC signal during non-standard replay. When the slow or fast motion facility is in operation, the rate at which the video is recovered may vary by several percent.

For a C format machine in stop, there is a loss of 3½ video lines per field. This is caused by the lack of tape movement, which means that part of each track is out of contact with the drum. These lines must be replaced in the TBC.

The change in the number of lines scanned implies a reduction in the velocity of the head across the tape, which in turn means a reduction in the rf frequency. Although the picture contrast is proportional to the deviation, the black level is proportional to the absolute frequency. The TBC must therefore make a correction to the brightness in non-standard play.

For a still frame, a single field or frame is replayed continuously. The PAL or NTSC is therefore destroyed, and the TBC must recreate it. The chrominance is stripped decoded before being recoded with a stable chrominance subcarrier. The phase angle may be adjusted in this process and may have to be manually set to account for machine and recording tolerances.

P Audemars MA
Senior Film & Video Editor, London Weekend Television

38 Film and Video Tape Editing

This section details and compares the various procedures involved in film and video tape editing, and examines the increasing inter-relation between the two in television production. The overall aim of each system is of course the same, as are the basic editing conventions or aesthetic principles. Similarly, the larger practical limitations in terms of available source material, allocated time and budget are equally applicable. However, the actual procedures of the editing process are different, and these are discussed separately (sections *38.1* and *38.2*) and compared in section *38.3*.

38.1 Film editing

38.1.1 Formats

38.1.1.1 16 mm reversal, combined magnetic and separate magnetic

For many years, 16 mm reversal film in either the combined magnetic format with a magnetic sound stripe on the edge of the picture or mute film with sound on separate 16 mm magnetic film was the mainstay of television news and current affairs film output. It is a system in which the camera original is processed directly into a positive transparent image, in the same manner as slide film in still photography. This processed camera original is then edited by cutting and joining and either transmitted direct from telecine or transferred to videotape as one element in a complete programme.

The advantages of this system are:

- *cost effectiveness*, the only laboratory charges being for the processing of the camera original (more complex systems incur a number of other costs as discussed in subsequent sections)
- *speed of operation*, as the time taken for the additional procedures of more complex systems is eliminated

It was common practice in news operations for film to be shot, processed, edited, dubbed and transmitted within a few hours, and for quite complex current affairs programmes to be completed within a few days.

The disadvantages lie in:

- *dangers* of working with the irreplaceable master throughout the editing, dubbing and transmission processes with inherent risks of damage
- *limitations* of editing when shots can be shortened but not lengthened once cut without the inclusion of a join in the middle of a scene
- *visibility* of these physical joins (even when occurring more normally between scenes) on transmission
- *inability to grade* scenes except on telecine
- *inability to use any transition* between scenes other than a straight cut

This system is now very little used, having been replaced almost entirely by lightweight video recording equipment giving the same portability as 16 mm film cameras and by the development of non-segmented recording which makes possible still-frame editing in the film manner.

Although the initial investment in a video operation is heavy, it does of course result in an enormous saving on stock costs, as not only are video tapes much cheaper than reversal film stock and processing, but the rushes tapes can be erased and re-used.

38.1.1.2 16 mm negative/positive, separate magnetic

This is the most common form of television film used today. The camera is loaded with negative film which after exposure is developed and printed, usually as a *one light* (i.e. a consistent grading for the whole roll rather than the grading of individual scenes) *rushes* print. The sound is recorded on quarter-inch tape and transferred to separate 16 mm magnetic film stock. The great advantage over reversal film is that the rushes prints can be edited and re-edited without limitation, while the master negative remains untouched.

When the final cut stage is reached, the negative is cut to match the completed work print, each shot is graded to achieve a consistency of colour and density throughout, and a showprint is made as a complete new roll of film without joins. The sound track is edited together with the picture, and after track-laying is dubbed onto a new roll of magnetic sound film.

Normally the showprint and dubbed master sound are then run together on a telecine machine as a direct transmission or recorded onto videotape for later transmission.

38.1.1.3 35 mm negative/positive, separate magnetic

The outline in section *38.1.1.2* applies equally to the use of 35 mm film (except that the original sound would be transferred

Figure 38.1 Steenbeck film editing machine (W Steenbeck & Co)

to 35 mm magnetic film stock). 35 mm film offers considerable advantages in both picture and sound quality over 16 mm, because of the approximately four times larger frame area of the picture, and the $2\frac{1}{2}$ times greater linear speed and increased track width of the magnetic sound.

This quality difference is most apparent in theatrical projection where the film image is enlarged many thousands of times, and where the sound reproduction must be adequate for a large auditorium. 16 mm is not capable of such a satisfactory result in these circumstances, but for television transmission, particularly with recent advances in camera stocks and printing techniques and also considerable advances in the capability of telecine machines, it is generally considered more than adequate.

The extra cost of 35 mm production is not usually justified in television production other than for commercials where complicated special effects are common or for large scale productions or co-productions which have overseas sales potential or the possibility of theatrical exhibition as well as television transmission.

38.1.1.4 16 mm and 35 mm combined magnetic and combined optical

In television, where telecines have the ability to run separate picture and sound in synchronization, film sound is normally

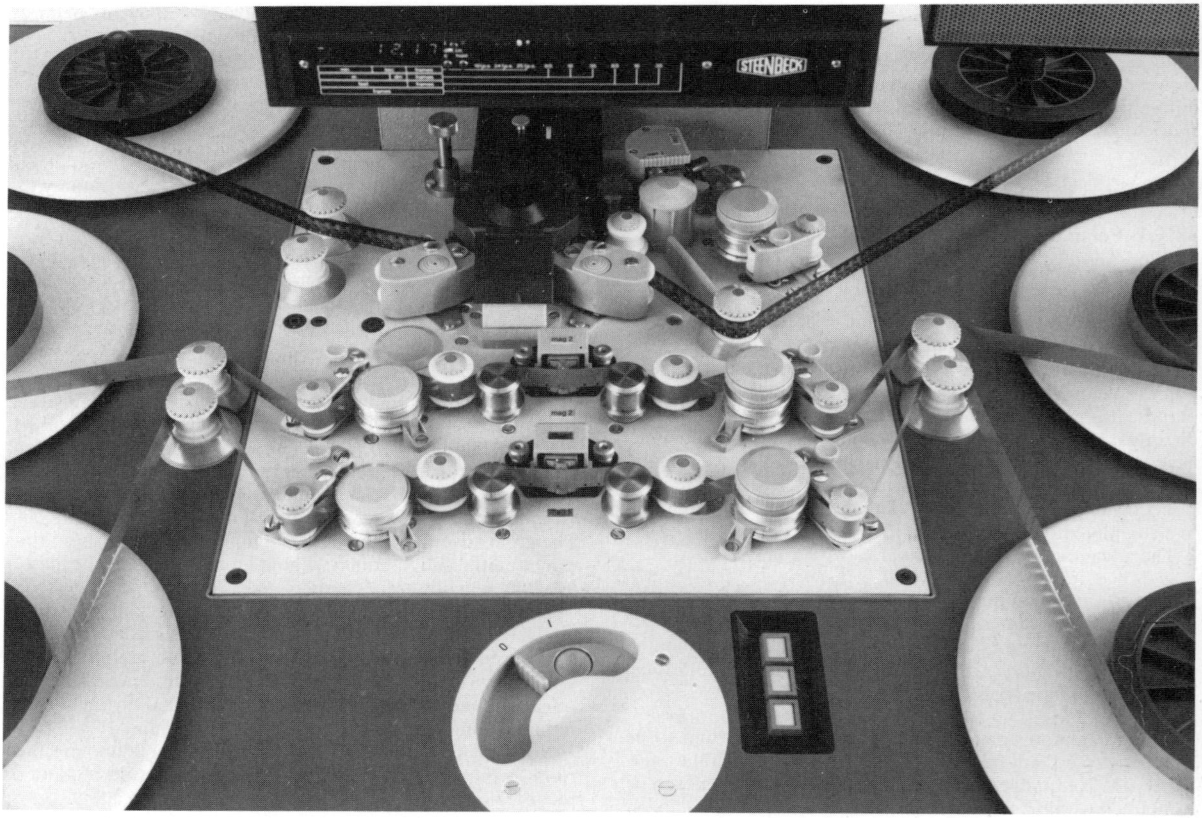

Figure 38.2 Passage of the film through the heads on a Steenbeck film editing machine (W Steenbeck & Co)

left separate, and the finished product is transmitted or recorded as a *double head*.

For theatrical exhibition using a conventional projector, however, a combined print is necessary. This is achieved either by transferring the final sound track onto a print that has been *striped* with a magnetic track or tracks outside the picture frame area, or by re-recording the final sound track optically onto a separate film negative. This is subsequently developed and printed onto the positive film as a separate process.

Modern film transport mechanisms tend to utilize a rotating prism to provide the intermittent illumination necessary to create the illusion of movement from a series of still frames, and the film itself moves continuously (see also section *39.5*). However, because earlier systems depended on an intermittent pull-down mechanism whereby the film stopped momentarily in the projection gate on each frame, and also because of the difficulty of locating the sound reproducing head level with the picture gate, it was necessary for the sound head to be positioned either before or after the picture gate at a point where the film is running continuously.

Allowance must therefore be made for this when making combined prints as follows:

Combined optical prints

16 mm	sound advanced 26 frames
35 mm	sound advanced 20 frames

Combined magnetic prints

16 mm	sound advanced 28 frames
35 mm	sound retarded 28 frames

38.1.2 Equipment

38.1.2.1 Editing machines

Film editing machines today are most commonly of the Steenbeck type, sometimes known as *editing tables*, where the picture roll runs flat between plates located on either side of the table through a rotating prism picture head that projects the picture via a series of lenses and mirrors onto a vertical screen. The sound roll or rolls follow a similar path via a magnetic sound head or heads.

This system has the advantage over earlier Moviola type machines (which depended on an intermittent motion through the picture gate) of being able to run at much higher speed when searching, and so it is far less likely to cause damage to the film during the editing process. This high speed access together with the large take-up plates removes the necessity to break down the picture and sound into individual takes, and thereby reduces film handling time. These factors made the use of reversal film in television much more viable.

Editing tables are most commonly either of the four-plate (picture plus one sound track) or six-plate (picture plus two sound tracks) types, although more complex machines are available. All machines have the ability to run picture in lock with sound, or separately at various speeds from frame by frame 'inching' to up to six times normal, or 'sound speed' either forwards or backwards.

38.1.2.2 Picture synchronizers

As the name implies, synchronizers are designed for the purpose of synchronizing picture and sound. Originally, synchronizers consisted of a number of toothed double wheels on a common shaft into which picture and sound sections could be located for synchronization during rushes syncing, assembly after individual shots had been cut on a Moviola type editing machine, or for track-laying. See section *38.1.3*.

With 35 mm film, the picture frame is of a sufficient size to locate and mark frames with the aid of a lightbox set in the editing bench beneath the synchronizer. However with 16 mm film, it was necessary to develop the picture synchronizer or *picsync* in which a rotating prism picture head is provided to project the picture onto a small screen.

Figure 38.3 Compeditor 720 picture synchronizer (Acmade International)

These machines subsequently became common in 35 mm format as well, and have gradually evolved into quite sophisticated devices incorporating motor drive at normal or sound speed as well as a declutchable fourth sound path for searching in addition to the traditional three sound paths locked to the picture. Special models offering twin picture heads or 16 mm/35 mm combinations are also available.

38.1.2.3 Ancillary equipment

The most important piece of ancillary equipment is the *joiner*. At one time, cement joiners were the only type available. In those, a scraper blade was used to remove a narrow strip of emulsion across the end of the first piece of film to be joined. This exposed the top of the film base which was then coated with a special film cement which dissolved the top layer of film base. The bottom of the film base of the next piece of film was then clamped on top involving a slight overlap, and the join would harden and fuse the two pieces of film together. This was a slow process that could be speeded by the addition of an electric heating element to the joiner.

The disadvantages of this system are considerable as far as the film editing process is concerned:

- it is time-consuming
- it leaves an overlap which can cause problems when running the edited film
- at least one frame is lost in the process, so that if the scene subsequently needs to be extended, spacing must be inserted to maintain sync with the sound

Now that the superior (from the film editing point of view) system of tape joining is available, the cement joiner is used only for negative cutting, where tape joins, if used, would show on the subsequent prints.

The most commonly used tape joiners consist of a base block into which is recessed a film guide path with locating pins. At the right-hand edge is a film cutting block with an attached blade. The two pieces of film to be joined are cut on the frame line in the appropriate place using the cutting blade, and then placed in the central film guide so that they lie flush end to end. A length of transparent joining tape is pulled away from the roll located at the front of the joiner and placed across the two pieces of film so that half the width of the tape is over each piece. The top plate is brought down to clamp the film together, and the operating lever depressed. This lowers a combined tape cutter which trims off the tape flush with the edges of the film and also perforates the tape to match the sprocket holes along the film's edge. The description takes far longer than the operation, which with familiarity takes only 2–3 s. There is no overlap, and the join can easily be unpeeled and remade an infinite number of times without losing any frames.

Figure 38.4 M3-16 mm 2T special splicer (CIR)

Film editing rooms are also commonly equipped with one or more trim bins which are used to hang pieces of cut picture and sound, with one or more rewind benches, with shelves or racks for storing film cans, and with various smaller items such as chinagraph pencils for marking film.

38.1.3 The editing process

There are many different ways of editing film, and various procedures are followed with regard to features or television drama, television or cinema commercials, television documentaries and current affairs programmes, news or magazine type programmes, and so on. The following paragraphs give mainly an outline of commonly used practices within television film editing.

38.1.3.1 Rushes syncing and edge numbering

As mentioned in section *38.1.1.2*, the cutting room is normally supplied with a *one light* (i.e. unvaried grading over the whole roll) or *rushes* print (a term shortened from *daily rushes* and derived from the practice of developing and printing overnight so that the material can be viewed the day after the shoot). The sound is recorded on a quarter-inch tape machine. Originally a pulse cable connected the camera and tape machine to ensure that the speed of the recorder was controlled by the speed of the camera and the two ran synchronously. Now, crystal controlled motors mean that camera and tape recorder can be operated completely independently of one another and still maintain exact synchronization. The quarter-inch tape is independently pulsed, and when it is transferred onto 16 mm magnetic film the speed can be checked by using an oscilloscope to read the pulse.

When the picture and sound rushes reach the film cutting room, the first task is to sync them together. This can be done either on a pic-sync or on an editing table.

Still the most common (because it is the most straightforward and reliable) method of sync reference is to use a *clapperboard* at the head of each scene. Scene and take number are marked on the clapperboard for visual reference, and a spoken announcement is made for aural reference. This enables each picture and sound take to be identified. The lever on top of the clapperboard is then brought down sharply onto the main part, making a sharp bang on the sound track. By matching the picture frame where the board is seen to close with the sound frame where the bang occurs, each take can be synchronized.

The identically spaced perforations of the picture film and sound film then ensure that picture and sound will remain in sync when run together through any mechanism in which the picture and sound transports are mechanically or electrically locked. Each take is trimmed to exactly the same length on picture and sound, and the next take is synced and joined on. By this process, the rushes can be assembled into convenient sized double-headed rolls.

The next task is to mark these rolls to aid subsequent identification and editing. This process is still often referred to as *Moy numbering* (after the manufacturers of the original machines) or as *rubber numbering* (after the early use of rubber numbering blocks which inked the numbers onto the film). At the present time, numbering machines no longer use ink, but rely instead on a narrow strip of adhesive tape through which a metal block prints onto the edge of the film, causing the tape to adhere to the film in the form of readable numbers. The numbering block advances by one digit on each revolution so that sequential numbers are printed at exact intervals.

By marking both picture and sound rolls with synchronous start marks, running first one roll (the order is unimportant) from the start mark, resetting the numbering block to the same start number, and then running the other roll from its start mark, both rolls will then be numbered in sync throughout, greatly facilitating identification and editing later, as even very short sections can still be matched.

In television documentary editing, it is usual to number rolls of film continuously, giving each roll a different prefix. Thus a simple numbering block might contain a choice of two letters (commonly A–K excluding I) which provides 100 possible combinations (AA,AB,AC,etc., followed by BA,BB,BC,etc., and so on). These letter prefixes remain constant throughout the roll, so that later on pieces of film and sound can be filed by roll letters. The numbers following (of which there are usually four in a simple block) increase by one digit every foot, so a roll might be numbered from AA0000 to AA0956. Having available up to 100 roll identifications, with each roll capable in theory of being numbered up to 9999 feet (longer than 3 km), is more than sufficient for most television productions.

In drama productions, where takes need to be more individually defined, it is usual to number each take separately. In this system, the block is more complex and is reset at the beginning of each take and lined up with the clapperboard. The scene and take identification is included in the number, e.g. scene 16 take 2 would be numbered 16T2 0000–16T2 0117. Obviously this is more precise but necessarily more time-consuming, as the numbering machine has to be stopped and the block reset many times during the course of numbering each roll.

38.1.3.2 Rushes viewing

When the rushes have been synced and numbered, it is usual to view them before any cutting takes place. At one time, this meant viewing all rushes at normal speed in a projection theatre (which is still highly desirable where time allows), but in many television situations where time is short, it is now more common to view rushes on a Steenbeck or a similar machine so that while some sections can be viewed at normal speed, other sections such as long interviews (or 'talking heads') can be viewed at high speed once it has been established that the overall picture and sound quality are satisfactory. At this stage, notes are made on the suitability of various scenes for inclusion in the first assembly, and a script may be marked up with selected takes.

38.1.3.3 Logging, breaking down, filing

It was once standard practice for all uncut film to be *logged*, which meant that a list was prepared with a note of every scene and take number available together with the first and last printed-on edge number (as described in section *38.1.3.1*), also the first and last number printed through from the negative (a guide mainly for negative cutting, discussed in section *38.1.3.12*), of each and every take. This log was subsequently used (and still is in many types of film editing) to identify trims, and was also useful when ordering reprints of picture or sound, when ordering dupe sections for optical effects, and for negative cutting queries.

However, now that 16 mm film is commonly numbered in 35 mm feet (i.e. with a number every 16 frames rather than every 40 frames), there should be no difficulty in locating any piece of film, however short, very quickly, without reference to a log sheet. This is provided that the cut sections that are so short as not to have printed numbers on them are marked with the adjacent number at the time of cutting, and that trims are filed on a peg marked with the appropriate prefix or put back in the relevant can.

It was usual to break down rushes (once viewed) into individual takes, because this made it easier to run and mark up on a Moviola type machine and subsequently cut and assemble on a synchronizer. However, with the introduction of high-speed Steenbeck type editing tables, this procedure is now largely unnecessary, and film tends to be left in larger rolls, often more or less as viewed.

Whether broken down or not, it is necessary to file rushes in such a way that they are easily accessible. A series of film cans marked with both lid labels and side tapes is the most common method, so that if for instance scene 27 take 3 is required a can marked *production name, production number, sync rushes, scenes 20–29, numbered AC 0000–AC 0567* can be instantly selected. If the rushes have been broken down, this will contain scenes 20–29 rolled into individual scenes with the picture and sound rolled in together, each with a tag attached reading 20-1, 20-2, 21-1, 21-2, etc., and 27-3 can be immediately retrieved and run on the machine. If the material has not been broken down, the can will contain one large picture roll and one large sound roll each marked *scenes 20–29* which can then be laced up on an editing table and run at high speed to the start of 27-3.

If working from individual scenes, after removing the appropriate piece for assembly, the front and end trims can be hung on a pin rack over a trim bin or rolled up and put back in the relevant can. If working from larger rolls, the appropriate piece is removed for assembly and the roll joined back together and returned to the relevant can. If the scene is subsequently shortened, the fine trims can be hung over the bin and later rolled in with the other trims of the same scene or joined back into the roll in the correct place. Whichever method or combination of methods is used, it is vital that everything is kept in its proper place at all times, and the methodical approach that takes a few minutes longer in the early stages can save hours of wasted time later on.

38.1.3.4 First assembly

Whenever a film or video editor tells someone outside the industry what his or her job is, the first comment is often: 'Oh, you're the person who goes through it and takes the bad bits out''.

This misconception stems of course from home movies where the editing process is often no more than this, and where the *cutting ratio* (the ratio between shot film in total and final edited length) is probably only 1½:1 or even less. In any professional production where the ratio is likely to be at the very least 10:1, and possibly 30:1 or 40:1 on a large scale documentary or even several hundred to one on a commercial, the process is much more akin (to use the same simple terms), not to 'cutting the bad bits out' but to 'cutting the good bits out' and assembling them as a separate roll.

In a documentary containing large numbers of interviews or talking heads, the most efficient way to proceed is to go through the script and from this mark all the pieces of selected interview (generally referred to as the *sync* even though in fact many other parts of the film involve synchronous sound) onto the transcripts of each individual interviewee. It is then a straightforward task to go through each roll of rushes (which ideally should not have one interviewee split onto more than one roll) and take out all the required sections of that particular interview as they occur, regardless of the running order of the script. This is much quicker than working directly from the script, which might call for a section of interviewee A, followed by B, followed by A again, then C, then D, then back to B, then back to A again, and so on. Taking the sections out in script order will involve much time-consuming lacing-up and rewinding of the same rolls over and over again. Using the first method, once all the required sections of A, all the required sections of B, and so on, have been removed and marked with the interviewee's name and the script number, they can be assembled quickly in the required order.

This stage is usually known as a *first assembly* or *sync assembly* and can give a very rough idea of the shape of the final programme, or at least one possible shape. If considered necessary, the length of the remaining sections can be roughly calculated from the length of the proposed commentary (if written at this stage) and added to the length of the sync assembly to give an approximation of running time, which at this stage is likely to be considerably over length.

38.1.3.5 Rough cut

To reach this stage, the first assembly is viewed, sections are shortened (and occasionally lengthened), sections are deleted and new ones added, and the running order might well be rearranged. At the same time, the illustrative sequences will be rough edited, i.e. they will be cut together in one possible way without too much regard for the final polishing of every edit, simply because there may well be drastic recutting later, which would render time spent in perfecting sequences at this stage as time wasted.

A typical rough cut will consist of interviews cut to length, presenter pieces to camera (if any), sequences illustrating the subject being dealt with (possibly with a rough commentary or with voice-over taken from the interviews), and perhaps such library or rostrum material (see section *38.1.3.6*) as is available at this stage.

The rough cut should give a fair idea of what the final programme might look like. It can also give a fair idea of what the programme might sound like, although at this stage the sound will be fairly rough, and will not contain any effects other than those shot at the time of filming, nor any music, and if there is commentary it is likely to be only a rough guide.

38.1.3.6 Fine cut and use of library film and rostrum material

After viewing the rough cut, it may well be decided that the general shape of the programme is misconceived, and that a completely new approach is called for, involving drastic rearrangement. At the other extreme, it may well be decided that the overall approach is completely correct, and proceeding to fine cut stage is only a matter of shortening by tightening up sequences and deleting redundant material. The rough cut is normally over length as sequences will have been cut loose and possibly alternatives will have been included. In practice, the necessary action will be somewhere between these two extremes.

Television documentaries often depend heavily on library material, and a film researcher will by this stage have obtained what is required from various archive sources. A duplicate negative of such material, compatible with the original camera negative of the programme, will have been made, and a duplicate print (with matching key or edge numbers, i.e. numbers printed through from the edge of the negative to the edge of the positive work print) supplied to the film editor. A television documentary is also likely to include rostrum material. Still photographs or artwork are shot on a rostrum camera so that such material can be included in the programme. This shooting can take the form of straight reproduction using a fixed shot or more likely will involve moves on the rostrum camera to give panning and zooming shots over still photographs or artwork or may involve stop-frame animation of the artwork to produce moving graphs or diagrams.

38.1.3.7 Optical effects and titling

To make optical effects as a separate process, it is necessary for the required sections of camera negative to be sent to the film laboratory. Duplicate positive prints will be made from these sections. These are then combined (effectively rephotographed) on an optical printer to produce a composite optical negative, from which an optical print can be struck (again with matching key numbers) and incorporated into the workprint or cutting copy. Today, the only commonly used optical effects in television film are dissolves from one scene to another and fades from picture to black and vice versa. These simple effects can be produced at the final printing stage provided that the negative used for printing is cut as an *A and B roll* (see section *38.1.3.12*). In television film, this is now almost universal practice.

Titling, whether opening titles, closing credits, or lower frame 'supers' or superimpositions (usually identifying a particular speaker, though also used for other information), can also be produced by optically combining the film background with a shot title, but is now usually added from a titles generator at the film to videotape transfer stage.

38.1.3.8 Commentary recording

The most satisfactory way to record commentary is to picture at a dubbing theatre. The picture is run in lock with the sound tracks and also in lock with blank magnetic film on a recording head. The commentator sits in a soundproof booth watching the picture through a window and listening to the sound on headphones. He is cued by a light in the booth, by cue marks on the picture, or by watching the footage counter beneath the screen, the relevant footages having been marked on his script. As the commentary is being recorded directly onto magnetic film which is running in lock with everything else, it can easily be retaken by rolling back the picture and tracks (this system, known as *rock and roll*, is now universal) and re-recording.

If the desired result can be obtained in this way, then the actual dub (mixing the just recorded commentary with the other sound tracks, assuming that these have already been laid) can proceed immediately. Otherwise the commentary track can be adjusted back in the cutting room, returning to the dubbing theatre later to do the final mix or dub as a separate operation.

It is possible to record the commentary in isolation from the picture, either subsequently cutting the film to match, or if the film is cut first, matching timings supplied by the film editor. Although these methods have their uses in certain situations, generally they are very restrictive. The commentary tends to lack expression and any feeling of involvement if it is recorded with no reference to the picture on the part of the commentator.

38.1.3.9 Post syncing

Apart from a possible *commentary*, a finished film sound track will also generally include *dialogue*, *effects* and possibly *music*. Dialogue will have been recorded at the shooting stage, and in television film editing it is common practice to use this original dialogue right through to the final stages of dubbing. It may be necessary to replace some or all of the magnetic film sound tracks with new transfers or *lifts* from the quarter-inch tape master recording because of wear during the editing process. The technique of *post syncing* is not common in television film editing, but is often used in features and commercials.

In post syncing, the picture is broken down into short sections and looped as is the original sound track to be used as a guide. These two loops are then run in lock with a loop of blank magnetic film on the recording head. An actor in the recording booth watching the screen and listening to the guide track through headphones can then record the lines again in controlled conditions in synchronization with the picture. This is done if the original location recording was unsatisfactory for any reason. The loops must of course all be exactly the same length or else drift will occur, and since each take will erase the previous one, any possibles are checked before proceeding.

It is best to use 35 mm film for this process, because it is much more robust than 16 mm and can withstand being run as a continuous loop for long periods. Also, with fully coated 35 mm magnetic stock, it is possible either to turn the recording loop round the other way and record an alternate take down the other edge or, with a triple recording head system, to record three alternatives running in the same direction. Additionally, 35 mm film affords the possibility of subsequently adjusting the sync by quarter frame increments once the loops have been transferred for tracklaying. This is because there are four sprocket holes to each frame of 35 mm film as opposed to only one per frame on 16 mm.

The same technique can also be used for post syncing footsteps or other effects and is commonly used for the whole dubbing process in commercials where the short total running time involved makes it quicker and more convenient to run picture and all tracks as loops rather than rewind to the beginning after each take.

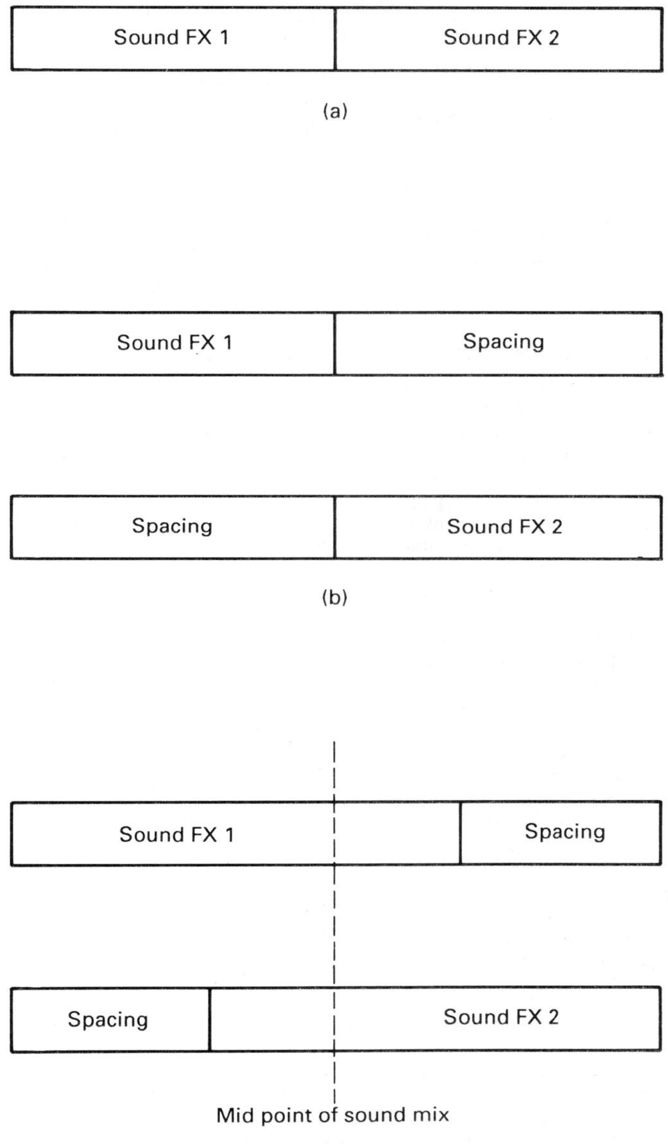

Figure 38.5 Basic tracklaying; (a) Butt joined sound, (b) offlaid sound, (c) offlaid sound with overlaps

An alternative method is the use of ADR (*automated dialogue replacement*), which consists basically of a high speed rock and roll dubbing system and a programmable locating device. Instead of looping the required sections, complete reels are put up, and the start and end footages of the first section required are entered. The system then rolls forward at high speed to the start of the first section to be post synced, plays the section at normal speed, reverses at high speed to the beginning of the section, plays it at normal speed again, and so on.

As the section is played over and over again, post syncing can be attempted and any possible good takes checked. It is, of course, still possible to use a triple track recording system if required.

Once an acceptable result has been achieved, the start and footages of the next section are entered and located, usually much more quickly than loops could be taken off and new ones put up (although this will depend on how much further on the next section is). A considerable amount of editing time can be saved by removing the necessity of making up and breaking down hundreds of loops.

38.1.3.10 Tracklaying and dubbing

The purpose of tracklaying is to enable all the required sound to be blended together. Sound that has been recorded on location will tend to be of varying levels and varying quality, and so

Picture	Dialogue 1	Dialogue 2	FX 1	FX 2
0 MS Two people	0 Man 1 dial		0 Interior cafe atmos	
27 CU Man 2	'later today' 27	27 Man 2 dial		
35 CU Man 1	35 Man 1 dial	'back soon' 35		
70 Man 2 walks into street	'get it now' 70		70 80	60 70 Exterior street atmos

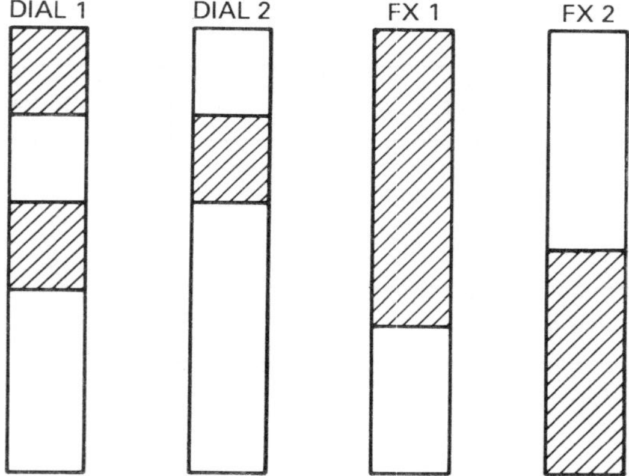

Figure 38.6 A simplified dubbing chart and the tracks it represents. On the extreme left of the chart are footages and picture cues. The dialogue is straight cut at the footages marked and the out words of each section are indicated. The background sound is to be mixed, the mid point of the mix is 70, and the tracks are overlapped 10 feet on either side

joining, for example, two sections of dialogue or two sections of effects together will be likely to give a jarring and hence unacceptable transition. If, on the other hand, these two sections are assembled onto separate tracks with spacing to maintain sync, it then becomes possible to re-record onto a new roll of magnetic film at the same time equalizing the tonal quality and adjusting the levels to match each other.

By overlapping the sound, it becomes possible to mix from one track to the other. This gives a smoother transition where required, the fader controlling the sound output from one track being progressively closed as the other one is progressively opened at the same rate.

Tracklaying also makes possible the inclusion of sound from different sources. Apart from the location sound, which might be split onto four or five separate tracks for the reasons explained, other effects or background atmospheres might be selected from sound effects libraries or especially recorded and assembled on further tracks, together with one or more music tracks (again either obtained from music libraries or especially recorded), the commentary if there is one, and also any post synced sound.

To enable the dubbing mixer to follow this multiplicity of tracks, a dubbing chart is prepared. This shows in diagrammatic form how the various tracks are assembled, what is on each track, and at which points they are available. Each track is laced up on a separate replay head, and they all run in lock with the picture and with a roll of magnetic stock on the recording head. The output of each replay head is fed to a separate channel in the mixing desk, and the combined output of the desk is recorded onto the new magnetic stock. The dubbing chart is marked in footages, and a footage counter beneath the screen enables the dubbing mixer to make the necessary adjustments at the appropriate time. Each fader on the desk will have been marked to correspond with the chart and to show which track it is controlling.

Although most dubbing theatres can run at least eight or ten tracks at once, there might well be more than this. For this reason, and also because even the most experienced mixer might well find it difficult to mix satisfactorily more than five or six tracks at once, *premixes* are often made. Perhaps four or five dialogue tracks might be combined into a *dialogue premix* and four or five effects tracks combined into an *effects premix* and then these two premixes combined with perhaps two music tracks and one commentary track to produce a *final mix* or *master*.

This system has the advantage that if a subsequent alteration is required on one of the tracks, it will be possible to remix just one of the premixes and then to do the final mix again (or possibly just a section of it). Similarly, in cases where it is likely that different versions (perhaps in foreign languages) will be made, it is usual to make an *M & E* (music and effects) track. This ensures that those elements of the sound track that can stay the same in the foreign version will not have to be mixed again, as the M & E can be combined directly with the new dialogue or commentary.

38.1.3.11 Optical sound transfer

In television film production, it is usual for the mixed master sound to remain as magnetic, so after the dub is completed it is necessary only to make a safety copy of the master, often done as a playback to picture so that the quality and balance of the mix as well as the sync can be checked at the same time.

For theatrical exhibition, however, it is necessary to provide an optical sound track. This involves re-recording the final mix photographically onto the negative film using either a shutter (moving in synchronism with the sound waves) to vary the amount of light reaching the film, or else a mirror (again moving in synchronism with the sound waves) to oscillate reflected light across a stationary slit.

The *shutter* method produces a track of constant width but variable density and is known as the *variable density* system.

The *mirror* method produces a track of constant density but variable area and is known as the *variable area* system.

The exposed negative is then developed and printed in the same way as picture film, and the positive print can then be played back for checking. If satisfactory, the negative can then be synchronized to the cut picture negative in the relationship described earlier and combined prints produced.

38.1.3.12 Negative cutting

When the picture fine cut is complete, it is sent either to the film laboratory or to a specialist negative cutting firm to be used as a guide for matching the original camera negative and any other negative to it. By means of the key numbers which were printed through onto the positive, the negative can be cut to match exactly with the picture fine cut. Tape joins overlap considerably into the picture area on either side of the join and would show on the print, so cement joiners are always used for negative cutting. With 35 mm film, there is sufficient space between the frames for the negative to be cut as a single roll without the joins showing on the print.

However, this is not the case with 16 mm film, so a system known as *A and B chequerboard* has been adopted. This involves cutting each succeeding piece of negative onto alternative rolls and keeping sync by matching lengths of black spacing through which the printing light does not pass. The joiner is turned so that the overlapping joins always occur into the spacing and not into the picture. This renders the joins invisible on the print.

It is now common practice in television film for simple optical effects to be incorporated at the printing stage, and the A and B negative cutting system makes this possible by overlapping the two picture negatives for a dissolve or by overlapping one picture negative and clear spacing (which will print black) to give a fade up or down. By overlapping the negatives in this way and by cueing the printer to fade up or fade down the light in the

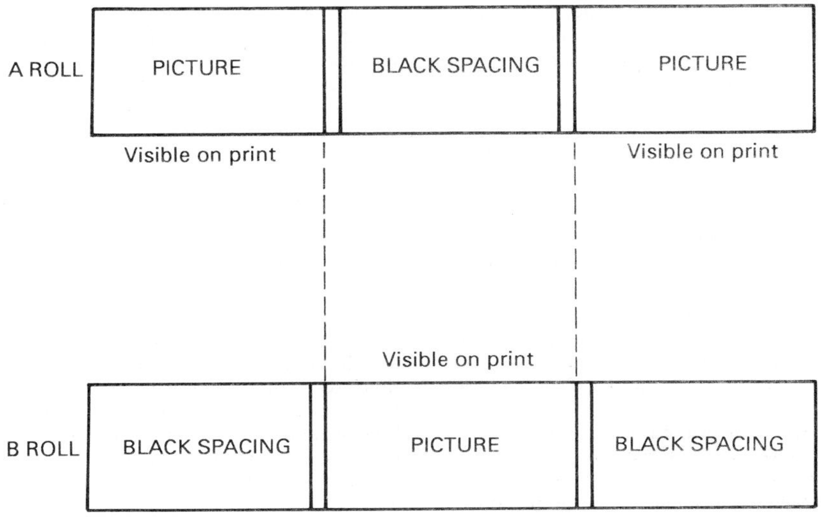

Figure 38.7 A and B roll chequerboard negative cutting

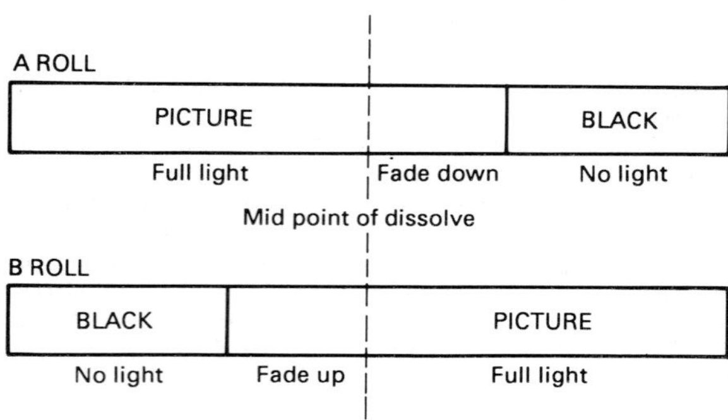

Figure 38.8 A and B roll dissolve

appropriate place, simple opticals can be achieved quickly and cheaply.

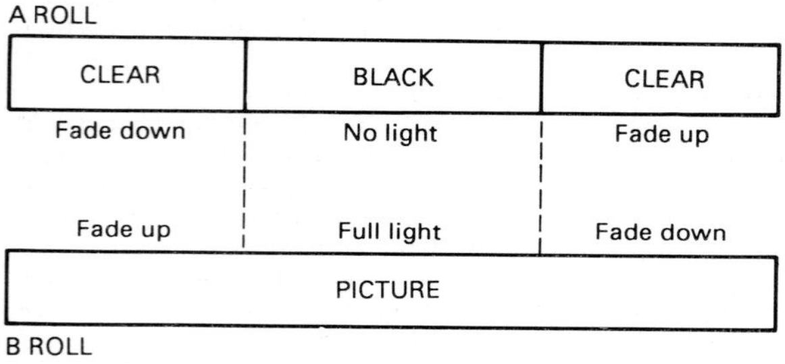

A ROLL
CLEAR | BLACK | CLEAR
Fade down | No light | Fade up
Fade up | Full light | Fade down
PICTURE
B ROLL

Figure 38.9 A and B roll fade up from black and fade down to black

38.1.3.13 Grading and printing

Rush prints are normally supplied as one light prints. However, once the film has been edited, scenes from many different rolls shot at different times under different conditions may well now be adjacent to each other. To maintain as much consistency as possible, the negatives of every shot must be *graded*, i.e. colour and density matched as closely as possible by varying the intensity and the colour balance of the printing light on each shot.

Graders used to depend on their experience to assess what a particular piece of negative would look like when printed at certain lights, or would grade each scene by reference to standard negative frames of varying densities each of which had a known printer light setting to give an acceptable result. Where doubt existed, tests would be run. But modern electronic analysis can now *reverse phase* the negative image so that it is seen as positive, and various printer light settings can be simulated to achieve the desired result. These settings are logged and passed to the printing department, where the printing machines can be cued to print each shot in the appropriate way.

Usually the A roll negative is printed first (shots 1, 3, 5, 7, etc.), the positive rewound and resynced with the B roll negative for a second pass (shots 2, 4, 6, 8, etc.). A third pass will then be required if an optical sound track is to be added. Once printing is complete, the positive can be developed.

The first printing from a graded cut negative is generally known as an *answer print*, and after viewing by the grader and usually the 'lab contact' (the liaison between the laboratory and the customer) it will, if satisfactory, be shipped to the customer for viewing by the editor and director. Suggested changes can then be incorporated in the next print, usually known as the *show print* or *transmission print* in television. While at one time several more attempts might well have been made at regrading, nowadays with the precision of electronic analysis it is rare that more than two gradings are required, and of course in television it is possible to further correct the print on telecine if necessary.

38.1.3.14 Preparation for telecine

It is usual for television film to be left as separate picture and sound for recording or transmission. Since telecine machines offer this facility, it makes good sense to utilize it, firstly to save the time and cost of making a combined print, and secondly and more importantly because the quality of sound reproduction is far superior from the magnetic master mix rather than from an optical track. This is mainly because of the superior inherent characteristics of magnetic tape and to a lesser extent because a further generation is avoided.

Basically, all that is required is for the telecine operator to be supplied with the show print and master sound, correctly marked up and leadered, together with the answer print and copy sound as a back up. A rehearsal normally takes place, and settings can be predetermined. On some machines instant corrections can be pre-programmed on shot changes. Film can be either transmitted directly from telecine, or more usually these days, transferred to videotape for later transmission.

It is now common practice either for the signal to pass via a presentation studio for the addition of titles before passing to the vtr machine, or for the signal to be recorded directly onto vtr and titles added at a subsequent vtr editing session. The second method offers more precision but adds a tape generation, but in television titles are now normally added in either of these ways rather than being made optically as part of the film editing process.

38.2 Video tape editing

38.2.1 Available formats

There are various formats currently in use, and the most important characteristic from the editing point of view is the distinction between segmented and non-segmented recording.

With the *segmented* format, such as the two-inch Ampex quadruplex or one-inch type B helical, each complete pass of the heads does not scan the entire video field, and therefore when the machine slows or stops there is no picture displayed (unless a frame store device is added). All editing is normally done *on the fly* (i.e. entering in and out points with the machine running).

With the *non-segmented* format, such as one-inch type C helical or smaller gauge systems, each complete pass of the heads does scan the entire video field, and therefore a picture can be displayed at slow speed or in a stationary mode making editing in the film manner possible. This ability to examine material at varying speeds forwards or backwards or frame by frame in the way taken for granted by film editors has been the most significant change of recent years. It has allowed, in addition to the previous system of making fairly minor alterations to material pre-edited by a vision mixer during a continuous multi-camera shoot, the development of a system of building a complete programme from 'rushes' recorded shot by shot (and not necessarily in order) using a single camera in the film manner.

38.2.1.1 Two-inch Ampex quadruplex

This system uses two-inch video tape running between open reels at a speed of 15 inches a second (381 mm/s). The quadruplex or 'quad' head assembly contains four video heads mounted at 90° angles in a rotating head wheel, and one complete field is scanned every four revolutions. The video tracks are recorded at a near 90° angle to the tape, and the one or two audio tracks together with the control and cue tracks are recorded longitudinally. For many years the mainstay of broadcast television, two-inch machines are now generally being replaced by one-inch machines on the grounds of comparable quality with lower costs, less bulk, more convenient tape handling and storage, and increased reliability and easier maintenance.

38.2.1.2 One-inch type B

This system uses one-inch video tape running between open reels at a speed of 9.65 inches a second (245 mm/s), and the supply and take-up reels use a *B wind* with the oxide facing outwards. The video head assembly is smaller than on a two-inch machine and incorporates two rotating video record/replay heads mounted opposite each other and two erase heads also mounted opposite each other, the four heads being at 90° in the assembly. The video tracks are recorded at a 45° angle to the tape, and the three audio tracks and one control track are recorded longitudinally. Like the two-inch quad machine, one-inch type B uses a segmented recording system and cannot reproduce picture information in slow motion or still-frame mode without the use of frame store devices. It is much less common in use than the non-segmented type C.

38.2.1.3 One-inch type C

This system also uses one-inch video tape running between open reels at a slightly slower speed of 9.61 inches a second (244 mm/s). The tape wraps fully around the video head assembly, which includes two record/replay heads. One is used to record and play back the visible part of the picture (the *video* head), and the other is used to record and play back the vertical interval part of the video signal (the *sync* head). There are also two erase heads, one for erasing the visible picture (the *video erase* head), and the other for erasing the vertical interval (the *sync erase* head).

Because the visible part of the video signal is recorded or replayed by only one head, each revolution of the video head assembly scans a complete video field. This non-segmented mode of operation means that non-standard speed or stop frame editing can be carried out. The video tracks are recorded at a shallower angle than on the B format. There are again three audio tracks, and the control track is recorded between the picture video and the sync.

38.2.1.4 Three-quarter-inch U-matic

In this system, the three-quarter-inch video tape is enclosed within a cassette and is threaded automatically on loading, forming a U pattern around the video head assembly. The tape runs at a speed of 3.75 inches a second (95.3 mm/s), and a non-segmented recording system is used. Two audio tracks and a control track are provided; originally, if time code was required, one of the audio tracks would carry this as in the systems described previously. However, with the BVU (*broadcast video U-matic*) system, an address track is provided which permits the recording of time code across the video recording. Because this is recorded along the unused area of vertical blanking, it does not affect the video recording and frees the second audio track for audio use.

38.2.1.5 Half-inch Betacam

Increasingly widespread use is being made of this cassette format which runs at 4.0 inches a second (101.5 mm/s), employs a non-segmented recording system, and provides an address track for time code in the same manner as BVU. The tape also carries two normal sound tracks and two fm tracks designed for high quality recording and reproduction. The fm tracks cannot normally be edited and, if used for field recording, have to be transferred if editing is required. They can, of course, be used for laying back a mixed track onto the edited master if suitable equipment is available.

When used in the composite mode, the picture quality of this system is similar to BVU, but in fact the system is designed to be used in the component mode, whereby the luminance, chrominance and sync information are recorded and reproduced separately, giving a greatly enhanced picture quality and being capable of more generations before unacceptable degradation occurs.

38.2.2 Control track and time code

All video tape recordings will carry a control track of continuous low frequency signal to stabilize the servo system of the vtr. Incorporated in this signal is a series of sharp pulses occurring at the beginning of each frame. It is possible to edit video tape electronically (as opposed to the earlier physical methods) by means of a system of counting these pulses (*control track editing*), but more sophisticated systems employ time code (*time code editing*).

Longitudinal time code (ltc) is recorded either on an audio track or on the address track of the BVU or Betacam systems. It is possible to edit satisfactorily using only this system, but from the editing point of view its disadvantages are its tendency to misread the code when running at slow speeds and its inability to read the code at all when stationary.

The alternative is to record *vertical interval time code* (vitc) in the vertical interval during the shoot, which can be read when stationary. However, vitc tends to become unstable at high speed, and for this reason it is preferable to have both systems on the field tapes. The pre-blacked master tape for editing onto (see section *38.2.3.*) will, of course, carry only ltc, as vitc must be inserted into the actual video signal at the time of recording.

38.2.3 Assembly editing and insert editing

In the *assembly edit* mode all information, video, audio, and control track are laid down at once onto blank tape or will over-record anything that is already on the tape (including any existing control track and time code). At the end of the edit, video break-up will occur as the new control track finishes or reverts to the pre-existing one. This break-up will be over-recorded by the in point of the second edit and then occur again at the end of the second edit. This process will continue to the end of the sequence where a black signal can be edited on to secure a clean finish.

It will be obvious that this mode of editing will not be satisfactory when it is necessary to insert a new edit into a previously recorded sequence. Here the *insert edit* mode must be used. Although it takes its name from its ability to insert new edits cleanly into previously recorded material, this mode is in fact almost always used for all broadcast editing. In this mode, the existing control track remains, and any combination of video and audio edits (video only, audio one only, video and audio two, etc.) can be made according to which recording heads are activated. Because the control track is left stable and continuous, break-up does not occur at the edit out point. It is also usual, especially with smaller format vtrs, to use a time base corrector on each machine to correct any remaining video signal instability.

Because, in the insert edit mode, the record vtr must lock into an existing control track when editing onto a new tape, this tape must be *pre-blacked*. This means that the tape which will become the master must be pre-recorded with colour reference burst (usually a black signal) and with a continuous control track. Continuous longitudinal time code will also be required if time code editing is to be undertaken. Use of the insert edit mode is essential in time code editing as the time code on the master must remain undisturbed throughout editing, re-editing or re-conforming, tracklaying, dubbing and lay-back.

38.2.4 Editing systems

There are currently three commonly used editing systems: control track, time code, and computerized time code. In the simplest of these systems, *control track editing*, the method employed involves the edit controller counting the frame pulses recorded on the control track signal. The *in* points of the edit must be entered on both the record and playback vtrs. Additionally, an *out* point can be entered on one or the other, or the edit left open ended and the recording stopped when sufficient has been laid down, the next edit being used to over-record from the appropriate point.

Once the *in* points have been entered, *preview* or *record* is selected, and the edit controller will pre-roll both machines back to a point sufficient to give time to run up to speed and lock together (usually 5 s before the *in* point). The machines then roll forward together, and the edit is performed at the entered point. The advantages of this system are low cost and simplicity of operation; the disadvantages are lack of frame accuracy (there can often be a two or three frame error even on a good, well-maintained system) and the lack of an EDL (*edit decision list*). This means that although edits can be modified at the preview stage, once recorded the corresponding information is discarded by the edit controller to enable it to accept the information necessary for the next edit, so in any subsequent re-editing the edits will have to be set up again manually.

A *time code editing* system is capable of reading the time code on both the playback and record vtrs. Edit points can be marked in with the machines stationary or running, or can be entered via a keypad. Once the time code information has been entered, the system will search for the entered points on both machines and perform the required edit to the frame, which means that edits can be accurately repeated or modified. On higher priced systems, an EDL can be stored and subsequently used for alterations either on the same system or on a compatible one. It is also possible to control accurately two or more playback vtrs (essential if any optical effects such as dissolves, wipes or keys are required).

A *computerized time code editing* system will be capable of controlling a larger number of sources (probably six or more) and will have two outputs for duplex recording. The sources need not necessarily all be vtr machines. Commonly they will include a vision mixer, a digital video effects system, and possibly a multi-track audio system. The computer will store a complete record even of complex edits, and will have the ability not only to reproduce edits accurately one by one but also to reproduce (or *reconform*) them automatically and continually. This has been a great influence in shaping current editing procedures.

38.2.5 Special effects

One of the advantages of video over film is the ability it gives to the editor to create his or her own effects within the edit suite. Rather than marking them up on a cutting copy and waiting days for the optical print to come back, and then quite often finding that the result isn't quite what was expected, with video editing, effects can be programmed, previewed, adjusted as necessary, and then recorded as part of the editing process.

Most edit controllers capable of running two or more playback vtrs have inbuilt facilities for simple dissolves, wipes, keys, etc. On the more complex computerized systems, the range of inbuilt effects can be quite extensive. Beyond this a vision mixer is normally used, so that in addition to performing effects automatically through the computer, it is also possible to add soft edges, coloured borders, drop shadows, and so on. It must be remembered that while effects performed entirely through the edit controller will be stored in the EDL and can therefore be easily reproduced at any time, effects performed manually using the vision mixer will not be stored in the same complete way and will have to be manually set up again during any subsequent re-editing session.

In addition to the special effects available through the edit controller and the vision mixer, widespread use of DVE (*digital video effects*) systems is now commonplace. In these various systems, the analogue video signal is converted into a digital signal and enters the memory network of a digital frame store. Each frame of information is held for a one frame duration and then re-converted to an analogue signal and sent to the output. During this $^1/_{25}$ s delay within the frame store, the signal can be manipulated in various ways by varying the output characteristics of the memory network. Of course, from the editing point of view, the one frame delay means that sync sound scenes will subsequently be one frame out of sync, and the sound must also be delayed by one frame at some stage to compensate.

Using these systems a wide variety of effects are possible, actually altering the image in terms of aspect ratio, shape, size, position and duplication. Although previously images could be combined or overlayed in various ways, the images themselves could not be altered.

38.2.6 The editing process

As in film editing, viewing of the available field tapes (still commonly referred to as 'rushes') and other material, such as transfers from library sources or rostrum material, must be the first step. This can be done in the edit suite, or to save edit suite time (and therefore cost), VHS copies with a 'burnt in' or 'window', time code (where the time code is transferred as part of the video information and is therefore permanently displayed in the picture area) can be viewed using an ordinary domestic vcr. The script can then be marked up with the reel numbers and time codes of selected scenes.

If editing directly *on-line* (i.e. to transmission standards) which is common practice within broadcasting organizations, the aim is normally (in contrast with film) to achieve as near as possible to a fine cut on the first version. It is also necessary (again in contrast with film) to edit the programme in a linear manner, starting with the first shot and going through in order to the last. This is because, in normal video editing, there is no random access into the material. It is not possible merely to change the length of a shot or to remove it altogether in the middle of a completed programme and then to join the programme back together again.

If a second generation master is required (as is normally the case with BVU editing), any alterations to the first completed version will involve re-editing the whole programme after the point of alteration (in the sense of re-recording every shot, although the edit decisions may remain the same).

The auto reconforming facility offered on computer systems makes this process far easier, although still time consuming. Using this system, a shot can be deleted from the memory (to use the simplest example), the next shot 'pulled up'' and the EDL 'rippled' to bring all the *in* and *out* points of every subsequent edit forward by a commensurate amount. This might take only a few seconds but, although the EDL has been altered, what is actually on the master tape has not. It is then necessary to re-record everything that has been altered in the EDL in the new position on the tape. With one-inch C format or half-inch Betacam used in the component mode, quality loss through additional generations is not such a problem. In this situation, it would be usual to make the necessary alterations while re-recording onto a third generation master, which means that the majority of the programme which remains the same can be transferred in real time as one edit.

38.2.7 Off-line and on-line

In order to avoid excessive reconforming or excessive generations, and therefore to save time in an expensive on-line facility, it is common practice where more complex programmes are involved to edit them off-line first. In the simplest process, small format copies can be made from the field tapes. These copies will carry burnt in time code as previously described. These copies can then be edited using a simple facility to produce a first cut version; this first cut can then be put in the player and used as a source together with the field tape copies to produce a second cut version; the second cut can be used to produce a third, and so on. The quality loss on each succeeding generation is irrelevant as long as the time code numbers can still be read, and as they form part of the picture they will be reproduced correctly on every version. The edits are then logged by hand from the final edited picture and entered by hand into the EDL of an on-line system, which can then duplicate the programme direct from the original field tapes to a master tape to give a second generation master on a broadcast standard format.

More advanced off-line systems will generate their own EDL, which can be off-loaded onto floppy disk and then reloaded into a compatible on-line system. Obviously if, for example, the final off-line is a fourth cut, the time codes on the final EDL will be mainly from the third cut, and not from the field tape copies. However, there are trace programmes available which will convert the final EDL back to the equivalent time codes of the field tape copies as long as each cut version has been assigned its own reel number (different from any of the reel numbers assigned to the field tape copies) before being used as a source tape.

38.2.8 Tracklaying and dubbing

Unlike film editing, where it is possible to lay a virtually unlimited number of sound tracks, each of which can be physically moved independently of one another, in video editing, track laying is usually done on a multi-track audio tape machine, in which the relationship between one track and another can be altered only by re-recording. Another limiting factor is the number of tracks available on the video tape, e.g. on a one-inch system, one of the three audio tracks will be carrying time code, leaving only two for audio. Even using the address track for time code on a BVU or Betacam system, there are still only two tracks available for sound editing.

It is common practice to lay as much sound information as possible onto the two audio tracks of the video master, to transfer this edited sound onto two tracks of a multi-track audio tape, then to add extra effects, music, commentary, etc., and mix down within the multi-track, and finally to lay the mix back onto the video tape. This does mean, however, that the first re-recording is from field tape to master tape (at least as far as the main sync tracks are concerned), the second is from master tape to multi-track, the third occurs during the mix down within the multi-track, and the fourth when the mixed sound is laid back onto the master tape. These four re-recording stages mean that the final sound is fifth generation, which can be of poor quality especially when using small format systems. Although this method provides ample tracks at the dubbing stage, it is the lack of tracks at the editing stage that is the real problem.

One answer is to slave, for example, a one-inch eight track audio recorder via the computer (which will then operate it in lock with the record vtr) and record directly from the field tapes onto the one-inch audio machine. With an eight track audio machine, one track will be carrying time code, one will probably be left blank as a guard band between the time code and the audio, and one will be required for the mix down. There will still be five tracks left for track laying. By going direct from field tapes to high quality multi-track, not only is a generation saved but a small format vtr sound generation is saved which greatly improves the final result. Extra tracks can be added direct to the multi-track machine, then a mix down takes place and the result is laid back onto the master vtr.

The principles of commentary recording, tracklaying and dubbing on video are similar to those for film editing. The main difference in practice is that the editor is unable physically to cut and lay original sound ready for the dubbing mixer to re-record. Each stage of editing and tracklaying involves the editor in re-recording the original sound. It is therefore essential to maintain the quality as far as possible throughout, as although with more sophisticated systems one can locate and re-transfer the original sound at the dubbing stage, this is obviously wasteful in terms of time and therefore money if done to excess.

38.3 Interrelationship between film and video editing

There have been a number of attempts in recent years to increase the integration of film and video editing, some more successful than others. One system which seems likely to achieve widespread use is the Profilm EFC System (*Electronic Film Conforming*) first used in Sweden by AB Film Teknik in 1982, and in the UK by Filmatic Laboratories in 1984.

This system can be used in many different ways, the real breakthrough being its ability to work with film edge numbers as well as with time code. A production shot on film can therefore be developed and rush printed in the normal way, the edge numbers logged into a computer, and the rush print transferred to a one-inch C format videotape known as a *source tape*.

Alternatively, if the negative is correctly exposed (to within half a stop either way), it is possible by reverse phasing the image to transfer directly from negative to one-inch tape, thereby saving the print generation and enhancing the final quality. Film editing can proceed as usual, and video wipes can be marked on the cutting copy, as can dissolves of between 4 and 254 frames, freezes and repeat shots. After film dubbing, the cutting copy is logged into the computer and the programme can be conformed from the one-inch source tape to a one-inch master tape, adding effects and titles and correcting colour and density as required.

The system allows for conventional film editing, which is much cheaper than video editing on a cost per hour basis, and also eliminates negative cutting and grading and printing, as well as the conventional transfer from film showprint to vtr and the subsequent vtr edit for the purpose of adding captions and titles.

Alternatively, a small format video transfer can be made from the one-inch source tape after the film has been transferred, and the programme can be edited as off-line video, or material originated on video tape can be supplied as a telerecording with burnt in time code or as a small format video tape for off-line editing. It is therefore possible to edit a film originated programme as film or video, or a video originated programme as video or film, or use any combination of

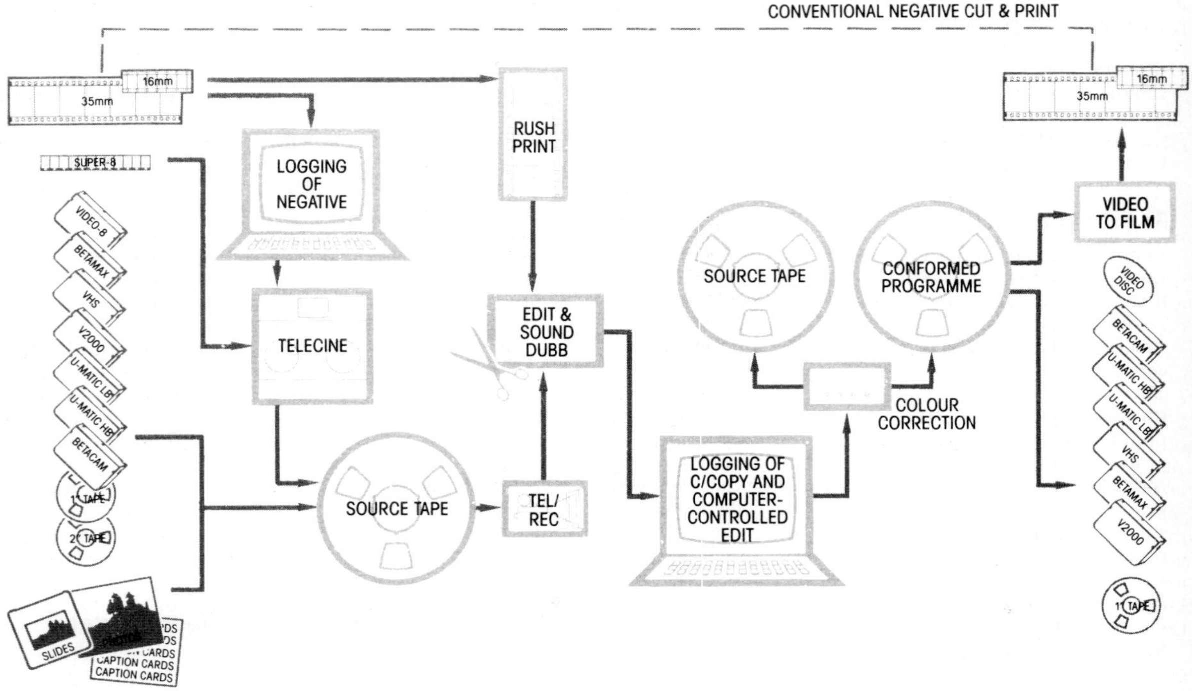

Figure 38.10 EFC process: film edit

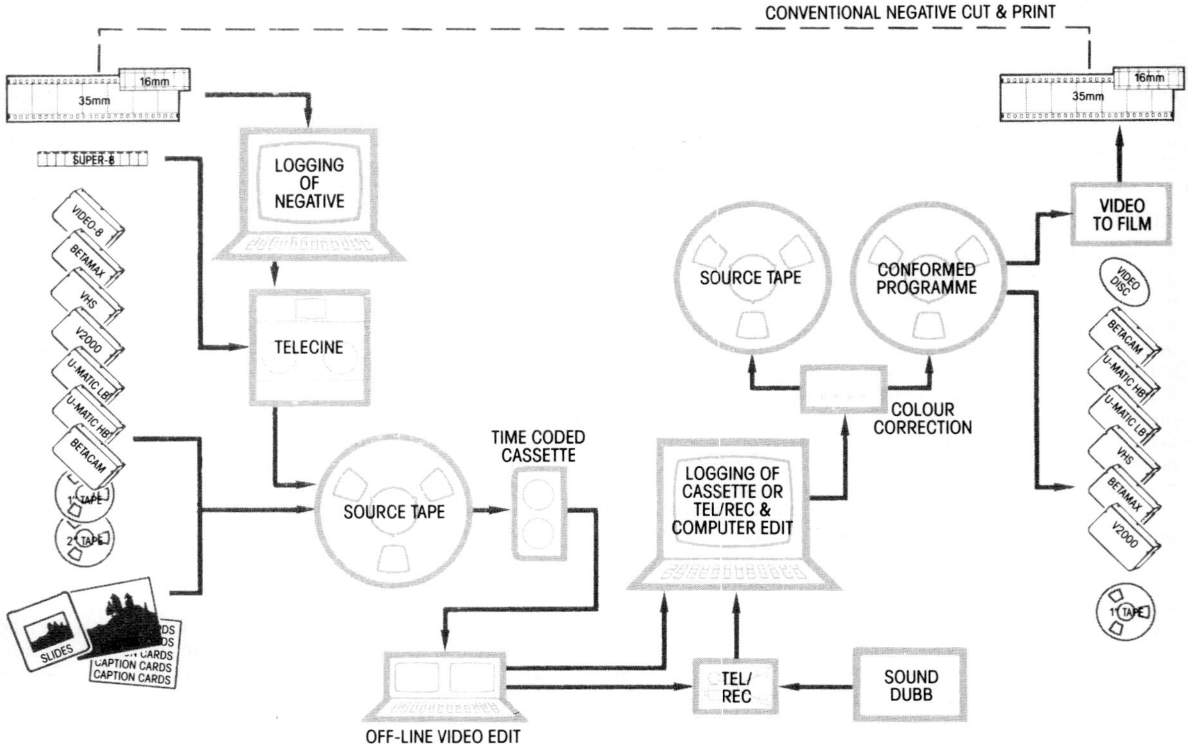

Figure 38.11 EFC process: off-line video edit

methods. This is useful when making a compilation programme from many different sources. Whichever method or combination of methods is used, the end result will be a one-inch master tape ready for transmission.

38.4 Future developments

The greatest disadvantage of video editing compared with film editing is the lack of random access into the material. Apart from the time taken to locate the required rushes material (which can be equally time consuming with film), once material has been re-recorded onto the master even the smallest alteration will involve either reconforming a substantial part of the master or going down a generation.

In the early 1970s, an attempt was made to overcome this problem with the RAVE system (*Random Access Video Editing*) developed by CBS and Memorex, which became the CMX 600 off-line system. This involved transferring either film or videotape to video discs from which any shot could be accessed immediately. The programme could then be edited and re-edited into the memory of the system without having to spend time running through rolls of film or videotape to reach the required points, in fact without touching the original film material or re-recording anything on tape. Following this, film negative cutting or on-line video editing could take place. Unfortunately, the cost of the large disc drives involved made the system uneconomic at the time.

However, the development of digital recording techniques enabling video and audio information to be stored in a non-linear mode could make random access possible in a more economic way. Digital disc systems are already widespread for creating special effects, and cheaper systems may in the future make random access video editing commonplace.

38.5 The role of the editor

The overall aim of both film and video editing must be to produce a polished product that will effectively achieve the task it sets itself, whether this task is to convey a dramatic story, to amuse or entertain, to inform, or to persuade the audience to watch a particular programme or to buy a particular product.

This section has described in general terms the more common practical techniques used in most areas of film and video editing, but the artistic or creative skills required vary enormously from one area to another, and consequently a lot of editors tend to specialize in one area only, such as drama, comedy, documentaries or commercials.

An editor's role is to bring to the rushes a fresh eye, unprejudiced by anything that may have happened in pre-production or on the shoot, and hopefully have the opportunity to make just one of the many creative inputs that make a successful programme, which after all is always a team effort.

There are always limitations in terms of time and money, and sometimes the schedule is so tight (particularly in television current affairs or news, and increasingly in other areas) that the main priority for the editor must be to get the job finished on time to a pre-determined brief. This inhibits any radical re-structuring. Achieving the required result in a very short space of time is in itself a considerable skill, although perhaps not so satisfying to exercise as the skill of gradually building an outstanding piece of work from many possible permutations.

Although always part of a team, an editor can have a considerable individual influence on the final product, depending on the circumstances of the production and particularly on the editor's relationship with the director and the rest of the production team. Once a good relationship has been established, the editor may well find himself increasingly left to make major creative decisions affecting the overall programme as well as the minor creative decisions affecting particular sequences which are part of the job on a day-to-day basis.

Bibliography

REISZ, K and MILLER, G, *The Technique of Film Editing*, 2 edn, Focal Press (1968)

WALTER, E, *Technique of the Film Cutting Room*, Focal Press (1969)

CRITTENDEN, R, *Manual of Film Editing*, Thames & Hudson (1981)

J D Millward B Sc, C Eng, MIEE
Head of Research, Rank Cintel Ltd

39 Telecines

39.1 Telecine types

Telecines can be grouped into three distinct types: photoconductive, flying spot and ccd.

The term *photoconductive* applies to those telecines that incorporate an electronic camera as the pickup device; a variety of camera tubes can be used such as plumbicon, vidicon, saticon, etc. The light source is usually a tungsten halogen lamp and the projector will have an intermittent film transport so that the image on the film frame can be transferred to the camera tube target while the film is stationary, a necessary requirement since the camera tubes exhibit storage characteristics. The photoconductive camera has three separate tubes to produce the red, green and blue signals, and the scans on these tubes have to be accurately matched to obtain colour registration.

The *flying spot* telecine uses a single flying spot cathode ray tube with an unmodulated raster as a light source. An image of the raster is projected onto the film by an object lens. Light passing through the film is modulated by the film image, and the light is then collected and directed through colour splitting optics to photomultipliers which produce the electrical signal.

Since the colour splitting process occurs after the imaging process, there is no requirement for colour registration. Flying spot telecines at present being manufactured use continuous motion projectors although there are a small number still in use which have a 16 mm projector with intermittent motion and a pull down time of 1.2 ms.

The *ccd* telecine is similar to the photoconductive telecine with a tungsten halogen light source, but with the three tubes replaced by three ccd linear arrays. A ccd (*charged coupled device*) consists of a single line of photosites adjacent to a charge coupled shift register, into which the charge is transferred from the photosites at frequent intervals. Then, while new charges are being generated in the photosites, the transferred charges are clocked out serially from the shift register to produce the electrical signal.

Since the ccd used in telecine has only one line of photosites, it normally scans the film only in the horizontal direction, and film movement provides the vertical scan. Therefore, as with flying spot telecines, continuous motion projectors are used, but there is a requirement for colour registration of the three ccds as in the case of the photoconductive telecine. Also, since this method of scanning the film produces a sequential or progressive scan, a sequential-to-interlaced converter is required before the signal can be transmitted.

39.2 Film formats

Although there are various film gauges, the vast majority of telecines handle only 35 mm and 16 mm cine film, and 50.8 mm square slides. The other film gauges, if required for television transmission, are normally transferred to 16 or 35 mm at the printing stage.

The standard aspect ratio (ar) of the transmitted television signal is 4:3, while many films produced for the theatre have a wider ar varying from 1.66:1 to 2.35:1. Also, although the displayed ar may be 2.35:1, the film frame may be 1.17:1; this is known as *anamorphic film*, the image having different magnification in the horizontal direction from that in the vertical direction. For photoconductive telecines, special anamorphic lenses would be required to transmit this film. To avoid this, the film is usually printed in the displayed aspect ratio. Flying spot and ccd telecines can transmit anamorphic film without the need for special lenses, the correction being performed electrically.

Having obtained the correct display aspect ratio for the particular film, it still has to be decided whether to transmit a wide screen format exactly as intended or as a 4:3 ar to match the home receiver. In the former case, part of the crt is not utilized and the picture is smaller than it could be; in the latter case, although all the screen is utilized, some film information is lost. To insure that important information is not lost, the part of the frame to be transmitted is selected frame by frame prior to transmission.

The former method of transmission is called *letterbox* and the latter *panscan*. Obviously the letterbox mode does not require editing, but the panscan mode does, and this editing is performed prior to transmission by pre-programming with the aid of a small computer and a memory. Two forms of panscan can be selected during pre-programming, either a *pan-cut*, i.e. an instantaneous shift during field blanking, or a *slow pan* with variable rates.

One big advantage that telecines have over optical projection is that negative film can be used directly by means of electronic inversion. This avoids several forms of degradation arising from the printing process and so provides better picture quality.

Two easily understood improvements are detail resolution, and vertical and horizontal stability, especially if the original camera negative is available.

A third important improvement, which is not immediately obvious, is grey scale resolution or linearity. Negative film is manufactured with a very low gamma so that it is very tolerant of a wide range of exposure variations without loss of information. When the negative is printed, the exposure can be adjusted to take account of the density of the negative, but since the print is normally intended for optical projection it must have a gamma of at least unity to reproduce the original scene. Therefore scenes which have a contrast range in excess of 100:1 require a print density of at least 2. Under these circumstances, with existing film dye technology, the highlights and shadows become compressed even when the print is correctly exposed. The negative in effect compresses all the information equally from highlight to shadow so that when the telecine electronically decompresses the information there is no loss of linearity from highlight to shadow.

39.3 Film transports

The film speed in the majority of telecines must be related to the television system in which they are operating for various reasons connected with the type of telecine. The two current television picture rate standards are 60 fields/30 pictures per second and 50 fields/25 pictures per second.

Prior to the advent of television, the standard film rate was 24 frames/second and this speed is still probably the most common. Obviously film produced at 24 frames/second could not be transmitted at 30 frames/second for that particular system because the increase in sound pitch and picture speed would have been unacceptable. Fortunately, there is a direct if not ideal relationship between 24 frames/second and 60 fields/30 pictures per second, that is, two and then three television fields per film frame alternately. This gives synchronization between the projector and the television system, but moving objects in the scene or camera panning produce just noticeable non-uniform movement under this arrangement, although telecines using photoconductive tubes with their inherent frame-to-frame lag tend to obscure this non-uniformity. Therefore, although most film shot for the 30 frame television standard is still 24 frames/second, some film has been produced at 30 frames/second.

The other television standard of 50 fields/25 pictures per second can manage by operating the film at 25 frames/second, which gives a more direct relationship between projector and television system and an increase in sound pitch and picture speed of 4 per cent which is quite acceptable to most observers. Nevertheless, film shot for this particular television standard has been produced at 25 frames/second which again is acceptable on the 30 frame system or for optical projection at 24 frames/second.

The introduction of frame stores and digital sound compression/expansion equipment which correct the sound pitch has enabled the use of a much wider range of film speed whilst still retaining a whole number of television fields per whole number of film frames.

39.3.1 Intermittent motion

Photoconductive pick-up tubes convert photons into an electronic charge which is stored on the target. When using tungsten halogen light sources, exposure times measured in milliseconds are required to produce sufficient charge on the target. To avoid image blur, the film image must be stationary during the exposure, and hence an intermittent film motion projector is required for photoconductive telecines.

Optical projectors also operate in the intermittent mode and have, therefore, been adapted for telecine application.

By far the most common type of intermittent mechanism is the Maltese cross or Geneva movement as described by Wheeler[1]. It is adapted for television by synchronizing the projector to the television synchronizing pulses and modifying the shutter to produce an equal number of light pulses per television field. For projectors operating at 24 frames/second on 60 fields/second television systems, a further adaption is required, and the Geneva movement is modified to move the film with an alternate two/three television field sequence.

Projectors operating at 25 frames/second on the 50 fields/second television system typically move the film frame in 10 ms with a stationary period of 30 ms, during which time the shutter produces two light pulses of equal width and spacing and synchronous with the television fields.

A small number of television stations in Europe still use a 16 mm intermittent projector with a pull down time close to 1.2 ms using a flying spot scanner. Fast pull down is required when using intermittent projectors with flying spot systems because there is no inherent storage of the picture in such systems.

Electronic picture storage, which is now a common feature of television studio equipment, could be applied to intermittent projector/flying spot systems, allowing longer pull down times.

It also solves many problems with continuous motion projectors. Consequently, since continuous motion projectors are preferred due to their smooth and silent operation, intermittent mechanisms for flying spot systems are no longer manufactured.

39.3.2 Continuous motion

Continuous motion projectors can be subdivided into three distinct types, sprocket driven, claw driven and capstan driven.

In *sprocket driven* projectors, the sprockets rotate at a uniform velocity synchronized to the television sync pulses to maintain the required film frame rate. As film shrinks with age, up to 3 per cent for nitrate base and 0.5–1 per cent for acetate base, the rotational speed of the sprockets and mean film velocity falls in proportion to the film shrinkage. Also, the pitch between the teeth on the sprockets must be greater than the film sprocket hole pitch at all times because of film shrinkage. This means that only one tooth, or pair of teeth with double sided sprockets, is driving the film at any particular instant. As the sprocket rotates, the drive transfers from tooth to tooth, the film slowing down during the transfer from one tooth to the next.

Thus the film motion is not perfectly smooth with sprocket drive and it is necessary to remove this velocity variation with *flywheel filters*. These consist of film rollers with flywheels attached, being driven by the film. When the projector starts, the acceleration has to be low to allow the film to accelerate the flywheel without slippage to avoid scratching the film.

The sound head is between 20 and 28 film frames from the picture gate and a sound flywheel roller is also required.

The *claw driven* projector can be distinguished from the sprocket driven type by two important characteristics:

- The film velocity is constant during the active picture period whatever the film shrinkage.
- The changeover from one claw to the next occurs during the picture blanking period and therefore the velocity change is not perceived.

Thus, no flywheels are required at the picture gate, although a sound flywheel is still necessary.

The *capstan driven* projector is now the most common, being applied to flying spot and ccd telecines. As there are no driving sprockets, the film motion is absolutely uniform with no need for picture or sound flywheels. This gives many advantages, including fast start, fast shuttle and multi-gauge projectors.

Assuming no errors in the filming and printing processes, vertical film stability in capstan and sprocket driven telecines is a function of two factors:

- uniformity of the film sprocket hole pitch which determines the spacing of the film frames,
- uniformity of the film velocity in the telecine.

Since timebase correctors cannot be applied to telecine to correct vertical stability, we depend on accurate control of these factors. The first is controlled by the film manufacturer and the second by the telecine manufacturer.

In the telecine, the capstan is the main controlling element, but because of the spacing between sound and picture gates and the need for constant velocity at both points, any intervening rollers must also be near perfect. The capstan is coated with synthetic rubber and has a diameter of about 50 mm, giving good traction with moderate surface pressure and film tension while eliminating the need for a pinch roller.

A typical capstan drive telecine layout is shown in *Figure 39.1*. Besides the capstan motor, there are two spooling motors, not coupled to the capstan or each other electrically or mechanically except by the film. The capstan is driven at a constant velocity to synchronize the film frame rate.

To maintain good film stability the capstan servo has two feedback loops: velocity and phase. The velocity loop is responsible for stability, and the phase loop determines the film frame rate. The reference for the velocity loop is an optical tachometer attached to the capstan shaft which supplies several thousand pulses per revolution.

The reference for the phase loop is a film sprocket hole detector. This usually takes the form of a sprocket driven by the film. It is mounted in free bearings and so does not impart velocity modulation as is experienced with sprocket driven telecines.

A mechanical sprocket detector is not absolutely necessary but has several practical advantages over a purely optical device. When 35 mm film is correctly laced on such a detector, it can be arranged to give one pulse per film frame correctly phased, whereas an optical detector has difficulty in determining which of the four sprocket holes provides the correct phase relationship. Also if a number of sprocket holes are damaged, the mechanical sprocket will continue to provide correctly phased pulses.

Each film spool motor is controlled by its own servo, to maintain constant film tension in all modes of operation. The tension rollers provide the tension reference for the servos.

Figure 39.1 shows the sound heads situated very close to the capstan to give the best possible wow and flutter performance. The capstan is able to handle 35 mm, 16 mm and 8 mm film formats with a common optical centre line, so that all film formats are supported over their full width. The optical sound head detectors are located in narrow slots inside the capstan which is machined from one piece.

The film path for 35 mm film includes one extra roller between the vision gate and the capstan to obtain correct sound synchronization. All rollers are triple gauge so that when changing from one film gauge to another it is necessary only to change the film spools or back plate and change the vision gate.

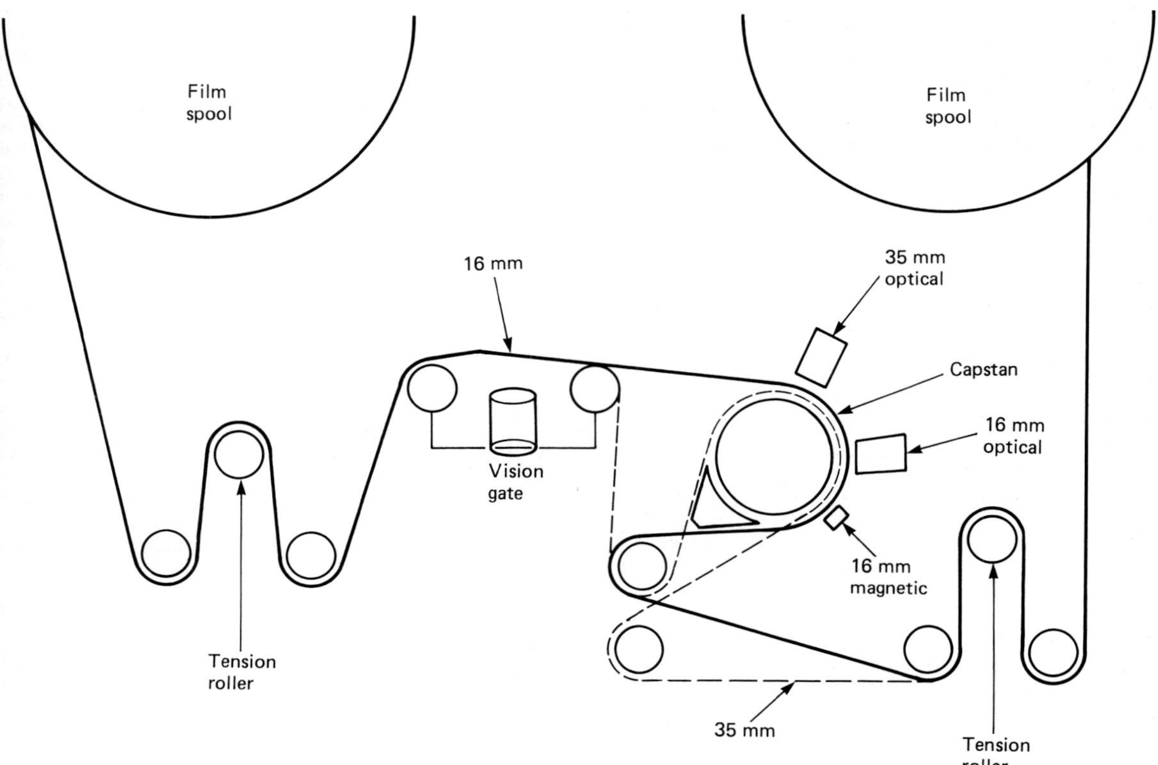

Figure 39.1 Capstan driven telecine

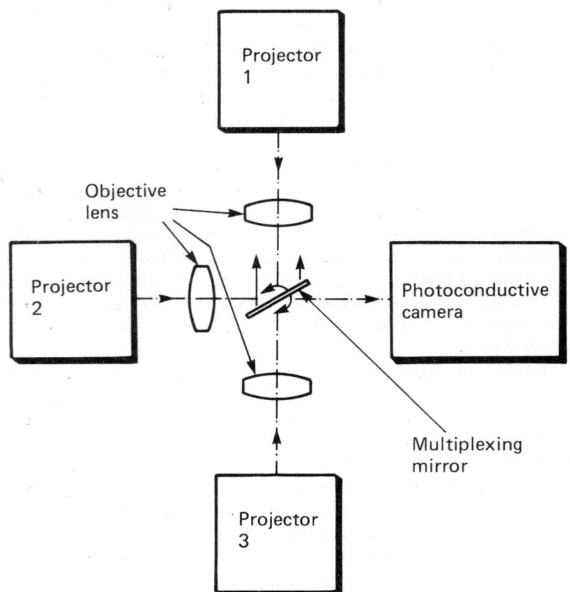

Figure 39.2 Photoconductive telecine

39.4 Sound reproduction

Until quite recently, telecines were equipped only for monophonic sound because television broadcasts were themselves only monophonic. Some countries are now broadcasting television sound in stereo and others have stereo under consideration. Video cassette players are now available with stereo, and programmes for this medium come mainly from film. Thus telecines now require stereo sound for some applications.

The sound tracks can be married to the film image or on separate sound only films.

Married sound tracks are most commonly 35 mm optical, 16 mm optical or 16 mm magnetic, heads for these tracks being fitted as standard as shown in *Figure 39.1*. The 35 mm optical sound track is 21 film frames in advance of the picture, the 16 mm optical sound track is 26 frames in advance and the 16 mm magnetic sound is 28 frames in advance.

Optical sound tracks can be variable in density or variable in area, the latter being by far the most common. The optical head consists of an illuminated slit, imaged by an objective lens onto the sound track. Light passing through and modulated by the film falls on a photocell to produce the electrical signal. 16 mm optical sound tracks can be on either side of the film base, and therefore the slit is focused midway between the two sides, so that after electrical compensation a flat frequency response is obtained from sound tracks on either side.

Sound tracks on separate film are reproduced on *sound followers*, i.e. film transports very similar to the telecine without the picture components. The sound follower is synchronized to the telecine by comparing the sprocket hole pulses from the two machines and adjusting the speed of the sound follower accordingly. Both machines need to be synchronized manually at the beginning of the programme.

39.5 Optical and scanning systems

39.5.1 Photoconductive

A simplified diagram of a multiplex photoconductive telecine is shown in *Figure 39.2*. Here one camera is linked optically to three projectors, which typically could be two 35 mm and one slide, or one 35 mm, one 16 mm and one slide.

Switching between the projectors is accomplished by means of the multiplexing mirror, and so a continuous programme can be maintained with one multiplex telecine.

The vision gate in a photoconductive telecine is normally flat rather than curved, and the active picture area of the film is unsupported to avoid scratching the image. Consequently, the

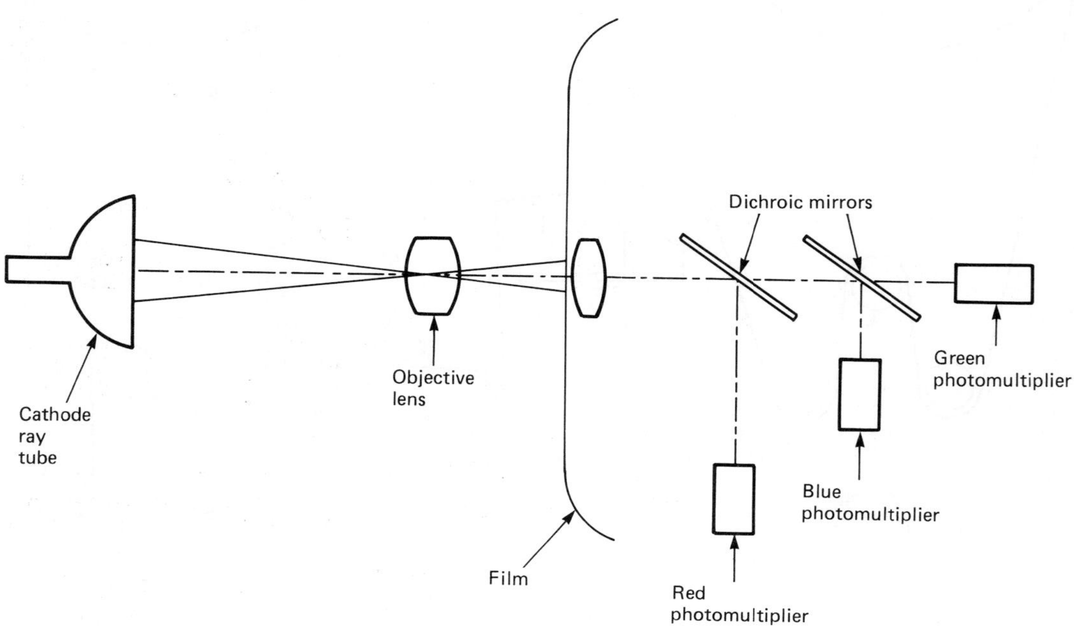

Figure 39.3 Twin lens telecin

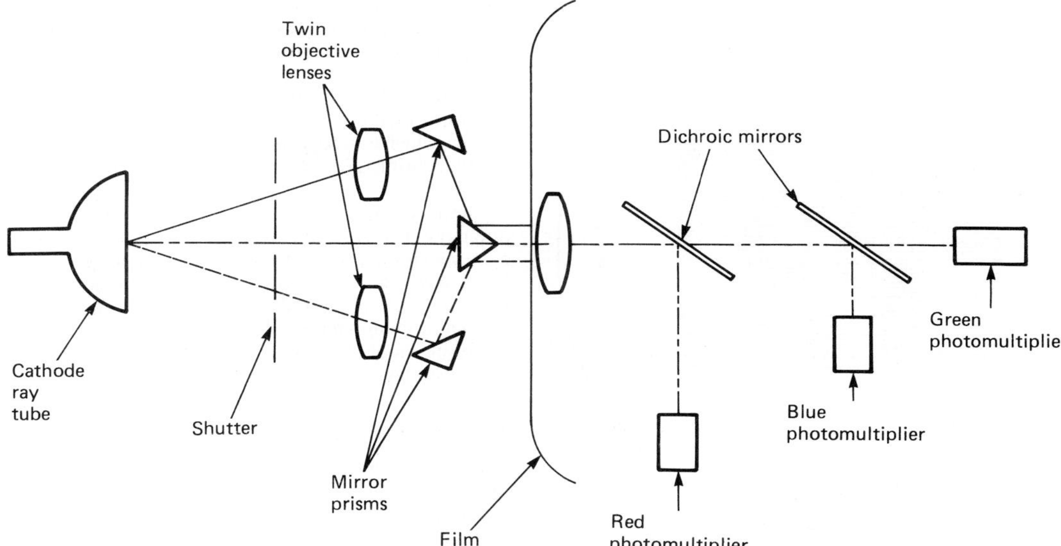

Figure 39.4 Jump scan and sequential scan telecine

film-to-objective lens spacing can fluctuate slightly depending on the state of the film and its previous storage conditions. Sufficient depth of focus is necessary in the objective lens to maintain resolution and this means operating at low lens apertures. Fortunately sufficient light can be obtained from tungsten halogen light sources to maintain a respectable signal/noise ratio.

39.5.2 Flipping mirror and mirror drum

Flipping or oscillating mirror telecines are still in active use in some countries and, as with the mirror drum, they freeze the image on the continuously moving film so that it can be scanned by an unmodulated raster on a flying spot crt. The light collecting, colour splitting and signal processing techniques are similar for all flying spot systems and are described in more detail in sections *39.6* and *39.7*.

39.5.3 Polygon

The polygon is a rotating prism with 30 facets, the film being moved in synchronism with the polygon, one film frame per facet. The polygon has an effective refractive index of 2 so that, viewed from the opposite side of the polygon from the film, the film appears to be at the centre of rotation and therefore stationary. A simple flying spot crt can scan this stationary virtual image, and as the film and polygon accelerate from rest, the film frames optically dissolve from one frame to the next.

These telecines can be operated at any speed from zero upwards on any television standard, and they are the only machine where the number of television fields per film frame does not need to be an integer. Although a few of these machines are still in use, they are no longer manufactured because of the difficulty in manufacturing polygons to an acceptable standard of quality.

39.5.4 Twin lens

The twin lens system was designed specifically for television standards and film speeds where two television fields from each film frame are required. This applies to the 625/50, 25 frames/second standard which covers more than half the world but cannot operate at 525/60, 24 frames/second.

The film moves at a constant speed through the vision gate as shown in *Figure 39.3*, the first field being scanned by the upper lens and the second by the lower lens, the continuously moving shutter selecting the appropriate lens. Sprocket driven projectors move the film at a constant speed. As the speed is affected by film shrinkage. the distance between the two objectives must be adjustable to ensure that the two television fields scanning one film frame are correctly registered in the direction of film motion.

This adjustment is normally performed automatically by mounting a film sprocket on a pivoted arm, which is under spring tension. The film is laced around this pivoted sprocket between two fixed sprockets so that the angle of the pivoted arm measures the film shrinkage and is mechanically linked to one of the objective lenses. Claw driven projectors do not require shrinkage correction because the film velocity is constant during the active picture period whatever the film shrinkage.

The magnification of the two objective lenses must be carefully matched to obtain field-to-field registration in both horizontal and vertical directions, and the light transmission of each lens also needs to be matched to avoid a 25 Hz brightness flicker.

The flying spot crt displays a raster with an 8:3 aspect ratio, i.e. half the vertical amplitude of the displayed picture, the other half of the vertical scan being provided by the film motion. The raster is interlaced and fixed, and therefore a 1 per cent geometry specification is adequate.

39.5.5 Jump scan and geometry correction

Figure 39.4 shows the simplest optical arrangement possible with no moving optical components or shutters and a single objective lens.

This simple system is used in the *jump scan* and *sequential scan* modes. (For the latter, see section *39.5.6*.)

To obtain two or three field scans of each film frame as the film moves through the projector in the jump scan mode, each

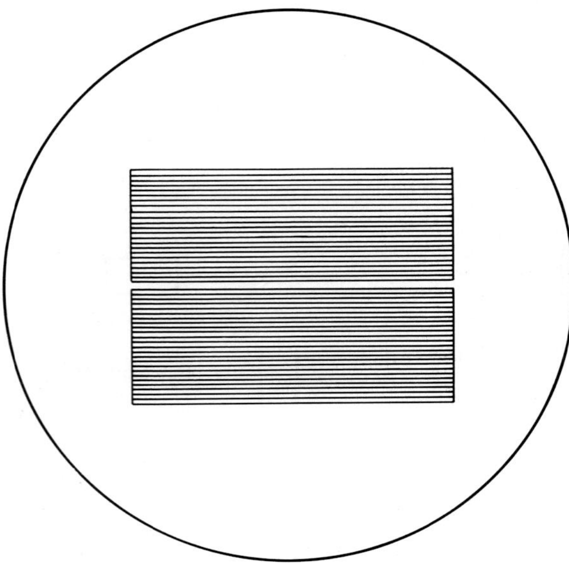

Figure 39.5 Jump scan 625/50 field

field scan is displaced vertically on the cathode ray tube to follow the film movement. *Figure 39.5* shows the two separate television fields in the 625/50 system. Each separate field has an 8:3 aspect ratio, the remaining vertical scan being provided by film motion.

The two separate field scans shown in *Figure 39.5* scan one film frame and therefore must be identical geometrically to within an accuracy of 0.1 per cent to meet the necessary field-to-field registration requirements.

Using known accurate methods of scanning, 0.5–1 per cent errors are normal; to obtain errors of less than 0.1 per cent the addition of a second correction deflector is necessary. This correction deflector has a maximum deflection capability of 2 per cent of the main deflector. Since the deflection errors in each corner differ, the tube face is divided into four quadrants and the correction deflection currents for the four quadrants are adjusted separately.

It is important to note that the total vertical scan using the jump scan system of 625/50 is twice that of the twin lens system, and its importance is related to the curved vision gate required for flying spot telecines.

The light source is the cathode ray tube and the light available is therefore limited. To obtain acceptable signal/noise ratios, wide aperture objective lenses are required. Wide aperture lenses have a short depth of focus, and a flat gate is unacceptable because the film position is unstable. Bending the film round a curve produces stiffness in the film even when only supported at the edges, and thus a horizontal line through the film at the centre of the optical axis is maintained flat. Either side of this central horizontal line the film is curving either towards or away from the objective lens. A compromise has therefore to be made regarding the radius of curvature of the gate, which affects the stiffness and stability of the film, the size of the vertical scan and the lens aperture which controls the depth of focus. Having made a compromise, an improvement in resolution can be obtained by including a rombic correcting lens in the vision gate close to the film which partially compensates for the curved gate.

It was noted in the twin lens system that the objective lens spacing was adjusted automatically to compensate for film shrinkage; for the same reason the spacing between the two field scans on the crt needs to be adjusted. A capstan drive telecine has no sprockets with which to measure the film shrinkage and therefore another accurate method is required.

This method depends on the accuracy of the crystal locked colour television sync generator and the capstan optical tachometer. The film frame rate is locked to the crystal sync generator by the capstan servo so that the capstan tachometer output, which is a frequency measurement, is directly related to the film shrinkage and is used to control the spacing between the two field scans.

Jump scan systems require much smaller shading errors than other systems, typically less than 1 per cent. The separation of each television field on the crt means that any shading error has frequency components equal to the field sequence rate which for 625/50 is 25 Hz and 525/60 is 12 Hz. Any shading errors with frequency components below 48 Hz can be objectionable and therefore should be of very low magnitude. Shading correction is discussed in more detail in section *39.7.4*.

39.5.6 Sequential scan

This involves scanning all the horizontal lines in one vertical sweep rather than two sweeps as in a 2:1 interlaced scan.

Sequential scanning was introduced to overcome the problems and reduce the alignment procedure associated with the jump scan system, while retaining its simple optical arrangement.

As each film frame is scanned only once, there is no problem with field-to-field registration or field-to-field shading, and no requirement for shrinkage correction.

It is necessary obviously to convert the sequential scan to interlaced for transmission purposes, and for this a frame store is used as described in section *39.7.10*. It was not possible to use sequential scan until frame stores became economically viable for general use in studio equipment.

39.5.7 CCD linear array

The optical arrangements of the ccd telecine are very similar to those of the photoconductive telecine except that, since the film moves continuously rather than intermittently, the vision gate is curved instead of flat. Also, as the ccd consists of only one horizontal line of photosites at the centre of the optical axis, the radius of curvature of the film in the vertical direction does not affect resolution in the vertical axis. Wide aperture lenses can therefore be used, which is necessary to obtain a good signal/noise ratio in the blue channel where not only is the tungsten halogen light source lacking in energy, but the ccd has poor sensitivity.

39.6 Colour response

Telecines, like electronic cameras, usually have three colour channels: red, green and blue. The main difference between them is the spectral width of each channel.

Electronic cameras need to be sensitive to all wavelengths in the visible spectrum and so require overlapping spectral characteristics. In the case of telecine, the film is sensitive to all wavelengths, and the telecine needs only to determine the density of each colour dye. For this purpose a single narrow line response for each dye would suffice. In practice, because of signal/noise considerations, the responses are usually made as wide as possible without overlapping. In general, the photoconductive responses tend to be narrower because more light is available.

Light can be split into three components by dichroic mirrors or dichroic prisms. Mirrors are the simplest arrangement but can only be used outside the imaging optics because of secondary reflections at the second glass/air interface of the mirror. Therefore dichroic mirrors are used only in flying spot systems. A typical arrangement is shown in *Figure 39.4*.

On the glass surface of dichroic mirrors and prisms there are several subwavelength glass layers of various refractive indices to give the required characteristic for a defined angle of incidence.

The dichroic glass/air interface reflects all light above or below a designed wavelength and transmits the rest with virtually no absorption. The change from total reflection to total transmission occurs over approximately 20 nm of wavelength. The wavelength at which half the light is reflected and half transmitted (known as the *50 per cent point*) is the design wavelength for the dichroic surface.

The 50 per cent point changes with angle of incidence, and therefore off-axis light rays from the edges of the picture suffer a shift of the spectral characteristic which is manifested as a change in amplitude resulting in colour shading. The rate of change of the 50 per cent point increases with increase of angle of incidence, so angles of incidence less than 45° are normally selected — typically between 25 and 35°.

Using just two dichroic surfaces will split the visible light into three components as shown in *Figure 39.4*, but the spectral characteristics for the three channels will be overlapping since there is very little absorption. To give some separation between the three channels, additional filters are added. It is also necessary to add an infrared filter in the red channel because even though there is normally very little sensitivity at these wavelengths, it is enough to cause red shadows with some films.

The ccd/quartz halogen light source combination is at its most sensitive in the infrared region, and this characteristic can be put to good use. Colour film dyes are transparent in the infrared region, and by adding a third dichroic surface and fourth ccd to give an infrared channel, dirt and scratches on the film can be detected. When dirt is detected, it is possible to reduce its visibility by replacing the signal with a more appropriate signal.

Figure 39.6 shows a typical spectral characteristic of a flying spot telecine. It can be seen that the blue response is narrower than green or red and is due to the falling response of the crt phosphor at the shortest blue wavelength. The blue channel in consequence has the poorest signal/noise ratio.

The falling response at the long wavelength end of the red response is a combination of the crt phosphor, the photomultiplier and the infrared filter.

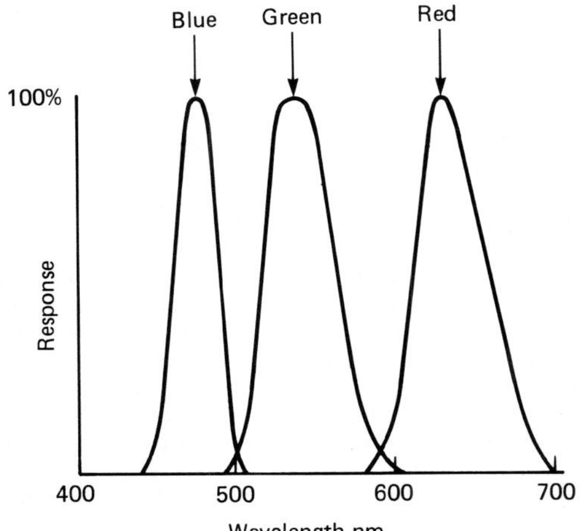

Figure 39.6 Telecine colour response

39.7 Signal processing

39.7.1 Noise sources and signal gain control

The maximum output current from a photoconductive tube at saturation is typically 300 nA. At this level of signal current, head amplifier noise is the predominant noise component. Therefore, to maintain the best signal/noise ratio the tubes are operated near saturation whenever possible. As the film density increases, the light is increased to maintain the same peak current, and this therefore is the mode of gain control with photoconductive telecines. The light is controlled by means of a variable neutral density disc so that a constant colour balance is obtained with varying light output.

As the noise is generated at the head amplifier, it is independent of the signal amplitude, i.e. it is the same at white or black. The human eye is more sensitive to noise in darker parts of the picture, so when aperture correction is added which boosts the resolution and the noise, it is made level dependent, i.e. it reduces to zero near black.

The signal in a flying spot telecine is generated in a photomultiplier and is between 50 and 100 μA, or some 300 times larger than the photoconductive signal. The head amplifier noise is therefore negligible, the noise being shot noise which is a function of the number of photons striking the photosensitive cathode and the number of electrons released in the photomultiplier. The noise is therefore a function of incident light and is proportional to the square root of the signal current, falling to zero at black level.

Gain control is obtained by varying the supplies to the photomultiplier to maintain a constant signal output as the light output decreases. The gain of the photomultiplier tends to increase more rapidly with increasing supply voltage. This nonlinearity does not affect the linearity of the light input/signal current output characteristic which is linear at all normal operating voltages. It affects the colour balance with gain change since the three photomultipliers are operating at different mean voltages due to the variation in the light output from the crt and colour splitting optics. To maintain colour balance as the master gain control is adjusted, it is necessary to preset the gain change of the red and blue photomultipliers to match the green.

The ccd produces a nominal voltage output of 1 V rather than current. Here again, any noise is generated in the ccd itself, and as this is a semiconductor, lf noise predominates, measuring approximately -60 dB. As the noise is so low and the signal high it is possible to use electronic means of gain control instead of light control. Electronic control is advantageous for rapid change from film frame to film frame.

39.7.2 Afterglow correction

This only applies to flying spot telecine and corrects for the phosphor persistence on the flying spot crt. The light output from the phosphor falls to 10 per cent of the excited amplitude after removal of the excitation in 120 ns. Unfortunately the afterglow curve of *amplitude* against *time* does not follow an exponential curve and therefore the corrector includes a number of capacitance/resistor networks with separate amplitude adjustments to assimilate the afterglow curve.

39.7.3 Burn correction

Burn correction is needed only with flying spot telecine where the raster on the crt will reduce the efficiency of the phosphor and faceplate with time. If the raster were always in a fixed position and of fixed amplitude, there would be no need for burn correction, but facilities such as zoom and variable film frame rate alter the size of the raster. Thus some parts of the phosphor screen are used more than others and so give less light output after a time. The burn corrector consists of a photomultiplier which measures the light output from the crt, and a reciprocal amplifier the output of which drives shading correctors in each colour channel.

39.7.4 Shading correction

Every type of telecine is affected by variation in brightness across the picture. Whether or not correction is necessary, and what amount, depends on the type of shading error.

If the shading is only in luminance, then the human eye can tolerate very large variations before it is evident that something may be wrong. The eye is much more sensitive to variations in colour shading and extremely sensitive to field-to-field shading which can cause flicker at 25 Hz and lower frequencies.

Shading correction is therefore essential for jump scan telecines, highly desirable in colour telecines, but not necessary in monochrome telecines.

Shading errors arise from the light source, objective lens and colour splitting dichroic mirrors and prisms. The errors are usually constant and can be corrected by developing waveforms from scan currents in the deflection system. These waveforms are adjusted individually in amplitude and shape and used to modulate the individual red, green and blue signal channels.

39.7.5 Stripe stripping

This particular process is only relevant to the ccd telecine. The term *stripe stripping* refers to vertical stripes that are seen if the errors in the ccd remain uncorrected.

A ccd consists of a number of separate photosites, and the variation in sensitivity between adjacent sites can be as much as 5 per cent and is quite random. As the film moves through the telecine, each photosite scans a vertical line of the film frame and therefore if all photosites are not equally sensitive, a vertical line pattern will appear in the picture. As the variations in photosite sensitivity are independent of temperature, a fixed read only memory (ROM) can be used to make corrections.

Alternatively, during the initial line-up procedure an electronic measuring feedback arrangement can correct errors, storing the corrections in a random access memory (RAM). This has the advantage that somewhat variable shading errors due to the tungsten halogen light source can also be corrected day by day.

39.7.6 Gamma correction

Like electronic cameras, telecines incorporate a gamma corrector which is a power law transfer characteristic to compensate for the power law transfer characteristic of the receiving crt. There is one corrector in each colour channel.

In telecine, the gamma corrector fulfils another function, i.e. the correction of colour film errors from scene to scene or film to film, errors which electronic cameras do not normally encounter.

Film colour balance errors are not normally objectionable when projected optically in a dark theatre because the human eye adapts its own colour balance to the predominant wavelength in the incident light[2]. When viewing a television receiver, the observer usually has a reference colour near the receiver and only a small proportion of the light incident on the observer's eye comes from the receiver. Therefore, colour balance errors in film shown on television need correction, and differential gamma variation between the colour channels is part of this correction (see section *39.7.8*).

Continuously variable gamma is obtained by using logarithmic and antilog amplifiers. If the amplifier gain between the log and antilog stages is unity, the channel will be linear, or in mathematical terms:

$$y = x^{\gamma}$$

By varying the gain between log and antilog stages, a range of gamma can be obtained.

39.7.7 Log masking correction

As one would expect, film dye characteristics can never be perfect, and the telecine spectral characteristics can never be ideally placed with respect to the film dyes since light sources and pickup device limit complete flexibility.

Consequently when combining two imperfect media, the film and the telecine, there will inevitably be some saturation and hue errors. Correcting these errors involves multiplicative crossmodulation between the channels, a function which can be realized with logarithmic masking[3], i.e. addition or subtraction of logarithmic signals.

The logarithmic matrix can be calculated[4], or alternatively a colour bar film can be placed in the telecine and the matrix resistors adjusted to simulate an electronic colour bar pattern[5].

39.7.8 Colour correction

Colour telecines are now provided with a whole array of controls for colour correction by popular demand.

There are *gain*, *gamma* and *lift* controls for each colour channel which could result in nine individual controls. However it is usual to combine these nine controls in three joysticks, one each for gain, gamma and lift. Rotation of the joystick spindle changes the master gain, gamma or lift with no colour change and movement of the joystick itself alters the colour balance in highlights, mid-greys or dark shadows with constant luminance.

There are also separate *hue* and *saturation* controls and secondary colour controls. The latter, twelve in all, enable the adjustment of hue and saturation over a limited wavelength of red, green, blue, magenta, yellow and cyan.

Obviously, adjustment of this number of controls 'on air' would be impossible. The film is therefore previewed, and at each scene where correction is necessary, the film is held stationary in the film gate while the necessary adjustments are made at leisure. When a satisfactory balance has been obtained, all the control settings are stored in RAM together with film frame numbers at the beginning and end of each section for which the control settings apply. Each scene is calibrated in this way, and when the film is transmitted, the control settings are extracted automatically from RAM, the change of control values occurring during frame blanking as instructed by the film frame counter.

For storage purposes, the control settings for a particular film are transferred to a floppy disk which can be stored with the film for subsequent use. For further replay, the control values are transferred back to RAM for rapid access.

There are occasions, e.g. in newsfilm, when there is insufficient time to preview the film and make the necessary corrections as described above. In this case an automatic colour corrector can be used, although it can never produce perfect results unless the film happens to agree with the rules and assumptions made by the corrector. There are therefore two modes of correction. If the film does not match the corrector assumptions, the corrector can be switched to the simplest mode.

In this least error prone mode, it is assumed that the majority of the darker shades in a picture are monochrome. Since black is monochrome, this assumption is more often correct. The peak blacks in the three channels are therefore monitored, and made equal to the lowest of the three, so ensuring that the darker shades are monochrome.

In the second mode, not only are the peak blacks equalized, the integrals of the red and blue signals are made equal to the integral of the green signal, sometimes called *integration to grey*['6,7]. This mode assumes that the average picture contains equal amounts of red, green and blue. Obviously some pictures with a predominance of colour would then be incorrectly portrayed; to reduce such errors, saturated colours are detected and clipped before measurement.

39.7.9 Aperture correction

Aperture correction is compensation for detail or resolution losses due to film camera lenses, printing processes, telecine lenses, and scanning spot size or distortion.

A television image consists of a number of horizontal lines. A fine pattern of parallel lines in that image produces a signal frequency that is much higher than the horizontal scanning frequency when the image pattern lines are at right-angles to the scanning lines, and lower than the scanning frequency when the image pattern lines are parallel to the scanning lines. To correct the resolution loss of patterns or edges at any angle, correction has to be applied horizontally and vertically at two quite different signal frequencies. Also, to maintain the symmetry of pulse edges after aperture correction, the correction network should have a linear phase response with frequency.

A transversal (non-recursive) filter fulfils the above requirements. It consists very simply of two delay lines in series, each with a delay T and suitable summing amplifiers to produce a correction signal:

$$V_T - \frac{V_{2T} - V_O}{2}$$

where V_T is the output of the first delay line, V_{2T} is the output of the second delay line and V_O is the input undelayed signal. In terms of frequency the correction signal is:

$$V_0 \left[1 - \cos \left(\frac{\pi f}{2f_O} \right) \right] \qquad (39.1)$$

where $T = 1/(2f_o)$ and f_o is the peak frequency.

The above expression gives a cosinusoidal response and the peaking frequency can be adjusted by varying the length of the delay lines.

Horizontal aperture correction requires delay lines between 50 and 200 ns depending on the film format being scanned. Some telecines incorporate two transversal correction filters with separate amplitude controls so that the peaking frequency can be controlled remotely or preprogrammed, to suit the particular film format in the telecine. Horizontal delays of 50–200 ns can be realized using lumped constant networks of inductance and capacitance.

The vertical aperture corrector requires delays of one horizontal line or multiples of one line to ensure that the vertical correction is only vertical and not a combination of vertical and horizontal. These delays are therefore in multiples of 64 μs and cannot be realized in lumped constant form.

Telecines without frame stores use glass ultrasonic delay lines which operate with an amplitude modulated 27 MHz carrier to obtain a 5.5 MHz bandwidth. Telecines with frame stores use digital one-line delays to develop the vertical correction signal.

The combined vertical and horizontal aperture correction signal can then be 'cored'. The coring process can best be explained by considering the nature of the correction signal. Expression 39.1 indicates that at zero frequency the correction signal is zero and there is no direct component. The correction signal therefore consists of positive and negative voltage excursions about zero. The coring process limits any correction output until the positive or negative excursion exceeds a preset level. Consequently small correction signals due to film grain and electronic noise are excluded.

As the chrominance signal is bandlimited in the transmission path, it is necessary to apply aperture correction only to the luminance signal, and in some telecines, particularly photoconductive, only the green channel is enhanced.

39.7.10 Frame stores

Frame stores were first introduced in telecine to convert sequential scan to interlaced for reasons described in sections *39.5.6* and *39.5.7*.

The telecine RGB output is normally coupled to a colour encoder when this output is interlaced, but when the telecine output is sequential there are at least three options for interfacing the store:

1 Colour encode the RGB sequential signal and follow this with a single channel store for conversion from sequential to interlaced.
2 Convert the RGB sequential signals to interlaced using three stores and then encode.
3 Convert the sequential RGB signals to sequential luminance and chrominance, which requires two stores for sequential-to-interlaced conversion before coupling to the colour encoder.

Option 1 requires a special sequential signal colour encoder, but uses the minimum amount of storage. 2 and 3 use standard colour encoders but require more storage and can provide a freeze frame without any extra storage.

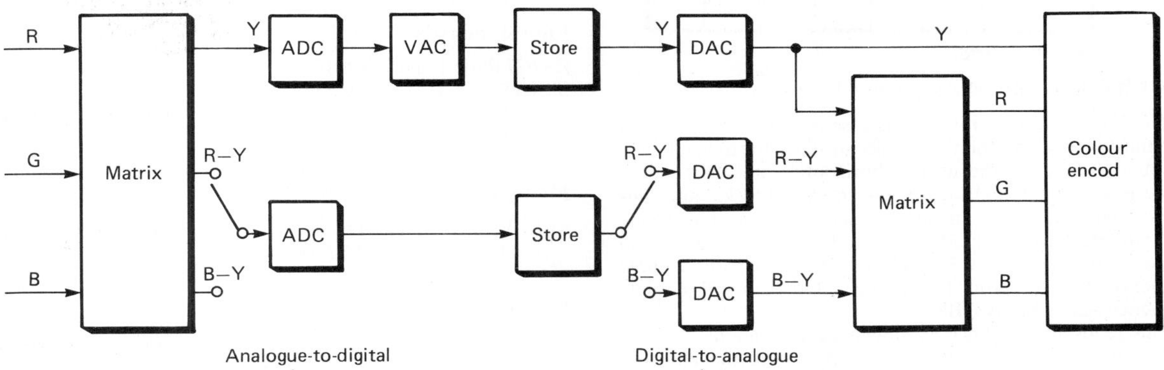

Figure 39.7 Frame store block diagram

Option 3 is most popular, being economical on storage, and because the digital studio standard specifies digital luminance and chrominance rather than RGB. *Figure 39.7* is a block diagram of this option. The analogue RGB signals are first matrixed into Y, R–Y and B–Y before digitization by analogue-to-digital converters (ADCs). The R–Y and B–Y analogue signals are connected to the chrominance ADC alternately by an analogue switch operating at half the ADC clock rate so that R–Y and B–Y digital words appear alternately at the output of the chrominance ADC. As the cost of ADCs fall, an alternative arrangement gaining favour is the use of half the luminance ADC clock frequency. The R–Y, B–Y digital words are then combined into one parallel bit stream at the luminance clock frequency.

The ADCs have eight bit resolution giving 256 levels of grey scale, which is universally recognized as adequate for gamma corrected signals. Linear signals would require 11 or 12 bit resolution.

The digital luminance signal then enters the vertical aperture corrector (VAC) where a vertical correction signal is generated as described in section *39.7.9* and added to the luminance signal.

The luminance and chrominance signals are then connected to two identical stores, but since the luminance signal is delayed in the vertical aperture corrector by one or two horizontal lines, the read address to the chrominance store is designed to delay the chrominance read-out by the appropriate number of lines to resynchronize the two signals.

When writing the incoming sequential signal into the store, it is normal to separate the lines into their correct fields at the writing stage. This is shown diagramatically in *Figure 39.8* where A and B represent the first and second fields in address space, two fields of storage being required in the simple arrangement for 525/60 operation. The first scanned line of film frame 1 is written into address space field A, the second in field B, the third into A, and so on alternately.

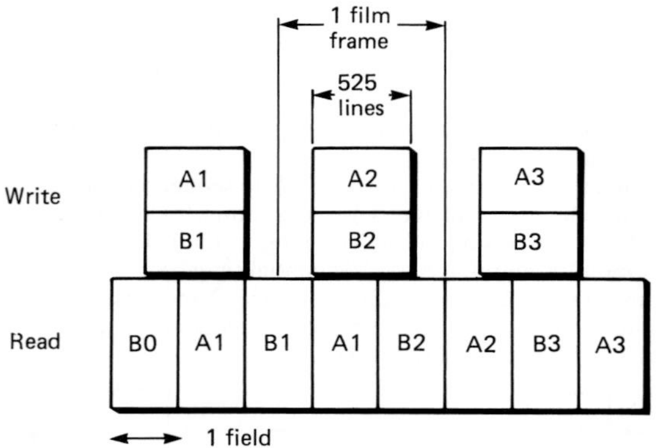

Figure 39.8 Timing diagram for sequential-to-interlace conversion

The horizontal scan rate in this case has been increased by one third so that 2 complete fields are stored in the time taken to output 1½ fields at the normal horizontal rate. This means there is a complete field gap between write blocks, and thus three complete fields can be obtained from one film frame without interfering with adjacent film frames. This three-field sequence is shown as A1, B1, A1, to be followed by a two-field sequence, B2, A2, from film frame 2 and then a three-field sequence again, B3, A3, B3.

The read-out sequence is shown as continuous. This is necessary to maintain a picture except during the blanking periods. The blanking periods are short compared with the storage time of dynamic RAMS and therefore, since the dynamic RAMs are continually exercised, they are automatically refreshed even when writing is inhibited.

The digital signal data rate is 13.5 MHz or higher to satisfactorily process analogue bandwidths of 5.5 MHz. Low cost, high density dynamic memories do not operate at 13.5 MHz and therefore the incoming 8 bit word to the store at 13.5 MHz is translated to a 64 bit word at 1.8 MHz by serial-in, parallel-out shift registers.

At the output of the stores, the 64 bit words are reconverted to 8 bits and the chrominance signal is also demultiplexed into R–Y, B–Y. The three digital signals are then converted to analogue to become Y, R–Y, B–Y, where they can be directly colour encoded.

Figure 39.7 shows that these signals are reconverted to RGB because in many cases the colour encoder is separate from the telecine and does not accept R–Y, B–Y inputs. Also, there are applications for the telecine as background pictures where a RGB signal is preferred.

References

1 WHEELER, L J, *Principles of Cinematography*, 3rd Ed., Fountain Press, London, 224 (1953)
2 HUNT, R W G, *The Reproduction of Colour*, 2nd Ed., Fountain Press, London, 68 (1957)
3 BURR, R P, 'The use of electronic masking in colour television', *Proc IRE*, **42**, 192 (1954)
4 GRIFFITHS, F A, 'The calculation of electronic masking for use in telecine', *BBC Research Dept Report* 1972/24 (1972)
5 HUNT, R W G, *The Reproduction of Colour*, 2nd Ed., Fountain Press, London, 426 (1957)
6 HUNT, R W G, *The Reproduction of Colour*, 2nd Ed., Fountain Press, London, 70 (1957)
7 WRIGHT, D T, 'An automatic tariff for news film', *BBC Research Dept Report* 1974/15 (1974)

Bibliography

Cinematography

WHEELER, L J, *Principles of Cinematography*, 3rd Ed., Fountain Press, London (1953)

Colour

HUNT, R W G, *The Reproduction of Colour*, 2nd Ed., Fountain Press, London (1957)

Film standards

BS 677 Parts 1 and 2 (1958)
SMPTE Standards, New York
DAVIES, H, Colloquium on Sound on Film, *IEE*, London (1966)

Flying spot telecine for 525/60

MILLWARD, J D, 'Flying spot scanner on 525-line NTSC standards', *Jour SMPTE* (Sept 1981)

Frame stores

JESSON, G S, 'A variable speed frame store for flying spot telecine', *Professional Video* (Sept 1982)

Logarithmic masking

BURR, R P, 'The use of electronic masking in colour television', *Proc IRE*, **42**, 192 (1954)

GRIFFITHS, F A, 'The calculation of electronic masking for use in telecine', *BBC Research Dept Report* 1972/24 (1972) JONES, A H, 'A theoretical study of the application of electronic masking to television', *BBC Research Dept Report* PH-16 (1968)

Twin lens telecine

NUTTALL, T C, 'The development of a high quality 35 mm film scanner', *Proc IEE*, **99**, Part II A No 17 (1952)

NUTTALL, T C, BOSTON, D W, ASKEW, G H and LOWRY, P, 'The Rank Cintel twin claw twin lens flying spot 16 mm film scanner', *BKSTS Jour*, **48**, 1, 2 (1966)

Part 8
Mobile and Portable TV Equipment and Operations

Frederick M Remley Jr FSMPTE, MBKSTS
Centre for Information Technology Integration
University of Michigan

40 Portable Television Cameras and Videotape Recorders

40.1 Background

Contemporary selections among portable, hand-held television cameras are attractive in terms of size, weight and performance. Similarly, designs for small television recorders, often attached to these cameras to produce combinations known commonly as *camcorders*, achieve high levels of performance and are compact and lightweight. There are good reasons to be confident that the trends toward smaller size and higher quality will continue as the technologies of camera and video recorder design advance.

Although portable professional quality video recording equipment has been available for more than 20 years, beginning with the Ampex V R 3000, a rather bulky, heavy, backpack-carried transverse track (quadruplex) recorder of the late 1960s, the trends toward compact cameras and recorders began in earnest in the middle 1970s.

Figure 40.1(a) Ampex VR-3000 backpack recorder (Ampex Corp)

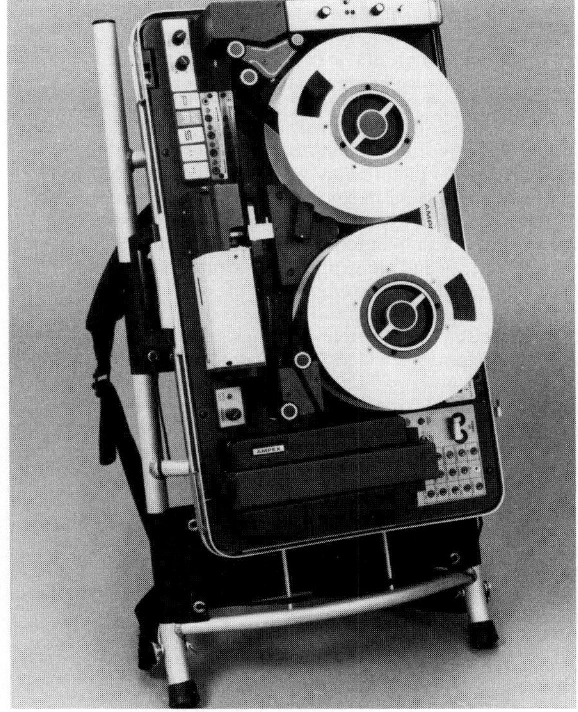

Figure 40.1(b) The VR-3000 with cover removed (Ampex Corp)

At that time a move by broadcasters to replace film equipment with video equipment for news gathering purposes began to gain momentum. In response, camera and video recorder manufacturers reacted to the firmly expressed needs of potential purchasers by finding new approaches to equipment design. ENG (Electronic News Gathering) equipment became available in quantity during the 1970s and eventually displaced 16 mm motion picture film as the dominant television news recording medium.

The advantages of using video equipment were, and are, increased ease of use and reduced cost of operation. The electronic camera/recorder produces material that can be reviewed as it is recorded, and is immediately available for electronic editing and broadcasting. By contrast, the 16 mm motion picture film process requires skilled exposure determination, chemical development at a later time with specialized film processing equipment and physical editing before it is ready for broadcast. In addition, the labour costs of a news gathering crew using photographic equipment tended, in the 1970s, to be higher than the costs associated with a crew using electronic systems. Such cost differentials would be much greater today because photographic systems have continued to increase in relative cost while electronic systems have become relatively less expensive.

The best of the early compact cameras had a similar physical size and shape to the 16 mm film cameras then in use. However, the power demands were rather high, and large, often unreliable, battery packs were required. Lead oxide pickup tubes were commonly used in these cameras although selenium-antimony-tellurium pickup tubes were also employed. Because of bulk and weight, the recording equipment usually was carried and operated by an assistant to the camera operator. Sometimes the recorder was rolled to the news scene on a small wheeled cart. This was usually the case when B format or C format recorders using 25 mm tape were used. In time, 19 mm U-format cassette recording systems became the most frequently used recorders for ENG work, and new designs appeared that were compact, battery powered and durable.

ENG endured some difficult problems when initially introduced. The picture quality of many ENG recordings, when using early U-format recorders, was inferior to that of well exposed motion picture film. Partially offsetting this problem was an improvement in sound quality, at least in cases where photographic news recordings had used optical soundtracks. Overall, the combination of somewhat primitive portable camera designs and conventional U-format recorders resulted in less than optimum picture quality.

No sooner had cameras and recorders of reasonable size been placed into the hands of television news people than a demand arose to use portable video equipment for kinds of productions unrelated to news gathering. Once again, the use of motion picture film in television programme production had been (and is to this day) commonplace, usually involving 35 mm film equipment in the case of studio productions. The use of video equipment in remote locations outside of the studio was proposed as a cost saving technique. Soon another term was coined: Electronic Field Production or EFP. As noted above, most early ENG equipment was not well suited for EFP work, since the quality of the video signal produced by early cameras and the quality of the recordings produced by early portable recorders sometimes did not meet programme production specifications. There grew, then, families of cameras that could be hand-held by a sturdy person, often derived from studio camera designs and using studio quality pickup devices, connected by cables to control equipment mounted remotely, perhaps in an OB vehicle. The recording equipment was located in the vehicle as well, and this recording equipment used transverse track (quadruplex), B format or C format, depending on the recording format chosen by the relevant production company.

40.2 Recording systems for portable use

The long standing limitations on picture quality, weight and size of video recorders used for high quality professional work have been largely solved with the introduction of component analogue recorders, and camcorder combinations, making use of 12.65 mm wide video tape packaged in cassettes. Improvements in 19 mm U-format cassette systems have also been achieved and the new versions of U-format equipment have been placed into service in some organizations. Even so, contemporary equipment choices made by most field production and news gathering organizations include cassette-based component recording systems exemplified by the IEC Type L and IEC Type M II systems. Just beginning to be discussed are the professional application requirements for future designs using recorders based on even narrower tape, most often predicted to be either 8 mm or perhaps 6.35 mm. It is generally assumed that these new, smaller designs will be based on digital component recording techniques and will use high coercivity tape, having either metal particle or evaporated metal magnetic coatings.

40.2.1 The design of portable recorders

Portable use of video recording equipment poses many problems for the equipment design engineer. The environment encountered by all portable equipment, television or other, tends to be quite hostile. Factors such as transportation from one place to another, exposure to moisture, dust, heat and cold while being transported and while in use, coupled with wide variations in power source stability, and so on, all offer challenges to an equipment designer. In addition, professional users set demanding requirements for factors such as reliability, quiet operation, minimum weight and ease of use. These requirements conflict with purchaser expectations of low purchase prices and operating costs and serve further to compound the design problems.

Figure 40.2 Betacam SP camera and recorder (Sony)

Early in the design of second-generation, post-U-format, portable video recording systems for ENG and EFP use, some potential users insisted that new equipment designs use tape cassettes identical to the types originally designed for domestic video recorders e.g. VHS format or Beta format. This was thought to be a necessary requirement despite the fact that ENG/EFP recorders were planned from the beginning to use new, professional-quality track formats for the actual recordings on tape. It seemed important to the pioneering broadcast news groups and to some equipment designers that users be able to purchase video recording cassettes from any convenient seller, anywhere in the world, whether or not such seller normally catered for broadcast clients or domestic clients. This rather idealized situation has not, in fact, prevailed. Professional users are more critical of video tape quality than are domestic users. Tape defects such as dropouts may not matter very much in the home but may be disastrous in professional ENG or EFP use. In addition, the consumer video cassette market is very price sensitive and video cassettes are consequently sold in several quality ranges. Too often professional users find that only lower-grade cassettes are available in locations remote from metropolitan areas. It is usually not obvious that these quality variations exist since no internationally agreed quality standards have been promulgated. Thus, the use of a cassette of unknown quality can lead to difficulty.

40.2.2 Professional tape cassettes

Given the factors noted above, a decision has been made by most broadcast organizations to abandon the use of consumer market cassettes for professional ENG and EFP applications. This has a great advantage in making it possible to use special magnetic tape formulations to improve the performance of professional cassette recording systems. However, accompanying this decision is an inevitable increase of cost for professional cassettes as compared with consumer cassettes. It remains to be seen if more sophisticated consumer cassette designs, for example the design internationally agreed to for 8 mm camcorders, will satisfy professional requirements.

Professional users normally opt, as was almost always the case when film was used for similar applications, to carry their blank recording medium with them and do not rely on purchases on location. Once this fact is accepted, it is possible to design portable recording systems to use magnetic tape of different, much improved performance characteristics; tapes that cater specifically for the needs of professional users. Professional quality tape cassettes may not function with domestic machines in some cases, since the improved tape may require different video recording characteristics or may not erase satisfactorily or may have different audio characteristics from tapes designed for domestic products.

40.3 Component video systems

Two major trends continue to emerge in the design of contemporary television equipment. The first is a trend towards digital signal processing. In consequence, standardized studio digital tape recorders are in use, as are facilities providing digital signal generation and manipulation for a variety of television studio uses, most especially special effects and graphics. Studio facilities using digital signal processing techniques will inevitably increase in number and complexity and present-day digital recorders and telecine equipments will be joined by new apparatus to permit fully digital studios.

The second trend is a strong move towards component video signal origination and distribution in both studio and field applications. Component-based video systems can be designed to operate with either analogue video signals or digitized video signals. We will look primarily at the application of analogue component techniques to small cameras and recorders.

The use of component video stands in contrast to the tradition of converting all internal camera and recorder signals, for example the red, green and blue camera pickup signals, to composite signals before they are routed through the studio plant. Composite signals are conveyed through a single coaxial or other transmission channel to other pieces of television apparatus. In component systems, by contrast, the convenience of a single video channel is abandoned in favour of picture quality improvements gained by retaining the video signal in an unencoded form of luminance (Y) and colour difference (R–Y, B–Y) signals through most of the origination, recording and processing steps of television production. The quality advantages are significant and the forward march of electronic technology has made such an approach eminently practical.

Most new portable EFP and ENG recording systems are designed to use component signal processing. The recorder portion of the camera/recorder combination records component video signals in track patterns (formats) unlike those used for consumer equipment and usually records these tracks on special, professional quality tape. Higher signal writing speeds are used and thus the tape consumption per unit time is increased. A professional cassette may record for only a fraction of the time that a similar cassette can record when used in a domestic video cassette recorder. This is a trade-off that is made in favour of increased video signal quality and does permit the use of special, high energy video tape while the cassette housing itself can be made in the same moulds used for consumer housings, and thus achieve manufacturing economies.

40.4 Sound recording features of compact video recorders

The performance of the typical longitudinal sound recording channels of professional analogue video recorders has never been fully satisfactory for the most demanding kinds of applications. This is true for all of the many professional formats that have been used in broadcasting and in professional television production: transverse track, Type B, Type C, U-format, Type L and Type M II. The deficiencies usually found include relatively low signal/noise ratio, relatively high distortion, and relatively poor frequency response.

The longitudinal tracks designated for audio recording in portable video recording systems are quite narrow, limiting the signal/noise ratio. Because of the rotary head principle used for the video portion, the longitudinal tape velocity is low, causing degradation of sound recording frequency response due to losses at short wavelengths. Domestic recordings suffer the most, since the very low tape velocities typical of Beta and VHS domestic recorders can result in disappointingly poor sound channel performance. In some consumer recording equipment the use of audio processing and noise reduction techniques has improved the sound quality obtainable from these formats, but the addition of processing equipment almost inevitably introduces compatibility and interchange problems for the user.

New equipment designs for both domestic and professional recording equipment address the matter of sound recording quality directly. A new approach has emerged from the design laboratories. It uses frequency modulated (fm) subcarriers, placed in a protected part of the video recording spectrum, to provide superior sound channels. These fm subcarriers are capable of excellent performance, with high signal/noise ratios, good frequency response and low distortion. The fm sound channels typically provide wow-and-flutter performance that is better by an order of magnitude when compared with direct longitudinal recording, thus providing higher quality sound in this regard as well.

40.5 Portable cameras

The design of colour television cameras intended for use in ENG and EFP conditions requires that detailed engineering consideration be given to several factors that are usually of secondary importance in studio camera designs. For example, the ENG/EFP picture is expected to be within specifications after only a few seconds of camera operation. This requirement must be met under adverse conditions of excess heat or bitter cold. Not only must the camera furnish almost immediately a fully acceptable image, but an accompanying video tape recorder must also function under extreme conditions as quickly as the camera. Both the recorder and the camera must be equipped with electronics, mechanics and power sources assuring such performance. Consequently, the design of all parts of the camera electronics and the recorder mechanics systems can be very challenging. The design of ENG/EFP battery packs is a speciality in itself.

The following factors must be taken into account when considering the specifications for a portable television camera:

- sensitivity and response characteristics of photosensors,
- colour temperature compensation range for a variety of light sources,
- satisfactory ambient temperature range for storage and for operation,
- humidity effects on cameras, lenses and recorders,
- lens availability in a choice of focal lengths and a choice of manufacturers,
- total system weight and weight distribution when hand-held,
- power requirements and available types of power packs,
- resistance to mechanical shock,
- resistance to invasion by dust and liquids.

Studio cameras using solid-state internal design appeared many years before lightweight portable cameras became available. The space occupied by the large number of discrete components needed in early solid-state studio cameras coupled with the relatively high power consumption typical of discrete device systems delayed the design of compact cameras until integrated circuit technology made some significant advances in increasing junction packing density. The first truly compact ENG cameras made pioneering use of medium scale integration (msi) semiconductor devices.

Even after the signal, control and deflection circuits were compressed into very small volumes through advanced semiconductor technology there remained a very real stumbling block along the road to the camcorder system. This involved the imaging devices used.

40.5.1 Photosensors

For many years most television camera imaging devices have been special forms of the vidicon, a member of a class of photosensors based on changes in the conductivity of a semiconducting material as a function of photons striking the sensing surface. The back of the semiconducting material is scanned by an electron beam to generate an electrical signal that corresponds to the optical image that the camera lens has focused on the front surface (see *Figure 40.3*).

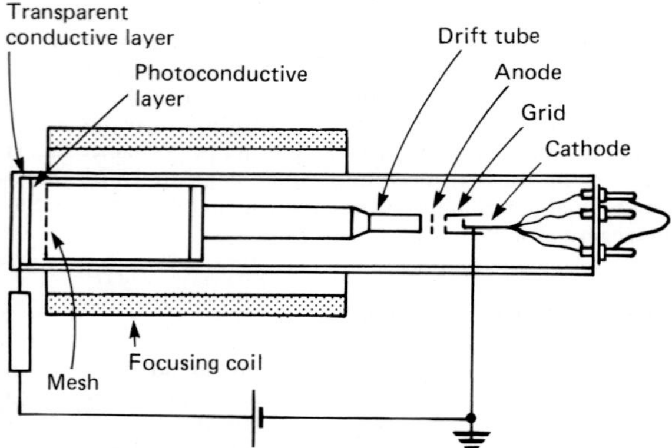

Figure 40.3 Vidicon (Saticon, Plumbicon) camera

Typically, lead oxide and selenium-arsenic-tellurium photoconductor constructions have been used. Trade names such as Plumbicon and Saticon have been applied to these devices. The virtues and faults of this technology are well known to television engineers. Photoconductive tubes provide excellent pictures under good lighting conditions, but tend to produce pictures with several kinds of defects under difficult lighting situations. Some of the problems seen in vacuum tube photoconductive sensors include:

- *picture lag*, a defect that causes images to have smeared colour trails following behind moving objects,
- *noise*, resulting from inadequate depth-of-modulation of the scanning electron beam under low light conditions,
- *overload* by bright light sources and overall limited contrast range,
- relative *fragility* and a tendency to vibration sensitivity under field conditions.

In spite of these real deficiencies of photoconductor vacuum tube sensors, the devices are capable of producing television pictures of the highest quality when lighting conditions are satisfactory. Other, newer technologies must be judged against this demonstration of excellent performance.

40.5.2 Semiconductor photosensors

Given the several problems noted above that affect cameras using photoconductive sensors, designers have for many years sought suitable semiconductor equivalents to the photosensor tube. To function in an ENG/EFP environment and to replace the tube sensor in field applications, semiconductor photosensors must exhibit most of the attributes of other complex semiconductors — low production cost, small size and light weight, low power consumption, reproducibility, durability, etc. When used in a portable television camera the following additional requirements are sought:

- ruggedness of basic sensor and of the camera optical system,
- good performance under all kinds of lighting conditions,
- good colour response, complementing the television system,
- good image response to rapid subject motion.

The charge coupled device photosensor (ccd) is used in many modern portable cameras. The ccd is a solid-state analogue image sampling device consisting of a complex pattern of photosensitive junctions that respond to light intensity. The pattern of these junctions is carefully arranged to minimize possible problems that sometimes affect sampled video systems, including moiré and aliasing.

Charge coupled devices were applied to domestic television camera designs before they were introduced to professional equipment. In a real sense, the amateur videographer accelerated the availability of professional ccd cameras, because the funding of much fundamental ccd design research was accomplished on the basis of large potential consumer sales. Most of the basic factors listed above as being important to professionals are also important to amateur users. However, professional camera designs have several added requirements:

- ability to tolerate wide exposure ranges, including specular reflections and uncontrolled brilliant scene objects, with little video overload and no sensor damage. The ccd should match the photoconductive pickup device in this if possible.

• a contrast range suited to the professional television system. Usually this implies a video s/n ratio of about 60 dB and good transfer characteristic ('gamma') correction.
• low noise, both electrical (thermal) noise and sampling noise,
• low dark current. Dark current is a defect which reduces the effective contrast range and makes gamma correction more difficult.
• freedom from manufacturing blemishes at reasonable production cost.

40.5.3 Characteristics of available ccd devices

CCD imaging devices consist of an array of light sensitive junctions together with a semiconductor structure for reading information contained in these junctions and transferring the information to the camera system. The fact that an array of individual junctions is used means that the output of the sensor is sampled, and each sample contains information from a small part of the camera image. Since each junction stores a charge that is related to the incident light falling upon it, the result is a sampled analogue system. What is read from the sensor is a series of samples of current that represent the scene brightness imaged onto the ccd surface at the junction points. These samples are related in time to the television scanning system in use and are combined together to form a continuous analogue video signal.

Several choices are possible in the design of a ccd. The light sensitive junctions may be simple photodiodes consisting of silicon p-n junctions. Other designs use mos (metal-oxide-semiconductor) transistor elements. The junctions may be arranged in various ways and the ancillary reading circuitry can scan the junctions in different fashions. For example, it is possible to store sequentially the information needed to generate a frame of video information and to read it during the television scanning vertical blanking interval in one pass of a decoding circuit. Many designs use this *frame transfer* method and perform well (*Figure 40.4*). However, rapidly moving subjects may exhibit a blur when imaged with a frame transfer ccd and it is often necessary for the camera designer to provide the system with a mechanical shutter, or its electronic analogue, in order to achieve sharpness in moving images.

In the frame transfer device, two similar areas of elements are provided, usually one above the other on the same silicon chip. One is the photosensitive (imaging) area and the other is a storage area. The number of elements is the same for both areas. During the television active scanning interval, the photosensitive area is exposed to light from the camera optical system and charge packets are formed in each ccd element

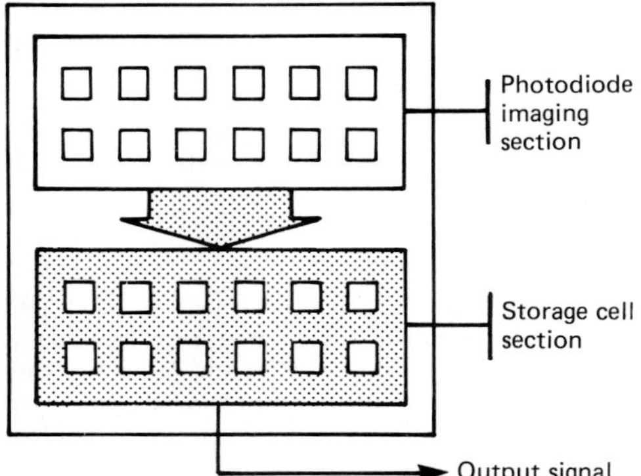

Figure 40.4 Frame transfer ccd device

proportional in charge to the image brightness. During vertical blanking, all elements of the photosensitive area are caused to transfer their charges to the corresponding elements in the storage area. The stored video signal is then read out during the period between blanking intervals.

Other designs use schemes where there are charge storage elements embedded in the active picture area of the imaging device. These are termed *interline* transfer devices and provide for transfer of image signal charge from the photodiodes to light-shielded vertical shift registers in the image area. The vertical shift registers are shifted line by line in turn and load their charges into a horizontal shift register and thence to the camera channel preamplifier (*Figure 40.5*).

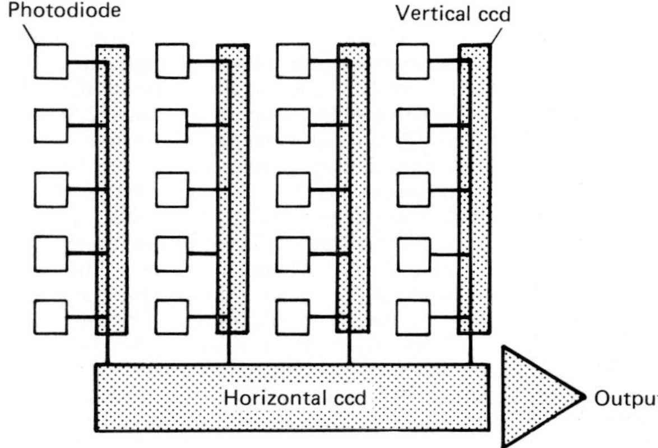

Figure 40.5 Interline transfer ccd

Designs have been introduced that incorporate both the frame transfer and the interline transfer design elements. By using very high clocking speeds for the interline transfer, it has proved to be possible to reduce greatly image smearing caused by subject motion that might occur while the charge transfers are being made (*Figure 40.6*).

Most ccd professional cameras use three photosensor devices. These may be arranged in the traditional pattern of

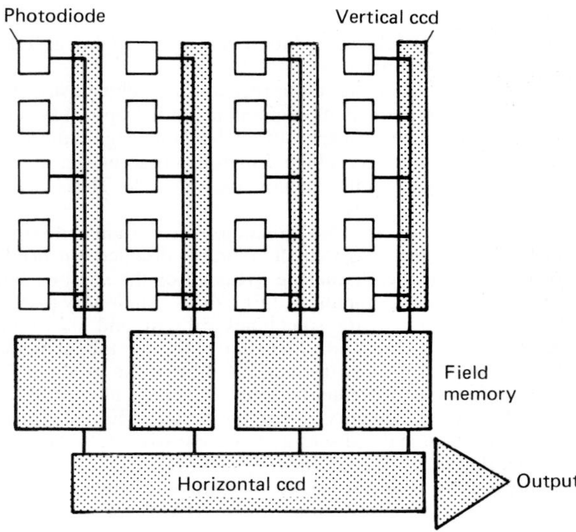

Figure 40.6 Combination ccd

three elements fastened to a prism assembly that provides three light output ports; red, green and blue (RGB output). Other designs use two ccds to sense the green channel, with improved equivalent green image sharpness resulting from an offset of one green ccd relative to the other by half a pixel. In this case, a third ccd having filter stripes applied to its front surface is used to sense the red and blue components at lower resolution. The output of the sensors is combined to form an RGB signal.

The resolution of a ccd imaging device is fundamentally determined by the number of sensing junctions provided in the ccd matrix as well as by the relative area of the junctions compared with the areas of the images formed by the lens system. Typically, 500–800 junctions are used in both the horizontal and vertical image planes and the sensitive area may be equivalent to that typical of the image area of a 17 mm photoconductive tube. Experimental ccd sensors having many more elements have been announced and demonstrated. For example, a ccd sensor built to operate with optics typical of a 25.4 mm photoconductive sensor camera contains 1280 horizontal elements and 970 vertical elements. Horizontal resolution is rated at 960 lines and is nearly good enough for high definition television camera use. It is expected that resolution performance will continue to improve at a rapid pace.

Overall system resolution is also affected by the fact that the ccd device is a sampling device. At high image spatial frequencies, moiré and aliasing disturbances are often produced and must be dealt with by means of electrical or optical filters. Optical low-pass filters may be used to band-limit the image that falls on the photosensor junction matrix. Electrical filtering may also be required. In either case, the system resolution, as plotted on a chart of modulation transfer function (mtf), will be significantly lower than would be indicated solely by considerations of photo junction density and area.

40.5.4 CCD camera system considerations

Because the ccd has uniform resolution over its entire photosensitive area, in contrast to the photoconductive tube which inherently produces less image edge resolution than centre-image resolution, a new kind of burden is placed on the lens designer who plans to provide objectives for modern ENG/EFP cameras. Lenses for professional ccd use must exhibit much improved image edge resolution and reduced image edge chromatic aberration.

CCD cameras have no colour registration problems since the ccd sensors are permanent components of the optical system. No aging of the sensors occurs through picture degradation and the overall stability is very high. Environmental changes, due to temperature or humidity changes, are negligible. These factors result in great reliability.

However, there are some limitations in certain ccd implementations. Some designs are temporarily disabled if struck by very high light levels, causing overload of the sensor itself. Improved designs minimize this effect. Stationary-pattern image disturbances are sometimes present and horizontal resolution is frequently inferior to that of the best tube cameras. Improvements in ccd performance tend to occur at about the same rate as improvements occur in semiconductor RAM (Random Access Memory) integrated circuits. The first generation of professional ccd sensors have approximately the same feature density as typical 256 kb dynamic RAM chips, but with a larger active area. Hence, ccd performance will continue to improve as 1 Mb and larger dynamic RAM chips move into mass production, because the semiconductor fabrication technologies are similar.

40.6 Implementations of compact recording system designs

40.6.1 The IEC Type L recording system

The IEC Type L system is a compact professional television recording system. It was originated by the Sony Corporation but is also available from other licenced manufacturers. This description covers the recorder and its recording format. Many cameras are available to work with the portable Type L recorder. The most compact of these are especially configured to attach to, and operate efficiently with, the portable recorder. Studio recorder/reproducers are available that permit editing and full-feature playback into the studio system of the Type L component video recordings.

The Type L format was introduced in 1982 as a system designed for ENG work. It proved to be successful and has been upgraded over the intervening years with new models of recorders and cameras. In 1987 the Type L SP system was introduced — an improved, upwardly compatible version of the system. Type L SP can make recordings suited to reproduction on older Type L equipment and similarly Type L SP can play recordings from older systems. Moreover, the SP version offers new features not available on the older Type L equipment.

Type L SP recorders can use a new cassette design that permits longer recording times and can use metal particle tape for improved signal/noise ratios. The newer system provides for two high quality fm audio channels, suitable for stereophonic use, in addition to two longitudinal audio channels. In addition, the luminance and chrominance video fm carriers have been moved upwards in frequency for improved video quality. Type L SP has been introduced in response to requests for improved picture and sound quality to meet more elaborate EFP production requirements.

40.6.1.1 Original system

True to the philosophy of early ENG equipment, the original Type L system made use of a cassette physically identical to the domestic Beta cassette. However, because of the increased tape speed used to improve the video performance of the Type L system, the standard Beta cassette provides just 30 minutes of Type L recording time.

40.6.1.2 Mechanical parameters

Figure 40.7 shows the basic Type L recording format. It is important to note that Type L is based on the use of component video processing and recording. The components recorded are luminance and multiplexed chrominance. The tracks are assigned in typical fashion, with the helical video tracks occupying most of the tape width. Azimuth recording is used, a technique more fully described later in the discussion of the Type M II format. Two longitudinal audio channels are placed at the top of the tape, and a control track and a time code track are placed at the bottom. The video tracks are 86 μm wide. The video scanner has a diameter of 74.5 mm and produces a video track 115 mm long.

40.6.1.3 Electrical specifications

Type L records simultaneously two parallel helical video tracks, one for luminance information and the other for chrominance. *Figure 40.8* shows the carrier frequencies and deviation bands for the relevant signals. The luminance (Y) signal is processed before recording to modify the horizontal synchronizing signal to add a timing reference pulse, to provide detail enhancement, video recording pre-emphasis and clipping. The chrominance (C) signal is composed of time-scale compressed R–Y and B–Y

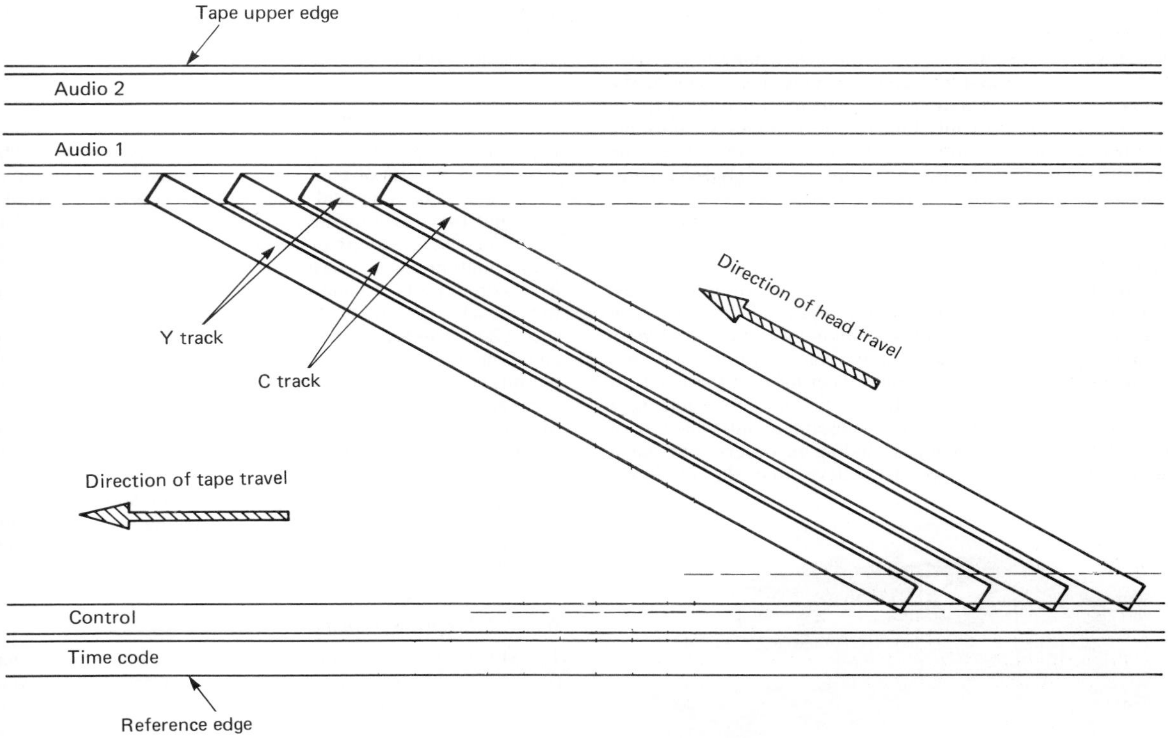

Tape upper edge
Audio 2
Audio 1
Y track
C track
Direction of head travel
Direction of tape travel
Control
Time code
Reference edge

Figure 40.7 Type L system tape format

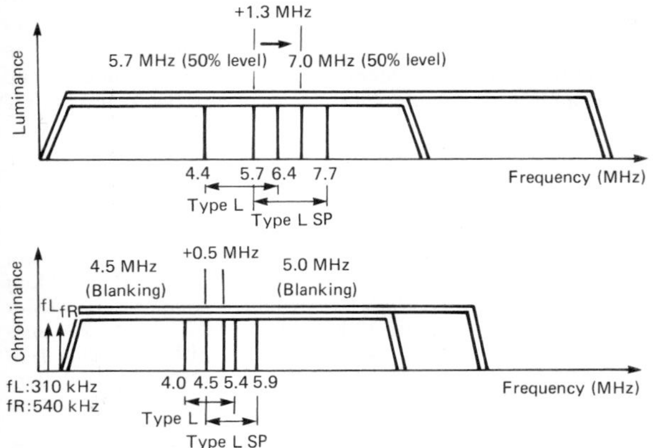

Figure 40.8 Type L carrier frequency allocations

signals multiplexed together so that both signals are recorded during a single horizontal line interval. In addition, the C signal has added to it a horizontal timing pulse, video pre-emphasis, detail enhancement and clipping.

40.6.1.4 Type L SP improvements

In introducing Type L SP the manufacturers took a stand for upward compatability. Type L recordings can be reproduced on Type L SP equipment and Type L SP equipment can record material for Type L use if the smaller cassette (two sizes can be used in the SP equipment) is used with ferric oxide (not metal particle) tape.

The improvements incorporated in Type L SP are intended to make the system useful in applications requiring high initial picture and sound quality and to allow for multiple generation copying and editing procedures. By adopting metal particle tape for Type L SP use, it was possible to raise the fm carrier frequencies used in recording the Y and C video components (see *Figure 40.8*). Higher carrier frequencies allow designers to improve channel bandwidths and to refine filter designs. The new system shows reduced moiré components, wider video bandwidth and improved K-factor response. In addition, it is possible to include two high quality fm audio subcarriers, suitable for stereophonic use, in the chrominance channel, recorded with the C channel rotary head, using carrier frequencies of 310 kHz and 540 kHz.

A new, larger cassette was introduced allowing up to 90 minutes of recording/playing time in the studio versions of Type L SP machines.

40.6.1.5. Other Type L components

This system has been designed from its inception for professional use. Accordingly, a full line of studio recorders, studio players, time-code generators and readers, editing controllers and other ancillary equipment is available from a variety of manufacturers. These items allow for systems to be designed to preserve the component-form video signals throughout the editing process or to be designed for immediate conversion of the component signals to composite analogue or digitized component or analogue signals.

Reproducing equipment is available for automatic playback of cassettes containing short messages.

40.6.2. The Type M system

The Type M format uses a cassette design based on the VHS consumer cassette. Two cassette sizes are offered, the larger of which offers 95 minute recording capacity. The most highly developed version of the Type M system is known as M II. This format is based on the use of metal particle tape having thin (13.5 μm) backing material and uses new technology in the design of the record and playback heads. In addition, the format offers a high quality fm audio recording system usable for stereophonic recording. The video system uses the chrominance time-compressed multiplex (ctcm) video recording technique (*Figure 40.9*).

The M II format resulted from a joint effort by NHK, the national broadcasting organization of Japan, and the Matsushita Electric Industrial Company to develop a universal component analogue video recording format. In other words, the project was directed towards developing a format based on well proven analogue recording technology that could serve in portable equipment as well as in the studio environment. It was hoped that this new format would meet most of the needs of broadcasters from its introduction to the end of the analogue studio equipment era, with the subsequent introduction of digital television equipment, including recorders, into most studio systems. Many experts believe that the all-digital television production facility will become dominant by the middle 1990s and that analogue video recording will diminish in importance within a similar time period.

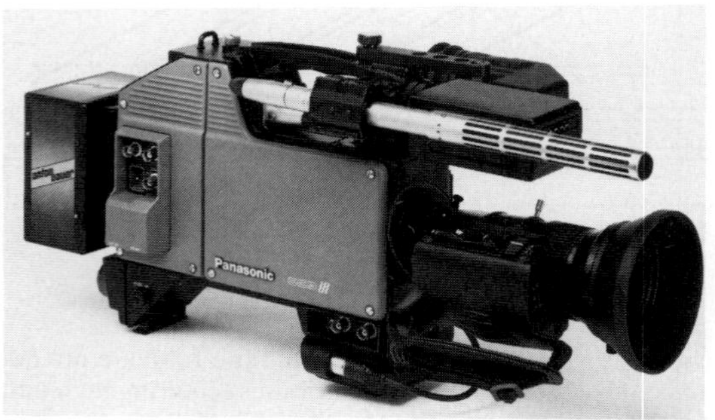

Figure 40.9 Type M II camera and recorder (Panasonic)

40.6.2.1 Mechanical format characteristics

Figure 40.10 shows the tape format used in the M II system. The mechanical aspects of the format are relatively conventional, with a time-code track and a control track at the bottom of the tape and the two longitudinal audio tracks at the top. The record/playback azimuth of the heads used to record the Y and C information on successive video tracks differs by 30°; the Y head azimuth is 15° counterclockwise from a line perpendicular to the track edge, and the C head azimuth is 15° clockwise from such a line. This arrangement, known as 'azimuth recording', is common in contemporary recorder design. Azimuth recording permits good isolation between adjacent video tracks with little or no guard bands between them. The scanner diameter for the M II system is 76 mm and the shortest recorded wavelength is about 1 μm. The video track width is 38 μm and the video track length is 118.3 mm.

To provide flexible editing facilities in studio recorder designs, the format is designed to accommodate automatic tracking playback heads and flying erase heads in the scanner assembly.

40.6.2.2 Electrical format characteristics

The M II format uses ctcm signal processing, which means that the colour component signals R–Y and B–Y are each stored in 1 H ccd analogue memories using a clock frequency of 455 times the horizontal scan frequency (f_h) and read back after a suitable delay at a clock frequency of 910 f_h. This results in a 2:1 time compression of the colour component signals. The compressed component signals are then suitably combined to produce the C signal which is made up of 0.5 H of R–Y and 0.5 H of B–Y. In order to improve the final jitter correction performance of the system, a pilot signal is added to the horizontal blanking interval (back porch) of the Y and both C

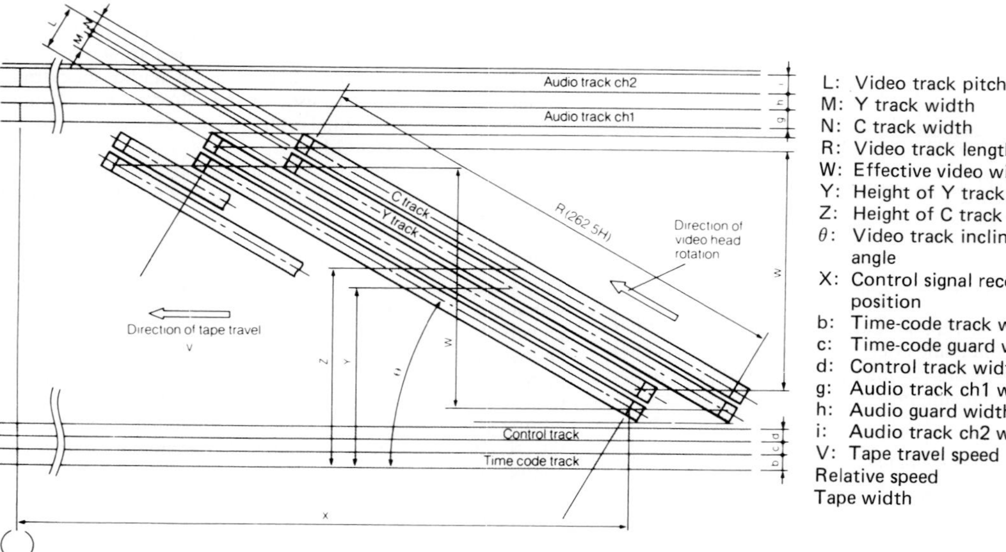

Figure 40.10 Type M II system tape format

component signals prior to recording. The pilot signals are used by the system timebase corrector to reduce signal jitter errors on playback. In addition, identification sync pulses are added to the R–Y signal before recording.

The fm audio carriers are placed on the chrominance track at base band frequencies of 0.4 and 0.7 MHz and the audio carrier deviation is ±105 kHz. The specifications of the fm sound system are of professional quality.

Figure 40.11 shows the fm frequency allocations for the Y and C tracks in the M II format. Note the audio subcarriers on the channel spectrum.

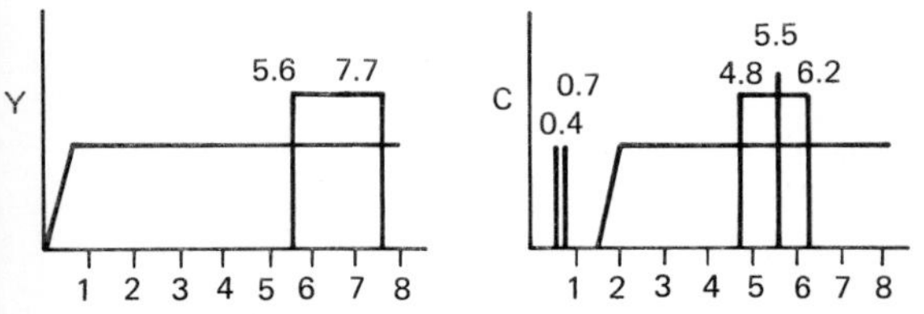

Figure 40.11 Type M II carrier frequency allocations

40.6.2.3. Other Type M equipment

A selection of production and post-production equipment is available to complement the camcorder equipment. It is normally expected that the editing of M II recordings will take place in the component form, but conversion of the component signal to composite form is possible at virtually any point in the system. Automatically loaded, multiple cassette machines are available to simplify the broadcast of short messages.

J T P Robinson
MVC Crow Ltd

41 Outside Broadcast Vehicles and Mobile Control Rooms

41.1 Evolution of OB vehicles and mobile control rooms

The ability to remove the microphone and the television camera from the studio environment and place them as witnesses in front of events as they happen has probably been one of the most significant steps in the development of television, and one that has provided some of its most dramatic moments.

This ability to televise outside events live was incorporated into the overall technical requirements when the foundations of the modern electronic scanning system were laid down in the 1930s by the EMI team of scientists and engineers led by Sir Isaac Schoenberg.

Plans were drawn up for a number of vehicles to be specifically constructed for this purpose. One vehicle housed the Emitron cameras and their associated control equipment together with simple vision and sound mixers with the synchronizing pulse generators. A second vehicle contained a transmitter for handling the video and audio signals from the 'scanner' vehicle and transmitting them on a carrier sufficiently far removed from the Alexandra Palace main transmitter frequency of 45 MHz (vision) to avoid mutual interference when both transmitters would be on the air simultaneously.

A Dennis fire brigade rescue ladder vehicle modified to take the transmitter aerial was provided so that the required height could be achieved for reaching the reception point in North London. A fourth vehicle was equipped with a diesel generator to provide the not inconsiderable power required for the scanner and transmitter vehicles when on site.

With this mobile fleet of equipment the complex technicalities of live outside broadcasts were mastered some 50 years ago. Their first notable occasion to witness was the coronation of King George VI and Queen Elizabeth in May 1937.

Since then, developments in camera tube technology have resulted in highly sensitive colour cameras of a size that enables them to be taken into areas and situations undreamt of before.

Similarly, the ability to record the video output of the camera for faithful playback at a later date has also expanded the outside broadcast horizon. The early 1960s saw the first Ampex machine using 2 in wide magnetic tape housed in a floor standing console of great weight. The present day helical scan machine is truly portable and has a performance that makes it difficult to distinguish between live material and a playback.

For on-site recording with an early quad head machine, a dedicated vehicle was required to house two machines together with their noisy air compressors. It is now common practice for one vehicle to house not only four or even six colour camera channels but also two helical scan vtrs.

Although perhaps the two most significant advances in television OB development have been in cameras and video recording, other areas where developments have been equally advanced but not so much in the public eye have been those that now enable broadcasters to achieve a programme subtlety coupled with technical reliability which was just not possible, even in the 1970s.

The advent of digital technology to vision mixers for special video effects and to the provision of on-board caption generators enables subtitles to take the place of caption cards perched on a makeshift easel placed in front of a hastily set up camera. The required captions can be typed in the comfort of the mobile control room prior to transmission, stored in a memory and called up as and when required. Furthermore they can be instantly changed to suit any unforeseen circumstances that may occur on transmission. This is just one example of how technology has influenced the evolution of OB vehicles.

Figure 41.1 Modern 16 tonne GVW chassis with TV OB body (MVC Crow Ltd)

The use of microprocessors in equipment such as synchronizing pulse generators, video measuring devices, test signal generators, video and audio switching matrixes and even for the co-ordination of facilities of large audio mixing consoles, has released broadcasting engineers from the two main problem areas which harassed their predecessors in the totally analogue era of television. These areas were equipment unreliability and signal parameter drift due to heat build up in valved equipment. In the 50s and 60s both of these accounted for a considerable expenditure of time and energy on the part of the operational staff engaged on OB work and called for a specific dedication in that field.

It is against this background and wealth of experience that the modern OB vehicle has evolved. It incorporates not only the technical excellence vital in today's competitive climate but equally important can house the operational staff in a comfortable working environment for many hours at a time, whether the vehicle is parked in Manchester, Moscow or Melbourne.

41.2 Vehicle design

The design of a modern OB vehicle depends on three key factors. These are:

- choice of chassis and overall dimensions,
- country of destination,
- extent of facilities.

Let us look at each of these factors in turn.

41.2.1 Choice of chassis and overall dimensions

The overall dimensions of a vehicle to travel legally on the roads of Europe are:

overall width (maximum)	2.50 m
overall height (maximum)	4.00 m
overall length (maximum)	
for a rigid chassis	11.00 m
for a combined tractor/trailer unit	15.50 m
for a draw bar trailer combination	18.00 m

There are also restrictions and relationships between wheelbase, rear overhang and axle weights, so a watch must be kept on the wheelbase measurements.

When considering the use of a particular chassis of whatever size for road use in any part of the world it is vital to ensure that the supplied unit with the bodywork will be able to be legally used on the road in the country of destination. It may well be worthwhile asking for the co-operation of the end customer or even hiring the services of a local consultant to make sure that the chosen chassis conforms in all respects to the road traffic laws of that country.

For instance, lighting regulations concerning lamp height, colour of lens, reflector positions and so on will vary considerably in different parts of the world, stemming a great deal from whether vehicles drive on the right or left hand side of the road.

Axle load limits also vary, with some countries legislating a weight limit on a rear axle, for example, that is well within the chassis manufacturer's maximum loading for the same axle.

Weight distribution within the frame of the vehicle must be borne in mind when choosing the make and model of the chassis. It is most important that the completed vehicle with all equipment, cables and camera accessories in position adopts a level attitude when on the road. It should not be nose down indicating excess front axle loading and equally it should not be tail heavy. It should also adopt an even lateral attitude and not list to either the offside or the nearside.

The centre of gravity of the vehicle should also be kept as low as possible by restricting overhead weight to essential items only and making full use of skirt lockers for housing compressor motors and similar heavy units.

It is thus necessary at the outset of the planning stage to know the weight not only of the technical hardware that is to be carried such as vtrs, mixers, camera control units, etc., but also of the air conditioning equipment, the cables and their connectors (including the permanent installed cables), batteries, chargers and so on.

Last but by no means least, the weight of the coachwork making up the body of the vehicle together with all its fittings and materials must be calculated and added to the technical equipment weight, to give the payload. It is this payload that will figure prominently in deciding upon which chassis the body is to be built.

Manufacturers rate their chassis carrying capacity in gross vehicle weight (GVW). This is the total weight allowed to bear upon the road surface which is in contact with the tyres. This figure therefore includes not only the payload mentioned above but the actual weight of the basic chassis as it left the factory. The tyres, springs and engine will be designed to support and propel that GVW with an appropriate safety factor.

For normal commercial use the all-up weight can be allowed to approach the GVW limit since the payload will vary from day to day or week to week depending upon the vagaries of the type of work involved. This variable duty will allow the suspension a degree of recovery since heavy abuse of the springs on one journey will be compensated by a lighter load or perhaps no load at all on the next journey.

However, a TV OB vehicle has a permanent load which, by the very nature of its role, will never or hardly ever vary. It is for this reason that care should be taken to make sure that the kerbside weight of the fully loaded vehicle is not much more that 80 per cent of the GVW. This will give a good handling characteristic to the driver and will allow some degree of liveliness from a medium sized engine in the range of chassis under consideration. It should also allow the suspension to still exercise its prime duty of shock absorption over bad road surfaces without bottoming.

For example, let us suppose that a rigid TV OB vehicle is required to house four colour cameras and two helical vtr machines complete with timebase correctors, and it is to be equipped to a high standard in terms of peripheral equipment. It is to be used in a Far East climate with seven operators, and is to be around 9 m in overall length. No on-board generator is required. The calculation is as follows:

Total weight of all electronic equipment, found from manufacturers' brochures, say: 1.5 tonnes (1500 kg)
Total weight of all interconnecting cables and connectors, say: 0.3 tonnes
Total weight of any cable and cable reels for external use, say: 0.5 tonnes
Total weight of air conditioning equipment, say: 0.5 tonnes
Total weight of body structure to house all the above, say: 5.0 tonnes

Adding up these five items gives the total payload as 7.8 tonnes which may be rounded up to 8 tonnes.

From chassis manufacturers' data we see that for a payload of 8 tonnes a chassis is required which is at least 13 tonnes GVW. However, the kerbside weight of this chassis without payload is 4.5 tonnes which for our example would give a margin of only 0.5 tonnes, arrived at by subtracting the all-up kerbside weight of the finished vehicle (12.5 tonnes) from the GVW (13 tonnes).

The resultant 96 per cent ratio of kerbside weight to GVW gives too low a safety margin for a TV mobile, and a higher

GVW chassis rating will be necessary to meet the criteria outlined earlier. To complete the illustration, a 16 tonne GVW chassis will have a chassis weight of 5 tonnes but its payload capacity is nearly 11 tonnes. Therefore if the example of body payload capacity of 8 tonnes given earlier is now added to the 5 tonnes of the basic chassis, a 13 tonne all up kerbside weight will result. This gives a kerbside weight/GVW ratio of 81 per cent which, coupled with a medium sized engine in this particular model range, would fulfill the necessary requirements.

Mention should be made here of the popularity of the panel van as a variant to the coachbuilt specialist body. Several major chassis manufacturers market a range of small panel vans which have proved popular with many broadcasters where the larger purpose built vehicle is not required and where the greater manoeuvrability and reduced parking demands are particularly important such as when a considerable amount of TV work in a city is necessary.

Figure 41.2 Typical TV OB vehicle based on panel van (MVC Crow Ltd)

Such panel vans are generally 6 m in overall length, have a high roof to give an internal height of 2 m and have a maximum GVW of 6 tonnes. The kerbside weight of the empty van is usually about 3 tonnes leaving a maximum of approximately 2 tonnes of payload if a figure of 80 per cent is used as the ratio of kerbside weight/GVW.

While 2 tonnes of payload may not seem a great deal, the advantage is that very little weight is added in the coachbuilding/conversion stage since the external shell already exists and is not included as part of the payload. If this factor is coupled with a sensible choice of equipment in terms of cameras, vtrs, etc., a very versatile OB unit can be built for a relatively modest cost, proved by the number of such vehicles giving excellent service in many parts of the world.

Before leaving the subject of vehicle body options, a development of interest to broadcasters and others has been the demountable body. This system allows the coachbuilt bodywork with all its technical equipment to be disengaged and raised above the chassis by means of inbuilt hydraulic or mechanical jacks placed at each corner of the body, allowing the chassis to be driven away from underneath. The chassis-less body is then lowered to near ground level to become a static studio control room.

This concept is of interest for the following reasons:

1 The unit can be taken to a site, demounted and left as a fully operational unit for days or weeks on end where continuous long term coverage of events is required. This system has even been used in a naval capacity where the demounted body has been slung aboard ship and taken to sea.
2 If two or more units are operated, then only one chassis with engine and driver is needed to deploy them.
3 For detailed chassis maintenance and inspection purposes, removal of the chassis away from the body creates greater access.

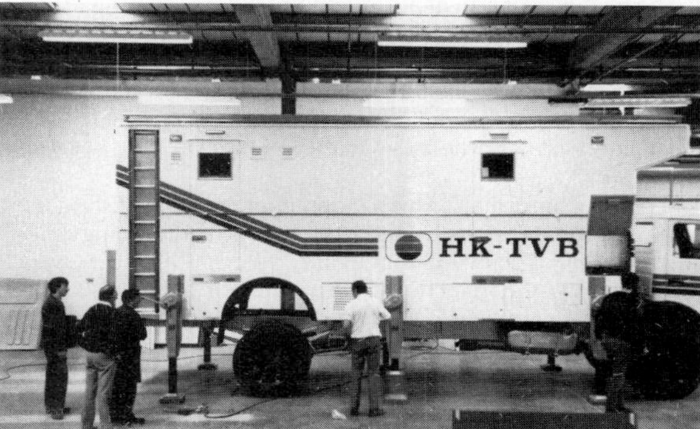

Figure 41.3 Demountable TV OB body. The body is raised from the chassis and the chassis driven out from underneath (MVC Crow Ltd)

4 Only the bodywork need be constructed at the factory. The chassis may be locally supplied by the end user and united with the body upon its arrival. This could have important tax and import duty advantages to the end user.
5 Where chassis delivery is protracted, the bodywork may be constructed independently thus cutting down on the overall delivery time.

41.2.2 Country of destination

The air conditioning requirements of the vehicle will be wholly determined by the climate in which it is to operate. The hotter the climate, the more on-board air conditioning equipment will be necessary for a given number of operators and for technical heat dissipation. The construction of the vehicle will affect the heat retention or loss.

However, it must be borne in mind that the greater the number of air conditioning units that are carried on a vehicle, the greater the payload penalty. Since one air conditioning unit having an output of 6000 Kcal/h (24 000 Btu/h) will weigh about 100 kg and with the typical number of such units for a larger vehicle being at least four then a weight penalty of between 400 and 500 kg can be expected.

To assess the necessary size of the air conditioning system, the following information is required to perform the heat calculations:

1 The structural heat gain of the bodywork. The structural materials and construction of the walls, roof and floor must be identified and allocated a U value.
2 The internal dimensions and areas of the vehicle where air conditioning is to be applied.
3 The setting of an acceptable interior temperature and humidity based upon the country of destination.
4 The heat dissipation of all equipment within the vehicle. This will be based mainly upon the technical equipment power consumption.
5 The occupancy level of the vehicle as an operational average.
6 The interior lighting heat dissipation.
7 The inclusion or otherwise of a roof platform.

From this information will emerge an overall sensible heat gain which has then to be matched to an air conditioning unit or units capable of rejecting that figure.

In practice the rejection figure will normally exceed the sensible heat gain figure by around 20 per cent in order to take care of latent heat gains due to deliberate external fresh air intake and heat gain from the opening of exterior doors, etc., during operation.

Item 7 (roof platform) in the above list is an example where occasionally two factors come together to assist each other rather than oppose. A roof platform may be called for in the specification to provide a high level vantage point for cameras, commentators or even for a microwave starter link. By spacing the platform off the actual roof of the vehicle by about 50 mm and allowing a free flow of air to circulate beneath it, a very real assistance is given to the overall efficiency of the air conditioning system since the platform then acts as a solar shield.

Having selected a suitable air conditioner to handle the interior heat gain, note should be taken of both its start-up current and normal running current for future calculations on required power intake to the vehicle. A useful guide is to note that generally the start current, lasting about one second, is about 3–4 times the normal run current. Because of these high peak currents upon switch on of the compressor motor, the practice is now to allow the motor to continue to run for as long as the main power is switched on. If this were not so, then continued heavy short duration pulses of intake current as the motors started up and stopped each time in response to the environmental demands would reflect back upon the vehicle technical power circuits, causing equipment supply voltage variations at the least.

Thus an alternative method has been developed which varies the cooling demand while allowing the compressor motor to run continuously. This is the *hot gas bypass system*.

In this system a solenoid valve is placed across the vehicle interior evaporator coil and external compressor and is opened by the interior thermostat when the required vehicle interior temperature has been achieved. This allows the hot gas from the compressor to be routed direct to the evaporator rather than going through the cooling process of the external condenser coil and thus allows the evaporator temperature and hence the interior temperature to rise, in turn creating a demand for cooling. The solenoid valve is then closed by the thermostat and hot gas is routed to the condenser coil, cooled, liquified and passed through the evaporator coil in the normal way where it absorbs heat from and therefore cools the vehicle interior.

This cycle continues with the compressor motor running continuously. In the hot gas bypass mode it is normal for the total cooling capacity of the system to be reduced by more than 60 per cent which, if the system has been sized correctly, will allow the interior temperature to build up and cause the cooling demand requirement to re-occur. It is therefore most important that the equipment selected for this cooling process is not excessively oversized. If there is over specification on the air conditioning, this cycle of allowing the interior heat build up will not occur and the complete system will eventually ice up and malfunction.

Where several air conditioners are installed in a vehicle, it is advisable for a logic system to control their switch on when power is first applied. Time delays will then ensure that only one compressor motor starts at a time, the next motor in the chain being delayed in starting until the first motor has established itself in a functional state. This will minimize the overall primary current in-rush on switch on.

Common use is made of the split system of configuration for deployment of air conditioners. Here the evaporator coil with its fan is physically separated from the condenser coil and compressor, the units being connected only by rigid pipework or, more usually for vehicles, a flexible pressure hose having a low effusion rate or loss for the refrigerant to be used (normally Freon, R22).

The advantage of a split system for vehicles is that the evaporator unit can be sited within areas inside the vehicle where it is able to perform efficiently in terms of providing conditioned air flow. The compressor/condenser unit, on the other hand, requires access to an external air flow for cooling purposes and in any case is a source of noise and vibration. It is therefore normally located external to the operational shell, usually in a side skirt locker, and suitably treated to reduce both of these factors.

To ensure efficient operation of the air conditioning system, all filters and radiator matrixes must be kept clean and free from a build up of dust. Once a system has been charged and commissioned under operational conditions there is little else requiring attention, but as with all things mechanical, regular inspection of the system should be scheduled at specified intervals.

It is important to consider the way that the conditioned air is delivered into the operational areas. A direct flow of cold air from the evaporator unit via front, side or overhead outlets is not to be recommended for operational areas as the occupants will raise objections to the resultant draughts at ear, eye and top of the head levels!

A more sophisticated approach is to deliver the conditioned air overhead but via a perforated ceiling. This ensures a very much more indirect delivery and satisfies the environmental criterion of moving a large mass of cold air relatively slowly.

Just as important as air delivery is air return. Any air delivered into an area must eventually find its way back to the evaporator otherwise the system will be starved of air circulation. It is also a prerequisite that a portion of fresh air is mixed with the circulating air to prevent stale air building up inside the vehicle.

Unless there are items of equipment on board which are specifically temperature conscious, the normal practice for interior air circulation is to first deliver the air to the operational areas and then take the return air via the electronic equipment, usually housed in 19 in (482 mm) equipment racks. This ensures that the slightly warmer air, or more accurately the less cold air, than that delivered directly from the evaporator unit is passed through the equipment last and helps to prevent undesirable condensation forming on the equipment. Thus the priority in terms of environment is that of treating the operational areas first and the technical equipment second. Not all that long ago the reverse would have been the case; this reflects the way that component manufacturing techniques have eliminated the heat problems of the 1950s and 1960s.

Hot daytime temperatures are often accompanied by low night temperatures and it is equally important to ensure that when the outside temperature falls below an acceptable minimum there is sufficient heat capability on board the vehicle to provide human comfort. This can be achieved by incorporating heater elements within the air conditioning units on the premise that heating and cooling will not be required simultaneously. However, this would probably result in heated air being discharged at roof level, perhaps by means of a perforated ceiling. This is contrary to natural thermal circulation and not conducive to a healthy environment.

A more acceptable alternative is to place electric fan heaters at low level within the vehicle or to use a natural fuelled system of warm air heating. Such units will run on a variety of fuels such as diesel, gas-oil, propane, etc., and will discharge the resultant warm air through purpose built ducting within the vehicle to low level outlets placed in the operational areas.

Since they can be operated from low voltage dc batteries, they are independent of mains electrical power and can therefore operate when such power is not available on site. If the main vehicle engine is a diesel, then quite often the warm air heater is chosen to be diesel operated and will take its fuel from the vehicle tank. By siting the take-off feed near the half capacity point, the heater will shut off when that point is reached, leaving enough fuel in the tank for the vehicle to reach its fuelling point safely.

For very cold climates, recourse is sometimes made to a centrally heated hot water system of distribution within large vehicles, the heated water circulating to domestic radiators placed in the operational areas.

41.2.3 Extent of facilities

The extent of the technical facilities placed on board the vehicle will dictate the overall size of that vehicle. The greater the number of cameras and vtrs, and the larger the sound desk, the more operators and engineering staff will be required demanding in turn a greater air conditioning capacity and a heftier GVW of the vehicle.

It is therefore important that sight is not lost of the original requirements during the constructional phase. It is very easy to add items of equipment to the system and even to alter the mode of the vehicle during the building such that the margins hopefully allowed for during the planning stage are steadily eroded away. In extreme cases, it is not unknown for an extra load bearing axle to have to be added at a late stage in construction with consequent disruption of the delivery time scale and endless arguments as to how the situation arose.

A useful way of monitoring the weight of the vehicle as construction proceeds is to place a load bearing pad under each wheel which gives a readout in kilogrammes. This will not only give an overall picture of axle loading as a percentage of the maximum but also indicate the lateral or side to side weight loading.

41.3 Outline of constructional techniques for TV mobiles

Whereas mobile OB units built in the 1950s used a considerable amount of hardwood for their framework following the trend of most commercial vehicle box bodies of that period, the 60s saw the move to aluminium for both framing and skinning of specialist vehicles, bringing with it a much needed reduction in body weight.

Figure 41.4 Aluminium framework prior to cladding (MVC Crow Ltd)

This trend has continued to the present day, with the only other real departure of note being the occasional use of a glass fibre or GRP sandwich construction for large expanses of side wall, supplemented by conventional construction for underskirt lockers, main entry doors and other similar areas.

The construction of an OB mobile starts off on the drawing board in the proposal stage, where consideration must be given to operational ergonomics, air conditioning requirements and technical equipment housing based upon a known chassis. If this proposal is accepted then more detailed dimensional drawings of vehicle layout plan and elevations are made. From these the various extrusion types and lengths, aluminium panelling and indeed all the many hundreds of items which go to make up the finished vehicle are listed. These comprise the material schedules which enable orders to be placed with suppliers. Since the timescale for the construction of a large TV mobile from placement of order to finished vehicle may be between four and five months, it is necessary to keep track of the various stages of construction for the benefit of the customer as well as the builder. To this end, use is made of the familiar bar chart, where the x axis represents the time element, usually calibrated in week numbers, and the y axis the 25 or so benchmarks of the construction from chassis modifications through to painting and signwriting.

The skeletal framework of aluminium is built upon aluminium transverse cross bearers, these in turn being bolted to the aluminium bearers running the length of the chassis and coincident with the vehicle chassis main steel members.

To provide a degree of 'give' between the steelwork of the chassis and the coachbuilt body a semi-resilient packer is sandwiched between the two. This 'give' factor is very important for specialist vehicles since it allows the various sections of the body to maintain their positions relative to each other as the vehicle settles down over its life.

It is important to choose construction materials that are not going to be problematical in tropical climates. Any hardwood used must be treated against attack by insects and a humid atmosphere whilst fabrics should be wholly man-made and not derived from natural sources. Corrosion of metal fittings and screws used on external surfaces is best prevented by the use of stainless steel wherever possible. Otherwise heavy chrome on brass will resist attack provided that the quality of plating is good.

When the detailed drawing is complete, the framework is in effect built from inside to out.

Thin aluminium sheet or exterior grade 6 mm plywood is fixed via a thermal break to the inside of the aluminium

Figure 41.5 Interior cladding in progress (MVC Crow Ltd)

framework with a 50 mm build up of very light but high thermal insulation material placed so that every cavity between the inner and the as yet absent outer aluminium skins will be completely filled. An alternative technique is to use a polyurethane foam sprayed into the cavity. Whichever method is used it is important to ensure that the result gives as high a U factor as possible to the body shell.

Similar treatment must be applied for the same reasons to the roof and the floor, but within these areas provision must be made for cable ducts in the floor and for delivery of conditioned air within the ceiling. If the preferred method of air delivery is via a perforated ceiling then a suitably sized duct must be formed immediately below the insulated roof, faced on its lower level by perforated sheet. Such sheet will be of aluminium and should be treated with some form of anti-condensate finish. To avoid air turbulence and hence noise within the ducts — there may of course be more than one such duct within the ceiling void, depending upon the air conditioning specification — they must be lined with a non-toxic polyurethane sheet foam.

Prior to the external skinning of the vehicle it will be necessary to provide electrical conduits within the walls for all mains services for general and operational lighting switches, power outlet points and also for the various dc services such as skirt locker lighting, emergency lighting, compressors for any pneumatically operated telescopic masts, fire, smoke and other warning systems and so on.

By now a roof platform, if called for, will have been constructed off line and then fixed to the roof to give a 50 mm air gap as described earlier. Equipment racks, partitions and console will have been constructed, and the air conditioning equipment will have been installed and internal finishes will have started. To give an acceptable aesthetic finish and at the same time provide a reasonably warm acoustic feel to the interior, a high quality man-made carpet is applied to the interior wall. The desk surfaces will have been treated with one of the many laminate finishes now available and apertures cut to accept the various sizes and shapes of the equipment to be positioned within them, edged with either a hardwood or an aluminium trim. Hardwood or soft cushioned edging will be fixed to the fronts of the various desks, and decisions on the type and positioning of fire extinguishers and first aid boxes will signal the end stages of coachbuilding.

The interior should be vacuumed out and the floor covering protected for the next stage of the construction. This will be the technical installation of all necessary cables and equipment in accordance with the schedules which will have been prepared in parallel with the coachwork construction.

Hopefully, the end of the constructional phase will coincide with the start of the installation phase, ensuring a gradual change of emphasis in the work content from mechanical construction to technical installation.

After this phase and preferably before commissioning and acceptance trials, the final stage in the coachbuilt activity is painting and signwriting. Sometime during the building phase, and forming one of the many bars on the bar chart, will be the need to obtain the customer's requirement for the paint finish in terms of colour and coachline positions as well as for signwriting of the station identification lettering and logos. Also, such things as tyre pressures, fuel identification, paint specification, vehicle length, height and width must all be determined and enscribed on or in the vehicle as appropriate.

If coachbuilding only is the contractual requirement, then it is necessary to drive the vehicle to a public weighbridge and take front and rear axle loadings plus overall loading. The weighbridge tickets should then be handed over with the vehicle to the customer for his own processing of the necessary documentation to register the vehicle for road use.

If the contract is a turnkey whereby both coachbuilding and the technical installation are carried out by the one contractor then weighing should be made of the finished coachbuilt vehicle before the technical installation and again at the point prior to delivery. This will provide a useful record for future use, but as pointed out earlier, if continual weight monitoring can be integrated into the constructional phase by the use of load pads then this is by far the best way to avoid embarrassing last minute weight problems.

I G Aizlewood
Managing Director, Continental Microwave Ltd

42 Microwave Links for OB and ENG

In fewer than 30 years, the outside broadcast link has evolved from three suitcases of electronics per terminal (weighing more than 100 kg, requiring power supplies of 1 kV or more and producing only mediocre monochrome performance) to a package easily carried by one man and with colour transmission capability rivalling some test equipment.

Frequency synthesis, now almost universally adopted, has eliminated the problems of arriving on site with the anticipation of a day's work to conquer the twin unknowns of aligning transmit/receive antennas whilst unsure of even maintaining a common frequency for the transmitter and receiver.

Here is one technology that has not only kept pace with market needs, but actually generated new programme production opportunities. With special applications transmitters now capable of being comfortably accommodated in a jacket pocket and powered from a few Nicad cells, no situation is safe from live transmission; no one can deny the emotive impact of live pictures from a formula 1 racing car or whilst 'accompanying' a parachutist departing from his aircraft.

The enormous spectrum of OB transmission equipment now available prevents detailed examination here, but we can explore the most popular current techniques in generation and reception of microwave power and evaluate some approaches adopted by manufacturers to meet the different priorities of point-to-point portables, ENG and special application links.

The aim is to present a practical guide to system and subsystem choice, discuss the relative merits of different technical approaches and look at a few performance calculations.

42.1 System concepts

42.1.1 Transmitters

Siting of the transmitter is largely dictated by location of the event to be televised and is therefore substantially outside the operator's choice. Consequently, the transmit terminal generally makes greater demands of the designer. Let us examine the main options available to the development engineer.

Two broad approaches are common: direct modulation and heterodyne.

42.1.1.1 Direct modulation

The generation of a uhf signal and its modulation and multiplication to the required output frequency is a technique generally termed *direct modulation*. Direct modulation has advantages of simplicity, small physical size and low cost.

The advent of modulatable synthesizers has eliminated a major difficulty in maintaining effective frequency stability while applying video modulation to the master oscillator, but there still remain significant disadvantages to this method of power generation, notably in multi-hop operation which is now often demanded of portable point-to-point equipment.

FM modulation is applied to a relatively low frequency oscillator and its modulated output subsequently multiplied to the desired final frequency. Deviation is therefore effectively multiplied by the same factor as the oscillator frequency. CCIR standards set the final deviation (usually 8 MHz peak-to-peak at the baseband crossover frequency), and it follows that deviation at the modulation frequency must be reduced by a factor F/N where N is the multiplier and F is the final frequency.

Unfortunately the dominant system noise emanates from, or is prior to, modulation and is therefore multiplied with the deviation. Ultimate noise performance is thereby limited, and the technique becomes less practical as the operating frequency band increases.

Direct modulation has a further disadvantage. When line of sight is not available for the required path, a repeater is necessary. Transferring the signal from a receiver to following transmitter without demodulation at a repeater is very attractive as non-linear distortions and noise associated with the demodulation and remodulation process are eliminated.

A receiver will generally have available a suitable if output to feed the ongoing circuit (usually 70 MHz), but compatability with direct modulation transmitters is not possible as no matching frequency is available to access for injection. The modulated master oscillator frequency and subsequent multipliers are determined by the designated transmission frequency and hence different in every case.

Recovery of at least composite video is therefore necessary at each intermediate station, followed by a repeat modulation process — giving rise to the term *remodulating system*.

Manufacturers' descriptions should be treated with caution when they infer if repetition but actually only achieve acceptance of an if input into the ongoing transmitter by incorporating a demodulator integral to it. Such transmitters may apparently offer an if input capability but confer none of the advantages of true if repetition.

Examination of the relative merits and deficiencies of remodulating systems suggests that they are best suited to miniature, special event, ENG and 'starter' situations.

42.1.1.2 *Heterodyne*

A more 'purist' solution is offered by the up-conversion or heterodyne philosophy. Here a separate video modulator, usually running at 70 MHz, is mixed with a suitable shf pump frequency so that one sideband provides the required shf channel frequency. FM deviation present at the modulator will be directly translated to the output frequency.

FM noise is inherently easier to control with this system concept as the pump generator largely responsible for generation of the microwave frequency is unmodulated and can have a narrow loop bandwidth. In addition, deviation is unaffected by the mixing process, and modulator noise contribution is hence not magnified (as with subsequent multiplication in the direct modulation system).

Perhaps the greatest advantage is yielded by the constancy of a 70 MHz modulator signal irrespective of the required shf output frequency, which can be independently adjusted by changing the pump frequency. Availability of a 70 MHz interface in the transmitter permits the local modulator to be replaced by a compatible signal derived from the if of a preceding receiver when required, so true non-demodulating repetition is now possible.

On the surface then, this transmission concept eliminates all the shortcomings of the direct modulation method for a modest increase in cost and physical size acceptable in point-to-point portable applications. However, a new disadvantage now arises which greatly limits operational versatility of the single conversion concept.

In realizing the shf frequency, the pump frequency and unwanted sideband will also be present at the mixer output, only displaced by 70 MHz and 140 MHz respectively from the wanted signal.

To remove the unwanted components, there exists the need for an output filter with very high rejection only 70 MHz from the wanted sideband, but with a passband of at least 20 MHz to pass the fm modulation. While this is realizable, it restricts transmitter operation to a single shf channel without filter change, a major deficiency in congested operating environments with multiple co-located systems and high rfi.

The ultimate solution is achieved by double up-conversion as shown in *Figure 42.1*.

The 70 MHz first if is retained to provide if inject facilities at a repeater, but the first pump oscillator translates this to a

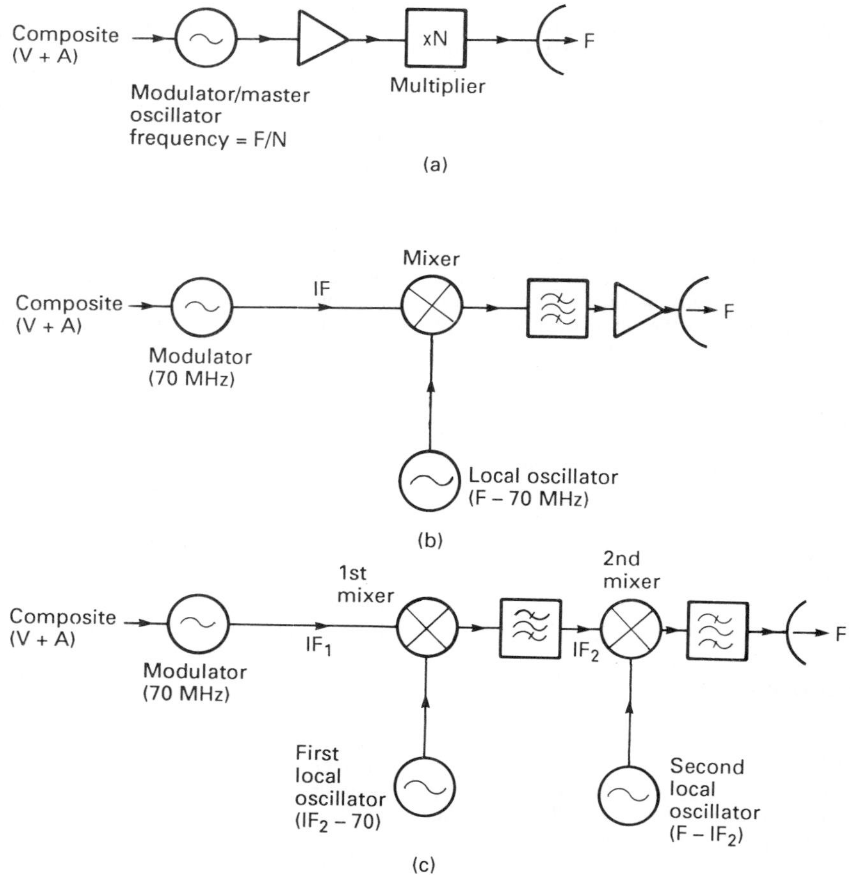

Figure 42.1 Simplified schematics of the transmission concepts: (a) remodulating, (b) single conversion heterodyne, (c) dual conversion heterodyne

second intermediate frequency rather than to the final shf channel. (Frequencies between 300 MHz and 1.5 GHz are popular.) The requirement remains for a filter at the second if frequency capable of rejecting the uhf pump frequency while not distorting the modulation band, but as the ratio of bandwidth to absolute frequency is now much greater, and since the second if is a fixed frequency, this filter is easy to realize and its presence not restrictive to final system frequency agility.

A second up-conversion then provides a modulated sideband at the required shf channel. Rejection of the pump frequency is again required, but this time filter parameters are less stringent as pump offset will be significantly greater because of the higher if. Indeed, it is quite possible to tune the second local oscillator over bandwidths up to 80 per cent of the uhf local oscillator frequency without the pump signal encroaching as a spurious output.

So now we have an elegant solution offering frequency agility, true if repetition and excellent noise performance.

42.1.2 Receivers

Single and double conversion solutions are common (*Figure 42.2*). Down-conversion direct to 70 MHz offers economy of design but, with an image frequency only 140 MHz removed from the wanted signal, precludes operation without a relatively narrow band input filter. It renders frequency agility minimal (a similar situation to single up-conversion transmitters).

Use of a higher if frequency would extend frequency agility of the receiver but at the expense of losing a 70 MHz output for if repetition.

The most versatile concept again therefore becomes the double superhet. Double down-conversion follows exactly the concept of the double up-conversion transmitter in reverse. Indeed, with at least one well-known manufacturer, the same filters, mixers, uhf and shf local oscillators are common to transmit and receive terminals — a very significant practical advantage to spares holdings and servicing.

42.2 Multiplexing

Operation of two or more independent links from a common antenna is now prevalent. Dual channels may be required for standby or for two separate vision channels, whilst a reverse channel for editing is also popular.

Several methods of *multiplexing* (*duplexing* or *diplexing*) multiple rf signals onto one antenna have been explored by manufacturers. The most common are: filter/circulator, hybrid and bipolar.

42.2.1 Filter/circulator multiplex

Filter/circulator multiplexing is traditional (derived from fixed system philosophy). It has one valuable advantage of positively protecting receivers from rf interference and a more dubious one of assumed low transmission loss.

However, this method also drastically restricts frequency agility, thus destroying the most popular feature of wide-band tuning now available on most modern links.

In basic form, filter/circulator multiplexing is shown in the block schematic of *Figure 42.3*. The filter is chosen to have a pass-band for T_1.

T_1 output passes through the filter influenced only by its insertion loss and any minor distortions dictated by limitations of filter bandwidth. Generally, a group delay equalized filter with bandwidth 28 MHz to the -1 dB points will not significantly affect a (video +4 audio) modulation to CCIR Rec. 405 (B). It is perfectly acceptable to increase the bandwidth of this filter to accommodate some variation in T_1 frequency, but at the expense of minimum channel spacing between T_1 and T_2.

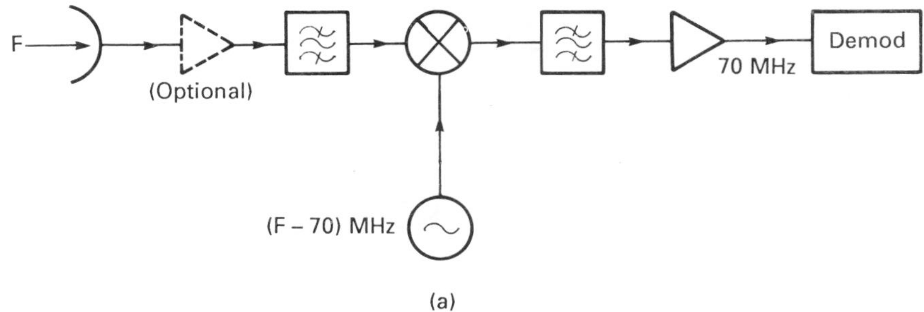

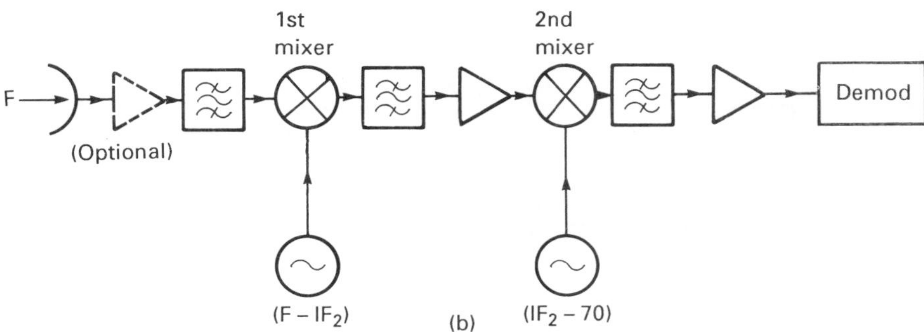

Figure 42.2 Simplified schematics of the reception concepts: (a) single conversion heterodyne, (b) dual conversion heterodyne

The reason for this becomes clear if the T_2 transmission path is examined. T_2 enters the circulator and passes clockwise to the next available port, which coincides with T_1 entry via F_1.

T_2 signal tries to exit from this port, but is reflected by F_1 filter (which will appear as a mismatch at F_2) and re-enters the circulator to emerge again at the matched antenna port. Undistorted T_1 signal flow relies on F_1 appearing as a true short-circuit to F_2 frequency (including its full modulation bandwidth). However, in practice T_1 and T_2 frequencies may not be displaced adequately to guarantee uniform reflection for F_2 modulated bandwidth from the filter F_1. In this case F_2 will reflect from the skirt of filter F_1 causing progressive phase distortion as the fm signal traverses the slope and encounters a varying return loss and phase effect.

Increasing the skirt slope of F_1 filter by adding sections will improve distortion on the bounced channel, but only at the expense of insertion loss to the T_1 transmission path.

A typical filter would employ five sections with 0.8–1.2 dB insertion loss. Increasing the filter to six sections would steepen the skirt slope but typically add 0.5 dB loss.

It will be noted from *Figure 42.3*(a) that the simplest multiplex circuit as illustrated is 'handed', i.e. it would not be possible to swap the receiver multiplexer to the transmitter terminal (or vice versa), without physically changing the filter to another circulator port.

In practice this would represent a severe operational limitation, and multiplexers therefore typically include filters in both ports, as shown in *Figure 42.4*. Positive interference rejection is given to both receivers by this arrangement and, when required, the same equipment complement can of course be operated bidirectionally. However, multiplex insertion loss on a full hop now rises to a practical minimum of two filters at, say, 1.0 dB each plus three circulator passes at 0.3 dB each, giving a total of 2.9 dB. These are the minimum likely losses and may

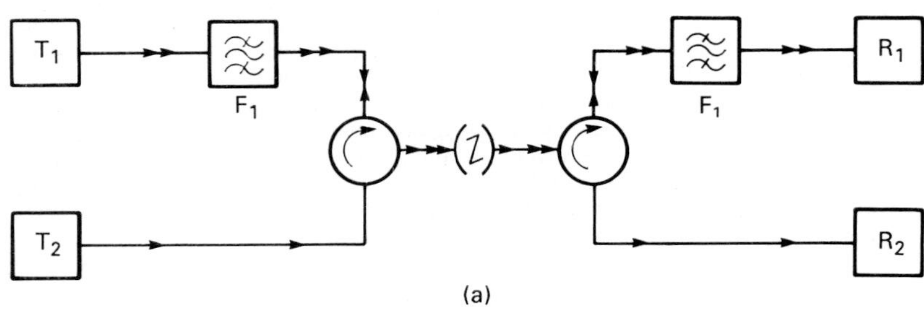

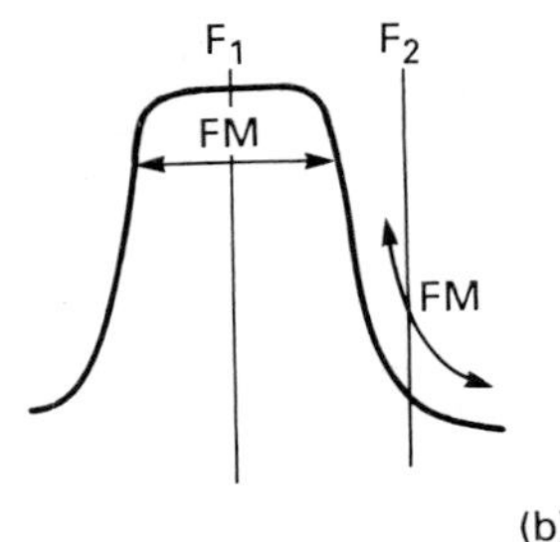

Figure 42.3 Filter/circulator multiplex schematic (a), showing the effect as the fm signal traverses the slope of the filter (b)

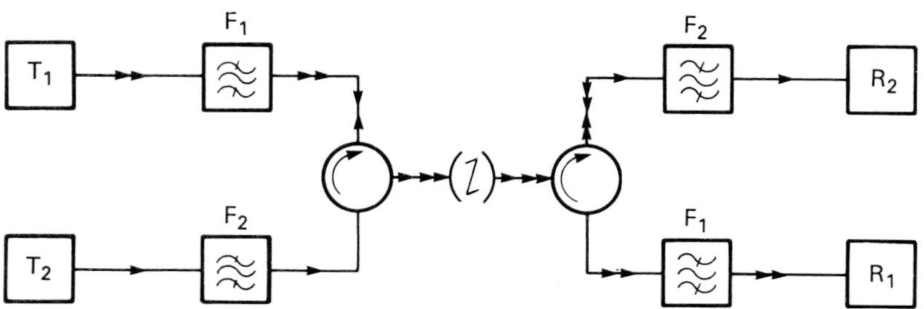

Figure 42.4 Schematic of multiplexes with filters in both ports

not be realizable in practice due, for example, to mechanical constraints on layout, waveguide components, etc.

So it will be seen that the presumed advantage of low loss in filter/circulator multiplex is not in practice very significant when compared with other methods.

In an attempt to regain some limited frequency agility, two 'variations' of filter/circulator multiplex are possible:

- Using the same multiplex arrangement, filter bandwidths are increased from one channel of, say, 30 MHz to encompass a wide sub-band. As filter bandwidth increases, skirts become proportionally more shallow for any given number of filter elements, so it is not possible to divide the available band directly into two usable sections without a 'protection' bandwidth of typically 15 per cent mid-band.
- Alternatively multi-channel combining arrangements are quite practical (subject to size constraints) and can be constructed with a mixture of channel and sub-band elements tailored to offer the best user versatility for a particular environment and available operating frequencies. Such multiplexers are reversible between transmit and receive, so any combination of go/return traffic can be accommodated.

42.2.2 Hybrid multiplex

Provided the receivers in use are of double conversion superhet design and may therefore be operated in reasonable rf interference environments without external channel filters (or alternatively, if the receivers are fitted with integral channel filters), hybrid combining and separation provides a cheap, wide-band and physically small solution.

There will of course be a minimum theoretical loss over a full path comprising two such multiplexers of 6 dB with this method (typically 7 dB). However, when judged against the practical performance achieved by the filter circulator multiplex solution, the additional loss is unlikely to jeopardize significantly the overall system noise except at extreme range.

Hybrid multiplexing may not offer inter-port isolation and would therefore be unsuitable for duplex (bidirectional) operation unless receivers are separately protected by filters to prevent front end damage from the high incident power of the adjacent transmitter.

42.2.3 Bipolar multiplex

Strictly, bipolar systems are not multiplexed at all. They operate from dual feeds which happen to share a common reflector and therefore behave as two single links. Bipolar advantages and disadvantages fall some way between filter/circulator and hybrid options.

The bipolar approach offers wide-band operation without any of the losses inherent in the 3 dB hybrid or filter/circulator multiplex solutions. Price is not dissimilar overall to a hybrid multiplexer, and without filter constraints no multiplexing distortions arise.

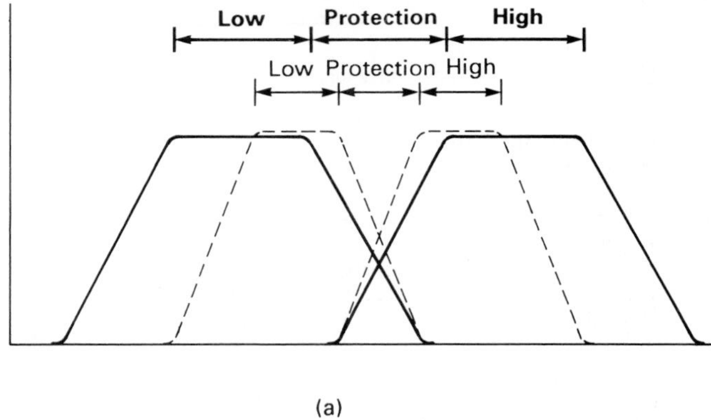

(a)

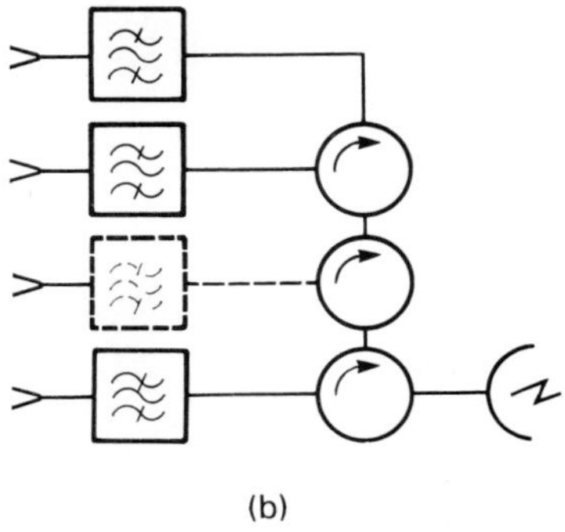

(b)

Figure 42.5 Variations of filter/circulator multiplex: (a) bandwidths increased to encompass a wide sub-band, (b) multi-channel combining arrangements

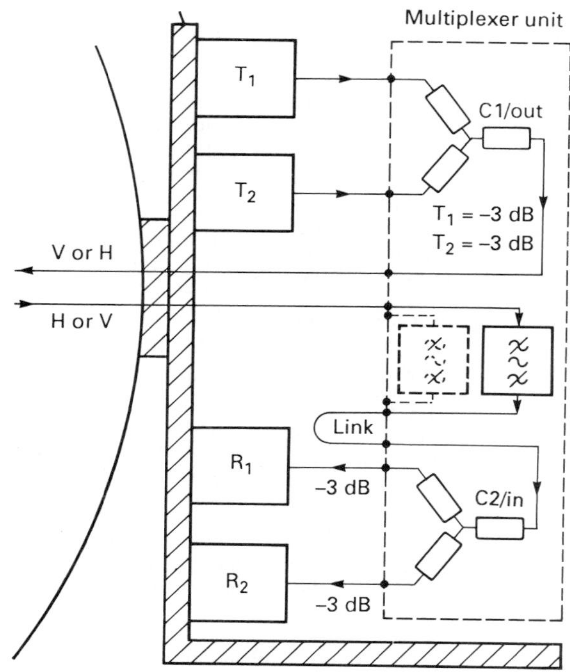

Figure 42.6 Hybrid/filter/bipolar multiplexer (arranged for 2x bidirectional channels)

However, bipolar operation is only conditionally able to support bidirectional operation without filter protection as antenna inter-port isolation will be limited even in the best designs to around 50 dB.

In practice, as system frequency increases, achievable transmission output power reduces, and the physical elements of the feed become smaller so that higher inter-port isolations are more realizable within the feed. The net effect of these parameters is to progressively reduce bleed energy from a transmitter into its adjacent receiver and to render bidirectional bipolar operation more practical in the higher frequency bands.

Type	*Advantages*	*Disadvantages*
Hybrid	Wideband Compact Cheap Close adjacent channel use without distortion	High loss Diplex only unless used in association with filters or bipolar antenna
Bipolar	Wideband No multiplexing loss No additional space required (integral to feed) Relatively cheap Close adjacent channel use without distortion	Conditional diplex operation unless receivers are filter protected
Filter/ Circulator	Can be designed for any combination of multiple or two way traffic with minimal interference risk	Restricted bandwidth operation relatively high cost Less compact Distortion when used with very close adjacent channels

Table 42.1 Advantages and disadvantages of multiplexing methods

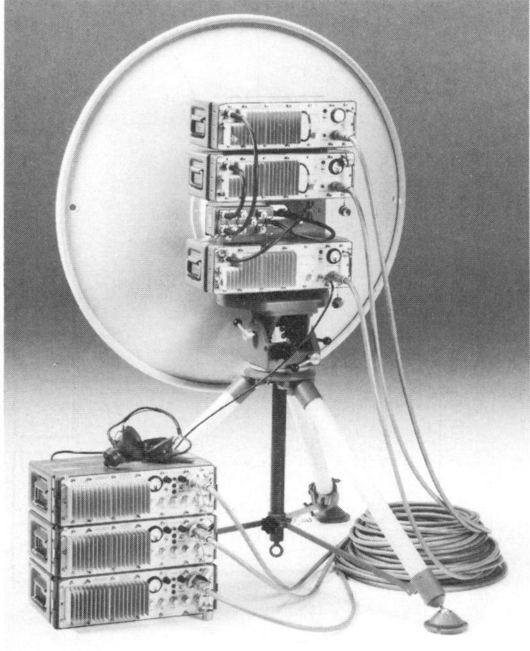

Figure 42.7 Three channel mobile with remote control; hybrid multiplex and bipolar feed, and 1.1 m antenna

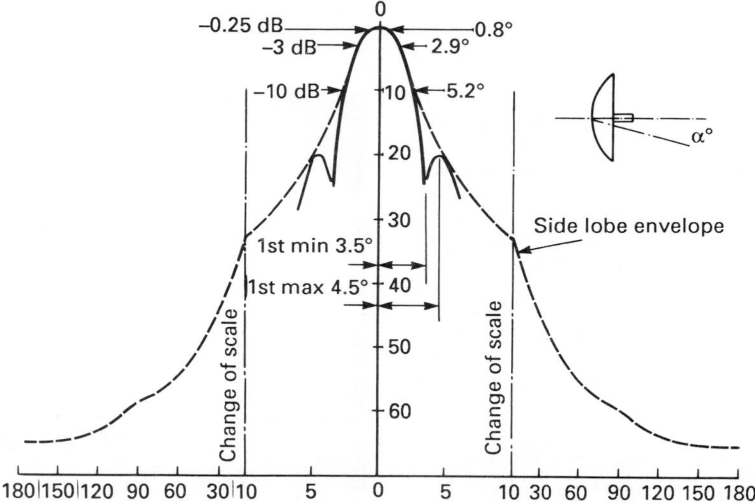

Figure 42.8 7 GHz 1.1 m antenna polar diagram

Antenna size ft	m	*Gain (dB) at frequencies 2-23(GHz)* 2	2.5	3.5	5.5	7	8.5	10	13	14.5	23
2	0.6	18.7	20.6	23.6	27.6	29.6	31.3	32.8	35.0	35.9	39.8
4	1.2	24.7	26.6	29.6	33.6	35.6	37.3	38.8	41.0	41.9	45.9
6	1.8	28.2	30.2	33.2	37.0	39.1	40.8	42.2	44.4	45.4	–
(8)	(2.4)	30.7	32.6	35.6	39.6	41.6	43.3	44.8	47.0	47.9	–
(10)	(3.0)	32.8	34.6	37.6	41.6	43.6	45.2	46.7	49.0	49.9	–
(12)	(3.7)	34.2	36.2	39.2	43.0	45.1	46.8	48.2	50.4	–	–

The frequencies chosen represent commonly used OB bands.
For portable equipment an illumination efficiency of 45 per cent has been assumed.
All performance is 'typical' and will vary slightly with antenna bandwidth, method of illumination, mechanical configuration etc.
Sizes in parentheses are given for information only and are not generally appropriate to portable operation.

Table 42.2 Parabolic antenna sizes versus gain for various frequency bands

Bidirectional bipolar operation without receiver filter protection always has an element of risk as antenna inter-port isolations can be significantly modified by the presence of local reflecting surfaces such as safety barriers or tower legs. Careful positioning of systems with respect to local obstructions is important in optimizing system performance.

Some advantages and disadvantages of the various multiplexing options discussed are summarized in Table *42.1*.

In practice, combinations of the options discussed above can provide highly versatile packages which may be configured quickly on site for multiple transmission requirements. An example of a hybrid/filter/bipolar combination is shown schematically in *Figure 42.6* and a practical realization of such a system in *Figure 42.7*.

42.3 Antennas

42.3.1 Point-to-point

This category of link is required to operate over very long ranges (typically in excess of 50 km). It may well be co-sited with several other systems at starter or repeater sites when a

Useful frequency band (GHz)	*WG Flange*	10dB gain Size (mm)*	15dB Gain Size (mm)*	20dB gain Size (mm)*
2.6-4.0	10	120 × 90 × 225	220 × 150 × 420	–
4,0-6.0	12	75 × 55 × 150	220 × 150 × 420	230 × 170 × 420
5.9-8.2	70 × 40 × 110	90 × 95 × 280	160 × 120 × 340	
8.2-12.4	16	40 × 30 × 70	70 × 50 × 150	130 × 95 × 270
12.4-18.0	18	25 × 20 × 55	40 × 30 × 95	75 × 55 × 160
18.0-26.5	20	20 × 16 × 40	30 × 22 × 70	50 × 37 × 110

* Sizes are given in the order: flair width × flair height × front-to-back length

Approximate beam widths to − 3dB points are:
10dB gain = 50°
15dB gain = 30°
20dB gain = 20°

Table 42.3 Horn antenna parameters

major event is to be televised and is therefore potentially subject to high rf interference.

In general therefore, point-to-point operation of mobile links demands high antenna efficiency and a good polar diagram at the expense of some portability.

Derivations of parabolic antennas are generally chosen to provide high gain with narrow beam width, low side-lobe radiation and best front-to-back ratio.

Several manufacturers prefer offset paraboloids which have the advantage of better clearance over local obstructions such as safety barriers, and lower aperture blocking if the feed is offset. However, it is doubtful if these advantages outweigh the relatively lower gain, increased mechanical complexity, storage difficulties and manufacturing cost over a simple centre-fed solution.

Since most mobile links operate in reasonable line of sight conditions, linear polarization is traditional and still generally favoured, although circular polarization is now sometimes used at lower frequencies.

Figure 42.9 Triple mobile with remote control. Hybrid multiplex and discrod antenna (2.5 GHz/21 dB gain)

Table *42.2* lists typical antenna size, frequency band and gain for use in the specimen path calculations in section *42.5*.

42.3.2 Horn

For short range point-to-point operation, simple horn antennas providing vertical or horizontal polarization can be a cheap and effective solution. These items are rarely seen in the commercial market, but can often be fabricated easily in the broadcaster's model shop.

Table *42.3* provides dimensions and gain for some useful frequency bands.

42.3.3 ENG systems

ENG systems are generally used in rapid deployment situations where the path is dictated by the origin of an event and a clear line of sight may not be available.

In order to provide real time television from such adverse environments, antennas need lightweight portability and the capability to utilize deliberate 'bounce' opportunities where line of sight is not achievable. Low installed wind resistance is also a requirement so that such antennas may be operated on pump-up towers, lightweight tripods, etc.

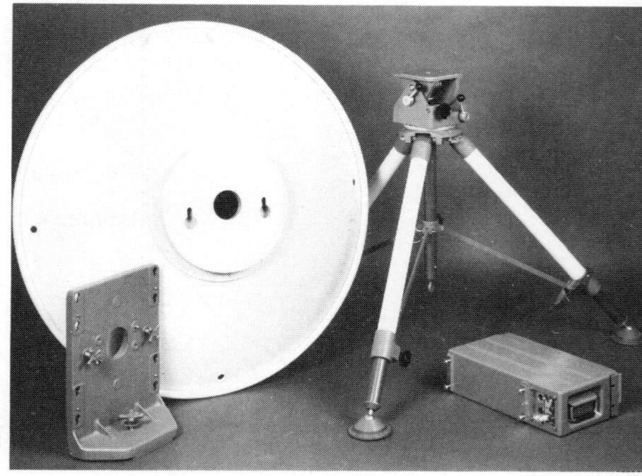

Figure 42.10 Broken down mobile mechanics set

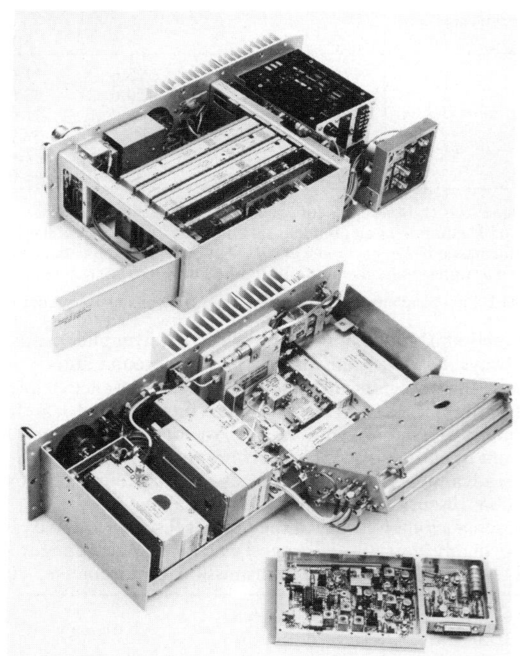

Figure 42.11 Typical interior layout of head and control electronics

Although small parabolic reflectors (0.3–0.6 m) are practical, most users seem to prefer helix or discrod solutions in single or combined form. Helix antennas have some nominal advantage of polar diagram over discrods, particularly in side-lobe performance. They are however restricted to a single pre-chosen direction of circular polarization. Discrods have the very real attraction of switchable polarization, any combination of right or left circular or vertical or horizontal linear being realizable.

Both helix and discrod antennas are relatively expensive due to their specialized, low volume nature, especially when supplied in multiple formats with integral combiners. Helices and discrods are practical to at least 8.5 GHz when required.

42.3.3.1 Transmission on the move

Effective antennas are a key component of any microwave transmission path, but nowhere is performance more critical than in the achievement of high quality colour transmission from a moving source.

Figure 42.12 Mini 1 W 2.5 GHz ENG terminal (video and two audio) on lightweight mechanics with dual discrod antenna (21 dB gain)

Two factors dominate the received signal quality:

- signal strength
- distortions due to multiple received signals (multi-path)

Signal strength is enormously reduced when the source is masked from the receiver by a reflecting or absorbing surface. Even one tree close to and in line with the transmitting antenna is sufficient to impose unmanageable path fade at frequencies of 2 GHz and above. Worse, reflecting surfaces within the transmit antenna bandwidth can often present the receiver with two or more signal sources. Since these are derived from a common transmitter, the receiver will readily accept them all.

However, the direct and secondary (reflected) signals will travel different distances and hence arrive at the receiver in different phases. Because a wavelength at 2 GHz is only about 15 cm, even minor relative movement between transmitter and receiver will transcribe multiple full cycle phase changes as seen by the receiver. When the multiple signals are in phase, enhancement will occur and the receiver will see a signal significantly in excess of that expected. Conversely at 180° phase conflict, signals will tend to cancel. It is quite possible in practice for microwave power to be reflected highly efficiently, and therefore for total signal cancellation to occur at the receiver.

Possibly the most objectionable effects occur when the receiver demodulator is presented with phase conflicts which translate into group delay distortion of the demodulated signals with disturbance of colour parameters in particular.

Several methods are in common use to alleviate distortions, notably:

- if agc characteristics with fast response times able to react to flutter speeds of several kilohertz
- powerful limiting for am suppression in excess of 50 dB
- video clamps
- chroma agc
- component transmission

The last is a technically superior option, but it is not practically realizable yet within the size constraints demanded of special event transmitters. However effective these measures are, none can ever equal tackling the problem at source.

It follows that the best way to cope with multi-path distortions is by prevention, or at least reduction of multi-path itself, and this is where specialized antennas with parameters optimized to operational requirements play a vital part.

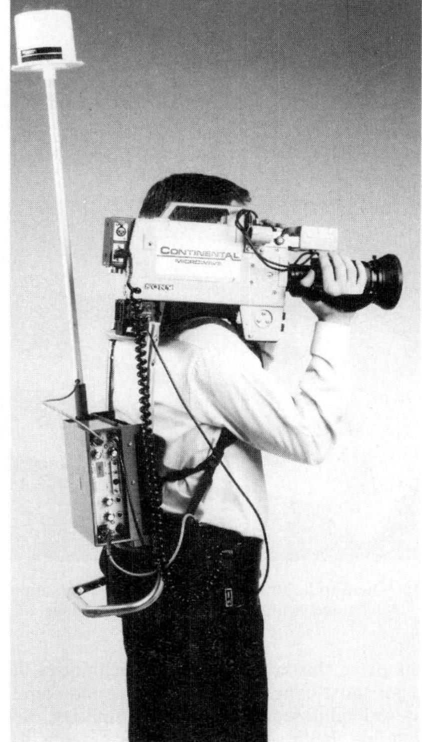

Figure 42.13 One man backpack operation, circular polarized omni antenna

Linear antennas give rise to reflections of the same polarization, and a receiving antenna is therefore equally receptive to the primary and any secondary signals. However, circular polarization (following the laws of light) reverses its direction of rotation on reflection, and the receiving antenna can therefore discriminate against a first reflection signal, typically by a factor in excess of 20 dB.

Circular polarization is not a total solution since many practical paths permit double reflections, where the direction of signal polarization of course reverts to the original. Generally however, signal strength of a double reflection is significantly less and does not present a major influence except in enclosed urban or sports stadium environments. For such situations, the use of semi-directional antennas to reduce the number of possible reflection opportunities, together with circular polarization, generally yields acceptable results.

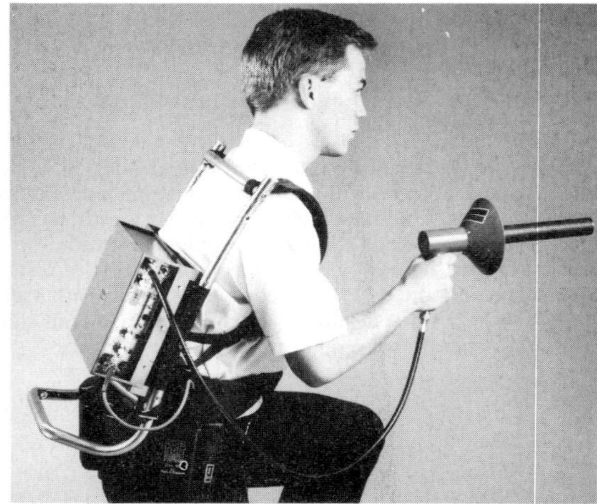

Figure 42.14 Backpack terminal with hand-held helix 'gun' antenna. 12 dB gain at 2.5 GHz; very effective in reducing multipath

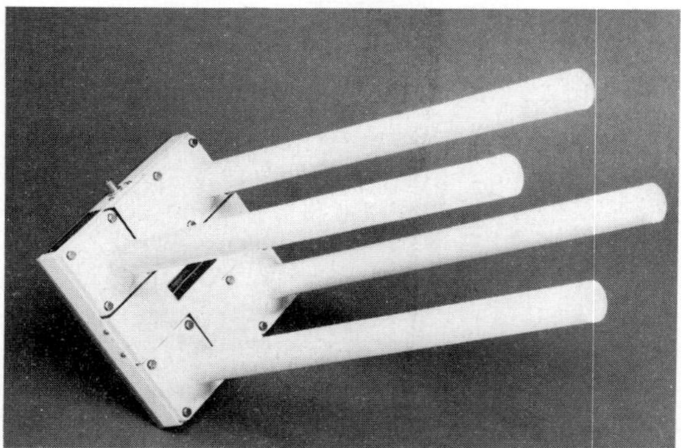

Figure 42.15 'Quad-rod': four discrods combined for approximately 23 dB gain. Advantage is high gain with low windage and circular polarization

Antennas using this combination of techniques have been developed for most conceivable situations and some popular types are described in section *42.4*. For simplicity, direction of transmission has been assumed from a moving to a fixed terminal. In reality the same commentary would apply for reversed transmission.

42.4 Central ENG

It is convenient to divide central ENG station considerations into *antenna* and *receiver* elements since maintenance factors usually dictate siting the receiver remote from its antenna in a more hospitable operational environment.

There are three fundamental approaches to remote mounted (masthead) central receiver antennas with 360° coverage: omni antennas, sector antennas operated with combiners, or switches and directional antennas with servo driven remote steering systems.

In general the options are described in order of ascending cost!

42.4.1 Omni antennas

Omni antennas provide a very effective solution where the possibility for mounting at the topmost point of the tower exists to give full azimuth coverage without shadow from the tower structure.

Omnis can be produced with any polarization mode but do not subsequently offer easy change of polarization. The choice of circular polarization generally offers superior multipath performance (see section *42.3.3.1*), and co-linear combining can yield useful gain. In fact, in such fixed applications, omni-direction gains of 11–14 dB in azimuth are quite achievable without antenna size becoming excessively cumbersome.

However, this solution is not ideal for long range operation or where local topography suggests a multipath risk.

42.4.2 Sector antennas

Sector antennas such as *quad horns* are useful where tower top mounting is not feasible. They are typically mounted on tower legs with each horn scanning a 90° sector. Choice of the relevant sector for any transmission is made by a remote switch.

Quad configuration generally permits selection of multiple polarizations: circular right or left or linear vertical or horizontal, giving a remote selection function of 1 from 16 to optimize the received signal. Horn gains of 14–16 dB are realizable, yielding a small but useful gain advantage over the simplest colinear omni solutions and significant multipath benefits.

42.4.3 Directional antennas

The most sophisticated central ENG antenna solution uses a narrow beam paraboloid (or offset paraboloid or cosec^2 variant) for optimum multipath rejection and maximum gain (range).

Because of the directional antenna properties, at least azimuth steering by remote servo control is necessary. The receiver is generally required to provide a remote output proportional to signal strength which can be used to position the antenna from a distant location, perhaps fed out on a reverse link or telephone line.

Some antennas are deliberately given a distorted radius profile to increase beam width in the vertical plane (cosec^2) and eliminate the complexity of elevation adjustment. Typical gains of 27–29 dB are realized offering a range almost quadruple other solutions for a given received signal level.

Several remotely steered paraboloid solutions are commercially available. However, the superior performance available from this antenna type must be weighted against initial capital cost and increased maintenance of the moving parts.

42.4.4 Low noise masthead pre-amplifiers

Because ENG transmission typically operates with small transmission antennas and less than perfect propagation conditions, received signal level is often close to threshold limits and any boost in signal/noise performance is welcome.

A low noise pre-amplifier (lna) situated adjacent to (or integral with) the receiving antenna is highly desirable in optimizing effective receiver noise factor by overcoming antenna feeder cable losses. This item, whilst almost universally of high reliability gallium arsenide fet design is nevertheless vulnerable to extremes of temperature, water ingress and lightning strikes, so dual configuration or bypass facilities are preferable.

Channel or at least sub-band/band filtering prior to the lna is usually included for out-band interference protection to avoid saturation or inter-modulation occurring within this low level device. Most central ENG receivers now are of double superhet concept, so wide-band noise from the lna output does not present a problem and lna output filtering is not required.

Some typical configurations are shown in *Figure 42.16*.

The isolator prior to the lna buffers filter return loss, and subsequent to the lna offers a good source impedance to the feeder to prevent reflections which could significantly affect colour performance.

42.4.4.1 Effective noise figure of masthead lna

It is often assumed that a masthead lna with 2 dB noise factor will provide 2 dB performance in a system. The reality is considerably different as every element in the chain will introduce some modification of the effective system noise figure.

Figure 42.17 shows a typical example. It assumes a masthead remoting cable such as LDF5-50A foam filled heliax and 100 m typical cable length. The loss over that length will be 6.6 dB. The receiver noise figure in this example is 9 dB, typical for a central ENG receiver without an integral lna.

So, moving towards the antenna, the noise figure measured at the antenna end of the cable, including the isolator, will become 9.0 + 6.6 + 0.3 = 15.9 dB. This will appear as the effective noise figure of the receiver as viewed from the lna. The noise figure as seen from the lna input will be 2 dB 'diluted' by the effective receiver noise figure 15.9 dB (because of lna transparency), and the dilution will be inversely proportional to the lna gain, i.e. with very low lna gain, the noise figure as viewed from the lna input will not yield the expected improvement.

Choice of lna gain is therefore largely based on the need to overcome the 'effective' receiver noise figure viewed down the antenna cable.

The actual effective noise figure can be deduced from:

$$N_{\text{effective}} = N_{\text{lna}} + \frac{N_{\text{post lna}} - 1}{\text{lna gain}}$$

In our example, let us calculate effective lna noise figures for lna gains of 8 and 16 dB (being those typically realizable from one and two stage pre-amplifiers).

With lna gain of 8 dB:

$$N_{\text{effective}} = 2 + \frac{15.9 - 1}{8} = 3.9 \text{ dB}$$

With lna gain of 16 dB:

$$N_{\text{effective}} = 2 + \frac{15.9 - 1}{16} = 2.9 \text{ dB}$$

As can be seen, the difference is a useful 1 dB improvement in noise figure and may be well worth the additional lna stage. However, in neither case was the 2 dB lna noise factor realized.

For both options, note that prior to the lna, the effective noise figure viewed from the antenna will include the cable, band filter and isolator losses, yielding a final system noise figure for our example of 2.9 + 0.3 + 1.2 + 0.3 = 4.7 dB.

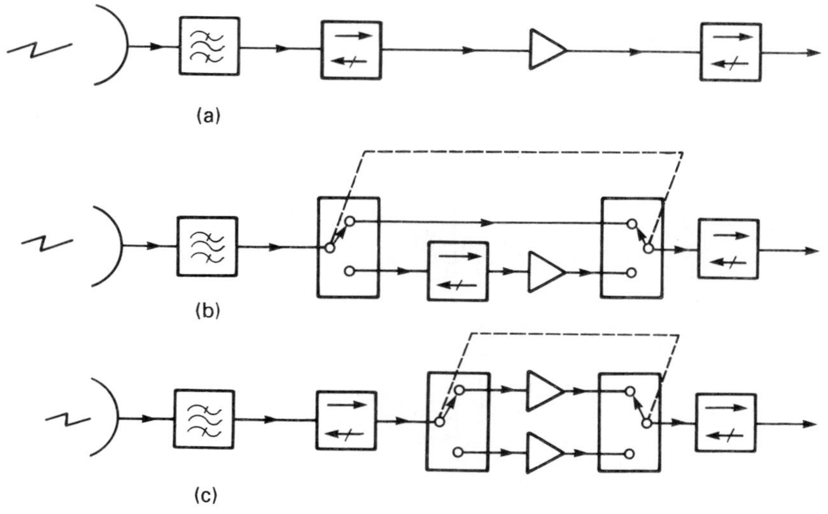

Figure 42.16 Some lna configurations: (a) unduplicated, (b) unduplicated with bypass, (c) fully duplicated

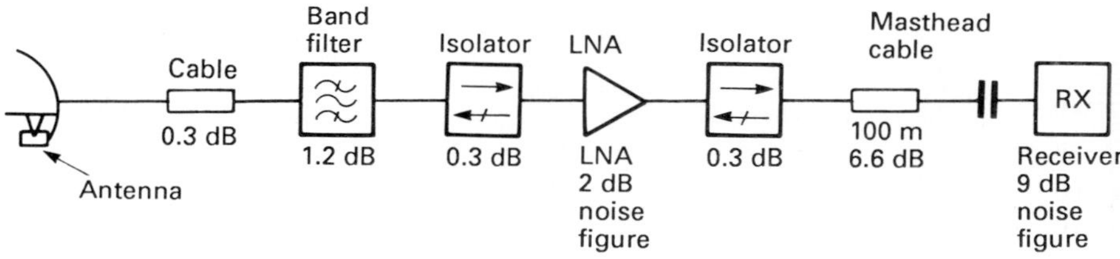

Figure 42.17 Typical calculation of effective noise figure. See text

42.5 System calculations

42.5.1 By computation

The example assumed is:

1 a frequency of 7.5 GHz
2 a path length of 40 km (25 miles)
3 an output power of 1 W (0 dBW/+30 dBm)
4 an antenna size of 0.6 m at the transmitter
5 an antenna size of 1.2 m at the receiver
6 a receiver with lna (which therefore approximates to the 4 dB noise figure assumed for the full calculation)
7 a receiver threshold of -112 dB

42.5.2 Short method using nomographs

The nomographs in *Figures 42.18* and *42.19*, and the graph in *Figure 42.20* provide a quick everyday method of calculating expected system performance to a first approximation.

It is of course necessary to know some basic system parameters and, for the sake of useful comparison, we shall repeat the full calculation of section *42.4.4.1* and assume again:

1 a frequency of 7.5 GHz
2 a path length of 40 km (25 miles)
3 an output power of 1 W (0 dBW/+30 dBm)
4 an antenna size of 0.6 m at the transmitter
5 an antenna size of 1.2 m at the receiver
6 a receiver with lna (which therefore approximates to the 4 dB noise figure assumed for the full calculation)
7 a receiver threshold of -112 dB

To a first order approximation, items 6 and 7 can be assumed similar for most current OB links.

1. Equipment type mobile link/integral antenna		
2. Frequency		7.5 GHz
3. Path length		25 miles
		40 km
4. Path loss (free space)		142.0 dB
5. Transmitter output power		1000 mW
		0 dBW
6. Transmitter losses: multiplex		1.1 dB
	antenna cable loss	0.3 dB
	TOTAL	1.4 dB
7. Transmitter antenna size		1 ft
		0.6 m
	gain	31.0 dB
8. Effective radiated power (Item 5-6 + 7)		29.6 dBW
9. Power to receiver antenna (Items 8-4)		−112.4 dBW
10. Receiver antenna size		4 ft
		1.2 m
	gain	37.0 dB*
11. Receiver losses: Antenna cable loss		0.3 dB
	Multiplex (filter + 2 circulator passes)	1.4 dB
	TOTAL	1.7 dB
12. Signal level to receiver (Item 9 + 10-11)		77.1 dBW

*Transmitter and receiver antenna gain obtained from tables or *Figure 42.18*.

Table 42.4 Calculation of received signal level.

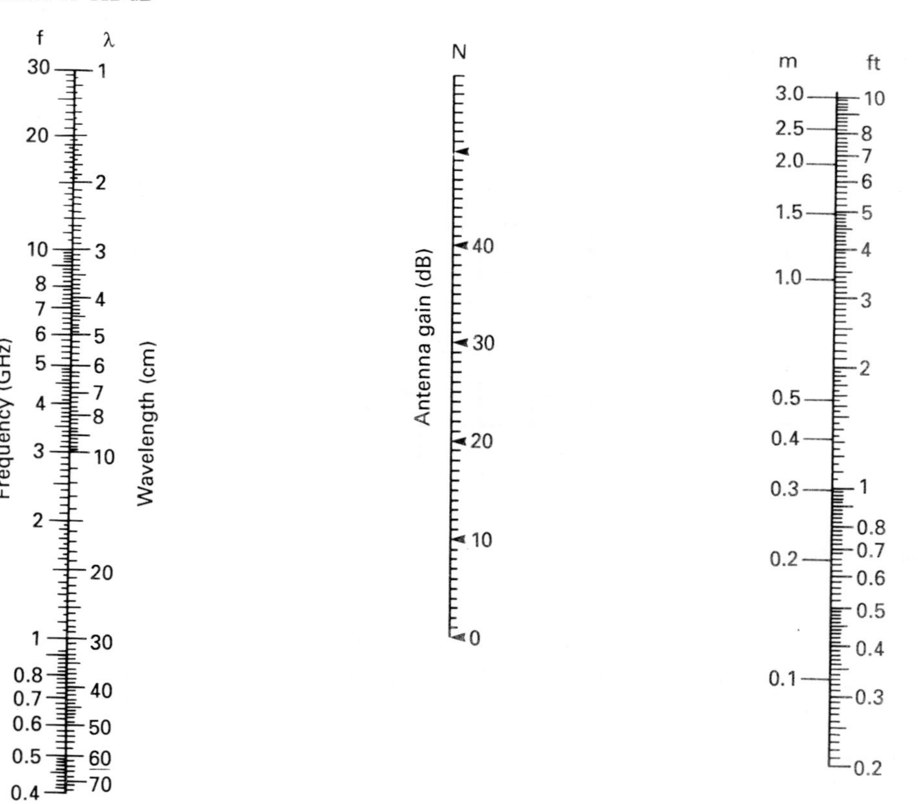

Figure 42.18 Antenna gain as a function of frequency and size

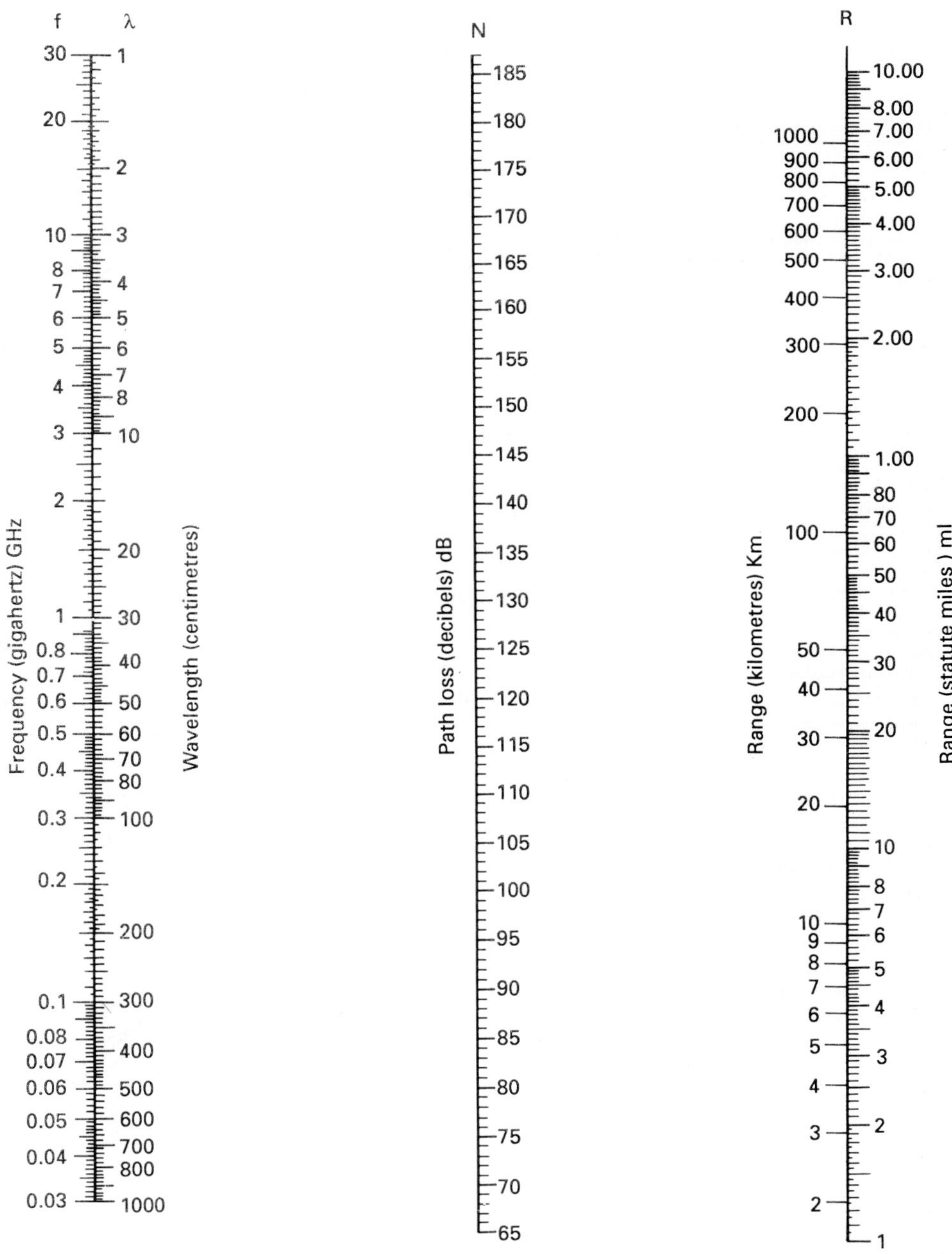

Figure 42.19 Path loss as a function of frequency and range

13. Receiver noise figure N	4.0 dB
14. Receiver bandwidth (between −3dB points)	28 MHz
15. Receiver noise power = K.T.+B+N	−125.0
where K.T. = Boltzmann's constant (K) × absolute temp. (300° kelvin) = 203.5dBW/Hz	−203.5
B = receiver bandwidth (10 $\log_{10}$ Hz)	74.5
N = receiver noise factor (dB)	4.0 dBW
16. Carrier/noise ratio (item 12-15)	47.9 dB
17. Receiver threshold	−112.0 dBW
18. Fade margin (free space) (12-17)	34.9 dB

Table 42.5 Calculation of carrier/noise ratio and fade margin

19. TOTAL fm improvement $= 10\log\frac{B}{2f} + 20\log\frac{SF}{f}{}_3 = 4.47 + 2.83$	7.3 dB	
where B = receiver bandwidth	(28 MHz)	
f = highest modulation frequency	(5 MHz) (for noise measurement)	
6F = peak deviation	(4 MHz)	
20. Pre-emphasis degradation = 0.5	−0.5 dB	
21. Video signal/noise ratio (Item 16 + 19 + 20)	54.7 dB	
22. Picture signal/rms noise ratio due to phase thermal noise only	60.6 dB	
Video s/n ratio + picture signal/rms signal ratio (Item 21 + 20 log 2 2 × 0.7)	(5.9)	
23. Picture signal/rms noise ratio due to equipment contribution	62.0 dB (dependent on equipment)	
24. Effective overall picture signal to rms noise ratio (Item 22 + 23)	58.2	58.4 dB
25. Weighting improvement for noise (CCIR Rec. 267) = 16.3	9.8 dB	
26. Picture signal/weighted rms noise ratio due to phase thermal noise only (Item 22 + 25)	70.4 dB	
27. Picture signal/weighted rms noise ratio due to equipment contribution	65.0 dB (dependent on equipment)	
28. Effective overall picture signal/weighted rms noise ratio (Item 26 + 27)	63.9	64.0 dB

Table 42.6 Calculation of video unweighted and weighted signal/ noise ratio

First, to find antenna gain use the graph in *Figure 42.18*. Place a straight edge between the frequency (assumed here at 7.5 GHz) and antenna size columns. Read off the gain from the centre column.

In the case being considered, the antenna gains are:

for the transmitter antenna of 0.6 m	31 dB
for the receiver antenna of 1.2 m	37 dB

Now to establish path loss, use the graph in *Figure 42.19*. With a straight edge from 7.5 GHz to 40 km, path loss is available from the centre column as 142 dB.

So received signal level equates to:

transmitter output power +:	0 dBW
transmitter antenna gain +:	31 dB
receiver antenna gain −:	37 dB
path loss:	−142 dB
Signal level to receiver=	−74 dBW

Since we know, for our example, that miscellaneous 'per hop' losses equate to 3.1 dB, we can now include these to improve correlation of our follow-on calculation.

So signal level to receiver = −74.0 − 3.1 dB = −77.1 dBW

Fade margin = received signal − threshold dBW = −77.1 − (−)112 dBW = 34.9 dB

Finally, for signal/noise ratio, refer directly to *Figure 42.20*, where −77.1 dBW received signal equates to 64 dB weighted s/n.

This result correlates closely with our earlier full calculation.

So, for all practical purposes, a graphical computation is quite adequate for mobile performance prediction.

Note that the curves in *Figure 42.20* also provide unweighted luminance and chrominance s/n figures with and without received lna.

If first approximation audio performance is also required, refer to *Figure 42.21*.

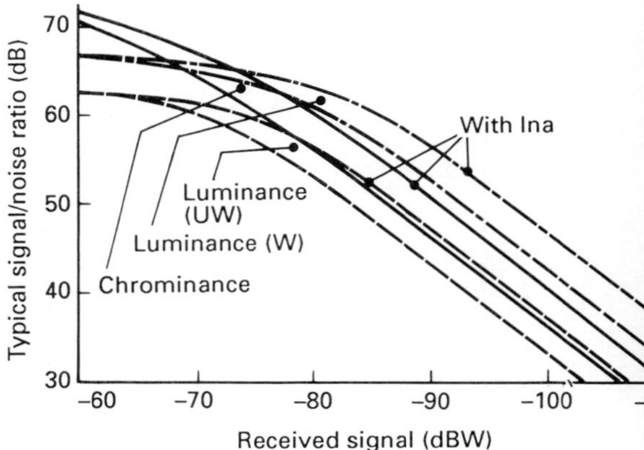

Figure 42.20 Typical video noise performance

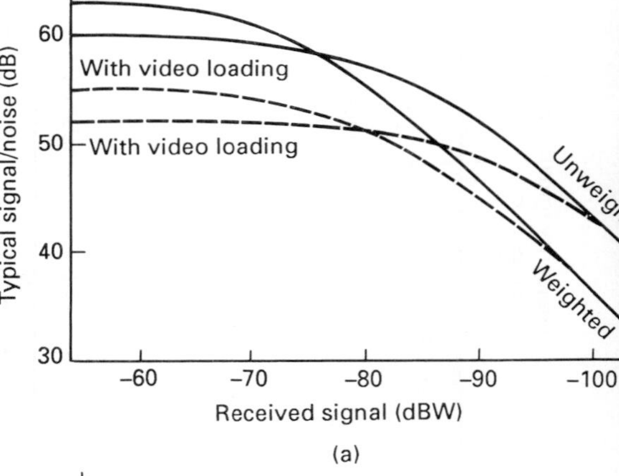

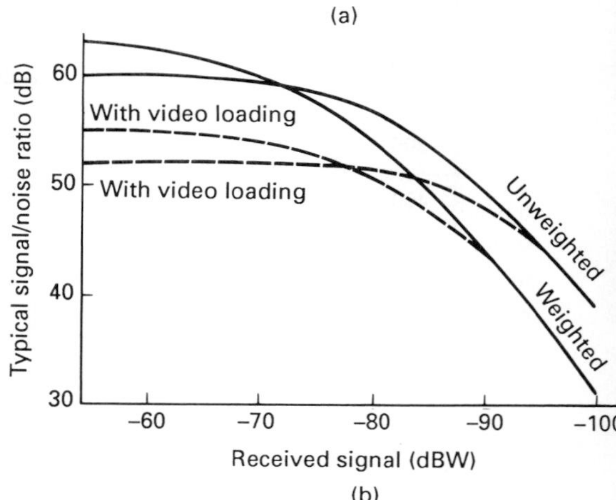

Figure 42.21 Typical audio noise performance: (a) with lna, (b) without lna. All measurements are with respect to O dBm audio loading, by ppm to CCIR 468-2; video loading is multi-burst

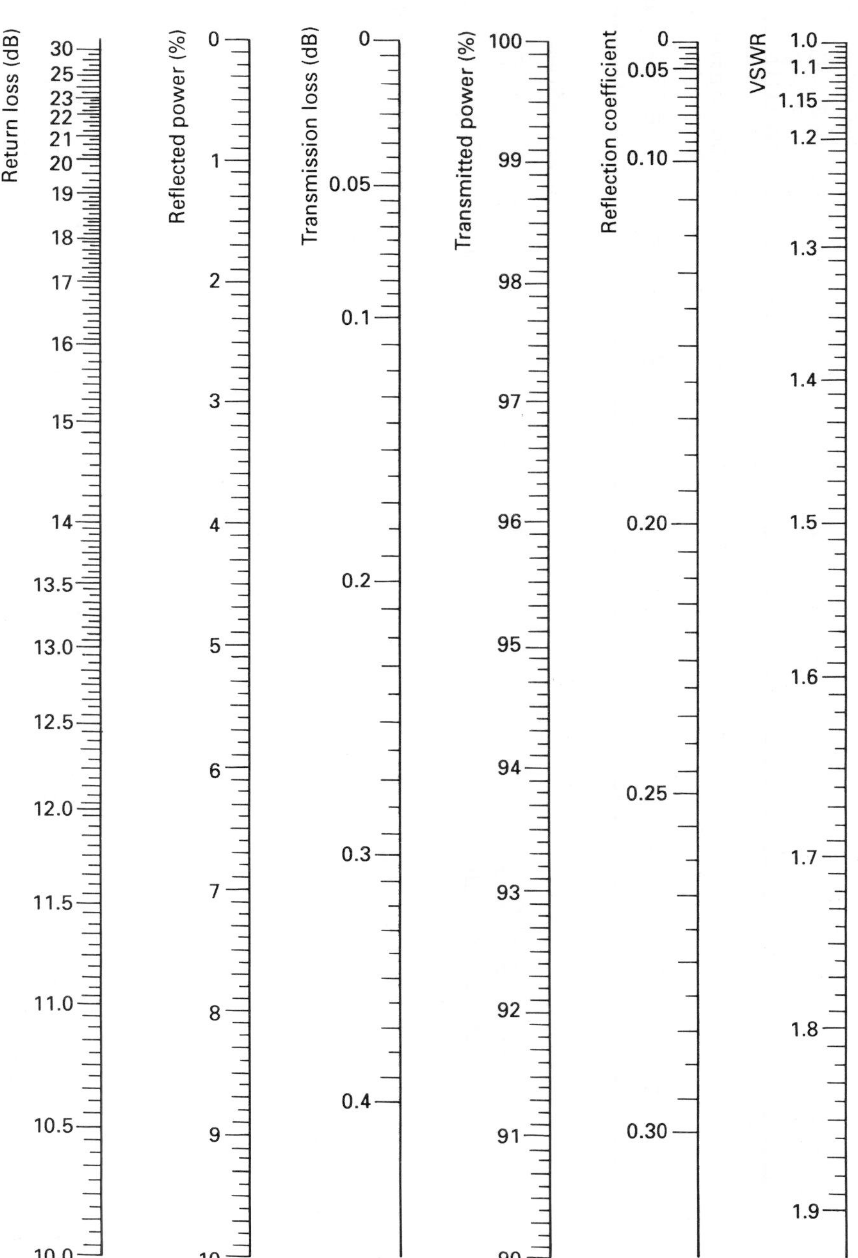

Figure 42.22 Return loss nomogram

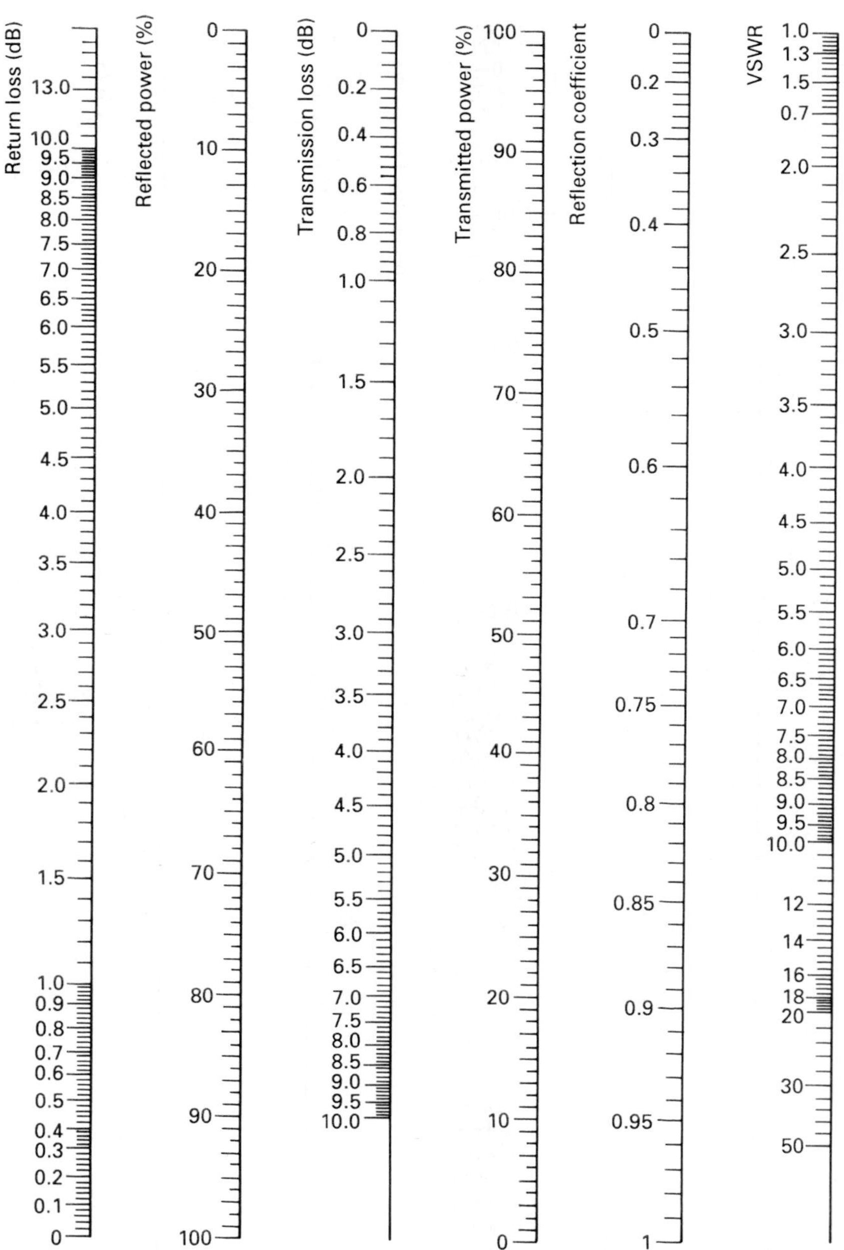

Figure 42.23 Return loss nomogram

A Todorović
Director, Televizija Beograd

43 Electronic News Gathering and Electronic Field Production

43.1 Electronic news gathering

43.1.1 Introduction

Preparing and transmitting news reports is certainly one of the most important tasks of broadcast organizations. Radio news reporting has a distinguished and brilliant history: in some of the crucial moments of the twentieth century, radio played an essential role. Today the proliferation of car radios, portable radios, miniature ones, etc., ensures that it is still a powerful news dissemination medium.

The introduction of television was the next step in the development of the medium. The availability of a picture affected the domain of newscasting. While for radio news the voice of the reporter was, more or less, sufficient, in the case of television news it was certainly not: the mandatory requirement for any television newscast is the picture of the event (or at least of the location where it happens). The history of television newscasting is, therefore, the history of the long fight against the clock for innumerable programme makers in their desperate attempts to secure a usable and meaningful picture. Photographic techniques were certainly available, the reversal film was a viable compromise between speed and quality, and OB vans were usually available. But television presupposes a *moving picture*. The use of film limits the timing of usable events (the film has to arrive at the laboratory about two hours before its planned use to permit all the processing and editing). An OB van requires even more time to be rigged and set to work, and then its presence on the spot may sometimes be more important than the event it was supposed to cover.

Obviously the answer to the programme makers' needs was a portable camera and portable recorder, but it took 30 years from the first all electronic camera to the development of a high quality portable black and white camera. Fortunately, it took less time to make a 'portable' studio quality recorder, weighing however over 20 kg. In a way, all that hardware arrived too late. By the time it was commercially available, television broadcasting was moving into colour, and a portable colour camera accompanied by an adequate recorder was the new goal.

At the same time in America national and international events attracted a large audience for the news programmes. Suddenly, these programmes became trump cards in the ratings game. One of the three networks, CBS, rightly assessed that the immediacy of pictorial news reporting was of utmost importance. They set their R and D departments on that task, collaborated with the manufacturers, and spearheaded the ENG revolution.

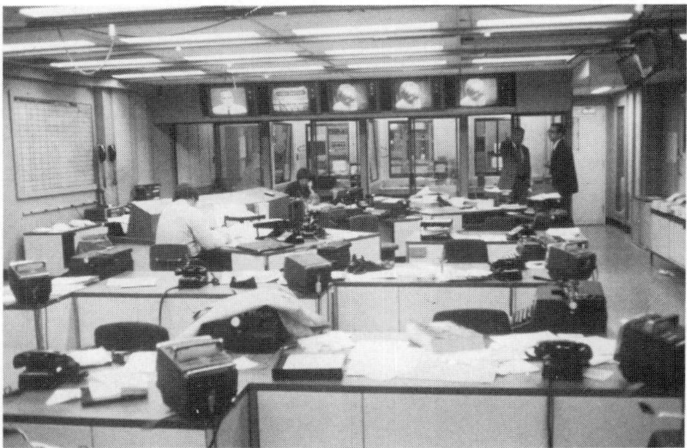

Figure 43.1 General view of the KMOX–TV newsroom; the first 'all-electronic' television station in the USA (CBS)

It was not long afterwards that the first 'portable' colour cameras became available. However, although the camera head was of an acceptable size and weight, it was attached by a short umbilical cord to a backpack full of electronic circuitry. Nevertheless a robust cameraman could carry all that hardware for a reasonable period. The camera was there, but what of the recorder? The existing 'portable' quad was too bulky, too heavy and too unreliable. The only alternative was a machine developed and used in a completely different field of industrial and educational application: the U-matic recorder. For the relatively low subcarrier of NTSC, that 'colour-under' recording format gave a quality considered by the Americans to be adequate for news reporting. The ENG was born. Courageously the same network took one of its stations (KMOX–TV in St. Louis) and transformed it into a so-called 'all-electronic station'. That meant that the processing labs were closed, all film cameras withdrawn and the whole system was based on electronic cameras and U-matic recorders (for the field as well as for the base). The full success of the KMOX–TV operation marked a real watershed in the USA.

European broadcasters followed with great interest the development of the new technology on the other side of the Atlantic. They also had some experience with black and white cameras and portable quad recorders, but unfortunately the rather high subcarrier in their PAL and SECAM systems prevented the successful use of the existing U-matics. For that reason the EBU pressed vtr manufacturers by formulating precise operational and quality goals for a new generation of small portable cassette video tape recorders. The result was the U-matic H developed by Sony.

At the same time cameras have been the object of extremely dynamic development. Only a couple of years after the first 'backpack portable' colour cameras, a new generation of compact ones (where all relevant parts are housed in a single hand-held package) engulfed the market. The road for the worldwide acceptance of ENG technology was now open.

During these dynamic years the majority of broadcasters saw the introduction of ENG only as the replacement of a film camera by an electronic one. An EBU report even states: "... perhaps nothing is more innovative in ENG than its name and acronym...". But we are often tempted to underrate what is happening in front of our eyes. Back in the 20s, C P Scott stated that "... nothing good will come out of television..."; in 1956 the vtr was greeted as a marvellous means of overcoming the time zone problem, etc. Similarly the ENG proved that the EBU report underestimated the tremendous potential of this new technology which so profoundly transformed the television news storytelling, and had also such an important impact on the development of television equipment in general.

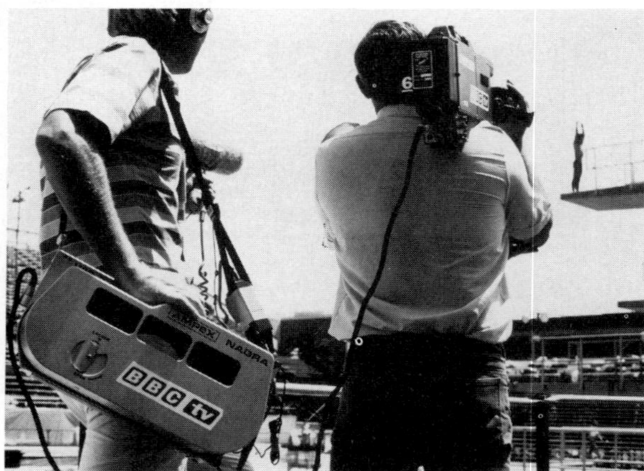

Figure 43.2 BBC crew at the Los Angeles Olympic games (Ampex Corp)

Portable colour cameras, equipped with high quality 18 mm pickup tubes, soon provided a very good picture, almost comparable to that obtained from the top of the range studio cameras, and programme makers were tempted to use them for programmes other than just 'hard news'. Fortunately, the newcomers in the field of video tape recording, the one-inch B and C formats, offered not only standard studio consoles but also real portable recorders. These recorders were heavier and less handy than the U-matic ones, but they provided one-hour recording, and, what is even more important, full broadcast quality recordings, playable and editable on standard studio machines. This combination of the top of the range portable colour camera and the one-inch portable recorder (in conjunction with the tremendous development of tape editing techniques) permitted the introduction of electronic field production (EFP).

It is difficult to find today a broadcasting organization that does not rely heavily on ENG, or has not even completely converted its news operation. A consequence of this demand together with continuing technological development is a large availability of all sorts of ENG equipment in the world broadcasting market. Many of the products which looked rather wishful futuristic dreams only a few years ago are now available on 90 days delivery.

That profusion has also its drawback. In the field of video tape recording the world is now crowded with non-compatible formats. Each has its advantages and consequently its supporters, but the happy old days of a single worldwide broadcast standard are gone. The visiting crews have now to inquire not only about the mains voltage, the scanning standard and the colour standard, but also about the local ENG recording format, and following some of the corollaries of Murphy's laws, nine times out of ten it proves to be a different one. It is safe to state that the war of the formats will be with us for many years to come.

43.1.2 What is ENG?

ENG, or electronic news gathering, is, as its name says, the collection of news stories intended for broadcast during different television newscasts. In fact since this acronym was coined, ENG has evolved into a technology of its own, taking care not only of 'gathering' in the strict sense of the term, but also of all post-production and airing of news items.

Although based completely on electronics, this technology owes a lot to film practice. When ENG equipment first became available, broadcasters started to replace their film crews in news departments by ENG crews, assigning them the same duties, and sending them to solve the same sort of problems film crews had fought for years. But, ENG immediately demonstrated important advantages over film.

Its first and most obvious advantage is its *speed*. The use of this technology permits dramatic shortening of the time which has to elapse between the actual event and the moment when the news story is ready for airing.

Second is *quality*. Although it is true that for some news stories the technical quality is not of prime concern, the overall quality expectation of the average viewer has considerably increased. If we compare the quality parameters of a 16 mm news reversal film with what can be obtained by broadcast quality ENG, we can see that ENG has several obvious advantages:

- The colorimetry of portable colour (three tube) cameras is better than that of fast films and provides a better match to the studio shots.
- Due to a higher sampling frequency the reproduction of movement is considerably better.
- The problem of frame steadiness is virtually nonexistent.

The better quality of the output and its speed of handling are supported by several other important operational advantages:

- In dim light situations, which are rather frequent in the news gathering business, the viewfinder on the colour camera facilitates focusing and framing.
- The same original, shot on location, can be re-used easily for several different versions of the same story (different newscasts, magazine programmes, etc).
- ENG equipment is considerably quieter than the corresponding film equipment.
- The recording can, in most cases, be checked on the spot.

• The cassette principle makes the change of cassettes at a convenient moment easy and fast.
• For important international gatherings, the ENG technology permits the visiting crew an almost complete independence from the local broadcaster (whose facilities are on such occasions usually overbooked).

All these advantages are backed by a favourable economic balance sheet. Although ENG equipment requires higher investment costs (a complete comparison is difficult to make, since film should include the building costs of a processing lab, etc.), its running costs are certainly lower. It requires less expenditure on material since the tape is re-usable, a camcorder crew can be smaller than the corresponding film crew, there are no chemical costs, tape copies are cheaper and easier to produce, and so on.

However, film technology still retains some advantages, and that situation will probably remain unchanged for many years to come. Film shooting equipment is simpler, requires less maintenance and consumes less power (in the extreme, a clockwork 16 mm camera is completely free from the availability of a power source).

43.2 Operating practices

There are many ways to organize and run an ENG operation. Essentially we need to look at two basic segments and one special case:

• the coverage of the news event and the forwarding of the programme material back to the main television station,
• the processing of the news material and its distribution to the viewers,
• the situation when all or most of the post-processing is done in an improvised facility near the place where the covered event takes place.

43.2.1 Field operation

There are two basic approaches to news covering by electronic means: the *recording* concept and the *microwave* concept. These two concepts are by no means exclusive. On the contrary, they are just two different ways of operating with the same ENG crew, and sometimes they can even represent two stages of the same operation.

The recording concept is certainly the most common way of covering news events by ENG crews. As it consists of recording an event on a portable recorder, the equipment required for such an operation is extremely reduced. The basic requirement is a portable camera and a portable recorder (or a camcorder), a microphone, batteries and perhaps a lighting source. Such equipment can be operated by an extremely reduced crew. In some situations it can consist of one person only (especially if a camcorder with a built-in microphone is used). The nature of the programme and the character of the event will dictate how much equipment is needed. If, for example, it is expected to cover only the arrival of high officials to a formal gathering, a single man crew with a camcorder and a built-in microphone for the ambience sound will suffice. But, if an indoor interview is likely, it will be necessary not only to bring lighting units and separate microphones, but also to have a camera tripod and even a sound recordist with a portable sound console. All that equipment and other ancillary items can be stored at the base and packed in a saloon car on reception of the assignment order, but it is also possible to conceive a specially equipped van, fitted with all that might be needed in any situation. There is however a danger of overequipping such a car and transforming a very handy operational tool into an institution.

A specially dedicated and equipped ENG van can be fitted also with a microwave transmitter, which leads us to the second concept of news coverage. It might happen, although not too frequently, that an event is taking place during the regular newscast. To achieve live coverage it is necessary either to establish a special microwave link between the television centre

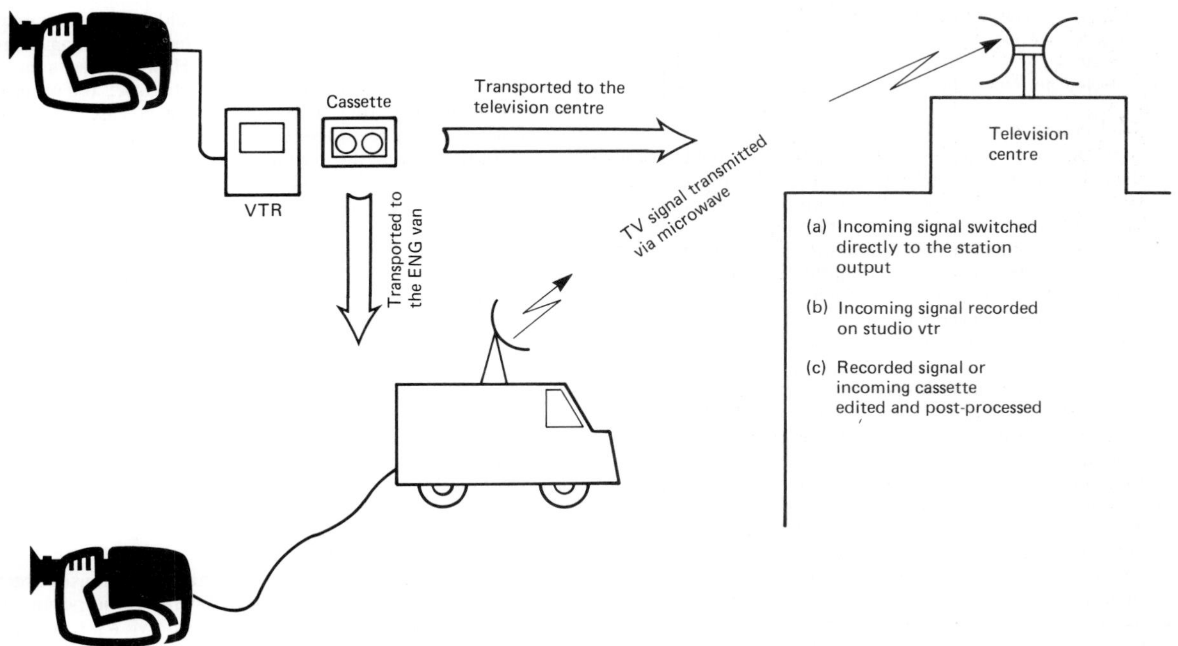

Figure 43.3 Alternative schemes of ENG operation

and the location where the event takes place, or to install permanently on a convenient spot a suitable 360° receiving point and to connect permanently (by cable or microwave) that point with the television centre. Such a receiving centre permits a fast operation, since the mobile crew, equipped with a microwave transmitter, has only to establish a liaison with the central receiving point. This is confirmed by the centre over a radio co-ordination line. The same radio co-ordination is used for cueing, but the crew in the field can also follow the regular programme on a portable off-air receiver and see when they are 'on air'. Although events usually refuse to happen during the scheduled time of newscasts, the live approach can still be used. A live picture in a newscast can always enhance its visual appeal ('latest news' live from a news agency, a weatherman live from the street under a heavy snowfall, etc.). On the other hand, a microwave connection to the studio will eliminate the usually painful journey through heavy traffic and will ensure that the post-production can start as soon as the event has ended.

A combination of these two operational methods, i.e. the recording of the event on a portable recorder and the microwave link between the field and the studio, permits adequate covering of specific situations. A small van equipped with a microwave transmitter, radiocommunication equipment, a video tape player and, if the format requires it, a timebase corrector, can be parked at a location which is at the same time suitable for an easy microwave connection with the studio (or the central receiving point) and reasonably near to the spot where the covered event takes place. Such an approach is appropriate only if it is vital to have the story back at the studio quickly, or if the journey back to the base may be delayed, and when it is not possible to establish easily a microwave connection between the spot in question and the studios' receiving point. In such a situation, the event will be recorded and the tape or cassette carried to the van where it will be played back and transmitted over microwave to the television centre.

Many broadcasting organizations have installed special 'injection points' at many of their accessible transmitter sites. These injection points can be accessed by travelling ENG crews; they are equipped with facilities that enable the crew to communicate with the base and obtain access to some contribution circuit. Any television centre linked to the national or international network can also be used as an injection point.

43.3.2 ENG operation at the base

Once the picture and the sound of an event are brought back to the television centre — the base — by any of the methods described earlier, they have to be post-processed in order to obtain a 'news item' in a suitable form for direct insertion in a given newscast. The term 'post-processing' encompasses editing, sound sweetening, adding of graphics and sometimes compiling on the same tape, in sequence, all items that are planned to be used in a given programme.

As in the gathering of the news, so in post-production we can develop a given facility from the simplest case of two video tape recorders, one of which should preferably be of the same format as the recorders used in the field, surrounded by the usual editing control console and audio and video monitors, up to a sophisticated editing suite including a video and audio mixing and special effects unit and a character generator, linked to other picture sources in the plant (electronic still store, etc.). The necessary sophistication of the post-production areas will depend on the way in which the programme material is presented. It is possible simply to assemble the news item in the editing suite and to add all voice-overs, graphics, etc., during the airing of the news programme. Such a procedure certainly shortens the editing of the news story, but makes the whole newscast less reliable, and should therefore be used only when it is the only way to be on time for the scheduled programme.

The use of current electronic post-production facilities makes it possible to assemble in a reasonable time very sophisticated programme items in the style of a particular organization. Further, the same camera material can be used for several different versions of the same event: a short version for an 'overcrowded' newscast, a larger one for a magazine programme, a different one without commentary for an EBU exchange of news, etc. In order to simplify the creation of the version intended for the exchange, as well as a later re-use of the same edited version in a different programme, it is advisable to record the commentary on a separate audio track (all current recording formats are able to offer two broadcast quality sound channels), and to mix that sound with all effects during the airing.

If there is no time constraint, such a recording with two audio tracks can be dubbed with both sounds mixed on a single track. However, this will represent an additional generation of the video tape recording, and might lead to an unacceptable degradation of the technical quality. It is obvious that all these aspects — the need for a version with only the international sound, the desire to simplify the airing of newscasts, the concern about the technical quality, etc. — have to be borne in mind when deciding how to run the ENG post-production in a television plant.

Experience has shown that the ENG editing or post-production suites should be located as close as possible to the newsroom; ideally they should be part of the same complex. Then it is possible for the news editors to preview all programme materials at any stage of the preparation of a given news programme. Also, there should be an additional station in the newsroom at the duty editor's desk for the radiocommunication link between the base and the crews in the field.

It is advisable to use on-line time-code editing which represents a good compromise in speed and editing sophistication. For a simpler concept it is possible to envisage a frame count editing method.

When a news item has been edited and shaped as desired, it has to be fitted into the framework of a news programme. The simplest and the most reliable way would be to assemble, adequately separated and cued, all items in their running order on the same reel of tape or video cassette. However, last minute events often perturb a carefully planned schedule , and require a different running order. The situation may be eased by using several video tape recorders, but this is not always the most efficient utilization of the existing facilities.

The ideal solution is a random access multi-transport video tape recorder. The simplicity of programming and reprogramming the sequence in which the events will have to be played back, the simplicity of replacing a given version of an item by another one, the flexibility of cueing the items, etc., make it perfectly suited for news operations. Unfortunately, we encounter again the recording formats problem. These machines are not available in some of the formats used in ENG, and if a user is orientated towards those formats he will have to rethink his whole post-production system in order to introduce a new format, or will have to dub all the edited items. The development of ENG technology will continue with formats which will, like Betacam, Betacam SP or M II, offer multi-transport machines in their range.

43.2.3 Creating a temporary base

Large international sporting or political occasions attract crowds of journalists and film and ENG crews who permeate the event, shooting the highlights, taping interviews, recording 'stand-ups', in short covering the event. Afterwards they all rush to the premises of the local broadcast organization asking for processing, dubbing, editing and broadcasting facilities.

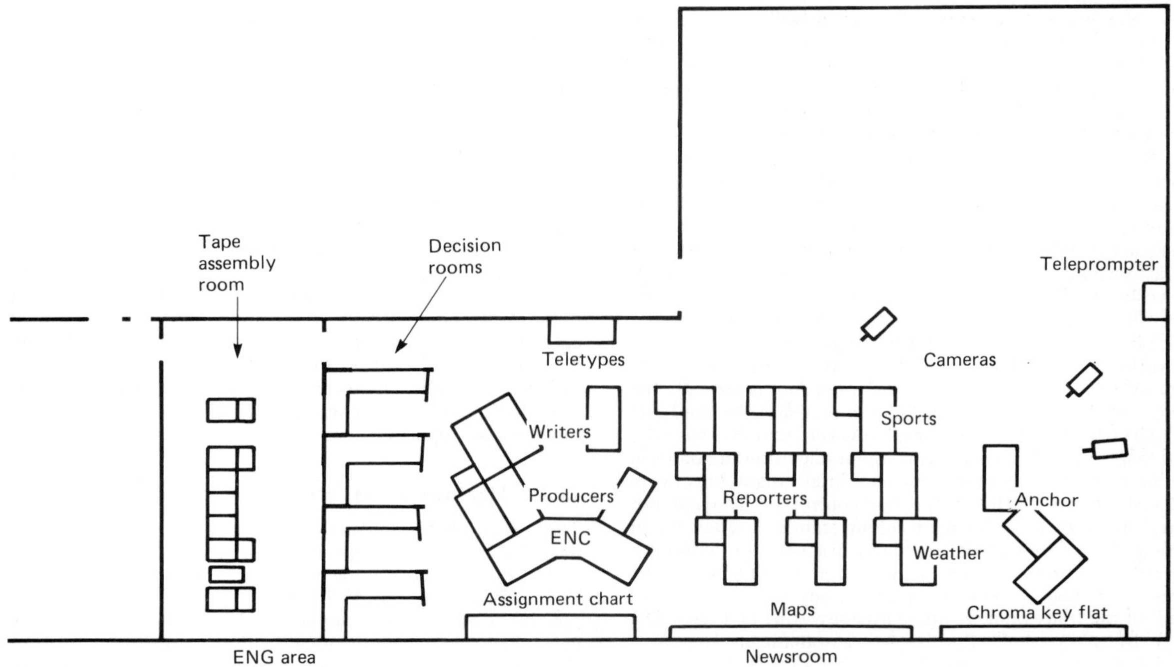

Figure 43.4 Layout of the newsroom at KMOX–TV, as equipped for electronic news gathering (CBS)

Whatever the size of that TV station, its post-production facilities will be limited, and they will all be required simultaneously. In such situations it might be good policy to equip the travelling crew not only with the portable equipment but also with all the necessary hardware to improvise a post-production suite at the site where the event takes place. The transportation costs for the additional equipment and the costs for the additional staff will be similar to the price usually charged for the use of post-production facilities.

It is usually required in such situations to provide a relatively simple post-production. The equipment for a 'mobile' post-production facility therefore is limited to:

- two video tape recorders of the same format as the portable unit(s), fitted with full editing facilities,
- two video monitors,
- one editing control console (capable of on-line time-code editing, or at least, frame count editing),
- additional microphone(s),
- portable small (five inputs?) audio mixer (optional).

If this improvised post-production facility is installed at the premises of the local broadcaster it might be worth ensuring that the product can be transmitted to the international network straight from that base. It will then be necessary to consider the playback characteristics of the recorders. If the format in question requires an external time-base correcter, this must be added to the list.

The EBU has issued a recommendation (No. R 24–1980) to its members advising certain minimum working conditions for the travelling ENG crews, such as:

- a room with benches,
- a connection to the ac power supply,
- connection to the master control room for access to the international or satellite circuit,
- telephone lines, as well as necessary co-ordination and control lines (with adequate terminals).

At the same time, in order to facilitate the interconnection of the equipment brought by a travelling ENG crew with the equipment of the local broadcast organization, the EBU have recommended (No. R 21–1979) a set of preferred connectors for the video and pulse connections:

- for video and pulse connections — BNC connectors,
- for audio connections — Cannon XLR three pole,
- for the battery connection — Cannon XLR four pole,
- for equipment having direct mains connection — Cannon XLR–LNE,
- for headphone monitoring — Cannon XLR three pole,
- for connection between transmitter and aerial — Type N.

43.2.4 Getting the signal back

When an ENG crew is operating in the immediate vicinity of its base, or within its own country, there are various ways of getting the signal back to the studio. This is also true when the crew is operating abroad. If exterior post-production of the recorded material is envisaged, it is best to find the safest and the most efficient way of dispatching the cassette(s) back to the home base. But if the assignment is linked with the daily newscasts it will be necessary to use international terrestrial or satellite circuits to transmit the signal as soon as practicable.

In the European zone it is always possible to use the Eurovision network, which can be ordered either by the organization to which the travelling crew belongs, or by the local broadcaster at the request of that same crew. The planning is done by the EBU Technical Centre, depending on available time and the assent of transmitting and receiving units. However, for the actual booking of the circuits, several factors have to be taken into consideration. If the travelling crew belongs to an active member of the EBU and intends to use

Eurovision permanent circuits, the booking will be through the EBU Technical Centre in Brussels. If the same circuits have to be used by a crew belonging to a non-member organization, the request for the circuits will have to be passed through the local PTT administration (or the local broadcasting organization if it owns the circuits), and the EBU Technical Centre will be asked to release the permanent circuits for the requested time. If other circuits than the permanent ones have to be used (regardless of the affiliation of the travelling crew), the request will have to be done through the local PTT administration or the local broadcasting organization if it owns the circuits.

East European countries are linked by their own Intervision network. That network has several connections with the Eurovision one, and its booking is to be done in a similar way, which makes it easy to transmit ENG signals through terrestrial networks all over Europe. If parts of both networks have to be used for the same transmission, the procedure is virtually the same, and the necessary co-ordination is ensured by the EBU and the OIRT Technical Centres. However, attention has to be paid to the fact that there are two colour television standards in use in Europe, and that some countries might have very reduced, if any, facilities for the control of a signal not complying with their adopted colour standard. They might consequently be reluctant to allow a transmission without a prior transcoding. This problem is even more acute if the scanning standard is also different. In any case the possible need for the transcoding or standard converting of a recording made by a travelling crew implies automatically that at a receiving end the signal will have to be converted (or transcoded) back, and such a cascading of conversion or transcoding processes will unavoidably harm the quality of the video signal.

Between continents, the transmission of an ENG recording back to the home studio will certainly be done via a communication satellite. In that case it is again advisable to act through the local common carrier who will investigate with Intelsat (or a corresponding organization) the availability of the circuits at the requested time. Intelsat will report that time is available and expect to receive a matching order from the PTT administration (or a common carrier) in the country from which the signal will originate. That matching order also confirms that between the place of origination of the signal (some television studio) and the earth station all needed facilities are available. It is necessary to bear in mind that the cancellation of booked satellite facilities may (particularly if late) incur considerable costs. For example, if the cancellation is made less than 24 hours from the reserved time, the whole originally booked time will be billed.

Within the United States, ENG recordings are beamed back in a similar manner, although the transmission network in the USA does not belong to the broadcasters, the PTT or telecom administrations, but to private agencies known as *common carriers*. There are more stringent requirements on bookings.

However, the size of the country and the availability of an impressive number of commercial communication satellites, linked with recent developments in technology, led American broadcasters to introduce Satellite News Gathering (SNG). Basically, SNG replaces the terrestrial microwave network and its multiple access points with a space segment of some of the communication satellites, and uses a mobile earth station or uplink for access. Such an approach is obviously very attractive for the long distance hauls one may encounter inside the USA or Canada, or for countries with particular geographical difficulties, such as Japan.

It can also be useful for communications between countries within a given region. In Europe, one can take advantage of a developed network of terrestrial microwave links, but the traffic over those lines is sometimes quite heavy and it can be difficult to ensure a transmission at a desired time. In other continents, such a terrestrial network is not yet fully developed and SNG would be a good means of increasing the flow of information.

There is still much to be done before this new technology can become routine. The mobile earth stations are still very large and bulky and they cannot be compared with the rest of the standard ENG equipment. Some existing regulations need to be changed and some new ones implemented. In a number of countries, broadcasters do not have access to receiving or sending satellite earth stations, which are the exclusive property of the PTT or telecom administrations. Elsewhere, it is still necessary to improve the booking procedure in order to simplify and facilitate access to the satellites. The full extent of this new technology will be possible only if international agreements and conventions permit ENG crews to travel freely around the world and give them fast access to communication satellites. When the hardware is miniaturized and regulations simplified, SNG will become a really flexible and invaluable tool for news reporting.

43.2.5 Equipping ENG crews

There are probably as many ways of equipping an ENG crew as there are different concepts in equipping a television studio or an OB van. We will outline here some possible schemes.

An ENG crew should be as mobile and as versatile as possible. Unfortunately, mobility and versatility are frequently in conflict, since versatility means ability to satisfy a number of different possible assignments and situations, which in turn means more equipment. The following list represents a minimum of hardware, which need not always be carried in the field, but will be at the disposal of the crew in the base. A careful analysis of the usual operational and programme practices of a given television organization may show that some of the equipment listed may be shared by several crews:

- portable colour camera,
- portable video tape recorder of a selected format,
- microphone(s),
- camera tripod and/or shoulder mount,
- portable battery light,
- portable (battery powered) audio mixer,
- ac adaptor,
- required batteries,
- cassettes (or tapes).

If a broadcaster chooses to use some of the analogue component recording formats, the two first items in the above

Figure 43.5 Single man operation with a camcorder (S Kragujevic Beograd)

list will probably be replaced by a camcorder. The number of microphones will depend on the foreseen tasks, but it is wise to equip a crew with a reduced but comprehensive selection of microphones (a general purpose cardioid dynamic microphone, a highly directional one, a lavalier or clip microphone with a reasonable length of cable, and, as an option, a low power wireless microphone system). A single portable battery light will certainly not offer ideal lighting conditions for shooting, but more lighting would unavoidably require additional crew members, more transportation space and the availability of an ac power supply. It would thus reduce the operational flexibility of the ENG operation.

The portable audio mixer will probably not be used too often and may be shared by several crews. The batteries are probably the weakest link in any chain depending on that sort of power supply. They are prone to leakage, temperature oversensitive, need charging and recharging with extreme care (a forced fast charging seriously reduces their lifespan) and have an unfavourable ratio of capacity to size and weight. For all these reasons, and bearing in mind that the batteries in some cameras and/or video tape recorders are designed to allow recording and viewing of only one single cassette, it is strongly recommended that no ENG crew leaves base without a set of spare batteries. A twofold increase in power supply weight is largely repaid by a multifold increase in operational reliability.

If an ENG crew has to travel abroad, or travel relatively far away from its base, the above list has to be complemented with:

- battery charger,
- portable colour monitor,
- portable video tape player,
- test charts,
- multimeter,
- tools, cleaning material, gaffer tape, etc.

The need for a battery charger is rather obvious, like the need for a colour monitor with sound. Since some recording formats do not have playback facilities incorporated in the portable recorder, it is advisable, when operating far away from the base, to ensure that the crew will be able to preview recorded tapes in the field. Test charts are needed for the realignment of the camera, and the multimeter to check the offered ac and dc supplies, etc. Tools and cleaning material will considerably enhance a crew's autonomy in the field. Some manufacturers offer special 'production boxes' fitted with all possible requirements, for example: two rolls of gaffer tape, a nylon rope, a package of plastic garbage bags, a folding umbrella, a tool kit, a soldering iron, a magnetic head cleaning kit, a bottle of alcohol, chamois cloth, paper towels, rolls of aluminium foil, work gloves, stop watch, felt tip pens and logging sheets, writing pad, white adhesive tape, a can of dulling spray, ac extension cable with a four way termination box, set of different connectors and plug adaptors.

If an assignment in the field is connected with the need to establish a temporary base, to process news stories before sending them back home, it will be necessary to add editing and playback equipment to the list.

The need for a specially equipped car has been a topic for debate since the early days of ENG. On one hand, a car or van containing all the necessary and back-up equipment, as well as all possible auxiliary items, may be of considerable help to the crews in the field. On the other hand, it will unavoidably reduce their flexibility. It seems that a car is especially useful in situations where ENG crews are mainly covering city news and extensively using microwave links to beam the signal back to the studio. Otherwise, the absence of a car directs one to the fastest (or the most practical) means of transportation, and encourages an appropriate selection of equipment to meet the requirements of the given assignment.

Some broadcasters have compromised with an estate car fitted with an appropriate fastening arrangement for standard ENG equipment cases and with a radio communication station. Such a car can be used for other purposes when the ENG crew does not require it.

43.2.6 Preparations and checking

Electronic news gathering is now well established and has inherited many practices present in any broadcasting organization. It is, therefore, rather difficult to recommend a single set of routine checks and preparatory actions, valid in all cases. Nevertheless, some general rules might be acceptable everywhere.

Although the everyday routine assignments do not need a special planning session, a special project, as well as an assignment involving foreign travel, should be preceded by a planning meeting. This meeting will make clear what equipment and accessories are needed. It is then necessary to check the local television standards and the formats used by the local broadcaster, as well as the mains voltage and type of plugs and sockets. If the planned assignment requires the use of wireless microphones or other rf equipment, it is mandatory to contact the local authorities (preferably through the local broadcaster). It might be advisable to hire such equipment locally, to avoid difficulties with frequency allocations and required permits.

Special attention should be paid to the environmental conditions. Although modern electronic equipment is designed to work properly in a wide range of ambient temperatures and humidity, extremes of cold or heat may have a pernicious effect. Protection against extreme conditions can be provided by appropriate equipment covers, or by other special measures.

For foreign travel it is essential that all paperwork and documentation is accurate and complete.

Before any complex assignment, or before the first daily assignment in the case of routine work, certain checks are advisable. The camera should be checked for picture, alignment, the functioning of all operational controls, registration, back focus, framing and grey scale tracking. Test recordings should be made on the video tape recorder of both video and audio signals. The contents of 'blank cassettes' should be checked to ensure that the right ones have been supplied. Some broadcasters bulk erase cassettes for portable recorders before handing them to ENG crews, while some authorize the erasure of the previously recorded material during the next recording.

To get the best from electronic equipment, it must be looked after and precisely aligned. All ENG equipment should therefore be sent at regular intervals to the maintenance department for a more detailed checking and optimizing. This will guarantee the quality of its performance and improve its reliability — of paramount importance in any news gathering operation.

43.3 Electronic field production

The development of high quality portable colour cameras, and especially the availability of portable video tape recorders capable of delivering the same sort of quality as top of the range studio recorders (one-inch format B or C), had its impact in another area — on-location production.

At the same time there developed an emphasis on post-production which became more innovative as video tape editing evolved. The cinematographic approach — basically a single camera production — had many attractive features, including an easier relationship with the performer, better lighting, etc. It was, however, less effective than the multi-camera approach typical of television. Electronic field production (EFP) tried therefore to establish a new operational technology somewhere in the grey area between classical television and film. On one extreme of that area appeared 'EFP vans'. These were, in fact, conventional OB vans of a smaller size and equipped with lightweight cameras. At the other extreme is 'electronic cinematography' (see below), where the production rules are almost identical to the cinematographic ones, and whose major aim is to achieve the 'film look'.

A possible distinction between a more or less 'classical' OB operation and an EFP one might be based on the proportion of production edits (cuts or junctions between different picture sources performed on a production video switcher during the recording process) to post-production edits. If the ratio of production edits to post-production edits is greater than one we are facing an OB production. If it is less, then we have an EFP production.

The programme content will dictate the choice of production technology. A documentary or high budget drama will most certainly be best served by a pure single camera technique. On the other hand a situation comedy or a light entertainment show may use a classical multi-camera approach or a specific cross-breed. For years the film industry used a multi-camera approach for some critical sequences (like the final court sequence in Hitchcock's *Witness for the Prosecution*) or for the fast production of television serials (mainly situation comedies). The 'cross-breed' approach involves replacing three film cameras by three electronic ones, linked to three video tape recorders (and sometimes also to the input of a production switcher whose output is connected in turn to a fourth video tape recorder). Such a programme is edited by synchronously running three tapes (synchronization achieved by means of the recorded time code) and routeing the reproduced signals to a production switcher. This 'rough cut' tape is then transferred to a more conventional editing system for the final re-editing.

Whatever EFP technique is used, it will be necessary in post-production not only to deal with a considerable number of edits, but also to handle more sophisticated transitions and most probably many more generations than with a conventional OB production. This requires *off-line* editing systems capable of providing flexible and time-sparing re-editing. These systems are usually based on the utilization of an auxiliary magnetic recording format (cassette or disc) for the whole editing procedure, and the *simulation* of edits. Simulation of the edits allows easy changes in the established 'edited' sequence, where deletions can be restored or changed. The final decisions are transferred to a storage medium that will be used to instruct the computer controlling the machines with the original recordings and the master machine. In such a process the sound is treated separately, using a dub of the edited final video version. With such video and audio post-production facilities, the single camera EFP approach uses the best of both film and television technologies.

Electronic cinematography originally meant a single camera production. Later, with broadcasters demanding a 'film look', the industry responded with specially conceived cameras. Meanwhile, the term became associated with the idea of producing cinema films by electronic means (with all production and post-production performed by electronic means and the final version transferred to 35 mm film for release). Equipment developed for the earlier concept of electronic cinematography was inadequate for such a use, in fact any equipment based on existing 525 or 625 scanning standards was inadequate for the electronic production of cinema films. Higher definition television is necessary. A new standard with at least twice the horizontal and twice the vertical resolution of the present ones is required (see section *63*).

Bibliography

BATTISTA, T A and FLAHERTY, J A, 'The all-electronic newsgathering station', *Journal SMPTE* **84**, No.12 (1975)

FLAHERTY, J A and NICHOLLS, W C, 'Editing systems for single-camera production', *121st SMPTE Technical Conference*, Los Angeles (1979)

HERICOURT, M F and Le BRETON, D J, 'Desirable operational and functional characteristics of EFP cameras', *EBU Review* No.188 (1981)

MORRISON, W A and BROGDEN, D, 'ENG' *Australian Film and Television School*, North Ryde, New South Wales (1981)

SIEDLER, R, 'ENG — Development in Europe in view of Swiss experiences', *11th International Television Symposium*, Montreux (1979)

STICKLER, M J, GRANT, D W and EDWARDSON, S M, 'ENG — Electronic news gathering: Part 1 The engineering planning', *BBC Engineering*, No.112, April (1979)

STOW, R L and WYLAND, G, 'Electronic cinematography; progress towards high definition', *EBU Review*, No.209 (1985)

TARNER, H C J, 'ENG — Electronic news gathering: Part 2 Operational practice and experience during the experimental first year', *BBC Engineering* No.112, April (1979)

EBU Report on Electronic News Gathering, 2ed, Technical Centre of the EBU, Brussels (1981)

Electronic News Gathering — An Insight, ITCA, London (1981)

ENG/EFP Planning Guide, 1st edn, NANBA, Toronto Canada (1983)

J T P Robinson
MVC Crow Ltd

44a Transportable Power Sources: Generators

44.1 On-board power generation

If the specification of an OB vehicle includes a requirement for generation of power, several factors relating to weight loading must be carefully considered in addition to those described in *section 41*.

The first is that of power capacity. If the generator is to supply the complete operational power requirement of the air conditioning and the technical load, then this must be assessed on the basis discussed in section *41.2.3*. This gives the net power requirement and represents the minimum generator capacity. However it is wise to increase this figure by 15 per cent to allow for future demands and then to calculate the gross power requirement by applying derating factors for the normal ambient temperature and average altitude that the generator will be exposed to in the country of destination.

For a naturally aspirated engine, both of these conditions together can account for a further 15 per cent increase in the net figure arrived at earlier if the vehicle is to operate in temperatures of 30° C at high altitude. Under high altitude conditions it would be wise anyway to ensure that the generator prime mover is turbo driven since the altitude derating factor is appreciably less than for a naturally aspirated engine.

The *derating factor* is the difference, expressed as a percentage, between the maximum output power of the generator when it is at an altitude of less than 150 m and/or a temperature of less than 30° C and that when it is operated at an altitude of more than 150 m and/or a temperature higher than 30° C. The maximum output of the generator is thus derated in accordance with a graph similar to *Figure 44.1* if the above criteria of altitude and temperature are met.

Generating set manufacturers publish a wealth of informa-

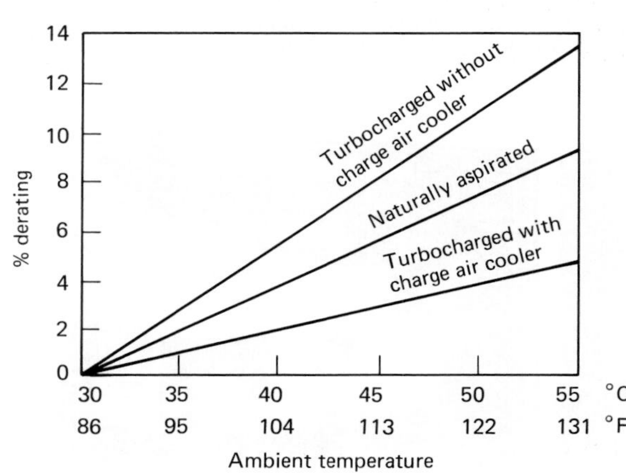

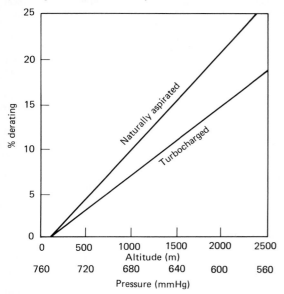

Figure 44.1 Typical derating graphs for diesel generators (G & M Power Plant plc)

tion on the use of their products and it is wise to consult them if there is any doubt concerning the application.

Most generator prime movers are diesel fuelled units rotating at 1500 rpm to give an output frequency of 50 Hz. For 60 Hz operation rotational speed is 1800 rpm. In both cases, the rotational stability is usually insufficient to ensure freedom from so called 'frame' or 'hum' bars and recourse is made to an electronic frequency comparator which maintains a closer control of generator speed than is possible with the conventional mechanical governor.

It is uncommon to find petrol engined generators having a power rating above 5 kW. Though they are free from the familiar low frequency diesel 'thump' and therefore more amenable to acoustic treatment, factors such as economy in running, ruggedness and long life all weigh heavily in favour of the diesel. Such engines will happily run at a constant speed for very long periods and drive alternators having power ratings in hundreds of kilowatts (*Figure 44.3*).

Figure 44.2 Typical 9 kW aircooled diesel generator prior to installation (Onan Corporation)

Because of the maximum width of 2.5 m imposed upon road vehicles, it is not possible to obtain a power output before derating of much more than 30 kW from a generator set mounted across a vehicle. This restriction is imposed by the dimension of the diesel engine when directly coupled to the alternator and when both are mounted on a common flat bed. In addition, the size of the acoustic housing for the generator set will add considerably to the overall dimensions. The limit could be raised by splitting the diesel engine from the alternator so that the former is mounted below the latter and connected by a belt and pulley system, but other problems would then arise not the least of which would be those associated with servicing. Thus if powers much above 30 kW are required in a mobile form, the set must be placed in line with the chassis. The vehicle then becomes a dedicated mobile power generating unit in its own right.

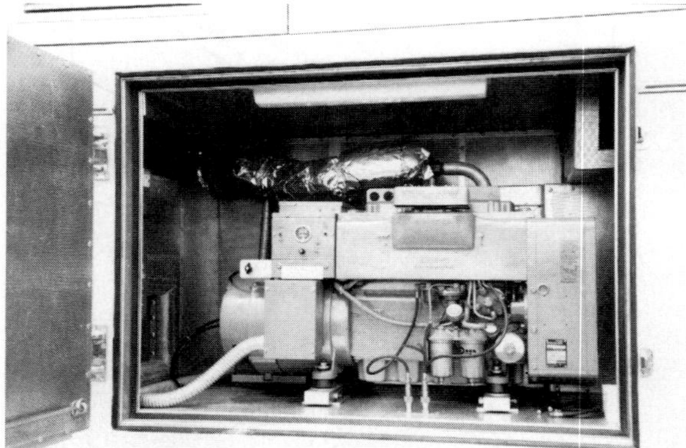

Figure 44.3 9 kW aircooled diesel generator installed in acoustic housing at rear of vehicle (MCV Crow Ltd)

With the diesel alternator set mounted conventionally at the rear of the vehicle and placed across its width, it is advisable to ensure that access can be obtained to the 'blind' side of the engine for maintenance. This can be done by building the alternator set on heavy duty telescopic slides so that the whole unit may be pulled out of its housing. For major servicing it must be possible to easily disconnect the generator from fuel, electrical and exhaust connections and lift the unit clear of the vehicle (see *Figure 44.3*).

Diesel engines can be cooled by either air or water but the higher output units will usually be water cooled, the break point being around 10–15 kW. If an air cooled unit is to be installed into a vehicle without an undue noise penalty inside and outside the vehicle, particular care must be taken to meet the manufacturer's air flow figures across the engine and alternator without creating undue air flow noise. This volume of air flow however must not be at the expense of the acoustic treatment and a careful balance must be reached between the two.

A water cooled unit installed in a vehicle will probably provide better noise figures than an equivalent air cooled unit. Such noise figures are expressed in acoustic decibels (dBA) and are taken within the operational area and external to the vehicle with the engine running on full load. It is important to recognize that the noise spectrum emitted by an alternator set will differ between full load and no load as it will between air and water cooling so that a weighting of the individual frequency readings as given by the noise meter must be made to arrive at the dBA figure. A noise figure of between 50 and 55 dBA in an operational area of a mobile control room and a figure of 65 dBA at 1 m from the air outlet external to the generator housing would be considered quite good for a medium sized vehicle with an on-board air cooled diesel generator.

Vibration must also be taken into account. Diesel engines, by the basic nature of their operation, will require more preventive treatment than will a petrol engine. The generating set manufacturer will have provided a degree of mechanical isolation, but further treatment will be needed during vehicle installation. This again is an area where specialized knowledge is required to obtain the right amount of treatment and to integrate this with the overall noise reduction.

Having decided upon the appropriate generator output for the conditions under which it is to be used, the generator weight as given by the manufacturer plus an allowance for fuel must be added to the calculated weight of acoustic treatment needed to meet the required noise figure. Since the material necessary to attenuate the basic generator noise will consist of steel, zinc sheet, hardwood and very dense rockwool, their use will add considerably to the overall weight of the vehicle. A useful rule of thumb is to double the generator set weight. Thus if the selected generator has a net weight of 300 kg, then a figure of 600 kg should be used as the guide weight for the on-board silenced generator. This figure is then added to the payload figure.

44.2 Trailer generators

The alternative to on-board power generation is for the alternator set to be housed inside an acoustic or semi-acoustic canopy mounted on either a single or a twin axle trailer according to the weight involved and towed behind the OB vehicle.

This arrangement has the following advantages:

1 The weight penalty is removed from the OB vehicle itself and is transferred on to wheels. This weight can now be towed behind the main vehicle and in most cases can be neglected for all but the smallest of vehicles. For example, a 16 tonne (16 000 kg) GVW vehicle can legally and safely tow an 8 tonne trailer if necessary. However, a 50 kW semi-silenced trailer generating set will not weigh more than about 2.5 tonnes and will present no problem to such a vehicle (*Figures 44.4* and *44.5*).
2 When on site the trailer generator can be detached from the main vehicle and parked away from the area where the televised action is to take place. This arrangement will provide a greater degree of noise immunity for the vehicle but care is required in parking the trailer in relation to the microphones.

Figure 44.4 50 kW trailer generator towed behind 16 tonne GVW OB vehicle

Figure 44.5 Interior of 50 kW trailer generator showing the prime mover on the left with the alternator/control and metering section on the right (G & M Power Plant plc)

3 The trailer generator, being an independent unit, can be used for other purposes requiring portable power.
4 Servicing and maintenance of the trailer generator is simpler than that of an on-board generator.

A trailer generating set has the disadvantage of being a separate unit that can never take the place of the integrated facility which travels on the vehicle and literally provides power at the press of a button. Certainly for an ENG vehicle engaged in news coverage or an EFP vehicle where constant stop and start recordings are made at widely different venues, the on-board generator is essential.

On the move recording and transmission productions using highly mobile OB units must also have on-board power facility but in this case recourse may be made to a lead/acid battery and an inverter.

D Hardy
Design Manager, PAG Ltd, and
C Debnam
Quality Assurance Manager, PAG Ltd

44b Transportable Power Sources: Batteries

With the increased use of EFP and ENG equipment, reliable lightweight rechargeable power sources are of paramount importance. Supplies have been obtained from sources ranging from dry primary batteries to heavy wet lead/acid secondary systems, but the broadcast industry has settled on the best compromise in the universal adoption of sealed nickel-cadmium battery systems (*Figure 44.6*).

44.3 Battery sources

Initially, ENG/EFP equipment consisted of separate camera, recorder and lighting equipment, all of which required power. The ENG system has evolved to the point where all functions including lighting can be mounted on-board the camera. A typical example is the Sony Betacam (camcorder combination) with a 100 W 12 V lamp mounted directly onto the camera carrying handle.

The current consumption of the latest camcorder combinations is as low as 2–2.5 A with all functions operating (such as the recorder and lens focusing). The total current drain at 12 V is approximately 10–10.5 A, including lighting with a power of 100 W.

In the ENG situation, all the equipment weight is generally borne on the shoulder of the camera operator, and due consideration must be given to balance and to keeping the weight of the battery to a minimum consistent with reasonable capacity. A 4 Ah battery has been adopted by most users in the form of the Sony BP90, PAG Master 90 or Anton Bauer Propak 90. All of these batteries conform to the same voltage, size and connector standards.

Problems have been encountered by adopting 12 V batteries as the standard due to the lock-out voltages (the voltage at which the camera automatically turns off) inherent in the new generations of camera/camcorder systems. Generally, the lock-out voltages are in the 11.3–11.7 V range. Consequently, the full range of the 12 V nickel-cadmium battery is not being utilized.

The useful range of the single nickel-cadmium cell is from fully charged (around 1.3 V) down to 1 V. However, in a 12 V (10 cell) battery, the capacity actually used on a modern ENG/EFP camera is only between 1.3 and 1.15 V per cell.

An approach worth considering is the adoption of an 11 cell (13.2 V) battery, given that camera lock-out voltages will remain in the 11.4–11.7 V band. The 11 cell battery will be utilized throughout the full cycle range with consequent improvement in camera running times.

Many broadcasters and professional video users employ the 13.2 V battery approach and are finding that the increase in camera/camcorder running times is worthwhile.

An alternative to the on-board power source is a *battery belt*. Power belts utilizing nickel-cadmium cells are available up to 30 V, with capacities up to 10 Ah.

Like all systems, battery belts have advantages and disadvantages. They are a convenient way of carrying weight, since the burden is borne around the wearer's hips. A small person may however find them uncomfortable or insecure. Many manufacturers offer belts in the range 12–30 V including Cine 60, Christie and PAG. Most belts have a built-in charger for overnight slow charging (14 h) from ac mains supplies.

While the camera system running time is an important factor for ENG/EFP crews, this must be balanced against the need for ease of portability. Aspects such as weight and bulk of the equipment, particularly when the camera is being used on the operator's shoulder, become vital. Tripod mounting of the camera does permit the use of larger capacity on-board batteries with a consequent increase in available running times. In the tripod mounted role, the camera can be fitted with a 7 Ah battery increasing the running time by 75 per cent.

44.3.1 Battery fitting systems

Battery manufacturers have devoted effort to the problems of increased running times and methods of attaching larger batteries to the camera systems currently available. The leaders in this field are Anton Bauer (USA) and PAG (Europe).

The Anton Bauer system (*Snap-on*) employs three mushroom lugs for mechanical fitting with 2.1 mm 'banana' pins as the electrical contacts. PAG offer the PAGLOK system which incorporates additional negative and positive self cleaning electrical contacts and a four lug locking mechanical fitting. Both systems are available for the majority of professional broadcast video cameras.

44.3.2 Integral batteries

The Sony NP1 style of battery (12 V, 1.7 Ah) has been adopted

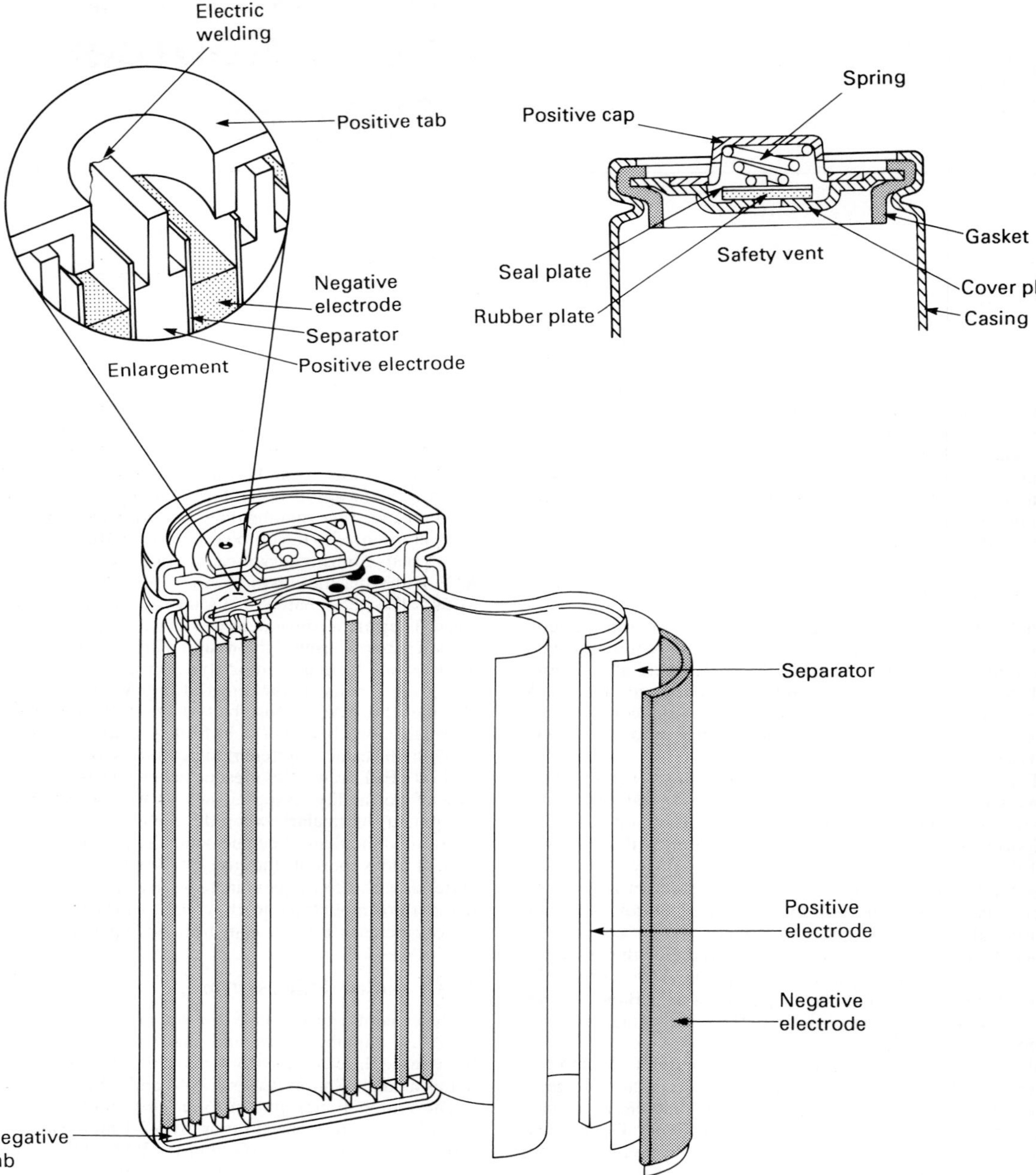

Figure 44.6 Cross-section of a typical sealed nickel-cadmium cell

as the integral battery for the professional broadcast video camera. It is small (185 × 70 × 24 mm) with flat electrical contacts that automatically connect when the battery is fitted into the camera housing. It has many imitators, and this style of battery is available from other manufacturers in 12.0, 13.2 and 14.4 V versions.

While the NP1 battery is small and light, it has one serious drawback — a comparatively low capacity. Despite the shorter running times achieved with NP1 style batteries, they are becoming popular with users.

New cameras tend to use less power than the preceding models. CCD cameras demand far less power than earlier tube versions. Future cameras may well require even less power, so making the smaller NP1 style battery a much more viable power source.

44.4 Battery charging systems

The nickel-cadmium battery is dependent upon good charging management to ensure an economic duty life. With adequate care and a controlled charging regime, it is not unusual to achieve up to 900–1000 cycles from a good quality nickel-cadmium battery. Nickel-cadmium batteries must be charged at a constant current and *not* at a constant voltage (as is specified for lead/acid batteries). Most of the cells used in broadcast standard batteries will accept charging at a fast (1 h) rate or at a slow (14 h) rate.

There are many chargers on the market to cover the overnight or 14 h rate charging requirement. Many users are content with 14 h charging at the C/10 (one-tenth capacity) rate, but the broadcast and professional video markets are increasingly converting to fast charging.

Slow charging is simple and reliable, since no sophisticated control is needed over the point of charge termination. Because the charge is normally confined to the C/10 rate, there is little chance of battery damage through excessive overcharge unless batteries are left connected for periods well in excess of 24 h.

In modern broadcast batteries from Japan, the USA and UK, the cells are specially selected for their fast charge characteristics coupled with high voltage hold up on discharge.

With these cells, fast charging is usually carried out using a charge current of between 2 and 4 A. The cells will accept charge at this rate, but will not tolerate long periods of overcharge without serious damage such as venting of electrolyte or swelling. Reliable charge termination is therefore of paramount importance. It should be noted that not all nickel-cadmium cells are designed to cope with fast charging (C rate or 1 h rate), and cell damage may occur if high charge currents are used. If in doubt, the battery manufacturer should be consulted.

Each of the main charger manufacturers, Anton Bauer (USA), PAG (UK) and Sony (Japan), employs a different method to cope with the need to fully charge the battery and to terminate charge safely.

The Anton Bauer concept uses temperature sensing as the basis for terminating charge and is reliant upon temperature sensors fitted inside the battery. This technique operates successfully but has the limitation that only Anton Bauer batteries can be fully charged on an Anton Bauer charger since the charge termination is dependent upon a third wire linking the battery sensors to the charger.

The PAG system for fast charging employs a specialist microcomputer to monitor the battery response to the applied charge current. This system is extremely flexible since it does not depend upon any third wire connection between battery and charger. The PAG fast chargers can successfully and safely fast charge any fast chargeable battery in the range of 2–10 Ah.

Sony tend to manufacture specialist fast chargers for specific batteries such as the BP90 (12 V, 4 Ah) and the NP1A (12 V, 1.7 Ah). The chargers are dedicated and are not recommended for charging batteries which do not conform exactly to the Sony standards with regard to battery voltage and internal battery protection circuitry.

44.5 Nickel-cadmium battery management

Appropriate care of sealed nickel-cadmium batteries will result in optimum performance and a long and useful life.

44.5.1 General treatment

Nickel-cadmium batteries are suitable for use in any orientation. They must be protected from severe shock and vibration.

Keep the batteries away from intense heat. Do not immerse them in water or expose them to driving rain, steam or high humidity.

Never short-circuit nickel-cadmium batteries: they are capable of delivering very large currents which could result in fire.

44.5.2 Maintenance

Nickel-cadmium batteries need little routine maintenance in normal use. If a battery should get wet, shake out any excess water and allow it to dry naturally in a warm dry place. Do not attempt to use the battery until it has fully dried.

If the battery fuse blows repeatedly, there is almost certainly a fault within the battery pack, the connectors, the equipment or the cables. These should all be checked for possible short-circuits. Never replace the fuse with one of a higher rating than recommended.

When repairing a nickel-cadmium battery do not replace cells indiscriminately. Battery performance is only as good as the worst cell.

44.5.3 Charging

Nickel-cadmium batteries should only be charged using a constant-current charger. This type of charger maintains the correct charging current irrespective of mains voltage and battery condition. Do not be tempted to use a charger intended for lead/acid batteries. The uncontrolled current could cause permanent damage to the battery cells.

44.5.3.1 Slow (overnight) charge

- Allow the battery to cool after heavy discharge. The ideal temperature for charging is around 20° C.
- Connect the battery to the charger ensuring correct polarity.
- Charge at C/10 rate for 14 h (i.e. 400 mA for a 4 Ah battery, 700 mA for a 7 Ah battery).

Nickel-cadmium batteries may be left on continuous charge at the C/10 rate for periods in excess of 24 h. Prolonged charging beyond this time is not recommended as this will lead to battery heating which can have a long term detrimental effect.

44.5.3.2 Fast charge

Always ensure that the batteries are *fully compatible* with the fast charger being used and observe the following:

- Allow the battery to cool after heavy discharge.

- Connect the battery to the charger ensuring correct polarity.
- Charge the battery until the charger indicates that it is charged.

Nickel-cadmium batteries must not be fast charged when they are colder than 0° C as this could lead to venting of hydrogen.

Do not attempt to fast charge nickel-cadmium batteries in *parallel*. The charging current will not be equally shared, and battery damage could result.

Do not attempt to fast charge batteries in *series*, because the batteries will not be sufficiently matched to ensure proper termination of fast charging.

Nickel-cadmium batteries benefit from occasionally being slow charged (C/10 rate) for 24 h. This balances the charge in the cells.

44.5.4 Battery usage

When using batteries to power equipment, start with a fully charged battery whenever possible.

Nickel-cadmium batteries should not be discharged too deeply. If the voltage of a battery pack falls below 75 per cent of its nominal voltage (i.e. 9 V for a 12 V pack), cells may be damaged by voltage reversal thus shortening the battery's life.

Long power cables are to be avoided if possible as their resistance causes voltage losses. Nickel-cadmium batteries should not be connected in parallel to achieve the effect of one large battery. A large current may flow from one pack to the other if they are mismatched.

44.5.5 Capacity testing

Measuring the voltage of a nickel-cadmium battery is not a reliable way to tell whether the battery is charged or discharged, in good condition or bad. Nickel-cadmium batteries have an output voltage that is essentially independent of their state of charge over most of their range. The only accurate way to evaluate the condition of a battery is to discharge it under a similar load to the equipment on which the battery is to be used.

44.5.6 Storage

Store batteries in a cool dry place. Ideally, the temperature should be from -20° to +35° C. High storage temperatures (above 45° C) will reduce the battery's life because of unwanted chemical reactions. Excessively low storage temperatures (below -30° C) are also to be avoided since the electrolyte will freeze and permanent cell damage will result.

After prolonged storage, do not fast charge the battery immediately. The cells should first be reformed and balanced by giving the battery a slow (C/10) charge for 24 h.

Batteries should be given a top-up charge even after one week in storage. At room temperature (20° C), nickel-cadmium batteries lose 1 per cent of their charge every day. The top-up charge needed is one hour at C/10 for each week of storage.

44.5.7 Memory effect

There is a great deal of misunderstanding about the so-called 'memory effect' in nickel-cadmium batteries, which is supposed to be caused by repeated shallow discharging. In practice, this effect is very difficult to produce and does not normally occur. It can be disregarded by most users.

Batteries with their cells seriously out of balance often give this appearance of 'memory effect'. It is particularly noticeable when such batteries are fast charged, where charging is liable to be terminated before all cells are fully charged and so exhibit a reproduceability of low capacity. A 24 h slow charge at the C/10 rate will rebalance a battery provided that the cells are themselves undamaged.

Part 9
Television Sound

M Talbot-Smith B Sc, C Phys, M Inst P
formerly BBC Engineering Training Department

45 Sound Origination Equipment

Equipment for sound origination can conveniently be divided into two categories: *primary* sources and *secondary* sources. Primary sources are those which are, as it were, right at the start of the chain and convert acoustic signals into electrical signals. Microphones are, of course, primary devices. Secondary sources are essentially devices which store the outputs of the primary sources, i.e. recording and reproducing equipment.

45.1 Primary sources

There are a number of basic features that should be present in any professional microphone. While compromises may be necessary in practice, the following items form a basic checklist:

- The *frequency response* should be as flat as possible, although it is desirable where a microphone is to be mounted in a boom or on a hand-held pole that there should be some bass cut below about 150 Hz to reduce the effects of rumble. On some microphones this bass cut is switchable.
- There should be good *transient response*, i.e. response to the important short-lived frequencies present in the first few milliseconds of a sound. This response can only be judged by ear.
- *Sensitivity* may be very significant. An approximate but useful guide is to see what the output is for normal speech at a distance of about 0.5 m. In such conditions a sensitive *electrostatic* microphone will produce a peak output of some 55–60 dB below a reference level of 0.775 V. This is of the order of 1 mV. A typical *moving coil* (dynamic) microphone gives around -70 dB, and a *ribbon* microphone or a low sensitivity moving coil may produce about -80 dB, i.e. about one tenth of the voltage of an electrostatic microphone.

(0.775 V is a standard audio reference voltage. It originated from the fact that it is the voltage across a 600 ohm resistance when 1 mW is dissipated in it.)

- The stated *polar diagram* should be well-maintained over the majority of the audio range.
- The microphone should not be unduly affected by *environmental influences* such as humidity, temperature, stray magnetic fields (which may cause 'hum' in the output), radio frequency pick-up, rumble and vibration.
- The electrical load into which the microphone will be connected should be compatible with existing equipment. It is the usual practice among manufacturers of professional microphones to give them an impedance in the region of 150–200 ohms, but specify that the electrical load should not be less than five or six times this value (around 1000 ohms or more). The microphone inputs on professional mixing consoles generally have an impedance of 1000–1500 ohms, for this reason.
- Electrostatic microphones need a *power supply*. This may be provided by batteries (in which case it is desirable that the batteries are readily available) or by a 'phantom power' system (see section *45.1.2.3.1*) and this also should be compatible with existing equipment.
- Susceptibility to wind noise, or *popping*, is very important if the microphone is to be used out of doors, or if it is to be used close to the mouth as with vocalists' hand microphones. Some microphones have built-in windshielding. Others may need a separate foam or gauze shield. This should not affect the quality of the microphone's output, but at the same time should do its job satisfactorily.
- Microphones used in professional work need to be as *robust* as possible. However carefully microphones are handled, there is always a risk of accidental damage. Therefore a good repair service by the maker is most desirable. This is most likely to be the case if the makers are well-known as long-standing suppliers to the professional studios.
- The availability of suitable *holders* such as clips (for stands), boom cradles, etc., must be looked into. This is not likely to be a problem with good manufacturers.
- There are sometimes advantages in having built-in *attenuators* or frequency correction circuits, selected by switches on the microphone body (see the first item). Some microphones have on/off switches. The latter are often more trouble than help as it can be quite easy to discover, after the microphone has been rigged in a place difficult of access, that the switch is in the off position!
- *Cost* is usually a factor, but not one for which much general advice can be given. It is, however, worth noting that good professional microphones are not cheap to buy but low cost microphones may prove more expensive in the long run.

45.1.1 Microphone sensitivities

Different manufacturers tend to use different units to describe the sensitivity of their products. *Table 45.1* shows with sufficient accuracy the relationship between some of the commonly used sets of units.

db rel 1V/0.1N/m²	*db rel 1V/N/m²*	*mV/μbar*	*mV/10μbar*
−40	−20	9.5	95
−45	−25	5.5	55
–	−30	3.0	30
−55	−35	1.8	18
−60	−40	1.0	10
−65	−45	0.55	5.5
−70	−50	0.3	3.0
−75	−55	0.18	1.8
−80	−60	0.10	1.0
−85	−65	0.055	0.55

Table 45.1 Relation between scales of units of sensitivity

The first column, showing decibels relative to 1 volt/newton/10(metre)^2, is included because average loudness speech at a distance of half a metre produces very roughly a sound pressure of 0.1 newton/m^2. Thus a microphone with a sensitivity of -60 dB rel 1 V/0.1 N/m^2 will give a peak output of approximately 60 dB below 1 V.

45.1.2 Microphone transducers

The transducer converts diaphragm movements into an electrical voltage. In principle there are many mechanisms that could be used. In practice professional microphones use no more than three: moving coil, ribbon and electrostatic.

45.1.2.1 Moving coil transducers

These are sometimes known as *dynamic*. *Figure 45.1* shows the simplified construction. The coil, typically some 20 turns of wire (often aluminium for the sake of lightness) moves in the magnetic field and an emf is generated. A typical impedance is of the order of 30 ohms, although, as stated in section *45.1*, the intended electrical load may be as much as 1000–1500 ohms.

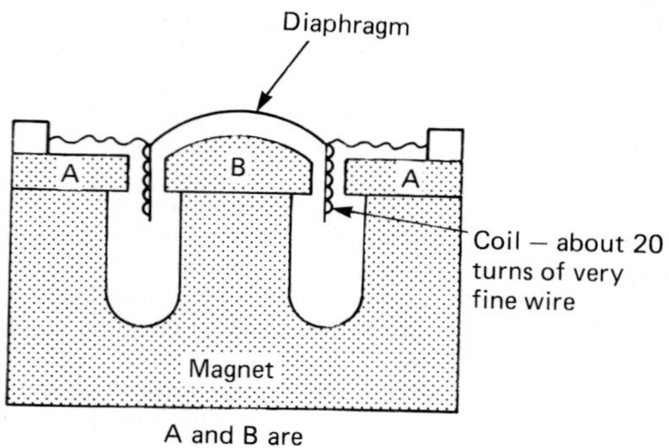

Figure 45.1 Moving coil transducer

Two main advantages of the moving coil transducer are:

- It is generally robust.
- No external power source is needed.

Three disadvantages are:

- While the quality of reproduction may be very satisfactory it is rarely as good as can be obtained with other systems. This is because, light though the diaphragm and its attached coil are, they are nevertheless not as light as those in electrostatic and ribbon microphones.
- The sensitivity is generally some 10–15 dB (or more) below that of an electrostatic microphone.
- The delicate work involved in assembling moving coil microphones makes them relatively expensive.

45.1.2.2 Ribbon transducers

As with the moving coil device, the emf results from the movement of a conductor in a magnetic field, but this time the conductor is linear, being formed of corrugated aluminium. The length of the ribbon is of the order of 2 cm, and it may be from 0.5 cm wide (*Figure 45.2*).

The impedance of the ribbon is very low, usually less than 1 ohm, but an integral transformer is used to produce an impedance at the output of something much higher than this. The emf is also raised, of course. A typical load impedance is, as with other transducers, 1000–1500 ohms.

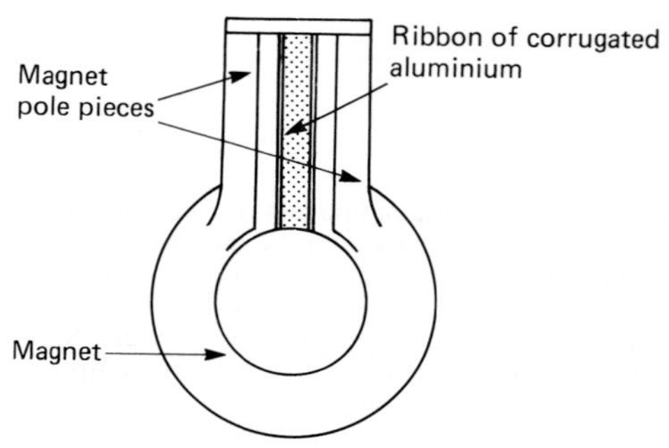

Figure 45.2 Ribbon transducer

Relatively few ribbon microphones are made now, and what are available tend to be expensive. Their main advantages are:

- It is easy to produce a very good figure-of-eight polar diagram (see section *45.1.3.2*).
- The lightness of the ribbon results in a very good response to sound transients.

The disadvantages of a ribbon transducer are:

- It has a relatively high cost.
- The sensitivity is low, being some 20–25 dB below that of an electrostatic microphone.
- The tendency is for it to be large and heavy.
- It is fragile.

Despite these drawbacks many professional organizations still find applications for ribbon microphones.

45.1.2.3 Electrostatic transducers

Essentially an electrostatic transducer consists of a very light circular diaphragm close to a metal plate. The two are typically 0.02 mm apart, insulated from each other and forming a capacitor whose value is generally of the order of 10 pF. The combination is known as the *capsule* (*Figure 45.3*).

Older types of electrostatic microphone (sometimes called *condenser* microphones) had a dc potential of around 50 V applied between the conductive diaphragm and the back plate. More modern microphones make use of *electret* materials, i.e. a permanent electrostatic charge is carried by either plate or diaphragm. In the former system the direct current is supplied series with a very high resistance (500 megohm is typical) so at the time-constant of the CR combination is sufficiently ng for the charge on the capacitor to be effectively constant. ith an electret microphone the charge is constant anyway.

om

$= CV$

ıere Q is constant and C, the capacitance, varies as a result of und wave pressures affecting the diaphragm, then V, the ltage across the capacitor, also varies. Because of the very gh impedances involved it is necessary for a pre-amplifier, ually an fet device, to be mounted as close as possible to the psule.

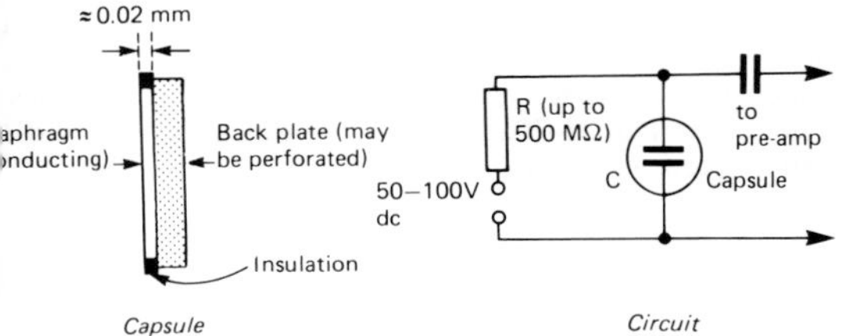

gure 45.3 Electrostatic transducer

Advantages of an electrostatic transducer are:

It generally has very good frequency response and good ansient response, because of the lightness of the diaphragm.
The sensitivity is normally high — around -35 dB relative to V/N/m².
It can be made very small.
It is fairly easy for electrostatic microphones to be designed have interchangeable capsules, etc.
Despite the apparent complexity, manufacture is often not s difficult as is the case with, say, moving coil microphones; ence costs are competitive.

The main disadvantage of the electrostatic transducer is:

Because of the very high insulations needed, proneness to umidity is sometimes a problem.

Condensation on the capsule and its associated components an occur if a cold microphone is brought into a warm studio. 'his usually manifests itself as a marked 'frying' on the output, 'hich will normally disappear after several minutes if the icrophone is put in a warm place.

5.1.2.3.1 Powering arrangements for electrostatic microhones

'he pre-amplifier for the capsule needs a little power for its peration. Some electrostatic microphones have a battery in a uitable case. More commonly a *phantom power* system is used *Figure 45.4*). The standard system uses three-core microphone able to carry an earth (ground) and two programme wires for he audio signal, while the programme wires both take +48 V o the microphone, the earth wire being the return for the ower.

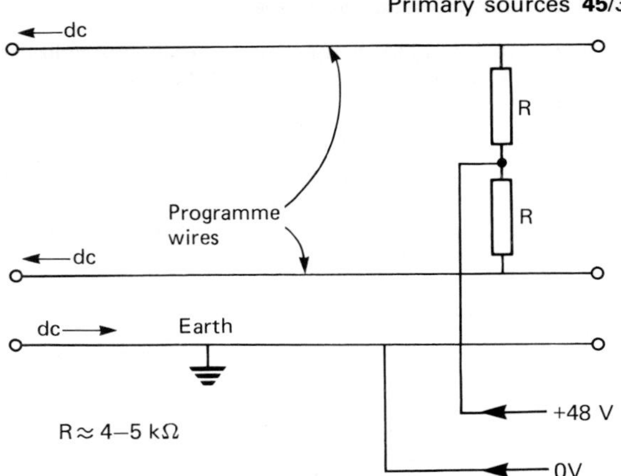

Figure 45.4 Standard 45 V phantom power system

In studio installations it is usual for one power supply to feed all the microphone sockets. Note that non-powered microphones can be plugged into these sockets without ill effects, although it is better to ensure that the appropriate faders are out if loud clicks or bangs in the loudspeaker(s) are to be avoided.

An alternative system, usually for feeding only one microphone at a time, is termed *modulation-lead* powering, or, more commonly, *A-B powering*. Here a low voltage, usually less than 12 V and derived from batteries, is carried to the microphone on the two programme wires. A drawback is that the microphone fails to work if there is an accidental phase reversal in the microphone cable.

45.1.2.4 RF/electrostatic microphones

The capsule is basically similar in design to that in a conventional electrostatic microphone described above. However it is not charged. Instead it forms part of a tuned circuit controlling a frequency modulation discriminator. The latter is fed with rf at several megahertz from a stable oscillator. Changes in capsule capacitance cause variations in the tuning control of the discriminator and consequently produce an audio output. Advantages are:

- high sensitivity,
- very good freedom from humidity problems.

RF electrostatic microphones are much used for outdoor news gathering, exterior filming, and so on.

A disadvantage can be:

- relatively high cost.

45.1.3 Acoustic characteristics of microphones

The polar response, i.e. the sensitivity of the microphone to sounds arriving at different angles, is usually of great importance to a sound engineer. This response is best represented by a *polar diagram* in which the microphone is considered to be at the centre and the distance from there to the graph is a measure of the microphone's sensitivity at each angle. Such diagrams are, of course, shown as two-dimensional plots, but it should be remembered that the polar response of a microphone is three-dimensional. Usually the three-dimensional response can be assumed to be the two-dimensional diagram rotated about the acoustic axis of the microphone. The polar response is determined by the acoustic characteristics of the microphone, in particular the way in which sound waves reach the diaphragm. The most important responses are set out below.

45.1.3.1 Omnidirectional response

The polar diagram is basically a circle (in reality, a sphere). In such a microphone, sound waves are allowed only to reach the front of the diaphragm. This mode of operation is termed *pressure operation* (*Figure 45.5*).

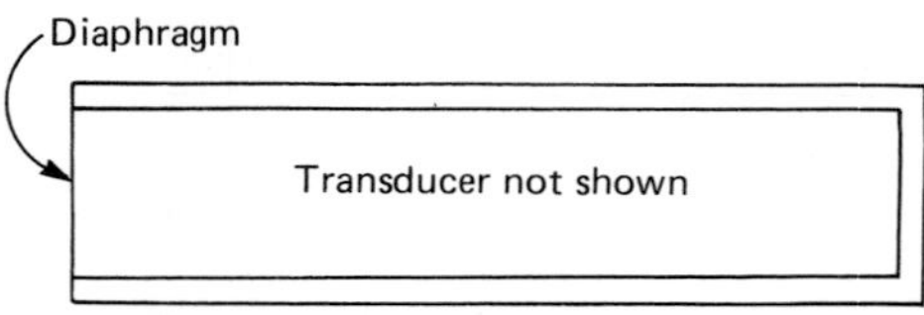

Figure 45.5 A pressure operated microphone (simplified)

It can be assumed that at low frequencies, when the wavelengths are much greater than the dimensions of the microphone, sound waves will diffract round onto the diaphragm no matter what their angle of incidence is. The microphone will thus be truly omnidirectional.

However at higher frequencies, when the sound wavelengths are similar to the diameter of the microphone, this diffraction process does not occur so effectively, and the response of the microphone to sounds from side and rear falls off. For a microphone 2 cm in diameter, the departure from truly omnidirectional behaviour starts to occur at around 3 kHz. For a 1 cm diameter microphone it will be about 6 kHz. It is a useful characteristic of omnidirectional microphones that they are freer from rumble and vibration effects than most other types.

Omnidirectional microphones have limited applications in studios because they tend to pick up unwanted sounds, excessive reverberation, and so on. However, they are perfectly satisfactory for small 'personal' microphones clipped to clothing, and they are also well-suited as hand-held microphones for interviews, when their lack of directionality and relative freedom from rumble can both be advantageous.

45.1.3.2 Figure-of-eight response

The shape of the polar diagram of a figure-of-eight (or *bi-directional* microphone) is self-explanatory. The effect is achieved by allowing sound to reach both sides of a diaphragm. The force on the diaphragm results from the difference in acoustic pressures on the two sides. This difference, in turn, is a consequence of the path difference travelled by sound waves reaching front and back of the diaphragm. The process is generally referred to as *pressure gradient operation*. The obvious type of transducer to permit this is the ribbon type, and indeed most ribbon microphones are figure-of-eight. Electrostatic transducers can, however, be used (see section *45.1.3.4*). In order to have a flat frequency response it is necessary for the diaphragm to have a low mechanical resonant frequency, and this makes figure-of-eight microphones prone to rumble. They also exhibit an effect know as *bass tip up* or *proximity effect*, which means that an excessive bass output results when the sound source is close (typically nearer than 0.5 m) to the microphone, when not only is there a phase difference in the sound waves striking the sides of the diaphragm but also an amplitude difference because of inverse square law effects.

Figure-of-eight microphones are used most commonly in sound studios, where the two 'dead' sides can be useful in reducing the effects of unwanted noises. A particular application is in some so-called *noise-cancelling* microphones. Used close to the mouth the otherwise excessive bass-rise due proximity effects is removed by a built-in bass-cut circu Distant noise, which will not have a bass-rise, is also reduced the low frequency end by the equalizer.

45.1.3.3 Cardioid response

The heart-shaped polar diagram of a cardioid microphone produced by allowing some sound to enter the microphone a reach the back of the diaphragm. Such microphones are usua recognizable by slots or other apertures to the rear of diaphragm. To achieve the cardioid pattern an acoustic lab rinth introduces a delay, or phase shift, into sounds that ha entered the aperture. Cardioid microphones are partly pre sure-gradient operated (see section *45.1.3.2*) and are thus apt exhibit a degree of proximity effect and also a tendency to rumble-prone. It is virtually impossible to design cardi microphones that maintain their polar diagram well over t whole audio range. The front/back ratio, which expresses decibels the difference between front and rear sensitivities, rarely better than 25–30 dB, and may be as little as 5–10 dB some frequencies (*Figure 45.6*).

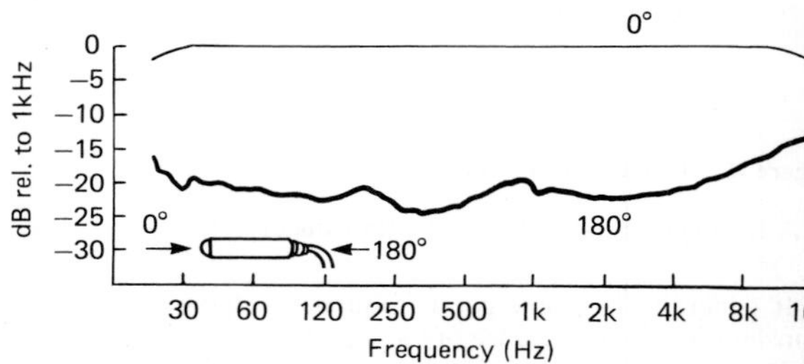

Figure 45.6 Typical front/back ratios for a high-grade cardic microphone

Cardioid microphones are probably the most widely us type in professional work because of their insensitivity sounds arriving from behind the microphone. Televisi booms, vocalists' microphones, musical instrument pickup, a so on, are frequently cardioids.

Their use out-of-doors can be limited unless well win shielded because the apertures behind the diaphragm can ma them vulnerable to wind effects.

45.1.3.4 Variable response

With microphones with a variable polar diagram, a range responses is available, typically omnidirectional, cardioi figure-of-eight and maybe *hypercardioid* (a pattern betwe cardioid and figure-of-eight). Selection of pattern is usually by switch on the body of the microphone, but remote selection sometimes available.

The operation depends on two electrostatic cardioid capsul mounted back-to-back. The overall polar pattern depends c how the outputs of the capsules are combined. Switching o one cardioid results in the response of the remaining cardioid; the two cardioid outputs are added, an omnidirection response is obtained; subtraction gives a figure-of-eight, whi partial addition of one cardioid to the other gives a hyperca dioid (*Figure 45.7*).

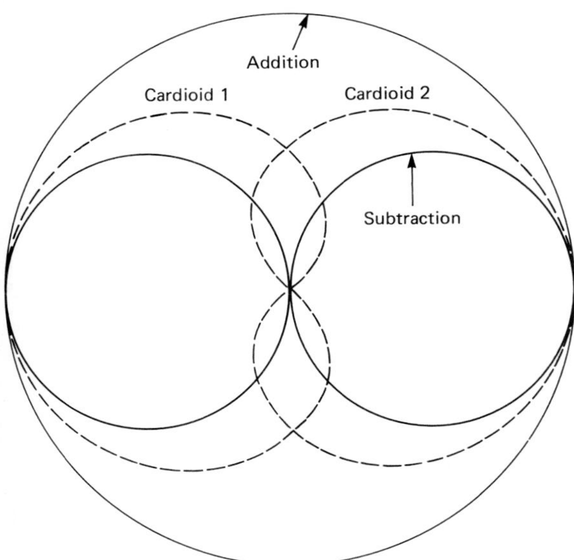

Figure 45.7 Addition and subtraction of cardioids

These microphones are useful for studio work where the ability to vary the polar response is an advantage. They tend to be expensive.

45.1.3.5 'Gun' or 'rifle' microphones

These consist of a slotted or perforated tube, usually about 0.5 m in length, behind which is a normal capsule, often cardioid with either an electrostatic or moving coil transducer. The slotted tube in front of the diaphragm has little effect on sound waves arriving from along its axis. Sounds arriving at an angle greater than about 20° tend to undergo phase cancellations inside the tube because their path length to the diaphragm depends on where they enter the tube. The result is that the microphone is very directional, provided that the sound wavelengths are comparable with, or less than, the length of the tube. At the higher audio frequencies, a good gun microphone will have a flat frequency response over a total front angle of about 30°. At low frequencies, gun microphones tend to be omnidirectional, but the use of a cardioid capsule can help to give some directional effect even then.

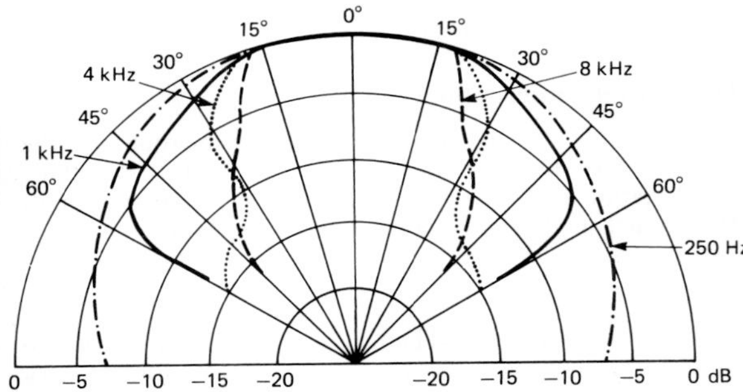

Figure 45.8 Polar diagrams for a typical gun microphone

Gun microphones are used mostly for exterior work such as location shooting and news gathering. For such purposes an rf electrostatic capsule (see section *45.1.2.4*) is to be preferred for its sensitivity and freedom from humidity effects. In a large studio, gun microphones are used for audience contributions, but it should be noted that directionality is generally poor in small rooms. Shorter tubes are often favoured for location shooting as they are lighter and less unwieldy for the operator. Their polar response is wider and approximates to a hypercardioid (*Figure 45.8*).

45.1.4 Specialized microphones

These are microphones basically embodying at least some of the features dealt with in previous sections but with applications that are such as to warrant extra attention.

45.1.4.1 Personal microphones

Personal microphones are very small microphones which can be clipped inconspicuously to clothing, or concealed inside it. The majority have electrostatic (electret) transducers and are pressure-operated (i.e. omnidirectional). Pressure operation is generally employed because such microphones can be made smaller more easily than can cardioids. Also, the relatively rumble-free character of the omnidirectional microphone is an advantage when the microphone is to be worn. Some personal microphones have a battery pack which also needs concealing in clothing. Care has to be taken in fitting personal microphones to artists as the nature of adjacent clothing can have a marked effect on the performance of the microphone.

If circumstances allow the microphone to be visible it is usually sufficient to position it so that it, and possibly its cable, does not rub against clothing. There can be problems, however, when it must be invisible. Man-made fibre materials can generate static electric charges which may discharge through the microphone, resulting in audible effects. Also some materials, especially heavy or closely woven ones, can attenuate the higher audio frequencies. It is therefore very important to carry out tests with the microphone in place at the rehearsal. A measure of equalization is often needed with these microphones, even if outside the clothing, as the microphone is off the high-frequency axis of the voice. It is interesting to note that it is sometimes possible for a conversation between two people, such as an interview, to be picked up with only one of the persons wearing a microphone of this type. This technique cannot, of course, be used in stereo and it is unlikely to be satisfactory for television purposes.

45.1.4.2 Pressure zone microphones

These are devices in which there is a fairly conventional transducer, usually electrostatic and pressure-operated, but which differ from conventional microphones in that the diaphragm is effectively in the plane of a hard reflective surface. In some types the transducer element is mounted above, but very close to, a metal plate so that the diaphragm faces the plate. In others the transducer is behind the reflective surface which may take the form of a wooden block; the diaphragm, suitably protected by a screen, faces upwards and is approximately level with the surface of the block.

Stated simply, the principle is that when sound waves are reflected from a hard surface there is an acoustic pressure at the surface which is greater than the maximum pressure in the sound wave in free field conditions. This pressure increase, occurring in the 'pressure zone', varies according to the size and shape of the reflecting surface, but is typically around double (i.e. 6 dB increase) for large surfaces and normal incidence. For grazing incidence there is no pressure increase. For random incidence an increase in sound level at the diaphragm of 3 dB is often quoted.

It is very important that the diaphragm is as close as possible to the plane of the reflective surface so that:

- it is in the increased pressure region,
- there is no significant phase cancellation causing degradation of sound output (*Figure 45.9*).

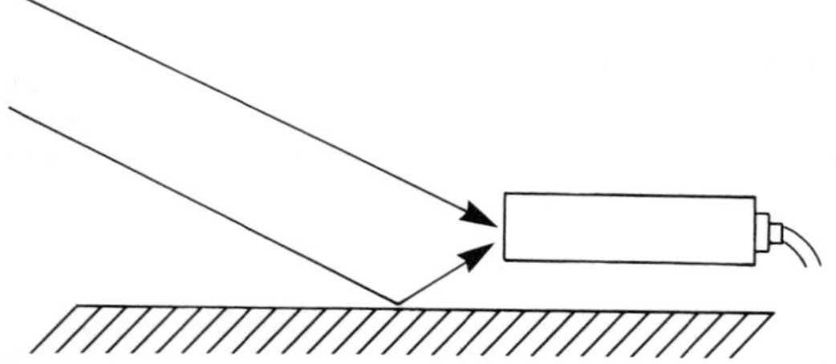

Figure 45.9 Phase cancellation

The reflective plate or surface supplied by manufacturers is of the order of 15–25 cm across; on its own this is too small to be effective except at audio frequencies above at least several hundred hertz. It is generally recommended that the plate is placed on the ground or some other large surface, e.g. a table top. In such conditions the polar response is a hemisphere above the surface.

These microphones can be effective where there is a need to pick up speech from several positions in their vicinity, such as in round table discussions. Within certain limits pick-up is relatively independent of proximity of the source to the microphone.

A reasonably effective pressure zone microphone can be created by taping a small personal microphone to the top of a table.

45.1.4.3 Stereo microphones

These and their use are a large subject and can only be dealt with here in outline. A *coincident pair* has a number of applications in sound recording and broadcasting. Essentially it consists of a pair of microphones mounted as close together as possible. This may be achieved in one of two ways:

- A specifically stereo microphone having two separate capsules, generally mounted one above the other. Their polar responses can be individually adjusted and one capsule can be rotated relative to the other.
- Two conventional mono microphones fitted onto a suitable mounting such as a 'stereo bar'. This is a metal bar some 20 cm long and fitted with clips or other means of attachment. The bar itself can be screwed onto a microphone stand. There may be adjustments so that the two microphones can have their spacing and the angle between them varied.

The choice of polar diagram affects very significantly the nature of the reproduced stereo image, particularly in respect of its width. If the two capsules have their axes mutually at right-angles, then to produce a stereo sound image which extends right across the region between the two reproducing loudspeakers, the sound stage should subtend the following angles (*Figure 45.10*):

- 90° for two figure-of-eight microphones,
- 130° for two hypercardioid microphones,
- 180° for two cardioid microphones.

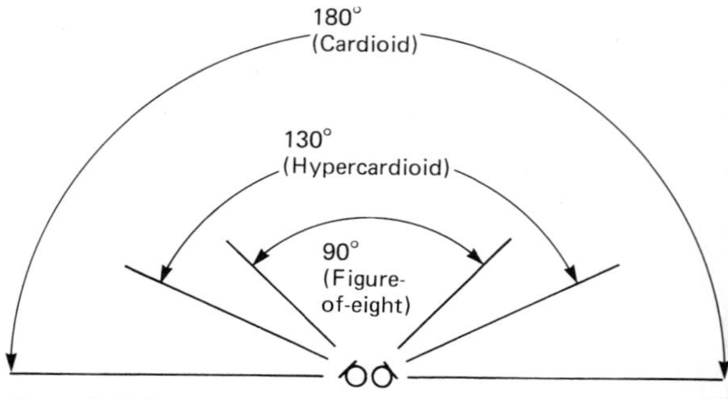

Figure 45.10 Stereo microphone angles of pickup for coincident pairs at 90° to each other

45.1.4.4 Direct injection boxes

These are included as alternatives to microphones. Very simply, a direct injection (DI) box is inserted somewhere between an electrical or electronic musical instrument and its loudspeaker(s). A socket on the box can then be used to feed the instrument's signal to the sound mixing console instead of placing a microphone near the loudspeaker.

DI boxes are not normally expensive but it is advisable to check on their internal electrical insulation properties. A power fault in a mains-powered instrument may in itself be serious enough, but there could be further unfortunate consequences if the output of the DI box were not suitably isolated.

The nature of the sound signal obtained with a DI box may be preferred to that picked up by a microphone in front of the instrument's loudspeaker; there is no risk of the sound of other instruments being picked up and consequently 'separation' is greatly improved. This can be a big advantage where musicians and their instruments are in a cramped space.

45.1.4.5 Soundfield microphones

A soundfield microphone is very complex. The microphone capsule assembly contains four capsules arranged in a regular tetrahedron and is inside a unit that is only slightly larger than many conventional microphones. The output goes to a control box. This may be connected directly to monitoring loudspeakers and line, or a four-track tape machine can be connected to it. The complete unit can be thought of as a coincident pair, the characteristics of which — angle, polar diagram, tilt, etc. — can be controlled on the box. If a four-track recording is made, it can be processed 'off-line', so that the parameters may be adjusted subsequently. The soundfield microphone can be used for ambisonic recordings, although this is of limited interest at present.

45.1.5 Radio microphones

There are two broad categories of radio microphone: the miniature type intended to be concealed in the artist's clothing or incorporated in a hand microphone, and the larger pattern which is strapped to a belt or worn as a backpack.

45.1.5.1 Miniature radio microphones

Usually the transmitter and battery compartment are built into the same housing. The aerial is typically a dipole formed from a thin wire plugged into the casing and the screen of the microphone cable, although the radio hand microphone usually has only an aerial protruding from the base. Where the microphone is separate from the transmitter it is normal to have a small personal microphone, although in principle other types can be used. There may be difficulties, however, if the separate microphone needs 48 V phantom power. An audio limiter is generally built into the system to avoid over-modulation. The receiver may be battery or mains powered. Some types are fitted with signal strength indicators, and there is often a headphone-listening jack as well as a microphone-level output intended to be connected directly into the microphone inputs of a mixing console.

Because of the small size of this type of radio microphone, the batteries which power them also have to be small and this means the radiated power is not great — typically of the order of a few tens of milliwatts. In turn the working range is limited. Under very good conditions, such as line-of-sight in the open air, a range of 300–400 m is possible, but may not be reliable. Inside a studio the range can be very much less, not just because of the dimensions of the studio but because the presence of reflecting metalwork can result in regions where there is interference between direct and reflected signals. A range of as little as 5–10 m is not unknown.

The problem of this interference can be greatly reduced by the use of *diversity reception*. More than one receiving aerial is used and these are placed a few metres apart. The diversity reception unit automatically switches to the output of whichever aerial system is receiving the best rf signal. Most of the miniature radio microphones use a frequency modulated carrier in the vhf bands. The audio performance of the microphone to receiver output link is normally good; the quality of the microphone capsule is often the limiting factor.

The most obvious advantage bestowed by a miniature radio microphone is freedom of movement for the artist, and for many television applications this is very valuable. It must, nevertheless, be recognized that a radio microphone is less reliable than a conventional microphone on the end of a cable. Also if more than one radio microphone is used in a studio, care has to be taken that each is on a different channel. Further, in some environments, there can be serious interference from nearby radio sources.

Careful checks of a radio microphone's performance in all the places where it is going to be used should be carried out before recording or transmission whenever possible. It should not be assumed that anywhere in a studio will be satisfactory. If difficulties are encountered it may help to change the position of the receiver aerial.

45.1.5.2 High power radio microphones

These are more in the nature of small radio link systems. A conventional microphone may be plugged into the transmitter, or, in some cases, a 'head-and-breast set' consisting of a small microphone on a rod (or *boom*) attached to the headphones is used. In the latter case the transmitter/receiver system is obviously a two-way arrangement. Batteries for these units are frequently in a separate pack and are then large enough to provide relatively high power. The rf output may be several watts, giving a range of up to 2–3 km. Radio frequencies used may be in Band 1 or they may be very much higher.

45.2 Secondary sources

45.2.1 Tape recording and reproducing systems (analogue)

The general principles of magnetic tape recording are dealt with in other books (see *Bibliography*), Particular points and a typical specification for a professional recorder will be given here.

45.2.1.1 Tape

The standard tape width for mono or stereo recordings is 6.25 mm, frequently referred to as *quarter-inch tape*. The base is polyester, commonly some 30 μm thick with a coating of suitable oxide of perhaps half this thickness. For stereo recording, the two tracks are usually 2.75 mm wide, with a guard track of about 0.75 mm between them. Twin-track recordings, where there may be totally different types of programme on the two tracks, have a wider guard track of about 2 mm, leaving the active tracks 2.1 mm each. There is then better separation between the tracks but noise figures are less good. For multi-track recording work, the standard tape width is *two inch* (50.8 mm). Standard tape speeds are:

- 38.1 cm/s (15 ips) for general recording,
- 19.05 cm/s (7.5 ips) where tape economy is important and the lower performance can be accepted; this speed is generally quite adequate for speech,
- 76.2 cm/s (30 ips) as a preferred alternative to the use of noise reduction systems at 38.1 cm/s; however standard NAB spools of tape carry only about 16 minutes of recording at this speed.

45.2.1.2 Bias

Satisfactory analogue recordings cannot be made unless a high frequency signal is mixed with the programme signal before it is applied to the record head. This bias frequency is not critical, but is usually in the range 150–200 kHz. The bias level however is critical. For optimum performance, the bias current should normally be slightly higher than is needed for maximum recording sensitivity (see *Figure 45.11*).

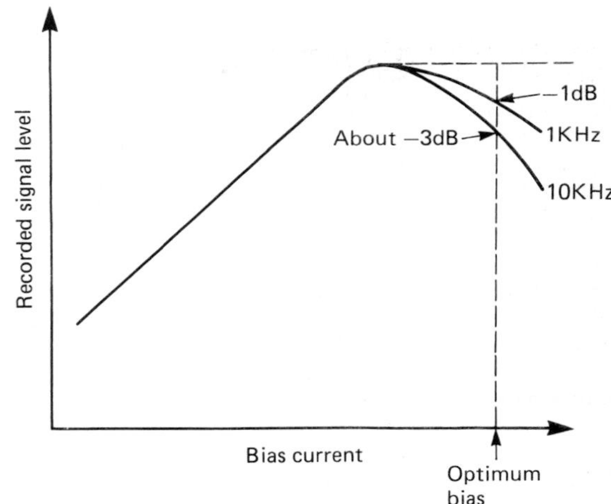

Figure 45.11 Optimum basis

45.2.1.3 Equalization

For further optimization of recording, equalization is needed. The exact characteristics vary; the IEC, for example, sets standards that are common in Europe. In North America, NAB standards are adopted. The IEC recording characteristic is flat up to a frequency which depends on tape speed, being 4.5 kHz for 38.1 cm/s, and then falls off at 6 dB/octave. The corresponding time constant is 35 μs.

Replay equalization, which compensates for the fact that the emf induced in the replay head is proportional to rate of change

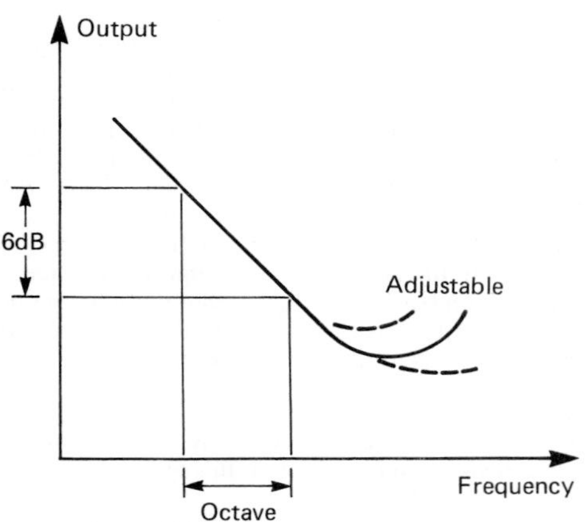

Figure 45.12 Replay equalization curve

of flux, and hence to frequency, is shown by a curve falling at 6 dB/octave (*Figure 45.12*).

45.2.1.4 Compact cassettes

Until relatively recently these cassettes and their machines had no place in the professional world. However, with modern recording and replay machines, marked improvements in tape materials and the use of noise reduction systems, cassettes can be useful, especially in broadcasting where the small size is useful for news gathering, for example. A major problem is not so much the poorer quality compared with full-size machines using 38 cm/s tape, but the virtual impossibility of editing cassettes. The width of cassette tape is 3.81 mm (0.15 in) and the track layout is shown in *Figure 45.13*.

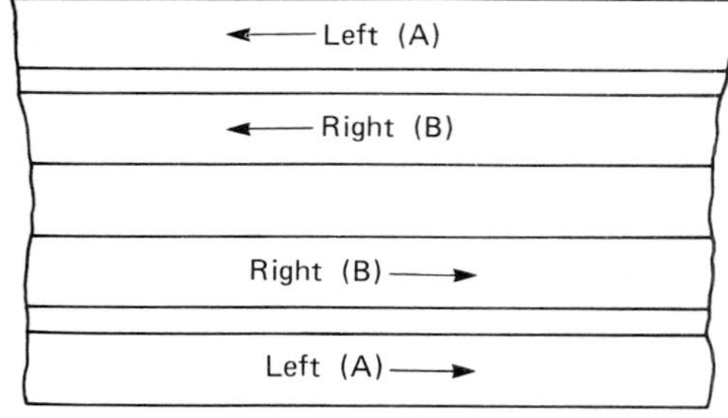

Figure 45.13 Cassette track layout

45.2.1.5 Typical specifications for a professional tape machine

A professional machine may have the following specifications:

Tape speed deviation	±0.2 %
Tape slip	0.1 %
Wow and flutter, weighted, at 38 cm/s	0.04 %
Effective start time	0.5 s
Rewind time for full NAB spool(approx.)	90 s
Input impedance	8–10 k
Input level	+22 dBm max
Output impedance	30 Ω
Load impedance	200 Ω min
Output level	+24 dBm max
Frequency response at 38 cm/s	
30 Hz–18 kHz	±2 dB
60 Hz–15 kHz	±1 dB
Signal/noise ratio, CCIR weighting at 38 cm/s, full track	56 dB
Stereo crosstalk rejection at 1 kHz	45 dB
Erase efficiency at 1 kHz	>75 dB
Erase and bias frequency	150 kHz

45.2.2 Digital tape recording

Brief generalizations cannot be made here as there are several different systems. Almost the only common factors are the sampling rates (44.1 kHz or 48 kHz)! Basically the systems can be divided into two broad categories: stationary head machines and rotary head machines.

Stationary head machines usually use 6.25 mm ('quarter-inch') tape running at 38 cm/s. A typical machine has several data tracks, including SMPTE time code, plus eight digital tracks carrying two 'music' channels. Each channel is thus spread over four digital tracks using very complex coding to reduce the effects of tape 'drop-outs' and also to record the very high bit rate (approaching 1 Mb/s).

Rotary head machines achieve high bit rate recording by using the principle of video tape machines. One popular low-cost system in fact uses a domestic video recorder in conjunction with a unit that converts the input stereo audio signals into digital form and then arranges the digital data into a format that simulates video signals and is thus acceptable to the recorder.

Advantages of digital tape recording are:

- very high signal/noise ratios — some 30 dB better than analogue recordings,
- multiple copying — up to 100–200 times compared with less than half a dozen for analogue,
- print-through and crosstalk between tracks virtually eliminated,
- tape speed constancy less of a problem with a digital system as small cyclic fluctuations can be corrected by storing the digital data and 'clocking' it out at the correct rate,
- bias and equalization unnecessary,
- rotary head machines relatively cheap.

Disadvantages are:

- with rotary head machines, only dub-editing possible,
- stationary head machines expensive.

Cut editing can, however, be carried out with a razor blade on stationary head machines because the digital data are 'scattered' around the tracks. Complex circuitry can be used to interpolate missing data in the region of the cut.

45.2.3 Disc systems

Although analogue discs are threatened by the advent of the compact disc (cd) they are likely to be used or produced by professional studios for some time yet.

Typical performances of digital and analogue systems are given in *Table 45.2*. This listing appears to suggest that the vinyl (analogue) disc is a very inferior alternative to the compact disc. In terms of quality of reproduction this may be true, although a new vinyl disc can produce excellent quality. However, broadcast studios often rely on discs being accurately cued-up for sound effects purposes in radio or television dramas as well as for musical programmes, and many operators feel that in some circumstances they have better control over vinyl discs, which they can see, than over compact discs. The relative invulnerability of compact discs over vinyl discs remains, however, a major point in their favour.

	digital	*analogue*
Frequency range	20Hz-20kHz 0.5dB	30Hz-18kHz 1dB
Signal/noise ratio	±90dB	±60dB
Seperation between channels	90dB	better than 25dB at 1kHz
Total harmonic distortion	0.005%	0.2%
Life expectancy of recordings	disc: probably infinite* Laser: several thousand playings	disc: about 100 playings

* The long-term effects of age on compact discs is not yet known. Shrinkage of the materials may occur, for example, although there seems to be no evidence at present of this.

Table 45.2 Comparison of digital and analogue systems

Bibliography

ROBERTSON, A E, *Microphones*, 2nd Ed., Iliffe (1963)Although published so long ago this is still one of the most exhaustive works on the subject

ALKIN, G, *Sound with Vision*, Newnes-Butterworths (1973)

NISBETT, A, *The Use of Microphones*, 2nd Ed., Focal Press (1983)

ALKIN, G, *Sound Recording and Reproduction*, Butterworths (1981)

AMOS, S W, (editor), *Radio, TV and Audio Technical Reference Book*, Newnes-Butterworths (1977)

M Talbot-Smith B Sc, C Phys, M Inst P
formerly BBC Engineering Training Department

46 Sound Mixing and Control

The majority of professional sound desks or mixing consoles are analogue, and are likely to be for some time. The basic principles of digital mixing are outlined in section *46.2*, but the emphasis here is on analogue principles. A very wide range of equipment is available with a correspondingly wide variation in detail. However, almost all have the same basic 'skeleton', and an understanding of this skeleton should make it possible to acquire a working knowledge of almost any type of analogue mixer very quickly.

46.1 Basic functions of a sound desk

The basic functions of a sound desk or mixing console are:

1 to provide amplification for low-level sources such as microphones, the outputs of which are generally of the order of a millivolt; to minimize the effects of noise, most processes in a mixer are carried out at or about zero level (nominally 0.775 V), and a total gain of 60–70 dB is needed
2 to allow processing of the audio signal by equalization (EQ), the use of automatic gain control devices such as limiters and compressors, and artificial reverberation ('echo'), etc.
3 to allow the operator to control the level of each source and then combine different sources in a way that is satisfactory both technically and artistically
4 to enable the sound signals to be monitored
5 to provide outputs suitable for recording equipment, transmission, and the exchange of signals with other studios, as well as for performers and staff in the studio
6 to provide communications within the studio and with associated areas

In this section, function 1, the provision of suitable amplifiers at points in the desk, will be taken for granted.

Function 2 is best introduced by looking at a typical set of channel facilities. A *channel* in this context means the path of a single source (e.g. the output of a microphone) up to the point where it may be combined with other sources.

46.1.1 Channels

Figure 46.1 illustrates a typical channel. Two input sockets are shown, one with attenuation making it suitable for 'high-level' sources such as tape machines. On some consoles, there may be only one input socket. The function of the *variable gain amplifier* is to bring all sources into a convenient operating range on the fader. Thus insensitive microphones can be given

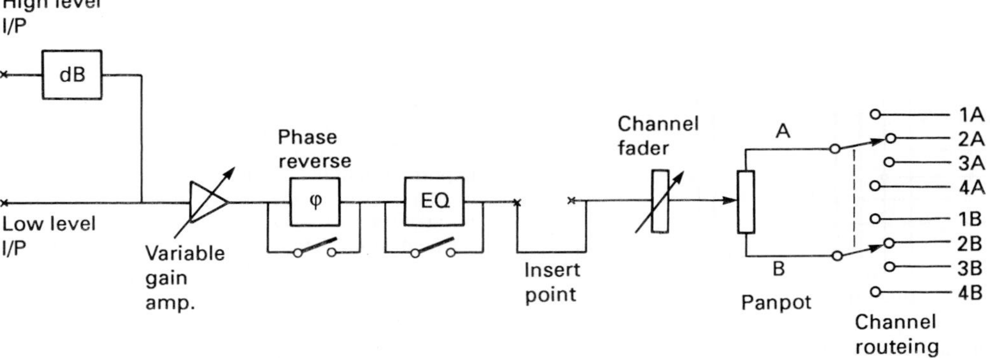

Figure 46.1 A channel on a mixing console (see text)

more gain than ones with a high output. Where there is only one input, there will normally be switching in the variable gain amplifier which produces an input impedance of around 1 k ohms (1.2 k ohms is common) for microphones, and a much higher impedance, often 6.8 k ohms, for high level sources.

The *phase reverse* in effect changes over the connections when incorrectly wired equipment is plugged in.

EQ which, like the phase reverse, can be bypassed, will normally provide a wide range of frequency correction devices, including steep (18 dB/octave) filters at both ends of the frequency range, conventional bass lift/cut, treble lift/cut and mid-frequency (*presence*) lift and cut, and parametric equalizers in which the frequency of, say, the presence lift can be continuously variable.

Insert points, which may be before or after the EQ, or maybe both, allow additional 'outboard' devices to be plugged in. These could include graphic equalizers, limiters or any one of a wide range of processing units. Insert points are usually on a separate jack field (*patch panel*) at the side of the main console.

The *channel fader* needs very little explanation. It meets part of function 3 in the earlier list. The vast majority of faders are *linear*, making use of conductive plastic tracks. However, on small mixers used for location recording with, for example, ENG (electronic news gathering) operations, the faders may be *rotary*. Linear faders are usually arranged so that pushing the slider forwards *fades up*, i.e. increases the level of the signal.

Positioning of a source in a stereo image is achieved by the *pan-pot* (panoramic potentiometer), which is normally a rotary control.

The final item in *Figure 46.1* is the *channel routeing*, shown for simplicity in the figure as a pair of rotary switches, although in practice pushbuttons are normal. Here the signal in the channel is sent to a *group*.

46.1.2 Groups

This simply means the facility of allowing the outputs of more than one channel to be combined together to pass through a *group fader*, which thus acts as a kind of sub-master. A further part of function 3 is then catered for. *Figure 46.2* illustrates grouping; for clarity only the A (left) leg of the stereo system is shown. The B leg is exactly similar.

At the left of the figure are the connections from the routeing switches of three channels, although in a practical desk there may be up to 50 channels and eight or more groups.

There then follow the *group busbars* which combine together the individual channel outputs as selected on the routeing. Each busbar is connected to a *group fader*. Note that some form of buffering is necessary at the busbars to avoid interaction between different channels. Small amplifiers are one way of achieving this. An often satisfactory and much cheaper method is to insert relatively high resistances in the channel feeds but keep the busbars themselves at low impedances.

Each group may contain many of the facilities that were shown in *Figure 46.1* for each channel: insert points, EQ, etc. A consideration of the use of groups will make clearer the need for some duplication of the channel facilities. A good example can be found in music recording where there can be large numbers of microphones, often one to each instrument. It may be very useful to an operator to have one group for all the upper string microphones, another for the lower strings, one for brass, one for woodwind, and so on. This enables the operator, having balanced the individual instruments in a group, then to use the group faders to balance one complete section against another.

If there is a need for, say, limiters in the brass section, it will generally be more convenient, less expensive and possibly artistically better to have one limiter in the group than several such units in the individual brass channels.

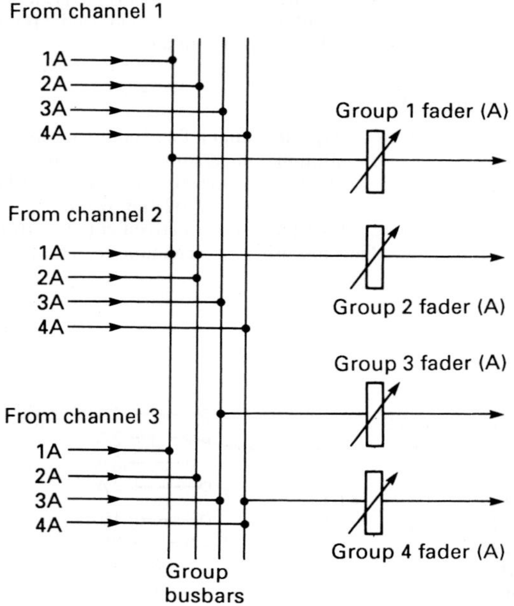

Figure 46.2 Grouping (just one half of the stereo system is shown, for clarity)

46.1.3 Main output

This deals with function 5 in the list. *Figure 46.3* shows a typical output stage in simplified form.

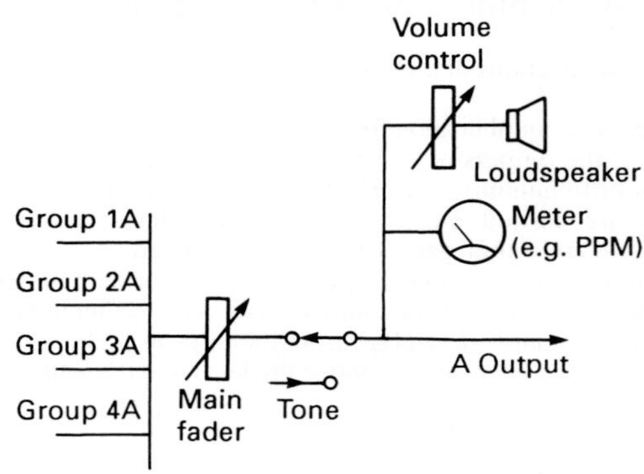

Figure 46.3 A typical output stage (only one half of the stereo path is shown)

The lines on the left represent the outputs from the group faders. (The stereo A signal path only is shown.) The most important item here is the *main* (or *master*) *fader*. This gives overall control of the desk output. It is this fader that will normally be used to *fade in* or *fade out* the studio, and it is also the point at which the operator will obtain a correct signal level, avoiding excessively high levels that may cause distortion in subsequent stages, and at the same time avoiding *under modulation*, which may carry the signal into the noise that is inevitably present sooner or later.

Because the main fader is in a critical position in the signal path and any failure would be catastrophic, there may be safety measures such as dual tracks and a bypass switch to select the second track if the first becomes faulty.

Figure 46.3 also shows the basic elements of *monitoring* (function 4 in the list).

It is often important for an operator to be able, at many stages in the desk, to monitor (i.e. check the signal level and the sound quality of) individual channels, the feeds to external equipment such as echo devices and so on. The illustration simply indicates the monitoring of the desk output. Switching, possibly fairly complex, is needed to allow satisfactory monitoring of all the other stages.

Monitoring needs to be done both by the operator's ears and by his eyes. *Visual monitoring*, using VU meters, PPMs or other display devices, is the only way of checking that the signal level is within its correct limits. Meters, however, can give no indication of the *quality* of the sound nor of the *balance* between different sources.

Loudspeaker monitoring is essential for checking balance and quality generally, including detecting any distortion. In some circumstances, headphones have to be used. It is probably fair to say that headphone monitoring is second best to the use of loudspeakers. This does not necessarily imply that the quality of reproduction of headphones is inferior. The reason is possibly partly psychological, arising from the isolation of the operator. Stereo imaging with headphones is apt to be very different from that perceived using loudspeakers.

Important features of a good stereo monitoring system include, for the loudspeakers, *volume control* (ganged dual controls), *balance*, to ensure equal sensitivity for both speakers, and *phase reverse*, operation of which is sometimes the only way of confirming the correct phase of material.

Mono on both speakers gives a central image which is necessary for setting the balance.

Mono on left simulates conditions for the mono listener better than mono on both. In principle, it clearly does not matter on which loudspeaker the mono signal is reproduced, but in the UK the left (A) speaker is generally preferred.

For visual monitoring, a pair of double pointer PPM meters are generally considered to give the best indications for stereo working. A stereo PPM has a pair of pointers on coaxial shafts rather like the hour and minute hands of a conventional clock. Two such meters side by side can indicate:

- *left signal* (A) and *right signal* (B)
- *M signal* (A + B) and *S signal* (A – B)
- *S + 20* (i.e. the S signal given 20 dB extra gain)

The last is particularly useful in some line-up operations where, to ensure close similarity between the levels of the two channels, adjusting for minimum difference is more accurate than aiming for equal meter settings.

It is usual to work with one meter indicating A and B and the other showing M and S. Either can be switched to S + 20 where needed.

A further feature shown in *Figure 46.3* is the provision for sending a line-up signal (*tone*) to *line*, i.e. the destination. This is usually a zero level (0.775 V) 1 kHz sine-wave, although other frequencies may be selectable. Its purpose is primarily to enable the destination to check that signal levels in the system are correct. Identification by the destination that the source is the correct one can be achieved by 'cutting' the tone at an agreed instant.

It is possible on many broadcast desks to send *tone to line* by injecting it after the monitoring. This allows studio rehearsals to continue while the tone is being sent.

46.1.4 Channel or group peripherals

Feeds to and from artificial reverberation units, public address (PA) and foldback (FB) facilities cover functions 2 and 5 in the list.

Figure 46.4 details the channel fader section of *Figure 46.1*. It shows how suitable signals for these peripherals can be derived. Note that there is generally a similar arrangement at the group faders.

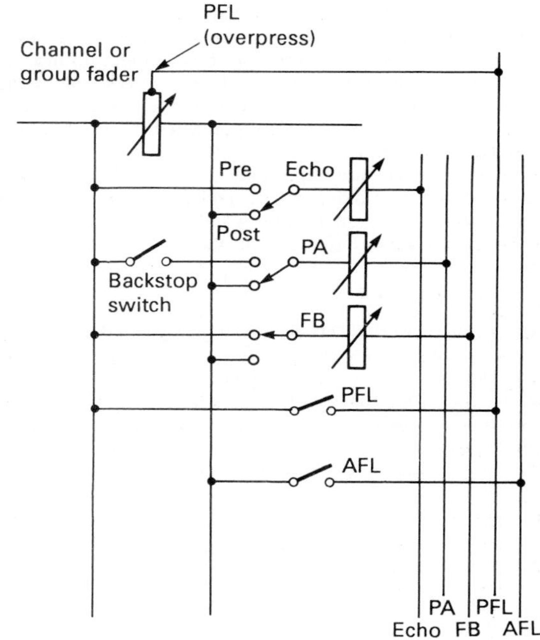

Figure 46.4 Channel fader section

The term *echo*, although strictly incorrect, is used almost universally to mean *artificial reverberation*. We shall use it here in that sense.

It will be seen in *Figure 46.4* that there are two busbars, one before and one after the fader. These allow for *pre-fader* or *post-fader* operation. The reasons for these two modes are outlined briefly in sections *46.1.4.4* and *46.1.4.5*.

46.1.4.1 Echo

A proportion of the channel or group signal is fed to an external artificial reverberation unit, the output of which is returned to be mixed with the 'direct' signal at a later date. Switching allows the choice of pre-fader or post-fader operation. Pre-fader echo derivation means that the fader setting has no effect upon the level of the *echo go* signal (or *echo send* signal — both terms are used). Post-fader selection results in the *echo go* signal being proportional to the direct signal.

From the point of view of the operator, there are advantages and disadvantages in each method. What is often more satisfactory is an *echo mixture* control (not shown), in which the direct signal is decreased as the *echo go* signal is increased and vice versa. The sum of the *direct* and *echo go* signals is thus approximately constant.

A small rotary control can be used to adjust the level of the *echo go* signal. Not shown in *Figure 46.4*, but common on many large sound desks, is the provision of EQ, insert points and monitoring in the echo circuits.

46.1.4.2 Public address (PA)

This output is taken to loudspeakers in the studio. These are positioned around areas where a studio audience, if any, is seated. The purpose is to provide the reinforcement which is almost always needed of the 'live' sound. It is important to note that the PA feed will normally have to be different from the main output of the desk, as some sources will need less reinforcement than others for the audience. The basic circuitry is similar to that for echo, and on some small desks it may be found that such circuits are called *auxiliaries*, or *AUX*. Auxiliary outputs can be used for echo, PA or foldback.

An additional feature sometimes found in the PA circuit is the *fader backstop switch*. In the pre-fader mode, a full level signal is sent to the PA loudspeaker. To prevent this being present at all times, whether needed or not, the PA feed is broken by a microswitch which is open when the fader is fully 'out' and closed as soon as the fader control is moved away from its 'out' position. The use of pre-fader PA ensures that the PA levels in the studio are independent of the fader setting, and this can have certain advantages. Post-fader PA, though, may produce a more artistic effect for the audience.

46.1.4.3 Foldback (FB)

Basically similar to PA, this is again a feed to loudspeakers in the studio, this time for the benefit of performers, when the loudspeaker output may carry such things as a backing track for vocalists, cues from film or videotape sound tracks, and so on. The studio microphones should not normally pick up any of the output of either PA or FB loudspeakers.

On a large desk, the PA and FB feeds will often have insert points where equalizers or other processing equipment can be plugged in. There will also be provision for monitoring.

46.1.4.4 Pre-fade listen (PFL)

This is sometimes called *pre-hear*. It is a facility by means of which an operator can monitor, both visually and via loudspeakers, a source before fading it up. A good example is in working with external contributions (i.e. an *outside source*, or *OS*) when it is vital to check that the contribution circuit is available and that the signal will be at a satisfactory level and of acceptable quality.

PFL may often be accessed in two ways: by operating a switch (usually a button) or by fader overpress. The latter is indicated diagramatically by a connection to the top of the fader symbol. Operation is by pressing the fader control against a sprung microswitch in the faded-out position. It may be necessary to set additional switching in the monitoring system to make the PFL signal appear on meters and/or loudspeakers.

46.1.4.5 After-fade listen (AFL)

The basic concept is simple. The facility enables an operator to check the quality and/or level of an already faded-up source but without affecting the output of the desk. An example is to check for distortion in a particular microphone during a music recording session. Additional switching arrangements are needed to feed the AFL signal to meters and loudspeakers in place of the normal (desk output) feed. An additional visual warning, such as a small light adjacent to the meters, is often arranged to come on when AFL is being used.

46.1.4.6 Echo return

The reverberant version of the signal that comes out of the echo device clearly has to be mixed with the direct signal at some stage. This is commonly near the group faders, where the returned echo signal passes through an *echo fader* (dual tracks for stereo) to join the direct signal either before or after the group faders. There can be operation advantages in each mode, and appropriate switching is available on the larger desks.

46.1.5 Clean feed (mix minus)

This feature is virtually essential on broadcast desks and may be useful on others. It is best described as a desk output from which one or more sources are missing. *Figure 46.5* shows it in a simplified form.

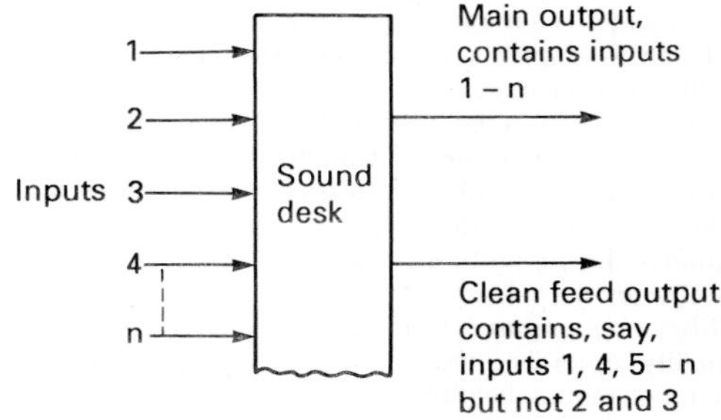

Figure 46.5 Clean feed

Two examples of the use of clean feed are:

- In a two-way interview with a remote studio. If, say, the presenter in the remote studio hears themself in the output of the main studio, this can be disturbing, especially with the delays which can be experienced on satellite links. If the remote presenter has a clean feed from the main studio the problem is overcome (*Figure 46.6*).

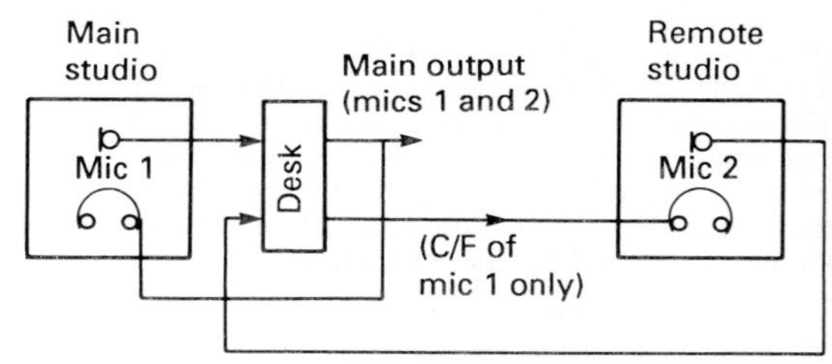

Figure 46.6 Clean feed for two-way interview with remote studio

- A frequent requirement in a music recording is for a vocalist or other musician to have a headphone feed of a particular section of the band. Since this feed would consist of only certain of the sources, it is in fact a clean feed, although it would probably be referred to as a *foldback*.

46.1.6 Multiway working (MWW)

This is a form of multiple clean feeds and is very widely used in broadcasting where a number of remote studios or locations all contribute to the main studio's programme. The basic requirement is that each location receives as an incoming signal a mix of all locations *except itself*. This calls for the use of a *multiway working matrix*, as shown in *Figure 46.7*.

Buffering amplifiers, etc., are omitted from this illustration. The connections ensure that each remote source does not receive its own output.

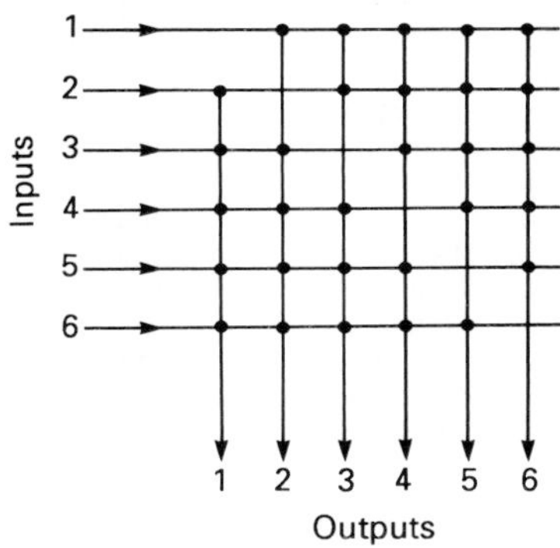

Figure 46.7 Multiway working matrix

46.1.7 Derivation of mono from stereo

It can be assumed that, as a general rule, the simple addition of the A/B (X/Y) stereo signals will give a reasonably satisfactory compatible mono signal, certainly in respect of the balance. There may, however, be problems with levels. If the left and right stereo signals are *coherent* (i.e. they are exactly equal), as is the case with a central sound source, the addition of these gives a mono signal which is 6 dB higher than either. *Non-coherent signals* (i.e. when the left and right bear no relation to each other) when added together give a signal the level of which will depend on circumstances but may be 0 to perhaps 3 dB higher than either left or right.

If the control of the stereo signals is based on the mono level, and the level of the mono signal is monitored to prevent over-modulation (e.g. PPM 6 is taken as the maximum), there is severe undermodulation of A and B with central sound sources. A compromise is therefore usually adopted by the insertion of a fixed attenuator as in *Figure 46.8*.

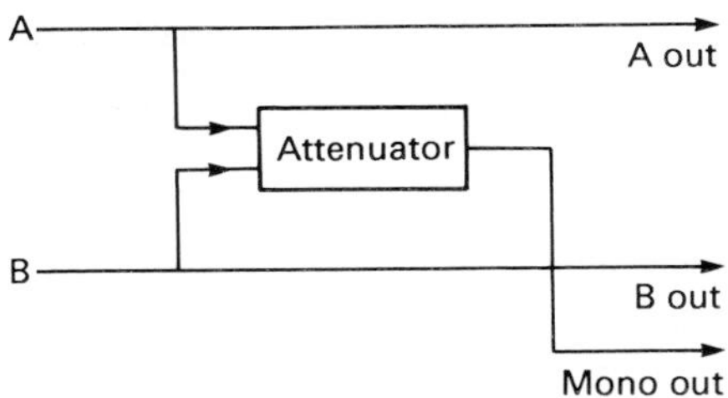

Figure 46.8 Addition of an attenuator to provide a mono signal from stereo

Unfortunately the amount of attenuation is not standard. In the UK, 3 dB, 4.5 dB and 6 dB are all used by various broadcasting organizations. There can thus be confusion when equipment such as a tape machine is being lined up.

46.1.8 Studio communications

Since talkbacks and similar communication channels are audio, it has become the practice in many, if not all, organizations for these to be associated with the sound desk and its peripheral apparatus. The most important studio communications are listed in sections *46.1.8.1* and *46.1.8.2*.

46.1.8.1 Sound-only studios

46.1.8.1.1 Talkback into the studio

This is normally via microphones available to the producer and sound operator, the microphones being operated by a switch or, for the operator, either a switch or a pedal. In *rehearsal* conditions, the talkback in the studio will usually be through a loudspeaker. In *transmission* conditions, i.e. when recording or on-air, the loudspeakers are muted and headphones are used instead. Talkback is frequently extended to other areas such as recording suites or remote contributing studios.

46.1.8.1.2 Reverse talkback

During rehearsals, communication from studio to control room can often be via one of the studio microphones. This may be inappropriate on occasions, perhaps when part of an orchestra is rehearsing, and reverse talkback using, for example, a microphone attached to a musical director's desk and linked to a loudspeaker in the control room, can be a more convenient form of communication.

46.1.8.2 Television studios

46.1.8.2.1 Production talkback

Production talkback is usually via an 'open' microphone in front of the producer/director, although other key personnel will generally have access to it through switched microphones. This talkback is normally fed to headphones worn by studio staff (cameramen, etc.) and to loudspeakers in other technical areas. Studio performers will normally have the producer's instructions relayed to them from a floor manager who will have a small radio receiver picking up the talkback.

46.1.8.2.2 Loudspeaker talkback

This talkback is fed to studio loudspeakers. It is less used for obvious reasons. Other persons besides the producer/director will have access to it.

46.1.8.2.3 Sound talkback

Sound talkback enables the sound desk operator to speak to members of the sound crew wearing headphones. A particularly important recipient is likely to be the boom operator(s) in a drama production.

46.1.8.2.4 Boom reverse talkback

A boom operator's headset normally is equipped with a microphone allowing communication with the sound desk operator.

46.1.8.2.5 Other talkbacks

These include communication from vision control staff to cameramen, with a camera reverse talkback.

46.2 Digital sound desks

46.2.1 Principles and processes

46.2.1.1 Faders

Any change of signal level in a digital system must involve multiplication of the number representing each sample by another number which will be less than unity in the case of attenuation. *Figure 46.9* illustrates this.

For example, multiplication of the sample by 2 is equivalent to a 6 dB increase in the analogue signal voltage. Very high computing rates are required in the processor, as will be quickly realized with a sampling rate of 48 kHz and the fact that there are 50 or more faders in a desk, all of which are typically scanned in $^1/_{25}$ s.

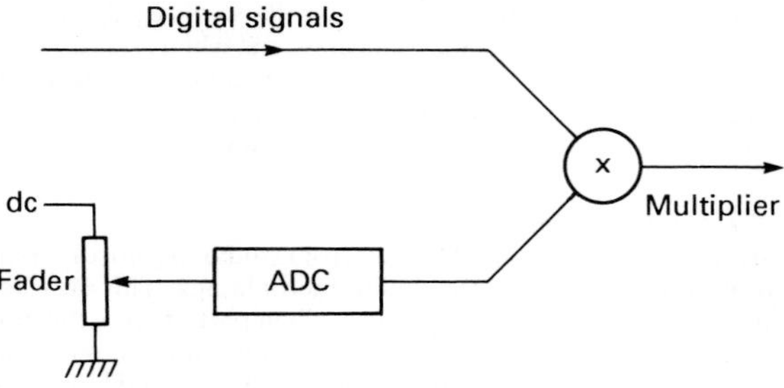

Figure 46.9 Change of signal level in a digital system

46.2.1.2 Mixing

This term is used in the sense of combining signals, as happens on the group busbars of an analogue desk. The digital process is one of addition (see *Figure 46.10*).

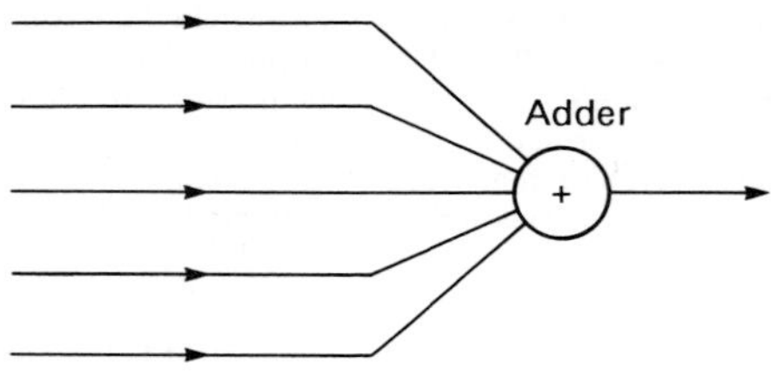

Figure 46.10 Digital mixing

46.2.1.3 Limiting and compression

Limiting and compression are essentially matters of attenuation and thus multiplication of digital samples is required.

46.2.1.4 Filters (EQ)

This is a more complex process, which basically involves delays and multiplication. In *Figure 46.11*, each sample enters at the left and moves along the chain of stores, being held for a sample period in each one.

The multiplying units operate on each sample value by coefficients which have been determined to give the desired EQ effects. After multiplication, the new sample values are added. Each sample value is operated upon a number of times as it passes along the line of stores. The final output from the adder will represent a waveform which is different from the original. In other words, it has been processed.

On a digital sound desk, operation of the EQ controls has the effect of changing the coefficients of the multipliers. In this way, all the EQ operations of an analogue desk can be performed in the digital mode.

46.2.2 Assignable controls

An increasingly important facility is that of assignability. In other words, the number of desk controls is kept to a minimum but any control's setting is stored in memory and then assigned to a channel, or possibly more than one. A fully digital desk clearly lends itself to assignability, but not all assignable desks are fully digital. In a non-digital desk, considerable use has to be made of *voltage controlled amplifiers* (VCA), as recall from memory and subsequent conversion from digital to analogue signals means that a voltage and not mechanical movement is needed to make an adjustment to a setting.

The desk controls themselves have to be different from a conventional desk. Shaft encoders, capable of continuous rotation, take the place of potentiometers. This is because, for example, a control may have been set to maximum for one control setting, and on assigning to another channel the existing setting may be low so that a further increase in the control's position is needed. This can only be satisfactorily achieved with shaft encoders. The problem of display then arises. The position of a pointer on a control knob is now without meaning, so separate displays, such as arrangements of leds perhaps in a vertical stack or in a ring around the control, are used.

A big operational advantage of assignable desks is that control settings may be stored on floppy disks and recalled if necessary at a later date, perhaps for a further editing session.

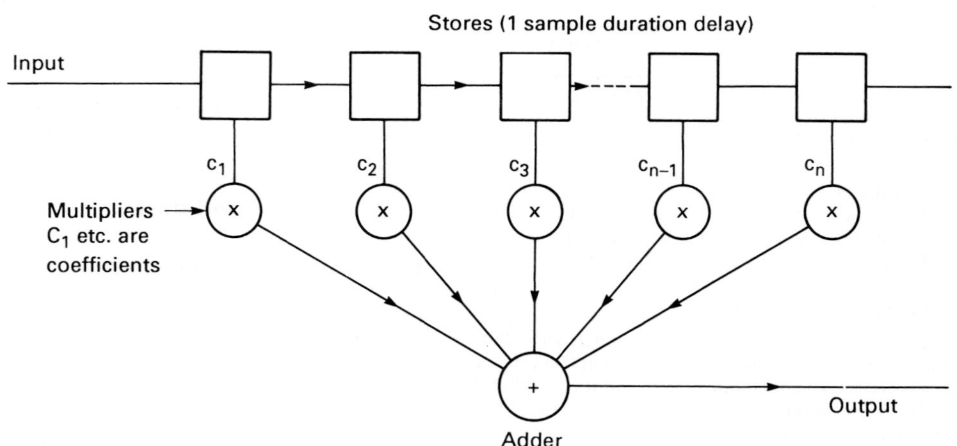

Figure 46.11 Digital filter

M Talbot-Smith BSc, C Phys, M Inst P
formerly BBC Engineering Training Department

47 Sound Recording Processes

47.1 Analogue recording

The impression is sometimes given that analogue recording is now an out-of-date technology; that digital recording is vastly superior and nothing less than digital quality should ever be considered. In some circumstances this view has validity – in mastering compact discs, for example. In other areas analogue tape still has an important part to play. It is easily edited, and for broadcasting, where impairments in quality introduced by transmission and reception are unavoidable, there may sometimes be little point in using digital recording – unless, of course, multiple copying is going to he carried out, when the progressive impairments with analogue systems mean that digital methods have to be used. On the other hand, there are noise-reduction systems now available which can make analogue recordings comparable in quality with digital ones.

Considered below are tape, the recording pròcess, erase, and the replay process. It probably need not be stated here that professional machines have three heads so that off-tape monitoring of the recorded signal is possible.

47.1.1 Tape

The magnetic material normally used on professional tapes is gamma ferric oxide – a form of Fe_2O_3 in the form of needle-shaped particles about 1 μm in length, and these are laid on a base of polyester or other similar material. Other substances are also present to bind the oxide to the base and these have to be chosen with some care (for example, to avoid problems arising from static electricity which, apart from producing audible clicks in the output can also affect the evenness of the tape when spooled fast). The total thickness of base plus oxide is 55 μm. A complete tape spool is 26.7 cm diameter (10½ in) carrying 732 m (2400 ft) of tape giving a playing time of 32 min at 38 cm/s (15 ips). The intention behind a playing time of 32 min is that this allows 30 min for recording plus up to 2 min for test tones and identifying signals.

Calibration of any tape machine is a requirement before it is used for recording. The stability of modern tape recorders is good and full calibration is generally needed relatively infrequently, unless the machine has been physically moved or maintenance work has been carried out on it. Consequently it is often permissible simply to check that record and replay levels are satisfactory. For this purpose, a 'line-up tape' is used. In the UK a common standard is to have such a tape carrying 1 kHz tone with a recorded flux of 320 nW/m (nanowebers/metre), but other standards exist for different purposes and in other parts of the world.

47.1.2 The recording process

The object is obviously to produce a magnetic signal on the tape which is virtually a replica of the original sound signal. The main problem is that the initial magnetization curve is non-linear (*Figure 47.1*) and serious distortion occurs if a correcting process is not used.

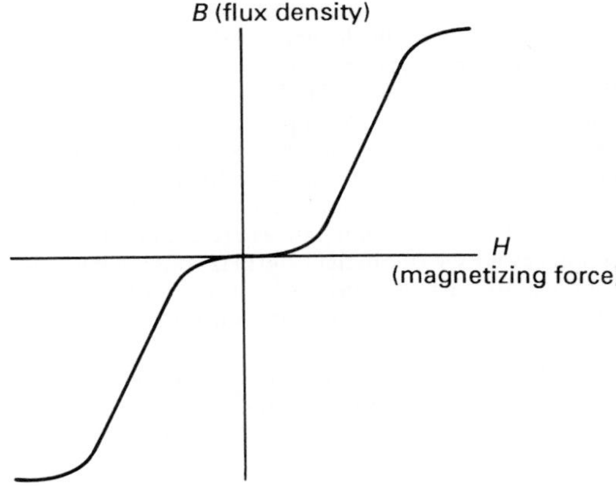

Figure 47.1 The initial magnetization curve.

A permanent magnet system has been used in the past to 'lift' the recorded signal on to a linear portion of the curve, but this is unsatisfactory, as it results in a background hiss which is unacceptable for serious recording. The use of an ac bias of suitably high frequency and at the correct level can, however, eliminate the non-linearity by a process which is complex and beyond the scope of this book to try to explain. The bias frequency is not critical – usually in the range 150–200 kHz – but the bias level needs careful adjustment. (See section *45* for more information and an illustrative diagram.)

The dimension of the record head gap is a matter of compromise. The gap needs to be wide enough for the flux to spread out into the oxide of the tape, but at the same time, if it is too wide, then an excessively large record current in the coil is needed to provide sufficient flux. A head gap of 20 μm (0.02 mm) is typical.

Some signal equalization is needed in the recording amplifier to compensate for high-frequency losses in the head. This takes the form of a more-or-less flat characteristic with high-frequency lift.

47.1.3 The erase process

The erase head, obviously, precedes the record head in the order in which the tape moves. The object is to give a traditional demagnetization operation by carrying the magnetic material on the tape through many hysteresis cycles of steadily diminishing amplitude (*Figure 47.2*).

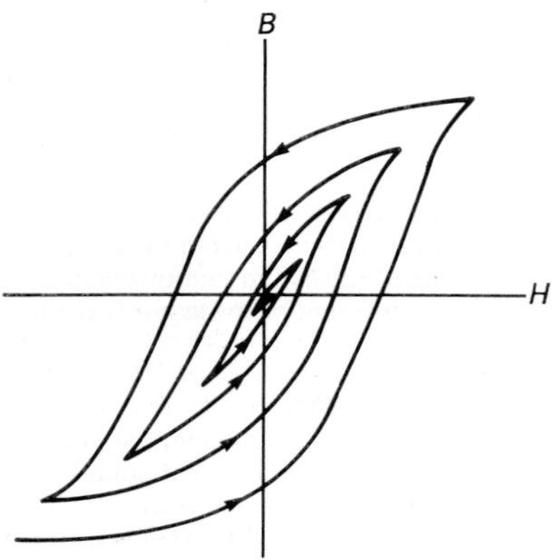

Figure 47.2 Demagnetization cycles

This is achieved by having a relatively wide head gap – around 100 μm – and using a high-frequency erase signal so that the amplitude of the signal is reduced as the tape moves through and away from the gap. A simple calculation shows that for tape moving at 38 cm/s, around 25 cycles of a 100 kHz frequency occur in the time taken for the tape to pass the erase head gap. The erase signal is commonly derived from the same oscillator as is used for bias, although where a particularly high bias frequency is used (more than 200 kHz) this is sometimes halved for the erase current. Erase currents are large in order to carry the magnetic material into saturation at the start of the cycle of events.

Neither the amplitude nor the exact frequency of the erase current is critical, so long as initial saturation is achieved and there are sufficient cycles of diminishing size to result in demagnetization.

47.1.4 The replay process

The replay head is simply a device with a coil into which an emf is induced by the varying flux produced as the tape moves past it. The gap width has to be chosen with some care. If it is too small then the induced emf is also small. The maximum size is limited, however, as it must be smaller than the shortest recorded wavelengths. (There will be no induced emf if the recorded wavelength equals the gap width, as there will then be no magnetic 'difference' at the ends of the gap. This corresponds to the 'extinction frequency', f_e (*Figure 47.3*).

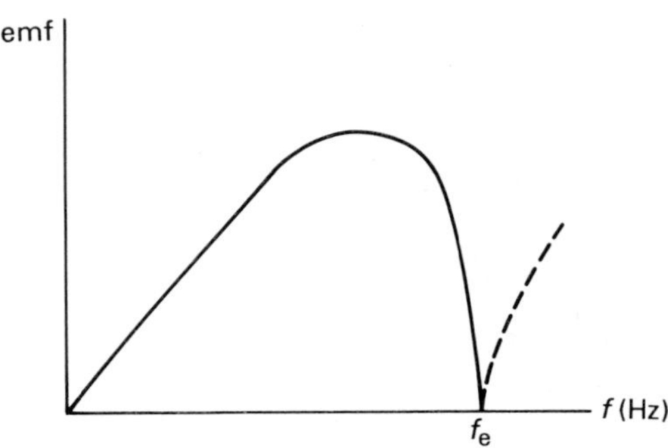

Figure 47.3 Extinction frequency

Gap sizes of about 5–10 μm are usual in studio tape machines (but much smaller in cassette machines – around 1 μm). Taking 8 μm as an example and given a tape speed of 38 cm/s (15 in/s), f_e can be found from (tape velocity/gap width), which in this case is (0.38 m/s/8.10^{-6}m) or approximately 48 kHz. A high value of f_e here is necessary because it is likely that the tape machine may be required to have 19 cm/s (7½ ips) as a selectable tape speed and with the given gap width f_e would then be only 24 kHz.

A further complication is that the emf induced in the coil is proportional to rate-of-change of flux. It is thus proportional to the frequency of the recorded signal (and also to tape speed). The reproduced emf is then very small at low frequencies and rises at 6 dB/octave. Equalization is needed to compensate for this 6 dB/octave rise and also, when necessary, for proximity to the peaks of the extinction frequency curves in *Figure 47.3*. (See section *45* for a typical replay equalization characteristic.)

It should be noted that any speed-change controls will normally automatically carry out the necessary switching of equalization circuitry.

47.1.5 Azimuth

This means the angular deviation of a head gap from the vertical (*Figure 47.4*). The problem which arises from an incorrect adjustment is that the effective head gap is increased.

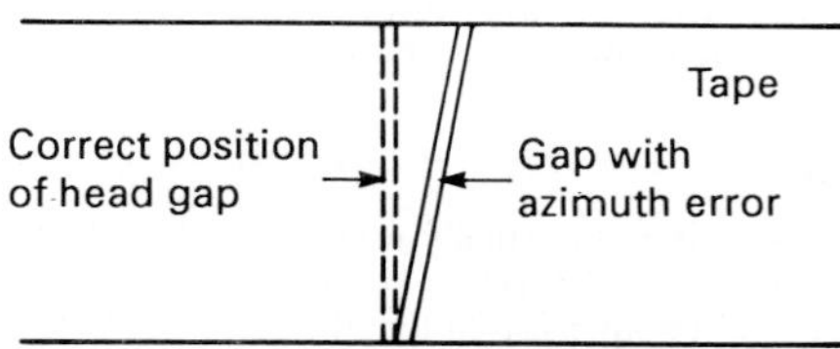

Figure 47.4 Head gap azimuth

On a record head this effect may not be too serious, but on a replay head the extinction frequency is lowered. In mono recordings this can manifest itself as a loss of high frequencies. On a stereo machine the loss of high frequencies *in stereo listening* may not be apparent. Each stereo track is narrow, but the mono signal derived from the stereo may be noticeably lacking in high frequencies.

It is worth noting that a very small azimuth error in angular terms can have serious effects. For example, taking a replay head gap of 8 μm used with 6 mm (quarter-inch) tape, then if the gap is only 4 minutes of arc out of the vertical the effective head gap is approximately doubled and the extinction frequencies are halved. There are potentially even greater errors with multitrack machines because of the wider tape. Consequently great care has to be taken that the azimuth is properly set, and on professional machines there are mechanical adjustments to correct it where necessary. Normally the azimuth on a studio machine should not alter but where a machine is mobile, as, for example, in outside broadcasts (OBs), regular adjustment may be needed. It is good practice also to check the alignment on *any* machine which has had to be transported from one place to another.

47.1.6 Tape speed control

Accurate tape speed is vitally important in any professional machine, avoiding either short-term fluctuations which will be heard as 'wow' or 'flutter' but also long-term variations which may only show up when a section at the end of a tape (for example, a 'retake') is edited into an early portion of the same tape, so that any pitch changes caused by a slow progressive speed change will be noticeable. Also consistency of timing between different machines is important, especially in broadcasting, where an error of only a few seconds in the stated duration of a programme could cause an embarrassing under- or over-run. Servo control of motors is necessary with a crystal-controlled oscillator as a reference. In machines with a 'varispeed' facility, giving a significant percentage change in the tape speed, the crystal control is replaced by a stable variable-frequency oscillator.

47.2 Noise reduction

A limitation in even high-quality analogue machines is that there is inherent noise caused by the fact that the tape coating consists of separate magnetic particles. A signal/noise ratio in the region of 55–60 dB is as much as can be expected. This can be much improved by the use of one or other of the standard noise-reduction systems which are outlined in sections *47.2.1–47.2.6*.

These work basically by reducing the dynamic range before entering the noise domain and expanding it afterwards using a *compander*, which is a *compressor* followed by an *expander*.

47.2.1 Dolby A

This system is widely used, especially in multi-track work where the mixing together of many tracks has an additive effect on the noise. In general, tape noise may be reduced by about 10–15 dB.

The system avoids many of the undesirable effects which a simple compander can introduce. These are:

1. Exact tracking with a simple compander is difficult – that is, expanding the dynamic range by exactly the same amount by which it has been reduced. The Dolby system uses the same circuitry for both processes since it works by subtracting or adding as appropriate a portion of the audio signal. It is thus a linear process.
2. A compander works on the higher levels of the audio signal, but the higher levels are those which will tend to mask the noise and therefore are those parts which need noise reduction least of all. The Dolby A system detects when the signal levels are low and raises these at compression and lowers them again in expansion. *Figure 47.5* illustrates this process.

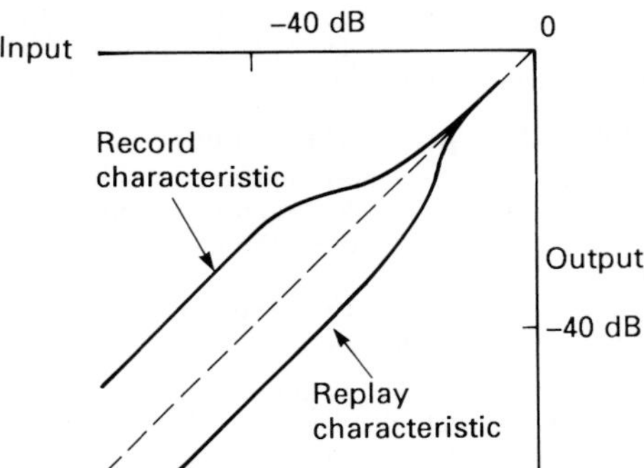

Figure 47.5 Dolby A process

3. Companding action over the entire frequency range may result in unwanted modulation effects. Thus, if a high-level signal and a low-level one occur at widely different frequencies the processing will take place because of the low-level one, although the other does not need any action. To avoid this, the Dolby A system divides the frequency range into four bands, each designed to help defeat particular types of noise:

Band	Frequency range	Type of noise targeted
I	below 80 Hz	Hum and rumble
II	80 Hz to 3 kHz	Cross-talk and print-through
III	3 kHz and above	Hiss and modulation noise
IV	9 kHz and above	Hiss and modulation noise

47.2.1.1 Dolby A line-up

One minor drawback of the Dolby A system is that accurate line-up of programme levels is very important: if the encoder introduces compression because the signal level is low then the decoder must receive the signals at exactly the right level if it is to expand correctly. For this reason, it is most important that all Dolby-encoded tapes are clearly marked. Exact details depend on individual installations, but it is worth pointing out here that 'Dolby Tone' – an easily identifiable warble produced by frequency changing (not level changing) – must always be put on the start of any tape. This does two things. It identifies for later operators that the tape is Dolby encoded and it also helps the signal levels to be correctly set. The Dolby units themselves contain the circuitry for producing the tone.

47.2.2 Dolby B

This system is a low-cost version designed to be used with cassette players where the tape noise, unless treated, is objectionable. It was originally intended for the domestic market but it is now being used professionally not only with audio cassettes but also with the soundtracks of various types of video equipment.

Dolby B is a much-simplified version of Dolby A, the principle of operation being exactly similar but using only one frequency band, namely from 2 kHz to 5 kHz and above. The lower frequency of the band is variable. (Note that tape noise – 'hiss'– is predominantly a high-frequency effect and falls largely into the band dealt with by Dolby B.)

47.2.3 Dolby C

This is a later and improved version of Dolby B, again originally intended for domestic audio cassettes but now appearing on professional equipment. It has two frequency bands and in a sense is therefore nearer to the A system. Most good-quality cassette machines incorporate Dolby C as well as B. About 20 dB of noise reduction can be obtained.

47.2.4 Dolby SR ('Spectral Recording')

The latest of the Dolby systems (it appeared first in 1986), this embodies concepts from the A, B and C systems. The word *spectral* refers to the spectrum of the audio range which is analysed in the processing.

To see how Dolby SR works we need to look at its action at low, medium and high signal levels:

1. *Low levels*. The idea here is to use a high recording level using a fixed equalization.
2. *Medium levels*. With an increase in signal level there could be a risk of overload, so some gain reduction is applied. Basically, there is a group of variable filters. Some are of variable bandwidth and others are of variable gain. The fixed-bandwidth ones have their *gain* electronically varied while the others, those with fixed gain, are adjusted to cover different frequency ranges. A control circuit thus creates, in effect, a very large number of filters which are appropriate to the programme signal.
3. *High levels*. Gain reduction takes place but only in the frequency region near to that of the high-level signal.

Overall, a noise reduction of up to 24 dB can be obtained with Dolby SR. This makes analogue recordings have noise figures comparable with digital recordings, and for this reason it is being increasingly used with multi-track recording. However, to equate SR recordings with digital neglects the fact that multiple-copying SR recordings, like any other analogue recordings, will deteriorate after relatively few copies.

47.2.5 dbx

This is a relatively straightforward compander system, using a 2:1 compressor and matching expander. The use of pre-emphasis, that is, high-frequency lift in recording which is subsequently corrected in replay, improves the performance by causing more compression at the high frequencies. Additional signals outside the audio band are used to match the expander to the compressor. One advantage is that line-up of levels is not critical and the noise reduction figures are impressive – around 25–30 dB is quoted. However, it is possible for the companding process to be audible, and also the effects of tape drop-outs are magnified by 2:1 on replay.

47.2.6 Telcom c4

This can be thought of as a combination of Dolby A and dbx in that four frequency bands are used (like Dolby A) and a compander system is employed as in dbx, one compander to each band. However, the companding is carried out at 1:1.5, not 1:2. The lower the ratio the better, in the interests of minimizing possible ill effects resulting from the processing. The four frequency bands are:

I Up to 215 Hz
II 215 Hz to 1.45 kHz
III 1.45 kHz to 4.8 kHz
IV 4.8 kHz and above

Disadvantages of Telcom c4 are that, as with dbx, tape drop-outs are magnified in the expansion process. Also it is the most costly system. Noise reduction of 15 dB or more can be obtained.

47.3 Digital recording

There are two main types of digital recorder currently in use. One, the DASH system (Digital Audio Stationary Head), is mainly found in multi-track recorders. Other DASH machines with two tracks have the advantage that razor-blade editing can be carried out on 25 mm (quarter-inch) tape which appears to be exactly similar to the tape used for conventional analogue recording. (It is, in fact, different in its characteristics.) DASH machines are currently very costly and this section deals only with the alternative system that is being widely used.

Originally, digital audio recorders were all expensive devices, but this situation changed with the use of portable video recorders fed with the outputs of analogue/digital converters, these signals being processed to be compatible with the format of video signals. A relatively cheap system was thus widely available. A minor disadvantage was that, normally, two units were required, one for the processing and one for the recording. A further (but in the circumstances acceptable) disadvantage was that tape usage was much greater than was strictly necessary, since the signal/noise ratios required for video recording are much greater than those needed for digital operations. The advent of the system known originally as R-DAT (Rotary head Digital Audio Tape recording), but now more commonly called simply DAT, has brought about a minor revolution in recording. Present-day DAT recorders combine low cost with small size – as little as 253 × 55 × 191mm (10 × 2¹/₄ × 7⁵/₈ in) with a weight of 2 kg – and, at the same time, provide digital audio quality with stereo capability.

The DAT system can be thought of as a small version of a video cassette recorder. A rotary head system is used and the cassettes are smaller than the standard compact audio cassette, the dimensions being approximately 73 mm × 54 mm × 11 mm. There have been general improvements in the design compared with the analogue cassettes. For example, a brake is applied to the hubs when the door to the cassette is shut, as it is when the cassette is not inserted in the DAT machine. The tape is very thin – the base is 10 μm thick and the oxide coating is about 3 μm, making a total of 13 μm. The tape width is 3.81mm, being only slightly wider than the tape in a compact cassette.

Several standards of tape speed and other parameters are available, some intended for the domestic market using pre-

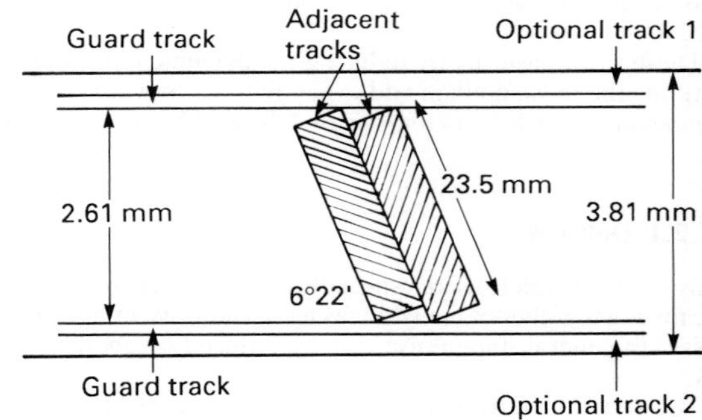

Figure 47.6 DAT track format

Sync	ID	Block address	Parity	Audio data
1 byte	1 byte	1 byte	1 byte	32 bytes
8 bits	8 bits	8 bits	8 bits	←256 bytes→

←288 bits→

Figure 47.7 Audio data block

recorded tapes. At present (1990) the domestic market has not developed and therefore the only figures given here are for professional applications:

Sampling rate	48 or 44.1 kHz
Quantizing	16-bit linear
Tape speed	8.15 mm/s
Play/record time	2 h max.
Drum speed	2000 rpm
Effective head/tape speed	3.133 m/s

The automatic lacing of the tape is simpler than in a video cassette machine as only a 90° wrap is needed. This is made possible by intermittent recording: the blocks of samples are fed into a store and then taken out a much higher rate. The gaps between these time-compressed blocks cover the time when a head is not in contact with the tape.

Figure 47.6 shows the essentials of the track format. Each track is about 23.5 mm long and is at an angle of just over 6° to the direction of travel of the tape. The two linear tracks at the edge act more as protection against damage to the tracks. Because of the low tape speed they are of little real value as audio tracks, the highest recordable frequency being in the region of 3 kHz.

The use of aximuth recording, in which each track is recorded with a 20° 'slant', alternate tracks having alternate 'slopes', reduces markedly the risk of crosstalk between tracks.

47.3.1 Blocks

The audio data block is shown in simplified form in *Figure 47.7*. In the 'preamble' to the 256 audio data bits, which include error correction, there are four bytes, the first of which is for synchronization. The second, the ID code, specifies, for example, the sampling frequency and the number of channels. The third byte states whether the block consists of digital audio signals or whether it is a subcode, while the fourth provides parity checks. There then follow the 256 data bits (32 bytes), making so far a total of 36 bytes.

A complete track is shown in *Figure 47.8*. It is made up of 196 blocks performing the following functions:

1. Margins at beginning and end (11 blocks each). These are effectively guard bands.
2. Subcodes 1 and 2 (11 blocks each). These carry additional information (running time, contents, etc.) rather in the manner of a compact disc.
3. ATF 1 and 2 (Automatic Track Following) (five blocks each). These blocks provide information for the servo systems which cause the heads to follow the tracks accurately.
4. Four spacing sets of three blocks each – the Inter Guard Bands.
5. 130 blocks of audio data, each of which is as in *Figure 47.7*.

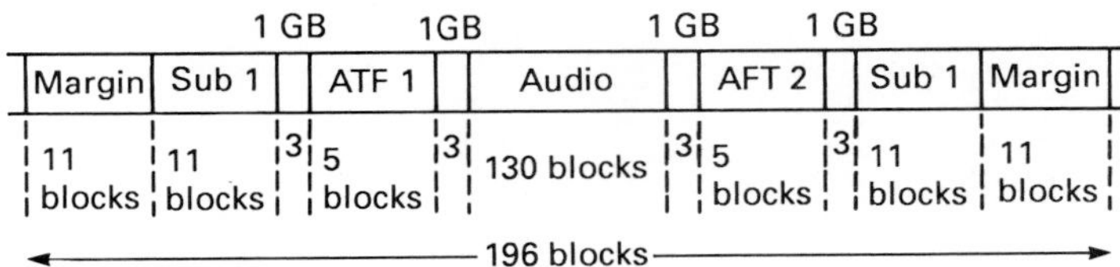

Figure 47.8 One audio track

S R Ely PhD, C Eng, MIEE
Head of Carrier Systems Section, BBC Research Dept

48 Multi-channel Sound Systems

Stereo or multi-channel sound in the cinema and in home hi-fi systems is taken more or less for granted these days. But television too benefits enormously from the addition of stereo sound, and the provision of sound signals in two or more different languages is an urgent requirement in many countries where there are two or more major ethnic groups in the population.

However, stereo or multi-channel sound with television has, until recently, not been available from terrestrial broadcast television signals in most countries. An impetus towards broadcasting multi-channel sound with television has come from its provision through media such as satellite, cable and pre-recorded video cassettes. Many of these new systems have offered stereo or multi-channel sound from the outset. In order to match the services offered by these new competitors, many television broadcasters have now added, or have plans to add, stereo or multi-channel sound to their broadcasts.

This survey of the more commonly used systems for broadcasting multi-channel sound with television is not exhaustive. The emphasis is upon the systems used in the UK and the rest of Europe. The major systems used in the USA, Australia, and Japan are, however, described in outline and references given where full details of these and other systems can be found.

48.1 Overview

48.1.1 Survey of systems

In 1990 at least eight different systems for broadcasting multi-channel sound with television were in regular use in various countries in the world (see Table *48.1*). This contrasts strongly with fm radio, where a single system, the pilot-tone stereo system developed by Zenith/General Electric (see section *48.2.2.1*) is used to broadcast stereo radio programmes in most countries in the western world.

Some of the reasons for this lack of standardization relate to the basic differences in the television systems used in various countries, while others relate to the functional requirements, constraints, and operating practices of individual broadcasters.

48.1.2 Functional requirements

48.1.2.1 Quality

Although the quality of the television sound signal delivered to viewers has often been constrained by limitations of receivers and their loudspeaker systems, the quality of the television sound signal at the source is usually good and deserves the same full audio bandwidth (at least 15 kHz), high signal/noise ratio, and low distortion that is available from fm radio services. This is especially so when stereo is presented, because the enhancement which stereo can provide gives a stimulus towards better receivers, loudspeakers and listening conditions.

A particular problem for analogue television sound systems, which does not afflict fm radio signals, is that of interference from the vision signal. This is a special problem for systems which use subcarriers (see sections *48.2.2.1* and *48.2.4*). Analogue companding systems are commonly used to attempt to overcome this problem. Digital multi-channel television sound systems (see section *48.3*) overcome the problem of vision on sound almost completely.

Separation (i.e. low crosstalk) between the audio signals is also important. Separation between the pair of sound signals in a stereo system need be only around 20–30 dB in the mid-frequency range and less at the high and low frequencies; indeed this is the best that could be achieved from a conventional vinyl gramophone record. However, where the sound signals are unrelated, as, for example, in bilingual broadcasts, much greater separation, typically 55 dB or more, is needed between the audio channels.

48.1.2.2 Compatibility

In the case of new programme services, especially those delivered via satellite, stereo or multi-channel sound can be an objective from the outset, and it can be delivered via a purpose designed integrated transmission system such as the MAC/packet family of systems (see section *29*).

In far more cases, however, multi-channel sound has been added as an overlay on existing monophonic programme services. This is in many ways similar to the conversion of monophonic fm radio services to stereo, and is analogous with the change from monochrome to colour television. In these circumstances, where it is proposed to add new signal components to an existing service, the new signals must not cause interference to reception of the picture or mono sound on existing receivers which were not designed with the new signals in mind.

System	Description	Where adopted	Date
FM-FM	FM subcarrier modulating existing mono fm carrier	Japan	1978
BTSC/MTS	AM subcarriers modulating existing mono fm carrier. Audio signals conveyed on subcarriers are compressed.	USA	1984
Dual Carrier A2	Additional fm sound carrier (conveys R signal for stereo)	Germany Australia Italy Netherlands Switzerland	1981
Dual Carrier	Additional fm sound carrier (conveys L-R signal for stereo)	South Korea	1984
NICAM 728	Additional digitally modulated carrier conveys two digitized NI companded sound signals	UK Sweden Finland Norway Denmark Honk Kong New Zealand Spain	1986
Wegener family	Low deviation fm subcarriers on fm satellite channels	Astra and other satellites. Marcopolo 1 (BSB), TVSat-2	
MAC/packet family	Digitally coded sound conveyed in tdm with analogue picture signal components	TDF-1 and Tele-X satellites	1990
B-MAC	Dolby adm encoded digital tdm with analogue picture signal components	Australian satellite service and communications satellites	1984

Table 48.1 Multi-channel sound with television systems

A further important compatibility consideration is that the new signals must fit into existing frequency allocation plans. In particular, the new signals must not cause increased interference to other services (perhaps broadcast across a national border in an adjacent country) operating in the same or adjacent channels.

A lesser, but still important, compatibility consideration is that, if possible, the signals of the multi-channel sound system should be compatible with the existing equipment and infrastructure owned and operated by the broadcaster. This is especially important for uhf terrestrial networks where transmitters at many hundred different sites are used to cover one country.

48.1.2.3 Ruggedness

The signals of the multi-channel sound system should be capable of being reliably received on suitably equipped sets wherever this is possible. Thus the system should be adequately robust against impairments to reception such as low field strength, multipath propagation, and interference from other signals that are broadcast in the same or adjacent channels.

Ruggedness and compatibility often conflict. For example, in a system which uses a subcarrier or additional carrier to convey the additional information needed for multi-channel sound, making the level of the additional signal too large may result in a compatibility problem such as patterning on the picture or interference to reception of the mono sound on existing receivers. Conversely, if the new signal is made too small, it may not be reliably received towards the fringes of the service areas and in other places where reception is already difficult.

48.1.2.4 Receiver cost

One of the most crucial factors for the general and rapid acceptance of a multi-channel sound system is that of receiver cost. Suitably equipped receivers need to be readily available at a price which the viewers are prepared to pay. However, the price paid by viewers for their receivers is, in general, not related in any simple way to the complexity of the circuits which they contain. Instead, it is determined by complex techo-economic factors. It is worth noting that, in those countries such as Germany where multi-channel television sound services are well established, multi-channel receivers have, except for small portable sets, become normal, and new monophonic sets of 59 cm or greater screen size are comparatively rare.

48.2 Analogue systems

48.2.1 Simulcasting

In the UK since 1972, occasional use has been made of the so called *simulcast system*[1] to provide stereo sound with television. In this system, the stereo sound associated with the television picture is broadcast via a stereo fm station. This service has been much appreciated by viewers, who can enjoy it without buying any special new equipment apart from a standard fm stereo radio receiver which they may already possess.

Simulcasting is not, however, convenient for the broadcaster because it occupies two separate programme services and requires close coordination of the schedules of a radio service with a television service. Furthermore, it can be used only for programmes where the sound signals on their own provide an acceptable programme for the radio listeners. Most television programmes are not suitable for broadcasting in this way. Even in those programmes, such as music, which are suitable for simulcasting, the ideal sound balance and stereo image-width for the television viewers may not be best suited to the perspective of the radio listeners, and compromises may therefore have to be made.

Simulcasting should be regarded only as an interim solution until a dedicated system for conveying multi-channel sound along with the associated television picture signal can be provided.

48.2.2 Subcarrier systems

48.2.2.1 Pilot-tone stereo

FM radio stations have broadcast in stereo since the early 1960s using the *pilot-tone* multiplex system developed by Zenith/General Electric in the USA. In this system, the left and right audio signals are fed via pre-emphasis networks (50 μs in Europe, 75 μs in the USA) to a matrix which produces a *sum* signal, (L+R)/2, and a *difference* signal, (L-R)/2. The (L-R)/2 signal is double-sideband suppressed carrier modulated onto a 38 kHz subcarrier and occupies the region 23–53 kHz in the multiplex spectrum applied to the modulation input of the fm transmitter (see *Figure 48.1*). A low-level pilot-tone at 19 kHz is broadcast as a reference to enable the stereo decoders to demodulate the suppressed subcarrier signal. The (L+R)/2 signal modulates the fm transmitter directly and therefore comprises components in the 0–15 kHz region of the broadcast multiplex spectrum. It thus provides a mono signal which can be received by receivers not equipped with a stereo decoder. This satisfies the compatibility requirement.

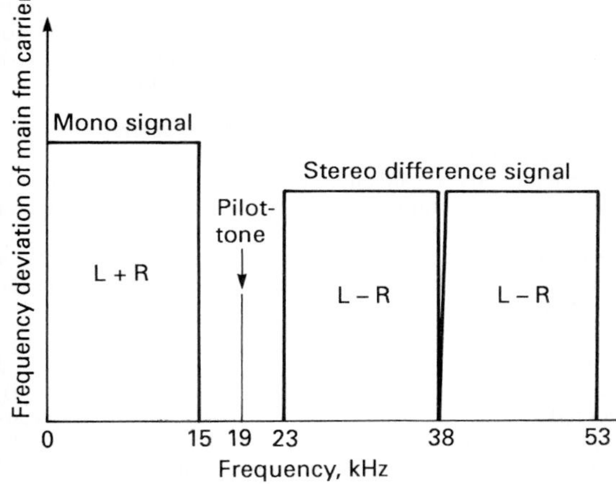

Figure 48.1 Spectrum of a pilot-tone stereo multiplex signal (vertical axis not to scale)

With this pilot-tone system, the bandwidth of the stereo multiplex signal at the output of the fm discriminator is 53 kHz, which is about 3.5 times (53/15) greater than the bandwidth of the mono signal. In an fm system, the noise at the output of an fm discriminator is 'triangulated', i.e. the voltage spectral density of the noise increases linearly with frequency. Therefore the power of the noise at the output of the fm discriminator increases as the cube of the bandwidth in which it is measured. Thus the signal/noise ratio of the stereo audio signals at the output of a pilot-tone stereo decoder is, in theory, under field-strength limited reception conditions, about 20 dB worse than that of the mono signal. In practice, with an adequate antenna system within the planned service areas of the broadcasts, stereo fm radio receivers can deliver stereo sound with a very good s/n ratio.

The noise penalty for pilot-tone stereo applies not only to random noise (*hiss*) but also to other interfering signals such as *buzz* or *whistles*. This is very significant if the pilot-tone system is applied to the fm sound signal associated with a broadcast television signal, because there are relatively large amounts of interference to the fm television sound signal from the vision signal.

Many separate mechanisms can cause this vision-on-sound interference[2,3], but a lot are due to the use of *intercarrier sound* in most mono television receivers. In this method of detecting an fm television sound signal, the fm sound signal is mixed at if with the recovered vision carrier to derive an fm signal centred on a frequency sound equal to the intercarrier spacing of the signals, i.e. the difference between the frequency of the broadcast vision carrier and that of the fm sound signal. These frequencies are 4.5 MHz, 5.5 MHz and 6.0 MHz for television systems M/N, B/G and I respectively. Intercarrier sound is popular because it eliminates (by making it common mode between the vision and sound signals) the effects of phase noise in local oscillators in the broadcast system and receivers. However, the penalty for intercarrier sound is that incidental phase modulation of the vision carrier is transferred to the intercarrier sound signal. Such incidental phase modulation of the vision signal occurs in transmitters and in receivers. It is an almost inevitable consequence of the vestigial sideband (vsb) filtering in the receivers.

Since the unwanted phase modulation of the vision carrier is generated by the vision signal, it has a similar spectrum to the video signal. Thus there are strong components at the field frequency (60 or 50 Hz) and at the line frequency (15 734 or 15 625 kHz) and its harmonics. In mono transmission, the harmonics of the field-rate components cause most trouble and are heard as low-frequency "buzz'. Buzz, however, is, especially with the limited bandwidth loudspeaker systems commonly used in television receivers, usually perceptible but not annoying. If conventional fm radio pilot-tone stereo is used, harmonics of line frequency become very significant. For example, if the 38 kHz difference signal were broadcast with a 625-line 50 Hz television signal, then 6.75 and 8.875 kHz whistles would result from the demodulation of interference components at twice and three times the line frequency (31 250 and 46 875 kHz) respectively.

Thus, to apply pilot-tone stereo to television, it is necessary to take steps to minimize interference from the harmonics of the line frequency of the video signal. Two practical systems which approach this problem in different ways are described in sections *48.2.2.2* and *48.2.2.3*.

1. *First sound channel (same as the monophonic channel):*	
Maximum frequency deviation of the main carrier	± 25 kHz
Audio frequency range	50 – 15 000 Hz
Pre-emphasis	75μs
Programme signals	First language (dual sound) L+R (sum signal) (stereo)
2. *The second sound channel:*	
Subcarrier frequency	$2f_H$ (f_H=15 734.264 H_3)
Frequency deviation of the main carrier by the subcarrier	± 15 kHz (dual sound) ± 20 kHz (stereo)
Audio frequency range	50 – 14 000 Hz
Pre-emphasis	75μs
Programme signals	Second language (dual sound) L–R (difference signal) (stereo)
In the case of stereo, the left-hand signal (L) produces a deviation in the same direction in both the subcarrier and the main carrrier.	
Compensation for receiver time delay (stereo)	20μs
3. *Control signal*	
Subcarrier frequency	$3.5f_H$
Modulation frequency	922.5 Hz (dual sound) 982.5 Hz (stereo)
Modulation	60% am
Maximum frequency deviation of the main carrier by the control signal subcarrier	± 2kHz

Table 48.2 Transmission characteristics of the fm-fm system, television system M

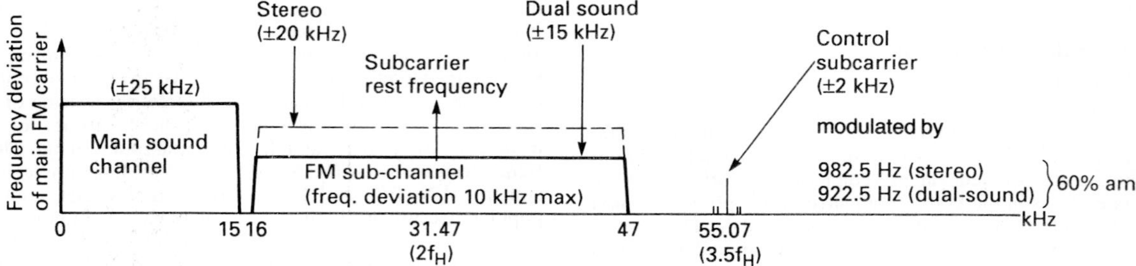

Figure 48.2 Baseband spectrum of the fm/fm system (vertical axis not to scale). Line frequency f_H = 15 734.264 KHz

48.2.2.2 The Japanese fm/fm system

The first multi-channel sound with television system to be put into regular service was the fm/fm system. This was devised by NHK of Japan who have used it with their system M NTSC broadcasts since 1978. In the fm/fm system[4], the second sound channel is provided by a frequency modulated subcarrier with a rest frequency of 31.469 kHz (chosen to equal the frequency of the second harmonic of the line frequency to reduce the audibility of interference from the video signal into the second sound signal). A control subcarrier at 55.070 kHz (3.5 times the line frequency) is amplitude modulated by tones to indicate to the receiver whether mono, stereo or dual sound programme signals are being broadcast. For stereo, the fm subcarrier is used to carry the (L-R)/2 difference signal. The principal transmission characteristics of the fm/fm system are given in reference 5 and summarized in Table *48.2* and *Figure 48.2*.

Because of the use of frequency modulation of the subcarrier, the signal/noise and interference ratios are improved compared with those obtained by amplitude modulation (as in the Zenith/GE pilot-tone system). Furthermore, the separation between the two audio channels is sustained better in the presence of receiver imperfections than with the Zenith/GE system. This enables NHK to use the system to convey bilingual broadcasts, and for other applications where two separate sound signals are conveyed.

But these improvements are obtained at the expense of other problems, of which *buzz-beat* distortion is regarded as the most serious. Buzz-beat distortion is caused by multiplicative disturbance between the fm subcarrier and interference related to the picture signal. The frequency of the buzz-beat tends to be equal to, and its amplitude proportional to, the instantaneous frequency deviation of the fm subcarrier. Thus, although there is no interference during silent passages, buzz-beat results in 'rough' sound to high frequencies such as harmonics of violin or piano music.

Furthermore, there is inherent harmonic distortion due to band-limiting of the fm subcarrier signal. Such band-limiting is necessary in the encoder to prevent the fm subcarrier signal interfering with the baseband audio signal ((L+R)/2 in the case of stereo), but the removal of the second and higher order sidebands of the fm subcarrier signal results in about 2 per cent harmonic distortion.

Despite these problems, the fm/fm system provides good compatibility with mono receivers and transmission networks and is relatively cheap to implement for both the broadcaster and the receiver manufacturer. Its use was therefore considered by several European broadcasters in the early 1970s. It was, however, found[5] that the two-carrier system described in section *48.2.3* was slightly better under conditions of multipath reception such as are found in mountainous regions. Problems were also found under conditions of co-channel interference, especially when the vision carriers were offset by certain amounts to minimize interference between the co-channel vision signals. Under some conditions the signal/noise ratio at the output of the fm/fm decoder was found to be severely impaired even when the picture signal was almost unimpaired. The fm/fm system has not been adopted outside Japan.

48.2.2.3 The USA BTSC multi-channel sound system

In the USA in late 1978, the Broadcast Television System Committee (BTSC) of the Electronic Industries Association formed a subcommittee to formulate standards for the broadcasting and reception of multi-channel television sound (MTS).

The objectives for the system were set to include:

- a mono signal compatible with existing receivers
- a good quality stereo signal
- a lesser quality second audio programme (SAP) subchannel for ancillary programme services such as a second language or commentary for the blind
- a third service for professional use, such as telemetry or electronic news gathering; this third service is referred to as the Non-Public Channel (NPC).

Three basic proposed systems, Zenith, Telesonics and EIA-J (the latter was a variant of the Japanese fm/fm system), emerged for consideration[6]. All of them used multiple subcarriers to convey the stereo difference, SAP and NPC channels, and in all cases the (L+R)/2 compatible mono signal deviated the main subcarrier by up to ±25 kHz, as for mono services. The new subcarriers added extra frequency deviation of the main fm sound carrier. Thus no main channel deviation was sacrificed to add the new signals. It was decided that noise reduction would be needed to obtain satisfactory signal/noise ratio on the stereo and SAP services. To maintain compatability with existing mono receivers, the noise reduction could not be applied to the (L+R)/2 mono signal but could replace the normal 75 μs pre-emphasis on the (L-R)/2 stereo difference and SAP signals. Three noise reduction analogue companding systems based on CBS-CX, dbx and Dolby C were considered.

After extensive testing and review by the industry, the Zenith am subcarrier system, with dbx-TV noise reduction was selected in 1983 by the BTSC subcommittee to be recommended to the Federal Communication Commission as the USA standard for MTS. The FCC opted not to standardize on any one MTS system, preferring to let the market choose a system. But protection was afforded to the 15.734 kHz pilot frequency of the BTSC system, and the FCC stipulated that if a pilot signal is transmitted at 15.734 kHz, then the BTSC system must be used.

The transmission characteristics of the BTSC system are summarized in *Figure 48.3* and Table *48.3*[7,8]. *Figure 48.3* shows the spectrum of the baseband signal with the system configured

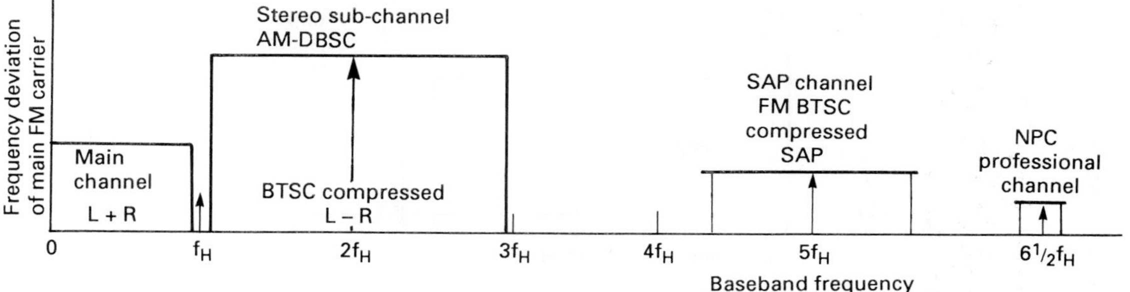

Figure 48.3 Baseband spectrum of the BTSC multi-channel sound system (configured for stereo, SAP and NPC) (vertical axis not to scale)

Service of signal	*Modulating signal*	*Modulating frequency range kHz*	*Pre-emphasis or Companding*	*Sub carrier frequency*	*Sub carrier modulation, kHz*	*Sub carrier Deviation*	*Sound carrier peak deviation, kHz*
Sum (mono)	L+R	0.05–15	75 μs				25*
Pilot-tone				f_H			
Difference (stereo)	L–R	0.015–15	dbx compression	$2f_H$	AM-DBSC	5	50*
Second audio programme (SAP)		0.05–10	dbx compression	$5f_H$	FM	10	15
Non-Public Channel (NPC)	Voice or data	0.3–3.4	150 μs	$6.5f_H$	FM or FSK	3	3
		0–1.5	–				
Total Deviation:							73 kHz

*Combined deviation due to sum and difference signals does not exceed 50 kHz
f_H = line frequency = 15 734.264 kHz

Table 48.3 Outline specification of the BTSC MTS system

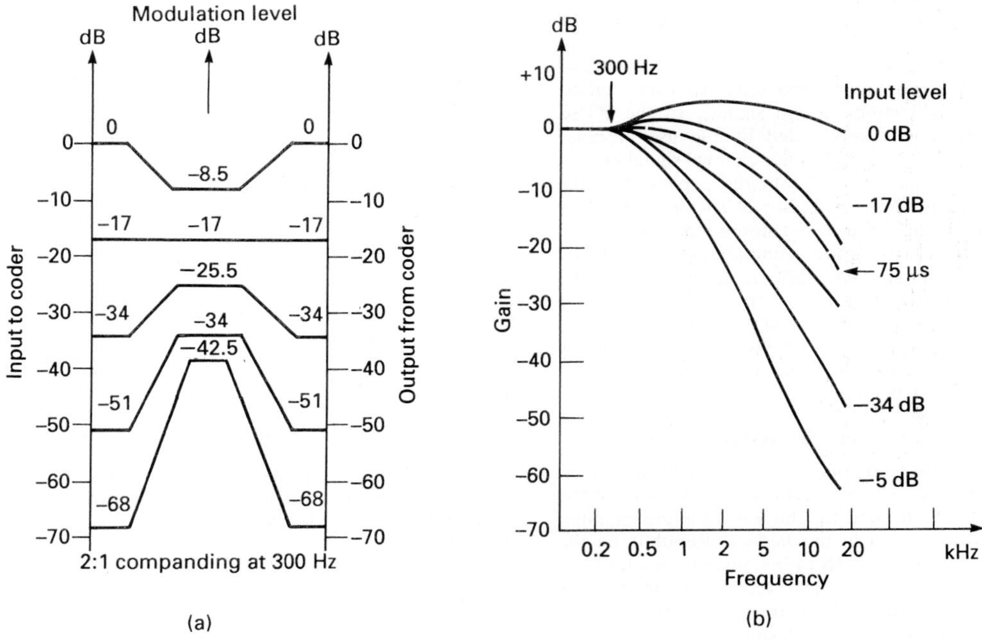

Figure 48.4 (a) BTSC dynamic gain compression, (b) Variable de-emphasis in the receiver

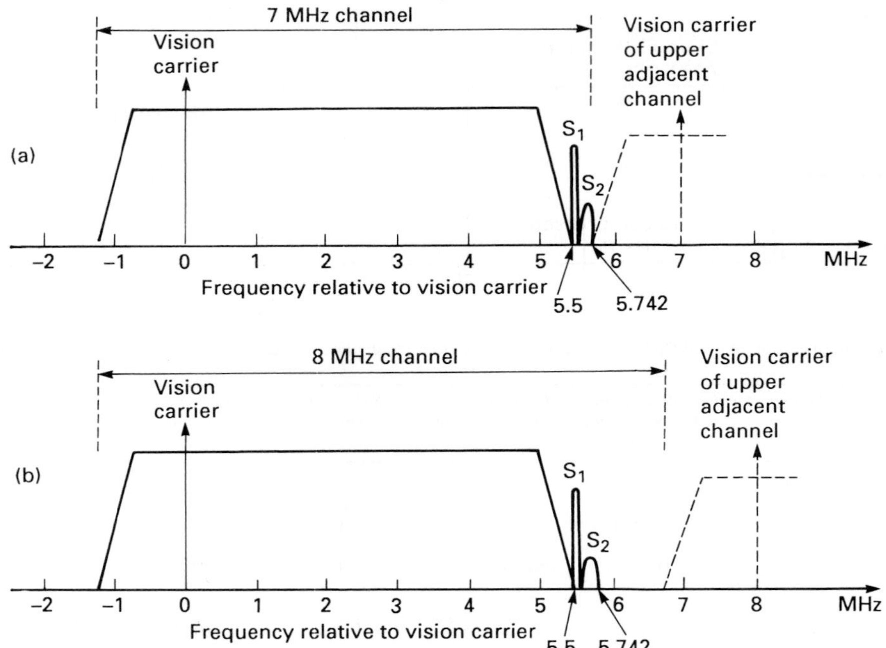

Figure 48.5 RF spectra of the dual carrier sound system (a) applied to television system B, (b) applied to television system G

to convey stereo and the SAP and NPC channels. Other configurations are possible, e.g. mono and SAP[7].

Figure 48.4 illustrates the dbx-TV companding algorithm which is applied to the stereo difference signal and the SAP channel. The dbx system may be seen to comprise a combination of wideband compression and fixed and variable pre-emphasis (known as spectral compression).

In any matrixed stereo system in which the sum and difference signals travel separately, care must be taken to maintain the phase and gain relationships between the sum and difference channels in order to preserve separation between the decoded left and right stereo sound signals. In the BTSC system, this problem is compounded by the use of noise reduction companding on the difference channel signal but not on the sum signal. This requires that the expander in the receiver closely tracks the compressor at the transmitter in order to avoid degrading seriously the stereo separation.

The BTSC system also requires complex closely matched low-pass filters in the sum and difference channels of the encoder[9]. In spite of these problems, the BTSC system has become well established in the USA and is in regular service on many networks including the NBC television network, who by 1987 were broadcasting over 30 hours a week in stereo from their 113 affiliated stations.

48.2.3 West German dual carrier system

Instead of using a subcarrier, the German dual carrier television sound system uses a separate fm carrier to convey the additional information needed for the second audio channel. This system was developed in the Federal Republic by the Institut für Rundfunktechnik (IRT) in conjunction with German receiver manufacturers[10,11]. The system was launched on-air on the ZDF network in Berlin at the 1981 Funkaustellung (International Radio and Television Exhibition).

The transmission characteristics of the dual carrier system are summarized in Table *48.4* and in the spectrum diagrams of *Figure 48.5*.

Frequency of the sound carriers		
Sound 1	5.5 MHz	Above vision carrier frequency
Sound 2	(5.5 + 0.2421875)(1) MHz	Above vision carrier frequency
Deviation of the sound carriers		
Sound 1	± 50 kHz peak	
Sound 2	± 50 kHz peak	
Pre-emphasis		
Sound 1	= 50 μs	
Sound 2	= 50 μs	
Identification subcarrier		
Frequency	54.6875(2) kHz	
Modulation	mono	= unmodulated
	stereo	= 117.5 Hz (3) 50% (4) am
	dual language	= 274.1 Hz (5)
Deviation of the second sound carrier by the subcarrier		= ± 2.5 kHz
Programme signals		
Sound 1	mono	= mono
	stereo	= (L+R)/2
	dual language	= primary languange
Sound 2	mono	= mono (same as Sound 1)
	stereo	= R
	dual language	= secondary language

(1) The frequency difference between both sound carriers is 15.5 × line frequency = 242.1875 kHz. Phase locking of both sound carriers with line frequency gives improvements, but is not abosolutely necessary.
(2) The frequency of the identification subcarrier is 3.5 × line frequency. The subcarrier and identification frequencies are phase locked with the line frequency.
(3) Line frequency/133
(4) The residual 50 per cent modulation is reserved for future identification of audio companding systems.
(5) Line frequency/57

Table 48.4 Transmission characteristics of the dual carrier system (for television systems B, G, H)

The difference (approximately 242 kHz) in the rest frequencies of the two fm sound carriers is carefully chosen to minimize the visibility of any patterning caused by intermodulation products produced by the effects of non-linear distortion in transmission of the two sound signals.

In the case of stereo broadcasts, although the primary sound carrier carries the (L+R)/2 sum signal, the second sound carrier *does not* (except in the case of the Korean variant (see below)) convey the (L-R)/2 difference signal. Instead it conveys the R (right-hand) signal only. This form of matrixing was chosen because, in the presence of coherent interference from the vision signal (buzz-on-sound), it ensures that similar correlated interfering signals appear equally (centre stage) in the recovered stereo L and R signals, rather than predominantly to one side and at a higher level.

This may be understood from the following matrix equations. Consider first a conventional stereo matrix used in the Zenith/GE pilot-tone stereo system for fm radio and assume that a coherent interfering signal, N, affects both the sum and difference signals equally. Then the conventionally dematrixed signals comprise:

$$\{(L+R)/2 + N\} + \{(L-R)/2 + N\} = L + 2N$$

and

$$\{(L+R)/2 + N\} - \{(L-R)/2 + N\} = R$$

Thus the interference appears in the left-hand channel only. Now with the matrix used in the IRT dual carrier system, the corresponding equations are:

$$2\{(L+R)/2 + N\} - (R + N) = L + N$$

and directly,

$$R + N = L + N$$

Thus the interference is divided equally between the two channels. Note that, in principle, the combined levels of the interference in the L and R channels are the same for both matrices, although in practice, because some phase dispersion is likely to occur due to the loudspeakers and room acoustics, the perceived level of *coherent* interference is expected to be about 3 dB lower with the IRT matrix.

The dual carrier system has three modes of operation: mono, stereo and bilingual. These are indicated using a subcarrier signal at 3.5 times line frequency (54.6875 kHz) which is 50 per cent am modulated by tones of 117.5 and 274.1 Hz for stereo and dual-language broadcasts respectively (see Table *48.4*). The unused 50 per cent modulation depth of the subcarrier is reserved to identify analogue companding (e.g. Dolby or HICOM). So far as is known, this companding option has never been used.

For bilingual operation, matrixing is not used; the primary (mono compatible) fm sound carrier conveys the main language, and the second sound carrier conveys the other language. In this mode, the separation of the sound signals is not limited by the precision of the matrices or by the relative fm deviation of the two sound carriers. This last factor is, however, critical for stereo operation, and special techniques have been developed to ensure the precise alignment of the fm deviations of the two carriers. Most dual channel receivers can simultaneously decode both languages of a bilingual broadcast, and provision is usually made to select one language to the loudspeakers and the other to a headphone socket. The listener can manually select which language goes to which output.

The dual carrier system provides much better separation in the bilingual mode than the fm/fm or BTSC subcarrier systems. But it is more expensive to implement both for the broadcaster and the receiver manufacturer.

At the transmitters, provision must be made to radiate the new sound carrier. This will require the bandwidth of the sound transmitter and sound/vision combiners to be increased, and it will need the sound transmitters to operate in a linear mode. With a single mono fm carrier, the sound transmitters can operate in class C to provide energy saving.

The chosen level of the two sound carriers (-13 dB and -20 dB for the first and second sound carriers respectively) relative to the associated vision carrier is a compromise between compatibility and ruggedness. If the amplitude of the sound carriers is too large, the intermodulation between them and with the picture signal components may result in patterning on the picture. If the amplitude of the sound carriers is too low, then the sound signals will be insufficiently rugged, and they will not be received reliably in places where the signal is weak or disturbed by multipath propagation caused by, for example, mountains or tall buildings.

A particular problem can arise under such conditions of multipath propagation, where selective fading can depress the level of one or both of the received sound carriers by up to 12 dB relative to the vision carrier. When this occurs, the received sound signal may be very noisy even though the picture signal is relatively unimpaired. Although the dual carrier system is more robust than the fm/fm system in this respect, care is needed in the design of the receivers to get the best from the system.

Significantly better performance can be obtained from the dual carrier system by using a *quasi-split sound* or *dual if* receiver[12]. In this, separate if filters, amplifiers and detectors are used for the vision and sound signals. The vision if is almost conventional and has the appropriate vestigial sideband filtering. The sound if, however, has a special "dual-hump' saw filter with two peaks in its response to pass the vision carrier and the sound carriers. Intercarrier detection is still used but, because the vision carrier used in the sound detector has been filtered by the special dual-hump filter, it has much less incidental phase modulation than in a conventional receiver. Quasi-split sound receivers have now become popular wherever the dual carrier system is used.

The German dual carrier system has been quite widely adopted in other countries, notably Australia, the Netherlands, Switzerland, Austria and Italy. A variant of it, in which the second sound carrier conveys the difference signal instead of the right-hand signal, has been adopted in South Korea.

A variant of the dual carrier system, adapted to system I PAL, was tested by the BBC in the UK in 1982. The results of these tests indicated[13] that while it would have been feasible to use it in the UK, the margin between compatibility and ruggedness problems would have been small with the different parameter values needed for system I. It was concluded that better prospects were offered by the then emergent digital techniques described in section *48.3*.

48.2.4 Wegener multi-channel systems for satellite services

All the systems described previously are intended to be used with terrestrial television broadcasts in which the picture signal is conveyed by vestigial sideband amplitude modulation. In terrestrial broadcasting, the composite picture and sound signals must be conveyed within the relatively narrow channels available (6, 7 or 8 MHz for systems M/N, B, G/I respectively).

When an NTSC or PAL signal is conveyed by satellite, for point-to-point communications (as in a contribution circuit) or point-to-multipoint (as in feeds to cable head ends or for direct to home (DTH) reception), the composite baseband picture and fm sound signals are conveyed by frequency modulation of the shf carrier. Multiple sound signals are usually conveyed as low-level subcarriers added to the baseband picture signal. This is in some ways analogous to the subcarriers used in the pilot-tone stereo systems described in section *48.2.2.1*, but in this case it is the fm sound carriers themselves, at typically 5–8 MHz above the vision carrier, which comprise the subcarriers.

A USA corporation, Wegener Communications Inc., devised a standardized band plan to use up to eight fm subcarriers which are added at low level to a normal NTSC or PAL signal for transmission via an fm satellite link[14]. The main principles of the band plan are that the modulation index (the ratio of the deviation of the main fm carrier to the frequency of the subcarrier) should be between 0.14 and 0.18, and that the subcarriers should be spaced by 180 kHz. Each of the fm subcarriers is deviated up to ±50 kHz by one audio signal.

By keeping the modulation index due to each subcarrier very low, the contribution to the total frequency deviation of the main shf carrier due to the subcarriers is small comparted with that produced by the picture signal. The subcarriers consequently have little impact on the bandwidth needed to convey the composite fm signal or on the signal/noise ratio of the received picture signal. The penalty for the low modulation index is that the signal/noise ratio in the audio signals recovered from the subcarriers is reduced compared with a conventional mono signal. This is usually not a problem on contribution circuits or feeds to cable heads where a large receiving dish can be used in order to provide a good carrier/noise ratio. It may be more of a problem for DTH receivers which use the Wegener system. To help this, analogue companding of the audio signals is sometimes used.

For NTSC signals, the subcarriers are typically placed between 5.2 and 8.5 MHz above the vision carrier frequency. For PAL broadcasts the spacing is in the range 6.3–7.92 MHz with 7.02 and 7.2 MHz the most commonly used for two-channel operation.

Each 180 kHz subcarrier band can be allocated to a 15 kHz audio signal or subdivided to convey two or more narrow-band audio signals. Alternatively, data may be conveyed using frequency shift keying or phase shift keying. Using the latter, sound signals digitized by the Dolby adm system have been conveyed using two adjacent 180 kHz slots to convey the 256 kbit/s needed for a single mono Dolby adm (see section *48.4.2*) encoded sound signal.

The Wegener 1600 series system was used by the BBC during the 1984 Olympic Games to provide multi-channel (up to nine channels) audio contribution circuits from Los Angeles to London.

The Wegener Panda 1 system has been in service for several years to convey stereo sound associated with the Music Box programme services on Eutelsat.

New variants, with different companding systems have been developed (Panda 2) and, for example, are used by Sky to convey the multiple sound signals of their programme services broadcast via the Astra satellite. These Sky Astra services are intended for direct-to-home reception and the cheaper Astra receivers decode only the Wegener subcarriers, ignoring the conventional 6 MHz mono carrier.

48.3 Digital systems

The advent of compact discs in 1983 heralded the introduction of digital sound systems into the home and provided much of the basic enabling technology, in the form of low cost components such as high quality digital/analogue converters, to make digital multi-channel sound with television feasible.

A system that conveys television sound signals in digitally encoded form offers a number of important advantages over an analogue system:

- improved signal/noise ratio and low distortion; interference from the vision signal is almost completely avoidable
- theoretically unlimited separation between the decoded sound signals; this is especially important for bilingual broadcasts
- uniform performance throughout the service areas of the transmitters; noise and distortion are not cumulative through the transmission system
- additional capacity to provide a low-rate data channel e.g. to carry encrypted access control data in a conditional access television system; when not needed to carry stereo or dual-language sound signals, the main digital channels may be used to convey data

There are several ways in which digitally encoded sound signals can be introduced into a television channel:

- frequency division multiplexing as in the NICAM 728 system
- time division multiplexing at baseband as in sound-in-sync and B-MAC, D-MAC and D2-MAC
- time division multiplexing at radio frequency as in C-MAC

48.3.1 The NICAM 728 digital system

The NICAM 728 system[15–23] for broadcasting digital two-channel sound with a terrestrial television system was developed by the BBC in the UK in consultation with the IBA and representatives of the British Radio and Equipment Manufacturers Association (BREMA). The system conveys two high-quality digitally coded sound signals along with the picture and mono fm sound signals of the UK television system I. NICAM 728 has also been adapted for use with television systems B and G and may also be suitable for use with other television standards.

The digital multiplex of NICAM 728 may be configured either as a single stereo channel, or as two independent mono channels suitable for dual-language broadcasts. There is also the option of transmitting binary data with, or without, a single mono digital sound channel, giving a data channel capacity of 352 kbit/s or 704 kbit/s respectively.

The joint BBC/IBA/BREMA specification[15] of the NICAM 728 system for use with system I broadcasts was approved by the UK administration in 1986, and with minor revisions in 1988.

The European Broadcasting Union[24] recommends that those members planning to introduce digital two-channel sound with terrestrial television systems B, G, H and I should base their choice on the NICAM 728 system.

The outline characteristics of the NICAM 728 system are given in Table *48.5*.

48.3.1.1 Conveying the digital information

The four basic ways in which digitally encoded sound signals can be introduced into a television channel have been listed. Radio frequency time division multiplexing (tdm) could not be introduced compatibly on existing terrestrial broadcasts. Baseband tdm was, however, a possibility using data inserted into television lines in the field blanking time or into the line blanking time of the video signals. Teletext systems, however, already occupy many of the available lines in the field blanking of UK broadcasts. And although sound-in-sync has been successfully used for many years to convey sound signals digitally in the BBC's point-to-point distribution circuits (see section *49*), data introduced into the line-blanking interval were found to be incompatible with some existing domestic receivers, especially under conditions of multipath interference (when the data signal becomes visible as crawling dots).

1. Specification of the digitally modulated carrier:	
Carrier frequency:	System I: 6,552 MHz above the vision carrier. Systems B & G: 5.85 MHz above the vision carrier level.
Carrier level:	-20 dB with respect to peak vision carrier level.
Modulation:	Differentially encoded quadrature phase shift keying.
Spectrum shaping: (overall with ideal receiver)	System I: 100% cosine roll-off, split equally between transmitter and receiver (overall bandwidth of digital signal approximately 728 kHz). Systems B & GL 40% cosine roll-off, split equally between transmitter and receiver (overall bandwidth of digital signal approximately 510 kHz).
2. Level of primary f.m. sound carrier	- 10 dB with respect to peak vision carrier level.
3. Overall bit-rate:	728 kbit/s
4. Sound coding characteristics:	
Pre-emphasis:	CCITT Recommendation J.17
Audio overload level:	System I: + 14.8 dBu0 at 2.0 kHz (0 dBu0 = 0.775 Vrms) Systems B & G: +22.0 dBu0 at 400 Hz. (equivalent to + 12.5 dBu0 at 2.0 kHz)
Sampling frequency:	32 kHz
Initial resolution:	14 bits per sample.
Companding characteristics:	Near-instantaneous (NI), with compression to 10 bits per sample in 32-sample (1 ms) blocks.
Coding for compressed samples:	2's complement.
Number of coding ranges:	5 } Signalled by 3-bit scale factor code.
Number of protection ranges:	7
Error protection:	One parity bit added to each 10-bit sample to check the six most significant bits (parity bit modified for scale factor signalling).
Scale factor signalling: 3 bits per sound coding block (two blocks per frame)	By modification of 9 parity bits per scale factor bit, detected by majority logic decision.
5. Bit interleaving:	44 × 16. Frame alignment word *not* interleaved.
6. Energy dispersal scrambling:	By modulo-two addition of a pseudo-random binary sequence of length 2^9 - 1 bits, synchronously with the multiplex frame. Frame alignment word *not* scrambled.
7. Frame format:	728 bits per (1 ms) frame with 8-bit frame alignment word.

Table 48.5 Summary of the characteristics of the NICAM 728 system for digital two-channel sound with terrestrial television

The suggestion of using a frequency division multiplex (fdm) system with an additional digitally modulated carrier to convey sound or data signals with system I television signals was made in 1978[25]. During 1983, a preliminary experimental system based on this proposal was built, and laboratory and field tests begun.

There are four main processes involved in preparing the sound signals for transmission via this digital system:

- analogue/digital conversion
- companding
- multiplexing
- modulation

In the following sections each of these processes is described in detail in the context of the NICAM 728 system.

48.3.1.2 Analogue/digital conversion

The process of analogue/digital conversion of the sound signals involves sampling at a frequency greater than twice that of the highest frequency signal components (see section *50*). A sampling frequency of 32 kHz was selected for this application because of its existing international use in point-to-point distribution circuits and in the MAC/packet family of systems (section *29*). The sound signals are therefore band-limited by 15 kHz low-pass filters at the input to the analogue/digital converter (ADC).

For high quality digital audio, an initial coding accuracy of at least 14 bits per sample is needed to represent the sound signals. If fewer bits per sample are used, the quantizing error will become audible. This usually sounds like random noise added to the signal, but when the signal is at very low-level the effect is to impart a 'gritty' sound to the reproduced signal. This is known as *granular distortion*.

48.3.1.3 Companding

With a sampling frequency of 32 kHz and an initial coding accuracy of 14 bits per sample, about 1 Mbit/s would be needed to convey the digitally coded sound signals directly (i.e. linear coding). This data rate could not easily be accommodated within the bandwidth available in the 8 MHz television channels. However, techniques for bit-rate reduction of digital sound signals are well established and are used to reduce the transmitted bit rate to manageable proportions.

Bit-rate reduction is achieved by using *companding*. Companding involves compressing the sound signals prior to transmission (or recording) and expansion at the receiving terminal (or on playback). It is useful for analogue as well as digital systems. For example, it is widely used with analogue tape recorders. In analogue systems, however, companding improves the signal/noise ratio, whereas the object of companding in digital systems is to reduce the bit-rate requirements of the signal and thereby permit economical use of the available channel capacity.

In analogue companding, a major problem is matching the expander characteristic in the decoder to that of the compressor in the coder. Mistracking will produce audible non-linear distortion. In a digital companding system, the compression and expansion are performed in the digital domain and therefore the coder and decoder can be exactly matched, thus avoiding any mistracking and consequent distortion.

The companding technique used in the NICAM 728 system is that of NI (near instantaneous) companding which was developed by the BBC in the early 1970s[27,28]. Indeed NICAM is an acronym for *near instantaneous companding and audio multiplex*. The same companding technique is used in two of the sound coding options available in the MAC/packet family of systems[26]. *Figure 48.6* illustrates this process. The 14-bit sound samples emanating from each ADC are coded in two's complement form (see section *50.1.1.4*). The samples from each channel are partitioned into separate blocks of 32 samples (1 ms) and the largest sample in each block found.

For blocks in which the largest sample is up to one-sixteenth of full amplitude, the only processing is to truncate the transmitted samples by suppressing the four most significant bits which are next to the *sign bit* (i.e. the bit of highest significance). The most significant bits (msbs), which are needed to identify the sign of the samples, are always transmitted whatever the amplitude of the signal being coded. Thus low amplitude signals are conveyed with the full 14 bits per sample initial coding accuracy. There is therefore no increase in granular distortion. Blocks of samples in which the largest is up to one-eighth full amplitude are processed by suppressing the

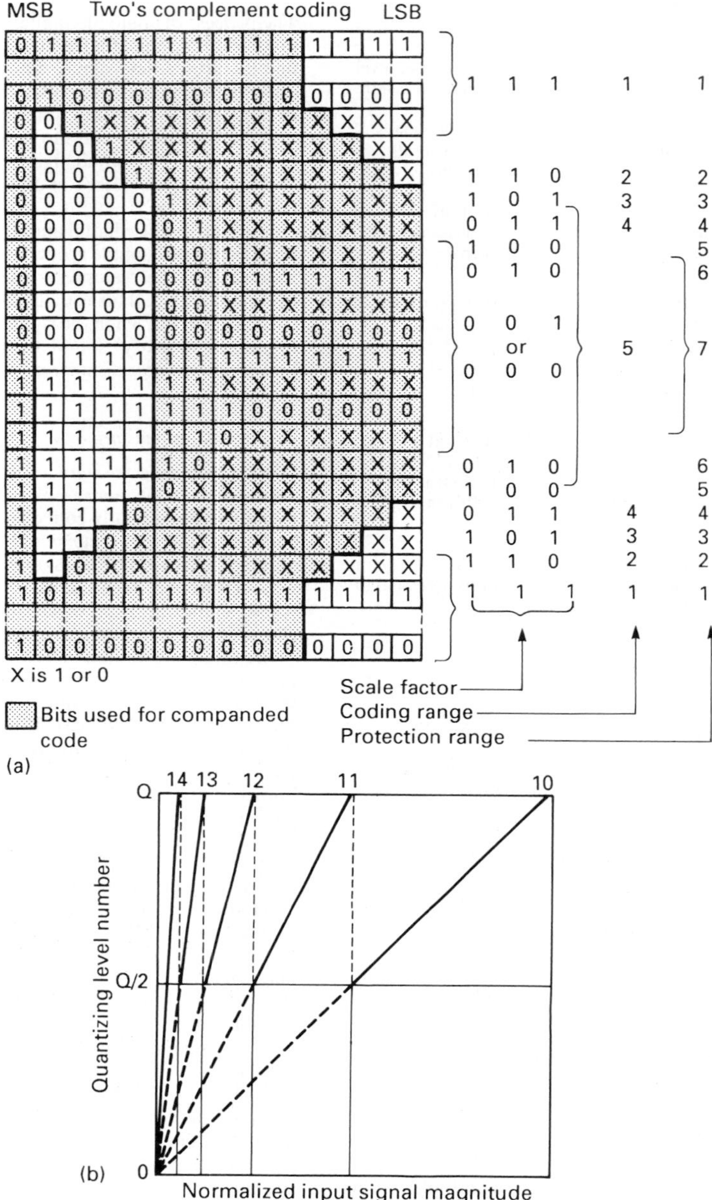

Figure 48.6 NI companding: (a) coding of the companding sound signals, (b) quantization of the signals

least significant bit (lsb) and three of the msbs. They are therefore transmitted effectively with 13 bit coding resolution.

Those blocks with samples up to one-quarter full amplitude are transmitted with 12 bit effective resolution, and so on to blocks containing samples greater than half full amplitude which are transmitted with an effective resolution of 10 bits per sample. These ranges of signal amplitudes are referred to as *coding ranges* (see *Figure 48.6*(b)).

The expander in the decoder must, of course, be told the coding range for each block so that it can expand it correctly. This requires the transmission of *scale factor* (or *range code*) information. With the five coding ranges of this system, three scale factor bits per 32 sample block must be transmitted.

All companding systems, whether analogue or digital, add programme modulated noise to the signal. Large audio signals are accompanied by more noise than small signals, and companding relies upon the fact that the programme signal masks the noise at the higher levels. *Figure 48.7* shows the signal/quantization noise characteristic of the NI companding system used in this application. Note the four 'gear-changes' of 6 dB in the signal/noise ratio as the system moves through the five ranges.

The use of CCITT Recommendation J.17 pre-emphasis[29] prior to compression, with the corresponding de-emphasis after

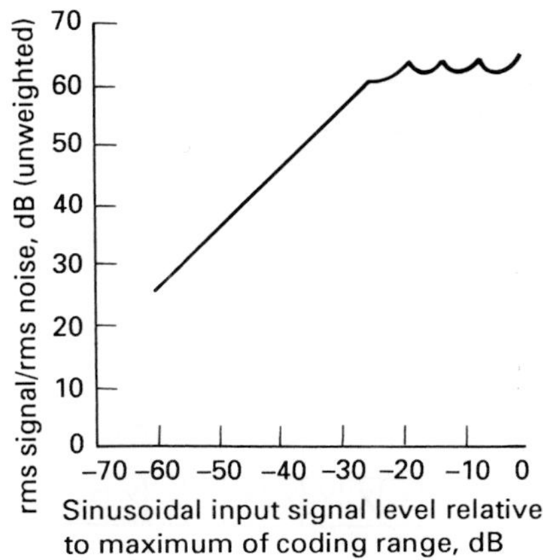

Figure 48.7 Quantization noise for 14/10 NI companded sound signal. No allowance has been made for the use of pre-emphasis or for the fact that, for low frequency sine-wave signals (<1 kHz), range changes will occur during each cycle

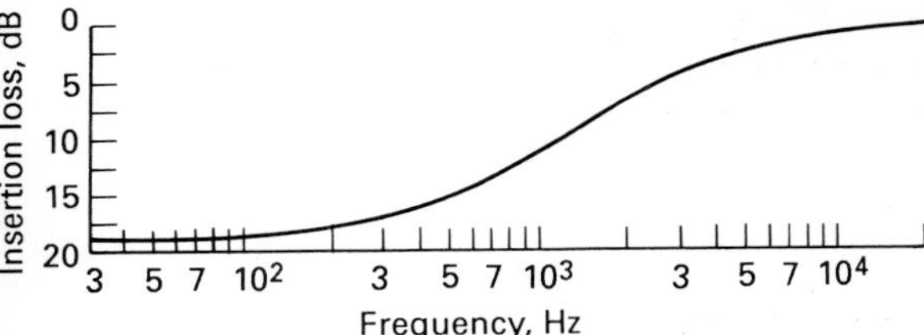

Figure 48.8 CCITT J-17

expansion, has been found to reduce significantly the audibility of programme modulated noise. This makes use of the fact that the spectrum of typical programme signals contains less power at high audio frequencies than at low frequencies.

48.3.1.4 Multiplexing

48.3.1.4.1 Frame structure

The transmitted serial data stream is partitioned into 728-bit frames, each of 1 ms duration, giving an overall data rate of 728 kbit/s made up as follows:

8-bit frame alignment word	8 kbit/s
5 bits for control information	5 kbit/s
11 bits for additional data	11 kbit/s
704 sound, parity or data bits	704 kbit/s

Diagrams of the frame structures for conveying stereo and mono sound signals are shown in *Figure 48.9*. The two different frame structures (one for stereo and the other for mono) are used to accord with those specified in the MAC/packet family of systems. There, the principle of packet multiplexing requires that information for different services (e.g. mono sound channels) is conveyed in separate packets. Information for the two channels comprising a stereo signal must, however, be carried in the same packet to maintain the phase relationship between the two sound signals. This is because, in a packet system, sampling is not synchronous between packets, and therefore there is an undefined delay in transmitting and receiving successive packets.

The 720 bits which follow the frame-alignment word form a structure identical with that of first-level protected, companded sound signal blocks in the systems of the MAC/packet family, so that decoding of the sound signals could be performed by using the same type of decoder that is used in these MAC systems.

In each 728 bit frame, the first eight bits comprise a frame-alignment word that is needed to enable the receiver/decoder to synchronize to the received frame structure and thence re-partition the data stream. The five bits of control information include three which identify the application of the 704 bit sound/data block. The eleven additional data bits may be used for ancillary applications yet to be defined.

48.3.1.4.2 Bit interleaving

Bit interleaving is applied to the 704 bit sound/data block to minimize the effect of multiple bit errors due, for example, to differential decoding of the received data signal (see section *48.3.1.7*). The specified interleaving pattern[15] places data bits which are adjacent in the frame structure of *Figure 48.9* in positions at least 16 clock periods apart in the transmitted bit stream. Conversely, bits which are adjacent in the transmitted bit stream are at least 44 bits apart in the frame structure (except across frame boundaries). This interleaving structure is convenient for decoder design because at least 44 bits comprise exactly four 11 bit companded samples (10 bits plus a parity bit), thus allowing easy management of the memory used in de-interleaving.

48.3.1.4.3 Energy dispersal scrambling

In order to ensure that the transmitted signal is as noise-like as possible (which is desirable to help compatibility with the picture and mono sound signals in the wanted and adjacent channels), the transmitted bit-stream is scrambled. This scrambling is achieved by adding, modulo-two, a pseudo-random sequence to the serial data-stream comprising the 728 bit frames. The frame alignment words are not scrambled because they are needed to synchronize the pseudo-random sequence generator used for descrambling in the receiver.

The specified pseudo-random sequence is only 511 bits long and the generator is re-initialized on the first scrambled bit of each frame; thus if the sound/data bits are the same from frame-to-frame, the transmitted bit pattern will repeat in each frame. A short scrambling sequence was used to simplify the receiver and to speed acquisition of frame-lock, and, in practice, no disadvantages have been found from using this relatively short scrambling sequence.

48.3.1.5 Error protection for sound signals

The sound samples are protected by a simple parity checking arrangement. The basic principle of this error protection is as follows.

One parity bit is added to each 10 bit sound sample, to allow checking in the decoder for the presence of errors in the six most significant bits. At the coder the six msbs are added together modulo-two (this is equivalent to the logical exclusive OR function). The result will be either 0 or 1. If 0, then the parity bit is 0; if 1, then the parity bit is 1. Thus the modulo-two sum of the six protected sample bits and the parity bit is always 0. That is, the total number of 1 symbols in the parity group (including the parity bit) is always even.

In the decoder, the parity check is performed by recalculating the modulo-two sum of the six protected sample bits and the

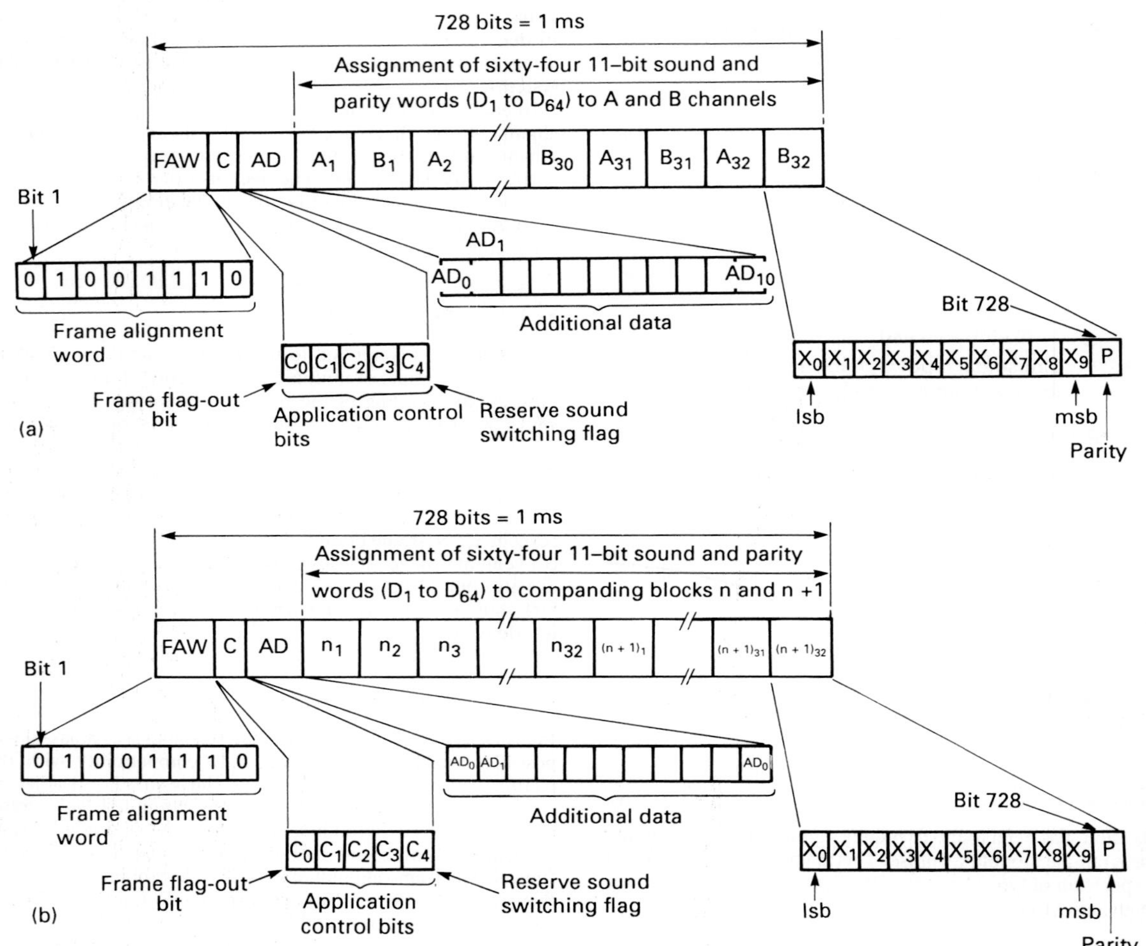

Figure 48.9 Structure of a 728-bit frame containing a sound signal (before interleaving): (a) stereo, (b) mono

parity bit. A single error in the group of six protected sample bits or the parity bit causes the modulo-two sum to become 1 instead of 0, and thus a single error can be detected. The decoder cannot, however, tell which particular bit in the group is in error, and double or any even number of bit errors are not detected by this simple parity check. However, with the bit interleaving described above, single errors are much more likely than multiple errors, so this simple parity check is effective.

If an error is detected in a sample, that sample may be replaced by an interpolated sample value formed from the previous and following correct sample values (see *Figure 48.10*). Of course, the interpolation is not always perfect, but subjectively it gives good results. For example, without concealment, the impairment resulting from random errors was judged to be Grade 3 (*slightly annoying*) on the CCIR 5-point quality scale at a bit-error ratio of about 2 in 10^5. With concealment, the same impairment does not occur until the bit-error ratio exceeds 1 in 10^3. In terms of carrier/noise ratio (see section *48.3.1.8*), error concealment yields an effective improvement of about 2 dB.

The parity check and error concealment is confined to the six msbs because listening tests conducted by a number of broadcasting organizations during the development of the MAC/packet family showed that to be the best compromise for companded sound systems of this type. Errors in the least significant bits are usually not annoying, and it is therefore better to leave them unprotected in order to afford better protection for the msbs. If more bits are protected, then there is a greater chance of an undetectable multiple error; if fewer bits are protected then the unconcealed errors become more annoying.

48.3.1.6 Transmission of scale factor information

The scale factor word for each 1 ms, 32 sample sound coding block comprises three bits signalling the five coding ranges. Since there are two sound coding blocks in each 728 bit frame, six scale factor bits must be transmitted in each frame. This is done without using any additional bits by using a technique known as *signalling in parity*[30]. This exploits some of the redundancy inherent in the parity bits and the fact that it is unlikely that the majority of samples in a small group will contain errors.

Signalling of the scale factor in parity operates as follows. (We shall consider the case of a 728 bit frame which contains stereo signals; the principle is the same for frames containing more signals, but, of course, due to the different frame format the details are different.)

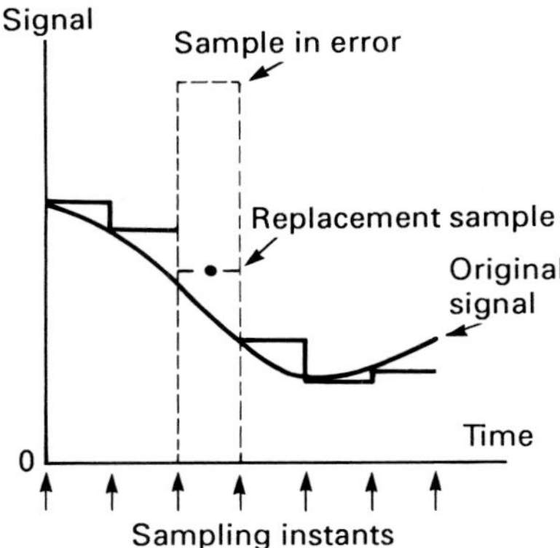

Figure 48.10 Error concealment by interpolation

The first 27 of the 32 sound samples in a 1 ms companding block are divided into three groups of nine samples. Each group is allocated to one of the three scale factor bits needed for that companding block. Each bit of the scale factor word is then signalled by allocating even parity to each sample of the group of nine if the bit to be signalled is 0, and odd parity if the bit to be signalled is 1.

In the receiver, the parity check for each sample is initially recalculated in the normal way but is not used directly. Instead, the parity checks over each of the nine sample groups are inspected. For those groups in which the majority of the samples have odd instead of even parity, it is taken that the scale factor signalled by that group is 1. The parity on all nine samples in that group is complemented and error concealment applied to replace any samples which then yield odd parity. For those groups in which the majority of the samples have even parity, it is taken that the scale factor signalled by that group is a 0. Error concealment is applied directly to any samples with odd parity in these groups.

Signalling in parity has the advantage of being very robust, because five or more of the nine samples in a group must be in error before a wrong decision on scale factor is taken by the decoder. For example, with a bit error ratio of 5 in 10^3, the mean periodicity of scale factor errors is 33 s[31]. This is important because scale factor errors can produce very annoying clicks.

The penalty to be paid for exploiting the redundancy inherent in the parity bits in this way is a small weakening of the protection they provide against sample errors. But for bit-error ratios better than 1 in 10^2, the consequent degradation is negligibly small[32].

There is one further form of protection against errors provided in the system. *Figure 48.6*(a) shows the way in which the three scale factor bits signal the five coding ranges using five of the eight possible binary codes. The remaining codes are used to signal two further 'protection' ranges. If the decoder is informed when the input signal to the coder is at a low level, less than half of the maximum amplitude of the lowest coding range, then it can make certain deductions about the msbs of the incoming sound samples. The samples are in two's complement code so if, for example, the maximum amplitude of the samples in a given sound coding block is less than $^1/_{128}$ of full amplitude, then the seven msbs should all be the same. Thus if the decoder is told that the amplitude of the signal lies within protection range seven (see *Figure 48.6*(a)), then errors in those seven msbs (of which three are transmitted) should all be identified even when left undetected (as a result of multiple bit errors) by the ordinary parity check. A majority decision can then be used to correct these errors.

48.3.1.7 RF characteristics

48.3.1.7.1 Frequency of digitally modulated carrier

In the system I version of NICAM 728, the digitally coded sound signals are conveyed on an additional carrier which is at a frequency 6.552 MHz above the vision carrier (see *Figure 48.11*). This frequency spacing was determined by the need on the one hand to avoid interference to or from the mono fm sound carrier of the wanted signal, and on the other to avoid interference to or from signals in the upper adjacent channel. The precise spacing of 6.552 MHz was chosen because it is numerically equal to nine times the bit rate of 728 kbit/s.

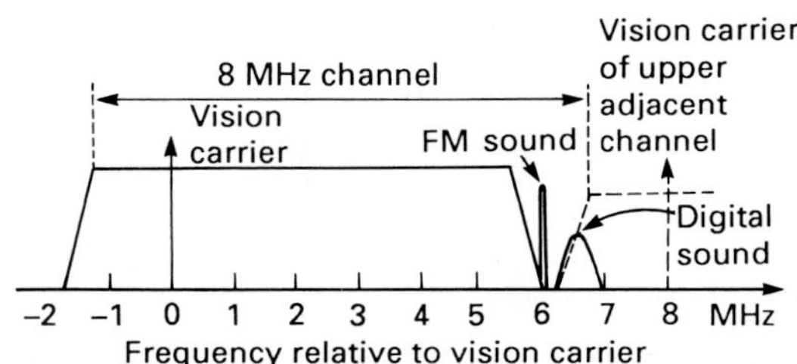

Figure 48.11 Spectrum of system I television signal with digital two-channel sound signal (vertical axis not to scale)

Most of Europe other than the UK uses television systems B and G. System B exists in 7 MHz vhf channels rather than the 8 MHz uhf channels of system I. There was therefore some question as to whether it would be possible to use a digital system in these narrower channels. However, the spectral gap available for the digital signal is the same in system B as it is in system I, because the narrower channels are compensated by a smaller intercarrier spacing between the vision carrier and the mono fm sound carrier (5.5 MHz instead of 6 MHz) and a narrower vestigial sideband for the picture signal.

Broadcasters in the Nordic countries had started independent tests with a 512 kbit/s digital system in 1984. This lower bit rate allowed the frequency of the digital carrier to be set at only 350 kHz above the fm sound signal, i.e. 5.85 MHz above the vision carrier (see *Figure 48.12*). Although encouraging results were obtained with this 512 kbit/s system, the reduced bit rate caused some loss of quality compared with the 728 kbit/s system. And the advantages of commonality with the UK system (and the MAC/packet family) were recognized by the Nordic broadcasters and the receiver industry. Consequently, in 1986, the Nordic broadcasters, with some help from the BBC, made strong efforts to adapt the 728 kbit/s UK system for use in systems B and G.

To avoid interference to receivers tuned to the signal in the upper adjacent channel, it was found to be essential to restrict the intercarrier spacing in system B to 5.85 MHz.

An intercarrier spacing of 5.85 MHz is also used in system G where, with 8 MHz wide uhf channels, there is plenty of space for the digital signal. This use of the same intercarrier spacing (and data signal spectrum shaping) in system G as in system B is desirable, even though system G would allow a wider intercarrier spacing, to avoid the extra complication of switching filters,

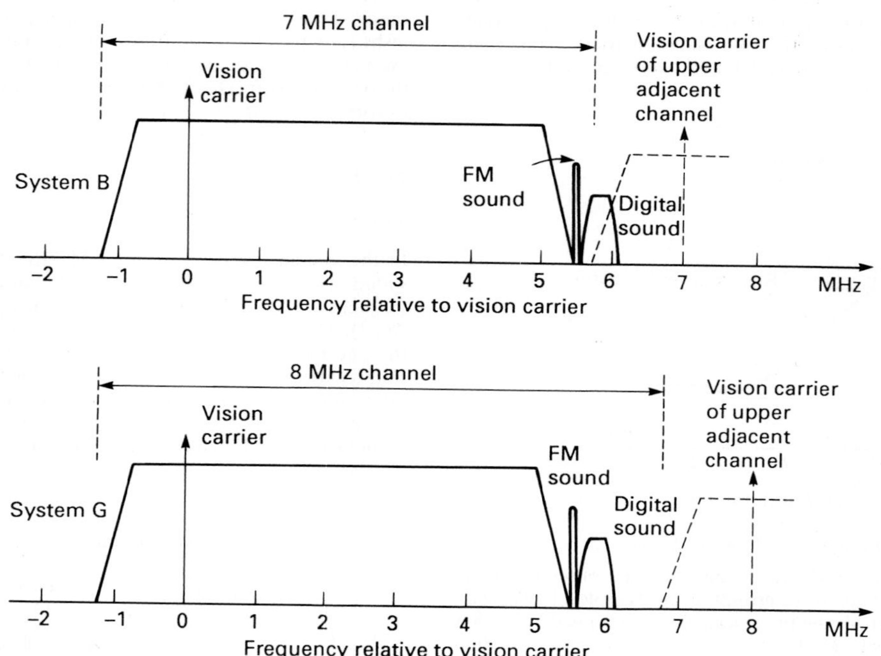

Figure 48.12 Spectra of digital two-channel sound signal (vertical axis not to scale)

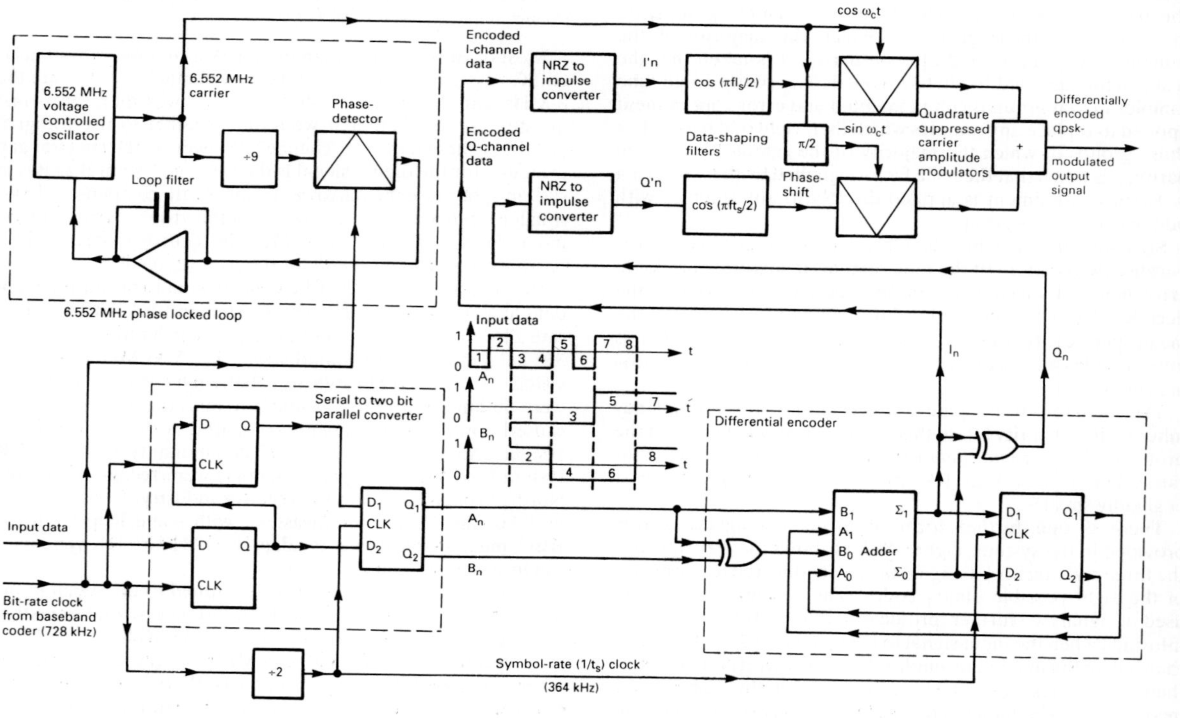

6.552 MHz voltage controlled oscillator
6.552 MHz carrier
÷9
Phase-detector
Loop filter
6.552 MHz phase locked loop
Encoded I-channel data
NRZ to impulse converter
I'n
cos ($\pi f I_s/2$)
cos $\omega_c t$
Encoded Q-channel data
Data-shaping filters
$\pi/2$
Phase-shift
−sin $\omega_c t$
Quadrature suppressed carrier amplitude modulators
+
Differentially encoded qpsk-modulated output signal
NRZ to impulse converter
Q'n
cos ($\pi f t_s/2$)
Input data
A_n
B_n
I_n
Q_n
Serial to two bit parallel converter
D Q CLK
D Q CLK
D_1 Q_1 CLK D_2 Q_2
Input data
Bit-rate clock from baseband coder (728 kHz)
÷2
Symbol-rate ($1/t_s$) clock (364 kHz)
Differential encoder
B_1 Σ_1 A_1 B_0 Adder A_0 Σ_0
D_1 Q_1 CLK D_2 Q_2

Figure 48.13 Block diagram showing the processes of differential encoding, data signal spectrum shaping and modulation at the transmitter (system I)

etc., in system B/G receivers. (Furthermore, in those countries which use systems B and G, transposers are often used which have a signal in a system B channel at their input and one in a system G channel at their output.)

48.3.1.7.2 *Level of digitally modulated carrier*

The level of the modulated signal is set at 20 dB below the peak vision carrier level by the need to balance very carefully the conflicting requirements of compatibility and ruggedness. If the level of the digital signal were set too high then a compatibility problem would result, with interference to the picture or mono sound on existing receivers; if it were set too low, then reception of the new digital service would be unreliable on the new sets developed to decode it.

48.3.1.7.3 *Modulation system*

Differentially encoded quadrature phase shift keying (dqpsk) modulation was selected as offering the best overall compromise between efficient use of the available spectrum space and the need for reliable reception with inexpensive receivers. With the specified data signal spectrum shaping, the overall bandwidth (to the -30 dB points) of the transmitted digital signal is about 728 kHz for system I and 510 kHz for systems B and G.

Figure 48.13 illustrates the processes of serial to two bit parallel conversion, differential encoding, data signal spectrum shaping and quadrature modulation used to produce a dqpsk signal as used on the system I version.

DQPSK is four state phase modulation in which each change of state conveys two data bits. The phase of the transmitted signal has four rest states 90° apart, as shown in *Figure 48.14*(a). An input bit-pair will shift the carrier phase into a different rest state by the amount of phase change assigned to that bit-pair. The transmitted phase changes and subsequent carrier rest

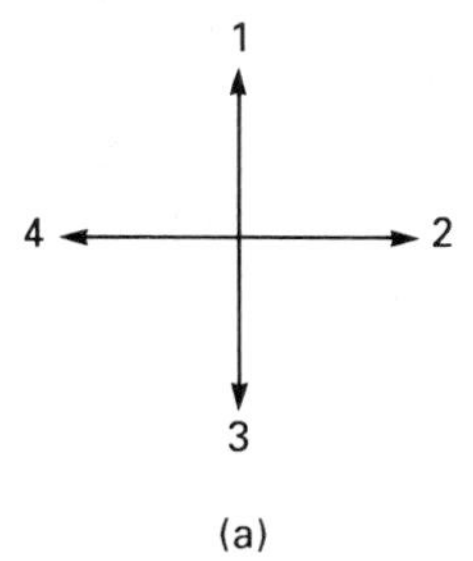

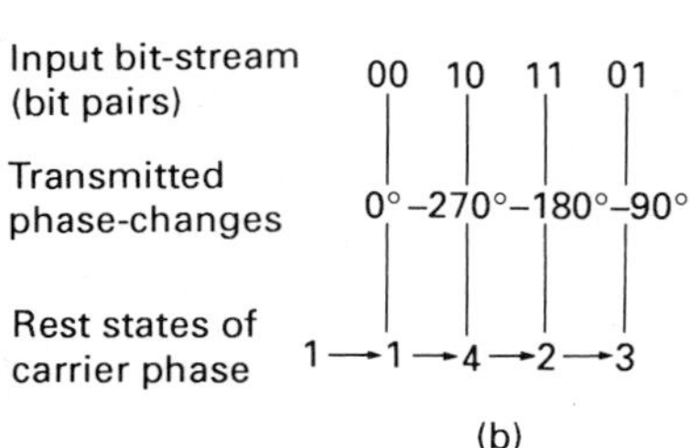

Figure 48.14 (a) The rest states of carrier phase 90° apart; (b) the transmitted phase changes and rest states of carrier phase for the input bit-pair sequence 00, 11, 01, assuming the carrier to be initially in rest state 1

states for the input bit-pair sequence 00, 10, 11 and 01 are illustrated in *Figure 48.14*(b).

Referring to *Figure 48.13*, the data to be transmitted are presented in pairs, after differential encoding and spectrum shaping, to the modulation inputs of two suppressed carrier modulators. The 6.552 MHz carrier inputs to these modulators are in quadrature. The dqpsk signal is formed by the linear addition of the outputs of these two quadrature modulators.

48.3.1.7.4 *Spectrum shaping*

In the system I version, the data signal spectrum shaping is designed to be 100 per cent cosine roll-off overall with the filtering split equally between the transmitter and receiver. In *Figure 48.13*, impulses at the symbol rate of 364 kHz are filtered by a low-pass filter with the following amplitude-frequency response:

$$H(f) = \cos\left(\frac{\pi t_s f}{2}\right) \quad \text{for } f \leq \frac{1}{t_s} \qquad t_s = \frac{1}{364}\ \text{ms}$$

$$= 0 \quad \text{for } f > \frac{1}{t_s}$$

This data signal spectrum shaping was chosen because the matching filter in the receiver is easy to implement and because it yields a wide data *eye* which is tolerant of errors in the timing of the sampling clock in the decoder. This spectrum shaping is also tolerant of errors in the overall amplitude/frequency and/or group delay/frequency responses.

(The term *eye* is used to describe the pattern produced when the demodulated data signal is displayed on an oscilloscope, the timebase of which is locked to a symbol rate clock. The vertical height of the eye at the sampling instant indicates the margin available against noise; the horizontal width of the eye indicates the margin available for errors in the timing of the sampling clock.)

In the system B/G version, the spacing of the digitally modulated sound signal from the fm sound signal is only about 63 per cent of that in system I. This is inadequate to accommodate the spectrum of the digital signal specified for the UK system, which is 728 kHz wide. However, by using narrower bandwidth data shaping filtering (see *Figure 48.15*), 40 per cent cosine roll-off overall instead of 100 per cent, the 728 kbit/s signal can be accommodated in system B with 5.85 MHz intercarrier spacing (see *Figure 48.12*).

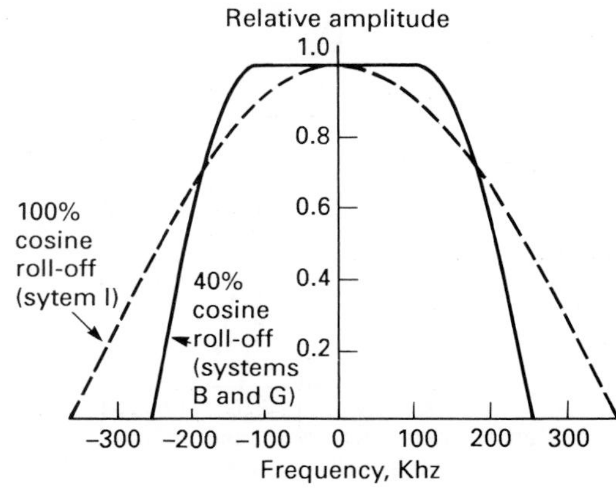

Figure 48.15 Spectrum shaping for system I and system B/G

This 40 per cent cosine roll-off data signal spectrum shaping for systems B and G is achieved (referring to *Figure 48.13*) by filtering impulses at the symbol rate of 364 kHz using a low-pass filter with the following amplitude-frequency response:

$$H(f) = \begin{cases} 1 & \text{for } f < \frac{1-k}{2t_s} \\ \cos\left[\frac{\pi t_s}{2k}\left(f - \frac{1-k}{2t_s}\right)\right] & \text{for } \frac{1-k}{2t_s} \leqslant f \leqslant \frac{1+k}{2t_s} \\ 0 & \text{for } f > \frac{1+k}{2t_s} \end{cases}$$

where $k = 0.4$ and $t_s = \frac{1}{364}$ ms.

There are small penalties for using the sharper roll-off filter characteristic: the width of the data eye is smaller (though the eye-height under ideal conditions is still 100 per cent) and consequently there is smaller tolerance than with 100 per cent cosine roll-off filtering for sample timing errors in the dqpsk demodulator/decoder. Furthermore, the data shaping filter in the receiver is slightly more difficult to implement, and the system is less tolerant of degradation in the overall amplitude/frequency and/or group delay/frequency response than with the 100 per cent cosine roll-off characteristic (including such degradation consequent upon multipath propagation). However, in practice, these are only marginal penalties, and are not significant.

48.3.1.8 Performance in field strength limited reception conditions

Under field strength limited reception conditions, the bit error rate may be calculated from the carrier/noise ratio at the input to the digital demodulator using the standard result for dqpsk with Gaussian white noise, as follows[32].

The bit error probability for coherently demodulated qpsk, ignoring differential encoding, is given by:

$$p_e = \frac{1}{2}\,\text{erfc}\,\sqrt{\frac{E_b}{N_o}}$$

$$p_e = \frac{1}{2}\,\text{erfc}\,\sqrt{\frac{C}{N}}$$

where E_b is the average energy per bit (joules), N_o is the power density of the noise (W/Hz), C is the average modulated carrier power (W), N is the noise power (W) measured in a bandwidth numerically equal to the bit rate, and

$$\text{erfc}\,(y) = \frac{2}{\sqrt{\pi}} \int_y^\infty e^{-z^2}\,dz \text{ for } y > 0$$

Note that the noise bandwidth of an ideal demodulator for this system is numerically equal to *half* the bit rate; a noise bandwidth of twice that is taken here, however, to accord with convention.

Allowing for the differential decoding needed for dqpsk (but still coherent *demodulation*), the bit error probability becomes:

$$p'_e = 2\,p_e(1 - p_e)$$

Figure 48.16 shows the results of this calculation and also the measured performance of an experimental demodulator.

Note the characteristically rapid failure with declining carrier/noise ratio. The system changes from a bit error ratio of 1 in 10^4 (which would yield almost unimpaired sound quality) to 1 in 10^2 (which would be unusable) over a range of only about 4 dB.

Relating the carrier/noise ratio of the digital signal to other parameters, such as the voltage applied at the antenna input of the receiver or the picture signal/noise ratio, is difficult because many factors, such as the noise factor of the receiver, have to be taken into account. However, in the BBC experimental receiver, a bit error ratio of 1 in 10^4 was obtained with a potential difference across the antenna input of approximately 35 μV rms (measured at the tips of syncs of the vision signal). The picture signal/rms unweighted noise ratio for this input voltage was measured to be about 15 dB, which corresponds to a very poor quality picture. It may be noted that, in theory, you can achieve the same bit error rate with a signal 2 dB weaker.

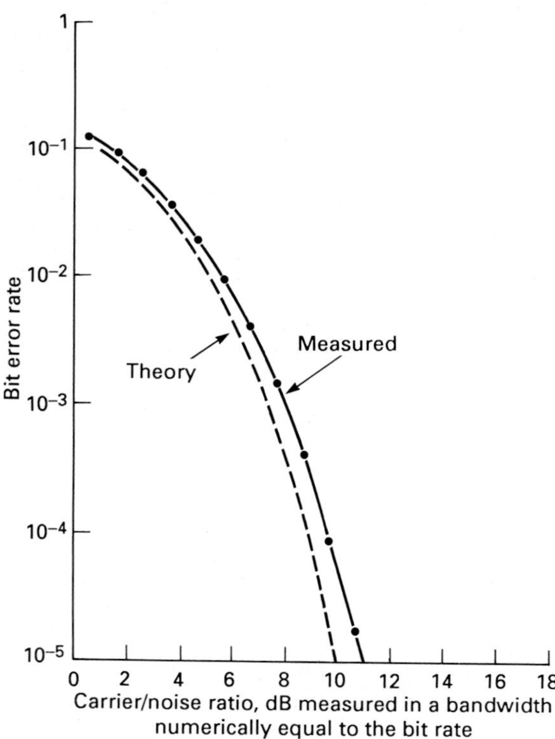

Figure 48.16 Bit error rate versus carrier/noise ratio

48.3.1.9 Performance in multipath reception conditions

Theoretical analysis of the performance of dqpsk in multipath reception conditions is given in reference 33. The effect of echoes is found to be critically dependent upon the precise delay of the echo(es) relative to the direct signal, but typically, for a single echo of about 10 μs delay, an echo of up to about 30 per cent of the amplitude of the direct signal can be tolerated without serious loss of performance.

Field tests of the system in Wales in 1983[34] proved its ruggedness against multipath. It is significantly more robust in this respect than teletext.

48.3.1.10 NICAM 728 broadcasts and receivers

Regular experimental broadcasts using NICAM 728 began from the BBC transmitter at Crystal Palace in London in 1986, and a NICAM 728 stereo service was expected to cover 75 per cent of the UK population in 1991.

The IBA and some of the independent television companies began NICAM 728 broadcasts in 1989[35].

Broadcasters in Finland (where the main application is dual language broadcasts) and Sweden started NICAM 728 services in 1988, and Denmark and Norway started in 1989.

Others in progress include New Zealand[36], Hong Kong, Spain, Belgium, France (on system L SECAM), Yugoslavia, Hungary (on system D/K SECAM), Germany (on cable networks), Italy (on satellite services) and China (on system D/K PAL).

The first commercial NICAM 728 receiver (actually implemented as part of a VHS vcr) was put on sale in the UK in 1987. Since then, even in the absence of a publicized service, most major receiver manufacturers have added NICAM 728 decoders to their more expensive receivers and vcrs.

48.4 Multiple sound signals in MAC systems

As described in section *29*, new television systems intended for transmission via satellite or cable have been developed based on multiplexed analogue component (MAC) coding of the vision signal components. Many different variants of the MAC format exist or are under development. In many cases, the principal differences between the variants relate to the way in which the digitally coded sound signals are multiplexed with the MAC vision signal components.

48.4.1 EBU MAC/packet family

The EBU MAC/packet family comprises[26–29]:

C-MAC: rf time division multiplexing of analogue picture signals and digitized sound or data signals conveyed in a 3 Mbit/s multiplex

D-MAC: baseband time division multiplexing with the 3 Mbit/s multiplex of data/digitized sound conveyed using duobinary coding

D2-MAC: as D-MAC but with only 1.5 Mbit/s data/digitized sound, yielding a reduced bandwidth signal that can be conveyed in a 7 MHz channel

Four different configurations of sound coding and error protection are defined in the MAC/packet system:

- Near instantaneous companding from 14 to 10 bits and error protection comprising one parity bit per sample covering the six msbs in each sample (first level protection). The scale factor information is conveyed by modification of the parity bits. This configuration is the same as that used in NICAM 728 (see section *48.3.1*).
- Near instantaneous companding from 14 to 10 bits and error protection using an 11,6 extended Hamming code in which the five parity bits enable the correction of one single error and detection of one double error in the six msbs of each sample (second level protection).
- 14 bits per sample linear coding and error protection comprising one parity bit per sample covering the eleven msbs in each sample (first level protection).
- 14 bits per sample linear coding and error protection comprising a 16,11 extended Hamming code in which the five parity bits enable the correction of one single error and the detection of one double error in the eleven msbs of each sample (second level protection).

In all four cases, for high quality sound, 32 kHz sampling is specified. However, in each configuration (but only in mono) it is possible to have two reduced bandwidth channels with 16 kHz sampling, giving a bit-rate which is half that of the high quality option. Such reduced bandwidth channels could, for example, be used to convey commentaries.

48.4.2 B-MAC and Dolby ADM

In the USA and Canada, a different variant of MAC, known as B-MAC, has been developed for both 525-line and 625-line television systems. B-MAC is also now used on several European communications satellites and on a satellite service in Australia[40].

As in EBU D-MAC, B-MAC uses baseband time division multiplexing of the analogue picture and digital sound and data components. In B-MAC, however, the clock frequencies are integer even multiples of the NTSC colour subcarrier frequency.

In B-MAC systems, Dolby adaptive delta modulation (ADM) sound coding is usually used. ADM is not, on its own, a multi-channel sound transmission system; it is a companding system applicable to broadcasting and recording. As such it is used in some multi-channel sound broadcasting systems, notably B-MAC, and was at one time considered as an alternative to NI companding for use in the digital multi-channel sound for terrestrial television system now known as NICAM 728.

Dolby ADM is founded on delta modulation which may be considered as a special case of differential pulse code modulation with a one-bit quantizer. The digital/analogue convertor in the decoder comprises a simple integrator circuit.

In simple linear delta modulation, slope overload for large amplitude, high frequency signals is a problem unless a very high bit rate is used. Adaptive delta modulation, in which the step size of the correction pulses is variable, has often been used to overcome this problem. Adaptive delta modulation is a form of digital companding system which operates, not according to the amplitude of the input signal (as, for example, does a NICAM system), but rather on the slope of the input signal. In digital adaptive delta modulation, the step size is constrained to a number of finite sizes, and a digital control signal is sent to the decoder to cause it to adjust its step size to track that of the encoder.

In common with all companding systems, noise modulation can be a problem. In the case of ADM it is high-slope signals which are most troublesome. "Gain-blipping' can also be troublesome because the step size control data are critical, and if an error happens to hit a critical bit which conveys the step size information there is a sudden change in gain.

Dolby Corporation improved the basic ADM method by applying the following techniques[41–44]:

- Before digitization, the audio signals are processed by the Dolby proprietary 'sliding band' pre- and de-emphasis. This reduces noise modulation without the penalty of reduced high frequency headroom or low frequency noise emphasis in the presence of predominantly high frequency programme material.
- The critical step size control information is conveyed as a separate low data-rate bit stream in which all bits have equal weight. The step size control is thus well protected against errors.
- The transient response of the system is improved by a delay line in the encoder which allows the low rate control signal to indicate a step size change ahead of an oncoming audio transient.

In subjective tests conducted by the BBC and Swedish Radio, one version of Dolby ADM, operating at 728 kbit/s to convey two high quality sound channels, was found to give performance that is similar to that of an NI companding system operating at the same overall bit-rate. On some kinds of programme material and at certain bit-error ratios, Dolby ADM was judged to be marginally subjectively better than the NI companding system, while for other kinds of programme material (notably speech) and at other bit-error ratios, the NI companding system was judged to be preferable.

With this close matching of performance of the two companding systems, the choice of NI companding for digital terrestrial multi-channel sound with television was made mainly because the receiver industry favoured commonality between terrestrial NICAM systems and one of the sound coding options of MAC/packet systems.

48.5 Multi-channel sound for HDTV

High definition television (HDTV) will bring new conditions under which viewers watch television and new challenges for the provision of sound. The prospect of larger, wider aspect ratio pictures and closer viewing distances relative to picture size suggests a change in viewing angle and consequently a change in the degree of listener head movement.

It seems reasonable to argue that a sound system for HDTV should match the improved picture size and quality with increased realism and a wide sound stage. Studies[45,46] suggest that two-channel stereo may be inadequate because localization errors, which are likely to be disturbing in this context, will occur with listener head movement and non-axial viewing. To improve this, three or more sound channels are likely to be needed for HDTV.

Acknowledgement

The author gratefully acknowledges the substantial contribution to this section made by his colleagues N H C Gilchrist, A P Robinson, A J Bower and A H Jones.

References

1 ANGUS, J, 'Simulcasting', *Broadcast Systems Eng.* (December 1985)

2 EDWARDSON, S M, 'Stereophonic and two-channel sound in terrestrial television broadcasting', IBC 1982, *IEE Conf Publ 222*, 276–281

3 EDWARDSON, S M, 'Stereophonic and two-channel sound in terrestrial television broadcasting', *Radio Electronic Eng.*, **53**, 11/12, 403–406 (November/December 1983)

4 NUMAGUCHI, Y, and HARADA, S, 'Multichannel sound system for television broadcasting', *IEEE Trans. Consumer Electronics*, **CE-27**, 3, 366–371 (August 1981)

5 CCIR, *Recommendations and reports of the CCIR, 1978. Volume X, Broadcasting service (sound)*, Report 795 XIVth Plenary Assembly, Kyoto (1978)

6 TINGLEY, E M, 'US multichannel television sound technical standards', IBC 1984, *IEE Conf Publ 240*, 308–311

7 EILERS, C G, 'TV multi-channel sound, the BTSC system', *IEEE Trans. Consumer Electronics*, **CE-31**, 1, 1–7 (February 1985)

8 KELLER, T B, 'Stereo audio in television: the BTSC multichannel sound system', *SMPTE Jour*, 1024–1027 (October 1985)

9 HOFFNER, R, 'Multichannel television sound broadcasting in the United States', *Jour Audio Eng Soc*, **35**, 9, 660 (September 1987)

10 DINSEL, S, 'Stereophonic sound and two languages in TV, the double-sound carrier method', IBC 1980, *IEE Conf Publ 191*, 207–211

11 DINSEL, S, 'Two carrier system in Germany since 1981', IEE Colloq Dual Channel TV Sound Terrestrial Broadcasting and Reception, April 1983, *IEE Digest 1983/39*, 1/1–1/5

12 LOOSER, D, 'Stereo TV sound', *Television*, 530–532 (August 1984)

13 JONES, A H, 'The two-carrier method for dual channel TV sound: over-air tests in the UK', IEE Colloq Dual Channel TV Sound Terrestrial Broadcasting and Reception, April 1983, *IEE Digest 1983/39*, 2/1–2/4

14 MOUNTAIN, N, 'US subcarrier system for Europe', *Cable and Satellite Europe*, 45–46 (June 1985)

15 *NICAM 728: Specification for two additional digital sound channels with system I television*, BBC/IBA/BREMA (1988)

16 ELY, S R, 'The UK system for digital stereo sound with terrestrial television', *Jour Audio Eng Soc*, **35**, 9, 653–659 (September 1987)

17 JONES, A H, 'Digital stereo sound with terrestrial television', *SMPTE Jour* (October 1985)

18 ELY, S R, 'Experimental digital sound with terrestrial television', IBC 1984, *IEE Conf Publ 240*, 312–317

19 JONES, A H, 'Digital two-channel sound with terrestrial television', *Electronics & Power*, 801–803 (November/December 1986)

20 BOWER, A J, 'Digital two-channel sound for terrestrial television', *IEEE Trans Consumer Electronics*, **CE-33**, 3, 286–296 (August 1987)

21 ELY, S R, 'Progress and international aspects of digital stereo sound for television', IBC 1986, *IEE Conf Publ 268*, 138–143

22 ELY, S R, 'The UK system for digital stereo sound with terrestrial television', *Communication & Broadcasting*, 29, 21–29 (March 1988)

23 JONES, A H, 'Survey of the twin-channel TV sound broadcasting situation', IEE Colloq Twin Channel Digital TV Sound for Terrestrial Broadcasting, Jan 1986, *IEE Digest 1986/2*, 1/1–1/3

24 EBU, 'Specification for transmission of two-channel digital sound with terrestrial television systems B,G,H and I', *SPB 424 3rd revised version* EBU Technical Centre Geneva (1989)

25 EATON, J L and HARVEY, R V, 'A two-channel sound system for television', IBC 1980, *IEE Conf Publ 191*, 212–215

26 EBU, 'Specification of the systems of the MAC/packet family', *Doc Tech 3258-E*, EBU Technical Centre, Brussels

27 OSBORNE, D W and CROLL, M G, 'Digital sound signals: bit-rate reduction using an experimental digital compander', *BBC Research Department Report No 1973/41*

28 CAINE, C R, ENGLISH, A R and O'CLAREY, J W H, 'NICAM-3: Near-instantaneously companded digital transmission system for high-quality sound programmes', *Radio & Electronic Eng*, **51**, 10, 519–530 (October 1980)

29 CCITT Red Book, Volume III, Fascicle III.4, *Transmission of Sound Programme and Television Signals, Recommendation J.17*, 'Pre-emphasis used on sound-programme circuits'

30 CHAMBERS, J P, 'Signalling in parity: a brief history', *BBC Research Department Report* 1985/15

31 OLIPHANT, A, 'The effect of transmission errors on sound signals in the MAC/packet family', *EBU Review — Technical 216* (April 1986)

32 BHARGAVA, V K, HACCOUN, D, MATYAS, R and NUSPL, P P, *Digital Communications by satellite*, Wiley, New York, 46–47 (1981)

33 KALLAWAY, M J, 'An experimental 4-phase dpsk stereo sound system: the effect of multipath propagation', *BBC Research Department Report No 1978/15*

34 ELY, S R, 'Experimental digital stereo sound with terrestrial television: field-tests from Wenvoe, October, 1983', *BBC Research Department Report No 1983/19*

35 GARDINER, P, 'NICAM Digital Stereo', *Electronics & Wireless World*, **95**, 1642, August 1989, 754–757 and **95**, 1643, September 1989, 920–923

36 INGHAM, J D, 'Multichannel sound for television', *ABU Technical Review*, 13–18 (November 1985)

37 MERTENS, H and WOOD, D, 'The C-MAC/packet system for direct satellite television', *EBU Review* 200, 172–185 (August 1983)

38 MERTENS, H, 'The overall structure and digital aspects of the EBU DBS system', IBC 1984, *IEE Conf Publ* 240, 180–184

39 LOTHIAN, J S and O'NEILL, H J, 'The C-MAC/packet system for satellite broadcasting', *IEE Proc*, **13**, Pt F, 4, 374–383 (July 1986)

40 SMITH, G I, 'Introduction of an operational B-MAC system for broadcasting and programme distribution', IBC 1986, *IEE Conf Publ* 268, 21–23

41 GUNDRY, K, 'An audio broadcast system using delta modulation', *SMPTE Jour* (November 1985)

42 TODD, C C and GUNDRY, K, 'A digital audio system for broadcast and pre-recorded media', 75th Convention of the Audio Engineering Society, *Jour Audio Eng Soc* (Abstracts), **32**, 480, preprint 2071 (June 1984)

43 GUNDRY, K J, ROBINSON, D P and TODD, C C, 'Recent developments in digital audio techniques', 73rd Convention of the Audio Engineering Society, AES preprint 1956 (March 1983)

44 TODD, C C and GUNDRY, K, 'A digital audio system for DBS, cable and terrestrial broadcasting', IBC 1984, *IEE Conf Publ* 240, 414–418

45 PLENGE, G, 'Sound design and sound transmission in a future HDTV system', 79th Convention of the Audio Engineering Society, AES preprint 2306 (October 1985)

46 KOMIYAMA, S, 'Subjective evaluation of angular displacement between picture and sound directions for HDTV sound systems', *Jour Audio Eng Soc*, **37**, 4, 210–214 (April 1989)

Bibliography

YONGE, M, 'Stereo sound for television', *Television: Jour RTS*,June 1986, 150–154 & August 1986, 188–194

LEWIS, G, 'Dual-channel TV sound systems', *Television*, January 1988, 203–206, February 1988, 269–273 & March 1988, 366–369

J G Sawdy B Sc, C Eng, MIEE
Independent Broadcasting Authority

49 Sound Distribution including Digital Sound-in-sync

Television networks have to carry two rather different signals, the *sound* and the *vision*. Until recently, the sound signal has usually been a single audio channel of 10–15 kHz bandwidth because the broadcasting standards used have only provided for a single monophonic audio channel to be transmitted to the viewer. However ways have now been found of broadcasting more than one audio channel. In Germany, a second audio channel has been provided by adding a second, frequency modulated, sound carrier, but more countries are now adopting the EBU recommended NICAM system[1] which adds a digitally modulated sound carrier capable of carrying two high quality sound channels in addition to the conventional analogue sound carrier (see section *48*).

This means that the networks that are used for programme production and to feed the transmitters must also be upgraded to carry two or even three audio channels. Where the digital system is used for broadcasting, it is highly desirable to use digital systems for networking to ensure that the broadcast sound quality is not limited by analogue distribution networks.

49.1 Methods of sound distribution

The way in which the television programme sound signal(s) are distributed is likely to be influenced by the telecommunications policy of the country in question. Where broadcasters are responsible for providing their own programme distribution networks, the programme sound signals are usually carried along with the vision signal. The way in which the signals are combined will depend on the type of system used.

For a system based on microwave links, combination is usually achieved most economically by the use of one or more frequency modulated subcarriers added to the video signal above the highest video frequencies. This system can also be used on coaxial cable and fibre optic systems using analogue transmission. Where digital techniques are used, the digitized sound channels, which require a relatively low bit rate compared to the vision signal, can be multiplexed with the vision to form a combined sound and vision bit stream.

In many countries, broadcasters' circuits are provided in part or whole by PTOs (Public Telecommunications Operators). PTO provided sound circuits are often based on extensions of the plant and principles used to provide the telephony networks, and are integrated with this service at a fairly low level. They may, for example, share pairs in a cable carrying telephony in a local distribution, or use several adjacent channels of bandwidth in an analogue telephony group, or perhaps 4–6 sound programme bandwidth audio channels will be multiplexed into a 2048 kbit/s tributary, in a 140 Mbit/s or 565 Mbit/s trunk route mainly carrying telephony and data.

In contrast, however, PTOs' vision circuits are usually based on a separate overlay network dedicated to television use, although sites and some types of hardware will usually be common to those used on intercity telephony trunk routes. PTO vision circuits are normally specified to cover just sufficient bandwidth to handle the highest frequencies in the video signal.

In PTO provided television networks consisting of separate sound and vision circuits, the sound and vision signals are therefore often carried quite separately on the two types of circuit, coming together only at network nodes and switching points.

49.2 Sound-in-sync

In order to reduce the cost of television signal distribution by eliminating the need for separate sound circuits, the BBC, in the 1960s, proposed a novel way of carrying the sound signal as part of the video waveform. They developed the first *sound-in-sync* (sis) equipment, the design of which was later licensed for manufacture by Pye TVT Ltd (now Varian TVT Ltd) of Cambridge, England.

As the name implies, the principle is to make use of the time available during the sync pulse periods of the television waveform to carry sound signals.

The sync pulse time is effectively 'spare time' on the transmission circuit. The sync pulse periods need to exist in the transmitted signal to allow sufficient time for flyback in domestic TV receivers, but they convey no information other than the timing reference provided by the pulse edge. Provided therefore that enough information is left to allow the eventual restoration of the original syncs, the sync pulse time can be 'borrowed' without any impairment to the video signal.

Carrying the sound in this way does not increase the bandwidth of the video signal, and therefore it should be possible to carry sis coded vision on any circuits suitable for normal video signals.

Sound-in-sync coding of the video signal does impose some constraints. If the sis coded video signal is to be monitored along its transmission path, then at each monitoring point either a blanker unit must be provided to blank out the data and restore the normal sync pulses, or the monitoring equipment used must employ a more elaborate type of sync separator circuit which will allow it to operate in the presence of the sound-in-sync pulses.

If the video circuit employs any clamp circuits, these must clamp to the black level porches of the video signal rather than the sync bottoms as well as being compatible with sound-in-sync signals.

Most importantly, perhaps, the use of sound-in-sync imposes greater constraints on the source video signal. Excessively noisy or jittery video levels, or signals containing other impairments such as spikes below black level or missing or corrupted sync pulses, can cause not only sound failure but also quite possibly additional corruption of the video signal.

Nevertheless, the BBC/Pye system of monophonic sound-in-sync has come into widespread use. In the UK, it has been used on both BBC television networks and was used from the outset on the major part of the IBA's Channel 4 network. In Europe, monophonic sound-in-sync is used on the EBU Eurovision terrestrial and satellite networks.

The audio sampling in monophonic sound-in-sync is done at exactly twice the television line rate (31.25 kHz) giving an audio bandwidth of 14 kHz. The audio signal is first pre-emphasized and then compressed by an analogue compander before being sampled by a 10 bit analogue/digital converter. A pilot tone is added at the compander input to assist correct decompanding at the decoder.

One of the two 10 bit words produced each line is complemented before being interleaved with the other in order to minimize variations in mean amplitude of the data burst with audio modulation which could otherwise cause sound to vision crosstalk. A marker pulse is added, making up the 21 bit binary signal that is inserted in each sync pulse.

Finally, the pulses are filtered to a sine squared pulse shape and inserted in the sync pulse period at 700 mV amplitude and at a data rate of 5.5 Mbit/s. The entire data burst is thus 3.82 μs long which is comfortably contained within the sync pulse width of 4.7 μs.

Because two audio samples have to be transmitted in every TV line period, a problem occurs during the field sync pulse time where, instead of normal line sync pulses, there are equalizing pulses half as wide as a line sync pulse. The system copes with this situation quite simply by encroaching outside the sync pulses and putting normal double audio sample pulse bursts in at line rate. This is normally of no consequence as a sis decoder restores the correct sync waveform, but is worth bearing in mind if blanker units are employed to remove the sound-in-sync data. Some of these simply blank all the data areas down to sync bottom level leaving alternate equalizing pulses broadened. While this is quite adequate for picture monitoring purposes, it will not be adequate where a full specification sync waveform is required.

49.2.1 The economics of sound-in-sync

There are several factors involved in assessing the economic advantages of using sound-in-sync in any given network situation. Some are obvious, others less so.

In the simplest case, comparing the cost of a single rented vision circuit equipped with sound-in-sync equipment with the cost of renting separate sound and vision circuits, the considerations will be the capital cost of the sound-in-sync equipment and any necessary spares against the recurring annual cost of a separate sound circuit. Such decisions, based on trading a one-off cost against recurrent payments are never straightforward, but techniques such as discounted cash flow analysis exist to reconcile the two alternatives.

In general, it is obvious that the use of sound-in-sync is more attractive the longer the route, because the cost of a separate sound circuit will normally be distance related while the cost of sis equipment is the same irrespective of distance.

In the case of a complete network, the total costs of sis equipment, any necessary blanker units and an allowance for spares together with the recurring costs of equipment maintenance, must be weighed against the alternative means of sound network provision.

It may appear that it would be most economic to use sis equipment on the longer routes of a network together with analogue circuits on the shorter ones. This solution may be suitable for relatively fixed networks, but often the inflexibility imposed by the inability to interconnect all circuits without the provision of sound-in-sync coders or decoders at intermediate points will make this solution less attractive than using sound-in-sync on all network routes.

49.3 Stereo sound-in-sync

The move towards the use of stereo or dual language sound with television has inevitably led to the development of modern versions of sound-in-sync equipment capable of carrying two channels of sound.

While the terms *stereo* and *dual channel* are often used interchangeably, there are small differences in the technical requirements of systems to carry stereo and dual language services. Basically, dual language systems require much better interchannel crosstalk performance than stereo ones, while stereo systems require better matching of the characteristics of the two channels than systems for dual language. In practice, all two channel sound-in-sync systems are engineered to cover both requirements.

The maximum rate of binary data that can be accommodated within a 5.5 MHz bandwidth vision circuit is about 7 Mbit/s (UK teletext, for example, uses a data rate of 6.9375 Mbit/s). Given that only about 4 μs in every 64 μs line is available in the sync pulses for data, this means that the maximum data rate available for sound-in-sync is about 440 kbit/s. In practice, it is advisable to use a somewhat lower data rate than teletext if reliable operation is required over long distance routes without regeneration.

A slightly higher data rate could be achieved by putting data in the broad field sync pulses as well as the line syncs, but the improvement is hardly worth the extra circuit complexity involved, especially as this would make the signals incompatible with equipment designed to blank mono sound-in-sync.

A data rate of around 400 kbit/s is inadequate to carry two channels of high quality audio unless very elaborate bit rate reduction techniques are used. Rather than use binary data, the BBC proposed a system[3] using four level data signals. Each four level 'symbol' can represent four possible states, the equivalent of two binary bits, so that the effective bit rate is double that of a binary system.

An eight level system has also been tried experimentally. This gives three times the bit rate of a binary system (about 1.2 Mbit/s), enough to allow two audio channels using uncompanded 16 bit linear coding with sampling at 32 kHz to be carried. This was found to be insufficiently robust for long distance use on vision circuits in the UK.

With a maximum available bit rate of about 800 kbit/s, some form of companding is still required. The BBC produced the first dual channel sound-in-sync system (also licensed to Varian TVT) which was based on NICAM 3 coding[4]. NICAM 3 is a system designed to carry six high quality audio signals on a 2.048 Mbit/s circuit, and a stereo pair is coded at 676 kbit/s. In this sis system, the 676 kbit/s bit-stream is asynchronously inserted into the sync pulses of the TV waveform so, unlike mono sound-in-sync, the audio sample rate is not locked to line frequency. This allows the audio coding to be done remotely from the point where the signal is inserted into the TV waveform.

However, another version of NICAM, NICAM 728 (so called because it uses a bit rate of 728 kbit/s to carry a stereo pair) has been adopted for terrestrial digital dual sound with television[1,2]. This version was adopted because it uses the same coding structure as one of the sound coding options of the MAC/packet family[5] so that decoding can be performed by the same type of decoder as used in MAC/packet systems.

There is considerable attraction in using this version of NICAM as the basis of a sound-in-sync system, because the digital signal recovered by the sis decoder at a transmitter can, with a minimum of digital processing, be used to modulate directly the quadrature phase shift keyed digital sound carrier.

The alternative of using another format, such as NICAM 3, for sound-in-sync distribution would require that the signal be either digitally transcoded into NICAM 728 for transmission or decoded to analogue and recoded in the NICAM 728 format. Either of these solutions is likely to be more complex and thus more costly and less reliable, as well as being unable to support the data transmission options of NICAM 728. Using an analogue connection is particularly unattractive as quality will be lost, and it is not easy to monitor that the system is functioning correctly.

Digitally transcoding between NICAM 3 and NICAM 728 is not simple because, although both systems use the same sampling frequency and 14 to 10 bit companding principle, NICAM 728 specifies that the two channels are simultaneously sampled while NICAM 3 specifies alternate sampling.

One advantage of NICAM 728 in comparison with NICAM 3 is that it uses a 1 ms frame period as opposed to the 3 ms used in NICAM 3. This means that the overall audio delay of a coder/decoder pair is less. Typical figures are about 5 ms for NICAM 728 and 13 ms for NICAM 3. This can be quite important in television systems, because if the build up of audio delay with respect to vision is too great, noticeable lip-sync problems arise.

One final attraction of using NICAM 728 as the basis of a sound-in-sync system is that, because of its use as a broadcast format, integrated circuit decoders for this system are available at low cost. Use of these devices can significantly reduce the cost and complexity of the sis decoder.

49.4 Stereo sound transmission using NICAM 728

As discussed in section *49.3*, the NICAM 728 system is the basis of the EBU recommended system for the transmission of two-channel digital sound channels with terrestrial television systems B, G and I. An overview of the system is given here; full details are available in references 1 and 2.

The system is based on the addition of a new carrier, which is digitally modulated with a continuous 728 kbit/s signal, to the existing television transmission.

This 728 kbit/s signal is made up of 1 ms frames each consisting of 728 bits transmitted continuously without gaps.

The first eight bits of each frame consists of an eight bit frame alignment word, 01001110, which identifies the start of the frame.

The next five bits, C0–C4, are control bits. C0, the *frame flag bit*, is automatically inserted by the NICAM 728 coder. It is alternately 1 for eight frames and 0 for the next eight, thus defining a 16 frame sequence. This sequence is used to synchronize changes in the data mode as defined by the next three bits.

C1, C2 and C3 are the *application control bits*. Between them they could define eight different states, but currently only four possible modes are defined and only C1 and C2 are permitted to vary. C3 is therefore always set to 0. Setting it to 1, as would be the case for the four as yet undefined modes, should cause first generation decoders to give no sound output and the receiver to revert to receiving the normal analogue sound transmission. The currently defined modes are:

C1	C2	C3	
0	0	0	stereo signal
0	1	0	two independent mono signals
1	0	0	one mono signal plus 352 kbit/s data
1	1	0	704 kbit/s data

The last control bit, C4, is the *reserve sound switching flag*. Its purpose is to indicate if the analogue sound carrier is carrying the same programme as the digital carrier (C4 set to 1), in which case a sophisticated receiver could select the analogue sound demodulator if the digital signal was being poorly received. With C4 set to 0, such a receiver should not switch to the analogue sound as it would not be carrying the same programme.

After the five application control bits come 11 further bits, AD0–AD10. These additional data bits are reserved for future applications and are not yet defined. In effect they provide an 11 kbit/s data channel carried continuously by the digital sound carrier.

The remaining 704 bits in each frame carry the sound samples or can be used in whole or part for data if this is signified by the application control bits. However, no format has yet been defined for data transmission.

All modes that convey audio use the same sampling and companding methods, so the system performance is the same regardless of mode. As well as the difference in control bits, the arrangement of the sound sample bits between stereo and single or dual mono modes is also quite different. This difference stems from the two different structures that NICAM 728 has inherited from the MAC/packet specification.

In the stereo mode, the 704 bit sound/data block of each frame is made up of alternate samples from each channel. Both channels are simultaneously sampled to an accuracy of 14 bits. These samples are digitally companded to 10 bits using the NICAM method described below, and a parity bit is added to each sample. Thus each frame contains 32 11-bit sample words for each channel, making 704 bits in total.

When two independent mono signals (M1 and M2) are being transmitted, the samples from each channel are contained in alternate *frames*. Odd numbered frames, as defined by the 16 frame sequence of C0 (the first frame with C0 = 1 being frame one), contain 64 samples from audio channel M1, while even numbered frames contain 64 samples from channel M2.

If a single audio channel is being transmitted along with 352 kbit/s of data, the audio coding is exactly the same as for M1 above with the data carried in the even numbered frames.

NICAM is an acronym for *Near Instantaneously Companded Audio Multiplex*, and the technique is used in NICAM 728 to reduce each 14 bit audio sample to 10 bits. The principle is called near instantaneous companding (*NI companding*) because the audio samples are companded in 1 ms groups as opposed to other digital companding systems, such as A-law, where each sample is individually converted to a new sample word containing fewer bits.

The 1 ms companding blocks are each frame in stereo mode. When independent mono signals are being carried, each frame, containing 64 samples from the same channel, is two companding blocks.

The principle of NICAM is simply to examine the 32 samples and to send the 10 most significant active bits. The same bits are sent from each of the 32 samples, so it is effectively the largest sample in the block that determines which bits are sent. Two's complement coding (see section *50.1.1.4*) of the audio samples is used, so the sign bit is always sent. The remaining nine bits can vary from the most significant, in which case the least significant four bits are lost, to the least significant, in which case the four most significant are regenerated in the decoder. (This *can* be done as by definition there is no activity in these more significant bits when the least significant ones are being sent.)

The decoder will need to be told which bits are being sent. There are five possibilities between all the most significant bits being sent (when the audio signal is within the top 6 dB of its dynamic range) and the least significant bits being sent (when the signal amplitude is more than 24 dB below maximum level). A three bit number is needed to represent these five states and is known as the *scale factor*.

There are no specific bits in the 728 bit frame corresponding to the scale factor bits, because a special technique is used to carry them. They are sent very securely without apparently using any bit space by modifying the parity bits accompanying each audio sample.

Firstly, the parity bits are calculated in the normal way on the six most significant bits of each audio sample. The parity is even, i.e. there is an even number of ones in the group formed by the six protected bits and the parity bit. Each scale factor bit is then used to modify a group of nine parity bits: if the scale factor bit is 1 the nine parity bits are complemented, if it is 0 the parity bits are unchanged.

In the decoder, the parity bits are re-calculated and compared in majority decision logic. If most or all of the parity bits in the group of nine come out even, then the scale factor bit is taken as 0, but if most or all of the parity bits are odd, then the scale factor bit is taken to be 1 and all nine parity bits are complemented again to restore them to their normal state. Once the scale factor bit has been calculated and the parity bits corrected if necessary, the parity bits can be used in the normal way to inhibit samples that contain an error.

As each frame containing audio samples contains two companding blocks (either one block each from left and right channels in stereo mode or two blocks from the same channel for a mono channel), there are six scale factor bits to be conveyed. Each scale factor bit modifies nine parity bits, a total of 54. The specification also permits the ten remaining parity bits to be similarly modified in two groups of five in order to convey securely two additional bits of information.

There is one further refinement in the companding system of NICAM 728 which gives greater protection against errors when low level audio signals are being transmitted. Although there are only five possible coding ranges to be conveyed by the three scale factor bits, these three bits can of course indicate eight possible states. When the system is in the fifth coding range where all the least significant bits are being sent, these additional states are used to define two further *protection ranges* indicating that either one or two of the most significant bits being sent (excluding the sign bit) is inactive. It would in fact be possible to define three additional protection ranges rather than two, but this is not done in order to maintain maximum compatibility with the MAC/packet specification which uses the scale factor 000 to indicate periods of silence during which buffer adjustments may be carried out.

The 704 audio samples in a frame are not transmitted in sequence but are interleaved with each other. This minimizes the disturbance when errors occur in bursts rather than being randomly spaced. There is a high chance that multiple bit errors in a single sample will not be detected by parity, but interleaving spreads such errors into single bit errors in several samples where the parity will detect them and enable the errors to be masked.

Finally, the 728 bit frames are scrambled to make the spectrum of the transmitted digital sound signal as noise-like as possible. This minimizes the likelihood of the digital sound carrier causing patterning interference to the analogue vision signal. The scrambling is achieved by inverting all the bits in the frame apart from the frame alignment word in accordance with a pseudo-random sequence, and is done synchronously with the frame sequence, i.e. the same pseudo-random inversion pattern is applied to each frame.

The audio channels in NICAM 728 are pre-emphasized to the CCITT J.17 characteristic[6]. In the UK, the level that gives a full amplitude signal in the digital domain is defined as +14.8 dBu at 2.0 kHz which permits the system to handle signal amplitudes of at least +8 dBu at all audio frequencies.

The 728 kbit/s signal produced by the above process is used to modulate digitally the additional sound carrier using the modulation system known as *differentially encoded quadrature phase shift keying*. This means that the carrier may occupy four phase states and each phase change conveys two data bits.

Thus, approximately every 2.75 μs (the period of two bits at 728 kbit/s) a carrier phase change may occur. A -90° phase change signifies the bit pair 01, -180° signifies 11, -270° signifies 10, while no change at all is taken to signify the bit pair 00.

In television system I, the carrier frequency of the digital signal is 6.552 MHz which is nine times the modulation bit rate. In some system I countries, the two frequencies are locked to each other, and both thus have the same frequency stability which is specified as ±1 part per million for the 728 kbit/s signal. In television systems B and G, the carrier frequency used is 5.85 MHz which is unrelated to the bit rate. In both cases, the phase change impulses are filtered by a constant group delay low-pass filter for spectrum shaping purposes.

The approximate power level of the modulated digital signal is 20 dB below the peak vision carrier level, while the approximate level of the conventional fm sound carrier below peak vision carrier level is 10 dB in system I or 13 dB in systems B and G. In some cases, this implies a slight reduction in the power of the analogue sound carrier when the digital sound signal is added.

References

1 *SPB 424 Specification for the Transmission of Two-channel Digital Sound with Terrestrial Television Systems B, G and I*, European Broadcasting Union, (2nd rev edn October 1987)

2 *NICAM 728: Specification for Two Additional Digital Sound Channels with System I Television*, jointly published by IBA, BBC and BREMA (August 1988)

3 HOLDER, J E, SPENCELEY, N M and CLEMENTSON, C.S., 'A two-channel sound in syncs transmission system', paper presented at IBC 84

4 CAINE, C R, ENGLISH, A R and O'CLAREY, J W H, 'NICAM 3: Near instantaneously companded digital transmission system for high quality sound programmes', *Radio Electr Eng*, **50** (10 October 1980)

5 *Specification of the System of the MAC/packet Family*, European Broadcasting Union Technical Document 3258 (1986)

6 CCITT Red Book, Volume III, Fascicle III.4, *Transmission of Sound-Programme and Television Signals, Recommendation J.17: Pre-emphasis used on sound-programme circuits*

E P Tozer B Sc(Hons)
Principal Lecturer, Sony Broadcast and Communications

50

Digital Audio Concepts and Equipment

To understand the advantages of processing audio digitally, it is necessary to look at the fundamental differences between analogue and digital signals (*Figure 50.1*).

An *analogue* signal can, within the limits of peak level and signal bandwidth, exist at any level and at any time. This means that if, during processing, there is any level variation, caused by distortion or the addition of noise, or any time variation, caused by wow and flutter, the new signal is a valid one.

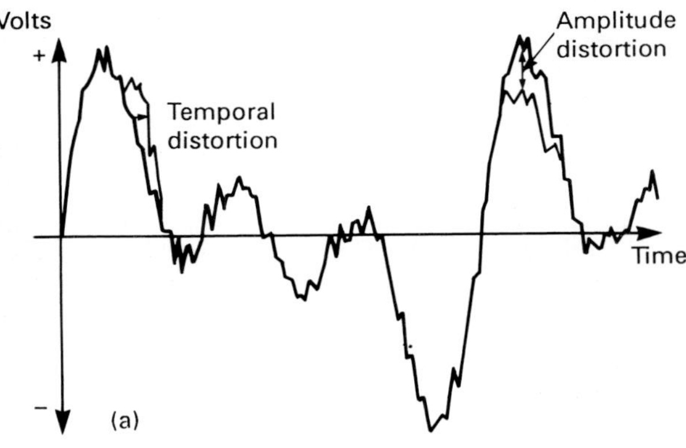

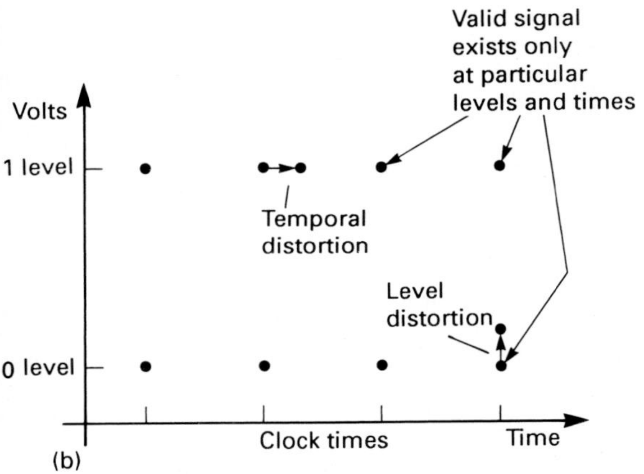

Figure 50.1 Signal distortion: (a) analogue. (b) digital

A *digital* signal differs from an analogue one in that it is constrained to be valid only at particular levels, generally one and zero, and particular times, clock intervals. Digital signals thus have an inherent immunity to change. So long as the temporal or amplitude variation is small they may be brought to the nearest allowable value. This perfect regenerating ability of digital signals means that digital audio may be duplicated or transmitted with zero degradation.

Figure 50.2 shows the path of an audio signal through a typical digital audio recorder which contains virtually all the different types of processing applied to a digital audio signal.

The layout of a typical mixing console is shown in *Figure 50.3*. Here the particular aspects of processing are not explicit, most processes being performed as software execution.

50.1 Digital audio concepts

50.1.1 Analogue/digital interface

50.1.1.1 Linear pre-emphasis

Linear pre-emphasis is used with digital audio processing in order to improve the overall signal/noise ratio of signals containing high frequencies at only low levels. It is used in precisely the same manner as for analogue systems, but is generally less useful for digital recording. Digital meters read even the shortest transient, so causing the operator to reduce input level to the system, in turn reducing any benefit obtained from the pre-emphasis. The most commonly used form of pre-emphasis is EIAJ (Electronic Industries Association of Japan) pre-emphasis (*Figure 50.4*).

50.1.1.2 Anti-alias filtering

The function of the anti-alias filter is to remove any audio signal in excess of half the sample rate. This filtering is required as a sampled signals spectrum is repeated to infinity at multiples of the sample frequency[1] (*Figure 50.5*).

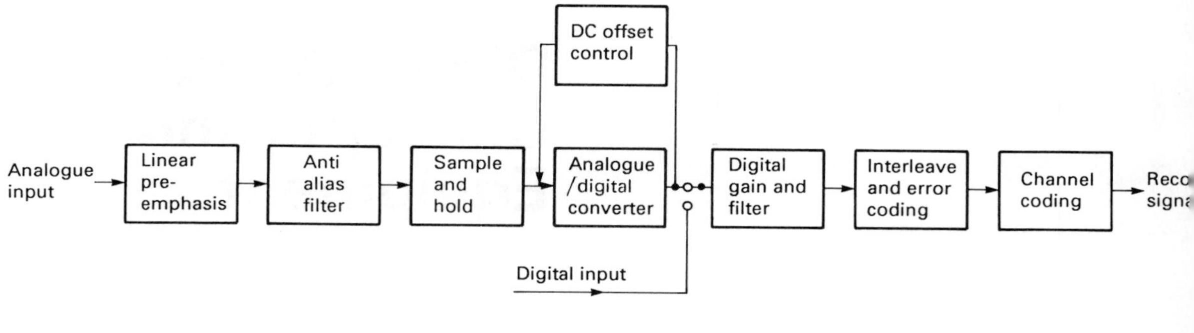

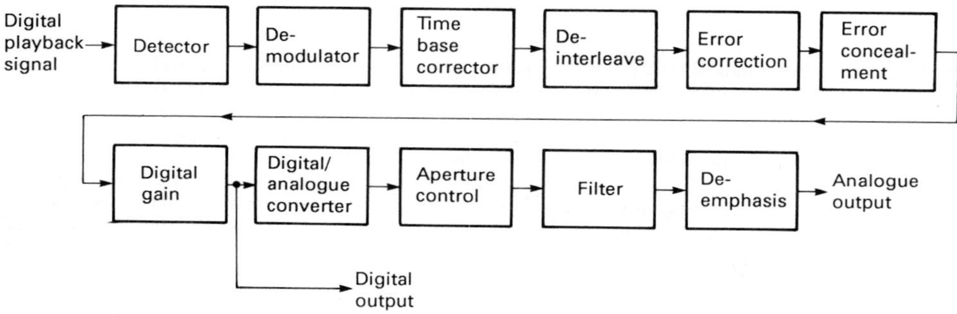

Figure 50.2 The path of an audio signal through a digital recorder

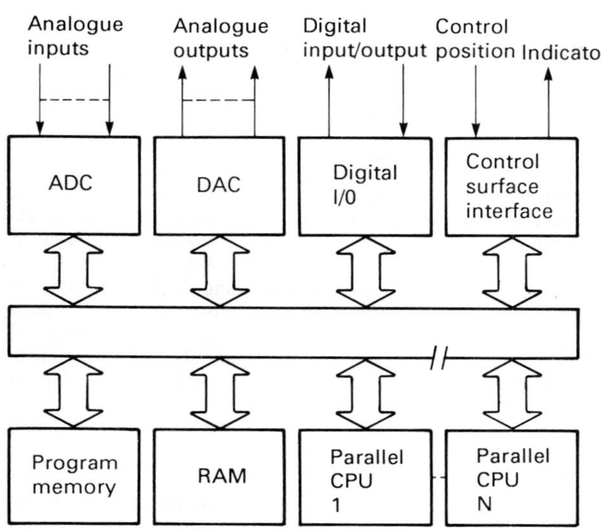

Figure 50.3 Electronic structure of a digital console

The repetition of the spectrum can be understood by considering the sampling process as equivalent to the multiplication of a signal by a sample rate train of impulses (*Figure 50.6*).

The spectrum of a train of impulses is a set of frequencies extending to infinity, spaced at the sample frequency. The signal spectrum is modulated about each of these carrier frequencies.

If the signal contains no components greater than half the sampling frequency, the sidebands will not interfere. However, if the signal contains components at frequencies in excess of half the sampling frequency, then aliasing (*overlap*) of the sidebands will occur. This overlap is equivalent to a folding of the audio spectrum about half the sample frequency (*Figure 50.7*).

The effect on the signal is that frequencies exceeding half the sample rate become lower than half the sample rate by the same amount. For example, in a 48 kHz sample rate system, a frequency of 30 kHz would be aliased to 24 kHz - (30 kHz - 24 kHz) = 18 kHz.

An anti-alias filter will be required to allow frequencies up to 20 kHz to pass unattenuated, and frequencies in excess of half the sample rate to be attenuated by around 90 dB. The frequency and phase response of a typical anti-alias filter is shown in *Figure 50.8*.

It can be seen that, although the frequency response is flat, there is severe phase distortion (*group delay*), caused by the high rate of amplitude roll-off. This phase distortion will cause an audible degradation of sound quality[2]. To overcome the

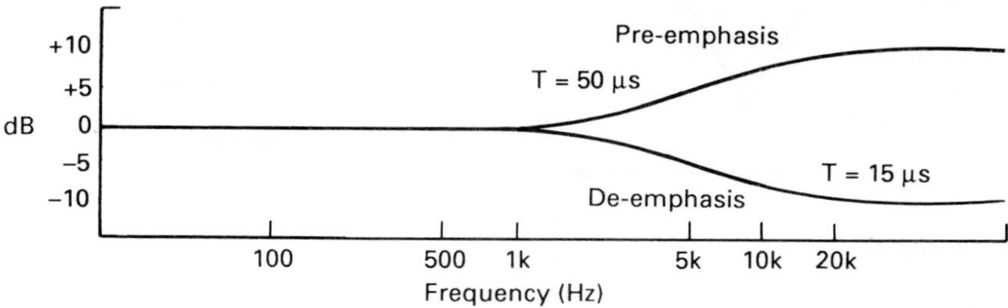

Figure 50.4 EIAJ pre-emphasis curve

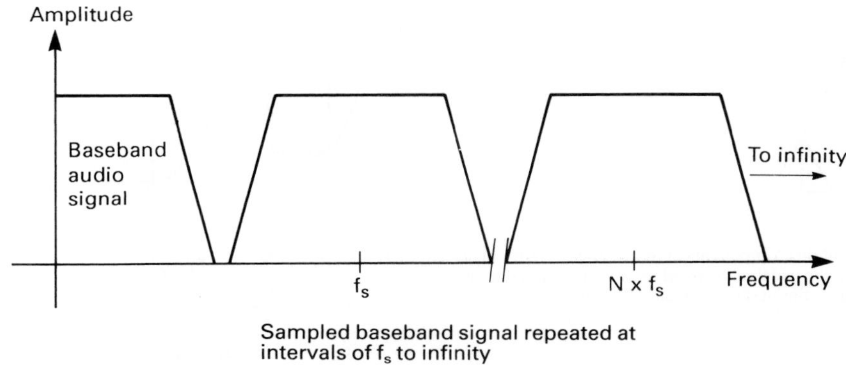

Figure 50.5 Sampled baseband signal repeated at intervals of f_s to infinity

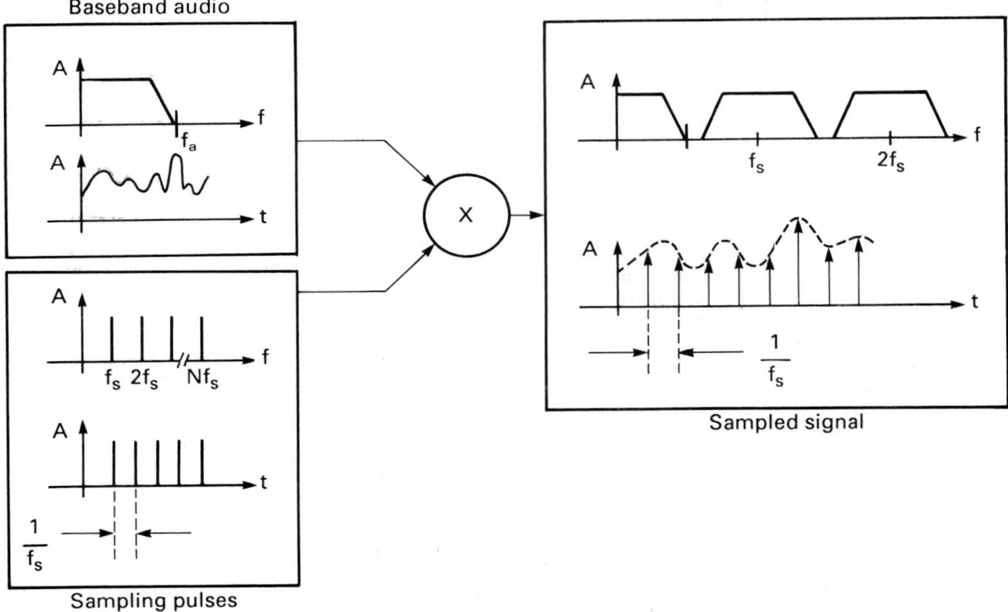

Figure 50.6 The sampling process may be considered equivalent to the multiplication of a signal by a sample rate train of impulses

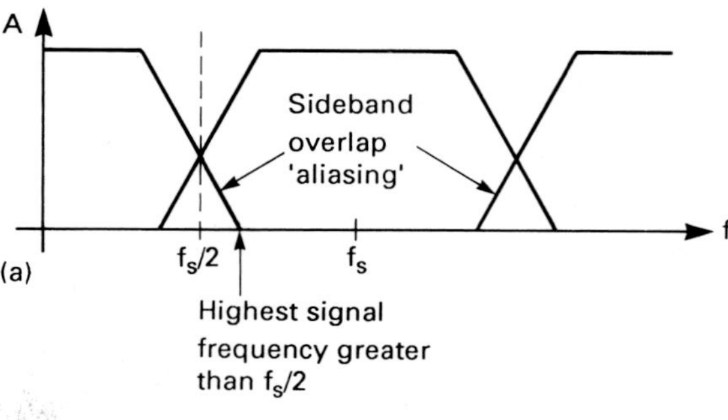

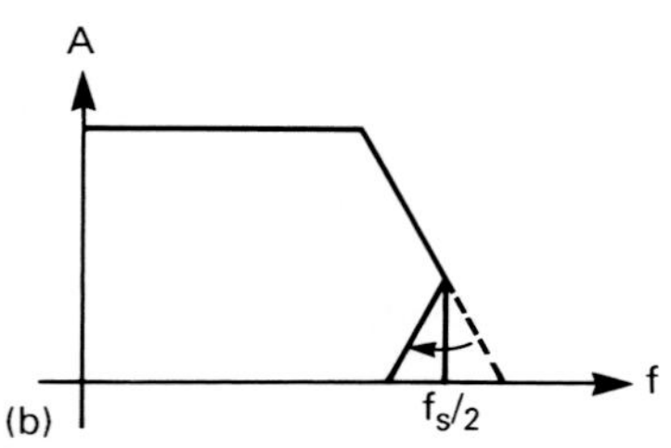

Figure 50.7 If a signal contains components at frequencies of more than half the sampling rate, the sidebands overlap (a). This overlap is equivalent to the folding of the audio spectrum about half the sample frequency (b)

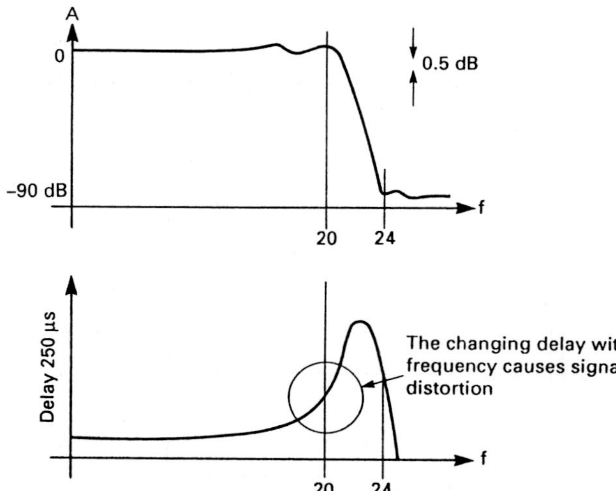

Figure 50.8 Typical amplitude and phase responses of an anti-alias filter used in a non-oversampling system

problems caused by group delay in the anti-alias filter, it is possible instead to use an oversampling system.

50.1.1.3 N bit systems

When dealing with a digital audio system, reference is made to an N bit system. N refers to the number of binary digits used to represent the digitized signal. N bits will allow 2^N discrete levels of signal to be represented.

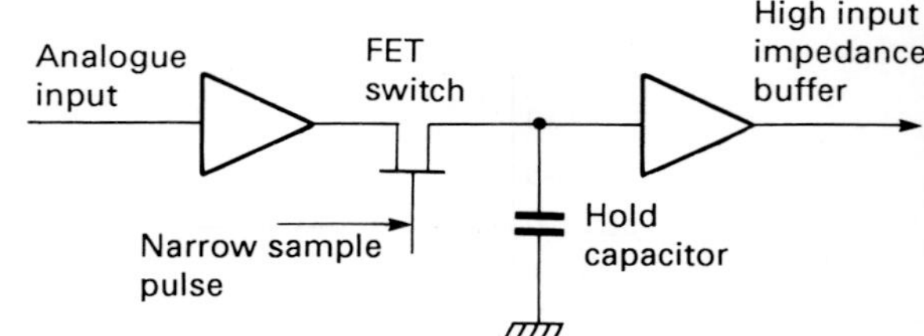

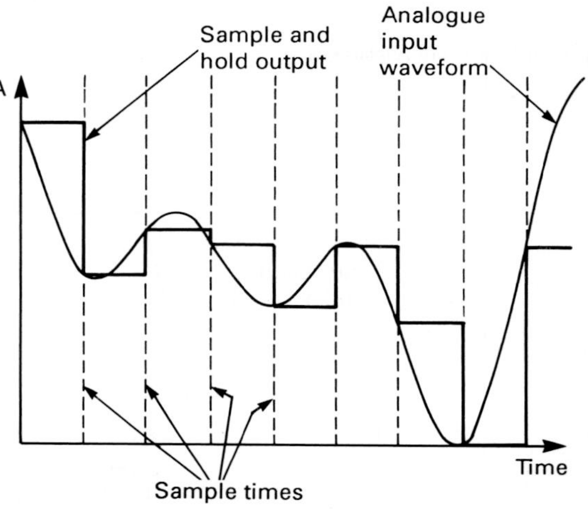

Figure 50.9 Sample and hold circuit and associated waveforms

Currently the most commonly used representation is a 16 bit system. 16 bits allow 2^{16} = 65 536 separate levels to be represented in binary form. Often a digital audio processing system will use 16 bit DACs followed by 20 bit or more 'internal processing' (*Figure 50.9*).

Representing the digital audio more accurately than its original conversion allows for many stages of audio processing, gain, filtering, equalization, etc., each adding rounding errors[3] (*rounding noise*) to the signal. The use of these extra 'internal' bits means rounding errors will always be smaller than the lsb of the original conversion. This process is analogous to performing calculations on a calculator which is accurate to a large number of digits, then finally rounding the result back to the accuracy of the original data.

50.1.1.4 Two's complement notation

Two's complement notation is the binary representation universally employed for digital audio. It is a method of representing binary numbers as a leading sign bit followed by a number of magnitude bits. For 16 bit digital audio, two's complement notation is a sign bit followed by 15 magnitude bits, allowing representation of numbers between -32 768 and +32 767, i.e. 65 536 possible numbers.

The convention for the sign bit is 0 for positive, 1 for negative. In order to make two's complement notation a mathematically useful system, the magnitude bits are inverted for negative numbers, and in order to eliminate two zero representations (+0 and -0), one is added to all negative numbers. Table *50.1* shows the 16 bit two's complement numbers around 0 and around plus or minus maximum.

As may be seen from Table *50.1*, a count between plus maximum and minus maximum involves only a one bit change,

Decimal	Two's complement Binary	
−32786	1000000000000000	1 bit overload
+32767	0111111111111111	+maximum
+32766	0111111111111110	
:	:	
:	:	
+ 2	0000000000000010	
+ 1	0000000000000001	
0	0000000000000000	
− 1	1111111111111111	
− 2	1111111111111110	
− 3	1111111111111101	
:	:	
:	:	
−32767	1000000000000001	
−32768	1000000000000000	−maximum
+32767	0111111111111111	−1 bit overflow

Table 50.1 Some 16 bit two's complement numbers

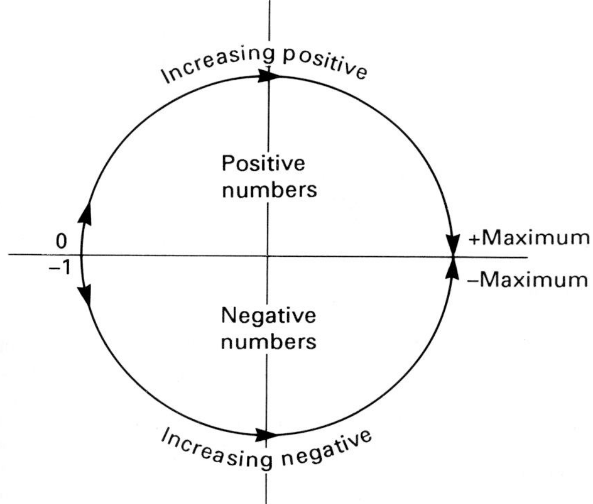

Figure 50.10 Two's complement ring

and a change between -1 and 0 involves only a one bit overflow. It is therefore convenient to envisage two's complement numbers as a ring of numbers (*Figure 50.10*).

50.1.1.5 Analogue/digital conversion

Analogue/digital conversion is the starting point of most current digital systems. The vast majority of digital audio signals start life as analogue and must therefore pass through an analogue/digital converter (ADC).

The ADC is the section of circuitry which causes the most problems for digitally processed audio in terms of noise and distortion. For digital audio, ADCs are often based around a comparator, counter and constant current source (*Figure 50.11*).

At the start of the conversion process, the counter is set to zero; the constant current source is turned on and starts discharging the hold capacitor of the preceding sample and hold gate; at the same time the counter is clocked at a high frequency. These processes continue until the hold capacitor is discharged, the comparator detects this and the counter stops. The count reached by the counter will be proportional to the time taken to discharge the hold capacitor, which in turn, as discharge is by a constant current source, will be proportional to the initial hold voltage.

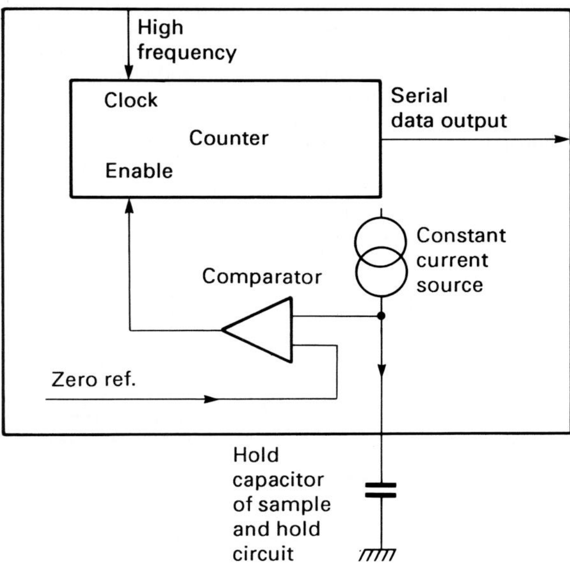

Figure 50.11 Simple analogue/digital converter

The clock frequency required for an actual system of this type is not practicable (for a 48 kHz 16 bit system, the counter must be capable of counting to 65 536, 48 000 times a second; this gives a clock frequency of 48 000 × 65 536 Hz = 3.15 GHz.

To reduce the required clock rate, a dual slope converter is used which splits the counter into two parts, an upper and a lower, each with its respective current source and comparator.

A typical circuit arrangement is seen in *Figure 50.12*. Firstly the upper 8 bit counter current source discharges the hold capacitor to within 1 lsb (*least significant bit*) of 0, followed by the lower counter current source discharging to exactly 0. The clock frequency is now reduced to 2 × 256 × 48 000 Hz = 24.6 MHz. If the counter clock is not locked to sample rate, low level audible beats may be heard, and the clock frequency should be trimmed for optimum audible effect.

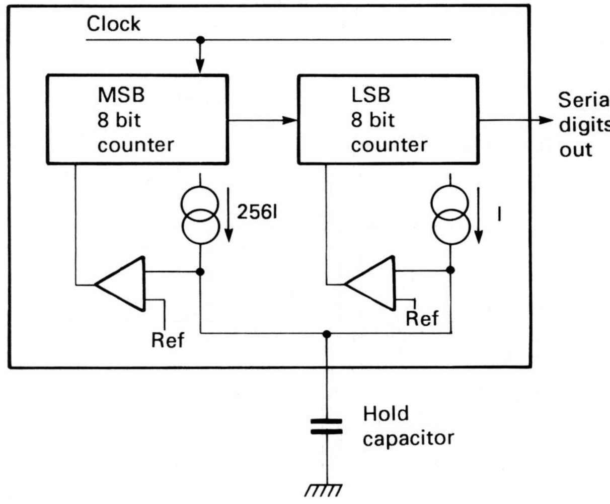

Figure 50.12 Dual slope analogue/digital converter

The two major problems associated with ADCs are distortion and dc offset. *Distortion* can be caused by a mismatch between the two current sources in the converter. This mismatch causes a non-monotonicity, and hence distortion in the conversion, as shown in *Figure 50.13*.

DC offset is caused by an offset between the range of audio input and the resulting digits (*Figure 50.14*). Its value is often dependent on the sample rate used, causing problems in variable sample rate systems.

Two methods are employed to eliminate variable dc offset: digital filtering and sign bit averaging. To eliminate the dc with a *digital filter*, it is necessary only to follow the ADC with a digital high-pass filter operating at, say, 0.5 Hz. In *sign bit averaging* (*Figure 50.15*), the converted audio, in two's complement form, has the sign bit extracted from the data stream and is applied to a low-pass filter.

The low-pass filtered sign bit is then subtracted from the analogue audio at the input to the ADC. The result of this process is that, on average, the audio being output from the ADC will have an equal number of positive and negative samples, the subtraction of the filtered sign bit being in opposition to any dc drift caused by the ADC. This sign bit averaging is only suitable for symmetric signals. If conversion of an asymmetric signal is attempted, the sign bit averaging will instead add a dc offset (*Figure 50.16*).

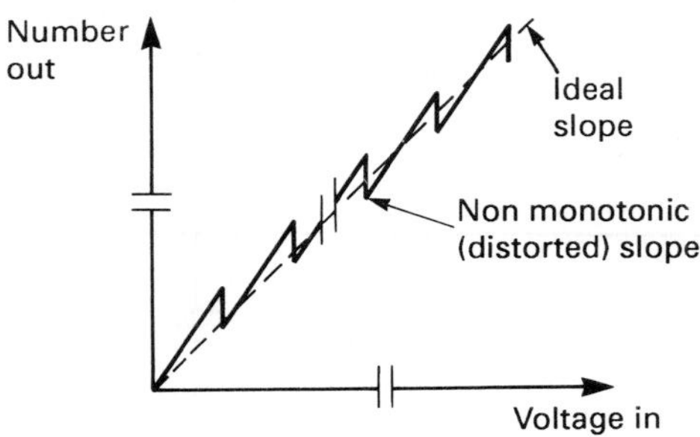

Figure 50.13 ADC distortion

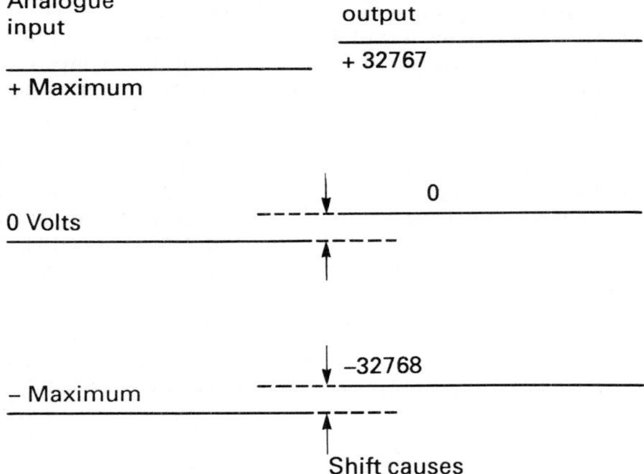

Figure 50.14 ADC dc offset

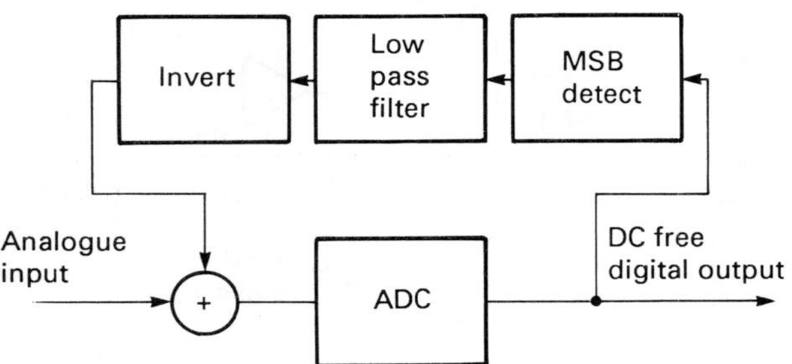

Figure 50.15 MSB averaging to remove dc

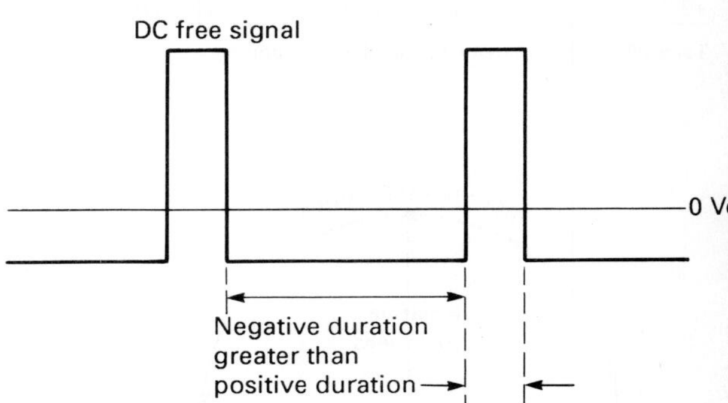

Figure 50.16 Asymmetric dc free signal

When converting from analogue to digital, a known dc offset is sometimes intentionally added to the analogue signal before conversion and subtracted in digital form after conversion. This dc offset prevents analogue noise causing switching between digital 0, two's complement 000...0, and digital -1, two's complement 1111...1. This switching between all 0s and all 1s could modulate the supply voltage of the converter, which in turn could cause modulation of the analogue signal, leading to increased noise levels.

50.1.1.6 Digital/analogue conversion

Digital/analogue converters (DACs) generally operate in a similar manner to ADCs, utilizing a counter and constant current source (see *Figure 50.17*). Again, unlocked clock sources may cause beat frequencies.

50.1.1.7 Sample and hold, and aperture effect

Sample and hold circuits are required for both ADC and DAC conversions. A simple form of sample and hold circuit and the waveforms associated with it are shown in *Figure 50.18*.

For use with an ADC, the sample and hold gate is required to hold the signal level constant while the ADC calculates the digital representation. When used with DACs, the sample and hold gate has a more complex function. The signal output from a DAC should be an infinitely narrow pulse, if the frequency response of the analogue signal is to be correct[2]. It is also necessary that the analogue signal be low-pass filtered to remove unwanted harmonies that might cause intermodulation distortion in later stages (*Figure 50.19*).

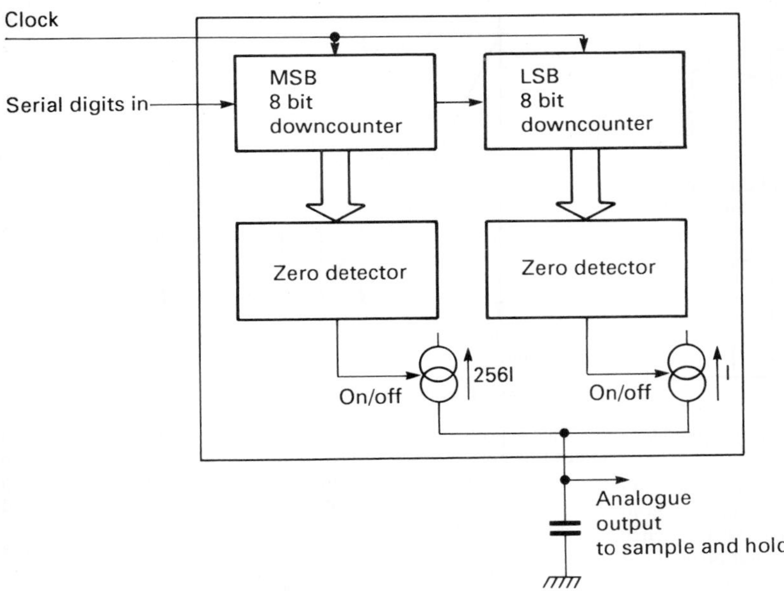

Figure 50.17 Dual slope digital/analogue converter (cf *Figure 50.12*)

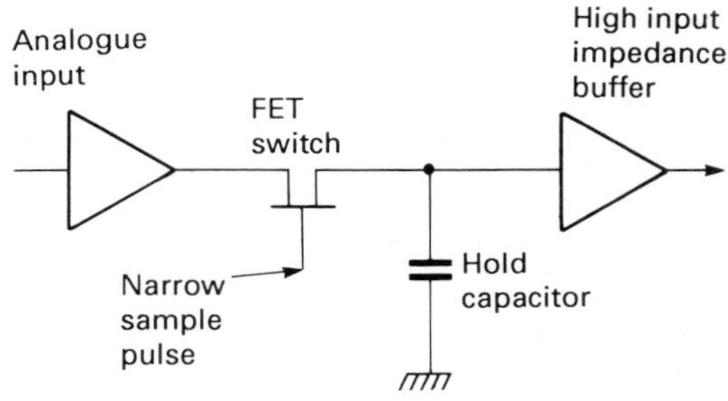

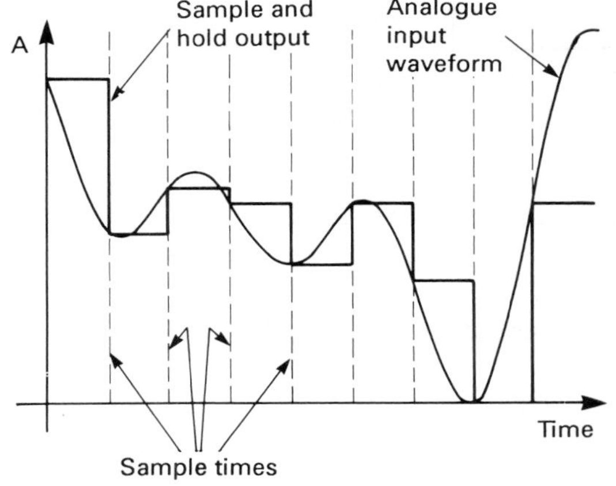

Figure 50.18 Simple sample and hold circuit and its associated waveforms

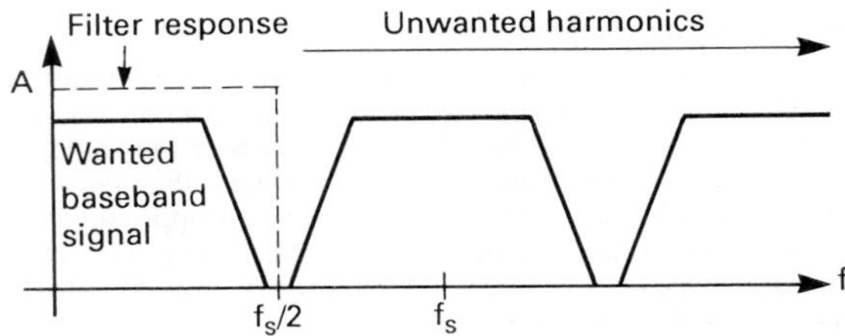

Figure 50.19 DAC output filtering

The signal from the DAC must therefore be held for some finite time before low-pass filtering so that there is a finite amount of energy to filter. The hold duration has an effect on the frequency response of the filtered signal as shown in *Figure 50.20*. This frequency response error is known as the *aperture effect*[3].

A hold for the full duration of the sample period will lead to a frequency response error of approximately -3.9 dB at half the sample frequency.

For most systems, the aperture effect causes no problems. It is of a known value and may be corrected with a simple RC

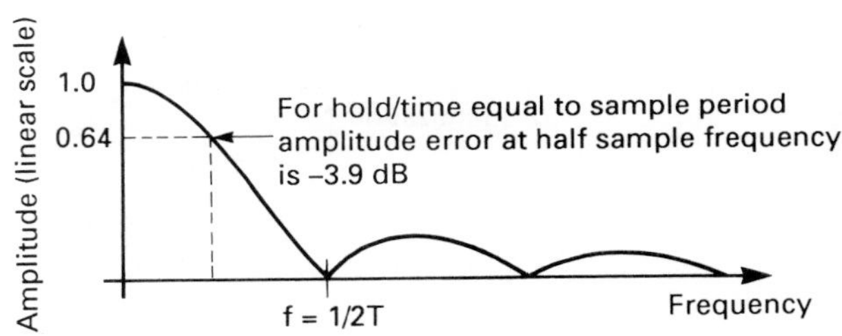

Figure 50.20 Aperture frequency response; T = hold time

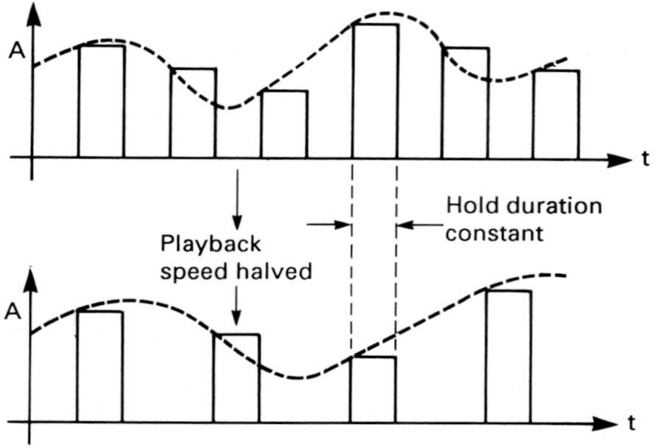

Figure 50.21 Constant hold duration in vari-speed playback leads to constant response errors

network. However, for systems capable of variable rate playback, e.g. vari-speed tape machines, full duration hold cannot be used, as the hold time and hence the frequency response will depend on the sample rate. The solution adopted for vari-speed systems is to use a constant hold time (constant aperture) which is less than the shortest sample period to be encountered (see *Figure 50.21*). This results in a frequency response error independent of sample rate.

50.1.1.8 Signal quantization

When a signal is converted from an analogue representation to digital, the continuously varying analogue signal is converted to a finite range of digits. The relationship between the analogue voltage and its digital representation can be either linear or non-linear. Whichever method is chosen, an approximation will be made by choosing the closest digital value to the actual analogue value (*Figure 50.22*). This approximation is known as the *quantization error*.

The quantization process is equivalent to adding to the signal a noise voltage that has the same form as the approximation.

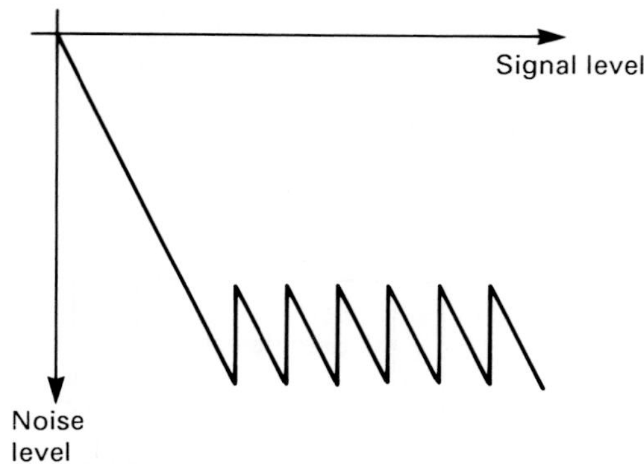

Figure 50.23 Noise level in a non-linearly quantized signal

This hypothetical noise is the *quantization noise*. The peak level of this noise voltage is plus or minus one-half the quantization level, which in an N bit linear quantization system leads to a quantization noise level of $6.02 \times N + 1.76$ dB[5]. In a 16 bit system, this is a noise level of −98 dB.

For non-linear quantization systems, the signal is first converted linearly and then converted to non-linear representation using a lookup table. In a non-linear quantization system, the noise level will be dependent on the signal level (*Figure 50.23*), and the noise floor for a given number of bits will be lower.

50.1.1.9 Dither

Dither is often used in conjunction with an ADC, to overcome problems of distortion when converting low level signals, and in digital signal processing to overcome similar problems caused by rounding errors.

Figure 50.24(a) shows the effect of applying ±1 lsb sinusoidal signal to an ADC. The low level sine-wave has been approximated to the nearest quantization levels, and the digital representation is now closer to a square-wave than a sine-wave.

Figure 50.24(b) shows the harmonic spectrum of the con-

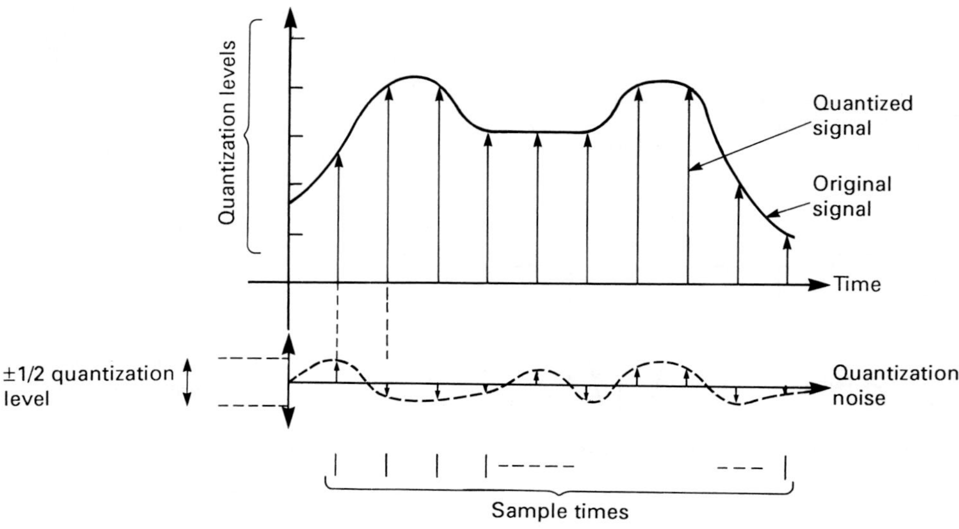

Figure 50.22 Signal quantization and associated quantization noise

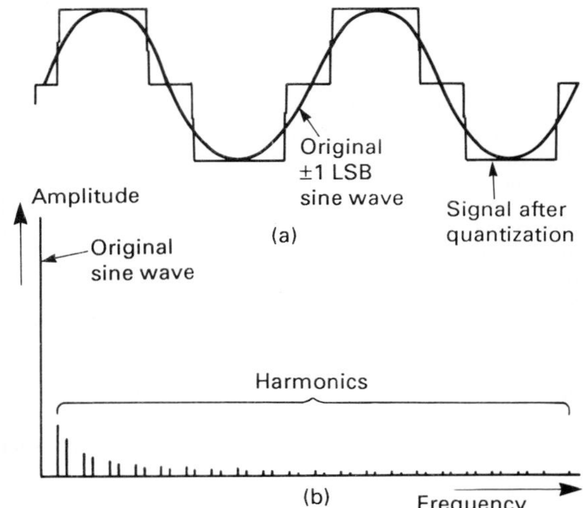

Figure 50.24 Quantization of low level sine-wave: (a) shows time domain, (b) shows frequency domain

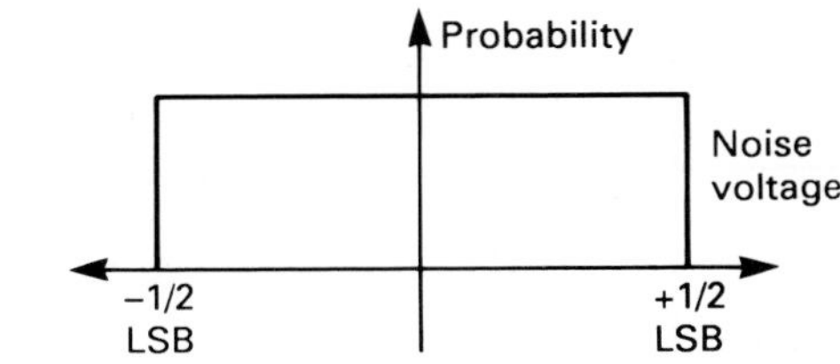

Figure 50.25 The probability density of square dither noise

verted sine-wave, which contains the fundamental plus a high level of audibly objectionable odd harmonics. This digital representation of the signal is a distorted version of the input signal. A second feature of this distortion is that the level of harmonic distortion varies suddenly as the number of quantization levels covered by the signal varies. This stepping of distortion is also audibly apparent.

To overcome the problem of quantization (*granulation*), noise dither is added to the analogue signal before quantization. Dither is most commonly a white noise signal of ±0.5 lsb amplitude. The probability density function (PDF) is shown in *Figure 50.25*.

The effect of adding dither before the ADC is shown in *Figure 50.26*(a). The quantization of the signal is randomized,

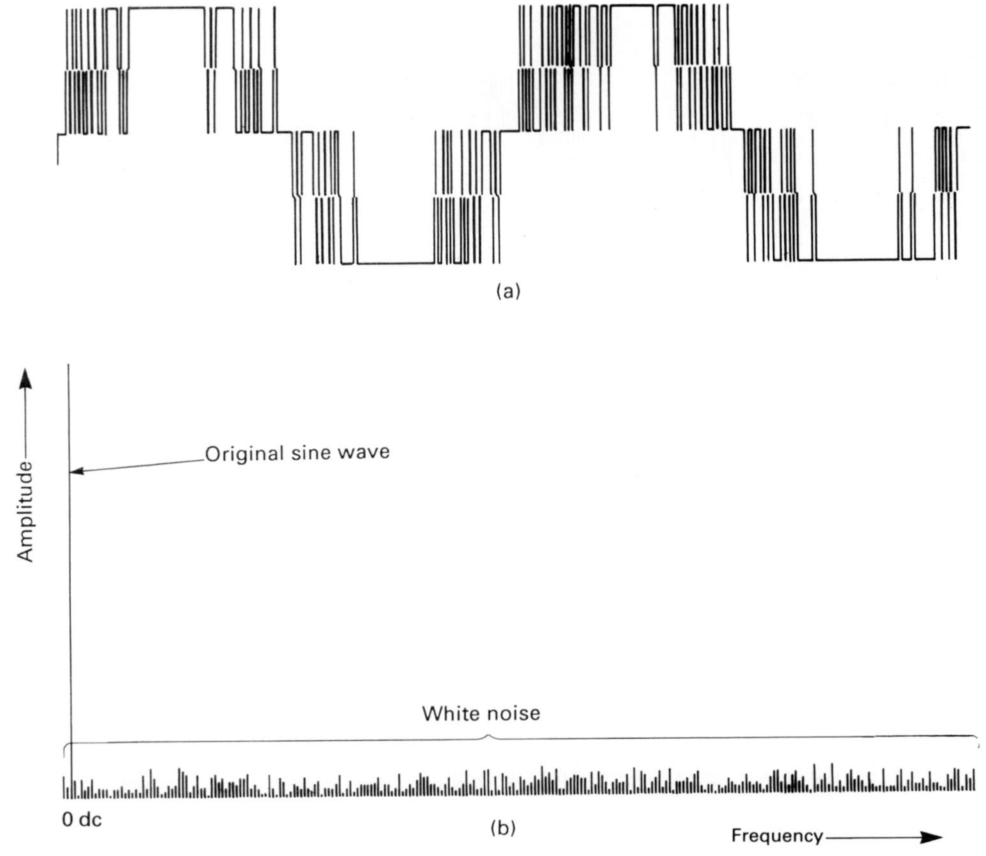

Figure 50.26 ± sb sine-wave ±1/2 sb square dither after quantization (a), and the spectrum of the signal (b)

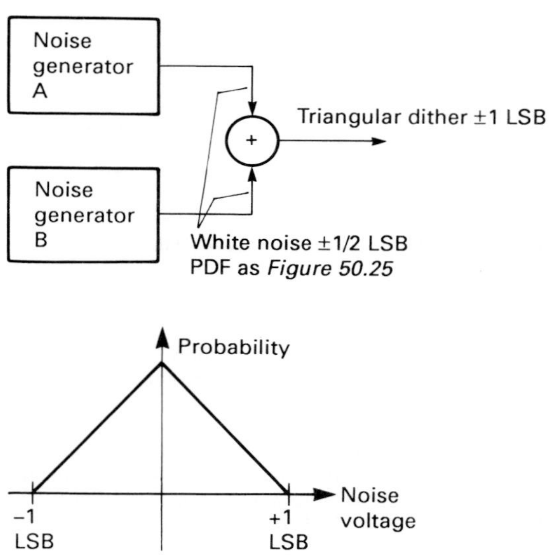

Figure 50.27 Triangular dither generator and associated PDF

causing an effect similar to random pulse width modulation on the converted signal. The dramatic effect on the harmonic spectrum is shown in *Figure 26*(b), the harmonic distortion having completely disappeared, and been replaced by the audibly more pleasant white noise floor.

This type of dither has the advantage that, in an inactive digital filter, ±½ lsb of dither will not alter the sample value, thus enabling cloning of material. It has the disadvantage of level modulation by the analogue signal.

To overcome the problem of noise modulation by the programme material, *triangular dither*[6] is employed. Triangular dither is generated by adding together the outputs of two white noise generators, to produce a dither signal with a white spectrum and PDF as in *Figure 50.27*.

The effects of adding triangular dither to a low level signal are shown in *Figure 50.28*. The effect is the same as for square dither, except that the noise level is 3 dB worse and the noise floor is no longer signal level dependent. However, ±1 lsb triangular dither has the disadvantage that in an inactive filter the sample values will be changed, making cloning impossible.

50.1.1.10 Oversampling conversion

Oversampling is used both to eliminate the group delay distortion caused by the anti-alias filter, and to improve the signal/noise ratio of the digitized signal. In an oversampling

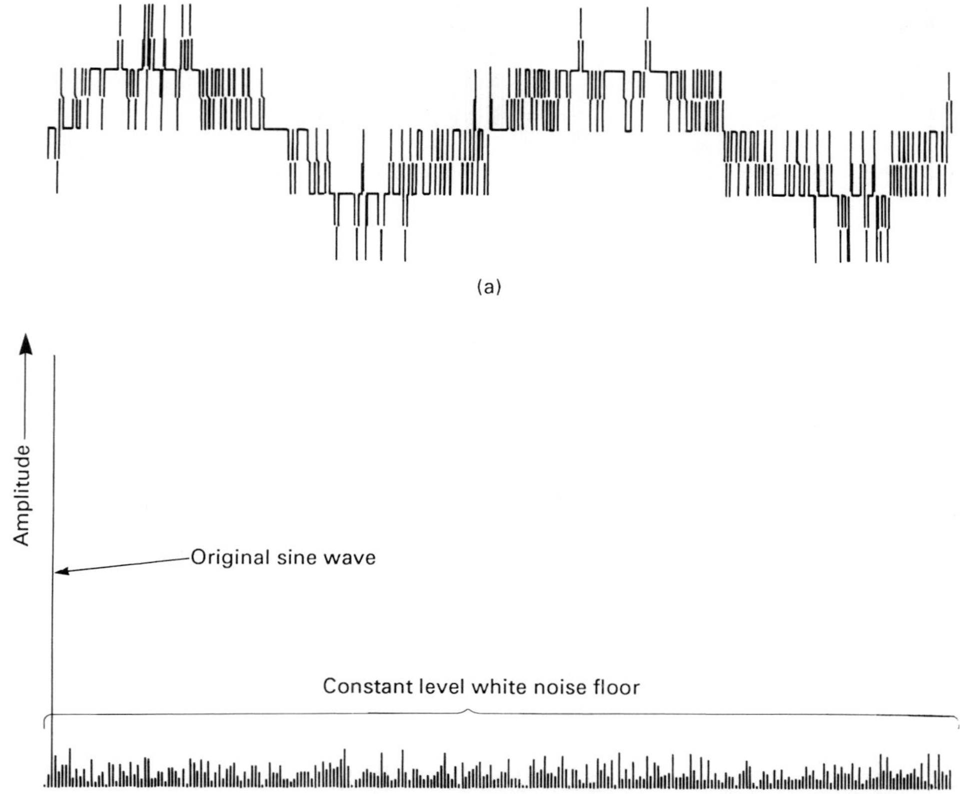

(a)

Amplitude
Original sine wave
Constant level white noise floor
0 dc
Frequency

(b)

Figure 50.28 ±1 sb sine-wave ±1 sb triangular dither after quantization (a) and the spectrum of the signal (b)

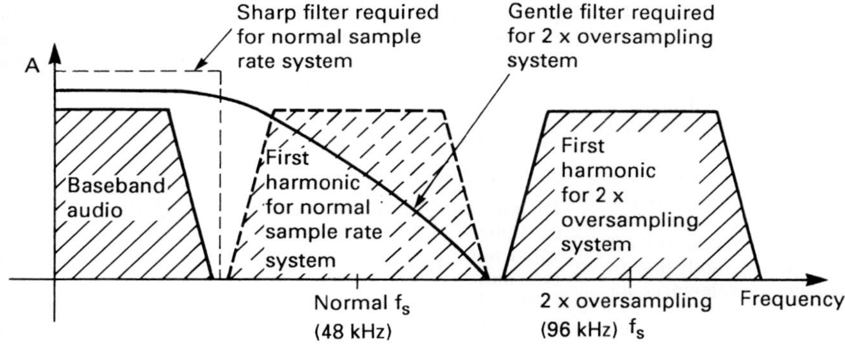

Figure 50.29 Comparison of filters required for a 2× oversampling system and a normal sample rate system

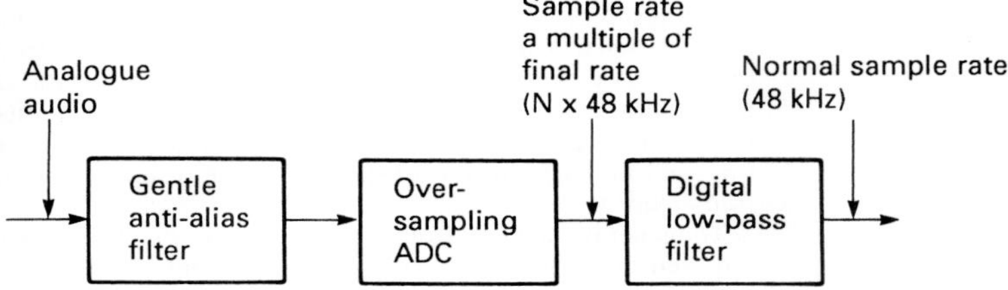

Figure 50.30 An oversampling system

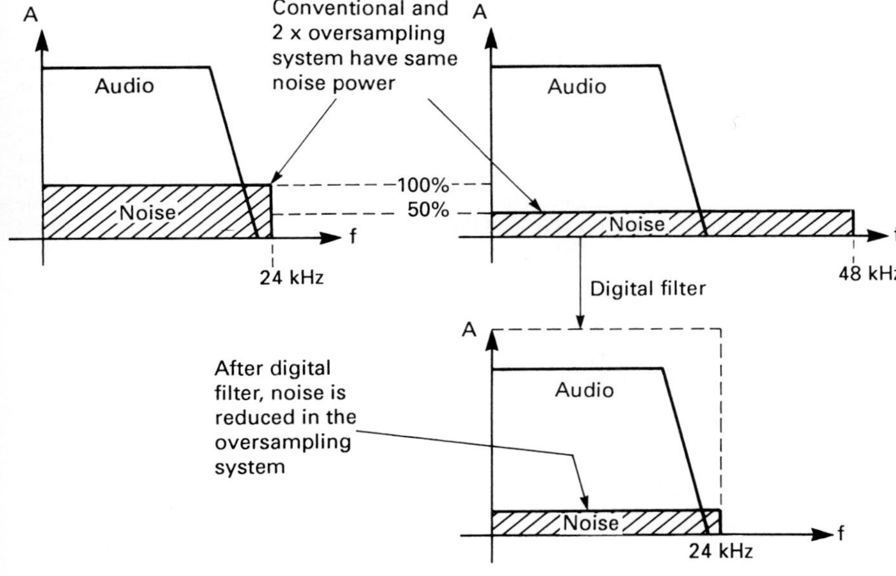

Figure 50.31 Comparison of noise levels in conventional and 2× oversampling system

system, the ADC is operating at a multiple, normally a power of two, of the actual sample rate (*Figure 50.29*).

The first image frequency of the converted signal is now centred around the higher sample rate, which means that the rate of roll-off of the anti-alias filter can be drastically reduced, to virtually eliminate phase distortion.

The sample rate of the oversampled signal is now higher than necessary, and must be reduced before the digitized signal is further processed. To do this a digital low-pass filter is used (*Figure 50.30*).

In an oversampling system, the bandlimiting of the signal to half the sample rate is moved from the analogue domain to the digital. The advantage of performing the filtering in the digital domain is that the low-pass filter can be designed for fast roll-off phase response errors much more easily and accurately than in the analogue domain, as digital filter component tolerances are zero.

A second advantage of oversampling is that the noise level of the converted signal is reduced. The noise caused by signal conversion in a 16 bit system is a white noise floor at approximately 96 dB below peak signal level. If the conversion is at, say, twice the sample rate, then this noise is spread over twice the bandwidth, and consequently only half the noise power is in the signal band. The digital filter following the oversampling ADC removes this out-of-band noise, consequently improving the s/n ratio and increasing the number of bits representing the signal (*Figure 31*).

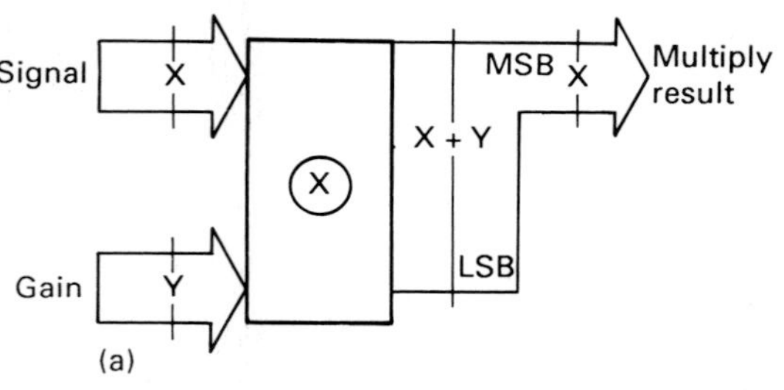

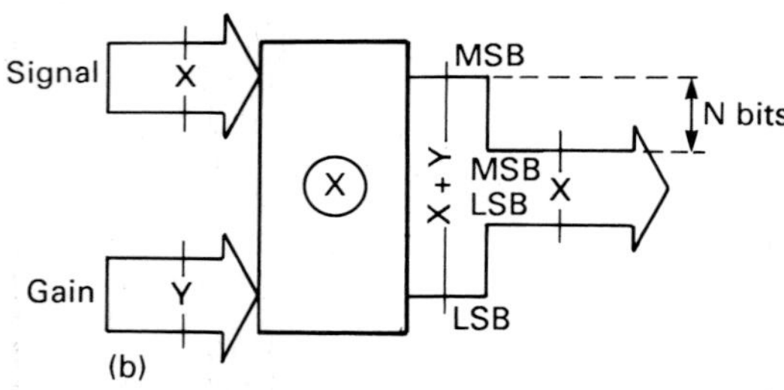

Figure 50.32 Multiplier systems having maximum gain (a) of 0 db, (b) of 6N db

50.1.2 Digital signal processing

50.1.2.1 Digital gain

Digital audio gain control is effected through the use of binary multipliers. Multiplying a signal by numbers greater than unity will achieve gain, whilst multiplying a signal by numbers less than unity will achieve attenuation.

Multiplying an x bit signal by a y bit coefficient will produce a result of $x + y$ bits. This means, for example, that when multiplying a 16 bit signal by a 16 bit gain coefficient, the true result will consist of 32 bits. If the result is to be returned to 16 bit format, the result must either be *truncated*, the least significant bits being thrown away, or *rounded*, using the nearest 16 bit number. Both truncation and rounding will distort the signal. This approximation distortion may however be masked by the use of dither.

The result of a multiplication consists of more bits than the original signal. Choosing which bits of the result are subsequently used for the signal will decide whether gain or attenuation is achieved; by using only the most significant bits, unity is the highest gain achievable. However, a gain of 6 dB is achieved for each downward skewing of the result by one bit (see *Figure 50.32*).

To generate a fade from unity to zero, the two inputs required to a multiplier are the signal and, as the gain coefficient input, a count from maximum (unity gain) to zero.

Figure 50.33 shows a typical fade arrangement, the down counter being formed from a subtractor constantly subtracting a fixed rate value from an initial preset to unity. To increase the rate of fade it is necessary only to increase the rate value, which in turn speeds the down count from maximum to zero. In order to change the fade law from being volts-linear to some other characteristic, a ROM based law lookup table can be placed between the counter and the multiplier.

For manual fades, there must be an input device which can either be a binary fader or a conventional fader followed by an ADC. The digital gain value is then applied to the fader law lookup table and hence to the multiplier as coefficient input.

If a binary fader is used, there may be only 256 steps to cover a range of +12 dB to $-\infty$ B, which will necessitate large steps of gain at the low gain end of the fader. This will result inaudible stepping of the level (*zipper noise*) at low gains. To overcome stepping, the lookup table is followed by a low-pass

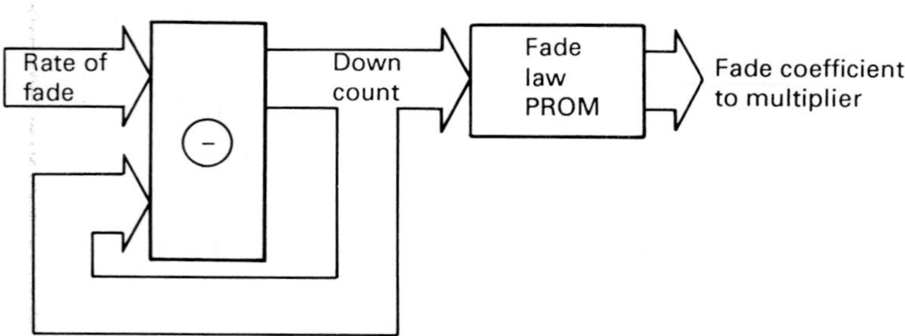

Figure 50.33 Fade coefficient generator. The subtractor is preset to maximum value, and a rate number is constantly subtracted

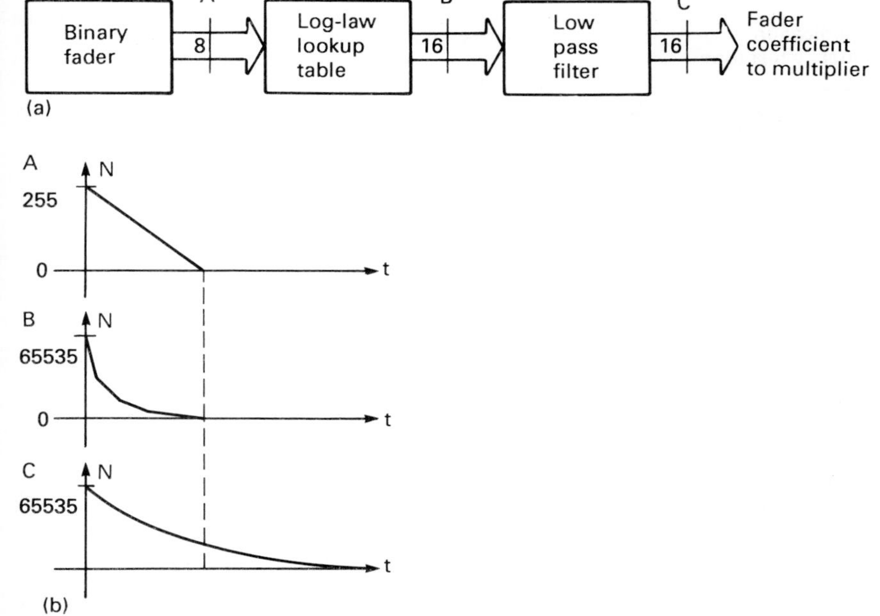

Figure 50.34 (a) Using a log-law lookup table and low-pass filter to generate a smooth logarithmic fader action, (b) shows the associated waveforms

filter to smooth out the gain changes. This approach will however lead to some lag if the fader is moved rapidly (*Figure 50.34*).

50.1.2.2 Digital audio clipping

In any system, if the gain is increased too far, overload will occur. In a two's complement system, overload generates a very unpleasant distortion of the signal, caused by the cyclic nature of two's complement numbers where a 1 bit overload of maximum positive value generates maximum negative value.

Figure 50.35 shows the effect of overload on a sine-wave in two's complement notation. To improve the subjective effect of overload in a two's complement system, it is normal to follow any process capable of supplying gain by a limiter, driven from an overload detector. When overload occurs, the limiter will supply either maximum positive or maximum negative value in place of the multiplier output, depending on the polarity of the input signal, returning the overall effect to normal clipping.

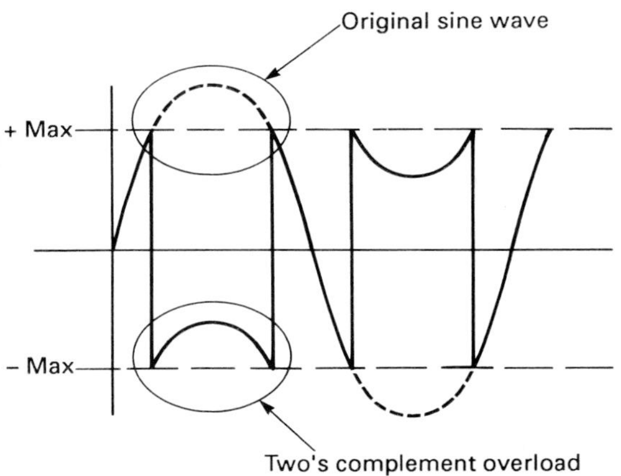

Figure 50.35 Overload effect with two's complement representation

50.1.2.3 Levels, metering and overlevel indication

Digital audio metering is performed in the digital domain using peak reading meters with zero attack time. A peak hold facility and overload indication are usually provided. The use of peak reading meters is essential for controlling levels in a digital audio environment as, unlike analogue tape machines where distortion gradually becomes worse as tape saturation effects occur, distortion in a digital recorder or system is negligible up to peak level, and from that point on the signal is clipped as the system overloads (*Figure 50.36*).

Overload indication in a digital audio meter is performed by detecting consecutive samples at peak level. A single sample at peak level cannot be considered as an overload, but consecutive samples at peak level will normally have been caused by signal clipping.

In a digital audio overload indicator, it is normal to be able to set the number of consecutive samples which cause overload indication. When metering low frequency peak level tone, as may be the case when playing a test tape or test disc, it is possible that overload indication occurs even though the signal

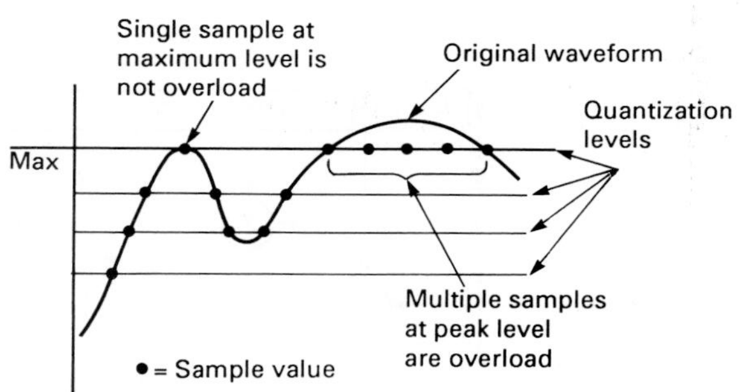

Figure 50.36 Overload detection in digital metering

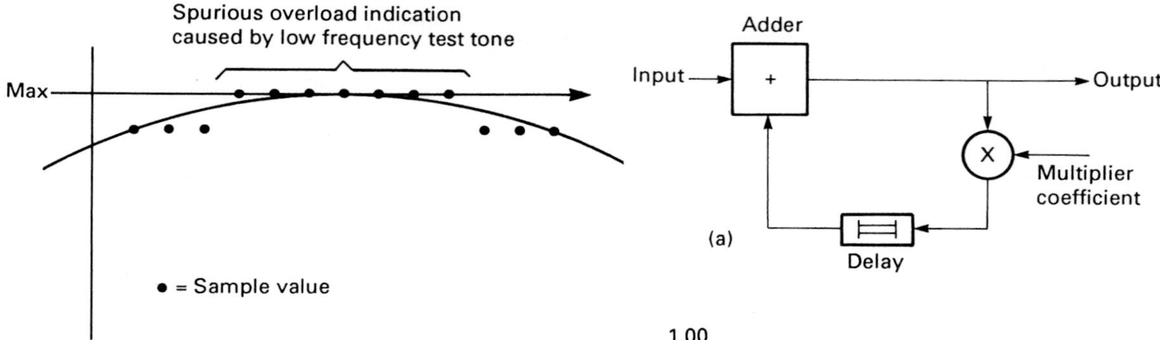

Figure 50.37 Spurious overload indication caused by low frequency test signal

is at maximum level and not over. This spurious overload indication is caused by full level low frequency signals having several consecutive samples at peak level (see *Figure 50.37*).

Standard peak analogue level in a digital audio system will normally be +24 dB for a studio environment (occasionally +18 dB will be used if the analogue mixing console is only capable of this level), or +14 dB for a broadcast environment.

50.1.2.4 Digital filtering

Digital filtering of signals is necessary for a number of reasons. Equalization, oversampling converters and dc removal all require the use of digital filters. Digital filters divide into two classes, infinite impulse response (IIR) and finite impulse response (FIR). IIR filters contain feedback paths while FIR filters have none. All filters consist of different arrangements of only three basic elements: the multiplier, the adder and the delay.

A basic form of *infinite impulse response* low-pass filter is shown in *Figure 50.38*. The input signal is fed through an adder to the output, a portion of the output, determined by the multiplier coefficient, being fed back to the adder through the delay.

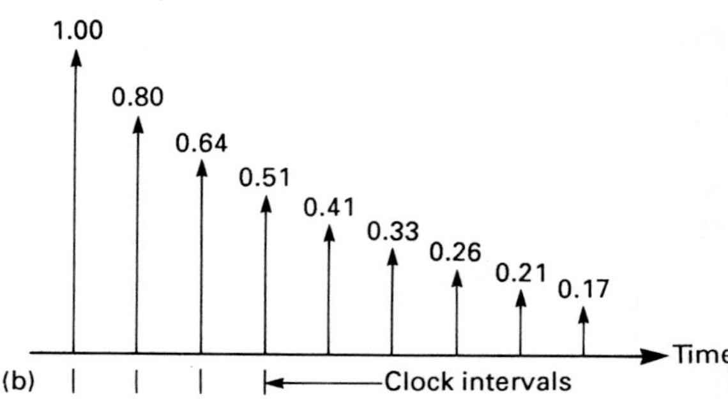

Figure 50.38 (a) Basic IIR and (b) the filter output for unity impluse input and coefficient of 0.8

Figure 50.38(b) shows the circuit output for an impulsive input, a delay of one clock cycle and a gain of 0.8. It is clear that, if the multiplier coefficient is set to 1, then the output will continue at the same level for all time. This is possible due to the feedback path in the circuit, thus the name of *infinite* impulse response.

A basic *finite impulse response* filter structure is shown in *Figure 50.39*. The response of any filter is uniquely determined by its impulse response[7]. For an FIR filter, this impulse response will be identical to the multiplier coefficients. The filter coefficients may therefore be determined by transforming the desired frequency to the time domain. The coefficients for a simple filter to generate a ramp response for an impulse input are:

$n_1 = 0.1$	$n_4 = 0.4$	$n_7 = 0.7$
$n_2 = 0.2$	$n_5 = 0.5$	$n_8 = 0.8$
$n_3 = 0.3$	$n_6 = 0.6$	$n_9 = 0.9$

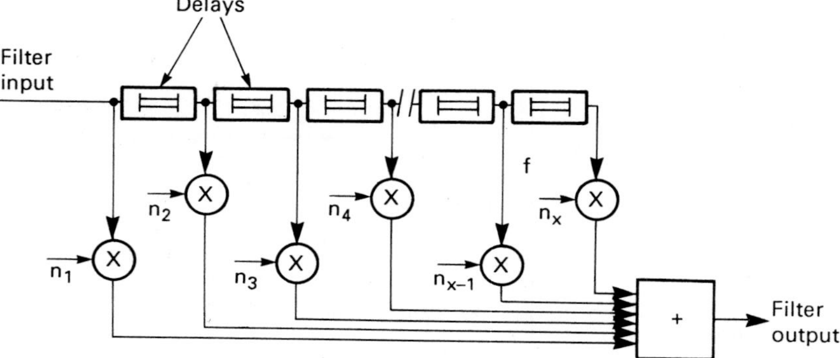

Figure 50.39 Basic FIR filter structure

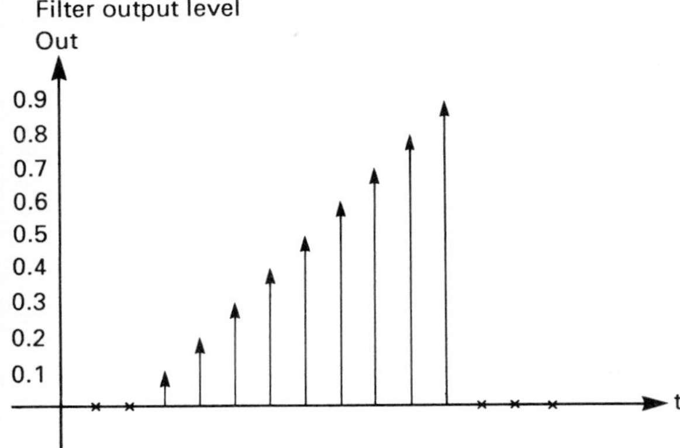

Figure 50.40 Nine element FIR filter impluse response for multiplier coefficients n_1–n_9 = 0.1–0.9

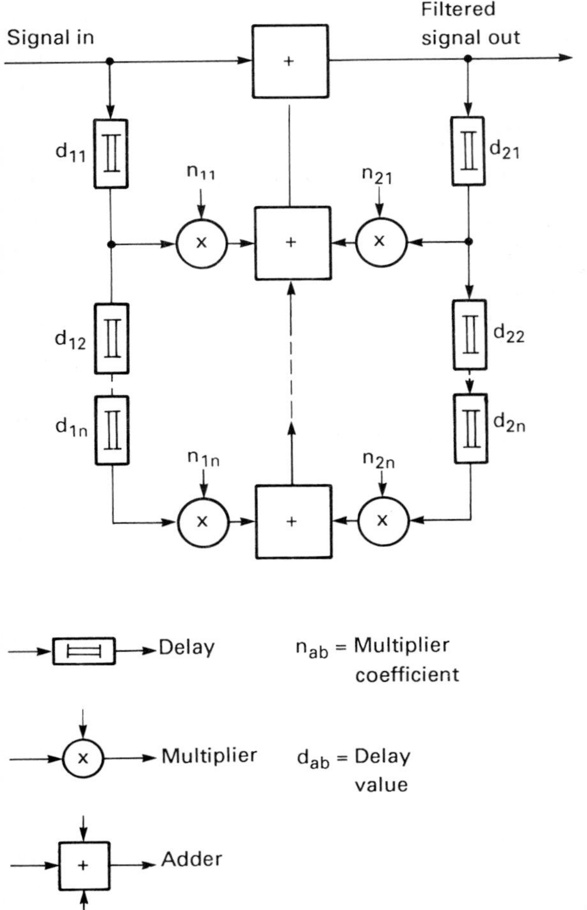

Figure 50.41 General form of digital filter

The resulting output for impulsive input is shown in *Figure 50.40*. As there is no feedback path in an FIR filter, the duration of any output, for impulsive input, cannot exceed the total delay time. Thus the output duration is finite. A realistic version of the filter would typically have 96 delay elements, the delay consisting of RAM. The filter coefficients would be stored as a lookup table in ROM, with the multiplier and adder time division multiplexed between filter sections.

A general structure for a complex digital filter is shown in *Figure 50.41*, the actual filter characteristics being determined by the multiplier coefficients and delay lengths[8].

50.1.2.5 Interleave and error coding

Interleave is used in digital audio record systems to:

- disperse the error bursts caused by dropouts or contamination
- allow splice (razor blade) editing of tapes

Error burst dispersal is necessary for the error correction systems to work more effectively. Should the error correction system become overwhelmed, it is also easier to interpolate from the remaining correct samples, to construct replacement samples, if the errors are dispersed.

Either block based interleave or convolutional interleave can be used.

Block based interleave is used in the Pro-Digi, Sony PCM 1630 and DAT formats. In a block based interleave, a fixed number of audio samples are read into a memory, reordered and then recorded on tape (*Figure 42*).

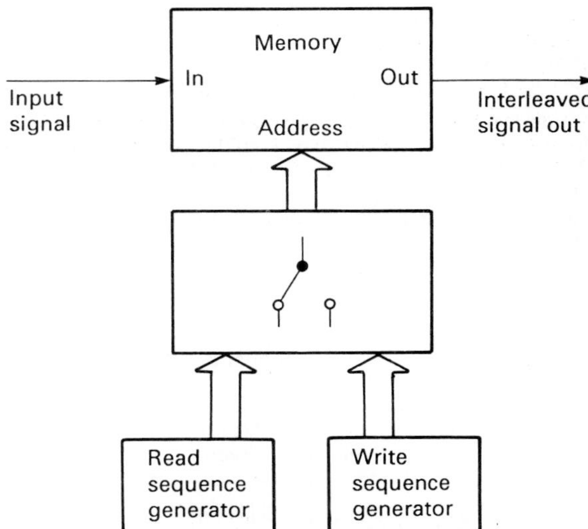

Figure 50.42 General arrangement for block interleave system

Editing a block based interleave system requires edit accuracy to be limited to block ends. Block based interleaves are convenient for rotary head systems as the interleave length can be related to integral numbers of rotations of the drum. For example, in DAT one interleave period is equal to one drum rotation. In PCM 1630 format, 14 interleave periods occur in rotation. In PCM 1630 format, 14 interleave periods occur in one drum rotation (one frame of video).

Convolutional interleave is a continuous one where different

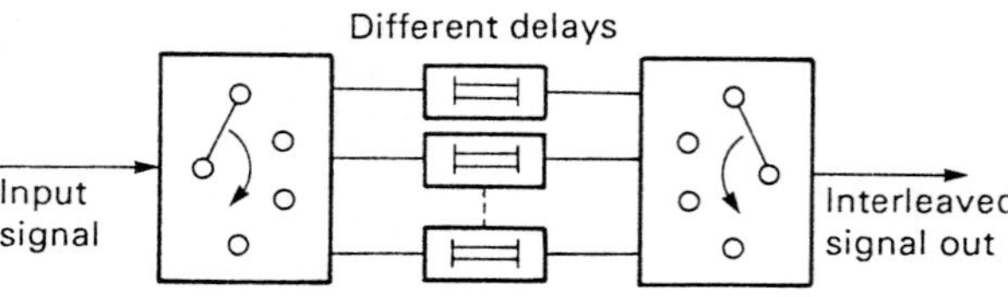

Figure 50.43 General form of convolutional interleave

audio samples are delayed by varying times. The longer the interleave distance, the further the errors will be spread and the better the error correction may be performed (*Figure 50.43*).

However, there is a conflicting requirement for the interleave system. This is its ability to deal with splice edits. As may be seen from *Figure 50.44*, as the interleave length increases so does the amount of damage done to the error correction system when a splice edit is performed. A compromise is therefore drawn between burst protection and splice ability.

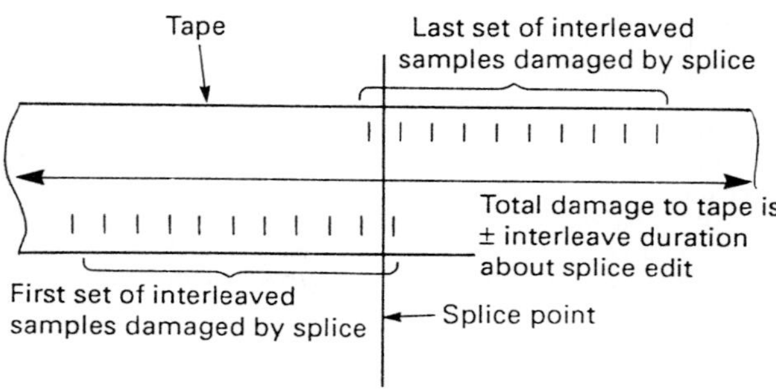

Figure 50.44 Disruption to interleave system caused by splice. Tape motion is from right to left

Interleave is utilized during splice editing to enable audio from both sides of the edit to be available simultaneously after de-interleave. Thus the signals from either side of the edit may be cross-faded. This type of interleave is achieved by separating alternate (odd and even) samples before recording (*Figure 50.45*).

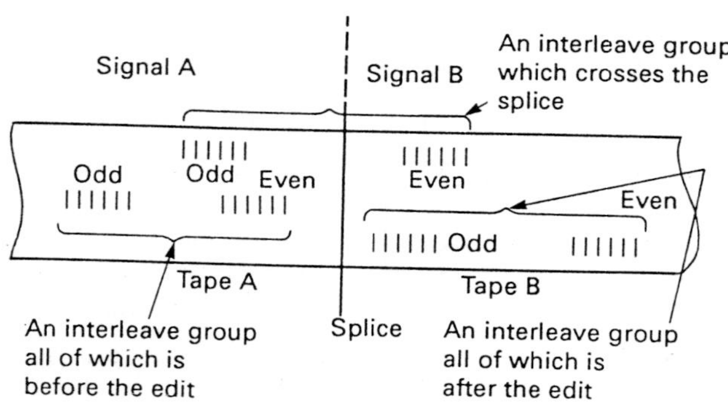

Figure 50.45 After de-interleave the group of data which crosses the splice will become alternate samples from either side of the splice

At a splice edit, alternate samples from the decoder will be from either side of the edit. If alternate samples from either side of the edit are available, both signals may be reconstructed by interpolation. Once both signals have been reconstructed, they can be cross-faded to give an equivalent effect to a 45° razor edit on analogue tape, but without any interchannel delay (*Figure 50.46*).

Interleave has an effect on electronic tape editing, due to the time delay required for decoding the playback audio and for encoding the record audio. To enable a cross-fade between playback and record audio at an electronic edit, it is normal to playback the audio from a head earlier in the tape path than the record head. This enables the playback audio to be decoded and re-encoded in the time it takes the tape to travel between the two heads.

The cross-faded audio is thus re-recorded on the tape in exactly the same physical position that it was played from. The record head downstream from the playback head is referred to as the *sync record* head (*Figure 50.47*).

50.1.2.6 Error correction

Error detection and correction systems are required in digital audio transmission and recording systems for detection and prevention of data corruption. All error correction systems use additional redundant information for error detection and correction, but there are many ways of implementing the redundancy[9].

The three commonly used forms of redundancy are CRCC (cyclic redundancy check code), parity and Reed-Solomon codes.

A *CRCC code* is a number generated from and appended to the digital data stream which is used to detect playback errors.

Parity words are generated by an exclusive-or (XOR) of data words and are used to correct errors indicated by a CRCC check circuit.

Reed-Solomon codes have recently become possible with the advent of galois field[10] processing lsi chips. Reed-Soloman codes are basically a hybrid between CRCC and parity codes and may be designed to both detect and correct errors. Overall error correction techniques result in vast improvements in error rates[11]. Random error rates of hundreds per second can be reduced to single errors per hundreds of minutes.

50.1.2.7 Channel coding

Channel coding[12,13] is used by record systems to change the binary, two's complement, signal to a form more suitable for recording on a particular medium.

A signal to be recorded has numerous requirements. Firstly, the recorded signal should have a strong clock content, so that it may be easily extracted on replay (on a multi-track tape, each channel will require its own clock signal as tape path instability will cause interchannel timing variations).

Secondly, tape record/replay systems have poor low and high frequency responses, and consequently any signal recorded on tape should have limited lf and hf content.

Thirdly, high density magnetic recording systems suffer from peak shift[14] distortion — a widening of narrow asymmetric pulses.

Problems with binary signals are numerous. For example, digital silence, a data stream of all zeros, has a strong lf content, no clock content, and very high asymmetry, rendering it unplayable. A channel coding system attempts to match the record signal to the medium.

All channel codes work in a similar manner, by converting one group of data into another more desirable set of record data. The codes break down into two basic types: bit or word substitution codes and convolutional codes.

In *substitution codes*, a group of bits, generally a byte, are converted into an alternative group of bits with more desirable characteristics. These are found by using a lookup table.

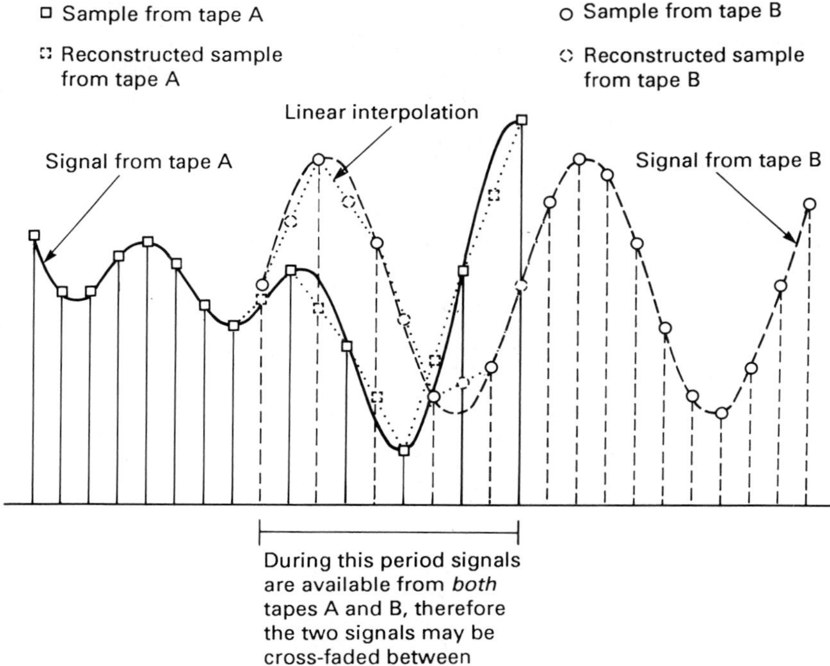

Figure 50.46 Signal reconstruction at a splice edit

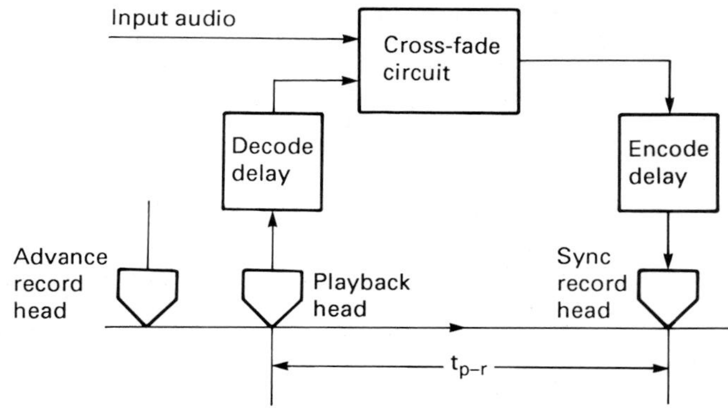

Figure 50.47 The use of a sync record head to enable electronic editing. The time taken for the tape to travel from the *play* head to the *record* head (t_{p-r}) must equal the sum of the encode and decode delays

Substitution codes are used by DAT, (8–10 modulation), Pro-Digi (4/6M), and compact disc (EFM).

Convolutional codes differ from substitution codes in that the channel code for a particular group of data bits depends not only on the data bits themselves, but also on the data that have gone before. This dependence on previous data leads to an extra requirement when designing a suitable code: error propagation in the code should be small.

Error propagation in convolutional codes comes about due to the prior dependent nature of the code. If a bit is corrupted, then as later bits depend upon its state, later bits will also be corrupted. A convolutional code HDM-1 is used by the DASH format.

50.1.2.8 Signal detection

The playback digital signal from tape will be significantly different from that recorded (see *Figure 50.48*). The playback signal is a differentiated version of the record current, and suffers from peak shift caused by the spreading of asymmetric record signals. Before the playback signal can be converted into a true digital signal for playback processing, it must undergo equalization to eliminate the peak shift, followed by integration to restore the wave shape, and finally slicing at a 1/0 decision level to recreate a true digital signal. (*Figure 50.49*)

50.1.2.9 Timebase correction

Timebase correction of the playback signal in a recording

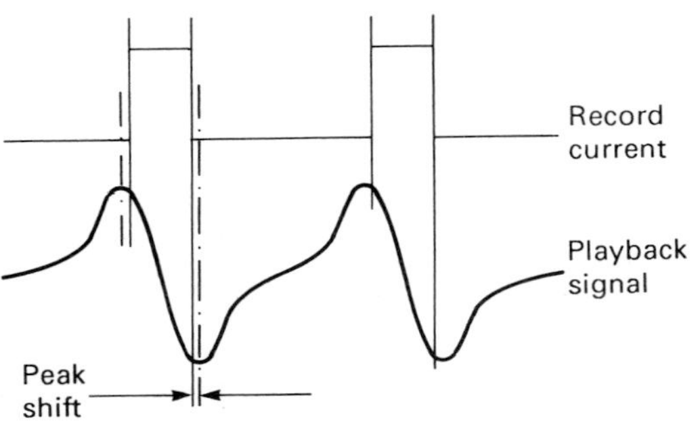

Figure 50.48 Playback signals suffer from peak shift effect. The playback signal is also a differentiated version of the record current

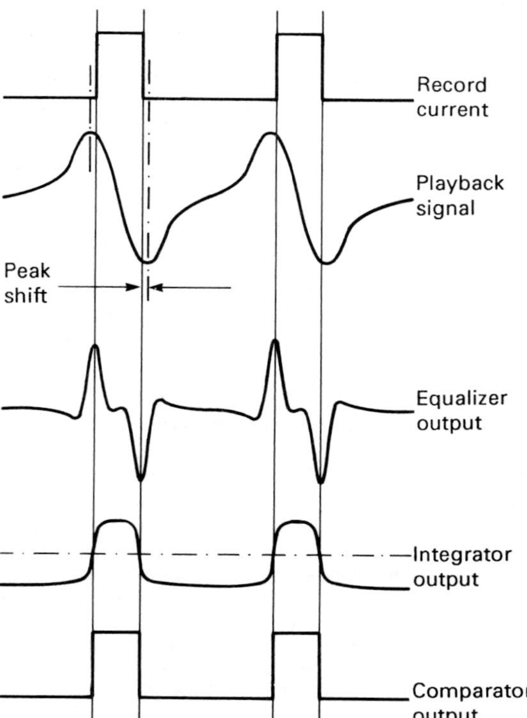

Figure 50.49 Processes required to reform the playback signal

system is necessary to eliminate playback timing instabilities. These instabilities are caused by a variety of problems such as servo lock errors or tape weave. Timebase correction is performed by extracting the timing information from the playback signal, then writing the playback data into a memory using the extracted clock (see *Figure 50.50*).

The data may then be read from the memory at a constant, stable rate using the reference clock of the system. The size of memory must be greater than maximum instability, i.e. if the maximum playback jitter encountered is ±40 samples, then an 80 sample memory will be required.

50.1.2.10 Error concealment

If random errors occur in a digital system, the audio samples will be replaced by random numbers. The analogue equivalent of a continuous stream of random numbers is full level white noise. It is therefore necessary to employ an error concealment system in case the error correction systems are overwhelmed and unable to correct corrupted data.

Error concealment systems normally operate in a number of different manners according to the severity of the error. The first level of concealment occurs when only a single sample is in error. The usual strategy for concealment is to replace the sample by an average of the preceding and the following samples (*Figure 50.51*(a)). This is also referred to as *interpolation*.

If multiple errors occur, the next level of concealment is hold followed by average (*Figure 50.51*(b)). Here the value of the last correct sample is repeated until the corruption is finished, the last corrupt sample being replaced by an average of the held value and the first uncorrupted sample.

For severe errors, the strategy of concealment by hold is inappropriate, as the many transitions between held data and good data will produce large numbers of transients which are audibly objectionable. For severe data corruption, the normal approach to concealment is to mute the output, replacing the corrupted data with a fade to and from zeros for a given duration. This has the advantage over hold that, if a single uncorrupted sample is output, then it will not produce a transient, audible as a click.

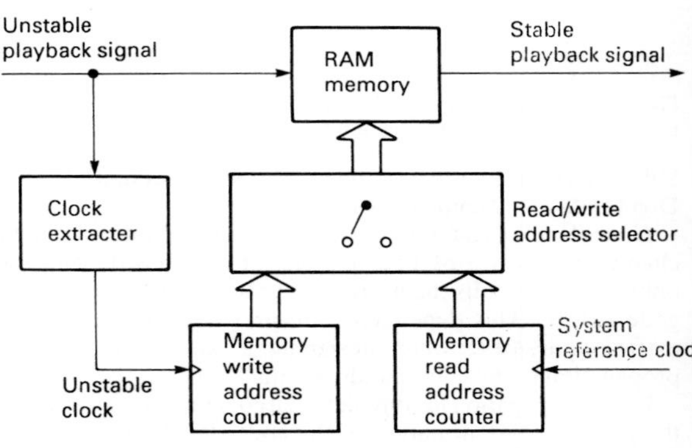

Figure 50.50 Time base corrector

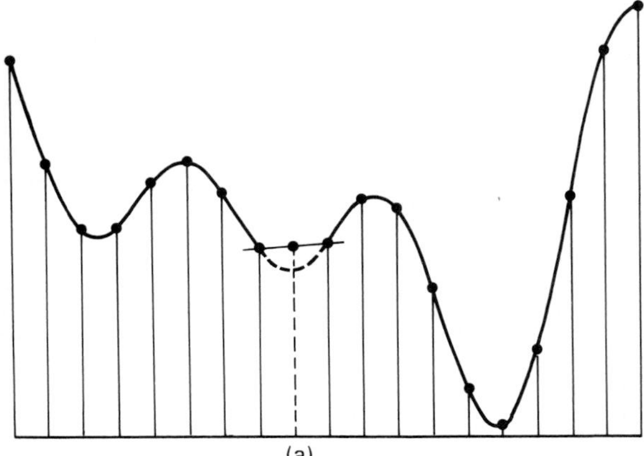

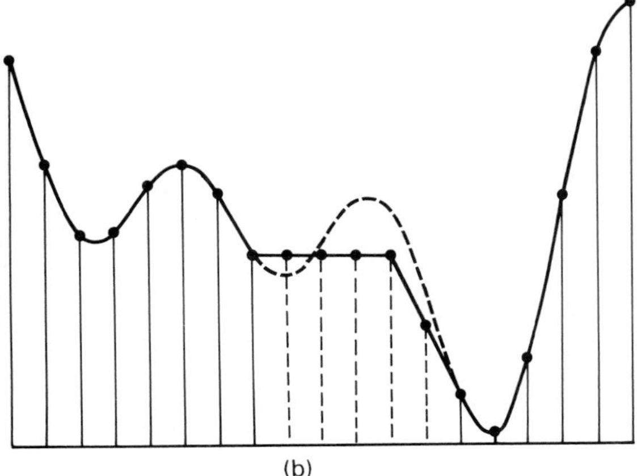

Figure 50.51 Concealment of errors: (a) concealment of single error by averaging (interpolation), (b) concealment of multiple errors by hold and average

50.2 Digital audio in application

50.2.1 Standard formats and their conversion

50.2.1.1 Sample rates

Sample rates for digital recording have had a much debated history. In 1981, 60, 54, 52.5, 50.4, 50.35, 50, 48, 47.25, 47.203, 45, 44.1, 44.056, 32, and 30 kHz were all considered as possible contenders for a digital frequency standard[15].

Now there are four rates in normal usage. 48 kHz is used by convention in sound recording studios and for digital sound on vtrs, this rate being a convenient multiple of both the 25 Hz and the 30 Hz television frame rates. 44.1 kHz is the standard for compact disc and its associated mastering equipment. 44.056 kHz evolved from 44.1 kHz and enables equipment designed for 44.1 kHz to be used with drop frame NTSC video (29.97 Hz frame rate). 32 kHz is used for PTT, TV sound and fm radio where the audio bandwidth is limited to 15 kHz.

50.2.1.2 Sample rate and format conversion

The use of vari-speed playback, along with the variety of sample rates and transmission formats in use, means that digital audio standards converters are required to convert between standards and sample rates.

Vari-speed operation with a digital radio recorder is a rather more complex operation than with an analogue machine. Not only does the playback speed and pitch of the programme vary, but so does the playback sample rate, becoming non-standard. Variations of the sample rate will not be a problem if the machine is being used only via its analogue inputs and outputs, as the digital machine is effectively being used as an analogue recorder.

Problems occur when digital audio is played back at non-standard speed and is required as input for a digital system operating at a standard rate. There are two possible solutions. The first is to forget that the operation is digital and connect the two systems via their analogue inputs and outputs. This rather inelegant solution will however degrade the signal quality. The second solution is to employ a digital sample rate converter to change rates without entering the analogue domain.

When using a digital sample rate converter, one must be aware of its limitations. When changing from a known high sample rate, say 48 kHz, to a lower rate, say 44.1 kHz, then in order to avoid aliasing of high frequencies, the audio must be low-pass filtered by the converter. As well as losing out-of-band frequencies, this filtering will lead to a small level change between input and output of the converter, caused by the filter characteristics.

A second problem arises when converting a variable rate source to a fixed rate. As the actual range of sample rate input may be very wide, there is the choice of either always filtering the audio to half the lowest allowable sample rate, or of not filtering the audio and producing alias signals when the input sample rate becomes lower than the output sample rate. The latter approach is normally chosen.

50.2.1.3 Transmission formats

There are four standard formats used to interconnect digital audio equipment. The transmission standards are single channel SDIF[16] (Sony Digital Interface Format), twin channel AES/EBU[17] format, its domestic variation SPIF[18](Sony Philips Interface Format) and multi-channel MADI[19] format (multi-channel digital audio interface).

SDIF is based on a transmission format of one audio channel per connection. It is not a self clocking signal and will therefore also require connection of a synchronizing signal (word clock, a square-wave sample rate) between the transmitter and receiver.

An SDIF connection is capable of handling a single channel of audio of up to 20 bits. Emphasis information is transmitted with the audio. The format also provides a facility for user data, although in practice this is unused. An SDIF connection uses 75 ohm coaxial cable with BNC connections.

AES/EBU transmission format sends one or two audio channels of up to 24 bits per connection along with considerable quantities of subcode information. The subcode indicates amongst other things sample rate, two channel mode and clock status. The AES/EBU format also transmits two additional bits of information with each audio word. The first is a *parity* bit, used to assess transmission link quality; the second is a *validity* bit intended to indicate whether the audio word is a genuine sample, or whether the sample has been generated by interpolating from other data.

The AES/EBU format is a balanced self clocking format designed to use existing audio cabling. The standard connection is via 110 ohm cabling with XLR connections. As AES/EBU format is self clocking, there is no requirement to send additional synchronizing information with the data. However, it is good practice to synchronize equipment using a dedicated clock, in preference to the data stream.[26]

SPDIF is an unbalanced variant of AES/EBU format, intended for domestic use, utilizing phono plugs for connection. The major difference between AES/EBU and SPDIF formats is in the subcode channel, the SPDIF standard having provision for copyright information in the form of ISRC data (International Standard Recording Code).

The MADI format is designed to connect up to 56 channels of digital audio using a single 75 ohm coaxial connection. MADI format is transparent to AES/EBU audio and subcode, which is, in effect, a subformat.

50.2.2 Digital audio equipment

50.2.2.1 Digital audio recorders

There is a wide variety of digital audio recording formats. Those currently in professional use are:

- DAT, a tape cassette based format
- Pro-Digi and DASH, two open reel formats available in a wide variety of tape speed, tape width and quantity of channel options
- Sony PCM 1630, which utilizes U-matic recorders and cassettes for recording and storage.

DAT[20,21] is a rotary head format, originally intended for domestic use. Two hours of stereo recording are available on a cassette measuring 73 × 54 × 10.5 mm. This combined with a track search speed of ×200, means the format excels at bulk storage and retrieval. DAT is capable of working at 48, 44.1 and 32 kHz. The format uses a 33 ms block interleave structure with Reed-Solomon error correction and 8–10 channel code.

The proposed professional DAT time code standard is base on an internal 33⅓ Hz rate recorded on the helical data track. This 33⅓ Hz time code must be converted to or from standard EBU or SMPTE format for input and output from the DAT machine. This inherent time code conversion leads to the possibility of DAT machines being compatible and synchronizable with both 25 Hz and 30 Hz video standards.

The advent of a time code format for DAT has led to the introduction of DAT editing systems, changing DAT from a simple storage medium to a full professional digital audio format.

Pro-Digi[22] and DASH[23] formats are both analogue machine replacement formats, both being designed for razor blade editing with operation in a manner familiar to analogue machine users. The range of machines available reflects the range of analogue machines available. Both Pro-Digi and DASH formats are capable of Vari-speed operation.

The Pro-Digi format is available in quarter-inch two channel, half-inch 16 channel and one-inch 32 channel variations. The DASH format is available in quarter-inch two channel, half-inch 24 channel and half-inch 48 channel variations. The main difference between the two formats is the choice of interleave structure.

Pro-Digi format groups eight digital audio tape tracks with two error correction tracks. This separation of data and error codes allows a complete dropout of a tape track to be fully corrected. However, it also means that if a single audio channel is to be recorded, all other tape tracks in the group must be re-recorded. Pro-Digi uses a block based interleave with Reed-Solomon error correction and 4/6M channel coding.

DASH format maintains a single track for a single audio channel. The interleave is convolutional, with parity and CRCC codes used for error correction. The channel code is convolutional HDM-1.

Sony PCM 1630/U-matic vtr[24] format is the accepted standard for compact disc mastering, the maximum cassette length of 75 minutes reflecting this use. The format records two channels of digital audio, modulated onto video carrier as the video signal on a 30 Hz U-matic vtr, along with time code and P and Q data (compact disc subcode data) on the vtr's linear tracks, which would normally be utilized for audio. Sample rates available are 44.1 and 44.056 kHz. There are two block interleave variations: standard, 210 sample, 35 TV line, for digital audio use, and extended interleave, 2940 sample, one TV frame, for CD-ROM applications. The error correction system uses parity and CRCC coding.

50.2.2.2 Record modes

With stationary head recorders there are two record modes:

- advance record, which uses a record head in advance of the playback head
- synchronous (sync) record, which uses a record head after the playback head (*Figure 50.52*)

Advance record mode is used for normal recording with confidence (off-tape) monitoring possible from the playback head. This is the same as *normal record* mode in a three head analogue machine.

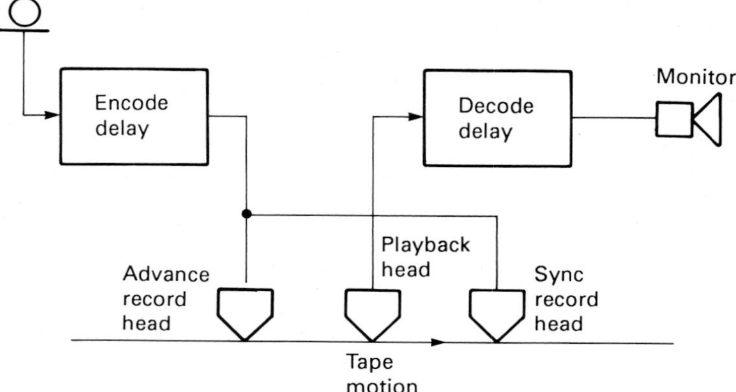

Figure 50.52 The use of sync record and advance record heads on a digital audio tape recorder. The sync record head allows recording to be made the same point on the tape from which the monitor signal was taken. The advance record head provides off-tape monitoring

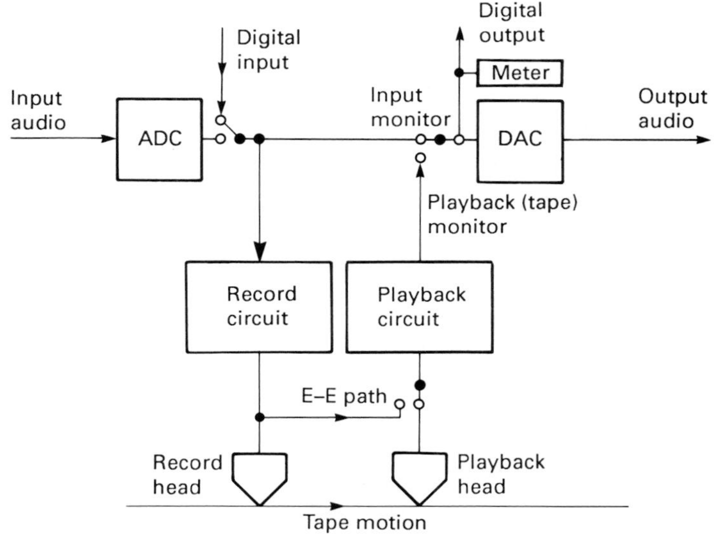

Figure 50.53 Monitor paths of a digital audio recorder

Synchronous record in a digital machine is the equivalent of analogue machine recording using the playback head, in order that audio monitored from other tape tracks is synchronous with the record signal. In a digital machine, the record head must be after the playback head to allow time for the playback audio to be decoded and the record audio to be encoded. The time taken for the tape to travel between playback and sync record heads must be exactly equal to the encode time plus the decode time.

The monitor structure of a digital audio recorder is shown in *Figure 50.53*. The input source, analogue or digital, is selected and fed both to the record path and to the monitor and meter path.

In a digital audio recorder there are three possible monitor paths from input to output. Firstly, as in a standard audio recorder, there is input monitor. In a digital machine this path will contain the analogue or digital input and output circuitry, which in the case of the analogue inputs and outputs contains the anti-alias filter, ADC and DAC. The input monitor path can therefore be used in the same manner as input monitor on an analogue machine and is additionally useful for isolating problems associated with the converters.

The second monitor path on a digital recorder has no real equivalent on an analogue audio machine, but is the equivalent of input monitor on a video tape recorder. This is the electronic to electronic (e-e) path which encompasses all the record and replay electronics with the exception of the record and playback head amplifiers and tape path.

The e-e path is not used during the normal operation of a digital audio recorder. Input monitor is provided, as in e-e mode the audio monitored at the output is delayed with respect to the input audio by the sum of the encoder and decoder delays. The real value of e-e monitoring is that it enables the user to isolate the tape path.

This is particularly useful in fault finding, when the majority of problems with digital audio recorders occur getting the signal on and off the tape.

The last monitor path is off-tape monitor, as in an analogue machine. In record mode, this monitor path will provide 'confidence playback' of the recorded signal. However, it must be borne in mind that this path will be considerably delayed with respect to the input signal, as not only has the audio been delayed by the time taken for the tape to pass between the heads, but also by the encode and decode time. The overall delay may be of the order of 200 ms.

50.2.2.3 Editing digital audio

50.2.2.3.1 Open reel editing

With the exception of razor blade editing, which is carried out in the same manner as for an analogue tape (but with a little more concern for care and cleanliness), there are two possible approaches; the assemble edit and the insert edit. Assemble and insert editing are the same as for video tape recorders, the difference between the two methods being the presence or absence of a previously recorded reference signal.

Assemble editing is used to build up a composite tape from start to end and does not require a tape with previously recorded signal (a pre-striped tape). An initial recording is made on the tape. A second portion of material is then assembled onto the first. The start of the second segment may be anywhere after the start of the first; the record machine will ensure that at the assemble point the new recording is correctly synchronized, also if under the recorders control, that the time code track is continuous (*Figure 50.54*).

This continuity of recording implies that the recorder has knowledge of the signals recorded on tape up to the edit point. This information is obtained during 'pre-roll', when the recorder plays back the recorded tape up to the edit point in order to synchronize tape speed and timing to the on-tape signal. The process of assembling further segments then continues until the recording process is complete.

Insert editing is only possible on a digital audio recorder when a tape has a previous recording on it, as insert editing requires an on-tape reference to synchronize the inserted portion correctly at both the edit-in and edit-out points. To insert edit a segment of material, it is necessary to inform the recorder of the edit-in and (possibly) the edit-out point; the recorder than pre-rolls the tape, in order to synchronize tape motion, and overwrites any previously recorded material on the chosen audio tracks (*Figure 50.55*).

50.2.2.3.2 VTR cassette editing

The edit resolution provided by recorders based on vtr formats is generally insufficient for normal usage, as edits are possible

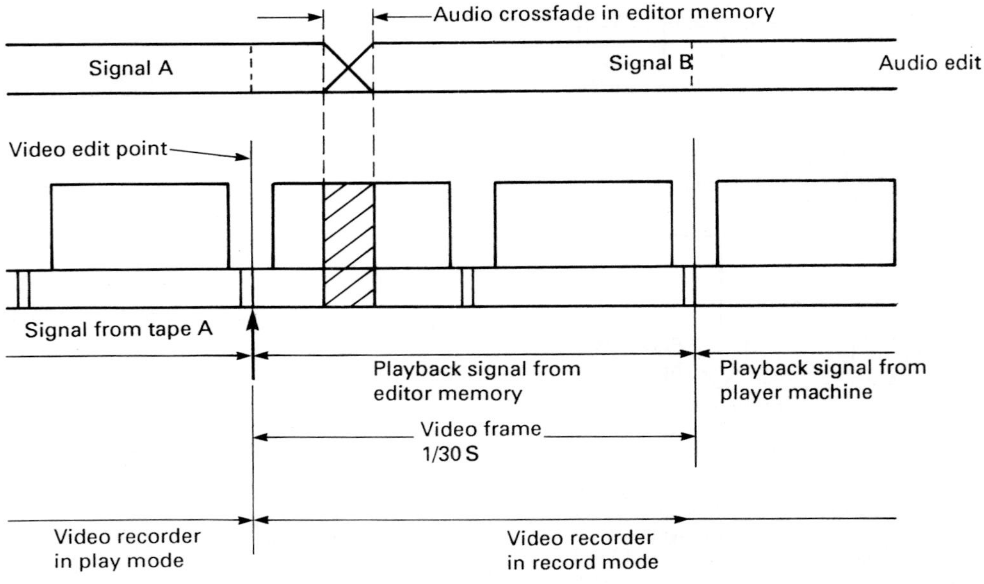

Figure 50.54 Assemble editing using vtrs and edit controller

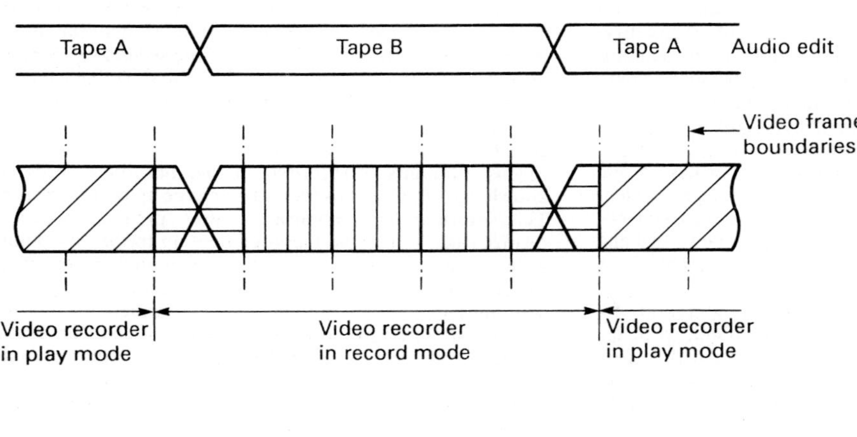

Figure 50.55 Insert editing using vtrs and edit controller

only at video frame boundaries, i.e. every 1/30 s. Only cut editing is possible. There is no possibility of a cross-fade between the two halves of the edit, so supplementary editors are used.

The digital audio editor makes editing possible within the video frame by storing audio from either side of the edit and, after rehearsal, performing the audio edit in memory. The editor assembles onto the recorder firstly the edit point from memory and secondly the subsequent audio from the player machine.

The process may be extended to insert editing. First, the edit-in and edit-out points are rehearsed in the editor memory. The editor will then insert the in point edit, the player playback material, and finally the out point to complete the audio insert. If the overall duration of the insert is not greater than the memory capacity of the editor, the whole insert can take place from the editor memory.

50.2.2.3.3 Hard disc editing

The final possibility for digital audio editing is to use a hard disc based system. When using a disc based editor, all the material to be edited is firstly placed into the editor and recorded onto hard disc. The editor will then provide a range of cut and splice, cross-fade, duplication and repeat facilities simulated in software.

The actual editing process is also simulated in software as the edited material comprises a list of markers to material and effects to be played back, rather than actual edited audio.

The advantages of using a disc based system are that the original material is left unaltered and that even complex editing can be very rapid. Access to edit points anywhere in the material is virtually instantaneous, there being no delays while tape winds forwards and backwards. The major disadvantage of disc based systems is that all material must first be played from the master tape into the editor and, once the editing has been completed, played out of the editor back onto tape. To store the edited material and edit decision list at the end of an editing session, the material must be dumped onto floppy disc or tape streamer.

50.2.2.4 Time code

Time code recording with a digital machine poses an extra set of problems as the time code and reference sync/word clock are both possible reference sources. If the digital audio recording is being used as though it were an analogue signal, i.e. only the analogue audio is being used, then it is of no consequence whether the time code and sample rate are locked together. However in more complex set-ups where, for example, the digital audio must be synchronized with video, then it is essential that the sample rate and time code are locked. This is normally done by locking the record time code to record sample frequency.

More complex situations can arise where, for example, a tape mastered at 44.1 kHz sample rate, locked to 30 Hz SMPTE time code, must be synchonized with 25 Hz PAL vtr pictures. Here some form of time code 'gearbox' must be used. In the Sony PCM 3402 (*Figure 50.56*), there are independent readers and generators for both tape and external time code all of which are converted to an internal standard form. This means that, for example, external reference time code can be different from on-tape time code, which can in turn be different from output time code, and all can remain locked.

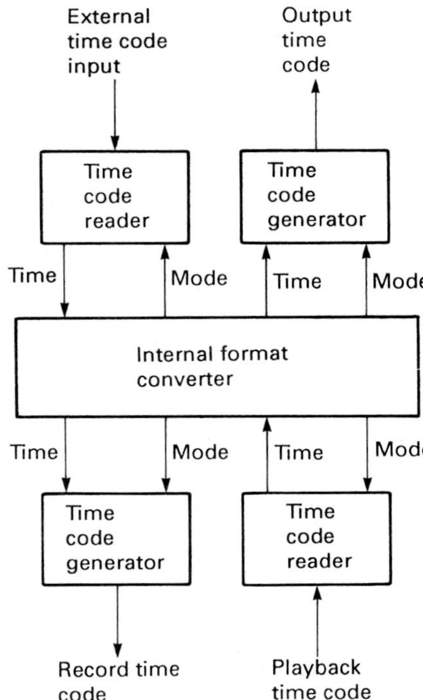

Figure 50.56 A multi-standard time code system which is capable of working with different time code standards simultaneously

50.2.2.5 Compact disc

Compact disc[25] is the most commonly known digital audio format. Two channels of digital audio data are recorded at a sample rate of 44.1 kHz for a maximum of 75 minutes, along with eight channels (P, Q, R, S, T, U, V, W) of subcode. The P and Q subcode data are used to indicate track running time. The R W channels are used in such applications as computer graphics, to accompany the recorded music.

Compact disc uses Reed-Solomon error coding with EFM (*eight to fourteen modulation* converts an 8 bit data byte to a 14 bit channel code, see section *50.1.2.7*).

50.2.2.6 Mixing consoles

The digital mixing console has one of two forms. Either the console appears, functions and behaves as a traditional analogue console, or the digital console has a truly innovative design. The differences between an analogue desk and its digital mimic are minimal. The desk has a fixed number of channels each with standard fixed functions, equalization, filters, gain, pan, etc. The alternative approach to the digital console is the *assignable* desk, where very few features or functions are fixed.

The assignable desk may have more audio channels than control channels. The audio channels to be worked on are assigned to the available control channels. If the audio channel is unassigned then all settings are maintained.

The controls of assignable consoles can also vary. A single control could be assigned as a high-pass filter or a low-pass filter. This variable console structure is made possible by the electronic design (*Figure 50.3*) which is not at all like its functional structure.

A feature of the computer based approach to console design is that, as the available processor power is shared between desk functions, the more equalization that is used the more processor power that is required. This allocation of processor power leads to the anomaly that the console may not be capable of supporting all its functions at the same time. However, the parallel processing approach allows the solution of adding more processing power if and when required.

50.2.2.7 Tape and disc analyzers

Digital tape and disc analyzers are used to assess the quality of a digital audio disc or tape, providing a printed reference sheet detailing the time, quantity and nature of any defects. An analyzer will only log defects in the reproduction occurring at the time of play so, if a defective recording is re-recorded onto a new tape or disc, any playback (hence loggable) defects will disappear, even though the audio defects are still present.

References

1 CARLSON, A B, *Communications Systems*, 299, McGraw-Hill
2 HOSHINO, Y and TAKEGAHARA, T, 'Influence of group delay distortion of low-pass filters on tone quality for digital audio systems', *Proc AES 3rd Inter Conf. Present and Future of Digital Audio*, 1985, 115
3 TAUB, H and SCHILLING, D, *Principles of Communication Systems*, 5.4–5.6, McGraw-Hill
4 RABINER, L and RADER, C (eds), *Digital Signal Processing*, B LUI, 'Effect of finite word length on the accuracy of digital filters — A review', IEEE Press, 361
5 POHLMAN, K, *Principles of Digital Audio*, 55, Howard W Sams
6 TAUB, H and SCHILLING, D, *Principles of Communication Systems*, 71, McGraw-Hill
7 CARLSON, A B, *Communications Systems*, 2.5, McGraw-Hill
8 RABINER, L and RADER, C (eds), *Digital Signal Processing*, IEEE Press
9 LIN, S and COSTELLO, D J Jr, *Error Control Coding: Fundamentals and Applications*, Prentice-Hall
10 BIRKHOFF, G and MACLANE, S, *A Survey of Modern Algebra*, 15, Macmillan
11 TOSHI DOI, 'Error correction for digital audio recordings', *AES Prem Conf Collected Papers* (1982)
12 BLESSER, B, LOCANTHI, B and STOCKHAM, T G Jr (eds), 'Digital audio', *AES Prem Conf Collected Papers*, 5 (1982)
13 DOI, T T, 'Channel coding for digital audio recordings', *AES 70th Convention* (1981)
14 JORGENSEN, F, *The Complete Handbook of Magnetic Recording*, 262, Tab Books
15 GIBSON, J J, 'A review of issues related to the choice of sample rates for digital audio', *SMPTE Digital Television Group* (1981)
16 POHLMAN, K, *Principles of Digital Audio*, 156, Howard W Sams
17 'AES recommended practice for digital audio engineering — serial transmission format for linearly represented digital audio data', *Jour Audio Eng Soc*, **33**, 975–984 (1985)
18 *Draft Standard for a Digital Audio Interface*, IEC (February 1987)
19 WILKINSON, J, EASTY, P, WARD, D.G. and LIDBETTER, P, 'Proposal for a Serial Multichannel Digital Interface', *AES and EBU paper*
20 The DAT Conference Standard, Digital Tape Recorder System June 1987, Electronic Industries Association of Japan Engineering Department
21 WATKINSON, J, *The Art of Digital Audio*, 8.10, Focal Press
22 ibid, 9.16
23 ibid, 9.4
24 ibid, 8.3
25 ibid, 13.1
26 'Synchronization of digital audio equipment in studio operations'

Part 10 Television Receivers

D G Thompson B Sc
Philips Components

51 Basic Receiver Design Principles

In most developed countries, the majority of television receivers are colour sets, with a minority of portable monochrome or colour receivers for use as second sets. In spite of the demand for a low cost product in a competitive market, the need to operate well under diverse conditions, including poor signals, varying supply levels, high and low temperatures and ambient illumination, has led to a highly sophisticated performance as the norm even in the monochrome field. For this reason, this survey of receiver design principles emphasizes the requirements for colour receivers, but is inclusive of most monochrome concepts.

The main colour concepts divide quite naturally into several levels of cost and sophistication. Although some manufacturers have produced a range of models based on a number of specific chassis designs, there is a strong tendency to begin with one basis chassis type covering a range of picture tube sizes, to which selected additional features such as a teletext decoder, remote control or electronic tuning may be added by incorporating additional components or modules.

At the upper extreme, receivers of the 'flagship' level are designed from the start with features such as large screen, 110° deflection with its proportionately low cabinet depth requirement, stereo sound, computer controlled tuning and teletext, and baseband peripheral television interfaces for video cassette recorders or home computers.

In continental Europe, a further requirement is for receivers to be used on signals from neighbouring countries having differing transmission standards and signal coding.

Provision for multi-standard operation may entail allowance for:

- transmission in other bands than local service
- different sound modulation methods
- reversal of vision and sound carrier positions in if spectrum
- different channel width and channel spacings
- different vestigial sideband overlap
- different group delay characteristics within the pass-band
- different frequencies for adjacent-channel traps
- positive or negative polarity of vision modulation
- different colour encoding system — PAL, SECAM or NTSC
- different scanning rates (625-line 50 Hz field or 525/60)

51.1 Receiver performance requirements and structure

For good picture and sound quality, an antenna signal greater than 1 mV emf is required, but the quality of the antenna installation may determine the effects of:

- ghosts or rings due to reflections from large physical objects interfering with the propagation path
- ghosts, rings or distorted frequency and phase response due to impedance mismatch of antenna downlead to tuner input
- signals injected into the receiver by routes other than the antenna input
- interference from electrical apparatus operating in the vicinity of the receiver and radiated directly or via the mains cable

Most receivers are designed to continue functioning on signals down to below 10 μV emf despite the probable degradation of picture quality due to noise, ghosting and spectral distortion which can affect sound and synchronization of some channels. In addition to the high sensitivity required in these conditions, great care is needed in the internal chassis design to avoid radiation from the receiver's own circuits being picked up by the antenna.

The functional diagram in *Figure 51.1* shows the essential parts of a television receiver. The signal passes through the front end, which converts the modulated rf signal into two paths: the demodulated baseband video signal which is processed by the colour and teletext decoders and used to modulate the picture-tube beam-current, and the am or fm sound-carrier signal which is demodulated, amplified and applied to the loudspeaker. The video signal is also applied to the synchronizing pulse separator which provides the basis for a correctly timed, stable raster, generated by the line and field timebases.

The high voltage and heater supplies for the picture tubes are usually obtained from the line output stage, while the power supply for the receiver provides a stable, smoothed high voltage supply for the line scanning circuit. The low voltage supplies for the small-signal and control circuits may be derived either from the main power supply or from the line output circuit.

51.1.1 Receiver front end

The front end selects the desired channel and reproduces the baseband video and audio signals from the carrier modulation. It comprises a vhf or uhf tuner, an intermediate frequency signal processor for the vision and sound signals, and a tuning control system. The tuning and amplification functions are optimized by automatic frequency control (afc) and automatic gain control (agc) systems associated with the vision demodulator.

51.1.1.1 The tuner

The tuner is a specialized sub-assembly built within a metal screening box. Its functions are outlined in *Figure 51.2*. It provides:

- rf amplification at the frequency of the received channel
- gain control to avoid overloading of subsequent stages
- selectivity to prevent cross-modulation and intermodulation effects from unwanted strong signals
- frequency conversion to apply the received signal to selective circuits having the precise responses required for demodulation of the vestigial sideband (vsb) vision signal and of the sound signal
- if amplification adapted to the insertion loss of the if selective circuits
- a frequency-divider circuit sampling the local oscillator frequency, for pll (phase locked loop), or frequency-synthesized electronic tuning systems

Because of the technical difficulty and development cost for the design and manufacture of tuners, covering many different

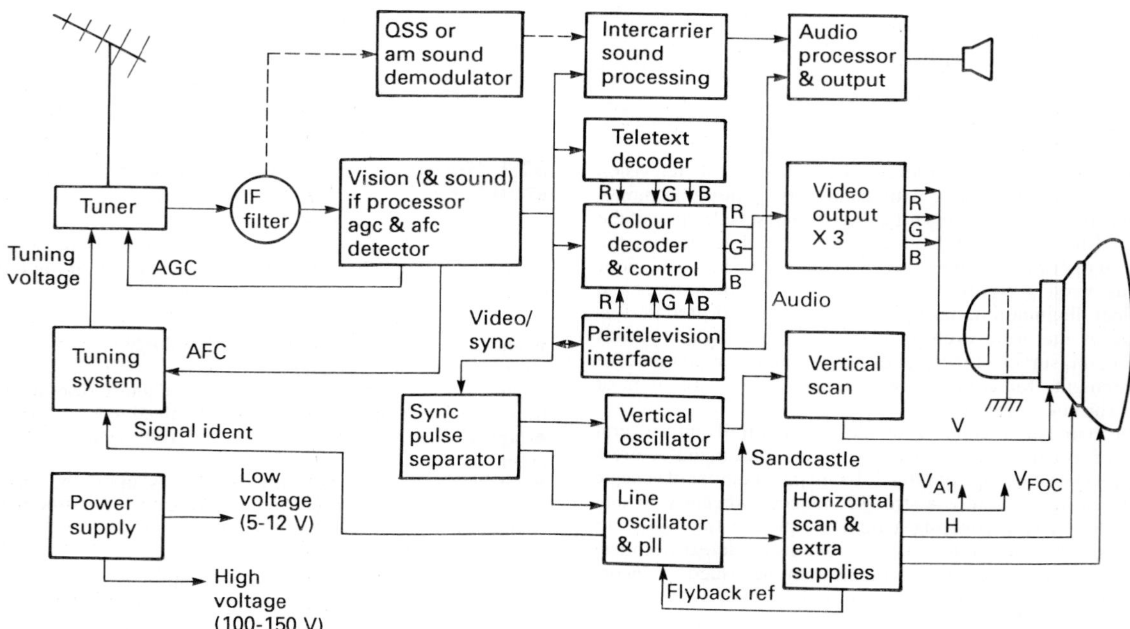

Figure 51.1 Functions of a television receiver

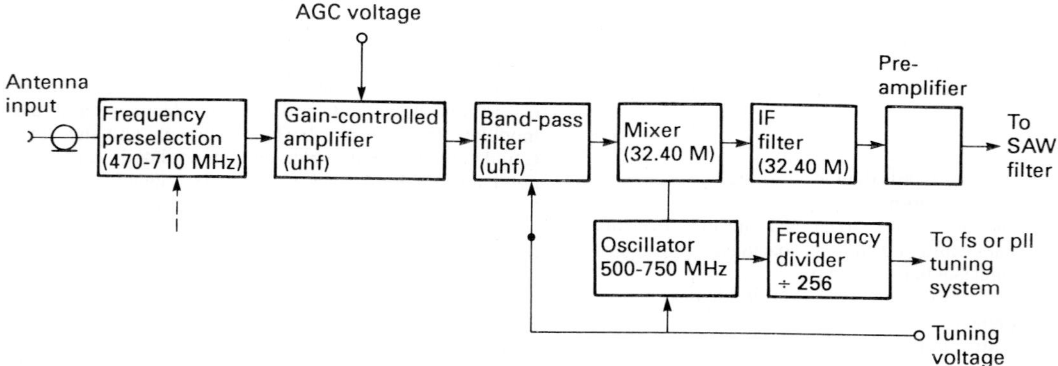

Figure 51.2 Outline of uhf television tuner with pre-scaler

requirements for good performance in different areas, most set manufacturers buy tuners from specialist manufacturers. The example shown in *Figure 51.3* uses surface mounted components on the copper side of its substrate board, and this requires no alignment or adjustment after mounting in the receiver. Channel tuning is effected by varicap diodes controlled by a tuning voltage derived from the tuning control system.

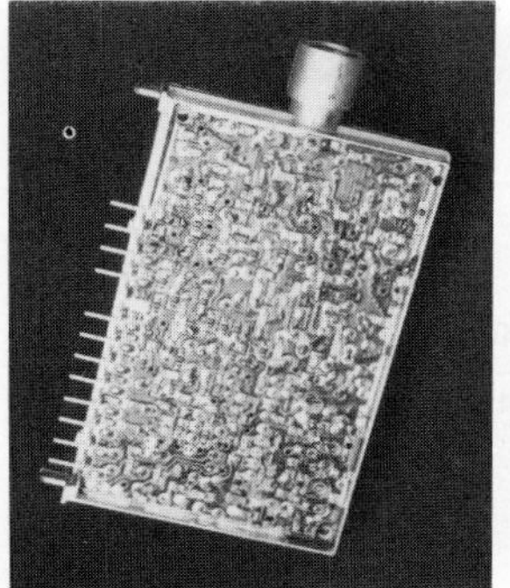

Figure 51.3 Constuction of a modern television tuner

For terrestrial television transmissions, the rf signal comprises an rf carrier which is amplitude modulated by the vision modulation signal, and a sound carrier which is placed outside the range of sidebands arising from the video modulation. To minimize the bandwidth required for each channel, the high frequency sidebands are filtered out from one side of the vision carrier frequency at the transmitter, leaving only a vestigial sideband on this side of the carrier. This vestigial sideband (vsb) is either 0.75 MHz or 1.5 MHz in width, depending on the national system specification. The higher value applies for systems H, I and L.

The application of frequency conversion to the rf signal is illustrated in *Figure 51.4*. As is usual, the local oscillator frequency is placed on the high side of the rf signal. The if signal, resulting from the difference-frequency components of the mixing process, appears as a vsb signal with the sound carrier and main picture components below picture carrier frequency.

In receivers designed for French system L, however, the local oscillator is usually placed below the wanted transmission frequency. This gives an if signal with the main components above picture carrier, since the French national receiving specifications place the vision and sound if carrier frequencies at 32.7 and 39.2 MHz respectively. But the lowest vhf channels in Band 1 occur just above the if band; on these channels, therefore, the local oscillator frequency can only be higher than the wanted transmission. So in France, the vhf Band 1 channels (only) are transmitted with the main components below the picture so that they can appear at if at the same frequencies as with the other channels. Thus the receiver's local oscillator operates above the received carrier in Band 1 and below on Bands 3, 4 and 5.

In a multi-standard receiver, for use around the French borders, normal System B and G vision intermediate frequencies and vision-to-sound orientation can be used for the high-frequency channels, with the local oscillator operating on the high side of the incoming signal. For the Band 1 case, however, the picture and sound carriers are now reversed. This problem may be resolved in three ways:

- using an additional transposing up-convertor unit which re-creates an if signal with the vision and sound carriers reversed
- providing a dual-purpose if processing channel suitable for vision carriers placed at the high side or low side of the response curve
- avoiding the problem by limiting the receiver to use in areas not covered by this format

The responses in *Figure 51.5* show the essential relationship between the signal spectrum (broken line) and the if selective response. The selective circuit (shown solid) has its 50 per cent transmission point set at vision carrier frequency (the -6 dB Nyquist condition). The upper and lower sidebands tend towards zero and 100 per cent transmission respectively. This ensures a flat frequency response for all modulation components over the transition from double to single sideband characteristics in the transmitted signal.

The mixer may comprise a bipolar transistor, a field effect transistor (fet), a Schottky diode, or be incorporated in a tuner IC. The output signal level is related to the if filter insertion loss; frequently an additional gain stage is included prior to this filter to ensure adequate signal/noise ratio in the if processor.

Tuners may be designed for vhf, uhf, hyperband (certain

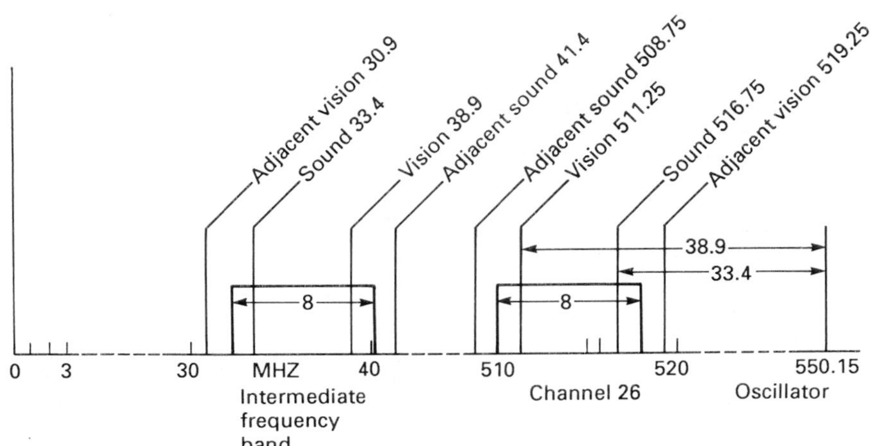

Figure 51.4 Frequency conversion applied to a television channel

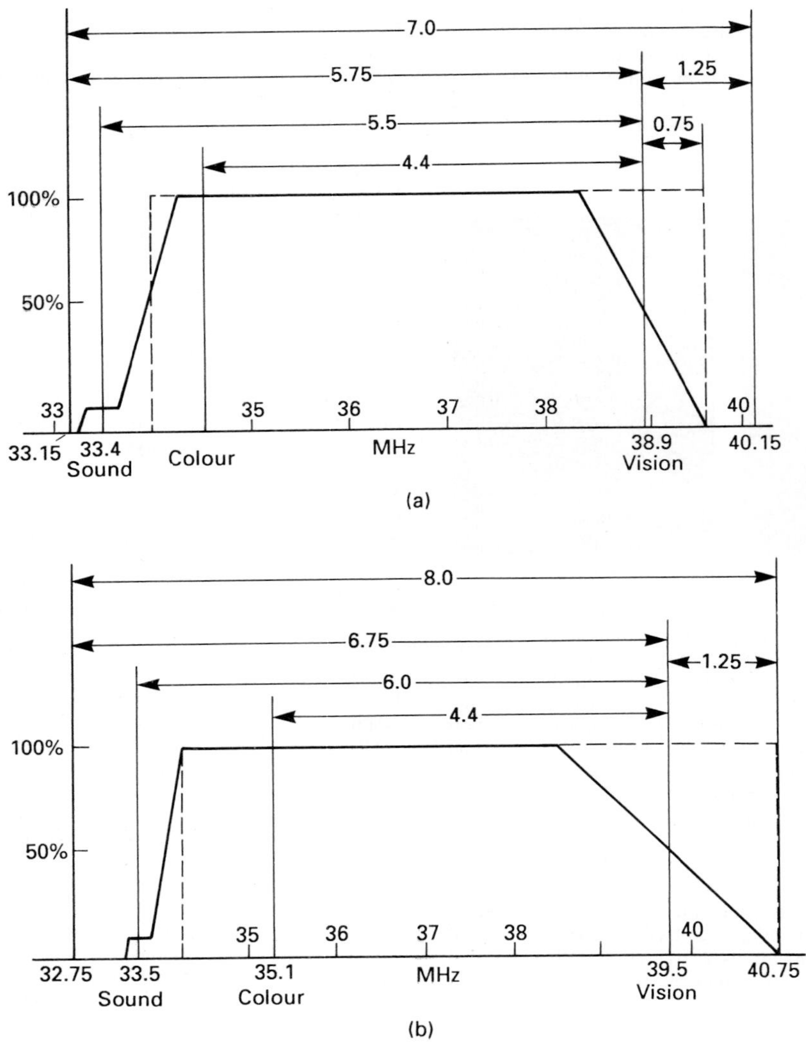

Figure 51.5 Nyquist responses for demodulation: (a) for a 0.75 MHz vestigial sideband, (b) for a 1.25 MHz vestigal sideband

channels allocated for cable TV), or a combination of these bands. Thus a receiver may incorporate a multi-band tuner or individual tuners with the option of separate antenna feeds. Multi-band tuners include circuits for splitting the antenna signal into the separate bands required by the separate internal front end sections, required for optimal processing of these signal frequencies. A frequency multiplexer may be provided at the antenna site to combine the signals picked up on different antennas onto a single downlead.

51.1.1.2 Tuning control system

The tuning control system may provide three outputs:

- the tuning control potential for varicap diodes which are the main tuning elements within the tuner
- band-switching outputs where required, to set the tuning range of the tuner
- a 'mute' signal to silence the sound, and also to disable the automatic frequency control (afc) during channel changing, setting up

Three types of system are used:

1 mechanical potentiometers, one for each channel, each associated with a band-selecting switch, both being activated as required by a set-front pushbutton or an electronic switch in the case of remote control
2 electronic voltage-synthesis (vs) systems in which the required tuning potential for each channel is converted from data stored in a digital non-volatile memory by a D/A convertor
3 electronic frequency-synthesis (fs) systems which measure and control the oscillator frequency with reference to a set of standard frequencies stored in a read only memory (ROM); a non-volatile memory loaded by the user relates a set of program numbers to the preferred channels

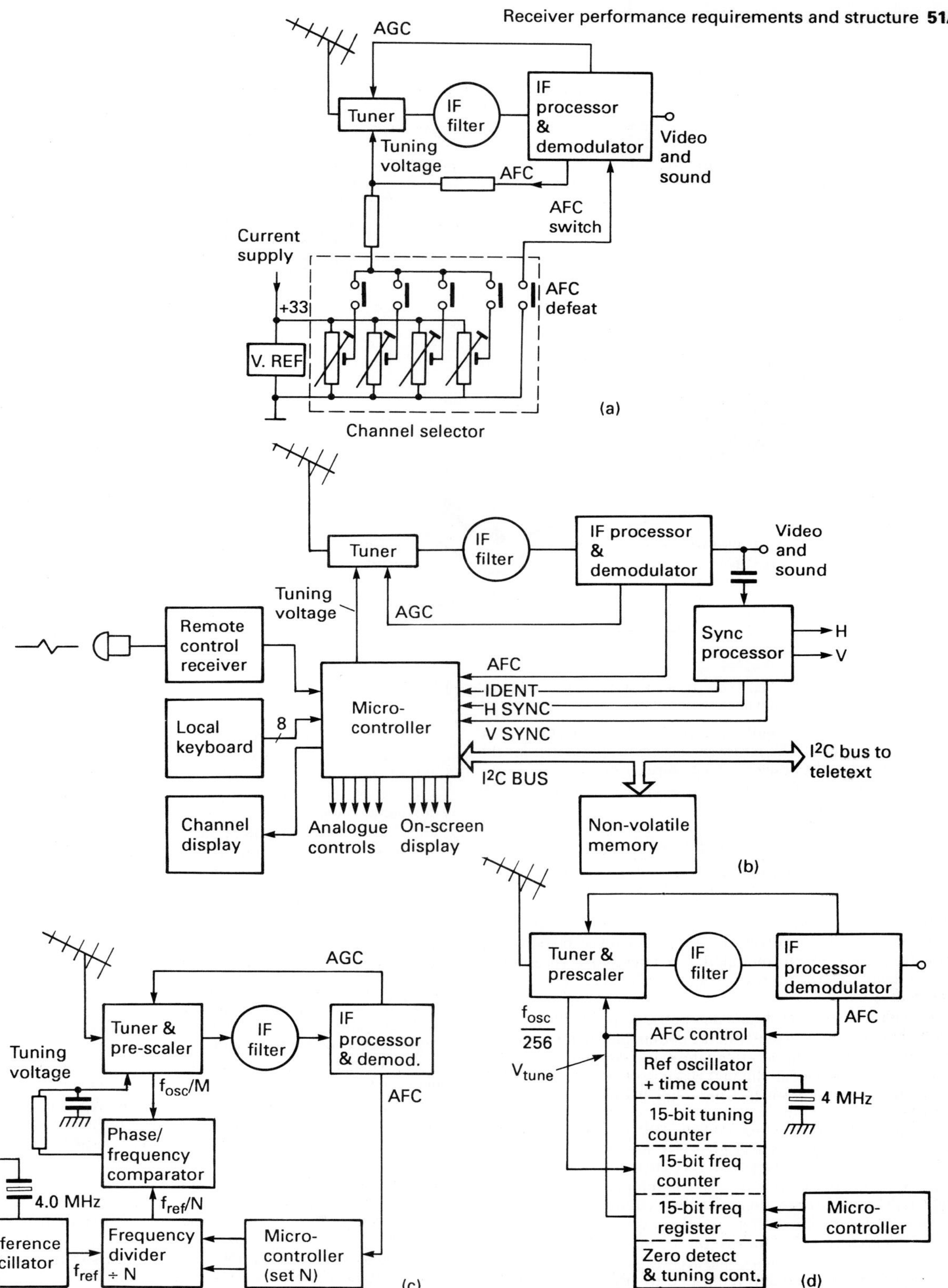

Figure 51.6 Electronic tuning systems: (a) switched potentiometer channel selction, (b) computer controlled voltage synthesis, (c) principle of pll frequency synthesis, (d) computer controlled frequency synthesis

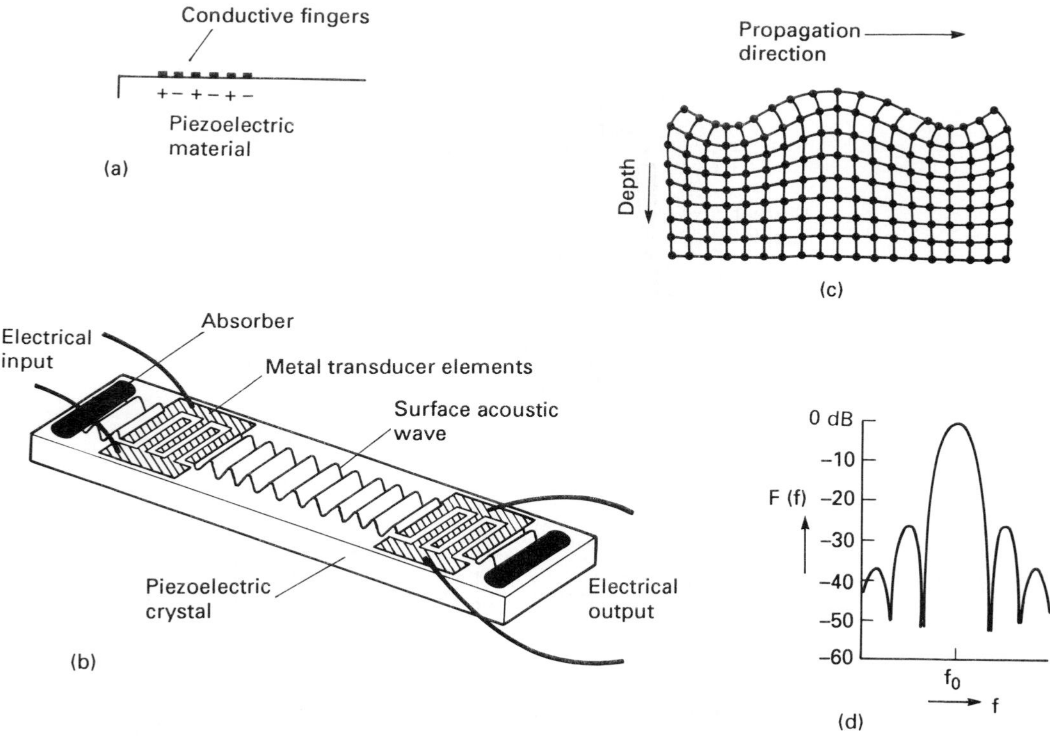

Figure 51.7 Principle of surface acoustic wave filter: (a) transducer principle, (b) construction, (c) wave motion in surface of crystal, (d) amplitude response of a SAW filter with uniform transducers

The first two types require initialization by the user using the transmitted signals. The fs system can be programmed for the desired signal in the absence of a picture carrier, except for the video cassette type of source which does not accurately conform to a standard frequency.

51.1.1.3 IF selectivity and demodulation

Prior to the main if amplification and demodulation, the principal selective circuit is interposed. Its requirements are:

- shaping of response in vision carrier region to conform to Nyquist condition for correct amplitude and phase response for low frequency video modulation components (prescribed group-delay characteristics vary with national system standards)
- providing a level amplitude and phase response through the entire pass-band particularly in the colour subcarrier region
- providing the required attenuation at if frequencies corresponding to the adjacent-channel vision and sound carriers and all other unwanted products from other channels

These requirements may be met with a block filter using lumped components or with a surface acoustic wave (SAW) filter which provides a response very close to the ideal, at the expense of greater insertion loss. The lumped filter requires a high-order configuration to realize the required responses, but in view of the precision required, necessitating alignment of many components by means of a complicated alignment procedure, it does not compete in accuracy with the SAW filter, which can be designed and replicated to any desired order of excellence.

The SAW filter consists of a piezoelectric substrate wafer on whose surface a mechanical wave motion produced by the input signal is set up by means of an array of transducing and wave shaping elements. The substrate material, its dimensions, the axis of cut of the wafer in relation to the crystallographic axes, and the pattern and dimensions of the transducer array all determine the precise frequency response, group delay characteristics and insertion loss of the filter.

The basic response produced by a uniform array can be modified to derive the amplitude and phase response characteristics required for television if processing by manipulating the number of fingers and their dimensional pattern, using computer aided design methods. The manufacturing process is analogous to integrated circuit technology. The resulting performance is highly predictable and so precise that no alignment adjustments are required.

SAW filters are also available for nearly all the world systems, and are easily matched to the ideal impedance requirements of the tuner and if signal processor.

The characteristics of SAW filters for system I from two manufacturers are shown in *Figure 51.8*. The logarithmic scale displays the performance in the important stop-band, and the accepted performance limits are shown to be met in both cases.

The group delay characteristics of the transmitted signal are prescribed by each national broadcasting authority, and should be matched in the receiver's if design. This requirement is readily satisfied by appropriate SAW filter design for a given country, but it requires some compromise in a multi-standard receiver.

The functions of a typical if signal processor are shown in *Figure 51.9*. The signal passes through a three stage gain-controlled amplifier into a synchronous demodulator. Its reference signal can be derived from a 'tank' circuit tuned to vision-carrier frequency, which is excited by the amplified

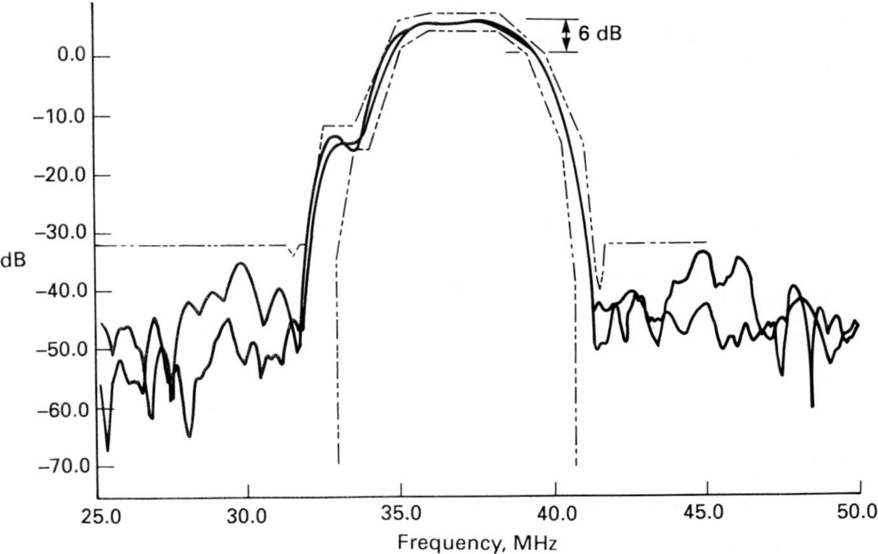

Figure 51.8 Frequency responses of two SAW filters for system I

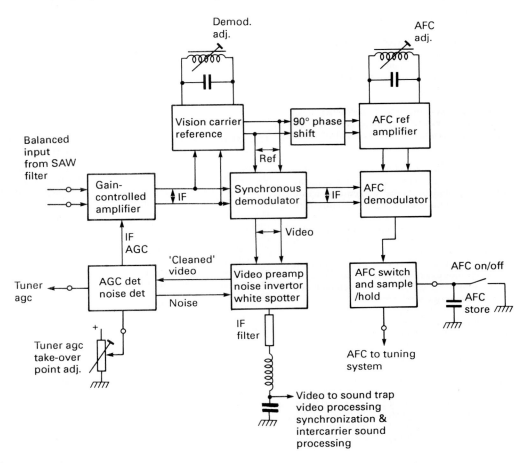

Figure 51.9 IF signal processor outline

signal. An amplitude limiter ensures that it operates constantly over the full range of carrier modulation levels.

The synchronous demodulation process consists of multiplication of the whole if signal with the derived carrier signal. This process is more sophisticated than the simple envelope detector used in the past and has performance advantages in the following areas:

- lower quadrature distortion resulting in improved transient performance, of particular benefit for colour and teletext
- improved signal/noise performance, 3 dB better than an envelope detector
- inherently good linearity particularly for low signal strengths
- lower differential gain and phase distortion

51.1.1.4 AGC, afc and signal polarity

To obtain consistent operation over the full range of signal strengths, the agc system detects the maximum carrier level, which is used to control the gain of the tuner and if amplifier stages. For negative modulation, the peak carrier amplitude occurs during the sync pulse tips. Peak detection of the sync tips is therefore a satisfactory method provided sporadic noise spikes are not permitted to disturb the detected level, leading to excessive gain reduction.

Three precautions are commonly applied:

- detection of noise spikes and their cancellation in the output video waveform
- time-gating the agc detector, using synchronized line keying pulses
- avoiding too fast a charge time-constant in the agc amplitude detector, thus avoiding excessive response to short duration pulses

For positive modulation systems, the peak carrier level corresponds to peak white, whose amplitude and duration depend on picture content. A choice of three types of system is available:

1 mean level agc in which the demodulated video signal is integrated, and the resulting dc potential used for agc; the resulting level is highly dependent on the momentary picture demodulation, leading to continuous changes in picture contrast and sync-pulse amplitude
2 peak white detection, which is less dependent on picture content but must operate on short-duration signals, and is vulnerable to noise spike interference
3 black-level detection, a means of maintaining a constant sync/pulse amplitude; it requires a reliable method for sampling the blanking level in the sync waveform from the demodulated video, and is therefore the most complex solution

The proportions of agc applied to tuner and if gain-controlled stages are arranged so that the maximum allowable signal is applied to the if processor before the tuner's agc comes into operation. This 'delayed' agc is developed within the if processor, and is subject to a specific adjustment in each receiver to accommodate spreads in tuner gain, SAW filter insertion loss and in the transfer characteristic of the processor. The relationship between signal levels, agc curves, and noise performance for a typical receiver is illustrated in *Figure 51.10*. Setting the tuner's agc take-over point to a high input level increases the signal/noise ratio, but this may reduce the ability of the agc system to prevent overload on strong input signals.

It is necessary for the frequency of the oscillator in the tuner to be maintained to within a tolerance of +0.2 MHz -2.0 MHz. This is to comply with statutory radiation requirements. For good colour and sound signal reception, a tolerance of ±100 kHz is accepted while a more severe requirement of ±50 kHz applies for receiving teletext. This entails the use of an automatic frequency control (afc) system, which operates by producing an error signal derived from a measurement of vision carrier frequency which is added to the tuning potential applied to the tuner.

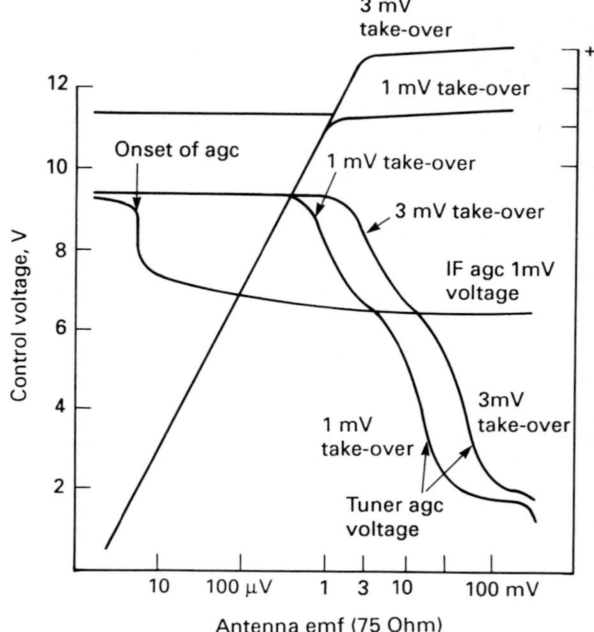

Figure 51.10 Automatic gain control for tuner and if amplifier. Effect of take-over setting is shown for 1 mV and 3 mV levels at antenna input

The afc signal is obtained from a second synchronous demodulator operating with a vision carrier reference derived from the vision demodulator via a frequency dependent phase shifting network whose output is in phase-quadrature at the correct vision carrier frequency. This demodulator produces no output when the carrier frequency is correct. A positive or negative output proportional to the frequency error is produced when the signal is mistuned.

Some forms of afc detector are responsive to the video modulation of the signal. This can be overcome by keying the detector to act only during the line blanking time, using a sample and hold circuit to store the control voltage.

The sense of afc voltage required by the tuner's varicaps depends on whether the oscillator operates above or below the vision carrier frequency.

51.1.1.5 Components for video baseband and sound processing

Whereas the SAW filter is well suited to the bandwidths and range of frequencies of the vision if, the requirements for sound intercarrier and chrominance processing are more readily met by discrete circuit elements using fixed or adjustable inductors in conjunction with close tolerance capacitors and resistors. Alternatively, pre-trimmed component assemblies or ceramic resonators may be employed.

The capacitors are available with temperature coefficients which may be specified to compensate for the inevitable behaviour of other components and to realize an overall circuit which is stable with temperature. Variable inductors use an adjustable magnetic core located within a screw thread, which varies the permeability of the magnetic path around the winding.

Ceramic filters make use of the piezoelectric effect in combination with mechanical resonances of a polycrystalline ceramic material which has been electrically polarized by a strong dc electric field applied during cooling and solidification.

Various modes of vibration are possible, and the required mode is selected in the mechanical construction such that the spurious responses corresponding to the unwanted modes are well removed from the required frequency range and may be suppressed by additional circuitry, or ignored.

Over its intended frequency range, the ceramic resonator provides a close equivalent of the formerly used discrete circuits, and it can be supplied with close enough tolerances to obviate alignment adjustments.

51.1.1.6 Sound signal processing

There are several distinct methods of sound modulation employed in the different national transmission standards. These require quite different techniques and architectures in the receiver's signal processing system.

Consideration here is limited to the terrestrial forms of monophonic and stereo transmission, because equipment for reception of direct broadcast by satellite (dbs) takes the form of set-top adaptors.

Monophonic signals may be carried by amplitude modulation (am) or frequency modulation (fm) of the sound carrier. For a dual sound or stereo service, the extra channel may be carried by additional modulation components on the sound carrier, or by introducing another sound carrier at a different frequency within the allocated channel space.

The systems which apply additional modulation to the sound carrier permit realization of stereo, dual language operation, a low capacity data system or an additional voice quality channel.

The most extended facilities are available with the use of an additional subcarrier with digital modulation, such as the NICAM 728 system.

The single carrier format is used with systems M and N in the USA and Japan, while additional sound carriers are used in the German Zweiton system and in the digital NICAM 728 systems. Details of these formats are given in section *51.5*.

The spacing between the monophonic sound carrier and the vision carrier is precisely fixed for each system, and one product of the vision demodulation process is a subcarrier at the intercarrier frequency which contains the sound signal modulation. For a simple monophonic fm sound modulation, this sound intercarrier is suitable for processing and demodulation. It may be extracted from the video signal by means of a selective circuit or ceramic resonator following the video demodulator. The signal is then amplified, limited and demodulated by the chosen form of fm demodulator (see *Figure 51.11*).

For am sound and for stereo fm subcarriers, the intermodulation requirements are much more stringent, and the sound channel processing is usually split off from the vision ahead of the if processor. An am sound demodulator will respond directly to all amplitude modulation of the carrier, and an intercarrier produced by mixing the vision carrier and sound signal would be heavily modulated with the video signal, unless the vision carrier has been very cleanly separated from its vision modulation components.

Direct demodulation of an am sound if signal, as shown in *Figure 51.12*, is customary but, because of its narrow bandwidth, very accurate tuning of the tuner's local oscillator is required.

An intermediate format, known as quasi-split sound (QSS), separates the sound signal from the vision if signal path ahead of the vision if processor by using an if filter which provides separate ports for the vision and sound processing channels. The response for the vision channel has the benefit of a high attenuation at sound carrier frequency. The vision carrier, minus its high frequency modulation components, is still required for the generation of the carrier reference used in the

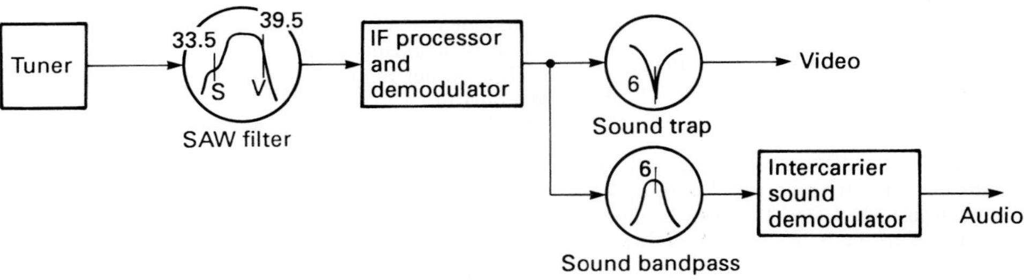

Figure 51.11 Conventional fm sound processing

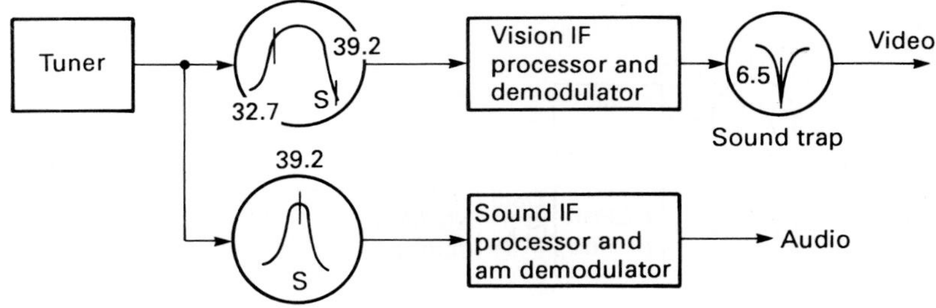

Figure 51.12 AM sound processing in system L

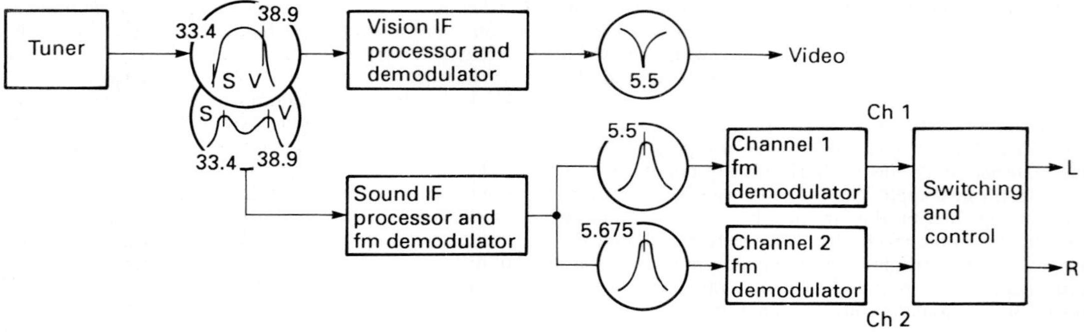

Figure 51.13 Quasi-split sound method applied to German dual-sound system

synchronous sound carrier demodulator, and so the sound channel response has symmetrical peaks around the sound and vision carrier frequencies.

A further feature of this system is that the vision carrier reference used in the sound carrier demodulator is in phase quadrature with that which would be required for vision demodulation, and this helps rejection of the lower frequency video components. It meets the requirements for the German stereo system and for NICAM digital stereo, but is not generally applied for am sound processing.

SAW filters are available in versions which meet the requirements for QSS for the various national systems, including some which cover the extended band-pass requirements for digital NICAM sound.

The quasi-split sound system provides the following benefits:

- filter selectivity reduces response to high frequency video modulation
- filter transfer characteristic is phase-linear in vision and sound carrier regions, which avoids detection of spurious noise or video modulation components arising from local oscillator frequency inability or 'pulling'
- intercarrier demodulator is made insensitive to vision carrier modulation by using an amplitude limited reference signal in quadrature with the vision carrier, thereby cancelling low frequency vision modulation which occupies the double-side-band part of the channel spectrum

For the fm sound systems, a 6 dB/octave de-emphasis is required at the demodulator output to correct for the pre-emphasis applied to the transmitted signal, having a time-constant of 50 μs for most systems. Systems M and N, however, apply 75 μs.

The NICAM decoder system shown in *Figure 51.14* operates in parallel with the conventional monophonic decoder, to allow for situations and sources in which the NICAM signal is not provided. It provides:

- band-pass selective circuit centred on 6.552 MHz in system I or 5.850 MHz in systems B and G
- differential quadrature phase-shift keyed (dqpsk) demodulator
- NICAM demodulator
- digital/analogue convertor
- channel selection switch

The band-pass circuit provides a high attenuation for video and fm sound intercarrier components to minimize interference to the digital symbol recovery. The dqpsk demodulator uses a phase-locked loop (pll), which synchronizes the sine and cosine reference carrier signals for quadrature demodulation of the dqpsk carrier input. Two parallel bit streams are produced in this process, and are converted into the NICAM serial data stream.

The NICAM decoder reformats the demodulated data, and performs the required de-scrambling to restore the action of energy dispersal applied to the transmitted signal, or to decrypt conditional access signals if required. Error correction and 10 bit to 14 bit expansion are also applied at this stage.

The output of the NICAM decoder is in standard stereo digital form, with a 32 kHz clock rate, and this is converted in a standard DAC as used in compact disc players. The filtered outputs are then fed to a selector switch, which is controlled by user instructions and by control bits received with the signal which identify the type of service being transmitted.

Audio signal processing functions may include:

- muting function when no valid signals are received
- source switching when peri-television sources are connected

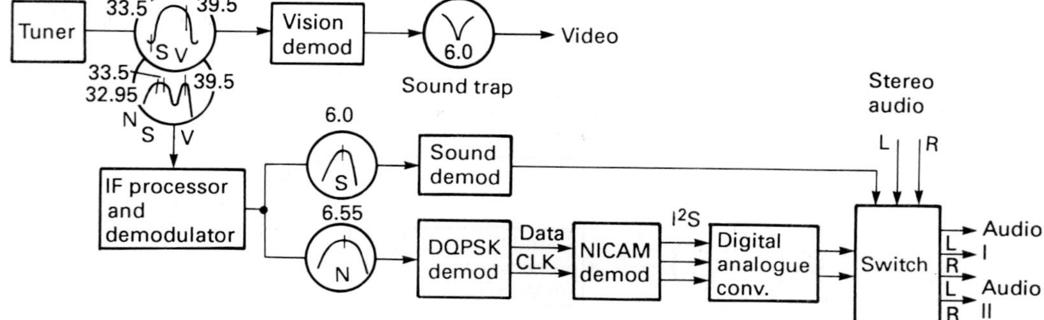

Figure 51.14 Processing system for NICAM digital sound

- stereo or multi-language channel selection
- dc controlled volume control, tone control and stereo controls for connection with remote control system

Some further requirements of the audio system in a television receiver are:

- power supply should have low common impedance with video and scanning circuits to avoid moving patterns on the screen due to supply loading
- loudspeakers should have low external field to avoid disturbance of picture geometry and colour purity

51.1.2 Picture tube drive requirements

Two essential features of a receiver depend on the picture tube driving conditions. These are maximum contrast and resolution of detail.

The *perceived contrast* depends on the ratio of maximum tube illumination to the residual illumination caused by ambient light when the beam is cut off. The maximum illumination depends on:

- maximum beam current
- picture tube screen size
- phosphor efficiency
- faceplate transmission factor

Residual illumination, which is an unwanted detraction from picture quality, is dependent on ambient lighting and picture tube design. Unfortunately, the factors which provide low residual illumination (good blacks) also reduce the wanted illumination, and therefore higher beam currents are demanded.

Apart from power considerations in the receiver, the maximum usable screen illumination is limited by the onset of visible flicker, which is strongly related to field frequency; the 60 Hz scanning rate gives a six-fold advantage over 50 Hz systems.

The characteristics of currently used picture tubes require 30–100 V of video drive on the cathodes for full modulation.

51.1.2.1 Customer controls

The functions of the 'monochrome' controls (brightness and contrast) are best understood from consideration of their ideal implementation. The *contrast* control sets the video drive amplitude in relation to a fixed picture black level. Deviations from the ideal black level for low key scenes or within the whole grey scale are corrected with the *brightness* control.

In principle, these adjustments should be made once and for all when the receiver is installed, but readjustments may be necessary from time to time for two reasons:

- Estimation of picture black is subjective; it may vary with different programme material, and is anyway specified differently in relation to blanking level on different transmission systems or signal sources (e.g. vcr machines connected to the receiver).
- The optimum condition for visual cut-off depends on ambient illumination which will vary from time to time.

Unwanted variations in black level may also occur in the receiver circuits if the dc component of the signal is not fully maintained. This may apply in some monochrome receivers, but in colour receivers one or more clamp circuits operate as dc restorers during the back-porch or line-blanking period.

The brightness control function essentially consists of varying the level to which the clamp circuit is referred, thereby moving the entire video waveform at the picture tube drive electrode. Unless high level clamps are applied at the tube drive-electrodes, part of the video stage drive capability must be reserved for this function.

Another method, applicable in monochrome receivers, is to vary the potential of a second tube drive-electrode on the picture tube.

Saturation control adjusts the relative amplitude of colour difference signal with respect to luminance. It operates by controlling the amplitudes of the demodulated R-Y and B-Y signals. For its effect to appear constant, a coupling of the contrast control setting is required so that the colour-difference signal tracks with the luminance amplitude as the contrast control is adjusted.

Hue control is applied in NTSC receivers to counteract subcarrier phase-errors and is applied in that case as a phase adjustment to the reference signal. If applied in a PAL or SECAM decoder, it can comprise a change in the colour drive ratios.

51.1.2.2 Beam current limiting

It is necessary to provide a supervisory control function to limit the picture tube drive current to a predetermined level to avoid several problems:

- excessive demand on the power supply and overstress of some components
- large area flicker
- local optical disturbances (possible defocusing or colour purity loss in some colour picture tubes)

To make the action of beam limiting inconspicuous, the control should be applied to the contrast control function. This does not disturb the tonal scale in the critical black area, nor produce colour changes in a colour set. As it accompanies the changes in picture content which provoke its action, the eye accepts it as natural. However, excessive beam current is also likely when the black level is set too high, and no amount of contrast reduction can then reduce the beam current adequately. Some receivers therefore use a two-stage beam limiting strategy whereby brightness reduction is applied in succession after contrast reduction has reached its limit.

Three methods of sensing beam current are:

1 measuring picture tube drive voltage
2 measuring picture tube cathode current
3 measuring eht supply current

Method 1 may be applied indirectly in the signal processing IC by applying contrast reduction whenever one of its outputs attempts to exceed a predetermined level. The operating conditions of the video circuits are then set so that this limiting level corresponds with the specified maximum.

51.1.2.3 Flashover protection and electromagnetic compatibility

Part of the general topic of electromagnetic compatibility (see section *66*), which impinges on many parts of the receiver circuitry, is the strategy for avoiding problems arising from mains spikes and sporadic flashovers in the picture tube. Flashovers produce transient disturbances on a tube electrode, comprising a high amplitude, fast rising pulse of short duration.

It is possible, and essential, to divert such transients away from the small-signal circuits, and particularly any circuits involving digital memories whose contents may alter the receiver's mode of operation. Precautions are also required to

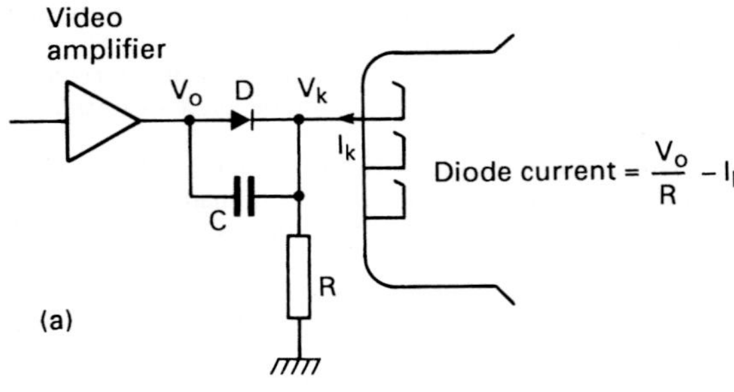

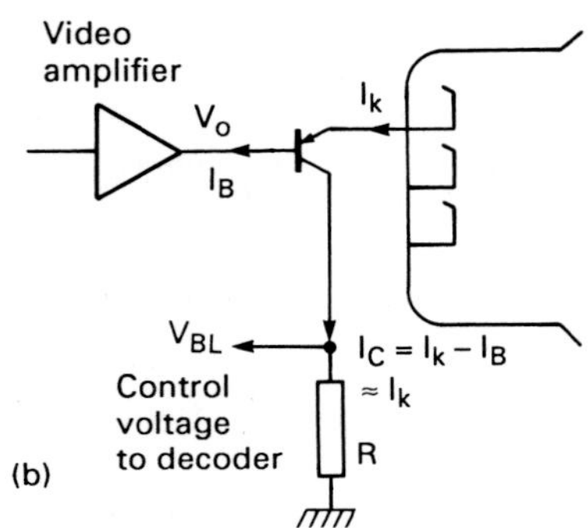

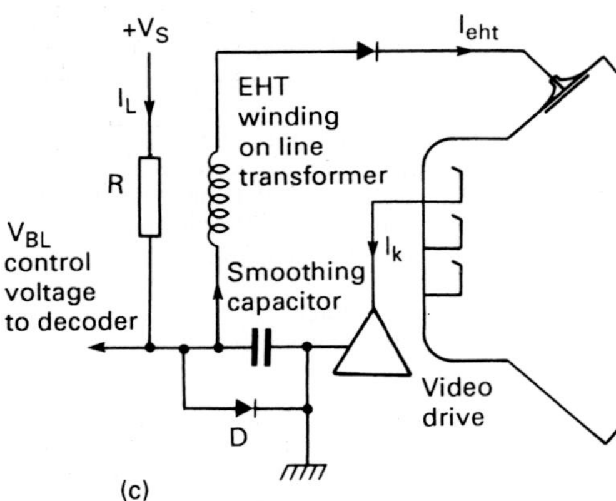

Figure 51.15 Three methods for beam current limiting: (a) diode current limiter. If V_0 is less than $I_{klim}R$, the diode is opene-circuited and mean current cannot increase. Each gun is limited separately so colour balance may be disturbed. (b) Transistor beam current sensing. A high voltage pnp transistor is place in the cathode current stream of each gun. Collector currents of three transistors are summed in resistor R. (c) Supply current to tube. V_{BL} goes negative when $I_{eht} > V_2/R$

protect any components connected to the picture tube elements. These include the use of spark gaps and decoupling capacitors returned to the picture tube aquadag and high voltage resistors in series with the tube elements to limit the transient current flow.

Flashovers become rarer after the initial 'burn-in' of the receiver. Appropriate protection ensures that there is no consequence for the life or performance of the receiver.

51.1.3 Synchronization

Display of a stable, correctly positioned picture depends on a synchronizing system that performs consistently for weak or strong signals, and tolerates sub-standard or non-standard signal formats. In general, the extremes to be allowed for cannot be tolerated by a single mode system, and various adaptive features may be incorporated to extend operation without apparent compromise.

51.1.3.1 Synchronizing pulse separation

The composite sync waveform, which incorporates all the required line and field synchronizing information, occupies the 'blacker than black' part of the video waveform. It is an apparently simple operation to strip this off the composite video waveform from the vision demodulator, but the following problems must be allowed for:

- Amplitude is uncertain, due to different signal sources, very weak or noisy signals, vcr sources via the antenna or peri-television connector; also the action of mean level agc produces changes of sync pulse amplitude with picture content.
- Input video waveform may have high frequency distortion caused by mistuning or propagation problems which makes recognition of sync pulses uncertain.
- Input waveform may have low frequency hum due to co-channel interference or field-rate level variation due to poor agc action or ac coupling from a video interface with poor low frequency response.
- Video waveform may be corrupted with noise spikes.
- Non-standard composite sync waveforms may be introduced by electronic games, vcrs in 'feature modes' (still pictures, fast search, or slow motion, all of which have a non-standard line count per field), video cassettes or other sources using an anti-copy process, etc.

Essentially the task of the sync separator is to find the level mid-way between sync tip and blanking level and identify the time and direction of all crossings of the video waveform with respect to this level. The best available circuits function correctly with input amplitudes from 1–2 V down to 50 mV.

51.1.3.2 Line flywheel

The line scanning circuit almost invariably uses a single or a dual pll (*phased locked loop*) flywheel. A pll consists of a voltage controlled oscillator whose frequency is controlled by the output of a phase detector. The phase detector compares the output of the oscillator with the input signal to which the system is to be locked. The control voltage is low-pass filtered, and it is this filtering that provides the desired noise immunity.

Use of a flywheel provides several advantages over a directly synchronized system:

- a symmetrical hold and catch which prevents picture tearing when sync pulses are missing and allows for long term oscillator drift in either direction
- drive to the timebase output stage may precede the line sync pulse; hole-storage delay in the switching circuits may be accommodated
- a low noise bandwidth may be achieved along with tight phase control, giving accurate and jitter-free positioning of the picture

The characteristics of the flywheel pll which are optimized to achieve the desired performance are the system gain, the filter response and the control range limits. Furthermore, these parameters may be varied automatically to adapt the system behaviour according to the signal conditions. A cyclic adaptation during the field blanking interval can also be introduced to avoid the top flutter effects caused by the field sync sequence.

This disturbance wiould be insignificant, if the line synchronization was based on the leading edge only of the line synchronizing pulses, which retain the correct timing throughout the field synchronizing sequence. However, this method is sensitive to noisy signals, and most receivers apply the complete line pulse to the line pll. The spurious response in the line flywheel may then be reduced by desensitizing the pll during the field blanking interval.

The anti-top-flutter feature is also valuable in vcr replay to provide a quick and invisible recovery from the line phase jump which occurs at the instant of head scan changeover at the end of each picture field. In addition, the phase detector may be keyed at line rate to eliminate disturbances, or random noise, occurring outside the time interval occupied by the sync pulse.

A dual-loop flywheel circuit uses a phase comparator fed by the first pll output waveform and the line flyback waveform to control the timing of the line drive waveform. It provides the following facilities:

- picture centring by adjustment of the second pll to set the relative phase of line scan with respect to the first loop oscillator without detuning, and thus preserving symmetrical catch and hold ranges
- a fast response to enable transient hole-storage delay variations in the line output transistor in response to fast load-current changes with picture content to be corrected; the noise immunity obtained with the first loop, which may be relatively slow acting for weak, noisy signals, is not sacrificed

51.1.3.3 Field synchronization

Field synchronization, in its simplest form, consists of an integration of the composite sync waveform obtained as described in section *51.1.3.1*.

To achieve satisfactory interlace of successive field scans, it is necessary to prevent line information from the timebase being coupled into the field timebase synchronizing circuit, and to choose an integrating circuit that provides an unambiguous pulse edge for triggering the field oscillator.

A directly synchronized oscillator will have a catching and hold range on the high side of the free-running frequency. The catch and hold range depends on the duration of scan in which the oscillator is responsive to the sync pulse. If this is made too large, the response to spurious noise pulses is enhanced. Thus, a limit must be set which just allows for ageing of the timing circuit and for the worst case frequency error arising from non-standard signals from a vcr in 'feature modes'.

Loss of a sync pulse will cause a picture roll at a speed determined by the difference between free running oscillator frequency and the field rate of the received signal.

A modified oscillator format that provides a symmetrical catch range with direct sync has been employed in some cases, but the best overall noise performance is provided by an indirect count down system that makes use of the high noise immunity achieved in the line flywheel, and provides several extra features:

- display is stable even when noise disturbances eliminate several field sync pulses
- there is no field oscillator and no field hold adjustment
- the system automatically recognizes 50 or 60 Hz picture standards and maintains correct scan amplitude and synchronism in either case
- use of twice line frequency input ensures accurate positioning of field drive waveform, giving perfect interlace
- accurate subsidiary timing waveforms are derived for signal blanking, and for flywheel parameter changes during the field blanking interval to overcome phase disturbances due to the field sync waveform, or head gap effects on vcrs

51.1.4 Scanning system

The scanning system utilizes magnetic deflection produced by scanning currents in deflection coils on the picture tube neck. This requires the scanning circuits to produce sawtooth scanning waveforms whose linear rise corresponds with the active picture information in the input video signal. Retrace of the beam takes place during the horizontal and vertical blanking periods.

51.1.4.1 Linearity and S-correction

In principle, the angle of deflection of the electron beam is proportional to the deflection current. A practical picture tube,

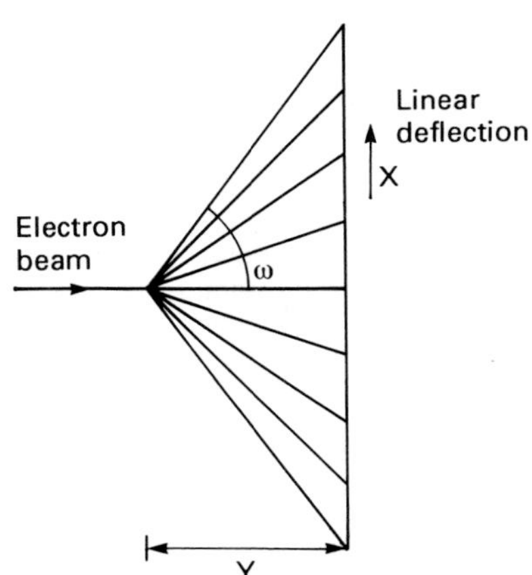

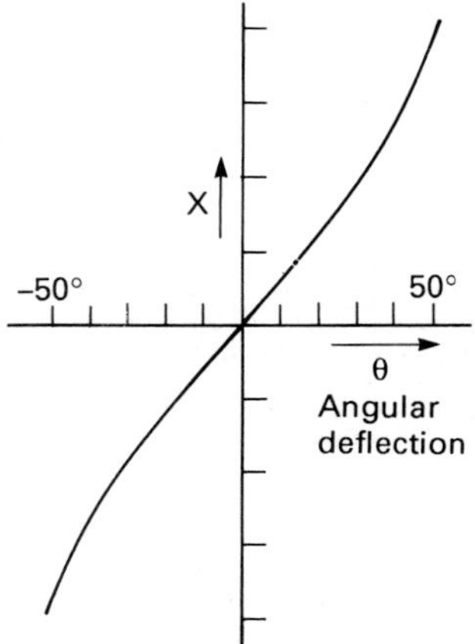

Figure 51.16 Scan angle and raster linearity. Equal deflection steps are subtended by diminishing deflection angles, according to tan ω = X/Y. for a uniform magnetic field, ω is proportional to scan current

however, has an almost flat faceplate, which means that the deflection sensitivity on its surface increases with deflection angle as illustrated in *Figure 51.16*.

This leads to a requirement for two forms of correction:

- *pincushion correction* in the N-S and E-W directions
- *S-correction* of the scan current waveform

Pincushion correction, which is required to vary the overall scan amplitude in vertical and horizontal directions, may be provided either in the scanning circuits or by special design of the scan coils to provide an optimum magnetic field distribution.

S-correction consists of a gradual flattening of the idealized sawtooth waveform at the ends of scan, to match the curve shown in *Figure 51.16*. Use of the field distribution method is well established for 90° *raster correction free* (rcf) colour or monochrome tubes. For 110° colour tubes, however, the rcf feature is not available for horizontal scan, and a modulation of line scan amplitude at field rate is required.

51.1.5 Vertical scanning

Vertical deflection requires a 50 Hz or 60 Hz sawtooth current waveform synchronized with the transmitted video waveform, providing two interlaced fields in each 20 ms or 16.7 ms frame period, respectively. The frame repetition rate is 25 Hz or 30 Hz, and the use of field interlace has greatly reduced large area flicker without sacrifice of vertical resolution. For most display requirements this standard is sufficient, but two exceptions have been realized:

- digitally generated characters or graphics may use a non-interlaced display, which avoids interline flicker on horizontal edges, but reduces vertical resolution by half; interline flicker is more objectionable on this type of display than for 'live' pictures, so non-interlace is preferred for full screen text displays
- a double scan-rate display with field rate of 100 Hz and line frequency at 32 kHz

The latter form of scanning provides a great reduction of large area flicker and in some versions, interline flicker is also removed. Standard signals may be processed in the receiver with a complex digital system incorporating digital field storage. Its advantages can apply to both picture and text displays, without sacrifice of vertical resolution.

The scanning circuit principles are the same whichever of these scanning rates is required. Signal processing and synchronization become more complex, and scanning energy is increased in the case of double rate scanning.

For field scanning, the required current waveform is derived by amplification of a small signal linear sawtooth which is initiated by the field sync system. It is produced by switched charge and discharge currents applied to a capacitor. The amplitude depends on the charging current during the scan time, and this current requires control to allow for two factors:

- If line scan current is varied during operation to counteract changes in its eht supply voltage with picture brightness (see section *51.1.7.4*), the field scan current should change in sympathy to preserve a constant aspect ratio, e.g. by deriving the capacitor charge current supply from the line timebase.
- If 50 and 60 Hz operation are both provided, a larger charging current is required in 60 Hz mode to achieve the correct scan amplitude within the reduced scan period; this condition may be provided automatically by using a 50/60 Hz sensing circuit.

The sawtooth voltage waveform requires amplification, linearity modifications and conversion at the output stage into the required current waveform. For field scanning, the load impedance comprises the self inductance of the coil in series with its winding resistance. The resistive component is the major component at field frequency, and any change in its value due to ambient temperature or self heating must be prevented from affecting the scan current and picture height. This condition may be met by providing a current feedback circuit, or by temperature compensation of the coil currents.

Another major component is the coupling capacitor which serves, not only as a dc blocking element where a single supply rail is used, but also as a source of S-correction via the feedback network. A typical basic arrangement is illustrated in *Figure 51.17*. The voltage waveform produced by the sawtooth current on the sampling resistor R_s is fed back to the differential amplifier and compared with the reference sawtooth waveform. The voltage at the top of the coupling capacitor C_B is also applied via a shaping network for S-correction and dc stabilization of the output voltage.

51.1.6 Horizontal scanning

Whereas generation of the vertical scanning current involves linear amplification, the line scanning process lends itself to a

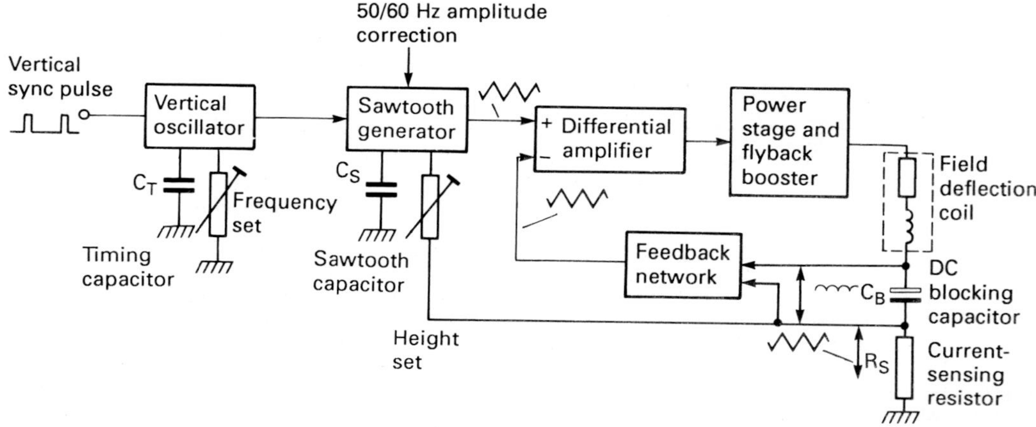

Figure 51.17 Outline of a field timebase

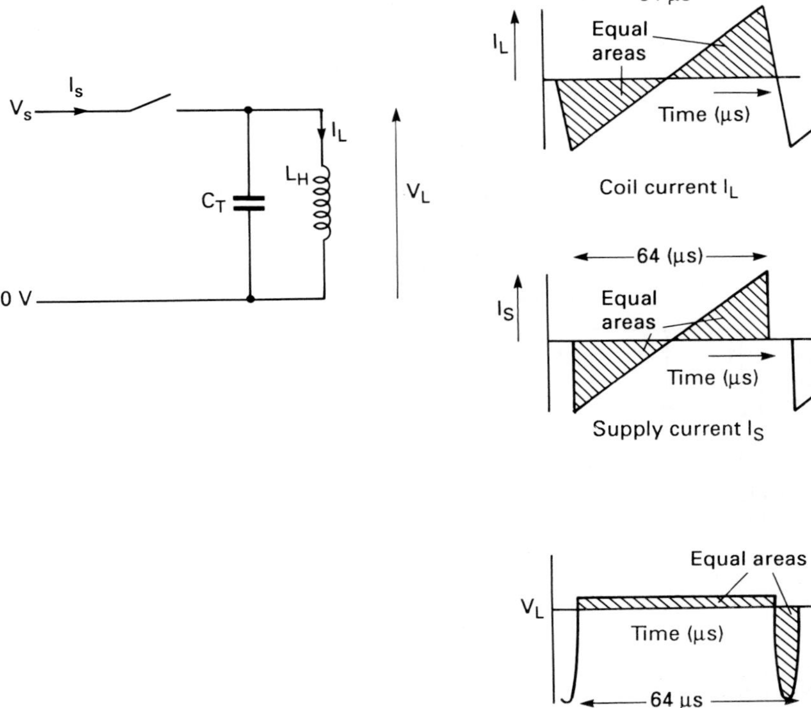

Figure 51.18 Principle of line deflection

switching principle, in which the scan current is generated and defined by imposing a fixed dc voltage across the scan coils during each scan period. The scan amplitude may be set by changing this dc voltage, or by means of a scan modulator circuit which dynamically varies the current in the scan coils without affecting the switching circuit.

An active circuit of very high efficiency is achieved, because the scan coil's impedance is almost purely inductive at line frequency, and the required voltage waveform is simply that shown in *Figure 51.18*. While a fixed voltage is applied to the coil inductance L_H, the current I_L increases linearly. At the end of scan, the switch is opened, and the coil current flows into the tuning capacitor C_T, commencing a sinusoidal oscillation with period given by the values L_H and C_T. After a half cycle the switch is again closed, and the coil current I_S, which is now reversed, flows back into supply V_S.

A simple practical circuit, using a bipolar transistor and a diode for recovering the energy developed in each scanning cycle, is shown in *Figure 51.19*. The diode current begins when the collector voltage V_C has fallen to the negative supply rail. The transistor is driven into conduction in time for the current reversal in the coil, at which time the diode ceases conduction.

The scan current is sourced in turn by the efficiency diode D and by the scan transistor. During flyback, the current flows entirely via the tuning capacitor until the voltage across the coil brings the diode back into conduction.

As the currents in the transistor and the diode charge and discharge the storage capacitor equally, there appears to be no net current demanded from the supply, V_S. In a practical circuit, there are losses caused by coil resistance, imperfect switching in the transistor and diode, and various other loads that are added to use the circuit as a dc/dc convertor. All this energy is taken from the supply.

The existence of some winding resistance in the coils leads to a spurious linear sawtooth waveform in series with the required fixed voltage across the coil inductance, which causes an assymmetrical loss of sensitivity across each line scan. Usually this is counteracted by means of a series connected inductor with a saturable magnetic core, which is biased by a permanent magnet so that the inductance falls as scan proceeds from left to right. The voltage drop across this component varies in the opposite sense to the resistive loss in the scan coils, thus providing the required correction.

The line output transistor is a very high voltage device; collector peak voltage during flyback may be more than 1 kV. Flyback begins when the switching transistor is turned off by the drive waveform. To overcome the switching transistor hole-storage effects, this drive waveform must quickly remove the charge built up by the large drive current required to support the end-of-scan collector current. A transformer coupling to the base of the transistor satisfies this requirement, making use of the negative voltage swing at the instant of switching off.

The flyback time is usually about 12 μs, which approximately equals the line blanking interval in the video waveform. Because the hole-storage delay time of the output transistor may amount to a few microseconds, the timing of the switch-off waveform must precede the start of line blanking. This condition is readily met by means of the pll system.

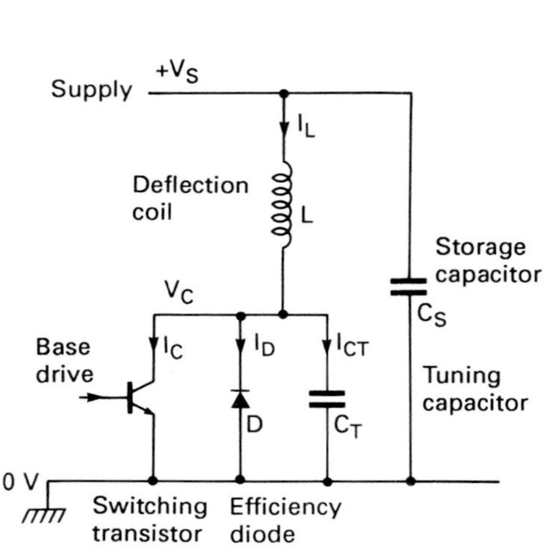

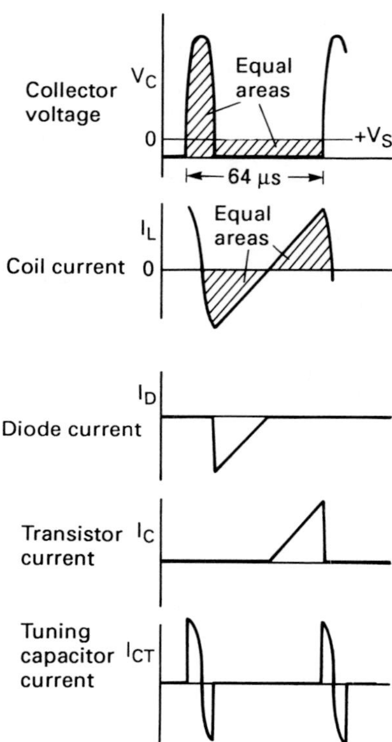

Figure 51.19 Transistorized line output circuit

51.1.6.1 Line output transformer

Ideally the scan coil is coupled to the switching device through a dc blocking capacitor, since the direct current component must not be added to the scan current which is symmetrical about zero. The dc requirement of the circuit is therefore applied via a separate inductor, which is normally extended into a multi-function line output transformer. The chosen dc blocking capacitor also provides the S-correction required by the particular tube and its deflection circuit.

The various secondary or tertiary windings may provide the following functions:

- eht and first anode supply generation
- impedance matching of scan coils if not directly connected as above
- power supply for field timebase
- power supply for video output stage
- reference waveform generator for second line flywheel loop
- line blanking waveform generation
- low voltage power supply for signal circuits
- heater supply for picture tube

E-W modulation for pincushion correction can be achieved in several ways:

- variation of supply voltage
- insertion of a saturable reactance transformer in series with scan coils
- use of a diode modulator circuit

The first is satisfactory only where the subsidiary power supplies are derived from elsewhere. The second is applicable for low energy tube concepts; the transformer becomes too expensive for 110° full performance colour sets.

The popular diode modulator circuit gives full control of raster shape and scan amplitude, while providing a constant load current and a constant flyback time. It is therefore suitable for systems in which the various secondary supplies are derived from the line output stage.

The line output transformer nowadays incorporates the eht rectification, and also the generation and preset adjustments of focus and first anode potentials.

51.1.7 Power supplies

A wide variety of power supply systems are in use, reflecting the degree of complexity of the receiver models and the level of performance required. The main determining factors are:

- range of mains supply voltage to be handled
- value and range of load currents
- requirement for interconnection between receiver and external equipment
- avoidance of interaction between scanning power supplies and signal circuits, most particularly the audio system

The voltage level for these supplies has been determined by the voltage and current capabilities of the major power devices. For the line output circuits, the optimal supply voltage lies in the range 100–180 V; for the small-signal parts, the range is 5–12 V. For medium power circuits, such as audio power, and field output stages it is 12–30 V.

51.1.7.1 Non-isolated power supplies

Until recently, a large proportion of receivers were non-isolated, the principal high voltage supply being derived from the mains by direct rectification. Stabilization is conveniently and efficiently applied by means of a phase controlled rectifier such as a thyristor, or by a switched-mode dc/dc convertor, supplied by a bridge rectifier fed from the mains input.

Thyristor power supplies using both half wave and full wave rectification systems have been widely employed, the latter providing a more acceptable current load waveshape for the mains. Although good stabilization against operational load current and supply voltage variations is provided, the inherent ripple due to the 50 Hz or 100 Hz (full wave) input requires heavy filtering, with large electrolytic capacitors and high dissipation resistors, and this detracts from the benefits achieved by the switching principle realized in the thyristor.

A switched mode dc/dc convertor is now used in most receivers. Its demands on filtering capacitance are significantly reduced, as its high control loop gain provides excellent stabilization and ripple reduction. If operated in series mode, however, some protection against short-circuit failure of the switching device is essential.

51.1.7.2 Isolated power supplies

Galvanic isolation of the receiver from the mains supply has advantages in two main respects:

- safe and simple interface via galvanic interconnection with external video equipment
- metal parts on isolated parts of the receiver may be touched safely; double insulation of entire chassis is no longer required

Monochrome and small screen colour receivers have been based on a mains transformer, which provides isolated high and low voltage supplies. However, the disadvantages of weight and stray magnetic field and the remaining requirements for stabilization and ripple filtering of each supply rail have limited its use in most applications.

A high frequency switched-mode power supply operating in the range of 15 kHz to several hundred kilohertz offers the following advantages:

- high basic efficiency due to the switching principle
- availability of a number of different stabilized supply voltages by means of taps and separate windings on the transformer
- simple filtering requirements, since mains frequency ripple is eliminated by stabilization, leaving only the switching frequency ripple
- lightweight transformer, especially for high frequency versions
- integrated circuit control systems provide sophisticated protection
- overvoltage malfunction cannot be produced by failure of major switching components
- standby condition is possible if control circuit is powered

The switching of the power supply is sometimes chosen to be synchronous with the line scan. This may reduce visibility of any residual ripple and permit simplifications of the drive circuitry which may be common to power supply and timebase. Great care is needed to prevent pick-up of the switching edges in the form of visible vertical lines.

If a higher switching frequency is chosen, the system may be based on smaller components, and radiation may be incoherent with scan and less noticeable.

51.1.7.3 Signal circuit power supplies

Ever since the introduction of semiconductors into television receivers, a low voltage supply of about 12 V has been used in the signal processing circuits, in conjunction with higher voltage supplies for the line and field timebases. The low voltage supply invariably requires an efficient conversion method to avoid a disproportionate power loss in a dropping resistor.

In most receivers, this lt supply is derived from a line frequency, or similar, switched source, and an additional regulator is used to stabilize it against load variations and ripple. If several ICs are incorporated, it is usually important that they are operated from exactly the same supply voltage so that their interconnections are made at voltage levels referred to one value. At the same time, ripple and impulsive disturbances can be imposed on the supply rails by signal currents and other IC functions. Accordingly, the supply pins of the IC are decoupled with a capacitor between supply and the substrate pin, and fed through a low value series resistance or inductor.

If the signal circuits are powered from the line output circuit, and only a high voltage supply is available from the mains source, the line oscillator and driver stage must be supplied directly from this supply to enable the receiver to start up, and thereafter sustain the operation of the other parts deriving their power from it.

51.1.7.4 Picture breathing

The field scan amplitude is required to track that of the line scan (see section *51.1.5.1*). Why should they vary at all? In general, the eht generator has a source impedance of 1 megohm or more, and the consequent reduction of eht voltage resulting from an increase of beam current increases the scan sensitivity. This effect, known as *breathing*, may be reduced by reducing scan current when beam current increases. This change must be duplicated by the field generator circuit.

In practice a 'breathing resistor' may be inserted in series with the line output stage, whose current demand increases in response to the increase in beam current. The resulting reduction in supply voltage reduces the scan amplitude, as required. It is necessary, in this case, that no other loads with variable demand, such as a class B audio circuit, should be supplied from the line output stage, as this would worsen the effects of sound modulation affecting the picture.

51.2 Colour decoding

Colour decoding requires separation of the chrominance signal from the composite luminance-plus-chrominance, and recovery of the colour-difference signals which are then recombined with the luminance to produce the RGB drives for the three guns of the picture tube.

The following paragraphs describe the process of demodulation for the three principal terrestrial colour transmission systems. The principles of decoding systems for the MAC satellite systems are described in Part *6*.

51.2.1 NTSC decoding

The NTSC and PAL systems use a single subcarrier that is phase and amplitude modulated. This complex modulation is implemented in the encoder at the transmitter by modulation of two subcarrier reference signals in phase quadrature using the prescribed colour-difference signals.

In the NTSC system, the subcarrier is modulated by two signals derived from the R-Y and B-Y signals in such a way that three requirements are met:

- the reference phase for the system lies along the axis of the B-Y signal

- the amplitudes of the two signals are scaled so as to produce equal peak amplitudes of the chrominance signal for fully saturated colours when the two signals are added in quadrature
- the axes chosen for the I and Q (in-phase and quadrature) signals permit advantage to be taken of the different acuity of the human eye for colours lying on different colour axes, by reducing the bandwidth of the less critical component

The derivation of these relationships is given in section *51.5.2*, and the relationship between phase and amplitude of the fully saturated colours as used for the standard colour bar pattern is shown in the graticule diagram of *Figure 51.20*. It indicates the vector locations of colours in the standard colour bar sequence. Each colour axis intercepts two fully saturated complementary colours on the circumference, with the saturation represented by the vector amplitude and hue dependent on the phase angle. The origin represents the reference white point for the system, and the colour burst lies at +180° from the B-Y reference axis.

51.2.1.1 Demodulator and matrix

The colour-difference signals E'_r-E'_y, E'_b-E'_y and E'_g-E'_y, as defined in section *51.5.2*, could each be recovered by synchronous demodulation along these axes at 0°, 90° and 250° as shown in *Figure 51.20*. It is preferable, however, to demodulate two axes; the third colour-difference signal may be derived by a matrix operation upon two colour-difference signals. Two such signals may be chosen, either the I and Q signals used in the transmitter encoder or R-Y and B-Y, which allow the simplest form of matrix to be used for recreating the RGB signals for video driving.

Some of the earliest NTSC receivers demodulated along the I and Q axes. This method is technically advantageous because it permits the receiver to complement the different bandwidths of the I and Q components in the encoded signal. This allows full recovery of the wide-band 1.4 MHz I channel modulation, while high frequency noise and crossstalk are reduced in the Q channel whose bandwidth is reduced to about 0.5 MHz.

This feature introduces a differential time delay between the I signal and the Q signal, and a wide-band delay circuit is required in the I channel to equalize the timing. Also, the encoded I and Q signals are unequal because of the scaling factors used in the NTSC colour equations; the correct amplitude ratio should be restored prior to the colour matrix operation.

The complication of providing this form of decoding is not usually acceptable to manufacturers, and the usual technique is to demodulate on the R-Y and B-Y axes, using a common compromise bandwidth in the two channels. The in-phase and quadrature subcarrier reference waveforms are produced directly by the pll system synchronized to the colour burst, and a relatively simple form of matrix is used for recreating the RGB signals for video driving.

The timing of luminance and chrominance signal components is made equal at the transmitter's encoder. However, in the receiver, the bandwidths of the chrominance filters and of the post-demodulator subcarrier-rejecting filters are much less than the luminance bandwidth. This delays the chrominance signal by approximately 0.5 μs, and this must be equalized with a broad-band delay of this value in the luminance channel, prior to recombining these components in the RGB matrix. This scheme, which incorporates only one delay line, is outlined in *Figure 51.21*.

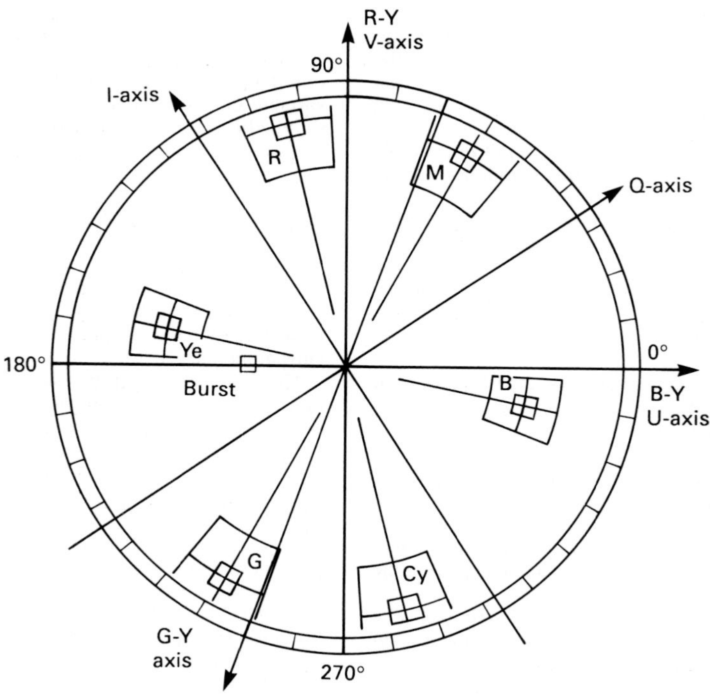

Figure 15.20 Colour vector diagram incorporating phase-amplitude representation of six principal colours from standard colour-bar test pattern. Black and white bars occupy the origin. This display is produced by connecting R-Y and B-Y outputs to the X and Y inputs of an oscilloscope. A dedicated instrument for analyzing the colour subcarrier in this way is called a *vectorscope*

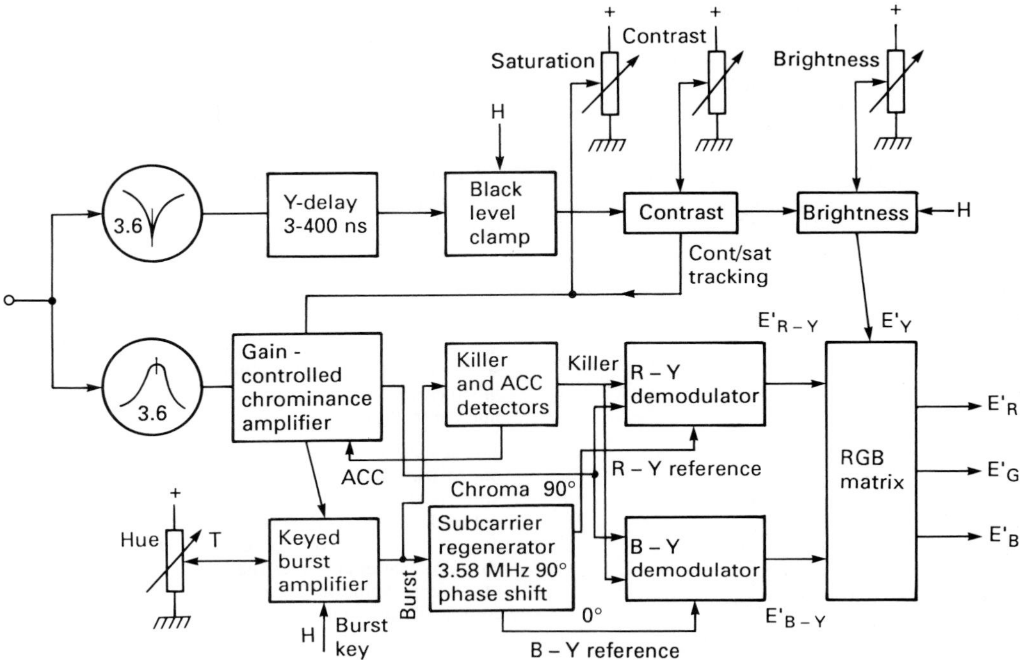

Figure 51.21 UV axis NTSC decoder

Another refinement frequently used in NTSC is to modify the demodulation axes by rotating the B-Y demodulating axis through about 30°. This variation has been introduced to improve perceived flesh tones. Where the demodulating axes cannot easily be modified, the modification may be implemented by an equivalent process which couples a portion of the demodulated B-Y signal into the R-Y channel.

51.2.1.2 Subcarrier regeneration

The quadrature reference signals required for demodulation are generated by a continuously running crystal oscillator within a phase locked loop. It is synchronized by the colour burst signal which is keyed by the standard timing waveform (*sandcastle*) and applied to a phase detector which comprises a synchronous demodulator detecting the burst signal along the B-Y axis.

In the NTSC system, any phase error in the reference subcarrier relative to burst is perceived as a change in hue of the picture. Such errors can arise both at the transmitter distribution system and in the receiver. Most NTSC receivers therefore provide a user adjustment to allow the warmth of the colour to be corrected as required.

The crystal oscillator may be designed to operate at twice subcarrier frequency and to drive a divide-by-two circuit to produce the required reference signal. This arrangement has two significant advantages:

- Two signals precisely in quadrature are derived from flip-flop dividers without the need for a precision 90° phase-shift circuit.
- Spurious coupling of the crystal oscillator, at twice subcarrier frequency, into the chrominance signal path prior to demodulation, produces no response at the demodulators' outputs. Pick-up of subcarrier frequency prior to demodulation can produce a colour cast.

51.2.1.3 Selectivity requirements

The luminance and chrominance signals are both present on the input composite video signal, and filters with characteristics such as those shown in *Figure 51.22* are required for separating these prior to demodulation. Some variations in these requirements may be applied to correct for response shortcomings of the front end, but this could degrade the nominally correct signals provided via a baseband interface.

The use of these filter functions limits the luminance bandwidth and does not prevent annoying cross-colour effects. These problems may be largely overcome by incorporating a *comb filter* system which takes advantage of the fact that, for a constant picture content in adjacent lines, the half-line offset incorporated in the subcarrier frequency ensures that the chrominance signal in vertically adjacent lines is in anti-phase. Thus, if a precise one line delay is introduced and its output is fed into a sum and difference network, the sum signal will contain luminance only, and the difference signal will contain chrominance with the luminance cancelled out.

This type of separation system avoids band limiting of the luminance and chrominance signal components as depicted in *Figure 51.22*, and avoids large area cross-luminance and cross-colour effects. At certain edges at which adjacent lines have different information, and at moving edges, some disturbances remain visible.

51.2.1.4 ACC and colour killing

The amplitude of the colour subcarrier relative to the luminance signal may vary with reception conditions, and the state of alignment of the receiver's selective circuits. It is therefore necessary to provide an automatic control of the chrominance signal amplitude (acc) applied to the colour demodulators. Estimation of the chrominance amplitude is based on the colour burst amplitude. This method includes the effect of noise on the

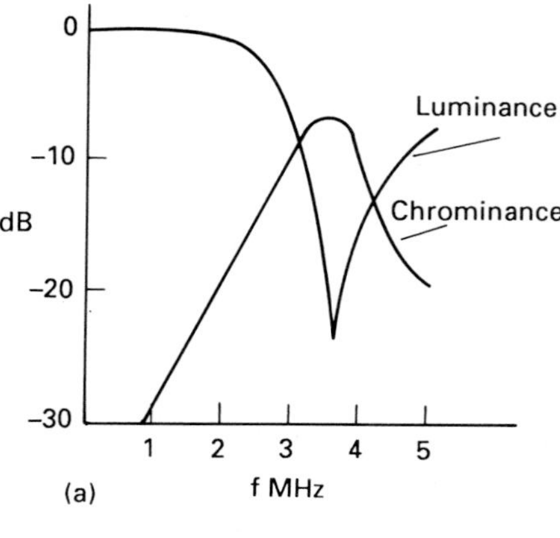

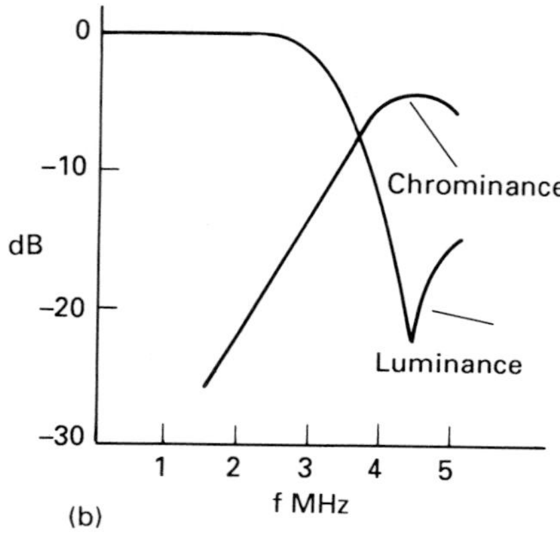

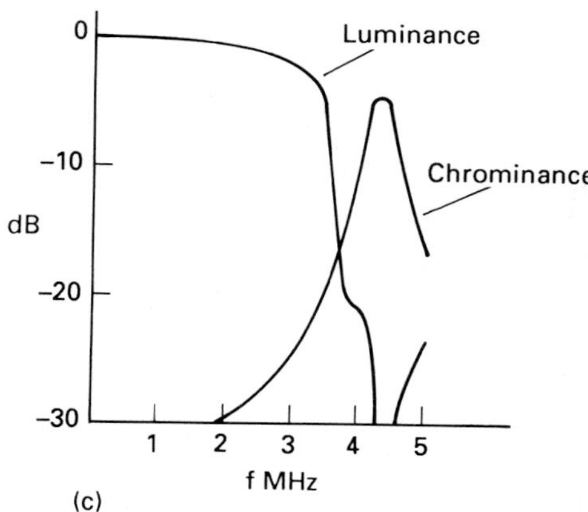

Figure 51.22 Video selectivity requirements for various colour systems: (a) American system M, subcarrier frequency 3.5756 MHz, (b) PAL systems B, G, H and I, subcarrier frequency 4.4336 MHz, (c) SECAM system subcarrier rest frequencies 4.25 MHz and 4.41 MHz, cloche filter tuned to 4.29 MHz

signal, so that the colour saturation is progressively decreased as the signal becomes noisy, which is less annoying than the appearance of high amplitude demodulated coloured noise.

For monochrome signals, or no-signal conditions, the demodulators should be inactive, and this *colour killing* operation is based on the synchronous detection of the burst amplitude along the B-Y axis. It therefore depends on the suppression of the colour burst at the transmitter.

51.2.2 PAL system decoding

The main features of the PAL encoded signal are outlined in section *51.5.3*. The PAL decoding system is based on the same principles as the NTSC decoder. It must provide for the following additional features:

- Only U and V axes are usable for demodulation because of phase reversal on alternate lines of the V axis signal.
- R-Y demodulation must be switched by H/2 square-wave to give 180° phase shift on alternate lines.
- Mean phase of burst is used to synchronize the pll; the alternating component is used for identification of H/2 phase switching circuit and colour killing.
- Use of PAL delay line eliminates effects of any phase errors in subcarrier regenerator or transmission path; no phase shift hue control is required.

The selectivity curves for chrominance band-pass and chrominance trap in the luminance channel shown in *Figure 51.22* apply also for a PAL decoder. The response shapes and group delay characteristics represent a conflict between the requirements for low cross-colour, resolution of luminance and chrominance detail, and freedom from rings, to which there is no unique solution. A variety of filter circuits of varying complexity are in use.

51.2.2.1 PAL delay line

While a PAL decoder could function without using a one line delay in the colour-difference channels, the advantages of PAL delay are so powerful that no receiver design known to the author has appeared without it. The delay line has generally taken the form of a glass block in which mechanical compression waves are generated at one end surface by a piezoelectric transducer. After propagation through the material via several criss-crossing paths defined by accurately ground reflecting surfaces, the delayed signal terminates in a matched receiving transducer on the final end face.

The transducer is designed for maximum efficiency at subcarrier frequency with the bandwidth required by the chrominance channel. The propagation path is machined to provide a path length that is accurate to within a fraction of a wavelength at this frequency. This form of delay line must be applied ahead of the colour demodulators.

When combined with the undelayed chrominance signal via a sum and difference network, the outputs are the U and V subcarrier signals respecticely. In addition to the preselection of the chrominance components, the delay line matrix also eliminates the effect of phase errors occurring anywhere in the signal path from the transmitter to the recovered U and V signals.

Correct functioning of the PAL delay line depends on precise matching of the amplitude and phase of the delayed and undelayed signals. Matching is optimized by a preset gain adjustment of the undelayed signal and a phase adjusting component terminating the delay line transducer.

It has long been expected that the PAL delay function would be realized with an electronic delay circuit using many delay elements in a charge coupled or similar array, operating at baseband on the R-Y and B-Y signals. This will achieve the same advantages as the glass delay line without the need for precise phase matching and amplitude adjustments.

51.2.2.2 Identification and colour killing

The PAL switch, which reverses the phase of the V signal on alternate lines, must be synchronized with that used in the encoder at the transmitter. The necessary information provided by the phase alternating colour burst is not used directly, because an occasional disturbance of the burst signal could then cause mis-identification over a period of some lines, which is unacceptable.

To preclude such an effect, the PAL switch is driven by a flip-flop triggered by the highly noise-immune line timebase pulses. The flip-flop output and the H/2 identification signal from the V axis demodulator during the colour burst are fed to a comparator, and the phase of switching is reset only if an error exists over a long term.

The presence of the H/2 signal from the V axis burst demodulator is a very reliable indication of the presence of a PAL colour signal, and its presence or absence is used for PAL system identification and for colour killing.

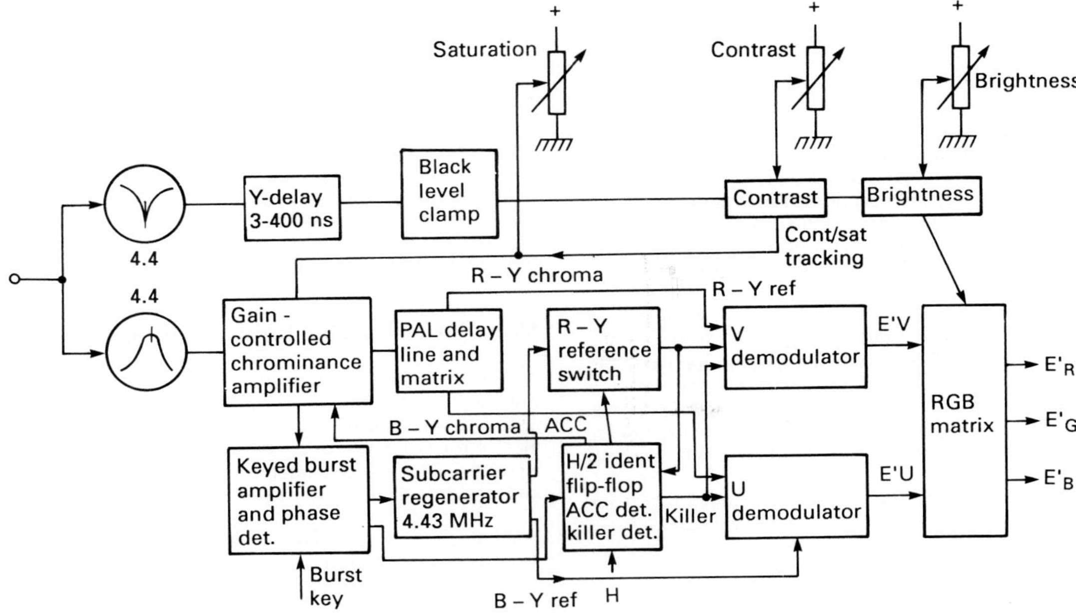

Figure 51.23 Outline of a PAL decoder

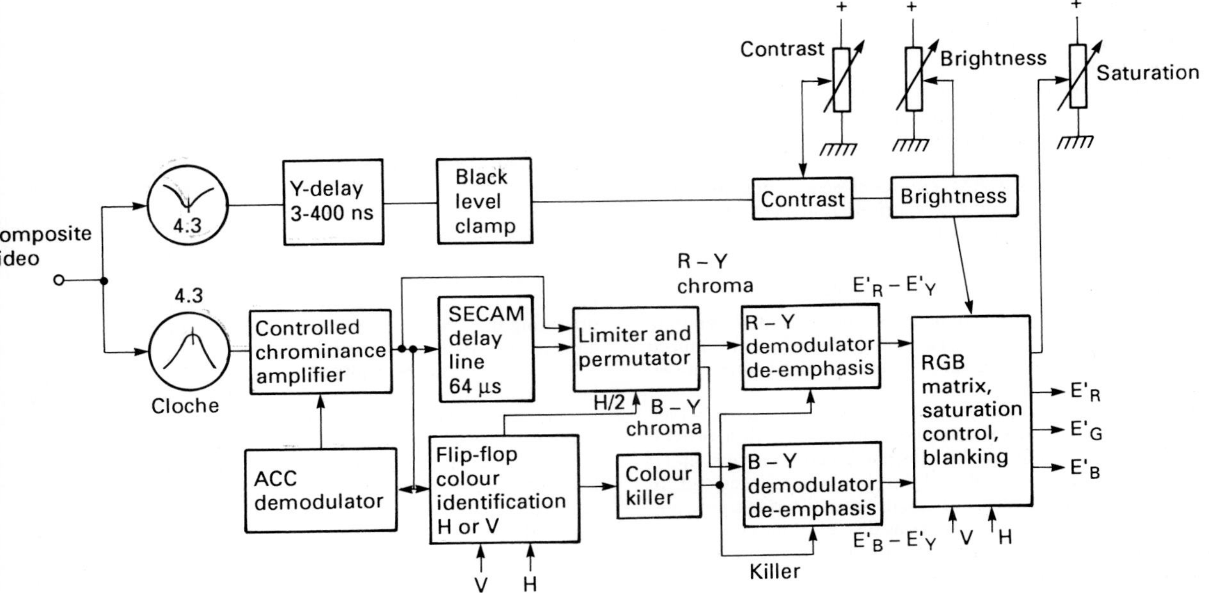

Figure 51.24 Outline of a SECAM decoder

51.2.3 SECAM system decoding

The SECAM system avoids the effects of subcarrier phase errors by using a frequency modulated colour subcarrier, which transmits the R-Y and B-Y information on alternate lines. The colour difference signals used in the SECAM system are weighted differently from those used in PAL and NTSC. The weighted signals are known as D_r and D_b. The weighting factors applied to these signals differ from those used for the PAL and NTSC systems because the two signals are transmitted alternately, and it is the peak amplitudes of the two signals that are normalized. The relevant equations of these signal components are given in section *51.5.4*.

The vertical resolution of colour detail is reduced by the same amount as for the PAL system, as every line of colour information is used twice.

The SECAM decoder (*Figure 51.24*) is structured in a way which complements the system signal specification. The spectral shaping circuit (cloche filter) prior to fm demodulation, and the de-emphasis filter after demodulation, provide the inverse frequency characteristics to those introduced at the transmitter in the interests of signal/noise ratio and low visibility of the colour subcarrier (cross-luminance).

The cloche filter requires careful alignment for avoidance of colour transient distortion, and it can be observed that the amplitude modulation of the chrominance signal applied to the demodulator with standard colour bar test signals is at a minimum when correctly tuned. The fm demodulators are adjusted so that the system rest frequencies for the D_r and D_b channels produce outputs equal to the levels without subcarrier.

Scale factors are also incorporated in the encoded subcarrier signals in the ratio $D_r/D_b = -1.9/1.5$, and the necessary inverse correction is applied ahead of the colour-difference matrix.

51.2.3.1 The SECAM delay line and commutator

The commutator switches alternate lines of input signal and one line delayed signal into the fm demodulators. As in the PAL system, an H/2 identification is required, and this is derived from the modulated subcarrier, either as nine lines in the field blanking interval (*field identification*), or as a burst in the back porch comprising the rest frequency for the following line (*line identification*).

The use of line or field identification depends on the system chosen by the national broadcaster. Line identification is now most widely favoured. Colour killing is derived from the identification circuit.

The delay line for SECAM requires close gain matching with the undelayed signal, and low spurious reflection levels, but highly precise phase accuracy is not a requirement.

51.2.4 Multi-standard receiver systems

The earliest multi-standard receivers incorporated separate colour decoders for each system option, based on specialized integrated circuits, but the most recent systems centre on two alternative formats.

The first utilizes a complete multi-standard decoder IC which includes full signal processing and system identification, providing automatic switching of the external selective circuits, and if required, an indicator display of the system being received. In localities in which the received signals have differing sound and vision transmission characteristics as discussed in section *51.1*, some further processing instructions may be stored in the memory of the tuning system.

The second format has arisen because of the frequent need for PAL receivers to accept SECAM encoded signals, whether from neighbouring countries' broadcasts or from imported video cassettes. For some manufacturers, this is a minority market requirement, and it is preferable to add SECAM decoding to a receiver model designed initially for PAL by means of an add-on SECAM panel using a transcoder IC which operates in conjunction with the PAL decoder. System recognition is automatic, and there are no serious performance limitations. Similar decoders are available which also provide NTSC decoding.

The transcoder may take two possible forms:

- operating independently of the PAL decoder, having its own subcarrier crystal and system identification, and generating a normal PAL encoded signal
- accepting the subcarrier reference from the PAL decoder, using the PAL identification provided by the PAL decoder, and generating a modified 'quasi-PAL' encoded signal

The former type is suitable for the most basic PAL decoders which do not provide access to the required identification signals, or for operation in a separate unit which is separate from the receiver.

51.3 Requirements for peripheral equipment

All but the most basic economy receivers make provision for the connection of external peripheral equipment to the receiver, with some adaptation to its special requirements. The most obvious feature to be added is the baseband interface, which provides galvanic interconnection between equipments and may provide the following benefits:

- elimination of rf signal processing elements with possibilities for distortion, interference, bandwidth limitation, and sub-optimal adaptation
- not limited by video sources and receivers designed for specific local broadcast system standards
- provision of stereo sound from a peripheral television source even where the local broadcast system does not provide it
- receiver and video equipment interconnected with a common control system via a data control line

51.3.1 VCR adaptation

The requirements in the receiver for correct handling of vcr signals place special constraints and limitations on the synchronizing and control parts of the receiver. This has a number of effects and implications.

The tape transport produces a timing jitter due to surface asperities in the tape, causing friction and eccentricities in the rotating parts. These disturbances extend over a number of line periods, and the picture stability is greatly improved by operating the line flywheel in a high gain, high speed mode to correct these dynamic disturbances.

The helical scan principle introduces a phase jump at the head changeover position in the video waveform, usually placed just ahead of field flyback in the undisplayed overscanned area at the bottom of the picture. It requires a fast recovery of correct line phase and elimination of line keying features which would delay recovery of synchronization and agc actions.

Anti-copy waveforms are present on commercial pre-recorded cassettes, comprising several lines with selected waveforms designed to disturb the recording circuits of an illegally connected second vcr. The receiver's agc, synchronization and system identification circuits must be made insensitive to this feature.

Non-standard line counts and head changeover effects occur during *fast forward* and *reverse* picture search operations. In most vcrs operating in these modes, the head changeover occurs several times in each field scan, where the 'anti-top-flutter' feature described in section *51.1.3.2* is of no effect.

The synchronizing and muting characteristics required for good receiver performance when receiving weak signals are in conflict with the above requirements for vcr operation, especially in the feature modes. Because a receiver must provide high performance in adverse reception conditions, it is essential for the rf signal from a vcr to be greater than the level at which the receiver is in its weak signal mode. The use of a passive multi-way antenna splitter may thus be unsatisfactory.

The vcr incorporates an rf modulator and directional coupler which outputs the antenna signal together with the vcr's own output signal. The latter signal is non-standard in two respects: it is double-sideband modulated with sound carriers above and below the vision carrier, and the vision carrier is not precisely located at one of the designated channel frequencies for vcr operation. The first presents no serious problem, but the second may affect some forms of fs tuning systems.

Most of the problems of adaptation identified in this section are removed with the use of a video interface, in which the receiver's operating mode may be defined by forced switching of the circuits which influence the quality of vcr adaptation.

51.3.2 IEC peripheral television connection

The SCART interconnection system has been incorporated in many receivers and vcrs on sale in Europe, particularly in France where its inclusion is mandatory. The standard fixes the pin allocations, although not all facilities are implemented in most receivers; in particular, the RGB inputs and the data lines are often omitted.

A variety of connecting lead formats is also available, some omitting the unused features, others providing interface with other connecting systems, such as BNC, RCA Phono, or the 6 pin DIN standard. A receiver may incorporate several IEC sockets, permitting several peripheral equipments to be connected via the receiver acting as a 'telephone exchange'.

For some purposes, such as externally received satellite or cable signals, it is valuable to retain the decoded video signals in component form rather than as a composite PAL or SECAM signal. The IEC socket permits an RGB component interface, but such linear analogue signals impose tight matching tolerances on the three channels to preserve an accurate grey scale and realistic flesh tones.

An interconnection using the YUV format is much more tolerant of amplitude matching errors, but no convenient interconnection system using a single connector has yet been agreed.

51.3.3 Y-C interconnection format

Another interconnection format, using a luminance and a PAL, SECAM or NTSC chrominance signal on two separate conductors, has been adopted by Japanese manufacturers in connection with the Super VHS and ED Beta formats. This is a simple video interface using a unique 4 pin socket, known as the *S connector*. The sound signals in stereo or mono can be routed via standard phono leads to the TV receiver or to a hi-fi system.

The Y-C interface offers the following advantages:

- pre-recorded video material reproduced totally free of cross-luminance and cross-colour effects
- luminance bandwidth not limited by chrominance trapping
- gain matching of luminance and chrominance channels uncritical since the receiver's acc function maintains correct amplitude of demodulated U and V components
- vcr able to utilize a sophisticated form of luminance/chrominance comb filtering, giving improved performance on off-air signals, which is not sacrificed by combining into a composite CVBS signal

The S connector with its associated switching circuitry can provide obvious performance advantages over previous methods, and is likely to be an accepted feature of a high proportion of receivers worldwide.

51.4 Remote control and I²C bus

Remote control systems transmit the required instructions via a serial data stream transmitted almost universally by means of an infrared light transmitter in a hand-held keypad, and received by an infrared light optical sensor on the front of the receiver. The handset is battery powered, and it is designed so that power is consumed only during transmission of a command. As a binary code is used, the number of distinct command messages available is a function of the number of bits in each message word. Thus a 5 bit system without error protection would provide 32 possible commands, a 6 bit system would provide 64, and so on.

A 64 command keypad represents a reasonable practical limit, both from the point of view of easy visual or tactile recognition of the keypad layout, and of limiting the message length to allow quick response of the system to a string of commands.

Where an increased menu of commands is called for, the same set of command code words may be used all over again by using a mode change command which causes the receiver to interpret subsequent messages according to a different command set. Such mode changes may be used to giver a manifold extension of the command capability, with one or more commands in each set being reserved for mode changing.

The chosen modes would normally be made logically cohesive, relating to different operating modes of the receiver, such as:

- teletext
- videotex services
- stereo audio controls
- satellite and cable TV adapter controls
- vcr and LaserDisc controls
- other peripheral equipment connected externally

Such remote control systems are so powerful and comprehensive that many receivers have limited the controls on the receiver front panel to the barest minimum required to obtain a wanted picture, with the remaining text, stereo sound parameters and so on accessible only through the remote control.

The most basic remote control systems use a set of instructions in which the codes have been predetermined by the IC manufacturer for the available modes. In this case, the decoding IC provides all the analogue and switching function control signals. In the case of teletext, the command codes are passed into a control bus which is active during text mode and these commands are decoded in the text processor.

For the more advanced forms of receiver, a much more flexible system is provided, in which the command structure, the channel-tuning algorithms, and many other aspects of operation are determined by a microcomputer for which the software, which determines the operating system, is designed or specified by the receiver manufacturer. The incoming remote control messages are processed within the microcomputer according to these rules, and its output ports provide direct control of the receiver functions in the way specified by each of the controlled circuits.

The use of individual leads to control each function addressable from a present-day remote control system leads to an excessive amount of wiring in the receiver as well as a high pin-count on the controlling IC. This problem is overcome by the adoption of the inter-IC bus (*I²C-bus*), which permits any of a large range of signal processing, tuning system, scanning control, and display generating ICs to be interconnected via a two-wire bus.

A wide range of ICs for direct control via the I²C-bus is now available. Each IC type responds only to messages addressed to it via the bus wires, and ignores all messages addressed to other destinations. In addition, it is possible for an IC to source information about its operating state via the bus, and the entire communicating system functions under the supervision of the microcontroller.

The I²C bus strategy is compared in *Figure 51.25* with the interconnection requirements of a typical receiver. Apart from the main signal paths, the control signals for each IC are conveyed on the bus data path comprising only two wires, and this greatly reduces the constructional complexity of the receiver. Furthermore, the system may be extended to other areas of control which have not been provided previously, such as scanning circuit alignment using an extended remote control set not accessible to the final user. This facility would permit some service alignments to be performed without removing the back of the cabinet.

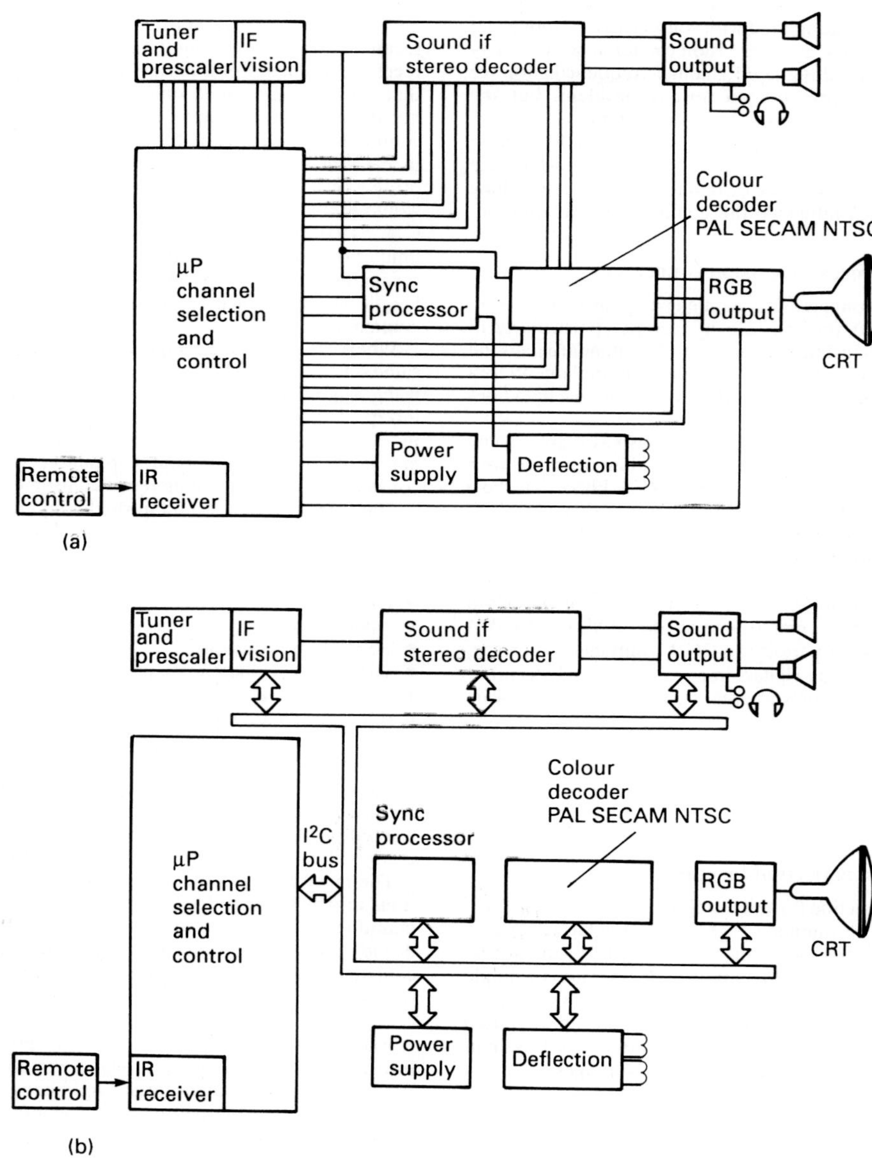

Figure 51.25 Application of a microcomputer and the inter-IC bus system: (a) interconnection wiring in a conventional TV set, (b) interconnections in a receiver using the I²C bus

Principal country	Main sound carrier	Main additional subcarrier	Extra facilities	
USA	4.500 MHz fm	2 × line frequency am (stereo difference channel)	5 × line frequency fm (second audio program)	6.5 × line frequency fm (narrowband voice or data)
Japan	4.500 MHz fnm	2 × line frequency fm (stereo difference or second sound)	3.5 × line frequency am (service identification)	

Table 51.1 Systems based on main sound carrier

Principal country	Main sound carrier	Additional carrier	Bandwidth of additional carrier	Extra facilities
Germany	5.500 MHz	5.742 MHz analogue	±80 kHz	Stereo, dual-language.
Scandinavia	5.500 MHz	5.850 MHZ digital NICAM 728	±60 MHz	Stereo, dual-language, data encryption.
United Kingdom	6.000 MHz fm	6.552 MHz digital NICAM 728	±360 MHz	Stereo, dual-language, data, encryption.
China	6.500 MHz	6.742 MHz	± kHz	Stereo, dual-language.

Table 51.2 Systems based on additional sound carriers

51.5 Basic data

51.5.1 Stereo and additional sound formats

Systems based on main sound carrier are defined in Table *51.1* and those based on additional sound carriers in Table *51.2*.

51.5.2 NTSC colour signal

The colour signal is derived from the red, green and blue camera outputs E_r, E_g and E_b after several basic processing steps:

Gamma correction. Assuming the voltage outputs of the camera to be linearly scaled to the light density of the scene, it is necessary then to provide gamma correction to complement the assumed characteristic of the display device. The system is balanced for a display device having primary colours with the following chromaticities in the CIE colour system:

	X	Y
red	0.67	0.33
green	0.21	0.71
blue	0.14	0.08

With equal drives $E'_r = E'_g = E'_b$, the display output should match illuminant C (x = 0.310, y = 0.316). The transmission of such gamma corrected signals reduces the visibility of noise which arises in weak signal conditions. The display device is assumed to have a *gamma* (transfer exponent) of 2.2 for each primary colour. Conventionally, the gamma corrected voltages are expressed using the primed symbol thus:

$$E'_r = E_r^{1/\gamma}$$
$$E'_g = E_g^{1/\gamma}$$
$$E'_b = E_b^{1/\gamma}$$

Separation of luminance and colour-difference components. The luminance signal is defined as:

$$E'_y = 0.30\,E'_r + 0.59\,E'_g + 0.11\,E'_b$$

and the colour-difference signals are:

$$E'_b - E'_y$$
$$E'_r - E'_y$$

The luminance signal conforms to all the requirements for reception by a black and white receiver, which also substantially ignores the presence of the chrominance modulation.

Conversion into in phase *and* quadrature *modulating signals*.

$$E'_q = 0.41\,(E'_b - E'_y) + 0.48\,(E'_r - E'_y)$$

$$E'_i = -0.27\,(E'_b - E'_y) + 0.74\,(E'_r - E'_y)$$

Modulation of colour subcarrier and addition of luminance component.

$$E_m = E'_y + \{E'_q \sin(\omega_{sc}t + 33°) + E'_i \cos(\omega_{sc}t + 33°)\}$$

Note that for frequencies below 500 kHz, this expression is reduced to:

$$E_M = E'_Y + \frac{1}{1.14}\left\{\frac{1}{1.78}(E'_b - E'_Y)\sin\varphi_{SC}t + (E'_T - E'_Y)\cos\omega_{SC}t\right\}$$

The subcarrier angular frequency is 2π times the specified chrominance frequency, 3.579 545 MHz, which is related to the line and field scanning frequencies thus:

$$f_H = 2f_{sc}/455$$
$$f_V = f_H/525$$

The Q-channel bandwidth is greater than 400 kHz at -2 dB; the I channel bandwidth is greater than 1.3 MHz at -2 dB.

51.5.3 PAL colour signal

The PAL colour signal coding is similar to the NTSC signal coding with various principal differences as follows.

The R-Y and B-Y axis signals are encoded with equal bandwidths, using modulating signals E'_u and E'_v derived from the basic colour-difference signals thus:

$$E'_u = 0.493\,(E'_b - E'_y)$$
$$E'_v = 0.877\,(E'_r - E'_y)$$

The total picture signal is defined as:

$$E_m = E'_y + E'_u \sin \omega_{sc}t \pm E'_v \cos \omega_{sc}t$$

The ± sign signifies an alternating phase where the positive sign applies during odd lines of the first and second fields, and during even lines of the third and fourth fields.

The display gamma is assumed to have the value 2.8, and the chromaticity for equal drives to be illuminant C for systems B, G and H and illuminant D6500 for system I (x = 0.313, Y = 0.329). System I also assumes the primary colours to be:

	X	Y
red	0.64	0.33
green	0.29	0.60
blue	0.15	0.06

The subcarrier chrominance frequency is specified at 4.433 618 75 MHz, which relates the line and field scanning frequencies thus:

$$f_H = 4f_{sc}(1135 + 4/625)$$
$$f_V = 2f_H/625$$

The colour burst comprises ten cycles at subcarrier frequency with an amplitude of 3/7 of the black to white amplitude of the signal and a phase, relative to the $+E'_u$ axis, as follows:

- +135° during odd lines of the first and second fields and on even lines of the third and fourth fields
- −135° during even lines of the first and second fields and on odd lines of the third and fourth fields

51.5.4 SECAM colour signal

The SECAM system derives the luminance and colour-difference signals in the same way as for NTSC, assuming the primary colours and illuminant C for equal drives. The colour subcarrier, whose spectrum lies within the spectrum of the luminance channel, is frequency modulated by the two colour-difference signals on alternate lines. The principal features of this modulation are as follows.

The colour-difference signals applied to the modulator are:

$$D'_r = -1.9\,(E'_r - E'_y)$$
$$D'_b = 1.5\,(E'_b - E'_y)$$

The colour-difference signals have a low frequency pre-correction with corner frequencies of 85 kHz and 255 kHz. The resultant pre-corrected signals are expressed by $D'_r{}^*$ and $D'_b{}^*$.

The equation for the modulated chrominance signal for given values of the signals $D'_r{}^*$ and $D'_b{}^*$ is:

$$m(t) = M \cos 2\pi\,(f_{oR} + D'_R{}^*\Delta f_{oR})t$$

or

$$m(t) = M \cos 2\pi\,(f_{oB} + D'_B{}^*\Delta f_{oB})t$$

on alternate lines, where M defines the amplitude of the subcarrier signal which, to minimize visibility, is modulated in response to its instantaneous frequency spectrum and also to the luminance signal value. Also, its phase is changed by 180° every third line.

Δf_{oR} and Δf_{oB} are the nominal deviations for the lines modulated by $D'_r{}^*$ and $D'_b{}^*$ respectively.

Their values are:

$$\Delta f_{oR} = 280$$
$$\Delta f_{oB} = 230$$

The nominal values for the frequencies corresponding to zero chrominance are:

$$f_{oR} = 282\ f_H = 4.406\ 25\ \text{MHz}$$

and

$$f_{oR} = 272\ f_H = 4.250\ 00\ \text{MHz}$$

E C Thomson
Mullard Application Laboratory

52 Picture Displays

The dominance of the cathode ray tube (crt) for use in domestic TV displays has not been seriously challenged by any of the alternative technologies. Small screen non-crt systems are available, but it is anticipated that the crt is likely to dominate for many years to come. This is partly due to the crt constantly being improved to stay in front of the competition.

Advantages of the cathode ray tube are:

- keeps pace with evolving markets
- versatile in application
- high resolution colour >1000 TV lines
- high resolution monochrome >2000 TV lines
- high brightness
- available in flat form
- available to MIL spec for ruggedness
- requires few external connections

Disadvantages are:

- conventional tubes have a high bulk
- requires an eht supply
- high currents required for scanning

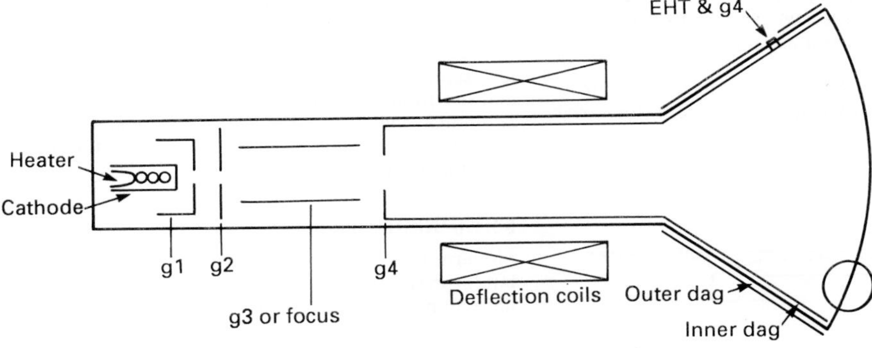

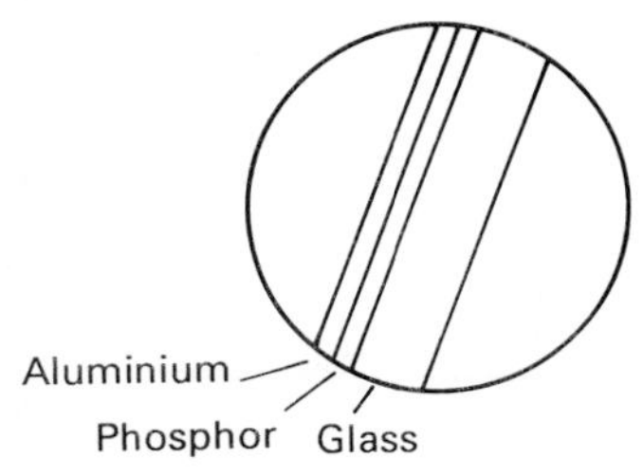

Figure 52.1 Cathode ray tube principle

- halo produced by light scattering in faceplate
- maximum size is limited
- flicker problems with interlaced scanning
- raster distortions produced by scan coil
- not directly compatible with digital circuits
- life currently suggested at 20 000 h

52.1 Cathode ray tube principles

The basic principles of the crt are shown in *Figure 52.1*.

52.1.1 Electron source

Electrons are emitted from a heated cathode. These pass through a small hole in the first grid attracted by the potential on the second grid. The intensity of the beam, and hence the spot brightness, is dependent upon the potential present on the grid g1. The more negative that g1 is biased, the smaller the beam current. Modulation for video may be applied to either g1 or the cathode.

52.1.2 Acceleration and focus

Having passed through the first grid, the electrons are accelerated by the second grid, sometimes called the first anode (A1), into the focus field generated by the third grid g3, and then further accelerated to the screen by grid g4 and the final anode. Two types of focusing are used, the *unipotential* gun and the *bipotential* gun (see *Figure 52.2*).

The focus electrode of the unipotential gun, commonly found in monochrome tubes, is intended to operate at about 250 V with a range from 0 to 350 V. The gun is known as *unipotential* because the electrodes on either side of the focus electrode are at the same potential. The advantage is that the gun is relatively insensitive to changes in focus potential. The disadvantage is that the gun can produce aberrations due to the beam passing through areas of highly distorted electrical fields produced by the unipotential electrodes, an effect which is normally reduced by shuttering the beam to prevent it passing through the areas of most distortion.

The focus electrode of the bipotential gun, used in most colour tubes, operates at approximately 18 per cent (for early tubes) to 30 per cent (for current tubes) of final anode voltage. It is known as *bipotential* because the electrodes on either side of the focus electrode are operated at different potentials. The advantage of the gun is that very little distortion is produced. The disadvantage is that the focus potential is high and requires to be stable, preferably tracking any changes in the eht.

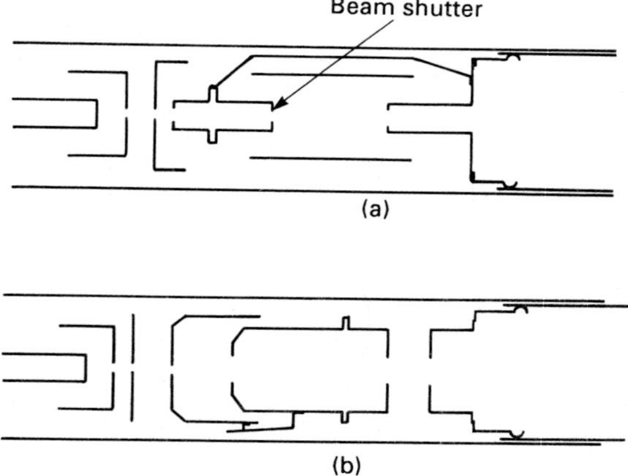

Figure 52.2 Focus grid construction: (a) unipotential electrode structure, (b) bipotential electrode structure

52.1.3 Deflection

Deflection can take two forms: electrostatic or magnetic. *Electrostatic* deflection, used only for small angles of deflection, consists of a pair of plates which when suitably charged attract and repel the beam to scan a line. A second pair of plates at 90° to the first completes the raster. *Magnetic* scan, used for most domestic tubes, uses the principle that the electron beam is deflected at 90° to a magnetic field. Two pairs of coils are used to generate the raster, one for horizontal or line-scan, the other for vertical or field scan.

52.1.4 Screen

Having passed the deflection field, the beam then accelerates towards the screen. Here the beam activates the phosphor which converts the electrons into light. Immediately before the phosphor, the beam passes through an aluminium backing film. The aluminium prevents ion burns and ensures that most of the light emitted from the phosphor is transmitted through the tube faceplate. (Ions are produced by the residual gas, left in the tube after sealing, being ionized by the electron beam. Ions are large and heavy and will burn the phosphor if allowed to strike it. The aluminium layer is thin allowing the electrons to pass while dissipating the energy of the ions.)

In a colour tube, colour selection is achieved by a shadow mask, which has the disadvantage that some 80 per cent of the total beam current is intercepted, in order to ensure that each of the beams activates only the phosphor for which it is intended.

52.1.5 Current return

To complete the current path of the electron beam, the internal surface of the tube bulb is coated with a graphite loaded conductive coating, joining the cavity connector to the aluminium screen backing and the g4 gun electrode. All electrons emitted from the gun are thus returned to the supply. An additional coating on the exterior of the tube provides a capacitor which may be used as a reservoir for the eht generated from the line flyback.

52.1.6 Raster shape

Modern styling requires the tube face to be as flat as possible but, as may be seen in *Figure 52.3*, a flat screen in the curved deflection field leads to *pincushion distortion* of the raster shape.

Correction may be applied by:

- winding the deflection coil to barrel distort the deflection field, thereby neutralizing the pincushion
- electronically modulating the deflection current, a technique suitable for use with a three gun colour system
- applying small permanent magnets to the periphery of the coil former or the tube bulb, effectively distorting the deflection field and correcting the picture shape

52.1.7 Modulation

We have seen that video modulation is achieved by adjusting the potential between cathode and first grid. In practice,

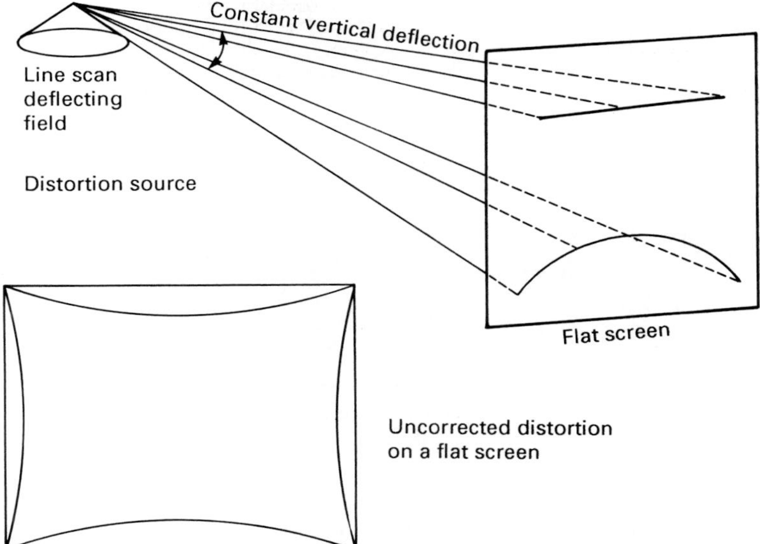

Figure 52.3 Inherent raster distortion on a flat screen cathode ray tube

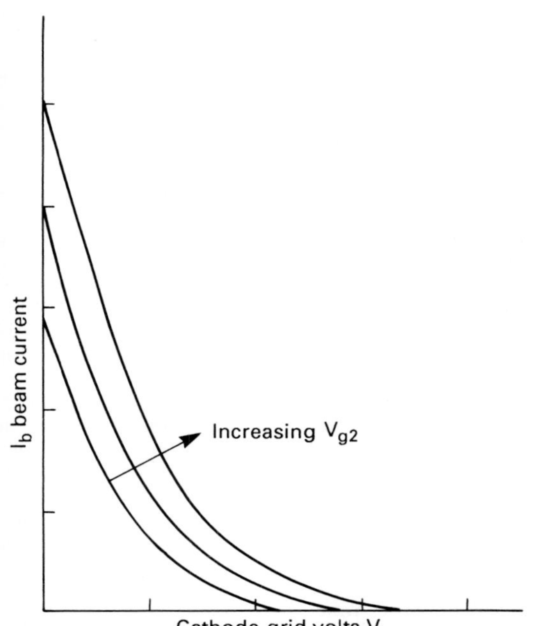

Figure 52.4 Cathode drive curve

modulation can be applied to either grid or cathode or both. Today cathode modulation is preferred. *Figure 52.4* shows the cathode drive characteristic for a typical tube. As may be seen, the relationship between the drive voltage and the light output is not linear.

The relationship between the beam current and the drive voltage is given by:

Cathode drive

$$I_{beam} = \frac{K(1+D)^3 V_{dr}^3}{[1+D V_{dr}/V_{co}]^{3/2} V_{co} 3/2}$$

Grid drive

$$I_{beam} = K(V_{dr}^3/V_{co} 3/2)$$

where D is the phosphor penetration factor, V_{dr} is the drive voltage, and V_{co} is the cut-off voltage.

The non-linearity is known as *gamma* and has a value of approximately 2.2 for monochrome tubes and 2.8 for colour. Signals produced from linear sources, e.g. the plumbicon camera tube, and the telecine require pre-distortion, gamma correction before transmission so that the original scene may be accurately reproduced. The above equations indicate that the gamma is slightly different for the two methods of driving and that the drive voltage amplitude is higher in grid drive than cathode drive.

In the early days of colour, a luminance drive voltage was applied to the cathodes and a colour difference to the grids. When large scale integrated circuits were introduced in which red, green and blue video signals could be easily generated, cathode-only drive became popular, reducing the number of video output stages. Colour tubes used in today's domestic receivers can only be driven this way as all the grids are connected in parallel between the three guns. Techniques for setting the grey scale with this gun construction are described in section *52.5.3*.

52.1.8 Cathodes

Traditionally, electrons are emitted from a heated cathode. This is normally a nickel tube with a coating of a barium rich material, heated by a tungsten filament. The temperature and coating composition are critical factors of tube life. Attempts have been made to extend the life of the cathode by providing a reservoir of barium compound to constantly refresh the active area. This increases the life by about three times, but the running temperature of this cathode is higher than normal which in itself tends to reduce the reliability.

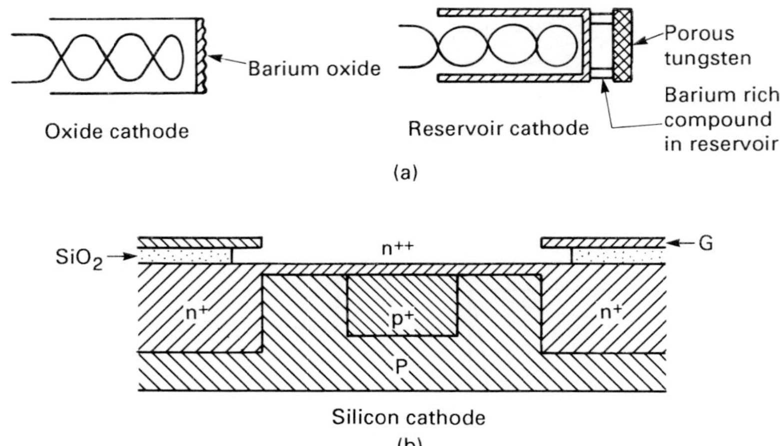

Figure 52.5 Cathode construction: (a) are heated cathodes, (b) is a cold cathode structure with a 10Unm emitting area

For some years intensive investigations into alternative electron emitters have been made. The *cold silicon cathode* is based on the principle that a pn junction will emit electrons at or near its surface when in a vacuum. Early work using silicon carbide materials were not successful because there was insufficient emission and the current density was too low. More recently an effect of electron injection from silicon in silicon dioxide layers, noticed because it degraded mos transistors, has been explored. The construction shown in *Figure 52.5*(b) is capable of high current densities and high injection efficiency. Cold cathodes in this form can be modulated directly. To change the diode current and hence the emission current requires only a few volts, thus the video amplifier may be simplified. A very wide bandwidth is easily achieved because the capacitance of the active area is small.

52.2 Monochrome tubes

Monochrome picture tubes used in television today differ little from the first commercial displays used for the first Marconi EMI 405-line transmissions.

Monochrome tubes for domestic use usually have a unipotential gun and a small neck (20 mm) to keep the overall power consumption down. No raster correction is applied to the winding of the deflection coil; instead small magnets are positioned on the periphery of the coil mounting which can be adjusted for optimum correction of the pincushion distortion.

Many different phosphors are available ranging from the medium short persistence white for TV ($x = 270$, $y = 300$ or 11 000 K), through green or orange for VDU displays, to the long persistence phosphors required for direct viewing radar.

52.3 Single beam colour systems

52.3.1 Beam penetron

This is a single gun system in which the colour selection is achieved by beam energy. Two types of phosphor are possible: voltage sensitive, and current sensitive.

Voltage sensitive phosphor is in use today, with two and three colour versions available. A multi-colour screen can be made up from phosphors laid successively on the screen, the individual colours being separated by layers of barrier material (see *Figure 52.6*). Alternatively the individual particles of phosphor can be constructed like an onion, with layers of different colour phosphors, again separated by a barrier material. Either way, colour selection is achieved by varying the eht, typically giving red at 10 kV, green at 14 kV and blue at 18 kV, the layers of barrier material assisting this selection.

Colour purity is reasonable due to the phosphors having a peak light output at a particular beam energy. Thus if red is the low energy phosphor, when the beam is accessing the green phosphor the output from the red will appreciably drop.

The main problem with this tube is that changing the eht changes the deflection sensitivity, so for each colour change the scanning amplitudes also have to be changed. This would be difficult to do on a pixel basis, so the tube is normally run in sequential field mode, so that adjustments to eht voltage and scanning amplitudes can be made during the field flyback time. To generate a suitable video waveform giving a flicker free display from composite video or RGB requires video field stores and high frequency scanning of the tube. This is thus an expensive display with very good resolution. *Current sensitive* penetron tubes can be made, but setting the grey scale is a problem. With the eht remaining constant, the beam energy is constant so that changes in colour must be accompanied by large changes in brightness (in the order of 1000:1). Pollution of the second colour by the first is much higher due to the efficiency of the first colour remaining constant. However, since the scanning amplitudes remain constant, colour changes can be made during a scan. Brightness variations could be compensated for by introducing modulated pulse width operation, reducing the duty factor for the high current phosphor. Thus this version could be cheaper to drive while maintaining the inherent high resolution of penetrons.

52.3.2 Beam index tube

A beam index, or *apple*, tube consists of a single beam colour display in which the colours are laid in vertical strips separated by guard bands. Line scanning produces continuous colour sequence (CCS), i.e. R,G,B,R,G,B,... Unlike most other display systems, the displayed colour is totally dependent on the instantaneous beam position. Variations in the line-scan current waveform and the stripe pitch require the instantaneous position of the beam to be obtained from the screen. This information can be used to form a suitable CCS composite signal to drive the single cathode.

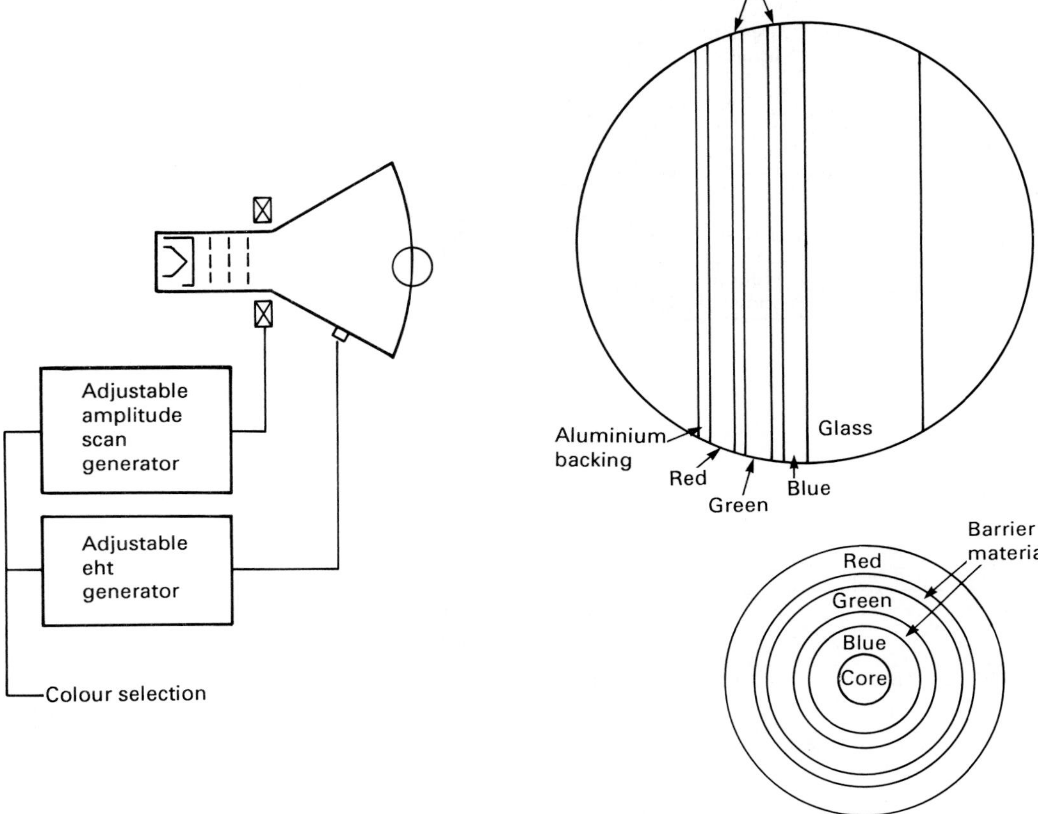

Figure 52.6 Beam penetron tube

One possible arrangement is shown in *Figure 52.7*. Index lines of ultraviolet light-emitting material are laid behind the screen aluminizing, in registration with the guard bands. The index signal, generated when the beam scans a line, is used to generate the CCS waveform. This signal, if its frequency and phase are constant, may be used to directly demodulate NTSC encoded colour signals. However screen non-uniformities and a line-scan generated by a conventional line timebase will produce frequency and phase variations which make the decoding less reliable. The index signal, reflecting the phase errors, may also be used to correct line timebase non-linearities and thus improve the system.

As the beam traverses the phosphor triplets, the frequency at which the beam switches between the colours is referred to as the *writing frequency*. To obtain colour purity, the spot width must not exceed one third of the triplet width (i.e. one phosphor stripe plus one guard band). To reduce the visibility of the individual phosphor lines, the phosphor structure must be as small as possible. A phosphor triplet pitch of 1.2 mm on a 21 inch tube operating at 625-lines would give a writing frequency of 8 MHz and a spot width of 0.4 mm maximum.

This is difficult to achieve with a circular spot shape since, at high beam currents, space charge effects cause the beam to spread. One solution is to make the spot shape elliptical with the major axis parallel to the phosphor lines. An elliptical hole in the control grid (g1) will form this beam shape. However, care must be taken to ensure that no beam rotation takes place in the focusing lens or by deflection field aberrations. To maintain a suitably shaped spot over the whole picture area, not only must the deflection coil be carefully designed but some form of dynamic focus may also be required at both line and field rates.

Detection of the ultraviolet index signal may be by use of a suitable detector positioned behind a window in the tube bulb. An alternative method of generating an index signal is to use lines which produce a high yield of secondary electrons. These electrons may be collected by a conductive coating inside the cone maintained several kilovolts higher than the screen.

As index circuits can have only a limited bandwidth, variations in the index signal frequency and phase are limited to a few per cent. The line-scan linearity and S correction must be tightly controlled and free from "ringing', and the eht must be stabilized against beam current variations. The stripe structure may be matched to the nominal line-scan to ease these stringent requirements.

It might at first appear that the simplest possible arrangement is to have one index stripe per phosphor triplet, so that the index frequency, f_i, is equal to the triplet frequency and

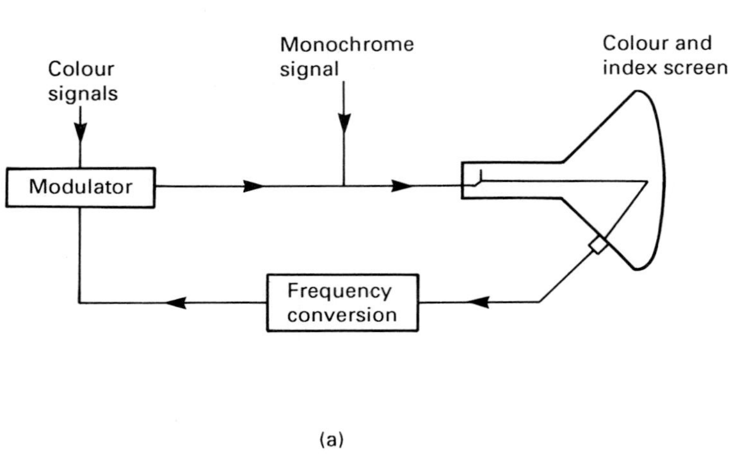

Figure 52.7 Beam index tube

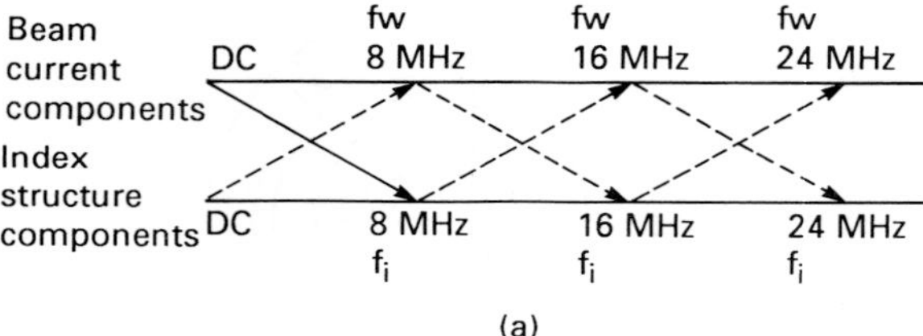

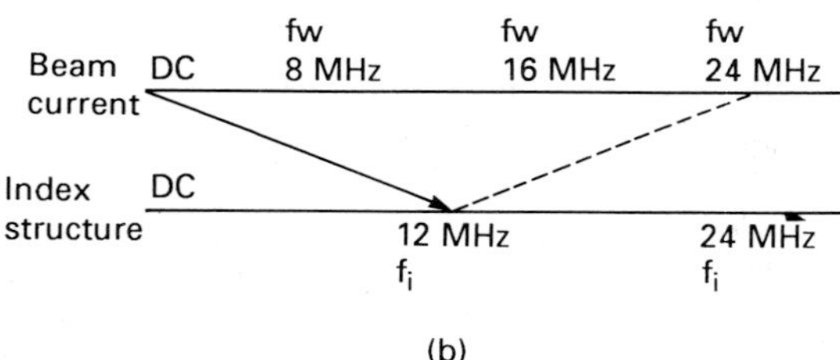

Figure 52.8 Crosstalk: (a) with $f_i = f_w$, (b) with $f_i = 3/2\ f_w$

therefore directly related to the writing frequency, f_w. *Figure 52.8*(a) shows the harmonic relationship of the index frequency and writing frequencies of this arrangement. Simply scanning the index stripes will give f_i plus harmonics. If the gun is modulated at f_w, when writing to the phosphors this will intermodulate directly with f_i producing phase errors and thus errors of hue in the final picture.

There are several solutions, the one in common use relating the index pitch to a non-integral multiple of the triplet pitch. The situation with $f_i = 1.5f_w$ is shown in (b). As may be seen, the crosstalk is present only at a wider spacing and is therefore easily filtered. As the index stripes now do not directly relate to the phosphor colour being accessed, it is common to have a run-in section of several index lines with a pitch of $f_w/2$ before the start of the phosphor stripes, so that the first line and therefore the correct sequence may be identified.

Advantages of the system are:

- single beam, therefore no convergence problems
- no shadow-mask to intercept a large proportion of the beam current

Disadvantages are:

- the beam current cannot be reduced to zero since the index signal will disappear
- stray light from the rear of the screen can interfere with the index signal
- line-scan has to be extremely linear to provide index signal at constant frequency

Sony and Sanyo have both developed small flat versions of the index tube with the beam accessing from the front of the phosphor. The semiconductor index detector is coupled to the stripes with a plastic light pipe, containing a phosphorescent frequency changing dye, allowing the detector to operate at peak efficiency. Sony have also described a beam index tube for use in a projection system.

52.3.3 Lawrence tube

The Lawrence tube or *chromatron* is a single gun tube constructed around the principle of post-deflection focusing. The original device as developed by Chromatic Television Laboratories Inc. had horizontal phosphor lines laid on a flat glass plate. A wire grid is positioned some 20 mm behind the phosphor. Alternate wires are connected together to form two interleaved grids. Wires from one grid are positioned behind the blue phosphor stripes and from the other grid behind the red, leaving the green unscreened. There are twice as many green phosphor stripes as red or blue.

In operation, the phosphor screen is maintained some 17 kV above the grid wires, giving considerable post-deflection focusing to the beam. If both grids are at the same potential then the beam is focused onto the green phosphor. By applying a potential difference of some 400 V between the two grids the beam can be deflected onto the blue or red phosphor stripes (see *Figure 52.10*). By driving the grids at a frequency of 3.58 MHz, each primary colour is selected for 0.28 μs giving dot sequential operation, allowing the possibility of directly decoding an NTSC signal. The grid structure has typical capacitance of some 3 nF, and may be resonated with a high Q inductance and driven by a class C output stage. The critical

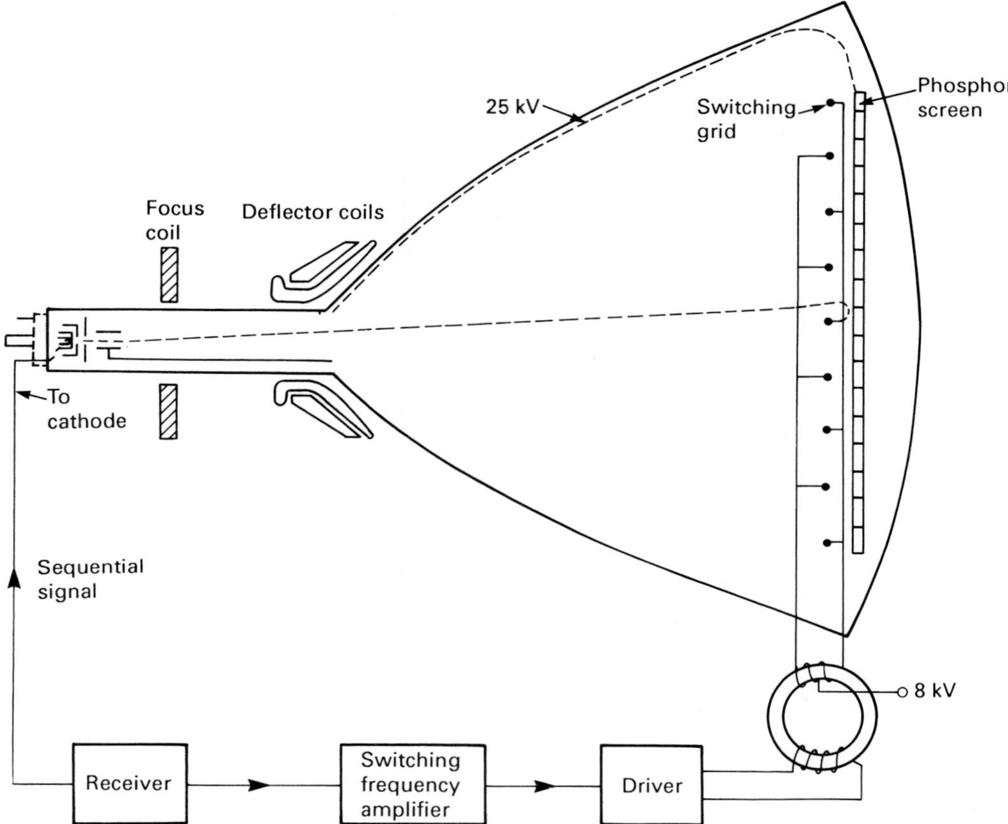

Figure 52.9 The Lawrence tube

components may be mounted inside the tube envelope to reduce losses and keep electromagnetic radiation to a minimum.

Later developments used a 22 inch 90° tube with the phosphor deposited directly on the faceplate. Red was chosen as the unswitched colour instead of green to compensate for the lower phosphor efficiency, and the number of stripes was doubled to 1000 red with 500 blue and green. A three-gun version was also developed in which a single vertical grid focused the beams onto the screen arranged as a repeating colour sequence. This tube with its single vertical aperture grid may be considered the forerunner of the Trinitron tube.

The Lawrence tube was the brightest in its day and would hold its own today but for a dramatic contrast reduction caused by secondary emission from the beam steering wires and poor resolution due to low beam energy prior to the wire grid, resulting in a large spot size. The system is still being considered and with modern materials could make a comeback.

52.4 Side gun and folded electron optical systems

52.4.1 Gabor and Aikin tubes

The main criticism of conventional crts is that they consume a lot of power and are bulky. Work to reduce both the power consumption and the size has been going on, somewhat intermittently, since the early 1950s. Aikin and Gabor, independently, showed the way to produce flat tubes, only a few centimetres thick.

The Aikin tube used a side gun and two orthogonal deflection systems, the first for line-scan and the second for field. Both scans used approximately ten plates, the line running at some 1 kV and the field at 12–15 kV. This high voltage switching proved difficult, and Aikin developed a multiple anode triode with variable-μ grid to provide the scanning voltages. The Gabor tube used a gun mounted vertically behind the screen, electrostatic line-scan plates, a lens system to turn the beam through 180°, and an array of field plates which had to be switched through some 15 kV to provide the field scan.

52.4.2 Mullard Slimscreen

Both the above systems suffered from the problem that full beam current had to be deflected through a tortuous route to the screen. The problem is less if the beam can be formed and scanned at low voltage and current and then the beam energy increased to activate the phosphor screen. Mullard developed a technique to increase the beam current, using a microchannel plate electron multiplier (mcp). The beam energy used is only 600 eV, allowing the raster to be generated relatively easily, and the beam current is multiplied to a level suitable to excite the phosphor.

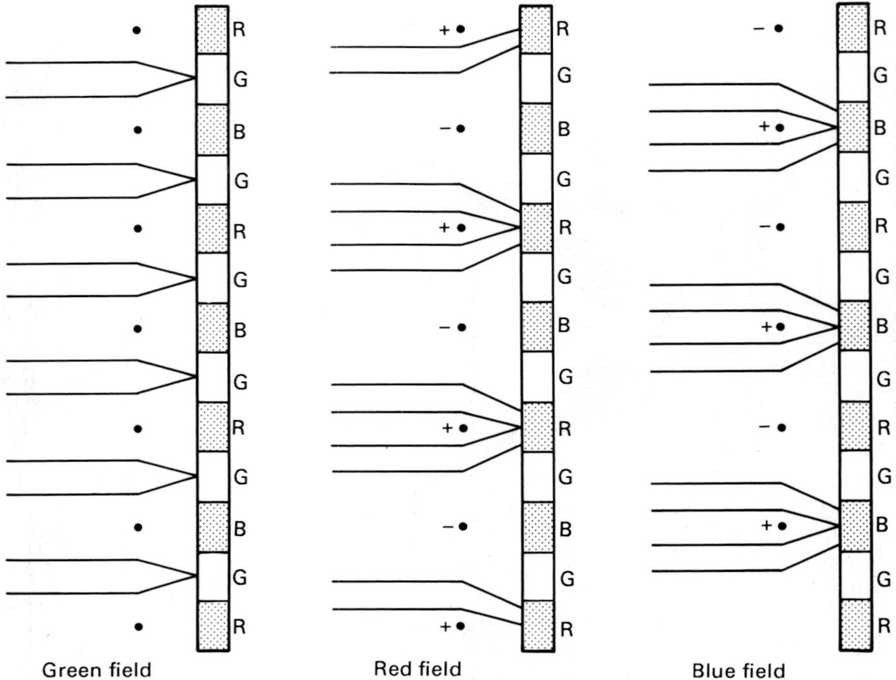

Figure 52.10 Switching operation of Lawrence tube

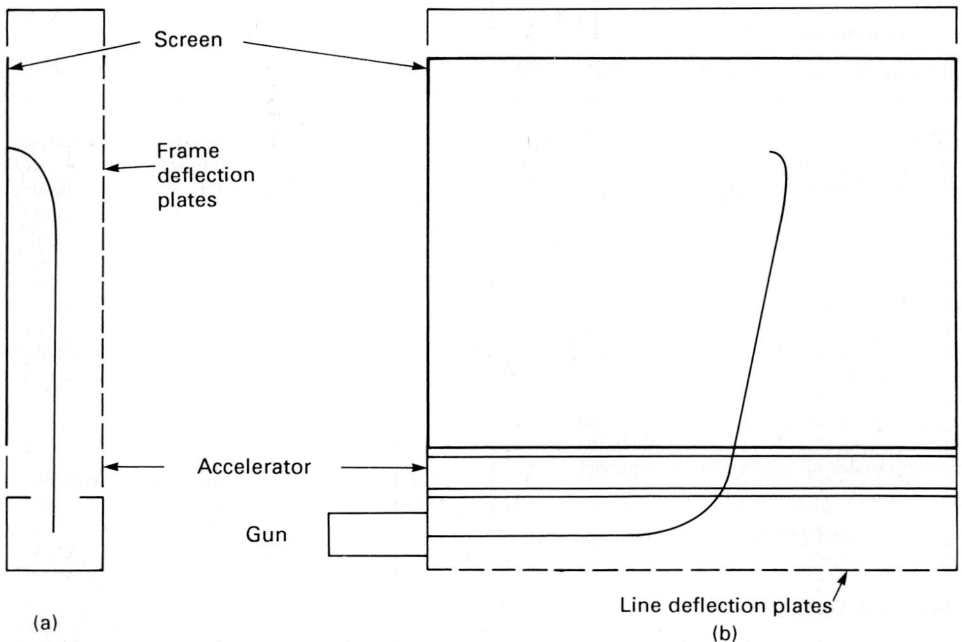

Figure 52.11 Aikin slim tube: (a) side view, (b) front view

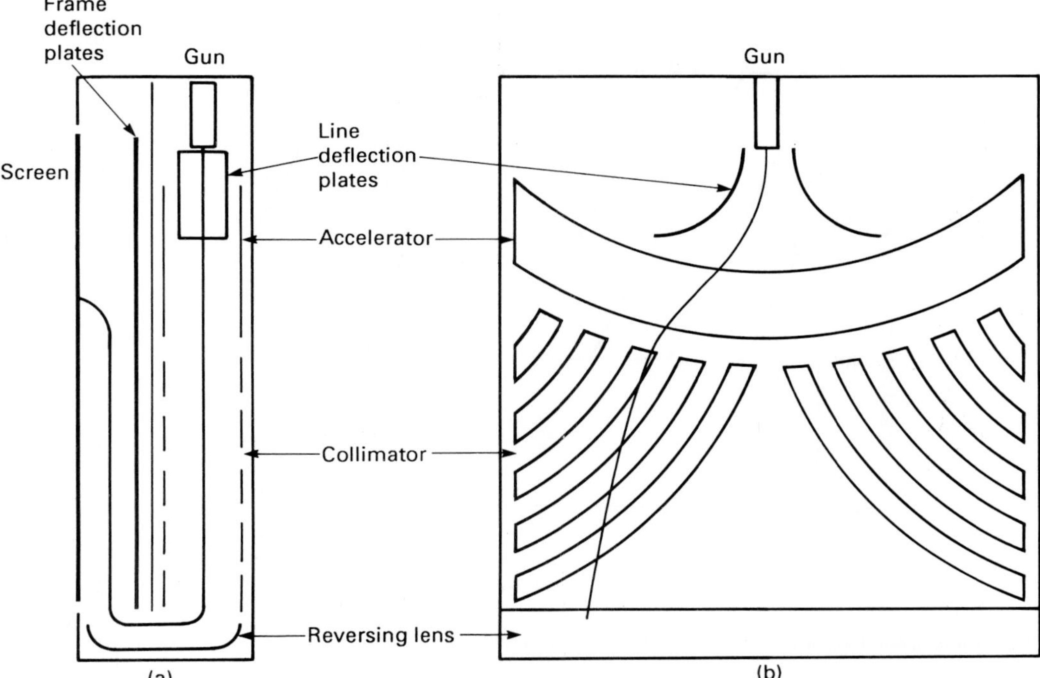

Figure 52.12 Gabor slim tube: (a) side view, (b) rear compartment

The Slimscreen tube with a 200 mm screen and about 50 mm deep is based on the Gabor system, with a vertical gun, 180° reversing lens and eleven plates to provide field scan. An electron gun mounted at the rear of the assembly directs an electron beam vertically downwards towards a reversing lens mounted in the lower part of the tube. The beam passes between a pair of conventional electrostatic deflection plates producing the line-scan. The beam is then turned through 180° emerging in the slot between the planar array of field deflection plates and the mcp. The beam is electrostatically deflected, by the field deflection plates, forwards onto the input of the mcp. Here the beam current is multiplied by approximately 1000, and the output is proximity focused onto the screen. The beam acquires sufficient energy to provide a highlight brightness of some 1200 cd/m^2.

The electron gun is conventional in construction, working at 600 V and a beam current of 200 nA maximum. With such a low beam current the cathode drive potentials are also low, thus a high frequency high resolution system may be easily achieved.

Prior to line deflection an extra pair of deflection plates, *Z plates*, steer the beam to the correct entry point in the reversing lens. Line deflection is performed by a pair of electrostatic plates, *X plates*, driven with a sweep potential of 120 V. Line deflection produces deflection defocusing of the beam which is corrected by a combination of dynamically changing the main focus and the mean dc potential of the X plates. Focusing thus has two aspects: the main focus electrode provides the overall beam shape and dynamic correction which compensates for deflection defocusing and astigmatism introduced by the line deflection plates. The beam then enters the reversing lens to be deflected into the zone behind the mcp.

Field deflection is accomplished by voltage ramps applied to the field deflector plates. With the mcp input potential and the field plates at 0 V, the beam is undeflected. Decreasing the

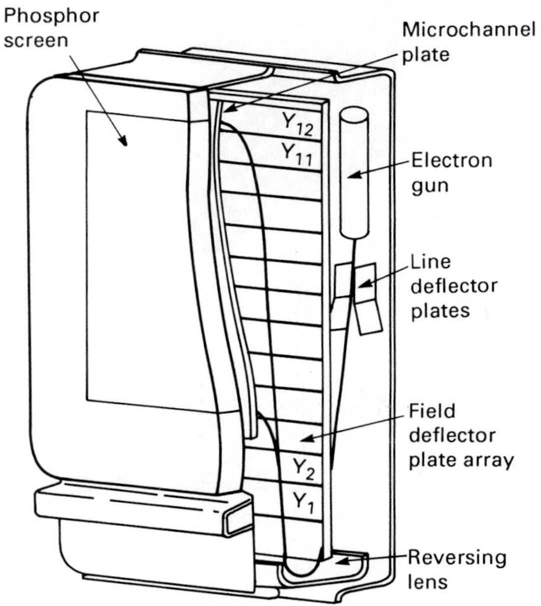

Figure 52.13 Mullard Slimscreen tube

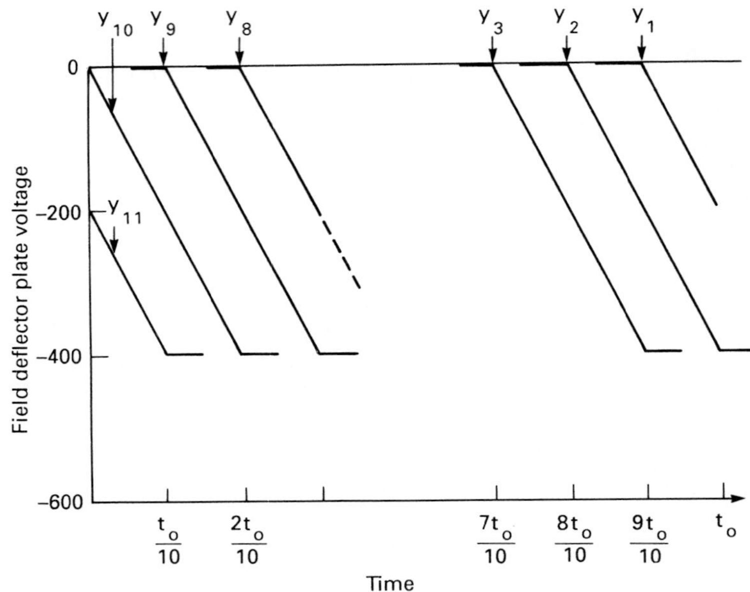

Figure 52.14 Field defection waveform of a Slimscreen tube. t_o = field display time

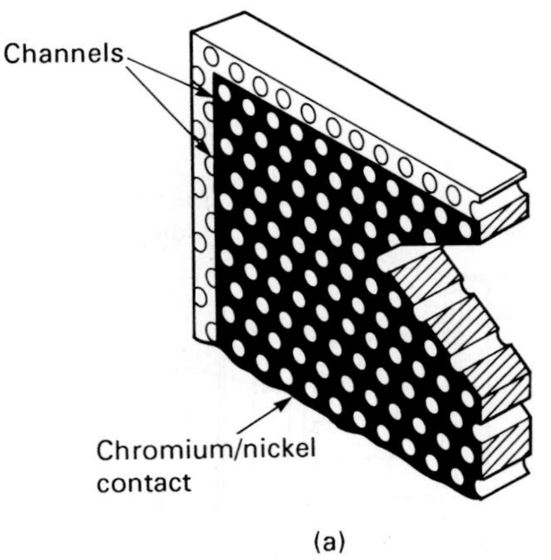

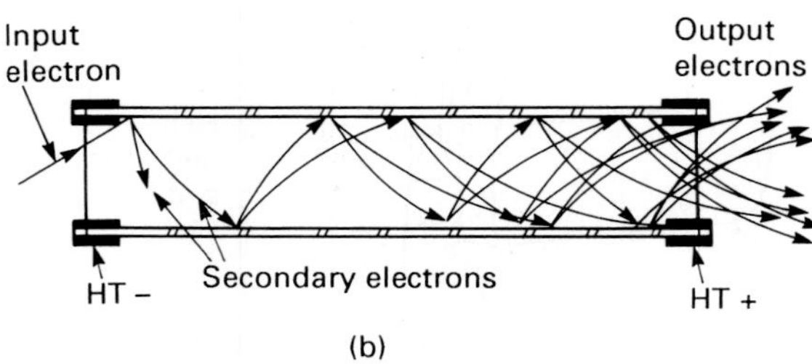

Figure 52.15 Microchannel plate assembly (a) and the electron multiplication mechanism in a single microchannel (b)

potential on any deflector plate will bend the beam towards the mcp input.

Field scan is achieved by applying sequential ramps to the field deflector plates. Two plates are sequentially ramped at any one time, the most negative waveform applied to the upper plate and an intermediate potential on the lower plate, as shown in *Figure 52.14*.

The mcp multiplies the low energy beam current to give sufficient beam current to excite the screen phosphor. The microchannel plate consists of a closely packed array of glass tubes fused together to form a thin plate. Electrodes are evaporated onto each face to allow a potential difference, typically 1000 V, to be applied between the ends of each tube or channel. The mechanism of current amplification within a single channel is illustrated in *Figure 52.15*(b). The glass is a special semiconducting type with a high secondary emission ratio.

When an electron strikes the input, a number of secondary electrons are produced. Each of these electrons is accelerated and strikes the wall farther down the tube, producing more electrons. This process is repeated down the tube producing a large current pulse at the output for each input electron. The amplified beam current is accelerated and proximity focused onto the screen to produce a picture with a full grey scale and a highlight brightness of up to 1200 cd/m^2.

The tube has been developed in monochrome form with a colour version under consideration using a penetron phosphor. The advantage with this type of tube when used with a penetron phosphor is that the scanning amplitudes do not require adjustment when the eht is changed, since it is only the proximity focused output from the mcp that is affected.

52.5 Three beam colour systems

52.5.1 RCA shadow-mask tube

The first practical colour tube to be produced had 90° deflection with a delta gun and a shadow-mask using a dot structure

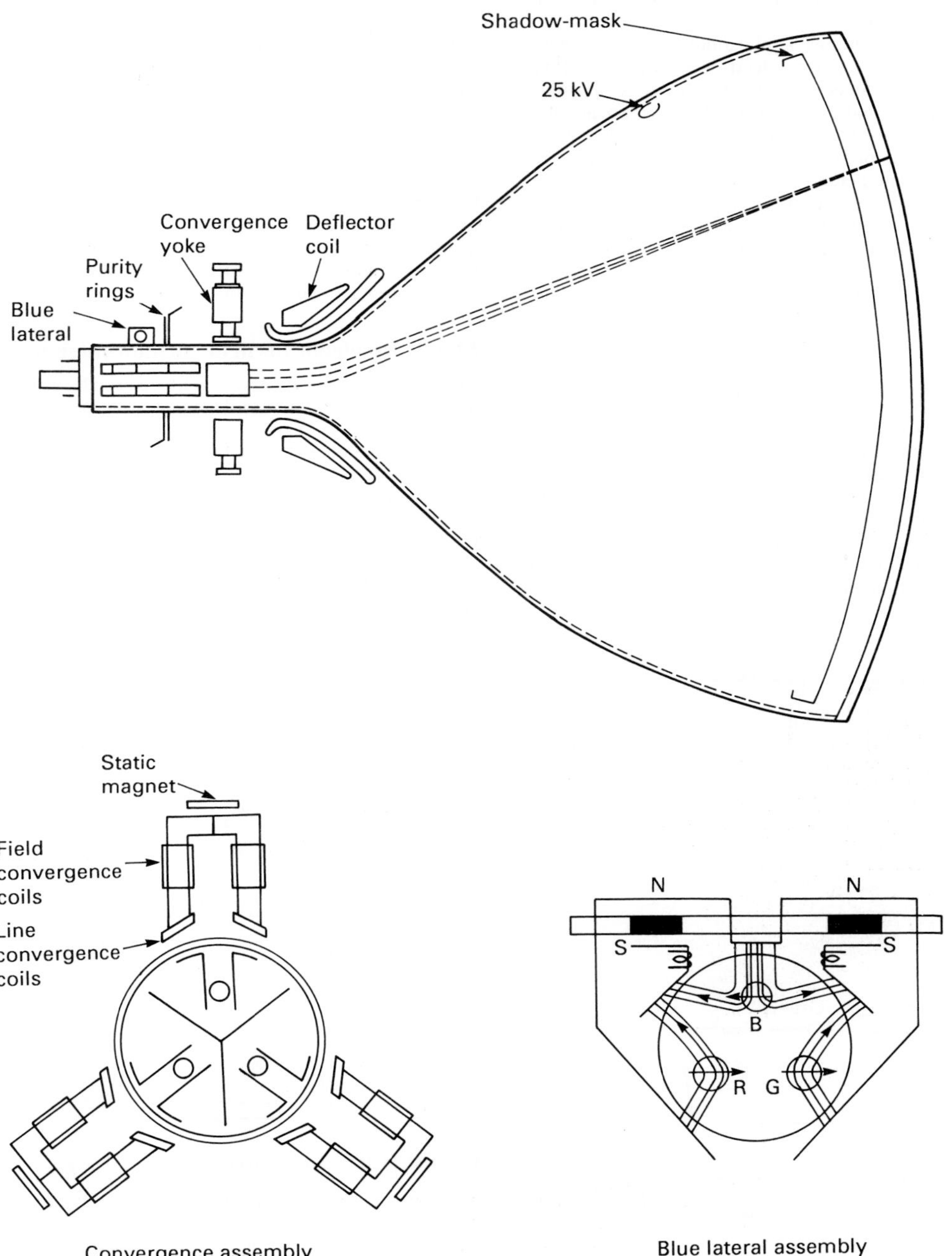

Figure 52.16 RCA shadow-mask tube

screen. While the basic principle of the tube is simple, the requirements of high definition, colour purity and registration pose some difficult problems. The fact that these tubes, and their derivations, have been manufactured in huge quantities is a tremendous achievement.

The principle of the RCA shadow-mask tube (shown in *Figure 52.16*) is as follows:

A single envelope contains a phosphor screen, a metal shadow-mask and three separate electron guns whose beams are scanned by a single deflection coil assembly. The phosphor screen is built up from small circular dots of red, green and blue light-emitting phosphors that are arranged to touch tangentially so that their centres lie on the vertices of equilateral triangles. Approximately 2.5 cm behind the screen is a metal plate, pierced with as many holes as there are triads of phosphor dots. The three electron guns are mounted in a delta configuration with the overall geometrical arrangement of the assembly such that each electron beam can only impinge on the phosphor colour for which it is intended. The other phosphors are shielded by the shadowing effect of the shadow-mask. If the three guns are now suitably modulated, a colour picture can be produced.

In order to obtain a high quality display allowing for production spreads, a number of additional beam steering devices are necessary.

For the electrons from each gun to strike only the intended phosphor they must pass through the approximate centre of deflection. This is achieved by a pair of two pole magnets, or *purity rings*, mounted approximately over the first anodes the field from which corrects the axial position of the beams prior to deflection. The deflection coil can be moved forwards and back along the tube neck to adjust the centre of deflection.

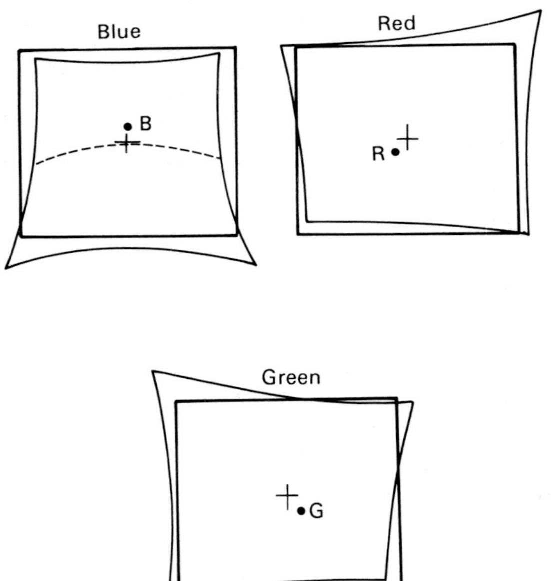

Figure 52.17 Raster distortion produced by delta gun

Having degaussed the tube, the setup procedure is to move the deflection coil fully back then, using the red gun only, adjust the purity rings for a pure red ball in the centre of the screen. Now move the coil forward until the pure centre expands to fill the screen. The other colours are then checked and, if necessary, small adjustments made to the rings and coil position.

The beams must converge at all points on the screen. Convergence is achieved by a combination of static and dynamic magnetic fields applied individually to each beam. *Figure 52.16* shows the general arrangement of the magnet assembly. There are four magnetic fields to steer the beams so that they converge at the tube centre. Three adjust the radial positiion of the beams. The fourth, a double magnet assembly, shifts the red and green with respect to the blue. These controls are known as *static convergence* with the last called the *blue lateral*.

The uncorrected raster shapes due to the offset guns are shown in *Figure 52.17*. These rasters may be converged using both line and field parabolic and sawtooth waveforms, applied to the dynamic convergence yokes. Note that the corrections available with domestic circuitry can only be made across the horizontal and vertical centres of the tube, and any residual corner misconvergence is due to coil errors.

52.5.2 Developments in the shadow-mask tube

The shadow-mask tube has developed as to be almost unrecognizable when compared to the early RCA tube.

- The deflection angle can be increased to 110°

Increasing the deflection angle to 110° reduces the depth of the tube and therefore the cabinet, giving designers an opportunity for an elegant slim profile. The penalty for the increase in deflection angle is power consumption and circuit complexity, notably the addition of east/west correction. The increase in angle has not been easy. Early tubes based on a delta gun had elaborate correction circuitry, including corner convergence, and north/south and east/west pincushion. Before the introduction of the diode modulator there were two line timebases: the main scan generator also developing the eht, etc., and a smaller line-scan generator used to modulate the main scan generator, providing east/west pincushion correction.

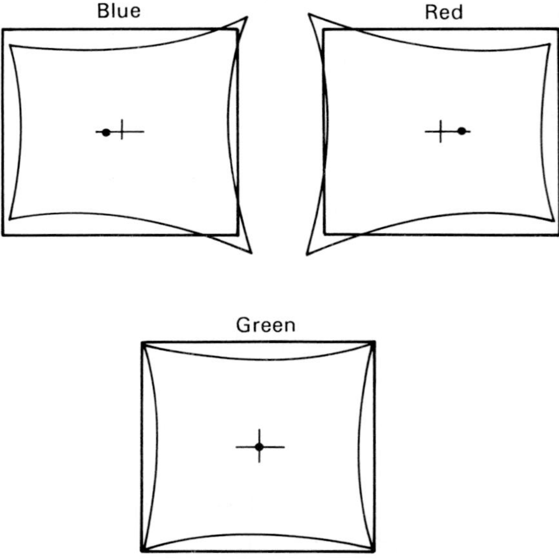

Figure 52.18 Raster distortion produced by in-line gun

- Guns are mounted in line

The early delta gun has been changed to an in-line gun. A comparison of *Figures 52.17* and *52.18* shows that convergence problems are mush eased, due to the symmetrical distortion of the raster shape produced by the two off-axis beams, which are now on the horizontal centre line.

With the introduction of fs (*flatter squarer*) tubes, Mullard reduced the neck diameter from 36 mm to 29 mm and redesigned the gun to suit. This gun uses less deflection power and has common grid and anode components to reduce the circuit complexity. An additional advantage is that inherent convergence errors may be reduced by tightly controlling the winding of the deflection coils. Toshiba have further reduced the neck diameter to 22 mm. This small diameter has the advantage that the deflection power can be further reduced. The beams are physically closer, making good convergence easier to achieve, but the colour selection angle is smaller reducing the colour reserve, making the positioning of the deflection coil more critical.

- Slot holes in the shadow-mask

Shadow-mask holes have evolved from round to rectangular slots with the long axis vertical. This improves the transmission of the shadow-mask, giving brighter pictures. When used with screen deposition techniques that allow for continuous vertical stripes of phosphor, the sensitivity to terrestrial magnetic effects is reduced, improving purity and allowing for a simpler method of degaussing.

- 90° tubes are raster correction free

90° tubes can have the gun structure slightly modified to allow the coil to be wound to give *raster correction free* (rcf) operation, a system which is in common use on tubes up to 21 inch. Toshiba have produced a tube with 110° deflection which requires no raster correction. This tube is expensive but shows the possibilities for such a tube.

- Only 110° tubes require east/west correction

The deflection coil for 90° tubes can now be wound to give a deflection field which also converges the beams, leaving equal raster distortion for all three beams, visible as vertical pin-cushion distortion of the picture, known as *east/west*. The line timebase can be modulated using a diode modulator to correct for this distortion.

- Magnet ring

With the introduction of the in-line gun, a magnet assembly was used on the tube neck which replaced the original two pole purity rings with the addition of a four pole pair of rings for statically converging red and blue, and a six pole pair for converging the green with red and blue. This assembly is prone to drifts and accidental misalignment. In many tubes these have all been replaced by an internal ring, magnetized automatically during production. The internal ring gives an improvement in reliability and reduces the manual component of setting up.

- The picture tube shape now has a 4:3 aspect ratio with square corners

The screen shape has changed from 5:4 ar to 4:3 ar (see *Figure 52.20*). As the transmitted picture is 4:3, more of it is visible. The change is a consequence of cabinet restyling and an evolution of line timebases to produce less raster distortions, ringing, etc., so there is extra useful picture area. The removal of the radiuses in the corners of the picture area and a straightening of the sides give a rectangular picture shape.

- The radius of the screen face has been doubled

As well as squaring the picture, the radius of the front plate has been doubled, giving a flatter picture. This gives the advantage of a wider viewing angle and, of course, modern styling. Unfortunately this screen flattening has proved difficult to implement. The flatter shadow-mask has proved to be thermally problematical. It is difficult to compensate for expansion of the mask, particularly local expansion in high contrast ratio pictures. A special corner mounting suspension has been designed to compensate automatically for overall thermal expansion which, together with techniques developed to compensate for the mask 'local doming' (an expansion of the

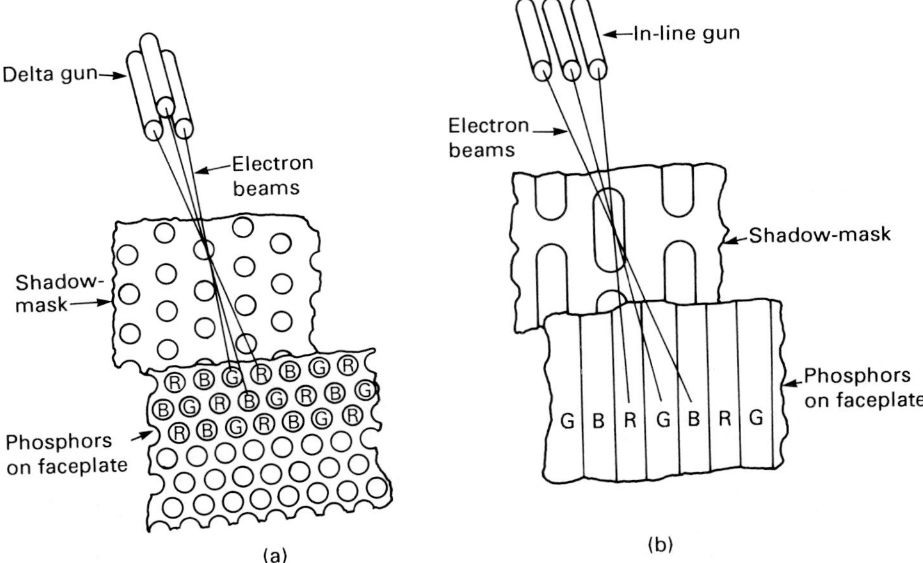

Figure 52.19 (a) Delta gun with dot shadow-mask, (b) in-line gun with slot shadow-mask

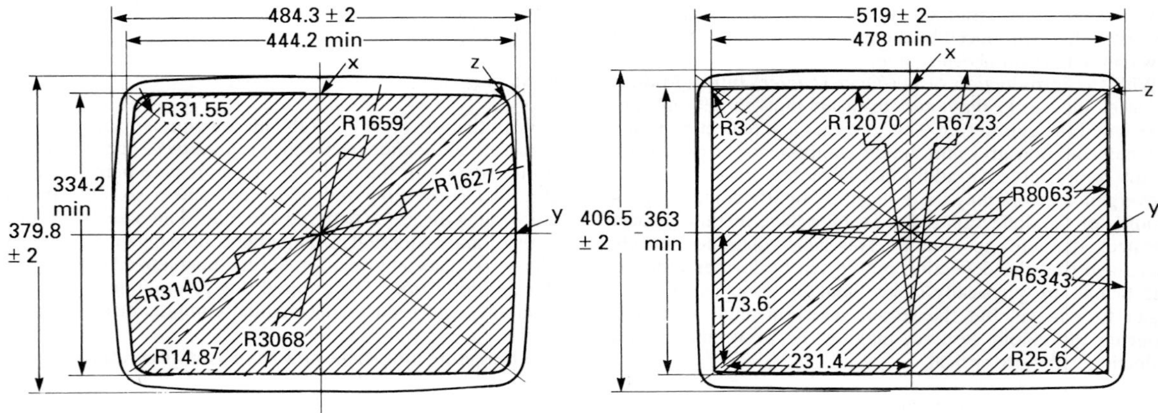

Figure 52.20 Comparison of traditional and fs bulb shapes

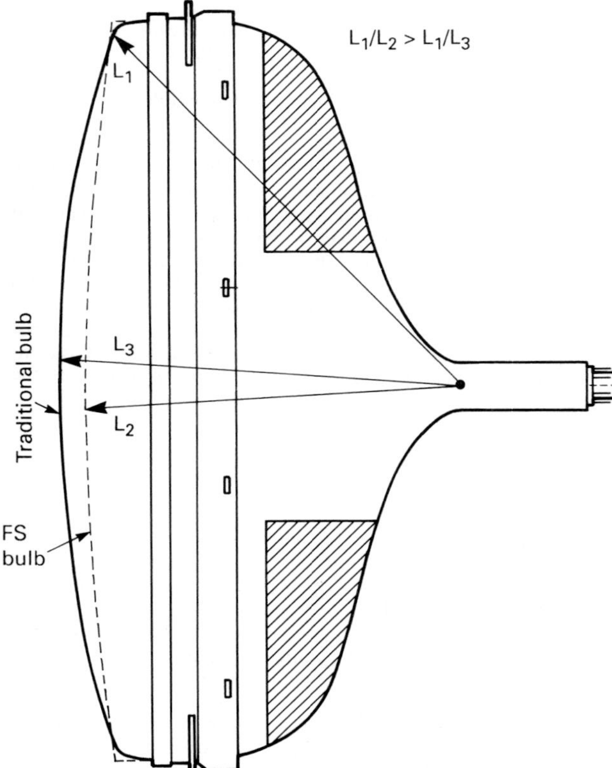

Figure 52.21 Curvature comparison

shadow-mask producing a small area of colour impurity with a high contrast ratio picture), gives a performance measurably better than the earlier tubes.

The gun requires a much tighter specification since the ratio of distances of gun to picture centre and gun to picture edge is higher than any predecessor tube (see *Figure 52.21*). The focus requirements are also more stringent with the spot size, astigmatism and haze all tightly controlled in the new gun to provide pictures with at least equal, and usually better, performance than the predecessors. In the receiver, the flatter screen requires more S correction for both line and field timebases. East/west pincushion is corrected in the usual way with a diode modulator. Larger tube sizes require inner pincushion correction at the 2/3 point (see *Figure 52.22*). An inner pincushion distortion of some 4 mm is present in a 660 mm tube which may be corrected using dynamic S correction.

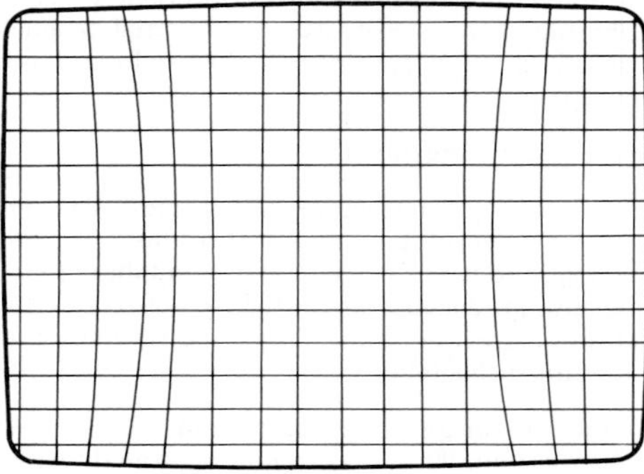

Figure 52.22 Inner pincushion on fs tube (exaggerated)

- Neck diameters have been reduced

The tube shape defined by the two previous items is known as a flatter squarer tube and has required the development of a new gun structure. Traditionally the 110° tubes had a 36 mm neck with three separate guns. This gun separation gives a good colour selection angle, and therefore good purity, but a deflection coil providing good convergence and raster shape on the flatter screen would be difficult to manufacture. The neck diameter was reduced to 29 mm which gives a suitable compromise for good convergence and a sufficient colour selection angle for purity. The smaller diameter requires less deflection power thus reducing the total power consumption of the set.

- The deflection coil is married to the tube in the factory

The introduction of smaller tube neck diameters with small gun spacing means that the positioning of the coil is becoming

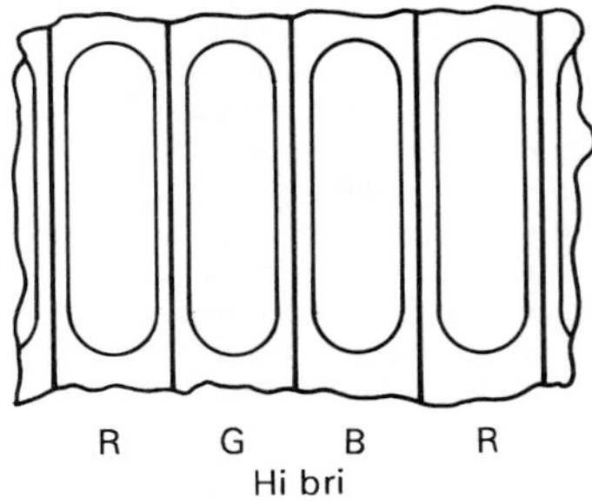

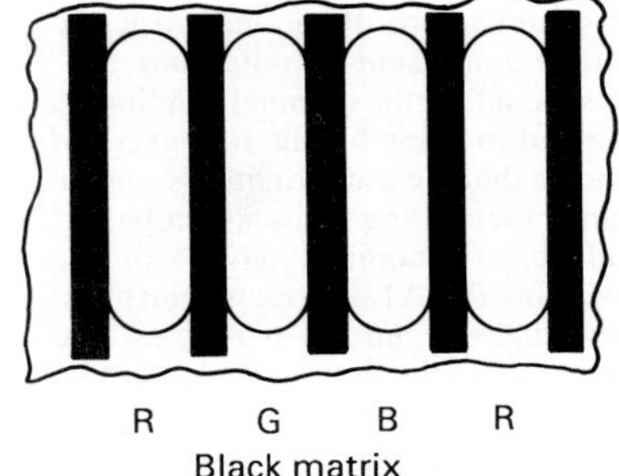

Figure 52.23 Black matrix screen construction

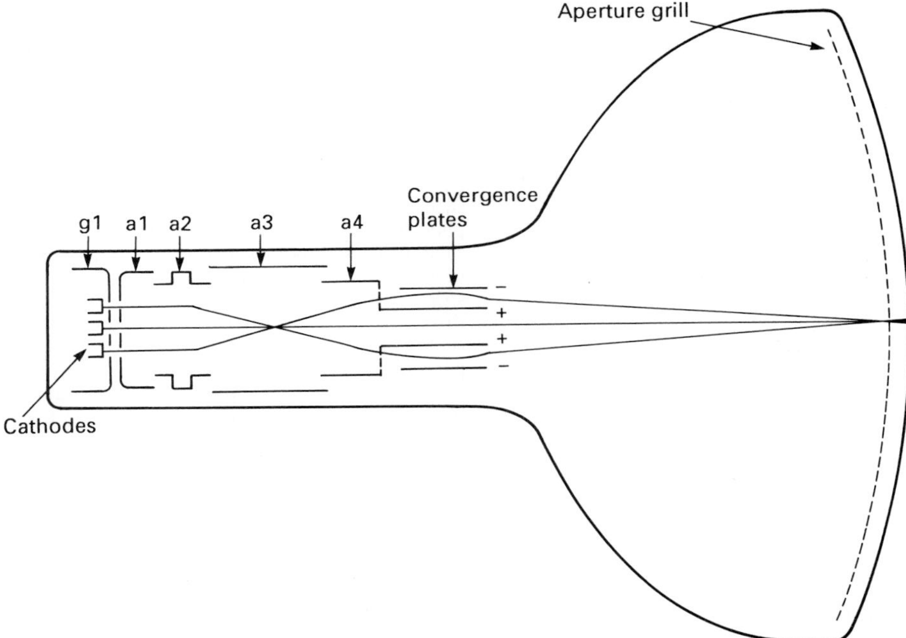

Figure 52.24 Trinitron tube

more critical. Not all tube/coil combinations are capable of providing the high standards required today. Attaching the deflection coils in the factory, correctly setting the axial position of the coil and the field strengths of alignment magnets (if any), should give optimum picture quality during the life of the tube.

- The maximum picture tube size has increased to more than 1 m

The achievable size of factory-produced picture tubes has gradually increased over the years from 25 inches (63.5 cm) to 42 inches (106.7 cm), bringing cinemascope into the home at a price. The commonly produced tubes range from 9 to 27 inches (22.9 to 68.6 cm) all in the flatter squarer format which is gradually replacing the traditional shape. Completely flat faced tubes have been publicly demonstrated.

- Black matrix

If an unlit screen is viewed in daylight it looks light grey, the normal colour of the phosphor material. When viewing a picture, the black part of the grey scale is represented by this unlit phosphor. In order to obtain a high contrast ratio, the illuminated parts of the screen must be very much brighter than the incident light reflected from the phosphor. A technique known as *black matrix* has been developed to increase the contrast ratio by reducing the light reflected from the screen. *Figure 52.23* shows the general arrangement of the screen.

In a screen with a triplet pitch of 0.6 mm the black stripes are 0.08 mm wide giving a total black width of 0.24 mm, i.e. 40 per cent of the total width of the triplet. Therefore 40 per cent of the incident light will be absorbed giving an apparent increase in contrast ratio. This has a secondary effect of providing guard bands between the phosphors giving a greater purity reserve and/or a reduction in sensitivity to terrestrial magnetic effects.

52.5.3 Grey scale setting procedure

Modern gun structures incorporating common g1 and g2 components require the grey scale to be set entirely from the potentials applied to the cathodes. The controls usually provided are red, green and blue background or cut-off, two or three drive controls and a screen or A1 control. There are tolerances associated with manufacture of the gun structure which result in a spread of the required cut-off potentials.

Initially all three cathodes are set for the nominal conditions, then these settings are modified to allow for the tolerances of the individual guns. First ensure that the user brightness control is set to the centre of its range then, using an oscilloscope, set the cut-offs to the manufacturer's nominal data with the background controls. Next, using the A1 or screen control set for raster extinction of the highest gun, then readjust the background controls to give raster extinction for the other two guns. Some receivers allow the field scan to be collapsed for this adjustment, since extinguishing a single horizontal line is more accurate.

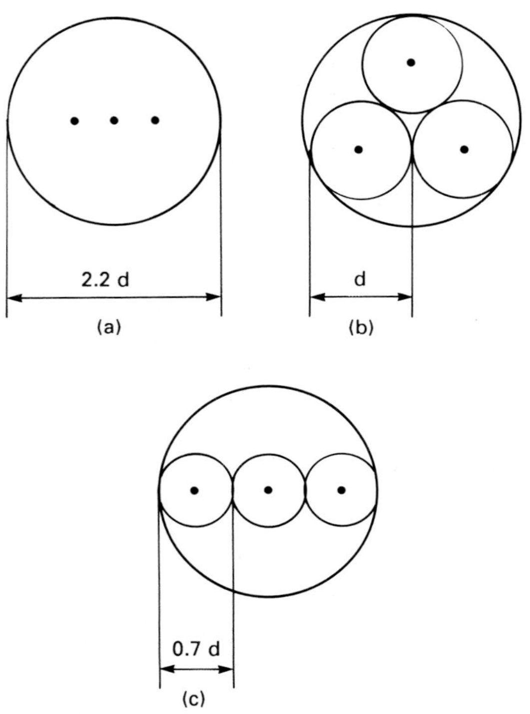

Figure 52.25 Comparison of effective electron lens diameters: (a) Trinitron, (b) three delta guns, (c) three guns in-line

Then the white point is set, using a white raster, and the drive controls are adjusted for the colour temperature as recommended by the set maker or according to personal preference. Ideally a colorimeter should be used, but a fluorescent tube of the correct colour may be employed as a grey scale reference, using the eye as the comparator, a job for which it is surprisingly accurate. Illuminant D at 6500 K ($x = 313$, $y = 329$) used to be the standard, but set makers are now tending to use higher colour temperatures of up to 8400 K.

52.5.4 Trinitron tube

A tube devised by Sony of Japan uses a shadow-mask in the form of vertical slots. The resulting phosphor pattern on the screen is a vertical stripe pattern with a triplet pitch of 0.55 mm.

To illuminate this screen Sony has developed a special gun, which consists of a single electron optical system containing the three beams, shown in *Figure 52.25*(a). The large diameter electron lens system thus formed presents a much more uniform electrostatic field to the electron beams, and produces fewer aberrations than the conventional three gun system. It is claimed that this gun produces a higher beam current and smaller spot size than conventional guns.

Figure 52.26 shows the development of the gun and the present system in which the three beams are produced from three independent cathodes feeding common grids and anodes. The anode two, anode three structure forms a weak lens converging the three beams so that they pass through the centre of the large focusing lens to reduce the aberration. The focusing lens is of the unipotential type not requiring a separate supply. As they leave the gun, the outer beams are converged by a pair of electrostatic prisms. The early tubes used a convergence waveform consisting of 350 V static, modulated by some 30 V dynamic, required to be biased close to the eht potential. Sony initially used a separate connection on the tube neck, then a coaxial eht connector. In a current tube, the coil combination has been wound for 'raster correction free' operation, and thus only static convergence is required. A thin film resistor is mounted alongside the gun terminating in a variable resistor on the base board marked *H-STAT*. The convergence prism is thus operated by a potential divider across the eht to ground, controlled by the potentiometer at the bottom of the chain.

Colour selection in this tube is achieved by use of an *aperture grill*. A conventional shadow-mask consists of a thin metal sheet pierced with many holes; for each hole three areas of phosphor are deposited on the tube face. The Trinitron shadow-mask, however, consists of a metal plate etched into vertical strips, and for each slot thus formed, three phosphor stripes are deposited on the faceplate. Advantages claimed for this grill structure are: higher transparency to the beams (20 per cent in the centre and 15 per cent at the edges, compared to 15 per cent and 12 per cent respectively for a conventional shadow-mask); a reduction in sensitivity to terrestrial magnetism, and freedom from Moiré patterning produced by the beating of the horizontal scanning lines with dot or block structure of the shadow-mask tube. A triplet pitch of 0.55 mm is employed with no guard bands.

The aperture grill and screen are cylindrical in shape instead of the usual spherical form used in a conventional picture tube. This makes the production of a stable grill possible, although additional support in the form of 3 μm wire spirals is used to damp out any microphonic resonance that may occur. A disadvantage of the system is that the grill could be destroyed under collapsed field scan conditions, a situation not assisted by the gun which is reputedly capable of producing a focused beam of at least twice the current of a conventional tube.

Advantages of the Trinitron system are:

- improved brightness and definition due to improved gun and increased grill transmission
- simple and more reliable convergence, only static convergence being required for outer beams; the electrostatic prisms for the convergence are built into gun structure
- freedom from Moiré patterning
- less sensitivity to terrestrial magnetic effects

52.6 Considerations when using crts

52.6.1 Flashover

A much overlooked aspect of tubes is that due to the high voltages involved, occasional internal breakdowns are to be expected. All tubes will exhibit this phenomenon to a greater or lesser extent. The flashover can occur from the final anode to any other electrode and have many causes, e.g. stray particles

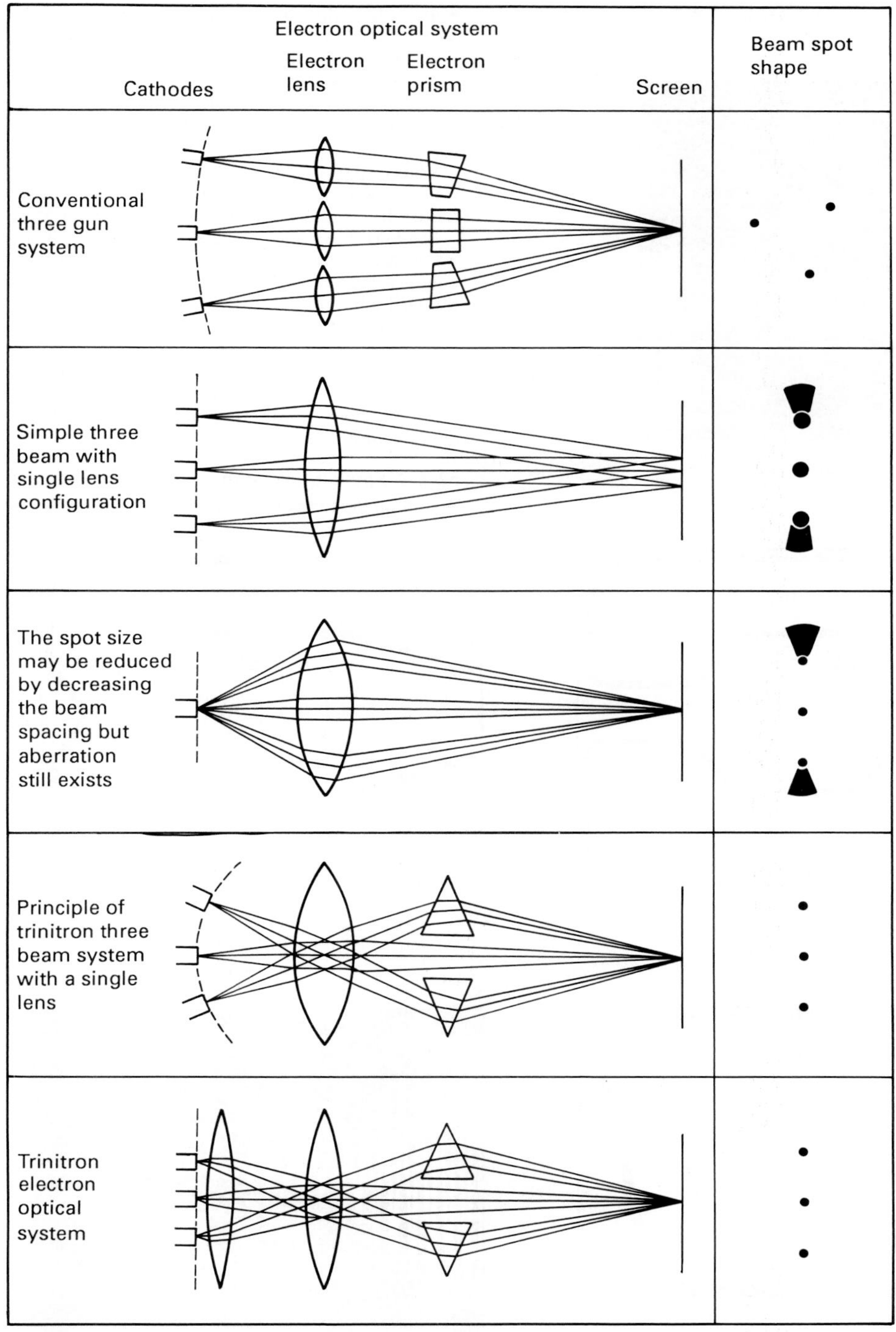

Figure 52.26 Development of Trinitron electron optical system

or electrode surface irregularities. Tubes have always flashed over, but the effect only became a serious problem when semiconductors replaced valves and unexplained failures of tuners and ifs were occurring. Much work was expended in investigating the effect and devising techniques for protecting semiconductor circuits. The resulting rules are simple:

Figure 52.27 Printed circuit spark gaps

- Use a good set of spark gaps on the tube base to provide a high voltage shunt in the event of a flashover.
- Provide a short thick lead from the spark gaps to a good contact with the aquadag (the outer coating on the tube) to return the energy to the tube.
- Resistively isolate the tube drive circuits where possible.
- Keep all printed circuit boards and wiring away from the tube and the rim band, preferably by a minimum of 30 mm.

Today spark gaps are built into the tube base connector which, when used with a direct connection to the aquadag, will return the current safely to the tube capacitance.

Most crts have undergone a change in that the inner coating, which was a similar coating to the outer aquadag, has had a quantity of iron oxide added to increase its resistivity. In the Mullard tube, this additive reduces the flashover current from some 900 A to only 60 A. The discharge current is thus dramatically reduced, as are various elements in the series current path. The components in the tube drive circuitry are therefore subjected to much less stress. This coating is called *soft flash* and is now used in one form or another by most tube manufacturers.

52.6.2 Spot suppression

A spot on the tube on switch off can damage the phosphor, and

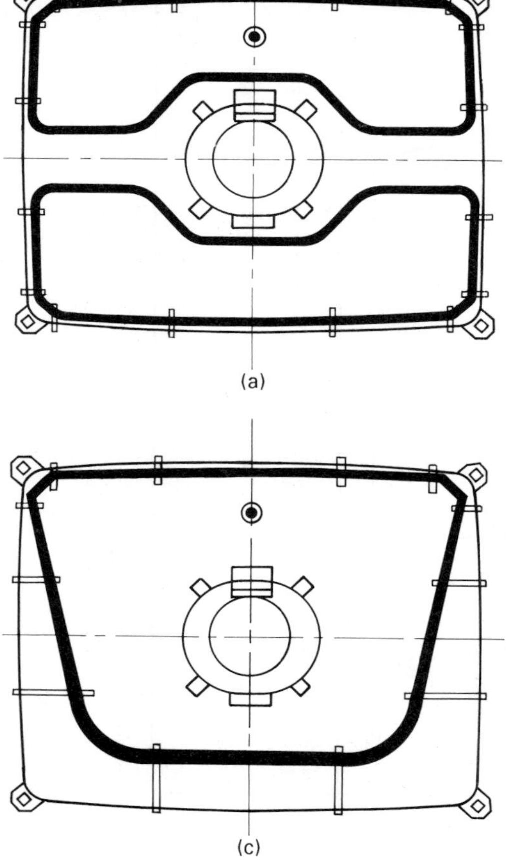

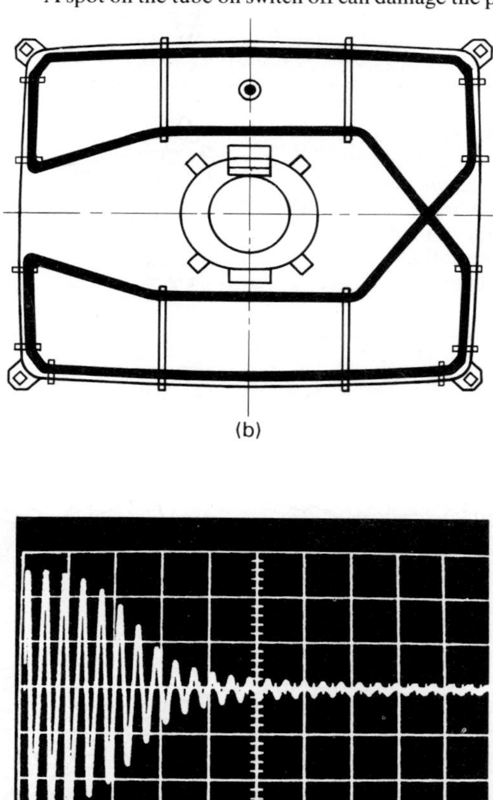

Figure 52.28 Degaussing: (a) double loop, (b) single twisted loop, (c) single loop, (d) decaying current

measures must be taken to prevent its appearance. Alternative methods are:

- Turn the gun hard on before the scanning has fully collapsed, thus discharging the eht capacitance. This may be assisted by holding the g2 voltage high, reducing the video amplifier supply voltage rapidly, and ensuring that the time base oscillators keep running while the scan is collapsing.
- Hold the gun off while the heaters cool down so that there is no emission after switch off. Problems with this technique are that the eht remains at full potential, a possible hazard to service personnel, and there is a high possibility of stray emission causing a pattern of illuminated phosphor on the screen, not dangerous but unacceptable to most people.

The first method is the preferred technique for reducing the spot burn problems.

52.6.3 Degaussing

Degaussing is a very important, and much misunderstood, aspect of shadow-mask crt operation. Consider the electron beam projected down the tube, aimed at a phosphor dot and passing through a hole in the shadow-mask. Any small magnetic field, especially random magnetization of the shadow-mask, will deflect the beam so that it lands on an adjacent phosphor producing an impure picture. To ensure colour purity, magnetic shielding is provided in the form of an internal shield which, with the shadow-mask, completes a magnetic circuit. If this structure is magnetized in an equal and opposite way to the magnetic fields acting on the tube, a field free zone is produced through which the beam can pass undeflected.

The procedure of magnetizing the shielding to produce the field free zone is called *degaussing*. To produce a suitable degaussing field, coils are positioned on the back of the tube (see *Figure 52.28*) so that their fields intersect the shielding components within the tube. A degaussing ac current is then passed through the coils, starting at 5 A and decaying to less than 100 mA in approximately 10 cycles. After 30 s the current is reduced to less then 5 mA. The peak field produced is only sufficient to compensate for the earth's magnetic field. If the tube becomes magnetized either during transport or by some strong magnetic device in the home, e.g. the bass unit of the hi-fi, then a powerful hand degausser is essential to restore pure colour pictures.

The decaying current shown in *Figure 52.28*(d) can be produced by using an ac supply and a dual positive temperature coefficient thermistor (ptc), where the two elements are in thermal contact. One element is in series with the coils which, as it passes the current, heats up and increases its resistance, thereby reducing the degaussing current. The second ptc element is in parallel with the supply and is constantly heated by the current flowing through it, which further heats the first ptc and thus keeps the residual current low.

52.6.4 Heater supply

Since the introduction of high performance low thermal inertia cathodes, the heater supply voltage has become more critical. The present heater voltage specification is 6.3 V +0 -5 per cent, and it is important to operate within these limits. If the heater is operated above the top limit then the tube life will be reduced. Operating below the lower limit will result in the heater taking a long time to heat up and can lead to degradation of the emissive material.

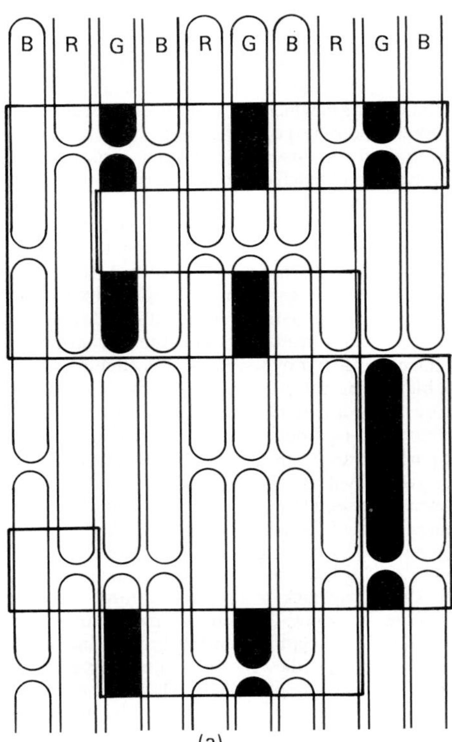

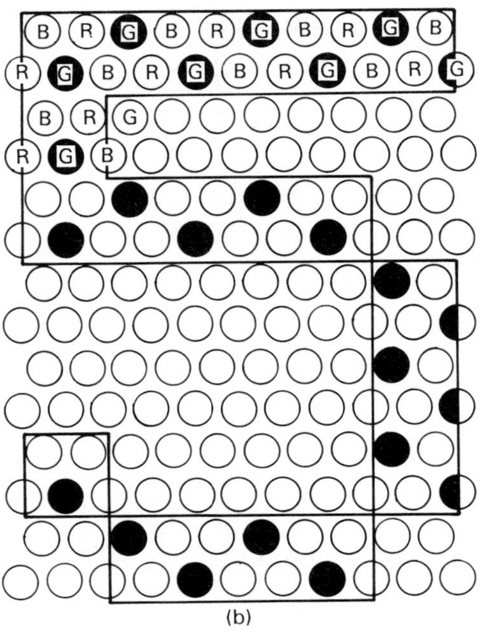

Figure 52.29 Resolution of characters (a) in a slot shadow-mask, (b) in a dot shadow-mask. The filled areas represent illuminated green phosphor

52.6.5 Shadow-mask pitch

Shadow-mask type tubes have the disadvantage that the maximum resolution is limited by the pitch of the phosphor triplets on the screen. A tube intended for use in a domestic TV set will have a pitch of some 0.8 mm which gives a theoretical maximum resolved frequency, on a 21 inch (510 mm) tube, of 5.5 MHz. This is adequate for an off-air picture but not for a computer generated display with 80 alphanumeric characters per line. For this application there are *medium resolution* tubes with a pitch of 0.42 mm, while for the highest performance computer displays, tubes with a dot structure pitch of 0.27 mm are available.

The dot structure tubes are used for the highest resolution tubes because, as may be seen in *Figure 52.29*, the vertical phosphor line structure is broken up and appears as a finer pitch when comparing pixels line to line. Unfortunately some light output is sacrificed when using this structure.

52.7 Projection systems

Projection television has been around from the early days of commercially available television sets, when Philips produced a monochrome back projection set with a 20 inch picture using a 2 inch tube run at 25 kV. When the direct view tube was made in larger sizes, the home projection systems declined.

Today, however, there are several colour systems available, generally front projection, using three 5 inch (12.7 cm) tubes and refractive optics to produce a 4 ft (1.2 m) picture. Two optical combining systems are in common use in addition to direct projection. Keystone raster correction is required for the two tubes not directly on the main optical axis. Over the years, several attempts have been made to construct optical combiners so that little or no raster correction is required and only one set of projection lenses.

Projection tubes are generally run at 30 kV or higher which can present an x-ray hazard when working on the system with the covers off. Otherwise the tubes are similar to normal monochrome tubes, using a large gun capable of providing a small well focused spot. A deflection angle of 70° is used, the tube displaying an underscanned very bright raster.

While these systems make a large screen size possible and a wide variety of tubes are available, they have the disadvantages that special high reflectance screens are used, which tend to be omni-directional, crt life is short at high brightness and there may be x-ray radiation under fault conditions.

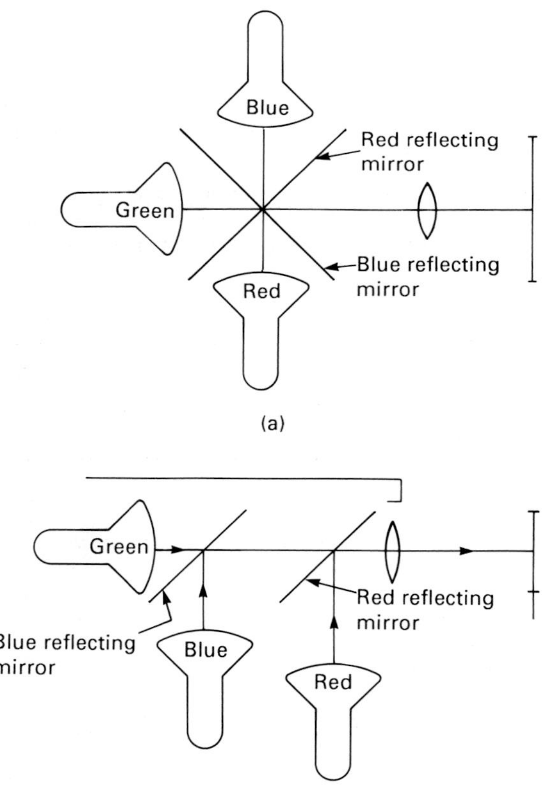

Figure 52.30 Three tube projection systems: (a) crossed mirror system, (b) parallel mirror system

52.8 Alternative technologies

There are three technologies competing for the opportunity to replace the crt as the display device of the future. The devices are based on gas plasmas, electroluminescence and liquid crystals. These devices tend to be used for specialized applications, and an outline of their principles is given here.

52.8.1 Plasma displays

These are based on gas discharge techniques with their roots in the trigger tube and Nixie display era. The display is matrixed but has the advantage that it can be self-scanned, dramatically reducing the external connections and keeping the costs low.

With all gas discharge devices, there is a problem of ensuring that the striking potential (the voltage at which a discharge can be started) of the tube is constant. A tube in which there has not been a discharge for some time will have a higher striking potential than a tube in which there has been a recent discharge. A small auxiliary discharge, or priming, can provide the free particles required to ensure a constant striking potential. Alternatively, a small patch of radioactive paint on the inside of the envelope will have the same effect.

The operation of the self-scan tube is shown in *Figure 52.31*. Every third cathode bar is connected together to form an interleaved grid. The discharge is transferred from cathode to cathode by a series of negative pulses applied to each cathode set in turn. The cathode bars are horizontal and represent field scan. Line scan is produced by moving the discharge along the length of the cathode bar by sequentially switching the anodes.

The problem of priming is overcome at the beginning of each line-scan by using auxiliary electrodes with a lower striking voltage which are used during the flyback time to ensure that sufficient particles are available to commence the scan. These electrodes are known as *keep alive*, and their discharge is not visible. As may be seen in *Figure 52.31*, one method of making the discharge visible is by perforating the cathode allowing the photons generated by the ionization to pass through.

A number of displays based on these principles have been marketed. One drawback is that the natural colour of the plasma display is orange, which is unacceptable for most analogue displays. Acceptable monochrome and full colour displays are possible, either by using alternative mixtures of gases or by incorporating phosphors which are excited by the gas discharge. Grey scale can be achieved by modulating the discharge current.

Later developments in plasma technology have incorporated a memory inherent to each cell or pixel to increase the apparent

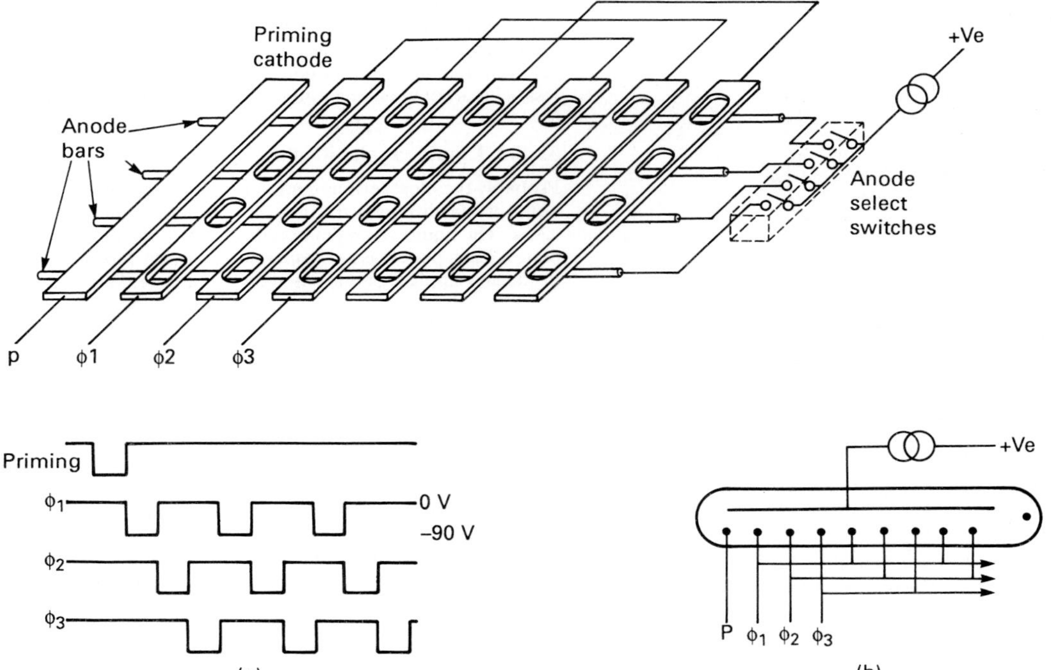

Figure 52.31 Principle of operation of the self-scan plasma tube. (a) is the timing diagram and (b) the equivalent circuit

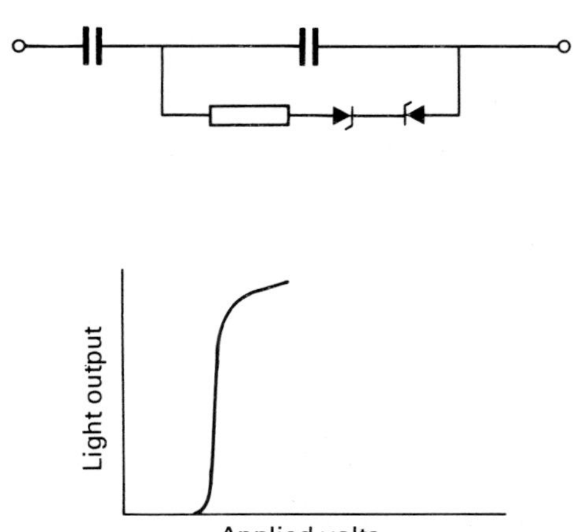

Figure 52.32 Equivalent circuit and characteristic for an ac thin film electroluminescent device

brightness. Thus only update information on picture changes is required on each scan.

This technology, commonly found in computer vectographic type displays, is waiting for a breakthrough to compete with the crt for more general-purpose displays.

52.8.2 Electroluminescent displays

These are known for their application in backlighting rather than displays. However matrix displays have been available since the early 1980s, mainly for use with computers. Unfortunately electroluminescent light generation is only fairly well understood and we lack a good model which would allow rapid improvement and development of the system. An explanation of the current theory for the ac thin film device relies on the use of a phosphor and an activator e.g. ZnS:Mn. *Figure 52.32* shows the accepted equivalent circuit and characteristic for this material.

As may be seen the curve is very steep, an indication of the deep levels to which the majority of the carriers tunnel into the conduction band. The carriers thus set up an interval field and, if a suitable external field is applied in phase with the internal field, there is sufficient energy to activate the electrons in the manganese atoms and thus emit photons of light.

When used as a 'thin film' lamp, a film of phosphor is trapped between a layer of aluminium and a transparent layer of tin or indium-tin oxide, forming a capacitor, driven from a supply of 100 V at 400 Hz – 5 kHz. Used as a matrix display, the thin film has a high transparency which, when used with two transparent electrodes, has the advantage that different colour pixels may be overlaid allowing the screen area to be used more efficiently.

Electroluminescent compatible phosphors can produce most colours. Red is probably the most difficult to make, providing sufficient light output to match that of blue and green. It has been found that the use of a fluorescent colour converter material overlaying the top electrode and activated by one of the more efficient phosphors, green or blue, will produce the best results.

Driving the matrix lines of a large area display is well within the capabilities of current video transistors since only about 100 V is involved. The greater problem is that ac thin films are difficult to use with matrix displays at the frequencies involved. An alternative phosphor/activator combination had to be found

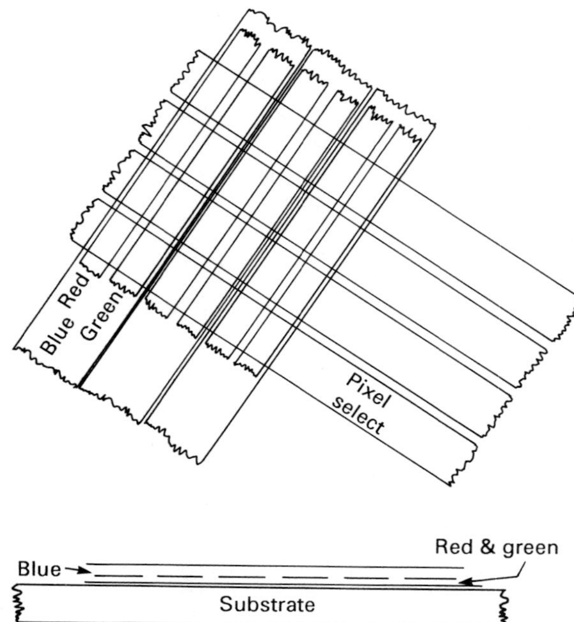

Figure 52.33 Construction of tri-colour electroluminescent display

which was compatible with the waveform found in the matrix driving circuits.

Lifetime is the major problem with these devices. A thin film lamp may be expected to last for only 1000–3000 h to the 50 per cent of initial light output point, and a matrix display for even less. This is not very long when compared with a crt at 20 000 h.

52.8.3 Liquid crystal displays

Probably the most serious contender for replacing the crt is the liquid crystal display (lcd). Its basic operation is shown in *Figure 52.34*. The liquid crystal material is trapped between two glass plates. Each plate has a transparent electrode deposited on the inside and a polarizer deposited on the outside. With zero volts applied to the cell (V_{ns}), no polarization twist occurs within the material and light can pass through the cell attenuated only by the polarizing layers. When a potential is applied (V_s), the material twists the light so that it is no longer parallel to the polarizers, and light no longer is able to pass through.

The cell may be operated in two ways: *transmission*, in which the cell is interposed between the light source and the observer, and *reflection*, in which the incident light passes through the cell, is reflected and returns through the cell to the observer.

Four to eight rows of 80 characters are available from several suppliers with displays capable of 24 rows of 80 characters (600 × 640 pixels) becoming available, and even 512 × 720 pixels.

In order to increase to this level of complexity, a new technique of activating the display was required, in the form of depositing a thin film transistor (tft) to control each pixel of the display. Now the matrixing is removed from the elements themselves, which creates problems of contrast ratio and

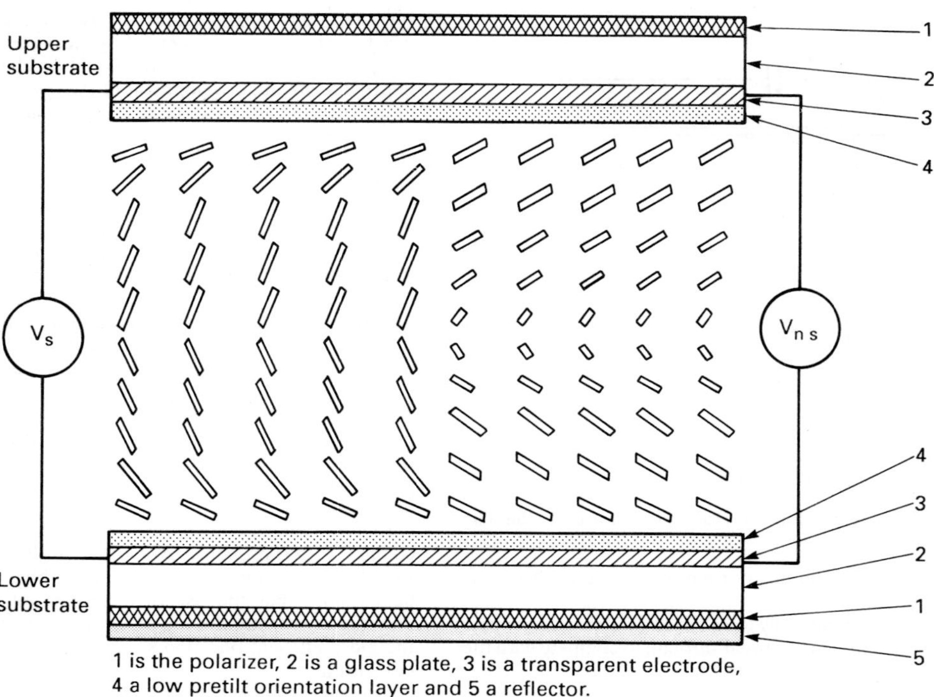

Figure 52.34 Operation of an lcd cell. 1 is the polarizer, 2 is a glass plate, 3 a transparent electrode, 4 a low pretilt orientation layer, and 5 a reflector. V_s = select voltage and V_{ns} = non-select voltage

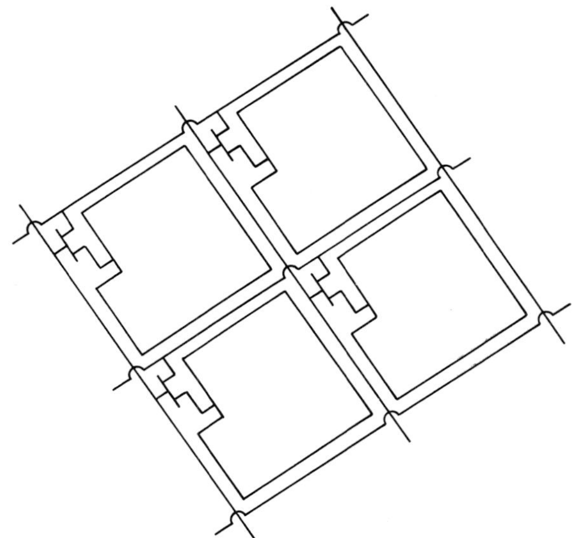
Figure 52.35 Connection of a tft transistor cell

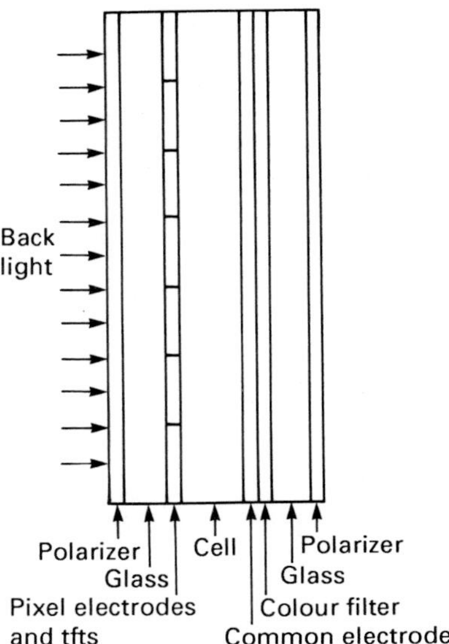

Figure 52.36 Construction of a colour lcd display

viewing angle, and activates the tft transistors which in turn select and sustain the voltage on the indium-tin electrode on the pixel. The matrix wiring passes between the pixels and need not be transparent.

This technology is sufficiently developed to allow production release of a 96 × 28 mm (6 inch diagonal) display with 640 × 600 pixels which, when used as a colour display, has 640 × 200 colour pixels. Colour is produced by placing colour filters in front of adjacent pixels as shown in *Figure 52.36*. Due to the high attenuation of the multiple layers, this display requires to be back lit. At present a fluorescent tube is preferred.

The disadvantage is that, although the display is low power, the back light consumes considerable power, and special measures have to be taken to ensure an even illumination. Another disadvantage is that this display has no grey scale with the pixels either *on* or *off*. The display is thus eminently suited to alphanumeric and graphic displays but not yet a true TV picture.

52.8.4 Operating the lcd display

The lcds and tfts operate from a 5 V display. The problem with the display is that we have 1240 connections to make from the matrix lines to the decoder/driver ics presently mounted on an ancillary pcb positioned behind the display. Several methods may be employed to connect the matrix lines to the pcb. The two most popular are:

- use of conductive silcoset rubber, consisting of a bar made up from conducting and non-conducting strips laid in parallel, placed such that the conductive strips connect the glass substrate to the pcb; the pitch of the rubber conductive strips is such that there is a minimum of three connections for each matrix line
- a flexible plastic substrate supporting carbon conductors which can be welded to the glass substrate and the pcb providing contacts between substrate and pcb

This connection system is a weak link in the reliability of the lcd system, and the next development is to place the decoder chips directly on the glass substrate thus reducing the number of connections and increasing the reliability.

Hitachi have produced an integrated circuit which can replace the 6845 crt controller capable of controlling displays up to 4096 × 1024 pixels and another which can take an RGB video signal and produce the appropriate drive signals for the lcd. This greatly eases the introduction of the lcd into graphic displays for computer outputs.

K Komada
General Manager, Toshiba Corp

53 Production Engineering and Reliability

53.1 Production engineering

The manufacturing process for colour television receivers is shown in *Figure 53.1*. It can be divided into three sections: printed circuit board (pcb) assembly, cabinet assembly and set assembly.

53.1.1 Assembly technology

53.1.1.1 Assembly of printed circuit boards

The technology for assembling electronic circuits has evolved as illustrated in *Figure 53.2*. The introduction of printed circuit boards made it possible to solder all the inserted components on to the pcb at once, instead of having to solder each component to terminals one by one with the aid of a soldering iron. This development was responsible for a reduction of assembly time by about 50 per cent. With the introduction of automatic insertion and integrated circuits, a further reduction of assembly time by 75 per cent was achieved.

Printed circuit board assembly requires the following equipment:

- axial insertion machine
- radial insertion machine
- special insertion machine
- manual insertion conveyor (other parts)
- inspection conveyor (insertion check)
- soldering equipment
- adjustment and inspection equipment

53.1.1.1.1 Automatic insertion
Components for automatic mounting are generally axial or radial in shape, but following recent developments other types of components also can be mounted automatically.

Automatic insertion of axial and radial components, which usually constitute 65–75 per cent of the total number of components on a printed circuit board, has become very common as the necessary machines have become widely available. In a highly automated production system, special components are also inserted by machines, to bring the number of components to be inserted automatically up to 90 per cent.

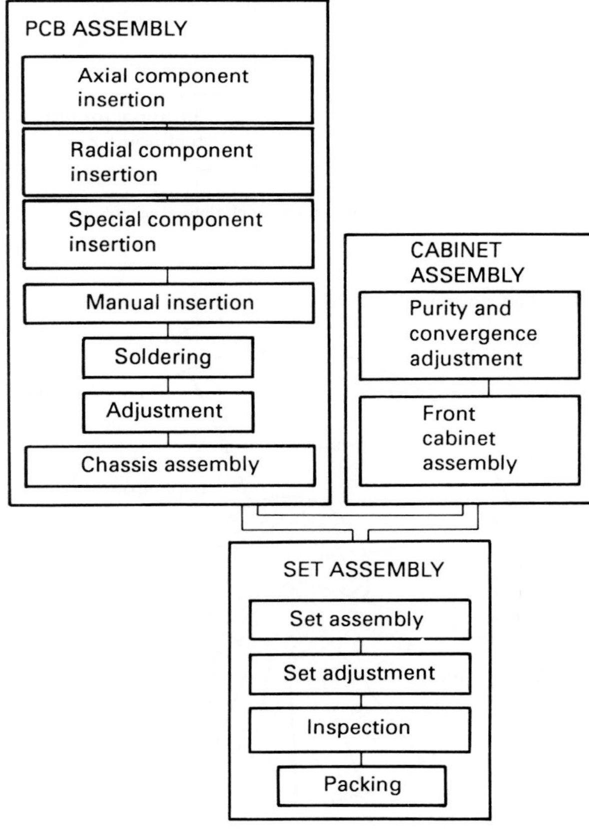

Figure 53.1 CTV production process

Many kinds of automatic insertion machines are available. They may be classified into *sequence* types and *in-line* types.

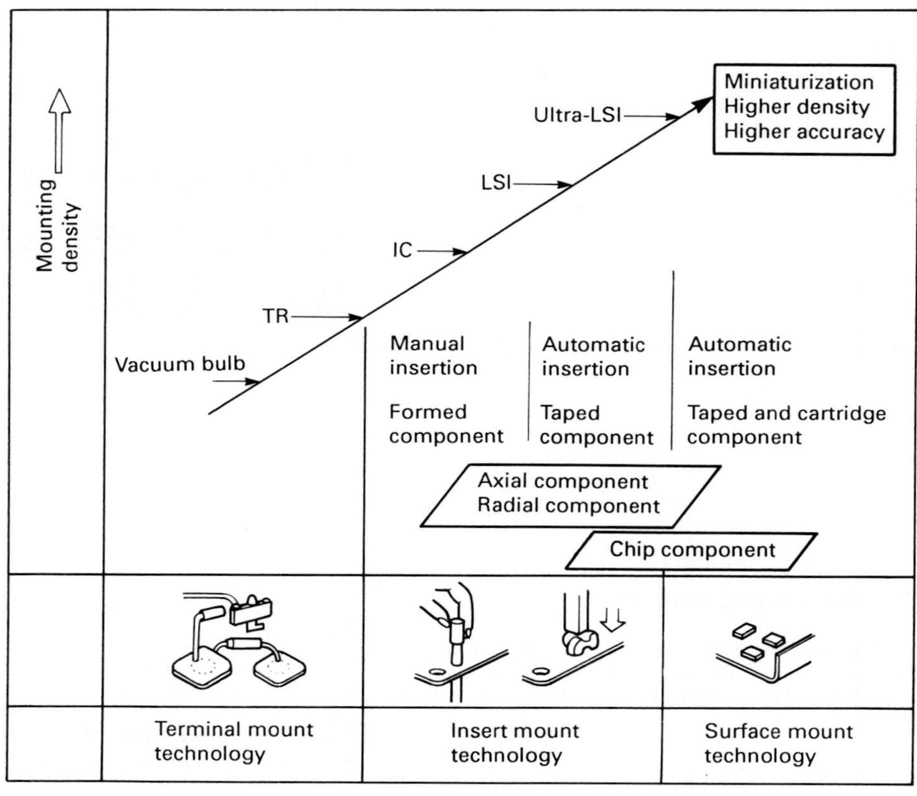

Figure 53.2 Development of component mounting technology

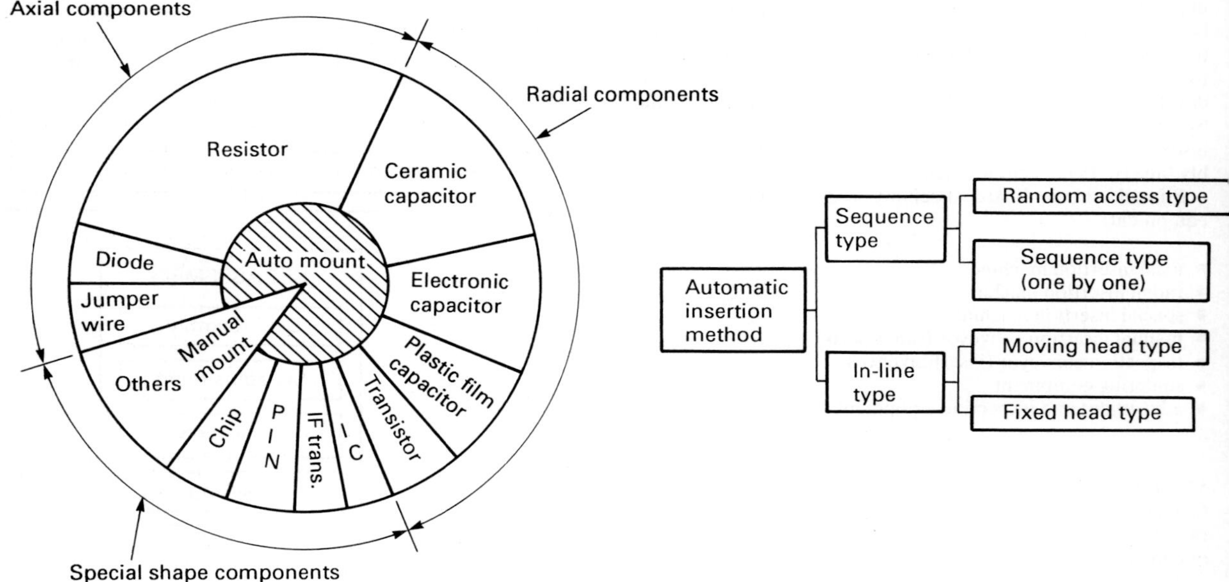

Figure 53.3 Constitution of automatic mounting components

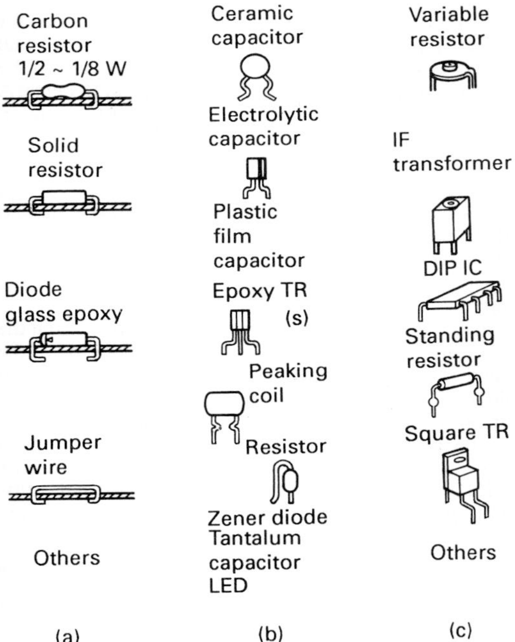

Figure 53.4 Components for automatic insertion: (a) axial, (b) radial, and (c) special shape

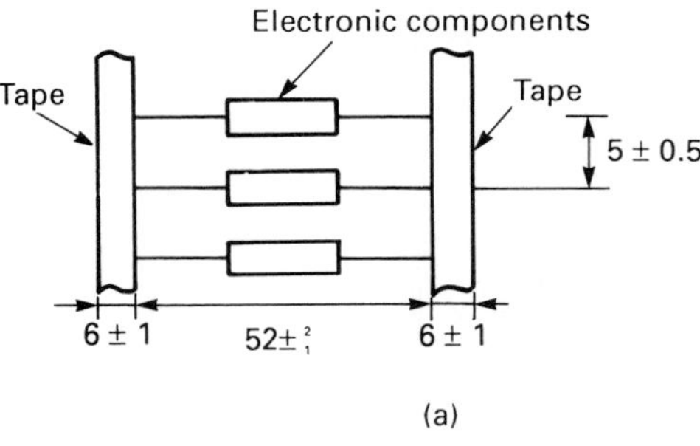

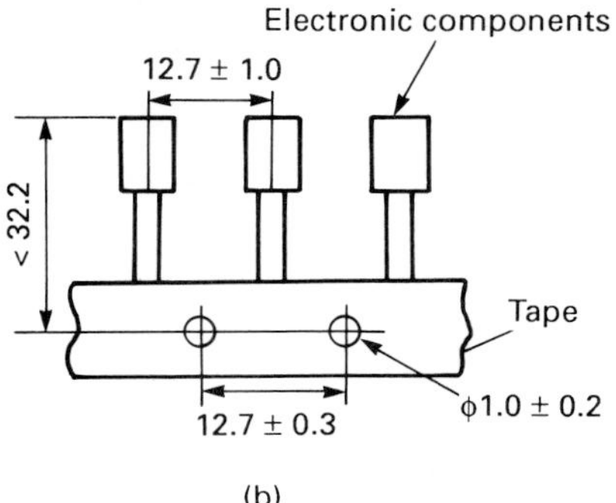

Figure 53.5 Taping components: (a) axial, (b) radial. The units are millimetres

Sequence type machines may be *random access* or *sequence* (one by one). In-line types may have *moving heads* or *fixed heads*.

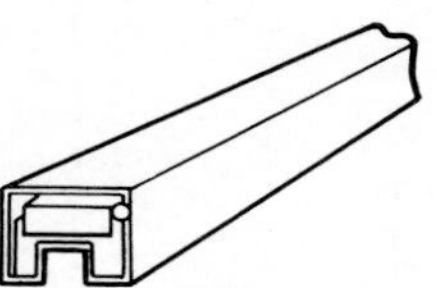

Figure 53.6 Stick magazine package

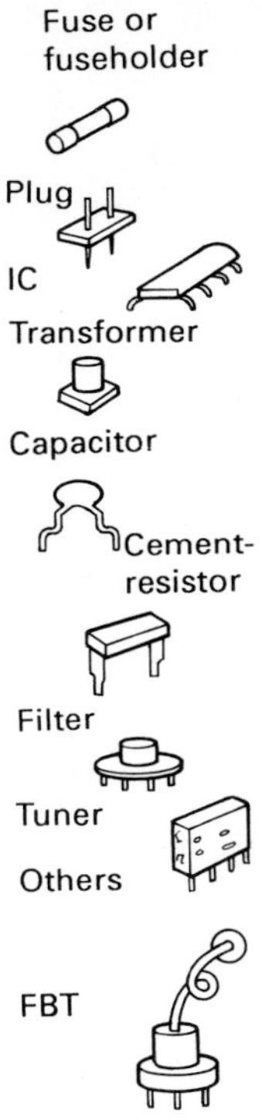

Figure 53.7 Manual insertion components

In a sequential system, components are inserted one by one according to the programme, whereas in an in-line system, basically one component is inserted with one head so that the number of machines required equals the number of components to be inserted.

Axial components for automatic insertion are shown in *Figure 53.4*(a). Typical specifications of a machine for inserting these are:

insertion speed per piece:	0.28–0.5 s
insertion pitch:	5–20 mm (fixed and variable)
kinds of components:	40–120

The components should be taped as shown in *Figure 53.5*(a).

Radial components for automatic insertion are shown in *Figure 53.4*(b). Typical specifications of a machine for inserting these are:

insertion speed per piece:	0.5–0.6 s
insertion pitch:	5 mm (fixed)
kinds of components:	40–80

The components should be taped as shown in *Figure 53.5* (b).*Special components* for automatic insertion are shown in

Figure 53.8 Robot line for component insertion

Figure 53.4(c). Typical specifications of a machine for inserting these are:

insertion speed per piece	0.9–1.2 s
kinds of components by shape	5–10

The components should be taped in the same way as shown for axial or radial components (*Figure 53.5*) or in a stick magazine (*Figure 53.6*).

53.1.1.1.2 Manual insertion of components

Components whose shapes are complicated or large cannot be inserted by the machines mentioned above. Components such as those illustrated in *Figure 53.7* are usually inserted manually. Recently, however, robots have been specially developed for some of these types of components (see *Figure 53.8*). In manual insertion processes, printed circuit boards are carried by conveyors. Each operator will insert between 7 and 15 items. A checking process is then required to confirm that the components are inserted properly.

53.1.1.1.3 Soldering

Soldering is the vital aspect in printed circuit board assembly. It was with the pcb assembly process that the soldering action was first automated. Subsequently, continuous development has resulted in improved equipment which minimizes faulty soldering.

Flux coating is first necessary to remove any oxides from the surface of lead wires of components and the copper foil of the pcbs, and then to coat them with flux to improve their ability to take solder.

It is important to keep the density of the flux constant in a specific gravity range of 0.82–0.86. Equipment has been developed to coat flux which can control specific gravity automatically. Compressed air is usually used to foam the flux. Fluxing is followed by *pre-heating*. This is required to avoid giving thermal shock to electrical components and to activate coated flux. The period and temperature of pre-heating depends on the specification of the printed circuit board and the number of electrical components to be soldered. Heating for one minute at around 100–120°C is desirable.

There are basically three different types of automatic soldering machines: dip system, flow system and flow-dip equipment.

The *dip system* is usually appropriate for pcbs with discrete components at average density. The printed circuit boards are dipped into the still surface of solder.

In a *flat dip* (or *drag*) the solder is molten in a suitably sized satinless steel bath. The pcbs are dipped in the bath for

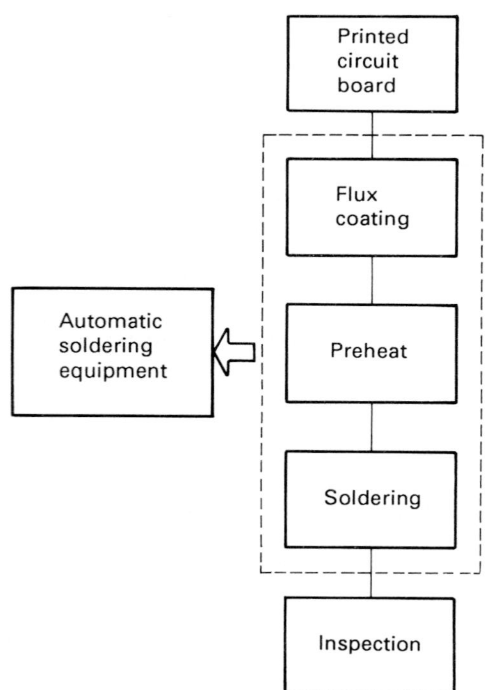

Figure 53.9 Soldering process

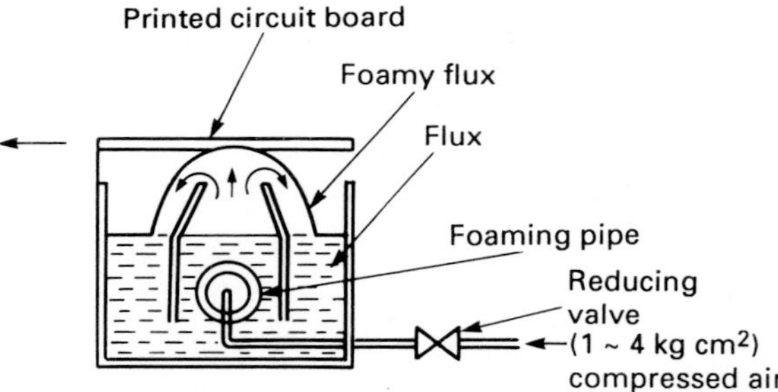

Figure 53.10 Equipment for foaming flux

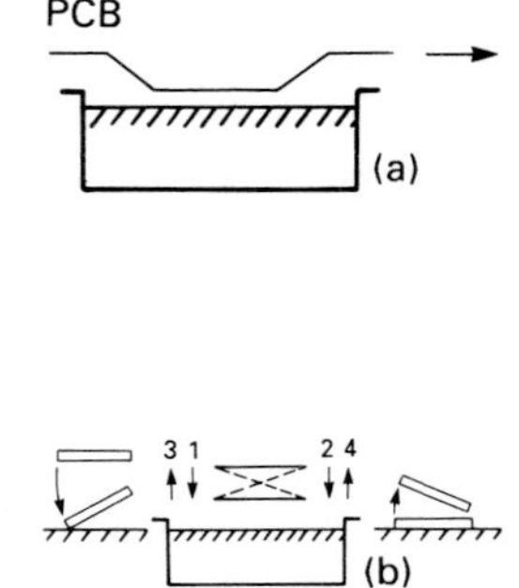

Figure 53.11 Examples of dip type soldering: (a) Flat dip, (b) vertical dip (movement seen from the direction of board travel)

soldering as shown in *Figure 53.11*(a). Oxidized solder slugs or carbonized flux are left on the surface of the solder, and these must be removed after each board is soldered.

In a *vertical dip*, the pcbs are slowly placed in a static surface solder bath beginning with one side, then withdrawn beginning with that side to complete soldering (see *Figure 53.11*(b)).A *flow system* is usually appropriate for hybrid use of discrete and chip components. In this system, printed circuit boards travel over a jet stream of molten solder horizontally or at a fixed angle.

A pump, a chamber and a jet nozzle are provided in the soldering bath. The pump is driven to jet the solder, the width of the jet being determined by the size of the pcbs.

In a *wave* variant, the solder is jetted into waves. This method is also known as *hollow wave*. Dual waves are opposed to each other to jet the solder. The solder jet height can be increased up to 50 mm, but a high-output motor is required to achieve this.

In an *inclined flow double wave*, flow waves and jet waves are combined to solder both chip and discrete components. Manufacturers have developed their own processes of jetting flux gases by delicately adjusting primary waves for soldering.

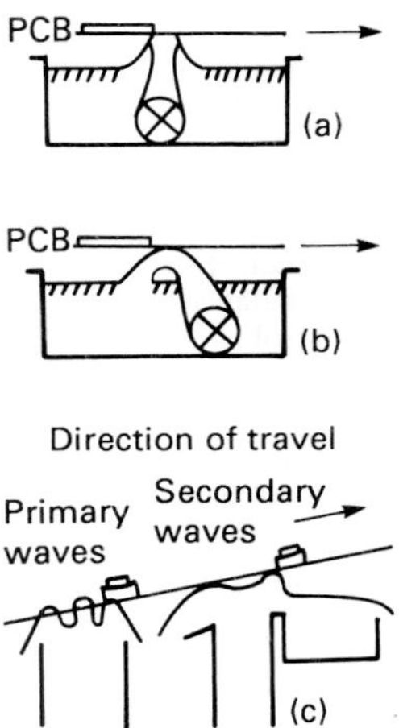

Figure 53.12 Flow type soldering: (a) simple flow, (b) wave, (c) inclined flow double wave. The primary waves provide preliminary soldering and the secondary waves finish soldering.

A *flow-dip system* is a combination of flow system and dip system and usually appropriate for printed circuit boards with discrete components at high density.

A *plain flow dip* keeps the sides of pcbs always clean with a jet pump without scraping off solder oxide films as with the dip type. It is suitable for soldering long component leads and provides stable jet solder sides.

A *flow and dip* method combines the flat dip with a flow mechanism to remove weaknesses of the dip method.

Soldering is as shown in *Figure 53.13*(b). Flow soldering takes a fraction of a second. The dip bath requires protection against residual flow waves and flow off of oxides. Because the solder bath surface is static, oxides must be removed after each pcb is soldered. A dip flow type operating in a reverse way has been used.

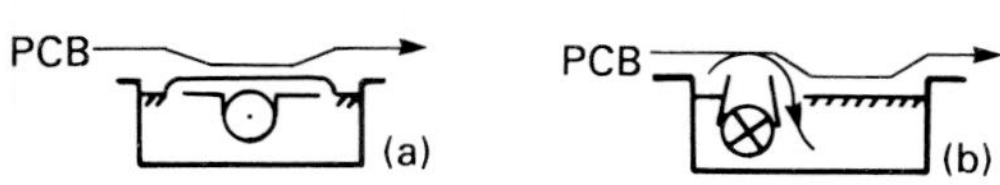

Figure 53.13 Mixed flow and dip type soldering

53.1.1.2 Cabinet assembly

The cabinet assembly line screws together the colour picture tubes and speakers. A television cabinet is the major design element in a set, and so cannot be standardized.

On the other hand, parts that are to be mounted onto these cabinets are big and heavy. It is very desirable, therefore, to automate the process. However, automation could not be accomplished until the recent development of an assembly line from an adaptation of an assembling robot.

53.1.1.3 Set assembly

On the set assembly line, cabinets with colour picture tubes and speakers mounted are fitted with chassis and wired. The sequence of the production process is:

- chassis assembly
- wiring
- crt adjustment and inspection
- receiving inspection
- installation of back cover

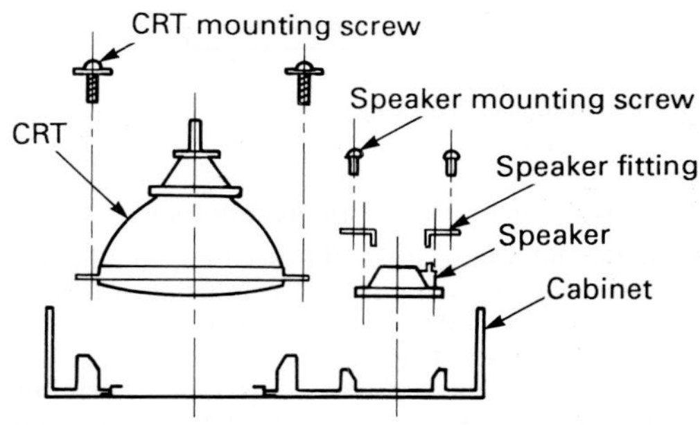

Figure 53.14 Cabinet assembly

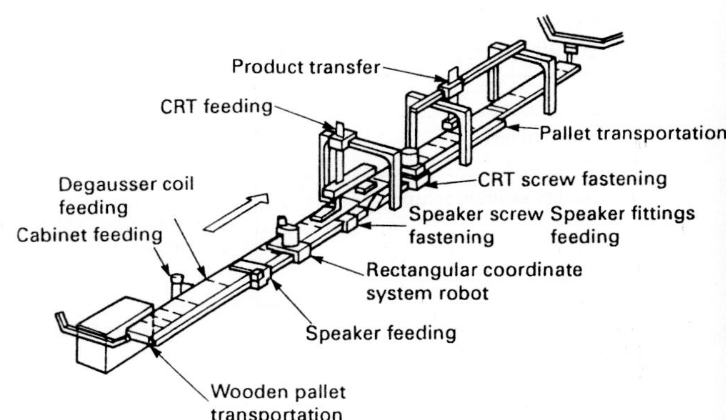

Figure 53.15 Outline of cabinet assembly

Packing is an operation that is necessary to get the products safely to users. A method of packing is required that will withstand impacts, vibrations and handling of freight during transportation. Automation of the packing process is progressing. The steps in the process are indicated in *Figure 53.16.*

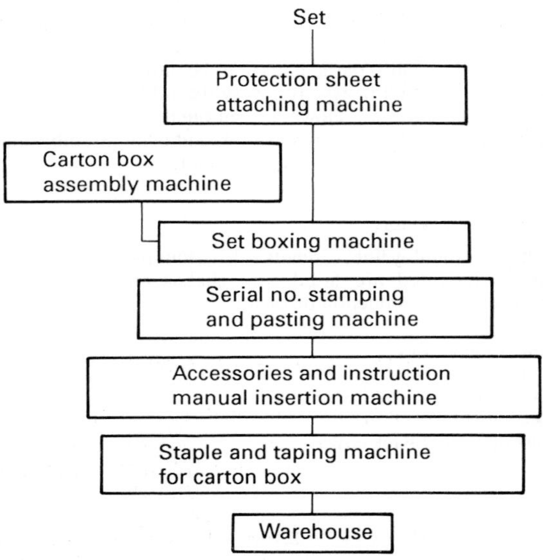

Figure 53.16 Automation packing process

53.1.2 Adjustment and testing technology

Adjustment and testing are vital to secure high reliability and high performance of the products. The function of *adjustment* is to limit deviation of performance, and that of *testing* is to secure reliability of the products.

Testing in itself does not create any value. Therefore, the number of testing points should be kept to the minimum necessary, and their position in the production process should be carefully chosen. Tests fall into four categories as shown in Table *53.1*.

53.1.2.1 Adjustment and testing of pcb assembly

The process of adjustment and testing of printed circuit board assembly includes the following steps:

- bare board
- components insertion
- inspection of insertion
- soldering
- inspection of soldering
- in-circuit test
- small signal circuit adjustment and inspection
- deflection circuit adjustment and inspection
- channel presetting
- chassis assembly
- receiving test

The inspection of components insertion is usually a visual check, as is the inspection of soldering.

The role of the in-circuit test is to find defects which cannot be easily detected visually. It checks the performance of the circuit by applying small signals before a dynamic test. It shows up defects of individual components, resistors, capacitors, coils, transistors, diodes, etc., faulty soldering and insertion of components, wrong components and missing components.

53.1.2.2 Adjustment and testing of receiving circuits

The adjustments and tests that can be done independently of the total set assembly, after pcb assembly, include:

Adjustments:

- picture intermediate frequency
- automatic fine tuning detection
- automatic fine tuning balance
- rf automatic gain control

Testing:

- frequency characteristics of picture if and video circuit
- amplification constant of video circuit

Other circuits are adjusted and tested in the same way, e.g.:

Adjustment:

- voltage of +B supply
- horizontal and vertical synchronizing

Testing:

- level of audio output

A receiving test is carried out to check the basic performance of the chassis before they are installed into cabinets.

Type	*Objectives*	*Methods*
Function test	To check the performance and function of the products	* Checking waveforms of the circuits * Receiving test
Quality assurance inspection	To check the assurance of value of the products considering environments and operating time	* Thermal shock test * Humidity test * Vibration test * Dropping test * Chemical test
Detection of failure	To detect any failures that escape preceding operation	* In-circuit test * Function test * Soldering inspection * Receiving test
Inspection of workmanship	To check defects caused by workmanship	* Checking missing components * Check clinch condition automatically inserted component

Table 53.1 Types of inspection

53.1.2.3 Adjustment of picture tubes

Colour picture tubes fitted with deflection yokes, purity magnets, and convergence magnets, need to have their purity, convergence and picture tilt adjusted. The process is shown in *Figure 53.17*. Equipment which automates this process is shown in *Figure 53.18*.

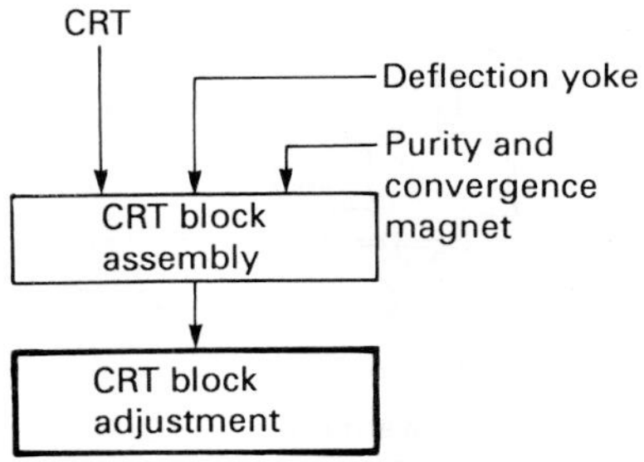

Figure 53.17 Adjustment and inspection process of cathode ray tubes

Figure 53.18 Equipment for automatic adjustments of colour picture tubes

53.1.2.4 Adjustment and testing of complete sets

The adjustment and test sequence after set assembly is:

- cabinet assembly
- crt adjustment
- receiving inspection
- dielectric strength test
- shipping inspection
- packing
- shipping

The pictures are adjusted after set assembly to their ideal, including white balance, focus, brightness and vertical amplitude.

After this adjustment, receiving tests are done to confirm the performance of the sets. These involve performance, picture quality and overall check of other items.

The dielectric tests involve an inspection of the dielectric strength of metal parts on the exterior of cabinets. A dielectric test and an insulation test is carried out. Finally, before shipping, a check is made on picture reception to reaffirm the performance and confirm that the sets can be released.

53.1.3 Production control system

The application of 'mechatronics' in production equipment and other technical developments have made it essential that the manufacturing operation should be systematic. At the same time, to keep control of expanded production, the introduction of computers has become necessary.The use of computers in production control has made it possible to provide management with timely production information. Yet management support is not the prime function of the production control system. The important items are:

- master production planning and daily scheduling
- capacity requirement planning
- manpower planning
- quality control information
- production process control
- machine group control
- shipping control

53.2 Reliability and quality assurance

53.2.1 Concept of quality assurance

The transition of the concept of quality assurance since the commencement of mass production of colour television receivers in Japan is shown in *Table 53.2*. In the early days of mass production, importance was put on sampling inspection; this was the era of *inspection*. Later, various methods of control were introduced into the manufacturing process; the era of *process control*, or *in-process quality control*.

The basic concept of in-process quality control is that, at all levels of management, every operational unit is responsible for both manufacture and checking, to avoid failures leaking to the next unit of operation. At individual operator level, each operative on the production line does both activities: manufacturing, e.g. soldering, and checking their soldering to avoid letting failures go to the next operator. At a process level, in the case of the printed circuit board assembly process, for example, various checking functions are built in the process as described in section *53.1.2*.

This concept reduces the importance of sampling inspection. For in-process quality control, it is essential to have an efficient information system which gives detailed data quickly. This enables an operator to take quick corrective action.

More recently, design activities have been highlighted. It is realized that a high level of reliability to meet the severest conditions under which the products may be used demands correct design work; this is the era of *design* or *reliability*. We currently see quality assurance in the terms described in Table *53.2*.

The *headstream control* applies *in-process control* to the headstream process of the total system, including the components production process and the design process of new products. This goes beyond the scope of the activities to produce colour television receivers in factories.

This concept of quality management requires the management of:

- the quality control of the production process by in-process quality control
- computer information of the quality covering the design, the production process, and the field
- review or approval at every stage from the new products plan to the initial shipment of finished products
- cooperation with suppliers of components to implement headstream control

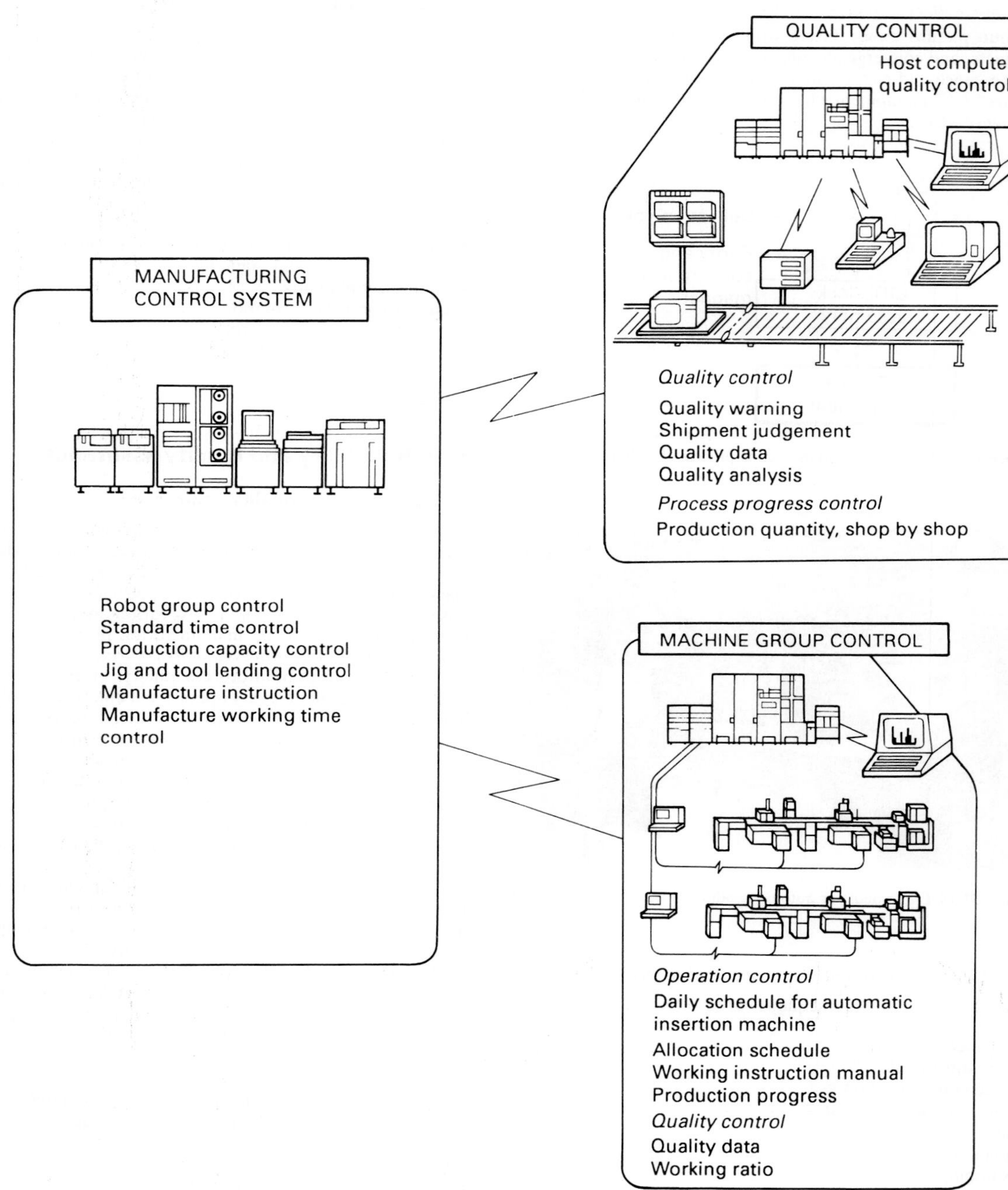

Figure 53.19 Production control system

- reliability, involving reliability prediction and evaluation test simulation at design stage of conditions under which products may be used

53.2.2 Computerized quality information

The system described is used in Japanese factories where some of the assembly processes are being subcontracted out. The total system is shown in *Figure 53.20*, and has five main features:

- There are some 300 input/output terminals at subcontractors and service stations throughout the country for on-line realtime data exchange.
- The host computer is provided for exclusive use for quality assurance, to manage the quality information of subcontrac-

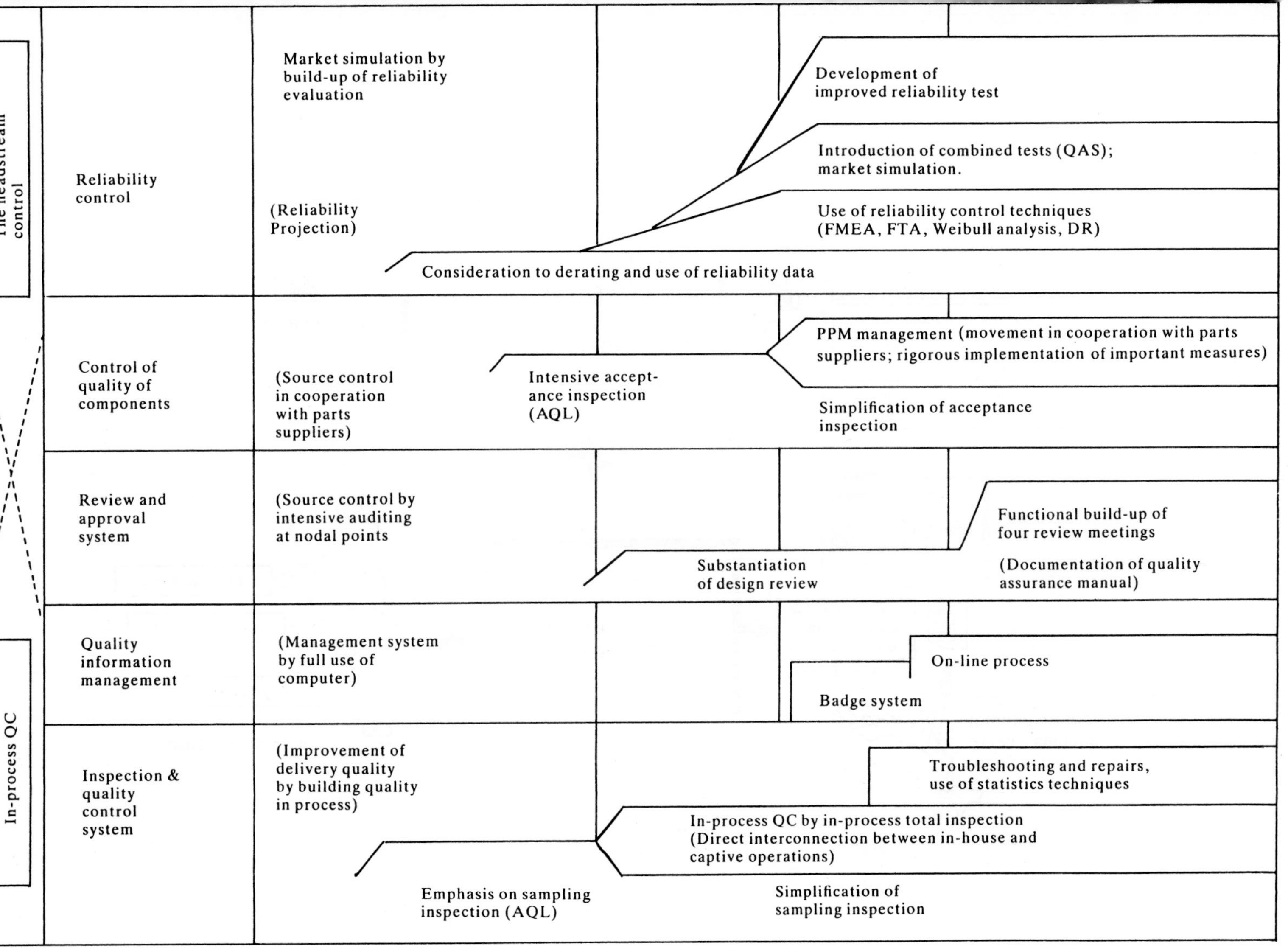

Table 53.2 The transition of concept of quality assurance

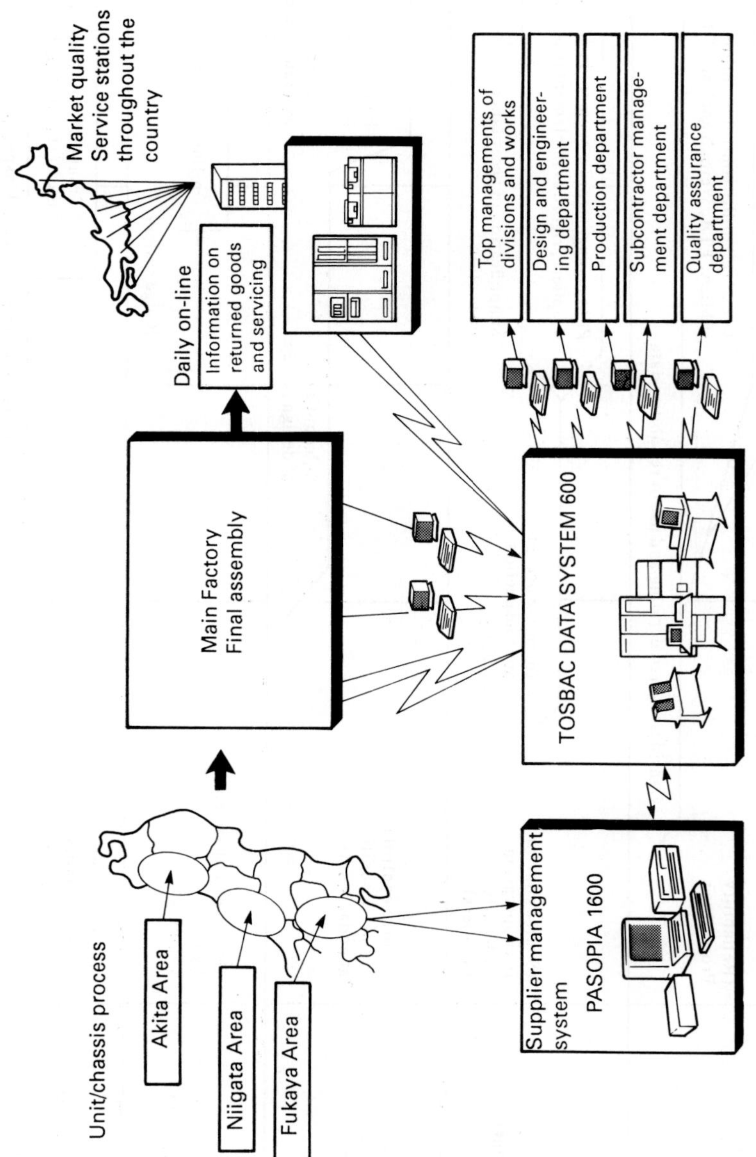

Figure 53.20 A schematic of a computerized quality information system

tors, the main factory and service stations in an integrated manner.

- Each key stage is monitored for abnormalities according to preset criteria. Upon detection of any abnormality, a warning is automatically given to the department responsible for it.
- Codes and languages are common to all key stages so that every operator can easily handle data in standardized formats, through interactive menu, input/output operations.
- Quality information can be interfaced with design and production plan information such as CAD and production schedules.

Objectives of the system, and key control points are:

- management of abnormality
- improvement through integrated analysis
- pass/fail judgement of shipment
- in-house reliability control
- reliability management

Management of abnormality involves the necessary corrective action when abnormality is detected by statistical analysis of the quality information from subcontractors, the main factory and the service stations.

Improvement through integrated analysis requires implementation of specific and viable measures formulated by integrated analysis of subcontractors, processes, models, symptoms, causes and various other factors.

Pass/fail judgement of shipment is made daily by evaluating quality information from subcontractors, set assembly processes, and sample inspection of products due for shipping.

In-house reliability control requires the formulation and implementation of improved control and management by analyzing the results of various simulated tests such as a 50 h ageing test, a 300 h ageing test, an aging test during processes, and an ageing test under special conditions.

Reliability management requires prediction of reliability by quality information from service stations, to achieve early detection of abnormalities, early detection of failures due to wear and tear, and development of reliability prediction at design stage. It also involves the development of improved simulation tests and calculation of the acceleration factors.

The system configuration includes:

- 1 host CPU DS600
- 285 on-line input/output terminals
- 20 subcontractor management terminals
- 20 hard disks
- 8 inspection data input terminals
- 5 plotters

53.2.3 Review and approval

It is essential to confirm that all the necessary actions have been carried out on new products, and that nothing has been overlooked at every stage from the very beginning of the system to the end, i.e. from planning of new products to initial shipment of finished products.

53.2.3.1 Approval of design plan

Plans of new products have to be approved before the actual design work is commenced taking the following factors into account:

- evaluation of prototype chassis in terms of performance, reliability, quality, serviceability and productivity
- evaluation of specification and appearance design with special regard to possible problems arising from past experience

53.2.3.2 Design approval

Design approval must be given before drawings are issued for production. This confirms that the design has been completed according to the approved design plan.

The main items to be confirmed are:

- level of performance, e.g. picture quality, sound quality
- productivity
- facilities for production
- quality and reliability
- availability of materials
- cost calculation

Investigation of these items is carried out by the department concerned, evaluating data and/or working samples which are prepared by the design department.

A conference is then called of all key members. Every investigation is discussed and confirmed by the conference, at the end of which the chairman gives approval or not.

53.2.3.3 Approval of pre-production

Before mass production starts, there is the pre-production of 50–100 units.

The principle objectives of this are to confirm:

- production facilities, jigs, instruction manuals for operators, etc.
- productivity: assembly, testing, etc.
- quality of new components
- design approval

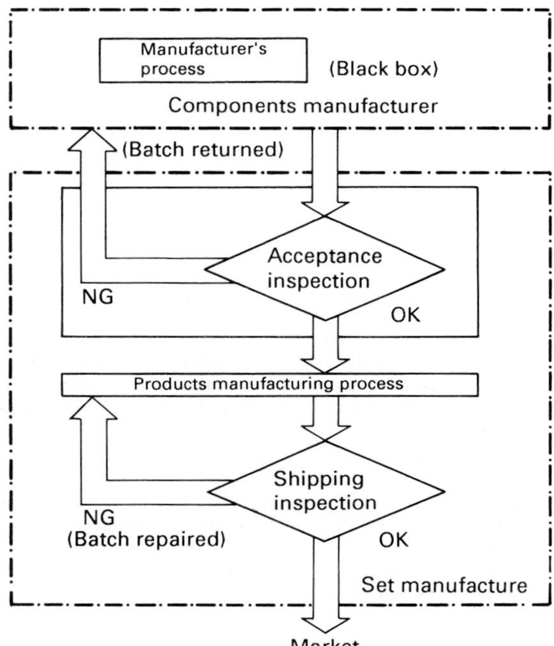

Figure 53.21 Conventional control system mainly depending on sampling inspection

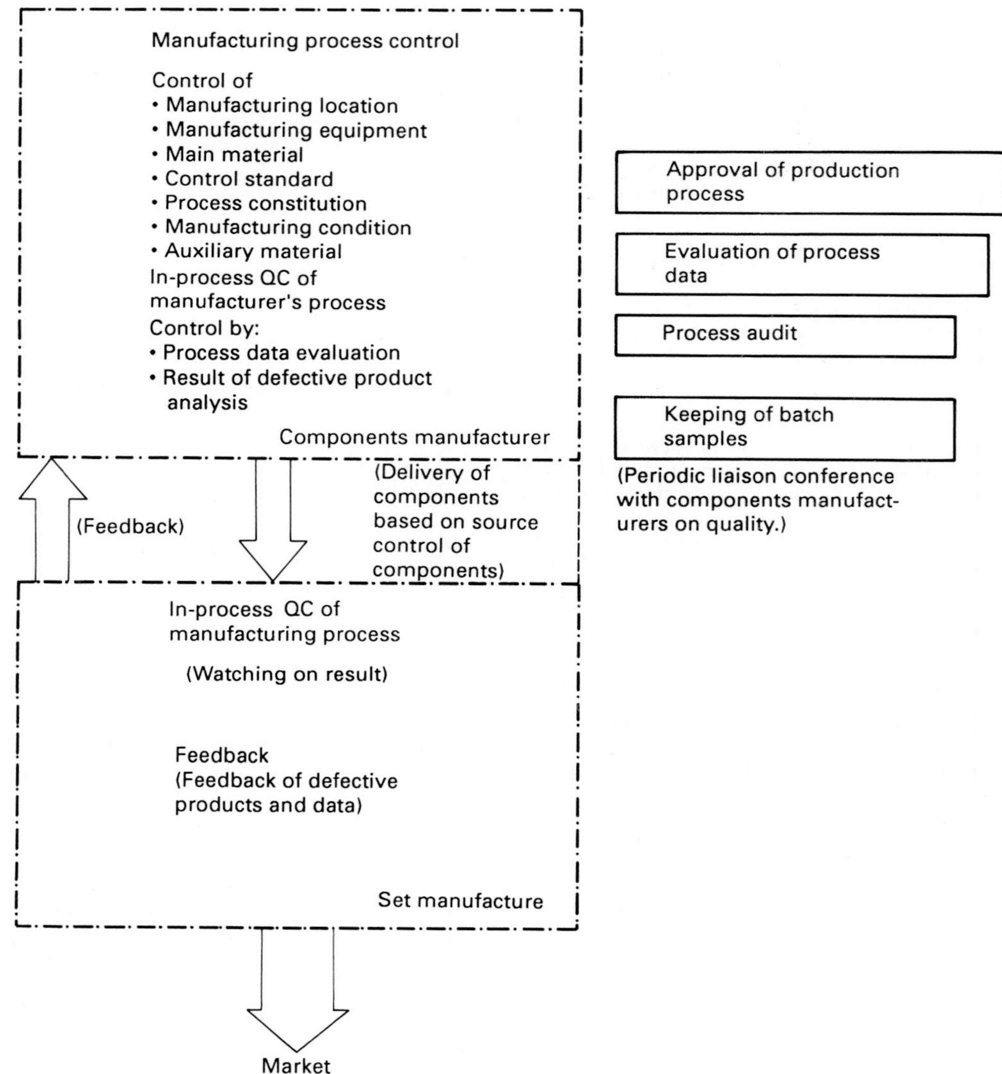

Figure 53.22 The main stages in headstream control of components

New components which are developed for new products must be investigated carefully. New components for experimental purposes are sometimes not manufactured in the same way as they would be for mass production. In the pre-production process, these components must be produced in the same way as they will be in mass production.

This difference in quality of components between those supplied for a trial and those used in mass production does sometimes mean a reconsideration of the design approval.

Following pre-production, the factory general manager calls a conference of all those involved to decide if the product can be released for mass production.

53.2.3.4 Approval of initial shipment

After completion of the first batch of mass production, the results of all the checks and quality control, including components, are investigated. The factory manager decides if it can be released.

53.2.4 Quality control of components

Quality control of components is based on *headstream control*. The concept of sample inspection is shown in *Figure 53.21* and the headstream concept in *Figure 53.22*. In sampling inspection, components were dealt with mainly by judgement in batches. In this system, components suppliers were considered as a black box. Set manufacturers were not concerned with the inside of the box and accepted defective components up to a given percentage.

In the headstream concept, components suppliers are considered as one of the manufacturing processes in the total system. Five key elements are:

- approval system for the production process

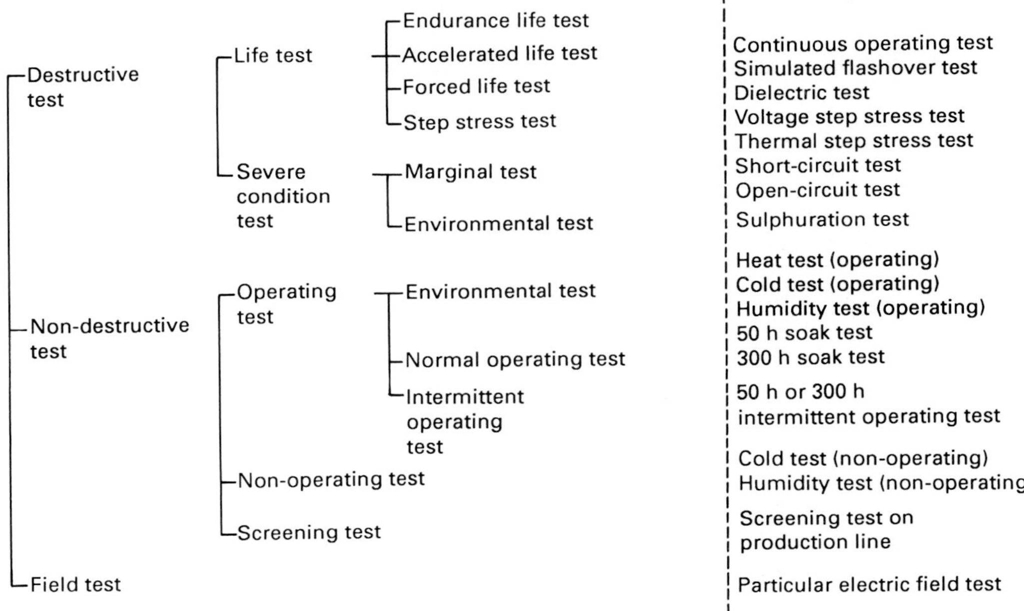

Figure 53.23 Reliability test

- evaluation system for quality information concerning the production of components
- inspection system for the production process
- batch sample system
- rating system, i.e. penalty or reward

The approval system for the production process of components by the set manufacturer is important for any new component. Items to be checked are factory, flow chart of the process, production facilities, various conditions of production, materials and quality control manuals.

The evaluation system of quality information for a batch to be shipped provides information that is transferred for every batch to be delivered.

The inspection system for production processes is carried out periodically by the set manufacturer to assess the status of management of quality control by the component manufacturers.

The sample batch is retained. This is important for critical components such as line output transformers.

The rating system evaluates the component supplier's performance over a pre-determined period.

This system cannot be successful unless the set manufacturer and the component manufacturer have common objectives and equal weight is given to both parties. All the items do not necessarily have to be implemented for all the components. The basis is good communication and cooperation between the set manufacturer and the components manufacturer especially in the area of quality and engineering. When there is a problem, a joint working team may be formed to solve it. This system allows a very high level of consistent quality to be achieved.

53.2.5 Control of reliability

Television receivers are used in various environments and by many kinds of people. With this in mind, safety and reliability must be assured.

53.2.5.1 Reliability test

To verify the reliability of television receivers, it is necessary to carry out various kinds of acceleration and simulation tests. These are shown in *Figure 52.23*.

53.2.5.2 Reliability prediction

High reliability of new models can be realized by using methods to control reliability such as FMEA (*failure mode effective analysis*), FTA (*fault tree analysis*), *Weibull distribution analysis*, etc., at design stage of new products. Reliability prediction is made according to the following steps:

1. Clearly define the system, the failure and the operating conditions.
2. Divide the system by function into different blocks and divide those blocks by failure mode into different sub-blocks to ease the calculation.

The most basic sub-blocks are parallel or series configuration. In the case of *parallel configuration*, the total failure rate λ_s is:

$$\lambda_s = \frac{1}{\sum_{i=1}^{m} (1/\lambda_i)}$$

where λ_i is failure rate of components or unit. In the case of *series configuration*, the total failure rate λ_s is:

$$\lambda_s = \sum_{i=1}^{n} \frac{1}{\lambda_i}$$

λ_i can be derived from past records and MIL-HDBK 217.

3. MTBF (*mean time between failures*) can be used as an index of reliability if the total failure distribution shows random failure mode.

In this case:

$$\text{MTBF} = \frac{1}{\lambda_s}$$

Failure rate after t hours, reliability, R_t, is:

$$R_t = e^{-\lambda st} = e - \frac{1}{MTBF}$$

As one example of predictions according to this method, the MTBF of a 14 inch colour television set which has 479 components is 51 080 h. On the other hand, the MTBF of the same model which is obtained by applying the data of a 300 h ageing test to the MTBF estimation from observed test data (which is shown in MIL-STD 781C) is 50 823. These two figures are very close.

Part 11 Television Receiver Installation and Servicing

R S Roberts C Eng, FIEE, Sen MIEEE
Consultant Electronics Engineer

54 Receiving Antennas

Electromagnetic energy is radiated into space in the form of a wave (see section *12*). This wave has two fields associated with it, related to each other and the direction of propagation as shown in *Figure 12.3*. The field magnitudes vary in an alternating fashion at the frequency of the current in the radiating antenna. In free space, the field strength falls inversely as the distance from the transmitting radiator. The induction fields are present round the radiating element, but their field strengths vary inversely as the square of the distance and consequently rapidly fall to insignificant values at moderate distances from the radiator.

54.1 Properties of antennas

54.1.1 Polarization

If a conductor is placed in the path of an oncoming wave, an alternating voltage will be developed in the conductor. This voltage will have a maximum value if the conductor is parallel to the *electric* flux lines, and will be zero if the conductor is parallel with the *magnetic* flux lines. *Polarization* is the term for the relationship between the radiated electric field and the receiving antenna. In television broadcasting, both *horizontal* and *vertical* polarization are used, and receiving antennas must lie in the appropriate plane for their area. Polarization may be used to provide some degree of isolation between two transmitters that operate on the same frequency; one can be vertically and the other horizontally polarized. This is a very useful facility in planning a service.

Polarizations other than vertical or horizontal are used for other services. *Slant* polarization may be used for vhf fm sound transmissions, so that a useful signal can be received with both vertical and horizontal antennas. *Circular* polarization is often used with satellite antenna operations; the electric field rotates continuously at the radio frequency, and considerable discrimination is provided between right-hand and left-hand polarization.

54.1.2 Reciprocity

The evolution and operation of a radiation system is described in section *12.2.2*, which looks at the properties of a radiating antenna. The same antenna, used as a receiving system, will exhibit the same properties. For example, an antenna used as a radiator will have an impedance, a bandwidth and directional properties whereby energy is radiated in such a manner that the field strength has maximum and minimum values in particular directions. If the antenna is used for reception, it will have the same impedance and the same bandwidth. The directional properties will be the same, i.e. the voltage developed in the antenna by an electromagnetic field will have maximum or minimum values when the field is arriving from the same angles as would apply to the antenna used as a radiator.

This is the *reciprocity* principle, and is helpful in considering antenna behaviour. It is often more convenient to look at the operation of an antenna as a radiator, knowing that the principles are directly applicable to the same antenna when used as a receiving device in which a voltage will be developed by an incident field. The principles derived here for a receiving antenna will therefore, by reciprocity, augment those established in section *12*.

54.1.3 Field strength

The field strength can be expressed in either electric field or magnetic field terms. The *electric field strength* is generally used because it is the most convenient for determining the voltage developed in the receiving antenna. For example, assume that a field sweeping past a receiving point has an electric field strength of 1 mV per metre. This means that the electric field voltage gradient is 1 mV between any two points 1 m apart in the field. If a conductor is erected in alignment with the electric flux, with an effective length of 2 m, a voltage of 2 mV will be developed between the ends of the conductor. The magnetic field intensity could have been used to determine the antenna voltage, but the calculation is more complicated.

54.1.4 Effective length

There are two qualifications to be applied to the simple calculation in section *54.1.3*. In practice, a receiving dipole is not of a random length, but is tuned to the operating frequency and optimized at a length of near half-wavelength for a maximum performance. Thus, for a given field strength, the voltage developed in a tuned dipole is proportional to the wavelength, and a uhf dipole will require a relatively high value of field strength for a useful value of voltage to be developed.

The second aspect concerns a difference between the *actual* length and the *effective* length. The current distribution along the length of a half-wave dipole is not uniform, but is sinusoidal, with a maximum value at the centre and zero at the ends.

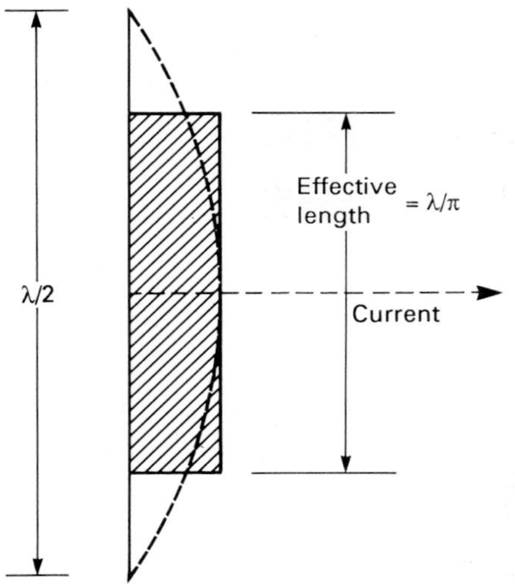

Figure 54.1 Current distribution along a λ/2 element is sinusoidal, with a maximum value at the centre and zero at the ends. An equivalent rectangle, with the maximum current value throughout its length, is shown shaded, and its effective length is λ/π

The receiving dipole will have a voltage developed in it, which will result in a current flow. Charge and discharge currents will be in phase at the centre, but will cancel at the ends of the conductor, producing the current distribution shown in *Figure 54.1*, where the value is maximum at centre and zero at the ends. The effective length is not λ/2, but is the length the conductor would have if the current were of a uniform value throughout its length.

The shaded rectangle in *Figure 54.1* has the same area as that under the sinusoidal curve for the same value of maximum current. The correction factor is 2/π, and the effective length of a λ/2 dipole is therefore λ/2×2/π = λ/π.

The voltage developed by an incident field is the product of the field strength and the effective length of the dipole.

54.1.5 Receiver input voltage

If the dipole of *Figure 54.1* is opened at the centre, the receiver input could be connected at this point. The current that results from the voltage derived from the field would then pass through the receiver input to be processed. The whole system then simplifies to an equivalent circuit (*Figure 54.2*). The voltage *e* is derived from the field, and is given by:

$$e = E.\ \frac{\lambda}{\pi}\ \text{volts} \qquad (54.1)$$

where *E* is the field strength in volts per metre. The impedance is usually about 70 ohms.

In practical terms, to transfer maximum power from the antenna to the receiver input, it is necessary for the receiver input impedance to have a value that 'matches' the impedance of the voltage source. This is usually about 70 ohms.

The receiver is usually connected to the antenna by a feeder cable that has an impedance of about 70 ohms. Such a cable is low-loss coaxial and has an impedance which is determined by the dimensions. In the complete system (*Figure 54.3*), the voltage *e* is derived from the field, the resulting power is transferred to the matching cable, and this in turn is matched to the receiver input. This can be reduced to the equivalent circuit shown in *Figure 54.4* where the cable/receiver system is represented by the matching load of 70–75 ohms.

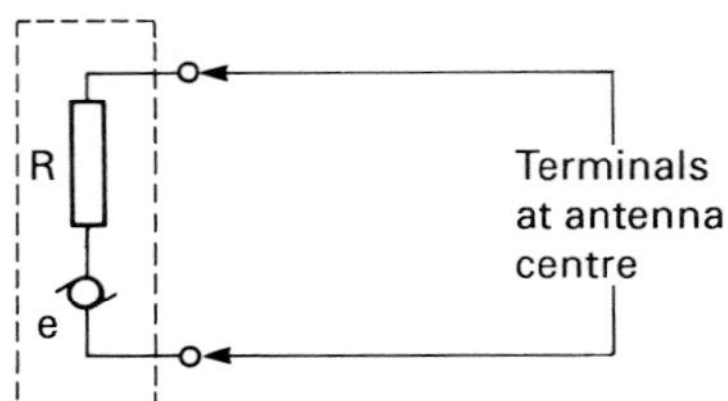

Figure 54.2 The antenna equivalent diagram with the antenna regarded as a signal source

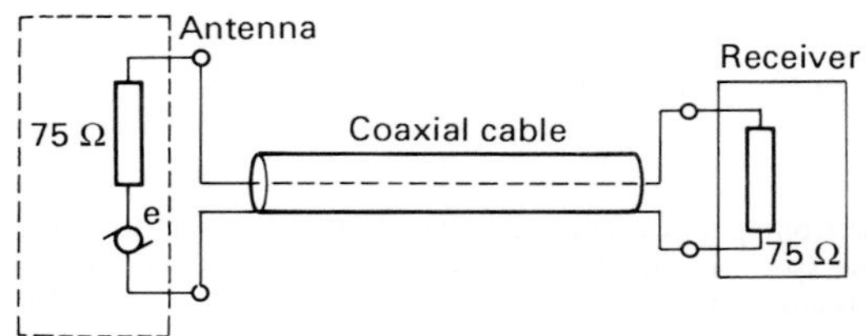

Figure 54.3 Connection from the antenna to the receiver requires two stages of impedance matching: antenna to cable and cable to receiver

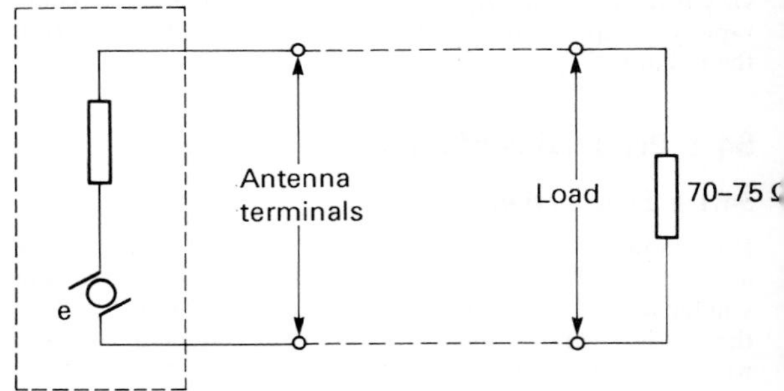

Figure 54.4 The load presented to the antenna is the cable/receiver system of *Figure 54.3*

Consider a system using a uhf dipole as an antenna operating on a frequency of 600 MHz at a site where the field strength is 2 mV/m. The wavelength is 0.5 m and, from equation (54.1), *e* will be 2 × 0.16 = 0.32 mV or 320 μV. From *Figure 54.4* we can see that the voltage delivered to the receiver input will be *e*/2, i.e. 160 μV. The connecting cable will not be perfect, and will have some losses that will reduce the receiver input voltage to a value less than 160 μV.

The uhf television receiver requires a minimum input voltage of the order of 2 mV for acceptable picture quality, and the above calculation shows that a uhf dipole would be a very poor receiving antenna. It would require unacceptably high transmitter power to generate the very large field strength needed.

The antenna for domestic reception needs to be considerably more sophisticated than a half-wave dipole. At the longer wavelengths of the vhf bands, the dimension λ/2 will be appropriately longer, and the voltage developed in the dipole by the field will be greater than that for uhf. For example, a receiving dipole operating in the European Band III on, say, 200 MHz (a wavelength of 1.5 m) would develop three times the antenna voltage for the same field strength at the antenna site considered above. At 50 MHz in Band I, the dipole becomes 3 m long, and the voltage *e* of our example would become nearly 4 mV.

54.1.6 Bandwidth

Requirements in the UK are for a group of four channels to be capable of excellent reception anywhere in the mainland. Four separate antennas could be used at each reception site, one for each channel, but such an unwieldy system would be costly and unacceptable. The obvious solution is a single antenna that operates efficiently over a bandwidth wide enough to cover the local four channels, but not so wide as to bring in interfering signals from adjacent four-channel groups.

Each UK channel is 8 MHz wide and, if the four channels were on adjacent frequencies, a single antenna with a bandwidth of 35 MHz would be adequate. However there are many reasons why adjacent spacing of the four channels cannot be used, and spacings are of the form n, $n+3$, $n+6$ and $n+13$, requiring an overall bandwidth of 88 MHz.

There are 44 channels available and, by using each channel frequency for several transmitters (relying on geographical and other features to avoid mutual interference), nine groups become available for national coverage.

Antenna manufacturers were faced with the problems of manufacture and stocking nine different types of antenna, each with a bandwidth of 88 MHz. This was simplified by the leading manufacturers who found, by further design work, that the overall bandwidth could be made 100 MHz, and the number of antenna types could be reduced to four. UHF antennas are now of four main types, classified as *groups*, identified by colour and letter codes as shown in Table *54.1*.

Table 54.1 UHF antenna groups

Channel Group	*Code letter*	*Colour code*
21–34	A	red
39–53	B	yellow
48–68	C/D	green
39–68	E	brown
21–68	W	black

Groups C and D, originally separate, are now combined
Group E has a wider bandwidth to suit a special region requirement
Group W is used for an antenna that covers the entire uhf Bands IV and V

The half-wave dipole is, essentially, a single frequency device, being tuned to one frequency. If the dipole is very thin, the voltage developed by an incident field would fall in value if the field frequency deviated from the antenna centre resonant frequency. If the dipole is of large diameter, the bandwidth becomes wider than it would be for a thin conductor.

A second aid to obtaining a wide bandwidth is to use folded dipoles. A folded dipole (see section *12.2.4*) has a self-compensating feature for *off-resonant* operation whereby the impedance tends to remain constant over a wide frequency band. A wide band of operation can be obtained by using an antenna design incorporating many elements. The elements are coupled to each other, and they can be tuned to different frequencies over the band. In practice, all available methods are used for design of wide-band antennas because a price has to be paid for a wide bandwidth; the gain is less than for a narrow bandwidth.

54.1.7 Directivity and beamwidth

Before any planning could be attempted for the uhf television service, it was necessary to know what type of receiving antenna would be made readily available by the manufacturers.

Unlike earlier television services in the vhf bands, uhf provided the first opportunity for a new broadcast service to be fully planned from the start. The BBC transmitted experimental tests in 1957 and 1958 for use by receiver and antenna manufacturers. All planning and design features were made on the basis that an outdoor antenna, 10 m high, would be used at the receiving location, and this assumption is still used. The receiving antenna must have all the required technical properties and, above all, it must be easy to erect and of reasonable cost.

It was agreed that the directional characteristics shown in curve A of *Figure 54.5* should represent the minimum directivity to be provided by the receiver antenna if a workable transmitter plan could be put into operation. As a result, the plan for national coverage in the UK uses the operation of transmitters on common channels. About 50 main, high-power stations provide the service, plus several hundred low-powered transmitters to 'fill-in' areas which are not adequately covered by the main transmitters.

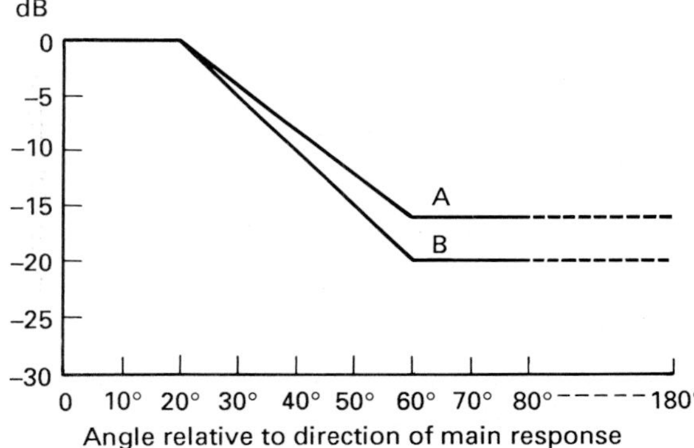

Figure 54.5 Minimum directivity requirements for receiving antennas

The antenna manufacturers found that curve A could be met relatively easily and, at a later date, curve B (*Figure 54.5*) was added. Curve A now represents a minimum for Band IV, local and medium ranges, and curve B should apply to Band V and antennas intended for use at fringes of service areas.

A most important feature of the receiving antenna performance is the *front/back* ratio which, on curve A, is shown as a minimum value of 16 dB. In practice, most antennas from reputable manufacturers have front/back ratios much nearer to 30 dB or more.

54.2 Types of antennas

54.2.1 Yagi antenna

The main requirements for receiving antennas can be met by a number of designs. None can provide all the features that are required, and compromises have to be made, e.g. a high gain can be provided if the bandwidth is not very wide. Of all the possible types of antenna, that first developed by Yagi and Uda has proved to be the most suited for reception. The principles of this antenna were first published by Yagi in 1928.

It is shown in section *12* that, for transmission, a second antenna element in close proximity to a dipole can be energized from the induction fields of the main driven dipole. Reciprocity considerations enable us to derive the contribution of a parasitic element to a main receiving dipole by an incident field. *Figures 12.10, 12.11* and *12.12* show how the omni-directivity of a single dipole can become unidirectional by the addition of a parasitic element.

Yagi found that if a parasitic element, with an inductive reactance, were spaced about λ/4 away from the main dipole, the radiated field would be distorted in the direction parasite-to-antenna, with an enhanced field strength as though the parasite were a *reflector*. He found that the same effect could be obtained if the parasitic element was positioned in front of the main dipole and had a capacitive reactance. In this case, the parasitic element is termed a *director*.

In practice, the reflector can be a single element, a multiplicity of elements or a conducting sheet, and the gain in the 'forward' direction can be of the order of 3–4 dB in such a two-element system. While only one reflector can be used, any number of parasitic elements can be added to improve the directivity and increase the gain in the forward direction.

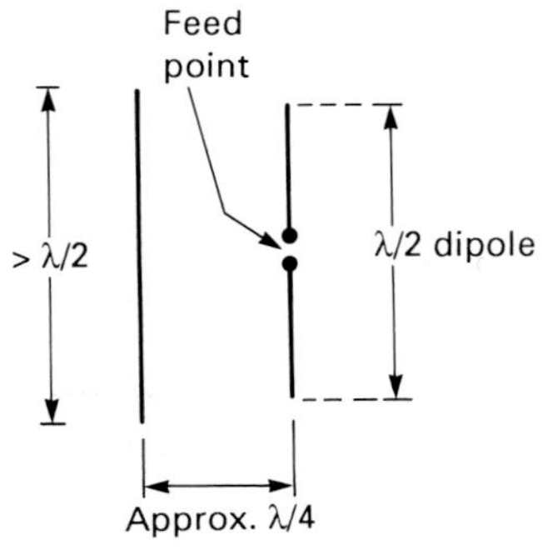

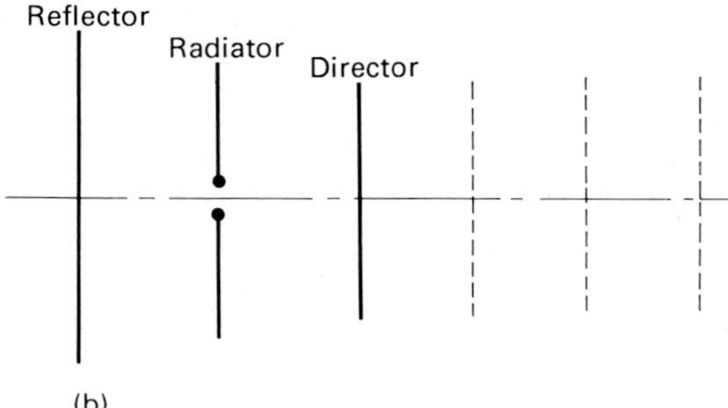

Figure 54.6 Addition of a reflector to a dipole. (a) The parasitic element is excited by the induction fields generated by the current in the dipole. (b) The addition of a director. Further directors may be added as shown dotted

A general purpose uhf antenna for use within the service area of a transmitter is shown in *Figure 54.7*. This antenna has ten elements, a wide-band dipole and a four-element reflector. The forward gain is just under 12 dB, the front/back ratio is 31 dB, and the beamwidth is about ±20° (see *Figure 54.8*).

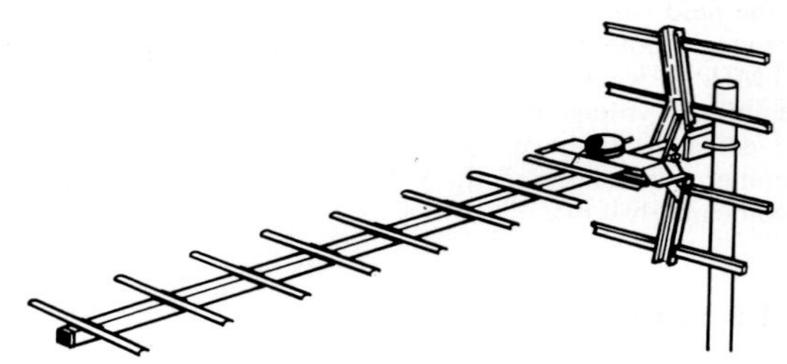

Figure 54.7 A general purpose ten-element antenna (Antiference Ltd)

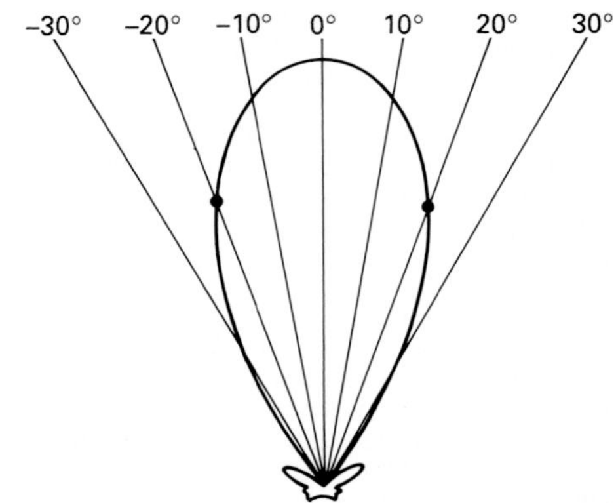

Figure 54.8 The beamwidth is the angle between the two points where the maximum values are reduced by 3 dB

The impedance of a dipole is reduced drastically by close coupling to other elements. The dipole and reflector of *Figure 54.7* has an impedance at the antenna terminals of about 140 ohms. This is reduced by the coupling of the directors to about 70 ohms.

The ability of the Yagi to increase the gain by increasing the number of directors is useful, but has some limitations. *Figure 54.9* shows how the number of directors and the gain are related.

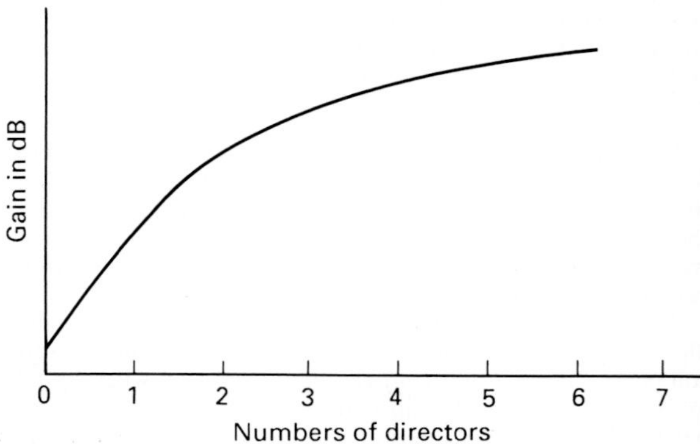

Figure 54.9 The yield of gain per director becomes smaller as more directors are added

Every additional director adds to the gain, but the increase of gain per director becomes less as more are added, and a practical limit is reached when it is no longer worthwhile to add more. *Figure 54.9* does not show any figures for the scale of gain. The actual gain figures will depend on the precise compromises that the designer has made for the various features that a particular antenna must have.

Figure 54.8 shows two further features of importance. Firstly, beamwidth refers to the included angle between two points on the polar diagram where the field strength falls from the maximum value by 3 dB, as determined by the length of the radials at any angle. Secondly, the small loops at the base of the diagram are referred to as *side-lobes;* if they are large they may seriously impair the directivity provided by the main lobe.

54.2.2 Log-periodic antenna

This antenna is similar to the Yagi in appearance, but it operates on a different principle. It is a frequency-independent antenna (i.e. wide-band), and is shown in *Figure 54.10*. It consists of an array of dipoles having lengths and spacings differing by a common ratio, and, unlike the Yagi, each element is connected to a transmission line.

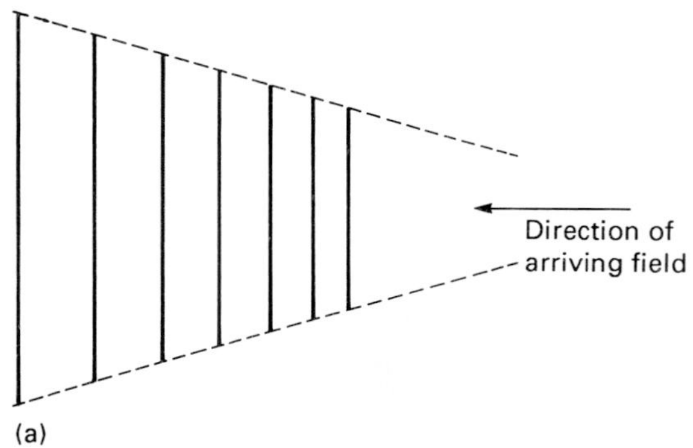

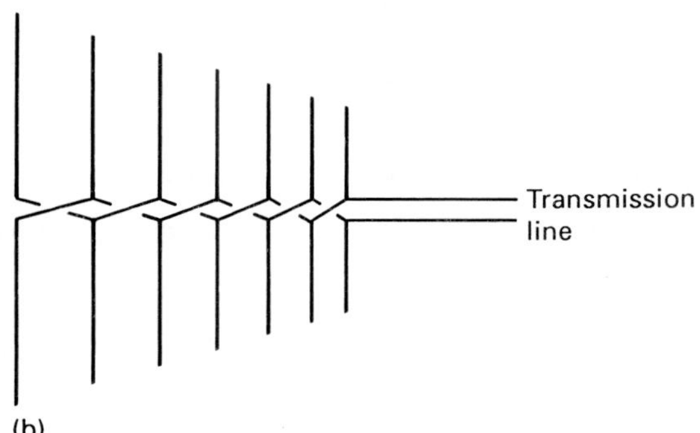

Figure 54.10 A log-periodic antenna, (a) the element lengths and spacings differ by a common ratio, (b) shows one method of obtaining correct phase for each element

The only constraints on the bandwidth are determined by the lengths of the longest and shortest elements. The lowest frequency is determined by the longest element, and the highest frequency by the shortest element.

The highest frequency end of the array is directed towards the transmissions that it is intended to receive, and the directivity comprises a polar diagram which is virtually free from side-lobes. The gain is low because different parts of the array are resonant to a different band of frequencies. At the high frequency part of the band, a few elements are functioning and the remainder of the array acts as a form of reflector. At the centre of the band, a few elements at the centre of the array come into operation. The lowest frequencies will energize the long elements. The antenna yields a modest gain of about 6– dB because only a few elements are in operation at any of the frequencies in the band, but it is used where no side-lobes are permissible if interference is to be avoided.

54.2.3 High gain antennas

The Yagi has practical limitations to the maximum gain that can be realized. An eighteen-element version of the antenna shown in *Figure 54.7* has a gain of about 14 dB, a front/back ratio of about 32 dB and a beamwidth of ±16°, but it is bulky and is probably at a limit for a Yagi design.

There are many locations where high gain is required, because severe shielding of the receiving site depresses the field strength. Another form of Yagi antenna is shown in *Figure 54.11*. Here the director elements are not single elements, but each director unit consists of four λ/2 elements. This provides a basic increase of gain for each director unit. For example, an eight-element version of this 'high gain' design can provide a gain of about 16 dB, compared with the ten-element conventional Yagi which has a gain of about 12 dB.

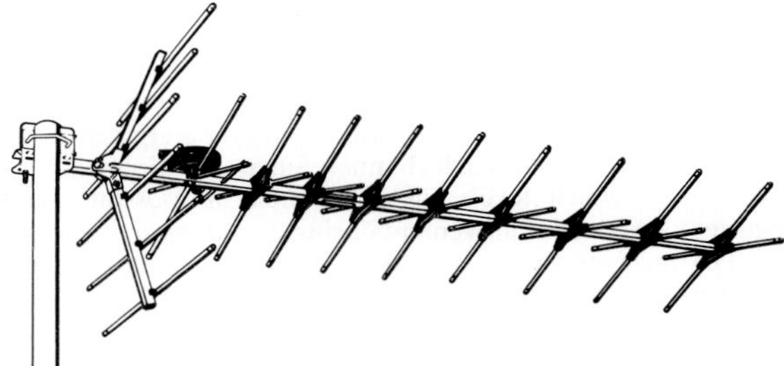

Figure 54.11 A 'higain' version of the ten-element Yagi shown in *Figure 54.7* (Antiference Ltd)

An interesting version of the 'high gain' is shown in *Figure 54.12*. In terms of Yagi designations for the elements, this has 23 elements, and its performance provides a gain of more than 20 dB, a front/back ratio of 33 dB and a beamwidth of ±13°.

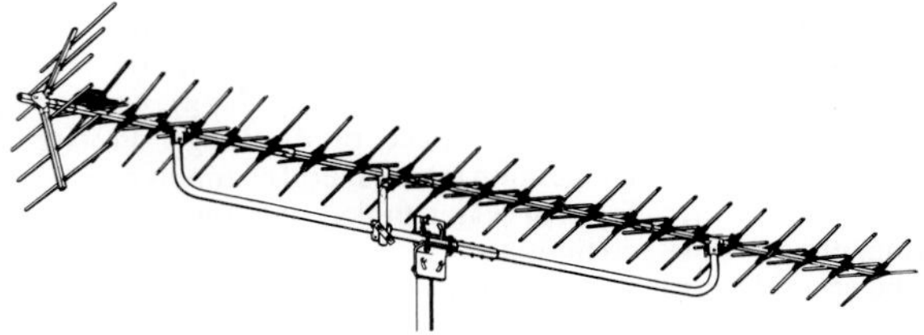

Figure 54.12 A 23 element 'higain' antenna (Antiference Ltd)

54.3 Antenna erection

The short wavelengths used for television give rise to effects similar to the behaviour of light. Shadowing effects produced by hills and high buildings in the path between the transmitter and receiving site can result in low signal strengths at the receiver input. Receiving sites at the bottom of a valley can have a low field strength compared with those on high ground.

Reflection from ground will always take place, however narrow the beamwidth may be.

Figure 54.13 shows a very simplified situation where P_1 is the direct path, and the ground is a simple reflector which reflects the path P_2. Thus, signals arrive at the receiving antenna by two routes, and the two fields will not be in phase. In the practical situation, signals reach the receiving antenna by many paths. In the horizontal plane many reflecting points may exist and, in the case of a high metallic structure such as a gasholder, for example, the reflected signal may be of comparable value to the signal arriving along the direct path.

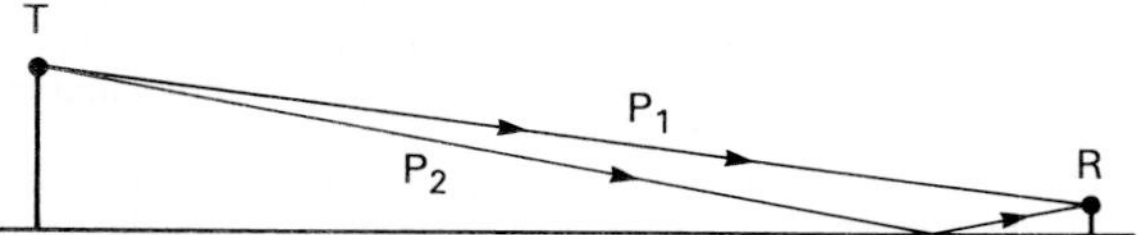

Figure 54.13 Signals received direct and reflected (not to scale). The angles involved are very small, and no practical antenna can have a beamwidth that will discriminate between P_1 and P_2

The difference between the path lengths may cause the reflected signal to have a considerable delay, and produce a second, *ghost*, image displaced to the right of the main image. An antenna with a narrow beamwidth will provide the required directional discrimination to minimize the effect of ghosts.

Reception of several channels will be more complicated because the position of reflection points are wavelength dependent. It is thus necessary to find a position for the antenna that will provide the receiver with approximately the same signal input voltage for each channel. Such a position will be a compromise which levels signals as near as possible, and requires a freedom of movement during the erection of about 1 m^3. The receiver has its own agc system which will level the compromise established during erection of the antenna.

54.4 Indoor antennas

Planning of the uhf service is based on the use of an outdoor antenna at a height of 10 m. There are many situations where outdoor antennas cannot be used, and some form of indoor antenna is necessary.

Any indoor antenna will have a performance which is inferior when compared with that at an outdoor site. However, at sites near transmitters, it may be possible to obtain an acceptable signal from an indoor antenna.

Indoor antennas are of two types. The standard uhf antennas for outdoor use are of small physical dimensions (e.g. the ten-element Yagi shown in *Figure 54.7* is about 1 m long), and can often be installed in the roof space of dwellings. Precise positioning for uniform reception from several channels may be difficult due to the effects of water tanks, plumbing, electric wiring, etc., in close proximity.

The second type is the room (or *set-top*) antenna. This may have a performance worse than an antenna in the roof space but, well within the service area, might provide a useful source of signals. The best position is not on top of the receiver, but has to be found by experiment.

Any room antenna must conform to BS 5373. This requires that the antenna must be isolated from the receiver input socket by means of a capacitor or a transformer. This is a safety precaution in case some electrical failure in the receiver connects the receiver input socket to the mains power supply. This is extremely unlikely with modern receivers, but a 'live' room antenna could be lethal. The standard is concerned only with safety, and not in any way with the performance of the antenna.

Over the years, room antenna design has progressed through many types but, more recently, appears to have settled down to a final form. There are three main problems associated with it:

- a dipole is not as much use as a uhf antenna, even at short distances from the transmitter
- extra elements are needed to provide as much gain as possible, which makes an overlarge device, unsuitable for room use
- aesthetically, it must be visually acceptable in any domestic decor

The log-aperiodic antenna has several features that can be embodied in the room antenna. It was explained in section *54.2.2* that, over a very wide frequency band, only a few elements of the array may be useful. By increasing the constant ratio of element length to spacing, and using elements of large cross-section, all the elements can become useful over the 2:1 bandwidth of the uhf broadcast band.

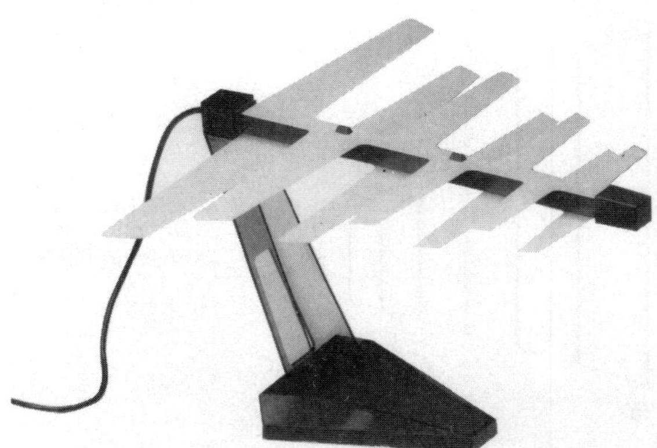

Figure 54.14 A log-aperiodic room antenna (Antiference Ltd)

The performance of the antenna shown in *Figure 54.14* is such that the forward gain is 5–7.5 dB, the front/back ratio between 16 and 22 dB, and the beamwidth between 32° and 35°. The total length is only 30 cm.

54.5 Standards and Codes of Practice

There are several standards concerned with receiving antennas, apart from BS 5373 referred to in section *54.4*.

There are difficulties associated with measurements of antenna performance, and a form of standardized measurement procedure has taken many years to reach publication.

The international IEC Standard is IEC Publication 597: *Aerials for reception of sound and television broadcasting in the frequency range 30 MHz to 1 GHz*. This is in four parts:

IEC 597-1. Part 1: *Electrical and mechanical characteristics*.
IEC 597-2. Part 2: *Methods of measurement of electrical performance parameters*.
IEC 597-3. Part 3: *Methods of measurement of mechanical properties, vibration and environmental tests*.
IEC 597-4. Part 4: *Guide for the preparation of aerial performance specifications. Detailed specifications sheet format*.

The British Standards Institution equivalent is BS 5640. The IEC and BSI specifications are virtually the same. As a result, BS 5640 is available in parts with the same titles as listed above.

Two Codes of Practice exist in the UK. The British Standards Institution Code is published as a specification, BS 6330: 1983 entitled *British Standard Code of practice for reception of sound and television broadcasting*.

The second code of practice is that published by the Confederation of Aerial Industries. This is *Code of practice for the installation of radio and television aerials*. This document is primarily concerned with the actual installation of antennas.

E Trundle MSERT, MRTS, MISTC
Chief Engineer, RNF Services Ltd

55 TV Maintenance and Servicing

Unlike most other equipment described in this book, domestic television receivers are mass produced for a high volume market. Their maintenance and servicing has to be based on a *high throughput at low unit cost* philosophy to remain viable. Since the technologies used in consumer products are similar to those in professional and broadcast equipment, similar levels of expertise, data and test equipment are required, but with heavy constraints on labour time and costs in the domestic sector. Fast fault diagnosis is the key to success in the TV servicing industry, and this in turn depends on the ready availability of service data, product training and suitable test equipment.

55.1 Test equipment

Sophisticated test gear is seldom required for fault diagnosis in TV receivers. The basic requirements are a sensitive multi-range testmeter, an oscilloscope, and occasionally a frequency counter. If (as is usually the case) video cassette recorder service is also envisaged, the counter will be more fully employed, and test jigs and alignment tapes are also required. Into a second category come such gear as pattern generators, logic probes and picture-tube testers.

55.1.1 Oscilloscope

An oscilloscope is the most useful tool in fault tracing. For TV work, a minimum requirement is a 10 MHz single-beam type with sensitivity of 2 mV per screen division. More versatility and utilization is possible if the oscilloscope has extra sensitivity down to 1 mV/division, greater bandwidth of 20 or 50 MHz, and dual-beam capability. The last two qualities, combined with a high PDA (*post-deflection acceleration*) enable the instrument to be used for servicing other consumer products like cd audio, video cassette recorders, etc.

To avoid circuit loading, use of a 10:1 probe at the Y-amplifier input is essential. This desensitizes the instrument by a factor of ten. For most applications, dc coupling of the oscilloscope is appropriate, giving at-a-glance readout of all characteristics of the signal at the test point.

Sweep and trigger requirements for TV service oscilloscopes are a maximum of 200 ns/division, and the provision of a sync separator for use with composite video input signals. Switch-selected trigger sources of TV line, TV field, mains rate and high/low filtered Y-channel signals are useful, selectable between Y1, Y2 and external sources. For reliable triggering, *external* synchronization from line-, field- or subcarrier-sections of the TV is recommended.

55.1.2 Multimeter

For testing voltage, current and resistance, the choice lies between analogue and digital meters. Modern digital multimeters (dmms) are accurate, robust and impose little loading on the circuit under test. A $3^1/_2$ digit type with lcd readout is suitable for TV service work. Battery operation gives the greatest convenience in use, and a basic accuracy of 1 or 0.5 per cent (dc voltage ranges) is quite adequate.

Many purpose-designed dmms also have facilities for testing semiconductor junctions and eyes-off 'bleep' continuity tests; some incorporate a tone generator whose pitch indicates the level being measured. A useful accessory is a high voltage probe with which to measure eht voltages in picture-tube circuits.

55.1.3 Frequency counter

A counter to give a readout of frequency or period of the applied waveform typically has a $7^1/_2$ digit scale length and is capable of counting to about 100 MHz. Pre-scalers are available to extend this upwards.

A highly accurate and expensive type is not justified for TV service work, as the instrument can be simply checked and recalibrated against the highly accurate timebase and subcarrier frequencies present in a TV tuned to a broadcast transmission. In use, beware of misleading readings caused by complex input waveforms and frequency drift during the measuring period.

55.1.4 Logic probe

Increasingly, microprocessor and digital techniques are being used in TV and allied equipment, and analysis of data is made easier by the use of a simple logic probe. The presence and nature of data on both parallel and serial buses is quickly shown by body-mounted leds. Best suited for TV and vcr work is a hand-held type with red, orange and green indicators and an 'eyes-off' tone generator.

Complementary to the logic probe, though not as useful in everyday servicing, is the *logic pulser* which generates single or continuous pulses for injection into the data line under test.

55.1.5 Pattern generator

Specialized patterns for convergence set-up and colour decoder adjustment are seldom required with modern self-converging tubes and one-chip colour decoders. Colour bars and plain colour fields are required for decoder/RGB fault-finding and display-purity adjustment. Other common requirements are crosshatch and 'multi-burst' patterns for focus and bandwidth tests, and edge castellations for centring and line-phasing checks. The use of test patterns in general is discussed in section *55.2*.

Versatility of output modes is important for a workshop-based generator. An instrument with multi-band coverage, calibrated and variable output levels, multi-system sound modulation, and video baseband and RGB outputs amply repays its cost over the years.

55.1.6 Picture-tube tester

The picture tube is the most expensive and wear-prone part of a TV receiver; replacement of a worn or faulty tube is seldom economically viable, even though virtually no setting-up adjustments are required. For these reasons, a comprehensive tester is a useful tool in proving the diagnosis. The best types have separate indicators for each of the three electron guns, a wide range of tests, primarily for cathode emission and inter-electrode leakage, and a comprehensive selection of base-socket adaptors.

Usually a *re-activation* facility is provided, with which the guns' cathodes can be electron-blasted to expose a clean emissive surface. Thus rejuvenated, old tubes are given a further lease of life in many cases.

55.1.7 Other test equipment

At times, other test equipment is used in servicing TV and allied equipment, most of which will be familiar to electronics engineers. Into this category come *variable power supplies*, *field strength meters*, *signal injectors*, *magnetic degaussers* and *component test bridges*.

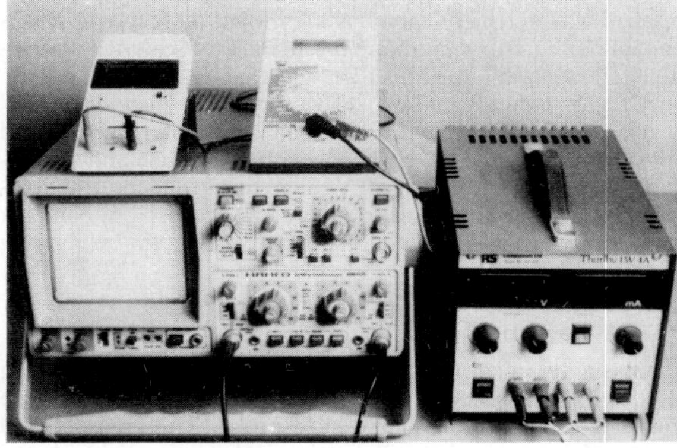

Figure 55.1 TV test instruments. Clockwise from right are a regulated lt power supply, wide-band oscilloscope, frequency counter and digital multimeter. The small instruments are battery powered

55.2 Setting up and performance appraisal

Many aspects of the performance of a TV set, and the quality of its input signal, can be judged by careful examination of a test card like the typical one shown in *Figure 55.2*.

Figure 55.2 BBC test card G. Based on the Philips PM5544 pattern, this one has 95 per cent saturation, 100 per cent amplitude colour bars. Other features are described in the text

The main features of this general purpose card are:

- *Border castellations*. Set scan amplitudes so that the edge blocks are half or two-thirds visible. In practice, the picture width is governed by the ht voltage setting. Off-centre and tilted pictures are revealed by the border castellations.
- *White grid and centre circle*. These are primarily for test and adjustment of scan linearity. The grid also provides a check of colour registration or convergence. Primary colour fringing should be zero at screen centre, less than 1 mm at edges, and less than 2 mm in extreme corners.
- *Colour bars*. The standard YCGMRB colours are displayed for subjective check of decoder, display and signal-path performance. The amplitude and saturation characteristics of the bars depend on the pattern and its originator.
- *Grey-scale step-wedge*. Towards the bottom of the centre circle, these graduated blocks, from black to white, represent equal increments of luminance in six steps. They check linearity of the video amplifiers, and become truly grey in colour when the grey-scale tracking is correctly adjusted.
- *Multi-burst*. Here just below screen centre, the progressively finer vertical gratings provide a check on receiver bandwidth and, together with the centre crosshatch, picture-tube focusing. For this type of broadcast test card, the gratings correspond to sinusoidal video waveforms of 1.5, 2.5, 3.5, 4, 4.5 and 5.25 MHz.
- *White/black rectangles*. In the top and bottom quarters of the centre circle, the white/black rectangles check lf response in the video circuits generally. Streaking or smearing indicates poor lf response. The black 'needle' in the upper block gives rise to ghost images to its right if multipath or reflection problems are present in the signal chain.
- *White/grey rectangles*. In a similar way, the white/grey rectangles above the colour bars check transient and mf frequency response. Streaking, smearing and overshoot should be absent from these.
- *Colour-fit pattern*. The bottom section of the circle contains a sandwich of red in yellow. Colour bleeding between the two, or an off-centre red block, indicates timing errors between

chrominance and luminance signals. Due to the low bandwidth of the chroma signal channel, some 'woolliness' is to be expected here, especially in receivers not using CTI (*colour transient improvement*).

- *Interlace check*. The card's horizontal centre line has a *frame* -rate component to check field scan interlace. If its thickness is different from that of the other horizontal lines, erratic field triggering is indicated. Problems of sync-separator performance or line-pulse breakthrough into the field circuit are usually responsible.
- *Colour-pattern 'brackets'*. On each side of the central circle, these contain specially encoded chroma signals to give a comprehensive check on PAL decoder performance. Characteristics of these signal components vary between test patterns. Specific details are available from the manufacturer or broadcaster involved. Similarly, the blocks between the brackets and edge castellations may contain decoder test signals which render colourless squares only if chroma demodulation phases are on the correct axes.

55.2.1 Picture-tube adjustments

Modern display tubes require very little in the way of adjustment. Purity and static convergence are governed by ring magnets on the neck, whose disturbance and adjustment should only be made after the set has been thoroughly degaussed by a hand-held coil, and then with reference to the set-maker's alignment instructions.

Some tubes have provision for fine adjustment of *dynamic convergence* (colour registration at screen edges and corners) by manipulation of the scan-yoke's front flare. One can pan horizontally to overlay red and blue lines adjacent to and parallel to screen edges, and tilt vertically to overlay the extremities of red and blue lines which pass through the centre of the screen.

An important aspect of colour reproduction is grey-scale tracking. There should be no coloration of a picture with no chroma signal content. The tube's gun-bias controls (d.c. operating points) should be adjusted for neutral tint in the darkest parts of the picture, then the RGB drive controls can be set for pure white highlights. In modern decoder IC designs, the black level of the three rasters is constantly monitored and adjusted in an *auto-grey-scale* system, wherein each gun's 'black current' is sampled once per field.

55.3 Fault diagnosis

Apart from the grey-scale set-up described in section *5.2*, virtually no routine maintenance is carried out on a TV set. With modern technology, design and construction, it should not be needed. In general, the receiver is presented for service only when it has actually developed a fault. The following text will examine each section of a TV receiver in turn, from the point of view of fault diagnosis.

For test and diagnosis, it is essential to break down the set into individual function blocks and isolate the fault to one of them by examining the signal, supply voltage and feedback conditions at the exchange points. Once the faulty block has been identified, individual sections within the block can be examined, narrowing down the field of search to one or more individual components. Bear in mind that a faulty or damaged component may be the victim of a fault elsewhere, and that *parametric* faults (subcarrier frequency drift, incorrect tuning, too high eht voltage, etc.) can give rise to *catastrophic*

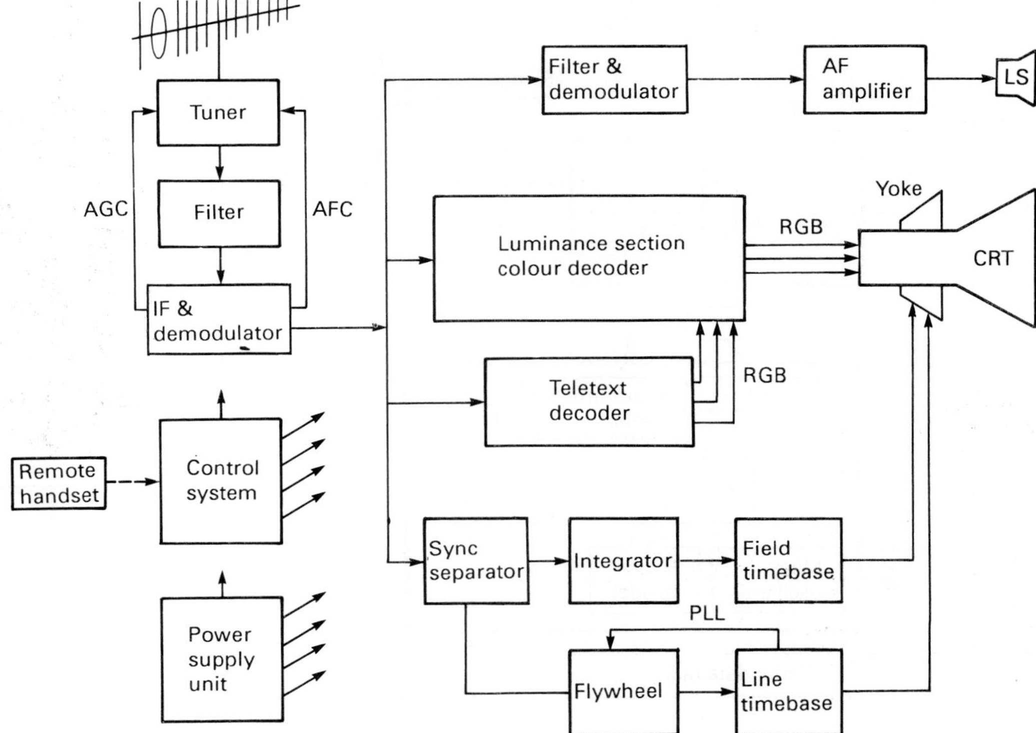

Figure 55.3 TV receiver block diagram. Each arrowhead represents a key test point

symptoms due to automatic protection, muting and 'killer' circuit features.

Figure 55.3 is a typical block diagram for a modern TV, showing the junction points at which the first diagnostic steps are taken.

55.3.1 Tuner and receiver

The varicap tuner is most often part of a phase locked loop (pll) in which the receiver is held on the required broadcast band by a frequency synthesis tuning loop based on a programmable divider. The reference for the system is a stable and accurate crystal. Station selection in the vhf or uhf bands is made by the user through a digital data link from a local or remote keypad.

An outline of the system is given in *Figure 55.4*. The most common operational fault is loss of signals, with the 'front-end' implicated by a lively display of noise (snow) on screen. Establish that the tuner has correct operating and agc voltages before attempting to isolate the fault to the tuner or its control system.

Check the varicap control voltage to the tuner, if necessary substituting a variable voltage from an external source. An absence of signals over an input range of 0–30 V implicates the tuner itself. If, alternatively, the 'steering' voltage from the control system is wrong or missing, check ICs such as the CITAC and memory to ensure they have an operating voltage and that their clock oscillator is running. Check also, if necessary, that data are coming from the control microprocessor. These are usually in serial form on a two-wire bus.

When peripheral components have been eliminated, the CITAC chip or its equivalent is usually responsible for tuning-control faults. Memory chip faults do not usually prevent station-search or direct entry of channel number. A situation in which the search sweep does not stop at each channel generally indicates an absence of feedback from a downstream section of the TV. Sometimes this is the vision demodulator section, more often it is the line oscillator IC, which sets up a 'flag' when it synchronizes to a transmission.

Tuning drift is rare in modern sets, especially those using fst (*frequency synthesis tuning*) control systems. The tuner itself can be at fault, but since it and much of the rest of the circuitry is within the control loop, a more likely culprit is the master crystal. Checking this and the input/output frequencies of the pre-scaler with a counter should reveal the trouble spot.

Low rf gain is indicated by excessive noise in the picture, manifest as 'snow' on the screen. First ensure that the input signal from the antenna is adequate (see section *54*). If it is, check the setting and condition of any tuner agc control that is fitted, and if necessary apply a suitable agc control voltage from an external source to the appropriate pin on the tuner. If no improvement is seen, the tuner is suspect: low-gain faults farther downstream (e.g. if stages or demodulator) result in low contrast rather than noise on picture. A faulty saw filter (used for if response shaping) can result in a 'noisy' picture, but it is usually accompanied by a ringing or "ghosting' effect.

The rest of the receiver section comprises the vision and sound if amplifier and demodulator. This is a relatively trouble-free section, invariably embodied in a single IC. As with all such, its internal circuits are dc-coupled, so pin voltages are a good guide to where any problem lies. Very often, such faults as are encountered lie in peripheral components such as electrolytic capacitors. This is particularly true of those used for supply-line and agc decoupling. AGC problems typically upset video output levels, tuner gain or field sync pulse integrity.

Many strange picture defects can be caused by misalignment of the carrier coil associated with the vision detector. For empirical adjustment, lowest picture brightness and best definition generally coincide at the optimum tuning point. The setting of this and the associated afc coil have a large effect on the quality of teletext reception. With suitable test equipment, adjust for best eye height of the text data pulses.

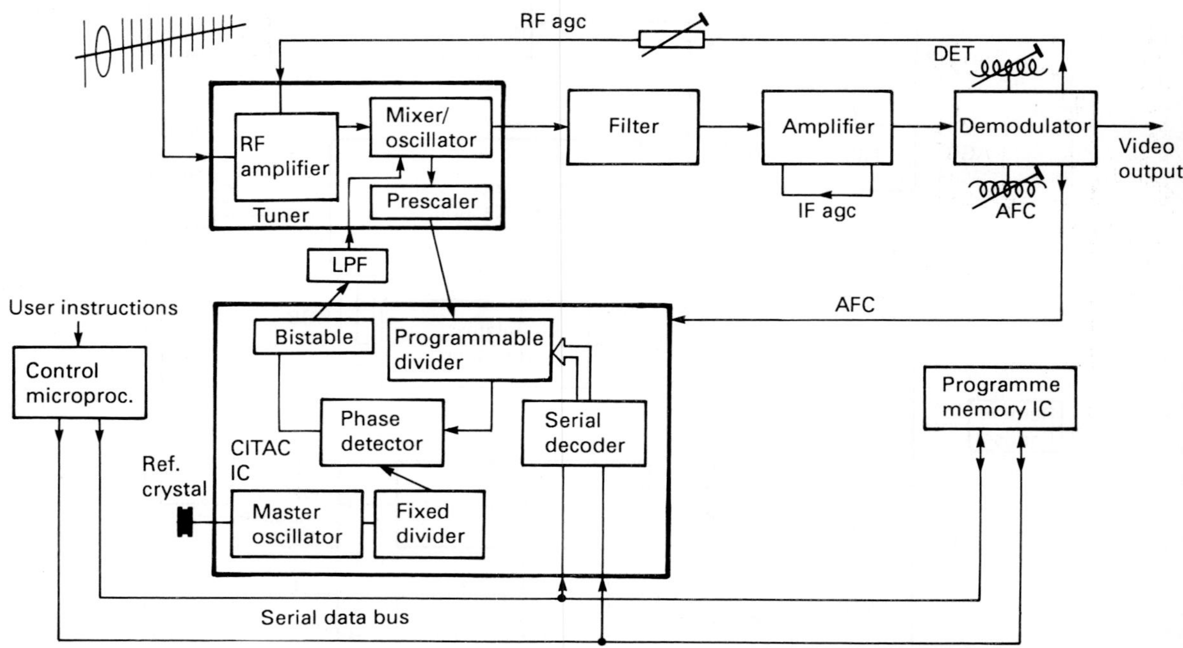

Figure 55.4 Tuning system and if amplifier. Station tuning is locked to the reference crystal in a phase locked loop. Division ratio is governed by the user's programme selector via the programmable divider

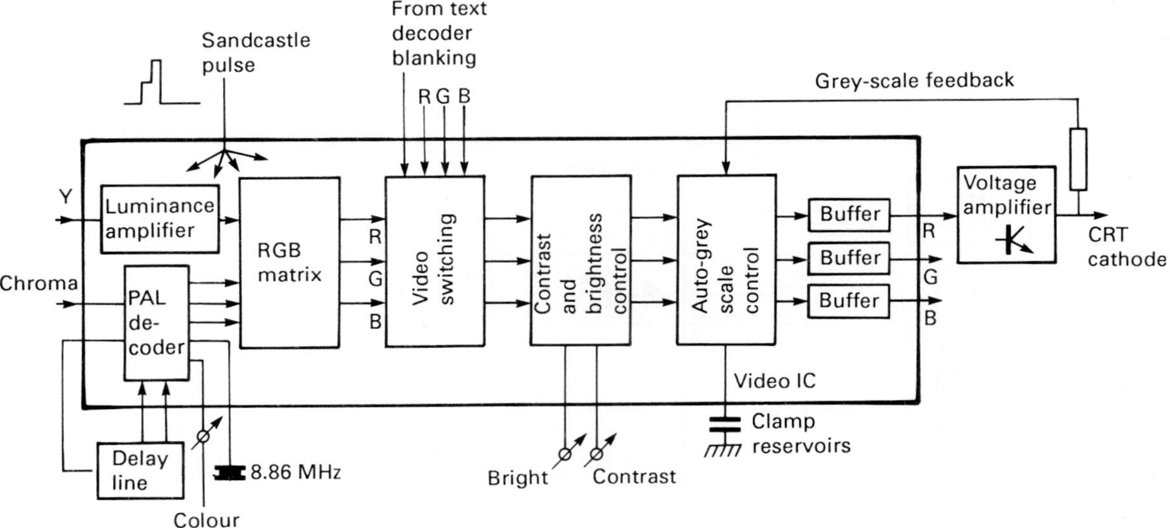

Figure 55.5 One-chip decoder system. The Y and chroma inputs come from the same source, via a delay line and a band-pass filter respectively

55.3.2 Colour decoder and video section

After demodulation, the video signal is filtered and passed on to the chroma decoder section. In most designs, a single IC is used here in a configuration like that of *Figure 55.5*. For servicing, the ideal input signal is from a local colour-bar generator, with dmm and oscilloscope as diagnostic tools.

A common symptom is *no picture*, with no visible raster on the picture tube. First ensure that the tube is correctly furnished with eht, focus and anode potentials and that its heaters are alight. If so, the guns are biased off by too high a voltage at the cathodes: they generally operate at a potential of about 125 V. At this point, investigation of the Y-RGB signal sections can begin. Again, dc voltage checks around the decoder chip should quickly lead to the cause of the fault. If the main disparity between the actual and expected voltages is at the RGB output pins, check any 'common' components in the emitter-to-ground paths of the RGB output stage transistors. Failure here can 'lift' all three emitters to cut off the output stages altogether.

Modern decoder ICs depend for internal clamping, keying and gating on a *sandcastle* pulse originating in the sync/timebase section. The level, shape and duration of all components of this composite pulse are crucial to correct operation. Any problem here generally leads to black-out of the picture. Similarly, the IC is very sensitive to incorrect conditions during the feedback phase of the auto-black-level measuring process. Thus a fault in one of the RGB output stages can cut off all three, for reasons that are not immediately obvious.

Most other no-picture fault culprits are revealed by dc voltage checks. When analysing control voltages to contrast- and brightness-setting IC pins, bear in mind the presence and influence of beam-current limiting, blanking and muting circuits; and that brightness control is governed by both the user's brightness control *and* the output from any remote-control decoder system.

Loss of one primary colour from the display, or suffusion of the screen with one colour, indicates incorrect bias on the gun concerned, or that the picture tube has internal leakage. This type of symptom must stem from the 'back-end' of the decoder/driver block, and since the three amplifiers and guns are identical, interchanging signal feeds (to tube or voltage amplifiers) can speed up the diagnosis where the fault is not due to some obvious cause like an open circuit load resistor or leaky output transistor. When interchanging signal feeds in a substitution test like this, beware of confusing the auto-gst (*automatic grey scale tracking*) section of the drive IC.

The next most common fault is probably that of *no colour*. It usually stems from a fault in the reference ('housekeeping') rather than the signal sections of the decoder system. Thus the first suspect is the subcarrier crystal; this has a frequency of 4.433 MHz in older designs and 8.86 MHz in later ones. First establish that it is running, then that its frequency is correct. If the regenerator pll comes out of lock, the automatic colour killer comes into operation to cut off the chrominance signal channels. Some decoder chip designs have provision for manual over-ride of the colour killer. With the killer over-ridden, unsynchronized colour suggests an absence of burst signal (check the chroma signal input) or line-rate keying pulses from sync or line-scan stages. If the crystal cannot be made to free-run at the correct frequency, check its two or three peripheral components (e.g. trimmer capacitor) before replacing it.

The PAL chroma alternations are removed in the decoder IC. Any amplitude or hue difference in adjacent pairs of scanning lines (*Hanover bars*) are more likely to be in the matrix circuit (consisting primarily of the 64 μs glass delay line and peripheral components) than in the IC itself. Waveform examination with an oscilloscope is the key here, though sometimes a substitution test on the delay line is quicker.

If the PAL switching is totally out of phase, the effect is reversal of reds and greens on screen, leading to a user complaint of 'green faces'. This symptom of ident failure is rare in modern IC decoder designs. A check of the line-rate steering pulse input is recommended before the chip is condemned.

Poor definition pictures may be due to a fault in the post-demodulator vision processing so long as the blurring is confined to vertical lines. Where no colour is present in the picture, check first the if amplifier and demodulator sections. This fault is not likely to be due to IC or semiconductor failure; capacitors and the luminance delay line circuit are main suspects. Where the lack of bandwidth is confined to one of the

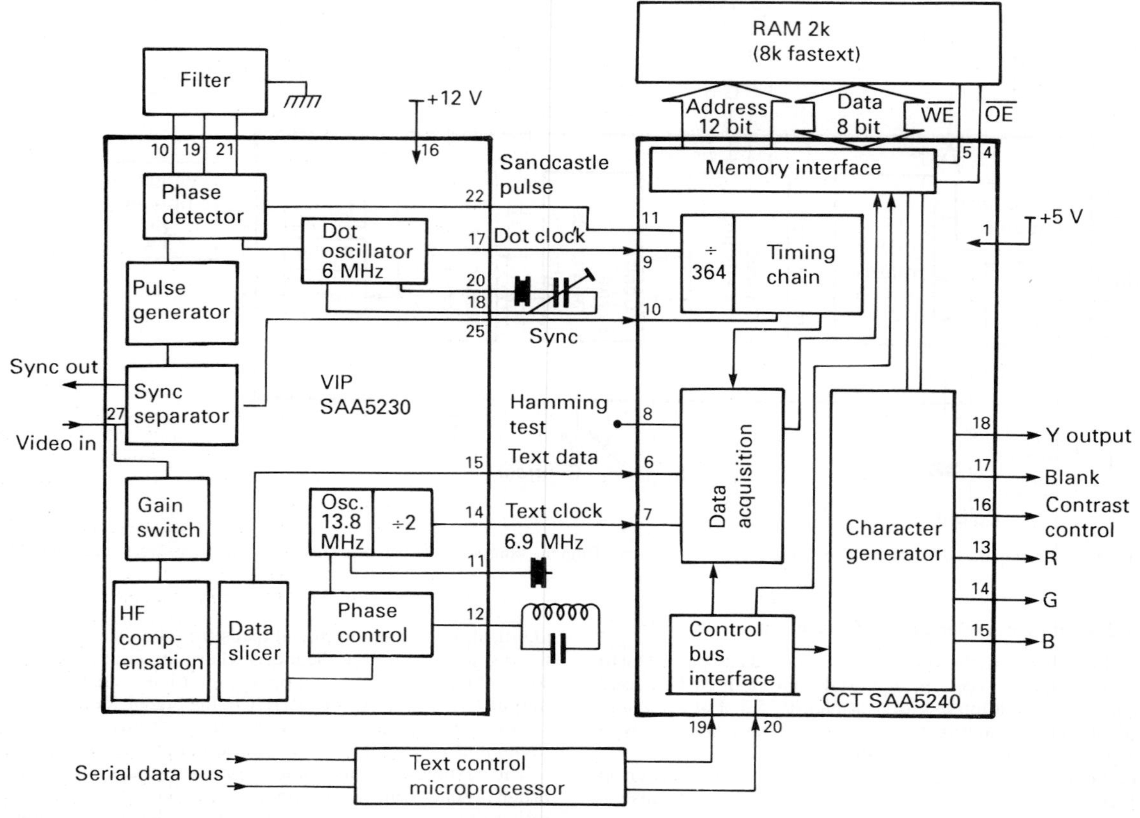

Figure 55.6 Teletext decoder block diagram. The outputs on the right interface with the type of IC shown in *Figure 55.5*. The Hamming test on pin 8 of the CCT is a useful indication of received data quality. It stays high in the absence of errors, and is reset one per line

three RGB output stages, the effect is rather like that of misconvergence, i.e. there is colour fringing. Again, this and poor lf response (smearing effect in one colour) are most likely to stem from faulty capacitors in coupling, decoupling or 'peaking' roles, or occasionally those which store a reference potential for clamping purposes.

While most cases of picture ringing and ghosting stem from faults in tuner and if stages or multipath signal reception, the luminance delay line can give rise to this symptom if its ground connection is broken by a dry joint or internal fault.

An occasional problem encountered in the baseband video processing section is picture shading, with a graduation in brightness and/or contrast from one side of the screen to the other, sometimes accompanied by smearing. The usual cause of this is a line-rate sawtooth waveform superimposed on the tube drive signals as a result of poor smoothing or decoupling of an lt or ht supply line from the line output or chopper transformer. Faulty electrolytic capacitors are usually responsible.

55.3.3 Text decoder and digital sections

An increasing trend in TV receiver design is the use of digital techniques in control and housekeeping sections, and in data transmission and display. See sections *3* and *61*. Our purpose here is to examine them from the point of view of fault diagnosis and repair.

The essence of a modern TV control system is a purpose designed central microprocessor whose inputs are user-commands from a local or (IR-linked) remote keypad, and whose outputs are distributed to all sections of the set on a two-wire serial bus system. In the most commonly used format (Philips I²C bus) the two lines are designated SCL (*clock*) and SDA (*data*). I²C is *inter-IC communication* (data bus in consumer equipment).

It is not possible with everyday equipment like a logic probe or oscilloscope to analyse the data on SCL and SDA lines, though either instrument will indicate the presence of data, which is usually sufficient. Data errors are rare, and except for the odd case of a faulty memory chip, if the data are present at all they are usually correct. If only one 'slave' chip is malfunctioning, it is more likely to be faulty than the main microprocessor.

Complete loss of response may typically be due to one or other of the data lines 'stuck' high or low due to leakage or short-circuit in one of the chips connected to it. Isolating each in turn is the most practical approach to this. IC failure may have been instigated by flashover in the high voltage (picture tube) section of the set.

In cases where all data are absent from the bus lines, the control microprocessor is the starting point for tests. Check that its clock is running, that a reset pulse is applied at switch-on, and that its operating voltage source (VDD/VSS) is present within 10 per cent of nominal level and free from hash and ripple. If these are in order, and the input signals and interfaces are in order, the microprocessor itself is suspect.

Failure of the remote control system, where local controls are working, should lead to a check first on the batteries and IR emission of the handset, then if necessary on the data stream through the photosensor and preamplifier stages in the TV, and its application to the control microprocessor.

The main application of digital processing in a conventional TV, however, is in the teletext decoder, which is also driven by the user via the serial control data bus. A functional block diagram of a modern text decoder is shown in *Figure 55.6*. It uses two lsi chips in the data processing section, with support from an additional RAM chip. The diagram has been prepared with special emphasis on the test points for signal and data interchange between ICs and main control system, for fault-finding.

If there is a complete absence of text display, first check the operating voltages: 12 V to VIP pin 16 and 5 V to CCT pin 1. Then confirm that a CVBS (composite chroma, video, blanking and sync) video signal is entering pin 27 of VIP. If so, the clock pulse feeds are the next test points: text (system) clock at 6.9 MHz from VIP pin 14 to CCT pin 7, and dot (character) clock at 6 MHz from VIP pin 17 to CCT pin 9. Failure of the viewer's commands to reach the CCT chip can cause no text display; where the rest of the control system is working, tests can be confined to the entry points of SCL and SDA on pins 19 and 20 of CCT, and the operation of any dedicated text control microprocessor interposed between here and the main control bus.

Perhaps the most common problem in teletext reception is character corruption, in which the screen display is 'garbled'. This may well arise from faults outside the decoder such as multipath reception or poor signal handling in the tuner/if demodulator sections. When it is proved (by substitution) that the fault does lie in the text module, either the decoding or the memory sections may be responsible. In the first case, the errors are random and tend to diminish as pages are refreshed during transmission; check the clock synchronization, VIP and peripheral components. In the second case, display errors have a pattern, unchanged during update, refresh or reselection of a page. Here the memory section (or occasionally CCT chip) is implicated, with a faulty RAM, or stuck data or address lines.

55.3.4 Field timebase and synchronization

The field timebase is less complex than its line counterpart. For fault-finding purposes it can be resolved into six blocks as shown in *Figure 55.7*. The driver and output sections are interdependent, with dc and signal feedback to stabilize operating conditions and maintain a full linear scan. In practice, the field timebase section is embodied in one or two linear ICs with (in some large-screen sets) a discrete output stage to drive the scan yoke.

Failure of the field timebase produces a horizontal white line across screen centre, though a blanking circuit for tube protection may render this invisible. In cases of field scan collapse, check first that operating power is available to all sections. This is typically derived from line or chopper transformer via a rectifier/capacitor set. Burnout of a fuse or safety resistor in this supply line usually signals short-circuit or leakage in the IC or transistor output stage.

Where supply voltage is available, oscilloscope checks should next be made to ascertain the presence and progress of the 50 Hz scanning waveform. It often saves time to start at the output pin of the IC, where a full amplitude square-wave may well be found. This indicates an open circuit in the scan yoke circuit. Seldom is the yoke itself responsible; faults in printed circuit board joints, coupling components and plug-socket joints are far more common. Forward tracing with the oscilloscope quickly reveals the culprit.

If no output is coming from the power amplifier IC, work back with the oscilloscope to check input and output waveforms of the driver and oscillator stages, which are usually passed in and out of the chip for waveform shaping and feedback addition purposes. Most set-makers provide internal block diagrams of the ICs in their products, from which the function of peripheral components can be deduced. In the low-power sections of the timebase, the only vulnerable passive components are electrolytic capacitors and resistive trimmers for presetting. Most often the IC itself will be responsible, diagnosed by analysis of IC pin voltages, also given in the service manual.

Beware of situations where a fault in the output section upsets dc feedback to the point where the driver (or even oscillator) stage ceases to function. Failure of synchronizing input will not prevent the field oscillator from running.

Short of complete failure of the field timebase, various parametric faults can appear. Insufficient picture height with good linearity suggests a problem in the height control circuit itself, or a failure of the negative feedback circuit around the field output stage. Most designs feed back a sawtooth voltage from across a low-value resistor effectively in series with the scan yoke. If the value of the resistor goes high, low height is the symptom. Conversely, gross vertical overscan (usually with non-linearity) is invariably due to failure of a feedback signal.

Poor field linearity, where the supply voltage to the timebase is adequate and well regulated, is also usually due to a problem in the field-scan feedback circuit, though coupling and decoupling components can also be responsible. Again, electrolytic capacitors are high on the list of suspects, and should be checked first (most easily by substitution) where they form part of an afflicted stage.

Because teletext and other data are broadcast during the field blanking interval, a too slow flyback period in the field timebase

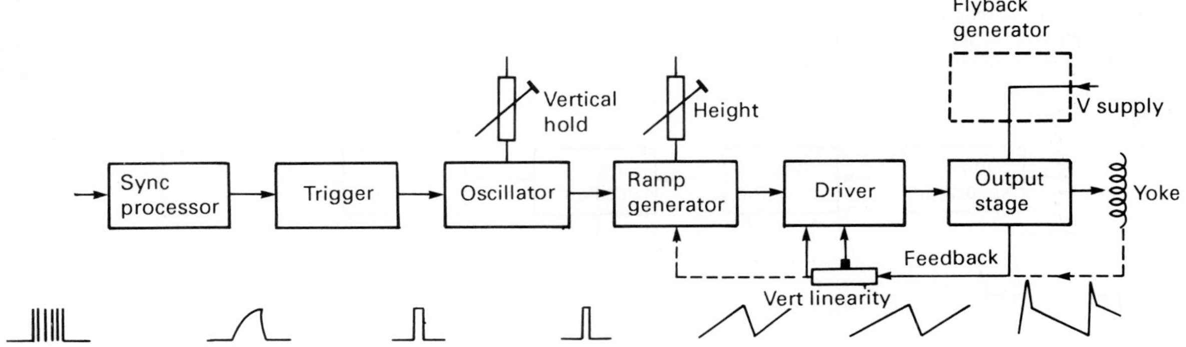

Figure 55.7 Stages in the field timebase, with normal waveforms. Most problems stem from driver and output stages

introduces the risk of superimposing twinkling dots and lines on the upper part of the picture. In modern designs, the flyback period is generally less than 1 ms. A slow or misshapen flyback waveform (oscilloscope check) should direct attention to the scan-output stage, and particularly the flyback-generator section which is a feature of most IC designs. It depends on a scan-charging circuit, usually based on a diode and capacitor external to the chip. Other symptoms arising from failure here may be cramping or foldover at picture top. Where linearity is perfect but text-line interference is present, the field-flyback blanking link to the luminance processing chip may be broken or faulty. An *over-bright* picture with field-flyback lines superimposed, on the other hand, is usually due to incorrect picture-tube bias, e.g. excessive first-anode voltage.

Incorrect field oscillator speed gives rise to rolling pictures. If the roll rate is fast (in either direction), the fault will lie in the field oscillator itself, whose timing components (R, C) should be checked. If, however, the picture is rolling slowly upwards, the cause is an absence of trimmer pulses. Check the condition and operation of the sync separator and field-pulse integrator sections.

Interlace faults are rare in modern TV designs. The effect is a coarse raster-line structure in which the lines of one field are not equally spaced between the lines of the other. The cause is uncertain firing of the field oscillator, and may be traceable to line-pulse interference with the sync signal (lead dressing, decoupling, common earth-paths), or a problem in sync separator or processor sections.

55.3.5 Line scan and eht generation

The line timebase, and particularly its output stage, accounts for a large proportion of faults and breakdowns in TV receivers. This stems from the relatively high scanning currents involved in wide-angle picture tubes, the need to generate high voltages to service focus and beam-accelerating requirements of the tube, the high rate of change of V and I during horizontal flyback, and the relatively high temperatures at which some components operate. The individual stages in the line timebase section are outlined in *Figure 55.8*. As with other fault diagnosis, the trouble should be tied down first to one block, then traced to component level within the block. The first two sections, sync separator and line oscillator, are invariably combined within a single IC, whose 64 μs period pulse output feeds the line driver, an intermediate power stage to provide a 'stiff' low-impedance feed to the line output transistor.

The line output transistor operates as a fast switch, alternately grounding and isolating the bottom of the line output transformer (lopt) primary winding. Linear sawtooth currents are thus generated in the windings of the highly inductive transformer and line scan yoke. Various auxiliary supplies for use in other parts of the set are tapped from the lopt via rectifier/capacitor sets. Tube heater energy is also drawn here.

Line oscillators free run at 15 625 Hz (in Europe) and are controlled by a reactance stage (or its digital equivalent) in a flywheel sync circuit. Incorrect line speed breaks the picture into diagonal or horizontal bars, and usually stems from a fault in the external (RC) time-constant components associated with the line oscillator block in the IC. Modern designs use a double-speed line oscillator from which a countdown circuit derives field sync pulses, but the principle remains the same. Normally a preset resistor is provided with which to set line speed. Sometimes there is a second preset to adjust line phase, in order to centre the picture within the raster. Check these and associated capacitors in cases of wrong line frequency and incorrect picture/raster phasing respectively.

Other faults associated with the line oscillator section are *no horizontal sync* (check video input, biasing and pin voltages at the sync separator section), *horizontal picture twitch* (check decoupling capacitors and chip-peripheral passive components), and *loss of output*, resulting in a 'dead' set symptom. For this, check supply voltage and oscillator peripherals before suspecting the chip.

The line driver stage generally consists of a transistor and a step-down transformer. Its main requirement is a fast transition between *on* and *off* states to switch cleanly (and thus prevent over-dissipation in) the line output transistor. Problems here are generally confined to transistor failure or dry joints at the driver transformer pins.

Figure 55.9 shows a skeleton circuit of a typical line output stage. TR1 is the switching transistor, T1 the line output transformer and C1 the flyback tuning capacitor. These and the clamping diode D1 are the essentials of this section.

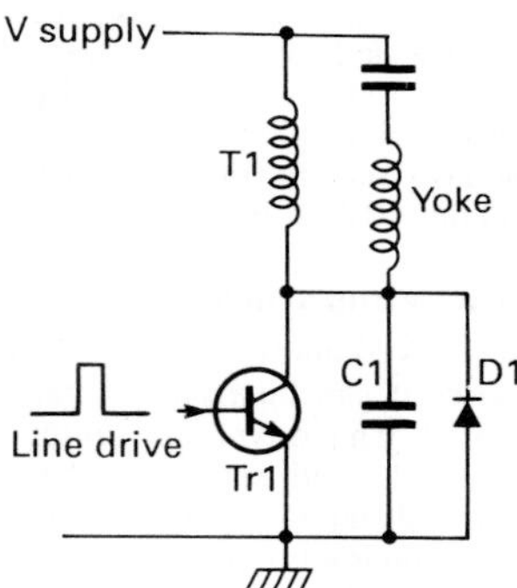

Figure 55.9 Main operational components in the line output stage

Complete failure to work, so long as dc operating supplies and drive pulses are available, is usually due to failure of the transistor or the joints in the primary transformer circuit. More often the symptom is excessive current consumption, blowing a fuse or invoking some overload protection device in the power

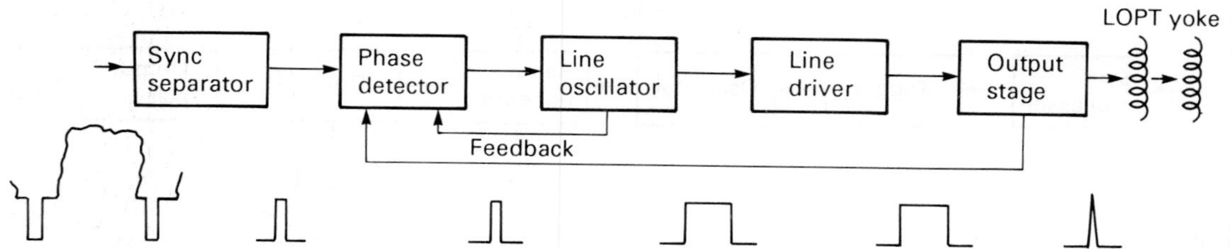

Figure 55.8 Components of the line timebase

supply section. First check the transistor and diode for internal short-circuit or leakage, then suspect that the inductance of the lopt is being damped by too heavy an electrical load.

While this may be due to short-circuit turns or insulation breakdown within the lopt itself, external components are more often responsible. Check scan and flyback rectifier diodes (and if necessary the circuits depending on them) for short-circuits and leakage; check associated inductors like scan yoke, feed choke and shift transformer for shorted turns; disconnect any external eht voltage-multiplier circuit; with diode-split lopts remove the crt anode connector, and so on. At the point where loading remains excessive with the lopt primary the only winding left in circuit, it may be assumed that the transformer has failed internally.

The picture width and eht voltage are closely dependent on the supply voltage to the top end of the lopt, and diagnosis of any picture size problems should begin with a check of this. The other main factor which affects picture width and eht voltage (and the relationship between them) is the flyback tuning capacitor, shown in *Figure 55.9* as C1. When the transistor cuts off, this and the lopt form an LC tuned circuit to generate a positive pulse during scan flyback. Too high a resonant frequency gives a dangerously large flyback pulse which can quickly break down insulation. If the capacitor is present and correct, the flyback period (measured with an oscilloscope) should be approximately 11 μs in duration. This test can be safely made with reduced ht voltage.

Horizontal scan collapse to a vertical white line on screen can only be due to a break in the scan-coil circuit itself in a conventional design like that of *Figure 55.9*, since eht is also generated in the lopt.

Some picture tubes need pincushion distortion correction, and this is provided in the line output stage by modulating the line scan current at field parabolic rate. The modulators are a pair of diodes (taking the place of D1 in *Figure 55.9*) and a medium-power driver transistor. These or the associated east-west transformer/choke may fail, to give the following symptoms: incorrect geometry at picture sides; incorrect width with eht voltage correct; heavy loading on the lopt.

Before we leave the line-scan section of the TV receiver, we must mention two less commonly encountered faults and possible causes. For picture ballooning with brightness changes, check regulation of ht and eht supply lines. For vertical striations at the left hand side of the picture, check the damping components (i.e. resistor) across inductors, particularly the line linearity control coil.

We have covered the most common faults encountered in TV line timebases in practice. As section *55.3.6* shows, there is interdependence between line-scan and power-supply sections in many designs.

55.3.6 Power supply section

Switch-mode power supply designs are used in mains operated TV sets. They consist of four basic elements (see *Figure 55.10*). A rectifier/capacitor set provides an unstabilized 320 V supply to the switch section. Each time the switch closes, energy is passed into an LC reservoir, on which the load (comprising the TV circuits) draws continually.

The mark:space ratio of the switch is governed by a control circuit culminating in a pwm (*pulse width modulator*) drive stage. The control circuit monitors output voltage and ensures that the required level is maintained in spite of mains voltage and beam current variations.

As an illustration of typical practice, *Figure 55.11* is a skeleton circuit of a blocking oscillator power supply unit (psu) using a control IC. Here Tr1 is the switch element, and the energy reservoirs are chopper transformers T1 and capacitors C5 and C6. Drive comes from IC1 pin 8, based on feedback information entering the chip at pins 2 and 3. This pulse drive is also conditioned by chip internal protection and soft-start systems.

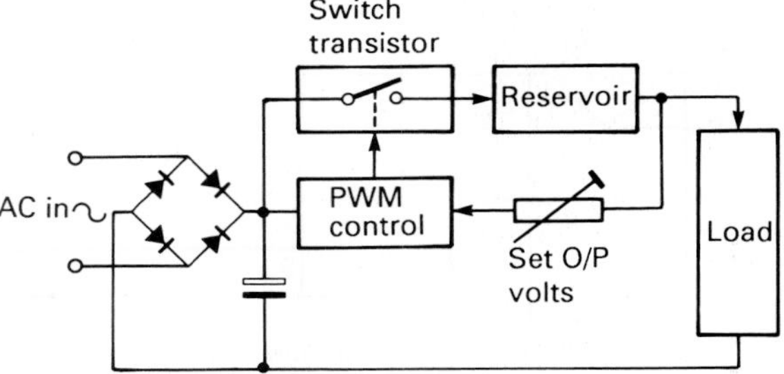

Figure 55.10 Principle of operation of a switch-mode power supply

In this type of psu, complete failure will often be found to stem from a blown and blackened mains input fuse. Occasionally the bridge rectifiers in BR1 are responsible, but more often the switch transistor Tr1 will be found to have gone internally short-circuit; if it is replaced without further investigation, the replacement may well fail at switch on. It is best to replace Tr1, IC1 and R4 together, and check C3 and R9 before restoring power.

Faults elsewhere in the psu or the load will trip a protection circuit within the IC, whereupon the operating frequency falls to about 1.5 kHz, discernable by a purr from T1 and a needle-pulse effect on oscilloscope inspection of the drive waveform. It is usually the result of a short-circuit in rectifiers D4–D5 or a shorted or heavily loaded line output stage, all easily checked with an ohmmeter. Failure of the sampling/feedback network D2–C2 or the resistive chain R5–R7 has the same effect. If all these points are in order, it may be found that pin 5 of the IC is below 2 V. This signifies that the mains input voltage has fallen below a satisfactory level or, more likely, that the pin is being pulled low by an external influence like the standby control line.

Other designs of psu work on similar principles, a numerous class being line synchronous to switch on the chopper transistor at 64 μs intervals, again with its duty-cycle being controlled by the demands made on the output lines. Often these incorporate protection circuits which trip continuously to give a 'pumping' effect. Faced with a tripping psu, the first step is to observe the symptoms; if there is a burst of sound and light on each 'pump cycle', the overvoltage trip is operating. If the set is dead with a 'tick' or 'thump' effect from the chopper transformer, it is likely that the overcurrent protection system has been invoked.

For overvoltage tripping, confirm with an oscilloscope that the psu voltage exceeds normal on each energy burst, then check the control circuit, specifically the reference (e.g. zener diode) and feedback (diode/capacitor/resistor) sections which determine the output voltage. Where sampling is carried out at the main (typically 160 V) output line, low capacitance in the reservoir capacitor may be responsible. Overcurrent tripping is due to a too heavy load, traceable either by seeing which output line rises *least* on each pump cycle, or by progressively unloading sections from the psu in the manner described in section *55.3.5* for an overloaded lopt. These strategems are necessary only when the culprit cannot easily be traced with simple ohmmeter checks.

Some psus will not operate properly without a load, and where there is doubt as to the fault area, the load on a psu can be simulated by using an ordinary 60 W or 100 W mains bulb. At half voltage it will dissipate one quarter of its rated power, and

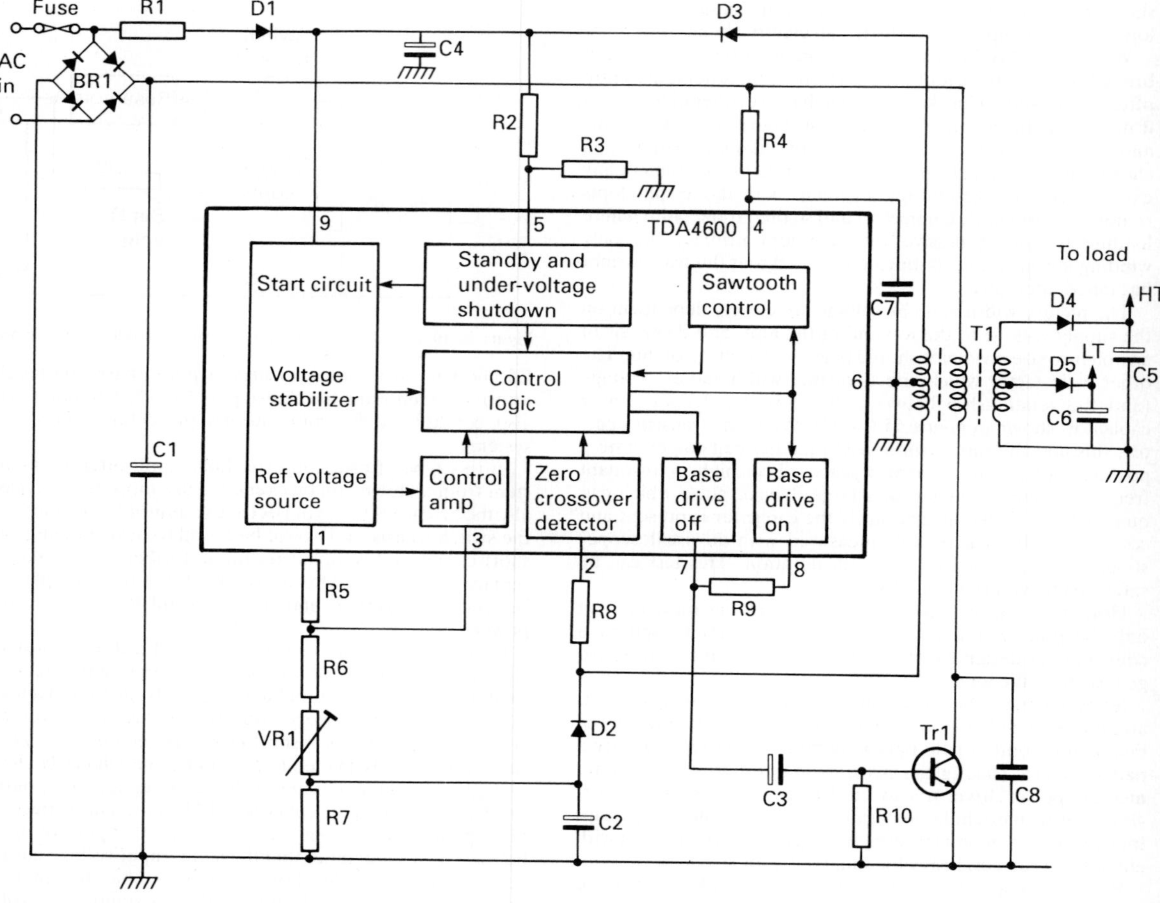

Figure 55.11 One form of switch-mode psu control system. It is based on a TDA4600 IC

the light emitted gives a good indication of the applied voltage. These bulbs can also be used in a protective role.

Where an obscure fault is present, or where the control loop fails to restrain the output of the psu, damage and semiconductor destruction may take place at the moment of switch on, rendering diagnosis difficult. In these circumstances a mains *variac* is a useful tool. It enables the applied mains voltage to be increased slowly while monitoring voltage and current. Some kick-start circuits will not operate with a smooth tail-on of mains power. To overcome this, switch on at an input level of 80–90 V rms, or (better) power the psu driver stages from an external and isolated source of lt as appropriate. Except in blocking oscillator designs (where it is not applicable at all), this latter method facilitates 'cold checking' of drive waveforms.

An alternative protection artifice, especially when the circuit's own protection systems have to be over-ridden for check purposes, is to use a 60 or 100 W mains bulb as a current-limiting resistor for the psu. It can be fitted either in one leg of the mains input, or effectively in series with the chopper element so long as there is some form of downstream decoupling. The same current-limiting effect can be used to advantage with a faulty line output stage.

Variations in psu output voltage are reflected in the picture size, and symptoms range from spasmodic jitter to poor regulation of the power line. As most possible culprits are within the stabilization loop, primary suspects are the reference and sampling networks concerned with setting the output voltage, and the presence of dry joints in the psu section, especially at heavy or hot components.

When working on mains psus, keep firmly in mind the risk of mains-to-earth shock, especially as most antenna and other signal input sources are at least partially grounded. Use of a 1:1 isolating transformer eliminates this risk. A rating of 250 VA is sufficient to operate modern colour receivers.

After completing repair work on psu circuits, ensure that all safety and protection devices are working, that fuses have been replaced with the correct type, and that output voltage levels are set to within 1 per cent of specification using a digital voltmeter. See also section *55.6*.

55.4 Replacement of components

The major aspect of TV servicing is *fault diagnosis*. Except in cases of miniaturized assemblies using surface-mounted devices (see section *5*), the removal and replacement of the faulty component(s) is a straightforward process. PCB-mounted components can be unsoldered with either a hand-held solder suction pump or with flux-impregnated copper braid while being heated with a soldering iron. Use a hot, clean soldering iron with 60/40 solder of 18 swg for ordinary work, and 22 swg for surface mounted devices, small components and ICs.

A common physical fault in TV equipment is a cracked printed circuit board. In a non-stressed part of the board it may be sufficient to bridge the cracks with solder, but a more reliable repair is made by using fine wire soldered across the break. 5 A fuse wire is adequate in most places, but where the board is broken, or heavily loaded physically, thicker wire is needed. To stop the crack in the pcb from spreading, drill a small hole at each end.

Surface-mounting technology is being increasingly used in TV circuits and sub-assemblies. Dealing with surface mounted devices is more difficult than work with conventional printed circuit boards. The main problem is in component removal, where all bonds between the component and the pcb must be broken without damaging the board. For a two- or three-tag component, desolder all connections with capillary braid, then gently twist the component; it can generally be removed without further heating.

Multi-pinned components like ICs present a greater challenge. Hot-air blowers and specially shaped desoldering bits are available, but a method which has proved effective in practice is to first remove as much solder as possible from the connection pins and board bands with capillary braid, then separate land and foot by a 'cheesewire' process. Thread a thin copper wire (e.g. 5 A fuse wire) under the lead-outs along one entire side of the IC or other device, then anchor its end with a solder blob at some convenient point on the board. Now, using a very dry, hot iron, heat each pin in turn as you progressively pull the cheesewire underneath, keeping it as close to the board surface as possible. Repeat along the other sides of the IC until all feet have been released.

Fitting replacement surface devices is easier. Anchor the device with a pinpoint of inert adhesive, then in the case of a multi-tag component very carefully centre it on the pcb lands. Solder opposite corners to establish it in position before soldering all lead-outs with a 1.5 mm bit and a minimum of 22 swg flux-cored 60/40 solder. When the job is complete, examine every joint with an 'eyepiece' ×8 magnifier to check for inter-pin short-circuits.

Keep surface mounted devices in their packaging until the last moment before fitting to ensure that they are protected from static discharge and correctly identified, since many are too small to have their ratings marked on the body.

55.5 Intermittent faults and scan testing

The most difficult and frustrating faults to diagnose are those which are intermittent or spasmodic in appearance. Generally they fall into one of three categories: physical, thermal or random.

Physical faults are the most common, and the easiest to tackle. They stem from pcb faults, dry joints, bad plug-socket connections or contact between adjacent conductors. By flexing and tapping printed circuit boards, wires, components, etc., a physical fault can generally be traced. Sometimes even close examination of the assembly does not reveal the fault, and especially on tightly packed boards it is necessary to get the equipment into fault state and trace the trouble by conventional means with meter and oscilloscope. In all but low-level signal circuits, a quicker route to diagnosis may be close examination of the interior of the set in *total darkness*. A spark or arc can be seen with surprisingly small currents at the trouble spot.

Thermal faults may appear or disappear as the equipment warms up. Two very useful diagnostic tools here are an ordinary mains powered hair dryer and an aerosol can of freezer spray. Assemblies and individual components can (with care) be temperature-cycled over a very wide range by these means. It is important to bear in mind here that semiconductor characteristics are temperature-dependent. Very small areas can be heated by holding them in the hot-air stream above the bit of a soldering iron.

Random faults are the most difficult of all to track down because they cannot be provoked by mechanical or thermal means. For these, the best approach is to leave diagnostic equipment hooked onto key test points during soak testing in order to gain as much information as possible while the fault *is* present. Some types of digital multimeter are designed to give an audible bleep when their input conditions suddenly change, a useful attention-caller to this type of fault. The general technique is to narrow down progressively the field of search on the basis of previous test results.

Many intermittent faults in the two last categories described can stem from incorrect settings or circuit conditions creating a 'borderline' situation, in which a small change in signal characteristics or environment triggers a threshold change. Examples are plls running close to the edge of their hold-in range, psu output voltage or current levels on the point of triggering protection circuits, and a wrongly biased sync separator whose pulse ouput is corrupted by certain types of picture content. Thus operating conditions and preset adjustments are worth checking where the cause of an intermittent fault is not obvious.

55.6 Safety in servicing

Although the onus for ensuring that equipments in Britain comply with BS 415 and BEAB requirements, and with corresponding standards in other parts of the world, rests with the manufacturer and retailer, the service technician has a duty to ensure that no action of his negates the safety features incorporated, and that no dangerous deterioration of insulation has taken place. In practice this means that all safety-crucial components (marked

in the circuit diagram and/or parts list) must be obtained from or approved by the set-maker, and mounted in the same way as the original.

In particular:

- all cable ties, clips and cleats removed during service must be restored
- an air gap of at least 6 mm must be maintained between live and isolated conductors
- shields and insulators must be maintained
- cabinet and covers must be checked to be whole, undamaged and securely fastened.

A test voltage of 1.5–2 kV applied for one minute between the mains input connections and user-accessible parts must not provoke any flashover or sparking, nor draw a current greater than 2 mA dc or 700 μA ac.

Bibliography

TRUNDLE, E., *TV and Video Engineer's Pocket Book*, William Heinemann (1987)

TRUNDLE, E., *Servicing TV and Video Equipment*, Heinemann Newnes (1989)

Part 12
Video and Audio Recording and Playback (Domestic)

R Watson C Eng, MIEE, MRTS, MBKS
Consultant

56 Video Cassette Recorders

The origins of current domestic video tape recorders can be traced back to the monochrome recorders of the late 1960s and early 1970s which used half-inch wide tape. Typical of these were the Japanese EIAJ-I standard recorders using half-inch (2.54 cm) tape on open reels.

Tape consumption and cost were however too high for the domestic market, and the threading of the tape by hand through the guides and around the head drum was liable to cause head or tape damage.

In Europe, Philips were developing a half-inch tape cassette machine and devised the *colour-under* method of carrying the chroma information. This method was also adopted and adapted by the Japanese to become the EIAJ-II standard.

At that time there was an agreement between Sony and Philips to produce a common half-inch video cassette standard, but when this did not materialize, Sony decided to develop the higher quality three-quarter-inch U-format for the industrial market. This early cooperation can be seen in the similarity of tape loading and transport methods between the Philips machines, the U-format and Beta cassette recorders.

56.1 Philips N1500 and N1700

The Philips N1500 was the first cassette machine, introduced in Europe in 1972, using a half-inch tape cassette with the two reels stacked on top of each other. A timer and tuner unit was included to enable broadcast programmes to be recorded, which gave the public the ability to 'time-shift' television programmes. The facility to record 'off-air' was an essential feature as it obviated the need for a large catalogue of pre-recorded tapes. The maximum playing time for the N1500 was one hour

The tape format is shown diagrammatically in *Figure 56.1*. The basic mechanical data for the N1500 system are:

drum diameter:	105 mm
speed, head disc:	1500 rev/min
scanning speed (relative speed, video head/tape):	8.1 m/s
tape/speed:	14.29 cm/s
video head gap length:	0.8 μm
video track width:	130 μm
distance between two video tracks (guard band):	57 μm
audio track width:	0.7 mm
sync track width:	0.3 mm

In 1977, Philips introduced the N1700 series which doubled the playing time. This increase in playing time was achieved by reducing the video track width from 130 to 85 μm and eliminating the guard bands. Crosstalk was prevented by angling the video recording heads at +15° and -15° (*azimuth recording*) and also slightly changing the video track angle to give a $1^1/_2$ line offset between adjacent tracks (*Figure 56.2*). Signal/noise ratio was maintained, despite the narrower track width, by the improvements in tape technology. Although the linear tape speed was halved to 6.56 cm/s, the video head writing speed was maintained at 8.1 m/s as the head drum diameter and rotation speed were not changed.

The basic mechanical data for the N1700 system are:

drum diameter:	105 mm
speed, head disc:	1500 rev/min
scanning speed (relative speed, video head/tape):	8.1 m/s
tape speed:	6.56 cm/s
video head gap length:	0.6 μm
video track width:	85 μm
distance between two video tracks (guard band):	0
audio track width:	0.7 mm
sync track width:	0.7 mm

The tape format is shown in *Figure 56.3*.

The launch of the N1700 in Europe was timely as it coincided with the introduction into the NTSC countries of the VHS and Beta systems by JVC and Sony respectively. Until then the only challenge to the Philips cassettes came from the National Panasonic cartridge machines (NV5120), using the EIAJ-II colour standard, with the tape on a single spool. A special stiff leader section enabled the tape to be loaded automatically. The inherent disadvantage of a cartridge system is that the tape can only be removed from the vtr when all the tape has been spooled back into the cartridge, unlike the cassette, which can be loaded or unloaded at any point.

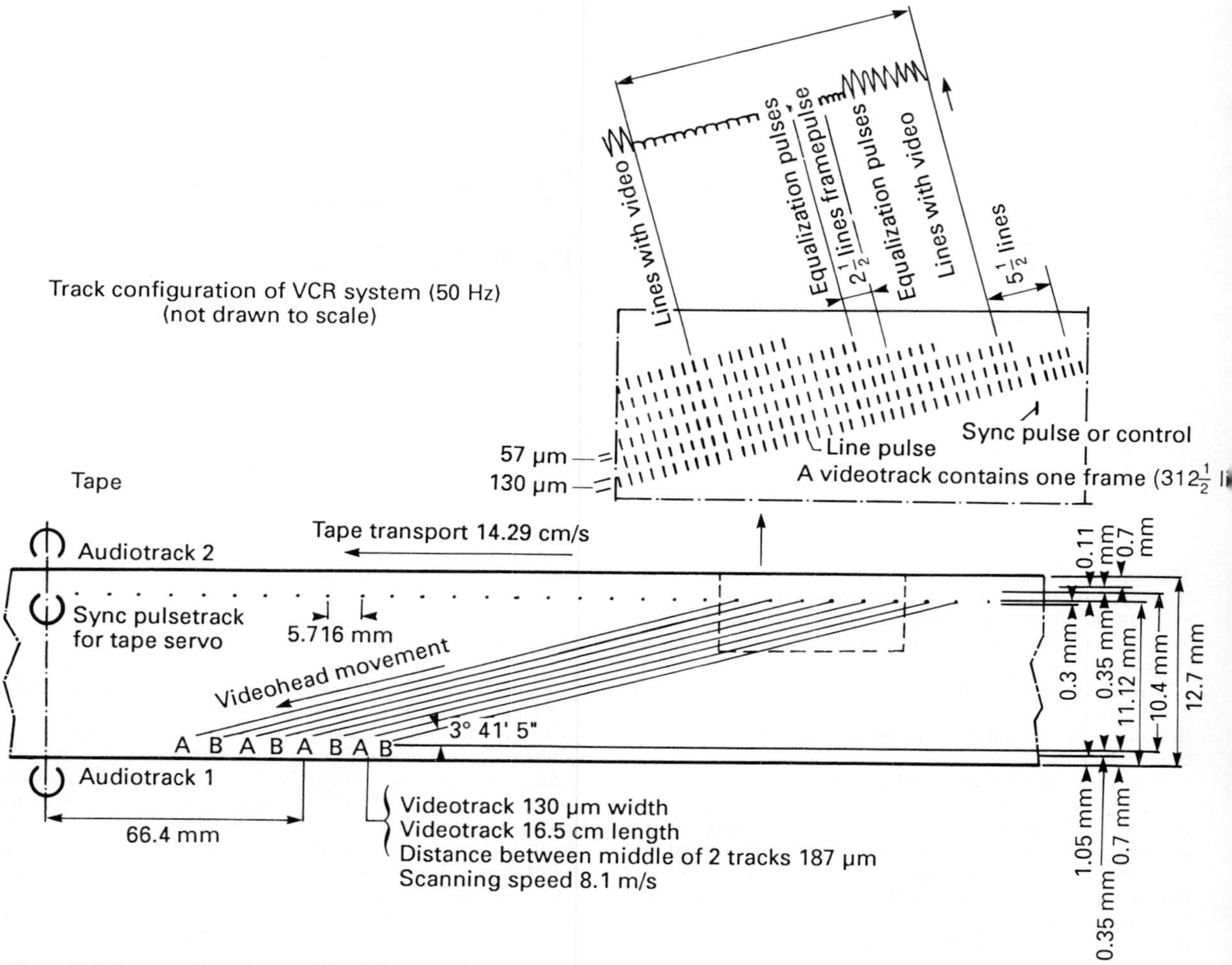

Figure 56.1 Track configuration of N1500 50 Hz system (not to scale)

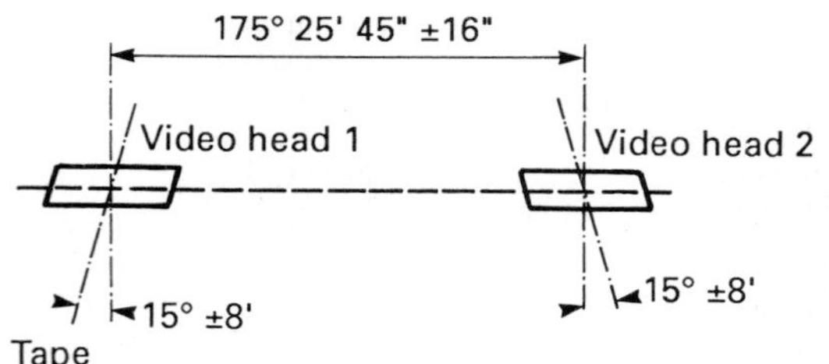

Figure 56.2 N1700 head azimuth

The first NTSC Beta and VHS machines played for two hours, but competition between the two systems stimulated the production of extended play versions with switchable playing times.

By the time the PAL versions were introduced into Europe in 1978, the playing time had been standardized at three hours, and we were spared the confusion of different recording speeds until 1984.

The increased recording time for VHS and Beta has been achieved by reducing the video track widths, with no guard band and luminance crosstalk eliminated by azimuthed heads (6° for VHS, 7° for Beta). The lower frequency colour-under crosstalk was cancelled by changing the phase between adjacent tracks for VHS, and changing the frequency for Beta.

56.2 Beta format

The NTSC Beta system was launched in 1977, just prior to the VHS, and competition between the rival systems stimulated the development of Beta machines with longer playing times. The PAL machines were single speed models, with the same three hour record/replay time as the VHS. The Beta and VHS formats are incompatible. The Beta tape wrap around the head drum is a modification of the U wrap, developed for the three-quarter-inch Sony U-Matic machines (*Figure 56.4*).

All Beta systems are compatible, including the smaller diameter head drum recorder used in the Beta Movie. See section *56.8.2*.

In the NTSC Beta system the chrominance signal is recorded on track A with every other line inverted in phase. Track B is recorded without any inversion.

In the PAL system, in addition to the frequency shift between tracks A and B detailed in Table *56.1*, a pilot burst is added to the video waveform. The amplitude is between 1.2 and 1.8

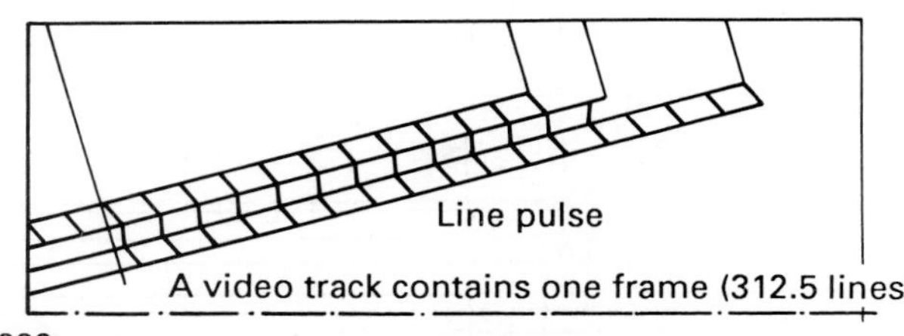

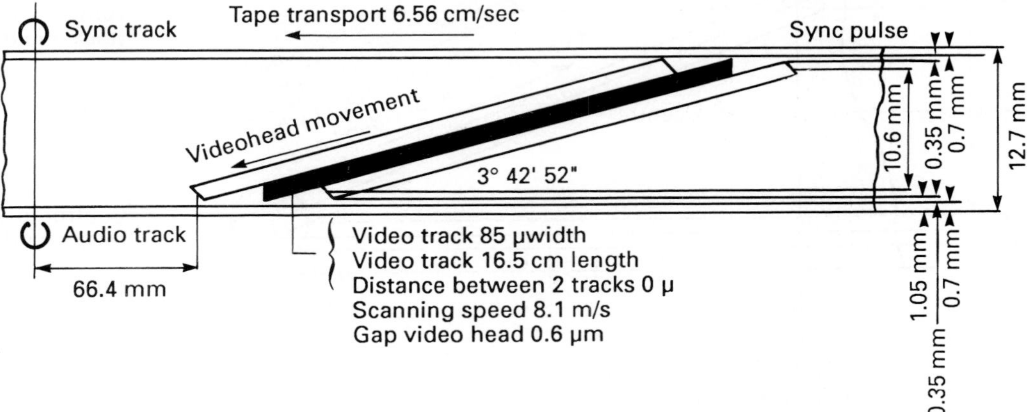

Figure 56.3 Track configuration of N1700 system

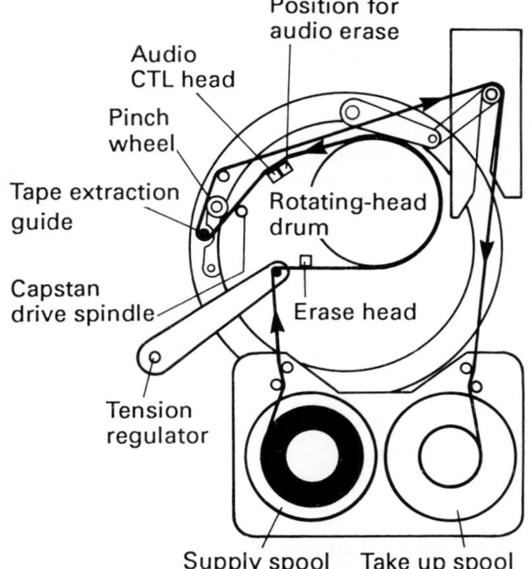

Figure 56.4 The Beta tape wrap is a modification of the U-matic

times the colour burst amplitude, and the phase must be 180 ±10° relative to the U-axis of the input signal. The pilot burst frequency is the same as the colour subcarrier of the video input signal.

In the SECAM system, in addition to the frequency shift between tracks A and B and the different colour-under frequencies listed in Table *56.1*, an identification signal is added during the vertical interval on track B, every other frame, to identify the delayed fields.

In all three systems, the recorded luminance to chrominance timing difference should not exceed 0.1 μs. The basic mechanical data for 625-line/50 field system are:

winding angle:	180°
drum diameter:	74.487 mm
lead angle:	5°00′00″
video head:	+7° 2 head
field number per 1 rotation:	2 fields
drum relative speed:	5.831 m/s
drum rotation:	25 rps
f_V:	50.00 Hz
f_H:	15.625 kHz
REC Y:	3.8–5.2 MHz
REC C:	f_{CA} 685.54 kHz
	f_{CB} 689.45 kHz

56.2.1 Audio

A mono audio signal should be recorded on both audio channels, or audio 1. Audio 2 may be used for time code.

Hi-fi audio signals on the NTSC system are recorded by the video heads as part of the recording envelope (between the chrominance and luminance signals). Four carrier frequencies are used, two by head A, and two by head B.

In the 625-line systems, separate hi-fi heads are mounted on the head drum, and carrier frequencies of 1.44 and 2.10 MHz are used in a similar manner to the VHS hi-fi system. Unlike the VHS hi-fi system, the Beta hi-fi track is almost as wide as the video track, but is offset by one and a half track widths. The audio track therefore spans across the upper half of one video track, and the lower half of the adjacent video track.

56.3 VHS format

The NTSC version of VHS was launched in 1977, and the PAL and SECAM models followed in 1978.

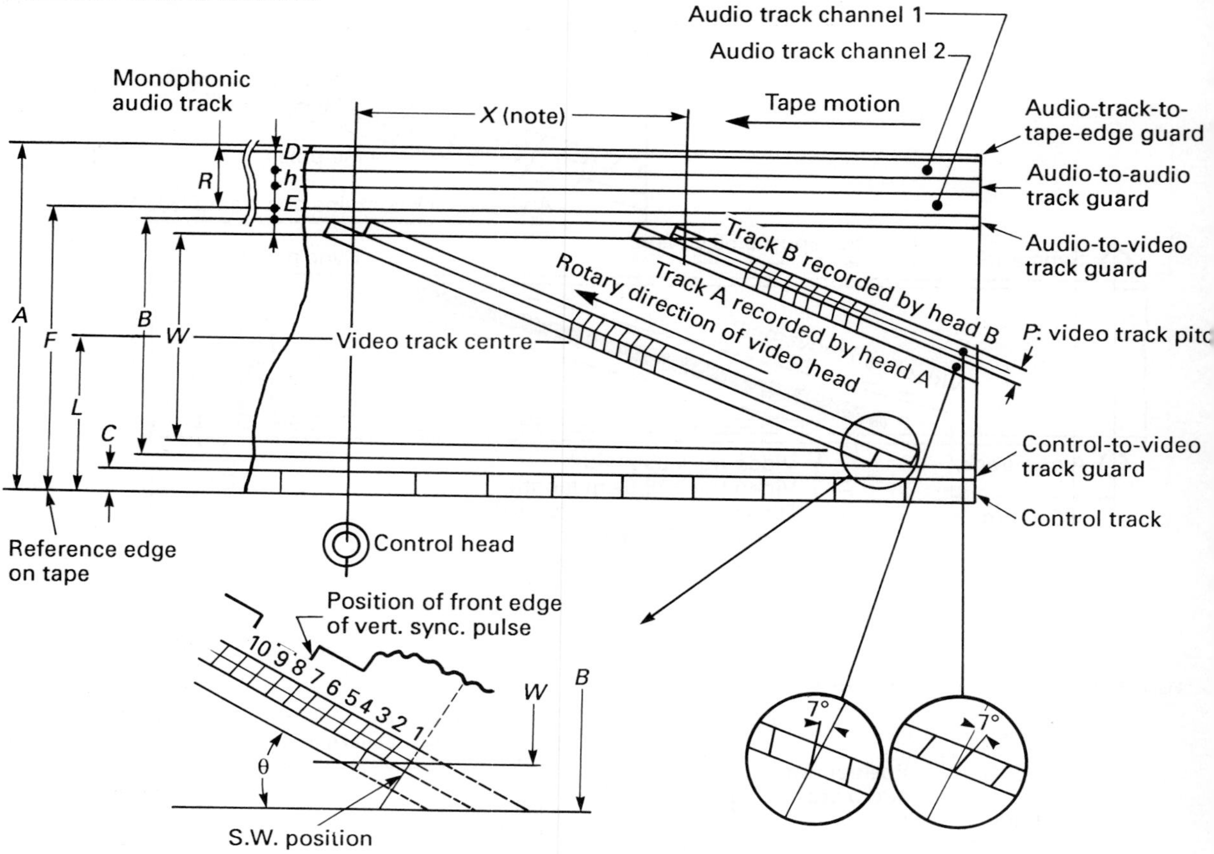

Figure 56.5 Beta tape configuration. The letters refer to dimensions listed in Table *56.2*

The JVC designed tape loading method departed from the traditional U format, and employed an M wrap. An ingenious system of locating the guide rollers into accurately positioned vee blocks ensured good tape interchange, which had been a problem with earlier systems. The tape path is shown in *Figure 56.7*.

This M wrap has been used throughout the VHS range, except for the camcorder and small portables where a smaller head drum is employed. These recordings are, however, compatible with the standard size head drum format.

56.4 Grundig SVR 4004

In 1978 Grundig introduced into the German market a four hour cassette machine (SVR 4004) based on the N1700 cassette, but with higher grade tape. Its basic mechanical data are:

drum diameter:	105 mm
speed, head disc:	1500 rev/min
scanning speed (relative speed, video head to tape):	8.2 m/s
tape speed:	39.5 mm/s
video track width:	51 μm
ref white level:	5.1 MHz
ref sync level:	3.1 MHz
video bandwidth:	3 MHz 6 dB

56.5 Technicolor VCV 212E

An interim development was the Technicolor CVC system using a quarter-inch wide tape cassette with a maximum playing time of 60 min. Conventional colour-under modulation was used, with two heads azimuthed at 11°. The basic technical data for the model 212E are:

tape speed:	37.4 mm/s
tape width:	6.25 mm (¼ inch)
resolution:	240 lines (-6 dB at 3.5 MHz mono)
audio:	200 Hz–8 KHz (42 dB s/n rate)
head azimuth:	±11° (2 heads)
fm deviation:	3.8–4.8 MHz
subcarrier for colour:	623 kHz, phase modulated

56.6 The second generation

The first generation of cassette machines used a central motor, with belts, pulleys and slipping clutches to drive the various functions. Operation was by 'piano keys' and mechanical linkage. Later models used individual motors for each function, microprocessor controlled, with touch button operation and consequently suitable for remote control.

The first machines were not able to reproduce still or slow motion. This was achieved in later models by using wider heads for *still* and *slomo* playback. In the more expensive industrial machines these heads were used only for this playback, but in the cheaper models they had also to double as the record head.

	Item	525-line/60 field			625-line/50 field
		Tape speed			
		40 mm/s	20 mm/s	13.3 mm/s	18.7 mm/s
B	Total video width	10.6	10.6	10.6	10.6
W	Effective video width (180°)	10.2	10.2	10.2	10.2
L	Video track centre from reference edge of tape	6.01	6.01	6.01	6.01
P	Video track pitch	0.0585	0.0292	0.0194	0.0328
C	Control track width	0.6 ± 0.1	0.6 ± 0.1	0.6 ± 0.1	0.6 ± 0.1
R	Audio track width (monophonic)	1.05 ± 0.1	1.05 ± 0.1	1.05 ± 0.1	1.05 ± 0.1
D	Audio track (channel 2) width (stereophonic-right)	0.35 ± 0.05	0.35 ± 0.05	0.35 ± 0.05	0.05 ± 0.05
E	Audio track (channel 1) width (stereophonic-left)	0.35 ± 0.05	0.35 ± 0.05	0.35 ± 0.05	0.35 ± 0.05
F	Audio track reference line	11.51 ± 0.05	11.51 ± 0.05	11.51 ± 0.05	11.51 ± 0.05
h	Audio-to-audio track guard width	0.35 ± 0.05	0.35 ± 0.05	0.35 ± 0.05	0.35 ± 0.05
O	Video track angle	5° 01′ 42″	5° 00′ 51″	5° 00′ 34″	5° 00′ 58″
O_o	Video track angle (tape stationary)	5°00′	5°00′	5°00′	5°00′
X	Position of audio and control head	68.00	68.00	68.00	68.00
					68.00
	fm carrier (MHz) peak white		4.8 (A)		5.2 (A)
		4.8 (A)	4.8 (B)	4.8 (A)	5.2 (B)
		4.8 (B)	+7.867 KHz	4.8 (B)	+7.812 KHz
			3.6 (A)		3.8 (A)
		3.5 ± 0.1 (A)	3.6 (B)	4.8 (A)	3.8 (B)
	sync tip	3.5 ± 0.1 (B)	+7.867 KHz	4.8 (B)	+7.812 KHz
	Colour-under	688 (A)	688 (A)	688 (A)	685.5 (A) } PAL
	subcarrier (kHz)	688 (B)	688 (B)	688 (B)	689.4 (B) } PAL
					685.5R (A) } SECAM
					841.8R (B) } SECAM
					689.4B (A) } SECAM
					845.7B (B) } SECAM
	Head A offset				
	equiv. to 95 lines delay	zero	zero	zero	17′ 17″
	Hi-fi audio head (L)				
	head B (R)				1.44 MHz
	(Deviation ± 200 kHz)				2.10 MHz
	Video head width (μm)	42	42	42	42
	Azimuth A	+7° ± 18′	+7° ± 18′	+7° ± 18′	+7° ± 18′
	Azimuth B	−7° ± 18′	−7° ± 18′	−7° ± 18′	−7° ± 18′
	Hi-fi head width (μm)				40
	Azimuth a				−30°
	Azimuth b				+30°

Table 56.1 Parameters of Beta track format. The dimensions are in millimetres

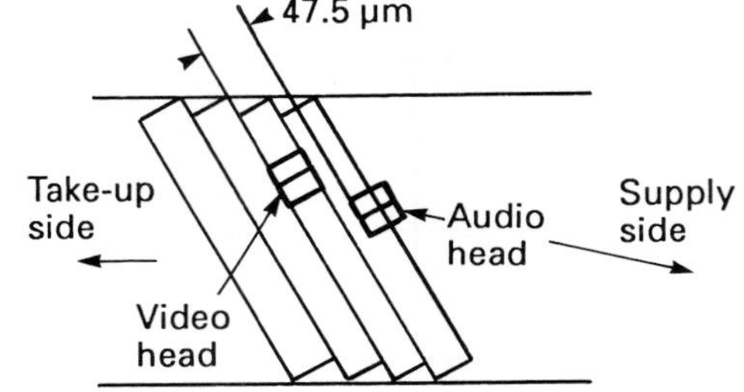

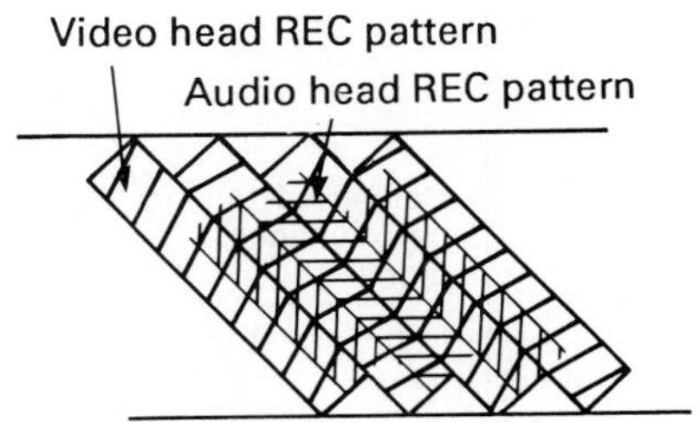

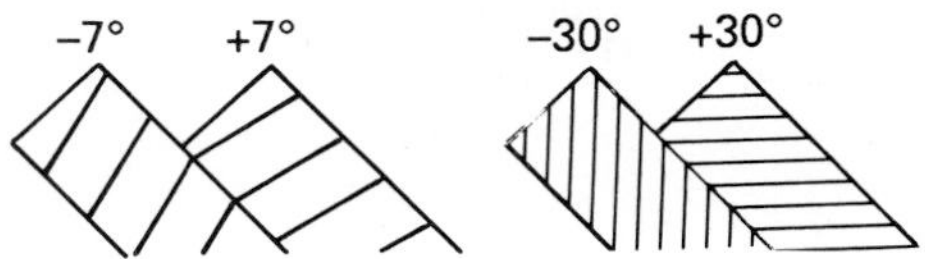

Figure 56.6 Audio and visual tracks in the Beta system

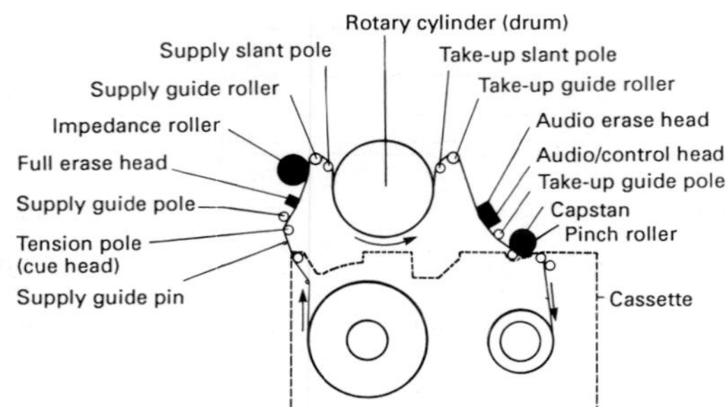

Figure 56.7 M wrap tape path of VHS

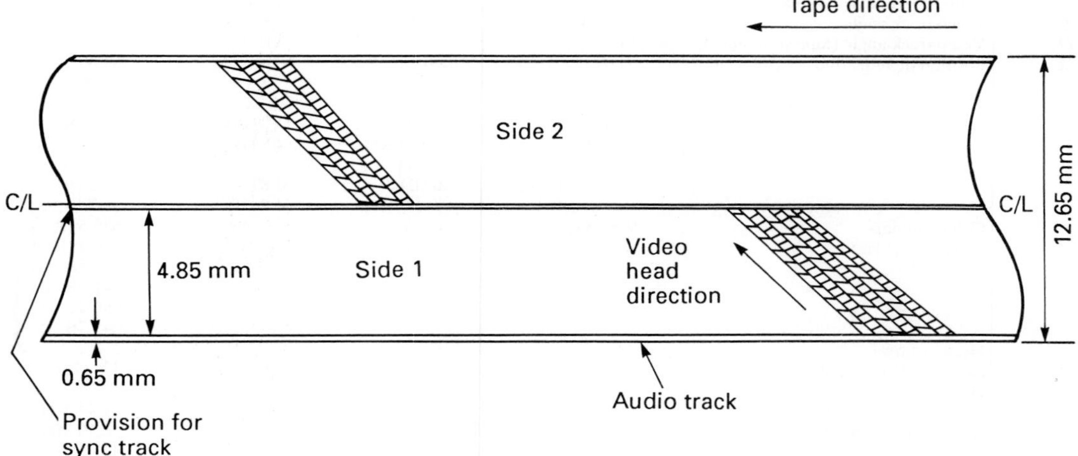

Figure 56.8 The V2000 tape format

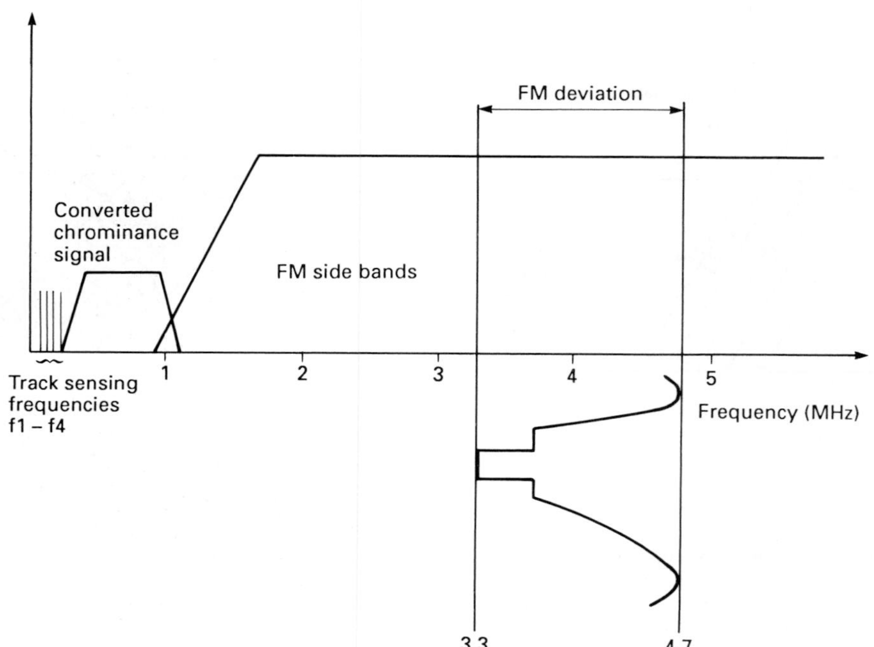

Figure 56.9 The V2000 rf envelope. The track sensing frequencies are f_1(ch1):102.187 kHz. The switching point from ch1 to ch2, etc., must be 8 ± 5 lines before the leading edge of the vertical sync pulse.

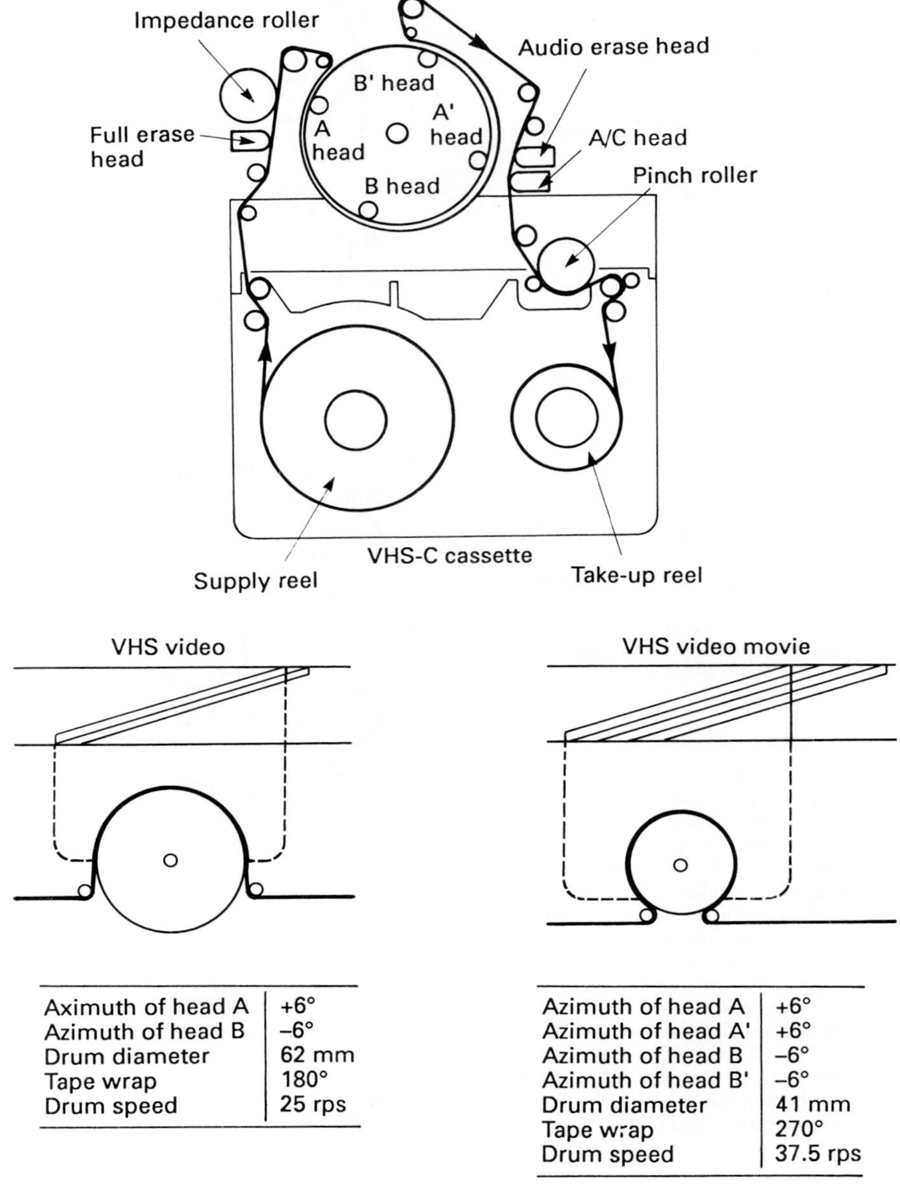

Aximuth of head A	+6°
Azimuth of head B	–6°
Drum diameter	62 mm
Tape wrap	180°
Drum speed	25 rps

Azimuth of head A	+6°
Azimuth of head A'	+6°
Azimuth of head B	–6°
Azimuth of head B'	–6°
Drum diameter	41 mm
Tape wrap	270°
Drum speed	37.5 rps

Figure 56.10 VHS-C heads and tape transport system

56.7 Philips V2000

The next important step was taken in 1980 when Philips introduced the V2000 series. This series of vcrs records the video signal on half the width of a half-inch tape enabling it to be turned over like a compact audio cassette. The recording time is four hours per side, i.e. eight hours per cassette. The video track is 22.5 μm wide with no guard bands. The video head, which is 23 μm wide, is kept on the correct track by *dynamic tracking* (*Figure 56.8*).

This dynamic tracking is achieved by mounting the video head on the head drum via 'actuators' made of piezoelectric crystals. Voltages are fed to these crystals to guide the head along the correct path. The head movement is sufficient to correct for the change in track angle during *still frame*, and the fields are scanned by their respective video heads to give a true still frame, rather than the same field that is reproduced twice.

An audio track is provided along the two outer edges of the tape, with the option of a *search-and-cue*track along the centre of tape between the two video tracks.

Another innovation in the V2000 is the absence of a conventional control track. This is because a series of test frequencies are recorded in the video rf envelope (see *Figure*

56.9), below the frequency of the colour under signal. These are used to provide the synchronization pulses as well as the signal to guide the dynamic tracking.

The basic mechanical data for the V2000 system are:

drum diameter:	65 mm
speed, head disc:	1500 rev/min
scanning speed (relative speed, video head/tape):	5.08 m/s
tape speed:	24.42 mm/s
video head gap length:	0.2–0.3 μm
video track width:	22.5 μm
distance between two video tracks:	0
audio track width:	0.65 mm
sync track width (cue track):	0.4 mm
video head width:	23 μm

56.8 Camcorders

The demand for a small portable camera and recorder combined into one unit, like an 8 mm cine camera, stimulated both JVC and Sony in 1984 to modify their VHS and Beta recorders to use smaller head drums, which would maintain compatibility with their original formats.

JVC also designed a compact cassette, VHS-C format, capable of 30 minutes recording time, which could be used in a standard machine when fitted into an adaptor.

56.8.1 VHS-C

The VHS head drum was reduced in diameter by one-third (from 62 to 41 mm), the tape wrap around the drum increased from 180° to 270° and the drum speed increased by 50 per cent (from 25 to 37.5 rps). To maintain compatibility, four recording heads were used with the standard azimuth of ±6°.

The VHS-C compact cassette enabled a very small, lightweight camera to be produced, but a number of portable cameras were designed to use the full size cassette. These cameras usually contained an insert-edit facility, which required an additional pair of 'flying-erase' heads to be mounted on the drum.

The flying-erase heads are usually the same width, and spaced just before the standard record/replay heads. They erase the video track just before the new information is recorded. This also erases the hi-fi track, but the longitudinal audio and the control tracks are not erased (*Figure 56.10*).

56.8.2 Beta Movie

In the Beta Movie system, a substantial saving in the space and weight of the recorder has been achieved by reducing the head drum diameter from 74.5 to 44.7 mm, and increasing the tape wrap from 180° to 300° (*Figure 56.11*). Only one recording head is used but, to provide Beta compatability, two head gaps are built into this head, azimuthed at +7° and -7° and spaced to provide a time difference equal to two lines. The head drum rotates at twice the normal speed, but to overcome the fact that no recording can be made during 60° (one-sixth of the time), the camera line rate is increased by one-sixth, and also picture overscanned by one-sixth in the vertical direction. Thus the camera runs at a non-standard rate, but the information recorded on the tape is correct.

This means that the Beta Movie recorder can only record with a modified camera and cannot be used as a stand-alone recorder. The basic mechanical data are:

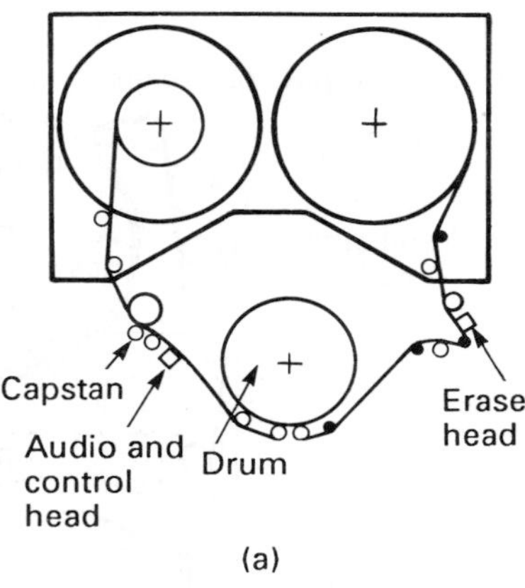

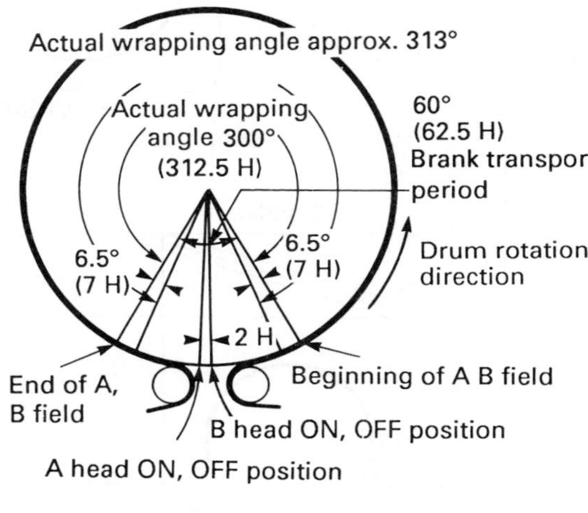

Figure 56.11 Tape transport system (omega wrap) and drum/head specifications for the Beta movie

winding angle:	300°
drum diameter:	44.671 mm
lead angle:	5°00′09″
video head:	double azimuth 1 head
field number per 1 rotation:	1 field
drum relative speed:	6.998 m/s
drum rotation:	50 rps
f_V:	50.00 Hz
f_H:	18.750 kHz
REC Y:	4.56–6.24 MHz
REC C:	f_{CA} 822.656 kHz f_{CB} 827.344 kHz

These figures should be compared with those for Beta format in section 56.2.

56.9 Hi-fi sound

In video cassette recorders sound has always been the poor relation, squeezed in on a narrow track at the edge of the tape.

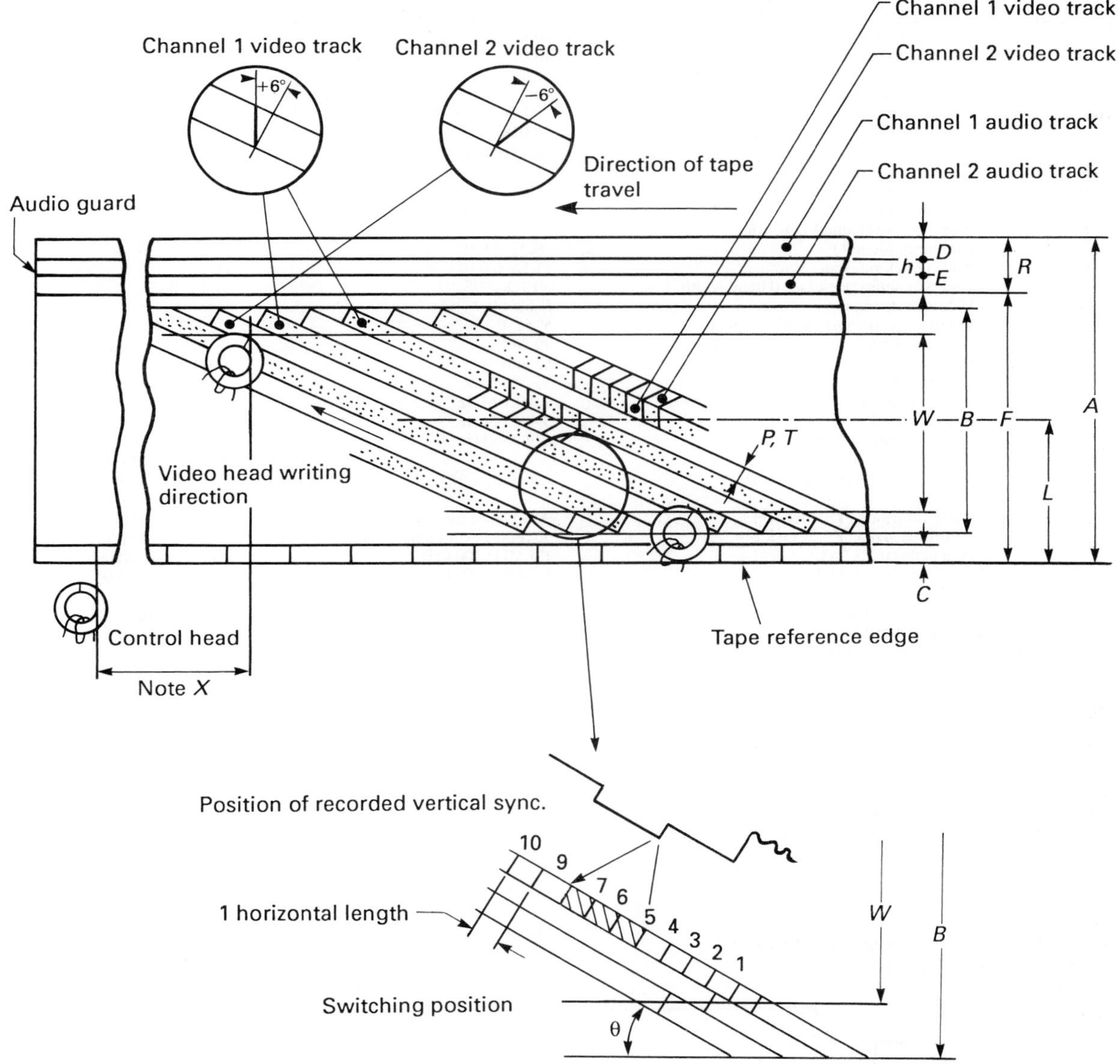

Figure 56.12 Beta tape format with two audio tracks. The letters refer to dimensions listed in *Table 56.2*. In the monaural version, audio tracks 1 and 2 are combined into a single track

The quality was further degraded when this track was split into two to provide stereo sound. The introduction of the audio compact disc in the mid 80s with its high quality provided a challenge that could not be ignored by the video tape recorder designers.

In 1984 a hi-fi audio track was added to the 625-line VHS and Beta cassette tape systems. This was achieved by mounting an additional pair of heads (half the width of the video heads, for VHS) onto the head drum. These audio heads, azimuthed at about 30°, lay down fm signals onto the video track, just prior to the recording of the video signal. The audio signals are partially erased by the video signal, but as the original recording level is high, sufficient signal remains to be detected on playback.

The hi-fi audio head has a wider gap than the video head in order to create a larger envelope for the magnetic flux. This larger field penetrates farther into the magnetic layer on the tape to produce a 'depth recording'. The wider head gap reduces the maximum frequency that can be recorded but is adequate for the 2 MHz audio carrier signal.

The video head, with the narrower gap, produces a smaller envelope which over-records a video signal on the surface of the tape, but leaves the audio modulation unaffected at the deeper level.

The audio fm carrier frequencies were chosen to sit between the luminance and chrominance envelopes, with stereo provided by two separate carriers. This produces an audio frequency response up to 20 kHz with a dynamic range of 80 dB, i.e. comparable to NICAM broadcast television sound.

	Item	*525-line/60 field*	*652-line/50 field*
B	Total video width	10.60	10.60
W	Effective video width (180°)	10.07	10.07
L	Video track centre from reference edge of the tape	6.2	6.2
P	Video track pitch	0.058	0.049
T	Video track width	0.058	0.049
C	Control track width	0.75 ± 0.1	0.75 ± 0.1
R	Audio track width (monophonic)	1.0 ± 0.1	1.0 ± 0.1
D	Audio track (channel 2) width (stereophonic-right)	0.35 ± 0.05	0.35 ± 0.05
E	Audio track (channel 1) width (stereophonic-left)	0.35 ± 0.05	0.35 ± 0.05
F	Audio track reference line	11.65 ± 0.05	11.65 ± 0.05
h	Audio to audio track guard band width	0.3 ± 0.05	0.3 ± 0.05
O	Video track angle	5° 58′ 09.9″	5° 58′ 09.9″
O_o	Video track angle (tape stationary)	5° 56′ 07.4″	5° 56′ 07.4″
X	Position of audio and control head	79.244	79.244
	Head drum diam.	62 mm	62 mm
	Tape speed	33.35 mm/s	23.39 mm/s
	Ref white level	4.4 0.1 MHz	4.8 0.1 MHz
	Ref sync level	3.4 0.1 MHz	3.8 0.1 MHz
	Colour-under subcarrier	629.36 kHz	626.953kHz (PAL) 1.08 MHz (SECAM)
	Subcarrier band pass filter (MHz)	3.58 0.5	4.43 0.5 (PAL) 4.32 0.8 (SECAM)
	Phase rotation of chrome signal — Track 1	90 advance every line	Remain at 0 PAL only
	Phase rotation of chrome signal — Track 2	90 retard every line	90 retard every line
	Hi-fi audio track 1 (L)	1.3 MHz 200 kHz	1.4 MHz 200 kHz
	Hi-fi audio track 2 (R)	1.7 MHz 200 kHz	1.8 MHz 200 kHz
	S-VHS		
	Ref white level	7.0 0.1 MHz	7.0 0.1 MHz
	Ref sync. level	5.4 0.1 MHz	5.4 0.1 MHz
	Colour-under subcarrier	as VHS	as VHS
	Hi-fi audio	as VHS	as VHS

Table 56.2 Parameters of VHS track format. The dimensions are in millimetres

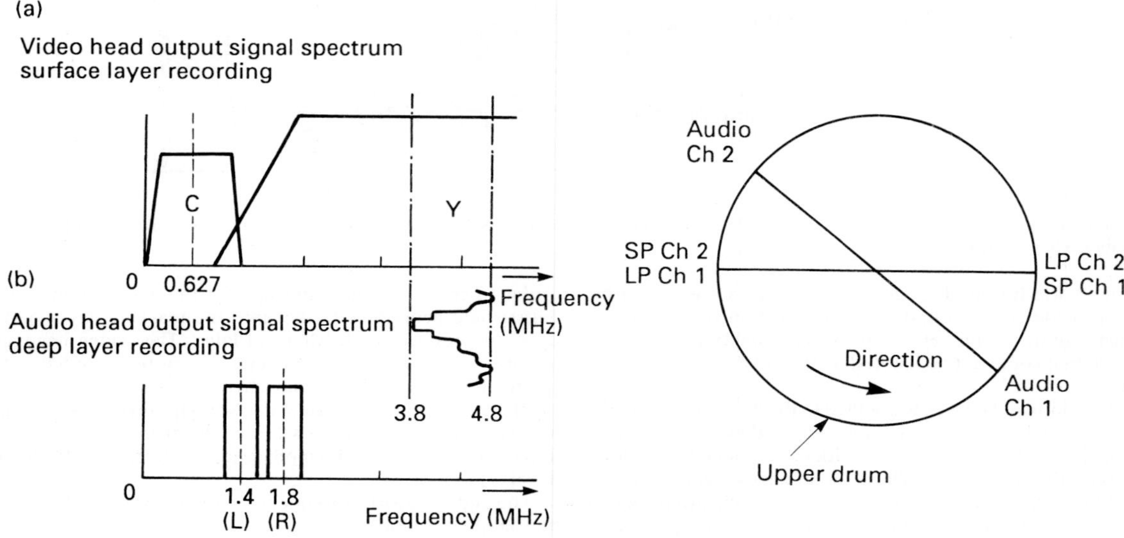

Figure 56.13 VHS record/replay hi-fi audio heads mounted on the head drum. The rf envelope is recorded by the video heads (a) and audio heads (b)

56.10 VHS long play developments

The Philips V2000 series with the capability of recording eight hours per cassette (four hours per side) could not go unchallenged by the VHS designers.

By 1982, the E240 VHS cassette had arrived, which by using thinner tape in the standard VHS cassette housing, had increased the maximum recording time from three to four hours.

In 1984, a half-speed PAL VHS recorder was introduced. This could be switched between the two speeds, (SP and LP) with auto sensing on playback on the more expensive models. With the E240 cassette, this provided eight hours of uninterrupted recording.

The half-speed recording required a narrower video track, provided by a video head that was 25 μm wide rather than the standard 49 μm, with an offset between adjacent tracks halved to 0.75 of a line.

This narrower video track reduces the quality of the video signal, but where a hi-fi audio head is used, the audio head width is unaltered and the audio response is virtually unaltered. The result is a relatively inexpensive machine that provides eight hours of continuous high quality audio reproduction.

On a two-speed machine the drum must carry two standard video heads, two half-width video heads and possibly two hi-fi heads (see *Figure 56.13*).

56.11 8 mm video

The introduction of the portable video colour camera and recorder, with its immediate playback and low tape cost compared to film, killed the 8 mm cine film camera market. The film giants such as Kodak pressed for a video replacement, and as a result an 8 mm video cassette standard has been produced. The standard was agreed with most of the photo film and video companies including Sony, who do not have a compact cassette in the Beta range.

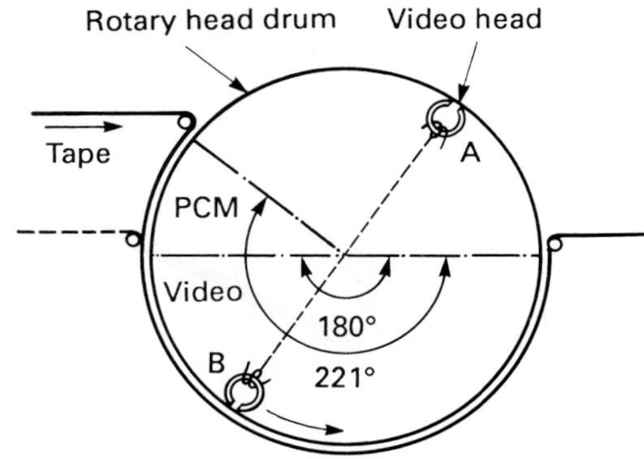

Figure 56.14 Relation between tape and head in 8 mm format

The 8 mm format was developed almost ten years after VHS and Beta and takes advantage of tape and format improvements. The specified tape is either metal powder (type A) or metal evaporated (type B) up to 90 minutes in length. The video recording method is similar to the Philips V2000 system, with dynamic tracking and no separate control track. The format includes pcm stereo sound in addition to a conventional longitudinal track and an additional longitudinal auxiliary cue track.

The format uses many of the features of the Philips V2000, but the full width of the 8 mm tape is used. The tape wrap around the head drum is 221°, with 180° being used for the video signal, and 36° for a pcm audio track. The pcm technique follows the method used for the compact audio disc. The head drum contains only two heads, spaced 180° apart, so that head A is recording the pcm audio track whilst head B is recording

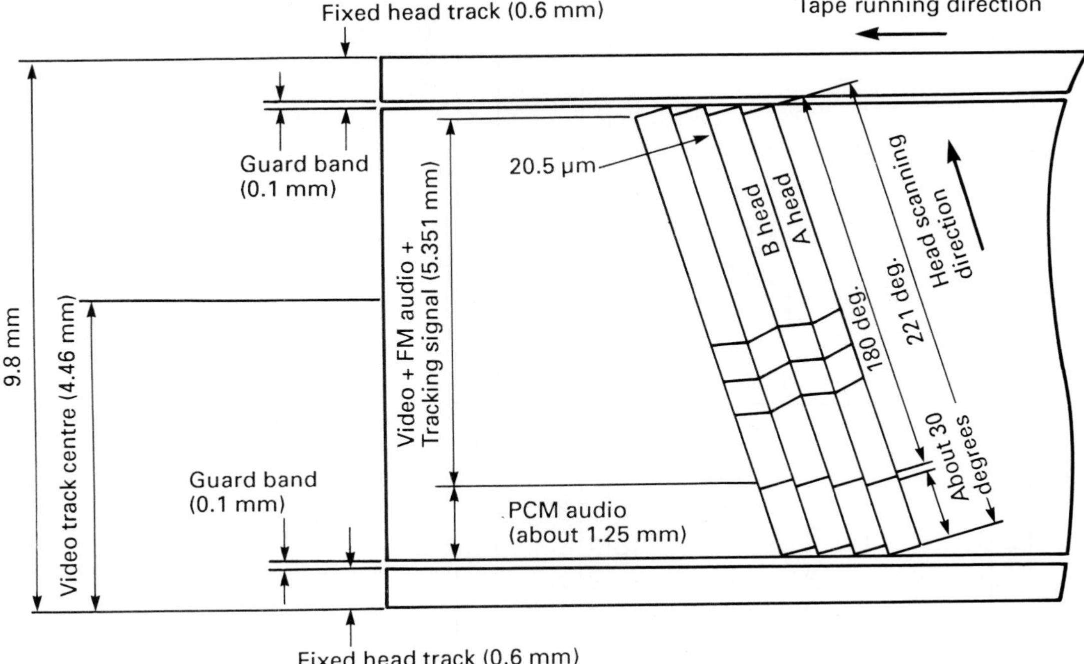

Figure 56.15 8 mm format tape configuration

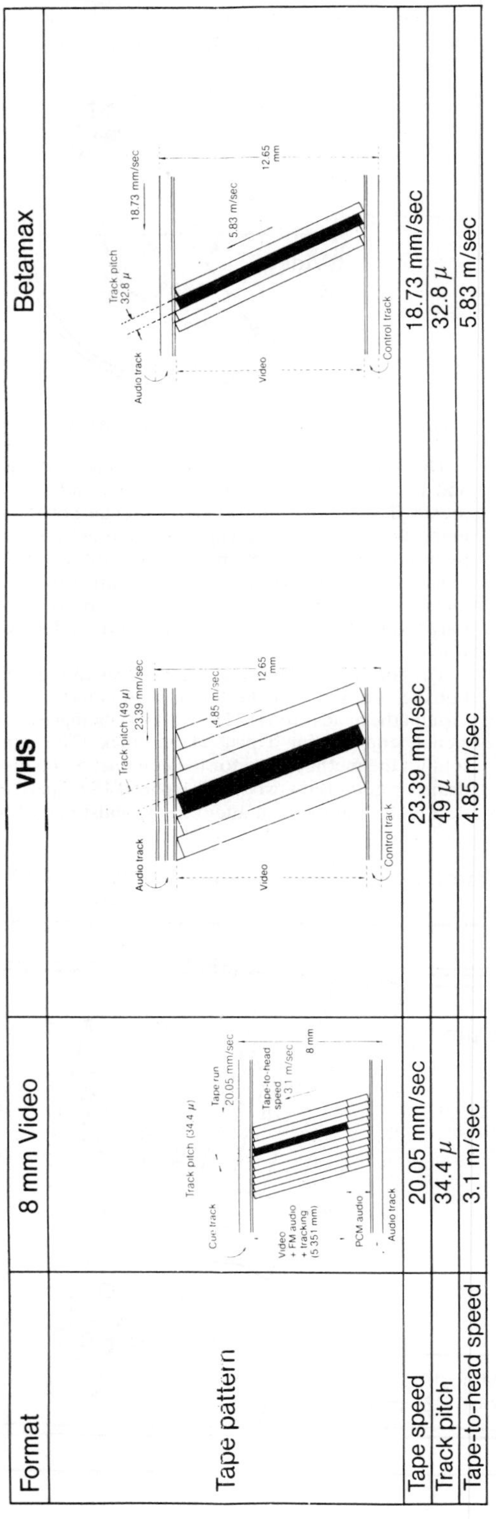

Format	8 mm Video	VHS	Betamax
Tape pattern			
Tape speed	20.05 mm/sec	23.39 mm/sec	18.73 mm/sec
Track pitch	34.4 μ	49 μ	32.8 μ
Tape-to-head speed	3.1 m/sec	4.85 m/sec	5.83 m/sec

Format	8 mm Video	VHS		Betamax	
Head drum		VideoMovie	Standard	Betamovie	Standard
Drum diameter	40 mm	41 mm	62 mm	44.7 mm	74.5 mm
Drum speed	25 rps	37.5 rps	25 rps	50 rps	25 rps

Figure 56.16 Comparison of specification among different formats (all 50 field formats)

the last part of the previous video field. The cassette incorporates five holes in the base for cassette identification and an autochange grip position. The tape path is shown in *Figure 56.14*. Tape speeds are 14.345 and 20.051 mm/sec for the 525/60 and 625/50 systems respectively. The 40 mm drum contains head CH1 azimuthed at +10° and head CH2 at -10°.

The tape recording format is shown in *Figure 56.15*. The fm audio carrier at 1.5 MHz is sandwiched between the luminance and colour carrier frequencies, and additional longitudinal audio track is provided along the bottom edge of the tape. A similar track exists along the top edge but is designated as a *cue-and-search* track. There is no control track.

Tracking pilot signals are continuously recorded, at frequencies below the chroma envelope, which serve the dual purpose of providing the monitoring signal for the dynamic tracking, and the cue track. Four separate pilot frequencies are used and switched in sequence to identify the four field sequence. The frequency change occurs just before the leading edge of the vertical sync pulse.

A fifth pilot signal, not exceeding 230 kHZ, may be used as a recording head positioning signal. This ensures equal track width and occupies three lines between the end of the pcm recording and the start of the video recording.

The pcm circuit provides a high quality stereo audio channel. The sampling frequency is twice line rate, providing an audio response of 10 kHZ. The recorded bit rate is 8, compared with 10 for a compact disc. The dynamic range is however expanded from the theoretical 48 dB by first quantizing the signal by 10 bits and then converting to 8 bits, depending on the signal amplitude. This provides a better dynamic range at the quieter levels, where noise would be objectionable, at the expense of a restricted range at the higher levels. A compandor system with a compression ratio of 1:2 is also used to double the dynamic range.

The pcm signal is recorded over a 30° arc compared with 180° for the video signal. The pcm signal must therefore be time compressed by 6:1. With error correction, this produces a transmission rate of 5.79 Mbits for the 525/60 system, and 5.75 Mbits for 625/50. It is also possible to record further pcm signals instead of the video information. A total of 6 pcm stereo tracks are then available.

56.12 S-VHS

In 1987 JVC countered the introduction of the 8 mm video cassette format by upgrading the VHS format to Super VHS (S-VHS). This was achieved by taking advantage of the higher grade tape currently available, and increasing the bandwidth of the luminance component of the VHS format signal.

The luminance peak white modulation frequency has been increased to 7 MHz compared to 4.8 MHz in the standard VHS format, and the deviation increased from 1 MHz to 1.6 MHz. The chrominance subcarrier frequency has not been changed but the bandwidth is wider, as the luminance sidebands no longer intrude into the chrominance envelope.

The S-VHS standard is claimed to be able to reproduce 400 lines per picture width and, when set to record at half speed, will have a performance similar to the original VHS standard.

The S-VHS tape is contained in the standard VHS cassette, but an additional hole is added to the cassette housing to identify the S-VHS standard.

It is possible to playback existing libraries of the original VHS recordings in the S-VHS machines and the quality will be of the original VHS standard. S-VHS cassettes cannot be used however in standard VHS machines.

56.13 Digital waveforms

The analogue video waveform is likely to be superseded by a digital waveform. This will be in the form of a brightness component (Y) and two colour difference signals (B-Y and R-Y). The B-Y and R-Y components will generally each require half the bandwidth of the Y signal. The brightness level will be determined by the digital bit code.

An analogue/digital converter will accept an analogue signal and convert it into digital form.

The resolution is determined by the sampling rate, i.e. a resolution equivalent to 6 MHz would require a sampling rate of 12 MHz.

The brightness information of each sample is provided by a series of digital pulses. The more pulses there are in the series, the better the level differentiation will be in the grey scale. An 8 bit sample will give a brightness gradation of 256 steps, whereas a 6 bit sample would provide only 64 steps. An 8 bit signal is used to broadcast video information but 16–20 bits are used for audio.

In order to correct errors in the digital signal caused by drop-out on the tape, 'parity' checks are made. This can be achieved, for example, by adding up all the numbers in a 'line'; the parity number is the total number that should be produced. This can be cross-checked by adding the numbers in a different sequence to obtain a different number. This would then be parity 2. Should these numbers not tally then error correction signals would be inserted.

It will be appreciated from the above that digital television requires considerably wider bandwidths than analogue waveforms, but in compensation, it is not necessary to assess the amplitude of the digital signal, but simply to determine if the signal is a '1' or a '0'. A digital vcr must be capable of recording and replaying the luminance, Y, signal and the colour difference signals, B-Y and R-Y (U and V). Coding methods will undoubtedly be devised to compress this information in an economical manner.

R Watson C Eng, MIEE, MRTS, MBKS
Consultant

57 Video Discs

It is interesting to conjecture whether the gramophone record would have succeeded if the audio tape recorder had existed for many years before the record was introduced.

In the 1970s and early 1980s, several companies developed video disc systems to the point where they would have been technically acceptable for the general public, but were restrained from marketing them by fears for their viability.

The advantage originally claimed for a video disc was the low manufacturing cost compared with a pre-recorded video cassette. This advantage is however insignificant when the programme material costs are included. The vcr is already well established in the home, and as it can also be used to record off-air there is little incentive to purchase a video disc machine for the domestic entertainment market.

A video disc has two major advantages over a video cassette: the facility to reproduce a *still frame*, and the very rapid *random access* time. These features make video discs particularly suitable for the 'data' market, including interactive teaching aids.

57.1 Early systems

In 1975 the TED or *Teldec* system was launched by a partnership between Telefunken in Germany and Decca in the UK. This system had teething troubles, but before they could be resolved the video cassette recorder had been introduced and was rapidly gaining in popularity. TED struggled on for two or three years before being abandoned.

TED used a 20 cm (8 inch) disc rotating at 1500 rpm for PAL and 1800 rpm for NTSC, playing for 10 minutes. The information was read off the disc by a diamond stylus and ceramic transducer, working on the hill-and-dale principle. The audio and video signals were fm coded on the same track.

Matshushita (National Panasonic) demonstrated in 1978 their *Visc* system which used a capacitance pickup on a 30 cm (12 inch) diameter disc. This rotated at 450 rpm (PAL) or 500 rpm (NTSC) with a playing time of one hour each side. Still frame was not possible, and the system was never marketed.

RCA developed a video disc system under the name of *Selectavision* (previously used for a video cassette system) which evolved into the CED system. This used a capacitance pickup with a groove to guide the stylus. The CED system was eventually abandoned by RCA in 1984.

57.2 Current systems

By the mid 1980s only two systems had survived. These were the JVC *VHD* systems using a capacitance sensor, and the Philips *LaserVision* employing an optical disc with no mechanical contact.

57.2.1 VHD

The VHD/NTSC system was first shown in 1979, but a 625-line version was never produced, and the NTSC production lapsed by the mid 1980s. This system used a 26 cm (10 inch) diameter disc with a playing time of 60 minutes each side. The video and two audio signals were carried on the same spiral track using different subcarrier frequencies on an fm carrier of 7.24 MHz. This information track comprised a series of pits in the surface of the disc (*Figure 57.1*). Tracking was provided, not by a mechanical groove, but by two pitted tracks either side of the information track. These pits were arranged in groups to produce different frequencies (508 and 711 kHz), but the spacings provided bursts of 'tone' with a repetition rate equal to the horizontal sync period.

The top surface of the disc was flat, and it was possible for the stylus to glide across the disc without damaging the information, which was contained below the surface of the disc. However the stylus was in physical contact with the surface of the disc, and so some wear did occur.

57.2.2 LaserVision

The Philips LaserVision system was introduced into the entertainment market in 1978 but has since developed more steadily as an information storage device.

The disc has only one information track, which carries the video and two audio channels on a single fm carrier similar to the VHD disc, but the storage is by variations of a reflective coating, protected by a transparent faceplate. The encoded information is scanned by a small lightspot from a laser source, and the reflected signal measured by a photocell.

The LaserVision disc is available in 20 cm (8 inch) or 30 cm (12 inch) diameters and scanning is from the centre, anticlockwise to the outside. A coding at the start of the disc identifies the disc diameter.

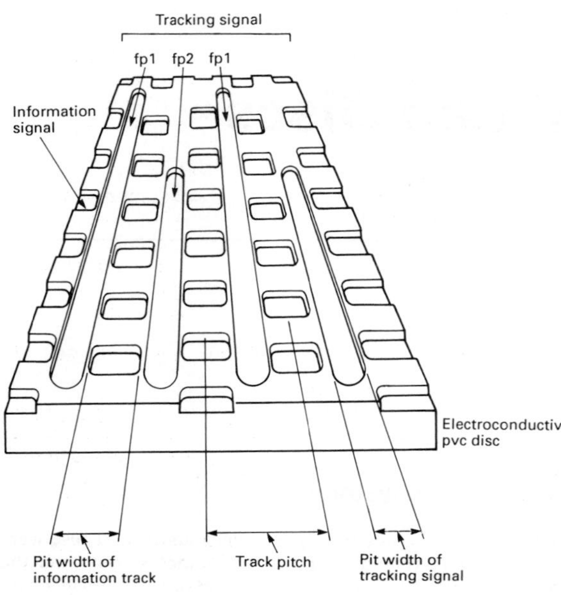

Figure 57.1 Information and tracking pits on the VHD disc

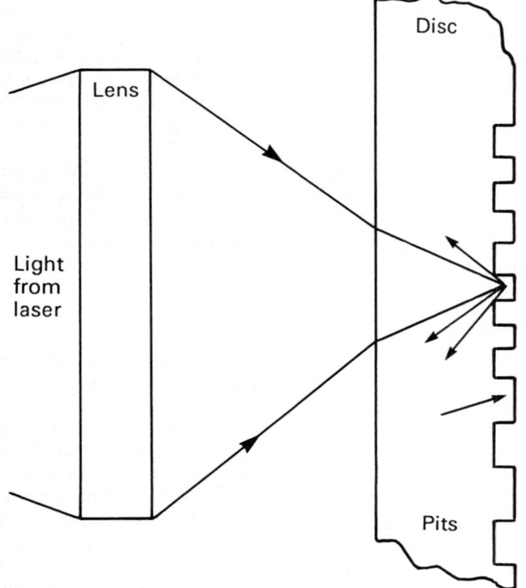

Figure 57.3 Laser beam focused onto information pits

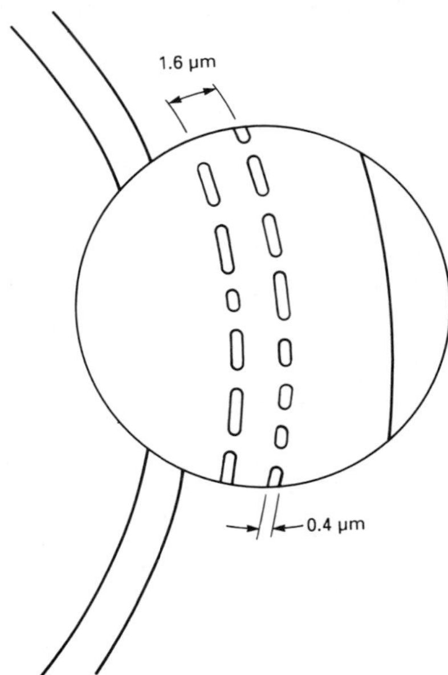

Figure 57.2 'Pits' on a LaserVision disc

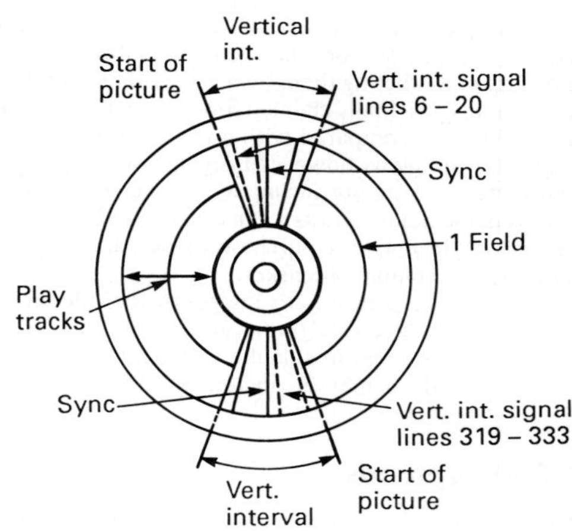

Figure 57.4 Information on a LaserVision disc

There are three methods of encoding the signal: constant angular velocity, constant linear velocity and constant angular acceleration.

Constant angular velocity (CAV) or *active play* runs at 1 revolution per TV frame, with facilities for freeze frame. The speed is 1500/1800 rpm (PAL/NTSC).

Constant linear velocity (CLV) or *long play* requires the rotational speed of the disc to be reduced as the sensor tracks from the centre to the outside. This provides a longer playing time, but freeze frame is impossible. This is not normally important for the entertainment market. The disc speed varies around 500 rpm.

Constant angular acceleration (CAA) is a hybrid of CAV and CLV. The angular velocity is kept constant for sections of the recordings, then changes velocity in steps instead of continuously each frame. This has the advantage of reducing horizontal sync pulse breakthrough between tracks.

The LaserVision player can switch between the three modes.

In all three formats the information is recorded on a continuous spiral track running anti-clockwise from the centre. The track consists of a series of 'pits' 0.4 μm wide and approximately 0.1 μm deep, spaced 1.6 μm between adjacent grooves.

A helium-neon laser with a wavelength of 0.633 μm is used to produce a spot of light which is focused onto the information track. The light will be reflected back to the photocell from the highly reflective surface, but will be scattered by the pits.

The frequency modulation is provided by varying the length and frequency of the pits (see *Figure 57.2*). With CAV coding, two fields, i.e. one frame of information, occupy 360° of rotation. The vertical interval blanking period is used to carry frame identification, and tracking instructions such as still frame (*Figures 57.3* and *57.4*).

CLV coding also starts off at the centre with one revolution per frame, but as the sensor moves away from the centre of the disc, each frame occupies a shorter arc. This provides a longer playing time as the speed of rotation decreases, to keep the linear reading rate the same. Still frame is not possible, so the vertical interval signal contains instructions to keep the machine in the CLV mode as well as providing a time code in hours and minutes to show the elapsed time.

The frequency spectrum shown in *Figure 57.5* is high enough to record a composite colour signal, making colour-under coding unnecessary.

The two audio channels can be used in any of the following four modes:

- stereophonic sound
- monophonic sound: two separate audio programmes
- monophonic sound: one audio programme on both channels
- one or both channels for control or cueing information

The BBC published their Doomsday project on two LaserVision discs. This contained a mixture of text and pictures, providing information obtained from many sources throughout the UK. The audio channels were used as a CD-ROM (see section *57.3*) to reproduce the digital data for the interactive element. In later versions of the LaserVision disc, the fm audio can be substituted for digital audio recording using the standard compact disc method. A coding signal at the start of the disc identifies the method used to record the audio component. The fm audio discs are coloured silver, and the digital audio are gold.

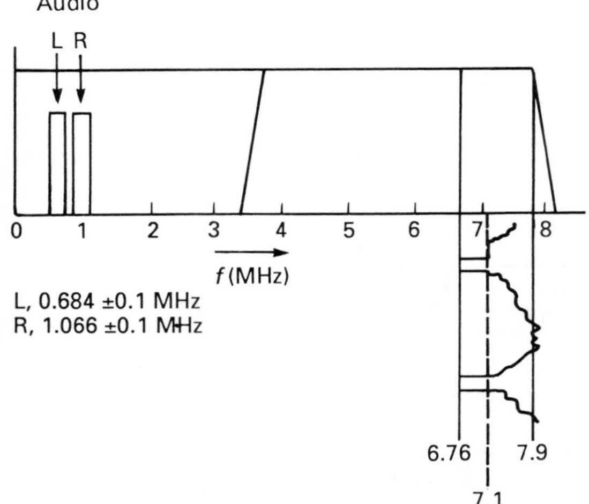

Figure 57.5 LaserVision modulation envelope (PAL)

Vertical interval control and address signals are:

	CAV format	*CLV format*
1	Lead-in	Lead-in
2	Lead-out	Lead-out
3	Picture numbers	Programme time code
4	Picture stop	CLV code
5	Chapter numbers	Chapter numbers
6	Programme status code	CLV picture number
7	Users code	Programme status code
8		Users code

Parameters of the LaserVision disc are given in Table *57.1*.

Table 57.1 LaserVision disc parameters

	625 PAL	525 NTSC
Disc diameter	20cm or 30cm	20cm or 30cm
Centre hole diam.	35mm	35mm
Thickness of disc	2.7mm	2.7mm
Speed of rotation		
CAV	1500 rpm	1800 rpm
CLV	1500-570 rpm	1800-700 rpm
Track width	0.4 μm	0.4 μm
Track depth	0.1 μm	0.1 μm
Track pitch	1.6 – 2 μm	1.6 – 2 μm
Laser wavelength	0.0633 μm	0.0633 μm
Numerical aperture	0.40	0.40
Carrier frequency:		
Sync tip	6.75 MHz	7.59 MHz
Black	7.10 MHz	8.10 MHz
Peak white	7.90 kHz	2.30 MHz
Audio I (left)	684 kHz	2.30 MHz
Audio II (right)	1066 kHz	2.81 MHz
100% mod. deviation	100kHz	100 MHz
Pilot burst freq.	3.75 MHz	colour burst
Vertical interval signals:		
a) Coding	24 bit biphase hexadecimal	24 bit biphase hexadecimal*
b) vert Int Test Sig (VITS)	Lines 19, 20 332, 333 (Lines 22 and 335 black)	19, 20, 282, 283
Address signals viz:	Lines 6 – 18 and 319 – 331	10 – 18 and 273 – 281
c) Picture numbers (CAV)	Lines 17 & 18 and 330 & 331	17 & 18 and 280 & 281
d) Picture numbers (CLV)	Lines 16 or 329	
e) Chapter number (CAV) when picture no. not used	Lines 17 & 18 and 330 & 331	17 & 18 and 280 & 281
f) Chapter number (CLV) when time code not used.	Lines 18 & 331	18 or 281
g) Time code (CLV)	Lines 17 & 18 or 330 & 331	17 & 18 or 280 & 281
h) CLV operation code to alternate with time code	Lines 330 or 17	280 or 17
i) Picture stop (CAV only)	16 & 17 or 329 & 330 (takes priority over chapter no.)	16 & 17 or 279 & 280

* In the NTSC disc, a 40 bit fm coded signal may also be used. On the CAV disc, the 40 bit codes provide picture number information. On the CLV disc, the code provides programme time and CLV information. Both CAV and CLV discs also include a first field white flag.

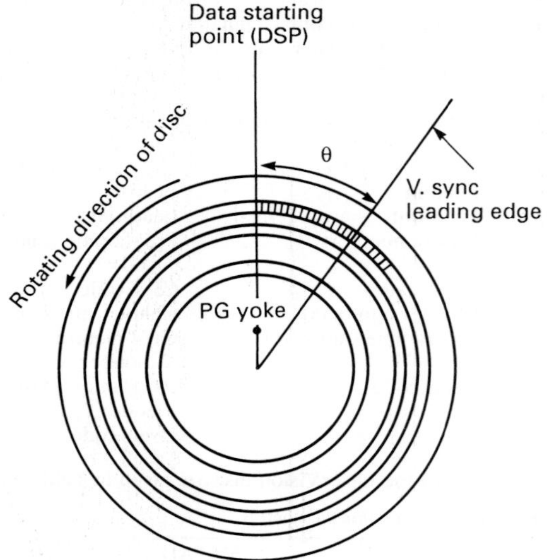

Figure 57.6 Floppy disc track allocation, showing the track pattern of the recorded side. For a 525-line, 60 field system, $\Omega = 7 \pm 2H(9°36' \pm 2°45')$. For a 625-line, 50 field system, $\Omega = 7 \pm 2H(8°4' \pm 2°18')$

Developments are in hand to produce an erasable and re-recordable LaserVision disc, but this is unlikely to be marketed to the general public.

57.3 CD-ROM

Compact discs are now available on the domestic market to reproduce high quality audio, using digital coding techniques. This method could be used to reproduce other forms of information such as a software program for a computer. Such a disc is then known as a CD-ROM (compact disc — read only memory).

57.4 Computer floppy disc

A 'floppy' disc or diskette is used in home computers and word processors and could now be classified as a domestic product.

Single sided discs are available, but double sided are more general. Each side has a magnetic coating which contains one cue track and 50 information tracks. Some disc drives can read both sides simultaneously. These 50 tracks may be used for computer data, but alternative standards exist for the recording of video information on both 525/60 and 625/50 line/field standards.

The track allocations on the disc are shown in *Figure 57.6*; the physical dimensions of the tracks are as follows:

Disk outer diameter:		47 +0.1 mm −0.2 mm
Rotating speed	525/60:	3600 rpm
	625/50:	3000 rpm
Rotating direction:		counter-clockwise
Radius of *n*th track:		20 - (*n*-1) × 0.1 mm
Radius tolerance of track:		±0.014 mm
Radius of 1st track:		20 mm
Radius of 52nd track:		14.9 mm
Track pitch:		0.1 mm
Track width:		0.06 + 0 mm − 0.006 mm
Data starting position 525/60:		0° ±1°22′
625/50:		0° ±1°9′
Azimuth angle:		0°
Azimuth tolerance:		±6′

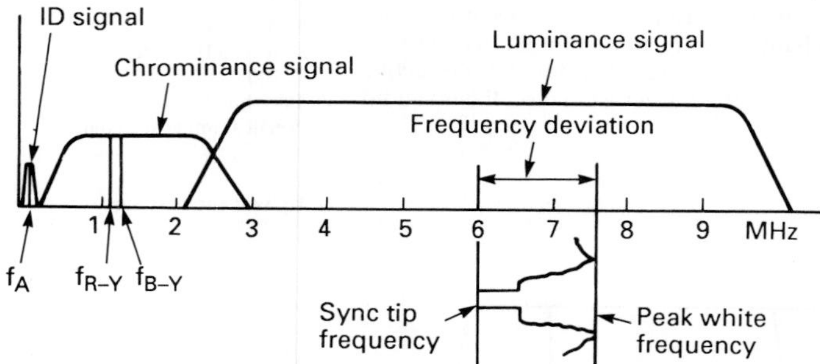

Figure 57.7 Frequency allocation of recording signals. f_A = carrier frequency for id; f_{R-Y} = fm R−y centre frequency, f_{B-Y} = fm B−Y centre frequency

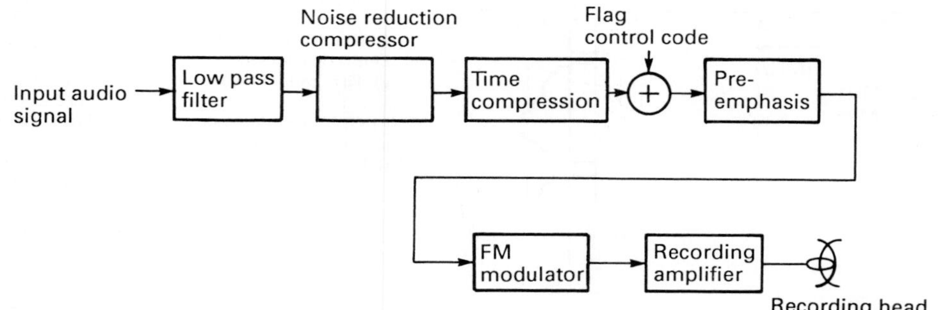

Figure 57.8 Audio recording circuit

It will be seen from *Figure 57.7* that the chroma information is recorded by the colour-under method, recording R-Y and B-Y as two separate signals, which are combined alternately into a line-sequential chrominance signal. The centre frequency of R-Y is 1.2 MHz and of B-Y is 1.3 MHz. Each track records one field, so that two consecutive tracks comprise one frame.

The id signal shown in *Figure 57.7* with a centre frequency of f_A is an identification code for logging data such as track numbering and dating. The carrier frequency f_A is fixed at 13 times the horizontal sync frequency i.e. 204.54 kHz for 525/60, and 203.13 kHz for 625/50 standards.

Audio signals may be recorded using a time-compressed analogue format. The block diagram of the recording circuit is shown in *Figure 57.8*, and the track format in *Figure 57.9*.

The time compression ratio for a 525/60 system is 320, 640 or 1280, and for a 625/50 system 272, 544 or 1088. Each track shown in *Figure 57.9* has four sectors, with five 'spaces'. The signal arrangement in each sector is shown in *Figure 57.10* and the carrier frequencies are as shown in Table *57.2*.

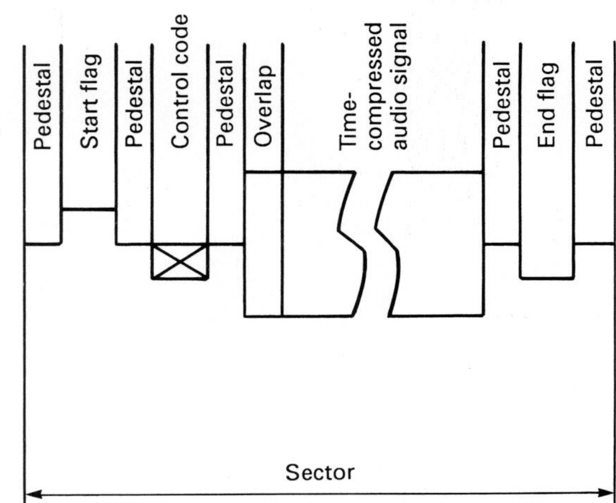

Figure 57.10 Signal arrangement for each sector. Each sector has an overlap portion which contains the same signal as the last part of the preceding sector in the sequence

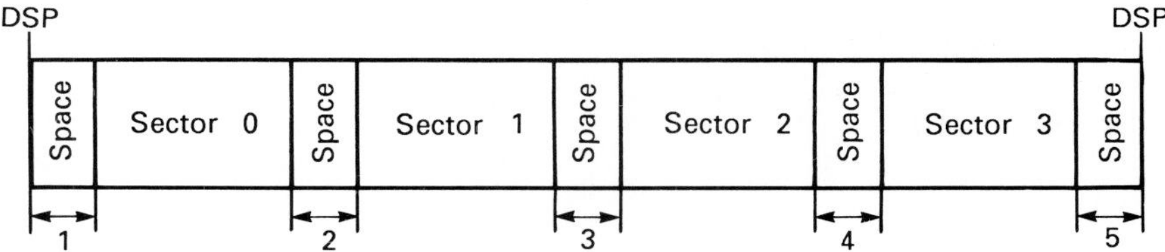

Figure 57.9 Track format

Signal portion	*525-line/60 field system*		*625-line/60 field system*	
Time-compressed audio signal				
Centre carrier	6 ± 0.15 MHz (f1)		5 ± 0.12 MHz (f2)	
Reference deviation (at 400 Hz)	± 2 MHz	(320)	± 1.7 MHz	(272)
	± 1.5 MHz	(640)	± 1.3 MHz	(544)
	± 1 MHz	(1280)	± 0.8 MHz	(1088)
	( ): Time compression ratio		( ): Time compression ratio	
Maximum deviation	± 3 MHz		± 2.5 MHz	
Pedestal	f1		f2	
Flag				
High level	f1 + 1 ± 0.05 MHz		f2 + 0.83 ± 0.04 MHz	
Low level	f1 − 1 ± 0.05 MHz		f2 + 0.83 ± 0.04 MHz	
Control code				
'1' level	f1		f2	
'0' level	f1 − 1 ± 0.05 MHz		f2 + 0.83 ± 0.04 MHz	
Space	6 ± 0.3 MHz		5 ± 0.25 MHz	

Table 57.2 FM carrier frequencies

R Watson C Eng, MIEE, MRTS, MBKS
Consultant

58 Electronic Cameras

The first range of non-broadcast cameras to be developed were for the surveillance market. This application still exists, and the pickup device used to convert the image into a television signal is usually similar to that used in the domestic market.

A *surveillance* camera is often mounted in a fixed position, and consequently there is no requirement for a viewfinder on the camera. This siting may also be inaccessible for a tape recorder, and a centrally located recorder may be used which will be capable of recording the inputs from a number of cameras, switched in sequence. The following description of the types of pickup devices used in domestic cameras will in general apply also to surveillance cameras.

58.1 Domestic cameras

The essential requirements of a *domestic* television camera are that it should be compact, lightweight and simple to use, it should have a built-in viewfinder, and be able to be coupled to a portable tape recorder.

The development of the portable video tape recorder, using half-inch wide tape, in the early 1970s stimulated a demand for an accompanying camera. This was met by the mid-1970s with the monochrome open reel EIAJ-I standard portable recorder and monochrome vidicon camera in the style of the 8 mm cine camera.

Since then development has been very rapid with the portable colour cassette recorder replacing the open reel recorder by 1980, with initially two-tube colour cameras.

The two-tube colour camera quickly disappeared as the performance of the single-tube camera improved, then the cameras became smaller as the one-inch diameter vidicon was replaced by the ⅔ inch and then half-inch tube.

An important innovation in 1983 was the VHS compact cassette. This still used the standard VHS recording format, but the cassette was considerably smaller, limited to 30 minutes playing time, using a high grade tape. This compact cassette could be played back in a standard VHS machine by placing it in an adapter. The obvious advantage of this smaller cassette was that the volume of the recorder could be substantially reduced to the point where in 1984 the recorder could be combined with the camera (JVC GR-C1). A similar combined *camcorder* was developed for the Beta system, but using the standard size cassette (Beta Movie). Both systems have an ingenious head drum assembly (see section *56.8*).

The mid 1980s saw the vacuum camera tube challenged by the solid-state chip. The advantages of the chip are its small size, enabling more compact camera/recorder combinations to be developed, an ability to handle a wide contrast range, and low power consumption and virtually infinite life. One of the early single chip domestic colour cameras was the Hitachi VK-C3000. This used a chip with 384 horizontal elements and 485 vertical elements.

Camera tubes were still popular until 1987 when a new range of camcorders were introduced using *interline charge-coupled device* (ilccd) sensors. These gave similar picture quality to the tube cameras, but were more compact and had very low power consumption.

Automatic light control, by the use of an automatic iris circuit has always been a feature of the domestic colour cameras. In the early 80s, auto-focus was available for the top-of-the-range cameras. Various methods of automatically focusing by means of triangulation optics, infrared and ultrasound radar were tried. Most domestic cameras use one of these methods, depending on the cost of the system. The ultrasonic method is not popular as it cannot function through a glass window.

The colour temperature of a light source, quoted in degrees kelvin, gives an accurate measurement of the ratio of red to blue energy in an apparently 'white' light. The human eye is very tolerant of the very wide change in the colour of an object when the illumination is changed from, say, daylight to artificial (tungsten filament) lighting. Photographic film and television cameras do not have this tolerance and they have to be corrected for the ambient lighting. In a photographic camera this is done by using the correct film (daylight or artificial) and/or using correction filters in front of the lens. In a television camera, filters can be used for major corrections, but precise correction can be achieved by adjusting the sensitivity of the red and blue colour channel amplifiers.

Daylight (5000°–6000° K) has roughly equal energy for red, green and blue whereas tungsten (2800°–3200° K) is rich in red but low in blue. A television camera balanced for daylight but used in tungsten lighting would reproduce very 'warm' pictures, i.e. they would be too red. This can be corrected in practice by changing the colour filter in front of the camera or changing the relative amplification of the red and blue camera channels. In most domestic cameras this can be accomplished by simply pressing the *white balance* button, with the camera focused onto a white card.

The television camera alone has virtually no capacity to store pictures. They can be fed directly from the camera to a picture monitor or receiver, but if reproduction at a later date is required, a video recorder must be used. The television signal can be conveniently stored on magnetic tape, but certain steps must be taken to ensure that the picture can be reproduced later in the original form. The signal from the camera contains synchronizing pulses, and these must be preserved in the recording. In addition, the tape must play back at the correct speed, and this is determined by the *control track* pulses on the tape. It is therefore necessary to maintain a precise relationship between the camera synchronizing pulses and the control track pulses.

Control track pulses on video tape can be likened to the sprocket holes on cine film. If either are missing, the reproduced picture will be unstable.

When a television camera is used with a recorder, either as a combined unit or two separate items cabled together, the television camera will be producing pictures continuously, but they will be recorded intermittently only upon a command from the cameraman. It is essential that these discrete recordings butt onto each other correctly if the resulting playback is not to suffer irregularities at each start-up point. A smooth transmission is achieved by the *back spaced edit* facility in the recorder. This is necessary because the recorder cannot instantly reach the correct speed at the moment of switch-on.

Back spaced editing is achieved by running the tape back for a second, or a fraction of a second, each time the recording stops. At the start of the next *record* command, the recorder starts in the *play* mode and reproduces the last second of the previous recording as it gathers speed, and falls into step with the previous control track pulses. At this point it is ready to switch from *play* to *record*, and lay down onto the tape the new picture and associated control track (see section *38*).

New cameras are introduced onto the market at a steady rate so that type numbers soon go out of date. However, any camera will probably continue to give satisfactory service for 5–10 years.

Many of these cameras use the same type of camera tube or chip, so a description of camera tubes and solid-state pickup devices is more appropriate than a description of particular cameras.

58.2 Camera tubes

58.2.1 Types

Single-tube colour cameras reproduce the colour by either a striped colour filter on the faceplate or a tri-electrode faceplate providing three separate colour signals when scanned by a single beam of electrons.

The material used on the faceplate to convert the light to electrical signals distinguishes the main tube types. The five principal ones are:

Vidicon This has its photosurface composed of antimony trisulphide. It is inexpensive to make, with a gamma of approximately 0.5, capable of handling a wide range of light levels by automatic adjustment of the target voltage. Its main disadvantages are the likelihood of burning the photosurface, if exposed to very bright light, and the *lag* characteristic. Lag is the residual storage in the photosurface which reproduces a low level signal for a number of frames after the illumination has ceased. The effect is to produce smeary pictures on moving objects, particularly at low light levels when the vidicon is working at high sensitivity.

Saticon This Hitachi tube has a photosurface composed of amorphous selenium-arsenic-tellurium. It is more expensive than the vidicon, with unity gamma. Contrast range is restricted to about 50:1, with no sensitivity control. Its main advantage over the vidicon is its ability to handle highlights with considerably less lag.

Chalnicon This is a Toshiba product with a cadmium selenium photosurface, 20 times more sensitive than the vidicon, with high resistance to 'burn-in'.

Newvicon This Matshushita tube is similar to the chalnicon, with 20 times the sensitivity of a vidicon.

Plumbicon This Philips tube with a lead oxide photosurface is also made by English Electric Valve Company under the trade name *Leddicon*. It has similar sensitivity to a Saticon, but reduced lag at low light levels. It is generally too expensive to be used for domestic single-tube cameras.

58.2.2 Colour

The tri-electrode faceplate did not find favour in the domestic market, as the striped faceplate was able to give adequate results at a lower cost. In the earlier tubes, the stripe filter was mounted on the outside of the faceplate of the camera tube, but as production techniques improved it was possible to deposit the filter on the inside, in close contact with the photosurface (*Figure 58.1*).

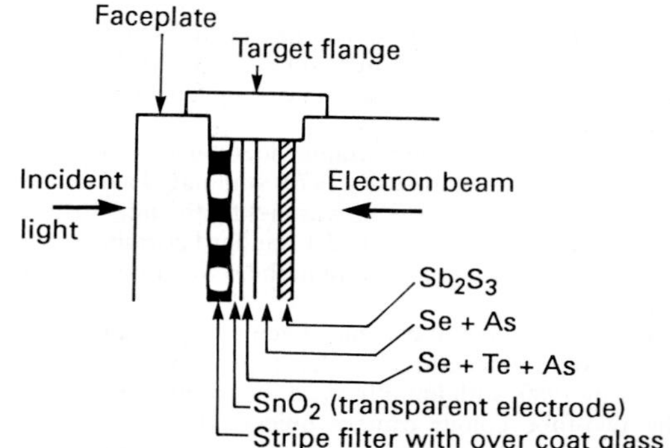

Figure 58.1 Structural diagram of target

Several different colour stripes have been used, but the following description is typical of a single-tube stripe filter decoding system.

The striped filter is positioned on the inside of the faceplate of the camera tube as shown in *Figure 58.2*. The stripes are laid at an angle of 25°24′ to the vertical as shown in *Figure 58.3*(a), using yellow filters that pass red and green wavelengths and cyan filters that pass blue and green. Where these filters cross, only green light is passed, and conversely, where no filter is

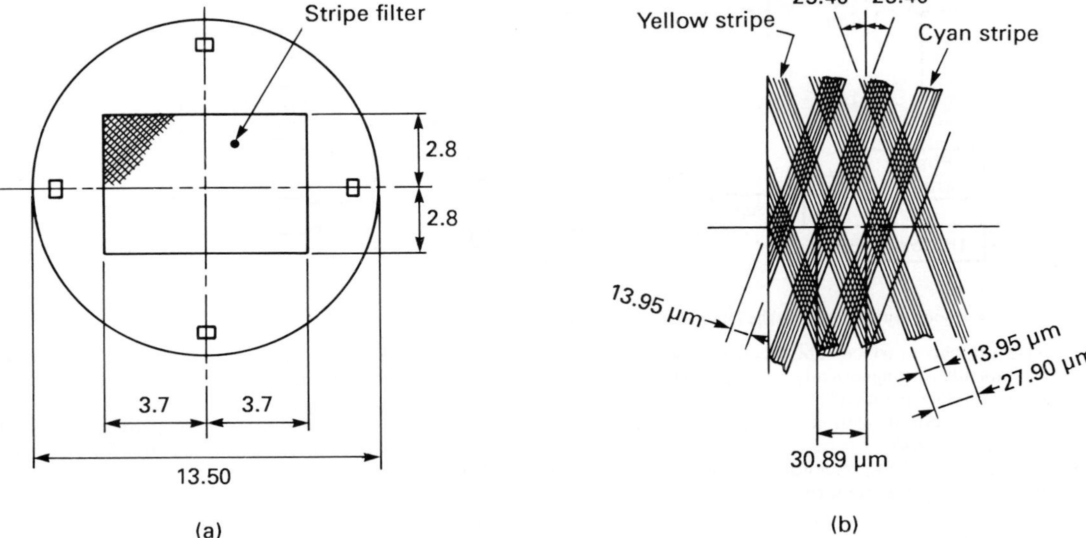

Figure 58.2 Half-inch Saticon with built-in stripe filter: (a) faceplate, (b) stripe filter

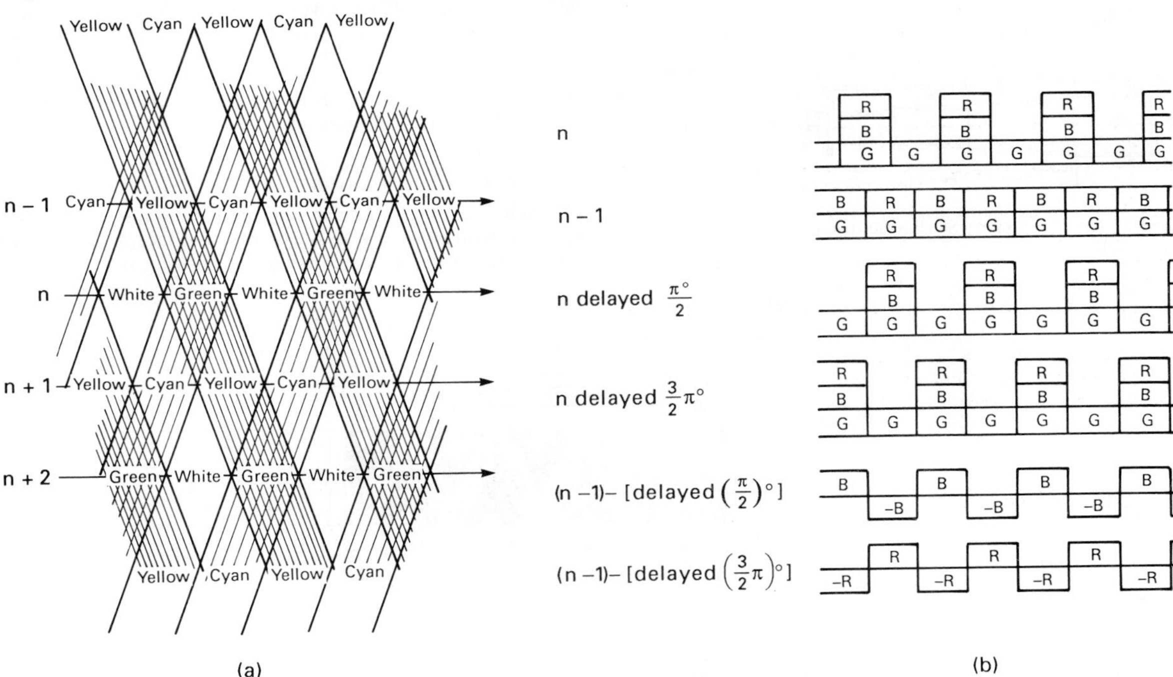

Figure 58.3 A striped filter as applied to the inside of the faceplate of a camera (a), and the resulting colours (b) that are read out by the camera tube beam

present, all three colours, red, green and blue, are available. The resulting colours that are read out by the camera tube beam are shown diagramatically in *Figure 58.3*(b).

These vary line by line and, by using delay lines and sum and difference circuits, it is possible to decode the red, blue and green signals.

The pitch of the striped filter is chosen to modulate the colour component at a frequency of 3.9 MHz, so that it is possible to separate the luminance from the colour component by passing the luminance signal through a low pass filter (lpf) as shown in *Figure 58.4*. This figure also shows the circuit blocks required to produce the standard composite colour signal. It will be noticed that a component output of Y and (R-Y)(B-Y) could also be produced. This component output signal is required in the S-VHS camcorder, for example.

58.2.3 Operation

The various types of photosurfaces described in section *58.2.1* all require the same form of electron beam to reproduce the

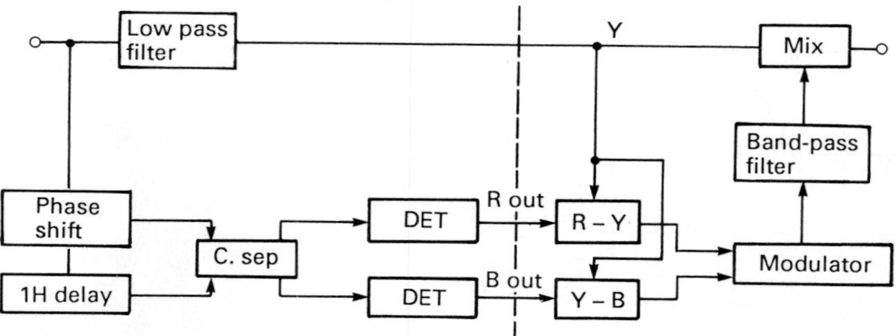

Figure 58.4 Separating the colour component from the luminance component

signal and discharge the photosurface. The electron beam can be focused electrostatically or magnetically, and also deflected across the photosurface electrostatically or magnetically. It is possible to use a hybrid method, with electrostatic focus and magnetic deflection or vice versa.

The advantage of *electrostatic* focus and/or deflections is that the external magnetic coil is not required, so saving power, space and weight. The disadvantages are that the camera tube is more complex, and possibly more expensive, and a better focus can usually be obtained using a *magnetic* focus coil.

Figure 58.5 shows a camera tube using electrostatic focus and magnetic deflection, and *Figure 58.6* shows one with magnetic focus and electrostatic deflection.

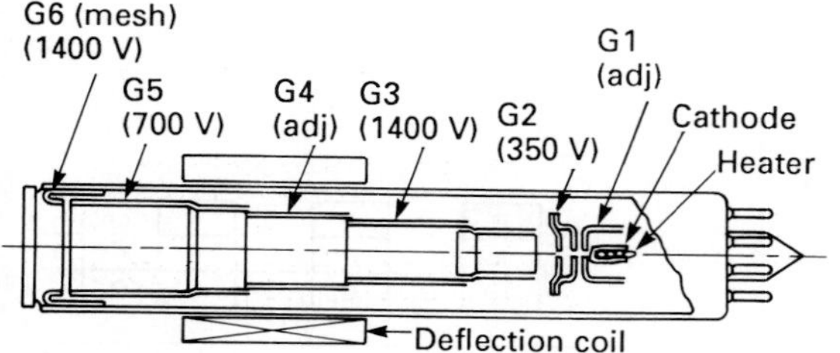

Figure 58.5 Cross-section of a Saticon tube using electrostatic focus and magnetic deflection

58.2.3.1 Focus modulation

When the beam scans the photosurface, there is a slight difference in the distance travelled between the centre and the four corners. This gives rise to a difference in focus between the centre and edges which can be corrected by applying a small parabolic correction waveform to the focusing electrode (usually G3 or G4). This improves the corner focus and consequently the colour reproduction.

58.2.3.2 Automatic beam control

Highlights, such as specular reflection off water and glass, produce small areas of gross overload on the faceplate that can lead to smearing or comet tailing on moving objects. This

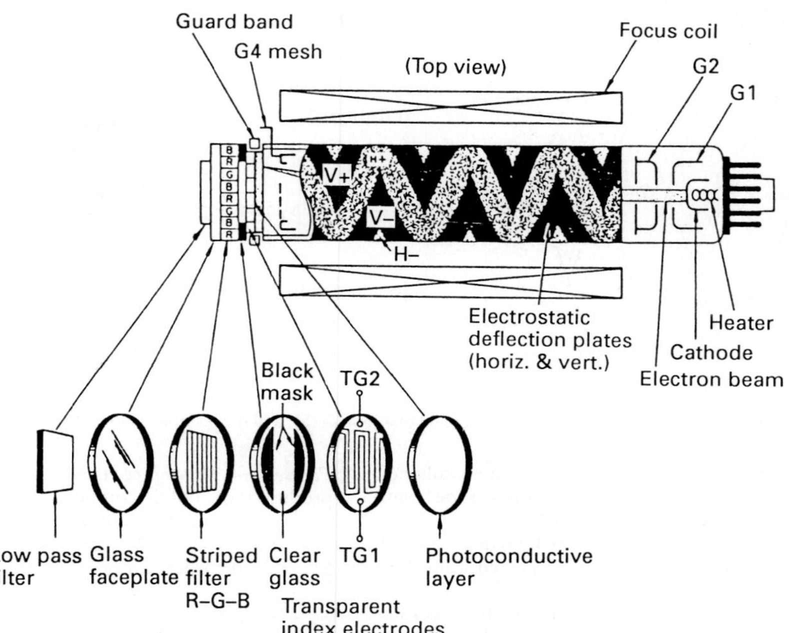

Figure 58.6 A conventional MF Trinicon with indirect heater

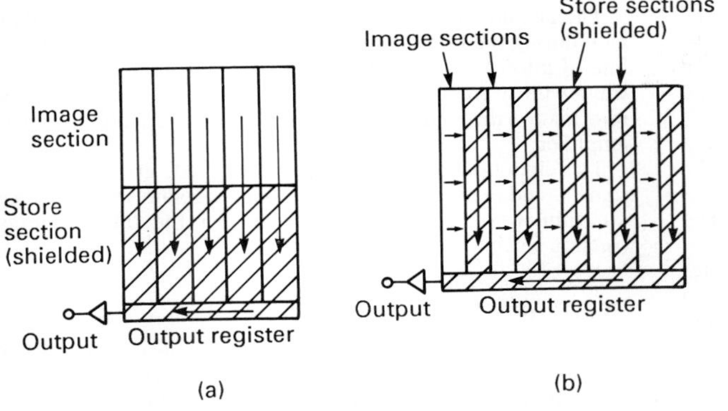

Figure 58.7 Frame transfer (a) and interline transfer (b) chip configurations

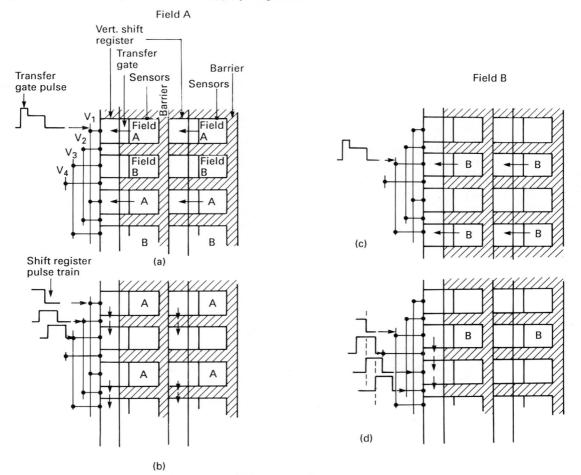

Figure 58.8 Sequence of interline transfer: (a) and (b) show field A, (c) and (d) field B

undesirable effect can be reduced or eliminated, by momentarily increasing the beam current to discharge the highlight. The beam focus will deteriorate, but this is usually not noticeable because the highlight obscures the resolution.

A circuit is included in most cameras to increase the beam current by applying a pulse to the grid G1 to discharge a highlight of eight times (three stops) peak white.

58.2.3.3 Scan failure protection

There is a risk of damage to the photosurface if the horizontal and/or vertical defection circuits fail. To prevent this damage, a safety circuit may be included to cut off the beam current should a scan failure occur.

58.3 Solid-state camera devices

Most domestic cameras use a ccd (charge-coupled device) chip now instead of a camera tube, as the chip requires less space and lower power consumption.

There is less difference between a tube and a chip camera than may be thought. Both are analogue devices, i.e. when the light falls on the photosurface the amplitude of the signal produced is directly proportional to the incident light. The sensitivity is also similar.

Charge-coupled device refers to the method used to transfer the charge on the photosensitive picture element to the video preamplifier. From then on the signal processing in both the chip and tube cameras is very similar. There are two basic variations in the configuration of ccd camera chips; *frame-transfer* (ft) and *interline-transfer* (il). The latter is generally used for single tube colour cameras (see *Figure 58.7*).

It will be noticed that, in the interline chip, the photosurfaces and transfer elements are arranged in alternate vertical lines. During the field vertical interval, the charge patterns on the photosurfaces are quickly transferred to the adjoining transfer element, discharging the photoelement ready to build up the new charge during the next field. *Figure 58.8* shows the construction of a pixel in more detail.

At the start of the vertical interval, the transfer gate 'opens' by being driven more positive than the photosurface and thus attracting all the photo-generated electrons through the transfer gate and into the adjoining vertical register store. A series of pulses applied to the vertical register 'directs' the store signals line by line to the video amplifier. Fields A and B both use the same shift register, but each field can be identified by the transfer gate pulse, which is applied to the V_1 electrode at the start of field A, and to the V_3 electrode at the start of field B.

The signals are transferred through the vertical shift register by a train of four pulses, each overlapping the previous pulse by about half a line, to cause the charge pattern to shift through the register without any dilution of the original charge.

The last line of the vertical register is switched by a further gate during the horizontal blanking period to load a complete line of information into the horizontal shift register. A pulse train with a frequency of at least 5 MHz is used to clock information through the horizontal shift register, into the video amplifier.

The circuits used to separate out the red, green and blue signals, and then adjust the amplitude to obtain *white balance* are similar to the circuits described for the tube cameras (section *58.2*).

58.3.1 Auto-focus

The *active* method uses a signal generated in the camera (either ultrasonic or infrared) and measures the distance of the subject using the reflected signal. The *passive* method relies solely on the light entering the camera from the subject.

Figure 58.9 shows an active system. Using an infrared beam, the distance of the subject is calculated by trigonometry; using ultrasonic pulses it is calculated by measuring the time taken by the reflected signal.

Passive systems are more complex, but are electronic equivalents of the split image focusing systems used in many still photographic cameras. In the viewfinders of these cameras, a small central area is split horizontally, and the vertical lines appear to move from left to right as the lens is focused. Correct focus is achieved when the upper and lower images are coincident.

In the television camera the same split image method is used, but the eye is replaced by a series of lenticular lenses and pairs of photodiodes. When the images are correctly aligned, the same signal level will be measured by each diode pair.

In the unfocused condition, the diode pairs will be unbalanced, and this condition will generate signals to drive the focus servo motor to seek a balanced signal, and consequently the optimum focus condition.

The sensor can be a true through-the-lens system as shown in *Figure 58.10*, or a separate coupled lens as in *Figure 58.11*.

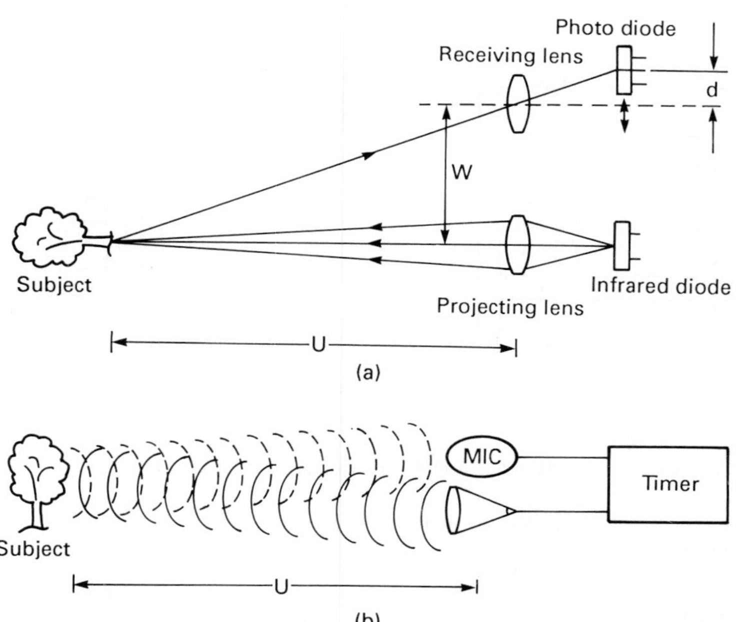

Figure 58.9 Active auto-focus system: (a) using an infrared beam, (b) using ultrasonic pulses. U is the distance of the subject from the lens, W is the distance between the transmitting and receiving lenses and d the displacement

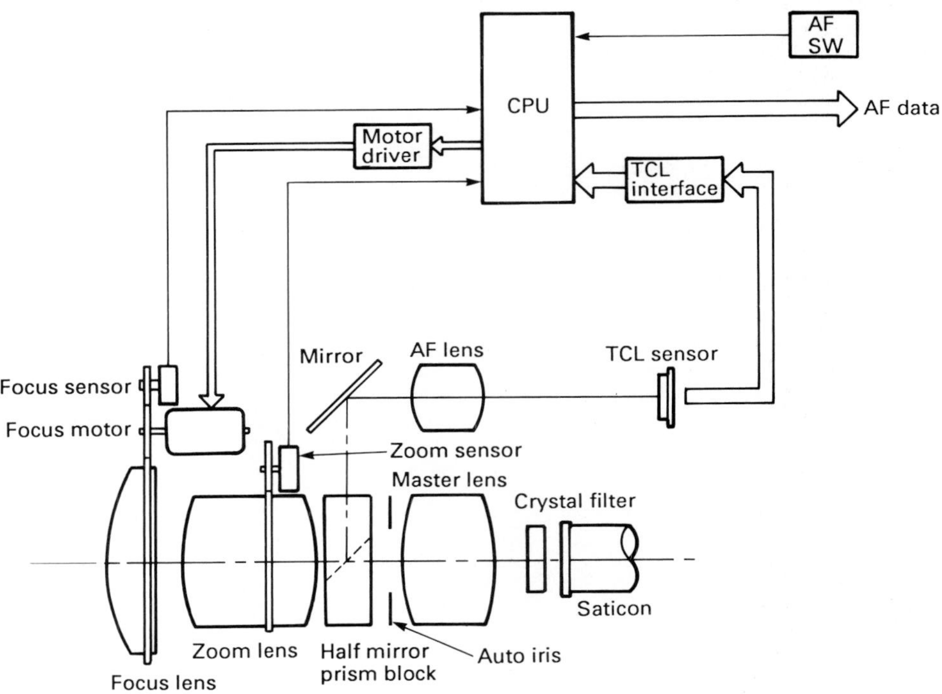

Figure 58.10 Through-the-lens auto-focusing

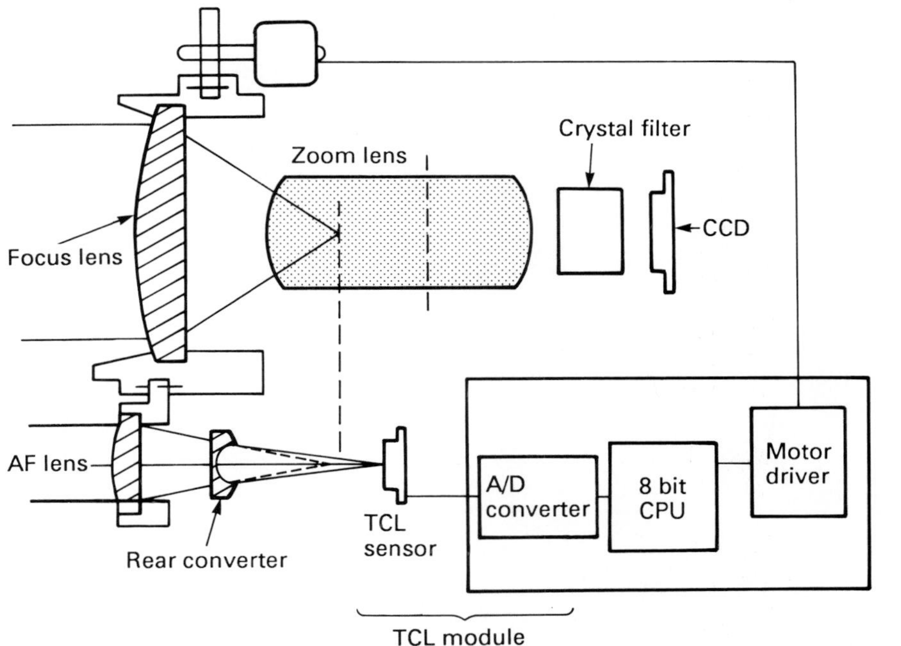

Figure 58.11 Auto-focusing with a separate coupled lens

R Elen
Creative Technology Associates

59

Audio Recording and Playback

The field of audio recording is an extensive and complex one, but the worlds of sound and vision are so interlinked, particularly since the advent of interactive multimedia systems such as LaserVision and Compact Disc Interactive (CDI), and the parallel development of audio compact discs and cd video, that it is important for a TV and video engineer to have at least an overview of the field.

Earlier types of equipment, specifically open-reel recorders, are essentially identical to those in professional use and discussed in Part *9*. Consumers utilizing open-reel machines today may well be using exactly the same equipment as may be found in the TV sound studio.

Consumer standards such as quarter-track on quarter-inch tape have fallen into disuse with the exception of the so-called *semi-pro* machines used in the 'home studio' environment, where such techniques are provided for multi-track recording. These are also to be found in the professional TV sound environment (see Section *45*).

The traditional audio sources in the home have been changing dramatically. The vinyl record has been overtaken rapidly by the digital compact disc which is now the largest-selling 'album format' medium. The newest source, for better or worse, is digital audio tape.

Digital audio, in the shape of pcm (pulse code modulation) digital recording as utilized in the compact disc audio and video formats, Video 8, and digital audio tape (DAT or R-DAT), is therefore the central focus of this section, and these applications will be considered in detail.

59.1 Background to consumer audio

Modern consumer audio recording and reproduction techniques need to be placed into an historical perspective, so that we can see the origin of all the current techniques, and note common ground where audio developments with professional implications share their origins with more prosaic consumer developments.

Strangely, audio recording has been possible for a good deal longer than audio playback. Nearly 150 years ago, it was discovered that sound signals could be recorded graphically by attaching a diaphragm to a suspended needle or brush which traced a track on a rotating drum covered in paper coated with lamp-black or carbon. But there was no way of replaying the information. The technique did a great deal to indicate the nature of sound, but it was not until Edison's invention of the phonograph (demonstrated first in 1877) that recording *and* reproduction of audio became possible.

The phonograph in principle used the same technique, but by inscribing the waveform with a needle or stylus in a surface such as wax (or in some cases metal foil), it became possible to track the recorded information with another stylus fitted with a diaphragm and amplifying horn attached. The waveform recorded by a phonograph was represented by changes in the *cutting depth* of the recording stylus into the medium. This required the medium to be of appreciable depth, but allowed the recording cylinder to be advanced by a constant amount per revolution on a lead screw of fixed pitch. This is known as 'hill and dale' cutting for obvious reasons. Had the recorded signal been captured by scoring grooves that altered laterally with amplitude, the recording time would have been reduced as the lead-screw pitch would have needed to have been greater to allow for wide excursions of the cutting stylus.

As use of the phonograph spread, mass-produced pre-recorded cylinders were made possible by the use of moulds made from the original cylinder. The pattern of ridges on the moulds was used to create an image of the original, in wax or later in shellac — a technique which is still used today, not only for vinyl records but also for compact discs.

Around 1900, Emil Berliner came up with his own novel method of recording audio signals — a flat disc-based system. His technique laid the foundations of the gramophone, and by World War I the flat disc was becoming pre-eminent, initially with one-sided but later two-sided lacquer records. With the greater surface area offered by a flat disc, some companies used lateral cutting while others retained the original 'hill and dale' or vertical technique. Both, however, retained fixed-pitch lead screws on the cutting lathes of the time. Domestic replay equipment was able to reproduce both types of disc thanks to the fact that the only method used to ensure that the disc was tracked correctly was the weight of the replay stylus (generally available in different grades for differing sound quality, and made of anything from steel to rose thorns) which dragged the whole replay assembly in towards the centre of the disc as it played.

These shellac discs are commonly called *78s*, and they are

generally regarded as having rotated at 78 rev/min. This standardization did not come quickly, however, and speeds of anything from 70 to 80 rev/min were common for several years, hence the speed control on many gramophones.

Early recordings were as likely to be recorded from the centre to the edge as the other way around, and although clockwise rotation was agreed relatively early on, standardization of recording from edge to centre came rather later. An interesting feature of the spiral groove on a record is that, as the stylus tracks the disc, the actual linear velocity of the track passing under the stylus changes, being higher at the edge and lower towards the centre. The higher the speed, the better is the audio quality, particularly in terms of background noise. When it came to recording longer works which required several sides, it became common practice in the broadcast industry to minimize the change in audio quality between sides by alternating between edge-to-centre and centre-to-edge. So the first disc would play inwards, for example, while the second played outwards, and so on.

Meanwhile, in Scandinavia, Valdemar Poulsen came up with a new method of recording and reproducing sound. He passed a steel wire past an electromagnet driven by a telephone-type microphone, and replayed the recording by passing the wire across the same *magnetic head* connected to an earpiece. The field of the head during recording made the previously randomized magnetic domains of the wire follow the changes of amplitude of the input signal, and these magnetic changes induced varying currents in the replay head similar to those used while recording. The *Telegraphone* system became used for instrumentation purposes for many years, but it was the development of the tape recorder that represented the major implementation of this technique.

In the 1920s and 1930s, at least two organizations began to experiment with multi-channel recording as a means of producing a spatial effect for the listener: Bell Laboratories in the USA and Alan Blumlein at EMI in Britain. Discrete multi-channel audio was also being developed for film use, for example by Disney for *Fantasia*, but the techniques in use there (based on optical encoding techniques) have never had any implications for the home audio environment.

Stereophony, as it came to be called, had in fact been demonstrated as early as 1881, where stereo pairs of telephone lines were used to relay the performances at the Paris Opera to listeners in the hall of a major exhibition of the time, but Blumlein developed the techniques to a height that has never been bettered. Using a pair of microphones located at the same point in space, he was able to represent a natural sound field very accurately, while the American researchers were using spaced microphones and achieving dramatic, but unnatural results. Developments of Blumlein's ideas into three dimensions in the early 1970s led to the introduction of the British *Ambisonic* surround-sound system, which is now used in a number of applications.

59.1.1 Disc recording

The researchers at Bell Laboratories were also in trouble over the means of recording and playing back stereo information on disc. Their approach was to use two cutting and playback styli, and cut two concentric grooves in the disc surface. This was not very satisfactory. Blumlein, however, developed a technique of encoding the two channels into the same groove by recording left and right channel data at right angles to each other. This technique, called *45-45 recording* for obvious reasons, is still in use today, although the earliest stereo discs issued used the *vertical-lateral* system where one channel was recorded vertically ('hill and dale') and the other horizontally. The trouble with the latter was that the rumble from the recording lathe or playback turntable was more noticeable in the vertical direction.

In the 1950s, the two major forces in US music and recording were the two major forces in American broadcasting at the time: the Radio Corporation of America (owners of NBC) and Columbia Broadcasting Systems (CBS). Both companies used new developments in plastics technology to upgrade the traditional disc. RCA used vinyl and narrower recorded grooves to produce a seven-inch diameter disc that had the same playing time as a twelve-inch 78 rev/min disc, assisted by the fact that the lower noise level of vinyl and other factors made it possible to use a lower rotational speed of 45 rev/min. CBS, meanwhile, developed *microgroove* techniques to a higher degree and introduced the 'long play' (lp) record — a twelve-inch disc with the ability to store over 20 minutes of information per side by rotating at 33⅓ rev/min.

The storage capacity of these new media was further enhanced by the development of *varicutting*. This changed the width of the cutting groove according to whether the cutting stylus required to make a wide or narrow excursion. It required a method of 'looking ahead' to see when a loud passage was coming, and opening the groove pitch to cope with it. This was (and is) done by using an additional tape head on the master replay machine during cutting, ahead of the main audio head. The overall amplitude read by this head is used to control the groove spacing.

Such techniques extended the possibilities of the vinyl disc medium and remarkable increases in quality became possible. The final development was the introduction of *direct metal mastering* (DMM) in the late 1970s. Instead of cutting on conventional lacquer-coated aluminium blanks, which were then coated with a conductive layer, electroplated and the resulting 'mothers' used to produce stampers for the presses, DMM uses a copper blank which is used to make stampers more directly, missing out several intervening stages. The result is much higher quality and a longer playing time (up to 45 minutes per side). This means that it is possible to create a recording that can be issued on one DMM disc but needs two single-sided compact discs (although with developments in cd technology resulting in longer playing times, this may not be the case for long).

59.1.2 Tape recording

Meanwhile, analogue tape recording had been developing. In the 1920s, experiments had been made with steel tape rather than Poulsen's steel wire. The resulting *Blattnerphone* steel tape recorders to be found in broadcasting institutions during the 1930s were placed in special booths. These huge machines ran tape at 60 inches a second and occasionally the tape snapped, and could kill. Tapes were joined by welding or riveting, and as a result the machines often had more than one set of heads. When one set became damaged by a passing 'splice', the new set could be used instead.

Most of the research on tape technology was going on in Germany, including the development of better tape materials. BASF developed a quarter-inch wide paper tape base that was coated with a black iron oxide powder. The early *Magnetophon* machines used no bias, and as a result the recordings were not of exceptionally high quality. To provide the best quality, tape must be *biased* to run in the linear part of its transfer curve. Initially this was done by applying a dc voltage to the *record* head. There is an apocryphal story that an engineer was one day trying to repair a dc-bias Magnetophon when the recording amplifier went into ultrasonic oscillation. Suddenly the recording quality increased tremendously! True or false, this is how analogue tape recording is performed today: the recorded signal is mixed with a high-frequency bias signal, the bias oscillator also generally being used to erase the tape before recording.

Following the introduction of consumer open-reel recorders in the early 1950s, the decade saw various attempts to get more and more audio storage into smaller and smaller spaces. The original tape recorders had used quarter-inch tape at speeds as high as 30 inches a second (76 cm/s); this soon came down to 15 inches a second (38 cm/s) for professional use, and then $7^1/_2$ inches a second (19 cm/s). The original specifications were in inches. Acetate and then polyester tape substrates came into use, making magnetic tapes more resilient and enabling more playing time to be offered. The exact length of *standard play* tape generally available on a single reel ($10^1/_2$ inches in diameter, NAB hub) varies from one manufacturer to another, but the playing time is almost always more than 30 minutes at 15 inches a second.

Today, professional recorders generally use 15 inches a second, with $7^1/_2$ inches a second used occasionally in broadcast applications and 30 in some multi-track recording environments. However, in the consumer field, speeds and reel sizes continued to fall. $3^3/_4$ inches a second became common, followed by $1^7/_8$ and even $^{15}/_{16}$, although these speeds were little good for anything other than speech. Reel diameters fell from $10^1/_2$ inches to 7, $5^1/_4$ and 3 inches.

At the same time, manufacturers attempted to increase the playing time on a given length of tape by reducing the track width. Original tape recorders had offered a single monaural audio track across the width of the tape. For stereo, this track was divided into two, with guard bands to reduce crosstalk. The introduction of *four-track* (more properly referred to as *quarter-track*) divided the quarter-inch tape width into four: these tracks could be used in an 'out-and-back' configuration, either in mono (so that by exchanging spools at the end of each of four passes you could get four times the playing time) or in stereo, where the user was able to obtain double the regular playing time in two passes.

Tape thicknesses were reduced, allowing longer lengths to be fitted on the same sized spools. Reels of *long play* tape gave $1^1/_2$ times the playing time of standard play, while *double play* offered twice the playing time. There were even triple and quadruple play tapes available. The disadvantage of tapes thinner than long play, however, was that the very thin tape was easily damaged.

The next step came in the early 1960s. Some manufacturers, aware of the fact that changing reels and threading tape into spools was a hassle for many users, designed tape cartridges which could be clipped on to a tape machine, containing the permanently threaded feed and takeup spools. Once installed on a machine, they performed exactly as conventional reels. The ultimate extension of this idea came from Philips, who in 1963 introduced the *compact cassette*, which went on to become the most important consumer audio medium for some years.

59.2 The compact cassette

The Philips compact cassette took many elements from conventional tape recorders and included them, in mass-produced form, in the cassette housing. The spools do not have flanges; instead the tape is kept in place by lubricated foils between the upper and lower tape edges and the cassette housing. Tape passes from right to left, as is generally the case, but the tape is wound with the oxide on the outside, as in European broadcast practice. This enables the tape heads to be positioned from the outside of the cassette rather than the inside. The tape leaves the feed spool and passes round a (generally rotating) tape guide, then across the *erase* head which fits into a cavity in the shell, tape-to-head contact being created by tape tension alone.

The tape then passes in front of the main *record/play* head assembly, usually but not always a single tape head. This again fits into a cavity in the cassette shell, but here tape-to-head contact is ensured not only by back tension but also by means of a sprung pressure pad mounted in the cassette housing. This has traditionally been one of the factors limiting cassette quality. Behind the pad assembly is a metal screen which serves to reduce hum induction in the same way as the head shield on a conventional reel-to-reel machine.

The tape width in a compact cassette is one-eighth of an inch (again non-metric). The original cassette recorders featured a mono recording configuration, with the tape track taking up half the tape width (including guard bands, etc.) allowing for two tracks to be recorded, one in each direction. A stereo configuration was later introduced, in which the stereo pair was recorded in the space previously taken up by a mono track, again offering two passes per cassette and also offering mono/stereo compatibility. This has become the standard configuration for all but the cheapest machines.

More recent modifications of the cassette system have included the introduction (at the semi-pro end of the recording industry) of true four-track cassette machines in which four tracks are recorded across the width of the tape, all in the same direction. While cassettes normally run at $1^7/_8$ inches a second, in these cases the speed is often increased to $3^3/_4$ inches a second for improved hf response. These systems, aimed at the 'home studio' marketplace, usually offer the facility to record one or more tracks while listening to the others. A cassette recorder is also available providing a full eight tracks on conventional cassettes.

59.2.1 Noise reduction

Initially, cassette machines were most suitable for recording speech, but with the introduction of stereo and Dolby noise reduction, the cassette recorder began to be seen as part of hi-fi systems. The principle of the original *Dolby system* is to compress the signal being recorded when its level drops below a certain threshold, and then expand it again by an identical amount on replay. To minimize side effects of the *compansion* process, such as 'pumping' and 'breathing', the signal is split into five frequency bands which are treated separately. This system was originally developed for professional recording systems and is referred to as *Dolby A*.

For consumer audio applications, the system was simplified so as to treat only one band, the upper frequency ranges which contain the majority of audible noise, the rest being masked to a greater or lesser extent by the signal. This simplification, *Dolby B*, enabled the system to be incorporated into a single ic with a minimum of support chips. Dolby B offers a worthwhile overall noise reduction of around 10 dB.

The disadvantage of these early Dolby systems was that they required a reference level to govern the amount of compansion and the level of the threshold above which no compansion takes place. In professional systems, this is achieved by means of a sophisticated alignment procedure and the recording of reference tones. While such adjustments were present on the very first consumer Dolby units (particularly those made in the UK by Kellar) they were, and are, omitted on almost all cassette recorders. This means that the setup is defined at the factory, for a specific type of tape and for one IEC type. As a result, the performance of Dolby units in cassette recorders is seldom optimum, the most usual problem being incorrectly adjusted replay thresholds resulting in loss of hf information on playback.

Several attempts have been made to circumvent this problem, including the utilization of other noise reduction techniques borrowed from the professional arena, such as the *dbx* system. dbx uses compansion too, but uses a linear compression/expansion curve with a fixed 2:1 ratio at all signal levels. As a result there is no problem with reference levels (there is no need for them) but in exchange there are increased problems with compansion side effects, and above all, there is stereo image instability as the compression on each channel is independent.

Experiments were also made, particularly by Philips, into the use of 'single-ended' noise reduction systems, notable their DNL (dynamic noise limiting) system. The term *single-ended* indicates that the system is applied solely on replay, there being no need for a 'mirror-image' encode/decode process. DNL functions through the use of a frequency-agile low-pass filter, the 'knee' frequency of which is defined at any instant by the highest frequency in the input signal exceeding a set threshold level. The effect is therefore to attenuate substantially any noise higher in frequency than the highest significant part of the wanted signal. As in Dolby B, this results in the most noticeable noise band (the highest) being reduced. An apparent reduction in noise of up to 30 dB can be achieved (apparent because the noise has not been removed across the whole spectrum, only at higher frequencies).

The disadvantage of DNL as it is usually used is that problems arise when there is a large gap between the highest frequency present in significant amounts in the input signal and the next significant frequency. The DNL system can only remove noise above the highest wanted frequency, so noise below it is unchanged. Thus a musical passage consisting of basses and a glockenspiel, for example, would suffer from the glockenspiel modulating the noise content, as noise below the pitch of the glock notes would be left alone when the glock was playing, and removed when it was not. A solution to this is to combine a frequency-agile lpf with an expander, both set to operate at the same threshold levels. Both devices have their problems, which fortunately virtually cancel out, producing a composite signal processing algorithm that can be very effective. While this combination has been used in the professional audio environment, in units such as the *Dynafex*, it has not been exploited in the consumer field.

59.2.2 Tape types

Major advances in tape technology led to the introduction of a range of different coatings. A range of IEC specifications was introduced, dividing cassette tapes into four types. The original tapes were based on various ferric oxides (IEC type I) which required relatively low levels of bias and 120 μs equalization. Type II tapes, characterized by a higher bias level requirement and benefiting from 70 μs eq, are usually made from chromium dioxide (CrO_2 or *chrome*) or various blends of oxides including cobalt. Type III tapes had an intermediate coercivity level due to their formulation of a combination of CrO_2 and iron oxide (known as *ferrochrome* or *FeCr*), but due to problems with the stability of such formulations they are now rare. The final IEC type, type IV, is reserved for metal particle tapes which have the highest coercivity, and arguably the best combination of frequency response and noise level. They also require 70 μs eq and offer superior headroom.

The major current tape types can be selected automatically by the machine by the use of a series of sensing 'fingers' similar to that used to detect the presence of the *record enable* tab at the rear of the cassette housing. Type II tapes have an additional notch adjacent to the record enable tab, while type IV tapes have the additional notch near the centre rear of the housing. The presence of an additional notch in an appropriate position can be sensed and used to set up bias levels, headroom indications and replay eq time constants automatically.

Commercial pre-recorded tapes, now largely recorded on chrome tape, still however require 120 μs eq, and as a result do not possess an extra notch. As far as the replay machine is concerned they contain type I tape.

Other advances contributed to the quality levels achievable on cassette, notably the development of compact drive mechanisms and dc servos which made it possible to produce small motor assemblies that were able to maintain speeds with a low proportion of short and long term variation (known as *flutter* and *wow* respectively). This development has continued with the growth of the personal stereo and in-car entertainment markets.

59.3 Other cassette systems

The potential problems of achieving high quality levels within the cassette format led to some attempts to develop other formats that would offer better performance. One of the problems with the cassette concept was that many of the elements of the tape path, particularly the guide rollers, that required the highest manufacturing accuracy had to be mass-produced as a part of the cassette itself. Both Sony and BASF, with the *Elcaset* and *Unisette* respectively, sought to overcome these problems either by improving the quality of those parts or by utilizing a cassette design which made them external to the cassette and put them back in the recorder, where they came from. Both manufacturers also used quarter-inch tape and higher recording speeds. Neither format was successful. The Elcaset disappeared completely while the Unisette had metamorphosed into a professional broadcast format almost before it was announced, where it made an unsuccessful attempt to supplant the NAB cartridge which is still in current use for broadcast applications.

The NAB cartridge itself spawned a consumer format in the shape of the *eight-track* cartridge. In this format, back-lubricated quarter-inch tape was used on an endless loop with four stereo pairs across the width of the tape (i.e. eight tracks) all recorded in the same direction. The recorded material was split into four 'programmes', and a stereo head was stepped up and down to read the tracks by means of a solenoid activated by a piece of metallic foil at the start of the loop. This system relied not only on mass-produced tape guides in the cartridge but, unlike the cassette or the NAB cartridge, also on a mass-produced pinch wheel. This, coupled with the use of a tape head stepped up and down mechanically, produced dire results, and the format only had serious success in the in-car field due to its convenience.

59.4 The digital audio disc

A major development in 1982 was the first successful consumer digital audio format: *compact disc*. Developed initially by Philips but completed jointly with Sony, the format is now the primary commercial format for pre-recorded material.

Philips started research into optical disc-based audio recording in 1974, and it soon became clear that the analogue method used in the VLP/LaserVision videodisc system would not be good enough. The answer lay in pulse code modulation: *pcm digital audio*. (In fact, the final specification for the NTSC version of LaserVision includes digital audio tracks, but this was not originally possible in the PAL version.) A provisional agreement was made with Sony in 1979 to continue the work, and the final results began to appear from both companies in 1982, with a commercial release in 1983 of the compact disc (cd) digital audio system.

In the LaserVision system, data are recorded in analogue form on a spiral track consisting of a series of 'pits' separated by 'lands'. Each transition from pit to land or vice versa marks a zero crossing of the modulated video signal. The standard 120 mm diameter cd uses a similar system, but here the data are stored digitally; the transitions represent binary 1 while all the channel bits in between are 0. The compact disc video (CDV) system is essentially a reworking of the LaserVision format and still stores the *video* information in analogue form.

The pattern of pits is tracked by an AlGaAs laser beam which also employs sophisticated servo systems to ensure perfect focus as the disc rotates. The laser beam is projected through the transparent polycarbonate substrate material (see *Figure 59.1*), generally from below, and is reflected from the pits thanks to the fact that the upper surface of the substrate is coated with a reflecting layer, generally of aluminium but occasionally of other substances such as gold. A protective layer of varnish is applied to the reflecting layer, and the label information is printed directly onto the varnish surface. The majority of cd pressing plants ensure that the silvering does not extend to the very edge of the disc, and that the varnish layer does, thus protecting the silvering from edge damage. The fact that some early discs did not include this protection led to a brief scare in 1987 that cds would have a limited life. This has proved to be unfounded, and is now almost universally avoided by a simple change of manufacturing practices.

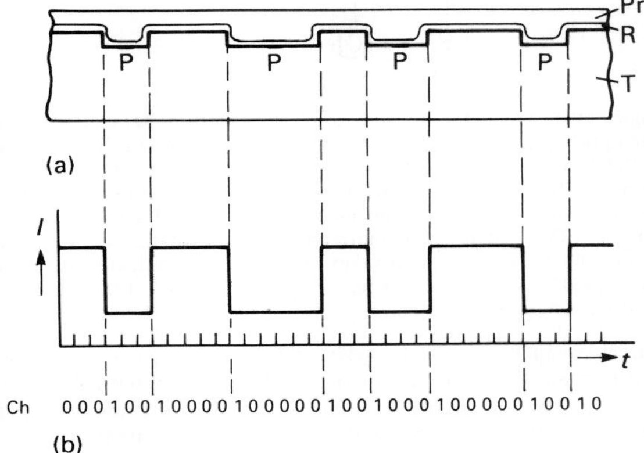

Figure 59.1 (a) Cross-section through a compact disc in the direction of the spiral track. T is transparent substrate material, R reflecting layer, P_r protective layer and P the pits that form the track. (b) *I* is the intensity of the signal read by the optical pickup plotted as a function of time. The digital signal derived from this waveform is shown as a series of channel bits (Ch)

Beam focusing is achieved by one of two methods in most players. The most effective, employed by Philips, is to use the main read beam to provide its own focus and tracking error signals by means of a beam splitter and photodiode array (see *Figure 59.2*). When the beam falls on a land area, almost all the light is reflected back to the photodiodes, while when it falls on a pit (whose depth is about one-quarter of the beam wavelength of 800 nm) interference significantly reduces the amount of light falling on the photodiodes. Summing the diode outputs therefore yields a fair approximation to the digital signal required. Analyzing the light pattern at the photodiodes also provides the tracking and focus error signals. Other manufacturers split the beam, typically into three, and use these to derive the focus and tracking error signals. This method is less effective as the focus and tracking signals can be affected by variations in the refractive index of the disc, e.g. as a result of stress patterns in the substrate.

CD uses a very high data storage density. A data bit takes up

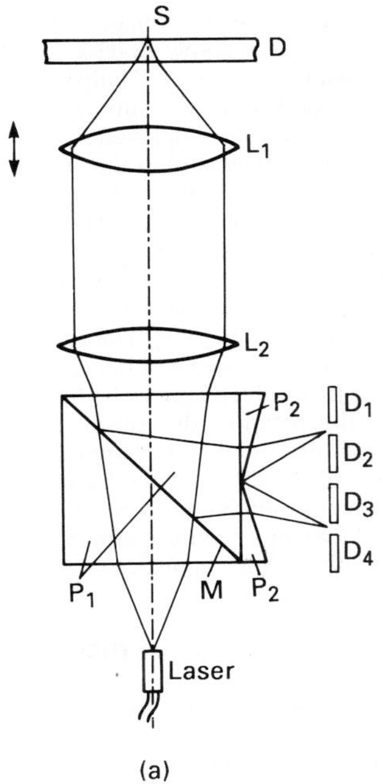

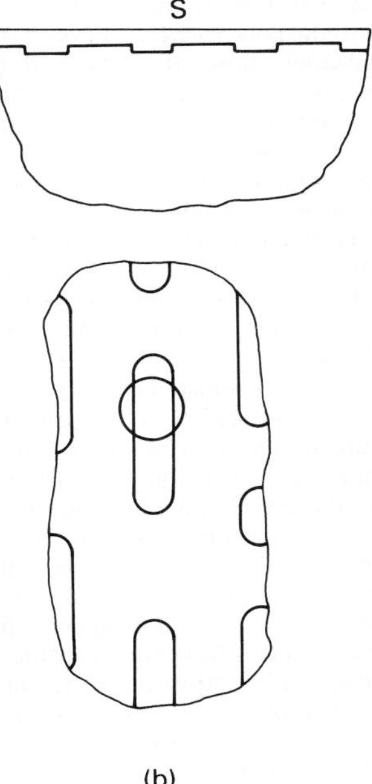

Figure 59.2 The optical pickup is shown in (a) where D is the radial section through the disc, S the laser spot, the image on the disc of the light-emitting part of the semiconductor laser, L_1 is the objective lens, adjustable for focusing, L_2 the lens for making the divergent laser beam parallel, M a half-silvered mirror formed by a film evaporated on the dividing surface of the prism combination, P_1 and P_2 beam-splitter prisms, and $D_1 - D_4$ photodiodes whose output currents can be combined in various ways to provide the output signal from the pickup and also the tracking-error signal and the focusing-error shown in (b). It can clearly be seen that the diameter of the spot (about 1 μm) is larger than the width of the pit (0.6 μm)

to 1 μm^2 on the disc, and the beam diameter is only 1 μm. The track pitch is 1.6 μm while the width is 0.6 μm and the depth 0.12 μm. The track is optically scanned from centre to edge at a constant linear velocity of 1.25 m/s. As a result, the speed of rotation of the disc varies from about 8 rev/s at the centre to about 3.5 rev/s at the edge.

The 16-bit stereo audio signal is recorded at a sample rate of 44.1 kHz (theoretically allowing for a frequency range from dc to 20 kHz). This results in a signal/noise ratio in excess of 90 dB. The net bit rate is $44.1 \times 10^3 \times 32 = 1.41 \times 10^6$ audio bit/s. The audio bits are grouped into frames each of which contains six samples. Each block of audio data has blocks of parity data added according to a system called *CIRC* (Cross-Interleaved Reed-Solomon Code). This system allows not only the detection of errors in the decoded signal but also their correction. The overhead for this operation is one additional CIRC bit for every three data bits.

The compact disc system has therefore three levels of error handling:

- *error correction*, where a missing data word can be correctly reconstructed from the combination of the misread data and CIRC
- *interpolation*, where the missing data values can be estimated from the context of adjacent correctly interpreted words
- *muting* in the case of the data being too garbled to read at all

In general use, correction is carried out frequently; interpolation occasionally; and muting virtually never. The primary cause of data loss is not physical damage (although this can affect replay quality if it is severe) but mistracking due to the player's inability to follow the disc surface accurately, either because of warping or because of incorrect seating on the spindle.

After CIRC calculation and addition, *control* and *display* bits are added. These can be used to display information to the listener as well as indicating to the player where a desired track or subsection of the programme material is located, thus enabling a rapid search by the user for desired material. These *subcodes* also carry other information. Next, the bitstream is modulated into a form suitable for disc encoding, using the efm code (*eight-to-fourteen modulation*). Using this scheme, blocks of 8 bits are translated into blocks of 14 bits, which are linked together by 3 'merging bits'. To enable synchronization, a synch pattern consisting of 27 channel bits per frame is added. This all adds up to a total data transfer rate of 4.32×10^6 channel bits/s. As the scanning speed is 1.25 m/s, this means that a channel bit on the disc occupies about 0.3 μm.

The optical section is about 45 × 12 mm and is mounted on a pivoted arm from where it can scan across the entire surface area of the disc. It is generally driven by a coil attached to the arm moving in the field of a permanent magnet, allowing the optics to be positioned under the disc in much the same way as an analogue milliammeter needle is positioned to indicate current flow. Correct tracking is ensured by a servo, controlled by an error signal from the photodiode array. A system not unlike that used in a loudspeaker (i.e. a coil moving in the field of a permanent magnet) is used to focus the beam according to a servo driven again by processing the output from the photodiode array. As the depth of focus of the optics is around 4 μm and the disc surface can move up to a maximum of 1 mm, this focus servo is vital to correct operation.

The signal processing on replay is as shown in *Figure 59.3*. First the digital signal is demodulated and reconstituted by the reverse of the efm encoding algorithm. The result is stored in a temporary memory buffer and fed to the *error detection and correction circuit*, ERCO. The parity bits are used here to correct errors or to detect them if correction is impossible.

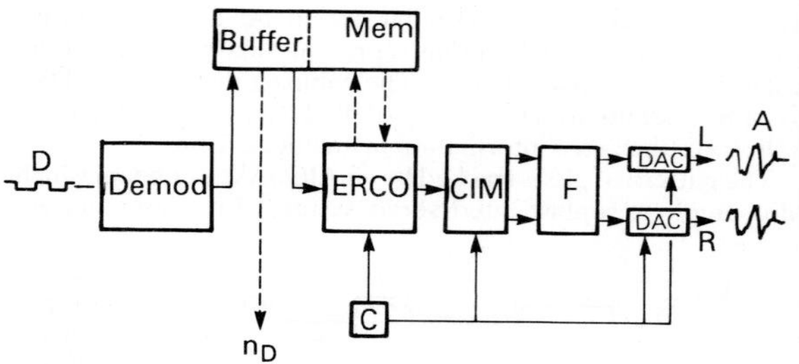

Figure 59.3 Signal processing in a cd player. D is the input signal read be the optical pick-up, A the two output analogue audio signals from the left (L) and the right (R) audio channels. DEMOD is a demodulation circuit, ERCO an error-correction circuit and BUFFER a buffer memory, forming part of the main memory MEM associated with ERCO. CIM is a concealment interpolation and muting circuit in which errors, that are only detected if they cannot be corrected, are masked or 'concealed'. F contains filters for interpolation. C is a clock generator controlled by a quartz crystal.

Interleaving of the CIRC codes ensures that the error rates do not exceed the capability of the ERCO to deal with them. A flaw on the disc such as a severe scratch will cause a 'burst error', and the ERCO can correct a burst of up to 4000 data bits, primarily because interleaving spreads them out.

If the error rate exceeds the maximum that can be handled by the ERCO, the errors detected are masked by the CIM (concealment: interpolation and muting) block. The corrected signal, now split into left and right channels, then passes to the digital-to-analogue conversion stage, consisting primarily of a digital/analogue converter and low-pass filter. Filtering carries out two related activities: firstly it 'smooths' the digital data by removing spurious 'steps' between adjacent sample values; secondly it removes any 'images' developed beyond the Nyquist limit (in this case, above 22.05 kHz) where the audio information is meaningless. In the Philips design and now some others, filtering is performed digitally. This is just as well, as the filter specification usually employed is a 'brick wall' roll-off at about 20 kHz. If implemented in the analogue domain, such a filter can cause severe problems, notably ringing and massive phase shifts all the way down the audio band.

Synchronization for the digital/analogue converters, the CIM and the ERCO are provided by a clock generator. This also by definition determines the rate at which data are read *out* of the temporary buffer memory. Data are read *in*, however, at a rate determined by the speed of the disc. Thus the amount of buffer memory that is filled at any one time depends on the clock rate and the disc rotation (thus the data acquisition) rate. The clock rate, and thus the data rate, must be fixed. However the disc rotation rate can be controlled, and this is done by ensuring that the buffer is only 50 per cent full at any given moment. The result is that all wow and flutter effects are removed from the signal without any sophisticated speed control requirements.

59.5 Related optical disc systems

In addition to the audio capabilities of compact disc, the same principles are utilized in the digital audio tracks of NTSC LaserVision and CDV. CDV discs are produced in 120 mm, 200 mm and 300 mm forms. The 120 mm format provides six minutes of video with digital audio plus 20 minutes of digital audio only. The 200 mm format provides 2 × 20 minutes of digital sound, plus video, while the 300 mm format (essentially the original LaserVision/VLP format) can contain 2 × 60 minutes of video and digital audio.

Additional coding can be used with the conventional cd to store such information as lyric sheets or still photographs, while a development of the technology, CD-ROM, uses the disc for data storage according to internationally agreed standards. There are also developments currently taking place regarding the use of the cd system in interactive multimedia (CD-I) combining sound, pictures and data storage. A recordable compact disc system has 'single' 80 mm diameter format.

59.6 Digital audio tape

Digital tape recording systems were the earliest forms of digital audio systems to appear, and they did so in the mid 70s with the advent of systems like the Sony *PCM* series of digital audio processors. These processors took a stereo audio input and converted it to digital form, but due to the wide bandwidth required to store the data effectively, they converted the data into a form of video signal (generally referred to generically as a *pseudo-video* signal) which could be recorded on a conventional video recorder, typically a U-matic video cassette machine in the case of professional systems. Such systems still offer a number of important benefits to professional users.

In the 70s, several Japanese manufacturers, led by Sony with their portable *PCM-F1* processor and companion *SL-F1* Betamax recorder, introduced consumer digital audio processors that would interface successfully with conventional consumer video recorders, either Beta or VHS. The Japanese introduced an EIAJ standard for this type of encoding scheme, although it allowed only 14-bit stereo digital recording, and this was utilized by all the major players. However, while in the professional field all the pseudo-video machines used NTSC as a video standard by default, to aid international exchange of master tapes, the consumer systems existed in versions to suit the major TV standards. This led to a good deal of confusion.

Sony decided to implement a 16-bit scheme based on EIAJ, but using two of the error-correction bits for data. This reduced the error-correction capability of the system, but gave it the same data level of fidelity as compact disc was later to offer. Sony offered Betamax recorders with their video error correction (usually based on line repetition) switchable so that it would not interfere with the EC system on the digital processors. Other manufacturers did the same thing with VHS recorders.

The theory was that as so many homes had video recorders, the chance to *record* as well as replay digital audio would become an important marketing point and that pseudo-video consumer digital audio systems could supersede both vinyl discs and cassettes. Pre-recorded digital audio cassettes would be available, and above all people would be able to record their own material, on a VCR they already owned, for the cost of a digital processor. The consumer market did not respond. However smaller professional users purchased the systems for low-cost digital mastering in budget-conscious studios. Digital converters were produced to transfer between the 16-bit so-called *F1-format* and the professional *1610* system.

The latest consumer audio development, digital audio tape or *DAT*, has its clear origins in the combination of digital audio processor and video recorder, and now looks as though it may suffer the same fate, that of being accepted readily as a low-cost professional system but remaining relatively unsuccessful in its original consumer market.

DAT was hardly the surprise challenger to the record industry that some pundits have labelled it. DAT was announced shortly after cd was released, in mid-1983, and the record industry knew at once that it would represent a prime medium for consumers to use to copy pre-recorded material onto their own tapes — a potential infringement of copyright. The opposition of the record companies has dramatically affected both the technical and marketing development of DAT.

Before 1983, all the systems proposed for digital audio specific cassette applications utilized fixed head systems. These generally required higher tape speeds (to obtain the necessary bandwidth) and thus larger cassettes with more tape. In 1983, however, Sony proposed a system based very much on their experiences with video recording and digital processors, utilizing helical-scan technology to increase the effective tape-to-head speed and offering the possibility of smaller cassettes and transports, slower tape speeds, longer recording times, and lower production costs. Within two years technical specifications for both S-DAT (stationary head) and R-DAT (rotary head) systems were standardized, but only R-DAT machines have been made available.

Record companies, with a heavy commitment both to compact disc and to copyright protection, have released virtually nothing in these formats. With a lack of pre-recorded software, the sales of DAT machines in the countries where they are available have been well below expectations, while in contrast cd exceeded sales forecasts almost from the beginning.

As in the case of the old F1-format digital audio processors,

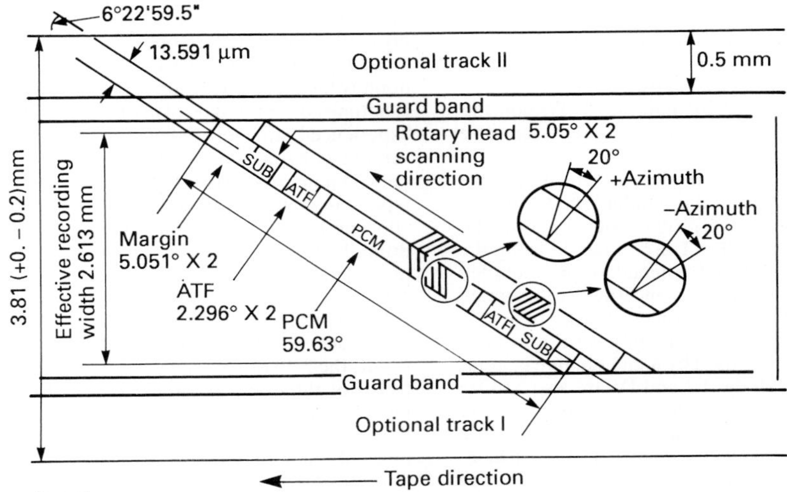

Figure 59.4 DAT tape format

Mode / Item	Mode I	Mode II	Mode III	Mode IV	Mode V	Mode VI
	Standard mode	Option 1	Option 2	Option 3	Pre-recorded tape	
		Compatible with satellite broadcasting A mode	Long-time mode	4-channel mode	One-to-one normal track	High speed wide track playback only
Available channels	2	2	2	4	2	2
Sampling frequency	48kHz	32kHz	32kHz	32kHz	44.1kHz	
Bit numeral of quantization	16-bit (linear)	16-bit (linear)	12-bit (non-linear)	12-bit (non-linear)	16-bit (linear	
Transmission speed	2.46Mbit/s	2.46Mbit/s	1.23Mbit/s	2.46Mbit/s	2.46Mbit/s	
Sub-code capacity	273.1kbit/s	273.1kbit/s	136.5kbit/s	273.1kbit/s	273.1kbit/s	
Modulation system	8–10 conversion				8–10 conversion	
Error correction system	Dual Reed Solomon				Dual Reed Solomon	
Redundancy	37.5%	58.3%	37.5%	37.5%	42.6%	42.6%
Tracking system	Area split ATF				Area split ATF	
Tape width	3.81mm				3.81mm	
Tape depth	$13\mu m \pm 1\mu m$				$13\mu n \pm 1\mu m$	
Tape is use	Metal powder				Metal powder	Oxide tape
Tape speed	8.15mm/s	8.15mm/s	4.075mm/s	8.15mm/s	8.15mm/s	12.25mm/s
Relative speed	3.133m/s	3.133m/s	1.567m/s	3.133m/s	3.133m/s	3.129m/s
Standard drum specs	Ø 30, 90° lap				Ø 30, 90° lap	
Drum revolution	2,000 rpm	2,000 rpm	1,000 rpm	2,000 rpm	2,000 rpm	
Track pitch	13.591μm				13.591μm	20.41μm
Track angle	6°22′59.5″				6°22′59.5″	6°22′59.5″
Head azimuth angle	±20°				±20°	
Recording time	120 min. (tape depth of 13μm)		240 min. (tape depth of 13μm)	120 min. (tape depth of 13μm	120 min. (tape depth of 13μm)	80 min. (tape depth of 13μm)
	180 min. (tape depth of 10μm)		360 min. (tape depth of 10μm)	180 min. (tape depth of 10μm)	180 min. (tape depth of 10μm)	120 min. (tape depth of 10μm)
Cassette size	73mm × 54mm × 10.5mm				73mm × 54mm × 10.5mm	

Table 59.1 Operating modes for digital audio tape

however, DAT has taken off as a new method of producing low-cost, high-quality digital audio masters for the cost-conscious recording studio. It offers 16-bit technology and levels of quality equivalent to those of compact disc.

A DAT cassette is completely sealed and measures 73 × 54 × 10.5 mm, weighing about 20 g. The tape in the cassette is 3.81 mm across, while the actual helical track (see *Figure 59.4*) is 13.591 μm in width and 23.501 mm long. Within this structure, each data bit is recorded in 0.67 μm, representing a data density of 114 Mbits per square inch.

DAT can operate in each of six modes, as shown in Table *59.1*. It will be noted that Mode I, the standard mode, utilizes 48 kHz sampling, the same as the professional studio digital format. However, studios using DAT are in fact utilizing the 44.1 kHz sampling rate of Mode V (designed for pre-recorded tapes only, as far as the consumer is concerned) to maintain compatibility with compact disc mastering systems.

Mode II is designed to offer compatability with the widely used 32 kHz digital audio paths used by many satellite systems and some terrestrial links. Modes III and IV are also 32 kHz sampling modes, Mode III offering double playing time and two channels and Mode IV offering four-channel audio. Mode VI would be difficult to record with a regular DAT machine, although it can be reproduced. This mode is designed to play back pre-recorded tapes (at 44.1 kHz) made by commercial tape-copying facilities (presumably using a 'contact-printing' type of copying process). The benefit of this mode is that oxide based tapes can be used, thereby reducing the cost. The other modes require metal tape (at least metal powder, and preferably metal-evaporated) to operate at the quoted error rates.

Many of the techniques used are common to compact disc, notably the error correction scheme, although here the system is optimized for the type of error bursts which will normally be experienced with tape. Other specifications are reminiscent of a vcr, but with some important differences. The drum diameter is only 30 mm, and a 90° wrap means that only a short length of tape is in contact with the drum, reducing tape damage and allowing high-speed search with the tape in contact with the drum. It also means that low tape tension can be used, ensuring increased head life and, if four heads are used, the heads can be separated by 90° to allow simultaneous monitoring off-tape.

In a typical two-head configuration, this means that a head is in contact with the tape for only 50 per cent of the time. As a result, the 2.46 Mbit/s signal to be recorded is compressed by a factor of three and recorded at 7.5 Mbit/s so that it can be recorded discontinuously. This has the major advantages that the transfer of data to and from the heads is much more efficient, and that the drum speed of 2000 revs a minute introduces a flywheel effect which increases mechanical and speed stability. Eight-to-ten conversion is used, converting 8 data bits into 10, unlike the 8–14 method used in cd. This reduces the range of wavelengths that must be handled: the maximum wavelength is four times longer than the minimum wavelength. It also permits overwriting, eliminating the need for a flying erase head; short wavelengths can overwrite previously recorded longer wavelengths.

In addition, high density recording is made practical by eliminating guard bands between each helical stripe across the tape. (There are, however, guard bands between the helical scan area and the two optional longitudinal tracks at the edges of the tape.) This is achieved by using heads 50 per cent wider than the track width.

Error correction is performed with double Reed-Solomon codes (see section *59.4*), and the blocks are interleaved across two tracks. Each block contains 288 bits, broken down into four 8-bit nibbles (sync, id code, block address and parity respectively) plus 256 data bits consisting of the pcm data plus parity, and configured as 32 8-bit symbols. One track contains 128 blocks (4096 symbols of which 1184 are used for error correction and 2912 for data). There is also a reserved subcode area.

Different areas of each track are allocated to a series of signals, via time-division. To avoid interblock crosstalk, interblock gaps are also included. According to the settings of these gaps, certain areas can be partially overwritten without affecting adjacent areas. The main pcm programme data areas are thus separated from the subcode areas used to write details of programme contents, start ids, etc. As a result, the latter can be edited and rewritten without affecting the main data areas.

Also provided are areas to enable *automatic track finding* (atf). ATF uses a pilot signal at 130.67 kHz, two sync signals at 522.67 and 784.00 kHz, and an erase signal at 1.568 MHz. When the head advances, the presence of an atf signal is detected by picking up either of the two sync signals. The adjacent pilot signals are then compared and a decision made as to the accuracy of tracking. The pilot signal is relatively low in frequency, sufficient to be unaffected by azimuth settings, so that crosstalk can be picked up from both sides. The comparison method is analogue (see *Figure 59.5*). The system neatly eliminates the need for a tracking control head as found on vcrs.

The high-speed search facility is also important and is made possible by the tape speed (8.15 mm/s) and the wrap angle on the drum. The recording format enables data to be read at high speed, as the heads can read a section of several adjacent tracks at a time. This is sufficient to pick up the address within the block format and means that, thanks to a fast wind speed, a desired cue point can be located quickly and accurately.

Subcode areas are also provided for in the block structure, at the rate of 273.1 Kbits/s. This is over four times the subcode capacity of compact disc. It is, however, used simply for id codes recording a number of useful parameters such as sample rate, channel number, tape speed, copy protection, pre-emphasis, etc.

59.7 Other systems

Another system worthy of brief consideration is the so-called ‘hi-fi’ system now available for audio recording on current consumer video cassette recorders: *VHS hi-fi* and *Beta hi-fi*. Both forms can offer audio quality approaching that of digital audio (80 dB dynamic range, for example, distortion at less than 0.3 per cent and wow and flutter in the 0.005 per cent region), and this is achieved by means of frequency modulation. The audio and video signals are recorded in two separate

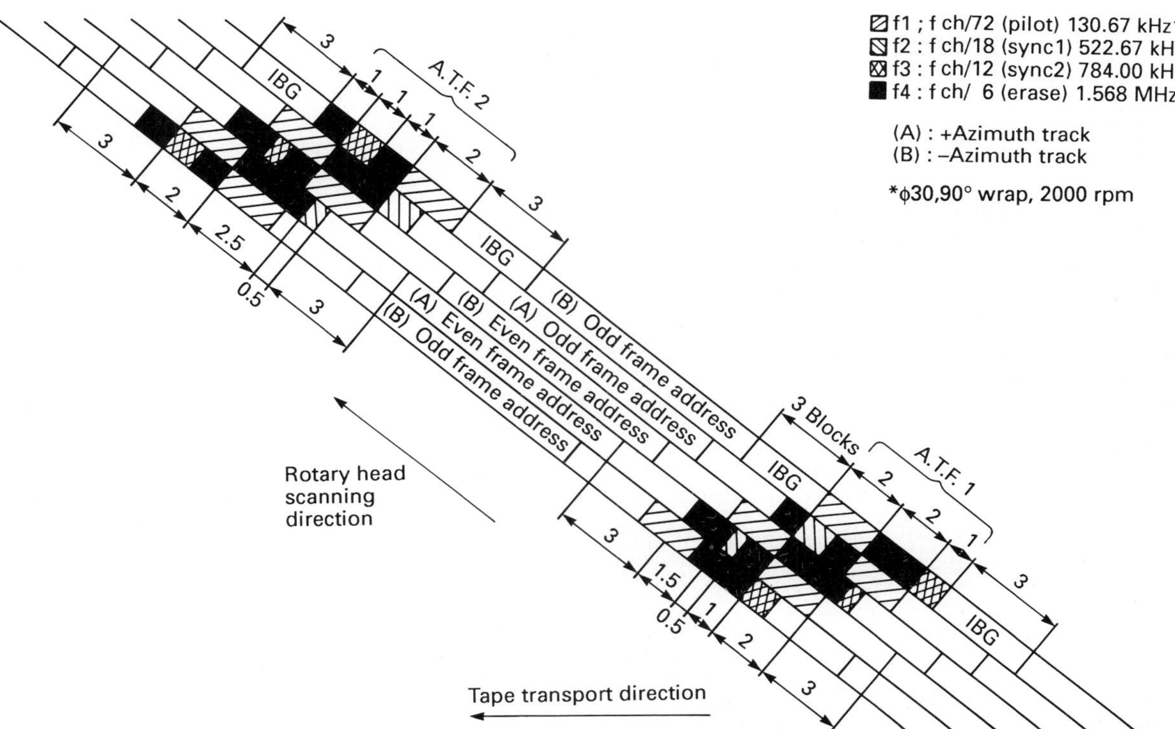

Figure 59.5 Track pattern for automatic track finding on DAT. Even frame address track 0.5 block sync. Odd frame address track 1 block sync

helical scans with different azimuth angles across the same part of the tape. This ingenious method allows a 4 MHz video signal to share the same physical space as a 20 kHz audio signal.

There are in fact two different versions of the system: *frequency multiplex*, which is used on the NTSC-format Beta hi-fi machines, and *depth multiplex* which is used in PAL Beta hi-fi and VHS hi-fi (both PAL and NTSC). In the latter system, a separate revolving head is used which records audio on a deeper level of the tape. Then a video head records the video signal on the same diagonal track on top of the existing audio information. Because the wavelength of the video signals is shorter, and the video signals are recorded primarily at the surface of the tape, there is no interference between the two.

Video 8 offers three audio recording systems, a *standard longitudinal system*, an *fm system*, and a *pcm digital system*. The fm system is not unlike that used in NTSC versions of Beta hi-fi, except that it offers only a single channel and has a single-frequency carrier of 1.5 MHz. The deviation is ±100 kHz. Because Video 8 uses metal evaporated tape, which has a thickness of less than 0.2 μm, depth multiplexing is impractical. Also, additional audio heads would have increased the cost. The wider the azimuth angle, the more the crosstalk decreases, but if the azimuth angle becomes too wide, the hf area of the video signal begins to deteriorate. Therefore the azimuth angle is set to 10°.

The pcm system used in Video 8 offers stereo audio and may be recorded simultaneously with the video or afterwards (the fm system can only be used for simultaneous recording for obvious reasons). Extra space is made for the pcm signal by increasing the wrap angle from 180° to 210°, enabling pcm blocks to be recorded at the edge of the scan. The system used employs only eight data bits which would normally limit the dynamic range to 48 dB. However a combination of non-linear data compression which effectively squeezes 10 bits into the space of 8, plus 2:1 compansion analogue noise reduction, achieves a claimed performance equivalent to 13 bit linear pcm. The sampling frequency is low, however, compared with cd or DAT, being 31.25 kHz PAL and 31.5 kHz NTSC. This limits the frequency range of the system to 15 kHz.

Error correction is performed by two interleaving error correcting codes on each data block every eight 8-bit words — a *cross interleave code*. 1250 words per field are recorded in PAL and 1050 per field in NTSC.

Normally, the pcm signal recorded on a Video 8 recorder is recorded during the final 30° of wrap, 180° being used for the video signal. An interesting possibility exists on many Video 8 recorders, however, i.e. the ability to record five extra pcm stereo pairs in place of the video signal. Machines with a *pcm audio only* mode can record stereo audio for, for example, 18 hours on a P5-90 tape in lp mode, each track offering three hours playing time.

Bibliography

THORN, E A , *Understanding Copyright: A Practical Guide*, Jay Books (1989)

Part 13
Teletext and Similar Technologies

P L Mothersole F Eng, C Eng, FIEE
VG Electronics Ltd

60 Broadcast Teletext Systems

60.1 The teletext data signal

Teletext is a system of transmitting digitally coded alphanumeric data in the vertical blanking period (vbi) of the television signal. The data are displayed on a television receiver as a page of information. To be compatible with a normal domestic receiver, a display format of 40 characters per row and 24 rows per page has been adopted. (Normal computer displays use 80 characters per row but are viewed with the operator sitting relatively close to the screen.)

The teletext pages are transmitted cyclically, i.e. one after the other in a continuous sequence. When the complete sequence or magazine of pages has been transmitted it is then repeated.

The first row of a page is referred to as the *header row*, or row 0. This row contains the magazine and page number, the service name, date and time. In order to capture a page for display on a receiver, the magazine and page number are keyed in by the viewer. When the requested page in the magazine is transmitted, it is then captured by the decoder and displayed on the screen.

In a teletext system the pages are transmitted sequentially, and it is unlikely that the page will be transmitted at exactly the time a viewer requests it. There is therefore a delay between the request and the moment when the page is captured and displayed. This delay is called the *access time* and is a critical parameter in a teletext system. It must be as short as possible and is related to the data rate, the data format, the number of pages being transmitted, and the number of vbi lines being used.

The nrz teletext data signal is carried in the vbi of a normal television signal, and the energy is concentrated at the high frequency video end of the band. The data pulse shape is chosen so that most of the energy is contained inside the normal video bandwidth of 5 MHz, 5.5 MHz or 6 MHz for 625-line and 4.2 MHz for 525-line systems. As a result of experience gained on video distribution networks feeding transmitters, a 100 per cent raised cosine filter was found to be optimum for pulse shaping and is now standard (*Figure 60.1*).

The data pulse amplitude is specified for World System Teletext (WST) as 66 per cent peak white. This amplitude was determined after extensive tests to avoid sound buzz and visibility of the data signal at the top of the picture on existing receivers (see *Figure 60.2*).

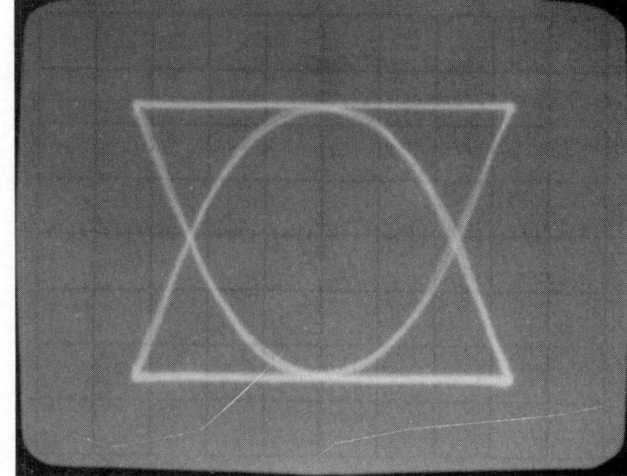

Figure 60.1 Eye display at data inserter

Teletext is a 'one way' transmission system utilizing television broadcast networks. Teletext information can therefore be rapidly transmitted to millions of receivers simultaneously. To maximize the transmission rate, it is therefore desirable to minimize the amount of information required for each page.

Various data formats have been specified for different teletext systems, but any of the specified systems can achieve similar results in terms of displayed information, the quality of the graphics, etc. However, there are significant differences between the systems in terms of transmission time because of the amount of data required to be transmitted for a given page and the complexity of the receiver decoders. Full field teletext systems, as the name implies, are systems where teletext data occupy all transmitted lines, there being no video information.

Teletext pages are normally transmitted so that any receiver with a suitable decoder can capture the pages for display purposes. Teletext pages may be encrypted so that they can be displayed only on receivers which incorporate decoders with suitable de-encryption circuits.

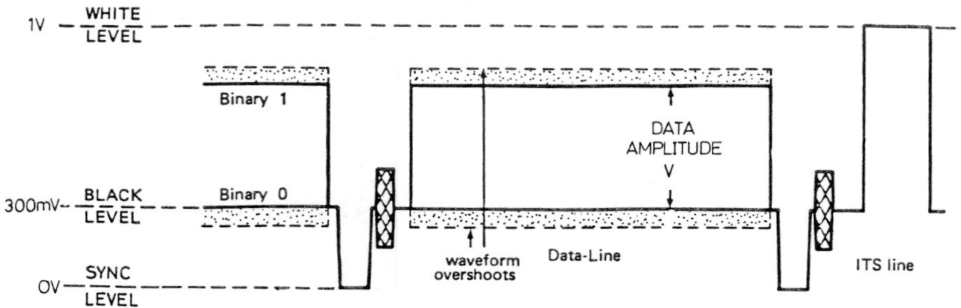

625 LINE WORLD SYSTEM TELETEXT T = 12 μs V = 462 mV
(Australian standard V = 490 mV)

625 LINE ANTIOPE T = 12·8 μs V = 700 mV

525 LINE WORLD SYSTEM TELETEXT T = 11·7 μs V = 500 mV.
ANTIOPE & NABTS.

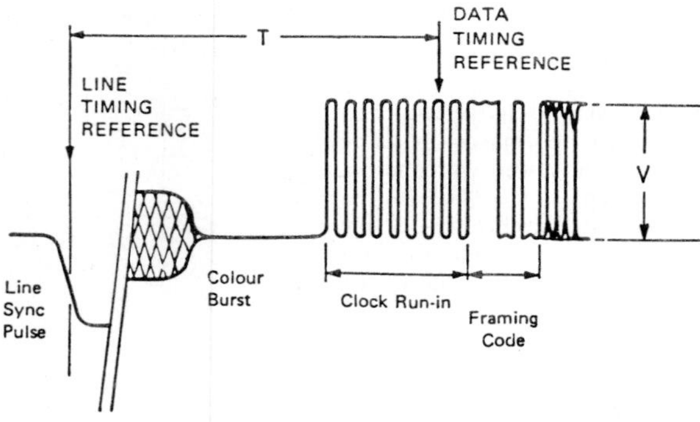

Figure 60.2 Data levels and timing

Encryption is normally undertaken at the data source or editing facility, prior to insertion into the teletext system. The degree of security needed and, when required, the control of specific teletext decoders from the source is therefore a function of the encryption facility.

The teletext system is transparent to the encrypted pages or data packets and acts as a carrier for the data. However when eight bit data code, as used with encrypted data or software packages, is to be transmitted, care must be taken to ensure the network is able to handle this without causing errors (see section *61.6*).

60.2 Teletext data format

There are basically two techniques for formatting pages of data that are to be transmitted in the vbi of television signal.

60.2.1 Variable data format

A page of data and graphics can be represented as a continuous string of digits using the normal *line feed* and *carriage return* code to signify the end of each line. (This is similar to a view data arrangement, see section *63*.) A typical teletext page might contain some 8 k bits of data which could then be divided up into a series of 50 μs data blocks. These data blocks could be added into the vbi lines of a television signal at 50 Hz intervals, i.e. one block of data per field. Each block of data would need to be preceded by a data clock run-in together with some additional bits for byte synchronization.

At the decoder, the blocks of data would be stored so that, several field periods later, a complete page of information would be held in the decoder store and could be processed for display.

With variable format of the data there is no relationship between where characters should appear on the screen and their position in the data block. Positional information must therefore be transmitted as part of the data stream for use by the decoder.

Any interference or distortion resulting in any loss of information during the transmission will result in errors in the received data. The errors would therefore show up as either incorrect characters in the page or information in the page being displaced and appearing in the wrong position. To guard against this, various forms of error protection must be employed; these significantly increase the amount of data required for one page of information. Furthermore, a processor is an essential part of the decoder to enable the complex data stream to be satisfactorily decoded and displayed.

Variable format is used in the French *Antiope* system, and in the *North American Broadcast Teletext Specification* (NABTS), that has yet to be implemented as a domestic service. An advantage of a variable format system is that the data can be readily supplied from external computers as there is no direct relationship to the displayed page. Disadvantages are that a very considerable amount of additional information is required for adequate data protection to avoid reception errors seriously disturbing the display, and a more complex decoder is required to process the additional data.

60.2.2 Fixed data format

The fixed format system, as used by *World System Teletext* on both 625-line and 525-line systems, exploits the regular and defined timings of the television signal that carries the data to ensure that the characters are always displayed in the correct position on the screen. No positional information has to be transmitted, and the decoder is correspondingly simplified.

One row of teletext data is transmitted in one television line period on the 625-line version. Each row starts with a data clock run in sequence to synchronize the decoder clock, followed by byte synchronizing bits for logic synchronization and the magazine and row number (see *Figure 60.2*).

These data are then followed by an eight bit code corresponding to each of 40 characters in one row of displayed information. When no character is present, a blank space character is transmitted so that one row of information always has code corresponding to 40 characters present in the data stream. The position of the character in the display is then directly proportional to its location in the data block.

The header row contains the magazine and page number and service name, date and time. Then row 1, row 2, row 3, etc., follow until the complete page has been transmitted, when the header row for the next page is transmitted. This is used as an indication to the decoder that a complete page has been received. There is therefore a direct one-to-one relationship between the position of the data bits corresponding to a particular character in the row and its position on the screen.

Each row of data for a particular page is transmitted after the header row and has its own row number. Blank rows of information need not be transmitted. The technique of transmitting only rows of information that contain data is called a *row adaptive* transmission. This feature is utilized in the transmission of subtitles, as these normally contain only one, two or possibly three rows of information, and of news flashes, which again normally contain only a few rows of information.

The video bandwidth of 525-line systems is less than that for 625-line systems, although the active line period is approximately the same (52 μs). The data rate is therefore reduced in approximately the ratio of 525:625. This means that only 32 characters per row can be transmitted in one TV line period, although 40 characters are required to be displayed. The gearing technique used to overcome this limitation of the fixed format system is to transmit the last eight characters for each of the previous four rows as a separate row. The decoder displays the rows correctly, and each page of 24 rows therefore requires 30 data lines. As the field frequency of 525-line systems is 60 Hz compared to 50 Hz for 625-line systems, the overall transmission rate for the systems is similar.

60.2.3 Control characters

Control characters are required to augment the normal alphanumeric display characters in order to achieve special effects. These include colour changes, flashing, double height or double width characters, or graphic symbols. In *variable format systems*, the location of the control characters in the data stream is not critical. In *fixed format systems* the control characters are normally inserted in front of the block of text to which they refer, and transmission of a control character occupies a character space. This does not normally impose limitations since the control characters are inserted in spaces at the start of sentences or prior to a graphics symbol. Control characters contained within the page are referred to as *serial attributes*.

The character code tables for use with fixed format pages, i.e. pages that contain all the information within the 24 rows, are limited to a character font of 96 characters. Certain languages, e.g. Spanish and Arabic, require additional characters. Such characters are transmitted on additional data rows that are not displayed. These rows were originally referred to as *ghost rows* but more recently they are known as *data packets* and have been assigned packet numbers.

60.2.4 Enhancements

Evaluation of various teletext systems has been undertaken by many broadcasting organizations, and the fixed format or World System Teletext (WST) has now been adopted by over 30 countries worldwide. At the same time, the system has been further developed to cater for high resolution graphics, the special characters required for different languages, and the transmission of still pictures. These enhancements have been incorporated without sacrificing the basic transmission ruggedness of WST afforded by the fixed data format which has been retained. The various enhancements have been specified as a number of *levels* as follows:

- Level 1 Basic 96 character font, double height, flashing, 8 colours and mosaic graphics.
- Level 2 Additional data packets (non-displayed rows) and pseudo pages for transmission of additional characters and non-spacing attributes, a wider choice of background and foreground colours, a smooth graphics capability and other display features.
- Level 3 DRCS (dynamically redefinable (down-loaded) character sets).
- Level 4 Alphageometric displays.
- Level 5 Alphaphotographic displays (still pictures).

The higher levels, in particular levels 3, 4 and 5, require increasing amounts of page storage and associated data processing so that decoders are progressively more expensive. Page creation also requires more complex digitizers and editing software. The additional data required for each page also increase the access time, which reduces the response time of the system.

60.3 Error protection

Characters are transmitted in eight bit data groups formed from seven-bit character codes plus one (odd) parity bit.

Addresses are transmitted in eight bit code groups, protected by the use of a *Hamming error correcting code*. The Hamming code words comprise four message bits and four protection bits. The use of four protection bits permits any single error in the received code group to be corrected and in any double or other even number of errors (except eight) to be detected.

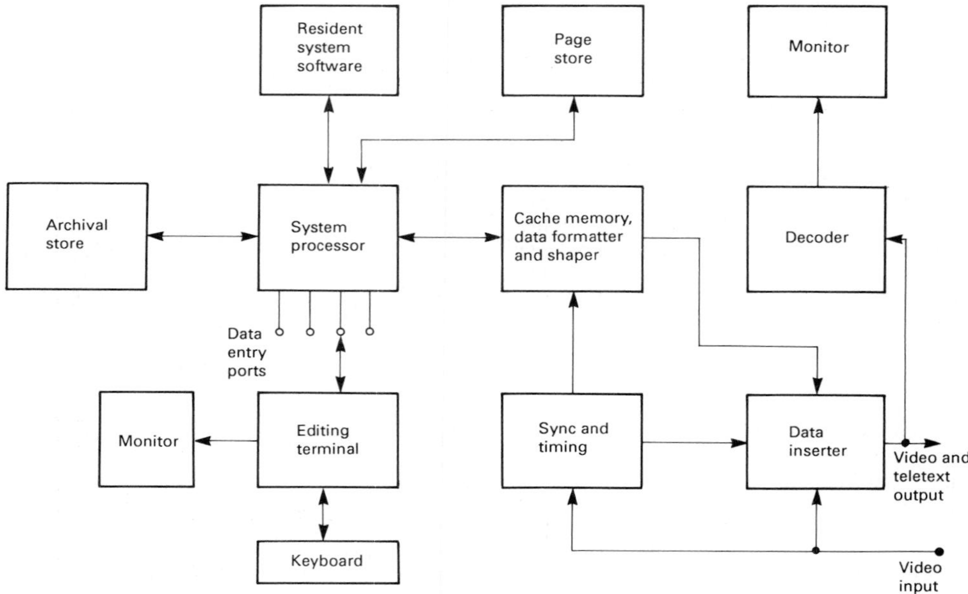

Figure 60.3 Basic teletext system

Hamming protection of the addresses, i.e. magazine and page numbers, ensures that decoders will receive the data for the individual pages. The parity bit check for individual characters normally ensures that, should an error be received, the character will not be displayed. In the event of two errors being received, the decoder will display an incorrect character. Since the teletext magazine of pages is transmitted cyclically, when the page is next received, characters that were received with an error on the first acquisition almost certainly will be received and displayed correctly the second time round. Should an error be received for a character that is already displayed correctly then that character will not be changed. In variable format systems, Hamming protection is essential for all positional and control information in addition to the page address.

60.4 Access time and data rate

As explained in section *60.1*, it is very desirable that access time should be as short as possible. To obtain the maximum throughput of data in a teletext system, the data rate must be as high as practical bearing in mind the constraints of the television channel. For 625-line systems, the video bandwidth is 5 MHz, (5.5 MHz for system I and 6 MHz for system D).

When the teletext data standard was being evolved, 625-line reception tests showed that the noise immunity of data transmitted over the television channel was not significantly less for data at a bit rate of just under 7 M bits per second than for data at a lower rate of 4.5 M bits/s.

A bit rate of 6.9375 M bits/s was chosen for the 625-line systems. This bit rate corresponds to 444 times the horizontal line frequency of 15.625 kHz, but the date are not locked to the television signal. The bandwidth of 525-line TV systems is normally 4.2 MHz, and a corresponding data rate of 5.727 272 M bits/s was therefore chosen which corresponds to 364 times the horizontal line frequency of 15.734 264 kHz. The teletext page is displayed in colour, but the teletext system and data signal are independent of the video colour standard (PAL, SECAM or NTSC). The video signal simply acts as a carrier for the data, the decoder providing RGB output signals.

The data rate chosen for the 625-line WST teletext standard enables one complete row of text to be carried in one TV line period. The page is normally made up of 24 rows, and it follows that approximately two pages are transmitted per second per data line used. When six data lines are used, some twelve pages per second are transmitted. The worst access time will therefore be a maximum of about 16 seconds with an average of eight seconds when 200 full pages are transmitted. 525-line WST systems have a similar access time as, although the data rate is slower, the field rate is higher.

60.5 Transmission systems

The essential elements of a teletext transmission system are shown in *Figure 60.3*. Teletext pages are created using an editing terminal. This consists of an editing keyboard connected to a processor with resident software and an associated bank of memory and a display monitor.

The software contained in the terminal would normally contain the usual word processor functions together with additional software to simplify creation of the teletext pages. It is normal practice for an editing terminal to contain sufficient memory to store several teletext pages and to have provision for swopping information between the pages. A floppy disk drive is often associated with editing terminals to allow editors to create a set of pages off line from the main system.

Extra data ports can be provided to enable the terminals to be connected to external data bases so enabling the editors to have direct access to additional material.

Graphic figures can be produced using the editing terminal and special keys. The simple mozaic graphics cell occupies the space normally required by a character, and the cell is divided into six segments having a 2 × 3 format. Any combination of the segments can be illuminated which allows 64 different combinations. *Figure 60.4* shows the 96 character codes set and corresponding characters and graphic shapes used for teletext broadcasting using the WST standard.

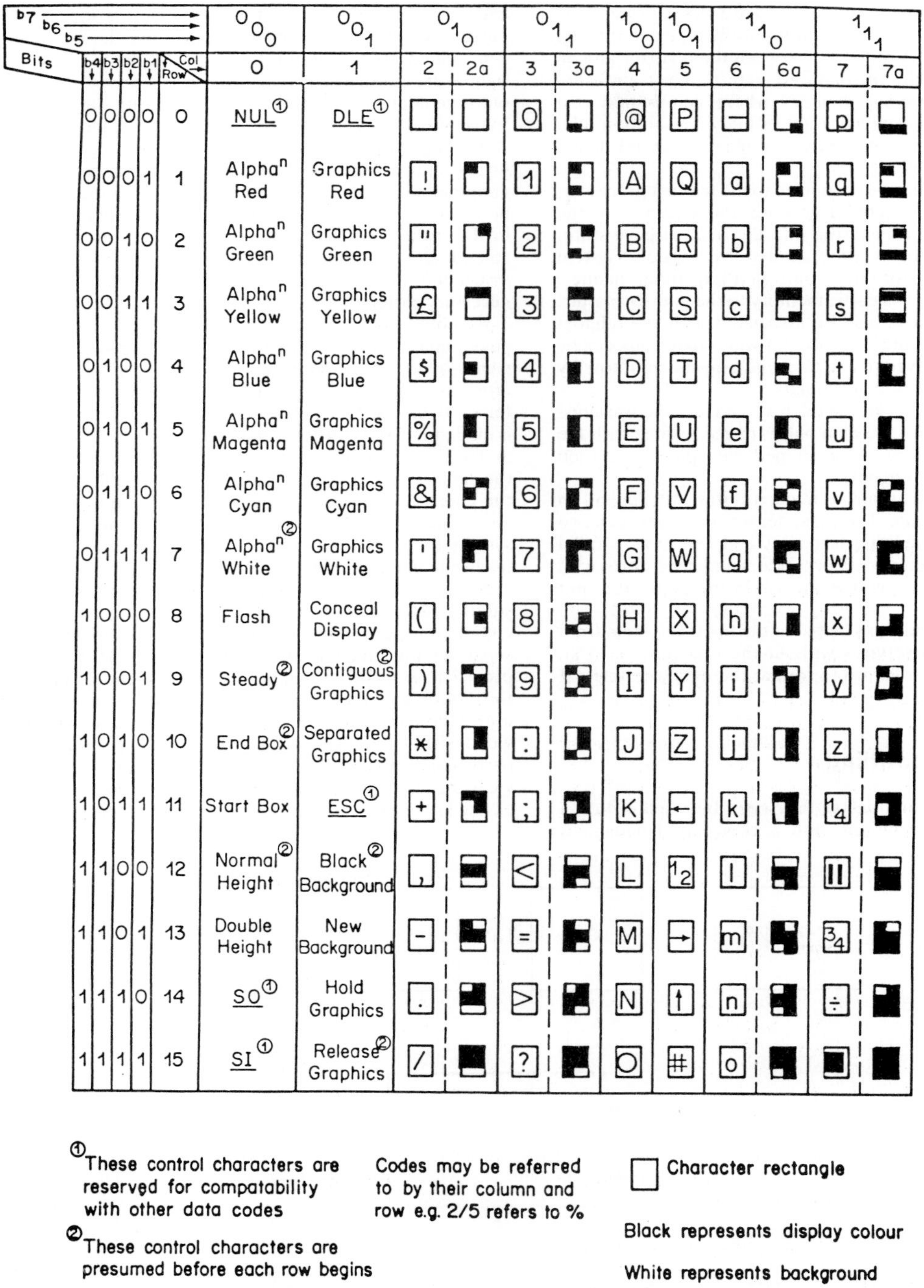

b7 b6 b5 →					000	001	010		011		100	101	110		111	
b4	**b3**	**b2**	**b1**	**Row \ Col**	**0**	**1**	**2**	**2a**	**3**	**3a**	**4**	**5**	**6**	**6a**	**7**	**7a**
0	0	0	0	0	NUL①	DLE①			0		@	P	—		p	
0	0	0	1	1	Alpha^n Red	Graphics Red	!		1		A	Q	a		q	
0	0	1	0	2	Alpha^n Green	Graphics Green	"		2		B	R	b		r	
0	0	1	1	3	Alpha^n Yellow	Graphics Yellow	£		3		C	S	c		s	
0	1	0	0	4	Alpha^n Blue	Graphics Blue	$		4		D	T	d		t	
0	1	0	1	5	Alpha^n Magenta	Graphics Magenta	%		5		E	U	e		u	
0	1	1	0	6	Alpha^n Cyan	Graphics Cyan	&		6		F	V	f		v	
0	1	1	1	7	Alpha^{n}② White	Graphics White	'		7		G	W	g		w	
1	0	0	0	8	Flash	Conceal Display	(		8		H	X	h		x	
1	0	0	1	9	Steady②	Contiguous② Graphics	)		9		I	Y	i		y	
1	0	1	0	10	End Box②	Separated Graphics	*		:		J	Z	j		z	
1	0	1	1	11	Start Box	ESC①	+		;		K	←	k		¼	
1	1	0	0	12	Normal② Height	Black② Background	,		<		L	½	l		‖	
1	1	0	1	13	Double Height	New Background	-		=		M	→	m		¾	
1	1	1	0	14	SO①	Hold Graphics	.		>		N	↑	n		÷	
1	1	1	1	15	SI①	Release② Graphics	/		?		O	#	o		■	

① These control characters are reserved for compatability with other data codes

② These control characters are presumed before each row begins

Codes may be referred to by their column and row e.g. 2/5 refers to %

□ Character rectangle

Black represents display colour

White represents background

Figure 60.4 96 English teletext character code table.

Diagrams and line drawings can readily be produced on editing terminals, but pictures, logos and maps are normally produced using a graphics digitizer. This consists of a monochrome television camera used to produce a video image of the required picture which is then automatically digitized directly into teletext format. A data link between the digitizer and an editing terminal enables the digitized picture to be transferred into the editing terminal memory where colour and text information are added. Generation of higher resolution teletext images can only be cost effective if an advanced digital graphics facility is used.

The editing terminals are connected to the teletext processor, or system, with its associated page store. Pages created by the editing terminals are transferred into the teletext page store. The system processor is programmed by resident system software to read the store in a controlled sequential manner, the output being fed via a cache memory to the data formatter. Additional memory can also be associated with the teletext processor for archival purposes. In this way a data bank can be created and pages moved to and from the transmission store by commands from the editing terminals.

The data formatter forms the teletext data into the transmission format and shapes the data pulses in accordance with the transmission specification. The shaped data pulses are then inserted into the video signal.

The complexity of a broadcast teletext system will depend on the service requirements. Typical systems employ between two and twelve or more editing terminals and can have parallel operated standby teletext data processors and page stores with automatic changeover facilities for use in the event of a unit failure.

Direct data feeds may also be fed to the teletext system from external information services without the intervention of an editor, using special software to format the data into teletext pages.

60.6 Teletext subtitling

Teletext is ideal for subtitles as they need be displayed on a receiver only when required, and a choice of language is possible. The subtitle is prepared by an editor, independent of the teletext system, using equipment as shown in *Figure 60.5*.

The programme to be subtitled is displayed on the monitor, and the subtitle is prepared with the editing keyboard. The position of the subtitle on the screen and the colour is also determined by the editor from the keyboard. Time coded signals are produced by the programme source and entered onto the floppy disk together with all the subtitle information. In this way the start time, the subtitle text and the stop time are stored sequentially. This method of subtitle preparation allows for maximum flexibility; for example, a number of TV programmes can be worked on concurrently and the subtitles can be viewed independently of the teletext system and previewed in the order that they will eventually be transmitted. If an amendment is found necessary to either the text or the associated time code, flexible editing facilities enable the necessary corrections to be made.

A subtitle page cannot simply be inserted according to the page number in the teletext magazine. If this were done, the page would be transmitted anywhere between 0 and 25 seconds after insertion (assuming a 100 page magazine using two TV lines). Even inserting the subtitle at the end of the current page is inadequate since the transmission delay is variable between 0 and 0.25 seconds. It is therefore necessary to start transmitting the data for the subtitle page on the first available data line after the cueing point rather than wait for the end of the current page. Special techniques are therefore necessary in the teletext system to enable the subtitle page to be inserted into the magazine in this manner.

An alternative arrangement is to use a separate vbi line for subtitle data. This enables the subtitle to be transmitted completely independently of the teletext system and therefore separates the function of subtitling and teletext information provision. A further advantage in the use of a separate vbi line for subtitling is that it enables regional stations or other users to utilize the subtitle information in the vbi in conjunction with the video signal independently of the teletext service. This is particularly important when the regional station is storing the video signal on tape for transmission at a different time.

The use of the separate vbi line enables the programme to be

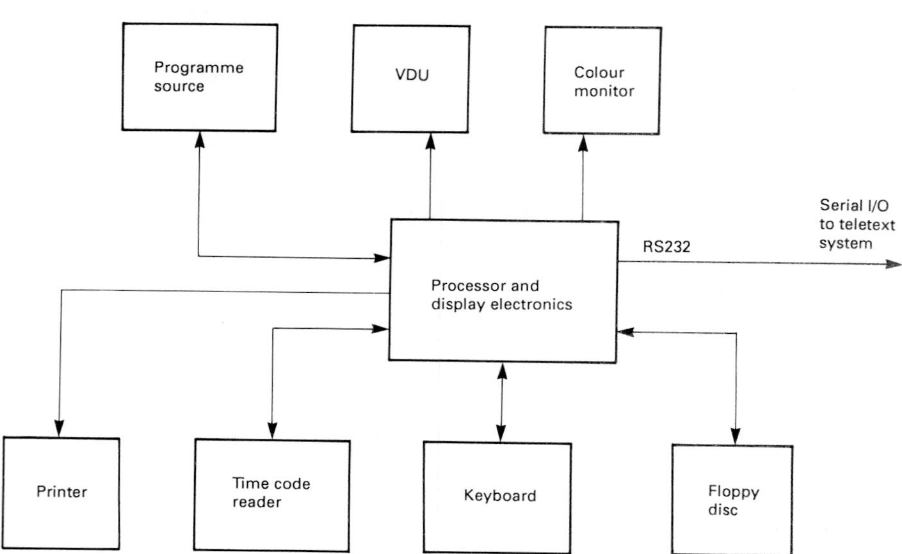

Figure 60.5 Subtitle preparation system

recorded complete with the subtitle, and when the programme is transmitted the subtitle is transmitted directly from the tape recorder with the video signal. The only additional data processing is regeneration of the teletext signal to compensate for shortcomings of the recorder (see section *60.8.2*).

60.7 Regional teletext services

Regional or local TV stations will often wish to use a teletext service available from a major network and insert their own local pages. Alternatively they may wish to add a complete local magazine.

If an additional vbi line is available, then the local station can insert its teletext signal onto this line, provided, of course, a different magazine number is chosen by the local station to those used by the main network. When two teletext services operate in parallel in this manner, the parallel magazine control bit must be set. The receiver decoder will then operate without confusion between the two teletext services.

An alternative arrangement is to use a teletext data store at the local station which is fed with the teletext pages from the incoming video signal. The locally created teletext pages (or separate magazine) can also be entered into the store from the local editing terminals. The system processor is programmed by resident software to read the store sequentially and feed the output to the teletext data formatter. The output from this unit will then be inserted onto the vbi of the regional video signal after erasing any existing teletext signals.

60.8 Teletext data networking

The teletext data signal parameters have been chosen and specified to enable the data to be inserted into the vertical blanking period of a normal television signal without disturbing the normal vision and sound signals. The data are shaped to be skew symmetrical about the half bit rate and to be substantially zero in amplitude at 5 MHz in 625-line systems, or 4.2 MHz in 525-line systems. The data signal, consisting of high speed pulses, is sensitive to amplitude, group delay and non-linear distortion. Such distortion causes overshoots and a deterioration in the separation between the 0 and 1 levels of the data signal.

When a television signal is fed into a widespread distribution network, it passes through various links and switching centres. Although the quality of the colour television signal is maintained to broadcast standards, the data signal can suffer some degradation. To provide the maximum teletext service area and to ensure the data signal does not disturb the reception of normal television sound or vision signals, it is essential that the data signal be radiated without degrading the pulse shape and that the amplitude is at its correct value.

The digital signal can be processed independently of the video signal. This enables the data to be regenerated or linked to other networks as required for the teletext service, without disturbing the normal television network.

60.8.1 Data bridging

Television networks often have many special arrangements for distributing video signals which can vary during the day depending on where programmes are originated and which transmitters are to radiate them. The teletext data signal may be generated centrally, but it must be fed to all the transmitters in the network. The data signal can be transferred between networks using a data bridge.

The data from the input video signal are stored in a buffer memory and read out under the control of the regional signal (*Figure 60.6*). The two video signals can be asynchronous, and the vbi lines on which the data are carried need not be the same. The data must be retimed, pulse shaped, and the video line erased prior to insertion to ensure the highest possible quality of the output teletext signal. A teletext signal can therefore be passed through very complex video television networks or it can be bypassed from studio centres as required, without any disturbances to the normal television operation.

The data bridge may be designed to incorporate a test page generator which is initiated whenever no data signal is received from the input network. The page can be programmed to contain an apology caption and also an identification code so

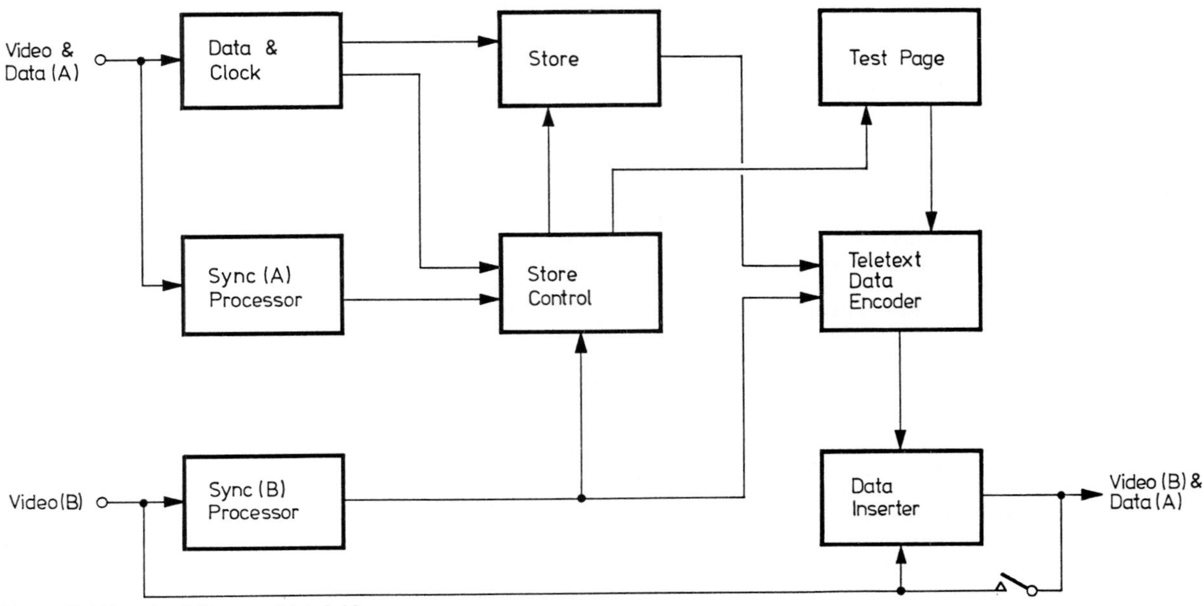

Figure 60.6 Functional diagram of data bridge

that its position in the video network can be readily ascertained remotely.

The line shifting facility of a data bridge enables several teletext programmes to be combined on one television channel and separated remotely. This facility enables various teletext services to be combined onto one video signal for distribution purposes. This technique is particularly useful when a satellite distribution system is used to feed a number of different networks or cable operators. It enables one satellite signal to carry a number of independent teletext channels which can be selected by different networks or operators.

60.8.2 Data regeneration

The transmission of television signals in a widespread network inevitably causes some distortions to occur. The specification for video links and the associated measuring techniques are designed to meet the needs of television signals to ensure that the link satisfies broadcast standards. Different criteria may apply to the requirements for fast data signals which are particularly sensitive to group delay and non-linear distortion that can occur in some rebroadcast links (rbl).

Provided the teletext data can be decoded correctly, its characteristics can be fully restored by regeneration. At the present time, regeneration of the data signal must be carried out at video. Regenerators can therefore be used only where the baseband signal is available such as at the input to a transmitter.

The regenerator strips the data from the video signal, completely reprocesses and retimes them, carries out band shaping and then inserts the data back onto the video signal one field (or one frame) later, after erasing the video line prior to insertion. The video inserter section of the regenerator must meet the full colour television broadcast specification as it is in the main programme path, and it must also be equipped with a physical bypass relay in the event of a malfunction.

In addition to their use at transmitting sites, regenerators are also necessary at the outputs of video tape machines to restore the recorded teletext data when subtitles are recorded as part of programme material.

60.9 Measurement of teletext signal quality

The teletext data signal contained in the vertical blanking interval is very difficult to measure accurately with a normal oscilloscope, particularly as the data are not locked to the video signal. To examine the data, the oscilloscope timebase must be triggered from the data clock run in at the start of the data line, but this requires a special trigger circuit sensitive only to the clock frequency.

The teletext signal is generated by passing the fast digital bit stream, which has identical rise and fall times, through a special band shaping filter. Degradation of this signal by noise, reflections, non-linear amplitude and group delay all contribute to a loss in the *decoding margin*, i.e. the difference in level between the worst 0 and 1 in the bit stream. The principal and most critical parameter of a teletext signal is the decoding margin, and this is the prime criterion of teletext signal quality.

A Lissajous figure may be displayed on an oscilloscope (*Figures 60.1 and 60.7*) by applying the data to the vertical deflection amplifier and using a sub-multiple of the data clock for the horizontal deflection (usually quarter clock frequency). A data line bright-up pulse is also necessary to exclude all television information.

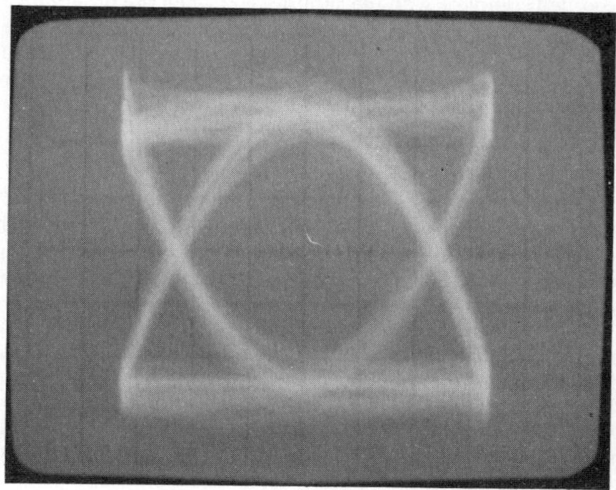

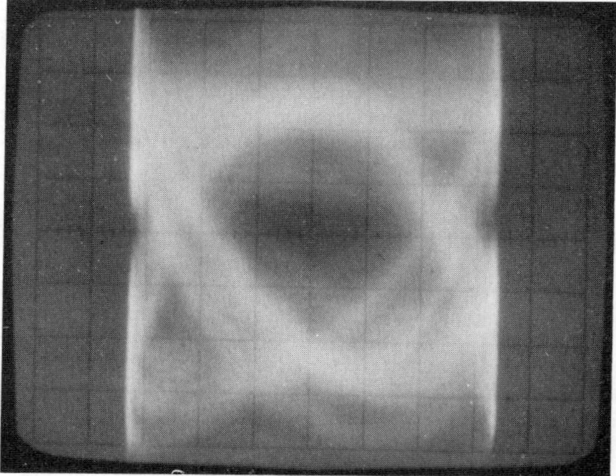

Figure 60.7 Eye displays of distorted data. The eye opening (the separation between the 0 and the 1) is reduced by noise and reflections. Clock jitter and non-linear distortion further reduce the eye opening (lower photo)

The resulting display resembles an 'eye' and the difference between 0 and 1 levels is referred to as *eye height*. The eye opening is normally expressed as a percentage of the true data amplitude. As the data signal is degraded, the eye height falls until, in the limit, decoding is virtually impossible.

The Lissajous figure method of measurement suffers a fundamental disadvantage in that the reference data clock must be extracted from the distorted data stream. Any phase jitter of the reference clock derived from the data stream will cause phase jitter to the horizontal waveform and falsely reduce the height of the eye diagram. This difficulty increases with low values of eye height when extraction of a jitter free clock reference becomes more difficult.

The data quality may be estimated by direct inspection of the data stream on a normal oscilloscope display of the *clock cracker* page (*Figure 60.8*). This test page is composed of alternate symbols having the minimum number of transitions. This page therefore imposes the most searching test of the decoder clock recovery circuits which are synchronized by the data transitions. It is also the least confused bit stream for visual estimation of eye height or decoding margin when using an oscilloscope. Errors on the displayed page of the text are of course very easily seen.

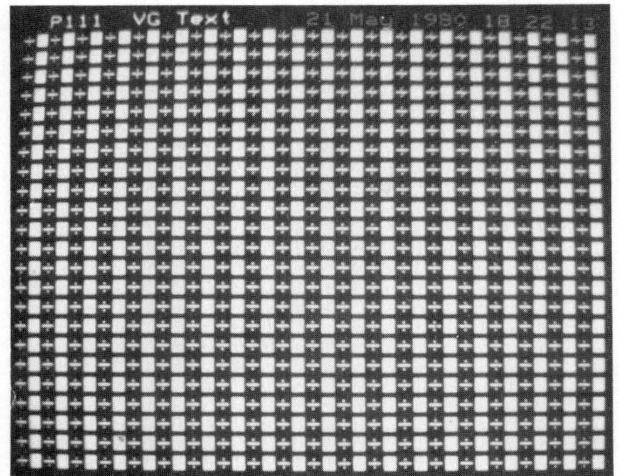

Figure 60.8 Clock cracker page

An alternative method of determining signal quality is to measure the decoding margin over many field periods. This measurement requires the use of a specially designed instrument which measures the effective separation between the worst 0 and 1 levels over typically 100 or 1000 data lines.

Bibliography

Broadcast Teletext Specification, Joint BBC/IBA/BREMA publication (1976)

North American Broadcast Teletext Specification (NABTS), CBS Television Network (1981)

World System Teletext and Data Broadcasting System, Technical Specification, BREMA (B.J. Rogers), Landseer House, 19 Charing Cross Rd, London WC2H 0ES.

Didon-Antiope Specifications Techniques, Télédiffusion de France (1984)

GECSEI, J, *The Architecture of Videotext Systems*, Prentice-Hall, New Jersey (1983)

HAMMING, R W, 'Error detecting and error correcting codes', *Bell System Tech Jour*, **29**, 2 (1950)

HUTT, P R and McKENZIE, G A, 'Theometrical and practical ruggedness of UK teletext transmission', *Proc IEE*, **126**, 12 (1979)

SHERRY, L A and HILLS, R C, 'The measurement of teletext performance over United Kingdom television network', *Radio and Electronic Engineer*, **50**, 10 (1980)

WITHAM, A L et al 'Developments in teletext', *IBA Tech Rev* (May 1983)

MOTHERSOLE, P L, 'Teletext and Viewdata — New information systems using the domestic television receiver', *Proc IEE*, **126**, 12 (1979)

MOTHERSOLE, P L and WHITE, N W, *Broadcast Data Systems – Teletext and RDS*, Butterworth Scientific (1990)

CHAMBERS, J P, 'Teletext: Enhancing the basic system', *Proc IEE*, **126**, 12 (1979)

CHAMBERS, J P, 'Datacast — auxiliary services using teletext technology', *IBC Conf Publication* (September 1986)

GREEN, N, 'Subtitling using teletext service: technical and editorial aspects', *Proc IEE*, **126**, 12 (1979)

HEDGER, J and EASON, R, 'Telesoftware: adding intelligence to teletext', *Proc IEE*, **126**, 12 (1979)

CROWTHER, G O and HOBBS, D S, 'Teletext and viewdata systems and their possible extension to the USA', *Proc IEE*, **126**, 12 (1979)

CROWTHER, G O, 'Adaptation of UK teletext system for 525/60 operation', *IEEE Trans on Consumer Electronics*, **CE-26**, 3 (1980)

STOREY, J R, VINCENT, A and FITZGERALD, R, 'A description of the broadcast Telidon system', *IEEE Trans on Consumer Electronics*, **CE-26**, 3 (1980)

KINGHORN, J R, 'New features in world system teletext', *IEEE Trans on Consumer Electronics*, **CE-30**, 3 (1984)

TARRANT, D R, 'Teletext for the world', *IEEE Trans of Consumer Electronics*, **CE-32**, 3 (1986)

TARRANT, D R, 'Data link using page format teletext transmissions', *IEEE Conf Publication* No 69 (1986)

P L Mothersole F Eng, C Eng, FIEE
VG Electronics Ltd

61 Teletext Decoders

61.1 Basic functions

The teletext decoder, when incorporated in a television receiver, consists of additional integrated circuits associated with the video processing section of the receiver. The decoder input circuit is normally fed with a video signal of about 2 V amplitude derived from circuits associated with the video detector. The decoder output normally provides RGB and blanking signals of similar amplitude which are fed into the low level RGB amplifiers of the receiver. Teletext decoders are only incorporated into receivers which have remote control, and the control signals for the decoder are therefore derived from the remote control circuits.

Various data formats have been specified for different teletext systems, but any of the specified systems can achieve similar results in terms of displayed information and quality of the graphics. The major differences between the systems are the transmission times, because of the amount of data required to be transmitted for a given page, and the complexity of the receiver decoder. (See sections *60.1–60.4* for details of the signal format and associated topics.)

The differences in the decoder are principally concerned with the processing that needs to be carried out on the data stream before the information is in a suitable form to be displayed. The decoder input and output interfaces with the video circuits, and the functions of the input and output circuits of the decoder, are similar. Decoders respond to teletext data on any line (or lines) in the vertical blanking interval (vbi), the acquisition circuits being designed to respond to the clock run-in and framing code to identify the teletext data.

61.1.1 Input circuits

The input, or *data acquisition circuit*, is used to process and select the required teletext data for a particular page so that it can be written into memory. The circuit therefore includes data slicing and timing functions. The data clock, which is locked to the incoming data stream, is also generated by the acquisition circuit. A crystal oscillator normally forms part of this circuit function, the crystal operating at a frequency corresponding to the data rate or a multiple of the data rate. Error correction circuits are also incorporated into the acquisition circuit. Hamming coded address words and the positional information required by free-format systems are checked, and those having a single bit error are corrected.

Alphanumeric characters forming the page content are transmitted in eight bit data groups formed from seven bit character codes plus one (odd) parity bit. The parity bit check for individual characters normally ensures that, should an error be received, the character will not be displayed. If two errors are received, the decoder will display an incorrect character. Since the teletext magazine of pages is transmitted cyclically, when the requested page is next received characters that were received with an error on the first acquisition almost certainly will be received correctly the second time round and therefore displayed correctly. Should an error be received for a character that is already displayed correctly, that character will not be changed.

The viewer selects pages using the remote control unit, and the data from the remote control system are applied to the data acquisition circuit. This control data will specify the page and magazine number for the page that is required to be displayed. A comparator system is incorporated into the circuit so that only requested data are processed by the error correction circuits and fed to the memory as parallel data.

61.1.2 Output circuits

The decoder output circuit contains a character generator ROM for converting the seven bit character data into a dot matrix pattern form, typically a 7 × 5 dot form for each character. The ROM contains the complete set of symbols to be displayed. A typical device would contain at least 96 symbols, which can be selected by means of the seven bit ASCII (American Standard Code for Information Interchange) code applied to the input circuit. Each character is formed from the appropriate dots in the 7 × 5 dot matrix.

The ROM is operated on a *row scan* system, which means that all the dot information in a horizontal row of a character is available simultaneously at the output. A second set of inputs, known as *row address input*, provides the vertical element of the character by determining which of the seven rows is supplied to the output. The dot information is then fed out from a serializing circuit at a rate determined by the write clock, typically about 6 MHz. This signal forms the video output, the

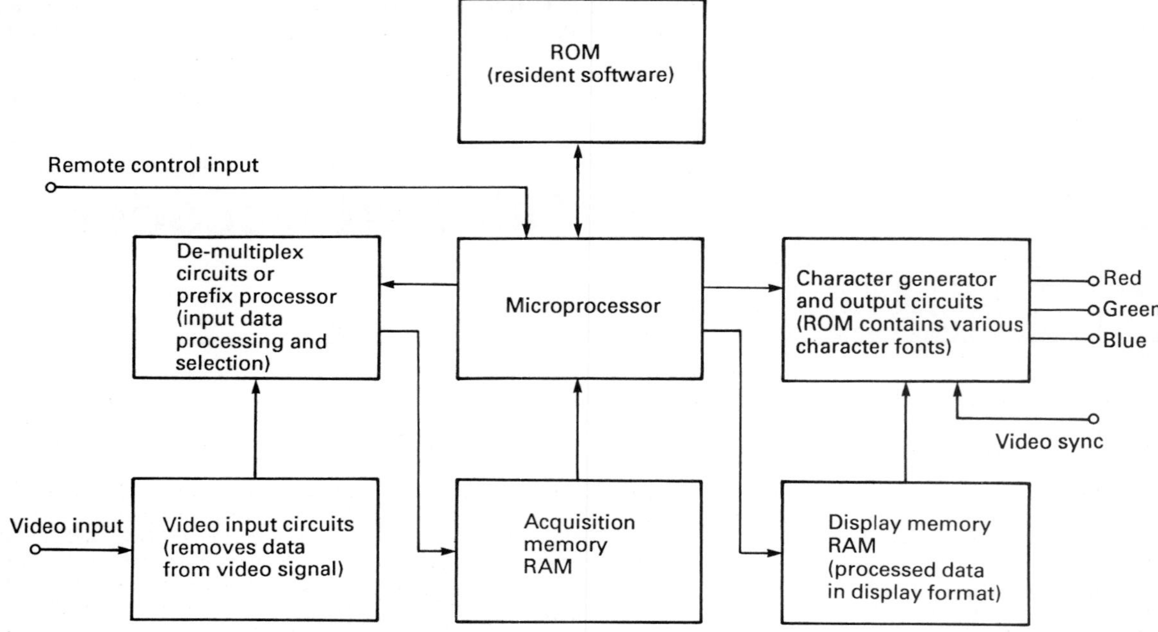

Figure 61.1 Functional diagram of variable data format decoder

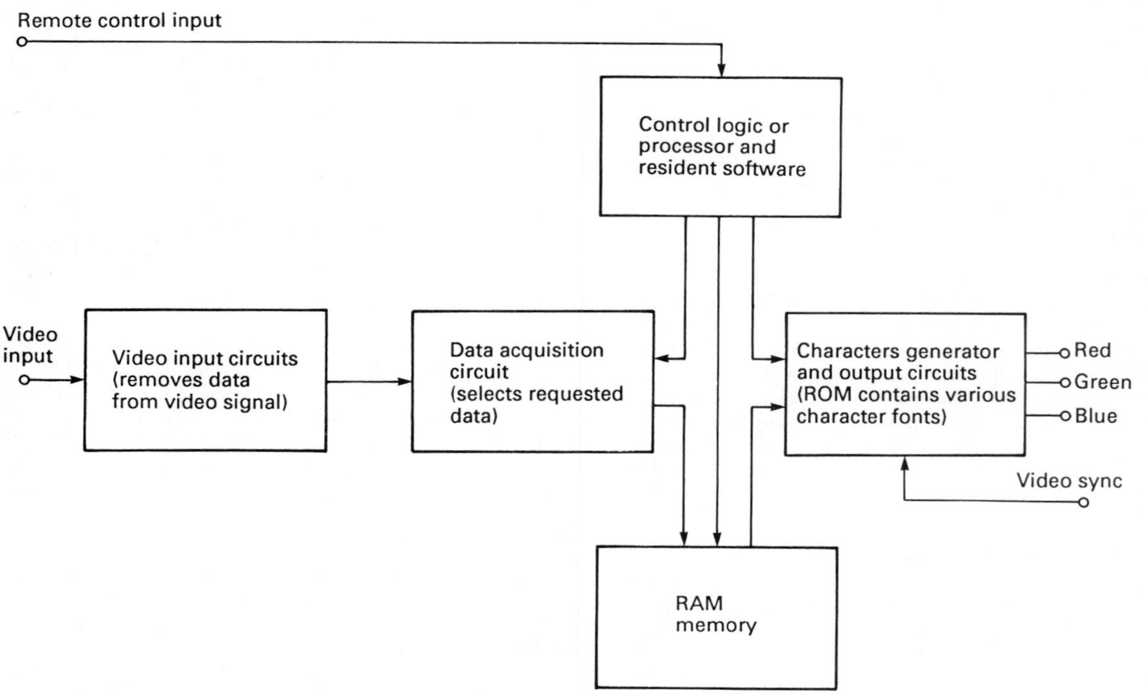

Figure 61.2 Functional diagram of fixed data format decoder

character colour being determined by the colour control circuit selecting the appropriate combination of RGB output signals.

Character rounding can be used to improve the resolution of the characters and makes use of the fact that all the letters, numbers and symbols in normal use are made up from a series of lines. The character rounding circuit modifies the video signal from the character generator when a diagonal line is being generated so that a smoother display is obtained. This does not require any increase in the bandwidth of the video amplifier in the receiver. In effect, the dot resolution of each character is doubled in both the horizontal and the vertical directions. A teletext character generated on the 7 × 5 dot matrix is displayed with a 14 × 10 dot resolution.

Display of alphanumeric data on an interlaced television raster gives rise to interline flicker, which can be objectionable. This effect is overcome by using a non-interlace display for teletext pages. While this effectively removes the interline flicker effects, it also prevents character rounding from being used.

Recently developed decoder output circuits enable the display to be switched for either interlaced or non-interlaced as a design option. At the same time the dot matrix in the character generator ROM has been increased to a 9 × 10 matrix to improve the horizontal definition. Interlace must be used for subtitles and news flashes which are inserted into the video picture.

Certain control functions are also undertaken in the decoder output circuit. These include the selection of graphics or alphanumerics and flashing of words or symbols. A blanking signal for use in the video circuits when a teletext page, or part of a page, has to be inserted into the video signal is also produced by the output circuit. This blanking signal is needed for news flashes and subtitles as these are normally displayed as boxes within the television picture; the header row (row 0) is not displayed. Timing signals are also fed to the output circuit to ensure that the RGB video and blanking signals are timed correctly with respect to the receiver display.

The ROMs contained in the output circuit must contain the complete character fonts needed for various languages that are required to be displayed. Recently developed output circuits also contain an area of programmable RAM. This enables special character shapes to be down-loaded from the editing system to enable the receiver to display high resolution graphics or the special characters required for languages such as Chinese.

Such displays require more information than that contained in a normal teletext page of 960 characters. Larger page memories are therefore required in the decoder, and the access time when such pages are being used becomes correspondingly longer (see sections *60.2–60.4*).

61.1.3 Variable data format page processing

Decoders for use with variable data format systems require two areas of memory together with a processor and the appropriate resident software. The functional diagram of a basic decoder is shown in *Figure 61.1*. The data from the de-multiplexing circuit are fed into the acquisition memory. The processor reformats these data in accordance with the received control and positional information to meet the display requirement and loads them into the display memory. The parallel output of the display memory feeds the character generator and output circuits.

61.1.4 Fixed data format page processing

Decoders for the World Standard Teletext (WST) fixed format system require only a single memory as the data from the acquisition circuit are already in page format form. The memory therefore feeds directly to the character generator and output circuits. The functional diagram of a basic decoder is shown in *Figure 61.2*.

This simplification of the decoder function means that the fixed format systems can display information virtually instantaneously as it is received, and the absence of the processor enables very simple decoder circuits to be achieved. The higher levels of WST teletext, as used for high quality graphics, require additional memory together with a processor. A processor is also used when multiple page stores are incorporated in the decoder.

61.2 Decoder performance

The performance of a teletext decoder or a complete teletext receiver is normally judged by the errors in the displayed picture. The teletext data signal is digitally coded alphanumeric data and, provided the 0 and 1 levels of the data stream are well separated, the decoder is able to decode the data without errors. Degradation of the received data will gradually reduce the separation between 0 and 1 levels until a point is reached where the decoder cannot make correct decisions, and errors will start to be produced. The address information, i.e. the page address and the positional information required by free format systems, is protected by Hamming code (see section *60.3*).

Hamming protection of the magazine and page numbers ensures that the decoder will receive the correct pages. When a page is first received, initial errors will show up as an incorrect character on the page of text. When the page is received for the second time, the decoder will correct these display errors. The teletext service can be considered at its limit when typically three or four errors occur when a page is first received.

As the signal impairment increases, the rate of errors rapidly increases making the page unusable. The service area for teletext, unlike that for normal colour television reception, is relatively abrupt, the transition from good performance to unusable occurring quite rapidly as is common with all digital systems.

The main signal degradation that causes the decoder to produce errors is that due to reflections which distort the pulse shape creating confusion between 1 and 0 levels in the data stream. The teletext data signal may also be degraded by noise and co-channel interference. Impulsive interference does not normally cause any problems unless it occurs during the period when the teletext data signal is being acquired by the decoder, i.e. during the vbi.

The accepted design target for receivers is to operate down to a decoding margin of about 25 per cent with three or four errors occurring when the page is first acquired. Allowing for degradation in the tuner and if circuits, it follows that the decoder would need to operate with a decoding margin of some 20 per cent at its input.

61.3 Signal path distortion

To minimize the distortion that occurs in the conventional television section of the receiver, close attention must be paid to the tuner, if amplifier and video demodulator to optimize teletext performance.

The use of surface wave filters (swaf) significantly improves the performance of if amplifiers compared to LC block filters. Receiver de-tuning has more detrimental effects on teletext reception that it does on the reception of the television video signal. A tuning accuracy of ±50 kHz is desirable, and so a high performance afc system or digital tuning is necessary.

The video demodulator is the main source of non-linear distortion. Simple diode demodulators, which respond to the envelope of the rf signal, introduce a high level of quadrature distortion. This seriously affects the data eye characteristics, causing a loss of eye height and symmetry (*Figure 61.3*). The use of vestigial sideband transmission requires a detector that responds only to the modulation component in phase with the carrier. Fully synchronous demodulators give the best eye height performance, but cost prevents their widespread use in domestic TV receivers.

Quasi-synchronous demodulators give excellent performance with both TV signals and teletext data.

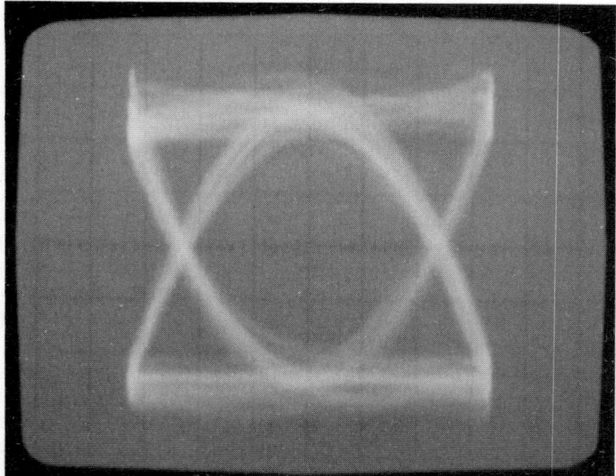

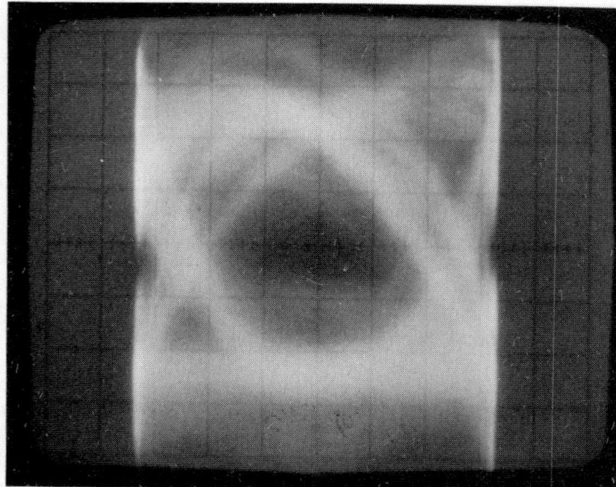

Figure 61.3 Eye displays of distorted data. The eye opening (the separation between the 0 and the 1) is reduced by noise and reflections. Clock jitter and non-linear distortion further reduce the eye opening (lower photo)

To measure and assess accurately the performance of teletext decoders requires a source of signals with controlled levels of distortion. A technique of simulating distortion caused by reflections, devised by IBA engineers, is DELPHI (Defined Eye Loss with Precision Held Indication). A calibrated distortion unit designed for decoder measurement employs this technique for controlling the eye height of the teletext data signal. Provision is also made for adding white noise and simulated co-channel interference in controlled amounts.

In domestic locations, teletext performance can easily be marred by poor antenna installations and reflected signals. Experience has shown that if an antenna is adjusted for good teletext performance, the resulting colour television picture is significantly improved.

61.4 Multipage decoders

Decoders capable of storing several teletext pages are now being fitted in WST receivers. A processor is also incorporated in such decoders to control which pages are to be acquired and stored as required. Various strategies are being employed to reduce the access time (see section *60.4*) and to make the decoder more user-friendly.

One arrangement that does not require any editorial involvement is to incorporate eight page stores in the decoder and arrange for the decoder to capture the seven subsequent pages to that requested by the viewer. This enables the viewer to increment through the magazine with a virtually instantaneous display of the requested page.

An alternative arrangement utilizes additional information provided by the editor. This is added to the page in the form of a data packet, i.e. a row of data that is not displayed by the decoder but is used to instruct the decoder which additional pages it should capture. This instruction is based on the editor's anticipation of the viewer's requirements. The decoder is also made more user-friendly by accepting a command from a single key press for both magazine and page number on the viewer's handset.

Special additional information is displayed on the bottom row of the page to provide simple instructions for the viewer's use. For example, the additional instruction row might contain four topics such as *sport*, *news*, *financial* and *travel* information. Each of these topics would be given a different colour background corresponding to coloured buttons on the viewer's handset. When the appropriate coloured button is pressed, the decoder immediately captures the pages corresponding to this topic, e.g. *sport*. The names displayed on the bottom row instruction line may then change to *football*, *cricket*, *tennis* and *swimming* by the decoder using data in the instruction packets. Again, when the appropriate coloured button is pressed, the pages associated with that particular sport would be captured ready for display. This technique has been developed as part of World System Teletext and is now called *Fastext*.

A system developed in Germany aimed at achieving similar results but without using data packets is called *TOPS*. In this system a special control page, or *table of pages*, which defines all the other pages, is also transmitted. Each 8 bit character position in this page defines the category of each of the other pages. This page is captured and held in the decoder memory and is used to instruct the decoder processor as to which pages should be aquired ready for display in conjunction with the viewer's instruction received via the remote control handset. The handset is fitted with four additional buttons to simplify operation. To guide the viewer, additional information is also transmitted on special pages and is displayed on the bottom row of the page.

61.5 Teletext adaptors

Teletext adaptors are designed so that they may be added to a conventional receiver to provide a teletext facility. The adaptor normally has its own tuner, if amplifier and demodulator driving a teletext decoder together with a remote control system. In fact, it incorporates virtually all the circuits contained in a TV receiver with the exception of the crt and the associated timebases. The RGB outputs from the teletext decoder are recoded into a conventional television signal

(PAL, NTSC or SECAM). This signal is then modulated onto an rf carrier so that it can be fed into the antenna socket of a conventional TV receiver. Such adaptors are not user-friendly in that when the television channel is changed, the decoder channel selector has to be changed in sympathy and a second remote controller is required for the adaptor.

The main limitation of teletext adaptors is the need to recode the RGB signals from the teletext decoder into a conventional colour coded television signal. Text signals are usually coloured and have fast edges. When such signals are converted into a coded colour television signal, the bandwidth of both the colour and the luminance components are limited causing a very significant loss of legibility in the final displayed picture. Teletext adaptors have proved unpopular due to their relatively high cost and limited performance.

More recent adaptors drive the receiver via the Scart socket with RGB and audio signals. Decoders are now being fitted into video recorders. The adaptor (or recorder) can then be used for television station selection and only one controller is necessary.

61.6 Encrypted teletext services and telesoftware

Digital data can readily be encrypted. When teletext data are transmitted in encrypted form, they can then only be used by receivers whose decoders are equipped with necessary de-encryption circuits. Such services are not normally for consumer use but are used in conjunction with a microcomputer facility having an RS232 input. The decoder for use with such systems therefore consists of a tuner, if amplifier and demodulator, similar to those used in a conventional television receiver, and a teletext decoder with serial data output providing an RS232 feed. In such units, the channel tuning and page address are often fixed and preselected.

The level of security required for such special encrypted services, and the arrangements for identifying the user, are functions of the encryption system; the television service provides the data transport mechanism. Teletext data format is used as this has been designed for data transmission in conjunction with a television signal. It also enables various integrated circuits, designed for consumer use, to be employed in the special decoder to improve the cost effectiveness. A decoder for encrypted services can be a stand-alone unit, rather like a teletext adaptor, but with an RS232 data output feed or incorporated into the microcomputer or end-user facility.

Software can be down-loaded to subscribers over the teletext network, but special precautions are necessary since a received error could corrupt the program. The data stream can be page formatted or sent as individual rows of data with a suitable address code. A cyclic redundancy check word (crc) is normally added to the page (or part of a page) or row to minimize the acceptance of errors. The principle of crc is to generate a check word from the data stream at regular intervals and to insert it into the data stream. The decoder is able to perform the inverse operation on the received data and check word whose composition is such that an output other than zero indicates an error.

The character code normally contains eight bit words, and hence a code could be all 0s or 1s, i.e. with no transitions occurring in the character code. The absence of such transitions, which are used to help synchronize the data clock, have a detrimental effect on the clock recovery circuits used in data bridges, regenerators and decoders. It is essential therefore that the teletext network is able to handle eight bit data code without creating errors within the network.

Bibliography

Broadcast Teletext Specification, Joint BBC/IBA/Brema publication (1976)

World System Teletext and Data Broadcasting System, Technical Specification, BREMA (B.J. Rogers), Landseer House, 19 Charing Cross Rd, London WC2H 0ES.

Didon-Antiope Specifications Techniques, Télédiffusion de France (1984)

RAYERS, D J, 'Telesoftware by teletext', *IBC Conv Publication* (September 1984)

ROGERS, B J, 'Methods of measurement on teletext receivers and decoders', *Proc IEE*, **126**, 12 (1979)

CHAMBERS, J P, 'BBC datacast', *EBU Review-Technical*, 222 (April 1987)

BUGG, R E F, 'Hardware minimization of videotext or teletext decoders of display redundancy', *IEEE Trans on Consumer Electronics*, **CE-30**, 3 (1984)

BUGG, R E F, 'Defined-format teletext moves to higher levels of display and yet fulfils its promise of compatability with first generation decoders', *IEEE Trans on Consumer Electronics*, **CE-31**, 4 (1985)

KINGHORN, J R 'Using extensions to world system teletext', *IEEE Trans on Consumer Electronics*, **CE-31**, 4 (1985)

KINGHORN, J R and GUENOT, A, 'Packet and page format data reception using a multistandard acquisition circuit', *IERE Conf Publication* No 69 (1986)

WITHAM, A L et al, 'Development in teletext', *IBA Tech Rev* (May 1983)

ARD/ZDF/ZVEI Directive 'TOP' System for Teletext, Intitut für Rundfunktechnic, Munich

MOTHERSOLE, P L and WHITE, N W, *Broadcasting Data Systems – Teletext and RDS*, Butterworth Scientific (1990)

C Dawkins BA (Cantab) and
C P Arbuthnot B Eng
British Telecom Research Laboratories

62 Interactive Videotex

The videotex concept was developed originally at the Post Office Research Centre at Dollis Hill in London (later to become the British Telecom Research Laboratories at Martlesham Heath) during the 1970s.

Videotex, then known as *viewdata*, grew out of early feasibility studies into possible picture/phone technology. Full television picture transmission over the telephone network proved impossible to achieve at that time, so viewdata was conceived as an achievable blend of television and telephone technology which would stimulate increased use of the telephone network outside busy periods.

It was first demonstrated publicly at the Eurokom conference in 1976, and after a market trial was launched, under the name *Prestel*, as a public service in the UK amidst considerable publicity in 1979.

Although originally conceived as a mass-market, residential service, centrally provided, in practice the most rapid take-up was in specialized business sectors, notably the travel trade. As well as the public Prestel service, a substantial market developed in private videotex systems (pvs) serving particular user groups.

Other public telecommunication operators throughout the world watched the early UK developments with interest, and many purchased copies of Prestel software to run their own market trials. In France, a more radical approach was taken, and rapid market growth was achieved by the (state owned) France Télécom supplying videotex terminals free of charge to telephone subscribers to enable them to access an electronic directory enquiry (edq) service. It was originally thought that this would make it possible to phase out both the printing of telephone directories and the provision of the operator assisted directory enquiry service. While this objective has not yet been completely realized, a very wide variety of application providers has emerged, providing services covering every aspect of business and residential life.

62.1 Fundamentals

62.1.1 Definition

Videotex, or viewdata, is the generic term given to text and graphic based computer services, accessed over telecommunication networks, designed for use by non-expert users. It brings together computer, telecommunication and television technologies.

It should be noted that the CCITT (International Telegraph and Telephone Consultative Committee) uses the term *videotex* to cover both *interactive videotex* and *broadcast videotex* (more commonly *teletext* – see section *60*).

62.1.2 Characteristics of a videotex system

The main characteristics of a videotex system that a user may perceive are:

- two way interactivity between the videotex terminal and service host computer,
- retrieval of a broad spectrum of information from one or more sources,
- relatively low cost of provision,
- provision of other services (messaging, closed user groups, processable data distribution, etc.),
- public or private system.

A typical videotex system comprises four main components (see *Figure 62.1*):

- *User terminals*. The videotex information received from the service is decoded and presented on a display device. Terminal control and information requests are entered using a keyboard. Typically, terminals are connected to the telecommunications network using a modem (device converting digital signals to analogue, and vice versa, in order that they may be transmitted within the audio bandwidth of the network).
- *Telecommunications network*. The terminal is connected to a service centre via a telecommunications network. Usually the public switched telephone network (PSTN) is used for local access to a national *packet switched* data network. The packet switched network routes the videotex information between the terminal and service centre more efficiently than using a direct PSTN connection.
- *Service centre*. This provides the access point for the user terminal to the videotex application services' databases and carries out certain control and management functions such as session set-up, billing, gateways to external services, etc.
- *Application service*. This holds the information on a database

in a structured order (commonly in logical units called *pages*) and may provide some data processing application functions. Application services and their databases may be situated locally or remotely from the service centre. Information is created using special editing terminals or directly from other sources such as stock exchange computers.

62.2 Terminals

62.2.1 User terminals

There are three main classes of user terminals: dedicated, adaptor and personal computer (PC) emulators.

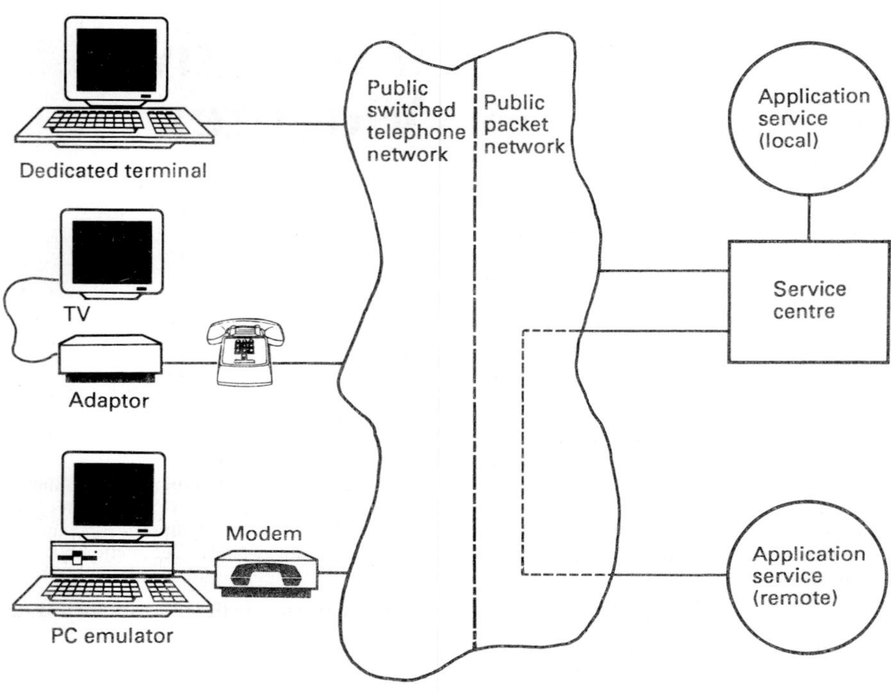

Figure 62.1 Typical videotex system using public telecommunication network

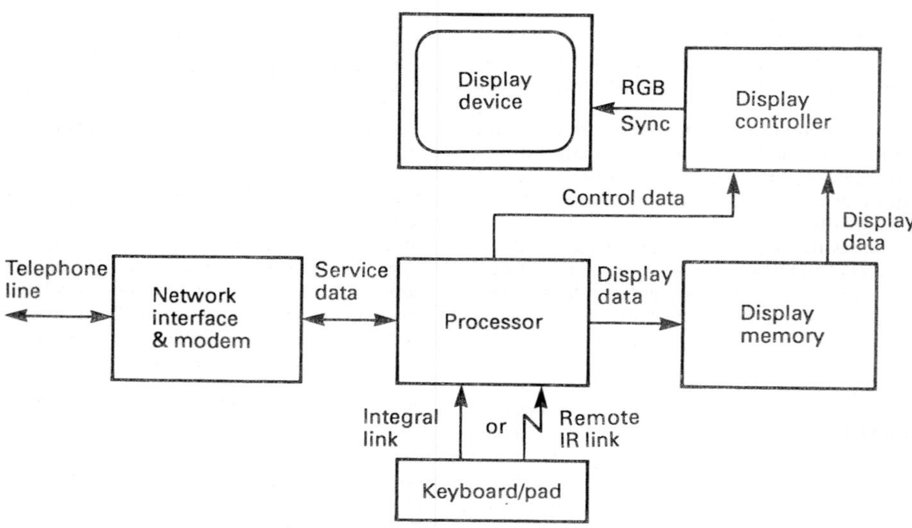

Figure 62.2 Components of a user terminal connected to the PSTN

Dedicated terminals are purpose designed terminals that range from low cost units with monochrome display and numeric keypad, handling text and simple graphics information, to elaborate units providing colour display, with an alphanumeric keyboard and capable of displaying geometric and photographic images.

Adaptors provide all the basic functionality of a dedicated terminal except for the display device which is usually a domestic television. The resulting display presentation is inferior due to the limited bandwidth of the television's demodulator and decoder stages, the low crt resolution and interlacing.

A *PC emulator*, utilizing the processing power and display capabilities of modern personal computers with suitable application software can usually provide a videotex terminal function, although a modem will also be required if accessing videotex services over the PSTN.

There are six main components of any user terminal connected via the PSTN (see *Figure 62.2*). These are:

- the *network interface* and *modem* for signalling and the transmission and reception of data by modulation and demodulation of voice band frequencies,
- the *processor* for managing the terminal's operation (e.g. modem control, keyboard polling, handling of peripheral ports for printers, etc.) — this decodes the videotex data for presentation, executes commands received from the service affecting terminal control and encodes information to be sent to the service centre,
- the *display controller*, which manages the display memory and handles the refreshing of the display device from the data held in the display memory by providing the necessary video signals (R, G, B, sync, etc.),
- the *display memory* for holding the decoded information for display refresh,
- the *display device* for presenting the information to the user (usually a raster based crt is used although flatscreen liquid crystal display (ldc) technology is becoming more widely used),
- the *keypad* (incorporating digits 0–9, * and #) or *keyboard* (alphanumeric and application function keys) for data entry and terminal control.

62.2.2 Editing and inputting terminals

Information for the videotex database can be created and edited by *information providers* (IPs) using special editing terminals. These terminals vary greatly in functionality depending mainly on the presentation techniques used (see section *62.3*). Good editing terminals allow information containing complex graphics to be quickly and accurately generated and manipulated by the operator and may utilize digitizer boards, pointing devices and video cameras.

The information pages created may then be uploaded onto the application service's database, often using a *bulk update* procedure. Alternatively, a few pages may be created or edited while on-line to the database.

62.3 Presentation techniques

62.3.1 Display characteristics

Most videotex systems present their information in *frames*. (A frame is a full display screen, and not to be confused with a television raster frame.) Frames are displayed within the *defined display area* (dda) which is a central rectangular area with an aspect ratio of typically 4:3 leaving a surrounding border area (which may be of a defined colour). This helps to ensure that all information is displayed despite any variation of video timing circuits (i.e. overscanning) that may exist from one display device to another.

For character based information, the frame format is normally 24 rows of 40 character cells (960 character cells per frame) but is dependent on the character cell pixel dimensions used, the raster resolution of the intended display device and the dda. In practice, for a 6 MHz 625-line raster display, a dda is set to 240 lines per field and an active line period of 40 μs which gives a character cell of 6 horizontal by 10 vertical pixels. This provides acceptable clarity for text display on a domestic television. For 525-line displays, the reduction in lines per field means that either the number of rows is reduced to typically 20, maintaining a vertical character cell dimension of 10 pixels, or for 24 rows the cell height is reduced to 9 pixels.

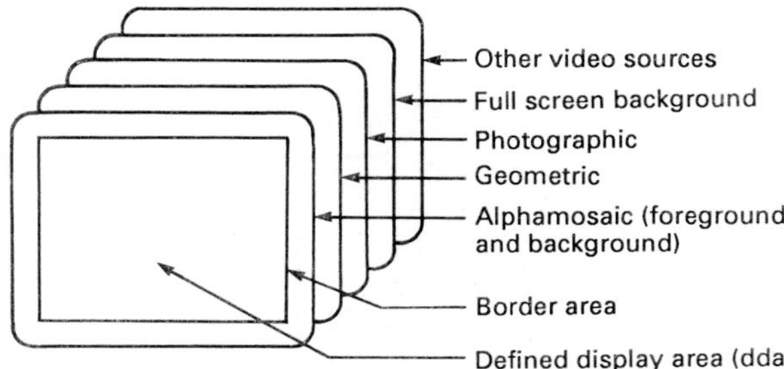

Figure 62.3 Theoretical layered structure of display

PC emulators often provide high resolution displays. This enables improved definition character cells but the frame format (e.g. 40 × 24) will still be the same for any one videotex system.

Many private videotex systems provide an 80 × 24 format using PC displays.

62.3.2 General presentation formats

The success of a videotex system depends on the effective presentation of information. With the rapidly evolving technologies used by videotex, newer techniques for information presentation are becoming available, such as animated geometrics, high resolution photographics and sound.

Present technology applied to videotex may be divided into four independent presentation entities (*modes*), increasing in graphics capability but also in complexity and hardware costs:

- *alphamosaic*; character based display of text and simple graphics with a range of attributes,
- *dynamically redefinable character sets (drcs)*; enhancement to the alphamosaic mode which allows character sets to be custom defined,
- *geometric*; display of graphic images constructed from primitive geometric shapes (the term *alphageometric* is used when combined with the alphamosaic mode),
- *photographic*; display of still photographic images using television techniques (the term *alphaphotographic* is used when combined with the alphamosaic mode).

Any combination of the above may be used in a videotex service. The theoretical structure of the display is normally in the following order of precedence (see *Figure 62.3*):

1. alphamosaic and drcs foreground (i.e. character shape)
2. alphamosaic and drcs background
3. geometric information
4. photographic information
5. full screen background
6. any other video source

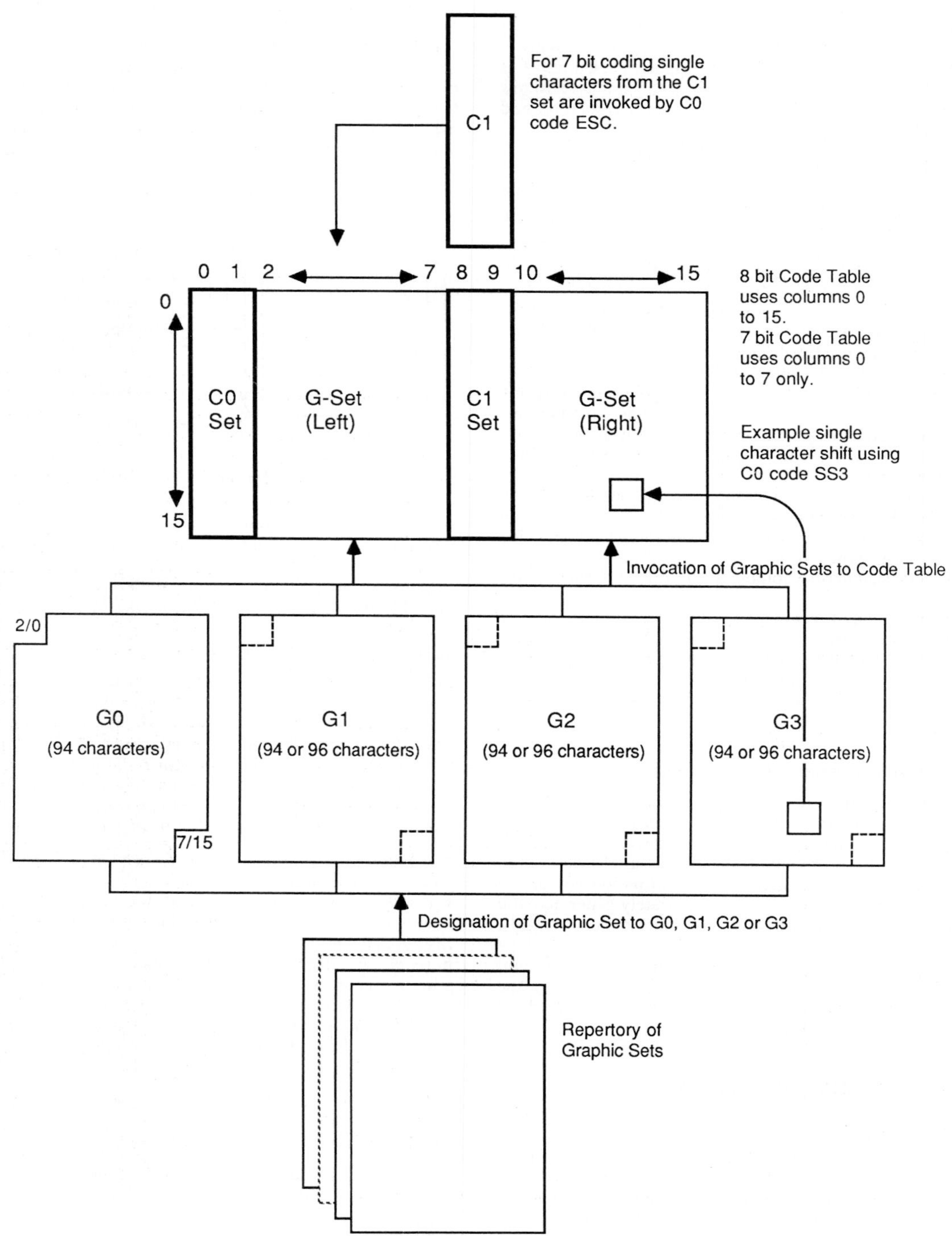

Figure 62.4 Principle of the code extension technique

However, in some systems, multi-mode frames are built up out of a chronological sequence of drawing commands with the effects of each superimposed over those of previous ones.

62.3.3 Graphic and control sets and their coding

Displayable characters and controls that effect the display and other terminal functions are respectively coded for storage and transmission as *graphic* (G) and *control* (C) sets. They are described in terms of 7 or 8 bit code tables which are stored in the terminal and used as defined by ISO 2022 (Code extension techniques).

The code table consists of either 128 or 256 positions, depending on whether 7 bit or 8 bit coding is employed, and is divided into 8 or 16 columns of 16 rows with each position defined by a unique 7 or 8 bit code respectively. The codes are expressed in the notation *column/row* with the range 0/0–7/15 for 7 bit codes and 0/0–15/15 for 8 bit codes. These codes are used to transmit the data between the service centre and the terminal. However, certain data types, e.g. photographic, use a transparent data mode of transmission (see section *62.3.4*).

A 7 bit code table retains columns 0 and 1 for a C-set and columns 2–7 for a G-set, whereas an 8 bit code table contains an additional C-set and G-set in columns 8 and 9 and 10–15 respectively. C-sets therefore comprise 32 positions and G-sets comprise either 94 or 96 positions. A 94 position G-set reserves positions 2/0 for the *space* character and 7/15 for *delete*, while a 96 position G-set allows 2/0 and 7/15 to take on other meanings.

At any one time four G-sets (G0, G1, G2 and G3) and two C-sets (C0 and C1) may be 'designated' from a larger repertory of sets (see examples in *Figure 62.5*).

The initial default sets normally correspond to the following:

- G0: *primary graphic set* compromising 94 alphabetic, numeric, punctuation and miscellaneous symbolic characters
- G1: *mosaic graphics set* or *geometric* primitives
- G2: *supplementary graphic set* containing accents, other characters and symbols
- G3: *line-drawing* and *mosaic graphic set*
- C0: *primary control set* including codes called *format effectors* that control the movement of the cursor (i.e. the screen position at which the next character is to be placed), and codes to clear the screen, invoke a new graphics (G) set, and control transmission and other special terminal functions
- C1: *supplementary control sets* containing the graphic attribute codes

The desired graphic sets having been designated, a G-set is 'invoked' into the code table by various C0 controls. These may either replace the complete G-set resident using a 'locking shift' C0 code (e.g. SI, SO, LSO), or carry out a single character shift, the latter only applicable to G2 and G3 sets using 'single shift' C0 codes (SS2 and SS3).

The single character shift is useful for encoding accented characters where the accent is transmitted first followed by a G0 character. This is referred to as the *composition technique*.

Additionally, in some videotex systems, a 96 position L-set is used which contains mosaic and upper-case alphabetic characters. This is invoked by certain C1 controls until a contradictory control or graphic character is received or the start of a display

	0	1
0	NULL: Has no effect	
1		CON: Cursor On
2		RPT: Repeat display of character
3		
4		COF: Cursor Off
5		
6		
7		
8	APB: Active Position Back	CAN: Cancel. Fills to end of row with space
9	APF: Active Position Forward	SS2: Invokes one G2 character
10	APD: Active Position Down	
11	APU: Active Position Up	ESC: Escape. Used for additional control
12	CS: Clear Screen	
13	APR: Active Position Return	SS3: Invokes one G3 character
14	SO: Shift Out (Invokes G1 Set)	APH: Active Position Home
15	S I : Shift In (Invokes G0 Set)	APA: Active Position Address

(a)

	2	3	4	5	6	7
0		0	@	P	`	p
1	!	1	A	Q	a	q
2	"	2	B	R	b	r
3	#	3	C	S	c	s
4	¤	4	D	T	d	t
5	%	5	E	U	e	u
6	&	6	F	V	f	v
7	'	7	G	W	g	w
8	(	8	H	X	h	x
9	)	9	I	Y	i	y
10	*	:	J	Z	j	z
11	+	;	K	[	k	{
12	,	<	L	\	l	|
13	—	=	M	]	m	}
14	.	>	N	^	n	‾
15	/	?	O	_	o	

(b)

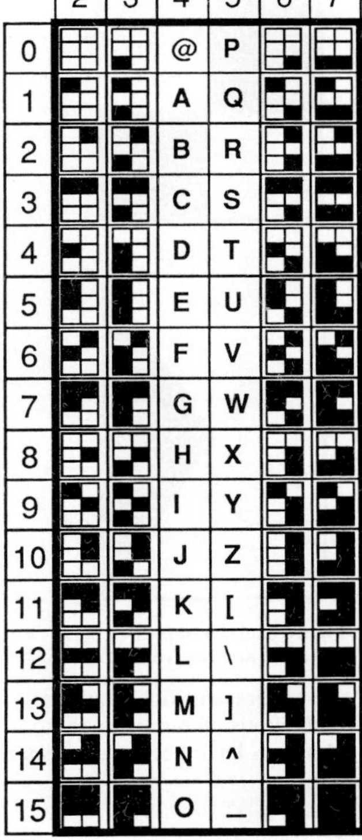

	2	3	4	5	6	7
0			@	P		
1			A	Q		
2			B	R		
3			C	S		
4			D	T		
5			E	U		
6			F	V		
7			G	W		
8			H	X		
9			I	Y		
10			J	Z		
11			K	[		
12			L	\		
13			M	]		
14			N	^		
15			O	_		

(c)

Figure 62.5 Typical control and graphic sets (as used in CCITT DS II): (a) primary control set (default CO), (b) primary graphic set (default GO), (c) supplementary mosaic graphics L-set

row is reached when the previously invoked G-set is re-invoked into the code table.

62.3.4 Transparent data coding

Transparent data mode is used for increased efficiency when certain presentation modes like geometric and photographic contain a large amount of random data which may contain what would normally be interpreted as control codes. Data are transmitted within a packet which has a header indicating that the following *n* data bytes are transparent and therefore should not be decoded in the normal manner.

62.3.5 Alphamosaic display

An alphamosaic display is a character based display using a subset repertoire of alphanumeric characters taken from the ISO 646 standard and additional mosaic characters (see *Figure 62.6*), examples of which are:

- *block mosaics* consisting of a combination of six rectangular elements (2 wide × 3 high) in the character cell giving an 80 × 72 element dda resolution for a 40 × 24 frame format,
- *smoothed mosaics* where the shapes are bounded by lines between the corners of the six rectangular elements or between the cell corners and the cell centre,
- *line-drawing graphics*, *jointive arrows* and miscellaneous graphics.

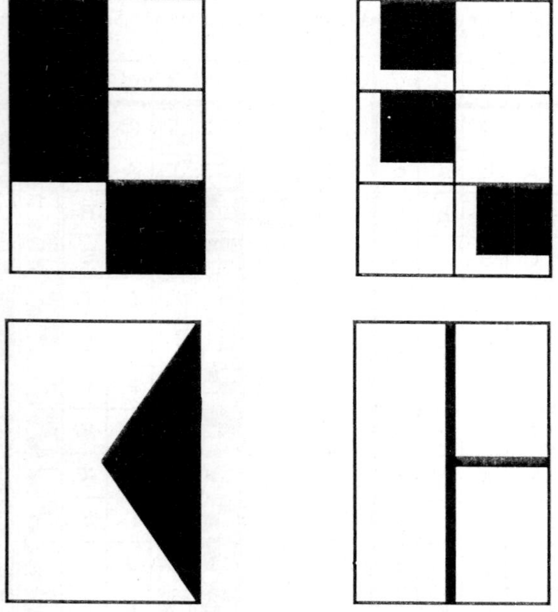

Figure 62.6 Examples of mosaic characters: (a) continuous, (b) separated, (c) smooth, (d) line drawing

Mosaics and line-drawing graphics may be displayed with contiguous elements or separated elements in which case the elements have a separation space usually on the left and lower edges.

62.3.5.1 Attributes

Attributes from a repertoire of control codes (usually C1) may be assigned to graphic characters. Some common attributes are:

foreground colour: colour of alphamosaic character shape
background colour: colour of remaining character cell area
lining: underlines alphanumeric characters and separates mosaic and line-drawing graphics
size: for double width, double height and double size characters
flash: characters displayed alternately in the prevailing foreground and background colours, or another background colour. (Flashing may have an even mark:space ratio or can be three-phase to give dynamic effects with rates of 1–2 Hz.)
conceal: characters displayed as spaces (i.e. prevailing background colour) until the user chooses to reveal them
invert: character foreground and background colours exchanged

A minimum of eight attribute colours are provided comprising the primary colours (red, green, blue), secondary colours (magenta, yellow, cyan), black and white.

Attributes may be applied to characters in either serial or parallel mode. *Serial attributes* are applied from the current cursor position to all successive character positions on the same row, until either a contradictory attribute control is reached or the end of the row. By storing the attributes in place of characters, very efficient use of display memory is obtained, requiring 960 bytes for a 40 × 24 frame format. Where a character position is used up by an attribute, a space character is implicitly displayed leading to obligatory spaces whenever a new attribute is required. In practice, this is rarely a problem with text where attributes can be applied between words, but it can place some constraints on intricate mosaic patterns such as maps.

Parallel attributes apply to all characters subsequently written to the display irrespective of their position until a contradictory attribute control is received by the terminal or certain cursor and display controls are received. This scheme requires at least two bytes per character position, i.e. a minimum of 1920 bytes for a 40 × 24 frame format, but overcomes the necessity of displaying a space when attributes are required.

62.3.6 Dynamically redefinable character sets (drcs)

Character patterns are down-loaded from the service to allow any shape character to be defined (including Chinese, Cyrillic and Greek) and enable fine graphic pictures to be constructed. A 40 × 24 frame format with a 6 × 10 pixel character size gives an overall dda pixel resolution of 240 × 240. Once a drcs (or part of it) is down-loaded, it can be designated as a G-set and then invoked into the code table in the usual way, allowing frames to contain both alphamosaic and drcs characters. It is important that the drcs character cell size is in accord with that of the terminal otherwise the data describing the drcs patterns will be incorrectly presented.

There are two types of drcs: basic and colour (see *Figure 62.7*). *In basic drcs*, only the character patterns are down-loaded, and the foreground and background colours depend upon attributes as with alphamosaic characters. *Colour drcs* defines the colour of each pixel in the character cell and when displayed they are not affected by colour attribute controls. The pixel colour may be one from a range of 2, 4, 8, 16 or more colours held within a colour map, which may be defined from a much larger palette of, for example, 4096 colours (see section *62.3.9.2*).

The creation of intricate drcs patterns can be time consuming without the aid of special editing terminals, and significant transmission delays may be incurred while down-loading extended colour drcs. Extra memory is required to store any drcs that is down-loaded. A 94 character drcs using the common 12 × 10 pixel character cell each described by 4 bits (i.e. 16 colours) requires 5640 bytes of additional memory.

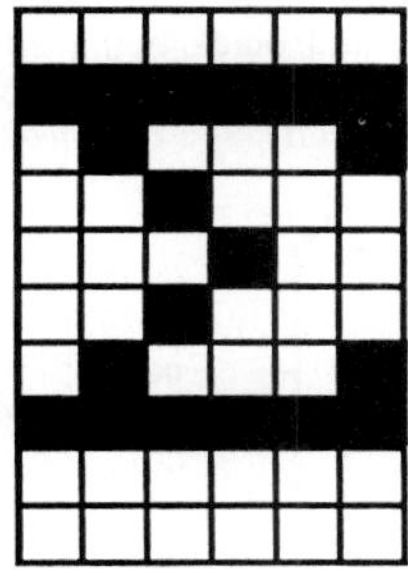

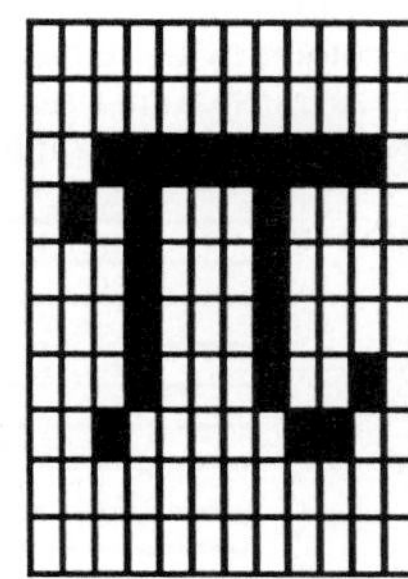

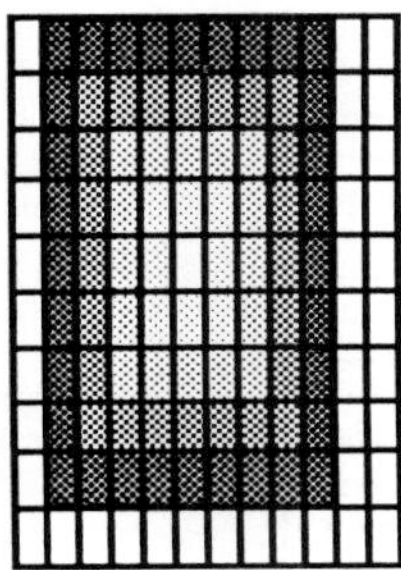

Figure 62.7 Examples of drcs characters: (a) 6 μ 10 basic, (b) 12 μ 10 basic, (c) 12 μ 10 colour

62.3.7 Geometric displays

For geometric coded displays, the dda is regarded as a continuous area on which any point may be defined using a pair of coordinates (horizontal and vertical distances from the lower left-hand corner of the dda) and may be specified as accurately as desired. A typical resolution would be 240 × 240 pixels with each pixel defined by 4 bits and therefore requiring over 28 kbytes of display memory.

Picture description instructions (pdis) or *primitives* are used to describe the pictorial information for display, being abstractions of basic actions that a geometric graphics device can perform. Examples are *point*, *line*, *arc*, *rectangle*, *polygon*, *incremental*, etc. The incremental primitive enables random shapes (e.g. maps and handwriting) to be rendered. Text may also be displayed with position, size and orientation being defined.

Attribute controls specify the characteristics of the primitives on the display and include *colour*, *brush*, *line width*, etc. A set of attributes associated with one primitive may be grouped together into a bundle. All or part of a display image may be rotated, scaled and translated.

Some videotex systems define their own geometric pdis, while others are based on geometric standards that have already been defined for other applications (e.g. computer aided design systems) such as the *graphical kernal system* (gks) defined by ISO 7942. Geometrics extend the graphic capability by providing flexibility and creativity in the composition of pictorial displays. They also offer to an extent wider compatibility between databases and terminals of, for example, different display resolutions.

62.3.8 Photographic displays

A still colour photographic image may be included within a videotex frame consisting of individually defined pixels with many grey/colour levels. This is sometimes called a *photo-videotex*. Colour television techniques are often used to define the image in digital form, based on the CCIR digital television studio standard (Recommendation 601) using luminance (Y) and chrominance (R-Y, B-Y) components, the latter, when scaled to fill the digital quantization ranges available, being termed U and V respectively.

The image sampling structure commonly employed shares the chrominance component between two adjacent pixels on a raster line (in accordance with CCIR Recommendation 601) leading to each pixel typically defined by 16 bits, 8 for the luminance component and 8 for alternate chrominance components.

The display resolution used with this sampling structure is either 2:1:1 or 4:2:2, using the CCIR nomenclature, where the frequencies for the three components Y, U and V are expressed in sequence and relative to 3.375 MHz. For example, 4:2:2, which is the CCIR studio standard, requires Y samples at 13.5 MHz and U, V samples at 6.75 MHz. For a dda with 40 μs active line period this gives 540 pixels per line ($40 \times 10^{-6} \times 13.5 \times 10^{6}$), and for a 625-line interlaced raster display a vertical resolution of 480 pixels (2 × 240), a total of 259 200 pixels. For 16 bits/pixel, a display memory of more than 506 kbytes is required.

Photographic images may be of a defined 'photo area' and position within the dda. A system may employ a restriction to the photographic area and a lower sampling structure in order to limit the display memory requirement of the terminal. For example, a photo area of one sixth of a 270 × 240 pixel dda (10 800 pixels) using 2:1:1 (6.75 MHz) sampling requiring 16 bits/pixel can be implemented with 22 kbytes of display memory.

A terminal providing a photographic display capability will usually use a separate photographic display memory (often referred to as a *frame store*) from that used by previously mentioned character based presentation modes. The latter will normally have a higher display priority (refer to section *62.3.2*) with the underlying photographic image being displayed wherever pixels of the higher order presentation entities have a colour assigned to them that signifies transparency (see section *62.3.9.2*).

Special coding algorithms are employed to reduce the amount of data needed to be stored by the database and also transmitted. Currently, these include *differential pulse-code modulation* (dpcm), *discrete cosine transform* (dct), and *modified Huffman coding*. DPCM stores and transmits the differences between adjacent digitally encoded sample values of a picture providing 50 per cent data compression and is relatively simple to implement, while dct can reduce the bits/pixel from 16 to between 1 and 2 through efficient coding of the spatial frequency components. Modified Huffman is a run length coding scheme (similar to that used for facsimile).

More recent work on an *adaptive dct* (adct) algorithm has provided compression ratios in the order of 16:0.75, reproducing images that are subjectively very close to the original. The technique offers either sequential or progressive picture build-up, the latter enabling a crude picture to be displayed more quickly using a 16:0.08 compression ratio (see *Figure 62.8*). Photographic data are transmitted in transparent data mode (see section *62.3.4*).

Photographic displays allow the full capabilities of the display device, the crt, to be used and provide a simple method of frame creation (e.g. capturing an image with a video camera) and editing.

Figure 62.8 (Top) Image data compressed by 16:0.75 (approximately 21:1). (Bottom) Image data compressed by 16:0.08 (200:1)

62.3.9 Display management functions

Several functions are often provided as part of a videotex system's data syntax to manage and regulate one or more of the presentation entities described earlier. These include macro, colour, format, reset, terminal facility identifier, timing control, scrolling and processable data.

62.3.9.1 Macro

The *macro* function provides the capability of encoding sequences of presentation entity codes to be executed upon command. A macro definition consists of an arbitrary string of locally buffered presentation entity code that is identified by a code from a macro G-set after designation and invocation. A macro code may be included within another macro definition, providing a nesting capability, and may be linked to a user input mechanism.

62.3.9.2 Colour

Colours used by any of the presentation entities (usually with the exception of photographic) are defined within a look-up table called the *colour map*. This holds actual colour values (in RGB format) in each of the colour map's ordinal addresses. The addresses are associated with each of the pixel values of the videotex frame stored in the terminals display memory. A typical colour map may have 16 or 32 addresses, normally the lower eight defaulting to black, blue, red, magenta, green, cyan, yellow and white and the following eight containing the same colours with reduced intensity, and black being interpreted as transparent (thus allowing lower priority presentation entities such as photographic to be displayed). See *Figure 62.9*.

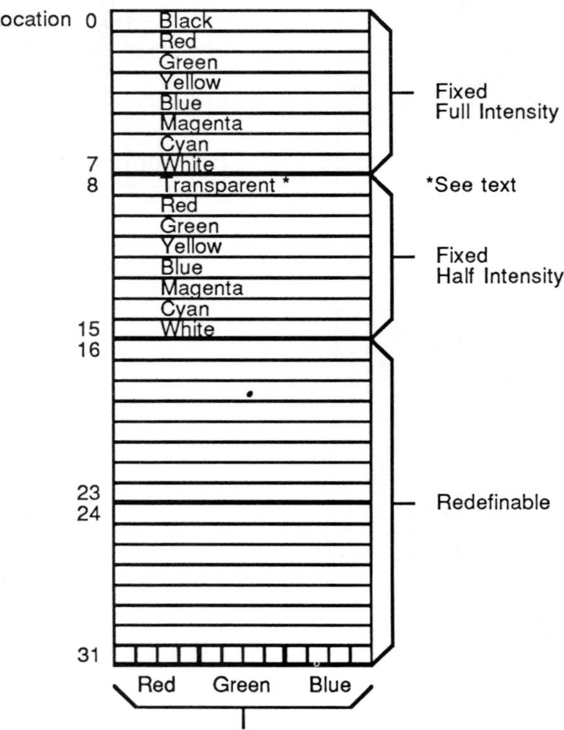

Figure 62.9 The colour map

This technique enables the relatively small number of colours that can be addressed at any one time to be selected from a large range of colours, the range being 2^n where n is the bit width of each colour map location. Thus a 12 bit wide colour map with 32 addresses provides 32 colours from a palette of $2^{12} = 4096$. Part of the colour map may be redefined by the service which may take effect immediately to give interesting dynamic results. Normally the lower 8 or 16 locations are fixed and only the upper locations may be redefined. This ensures that certain service messages will be displayed in a legible form at all times.

62.3.9.3 Format

The *format* function defines the number of columns and rows displayed within the dda for an alphamosaic and drcs display. Common defaults are 40 × 25 and 40 × 20.

62.3.9.4 Reset

The *reset* function allows the service to set the terminal to a predefined state thereby synchronizing the terminal to the service. There may be different forms and levels of reset, typically:

general reset: brings the terminal to its 'power-up' state
partial reset: re-establishes the G- and C-sets to their default states
service row reset: temporarily resets the terminal to alphamosaic mode to enable the service to display a message on a particular row, and then returns it to the previous state

A reset function normally takes priority over everything else.

62.3.9.5 Terminal facility identifier

The *terminal facility identifier* (tfi) enables a service that may support a number of different types of terminals to ascertain the type or a specific capability it may have (e.g. geometric or photographic display).

62.3.9.6 Timing control

The *timing control* allows a specified delay to be introduced in the execution of commands by the terminal which enables display build-up to be controlled for special effects.

63.3.9.7 Scrolling

The *scrolling* function provides a method of extending the amount of information held on a videotex frame. The scrolling area may be the complete dda or a designated number of rows. Some systems provide scrolling from the lead frame or from the second frame, with high and low scrolling rates and controls to stop and start the scrolling.

62.3.10 Processable data

The *processable data* facility enables reliable down-loading of data files between the service centre and terminal, more commonly in the terminal direction only. The data files may contain computer software (often referred to as *telesoftware*), information for the control of terminal peripherals such as printers and video disc players, or any other data that may require further processing once received.

62.4 International videotex standards

The majority of videotex systems used worldwide conform to some extent to one of three recognized standards that have been drawn together by the CCITT as Recommendation T101 and are known as *Data Syntax* (DS) *I*, *II* and *III*. These generally correspond to systems adopted by Japan, Europe and North America respectively.

Each data syntax describes the formats, rules and procedures for encoding textual, pictorial and, for DS I, sound information for videotex applications. They conform to the architecture defined in ISO 7498 and CCITT's seven layered reference model for open systems interconnection (osi), specifying mainly the data syntax for the presentation layer (responsible for the encoding of text, graphic and display control information).

Each data syntax defines one or more service reference models (srms) that specify the minimum implementation of a receiving device (i.e. terminal) and the maximum implementation to be assumed by the application service. An overview of the three data syntaxes is given in sections *62.4.1–62.4.3*. Additionally, T101 defines a data syntax to enable some degree of interworking between DS I, II and III.

62.4.1 Data Syntax I

DS I presentation layer data syntax provides coding for alphanumeric, Japanese Kanji (3657 characters) and other forms of Japanese characters, mosaic graphics (two sets, one of which is compatible with DS II), basic drcs, geometric (compatible with DS III) and pseudo-photographic display. Additional capabilities include a colour map, macros, scrolling and variable character size. Operation is accommodated in both 7 bit and 8 bit character coding environments except photographic that uses a transparent data mode. Modified Huffman coding is used for photographic data compression.

Optionally available are musical note encoding and animated picture coding. *Music encoding* uses a 2 byte musical tone character set to represent pitch and duration for each note, and a musical control set of 12 codes to define the start and end of a music code sequence, set the sound level, change the timbre, etc.

A *move* instruction enables simple animated picture presentation by causing a specified frame to be moved virtually with reference to other frames. In this context a frame defines a number of bits/pixel from the total number of available bits/pixel of the display memory configuration.

61.4.2 Data Syntax II

Data Syntax II corresponds to the European CEPT Recommendation (T/TE 06 series) presentation layer data syntax used by most European countries operating videotex services. DS II presents within its srm a conformance section that lists a selection of specific facilities taken from the main technical annex. This helps to provide harmonization and ease interworking between different services. The degree to which a service implements these facilities is determined by the service profile. Four service profiles (1, 2, 3 and 4) are recognized by the standard within DS II.

Profile 1 uses a bit coding and allows both parallel and serial attribute controls. It provides alphamosaic, basic and colour drcs display with a 32 location colour map from a palette of 4096 colours. Formats of 40 columns × 24 or 20 rows are provided, as well as various modes of reset and a terminal facility identifier.

Profile 2 uses 7 bit coding and parallel attribute controls. It provides only an alphamosaic display with a format of 40 × 24.

Profile 3 uses 7 bit coding, serial attribute controls and a limited range of display format effectors. It provides only an alphamosaic display with a format of 40 × 24.

Profile 4 is identical to profile 3 except that all format effectors defined are recognized and executed.

On top of any of these four service profiles, optional enhanced facilities may be provided by a service application; examples are geometric and/or photographic displays (the latter employing dpcm or adct data compression) and processable data. The extent of implementation is determined by the srm conformance section. The geometric display defined by DS II is based on gks and the photographic display is based on CCIR digital television techniques.

62.4.3 Data Syntax III

Data Syntax III generally corresponds to the *North American Presentation Level Protocol Syntax* (NAPLPS). Operation in 7 and 8 bit environments is accommodated. Alphamosaic, basic DRCS and geometrics are provided with additional capabilities which include a colour map, macros and partial screen scrolling.

DS III goes some way to providing a coding scheme by which specific information may be conveyed and correctly interpreted regardless of the hardware dependencies of particular receiving device configurations (e.g. display resolution).

62.4.4 Interworking Data Syntax (IDS)

T101 also recommends methods of videotex interworking between the three data syntaxes used by different countries. Basically it states the following for videotex interchange:

- If two countries implement the same DS, then interworking

can use the same DS.

- If two countries implement two different DSs, then interworking can use either the CCITT IDS, or any one of the three DSs and convert directly between DS I, II and III.

If the IDS is used, then the service provider in the country in which the database is located is responsible for the conversion into IDS, and the service provider in the country in which the user terminal is located executes the conversion from IDS.

62.5 Functionality

A very wide variety of different application services can and have been implemented using videotex technology. As well as using the common display and presentation techniques described in section *62.3*, they typically employ a number of common dialogue and service functions, so as to present the user with a consistent interface style, and to facilitate the production of standard application packages for use by application providers.

Functions found in many videotex systems include: user access control, closed user groups, database retrieval, keyword access, standard commands, database update, form filling, mailbox, realtime dialogue and billing.

62.5.1 User access control

Most videotex systems implement some form of user identification, both to prevent access from unauthorized terminals and to identify users for billing and accounting purposes. Typically a user must identify himself to the system by a user identity and a password. Sometimes the system also interrogates the terminal for its terminal identity and cross checks this against the user identity. More sensitive applications may implement further mechanisms, such as the use of transaction numbers, or employ additional hardware at the terminal end to implement data encryption, often in association with a badge or smart card reader to provide further user identification.

62.5.2 Closed user groups

The user identification system above may be extended to provide a closed user group (cug) facility. This allows groups of users to be defined with particular access rights. In a public system, for example, certain areas of the database may be restricted to users who have paid a subscription. In a private system, such as one used within a business, cugs may be used to control access by different business departments. More complex cug systems allow for a hierarchy of access arrangements.

62.5.3 Database retrieval

Data are usually arranged in the form of pages or frames — single screenfuls of information which the user can retrieve one at a time.

The simplest method of page selection is by use of menus. A menu is simply a list of options for the user to select, by keying a one or two digit choice. By arranging menus in an hierarchical fashion, or tree structure, the application provider can guide the user in more and more detail until the required page of data is obtained. The design of such hierarchical or tree-structured menu structures is difficult, and needs to achieve an acceptable balance between speed of access and clarity of choice for the user. *Figure 62.10* gives an example of a tree-structured menu.

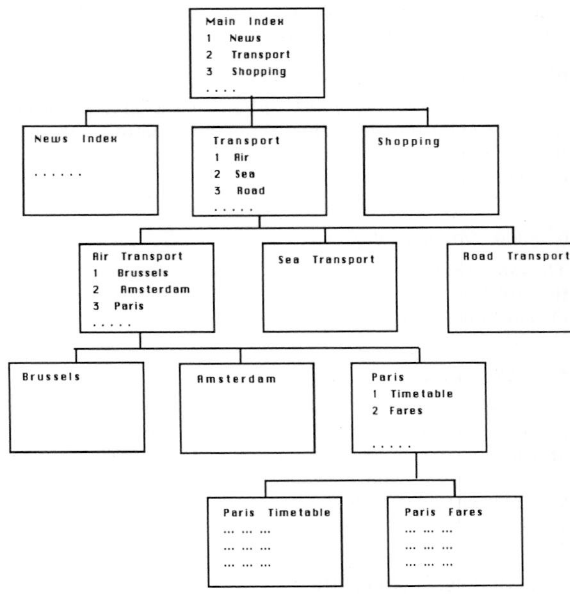

Figure 62.10 Example of a tree-structured menu

Some form of direct page access is usually also provided, so that the user may go directly to a previously accessed page, by keying in a page number. Some systems provide a method of page marking whereby the user can mark a particular page and return to it later by a simple command. Sometimes a printed directory of useful page numbers is provided, either by the videotex service provider or an independent publisher.

62.5.4 Keyword access

More advanced systems provide keyword access facilities, so that a user can retrieve a page directly without needing to progress through a series of intervening menu pages. Keyword systems typically need to be able to cater for:

synonyms: different keywords with the same meaning, e.g. *car* and *vehicle*
mispellings: where the meaning is unambiguous, e.g. *thetre* for *theatre*
combinations: several keywords combined to make a more precise search, e.g. (*travel* and *air* and *France*), or (*cars* or *vans*)
prompts: where a keyword is insufficient to identify a single page, e.g. *travel* may prompt *air* or *sea* or *land*

62.5.5 Commands

A number of standard user commands are available for frequently used functions, such as: *go back to previous page*, *go back to start page*, *redisplay page*, *help*, *leave the system*, *display current bill*.

Such commands may be keyed in a simple string of characters by the user, e.g. *0#, *#, *09, etc. In some systems dedicated function keys are provided on the terminal for this purpose; the French Minitel terminal has eight such keys labelled *help*, *index*, *previous*, *next*, *send*, *cancel*, *delete*, *repeat*.

62.5.6 Database update

Various means have been implemented for updating the database, depending on the type of host system and network.

Many videotex systems provide a simple editing system whereby an information provider (IP) can use a standard terminal to create and amend pages on the database directly. Such an editor usually makes use of standard host functions such as form filling (see section *62.5.7*) in its dialogue with the IP, who must remain logged onto the host throughout the editing session.

More sophisticated editing terminals allow an IP to create and amend pages locally before establishing a call to the host, and often provide many more complex editing features. Once a set of pages has been created to the IP's satisfaction, the terminal is instructed to 'download' them to the host system, and no further operator intervention is required. A special protocol, known as *bulk update*, is used to control the interaction between the host and such editing terminals. No formally established standard exists for a bulk update protocol, since different host systems require different functions.

The bulk update protocol can also be used to allow an IP with his own computer system to update a videotex host. This technique is often used where an IP has his own database system for internal use, and wishes to copy a subset of it to a public videotex system. A number of software packages are available for reformatting information from various proprietary database systems into videotex pages. As well as simple reformatting, such packages are also able to construct menu pages and indexes to assist the eventual user to access the data.

Where such copying is relatively infrequent and not time-critical, it is often more economical to use an off-line transfer medium, such as magnetic tape or disk, instead of a communications link. Again, there are no formally established standards for the format or content of such tapes.

62.5.7 Form filling

A technique used frequently in videotex applications is to present the user with a form, e.g. to order goods. A form is a frame containing one or more fields to be entered by the user. Many videotex systems provide standard routines for form design and filling, so that a common interface is presented to the user across many applications. Such routines allow for fields to be filled and maybe amended by the user, and checked by the system (e.g. numeric characters only may be required in particular fields) before being forwarded to the IP. It may also be possible for the IP to provide default values in certain fields for the user to accept with a single keystroke. Special system fields may also be defined by the IP, which are filled in automatically by the host system and not accessible to the user. Such fields may be used to contain, for example, the date and time of a transaction, the user's name and address, etc.

62.5.8 Mailbox

Many videotex systems provide some form of simple electronic mail service, usually known as *mailbox*. The sending user fills in a mailbox form, typically using standard form filling routines, and when the recipient next logs onto the system he is informed that a message is waiting. Messages may be pre-addressed, addressable, pre-formatted or free format.

In *pre-addressed* messages (also known as *response pages*), the recipient is pre-defined by the application provider when the page is first created and cannot be changed by the user. Response frames are typically used in applications such as ordering goods or services, requests to an IP, or data collection.

The pages of *addressable* messages contain a field to be entered by the user identifying another user to whom the message is to be sent.

The information to be filled in by the sender of *pre-formatted* messages is restricted to a few fields in a pre-defined form.

Free format messages permit the sender to enter text in any format or position on the page.

A good mailbox system will also provide extra facilities, such as *message forwarding*, *distribution list management* and a *filing system*. It may also have a directory of users and a library of standard message pages.

62.5.9 External mailbox

The mailbox service can be used as a mechanism by which a user may exchange messages with other systems outside the videotex system. For example in Prestel and also in the German Bildschirmtext service, a user may initiate a telex message by sending a standard videotex mailbox message to a *telex interface*. The interface software takes the videotex message, reformats it and transmits it to the specified telex number, anywhere in the world. Incoming telex messages are similarly reformatted into videotex mailbox messages and sent to the appropriate user. It is of course necessary to ensure that the incoming telex message clearly identifies the user so that it can be correctly addressed on the videotex system. Similar links have been established with other electronic mail systems; an international standard (CCITT Recommendation X400) has been established to facilitate such links.

62.5.10 Realtime dialogue

As well as the mailbox services described in sections *62.5.8* and *62.5.9*, some videotex services provide a realtime dialogue service, allowing users to 'talk' directly to one another. Such devices have proved very popular in the French Teletel system, where they are known as *messageries*. Usually such systems enable users either to hold a one-to-one private conversation or else a more general conference in which many users may converse at the same time.

62.5.11 Billing

Public videotex service operators often operate a billing system on behalf of the application providers, so that the users receive only a single bill, regardless of how many applications they have accessed. The income received from users is distributed by the service operator to the relevant application providers, according to the overall use of the various services. There are three main ways in which users may be charged: by time, by page or by transaction.

62.5.11.1 Time charging

The user is simply charged for the time he is logged onto the system, in units of (say) a minute, in a similar way to telephone calls. Different charge rates may be used at different times of the day, for different areas of the database, or for different groups of users.

One method of collecting a time charge is by use of the *premium charge* facility which is commonly available in the public telephone network. This is a 'revenue sharing' arrangement, whereby the user is charged a higher rate for his telephone call, and the application provider is paid a proportion of the extra revenue collected by the telephone company. This technique is used a great deal in the French Teletel system, where the 'premium rate' part of the service is known as *Kiosque*.

62.5.11.2 Page charging

A charge (typically a few pence) is raised each time the user accesses a page. Different pages in the system may bear different charges, at the discretion of the information provider. Many pages, such as menu pages or help pages, may be free. In some countries, local regulations require that the user is informed in advance before a chargeable page is displayed.

62.5.11.3 Transaction charging

A charge is raised every time a user makes a logical transaction, for example each time an item is purchased from a shopping service, or for each operation performed in a banking service. A transaction may involve several page accesses, and take any length of time. The application service provider is responsible for defining what constitutes a 'transaction'.

The basic charging methods are not mutually exclusive. Prestel, for example, supports page charging and variable time charging, allowing application providers considerable flexibility in the way they charge their users. Teletel supports only time charging, via the Kiosque mechanism; but application services are free to make other charging arrangements outside the Teletel system.

62.6 Network topologies

Section *62.1.2* identified the four main elements of a typical videotex system: terminal, network, service centre and application service. In different systems round the world, these elements have been put together in a variety of arrangements or *topologies*, according to particular local requirements. Factors which affect the topology chosen include:

- size of user population; how many users are there, both in total registrations and simultaneously active in the system?
- user sophistication; are the users to be aware of the topology, or are they to be shielded from it?
- telecomms environment; is the system contained within a private network domain, such as a single company, or are public telecommunication network(s) involved?
- telecomms regulatory regime; what data are permitted on the network and what methods of connection are available?
- geography; are the users located in a small area, e.g. a single building or site, or are they widely distributed?
- application services; how many are there, and how complex (simple data retrieval, or interactive transaction processing)?
- database size and volatility; how large is the database and how frequently is it to be updated?
- user registration requirements; do users need to be individually identified by the system, for security or billing purposes?
- accounting arrangements; how is revenue to be collected from and for users, application services providers, network providers?

Some of the topologies which have been employed are described in sections *62.6.1–62.6.8*.

62.6.1 Simple network

Terminals are connected directly to a single host computer via the PSTN or local connection. The host computer performs service functions, such as access control, and application functions, such as database management. This topology is typically used for small scale private videotex systems, where the number of terminals does not exceed about 100, and where a single host computer has sufficient power to support all the required application services. *Figure 62.11* illustrates this simple network.

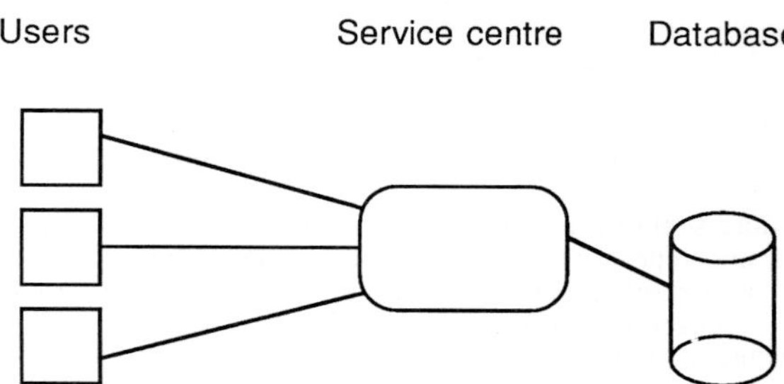

Figure 62.11 Simple network

For the PSTN connection, standard modems are used, often built into the terminal. Modem standards and data rates are defined by a number of CCITT recommendations, and different videotex systems support a variety of different standards. Those in most common use include V23 (1200 receive/75 baud transmit), V22 (1200/1200), V21 (300/300), and in more recent systems V22bis (2400/2400) or higher. Many systems support more than one standard, and modern modems often provide *autobauding* whereby the host computer modem automatically adapts itself to the appropriate rate for the calling terminal.

62.6.2 Replicated database

If the user population is too large to support economically on a single host computer, it may simply be replicated. Separate copies of the system may be geographically separated, to provide local access for different user populations. Users obtain service and applications from their 'home' system, and any database updates must be physically transported between the separate systems.

62.6.3 Replicated database with central update

This was the approach initially adopted by Prestel in the UK. The simple replicated *user service centres* (usc) of section *62.6.2* are all connected to a central 'hub' system or *update centre* (udc), which holds a master copy of the complete database. Users obtain service from their local usc, while information providers access the update centre directly when they wish to update their part of the database. The udc automatically distributes updates to the uscs, via high speed dedicated links.

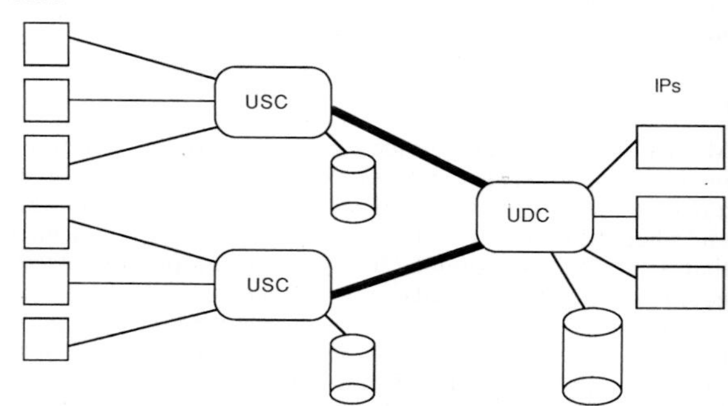

Figure 62.12 Replicated database with central update

A more advanced version of this idea is for the uscs to hold only a subset of the complete database, updated dynamically from the udc in response to user requests. Sophisticated update algorithms ensure that the usc keeps a local copy of the most popular parts of the database, so that a high proportion of user requests can be satisfied without needing to refer to the udc. This topology is illustrated in *Figure 62.12*.

62.6.4 External application services

If the number or type of application services makes it impractical or otherwise undesirable to centralize on a single (or replicated) host computer, one or more application services may be separated and hosted on a separate *application service computer* (asc), connected to the usc by a high speed link, typically a packet switched network. Such a connection is often called a *gateway*. The use of a separate asc gives an application provider more direct control over the database and user dialogue, and is often used where the videotex system forms only part of an application provider's total system.

In the replicated topology of sections *62.6.2* and *62.6.3*, a single asc would usually be connected to more than one usc. Not all application services need to be separated in this way. In many videotex systems, some applications are supported directly on the uscs, while others—typically the more complex, transaction processing type of applications—are supported via gateways. Indeed, in some cases a 'logical' service — the total application package as seen by the user — is implemented partly in the usc and partly in an asc. The user is not generally aware of the boundary between the two systems. The terms *local host* (ld) and *external host* (eh) or *external computer* (ec) are sometimes used for usc and asc respectively.

No internationally agreed standards have been established for the communications protocols used between usc and asc. Two main approaches have been taken:

- Many videotex systems support a simple protocol based on CCITT Recommendation X 29. This allows the asc to pass data directly to the user without reformatting by the usc, and gives the asc total responsibility for the user dialogue.
- Many videotex systems also support a more complex *gateway protocol*, providing higher level functionality such as billing, user identification, format conversion, etc., and relieving the asc of much of the detailed user dialogue processing. Examples are the Prestel gateway protocol (PGP) and the German EHKP4.

62.6.5 Replicated databases with central services centre(s)

This topology extends the 'centralized update' concept of section *62.6.3* by centralizing other service functions, such as access control, billing and message services. This, together with many external application service computers as in section *62.6.4*, is the topology currently adopted by Prestel in the UK and Bildschirmtext in Germany. In practice, the 'central' functions may be hosted on several computers, which are connected by a high speed dedicated network, and not necessarily physically co-located. Prestel, for example, currently has three centralized computers, an *update centre*, an *admin centre* and a *message centre*. This topology is illustrated in *Figure 62.13*.

62.6.6 Access network

As well as the 'back end' network of service centres, videotex systems often provide a front end access network to facilitate connection of users to their local usc. In the simplest case, this network is just the PSTN. Often this is supplemented by the user of remote multiplexer links to provide customers with local call access to a distant usc. Multiplexer networks can become quite complex and provide sophisticated network management facilities such as traffic concentration, rerouting in case of congestion, traffic monitoring, etc.

62.6.7 Distributed services

The topology adopted by Teletel in France has no centralized functions, but instead provides considerable functionality in the access network. User terminals are connected (via the PSTN) to simple *videotex access points* (vaps) which provide access to the *packet switched service* (pss) network. The user chooses the service he requires by typing a short (alphabetic) name, which is used by the vap to select the appropriate pss address, and set up a pss call to the required asc. The vap performs only this routing function; user registration, access control and billing, if required, must be performed by the individual ascs. The user must register separately on all the different services, and receives separate bills from each service.

To overcome this inconvenience, many application services make use of premium rate telephone call charging. The asc does not bill the user directly, but recovers income via the telephone bill. The user's telephone call is charged at a higher rate than standard, and a proportion of the extra income is paid to the application service provider by the network provider (usually the public telecommunication operator). In the French Kiosque service, a series of different charge rates may be applied by application services, by using different telephone numbers. With the advent of all-digital telephone exchanges, this concept can be extended further, so that the rate of charge can actually be adjusted by the asc during the call.

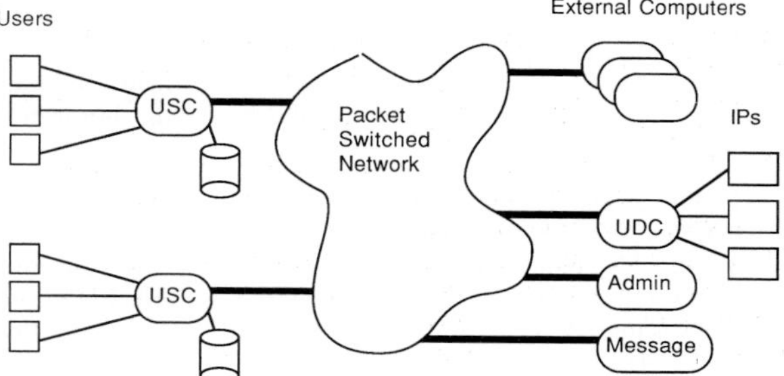

Figure 62.13 Replicated database with centralized service centres

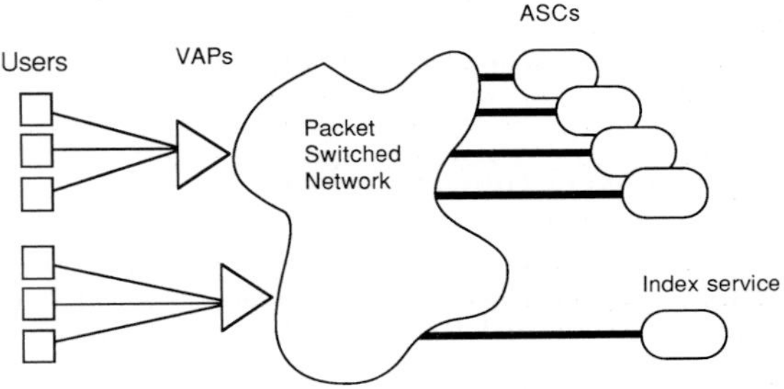

Figure 62.14 Distributed system

This topology does not, in concept, require any centralized services, other than the transmission networks and some method of keeping the vaps' name/number lookup table up to date. In practice a 'special' asc is normally set up by the administration, namely an index service, a directory of all the application services which the user may access. As far as the network is concerned, this service operates in exactly the same way as any other asc, and is selected by the user from the vap just like any other asc. On the Teletel network, the index is called the *Minitel Guide des Services* (MGS). *Figure 62.14* illustrates a distributed system topology.

62.6.8 Hybrid networks

The systems discussed above use conventional telephony and data networks in various combinations. Some interesting systems have been developed which use other transmission technologies, notably broadcast and cable television systems. The high data bandwidth available on such networks makes practical the provision of more advanced display modes, such as photovideotex.

62.6.8.1 Videotex via cable television

Cable television (ctv) systems may provide videotex services as well as standard TV channels. Such services may be local, specific to the particular cable network, or remote, consisting of a gateway to a public service such as Prestel. Two possible transmission methods have been used, depending on the capabilities of the cable network:

- One uses standard teletext technology. The videotex host, via the cable TV headend, inserts the videotex data requested by the user in the teletext slot of a TV channel. The user, who must of course possess a suitable teletext receiver, simply selects a designated teletext page (and possibly subpage) to view the data. Other users could 'eavesdrop' on the videotex session, by selecting the same page and subpage, but this disadvantage can be minimized by using a different page for each session. Transmission of data back from the user to the cable headend is achieved by using spare bandwidth on the network and specialized interface hardware in the user's premises.

- More advanced cable TV systems, such as British Telecom's switched star system in Westminster, London, can provide a dedicated TV channel unique to a given user. In this case, the videotex data can be decoded and formatted by the headend (or at some other point in the cable network, according to the technology) and transmitted to the user as a TV signal. The user need not use a teletext receiver, but simply tunes his TV set to the designated channel. Since the channel is switched uniquely to the given user, no other users can eavesdrop on the session. This cable network also provides a dedicated data path back from the user to the headend for other purposes, so videotex data can easily be carried by the same route.

62.6.8.2 Videotex via broadcast television

Where geographical and regulatory conditions allow, it is possible to integrate broadcast and telephone technology to achieve a very high bandwidth from host to user terminal while retaining the security and flexibility of PSTN transmission for data from user terminal to host. In the Singapore *Teleview* system, specially developed terminals are used which initially make contact with the videotex host via the PSTN, but may receive data back from the host via a dedicated TV channel. In this system the full bandwidth of the TV channel is used, and the data are encoded as 'full-field' teletext. Multiple presentation modes are available, including alphamosaic, geometric, photographic, and a specially developed Chinese character encoding. The transmission path to be used for any given piece of data is selected by the host, on the basis of size of data, TV reception conditions and security requirements. In general, the user is not aware which path has been used. A high level of security is achieved by dynamically altering the page/subpage used during a session.

Acknowledgement

Acknowledgement is made to the Director of Communications Systems Technology of British Telecom Research Laboratories for permission to make use of the information contained in this section.

Bibliography

CCITT Recommendations:

F300 *Videotex service*
T101 *International interworking for videotex services*
Annexe A *Interworking Data Syntax*
Annexe B *Data Syntax I (CAPTAINS Japan)*
Annexe C *Data Syntax II (CEPT Europe)*
Annexe D *Data Syntax III (NAPLPS North America)*
V21, V22, V22bis, V23 *Modem recommendations*

CEPT Recommendations:

T/SF 59 *International videotex service*
T/TE 06.01 *Videotex presentation layer Data Syntax*

ISO Standards:

ISO 646 *7 bit coded character set for information processing interchange*
ISO 2022 *Information processing; ISO 7 bit and 8 bit coded character sets; code extension techniques*
ISO 6429 *Information processing; ISO 7 bit and 8 bit coded character sets; additional control functions for character-imaging devices*

Terminal guides:

Prestel Terminal Technical Guide, British Telecom/Dialcom, London (1985)
Minitel 1B Videotex Terminal Specification, Intelmatique, Paris (1987)

Part 14 High Definition Television

Section 63a High Definition Television and Electronic Production of Movies – International Background

L. Strashun, Sony Broadcast Ltd

Section 63b High Definition Television and Electronic Production of Movies – European Standards

N. Wassiczek, G. T. Waters and D. Wood, European Broadcasting Union

L Strashun M Sc, C Eng, MIEE, MBKS, MRTS
Senior Manager, Sony Broadcast Ltd

63a High Definition Television and Electronic Production of Movies – International Background

63.1 Introduction

The term *high definition television* was first used in the mid 1930s. At the time the Selsdon Committee in Britain proposed that any system using more than 180 horizontal scan lines should be termed 'high definition'. Television engineers and inventors were dreaming then, with boundless enthusiasm, of 240 lines as 'true' high definition.

They had good reason for this. In the 1920s, the transmission of any image, even a shadow or a silhouette, was considered an achievement. Television was based on the principle of mechanical image scanning, first proposed by Dr Paul Nipkow in 1884 for a definition of 18 lines! This was developed and refined by Dr H E Ives, C F Jenkins and John Logie Baird to 24 lines, then 30 lines and finally 40 lines. Baird's method became the first in the world to be used in a regular TV service over the BBC in 1929, achieving 30 lines of definition with a Nipkow disk. Baird's *Televisor* receiver was successful enough to have sold 10 000 units by 1932.[1]

However, definition could not be improved much further by mechanical means. The future of television was all-electronic. The first major link in the electronic television chain was the iconoscope camera tube patented in 1926 by Vladimir K Zworykin. The concept was a practical extension of the theoretical work of A A Campbell Swinton and incorporated the principle of image storage. In 1933 and all-electronic system achieved a resolution of 240 lines. 'High definition' became a reality. Further accomplishments came rapidly and in 1937 the world's first standard of 405 lines was adopted in Britain.

Today colour comes in three major standards: PAL, SECAM and NTSC with either 625/50 or 525/60 lines/fields. The pictures are spectacular compared with previous systems and meet most viewers' requirements. There is little viewer complaint against picture quality as it appears on current receivers with a maximum screen size of 67 cm.

During the lifetime of television broadcasting, electronic technology has advanced at an enormous pace. With the demand for information growing at an equal rate, the broadcast industry has steadily expanded in volume and diversity of the information required, including new forms such as teletext, graphics and so on.

At one point in the history of cinema, at the beginning of the 1950s, Hollywood started to develop wide screen techniques which were capable of producing breathtaking and impressive images. The cinema audiences were made to feel as if they were taking part in the action, not merely spectators. Since then there has been a steadily growing demand for systems that can provide increasingly realistic images.

In Paris in 1986 what was then the biggest screen in the world was inaugurated. 'La Géode' is part of the Cité des Sciences et de l'Industrie and offers a 10 000 ft^2 screen surrounding the spectator with a complete hemisphere, exceeding the normal field of vision. The technique is Omnimax, developed in Canada from a system invented by an Australian, Ron James. Omnimax uses a 70 mm film and a 25 mm fish eye lens with an angle of view of 172°. To cover such a vast screen area from a single projector requires a massive light source; Omnimax uses a 15 kW water-cooled Xenon lamp. As extra attractions, lasers and holograms are combined with the projected film image.

Close examination shows that today's television image is not as good as optical film, even the 35 mm variant of it. The scanning standards for black and white television were developed during the 1940s, while the coding methods for colour television were standardized during the 1960s. The image quality is spoilt by impairments and defects known as *artefacts* (unintended image defects produced by some aspect of a television system): it flickers (especially in Europe's 50 Hz systems), the fine details often generate false patterns of shimmering colour, and moving objects create patterns not in the original image (*aliasing*), due to the standards of image acquisition and display and the limitations of the transmission channel.

The four main impairments of current European standards are described as follows:[2]

- *Large area flicker* occurs at 50 Hz and is perceived primarily in peripheral vision. It is aggravated with brightness increase because the response time of the eye is reduced and perceived flicker becomes worse. For a given brightness it is more annoying as the screen size increases or the viewing distance decreases.
- *Interline twitter* occurs at 25 Hz and is caused by the 2:1 interlace. It arises in areas of high amplitude vertical details with spatial frequencies approaching the Nyquist limit of 312.5 cycles per picture height. Perception of twitter is a function of display contrast because of an ocular property similar to the

brightness/response-time relationship. It may also be dependent on viewing distance or size of display, because of peripheral vision, and on the scanning spot size in the studio source.

- *Line crawl* is an unavoidable effect resulting from the interlaced scanning process. It is most readily seen when vertical motion in the scene forces the eye to scan the image at or near a critical rate of 11.5 seconds per picture height. When line crawl is apparent, the viewer perceives a 312.5 line picture, which halves the vertical resolution.
- *Static raster*. The individual lines of a picture can be seen when viewed on modern high brightness displays at moderately close viewing distances. This impairment would be aggravated by increases in screen size and brightness if the ratio of scanning spot size to picture size is maintained or reduced.

The viewed picture is itself limited. Because of the small screen and relatively narrow aspect ratio (ar) of 4:3, the image occupies only a small angle of view. Peripheral vision is not involved, and there is no need to move the eyes or the head to follow the action. Television therefore does not provide the feeling of involvement enjoyed in the cinema.

So, today, when it seems likely that what may be called an 'image-orientated society' of the future will be formulated on the basis of continually evolving television technology in an 'information-orientated age', there is a great need for the development of higher precision and higher quality television technologies. A wider screen high-fidelity television system seems necessary to meet the demand for instant information such as news, weather, finance and other time-variant data (teletext, viewdata, etc). It might be a major structural, not just sectoral, change in our way of life.

Wide ranging studies of picture quality, picture format (including aspect ratio and picture size), signal standards and broadcasting systems are being carried out in many countries, and novel equipments for all three links of the television chain — production, transmission and display – have been produced by manufacturers.

The Japan Broadcasting Corporation (NHK) inaugurated a high definition 1125 lines, 5:3 aspect ratio television service through satellite. High definition television and high definition electronic production are no longer a technology issue. They are now a business issue.

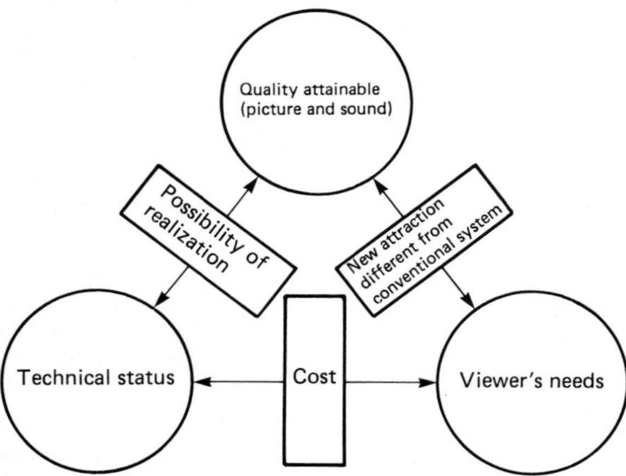

Figure 63.1 Requirements for HDTV broadcasting

63.2 Desirable features of HDTV

The ideal of television may be regarded as the reproduction of a remote image identical to an original scene, eventually stereoscopic television by holographic means. There is a certain relationship between the three main factors of an HDTV broadcast, namely the quality attainable, the viewer's needs or wishes and the available technology or technical status as shown in diagrammatic form in *Figure 63.1*.[3] It is certain that HDTV would involve additional expense.

To make it acceptable to the public at large, HDTV should be as visibly distinctive from current standards as colour TV was from monochrome TV. At the same time it must be technically feasible and the receiving sets and programmes must be available at an appropriate cost. The prime feature of HDTV is visual impact, providing a feeling of realism. For this, increases in the size of the screen and in the aspect ratio contribute even more than improvements in resolution.

63.2.1 Picture size

It is commonly acknowledged that a grand spectacle seen on a cinema screen loses its impressiveness when seen on a home receiver. Investigations conducted under the auspices of SMPTE by the HDTV group into the visual conditions in cinema theatres show[4] that typical mean values of viewing angles are: 58° in a front seat where the viewing distance is 0.9 times the picture height; 17.5° in a central seat at a viewing distance of 3.3 times the picture height and 9.5° in a rear seat at 6.0 times the picture height.

In domestic conditions with a set having a 51 cm diagonal (21 in) or 32 cm picture height and a viewing distance of about 2 m, or 6.3 times the picture height, the viewing angle is 9.2°.

Because human vision extends about 150° in the vertical plane and about 200° in the horizontal plane, the proportion of the visual field occupied by the picture area (which is closely related to the visual impact) is 25 per cent for a front seat in a cinema theatre but only 0.4 per cent in the home.

It is also significant that a large picture viewed from a distance seems more impressive than a small picture viewed from close up even if the viewing angles are the same.

63.2.2. Resolution

When picture size is large, sharpness (the subjective equivalent of definition) is an important factor of picture quality. Resolution, brightness and contrast all contribute to the sharpness of a television picture.

The resolution necessary for a television system is closely related to visual acuity and viewing distance. Vertical resolution depends on the number of scanning lines, so that if the viewing distance is three times the picture height ($3H$) more than 1000 scanning lines (for sequential scanning) are desirable, so that adjacent scanning lines become indistinguishable.

As the spatial frequency characteristics of human vision are similar to those of a low pass-filter, it follows that if viewing distance of a television picture is given, the general standard for an appropriate television system may be defined.

It is now accepted that for TV or cinema pictures with rapid movements a viewing distance that produces no visual fatigue is about four times picture height ($4H$). Viewers, however, tend to move closer if movement on the screen is not rapid. It is now widely accepted that the picture quality of future television should be good enough to allow a viewing distance of $3H$.[5]

Starting from this assumption, the number of scanning lines and therefore the signal bandwidth for HDTV can be calculated. As shown in *Table 63.1* this works out at about 1240 lines; after 1125 lines, however, the subjective gains are minimal.[6]

Viewing distance (dH)	$4H$	$3.3H$	$3H$	$2.5H$
n (lines)	935	1125	1241	1481
f_o (MHz)	4.5	6.5	7.9	11.4
Visual angle (deg)*	23.5	28.3	31.0	36.9
n	525†	951	1351	1601
dH	$7.2H$	$3.9H$	$2.8H$	$2.3H$
f_o (MHz)	1.1	4.7	9.4	13.3
Viewing angle (deg)	10.7	23.9	33.7	39.7
Standard	Interlace ratio (R_i) 2:1			
	Aspect ratio (ar) 5:3			
	Frame frequency (f_f) 30Hz			

f_o = cutoff frequency of noise weighting function
H= picture height
* = visual angle in horizontal direction viewed from optimum distance of the system
† = aspect ratio 4:3

Table 63.1 Television standards required as a parameter of viewing distance

63.2.3 Aspect ratio

The results of subjective evaluation tests relating picture size and aspect ratio (conducted by NHK) are given in *Figure 63.2*.

In general, an aspect ratio of 5:3 is preferred to the 4:3 ratio of current television systems. As the picture becomes larger the 5:3 ratio becomes even more desirable. For sports and landscapes, however, a 2:1 ratio is preferable. Inasmuch as cinema films would be one of the most important programme sources for HDTV and will even be produced by High Definition Electronic Production (HDEP) means, a 2:1 aspect ratio is considered desirable if problems of display can be solved.

The ATSC (Advanced Television Standards Committee) in the USA is backing a technically easier attainable aspect ratio of 5.33:3 as a compromise solution.

After broad agreement on desirable picture size, viewing distance, resolution and aspect ratio has been reached and after evaluation tests have clearly shown that the greatest gains in visual impact are obtained by increasing the number of scanning lines to about 1125 and the ar to 5:3, NHK have adopted this as an experimental standard (See *Figure 63.3* and *Table 63.2*).[5]

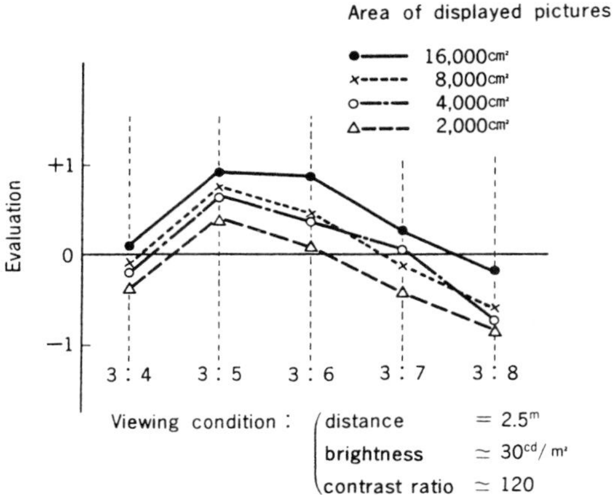

Figure 63.2 Subjective evaluation of picture quality as a function of aspect ratio

Figure 63.3 The proposed HDTV format (5:3) compared with normal format (4:3) (The British Tourist Authority)

Number of scanning lines (n)	1125
Aspect ratio (ar):	5:3
Line interlace ratio:	2:1
Field repetition frequency (f_v)	60 Hz
Video frequency bandwidth	
Luminance (Y) signal:	20 MHz
Chrominance (C) signal:	
Wideband (C_w):	7.0 MHz
Narrow-band (C_n):	5.5 MHz

C_w and C_n correspond to the wide-band axis and the orthogonal narrow-band respectively (see *Figure 63.7*)

Table 63.2 Provisional standards for a HDTV proposed by NHK

63.3 Technical considerations for future standards

At present there is very little possibility of a solution that would be compatible with existing standards. All recommendations for HDTV are based on different aspect ratios, different numbers of lines, different bandwidth for luminance and for chrominance, different coding solutions of these signals and finally a different number of frames/second in the non-NTSC countries.

However, it is generally believed that very soon the home receiver/display will work as a visual terminal, switchable through appropriate interfaces to numerous local and remote video and data sources working on different standards. Fundamental differences between the main characteristics of HDTV and present television systems are unavoidable, but there is a possibility of partial compatibility.[7] This means a choice of HDTV parameters so that economical interfaces for reciprocal compatibility may be achieved.

The television chain comprises three main links: production source, transmission and display. All three links have operated and do operate on the same standard and it has always been assumed that there is no alternative. This is because the costs of signal processing for even simple standards conversion has always been high and certainly prohibitive in a domestic receiver. While this remains true, priorities are changing. For the transmission link for instance, where frequency allocations are scarce, one can say that the electronics for signal processing are very expensive but the necessary additional bandwidth is non-existent!

The current approach to HDTV has been based mainly on an extension of existing standards with the video bandwidth increased by 4–6 times. There is a general consensus not to share the video spectrum between luminance and chrominance signals in order to avoid the present-day picture impairments.

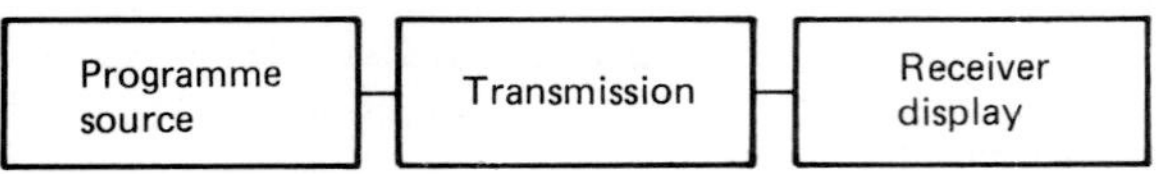

Figure 63.4 The television chain

This brings the necessary video bandwidth to over 30 MHz and the transmission bandwidth to over 80 MHz after modulation. Even though this was used in the experimental transmissions over satellite in Japan it is not acceptable in the long term.

Extensive research is being conducted in many countries. The MUSE (multiple sub-nyquist sample encoding) bandwidth compression technique for transmission[8] which brought down the necessary transmission bandwidth to no more than 27 MHz, the second generation HDTV standards converter from 60 Hz to 50 Hz[9] and the all digital receiver show that there is no longer reason to use identical codes for the video signals in programme production, transmission, satellite broadcasting or network distribution.

The optimized conditions for digital coding, for local memory devices and for bit rate compression can be specific for each link of the television chain — as happens in biological systems. All these developments will help in reducing the bandwidth necessary for transmission and in the adoption of optimum standards for each link of the chain, as well as improving compatibility. There is a high probability that in a few years, when HDTV domestic services will be in their initial stages, the video signals will no longer be analogue but digital, conforming to CCIR Recommendation 601, and frame and multi-field memories will be cheap enough to be used in domestic receivers to further enhance the quality of the image.

63.4 The HDTV elements design consideration

Experimental HDTV system hardware with 1125 scanning lines and an ar of 5:3 has been developed ranging from high performance cameras, to vtrs, switchers, telecines, standard convertors, crt displays and projectors. They all produce a picture of far superior quality to current standards.

The HDTV or HDVS (High Definition Video System) — as it is called by Sony — has reached a stage where it can be immediately utilized for film production and even for broadcasting if and when an adequate transmission standard and frequency band is found and allocated to it.

We will summarize the considerations for some of the elements and techniques of the HDTV chain.

63.4.1 Cameras

The fundamental problem is to achieve high resolution while maintaining a good signal/noise ratio (subjectively noiseless pictures) at lighting levels usually prevailing in the studio, as well as precise registration of the three scanning rasters. This in turn leads to:

- the development of a lens system with high resolution and low aberration,
- the development of a pickup tube of high resolution and methods of using it to obtain accurate registration,
- the requirement for wide bandwidth circuitry that does not seriously degrade the signal/noise and provides high stability.

HDTV camera lenses and prism systems must be capable of twice the resolution of conventional lenses and optics used for today's TV standards, and produce at least 35 mm film colour quality on a large screen with a minimum ar of 5:3. The large picture size and aspect ratio require a much higher degree of sharpness to attain the colour fidelity. This dictates that the optical system produces consistently high quality pictures regardless of aperture and focal length. Variations or inconsistencies between R, G and B tracks can significantly alter colour definition, resolution and contrast. Therefore the individual RGB channels have to track precisely over the entire zoom and lens aperture range.

A special one-inch tube having magnetic focus, electrostatic deflection, a Saticon layer and using a diode gun was developed in 1980 and then improved in 1982. It has a small aperture size of 16 μm and a mesh voltage (EG_4) of 750 V. The resolution obtained is related to the smallness of beam spot size but this also needs to be large enough to ensure that the target is sufficiently discharged during each scan. A beam spot size of 16 μm is thought to be optimum for an image size of 13.7 mm $\times$ 8.23 mm in a one-inch tube which is scanned by 1125 lines. Reducing the spot size to 12–13 μm causes a significant increase in lag time. To obtain an increased dynamic range a gun was designed which allows a beam current of eight times the normal level controlled by an automatic beam control (ABC).

To obtain a high signal/noise ratio with a wide bandwidth of about 30 MHz is a major problem. An increased number of scanning lines and a larger display screen cause the noise spatial frequencies to become lower and therefore more annoying.

The noise generated in the camera is strongly dependent on the performance of the pre-amplifier (head amplifier) and generally is proportional to the output capacitance of the tube. The output capacitance of the tube used in the Sony HDC-100 camera is less than 5.0 pF and matched to the input capacitance of the pre-amplifier (an fet) located at the head of the tube to keep stray capacitance to a minimum. A Percival coil is used to improve the s/n ratio.[10]

Special amplifier blocks with negative feedback, requiring no adjustment, have been developed for a bandwidth of 60 MHz. Wide-band gamma correctors have also been developed. The circuits feature image enhancing in order to compensate for the falling off of modulation depth in both vertical and horizontal directions. A conventional glass delay line in a cosine type of equalizer is not suitable for high definition, not only because of the requirement for 30 MHz of bandwidth, but also because of the need for better phase error and delay time characteristics. The s/n ratio may also be degraded.

In the HDC–100 camera a digital circuit is used with the following parameters:

sampling frequency	64.7 MHz
ADC	RGB 8 bits
DAC	RGB 8 bits
detail	10 bits

Other features of the Sony HDC–100 camera are:

- the one-inch electrostatic deflection tubes provide high resolution (1200 TV lines) and high picture quality,
- registration error is greatly reduced to 0.025 per cent at centre,
- adoption of the double optical filter wheels permit shooting under various conditions,
- an optical seven-inch monochrome viewfinder is available for studio use in addition to the standard 1.5-inch viewfinder,
- the high quality camera image enhancer (HDIE–100) increases sharpness and makes the picture clear and natural,
- full auto setup function gives comprehensive fine adjustment.

The distance between the camera and the ccu can be extended up to 200 m with multicore cable and up to 1 km with optical fibre cable.

The camera can be used with the HDCU–100 camera control unit which provides fine adjustment capability for enhanced colour fidelity and minimum registration error (*Figure 63.5*).

Figure 63.5 The HDC 100 camera (Sony)

The whole camera system is shown in *Figure 63.6*. The requirements for the HDTV camera are not very special or unusual: simply the degree of necessary quality is greater.

63.4.2 Scanning systems

Considerable interest exists in the possibility of achieving a common worldwide standard for HDEP studios.[12]

In spite of its many variations in format and aspect ratio, film has served and will continue to serve as a major worldwide exchange standard, not encumbered, as video tape formats are, by the need for degrading standards conversion. 35 mm optical film serves as a benchmark of the necessary image quality for the high definition standard.

The importance of the worldwide market places a high priority on the need for a single HDEP standard to augment and eventually replace the 35 mm optical film at 24 frames/second.[13] One question is whether this standard should use interlaced or sequential (progressive) scanning.

The scanning process is fundamental in television. It is the means by which two-dimensional images are converted into a continuous one-dimensional electrical signal. Similarly, scanning is used in the subsequent reconversion of the electrical signal into an image for display.

Scanning can be represented as a sampling procedure.[14] The electron beam scanning of a scene imaged onto a television camera target provides information about the scene centred at the discreet positions in the vertical direction that were scanned. Thus a scan that has 575 active lines produces a sampled image with a vertical sampling frequency of 575 cycles/picture height. If the same vertical position on the photoconductive layer is rescanned every $1/25$th of a second, this represents sampling in time with a frequency of 25 Hz.

The sampling action of scanning is not ideal, since the scanning aperture is not infinitely small in either the vertical or the temporal direction, but it is nevertheless a sampling process.

With the emergence of digital signal processing for television, it is becoming more common to consider the television image as an array of picture elements or pixels. Scanning and analogue/digital conversion procedures can then be combined conceptually as a means of converting a continuous real image into a discreetly sampled electrical version of the image. Digital/analogue conversion and rescanning for display provide the means for reconversion to a continuous image.

The 'pixel array' approach highlights the sampling nature of scanning. It is especially relevant to solid-state cameras and displays in which the scanning process involves addressing a pixel array.

The allocation of bandwidth or resolution capability in a television system can be specified by the scanning/sampling procedure.

There is generally unanimous agreement that transmission standards for HDTV should be interlaced, for the bandwidth efficiency offered. The merits of using sequential scanning in television displays have been well documented.[15] The choice of interlaced or sequential scanning is affected to some extent by the field rate used, since the artefacts introduced by interlaced scanning are less objectionable in the 60 Hz countries than in Europe.

In the case of cameras used for production, the only temporal pre-filtering, so important in reducing aliasing, is that due to the integration time of the tubes. This would be $1/25$ s in Europe and $1/30$ s in 525/60 Hz countries, if the reading out of the lines of one

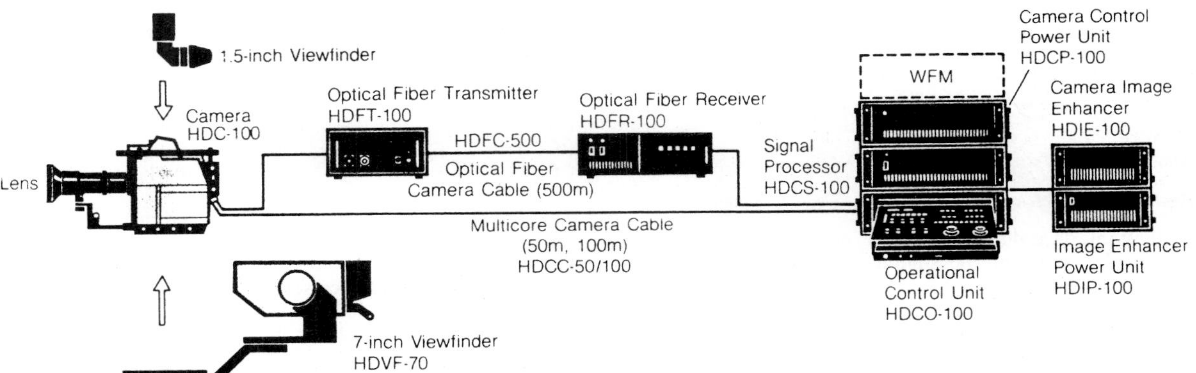

Figure 63.6 HD camera system (Sony)

field from the tube target left the lines of the second, interlaced, field unaffected. In practice things are different. Measured tube lag characteristics show that less than 10 per cent of the charge on the second (unscanned) field is left after the first field has been read out. Thus the integration time of cameras is nearer to $^{1}/_{50}$ s (20 ms) than to $^{1}/_{25}$ s. This implies that the camera may not be achieving its full degree of vertical resolution. (When two lines are read out at a time, the effective vertical profile of the scanning spot is twice as large as predicted.)

The use of sequential, rather than interlaced, scanning in the camera may overcome this and also allow better vertical aperture correction characteristics.

In a sequential scan system, vertically adjacent pixels are nearly simultaneous in time. Frames can thus be processed as complete pictures providing advantages in many post-production processes such as:

- image manipulation and special effects,
- editing ('one-field' edits),
- chroma key,
- slow motion,
- standards conversion,
- film recording.

On the other hand, the major disadvantage of sequential scanning is higher bandwidth and a reduction in camera sensitivity. It should be made clear that bandwidth does not double as conventional wisdom might lead one to believe. Jesty[16] defined as long ago as 1958 the concept of *interlace factor*. This is a multiplier that must be applied to the number of lines in addition to the *Kell factor*[17,18] (K) which indicates the extent to which practical systems fall short of their theoretical capabilities. (For example, a Kell factor of unity would correspond to a perfect system, while K = 0.5 means that only half the theoretical number of lines or horizontal black and white bars — on static pictures — can be resolved.)

According to Jesty an interlaced signal provides a vertical resolution subjectively equivalent to only 20–30 per cent greater than that of a single field rather than the expected 100 per cent. In other words, for equivalent subjective sharpness and equal horizontal resolution, the bandwidth of a sequential scan system must be only about 1.3 times or some 30 per cent greater than that of an interlaced system (*Figure 63.7*).

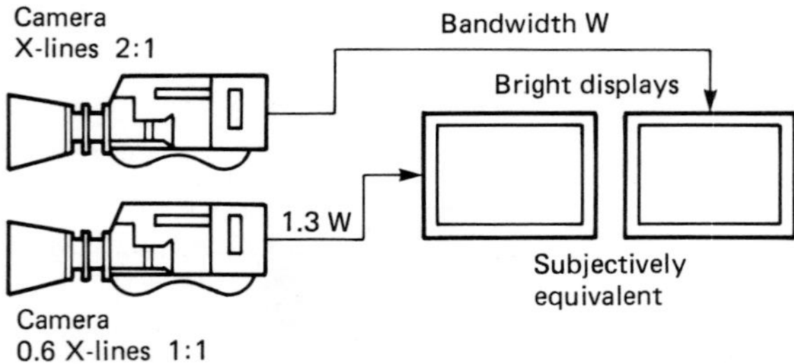

Figure 63.7 Subjective equivalence: interlace versus sequential

Many interlace due artefacts can be reduced with picture storage, but only for the stationary portions of a picture. Stationary pictures are very rare and usually either the camera or the objects in the scene are moving. Human vision is able to eye-track much of the motion and easily spots lack of definition of small moving objects such as a tennis ball. So far, the motion related artefacts of interlace have mostly been masked by the lag characteristics of camera tubes, but the advent of lag-free pickup devices such as ccd cameras will clearly single them out. The increasing demand for sharp slow-motion effects that such cameras can provide will further highlight the dynamic resolution limitations of interlaced scanning and the need for improvement.

This indicates that there is a certain advantage to be gained in using sequential scanning in an HDTV camera, but much experimental work remains to be done to prove the eventual source imaging and post-production values of the sequential scanning. Therefore an eventual standard must be studied by fully considering the future direction of technological advances. The urgency of achieving worldwide agreement on standards should not be allowed to stand in the way of achieving the best technical standard, because broadcasting standards cannot easily be changed once determined.

63.4.3 Field frequency

Field frequency depends on considerations of flicker and smoothness of movement. If it were possible to keep the luminescence of each pixel or picture element constant for a period of one field, the field frequency could be as low as 40 Hz, which is the minimum necessary to give the illusion of smooth movement. However, for an ordinary variable luminescence display with a picture brightness of 159 cd/m^2, a field frequency of about 60 Hz is appropriate if flicker is to be imperceptible.

As the field rate is raised, the quality differential observed between sequential and interlaced scanning reduces. Therefore the problem is much more acute in Europe and it was taken into account by NHK when choosing interlace.

63.5 HDTV formats and transmission

NHK have suggested over the years at least a dozen different formats for the HDTV signal. The format proposals differ in the way they handle the various signal components. These are the luminance Y and the two colour primaries, different from those used in PAL or NTSC and called C_w and C_n, to denote the wider and the narrower bandwidth components.

Extensive subjective tests on spatial frequency response of the human vision to chromaticity have shown that:

- the peak sensitivity to chromaticity of human vision is at about 625 nm (red) and 492 nm (green),
- the eye is less sensitive to wavelengths close to 565 nm (yellow) and 440 nm (blue).

Further, it is thought[5] that chrominance primaries for the HDTV should be chosen to correspond to colour axes crossing orthogonally on the chromaticity diagram in order that effects resulting from chrominance signal errors are uniform in all colour areas (*Figure 63.8*). Accordingly a provisional standard for transmission signals was chosen as follows:

$$\begin{pmatrix} Y \\ C_w \\ C_n \end{pmatrix} = \begin{pmatrix} 0.30 & 0.59 & 0.11 \\ 0.63 & -0.47 & -0.16 \\ -0.03 & -0.38 & -0.41 \end{pmatrix} \begin{pmatrix} R \\ G \\ B \end{pmatrix}$$

The video bandwidth for HDTV is large, and shf and ehf bands are being considered for satellite broadcasting and cable distribution including fibre optics. At the 1979 World Administrative Radio Conference several bands were allocated for satellite broadcasting services. Inside those bands the channel spacing was set to 19.18 MHz and the bandwidth to 27 MHz. Taking into account that rain attenuation is very large in the 42 GHz and 85 GHz bands, NHK have chosen for experimental satellite transmission the 12.50–12.75 GHz band and will expand into the 22 GHz band as soon as technological advances permit it.

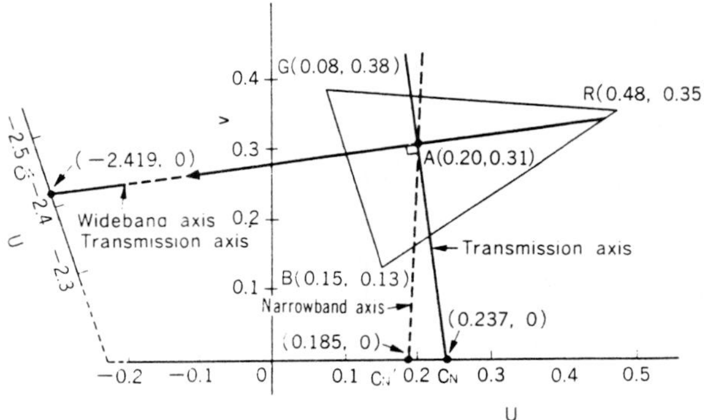

Figure 63.8 Wide-band and narrow-band chromacity axes and transmission primaries

63.5.1 Terrestrial broadcasting: The HLO–PAL format

Transmission standards depend on available power and bandwidth. They should be different for terrestrial and satellite broadcasting. For terrestrial broadcasting when sufficient power is available but bandwidth is insufficiently wide, vestigial sideband (vsb)–am transmission of a colour composite signal is most beneficial and NHK have developed a format called HLO–PAL (half-line offset, phase-alternation-by-line) using a colour subcarrier at 24 MHz[3,5,19] as best for bandwidth use efficiency.

The HLO–PAL composite format uses the same principles as PAL or NTSC but scaled up to a wider bandwidth. The baseband frequency spectrum is shown in *Figure 63.9*. The colour components are modulated onto a carrier of 24.3 MHz, which lies outside the Y signal bandwidth, to minimize cross colour and cross luminance effects. The system uses phase reversal of the C_w component on alternate scanning lines in a similar way to PAL. The Y and C primary signals are frequency-division multiplexed in their spatial and frequency domains as in conventional television. In this system signal processing can be simplified, especially the decoding at the receiving end. But chroma noise interference is visible on the picture.

63.5.2 Satellite broadcasting: The tci and MUSE formats

Satellite broadcasting is certainly the most effective means of distributing television widely at relatively low cost. The technologies of signal transmission for HDTV can be divided into two categories: one includes signal transmission between broadcasting stations (inter-station exchange), while the other includes broadcasting for the general public.

Inter-station exchange transmissions are required to have extremely high picture quality. This can be acheived only with considerable bandwidth. On the other hand, for public broadcasting, the bandwidth must be limited and narrow.

HDTV requires several times as much information as current television in order to offer features which current television lacks. But the 12 GHz band which was allocated for broadcasting offers no more than 27 MHz per channel. Satellite broadcasting technology in the 22 GHz frequency band is not yet commercially feasible. For this reason, systems for distribution and broadcast of HDTV signals require technologies to compress the signal bandwidth.

The bandwidth that is finally needed for HDTV signals distributed to homes depends on the display standards. Provided a complex signal processing system is allowed at the receiver, signals of narrow bandwidth can be transmitted.

One such method of bandwidth compression, called MUSE[8,21], was announced by NHK in 1984. This technique enables a satellite transmission channel of 27 MHz or 24 MHz to carry a 1125-line HDTV signal with an ar of 5:3, accompanied by digitally coded signals for dual sound and additional data necessary for reception. MUSE is a three-dimensional phase-alternating subsampling system (3D-pass) with motion compensation. The sampling is of a multiple dot-interlace type, and the cycle of the sequence is a period of four fields. The sampling pattern is shown in *Figure 63.10*.

For a stationary picture area, temporal interpolation is used with samples taken from all four fields. For moving picture areas, spatial interpolation is used to reconstruct the final picture, and all samples are taken from the same field in order to avoid multi-line blur. In this case, the transmitted band of frequencies is narrowed producing some blurring of pictures with an uncovered background. However, this degradation of quality is not serious, because human perception of sharpness is less sensitive to blur in moving portions of the picture.

In the case of camera movement caused by panning or tilting, the blur is more noticeable. To avoid this effect of spatial

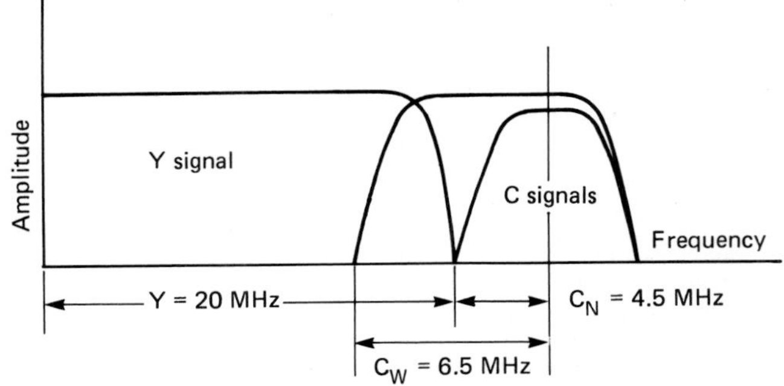

Figure 63.9 HLO–PAL frequency spectrum

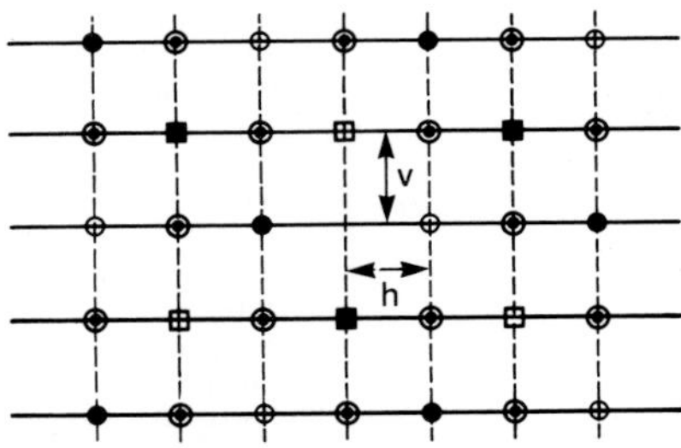

Figure 63.10 MUSE sampling pattern

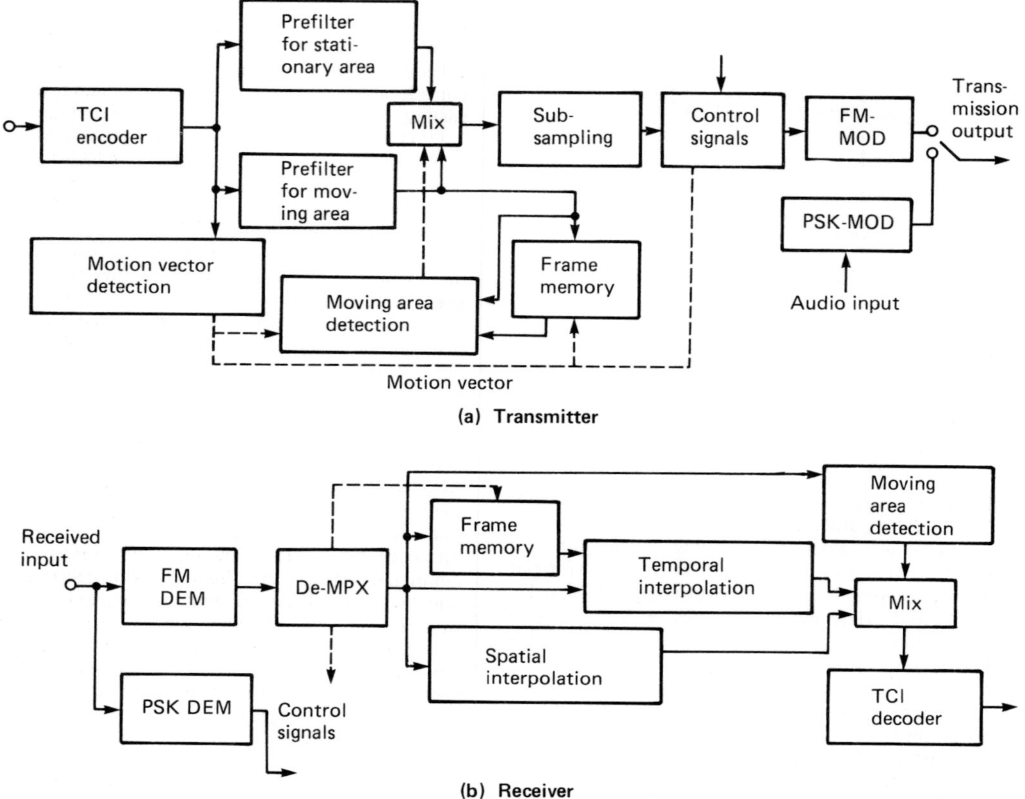

Figure 63.11 Block diagrams of MUSE transmitter and receiver

System:		Motion-compensated multiple subsampling system (Multiplexing of C signal is tci format)
Scanning:		1125/60 2:1
Bandwidth of transmission baseband signal:		8.1 MHz (−6 dB)
Re-sampling clock rate:		16.2 MHz
Horizontal bandwidth:	(Y)	20 22 MHz (for stationary portion of the picture) 12.5 MHz* (for moving portion of the picture)
	(C)	7.0 MHz (for stationary portion of the picture) 3.1 MHz (for moving portion of the picture)
Synchronization:		Positive digital sync
Audio and additional information		PCM multiplexed in VBLK using 4ø DPSK (2048 Kb/s)

*Values of a prototype receiver: these values should be 16 MHz and 4 MHz, if a perfect digital two-dimensional filter could be used

Table 63.3 Characteristics of the MUSE system

interpolation, motion-compensation is introduced. A vector representing the motion of a scene is calculated for each field at the encoder, and a vector signal is multiplexed in the vertical blanking period and transmitted to the receiver. In the decoder, the read-out address of sampled pixels of the preceding field in the frame memory are then shifted according to the motion vector. Together with this motion-compensation, temporal interpolation can be applied, handling data as in still-picture mode with no resultant blur.

The characteristics of the MUSE system are given in *Table 63.3*. Block diagrams of a MUSE transmitter and receiver are shown in *Figure 63.11*.

The MUSE system allows broadcast of HDTV signals on one standard channel of the 12 GHz band of 24/27 MHz so that an HDTV receiver will tune to one single channel exactly as happens with current TV standards. The total system concept is illustrated in *Figure 63.12*

In this system a phase-alternating sub-nyquist sampling method and motion compensated interframe coding have been combined and applied to bandwidth reduction. Luminance and colour information are first processed by the time compressed integration (tci) technique[20] in which the two chrominance primaries are encoded and temporarily compressed at a ratio of 4:1 to be multiplexed in the horizontal blanking period, not in the form of a 2:1 interlace but line sequentially as in *Figure 63.13*.

The synchronizing signal is composed of horizontal driving (hd) and frame synchronizing signals. The control signals and audio/extra information signals are multiplexed in the vertical blanking period. The control signals which include motion vectors are multiplexed to the video signal in the baseband. The audio signals are digitally coded in a format compatible with the Japanese BS–2 system, which can transmit two or four audio signals and additional information at a deviation of about 11 MHz. The system performance is summarized in *Table 63.4*.

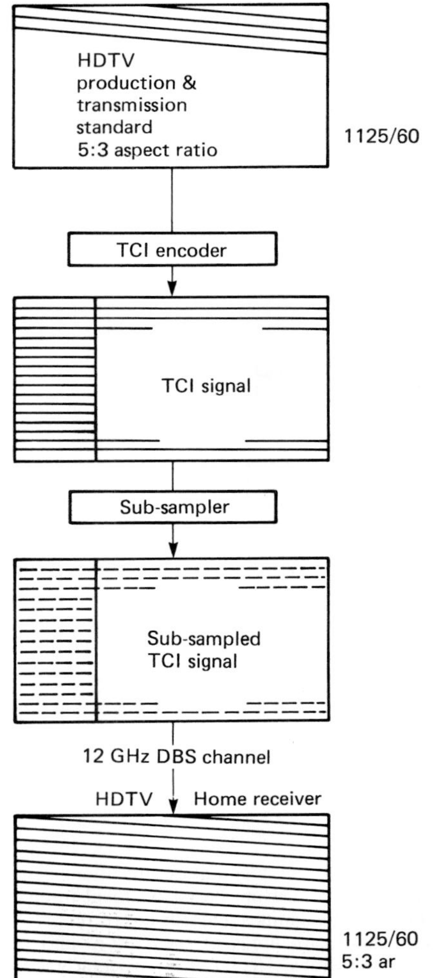

Figure 63.12 The one channel DBS HDTV system

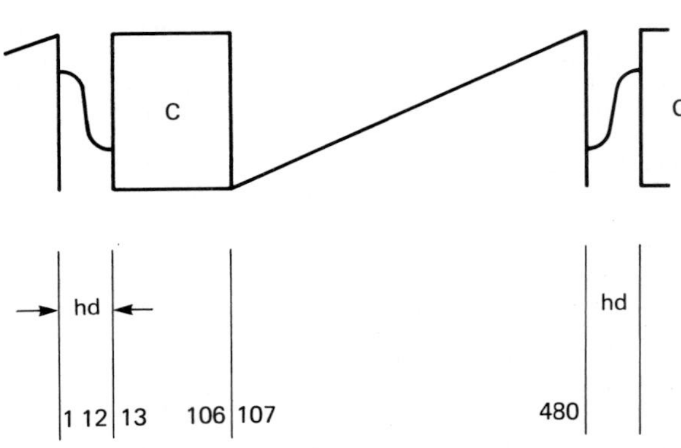

Figure 63.13 Transmission signal formatl

System:	Motion-compensated sub-sampling system
Input and output video signals:	R, G and B signals of HDTV system having 1125 lines, 2:1 interlaced 60 fields per second
Bandwidth of the reconstructed signal:	Luminance signal: 25 MHz for stationary portion of the picture, and 15 MHz for moving portion of the picture. Colour-difference signals: 7.0 MHz for stationary portion of the picture, and 4.5 MHz for moving portion of the picture.
Video multiplexing:	a simplified version of tci (time compressed integration) signal. Line sequential colour difference signals.
Audio multiplexing:	Two companded digital audio signals are multiplexed in the field blanking period 12 bits/sample, 22 750 samples/second
Synchronizing signal:	A positive synchronizing signal
Auxiliary signals:	180 bits/second of highly protected signal and 7200 bits/second of signal for general use.
Bandwidth of baseband signal for transmission:	8 MHz (−3dB bandwidth of a sharp cut-off filter)

Table 63.4 Parameters of the MUSE system

When MUSE is used in conjunction with a satellite transmitting power of 100 W and the diameter of the receiving antenna is 1 m, the s/n ratio is about 41 dB, which is quite satisfactory for a HDTV picture. The cost of receivers for HDTV is reducing due to extensive development of LSI fabrication and reduction of memory costs. A receiver for the MUSE system HDTV needs some 10 Mb of memory.

63.5.3 Standards conversion

The difference between field rates in the 50 Hz and 60 Hz countries constitutes the most serious obstacle to the spread of the NHK HDTV format into Europe. But in 1985, NHK announced[9] a second generation standards converter which offers the best hopes yet for a worldwide HDTV system. This is

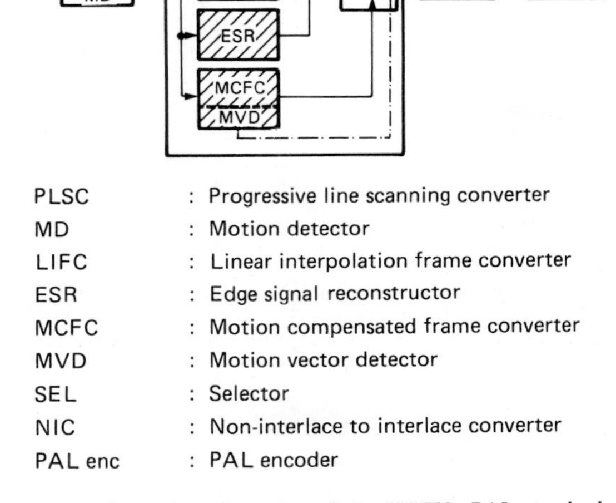

Figure 63.14 General configuration of the HDTV—PAL standards converter

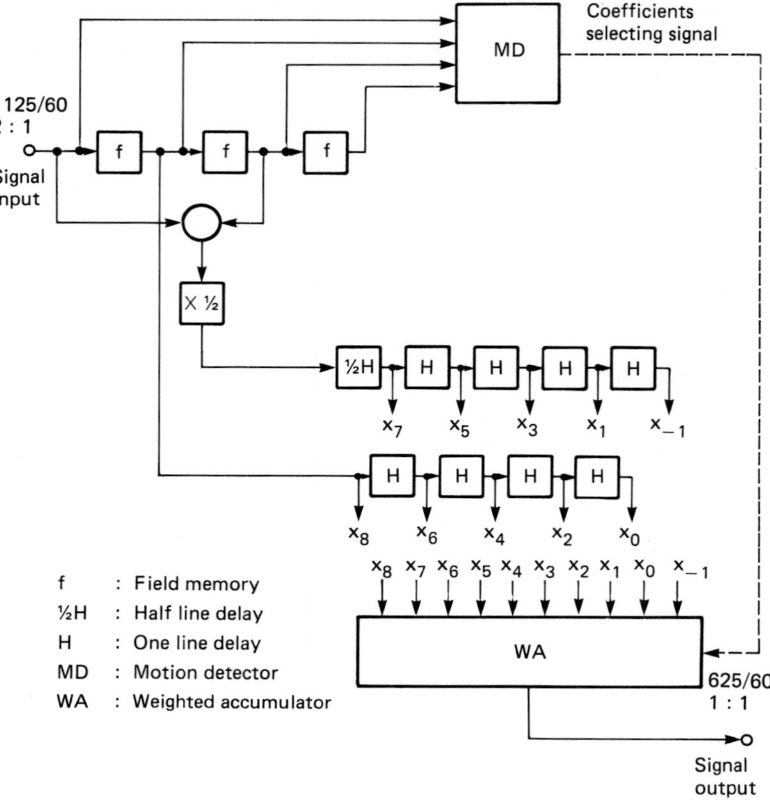

Figure 63.15 Block diagram of the line scanning converter

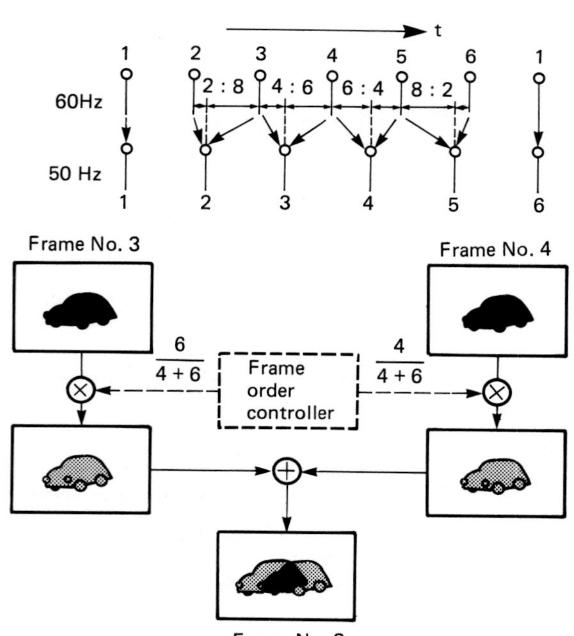

Figure 63.16 Linear interpolation

based on motion adaptive techniques and offers significantly better quality than converters based on fixed interpolation.

The problem with existing standards converters is that they cause blur in stationary areas of the picture and blur and jerkiness in moving areas. The latter is particularly noticeable.

The proposed standards converter uses a 1125-line to 625-line sequential scanning conversion with two-field storage (frame store) and then a 60 Hz to 50 Hz motion compensated frame conversion with an edge signal reconstructor to prevent the blur in moving areas (see *Figure 63.14*).

In the scanning line conversion, two operations are carried through: the conversion from 1125/60 to 625/60 and the conversion from interlaced scanning 2:1 to sequential scanning 1:1. The sequential scanning makes possible a more accurate detection of the motion vectors because scanning lines are in the same vertical position every $^1/_{60}$ s (see *Figure 63.15*). The optimum characteristics of filters used to prevent aliasing distortions are selected according to picture status: stationary or moving areas. The motion detector works out the inter-frame differences, using four successive fields for comparisons.

The most interesting part of the device is the 60 Hz to 50 Hz converter. This consists of a linear interpolation frame converter, an edge signal generator and the motion compensated frame converter.

The linear interpolation circuitry makes the weighted average of the current frame and the previous frame signals, as shown in *Figure 63.16*. Then an edge signal compensator determines the value of motion and generates a signal to compensate for the eventual blur. This signal determines the polarity and level of a differential signal used to compensate the final picture as shown in *Figure 63.17*.

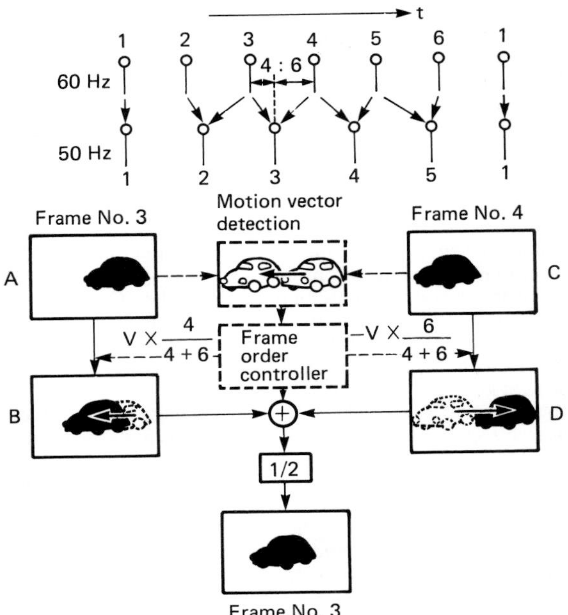

Figure 63.17 Positional interpolation

In the newly developed converter, four motion vectors are used to achieve high quality conversion as shown in *Figure 63.18*. The principle used is to select the pixels with a minimum value of difference by means of minimum inter-frame difference between two pictures, each of which is interpolated from previous and current frame pictures. Then the whole new picture at 625/50 is reconstructed pixel by pixel.

Finally the signals undergo another conversion from sequential scanning to interlaced and are encoded into PAL signals.

The output picture quality of this converter was found to be good enough to meet HDTV quality standards.

The same conversion algorithm can be applied to improve the picture quality of the transfer operation from HDTV to optical film.

63.5.4 HDTV receiver displays

In a high definition television system the final interface with the viewer is the displaying equipment. The performance of the receiver display is the main determinant in the evaluation of the whole system, because it provides the end product.

There are three main categories of display devices:

- direct view tubes,
- projection systems,
- flat panels.

Basically HDTV requires a wider viewing angle to be

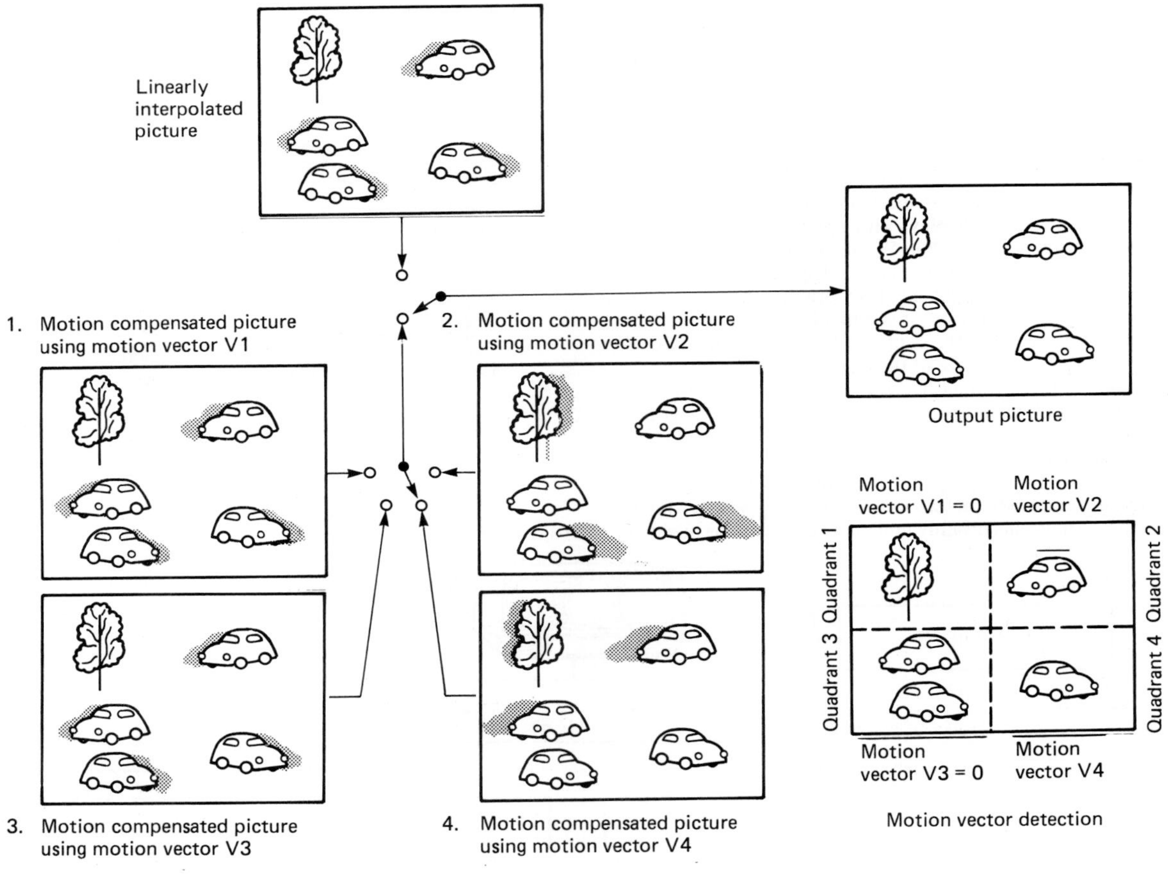

Linearly interpolated picture

1. Motion compensated picture using motion vector V1

2. Motion compensated picture using motion vector V2

Output picture

Motion vector V1 = 0
Motion vector V2
Quadrant 1
Quadrant 2
Quadrant 3
Quadrant 4
Motion vector V3 = 0
Motion vector V4

Motion vector detection

3. Motion compensated picture using motion vector V3

4. Motion compensated picture using motion vector V4

Figure 63.18 Motion compensation with four vectors

effective, and if one assumes that viewing distances in the home are not likely to change significantly, it follows that larger screen sizes are required. They are required not only by the television industry, but even more so by users of microcomputers. Demand for crt monitors with wider bandwidth, increased colour options and higher resolution is a reflection of the host-equipment manufacturer's demands for sharper and flicker free visual display units (VDUs) with improved text and graphics presentation features. The main impetus for technical improvements and innovation is coming from manufacturers of graphic terminals.

There is a worldwide display market incorporating television and computer industries *(Figure 63.19)*. Flat panel display technologies should make their presence known in the market place in the next few years[22] but the crt will continue to be the dominant display device because of its low cost per picture element.

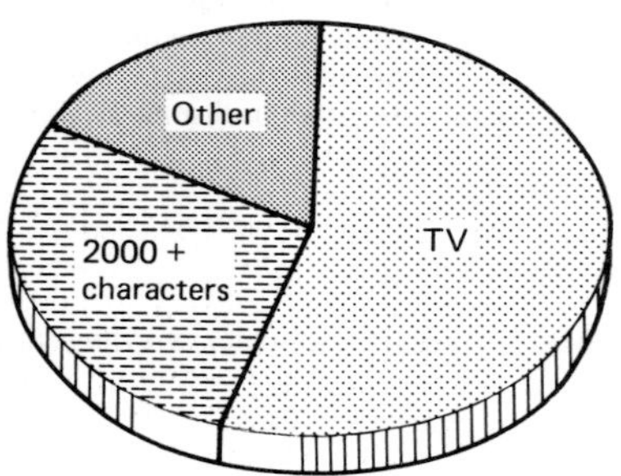

Figure 63.19 Worldwide display market

The development of VDUs and high definition television display technologies will, unavoidably, merge in the next few years and it is reasonable to expect more computer technology to penetrate the television industry and become part of it.

Back in 1985, several Japanese manufacturers had high definition *direct view* crt displays with a picture height of 40 cm and a screen width of more than 50 cm. Because the deflection angle is 90°, these direct-view receiver displays are very large and need a lot of driving power. They are very bright, with a marvellous picture quality but it is difficult to imagine them in an ordinary lounge. Thus the depth, the weight and the cost of crt tubes are limiting the size of the displayed picture.

Projection systems can easily service a screen of 5 m giving crisp pictures with the impact and realism of larger than life video. They can accept the output of a vtr, a video disc, a camera or a TV tuner *(Figure 63.20)*.

High-definition video projectors ensure a resolution of 2000 computer generated characters and pictures as good as a 35 mm optical film. They will certainly find many applications in video theatres, auditoriums and as background in concert halls. But in a normal home it is desirable that the equipment should not form an obstruction when not in use. Ideally, it should be a self-contained box. This means folded light paths and rear projection, which do not permit good resolution.

Ever since television sets started appearing in the domestic environment, there has been an interest in large screen *flat panel displays*.

The main considerations for flat panel displays are panel size and thickness as well as weight and power consumption. The luminous efficiency of the display should be greater than 2 lm/W in order to limit the power consumption.

There are several proposals for flat picture tubes. One is a flat crt using a localized source of electrons (point or line) and electron beam guides to distribute modulated current to each picture element or pixel. The beam guide is analogous to a fibre optic light guide.

The tube itself is made of two flat pieces of glass and four side walls (see *Figure 63.21*). The back of the tube is patterned with electrodes and the faceplate is coated with phosphors. Inside the box are electron sources, electron beam guides and an internal support structure. The electron sources and modulators are typically located along the bottom of the tube and inject current into vertical beam guide channels *(Figure 63.22)*.

A preferable technique is to utilize multiple electron sources with a discrete modulation point for each vertical beam guide (see *Figure 63.23*).

There is not yet a commercially available flat panel display suitable for HDTV. But there is a flurry of activity in the computer industry, where flat panel displays featuring

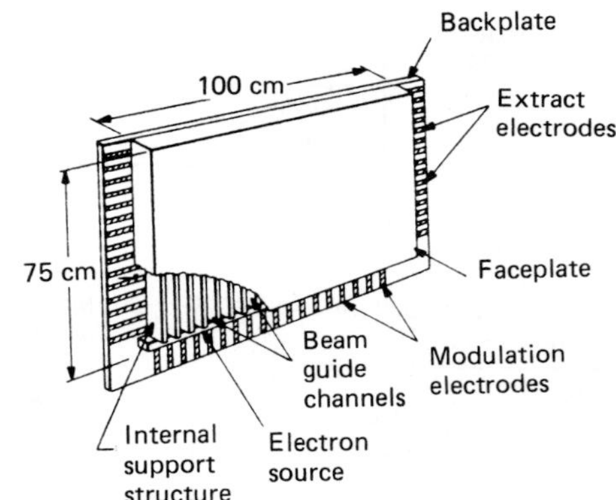

Figure 63.21 Flat panel display

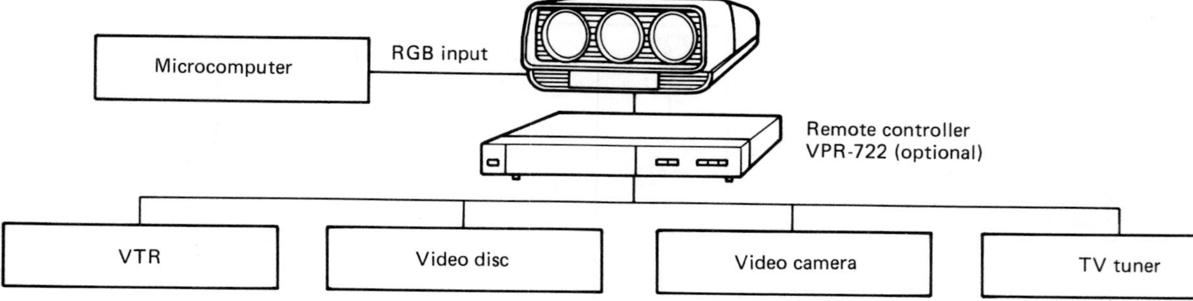

Figure 63.20 Possible connection to a projection screen

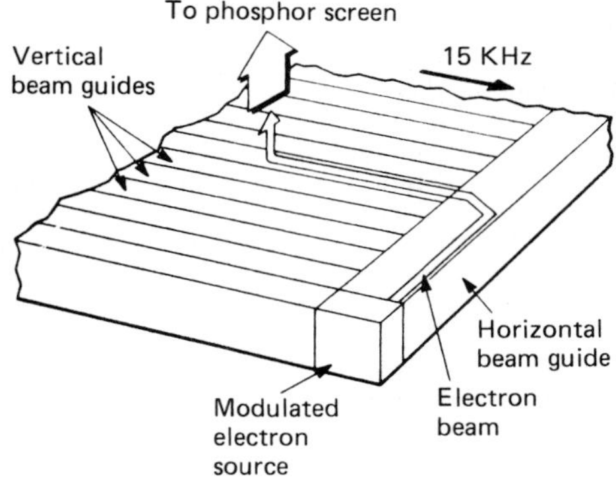

Figure 63.22 Detail of flat panel construction

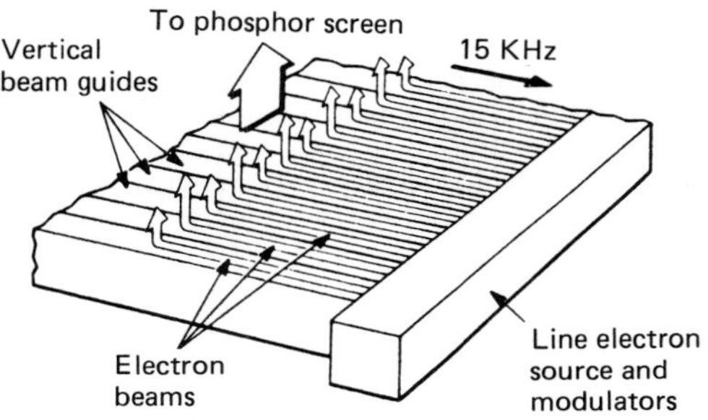

Figure 63.23 Multiple beam panel

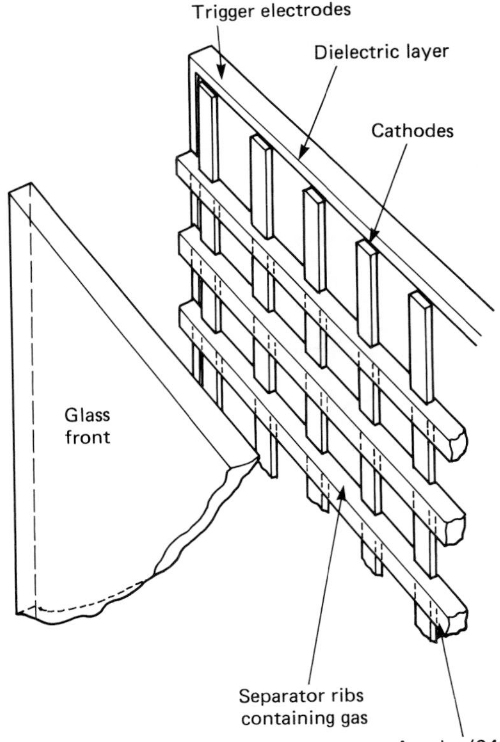

Figure 63.24 Dixy's prototype plasma panel

multiplexed fluid crystals, active matrix fluid crystals, plasma and vacuum fluorescence are being developed. The liquid crystal panels have very good prospects because of low production cost but they have serious drawbacks and limitations of which response time and contrast ratio are the most important.

Developments continue in plasma panels. For instance the Dixy prototype is a dc kind in which a gas mixture is sandwiched between two glass panels in such a way that pockets of gas are isolated one from one another — unlike the ac version where there is no isolation. They are located at the intersection of vertical and horizontal anodes and cathodes deposited on the glass panels. One of the difficulties with this kind of display is in switching the high voltage needed to break down the gas, necessitating some arrangement to keep the discharge 'alive' in order to reduce the switching voltage. This has been solved by using a capacitively coupled trigger electrode which is deposited on the rear glass plate, together with a dielectric layer situated underneath the cathode using a thick-film printing process *(Figure 63.24)*.

Data for the vertical electrodes are grouped in eight-bit packages and are shifted onto the display by clock-pulses. The horizontal electrodes are sequentially driven from top to bottom.

The manufacturers claim a luminance of 50 cd/m^2, a contrast ratio of 10:1 and a 120° viewing angle.

Matsushita have developed a prototype panel involving 'matrix drive and deflection'. This panel has the equivalent of 3000 electron guns (200 horizontally by 15 vertically) that excite primary phosphors, a contrast ratio of 50:1 and 0.5 mm element pitch and retains the luminance of a direct view crt display (see *Figure 63.25*).

The electrode structure has yet to be optimized but the secret

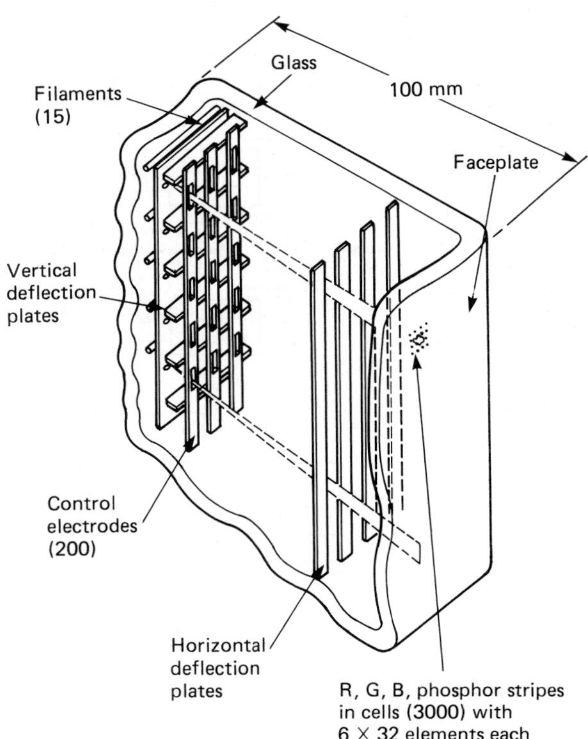

Figure 63.25 Matsushita's vacuum fluorescent display panel

to its construction is the technology required to cement, evenly and alternately, 0.1 mm grid electrodes with insulating plates. Signal processing and drive for the panel is digital.

Today most research is carried out in the computer industry where funds are available, but the eventual fall-outs are bound to benefit HDTV tomorrow.

63.6 High definition electronic production of movies

63.6.1 HDTV laser beam recording on 35 mm colour film

NHK have developed a laser-beam recorder, which records directly on colour film using laser beams that have their intensity modulated by video signals. This apparatus is used for video-to-film transfer of high definition television signals without any serious degradation of image quality.

Laser beams have high brightness, sharp directivity and good monochromacity. This is very convenient for electro-cinematography because it helps in obtaining:

- high resolution,
- high brightness,
- good colour reproduction,
- low flare light and excellent grey scale reproduction,
- low distortion when using mechanical deflectors.

The recorder is shown diagramatically in *Figure 63.26*. It consists of red, green and blue laser light sources, intensity modulated by optical modulators in accordance with the high definition RGB signals. All three beams are combined coaxially by dichroic mirrors into a single beam. The laser beam is then deflected horizontally by an optical deflector and focused on a 35 mm film in a continuously moving recording camera. The three laser sources are an He-Ne laser (50 mW) for the red line (632.8 nm), an A^+ laser (800 mW) for the green line (514.5 nm) and an He-Cd laser (15 mW) for the blue line (441.6 nm).

The intensity fluctuations and noise of the laser beams are reduced by special circuits in which the low-frequency component of detected noise signals is used to stabilize the intensity fluctuations in the feedback loop.

There are two types of optical modulator (AOM). The first has a centre frequency of 40 MHz and a modulation bandwidth of about 2 MHz, and is used for intensity modulation in accordance with the video signals and the cancellation of the laser noise.

The rotating mirror deflector is a 25 sided polygon with a diameter of 50 mm and a three-phase synchronous motor with aerostatic bearings, rotating at 81 000 rpm. The mirror is machined with great precision to avoid irregular scanning intervals. The relay lenses compensate for any remaining imprecision.

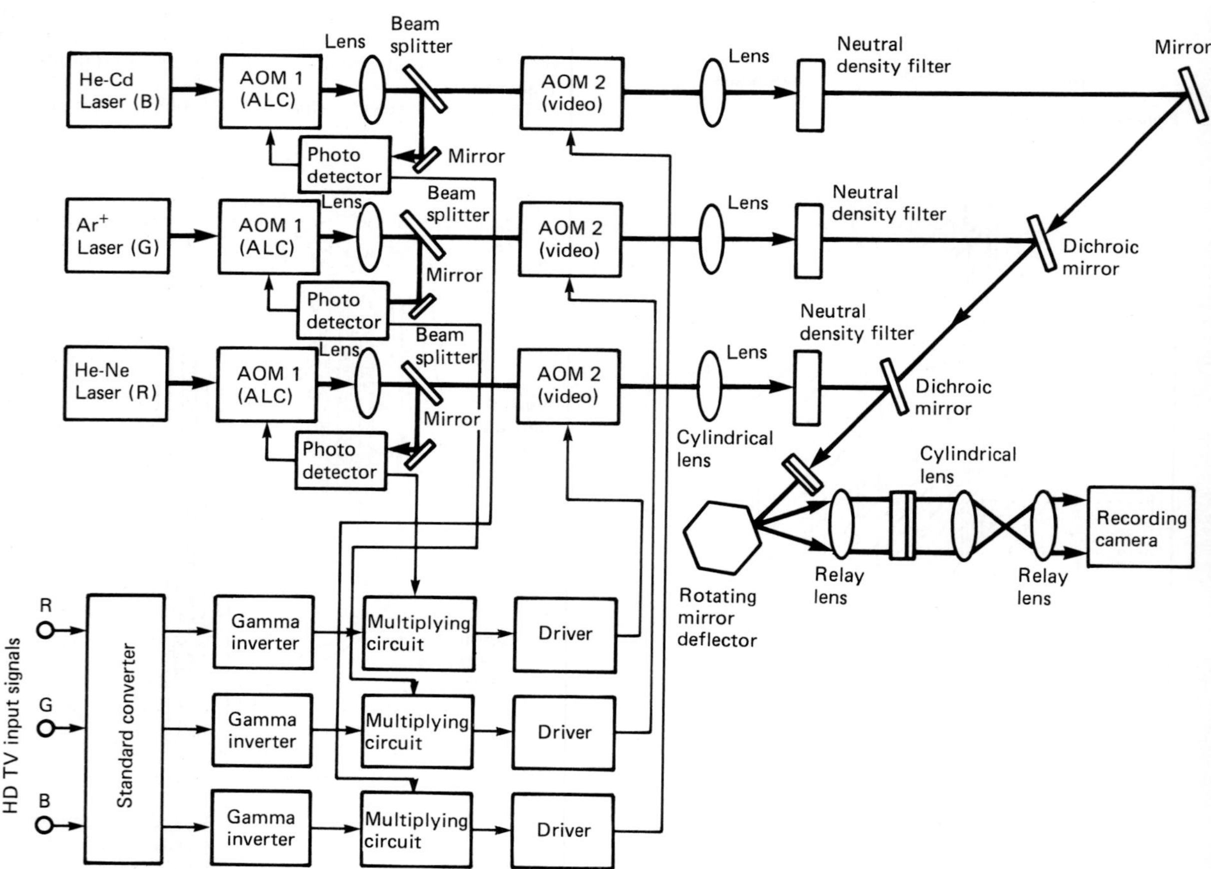

Figure 63.26 Configuration of the NHK laser beam recorder

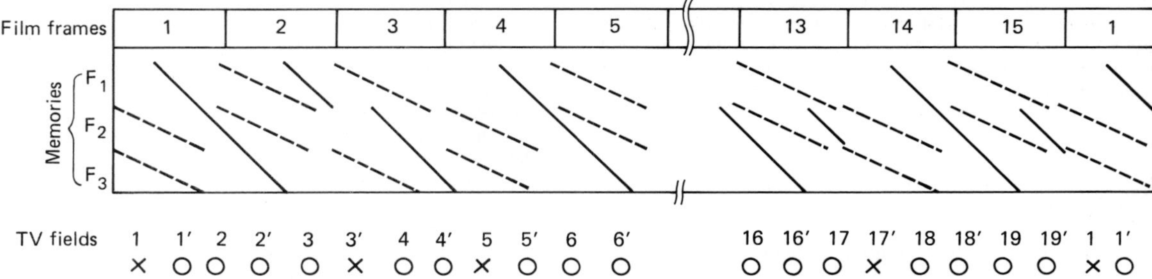

Figure 63.27 Principle of standards conversion using field memories. Solid lines indicate input to memories; broken lines are output from memories. The O and X sign show television fields which should or should not be stored in the field memories. Scanning conversion is accomplished by reading signals from two field memories alternately

The recording camera with its continuously moving mechanism needs a standards converter to convert the scanning signals from interlace to sequential mode. In this way uniform scanning lines are recorded.

The recording speed is 23.68 frames/second (slightly below the 24 frames/second of a conventional camera with intermittent movement). The picture size is 21.9 mm × 13.2 mm with a 5:3 ar.

The standards converter achieves scan conversion from interlace to sequential and simultaneously a frame conversion, according to the principle shown in *Figure 63.27*. It is important to match the spectral sensitivity curves of the film to the wavelength of the laser beams.

63.6.2 Comparison of optical film and video/film production systems

Film and television have been partners for half a century. During this time great improvements have been made in both media, with the public receiving most of the benefits.

In the production of motion pictures for cinema audiences and for television broadcasting, substantial flexibility and cost savings in the post-production process could result from the use of HDTV cameras for shooting and from the use of video tape rather than optical film for picture storage. The video tape should be equivalent, from a system point of view, to the dupe negative in the movies system.

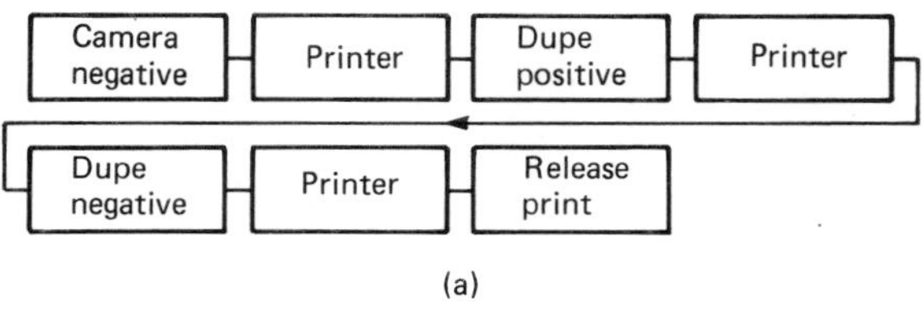

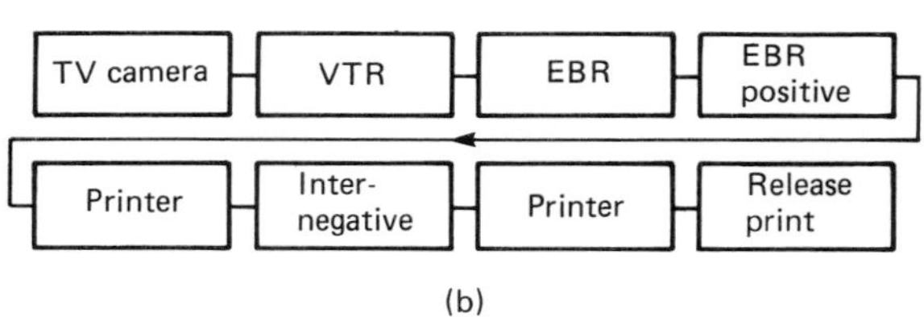

Figure 63.28 Production stages of (a) an optical film, (b) HDEP of movies

The production experience of RAI has shown that existing technology offered by Sony's HDVS and used to produce the feature film *Oniricon* (shown at Montreux in 1985) can provide the resolution capability necessary to match the 35 mm film system under similar lighting conditions and with almost the same ease of movement on site and in the studio.

In this film, video processes replaced most of the photographic processes and the immediate availability of the 'dupe negative' and the sophistication of the electronic post-production means facilitated the whole production.

To be able to make quantitative performance comparisons between film and HDEP one should analyse the equivalent production components[24]. The equivalent stages for optical film production and HDEP of movies are shown in *Figure 63.28*. There are seven stages in cascade in the optical film production and eight in the HDEP.

In the case of optical film, each stage causes a loss of resolution and introduces luminance fluctuations due to the grain structure of the film. These fluctuations (due mainly to the camera negative and the dupe negative) are transferred from one stage to another via the printing process. The luminance fluctuations and the noise seen in the release copy are a combination and a result of the granularity of each of the component films. The modulation transfer function (mtf) and the point gamma of the release print is a combined product of intermediate stages mtf and the point gammas of the component films. The noise from each of the four film sources is fed through a series of low-pass filters (printers, developed films, lenses) and, in each step, through a non-linear transfer characteristic, the incremental gain of which is the point gamma of the films through which the grain is being transferred.

In the case of the video/film system the noise sources are the scanning beam and the head amplifier in the camera, the video tape recorder, and the beam and film in the electronic beam recorder (EBR), the inter-negative and the optical release print.

The electrical noise signals are transferred through the video system in the same manner as grain noise is in the film system. But the significant difference here is brought about by means of electronic correction which narrows the noise bandwidth, corrects the final point gamma characteristic and compensates for most of the resolution loss. The amount of correction needed to achieve this is in the acceptable range of 4.5–6 dB.

The result is that the HDEP system allows a slightly better overall performance than optical film, at the same viewing distance.

63.6.3 HDEP TV system consideration

As far as the HDEP of movies is concerned things could be greatly simplified if the frame rate of a TV system specially designed for this purpose could be made the same as that of existing motion picture systems, i.e. 24 frames/second. On the other hand, there is no need for interlace, since film is the final output and the EBR could be made simpler. The other parameters might be better chosen if this principle were adopted.

References

1 FREEMAN, J P, 'The evolution of high definition television', *SMPTE Journal*, 492-501 (1984)

2 SANDBANK, C P and MOFFATT, M E B, 'High definition television and compatability with existing standards', *SMPTE Journal*, 552-661 (1984)

3 HAYASHI, K, 'Research and development on high definition in Japan', *SMPTE Journal*, 178-186 (1981)

4 FINK, D G, 'The future of high definition television', *SMPTE Journal*, 89-94, 153-161 (1980)

5 FUJIO, T, 'High definition wide screen television for the future', *IEEE Transactions on Broadcasting*, **BC-26**, No 4, 113-124 (1980)

6 FUJIO, T, 'The NHK high resolution wide screen television system', *Television Technology in the 80s, SMPTE 1981*, 146-150

7 PLONSKY, J, 'Questions on the orientation of research in HDTV in the 80s', *Television Technology in the 80s, SMPTE 1981*, 146-150

8 FUJIO, T, SUGIMOTO, M and NINOMIYA, Y, 'HDTV transmission method (MUSE) *14th International Television Symposium* (1985)

9 SUGIMOTO, M, FUJIO, T and NINOMIYA, Y, 'Second generation HDTV standard converter', *14th International Television Symposium* (1985)

10 NAKAMURA, T and SUZUKI, Y, 'Design considerations for a HDTV camera', *13th International Television Symposium* (1983)

11 SONY CORPORATION, 'HDVS — High definition video system', *Sony Catalogue* (1985)

12 CHILDS, I and TANTON, N E, 'Sequential and interlaced scanning for HDTV sources and display', *14th International TV Symposium* (1985)

13 POWERS, K H, 'High definition production standards interlace or progressive', *19th Annual TV Conference, SMPTE* (1985)

14 TONGE, G J, 'The television scanning process', *18th Annual TV Conference, SMPTE* (1984)

15 ROBERTS, A, 'The improved display of 625 line television pictures', *BBC Research Department Report 1983/8*

16 JESTY, L C, 'The relation between picture size, viewing distance and picture quality', *Proc. IEE* **105**, (Part B), 425 (1958)

17 KELL, R D, BEDFORD, A V and TRAINER, A M, 'An experimental television system' *Proc. IRE* **22**, 1246-1265 (1934)

18 KELL, R D, BEDFORD, A V and FREDENDALL, G L, 'A determination of optimum number of lines in a television system', *RCA Rev.* **5**, 8-30 (1940)

19 FUJIO, T et al, 'High definition television system', *NHK Laboratories Note No. 239* (1979)

20 FUJIO, T and SHIGETA, S, 'Technical problems in the broadcast distribution of high definition television', *13th International TV Symposium* (1983)

21 KIMURA, E and NINOMIYA, Y, 'High definition television broadcast system by a satellite', *Television*, 202-225 (1984)

22 EDITORIAL, 'Video displays', *Electronics and Wireless World*, 17-25 (July 1985)

23 CREDELLE, L, 'Large screen flat-panel' *Television*, 21-24 (1982)

24 SUGIURA, Y, NOJIRI, Y and OKADA, K, 'HDTV laser beam recording on 35 mm colour film and its application to electro cinematography', *SMPTE Journal*, 642-651 (1954)

25 POURCIAU, L L, 'High resolution TV for the production of motion pictures', *SMPTE Journal*, 1112-1170 (1984)

N Wassiczek, G T Waters, D. Wood
European Broadcasting Union

63b High Definition Television and Electronic Production of Movies – European Standards

The essence of the main debate in HDTV which is taking place in many international bodies, including the CCIR, is essentially about the number of lines, and fields/frames that should be used.

The chain from production studio to viewer can be considered in several parts. These include the system used for making programmes (the *studio* standard) and the system used to carry the HDTV over the airwaves to the viewer (the *broadcast* standard). Here we attempt to analyse the current situation regarding the studio standard and the broadcast standard, as seen with European eyes.

63.7 HDTV standards

The HDTV standards' debate has been the subject of many technical and techno-economic papers published in recent years,[1,2,3], and is summarized in section *63a*.

NHK, the Japanese Broadcasting Corporation, developed a range of HDTV studio equipment in the 1970s. In the early 1980s, broadcasters throughout the world recognized the importance of HDTV for the future of broadcasting. They also agreed that the objective should be unique standards throughout the world for the production and broadcasting of HDTV.

Everyone accepted the general principles of what HDTV should provide, in terms of picture quality. But they also recognized that the most difficult problem in achieving unique worldwide standards would be the choice of field rate.

About three-quarters of the world currently uses a 50Hz field rate and one-quarter of the world uses 60Hz. There are large infrastructures of television broadcasting built up, over the years, based on 50Hz and 60Hz. Using a field rate other than the locally used one would be certain to call for field rate standards conversion, even if just for use in normal television broadcasts.

It became clear after a relatively short time that the pressures on HDTV broadcasting (emission) would be to adopt a field rate compatible with conventional field rates. There remained however a relatively independent case for a unique studio production standard.

63.7.1 Studio standards

NHK put forward for discussion a potential parameter set for the HDTV production standard using 1035 active lines/frame, 60Hz fields/second, and interlaced scanning. A number of Administrations supported this parameter set in the CCIR as a unique worldwide standard.

The EBU initially put forward the idea of a unique production standard based on an 80Hz field rate. The argument in its favour was that such a field rate could be converted with relatively high quality, equally to both existing field rates 50Hz and 60Hz, without motion adaptive standards conversion. This proposal was not acceptable to North America, and the EBU therefore made a detailed study of the possibilities of using 60Hz in the European 50Hz environment.

NHK developed an 1125/60 to 625/50 PAL and YUV motion adaptive standards converter, and subjective assessments were made of the resulting picture quality. There were still some small artifacts in the converted picture in critical cases, but the work suggested that, with further development, virtually impairment-free field rate stndards conversion would be possible.

The EBU endeavoured to analyze other technical issues associated with the use of 60Hz in a currently 50Hz enviroment. These included the effects of choice of field rate on motion protrayal, lighting flicker, etc. An analysis was also made of the operational and economic consequences of operating with 60Hz in a 50Hz enviroment.

Although opinions were divided, the results of the analyses contributed to a reluctance on the part of several European authorities to accept 60Hz for HDTV production. Firstly, it was argued that using 60Hz would bring the disadvantage of systematic standards conversion. For the production community, this could constitute simply an expenses, for which the only return would be a slight deterioration in quality. Secondly, there appeared to be no major quality advantages in using 60Hz compared to 50Hz.

In the late 1980s, a group of European Administrations put forward to the CCIR the proposal that there should be a single worldwide standard based on 50Hz and progressive scanning.

However true the conclusions about using 60Hz in a 50Hz environment are, it can be argued that there are similar, perhaps even larger, reasons why a 50Hz standard could not be accepted in a 60Hz environment. It seems to the 60Hz world to be a contradiction in terms to move to a higher quality standard by accepting a lower picture rate. Equally, the broadcasting infrastructure in the USA is even more decentralized than in

Europe, so the costs of standards conversion become even more of a problem.

The position reached was thus that for both the 50Hz and 60Hz communities, accepting the other field rate for HDTV appeared to mean more losses that gains. The prospect of using a field rate which is equally inconvenient for both 50Hz and 60Hz remains not universally acceptable. We can conclude that the wish for a single standard extends only to the point where it will incur little extra cost.

Therefore, we have to make the best of the second best situation, i.e. accept the fact that there will be two standards, and try to minimize the difference between them.

63.7.2 Dual mode stanards

Possible objectives for the dual mode standard include:

- making possible switchable equipment, or changeable equipment modules
- aiming for highest quality and lowest cost standards conversion
- choosing standards most likely to lead to convergence on a single standard in future

The EBU can claim to be the first organization to encourage the CCIR to take up this matter, and a number of possible dual mode parameter set pairs have been put forward for discussion.

63.8 HDTV formats

The original EBU contribution[4] pointed the way forward in two general directions. These are now termed the *Common Image Format* (CIF) and *Common Data Rate Format* (CDR).

63.8.1 Common Image Format

In the common image format (CIF) approach, the number of active lines per frame of each of the 50Hz and 60Hz partners is arranged to be the same. In the digitally sampled versions of the two standards, the number of active samples per line is also arranged to be the same. A sub-set of this approach arranges for the number of active samples per line to be related to the number of active lines by a factor which is also the picture aspect ratio. This is called a *square sample distribution*

63.8.2 Common Data Rate Format

In the common data rate (CDR) approach, the sampling frequency of each of the 50Hz and 60Hz partners is arranged to be the same. As a consequence of this, the bit rate for the two partners can be arranged to be the same. A square sample distrubution is also possible in a CDR approach.

63.8.3 Arguments for the two approaches

The CDR approach will help reduce the cost of switchable equipment, or allow interchangeable modules, for equipment which involves signal processing, because of the common sampling frequency.

The CIF approach will help to make possible common solid-state cameras and displays. It will also help the eventual achievement of a completely unique standard, because there will be fewer scanning parameters remaining which are different.

The two preceding paragraphs are simplifications of complex ranges of technical factors, which, in some cases, bring advantages for the CIF approach, and in other cases bring advantages for the CDR approach.

Once again, we are faced with the very difficult task of weighing different sets of advantages in choosing an optimum. These are also of different importance for different regions of the world, and so even agreeing on appropriate criteria and priorities is difficult. This has been evident from the discussions in the CCIR.

To illustrate the difficulty, we could examine the claim that the CIF approach will remove the need for line standards conversion. In a CIF system, a frame will be a same, whatever the field rate. Or will it? In practice, in an interlace system, while there will be the right number of lines for both field rates, the information on the lines will not be correct, because the image will have moved to a lesser or greater extent between fields, unless the picture is stationary. The frames will not be the same. Their content will be different, depending on the field rate. So line conversion (*spatial standards conversion*) will still be needed. The 'common image format' will, strictly speaking, exist only in a progressively scanned system.

Manufacturers have different views about the impact on equipment cost and convenience of the CIF and CDR approaches. Some argue that CDR will result in less expensive equipment. Other argue that this would not necessarily apply to systems which temporal processing. Some argue that the DVTR will inevitably use bit-rate reduction, and therefore the choice of approach will not have a major effect. Others argue that bit-rate reduction will not be used for production equipment, and the choice of approach is a vital determinant of cost.

There are also 'instinctive' factors, which can possibly never be resolved by technical logic. A CDR format may always give a slight quality lead to the 50Hz partner, because the greater number of active lines which it gives rise to, may be more of an advantage than the extra fields. A CIF format may always give a slight quality lead to the 60Hz partner, because of the greater temporal headroom. Nobody wants to come in second.

There seem to be a number of constraints on the choice of parameters.

The most important need for European administrations appears to be that their partner should have no less than 1152 active lines per frame, because of the need to feed the HDTV emission system, HD-MAC (see section *63.9*). The best quality HD-MAC pictures for the viewer will result if the studio standard is 50Hz, progressively scanned, and has at least 1152 active lines. HD-MAC will need all the help it can get, because it is likely that interlace equipment would be used for an interim period.

The most important need for North America seems to be to have a common image format and a square sample distribution. From the European perspectives, these do not seem of overriding imprtance, but it has to be accepted that others may see things differently.

A system which brings together these two sets of primary requirements (European and North American) would be a CIF system with 1152 active lines/frame, and 2048 active samples per line.

However, this system does not seem acceptable to Japan. Japanese organizations have already invested considerably in 1125/60 equipment, and to increase the number of active lines per frame beyond about 1080 would mean, for them, only capital outlay for no benefit in return.

A system which may be a better match to the Japansese situation may be a CDR system, the 60Hz partner for which is compatible with existing 1125 equipment.

A new proposal, the *Common Image Part* (CIP), has been made, in which the active sample lattice of a 1125 system forms a sub-set of a 1250 active sample lattice. This may well be an acceptable solution to all sides.

European manufacturers began many years after Japanese manufacturers, and thus their equipment is at an earlier stage of maturity. European equipment is largely based on 1250/50 operation. This equipment can be used as sources for programme making, to familiarize production staff with HDTV. Hopefully, in the very near future the parameters of a dual mode standard will be agreed, and this will take us to the point where large investments can safely be made, in preparation for HDTV motion picture production and HDTV broadcasting.

At the CCIR meetings of Study Group 11 in 1989, a useful degree of progress was made in HDTV studio standardization. It was possible to draw up a draft recommendation for those parameter values which are common to the two proposals (Japanese and European), leading to agreement on interim parameter values for certain other aspects of the standard, including colorimetry and transfer characteristics[5].

63.9 HDTV broadcasting

There is a range of alternative ways by which HDTV programmes could eventually be brought to the public. Apart from distribution via tape or disc, HDTV might conceivably be broadcast in existing or future broadcast bands.

The two principal possibilities for broadcasting in Europe are both via dbs services. Systems are being studied which will allow satellite broadcasting in the bands above 12.7GHz. Efforts are being made by European braodcasters to analyze and promote the case for the allocation of a band, perhaps at about 20GHz, to carry digital HDTV broadcasts. Such systems are likely to have a net bit rate of 70-140MHz, and should give impairment free transmission of the HDTV studio standard.

More progress has been made on so-called *narrow-band* HDTV emission systems.

63.9.1 Narrow-band dbs HDTV emission systems

An HDTV emission system needs to provide the signals which will allow a successful consumer product. EBU specialists have examined what this is likely to mean, in terms of factors which will be influenced by HDTV system design parameters.

Traditional wisdom has it that, to be successful, any new system or service needs to provide a significant improvement on what people have already. No-one is sure how exactly to quantify this, but it can be provided by picture quality, convenience, software, etc.

A number of studies have been made in narrow-band HDTV emission systems in different parts of the world. In Europe the concept of a narrow-band HDTV system is one which will fit into a single WARC 1977 dbs emission channel. The main constraints here are the channel bandwidth and the permissile levels of interference to other services. The precise maximum base bandwidth permissible depends on spectral occupancy, but tests have shown that it will be in the region of 11-13MHz for vision type singnals.

In general, a European narrow-band HDTV emission system, derived from an appropriate HDTV studio standard, will require 4:1 or more bandwidth compression.

In any such system, even with the most sophisticated receiver processing, the quality available in the home will be inferior to some degree to that available from the studio standard. Studio standard parameters may, in any case, have been selected to provide a degree of post-processing headroom.

However, at minimum, the quality of pictures, as viewed on a large screen in the home, of greater than 1 square metre, must provide sufficient added value to justify the HDTV production and receiver development costs. In the European case, this means that the quality must be significantly better than that provided by normal 625/50 MAC.

Basic quality is not the only factor to be considered. For example, there should ideally be receiver future development possibilities. The system must comply with the WARC 1977 plan. The system must be transmittable over cable, suitable for home recording, and have appropriate sound and data capacity.

Some of these factors are considered in sections *63.9.1.1–63.9.1.5*.

63.9.1.1 Basic quality

As a fundamental premise, a European HDTV system needs to be a clearly noticeable step forward in quality from MAC, and any temporal artifacts created by the bandwidth reduction system must be sufficiently unobtrusive not to detract from the original quality of the programme source.

To evaluate the system, methods recommended by CCIR JIWP 10-11/6 should be used[6,7]. The degree of acceptability will depend, as always, on the test material used. But the following has been suggested: as a generality, for material which is 'critical but not unduly so', the quality of the test pictures should never fall more than about 1 grade (25 per cent on DSCQS scale) below source quality. The overall quality for a programme as a whole should be very close to source quality.

It is interesting to note that some studies[8] suggest that picture *sharpness* rather than *resolution* is the most critical quality factor. Sharpness is not exclusively linked to bandwidth and can be artificially enhanced in the receiver.

63.9.1.2 Receiver cost and complexity

The system design should be such that, given the normal pattern of consumer electronics development, narrow-band HDTV receivers are within financial reach of the majority of the general public, after a relatively short start-up period. Further, the system should allow the display performance to evolve over time, to maintain a differential between HDTV and convertional trelevison. The latter will probably benefit from a number of future receiver enhancements.

63.9.1.3 dbs coverage and WARC 1977 compliance

Systems developed for the 12GHz band dbs receivers are rather more sensitive than those foreseen in the WARC 1977 plan; and, in addition, studies suggest that the use of non-linear pre-emphasis may considerably improve the failure characteristics.

In evaluating the limit of the coverage area, the same quality criteria specified by the WARC 1977 plan should be used for noise and interference. This time, however, it should be assessed at a viewing distance of three times rather than six times picture height. The quality criterion for interference for PAL and SECAM is CCIR impairment grade 4.5 or better compared to reference quality. The criterion for noise is impairment grade 3.5 or better for at least 99 per cent of the worst month of the year.

63.9.1.4 Sound performance

Sound quality should be at least as good as that available from the MAC/packet system, and the sound failure characteristic should be such that sound fails after vision.

The improvement in 'realism' in vision performance with HDTV calls for a re-examination of the 'realism' in the accompanying sound system also. This may call for a multi-dimensional sound system.

As a starting point for discussion[9], the EBU specialists propose that the HDTV sound system should allow localization across a given sound image field. This could be about twice the image field. The localization properties should be maintained in a given listening area. This should be a trapezium beginning at about 2.5 times picture height and ending at about 4.5 times picture height. Within the listening area, it should be possible to co-locate the face and voice of a speaker on the screen.

Localization and technical fidelity are not the only characteristics of a multi-dimensional sound system. There are other quality factors such as naturalness, agreeableness and sense of reality. The criteria for these will depend on the method used to evaluate them. Currently there are no universally agreed methods, although studies are in hand in CCIR JIWP 10-11/6.

63.9.1.5. Compatible picture quality for MAC systems

The HD-MAC system, being developed in the Eureka 95 project is intended to allow simultaneous reception by conventional MAC receivers.

MAC receivers will also have a PAL/SECAM output if they are of the set-top type. Ideally, no impairments should be preceptible on a MAC (or PAL) receiver. However, to be realistic, narrow-band HDTV systems must use temporal processing and motion related sampling, which inevitably result in some visible impairments.

To some extent, the HDTV quality can be at the expense of the compatible picture or vice versa. This has to be borne in mind when specifying the compatible picture quality. In the long term, the purpose of the system is HDTV. Therefore, if there are sacrifices to be made which are unavoidable, they should be made in the compatible picture, rather than in the HDTV picture.

As a target, the general quality level of the compatible picture should not fall below that associated with PAL quality. This might mean that, for material which is 'critical but not unduly so', the compatible-MAC test picture should never fall more than about one grade (25 per cent on the DSCQS scale) below the quality of an equivalent direct MAC picture.

63.9.2 Provison of higher quality via dbs

In 1982-83, the EBU developed the MAC/packet dbs broadcast system[10]. The intention was to develop a system which would, at the same time, provide an improvement in vision and sound compared with the current PAL/SECAM systems, and provide Europe with a unique broadcast standard for dbs. This was intended to rectify, for dbs, the cultural, convenience, and cost problems, of the existing PAL/SECAM divergence.

The received MAC/packet signal has a potential bandwidth of 5.75 MHz. Conventional PAL receivers usually have an effective bandwidth, after a notch filter, of about 3.5 MHz. MAC/packet signals provide a step forward in quality, which is somewhere between conventional composite singnals, and expectations for HDTV. The MAC/packet system was also designed to operate with a standard or wide aspect ratio.

In 1982-83, it seemed reasonable to standardize on the MAC/packet system for dbs, because it would lend itself to enhancement to HDTV to a greater extent than PAL/SECAM. The adoption of the MAC/packet system for dbs was thus an insurance policy, or an investment for the future, against the day when HDTV would be called for.

HDTV itself was thought to be many years away, largely because of the technological development needed to arrive at large screen flat panel displays. HDTV resolution is only of benefit if the screen and the viewing angle are large. If the display is only of today's size, the additional definition which HDTV gives would largely pass un-noticed. MAC vision signals provide sufficiently high quality for moderate size screens.

In 1985-86, a major European research and development programme, Eureka 95, was set up to devleop an HDTV emission system which is a compatible, enhanced version of MAC. This was in line with the concepts considered when the MAC/packet system was developed, but the plan has not been entirely as originally foreseen by the EBU, for two reasons. First, the impetus for the development of an HDTV version of MAC came rather sooner than expected. It did not wait for the development of solid-state displays. Secondly, the projections for dbs services starting in Europe have proved to be over-optimistic. There has been a long series of delays in satellite launches and receiver devleopment.

63.10 DBS prospects in europe

There are thus currently several possible scenarios for European prospective dbs broadcasters.

The first is to broadcast in the MAC/packet system, either with 4:3 ar or 16:9 ar, and at some future date to up-grade the service to the HDTV version HD-MAC.

The second scenario is to commence directly with HD-MAC. In this case MAC/packet receivers would provide low cost receivers for the same service. These could be second sets, or receivers for the less financially able. Current cost projections are that they will be a fraction of the cost of HDTV receivers.

63.10.1 HD-MAC system

The main elements of the Eureka 95 HD-MAC coding algorithm were agreed in 1989, and a demonstration of a hardware system was given in September 1989. The system is based on the MAC/packet transmission system. For the MAC system, the luminance and colour difference signals are included on each line.

The first 10μs of each active scanning lines is given over to a packet of digital sound or data. The MAC/packet scanning system, 625/50/2:1, is the same as for PAL and SECAM. The conversion to PAL and SECAM can be done inexpensively in a set-top receiver if it is needed.

The coding and decoding of the MAC signal is done in the digital domain at the source and in the receiver. There are taken to be about 700 active luminance samples per line and 350 active colour-difference samples per line with the MAC/packet system. The aspect ratio of the MAC signal can be either 4:3 or 16:9.

For HD-MAC, the aspect ratio is taken to be always 16:9, and the number of samples per line is doubled to about 1400 for luminance (with half of this for each colour-difference signal), and the number of active scanning lines is also doubled. The image is then sub-divided into a large number of blocks of 16 × 16 samples horizontally and vertically. For each block, supplementary data are derived about the content, which is transmitted and used by the HD-MAC receiver to process the incoming picture in a number of different ways.

The system thus uses motion-adaptive processing, based on a knowledge of the general contents of each block. The data signal describing the block content is called DATV (*Digitally Assisted Television*).

For the luminance signal, the system is as follows: If the picture content is stationary, or moving very slowly (0 – 0.5 samples/40ms), a first *mode* (or signal path) is adopted. In this mode, the definition in the picture is built up over four successive fields. The true picture repetition frequency is 12.5 Hz, although the pictures are displayed at 50Hz (or 100Hz) in the receiver.

The picture that is built up is actually a quincunx sub-sampled version of the orthogonally sampled source. The Nyquist limit for resolution (with a quincunx sampling pattern) in the diagonal direction is half that in the original source, but the Nyquist limit in the horizontal and vertical directions is the same. Therefore, if the content of the block is stationary or nearly stationary, its contents are transmitted with a quality almost the same as the source. The quincunx sampling is not completely transparent to the source quality, but it is very nearly so.

If the content of the block is moving slowly (0.5 – 12 samples/40ms) a second mode is adopted. In this mode, the definition in the picture is built up over two successive fields. The true picture repetition frequency is thus 25 Hz. In this case the sub-sampling structure is arranged such that the vertical definition is half that in the first mode. The horizontal definition is the same as in the first mode. The frequency response is triangular.

This second mode also employs the technique of motion compensation in the receiver. In this technique, the 25 missing movement phases per second are calculated in the receiver and displayed. That is to say, the receiver generates, itself, every other displayed field by physically moving the displayed object to a point between where it lies in the two adjacent fields. The receiver calculates the missing field by using a motion vector which is transmitted with the picture. The motion vector is a binary number which signals in which direction the content of the block is moving.

If the content of the block is moving relatively quickly (more than 12 samples/40ms), a third mode is adopted. In this mode, the definition in the picture is only that associated with one field. The picture repetition frequency is thus 50Hz. In this case the sub-sampling is arranged such that the vertical and horizontal definitions are both half that in the stationary mode.

There are three similar modes for the colour-difference signal, although the motion compensation is not used in the second mode.

Finally, before transmission, it is necessary to arrange the signal to look like a 625-line conventional MAC signal. This is done by the technique of *line sample shuffling*. The samples from adjacent pairs of lines are moved together so that they lie on one line. They do not fall on the same places on their respective lines, so combining them in this way is possible. The final signal thus appears to be a MAC signal to a MAC receiver. An HDTV receiver will take adjacent samples from each line and put them back on two adjacent lines, as they were before encoding.

In summary, the HD-MAC system achieves a 4:1 bandwidth reduction by a motion-adaptive sub-sampling system. The vertical definition is equivalent to an 1150 active line/1440 sample per line signal, if the picture is near stationary. This definition is progressively reduced, the faster an object in the picture is moving.

The human psycho-physical system works generally in a way compatible with this. Definition is perceived as long as the eye can track an image. Beyond that it becomes more and more redundant. However, exactly how good an match HD-MAC is, and the extent to which mode transitions can be perceived, will be shown by subjective assessments of the final system.

63.10.2 Evaluations of HD-MAC system

EBU specialists have made proposals for the tests appropriate to evaluate the HD-MAC system to assess its performance in relation to the requirements mentioned in section *63.9.1*. The areas to be evaluated include:

- *Basic quality* – quality of the received pictures at high c/n in dbs environment, and high s/n in am/vsb cable environment
- *Impairment versus c/n* – assessment of the picture quality in the presence of falling c/n, from 20dB to 8 dB
- *Impairment versus ber in DATV channel* – assessment of the picture quality in the presence of increasing bit error ratio, from 10^6 to 10^3
- *Impairment versus if bandwidth* – determination of the influence of if bandwidth on the vision impairment
- *Impairment versus sampling time error* – impairment of the picture quality as a function of sampling time error
- *Impairment versus c/i ratio* – assessment of susceptibility to CCI and ACI interference.
- *Impairment versus receiver mistuning* – check on the afc accuracy of the receiver
- *Performance of am/vsb cable transmission* – assessment of the picture impairment, sound impairment and data threshold as a function of echo, noise and interference in cable channels

63.11 Conclusions

There is no doubt that HDTV will be the most important medium of the next century. It will have an impact on a wide area of life: broadcasting, computing, medicine, printing, etc. It is certainly well worth while taking considerable care in the choice of parameters. They will be with us for at least the next 50 years. However, it is also true that a sub-optimum standard is very much better than no standard at all. There will come a point at which industries will have invested too much money to change their ideas. At this point standardization by compromise will no longer be possible.

The point at which HDTV broadcasting will begin in Europe will depend on a range of technical and economic factors. Nevertheless we are clearly on the threshold of the most exiting phase of televison broadcasting since the advent of colour.

Acknowledgements

The opinions expressed in this paper are those of the authors, but the work described was undetaken by the EBU Technical Committee, its working parties, sub-groups, an specialist groups. The work of the chairman and members of these groups is gratefully acknowledged. The enormous contribution to European HDTV technology made by participants in the Eureka 95 project is also acknowledged.

References

1. Sandbank C and Wood D, 'EBU Studies in High Definition Televison'. *Intern. Con. Telecommunications*, Liege, Belguim (November 1983.
2. Haberman W and Wood D, 'Images of the Future – the EBU's part to date in HDTV system standardization,' *EBU Review – Technical* (October 1986)
3. Wassiczek N, Waters, G T and Wood D, 'European Perspectives on HDTV Studion Production Standards,' *IEEE Trans Broadcasting* (September 1989
4. *HDTV Studio Standards Derived from CCIR Rec. 601* EBU contribution 11/6–2010, 11/7–182 (June 1088)
5. *Basic Parameter Values for the HDTV Standard for the Studion and for Internationa Programme Exchange* Recommendation XA/11, CCIR document 11/719 rev. CCIR Study Period 1986–1990.
6. *Subjective Assessment of HDTV Picture Quality*, Recommendation XB/11. CCIR Study Period 1986–1990.
7. *The Subjective Assessment of HDTV Pictures*, Report AT/11, CCIR document 11/641(revl). CCIR Study Period 1986–1989.
8. Lupker S, Allen N, and Hearty P. 'The North American High Definition Television Demonstrations to the Public: The Detailed Survey Results.' DOC Canada.
9. *EBU Requirements for HDTV Sound Systems*, EBU Contribution IWP 10/12-2. CCIR Study Period 1990–1994.

10. *Specification of the Systems of the MAC/Packet Family*, EBU Tech. document 3258 (October 1986)

11. Vreeswikj F W P, 'HDMAC Coding for MAC compatible broadcasting of HDTV' *Third Inter. Workshop HDTV*, Torino, Italy (August 1989)

Part 15
Industrial, Commercial, Medical and Defence Applications of Television

E A Jones B Eng, AMIEE
Assistant Technical Support Manager, Philips Communications and Security

64 Industrial, Commercial, Medical and Defence Applications

64.1 Industrial applications

64.1.1 Robot vision

In areas of manufacturing such as car assembly where automation is the key to low cost production, substantial investments are made to incorporate robots on the production lines. For simple tasks such as component manufacture, numerical-control machines are used. Component manufacturing information is programmed into the machine memory and the system produces components for as long as it is fed with raw materials. When the components are brought together for final assembly, robots carry out positioning and welding with a high degree of precision.

The availability of solid-state ccd cameras makes the provision of machine vision for robots increasingly attractive. A system is in use where car windscreens are mechanically handled by a robot that has cameras as sensors. The cameras are used to determine the edges of the windscreens as they are brought close to the aperture, and their signals are fed into an image processing system. An algorithm is applied which causes the robot to automatically fit the screen into the aperture. The screen or surround would have sealant applied prior to fitting.

In this example of assembly on a mechanized line, the component is reasonably well defined in terms of supply, i.e. the windscreens would be supplied in an organized stacked stockpile.

Where components are supplied as loose items, systems employing bowl feeders are used to bring them to a position where feeder mechanisms load the parts for processing. Bowl feeders, however, depend on vibration to align the parts. If these parts are delicate the robots need to be able to pick them up without pre-alignment.

Cameras scanning the components connect to robot arms via image processing equipment. The imaging equipment looks for edges generating grey level changes and a match with an image stored in memory viewed from various angles. The system then produces an algorithm whereby the robot arm is adjusted so as to grasp the component in its view.

By using a similar evaluation procedure to this pick and place equipment, a quality control function can be realized using closed circuit television (CCTV) cameras. The image provided by the camera can be checked against a master image held in memory. Defects in components which in the past would be detected by visual inspection can be picked out by vision systems and thus release a human operator from a tedious task.

Typical applications here might be the checking of bare printed circuit boards, metal pressings, etc. Defects of components may include variations in dimensions. TV images can be checked by applying a grating or crosslines in horizontal and/or vertical positions. When the component edges cross the predetermined boundaries, the component is over or under size. The operator is able to monitor the system by viewing a screen, or change parameters to suit modified or new components.

64.1.2 Food production

The availability and relatively low cost of colour cameras have opened up new possibilities in many areas of industrial control systems. Food production and processing is one such area. Consumers are faced with a bewildering array of products as producers vie for a share of the market. Choice of product is made not only on the basis of low cost, but also on the basis of perceived quality.

Many items such as fresh fruit present problems for quality control because of the wide range of input quality. Size grading can be carried out by passing through various mesh sizes. However, items such as fruit need to be graded in terms of ripeness and freedom from surface blemishes. Colour cameras monitoring the passage of produce on a conveyor belt are connected to image processing equipment. The surface of the product is scanned by the camera, and its colour is compared with a predetermined level held as a reference in memory.

Where the product is fruit, the conveyor belt causes it to roll so that all its surface is exposed to the camera. Each item can be located on screen coordinates and, if below standard, the image processing equipment activates a mechanical flap arrangement to reject the item.

Items which undergo heat treatment are checked for finished colour to ensure that precisely the right product is achieved consistently. In this category would be included items such as biscuits, cakes, cereals, etc.

64.1.3 Steelworks

Automation in heavy industry such as steelworks has not only enabled financial saving to be made, but also provided a uniform quality level in the product and increased the safety level of the plant. As the plant requires fewer human operators, CCTV cameras enable the central control point to monitor all aspects of the production line.

At steel strip production plants, the raw material is stored as ingots which are several metres long and consequently extremely heavy. The special trucks required to transport the raw material to the production lines are so large that TV cameras are required to enable the driver to manoeuvre and pick up the ingots safely.

The dumper trucks carry the ingots to a walking beam system that carries the ingots into a reheat furnace. Cameras monitor this point as blockages need to be quickly dealt with. The ingots enter the reheat furnace on the walking beam where they are heated progressively to the correct temperature for rolling. A furnace viewing camera monitors the progress of the ingots in the furnace.

The ingots emerge from the furnace to be fed through several sets of rollers, thereby becoming thinner and longer at each successive set. Cameras are used to monitor the passage of the strip as it proceeds through the rollers and controlled cooling stages. As the leading edge of the strip nears the end of the line, it is forced into a coiling mechanism which feeds the strip into a convenient form for transport to the customer. Cameras are used to monitor the end-of-line bays where the strip cools down prior to removal to storage areas.

The control point is at all times aware of the state of all points of the production line by the use of the TV system and is able to maximize the speed and therefore productivity of the plant.

64.1.4 Furnace viewing cameras

The need to view the inside of an operational furnace arises for a number of reasons:

- to ensure the correct ignition of burners,
- to determine the correct fuel usage for peak efficiency,
- to check burner alignment for good temperature distribution,
- to check the furnace for carbonization and therefore need for maintenance.

Cameras for this kind of environment obviously require protection from the high temperatures. Two kinds of furnace viewing camera are employed in industry. These are air cooled cameras for flame viewing, and air/water cooled cameras for furnace monitoring. The latter are usually located in very high ambient temperatures.

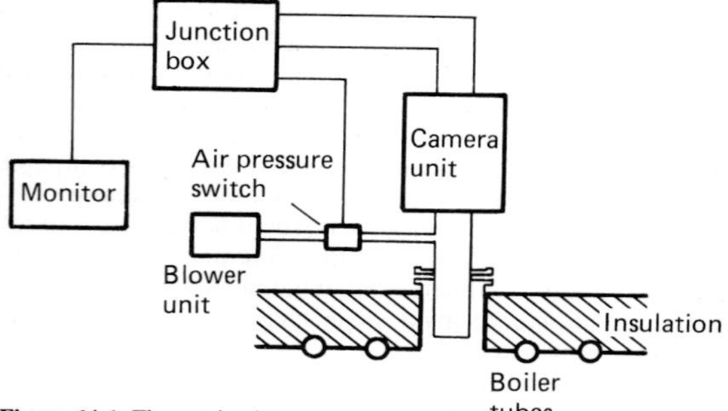

Figure 64.1 Flame viewing

When using an *air cooled camera*, the viewed object only is at high temperature but the camera is located in an ambient temperature within its specification. To protect the camera tube from the energy of the flame, a special lens system is required. This lens system consists of a combination of quartz lenses and heat reflecting filters. The infrared energy is absorbed by the lens system which is continuously purged by a flow of cooling air. The air vents into the furnace around the lens surface which helps to keep it free of deposits.

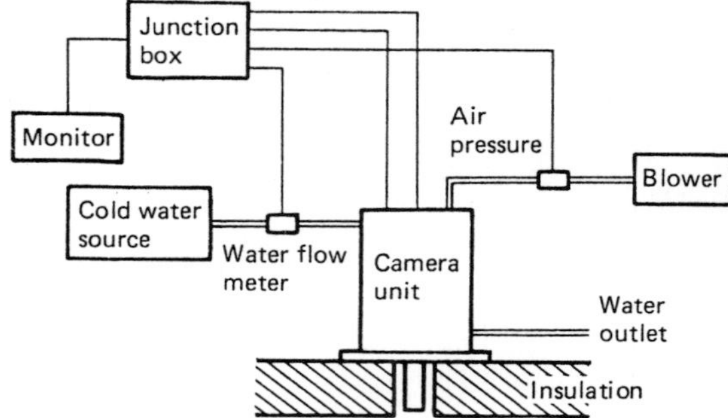

Figure 64.2 Furnace monitoring

Where the camera has to operate in a high ambient temperature, an *air/water cooled camera* is used. The body is contained within a dual skinned stainless steel housing carrying circulating cooling water. The special quartz lens system is incorporated to protect the camera tube. Sensors are fitted to the assembly to monitor air pressure, cooling water flow and internal temperature so that the camera can be withdrawn quickly to prevent camera damage. Flame viewing cameras have been designed that incorporate automatic withdrawal of lens and camera in the event of failure.

64.1.5 Coal mining

The mining industry is one in which the use of CCTV equipment adds greatly to the safety of the personnel and efficiency of the plant. Cameras for use underground are designed to be used in areas where explosive gases may be present. In coal mines, the seams are blasted to break up the coal. This then has to be moved up to the surface for treatment.

When the coal is ready to be transported away from the face, the miners use cameras fitted to the loaders to control the loading of the coal which ultimately ends up on conveyors.

The coal processing plant includes washing stations where coal and dirt are separated. Cameras monitoring this type of process ensure that blockages or other problems are quickly located at a central monitoring point. The alternative to cameras would be large numbers of employees having to monitor operations under uncomfortable conditions.

One interesting application of the use of TV cameras arose from filling in old mineshafts. These shafts make ideal repositories for waste material, while filling them removes the hazard that they present. An inexpensive camera and an electric bulb are located in the shaft and filling commences. The camera shows when the shaft is full and the cables are cut and left in situ.

In areas where conveyor systems are used to transport large volumes of material, e.g. coal fired power stations, aggregate production units and quarries, CCTV cameras facilitate monitoring of the conveyor transition points. A transition point is an area where the conveyor changes direction and is often where blockages occur. A single operator may be sufficient to monitor all conveyors on a large site. The operator is able to spot problem areas quickly and this ensures that the down time in the material supply is kept to a minimum in a cost effective way.

64.1.6 Remote inspection

The ability to inspect areas which are normally inaccessible is important in many types of industry. Such areas may include sewer pipes, heating circuits, box girders, processing plant tubes and many others. Problems such as scale build up in water carrying pipes, or corrosion, weld cracking, etc., must be detected at an early stage before major repair costs are incurred.

Cameras mounted on powered carriers enable inspection through horizontal pipes over several metres. The cameras are fitted with prisms or mirrors which can be rotated through 360° to enable viewing of the whole of the internal surface. Lighting units are attached to the camera and the complete unit is connected via an umbilical cable to the controller.

Where vertical pipes are to be inspected, the camera is lowered manually. Cameras used in the nuclear power industry are generally encased in stainless steel, and the inspection window is constructed from radiation resistant glass.

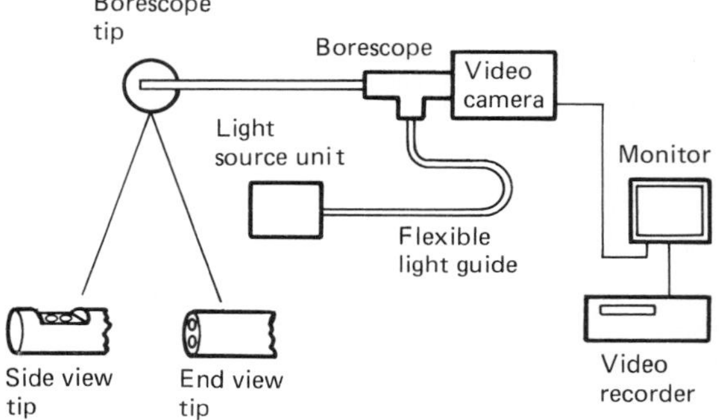

Figure 64.3 Rigid borescope

Remote inspection of areas which are not pipelines may be carried out using *borescopes*. Borescopes are basically optical instruments based on fibre optic techniques that enable a light source to be transmitted via a fibre bundle so that the area to be viewed is illuminated. A second fibre bundle carries the visual information back to an eyepiece. A TV camera may be attached to the borescope so that a video recording may be made and/or several persons are able to watch simultaneously.

Typical uses of borescopes are for inspecting the condition of the combustion chambers of engines, looking at impeller blades of gas turbine engines, checking gearboxes for wear and many others.

64.1.7 Positioning

Automation in industry often depends on components passing through workstations being correctly positioned so that processing always begins from a reference point. The signal from a TV camera monitoring the component is digitized by an image processing system and the component position required by the workstation is also held in memory as a reference point. The system must then energize actuators to reduce the difference between actual and desired positions to zero to allow processing to commence.

Typical applications here are for parts requiring machining, drilling, etc. Accuracy of position depends very much on two related factors, i.e. the lens system and the resolution capabilities of the camera. Lighting levels need to be adequate, not only for illumination of the component but to define the boundaries of the component clearly enough for the image processing system.

Positioning of components can be carried out in other ways. The object can be provided with markers. The camera viewing the object lines up the marker with a graticule on the camera. Camera graticules may be implemented in two ways. The simplest is to etch a graticule onto the surface of the camera tube or sensor (ccd chip). A more elegant solution is to superimpose an electronically generated graticule onto the video signal. Units are available for this purpose so that no modification to the camera is required.

One interesting application of positioning in the photographic industry arose from the need to align a film in complete darkness. Infrared light emitting diodes provided sufficient illumination for an infrared sensitive camera to be able to view the marker against the electronic graticule, while the film remained unexposed by the infrared.

64.2 Commercial applications

64.2.1 Baseband transmission

Transmission of video signals at baseband frequencies of up to 8 MHz can be implemented in the following ways:

- unbalanced (coaxial cable),
- balanced,
- freespace: radio, laser, IR, etc.,
- fibre optic cable,
- slow rate.

The most widely used transmission medium for baseband video is *coaxial cable*. The cable has a nominal impedance of 75 ohms and generally produces little deterioration in picture quality in lengths up to about 400 m for medium quality cable. Above this length the capacitive effect of the cable attenuates the higher frequencies of the video signal so reducing the resolution or sharpness of the picture. Frequency dependent amplification can be used to correct this attenuation, so that operation with coaxial cable can be extended to 1 km and more.

Balanced video transmission is increasingly being used mainly due to the reduction in cost of twisted pair cable compared with coaxial cable over long distances. The nominal impedance of balanced cable is 120 ohms. Where remotely controlled cameras are used in systems, quad cables, i.e. two twisted pairs, carry the telemetry signals to the camera on one pair and return the video on the other pair, so reducing the cost of cable installation. For the balanced signal, a video correction unit produces an unbalanced output suitable for display on most monitors.

Free space transmission is generally used where the cost of providing balanced or unbalanced cables is prohibitive. Modulation of microwave beams is the ideal solution, and transmission distances of many kilometres can be achieved provided line of sight is available. For the telemetry signal, modulation of a lower frequency radio channel is suitable. An alternative to microwave transmission is lasers. The range of the laser modulator is lower than that of the microwave beam, being typically 1 km. Lasers also require line of sight, but are affected by atmospheric conditions that limit visibility such as fog, rain, snow, etc. Modulation of an infrared beam may be used but is subject to the same limitations as the laser — generally with a shorter range.

Fibre optic cables enable transmission of video over long distances. The transmitter and receiver are galvanically isolated, so that hum is eliminated. With the transmission being by light, the problem of crosstalk is also eliminated. The basic system incorporates an infrared emitting diode which is current modulated by the video signal, the fibre line, and a PIN diode receiver/demodulator circuit. More sophisticated systems employ multiplexing techniques to reduce the number of fibres used.

Slow rate transmission is basically a technique of utilizing lines designed for a bandwidth of 3 kHz for video with a bandwidth of 8 MHz. The picture is sent as a modulated audio signal and the received picture reconstructed over a period of time line by line.

64.2.2 Security

Security is perhaps the most widely known application of CCTV. Theft, vandalism, wilful damage, minority extremist groups, both political and environmental, have all combined to create a large demand for security equipment and services.

Prices for CCTV cameras are now low enough for use outside domestic residences where they are fitted so that the owners are able to see what is happening outside while remaining safely indoors. Similar systems can be used in conjunction with intercom systems allowing the resident to see and speak to the caller prior to opening the door.

Security cameras installed in shopping arcades and centres provide a means of monitoring busy public areas. Patrolling security guards are in contact with the camera control point via two-way radio, and they can be directed to any source of trouble.

Shopping centres often have loading bays where large vehicles may congregate during peak delivery times. Cameras monitoring these bays allow the operators to activate traffic signals to restrict vehicular flow and reduce congestion.

Industrial areas that use TV cameras are often so large that to provide for site coverage by cameras with fixed angle lenses would make the system extremely expensive. The alternative to fixed cameras is a remotely controlled camera capable of being panned through 360° and tilted through 180°. Remotely controlled cameras usually have motorized zoom lenses fitted to vary the field of view. They can also be remotely focused.

A camera may be fitted with one of many tube types, e.g. vidicon for general purpose, newvicon for low light level, extended red newvicon for operation with infrared lighting and use in covert surveillance, and silicon diode for resistance to image burn-in. Solid-state cameras are now beginning to make advances in the security market because of their low cost of ownership. Unlike its tube counterpart, the ccd camera does not suffer from image burn, lack of emission, etc., and consequently the maintenance requirement is low. A low visual profile makes them eminently suitable for the security market.

64.2.3 Remote control

The simplest and most economical way of remotely controlling cameras is by *hardwire* control. This involves connecting a multicore cable between the camera and control. Hardwiring is generally reserved for simple systems and is satisfactory for short distances only. For long distances, other means must be used.

Analogue signals, in the form of *audio tones* generated by the control transmitter, provide a different tone for each camera function. Several tones may be transmitted simultaneously via a single pair cable to the camera location where tone decoder circuits regenerate the required functions. High stability circuits are required to overcome problems of frequency drift that can occur with extremes of temperature.

As an alternative to audio tones, *digital code transmission* may be used. Camera functions are encoded in a data word and are transmitted serially over a balanced line. Circuitry at the camera receives the data word, decodes it, and executes the functions in parallel form.

Frequency shift keying (FSK) is widely used in data communication between computer networks. Camera functions are encoded into a data word as for the digital system, but the data bits are converted into one of two frequencies and transmitted as tones. The tones fall within the audio speech band of telephone networks i.e. 300–3400 Hz.

Both digital and FSK systems can be configured for multidrop or star connections and the data word must contain the address of the camera to which the functions belong.

Remote control of cameras is also achieved via the *signal cable*. Camera control data are transmitted on the video cable during the blanking period of the frame. The data are decoded by the camera receiver circuitry which must combine its receiving function with the video transmission function.

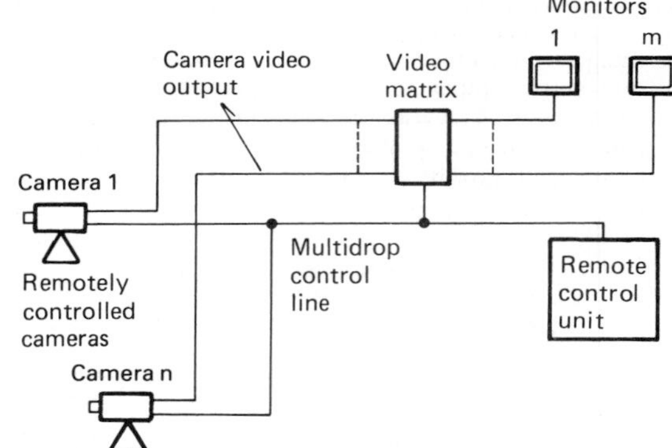

Figure 64.4 Typical multidrop camera system

Cameras operated by remote control offer several advantages over fixed cameras. Most important is that the operator, often a security guard, is able not only to survey a large area but also to alter the focal length of the lens to gain a better view of any intruder. The intruder may be tracked with the camera joystick controls, and if the system is fitted with audio facilities, he or she can be warned that detection has taken place. If the camera equipment is fitted with infrared lighting for covert surveillance, correction will need to be made to the lens focus and only remotely controlled cameras offer this capability.

The pan and tilt platforms used for camera adjustment may be fitted with preset positions so that, on activation of a command signal, the camera proceeds automatically until the selected field of view is reached. The preset position can also be applied to the zoom and focus motors. A typical application may be for monitoring several doorways or entrances with one camera. If each entrance is fitted with detection equipment such as magnetic contacts or trembler switches, then unauthorized entry activates the detector, brings the camera to the scene, and alerts the security guard via the alarm system.

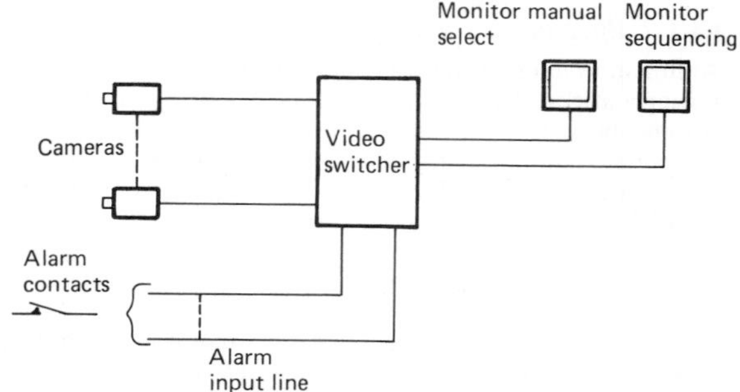

Figure 64.5 Basic integrated security system

On large sites where many cameras are installed, it may become difficult for a guard to determine which camera is being displayed on a monitor since video switching matrices will be used, (see section *64.2.4*). To overcome this difficulty, the camera is fitted with an identification circuit which displays a caption on the video output signal in the form of numbers, characters or a combination of both.

Recording information of intrusion from video cameras can be important, particularly if it is to be used in a court of law. However, before it can be used in court it is necessary to include an indication of when the recording took place. Time and date generators may be used for this purpose. Video recorders which provide up to 8 h of real time recording, and time lapse recorders providing up to 240 h from a standard 4 h tape, are in common use. Computer discs can be used to store video information provided the picture is digitized.

As an alternative to tape recording, there is available a means of producing a hard copy printout from video cameras. The video printer operates by storing a video frame in a random access memory. The frame is digitized into a number of pixels, each having a number of grey shades, by an analogue/digital converter. The print head converts the stored information into a series of dots of varying grey shades onto thermal paper. Colour video printers are also available using the same technique as the monochrome versions. The memory requirement is, however, much higher.

Cameras being used for surveillance and monitoring in areas where inflammable vapours may be present, i.e. petroleum processing plants or chemical works, must be installed in flameproof enclosures. Any camera generated sparking must be contained within the housing to prevent vapour ignition. All other components in the installation such as the cabling, pan and tilt platform, and wiper if fitted, must be classed as flameproof.

64.2.4 Video matrix

In large camera systems, it is generally impractical to connect a monitor to each camera as the numbers could run into hundreds. A more effective arrangement is to provide a relatively small number of monitors and arrange for the camera outputs to be switched through as required to each monitor. This simple sounding task is accomplished by the use of a video matrix.

Video matrices vary enormously in their complexity and cost. The simplest matrices take the form of video switchers which can have up to 12 camera inputs and 2 outputs, one providing a manually selectable camera and the other automatically sequencing through all the cameras. When several monitors are required in systems, a means of selecting desired video signals for display requires a control philosophy based on microprocessors. In large systems a monitor might be asked to display:

- set sequences of cameras,
- variable sequences of cameras,
- individual cameras for use with manual control,
- alarm pictures selected by operation of other detection equipment.

The requirements of the matrix are that it should have sufficient bandwidth to accept monochrome or colour signals, 12 MHz being satisfactory. The crosspoint density should be high enough to provide for a compact unit yet not cause the crosstalk or signal/noise ratio to be unacceptable. Video inputs need to be selectable between terminated and high impedance to allow looping through. Selection of crosspoints would be under computer control since large numbers are involved; thus a standard communications protocol is essential.

Some video switchers include alarm interfaces that can halt an automatic camera sequence and display the video associated with the alarm along with an audible signal. This forms the basis of an integrated security system.

64.2.5 Slow rate transmission

One of the greatest problems encountered when transmitting video signals over long distances is the choice of medium itself. Losses in coaxial cable become excessive, and even with equalization, signal/noise ratios deteriorate over a few kilometres. A means of transmission exists in the form of the telephone network, but here the bandwidth is limited to the audio frequencies of approximately 300 3400 Hz.

Slow rate transmission is a technique used to overcome this bandwidth limitation by converting the normal video bandwidth of 8 MHz to audio tones. The video signal is converted to digital form by a high speed analogue/digital or flash converter. The video frame is stored in a memory as a number of pixels, each pixel having sufficient bits to define a number of grey shades. The memory can be transmitted serially along voice grade lines using modulated audio tones.

At the receiver, the audio tones are converted back to digital form by a digital/analogue converter to form a video frame, line by line. As the telephone line is voice grade, data rates are insufficiently high to allow real time video to be transmitted. Frames are built up over a period of seconds. For high resolution pictures, the time taken to build up a frame will be longer than for a low resolution picture.

With the advent of broadband non-switched digital links, frame update times will decrease considerably bringing facilities such as videoconferencing into wider use.

64.2.6 Movement detection

Video cameras have been used to detect intruders with a limited amount of success in the security market. The method of achieving the alarm varies with the equipment but the basic technique is the same. The camera provides pictures which are sampled and memorized as a reference.

When movement occurs in the real time picture, the luminance levels of parts of the picture change. It is the change of luminance levels compared to the reference that causes the alarm to be generated. Obviously several other factors affect the system, e.g. the speed of the contrast change , the area of coverage and the relative brightness levels involved.

Where monitored scenes are unchanging in terms of illumination levels, e.g. lit corridors in buildings, any movement is quickly detected. Where the camera is located outdoors, several factors can cause the equipment to generate nuisance alarms. Trees blowing in the wind, snow particles, heavy rain, etc.

To overcome some of these problems, movement detection systems have been designed which allow certain areas of the scene to be excluded from the alarm generating reference picture and therefore the alarms are tailored to certain areas in the picture such as doorways, windows or gates. Digital frame stores allow the alarmed picture to be 'frozen' and are used when the alarmed scene has to be examined closely.

Alternatively the alarmed picture may be stored on a videotape recorder or computer disc.

64.2.7 Traffic monitoring

The availability of television camera systems has provided traffic engineers with the means to monitor and control traffic flow not only over large sections of motorway, but also in city centres, tunnel entrances or anywhere potential bottlenecks occur.

Cameras used for traffic control applications must meet demanding operational requirements. Not only do the cameras need to be resistant to water ingress and be able to operate over large temperature variations, but they have also to operate in areas where vibration caused by traffic spans a relatively wide frequency spectrum. Traffic cameras must operate in lighting levels that vary between bright daylight and starlight. Normally such a wide control range would be handled by an auto-iris lens system. In traffic applications however, highlights such as vehicle headlights or roadlighting would cause the iris to close down thus limiting the scene information. To overcome this problem, a peak white inverter can convert a highlight on the scene into a preset grey shade. The auto-iris lens then sees a lower average white level and consequently opens up.

Cameras designed for traffic applications tend to operate at sites far from the control centre. Frequently, the expense of installing the transmission system for video and control constitutes a large part of a total system cost. Fibre optic systems are becoming more widely used because of the freedom they give from crosstalk and hum induction. The installed cost however is generally justifiable only for relatively long transmission distances.

At motorway intersections, traffic build-up can occur for several reasons. Holiday traffic, peak time traffic occurring in areas where motorways are located near urban or industrial areas, and lane closures for either maintenance work or because of accidents, can all contribute to lane congestion. Traffic approaching the area must slow down in good time to avoid collision. Cameras looking at potential trouble spots indicate to the control centre that speed restrictions are necessary to warn oncoming vehicles. Computers are then used to transmit signals to gantry signs in the relevant road section.

The monitoring of traffic using camera systems is often assigned to more than a single control station. For instance, traffic control centres may be allocated primary control with police personnel allocated secondary control. The system configuration needs to allow for priority levels of access and control.

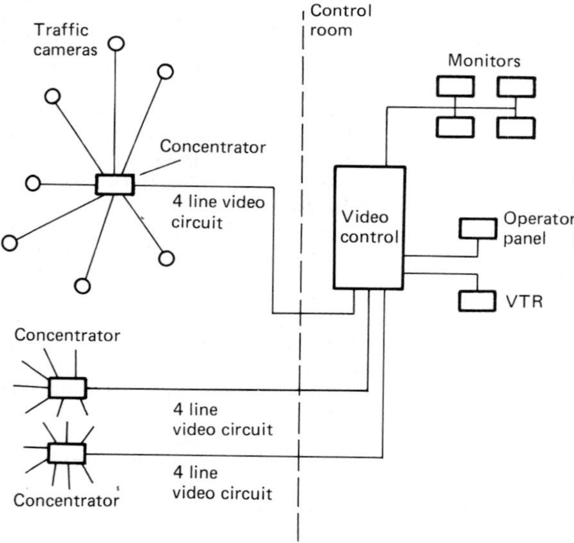

Figure 64.6 Traffic outstation concentrator

Camera systems for urban traffic control often require compromises to be made at the design stage because of the costs of rented video transmission lines. The number of video lines back to the control point may be effectively reduced by the use of video concentrators at the outstations. Cameras grouped in areas are connected to a concentrator which is then linked back to the control via the rented video circuits.

Not all cameras will transmit simultaneously, as only a limited number of video circuits will be used. A concentrator may group cameras into set cycling sequences, each group then sending one video signal back to control. Alternatively, several cameras may be included in a cycle with the remaining video channels dedicated to those cameras being operated by the remote control. With this method, sufficient monitoring information is gained by using the video circuits efficiently.

64.2.8 Split screen

When the design of CCTV systems requires monitors to display more than one picture, it becomes practical to use split screen equipment. In split screen pictures, only half of each video signal is displayed. Several split screen units offer choice in the resultant display and can be horizontally or vertically split. The ratio of the pictures can be altered in both directions and fade in/out can be used.

In systems where the number of monitors is limited by cost or space, split screen units can provide a cost effective proposition. In analogue split screen systems, cameras need to be synchronized so that the timebases are in phase. In digital systems, no synchronization is generally required as the video signal is converted by an analogue/digital converter and stored in a frame store. The resultant picture is a digital/analogue conversion of the frame store derived from both cameras.

Using digital techniques, four camera pictures may be accommodated on one screen. Unlike analogue systems which only provide part of each picture, digital systems provide complete pictures, but one quarter of the size.

64.2.9 Underwater television

With the increase in offshore oil production, CCTV has proved a useful tool for many operations. Cameras can be used to inspect the support struts of oil and gas production rigs to detect the onset of corrosion or cracks. This inspection procedure could be carried out by professional divers, but conditions are often too hostile for human life to be risked.

Cameras for subsea use require a unique set of design parameters. The housing of underwater cameras must be able to withstand the pressure at the operating depth. Housings for cameras used at depths of 1 km are now commonplace and depths between 7 and 10 km have been reached. Not only do the housings need to be capable of operation at great depth, the housing material needs to be able to withstand the corrosive nature of sea water and is therefore constructed of stainless steel or heavy-duty cast aluminium with a suitable surface finish.

The temperature of sea water drops with increasing depth, so the housing requires some form of heating. Seals between housing and window sections must be so designed that increasing external pressure increases the sealing effect. Finally, the electrical connections to the housing and the umbilical cable itself need to be suitable for the environment.

Ideally, the lens fitted to the camera would need a correction factor to be implemented to minimize the distortion generated by the differing refractive indices of water and air between object and sensor.

Cameras used under water often need to be stabilized against currents that would cause the picture to shift. For this purpose, the camera can be mounted on a submersible vehicle which is manoeuvred by motors controlled via the connecting umbilical cable. The camera also has the ability to be panned and tilted on the submersible via the same control cable.

Visibility decreases with increasing depth, so the camera needs to be a low light level type such as a newvicon. Supplementary lighting can be used with the option of being able to vary the output to suit the conditions. Alternatively, manually opening the iris has a similar effect with fixed lighting.

Hand-held cameras have been utilized in conjunction with video recorders to enable inspection of structure condition by many personnel. Other uses of undersea cameras are for carrying out seabed surveys prior to laying communications cables and pipelines, and also inspecting them should they become entangled by fishing equipment and anchors.

64.3 Medical applications

64.3.1 Patient monitoring

In hospital departments where intensive care is required, patient monitoring is on a 24 h basis. Areas in this category include paediatric wards, premature baby incubator wards, post-operative recovery rooms and cardiac wards. Patients are often linked to several monitoring instruments. CCTV cameras, however, are used to observe the patient so that the slightest deterioration in condition or any distress displayed by the patient are noted immediately.

One nurse is able to watch several patients by using cameras linked to a control station. The patient monitoring system interfaces to the camera system so that activated alarms cause the correct camera signal to be switched to a predetermined monitor at the nurse's station and possibly activate an audio alarm. The nurse is able to select any camera for observation or allow the system to sequence through all cameras. The camera used in this sort of system is required to operate down to the lighting levels encountered at night while still producing clear pictures.

Cameras provide a means of observing certain areas within a hospital such as the pharmacy where powerful drugs may be kept, or ambulance and emergency entrances which have to be kept clear.

64.3.2 X-rays

The use of x-rays in the diagnosis of medical problems is well known. They were originally used to provide information on photographic plates and today this facility is still widely used. Any x-rays passing through the body are partially absorbed, with denser areas such as bones, absorbing the most. Bones therefore appear as dark areas on the film. To provide sufficient information on a photographic plate, a dose rate was required that tended to have a destructive effect on cells in the irradiated area.

To protect the cells in the relevant area, the dose rate has to be reduced, but the result of this action is to reduce the information available on the photographic plate. To cater for the conflicting requirements of high sensitivity and low dose rate, the technique of image intensifier fluoroscopy can provide real time pictures. The x-rays pass through the body and are projected onto a screen which fluoresces under their influence. The fluorescence of the screen is then amplified by an image intensifier unit.

Closed loop control systems enable the output from the intensifier to remain constant. This is done by varying the dose rate from the x-ray source with a collimator controlled by the intensifier output. The intensifier window has to be able to allow as much x-ray absorption and as little scatter as possible. The larger the area of the fluorescent screen, then the bigger is the area of the body that can be examined. Correspondingly, the TV camera used at the end of the intensifier chain requires to be fitted with a tube of high resolving ability, i.e. a plumbicon type.

The camera which generates a video signal of the fluorescence on the screen can be operated alongside a photographic camera for hard copy records of the x-ray. The signal from the video camera can also be fed into image processing and analysis systems which provide a powerful tool for diagnosis. Images may be highlighted, contrast added, or colour superimposed to accentuate detail and many other features to extract the maximum amount of information from the x-ray.

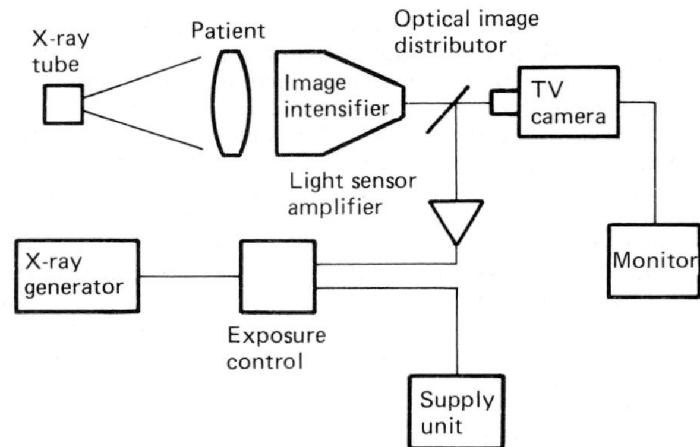

Figure 64.7 X-ray intensifier system

There are several benefits in acquiring the x-ray pictures as a video signal. Images may be digitized by computers and stored on magnetic tape or discs. This means that large quantities of film need not be stored. Time is also saved by removing the need to process film. Using slow rate television techniques, the x-ray image may be sent via telephone line to any part of the world. This is useful where a doctor may need to seek the opinion of an expert located outside the hospital.

64.3.3 Medical training

In the field of medicine, the need to train doctors and nurses in surgical techniques requires them to be able to watch operations as they occur. It is often impractical to have large numbers of students around the patient, and in pre-television days, galleries above the operating table allowed a certain amount of observation. The view was generally obscured by the lighting unit and therefore the need for a better system has been met by CCTV. Originally, the cameras were monochrome and proved to be of limited value, but with the advent of colour models, it has become feasible to mount a camera into the lighting unit to provide an excellent view of the operation.

Lighting sets used in operating theatres are precisely balanced units and require little effort to adjust the position. Fitting a camera to the unit must not alter this balance, so ccd cameras, being physically small, are a natural choice. As the lighting unit is mounted directly over the patient, only a limited amount of remotely controlled camera movement is necessary. A zoom lens provides close-up detail. Using cameras in this way means that the operation can not only be watched by students as the operation proceeds, but can also be recorded on video tape and used for teaching.

Student training in the field of psychoanalysis is facilitated by the use of CCTV. During treatment sessions, doctor and patient are necessarily alone. A TV camera monitoring the patient is one way for the student to be able to 'sit alongside the doctor'. The patient would be made aware of the fact that a camera and microphone are being used for recording the session for the purpose of student training. Generally, it is found that this is less inhibiting than having a group of students present.

64.3.4 Microscopy

Microscopy is one area where the application of a TV camera has produced many benefits. Microscopes can be fitted with an adaptor that allows the camera to view the image alongside a purely optical viewer. Colour cameras convey more information than monochrome models. The main advantage of combining TV and microscopes is that the display is not limited to viewing by one person. The picture can be transmitted around many areas and be made available to many persons. In addition, all viewers can see the effects of contrast, brightness and magnification.

Where the camera is fitted with an auto-iris lens, any change in lighting provided by the microscope system is handled by the camera. Detail on the slide may be referred to using an electronically generated pointer. Scales and grids may be superimposed on the screen in the same way, to carry out measurements. Magnification provided by the microscope is increased electronically by the camera.

Images from the camera may be stored on video tape, photographed directly from the monitor or digitized by microcomputer and stored on magnetic tape or disc. The image on the slide may be transmitted to a digital processing system which enables counting or size grading of cells when investigating growth rates of cultures.

64.3.5 Endoscopes

Closed circuit TV cameras have been used successfully in conjunction with endoscopes. These are instruments that allow observation of locations not easily accessible to direct view. They are used extensively in industry. In medicine, it is often necessary to view internal parts of the body and the alternative to surgery is the use of an endoscope.

The endoscope consists basically of two fibre optic channels, one carrying the light to the observation tip, and the other carrying the image back. Many endoscopes are fitted with eyepieces, allowing a single operator to observe, but when many medical staff wish to see the proceedings, or a recording is required, then a TV camera is attached. Generally a ccd camera will be used because of its small size, and invariably a colour model conveys more information than a monochrome unit. The camera, endoscope and ancillary equipment need to be constructed in a material resistant to the disinfecting fluids used in medical sterilization procedures.

The fibre optic portion is passed into the body via a tube whose end can be manoeuvred to place the sensing tip in the area required. Instruments may be sent via the same tube to remove small pieces of tissue from organs so that a biopsy may be carried out thus removing the need for major surgery.

64.3.6 Angiography

The application of x-rays to display the flow of blood in veins and arteries is known as *angiography*. A substance that causes the blood to become opaque to x-rays is injected into the bloodstream. The use of a low dose rate x-ray system allows doctors to monitor the flow of blood over relatively long periods. The TV camera scans the output of the intensifier and provides real time pictures of blockages and diverted flow on a video monitor.

By digitizing the video signals and using digital subtraction techniques, background areas in the picture may be removed to provide only the parts of interest to the viewer. These can then be electronically magnified, or the contrast and colour altered, to emphasize the information available.

64.4 Defence applications

64.4.1 Target tracking

Target monitoring is an important aspect of the use of television in a defence environment. Targets may be hostile, or they may be used for weapons testing. Cameras mounted on a servo-controlled platform (director units) allow for fast precision tracking. The type of camera used will be dependent on the lighting conditions encountered — normal daylight, low level lighting or haze, smoke, mist or total darkness. Where the light level is too low for TV cameras, a thermal imaging camera may be used.

Target tracking systems have been developed to the stage where signals from director mounted cameras, as described above, are used in automatic optical trackers. Monitoring of the speed and direction of the target ensures that if target lock is lost due to masking by clouds, then the stored date is used to compute the position of the target until lock is re-established. The system operates by digitizing the picture, storing the result, and activating the director unit in the direction that maintains the object in the screen centre. The system relieves the operator from having to manually track objects such as fast moving aircraft, missiles, etc.

Target recognition systems have taken the monitoring technique to the stage where signals, when digitized, are analysed by a computer. The information is compared with stored data and, when a match is found, the computer indicates the target type along with relevant facts. Obviously the target must have a minimum screen size to be able to determine its shape. Target tracking information may be fed into fire control systems which accept date from other sensors measuring wind speed, direction, etc., and used to compute the elevation and direction for weapons systems.

64.4.2 Remote inspection on the ground

In defence applications, protection of personnel is of paramount importance and this is most difficult in urban areas where terrorists use many diverse ways to inflict injury by means of concealed bombs, booby traps, etc. When an object is identified as a threat, military personnel are called in to determine the hazard level and possibly defuse the explosive.

The task of the military is made safer by the introduction of a remotely controlled vehicle designed specifically for remote inspection. The vehicle is equipped with tracked drive and a television camera fitted to a pan and tilt platform so that the controller can guide the vehicle and also closely inspect the explosive package.

Telescopic arms on the vehicle allow a certain amount of extension to the reach of the unit so that the camera can be manoeuvred into tight spaces. A rifle mounted on the arm allows the operator, using the camera for guidance, to detonate explosive from a safe distance. The video signal along with the controlling signals are carried by an umbilical cable which can reach distances of 100 m or more. Alternatively, radio control allows complete freedom of movement.

Some remotely controlled vehicles are equipped with two cameras so that perception of depth is gained whilst in operation. Cameras used on the unit are colour cameras and typically ccd for small size and low power, fitted with a zoom lens.

This type of vehicle with its vision facility has applications in areas other than military, e.g. checking the state of hazardous materials or operating in an area contaminated by toxic substances such as radioactive or chemical spillages. Where the unit is self-powered, freedom from the restrictions of cables allow the ability of perimeter patrolling on military sites. Here, the camera would be a low light level newvicon type, fitted on a pan and tilt platform enabling detailed inspection of the boundary fence. The task of the patrolling guard is therefore made less hazardous. This type of patrol vehicle can be equipped with listening devices and possibly public address equipment to ward off would-be intruders.

64.4.3 Remote inspection from the air

In order to be able to direct ground-based forces effectively, it is necessary to be able to locate the enemy positions. To do this, an elevated position above the battlefield is desirable. An observer standing on high ground with binoculars is probably the simplest way but not always the most effective. A TV camera mounted on a telescopic pole is another solution. Such units may be fitted to vehicles to increase effectiveness in the field. The camera could be a daylight, low light level or thermal imaging type or a combination to provide good pictures in all conditions. The only drawback is the need to stabilize the mast during strong winds especially where the camera has to be high as in wooded areas.

An alternative to a pole mounted camera is a remotely piloted vehicle (RPV) which is capable of carrying a camera. RPV designs have been based mainly on aeroplanes and helicopters. As they are generally quite small, they produce a very low radar signature and can be deployed deep into enemy territory.

The unit is self-contained with enough fuel on board to keep it in the air for sufficient time to observe the relevant area, which may be up to a few hours. Video signals are transmitted by the RPV to the control point by modulated rf. A radio link from the control to the RPV ensures that the machine is directed to the required areas. The radio link in both directions has to be able to withstand interference and jamming. Increasing the height of surveillance to minimize the possibility of the RPV becoming a target requires the lens system to have zoom function. At narrow viewing angles, stabilization of the camera platform is necessary to produce clear pictures.

An alternative to an RPV is the microlight type of aircraft. Again, its small size is an advantage over a helicopter or aeroplane and its lack of dependence on radio links may be considered an advantage over an RPV. Helicopters are able to carry a large payload but TV cameras need to be fitted to stabilized gyro mounts to produce stable pictures because of the extreme vibration of the airframe.

64.4.4 Coastal defence

Television cameras can play an important part in coastal or border defence. Cameras are set up to monitor a stretch of coast which may be impractical to cover by humans alone. Low light level cameras and thermal imaging cameras equipped with efficient lens systems are able to scan areas of coast that the eye could not possibly cover.

Many camera stations spaced so as to cover several kilometres of coastline run remotely with a preset scanning angle, each camera providing an overlap to its neighbour to completely cover the monitored section. The video signals from each camera are transmitted to a control point operated by a few personnel.

To provide further information of an intruding target, a laser rangefinder can, in conjunction with the camera, give precise position for interception by ships or helicopters. The cameras are mounted on precision director units which, being servo controlled, respond rapidly and accurately to directional changes. The operator at the control station is able to allocate control of the camera to an automatic video tracking system. The target shape is memorized by the system. The control system then adjusts the director unit so that the target shape is kept on screen at all times.

In a multi-station system, auto-tracking can be passed from station to station depending on the path of the target.

To provide for a greater range, camera stations can be shipborne, in which case stabilization of the platform is required to counter the normal movements of the ship. Transmission of video via microwave links to a shore-based control station provides for an effective monitoring network.

64.4.5 Naval television

Modern naval weapon systems incorporate missile launchers and big guns that are remotely controlled. In order to fire the charges, data are required from a fire control system. The fire control system acquires data from sensors that provide information on parameters which affect the trajectory of the shell or missile. Wind speed, direction, air temperature, etc., as well as ship pitch and roll, are fed into the computer, and the whole system calculates the required elevation and direction to maximize on-target strike.

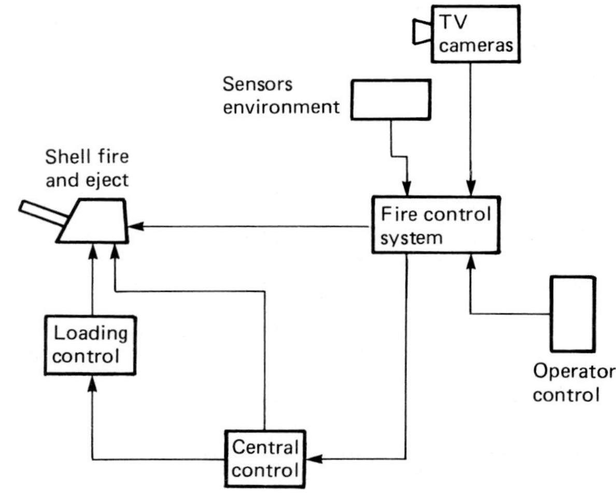

Figure 64.8 Naval gun control

Director mounted electro-optic sensor heads incorporate low light level TV cameras, thermal imaging cameras and laser or radar rangefinders to provide coordinates for target ranging. TV cameras have the advantage of being relatively immune to jamming by chaff, which could confuse radar, or flares, which would affect thermal imagers. Also, TV cameras are used where low angle targets such as sea skimming missiles produce degraded radar tracking. A disadvantage of using television for sighting is its poor visibility during adverse weather conditions.

The TV cameras generally have a remotely controlled field of view, and are used mainly for long-range operation. Guided missile systems used by ground forces have been designed so that TV guidance maintained by the aimer is transmitted to the missile. Alternatively the target is held on an automatic tracking system. Guided missile launchers may be static or mobile on tracked vehicles. Camera systems used on tracked missile launchers must be robust to withstand the extreme vibration and naturally are ccd units.

Submarines also use television systems. Periscopes have advanced far from the simple optical units of World War 2. Although submarines have for years been equipped with passive and active sonar detection systems for underwater sensing, the periscope has been enhanced by the addition of low light level TV cameras, thermal imagers and laser rangefinders. An attack periscope houses various combinations of optronic sensors, while a search periscope might offer a TV camera with an artificial horizon sextant. Recordings of seaborne operations may be made for subsequent evaluation back at base.

Part 16
Performance Measurements and Electromagnetic Compatibility

L E Weaver B Sc, C Eng, MIEE
Formerly Head of Measurements Laboratory,
BBC Designs Department

65 Television Performance Measurements

65.1 Introduction

The field of television measurements is vast, since it concerns every aspect of signal generation, recording, distribution, radiation and reception. Each of these requires appropriate test waveforms and measurement techniques, which may further be a function of the TV standard in use. These techniques are not even sufficient in themselves, since permissible tolerances must be laid down for each type of picture impairment, and consideration given to the way in which distortions add along a given signal path. Thought also needs to be given to measurement techniques for new developments in television.

It is proposed to concentrate here upon a discussion of the most common signal impairments with respect to internationally agreed test waveforms, supplemented by references to more detailed texts. The insertion test waveforms have been specifically designed to cover the most important analogue signal impairments, and some have a role in the testing of digital and component video systems.

The waveform photographs that follow may lack a little in clarity compared with line drawings, but provide a much better impression of operational conditions.

65.2 Insertion test signals

Insertion test signals (ITS), or in North America *vertical interval test signals* (VITS), take the form of ingeniously devised groups of waveforms which are carried by selected lines in the field blanking interval. Those to be discussed here are recommended by the international standardizing bodies, the CCIR and CMTT, for analogue signal transmission over long distances. The positions of the ITS and VITS in the field blanking interval are also laid down by these bodies[1].

Their advantages can be summarized as follows:

- Measurements can be made at any desired time and point in a network, even during 'in-service' conditions. Where automatic measurement equipments (AMEs) are installed, checks can be made at selected intervals throughout each day, giving advance warning of incipient failure and immediate notice of actual failure.
- Experience has shown that the most important signal impairments can be measured in this way to a satisfactory degree of accuracy, perfectly adequate for the control of programme exchanges over thousands of kilometres[2], or in studios[3]. They are equally suitable for repetitive 'out-of-service' measurements; indeed they have some advantages used in this way.
- If the ITS or VITS are inserted at the point where the programme is finally assembled and not touched subsequently, measurements at any point along the signal path will indicate the total distortion up to that point. Alternatively, the ITS or VITS can be inserted at any intermediate location for fault-finding purposes.
- Because this type of test signal is internationally standardized, equipment for generation and measurement is available from many sources, and has been tried and proved through operational experience.

Disadvantages are:

- As a result of the need to include as many waveforms as possible within a limited duration, some of the component test signals are not in the ideal format, although experience has shown that this restriction is less important than might be supposed.
- The position of the test waveform on the transfer characteristic is determined by the average picture level (apl). This is defined as the signal amplitude with respect to black level during the active portion of all lines, averaged over a field, and expressed as a percentage of white level, giving 0 per cent for a black picture and 100 per cent for one which is completely white. Any non-linear distortion is consequently a function of the apl, which during in-service conditions is governed by the picture content. For typical scenes, the apl is likely to be just less than 50 per cent. This inability to fix the apl does not, however, refer to out-of-service testing where it can very easily be varied.

It should be noted that there is no restriction on any individual broadcaster as regards test signals or positioning in the field blanking interval, provided no international programme exchange is intended, although it is normal to adhere

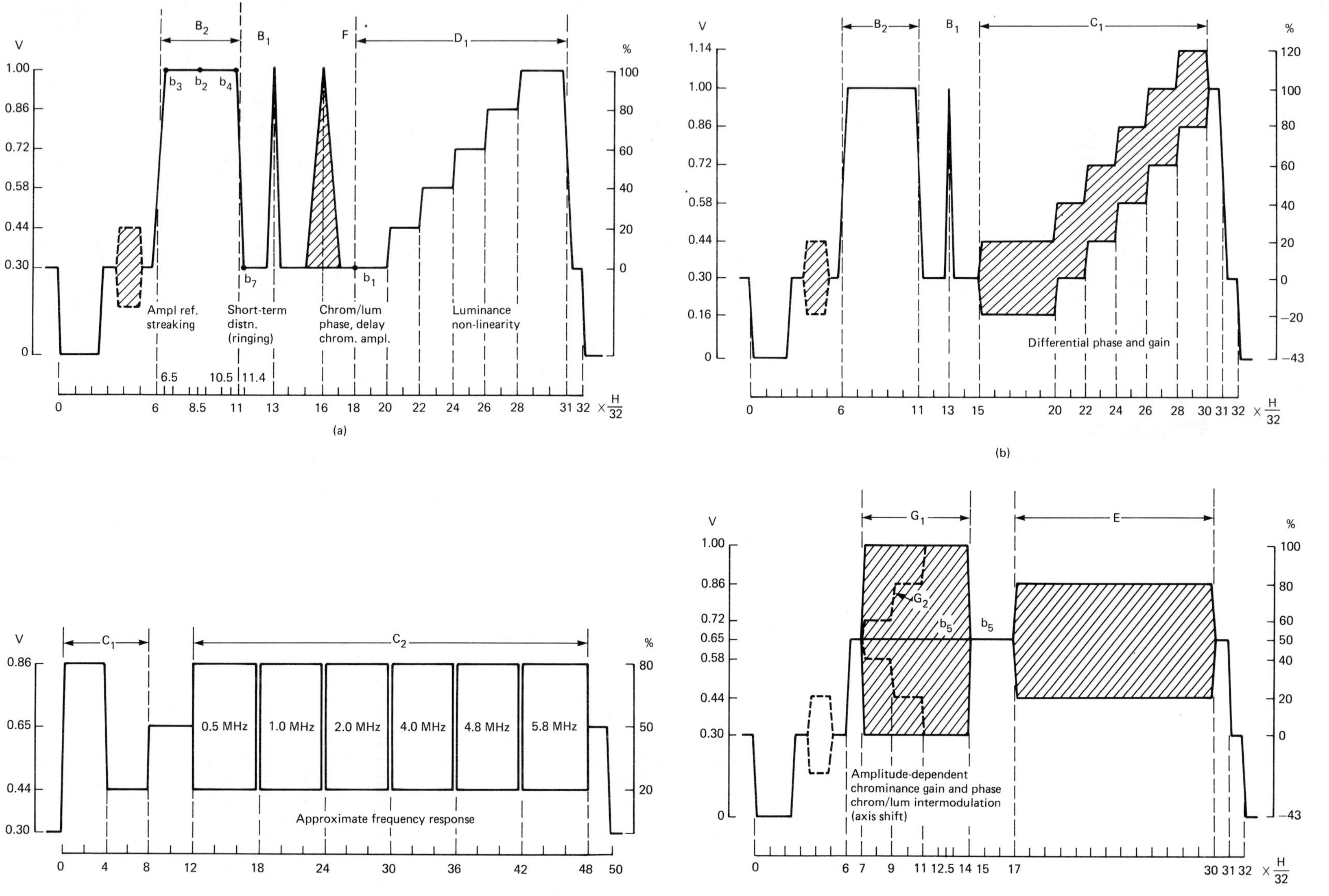

Figure 65.1 International 625-line ITS: (a) Line 17, (b) Line 330, (c) Line 18, (d) Line 331

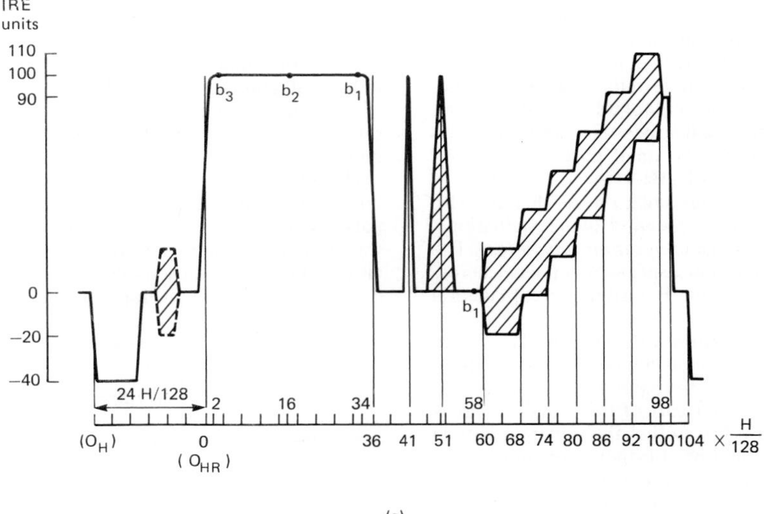

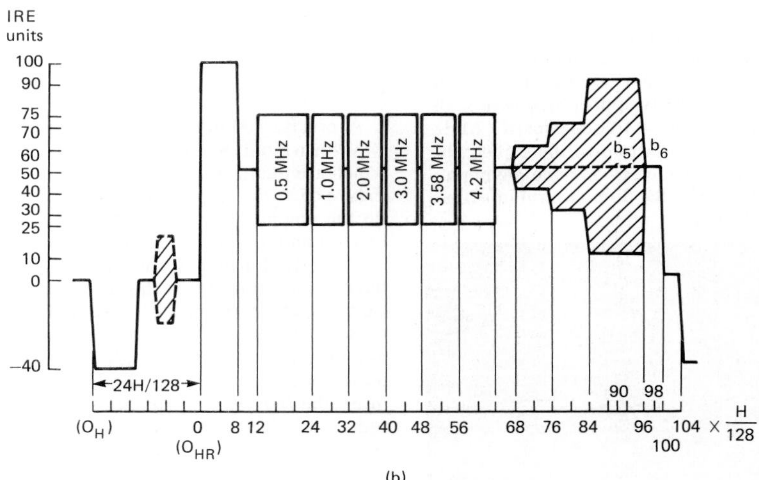

Figure 65.2 International 525-line VITS: (a) Line 17, Field 1, (b) Line 17, Field 2

as far as possible to the lines allocated for internal use with 625-line systems, i.e. 19, 20, 332 and 333. 525-line systems use line 17 in field 1 for one VITS and line 17 in field 2 for the other for all purposes (the line count in NTSC starts afresh for each field). In practice, the ITS and VITS do not differ significantly from the international recommendation in this instance, except for any special waveforms required for testing transmitters, teletext performance, etc.

It is essential, however, to avoid the insertion of any waveform on the so-called 'quiet lines', i.e. 22 and 335 for 625-line standards, and 10–16 for 525-lines. These are reserved for the measurement of random noise, and any tampering with them would make such measurements untrustworthy.

The international ITS is illustrated in *Figure 65.1* and the international VITS in *Figure 65.2*. Since the former is the more complex, and in any case the component waveforms are very similar in the two insertion signals, the discussion will be limited to the ITS. Any points of difference will be noted as they occur.

65.3 Measurement techniques

The principal linear and non-linear transmissions of analogue signals will be described in terms of the corresponding component waveforms of the ITS. A synoptic view of these is provided in *Figure 65.1*. This does not exclude measurements made on a line-repetitive basis for the optimum in accuracy, e.g. for acceptance testing, because the waveforms for this purpose are fundamentally the same, the principal difference being an extended duration in some instances. These analogue techniques are also employed where 'digital islands' exist in studio complexes, and with appropriate modifications will later form the basis of the measurement of digital transmission systems. The nomenclature of the distortions will be that of the CCIR[1] unless some special point needs to be made.

65.3.1 Signal amplitude

The importance of this measurement is too often underestimated. Signal amplitudes have to be maintained within very close limits for a number of reasons. In a long distribution system, errors can accumulate which may result in too high or too low a value at some point. The former may cause distortion, and the latter will result in a worsening of the signal/noise ratio. Within studio complexes it is essential for all originated signals to have amplitudes as nearly equal as possible, since the eye is remarkably sensitive to even very small differences. Good practice tries to limit these to ±0.1 dB[4,5].

It must be borne in mind that since one is dealing with equipment which is mostly used between 75 ohm terminations, the measurement ought to be that of *insertion gain* or *loss*. The difference between this and a measurement of level across terminations can be very significant. For a detailed explanation see Weaver[6,7].

A complicating factor is *return loss*. When the source impedance is not equal to the input impedance of an equipment, or the output impedance of the equipment is not equal to the terminating impedance, then part of the signal energy is reflected in the form of an echo, so that the signal amplitude passed on is less than it should be. Furthermore, since these impedances may be a function of frequency, amplitude and phase distortions can occur. The ratio of the signal amplitude under ideal matching to that of the reflection, expressed in decibels, is the return loss[6,7,8].

Another very frequently overlooked factor is the behaviour of coaxial connecting cables. It is usually assumed that these always have an impedance of 75 ohms, or whatever the nominal value may be. In fact, this value is only approached with high quality double-sheathed flexible cables at frequencies of 1 MHz and above. Moreover, inevitable variations in the cable constants can result in only a very few metres of cable having an intrinsic return loss of as little as 30 dB. Some useful information is given in Whalley[9].

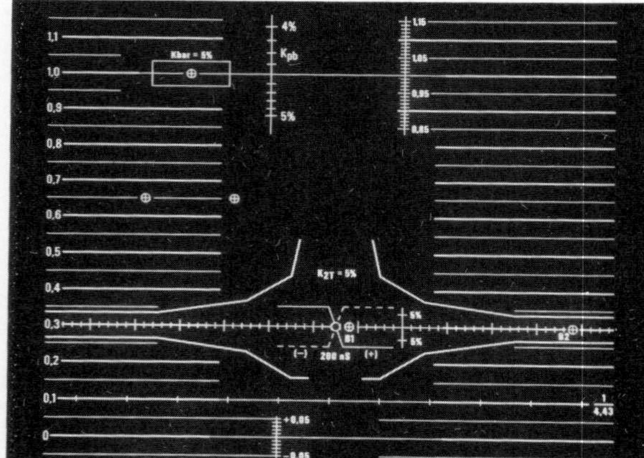

Figure 65.3 Composite ITS graticule (Tektronix Inc)

The 10 μs wide bar which is the first element in line 17 of the ITS (*Figure 65.1(a)*) is set very precisely to 700 mV in the insertion signal generator, and the difference between the centre of the bar top and black level is the standard measure of amplitude. Since disturbances may occur on black level, experience has shown that point b_1 is the most reliable black reference, and the desired amplitude is thus between points b_2 and b_1.

In the 525-line VITS the white-black amplitude is 714 mV, due to the pedestal ('set-up'), and this is defined as 100 IRE units.

The most usual technique for measuring the signal amplitude is to utilize the calibrated square wave provided by the waveform monitor to standardize its gain by ensuring that the top and bottom of the square wave precisely coincide with the 1.0 and 0.3 lines on the ITS measuring graticule, which will resemble that shown in *Figure 65.3*. Because of its convenience, this type of compound graticule is usually preferred to a set of individual graticules, even though each function has to be 'skeletonized'. When the ITS is viewed, the black level point is set to the 0.3 line, and the height of the bar is read from the scale on the left-hand side.

However, this procedure suffers from a number of errors not only possibly from the waveform monitor calibration, but from parallax and those introduced by the need to replace one waveform by another. A preferred alternative which eliminates all but the first of these is illustrated schematically in *Figure 65.4*. This has been used for many years in the BBC, and also commercially.

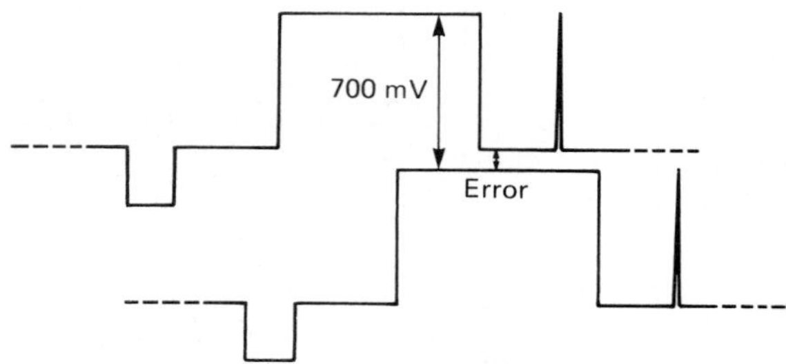

Figure 65.4 DC offset signal amplitude measurement method

A precise 700 mV square wave is added internally in the waveform monitor to the displayed waveform, giving two traces which are identical but displaced vertically. In the example of *Figure 65.4*, the signal amplitude is too low. A vernier control on the amplitude enables the two measurement points to be located on the same horizontal line, and the error can be read off directly. Further details are provided in Weaver[6,7]. Tests have demonstrated that this *dc offset* method has very significant advantages over others. The findings of Smith[10] agree with those of the writer that, provided the ITS general distortion is not unduly severe, an accuracy of ±0.05 dB can be attained. A useful practical tip if the displayed waveform is noisy, is to insert the *IRE filter* or similar network which can substantially reduce the 'fuzz' and thereby improve the setting accuracy. The synchronizing pulse amplitude may be found in the same way, except that the standardizing square wave amplitude is made 300 mV.

The measurement described above concerns the luminance component of the signal only. The difference between the chrominance and luminance component amplitudes is derived by a separate procedure (see section *65.3.2.6*).

65.3.2 Linear waveform distortion

The very convenient CCIR convention of classifying these distortions into groups according to the rough duration of the disturbance on the waveform (i.e. short-time, line-time, field-time, long-time) will be followed here.

65.3.2.1 Short-time waveform distortion

The length of a short-time disturbance is about 1 μs, taking the form of 'rings', i.e. overshoots on transitions. The method used is the *sine-squared pulse and bar* technique, which revolutionized ideas on television testing when it was introduced in the early 1950s. This can only be described here in the briefest outline, but further information is available in Weaver[6,7], as well as the original paper by Lewis[11]. The sine-squared measurement principle, in fact, takes in linear waveform distortions of other durations up to field rate. These will be dealt with below under the relevant heading.

The basic principles can be summed up very briefly as follows:

- The test waveform must as far as is practicable be representative of normal picture content. To this end it takes the form of a rectangular bar, corresponding to large areas of tone, and a narrow pulse which represents fine picture detail. These are the first two components of lines 17 and 330 of the ITS, and of the field 1 VITS (*Figures 65.1* and *65.2*).
- The energy in the pulse and bar should as far as possible be confined to the nominal video bandwidth which is taken to be 5 MHz for all 625-line systems and 4.0 MHz for 525-lines and the PAL systems using a 60 Hz field rate, e.g. PAL M. This nominal bandwidth is kept even for PAL I and SECAM with their wider bandwidths.
- The test waveform should be as well-shaped as is possible so that distortions are easily recognizable, and it should be capable of generation to a very high degree of consistency.

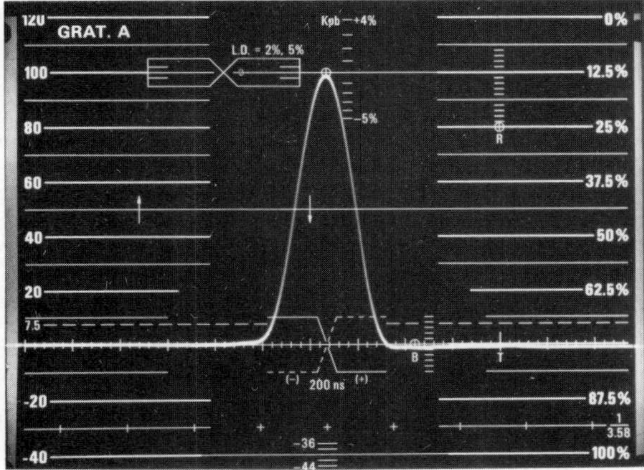

Figure 65.5 Generated sine-squared pulse

The final requirement was achieved by generating the pulse and bar initially with very rapid transitions, and applying them

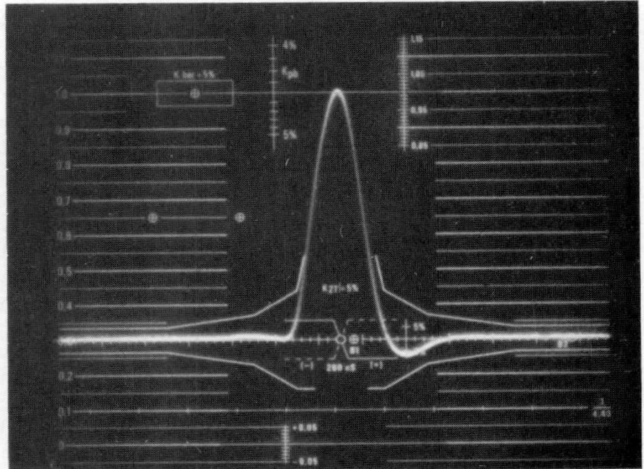

Figure 65.7 K_{2T} measurement

to the input of a special low-pass filter known as a 'Thomson network'[1,6,7] half-amplitude duration (had) of the pulse is defined in terms of a constant T = 0.1 μs, where its effective bandwidth is the reciprocal of this had. The most usual 625-line pulse is the 2T (had = 0.2 μs) giving a bandwidth of 5.0 MHz,

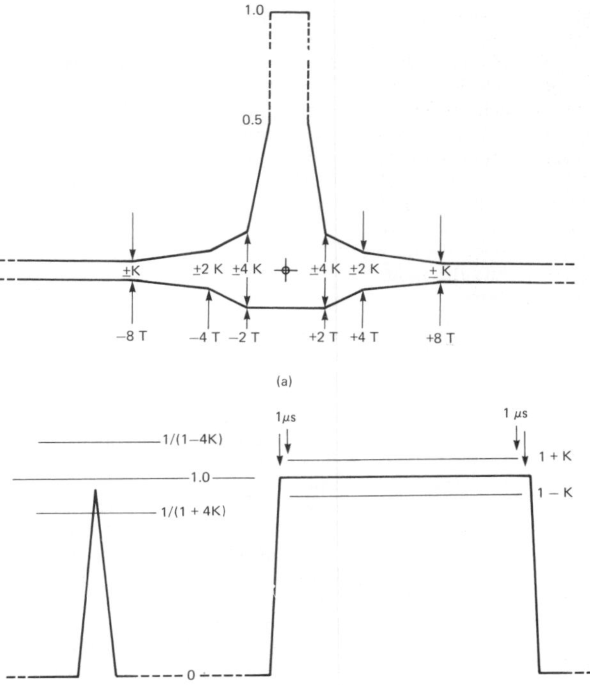

Figure 65.6 Basic tolerance schemes for pulses and bar rating: (a) 2T sine-squared pulse rating diagram, (b) Pulse/bar and bar ratings

but 1T pulses are also used for special purposes, and 2.5T for 525-line work.

The pulse shape agrees very closely with $E = \sin^2 (\pi t/2H)$ where H is the half-amplitude duration, hence its name. *Figure 65.5* demonstrates how well-shaped the practical pulse is; the only defect is a small and quickly damped overshoot on the right-hand side, which is tolerated. The bar transitions are virtually identical to those of the pulse.

Lewis and his colleagues[11] also took the very important and innovative step of rating the distorted pulse in terms of the equivalent subjective picture impairment by using the *paired-echo* principle of Wheeler[12,13]. For the 2T pulse this yielded the tolerance diagram of *Figure 65.6(a)*, where it is clear that a 4 per cent echo at a spacing of 2T is equivalent to a 1 per cent echo at 8T, and both would be given a 2T K-rating of 1 per cent. The diagram for the NTSC 2.5T pulse is identical except for the appropriate horizontal scaling. This particular K-factor is only one of several, and should more properly be denoted by K_{2T}.

For the practical measurement of K_{2T} the waveform monitor horizontal speed is first set to match the time scale of the graticule in use. Usually, this is preset. Then the pulse is adjusted in height and position so that its baseline lies on the central line of the tolerance diagram (see *Figure 65.3*), while the peak of the pulse coincides with the 1.0 graduation. K_{2T} is then given by the set of tolerance lines which would just contain the pulse overshoots. This usually has to be done by visual interpolation. In *Figure 65.3*, only K_{2T} = 5 per cent tolerance lines are given for clarity. Sometimes 2 per cent and 4 per cent are chosen but more sets of lines than two are confusing. A photograph of a K_{2T} measurement is shown in *Figure 65.7*.

K_{2T} is a unique and invaluable measure of the transient distortion associated with an extremely small item of picture detail in terms of the picture impairment it causes, but further information can be derived from the 2T pulse. Any bandwidth limitation gives rise to an increase in the transition time of transients, as is well known, which must increase the half-amplitude duration of the 2T pulse. However, the energy in the pulse is not changed, so its amplitude must consequently decrease. This can be recognized and measured by comparison with the amplitude at the centre of the bar top, which will not be modified, and yields a further K-factor K_{pb}, which in general terms is an indication of the resolution of the system. *Figure 65.6(b)* shows another set of tolerance lines, again related to the subjective impairment.

The practical measurement is made by setting the centre of the top of the bar to the 1.0 line in a graticule such as *Figure 65.3*, then moving the peak of the pulse to an upper auxiliary scale. The left-hand scale gives K_{pb} directly, while the right-hand scale provides the pulse/bar ratio in linear terms.

Before leaving the topic of short-time waveform distortion, two important points must be raised. The first concerns the tolerance scheme of *Figure 65.6*. Later and more refined work

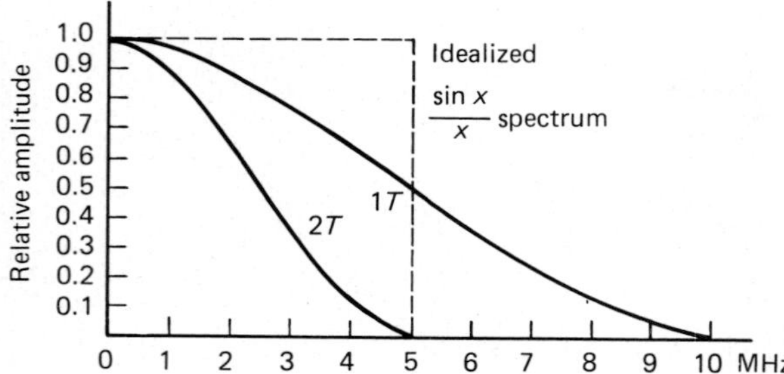

Figure 65.8 Comparison of spectra of 2T, 1T, and sin x/x waveforms showing the response of 1T and 2T shaping networks for 625-lines

by Allnatt[12] has demonstrated that this is not entirely correct. However, the errors are not thought to be serious enough to warrant modifying a technique which is in worldwide use, all the more since long experience has shown how valuable it is in practice, and how little the errors seem to matter operationally.

The second point can best be explained by reference to *Figure 65.8*. The spectrum of the 2T pulse falls to 0.5 of its initial value at 2.5 MHz, and to essentially zero at 5.0 MHz (2.0 MHz and 4.0 MHz for 525-lines). This implies that the upper part of the video band receives a much lower weighting than the lower. This is not as serious as it seems at first sight, first because distortions at the higher frequencies produce a smaller effect on the picture than the same distortions at the lower frequencies. Also, the ITS and VITS contain chrominance test waveforms which supplement to some extent the information derived from the 2T pulse.

However, especially in studios where the available bandwidth is often much higher than the nominal, it often seems desirable to know more precisely what effects are occurring towards the upper end of the nominal video band. This was not overlooked in the original work[11], where mathematical manipulation was proposed to remove the redundant and misleading transients which may occur when a 1T pulse (i.e. 0.1 μs half-amplitude duration) replaces the normal 2T pulse. This effect is especially serious if there is any bandwidth limitation in the system, as may be judged from the large amount of energy above 5 MHz in the 1T pulse demonstrated in *Figure 65.8*.

This method, although sound in itself, is tedious to implement and has only rarely been employed. Nevertheless, the coming into operational use of microprocessor-based automatic measurement equipment (see section *65.3.4*) might well provide a means for reviving this technique in an operationally convenient way.

A device extensively employed by some broadcasters is based on the use of one of the transitions of a 1T bar, i.e. the normal 2T Thomson filter in the generator is replaced by a 1T version. This has a spectrum which rolls off more rapidly than that of the 1T pulse, so that transients resulting from amplitude and phase effects above 5 MHz are much reduced. No K-rating is possible, and the output 1T transition is judged by comparison with an empirically derived tolerance graticule. Under conditions where high-frequency distortions tend to be small in any case, the 1T pulse is often employed as a means of comparing signal paths, since one is then only looking for significant differences.

65.3.2.2 Line-time waveform distortion

Even very small phase-angle errors at frequencies below a few hundred kilohertz produce a slope on horizontal areas such as the top of the ITS bar. The corresponding amplitude error is usually too small to be measured. A single distortion of this type is really exponential in shape, although frequently only the initial straight part of the curve is seen. More than one such distortion will give the bar top a complex shape.

When the effect is very severe (see *Figure 65.9*), the fact that the slope is an exponential causes interference between successive lines. For this reason, the CMTT has recommended[13] that the line immediately preceding the first ITS line should always contain a 50 per cent amplitude signal, which minimizes the interaction. The corresponding picture impairment is particularly serious due to the dragging-out of picture information into adjacent areas. Sometimes this appears as ‘stressing’ following transitions, sometimes as a general loss of contrast and definition. This, of course, applies to all standards.

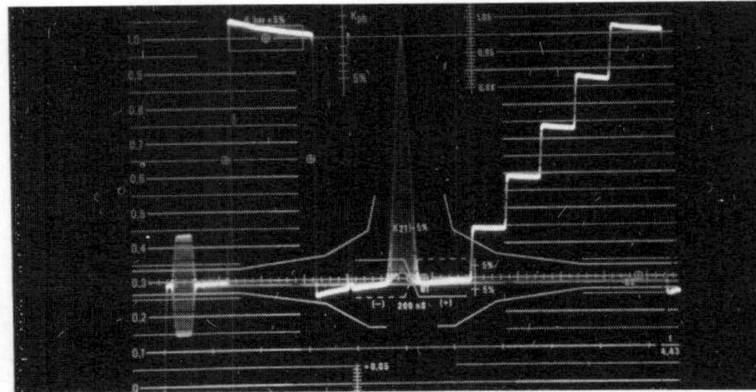

Figure 65.9 Example of line tilt

For ITS purposes, the bar slope is defined as the difference in level between points on the top of the bar 1 μs from each of the transitions, expressed as a percentage of the bar height. The first and last 1 μs intervals are omitted because they may contain short-time distortions which are irrelevant in this instance.

The process is simplified when using graticules such as that of *Figure 65.3*. The bar is located with its baseline on the 0.3 line, and the centre of the bar top passing through the mark in the middle of the K_{bar} 'box'. The transitions of the bar will pass through the half-amplitude marks. Since the width of the 'box' is 2 μs less than the bar width, i.e. 8 μs, it is simple to estimate the amplitude difference between the points where the bar top cuts the 'box' ends.

This shows up a difficulty with the ITS and VITS. 8 μs is a barely sufficient period over which to make this measurement, but a longer duration cannot be had. For the most searching test, a line-repetitive waveform is preferable, where the bar is about half the active line length, i.e. 25 μs. The procedure is similar, but in this instance the greater of the two level differences between the ends of the bar, again omitting 1 μs each end, and the bar centre gives K_{bar}.

65.3.2.3 Field-time distortion

Field-time distortion cannot be measured accurately with the ITS or VITS although some indication can be derived from the difference in level between corresponding points on the 10 μs bars in two successive fields. The best is the waveform given in CCIR[1], consisting of a series of lines with a total duration of 10 ms, each carrying white level bars for the whole of the active line duration. Each such 10 ms group is followed by another at black level, resulting effectively in a 50 Hz square wave with line synchronizing pulses. Field syncs may also be added if desired. For 525-lines the half-duration of the square wave is 8.33 ms. The composition of this waveform and its transition are shown in *Figure 65.10*.

The field-time distortion is defined as the greater of the two deviations of the bar top from the centre, expressed as a percentage. The first and last 250 μs are omitted. It may be necessary to modify the generator frequency in order to distinguish between true lf, slope and hum. In the original sine-squared pulse and bar method this was K_{50}, but on long circuits the measurement can be confused by other types of distortion, and it is now most often used in acceptance tests.

Figure 65.10 50 Hz test waveform

65.3.2.4 Overall K-rating

The rating method put forward by Lewis[11] advocated selecting the largest individual K-factor of the group as the overall K-rating. This has now been abandoned, not because it is a poor idea, but because it has been found more useful to record the separate values for informational and statistical purposes.

65.3.2.5 Long-time waveform distortion

Long-time distortion is a phenomenon of very long circuits which effectively contain a large number of series CR networks used for dc isolation. In an ac coupled condition, a sudden change in the average picture level (apl) is equivalent to the addition of a dc transient, which can be up to 700 mV in amplitude. This initiates a damped oscillation whose amplitude initially can be around 40 per cent theoretically[14], although work by the writer suggests that even a very small rise in amplitude at the lowest video frequencies can make that figure much larger. Its total duration may be as long as tens of seconds, during which time the signal can suffer severe non-linear distortion.

The test waveform resembles somewhat the 50 Hz square wave of *Figure 65.10*, but it is usual to reduce the total excursion to either 10–90 per cent or 12.5–87.5 per cent to approach practical conditions more closely. The waveform duration is preferably variable to suit the conditions of measurement.

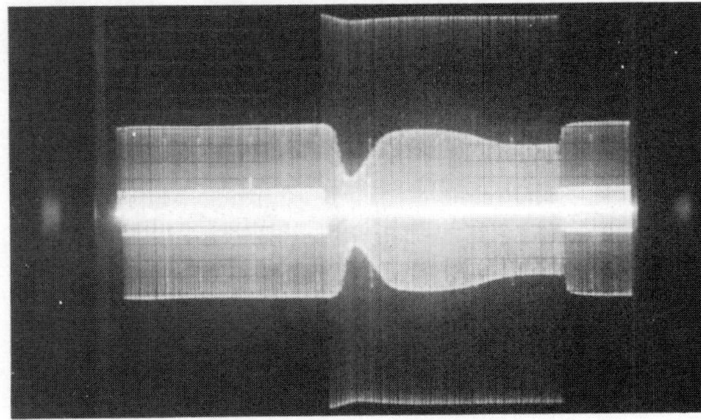

Figure 65.11 Long-time distortion at end of international circuit (differentiated)

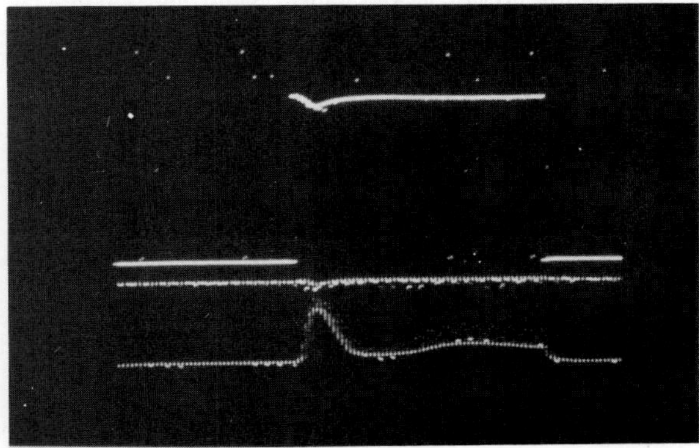

Figure 65.12 Long-time distortion at end of international circuit (sampled) (Tektronix Inc)

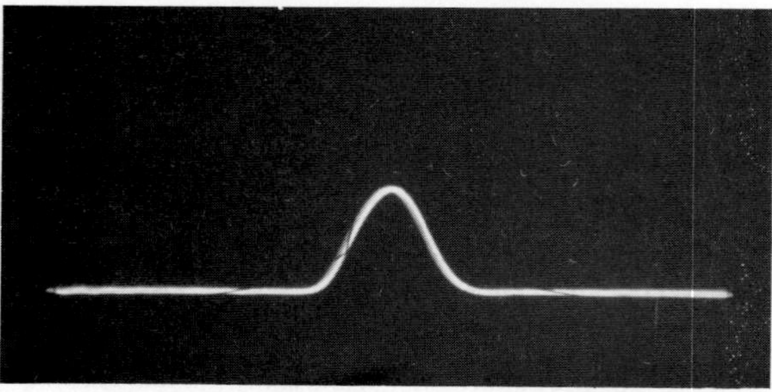

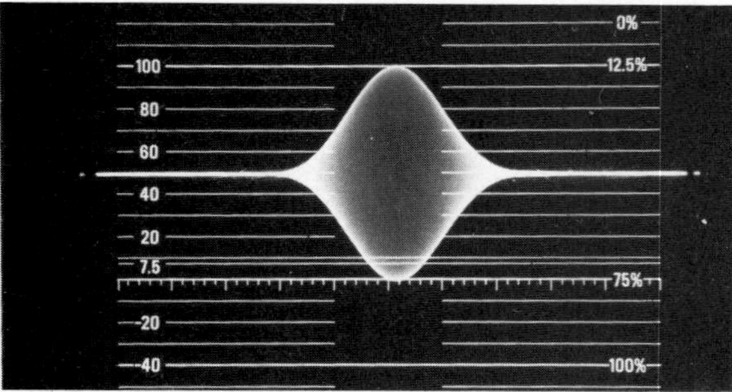

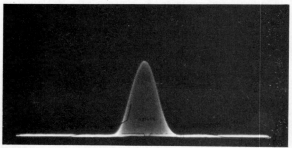

Figure 65.13 Formation of composite chrominance pulse: (a) Luminance component, (b) Chrominance modulated, (c) Composite pulse from addition of (a) and (b)

Since the process is so slow, direct viewing is not possible without either a storage or a digital sampling oscilloscope. Photography is the more usual solution, and both transitions must be measured. In order to reduce the total transient swing one may differentiate by using a high-pass filter (*Figure 65.11*). Alternatively an option for a well-known professional waveform monitor prefers a wide-range clamp and a sampling method for displaying the output, which has the advantage of maintaining the brightness roughly the same for most measurements. A warning light shows when the camera shutter has to be opened. The result, for the same waveform as *Figure 65.11*, is given in *Figure 65.12*, and is obviously much easier to analyze. In this instance, the duration of the 'bounce' was 12 seconds, during which time the sync pulse amplitude was for a time reduced to about one-third of its normal amplitude,and this is by no means an extreme example.

65.3.2.6 Luminance-chrominance inequalities

Ideally, the ratio of the amplitudes of the luminance and chrominance components of the video waveform should always remain unaltered (except with SECAM), otherwise the colour saturation is modified. The two channels must also arrive without any displacement in time, which could correspond to a registration error.

The test waveform is the composite chrominance pulse which is the third element in line 17 of the ITS and the field 1 VITS. Its generation is clearly shown in *Figure 65.13*. A luminance sine-squared pulse of 50 per cent amplitude is also modulated 100 per cent onto a subcarrier. When these two waveforms are added with precisely identical amplitudes and delays, a composite chrominance pulse is formed.

Any change in either the amplitude or delay, or both, will upset this delicate balance. If only the gains change, then the baseline acquires the shape of a half-period of a sinusoid, either above the baseline as in *Figure 65.14(a)* for a chrominance loss, or below for a chrominance gain. This effect is independent of the pulse duration, which for international purposes is 20T for 625-lines, and 12.5T for 525-lines. Internally, PAL I uses 10T.

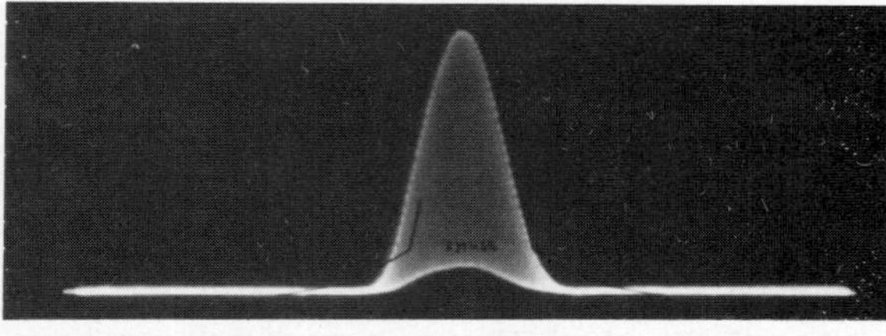

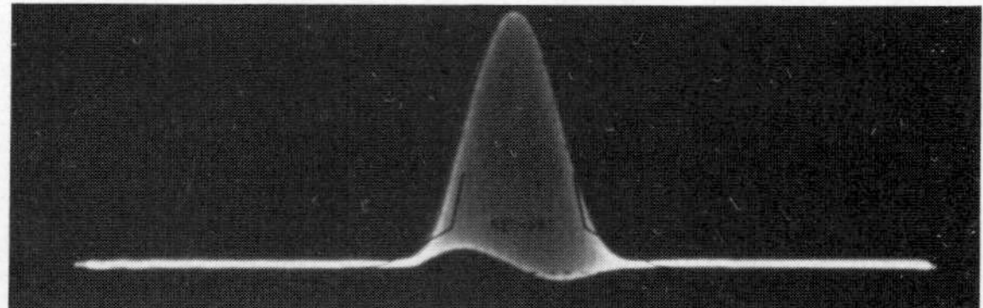

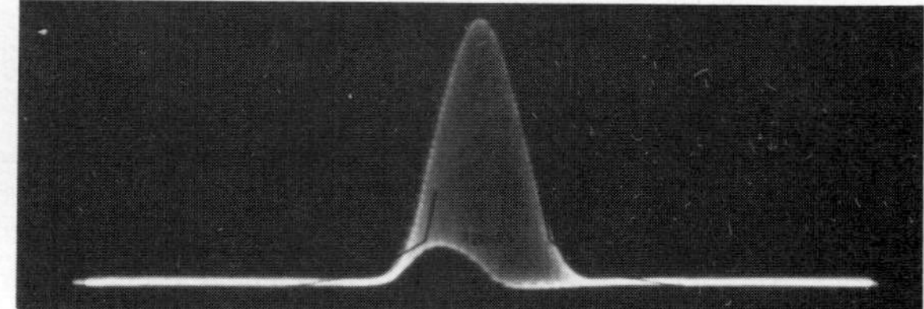

Figure 65.14 Distorted composite chrominance pulse: (a) Low chrominance gain, (b) Chrominance lagging, (c) Simultaneous gain and delay inequalities

When the relative delays differ, but the gains are equal, then the baseline takes the form of a complete period of a sinusoid. When the left-hand half period is positive, as in *Figure 65.14(b)*, the chrominance is lagging. More usually, both types of error are present at the same time, giving an asymmetrical baseline as

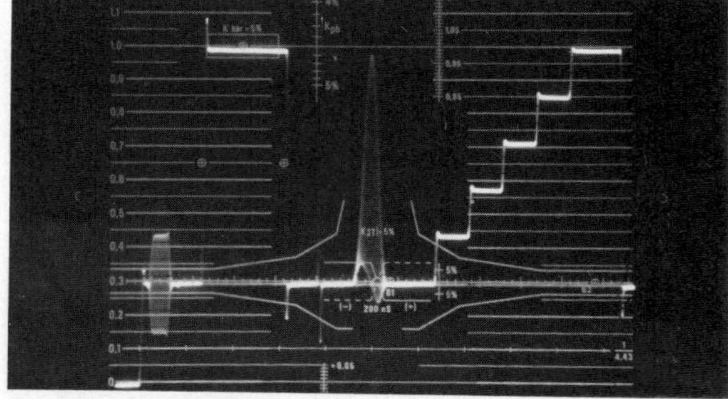

Figure 65.15 Estimation of 200 ns chrominance lag from graticule

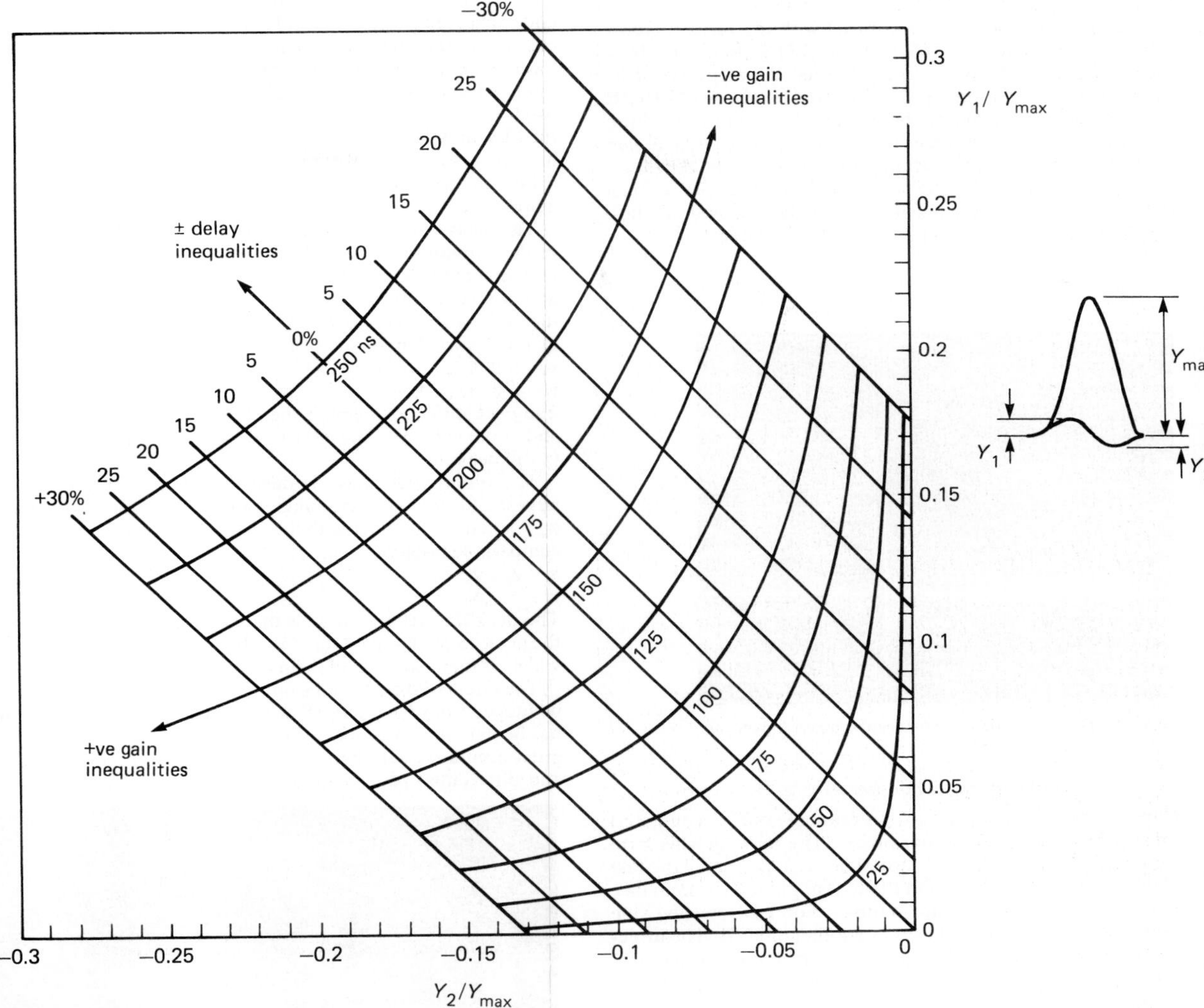

Figure 65.16 Nomogram for the 10T pulse (double delay readings for 20T)

in *Figure 65.14(c)*. The convention is that delay errors are in nanoseconds, and amplitude errors are percentages. For reporting purposes a chrominance loss and a lag are taken to be both positive, since the errors most often occur in that form.

In those instances where only one inequality is present at a time, the error is easily measured from the waveform. The signal amplitude is first standardized by setting the bar amplitude to 100 per cent (or 100 IRE); then if the peak amplitude of the baseline lobe is y, the amplitude error percentage is $a = 2y$. The delay error (in nanoseconds) is given approximately by $d = (2T_c.y)/100\pi$ where T_c is the half-amplitude duration of the composite pulse in nanoseconds. The graticule of *Figure 65.3* has a 'box' in the centre of the baseline for such an approximate measurement, used in *Figure 65.15* to estimate 200 ns chrominance lag.

When both inequalities are present at the same time, the calculations from the waveform become difficult and unsatisfactory. The full expressions can be found in Rosman[15] and Mallon and Williams[16]. Nomograms have been devised, and an example of the Rosman type is to be seen in *Figure 65.16*, but the accuracy is low and the process tedious.

A very much more satisfactory method is the use of a gain and delay tester. Originally a BBC design, this has been commercially available for a long time. It separates the two components of the composite pulse, whose amplitudes and relative delay can be varied by means of controls. The output waveform is observed, and when the baseline is seen once more to be flat, the errors are read directly from the controls. It is clear from the approximate expression above that the size of the lobe with delay errors is a function of the pulse width, which is the reason for preferring a 10T pulse in PAL I.

An excellent and more extensive discussion of chrominance-luminance inequalities can be found in D'Amato[17].

65.3.3 Non-linearity distortions

Non-linear distortions have no unique value since they are very much a function of the apl. During 'out-of-service' measurements it is usual to measure at apls of 10 and 90 per cent, or alternatively 12.5 and 87.5 per cent, with the larger of the two values in either case being regarded as the determining error. With 'in-service' testing one is forced to accept the long-term mean apl of a little less than 50 per cent.

65.3.3.1 Luminance non-linearity

Luminance non-linearity is measured with the plain staircase waveform on both the ITS and the VITS. Because direct measurement of the differences between the steps would be inaccurate, the waveform is differentiated by means of a CCIR defined network[1], which is built into professional waveform monitors. This converts the steps into a series of pulses (*Figure 65.17*) the height of each being proportional to the step from which it was derived. The luminance non-linearity is then defined as the difference between the largest and smallest amplitudes, expressed as a percentage of the largest.

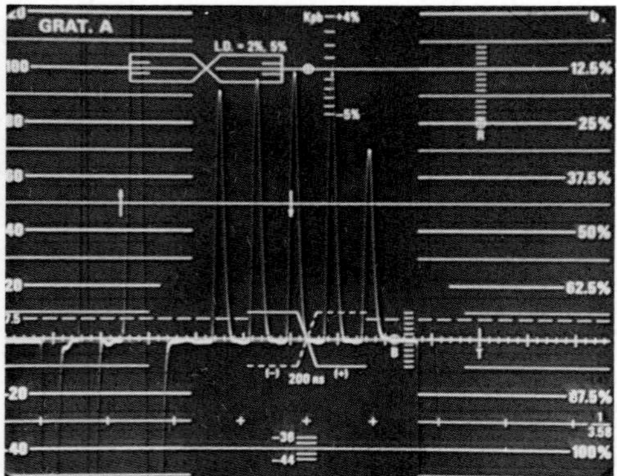

Figure 65.17 Differentiated luminance staircase waveform

65.3.3.2 Chrominance gain non-linearity

Chrominance gain non-linearity is normally only encountered on very long links and in transmitters. The relevant waveform for the ITS is the three-level chrominance step forming the first element in line 331. This, and the full-amplitude chrominance bar (see *Figure 65.1(c)*) are options. For EBU testing, the multiburst of line 18 is replaced by another line 331. The three-level chrominance step does not form part of the international VITS, but it is also employed, usually in a line which also contains a multiburst.

Although it is possible to measure this quantity on a waveform monitor using the inbuilt bandpass filter, the associated quantity, chrominance phase non-linearity, must be measured on a special instrument such as the vectorscope (see section *65.5.1*) which will provide both readings more conveniently.

With the amplitudes of the three bursts normalized to that of the central burst, the non-linearity is the larger of two possible differences in amplitude, expressed as a percentage of the central burst.

65.3.3.3 Chrominance phase non-linearity

Chrominance phase non-linearity is always measured in association with chrominance amplitude non-linearity, usually with a vectorscope, since the measurement of a subcarrier phase is needed. The magnitude of the distortion is defined as the largest phase-angle difference between the measurements on the three subcarrier bursts.

65.3.3.4 Differential gain and phase

Differential gain and phase are the result of a form of intermodulation between the luminance and chrominance channels. Ideally, they would be the changes in subcarrier gain and phase respectively resulting from differentially small changes in the luminance amplitude, and consequently could have an infinity of values over the black–white range.

In practice, the five-step staircase waveform of line 330 is used, corresponding to the luminance staircase of line 17, for the ITS, and the similar waveform in field 1 of the ITS for 525-line work. Both are overlaid with subcarrier, the ITS having ±140 mV and the VITS ±20 IRE units of a subcarrier amplitude.

The measurement of differential gain requires a comparison of the amplitudes of the subcarrier levels on the various steps with that at black level to find the differences. If the maximum and minimum differences are A_{max} and A_{min} respectively, and the subcarrier amplitude at black level is A_0, then two quantities are defined:

$x = 100(A_{max}/A_0{-}1)$;

$y = 100(A_{min}/A_0{-}1)$.

The *peak differential gain* is then numerically the larger of $+x$ and $-x$ per cent. The alternative *peak-to-peak differential gain* is $(x+y)$ per cent.

The differential phase is derived similarly by finding the largest and smallest phase-angle differences between the steps $ø_{max}$ and $ø_{min}$, together with the angle $ø_0$ at black level. Then in comparable fashion:

$x = ø_{max}ø_0$;

$y = ø_{min}ø_0$.

The peak differential phase is the larger of $x°$ and $y°$, whereas the peak-to-peak value is $(x + y)°$. It is purely a matter of choice which of these two sets of definitions is chosen.

The effect of differential gain on the staircase waveform can be judged from *Figure 65.18*, where the luminance component has been removed by means of a bandpass filter. Evidently no great accuracy could be expected from such a measurement, and differential phase cannot be displayed so simply. Many

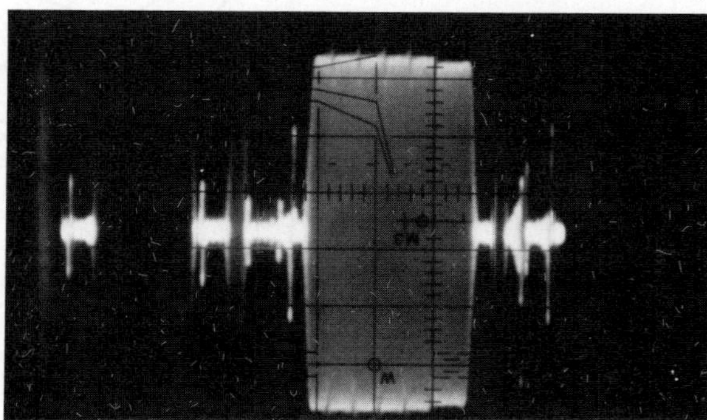

Figure 65.18 Luminance component of staircase removed to measure differential gain

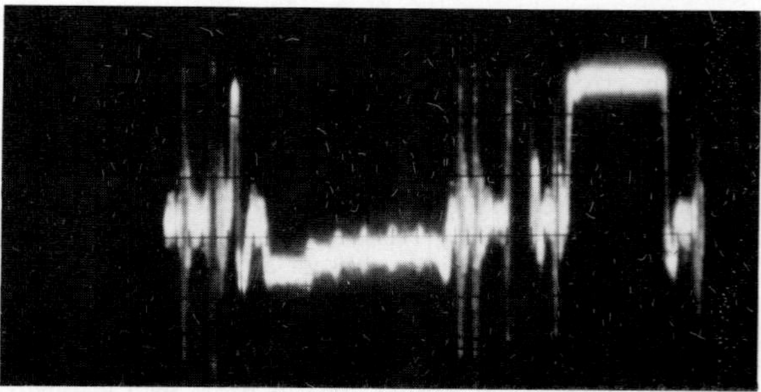

Figure 65.19 Differential gain display of off-air signal

excellent instruments are available for the measurement of both differential gain and phase. A display of differential gain from one of these is given in *Figure 65.19*.

A frequent alternative instrument is the vectorscope since this also serves other purposes. One of the displays of differential phase available from one commercial version is shown in *Figure 65.20*; the differential gain waveform would be rather similar. Although this is not the primary purpose of the vectorscope, the accuracy attainable can be really excellent.

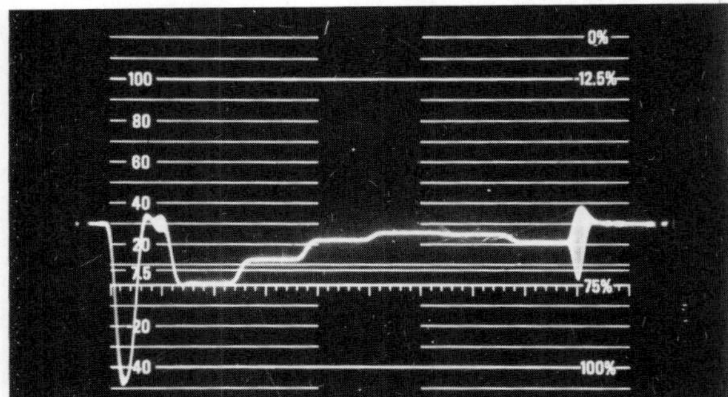

Figure 65.20 25° differential phase on vectorscope display

A word of warning is needed. It is sometimes believed that larger subcarrier amplitudes can improve the measurement accuracy. This is a fallacy. Experience shows that the amplitudes given above are the largest that will give reliable results. For studio work, where random noise is less of a problem, the subcarrier amplitude can be reduced by 50 per cent with advantage.

As far as picture impairment is concerned, the eye is very tolerant of differential gain, and of luminance non-linearity. Differential phase in NTSC is a great problem, since it gives rise to hue changes with luminance level, especially noticeable in skin tones, even for very small amounts of distortion. PAL and SECAM are very much more tolerant, since they were devised very largely with this in mind. PAL with delay line decoding converts differential phase into differential gain, and SECAM is affected in transitions only, where the impairment is visible mostly only to the skilled eye[18].

65.3.3.5 Chrominance-luminance intermodulation

Whenever amplitude non-linearity is present on a transmission channel such that the gains are unequal for the positive and negative half-periods of the subcarrier, rectification takes place with the result that a dc component is added to the waveform. The common term for this, *axis shift*, describes the effect clearly.

The chrominance bar of line 331 of the ITS (or the three-step alternative), and the three-step chrominance waveform of the VITS, are all suitable. A low-pass filter is used to remove the chrominance component, giving the effect shown in *Figure 65.21*, where the positive dc step in the area where the chrominance was situated is very visible. This dc step (or that produced by the largest of the three steps) is measured and expressed as a percentage of the amplitude of the 50 per cent bar in the chrominance bar waveform (not of the 100 per cent bar!). When the step is upwards, as in *Figure 65.21*, the distortion is recorded as positive.

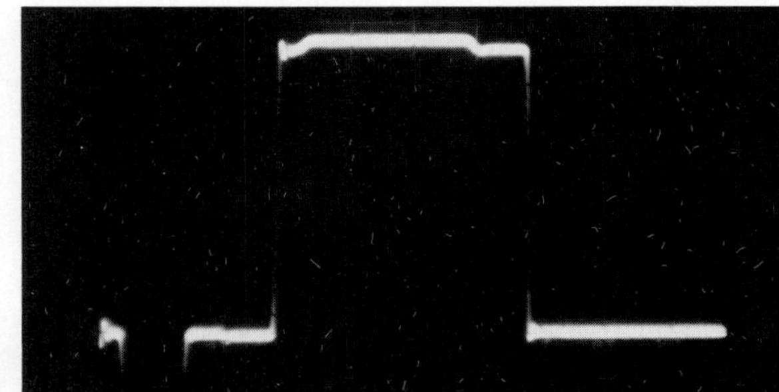

Figure 65.21 Chrominance-luminance intermodulation (chrominance removed)

Note that this distortion can give rise to large errors in the measurement of chrominance-luminance inequalities from the waveform monitor display, but not when the recommended specialized test set is used.

65.3.3.6 Multiburst

The multiburst waveform was introduced in the early days of video measurements with the aim of displaying the amplitude-frequency response in quantized form. It is found on line 17 of field 2 of the VITS, and line 18 of the ITS (but not for EBU purposes!). It takes the form of a white level reference pulse, followed by a sequence of rectangular frequency bursts between 0.5 and 5.8 MHz for the ITS, and 0.5 and 4.2 MHz for the VITS. The measurement is made by comparing the final burst amplitudes with that of the reference pulse.

In practice, the measurement turns out to be very unsatisfactory. The bursts do not represent a discrete frequency, but contain wide-spreading sets of sidebands. Not only is the response measured over an area, but there is usually an overlap between the sidebands. Further information can be found in Weaver[6]. Furthermore, the waveform is extremely susceptible to chrominance-luminance intermodulation, which is usually frequency dependent, and transient effects cause a lack of flatness of the tops of the bursts. In order to reduce the intermodulation, the burst amplitude is reduced to 75 per cent in the VITS, but this is only a palliative and not a cure.

A much more useful, and more soundly devised, version of the multiburst has been used in recent years by the BBC (section *65.5.2*).

65.3.3.7 ITS line 331

The perceptive reader may have wondered what purpose is served by the chrominance burst forming the second element on line 331 (and line 18 for EBU purposes). It was originally introduced to form a reference for subcarrier regeneration for differential phase measurements, but with advances in instrumentation it has now become redundant. The BBC have taken advantage of this by replacing this burst by a special teletext test waveform[35] for signals radiated within Britain.

65.3.3.8 Noise and interference

65.3.3.8.1 Random noise

True random noise consists of an assembly of pulses with amplitudes and times of occurrence which are known only statistically. In theory, these amplitudes can range from zero to infinity, but in practice circuit conditions impose obvious restraints. Random noise is a consequence of the discontinuous nature of matter, and can never be eliminated. The only measurable quantity associated with it is a mean power when averaged over a sufficiently long period, and the corresponding rms voltage. The distribution of the power with frequency is also significant.

There are two basic types of noise spectra, *white noise* where the rms voltage is constant with frequency, and *triangular noise* where it is linearly proportional to the frequency. This latter is important because it arises from the demodulation of a pure frequency modulated signal, although in practice it is modified by the pre-emphasis and other signal processing. Very often on long circuits the noise is *hypertriangular*, i.e. its spectrum rises more steeply with frequency than in the triangular case.

The definition of the signal/noise (s/n) ratio in television is the ratio of the white amplitude to the rms voltage, expressed in decibels. The justification for taking the rms voltage is that since the noise amplitude distribution is Gaussian, or a near approximation, the rms voltage is also the most probable.

Certain precautions must be taken before a noise measurement:

- Out-of-band noise must not be included, so the signal to be measured must be band-limited by a special filter[1,6,7] to 5.0 MHz (625-lines) and 4.0 MHz (525-lines).
- Power supply hum and other lf interference must be removed before a measurement. The CCIR recommends a high-pass filter which can be used in conjunction with the low-pass band-limiting filter. Residual subcarrier can be another problem, but this is readily removed by means of a notch network.
- The subjective picture impairment of random noise is dependent upon its spectrum. This uncertainty is removed by means of a weighting network[6,7] through which the random noise is passed before measurement. The CCIR recommends the 'unified' network[1] which is claimed to be suitable for both 625- and 252-line signals. In fact, it can be in error under certain conditions[12], but its use is still standardized.

A general difficulty with colour television signals is the simultaneous presence in a signal of the luminance and chrominance channels, which are utilized differently in the receiver. For example, one must consider in PAL and NTSC that although the chrominance bandwidth is so much lower than the luminance, the noise components falling into that region are demodulated down to become low-frequency noise in the chroma channel. The problem is even more complex in SECAM due to the considerable noise-weighting already built into the system[18] and the fm demodulation.

The method internationally agreed for PAL I[7,19] is the use of a specially designed bandpass network that gives the same degree of weighting to the chrominance channel as the wide-band network does to the luminance. Allnatt and Prosser[20] showed by subjective tests that this is true, and that a true measure of the overall s/n ratio can be obtained by combining the output of the luminance weighting and the chrominance weighting networks, with a 6 dB pad in series with the latter.

Although the use of the two weighting networks is preferred in the UK, it has nevertheless become common to accept the unweighted s/n ratio, in spite of the objections this raises, for general monitoring of circuits. Very often the weighted value is also measured and used together with the unweighted, since this provides additional information.

For the practical measurement of random noise, one of the 'quiet' lines is normally selected, e.g. 22 for 625-lines and 10 for 525-lines. The primitive procedure, once the normal method, is to measure the apparent peak-to-peak amplitude of the noise on a waveform monitor, and then to add a correction factor for conversion to rms voltage. The fact that this latter varies, according to different authorities, between 14 and 18 dB, gives some impression of the very poor accuracy. This is discussed in detail in Weaver[6,7].

Two of the superior visual methods are the *tangential* of Garuts and Samuel[21] and the *inserted noise burst*. In the latter, a small gap is made in the centre of the quiet line, into which is inserted a variable burst of random noise from a local generator. The inserted noise pedestal and amplitude are then varied until the gap disappears; then the s/n ratio is read from the instrument. *Figure 65.22* gives a deliberately incomplete adjustment to illustrate the process.

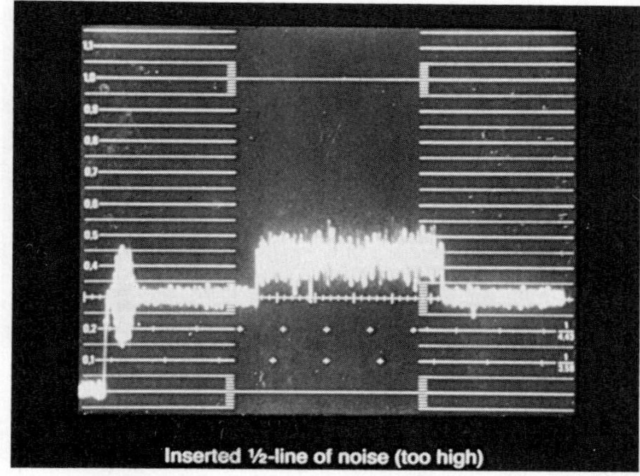

Figure 65.22 Inserted burst random noise measurement (with the burst deliberately offset) (Tektronix Inc)

Objective measurements, however, are very much to be preferred. One typical and very successful instrument is described by Holder[22]. In this, a narrow burst of the random noise is sampled from the line; after processing, its rms voltage is found. The instrument is so arranged that the s/n ratio is read off directly. Such equipment is capable of ample accuracy for practical purposes over a wide range of s/n ratios. Especially refined techniques are possible with digital automatic measuring equipment (see section *65.3.4*).

65.3.3.8.2 Interference

The term *interference* covers a wide range of unwanted phenomena which impair a television signal. The one thing they have in common is that they are measured on a peak-to-peak basis on a waveform monitor. When they are periodic or quasi-periodic in nature, it is usually possible with skill to trigger the monitor so as to lock the interference for long enough for a measurement to be made. When it is erratic, recourse may have to be made to a digital sampling or storage oscilloscope. Where more than one form of interference is present on a video signal, the best indication of their combined effect seems to be obtained by the quadratic addition of the individual amplitudes.

Moiré is a special form of interference which arises from the frequency modulation process in a video tape recorder. It is measured on a spectrum analyzer on a 100 per cent colour bar waveform (see section *65.5.1*) from a pre-recorded tape. The moiré components must be measured on each of the colour bars, since the effect is a function of hue. The total interference is again expressed in terms of the root-sum-square value[5,23].

65.3.3.8.3 Crosstalk

Crosstalk is the leakage of a signal in one path into another path by electromagnetic or electrostatic coupling, and occurs mostly in switching matrices. The number of possible combinations of disturbing and disturbed circuits is immense, so it is usual to concentrate attention upon those in close physical proximity. The test waveform may well be ITS line 17 or VITS field 1. In any case, high amplitude chrominance must be present. The disturbed path should preferably carry a signal with, say, a black line in the position of the disturbing waveform, otherwise the crosstalk cannot be identified easily. The crosstalk is defined as the ratio of white level to the peak-to-peak induced voltage, expressed in decibels[24].

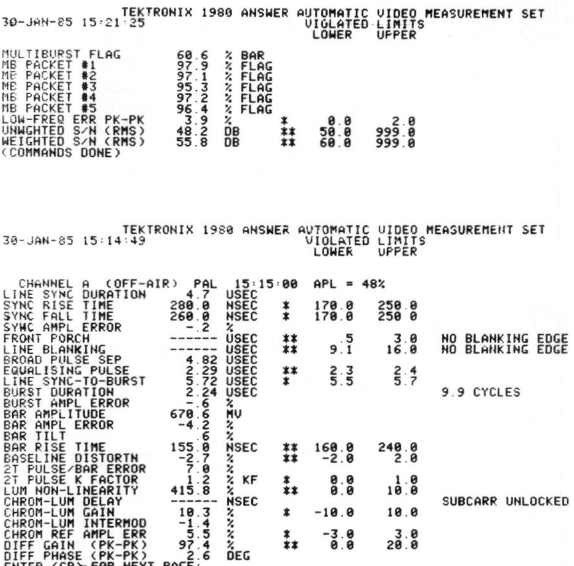

```
               TEKTRONIX 1980 ANSWER AUTOMATIC VIDEO MEASUREMENT SET
30-JAN-85 15:21:25                             VIOLATED LIMITS
                                                LOWER   UPPER

MULTIBURST FLAG        60.6  % BAR
MB PACKET #1           97.9  % FLAG
MB PACKET #2           97.1  % FLAG
MB PACKET #3           95.3  % FLAG
MB PACKET #4           97.2  % FLAG
MB PACKET #5           96.4  % FLAG
LOW-FREQ ERR PK-PK      3.9  %         *     0.0     2.0
UNWGHTED S/N (RMS)     48.2  DB        **   50.0   999.0
WEIGHTED S/N (RMS)     55.8  DB        **   60.0   999.0
(COMMANDS DONE)

               TEKTRONIX 1980 ANSWER AUTOMATIC VIDEO MEASUREMENT SET
30-JAN-85 15:14:49                             VIOLATED LIMITS
                                                LOWER   UPPER

  CHANNEL A  (OFF-AIR)  PAL  15:15:00  APL = 48%
LINE SYNC DURATION      4.7  USEC
SYNC RISE TIME        280.0  NSEC      *   170.0   250.0
SYNC FALL TIME        260.0  NSEC      *   170.0   250 0
SYNC AMPL ERROR        -.2   %
FRONT PORCH          ------- USEC      **     .5     3.0   NO BLANKING EDGE
LINE BLANKING        ------- USEC      **    9.1    16.0   NO BLANKING EDGE
BROAD PULSE SEP        4.82  USEC
EQUALISING PULSE       2.29  USEC      **    2.3     2.4
LINE SYNC-TO-BURST     5.72  USEC      *     5.5     5.7
BURST DURATION         2.24  USEC                          9.9 CYCLES
BURST AMPL ERROR       -.6   %
BAR AMPLITUDE         670.6  MV
BAR AMPL ERROR         -4.2  %
BAR TILT                 .6  %
BAR RISE TIME         155.0  NSEC      **  160.0   240.0
BASELINE DISTORTN      -2.7  %         **   -2.0     2.0
2T PULSE/BAR ERROR      7.0  %
2T PULSE K FACTOR       1.2  % KF      *     0.0     1.0
LUM NON-LINEARITY     415.8  %         **    0.0    10.0
CHROM-LUM DELAY      ------- NSEC                          SUBCARR UNLOCKED
CHROM-LUM GAIN         10.3  %         *   -10.0    10.0
CHROM-LUM INTERMOD     -1.4  %
CHROM REF AMPL ERR      5.5  %         *    -3.0     3.0
DIFF GAIN  (PK-PK)     97.4  %         **    0.0    20.0
DIFF PHASE (PK-PK)      2.6  DEG
ENTER <CR> FOR NEXT PAGE:
```

Figure 65.23 Printout from digital automatic measurement equipment (Tektronix Inc)

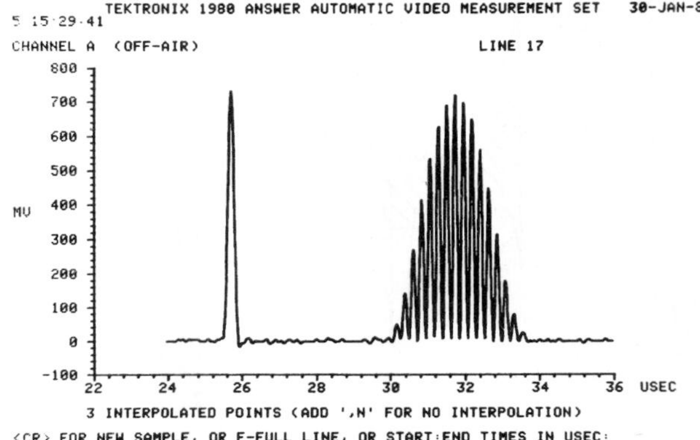

Figure 65.24 Waveform printout from digital automatic measurement equipment (Tektronix Inc)

65.3.4 Automatic measuring equipment

The ever-increasing size and complexity of television networks led long ago to the realization that most if not all of the routine measurements could advantageously be performed by automatic equipment (AME). This could tirelessly repeat sequences of measurements, even in locations difficult for staff to reach, issue alarms for actual or impending fault conditions, send results to a centre for statistical processing, and so on.

The earliest equipment used analogue techniques[25,26], although one early proposal for a digital system was made by Vivian[27]. Analogue AMEs have through long practical experience been developed to a high standard of performance and reliability. However, it must be admitted that they have a drawback in that the techniques used cannot always mimic those of an engineer, and it has been necessary for the CCIR and EBU to allow some modifications and relaxations[28,29]. In consequence, manual and automatic measurements cannot always be reconciled.

Digital methods are now in widespread use and offer great promise for the future from their reliability and versatility[30,31]. In the briefest possible terms, the test waveform is sampled at well above the Nyquist rate, and the data samples are stored, in one well-known instrument for 32 successive fields, permitting a noise reduction of 15 dB. These form a matrix of data values which can be processed entirely under the control of software. Not only can measurements be carried out by methods exactly analogous to those used manually, but they can be carried even further when necessary. For example, in the measurement of random noise, a fast Fourier transform can be applied to detect and remove periodic interference. A good impression of the range of possible measurements is provided by *Figure 65.23*. *Figure 65.24* demonstrates how the test waveform components can be reconstructed if required for information or record.

Criticism is sometimes levelled at the accuracy of AMEs, compared with manual methods. In the author's opinion, the precision of the latter is very often overestimated. One can be sure with modern AMEs that the accuracy is perfectly adequate for routine operational purposes, and possibly very much better than can be achieved by an engineer unless he is equipped with all of the specialized equipment needed for some of the measurements. Also, the AME is tireless, and is capable of functioning efficiently in remote or inaccessible situations for 24 hours a day, seven days a week.

65.4 Measurement tolerances

Measurements are insufficient in themselves. It is also necessary to know what errors are permissible at each point in the television chain. These must be determined globally from the subjective impairment experienced by viewers confronted by actual pictures with known values of the various distortions. This is a difficult but highly important subject, combining the techniques of psychophysics and statistics. A remarkably comprehensive account is given by Allnatt[12].

Once the total allowable error has been found, it must be shared between the viewer's receiver and the chain of equipment between the picture source and the transmitter. It might be thought that in a long chain the distortions would add according to a Gaussian distribution, i.e. root-sum-square, but experience shows that this is by no means always the case. Moreover, the situation is complicated by the fact that the errors are also a function of time[32,33,34].

There is yet another factor to be considered. The perceived picture impairment must be due to the effect of all the distortions present on the signal, and not to just an individual distortion. This problem seems to have been solved by Allnatt and his colleagues, who have shown that a quantity can be

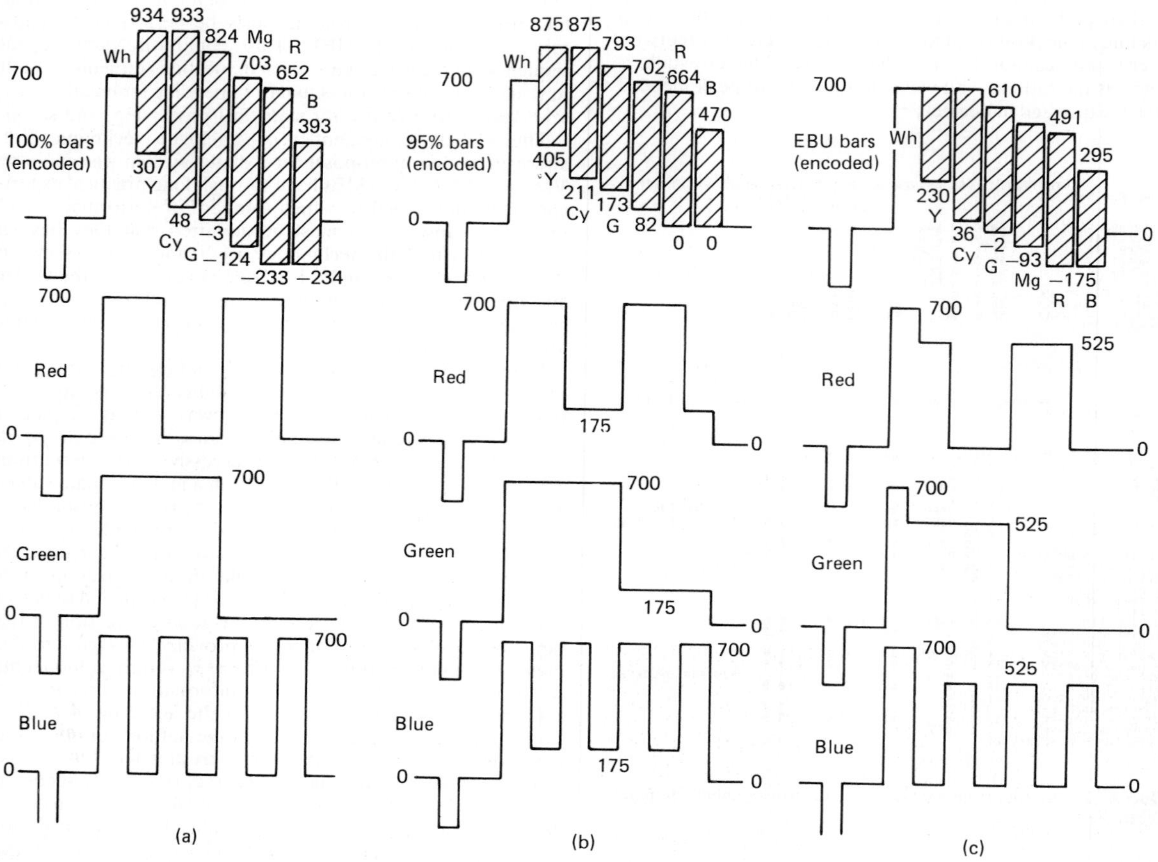

Figure 65.26 Composition of 625-line colour bar waveforms. All figures are in millivolts

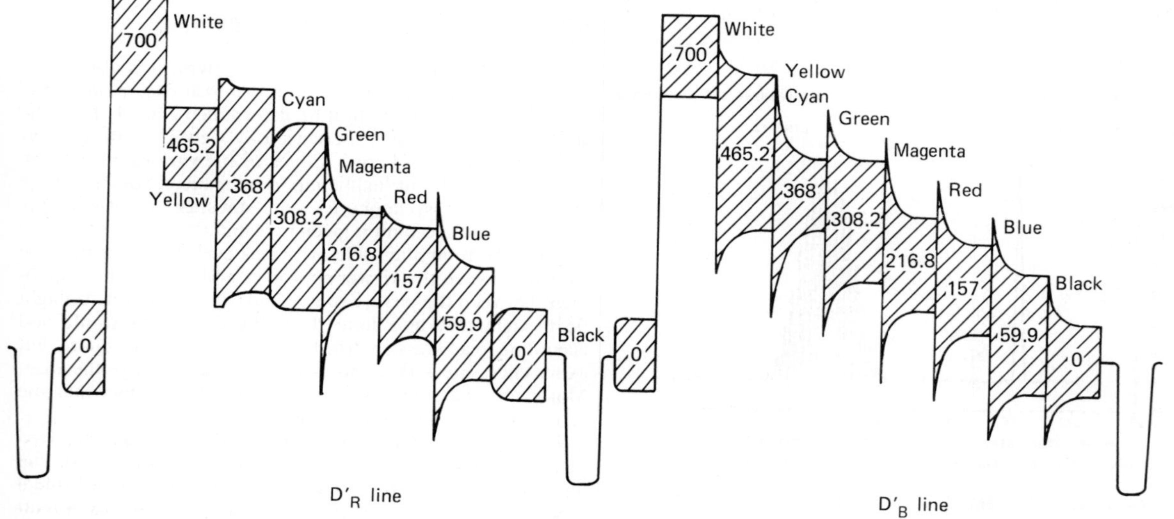

Figure 65.27 Standard SECAM colour bar waveforms. The figures within the bars are luminance levels in millivolts

derived from subjective tests, called an *imp* (impairment unit), which has the property of summability between unrelated distortions. This has thrown a very significant new light on an old problem[12,34].

Practical values for tolerances along the signal path can be found in references 1,5,7,34 and 35.

65.5 Teletext

Teletext is an information service, including subtitles for the hard of hearing, carried on lines of the field blanking interval by non-return-to-zero data pulses. The two principal systems are the British *Ceefax/Oracle*[36] and the French *Antiope*[37]. Since the data pulses are as short as possible to include the maximum information, transient distortion will cause confusion between pulses and lead to incorrect decoding in the receiver, so a further and rather stringent condition is imposed upon signal quality, especially as regards transmitters, for countries using teletext. A general survey of the problems is given in reference 38.

A fundamental criterion for teletext quality is the 'eye-height', i.e. the display obtained on a waveform monitor triggered so as to overlay a series of pulses (*Figure 65.25*). The eye-height is the maximum clear height within the pattern, in this case about 68 per cent of the possible value. The aim is to include the effect of noise as well, to obtain the decoding margin. For the definition and relationship to eye-height see references 39 and 40.

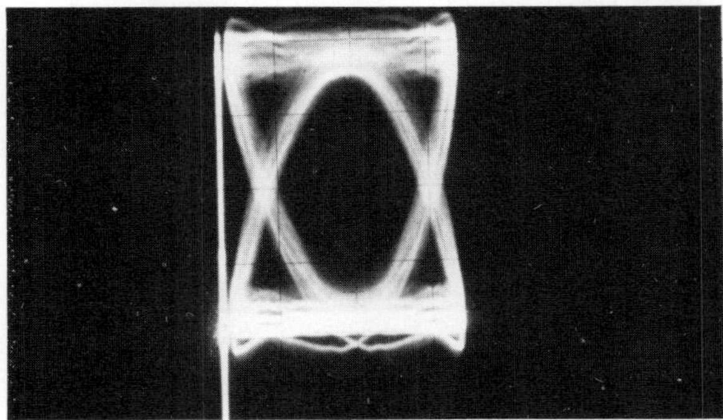

Figure 65.25 Teletext eye height display (BBC)

65.6 Specialized test waveforms

65.6.1 Colour bars

Whatever system is in use, good pictures are impossible unless the encoders are correctly adjusted at the points of signal origination. The colour bar waveform is the standard test signal for this purpose. It consists of a white reference pulse followed by six colour bars of the primaries red, green, blue, together with their complements yellow, cyan and magenta. By convention they are in a sequence of descending luminance values. SECAM bars must consist of a pair, which is also needed in PAL for delay-line decoding. The great virtue of this waveform is that it can be generated to a very high degree of accuracy and consistency.

The three most common PAL colour bars are illustrated in *Figure 65.26* and the SECAM bar pair in *Figure 65.27*, the former showing the colour separation components. The strange shape of the latter arises from the effects of the lf and hf pre-emphasis circuits. The 525-line standard closely resembles that of the EBU ((c) in *Figure 65.26*) but with the luminance bar at 75 per cent amplitude, and the addition of the 7.5 per cent pedestal. The 100 per cent bars are frequently used in studio practice since they correspond to the locus of 100 per cent saturated colours, but the '95 per cent' bars with their lower total amplitude range may often be preferred.

PAL and NTSC bars are universally measured with the vectorscope[6], which is a polar display formed, as shown clearly in *Figure 65.28*, from the (R–Y) and (B–Y) (or in NTSC, I and Q) colour difference signals. The important point is that for correct encoder adjustment the tips of the colour vectors must be located at predetermined points, which are given tolerance 'boxes' on the vectorscope graticule to allow a very rapid estimate of the encoder quality. The central dot is the *white point*, which again must be correctly located. Its shape is also significant.

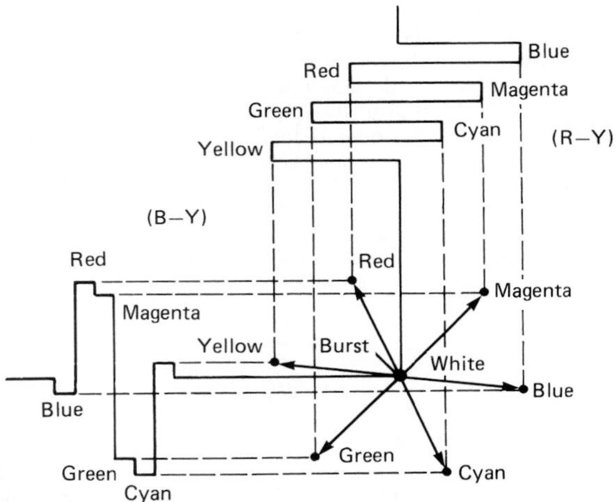

Figure 65.28 Formation of vectorscope display

The display of *Figure 65.28* corresponds to PAL with the normal V-axis switching. It is also possible to disable the switching, producing a display in which ideally the vectors of *Figure 65.28* are perfectly mirror-imaged about the U (horizontal) axis. This is invaluable for diagnostic purposes since the behaviour of each individual line is visible independently. This facility, of course, is not available with NTSC, where the display resembles *Figure 65.28* except that the burst vector lies on the horizontal axis.

The standard approach with SECAM has always consisted in measuring the colour bars in terms of the deviations of the two subcarriers, and extremely effective instruments exist using that principle. However, a vector display is also feasible and useful. At least one range of high-class picture monitors for SECAM provides outputs of (R–Y) and (B–Y) to drive an X-Y monitor and so produce a display closely resembling *Figure 65.27* except for the presence of the identification signals.

Some versions of the vectorscope are fitted with facilities for the measurement of differential phase and gain, which considerably increase the usefulness of this very versatile instrument.

65.6.2 Multipulse

The multipulse waveform was originally devised by the BBC for the measurement of the amplitude and delay responses of transmitters[41], and has since found applications in the USA. As shown in *Figure 65.29*, it consists of a series of ten pulses across the duration of a single line, making it suitable for use as an ITS. Each pulse is formed precisely in the same manner as the composite chrominance pulse of section *65.3.2.6* except

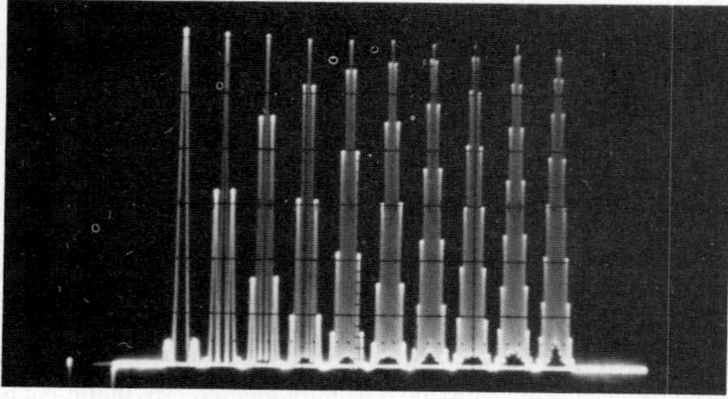

Figure 65.29 Multipulse waveform (Tektronix Inc)

that the modulation is performed with a sequence of frequencies spaced across the video band.

Exactly as with the chrominance pulse, the baseline distortion will yield not just the amplitude, but also the delay error (cf *Figure 65.14*). Moreover, the spread of the pulse sidebands is minimized by the sine-squared shaping, thus overcoming one of the objections to the multiburst waveform (section *65.3.3.6*).

Operational experience has demonstrated that this waveform is a powerful tool, especially for the delay correction of transmitters as well as distribution circuits, so it is very surprising that its use has not become more widespread. It is much preferable to the multiburst.

65.6.3 Sin x/x pulse

It was pointed out by Heller and Schuster[3] that test waveforms of very great accuracy and stability can be generated from binary numbers stored in read only memories. Since then, it has been realized that the same technique is capable of producing waveforms that would otherwise be extremely difficult or even impracticable by the hitherto conventional methods. A case in point is the sin x/x pulse, which can be considered as the result of passing an infinitely narrow pulse through an ideal low-pass filter.

The unique property of the sin x/x waveform is that its spectrum is flat over the whole band up to the cutoff frequency. Hence, as is clear from *Figure 65.8*, its use would enable an equal weighting to be given to all parts of the video band, unlike the sine-squared pulses whose sensitivity falls off rapidly towards the band limit. Its use for testing video systems was foreseen by Lewis[11], who proposed the derivation of the sin x/x response from the 1T response. However the technique is cumbersome, and the computational aids at the time were inadequate, so the method seems never to have been pursued.

Now it would seem that the sin x/x pulse is no more than a mathematical abstraction since the accompanying 'rings' ideally extend to plus and minus infinity. However, one prominent American manufacturer has put on the market a digitally generated waveform which, by limiting the number of 'rings' to 44 each side, together with a slight shaping, provides a close practical approximation to the ideal. The pulse is shown in *Figure 65.30*, and its spectrum in *Figure 65.31*. When the sidebands due to synchronization and pedestals are ignored, it becomes clear that the principal difference from the ideal response occurs in the close proximity of the band limit. This is a very interesting development which awaits further applications.

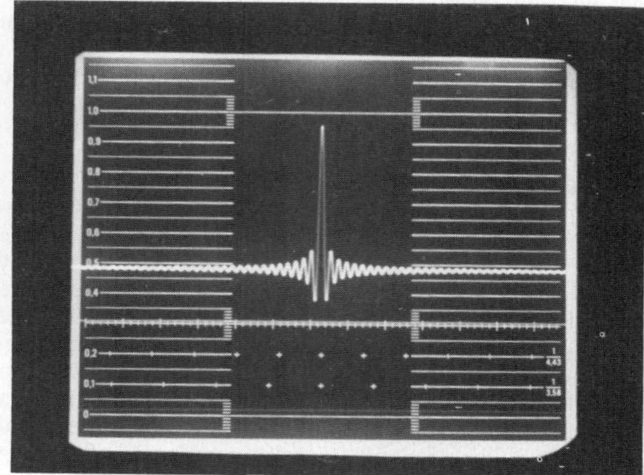

Figure 65.30 Synthesized sin x/x waveform (Tektronix Inc)

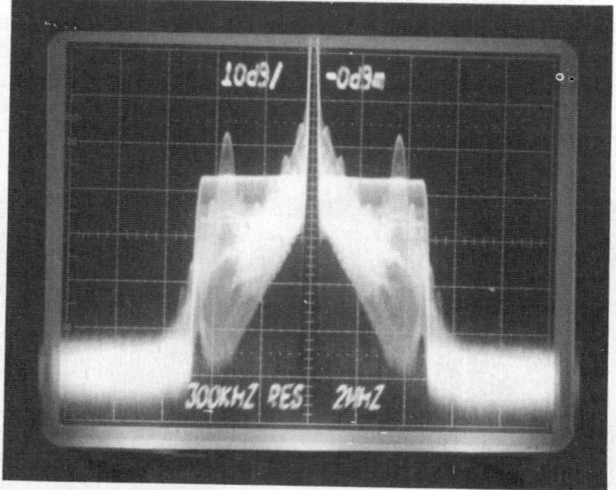

Figure 65.31 Spectrum of synthesized sin x/x waveform (Tektronix Inc)

65.6.4 Zone plate test pattern

In the search for better picture quality and advanced high-definition systems, it is now common to carry out filtration of the generated picture not merely in both horizontal and vertical directions, but also temporally. Other types of processing arise during digitization, e.g. sub-Nyquist sampling. This gave rise to a need to be able to identify picture impairments which the eye can only recognize as a degradation in quality, and since they are a function of the display cannot be revealed by either spectral or waveform analysis.

The *zone plate*, known for many years in physical optics, is prepared by drawing a set of circles whose radii are proportional to the square roots of the natural numbers; each alternate ring is then blacked out (see *Figure 65.32*). It was originally a demonstration of Fresnel's zone theory of diffraction, whence the name, and was first proposed by Mertz and Gray[42] in a classic paper on the theory of scanning to demonstrate the formation of alias components. Fairly recently it was revived by the BBC[43], who then designed a digital generator for both the circular and hyperbolic zone plate patterns[44], capable in addition of changes of scale and movement. This generator is commercially available, and its use is widespread, since there is no other equivalent.

Figure 65.32 Zone plate pattern (BBC Research Dept)

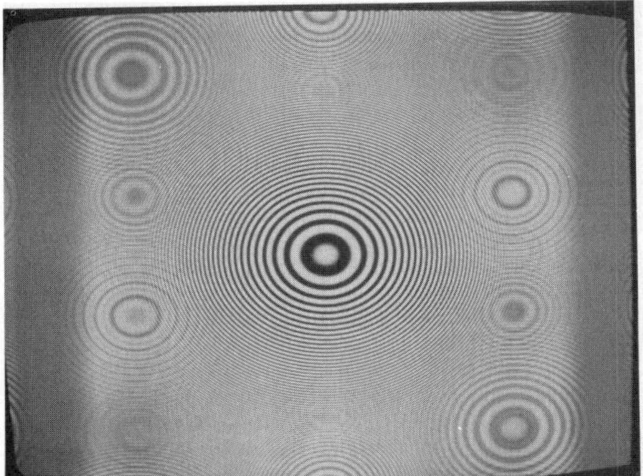

Figure 65.33 Displayed zone plate signal (luminance channel only) (BBC)

In the simplest possible terms, the zone plate may be considered as a linear two-dimensional frequency sweep. *Figure 65.33* shows the generated waveform displayed in luminance only on a waveform monitor. In this simplest instance the result is easily predictable. The dark portions at the left and right of the image are caused by the limited bandwidth of the video amplifier of the monitor. However, in the vertical direction the waveform is quantized by the scanning lines, whose number is equivalent to a spatial frequency too low for that of the display. The effect is the same as sub-Nyquist sampling, and as theory predicts, alias images of the zone pattern appear above and below.

Figure 65.34 is a monochrome picture of a zone pattern which has undergone PAL coding and decoding. The number of spurious images has not only increased considerably, but in reality they have characteristic colours and repetition frequencies from which the impairments may be determined. It must be said that this requires considerable skill and knowledge, but it is nonetheless possible for an expert. The temporal characteristics of the image can also be determined by introducing movement into the pattern.

Apart from research work into digital and high definition television, zone test patterns are also being utilized for the checking of digital processing equipment, standards convertors, decoders, picture monitor displays, and others too numerous to mention individually.

Figure 65.34 Displayed zone plate signal showing spurious waveforms resulting from PAL coding and decoding (BBC)

65.7 New TV generation and distribution methods

65.7.1 Digital signals

In principle, digitally coded signals should be distortionless, but ADCs and DACs contain analogue circuitry, notably low-pass filters, which can degrade the quality, and equipment failure of a number of kinds is always possible.

Signal impairments due to the encoding/decoding process and interference between the luminance and chrominance channels should not occur except at the transmitter or cable system where distribution takes place to the viewer. It appears that linear test waveforms such as the sine-squared pulse, possibly with a different width, will still be needed for overall quality checking. For fault-finding, it seems that a portable high-quality oscilloscope is the most generally useful tool. This will probably be supplemented by a decoder and picture monitor to pick up those impairments which are not revealed by test signals. Finally, a very powerful test waveform for ADCs and DACs is the simple ramp (sawtooth), which is unequalled for detecting quantization errors such as 'missing bits'.

A situation more likely to be encountered at present is a mixture of analogue and digital equipment in a studio area. The latter will consist of frame synchronizers, special effects generators, and so on. In this instance, all of the normal analogue measurements must be carried out.

A fundamental difficulty is encountered here with measurements of differential gain and phase, as well as noise, arising from the quantization of the composite, encoded signal. Highly misleading results are obtained because a signal amplitude may differ from its true value by $\pm 1/2$ least-significant bit. A detailed analysis is given by Felix[45], who furthermore shows that chrominance-luminance crosstalk, and chrominance gain and phase errors also occur, as well as quantization noise. Similarly, the same effects can also take place wherever a composite, encoded signal is digitized.

The remedy proposed by Felix is the doubling of the amplitude of the superimposed subcarrier (see section *65.3.3.4*) and making its phase a random quantity. More preferable is the addition to the test signal of a 'dither' waveform in the form of a small amplitude ramp at, say, line rate[46].

65.7.2 Analogue component distribution

For some very good reasons, there is now a strong swing towards the distribution of signals within a studio area in component form. This may take the form of R,G,B, and Y and linearly modified (R–Y), (B–Y) (C_R,C_B), or a time division multiplexed waveform derived from the MAC systems proposed for satellite transmission[47]. This latter has the advantage of single-wire transmission with luminance and chrominance on one line, but requires a rather complex decoder.

While three-wire methods, R,G,B and Y,C_R,C_B, avoid some of the impairments of composite encoded signals, they are especially sensitive to errors in relative gain and delay, and non-linearity since each of these gives rise to chrominance distortions. Crosstalk is another possible problem.

No standards for test waveforms have yet emerged. A comprehensive proposal is given by James and Marshall[48], and an interesting comparison of the ideas of several different authorities can be found in IEE[49]. Some commercial test equipment has already appeared, including a quasi-vectorscope type of display described in reference 49, but these play for safety in the sense that a sufficiently wide range of test waveforms is included for some to survive as the most useful and practicable.

References

1 CCIR 'Transmission of circuits designed for international connections', *Recommendation 567* (1978)

2 DOUGLAS, J N, 'International quality control through systematic measurements on ITS', *IERE Conf Proc*, 42 (1978)

3 HELLER, A and SCHUSTER, K, 'Application of ITS in TV studios and new methods for generating ITS', *IERE Conf Proc*, 42 (1978)

4 DARBY, P J and TOOMS, M S, 'Colour TV studio performance measurements', *IBA Tech Rev*, **1** (1972)

5 IBA, 'Code of practice for TV studio centre performance', *IBA Tech Rev*, **2** (1972)

6 WEAVER, L E, *Television measurement techniques*, IEE Monograph Series 9, Peter Peregrinus, London (1971)

7 WEAVER, L E, *Television video transmission measurements*, 2ed, Marconi Instruments Ltd (1978)

8 WHALLEY, W B, 'Colour TV coaxial termination and equalization', *Jour SMPTE*, **64** (January 1955)

9 THIELE, A N, 'Measurements of return loss at video frequencies', *Proc IERE (Austr.)* (June 1965)

10 SMITH, V G, 'TV waveform measurement', *Marconi Instrumentation*, **15**, No 4 (1977)

11 LEWIS, N W, 'Waveform responses of television links', *Proc IEE*, **101**, Part III, 258 (1954)

12 ALLNATT, J, *Transmitted picture assessment*, John Wiley (1983)

13 CMTT, *Document CMTT/124* (1976)

14 COMBER, G, and MACDIARMID, I F, 'Long-term step response of a chain of ac-coupled amplifiers', *Electronics Letters*, **8**, No 16 (1972)

15 ROSMAN, G, 'Interpretation of the waveform of luminance-chrominance pulse signals', *Electronics Letters*, **3**, No 3 (1967)

16 MALLON, R E and WILLIAMS, A D, 'Testing of transmission chains with vertical interval test signals', *Jour SMPTE*, **77** (August 1968)

17 D'AMATO, P, 'Study of the various impairments of the 20T pulse', *EBU Document Tech 3099-E* (March 1973)

18 WEAVER, L E, *The SECAM color television system*, Tektronix Inc, Oregon (1982)

19 CCIR *Recommendation 451-1* (1970)

20 ALLNATT, J, and PROSSER, R D, 'Subjective quality of colour television pictures impaired by random noise', *Proc IEE*, **113**, No 4 (1966)

21 GARUTS, V and SAMUEL, C, 'Measuring conventional oscilloscope noise', *Tekscope* (Tektronix Inc), **1**, No 1 (1969)

22 HOLDER, J E, 'An instrument for the measurement of random noise', *IEE Conf Report*, 5 (1963)

23 IEC, 'Measuring methods for television tape machines', publication 698 (1981)

24 DARBY, P J and TOOMS, M S, 'Colour television studio performance measurements', *IERE Conf Proc*, 18 (1970)

25 WILLIAMSON-NOBLE, G E and SEVILLE, R C, 'The television automatic monitor major', *IEE Conf Publication*, 25 (1966)

26 SHELLEY, L J and WILLIAMSON-NOBLE, G E, 'Automatic measurement of insertion test signals', *IERE Conf Proc*, 18 (1970)

27 VIVIAN, R H, 'Some methods of automatic analysis of television test signals', *IBA Tech Rev*, 1 (1972)

28 CCIR, 'Definitions of parameters for automatic measurement of television insertion test signals', *Recommendation 569* (1979)

29 EBU, 'Recommended definitions for parameters to be automatically measured on insertion test signals', *Document Com. T(T3)218* (1974)

30 RHODES, C W, 'Automated and digital measurement of baseband transmission parameters', *Jour SMPTE*, **86**, 832-835 (1977)

31 WATSON, J B, 'Digital automatic measurement equipment', *IRE Conf Publication*, 145 (1976)

32 D'AMATO, P, 'The determination of tolerances for chains of Television circuits', *EBU Rev Tech*, 156 (1976)

33 LARI, M, MORGANTI, G and SANTORO, G, 'The statistical addition of distortions in transmission systems', *EBU Rev Tech*, 143 (1974)

34 MACDIARMID, I F and ALLNATT, J, 'Performance requirements for the transmission of the PAL coded signal', *Proc IEE*, **125**, 6 (1978)

35 DEPARTMENT OF TRADE AND INDUSTRY, 'Specifications of television standards for 625-line system I transmissions in the United Kingdom' (1984)

36 BBC/IBA/BREMA, 'Specifications of standards for information transmission by digitally coded signals in the field blanking interval', *IEE Colloq Broadcast and Wired Teletext* (1976)

37 MART, B and MAUDIT, M, 'Antiope, service de télétexte'. *Radiodiffusion-Télévision*, 40 (1975)

38 IBA, 'Developments in teletext', *IBA Tech Rev*, 20 (1983)

39 SPICER, C R and TIDEY, R J, 'An automatic instrument for the measurement of teletext decoding margin', *IRE Conf Proc*, 42, 277-285 (1979)

40 HUTT, P R and DEAN, A, 'Analysis, measurement and reception of the teletext data signal', *IBC Conf Publication*, 166, 258-261 (1978)

41 HOLDER, J E, 'A new television test waveform', *Electronics Letters*, **13**, 9 (1977)

42 MERTZ, P and GRAY, F, 'Theory of scanning', *Bell System Tech Jour*, **13**, 464-515 (1934)

43 DREWERY, J O, 'The zone plate as a television test pattern', *IERE Conf Proc*, 42, 165-174 (1979)

44 WESTON, M, 'The electronic zone plate and related test patterns', *IBC Conf Publication*, 191 (1980)

45 FELIX, M O, 'Differential phase and gain measurements in digitized video signals', *Jour SMPTE*, **85**, 76-79

46 WILCOCK, P E, 'Analogue component video systems, measurement instrumentation and alignment', *IEE Conf Publication*, 240, 392-394 (1984)

47 CCIR, 'Satellite transmission of multiplexed analogue component (MAC) signals', *CCIR Draft Report AB/10-11* (1982-1986)

48 JAMES, A and MARSHALL, P J, 'Measurements in a television component environment', *IEE Conf Publication*, 240, 383-391 (1984)

49 IEE, Colloquium 'Component TV measurements and their relevance', *Electronics Division Digest*, 1985/23 (1985)

C A Marshman B Tech, C Eng, MIEE
York Electronics Centre, University of York

66 Electromagnetic Compatibility

In order to achieve electromagnetic compatibility between electrical/electronic equipment, it is necessary to control:

- unwanted emissions from equipment
- the level of immunity of equipment to externally generated interference

These objectives are achieved by using standards as guide lines, enforceable by regulations. Most countries develop their own standards and assign enforcement to a regulatory body. In the UK, standards are drafted by the British Standards Institute (BSI). Examples of regulatory bodies are the Federal Communication Commission (FCC) in the USA and the Fernmeldetechnisches Zentralamt (FTZ), the German Ministry of Post and Telecommunications.

Historically, electromagnetic compatibility has only been considered when interference prevents a system from functioning as required. Awareness is improving, which is reflected by the introduction and development of standards. In particular, the European Community agreed an all embracing directive in May 1989, to apply from 1 January 1992.

Regulations to control levels of electromagnetic interference have existed since the 1950s and are primarily concerned with interference to radio and television receivers. With the advent of microprocessor and microcomputer based systems operating at clock frequencies of several megahertz, increasing problems of interference in a wide range of commercial and industrial environments have been identified, resulting in more specific requirements and recommendations being produced by national and international committees. Many of these have been adapted from military specifications where the problem of achieving electromagnetic compatibility has been given greater emphasis.

Most standards are derived from the recommendations published by CISPR (International Special Committee on Radio Interference). CISPR makes recommendations for emission limits, immunity levels and test procedures and is a sub-committee of the IEC (International Electrotechnical Commission).

UK based companies have tended to design equipment for electromagnetic compatibility (EMC) only if their equipment's function is affected, if it is covered by the Wireless Telegraphy Act, or if it is for export to the USA and Germany, where it must meet respectively FCC requirements or the FTZ enforced VDE standards. However, from January 1992 all electrical/electronic equipment to be marketed or taken into service within Europe, including the UK, is legally required to comply with the EC Directive on EMC.

66.1 European Community Directive

From 1 January 1992, all electrical and electronic equipment 'placed on the market' or 'taken into service' must comply with the objectives of the European Community EMC Directive. This applies to both new and existing designs being manufactured and marketed after this date.

This directive is an essential precursor to the establishment of the single European market and is intended to provide an environment for the reliable operation of all electrical/electronic equipment. The objectives defined by the directive are mandatory, while standards are not themselves binding and are only defined as a means of demonstrating that compliance with the objectives has been achieved. These can therefore be adapted to take account of technological progress, ensuring that development is not stifled.

66.1.1 Objectives

The essential protection requirements are:

- equipment must be constructed to ensure that any electromagnetic disturbance it generates allows radio and telecommunications equipment and other apparatus to function as intended
- equipment must be constructed with an inherent level of immunity to externally generated electromagnetic disturbances

66.1.2 Scope

All electrical and electronic equipment, together with equipment and installations containing electrical/electronic components, is without exception deemed to be within the scope of the directive. Previous directives and associated legislation covering domestic equipment and luminaires are absorbed into it.

Likewise, the definitions of electromagnetic disturbances are all embracing, covering conducted and radiated emissions, conducted and radiated immunity, mains disturbances, electrostatic discharge (esd) and lightning induced surges.

The directive excludes equipment covered by other directives with EMC provisions. This includes vehicle spark ignition systems and non-automatic weighing machines. It should be noted that where these separate provisions exist, but cover only certain aspects of electromagnetic disturbances (e.g. immunity to radiated interference), equipment is still required to comply with the EMC directive in respect of the other aspects (e.g. radiated emissions).

Also excluded are amateur radio equipment that is not commercially available, and kit-built electronics.

66.1.3 Compliance with protection requirements

Manufacturers, and importers from outside the European Community, are required to provide a declaration that their equipment complies with the objectives of the directive. However, for demonstrating compliance, there is a choice of two available routes.

66.1.3.1 Self-certification

The simplest method, allowing self-certification, is by satisfying *relevant standards* either by in-house tests or contracting the tests to an independent test house. The directive delegates responsibility for standards to CENELEC (European Committee for Electrotechnical Standardization), which is required to produce standards in the form of European Standards (EN). These generally follow the recommendations of CISPR and are defined as *relevant standards*. Each national standards body is required to produce standards harmonized with the appropriate Euro Norm.

In the absence of a European standard, compliance with an existing national standard will suffice, if the particular standard is accepted by the Commission and published in the Official Journal of the European Communities. However, this is likely only to be an interim measure. If harmonized standards or approved national standards appropriate to a particular equipment are not in place when the directive becomes legally binding, then existing national arrangements remain in force until 31 December 1992. This provision is subject to review and the transitional period may well be extended in view of the extremely large task faced by CENELEC in producing standards for all types of equipment.

66.1.3.2 Technical file

The alternative method of demonstrating compliance is to produce and hold a *technical file*, to be available for inspection by the national body responsible for policing the directive. This form of certification implies that the technical file should demonstrate conformity with the objectives of the directive. The technical file should contain a description of the equipment and the EMC provisions made; it must also include a technical report from a 'competent body'. This may be based on a theoretical study and/or appropriate tests.

The manufacturer (or agent) is required to hold the technical construction file at the disposal of the enforcement authorities from any member state for a period of ten years.

This route for claiming compliance is obligatory after 1 January 1993 if there is no appropriate relevant standard.

In the UK, a competent body is likely to be a NAMAS approved laboratory. NAMAS, the UK National Measurement Accreditation Service, is a division of the National Physical Laboratory (NPL), whose purpose is to assess and accredit laboratories that have demonstrated their competence to perform defined measurements within prescribed limits of uncertainty. NAMAS was formed by an amalgamation of the British Calibration Service (BCS) and the National Testing Laboratory Accreditation Scheme (NATLAS).

A manufacturer may obtain accreditation for his own test facilities if these satisfy NAMAS requirements. In this instance, it would be necessary to demonstrate that the testing facility is not compromised by pressures from the production side of the manufacturer's operations. For example, the management 'tree' should indicate the independence of the testing facility and show no direct line of command from the control of production.

In the absence of appropriate European or national standards, it may be necessary for a manufacturer or the competent body to consult with the national authority in order to carry out testing to standards which are not specific for the application, but which the authority will accept in a technical file. Telecommunications terminal equipment (i.e. equipment directly or indirectly connected to a public telecommunications network to send, process or receive information) must be assessed by a *notified body* which will issue an EC type examination certificate. This also applies to radio transmitters excepting those used by radio amateurs.

A *notified body* will to all intents and purposes be the same as a *competent body*, and for the UK accreditation will also be the responsibility of NAMAS.

It should be noted that the manufacturer is required to comply with the objectives of the directive, not with particular standards, and the directive specifically refers to possible inadequacies of standards.

66.1.4 EC declaration of conformity

A manufacturer or import agent must hold an *EC declaration of conformity* for equipment to be placed on the market. This declaration must contain:

- a description of the apparatus to which it refers
- the specifications under which conformity is declared
- identification of the *signatory* empowered to bind the manufacturer or agent
- where appropriate, reference to the EC type examination certificate issued by the notified body

The declaration must be kept available to the enforcement authorities for ten years after the equipment to which it refers has been placed on the market.

The manufacturer or agent must apply the CE mark to the equipment or else to its operating instructions, or the guarantee certificate or its packaging. Alongside the CE mark should be the year in figures for which compliance was first claimed and, if appropriate, the distinctive letters of the notified body issuing an EC type examination certificate. Use of the CE mark indicates that the equipment complies with all the EC directives applying to it, e.g. electronic toys must comply with both the toy and EMC directives.

66.1.5 Responsibilities of member states

Apparatus complying with the objectives of the directive, i.e. bearing the CE mark, must not be impeded from being placed on the market. However, if an administration finds that apparatus does not comply, then the apparatus must be withdrawn from the market or its free movement restricted. The European Commission is then immediately informed, which in turn, assuming the action is justified, will inform all the national administrations. This effectively will 'ban' the equipment throughout Europe.

66.1.6 UK implementation

In the UK, the Wireless Telegraphy Act exists to preserve the quality of radio communications. Changes have been implemented to update the regulations for CB radios, portable tools and fluorescent lighting, to take account of earlier European directives. This existing legislation is responsive, whereas the EMC directive, which controls the conditions for apparatus to be placed on the market, is preventative. Thus, changes have been necessary in the legislation.

The DTI's Radio Investigation Service has been responsible for enforcement of the Wireless Telegraphy Act.

66.2 Relevant standards

Standards which are described as *relevant* for claiming compliance with EC directives are designated *Euro Norm* or EN. For the EMC directive, these are drafted by CENELEC and derived from CISPR or other IEC publications. It is necessary for the European Community member states to harmonize their own national standards with the appropriate EN. An example of this is BS 6527:1988, which is fully harmonized with EN 55022. It should be noted that EN 55022 itself refers to CISPR 22 and only lists modifications to it, whereas BS 6527:1988 contains the complete text of CISPR 22 with the appropriate modifications and thus provides the complete text for EN 55022.

Approximately equivalent standards are shown in Table *66.1*. EMC related British Standards are listed at the end of the section. Table *66.1* shows which of these will become relevant standards. However, the process of harmonization with the appropriate EN is not complete.

CENELEC have recognized that it is not possible to produce dedicated emission and immunity standards for each distinct electrical and electronic product type, given the timetable for the directive to become legally binding. Accordingly, with the European Commission's approval, CENELEC has commenced the development of generic EMC standards which may be applied to a broad range of products. In order to achieve a rapid introduction of these new standards, they must be based on existing accepted standards.

Generic standards for emissions and immunity are required. The former will be based on EN 55022 which was originally formulated for measuring the emissions from information technology equipment, while the latter will be based on IEC 801, originally developed for testing industrial process control and instrumentation equipment, but which has been increasingly applied to other types of equipment due to the lack of appropriate standards.

Some of the details of EN 55022 and IEC 801 are considered to illustrate the difficulties faced by manufacturers in achieving compliance with the directive.

66.2.1 EN 55022

The permitted levels of radiated and conducted emissions from information technology equipment (ITE) are defined by EN 55022 (BS 6527:1988). This may be interpreted as applying to all electronic equipment employing microprocessor or microcomputer devices and therefore has far reaching consequences.

Class A and class B limits are specified as follows:

> *Class A* equipment is information technology equipment which satisfies the class A interference limits but does not satisfy the class B limits. In some countries, such equipment may be subject to restrictions on its sale and/or use.
>
> *Class B* equipment is information technology equipment which satisfies the class B interference limits. Such equipment should not be subject to restrictions on its sale and is generally not subject to restrictions on its use.

It is actually necessary to look at the *Notes* to ascertain that class B limits are for domestic and class A for commercial establishments! So, in effect, the class definitions are similar to the FCC classifications, but the possibility to misinterpret the standard exists, whereas the FCC are quite precise in their definitions of class A and class B equipment and also of what is a computing device (i.e. equipment with a clock rate in excess of 10 kHz). The FCC also list equipment which is excluded from the regulations, e.g. transportation equipment and domestic appliances.

Table *66.2* shows the limits specified by EN 55022 for radiated field strength measured using an open field test site. The average and quasi-peak measurements are specified for conducted emission limits. The equipment under test (EUT) is required to meet both limits. Should the average limits be attained when using a quasi-peak detector, the EUT is deemed to meet both limits.

Full details of the measurement techniques are also specified.

Subject	*USA* FCC	*Germany* VDE/DIN	*British Standards* BS	*CISPR Pub.*	*CENELEC* EN
Emissions:					
Industrial Scientific and Medical (ISM)	47CFR Pt.18	0871	4809	11	55011
Ignition		0879	833	12	–
Radio and TV	47CFR Pt.15	0872	905-1	13	55013
Household Appliances		0875	800	14	55014
Luminaires		–	5394	15	55015
Information technology	47CFR Pt.15	0871	6527	22	55022
Immunity:					
Radio & TV			905-2	20	55020
Industrial Process Control	SAMA PMC33.1		6667	IEC 801	HD481*
	ANSI 63.12				

*HD is a harmonization document

Table 66.1 Equivalent standards

Of particular relevance to a manufacturer is the requirement to use an open field test site for radiated emission measurements.

The general arrangement of an open field test site is shown in *Figure 66.1*. The equipment under test is placed at a defined height above the ground plane with its power cables fed through outlets on the ground plane (with defined rf impedance). The measurement antenna is placed at a specified distance from the EUT and must be able to move vertically over a defined range. The common EUT to antenna distances specified are 3 m, 10 m and 30 m. The frequency range of the site is 30 MHz–1000 MHz.

The size of the EUT limits the minimum size of the site that can be used. BS 6527 specifies that the antenna to EUT spacing must always be greater than $2D^2/\lambda$, where D is the maximum dimension of the EUT. Essentially, this means that the larger the EUT, the greater the EUT to antenna distance, and hence the larger the site required.

For measurements to be repeatable, the reflection coefficient of the ground must be constant. This is achieved by using a metal ground plane beneath the EUT and antenna. *Figure 66.2* shows the minimum ground plane specified by BS 6527. If a ground plane is not used, then the ground reflection is dependent mainly upon the ground conductivity which is determined by its water content. On sites without ground planes, considerable performance variations can be expected due to changes in the local climate.

Reflecting objects close to the site will also affect the site performance. BS 6527 defines the minimum clear area for the site. This elliptical area is shown in *Figure 66.3*. The elliptical clear area minimizes the effect of reflections from outside the ellipse, so that they are small compared to the direct and ground reflected waves.

Often the radiation from the EUT may originate from the power or signal cables connected to it. The specifications generally require that the cables should be laid out either to represent the operation of the unit or to give the worst case radiation.

While each test site may be constructed in accordance with

Radiated field strength limits

Frequency range MHz	Test distance, m		Quasi-peak limit dBμV/m	
	Class A	Class B	Class A	Class B
30 – 230	30	10	30(39.5)*	30
230 – 1000	30	10	37(46.5)	37

Conducted emission limits

Frequency range, MHz	Limits			
	Class A		Class B	
	Quasi-peak, dBμV	Average, dBμV	Quasi-peak, dBμV	Average, dBμV
0.15 – 0.5	79	66	Decreasing linearly with the logarithm of the frequency from 66–56	56–46
0.50 – 5	73	60	56	46
5 – 30	73	60	60	50

*values in parentheses are 10m equivalent values for class A

Table 66.2 Limits specified in BS 6527:1988

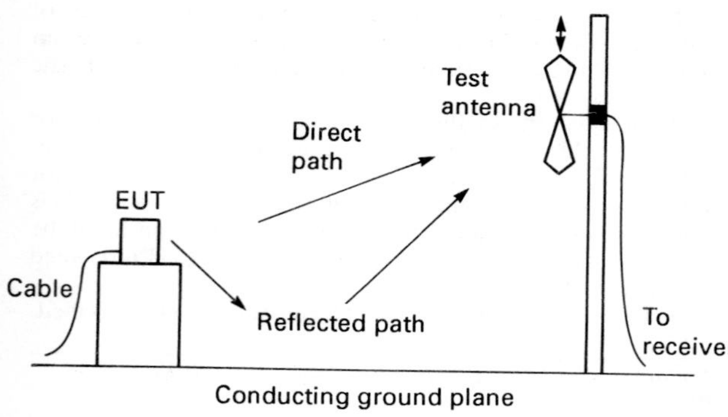

Figure 66.1 Open field test site

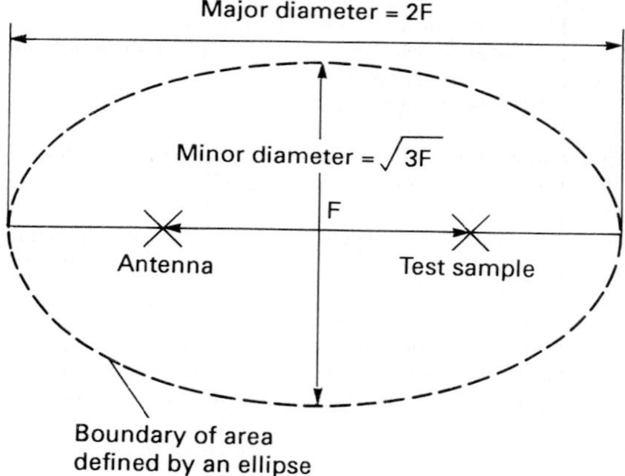

Figure 66.3 Plan of open field test site, showing volume above earth to be free of reflecting objects

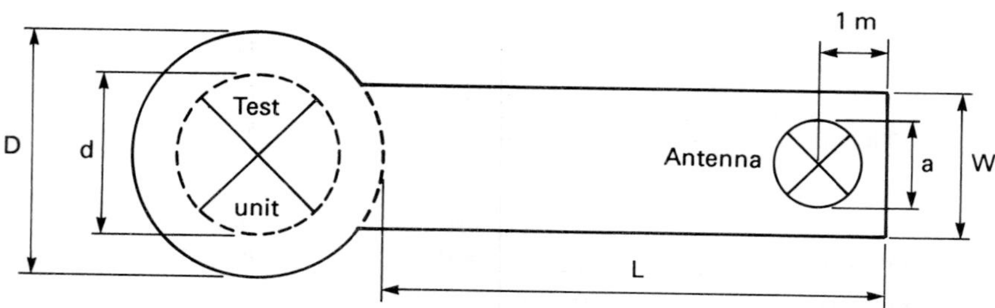

Figure 66.2 Minimum size of metal ground plane for radiation testing according to BS 6527. $D = d + 2$ metres, where d is the maximum test unit dimension; $W = a + 1$ metre, where a is the maximum antenna dimension; $L = 3$, 10 or 30 m

the specification, they will, in practice, differ slightly. Site evaluation methods are therefore required. The method adopted by the FCC is described in document OST 55 and is called *site attenuation*. The test results for a given site are required to be within typically ±3 dB of the ideal site.

The major disadvantage of open field test sites is that they are not isolated from the electromagnetic ambient. This may mean that at certain frequencies, occupied by say broadcast services, emissions from the EUT are masked. As the broadcast services themselves are to be protected by the emission specification, it is ironic that they should interfere with the emission measurement.

66.2.2 IEC 801

Few commercial specifications exist for the measurement of immunity to unwanted electromagnetic radiation. US Mil-Std-461C and DEF Std 59/41 specify immunity levels and measurement techniques for various categories of military equipment.

Examples of standards which apply to commercial equipment are part of American National Standards Institute (ANSI) C63, IEC 801 and its British Standard equivalent, BS 6667:1985.

IEC 801 consists of four parts covering EMC for industrial process measurement and control equipment. *Part 1* is an introduction; *Part 2* details evaluation of immunity to ESD; *Part 3* covers immunity to radiated electromagnetic energy, and *Part 4* immunity to bursts of fast transients. Parts of this standard have also been adopted for testing other types of equipment, e.g. Part 3 of BS 6667 is also incorporated in CEGB Standard DN5. Currently, Part 4 of IEC 801 has not been adopted by BSI, and Parts 5 and 6, covering *Disturbances in power supply environment* and *Conducted immunity requirements* have not been finalized by the IEC.

Part 3 of IEC 801, covering immunity to radiated electromagnetic fields, will be considered in detail. Immunity levels and measurement techniques are specified by this standard.

For severity levels 1, 2, 3 and X, the test field strength at frequency band 27-500 MHz is specified as follows:

1:	1 V/m
2:	3 V/m
3:	10 V/m
X:	special

The problem faced by the manufacturer is to choose an appropriate level of severity to test his equipment against. Here it is necessary to consider the threat level likely to exist within the operating environment for the equipment. The manufacturer also has to consider the level of performance degradation that can be tolerated. Here there may be differences in what is considered acceptable between the manufacturer and the end user. For example, is a temporary flicker on a VDU acceptable? It should be noted that severity level X is determined by agreement between manufacturer and user.

The test methods described in IEC 801-3 use shielded enclosures (screened rooms), anechoic chambers and the *stripline circuit* shown in *Figure 66.4*. When using a screened room, the EUT is illuminated with electromagnetic energy of the appropriate field strength by using antennas positioned 1 m away from the EUT. When a stripline circuit is used, the EUT is placed in the centre of the 'cube' part of the stripline on a support of foam. The aim of the stripline is to provide a uniform electric field in the cube position.

The stripline method can at best be described as giving a repeatable test result. However, it does not guarantee that the equipment will perform correctly when installed.

In practice, the stripline described in the standard is found (depending on the input impedance of the EUT) to be resonant in the frequency range 50–60 MHz, and hence the induced currents in the EUT connecting cables are significantly higher in this frequency band. Therefore, an EUT tested in accordance with IEC 801-3 and meeting its requirements at a nominal 10 V/m field strength, may actually have been tested at the

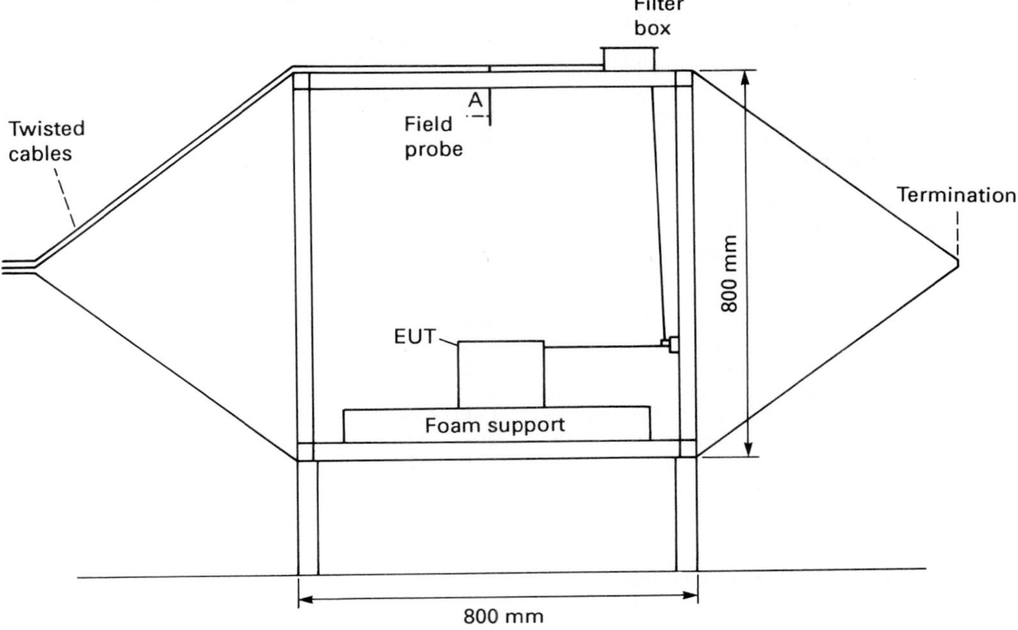

Figure 66.4 Stripline circuit specified by BS 6667 Part 3

equivalent of a significantly higher field strength. Also an equipment when installed may fail in practice because the length of the connecting cable may represent a significant fraction of a wavelength to an exciting frequency present in the environment and therefore be resonant. The effective field strength will therefore be higher than the 10 V/m at which the equipment was tested.

Alternative methods for immunity measurements use direct rf current injection techniques over which a far greater degree of control can be exercised. It is apparent that the standard itself is open to misinterpretation particularly in respect of choice of severity levels to be used for testing, and also that there are limitations on the test methods described and which therefore must be used with caution, since the directive requires compliance with the essential protection requirements not with a particular standard. Immunity standards are coming under severe scrutiny, and it is hoped that more meaningful test procedures will be established.

66.3 Conclusions

The aims of the directive are laudable in attempting to provide an environment for the reliable operation of all electrical/electronic equipment without interference to bone fide spectrum users. It removes any existing EMC regulations within the EC which may be used as a barrier to trade. It attempts to define, by virtue of standards, levels of emissions and immunity for manufacturers to attain.

Some manufacturers may not possess the necessary expertise required to produce the documentary support to claim compliance for their products. However, the onus will be upon companies to evaluate the electromagnetic performance of their products accurately (and inevitably incur extra costs). 12 months grace has been allowed after 1 January 1992, during which time existing national regulations should be observed if a 'relevant' standard has not been agreed. Of particular concern is:

- the lack of immunity standards, with the exception of specific areas, e.g. IEC 801 and the equivalent BS 6667, which is likely to be implemented in other areas as a generic standard despite known shortcomings
- the lack of EMC standards relating to physically large systems, or installations, such as standby gensets, telephone exchanges or rail traction equipment

A manufacturer is required to meet the objectives of the directive and is therefore faced with a dilemma. To protect himself legally, he can only consider testing against known standards; as has been illustrated, these may be open to misinterpretation, possess shortcomings, or there simply may not be an appropriate standard available.

Since the directive is effective for all products from January 1992, it is essential that manufacturers design all new products with EMC in mind. They must also assess which current designs remain in manufacture when the directive is effective and take appropriate steps to ensure that they comply. Despite the uncertainties of how the directive will be policed, and the availability of standards, action must be taken now to confer a 'reasonable' level of EMC to products.

It is suggested that the following action is taken:

- Define existing products which it is intended to market after 1 January 1992. Assess the electromagnetic performance of these using applicable standards where possible and establish whether or not a problem for compliance with the directive exists.
- For new developments, establish an EMC project strategy. Design with EMC in mind to avoid costly retrofits and evaluate the electromagnetic performance at prototype and pre-production stages.
- Where a lack of in-house expertise exists, use third party consultancy and/or test facilities. The directive is legally binding.
- Ensure that perceived problems can be considered by the standards making machinery, either by representation through trade associations or directly through BSI. If applicable standards are not available, it may be appropriate to devise suitable guide lines in conjunction with or through a trade association which may be submitted to form the basis of a harmonized European standard.

Bibliography

'The EC directive on electromagnetic compatibility 89/336/EC', *Jour European Communities*, L139, 19–26 (May 1989)

Electromagnetic Compatibility, HMSO Dd8221558INDYJO868NJ, DTI and COI (July 1989)

Electrical Interference: A Consultative Document, DTI/PUB207/10K, DTI Radio Communications Div (October 1989)

BS 6527:1988 (EN 55022), *Specification for Limits and Measurement of Spurious Signals Generated by Data Processing and Electronic Office Equipment* (CISPR 22)

BS 6667:1985 Pt 3 (IEC 801 pt 3), *EMC Requirements for Industrial Process Control Instrumentation* (IEC 801)

'Achieving compliance with the European Community EMC Directive', *IEE Colloq.*, Digest No 1989/126 (November 1989)

BSI publications relating to EMC

BS 613:1977 *Specification for Components and Filter Units for Electromagnetic Interference Suppression*

BS 727:1983 *Specification for Radio Interference Measuring Apparatus* (CISPR 16)

BS 800:1988 (EN 55014) *Specification for Radio Interference Limits and Measurements for Household Appliances, Portable Tools and other Electrical Equipment Causing Similar Types of Interference* (CISPR 14)

BS 833:1970 (1985) *Radio Interference Limits and Measurements for the Electrical Ignition Systems of Internal Combustion Engines* (CISPR 12)

BS 905 *Sound and Television Broadcast Receivers and Associated Equipment: Electromagnetic Compatibility*

Pt 1: 1985 *Specification for Limits of Radio Interference* (CISPR 13)

Pt 2: 1988 *Specification for Limits of Immunity* (CISPR 20)

BS 1597:1985 *Specification for Limits and Methods of Measurement of Electromagnetic Interference Generated by Marine Equipment and Installations*

BS 2316 *Radio-frequency cables*

Pts 1 and 2 1968 (1981) *General Requirements and Tests*

BS 4727 *Glossary of Electrotechnical, Power, Telecommunications, Electronics, Lighting and Colour Terms*

Pt 1: Group 09: 1976 *Radio Interference Technology* (IEC 50: Chapter 902)

BS 4809:1972 (1981) *Radio Interference Limits and Measurements for Radio Frequency Heating Equipment* (CISPR 11)

BS 5049:1973 (1981) *Methods of Measurement of Radio Noise from Power Supply Apparatus for Operation at 1 kV and Above* (CISPR 18)

BS 5260:1975 (1981) *Code of Practice for Radio Interference Suppression on Marine Installations*

BS 5394:1988 *Specifications for Radio Interference Limits and Measurements for Luminaires Using Tubular Fluorescent Lamps and Fitted with Starters* (CISPR 15)

BS 5406:1976 (1985) *The Limitation of Disturbances in Electricity Supply Networks Caused by Domestic and Similar Appliances Equipped with Electronic Devices*

BS 5602:1978 *Code of Practice for Abatement of Radio Interference from Overhead Power Lines* (CISPR 18)

BS 5783:1984 *Code of Practice for Handling Electrostatic Devices*

BS 6201 *Fixed Capacitors for Use in Electronic Equipment*
Pt 3:1982 *Specification for Fixed Capacitors for Radio Interference Suppression. Selection of Methods of Test and General Requirements* (IEC 384-14)

BS 6299:1982 *Method for Measurement of Radio Interference Terminal Voltage of Lighting Equipment* (CISPR 15)

BS 6527:1988 *Specification for Limits and Measurement of Spurious Signals Generated by Data Processing and Electronic Office equipment* (CISPR 22)

BS 6651:1985 *The Protection of Structures against Lighting*

BS 6656:1986 *Prevention of Inadvertent Ignition of Flammable Atmospheres by Radio Frequency Radiation*

BS 6657:1986 *Prevention of Inadvertent Initiation of Electro-explosive Devices by Radio Frequency Radiation*

BS 6667:1985 *EMC Requirements for Industrial Process Control Instrumentation* (IEC 801)

BS 6839:1987 *Mains Signalling Equipment*
Pt 1: *Specification for Communication and Interference Limits and Measurements*
Pt 2: *Specification for Interfaces*

DD 158:1987 *Filters for Mains Signalling Systems*

PD 6485:1980 *Limits of Radio Interference and Leakage Currents According to CISPR and National Regulations* (CISPR 9)

3G 100 *Specification for General Requirements for Equipment for Use in Aircraft*
Pt 4: Section 2:1980 *Electromagnetic Interference at Radio and Audio Frequencies*

Index